CELL PHYSIOLOGY
SOURCEBOOK
A Molecular Approach

THIRD EDITION

CELL PHYSIOLOGY SOURCEBOOK

A Molecular Approach

THIRD EDITION

Edited by

NICHOLAS SPERELAKIS

Department of Physiology and Biophysics
College of Medicine
and
Division of Pharmaceutical Sciences
College of Pharmacy
University of Cincinnati
Cincinnati, Ohio

ACADEMIC PRESS

A Harcourt Science and Technology Company

San Diego   San Francisco   New York   Boston   London   Sydney   Tokyo

Cover image: Superposition of the average structures of different
forms of *ras*-p21 on one another. In each of the three superpositions,
the normal, GDP-bound p21 is the lighter trace. Superimposed on the
normal protein are left, Val 12-p21; middle, GTP-p21 (bold); right,
Leu 61-p21. For more details, see Chapter 2, Figure 16.

This book is printed on acid-free paper.

Academic Press
A Harcourt Science and Technology Company
525 B Street, Suite 1900, San Diego, California 92101-4495, USA
http://www.academicpress.com

Academic Press
Harcourt Place, 32 Jamestown Road, London NW1 7BY, UK
http://www.academicpress.com

Library of Congress Catalog Card Number: 00-111100

International Standard Book Number: 0-12-656977-0 (pb)
International Standard Book Number: 0-12-656976-2 (case)

PRINTED IN THE UNITED STATES OF AMERICA
01 02 03 04 05 06 EB 9 8 7 6 5 4 3 2

This third edition of *Cell Physiology* is dedicated to planet Earth and its animal and plant inhabitants. Destruction of the environment and toxic pollution of the air, water, and soil are still occurring at an alarming rate. Our wilderness, forests, parks, and farmland are under increasing pressure. There is an urgent need for global population stabilization and global human rights. Our main salvation may be those numerous national and international nonprofit organizations that are dedicated to halting the ruthless destruction of our planet and the inhumane treatment of animals and humans. These organizations are concerned about the environment, wildlife, forests, farmland, overpopulation, animal rights and welfare, and human rights. Such deserving organizations are in desperate need of support from all of us. I urge all readers of this book to express their serious and urgent concern for the well-being of planet Earth and all of its inhabitants, and to help educate the public and governments worldwide.

Diagram courtesy of Richard S. Babb
(using Microsoft Office 97 and Windows 95 software)

"The Foundation of Every State is Education of its Youth."
Diogenes

In Memorium: Three of the contributors to the third edition of this book have recently passed away: Professor Shirley H. Bryant, Professor William J. Larsen, and Professor Earl T. Wallick. They were outstanding scientists, leaders in their fields, and men of great international stature. They will be sorely missed by the world scientific community.

Contents

SECTION I
Biophysical Chemistry, Metabolism, Second Messengers, and Ultrastructure

SECTION II
Membrane Potential, Transport Physiology, Pumps, and Exchangers

SECTION III
Membrane Excitability and Ion Channels

SECTION IV
Ion Channels as Targets for Toxins, Drugs, and Genetic Diseases

SECTION V
Synaptic Transmission and Sensory Transduction

SECTION VI
Muscle and Other Contractile Systems

SECTION VII
Protozoa and Bacteria

SECTION VIII
Plant Cells, Photosynthesis, and Bioluminescence

Contributors

Numbers in parentheses indicate the pages on which the authors' contributions begin.

Hugués Abriel (643), Department of Pharmacology, College of Medicine, Columbia University, 630 W 168th Street, New York, New York 10032

Francisco J. Alvarez-Leefmans (301), Department of Physiology and Biophysics, Wright State University School of Medicine, Room BS 158, 3640 Colonel Glenn Highway, Dayton, Ohio 45435

Juan Manuel Arias (1171), Department de Bioquímica, Centro de Investigación y de Estudios Avanzados del I.P.N., 07000 México, D.F., Mexico

David M. Balshaw (261), Department of Biochemistry, University of North Carolina, Chapel Hill, North Carolina 27599-7260

David W. Barnett (725), Departments of Medicine and Cell Biology/Physiology, Jewish Hospital, Washington University Medical Center, St. Louis, Missouri 63110

Clive M. Baumgarten (319), Department of Physiology, Medical College of Virginia, Virginia Commonwealth University, Box 551 MCV Station, Richmond, Virginia 23298-0551

Michael M. Behbehani (479), Department of Molecular and Cellular Physiology, University of Cincinnati College of Medicine, Cincinnati, Ohio 45267-0576

Kenneth M. Blumenthal (625), Department of Molecular Genetics, University of Cincinnati College of Medicine, Cincinnati, Ohio 45267-0524

John H. B. Bridge (283), Department of Physiology, University of Utah, Nora Eccles Harrison Building, Salt Lake City, Utah 84112

Shirley H. Bryant (653), Department of Pharmacology and Cell Biophysics, Universitiy of Cincinnati College of Medicine, Cincinnati, Ohio 45267-0575

Alma D. Chávez (1135), Departamento de Bioenergética, Universidad Nacional Autónoma de México, Instituto de Fisiologia Celular, Apartado Postal 70-243, 04510 México, D.F., Mexico

Alberto Darszon (509), Departmento Genética y Fisiologia Molecular, Instituto de Biotechnología, Universidad Nacional Autónoma de México, Apartado Postal 510-3, 62250 Cuernavaca, Morelos, Mexico

John R. Dedman (167), Department of Molecular and Cellular Physiology, College of Medicine, University of Cincinnati, Cincinnati, Ohio 45267-0576

Louis J. DeFelice (539), Department of Pharmacology, School of Medicine, Vanderbilt University, Nashville, Tennessee 37232-6600

Guy Droogmans (485), Department of Physiology, KU Leuven, Campus Gasthuisberg, Herestraat, B-3000 Leuven, Belgium

Istvan Edes (271), Department of Cardiology and Pulmonary, Medical University of Debrecen, 4004 Debrecen, Hungary

Joseph J. Feher (319), Department of Physiology, Box 551, Medical College of Virginia, Virginia Commonwealth University, Richmond, Virginia 23298-0551

Donald G. Ferguson (95), Department of Anatomy, College of Medicine, Case Western Reserve University, Cleveland, Ohio 44106-4930

Harvey M. Fishman (857), Department of Physiology and Biophysics, University of Texas Medical Branch, Galveston, Texas 77555-0641

Darrell E. Fleischman (1097), Department of Biochemistry and Molecular Biology, Wright State University, 3640 Colonel Glenn Highway, Dayton, Ohio 45435

Michael S. Forbes (95), Merry Oaks, Rt. 1, Box 245, Troy, Virginia 22974-9741

Jeffrey C. Freedman (3), Department of Physiology, SUNY Health Science Center at Syracuse, 766 Irving Avenue, Syracuse, New York 13210

Andrew S. French (761), Department of Physiology and Biophysics, Dalhousie University, Halifax, Nova Scotia B3H 4H7, Canada

Akikazu Fujita (601), Department of Pharmacology II, Faculty of Medicine, Osaka University, 2-2, Yamadoaka, Suita, Osaka 565, 0871 Japan

Nicole Gallo-Payet (705), Endocrine Service, Department of Medicine, University of Sherbrooke, Sherbrooke, Quebec, J1H 5N4, Canada

Keith D. Garlid (139), Department of Biochemistry and Molecular Biology, Oregon Graduate Institute of Science and Technology, 20,000 NW Walker Road, Beaverton, Oregon 97006-8921, or P.O. Box 91000, Portland, Oregon 97291-1000

W. Gibson Wood (81), Department of Physiology and Pharmacology, Texas A&M University, College Station, Texas 77843-4466

Diana González (1135), Departamento de Bioenergética, Universidad Nacional Autónoma de México, Instituto de Fisiologia Celular, Apartado Postal 70-243, 04510 México, D.F., Mexico

Hugo Gonzalez-Serratos (865), Department of Physiology, University of Maryland at Baltimore School of Medicine, 655 West Baltimore Street, Baltimore, Maryland 21201-1559

Thomas Gorrell (1041), Haskins Laboratories, Pace University, 41 Park Row (at Pace Plaza), New York, New York 10038-1598

Kathleen Heppner Goss (1161), Department of Biochemistry and Molecular Genetics, College of Medicine, University of Cincinnati, Cincinnati, Ohio 45267-0524

Anthony L. Gotter (1025), Laboratory of Developmental Chronobiology, Children's Service GRJ 1226, Massachusetts General Hospital, Boston, Massachusetts 02114

Michael S. Grace (832), Department of Biological Sciences, Florida Institute of Technology, Melbourne, Florida 32901

Steven M. Grassl (249), Department of Pharmacology, SUNY Health Science Center, 750 East Adams Street, Syracuse, New York 13210

Lawrence Griffing (1079), Department of Biology, Texas A & M University, College Station, Texas 77843

Joanna Groden (1161), Department of Biochemistry and Molecular Genetics, College of Medicine, University of Cincinnati, Cincinnati, Ohio 45267-0524

Dennis W. Grogan (1063), Department of Biological Sciences, University of Cincinnati, Cincinnati, Ohio 45267-0006

Augustín Guerrero (1171), Department de Bioquímica, Centro de Investigacion y de Estudios Avanzados del I.P.N., 07000 México, D.F., Mexico

Andrés A. Gutiérrez (1135), Departamento de Bioenergética, Universidad Nacional Autónoma de México, Instituto de Fisiologia Celular, Apartado Postal 70-243, 04510 México, D.F., Mexico

Eric J. Hall (1185), Center for Radiological Research, College of Physicians and Surgeons, Columbia University, New York, New York 10032

J. Woodland Hastings (1115), Biological Laboratories, Harvard University, 16 Divinity Avenue, Cambridge, Massachusetts 02138

Hiroshi Hibino (601), Department of Pharmacology II, Faculty of Medicine, Osaka University, 2-2, Yamadoaka, Suita, Osaka 565, 0871 Japan

Christopher D. Heinen (1161), Department of Biochemistry and Molecular Genetics, College of Medicine, University of Cincinnati, Cincinnati, Ohio 45267-0524

Judith A. Heiny (911), Department of Molecular and Cellular Physiology, College of Medicine, University of Cincinnati, Cincinnati, Ohio 45267-0576

Kent R. Hermsmeyer (899), Dimera LLC, 2525 NW Lovejoy, Suite 401, Portland, Oregon 97210

Nelson Horseman (191), Department of Molecular and Cellular Physiology, University of Cincinnati College of Medicine, Cincinnati, Ohio 45267-0576

Ching-hsien Huang (43), Department of Biochemistry and Molecular Genetics, University of Virginia School of Medicine, Box 440, Charlottesville, Virginia 22908-0001

Atsushi Inanobe (573), Department of Pharmacology II, Faculty of Medicine, Osaka University, 2-2, Yamadoaka, Suita, Osaka 565, 0871 Japan

Marcia A. Kaetzel (167), Department of Molecular and Cellular Physiology, College of Medicine, University of Cincinnati, Cincinnati, Ohio 45267-0576

Edna S. Kaneshiro (959), Department of Biological Sciences, University of Cincinnati, Cincinnati, Ohio 45221-0006

Robert S. Kass (643), Department of Pharmacology, College of Medicine, Columbia University, 630 W. 168th Street, New York, New York 10032

James G. Kereiakes (1185), Department of Radiology-Medical Physics, College of Medicine, University of Cincinnati, Cincinnati, Ohio 45267-0579

Ann B. Kier (81), Department of Physiology and Pharmacology, Texas A&M University, College Station, Texas 77843-4466

Evangelia G. Kranias (271), Department of Pharmacology and Cell Biophysics, College of Medicine, University of Cincinnati, Cincinnati, Ohio 45267-0575

Yoshihisa Kurachi (573), Department of Pharmacology II, Faculty of Medicine, Osaka University, 2-2, Yamadoaka, Suita, Osaka 565, 0871 Japan

William J. Larsen (523), Department of Anatomy and Cell Biology, College of Medicine, University of Cincinnati, Cincinnati, Ohio 45267-0521

Harold Lecar (441), Department of Molecular/Cell Biology, 102 Donner Lab, University of California at Berkeley, Berkeley, California 94720

Michael Levandowsky (1041), Haskins Laboratories, Pace University, 41 Park Row (at Pace Plaza), New York, New York 10038-1598

Simon Rock Levinson (455), Department of Physiology, Health Science Center, University of Colorado, 4200 East Ninth Avenue, Denver, Colorado 80262

Michael A. Lieberman (119), Department of Molecular Genetics, College of Medicine, University of Cincinnati, Cincinnati, Ohio 45267-0524

Arturo Liévano (509), Departmento de Bioquímica, Instituto de Biotechnología, Universidad Nacional Autónoma de México, Apartado Postal 510-3, 62271 Cuernavaca Morelos, Mexico

Alister G. Macdonald (1003), Department of Biomedical Sciences, c/o Zoology Building, University of Aberdeen, Tillydrone Avenue, Aberdeen AB24 2TZ, United Kingdom

Daniel C. Marcus (775), Departments of Anatomy and Physiology, Kansas State University, 1600 Denison Avenue, Manhattan, Kansas 66506

Francisco Martínez (1135), Departamento de Bioenergética, Universidad Nacional Autónoma de México, Instituto de Fisiología Celular, Apartado Postal 70-243, 04510 Mexico, D.F., Mexico

James Maylie (653), Oregon Health Sciences University, 3181 SW Sam Jackson Park Road, Portland, Oregon 97201

Michele Mazzanti (539), Dipartimento di Fisiologia e Biochimica Generali, Laboratorie di Erettrofisiologia, Universita degli Studi di Milano, Via Celoria 26, 1-20133 Milan, Italy

Gerhard Meisner (927), Department of Biochemistry, Biophysics, and Physiology, University of North Carolina, Chapel Hill, North Carolina 27599

Lauren A. Millette (261), Department of Pharmacology and Cell Biophysics, College of Medicine, University of Cincinnati, Cincinnati, Ohio 45267-0575

Stanley Misler (725), Departments of Medicine and Cell Biology/Physiology, Jewish Hospital, Washington University Medical Center, St. Louis, Missouri 63110

Catherine E. Morris (745), Departments of Biology and Medicine, Loeb Institute, University of Ottawa, Ottawa Civic Hospital, 1053 Carling Avenue, Ottawa, Ontario K1Y 4E9 Canada

Edward F. Nemeth (179), NPS Pharmaceuticals, Inc., 420 Chipeta Way, Salt Lake City, Utah 84108-1256

John New (839), Biology Department, Loyola University, 6525 North Sheridan Road, Chicago, Illinois 60626

Bernd Nilius (485), Department of Physiology, KU Leuven, Campus Gasthuisberg, Herestraat, B-3000 Leuven, Belgium

Denis Noble (xix), Department of Physiology, University of Oxford, Parks Road, Oxford OX1 3PT, United Kingdom

Richard J. Paul (941), Department of Molecular and Cellular Physiology, College of Medicine, University of Cincinnati, Cincinnati, Ohio 45267-0576

Marcel D. Payet (705), Department of Molecular and Cellular Physiology, Faculty of Medicine, University of Sherbrooke, Sherbrooke, Quebec J1H-5N4, Canada

J. Wesley Pike (191), Department of Molecular and Cellular Physiology, College of Medicine, Universiity of Cincinnati, Cincinnati, Ohio 45267-0576

Matthew R. Pincus (19), Department of Pathology and Laboratory Medicine, Veterans Administration Medical Center, Brooklyn, New York 11209

David M. Pressel (725), Departments of Medicine and Cell Biology/Physiology, Jewish Hospital, Washington University Medical Center, Street Louis, Missouri 63110

Raymund Y. K. Pun (441), Department of Molecular and Cellular Physiology, College of Medicine, University of Cincinnati, Cincinnati, Ohio 45267-0576

Robert W. Putnam (357), Department of Physiology and Biophysics, School of Medicine, Wright State University, 3640 Colonel Glenn Highway, Dayton, Ohio 45435

Ilaria Rivolta (643), Department of Pharmacology, College of Medicine, Columbia University, 630 W 168th Street, New York, New York 10032

Stephen D. Roper (815), Department of Physiology and Biology, University of Miami School of Medicine, R430, P.O. Box 016430, Miami, Florida 33101

Nancy J. Rusch (899), Department of Physiology, Medical College Wisconsin, 8701 Watertown Plank Road, Milwaukee, Wisconsin 53226-0509

Michael Sanderson (959), Department of Physiology, University of Massachusetts Medical Center, 55 Lake Avenue, Worcester, Massachusetts 01655-0127

William T. Sather (455), Department of Pharmacology, University of Colorado Health Sciences Center, Denver, Colorado 80262

Friedhelm Schroeder (81), Department of Physiology and Pharmacology, Texas A&M University, College Station, Texas 77843

Richard G. Sleight (119), Yale University Graduate School, P.O. Box 208236, 320 York Street, New Haven, Connecticut 06520-8236

Nicholas Sperelakis (209), Department of Molecular and Cellular Physiology, College of Medicine, University of Cincinnati, Cincinnati, Ohio 45267-0576

Kira Steigerwald (1161), Department of Biochemistry and Molecular Genetics, College of Medicine, University of Cincinnati, Cincinnati, Ohio 45267-0524

Richard G. Stout (1079), Department of Biology, College of Letters and Science, Montana State University, Bozeman, Montana 59717-0001

Janusz B. Suszkiw (689), Department of Molecular and Cellular Physiology, College of Medicine, University of Cincinnati, Cincinnati, Ohio 45267-0576

Nortisugu Tohse (585), Department of Physiology, College of Medicine, Sapporo Medical University, South 1 West 17, Chuo-ku Sapporo 060, Japan

Paivi H. Torkkeli (761), Department of Physiology and Biophysics, Dalhousie University, Halifax, Nova Scotia B3H 4H7, Canada

Timothy C. Tricas (839), Department of Zoology, University of Hawaii at Maroa, 2538 McCarthy Mall, Edmonson Hall, Honolulu, Hawaii 96822

Richard D. Veenstra (412), Department of Pharmacology, College of Medicine, State University of New York, 750 East Adams Street, Syracuse, New York 13210

Gordon M. Wahler (559), Department of Physiology, Midwestern University, 555 31st Street, Downers Grove, Illinois 60515

Earl T. Wallick (261), Department of Pharmacology and Cell Biophysics, College of Medicine, University of Cincinnati, Cincinnati, Ohio 45267-0575

Gary L. Westbrook (675), Vollum Institute and Department of Neurology, Oregon Health Sciences University, 3181 SW Sam Jackson Park Road, L474, Portland, Oregon 97201-3098

George Witman (959), Department of Cell Biology, University of Massachusetts Medical School, 55 Lake Avenue N., Worcester, Massachusetts 01655

Hisashi Yokoshiki (585), Department of Cardiovascular Medicine, Hokkaido University School of Medicine, Kita-15, Nishi-7, Kita-ku, Sapporo 060 Japan

Anita L. Zimmerman (807), Department of Physiology, Brown University, Box G-B329, Providence, Rhode Island 02912

Foreword to the First Edition

It was kind and generous of my friend Nicholas Sperelakis to relate this excellent book so closely to my own, *A Textbook of General Physiology*. In the preface to the first edition of my book, I had expressed the hope that it might be compared with Bayliss' *Principles of General Physiology*. If this comparison is valid, it is very appropriate that the present book be organized and partly written by one who is, along with Sir William Bayliss and myself, associated with University College London (Professor Sperelakis having spent a sabbatical year there). It is a pleasure to recall that it was there that I first met Nicholas, and I remember discussing his pioneering study on the potentials across the crystalline lens of the eye.

For reasons that I think bear no relation to its scope, this book has a different title; I presume it is because the distinction between "ordinary" and "general" physiology has become sufficiently blurred to demand something more appropriate. The definition of general physiology that I had proposed in the preface to the first edition of my textbook was "the study of those aspects of living material that show some immediate prospect of being described in terms of the known laws of physics and chemistry." Later, I had misgivings as to the narrowness of this definition, and I then suggested that it might be replaced by "the study of those features of life that appear to be common to all forms." Whatever definition we choose, however, it is of immense satisfaction to me that this new book, essentially, has been fitted into the same sectional headings that I employed in my own book.

If I may be permitted to reminisce further, I have wondered frequently how a single scientist, actively engaged in research, could write a new book of such wide scope. The answer is that I wrote the book during the 2 years immediately following the end of World War II. Thus, for several years—in Great Britain for as many as 10 years—very little original academic physiology and new research had been published. This made it possible to survey the original literature of a lengthy period without being overwhelmed by a rapid succession of new discoveries that would have rendered my task nearly impossible, a fate similar to that of Sisyphus. Today, this task would be impossible, and it only surprises me that Nicholas has been able to produce this magnificent book with so few collaborators.

Hugh Davson
1995

Foreword to the Second Edition

In his Foreword to the first edition of the *Cell Physiology Source Book,* Hugh Davson established it as the lineal descendent of his own well-known and highly respected work *The Textbook of General Physiology.* The second edition of the *Cell Physiology Source Book,* again edited by Nicholas Sperelakis, continues in this same tradition. Although the first edition was enthusiastically received by the cell physiology community because of its depth and breadth of coverage, considerable important progress has been made in this rapidly developing area since its publication. The second edition deals with these new developments by a thorough reworking of topics and by the inclusion of new chapters in all sections covered in the first edition. The new topics introduced into the various sections include lipid structure, mitochondrial physiology, cell responses to hormones, red blood cell transport, neuron physiology, developmental changes in ion channels, sonotransduction, excitation-contraction coupling, and electroplax cells. In addition, the scope of the new edition has been valuably broadened by the inclusion of two entirely new sections. One titled *Protozoa and Bacteria* covers the physiology of these organisms in two chapters. In the other, *Cell Division and Programmed Cell Death,* there are chapters on the regulation of cell division, the cancer cell, apoptosis, and the effects of ionizing radiation. The extensive revisions and the new material in the second edition raise it to a new level.

Cell physiology, an area of central importance in biology, has grown out of a number of more traditional fields, and as a result, the literature continues to be widely dispersed. The great value of the *Cell Physiology Source Book* is that it gathers together under a single cover a broad range of up-to-date chapters that, taken together, define the field. The various chapters exhibit a uniformity of style and level of presentation that are a credit to the editor. Because of this and the scope and clarity of the presentations, this book can serve exceptionally well as an advanced undergraduate- or graduate-level text for cell physiology courses. The broad coverage of this second edition also makes it very attractive for use in cell biophysics, membrane biology, and biomedical engineering courses. It can serve equally well as a textbook for introductory courses in ion channel structure and physiology.

I was pleased, and indeed proud, to be asked by my colleague Nicholas Sperelakis to contribute the Foreword to the second edition of the *Cell Physiology Source Book.* This book clearly sets a new standard of excellence.

Thomas E. Thompson
1997

Foreword to the Third Edition

When Nicholas Sperelakis kindly invited me to write this foreword, I wondered how I could possibly follow in the footsteps of Hugh Davson. I studied physiology at University College London (UCL) at the time when his monumental *Textbook of General Physiology* was published. UCL was an extraordinary place in which to be a student at that time. Not only was Hugh Davson laying his particular cornerstone of the subject, but Leonard Bayliss was also re-working his father's famous *Principles of General Physiology.* These were the books that convinced me to become a research physiologist and that cell physiology was the place to begin. Between them, Davson and Bayliss were responsible for seducing generations of medical students to discover the challenge and delights of physiological research. Later, as a young lecturer at UCL, I remember asking Hugh Davson why he didn't lecture very much to the students. He simply replied: "Denis, I've written it all. Tell them to read!" Actually, that remark inspired me to reread his books. My copies are still in my library, and they are well and truly thumbed.

It will be the greatest tribute to the work of Nicholas Sperelakis and his colleagues that the pages of their book will also become the well-thumbed bible of a new generation of physiologists. They will be entering the discipline at an exciting time, for cell physiology is the base on which our understanding of all aspects of integrative and systems physiology must rest. This book, therefore, will be an essential source for all of us who aspire to understand the "logic of life," which is, after all, the meaning and origin of the word "physiology." Life has no single "logic," of course. Unraveling the physicochemical mechanisms that nature has discovered, fashioned, combined, and interwoven into the tangled skein of processes that form the function of even the simplest cell is a process full of surprises as we discover with awe the audacity with which nature's molecular mechanisms are reused again and again in different contexts. One might call this audacity the "dance of the genes" were it not for the inanimate and unthinking nature of these bits of code that transmit the logic from one generation to another. The logic is physiological, not genetic. The molecular biologist Sydney Brenner put the point succinctly when he wrote recently (in a Novartis symposium titled The Limits of Reductionism in Biology) that "Genes can only specify the properties of the proteins they code for, and any integrative properties of the system must be 'computed' by their interactions." We are approaching the point at which our understanding of those interactions is deep enough for physiology to aspire to become the quantitative analytical discipline it needs to be to solve the problems it tackles. Brenner went on to remark, significantly, "This provides a framework for analysis by simulation." It is a sign of the maturity of our science that simulation is indeed also becoming an essential tool of analysis.

It is further such sign that the scope of cell physiology is now so wide. The breadth of this book is therefore one of its greatest strengths. The chapters encompass the great majority of important molecular and cell systems, so that it does indeed justify its title as a *source* book.

Denis Noble
1999

Preface to the First Edition

Hugh Davson's original textbook on cell physiology, *A Textbook of General Physiology,* was a true classic and a huge success. The first edition was published in 1959, and the fourth and final edition in 1970. In the past two decades, cell physiology has advanced by leaps and bounds, as, for example, in the area of ion channels and their regulation. The present book attempts to fill the void left by the discontinuation of Davson's monumental book. This book is dedicated to Professor Davson in recognition of the great impact he has had on the dissemination of the principles of cell physiology.

At present, there is a need for a good text on cell physiology. Several good texts on cell biology are available, but these generally treat cell physiology in a superficial and incomplete manner. Therefore, it was our intent to prepare a new work on cell physiology that is high quality, is comprehensive, and covers recent developments (the chapter authors were asked to limit their bibliography to no more than 40 key papers and review-type articles). We hope that the reader will find this book clearly written, thorough, up to date, and worthy of being the successor to Professor Davson's book.

This book focuses on physiology and biophysics at the cellular level. Organ systems and whole organisms are essentially not covered, except for unicellular organisms. The major topics covered include ultrastructure, molecular structure and properties of membranes, transport of ions and nonelectrolytes, ion channels and their regulation, membrane excitability, sensory transduction mechanisms, synaptic transmission, membrane receptors, intracellular messengers, metabolism, energy transduction, secretion, excitation-secretion coupling, excitation-contraction coupling, contraction of muscles, cilia and flagellae, photosynthesis, and bioluminescence. These topics are covered in a comprehensive, but didactic, manner, with each topic beginning in an elementary fashion and ending in a sophisticated and quantitative treatise.

This book is intended primarily for graduate and advanced undergraduate students in the life sciences, including those taking courses in cell physiology, cell biophysics, and cell biology. Selected parts of this book can be used for courses in neurobiology, electrobiology, electrophysiology, secretory biology, biological transport, and muscle contraction. Students majoring in engineering, biomedical engineering, physics, and chemistry should find this book useful to help them understand the living state of matter. Postdoctoral scholars and faculty engaged in biological research would also find this text of immense value for obtaining a better understanding of cell function and as a reference source. Medical, dental, and allied health students could use this book as a valuable comparison text to complement their other textbooks in medical/ mammalian physiology. The latter texts often strongly emphasize organ system physiology to the virtual exclusion (in some cases) of cell physiology, membrane biophysics, and mechanisms underlying homeostasis of the entire organism. The chapter authors were asked to include the following aspects wherever appropriate: comparative physiology, developmental changes, pathophysiology, membrane diseases, and molecular biology.

Considering the tremendous amount of new information gained during the past two decades using sophisticated new instruments and techniques, I have recruited a number of outstanding researchers, who are leaders in their respective fields, to contribute to this undertaking. I am delighted that they agreed to participate. It has been my great pleasure and honor to work with them on this important and timely project.

Nicholas Sperelakis
1995

Preface to the Second Edition

The first edition of the *Cell Physiology Source Book* was conceived to serve as the replacement for Davson's classic textbook of cell physiology, which was discontinued after the fourth edition was published in 1970. There was a pressing need for a comprehensive and authoritative textbook on cell physiology because the available cell biology textbooks treated cell physiology in a superficial and incomplete manner. To accomplish this Herculean task, I recruited a number of outstanding researchers who were leaders in their respective fields to contribute to the undertaking. The result was a comprehensive source book (738 pages, hardcover) that appeared in February 1995 and rapidly became successful. Within a year, the book had to be reprinted in hardcover, and a softcover version was printed. Book reviews were complimentary to the book, and I received numerous comments from contributors and other scientists about the excellence of the *Source Book.* One of my favorite comments was made by an Israeli physical chemist at a conference in Prague where the book was on display. He said, "Nick, this is the type of book that *speaks* to me." The first edition was selected as one of the outstanding academic books for 1996 by *CHOICE* (current reviews for academic libraries published by the Association of College & Research Libraries, a division of the American Library Association).

I began to formulate plans for the second edition almost immediately after the appearance of the first edition. I believed that it was important for the second edition to appear within 3 years after the first edition. It was especially important to have a relatively short interval because the field is advancing so rapidly. I also wanted to take the opportunity to include some new chapters on topics quite relevant to the discipline of cell physiology. In addition, several of the original chapters were expanded, reorganized, and restructured, and several other chapters were shifted in location, with the goal of making the *Source Book* more useful as a textbook. A total of 18 new chapters were added, and two new sections were organized: Section VII, Protozoa and Bacteria; and Section IX, Cell Division and Programmed Cell Death. The new chapters include lipid structure, physiology of mitochondria, responses to hormones, transport in erythrocytes, developmental changes of ion channels, calcium receptors, cytoskeletal modulation of ion channels, neuronal cells, sonotransduction, excitation–contraction coupling, electroplax cells, electroreceptors, protozoa, bacteria, cell division, cancer cells, apoptosis, and ionizing radiations.

These additional chapters should make the *Source Book* more complete and more useful to graduate students in neuroscience, biology, cell biology, transport physiology, and electrophysiology. Selected chapters in the *Source Book* can be used for courses in cell physiology, cell biophysics, electrophysiology, electrobiology, transport physiology, secretory biology, sensory physiology, neurobiology, and contractile systems. Students majoring in the physical sciences should find this book helpful in understanding the living state of matter. This book is intended primarily for graduate students and advanced undergraduate students majoring in the life sciences.

A new edition also allows the individual contributor the opportunity of adjusting the level at which his or her chapter is pitched (i.e., to aim it at the median level at which the majority of the chapters are pitched). Thus, parts of some chapters were expanded, whereas parts of others were constricted. Certain figures were improved. The subject index was improved by limiting entries to first order and second order.

I sincerely hope that the student will find this book to be clearly written, thorough, up to date, and worthy of being the successor to Davson's monumental book. The book contains several opening chapters and an appendix to help the student review the physical chemistry of solutions, protein structure, lipid structure, and electricity. Hence, within the covers of this one book, the student should have all the information needed to understand the various topics that constitute the discipline of cell physiology. The contributors, publisher, and I want to establish this book as the leading textbook of cell physiology.

Nicholas Sperelakis
1997

Preface to the Third Edition

The first edition and second edition of *Cell Physiology Source Book* have done very well in terms of sales and usage. In addition, each edition was selected as one of the outstanding academic books [for 1996 (1st ed.) and 1998 (2nd ed.)] by *CHOICE* of the American Library Association. This is indeed an honor for the editor, contributors, and publisher. Therefore, this textbook/sourcebook is now well established, and it is probably the leading book in the field of cell physiology and biophysics.

The second edition was substantially expanded from the first edition, and the third edition is somewhat expanded from the second edition. The new chapters include Membrane Structure (Chapter 4) and Intracellular Chloride Regulation (Chapter 20), an appendix on infrared detectors (Chapter 49), and an appendix on mechanism of electroreceptors (Chapter 50). Several chapters have new authors and new focus, including Transport of Ions and Nonelectrolytes (Chapter 16), Direct Regulation of Ion Channels by G Proteins (Chapter 34), Sensory Receptors and Mechanotransduction (Chapter 45), and Physiological Effects of Pressure on Cell Function (Chapter 59). In addition, some new authors were brought in to help update several chapters, including Diffusion and Permeability (Chapter 13), Regulation of Ion Channels by Phosphorylation (Chapter 33), Ion Channels as Targets for Disease (Chapter 39), Skeletal Muscle Action Potentials (Chapter 51), Smooth Muscle Action Potentials (Chapter 53), Amoeboid Movement, Cilia, and Flagella (Chapter 57), and Plant Cell Physiology (Chapter 63). One chapter was dropped due to time constraints (Polarity of Cells and Membrane Regions; old Chapter 21).

These changes, along with the updating of all chapters, should make the *Sourcebook* more complete and more useful to students and faculty. Graduate students and advanced undergraduate students majoring in the life sciences should find this book very useful. Students majoring in physics, chemistry, and biomedical engineering will find this book helpful in understanding the *living state* of matter. Selected chapters in the *Sourcebook* can be used for academic courses in cell physiology and biophysics, electrophysiology, transport physiology, sensory physiology, and neurobiology. The *Sourcebook* can be used as an adjunct text for courses in neuroscience, cell biology, and membrane biophysics.

The contributors and I hope that the student will find this *Sourcebook* to be clearly written, comprehensive, up to date, and worthy of being the successor to Hugh Davson's classic book on cell physiology. The book contains several opening chapters to help the student review the physical chemistry of solutions, protein structure, and lipid structure. The Appendix to the book provides a review of electricity relevant to biology. It is my pleasure, as editor, to work with a distinguished group of scientists who are experts in the various topics included in this *Sourcebook*. The contributors and I have put much effort into this enormous task, and we are very proud of our accomplishment.

Nicholas Sperelakis
2000

SECTION

I

Biophysical Chemistry, Metabolism, Second Messengers, and Ultrastructure

Jeffrey C. Freedman

1

Biophysical Chemistry of Physiological Solutions

I. Introduction

All living cells contain proteins, salts, and water enclosed in membrane-bounded compartments. These biochemical and ionic cellular constituents, along with a set of genes, enzymes, substrates, and metabolic intermediates, function to maintain cellular homeostasis and enable cells to replicate and to perform chemical, mechanical, and electrical work. **Homeostasis**, a term introduced by the physiologist Walter Cannon in *The Wisdom of the Body* (1932), means that certain parameters, including cellular volume, intracellular pH, the transmembrane electrical potential, and intracellular concentrations of salts are maintained relatively constant in resting cells. Cellular homeostasis depends on a relative constancy of the extracellular fluids that bathe cells. The extracellular fluid compartment was termed the *milieu intérieur*, or internal environment, by Claude Bernard, who recognized around 1865 that "*La fixité du milieu intérieur est la condition de la vie libre*," that the constancy of the internal environment is the condition for independent life. In *An Introduction to the Study of Experimental Medicine* (see 1949 translation), Bernard wrote that "only in the physico-chemical conditions of the inner environment can we find the causation of the external phenomena of life." The development of cell physiology was greatly influenced by the Bernard-Cannon theory of physical-chemical homeostasis.

Biophysical chemistry concerns the application of the concepts and methods of physical chemistry to the study of biological systems. Physical chemistry includes such physiologically relevant subjects as thermodynamics, chemical equilibria and reaction kinetics, solutions and electrochemistry, properties and kinetic theory of gases, transport processes, surface phenomena, and molecular structure and spectroscopy. Throughout this book, it is seen that many cellular physiological phenomena are best understood with a rigorous and comprehensive understanding of physical chemistry. Physical chemistry texts that specifically emphasize biological applications include those by Eisenberg and

Crothers (1979) and by Tinoco *et al.* (1994). During the past fifty years, outstanding monographs on biophysical chemistry have also been available: Höber (1945), Edsall and Wyman (1958), Tanford (1961), Cantor and Schimmel (1980), Silver (1985), van Holde (1985), and Bergethon and Simons (1990). To help the reader begin to understand how cellular homeostasis is achieved, this chapter and the following one will introduce some of the conceptual underpinnings of cell physiology by describing certain physicochemical properties of water, electrolytes, and proteins that are relevant for understanding the structure and function of living cells.

II. Structure and Properties of Water

Biological cells contain a large amount of water, ranging from 0.66 g H_2O/g cells in human red blood cells to around 0.8 g H_2O/g tissue in skeletal muscle. The amount of water in cells is determined by osmosis (for review, see Dick, 1959, and Chapter 21). Liquid water is a highly polar solvent, with a structure stabilized by extensive intermolecular **hydrogen bonds** (Fig. 1). The hydrogen bonds between adjacent water molecules are linear (O–H···O) but are able to bend by about 10°. From x-ray diffraction studies of ice crystals, it is known that each of the two covalent O–H bonds in water is 1 Å in length, that each of the two hydrogen bonds is about 1.8 Å in length, and that these occur around each oxygen atom at angles of 104°30′ in a tetrahedral array.

The energy required to break hydrogen bonds in liquid water is 5 to 7 kcal/mol, much less than the 109.7 kcal/mol required to break the covalent O–H bond. A calorie is a unit of work equal to 4.18 J in the SI system of units, where 1 joule (J) equals 1 Newton·meter. The newton ($N = kg·m/s^2$) is the unit of force in the International System of Units (SI); the dyne ($dyn = g·cm/s^2$) is the corresponding unit of force in the older cgs system ($1 N = 10^5 dyn$). Recall that, according to Newton's first law, the force (F) is mass times acceleration, and that work is defined as force times distance. Because of the strength of hydrogen bonds, liquid water has unique physicochemical properties, including a high boiling point (100 °C), a high **molar heat capacity** (18 cal/mol·K), a high **molar heat of vaporization** (9.7 kcal/mol at 1 atm), and a high **surface tension** (72.75 dyn/cm). As discussed by Bergethon and Simons (1990), all of these thermodynamic parameters are considerably higher for H_2O than for the analogous compound H_2S, which boils at -59.6 °C. The outer electrons of the S atom shield its nuclear charge and reduce its electronegativity, making the S–H bond weaker, longer, and less polar than the O–H bond. Also, the bond angle of H_2S is only 92°20′, an angle that does not form a tetrahedral array, and thus H_2S does not form a hydrogen bonded network like water.

The extensive intermolecular association in liquid water is due to the geometry of the water molecule, and also to its strong permanent **dipole moment**. Water molecules have no net charge, yet possess permanent charge separation along each of the two O-H bonds. Since oxygen is more **electronegative** than hydrogen, its electron density is preferentially greater. Oxygen thus acquires a partial negative charge (δ^-), leaving the hydrogens with a partial positive charge (δ^+) (Fig. 1C). The magnitude of the dipole moment (μ) of a chemical bond is computed as the product of the separated charge (q) at either end, and the distance (d) between the centers of separated charge ($\mu = q \cdot d$). The dipole moment of a molecule is the vector sum of the dipole moments of each bond. For water in the gaseous phase, the dipole moment is 1.85 debye, where 1 debye = 10^{-18} esu·cm. One esu (electrostatic unit), or statcoulomb, equals 3.336×10^{-10} C, the coulomb (C) being a unit of electric charge. In liquid water, the dipole moment becomes even larger (2.5 debye) due to association with other water molecules. In comparison, the linear molecule carbon dioxide (O=C=O) also has a separation of charge, with a partial negative charge on each oxygen atom and a partial positive charge on the carbon atom. In this case, the oppositely directed dipole moments of the two C=O bonds sum to a zero dipole moment for the CO_2 molecule. In H_2S, the S–H bond has a smaller dipole moment (1.1 debye) than the O–H bond in water because sulfur is less electronegative than oxygen. Electrostatic attractions between opposite charges in adjacent water dipoles stabilize the hydrogen-bonded structure of liquid water.

Water strongly affects the forces between ions in solution by virtue of its high **dielectric constant**. The dielectric constant (ϵ) of a medium known as a **dielectric** is defined as the ratio of the coulombic force (F_{coul}) between two charges in a vacuum to the actual force (F) between the same two charges in the dielectric medium.

$$\epsilon = \frac{F_{coul}}{F}$$

In a vacuum, the coulombic force, F_{coul} (newtons, N), between two ions with charges q^+ and q^- (C) separated by a distance d (meters, m) is given by **Coulomb's law**:

$$F_{coul} = \frac{q^+ q^-}{4\pi\epsilon_0 d^2}$$

where ϵ_0 is the **permittivity constant** (8.854×10^{-12} C²/N·m²). Coulomb's law states that the force between two point

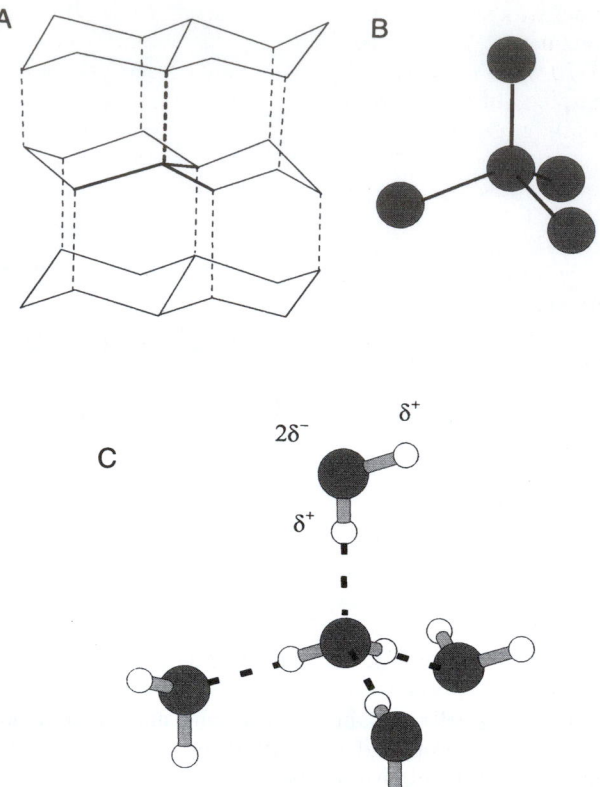

FIGURE 1. Hydrogen bonds in ice and liquid water. (A) Crystal structure of ice, showing the puckered hexagonal rings of oxygen atoms. (B) Each oxygen atom is connected to four others in a tetrahedron. (C) The linear hydrogen bonds are indicated by dotted lines between adjacent water molecules (O–H···O). Two covalent bonds (O–H), indicated by solid lines, and two hydrogen bonds occur around each oxygen atom in a tetrahedral array. The partial charge separation of the O–H dipoles is indicated on the upper water molecule.

charges is directly proportional to the magnitude of each charge, and inversely proportional to the square of the distance between the charges. Substituting F_{coul} into the above expression and rearranging yields the following.

$$F = \frac{q^+ q^-}{4\pi\epsilon_0 \epsilon d^2}$$

The dielectric constant of a vacuum is unity, and that of air is close to unity ($\epsilon_{air} = 1.00054$). Increasing the dielectric constant of the solvent decreases the attractive force between oppositely charged ions. The force felt by a distant ion is reduced in a dielectric medium such as water, as compared with a vacuum, because part of the interaction energy is spent aligning the intervening water dipoles and distorting their polarizable electron clouds. The dielectric constant of water at 25 °C is 78.5, much greater than methanol ($\epsilon = 32.6$), ethanol (24.0), or methane (1.7). In liquid water, the high dielectric constant weakens the coulombic attractive forces between oppositely charged particles, and thus promotes dissociation and ionization of salts.

Although the thermodynamic properties of liquid water are explicable in terms of extensive hydrogen-bonding, the actual structure of water is still unknown (see Eisenberg and Kauzmann, 1969). In the flickering cluster model, groups of 50–70 water molecules, resembling a slightly expanded broken piece of the ice lattice (icebergs), are continuously associating and dissociating on a picosecond time scale. This dynamic model contrasts with the extended order of solid ice. At a given instant, some water molecules are unattached to the clusters and are located in the interstitial regions of the network, but may attach and detach as the clusters continuously form and break down. Other theories treat water as a mixture of distinct states, or as a continuum of states, with considerable short-range order, characteristic of the crystalline lattice of ice.

Despite these uncertainties regarding the structure of liquid water and the influence of macromolecules on its properties in cells (see Cooke and Kuntz, 1975), the ability of water to act as a solvent inside cells closely resembles that of extracellular water. Thus, a variety of permeant, hydrophilic, nonmetabolized nonelectrolytes distribute at equilibrium across human red blood cell membranes with ratios of intracellular to extracellular concentrations that deviate from unity by less than 10% (Gary-Bobo, 1967). In mouse diaphragm muscle, C. Miller (1974) found that several alcohols, diols, and monosaccharides exhibit distribution ratios within 2% of unity, whereas certain other sugars appear to be excluded from membrane-bounded intracellular compartments. The amount of solute that dissolves at equilibrium is also nearly normal in water that is constrained in **gels**, which are cross-linked networks of fibrous macromolecules. The diffusion of solutes within gels, however, may be hindered by collisions and interactions with the macromolecules, and by the tortuosity of the diffusion paths.

Viscosity, another property of water, contributes to the resistance to flow. A fluid with a greater viscosity exhibits less flow under the influence of a given pressure gradient than a fluid with a lesser viscosity. For example, molasses, a concentrated sugar solution, is much more viscous than pure water. During **laminar flow**, a frictional force develops between adjacent layers (laminae) in the fluid, and this force impedes the sliding of one lamina past its neighbor. In a **Newtonian fluid**, the frictional force per unit area, or shear stress (τ, dyn/cm^2), is proportional to the velocity gradient, or rate of strain (dv/dy, s^{-1}) between laminae,

$$\tau = \eta \cdot (dv/dy)$$

where the viscosity (η) is the proportionality constant with units of poise (= 1 dyn·s/cm^2), named after Poiseuille. The viscosity of H$_2$O at 20.3 °C is 0.01 poise, or 1 centipoise (cp). In cultured fibroblasts, the fluid-phase cytoplasmic viscosity, as determined from rotational motions of fluorescent probes on a picosecond time scale, is only 1.2–1.4 times that of pure water (Fushimi and Verkman, 1991). Fluorescence studies also show that the viscosity is the same in the cytoplasm and nucleoplasm, and is unaffected by large decreases in cell volume or by disruption of the cytoskeleton with cytochalasin B. The fluid-phase viscosity, as determined from fluorophore rotational motions, is not nearly as affected by macromolecules as is the bulk viscosity, and it thus provides a more accurate view of the physical state of the aqueous domain of the cytoplasm. These and other studies (e.g. Schwan

and Foster, 1977; Horowitz and Miller, 1984) set limits on the extent to which the physicochemical properties of intracellular water differ from those of extracellular water.

III. Interactions Between Water and Ions

Ions in solution behave as charged, hard spheres that interact with and orient water dipoles. When crystals of sodium chloride are dissolved in water, the electrostatic attractive forces between water dipoles and ions in the crystal lattice overcome the interionic attractive forces between oppositely charged ions in the crystal. The dissociated ions then acquire the freedom of translational motion as they diffuse into the solution accompanied by a layer of tightly associated **hydration water** (Fig. 2). **Ion** is the Greek word for "wanderer." The strength of the attraction between ions and water dipoles depends to a large extent on the ionic charge and radius. For the alkali metal cations, the force of attraction, and the energy of interaction of ions with water, decreases according to the following series:

$$Li^+ > Na^+ > K^+ > Rb^+ > Cs^+$$

Li$^+$, being the smallest of the alkali metal cations, has the strongest interaction with water because its positively charged nucleus can approach most closely to the negative side of neighboring water dipoles. As the ionic radius increases with increasing atomic number in the alkali metal series, the filled outer shells of electrons effectively shield the cationic charge and reduce the distance of closest approach to water molecules. The smallest ion thus acquires the greatest degree of hydration and has the largest hydrated ionic radius.

According to Frank and Wen (1957), the orienting influence of ions on water dipoles results in three regions of water structure (Fig. 3). The electric field of an ion is sufficiently strong to remove water dipoles from the bulk water clusters and to attract to itself the oppositely charged ends of from one to five water dipoles. A certain number of these water dipoles, called the **hydration number**, then become

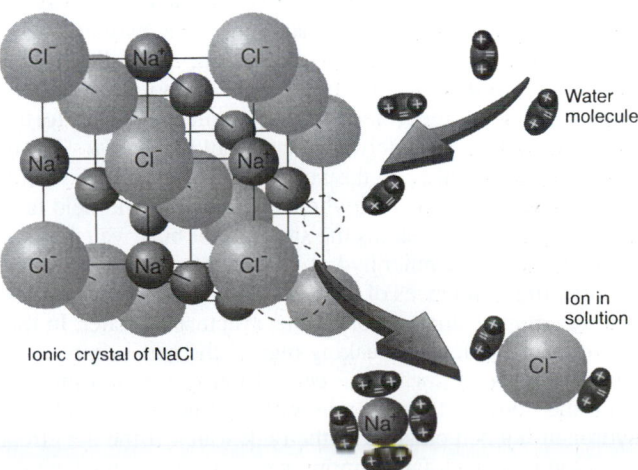

FIGURE 2. Dissociation of NaCl in water into hydrated Na$^+$ and Cl$^-$ ions.

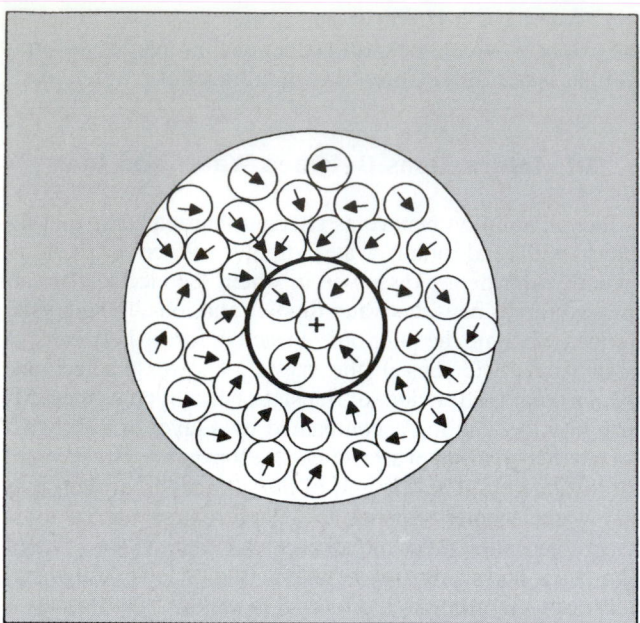

FIGURE 3. Three regions of water structure near an ion. A central cation is shown with a primary hydration shell of four oriented water dipoles, surrounded by a partially oriented secondary sheath. The gray region indicates the hydrogen-bonded structure of the bulk water.

TABLE 1 Radii, Enthalpies of Hydration, and Mobilities of Selected Ions

Ion	Nonhydrated radius[a] (Å)	$\Delta H^0_{\text{hydration}}$[b] (kcal/mol)	mobility[c] $10^{-4}\frac{\text{cm/s}}{\text{V/cm}}$
H^+	—	−269	36.25
Li^+	0.60	−131	4.01
Na^+	0.95	−105	5.19
K^+	1.33	−85	7.62
Rb^+	1.48	−79	8.06
Cs^+	1.69	−71	8.01
Mg^{2+}	0.65	−476	2.75
Ca^{2+}	0.99	−397	3.08
Sr^{2+}	1.13	−362	3.08
Ba^{2+}	1.35	−328	3.30
Cl^-	1.81	−82	7.92

[a] Radii are from Pauling (1960).
[b] Standard enthalpies of hydration at 25 °C are from Edsall and McKenzie (1978).
[c] Mobilities in water at 25 °C are from Hille (1992, p. 268).

trapped and oriented in the ion's electric field. The inner hydration shell includes water molecules that are aligned by the force field and in direct contact with the ion, 5 ± 1 for Li^+, 4 ± 1 for Na^+, 3 ± 2 for K^+ and Rb^+, 4 ± 1 for F^-, 2 ± 1 for Cl^- and Br^-, and 1 ± 1 for I^-. Thus in physiological saline at 0.15 M NaCl in water (55.5 M), about 1.6% (= 100 × 0.15 × 6/55.5) of the water is located in the inner hydration shells of Na^+ and of Cl^-. Water in the inner hydration shells of Na^+, K^+, and Ca^{2+} rapidly exchanges with bulk water on a nanosecond time scale, but in contrast, the smaller divalent cation Mg^{2+}, with its high charge density, is some four orders of magnitude slower in exchanging its inner hydration water. The inner hydration sheath of water molecules moves together with the ion as a distinct and single kinetic entity.

The mobilities of the alkali cations in water decrease as the nonhydrated ionic radius decreases, but as the hydrated ionic radius increases (Table 1; for discussion, see Hille, 1992). The ionic **mobility** (u) is defined as the proportionality constant that relates the velocity (v) of ionic migration to the force exerted by an external electric field E ($v = u \cdot E$); in other words, mobility is the velocity per unit electric field. Farther away from the ion, where the ion's electric field falls towards zero, water retains the structure of bulk water. In the region between the inner hydration sheath and the bulk water, the orienting influences of the ion and the bulk water network tend to compete and to disrupt the structure of water. In this intermediate structure-breaking region, the water dipoles are partially oriented toward the central ion, yet do not migrate with the ion, and, they join only infrequently with the hydrogen-bonded clusters of the bulk water. If the net effect of an ion is to disorganize more water in the intermediate region than is found in the primary hydration shell, then the ion is termed a *structure-breaker*. Conversely, water may form a

variety of hydrogen-bonded clathrate structures around apolar protein sidechains, whose action resembles that of the class of solutes termed *structure-makers* (see Klotz, 1970, for review).

The **enthalpy of hydration** of an ion is a measure of the strength of the interaction between ions and water, and is defined as the increase in enthalpy when one mole of free ion in a vacuum is dissolved in a large quantity of water. The enthalpy of hydration may be estimated from the heat released upon dissolving salts in water, usually less than 10 kcal/mol, taking into account the energy needed to dissociate the salt crystal and then to hydrate the ions. The enthalpy of hydration may be calculated by using either the Born charging method, which estimates the energy needed to transfer a rigid charged sphere into a structureless continuum, or, alternatively, by the Bernal-Fowler structural method, which takes into account the dipolar structure of water (see Bockris and Reddy, 1970). More accurate results are obtained if water is considered to be an electric quadrupole with four centers of charge, two partial positive charges near the hydrogen nuclei and two partial negative charges on the nonbonded electron orbitals near the oxygen nucleus. A further improvement in the calculated enthalpy of hydration is obtained when the polarizability (α) of the water dipole by the ion is taken into account. The electric field of the ion distorts the electron cloud of the hydration water along its permanent dipole axis, thus inducing an additional increment of charge separation and increasing the dipole moment. For small fields, this **induced dipole moment** (μ_{ind}) is proportional to the electric field strength (E), and the constant of proportionality is the **polarizability** ($\mu_{\text{ind}} = \alpha \cdot E$). In Fig. 4, the measured enthalpies of hydration for alkali metal cations and halides are compared with

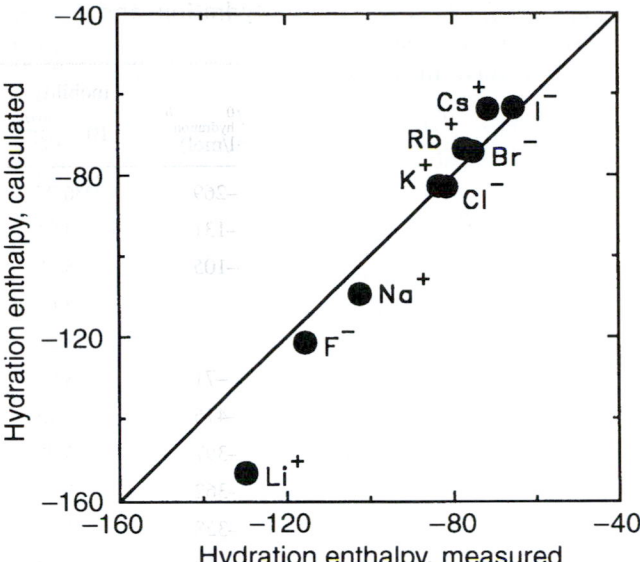

FIGURE 4. Comparison of measured enthalpies of hydration of alkali metal cations and halides in water (abscissa), with those calculated (ordinate) taking into account the Born charging energy, ion-dipole interactions, ion-quadrupole interactions, and ion-induced dipole interactions (data from Bockris and Reddy, 1970, p. 107).

calculated values; the impressive agreement demonstrates the primary importance of electrostatic forces in determining the solvation of ions in water. The enthalpy of hydration for the smallest alkali metal cation Li^+ is quite large at -131 kcal/mol. With increasing ionic radius, the enthalpy falls progressively for Na^+, K^+, and Rb^+, reaching -71 kcal/mol for Cs^+, the largest nonhydrated but smallest hydrated ion in the series (see Table 1). For the divalent, alkaline earth metal ions Mg^{2+} and Ca^{2+}, the enthalpies of hydration also follow the ionic radius but are considerably larger at -476 kcal/mol and -397 kcal/mol, respectively. As pointed out by Hille (1992), the magnitude of ionic hydration enthalpies approximates the cohesive strength of the ionic bonds in a crystalline salt lattice.

In 1929, the physical chemist Peter Debye explained the dielectric constant, a macroscopic property of water, in terms of its molecular properties—its dipole moment and polarizability. In a gas consisting of dipoles at concentration n, the macroscopic dielectric constant (ϵ) is related to the polarizability (α) and the molecular dipole moment (μ) at temperature T by the Debye equation:

$$\epsilon - 1 = 4\pi n\alpha + \frac{4\pi n\mu^2}{3kT}$$

where k is Boltzmann's constant (1.38×10^{-23} J/K). In this equation, the first term represents the effect of the polarizability (α) of the electron clouds, and the second term represents the effect of reorienting the permanent dipoles, with moments μ, in an electric field, as opposed by the randomizing influence of thermal energy (kT). Thus, increasing the temperature reduces the dielectric constant by disorienting the dipoles. In a condensed polar dielectric such as water, the Kirkwood equation for the dielectric constant takes into

account the formation of molecular groups with g nearest-neighbors linked to the central molecule. These groups, which could be a tetrahedral group of water molecules, orient as a unit in the specific local electric fields, as distinct from an externally applied field. The Kirkwood equation is

$$\frac{(\epsilon - 1)(2\epsilon + 1)}{9\epsilon} = \frac{4\pi n}{3}\left[\alpha + \frac{\mu^2\left(1 + g\,\overline{\cos\gamma}\right)^2}{3kT}\right]$$

where $\overline{\cos\gamma}$ is the average of the cosines of the angles between the dipole moment of the central water molecule and those of its bonded neighbors. In small spaces that restrict the formation of tetrahedral water clusters in such a way that g is reduced, the Kirkwood equation predicts that the dielectric constant will also be reduced, and electrostatic forces between ions will then be correspondingly increased. In the primary hydration shell of ions, the oriented water is polarizable but cannot be reoriented by applied fields, and so the dielectric constant is reduced from its bulk value of 78 to a value of about 6, with intermediate values in the partially oriented structure-breaking region between the primary hydration shell and the bulk water. For this reason, the dielectric constant of a solution decreases with increasing salt concentration, but in protein-free solutions at physiological salt concentrations, the extent of the decrease is small due to the small fraction of hydration water.

IV. Protons in Solution

The hydrogen-bonded structure of liquid water also contributes to high proton mobility by a mechanism that differs fundamentally from the migration of other hydrated ions. The proton is a highly reactive, positively charged hydrogen nucleus, devoid of electrons. Whereas ions with electron shells typically have diameters in angstroms, the diameter of the proton is only about 10^{-5} Å. With such a small size, the proton has a strong attraction to electrons, as indicated by the high **ionization energy** of 323 kcal/mol needed to remove an electron from a hydrogen atom to form a proton. By comparison, the ionization energies for the alkali cations decrease with increasing atomic number in the series as follows:

Li^+(124 kcal/mol) $> Na^+$ (118 kcal/mol) $> K^+$ (100 kcal/mol) $> Rb^+$ (96 kcal/mol) $> Cs^+$ (90 kcal/mol)

As more filled electron shells separate and shield the outer shell from the positively charged atomic nucleus, it becomes easier to form a cation by removing an electron. The high affinity of the proton for electrons, and its small size, explain its tendency to form hydrogen bonds with the unshared electrons of oxygen in water. Infrared spectra indicate that protons in solution exist predominantly in the form of **hydronium** ions H_3O^+. Nuclear magnetic resonance (NMR) data are consistent with a flattened trigonal pyramidal structure, with O–H bond lengths of 1.02 Å and H–O–H bond angles of 115°, a structure resembling that of NH_3 but with a radius similar to that of K^+. The free energy of formation of H_3O^+ from H_2O and H^+ is about -170 kcal/mol, which corresponds to a hypothetical concentration of free protons in solution at room temperature of about 10^{-150} M—"about as zero as one can get" say Bockris and Reddy (1970). The heat of hydration of H_3O^+ is an additional −90 kcal/mol. The change in the density of water with temperature is consistent with the hydration of H_3O^+ with an additional three water molecules, forming the tetrahedral cluster $H_9O_4^+$ (Bockris and Reddy, 1970). The proton mobility in water is 36×10^{-4} cm²/s · V, much slower than expected on the basis of hydrodynamic theory, if free protons were to carry the current, but curiously about seven times as fast as expected if H_3O^+ migrates like K^+. The abnormally high proton mobility in water (and also in ice) is consistent with a **Grotthus chain mechanism** (Fig. 5) in which the successive breakage and formation of hydrogen bonds, accompanied by proton jumps between neighboring water molecules, effectively results in the passage of H_3O^+ along a chain. The transport process is rate-limited by the time needed for each successive acceptor water molecule to rotate and reorient its nonbonded orbital into a suitable position to accept the donated proton. The Grotthus mechanism accounts for about 80% of proton mobility, with the remaining 20% due to H_3O^+ itself, as a single kinetic entity, undergoing a translational migratory movement through the solvent like other ions.

V. Interactions Between Ions

Interactions between ions may be weak or highly selective. Ions in solution attract ions of the opposite charge, in accordance with Coulomb's law. Since the long-range attractive forces are inversely proportional to the square of the distance between charges, the interaction energies are greater in more concentrated solutions. In a uni-univalent salt solution, defined as having monovalent cations and anions, of concentration c (mol/L), the average distance (d, angstroms or Å) between any two ions is given by

$$d = \left(10^{27}/2N_A c\right)^{1/3}$$

where N_A is Avogadro's number (6.023×10^{23}). The factor 2 is included because the anions and cations are both counted, and the factor 10^{27} is Å³/L. For 0.15 M NaCl, the average distance between Na^+ and the nearest Cl^- is 17.7 Å, at which range coulombic forces are highly significant. At pH 7.4, where the concentration of "protons" (really hydronium ions) is only 40 nM, the average distance between "protons" is 0.28 μm.

Attractive ion-ion interactions are stabilizing, and they lower the chemical potential of an ion from its value in an ideal, infinitely-dilute solution. The **activity** (a) of an ion is defined in terms of its **chemical potential** (μ) as follows:

$$\mu = \mu^0(T,P) + RT \ln a$$

where $\mu^0(T,P)$ is the **standard-state chemical potential**, or the chemical potential when the activity is 1. Increases in

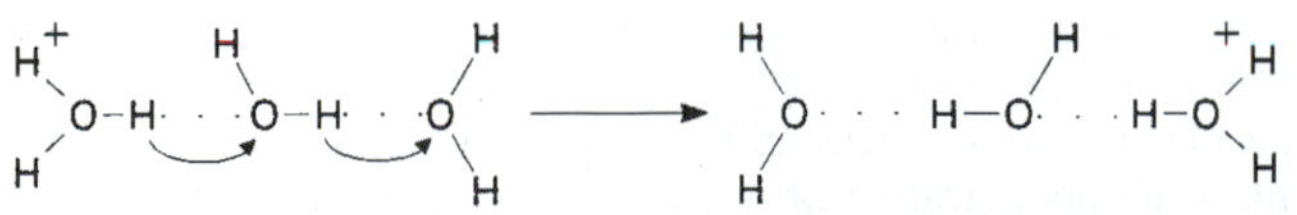

chemical activity increase the chemical potential. The chemical potential of a solute in an ideal dilute solution, where the activity equals the chemical concentration (c), is

$$\mu_{\text{ideal}} = \mu^0(T,P) + RT \ln c$$

In order to describe the properties of more concentrated nonideal solutions, G. N. Lewis introduced the **activity coefficient** (γ), such that the activity is the product of the concentration (c) and the activity coefficient ($a = \gamma \cdot c$). The chemical potential of an ion in a nonideal solution is

$$\mu = \mu^0(T,P) + RT \ln c + RT \ln \gamma$$

where the term $RT \ln \gamma$ includes the effect of ion-ion interactions on the chemical potential.

When a salt ($c_+ c_-$) is added to water, the cations and anions, at concentrations c_+ and c_- respectively, both contribute to the free energy of the solution. The chemical potentials of the cations and anions (μ_+ and μ_-, respectively) are given by

$$\mu_+ = \mu_+^0 + RT \ln c_+ + RT \ln \gamma_+$$
$$\mu_- = \mu_-^0 + RT \ln c_- + RT \ln \gamma_-$$

Adding these two expressions, and taking the average, gives

$$\frac{\mu_+ + \mu_-}{2} = \frac{\mu_+^0 + \mu_-^0}{2} + RT \ln\left(c_+ c_-\right)^{1/2} + RT \ln\left(\gamma_+ \gamma_-\right)^{1/2}$$

The **mean ionic activity coefficient** $\gamma_\pm$ is defined as

$$\gamma_\pm = \left(\gamma_+ \gamma_-\right)^{1/2}$$

Mean ionic activity coefficients of salts have been estimated by the **Debye-Hückel theory**, which supposes that central ions attract oppositely-charged ions (or **counterions**) in a diffuse and structureless ion cloud (or **atmosphere**). The forces of coulombic attraction between the central ion and its cloud of counterions are opposed by the randomizing influence of the thermal motion of the ions. In dilute solutions, in which the theory treats central ions as point charges relative to the size of the ion cloud, the mean ionic activity coefficient ($\gamma_\pm$) for salt ions with charges z_+ and z_-, is given by the limiting law as follows:

$$\log \gamma_\pm = -A\left|z_+ z_-\right| I^{1/2}$$

where I is the **ionic strength**, defined as

$$I = \frac{1}{2}\sum_i c_i z_i^2$$

The constant A, which equals 0.5108 kg$^{1/2}$ · mol$^{-1/2}$ is given in terms of B, which equals 0.3287×10^8 kg$^{1/2}$ · mol$^{-1/2}$ · cm^{-1}, both at 25 °C, as follows:

$$A = \frac{1}{2.303} \frac{N_A e_0^2}{2\epsilon RT} B$$

$$B = \sqrt{\frac{8\pi N_A e_0^2}{1000\epsilon kT}}$$

where N_A is Avogadro's number (6.023×10^{23} ions/mol), e_0 is the charge on an electron (4.80×10^{-10} statcoul $= 1.60 \times 10^{-19}$ C), ϵ is the dielectric constant, R is the gas

FIGURE 5. Grotthus chain mechanism for proton mobility in water.

constant (8.32 J/mol·K), T is the absolute temperature (K), and $k (= R/N_A)$ is Boltzmann's constant (1.38×10^{-23} J/K).

The derivation of the Debye-Hückel limiting law (see Bockris and Reddy, 1970) begins with Gauss's law, one of the four fundamental Maxwell equations of electromagnetic theory, and computes the spherically symmetric electric field around an ion, leading to Poisson's differential equation relating the charge density (ρ_r) of the ionic cloud to the electrostatic potential (ψ_r) at a distance r from the central point charge. The counterions, at concentration n_i for ionic species i at a distance r from the central ion, are considered to distribute in the electric field according to a **Boltzmann distribution**

$$n_i = n_i^0 e^{-z_i e_0 \psi_r / kT}$$

where n_i^0 is the bulk ion concentration (ions/L). By linearizing the Boltzmann equation, using a Taylor series expansion and retaining only the first two terms, which assumes that $z_i e_0 \psi_r \ll kT$, the linearized and integrated Poisson-Boltzmann equation shows how the electrostatic potential ψ decreases as a function of distance (r) from the central ion (Fig. 6A):

$$\psi_r = \frac{z_i e_0}{\epsilon} \frac{e^{-\kappa r}}{r}$$

where κ is given by

$$\kappa = \sqrt{\frac{4\pi}{\epsilon kT} \sum_i n_i^0 z_i^2 e_0^2}$$

The charge density (ρ_r) of the ionic cloud also decreases with increasing distance from the central ion (Fig. 6B)

according to

$$\rho_r = -\frac{z_i e_0}{4\pi} \kappa^2 \frac{e^{-\kappa r}}{r}$$

The amount of charge dq contained in a concentric spherical shell of thickness dr located at a distance r from the central ion (Fig. 6C) is given by

$$dq = -z_i e_0 e^{-\kappa r} \kappa^2 r \, dr$$

Furthermore, the maximum amount of charge contained in such a spherical shell occurs at a distance κ^{-1}, known as the **Debye length**, which defines the effective radius of the counterion atmosphere. It can also be shown that the Debye length is the distance away from the central ion where an ion of equal and opposite charge would contribute to the electrostatic field by an amount equivalent to that of the dispersed cloud of counterions (Bockris and Reddy, 1970). The Debye length decreases as the ionic concentration increases (Fig. 6D).

The Debye-Hückel limiting law predicts activity coefficients accurately only to concentrations of about 0.01 M. In more concentrated solutions, the finite radius (a, in cm) of the ion is taken into account in the **extended Debye-Hückel equation**, given by

$$\log \gamma_\pm = -\frac{A |z_+ z_-| I^{1/2}}{1 + Ba \, I^{1/2}}$$

where the constants A and B are still 0.5108 kg$^{1/2}$·mol$^{-1/2}$ and 0.3287×10^8 kg$^{1/2}$·mol$^{-1/2}$·cm^{-1}, respectively, for water at 25 °C. The ion-size parameter (a) is adjustable, but reasonable values extend the range of concentrations for which the Debye-Hückel theory accurately predicts ionic activity coefficients (Fig. 7).

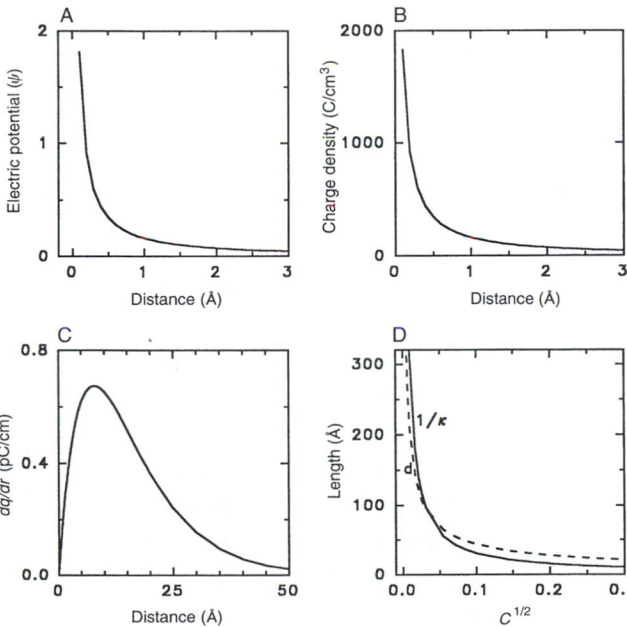

FIGURE 6. Debye-Hückel theory. (A) Electrostatic potential (ψ), (B) charge density (ρ) of the counterion cloud, and (C) the charge (dq) enclosed in a spherical shell of thickness (dr), each calculated and plotted versus distance d from the central ion. (D) Debye length κ^{-1} (solid line) and average distance d between ions (dashed line) versus square root of ion concentration.

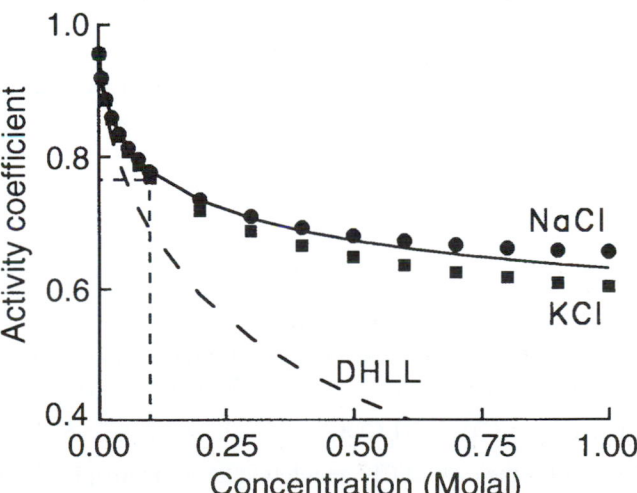

FIGURE 7. Activity coefficients of NaCl (circles) and KCl (squares). Data are from Robinson and Stokes (1959). The dashed curve represents the prediction of the Debye-Hückel limiting law. The solid line represents the prediction of the extended Debye-Hückel equation with an ion size parameter a of 4.7 Å. The dashed vertical line indicates that at 0.1 M, the activity coefficients of NaCl and KCl are 0.77 and 0.76, respectively.

At still higher salt concentrations above about 0.7 M, the activity coefficient stops decreasing and instead begins to increase with increasing salt concentration. This behavior has been explained by the increasing fraction of hydration water, which both reduces the amount of free water in the solution and raises the effective concentration, and therefore the activity, of the dissolved ions. Corrections for hydration enable prediction of the activity coefficient as a function of salt concentration over the full range of salt concentrations reaching to several molar.

The activity coefficients calculated by the Debye-Hückel theory indicate that the forces between ions in dilute solutions are weak and nonselective. The predicted activity coefficients depend on the ionic charge and the ionic strength of the solution, and they are largely independent of the specific ion within, say, the alkali metal series. A very small degree of selectivity is introduced with the ion size parameter. The mean ionic activity coefficients for NaCl and KCl as a function of salt concentration are shown in Fig. 7. At 0.1 M concentrations, the activity coefficient of Na^+ is 0.77, whereas those of K^+ and Cl^- are both 0.76. These relatively high activity coefficients correspond to ion-ion interaction energies of only about -0.3 kcal/mol (see Hille, 1992).

Agreement between theory and data, such as seen in Fig. 7 for the Debye-Hückel theory, does not necessarily imply the correctness of the theory, or of the model on which the theory is based. A fundamental criticism of the Debye-Hückel theory concerns the smeared charge model for the cloud of counterions (Frank and Thompson, 1960). Below a concentration of about 0.001 M, the Debye length κ^{-1} is greater than the average distance between ions in the solution (Fig. 6D), in which case the counterions could reasonably appear as a cloud of smeared charge from the vantage point of the central ion. Above 0.001 M, however, the Debye length incongruously falls below the average distance between ions in the solution. Moreover, at only 0.01 M, just one ion is needed to account for 50% of the effect of the counterion cloud on the central ion, yet this ion must be smeared around the central ion in a cloud of spherically symmetric charge distribution located only 25 Å away from the central ion (Bockris and Reddy, 1970). Thus, well below physiological salt concentrations, the assumption of smeared charge in the Debye-Hückel theory is not in accord with the coarse-grained structure of salt solutions, despite the impressive ability of the theory to predict activity coefficients. A quasi-lattice approach to salt solutions minimizing the free energy using modern computational power would seem to be a preferable alternative (see Horvath, 1985, for references).

Specific effects of small anions on the solubility, aggregation, or denaturation of proteins often follow the **Hofmeister (lyotropic) series** in the following order of effectiveness:

$$F^- < PO_4^{3-} < SO_4^{2-} < CH_3COO^- < Cl^- < Br^- < NO_3^- < I^- < SCN^-$$

The ions to the right of Cl^- are referred to as **chaotropic**, since they tend to destabilize proteins (for review, see Collins and Washabaugh, 1985). Ions may also associate with sites on proteins, as exemplified by the binding to albumin of Ca^{2+} (Katz and Klotz, 1953), and of Cl^- (Scatchard, Coleman and Shen, 1957; see ch. 8 in Tanford, 1961). Certain enzymes, ion channels, and membrane transport proteins interact with the alkali cations with a high degree of **ionic selectivity**. Considering the

five alkali metal cations, there are 5! ($= 120$) possible orders of selectivity that might arise. However, only 11 cationic selectivity orders commonly occur in chemical and biological systems, as listed in Table 2. G. Eisenman (1967) predicted these selectivity orders by calculating the ion-site interaction energies. If the field strength of a negatively-charged site is weak, then an associated ion remains hydrated. In this case, Cs^+, having the smallest hydrated ionic radius, is favored (Sequence I) because it can approach most closely to the site, and thus has the strongest coulombic force of attraction. If the field strength of the site is strong, and the interaction energy between the ion and the site is stronger than the energy between the ion and water dipoles, then the ion loses its associated water and becomes dehydrated. In this case, Li^+, having the smallest nonhydrated ionic radius, becomes favored (sequence XI). With intermediate field strengths, ions partially dehydrate, giving rise to the intervening selectivity sequences. Other sequences are possible when the sites are assumed to be polarizable (for review, see Eisenman and Horn, 1983).

Some ion-selective membrane channels are sufficiently narrow that ions and water permeate by a process of **single-file diffusion** in which water and ions cannot pass each other, at least in the narrowest part of the channel (for review, see Finkelstein and Andersen, 1981, and Chapter 12). In narrow channels, where the diameter of the permeating ion may approximate that of the channel itself, the hydration water is stripped from the ion and is probably replaced by dipolar groups of the proteins lining the channel walls, thus providing electrostatic stabilization of the permeating ion.

VI. Cell Cations

The predominant biological cations are potassium and sodium (see Ussing, 1960, for review), calcium (see Campbell, 1983, for review) and magnesium. In excitable cells such as

TABLE 2 Eisenman's Selectivity Sequences for Binding of Alkali Cations to Negatively Charged Sites

Highest field strength of site	
Li > Na > K > Rb > Cs	Sequence XI
Na > Li > K > Rb > Cs	Sequence X
Na > K > Li > Rb > Cs	Sequence IX
Na > K > Rb > Li > Cs	Sequence VIII
Na > K > Rb > Cs > Li	Sequence VII
K > Na > Rb > Cs > Li	Sequence VI
K > Rb > Na > Cs > Li	Sequence V
K > Rb > Cs > Na > Li	Sequence IV
Rb > K > Cs > Na > Li	Sequence III
Rb > Cs > K > Na > Li	Sequence II
Cs > Rb > K > Na > Li	Sequence I
Lowest field strength of site	

nerve and muscle, the inwardly-directed concentration gradient of Na^+, and the outwardly-directed gradient of K^+, both created by the Na^+,K^+-ATPase (see Chapter 17), are reduced by a very small extent during each action potential, during which Na^+ and K^+ move down their concentration gradients through voltage-gated ion channels (see Chapters 14 and 25). K^+ also functions as a specific cofactor for the glycolytic enzyme pyruvate kinase. The maximal catalytic velocity follows Eisenman sequence IV: $K^+ > Rb^+ >> Cs^+ \geq Na^+ > Li^+$ (Kayne, 1973). During *in vitro* protein synthesis, K^+ maintains an active conformation of 50S ribosomal subunits that catalyze peptide bond formation. The order of selectivity for this effect is Eisenman sequence III: $NH_4^+ \geq Rb^+ > K^+ > Cs^+$; Na^+ and Li^+ are ineffective (Miskin, Zamir and Elson, 1970).

The widespread importance of calcium (Ca^{2+}) in cellular physiology was first emphasized by L. V. Heilbrunn (see e.g., *The Dynamics of Living Protoplasm,* 1956). Injections of various salts into frog skeletal muscle fibers revealed that Ca^{2+} is the only intracellular ion that induces muscle contraction (see Chapters 55 and 56). Normally, release of Ca^{2+} from the sarcoplasmic reticulum (SR) initiates muscle contraction. During relaxation, Ca^{2+} is pumped out of the sarcoplasm back into the SR, and also across the plasma membrane into the extracellular solution by Ca^{2+}-ATPases (see Chapter 18). Using "skinned muscle fibers," in which the sarcolemma has been removed with fine needles, J. Gulati and R. J. Podolski (1978) and others have studied how the tension exerted by myofilaments is directly proportional to the concentration of free Ca^{2+} in the sarcoplasm. Pumping of Ca^{2+} from the cytoplasm of muscle, red cells, and other cells across the plasma membrane normally results in a submicromolar steady-state intracellular concentration of Ca^{2+}, whereas plasma Ca^{2+} is around 2.5 mM. The ratio of extracellular to intracellular Ca^{2+} concentrations is thus more than 1000, much greater than the ratio of around 25 for Na^+. In human mammary glands, Ca^{2+} is secreted by exocytosis via the Golgi system into milk to a total concentration of around 10 mM (Neville *et al.,* 1983).

The classic experiments of S. Ringer (1882) established that plasma Ca^{2+} is essential for the sustained beating of isolated hearts. An inward Ca^{2+} current down its concentration and electrical gradients through an ion channel constitutes a major component of the cardiac action potential (see Chapter 52). Transient elevations of intracellular Ca^{2+} also occur in such processes as fertilization, cell division, exocytosis during neurotransmitter release, activation of platelets, neutrophils, and lymphocytes, and the hormonal activation of cells. Ca^{2+} is an essential cofactor in blood clotting and in the activation of complement, and also has a structural role in membranes and in the mineralization of bone, teeth, and other skeletal structures.

In multicellular animals, the concentration of magnesium ion in both the extracellular and intracellular fluids is around 1 mM. Consequently, unlike Na^+, K^+, and Ca^{2+}, Mg^{2+} does not act as a carrier of ionic current across cell membranes or as a trigger for the initiation of cellular activities. Mg^{2+} ion, with a divalent positive charge and a nonhydrated diameter of 1.3 Å (see Table 1), has the highest charge density of all of the ions found in cells. Consequently, Mg^{2+} binds readily to anionic sites, particularly to polyphosphates such as ATP. MgATP is the substrate for all kinases and for some phos-

phatases such as the Na^+,K^+-ATPase and the Ca^{2+}-ATPase. Mg^{2+} is also a cofactor for the glycolytic enzyme enolase, for glutamine synthetase, and for other enzymes. Mg^{2+} is also tightly bound to porphyrin in chlorophyll, the primary biological molecule for capturing light energy (see Chapter 64).

VII. Cell Anions

The most abundant permeant physiological anion is chloride, whose concentration is typically less within cells than in the extracellular solution. Macromolecular cellular constituents including nucleic acids and most cytoplasmic proteins carry a net negative charge at physiological pH. Phosphorylated metabolic intermediates and organic acids are also negatively charged. The **condition of macroscopic electroneutrality** requires that the total charge-weighted concentrations of cations and anions be equal in each cellular compartment. However, a very small amount of charge separation occurs across membranes, giving rise to the **transmembrane electrical potential.** The intracellular cationic charge, primarily due to Na^+ plus K^+, is neutralized by intracellular protein, nucleic acids, organic anions, and by Cl^-, bicarbonate (HCO_3^-), phosphate, and sulfate. HCO_3^- and phosphate, along with proteins, are the principal buffers that regulate intracellular and extracellular pH (see Chapter 22). Whereas Cl^- functions primarily to maintain bulk electroneutrality, some cells also use Cl^- to carry ionic current. Halobacteria, for example, possess a light-driven Cl^- pump (Oesterhelt and Tittor, 1989).

VIII. Trace Elements

In addition to the predominant electrolytes found in the intracellular and extracellular solutions, other ions are tightly associated with certain proteins and enzymes, known as **metalloproteins.** The transport of oxygen (O_2) to all of the cells of the organism by red blood cells depends on the reversible association of O_2 with ferrous ions (Fe^{2+}) in the porphyrin groups of intracellular **hemoglobin.** Heme is an iron porphyrin whereas chlorophyll is a magnesium porphyrin. Binding of O_2 to Fe^{2+} in the muscle heme protein **myoglobin** provides a reservoir of O_2 for muscular work. **Catalase** is another Fe-heme protein which protects cells by converting hydrogen peroxide into water and oxygen. Iron is stored in cells in the cavity of the protein **ferritin,** which can accommodate as many as 4500 ferric (Fe^{3+}) ions per molecule. Iron (Fe^{2+}, Fe^{3+}) and copper (Cu^+, Cu^{2+}) are essential cofactors in **cytochromes** in the respiratory chain of mitochondria, and in the photosystems of chloroplasts. Manganese ions are essential in Photosystem II in plant cells. Selenium is a cofactor in **glutathione peroxidase,** an enzyme that reduces hydrogen peroxide and organic peroxides, and helps to prevent oxidative damage in cells. Zinc (Zn^{2+}) and copper are cofactors in **superoxide dismutase,** an enzyme that scavenges superoxide anion (O_2^-) and helps to prevent oxidative damage from toxic free radicals. Zn^{2+} also functions as a cofactor in some proteolytic digestive enzymes, the zinc proteases, and stabilizes structures known as "zinc fingers" in certain DNA-binding proteins. A notable example of the binding of an ion to a non-protein compound is that of tervalent cobalt (Co^{3+}) in vitamin B_{12} and the B_{12} coenzymes.

IX. Solute Transport: Basic Definitions

In the intracellular or extracellular solutions, a solute is said to be at **equilibrium** when its concentration is constant in time without requiring the continuous input of energy from metabolism or other sources. In human red blood cells, for example, Cl^-, HCO_3^-, and H^+ are at thermodynamic equilibrium, i.e. they are passively distributed. A solute is said to be at **steady-state** when its concentration is constant in time, but is dependent on the continuous input of energy from metabolism or other sources. **Active transport** utilizes energy and results in steady-state distributions that represent a deviation from equilibrium. In biological cells, the concentrations of Na^+, K^+, and Ca^{2+} are at a steady-state. Their concentrations are constant in time, but are dependent on the continuous hydrolysis of ATP by the Na^+-K^+ pump (for review, see Glynn, 1985, and Chapter 17) and the Ca^{2+} pump (see Chapter 18). With isolated red cell membranes ("ghosts"), active transport of Na^+ and K^+ against their respective concentration gradients has been observed directly by measurements of net fluxes (Freedman, 1976). The active production and maintenance of ion concentration gradients by membrane pumps, such as the Na^+,K^+-ATPase and the Ca^{2+}-ATPase, represent **chemical work** and **electrical work** done by the cell. The energy required for continuous pumping of Na^+ by resting frog sartorius muscles has been estimated to represent some 14–20% of the energy available from the hydrolysis of ATP, and the value is similar in red blood cells.

Passive transport is the movement of solutes toward a state of equilibrium. Passive transport of **hydrophobic** substances across cell membranes usually occurs directly by diffusion across the lipid bilayer. Passive transport of **hydrophilic** substances is usually mediated by specific membrane proteins via a process known as **facilitated diffusion.** Passive ion transport may occur through pores or ion channels under the influence of concentration gradients and electrical forces by a process known as **electrodiffusion.** The passive flow of ionic currents down their concentration gradients through specific ion channel proteins constitutes negative work done by the cell. The changes in free energy associated with ion transport and the Nernst equilibrium equation are described in the Appendix.

X. Measurement of Electrolytes and Membrane Potential

Ion concentrations in biological fluids may be expressed as millimoles per liter of solution (millimolar, or mM), or as millimoles per kilogram of water (millimolal). When a solution containing metallic ions is aspirated into a flame, each type of ion burns with a characteristic color, Na^+ giving a yellow flame, K^+ giving a violet flame, and Ca^{2+} giving a red flame. In the technique of **flame photometry,** the inten-

sity of the emitted light, in comparison with that produced by solutions containing known concentrations of ions, provides a convenient measure of ion concentration in extracellular fluids and in acid extracts of cells. With a uniform rate of aspiration, a flame photometer accurately measures the intensity of the emitted light, which is related linearly to the cation concentrations in suitably diluted standards and unknowns (see e.g., Funder and Wieth, 1966). **Atomic absorption spectroscopy** is an alternative technique that measures the light absorbed by ions during electronic excitation in a flame. Flame photometry and atomic absorption spectroscopy both measure total ionic concentrations in cell extracts irrespective of any intracellular compartmentation, and they are sensitive in the millimolar range of cellular concentrations. K^+, Na^+, Ca^{2+}, and other elements in single cells, or even in single cell organelles, may be measured by **electron probe microanalysis,** a technique that utilizes an electron beam to excite the emission of x-rays with energies characteristic of the various elements in cells.

The development of **ion-specific glass microelectrodes** by G. Eisenman (1967), and then of selective **liquid ion-exchange microelectrodes,** made possible the direct determination of intracellular cation activities. L. G. Palmer and M. M. Civan (1977) found that for Na^+, K^+, and Cl^- of *Chironomus* salivary gland cells, the ion activities are the same in the nuclear and cytoplasmic compartments. In a related study, Palmer, *et al.* (1978) found that during development of frog oocytes, the ratio of the cytoplasmic concentration of Na^+ to K^+ increases, while the corresponding ratio of ion activities decreases. This observation could reflect the development of yolk platelets and intracellular vesicles that contain ions at differing concentrations and activities than does the bulk cytoplasm. In frog skeletal muscle, the sarcoplasmic reticulum contains a solution resembling that of the cytoplasm, whereas the ionic composition of the solution in the t-tubules is extracellular. The extracellular space, however, consists of the interstitial space between the muscle fibers as well as the vascular space; solutes leave these two extracellular compartments with differing rate constants, thus considerably complicating the interpretation of experiments that assess the rate of membrane transport with radioactive isotopes of Na^+, K^+, and other solutes (Neville, 1979; Neville and White, 1979).

To measure transient changes of intracellular Ca^{2+} in the micromolar and submicromolar range, **fluorescent chelator dyes** such as Quin-2, Fura-2, Indo-1, and Fluo-3 have been developed (Tsien, 1988). Quin-2, Fura-2, and Indo-1 are fluorescent analogs of ethylenediaminetetraacetic acid (EDTA), which contains four carboxylate groups that specifically bind two divalent cations. EGTA is a nonfluorescent analog with a higher binding affinity for Ca^{2+} as compared with its affinity for Mg^{2+}, and it is thus quite useful in experiments where the extracellular concentration of Ca^{2+} is systematically varied. Fluo-3 is a tetracarboxylate fluorescein analog that exhibits a shift in the emission spectrum upon binding Ca^{2+}; in contrast, Fura-2 undergoes a shift in its excitation spectrum. By measuring the ratio of Fura-2 fluorescence upon excitation at two exciting wavelengths, changes in the concentration of Ca^{2+} may be monitored. The cells are incubated with a permeant ester form of the dye to enable the dye to permeate into cells; intracellular esterases then release the Ca^{2+}-sensitive chromophore. With video microscopy of cells stained with fluorescent Ca^{2+} indicators, it is also possible to obtain time-resolved and spatially-resolved light microscopic images of the changes in intracellular Ca^{2+}. Another fluorescent probe (SPQ), developed by Helsey and Verkman (1987), has been used to measure intracellular Cl^-, and to study its transport across cell membranes.

The transmembrane electrical potential is usually measured by means of open-tipped microelectrodes such as those developed and used by G. N. Ling and R. W. Gerard (1949) to obtain accurate and stable measurements of the membrane potential of frog skeletal muscle. With human red blood cells, stable potentials have not been achieved with microelectrodes. As an alternative technique, J. F. Hoffman and P. C. Laris (1974) utilized fluorescent cyanine dyes to monitor and measure red cell membrane potentials. Fluorescent cyanines, merocyanines, oxonols, sytryls, rhodamines, and other dyes have since been used in numerous electrophysiological studies of red blood cells, neutrophils, platelets, and other nonexcitable cells and organelles that are too small for the use of microelectrodes (for review, see Freedman and Novak, 1989a). The equilibrium distribution of permeant, lipophilic, radioactively labeled ions such as triphenylmethylphosphonium ($TPMP^+$) may also be used to assess the membrane potential using the Nernst equilibrium equation (see Freedman and Novak, 1989b).

XI. Ionophores

Ionophores are a class of compounds that form complexes with specific ions and facilitate their transport across cell membranes (for review, see Pressman, 1976). An ionophore typically has a hydrophilic pocket (or hole) that forms a binding site specific for a particular ion. The exterior surface of an ionophore is hydrophobic, allowing the complexed ion in its pocket to cross the hydrophobic membrane. A list of ionophores showing the ion specificity of each is given in Table 3. Ionophores are useful tools in cell physiology experiments. **Nystatin** forms a channel in membranes for monovalent cations and anions, and has proved useful for altering the cation composition of cells. **Gramicidin** forms dimeric channels specific to monovalent cations. **Valinomycin** carries K^+ across membranes with a high selectivity, and has been used extensively to impose a high K^+ permeability on cell membranes. **Monensin** is a carrier with specificity for Na^+. **Hemisodium** is a new synthetic Na^+ ionophore with an even greater degree of selectivity for Na^+ (Kaji, 1992). The **Ca^{2+} ionophore A23187** has been used extensively to permit entry of Ca^{2+} into cells, which normally have a low native permeability to Ca^{2+}, and thereby to activate a variety of cellular processes that are regulated by Ca^{2+}. **Nigericin** exchanges K^+ for protons, and has been used in many studies of mitochondrial bioenergetics to alter electrical and chemical gradients for protons (for review, see Harold, 1986). Ionophores such as **FCCP** and **CCCP** are specific for protons. Study of the mechanism of membrane transport mediated by ionophores has provided important conceptual insights (e.g. Stark and Benz, 1971;

TABLE 3 Ionophores and Their Ion Selectivities

Conductive carriers	
Valinomycin	K^+
Hemisodium	Na^+
FCCP, CCCP	H^+
Electroneutral exchangers	
A23187	Ca^{2+}/Mg^{2+}; $Ca^{2+}/2H^+$
Nigericin	K^+/H^+
Channels	
Gramicidin	$H^+ > Cs^+ \approx Rb^+ > K^+ > Na^+ > Li^+$
Nystatin	monovalent cations and anions

Finkelstein and Andersen, 1981) relevant to the understanding of ion transport mediated by native transport proteins.

XII. Summary

This chapter describes the hydrogen-bonded structure of liquid water and its dipolar and dielectric properties. In salt solutions, water exhibits three regions of structure: oriented dipoles near ions, an intermediate structure-breaking region, and flickering clusters with short-range order characteristic of ice. Electrostatic interactions between ions and water based on Coulomb's law account for the enthalpy of hydration, as well as for the sequence of ionic mobilities of the hydrated alkali cations. Intracellular water exhibits thermodynamic properties similar to extracellular water. Salts and nonelectrolytes dissolve in cell water with nearly the same solubility and activity as in extracellular water; even the viscosities are similar. Protons in solution are hydrogen-bonded to water to form H_3O^+, which is further hydrated due to electrostatic forces. Protons migrate through water by a Grotthus chain mechanism; in contrast, other hydrated ions migrate through water as hard spheres that interact with water dipoles according to the ionic radius and charge. Nonspecific ion-ion interactions reduce the activity coefficients of ions, as described by the Debye-Hückel theory, which conceptualizes a central ion surrounded by a cloud of counterions with smeared charge. Despite the predictive ability of the Debye-Hückel theory, a quasi-lattice theory of salt solutions is preferable at physiological salt concentrations. Selective interactions of ions with sites on enzymes or on channel proteins occur in certain predictable specific patterns (Eisenman sequences) that depend on the energy of interaction between the ions and water dipoles, relative to that between ions and their binding sites.

Next described are the functions and properties of the predominant cellular electrolytes—K^+, Na^+, Ca^{2+}, Mg^{2+}, Cl^-, and HCO_3^-. Steady-state concentrations of K^+, Na^+, and Ca^{2+} are established by the Na^+,K^+-ATPase and Ca^{2+}-ATPase, which utilize metabolic energy to pump ions against their electrochemical gradients. The passive flow of K^+, Na^+, and Ca^{2+} down their electrochemical gradients through voltage-gated ion channels constitutes the ionic currents that form action potentials in excitable cells such as nerve and muscle. Intracellular K^+ also activates the glycolytic enzyme pyruvate kinase and is required for peptide bond formation in protein synthesis. Changes in cell Ca^{2+} concentration initiate and modulate a variety of cellular functions, but too much Ca^{2+} inside cells can be harmful. Mg^{2+}, with its high charge density, binds tightly to polyphosphates and is a cofactor in all kinases and some phosphatases, including the Na^+, K^+-ATPase. HCO_3^- and phosphate, along with proteins, are the principal biological buffers that regulate pH. Cl^- mainly serves to maintain bulk electroneutrality. Trace ions, including those of iron, copper, manganese, zinc, selenium, and cobalt, are important cofactors in nutritional and bioenergetic pathways, and in enzymes that protect cells from oxidative and peroxidative damage.

All of the solutes in a cell contribute to the free energy of the intracellular solution. The Gibbs equation, which is derived in the Appendix by combining the first and second laws of thermodynamics, enables estimation of the changes in free energy when solutes cross cell membranes. The change in free energy is zero at equilibrium and negative for spontaneous processes. Solutes will redistribute until the electrochemical potential is the same in every compartment to which that solute has access. The Nernst equation (see Appendix), which follows from the Gibbs equation, relates the ratio of intracellular to extracellular ion concentrations at equilibrium to the membrane potential, and can be used to test whether a solute is in electrochemical equilibrium across a cell membrane. Cell cations are measured by means of flame photometry, atomic absorption spectroscopy, ion-specific electrodes, electron probe microanalysis, and fluorescent chelator dyes. Membrane-bounded compartments inside cells may contain different concentrations of solutes than does the bulk cytoplasm. Ionophores with a high degree of ion selectivity are useful tools in cell physiology experiments. The study of ion transport mediated by ionophores has provided instructive models for understanding the mechanisms of membrane transport.

Bibliography

Bergethon, P. R., and Simons, E. R. (1990). "Biophysical Chemistry. Molecules to Membranes." Springer-Verlag, New York.

Bernard, C. (1949). "An Introduction to the Study of Experimental Medicine." (H. C. Green, translator). Schuman, New York.

Bockris, J. O'M., and Reddy, A. K. N. (1970). "Modern Electrochemistry," Vol. 1. Plenum Press, New York.

Campbell, A. K. (1983). "Intracellular Calcium. Its Universal Role as Regulator." Wiley, New York.

Cannon, W. B. (1932). "The Wisdom of the Body." Norton, New York.

Cantor, C. R., and Schimmel, P. R. (1980). "Biophysical Chemistry, Parts I, II, III." W. H. Freeman, San Francisco.

Collins, K. D., and Washabaugh, M. W. (1985). The Hofmeister effect and the behavior of water at interfaces. Q. Rev. Biophys. 18, 323–422.

Cooke, R., and Kuntz, I. D. (1975). The properties of water in biological systems. Ann. Rev. Biophys. Bioeng. 3, 95–126.

Debye, Peter. (1929). "Polar Molecules." Dover Publications, New York.

Dick, D. A. T. (1959). Osmotic properties of living cells. Internat'l. Rev. Cytol. 8, 387–448.

Edsall, J. T., and McKenzie, H. A. (1978). Water and proteins. I. The significance and structure of water; its interaction with electrolytes and nonelectrolytes. Adv. Biophys. 10, 137–207.

Edsall, J. T., and Wyman, J. (1958). "Physical Biochemistry." Academic Press, New York.

Eisenberg, D., and Crothers, D. M. (1979). "Physical Chemistry with Applications to the Life Sciences." Benjamin-Cummings, Menlo Park, California.

Eisenberg, D., and Kauzmann, W. (1969). "The Structure and Properties of Water." Oxford University Press, Oxford.

Eisenman, G. (1967). "Glass Electrodes for Hydrogen and other Cations." M. Dekker, New York.

Eisenman, G., and Horn, R. (1983). Ionic selectivity revisited: the role of kinetic and equilibrium processes in ion permeation through channels. J. Membr. Biol. 76, 197–225.

Finkelstein, A., and Andersen, O. S. (1981). The gramicidin channel: a review of its permeability characteristics with special reference to the single-file aspect of transport. J. Membr. Biol. 59, 155–171.

Frank, H. S., and Thompson, P. T. (1960). A point of view on ion clouds. In: "The Structure of Electrolyte Solutions" (W. J. Hamer, Ed.), pp. 113–134. John Wiley & Sons, New York.

Frank, H. S., and Wen, W.-Y. (1957). Structural aspects of ion-solvent interaction in aqueous solutions: a suggested picture of water structure. Disc. Faraday Soc. 24, 133–140.

Freedman, J. C. (1976). Partial restoration of sodium and potassium gradients by human erythrocyte membranes. Biochim. Biophys. Acta 455, 989–992.

Freedman, J. C., and Novak, T. S. (1989a). Optical measurement of membrane potentials of cells, organelles, and vesicles. Meth. in Enzymol. 172, 102–122.

Freedman, J. C., and Novak, T. S. (1989b). Use of triphenyl-methylphosphonium to measure membrane potentials in red blood cells. Meth. in Enzymol. 173, 94–100.

Funder, J., and Wieth, J. O. (1966). Determination of sodium, potassium, and water in human red blood cells. Elimination of sources of error in the development of a flame photometric method. Scand. J. Clin. Lab. Invest. 18, 151–166.

Fushimi, K., and A. S. Verkman (1991). Low viscosity in the aqueous domain of cell cytoplasm measured by picosecond polarization microfluorimetry. J. Cell. Biol. 112, 719–725.

Gary-Bobo, C. M. (1967). Nonsolvent water in human erythrocytes and hemoglobin solutions. J. Gen. Physiol. 50, 2547–2564.

Glynn, I. M. (1985). The Na^+,K^+-transporting adenosine triphosphatase. In "The Enzymes of Biological Membranes" (A. N. Martonosi, Ed.), 2nd ed., Vol. 3, pp. 35–114. Plenum Press, New York.

Gulati, J., and Podolski, R. J. (1978). Contraction transients of skinned muscle fibers: effects of calcium and ionic strength. J. Gen. Physiol. 72, 701–716.

Harold, F. M. (1986). The Vital Force: a Study of Bioenergetics. W. H. Freeman, New York.

Heilbrunn, L. V. (1956). "The Dynamics of Living Protoplasm." Academic Press, New York.

Helsey, N. P., and Verkman, A. S. (1987). Membrane chloride transport measured using a chloride-sensitive fluorescent probe. Biochem. 26, 1215–1219.

Hille, B. (1992). "Ionic Channels of Excitable Membranes," 2nd ed. Sinauer Associates, Sunderland, Mass.

Höber, R. (1945). "Physical Chemistry of Cells and Tissues." The Blakiston Company, Philadelphia.

Hoffman, J. F., and Laris, P. C. (1974). Determination of membrane potentials in human and *Amphiuma* red blood cells by means of a fluorescent probe. *J. Physiol. (London)* **239**, 519–552.

Horowitz, S. B., and Miller, D. S. (1984). Solvent properties of ground substance studied by cryomicrodissection and intracellular reference-phase techniques. *J. Cell. Biol.* **99**, 172s–179s.

Horvath, A. L. (1985). "Handbook of Aqueous Electrolyte Solutions. Physical Properties, Estimation and Correlation Methods." John Wiley and Sons, New York.

Kaji, D. (1992). Hemisodium, a novel selective Na ionophore. *J. Gen. Physiol.* **99**, 199–216.

Katz, S., and Klotz, I. M. (1953). Interactions of calcium with serum albumin. *Arch. Biochem.* **44**, 351–361.

Kayne, F. J. (1973). Pyruvate kinase. *In* "The Enzymes" (P. D. Boyer, Ed.), pp. 353–382. Academic Press, New York.

Klotz, I. M. (1970). Water: Its fitness as a molecular environment. *In* "Membranes and Ion Transport" (E. E. Bittar, Ed.), Vol. 1, pp. 93–122. Wiley, New York.

Läuger, P. (1991). "Electrogenic Ion Pumps." Sinauer Associates, Sunderland, Mass.

Ling, G. N., and Gerard, R. W. (1949). The normal membrane potential of frog sartorius fibers. *J. Cell. Comp. Physiol.* **34**, 383–396.

Miller, C. (1974). Nonelectrolyte distribution in mouse diaphragm muscle. I. The pattern of nonelectrolyte distribution and reversal of the insulin effect. *Biochim. Biophys. Acta.* **339**, 71–84.

Miskin, R., Zamir, A., and Elson, D. (1970). Inactivation and reactivation of ribosomal subunits: the peptidyl transferase activity of the 50s subunit of *Escherichia coli. J. Mol. Biol.* **54**, 355–378.

Neville, M. C. (1979). The extracellular compartments of frog skeletal muscle. *J. Physiol.* **288**, 45–70.

Neville, M. C., and White, S. (1979). Extracellular space of frog skeletal muscle *in vivo* and *in vitro:* relation to proton magnetic resonance relaxation times. *J. Physiol.* **288**, 71–83.

Neville, M. C., Allen, J. C., and Watters, C. (1983). The mechanisms of milk secretion. *In* "Lactation. Physiology, Nutrition, and Breast-Feeding" (M. C. Neville and M. R. Neifert, Eds.). Plenum Press, New York.

Oesterhelt, D., and Tittor, J. (1989). Two pumps, one principle: light driven ion transport in Halobacteria. *Trends in Biochem. Sci.* **14**, 57–61.

Palmer, L. G., and Civan, M. M. (1977). Distribution of Na^+, K^+, and Cl^- between nucleus and cytoplasm in *Chironomus* salivary gland cells. *J. Membr. Biol.* **33**, 41–61.

Palmer, L. G., Century, T. J., and Civan, M. M. (1978). Activity coefficients of intracellular Na^+ and K^+ during development of frog oocytes. *J. Membr. Biol.* **40**, 25–38.

Pauling, L. (1960). "The Nature of the Chemical Bond and the Structure of Molecules and Crystals; An Introduction to Modern Structural Chemistry," 3rd ed. Cornell University Press, Ithaca, NY.

Pressman, B. C. (1976). Biological applications of ionophores. *Ann. Rev. Biochem.* **45**, 501–530.

Ringer, S. (1882). Concerning the influence exerted by each of the constituents of the blood on the contraction of the ventricle. *J. Physiol. (London).* **3**, 380–393.

Robinson, R. A., and Stokes, R. H. (1959) "Electrolyte Solutions," 2nd ed. Butterworths, London.

Scatchard, G., Coleman, J. S., and Shen, A. L. (1957). Physical chemistry of protein solutions. VII. The binding of some small anions to serum albumin. *J. Amer. Chem. Soc.* **79**, 12–20.

Schwan, H. P., and Foster, K. R. (1977) Microwave dielectric properties of tissue. Some comments on the rotational mobility of tissue water. *Biophys. J.* **17**, 193–197.

Silver, B. L. (1985) "The Physical Chemistry of Membranes." Allen and Unwin, Boston.

Stark, G., and Benz, R. (1971). The transport of potassium through lipid bilayer membranes by the neutral carriers valinomycin and monactin. *J. Membr. Biol.* **5**, 133–153.

Tanford, C. (1961). "Physical Chemistry of Macromolecules." Wiley, New York.

Tinoco, I., Sauer, K., and Wang, J. C. (1994). "Physical Chemistry: Principles and Applications in the Biological Sciences," 3rd ed. Prentice-Hall, Englewood Cliffs, New Jersey.

Tsien, R. Y. (1988). Fluorescence measurement and photochemical manipulation of cytosolic free calcium. *Trends in Neurosci.* **11**, 419–424.

Ussing, H. H. (1960). The alkali metal ions in isolated systems and tissues. *In* "The Alkali Metal Ions in Biology, Handbuch der Experimentellen Pharmakologie, Ergänzungswerk" (O. Eichler and A. Farah, Eds.), Vol. 13, Part 1, pp. 1–195. Springer-Verlag, Berlin.

van Holde, K. E. (1985). "Physical Biochemistry," 2nd ed. Prentice-Hall, Englewood Cliffs, New Jersey.

Appendix: Thermodynamics of Membrane Transport

I. Free Energy

When ions cross cell membranes, changes in free energy are involved. Consider a membrane permeable only to cations separating two solutions of KCl of differing concentrations, c_i and c_o, where the subscripts i and o represent the intracellular and extracellular compartments, respectively (see Fig. A-1). If the intracellular concentration (c_i) is greater than the extracellular concentration (c_o), then potas-

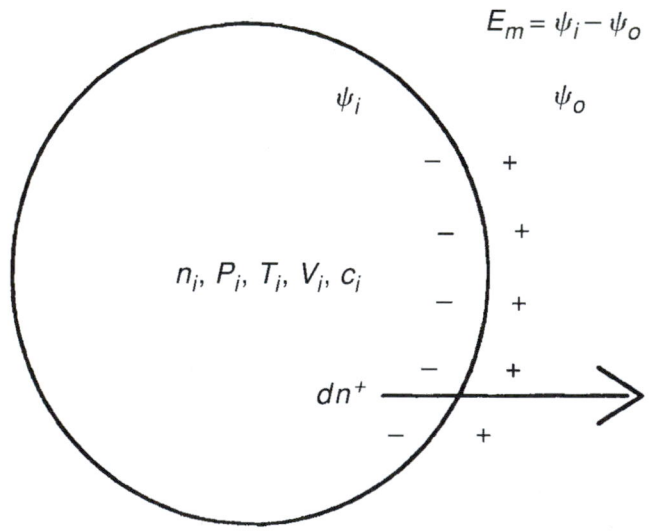

$$E_m = \psi_i - \psi_o$$

ψ_i

ψ_o

n_i, P_i, T_i, V_i, c_i

dn^+

n_o, P_o, T_o, V_o, c_o

FIGURE A-1. Free energy of ion transport. A cell membrane separates the intracellular (subscript i) and the extracellular (subscript o) compartments. Each compartment is at pressure P and temperature T, has volume V and electrical potential ψ, and contains solutes at concentration c. A free energy change is involved when dn^+ moles of cation leave the cell. The membrane potential E_m is $\psi_i - \psi_o$.

sium will tend to diffuse out of the cell down its concentration gradient. When only a very slight amount of potassium crosses the membrane (which for this example is assumed to be impermeable to chloride), the separation of charge creates a transmembrane electrical potential (E_m) that is negative inside, relative to outside. The resultant electrical force retards further efflux of potassium. An equilibrium is reached when the diffusion force favoring efflux of potassium exactly balances the electrical force, preventing efflux of potassium.

Prior to the attainment of equilibrium, the process of moving dn moles of K$^+$ ions out of the cell from compartment i to compartment o involves a change in **free energy** (dG) of the system. According to the laws of thermodynamics, the process will occur spontaneously only if $dG < 0$, and the system will be at equilibrium if $dG = 0$. Thus in order to compute the change in free energy (dG), we need to review certain basic thermodynamic principles.

The **first law of thermodynamics,** also known as the law of conservation of energy, states that for any system, the increase in **energy** (dE) is the gain in **heat** (dQ) minus the **work** (dW) done by the system.

$$dE = dQ - dW$$

For reversible processes, the **second law of thermodynamics** defines the change in **entropy** (dS) of a system in terms of the heat gained (dQ) and the absolute temperature (T) as follows:

$$dS = \frac{dQ}{T}$$

Combining the first and second laws gives

$$dE = TdS - dW$$

Since the difference in pressure (dP) between the internal and external solutions of animal cells is negligible, the work (dW) done by the system includes **pressure–volume work,** PdV, and **electrical work,** $z\mathcal{F}\epsilon_m dn$, summed as follows:

$$dW = PdV + z\mathcal{F}\epsilon_m \, dn$$

where P is the pressure, V is the total volume of the system, z is the ionic valence (eq/mol), $\mathscr{F}$ is the Faraday constant (= 96 490 C/eq), and dn is the number of moles of solute crossing the membrane and leaving the cell. The **transmembrane electrical potential** (ϵ_m) is the difference in electrical potential (ψ, in volts or J/C), between the two compartments.

$$\epsilon_m = \psi_i - \psi_o$$

The electrical potential (ψ) represents the work done in moving a unit positive test charge from infinity to a point in the solution. Note that when the transmembrane electrical potential (ϵ_m) is negative inside, the **electrical** work done by the system is negative whenever Na^+ or K^+ leave the cell, but positive when K^+ or Na^+ enter the cell, irrespective of the mechanism of transport.

According to the first and second laws, the change in energy (dE) of the system is

$$dE = TdS - PdV - z\mathscr{F}\epsilon_m dn$$

For systems of variable chemical composition, the **free energy** (G) is by definition

$$G = H - TS + \sum \mu_j n_j$$

where the **enthalpy** (H) is defined as

$$H = E + PV$$

The **chemical potential** μ_j of the jth solute in the system at constant T, P, and n_k is defined by

$$\mu_j = (\partial G/\partial n_j)_{T,P,n_k}$$

where n_j and n_k are the number of moles of the jth and kth solutes, respectively. The chemical potential, also called the **partial molar free energy,** represents the incremental addition of free energy to the system upon incremental addition of a solute. All solutes contribute to the free energy of a solution. While the free energy itself is a parameter of state for the whole system, the chemical potential refers to a particular solute. The free energy is thus given by

$$G = E + PV - TS + \sum \mu_j n_j$$

Differentiating yields

$$dG = dE + PdV + VdP - TdS - SdT + \sum \mu_j dn_j$$

Substituting the expression for dE given previously and simplifying yields a form of the **Gibbs equation,**

$$dG = -SdT + VdP - z\mathscr{F}\epsilon_m dn + \sum \mu_j dn_j$$

which states that the free energy of a system of variable chemical composition is a function of the temperature, the pressure, and the number of moles of each component in the mixture, or $G = G(T, P, n_j)$. For processes that occur at constant temperature and pressure, where $dT = dP = 0$, the Gibbs equation simplifies to

$$dG = -z\mathscr{F}\epsilon_m dn + \sum \mu_j dn_j$$

which states that the increase in free energy of a system is equal to the sum of the electrical work done on the system plus the total change in free energy due to changes in chem-

ical composition. Furthermore, when the system is at equilibrium, dG must equal zero. The second law of thermodynamics also implies that the change in free energy (dG) is negative for all spontaneous processes. Thus, the first and second laws of thermodynamics, when combined with the definitions of free energy and enthalpy, result in the Gibbs equation, which is the fundamental equation for the estimation of free energy changes when water, ions, or other solutes cross cell membranes.

II. Nernst Equilibrium

The **Nernst equation** describes the relationship between voltage across a semipermeable membrane and the ion concentrations at equilibrium in the compartments adjacent to the membrane. The Nernst equation provides a simple method of testing whether or not a particular solute is at equilibrium.

Considering the example described in the previous section, and using the Gibbs equation, the change in the free energy of the system that occurs when dn moles of K^+ ions move from compartment i containing K^+ at activity a_i to compartment o containing K^+ at activity a_o is given by

$$dG = -z\mathscr{F}\epsilon_m dn + \mu_i dn_i + \mu_o dn_o$$

where the change in free energy has been summed for the two compartments. For this process, the decrease in the number of moles of solute inside the cell ($-dn_i$) equals the increase outside the cell (dn_o), so that both may be represented by dn

$$-dn_i = dn_o = dn$$

and therefore

$$dG = \left[-z\mathscr{F}\epsilon_m - \left(\mu_i - \mu_o\right)\right] dn$$

The chemical potential in each solution is

$$\mu_i = \mu_i^0 + RT \ln a_i$$

$$\mu_o = \mu_o^0 + RT \ln a_o$$

For the system under consideration, the intracellular and extracellular standard-state chemical potentials are assumed to be identical

$$\mu_i^0 = \mu_o^0$$

and thus

$$dG = \left[-z\mathscr{F}\epsilon_m - RT \ln\left(a_i/a_o\right)\right] dn$$

At equilibrium, $dG = 0$, and since $dn \neq 0$, therefore

$$-z\mathscr{F}\epsilon_m - RT \ln\left(a_i/a_o\right) = 0$$

Rearranging yields the Nernst equation for a cationic concentration cell,

$$\epsilon_m = -\frac{RT}{z\mathscr{F}} \ln \frac{a_i}{a_o}$$

Converting the natural logarithm to base 10 yields

$$\epsilon_m = -2.303 \frac{RT}{z\mathscr{F}} \log \frac{a_i}{a_o}$$

The value of $2.303RT/\mathscr{F}$ is 58.7 mV at 23 °C, and 61.5 mV at 37 °C.

For charged solutes, the **electrochemical potential** ($\mu_j = dG/dn_j$) of the jth solute is defined as the sum of a chemical and an electrical component. The electrical contribution to the electrochemical potential is $z\mathscr{F}\psi$, and the chemical contribution is $RT \ln a$.

$$\mu_j = \mu_j^0(T,P) + RT \ln a_j + z\mathscr{F}\psi$$

Note that a fundamental condition of equilibrium is that the difference in electrochemical potential between compartments i and o is zero. The electrochemical potential of the solute is the same in each compartment to which that solute has access.

The Nernst equation is independent of the mechanism of transport, and is often used for ascertaining whether or not an intracellular ion is at electrochemical equilibrium. If so, then

$$a_i = a_o\, e^{-z\mathscr{F}\epsilon_m/RT}$$

Another way of understanding the Nernst equation is that at equilibrium, ions distribute across the membrane electric field in accordance with a Boltzmann distribution.

If $z = 0$, as for nonelectrolytes, then $a_i = a_o$, and the activities of the solutes will be the same at equilibrium on both sides of the membrane, as was found to be nearly the case for the distribution of nonelectrolytes across the membranes of human red blood cells (Gary Bobo, 1967). If $z \neq 0$, as for electrolytes, then at equilibrium each permeant monovalent ion will reach the same ratio of intracellular to extracellular activity. Such is the case for the passive distribution of Cl^-, HCO_3^-, and H^+ in human red blood cells.

$$r = \frac{[Cl^-]_i}{[Cl^-]_o} = \frac{\left[HCO_3^-\right]_i}{\left[HCO_3^-\right]_o} = \frac{[H^+]_o}{[H^+]_i} = e^{\mathscr{F}\epsilon_m/RT}$$

In red blood cells, Na^+, K^+, and Ca^{2+} deviate from this ratio due to the action of the Na^+-K^+ pump and Ca^{2+} pump. In skeletal muscle, K^+ and Cl^- have Nernst equilibrium potentials that are close to the actual measured resting potential, while Na^+ is far from equilibrium. In squid axons, the Nernst equilibrium potential for K^+ is closest to the resting potential, Cl^- is somewhat removed, and, as in muscle and red blood cells, Na^+ is far from equilibrium (see Chapter 13).

Matthew R. Pincus

2

Physiological Structure and Function of Proteins

I. Molecular Structure of Proteins

Proteins are biopolymers that are essential for all plant and animal life. All antibodies, enzymes, and cell receptors are proteins. The basis for all connective tissue is the fibrous proteins such as elastin and collagen. The entire genetic apparatus of every cell, regardless of how simple or complex its functions, is dedicated to the synthesis of proteins. The building blocks for proteins are α-amino acids that are linked to one another by peptide (or amide) bonds. The twenty naturally occurring amino acids (Fig. 1) all have the same **backbone** structure, the N–C^α–C=O unit, but differ from one another in that they have different side chains attached to the α-carbon, as shown in Fig.1. Also, all of these amino acids are chiral—optically active and in the L-configuration—except for the simplest amino acid, glycine, which has two H atoms attached to the α-carbon. Note that cysteine has a sulfhydryl group, giving it the capacity to form disulfide bonds (Fig. 1). Disulfide bonds are important in stabilizing the three-dimensional structures of proteins, as discussed in Section I.C.

Topographically, proteins are divided into the **fibrous proteins,** which are long repeating helical proteins, and the **globular proteins,** which tend to fold up into spherical or ellipsoidal shapes. As we will show, how proteins fold is governed uniquely by their linear sequences of amino acids.

A. Primary Amino Acid Structure

All proteins are composed of linear sequences of amino acids that are unique to each protein and are covalently bound to one another by peptide bonds (Fig. 2). This peptide bond is almost always arranged so that the C^α atoms of two successive residues are *trans* to one another and the atoms attached to the C' and N of the peptide all lie in the same plane since the C'–N bond has double-bond character, namely, the *trans*-planar peptide bond described long ago by Linus Pauling. The linear sequence of amino acids is called the **primary structure.**

Note that the amino acids in Fig.1 are classified as **polar, nonpolar,** and **neutral** depending on the nature of the side chain. Thus, for example, leucine and valine are nonpolar amino acids with aliphatic side chains that are hydrophobic, or nonmiscible with water and other polar solvents. Amino acids like lysine and glutamic acid, on the other hand, contain charged side chains that are highly soluble in water and are completely miscible in polar solvents. Amino acids such as glycine and alanine do not contain groups that predispose them to be classified as either polar or nonpolar, so they are referred to as neutral.

B. Regular or Secondary Structure in Proteins

The three-dimensional structures of over 300 proteins have been determined by x-ray crystallography. One striking feature in all of these proteins is the presence of recurrent regular structures, in particular, α-helices, β-sheets, and reverse turns (Fig. 3). The existence of the first two of these structural types was predicted by Linus Pauling in the early 1950s on the basis of x-ray diffraction patterns of model poly-α-amino acids.

Note that in the α-**helix** in Fig. 3A, the residues form a spiral such that at every 3.6 residues, the spiral or helix makes one complete turn. The repeat distance of this helix is 5.4 Å. This structural motif is observed in virtually all proteins, and some proteins like myoglobin are virtually completely α-helical. One very important interaction for stabilizing helices is the i-to-$i+4$ hydrogen-bonding scheme shown in Fig. 3A. The C=O of the ith residue accepts a hydrogen bond from the N–H of the $i+4$th residue. The hydrogen bond can provide up to 3 kcal/mol in stabilization energy. For an amino acid residue in the middle of an α-helix, two hydrogen bonds form to its N–H and C=O groups. This occurrence tends to cause helices to propagate because the more such double hydrogen bonds form, the greater the stabilization energy provided. Two other long-range factors stabilize α-helices. The

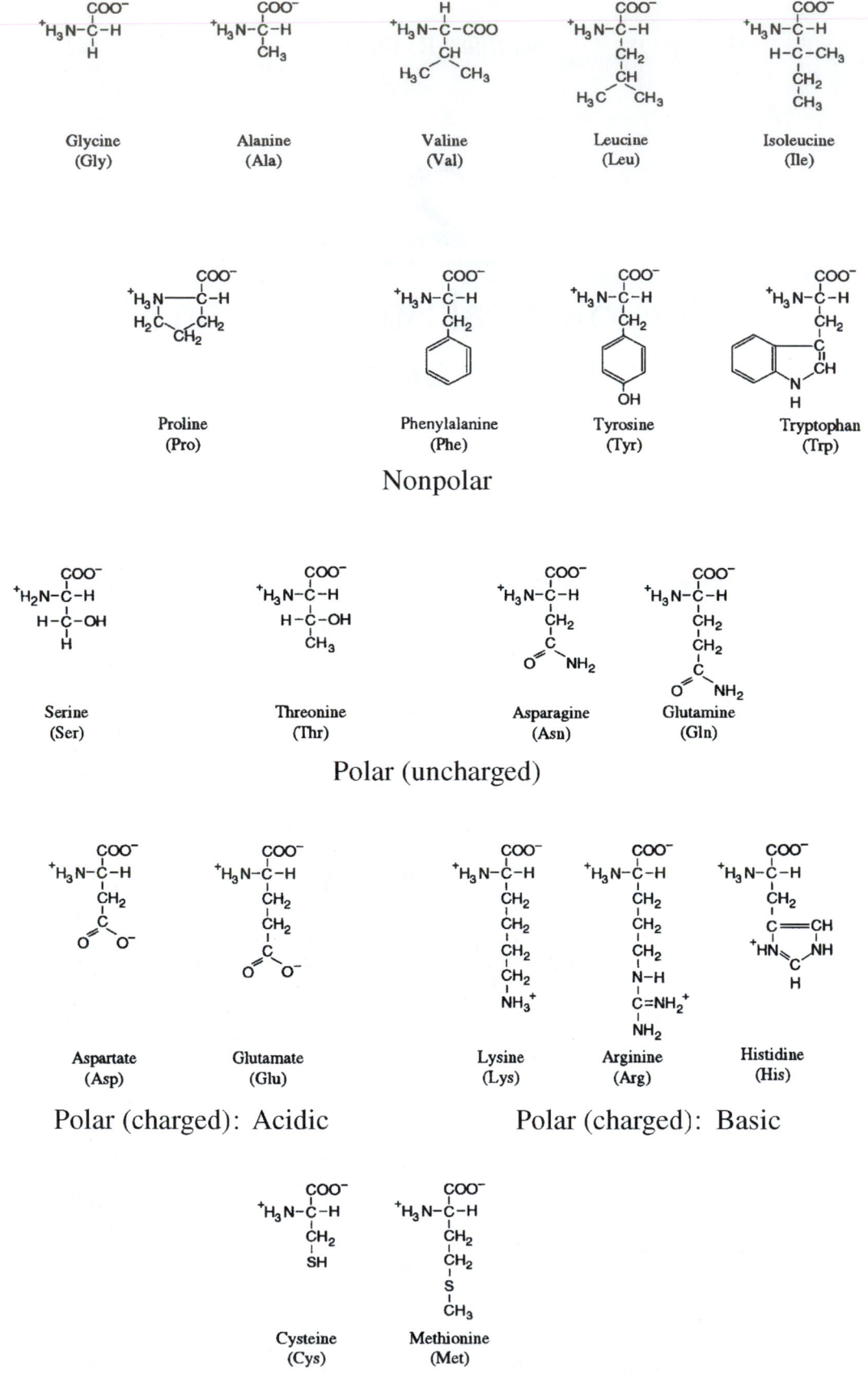

FIGURE 1. The 20 naturally occurring amino acids grouped by type. Note that Cys residues can form disulfide bonds in the following reaction: R–SH + R–SH → R–S–S–R, where R is the backbone atoms of Cys + the side chain CH_2 group.

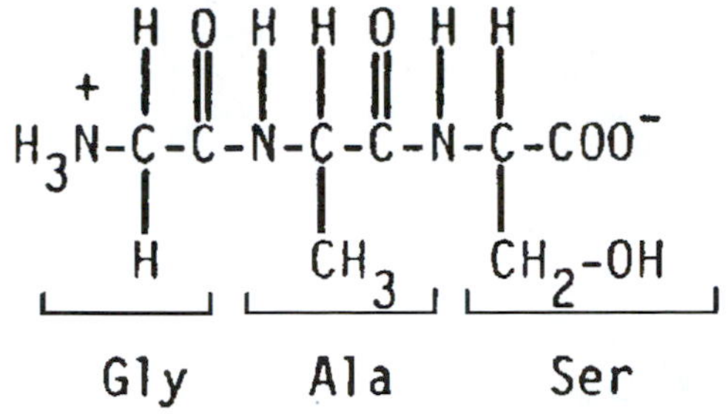

FIGURE 2. A typical tripeptide, Gly-Ala-Ser, showing the CO–NH peptide bond linkage.

first is the presence of negatively charged amino acid residues on the amino terminal end of a helix and positively charged amino acids on the carboxyl terminal end of the helix. This stabilization results from the dipole moment of an α-helix, where the center of positive charge is on the amino terminus of the helix and the center of negative charge is on the carboxyl terminus of the helix. The presence of the oppositely charged amino acids stabilizes the helical dipole.

The second stabilizing factor is **amphipathicity.** This refers to the i-to-$i+4$ side chain–side chain interactions such that nonpolar or hydrophobic amino acids contact one another on one face of the helix while other polar i-to-$i+4$ interactions occur on the opposite side of the helix. This segregation of polar and nonpolar faces allows for hydrophobic clustering to occur on one face and hydrogen bonding and charge neutralization to occur on the opposite face. This motif is vital in stabilizing α-helices in membrane proteins such as ion channel proteins.

α-helices are extremely important in membrane proteins, in the transmembrane domains of cellular receptors, and in ion channel proteins, as we discuss in Section IV.

In contrast, the **β-sheet** is composed of alternating residues of amino acids that are "flipped" 180° with respect to their nearest neighbors, as shown in Fig. 3B. We may regard the structure of single strands in these sheets as a flat helix whose repeat is every two residues. Note that within each strand of a β-sheet are minimal interactions between the backbone atoms of different amino acid residues. The hydrogen-bonding stabilization comes from other strands that lie in proximity to one another. As shown in Fig. 3B, the arrangement of the strands is either **parallel** or **antiparallel.** Regardless of the arrangement of the two interacting strands, it should be noted from Fig. 3B that two hydrogen bonds between the backbone atoms form for **alternate** residues; every other residue forms no hydrogen bond to the neighboring strand. Thus there are twice the number of hydrogen bonds in long α-helices as in long β-sheets. Evidently, other interactions are important in stabilizing β-sheets besides backbone-backbone hydrogen bonding.

Another type of regular structure is the **reverse turn,** the prototype of which is shown in Fig. 3C. In this structure, the polypeptide chain reverses its direction. A minimum of two amino acids is needed to form a reverse or hairpin turn, which is also called a β **bend.** Usually three amino acids are involved, however, because the C=O of the first residue can form an i-to-$i+3$ hydrogen bond with the N–H of the $i+3$rd residue, as shown in Fig. 3.

Of the three regular structures described, α-helices tend to propagate most strongly due to the double hydrogen bonding of the NH and C=O groups of central residues in the helix discussed above and illustrated in Fig. 3D. This type of "medium-range" interaction is not present for the two other types of regular structures.

The specific sequences of amino acids in proteins will adopt specific structures such as the three basic regular structures just described. Thus the amino acid sequences of proteins determine their respective three-dimensional structures. This conclusion has been verified as we now describe.

C. Tertiary or Three-Dimensional Structure of Proteins

1. The Amino Acid Sequences of Proteins Determine Their Three-Dimensional Structures

In a classic series of experiments, Anfinsen and his coworkers at the National Institutes of Health showed in the early 1960s that the primary structure (linear sequence of amino acids) of a protein determines its unique three-dimensional structure (Anfinsen *et al.*, 1961). These investigators took the protein ribonuclease A, which hydrolyzes RNA and whose amino acid sequence had just been determined and was known to contain four disulfide bonds, and denatured it in the denaturing agent 6 M guanidine hydrochloride. This agent and 8 M urea are both known to disrupt the hydrogen bonds in proteins and destroy their three-dimensional structures. In addition, using β-mercaptoethanol, they reduced the four disulfides to eight sulfhydryl (SH) groups, thereby destroying all of the major determinants of the three-dimensional structure. The denatured protein was completely inactive towards RNA hydrolysis and possessed spectral properties that differed greatly from those of the native protein. They then allowed the sulfhydryl groups to reoxidize to disulfides and dialyzed the 6 M guanidine hydrochloride. Within two hours, virtually all of the native enzymatic activity returned, and the spectral properties of the refolded protein were identical to those of the undenatured protein. In addition, all of the disulfides were paired exactly as they were in the native protein.

This experiment illustrated that the primary sequence of amino acids in a protein determines its three-dimensional structure. Further, it demonstrated that the interactions that govern the correct folding of a protein must be highly specific and strong because, of the vast number of possible structures that the polypeptide could have adopted, only one such structure actually ultimately formed. This conclusion is reinforced by the following consideration. Once the protein is reduced to the eight-sulfhydryl state, there are many ways in which these eight sulfhydryls can pair to give four disulfides. In particular, there are seven ways in which the first disulfide can form, five for the second, three for the third, and one for the fourth, or $7 \times 5 \times 3 \times 1 = 105$ possible ways. Of this total, only one pairing scheme, the native pairing, ultimately was found to form in this experiment.

Similar results have been obtained for a number of different proteins. In every case, the native structure is regenerated. Such dramatic results imply that, given the amino acid

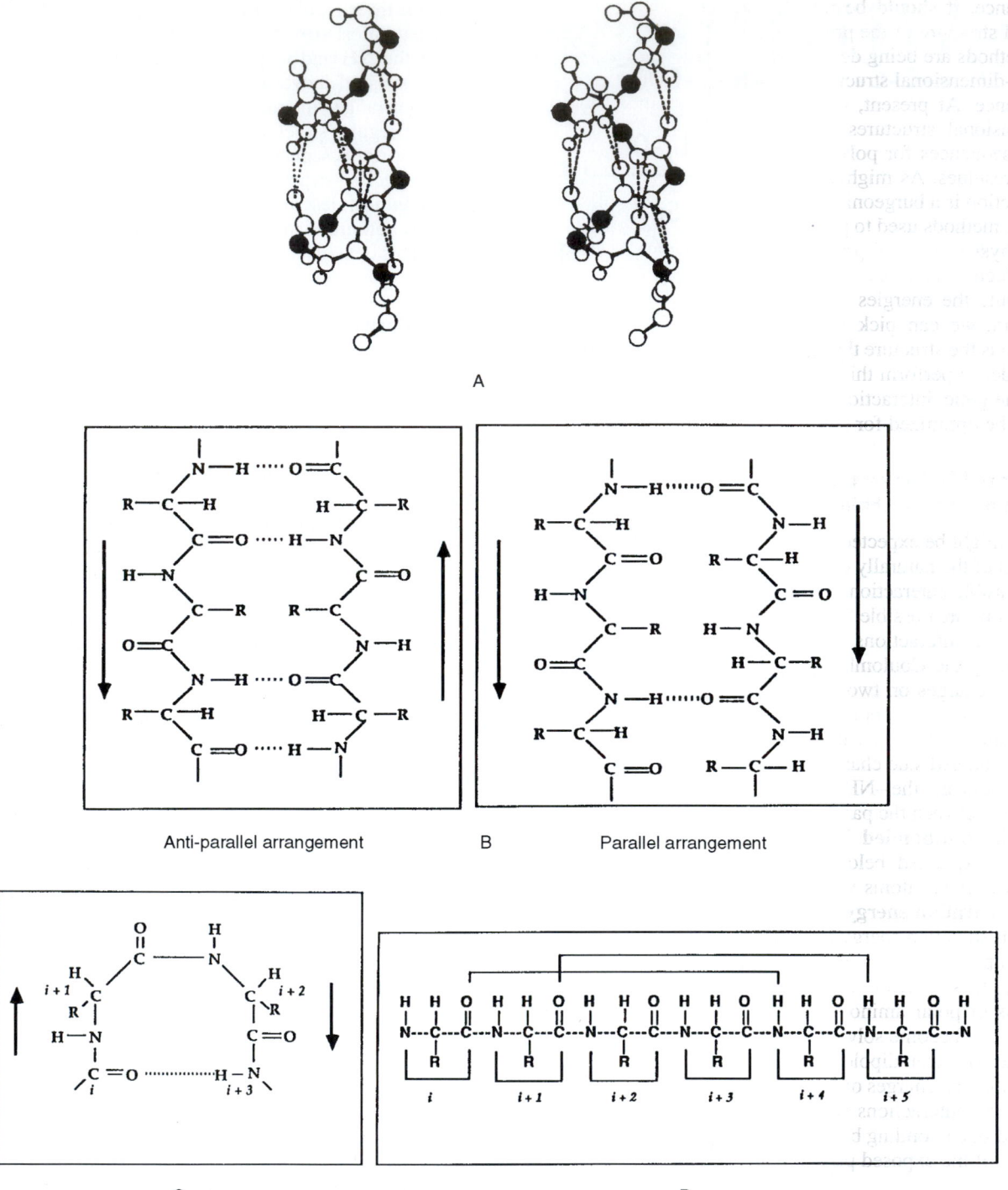

FIGURE 3.   Three types of regular (secondary) structure. (A) Stereo view of an α-helix for a polypeptide backbone showing the i-to-$i+4$ hydrogen bond. The filled circles are α-carbons; double dashed lines are drawn from C=O to NH (shorter dashed line) and from C=O to N (longer dashed line). (B) Two types of β-sheets. The left side shows the antiparallel arrangement while the right side shows the parallel arrangement, as indicated by the arrows in both figures. (C) The arrangement of backbone atoms in a β-bend showing how an i-to-$i+3$ hydrogen bond stabilizes the reverse turn structure. (D) Illustration of an extended α-helical chain showing how internal residues have both their NH and CO groups involved in hydrogen bonds.

sequence, it should be possible to infer the three-dimensional structure of the protein.

Methods are being developed that allow us to **predict** the three-dimensional structure of a protein from its amino acid sequence. At present, it is possible to compute the three-dimensional structures of polypeptides from their amino acid sequences for polypeptides of up to about 100 amino acid residues. As might be imagined, the field of structure prediction is a burgeoning one that is developing rapidly. All of the methods used to predict protein structure are based on the physicochemical principle that the observed structure of a molecule is the one with the lowest free energy. If we can compute the energies for the possible conformations of a protein, we can pick out the structure of lowest energy, which is the structure that the protein is most likely to adopt. In order to perform this task, we must understand the types of energetic interactions within a polypeptide chain that must be optimized for the protein to fold correctly.

D. Possible Interactions between Amino Acids in a Protein Chain

As might be expected from the differing nature of the side chains of the naturally occurring amino acids, a vast number of possible interactions between different side chains and backbone are possible for given amino acid sequences. These types of interactions include **electrostatic interactions** (given by the Coulombic potential $q_1 q_2 / DR_{12}$, where the qs are the charges on two different interacting atoms, R is the distance between them, and D is the dielectric constant of the medium) between positive and negative charges of oppositely charged side chains, such as the $-COO^-$ group of glutamic acid and the $-NH_3^+$ group of lysine as well as the interactions between the partial charges on all of the atoms of the protein; **nonbonded** interactions between the individual atoms, explained below; **hydrogen-bonding interactions** between polar atoms with H atoms and other polar atoms; and a **solvation energy** for most proteins.

This solvation energy is such that the nonpolar side chains of nonpolar amino acids tend to avoid water and therefore "bury" themselves in the core of the protein, while the side chains of polar amino acids tend to interact strongly with water and become solvated. This favorable solvation energy is due to water dipole interactions with charged residues such that the charges on the side-chain atoms are reduced by favorable interactions with the water dipoles. It is also due to hydrogen bonding between water molecules and the polar atoms of the exposed polar side chains.

1. Specific Interactions: Hydrogen-Bonding, Nonbonded, and Hydrophobic Interactions

In the preceding paragraph, we noted that hydrogen-bonding interactions between polar atoms in a protein can greatly stabilize the structure of a protein. These interactions are due to the sharing of a hydrogen atom between the donor atom to which the H atom is covalently linked and the receptor polar atom as in Fig. 4. This interaction stabilizes the distribution of charges over the three atoms involved in the hydrogen bonding system (Fig.4). For a given protein sequence, hydrogen bonding patterns can be arranged in many possible

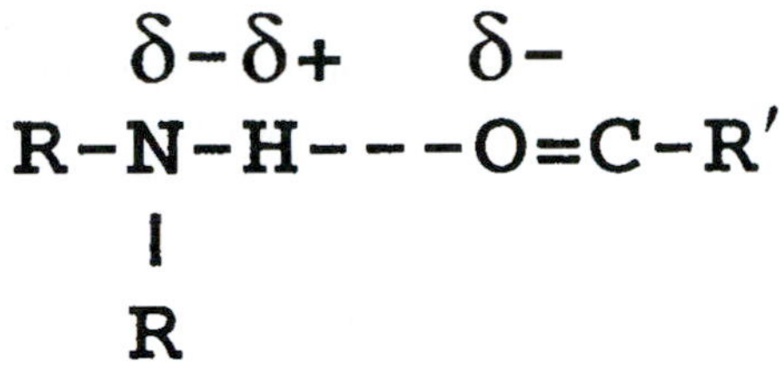

$$\overset{\delta-\ \ \delta+\ \ \ \ \ \delta-}{R-N-H---O=C-R'}$$
$$\overset{|}{R}$$

FIGURE 4. Illustration of how hydrogen bonds are stabilized by partial sharing of a polar H atom between two polar heavy atoms (N and O in this case).

ways. However, in every protein whose three-dimensional structure is known, only one or a few hydrogen-bonding schemes are observed, many of these occurring between backbone atoms (NH---O=C) and others occurring between the polar atoms of side chains.

In addition to the hydrogen-bonding interactions between the atoms of a protein, all pairs of atoms in a protein interact with one another so as to attract each other weakly until they become sufficiently close that their electron clouds repel one another. The weak attractive interactions between a pair of atoms are due to the polarization of charges in each atom between the negatively charged electrons and the positively charged nucleus that is induced by the other interacting atom. This induced-dipole–induced-dipole interaction is attractive because the positive end of the dipole of the first atom interacts attractively with the negative end of the dipole of the other atom of the pair and *vice versa*. The overall attractive energy for this interaction can be simply expressed as $-B_{ij}/R_{ij}^6$, where B_{ij} represents a constant that characterizes the magnitude of the attractive energy between any two interacting atoms and R_{ij} is the distance between the two interacting atoms. As shown in Fig. 5, as two interacting atoms approach one another, the attractive energy becomes progressively stronger until the electrons in the outer shells repel one another. This repulsive energy increases rapidly as the internuclear distance between the atoms decreases.

Most proteins fold in an aqueous environment. Changing the solvent, such as by adding nonpolar solvents to water, is known

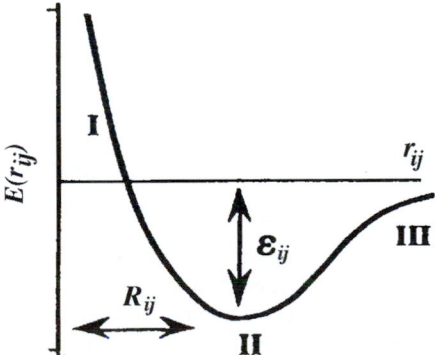

FIGURE 5. Interatomic van der Waals energy as a function of distance. Notice that two atoms can approach one another with decreasing energy (regions II and III) until the energy reaches a minimum (ϵ_{ij}) at a distance of R_{ij}. If the atoms become closer, they then repel one another so that the energy rises steeply (region I).

to denature proteins so that they cannot function properly. Evidently, water provides important interactions with proteins, such that they fold correctly to their native conformations.

The three-dimensional structures of over 300 proteins have been determined mostly by the technique of x-ray crystallography (see Section II.B). In all of these proteins, the nonpolar amino acid residues tend to "pack" in the interior of the protein, while the polar residues tend to interact with water on the surface of the protein. Avoidance of interaction of the nonpolar residues with water is due to the **hydrophobic (water-avoiding) effect.** This behavior is caused predominantly by an entropy effect. Water molecules tend to become highly ordered around the side chains of nonpolar or hydrophobic amino acid residues. On the other hand, water molecules can interact in a large number of low-energy complexes with polar side chains and so are less constricted structurally. This ordering of water molecules by nonpolar residues causes the latter to "pack" into the interior of the protein, allowing the water to be more disordered on the outside of the protein.

If one measures the free energy of transfer, ΔG, of nonpolar compounds (such as benzene or *n*-hexane) from nonpolar to polar solvents, it is found that the enthalpy of transfer, ΔH, is small whereas the entropy of transfer, ΔS, is a large negative value. From the Gibbs free energy expression,

$$\Delta G = \Delta H - T\Delta S \qquad (1)$$

it can be seen that a large negative entropy makes the overall process unfavorable; in other words ΔG becomes more positive. This effect can be attributed to the water structure effect (see Scheraga, 1984).

All of the above interactions have been taken into account in a variety of computer programs that generate the conformations of polypeptides and proteins and calculate their conformational energies. One such program is ECEPP (Empirical Conformational Energies of Peptides Program) developed in the laboratory of Professor Harold A. Scheraga of Cornell University. The energy parameters, such as the constants for the nonbonded interaction energies, have been determined experimentally. This program has been used to compute the structures of many polypeptides and proteins with excellent agreement between predicted and experimentally determined structures (Scheraga, 1984). We give examples below of how these methods have been used to predict protein structure and provide insight into the relationship between the structure of a protein and its function. First, we describe the properties of the structures of proteins.

E. Properties of the Structures of Proteins

1. Geometry and Dihedral Angles

Among the many protein structures that have been determined by crystallography, each of the amino acid residues that compose the protein have geometries that are remarkably constant throughout the protein and when compared between any two proteins. Remarkably, this geometry is essentially the same as that found in single crystals of the individual amino acids. Geometry refers to the **bond lengths** and **bond angles** of the individual amino acid. An example of the geometry of an amino acid is illustrated for glycine in Fig. 6. The basic vari-

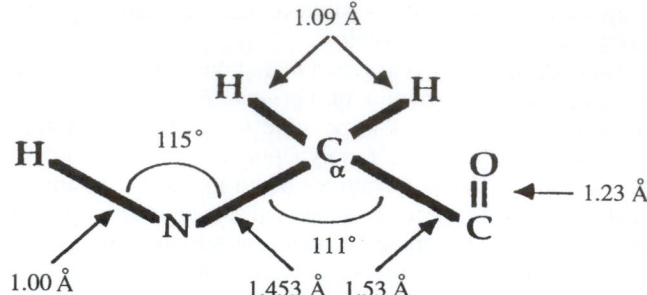

FIGURE 6. The geometry of a glycine residue showing typical bond lengths and bond angles.

ables that allow for changes in chain conformation are the **dihedral angles,** defined as the angle made between two overlapping planes. For example, in Fig. 7, the angle between the planes determined by the N–C^α–C' atom and the C^α–C'–N atoms of the two amino acid residues shown is a dihedral angle. These angles are generated by rotation around single bonds. As shown in Fig. 7, for the backbone of a single amino acid residue, three dihedral angles exist: Φ, Ψ, and ω.

The dihedral angle Φ is determined by the angle between the planes of C'–N–C^α and N–C^α–C' whereas the dihedral angle Ψ is determined by the angle between the planes of N–C^α–C' and C^α–C'–N. To determine Φ, one merely has to look down the N–C^α axis using the C'–N bond as a reference and determine the angle made between the C'–N bond and the C^α–C' bond as shown in Fig. 7. To determine Ψ, one uses the N–C^α bond as a reference and measures its angle with the C'–N bond. The dihedral angle ω is the interpeptide dihedral angle, which is determined by rotation around the C'–N bond shown in Fig. 7. It can be measured by determining the angle made between the C^α–C' bond and the N–C^α bond when sighting down the C'–N bond. This angle is almost always close to 180°, the *trans*-planar peptide conformation.

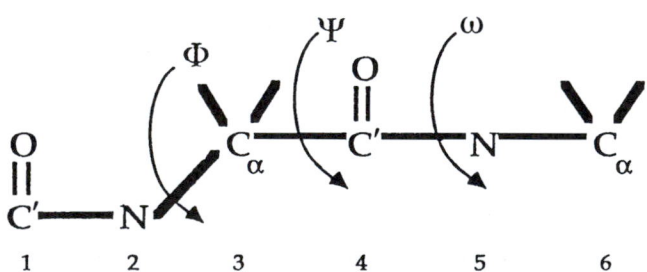

FIGURE 7. Dihedral angles, Φ, Ψ, and ω, for the backbone of an amino acid. Φ is determined by atoms 1, 2, 3, and 4. Clockwise rotations of the 3–4 bond relative to the 1–2 bond are considered positive while counterclockwise rotations are considered negative. 0° is the conformation in which the 3–4 bond is eclipsed relative to the 1–2 bond, while 180° is the conformation in which the two bonds are *trans* relative to one another. Ψ is determined by the 4–5 bond relative to the 2–3 bond. The same considerations apply to this dihedral angle as to those for Φ. ω is determined by the angle between the 5–6 bond and the 3–4 bond. This dihedral angle is almost always very close to 180° (i.e., the 5–6 bond is *trans* to the 3–4 bond).

Given the geometry and a complete set of dihedral angles for a given sequence of amino acids in a polypeptide, a unique three-dimensional structure can be generated, called the **conformation** of the protein. All regular structures described above have repeating values for Φ and Ψ from residue to residue. All α-helices have values for Φ and Ψ that are close to $-60°$ while, for the individual extended chains of β-sheets, the values for both of these dihedral angles lie close to $180°$.

In fact, the α-helical or A and extended or E conformations are minimum-energy conformations for all of the naturally occurring amino acids. Using conformational energy calculations based on ECEPP, all of the low-energy minimum conformations for the 20 naturally occurring amino acids have been computed as a function of the dihedral angles Φ and Ψ. In Fig. 8, these energies are plotted as isoenergetic contours in the same way that different elevations are plotted on contour maps. All of the amino acids have at least seven basic low-energy minima as indicated by the dots in Fig. 8.

Note in this figure that large regions of the map are energetically forbidden and that the allowed (low-energy) regions of the map are relatively restricted (Fig. 8). Analysis of the conformations of the individual amino acids in proteins whose three-dimensional structures are known reveals that virtually all of these amino acids adopt one of these eight basic conformational states, consistent with the results of the energy calculations.

Using programs such as ECEPP, systematic methods, reviewed extensively in Vasquez *et al.* (1994), have been developed that generate representative sets of the low-energy structures for a given polypeptide chain. These methods include the chain build-up procedure (Pincus, 1988), molecular dynamics (Karplus and McCammon, 1986), Monte Carlo procedures, and more specialized methods that allow for "jumping" potential energy barriers between different conformations of the given polypeptide chain (Scheraga, 1989). All of these methods take advantage of the energy map shown in Fig. 8, since the backbone of each amino acid is constricted to adopting one of the low-energy states shown in this figure. Once the low-energy conformations for a given polypeptide are generated and their energies computed, the lowest energy structure is selected and this structure is the one that should be observed experimentally.

Successful ECEPP computations of the three-dimensional structures of a number of polypeptides and proteins have been performed. Examples are shown in Fig. 9 for the cyclic decapeptide gramicidin A, collagen, and melittin (Pincus and Scheraga, 1985). There is close agreement between these computed structures and the experimentally determined ones. Excellent agreement between theory and experiment has also been achieved for other proteins such as avian pancreatic polypeptide and bovine pancreatic trypsin inhibitor (Scheraga, 1989), and a 100-residue segment of human leukocyte interferon (Gibson *et al.*, 1986). The experimental structures of these polypeptides and proteins have been determined by specific methods, namely, x-ray crystallography, two-dimensional nuclear magnetic resonance spectroscopy (NMR), or circular dichroism.

II. Techniques for the Determination of the Structures of Proteins

A. Regular (Secondary) Structure: Circular Dichroism

One of the most widely used techniques for determining regular structure is **circular dichroism,** which is based on the principle that molecules with asymmetric structures absorb light asymmetrically. Light is electromagnetic radiation which can be plane-polarized to the right or to the left. Normally light consists of both types of plane-polarized waves. When a molecule with an asymmetric structure absorbs light, it preferentially absorbs either the left- or rightpolarized light wave. The amount of light absorbed, A, is equal to $E \times C$, where E is the molar extinction coefficient and C is the concentration of the molecules. For molecules with asymmetric structures, the E values for left- and right-plane-polarized light differ. The difference $E_L - E_R$ depends on the wavelength of the incident light. A plot of $E_L - E_R$ (called the **molar ellipticity**) versus wavelength is referred to as the circular dichroism or CD spectrum. Regular structures in proteins are asymmetric both because the amino acids in these structures are themselves asymmetric and because these structures have a **handedness** or a twist-sense as in the case of the α helix, which is right handed. (See van Holde, 1985, for a complete treatment of circular dichroism.) Since the peptide bond has strong absorption in the far-UV wavelengths from 230 nm down to about 190 nm, CD spectra for each different

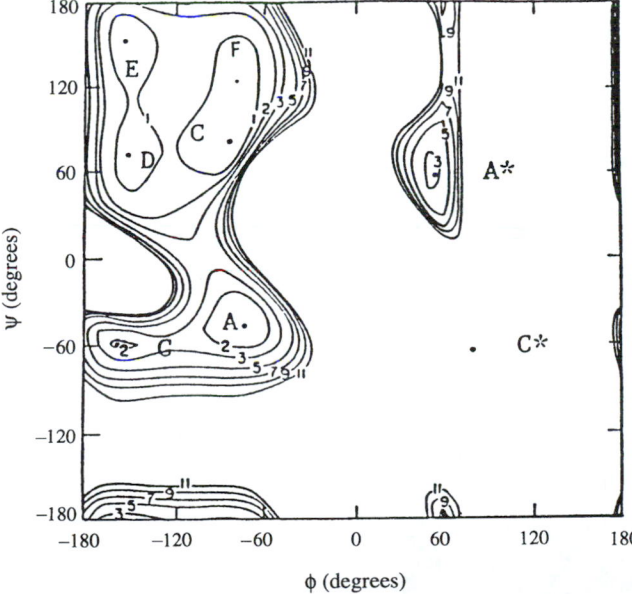

FIGURE 8. Contour map for Φ and Ψ for N-acetyl-alanine-N'-methyl amide as computed by ECEPP (Scheraga, 1984). Note that the actual energy minima are denoted by dots on the energy contour map. Seven low-energy minima are shown: A (α-helical), C, D (both of which are seen in bends), G, E, and F (both of which are seen in β-sheets), A* (left-handed α-helix), and C*, which is energetically forbidden for all L-amino acids except glycine. Note also that some of these regions are "split" (i.e., more than one minimum can exist for a given conformational region).

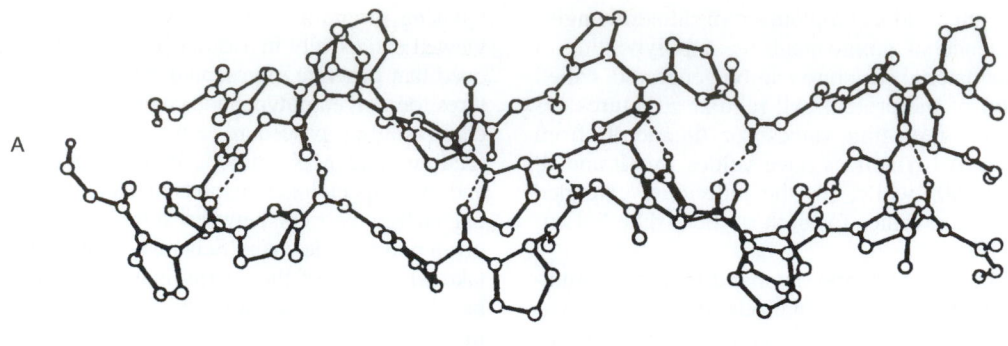

A

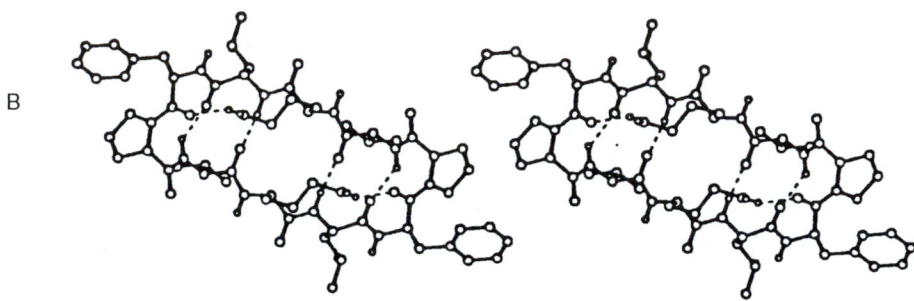

B

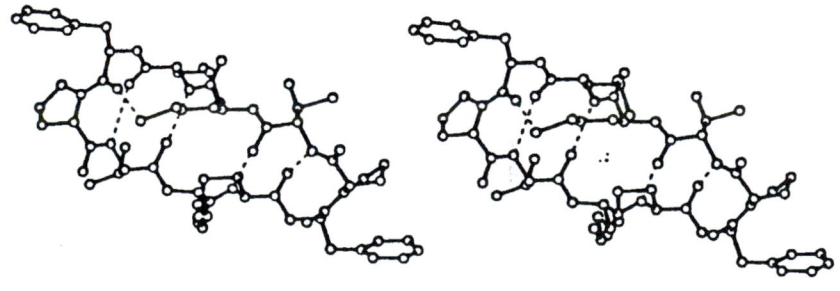

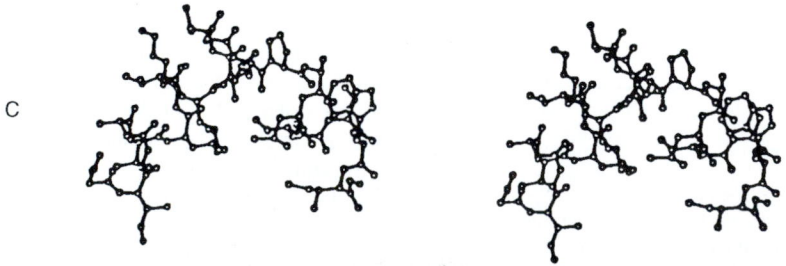

C

FIGURE 9.   Views of computed structures. (A) The structure of a section of collagen model polypeptide, (Gly-Pro-Pro)$_{20}$, (Miller and Scheraga, 1976), which was directly confirmed by single-crystal x-ray crystallographic studies subsequent to the computations of the structure of this molecule. (B) The computed structure for the cyclic decapeptide gramicidin A (upper part) and the x-ray structure of the same polypeptide (lower part) show the close agreement between the predicted structure and the x-ray crystal structure (Dygert *et al.*, 1975). (C) The computed structure for the membrane-active protein melittin, which consists of two

type of regular structure should be unique in this range of wavelengths. The patterns in the CD spectra found for each structure type are shown in Fig. 10. In this figure, the term *random coil* is a misnomer: the term refers collectively to all structures that are not α-helices or β-sheets, but these other structures are unique, not random.

For proteins of unknown three-dimensional structures, a knowledge of the contents of regular structures (α-helices and β-pleated sheets) can be obtained by taking their CD spectra. Since patterns for α-helix, β-sheet, and random coil structures are known, these patterns may be combined to reproduce the observed CD spectrum of a particular protein by curve-fitting. The combination of regular structure that best fits the observed CD curve is the one most likely to exist in the protein. This methodology has been tested on proteins of known three-dimensional structures (so that the percentage of regular structures can be determined) and found to reproduce the regular structure quite satisfactorily.

B. Three-Dimensional Structure of Proteins: X-Ray Crystallography

This technique is based on the principle that the electron clouds around atoms diffract incident light in a predictable

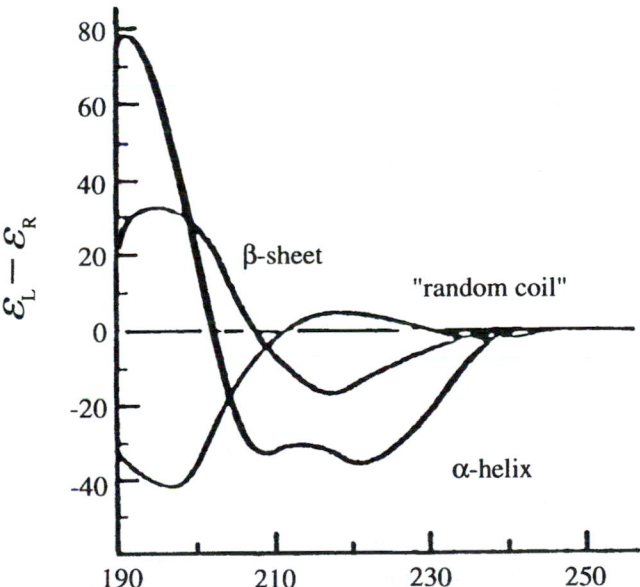

FIGURE 10. CD spectra for regular structures: α-helix, β-sheet, and so-called random coil. Note that α-helices have a characteristic "dip" in the region around 210 nm. For a given protein, one can compute the amount of each such structure from its CD spectrum by curve-fitting, using different amounts of each structure type and assuming that the resulting spectrum is the sum of the individual spectra for each structure type.

way. If the atoms are parts of an oriented molecule, in other words, the protein molecules are all oriented in a **crystal**, the same atoms of different molecules diffract the light in the same way. The closer two atoms are in the molecule, the farther apart the diffraction pattern will be. Conversely, the farther away two atoms are in the molecule, the closer the diffraction patterns from each atom will be. Thus the diffraction pattern is the reciprocal of the atomic pattern. The diffraction pattern can be analyzed with respect to the amplitude of diffraction and the relative positions of the scattering pattern. The light diffracted must be x-rays because the wavelength of light used must be of the order of single bond lengths, around 1.5–2.0 Å. Only x-rays have this range of wavelengths.

Once a diffraction pattern has been obtained, the problem is to relate the positions of the atoms to one another from the diffraction pattern. The relative positions of two atoms can be computed from their (diffraction) amplitudes and a phase factor that is related to the angular displacement of one atom relative to the other. This phase factor cannot be determined from the diffraction pattern. Since the position of at least one atom is required, in the x-ray diffraction of proteins, a heavy atom bound to the protein is used in the so-called heavy atom replacement method. Once the phase factor is determined, the individual atomic positions can be computed from the diffraction pattern as a Fourier transform that converts from reciprocal (diffraction) space to real (three-dimensional) space.

C. Two-Dimensional High-Resolution Nuclear Magnetic Resonance Spectroscopy (NMR)

In one of the most exciting developments in protein chemistry, this technique is being applied to solving the structure of proteins **in solution** without requiring them to be oriented in a crystal. NMR is based on the principle that most nuclei of atoms have spins. If a magnetic field is applied to the atom, the nuclei tend to align their spins with the field. The energy difference between the spins in the absence and the presence of the field is proportional to the magnetic field strength and is given as

$$\Delta E = h\nu = g\beta H \qquad (2)$$

where ΔE is the energy difference, h is the Planck constant, ν is the resonance frequency corresponding to the energy absorbed by the spinning nuclei, $g\beta$ is a constant, and H is the magnetic field strength. Nuclei like hydrogen exist in a protein in different environments so that some are more shielded by electrons than others, thereby requiring higher ν values or field strengths to excite them. The frequencies corresponding to the transitions of the spins of individual nuclei to align themselves with the field is in the radiofrequency range. Because different H atoms absorb at different field strengths or values, one can obtain an absorption spectrum for a given compound or protein,

---

α-helical rods, separated by a bend in the middle of the chain (Pincus *et al.*, 1982). These features have been described in the x-ray crystal structure of this protein. Each polypeptide represents a type of structure seen in many proteins, viz., the collagen helix (A) in fibrous proteins, β-sheets with reverse turns (B), and α-helices (C). (See Pincus and Scheraga, 1985, for a further description of these structures.)

where for a given field strength, *H*, different radio frequencies excite the nuclei to change their spins to orient with the field. When two H atoms from different parts of a protein approach one another, their spins either add to or subtract from one another and affect the intensity of each other's absorption in a phenomenon known as the **nuclear Overhauser effect** or NOE. This change in intensity is inversely proportional to the sixth power of the distance between the two H atoms. If the individual resonant frequencies for the two interacting H atoms are known, the distance between them can be calculated.

The specific H atoms that give rise to different NOE values in a protein can be identified by an elegant technique called two-dimensional NMR (2D-NMR) (Wuthrich, 1986). By irradiating the nuclei at two different frequencies, it is possible to construct a correlation map between the different interacting atoms. A large number of inter-H distances allows us to fit well-defined structures that have these distances so that a class of structures can be directly determined for the given polypeptide or protein.

Once a protein has adopted its final folded form, it possesses a unique size and shape that confer on it certain bulk properties, among the most important of which are electrical properties, which we now discuss.

III. Bulk Properties of Proteins: Proteins as Polyelectrolytes

Folded proteins will, in general, have different shapes and sizes depending on their specific sequences of amino acids. Virtually all folded proteins have their nonpolar groups buried in the interior of the protein as a result of hydrophobic interactions and their polar and charged groups on the surface of the protein interacting with the solvent. This distribution of charged groups on the surfaces of proteins strongly affects the manner in which these proteins will be oriented and migrate in electric fields that are either applied in electrophoresis experiments or that result from the transmembrane potential in cells. Proteins with net charges also contribute to the fixed charge inside cells. How proteins migrate in electric fields gives much information about their sizes, shapes, and molecular masses. We must therefore consider the properties of proteins as polyelectrolytes dissolved in solution.

A. Acid-Base Properties of Amino Acids

All amino acids have free α-amino (NH_2) groups and carboxyl (COOH) groups. Amino groups are **basic**, that is, **take up** hydrogen ions (H^+), while carboxyl groups are **acidic,** that is, **give up** hydrogen ions. Using glycine as an example, at neutral pH (7.0), the only form to exist is the dipolar, or zwitterionic, form, $^+H_3N–CH_2–COO^-$. Figure 11 shows a prototypical titration curve for a dibasic acid such as glycine over the pH range of 1–10. In this figure, note that there are two regions of the titration curve where the change in pH is minimal for added base: at pH of around 2.0 and 9.0, the pK_a values for the COOH and NH_2 groups, respectively. At pH 2.0 the buffering capacity of the COOH group is maximal, while at pH 9.0, the buffering of the NH_2 is maximal. The buffering capacity is governed by the Henderson-Hasselbalch equation,

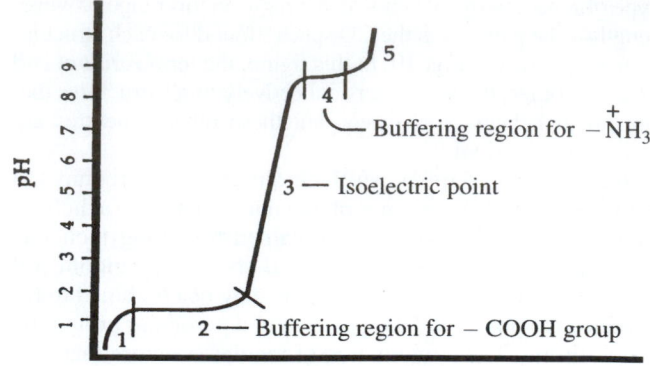

(1) $\overset{+}{H_3N} - CH_2 - COOH$

(2) $\overset{+}{H_3N} - CH_2 - COOH \longleftrightarrow \overset{+}{H_3N} - CH_2 - COO^-$

(3) $\overset{+}{H_3N} - CH_2 - COO^-$

(4) $\overset{+}{H_3N} - CH_2 - COO^- \longleftrightarrow H_2N - CH_2 - COO^-$

(5) $H_2N - CH_2 - COO^-$

FIGURE 11. Titration curve for a dibasic acid like glycine to show the two buffering regions for the COOH and NH_3^+ groups and the isoelectric point that can be computed as $(pK_{a1} + pK_{a2})/2$. A protein has multiple buffering regions, characteristic of each residue type that is involved in a prototropic dissociation.

$$pH = pK_a + \log \frac{\left[\text{conjugate base form}\right]}{\left[\text{acid form}\right]} \quad (3)$$

where the pK_a is the negative logarithm of the proton dissociation constant for the acid. For the acid segment of the titration curve in Fig. 11, these two forms are $H_3N^+–CH_2–COO^-$ and $H_3N^+–CH_2–COOH$, respectively; for the high-pH portion of the curve, these forms are $H_2N–CH_2–COO^-$ and $H_3N^+–CH_2–COO^-$, respectively. Note also that the following equilibria exist over this range:

$$H_3N^+–CH_2–COOH \rightleftharpoons H_3N^+–CH_2–COO^- + H^+$$

$$H_3N^+–CH_2–COO^- \rightleftharpoons H_2N–CH_2–COO^- + H^+ \quad (4)$$

We can write two separate sets of equilibrium conditions for the two reactions in Eq. 4 as follows:

$$K_1 = \frac{\left[H_3N^+–CH_2–COO^-\right]\left[H^+\right]}{\left[H_3N^+–CH_2–COOH\right]}$$

$$K_2 = \frac{\left[H_2N–CH_2–COO^-\right]\left[H^+\right]}{\left[H_3N^+–CH_2–COO^-\right]} \quad (5)$$

Solving both of these equations for the common zwitterionic species, $H_3N^+-CH_2-COO^-$, we obtain the expression

$$K_1 K_2 = \frac{\left[H_3N^+-CH_2-COOH \right]}{\left[H_2N-CH_2-COO^- \right]} \times \left[H^+ \right]^2 \qquad (6)$$

Taking the negative logarithms of both sides of Eq. 6, we obtain the expression

$$pH = \frac{pK_1 + pK_2}{2} + \frac{1}{2} \log \frac{\left[H_2N-CH_2-COO^- \right]}{\left[H_3N^+-CH_2-COOH \right]} \qquad (7)$$

where pK_1 and pK_2 are the negative logarithms of K_1 and K_2. This equation is the same as the Henderson-Hasselbalch equation for any acid-base equilibrium except, in this case, it is written for two equilibria. The **isoelectric point**, pI, is defined as the pH at which the total negative charge of the molecule is equal to its total positive charge (i.e., where only the zwitterionic species exists). In terms of Eq. 7, this condition is met if the numerator and denominator are equal. In this case,

$$pH = \frac{pK_1 + pK_2}{2} \qquad (8)$$

For amino acids with charged side chains, the isoelectric point is easily shown to be that pH equal to $(pK_1 + pK_2 + pK_3)/3$. In general, for multiple prototropic dissociations for a whole protein, the isoelectric point of the protein is simply that point on the titration curve for the whole protein where

$$pH = \sum pK_i / N \qquad (9)$$

where the sum is taken over all proton-dissociating groups, and N is the number of these groups in the protein.

B. Protein Charge and Solubility

Since all proteins are charged because of charged side chains and the α-NH_3^+ group and the $-COO^-$ carboxyl terminal group, there is generally a **net** charge on the protein. These net charges tend to be solvated and interact with counterions in solution. The solvation of these charges makes the protein soluble in H_2O. However, interactions of the charged side chains of the protein with surrounding ions and aqueous solvent become minimal at the isoelectric point, where charges on the protein exactly balance one another. At its isoelectric point, the solubility of the protein in water is minimal, and it will tend to precipitate from solution. Different proteins have different sequences and different charges and isoelectric points so that their tendencies to precipitate will differ from one another and depend on the pH of the solution. Thus, by changing the pH of protein solutions, it is possible to effect differential precipitation of these proteins. A more sophisticated version of isolation of proteins based on their isoelectric points is **isoelectric focusing**, an electrophoretic method discussed in Section III.D.

1. Isoionic Point

This term is used when proteins are dissolved in aqueous solutions where the only counterions are H^+ and OH^-; in other words, the counterion for the anionic groups is H^+, and

the counterion for the cationic groups is OH^-. Since the total number of positive charges for the system must equal the total number of negative charges for the system,

$$(H^+) + \text{(total positive charges on protein)}$$
$$= (OH^-) + \text{(total negative charges on protein)} \qquad (10)$$

From Eq. 10, if the isoionic point of a protein is at pH 7 (i.e., $H^+ = OH^-$), **the isoelectric point is the same as the isoionic point** since from Eq. 10, if $H^+ = OH^-$, then total positive charges on the protein equal total negative charges on the protein.

C. Titration of Proteins

Because of their large number of exposed acidic and basic groups, proteins behave as buffers. Generally, the pH of the buffering region will depend on the relative numbers of acidic and basic groups on the protein surface. Binding of protons to proteins may be considered to be formally the same as the binding of ligands (small molecules) to proteins. To understand the behavior of proteins as buffers, it is important to understand ligand binding theory.

1. Ligand Binding Theory

Suppose a protein has n equivalent sites for binding to a ligand and each site is completely independent of any other site. The equilibrium for the binding of a ligand, A, to any one site of the protein can be written as

$$A + P \rightleftharpoons AP \qquad (11)$$

where A is the free ligand, P is the unbound protein, and AP is the ligand bound to protein. The equilibrium association constant, K, for this process is

$$K = \frac{[A][P]}{[AP]} \qquad (12)$$

where [A] is the free ligand concentration, [P] is the concentration of unbound protein, and [AP] is the concentration of ligand bound to protein. If there are n equivalent binding sites per protein molecule, then the total concentration of sites is $n[P]_o$, where $[P]_o$ is the concentration of total protein and is equal to the sum of the concentrations of free and bound protein:

$$[P]_o = [P] + [AP] \qquad (13)$$

Combining Eqs. 12 and 13, we find that

$$[P]_o = [P] + K[A][P] \qquad (14)$$

Solving for [P], we obtain

$$[P] = \frac{[P]_o}{1 + K[A]} \qquad (15)$$

Since from Eq. 12, $[AP] = K[A][P]$, the ratio of bound protein to total protein is

$$\frac{[AP]}{[P]_o} = \frac{K[A]}{1 + K[A]} \qquad (16)$$

If there are n equivalent binding sites per protein molecule, the fraction of sites bound from Eq. 15 is

$$\frac{[AP]}{n[P]_o} = \frac{K[A]}{1 + K[A]} \qquad (17)$$

This equation can be linearized. Defining $R = [AP]/[P]_o$,

$$R/[A] = nK - KR \qquad (18)$$

Plots of $R/[A]$ versus R (called **Scatchard plots**) should give a straight line whose slope is K and whose intercept is nK. Thus both K and n are readily determined.

For n equivalent proton binding sites on a protein, we can write Eq. 16 as

$$R = \frac{K[H^+]}{1 + K[H^+]} \qquad (19)$$

In this case, $R = [PH^+]/n[P]_o$. If we use the dissociation constant $K' = 1/K$, Eq. 19 may be recast as

$$[PH^+]K' = \left(n[P]_o - [PH^+]\right)[H^+] \qquad (20)$$

Rearranging and taking logs of both sides of Eq. 20, we obtain the equation

$$pH = pK_a + \log \frac{n[P]_o - [PH^+]}{[PH^+]} \qquad (21)$$

where pK_a is $-\log (K')$. Note that Eq. 21 is the Henderson-Hasselbalch equation for multiple dissociations. The denominator in the logarithmic term in Eq. 21 represents the concentration of H^+-bound protein while the numerator represents non-proton-bound protein. This equation is useful for titration of proteins provided that the pK_a of each group is independent of that of any other group for a given set of groups, such as the carboxyl groups of Glu and Asp.

For proteins, it is not generally correct to assume that the above condition holds because, as each carboxyl group dissociates from a protein, the protein becomes progressively negatively charged, so that it becomes progressively difficult for the next proton to dissociate. In fact, the protein behaves as a polyelectrolyte wherein proton dissociations are strongly dependent on the electric field of the protein.

2. The Electrostatic Field Effect (Bull, 1943)

To account for the electrostatic field effect of the charged protein on the dissociation process, we note that the free energy of dissociation per mole of acid group, ΔG, can be written as

$$\Delta G = \Delta G_i + \Delta G_e \qquad (22)$$

where ΔG_i is the intrinsic dissociation free energy for the group and ΔG_e is the electrostatic free energy for dissociating a proton from the acid group in the presence of the field of the protein. From electrostatic theory,

$$\Delta G_e = eU \qquad (23)$$

where $e = 1$ electrostatic charge unit and U is the electrical potential from the protein. Since the standard free change for dissociation, ΔG, equals $-2.303 \times RT \log K$, and ΔG_i equals $-2.303 \times RT \log K_i$, (where R is the gas constant and T is the temperature in K), from Eq. 22, we obtain the relation

$$pK = pK_i + eU/2.303RT \qquad (24)$$

From Eq. 21, we have

$$pH - \log \frac{n[P]_o - [PH^+]}{[PH^+]} = pK_i + eU/2.303RT \qquad (25)$$

It is of obvious importance to obtain an explicit expression for U since it strongly influences the dissociation of protons from proteins.

3. Computation of the Electrical Potential

To compute the electrical potential U, it is necessary to use a model for the interaction of protein charges with those of small ions. The simplest model is to assume that the protein is a charged sphere and that the counterions surrounding it, like protons, form a spherical shell around the central charged protein sphere as represented in Fig. 12. The overall arrangement is that of two concentric spheres of radii R_1 and R_2, the first term being the distance from the center of the protein to its surface, and the second term being the distance from the center of the protein to the center of mass of the ions surrounding the protein. It is clear from this figure that the interactions of protons, and/or any ions, with the central, charged protein molecule depend on the distance of the centers of these ions from the center of the protein ion. The distribution of ions around the central charged protein ion is given by the Boltzmann distribution,

$$\Psi = \sum e n_i Z_i \exp -(e n_i Z_i U/RT) \qquad (26)$$

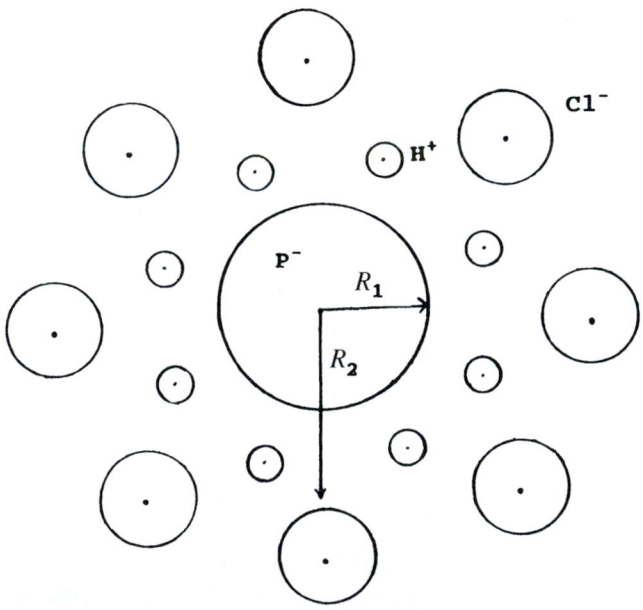

FIGURE 12. The model for the effect of the charge on a protein (p^-) on the dissociation of protons from groups on the protein involved in prototropic dissociations. R_1 is the radius of the protein, assuming that it has a spherical shape. R_2 is the distance from the center of the protein sphere to the mean center of the ions surrounding the protein. In this figure, H^+ ions are shown as the positive ions and Cl^- as the negative ions. Their distributions around the protein will depend upon the charge on the protein and the Boltzmann factors for each surrounding ion type. R_2 is therefore a mean distance from the center of the protein to the mean center of the surrounding ions.

where Z_i is the valence of the ith ion. We can expand the exponential term in a Taylor series as follows:

$$\Psi = \sum n_i Z_i e - \sum n_i Z_i e \left(e Z_i U / RT \right) \\ + \left(\sum n_i Z_i e / 2 \right) \left(Z_i e U / RT \right)^2 + \dots \tag{27}$$

We make the assumption that the eZU term is much less in value than RT, and that electrical neutrality must be preserved (i.e., the first sum on the right of Eq. 27 is 0). Then,

$$\Psi \approx - \sum n_i Z_i^2 e^2 U / RT \tag{28}$$

This charge distribution must satisfy the Poisson charge distribution equation,

$$\nabla^2 U = -4\pi \Psi / D \tag{29}$$

where $\nabla^2 U$ is the second derivative of the potential with respect to the sum of the coordinates of the system, Ψ is the charge distribution, and D is the dielectric constant of the medium, water. Combining Eqs. 28 and 29, we obtain,

$$\nabla^2 U = \left(4\pi e^2 \right) / (DRT) \times \sum n_i Z_i^2 U \tag{30}$$

Solution of this equation for U as a function of the coordinates of the system constitutes the Debye–Hückel theory, and it depends on the assumed configuration of small ions around the central large protein ion, assumed here to be spherical in shape as in Fig. 12.

Solution of Eq. 30 for this circumstance at R_1 gives

$$U = Ze \left(R_2 - R_1 \right) / DR_1 R_2$$

where $Z = \sum n_i Z_i$. Letting $1/\chi = R_2 - R_1$,

$$U = \frac{Ze}{DR_1 \left(\chi R_1 + 1 \right)} \tag{31}$$

Combining Eqs. 24 and 31, we obtain

$$pK = pK_i + \frac{Ze^2}{2.303RTD} \times \frac{1}{R_1 (\chi R_1 + 1)} \tag{32}$$

Setting

$$W = \frac{e^2}{2.303RTD} \times \frac{1}{R_1 (\chi R_1 + 1)} \tag{33}$$

Eq. 25 may now be written as

$$pH - \log \frac{n[P]_o - [PH^+]}{[PH^+]} = pK_i + 0.868WZ \tag{34}$$

This equation now takes into account the field effect of protein and surrounding ion charges provided that the system conforms to the assumptions made for this model: that $U \ll RT$, that the protein is spherical in shape, and that the ions around the protein form a spherical shell around it. Plots of the left side of Eq. 34 versus Z, the overall protein charge, should yield a straight line whose slope is $0.868W$. This term may be regarded as a structure term that depends on the radius of the protein and the position of the center of mass of the surrounding ion cloud.

4. More Explicit Models Based on Protein Structures

The above theoretical expression is based on several critical assumptions that do not always apply to proteins, namely, that the charge distribution around the protein is uniform and spherical and that its electrical potential is such that the ZU term $\ll RT$. We know from the many x-ray crystallographically determined protein structures that protein surfaces are far from spherical and that the charge distributions are not uniform. Thus, to understand the effects of protein charge and the effect of the medium on proton dissociations from the protein, more explicit and detailed models are needed. One treatment is based on conformational energy calculations on proteins surrounded explicitly by solvent molecules. These calculations have thus far been aimed at determination of the intrinsic pK_as of specific groups on protein surfaces.

In this approach (Russell and Warshel, 1985), the energy of dissociation of a particular acid group, AH, is computed for the whole protein surrounded by water molecules placed on a grid around the protein. The energy of interaction of the atoms of the protein with one another is computed using energy functions similar to those discussed in Section I.D. Included in this energy is the effect of the permanent dipoles of the atoms of the protein that induce dipoles on the other atoms of the protein. The magnitude of this energy depends on the polarizability of the interacting atoms. Further account is taken of the effect of the interaction of the dipoles of the atoms of the protein with water molecules which themselves are represented as dipoles. The calculations are then applied to the dissociation of a proton from a particular group, for example, Asp 3 and Glu 7 in bovine pancreatic trypsin inhibitor (BPTI) whose structure has been determined at high resolution by x-ray crystallography (see Section II.B).

In this procedure, a thermodynamic cycle is employed in which the free energy difference of the system is computed for the acid group in the AH form in the protein and for the AH group dissolved in water (process 1). The group is then allowed to dissociate in water at a given pH (process 2). The free energy difference between AH and A$^-$ in water is known (for Asp residues, this would be the free energy of the dissociation of acetic acid in water at a given pH) and is equal to $2.303RT(pK_a - pH)$. Finally, the free energy difference between solvated A$^-$ in water and A$^-$ in the protein (process 3) is computed. The overall free energy for dissociation is then computed as the difference between the free energy for process 1 and the free energy for process 3 plus a constant (free energy for dissociation of AH in water).

These calculations have been carried out on a supercomputer using molecular dynamics and iterative procedures (for the polarization energies) and have reproduced ionization constants satisfactorily. The method, which is computer intensive, includes the effects of both solvent and the atoms of the whole protein on the ionization process. Thus the approach appears to be a fruitful one in the investigation of prototropic dissociations in proteins.

D. Protein Charge and Electrophoresis

Different proteins have different net charges as discussed above. If these proteins are placed in an electric field, positively charged proteins migrate towards the negatively charged pole, the cathode, while negatively charged proteins migrate to the positively charged pole, the anode. If a protein is present at a pH equal to its isoelectric point, it will generally not move toward either pole. The electrostatic force on a protein present in an electric field is equal to ZeU, where U is the electric field, e is the charge of an electron (or a proton), and Z is the number of charges present. There is a drag force on the proteins that is proportional to the velocity at which they move towards either pole, fv, where v is the velocity of the protein moving in the electric field and f is a proportionality constant. This force balances the electrostatic force so that

$$ZeU = fv \qquad (35)$$

and

$$v = ZeU / f \qquad (36)$$

The value of f generally increases with increasing size and molecular weight so that, for proteins of similar charge but different molecular weights, the higher molecular weight proteins move more slowly than the lower molecular weight proteins.

There are several types of electrophoresis, all of which are extremely effective not only in separating proteins but in directly determining their molecular weights.

1. Slab Gel Electrophoresis

Proteins are separated on gels made up of such polymers as starch, polyacrylamide, and agarose. The separation is based mainly on charge, and, to an extent, on size.

2. Sodium Dodecylsulfate (SDS) Electrophoresis

SDS is a detergent that contains a long aliphatic chain and a sulfate group. This detergent interacts with denatured proteins to form a strongly negatively charged complex (the negative charge arising from the SO_4^{2-} groups of SDS). The proteins are first denatured by heat, then the SDS is added in large excess. The SDS-protein complexes all contain about the same negative charge because the SDS swamps out all of the protein charges. Since the charges are all the same, the proteins all separate from one another strictly on the basis of their sizes. From Eq. 36, the larger polypeptide chains of higher molecular weight migrate the most slowly while the lower molecular weight proteins migrate more rapidly towards the anode. Proteins of known molecular weight can be subjected to this procedure and used as "markers." A plot of the molecular weights versus the log of the distance traveled from the point of application yields a straight line. The log of the distance of migration of a protein of unknown molecular weight can then be plotted on this line and the molecular weight directly determined.

3. Isoelectric Focusing

In this elegant technique, a polymer is used in which acidic and basic groups change in density from one end of the polymer to the other. When this polymer is placed in solution, a continuous pH gradient is established along the polymer. A mixture of proteins is then applied to the polymer in a weak buffer. When the proteins migrate in the electric field, they experience local differences in pH on the polymer and eventually reach a local pH equal to their individual pIs. At the pI, the protein no longer migrates in the electric field. This method of separation depends only on the presence of proteins with different pI values.

4. Two-Dimensional Gel Electrophoresis

Proteins can be separated from one another in one lane as described in the preceding three sections; however, these approaches may not meet the conditions for separating all of the proteins. For example, on agarose gel electrophoresis, two proteins may not separate well at the pH of the buffer used. A buffer of different pH may then be added and the electrophoresis carried out at right angles to the original direction of migration to allow further separation of the proteins.

5. Western Blots (Immunoblots)

All proteins are antigens and can provoke the production of antibodies against them when injected into animals. The antibodies are immunoglobulins which are proteins that bind very specifically to and with high affinity for specific antigens. Antibodies and, in fact, all proteins can be conjugated covalently to fluorescent dyes or to enzymes. If we wish to identify a particular protein on an electrophoretogram, we can use a "tagged" antibody to the protein and reveal its presence on the electrophoretic gel. The antibody is "tagged" with a covalently labeled enzyme that catalyzes a reaction that produces a chromophoric (colored) reaction product. For example, alkaline phosphatase catalyzes the hydrolysis of p–nitrophenol phosphate, which is colorless, to p–nitrophenol, which is bright yellow. Since antibody cannot be added directly to the gel from electrophoresis, the gel itself is blotted onto nitrocellulose or another suitable membrane that contains all of the separated bands of proteins as they were on the original electrophoretic gel. This nitrocellulose strip is then incubated in a solution containing the antibody. The antibody itself may be conjugated to an enzyme like alkaline phosphatase, and the band can be identified, or antibodies to the primary antibodies, such as goat anti-rabbit IgG, are conjugated to the enzyme, and the blot is incubated with these secondary antibodies after being treated with the primary antibody. The secondary antibodies are then markers for the desired protein band.

IV. Relationship of Protein Structure to Function

In this section, we discuss some of the relationships of the conformations of proteins to their functions. This is a vast field, and so we concentrate on one aspect of fundamental importance to cell physiology, membrane polypeptides and proteins.

A. Membrane Polypeptides and Proteins

Membrane proteins have exceptional physiological importance in a number of different ways. Virtually all extracellular receptors have a transmembrane domain that is essential to the functioning of the receptor protein. Other membrane polypeptides are vital in the transport of secreted proteins across cell membranes. Still other membrane proteins, such as the components of the complement system, are involved in intercalating into the cell membrane, causing cell lysis. Yet other membrane proteins form ion channels which allow for selective entry or exit of specific ions to and from the cell. In this section, we discuss the relationship of the structures of some of these membrane proteins to their functions. It should be emphasized that this entire field of structure-function relationships of membrane proteins is quite new. Most of our knowledge in this field comes from experimental cell biological data on specific proteins and from the results of computations on membrane and membrane-associated protein structure.

1. Structure and Function of Leader Peptides

Each protein, after it is synthesized on the ribosome, must be transported across the rough endoplasmic reticulum (RER) membranes. If the protein is to be secreted from the cell, or to intercalate into the cell membrane, it must undergo transport across and/or into the cell membrane. For virtually every protein that is synthesized in the cell, there is a polypeptide segment beginning at the N-terminal methionine residue that consists of about 20–30 amino acid residues. Typically, these sequences consist of several hydrophilic amino acids followed by a long stretch of hydrophobic amino acids. When the protein is secreted across the RER membrane, this **leader sequence** is cleaved off, presumably by intracellular proteases. Absence of these leader sequences results in the inability of the protein to traverse the RER membrane, resulting in ultimate intracellular degradation of the protein. Thus, these leader sequences are of vital importance to cellular protein function in general.

The regulation of transmembrane transport of newly-synthesized proteins is delicate. It has been shown in *in vitro* systems that newly synthesized proteins that are secreted across reticulocyte membranes do not undergo leader sequence cleavage. These proteins are nonfunctional because they do not fold correctly. Thus, leader sequences must be attached to the polypeptide chain to enable protein secretion, but must be cleaved off to allow correct protein folding.

To explain how proteins are secreted across membranes, Engelman and Steitz (1981) have proposed the so-called **helical hairpin hypothesis** in which the leader sequence adopts an α-helical conformation. At the end of this segment there is a hairpin turn followed by another α-helix involving 30–40 residues of the protein itself. Both the amino terminal end of the leader sequence and the carboxyl terminal end of the growing polypeptide chain lie on the same side of the membrane while the hairpin turn lies on the opposite side of the membrane. Because the leader sequence helix and the succeeding helical sequence are both independently stable, they interact minimally with one another so that the growing polypeptide chain can slide past the hydrophobic leader sequence in the membrane. As the nascent polypeptide chain pushes through the membrane, anchored by the leader sequence, it begins to fold in the aqueous environment of the cytoplasm on the opposite side of the RER membrane. The leader sequence, however, remains in the membrane because of its hydrophobic character. The hairpin connection between the two helices is a signal for intracellular proteases that ultimately results in cleavage of the leader sequence.

Pincus and Klausner (1982), using ECEPP, have computed the low energy structures for the leader sequence of the κ–light chain immunoglobulin. This sequence, which contains 16 amino acids, is Asp-Thr-Glu-Thr-Leu-Leu-Leu-Trp-Val-Leu-Leu-Leu-Trp-Val-Pro-Gly. These investigators found that the first four polar residues tend to adopt an extended conformation followed by a long α-helix that ends in a hairpin turn at the carboxyl terminal Val-Pro-Gly sequence. This structure was lower in energy than any other competing structure by at least 10 kcal/mol. Thus, these calculations confirm the essential features of the Engelman-Steitz hypothesis, and have been further corroborated experimentally. For example, the CD spectrum of synthetically prepared leader sequences shows high α-helical content when the sequence is dissolved in hexafluoroisopropanol, a nonpolar solvent that simulates the low dielectric medium of the membrane. (Rosenblatt *et al.*, 1980).

In site-specific mutagenesis experiments, Thomas Silhavy and his coworkers at the National Cancer Institute have placed proline residues in the middle of the hydrophobic transmembrane domain of a bacterial leader sequence (Emr and Silhavy, 1982). Proline residues tend to disrupt α-helices. These investigators found that if they introduced two proline residues close together in the sequence, the protein to which the leader sequence was attached was no longer secreted. This result strongly corroborates the conclusion that leader sequences must adopt α-helices in the membrane.

In a series of elegant experiments, Blobel and his coworkers (1979) have shown that specific sequences in proteins will insert into membranes but will not lead to cleavage of the inserted sequence from the rest of the protein. These specific sequences contain what are referred to as "stop signals" that do not permit cleavage. Because the transmembrane domain is hydrophobic, it remains inserted in the membrane and there is no impetus for it to move through the membrane, resulting in secretion. This mechanism is extremely important for the function of receptor proteins, as will be discussed in Section IV.A.3.

2. Melittin

Containing 26 amino acid residues, melittin is perhaps the simplest protein that folds spontaneously. Interestingly, unlike most proteins, it folds in nonpolar environments and actually becomes denatured in water. It is the major component of bee venom and binds to cell membranes, causing lysis and resulting in extrusion of the intracellular contents. The sequence of melittin is $^+$H-Gly-Ile-Gly-Ala-Val-Leu-Lys-Val-Leu-Thr-Thr-Gly-Leu-Pro-Ala-Leu-Ile-Ser-Trp-Ile-Lys-Arg-Lys-Arg-Gln-Gln-NH$_2$. Inspection of this sequence reveals that this protein contains six positive charges and **no** negative charges. At least half of the residues are hydrophobic, as would be expected for a membrane-intercalating protein.

As noted in Section I.E, the structure of melittin has been computed (Fig. 9) and determined by x-ray crystallography with good agreement between the two structures. The structure of this protein may be thought of as a bent α-helical rod with α-helices from Gly 1 to Thr 10, a reverse turn at Thr 11 to Gly 12, followed by another α-helix from Pro 14 to Gln 26. Leu 13 can adopt an energetically favorable conformation, a D state (see Fig. 8) that results in a compact structure for the monomeric state. In the x-ray structure of tetrameric melittin, this residue adopts an energetically less favorable α-helical conformation, causing it to adopt a more open, "straight" structure. In both structures, melittin forms an amphipathic helix in which the nonpolar groups contact one another on the inside of the structure while the polar groups protrude out towards the solvent.

Inspection of the structure for the tetrameric protein reveals that packing between the monomeric units is such that the positive charges of each lie as far away from one another as possible while the hydrophobic cores pack as tightly as possible with one another. This can be accomplished if the monomeric units become slightly less compact by allowing Leu 13 to adopt the α-helical conformation. The sacrifice in conformational energy is about 3 kcal/mol.

Multiple theories about the mechanism by which melittin causes cell lysis have been proposed. One is that melittin acts like a detergent because it is amphipathic, containing both polar positively charged groups and hydrophobic groups, which allows it to interact favorably with water and the hydrophobic membrane lipid bilayer, respectively, resulting in solubilization of the membrane causing its breakdown. This theory does not account for the fact that virtually all biological membranes contain lysolecithins and like fatty acid derivatives that contain nonpolar and positively charged headgroups that actually stabilize membrane structure. Another theory maintains that melittin forms tetrameric complexes in the membrane, forming pores that allow movement of the intracellular contents into the outside environment.

A third theory (Kempf et al., 1982) proposes that melittin undergoes a conformational change in the presence of a transmembrane electric field such that it adopts a less compact structure and becomes a more rigid rodlike structure that spans the whole membrane and disrupts the lipids in the lipid bilayer. In this theory, monomeric melittin would form pores or channels. Evidence for this theory comes from experimental work in which melittin was added to artificial bilayers. Proteolytic enzymes (pronase) were added to the aqueous phase under the bilayer. If the transmembrane potential was 0 or such that the outside was negative relative to the inside of the bilayer, no proteolysis of melittin occurred. However, if the inside of the membrane was made negative with respect to the outside such that the transmembrane potential was −60 mv or less, proteolysis of melittin occurred. A potential of −60 mv, found in most resting cells, corresponds to an energy of 2–3 kcal/mol, sufficient to convert compact monomeric melittin to a straightened structure that then spans the membrane and encounters the proteolytic enzymes. This example illustrates the great biological relevance of the relationship of charges on the protein surface to their interactions with electric fields, as in electrophoresis, and to the effects of an electric field on the conformations of proteins. This relationship is vital in understanding the functioning of ion channel proteins as described in Section IV.A.4.

3. The Function of Transmembrane Proteins

Currently, there appear to be three classes of transmembrane proteins: intercellular adhesion molecules, receptor proteins, and ion channel proteins. Studies on the first category of these proteins are incipient whereas the second and third categories of proteins have been better studied. Generally, receptor proteins contain three specific domains: an extracellular, an intramembrane, and an intracytoplasmic domain. This motif has been found to exist for several growth factor receptors and for the T-cell receptor. While the detailed structures of these proteins have not been elucidated, certain features of their functioning are known.

As an example, the sequence for the growth factor receptor called *neu* or *her-2*, which is a protein of molecular weight 185 kDa, or p185, is known (Padhy et al., 1982). This growth factor receptor is extremely important in the control of mitogenesis in epithelial cells (Padhy et al., 1982). Defects in this receptor protein can result in unregulated cell division and have been found to be associated highly with breast cancer (King et al., 1985; Slamon et al., 1987). It is known that activation of this protein requires that its growth factor bind simultaneously to two receptors (Ben-Levy et al., 1992), thus cross-linking them. As a result of this cross-linking process, tyrosine kinases become activated, resulting in phosphorylation of critical target proteins in a mitogenic signal transduction pathway (Bargmann and Weinberg, 1988).

It is known that substitution of single amino acids within the transmembrane domain of this protein activates the protein, presumably by facilitating dimerization in the membrane (Bargmann and Weinberg, 1988). The essential segment of the transmembrane domain contains the sequence (residues 650–683) -Glu-Gln-Arg-Ala-Ser-Pro-Val-Thr-Phe-Ile-Ile-Ala-Thr-Val-XXX-Gly-Val-Leu-Leu-Phe-Leu-Ile-Leu-Val-Val-Val-Val-Gly-Ile-Leu-Ile-Lys-Arg-Arg-. The XXX amino acid residue can be Val, His, Tyr, Lys, or Gly for normal proteins and Gln or Glu for transforming proteins.

Recent calculations (Brandt-Rauf et al., 1989) on the structure of the transmembrane domain of the *neu* protein indicate that the entire transmembrane domain, most of which is hydrophobic, can exist in two states, as shown schematically in Fig. 13. The normal nonmutated transmembrane domain is a bent α-helix in which two helices are separated by a β-turn at residues 664–665. (These conformations are defined in Section I above.) In the "off" state for the normal protein, not bound to ligand, the α-helices are bent and are not favorably disposed to interact with one another. Amino acid substitutions at critical positions around where the bend occurs that obliterate the bend between the helices result in the formation of two regular α-helices that can easily associate with one another.

A complete analysis (Brandt-Rauf et al., 1990) of the Boltzmann distribution of the low-energy conformations of the p185 protein shows that 90% of the protein is in the bent helical state while 10% is in the all-helical state for all normal proteins. For the transforming proteins with Gln or Glu at position 664, this distribution reverses so that 90% of the protein is in the all-helical conformation while 10% is in the bent-helical conformation. These results predict therefore that if the normal protein is overexpressed by about a factor of 10, then sufficient all-helical forms would be present in the cell to cause cell transformation. This computed result has been confirmed in experiments where overexpression of the normal protein by factors of 1–4 causes no effect on the cell (Hudziac *et al.*, 1987) while ten-fold overexpression of the normal protein has been found to cause cell transformation (DiFiore et al., 1987). They have been further directly confirmed in ligand-binding experiments using Scatchard plots (see Section I C.1.) (Ben-Levy *et al.*, 1992), in which it was found that, for the normal cellular protein with Val at position 664, 90% of the protein was in a low-affinity state for the cross-linking ligand and only about 10% of the protein was in a high-affinity state. However, with Glu at position 664, a transforming substitution, the reverse was found: over 90% of the receptors were in the high-affinity state, or readily cross-linked.

Thus a change in conformation from a bent to a straight helix causes major changes in the functioning of a transmembrane protein, much like the proposed change in structure of melittin in the presence of an electric field. It is clear from the above examples that to understand cell function it is necessary to understand the conformational properties of proteins, the basis of which have been presented in this chapter.

4. Ion Channels (Caterall, 1995)

As discussed in the preceding section on melittin and as explained in detail in several later chapters of this textbook, cell membranes are polarized, resulting in a voltage change across them such that the inside of the cell is negative with respect to the outside. Furthermore, in the resting state of most cells, the extracellular sodium ion concentration is much higher than the intracellular sodium, while the reverse is true for potassium ions. During action potentials in the electrically excitable axons and dendrites of nerve cells, the membrane becomes depolarized. During this period, there is a large influx of sodium ions, followed by an almost equivalent efflux of potassium ions. These fluxes are then rapidly reversed, during which time the cell becomes refractory to further excitation.

The change in the permeability of the cell to sodium and potassium ions is caused by changes in voltage across the cell membrane. Thus the motion of these ions across the membrane is said to be **voltage-gated**. During action potentials, there is a **voltage-gated** increase in the conductance of the membrane to sodium, followed by a rapid and then a slower inactivation of sodium conductance.

All of these changes in conductance and perm-selectivity of the membrane are mediated by sodium, potassium, and calcium channel proteins, all of which have recently been cloned, purified, and expressed in different cell lines and in artificial lipid bilayers (Caterall, 1995). All of these channel proteins have remarkably similar linear amino acid sequences with differences that cause each to bind selectively to different ions.

All of the channel proteins are composed of an α-subunit of molecular mass 260 kDa that is heavily glycosylated. Sequencing of the gene for this protein reveals four domains that contain six repeating sequences (called S1–S6). These repeats are characterized by regular spacing in the sequence between hydrophobic amino acid residues and positively charged residues. As with leader sequences, melittin, and the transmembrane domain of the *neu/her*-2 protein, the six repeat sequences are highly hydrophobic. Therefore, they are thought to be transmembrane domains and have been postulated to be α-helical. The basic arrangement of the four domains is shown schematically in Fig. 14. In three dimensions, these four domains would be arranged in a cylindrical fashion and would surround a pore. The α-subunit in the sodium channel protein is noncovalently linked to a heavily glycosylated β-1 subunit of molecular mass 36 kDa and by a disulfide link to a β-2 subunit.

a. Voltage Activation. The actual voltage at which activation of the sodium channel occurs has recently been shown to depend on positively charged and hydrophobic amino acid residues in the S4 amphipathic helical segment of each transmembrane domain. As shown schematically in Fig. 15, seven critical positively charged Arg and Lys residues are interspersed among many hydrophobic residues. Each of these positive charges must interact with a negative charge from an Asp or Glu residue from another subunit.

In site-specific mutagenesis experiments, each of the positively charged residues has been replaced with neutral amino acid residues. Neutralization of positively charged residues 1, 3, 5, and 7 results in shifting the activation transmembrane potential to more positive values so that the influx of sodium ions begins to occur at more depolarized states. Neutralization of charged residues 2 and 4 produces the opposite effect. Thus residues 1, 3, 5, and 7 are involved with activation of sodium influx while residues 2 and 4 are involved with pore closing and inactivation of sodium influx.

To explain how the S4 segment causes activation of sodium influx, it has been hypothesized that at a certain level of depolarization, the S4 helix rotates such that positive charges on the helix change negatively charged partners in an upward spiral from the inner to the outer membrane, resulting in effective transfer of a positive charge to the outside of the membrane. Since, as indicated in the above discussion, not all positive charges perform the same function, this hypothesis may be oversimplified.

Another explanation for charge transfer across the membrane is that, at a critical transmembrane voltage, the S4 helix unfolds into a fully extended conformation, allowing for the shifting of positive charges towards the outer membrane surface. One problem with this explanation is that the energy required to convert an all-helical protein into an all-extended form requires the disruption of a large number of hydrogen bonds, each about 1–2 kcal/mol in energy. Unless the fully extended form can make hydrogen bonds with other segments in a β-pleated sheet arrangement, the energy barriers for this conformational transition would be very high. Nonetheless, it is clear that the change in the transmembrane potential causes a change in the conformation of the S4 helix,

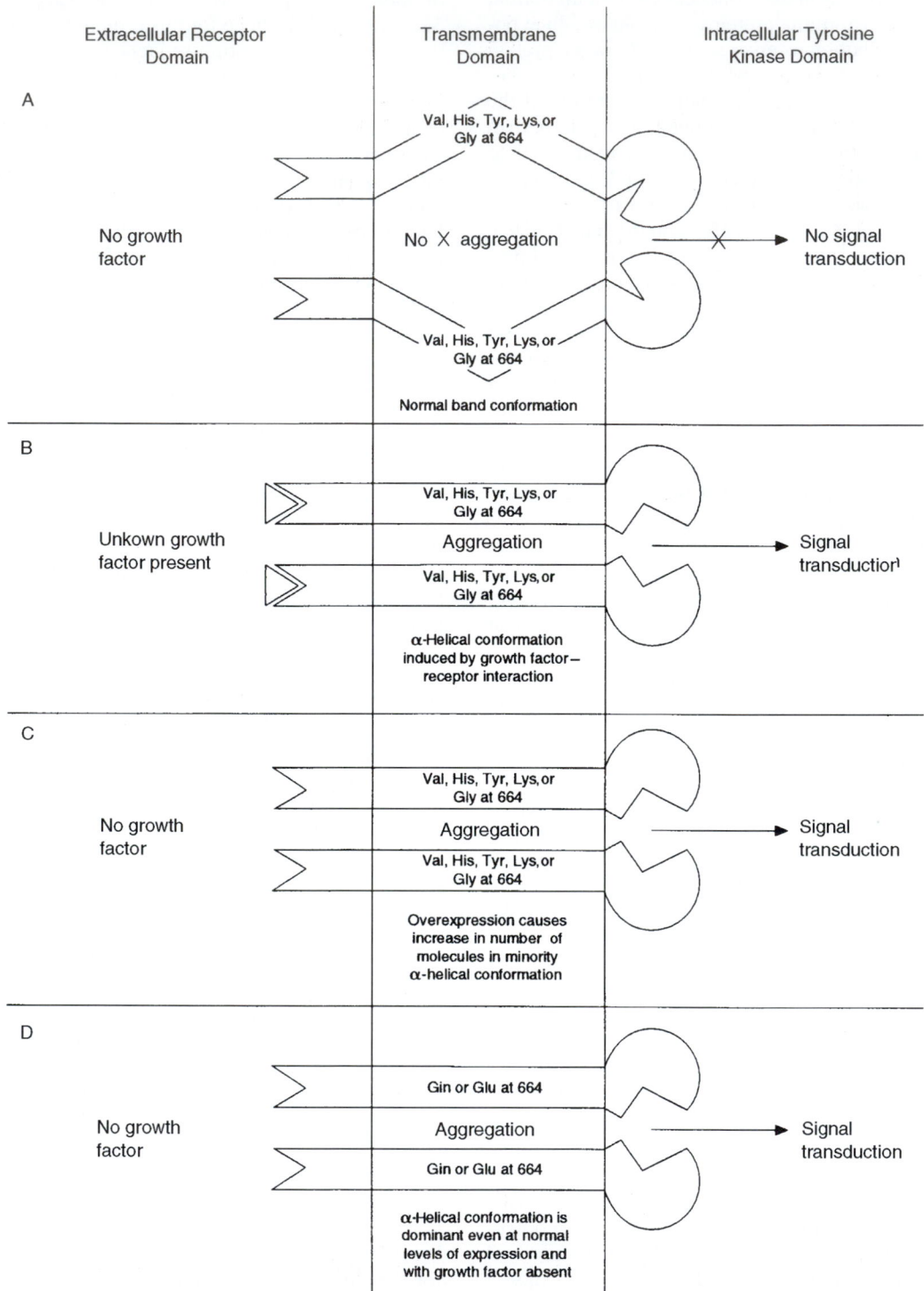

FIGURE 13. The effects of transmembrane domain structure of trans-membrane proteins on their function. The *neu*-onco-gene-encoded p185 protein is a transmembrane protein that has three distinct domains: an extracellular ligand-binding domain (left), a transmembrane domain (middle), and an intracellular signaling domain (right). To get signal transduction of a mitogenic signal to the nucleus, two p185 molecules must associate by the binding of their transmembrane domains to one another. This can be accomplished in a variety of ways as shown in this figure. (Modified from Brandt-Rauf *et al.*, 1990.) (A) The normal protein has Val, His, Tyr, Lys, or Gly at position 664 in the amino acid sequence for the normal protein. The structure of the transmembrane domain is a bent α-helix, preventing the two domains from associating with one another. (B) When a growth factor binds to the extracellular domain of the protein, the bend is removed from the middle of the helix, allowing two straight helices to associate. This dimerization allows for activation of mitogenic signaling elements via the intracellular domain in the pathway ending in the nucleus. (C) From the calculations of the distribution of conformations for the normal transmembrane domain (Brandt

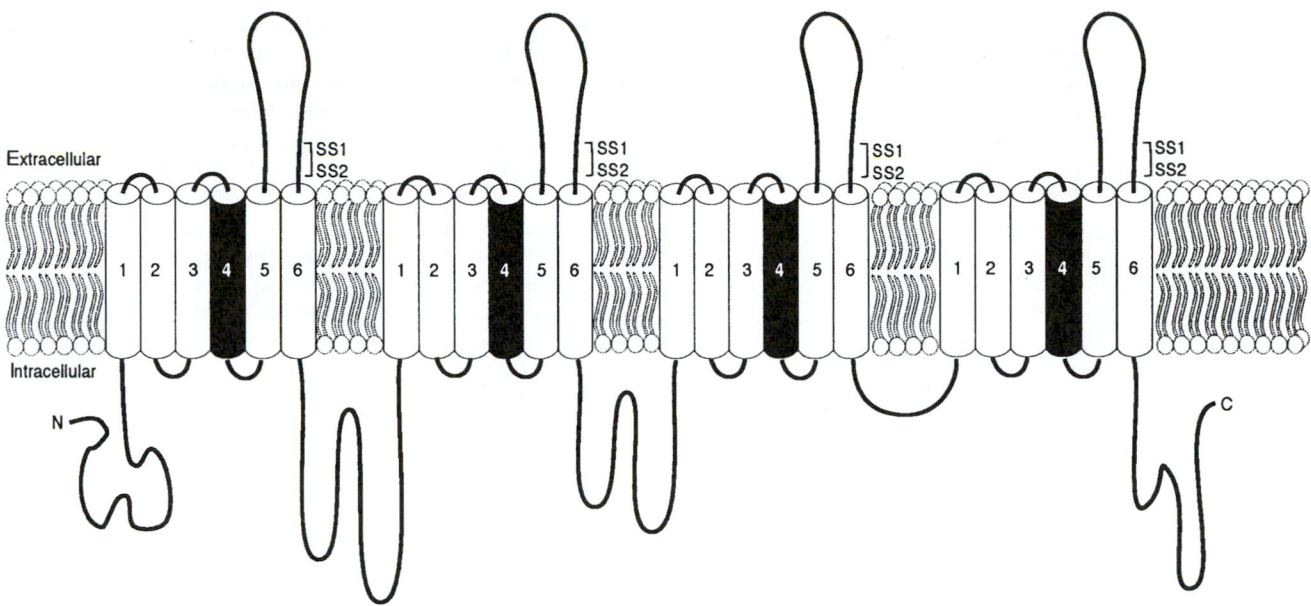

FIGURE 14. Model for the arrangement of the domains in the α-subunit of the sodium ion channel protein. The four domains are labeled 1–4. Each domain is shown to contain six transmembrane helical segments labeled 1–6. Subdomain 4 is highlighted in each domain because it contains the critical residues for voltage-dependent gating. The short segments 1 and 2 that are critical for ion selectivity are labeled SS1 and SS2, respectively. This region contains the critical residues Glu 387 and Phe/Tyr or Cys 385. (This model is modified from Hall, 1992, p. 109.)

resulting in the opening or closing of the sodium channel. This effect is strongly analogous to the proposed effect of the transmembrane potential on the conformation and function of melittin discussed above. Pore formation is clearly influenced by transmembrane voltage and results from a conformational change in the transmembrane protein.

b. Perm-Selectivity. Neural toxins, such as from snake venom, have long been known to bind with high affinity for and selectivity to specific ion channels. The toxin tetrodotoxin is known to bind to the sodium channel pore at its outermost surface. Localization of the tetrodotoxin (and sodium channel pore) binding site has recently been accomplished. Between the S5 and S6 helical segments are two short segments, named short segments 1 and 2 (SS1 and SS2), shown in Fig. 14. In SS2, neutralization of Glu 387 by site-specific mutagenesis results in a ten-thousand-fold decrease in the binding of tetrodotoxin to the mutated sodium channel. In similar experiments, it has further been found that additional residues are vital to perm-selectivity of the sodium ion channel. These include negatively charged residues in domains 1 and 2, a positively charged amino acid residue in domain 3, and a neutral amino acid in domain 4.

Position 385 is also a vital amino acid residue in SS2. In brain and skeletal muscle sodium channels, this residue is Phe or Tyr; in cardiac muscle, it contains the nonconservative substitution, Cys. The cardiac muscle sodium channel has a several hundred-fold lower affinity for tetrodotoxin than does brain or skeletal muscle channel protein. If the Cys residue of cardiac muscle channel protein is changed to Tyr or Phe, the high affinity for tetrodotoxin is restored. Thus Tyr/Phe 385 appears to be vital in tetrodotoxin binding and ion selectively.

Sodium and calcium channel structure are similar to one another. However, calcium ions bind with only about one-tenth the affinity to the sodium channel protein as that for sodium ions. However, mutation of Lys 1422 and Ala 1714 to Glu residues completely reverses the order of affinity of these two ions to the mutated sodium channel protein. These two residues are therefore implicated as being present at the mouth of the pore of the sodium channel.

c. Inactivation. Normally, after the voltage-gated increase in sodium ion permeability, there is a rapid inactivation of the sodium conductance. This inactivation process can be blocked by treatment of the intracellular domain of the sodium channel

-Rauf *et al.*, 1990) of p185, 90% of the molecules exist in the bent helix state, while 10% are straight helices. Overexpression of the normal protein by a factor of 10 was therefore predicted to produce sufficient straight helices to cause mitogenic cell signaling, leading to cell transformation. (D) Finally, substitution of Glu or Gln at position 664 reverses the distribution of bent and straight helices directly so that 90% of the molecules are straight helices while only 10% are bent helices. This results in significant levels of dimerization, resulting in permanent mitogenic cell signaling and oncogenic transformation of the cells.

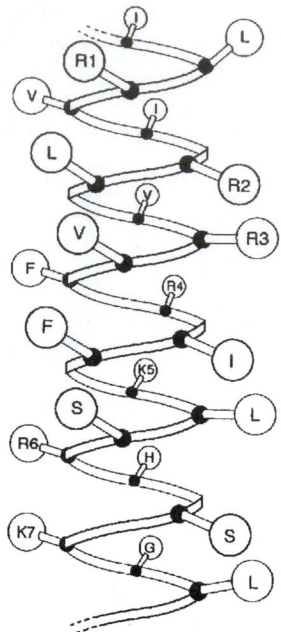

FIGURE 15. Helical model for the critical segment of the S4 subdomain of the sodium channel. The critical positively charged residues are labeled in numerical order. (Modified from Caterall, 1995, Fig. 3, p. 515.)

protein with proteases and by a monoclonal antibody directed against a segment of the protein connecting domains 3 and 4. Further studies involving site-specific mutagenesis have identified three critical amino acid residues as being vital to the inactivation process: residues 1488–1490, the sequence of which is Ile-Phe-Met, all hydrophobic. Mutation of Phe 1489 to Gln completely blocks inactivation of sodium channel conductance.

To explain the inactivation process, the three-residue segment has been proposed to constitute a "hinged lid," which moves to form a block in the sodium conductance channel after the conformational changes occur that increase sodium ion conductance. Phe 1489 would bind to another as-yet-unidentified pocket in the protein, blocking further movement of sodium ions.

Similar experiments on the potassium channel protein have identified a completely different segment of this protein involved in the inactivation of potassium conductance. This constitutes the amino terminal domain consisting of hydrophobic (especially Leu 7) and hydrophilic amino acid residues.

The model for inactivation for the potassium channel protein is the "ball and chain" model in which the amino terminal arm of the protein swings into the pore, blocking further potassium conductance. Convincing evidence for this model has been provided by experiments in which native amino terminal peptides hybridized with mutant potassium channel proteins, which do not inactivate potassium ion conductance, completely restore their abilities to inactivate potassium ion conductance. These results compared with those for the sodium channel suggest that inactivation of sodium channels occurs on the intracellular surface whereas that for the potassium channels occurs on the extracellular surface.

Our knowledge of the functioning of ion channels is incipient but has been greatly advanced by the cloning of the sodium, potassium, and calcium channel proteins and the site-specific mutagenesis work. These studies all suggest that the functioning of all of these proteins is based on the principles discussed earlier in this chapter. Selective binding of ions is caused by specific ionic (electrostatic) interactions; changes in conductance of the channel proteins is caused by changes in the three-dimensional structure of the proteins induced by changes in the electric field across the cell membrane as proposed for the simpler melittin protein. The basic structure of the channel protein consists of a series of amphipathic α-helices containing large numbers of hydrophobic amino acid residues. As found for the transmembrane domain of the *neu/her*-2 receptor protein, the interaction of the helices with one another is responsible for the functioning of the channel protein.

5. Effects of Amino Acid Substitutions on the *ras*-p21 Protein (Barbacid, 1987)

One of the most striking discoveries over the past two decades has been the finding that a single base change at codon 12 in the human *ras* gene, which encodes a protein of Mr 21 kDa called the p21 protein (with 189 amino acids in its sequence), results in the malignant transformation of quiescent (NIH 3T3) cells in culture. The encoded p21 protein contains a Val in place of the normally occurring Gly at position 12 in its amino acid sequence. Transfection of the oncogene, but not its normal counterpart proto-oncogene, into NIH 3T3 cells in culture results in their malignant transformation. Microinjection of cloned, purified oncogene-encoded (Val 12–containing) p21 protein, but not its normal counterpart protein, into NIH 3T3 cells also causes these cells to undergo cell transformation. Thus the transforming agent is the *ras*-gene–encoded p21 protein itself. Similarly, oncogenic but not normal p21 induces maturation (meiotic division) of frog (*Xenopus laevis*) oocytes. The biochemical changes induced by oncogenic p21 in all of these cells are very similar, suggesting that *ras*-p21 induces stereotypical events in each cell line. Incredibly, oncogenic *ras* genes have been identified overall in about one out of three common human cancers, in over 90% of human pancreatic and 70% of human colon cancers (Almoguerra *et al.*, 1988). Oncogenic p21 protein has also been found to occur at elevated concentrations in the sera of patients with specific types of malignant neoplasms.

It is now known that there are other critical positions in the polypeptide chain of p21, such as at Gly 13, Ala 59, and Gln 61, where single amino acid substitutions result in a transforming protein. By far the most common site for oncogenic substitution is Gly 12. Since 18 out of 20 amino acids that occur at position 12 cause the protein to become oncogenic, it seems reasonable to conclude that these substitutions induce changes in the three-dimensional structure of the protein, resulting in its being permanently activated.

a. Activation of ras-*p21 (Pincus et al., 1999). ras*-p21 is a **G protein** that becomes activated when GDP, bound to the protein in its inactive state, is exchanged for GTP. This process of activation is set in motion when a specific growth factor such as epidermal growth factor (EGF), itself a polypeptide, binds to its receptor. As discussed under the *neu/her*-2 growth factor receptor section (Section IV.A.3), the transmembrane receptor contains an extracellular growth factor binding domain, a transmembrane domain, and an intracytoplasmic domain which, for *neu* and EGF receptors, contains a tyrosine kinase domain. This kinase becomes activated when the growth factor receptor dimerizes as a consequence of the binding of the growth factor to it.

As a result, the intracytoplasmic domain binds to an "adapter" molecule, called *grb*-2, which simultaneously binds to a protein called SOS that, in turn, binds to p21, promoting the exchange of GDP for GTP. Modulating the effect of GTP on activation of p21 is the GTPase activating protein (GAP), which binds to activated p21 and induces the hydrolysis of GTP to GDP, thereby resulting in the inactivation of *ras*-p21.

All of these events occur in or near the inner cell membrane. *ras*-p21 is attached to the inner cell membrane by a covalent link at Cys 186 (Willumsen *et al.*, 1984) in which its sulfur atom is in thioether link to the farnesyl moiety, a lipophilic hydrocarbon that intercalates in the lipid medium of the cell membrane. Thus, like melittin, *ras*-p21 is active as a membrane-associated protein.

Thioether synthesis is catalyzed by the enzyme farnesyl transferase. In site-specific mutagenesis experiments, if Cys 186 is replaced by a Ser residue, *ras*-p21 not only is inactive, even if it contains oncogenic amino acid substitutions, but it competes with its cell–membrane associated counterpart protein so as to **inhibit** mitogenesis. The double inactive mutant of p21 with the Val-for-Gly 12 and Ser-for-Cys 186 substitutions, is therefore called a dominant negative mutant of *ras*-p21. Much work is currently being performed in the pharmaceutical industry to devise inhibitors of farnesyl transferase that will block insertion of p21 into the membrane and allow the cytosolic p21 to compete with its membrane-bound counterpart protein, resulting in the slowing down or halting of mitogenesis in *ras*-induced cancers (James *et al.*, 1993).

Signal transduction induced by G proteins is a generalized phenomenon that explains the action of many hormones and external stimuli. For example, polypeptide hormones such as ACTH, growth hormone, calcitonin, cholecystokinin, and insulin all activate G proteins that stimulate phosphorylation cascades. Light induces electrical signals on the retina; conversion of light to electrical signals traveling to the brain is performed by the G protein transducin in the retina. Much pharmacotherapy is currently aimed at interfering with the signal transduction events that occur after the hormone has bound to its receptor.

b. Function of ras-*p21.* A natural question that arises from the foregoing discussion is that, if *ras*-p21 is membrane bound, what is the mechanism by which it induces mitosis, when the latter process occurs in the nucleus? The answer to this question appears to be that activated *ras*-p21 induces a number of phosphorylation cascades in the cytosol in which successive proteins become activated until one or more of these proteins directly activates nuclear proteins that are in-

timately involved in DNA synthesis. These successive activation cascades are referred to as signal transduction pathways.

Activated *ras*-p21 binds directly to a protein called *raf*-p74 (molecular weight, 74 kDa). When activated by *ras* or other intracellular proteins, *raf* binds to and phosphorylates an extracellular **mitogen-stimulated kinase** called MEK, which, in turn, binds to and phosphorylates a vital protein, MAP kinase (mitogen-activated protein kinase). This protein is critical in catalyzing cytoskeletal rearrangements in mitosis and in activating the highly important nuclear transcriptional activating factor, *fos*. To be active, *fos* forms a heterodimeric complex with another protein, *jun*. Together these two activated proteins bind to promoter regions of the genome and induce the transcription of mRNA for a number of mitogenesis-promoting proteins such as the cyclins. The factor *jun* is activated by another pathway in which it is activated by the critical kinase, *jun* kinase (JNK). Further, oncogenic (Val 12–containing) *ras*-p21 interacts preferentially with both JNK and *jun* proteins, providing a "short circuit" pathway for activating direct nuclear transcription.

Remarkably, the above-described cascades occur in a wide variety of different cells with differing effects depending on the cell type. Thus, for example, in frog oocytes, oncogenic *ras*-p21 induces meiosis and cell **maturation**, not malignant transformation. In cardiac muscle that has suffered injury, infarction, or hypertrophy, the above cascades also become activated.

c. Relationship of Structure to Function of ras-*p21.* The fact that arbitrary amino acid substitutions at critical positions in the polypeptide chain cause the *ras*-p21 protein to become oncogenic suggests that critical changes in its three-dimensional structure occur in response to these substitutions. Another point of view is that these substitutions affect the ability of GAP to induce an increase in the GTPase activity of p21. There is an excellent correlation of the abilities of some oncogenic p21 proteins to induce cell transformation with their diminished levels of GAP-induced GTPase activities and with their binding affinities to GAP. On the other hand, there are striking exceptions to this correlation. For example, a Glu-for-Asp 38–substituted p21 protein binds tightly to GAP and has low GTPase activity but does **not** transform cells. Some substituted proteins efficiently hydrolyze GTP to GDP but transform cells. Also there are triply substituted forms of p21 that do not bind at all to any nucleotide but are permanently activated and strongly induce cell transformation. Thus it appears that for many substituted oncogenic p21 proteins, permanent activation is related to structural changes induced by the oncogenic amino acid substitutions themselves.

The x-ray structures of wild-type and oncogenic forms of p21 bound to GDP and GTP and of wild-type p21 bound to the *ras*-binding domain of *raf*, to SOS, and to GAP proteins have all been determined. Comparison of oncogenic and normal unbound p21 structures reveals some conformational changes in an effector loop from residues Tyr 32–Asp 47, which is involved in the binding of p21 to GAP, *raf*, and SOS. Because the protein occurs in a crystal lattice, wherein interactions between protein molecules limit conformational flexibility, the x-ray structures may not identify other regions that undergo conformational changes.

d. Use of Conformational Energy Calculations to Identify Effector Domains of p21. To identify other regions of the protein that may change conformation when the oncogenic form is compared with that of the wild-type protein, calculations of the low-energy conformations of these proteins have been carried out. These calculations are based on the principle, discussed in Section I.E.1, that the observed structure of a protein is the one of lowest energy; there may be a group of lowest energy structures for a protein that have the same overall chain fold but which differ from one another in the conformation of local regions.

In these calculations, using sampling methods such as molecular dynamics (discussed in Section I.E.1), the x-ray crystal structure is subjected to energy minimization and then allowed to move so that all of the low-energy structures that lie close to this structure, but which may differ from it regionally, are sampled. The average structure is the average of these low-energy structures. If oncogenic amino acid substitutions induce permanent, regional conformational changes in the three-dimensional structure of the protein, these will be reflected in differences in the average structures for the oncogenic proteins compared with that of the normal (wild-type) protein. We might expect that synthetic segments of p21 corresponding to these regions may inhibit the functioning of the oncogenic protein.

Results of these calculations on the wild-type protein bound to GDP and GTP and of the Val 12– and Leu 61–substituted proteins bound to GTP are shown in Fig. 16 (Monaco *et al.*, 1995). In this figure's superposition of the C$^\alpha$ tracings for the average structures of these proteins on one another, it can be seen that the overall fold of the proteins is virtually identical, but local regions differ in conformation between oncogenic and the inactive (GDP-bound) protein. There are six such regions: residues 4–20, 35–47, 55–71, 81–93, 96–110, and 115–126. Peptides corresponding to each of these regions have been synthesized and tested for their abilities to alter the ability of *ras*-p21 to induce mitogenesis.

e. Test Cell System for Identification of Functional Domains of ras-p21. As it happens, a very convenient cell system for measuring the abilities of agents to inhibit *ras*-p21 is oocytes. Oocyte maturation induced by a particular agent, such as oncogenic *ras*-p21, is easy to evaluate because of the large structural changes, like disappearance of the brown-colored animal pole, that each matured oocyte undergoes. Maturation of all oocytes in a given experiment occurs in a relatively short time, within 1–2 days after injection or treatment of the oocytes with a maturation-promoting agent. Finally, oocytes contain an abundance of insulin receptors; insulin is a growth factor that induces oocyte maturation by activating normal cellular *ras*-p21.

Proof that insulin acts through *ras*-p21 was provided by the finding that if oocytes are first injected with an inactivating antibody against *ras*-p21 and then incubated in insulin, they do not undergo maturation. Prior injection of other, unrelated antibodies has no effect; insulin induces maturation in these oocytes. Thus oocytes are ideal for studying the specificity of anti–*ras*-p21 agents to study whether they have a preferential effect on the oncogenic form versus the normal form of p21 by either co-injecting the agent with oncogenic p21 or by injecting the agent into oocytes that are then incubated with insulin.

f. Effects of p21 Peptides from Effector Domains Identified by Conformational Analysis (Pincus, 2000). Microinjection into oocytes of each of the synthesized peptides from the *ras*-p21 domains identified from the conformational energy calculations mentioned above, either together with oncogenic *ras*-p21 or into oocytes subsequently incubated with insulin, has been performed. Three peptides, 35–47 from the previously identified effector domain, 96–110, and 115–126, all strongly block the ability of oncogenic p21 to induce oocyte maturation. Significantly, these peptides, especially peptides corresponding to residues 96–110, and 115–126, only minimally affected the ability of insulin, hence activated normal

FIGURE 16. Superposition of the average structures of different forms of *ras*-p21 on one another. In each of the three superpositions, the normal, GDP-bound p21 is the lighter trace. Superimposed on the normal protein are left, Val 12-p21; middle, GTP-p21 (bold); right, Leu 61-p21.

p21, to induce maturation. This finding suggests that oncogenic and normal p21 induce mitogenesis by differing signal transduction pathways. As it happens, the latter two peptides have been found to interfere in the interaction of oncogenic p21 with JNK and *jun* proteins; their abilities to inhibit oncogenic p21-induced oocyte maturation strongly correlate with their abilities to inhibit this interaction.

The above results suggest that activated normal and oncogenic *ras*-p21 stimulate different pathways; the oncogenic form can preferentially activate JNK and *jun* directly, leading to unregulated nuclear transcriptional processes. By breaking this "short circuit" with effector peptides, selective inhibition of oncogenic *ras*-p21 can be achieved. This result has important implications for the design of new selective chemotherapeutic agents.

Thus, use of the methods of conformational analysis based on the principles of protein structure described in this chapter has resulted in our being able to identify critical residues of an oncogenic protein that are vital to their function, to design peptides that selectively inhibit oncogenic *ras*-p21, and to detect differences in signal transduction pathways induced by oncogenic and normal p21 protein.

V. Summary

We have seen in this chapter that the three-dimensional structures of proteins are dictated by their linear sequences of amino acids. A protein's structure is the one of lowest free energy for the given polypeptide chain, the detailed conformation of which is determined by the dihedral angles of the backbone and side chains. Proteins are structured in such a way that their polar amino acids tend to be directed towards the aqueous solvent while the nonpolar or hydrophobic residues tend to point towards the interior of the protein. The three-dimensional structures of proteins can be determined by a variety of techniques, which include circular dichroism (for regular structure), x-ray crystallography, two-dimensional NMR, and theoretical techniques.

The three-dimensional structures of proteins determine bulk properties of these proteins such as their behavior towards titration with acids or bases, their isoionic points, and their migration in electric fields, as in electrophoresis. The ionization of charged groups on the surfaces of proteins can be treated, and thus their titration curves can be predicted using ligand-binding theory combined with electrostatic field theory (using the Debye-Hückel formulation) or using more explicit models based on the actual three-dimensional structure of the protein surrounded by water dipoles.

The structure of a protein and its functioning have a well-defined relationship. Membrane proteins, for example, tend to fold into α-helices. This structure is of critical importance to the functioning of leader peptides that allow for proteins to be secreted both within the cell and outside of the cell. The ability of melittin to intercalate into cell membranes and to cause cell lysis depends on its ability to adopt an α-helical conformation and to change the angle of these helices. Certain large proteins, like *neu*/HER-2, contain major transmembrane domains, which can adopt several different conformations, some of which allow them to dimerize and thus initiate cellular signal transduction. This process is especially critical for oncogene-encoded proteins that are involved in mitogenic signal transduction to the nucleus, causing cell division. We have begun to understand the structural basis for this process using conformational analysis as described in this chapter. In this regard, conformational analysis has now been extended to the identification of the regions of oncogenic proteins involved in activation of downstream target proteins and has opened new possibilities in the field of drug design. Voltage-dependent conformational changes in the transmembrane domains of ion channel proteins are vital to their regulated functioning. The basic functioning of cells is therefore controlled by the proper folding and functioning of cellular proteins, making vital the understanding of protein structure.

Bibliography

Almoguerra, C., Shibata, D., Forrester, K., Martin, J., Arnheim, M., and Perucho, M. (1988). Most human carcinomas of the endocrine pancreas contain mutant c-K-*ras* genes. *Cell* **53**, 813–815.

Anfinsen, C. B., Haber, E., Sela, M., and White, F. H., Jr. (1961). The Kinetics of formation of native ribonuclease during oxidation of the reduced polypeptide chain. *Proc. Natl. Acad. Sci. USA* **47**, 1309–1314.

Barbacid, M. (1987). *ras* genes. *Ann. Rev. Biochem.* **56**, 779–827.

Bargmann, C. I., and Weinberg, R. A. (1988). Oncogenic activation of the *neu*-encoded receptor protein by point mutation and deletion. *EMBO J.* **7**, 2043–2052.

Ben-Levy, R., Peles, E., Goldman-Michael, R., and Yarden, Y. (1992). An oncogenic point mutation confers high affinity ligand binding to the *neu* receptor. *J. Biol. Chem.* **267**, 17304–17313.

Blobel, G., Walter, P., Chang, C. N., Goldman, B. M., Erickson, A. H., and Lingappa, R. (1979). Translocation of proteins across membranes: the signal hypothesis and beyond. *Symp. Soc. Exp. Biol.* **33**, 9–37.

Brandt-Rauf, P. W., Pincus, M. R., and Chen, J. M. (1989). Conformational changes induced by the transforming amino acid substitution in the transmembrane domain of the *neu*-oncogene-encoded p185 protein. *J. Protein Chem.* **8**, 749–755.

Brandt-Rauf, P. W., Rackovsky, S., and Pincus, M. R. (1990). Correlation of the transmembrane domain of the *neu*-oncogene-encoded p185 protein with its function. *Proc. Natl. Acad. Sci. USA* **87**, 8660–8664.

Bull, H. B. (1943). "Physical Biochemistry." John Wiley and Sons, London.

Caterall, W. A. (1995). Structure and function of voltage-gated ion channels. *Annu. Rev. Biochem.* **64**, 493–531.

Di Fiore, P. P., Pierce, J. H., Kraus, M. H., Segatto, O. S., King, R., and Aaronson, S. A. (1987). *erb*B-2 is a potent oncogene when overexpressed in NIH/3T3 cells. *Science* **237**, 178–182.

Dygert, M., Go, N., and Scheraga, H. A. (1975). Use of a symmetry condition to compute the conformation of gramicidin S. *Macromolecules* **8**, 750–761.

Emr, S. D., and Silhavy, T. J. (1982). Molecular components of the signal sequence that function in the initiation of protein export. *J. Cell. Biol.* **95**, 689–696.

Engelman, A. M., and Steitz, T. A. (1981). The spontaneous insertion of proteins into and across membranes: the helical hairpin hypothesis. *Cell* **23**, 411–422.

Gibson, K., Chin, S., Pincus, M. R., Clementi, E., and Scheraga, H. A. (1986). Parallelism in conformational energy calculations on proteins: partial structure of interferon. *In* "Montreal Symposium on Supercomputer Simulation in Chemistry. Lecture Notes in Chemistry" (M. Dupuis, Ed.), Vol. 44, pp. 198–213.

Hall, Z. W. (1992). Ion channels. *In* "An Introduction to Molecular Neurobiology." (Z. W. Hall, Ed.), pp. 81–118. Sinauer Associates, Inc., Sunderland, MA.

Hudziak, R. M., Schlessinger, J., and Ullrich, A. (1987). Increased expression of the putative growth factor reception p185^{HER2} causes transformation and tumorigenesis of NIH/3T3 cells. *Proc. Natl. Acad. Sci. USA.* **84,** 7159–7163.

James, G. L., Goldstein, G. L., Brown, M. S., Rawson, T. E., Somers, T. C., McDowell, R. S., Crowley, C. W., Lucas, B. K., Levinson, A. D., and Marsters, J. C., Jr. Benzodiazepine peptidomimetics: potent inhibitors of *ras* farnesylation in animal cells. *Science* **260,** 1937–1942, 1993.

King, C. F., Kraus, M. H., and Aaronson, S. A. (1985). Amplification of a novel v-*erb*B-related gene in a human mammary carcinoma. *Science* **229,** 974–976.

Karplus, M., and McCammon (1986). The dynamics of proteins. *Sci. Amer.* **254,** 42–52.

Kempf, C., Klausner, R. D., Weinstein, J. N., van Renswoude, J., Pincus, M. R., and Blumenthal, R. (1982). Voltage-dependent transbilayer orientation of melittin. *J. Biol. Chem.* **257,** 2469–2475.

Miller, M. H., and Scheraga, H. A. (1976). Calculation of the structure of collagen models. Role of interchain interactions in determining the triple helical coiled coil conformation. *J. Polymer Sci. Polymer Symposia,* No. 54, 171–200.

Monaco, R., Chen, J. M., Friedman, F. K., Brandt-Rauf, P. W., and Pincus, M. R. (1995). Structural effects of the binding of GTP to the wild-type and oncogenic forms of the *ras*-gene-encoded p21 proteins. *J. Protein Chem.* **14,** 721–730.

Padhy, L. C., Shih, C., Cowing, D., Finkelstein, R., and Weinberg, R. A. (1982). Identification of a phosphoprotein specifically induced by the transforming DNA of rat neuroblastomas. *Cell* **28,** 865–871.

Pincus, M. R., and Klausner, R. D. (1982). Prediction of the three-dimensional structure of the leader sequence of murine pre-kappa light chain, a hexadecapeptide. *Proc. Natl. Acad. Sci. USA* **79,** 3413–3417.

Pincus, M. R., Klausner, R. D., and Scheraga, H. A. (1982). Calculation of the three-dimensional structure of the membrane-bound portion of melittin from its amino acid sequence. *Proc. Natl. Acad. Sci. USA* **79,** 5107–5110.

Pincus, M. R., and Scheraga, H. A. (1985). Conformational analysis of biologically active polypeptides, with application to oncogenesis. *Accnt. Chem. Res.* **18,** 372–379.

Pincus, M. R. (1988). The chain build-up procedure in computing the structures of biologically active polypeptides and proteins. *Int. J. Quantum Chemistry: Quantum Biology Symposium* **15,** 209–220.

Pincus, M. R., Brandt-Rauf, P. W., Michl, J., and Friedman, F. K. (2000). *ras*-p21-induced cell transformation: unique signal transduction pathways and implications for the design of new chemotherapeutic agents. *Cancer Invest.* **18,** 39–50.

Rosenblatt, M., Beaudette, N. V., and Fasman, G. D. (1980). Conformational studies of the synthetic precursor-specific region of preproparathyroid hormone. *Proc. Natl. Acad. Sci. USA* **77,** 3983–3987.

Russell, S. T., and Warshel, A. (1985). Calculations of electrostatic energies in proteins. The energetics of ionized groups in bovine pancreatic trypsin inhibitor. *J. Mol. Biol.* **185,** 389–404.

Scheraga, H. A. (1984). Protein structure and function, from a colloidal to a molecular point of view. *Carlsberg Res. Commun.* **49,** 1–55.

Scheraga, H. A. (1989). Calculations of stable conformations of polypeptides, proteins, and protein complexes. *Chimica Scripta* **29A,** 3–13.

Slamon, D. J., Clark, G. M., Wong, S. G., Levin, W. J., Ullrich, A., and McGuire, W. L. (1987). Human breast cancer: correlation of relapse and survival with amplification of the *her*-2/*neu* oncogene. *Science* **235,** 177–182.

van Holde, K. E. (1985). "Physical Biochemistry." Prentice Hall, Inc., Englewood Cliffs, NJ.

Vasquez, M., Nemethy, G., and Scheraga, H. A. (1994). Conformational energy calculations on polypeptides and proteins. *Chem. Rev.* **94,** 2183–2239.

Willumsen, B. M., Christensen, A., Humbert, N. L., Papageorge, A. G., and Lowy, D. R. (1984). The p21 *ras* C-terminus is required for transformation and membrane association. *Nature* **310,** 583–586.

Wuthrich, K. (1986). "NMR of Proteins and Nucleic Acids." John Wiley and Sons, New York.

Ching-hsien Huang

3

Structural Organization and Properties of Membrane Lipids

I. Introduction

Lipids are of fundamental importance as they are a basic structure component of all cell membranes. In addition, many lipids in eukaryotic cell membranes are precursors of lipid-derived second messengers; hence, they are also functionally important. Topologically, membrane lipids are exposed extracellularly as well as intracellularly to an aqueous environment. Consequently, it is relevant and important to know how different membrane lipids are assembled structurally in an aqueous environment and how they interact energetically with each other within the fully hydrated assembly. In this chapter, we first discuss the classification of membrane lipids and the molecular structure of a representative lipid species within each class. We then describe the various organized assemblies of membrane lipids in excess water, with special emphasis on the different forms of lipid bilayers. Subsequently, we turn our attention to the thermally induced polymorphism of the lipid bilayer composed of a single type of phospholipid molecule. In particular, changes in the thermodynamic and conformational properties of the lipid bilayer as it undergoes the gel-to-liquid-crystalline phase transition are illustrated. Finally, in the last part of this chapter, the mixing behavior of two different phospholipids in the bilayer at various temperatures is considered in terms of the temperature-composition phase diagram. Specifically, basic thermodynamic equations that are applied for simulating the phase diagram for a binary lipid system in excess water are presented. Based on these equations, the simulated and experimentally determined phase diagrams can be compared and matched to determine the nonideality parameters of mixing for the two lipid species, thus leading to a quantitative estimation of the energetics involved in the lipid-lipid interactions in the two-dimensional plane of the lipid bilayer.

II. Classification and Structures of Membrane Lipids

Membrane lipids are amphipathic molecules composed of polar (or hydrophilic) and nonpolar (or hydrophobic) moieties. From a chemical structure point of view, these amphipathic lipid molecules can be broadly divided into three large groups: (1) **glycerophospholipids,** also loosely called **phospholipids;** (2) **sphingolipids;** and (3) **sterols.** Each membrane lipid group has an enormously wide range of chemically different species; however, a distinctive structural feature is common to all lipid species within each group, regardless of their bewildering diversity. All glycerophospholipids, for instance, are well known to contain a common glycerol backbone with L-configuration and a phosphate group ester-linked to the *sn*-3 carbon of the glycerol backbone. The structural feature common to all sphingolipids is the sphingosine backbone. Sterols, however, have four fused rings. The structural features common to all glycerophospholipids and sphingolipids are illustrated in Figs. 1A and 1B, respectively.

A. Glycerophospholipids

Glycerophospholipids, also known as phosphoglycerides, are classified on the basis of the polar alcohol esterified with the phosphate group at the *sn*-3 position of the glycerol backbone. The common polar alcohols are choline, ethanolamine, inositol, glycerol, and serine. Each class is further divided into two distinct subclasses based on the type of linkage between the *sn*-1 carbon atom of the glycerol backbone and the nonpolar aliphatic chain that is covalently attached to the *sn*-1 carbon atom. One subclass, called **diacyl phospholipids,** has an ester linkage at the *sn*-1 position. The chemical formulas of several commonly occurring diacyl phospholipids are presented in Fig. 2. The other subclass is characterized by an ether linkage at the *sn*-1 carbon

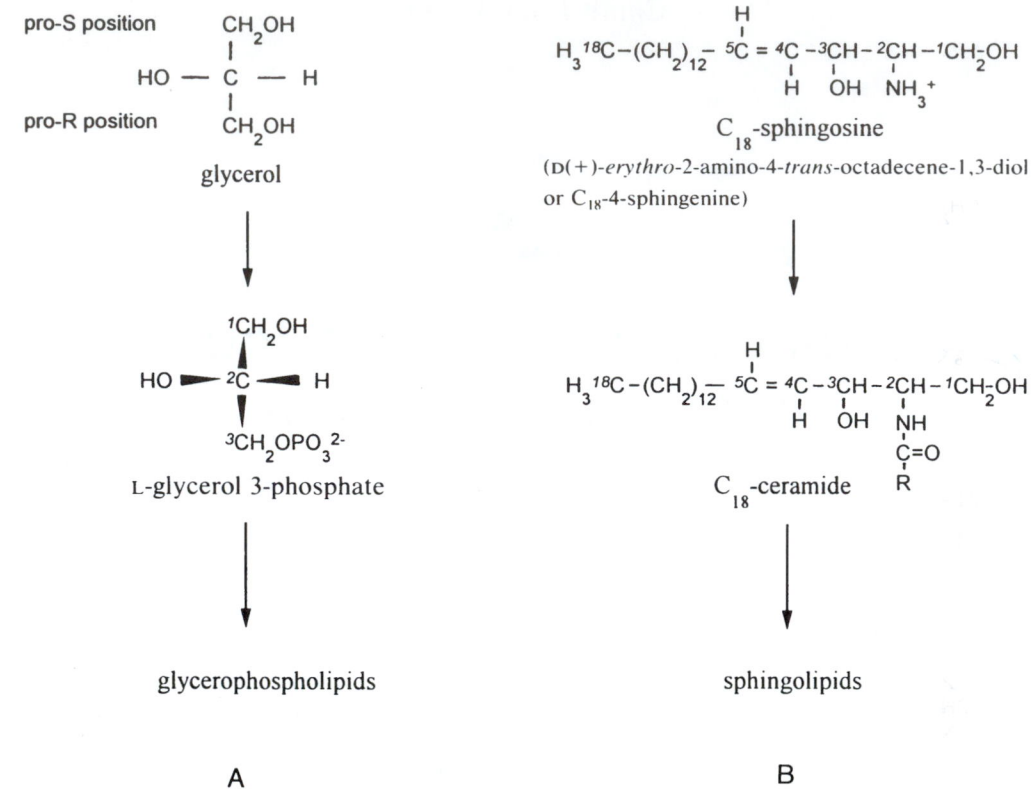

FIGURE 1. Structural formulas of glycerol and C_{18}-sphingosine. (A) All glycerophospholipids have an L-glycerol-3-phosphate backbone. (B) The sphingolipids are built from sphingosine base via ceramide.

of the glycerol backbone (Fig. 3). **Plasmalogens** and **platelet-activating factors** are examples of ether phospholipids belonging to this subclass. In particular, plasmalogens have a vinyl ether linkage, while platelet-activating factors have an alkyl ether linkage. In most glycerophospholipids, the aliphatic constituent at the *sn*-2 position is derived from a fatty acid, and this fatty acid is ester-linked to the glycerol backbone (Fig. 3).

In most mammalian cells, diacyl phospholipids constitute the major components of plasma membranes, although ether-linked phospholipids can be present in abundance in some subcellular membranes. A number of other membrane lipids, which are minor constituents of biological membranes, do not belong to either of the subclasses just discussed. These include dialkyl phospholipids, monoacyl phospholipids (e.g., lysophosphatidylcholines), and free fatty acids. The focus of this section will be upon the structures of various diacyl phospholipids that are found most abundantly in biological membranes. Information about the isolation, purification, and identification of these phospholipids from biological membranes can be found in Hanahan (1997).

Diacyl phospholipids isolated from biological membranes of animal cells are primarily mixed-chain lipids with the following three general features: (1) With the possible exceptions of mammalian lung and nerve endings, which contain large amounts of dipalmitoylphosphatidylcholine, identical fatty acids esterified at both the *sn*-1 and *sn*-2 positions of

the glycerol backbone occur rarely. Instead, the two esterified fatty acids have a different even number of carbon atoms ranging from 14 to 22. The fatty acids most commonly found in diacyl phospholipids are listed in Table 1. (2) The fatty acids in diacyl phospholipids are nonrandomly distributed. The one that is esterified with the *sn*-1 glycerol hydroxyl group, the *sn*-1 acyl chain, is often derived from a saturated fatty acid, whereas the other, the *sn*-2 acyl chain, is predominantly an unsaturated fatty acid. (3) The unsaturated fatty acids hydrolyzed from the *sn*-2 acyl chains have the structural formula $CH_3-(CH_2)_\alpha-[CH=CH-CH_2]_\beta-(CH_2)_\gamma-COOH$, where the subscripts are $\alpha = 1, 4, 5,$ and 7; $\beta = 1–6$; and $\gamma = 2–7$; and the *cis* double bonds ($\beta = 1–6$) are always separated by one methylene group (Kunau, 1976). Because of the large number of possible ways that permutations of chain length and unsaturation may occur, one can expect that a significant number of molecular species of diacyl phospholipid may be present in any given cell type. Indeed, it has been estimated that there are more than 1000 distinct molecular species of phospholipid in eukaryotic membranes (Raetz, 1986).

Structural features common to all diacyl phospholipids, as shown in Fig. 4, involve the following three lipid moieties:

1. The phosphate group ($pK_a \approx 1.3$) in a diacyl phospholipid is completely dissociated at physiological pH and therefore bears a negative charge. The four oxygen atoms bonded to the phosphorus atom are tetrahedrally

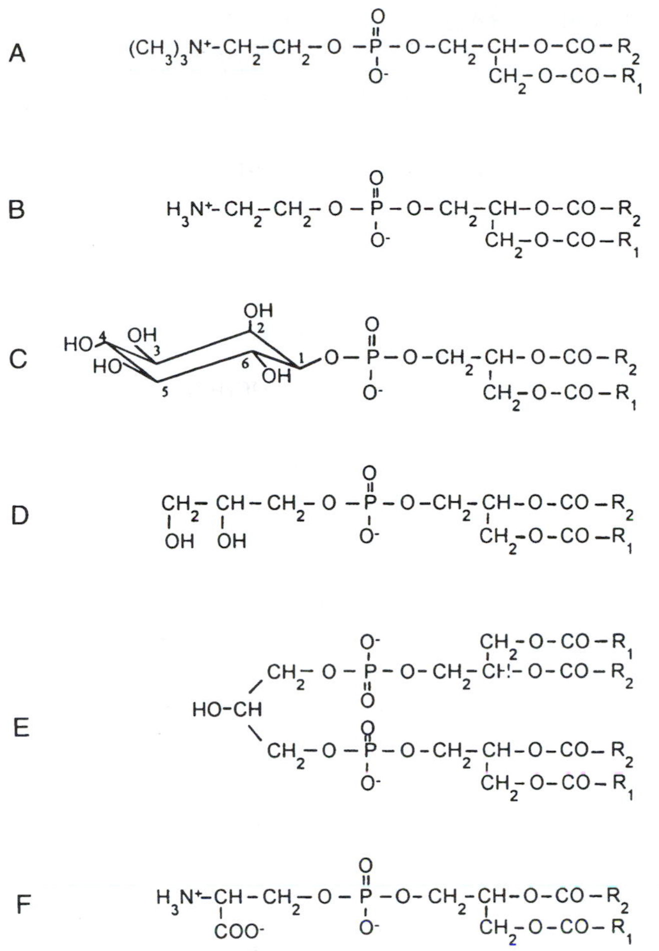

A $(CH_3)_3N^+-CH_2-CH_2-O-\overset{\overset{O}{\|}}{\underset{\underset{O^-}{\|}}{P}}-O-CH_2-\underset{\underset{CH_2-O-CO-R_1}{|}}{CH}-O-CO-R_2$

B $H_3N^+-CH_2-CH_2-O-\overset{\overset{O}{\|}}{\underset{\underset{O^-}{\|}}{P}}-O-CH_2-\underset{\underset{CH_2-O-CO-R_1}{|}}{CH}-O-CO-R_2$

C

D $\underset{\underset{OH\ \ OH}{|\ \ \ |}}{CH_2-CH-CH_2}-O-\overset{\overset{O}{\|}}{\underset{\underset{O^-}{\|}}{P}}-O-CH_2-\underset{\underset{CH_2-O-CO-R_1}{|}}{CH}-O-CO-R_2$

E

F $H_3N^+-\underset{\underset{COO^-}{|}}{CH}-CH_2-O-\overset{\overset{O}{\|}}{\underset{\underset{O^-}{\|}}{P}}-O-CH_2-\underset{\underset{CH_2-O-CO-R_1}{|}}{CH}-O-CO-R_2$

FIGURE 2. The chemical formulas of diacyl phospholipids that are commonly found in biological membranes. (A) Phosphatidylcholine (PC); (B) phosphatidylethanolamine (PE); (C) phosphatidylinositol (PI); (D) phosphatidylglycerol (PG); (E) diphosphatidylglycerol (cardiolipin); (F) phosphatidylserine (PS). R_1 and R_2 refer to hydrocarbon chains of fatty acids at positions 1 and 2 of the glycerol backbone in diacyl phospholipids.

ity, the two nonesterified phosphate oxygens are able to serve simultaneously as acceptors in two hydrogen bonds.

2. The two fatty acids esterified with the glycerol backbone of a phospholipid molecule at the *sn*-1 and *sn*-2 positions yield the primary and secondary ester linkages, respectively. The five atoms in the immediate neighborhood of each of the two ester bonds, $[C-O-C_1(=O)-C_2]$, are rigid and planar due to the partial double-bond character of the ester bond $(O-C_1)$ arising from resonance (Fig. 4B), where C_1 and C_2 are the first and second carbon atoms of the fatty acid, and C denotes the glycerol carbon atom. As a result, the $C-O$ and C_1-C_2 bonds adopt a *trans* configuration at physiological temperature. Moreover, the ester and the carbonyl oxygens bear partial positive and partial negative charges, respectively (Fig. 4B). The electronegative nature of the carbonyl oxygen enables it to serve as a H-bond acceptor.

3. The fatty acid ester-linked at the *sn*-1 position of the glycerol backbone in a diacyl phospholipid usually contains a long polymethylene chain. At low temperature (< 0 °C) and at thermal equilibrium, the long polymethylene chain adopts the minimum potential energy conformation that corresponds to the fully extended all-*trans* configuration. Consequently, the consecutive carbon-carbon single bonds along the polymethylene chain are aligned in a zigzag manner (Fig. 4C), with a separation distance of 1.27 Å between two neighboring carbon atoms along the long-chain axis. The two-dimensional plane occupied by the polymethylene chain with a zigzag conformation is termed the **zigzag plane.** If the zigzag plane is rotated 90°, the projected successive C–C bonds appear as a straight line with evenly spaced C–H bonds pointing above and below the straight line (Fig. 4D).

Before we turn our attention to the molecular structures of some diacyl phospholipids, it should be mentioned that the basic structure of a diacyl phospholipid molecule has customarily been regarded to consist of three regions: the **polar headgroup,** the **interfacial region,** and the **hydrophobic tail.** The polar headgroup refers to one end of the lipid molecule which is constructed from two polar groups, namely, the alcohol and phosphate groups. In the case of the phosphatidylcholine molecule shown in Fig. 2A, the quaternary base choline (an amino alcohol) is linked to the phosphate group by a phosphoester bond. The resulting

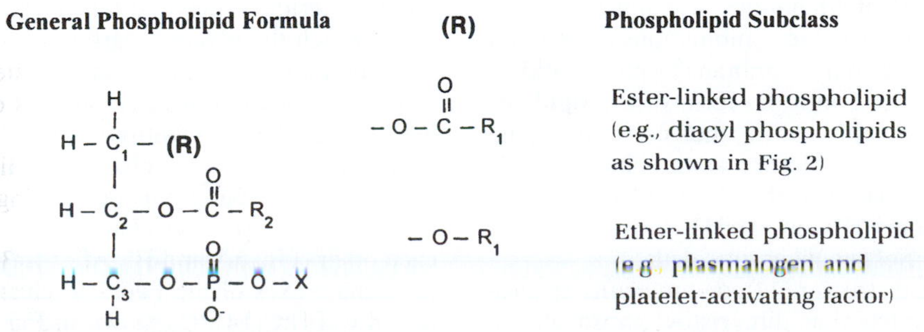

arranged. The two oxygens that do not participate in phosphodiester bonds, the pro-R and the pro-S phosphate oxygens, are electrostatically equivalent, each carrying a partial negative charge due to the resonance hybrid effect (Fig. 4A). Because of the electronegativ-

General Phospholipid Formula	(R)	Phospholipid Subclass
	$-O-\overset{\overset{O}{\|}}{C}-R_1$	Ester-linked phospholipid (e.g., diacyl phospholipids as shown in Fig. 2)
	$-O-R_1$	Ether-linked phospholipid (e.g., plasmalogen and platelet-activating factor)

FIGURE 3. The structural formulas of two subclasses of glycerophospholipids.

TABLE I   Some of the Common Saturated and Unsaturated Fatty Acids Found in Membrane Phospholipids

Common name	Systematic name	Abbreviated notation
Myristic acid	Tetradecanoic acid	C(14) or 14:0
Palmitic acid	Hexadecanoic acid	C(16) or 16:0
Stearic acid	Octadecanoic acid	C(18) or 18:0
Arachidic acid	Eicosanoic acid	C(20) or 20:0
Behenic acid	Docosanoic acid	C(22) or 22:0
Lignoceric acid	Tetracosanoic acid	C(24) or 24:0
Palmitoleic acid	cis-9-Hexadecenoic acid	$C(16{:}1\Delta^9)^a$ or $16{:}1(n-7)^b$
Oleic acid	cis-9-Octadecenoic acid	$C(18{:}1\Delta^9)$ or $18{:}1(n-9)$
Vaccenic acid	cis-11-Octadecenoic acid	$C(18{:}1\Delta^{11})$ or $18{:}1(n-7)$
Gadoleic acid	cis-9-Eicosenoic acid	$C(20{:}1\Delta^9)$ or $20{:}1(n-11)$
Erucic acid	cis-13-Docosenoic acid	$C(22{:}1\Delta^{13})$ or $22{:}1(n-9)$
Nervonic acid	cis-15-Tetracosenoic acid	$C(24{:}1\Delta^{15})$ or $24{:}1(n-9)$
Linoleic acid	All-cis-9,12-octadecadienoic acid	$C(18{:}2\Delta^{9,12})$ or $18{:}2(n-6)$
α-Linolenic acid	All-cis-9,12,15-octadecatrienoic acid	$C(18{:}3\Delta^{9,12,15})$ or $18{:}3(n-3)$
Mead acid	All-cis-5,8,11-eicosatrienoic acid	$C(20{:}3\Delta^{5,8,11})$ or $20{:}3(n-9)$
Arachidonic acid	All-cis-5,8,11,14-eicosatetraenoic acid	$C(20{:}4\Delta^{5,8,11,14})$ or $20{:}4(n-6)$
Adrenic acid	All-cis-7,10,13,16-docosatetraenoic acid	$C(22{:}4\Delta^{7,10,13,16})$ or $22{:}4(n-6)$
Timnodonic acid	All-cis-5,8,11,14,-17-eicosapentaenoic acid (EPA)	$C(20{:}5\Delta^{5,8,11,14,17})$ or $20{:}5(n-3)$
Clupanodonic acid	All-cis-7,10,13,16,19-docosapentaenoic acid (DPA)	$C(22{:}5\Delta^{7,10,13,16,19})$ or $22{:}5(n-3)$
Cervonic acid	All-cis-4,7,10,13,16,19-docosahexaenoic acid (DHA)	$C(22{:}6\Delta^{4,7,10,13,16,19})$ or $22{:}6(n-3)$

[a] The delta (Δ) refers to the unsaturated double bond and the superscript refers to the position of the double bond from the carboxyl end of the molecule.

[b] The positions of the double bonds are given relative to the methyl end of the molecule. Thus, $18{:}1(n-9)$ indicates that the double bond occurs at the ninth-from-last carbon atom.

phosphorylcholine is thus the polar headgroup of the phosphatidylcholine molecule. Phosphatidylinositol, illustrated in Fig. 2C, differs from phosphatidylcholine in that the polar headgroup has a polyol (inositol) in place of an amino alcohol (choline). The hydrophobic tail refers to the sn-1 and sn-2 acyl chains excluding the carbonyl groups. Geometrically located between the polar headgroup and the hydrophobic tail is the third region of the diacyl phospholipid molecule, the interfacial region. Within this region, two rigid and planar elements around the primary and secondary ester bonds exist.

1. Phosphatidylcholines

Phosphatidylcholines (1,2-diacyl-sn-glycero-3-phosphocholines or **lecithins,** often abbreviated as PC) are a structurally diverse class of diacyl phospholipids that are found most abundantly in cell membranes of higher organisms. The repertoire of phosphatidylcholines is extensive, originating from the numerous possible combinations of sn-1 and sn-2 acyl chains. In plasma membranes, phosphatidylcholine molecules aggregate into the form of the **lipid bilayer** due to their amphipathic nature, thus constituting the basic structural matrix. In addition, some phosphatidylcholine molecules serve as the metabolic precursors of intrinsic signaling elements (Exton, 1994), thus conferring some regulatory properties on eukaryotic cells.

The three-dimensional structure of one molecular species of saturated phosphatidylcholine, dimyristoyl phosphatidylcholine or C(14):C(14)PC, has been determined by x-ray

crystallographic approaches (Pascher *et al.,* 1992; Pearson and Pascher, 1979). In this chapter, the saturated PC is abbreviated as C(X):C(Y)PC, where the C(X) preceding the colon refers to the sn-1 acyl chain with X carbon atoms, and the C(Y) succeeding the colon gives the total number of carbon atoms (Y) in the sn-2 acyl chain. The single-crystal structure and the energy-minimized structure of C(14):C(14)PC (Huang and Li, 1996) are illustrated in Fig. 5A–D. Several conformational characteristics are revealed by the single-crystal structure. Firstly, the zwitterionic headgroup adopts a bent-down orientation so that the dipole axis or the P–N vector is inclined towards the interfacial region. Secondly, the primary and secondary ester planes are virtually perpendicular to each other, with the secondary ester plane, $C(2)$–O–C_1(=O)–C_2, running nearly perpendicular to the long molecular axis. Here, $C(2)$ denotes the sn-2 carbon atom of the glycerol backbone; C_1 and C_2 are the first and second carbon atoms of the sn-2 fatty acyl chain. Thirdly, the diglyceride moiety exhibits a roughly h-shaped geometry in which the glycerol carbon C_3, the primary ester oxygen, and all carbons in the sn-1 acyl chains are arranged in a fully extended conformation, and the sn-2 acyl chain is bent 90° at the C_2 position. The fully extended sn-1 acyl chain and the sn-2 acyl chain beyond C_2 are aligned in the same direction but nonparallel; however, the two zigzag planes of the two acyl chains are nearly perpendicular to each other (Fig. 5A and B). After force field refinement, the long-chain axes of the two acyl chains in the energy-minimized C(14):C(14)PC, shown in Fig. 5C and D, are nearly parallel with each other. It should be emphasized that al-

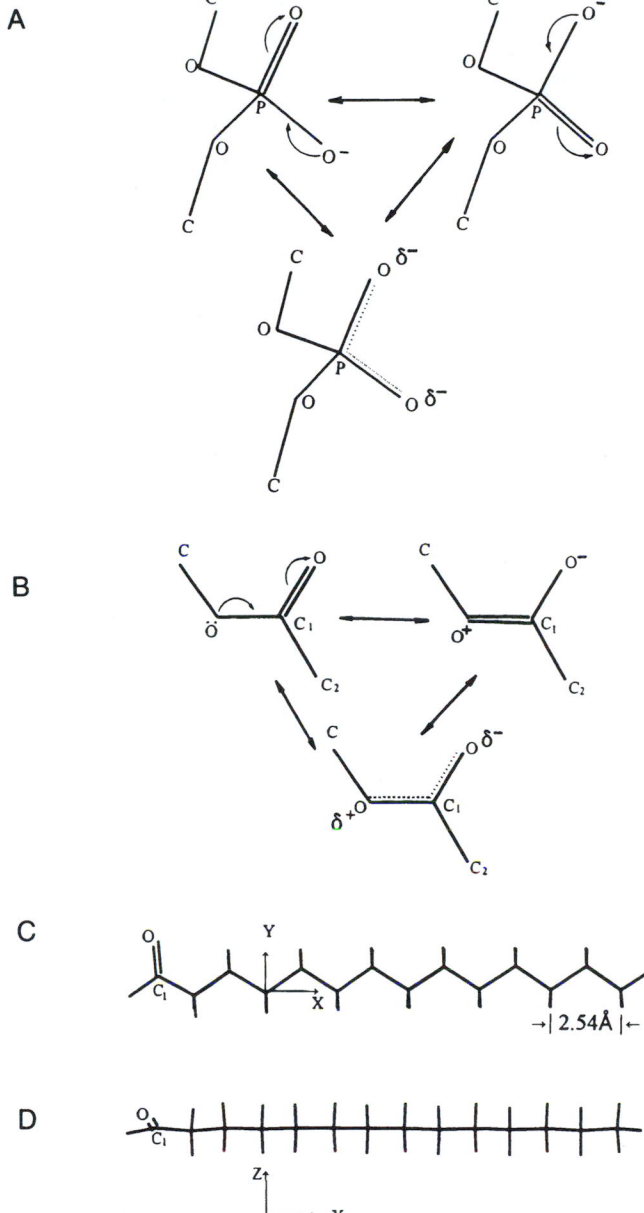

FIGURE 4. Structural features common to all diacyl phospholipids. (A) The three resonance states of the tetrahedrally arranged phosphate group. (B) The partial double-bond character of the ester bond as represented by the resonance hybrid effect. (C) The zigzag plane of the acyl chain viewed on the x–y plane. (D) The same zigzag plane shown in (C) viewed on the x–z plane.

2. Phosphatidylethanolamines

Phosphatidylethanolamines (1,2-diacyl-sn-glycero-3-phosphoethanolamines; PE) are also a major zwitterionic lipid component of biological membranes. They usually occur in lesser amounts in higher organisms than do phosphatidylcholines. However, phosphatidylethanolamines are often the principal diacyl phospholipids of microorganisms. Interestingly, phosphatidylethanolamines containing polyunsaturated fatty chains such as docosahexaenoic and arachidonic acids are found at extremely high levels in the retinal rod outer segments.

1,2-dilauroyl-DL-phosphatidylethanolamine or racemic C(12):C(12)PE is the first molecular species of diacyl phospholipid with which x-ray crystallographic analysis has been performed to determine the structure at the atomic level (Hitchcock et al., 1974). The single-crystal structure, however, was determined in the presence of acetic acid (Fig. 5E and F). The stereochemical features of the polar headgroup may therefore be affected by the contact with acetic acid. Nevertheless, the overall molecular structure of crystalline PE is very similar to that of C(14):C(14)PC shown in Fig. 5A–D. In particular, the diglyceride moiety of C(12):C(12)PE is also characterized by a roughly h-shaped geometry, with a sharp 90° bend occurring at C_2 of the sn-2 acyl chain. However, there are two notable differences. First, the zwitterionic headgroup of phosphatidylethanolamine has a smaller volume; it bends down slightly more than does the zwitterionic headgroup of phosphatidylcholine. Second, the zigzag planes specified by the long sn-1 and sn-2 acyl chains of C(12):C(12)PE are nearly parallel to each other, whereas the corresponding zigzag planes are nearly perpendicular in C(14):C(14)PC. It is also worth mentioning that the hydrogen atoms in the terminal $-NH_3^+$ group ($pK_a \approx 9.6$) of a phosphatidylethanolamine are able to form H-bonds with the nonesterified phosphate oxygens of an adjacent phosphatidylethanolamine. Such an H-bonding capability is absent in phosphatidylcholine.

3. Phosphatidylinositol

Phosphatidylinositol (1,2-diacyl-sn-glycero-3-phosphoinositol; PI) is an ester of a cyclic myo-inositol with a net negative charge. It is widely distributed among many organisms including plants, but constitutes only about 2–8% of all diacyl phospholipids in animal cell membranes. The biological function of phosphatidylinositol has been the subject of intense research in recent years. In eukaryotic cells, phosphatidylinositol on the outer leaflet of the plasma membrane, for instance, can be glycosylated to form glycosylphosphatidylinositol, which is capable of anchoring a large number of proteins to the cell surface (Low, 1989). In addition, two classes of glycosylphosphatidylinositol containing D-chiro-inositol and myo-inositol, respectively, have been demonstrated to play an important role of mediator in the action of insulin (Huang and Larner, 1993). On the other hand, phosphatidylinositol situated on the inner leaflet of the plasma membrane, which faces the cytosol, can be phosphorylated to form phosphatidylinositol bisphosphate. The phosphorylated PI, in turn, can be hydrolyzed by specific enzymes to yield inositol 1,4,5-trisphosphate and 1,2-diacylglycerol (Nishizuka, 1984). These

though sn-1 and sn-2 acyl chains of C(14):C(14)PC have the same total number of methylene units, there is an effective chain length difference between the sn-1 and sn-2 acyl chains within the lipid molecule. This difference is evident from the clear separation of the two methyl terminal groups along the long molecular axis. In fact, this effective chain length difference is one of the structural parameters that can be used to characterize the bilayer phase-transition behavior, and we shall discuss it in Section IV.C.

C(14):C(14)PC, crystal structure   C(14):C(14)PC, energy-minimized structure   C(12):C(12)PE, crystal structure

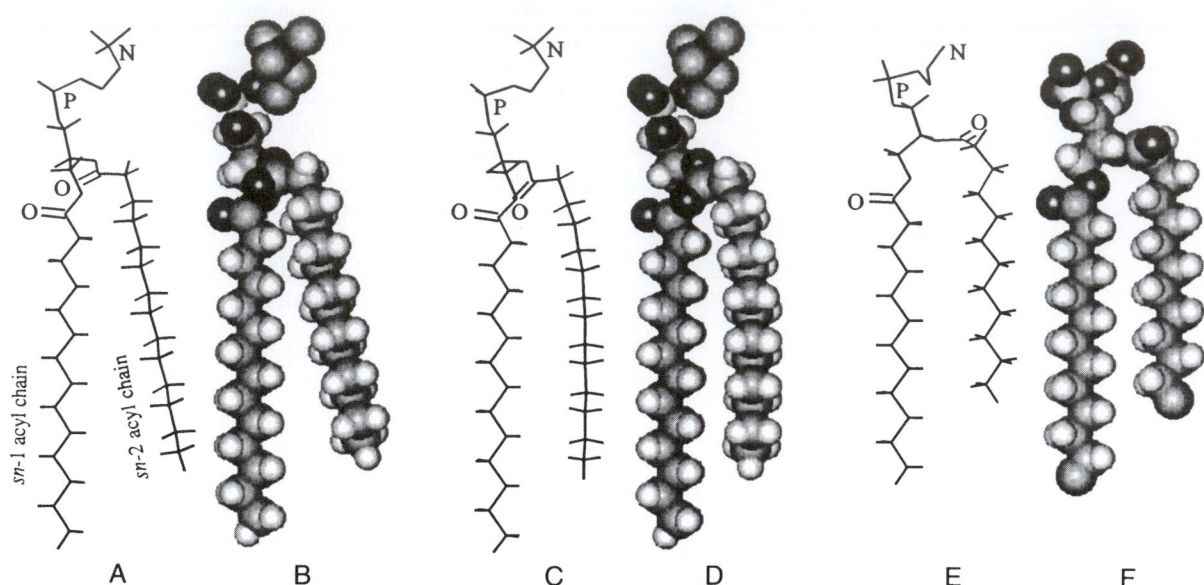

FIGURE 5.   The x-ray single-crystal structure of C(14):C(14)PC is shown in (A) and (B) as represented by wire and sphere models, respectively. The refined or energy-minimized structure of C(14):C(14)PC is shown in (C) and (D). The refined structure was obtained by molecular mechanics simulations using the x-ray crystallographic data as the starting data input. Two representations of the x-ray single-crystal structure of C(12):C(12)PE obtained in the presence of acetic acid are shown in (E) and (F).

hydrolytic products act as second messengers to activate intracellular cascades, thus leading to separate hormonal signaling pathways. From a functional point of view, phosphatidylinositol is an extremely important group of diacyl phospholipids.

The orientations of the rigid inositol ring structures of dimyristoyl phosphatidylinositol (C(14):C(14)PI) and dimyristoyl phosphatidylinositol 4-phosphate (C(14):C(14)PI-4P) in the bilayer of C(14):C(14)PC have been determined by neutron diffraction (Bradshaw *et al.*, 1996). According to the neutron diffraction results, the headgroup orientations of these two lipids are strikingly different. Specifically, the inositol ring of C(14):C(14)PI extends almost vertically from the bilayer surface. In addition, the hydroxyl groups at ring positions 2 and 6 are topologically located within H-bond distance from the two partially charged oxygens of the phosphate group (Fig. 6A and B). In the case of C(14):C(14)PI-4P, the rigid ring is inclined at an angle to the bilayer surface (Fig. 6C and D); moreover, the intramolecular H bonds are absent. At present, it is uncertain how the change in the headgroup orientation upon phosphorylation of the ring hydroxyl group is related to the variation of the function of phosphatidylinositol.

4. Phosphatidylglycerols

Phosphatidylglycerols (1,2-diacyl-*sn*-glycerol-3-phosphoryl-1-*sn*-glycerol or 1,2-diacyl-*sn*-3-phosphatidyl-*sn*-1'-glycerol; PG) are negatively charged diacyl phospholipids that are present as minor lipid components in animal cell membranes (largely mitochondrial membranes). It is also

the biosynthetic precursor of **cardiolipin,** shown in Fig. 2E, which is found primarily in mitochondrial membranes. Interestingly, the enzymatic activity of mitochondrial enzyme cytochrome *c* oxidase is stimulated specifically by cardiolipin. Phosphatidylglycerols are major constituents of chloroplast membranes of higher plants and gram-positive bacterial membranes. In a variety of plant cells, saturated identical-chain phosphatidylglycerols are found in chloroplast membranes. The amount of the saturated molecular species of phosphatidylglycerol appears to correlate positively with the susceptibility of the chloroplast membranes to chilling or low temperature–induced injury (Somerville, 1995).

The polar nonacylated glycerol of naturally occurring phosphatidylglycerol has a stereochemical configuration (*sn*-1-glycerol phosphate) that is opposite to the backbone glycerol (*sn*-3-glycerol phosphate). The single-crystal structure of dimyristoyl phosphatidylglycerol, C(14):C(14)PG, has been determined by x-ray diffraction (Pascher *et al.*, 1987; Pascher *et al.*, 1992). In addition, the energy-minimized structure of C(14):C(14)PG, calculated based on the single-crystal coordinates, is also known (Jin *et al.*, 1994). The diglyceride moiety of the single-crystal structure, shown in Fig. 7A and B, is essentially identical to that of the energy-minimized conformation (Fig. 7C and D). However, the packing pattern of the two acyl chains in C(14):C(14)PG is distinctly different from that of C(14):C(14)PC or C(12):C(12)PE. Specifically, the *sn*-1 acyl chain of C(14):C(14)PG is initially orientated perpendicular to the long molecular axis of the lipid molecule and then makes a

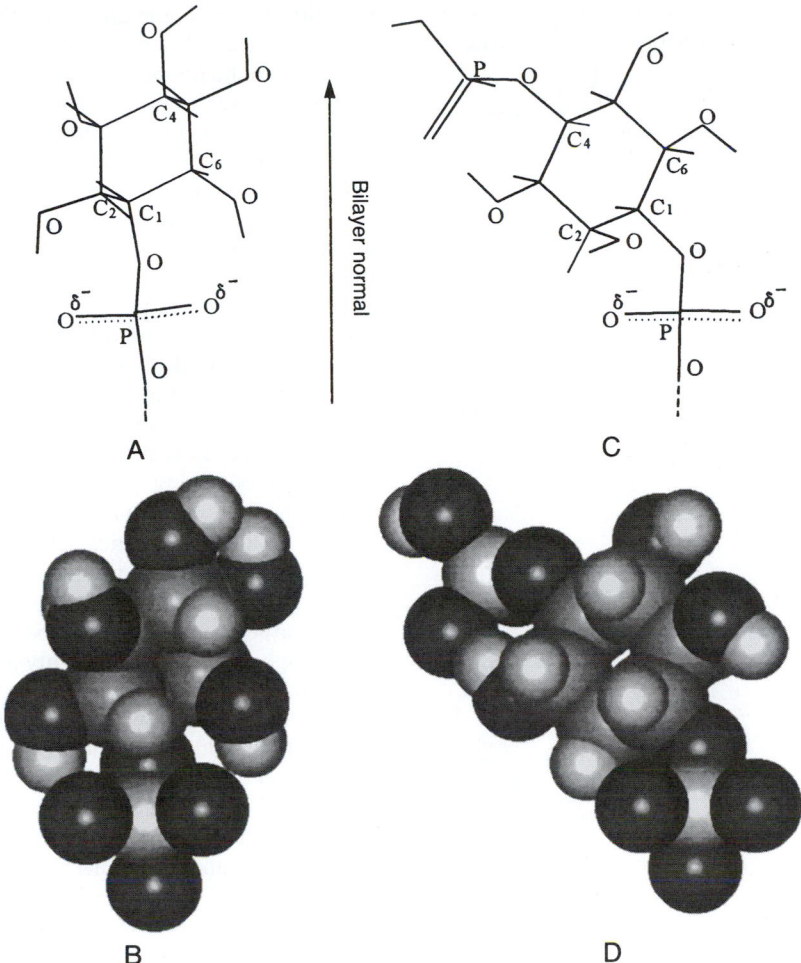

FIGURE 6. The orientations of the inositol ring structures of dimyristoyl phosphatidylinositol and dimyristoyl phosphatidylinositol 4-phosphate as determined by neutron diffraction are presented in the wire and sphere models as shown in A–B and C–D, respectively.

90° bend at C_2 so that the rest of the *sn*-1 acyl chain axis runs in parallel with the axis of the all-*trans sn*-2 acyl chain (Fig. 7). The overall acyl chain conformation can thus be considered to be opposite to that of C(14):C(14)PC or C(12):C(12)PE, in which the chain bend occurs at the C_2 in the *sn*-2 acyl chain (Fig. 5).

5. Phosphatidylserine

Phosphatidylserine (1,2-diacyl-*sn*-glycero-3-phospho-L-serine; PS) is a widely distributed, but minor, class of diacyl phospholipids. It is a negatively charged lipid and is an important constituent of brain cell membranes. In addition to serving as a structural component of membranes, phosphatidylserine is also an activator for all isozymes of protein kinase C. At present, the single-crystal structure of phosphatidylserine is not yet available. It should be mentioned that the serine moiety of phosphatidylserine is the carboxylated ethanolamine; hence, one would expect that the structure of phosphatidylserine may be somewhat similar to that of phosphatidylethanolamine. However, based on the data obtained with ^{31}P-NMR, Browning and Seelig (1980) have

suggested that the headgroup of phosphatidylserine is more rigid than that of phosphatidylethanolamine.

B. Sphingolipids

The second major group of membrane lipids, the sphingolipid, is derived from **sphingosine** (4-sphingenine), a long-chain (18 or 20 carbons in length) amino alcohol with a *trans*-double bond at C_4 (Fig. 1). Sphingosine has an amino group on C_2; this amino group is joined in amide linkage with a long-chain fatty acid to form **ceramide** (Fig. 1). The fatty acyl moiety of ceramide can be saturated hydrocarbons containing 14–26 carbons, and may be monounsaturated and α-hydroxylated. Unlike the sphingosine base, the monounsaturated C–C double bond in the fatty acyl chain has a *cis* configuration. The α-hydroxylation of the fatty acyl chain refers to an OH group covalently linked to the α-carbon of the fatty acyl chain. Ceramide can be further extended to form various kinds of sphingolipids by the addition of different polar headgroups to its terminal hydroxyl group at the C_1 atom. For instance, if a polar phosphorylcholine molecule is

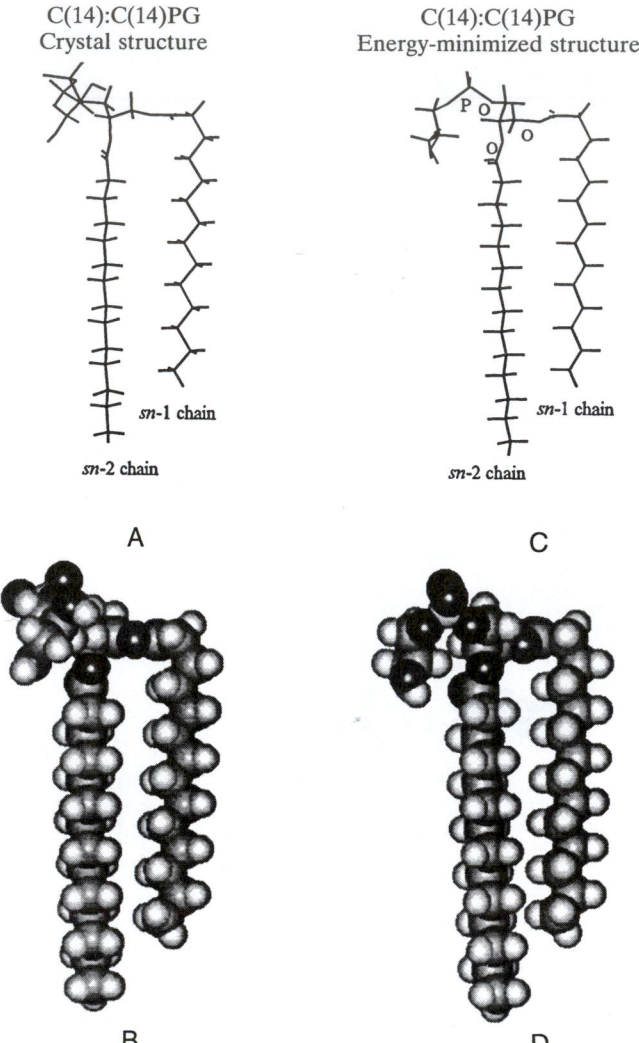

C(14):C(14)PG
Crystal structure

C(14):C(14)PG
Energy-minimized structure

sn-1 chain

sn-2 chain

sn-1 chain

sn-2 chain

A

C

B

D

FIGURE 7. The wire and sphere models of the single-crystal (A and B) and the energy-minimized (C and D) structures of C(14):C(14)PG.

linked to the C_1 hydroxyl group of ceramide by an ester bond, the resulting molecule, ceramide-1-phosphorylcholine, is **sphingomyelin,** the most commonly occurring sphingolipid. If, on the other hand, the reducing end of a carbohydrate residue is linked to the terminal hydroxyl group on the C_1 of ceramide by a β-glycosidic bond, then a **glycosphingolipid** (GSL) is formed. **Galactosylceramide** (or galactocerebroside), for example, is a well-known GSL; the molecule contains a C_{18}-sphingosine, a long-chain fatty acid, and a carbohydrate molecule of the hexose galactose. Another example is glucosylceramide (or glucocerebroside), in which the C_1 hydroxyl group of ceramide is ester-linked via a β-glycosidic bond to the C_1 of glucose. Additional sugars can be attached to the glucose moiety of glucosylceramide to yield oligosaccharide-containing GSLs such as negatively charged **gangliosides.** Table 2 lists the general structure and the nomenclature of a few representative GSLs. It should be emphasized that these glycosphingolipids are

ubiquitously present in animal cell membranes. For instance, galactosylceramide and its sulfate ester, **sulfogalactosylceramide,** are the major lipid species of the myelin sheath surrounding nerves in the central nervous system. An excellent review of the structure and the physicochemical properties of individual GSLs is given by Maggio (1994).

Let us now consider the structure of sphingomyelin, the most abundant sphingolipid. In this amphipathic molecule, the hydrophobic chain of the sphingosine backbone has a fixed length of 15 (or 17) carbons which can be considered as one of the two aliphatic chains of the hydrophobic tail. The fatty acid N-acylated to the sphingosine backbone constitutes the second chain of the hydrophobic tail. Typical fatty acids N-linked to the sphingosine backbone are stearic, behenic, lignoceric, and nervonic acids, with the lignoceric acid being the most prevalent one. Consequently, the most commonly occurring sphingomyelin is characterized by a marked hydrocarbon chain asymmetry (Fig. 8A).

Because of the identical phosphorylcholine headgroup, sphingomyelin appears structurally to resemble phosphatidylcholine. In fact, these two membrane lipids are quite different in other aspects. As discussed above, the most commonly occurring sphingomyelin is a highly asymmetrical molecule. In addition, the secondary hydroxyl group on C_3 of the sphingosine backbone is an allyl alcohol; hence, it can serve as an excellent H-bond donor. Furthermore, the NH and CO groups of the amide bond linked to C_2 of the sphingosine backbone can serve as H-bond donor and acceptor, respectively. These groups can, therefore, cause sphingomyelins in the bilayer to undergo intra- and intermolecular interactions mediated by H bonds. Phosphatidylcholine has no H-bond donor within the lipid molecule; hence, the intramolecular H bond and direct intermolecular H bonds between phosphatidylcholine molecules cannot occur.

The single-crystal structure of sphingomyelin is not known. The conformation of crystalline galactosylceramide, however, has been reported by Pascher and Sundell (1977). In addition, the conformation of the polar headgroup of sphingomyelin can be approximated by the same headgroup of phosphatidylcholine. A plausible model for the molecular structure of sphingomyelin can thus be simulated based on the crystal structures of galactosylceramide and phosphatidylcholine. This simulated molecular structure of N-lignoceryl sphingomyelin is illustrated in Fig. 8A. For comparison, the molecular structure of 1-palmitoyl-2-oleoyl- phosphatidylcholine or $C(16):C(18:1\Delta^9)$PC, obtained by molecular mechanics modeling (Huang and Li, 1996), is presented in Fig. 8B. Parenthetically, 1-palmitoyl-2-oleoyl-phosphatidylcholine is one of the most abundant membrane lipids found in animal cells.

C. Cholesterol

The third major group of membrane lipids, the sterol, can best be represented by cholesterol. Cholesterols are important components of animal cell membranes and plasma lipoproteins. In addition, cholesterol is also the precursor of many steroid hormones (e.g., testosterone, progesterone) and bile acids (e.g., cholate, taurocholate). In plasma membranes, cholesterol concentration is quite high. For instance,

TABLE 2 Name and Structure of Some Common Glycosphingolipids

Lipid	Chemical structure[a]	Abbreviated nomenclature (IUPAC-IUBS)	Common name
Galactosylceramide	Galβ1→1' Cer	GalCer	GalCer
Sulfogalactosylceramide	SO$_4$3Galβ1→1' Cer	SO$_4$3GalCer	SO$_4$3GalCer
Glucosylceramide	Glcβ1→1' Cer	GlcCer	GlcCer
Lactosylceramide	Galβ1→4Glcβ1→1' Cer	LacCer	LacCer
Gangliotriaosylceramide	GalNAcβ1→4Galβ1→4Glcβ1→1' Cer	Gg3Cer	Asialo G$_{M2}$
Gangliotetraosylceramide	Galβ1→3GalNAcβ1→4Galβ1→4Glcβ1→1' Cer	Gg4Cer	Asialo G$_{M1}$
Globotetraosylceramide	GalNAcβ1→3Galα1→4Galβ1→4Glcβ1→1' Cer	Gb4Cer	Globoside
G$_{M3}$ ganglioside	NeuAcα2→3Galβ1→4Glcβ1→1' Cer	II^{3}NeuAc-LacCer	Hematoside
G$_{M2}$ ganglioside	GalNAcβ1→4Gal(3-2αNeuAc)β1→4Glcβ1→1' Cer	II^{3}NeuAc-Gg3Cer	Tay Sachs ganglioside
G$_{M1}$ ganglioside	Galβ1→3GalNAcβ1→4Galβ(3-2αNeuAc)β1→ 4Glcβ1→1' Cer	II^{3}NeuAc-GgOse$_3$Cer	G$_{M1}$

[a]Cer, ceramide (acylsphingosine); Gal, galactose; Glc, glucose; GalNAc, N-acetylgalactosamine; NeuAc, N-acetylneuraminic acid (sialic acid); SO$_4$3Gal, a sulfate ester of galactose with the HSO$_4$ ester-linked at the C$_3$ hydroxyl of galactose.

in red blood cell membrane, the molar ratio of cholesterol/phospholipid is 0.8–1.0. In contrast, mitochondrial membranes contain little cholesterol.

Cholesterol is characterized structurally by a quasi-planar and frayed conformation. It consists of a fused tetracyclic ring system (the steroid nucleus) and a branched isooctyl side chain. Carbon atoms in the cholesterol molecule are numbered as shown in Fig. 9A. The four fused rings are named A, B, C, and D, with the A–B, B–C, and C–D rings being fused in the *trans* configuration. For instance, the C$_{18}$-angular methyl group is on the opposite side of the hydrogen atom attached at C$_{14}$.

The α face or underface of the fused tetracyclic ring system is relatively flat (Fig. 9B), because the seven axial hydrogen atoms at C$_1$, C$_3$, C$_7$, C$_9$, C$_{12}$, C$_{14}$, and C$_{17}$ are approximately coplanar. The β face, however, has greater relief due to the presence of two angular methyl groups (C$_{18}$ and C$_{19}$) attached at C$_{13}$ and C$_{10}$. In contrast to the alignment of the two angular methyl groups, which are perpendicular to the plane of steroid nucleus, the projected C$_{21}$ methyl group lies nearly parallel to the C$_{16}$–C$_{17}$ carbon-carbon single bond in the crystal structure of cholesterol monohydrate (Craven, 1976).

As a membrane lipid, cholesterol is unique due to the fact that the size of the headgroup, the β-hydroxyl group at C$_3$, is smaller than that of a water molecule. Consequently, cholesterols are too weakly amphipathic to self-assemble into the bilayer on their own in water. However, cholesterols can readily dissolve into the lipid bilayer of diacyl phospholipids.

III. Structural Organizations of Membrane Lipids

Since an individual diacyl phospholipid has two long acyl chains, it thus exhibits very limited solubility in water with a critical micellar concentration of about 10^{-10} M (Tanford, 1980). Above the critical micellar concentration, diacyl phospholipids self-assemble in water into numerous large organized structures, each characterized by a hydrophobic interior comprised of the acyl chains and a hydrophilic surface comprised of the headgroups. The driving force for forming spontaneously such a large organized structure is entropic. Specifically, bound water molecules are released into the bulk water from the acyl chains as these hydrophobic chains become sequestered in the interior of the organized structure. Since the motions of bound water are highly restricted, the release of bound water results in a substantial increase in entropy, thus favoring the formation of large organized structures. There are several forms of large organized structures for diacyl phospholipids in water: the micelle, bilayer, inversed hexagonal H$_{II}$ phase, and inversed bicontinuous cubic Pn3m phase. At physiological temperature and ambient pressure, the majority of diacyl phospholipids can self-assemble into **lamellae** or **bilayers** in excess water. The basic feature of a bilayer or lamella is a two-dimensional, bimolecular sheet consisting of two opposing leaflets. In this structure, the hydrocarbon chains of the aggregated diacyl phospholipids from the two opposing leaflets are submicroscopically aligned in a nearly parallel manner to form a microscopically two-dimensional hydrocarbon core, and the headgroups are layered on both sides of the hydrocarbon core in a sandwich arrangement.

Because lipid bilayers are structurally the fundamental building matrix of biological membranes (Danielli and Davson, 1935; Blaurock and Wilkins, 1972), studies of lipid bilayers composed of one or two types of pure diacyl phospholipids can provide insights into the properties of these chemically well defined lipid species arranged in two-dimensional arrays of lamellae. Consequently, the roles of these lipids played in biological membranes can be inferred. Two broadly defined model systems for the lipid bilayer have been widely used in recent years: the planar lipid bilayer (or black lipid membrane, BLM), originally developed by Mueller *et al.* (1962), and the liposome, admirably pioneered by Bangham *et al.* (1965).

The **planar lipid bilayer** is formed by applying a solution of lipids in an organic solvent such as decane to a small aperture in a hydrophobic sheet of Teflon or

N-Lignoceryl sphingomyelin

1-Palmitoyl-2-oleoyl phosphatidylcholine

A

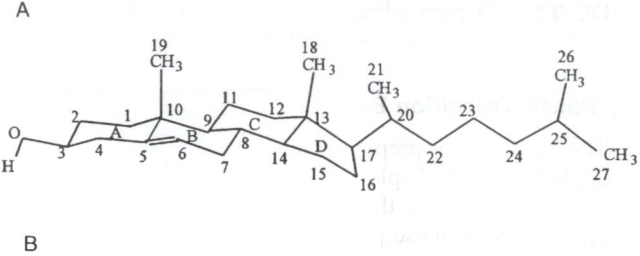

B

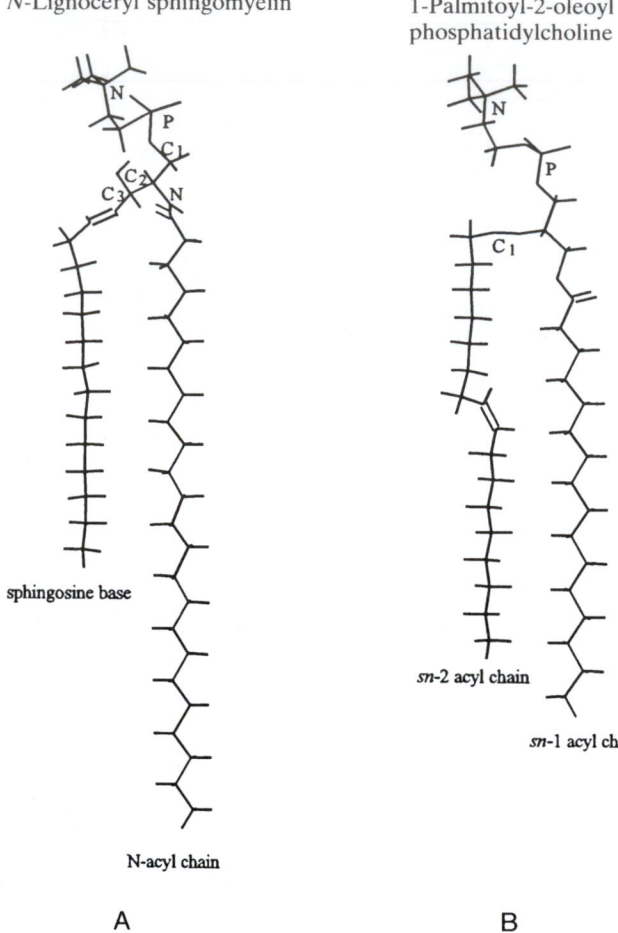

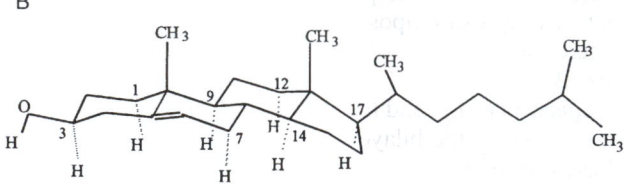

FIGURE 9. Conformation of cholesterol molecule. (A) The four fused rings of the steroid nucleus, designated as A, B, C, and D rings, are jointed in a *trans* configuration. The 18 and 19 angular methyl groups at positions 13 and 10 are located above the β face of the steroid nucleus. The polar β-OH group is linked to ring A at position 3. (B) The α face of the steroid nucleus is relatively flat, since the seven axial H atoms at positions 1, 3, 7, 9, 12, 14, and 17 are approximately coplanar.

sphingosine base

sn-2 acyl chain

sn-1 acyl chain

N-acyl chain

A   B

FIGURE 8. A structural comparison between L-lignoceryl sphingomyelin (A) and 1-palmitoyl-2-oleoyl-phosphatidylcholine (B). In sphingomyelin, C_1, C_2, and C_3 denote the first three carbon atoms in the sphingosine base as indicated in Figure 1B. In phosphatidylcholine, C_1 identifies the carbonyl carbon atom of the *sn*-2 acyl chain. N and P denote the nitrogen and phosphorus atoms, respectively.

polyethylene located between two chambers. This aperture is immersed in an aqueous medium. As the lipid bilayer forms spontaneously from the lipid solution under the aqueous medium, it becomes nonreflective to light; hence, it appears as a "black" film. This planar bilayer is an ideal model system for studying simultaneously the electric and transport properties of lipid membranes. An excellent book discussing the water transport across the planar lipid bilayer can be found in Finkelstein (1987). Many planar bilayer models using different methods of preparations are also available (Montal and Mueller, 1972; Schindler, 1980). Some can be readily used to incorporate ion channel–forming proteins to allow measurements of electrophysiological events associated with the bilayer (Hartshorne *et al.*, 1985).

Liposomes are formed spontaneously by allowing dry powders of phospholipids to undergo hydration in excess aqueous solution at a given temperature. If the temperature

is above the main phase transition temperature (T_m) of the phospholipid to be used, the dry lipid powders in solution can swell rapidly by vortexing to form concentrically enclosed multilamellar structures like hollowed-out onions, which are heterogeneous in size and shape. Multiple layers of aqueous solutions are partitioned into the multilamellar space between the polar surfaces of enclosed neighboring bilayers. These onion-like liposomes, resembling the myelin sheath of nerve cells, are often called **multilamellar vesicles** (MLV). If the milky liposome solution is subjected to ultrasonic irradiation under an N_2 atmosphere, it becomes a translucent one in which MLV with diameters up to several µm are transformed into small spherical vesicles. Upon gel filtration, small vesicles with an average diameter of about 200 Å can be separated, and each sized vesicle is a single-layered enclosed lamella (Huang, 1969). These are often referred to as **small unilamellar vesicles** (SUV). Because of the small radius of curvature, diacyl phospholipids packed in the inner and outer leaflets of SUV are subject to different constraints imposed by the different local surface curvatures; hence, they exhibit distinctly different packing properties. If the large liposomes are subjected to extrusion several times under pressure through a polycarbonate filter of well-defined pore size, single-layered large vesicles with an average diameter of about 0.1–1 µm, called **large unilamellar vesicles** (LUV), can be formed (Hope *et al.*, 1985). LUVs can also be formed by first converting the SUVs of saturated diacyl phosphatidylcholines into interdigitated lipid sheets in the presence of ethanol at low temperatures. After removing ethanol, these fused lipid sheets can be spontaneously vesicularized into LUVs by raising the solution temperature above a critical temperature (Ahl *et al.*, 1994). These LUVs have the potential to serve as drug carriers due to their large internal volumes.

IV. The Thermodynamic and Conformational Properties of Bilayers

A. Phase Transition Behavior

The lipid bilayer prepared from a single species of diacyl phospholipids can display interesting and unique properties in excess water. The thermotropic phase behavior, for instance, is one of these properties. This behavior can be determined by using liposomes as the bilayer system and the differential scanning calorimetry (DSC) as the detecting device (Chapman, 1993). Specifically, the **phase transition temperature** (T_m) and the **transition enthalpy** (ΔH) can be determined as the bilayer undergoes the thermally induced phase transitions.

In the case of liposomes prepared from dipalmitoyl phosphatidylcholines or C(16):C(16)PC that have been preincubated at 0 °C for a long time, the first DSC heating scan clearly shows three endothermic phase transitions (Fig. 10). The lower temperature peak is rather broad, centered at 21.5 °C (T_m), with a ΔH value of about 6 kcal/mol. Here the T_m value corresponds to the transition-peak position at the maximal height, and the ΔH value corresponds to the integrated area under the peak divided by the lipid concentration. This lower temperature transition is called the **subtransition**. The small middle temperature peak, $T_m = 35$ °C and $\Delta H = 1.0$ kcal/mol, is designated as the **pretransition**. The large and sharp upper temperature peak is the **main**

transition with $T_m = 41.5$ °C and $\Delta H = 8.9$ kcal/mol. Because of the three discernible transitions, four lamellar phases designated as L_c, $L_{\beta'}$, $P_{\beta'}$, and L_α can be defined for the lamellar C(16):C(16)PC within the temperature range of 0 to 50 °C (Fig. 10). The sub-, pre-, and main transitions can thus be ascribed to the $L_c \rightarrow L_{\beta'}$, $L_{\beta'} \rightarrow P_{\beta'}$, and $P_{\beta'} \rightarrow L_\alpha$ phase transitions, respectively. Upon immediate reheating, the subtransition is abolished; the pretransition is shifted down by about 1 °C; the main transition is the only one that is reproducible upon repeated heatings. Since the pretransition is very small, the main transition is often called the **gel-to-liquid-crystalline phase transition,** referring to the overall $L_{\beta'} \rightarrow L_\alpha$ transition.

Among the multiple phases exhibited by fully hydrated C(16):C(16)PC, the $L_{\beta'}$ and L_α phases have been studied most extensively by x-ray diffraction methods. For instance, in the wide-angle region, a sharp reflection at 4.2 Å with a shoulder at about 4.1 Å is observed for C(16):C(16)PC in the $L_{\beta'}$ or gel phase, whereas the L_α or liquid-crystalline phase is characterized by a diffuse reflection centered at 4.6 Å. These x-ray results thus indicate that in the $L_{\beta'}$ phase the acyl chains of C(16):C(16)PC are highly ordered with a tilt orientation relative to the axis perpendicular to the bilayer surface; however, the chains are melted at $T > T_m$, resembling simple liquid paraffins. More recently, refined x-ray analyses of fully hydrated C(16):C(16)PC in the $L_{\beta'}$ (20 °C) and L_α (50 °C) phases were reported by Nagle et al. (1996). Their analyses show that the bilayer thickness decreases from 47.8 Å at 20 °C to 39.2 Å at 50 °C. In addition, the surface area of C(16):C(16)PC expands from 47.9 to 62.9 Å² over the same temperature interval. Clearly, the lipid molecule as a whole has increased its volume as it undergoes the thermally induced main phase transition.

B. *Trans* to *Gauche* Isomerizations About the C–C Bonds in Lipid Acyl Chains

The structural changes of the lipid bilayer associated with the gel-to-liquid-crystalline phase transition can be attributed fundamentally to the following sequence of events: at $T < T_m$, the long acyl chains of C(16):C(16)PC adopt an all-*trans* conformation. Upon heating to the characteristic temperature of T_m, the long acyl chains are "energized"; the tight packing of the long acyl chain will loosen. This thermally induced loosening of the molecular packing is accompanied by a molar volume increase, arising mainly from the *trans* → *gauche* isomerizations of carbon-carbon single bonds along the acyl chains within each phospholipid molecule.

In Fig. 11A–C, the convention for designating the *trans* and *gauche* conformations of C–C single bonds in an acyl chain is illustrated. The *trans* conformation has a torsion angle of 180°, as shown in Fig. 11A, and is abbreviated as *t*. The two *gauche* conformations shown in Fig. 11B and C are *gauche* (+) with a torsion angle of 60° and *gauche* (−) with a torsion angle of −60°, abbreviated as g^+ and g^-, respectively. It should be mentioned that *t*, g^+, and g^- conformations are positioned in three different minimal troughs in the potential energy diagram, indicating that they all are stable conformations. The *t* conformation is of lower energy than g^+ or g^- by about 0.89 kcal/mol (Huang and Li, 1996); hence, *t* is more

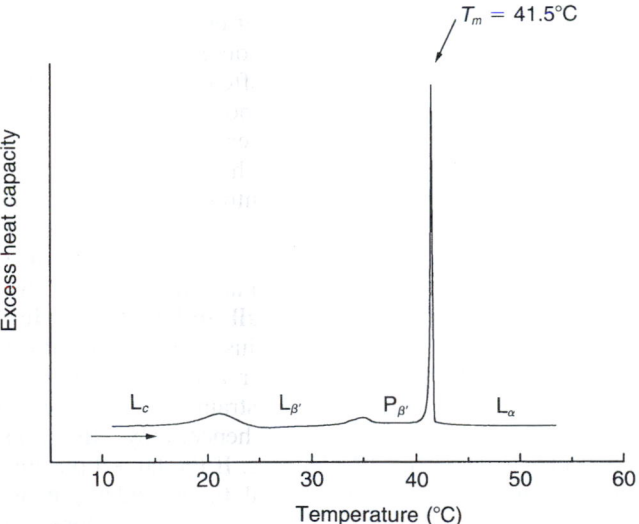

FIGURE 10. A representative DSC (differential scanning calorimetry) heating curve for the aqueous dispersion of C(16):C(16)PC. Prior to the DSC scan, the sample has been preincubated at 4.0 °C under an N_2 atmosphere for about a year. The DSC experiment was performed with a high-resolution MC-2 differential scanning microcalorimeter (Microcal, Inc., Northampton, MA) running at a constant scan rate of 15 °C/h. The buffer solution used for preparing the lipid sample was 50 mM NaCl solution containing 5 mM phosphate buffer and 1 mM EDTA at pH 7.4, and the final lipid concentration in the aqueous dispersion was 4.3 mM. T_m is the main phase transition temperature. L_c, $L_{\beta'}$, $P_{\beta'}$, and L_α are the four lamellar phases described in the text. The arrow indicates the ascending temperature direction of the DSC scan.

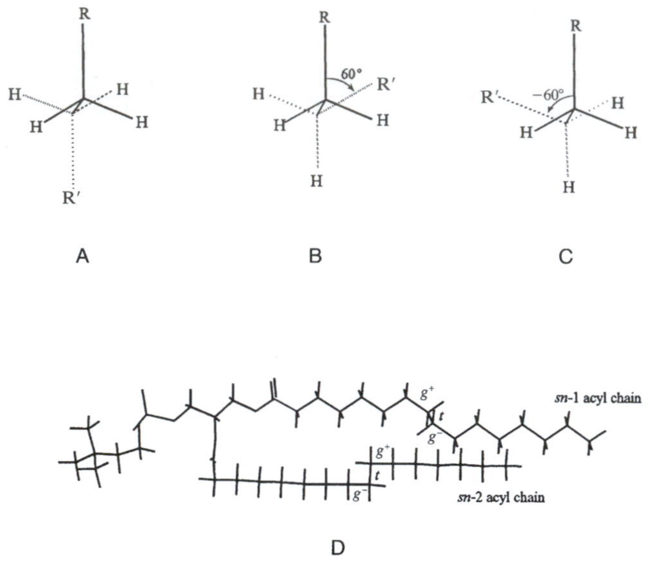

FIGURE 11. The conformations of *trans, gauche* (+), and *gauche* (−) bonds and g^+tg^- kinks observed in lipid acyl chains. The torsion angle around the C–C' bond shown in (A–C) is defined by the relative positions of R, C, C', and R' atoms. Looking down the C–C' bond axis, atom R is placed at 12 o'clock, and atom R' measures the torsional angle, plus if clockwise and minus if counterclockwise. The *trans* or *t* conformation is characterized by a torsion angle of 180° as shown in (A); the *gauche* (±) or $g^±$ conformations have torsion angles of ±60° as indicated in (B) and (C), respectively. Energetically, *gauche* (±) conformations are 0.89 kcal/mol higher than the *trans* conformation. In (D), coupled g^+tg^- kinks are shown in the *sn*-1 and *sn*-2 acyl chains of a PC molecule.

favored to exist at low temperatures. Upon heating, the conformational changes of the lipid acyl chains in the bilayer occurring abruptly at T_m are thus most likely to involve *trans* → *gauche* isomerizations. Since the *trans* → *gauche* isomerization proceeds with rotations of carbon atoms about the C–C single bond, different rotational isomers (t, g^+, and g^-) resulting from the rotated bond are also called **rotamers** or **rotomers.** In this section, some spectroscopic studies will be described, suggesting the simultaneous occurrence of coupled and isolated *trans* → *gauche* isomerizations about C–C bonds in lipid acyl chains at the phase transition temperature.

When a single g^+ or g^- rotamer is formed, the chain will make a 120° bend. If this pronounced bend is introduced into the center of an acyl chain in the bilayer, it will be inhibited sterically by the neighboring chains. Hence, the single g rotamer is most likely to occur, at $T > T_m$, near the methyl ends of the two acyl chains in a C(16):C(16)PC molecule where the steric hindrance due to neighboring chains is small. If, on the other hand, a sequence of two *gauche* bonds separated by a *trans* bond (g^+tg^- or g^-tg^+) is formed, the resulting acyl chain is then kinked in the shape of a crankshaft. The presence of such a kink leads to a largely parallel packing of the acyl chains in the bilayer. Consequently, the coupled g^+tg^- or g^-tg^+ kink is favored in the center of the acyl chain at $T > T_m$. In comparison with an all-*trans* chain, the overall length of a kinked chain is shortened by 1.27 Å along the long-chain axis. In Fig. 11D, this type of kink is graphically illustrated in each of the two acyl chains of a C(16):C(16)PC molecule.

One spectroscopic technique that has been elegantly used to examine the lipid chain order in terms of the *trans* → *gauche* isomerizations about the C–C bond is ^{2}H-NMR (Seelig, 1977; Davis, 1983). Specifically, the methylene and the terminal methyl groups in the acyl chain are substituted by the CD_2 and CD_3 groups, respectively; the congested powder patterns and the corresponding de-packed spectrum with a larger number of resolvable quadrupolar splittings can be obtained from a single ^{2}H-NMR experiment at $T > T_m$ (Sternin *et al.*, 1983). The experimentally determined quadrupolar splittings can be converted to the **order parameter** (S_{CD}) of the C–D bond (Seelig, 1977). Since the order parameter, $S_{CD} = 0.5 <3 \cos^2 \beta - 1>$, is a time-averaged parameter of the angular fluctuation (β) of the C–D bond relative to the bilayer normal on the ^{2}H-NMR time scale ($< 10^{-5}$ s), the quantity $|S_{CD}|$ can be used to infer the possible presence of *gauche* rotamers in the acyl chain at $T > T_m$. For instance, if the acyl chain is assumed to adopt an all-*trans* configuration and is simply reorienting about the long-chain axis, then $\beta = 90°$ and $|S_{CD}| = 0.5$. If, on the other hand, the chain is assumed to undergo an isotropic motion over all space, this motional averaging will then lead to $|S_{CD}| = 0$. The *trans* → *gauche* isomerizations of the CD_2 groups around the C–C bonds will undoubtedly result in a partially disordered orientation of the acyl chain, which can be expected to give rise to $|S_{CD}|$ values in between the two limiting cases (0.5 and 0). The $|S_{CD}|$ values of the C–D bonds along the acyl chain of C(16):C(16)PC at $T > T_m$ are found experimentally to be approximately constant and are, indeed, equal to about 0.2 over the initial segment near the interface; this part is usually referred to as the **plateau region.** A rapid decrease in $|S_{CD}|$ is observed near the methyl terminus, indicating that the motional averaging of the quadrupolar interactions increases towards the middle of the bilayer. The shape of this orientational order profile can be rationalized by a structural model of the acyl chain at $T > T_m$ as follows: the coupled $g^±tg^∓$ kinks are allowed in the plateau region, whereas the isolated $g^±$ rotamers exist predominantly near the terminal methyl end.

The relative ratios of *trans* and *gauche* rotamers in the lipid acyl chains at various temperatures can be detected directly by Raman spectroscopy (Levin, 1984). This method has indeed demonstrated that the gel → liquid-crystalline phase transition exhibited by the bilayer of C(16):C(16)PC is accompanied by an abrupt increase in the ratio of *gauche/trans* at the onset of T_m (Gaber and Peticolas, 1977). Moreover, single $g^±$ rotamers as well as g^+tg^- and g^-tg^+ kinks have been detected by FT-infrared spectroscopy in the bilayers of C(16):C(16)PC at $T > T_m$ (Mendelsohn and Senak, 1993; Casal and McElheney, 1990). Most interestingly, these single $g^±$ rotamers and coupled $g^±tg^∓$ kinks are also observed in the computer-simulated C(14):C(14)PC bilayer, in excess water, at $T > T_m$ using the molecular dynamics method (Chiu *et al.*, 1995). Moreover, the $g^±tg^∓$ kinks shown by molecular dynamics simulations are not fixed at any given preferred positions along the *sn*-1 and *sn*-2 acyl chains of C(14):C(14)PC molecules in the liquid-crystalline bilayer. Based on ^{2}H-NMR data, it has been suggested earlier that the coupled $g^±tg^∓$ kinks migrate up and down along the plateau regions of lipid acyl chains in bilayers at $T > T_m$ (Seelig and Seelig, 1980).

Based on the spectroscopic and computational results just discussed, it is obvious that the thickness of the lipid bilayer composed of saturated identical-chain phosphatidylcholines must be reduced at $T > T_m$ due to the presence of $g^{\pm}$ rotamers and $g^{\pm}tg^{\mp}$ kinks along the acyl chain. In addition, the lateral chain-chain van der Waals interactions must also be diminished at $T > T_m$ due to the random distribution and the dynamic nature of the $g^{\pm}tg^{\mp}$ kinks. Consequently, the average rate of **lateral diffusion** for saturated identical-chain phosphatidylcholine molecules in each of the two monolayers within the two-dimensional plane of the lipid bilayer can be expected to increase appreciably as the bilayer undergoes the thermally induced gel-to-liquid-crystalline phase transition. Indeed, it has been shown by using fluorescent lipid molecules that an averaged lateral diffusion coefficient of about 10^{-8} cm²/s is observed in the two-dimensional plane of the liquid-crystalline bilayer, and a much smaller value on the order of 10^{-11} cm²/s is estimated for the same fluorescent lipid molecules in the gel-state bilayer (Clegg and Vaz, 1985).

C. Factors Affecting the Gel-to-Liquid-Crystalline Phase Transition Temperature

The phase transition behavior of lipid bilayers composed of a given species of pure diacyl phospholipids can be influenced by many internal and external factors. Internal factors refer to changes in chain length, chain asymmetry, unsaturation, and headgroup structure. External factors include pressure, pH, and chemical composition of the medium. Of the two thermodynamic parameters (T_m and ΔH) that are associated directly with the main phase transition of the lipid bilayer, the value of T_m can be determined with greater accuracy by high-resolution DSC. We shall examine how some of the internal and external factors affect the T_m value associated with the thermally induced main phase transition of diacyl phospholipids in excess water.

Before we discuss the effects of chain length and chain asymmetry on T_m, let us first define some **structural parameters** (ΔC, $\Delta C/CL$, and N) underlying saturated diacyl phosphatidylcholines or C(X):C(Y)PC packed in the gel-state bilayer. To proceed, we return for a moment to Fig. 5C in which the effective chain-length difference between the sn-1 and the sn-2 acyl chains of an identical-chain phosphatidylcholine in the crystalline state, ΔC_{cry}, can be clearly identified. This value of ΔC_{cry} is seen in Fig. 5C to be 3.68 C–C bond lengths along the chain. For the same identical-chain phosphatidylcholine molecule packed in the gel-state bilayer, the effective chain-length difference is reduced to about 1.5 C–C bond lengths, and we designate it as the chain-length difference of the reference state (ΔC_{ref}). For a saturated mixed-chain C(X):C(Y)PC such as C(18):C(14)PC in the gel-state bilayer, the effective chain-length difference ΔC is related to X and Y as follows (Huang et al., 1994): $\Delta C = |X - Y + \Delta C_{ref}| = |18 - 14 + 1.5| = 5.5$ C–C bond lengths. Schematically, this structural parameter is presented in Fig. 12A. Here, a crystalline structure is conveniently drawn for C(18):C(14)PC; actually, all structural parameters of C(X):C(Y)PC are defined for lipid molecules packed in the gel-state bilayer with shorter chain lengths. A second structural parameter, CL, refers to the effective chain length of the longer of the two acyl chains in C–C bonds, and is defined as CL = X − 1 if the sn-1 acyl chain is the longer chain or as CL = Y − 2.5 if the sn-2 acyl chain is the longer chain. In the latter case, the effective chain length extends from the point corresponding to the position of carbonyl carbon of the sn-1 acyl chain to the methyl terminus of the sn-2 acyl chain. This structural parameter, CL, is also illustrated schematically in Fig. 12A. For C(18):C(14)PC, the CL value is 17 C–C bond lengths. The normalized chain-length difference or asymmetry is the ratio of $\Delta C/CL$. For C(18):C(14)PC, the $\Delta C/CL$ value is 0.32. Another structural parameter, N, is taken to be the distance, in C–C bond lengths, between the two carbonyl oxygens of the sn-1 acyl

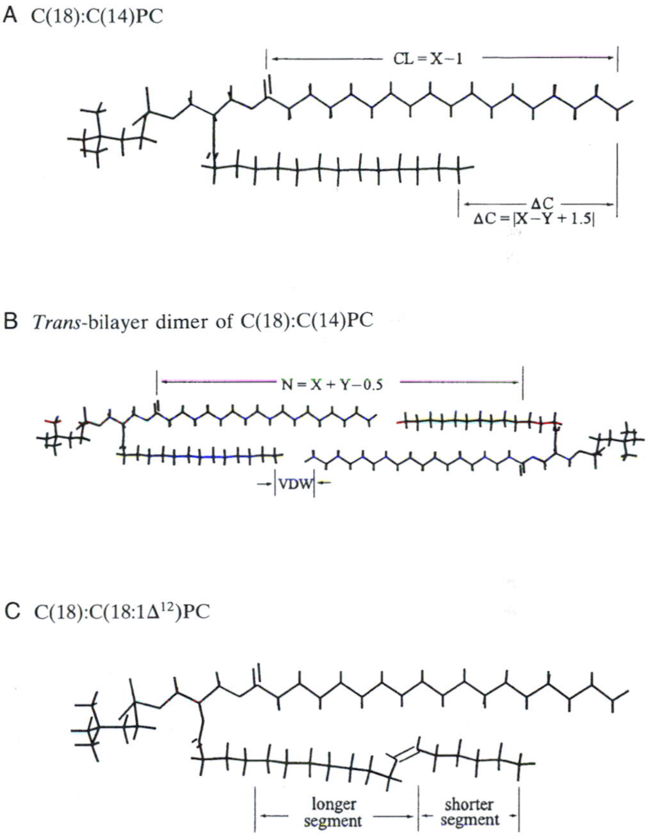

A C(18):C(14)PC

B *Trans*-bilayer dimer of C(18):C(14)PC

C C(18):C(18:1Δ^{12})PC

FIGURE 12. Molecular graphics representations of phospholipids with various structural parameters. (A) A monomer of saturated C(18):C(14)PC. CL is the effective chain length of the longer of the two acyl chains, and ΔC is the effective chain-length difference between the two acyl chains along the long-chain axis. For C(X):C(Y)PC in the gel state, the relations between CL, ΔC, and X, Y are indicated. In the case of gel-state C(18):C(14)PC, CL and ΔC are 17 and 5.5 carbon-carbon bond lengths, respectively. (B) A molecular graphics drawing to illustrate a transbilayer dimer of C(18):C(14)PC with a partially interdigitated packing motif. N is the hydrophobic thickness of the dimer, corresponding to the distance separating the two carbonyl oxygens of the two opposing sn-1 acyl chains. VDW is the van der Waals contact distance between two opposing methyl groups in the bilayer interior. (C) The monomeric C(18):C(18:1Δ^{12})PC. The sn-2 acyl chain is shown to adopt the crankshaftlike motif; hence it consists of two segments. The chain-chain lateral interaction is assumed to be energetically more favorable between the sn-1 acyl chain and the longer segment of the sn-2 acyl chain.

chains in a trans-bilayer dimer of C(X):C(Y)PC along the long molecular axis as shown schematically in Fig. 12B. The N value of a C(X):C(Y)PC dimer is related to X and Y as follows: $N = (X - 1) + \text{VDW} + (Y - 2.5) = X + Y - 0.5$, where VDW is the van der Waals separation distance between the two opposing terminal methyl groups in the trans-bilayer dimer (Fig. 12B), which is taken to be 3 C–C bond lengths in the gel-state bilayer.

When the normalized chain asymmetry ratio, $\Delta C/\text{CL}$, is less than 0.42, saturated C(X):C(Y)PC molecules in excess water at $T < T_m$ can self-assemble into the partially interdigitated bilayer, in which the methyl end of the longer acyl chain of one lipid molecule from one leaflet packs end-to-end with the methyl end of the shorter acyl chain of another lipid molecule from the opposing leaflet and *vice versa* (Li *et al.*, 1993). This partially interdigitated packing motif is diagrammatically shown in Fig. 12B. If the $\Delta C/\text{CL}$ value of a phosphatidylcholine molecule is greater than 0.42, this highly asymmetric lipid tends to self-assemble into the mixed interdigitated bilayer at $T < T_m$ (Li *et al.*, 1993). Phosphatidylcholines with $\Delta C/\text{CL}$ values greater than 0.42 are rare in biological membranes; hence, they are not discussed in this chapter.

At $\Delta C/\text{CL}$ value less than 0.42, DSC measurements on saturated C(X):C(Y)PC have demonstrated that the T_m value of the main phase transition is related to the structural parameters ΔC and N according to the following equations (Huang *et al.*, 1994; Huang and Li, 1999):

$$T_m = 161.75 - 3706.06\frac{1}{N} - 278.75\frac{\Delta C}{N} + 239.94\frac{\Delta C}{N + \Delta C} \quad (1)$$

for lipids with a longer effective *sn*-1 acyl chain, and

$$T_m = 155.11 - 3534.31\frac{1}{N} - 245.78\frac{\Delta C}{N} + 199.04\frac{\Delta C}{N + \Delta C} \quad (2)$$

for lipids with a longer effective *sn*-2 acyl chain. Implicitly, these two equations indicate that if ΔC is constant, the T_m value increases with increasing N; however, the increase tends to level off towards an asymptote. In contrast, the T_m value decreases with increasing ΔC if N is constant; as a result, ΔC counteracts the N-promoted increase in T_m. Thus, it is the delicate balance between the two structural parameters (N and ΔC) underlying the dimeric C(X):C(Y)PC molecules in the gel-state bilayer that determines the unique T_m value of the fully hydrated lipids. Equations 1 and 2 not only show the fundamentally antagonistic effect between N and ΔC in determining the T_m value, but also can be employed to predict the unknown value of T_m based on the chemical formula of C(X):C(Y)PC (Huang *et al.*, 1994; Huang and Li, 1999).

Table 3 lists some representative T_m values obtained calorimetrically with multilamellar vesicles containing diacyl phospholipids. The values of ΔC and N calculated according to $\Delta C = |X - Y + 1.5|$ and $N = X + Y - 0.5$ are also listed for those saturated lipid species. For the homologous series of saturated identical-chain phosphatidylcholines from C(14):C(14)PC to C(20):C(20)PC with a common ΔC value of 1.5 C–C bond lengths, the T_m value is seen to increase progressively with a step-wise increase in N which, in fact, is directly related to the chain length. This raises the in-

teresting question of why the T_m value increases progressively with an increase in the chain length of the saturated lipid molecule. The answer lies in the strength of the lateral chain-chain van der Waals interaction in the gel-state bilayer. This energetic term is directly proportional to the chain length. Specifically, the lateral van der Waals interaction between two all-*trans* saturated chains (E_{VDW}) is related to the number of methylene units in the chain (N_m) according to the following equation (Salem, 1962):

$$E_{\text{VDW}} = A\frac{3\pi N_m}{8lD^5} \quad (3)$$

where l is the distance between two methylene units along the long-chain axis (1.27 Å), D is the lateral distance separating the two all-*trans* chains, and A is a constant (-1340 kcal/Å^{-6} mol). If D is taken to be 4.8 Å, then $E_{\text{VDW}} = -0.49$ kcal/mol per methylene unit. Clearly, the longer the acyl chain or the larger the number of the methylene units (N_m), the greater is the attractive van der Waals interaction between the two saturated acyl chains. As a result, a larger thermal energy and a higher T_m are required to promote the *trans* $\rightarrow$ *gauche* isomerizations within the longer chains as the phospholipid molecules in the bilayer undergo the thermally induced gel-to-liquid-crystalline phase transition.

In Table 3, C(16):C(16)PC, C(14):C(18)PC, and C(18):C(14)PC are shown to have a common N value of 31.5 C–C bond lengths. The ΔC and T_m values of these saturated lipids are also shown to be 1.5, 2.5, and 5.5 C–C bond lengths and 41.5, 39.2, and 31.2 °C, respectively. This set of saturated C(X):C(Y)PC thus serves to demonstrate that the T_m value decreases with increasing ΔC when the N value is held constant. Similarly, another pair of the position isomers, C(16):C(18)PC/C(18):C(16)PC, exhibits the same inverse relationship between T_m and ΔC (Table 3). Fundamentally, the structural parameter ΔC represents the chain asymmetry. It is also a perturbation term that affects the closest lateral chain-chain interaction in the gel-state bilayer. This perturbation stems from the bulky methyl terminal groups.

Most of the naturally occurring phospholipids in eukaryotic cell membranes have a saturated *sn*-1 acyl chain and an unsaturated *sn*-2 acyl chain with different numbers and positions of *cis* double bonds. It is very convenient to abbreviate the names for these enormously complicated *sn*-1 saturated/*sn*-2 unsaturated phospholipid species according to the chemical natures of their two acyl chains. These abbreviated names, as specified in Table 3, are used throughout the rest of this chapter. Now, we can proceed to discuss the effects of unsaturation of the *sn*-2 acyl chain on the gel-to-liquid-crystalline phase transition temperature of phospholipid bilayers as monitored by high-resolution DSC. First, the introduction of a single *cis* double bond into the *sn*-2 acyl chain markedly reduces the T_m value of lamellar phosphatidylcholines. A good example is demonstrated by the bilayer composed of 1-stearoyl-2-oleoyl-phosphatidylcholines or C(18):C(18:1Δ^9)PC, which exhibits calorimetrically the T_m value of 5.6 °C. In contrast, the T_m value of 55.3 °C is detected for the bilayer composed of the saturated counterparts or C(18):C(18)PC. Second, the T_m-lowering effect of acyl chain monounsaturation depends critically on the position of

TABLE 3 The Main Phase Transition Temperature for Some Diacyl Phospholipids

Lipid[a]	ΔC	N	T_m (°C)	Lipid[a]	T_m (°C)
Saturated phospholipids					
C(14):C(14)PC	1.5	27.5	24.1	C(14):C(14)PE	49.6
C(16):C(16)PC	1.5	31.5	41.5	C(16):C(16)PE	63.2
C(18):C(18)PC	1.5	35.5	55.3	C(18):C(18)PE	74.0
C(20):C(20)PC	1.5	39.5	66.4	C(20):C(20)PE	82.5
C(14):C(18)PC	2.5	31.5	39.2	C(14):C(18)PE	61.6
C(18):C(14)PC	5.5	31.5	31.2	C(18):C(14)PE	54.9
C(16):C(18)PC	0.5	33.5	48.8	C(16):C(18)PE	70.8
C(18):C(16)PC	3.5	33.5	44.4	C(18):C(16)PE	66.0
C(20):C(18)PC	3.5	37.5	57.5	C(20):C(18)PE	76.3
Unsaturated phospholipids					
C(16):C(18:1Δ^6)PC			18.8	C(16):C(18:1Δ^6)PE	39.1
C(16):C(18:1Δ^9)PC			−2.6	C(16):C(18:1Δ^9)PE	26.1
C(18):C(18:1Δ^6)PC			24.8	C(18):C(18:1Δ^6)PE	42.6
C(18):C(18:1Δ^7)PC			16.7	C(18):C(18:1Δ^7)PE	35.4
C(18):C(18:1Δ^9)PC			5.6	C(18):C(18:1Δ^9)PE	31.5
C(18):C(18:1Δ^{11})PC			3.8	C(18):C(18:1Δ^{11})PE	29.8
C(18):C(18:1Δ^{12})PC			9.1	C(18):C(18:1Δ^{12})PE	32.3
C(18):C(18:1Δ^{13})PC			15.9	C(18):C(18:1Δ^{13})PE	35.4
C(18):C(18:2$\Delta^{9,12}$)PC			−14.6	C(18):C(18:2$\Delta^{9,12}$)PE	4.4
C(18):C(18:2$\Delta^{9,12,15}$)PC			−12.2	C(18):C(18:2$\Delta^{9,12,15}$)PE	—

[a]Saturated phosphatidylcholines and phosphatidylethanolamines are abbreviated as C(X):C(Y)PC and C(X):C(Y)PE, respectively. X denotes the total number of carbons in the sn-1 acyl chain and Y denotes the total number of carbons in the sn-2 acyl chain. sn-1 saturated and sn-2 monounsaturated acyl chains with the cis-double bond (Δ) at the position n from the carboxyl end abbreviated as C(X):C(Y:1Δ^n); similarly, lipids with sn-1 saturated and sn-2 diunsaturated acyl chains are abbreviated as C(X):C(Y:2$\Delta^{n,m}$), where m is the second cis-double bond position at the mth carbon atom from the carboxyl end.

the cis carbon-carbon double bond (Δ^n) in the sn-2 acyl chain. As shown in Table 3, the minimal T_m for a homologous series of C(18):C(18:1Δ^n)PC occurs at $\Delta^n = \Delta^{11}$ or when the cis double bond is positioned near the center of the linear segment of the sn-2 acyl chain. Moreover, the T_m value increases progressively as the single cis double bond moves from Δ^{11} toward either end of the acyl chain. Third, the progressive increase in the level of acyl chain unsaturation has a nonadditive effect on the T_m. Based on the T_m values of C(18):C(18)PC, C(18):C(18:1Δ^9)PC, C(18):C(18:2$\Delta^{9,12}$)PC, and C(18):C(18:2$\Delta^{9,12,15}$)PC shown in Table 3, it is evident that the introduction of the first cis double bond near the chain center gives rise to a drastic decrease of 49.7 °C in T_m. Although the introduction of a second cis double bond produces a substantial decrease of 20.2 °C in T_m, this decrease is not nearly as large as the first one. Most interestingly, the successive incorporation of a third cis double bond at Δ^{15} results in a slight increase of 2.4 °C in T_m.

Let us now consider a molecular model that has been invoked to interpret the large T_m-lowering effect of acyl chain monounsaturation and the characteristic dependency of T_m on the position of the cis double bond along the acyl chain (Wang et al., 1995). This molecular model makes three basic assumptions. (1) The monoenoic sn-2 acyl chain in the sn-1 saturated/sn-2 monounsaturated phospholipid molecule is assumed to adopt, at $T < T_m$, an energy-minimized crank-shaft-like kink motif as shown in Fig. 12C. The sn-2 acyl chain thus consists of a longer segment and a shorter segment separated by the Δ^n-containing kink. (2) The longer segment and the neighboring sn-1 acyl chain run in a nearly parallel manner with favorable van der Waals attractive distance between them. (3) The shorter segment is considered to be partially disordered at $T < T_m$, thus playing a perturbing role in the lateral chain-chain interactions in the gel-state bilayer. With this molecular model, we are now in a position to interpret first the drastic T_m-lowering effect of a single cis double bond. Because the shorter segment is a perturbing element, it behaves similar to the structural parameter ΔC to counteract the effect of N, thus lowering T_m. Moreover, since this short segment is already partially disordered at $T < T_m$, it cannot contribute significantly to the conformational disordering process of trans → gauche isomerizations involved in the phase transition. As a result, the total number of trans → gauche isomerizations during the phase transition is decreased appreciably for the bilayer composed of cis-monoenoic phosphatidylcholines relative to that of the saturated counterparts, leading to a significantly lower T_m value.

We can use the same molecular model to see how the T_m value varies as the single cis bond migrates along the sn-2 acyl chain. Since the longer segment of the kinked chain is assumed to undergo a favorable van der Waals contact interaction with the *sn*-1 acyl chain, this contact interaction energy must then depend on the length of the longer segment. When the single *cis* double bond is positioned at the center of the *sn*-2 acyl chain, the longer segment of the kinked chain has a minimal length, which is almost equal to that of the shorter segment. Hence, the van der Waals interaction with the *sn*-1 acyl chain is also minimal. As the single *cis* double bond migrates away successively from the chain center towards either end, the length of the longer segment is progressively increased, leading to a proportionally increased van der Waals interaction and hence a gradual increase in T_m. The characteristic change of T_m as a function of the location of the *cis* double bond, shown in Table 3, can thus be explained by the structural model.

The phospholipid headgroup can also exert a profound effect on the phase transition behavior of the lipid bilayer. For instance, when the headgroup's choline moiety in a saturated identical-chain phosphatidylcholine molecule is replaced by ethanolamine, the resulting phosphatidylethanolamine can self-assemble into the gel-state bilayer in excess water after incubating at $T < T_m$ for a brief time. Upon heating, the gel-state PE bilayer transforms directly into the liquid-crystalline (L_α) phase without going through the intermediate $P_{\beta'}$ phase that is commonly observed for fully hydrated PC with identical chains. Hence, the main phase transition of saturated identical-chain PE corresponds to the $L_\beta \to L_\alpha$ phase transition, where L_β denotes the gel phase with the saturated acyl chains orienting perpendicular to the bilayer surface. For comparison, the second DSC heating scans obtained with aqueous dispersions of C(16):C(16)PC and C(16):C(16)PE are shown in Fig. 13. Clearly, the T_m value is significantly higher for PE than for PC with the same saturated fatty acyl chain composition, although their ΔH values are comparable. In Table 3, the T_m values for nine species of saturated diacyl PE as well as PC are given. The T_m values for saturated PE are consistently higher than those for PC with the same fatty acyl chain composition. However, for the homologous series of identical-chain phospholipids from C(14):C(14)PE/PC to C(20):C(20)PE/PC, the difference in T_m between PE and PC (ΔT_m) decreases with increasing acyl chain length (Table 3). Figure 13 also shows the second DSC heating scans for aqueous dispersions of C(16):C(18:1Δ^6)PC and C(16):C(18: 1Δ^6)PE; here, the T_m value of PE is 20.3 °C higher than that of PC. In Table 3, the T_m values of a large number of *sn*-1 saturated/*sn*-2 unsaturated PC/PE are presented. Clearly, the T_m values of unsaturated PE are also consistently higher than those of unsaturated PC with the same fatty acyl chain composition. These higher T_m values can thus be taken as evidence to suggest that, in general, the lateral chain-chain interactions in the PE bilayer are stronger than those in the PC bilayer at $T < T_m$. This stronger lateral chain-chain interaction may be attributed to many factors such as the smaller PE headgroup, the interlipid H-bond networks on the two surfaces of the gel-state PE bilayer, the smaller hydration of the PE bilayer, and so on. At the present time, however, the quantitative contribution of each factor to the overall difference in T_m between PE and PC is not well understood and deserves further investigation.

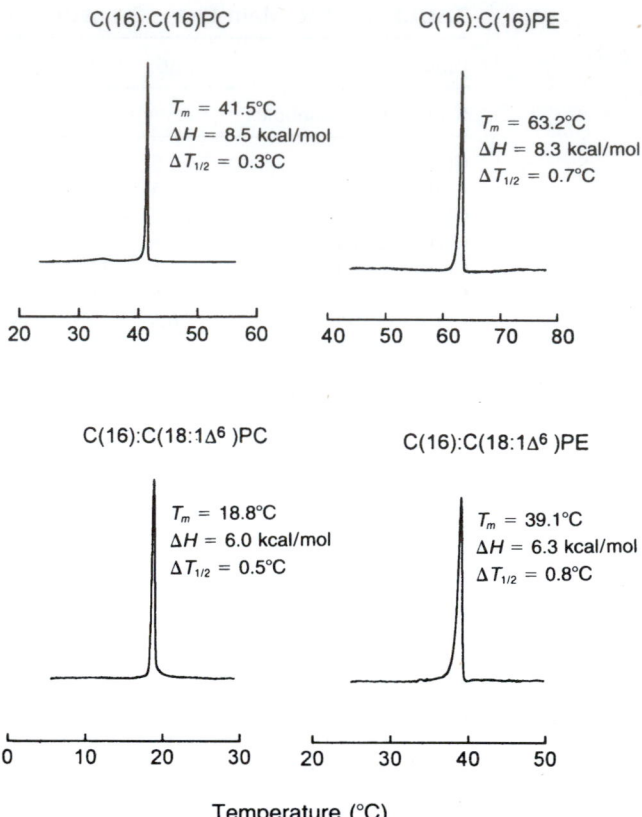

FIGURE 13. Representative DSC second heating scans for aqueous dispersions of PC and PE with the same saturated and monounsaturated acyl chain compositions. The T_m values of PE bilayers are observed to be consistently higher than those of PC bilayers with the same fatty acyl chain compositions irrespective of acyl chain unsaturation. These DSC experiments were performed with a Microcal Model MC-2 differential scanning microcalorimeter (Northampton, MA). Scan rate: 15 °C/h.

Based on experimental data obtained with 33 molecular species of C(X):C(Y)PE, two T_m equations similar to Eqs. 1 and 2 for C(X):C(Y)PC have also been derived. As a result, for PC and PE with the same *sn*-1 and *sn*-2 acyl chains, the T_m value of one lipid can be converted to the other via the common structural parameters. Specifically, the difference in T_m between PE and PC (ΔT_m) can be calculated based on their common values of N and ΔC according to the following equations (Huang and Li, 1999):

$$\Delta T_m^{PE-PC} = T_m^{PE} - T_m^{PC}$$
$$= -2.05 + 719.03\frac{1}{N} - 39.49\frac{\Delta C}{N} + 62.22\frac{\Delta C}{N+\Delta C} \quad (4)$$

for lipids with an effective longer *sn*-1 acyl chain, and

$$\Delta T_m^{PE-PC} = T_m^{PE} - T_m^{PC}$$
$$= 2.77 + 584.4\frac{1}{N} - 110.46\frac{\Delta C}{N} + 151.01\frac{\Delta C}{N+\Delta C} \quad (5)$$

for lipids with an effective longer *sn*-2 acyl chain.

Using Eqs. 4 and 5, the T_m value of a given C(X):C(Y)PC bilayer, T_m^{PC}, can be calculated from the experimental T_m

value of the corresponding C(X):C(Y)PE bilayer, T_m^{PE}. Likewise, the unknown T_m^{PE} value can also be calculated, provided that the value of T_m^{PC} is known experimentally. The two structural parameters, N and ΔC, associated with C(15):C(15)PC or C(15):C(15)PE packed in the gel-state bilayer, for example, are 29.5 and 1.5 C–C bond lengths, respectively. With these numbers, the ΔT_m^{PE-PC} term in Eq. 4 can be calculated to be 23.3 °C. The T_m value of 34.0 °C is known calorimetrically for the C(15):C(15)PC bilayer; consequently, the phase transition temperature underlying the gel-to-liquid-crystalline phase transition for lamellar C(15):C(15)PE can be estimated as follows: $T_m^{PE} = \Delta T_m^{PE-PC} + T_m^{PC} = 57.3$ °C. This calculated value has indeed been verified by DSC experiments.

One external factor that has been extensively studied is the pressure (Wong, 1986). Take the C(16):C(16)PC dispersions at 47 °C as an example. The lamellar PC in the L_α phase can undergo two isothermal phase transitions upon elevation of pressure up to 1.4 kbar. The first pressure-induced isothermal transition occurs at the critical pressure of 0.17 kbar, and the second occurs at 0.60 kbar. These two phase transitions correspond to the cooling-induced main and pretransitions, respectively. Clearly, an increase in pressure acts in a manner similar to a decrease in temperature by exerting an ordering effect on the acyl chain packing in the bilayer.

V. Binary Phospholipid Mixtures

Thus far the structure and the phase behavior of a limited number of one-component phospholipid systems have been discussed. Attention is now directed to the mixing behavior of two-component (A and B) phospholipid systems at ambient pressure. The mixing behavior of two component lipids in the bilayer is expected to depend on the difference in the pair-interaction energies between the mixed pairs (A-B) and the like pairs (A-A and B-B) which, in turn, depends on the extent of structural similarity between the two component lipids. Furthermore, the structural characteristics of A and B in the bilayer may change abruptly and significantly as the two-component bilayer undergoes the respective phase transition of components A and B. The relative pair-interaction energy may thus be quite different in the gel and liquid-crystalline states. For instance, the lateral phase separation of a given two-component system may occur in the gel phase due to their structural dissimilarity, whereas a complete miscibility between the same two components may take place in the L_α phase as a result of conformational changes of the component lipids occurring at the T_m. The mixing behavior and the interaction energies related to the component lipids in the gel and liquid-crystalline phases are fully contained in the **temperature-composition phase diagram** for binary lipid systems. In this section, we consider three of the most commonly observed temperature-composition phase diagrams for binary lipid systems. These phase diagrams are shown diagrammatically in Fig. 14A–C. In particular, each phase diagram is characterized by its position and shape, which are determined by the phase boundaries called the **solidus** and **liquidus** as shown in Fig. 14A. Experimentally, these phase boundaries are commonly constructed from the

DSC heating curves for a given binary phospholipid system at various relative compositions. Basically, the onset and completion temperatures for each transition peak, after appropriate correction, are plotted as a function of the mole fraction of the higher melting component lipid (X_B). These corrected onset and completion temperature points form the bases for defining the solidus and liquidus, respectively, of the temperature-composition phase diagram (Mabrey and Sturtevant, 1976).

Of all the known phase diagrams constructed on the basis of calorimetric data obtained with binary phospholipids, the mixing behavior of the two lipid components in the bilayer plane is found to be always nonideal. This implies that the phase transition of the component A (or B) is affected by the

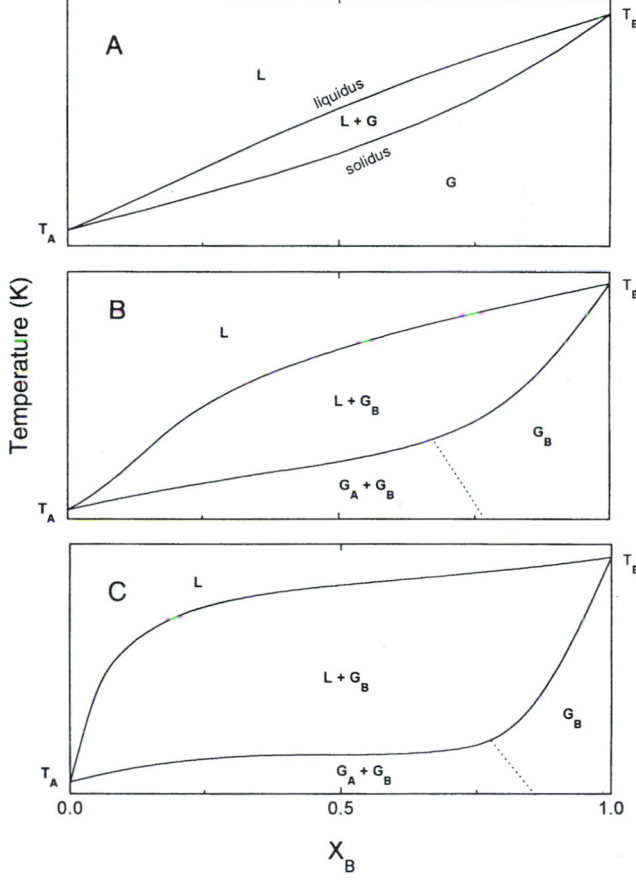

FIGURE 14. Three commonly observed phase diagrams determined calorimetrically for binary lipid systems. (A) The phase diagram for an isomorphous system as represented by the binary lipid system of C(14):C(14)PC/C(16):C(16)PC. (B) The phase diagram for the binary lipid system of C(14):C(14)PC/C(18):C(18)PC. (C) The peritectic phase diagram of the C(14):C(14)PC/C(20):C(20)PC system. X_B denotes the mole fraction of the higher melting component B. G and L denote the gel and liquid-crystalline phases, respectively. The subscripts A and B represent the lipid components A and B, respectively, in the binary lipid system. The values of T_A and T_B, in kelvin, for each of the three binary A/B systems are given in Table 4. The simulated values of the nonideality parameters of mixing (ρ^L and ρ^G) for the three phase diagrams are also shown in Table 4.

presence of the second component B (or A) in the binary mixture. The basic thermodynamic equations—derived from the regular solution theory (or Briggs-Williamson approximation), which describe phenomenologically the phase diagram for the nonideal system of binary mixtures of phospholipids (A and B)—are given by (Lee, 1977):

$$\ln \frac{X_B^G}{X_B^L} = \frac{\Delta H_B}{R}\left(\frac{1}{T}-\frac{1}{T_B}\right) + \frac{1}{RT}\left[\rho^L\left(1-X_B^L\right)^2 - \rho^G\left(1-X_B^G\right)^2\right] \quad (6)$$

$$\ln \frac{1-X_B^G}{1-X_B^L} = \frac{\Delta H_A}{R}\left(\frac{1}{T}-\frac{1}{T_A}\right) + \frac{1}{RT}\left[\rho^L\left(X_B^L\right)^2 - \rho^G\left(X_B^G\right)^2\right] \quad (7)$$

where X_B^G and X_B^L, the composition variables, are the mole fraction of the high-melting component B in the gel and liquid-crystalline phases, respectively; ΔH_A and ΔH_B are the transition enthalpies (kcal/mol) of pure components A and B, respectively; T_A and T_B are the phase transition temperatures (Kelvin) of the bilayers composed of the pure components A and B, respectively; ρ^G and ρ^L are two parameters having units of energy (kcal/mol) that describe the nonideality of mixing between components A and B in the gel and liquid-crystalline phases of the bilayer, respectively; and R is the gas constant. The nonideality parameter, ρ, is related to the pair-interaction energy between mixed pairs and between like pairs in the two-dimensional plane of the bilayer as follows: $\rho = Z[E_{AB} - 1/2(E_{AA} + E_{BB})]$, where Z is the number of nearest neighbors in the plane of the bilayer and E_{AB}, E_{AA}, and E_{BB} are the molar interaction energies of A-B, A-A, and B-B pairs of nearest neighbor pairs, respectively. If the value of ρ is positive, then there is a tendency for like lipid species to cluster together leading to lateral phase separation; $\rho < 0$ reflects the tendency of compound formation; $\rho = 0$ corresponds to ideal lateral mixing between components A and B in the bilayer.

Equations 6 and 7 allow the position and the shape of the phase diagram for a given binary lipid mixture to be determined, provided that the values of ρ^L and ρ^G and other experimental values (T_m and ΔH) are known. On the other hand, if the position and the shape of the phase diagram have already been determined experimentally, Eqs. 6 and 7 can be used to estimate the ρ^L and ρ^G values by matching the experimental and the simulated phase diagrams with a computer program (Brumbaugh and Huang, 1992). The mutual miscibilities of the two lipid components in the bilayer can thus be compared quantitatively based on the values of ρ^L and ρ^G.

A. Gel-Phase Miscibility and Liquid-Crystalline Phase Miscibility

The simplest phase diagram for binary lipid systems is the **isomorphous system,** in which the two components A and B are mutually miscible over the entire composition range in both the gel and liquid-crystalline states. Each binary mixture with a given X_B value in this isomorphous system exhibits calorimetrically a single endothermic phase transition, corresponding to the gel-to-liquid-crystalline phase transition. The transition curves obtained with various binary mixtures are broadened somewhat in comparison with those of the pure component lipids, particularly in the X_B range of 0.4–0.6. The solidus and liquidus, constructed on the basis of the onset and completion temperatures of these transition curves, are thus smooth curves, giving rise to a **lens-shaped phase diagram** as diagrammatically represented in Fig. 14A. Specifically, below the solidus is the single-phase gel region (G); above the liquidus is the single-phase liquid-crystalline region (L); between the solidus and the liquidus is a two-phase region in which the G and L phases are in dynamic equilibrium.

Binary mixtures of different saturated phosphatidylcholines, in which one component differs from the other by only two or fewer methylene units in each of their long acyl chains, exhibit complete miscibility in both gel and liquid-crystalline bilayers. The experimentally determined phase diagrams of these systems are characterized by the lens-shaped phase boundaries. A good example is provided by the binary lipids of C(14):C(14)PC/C(16):C(16)PC (Mabrey and Sturtevant, 1976). The nonideality parameters ρ^G and ρ^L for the lens-shaped phase diagram of C(14):C(14)PC/C(16):C(16)PC are 0.35 kcal/mol and 0.16 kcal/mol, respectively (Table 4). The value of ρ^G is, in fact, smaller than the average kinetic energy of the thermal environment at room temperature (0.58 kcal/mol); hence, the two lipid components, C(14):C(14)PC and C(16):C(16)PC, are mixed nearly ideally in the gel state at room temperature. It should be noted that the value of ρ^L is smaller than that of ρ^G; consequently, in the two-dimensional plane of the bilayer, these

TABLE 4 Values of Nonideality Parameters of Mixing (ρ) for Various Binary Lipid Systems[a]

Binary A/B system	ρ^L	ρ^G	T_A/T_B	$\Delta H_A/\Delta H_B$
C(14):C(14)PC/C(16):C(16)PC	0.16	0.35	297.2/314.6	6.0/8.5
C(14):C(14)PC/C(18):C(18)PC	0.32	0.85	297.2/328.4	6.0/10.9
C(14):C(14)PC/C(20):C(20)PC	0.60	1.21	297.2/339.5	6.0/13.2

[a]The units for ρ, T (phase transition temperature), and ΔH (transition enthalpy) are kcal/mol, K, and kcal/mol, respectively; the superscripts L and G denote the liquid-crystalline and gel phases, respectively; the subscripts A and B denote the lipid components A and B in the binary A/B system.

different lipid molecules can mix even better at higher temperatures in the L_α phase. In general, lens-shaped phase diagrams with small values of ρ^L and ρ^G are commonly observed for binary mixtures of different saturated phospholipids with the same headgroup and with a small difference in acyl chain lengths.

B. Gel-Phase Immiscibility and Liquid-Crystalline-Phase Miscibility

We have just seen the lens-shaped phase diagram exhibited by the binary A/B system of C(14):C(14)PC/C(16):C(16)PC. If the higher melting component, C(16):C(16)PC, in the binary system is replaced by C(18):C(18)PC, then the chain-length difference between A and B is increased from two to four methylene units. Upon heating, the DSC curve containing two overlapped or separated phase transitions is noticeably observed over a certain range of X_B, indicating that a lateral phase separation of G_A and G_B occurs in the C(14):C(14)PC/C(18):C(18)PC system. Here, G_A and G_B refer to the gel phases of A and B components, respectively. The phase diagram for the binary C(14):C(14)PC/C(18):C(18)PC system constructed from the DSC transition curves is illustrated in Fig. 14B, which deviates considerably from the lens-shaped phase diagram observed for C(14):C(14)PC/C(16):C(16)PC (Fig. 14A). Simulations of the phase boundaries yield the values of ρ^G and ρ^L to be 0.85 and 0.32 kcal/mol, respectively (Table 4). The ρ^G value is somewhat larger than the thermal energy of the environment; hence, a partial phase separation is expected between C(14):C(14)PC and C(18):C(18)PC in the gel state bilayer. On the other hand, the same two components are mixed nearly ideally in the liquid-crystalline state as indicated by the ρ^L value of 0.32 kcal/mol.

If the chain-length difference between the two saturated identical-chain phosphatidylcholines (A and B) is further increased from four to six methylene units, the increased mismatch will affect the lateral lipid-lipid interactions markedly, leading to $E_{BB} \ll E_{AB} \cong E_{AA}$ according to the Salem equation (Eq. 3) discussed earlier. This larger dissimilarity can thus be expected to cause a lower miscibility of C(14):C(14)PC and C(20):C(20)PC in the bilayer, particularly the gel-state bilayer. Indeed, the phase diagram for the C(14):C(14)PC/ C(20):C(20)PC system is seen in Fig. 14C to deviate drastically from the lens shape. Certain features of the mixing behavior between C(14):C(14)PC and C(20):C(20)PC are evident from this so-called **peritectic phase diagram**. Over a wide X_B range, the solidus is flat, indicating the coexistence of two immiscible gel phases (G_A and G_B) below the flat line. The liquidus is not flat, indicating that no obvious liquid-crystalline immiscibility of C(14):C(14)PC and C(20):C(20)PC occurs. However, the shape of the liquidus clearly suggests some degrees of nonideality of mixing in the liquid-crystalline state. Furthermore, the two-phase region enclosed by the liquidus and the solidus is considerably larger than the corresponding region in the lens-shaped phase diagram shown in Fig. 14A. The simulation of the phase diagram for C(14):C(14)PC/ C(20):C(20)PC yields $\rho^G = 1.21$ and $\rho^L = 0.60$ kcal/mol (Table 4). The relatively larger ρ^G value is indicative of more

extensive gel-phase immiscibility, which is, of course, consistent with the observed extended flat line of the solidus shown in Fig. 14C.

The experimental curves illustrated in Fig. 14A–C and the computational results given in Table 4 may serve as a paradigm to show that the degree of mutual miscibility between two chemically distinct lipid components (A and B) in the bilayer depends on the relative difference in the pair-interaction energies between the mixed pairs (A-B) and the like pairs (A-A and B-B) which, in turn, depends critically on the extent of structural similarity between A and B. In general, the structural similarity should also include the headgroup and the acyl chain unsaturation. If the headgroup of lipid B is different from that of A, this structural dissimilarity is expected to affect the mixing of A and B in the bilayer even if the acyl chain compositions of A and B are identical. This expectation is indeed borne out by the experimental observation that a peritectic phase diagram is detected for the binary lipid system of C(16):C(16)PC/C(16):C(16)PE. This peritectic phase diagram can be attributed to the stronger headgroup-headgroup interactions among PE molecules, arising mainly from hydrogen bonds between the amine hydrogens of the PE headgroups and the nonesterified phosphate oxygens of the neighboring PE headgroups in the gel-state bilayer.

We have discussed earlier that when a *cis* carbon-carbon double bond is incorporated into the *sn*-2 acyl chain of a phosphatidylcholine molecule, the shorter segment of the monounsaturated chain acts as a perturbing element in the gel-state bilayer. The disordered segments of monounsaturated chains esterified at the *sn*-2 position of lipids (B) can, therefore, be expected to hinder the close encounters of the highly ordered saturated phospholipids (A) in the gel-state bilayer composed of A/B, thus promoting the gel-gel phase separation or immiscibility. Indeed, the experimentally determined phase diagram for the binary lipid system of C(16):C(16)PC/C(16):C(18:1Δ^9)PC exhibits a shape that is very similar to the one observed for C(14):C(14)PC/C(20):C(20)PC, although the actual chain-length difference between the saturated and the unsaturated PC is less than two C–C bond lengths. Specifically, the shape of this phase diagram is characterized by an extended flat line in the solidus, indicating an extensive region of immiscibility between C(16):C(16)PC and C(16):C(18:1Δ^9)PC in the gel-state bilayer.

In summary, the mixing behavior of two different membrane lipids in the bilayer at various temperatures is rather diverse. This diversity arises from the fact that membrane lipids are characterized intrinsically by many different features including the acyl chain length, headgroup structure, and acyl chain unsaturation. Nevertheless, basic information regarding the miscibility and/or immiscibility of two different lipids in the bilayer can be obtained from the position and shape of the temperature-composition phase diagram. The construction and the simulation of phase diagrams for various binary lipid systems may thus be very useful in establishing the database for characterizing different binary lipid-lipid interactions, from which the more complex multiple lipid-lipid interactions in the two-dimensional plane of cellular membranes may be assessed.

VI. Summary

In this chapter, we have first discussed the classification and structures of membrane lipids with special emphasis on phospholipids. It is important to realize that within each class of phospholipid, a bewildering diversity of molecular species is present due to the complex composition of the fatty acyl chains. Nevertheless, all membrane phospholipids exhibit two common properties: (1) They are amphipathic molecules with a hydrophilic (polar) moiety and a hydrophobic (nonpolar) moiety. (2) These molecules, when hydrated, do self-assemble spontaneously into a variety of highly organized structures including the most commonly observed lipid bilayers. The structural polymorphism of the lipid bilayer composed of a single species of phospholipid as a function of temperature has been illustrated. Specifically, the thermally induced phase-transition behavior of the lipid bilayer, and the factors that influence this behavior, were presented. The main phase transition temperatures of bilayers prepared from various phosphatidylcholines and phosphatidylethanolamines were correlated with structural parameters underlying each lipid species. Finally, the mixing behavior of two different phospholipids in the bilayer at various temperatures was described in terms of phase diagrams. Moreover, simulations of the phase diagrams based on the Briggs-Williamson approximation were also presented. Although lipid bilayers composed of binary lipids are highly simplified compared with the bilayer matrix of biological membranes, the basic information regarding the lipid-lipid interactions obtained with the construction and the simulation of phase diagrams for binary lipid systems may be useful in assessing the more complex lipid-lipid interactions in biological membranes.

Bibliography

Ahl, P. L., Chen, L., Perkins, W. R., Minchey, S. R., Boni, L. T., Taraschi, T. F., and Janoff, A. S. (1994). Interdigitation-fusion: a new method for producing lipid vesicles of high internal volume. *Biochim. Biophys. Acta* **1195**, 237–244.

Bangham, A. D., Standish, M. M., and Watkins, J. C. (1965). Diffusion of univalent ions across the lamellae of swollen phospholipids. *J. Mol. Biol.* **13**, 238–251.

Blaurock, A. E., and Wilkins, M. H. F. (1972). Structure of frog photoreceptor membranes. *Nature* **236**, 313–314.

Bradshaw, J. P., Bushby, R. J., Giles, C. C. D., Saunders, M. R., and Reid, D. G. (1996). Neutron diffraction reveals the orientation of the headgroup of inositol lipids in model membranes. *Nature Structural Biol.* **3**, 125–127.

Browning, J., and Seelig, J. (1980). Bilayers of phosphatidylserine: a deuterium and phosphorus nuclear magnetic resonance study. *Biochemistry* **19**, 1262–1270.

Brumbaugh, E. E., and Huang, C. (1992). Parameter estimation in binary mixtures of phospholipids. *Methods Enzymology* **210**, 521–539.

Casal, L., and McElheney, R. (1990). Quantitative determination of hydrocarbon chain conformational order in bilayer of saturated phosphatidylcholines of various chain lengths by Fourier transform infrared spectroscopy. *Biochemistry* **29**, 5425–5427.

Chapman, D. (1993). Lipid phase transitions. *In* "Biomembranes: Physical Aspects" (M. Shinitzky, Ed.), pp. 29–62. Balaban Publishers, Weinheim.

Chiu, S-W., Clark, M., Balaji, V. Subramanian, S., Scott, H. L., and Jakobsson, E. (1995). Incorporation of surface tension into molecular dynamics simulation of an interface: a fluid phase lipid bilayer membrane. *Biophys. J.* **69**, 1230–1245.

Clegg, R. M., and Vaz, W. L. C. (1985). Translational diffusion of proteins and lipids in artificial lipid bilayer membranes. A comparison of experiment with theory. *In* "Progress in Protein-Lipid Interactions" (A. Watts, Ed.), pp. 173–229. Elsevier, Amsterdam.

Craven, B. M. (1976). Crystal structure of cholesterol monohydrate. *Nature* **260**, 727–729.

Danielli, J. F., and Davson, H. A. (1935). A contribution to the theory of permeability of thin films. *J. Cell Comp. Physiol.* **5**, 495–508.

Davis, J. H. (1983). The description of membrane lipid conformation, order and dynamics by ^{2}H-NMR. *Biochim. Biophys. Acta* **737**, 117–171.

Exton, J. H. (1994). Phosphatidylcholine breakdown and signal transduction. *Biochim. Biophys. Acta* **1212**, 26–42.

Finkelstein, A. (1987). "Water Movement Through Lipid Bilayers, Pores, and Plasma Membranes: Theory and Reality." John Wiley and Sons, New York.

Gaber, B. P., and Peticolas, W. L. (1977). On the quantitative interpretation of biomembrane structure by Raman spectroscopy. *Biochim. Biophys. Acta* **465**, 260–274.

Hanahan, D. J. (1997). "A Guide to Phospholipid Chemistry." Oxford University Press, New York.

Hartshorne, R. P., Keller, B. V., Talvenheimo, J. A., Catterall, W. A., and Montal, M. (1985). Functional reconstitution of the purified brain sodium channel in planar lipid bilayers. *Proc. Natl. Acad. Sci. USA* **82**, 240–244.

Hitchcock, P. B., Mason, R., Thomas, K. M., and Shipley, G. G. (1974). Structural chemistry of 1,2 dilauroyl-DL-phosphatidylethanolamine: molecular conformation and intermolecular packing of phospholipids, *Proc. Natl. Acad. Sci. USA* **195**, 3036–3049.

Hope, M. J., Bally, M. B., Webb, G., and Cullis, P. R. (1985). Production of large unilamellar vesicles by rapid extrusion procedure. Characterization of size distribution, trapped volume and ability to maintain a membrane potential. *Biochim. Biophys. Acta* **812**, 55–63.

Huang, C. (1969). Studies on phosphatidylcholine vesicles. Formation and physical characteristics. *Biochemistry* **8**, 344–349.

Huang, C., and Li, S. (1996). Computational molecular models of lipid bilayers containing mixed-chain saturated and monounsaturated acyl chains. In "Handbook of Nonmedical Applications of Liposomes" (D.D. Lasic and Y. Barenholz, Eds.), pp. 173–194. CRC Press, Boca Raton, FL.

Huang, C. and Li, S. (1999). Calorimetric and molecular mechanics studies of the thermotropic phase behavior of membrane phospholipid. *Biochim. Biophys. Acta* **1422**, 273–307.

Huang, C., Wang, Z., Lin, H., Brumbaugh, E. E., and Li, S. (1994). Interconversion of bilayer phase transition temperatures between phosphatidylethanolamines and phosphatidylcholines. *Biochim. Biophys. Acta* **1189**, 7–12.

Huang, L. C., and Larner, J. (1993). Inositol phosphoglycan mediators of insulin action: the special role of chiroinositol in insulin resistance. *Adv. Prot. Phosphatases* **7**, 373–392.

Jin, A. Y., Benesch, L. A., and Weaver, D. F. (1994). Computational conformational analysis of neural membrane lipids: development of force field parameters for phospholipids using semi-empirical molecular orbital calculations. *Can. J. Chem.* **72**, 1596–1604.

Kunau, W-H. (1976). Chemistry and biochemistry of unsaturated fatty acids. *Angew. Chem. Int. Ed. Engl.* **15**, 61–122.

Lee, A.G. (1977). Lipid phase transitions and phase diagrams. II. Mixtures involving lipids. *Biochim. Biophys. Acta* **472**, 285–344.

Levin, I. W. (1984). Vibrational spectroscopy of membrane assemblies. *In* "Advances in Infrared and Raman Spectroscopy" (R.J.H. Clark and R.E. Hester, Eds.), Vol. 11, pp. 1–48. Heyden, London.

Li, S., Wang, Z-q., Lin, H-n., and Huang, C. (1993). Energy-minimized structures and packing states of a homologous series of mixed-chain phosphatidylcholines: a molecular mechanics study on the diglyceride moieties. *Biophys. J.* **65**, 1415–1428.

Low, M. G. (1989). The glycosyl-phosphatidylinositol anchor of membrane proteins. *Biochim. Biophys. Acta* **988**, 427–454.

Mabrey, S., and Sturtevant, J. M. (1976). Investigation of phase transition of lipids and lipid mixtures by high-sensitivity differential scanning calorimetry. *Proc. Natl. Acad. Sci. USA* **73**, 3862–3866.

Maggio, B. (1994). The surface behavior of glycosphingolipids in biomembranes: a new frontier of molecular ecology. *Prog. Biophys. Mol. Biol.* **62**, 55–117.

Mendelsohn, R., and Senak, L. (1993). Quantitative determination of conformational disorder in biological membranes by FTIR spectroscopy. *In* "Biomolecular Spectroscopy" (R.J.R. Clark and R.E. Heister, Eds.), pp. 339–380. Wiley, New York.

Montal, M., and Mueller, P. (1972). Formation of biomolecular membranes from lipid monolayers and a study of their electrical properties. *Proc. Natl. Acad. Sci. USA* **69**, 3561–3566.

Mueller, P., Rudin, D. O., Tien, H. T., and Wescott, W. C. (1962). Reconstitution of excitable cell membrane structure *in vitro*. *Circulation* **26**, 1167–1171.

Nagle, J. F., Zhang, R., Tristram-Nagle, S., Sun, W., Petrache, H. I., and Suter, R. M. (1996). X-ray structure determination of fully hydrated L_α phase dipalmitoylphosphatidylcholine bilayer. *Biophys. J.* **70**, 1419–1431.

Nishizuka, Y. (1984). The role of protein kinase C in cell surface signal transduction and tumour promotion. *Nature* **308**, 693–698.

Pascher, I., Lundmark, M., Nyholm, P. G., and Sundell, S. (1992). Crystal structures of membrane lipids. Biochim. Biophys. Acta **1113**, 339–373.

Pascher, I., and Sundell, S. (1977). Molecular arrangements on sphingolipids. The crystal structure of cerebroside. *Chem. Phys. Lipids* **20**, 175–191.

Pascher, I., Sundell, S., Harlos, K., and Eibl, H. (1987). Conformation and packing properties of membrane lipid: the crystal structure of sodium dimyristoylphosphatidylglycerol. *Biochim. Biophys. Acta* **896**, 77–88.

Pearson, R. H., and Pascher, I. (1979). The molecular structure of lecithin dihydrate. *Nature* **281**, 499–501.

Raetz, C. R. H. (1986). Molecular genetics of membrane phospholipid synthesis. *Annu. Rev. Genet.* **20**, 253–295.

Salem, L. (1962). The role of long-range forces in the cohesion of lipoproteins. *Can. J. Biochem. Physiol.* **40**, 1287–1298.

Schindler, H. (1980). Formation of planar bilayers from artificial or native membrane vesicles. *FEBS Lett.* **122**, 77–79.

Seelig, J. (1977). Deuterium magnetic resonance: theory and application to lipid membranes. *Q. Rev. Biophys.* **10**, 353–418.

Seelig, J., and Seelig, A. (1980). Lipid conformation in model membranes and biological membranes. *Q. Rev. Biophys.* **13**, 19–61.

Somerville, C. (1995). Direct tests of the role of membrane lipid composition in low-temperature-induced photoinhibition and chilling sensitivity in plants and cyanobacteria. *Proc. Natl. Acad. Sci. USA* **92**, 6215–6218.

Sternin, E., Bloom, M., and MacKay, A. L. (1983). De-Packing of NMR spectra. *J. Magn. Reson.* **55**, 274–282.

Tanford, C. (1980). "The Hydrophobic Effect: Formation of Micelles and Biological Membranes." 2nd Ed., John Wiley and Sons, New York.

Wang, G., Lin, H-n., Li, S., and Huang, C. (1995). Phosphatidylcholines with *sn*-1 saturated and *sn*-2 *cis*-monounsaturated acyl chains: their melting behavior and structures. *J. Biol. Chem.* **270**, 22 738–22 746.

Wong, P. T. T. (1986). Phase behavior of phospholipid membranes under high pressure. *Physica* 139–140B, 847–852.

Jeffrey C. Freedman

4

Cell Membranes and Model Membranes

I. Membrane Structure

The term *plasma membrane* derives from the German *Plasmamembran,* a word coined by Karl Wilhelm Nägeli (1817–1891) to describe the firm film that forms when the proteinaceous sap of an injured cell comes into contact with water. The physiologist L. V. Heilbrunn referred to this and similar phenomena as the "surface precipitation reaction," as described in his book "The Dynamics of Living Protoplasm" (1956). "Protoplasm" was the old term for the substance inside cells, and was in general use before the techniques of electron microscopy and differential centrifugation helped to elucidate the detailed structures and specific functions of cell organelles as discrete entities. The involvement of what we know today as the cell membrane, and the biochemistry of the formation of a surface film, in response to cell injury have not subsequently been explained. The original usage of the term "plasma membrane," therefore, has an obscure relationship to its current meaning.

Much knowledge concerning membrane structure and function derives from studies of red blood cells (see Chapter 23), shown as a scanning electron photomicrograph in Fig. 1. Red cells are highly differentiated, and are specialized for the transport of oxygen and carbon dioxide in the blood. They consist essentially of a plasma membrane surrounding a concentrated solution of hemoglobin, and lack a nucleus and other intracellular organellar compartments. About 100 years ago, extensive osmotic and permeability studies of red cells by Hamburger, of plant cells by de Vries, and of many living cells by Overton implied that a lipoid membrane surrounded cells. Substances that are lipophilic and easily dissolve in lipids (i.e., fat soluble) crossed into cells easily, whereas water-soluble substances entered cells more slowly, if at all. Overton recognized a correlation between the oil-water partition coefficient and membrane permeability (see Chapter 13); nevertheless, in none of these early studies was a membrane postulated as a distinct structural entity to explain the findings (see Jacobs, 1962).

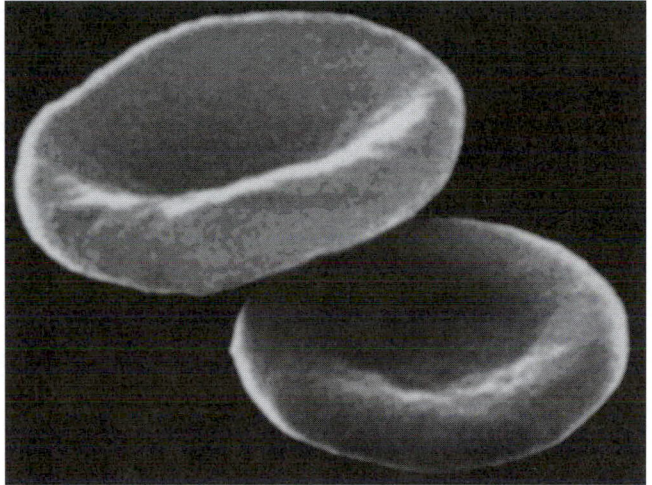

FIGURE 1. Scanning electron photomicrograph of human red blood cells. The biconcave discoid cells are 8 μm in diameter and 2.4 μm thick at the rim and 1.0 μm thick at the center. (From M. Bessis. (1974). "Corpuscles. Atlas of Red Blood Cell Shapes." Springer-Verlag, New York, Fig. 1, with permission.)

In 1925, Gorter and Grendel used acetone to extract lipids from a known quantity of red blood cells and, after evaporating the solvent, measured the area that the extracted lipids occupied as a monomolecular film at an air-water interface in a **Langmuir trough** (Fig. 2). From the area of the film of extracted lipids, and from the surface area of the red cells as estimated by light microscopy, they concluded that "It is clear that all our results fit in well with the supposition that the chromocytes are covered by a layer of fatty substances that is two molecules thick" (Gorter and Grendel, 1925, p. 443). Some 40 years later, however, it was pointed out that the red blood cell surface area is actually 50% greater; also, the acetone extraction had left about 30% of the lipid remaining in the ghosts. Fortunately, these two errors tended to offset each other (Bar, Deamer, and Cornwell, 1966), showing that on infrequent occasions in science, you can be

A

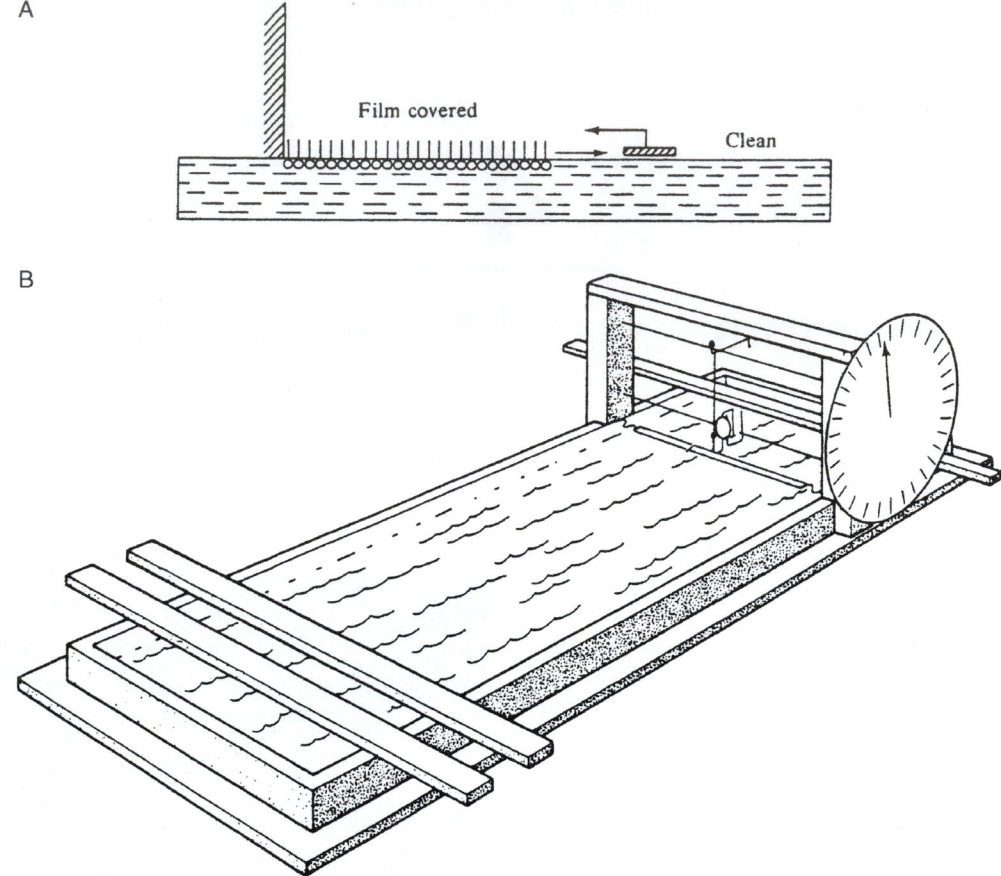

Film covered

Clean

B

FIGURE 2.   Langmuir trough for the study of lipid monolayers. (A) A monolayer of lipid on the surface of the aqueous trough. (B) The surface tension is measured by the force exerted against a spring balance. (From R. M. Dowben. (1969). "General Physiology, A Molecular Approach," Harper and Row, New York, p. 362, with permission.)

right for the wrong reasons. The bimolecular lipid leaflet, 75–100 Å thick, first proposed by Gorter and Grendel as a model for the cell membrane (Fig. 3), still forms the basis for contemporary thinking about the structure of cell membranes.

For pure lipids, the expected **surface tension,** as measured in the Langmuir trough, is about 9 dyn/cm, but the surface tension of marine eggs and other types of cells is about 50–100 times lower at only 0.1–0.2 dyn/cm. The surface tension can be visualized as the force necessary to close a slit in the surface of the membrane. Danielli and Harvey found that egg contents can lower the surface tension of an oil-water interface to about 0.6 dyn/cm, leading Davson and Danielli (1943) to postulate the presence of a film of protein, including possibly denatured protein, associated with the polar head groups at each side of the bimolecular lipid leaflet, a model that became known as the **Davson-Danielli paucimolecular membrane** (Fig. 4). Presumably, the protein strengthened and stabilized the thin lipid film. *Paucimolecular* means that this model included just a few molecules: a bimolecular lipid leaflet with adhering protein films on the inner and outer surfaces.

Using electrophysiological techniques, the electrical resistance of cell membranes was measured and found to be very

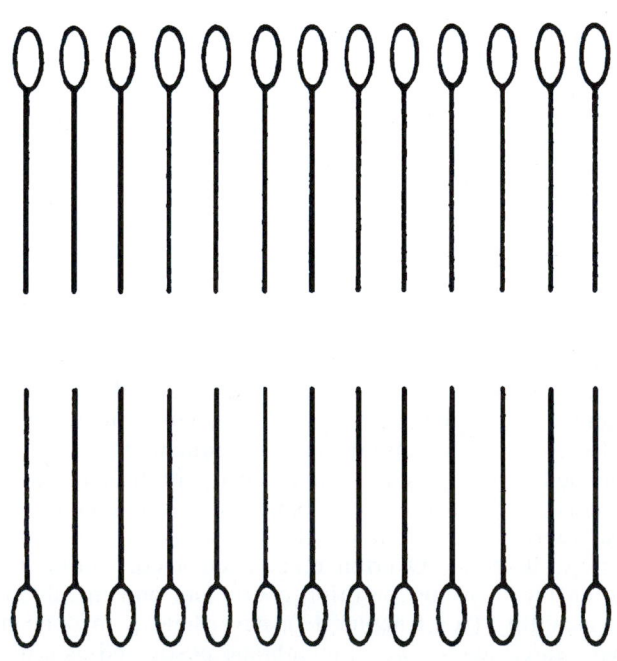

FIGURE 3.   Model of bimolecular phospholipid membrane as proposed by Gorter and Grendel (1925).

Exterior

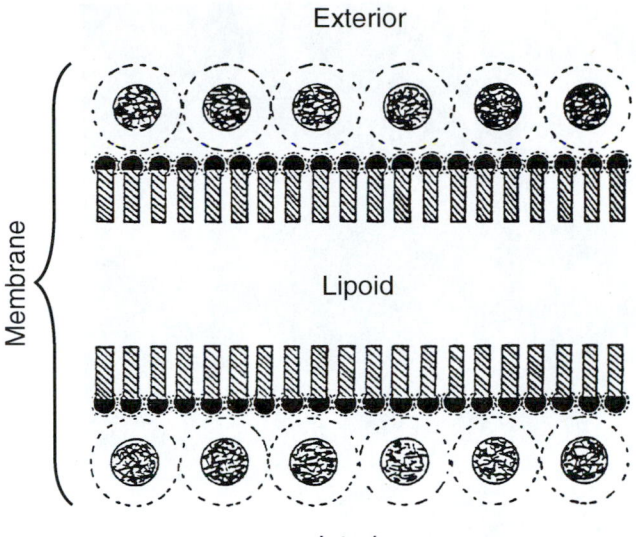

Membrane

Lipoid

Interior

FIGURE 4. Paucimolecular model of membrane structure. (From J. F. Danielli and H. A. Davson. (1935). A contribution to the theory of permeability of thin films. *J. Cell. Comp. Physiol.* **5,** 495–508, p. 498, reprinted by permission of Wiley-Liss, Inc., a subsidiary of John Wiley and Sons, Inc.)

high, a finding that was also consistent with the proposal of a lipid membrane surrounding cells. Other microscopical studies of membranes utilized a technique called **birefringence.** Birefringence is a specialized microscopic technique in which the specimen is placed between crossed Polaroids. A Polaroid film only passes light that has its electric vector parallel with the axis of the Polaroid film; two Polaroid films, crossed at right angles and held up to a light, appear black. But if a crystal, which has oriented material, or some other substance in which the molecules themselves are oriented, is placed between the crossed Polaroids, the transmitted light is circularly polarized, and the specimen looks beautifully bright. Any specimen that appears bright when placed between crossed Polaroids is said to be birefringent. **Intrinsic birefringence** is due to the oriented nature of individual molecules, such as filamentous proteins, whereas **form birefringence** is due to the oriented arrangement of molecules in an array, such as occurs with the parallel packing of actin and myosin filaments in muscle sarcomeres. When red cell membranes were viewed with a polarization microscope, the lipids contributed to birefringence, as did the underlying cytoskeleton, in a fashion consistent with the paucimolecular model of membrane structure.

High-resolution electron micrographs of the **unit membrane** also supported the Davson-Danielli paucimolecular model. The term *unit membrane* refers to the ubiquitous trilaminar structure, 75–100 Å thick, seen in electron micrographs of thin sections of cells and organelles. The image appears as two dark lines, each about 25–30 Å thick, sandwiching a lighter zone, and is especially well-resolved in samples fixed with potassium permanganate (for a view of the red cell plasma membrane, see Fig. 5). Virtually the

same trilaminar structure was seen not only at the surface of red cells but also in muscle cells, nerve cells, epithelial cells, plant cells, bacterial cells, and in just about every membranous cell organelle that was examined. In situations where two cells were tightly apposed, two trilaminar structures constituted a **double membrane.** In the myelin sheath that surrounds nerve cells, a series of trilaminar structures was seen in a spiral arrangement, consistent with the envelopment of nerve axons by the membrane of Schwann cells (see Fig. 6A). Since the detailed chemical reactions of potassium permanganate with tissue are unknown, some uncertainty remained concerning the basis of the image that was observed. The universal occurrence of the unit membrane, however, was taken as strong evidence in support of the Davson-Danielli paucimolecular model. Further studies of the multilamellar myelin sheath of nerve axons by x-ray diffraction of sedimented unfixed and unstained membranes yielded electron density profiles that were also consistent with the Davson-Danielli paucimolecular model (Worthington and McIntosh, 1973). As seen in Fig. 6B, the electron density is low in the hydrophobic core of the membrane and high in the polar regions of the phosphate groups. Moreover, the thickness of the membrane, as determined from these computed electron density profiles of unstained myelin sheath, quantitatively agrees with that seen as the unit membrane in thin sections stained with potassium permanganate—an observation constituting impressive evidence for the proposal of a bimolecular phospholipid leaflet. Thus, the evidence for the paucimolecular membrane model consisted of studies of permeability, electrical resistance, and microscopical observations of birefringence with the light microscope, as well as high-resolution images with the electron microscope. These convincing arguments were summarized in the classic monograph entitled "The Permeability of Natural Membranes" written by Davson and Danielli (1943), a book that greatly influenced subsequent development of cellular and membrane physiology.

The ability to separate intact red blood cell membranes from the rest of the cell by hypotonic hemolysis (Dodge *et al.*, 1963) was further evidence for the existence of a distinct membrane, and enabled detailed studies of its chemical composition. After centrifugation of red blood cells, followed by resuspension in distilled water, the water enters the cells by **osmosis** (see Chapter 21), and then the cells swell and burst. Following centrifugation and washing of the pellet by further resuspension and centrifugation at the optimal pH and ionic strength to remove hemoglobin, the resultant pellet is pure white. When these isolated washed membranes are examined by the technique of negative staining with an electron microscope (a negative stain is excluded from biological structures in contrast with positive stains, which bind more or less specifically to various structures), the image resembles an empty bag (see Fig. 7). In red cell physiology, the isolated membranes are known as **ghosts,** which can be either pink or white depending on the extent and conditions of washing. Red blood cell membranes were the first to be isolated and remain the best characterized.

The mass ratio of protein to lipid ranges from about 1:4 to 4:1 in various membranes. Myelin, which electrically insulates nerve axons, has only 18% protein by mass. In contrast,

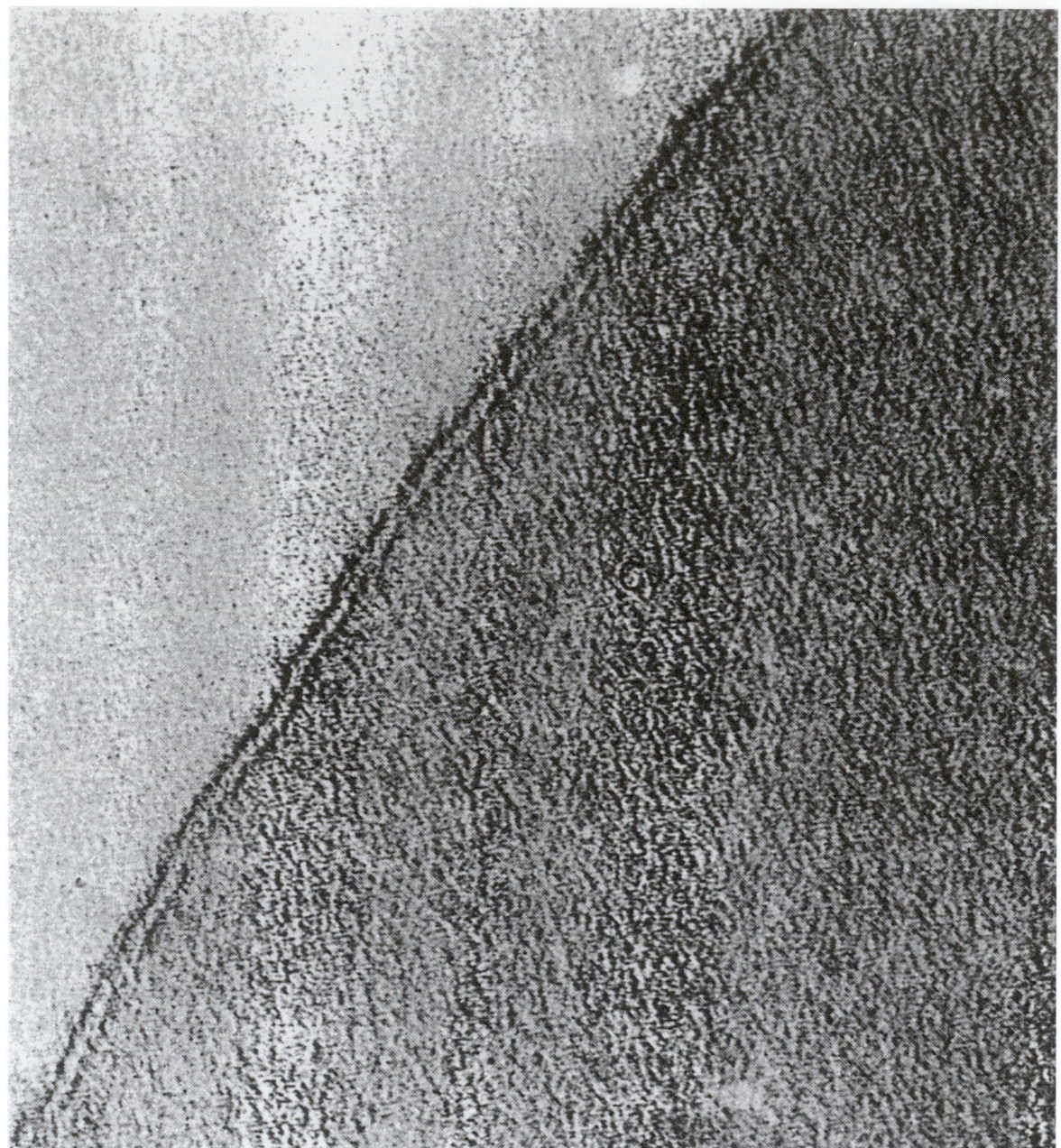

FIGURE 5.   Thin-section electron micrograph of the unit membrane of a red blood cell. (Photomicrograph by J. D. Robertson from R. D. Dyson. (1974). "Cell Biology. A Molecular Approach," Allyn and Bacon, Boston, with permission.)

the energy-transducing inner membranes of mitochondria and chloroplasts, where ATP is synthesized by oxidative or photophosphorylation, respectively, are densely packed with 75% protein. The chemical composition of isolated red cell membranes is summarized in Table 1, which indicates that these membranes, like many other membranes, are about half protein and half lipid by mass, with an additional fraction of carbohydrate. The carbohydrate exists as sugars on glycoproteins, glycolipids, and on sialic acid; the sugars always protrude from the outer membrane surface. Due to the sialic acids on **glycophorin,** an integral membrane glycoprotein, the cell surface is negatively charged. The external surface of

the red cell membrane is coated with adsorbed albumin and also with some plasma globulins. Since the mole ratio of phospholipid to cholesterol is 0.8 in red cells, the paucimolecular model needed modification to include cholesterols interspersed among the phospholipid fatty acyl chains in parallel with the phospholipids. The plasma membranes of eukaryotic cells are rich in cholesterol, whereas organellar membranes have less of this neutral lipid.

The proportions of phospholipids found in human red blood cell membranes are shown in Table 2. The different classes of lipids that are found in cells, and the properties of these lipids, were described in Chapter 3. The fatty acyl

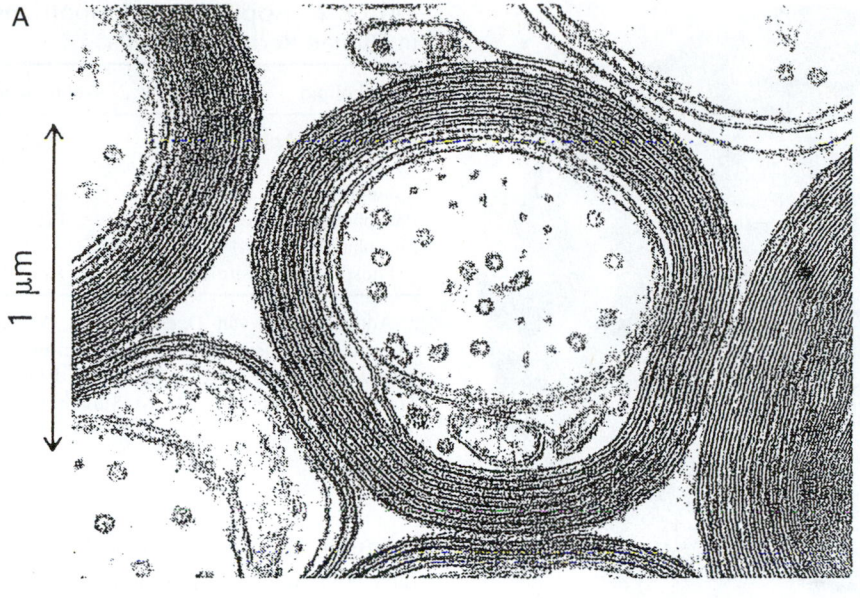

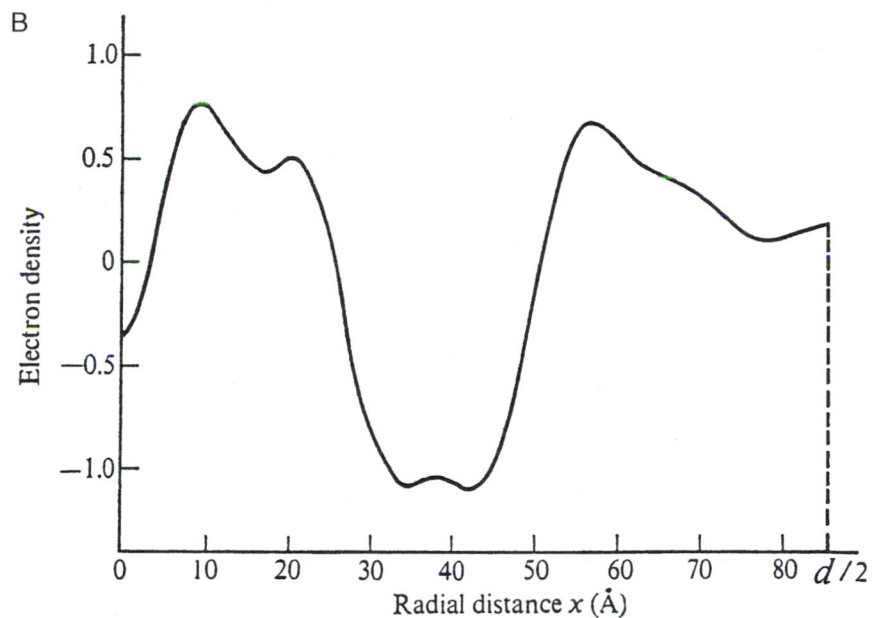

FIGURE 6. (A) Myelin sheath of spinal cord axon (courtesy of Dr. Cedric Raine.) (B) Electron density profile of frog sciatic nerve. (From C. R. Worthington and T. J. McIntosh. (1973). Direct determination of the electron density profile of nerve myelin. *Nature-New Biology* **245,** 97–99, p. 99. Reprinted by permission from *Nature* copyright 1973 Macmillan Magazines Ltd.)

chains in the phospholipids usually contain an even number of carbons between 14 and 24; the 16- and 18-carbon fatty acids are the most common. In red cells the major fatty acids are palmitic (16:0), oleic (18:1), linoleic (18:2), stearic (18:0), and arachidonic (20:4) (Table 3), which together account for some 92% of the total fatty acid content. Different membranes differ in their **lipid signature,** a term that refers to the characteristic pattern of phospholipids, cholesterol, and fatty acids that comprise the plasma membranes of different cells and of specific organelles. Lipid composition and content are usually measured by chromatographic meth-

ods, and have been measured and cataloged for many different membranes (Ansell *et al.*, 1973). The functional significance, however, of the different lipid signatures that occur in various specialized cells and organelles is largely not understood.

With two hydrophobic acyl chains, individual phospholipids have very low solubility in water, but form micelles at a critical micellar concentration of about 10^{-10} M. A **micelle** is a globular structure, usually less than 200 Å in diameter, with the polar head groups on the outside, and the hydrophobic fatty acyl chains buried in the interior. Salts of

FIGURE 7. Isolated red cell membrane negatively stained with phosphotungstic acid. (From P. N. McMillan and R. B. Luftig. (1973). Preservation of erythrocyte ghost ultrastructure achieved by various fixatives. *Proc. Natl. Acad. Sci. USA* **70**, 3060–3064, p. 3061, with permission.)

fatty acids readily form micelles, whereas phospholipids tend to form bilayers. The **self-association** of phospholipids to form a bimolecular leaflet depends on the **hydrophobic effect:** the release of structured water molecules from around the hydrocarbon tails of the fatty acids results in an overall gain of entropy that serves as the driving force for self-assembly of phospholipids into bimolecular membranes. The membrane is further stabilized by lateral van der Waals attractive forces between the fatty acyl chains, by electrostatic forces, and by hydrogen bonding between adjacent polar headgroups. Depending on the method of preparation, membranes can form as planar bilayers, multilamellar vesicles (MLVs), and as small (200–500 Å) or large (0.1–1 μm) unilamellar vesicles (SUVs and LUVs) (Fig. 8). In each structure, the polar, or hydrophilic, headgroups have

TABLE I   Composition of Red Blood Cell Ghosts

Substance	Mass percent
Protein	49%
Phospholipid	33%
Cholesterol	11%
Lipid (total)	44%
Carbohydrate (total)	7%

(Adapted from Rosenberg, S. A., and Guidotti, G. J. (1968). The protein of human erythrocyte membranes. I. Preparation, solubilization, and partial characterization. *J. Biol. Chem.* **243**, 1985–1992.)

TABLE 2   Approximate Proportions of Phospholipids in Human Red Blood Cells

Phospholipid	Abbreviation	Mass %
Phosphatidylcholine (lecithin)	PC	35%
Lysolecithin	lyso-PC	2%
Sphingomyelin	SM	23%
Phosphatidylethanolamine	PE	30%
Phosphatidylserine + Phosphatidylinositol	PS + PI	10%

Adapted from van Deenen, L. L. M., and de Gier, J. (1964). Chemical composition and metabolism of lipids in red blood cells of various animal species. *In* "The Red Blood Cell, A Comprehensive Treatise." (C. Bishop and D. M. Surgenor, Eds.) Academic Press, New York, p. 266.

an affinity for water whereas the nonpolar, or hydrophobic, fatty acyl chains avoid water.

When isolated red cell membranes are first solubilized in the detergent sodium dodecyl sulfate (SDS), then subjected to electrophoresis on polyacrylamide gels containing SDS, followed by staining with Coomassie blue, seven major bands are found (Fig. 9 and Table 4). In addition, stains for carbohydrate reveal glycophorin and other glycoproteins. Actually, more than 10 bands can be counted, and many more spots are seen when two-dimensional chromatography is employed. Specific proteins mediate the distinctive functions of membranes; for example, membrane proteins serve as ion channels, pumps, receptors, energy transducers, and enzymes. Membranes with different functions show different banding patterns of proteins on SDS gels. Whereas the

TABLE 3   Fatty Acids in the Membrane Phospholipids of Human Red Blood Cells

Fatty acid	$X:Y^a$	Mass%
Lauric	12:0	0.3
Myristic	14:0	1.0
	15:0	0.3
Palmitic	16:0	27.1
	16:1	3.4
	17:0	0.6
Stearic	18:0	9.4
Oleic	18:1	19.5
Linoleic	18:2	16.5
	18:3	0.5
	20:0	0.2
	20:3	1.4
Arachidonic	20:4	19.5

[a]In $X:Y$, X refers to the number of C atoms in the acyl chain and Y refers to the number of double bonds. (Data from de Gier, J., van Deenen, L. L. M., Verloop, M. C., and van Gastel, C. (1964). Phospholipid and fatty acic characteristics of erythrocytes in some cases of anaemia. *Brit. J. Haematol.* **10**, 246–256, p. 251.)

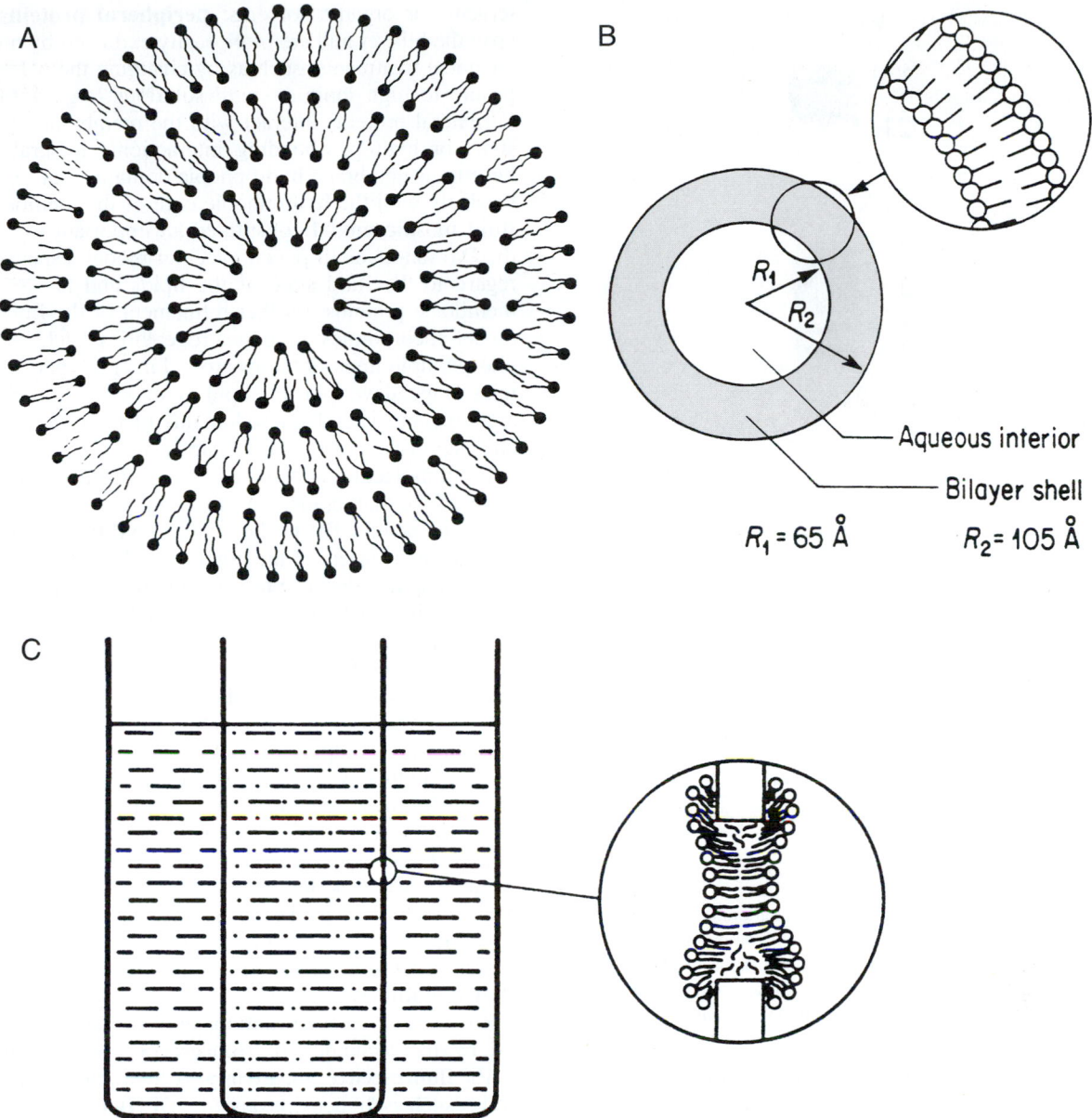

FIGURE 8. Model membranes: (A) Liposomes. (B) Vesicles. (C) Planar bilayers. (From P. L. Yeagle. (1993). "The Membranes of Cells," Second ed. Academic Press, NY, pp. 48, 52, and 54, with permission.)

membrane lipids form a permeability barrier to most polar molecules and also form closed compartments around and within cells, specific proteins mediate most other membrane functions.

When a cell membrane, such as the human red blood cell, is first rapidly frozen in liquid nitrogen and then fractured with the impact of a microtome blade, the cleavage plane can pass along the outer surface of the cell, and then slice through the hydrophobic core of the membrane. **Freeze-fracture electron micrographs** thus provide a view of the unfixed interior of the membrane. After shadowing with carbon and platinum, a replica of the membrane interior is obtained. A variation of this technique is **freeze-etching,** in which some of the ice is sublimed away before the shadowing. With this technique, globular intramembranous particles (75 Å) are seen to lie in the plane of the membrane (Fig. 10). In red cells, the surface density of these particles (10^6 per cell) approximately corresponds to the number of copies of the major membrane protein known as band 3, or capnophorin. In contrast, the freeze-etch images of myelin are relatively poor in intramembranous particles, and images of pure lipid membranes are devoid of particles.

Although spectroscopic studies had provided evidence that lipids and proteins have some degree of rotational and translational mobility in the plane of the membrane, the review article entitled "The fluid mosaic model of the struc-

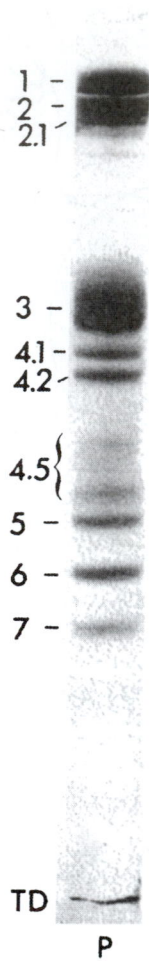

FIGURE 9. SDS gel of red blood cell membrane proteins. (From T. L. Steck and J. Yu. (1973). Selective solubilization of proteins from red blood cell membranes by protein perturbants. *J. Supramolec. Struct.* 1, 220–232, p. 223, Copyright © 1973. Reprinted by permission of Wiley-Liss, Inc., a subsidiary of John Wiley & Sons, Inc.

tergents or organic solvents; **peripheral proteins** do not span the bilayer and are more easily extracted by more mild chemical treatments, such as by changing the pH or by exposure to high ionic strength solutions (e.g., 1 M NaCl). Peripheral proteins interact with the membrane by electrostatic or hydrogen-bonding interactions; integral proteins have transmembrane hydrophobic domains that interact extensively with the hydrophobic core of the membrane. The **fluid mosaic model** constituted an important extension of the Davson-Danielli paucimolecular model, especially with regard to the fluid state of the lipids and the location of membrane proteins. Further refinement of the fluid mosaic model incorporates lipid heterogeneities such as annular phospholipid microdomains around integral membrane proteins, transbilayer phospholipid asymmetry, lipid phase separations, and lateral microdomains of cholesterol-rich and cholesterol-poor pools (see Chapter 5).

Lipid molecules and some proteins diffuse rapidly in the plane of the membrane, but rarely flip across the membrane from one hemileaflet to the other. The diffusion coefficient of lipids in a variety of membranes is about 10^{-8} cm^2/s, corresponding to a distance traveled of about 2 μm in 1 s. From the diffusion coefficient it is estimated that the viscosity of the membrane lipids is about 100 times that of water, similar to olive oil. Cell fusion experiments first demonstrated that fluorescent labeled antibodies attached to human and mouse cell surface proteins could diffuse a distance of several microns in approximately 1 min (Frye and Edidin, 1970). Another way of measuring the lateral translational mobility of membrane lipids and proteins is the technique of fluorescence recovery after photobleaching, as illustrated in Fig. 11. In human red blood cells a portion of band 3 protein is mobile, whereas another fraction is immobile due to attachments to the cytoskeletal protein matrix adjacent to the inner surface of the membrane. Proteins in different cells, and different proteins in a single cell, vary markedly in their **lateral mobility.**

Rotational molecular motions in membranes may be studied by the technique of **fluorescence polarization,** also called **fluorescence anisotropy.** A hydrophobic fluorescent probe, such as diphenylhexatriene (DPH), is first added to the membrane. The fluorescence is excited by light that is polarized, say in the vertical direction as illustrated in Fig. 12. Vertical exciting light only excites those probe molecules

ture of cell membranes" (Singer and Nicholson, 1972) popularized the view that the overall structure of the membrane is a two-dimensional solution of globular proteins embedded in a fluid lipid matrix (see Fig. 2 in Chapter 6). **Integral proteins** span the membrane and can be extracted with de-

TABLE 4   Protein Composition of Human Red Blood Cell Membrane

Band	MW	%	Position[a]	Name
Band 1	240 000	15	P	Dimeric spectrin
Band 2	215 000	15	P	Monomeric spectrin, myosin
Band 3	101 700	24	I	Capnophorin, Na$^+$,K$^+$-ATPase
Band 4.1	78 000	4.2	P	Junctional protein
Band 4.2	72 000	5.0	P	Junctional protein
Band 4.9	50 000			Dematin (actin-bundling protein)
Band 5	43 000	4.5	P	Actin
Band 6	35 000	5.5	P	Glyceraldehyde-3-phosphate dehydrogenase
Band 7	29 000	3.4		

[a]P refers to a peripheral, and I to an integral, membrane protein. (Data from Stack, T. L. (1974). The organization of proteins in the human red blood cell membrane. A review. *J. Cell. Biol.* **62,** 1–19, Table 1, p. 4.)

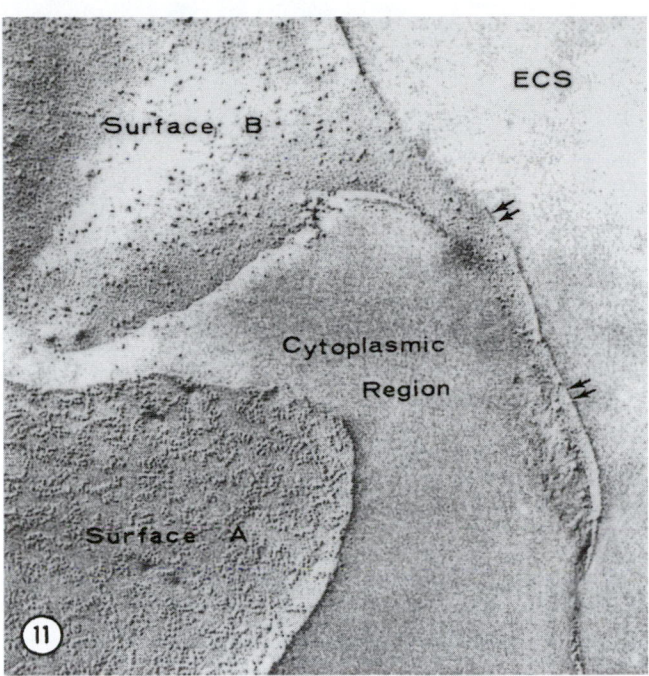

FIGURE 10. Freeze-etch image of red cell membrane showing intramembranous particles. (From R. S. Weinstein. (1969). Electron microscopy of surfaces of red cell membranes. *In* "Red Cell Membrane. Structure and Function." (G. A. Jamieson and T. J. Greenwalt, Eds.), J. B. Lippincott Company, Philadelphia, p. 54, with permission.)

that are oriented in the same direction as the polarized excitation beam, a phenomenon known as **selective excitation.** If the probe molecule is immobilized during the time interval between excitation and emission, all of the emitted light would remain polarized in the vertical direction. However, if the probe molecule can rotate before light is emitted, then the emitted light will no longer be vertically polarized, but instead will contain a mixture of light that is polarized in all directions. By detecting and measuring the intensity of the emitted light that is polarized in the vertical and horizontal planes, one obtains a measure of the rotational mobility of the probe. By using the equations of motion, one further obtains a measure of the **microviscosity** of the fluid membrane. The inverse of microviscosity is termed **fluidity,** and many studies have been performed to attempt to understand what determines membrane fluidity. Lipid composition, the length and degree of unsaturation of fatty acyl chains, the cholesterol content, and the temperature all have profound effects on membrane lipid fluidity (see Chapter 3).

II. Planar Lipid Bilayers

Planar lipid bilayers, also called **black lipid membranes** (BLMs), were first made by Mueller and colleagues using the painting technique in 1963. In this technique, a small aperture of about 1 mm diameter in a Teflon or polyethylene septum separating two aqueous salt solutions is coated by means of a beveled Teflon rod with a so-

lution of pure lipids that are usually dissolved in decane. The thick lipid film covering the aperture then spontaneously thins to a bimolecular lipid leaflet surrounded by a thicker annular torus (Fig. 13A). The bimolecular portion of the film appears black because light reflected from the front surface undergoes a phase shift of 180° and destructively interferes with light reflected from the back surface. The film thickness of about 100 Å is small compared to the wavelength of visible light (3800–7600 Å). The lipids used to make the film are usually commercially purchased phospholipids—phosphatidylethanolamine (PE), phosphatidylcholine (PC), phosphatidylserine (PS), cardiolipid—and are sometimes supplemented with cholesterol for added stability. The proportions of the various lipids in the bilayer are set by dissolving the pure lipids in chloroform/methanol, mixing in desired proportions, evaporating the solvent with a stream of nitrogen, and then redissolving the mixed lipids in decane.

The first attempts at reconstituting native ion channels into planar lipid bilayers were made by adding cell membrane extracts or partially purified membrane proteins into the aqueous solutions. Increases in electrical conductance were noted, often followed by breakage of the membrane; these effects were often due to adsorption of protein onto the bilayer and a generalized nonspecific disruption of the membrane associated with increased leakage of current. Some success was achieved by adding membranes directly to the lipid-forming solution. Subsequently, techniques were developed for fusing vesicles, either native vesicles isolated from cell membranes or, alternatively, vesicles reconstituted with specific ion channels, into the planar lipid bilayer (see Miller, 1986, 1987). Tip-dipping is an alternative technique involving dipping the tip of a micropipette into a pure lipid film twice to form a bilayer covering the tip of the pipette so as to form a tight electrical seal, usually 1–5 GΩ (1 GΩ = 10^{15} Ω). Tip-dipping improves the time resolution and reduces noise due to the smaller area of membrane that is being examined; this technique allows the study of asymmetric bilayers and does not require solvents if the monolayers are formed from dried lipids (Coronado and Latorre, 1983).

The planar lipid bilayer technique is an excellent method for studying the transport kinetics of lipid-soluble ionophores, such as the potassium-specific ionophore valinomycin, or the channel-forming antibiotic gramicidin. With planar lipid bilayers, one can easily vary the membrane lipid composition, and change the solutions and add reagents on either side of the membrane. Purified membrane channels may be studied apart from the regulatory mechanisms of intact cells, and the resultant information should complement that obtained from patch-clamp studies of intact cells. Planar lipid bilayers can also be used as an assay for channel purification. The planar lipid bilayer technique has several disadvantages. One is that the membranes tend to break. Once observed, the channels may disappear during the period of observation by diffusing into the thick annulus. The organic solvent, when present, may alter the properties of ion channels. The conductance increases seen may not always be physiologically relevant because the conductance properties may be altered during isolation of the vesicles or during purification of the channels, and the conductances themselves may originate from small amounts of impurities or from contaminating cells. Some channels incorporate into bilayers

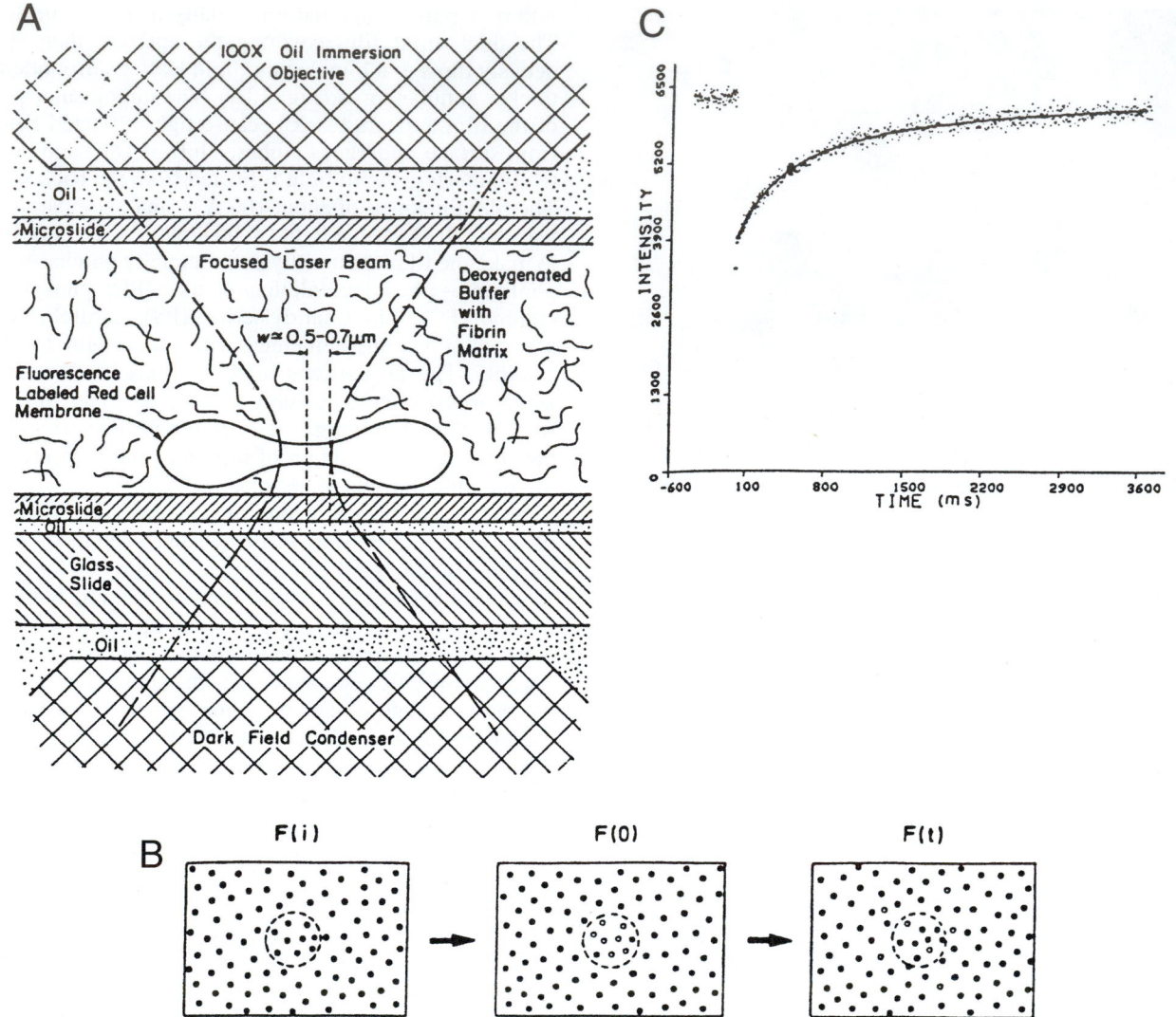

FIGURE 11. Lateral mobility determined by fluorescence recovery after photobleaching. (A) A fluorescence-labeled red blood cell is shown in a microslide, through which a focused laser beam is passed. (B) The fluorescence intensity at a spot on the cell is bleached, but then recovers as labeled lipids or proteins diffuse back into the bleached spot. (C) The diffusion coefficient of the labeled substance is computed from the time course of the return of the fluorescence. (From J. A. Bloom and W. E. Webb. (1983) Lipid diffusibility in the intact erythrocyte membrane. *Biophys. J.* **42**, 295–305, Figs. 1 and 4, with permission of the author and of The Biophysical Society.)

more easily than others, so there may be long periods of time when nothing happens, and the channel ultimately studied may not be the channel intended.

III. Ion Channel Properties in Planar Lipid Bilayers

Studies of ion transport through channels in planar bilayers revealed that ion channels are characterized by high transport rates and low temperature coefficients. Observations of single ion channels show that the current passing through a channel is on the order of picoamps (10^{-12} A). In order to convert 1 pA to the flux J in ions/s, note that the flux through a channel can be estimated from the single channel current i

(C/s), using Avogadro's number N_A (ions/mol), the Faraday constant $\mathscr{F}$ (C/eq), and the ionic valence z (eq/mole).

$$J = \frac{iN_A}{\mathscr{F}z}$$

So for 1 pA of current carried by a univalent ion, we have

$$J\,(\text{ions/s}) = \frac{\left(10^{-12}\,\text{C/s}\right)\left(6 \times 10^{23}\,\text{ions/mol}\right)}{\left(9.6 \times 10^{4}\,\text{C/eq}\right)\left(1\,\text{eq/mol}\right)}$$

$$= 6 \times 10^{6}\,\text{ions/s}$$

or 6 million ions/s. By comparison, enzymes and membrane transporters that are not ion channels have turnover rates that are less than 10^{5}/s. For example, the Na$^+$,K$^+$-ATPase pumps

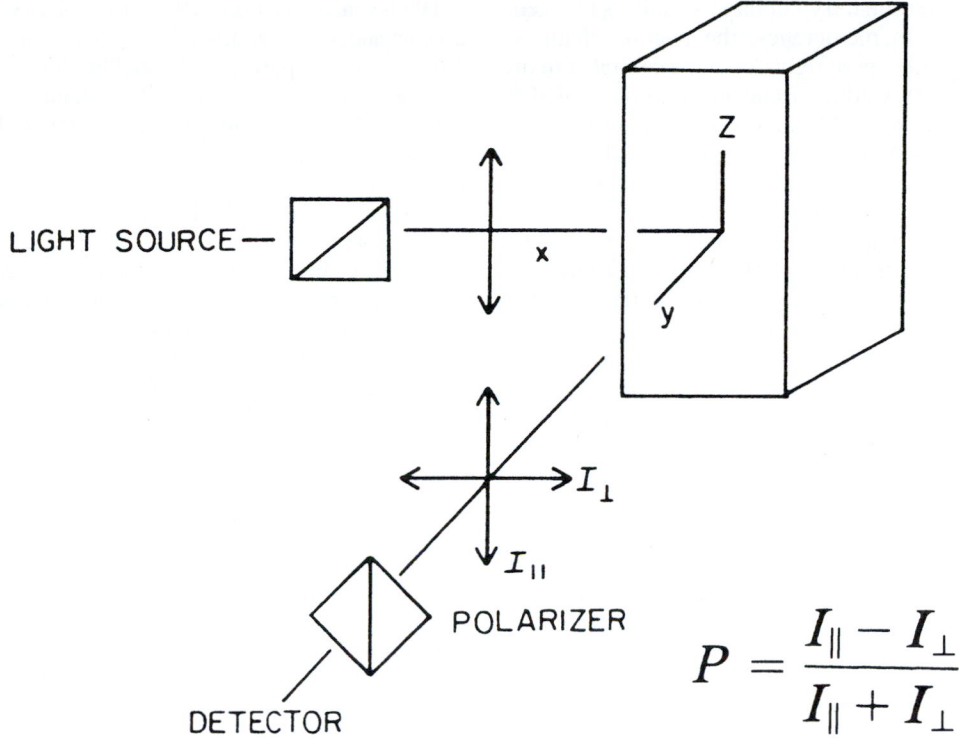

FIGURE 12. Membrane lipid fluidity determined by fluorescence polarization. (From J. R. Lakowicz. (1983). Fluorescence polarization. *In* "Principles of Fluorescence Spectroscopy," Plenum Press, New York, p.113, with permission.)

$$P = \frac{I_{\parallel} - I_{\perp}}{I_{\parallel} + I_{\perp}}$$

at a maximal rate of about 100 ions/s. The fastest non-channel transporter is the $Cl^--HCO_3^-$ exchanger of red blood cells, which has an exchange rate of 10^4 ions/s at 25 °C. The enzymes carbonic anhydrase and acetylcholinesterase have turnover rates of about 10^5/s. Thus, ion channels typically show high transport rates, one to several orders of magnitude faster than the fastest enzymes or non-channel transporters.

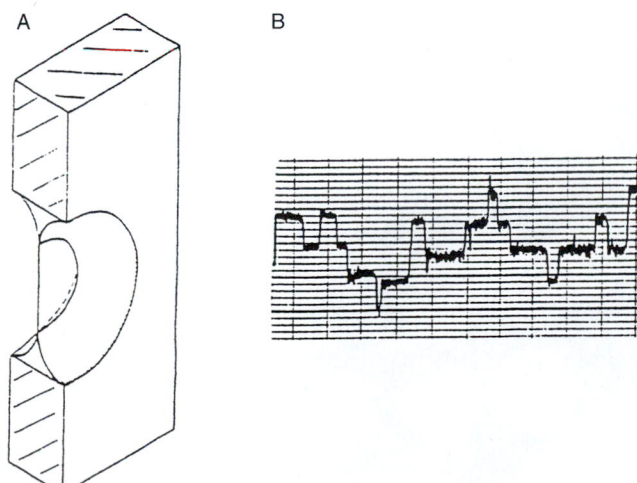

FIGURE 13. (A) Septum with planar lipid bilayer. (B) Traces showing currents through four dimeric gramicidin channels associating and dissociating in a black lipid membrane, as determined in the author's laboratory.

Ion channels were also found to have low temperature coefficients. Q_{10} is defined to be the change in the rate of a reaction when the temperature is increased by 10 °C. Hodgkin and Huxley found that Q_{10} for Na^+ and K^+ currents in squid axon is 1.2–1.4, which is comparable to that for unrestricted diffusion of ions in free solution, a finding consistent with these ions moving through ion channels in nerve membranes. From thermodynamics, the Q_{10} is related to the enthalpy of activation for the reaction. The Q_{10} for Na^+ and K^+ currents in squid axon corresponds to an enthalpy of activation of only 5 kcal/mol. Enzymes and non-channel transporters have Q_{10}s of 3–4 or higher, corresponding to higher enthalpies of activation.

IV. Gramicidin

One of the best-studied ion channels in planar lipid bilayers is gramicidin, an antibiotic synthesized by *Bacillus brevis* against gram-positive bacteria (hence the name). Gramicidin is commercially used as a topical bacteriostatic agent. The primary sequence of gramicidin A is a linear pentadecapeptide consisting of 15 alternating D- and L-amino acids. Natural sequence variations occur at position 11 with substitution of Trp with Phe (gramicidin B) or Tyr (gramicidin C). The natural mixture is termed gramicidin D, and contains about 80% gramicidin A. All of the amino acids in gramicidin are hydrophobic with no free charges. Gramicidin is thus virtually insoluble in water. Both end groups are blocked: the N-terminal valine (the head) is blocked by a formyl group; the C-terminal tryptophan (the tail) is blocked by an ethanolamine. In membranes, the peptide chain is wound in a β-helix with a pore right down the central core of the molecule (Fig. 14). Carbonyl and imino groups of the peptide bonds line the pore. All amino acid side chains extend away from the pore into the membrane lipid. The β-helix is stabilized by $-NH\cdots O-$ hydrogen bonds extending parallel to the pore axis. The aqueous pore is about 25 Å long and 4 Å in diameter. Gramicidin assumes different conformations in organic solvents; the double helical structure deduced from spectroscopic studies does not pertain to the channel conformation in membranes. The conducting pore is formed by a head-to-head dimer linked transiently by six hydrogen bonds. Evidence for a head-to-head dimer is that chemical modifications at the N-terminus (the head) drastically affect channel formation, whereas similar modifications at the C-terminus (the tail) do not. With fluorescent analogs of gramicidin, it was possible to measure simultaneously the conductance and the concentration of gramicidin in the membrane. The conductance was proportional to the square of the gramicidin concentration, and the equilibrium constant for dimerization was determined (Veatch *et al.*, 1975). Since gramicidin is synthesized nonribosomally *in vivo*, and contains D-amino acids that are not normally genetically coded, site-directed mutagenesis using recombinant DNA methods is not possible. Chemical synthesis has permitted alterations of the primary sequence for studies relating chemical structure to ion transport function.

Hladky and Haydon (1972) first observed single-channel conductances with gramicidin in planar lipid bilayers (see Fig. 13B for an example). The channel lifetimes are on the order of a second (compared to ms for Na^+ channels). With symmetrical 0.1 M NaCl, the single-channel current is 1.0 pA at 200 mV, corresponding to a single-channel conductance of 5 pS, and a flux of 6.3×10^6 ions/s. The high flux is consistent with a channel mechanism. The highest conductance so far reported is 107 pS at 23 °C with 3 M RbCl solutions in a neutral membrane made from glycerylmonooleate-hexadecane mixtures. The mean single-channel lifetime depends on lipid dynamics. For example, channels in thicker membranes, where pinching of the membrane may be required for dimer formation, have shorter lifetimes. Greater interfacial tension shortens channel lifetime.

Despite the absence of fixed negative charges in the pore, the gramicidin channels are cation selective (Myers and Haydon, 1972). The permeability ratios are determined from bi-ionic potentials, while the conductance ratios are determined in symmetrical salt solutions. The channels are ideally selective to cations. With a gradient of monovalent chloride salt, the reversal potential equals the Nernst potential for the cation, implying that the permeability to Cl^- is negligible. The channel is also impermeant to divalent cations. The selectivity order corresponds to Eisenman's sequence I, indicative of a weak field strength interaction between the pore and the ions.

The selectivity for $K^+:Na^+$ is only about 4:1—much less than that of the delayed rectifier K^+ channel (20:1 to 100:1),

FIGURE 14.　Space-filling model of gramicidin. (From D. W. Urry. (1972) A molecular theory of ion-conducting channels: a field-dependent transition between conducting and nonconducting conformations. *Proc. Natl. Acad. Sci. USA* **69,** 1610–1614, 1972, with permission.)

which is primarily responsible for resetting the resting potential of nerve and muscle following activation, and very much less than that of the carrier valinomycin (more than 1000:1).

The gramicidin channel is blocked by divalent cations (e.g., Ca^{2+}, Ba^{2+}, and to a lesser extent Mg^{2+} and Zn^{2+}). These ions probably block by binding near the channel mouth and occluding entrance into the channel. Flicker block may be observed with iminium ions; this transient block is thought to result from transient association of these ions with the channel wall during permeation.

Passage of ions or water through gramicidin is by means of **single-file transport.** The high proton conductance, and the known geometrical dimensions of the pore, suggest that the pore is filled with a continuous column of hydrogen-bonded water molecules (2.8 Å). The diffusive water permeability through gramicidin in bilayers has been measured with tracers, and is about 10^8 water molecules per second at low ionic strength. Water permeation is probably by means of a Grotthus, or "hopping" mechanism (see Chapter 1). The pore diameter is too small to permit passage of urea (5 Å diameter) or larger nonelectrolytes. Consequently, a **streaming potential** develops when nonelectrolyte is added to the salt solution on one side of the bilayer to make an osmotic gradient. In another type of study, an **electroosmotic** volume flow occurs when an ionic current is passed across the membrane. For single file transport, the number N of water molecules in the channel is obtained by dividing the water flux ϕ_w (molecules/s) by the ion flux J (ions/s):

$$N = \frac{\phi_w}{J}$$

For gramicidin channels, the number N of water molecules in the pore is 5–6. Quantitative analysis of streaming potentials and electroosmosis indicate that an ion passing through a gramicidin channel drags with it a column of about six water molecules in single file.

The channel is too narrow for water molecules to slip past each other, nor can ions and water molecules pass each other in the single-file part of the channel. However, sodium ion occupancy does not depress water permeability. The water permeability of a channel at high $[Na^+]$, when the channel always contains a sodium ion, is essentially the same as that of a channel at low $[Na^+]$, when the channel never contains a sodium ion. Thus water can pass the sodium ion somewhere in the channel, presumably near an ion binding site at the end of the channel.

The single-channel conductance for sodium increases as a function of Na^+ concentration, but then saturates at around 1 M NaCl. Half-maximal conductance is reached at 0.31 M NaCl, at which concentration half the channels have one Na^+ ion and half are empty. The curve of conductance g versus $[Na^+]$ fits a single-ion occupancy model, given by

$$g_{Na} = \frac{g_{max}[Na^+]}{K + [Na^+]}$$

where K is the concentration of Na^+ that gives half-maximal conductance. The single-ion occupancy model indicates that an ion binding site, or energy well, exists near the end of the channel. Since the dimeric structure is symmetrical, there

must be two binding sites, one at each end, but only one at a time is occupied by sodium. Additional compelling evidence for single-ion occupancy for sodium is that the flux ratio equation is satisfied at all sodium concentrations ranging from 0.1 to 5 M. Given that single-file transport occurs, a net flux in one direction would inhibit the unidirectional flux in the opposite direction, thus producing deviations from the flux ratio equation if the two Na^+ binding sites were simultaneously occupied. Consequently, double occupancy must not occur for Na^+. These two lines of evidence imply that there is never more than one sodium ion at a time in the gramicidin channel.

The dehydration energy for ions is of the order of 100 kcal/mol; the selectivity sequence indicating a weak field strength interaction with the pore suggests that ions do not dehydrate during passage through the gramicidin pore. The hydration shell of water around the permeating ions is partially replaced by hydrophilic groups lining the channel wall. The dielectric constant is lower in the middle of the pore than near the ends of the pore. Thus the electrostatic energy of the ion in the middle of the pore is greater than at either end of the pore. The actual flux of Na^+ at high $[Na^+]$ is within a factor of 5 of the maximal possible flux of Na^+, given the water permeability of an ion-free channel. This means that there is no significant electrostatic energy barrier to Na^+ movement through the channel. The rate of Na^+ transport is largely determined by the necessity for six water molecules to be moved along with the ion. Thus two energy wells near the ends of the channel are separated by a low-energy barrier (or a series of small barriers) in the middle of the channel. Strict single-filing of water and ions occurs between the two wells, whereas water can pass a sodium ion sitting in either well.

Evidence that gramicidin A can be occupied simultaneously by at least two monovalent cations other than Na^+ is as follows: (1) curves of single-channel conductance versus ion activity for Cs^+ are not monotonic but pass through a maximum and then decline at high aqueous salt concentrations; (2) at low ion concentrations, the ascending branch of the curve plotting single-channel conductance versus ion activity has an inflection point, or a curvilinear Eadie-Hofstee plot, indicating the existence of at least two interacting binding sites in the channel; (3) the exponent n of the generalized flux ratio equation

$$\frac{J_o}{J_i} = e^{-n\mathcal{F}E_m/RT}$$

as shown later, can be larger than unity (however, there is no direct correlation between the magnitude of n and the number of ions in the channel; kinetic models permit the fitting of the variation of n with ion activity); and (4) the permeability ratios, obtained from bi-ionic potentials, are functions of permeant ion aqueous activities. Two models were proposed to account for these observations. Model 1 assumes that the major barrier to ion movement is the translocation step through the middle of the channel, with the ion binding sites always in equilibrium with the aqueous phases. This model requires a minimum of four binding sites, two at each end of the symmetrical channel, to account for the experimental observations (Sandblom *et al.*,

1977). Model 2 assumes that the major barrier to ion movement is the dissociation step between the ion and the channel; therefore, ion binding sites are not in equilibrium with the aqueous phases (when there is a net flux of ions through the channel). In this model, only two (single-file) binding sites in the channel are required to account for the experimental observations (Hille and Schwarz, 1978; Urban and Hladky, 1979). Fortunately, the shape of current-voltage curves can differentiate between the two models. Model 1 predicts single-channel current-voltage curves to be superlinear and invariant to permeant ion activity, whereas Model 2 predicts single-channel current-voltage curves to be sublinear at low ion activities, and superlinear at high ion activities. Experiments showed that the current-voltage curve is approximately linear and nonrectifying up to about 150 mV. There is no voltage at which the channel is closed. There is some effect of voltage on channel properties, but much less than in the case of voltage-gated sodium or potassium channels. When the voltage range was extended to 500 mV, the data supported Model 2 (Andersen, 1983). In these experiments, there is some depression of conductance when the osmotic strength is raised with sucrose. However, the osmotic effects are small and alter the results quantitatively but not qualitatively. Thus there is no compelling need to assume more than two cation binding sites in the channel. The kinetic equations describing single-file transport are complex and beyond the scope of this text.

Cation transport through gramicidin also involves channel motions, **subconductance states,** and **channel flickering.** The gramicidin channel is not rigid, but instead motion of the peptide appears to be essential to its function. Solid-state NMR measurements of ^{15}N-labeled Ala 3 and Leu 4 indicate rocking motions of $\pm8°$ and $\pm15°$, respectively, for these sites in the absence of ions. Molecular dynamics simulations confirm local distortions in gramicidin structure during ion transport. Peptide rotation is not necessary for function since the channel is active in gel-phase lipid. Subconductance states, or "mini-channels," are frequently observed as an intermediate in the opening and closing of normal channels. These may constitute from 5 to 40% of the channel events in a single-channel recording, with lifetimes similar to those of normal gramicidin channels. The low-conductance state may correspond to less common side-chain conformations with altered coordinating ability in the conducting pore. Channel flickering to a low-conductance state with lifetimes ranging from 20 μs to about 1 ms is also observed. Rapid flickering may correspond to a state in which the dimer is partially dissociated. An increase in membrane thickness increases the frequency of these low-conductance states.

V. Summary

This chapter described the evidence for models of the structure of cell membranes, starting with the Gorter and Grendel bimolecular phospholipid leaflet, and proceeding through the Davson-Danielli paucimolecular model and the Singer-Nicholson fluid mosaic model. The succeeding chapter describes extensions of the fluid mosaic model incorporating heterogeneities of lipid phases and distributions. The lipid and protein composition of red blood cell membranes was summarized as an example applicable to other plasma membranes. Planar lipid bilayers made from synthetic lipids, and incorporating ionophores such as valinomycin or gramicidin, provide a useful tool for the study of membrane transport. Fusion of vesicles from native cell membranes into planar lipid bilayers provides information concerning ion channels that complements what is available from patch-clamp studies. Selective ion channels in planar lipid bilayers display high transport rates and low temperature coefficients, characteristics that distinguish ion channels from other modes of mediated transport. Finally, the features of cation transport by gramicidin in planar lipid bilayers were summarized. Detailed kinetic studies have shown that transport of cations and water through gramicidin occurs by means of single-file diffusion in which the ions and the waters cannot pass each other in the narrow part of the pore. The cation binds at a site near either end of the pore, and then is driven over an energetic barrier in the pore by the electrochemical gradient. The rate-limiting step, however, is the dissociation of the ion from the channel.

Bibliography

Membranes

Ansell, G. B., Hawthorne, J. N., and Dawson, R. M. C. (1973). "Form and Function of Phospholipids." Elsevier Scientific Publishing Company, New York.

Bar, R. S., Deamer, D. W., and Cornwell, D. G. (1966). Surface area of human erythrocyte lipids: reinvestigation of experiments on plasma membrane. *Science,* **153,** 1010–1012.

Bloom, J. A., and Webb, W. E. (1983). Lipid diffusibility in the intact erythrocyte membrane. *Biophys. J.* **42,** 295–305.

Davson, H., and Danielli, J. F. (1943). "The Permeability of Natural Membranes." Hafner Publishing Company, Darien, CT.

Dodge, J. T., Mitchell, C., and Hanahan, D. J. (1963). The preparation and chemical characteristics of hemoglobin-free ghosts of human erythrocytes. *Arch. Biochem. Biophys.* **100,** 119–130.

Frye, C. D., and Edidin, M. (1970). The rapid intermixing of cell surface antigens after formation of mouse-human heterokaryons. *J. Cell Sci.* **7,** 319–335.

Gorter, E., and Grendel, F. (1925). On bimolecular layers of lipoids on the chromocytes of the blood. *J. Exp. Med.* **41,** 439–443.

Heilbrunn, L. V. (1956) "The Dynamics of Living Protoplasm." Academic Press, New York.

Jacobs, M. H. (1962). Early osmotic history of the plasma membrane. In "Symposium on the Plasma Membrane, New York Heart Association, Inc." *Circulation* **26,** 1013–1021.

McMillan, P. N., and Luftig, R. B. (1973). Preservation of erythrocyte ghost ultrastructure achieved by various fixatives. *Proc. Natl. Acad. Sci. USA* **70,** 3060–3064.

Singer, S. J., and Nicholson, G. L. (1972). The fluid mosaic model of the structure of cell membranes. *Science,* **175,** 720–731.

Weinstein, R. S. (1969). Electron microscopy of surfaces of red cell membranes. *In* "Red Cell Membrane. Structure and Function" (G. A. Jamieson and T. J. Greenwalt, Eds.), J. B. Lippincott Company, Philadelphia.

Worthington, C. R., and McIntosh, T. J. (1973). Direct determination of the electron density profile of nerve myelin. *Nature-New Biology* **245,** 97–99.

Yeagle, P. L. (1993). "The Membranes of Cells," Second edition, Academic Press, New York.

Planar Lipid Bilayers

Andersen, O. S. (1983). Ion movement through gramicidin A channels. Single-channel measurements at very high potentials. *Biophys. J.* **41,** 119–133.

Coronado, R., and Latorre, R. (1983). Phospholipid bilayers made from monolayers on patch-clamp pipettes. *Biophys. J.* **43,** 231–236.

Ehrlich, B. (1992). Planar lipid bilayers on patch pipettes: bilayer formation and ion channel incorporation. *Meth. in Enzymol.* **207,** 463–470.

Finkelstein, A. (1974). Bilayers: formation, measurements, and incorporation of components. *Meth. in Enzymol.* **32B,** 489–501.

Labarca, P., and Latorre, R. (1992). Insertion of ion channels into planar lipid bilayers by vesicle fusion. *Meth. in Enzymol.* **207,** 447–463.

Miller, C. (1986). "Ion Channel Reconstitution." Plenum Press, New York.

Miller. C. (1987). How ion channel proteins work. *In* "Neuromodulation. The biochemical control of neuronal excitability" (L. K. Kaczmarek and I. B. Levitan, Eds.), Oxford University Press, New York, pp. 39–63.

Montal, M., and Mueller, P. (1972). Formation of bimolecular membranes from lipid monolayers and a study of their electrical properties. *Proc. Natl. Acad. Sci. USA* **69,** 3561–3566.

Mueller, P., Rudin, D. O., Ti Tien, H., and Wescott, W. C. (1963) Methods for the formation of single bimolecular lipid membranes in aqueous solution. *J. Phys. Chem.* **67,** 534–535.

Woodbury, D. J., and Hall, J. E. (1988). Role of channels in the fusion of vesicles with a planar bilayer. *Biophys. J.* **54,** 1053–1063.

Woodbury, D. J., and Miller, C. (1990). Nystatin-induced liposome fusion. A versatile approach to ion channel reconstitution into planar bilayers. *Biophys. J.* **58,** 833–839.

Gramicidin

Finkelstein, A., and Andersen, O. S. (1981). The gramicidin A channel: a review of its permeability characteristics with special reference to the single-file aspect of transport. *J. Memb. Biol.* **59,** 155–171.

Hille, B. (1992). "Ionic Channels of Excitable Membranes" Second ed. Sinauer Associates Inc., Sunderland, MA. Chapters 11, 14.

Hille, B., and Schwarz, W. (1978). Potassium channels as multi-ion single-file pores. *J. Gen. Physiol.* **72,** 409–442.

Hladky, S. B., and Haydon, D. A. (1972). Ion transfer across lipid membranes in the presence of gramicidin A. I. Studies of the unit conductance channel. *Biochim. Biophys. Acta* **274,** 294–312.

Myers, V. B., and Haydon, D. A. (1972). Ion transfer across lipid membranes in the presence of gramicidin A. II. The ion selectivity. *Biochim. Biophys. Acta.* **274,** 313–322.

Sandblom, J., Eisenman, G., and Neher, E. (1977). Ionic selectivity, saturation and block in gramicidin A channels: I. Theory for the electrical properties of ion selective channels having two pairs of binding sites and multiple conductance states. *J. Memb. Biol.* **31,** 383–417.

Urban, B. W., and Hladky, S. B. (1979). Ion transport in the simplest single file pore. *Biochim. Biophys. Acta* **554,** 410–429.

Veatch, W. R., Mathies, R., Eisenberg, M., and Stryer, L. (1975). Simultaneous fluorescence and conductance studies of planar bilayer membranes containing a highly active and fluorescent analog of Gramicidin A. *J. Mol. Biol.* **99,** 75–92.

Wooley, G. A., and Wallace, B. A. (1992). Model ion channels: gramicidin and alamethicin. *J. Memb. Biol.* **129,** 109–136.

Friedhelm Schroeder, W. Gibson Wood, and Ann B. Kier

5

Lipid Domains and Biological Membrane Function

I. Introduction

The most distinctive morphometric feature of all living cells is that they are enclosed by a membrane. In eukaryotes the cell interior is further subdivided into membrane-enclosed intracellular compartments, many of which communicate/interact via small membrane vesicles (Fig. 1). The major chemical entities responsible for membrane formation are lipids. Nearly 80 years ago, it was recognized that lipids spontaneously form membranes in water (Table 1). The precise structure adopted by lipids in biological membranes has significantly evolved into the **fluid mosaic model** wherein a fluid lipid bilayer serves as a matrix for embedded proteins functioning as ion channels, receptor-effector coupled systems, transporters, and so on (rev. in Yeagle, 1987). It is now accepted that lipids as well as proteins are organized into highly structured macro- and microdomains (rev. in Schroeder *et al.,* 1996; Schroeder *et al.,* 1998). The precise relationship between lipid and protein functional domains is the subject of intense interest.

II. General Structure of Biological Membranes

A. Cell Surface Plasma Membrane

The plasma membrane has a thickness of about 80 Å, with <50 Å of this due to the lipid bilayer and the remainder due to other molecules. The latter comprise the glycolipids and glycoproteins extending from the exofacial side and the cytoskeleton extending from the cytofacial side of the plasma membrane. Plasma membrane lipids are especially enriched in cholesterol, such that the cholesterol/phospholipid molar ratio ranges from 0.4 to 1.0 (rev. in Schroeder *et al.,* 1996). Cholesterol orders and rigidifies the membrane phospholipids and condenses their surface area (see Chapter 3), thereby making the plasma membrane lipid bilayer the most rigid in the cell. High choles-

terol and resultant low fluidity minimize the permeability of the cell surface lipid bilayer in order to limit diffusion of substances into and out of the cell (see Chapter 12). The relatively poor conductivity of lipids further helps to confer some of the basic electrophysiological properties on the cell membrane (see Chapters 13 and 14). The insertion of specialized proteins (receptors, channels, effectors, etc.) into the plasma membrane lipid bilayer confers on the cell the ability to control the intracellular ionic milieu, membrane potential, supply of basic building blocks (sugars, amino acids, nucleotides, lipids, etc.), intercellular communication, adhesion, and immunogenicity.

B. Intracellular Membranes

In contrast to prokaryotes, eukaryotic cells contain a highly developed intracellular membrane organization (e.g., endoplasmic reticulum (ER), Golgi complex, lysosome, peroxisome, mitochondria, nuclear membrane). These intracellular organelles as well as endocytic and recycling membrane vesicles functionally compartmentalize the eukaryotic cell interior for macromolecule synthesis, secretion, digestion, fatty acid oxidation, ATP production, genetic replication, and membrane biogenesis (Fig. 1). This compartmentation avoids futile metabolic cycles and allows subcellular specialization not possible in prokaryotes. The remainder of this chapter focuses on eukaryotic membranes.

Intracellular membranes differ significantly from the cell surface plasma membrane: they are less thick and more fluid with less carbohydrate and lower cholesterol. For example, the cholesterol/phospholipid ratio of ER, lysosomal, nuclear, and inner mitochondrial membrane is 0.08–0.2, 0.4–0.6, 0.1–0.2, and 0.01–0.12, respectively (rev. in Schroeder *et al.,* 1996). The nuclear envelope, which separates the DNA from the rest of the cell, is actually comprised of an inner and outer membrane. The outer nuclear membrane has physical continuity with the ER.

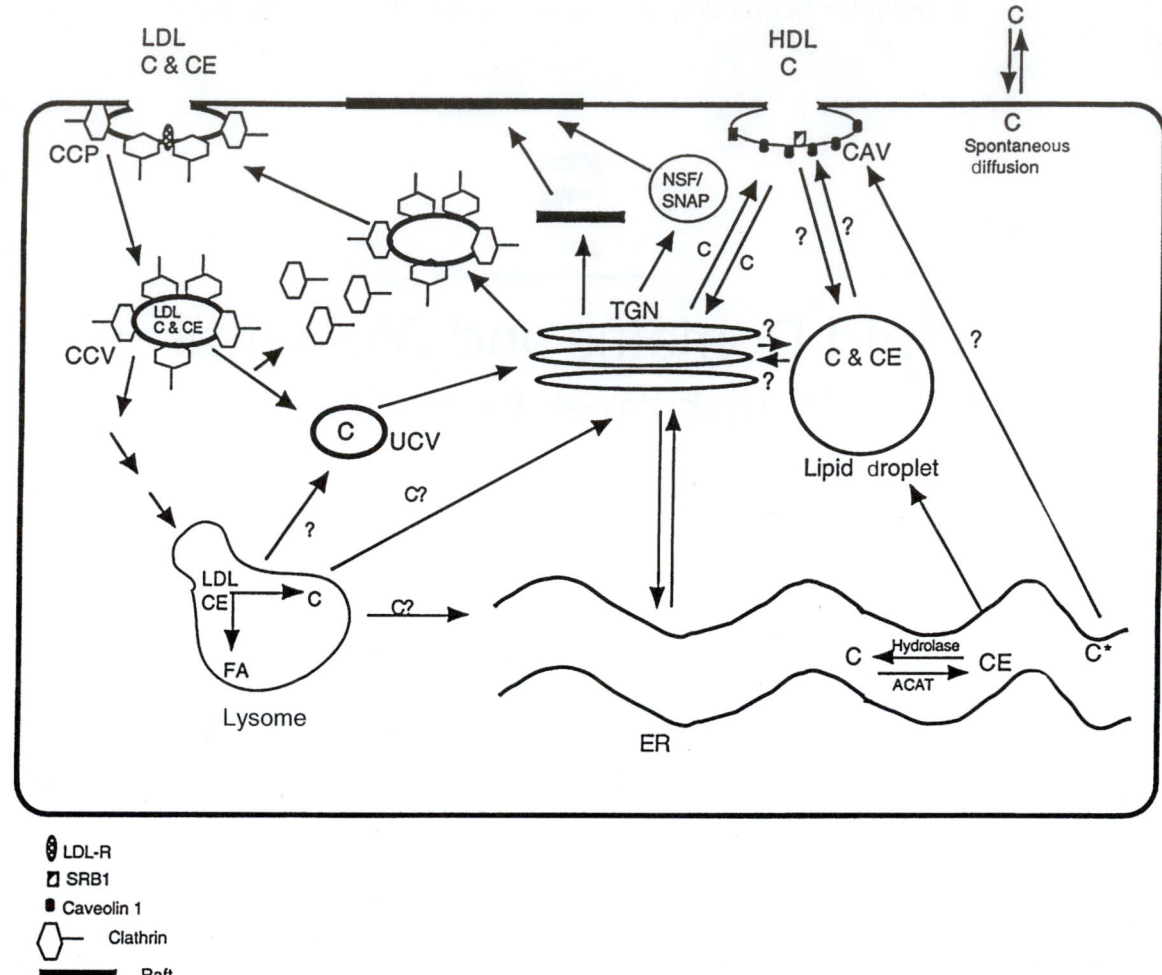

FIGURE 1. Plasma membrane domains, endocytic vesicles, and proposed intracellular cholesterol trafficking pathways: LDL-receptor–mediated clathrin-coated pit (CCP) pathway (upper left); rafts (heavy black lines, upper middle); HDL receptor (SRB1)–mediated caveolar (CAV) pathway (upper right); spontaneous diffusion (upper far right). Other symbols are as follows: TGN, trans–Golgi network; ER, endoplasmic reticulum; CCV, clathrin-coated vesicle; UCV, uncoated vesicle; ACAT, acyl CoA:cholesterol acyltransferase; C*, *de novo* cholesterol synthesis; C, cholesterol; CE, cholesteryl ester; FA, fatty acid.

Due to the fusion of inner and outer nuclear membranes, the nuclear envelope has pores of about 65–75 Å in diameter to allow passage of RNA and other small molecules of molecular weight less than 1000 (see Chapter 31). The ER is abundant in the cell and functions in macromolecule biosynthesis, fatty acid desaturation, detoxification, and some electron transport functions. The principal function of Golgi complex is the posttranslational modification of proteins to glycoproteins and vesicular secretion/trafficking. As such, it may also be important in membrane biogenesis (see Section IX.B). Because lysosomes are part of the phagocytic and degradative machinery of the cell, the

TABLE 1   Progression of Membrane Models

Year	Author	Technique	Concept
1925	Gorter and Grendl	Monolayer area	Plasma membrane is a lipid bilayer
1935	Davson and Danielli		Surface protein sandwiches the lipid bilayer
1960	Robertson	Electron microscopy	Bilayer phospholipid headgroups visualized
1970	Frye and Edidin	Fluorescence	Membrane components move laterally in bilayer plane
1972	Singer and Nicholson		Fluid mosaic membrane with integral and surface proteins

lysososomal membrane surrounds a variety of degradative enzymes that function optimally in the acidic pH of the lysosomal compartment. The very small organelles, peroxisomes, contain degradative enzymes—fatty acid, branched chain fatty acid (phytanic acid), bile acid, amino acid, and certain xenobiotic oxidations—as well as synthetic enzymes—dolichol, cholesterol, isoprenoid (farnesol and geraniol), and plasmalogen biosynthesis (rev. in van den Bosch *et al.*, 1992).

Mitochondria, which function in lipid oxidation and in ATP production (see Chapters 7 and 8), are surrounded by outer and inner membranes. The lipid composition of the outer membrane resembles that of the ER and contains cholesterol, while the inner membrane is essentially devoid of cholesterol. The rate-limiting step in steroidogenesis is the transfer of cholesterol from the outer to the inner mitochondrial membrane (rev. in Schroeder *et al.*, 1996; Schroeder *et al.*, 1998). Although there appear to be contact sites between the inner and outer mitochondrial membranes, the lipid complement of the two membranes is distinct.

C. Structure of Biomembrane Lipid Bilayer

The observation that synthetic lipids as well as lipids extracted from biomembranes spontaneously self-assemble to form membrane bilayers with a relatively random intramembrane distribution had a dramatic impact on our early understanding of biomembrane structure. It led to the depiction of vectorially organized membrane proteins as being localized in a sea of randomly structured lipids (Fig. 2) (Singer and Nicholson, 1972). Despite the nearly simultaneous discovery that biomembranes do not have a random distribution of lipid (Bretscher, 1972), it took nearly two decades before this transformation of the lipid bilayer concept became generally accepted (rev. in Schroeder and Nemecz, 1990; Schroeder *et al.*, 1996). The remainder of this chapter focuses on the structural and functional nature of these membrane lipid domains.

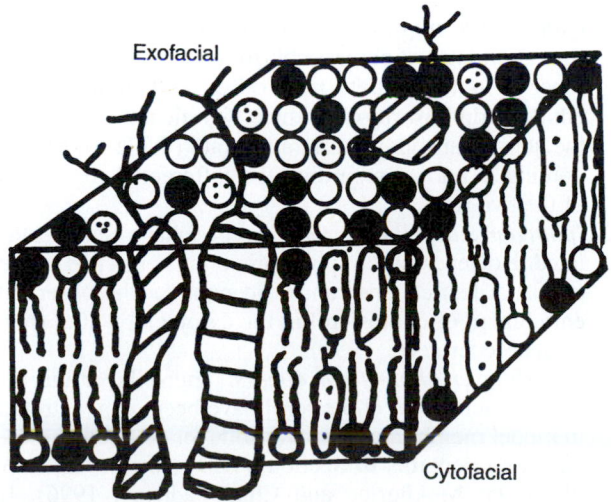

FIGURE 2.   Random lipid bilayer model of membrane structure.

III. The Plasma Membrane Lipid Bilayer: Transbilayer Lipid Distribution

A. Macroscopic Domains

Macroscopic domains are large domains that typically can be readily isolated and characterized. Large-scale (macroscopic) transbilayer lipid domains have been demonstrated in plasma membranes of epithelial cells from liver, intestine, and kidney, and blood sinusoidal, contiguous (basolateral), and brush border (microvillar, bile canalicular) membranes (rev. in Schroeder *et al.*, 1991). Within these domains the transbilayer lipid distribution is not random, but the two leaflets have evolved distinct lipid distributional patterns: the **exofacial (outer) leaflet** is enriched in neutral zwitterionic lipids such as phosphatidylcholine (PC) and sphingomyelin (SM) as well as glycolipids; the **cytofacial (inner) leaflet** is enriched in anionic phospholipids such as phosphatidylethanolamine (PE) and phosphatidylserine (PS) as well as cholesterol (Table 2). Furthermore, within a single cell type such as the hepatocyte, the three major macroscopic plasma membrane domains differ in transbilayer lipid distribution: PC is enriched in the exofacial leaflet of the blood sinusoidal plasma membrane segment; PE is nearly absent in the exofacial leaflet of the contiguous plasma membrane segment; and SM is more enriched in the exofacial leaflet of bile canalicular plasma membrane. Individual leaflets of the membrane bilayer can independently regulate permeability (Negrete *et al.*, 1996).

B. Microscopic Domains

Even when plasma membrane macroscopic domains (segments) cannot be isolated, the plasma membrane has an asymmetric transbilayer distribution of lipids that can be detected by microscopic or other techniques.

1. Phospholipid Transbilayer Domains

Phospholipids are asymmetrically distributed across the plane of the plasma membrane lipid bilayer (rev. in Wood and Schroeder, 1992). With the exception of PS, which appears to be localized exclusively in the cytofacial leaflet, the transbilayer distribution of phospholipids is not absolute. The exofacial leaflet is relatively enriched in the neutral zwitterionic phospholipids having large polar headgroups such as PC and SM (Table 2). In contrast, the cytofacial leaflet is enriched in phospholipids that have smaller polar head groups and are either neutral (PE) or acidic (PS, PI)

TABLE 2   Plasma Membrane Leaflet Selectivity of Lipids

Leaflet	Lipid enriched	Charge
Exofacial	Phosphatidylcholine	Zwitterionic
	Sphingomyelin	Zwitterionic
	Glycolipids	Neutral or anionic
Cytofacial	Phosphatidylserine	Anionic
	Phosphatidylinositol	Anionic
	Phosphatidylethanoalamine	Neutral
	Cholesterol	

(Table 2). The localization of phospholipids with smaller polar headgroups in the cytofacial leaflet provides for less geometric constraint in this leaflet in highly curved membrane areas (e.g., plasma membrane endocytic vesicles, plasma membrane spicules, microvillar tips, budding viruses). In addition, the distribution of anionic phospholipids primarily towards the cytofacial leaflet and zwitterionic neutral phospholipids towards the exofacial leaflet establishes a transbilayer charge gradient such that the cytofacial leaflet is more negative than the exofacial leaflet. This charge gradient may impact on the electrophysiology of the plasma membrane and on intracellular signaling, as in the release of cytofacially bound divalent metal counterions.

2. Cholesterol Transbilayer Domains

An asymmetric transbilayer distribution of cholesterol has been observed in plasma membranes from both normal (brain synaptosome, sperm, erythrocyte) and tumorigenic cells (L-cell). In each case, 3 to 7 times more cholesterol distributes toward the cytofacial leaflet than the exofacial leaflet (rev. in Schroeder and Nemecz, 1990; Parihar *et al.*, 1990). Plasma membrane transbilayer cholesterol domains contribute to the transbilayer fluidity gradient, the exposure of antigens in the exofacial leaflet, Ca^{2+}, Mg^{2+}-ATPase activity, Na^+, K^+-ATPase activity, and apoE and LDL-receptor activity (rev. in Wood *et al.*, 1999).

3. Sphingolipid and Glycolipid Transbilayer Domains

The sphingolipids and glycolipids are localized solely in the exofacial leaflet of the plasma membrane. Sphingolipid-cholesterol complexes are hypothesized to form membrane microdomains differing in thickness from other membrane regions (Smaby *et al.*, 1996). The intracellular trafficking, sorting, and targeting of glycosylphosphatidylinositol-linked proteins is associated with membrane microdomains rich in cholesterol, glycosylceramides, and sphingomyelin (Parton and Simons, 1995).

4. Fatty Acid Transbilayer Domains

The fatty acids associated with cytofacial leaflet phospholipids, PE and PS, from L-cell fibroblasts and brain synaptosomes are enriched with unsaturated fatty acids. In contrast, the phospholipids found in the exofacial leaflet are more saturated (rev. in Fontaine *et al.*, 1980). However, this transbilayer distribution of unsaturated fatty acyl groups is not universal. For example, in the small intestinal brush border membrane, the phospholipid in the outer leaflet contains more unsaturated fatty acids than the inner leaflet (Dudeja *et al.*, 1991).

5. Regulation and Pathophysiology

Spontaneous transbilayer migration of phospholipids is a remarkably slow process with a half-time of days to weeks. However, in the plasma membrane the transbilayer distribution of phospholipids is maintained by two plasma membrane integral protein systems: a **phospholipid translocase** and a **scramblase** (rev. in Williamson *et al.*, 1995; Devaux and Zachowski, 1994). The translocase is specific for anionic phospholipids, and rapidly translocates them from the exofa-

cial leaflet to the cytofacial leaflet. By exclusion, PC and SM largely remain in the exofacial leaflet. The scramblase protein, in contrast, has no specificity for phospholipid polar headgroups, and so randomizes all phospholipid classes across the bilayer when activated. Under normal circumstances, the transbilayer distribution of phospholipids appears to be remarkably stable. However, activation of platelets with thrombin and thapsigargin or with Ca^{2+} and ionophore induces a complete redistribution of all phospholipids across the lipid bilayer (Williamson *et al.*, 1995) (Table 3). In addition, lipid peroxidation, radiation (Schroeder, 1984), and ATP depletion (Devaux and Zachowski, 1994) can lead to loss of transbilayer phospholipid asymmetry. Alterations in phospholipid asymmetry have been observed in human sickle cell erythrocytes. Chronic ethanol exposure elicits a redistribution of phosphatidylcholine across the bilayer.

The transbilayer distribution of fatty acids occurs by a complex process in which phospholipids appear at the cytofacial leaflet with esterified fatty acids, which may also undergo fatty acid remodeling. The fatty acids, esterified or remodeled triphospholipids, are then translocated across the bilayer.

The transbilayer distribution of cholesterol is not fixed or immobile, but can be modified in intact cells or *in vivo* (rev. in Schroeder *et al.*, 1996; Wood *et al*, 1996; Wood *et al.*, 1999). Plasma membrane equilibrium transbilayer sterol distribution is determined by several factors: (a) Unsaturated fatty acids dramatically alter the plasma membrane transbilayer sterol distribution. (b) Chronic exposure of LM cell fibroblasts or mice to ethanol reduced the plasma membrane and synaptosomal plasma membrane transbilayer sterol asymmetry, such that two- to three-fold more sterol was localized in the exofacial leaflet, without any change in the total plasma membrane cholesterol/phospholipid ratio. (c) Aging dramatically altered mouse brain synaptic plasma membrane transbilayer sterol distribution, such that 2.4-fold more sterol was present in the exofacial leaflet. Although the spontaneous transbilayer migration rate of sterols is fast—on the order of minutes (rev. in Schroeder and Nemecz, 1990)—this cannot account for an asymmetric transbilayer sterol distribution. Several potential mechanisms regulating the equilibrium transbilayer sterol distribution have been proposed:

1. *Transbilayer migration rate (flip-flop).* The transbilayer migration rates of cholesterol from outer to inner versus inner to outer leaflets may differ. Although chronic ethanol administration to mice significantly increased the sterol transbilayer migration rate (Table 3), almost nothing is known regarding the differential regulation of the two rates.

2. *Sphingomyelin-cholesterol complexes* have been postulated. However, since sphingomyelin is enriched in the exofacial leaflet, this cannot account for the observed enrichment of cholesterol in the cytofacial leaflet (rev. in Schroeder *et al.*, 1996).

3. *Transbilayer cholesterol dimers.* Transbilayer tail-to-tail interactions of cholesterol have been demonstrated in model membranes (see transbilayer coupled cholesterol-rich domains in Section VII.D) (rev. in Harris *et al.*, 1995; Mukherjee and Chattopadhyay, 1996). If transbilayer coupled cholesterol dimers represent the

TABLE 3 Plasma Membrane Transbilayer Domain Regulation.

Parameter	Lipid	Plasma membrane	Effector
Transbilayer migration rate	Cholesterol	Synaptosome	Ethanol
Transbilayer distribution	Cholesterol	L cell	Ethanol
	Cholesterol	L cell	Polyunsaturated fatty acids
	Cholesterol	L cell	Sterol carrier proteins
	Cholesterol	L cell	Lipid peroxidation
	Cholesterol	L cell	Oxidized cholesterol
	Cholesterol	Synaptosome	Aging
	Cholesterol	Synaptosome	Ethanol
Transbilayer distribution	Phospholipid	Platelet	Activation
	Phospholipid	Erythrocyte	ATP depletion
	Phospholipid	Erythrocyte	Sickle erythrocyte
	Phospholipid	Erythrocyte	Ethanol
Transbilayer fluidity		Erythrocyte	Anesthetics
		Synaptosome	Anesthetics
		Intestinal microvillus	Anesthetics
		L cell	Anesthetics
		L cell	Polyunsaturated fatty acids
		L cell	Lipid peroxidation
		Synaptosome	Aging
		Synaptosome	Ethanol
		Synaptosome	Systemic lupus erythematosus

majority of cholesterol in plasma membranes, they would result in a symmetric transbilayer distribution of sterols. This is not observed, however, as stated above.

4. *Plasma membrane integral cholesterol-binding proteins.* Such proteins, if asymmetrically distributed across the membrane bilayer, could modulate transbilayer sterol distribution (rev. in Schroeder and Nemecz, 1990). While most membrane proteins are localized in cholesterol-poor areas (rev. in Schroeder *et al.*, 1996), caveolin is a cholesterol-binding protein localized to the cytofacial leaflet of plasma membrane microdomains called caveolae (rev. in Schroeder *et al.*, 1998; Smart and van der Westhuyzen, 1998).

5. *Cytoplasmic sterol carrier proteins.* Multiple cholesterol-binding proteins with nano- to micromolar K_d and 1:1 stoichiometry for cholesterol exist in the cytosol and some of these (e.g., sterol carrier protein 2, liver fatty acid-binding protein, caveolin) can elicit intermembrane sterol transfer to and from plasma membranes as well as to and from intracellular membranes (Fig. 1) (Schoer *et al.*, 2000; Gallegos *et al.*, 2000). Furthermore, by redistributing cholesterol within the cell, some of these proteins also altered plasma membrane transbilayer sterol distribution. The expression of sterol carrier protein-2 as well as its interactions with cholesterol and/or other lipids are altered by chronic or acute ethanol treatment of mice or rats (rev. in Wood *et al.*, 1999).

IV. Lateral Lipid Microdomains in Membranes

A. Plasma Membrane Lateral Microdomains of Sphingolipids and Glycolipids

Glycosphingolipids generally have high phase transition temperatures near 50–85 °C. In model membranes consist-

ing of stearoyl-oleoyl-phosphatidylcholine, glycosphingolipids phase separate into a solid-phase, glycolipid-rich domain and a stearoyl-oleoyl-phosphatidylcholine–rich fluid-phase domain at greater than 20% glycosphingolipid (Morrow *et al.*, 1992). However, even at low glycolipid concentration the glycosphingolipid and phospholipid exhibit solid-phase immiscibility. Likewise, sphingolipids cluster in the luminal leaflet of the trans-Golgi network of MDCK cells, where they form the budding site of apical membrane vesicles (rev. in Parton and Simons, 1995).

B. Cholesterol Lateral Microdomains in Plasma Membranes

Electron microscopy, cholesterol oxidase, electron spin resonance, and fluorescence techniques, as well as radiolabeled or fluorescent sterol exchange, have demonstrated the coexistence of cholesterol-rich and cholesterol-poor domains in cell surface membranes of red blood cells, platelets, fibroblasts, the renal brush border, and human immunodeficiency virus (rev. in Schroeder *et al.*, 1996; Schroeder *et al.*, 1998; Wood *et al.*, 1999). Multiple types of cholesterol-rich or cholesterol-poor lateral domains are now recognized in plasma membranes:

1. **Desmosomes** are intercellular junctions at the plasma membrane with a cholesterol/phospholipid ratio of 1.1 (Skerrow and Matoltsy, 1974).

2. **Clathrin-coated pits** are the site of LDL receptor–mediated endocytosis (rev. in Fielding and Fielding, 1997; Pearse, 1976; Severs and Robenek, 1983) (CCP in Fig. 1). Based on the lack of filipin staining, it was originally concluded that clathrin-coated pits do not contain cholesterol (rev. in Severs and Robenek, 1983). However, removal of the clathrin coat enables

filipin staining of cholesterol (Steer *et al.*, 1984), while chemical analysis of isolated coated and uncoated vesicles showed a cholesterol/phospholipid molar ratio of 0.3 to 0.36 (Steer *et al.*, 1984; Pearse, 1976). Thus, clathrin-coated pits are slightly poorer in cholesterol than the bulk plasma membrane, but are definitely not cholesterol-deficient domains.

3. Recent exciting discoveries indicate that plasma membrane **caveolae** are very cholesterol-rich, sphingomyelin-rich domains (three- to four-fold more than the bulk plasma membrane) containing the protein caveolin (CAV in Fig. 1). Caveolae are also enriched in lipidic signaling molecules (PIP_2, ceramide, diacylglycerol) as well as in proteins involved in signaling (eNOS, ras, EGF, TFPI, scr), transcytosis, and reverse cholesterol transport (SRB1) (rev. in Smart and van der Westhuyzen, 1998; Fielding *et al.*, 1998). Scavenger receptors (SRB1) interact with HDL to mediate cholesterol cellular efflux or HDL-cholesterol influx (Acton *et al.*, 1996; Frolov *et al.*, 2000; Atshaves *et al.*, 2000). In the absence of serum HDL, cholesterol entering the cholesterol-rich caveolar domain diffuses laterally to other microdomains within the plasma membrane.

4. **Rafts** (Fig. 1) are flat cholesterol-rich domains lacking clathrin or caveolin (rev. in Fra *et al.*, 1994; Bohuslav *et al.*, 1993).

5. Crystalline, pure cholesterol domains form in arterial smooth muscle plasma membranes (Tulenko *et al.*, 1998).

C. Regulation and Pathophysiology of Plasma Membrane Lateral Domains

The normal regulatory mechanism(s) that determine the lateral lipid domains are not known. The membrane content of sphingomyelin is directly correlated with the amount of nonexchangeable cholesterol domain in model and plasma membranes. Because sphingomyelin is particularly rich in the plasma membrane, the major portion (50–80%) of plasma membrane cholesterol is nonexchangeable (rev. in Schroeder *et al.*, 1996). Interaction of specific integral membrane proteins with cholesterol may cause formation of cholesterol domains. Cytosolic proteins (e.g., sterol carrier protein-2 and the brain fatty acid–binding proteins) alter the size of the sterol domains in plasma membranes *in vitro* (Table 4). Proteins containing basic regions (protein kinase C, myristoylated alanine-rich C kinase substrate, pp50[SRC]) bind to membranes and cause formation of negatively charged phospholipid domains. Lateral sterol domains in biomembranes are sensitive to temperature and ionic composition, as well as certain drugs and peptides (Table 4).

Relatively little is known about alterations in plasma membrane lateral lipid domains in pathological states. Pathogens such as Semliki Forest virus require cholesterol and/or specific cholesterol domains in target membranes for binding (rev. in Schroeder *et al.*, 1996; Smaby *et al.*, 1996). Influenza hemagglutinin and Alzheimer's amyloid peptide fragments induce formation of negatively charged phospholipid domains in membranes. Bacterial cytolysins (hemolysins, (θ-toxins) bind to specific cholesterol lateral domains and reveal multiple cholesterol domains in erythrocyte, lymphoma B, and BALL-1 cell surface plasma membranes, as well as in model membranes. Lateral sterol domains appear to be altered in sickle cell erythrocyte membranes in response to deoxygenation. Chronic ethanol treatment increases the half-time of exchange for the exchangeable sterol domain in brain synaptosomes. One of the most exciting aspects of plasma membrane lateral cholesterol domains is the possibility that their properties modulate the extent and/or rate of cholesterol efflux from cells. This process, called **reverse cholesterol transport**, is mediated by the cholesterol-rich caveolae. The finding that oxidized cholesterol disrupts caveolae may be of importance in atherosclerosis, diabetes, and other lipid disorders (rev. in Smart and van der Westhuyzen, 1998).

V. Transbilayer Asymmetry in Fluidity

The primary determinants of membrane lipid fluidity include phospholipid composition, fatty acid unsaturation, and cholesterol/phospholipid ratio. Phospholipids, fatty acyl chains esterified to the phospholipids, and cholesterol are asymmetrically distributed across the plasma membrane bilayer. Since each of these components has different effects on fluidity, it is not possible to accurately predict the net effect on

TABLE 4   Membrane Lateral Domain Regulation

Lipid	Cell	Membrane	Effector
Cholesterol	L cell	Plasma Membrane	Sterol carrier protein
	—	Model Membrane	Cytolysins
	Erythrocyte	Plasma Membrane	Cytolysins
	Lymphoma	Plasma Membrane	Cytolysins
	BALL-1 cell	Plasma Membrane	Cytolysins
	Erythrocyte	Plasma Membrane	Deoxygenation
	—	Model Membrane	Protein kinase C
Phospholipid	—	Model Membrane	MARCKS
	—	Model Membrane	pp 50_{SRC}
	—	Model Membrane	VSV
	Cell	Plasma Membrane	VSV
	—	Model Membrane	Alzheimer's amyloid

plasma membrane leaflet fluidity. A rough expectation can be based on the following: PC and SM are generally more fluid phospholipids than PE and PS; unsaturated fatty acids are more fluid than saturated fatty acids; cholesterol orders the membrane structure. Large-scale transbilayer lipid domains have been demonstrated in plasma membranes of epithelial cells from liver, intestine, and kidney (rev. in Schroeder *et al.*, 1996). Based on the transbilayer distribution of phospholipid and fatty acid in intestinal microvillus membrane one might predict that the cytofacial leaflet would be more fluid than the exofacial leaflet. On the contrary, the microvillus plasma membrane segment from intestine has a much less fluid cytofacial leaflet than exofacial leaflet (Dudeja *et al.*, 1991), which suggests that the cytofacial leaflet is enriched with cholesterol in the microvillus. If the transbilayer distribution of cholesterol is the major determinant of plasma membrane individual leaflet fluidity, one might predict that the cholesterol-rich leaflet is less fluid. This is indeed the case for L cell, brain synapse, spermatozoa, and erythrocyte plasma membranes. Thus, cholesterol transbilayer distribution is closely associated with transbilayer fluidity gradients in plasma membranes (rev. in Schroeder *et al.*, 1996; Wood *et al.*, 1999).

Transbilayer fluidity gradients in the plasma membrane are very sensitive to a variety of factors (Table 3). Acute ethanol administration fluidized the exofacial leaflet whereas anesthetics were leaflet-selective plasma membrane fluidizers: exofacial-leaflet selective (pentobarbital, phenobarbital, benzyl alcohol, ethanol) or cytofacial-leaflet selective (prilocaine, 2-[(2-methoxy-ethoxy)ethyl]-*cis*-8-(2-octylcyclopropyl)octanoate). In contrast, unsaturated fatty acids, aging, chronic ethanol administration, and systemic lupus erythematosus essentially abolished the plasma membrane transbilayer fluidity. The exofacial fluidizing drugs affected Na$^+$,K$^+$-ATPase and leucine aminopeptidase activity, while cytofacial fluidizing drugs affected Na$^+$-dependent D-glucose uptake and Na$^+$, K$^+$-dependent L-glutamic acid uptake.

VI. Lateral Plasma Membrane Lipid Domain Fluidity

A. Macroscopic Lateral Lipid Domains

The fluidity of large-scale (macroscopic) plasma membrane lipid domains has been demonstrated in plasma membrane segments of epithelial cells from liver, intestine, and kidney (rev. in Schroeder *et al.*, 1991; Schroeder *et al.*, 1996). These lateral lipid domains (sinusoidal, basolateral, and brush border or canalicular) are separable by centrifugation techniques. Even from the same cell, each different plasma membrane subfraction is characterized by unique enzymes, transport proteins, receptor/effector coupling systems, lipid composition, and physical structure. It is not understood how lateral intermixing of the lipid components of these macroscopic domains within the same cell is prevented. Lipids are free to diffuse laterally in the plane of the membrane. In the plasma membrane cytofacial leaflet, these lipids freely diffuse laterally across macrodomain boundaries. Tight junctions prevent the lateral diffusion of

lipid molecules in the exofacial leaflet across macrodomain boundaries. How the tight junctions accomplish this sorting of lipids (and proteins) in the exofacial leaflet is not known. Macroscopic lipid domains can also be separated by centrifugation as vesicles shed from cells. Depending on the cellular type, shed vesicle membranes can represent macroscopic cholesterol-rich or cholesterol-poor domains. Shed retinal rod outer–segment membrane vesicles are also macroscopic cholesterol-poor domains.

B. Microscopic Plasma Membrane Lateral Domains

Microscopic lipid domains are not readily separated by techniques such as centrifugation. Early work suggested that their existence can be inferred by histochemical staining with filipin to visualize filipin-cholesterol complexes (rev. in Schroeder *et al.*, 1991; Schroeder *et al.*, 1996). However, results obtained with histochemical stains such as filipin must be viewed cautiously. Proteins may interfere with filipin's access to cholesterol. The major breakthroughs in localizing microscopic lateral lipid domains have come through a variety of probe approaches as well as noninvasive physical techniques. A number of methods, based on cholesterol exchange, cholesterol oxidase, spin-labeled sterol, fluorescent sterol, and freeze-fracture, have provided evidence that such microdomains exist in both model membranes and biomembranes.

C. Structural Consequences of Lipid Lateral Domains in Model Membranes

Lateral segregation of membrane lipids has important structural consequences. Segregation of phosphatidylcholines differing by more than four methylenes (dimyristoylphosphatidylcholine and distearoylphosphatidylcholine) results in coexistence of fluid and solid phospholipid domains. Segregation of anionic (phosphatidylserine) from neutral zwitterionic (phosphatidylcholine) phospholipids results in lateral charge separation. Cholesterol segregates into cholesterol-poor and cholesterol-rich domains that represent coexisting immiscible fluid phases, with the former relatively less ordered than the latter. In short, lateral lipid domains can differ in phase, in charge, in glycolipid or sphingolipid content, or in cholesterol content. Structural consequences of lipid lateral domains have not been determined, although some evidence indicates that the cholesterol-poor regions of plasma membranes are more fluid than the cholesterol-rich regions.

VII. The Plasma Membrane Lipid Bilayer: Protein Distribution

A. Coupling of Protein Function to Transbilayer Lipid Distribution

Transbilayer lipid asymmetry is coupled to a variety of plasma membrane functions.

1. *Receptor-effector coupling.* The plasma membrane transbilayer cholesterol distribution is the primary determinant of transbilayer fluidity gradients. These gradients

and alterations therein thereby modulate the functions of plasma membrane proteins (reviewed in Schroeder and Wood, 1995; Schroeder and Sweet, 1988; Sweet and Schroeder, 1988) such as the glucagon receptor-adenylate cyclase coupled system located in exofacial and cytofacial leaflets, respectively.

2. *Ion transporter coupling.* Ion transporters that are composed of subunits function optimally in an associated/aggregated state. Plasma membrane leaflet cholesterol content and the associated transbilayer fluidity gradient may define specific ion transporter subunit association and activity as shown for Ca^{2+},Mg^{2+}-ATPase and Na^+,K^+-ATPase (rev. in Wood *et al.*, 1999).

3. *Cellular cholesterol influx.* Although the major mechanism for cholesterol uptake is via the LDL receptor–mediated endocytic pathway, some LDL free cholesterol may enter by an exchange pathway. This exchange may be facilitated by the low cholesterol content of the plasma membrane exofacial leaflet. Alternately, cholesterol may enter via the HDL-SRB1 pathway mediated via caveolae. Although the size of the exchangeable cholesterol domain and its transbilayer distribution in caveolae are not known, the transbilayer migration rate of cholesterol through caveolae is very rapid ($t_{1/2}$ of min).

4. *Cellular reverse cholesterol transport.* The reverse process of cholesterol influx is efflux and occurs by molecular cholesterol exchange. This process is very important to the HDL-mediated reverse cholesterol transport for the removal of cholesterol from the cell. The potential effects of cholesterol domain size and transbilayer migration rate are especially significant here.

5. *Translocation of proteins across the plasma membrane.* Protein translocation requires not only a leader sequence in the protein to be secreted and a protein translocase enzyme(s), but the lipid phase is also essential. The fatty acid composition, fluidity, and phospholipid charge are all vital, if poorly understood, factors contributing to protein translocation across membranes (Schatz and Dobberstein, 1996).

B. Transbilayer Coupling of Plasma Membrane Lateral Microdomains: Relation to Protein Function

The effect of the structural properties of lipids in one leaflet of the membrane on the structural properties of lipids in the apposing leaflet must be considered. Several points are especially pertinent to lateral domains in opposing leaflets of the membrane bilayer. At least three possibilities can give rise to coupling of laterally segregated lipids across the bilayer (Fig. 3). (1) Some lipids such as dolichol or dolichol phosphate are extremely long and extend across the bilayer (Fig. 3A). Dolichols are long-chain isoprenoid alcohols with chain lengths of generally 55–115 methylenes. These molecules are extremely potent in fluidizing membranes and in altering Na^+,K^+-ATPase activity in synaptosomal membranes. (2) Phospholipid species with fatty acids

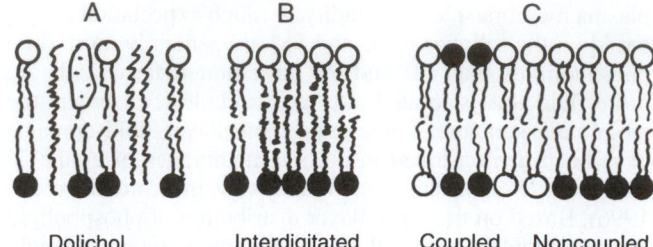

FIGURE 3. Transbilayer coupling of lipid domains in membranes. (A) Long-chain dolichol spans the entire bilayer and couples the two leaflets. (B) Phospholipids, glycolipids, and sphingolipids whose acyl chains are dissimilar in length may interdigitate into opposing leaflets. (C) Like lipids may form lateral domains that have identical lipids in the opposing leaflet (coupled) or in different lipids in the opposing leaflet (uncoupled).

esterified to C_1 and C_2 that differ in chain length by four or more methylenes, as well as many sphingomyelin species, can interdigitate (Fig. 3B) in the apposing leaflet acyl chains to varying degrees. (3) A fluorescent probe approach was recently used to examine the physical state of laterally segregated lipids coupled across apposing leaflets (Almeida *et al.*, 1992) (Fig. 3C). A membrane-spanning NBD-labeled phosphatidylethanolamine in a membrane bilayer containing lignoceryldihydrogalactosylceramide was used to show that the distribution of solid domains in one monolayer of the bilayer was independent of the distribution of solid domains in the apposing bilayer. In contrast, in membrane bilayers lacking lignoceryldihydrogalactosylceramide, the solid domains exactly superimposed upon solid domains in the apposing bilayer. These superimposed solid phases were very stable, with half-times of relaxation in excess of 7 h (Almeida *et al.*, 1992). Such transbilayer coupled and uncoupled lateral domains may be highly significant to the functions of transmembrane proteins (ion channels, pumps, receptor-effector coupling, etc.).

C. Annular Lipid Microdomains

Interfacial or annular lipid exists in the regions between lateral lipid domains or around integral proteins in cell membranes. At the boundaries of lipid domains such interfacial lipid (Fig. 4) is thought to be highly disordered and important to the actions of phospholipase enzymes. Integral membrane proteins are primarily localized in fluid, cholesterol-poor lipid domains (Houslay and Stanley, 1982). Moreover, in the fluid lipid microdomains the proteins are solvated by lipids. The lipids immediately adjacent to the embedded protein are referred to as the lipid annulus, boundary-layer lipid, or halo lipid. Although the annular lipid may differ from bulk lipid, it exchanges with neighboring lipids much more slowly (10^{-4} to 10^{-6} s) as compared to the extremely rapid hopping frequency, 10^{-7} s, of bulk lipid. The important concept here is that the annular lipid is not static and does turn over, albeit 10 to 1000 times more slowly than bulk lipid. As for interfacial lipid, the annular lipid is highly disordered. One of the most well studied examples is the annular lipid surrounding the Ca^{2+}-ATPase from sarcoplasmic reticulum (rev. in Houslay

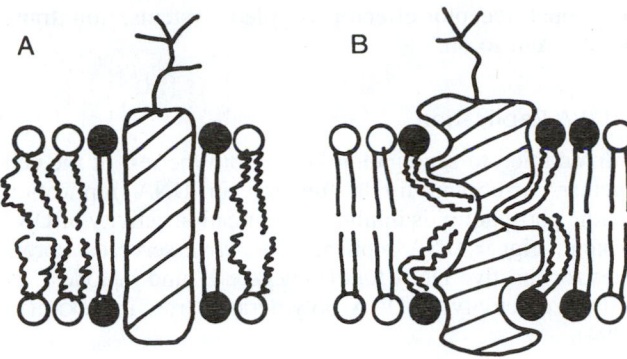

FIGURE 4. Annular lipid domains. Annular lipid is the first "shell" of lipid surrounding integral membrane proteins. (A) Annular lipid "ordering" by protein. (B) Annular lipid "disordering" by protein.

and Stanley, 1982). This protein spans the membrane and has an annulus of 15 phospholipid molecules in each leaflet of the bilayer. The annulus is essential for activity, since Ca^{2+}-ATPase activity is lost if less than 30 mol of lipid remain associated with the protein. When faced with a choice of phospholipids present in a mixed membrane (inhibitory and stimulatory lipids) the Ca^{2+}-ATPase segregates lipids necessary for proper function into the lipid annulus. Other examples of integral membrane proteins that segregate annular lipid include cytochrome oxidase, β-hydroxybutyrate dehydrogenase, glycophorin, Na^+,K^+-ATPase, rhodopsin, acetylcholinesterase, and protein kinase c (rev. in Houslay and Stanley, 1982; Nichols and Roufogalis, 1991.) Specific lipid species appear to be selected in the annulus surrounding these proteins: β-hydroxybutyrate dehydrogenase and phosphatidylcholine; 5'-nucleotidase and sphingomyelin; Ca^{2+}-ATPase and nonacidic phospholipids; cytochrome oxidase, glycophorin, Na^+,K^+-ATPase, and rhodopsin with acidic phospholipids. Acetylcholinesterase specifically requires associated annular phosphatidylinositol, whose fatty acyl groups differ from those of nonannular phosphatidylinositol.

Cholesterol in the annulus of integral membrane proteins may have both inhibitory and stimulatory roles. Cholesterol in the annulus of Ca^{2+}-ATPase is inhibitory (Houslay and Stanley, 1982). The annulus of the acetylcholinesterase receptor has both annular and nonannular cholesterol sites around the protein. The nonannular cholesterol sites have a 20 times greater affinity for cholesterol than the annular sites. The nonannular sites appear to be necessary for stabilizing α-helical structures in the acetylcholinesterase receptor, which are necessary to support a functional ion channel. However, other neutral lipids such as α-tocopherol, squalene, or cholestanol can substitute for cholesterol. Cholesterol enrichment also occurs in the immediate lipid layer surrounding the acetylcholine receptor, the oxytocin receptor (Gimpl et al., 1997), and vesicular stomatitis virus proteins G and M (Luan et al., 1995).

D. Lateral Coupling of Plasma Membrane Lipid Domains to Protein Function

The function of lateral lipid domains in membranes is not known. Lipids can influence the function of integral membrane proteins through the lipid annulus, lateral diffusion of the proteins, transbilayer domains, and lateral microdomains. Changes in transbilayer or lateral lipid domains can elicit conformational changes as well as vertical or lateral displacement of proteins (Schroeder et al., 1991; Schroeder et al., 1996; Bergelson, 1992). It is believed that lateral microdomains can buffer the integral membrane proteins embedded therein from structural changes occurring elsewhere in the membrane (rev. in Gimpl et al., 1997). Moreover, insertion of a membrane protein and refolding in the lipid bilayer requires lipid microdomains to be fluid. The spontaneous organization of cholesterol into domains may even be responsible for lateral segregation of proteins into specific cholesterol-rich or cholesterol-poor regions in the membrane (Bretscher and Munro, 1993). Lateral lipid domains, especially cholesterol, are coupled to a variety of biological processes.

1. Sperm Capacitation

In order to complete the acrosomal reaction to fertilize an oocyte, sperm must be capacitated. This requires shedding of cholesterol from the periacrosomal region of the sperm plasma membrane headpiece. Serum lipid transfer protein-I enhances sperm capacitation by increasing this cholesterol loss.

2. Cholesterol Absorption and Efflux

An intrinsic microvillar plasma membrane protein is required to potentiate cholesterol uptake by intestinal microvillus membranes (Schulthess and Hauser, 1995). The extent and direction of this process may be regulated by the size of the exchangeable cholesterol domains and/or the half-time of sterol transfer through the exchangeable domains. Reverse cholesterol transport (rev. in Schroeder et al., 1996; Smart and van der Westhuyzen, 1998; Fielding et al., 1998) is important in atherosclerosis and other serum lipoprotein disorders.

3. Membrane Protein Function

Both integral as well as soluble proteins can modulate the composition and properties of lipid domains. Integral membrane proteins are not randomly distributed in the lateral lipid domains. Lipids form the microenvironment of the proteins and the immediate annular lipids surrounding the proteins are relatively disordered with respect to bulk lipids. When ligands, agonists, antagonists interact with proteins, conformational changes occur in the proteins. The binding of hormones, antibodies, or lectins (prostaglandins, ricin, muscarinic agents, antibodies) to membrane-associated receptors elicits changes in the fluidity of specific membrane lateral microdomains in which these integral proteins are embedded (Sweet and Schroeder, 1988; Bergelson, 1992). Muscarinic agents decrease the fluidity of the microdomain wherein anthroylvinyl-phosphatidylcholine resides in rat brain membranes. One consequence of such lipid microdomain structure changes is that the activities of other proteins that cosegregate into the same lipid microdomain as the receptor protein may be modulated. For example, the acetylcholinesterase receptor, necessary for cholinergic synaptic transmission, requires both anionic phospholipid and cholesterol for optimal activity (Sunshine and McNamee, 1992). Binding of acetylcholine to the receptor is coupled to the opening of a cation-specific channel. The rate of desensitization of the receptor depends on the type of anionic and neutral lipid surrounding the receptor.

Soluble proteins such as the cytosolic cholesterol carrier proteins (SCP-2 and SCP/L-FABP) can also modulate lipid domains (rev. in Schroeder et al., 1991; Schroeder et al., 1996). In vitro studies with model membranes indicate that SCP-2 but not SCP/L-FABP alters both the size and kinetics of the cholesterol domains. In contrast, in vitro studies with biological membranes indicate that both SCP-2 and SCP/L-FABP can modulate sterol domains. These results have been interpreted as being consistent with SCP-2 interacting directly with membrane anionic phospholipids (rev. in Huang et al., 1999) while SCP/L-FABP interacts with a glycoprotein/glycolipid membrane receptor to elicit changes in cholesterol domain structure (Ishibashi and Bloch, 1981). These observations with SCP/L-FABP have been confirmed through molecular biological studies performed in vivo with intact cultured cells transfected with the cDNA encoding SCP/L-FABP. These alterations in cholesterol lateral microdomain structure dramatically influenced the plasma membrane content and proportion of Na^+,K^+-ATPase subunits, as well as the specific activity of the enzyme. The size of the exchangeable domain(s) may relate to the size of cholesterol-poor regions of the cell surface membrane. Lipid microdomains appear to regulate in part the activities of transport proteins including the Ca^{2+}-ATPase, Na^+,K^+-ATPase, Na^+-H^+-antiporter, and Na^+ dependent glucose transport. Other functions regulated by lipid lateral microdomains include the closure of the exocytotic fusion pore (Oberhauser et al., 1992) and protein kinase c activation/inhibition (Epand et al., 1992). Neutral phospholipids such as phosphatidylcholine as well as phosphatidylserine participate in the activation and differential regulation of the protein kinase c (Chen et al., 1992). It has been shown that autophosphorylation of protein kinase c may require a high order of protein-phospholipid aggregates. The lateral distribution of cholesterol into cholesterol-poor and cholesterol-rich domains may promote the aggregation of proteins into the cholesterol-poor domains. The size of the cholesterol-poor domain in membranes is usually small, and therefore regulation of the size of this domain by ions, drugs, and cytosolic proteins can have important consequences for integral membrane proteins that are activated in the aggregated state.

Just as lipid domains are modulated by proteins, protein functions themselves are modulated by lipid lateral domains. Most proteins reside in relatively cholesterol-poor lateral domains (rev. in Schroeder et al., 1991; Schroeder et al., 1996). Consequently, the size of the exchangeable or sterol-poor lateral domain may determine protein concentration in that domain, which may in turn determine the association or aggregation state of protein subunits to form functional receptor-effector coupled systems, ion transporters, and so on.

4. DNA Replication

In addition to segregating DNA from the rest of the cell, nuclear membrane lipids function in DNA replication. Acidic phospholipids inhibit the replication activity of DNA protein, trigger DNA replication in vivo, serve as storage sites for inactive DNA topoisomerase I, bind histones, and influence activity of DNA polymerases (rev. in Sekimizu, 1994).

5. Epithelial Cell Macrodomain Function

Epithelial cells from a variety of tissues (liver, kidney, intestine) have plasma membranes that differ markedly depending on their orientation (basolateral, serosal, microvillar). These different regions of the cell membrane have lipids and other components that are specifically segregated. For example, 5'-nucleotidase, leucine naphthylamidase, and alkaline phosphatase are localized predominantly in the bile canalicular region. Na^+,K^+-ATPase is located primarily in the contiguous region while glucagon-activated adenylate cyclase, glutamyl transpeptidase, and CMP-neuraminic acid hydrolase activities are primarily sinusoidal (rev. in Sweet and Schroeder, 1988).

VIII. Intracellular Membranes

A. General Properties

Intracellular membranes generally contain much less cholesterol, sphingomyelin, and glycolipid than the cell surface plasma membrane. The cholesterol/phospholipid ratios in nuclear, microsomal, lysosomal, and mitochondrial membranes are five- to ten-fold, four- to five-fold, one- to two-fold, and four- to thirty-fold, respectively, lower than in plasma membranes.

B. Organelle Lipid Domains

In general, the cytofacial-facing leaflet of those few intracellular membranes that have been examined appears to be enriched with acidic phospholipids (rev. in Buton et al., 1996). However, the mitochondrial outer membrane has a symmetric transbilayer distribution of phosphatidylcholine (Dolis et al., 1996). Transbilayer motion of glycerolipids across intracellular membranes is protein mediated and extremely fast, with a half-time near 25 s and 2 min for endoplasmic reticulum and mitochondrial outer membrane, respectively. With the exception of cholesterol domains, almost nothing is known regarding subcellular organelle membrane lateral lipid domains (rev. in Schroeder et al., 1996; Schroeder et al., 1998; Schoer et al., 2000; Gallegos et al., 2000). Similar to plasma membranes from a variety of tissues, fibroblast microsomal membranes have a large nonexchangeable sterol domain comprising 59% of total sterol. In contrast, mitochondria and lysosomes have a much larger nonexchangeable lateral sterol domain, 84% and essentially 100%

of total sterol, respectively. At least two and possibly three cholesterol pools exist in adrenal mitochondria.

C. Protein Functions

To date very few protein functions have been correlated with intracellular membrane lipid domains. One such process is mitochondrial steroidogenesis (rev. in Schroeder *et al.*, 1996). In mitochondrial steroidogenesis, the rate-limiting step is the transfer of cholesterol from the outer to the inner mitochondrial membrane. How this transfer occurs is not known, but possible mechanism(s) include: (1) A protein such as the steroidogenic acute regulatory protein (rev. in Schroeder *et al.*, 1998) that may alter the cholesterol transbilayer migration rate and/or equilibrium transbilayer distribution in the outer mitochondrial membrane. (2) A protein such as SCP-2 that stimulates cholesterol transfer from extramitochondrial sources to mitochondria and/or from outer to inner mitochondrial membranes. (3) A protein that allows cholesterol to migrate laterally from the outer to the inner mitochondrial membrane via contact sites between the two membranes.

IX. Membrane Biogenesis

In the past, the topic of membrane biogenesis used to engender thoughts on the origin of life: the primordial membrane enclosing the first protein/DNA, the entrapment of a primordial bacterium to become the progenitor of mitochondria in eukaryotes. In the past two decades attention has moved away from such global questions to a more pragmatic elucidation of how living cells make membranes. At least two processes are recognized: vesicular-mediated pathways and protein-mediated pathways.

A. Vesicular Membrane Lipid and Protein Trafficking

Vesicular pathways are important to receptor-mediated endocytosis, phagocytosis, secretions, protein sorting, protein insertion, and membrane partitioning during cell division (rev. in Morris and Frizzell, 1994; Warren, 1993; Pryer *et al.*, 1992; Schatz and Dobberstein, 1996; Schekman and Orci, 1996; Bogdanov *et al.*, 1996). Phospholipids (e.g., phosphoinositides) may be important regulators of vesicular membrane trafficking in the cell. The role of transbilayer or lateral lipid membrane domains in vesicular transport is at present unknown.

B. Protein-Mediated Membrane Lipid Trafficking

In contrast to the paucity of information on the involvement of lipid domains in vesicular trafficking, recent data on intracellular membrane cholesterol domains show that certain cytosolic proteins are capable of interacting with intracellular membrane lipid domains to elicit directional lipid trafficking, in certain instances against the cholesterol concentration gradient (rev. in Schroeder *et al.*, 1996; Schroeder *et al.*, 1998; Schoer *et al.*, 2000; Gallegos *et al.*, 2000).

1. *Spontaneous intermembrane sterol transfer between organelles*. The initial rate of spontaneous sterol transfer between plasma membranes, microsomes, and mitochondria varies over a 60-fold range, depending on the specific donor/acceptor pair. In addition, the initial rate of spontaneous sterol transfer varied up to 4-fold to 13-fold depending on the direction of sterol transfer examined.

2. *Role of sterol domains in sterol transfer between organelles*. Different acceptor organelles significantly affect the size of the exchangeable sterol domain in plasma membrane, microsomal, mitochondrial, and lysosomal donor membranes. For example, substitution of microsomal acceptor membranes by plasma membrane acceptor membranes decreased the size of the exchangeable sterol domain in donor microsomes by half. Thus, the nature of the acceptor membrane contributes to both the kinetics of spontaneous sterol trafficking and also the domain size in the donor membrane.

3. *Protein-mediated intermembrane sterol transfer between organelles*. SCP-2 (also called nonspecific lipid transfer protein), the liver sterol carrier protein (also called fatty acid–binding protein, L-FABP, or sterol and squalene carrier protein), brain fatty acid–binding protein (B-FABP), heart fatty acid–binding protein (H-FABP), and caveolin all stimulate intermembrane sterol transfer, but by different mechanisms (rev. in Schroeder *et al.*, 1996; Schroeder *et al.*, 1998). SCP-2 stimulated sterol transfer between all intracellular organelles, while L-FABP was more restricted to certain membranes. L-FABP but not SCP-2 decreases the size of the nonexchangeable domain. In contrast, neither brain fatty acid–binding protein (B-FABP) nor heart fatty acid–binding protein (H-FABP) enhance the initial rate of sterol transfer, but instead increase the size of the exchangeable sterol domain. Transfer of cholesterol mediated by SCP-2 and L-FABP required the interaction of the protein with the membrane and the presence of a sterol binding site in the protein. In the case of B-FABP and H-FABP, a sterol binding site was not evident. SCP-2 stimulated the initial rate of sterol transfer from microsomes to plasma membranes 5-fold *in vitro* and in intact cells, suggesting that SCP-2 may participate in reverse cholesterol transport. SCP-2 stimulated the initial rate of sterol transfer from plasma membrane to microsomes 26-fold, while in the reverse direction it stimulated the initial rate of sterol transfer from microsomes to plasma membranes only five-fold. Similarly, SCP-2 stimulated the initial rate of sterol transfer 12-fold from plasma membranes to mitochondria, but only 4-fold in the reverse direction from mitochondria to plasma membranes. SCP-2's effect on interorganelle sterol transport was vectorial and, depending on the donor/acceptor pair, it was also against the cholesterol gradient. Although caveolin is primarily a plasma membrane protein, it is also associated with the trans-Golgi, and a soluble complex of caveolin with cholesterol and several chaperone proteins has

also been observed. Thus, caveolin may play a central role in mediating cholesterol traffic from Golgi and ER to the plasma membrane caveolae for subsequent efflux to SRB1-bound HDL (reverse cholesterol transport). Alternately, caveolin may modulate influx of cholesterol into the cell by a pathway independent of the LDL receptor.

X. Summary

The landmark fluid lipid bilayer hypothesis of Singer and Nicholson became the model (more than two decades ago) for understanding lipid structure and the way proteins might interact with lipids. However, protein biochemists, biophysicists, and physiologists quickly found that this simplistic model did not consistently fit with their experimental data, especially on protein function. Unfortunately, rather than modify the model, some investigators concluded that lipids might not be that important to the structure and function of integral membrane proteins. Like a phoenix rising from the ashes, however, the discovery that the lipid bilayer comprised heterogeneously distributed lipids opened new areas of investigation (Bretscher, 1972). In fact, the growing awareness of lipid domains such as clathrin-coated pits and caveolae became a beacon for membranologists willing to embrace a modification of the basic bilayer theory to include the existence of an asymmetric distribution of lipids, or lipid domains. Although space limitations do not allow recognition of many of the original contributions in this field, the reader is referred to in-depth citations in the many reviews listed in the references.

In the coming millennium, membranologists find themselves with the task of attempting to understand the origin, regulation, and function of lipid domains in cells. Such lipid domains comprise macroscopic areas of epithelial cells, microscopic domains in essentially all membranes, and interfacial/boundary lipid in the protein-lipid interface. The challenge for structural biologists will be to find specific probes to explore not only how these domains regulate the tertiary structure and function of proteins, but also how they might regulate lipid trafficking within the cell and lipid efflux/influx from/to the cell. Biochemists will have to elucidate the pathways whereby lipid domains are regulated and how they interact with other membrane functions such as reverse cholesterol uptake, signaling, and endocytosis. The exciting progress made on plasma membrane lipid and functional domains (e.g., clathrin and caveolin) needs to be extended to characterize additional integral proteins that may regulate membrane lipid domains in plasma membrane as well as endocytic vesicles and intracellular organelles. Since lipids are synthesized in specialized subcellular areas of the cell, a major focus must be how lipids are sorted and targeted towards specific membranes and membrane domains. Physiologists will have to modify their concepts of how channels, transport proteins, and receptors function in, and may be regulated by, specific lipid domains in the membrane. The molecular biologist must begin identifying specific genetic sequences coding for the proteins that regulate lipid domains in membranes and for proteins that direct lipid trafficking from intracellular sites of synthesis. Contributions from all of these fields are necessary if our understanding of membrane protein function (transport, signaling, etc.) and lipid metabolism in normal physiology and disease states is to advance.

Bibliography

Acton, S., Rigotti, A., Landschulz, K. T., Xu, S., Hobbs, H. H., and Krieger, M. (1996). Identification of scavenger receptor SRBI as a high density lipoprotein receptor. *Science* **271,** 518–520.

Almeida, P. F., Vaz, W. L., and Thompson, T. E. (1992). Lateral diffusion and percolation in two-phase, two-component lipid bilayers. Topology of the solid-phase domains in-plane and across the lipid bilayer. *Biochemistry* **31,** 7198–7210.

Atshaves, B. P., Starodub, O., McIntosh, A. L., Petrescu, A, Roths, J. B., Kier, A. B., and Schroeder, F. (2000) Sterol carrier protein-2 alters HDL-mediated cholesterol efflux. *J. Biol. Chem.* **275,** 86 852–86 861.

Bergelson, L. D. (1992). Lipid domain reorganization and receptor events. Results obtained with new fluorescent lipid probes [letter]. [Review]. *FEBS Lett.* **297,** 212–215.

Bogdanov, M., Sun, J., Kaback, H. R., and Dowhan, W. (1996). A phospholipid acts as a chaperone in assembly of a membrane transport protein. *J. Biol.Chem.* **271,** 11 615–11 618.

Bohuslav, J., Cinek, T., and Harejsi, V. (1993). Large, detergent resistant complexes containing murine antigens Thy-1 and Ly-6 and protein tyrosine kinase p56[lck]. *Eur. J. Immunol.* **23,** 825–831.

Bretscher, M. S. (1972). Asymmetrical lipid bilayer structure for biological membranes. *Nature New Biology* **236,** 11–12.

Bretscher, M. S., and Munro, S. (1993). Cholesterol and the Golgi apparatus. *Science* **261,** 1280–1281.

Buton, X., Morrot, G., Fellman, P., and Seigreuret, M. (1996). Ultrafast glycerophospholipid selective transbilayer motion mediated by a protein in the endoplasmic reticulum membrane. *J. Biol. Chem.* **271,** 6651–6657.

Chen, S., Kulju, D., Halt, S., and Murakami, K. (1992). Phosphatidylcholine-dependent protein kinase C activation. Effects of *cis*-fatty acid and diacylglycerol on synergism, autophosphorylation and Ca(2+)-dependency. *Biochem. J.* **284,** 221–226.

Devaux, P. F., and Zachowski, A. (1994). Maintenance and consequences of membrane phospholipid asymmetry. *Chem. Phys. Lipids* **73,** 107–120.

Dolis, D., de Kroon, A. I. P. M., and de Kruijff, B. (1996). Transmembrane movement of phosphatidylcholine in mitochondrial outer membrane vesicles. *J. Biol. Chem.* **271,** 11 879–11 883.

Dudeja, P. K., Harig, J. M., Wali, R. K., Knaup, S. M., Ramaswamy, K., and Brasitus, T. A. (1991). Differential modulation of human small intestinal brush-border membrane hemileaflet fluidity affects leucine aminopeptidase activity and transport of D-glucose and L-glutamate. *Arch. Biochem. Biophys.* **284,** 338–345.

Epand, R. M., Stafford, A., Wang, J., and Epand, R. F. (1992). Zwitterionic amphiphiles that raise the bilayer to hexagonal phase transition temperature inhibit protein kinase C. The exception that proves the rule. *FEBS Lett.* **304,** 245–248.

Fielding, C. J., Bist, A., and Fielding, P. E. (1998). Selective uptake of lipoprotein free cholesterol and its intracellular transport—role of caveolin. *In* "Intracellular Cholesterol Trafficking" pp. 273–288. (T. Y. Chang and D. A. Freeman, Eds.) Boston, Kluwer Academic Publishers.

Fielding, C. J., and Fielding, P. E. (1997). Intracellular cholesterol transport. *J. Lipid Res.* **38**, 1503–1521.

Fontaine, R. N., Harris, R. A., and Schroeder, F. (1980). Aminophospholipid asymmetry in murine synaptosomal plasma membrane. *J. Neurochem.* **34**, 269–277.

Fra, A. M. W .E., Simons, K., and Parton, R. G. (1994). Detergent insoluble glycolipid microdomains in lymphocytes in the absence of caveolae. *J. Biol. Chem* **269**, 30 745–30 748.

Frolov, A., Petrescu, A., Atshaves, B. P., So, P. T. C., Gratton, E., Serrero, G., and Schroeder, F. (2000). High density lipoprotein-mediated cholesterol uptake and targeting to lipid droplets in intact L-cell fibroblasts: a single- and multiphoton fluorescence approach. *J. Biol. Chem.* **275**, 12 769–12 780.

Gallegos, A.M., Schoer, J. K., Starodub, O, Kier, A. B., Billheimer, J. T., and Schroeder, F. (2000). A potential role for sterol carrier protein-2 in cholesterol transfer to mitochondria. *Chem. Phys. Lip.* **105**, 9–29.

Gimpl, G., Burger, K., and Fahrenholz, F. (1997). Cholesterol as modulator of receptor function. *Biochemistry* **36**, 10 959–10 974.

Harris, J. S., Epps, D. E., Davio, S. R., and Kezdy, F. J. (1995). Evidence for transbilayer, tail-to-tail cholesterol dimers in dipalmitoylglycerophosphocholine liposomes. *Biochemistry* **34**, 3851–3857.

Houslay, M. D., and Stanley, K. K. (1982). "Dynamics of Biological Membranes." John Wiley and Sons, New York.

Huang, H., Schoer, J., Ball, J. M., Billheimer, J. T., and Schroeder, F. (1999). The sterol carrier protein-2 amino terminus: role in membrane interaction and sterol transfer. Submitted.

Ishibashi, T., and Bloch, K. (1981). Intermembrane transfer of 5 alpha-cholest-7-en-3 beta-ol. Facilitation by supernatant protein (SCP). *J. Biol. Chem.* **256**, 12 962–12 967.

Luan, P., Yang, L., and Glaser, M. (1995). Formation of membrane domains created during the budding of vesicular stomatitis virus. A model for selective lipid and protein sorting in biological membranes. *Biochemistry* **34**, 9874–9883.

Morris, A. P., and Frizzell, R. A. (1994). Vesicle targeting and ion secretion in epithelial cells: implications for cystic fibrosis. [Review]. *Annu. Rev. Physiol.* **56**, 371–397.

Morrow, M. R., Singh, D., Lu, D., and Grant, C. S. (1992). Glycosphingolipid phase behaviour in unsaturated phosphatidylcholine bilayers: a 2H-NMR study. *Biochim. Biophys. Acta* **1106**, 85–93.

Mukherjee, S., and Chattopadhyay, A. (1996). Membrane organization at low cholesterol concentrations: a study using NBD-labeled cholesterol. *Biochemistry* **35**, 1311–1322.

Negrete, H. O., Rivers, R. L., Gough, A. H., Colombini, M., and Zeidel, M. L. (1996). Individual leaflets of a membrane bilayer can independently regulate permeability. *J. Biol. Chem.* **271**, 11 627–11 630.

Nichols, C. P., and Roufogalis, B. D. (1991). Influence of associated lipid on the properties of purified bovine erythrocyte acetylcholinesterase. *Biochem. Cell. Biol.* **69**, 154–162.

Oberhauser, A. F., Monck, J. R., and Fernandez, J. M. (1992). Events leading to the opening and closing of the exocytotic fusion pore have markedly different temperature dependencies. Kinetic analysis of single fusion events in patch-clamped mouse mast cells. *Biophys. J.* **61**, 800–809.

Parton, R. G., and Simons, H. (1995). Digging into caveolae [comment]. *Science* **269**, 1398–1399.

Pearse, B. M. F. (1976). Clathrin: a unique protein associated with intracellular transfer of membrane by coated vesicles. *Proc. Natl. Acad. Sci. USA* **73**, 1255–1259.

Pryer, N., Wuestehube, L., and Schekman, R. (1992). Vesicle-mediated protein sorting. [Review]. *Ann. Rev. Biochem.* **61**, 471–516.

Schatz, G., and Dobberstein, B. (1996). Common principles of protein translocation across membranes. *Science* **271**, 1519–1526.

Schekman, R., and Orci, L. (1996). Coat proteins and vesicle budding. *Science* **27**, 1526–1531.

Schoer, J.K., Gallegos, A. M., McIntosh, A. L., Starodub, O., Kier, A. B., Billheimer, J. T., and Schroeder, F. (2000). Lysosomal membrane cholesterol dynamics. *Biochemistry* **39**, 7662–7677.

Schroeder, F. (1984). Role of membrane lipid asymmetry in aging. *Neurobiol. Aging* **5**, 323–333.

Schroeder, F., Frolov, A., Schoer, J., Gallegos, A., Atshaves, B. P., Stolowich, N. J., Scott, A. I., and Kier, A. B. (1998). Intracellular cholesterol binding proteins, cholesterol transport and membrane domains. *In* "Intracellular Cholesterol Trafficking" (D. Freeman and T. Y. Chang, Eds.) Kluwer Academic Publishers, Boston, pp. 213–234.

Schroeder, F., Frolov, A. A., Murphy, E. J., Atshaves, B. P., Jefferson, J. R., Pu, L., Wood, W. G., Foxworth, W. B., and Kier, A. B. (1996). Recent advances in membrane cholesterol domain dynamics and intracellular cholesterol trafficking. *Proc. Soc. Exp. Biol. Med.* **213**, 150–177.

Schroeder, F., Jefferson, J. R., Kier, A. B., Knittell, J., Scallen, T. J., Wood, W. G., and Hapala, I. (1991). Membrane cholesterol dynamics: cholesterol domains and kinetic pools. *Proc. Soc. Exp. Biol. Med.* **196**, 235–252.

Schroeder, F. and Nemecz, G. (1990). Transmembrane cholesterol distribution. *In* "Advances in Cholesterol Research" (M. Esfahami and J. Swaney, Eds.) Telford Press, Caldwell, NJ, pp. 47–87.

Schroeder, F. and Sweet, W. D. (1988). The role of membrane lipid and structure asymmetry on transport systems. *In* "Advances in Biotechnology of Membrane Ion Transport" (P. L. Jorgensen and R. Verna, Eds.) Serono Symposia, New York, pp. 183–195.

Schroeder, F. and Wood, W. G. (1995). Lateral lipid domains and membrane function. *In* "Cell Physiology Source Book" (N. Sperelakis, Ed.) Academic Press, New York, pp. 36–44.

Schulthess, G., and Hauser, H. (1995). A unique feature of lipid dynamics in small intestinal brush border membrane. *Mol. Memb. Biol.* **12**, 105–112.

Sekimizu, K. (1994). Interactions between DNA replication–related proteins and phospholipid vesicles *in vitro*. [Review]. *Chem. Phys. Lipids* **73**, 223–230.

Severs, N. J., and Robenek, H. (1983). Detection of microdomains in biomembranes: an appraisal of recent developments in freeze fracture cytochemistry. *Biochim. Biophys. Acta* **737**, 373–408.

Singer, S. J. and Nicholson, G. L. (1972). The fluid mosaic model of the structure of cell membranes. *Science* **175**, 720–731.

Skerrow, C. J., and Matoltsy, A. G. (1974). Chemical characterization of isolated epidermal desmosomes. *J. Cell Biol.* **63**, 524–530.

Smaby, J. M., Momsen, M., Kulkarni, V. S., and Brown, R. E. (1996). Cholesterol-induced interfacial area condensations of galactosylceramides and sphingomyelins with identical acyl chains. *Biochemistry* **35**, 5695–5704.

Smart, E. J., and van der Westhuyzen, D. R. (1998). Scavenger receptors, caveolae, caveolin, and cholesterol trafficking. *In* "Intracellular Cholesterol Trafficking" (T. Y. Chang, and D. A. Freeman, Eds.) Kluwer Scientific Publishers, Boston, pp. 253–272.

Steer, C. J., Bisher, M., Blumenthal, R., and Steven, A. C. (1984). Detection of membrane cholesterol by filipin in isolated rat liver coated vesicles is dependent upon removal of the clathrin coat. *J. Cell Biol.* **99**, 315–319.

Sunshine, C., and McNamee, M. (1992). Lipid modulation of nicotinic acetylcholine receptor function: the role of neutral and negatively charged lipids. *Biochim. Biophys. Acta* **1108**, 240–246.

Sweet, W .D., and Schroeder, F. (1988). Lipid domains and enzyme activity. *In* "Advances in Membrane Fluidity: Lipid Domains and the Relationship to Membrane Function" (R. C. Aloia, C. C. Cirtain, and L. M. Gordon, Eds.) Alan R. Liss, Inc., New York, pp. 17–42.

Tulenko, T. N., Chen, M., Mason, P. E., and Mason, R. P. (1998). Physical effects of cholesterol on arterial smooth muscle membranes: evidence of immiscible cholesterol domains and alterations in bilayer width during atherogenesis. *J. Lip. Res.* **39**, 947–956.

van den Bosch, H., Schutgens, R. B .H., Wanders, R. J. A., and Tager, J. M. (1992). Biochemistry of peroxisomes. *Annu. Rev. Biochem.* **61,** 157–197.

Warren, G. (1993). Membrane partitioning during cell division. [Review]. *Annu. Rev. Biochem.* **62,** 323–348.

Williamson, P., Bevers, E. M., Smeets, E. F., Comfurius, P., Schlegel, R. A., and Zwaal, R. F. (1995). Continuous analysis of the mechanism of activated transbilayer lipid movement in platelets. *Biochemistry* **34,** 10 448–10 455.

Wood, W. G., and Schroeder, F. (1992). Membrane exofacial and cytofacial leaflets: a new approach to understanding how ethanol alters brain membranes. *In* "Alcohol and Neurobiology: Receptors, Membranes, and Channels" (R. R. Watson, Ed.) CRC Press, Boca Raton, FL, pp. 161–184.

Wood, W. G., Schroeder, F., Avdulov, N. A., Chochina, S. V., and Igbavboa, U. (1999). Recent advances in brain cholesterol dynamics: transport, domains, and Alzheimer's disease. *Lipids* **34,** 225–234.

Yeagle, P. L. (1987). *The Membranes of Cells.* Academic Press, New York.

Michael S. Forbes and Donald G. Ferguson

6

Ultrastructure of Cells

I. Introduction: The Plasma Membrane as the Basis of Cellularity

To understand the fundamental mechanisms that constitute the normal activities of any organism, it is essential to consider the structural correlates that underlie these activities. Among the earliest observations that form the foundation of our current understanding of morphology was that of Robert Hooke, who in the 1660s examined thin slices of cork with a primitive light microscope and realized that the cork was composed of small compartments, which he termed "cells." These original observations were enlarged upon and refined over the next two centuries through the work of a great number of light microscopists. By the late 1830s, it had been independently postulated—by Schleiden in plants and by Schwann in animals—that all living organisms are composed of cells. This postulation has come to be recognized as the **cell theory**. Yet the cellular nature of two major tissues—nerve and muscle—was still subject to challenge around the beginning of the 20th century. Perhaps driven in part by this question, of cells versus syncytial conglomerates, a series of substantial advances in specimen preparation and staining techniques occurred during this same period of time, as well as improvements in the capabilities of microscopes themselves. This surge of research and development eventually won the day for the proponents of the cell theory. That being resolved, another important question arose which revolved around the nature of the barriers that separate cells from one another and from the surrounding environment. A major breakthrough in resolving this question was made independently by Gorter and Grendel in the 1920s and Davson and Danielli in the 1930s. The latter authors examined the behavior of phospholipids in aqueous environments and determined that the most favorable energetic configuration was a bilayer having the hydrophilic polar headgroups located at the water interface and the hydrophobic nonpolar tails apposed to each other. Such a cell membrane ("plasmalemma"), formed by a phospholipid bilayer, would be impermeable to fluids and charged species

such as ions. Thus it would isolate the cell contents from the extracellular environment.

Proteins were recognized as being closely associated with the phospholipid bilayer, and early models proposed that they form a layer on either side of the membrane (i.e., a protein-lipid sandwich). This model remained hypothetical until the introduction and refinement of the electron microscope (EM) in the 1930s and 1940s, for which Ernst Ruska was awarded the Nobel Prize. The EM, which produces an image with a beam of electrons rather than electromagnetic radiation in the visible spectrum, permitted the examination of tissues at much greater magnifications, and with far better resolution than had been possible with even the best light microscopes. In fact, resolution was so much improved that the morphological details observed with the EM came to be referred to as the **ultrastructure** (literally "beyond structure") of the cells. In the EM, it was apparent that all cells were bounded by a trilaminar membrane, 70 Å (= 7 nm) thick, that to some observers resembled a "railroad track" consisting of two darker lines separated by a relatively clear inner region (see Fig. 1A). This structure was termed the **unit membrane** by J. D. Robertson in the early 1950s. It had already been recognized by Davson and Danielli that many of the proteins associated with the cell membrane were tightly attached, since they were not extracted by changing the ionic strength of the bathing solutions or even by mild detergent treatment. Thus, the notion of a phospholipid bilayer with a coating of protein had evolved, by the 1960s, into the fluid mosaic model proposed by Singer and Nicholson (Fig. 2). In this model, there were both peripheral (or extrinsic) and integral (or intrinsic) membrane proteins. **Peripheral proteins** are associated with the inner and outer surfaces of the membrane, but are not anchored in it. **Integral membrane proteins** are actually embedded in at least one leaflet of the phospholipid bilayer (Fig. 1B, C; Fig. 2), and some traverse the entire thickness of the membrane.

The fluid mosaic model was more consistent with functional studies, which clearly demonstrated that cell membranes were not totally impermeable, but rather were

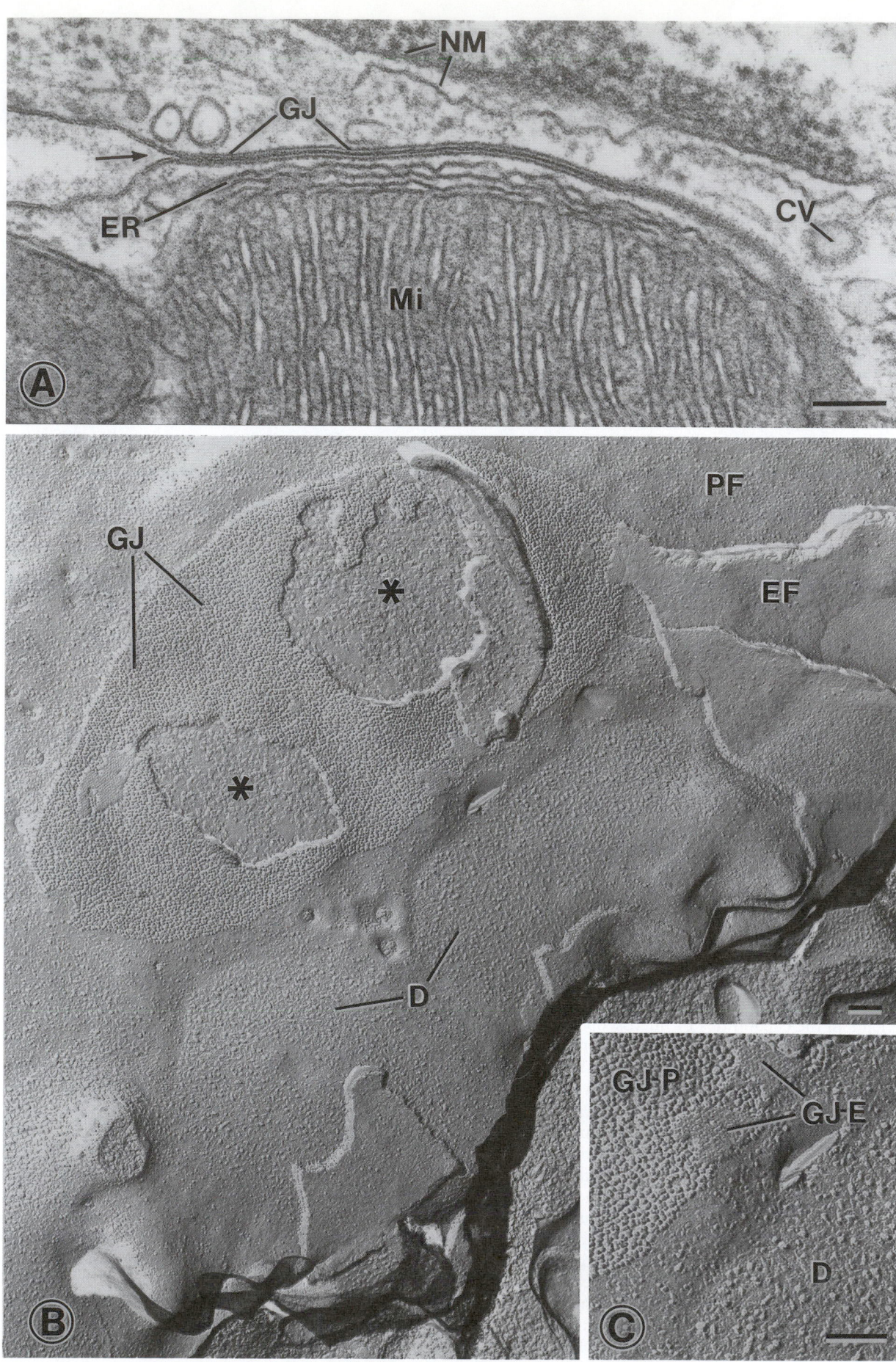

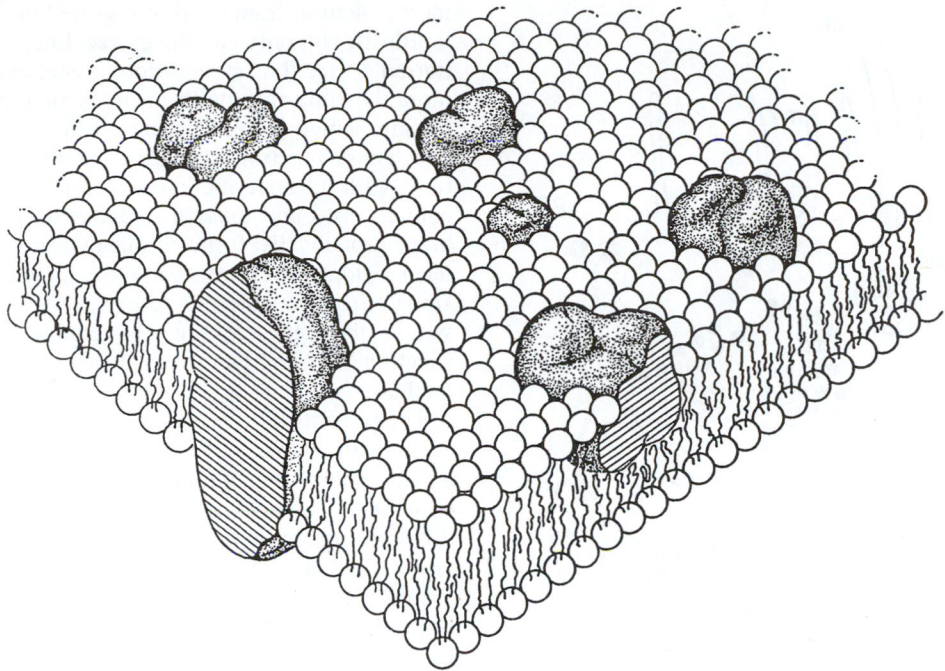

FIGURE 2. Fluid mosaic model of cell membrane. This diagram illustrates the organization of the phospholipid bilayer with the polar headgroups (depicted as spheroids) facing the fluid compartments of the cytoplasm and the extracellular space. The nonpolar tails of the phospholipids form the inner part of the bilayer. Integral membrane proteins are embedded in the bilayer, and some span the entire thickness of the membrane. (Reproduced with permission from S. J. Singer and G. L. Nicholson (1972). The fluid mosaic model of the structure of cell membranes. *Science* **175,** 720–731. Copyright 1972 American Association for the Advancement of Science.)

semipermeable. The integral membrane proteins included ion channels that acted as low-resistance pores, as well as energy-dependent pumps. This model better accounted for the establishment of ionic gradients and the influx or efflux of ions under various conditions.

As mentioned, the extrinsic proteins are **membrane-associated** rather than actually embedded in the lipid bilayer. The **extracellular matrix** exists on the outer cell surface as a layer of varying thickness, depending on the cell type. Historically, this coating was considered to be an

FIGURE 1. Unit membranes of cells. (A) Cell border formed by two cardiac muscle cells. Unit membranes are the basis of most of the structures shown here, including the double membrane of the nucleus (NM), the outer envelope and inner membrane shelves (cristae) of a mitochondrion (Mi), and several vesicular bodies, including a coated vesicle at the right of the picture (CV). The membrane-based, multilayered structure that dominates this view is a specialized adhesion known as a **gap junction** (GJ), formed between the two cells where the outer faces of the two cells' plasmalemmata come into close apposition. At the *arrow*, the relatively unspecialized cell surface membranes are clearly continuous with the membranes that form the gap junction. Note the membrane-based tubule of endoplasmic reticulum (ER) sandwiched between the mitochondrion and the gap junction. (B) and (C) show cardiac cell membranes as revealed by the technique of freeze-fracture, in which tissue is quickly frozen and broken, and a thin (ca. 2–3 nm) coating of metal is evaporated onto the specimen and stabilized with a thicker carbon film (similar techniques were used to produce the replicas depicted in Figs. 5D, 6D, 13B, and 13C). The tissue is then digested away, leaving only the metal-carbon replica of the fractured surfaces. The fracturing process specifically cleaves unit membranes to expose their inner "faces," thus showing the intramembranous particles (IMPs) that correspond to a variety of functionally important proteins. This also demonstrates the characteristic asymmetry inherent within a unit membrane; typically there is a greater density of larger particles on the **P face**, the face adjacent to the protoplasm (cytoplasm), whereas the **E face** (that closest to the "ectoplasm," i.e., the extracellular space) is characteristically smoother. Compare, for example, the regions labeled PF and EF at the upper right corner of B. This region shows the relatively unspecialized plasmalemmata of two apposed cells (cf. the cell surface membranes in A). The large oval body is a gap junction between the two muscle cells, broken away in two regions (*) to expose underlying intracellular structures, probably mitochondria (a similar association is shown in the thin section in A). The detail shown in panel C shows that though this gap junction replica consists mainly of large P-face particles (GJ-P), in some areas small regions of the junction's E face (labeled GJ-E) have adhered, which are characterized by a finer topography of pitlike depressions, into which the P-face particles fit in the intact gap junction. In B, two smaller ovoid membrane regions (D) appear which correspond to another type of intercellular junction, the **desmosome** (also see Fig. 13). The P-face particles belonging to desmosomes (also shown at higher magnification in C) are larger and more coarse-appearing than those of either the gap junction or the unspecialized plasmalemma. Scale bars in all micrographs = 0.1 μm.

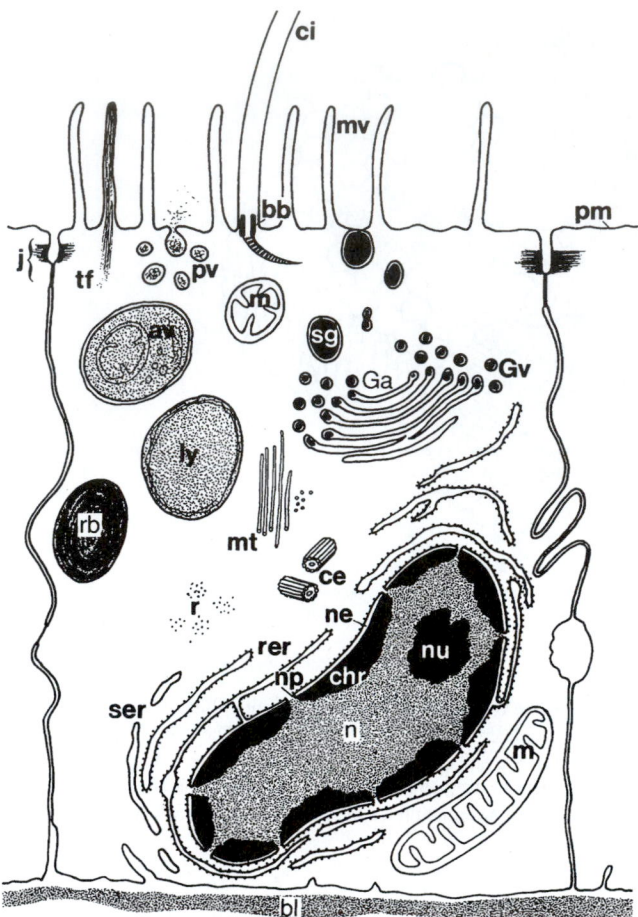

FIGURE 3. Diagram of the ultrastructure of a typical cell, in this case an epithelial cell. This picture provides a two-dimensional representation of the major organelles, based on their appearance in electron micrographs. The cell's contents, the cytoplasm and organelles, are enclosed by the plasma membrane (pm). The central nucleus (n) contains a nucleolus (nu) and clumped (hetero)chromatin (chr). The nucleus is bounded by a double bilayer membrane that is punctuated by discontinuities called nuclear pores (np). The outer nuclear membrane is studded with small dots, representing membrane-attached ribosomes, and is continuous with tubules of the rough endoplasmic reticulum (rer). Free ribosomes (r) also appear in the cytoplasm. Elements of smooth endoplasmic reticulum (ser) are observed as membranous profiles without attached ribosomes. An example of the Golgi apparatus (Ga) with its associated vesicles (Gv) and secretory granules (sg) is located just above the nucleus. Several types of lysosomes are also depicted: a primary lysosome (ly), an autophagic vacuole (av), and a residual body (rb). Several pinocytotic vesicles (pv) are shown at the apical surface of the cell. Two mitochondria (m) are drawn, composed of two layers of membrane. Cytoskeletal elements include actin-containing thin filaments (tf) that form the core of the apical membrane specializations called microvilli (mv). Microtubules (mt) are present both as cytoskeletal elements and as specialized microtubule organizing structures: the centriole (ce) and the basal body (bb) or a single cilium (ci) at the apical surface of the cell. A complex of intercellular junctions (j) that restrict movement of ions and solutes between adjacent cells is depicted at the lateral border of the cell near the apical surface. The basal lamina (bl) is a layer of fibrous material that underlies the cell. It is part of the extracellular matrix that regulates ion flow and provides strong connections to the deeper layers of connective tissue. Electron micrographs of these organelles are provided in subsequent figures. (Adapted with permission from A. Bubel. (1989). "Microstructure and Function of Cells," Fig. 1a. Ellis Horwood Ltd., Chichester, UK.)

inert structural framework composed of collagen and elastic fibrils. Not only can the extracellular matrix act as a filtration device, but, in addition, recent evidence points to a more dynamic role, with proteins such as fibronectin and laminin active in stimulating a number of cellular activities. These functions are likely mediated through interactions with intrinsic membrane proteins, which in turn are connected to and exert influences on the extrinsic membrane proteins that form subplasmalemmal networks on the inner (cytoplasmic) surface of the cell membrane. Ultimately, signals from outside the cell may be communicated deep within the cell via the elements of the cytoskeleton (Section VII) and can even influence gene activity in the nucleus.

The contents of the cell enclosed by the plasmalemma are collectively known as the **cytoplasm**. The cytoplasm is not an amorphous soup, but in fact contains a number of distinct structural elements known as **organelles** (by analogy to the different organs of the body). Figure 3 is a diagrammatic representation, and Fig. 4 a complementary electron micrograph, of a characteristic animal cell and its organelles. The organelles serve to compartmentalize different activities of the cell, just as the organs perform different functions within the body.

II. Nucleus

The most prominent organelle within the cell is readily visible in the light microscope, and since it is most often located at the cell center, it came to be called the **nucleus**. The nucleus contains most of the cell's DNA and its genetic apparatus as well. In most cells, the nucleus is an ovoid body, but it can assume a variety of shapes and sizes (Fig. 5A). Multiple nuclei are characteristic of certain cell types (e.g., skeletal myocytes). Early EM observations discovered the nucleus to be bounded by a double bilayer, the **nuclear envelope**, and noted as well a dense intranuclear body, the **nucleolus**. The nucleus was characterized in addition by masses of darker material, called **heterochromatin**, found dispersed throughout the nucleoplasm and also associated with the inner surface of the nuclear envelope. Beyond these initial descriptions, the regional specializations of the nucleoplasm remain rather poorly understood. Recent studies with antibody localization of nuclear-specific proteins demonstrate the presence of domains within the nucleoplasm; these are not readily apparent even with ultrastructural inspection, but appear to be organized through attachment to the filamentous proteins of the nuclear matrix. On the basis of compartmentation of DNA polymerases and RNA splicing factors, it has been suggested that discrete domains are involved in different processes, such as transcription and translation of the genetic material.

The nuclear envelope has a proteinaceous lamina at its inner surface, which is composed of cytoskeletal proteins called **lamins**; the major portion of the envelope is the prominent double membrane itself (Fig. 1A), which is interrupted at periodic intervals by circular discontinuities, the **nuclear pore complexes** (Fig. 5B–D). The nuclear lam-

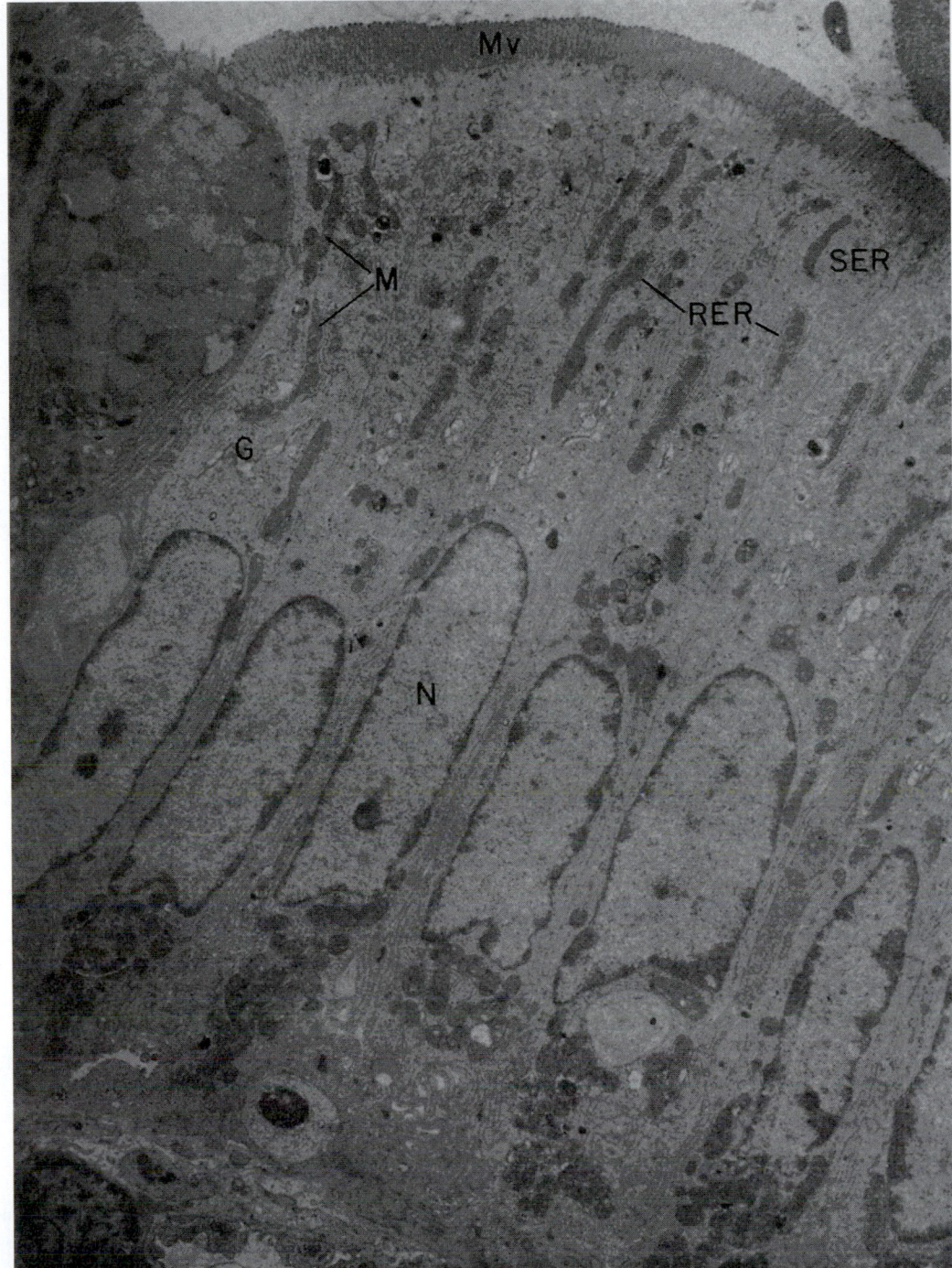

FIGURE 4. Electron micrograph of epithelial cells lining the lumen of the intestine. This low-magnification view illustrates the appearance of epithelial cells such as the one diagrammatically portrayed in Fig. 3. Each nucleus (N) is located toward the basal end of the cell. Also labeled are profiles of the rough endoplasmic reticulum (RER), smooth endoplasmic reticulum (SER), mitochondria (M) and microvilli (MV), these last being the characteristic apical membrane specializations of such cells. (Reproduced from R. R. Cardell, Jr., et al. (1957) *The Journal of Cell Biology* 34, 123–155, by copyright permission of The Rockefeller University Press.)

ina is not merely a scaffolding that supports the inner membrane surface. Rather, it acts as an attachment site for chromatin, and appears to be responsible for regulating the formation of the various nuclear domains. Detailed studies have demonstrated that the nuclear pore complexes are complicated structures indeed, incorporating an eight-subunit, ca. 133-nm-diameter annulus about a central aqueous channel, ca. 42 nm in diameter, which allows passive diffusion of small molecules and ions. The complex is composed in the main of highly conserved proteins known as **nucleoporins**, and its structure consists of a central spoke assembly with several arms and two peripheral rings, one on the cytoplasmic

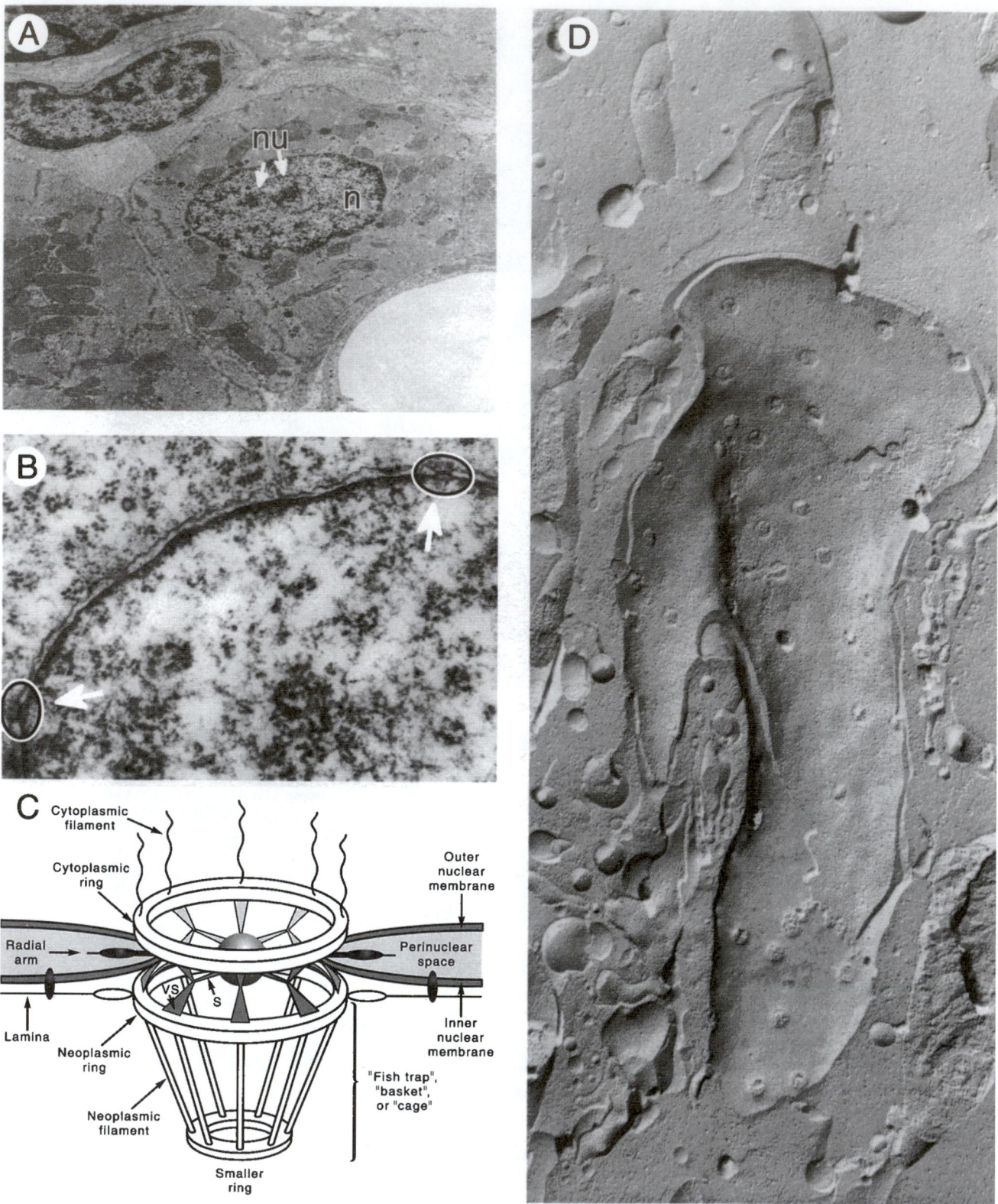

FIGURE 5.   The nucleus. (A) The centrally located nucleus (n) is one of the most obvious features of the majority of cells, with one or more nucleoli (nu) and chromatin distributed through the nucleoplasm and concentrated at the inner surface of the nuclear membrane. (B) A higher magnification view of the nuclear envelope illustrates the double membrane that encloses the perinuclear compartment, and also shows two circled nuclear pores (arrows). (C) This diagram depicts the details of the nuclear pore complex. Cytoplasmic and inner (neoplasmic) proteinaceous rings are attached to one another by eight vertical supports (VS). The central part of the pore is not a simple opening, but contains a central transporter assembly supported by spokes (S) that are attached to each of the eight vertical supports. The central transporter regulates translocation of proteins

side and one on the nucleoplasmic side, with a smaller ring deeper within the nucleoplasm (Fig. 5C). The central channel is attached to the spokes. The nuclear envelope is not a complete barrier, because it is necessary for RNA, proteins, and ions to pass in and out of the nucleus; the system of nuclear pore complexes regulates the trafficking of these substances.

As noted earlier, the nuclear envelope is a double bilayer. The outer membrane, often decorated with spherical RNA-containing particles called **ribosomes**, is continuous in places with a network of cytoplasmic membranes known collectively as the **endoplasmic reticulum** (ER).

III. Endoplasmic Reticulum

The ER was originally described by light microscopists in the late 1800s, who used special preparative procedures that selectively infiltrated this organelle system with a variety of staining moieties. Understanding of the structure of the ER was greatly advanced in the late 1940s and early 1950s by the elegant EM work of Porter and Palade, who demonstrated that the ER was an extensive network of interconnecting tubules and cisterns. More recently, the ER has been shown to be organized on a framework of microtubules, cytoskeletal elements that are described in Section VII.B.

Ribosomes (Fig. 6A) are compact rounded particles, approximately 300 nm in diameter, that dissociate in the presence of low Mg^{2+} into two smaller units. They consist in large part of ribosomal RNA, but in eukaryotes as much as 50% of the ribosomal mass can be composed of associated proteins. Ribosomes are often found organized into strands or rosettes called **polyribosomes** or simply "polysomes." These are aggregations of ribosomes active in the messenger RNA–directed linkage of amino acids to form peptide chains. Protein synthesis is initiated in the cytoplasm on ribosomes that subsequently bind to the ER to form a complex known as **rough endoplasmic reticulum** because of the beaded nature of the decorated membranes (Fig. 6). The rough ER is presumed to be involved virtually exclusively in protein synthesis. In contrast, ER that does not have attached ribosomes is known as **smooth endoplasmic reticulum** (Fig. 6B) and has been implicated in a number of different functions depending on the particular cell type. In steroid hormone–secreting cells, for example, smooth ER is associ-

ated with the production of secretory products. In liver cells, the smooth ER has been shown to be active in glycogen metabolism. In muscle, the smooth ER is better known as the **sarcoplasmic reticulum** (SR) (Fig. 6C, D), and SR membrane proteins are responsible for the uptake, sequestration, and release of Ca^{2+} during the cycles of excitation-contraction coupling and relaxation. There is now evidence that at least some ER in nonmuscle cells is also involved in Ca^{2+} handling.

IV. Golgi Apparatus

Synthesis of membrane proteins and secretory products is completed in another organelle, which is at least functionally connected with the ER, and is called the **Golgi apparatus**. This organelle was named for Camillo Golgi, a nineteenth-century microscopist-histologist who developed numerous staining techniques and described many of the light-microscopic features of cells. The Golgi apparatus does not stain with the dyes conventionally used for light microscopy, and thus appears as a clear area or "negative" image located close to the nucleus. In the electron microscope, this organelle was found to be composed of flattened disc- or saucer-shaped membranous elements known as **cisternae**. In each Golgi apparatus, numbers of these cisternae are organized into stacks, with individual cisternae interconnected through tubular channels. As secretory products pass through the Golgi apparatus, they undergo chemical modifications such as glycosylation, proteolytic cleavage, phosphorylation, and sulfation. Other functions performed by the Golgi apparatus include carbohydrate metabolism, targeting of plasmalemmal proteins (pumps, channels, and receptors), and the condensation of secretory materials.

Although a variety of terminologies have been used to identify different regions of the Golgi apparatus, four morphological subsections are generally recognized (see diagram in Fig. 7 and electron micrographs in Fig. 8). The cisternae are slightly dished in conformation, which confers both concave and convex faces to the Golgi complexes. The convex or *cis* face is also known as the forming or immature face, since it is the level at which newly synthesized proteins enter the Golgi. The medial or intermediate region comprises the varying numbers of cisternae in the middle of

and ribonucleic acids across the nuclear envelope. The nucleoplasmic ring is attached by filaments to an inner, smaller proteinaceous ring, and is also believed to be attached to the inner nuclear lamina. Intermediate filaments are associated with the outer cytoplasmic ring. The edges of the nuclear pore complex have radial arms that often appear to extend into the perinuclear space between the two bilayers of the nuclear envelope. (Reproduced with permission from Figure 2b, L. E. Maquat (1991). *Curr. Opin. Cell Biol.* **3**, 1004–1012.) (D) This image of the nuclear envelope was produced by freeze-fracture. As described in more detail in the legend for Fig. 1, this technique involves freezing tissue, cracking the frozen cells open to expose their interiors, and evaporating a heavy metal such as platinum to form a thin replica of the exposed surfaces. This approach is particularly effective for examining the ultrastructure of membranes and their integral membrane proteins. In this image, the nuclear pores appear as round structures distributed over the surface of the nucleus. In the lower part of the panel, the fracture plane passes through the outer nuclear envelope, and the pores appear to project toward the viewer. In the upper portion of the nucleus, the fracture plane is within the inner nuclear membrane, and the pores appear as depressions.

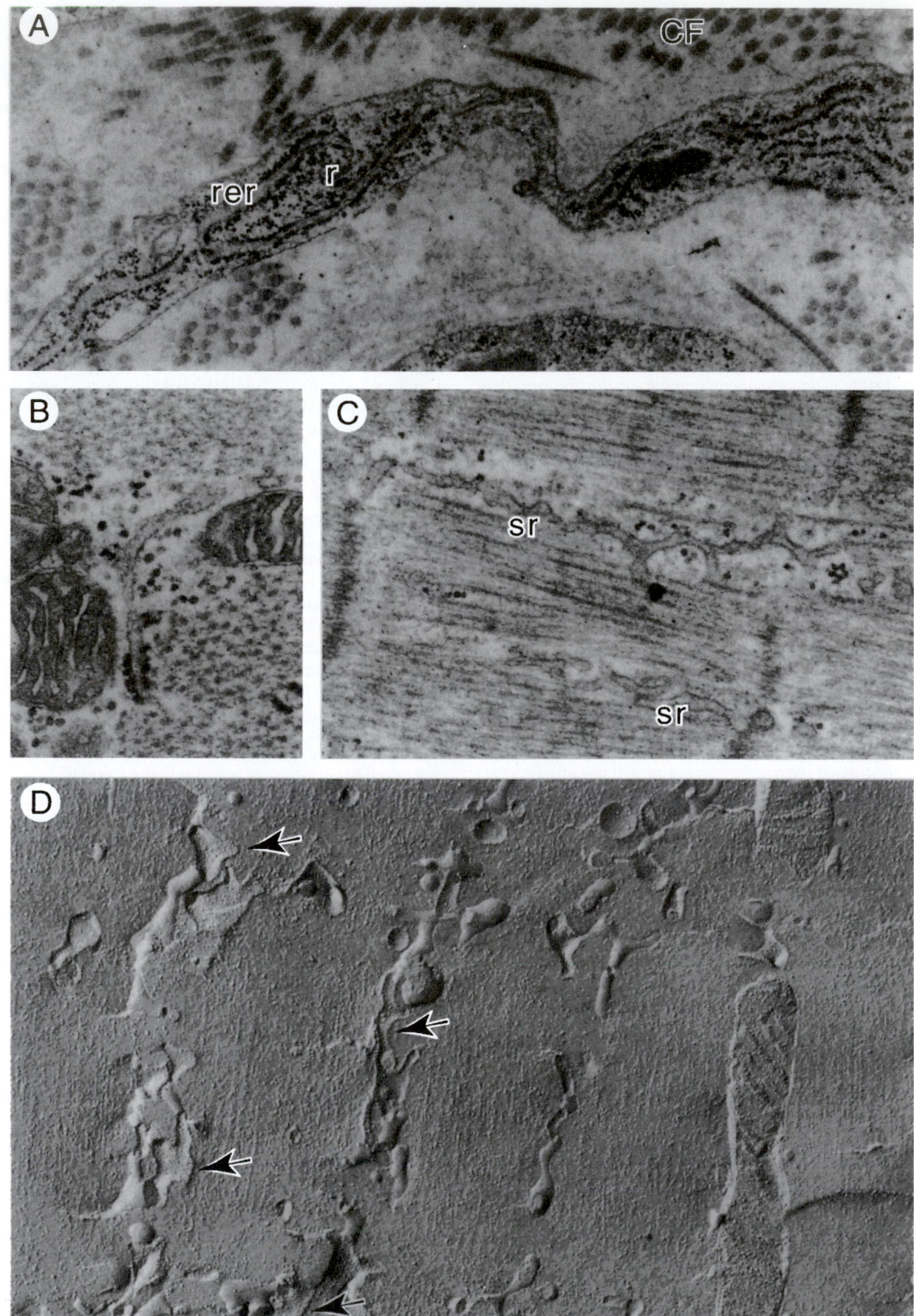

the stack. The concave or *trans* face is also called the mature or secretory face, since it is the site from which large secretory vesicles bud off after their contents have been modified for export. The **trans-Golgi network** is a reticulum of tubules emanating from the *trans* face, and is thought to be associated with the lysosomal system (Section V).

This general scheme is applicable to all Golgi complexes, and transport within the Golgi is generally vectorial, going from the *cis* to the *trans* face. However, more recent evidence suggests that maturation and pinching off is not totally restricted to the *trans* face. There is likely much more intra-Golgi shuttling back and forth of material than was originally believed. The numerous vesicles of various shapes and sizes in the immediate vicinity of the Golgi complex (Fig. 8A) may be transporting proteins along the pathway during maturation and processing. There is also evidence for some retrograde transport from the Golgi apparatus to the ER. Functional continuity of the Golgi complex with the ER has been clearly established, but whether this occurs through a direct connection or by way of vesicular traffic, or both, is not yet certain.

V. Lysosomes

In addition to packaging and modifying secretory products, the Golgi apparatus plays a role in the formation of another important cellular organelle, the **lysosome** (Fig. 8B, C). In fact, the apparent continuity of the ER, the Golgi apparatus, and the lysosomes has led to the concept of the GERL (Golgi–endoplasmic reticulum–lysosome) system (see Fig. 7). Lysosomes are a heterogeneous population of membrane-limited vesicles, differing in size and density, that were identified and characterized by De Duve in the mid-1950s. Lysosomes contain hydrolytic enzymes that can break down virtually every form of biological material, and thus these organelles act, in effect, as the cell's digestive system. Lysosomal enzymes include proteases (e.g., collagenase, acid phosphatases, and cathepsins), lipases (e.g., esterases and phospholipases), glycosidases (e.g., hyaluronidase, lysozyme, and galactosidases), and nucleases (e.g., ribo- and deoxyribonucleases). Many of the lysosomal enzymes are more effective at low pH, and the lumen of the lysosome accordingly is

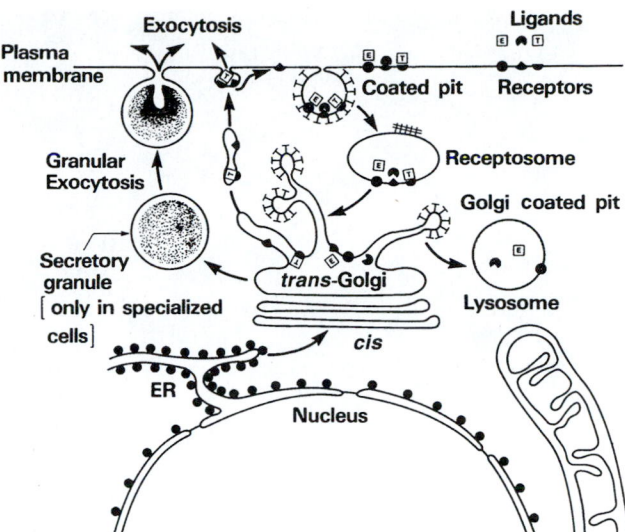

FIGURE 7. Diagram illustrating involvement of the Golgi apparatus–endoplasmic reticulum–lysosomal system (GERL) in exocytosis and endocytosis. The relationship between the Golgi apparatus, the nuclear envelope, and the endoplasmic reticulum (ER) is also depicted. The Golgi apparatus modifies secretory products, packages them in vesicles, and exports them by way of exocytosis. The Golgi also generates lysosomes that are involved in the process of receptor-mediated endocytosis. Ligands bind to receptors at the cell surface, and the receptor-ligand complexes migrate to specialized coated pits that are internalized and fuse with lysosomes for digestion. (Reproduced with permission from I. Pastan and M. C. Willingham. (1985). "Endocytosis," Chapter 1, Fig. 1. Plenum Press, New York.)

more acidic than the cytoplasm. The lysosome acts to break down and recycle intracellular components such as cytoskeletal proteins and defunct organelles such as mitochondria (by a process known as **autophagy**), as well as to digest extracellular material that has been trapped in phagocytic vesicles formed by internalized plasmalemma (**heterophagy**). Viewed in the electron microscope, newly formed, "primary" lysosomes are more uniform in size, with an amorphous electron-opaque content. After fusing with phagocytic vesicles or other cell contents, they form "secondary" lysosomes, whose size and internal density is more variable (Fig. 8B). Lysosomes containing indigestible material often remain in the cytoplasm

---

FIGURE 6. The endoplasmic reticulum. (A) Sections through processes of fibroblasts, one of whose major activities is the synthesis and export of large quantities of collagen, which extracellular matrix protein assembles into collagen fibrils (CF) that form the robust framework between cells in connective tissue. As might be predicted for cells that are so synthetically active, fibroblasts are rich in free ribosomes (r) and rough endoplasmic reticulum (rer). The RER here appears as membranous profiles with a beaded surface derived from the presence of attached ribosomes. (B) Although Fig. 3 and many other diagrams may suggest that the rough and smooth endoplasmic reticulum are distinct compartments, it is evident from this micrograph that elements of rough and smooth can be continuous with one another. (C) Smooth endoplasmic reticulum (SER) can perform different roles depending on the cell type in which it is found (see text); this example of SER is found in cardiac muscle, where it is better known as **sarcoplasmic reticulum** (SR). (D) A freeze-fracture replica of cardiac SR demonstrates that the sarcoplasmic reticulum is not merely a lipid bilayer, but a complex membrane packed with integral membrane proteins. The particles in the P-face leaflet of the SR (arrowheads) have been shown to be the Ca^{2+} pump protein that is responsible for removing Ca^{2+} from the cytoplasm during relaxation.

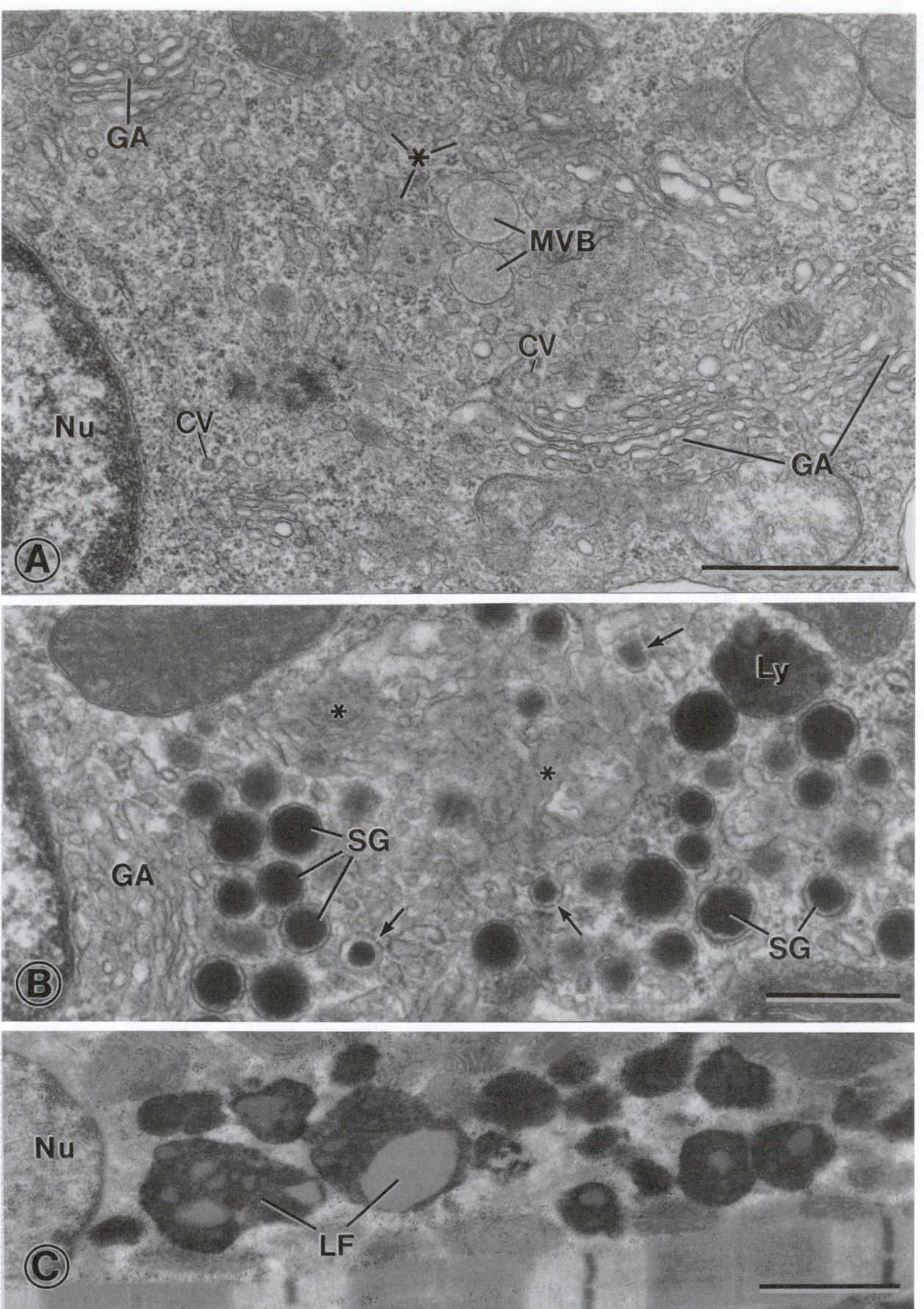

as residual bodies, also variously known as **aging pigment** or **lipofuscin** (Fig. 8C).

There are several different mechanisms whereby extracellular contents can be taken up by cells and directed to the lysosomal compartment. During **pinocytosis** ("cell drinking"), the plasmalemma pinches off small (0.1–0.2 μm diameter) vesicles containing fluid rather than solid material. **Phagocytosis** refers to the internalization of solid material from the extracellular space, and is often mediated by receptor proteins in the plasmalemma itself. This particular process is known as **receptor-mediated endocytosis,** and can be accomplished for a wide variety of receptor-ligand combinations. The ligands include nutrients too large to diffuse across the plasmalemma (e.g., iron or cholesterol) and extracellular matrix components. Additional ligands include hormones, growth factors, viruses, toxins, and immune complexes. Once receptors bind their specific ligand, they migrate to specialized flask-shaped "coated" pits in the plasmalemma that are lined with a unique 180-kDa protein called **clathrin**. Coated pits then pinch off to form **endosomes**, whose acidic lumina bring about dissociation of the receptors and ligands. A certain amount of sorting may be accomplished in the endosomes, and receptors may be recycled to the plasmalemma so as to be again available for binding additional ligand. Under certain conditions, receptors may be routed to the lysosomal compartment, which effectively down-regulates any activity that may be stimulated by that particular ligand-receptor interaction. Although most cells are capable of endocytosis, this activity is most prevalent in specialized immune cells such as macrophages, and in blood leukocytes such as neutrophils. Other cells showing higher endocytotic activity include those exposed to higher levels of foreign material, such as hepatocytes and epithelial cells.

VI. Mitochondria

Aside from the nucleus, the most obvious feature of the majority of cells is the collection of organelles called **mito-chondria** (Fig. 9). The term "mitochondrion" (literally "thread-grain") was introduced at the turn of the twentieth century based on the appearance of these bodies, under the light microscope, as elongated granules. In the EM, each mitochondrion is seen to be bounded by a double bilayer membrane (cf. Fig. 1A). The outer bilayer is smooth in contour and forms the boundary around the entire mitochondrion. The inner membrane has a variable number of infoldings, **cristae**, which increase the internal surface area substantially (Fig. 1A, Fig. 9). At high resolution, cristal membranes are seen to be studded with small raised structures known as **elementary particles**.

It is now clear that the mitochondria are sites of cellular respiration, and thus are the center of energy production in the cell. Elucidation of the various functions of different regions within the individual mitochondria has depended largely upon the development of procedures that isolate intact mitochondria, or that separate mitochondria into distinct fractions containing either outer or inner membranes. Using such preparations, it has been determined, for example, that the outer mitochondrial membrane contains large numbers of **porin** molecules, which form relatively nonselective channels that allow the passage of ions and small molecules having masses <10 kDa. This establishes a compartment—an **intermembrane space**—between the inner and outer membranes, which has essentially the same ionic characteristics as the cytoplasm. In contrast, the folded inner membrane is comparatively impermeable to ions, and encloses the innermost cavity of the mitochondrion, referred to as the **mitochondrial matrix**. The inner mitochondrial membrane contains abundant transport proteins responsible for establishing a proton gradient between the intermembrane space and the matrix. The protonmotive force created by this gradient is essential for driving many of the reactions of **oxidative phosphorylation**, a fundamental metabolic process occurring within the inner compartment of the mitochondrion. Oxidative phosphorylation generates much of the cell's adenosine triphosphate (ATP), the high-energy compound that is the source of energy for

FIGURE 8. The Golgi system, secretory granule formation, and lysosomes. Depending upon the type of cell and its metabolic state, there may be only a single Golgi apparatus consisting of a few stacked cisternae, or, as shown in A and B, there may exist a Golgi system, comprising multiple cisternal stacks in close specific association with other structures. (A) shows the perinuclear zone of a steroid-producing cell in the adrenal cortex. In addition to several profiles of Golgi apparatus (GA), this zone contains multivesicular bodies (MVB), likely related to lysosomal activity, along with a multitude of vesicles, both smooth-surface and coated (CV), tubular profiles (*), and scattered mitochondria. If traced in multiple serial sections, the seemingly isolated sets of Golgi membranes would likely be found to be connected into one large contiguous mass of tubules and cisternae. (B) demonstrates the typical appearance of the Golgi system in endocrine cells. A view of the perinuclear region of, say, a pituitary secretory cell producing ACTH or somatotrophin would be indistinguishable from this view, which in fact depicts an atrial myocardial cell. It doubles as both a contractile entity and a genuine endocrine cell, its particular hormonal product being ANP (atrial natriuretic peptide), which affects vascular tone and blood pressure. Endocrine cells often store their products in highly compacted forms known as **secretory granules** (SG). As these bodies develop, they first appear as smaller "immature" granules (arrows), forming as dense cores within vesicular envelopes and gradually enlarging to fill them. The plane of section in one place (*) passes tangentially through a Golgi cisterna, revealing it as a sort of perforated saccule. Lysosomes (Ly) are frequently found in the neighborhood of the Golgi. (C) Perinuclear region of cardiac muscle cell from an aged monkey. As an animal gets older, the heart and other organs increasingly accumulate waste products through the continuing action of lysosomes on various organelles such as mitochondria, which eventually leads to the production of numbers of large secondary lysosomal aggregates known as **lipofuscin** (LF). Eventually such masses of lipofuscin can impart a grossly darker color to the entire organ. Scale bars in A and C = 1 μm; in B scale bar = 0.5 μm.

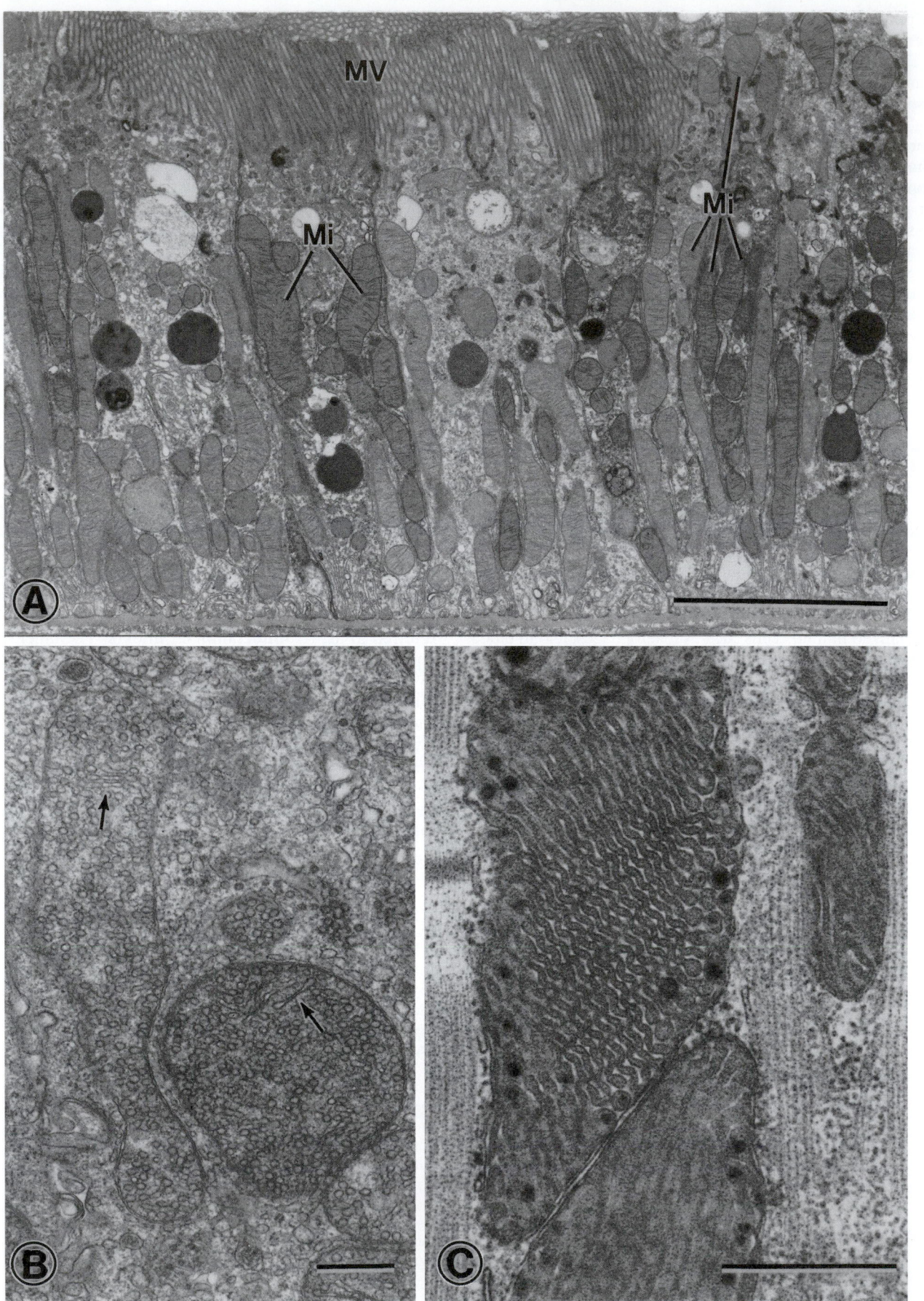

most cellular activities (for more details on the mechanism of this process, refer to Chapters 8 and 9). The enzymes responsible for the tricarboxylic acid cycle, one component of oxidative phosphorylation, are located in the mitochondrial matrix, and the enzymes involved in electron transport, another major function of the oxidative phosphorylation process, are associated with the inner mitochondrial membrane. In fact, the elementary particles on cristae have been identified as the molecules of the ATPase-synthase complex.

As would be expected, mitochondria concentrate at sites within the cell that require high energy utilization, as well as being more abundant in cells that exhibit a high degree of metabolic activity. For example, heart muscle tissue has one of the higher measured metabolic rates; accordingly, in myocardial cells, mitochondria constitute up to 40% of the total cell volume, and are sandwiched in between the collections of contractile proteins (myofibrils) (Fig. 9C).

Although mitochondria are responsible for the bulk of energy production in animal cells, plant cells contain an additional type of organelle, the chloroplast, considered in detail in Chapter 64.

VII. Cytoskeleton

As pointed out earlier, the cytoplasm is not just a watery bag in which organelles move about at will. In fact, cell contents are to one degree or another compartmentalized, with certain cytoplasmic regions specialized to subserve particular functions. Underlying this compartmentalization is the framework of fibrillar structures that make up the cytoskeleton of each cell. Normally 20–35% of the total protein of a cell is tied up in the cytoskeleton, although this proportion can vary, being considerably greater in muscle, where cytoskeletal proteins also form part of the extensive contractile apparatus. In all cells, cytoskeletal elements are involved in intracellular motility (such as migration of chromosomes during mitosis, translocation of organelles, and cytoplasmic streaming), cell locomotion, and maintenance of cell shape. More specialized functions, such as muscle contraction and ciliary/flagellar movement, also are supported by cytoskeletal elements.

A. Filaments

Filaments are composed of backbones of single proteins, polymerized to form long, slender fibrillar structures. Such filaments are both helical and polar, and frequently interact with one another. There are three general filament categories in cells: **microfilaments**, also known as "thin filaments," which are primarily composed of the protein **actin**, and range from 6–8 nm in diameter; **thick filaments**, consisting mainly of **myosin** and attaining 12–15 nm in average diameter; and **intermediate filaments**, so called because their customary 10-nm average diameter is intermediate between that of thin and thick filaments (thus also known as 10-nm filaments). Intermediate filaments may be composed of a variety of related proteins, depending on cell type.

1. Microfilaments

As noted above, the backbone of microfilaments is composed of the cytoskeletal protein actin; in most cell types, however, additional regulatory proteins (e.g., troponin) and structural proteins (e.g., tropomyosin) are intimately associated with the actin filaments. Single molecules of G (globular)-actin have a molecular mass (M_r) of 45 kDa, and are approximately 55 nm in diameter. Detailed structural studies have shown that the shape of G-actin molecules is not strictly globular, but instead consists of a bilobate entity shaped like a dumbbell. G-actin polymerizes to form filaments, and in this form is known as F (filamentous)-actin. The structure of the thin filament has been determined by the combination of special preparative techniques and electron microscopic examination. Generally, the thin filament is depicted as a double α-helix, 6–7 nm in diameter and of variable length, depending in part on the cell type and the type of cytoskeletal formation in which it participates. However, there is no evidence that the two strands of this double helix can dissociate or grow independently of one another. A number of associated proteins may regulate initiation and growth of a filament, as well as its final length. Tropomyosin, for example, stabilizes thin filaments of muscle cells.

In different cells, and in different tissues, thin filaments are associated with a variety of recognizable intracellular formations. Bundles of filaments, all having the same polarity, form the cores of microvilli in epithelial cells; actin filaments are also evident in smooth muscle and striated muscle cells as intrinsic parts of the contractile apparatus (Fig. 10). Bundles of thin filaments having opposite polarities are also found, located on opposing sides of the Z discs in skeletal and cardiac muscle, across from one another at dense bodies in smooth muscle, and within the stress fibers of non-

---

FIGURE 9. Mitochondria. (A) Low-power electron micrograph showing the wall of a kidney tubule. The columnar epithelial cells are closely packed together, and their apical surfaces are thrown into numerous microvilli (MV). In this section, the epithelial cytoplasmic contents consist in great part of elongated mitochondria (Mi), which though large and numerous are rather simple in construction, with relatively few internal membranes (cristae). (B) Mitochondria in cell from adrenal cortex. Steroid-secreting cells often are characterized by spheroidal or elongate mitochondria packed with cristae that take the form of fine tubules (arrows) or chains of tiny vesicles. (C) Mitochondria in cardiac muscle. In addition to being sandwiched between the myofibrils (collections of contractile filaments), cardiac mitochondria are densely populated with elaborate pleated shelves of internal cristae. Opaque granules often are present within the mitochondrial matrix. Scale bars: A, 5 μm; B and C, 0.5 μm.

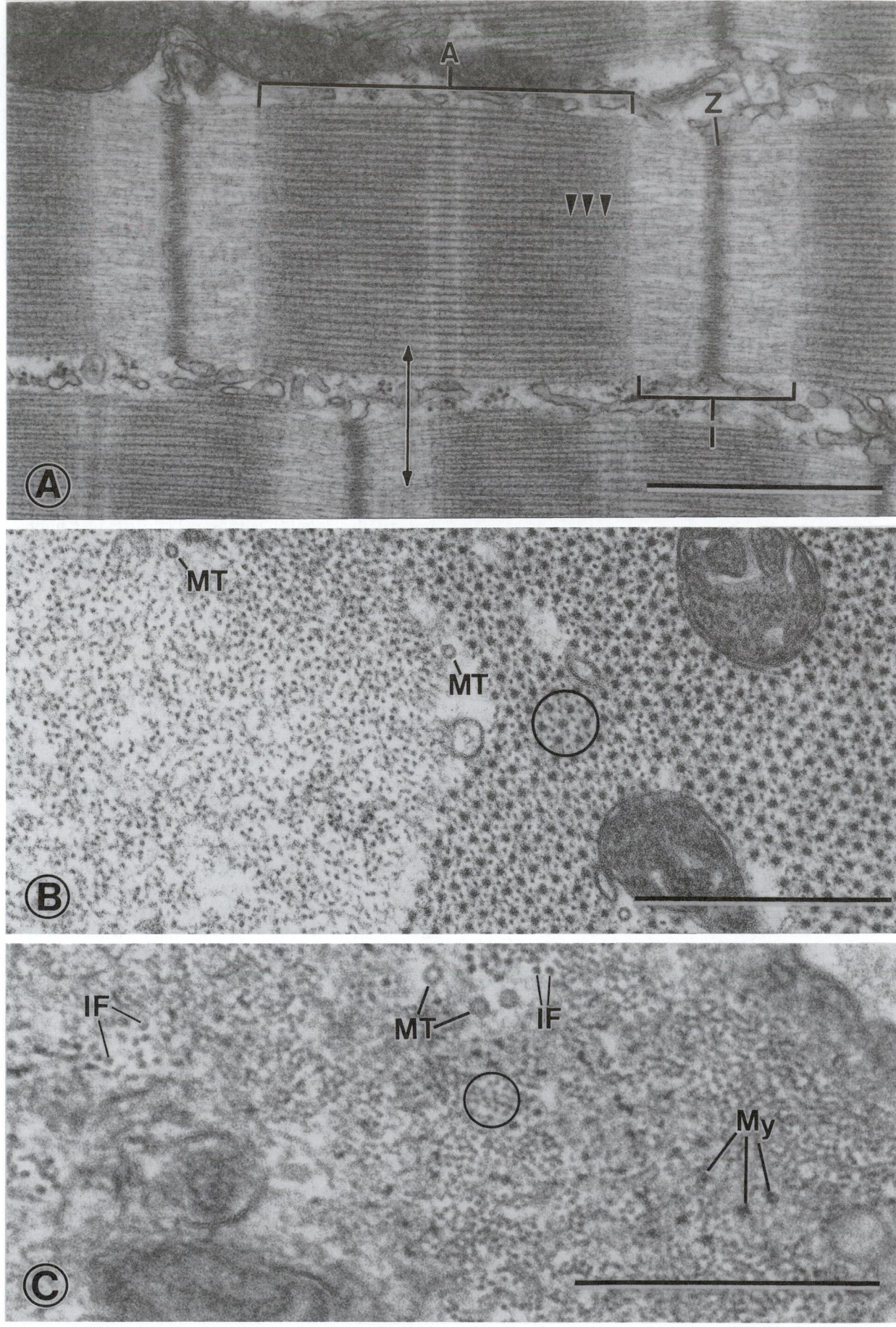

muscle cells. Three-dimensional networks of filaments are found in subplasmalemmal (cortical) arrays in many cells.

2. Thick Filaments

The most detailed structural information regarding myosin-containing filaments comes from the study of the particular myosin type (myosin II) found in striated muscle thick filaments (Fig. 10A, B). Single myosin molecules consist of dimers of two heavy chains. Each heavy chain has a globular head portion and a long slender tail. In myosin II, two different Ca^{2+}-binding light chains are associated with the head portion of each heavy chain. Thus, the myosin complex in striated muscle is a molecule having two globular heads and a tail approximately 150 nm in length. The tails of each myosin molecule are highly organized and form the backbone of each thick filament. The heads of the molecules project out at periodic intervals (Fig. 10A). These multistranded (three in vertebrate, four in insect striated muscle) α-helical filaments are approximately 15 nm in diameter. Thick filaments in smooth muscle and nonmuscle cells have not been so well characterized. Smooth muscle cells clearly contain thick filaments (Fig. 10C), but the exact organization of the molecules within the filaments is still controversial. It is likely that nonmuscle cells also contain thick filaments, but that they are shorter and less ultrastructurally obvious.

It is generally accepted that thick filaments are required for movement and motility, which are carried out through interactions with actin-containing thin filaments. The details of this interaction are dealt with in considerable detail in Chapter 56.

3. Intermediate Filaments

Intermediate filaments (IFs) actually comprise a related family of proteins found in virtually all cell types; lamins are IF proteins that form the internal lamina of cell nuclei, for example. However, there are also tissue-specific IF proteins. Vimentin is a 53-kDa protein found in mesenchymal tissue (connective tissue, bone, blood, cartilage); desmin is 52-kDa, and is localized in muscle; neurofilament protein, or neurofilamin (NF), is found in neurons and occurs in three molecular species: NF light (65 kDa), NF medium (105 kDa), and NF heavy (135 kDa). GFAP (for "glial fibrillar acidic protein") is a 50-kDa IF protein found in glial cells of the nervous system. Finally, keratins (including both acidic and basic categories) are IF proteins that are abundant in epithelial cells.

All the IF proteins have common features related to their molecular structure, which gives rise to 10-nm filaments when assembled as polymers. Pairs of molecules are the basic subunits that are joined as "coiled-coil core" dimers. The dimers are assembled into tetrameric complexes, about 2 nm in diameter, which then join end-to-end as **protofilaments**, which in turn assemble laterally into 4.5 nm **protofibrils**, four of which then join to form cylindrical 10-nm-diameter IFs.

IFs are often observed as loose three-dimensional networks, intermixed with other cellular components (Fig. 10C), but can form dense skeins that consist primarily of 10-nm filaments (see Figs. 12B, 13F). As pointed out above, IFs form the framework of the inner nuclear membrane, the nuclear lamina. IFs are also associated with subplasmalemmal plaques adjacent to the plasma membrane at specialized cell-cell contacts (desmosomes) (Fig. 13D, F) and are disposed in muscle in transverse meshworks that link and align myofibrils to one another, to the sarcolemma, and perhaps also to the nucleus. Thus, IFs are an internal scaffolding that can form networks to link peripheral and central components of the cell as a mechanically integrated complex.

B. Microtubules

Microtubules are elongate, hollow cylinders of notably larger diameter (ca. 25 nm) than other cytoskeletal fibrils (Figs. 10, 11, 12A) and are composed of the protein **tubulin**.

FIGURE 10. Fibrillar elements in muscle cells. (A) Longitudinal section through a cardiac muscle cell, showing the characteristic banded pattern of the ordered masses of contractile filaments known as **myofibrils**. Between myofibrils, other cell organelles such as mitochondria and endoplasmic (sarcoplasmic) reticulum are sandwiched. The units of the banding pattern are known as **sarcomeres**, and the major filament categories that contribute to their formation are thin (actin) filaments and thick (myosin) filaments. The so-called A band (A) is a composite of interdigitated actin and myosin, and the subtle crosshatching pattern derives from portions of the myosin filaments that form crossbridges (several shown by arrowheads) with the adjacent actin filaments. The I band (I) is occupied only by thin filaments, which insert into the dense material of the Z bands (Z), which are primarily composed of the protein α-actinin. Adjacent Z bands define a sarcomere unit, each of which covers a distance of ca. 2.2 μm in heart. Note in this panel that adjacent myofibrils are not always aligned with one another in terms of the levels of their sarcomere patterns. (B) A transverse section through two cardiac myofibrils, its plane of sectioning corresponding to the double-headed arrow in A. At the left of the picture, the section passes through the I-band region, and therefore reveals only actin filaments, whereas at the right the A band is cut, showing the characteristic hexagonally arrayed myosin filaments, each surrounded by actin filaments, usually in the "six-around-one" pattern circled. Mitochondria are enmeshed by the filaments in one myofibril, and additional fibrillar components, namely microtubules (MT), appear in the same orientation as the muscle filaments. (C) Cross-section through a vascular smooth muscle cell in the wall of a coronary artery. Like the cardiac muscle shown in the other panels, the smooth muscle contains examples of actin (circled) and myosin (My), but a geometric arrangement similar to that of cardiac or skeletal muscle is not evident. Cytoskeletal fibrils, including microtubules (MT) and intermediate filaments (IF), also are present in the same orientation as the contractile actin and myosin. Scale bars: A = 1 μm; B and C = 0.5 μm.

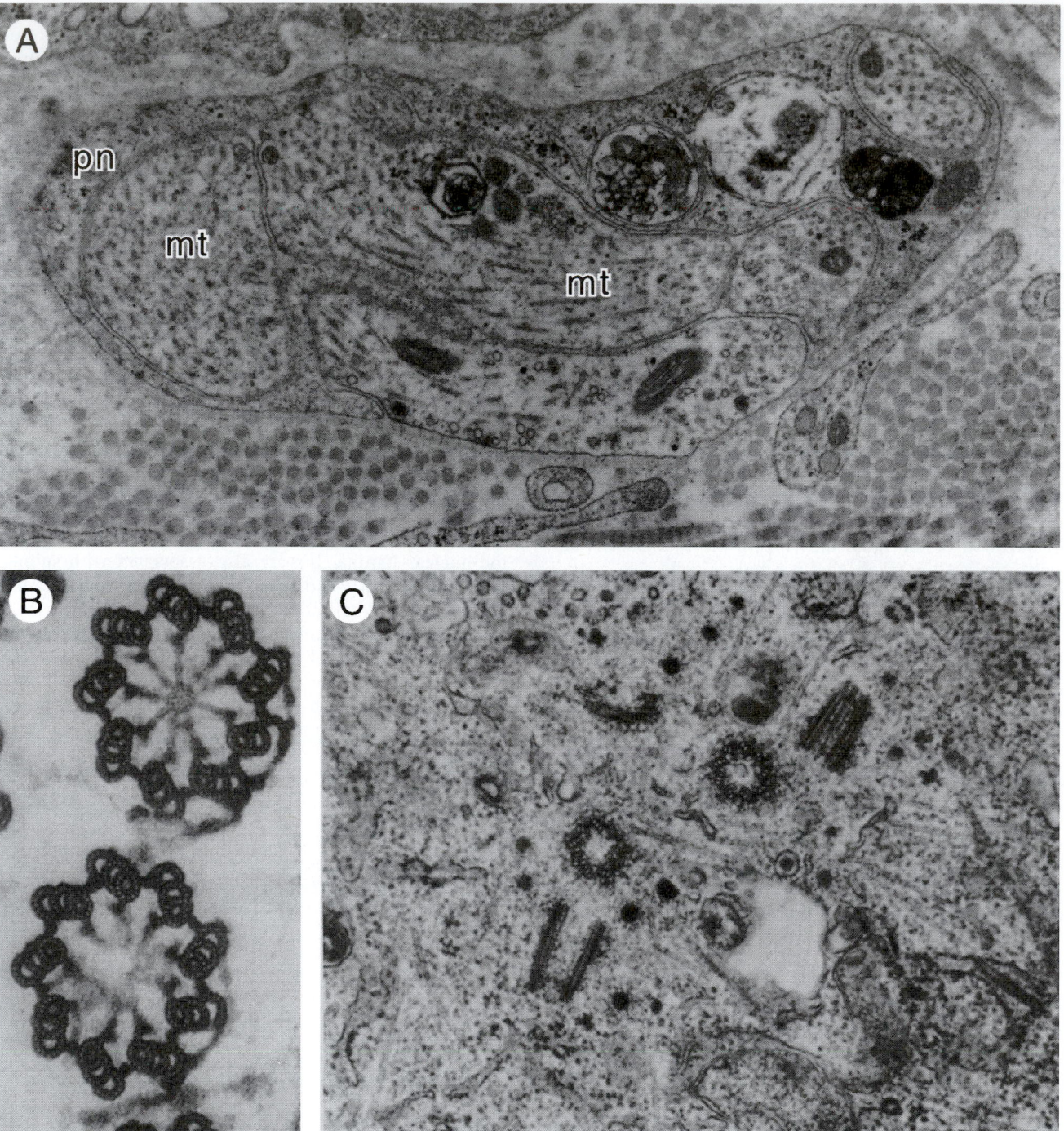

FIGURE 11. Microtubules. (A) This cross-section of a small peripheral nerve bundle (pn) contains profiles of several axons. The axons are filled with longitudinally oriented microtubules (mt) seen both in cross-section and in oblique view because of the different orientation of the different axon profiles. In neurons, microtubules are involved in transport of organelles from the nerve cell bodies out to the relatively distant synaptic terminal regions. (B) Two cilia in transverse section. Each profile is composed of nine units, each unit consisting of three microtubules; these structures form the support for the cilium, and sliding of the microtubules with respect to one another causes the cilia to bend and thus causes ciliary motility. (Reproduced with permission from B. Alberts *et al.,* "Molecular Biology of the Cell," 2nd ed., Figure 11-59a. Garland Publishing, New York.) (C) This electron micrograph shows two **centrosomes**. Each centrosome consists of a pair of centrioles oriented at right angles to one another. Centrioles are formed from microtubules organized in the same nine triplets observed in cilia and flagella. Centrioles act as microtubule-organizing centers: they give rise to the basal bodies of cilia and flagella and also serve to direct the orientation of the microtubules of the mitotic spindle (see Fig. 12A). (Reproduced with permission from B. Alberts *et al.,* "Molecular Biology of the Cell," 2nd ed., Figure 11-60. Garland Publishing, New York.)

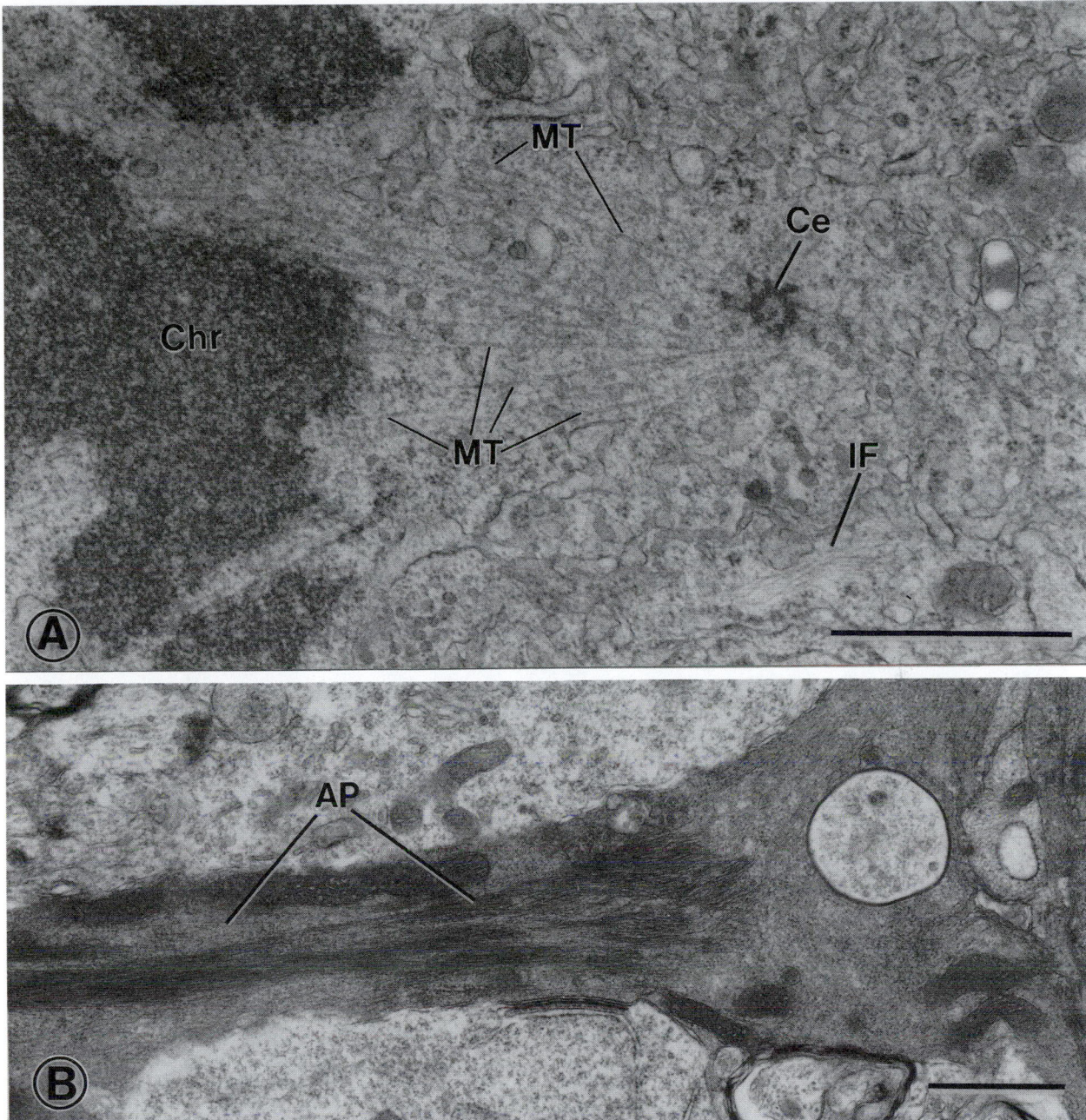

FIGURE 12.   Cytoskeletal fibrils. Although the two major cytoskeletal support elements, microtubules and intermediate filaments, are often found intermingled with one another (see Figs. 10C, 14B), under some circumstances and in certain cells one or the other type of fibril takes on a predominant role. (A) A portion of a mitotic endothelial cell in aorta. The condensed chromosomal material (Chr) is obvious at the left of the picture. The movement of these chromosomes is controlled in part by attached microtubules (MT) that make up the **mitotic spindle**. The spindle microtubules emanate from the centrosomal area of the cell, which also contains the centrioles (one shown at Ce). Intermediate filaments are present in these cells, (IF), but do not appear to take an active role in chromosomal translocation. (B) Concentrations of longitudinally oriented intermediate filaments support an elongated astrocyte process (AP) in brain. This particular type of IF is composed of GFAP (glial fibrillary acidic protein). Scale bars in A and B = 1 μm.

Tubulin has α and β isoforms that have an M_r of ca. 50 kDa. Each molecule is an α-β heterodimer of ca. 100 kDa. The walls of the microtubules are formed from 5 nm protofilaments composed of linearly arranged pairs of alternating α and β subunits. Protofilaments are aligned parallel to the long axis of the microtubule. The number of protofilaments that make up the microtubule cross-section varies, but 13 is the most frequent complement. The protofilaments are polar structures, and are assembled so that they confer polarity to the microtubule. The conditions that induce polymerization of tubulin protofilaments include the presence of certain **microtubule-associated proteins (MAPs),** GTP, Mg^{2+}, and low Ca^{2+}.

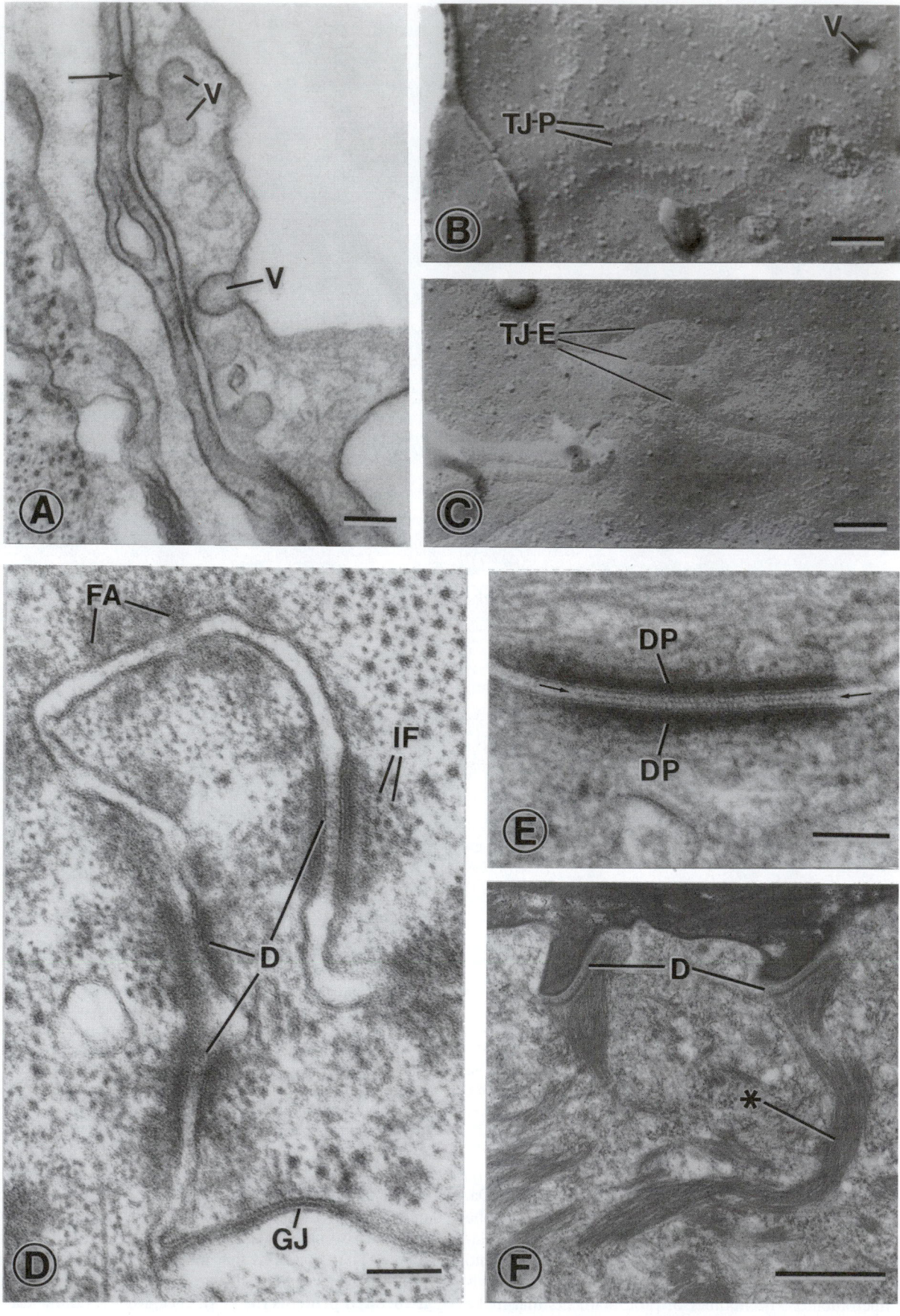

There are several different MAPs, including MAP1, MAP2, and tau. The MAPs serve to stabilize the microtubules and are the targets of regulatory signals. One of the most functionally important MAPs is **dynein**, a multimolecular complex with an M_r of 1000–2000 kDa. Dynein forms side-arms on the microtubules found in the cores of cilia and flagella. These side-arms enable microtubules to slide past one another, allowing bending of the whole structure and accounting for their basic motility (for further information, see Chapter [57]).

In addition to being the fibrillar component of cilia and flagella, microtubules are the basis of the so-called MTOCs (microtubule-organizing centers) such as centrosomes, basal bodies, and kinetochores (Fig. 11C). Microtubules are involved in movement of chromosomes during cell division (Fig. 12A), and perform a role in translocation of organelles within the cytoplasm. In some cases, microtubules also can form the framework on which some of the cell compartments are organized (as noted earlier, the ER is also involved in a similar function), and it has even been shown that the orientation of IFs depends, in some instances, upon the integrity of the microtubules in the same cell.

VIII. Cell Junctions

Where they are located in tissues and organs, cells do not exist in isolation, but rather exert multiple influences on one another. When they reside in different parts of the organism, the influences are usually realized through an intermediary milieu such as the bloodstream. Where neighboring cells abut one another, however, specialized structures or **intermembranous junctions** connect them. The electron microscope has proven a particularly apt tool for detecting such connections, since they usually incorporate some ultrastructurally visible form of modification of the apposed cell plasmalemmata.

Four types of intercellular junctions are commonly encountered in vertebrate tissue: (1) tight junctions; (2) gap junctions; (3) intermediate or **adherens** junctions; and (4) desmosomes. Epithelial layers such as seen in small intestine display a rather stylized "junctional complex" (cf. Fig. 3) that consists of the four junctional types arranged in a sequence from the luminal surface down to the basal surface; in order, the complex goes tight junction/adherens junction/desmosome/gap junction.

Because of the extreme thinness of sections used in the electron microscope, the views of cells and their constituents are essentially two-dimensional; in such views, the casual observer can easily fail to appreciate the true shape and extent of any particular profile. This is certainly the case for specialized intercellular junctions. More stringent observations, including examination of multiple sections of cells in various orientations, as well as use of serial sections, has shown—in epithelium as well as other tissues—a tendency for some categories of junctions to occupy continuous bands of plasmalemma (**zonulae**) that form rings about the lateral cell surface. Other junctions are more strictly delimited, in three dimensions occupying round, ovoid, or sometimes irregular patches of plasmalemma in each of the joined cells (e.g., Fig. 1B).

Tight junctions, viewed in thin sections, are identified—as the name would imply—by plasmalemmal appositions that come into such close contact as to obliterate the intercellular space between cells. Often multiple points of leaflet fusion are visible along short stretches of apposed membranes (Fig. 13A). Replication of membrane surfaces by freeze-fracture techniques (Fig. 13B, C) has revealed that

---

FIGURE 13. Cell-to-cell appositions, also called **intercellular junctions**. A, B, and C show endocardium (the endothelial lining of the heart) both in thin section (in A) and freeze-fracture replicas (B and C). (A) The plane of section passes vertically through two overlapping endocardial cells, which are decorated with cytoplasmic vesicles (V) that support transport of materials in and out of these cells. The plasma membranes of the two endocardial cells come into close proximity at several points, at one appearing to touch or fuse (arrow). Such junctions are known generally as "tight junctions," and form barriers to the penetration of certain materials past them in the intercellular spaces. (B) A freeze-fracture replica of the endocardial membrane's P face (cf. Fig. 1B) shows vesicular openings (V) as well as linear arrays of particles that correspond to the tight junction (TJ-P). (C) These correspond in replicas of the E face to linear indentations (TJ-E) into which the corresponding P-face particles fit to form an interlocking junctional barrier. D and E show junctions from the myocardial cell-to-cell junctional complex (the "intercalated disc"), which consists of interdigitated muscle cell tips that bear a variety of interspersed types of junctions, including gap junctions (GJ, also see Fig. 1). (D) At the top of the micrograph there appears a zone of **fascia adherens** (FA), the points at which thin actin filaments of the myofibrils insert into the intercalated disc. The adherens-type junctions are characterized primarily by accumulations of intracellular dense material that face one another across the intercellular space, but which incorporate no particular extracellular specializations. In contrast, the adhesive junctions known as **desmosomes** (three of which appear in this picture, D) not only bear intensely opaque intracellular densities closely associated with intermediate filaments (IF), but also contain a distinct component located in the extracellular space (also see E). (E) A desmosome is shown at high magnification. The intracellular material is concentrated into desmosomal "plaques" (DP) which, together with the associated plasmalemmal unit membranes, are arranged in a striking parallel array. The extracellular material, known as a "central lamella" (between arrows), is apparently responsible for maintaining the register of the apposed membranes, having toothed projections from a center linear component that anchor in the adjacent plasma membrane leaflets. (F) Keratinocytes in the epithelium of skin. Here the major filament type is **keratin**, considered to be one of the family of intermediate filaments. Keratin is concentrated into fibrous cords (*) that insert onto the intracellular plaque material of the numerous desmosomes (D) that connect the cells of this layer of the skin. Scale bars in A–E = 0.1 μm; scale bar in F = 1 μm.

tight junctions actually are composed of intramembranous particles arranged to form networks of "tongue-and-groove" interdigitations. The effect of such an architecture is to provide a barrier largely impenetrable to most particles and fluids. Tight junctional arrays prevent direct diffusion of various products through the extracellular fluid space, thus forcing their passage "through proper channels," so to speak, namely via mechanisms such as vesicular transport through the cytoplasm. Tight junctions are also found in abundance in pancreatic acini, as well as between endothelial cells in certain segments of blood vessels. To some degree, tight junctions are likely responsible for maintenance of the blood-brain barrier.

Gap junctions, also known as **nexuses**, were in early studies mistaken for tight junctions, but it was realized through the use of special staining procedures and particle tracer techniques that these complexes were different in their overall thickness, and furthermore allowed particles below a certain size to penetrate the extracellular space at their level. Gap junctions appear in thin EM sections to be composed of seven layers (Figs. 1A, 13D), the middle one presenting as a 2–4 nm "gap," explaining the common name for this type of junction. Closer examination has revealed that the gap junction is an assemblage of hexagonally packed "connexons" (as their individual intramembranous particles are called; cf. Fig. 1B, C), each of which contains a tiny pore at its center. It is thought that such pores are aligned to form thin channels, allowing intracytoplasmic passage of ions (e.g., Ca^{2+}) and other substances (such as ATP and amino acids) directly from the cytoplasm of one cell into the interior of the other, a distance calculated to be only about 2.5 nm. In this way, it is postulated, neighboring cells can be "coupled" both electrically and metabolically.

Adherens junctions, though often descriptively linked with desmosomes, can be distinguished from them structurally on the basis of the adherens junctions being more diffuse-appearing and more widely distributed than desmosomes (Fig. 13D). In addition to this morphological disparity are fundamental differences in the species of proteins associated with and composing the two types of junctions.

Adherens junctions are frequently associated with microfilaments, whereas desmosomes instead are accompanied by intermediate filaments. A well-described example of this dichotomy is found in cardiac muscle in the intercalated discs, the zones of extensive junctional attachment between adjacent cardiac muscle cells (Fig. 13D, E). Here at the cell tips, the actin filaments of the myofibrils terminate in the amorphous dense zones of the fascia adherens portion of the disc. In contrast, myofibrils do not come into contact with desmosomes. Furthermore, desmosomes exhibit a far more distinct and layered ultrastructure (Fig. 13E), including a central lamella lying in the extracellular space between cells and intensely opaque intracellular, subplasmalemmal plaques, within whose substance profiles of intermediate filaments can often be detected. In epithelial cells, both desmosomes and hemidesmosomes—which are essentially "half" desmosomes confined to the cells' basal regions—exhibit subplasmalemmal plaques that are the insertion points for cords of keratin-type intermediate filaments (Fig. 13F). As mentioned, distinctly different types of proteins are specific to each type of junction, and so—according to the philosophy of form following function—their structural resemblance to one another is related largely to their similar adhesive roles in cells.

IX. Special Tissues, Specialized Ultrastructure

Although a chapter such as this one necessarily deals in generalities, in reality few researchers work on generic "cells." Instead, certain types of cells, for one or many reasons, are of special interest to individual investigators. It is therefore useful to view cell ultrastructure as it is specifically arranged to form a cell type that is particularly appealing to physiologists: the nerve cell or **neuron**. This cell type not only generates and transfers electrochemical signals, it is also heavily devoted to synthetic activity. Though neurons vary considerably in size and shape in their various locations in the central, peripheral, and enteric nervous systems, as a group they are characterized by their ability to exercise their functions at points far removed from the central cell region that contains the nucleus. This action-at-a-distance ability can be directly correlated with ultrastructural features of neurons. The cell body or **soma**, which contains a single prominent nucleus, also houses the bulk of the synthetic machinery of the cell (Fig. 14A), including extensive, often multiple Golgi ap-

FIGURE 14. Electron micrographs illustrating specialized ultrastructure of neurons. (A) A portion of the cell body, or soma, that contains the nucleus (Nu). The cytoplasm in this region is filled with membrane-based synthetic organelles, including multiple profiles of the Golgi apparatus (GA) and cytoplasmic islands of rough endoplasmic reticulum and polyribosomes (known collectively as "Nissl substance," NS). Numerous small mitochondria are scattered throughout the soma, and other examples of typical cell organelles such as lysosomes (Ly) are represented as well. (B) High-magnification view of axons (Ax), the elongated cytoplasmic processes of neurons. These are cut transversely, and contain individual mitochondria (Mi) and numerous cross-sectioned cytoskeletal elements, including both microtubules (MT) and intermediate filaments (IF). (C) Axodendritic synapse (i.e., a nerve-to-nerve contact made by an axon with a dendrite, Den). The presynaptic axon terminal is packed with neurosecretory vesicles and also contains mitochondria. The hallmark of this complex is the synaptic density (arrows) which both attaches the two apposed nerve elements and delineates the region in which transfer of neurotransmitter will occur. (D) Motor endplate, a contact made by an axon terminal of the peripheral nervous system with a skeletal muscle fiber. Although much larger than the CNS terminal (cf. scale bars in B, C, D, and E), like it, the endplate terminal is filled with vesicles; instead of a dense plaque at the point of contact, however, there is a deep depression in the skeletal muscle into which the terminal nestles, together with elaborate secondary folds formed by the muscle cell membrane (*). (E) An afferent or sensory nerve terminal (Aff) in heart. This axon termination is considerably larger than most efferent terminals (cf. C and D) and contains a variety of inclusions such as mitochondria, clear and dense vesicles, and other structures. Scale bars in A, D, and E = 1 μm; bars in B and C = 0.5 μm.

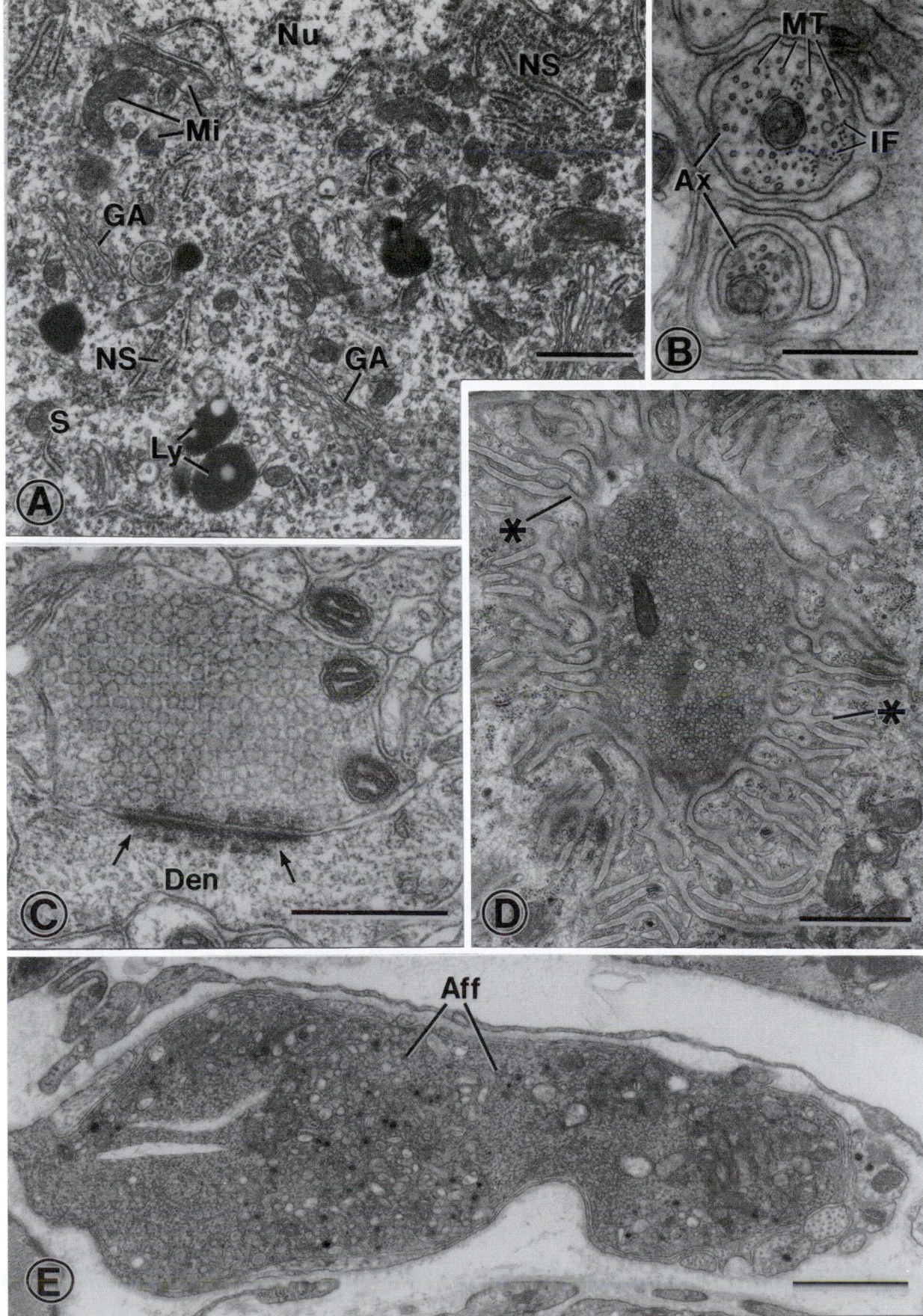

paratiuses and groupings of rough endoplasmic reticulum cisternal profiles, known collectively as "Nissl substance," because of the staining properties of their contained ribonucleoprotein. "Free" ribosomes (i.e., not attached to membranes) and polysomes abound as well within the neuronal soma.

It is the cytoplasmic processes of neurons, their **dendrites** and **axons**, that most effectively attest to their specialized functions. While dendrites, which generally bring signals toward the cell body, are usually limited to no more than several micrometers in length, the points at which the axons terminate may be located several millimeters away (or even farther, depending upon such factors as the size of the animal, type of neuron, and its particular location in the body). Both dendrites and axons, as would be expected, are well endowed with cytoskeletal elements, oriented preferentially along the processes in a configuration that best supports their attenuated shapes. Electron microscopic observation demonstrates that microtubules and intermediate filaments are the most prominent constituents of axons (Figs. 11A, 14B).

As pointed out, neurons, though organized along certain unified architectural lines, can vary markedly in their ultrastructure. Nowhere is this more pronounced than at their most distal portions, the tips of the axons, or **axon terminals**. These termini may have as their target other neurons, or may abut contractile cells, such as smooth muscle (both in viscera and blood vessels), cardiac muscle cells, or skeletal myocytes. Axon terminals are usually characterized by concentrations of vesicles of one type or another. Where these axon terminals occur, they may lack any specialized structure that attaches them to the target cell. Often, though, where neurons terminate on one another, a distinct appositional complex known as the **synapse** is found.

The axon terminal (also known as a "bouton") constitutes the presynaptic element, and may terminate on a neuronal soma or another axon, but in the central nervous system (CNS) more often is found in contact with a dendrite (Fig. 14C). In addition to the neurosecretory vesicles that fill the axonal bouton, there is a prominent **synaptic density** whose function appears similar to that of an adherens or desmosomal type of adhesive junction. The synapse functions through a process of vesicle fusion with the presynaptic membrane, with subsequent exocytosis of the transmitter chemical substance (which can be either excitatory or inhibitory) into the synaptic cleft, where it acts on the postsynaptic membrane side of the complex (i.e., the dendritic membrane in Fig. 14C), changing its electrical potential and thus generating a signal (thence the term "chemoelectric" for such an event). Where no synapse is formed, as for example in axon terminations near blood vessel walls, diffusion of neurotransmitter is sufficient to modulate the contractile activity of the vascular smooth muscle cells there.

In the case of vertebrate motor neuron axons, which terminate on skeletal muscle fibers, considerable specialization characterizes both the axon terminal and the adjacent muscle cell. This sort of "synapse" is better known as a **motor endplate** (Fig. 14D). The axon tip here is considerably larger than the typical boutons found in the CNS, and accordingly contains a considerably greater number of neurosecretory vesicles. The underlying muscle cell plasmalemma ("sarcolemma"),

furthermore, has itself become specialized, containing concentrations of membrane-associated receptors (usually for acetylcholine) and forming both a large depression into which the axon fits (the synaptic "gutter") and numerous secondary infoldings that increase the amount of excitable muscle membrane available to the released neurotransmitter.

The terminal axons discussed thus far are **efferent**, that is, they convey outbound signals from the nerve cell body to its distant target. Also found are **afferent**, or sensory, terminals, which function to send stimuli such as pressure or pain back to the soma. The contents of such terminals usually greatly vary, including not only vesicles of different shapes and sizes, but also numerous mitochondria, granules, and whorled membranous bodies (Fig. 14E).

X. Summary

Cell ultrastructure is studied to provide insights into the morphology that underlies the activities of the cell. These activities are often carried out in distinct regions, and may be compartmentalized as well by the cell's various organelles. In most cells, the nucleus is roughly spherical and centrally located, containing the cell's genetic machinery and initiating RNA transcription for eventual protein synthesis in the cytoplasm. Except during that time in the cell cycle devoted to division, the nucleus is contained in a double membrane, the nuclear envelope. Pores in the nuclear envelope regulate bidirectional transport between nucleoplasm and cytoplasm. The outer unit membrane of the nucleus is continuous with the system of cytoplasmic membranes known as the endoplasmic reticulum, which may exist as either "rough" (having ribosomes attached) or "smooth" ER. The rough ER—largely through the action of its ribosomes—is involved in the assembly or synthesis of polypeptides and proteins from component amino acids, and participates in the insertion of the resulting proteins into membranes or into packages for export out of the cell. Smooth ER has a variety of functions; in muscle cells in particular, it is involved in the regulation of free Ca^{2+} levels in the cytoplasm, and may perform a similar role in nonmuscle cells as well. The Golgi complex, which is functionally related to the ER, is a stack of disclike membranous cisternae; it receives synthesized proteins from the ER, further processes them in a variety of ways, and then directs them to the appropriate intracellular compartment, or compacts them into some form of secretory vesicle or granule for eventual export out of the cell. Lysosomes are spherical, membrane-limited organelles that contain a wide variety of digestive enzymes; their function is to break down cellular components, usually after those components have reached the end of their useful life. Once lysosomes have merged with cellular material, they take on a variegated appearance and are known as secondary lysosomes. Lysosomes can also process extracellular substances, including foreign ones, that have been taken into the cell by endocytosis. Mitochondria are generally numerous in the cytoplasm, and may be spherical or rodlike in appearance. They are composed of an outer limiting membrane and a system of inner convoluted membranes; this compartmentation of the mitochondrion plays an important role in metabolic processes such as oxidative phosphorylation. The mitochondria in fact are the cellular sites that produce the cell's major energy source, ATP. The framework of the cell, the cytoskeleton, is composed of several fibrillar structures: microfilaments, thick filaments, intermediate filaments, and microtubules, each with its specifically associated regulatory proteins. The cytoskeletal components collectively provide the scaffolding that underlies the compartmentation of cell structure and activity, and also are active in cell motility and organelle translocation. The cell surface membrane, the plasmalemma, is a phospholipid bilayer containing intrinsic proteins that can act, among other things, as ATP-dependent pumps or ion channels that maintain the ionic gradients and membrane potentials essential to the cell's activities.

Bibliography

Alberts, B., Bray, D., Lewis, J., Raff, M., Roberts, K., and Watson, J. D., Eds. (1989). "Molecular Biology of the Cell." Garland Publishing, New York.

Andre, J. (1994). Mitochondria. *Biology of the Cell* **80**, 103–106.

Bershadsky, A. D., and Vasiliev, J. M. (1988). "Cytoskeleton." Plenum Press, New York.

Block, S. M. (1996). Fifty ways to love your lever: myosin motors. *Cell* **87**, 151–157.

Clermont, Y., Rambourg, A., and Hermo, L. (1995). Trans-Golgi network (TGN) of different cell types: three-dimensional structural characteristics and variability. *Anat. Rec.* **24**, 289–301.

Damke, H. (1996). Dynamin and receptor-mediated endocytosis. *FEBS Lett.* **389**, 48–51.

Davis, L. I. (1995). The nuclear pore complex. *Annu. Rev. Biochem.* **64**, 865–896.

Dessev, G. N. (1992). Nuclear envelope structure. *Curr. Opin. Cell Biol.* **4**, 430–435.

dos Remedios, C. G., and Moens, P. D. (1995). Actin and the actinomyosin interface: a review. *Biochim. Biophys. Acta* **1228**, 99–124.

Easterwood, T. R., and Harvey, S. C. (1995). Modeling the structure of the ribosome. *Biochem. Cell Biol.* **73**, 751–756.

Egleman, E. H., and Orlova, A. (1995). Allostery, cooperativity, and different structural states in F-actin. *J. Struct. Biol.* **115**, 159–162.

Fawcett, D. W. (1981). "The Cell," Second ed. Saunders, Philadelphia.

Gerace, L. (1992). Molecular trafficking across the nuclear pore complex. *Curr. Opin. Cell Biol.* **4**, 637–645.

Gonatas, N. K. (1994). Contributions to the physiology and pathology of the Golgi apparatus. *Am. J. Path.* **145**, 751–761.

Hauri, H. P., and Schweizer, A. (1992). The endoplasmic reticulum–Golgi intermediate compartment. *Curr. Opin. Cell Biol.* **4**, 600–608.

Hughes, T. A., Pombo, A., McManus, J., Hozak, P., Jackson, D. A., and Cook, P. R. (1995). On the structure of replication and transcription factories. *J. Cell Sci.* Suppl. **19**, 59–65.

Kornfeld, S., and Mellman, I. (1989). The biogenesis of lysosomes. *Annu. Rev. Cell Biol.* **5**, 483–525.

Lee, C., Ferguson, M., and Chen, L. B. (1989). Construction of the endoplasmic reticulum. *J. Cell Biol.* **109**, 2045–2055.

Lodish, H., Baltimore, D., Berk, A., Zipursky, S. L., Matsudaira, P., and Darnell, J. Eds. (1995) "Molecular Cell Biology." W. H. Freeman and Co., New York.

Mandelkow, E., and Mandelkow, E. M. (1995). Microtubules and microtubule-associated proteins. *Curr. Opin. Cell Biol.* **7**, 72–81.

Mellman, I., and Simons, K. (1992). The Golgi complex: *in vitro veritas? Cell* **68**, 829–840.

Nickerson, J. A., Blencowe, B. J., and Penman, S. (1995). The architectural organization of nuclear metabolism. *Int. Rev. Cytol.* **162A**, 67–123.

Novikoff, A. B. (1976). The endoplasmic reticulum: a cytochemist's view. *Proc. Natl. Acad. Sci. USA* **73**, 2781–2786.

Nunnari, J., and Walter, P. (1992). Protein targeting to and translocation across the membrane of the endoplasmic reticulum. *Curr. Opin. Cell Biol.* **4**, 573–580.

Palade, G. E. (1956). The endoplasmic reticulum. *J. Biophys. Biochem. Cytol.* **2**, 85–98.

Pante, N., and Aebi, U. (1996). Molecular dissection of the nuclear pore complex. *Crit. Rev. Biochem. Mol. Biol.* **31**, 153–199.

Pavelka, M. (1987). Functional morphology of the Golgi apparatus. *Adv. Anat. Embryol. Cell Biol.* **106**, 1–94.

Peters, A., Palay, S. L., and Webster, H. deF. (1991) "The fine structure of the nervous system. Neurons and their supporting cells," Third ed. Oxford University Press, New York.

Peters, C., and von Figura, K. (1994). Biogenesis of lysosomal membranes. *FEBS Lett.* **346**, 108–114.

Robinson, M. S., Watts, C., and Zerial, M. (1996). Membrane dynamics in endocytosis. *Cell* **84**, 13–21.

Ruska, E. (1987). Nobel lecture. The development of the electron microscope and electron microscopy. *Biosci. Rep.* **7**, 607–629.

Schafer, D. A., and Cooper, J. A. (1995). Control of actin assembly at filament ends. *Annu. Rev. Cell Dev. Biol.* **11**, 497–518.

Seaman, M. N., Burd, C. G., and Emr, S. D. (1996). Receptor signalling and the regulation of endocytic membrane transport. *Curr. Opin. Cell Biol.* **8**, 549–556.

Shay, J. W. (1986). "Cell and Molecular Biology of the Cytoskeleton." Plenum Press, New York.

Spirin, A. S. (1986). "Ribosome Structure and Protein Biosynthesis." Benjamin/Cummings Publishing Co., Menlo Park, CA.

Tyler, D. (1992). "The Mitochondrion in Health and Disease." VCH Publishers, New York.

Tzagaloff, A. (1982). "Mitochondria." Plenum Press, New York.

Michael A. Lieberman and Richard G. Sleight

7

Energy Production and Metabolism

I. Introduction

Energy must be generated for all living organisms to grow and function. In animals, this energy is derived from the food consumed. The overall function of metabolism is to alter the food into chemical components that aid in cell growth and energy utilization. Two of the major functions of metabolic pathways are to provide energy to tissues when needed, and to store molecules as *potential energy* in times of energy excess. The pathways that **generate** energy are the **degradative**, or **catabolic**, pathways. The pathways that **store** energy are the **synthetic**, or **anabolic**, pathways. The anabolic pathways also produce many of the compounds required for cells to grow and divide, such as proteins, lipids, and nucleic acids. The synthesis of these compounds requires energy, which is provided by the catabolic pathways.

Metabolism can be viewed as two opposing sets of pathways, one leading to the biosynthesis of required compounds, the other to the degradation of these compounds. The catabolic pathways are used primarily under three sets of conditions. The first is when food supplies are plentiful. The food is degraded within the digestive tract into basic building blocks, and then converted into energy storage components (using the anabolic pathways). Degradative pathways are also used in the normal process of turnover of cell components. Many cellular constituents are degraded as they age in order to protect against the accumulation of damaged material within the cell. These constituents are then replaced by newly synthesized material. The third condition in which degradation is favored is when food supplies are scarce, such as during dieting and starvation. The degradative pathways are used to provide fuel for energy utilization. Because the biosynthetic pathways oppose the degradative pathways, they will be inactive when the degradative pathways are active. The major catabolic and anabolic pathways are outlined in Fig. 1.

Metabolism will be considered in the following manner. First, proteins are discussed in terms of their role as the **catalysts** of biochemical reactions. The regulation of **enzyme** activity will be introduced, along with a discussion of enzyme kinetics. This will lead to a more detailed discussion of enzyme regulation, before specific energy-generating or storage pathways are covered. The chapter concludes with a discussion of the metabolic changes seen during starvation, including specific changes within the liver, muscles, and fat cells.

II. Protein Enzymes

Protein molecules are composed of many amino acids linked together through covalent bonds. The sequence of amino acids within the protein determine both its shape and its function. Proteins perform many functions. Many proteins act as **catalysts** for the metabolic reactions that occur within a cell. These protein catalysts are known as **enzymes**. All metabolic reactions are catalyzed by enzymes. Enzymes enhance the rate at which metabolic reactions proceed without altering the equilibrium point of the reaction. For a reaction to proceed to completion, energy is required. Some of this required energy is utilized to correctly orient the molecules, so that the reactive groups are adjacent. Achieving the proper **spatial orientation** of the substrates during a reaction is known as reaching the **transition state** of the reaction. Once the transition state has been reached (see Fig. 2), the reaction will proceed. If a large amount of energy is required for the substrates to reach this transition state, the reaction will proceed very slowly, if at all. Enzymes alter the rate at which reactions proceed by **reducing the amount of energy required** to reach the transition state. This means that the rate at which the reaction will proceed is enhanced by the actions of the enzyme.

Because all of the reactions in metabolic pathways are catalyzed by enzymes, **regulation** of the pathways occurs at the enzyme level. The activity of an enzyme can be regulated in both positive and negative ways by specific **modifiers**. Understanding metabolic regulation is the key to understanding metabolism. There are several different levels of regulation. The first is known as **allosteric modification**, and requires that the modifier bind to a distinct site on the enzyme and alter enzyme activity. Table 1 lists examples of such enzymes, and indicates the **positive and negative regulators**. A second level of regulation is by **covalent modification**, in which one of the

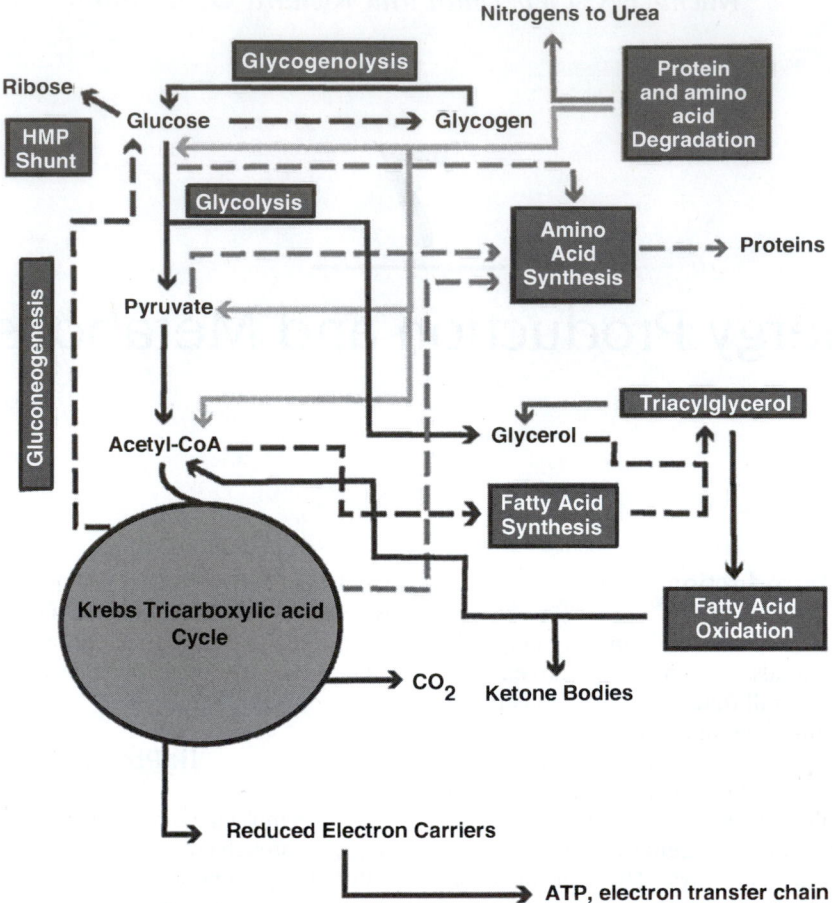

FIGURE 1.   The major pathways of metabolism. The pathway of **glycolysis** converts glucose to pyruvate, and is the entry of all sugars into metabolism. Intermediates along the pathway are used for **amino acid** biosynthesis and **triacylglycerol** formation. A key step of metabolism is the conversion of pyruvate to acetyl-CoA. Once acetyl-CoA is formed, it can be oxidized by the **Krebs tricarboxylic acid cycle** (TCA cycle) in order to produce reduced electron carriers. The reduced electron carriers then generate energy, in the form of ATP, via **oxidative phosphorylation** through the **electron transfer chain.** Intermediates of the TCA cycle are also used for amino acid biosynthesis. Excess glucose is stored in the form of **glycogen.** When energy is required, glycogen is degraded to glucose via **glycogenolysis.** Glucose can be synthesized from glycolytic and TCA-cycle intermediates via **gluconeogenesis.** Protein degradation leads to the generation of free amino acids, which are further metabolized to form intermediates of the TCA cycle and glycolysis. The major energy storage form of the cells is **triacylglycerol,** which is formed from glycerol and fatty acids. The glycerol is obtained from an intermediate of glycolysis, and the fatty acids are synthesized from acetyl-CoA. The pathway of triacylglycerol degradation leads to the formation of glycerol and acetyl-CoA, and derivatives produced from acetyl-CoA are known as **ketone bodies.** Glucose can be metabolized either through glycolysis or through the **hexose monophosphate shunt (HMP shunt)** pathway to produce the five-carbon sugar ribose. The HMP shunt is necessary to provide sugars for the biosynthesis of nucleotides, which are the building blocks of the nucleic acids **DNA** (deoxyribonucleic acid) and **RNA** (ribonucleic acid). In this diagram, the solid lines represent catabolic pathways, and the dashed lines anabolic pathways.

amino acid residues of the enzyme is chemically modified. This can occur in a number of ways, and examples of this type of modification are also presented in Table 1. In addition to natural modifiers of enzyme activity, many drugs act by inhibiting enzyme action. Thus, to understand drug action, it is important to understand how enzymes work and are regulated.

III. Enzyme Kinetics

The rate at which an enzyme works (Fersht, 1985) can be described by a value known as the **maximal velocity (V_{max}).** The V_{max} value is obtained by measuring the rate at which

the enzyme can form its products at various substrate concentrations. The V_{max} indicates how fast the enzyme can proceed when all of the available enzyme has bound substrate and is participating in the reaction.

Another kinetic parameter is the **Michaelis constant** (which will be discussed further), abbreviated as K_m. The K_m value is defined as the substrate concentration at which one-half of maximal velocity occurs. The K_m value reflects the **affinity** of the enzyme for its substrate, although it is not a true measure of that affinity. When a K_m value is low, it is an indication that the substrate **binds tightly** to the enzyme, and that half-maximal velocity can be obtained at low substrate concentrations. Conversely, a high K_m value reflects a reduced affinity of the

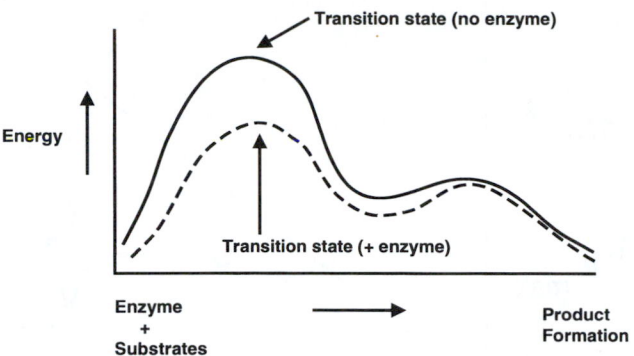

FIGURE 2. The energy required to reach the transition state of a reaction. The solid line indicates the energy required for the substrates to reach a transition state. The dotted line indicates how this energy requirement is reduced in the presence of an enzyme. By requiring less energy to reach the transition state, the reaction can proceed more rapidly, and the rate of the reaction will be enhanced.

enzyme for the substrate, and that a high substrate concentration is needed to reach half-maximal velocity. A typical enzyme kinetic plot, in which the initial velocity V of the reaction is measured as a function of increasing substrate concentration [S], is shown in Fig. 3.

Before one can understand how to mathematically represent the curve shown in Fig. 3, a number of terms need to be introduced. If one considers the reaction

$$E + S \underset{k_2}{\overset{k_1}{\rightleftharpoons}} ES \overset{k_3}{\rightarrow} E + P \qquad (1)$$

where E represents the enzyme, S represents the substrate that is being acted upon by the enzyme, P represents the product of the reaction, and ES represents an **enzyme-substrate complex**, which must form for the reaction to proceed. The velocity of this reaction is the rate at which the product is formed per unit time, such as moles of substrate per minute. The initial velocity of the reaction is the rate of product formation per

unit time when the reaction is first initiated. Under initial velocity conditions, there is virtually no back-conversion, i.e., conversion of P to S, due to the very high levels of S and the very low levels of P. This is what is represented by Eq. 1.

Two assumptions are made when one derives the mathematical equation that describes the curve shown in Fig. 3, and represents the reaction shown in Eq. 1. The first is that the reaction rapidly reaches equilibrium, or a steady state. This is true for the majority of biochemical reactions studied. The second assumption is that the product of the reaction does not go back to the starting material. This is not usually true for biochemical systems, but can be approximated by using the initial velocity of the reaction. The curve shown in Fig. 3 can be represented mathematically as

$$V = \frac{V_{max}}{1 + \dfrac{K_m}{[S]}} \qquad (2)$$

where V represents the initial velocity of the reaction, which is measured at different substrate concentrations [S] with a constant enzyme concentration. This equation is known as the **Michaelis-Menten** equation, named after the two scientists who initially derived it.

The curve shown in Fig. 3 is not very useful, as extrapolations need to be made to determine both the V_{max} and the K_m. Equation 2 can be manipulated algebraically into a form in which straight lines are generated. The equation which is derived is known as the **Lineweaver-Burk** plot. It is a **double reciprocal** plot, and is represented mathematically as

$$\frac{1}{V} = \frac{1}{[S]} \frac{K_m}{V_{max}} + \frac{1}{V_{max}} \qquad (3)$$

When $1/V$ on the y-axis is plotted against $1/[S]$ on the x-axis, a straight line is obtained, as shown in Fig. 4. The **slope** of the line represents the factor K_m/V_{max}, and the **y-intercept** represents the value $1/V_{max}$. It is much easier and more accurate to determine both the K_m and V_{max} from the Lineweaver-Burk plot than from the Michaelis-Menten plot shown in Fig.

TABLE 1 Examples of Allosteric Modifiers of Key Enzymes in Glycolysis and Glycogen Metabolism

Enzyme	Positive modifier	Negative modifier	Covalent modification
Phosphofructokinase-I (glycolysis)	AMP, fructose-2,6-bisphosphate	ATP, citrate	None
Pyruvate kinase (glycolysis)	Fructose-1,6-bisphosphate	Alanine	Phosphorylation inhibits in the liver
Phosphorylase kinase (glycogen metabolism)	Calcium	None	Phosphorylation activates
Phosphofructokinase-II (glycolysis)	Fructose-6-phosphate	Fructose-1,6-bisphosphate	Phosphorylation inhibits
Hexokinase (glycolysis)	None	Glucose-6-phosphate	None

This is only a partial list of the enzymes that can be modified either allosterically or by covalent modification, in this case by phosphorylation. Many other enzymes are also regulated in this way.

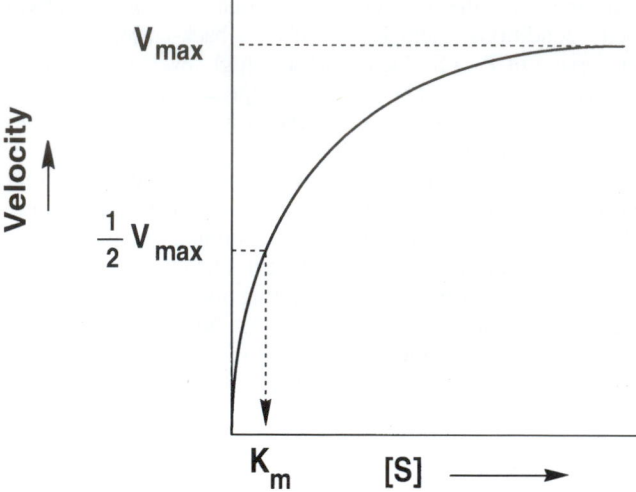

FIGURE 3.   A kinetic plot of reaction velocity V versus substrate concentration [S]. At saturation, the addition of more substrate does not increase the rate of the reaction, and the maximal velocity (V_{max}) is obtained. Once the V_{max} has been determined, the K_m can be determined, which is the substrate at which one-half maximal velocity occurs.

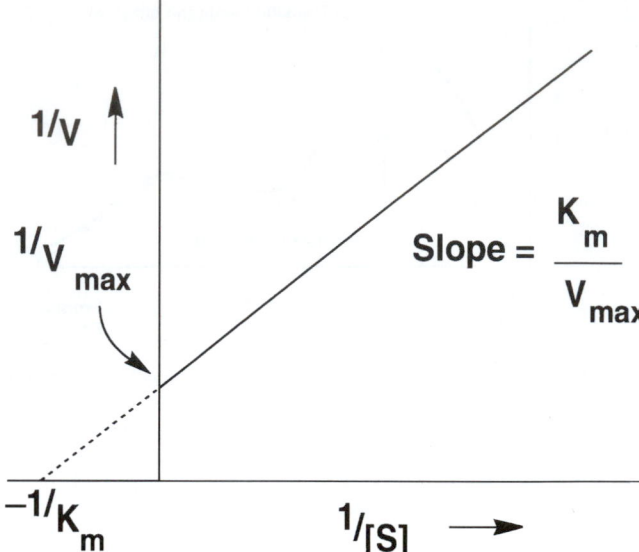

FIGURE 4.   An example of a Lineweaver-Burk plot. This is a more informative plot for analyzing kinetic data than the one shown in Fig. 3.

3. The Lineweaver-Burk plot is particularly useful for studying the effects of inhibitors on enzyme function, which is discussed in the next section.

IV. Enzyme Inhibitors

Since enzymes carry out the reactions of the metabolic pathways, regulation of the enzymes is very important physiologically. Enzymes can be activated or inhibited by various factors. In addition, many drugs act by inhibiting the action of enzymes. Aspirin, for example, inhibits the enzyme cyclo-oxygenase, which produces mediators of the pain response. Methotrexate, a potent anticancer agent, works by inhibiting the enzyme dihydrofolate reductase, which is needed to produce precursors for DNA synthesis. Inhibitors can work in many ways, but the two most common are termed **competitive inhibition** and **noncompetitive inhibition**. A competitive inhibitor can be overcome by adding excess substrate, as competitive inhibitors compete with the substrate for binding to the enzyme. Since there is competition for the substrate binding site, in the presence of a competitive inhibitor, the amount of substrate required to reach one-half maximal velocity will be increased. This will result in a greater K_m exhibited by the enzyme in the presence of inhibitor, as compared to in its absence.

A noncompetitive inhibitor does not affect substrate binding, but binds to the enzyme at a different site, reducing the ability of the enzyme to catalyze the reaction. A noncompetitive inhibitor, therefore, reduces the V_{max} of the enzyme. The K_m will not be affected by a noncompetitive inhibitor. These kinetic differences help to distinguish competitive from noncompetitive inhibitors.

The inhibition patterns exhibited by inhibitors are best represented graphically using the double reciprocal plot established by Lineweaver and Burk. In the presence of an inhibitor, Eq. 3 is altered. As a competitive inhibitor will also bind to the

enzyme, in competition with the substrate, the term K_i is used to indicate the **dissociation constant** of the inhibitor from the enzyme. This dissociation constant is a measure of the rate at which bound inhibitor will dissociate from the enzyme. If the effects of the inhibitor (I) are considered in the derivation of the Michaelis-Menten equation, Eq. 2 becomes

$$V = \frac{V_{max}}{1 + \frac{K_m}{[S]}\left(1 + \frac{[I]}{K_i}\right)} \tag{4}$$

In this equation, the concentration of inhibitor is indicated by [I], and the dissociation constant of the inhibitor from enzyme by K_i. Comparing Eq. 4 with Eq. 2 shows that the two expressions are very similar, except that the $K_m/[S]$ term is increased by the factor $(1 + [I]/K_i)$.

In the absence of inhibitor, the expression $(1 + [I]/K_i)$ would be 1, and Eq. 4 would be equivalent to Eq. 2. As the concentration of inhibitor is increased, the $[I]/K_i$ term increases, and the overall velocity of the reaction is reduced. When the reciprocal of Eq. 4 is calculated, the Lineweaver-Burk equivalent for the case of competitive inhibition is obtained as

$$\frac{1}{V} = \frac{1}{V_{max}} + \frac{K_m}{V_{max}}\left(1 + \frac{[I]}{K_i}\right)\frac{1}{[S]} \tag{5}$$

If $1/V$ is now plotted versus $1/[S]$, the y-intercept is $1/V_{max}$. and the slope of the line is given by

$$\text{slope} = \frac{K_m}{V_{max}}\left(1 + \frac{[I]}{K_i}\right) \tag{6}$$

Compared with the case in which no inhibitor is present, the slope has now been increased by the quantity $(1 + [I]/K_i)$. Thus, when the inhibitor concentration is zero (I is not present), the slope of the line is the same as for a non-inhibited

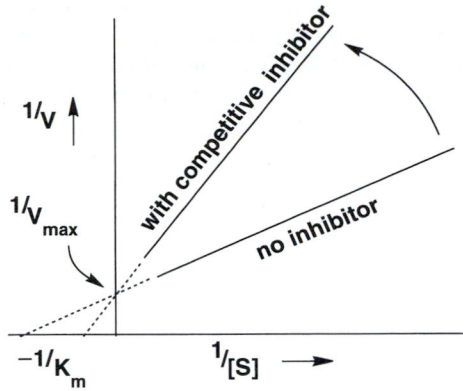

FIGURE 5. The kinetics of competitive inhibition. Note that the slope of the line will increase as the inhibitor concentration is increased. The lines will intersect on the y-axis, indicating that V_{max} is not altered in the presence of the inhibitor. As compared to the line in the absence of inhibitor, the slope of the line has been increased by the expression $(1 + [I]/K_i)$.

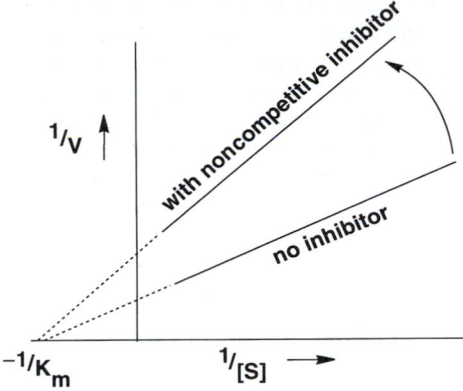

FIGURE 6. The kinetics of noncompetitive inhibition. Note that as in the case of competitive inhibition (Fig. 5), the slope of the line will increase in the presence of the inhibitor. In the case of noncompetitive inhibition, the lines will intersect on the negative x-axis, indicating that the binding of substrate to the enzyme is not inhibited by the inhibitor. However, the V_{max}, as represented by the y-intercept, decreases as the inhibitor concentration increases.

enzyme, namely K_m/V_{max}. However, as the inhibitor concentration is increased, the slope of the line increases, which results in an increase in the K_m. The kinetics of competitive inhibition are illustrated in Fig. 5.

As opposed to a competitive inhibitor, a noncompetitive inhibitor does not alter substrate binding to the enzyme. Rather, the inhibitor decreases the rate at which the enzyme can catalyze the reaction. Thus, K_m will not be altered, but V_{max} is decreased. The factor by which the velocity is reduced is equal to $(1 + [I]/K_i)$, the same as seen with competitive inhibition. The kinetics of noncompetitive inhibition is shown in Fig. 6. Notice that both lines (one in the presence of inhibitor, the other in the absence of inhibitor) intersect on the x-axis. In contrast, for competitive inhibition (Fig. 5), the two lines intercept on the y-axis. This reflects the fact that competitive inhibitors do not alter the V_{max}, which is **determined at the y-intercept**, whereas noncompetitive inhibitors do not alter the K_m, which is determined at the **extrapolated x-intercept**.

Regulation of metabolic pathways occurs primarily at the enzyme level. Enzymes can be regulated by molecules known as **allosteric effectors**, which bind to the enzyme and alter its activity. The binding of the effector leads to a shape change in the protein. The shape change then either makes it **easier** (in the case of an activator) or **more difficult** (in the case of an inhibitor) for the enzyme to catalyze a reaction. The kinetics of such regulated enzymes are shown in Fig. 7. The curves obtained are complex, but the end result is that sigmoidal kinetics are observed in a typical [S] versus V plot. In the presence of an inhibitor, the entire curve gets shifted to the right, indicating that higher levels of substrate are required to reach an equivalent level of activity. In contrast, activators shift the curve to the left, indicating that less substrate is required to obtain maximal activity. Another way of stating this is that activators reduce the K_m of the enzyme, whereas inhibitors increase the K_m. **It is important to realize that when a regulated enzyme is inhibited, it means that a higher concentration of substrate is required be-**

fore the reaction can proceed rapidly; at normal substrate levels, the reaction will proceed slower than in the absence of the inhibitor. Other means of regulating enzyme activity will be discussed later.

V. Metabolic Pathways

The major biochemical constituents of a biological cell are **amino acids, carbohydrates, nucleic acids,** and **lipids**. Examples of these four types of compounds are shown in Fig. 8. There are pathways to synthesize and degrade each of these key components. The major function of the degradative

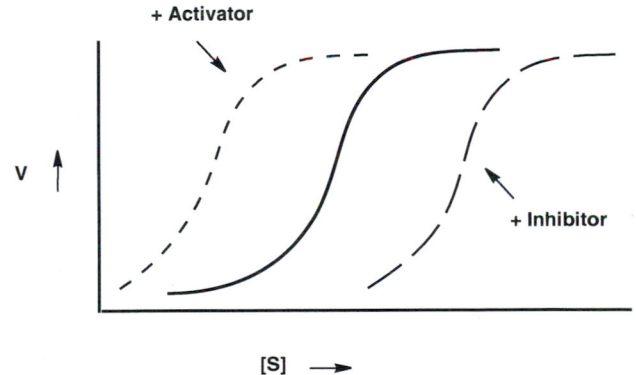

FIGURE 7. Kinetics of regulated enzymes. Regulated enzymes typically display sigmoidal kinetics, in which the activity of the enzyme increases dramatically over a small substrate concentration range. An inhibitor of the enzyme will shift the entire curve to the right, thereby increasing the level of substrate required to obtain one-half maximal velocity. An activator, however, will shift the curve to the left, in effect decreasing the level of substrate required to reach one-half maximal velocity. The effect of activation, therefore, is to allow the reaction to proceed more rapidly at lower substrate concentrations than in the absence of the activator.

pathways is to generate energy. When these pathways are activated, certain energy storage molecules are degraded to obtain energy. The energy that is obtained from the storage molecules is predominantly in the form of compounds containing **high-energy bonds**. A high-energy bond is one which will release a large amount of energy when the bond is broken. The key energy-containing molecules within the cell are nucleotide triphosphates, such as **adenosine triphosphate (ATP)**. The structure of ATP is shown in Fig. 9. Note how this molecule consists of a five-carbon sugar (ribose), three phosphates, and a nitrogenous base, which are the same components as the nucleic acids. The nucleic

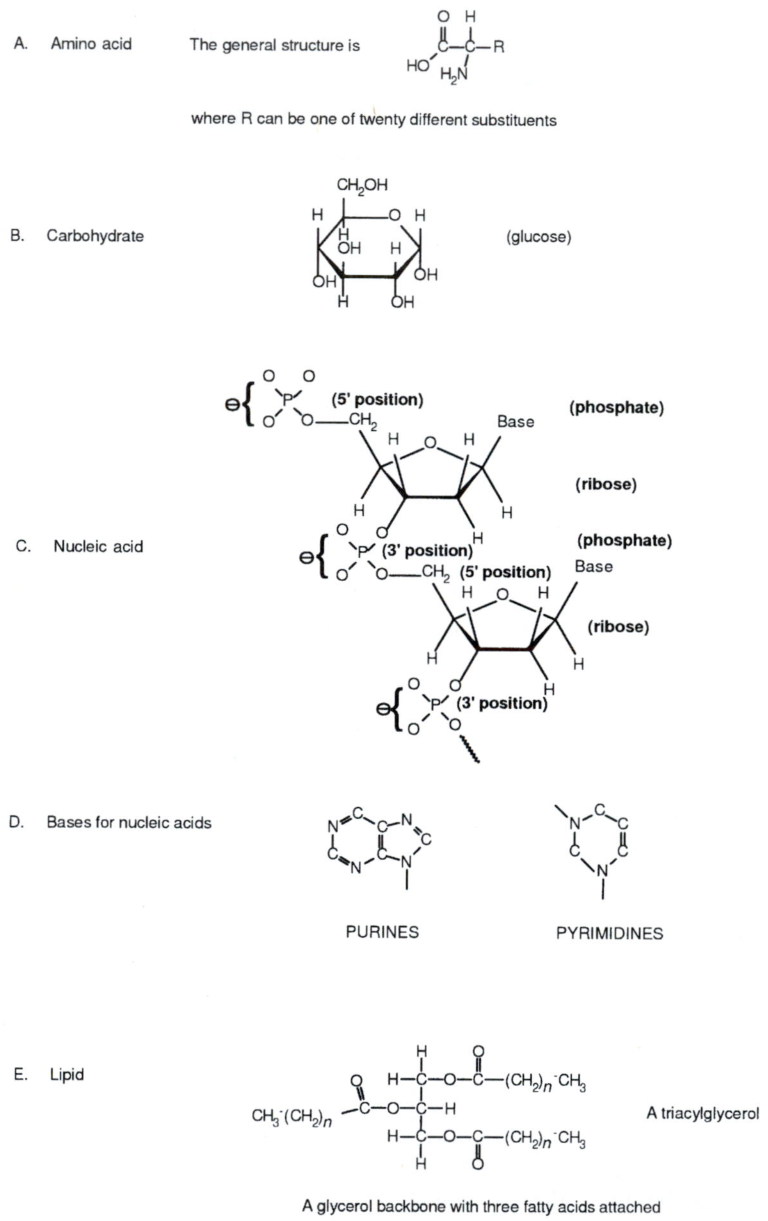

FIGURE 8.   Structures of the major chemical constituents of a cell. (A) The general structure of an amino acid. There are 20 different amino acids found in proteins, each differing in the R group attached to the central carbon atom. The R group ranges from components as simple as a proton (for glycine) to components as complex as a conjugated indole ring (for tryptophan). Peptide bonds are formed between the amino group of one amino acid and the carboxyl group of a second amino acid. (B) The structure of glucose as a representative carbohydrate. The molecule consists of a carbon chain with hydroxyl groups on the majority of carbons. (C) A schematic of a nucleic acid. Nucleic acids, such as DNA and RNA, are complex structures consisting of a phosphate, a nitrogenous base, and the five-carbon sugar ribose (for RNA) or deoxyribose (for DNA). The phosphates link different ribose-base units together to form a linear structure. The bases (D) are classified as either purines or pyrimidines based on their general structure. (E) The structure of a triacylglycerol. There is a glycerol backbone to which is esterified three fatty acids. Fatty acids consist of long hydrocarbon chains with a terminal carboxylic acid group.

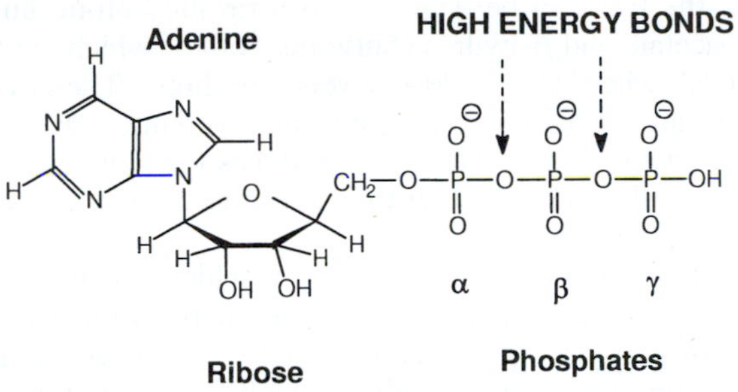

FIGURE 9. The structure of the high-energy compound ATP. ATP consists of the purine base adenine, the sugar ribose, and three phosphates linked in series to the number 5 carbon of the ribose. The high-energy bonds are indicated by the arrows. ATP and the other nucleotide triphosphates serve as the precursors of the nucleic acids, in addition to ATP's role as the major energy carrier of the cell.

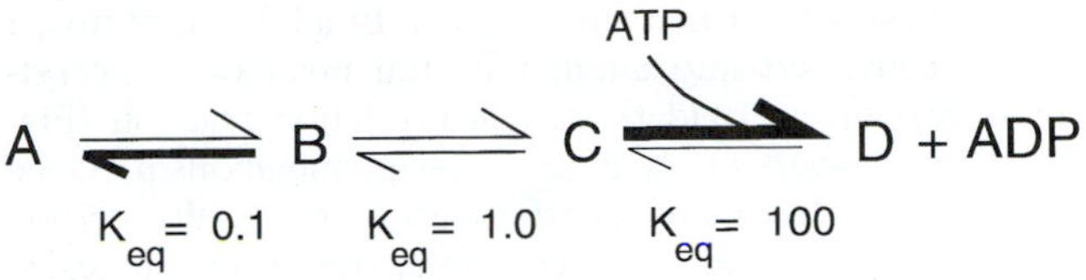

FIGURE 10. Linking an unfavorable reaction with a favorable reaction to create a pathway. In the example shown, the reaction A→B is unfavorable, such that at equilibrium 90% of A has not reacted. The reaction B→C has an equilibrium constant of 1, indicating that it is a freely reversible reaction. This means that at equilibrium 50% of C will have been converted to D. The reaction C→D, however, is essentially irreversible due to the release of energy when a high-energy bond of ATP is hydrolyzed. This means that at equilibrium only 1% of the original C will be present, since the other 99% will have been converted to D. However, as C is being converted to D, this affects the equilibrium between B and C. As C decreases in concentration, more B will be converted to C to maintain the equilibrium constant of 1.0. This has the net effect of decreasing the level of B within the reaction system. As the concentration of B drops, this will affect the equilibrium between A and B. Even though that is an unfavorable reaction, as the concentration of B drops, more A will be converted to B to maintain the A–B equilibrium. The net effect of coupling these three reactions is to push the unfavorable first reaction to completion, due to the hydrolysis of ATP and release of energy at the third step of the pathway.

acids are, in fact, synthesized by the condensation of the nucleotide triphosphates. The high-energy bonds within ATP are between the α and β phosphates and between the β and γ phosphates. Hydrolysis of one of these bonds results in the release of **7.3 kilocalories per mole** of energy, which for a biological system is a large amount. The high-energy bonds of ATP are also found in the other nucleotide triphosphates, such as GTP, CTP, TTP, and UTP.

How is the energy in ATP and other high-energy intermediates used? The primary role of these high-energy compounds is to release the energy at appropriate steps in a pathway to allow reactions that require energy in order to reach their transition state to proceed. One example of this is muscle contraction (Alberts *et al.*, 1989). Energy is required for muscles to contract. ATP is used to allow molecules (in myofilaments) to slide past one another, which brings about contraction. When ATP levels are low, muscle contraction is inhibited. Another example of energy use is to couple a number of enzyme-catalyzed reactions together, in the form of a pathway, in order to link thermodynamically-unfavorable reactions with those that are highly favorable. This is illustrated in Fig. 10. Notice how, by coupling an unfavorable reaction with one that releases a large amount of energy, a favorable sequence of reactions occurs despite the presence of the unfavorable reaction.

The key pathways of metabolism are outlined in Fig. 10 (Lehninger *et al.*, 1993). These pathways represent one of the unities of life, as they are common in both bacteria and humans. As stated previously, there are both anabolic and catabolic pathways. One of the major catabolic pathways is **glycolysis**, which is the major pathway for the conversion of carbohydrates (such as glucose and fructose) to pyruvate, a three-carbon intermediate. This pathway does not require oxygen, and generates a small amount of energy in the form of ATP during the reaction sequence. Glycolysis is the major pathway of sugar metabolism, as all carbohydrates eventually feed into this pathway.

The metabolism of five-carbon sugars (such as ribose) is connected to glycolysis at a number of points, through a pathway known as the **hexose monophosphate shunt (HMP shunt)**. The HMP shunt pathway performs two major roles. The first is the conversion of five- and six-carbon sugars. Five-carbon sugars are required for the synthesis of nucleoside

triphosphates, which are then used for nucleic acid biosynthesis. The second role of the HMP shunt is to generate a molecule known as reduced **nicotinamide adenine dinucleotide phosphate (NADPH)**. NADPH is an **electron carrier** required by many biosynthetic pathways. It is used to **donate electrons** for **oxidation-reduction reactions**, as illustrated in Fig. 11. Another major electron carrier is reduced **nicotinamide adenine dinucleotide (NADH)**, which has the same basic structure as NADPH, but does not contain the phosphate group. As will be seen shortly, the electrons carried by NADH or NADPH can be donated to oxygen to form both ATP and water.

The three-carbon molecule pyruvate plays an important central role in metabolism. It represents a key point in the determination of the eventual fate of the carbon atoms. Pyruvate can be converted into glucose through the process of **gluconeogenesis**. It can also be used as a precursor for

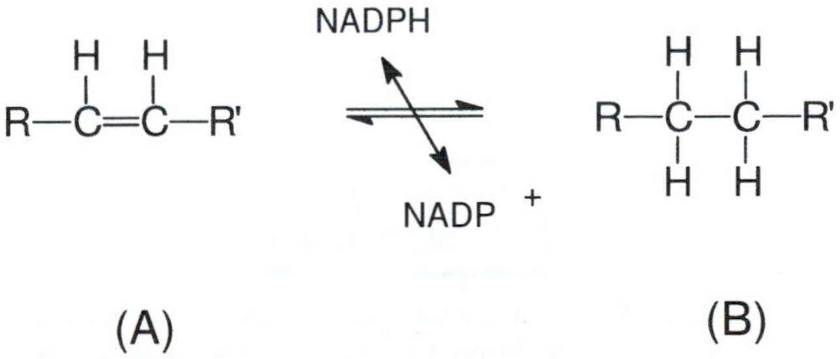

FIGURE 11. An oxidation reaction. Reactant A is oxidized, as compared to B. NADPH can donate its electrons, in the form of a hydride ion, to A to reduce the double bond to a single bond. In the course of this reaction, A is reduced to produce B, and the NADPH is oxidized to form NADP$^+$. Compounds that gain electrons are reduced, those that lose the electrons are oxidized. Oxidation-reduction reactions are common in biochemical pathways, and almost all require the participation of an electron carrier.

the synthesis of various amino acids. In addition, pyruvate can be converted into another central metabolite, **acetyl-CoA**, through an oxidative decarboxylation reaction (Fig. 12). This conversion is a key step in metabolism. Once pyruvate has been converted to acetyl-CoA, the carbon atoms can only be used to generate energy or to synthesize fatty acids, ketone bodies, or steroids. The carbons of acetyl-CoA cannot be used for amino acid or carbohydrate synthesis. If acetyl-CoA is used for energy production, it is oxidized to carbon dioxide (CO_2) and water through the **Krebs tricarboxylic acid cycle (TCA cycle)**, a pathway that requires oxygen to function. The TCA cycle generates a large amount of NADH, which donates its electrons to the electron transport chain. Through a series of electron transfers, the electrons are used to **reduce molecular oxygen (O_2) to water (H_2O)**, and during this process, energy is generated in the form of ATP. The **electron transport chain** is the major energy-producing pathway of the cell, and the mechanism of how energy is produced by electron transfer is discussed later in this chapter.

Fatty acids (FAs) can be synthesized from acetyl-CoA. Fatty acids play a number of roles within the cell. Their main role is as an energy source. In this role, fatty acids are stored in the form of triacylglycerol, which consists of a glycerol molecule to which three fatty acids have been esterified (refer to Fig. 8). A secondary role of fatty acids is as a **second messenger** (Lands, 1991) in the form of prostaglandins, leukotrienes, and prostacyclins (more information on this topic is provided in Chapter 9). Another secondary role of fatty acids is as an essential component of cell membranes (in the form of phospholipids), in which two aqueous compartments are separated by a hydrophobic barrier.

The energy from fatty acids can be obtained by their oxidation and conversion to acetyl-CoA, which is then further oxidized to CO_2 and water through the TCA cycle and electron transport chain. In addition, under specialized conditions, the FAs can be degraded to form the **ketone bodies** acetoacetate and β-hydroxybutyrate, both of which are synthesized when acetyl-CoA levels are high. These compounds are made under conditions in which the body has reduced energy stores. The liver produces the ketone bodies, which are then sent to other tissues for use as alternative energy supplies.

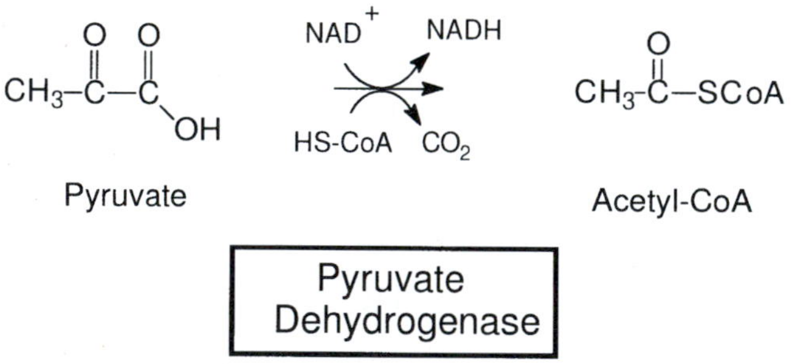

Diabetes mellitus is a disease in which glucose metabolism is altered, such that adequate energy cannot be derived from dietary glucose. This can occur when the pancreas stops producing **insulin**, which is required for the proper metabolism of glucose. Because of this problem, untreated diabetics produce excessive levels of ketone bodies in order to supply energy to all tissues of the body. As the ketone bodies are acids, excessive ketone body production can lead to **diabetic ketoacidosis** (Foster and McGarry, 1983), a condition that leads to a reduced pH in the blood and urine. Immediate treatment of this condition is required to prevent coma and death.

Glycogen is the carbohydrate storage form used by the cell. It consists of many glucose molecules organized in a branched structure. Glycogen is synthesized when the carbohydrate content of the cell is high, and is degraded when the cell or body requires more glucose. The pathways of glycogen synthesis and degradation are different, although the manner in which these pathways is regulated is similar. Regulation of these pathways will be discussed later in this chapter.

Amino acids, in addition to being the building blocks for proteins, can also be used for energy production via degradative pathways. When amino acids are degraded, they can be classified as either glucogenic or ketogenic. **Glucogenic amino acids** can give rise to glucose, through the pathway of gluconeogenesis. **Ketogenic amino acids** can only give rise to acetyl-CoA and ketone bodies, which can generate energy, but not glucose. Upon degradation, the amino acids are converted into acetyl-CoA or intermediates of either glycolysis or the TCA cycle. Humans can only synthesize 12 of the 20 required amino acids, because key enzymes of the biosynthetic pathways found in other organisms are not expressed in humans. The eight amino acids that cannot be synthesized are referred to as the **essential amino acids**, since they can only be obtained from the diet.

When amino acids are degraded, the amino group requires special disposal. Excessive levels of ammonia are toxic in all species, and different species have devised different methods for nitrogen disposal. Earthworms secrete ammonia directly, which diffuses away in the soil, whereas birds produce uric acid for nitrogen disposal. In mammals, the nitrogen is disposed of in the form of urea, which is synthesized in the liver via the **urea cycle**. Through use of this cycle, toxic levels of ammonia do not accumulate, and are instead converted into a nontoxic form, urea.

FIGURE 12. The oxidative decarboxylation reaction in which pyruvate is converted into acetyl-CoA. CoA represents coenzyme A, a vitamin-derived cofactor that activates the carboxyl group of acetate to allow it to participate in the subsequent reactions. The enzyme that catalyzes this reaction is pyruvate dehydrogenase. The reaction itself involves both a decarboxylation of pyruvate (loss of carbon dioxide) and an oxidation of pyruvate to form acetyl-CoA. Since the pyruvate is being oxidized, another molecule must accept the electrons that are being lost, and that is done by NAD^+ to form NADH (reduced nicotinamide adenine dinucleotide).

VI. Generation of Energy: Mitchell Chemiosmotic Hypothesis

One of the key functions of the catabolic pathways is the generation of energy, usually in the form of ATP. Although small amounts of ATP are produced through the glycolytic pathway, most of the ATP is produced through **oxidative phosphorylation**. These reactions occur within the **mitochondria**, whereas glycolysis occurs in the cytosol. Oxidative phosphorylation refers to the coupling of the energy produced by the energetically favorable reduction of oxygen to the phosphorylation of ADP to form ATP. The electrons used to drive this process are obtained from the TCA cycle through the oxidation of acetyl-CoA to CO_2. Electrons are obtained via electron carriers, such as NADPH, NADH, and $FADH_2$. $FADH_2$ is an electron carrier that has not been previously discussed, but is primarily generated during both the oxidation of fatty acids and in one reaction of the TCA cycle. As with NADPH and NADH, $FADH_2$ can donate its electrons to the electron transport chain to generate energy. The process of oxidative phosphorylation is shown schematically in Fig. 13.

Mitochondria are intracellular organelles that contain two membranes. The outer membrane is porous to molecules of less than 5000 daltons, and will allow the small molecules (such as electron carriers, ADP, and oxygen) required for oxidative phosphorylation to reach the inner mitochondrial membrane. The electron carriers of the electron transport chain are located within the inner mitochondrial membrane. The inner mitochondrial membrane is impermeable to virtually everything, including protons. The only compounds that can traverse this membrane are those for which specific carriers exist within the membrane, and thereby can transport the molecule into the lumen of the mitochondria.

Upon acceptance of electrons from either NADH or $FADH_2$, the electrons are transferred in an orderly fashion through the carriers to the terminal **electron acceptor** oxygen, which accepts the electrons and becomes reduced,

forming water. As the electrons are transferred from carrier to carrier within the membrane, protons are transferred from the inside to the outside of the inner mitochondrial membrane. As protons are positively charged, the active extrusion of protons during electron transport leads to the loss of positive charge from within the mitochondrion. This leads to both a difference in pH across the inner mitochondrial membrane, as well as a difference in charge. These H^+ differences across the membrane lead to a **proton-motive force** (PMF) across the membrane. This force consists of two components, ΔpH (change in pH across the membrane) and $\Delta\psi$, the difference in membrane potential (charge) across the membrane. The PMF (whose units are kcal/mol) can be expressed as

$$PMF = 2.3RT \ \Delta pH + z \ \mathscr{F}\Delta\psi \quad (7)$$

where R is the **gas constant** ($1.98 \ cal \cdot mol^{-1} \cdot K^{-1}$), T is the absolute temperature in kelvins (K), z is the charge on the particle (which in this case is 1, as the particle is a proton), and $\mathscr{F}$ is the **Faraday constant** ($23 \ kcal \cdot mol^{-1} \cdot V^{-1}$). The magnitude of the PMF is directly proportional to both the change in pH across the membrane and the difference in electrical potential across the membrane. By establishing a **proton gradient** in this manner, it becomes energetically favorable for protons to return to the mitochondria from the cytoplasm. This is because the protons will be traveling down both a concentration gradient and an electrical gradient, entering the compartment that contains fewer protons and is negatively charged. These two gradients are often lumped together as the **electrochemical gradient**.

Because the entry of protons into the mitochondria is energetically favorable (i.e., downhill) when the PMF is established, energy can be generated from proton entry. Protons are normally impermeable to the mitochondrial membrane. If this were not the case, a proton gradient could never be established by the electron transfer chain in the first place. Thus, for protons to enter the mitochondria, enzymes are required to facilitate the transport. There exists in the inner mitochondrial membrane an enzyme called the **proton translocating ATP synthase** (ATPase). This enzyme catalyzes the reversible reaction shown as

$$ADP + P_i \rightleftharpoons ATP + H_2O \quad (8)$$

The energy needed to synthesize ATP is derived from the inward transport of the proton down its chemical and electrical gradients. When the enzyme degrades ATP, protons are excluded from the mitochondria to contribute to the generation and maintenance of the PMF. This process **couples oxidation to phosphorylation**. The process of oxidation (converting oxygen to water) generates a PMF, and this leads to phosphorylation using energy generated by the inward movement of protons down their electrical and chemical gradients, to form a high-energy bond of ATP.

The elucidation of oxidative phosphorylation, and the demonstration of the PMF, was done by Peter Mitchell in the 1960s and 1970s, and the theory is frequently referred to as **Mitchell's chemiosmotic hypothesis** (Mitchell, 1963). Although ridiculed when first proposed, the theory has now gained wide acceptance. All data pertaining to the function of the electron transfer chain and the ATPase can be explained

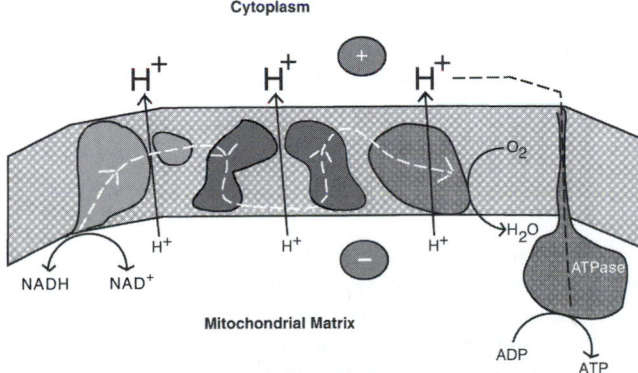

FIGURE 13. Generation of ATP by the Mitchell chemiosmotic hypothesis. In this model the asymmetric orientation of the membrane-bound electron transport chain results in the translocation of protons from the inside of the mitochondria (matrix) to the cytoplasm of the cell as the electrons funnel through the chain. The translocation of protons leads to both a proton and electrical gradient being established across the mitochondrial inner membrane. The end result of these gradients is that it is now energetically favorable for a proton to go down both its electrical and chemical potentials and enter the mitochondrion. This energy is harnessed by the proton-translocating ATPase to synthesize ATP from ADP and inorganic phosphate.

by this theory. Peter Mitchell was honored for his achievement with the Nobel Prize in Medicine in 1979. Oxidative phosphorylation is the major energy-generating pathway of the cell, and it requires O_2. In the absence of O_2, ATP is generated only through the glycolytic pathway, at a much reduced efficiency as compared to oxidative phosphorylation. However, even in the absence of O_2, a PMF can still be established by pumping protons out of the cell using the proton-translocating ATPase. This is important because a number of systems, including the transport of carbohydrates and amino acids into intestinal epithelial cells, require the energy generated by the inward movement of protons down their concentration gradient.

VII. Food and Energy

Because mammals do not eat continuously, mechanisms have been developed to allow energy derived from one meal to be stored and used as needed. There are two major energy storage molecules in the body. These are **glycogen**, which is a carbohydrate-based storage medium, and **triacylglycerol**, which is a lipid-based storage medium. Glycogen is stored in the tissues that synthesize it, which are primarily the liver and muscle. Glycogen is stored in small granules, along with the enzymes that both synthesize and degrade it. There are no specialized cells for the storage of glycogen. Lipids, which are stored in the form of triacylglycerol, are stored in specialized cells known as **adipocytes**. As more energy is stored, the size and fat content of the adipocyte increases. The roles of these two storage molecules are different, as will be outlined shortly.

The energy that is used for cell function and growth is obtained from food in the current diet, or from storage molecules synthesized when excess food was available. How is food processed so that energy can be obtained from it? In multicellular organisms, this task is carried out by specialized cells found within the digestive tract. In mammals, the food enters the stomach, where digestive enzymes from the saliva, stomach, and pancreas act to reduce it to smaller particles. This allows the smaller components to be absorbed by the cells lining the gut and intestines. Certain compounds, such as ethanol, can be absorbed directly by the stomach, but the majority of nutrients are absorbed within the intestine. Once absorbed by the **intestinal epithelial cells**, the small components may or may not be metabolized before being transported across the cell into either the blood or lymphatic system, which carries the nutrients to other organs and cells. This process is known as **vectorial transport** (see Fig. 14). During vectorial transport, the nutrients, which are at a high concentration in the lumen of the intestine, are transported down their concentration gradient into the epithelial cell. As the concentration of the nutrients increases in the epithelial cell, they are then transported out of the cell into the blood or lymph down their concentration gradient. The transport event can be either **active** or **carrier-mediated facilitated diffusion**. Active transport refers to the process of actively concentrating material within a compartment. Active transport can occur even when the concentration of the nutrients is greater inside the cell than outside. In order to accomplish active transport, energy must be expended. Carrier-mediated facilitated diffusion is not energy dependent, and refers to

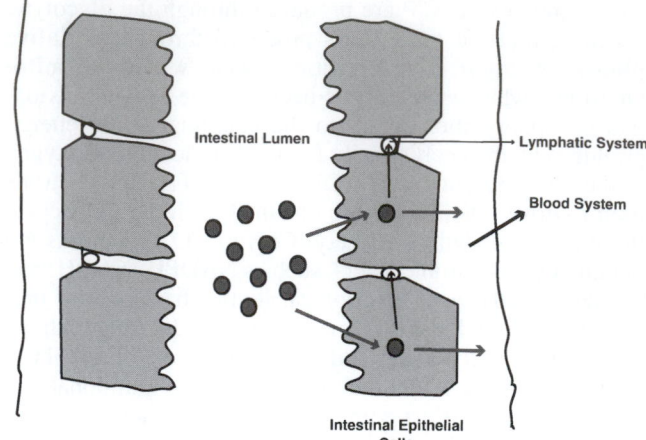

FIGURE 14. Vectorial transport by the intestinal epithelial cell. The required nutrient (gray circle), is transported into the cell from the intestinal lumen, down its concentration gradient. Once inside the cell, the nutrient may or may not be metabolized, and then it, or its metabolic product, is transported across the other side of the cell into either the blood or the lymph. Removing the nutrient from the cell ensures that the concentration of the nutrient within the epithelial cell will be lower than the concentration of the nutrient in the intestinal lumen, and that as much of the nutrient as possible can be absorbed from the lumen.

the use of membrane-bound carriers to transport nutrients down their concentration gradient. Both mechanisms of transport are used during intestinal absorption of nutrients.

Once food that has been ingested is broken down into constituent sugars, amino acids, and fatty acids, it is transferred to the blood or lymphatic system. Other tissues of the body will utilize these compounds. They will be used either for storage, so that energy can be obtained at a later time, or for the immediate generation of energy. **These two opposing pathways are never operative at the same time.** In order to understand why this is so, the basic principles of metabolic regulation need to be addressed.

VIII. Basic Pathways That Need to be Regulated

Of key importance is the regulation of pathways that either generate or store energy. The structures of the energy storage molecules glycogen and triacylglycerol are shown in Fig. 15. Glycogen synthesis occurs when blood sugar levels are elevated, indicating that adequate energy is available. Glycogen is stored within the tissues that make it, primarily the liver and muscle. As will be seen shortly, the role of glycogen is different in these two organs. Triacylglycerol is the primary fuel storage compound of mammals. Triacylglycerol, unlike glycogen, is stored in specialized cells known as adipocytes, which have an almost unlimited capacity to store fat in the form of triacylglycerol. Similar to glycogen, triacylglycerol is produced when blood glucose levels are high, to store energy for later use. The building blocks of triacylglycerol, fatty acids, are synthesized from acetyl-CoA, which is obtained from pyruvate.

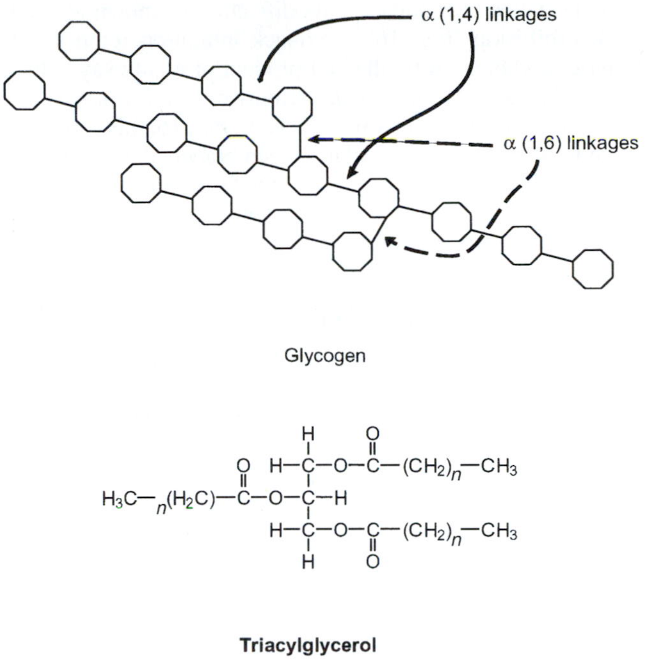

Glycogen

Triacylglycerol

FIGURE 15. The structures of the major energy storage forms in the cell, glycogen and triacylglycerol. The top structure is a schematic drawing of glycogen, which is a branched polymer of glucose (each hexagon represents a separate glucose molecule). The basic linkage between glucose residues occurs between carbons 1 and 4 of adjacent glucose molecules. The branches are formed by linkages between the 1 and 6 carbons of adjacent glucose molecules. The bottom structure is of triacylglycerol, which is composed of a glycerol backbone to which three fatty acids are esterified.

Fatty acid degradation also leads to the production of acetyl-CoA. As acetyl-CoA cannot form glucose, the oxidation of fatty acids can only lead to energy production, and not gluconeogenesis. This has important complications for certain physiological conditions, which will be discussed later.

A. Synthesis versus Degradation: Regulatory Principles

The metabolic pathways that synthesize the basic compounds are different from those responsible for their degradation. For example, the pathways of fatty acid synthesis and degradation are different and are compartmentalized. The rationale for this is to prevent the occurrence of **futile cycles**, in which a compound will be synthesized and then immediately degraded by the degradative pathway. Such futile cycles lead to no net gain of material and a useless loss of energy. By having alternate pathways for both the synthesis and degradation of a compound, futile cycles can be avoided by coordinate regulation of the pathways. **Typically, conditions that activate biosynthetic pathways simultaneously inhibit the degradative pathways.**

There are a number of different ways in which pathways can be regulated. Some of these are listed in Table 2. Regulation of pathways occurs by altering the activity of one or more enzymes specific to that pathway, such as controlling

TABLE 2 Different Methods for Regulating Enzyme Activity

Method of Regulation	Example
Change amount of enzyme through alterations in gene transcription	Enzymes required for fatty acid biosynthesis
Allosteric modifications	Regulated enzymes of glycolysis
Covalent modifications	Regulated enzymes of glycogen metabolism

the amount of enzyme present. Upon activation of the pathway, specific gene transcription is induced, which leads to more enzyme being made by the protein synthesis machinery of the cell. This is known as **up-regulating** the enzyme. By producing more enzyme, increased levels of substrate can be acted on per unit time, and the pathway reaction will proceed at an overall faster rate. Conversely, if a certain pathway is not used for an extended period of time, the enzymes of that pathway are **down-regulated**, less enzyme is made, and the pathway will be less active. Because this type of regulation requires alterations at the genomic level, they do not occur rapidly, but rather require four to seven days before the changes in enzyme levels become apparent. Such regulation of enzyme activity is known as **long-term adaptation**. Examples of enzymes regulated by long-term adaptation are the five enzymes necessary for synthesizing fatty acids. When food supplies are plentiful over a four- to seven-day period, the synthesis of these enzymes is increased; under conditions of minimal food intake, the synthesis and levels of these enzymes is decreased.

Long-term adaptation to environmental stimuli, while an important regulatory control, does not provide for a rapid means of altering enzyme activity. This is frequently desired, particularly under conditions of stress and exercise. In order to rapidly regulate pathways, other control mechanisms are utilized. These are (a) **allosteric modification** of enzyme activity and (b) **covalent modification** of the enzyme, which leads to an alteration in enzyme activity. The kinetics of allosteric enzymes has been discussed previously (Fig. 7). These enzymes have their activities either enhanced (activated) or decreased (inhibited) by small allosteric effectors. Activators tend to decrease the K_m of the enzyme, and inhibitors tend to increase the K_m of the enzyme. An example of an enzyme regulated in this manner is **phosphofructokinase-1 (PFK-1)** (Pilkis *et al.*, 1990) in the key regulatory step of glycolysis.

PFK-1 is **activated** by either **AMP** or **fructose-2,6-bisphosphate (F2,6BP)**, and is **inhibited** by either **citrate** or **ATP**. To understand why these compounds are used as regulators, it is necessary to comprehend the role of the pathway being regulated. The major role of glycolysis is to produce energy. This occurs in two ways. The first is anaerobic production of **ATP** by the pathway. The second is to rapidly produce **pyruvate**, which will be converted to acetyl-CoA, which when oxidized by the TCA cycle will generate energy through oxidative phosphorylation. Thus, under conditions in which energy is required, glycolysis is activated. High levels of **ATP** indicate that adequate energy levels are available, and glycolysis should not be activated. Thus, ATP inhibits PFK-1 and reduces the glycolytic rate. Alternatively, when ATP levels are low, and **AMP** levels elevated, it is an indication that energy is required by the cell. AMP will therefore activate PFK-1 in order to increase ATP levels. This activation continues until sufficient ATP is produced, at which point the newly-synthesized ATP will inhibit enzyme activity. As AMP is a precursor to ATP, the increase in ATP will also lead to a decrease in the levels of AMP, thereby simultaneously decreasing the concentration of an activator of the enzyme.

Citrate is produced in the mitochondria as an intermediate of the Krebs TCA cycle, whose major function is to produce energy. High energy levels will inhibit the rate at which this cycle functions. When energy levels are high within the mitochondria, the TCA cycle slows down, and this leads to an increase in the concentration of citrate. Thus, high levels of citrate indicate adequate levels of energy. Citrate in the mitochondria can diffuse across the mitochondrial membrane and into the cytoplasm. Once in the cytoplasm, the citrate binds to PFK-1 and inactivates it in a manner similar to ATP. Remember that for citrate levels to increase in the cytoplasm, adequate energy levels must be available. If adequate ATP is available, the rate of glycolysis should decrease.

The role of F2,6BP is more complex. The levels of F2,6BP increase in response to **hormones** that monitor blood glucose levels and indicate to the liver whether energy needs to be created or stored. It is through the modulation of F2,6BP levels that hormonal signals can regulate glycolysis. The enzyme that produces F2,6BP is activated as the result of specific hormone binding to the cell surface (in this case the hormone is **insulin**). This enzyme is inhibited when a different hormone (such as **glucagon**) binds to the cell. When the enzyme is activated, F2,6BP levels increase and PFK-1 is activated, thereby enhancing glycolysis. Hormonal control of metabolic pathways is very important, and will be re-emphasized when the regulation of glycogen metabolism is discussed.

A third means of regulating enzyme activity is through covalent modification of the enzyme. An example of this is the enzyme **pyruvate kinase (PK)**, another regulated enzyme in the glycolytic pathway. Pyruvate kinase catalyzes the reaction that produces pyruvate and ATP from phosphoenolpyruvate. Upon appropriate hormonal stimulation, an enzyme that can phosphorylate PK in the liver is activated. Enzymes that phosphorylate other proteins are known as **kinases**. When PK is phosphorylated, its activity in liver is reduced as a result of the covalent modification. A reduction in PK activity results in less pyruvate being produced, and thus leads to an overall reduction in the glycolytic rate. These changes correlate with the regulation of PFK-2 by F2,6BP. It was previously stated that the production of F2,6BP is hormonally regulated. The changes in PK activity and F2,6BP production are **coordinately regulated**. The hormones that lead to increases in F2,6BP production (such as insulin) also lead to activation of PK. This ensures that glycolysis is maximally active when insulin levels are high. Thus, hormonal signals can be transmitted into alterations in metabolic pathways through the regulation of a number of key enzymes in the pathway.

A special case of allosteric modification is known as **feedback inhibition** (Fig. 16). Feedback inhibition refers to allosteric modification by the end product of a pathway, which will feed back and modulate the activity of an enzyme that catalyzes a step of that pathway. As seen in Fig. 16, this can occur in a number of ways, particularly when branched pathways are considered. While feedback inhibition is not universal for all pathways, it does generally occur in biosynthetic pathways to ensure that a particular product is not overproduced.

B. Hormone Binding to Cells

It is apparent that hormone binding to cells is important in regulating biochemical pathways. A cell will be responsive to a particular hormone if the cell expresses a **receptor** specific for that hormone. For the case of peptide hormones, the receptor is on the cell surface. For steroid hormones, the receptor is intracellular. Hormone binding to the receptor can be analyzed in a manner similar to enzyme kinetics. From these analyses one can determine the affinity of the hormone for the receptor, known as the **dissociation constant, K_d**. In addition, the total number of hormone receptors on the cell surface, known as R_t, can be determined. This analysis was first worked out in 1949 by Scatchard, who derived the equation

$$\frac{B}{F} = -\frac{1}{K_d} B + \frac{R_t}{K_d} \tag{9}$$

where B **represents the concentration of hormone bound to the receptor** and F **represents the concentration of hormone present at equilibrium which is not bound to receptor.** Equation 9 is in the form of a straight line, in which the y-axis represents the value of the bound/free (B/F) ratio, and the x-axis represents the bound (B) material. The slope of the line defines the dissociation constant ($-1/K_d$), and the y-intercept gives R_t/K_d. Extrapolating to the x-intercept gives R_t, and al-

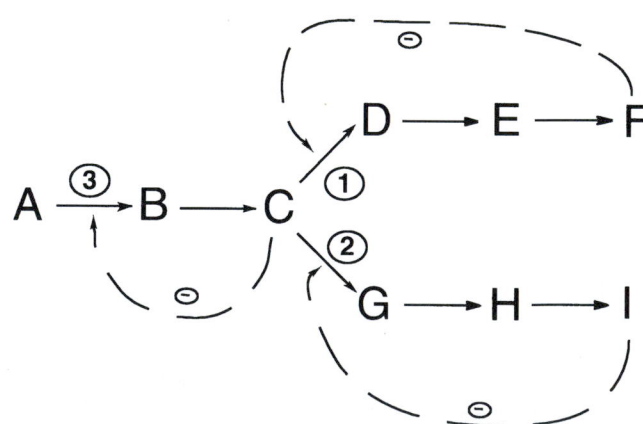

FIGURE 16.   Examples of feedback inhibition. End products F and I can inhibit the enzymes at the branch points from C (enzymes 1 and 2), whereas product C can inhibit enzyme 3. Product C will only accumulate when both enzymes 1 and 2 are inhibited (i.e., the end products E and F are present at high concentration). This type of feedback inhibition mechanism ensures that adequate quantities of all required compounds are available before any step of the pathway is inhibited.

lows the total number of receptors to be obtained. An example of a **Scatchard plot** is shown in Fig. 17. The Scatchard equation, while it does have certain limitations, has proven useful for the determination of both binding affinities and types of receptors on a particular cell. By knowing the affinity of a hormone for its receptor, the concentration of hormone required for a cell to respond to that hormone can be determined.

C. Regulation of Glycogen Metabolism

The degradative and biosynthetic pathways of glycogen are reciprocally regulated. When one pathway is activated, the other pathway is inhibited (Johnson, 1992). The regulation of these pathways is shown in Fig. 18, and is summarized in the next paragraph.

Glycogen synthesis is catalyzed by the enzyme **glycogen synthase**. The degradation of glycogen is catalyzed by **glycogen phosphorylase**. There are two major ways in which these enzymes are regulated. The first is by covalent modification, and the second is by allosteric modification. We will only consider the effects of covalent modification in this discussion. Before discussing the role of covalent modification, it is necessary to understand the role of glycogen in metabolism. As stated previously, glycogen is an energy storage molecule found primarily in the liver and muscle. The function of muscle glycogen is to present the muscle with an immediate source of glucose for energy production. The major function of liver glycogen is to provide glucose to all of the other organs of the body via the circulation. Thus, when blood glucose levels are low, liver glycogen will be degraded into glucose, and the glucose will be released into the circulation for use by other tissues.

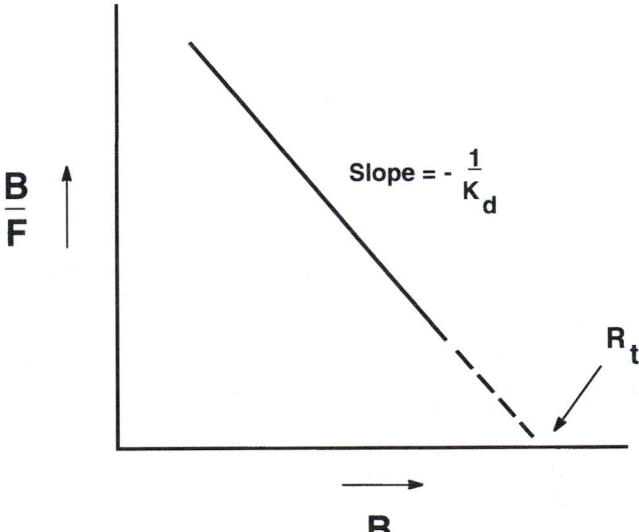

FIGURE 17. An example of a Scatchard plot. The slope of the line corresponds to $-1/K_d$, and the *x*-intercept will yield the total number of binding sites. If it is known how many binding sites there are per receptor, then the total number of receptors can also be estimated from this analysis.

Despite the different functions of glycogen in muscle and liver, the regulation of the metabolic pathways is the same in both tissues. **Both sets of enzymes are regulated by phosphorylation.** Glycogen synthase is inactive when phosphorylated. Glycogen phosphorylase is active when phosphorylated. This is a simple mechanism to ensure that both pathways are not active at the same time. This leads to the question, what activates and deactivates the phosphorylation system? The answer is circulating hormones. We have already seen that the major regulatory enzymes of the glycolytic pathway are controlled by hormones. The same hormones that regulate glycolysis also regulate glycogen metabolism. The next section describes how hormones regulate both glycogen metabolism and glycolysis.

D. cAMP/Phosphorylation Cascade

For a peptide hormone to initiate a cellular response, it must first bind to a specific receptor on the cell surface. Hormones are synthesized by specific tissues and released into the circulation. Once in the blood, hormones bind to specific receptors on the surface of appropriate target cells. Of prime importance for our discussion of glycogen metabolism are the hormones **insulin** and **glucagon**. These hormones regulate the level of glucose within the circulation. Specific cells within the pancreas can measure the level of glucose in the blood and release the appropriate hormone. When blood glucose levels are low, the pancreas releases glucagon, which activates glycogen degradation in the liver in order to raise blood glucose levels. When the blood glucose levels are high, the pancreas releases insulin, which stimulates both the muscle and liver to synthesize glycogen in order to store excess glucose.

Once a hormone binds to its receptor, the receptor alters its conformation and activates intracellular enzymes in order to signal the cell. In the case of glucagon secretion, caused by low blood glucose levels, hormone binding will activate the enzyme **adenylate cyclase**, which converts ATP to **cyclic AMP** (cAMP, Fig. 19). The increase in intracellular cAMP levels activates a **cyclic-AMP-dependent protein kinase**, which will phosphorylate various enzymes (Sutherland, 1972). Two of the enzymes that are phosphorylated are **glycogen synthase** and **glycogen phosphorylase kinase**. Phosphorylation of glycogen synthase inactivates the enzyme. Phosphorylation of glycogen phosphorylase kinase activates this second kinase. The primary substrate of glycogen phosphorylase kinase is **glycogen phosphorylase**, which will be activated upon phosphorylation. Thus, under conditions of glucagon release, the enzyme that makes glycogen is inactivated and the enzyme that degrades glycogen is activated. This enables the liver to degrade its glycogen in order to produce glucose to raise blood glucose levels. **Muscle does not respond to glucagon, as muscle does not express cell surface receptors for glucagon.**

The release of insulin indicates that blood glucose levels are high. Both muscle and liver express insulin receptors and will therefore respond to this hormone. Insulin binds to its specific receptor and **reduces intracellular cAMP levels.** When this occurs, the cAMP-dependent protein kinase is inactivated, and phosphorylation of the key enzymes no longer

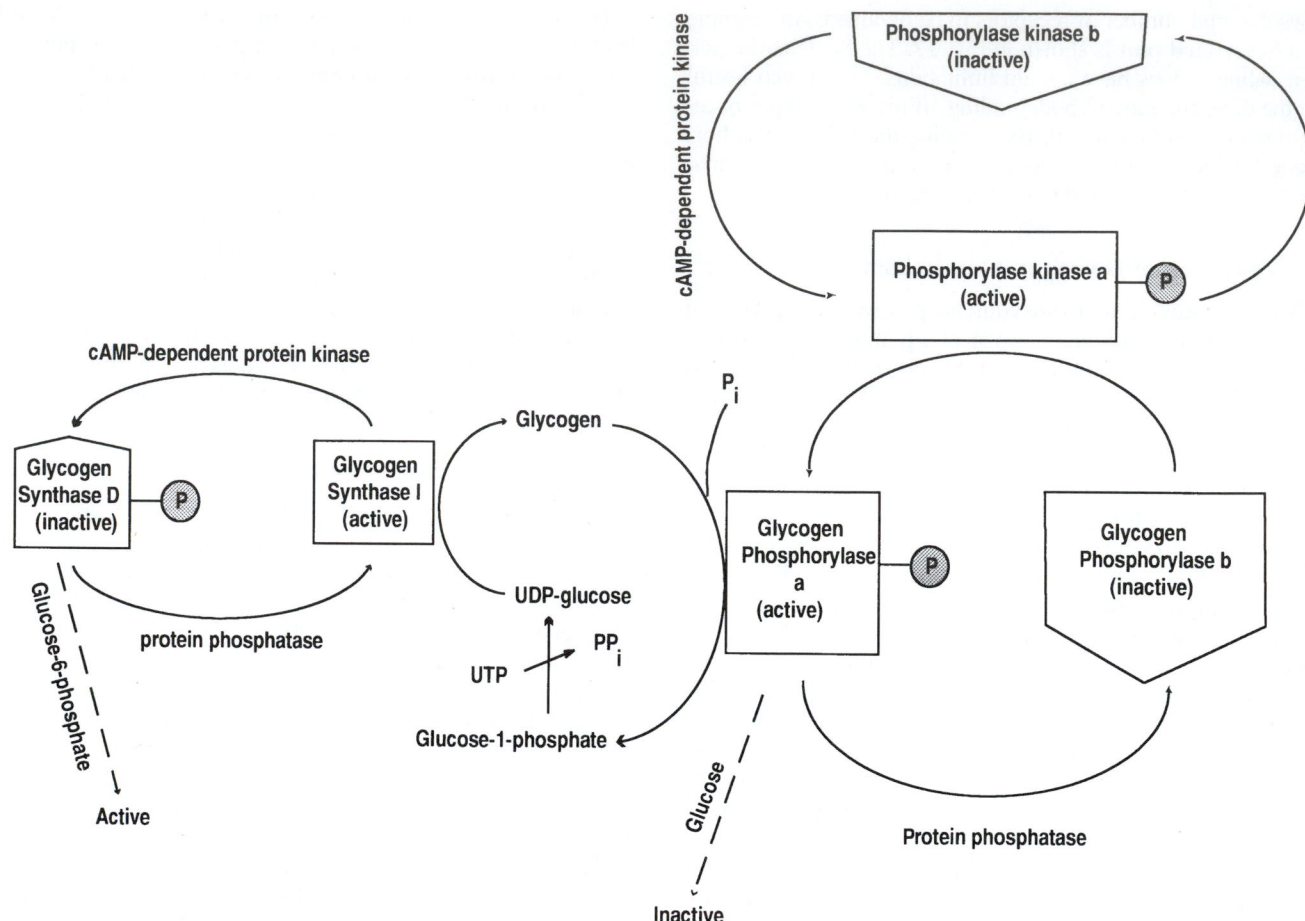

FIGURE 18. Regulation of glycogen metabolism. The key enzymes involved in degrading (glycogen phosphorylase) and synthesizing (glycogen synthase) glycogen are reciprocally regulated by phosphorylation. In addition, the enzymes responsible for phosphorylating these two key enzymes are themselves regulated, either by phosphorylation (phosphorylase kinase) or by allosteric activation (cAMP-dependent protein kinase). This system of regulation ensures that a futile cycle is not established, since only one pathway is active at any given time. The regulation indicated by the dotted lines is allosteric regulation of the two key enzymes. Allosteric regulation ensures that the enzymes can be activated or inhibited even in the absence of the phosphorylation controls.

occurs. This allows enzymes known as **phosphatases** to remove the phosphates from the enzymes, which activates glycogen synthase and inactivates both glycogen phosphorylase kinase and glycogen phosphorylase. Thus, under conditions of insulin release, glycogen synthesis is favored in both muscle and liver, and glycogen degradation is inhibited in the liver.

E. Tissue-Specific Regulation

The major role of liver glycogen is to act as a glucose reservoir for the bloodstream. When blood glucose levels drop, liver glycogen is degraded to glucose for use by other tissues. In order for this to occur, the pathway of liver glycolysis needs to be inhibited when liver glycogen degradation is activated. If it were not, liver glycolysis would utilize the glucose, and less glucose would be available for export to other tissues. Liver glycolysis is inhibited because two key enzymes of liver glycolysis (one of which is PK) are phosphorylated, and inactivated, by the activated cAMP-

dependent protein kinase. This reduces the activity of the glycolytic pathway to a very low level. While the liver is degrading glycogen and releasing glucose for other tissues, the muscle is also degrading glycogen for its own use. Since the muscle is using the glucose it is producing, **muscle glycolysis needs to be active during periods when liver glycolysis is inhibited.** Thus, **tissue-specific regulation** occurs. In the case of glucagon release, it is relatively clear why muscle glycolysis can continue when liver glycolysis is inhibited. It is because muscle does not contain glucagon receptors. This is actually one way in which differential tissue regulation is accomplished. Thus, when glucagon is released, the cAMP-dependent protein kinase is not activated in muscle.

There is another hormone, however, that will simultaneously activate the cAMP-dependent protein kinase in **both** liver and muscle. This hormone is **epinephrine**. Epinephrine is released during times of stress or exercise, and indicates that the muscles will be doing a lot of work, and that the muscle will need a lot of energy in the form of glucose.

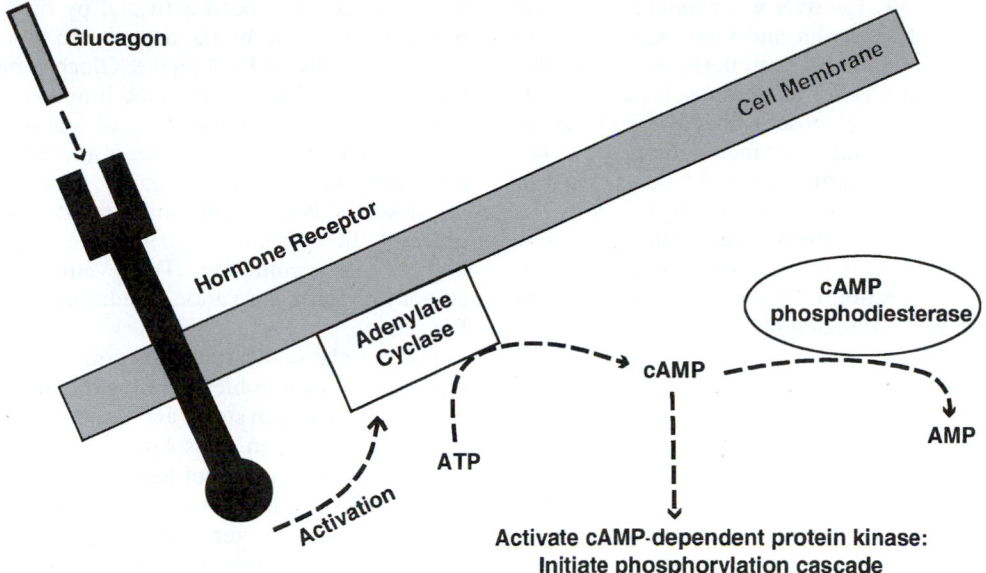

FIGURE 19. Initiation of the phosphorylation cascade in cells containing the glucagon receptor. Glucagon is released from the α cells of the pancreas in response to low levels of blood glucose and binds to specific cell surface receptors on the membrane. This binding activates the receptor and a signal is transmitted to the inside of the cell. The nature of this signal is discussed in Chapter 9. This signal activates the enzyme adenylate cyclase such that cAMP is produced. The cAMP produced activates the cAMP-dependent protein kinase, which initiates the regulatory cascade. The cAMP is degraded by a phosphodiesterase, which is also activated by opposing hormone action.

However, even when epinephrine is released, liver glycolysis is inactive while muscle glycolysis remains active. This occurs because muscle contains slightly different forms of the two glycolytic enzymes that are inhibited by phosphorylation in the liver. The muscle enzymes catalyze the same reactions as are catalyzed in the liver, but the muscle enzymes are not inhibited by phosphorylation (Pilkis *et al.,* 1990). This is an example of **isozymes**, two proteins with different structures that catalyze the same reaction. In most cases, the regulation of isozymes is different, even though the reaction catalyzed by each protein is the same. In muscle, phosphorylation of these key glycolytic isozymes does not inhibit their activity. Thus, the muscle isozymes have evolved to be regulated differently than the liver isozymes, and lead to tissue-specific regulation.

F. How Does Fat Storage Fit In?

The major energy storage form in the body is triacylglycerol. Energy can be derived from triacylglycerol by converting it to **glycerol** and **free fatty acids.** The glycerol enters the glycolytic pathway to generate energy, and the fatty acids generate energy by being converted to acetyl-CoA, which is then oxidized by the TCA cycle. Fatty acid oxidation generates considerable energy and is a very efficient energy source. In a manner similar to glycogen metabolism, the pathways of fatty acid synthesis and degradation are reciprocally controlled. Under conditions in which the body requires energy (such as glucagon or epinephrine release) the enzyme that initiates the synthesis of fatty acids (**acetyl-CoA carboxylase**) is phosphorylated and inactive. This results from the activation of the cAMP-dependent protein kinase by the hormone, which then directly phosphorylates acetyl-CoA carboxylase. The enzyme that initiates the degradation of triacylglycerol molecules from the adipocyte is also phosphorylated by the cAMP-dependent protein kinase, and is activated. This enzyme is known as **hormone-sensitive lipase.** This enzyme catalyzes the release of fatty acids from the adipocyte into the circulation, where they travel to other tissues to be used as an energy source. It is important to remember that simultaneous with fatty acid release the liver has also switched to the degradation of glycogen to provide glucose for the blood. Both of these pathways have been activated by the same mechanism.

Fatty acids cannot be converted to glucose. Thus, even though there is more energy available from fatty acids than from glycogen stores in the average individual, and fatty acid breakdown is also activated by glucagon release, the liver must still produce glucose from glycogen if blood glucose levels are low. This is important because the brain has an absolute requirement for glucose, and can not use fatty acids for energy. This metabolic oddity leads to difficulties in times of prolonged fasts or starvation. Under conditions in which food intake is drastically reduced, blood glucose levels are low for extended periods of time. Unfortunately, the liver only contains sufficient glycogen to keep blood glucose levels high for approximately one day. After the liver glycogen stores are depleted, the liver will still produce glucose, but will do so via the pathway of gluconeogenesis using amino acids, lactate, and glycerol as a source of carbon. The liver will also start to produce energy sources known as **ketone bodies**, which are produced from high levels of acetyl-CoA. The liver will produce the ketone bodies from fatty acid oxidation and then export them to other tissues to use as an alternative energy source to glucose. As the other tissues (such as the muscle) use the ketone bodies, there is more glucose available for the brain. The brain can also switch part of its metabolic needs to ketone body oxidation. Thus, fatty acid oxidation is used to produce ketone bodies, to help reduce the overall body demand for glucose. However, the ketone bodies are acids, and too large a buildup of ketone bodies can be detrimental, and may lead to acidification of the blood and urine (**ketoacidosis**), which is life-threatening.

G. A Starving Situation

As a unified example of whole-body energy metabolism and regulation, the case of a starving individual will be considered. The role of each organ during starvation will be considered before integrating all effects. The **major function** of the liver in a starvation situation is to produce glucose for use by other tissues. To do this, the liver will mobilize glucose from glycogen, produce glucose via gluconeogenesis, and, in order to spare glucose consumption by tissues other than the brain, produce ketone bodies from fatty acids. These processes can occur simultaneously, although they usually occur sequentially. **The overriding signal for a switch to these pathways is a reduction in the insulin:glucagon ratio.** A decreased insulin:glucagon ratio brings about the activation of the cAMP-dependent protein kinase. The cAMP-dependent protein kinase will activate glycogen phosphorylase and inactivate glycogen synthase. This results in a rapid mobilization of liver glucose derived from liver glycogen. However, after 12 to 24 hours of continual glycogen degradation, the stores of liver glycogen are depleted. At this point, the liver will switch to gluconeogenesis to maintain blood glucose levels. The pathway of gluconeogenesis has become favored over glycolysis because of the action of the cAMP-dependent protein kinase. This enzyme has been activated by glucagon binding to its receptor, and has phosphorylated two key glycolytic enzymes, both of which are inhibited upon phosphorylation. The liver uses glycerol, lactate, and amino acids as substrates for gluconeogenesis. The glycerol is provided by adipocytes from the degradation of triacylglycerol to glycerol and fatty acids. Lactate is derived from erythrocyte metabolism. The amino acids are derived from protein turnover and degradation.

The liver also begins to oxidize fatty acids, which have been released from the adipocytes by the activation of hormone-sensitive lipase. The liver is not synthesizing fatty acids because the enzyme needed to do this, acetyl-CoA carboxylase, is inactivated when the cAMP-dependent protein kinase is activated. The oxidation of fatty acids leads to an increase in acetyl-CoA levels in the liver, and this can lead to ketone body formation. The ketone bodies produced are then exported to other tissues to use as an alternative energy source to glucose. The events that occur in the liver upon starvation conditions are summarized in Fig. 20.

The role of the **adipocyte** during starvation is to degrade triglycerides to fatty acids and glycerol. This occurs through activation of the hormone-sensitive lipase by the cAMP-dependent protein kinase. The cAMP-dependent protein kinase has been activated by the release of glucagon, brought about by the decrease in blood glucose levels due to the reduced food intake. Glucose metabolism in the adipocyte is reduced, due to the limited availability of glucose. As excess accumulation of ketone bodies leads to acidification of the blood, there is a feedback mechanism to try to keep ketone body formation under control. High levels of ketone bodies will inhibit the degradation of triacylglycerol, thus partially regulating the level to which ketone bodies can accumulate. The events that occur in the adipocyte under starvation conditions are summarized in Fig. 21.

The muscle, under starvation conditions, still needs to do work, but it has a problem. During the first day of starvation, the muscle glycogen stores are utilized through allosteric activation of glycogen phosphorylase. This is necessary because the muscle does not have glucagon receptors, and so the cAMP-dependent protein kinase has not been activated in the muscle. However, since blood glucose levels are low, and insulin is not present, the glucose that is in the blood is not entering the muscle at a rapid pace. After 12 to 24 hours, the muscle glycogen stores have been depleted, and the muscle will switch to fatty acid and ketone body oxidation for energy. Under conditions of extreme starvation, the muscle will begin to degrade its own proteins to generate amino acids, which will be sent to the liver for conversion to glucose. This leads to the wasting observed in starved individuals. It should be noted that glycolysis is active in muscle due to allosteric activation. The interactions between all of the organ systems under starvation conditions is shown in Fig. 22.

IX. Energy Forms Revisited

How is energy stored in the tissues? Energy is stored in the form of high-energy phosphate bonds in nucleoside triphosphates, such as ATP. The last two phosphoanhydride bonds in ATP, when hydrolyzed, release 7.3 kcal/mol of energy, which for biological systems is a large amount. As previously discussed, ATP is synthesized either through substrate-level phosphorylation (as in glycolysis), or through oxidative phosphorylation. In the muscle, high-energy phosphates are also stored as **creatine phosphate** (the structure and synthesis of creatine phosphate is shown in Fig. 23). Creatine phosphate is generated by the action of **creatine kinase**, in which creatine is phosphorylated by ATP to produce ADP and creatine phosphate, with the high energy of the phosphate bond preserved in the creatine phosphate. The ADP produced can be regenerated into ATP by the action of the enzyme **myokinase**, which will convert two ADP molecules into one ATP molecule and one AMP molecule. The ATP can then be used to produce more creatine phosphate, which is used to store energy in muscle. Both ATP and creatine phosphate are energy-rich by virtue of their phosphate-oxygen or phosphate-nitrogen bonds, which when broken release a large amount of energy (7.3 kcal/mol released for ATP, 10.3 kcal/mol released for creatine phosphate). Subsequent metabolic reactions will utilize this energy to direct a series of linked reactions, such as a pathway, in the forward direction.

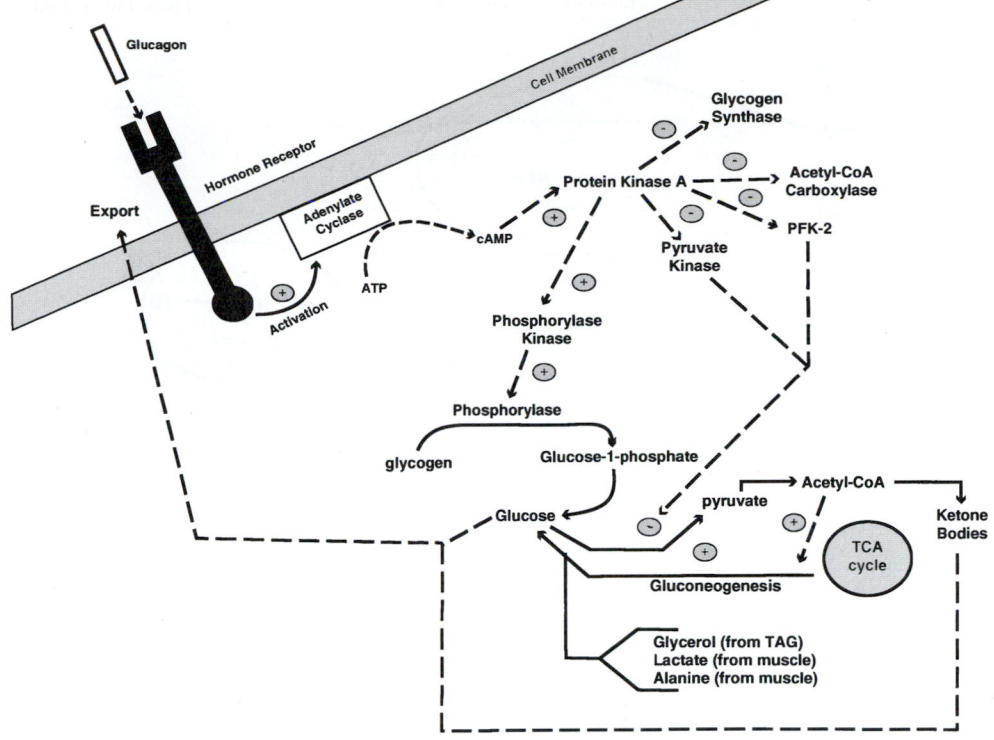

FIGURE 20. Regulation of metabolic pathways in the liver under starvation conditions. All of the metabolic changes are the result of the activation of the cAMP-dependent protein kinase by glucagon binding to its receptor. Once activated, the cAMP-dependent protein kinase phosphorylates the following enzymes: two enzymes of glycolysis (PFK-2 and PK), which are inactivated; one enzyme required for fatty acid synthesis (acetyl-CoA carboxylase), which is inactivated; and two enzymes involved in glycogen metabolism, activating one (phosphorylase kinase) and inactivating the other (glycogen synthase). These enzyme activity changes alter the overall flow of metabolism in the liver to one of energy export, as glycogen is degraded to glucose, and fatty acids are converted to ketone bodies, both of which are exported. Gluconeogenesis is enhanced by the inhibition of glycolysis, allowing glucose to be produced from amino acids and other glucogenic degradation products.

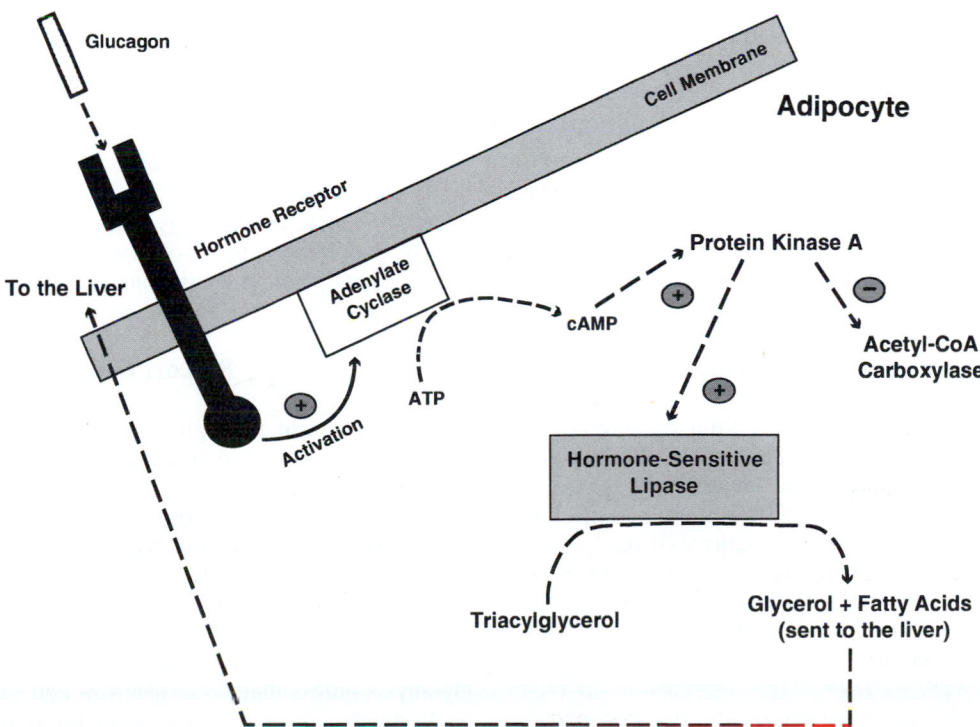

FIGURE 21. Events that occur in the adipocyte under starvation conditions. As with the liver, all events are initiated by the activation of the cAMP-dependent protein kinase (protein kinase A), and activation or inhibition of the appropriate enzymes within the cell.

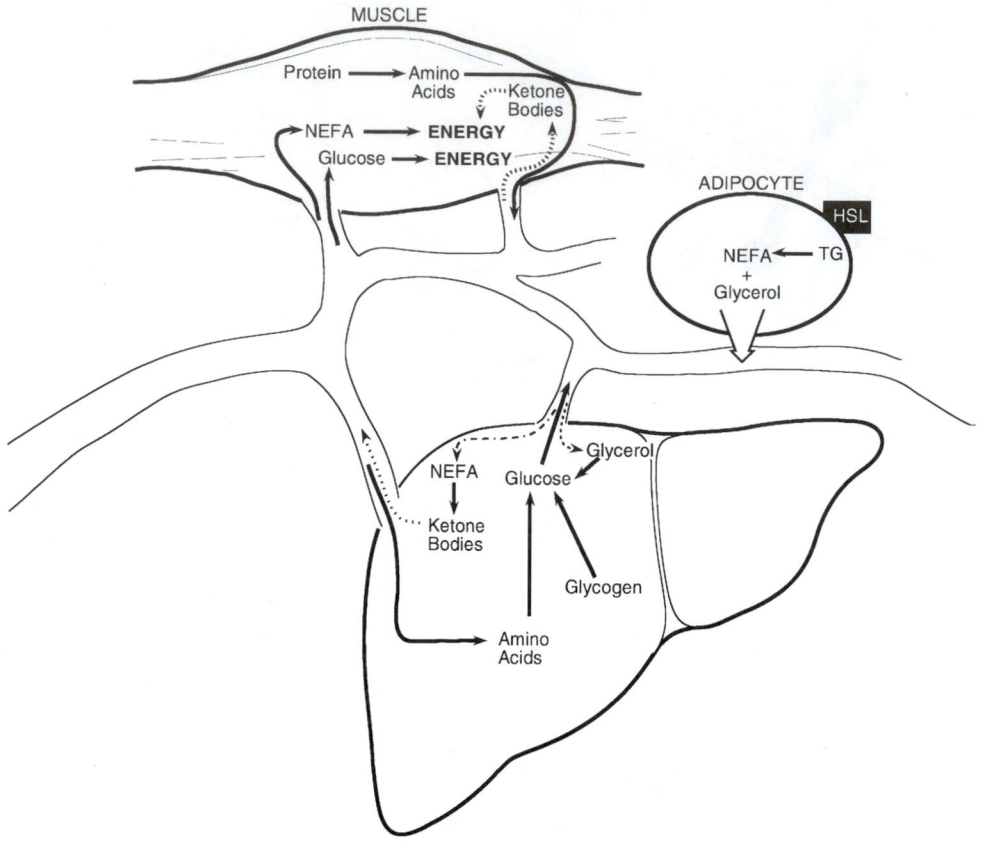

FIGURE 22.   Metabolic interactions between the liver, adipocyte, and muscle under starvation conditions. The liver is producing glucose from glycerol (obtained from the adipocyte) and amino acids (obtained from the muscle) by gluconeogenesis in order to supply glucose to the brain. The adipocyte is releasing fatty acids into the circulation for use by the muscle and liver as energy sources. The liver is also converting fatty acids to ketone bodies for use by other tissues as an alternative energy source. HSL stands for the enzyme hormone-sensitive lipase, specific for the adipocyte, and NEFA stands for nonesterified fatty acids, that is, fatty acids not attached to glycerol.

Why does muscle have this alternative energy source? The use of creatine phosphate as an energy source permits the muscle to store only small amounts of ATP. This is important because ATP is often used as a regulator of enzymes involved in the production of ATP. By keeping the levels of ATP low, and storing energy in the form of creatine phosphate, the enzymes that produce ATP will continue to make ATP, and the muscle can therefore store more high-energy compounds than if only ATP itself were the storage form. What happens when the muscle requires energy rapidly? The muscle creatine kinase catalyzes its reverse reaction and produces ATP from creatine phosphate and ADP. The muscles store creatine phosphate so that when bursts of energy are required there is sufficient energy available to carry out the required task. In fact, during a sprint, which generates energy through anaerobic glycolysis, the initial burst of activity is fueled through the stored creatine phosphate levels, as the amount of ATP within the muscles is only sufficient for a two-second burst of activity.

Lower invertebrates use a variety of different phosphorylated compounds to store energy in place of creatine phosphate. Phosphoarginine is the compound of choice for arthro-

pods (insects) and echinoderms (star fish, sea urchin), whereas the annelids (such as worms) utilize N-phosphorylated guanidinoacetic acid derivatives (White *et al.*, 1978). The structures of these alternate high-energy compounds are shown in Fig. 24. The important thing to notice with these compounds is that they all contain a high-energy nitrogen-phosphorus bond.

X. Cori Cycle

The end product of glycolysis in muscle is pyruvate. When muscle needs to work fast, and generate ATP at a rapid pace, this is best accomplished using glycolysis in the absence of oxygen. The problem that results from this is that glycolysis needs a continual supply of an oxidized electron carrier, NAD^+. During glycolysis the NAD^+ is converted to NADH, and usually the NADH is reconverted to NAD^+ via the electron transport chain. In the absence of oxygen, the electron transport chain is inoperative. This will slow down glycolysis unless there is a means of converting the NADH back into NAD^+. This does occur through a reduction of pyruvate to lactate, using as the electron donor NADH. The enzyme that catalyzes this reaction is **lactate dehydroge-**

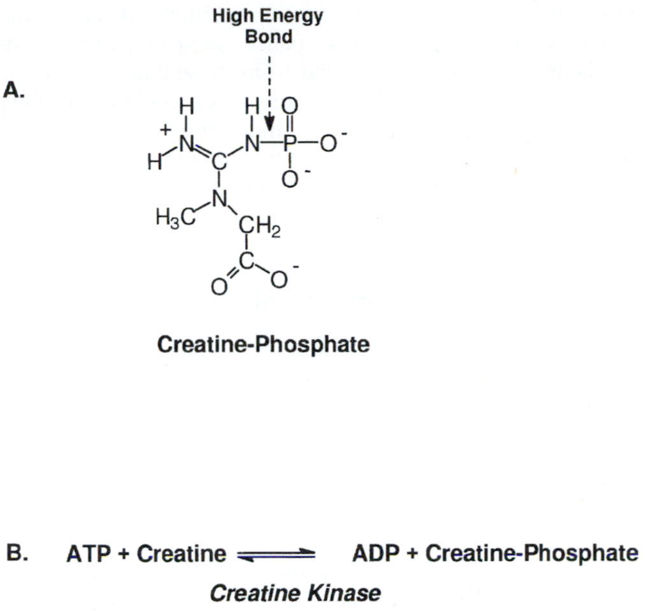

A.

High Energy Bond

Creatine-Phosphate

Phosphoarginine

Phosphoguanidinoacetic acid

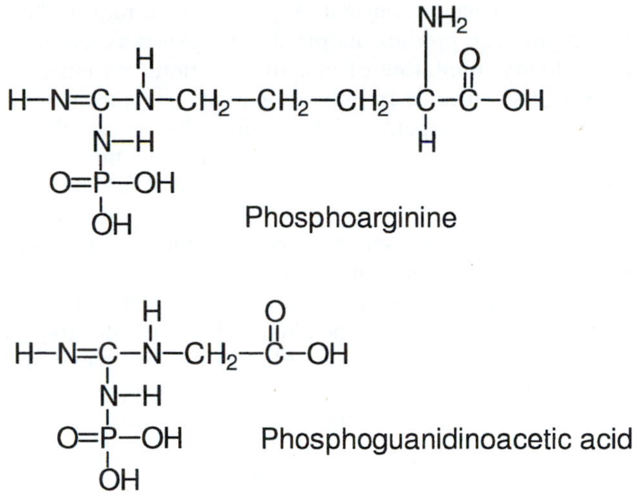

FIGURE 24. The structures of alternative high-energy compounds used in place of creatine phosphate in invertebrates.

B.   ATP + Creatine ⇌ ADP + Creatine-Phosphate

Creatine Kinase

FIGURE 23. The structure and biosynthesis of creatine phosphate. (A) The structure of creatine phosphate and the location of the high energy bond. This bond contains even more energy than the one in ATP, and will release 10.3 kcal/mol of energy when hydrolyzed. (B) The synthesis of creatine phosphate from ATP, catalyzed by creatine kinase. Creatine kinase catalyzes a reversible reaction; under conditions of strenuous exercise it will generate ATP, and under conditions of rest the synthesis of creatine phosphate will be favored.

nase. The lactate that is produced is then sent from the muscle to the liver for conversion to glucose. If the lactate were to remain in the muscle, the muscle pH would drop because lactate is an acid. A drop in muscle pH would lead to impaired performance, which is not desirable. Once the liver converts the lactate to glucose, it is sent back to the muscle as an energy source. This cycle is known as the **Cori cycle** and is shown diagrammatically in Fig. 25. Erythrocytes do not contain any mitochondria. Thus, all of their energy is derived from glycolysis. In order to allow glycolysis to con-

tinue, the erythrocyte regenerates NAD$^+$ using pyruvate and lactate dehydrogenase, producing lactate. The lactate is secreted by the erythrocyte into the blood, then travels to the liver, where it is used as a substrate for gluconeogenesis.

XI. Summary

All living organisms require energy, which is obtained from the sun and the food that is ingested. The food that is consumed is degraded into smaller components by the action of specific protein molecules in the digestive tract known as enzymes. Enzymes are proteins that catalyze specific reactions, primarily by reducing the energy required for the reactants to reach their transition state. Enzyme activity can be studied through enzyme kinetics, which measures the rate at which enzymes catalyze reactions. The rate at which enzymes work is a function of binding to the substrate, and this parameter can be approximated by a kinetic parameter

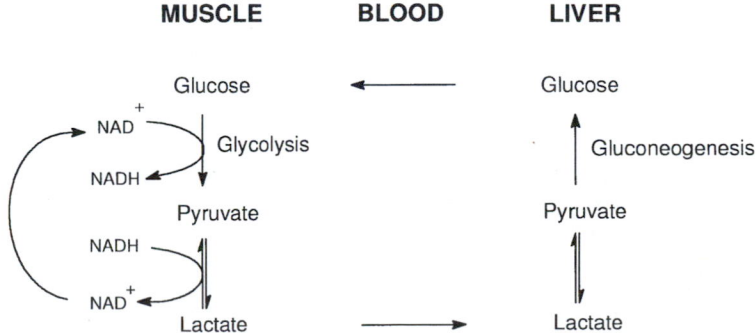

FIGURE 25. The Cori cycle. Under conditions of rapid muscle glycolysis, in the absence of oxygen, the supply of the required glycolytic cofactor NAD$^+$ becomes limiting. NAD$^+$ can be regenerated by converting pyruvate to lactate. This, however, leads to an increase in lactate levels within the muscle. This will reduce intracellular pH and inhibit muscle enzyme activity. To overcome the acidification of the muscle, lactate is sent to the liver, where it can be converted to glucose, which can be sent back to the muscle as an energy source.

called the Michaelis constant, K_m. The fastest rate at which the enzyme can produce its product is known as the maximal velocity. Inhibitors of enzyme reactions act either by blocking the substrate from binding to the enzyme, in a process known as competitive inhibition, or by altering the activity of the enzyme without affecting substrate binding, via noncompetitive inhibition. Certain enzymes can be regulated by small effectors. Regulated enzymes display a complex kinetic picture, which is best characterized by sigmoidal kinetics. Regulators either increase or decrease the level of substrate required to reach maximal velocity.

Metabolism consists of the chemical conversions used in living cells to generate energy. Major pathways required to accomplish this include glycolysis, gluconeogenesis, the TCA cycle, fatty acid synthesis and degradation, glycogen synthesis and degradation, and oxidative phosphorylation. Glycolysis is the process of converting sugars to the three-carbon intermediate pyruvate. Gluconeogenesis refers to the production of glucose from other components, such as amino acids, glycerol, and lactate. The TCA cycle is designed to oxidize acetyl-CoA to CO_2 and water, with the generation of a large number of reducing equivalents. The reducing equivalents then donate their electrons to the electron transfer chain in order to generate energy through the generation of the proton motive force and oxidative phosphorylation. There are two primary energy storage forms, glycogen (carbohydrate-based) and triacylglycerol (fatty-acid-based). Both glycogen and fatty acids are synthesized when energy stores are high, and degraded when energy stores are low. Regulation of these opposing pathways is critical.

It is important to regulate the metabolic pathways so that futile cycles are not maintained, and to accomplish required tissue-specific tasks. Multiple levels of regulation are evident, and include regulation of gene transcription, allosteric regulation of enzyme activity, and covalent modification of specific enzymes. Tissue-specific isozymes, which catalyze the same reaction but are subject to different regulatory controls, are also important for providing tissue specificity in the regulatory process. As a general rule, pathways involved in biosynthesis are reciprocally regulated from those that catalyze the degradation of their products. Thus, under conditions of fatty acid and glycogen biosynthesis, the pathways of fatty acid degradation and glycogen degradation are inhibited. By maintaining these rules the metabolic pathways are directed to accomplish their goals with a minimum of wasted energy.

Bibliography

Alberts, B., Bray, D., Lewis, J., Raff, M., Roberts, K., and Watson, J. D. (1989). "Molecular Biology of the Cell." Second edition. Garland Publishing, Inc., New York, pp. 613–629.

Fersht, A. (1985). "Enzyme Structure and Function." W. H. Freeman and Co., New York, pp. 47–120.

Foster, D. W., and McGarry, J. D. (1983). The metabolic derangements and treatment of diabetic ketoacidosis. *New England J. Med.* **309,** 159–169.

Johnson, L. N. (1992). Glycogen phosphorylase: control by phosphorylation and allosteric effectors. *FASEB J.* **6,** 2274–2282.

Lands, W. E. M. (1991). Biosynthesis of prostaglandins. *Ann. Rev. Nutrition* **11,** 41–60.

Lehninger, A. L., Nelson, D. L., and Cox, M. M. (1993). "Principles of Biochemistry." Second edition. Worth Publishers, New York, pp. 134–159, 198–239, 364–399, and 736–788.

Mitchell, P. (1963). Molecular, group and electron translocation through natural membranes. *Biochemical Society Symposia* **22,** 142–169.

Pilkis, S. J., el Maghrabi, M. R., and Claus, T. H. (1990). Fructose-2, 6-bisphosphate in control of hepatic gluconeogenesis. From metabolites to molecular genetics. *Diabetes Care* **13,** 582–599.

Scatchard, G. (1949). The attractions of proteins for small molecules and ions. *Ann. N.Y. Acad. Sci.* **51,** 660–672.

Sutherland, E. W. (1972). Studies on the mechanisms of hormone action. *Science* **177,** 401–408.

White, A., Handler, P. H., Smith, E. L., Hill, R. L., and Lehman, I. R. (1978). "Principles of Biochemistry." Sixth edition. McGraw Hill, New York, p. 1098.

Keith D. Garlid

8

Physiology of Mitochondria

I. Introduction

ATP is the common energy currency of the cell, where it is used to fuel the work of muscle contraction, protein synthesis, and active transport of ions. Most cellular ATP is synthesized by oxidative phosphorylation, in which diverse forms of chemical energy derived from food and body stores are transduced by oxidative phosphorylation into ATP. The complex of enzymes that couples substrate oxidation to ATP synthesis is located in mitochondria, as we learned in Chapter 7. The internal aqueous compartment of mitochondria, called the *matrix,* contains the enzymes of the Krebs tricarboxylic acid cycle. The matrix is enclosed by a highly folded, insulating membrane called the *inner membrane,* and this structure is separated from the cytosol by a more permeable *outer membrane.*

The inner membrane contains the anion carriers responsible for shuttling metabolic substrates between matrix and cytosol. It also contains the cation carriers and channels that regulate mitochondrial volume and matrix calcium. In brown fat cells, the inner mitochondrial membrane contains a unique protein whose function is to dissipate energy both for thermogenesis and to avoid obesity. The vectorial transport enzymes responsible for substrate oxidation, electron transport, and ATP synthesis are all inner membrane proteins.

As one might expect of a membrane-bounded organelle, the mitochondrion has its own transport physiology, which governs its survival and performance within the cytosolic environment. The physiological setting in which these transport cycles operate is provided by the chemiosmotic theory.

II. Chemiosmotic Theory

The chemiosmotic theory describes with elegant simplicity the mechanism by which substrate oxidation is coupled to ATP synthesis. Working far in advance of experimental evidence, Mitchell (1961) postulated that Nature uses protonic batteries to drive ATP synthesis and proposed that biological energy conservation is essentially a problem in membrane transport. The chemiosmotic theory consists of four postulates, each of which has been validated by experiment.

1. The inner membrane contains electron transport enzymes which are vectorially oriented so that the energy of electron transport drives ejection of protons outward across the membrane. The energy of substrate oxidation is thereby stored as a proton electrochemical potential gradient called the **proton-motive force.**
2. The inner membrane contains the F_1-F_0-ATPase, which is also vectorially oriented so that the energy of ATP **hydrolysis** will drive protons outward across the inner membrane. The ATPase is reversible, so that protons driven inward through the ATPase by the redox-generated proton-motive force will cause ATP synthesis.
3. The inner membrane must have a low diffusive permeability to protons and ions generally. Otherwise, ion leaks would short-circuit the proton-motive batteries, and ATP would not be synthesized.
4. The inner membrane must contain anion exchange carriers to enable substrates to reach their enzymes in the matrix, and it must contain cation exchange carriers to remove cations that entered the matrix by diffusion down the very large electrical gradient caused by outward proton pumping.

The consequence of the first three postulates, diagrammed in Fig. 1, is the generation of a proton-motive force (Δp), which is defined as the electrochemical proton gradient divided by the Faraday constant ($\Delta\mu_{H^+} / \mathscr{F}$):

$$\Delta p = \Delta\Psi - Z\,\Delta pH \qquad (1)$$

where $Z \equiv (RT \ln 10)/\mathscr{F} = 59$ mV at 25 °C, and $\Delta\Psi$ is the membrane potential (inside minus outside). The first measurement of Δp was made by Mitchell and Moyle (1969), who obtained a value of -230 mV in the absence of phosphorylation (State 4). Subsequent measurements have yielded values of about -200 mV in isolated rat liver mitochondria respiring in the nonphosphorylating state.

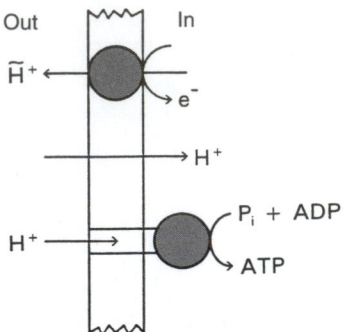

FIGURE 1. Chemiosmotic energy coupling of electron transport to ATP synthesis. The vectorial enzymes of electron transport and ATP synthesis are coupled indirectly via the protonmotive force. Efficient transfer of protonmotive energy from the redox chain to the ATP synthase is ensured by the low permeability of the inner membrane to protons.

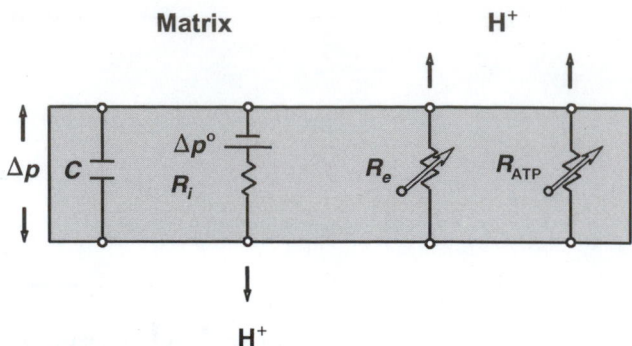

FIGURE 2. Circuit diagram of the mitochondrial electron transport chain. The figure contains an electrical diagram of the electron transport system (ETS). Δp^0 is the emf[0] of the overall system, corresponding to the free-energy drop from substrate to oxygen. Two electrons pass from the input redox couple (e.g., NADH-NAD$^+$) to oxygen with a redox span of ΔE_H, and the ETS simultaneously pumps n_H protons. Therefore, $\Delta p^0 = (2/n_H) \Delta E_H$ (Nicholls and Ferguson, 1992). The chemical reactions of the ETS will cause some losses due to frictional coefficients, and this is represented by the overall *internal* resistance, R_i. It can be shown that $R_i = r_i/N$, where r_i represents the internal resistance of each pathway, and N is the number of parallel ETS pathways.

C is the capacitance of the membrane, and R_e and R_{ATP} are resistances of the proton back-flux pathways. R_e is the sum of cation and proton leak resistances and those of the futile cation cycles necessary to regulate mitochondrial physiology, described in subsequent sections. Proton back-flux through the F_1-F_0-ATPase is designated by the element containing R_{ATP}. Thus, when ADP and phosphate are not present, or when the ATPase is inhibited, R_{ATP} is infinite. If conductance through both of these pathways were zero, there would be no respiration, no proton pumping, and Δp would equal Δp^0, the open-circuit voltage.

The fourth postulate demonstrates a deep insight into the physiological hazards posed by chemiosmotic coupling. As Mitchell (1966) has written:

> While the introduction of the foregoing sophistication [the chemiosmotic theory of energy coupling] solved one problem, it created another, for the membrane potential that would now be required to reverse the ATPase reaction would cause the ions of opposite sign of charge to the internal aqueous phase to leak in through the coupling membrane. To prevent swelling and lysis, the ion leakage would have to be balanced by extrusion of ions against the electrical gradient. It was therefore necessary to postulate that the coupling membrane contains exchange diffusion systems . . . that strictly couple the exchange of anions against OH$^-$ ions and of cations against H$^+$ ions.

It is noteworthy that the proposal for electroneutral exchange carriers was made at a time when there was no experimental evidence for the existence of ion exchange carriers in any membrane. (Indeed, the first evidence for the existence of Na$^+$-H$^+$ exchangers in biological membranes was obtained in Mitchell's laboratory.) Rather, the ion exchange carriers were born out of physiological necessity in an elegant application of the scientific method.

III. What Determines the Respiration Rate in Cells?

Mitchell correctly identified the electron transport system (ETS) as a collection of electromotive cells. The behavior of such systems is well-known from elementary physics and is described by the simple circuit diagram in Fig. 2. We note in particular that respiration is driven by the free energy contained in the redox drop, however, it is controlled by the proton back-flux through leak pathways and the ATP synthase. We note four salient aspects of this circuit: (1) The current through the system is determined entirely by the external resistances. (2) The battery will deliver increased current only when R_e or R_{ATP} is decreased. (3) The battery will respond the same whether current is drawn through R_e or R_{ATP}. (4) As current drawn from the battery is increased, the voltage will drop due to the internal resistance.

These features are illustrated in Fig. 3, which shows how Δp varies when electron current (measured as respiration rate) is progressively increased by addition of a protonophore that decreases external resistance, R_e. The decreased resistance to electrophoretic proton back-flux draws additional current from the ETS, and the resulting increased respiration causes Δp to fall gradually until, eventually, the V_{max} of the ETS is reached. What is being measured in such experiments is evident from Fig. 2:

$$\Delta p = \Delta p^0 - R_i V_o \qquad (2)$$

where V_o is the respiration rate. The constant slope of the curve, R_i, is the internal resistance of the ETS, representing the weighted sum of frictional coefficients of all the reactions leading to proton ejection. The intercept is Δp^0, the theoretical open-circuit voltage of the system; Δp^0 is never achieved experimentally, because there is always proton leakage across the inner membrane. Proton back-flux can be substantially increased by ionophores, by uncoupling protein, by futile Ca^{2+} or K$^+$ cycling, or by adding ADP and phosphate so that current is drawn via the ATPase. Careful measurements show that all methods of increasing electrophoretic proton back-flux yield points that fall on the same battery curve as illustrated in Fig. 3 (Nicholls and Ferguson, 1992). Thus, all four criteria described above are met by the mitochondrial ETS.

The same considerations apply to respiration within the cell, but here we must also take into account the fact that Eq. 2 con-

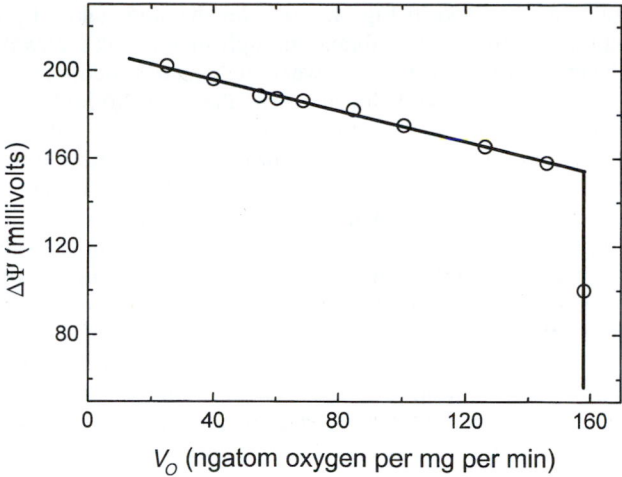

FIGURE 3. Dependence of proton-motive force on electron transport rate. Membrane potential ($\Delta\Psi$) of rat liver mitochondria is plotted versus respiration rate (V_o), which was varied by adding the protonophoretic uncoupler, CCCP. Respiration was measured using a standard Clark electrode. $\Delta\Psi$ was determined from the distribution of tetraphenylphosphonium cation. The slope of the curve, R_i, is 0.35, the intercept is 210 mV, and V_{max} is about 160 ngatom O/min·mg. These are typical values for rat liver mitochondria respiring on succinate. The pH gradient was 0.3 and assumed to be invariant with increased respiration. Therefore, Δp^0 is about 228 mV.

tains two parameters that could be varied in predictable ways by regulation. Thus, it is observed that R_i is increased by inhibitors of the ETS, which effectively reduce the number of elements in the redox chain. In principle, R_i should be decreased by activation of one or more individual steps in the ETS. This could occur by mobilizing additional ETS elements or by bringing individual elements closer together, thereby reducing the frictional losses. No convincing demonstration of this type of activation has been reported. For example, Δp^0 should be reduced by lowering the free energy of the redox couple by substrate restriction or hypoxia (Gnaiger *et al.*, 1995); Δp^0 should be increased by increasing the ΔG of the redox couple, resulting in higher $\Delta\Psi$, increased ion leaks (next section), and increased respiration. This is the cause of increased respiration observed in cells treated by glucagon (Yamazaki, 1975). The stimulatory effect of glucagon and other hormones is thought to be due largely to calcium-activation of matrix pyruvate dehydrogenase (Hajn'oczky *et al.*, 1995), which raises the NADH/NAD ratio and the overall free energy of the redox reaction.

It should be clear from the foregoing that the mitochondrial battery curve (Eq. 2) is a fundamental descriptor of mitochondria. Thus, respiration is controlled primarily by the rate of proton back-flux through R_e and R_{ATP}. In the absence of ATP synthesis (known as State 4 respiration), it is controlled entirely by the rate of ion leaks through R_e. (In principle, slip in the ETS machinery could also contribute to respiration; however, the contribution of slip to respiration appears to be quantitatively insignificant in mitochondria (Garlid *et al.*, 1992)). This explanation of the control of respiration is complete, insofar as the ETS is concerned, and

we may now turn to the important issue of what controls the rate of proton back-flux via the ATPase.

IV. What Determines the Rate of ATP Synthesis in Cells?

ATP synthesis by mitochondria must not only match the rates of ATP consumption by cellular ATPases, but it must also do so without compromising the need of these ATPases for relatively high ATP/ADP ratios. This has been a problem of intense investigation and considerable controversy over the years, but the problem has largely been solved for heart and skeletal muscle by Saks and coworkers (Saks *et al.*, 1994, 1998). The problem is this: the heart is capable of an 8–10-fold span of respiration and ATP synthesis rates, and even greater spans have been reported for skeletal muscle. How do heart mitochondria know that they must increase ATP synthesis by such amounts?

It is clear that ATP synthesis is primarily under thermodynamic regulation, although kinetic regulation has also been proposed. This means that the rate of ATP synthesis is determined by the phosphorylation potential, ΔG_P, in the mitochondrial matrix. Thus, when ADP production is increased at the myofibrillar ATPase, this signal must be transmitted to the matrix in the form of a lower ΔG_P. Indeed, the signal is transmitted very efficiently, because there is a strictly linear relationship between respiration and cardiac work (Williamson *et al.*, 1976). The problem, however, is that many workers have found that calculated ADP levels (based on equilibrium of the creatine kinase reaction: creatine + ATP ⇌ phosphocreatine + ADP) or average ADP levels in heart did not correlate with workload (Balaban *et al.*, 1986; Wan *et al.*, 1993). The tentative conclusion was reached that respiratory control in the heart cannot be determined by matrix levels of ADP. That this conclusion is incorrect is self-evident: even over short times, the rate of ADP produced in the cytosol by contraction must exactly equal the rate of ADP consumption in the matrix by the F_1-F_0-ATPase. A purely kinetic signal, such as calcium activation of the F_1-F_0-ATPase, cannot overcome this requirement (Saks *et al.*, 1998). Finally, most of the cytosolic ATPases require high ΔG_P, so there is a good reason for the cell to maintain low ADP levels, even in the face of high work. This brings us back to the primary question, how is peripheral ADP production transmitted so efficiently to the mitochondrial matrix?

The answer, developed over several years by Saks and coworkers (1994, 1998; Vendelin *et al.*, 1999), is that the signal is delivered by a combination of group transfer between myofibrils and mitochondria, and metabolic channeling within the intermembrane space of mitochondria. In heart and skeletal muscle, creatine kinase exists as two isoforms, one associated with the myofibril (cytoCK) and one in the intermembrane space (mitoCK). CytoCK efficiently converts ADP to creatine, which is present at high concentrations and is competent as a carrier of the "low energy" charge. Creatine crosses the outer membrane and reacts with mitoCK. MitoCK is self-associated (probably as a dimer) and also functionally associated with the ATP/ADP translocator. Thus, creatine is phosphorylated by mitoCK with simultaneous production of

ADP, which is taken up immediately into the matrix and phosphorylated. Phosphocreatine, simultaneously produced from exchanged ATP, shuttles back to the myofibril. In this manner, the ADP produced by the myofibril is efficiently transmitted to the mitochondrial matrix.

A large body of evidence supports this mechanism, and a major new piece of evidence from the transgenic mouse model puts the icing on the cake: Wieringa and collaborators have produced a mitoCK knockout mouse (van Deursen *et al.*, 1993). These mice develop normally, but knockout of both cytoCK and mitoCK profoundly impairs the power and work output of muscles and alters the cell architecture. Very similar results were obtained for the heart when CK was inhibited by iodoacetamide or inactivated by substituting guanidino propionate (GPA—its phosphorylated form is not used by CK) for creatine. The hearts from these animals exhibit the same linear dependence of respiration on cardiac work as those from normal mice, with one profound difference—they can only achieve 40–50% of the normal span of cardiac work. This outcome is exactly as predicted: when metabolic channeling through mitoCK is prevented, the cardiac myocyte must rely solely on direct transmission of the signal via ADP diffusion to mitochondria. Consequently, the mitochondria are no longer able to keep up with high rates of ATP consumption (data summarized by Saks *et al.*, 1994, 1998 and Vendelin *et al.*, 1999). Moreover, the tissue attempts to adapt to loss of CK channeling by increasing many-fold its phosphoryl group transfer via the adenylate kinase system, thereby reducing the necessity for free diffusion of ADP (Dzeja *et al.*, 1998).

The rate of ATP synthesis in muscle cells is therefore controlled primarily by ΔG_P; however, the transmission of this signal involves metabolic channeling through mitoCK. The mitoCK assembly will be sensitive to factors that perturb the intermembrane space, a matter that will be discussed later.

V. Ion Leaks in Mitochondria

Despite the low permeability of the inner membrane to ions, cation leaks occur at significant rates in respiring mitochondria, because $\Delta\Psi$ is maintained at very high levels. Ion leaks in mitochondria are physiologically important: inward potassium leak causes matrix swelling. Inward proton leak dissipates energy and contributes to the basal metabolic rate.

Diffusive transport of ions does not differ fundamentally from transport of polar nonelectrolytes across thin membranes. Thus, the rate of transport is proportional to the concentration difference, and the proportionality constant (the permeability coefficient) is a function of the energy barrier that must be overcome in order to extract the ion from the aqueous phase. The main complication associated with ionic charge movements derives from long-range effects of the imposed electric field on the local free energy of the diffusing particles.

All of the essential features of current voltage relationships in biomembrane systems can be derived from a simple model, described by Garlid *et al.* (1989), in which a single, sharp energy barrier is located at the center of the mem-

brane, as depicted in Fig. 4. Consider the steps taken by a cation crossing the membrane through such a leak pathway. The cation first partitions between the aqueous medium and a surface energy well, located near the phospholipid head groups. The ion must next overcome an extremely unfavorable Gibbs energy of transfer in order to move into the hydrophobic interior of the membrane. Only those ions having sufficient energy to reach the barrier peak (center of the membrane) will cross to the energy well on the opposite side and, thence, into the aqueous medium. Net flux will be proportional to the differential probability of getting to the peak from either side.

The major consequence of the high $\Delta\Psi$ in mitochondria is that inward cation flux is no longer ohmic but rather exhibits an exponential dependence on $\Delta\Psi$. To see why this is so, consider the probability of an ion in water having sufficient energy to move to the center of the membrane. This is given by an exponential Boltzmann function, $e^{-\Delta\mu_p/RT}$, where $\Delta\mu_p \equiv \mu_p - \mu_{aq}$ is the Gibbs energy of the ion at the peak (p) relative to its value in the aqueous energy well at the surface of the membrane. It is simple to show (see Garlid *et al.*, 1986) that ion leak across biomembranes is given by

$$J = P\left(C_1 e^{u/2} - C_2 e^{-u/2}\right) \tag{3}$$

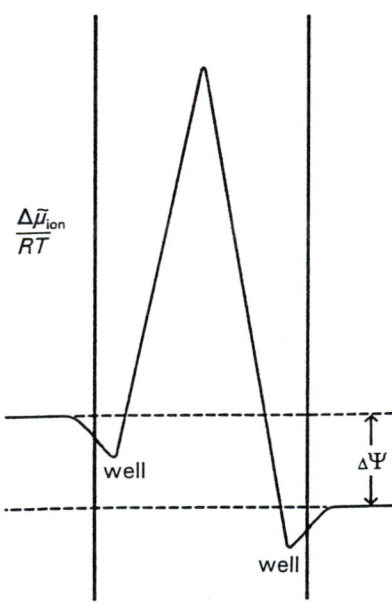

FIGURE 4. The energy barrier to ion leak across the mitochondrial inner membrane. Experimental data (see Fig. 5) confirm that the energy barrier to cations is best represented by a sharp peak at the center of the membrane. Cations must first partition into the surface energy wells created by the phospholipid head groups. Those ions with sufficient energy to overcome the unfavorable free energy of transfer from the well to the peak will cross to the other side. Protons and hard cations partition into the same interfacial energy well; however, protons go on to partition into a deeper energy well at the acylglycerol linkages of the phospholipids, where their concentration is near saturation. These surface features cause proton leak to differ **quantitatively** from hard cation leak; however, as shown in Fig. 5, leaks of protons and cations are **qualitatively** identical.

where $u \equiv -z\mathscr{F}\Delta\Psi/RT$, and P is the permeability constant, given by

$$P \equiv k e^{-\Delta\mu_p^0/RT} \qquad (4)$$

The energy barrier at the center of the membrane splits $\Delta\Psi$ in half when we make the customary assumption that the potential gradient is linear within the hydrophobic interior (the Goldman constant field assumption). This is the origin of the factor 1/2 in the exponents of Eq. 3. Note in Eq. 4 that $\Delta\mu_p^0 \equiv \mu_p^0 - \mu_{aq}^0$ is the standard Gibbs energy of transfer of the ion to the hydrophobic center of the bilayer; $\Delta\mu_p^0$ is the activation energy, and $\Delta\mu_p^0/RT$ represents the height of the barrier that must be overcome in order for the ion to traverse the membrane at equilibrium.

As is true of all flux equations, C_1 and C_2 do not refer to the bulk aqueous phase concentrations, but rather to concentrations adjacent to the energy barrier—in this case, to the concentrations in the energy wells. We may define a partition coefficient, $f = C_1/C_{10} = C_2/C_{20}$, where C_{10} and C_{20} are aqueous concentrations, and we assume the membrane to be symmetric with respect to surface potentials and ion extraction. This results in a practical flux equation.

$$J = fP\left(C_{10} e^{u/2} - C_{20} e^{-u/2}\right) \qquad (5)$$

It is instructive to consider special cases of Eq. 5. When $z = 0$ (transport of a nonelectrolyte) or $\Delta\Psi = 0$ (transport of an ion in the absence of an electric field), Eq. 5 reduces to Fick's law.

$$J = fP\left(C_{10} - C_{20}\right) \qquad (6)$$

When $\Delta\mu$ is sufficiently small, and $C_{10} = C_{20}$, Eq. 5 reduces upon expansion of the exponentials to

$$J = -fPC\left(\mathscr{F}\Delta\Psi/RT\right) \qquad (7)$$

Equation 7 describes an ohmic (linear) flux-voltage relationship and reveals the important point that Ohm's law is an approximation which is valid only when potential gradients are sufficiently small. Thus linear flux-voltage dependence should not be viewed as the expected behavior for ion leak. The electrical gradients across the mitochondrial inner membrane greatly exceed the ohmic limit, and flux is exponential with voltage.

The approximation of greatest practical application to mitochondrial bioenergetics is achieved by dropping the second term in Eq. 5. This term represents back-flux of cations from the matrix and becomes negligible at the high values of $\Delta\Psi$ maintained by respiring mitochondria. Thus, Eq. 5 reduces to a simple exponential function of $\Delta\Psi$:

$$J = fPC_{10} e^{u/2} \qquad (8)$$

Measurements of cation leak in mitochondria are in good agreement with the predictions of Eq. 8, as shown by the flux-voltage plots for H^+ and TEA^+ (tetraethylammonium ion) in Fig. 5.

A great deal of attention has been paid to two "anomalies" of proton leak across lipid bilayers and biomembranes (Deamer and Nichols, 1989). First, the rate constant (flux divided by concentration) for proton transport is about 10^6 higher than that for TEA^+ or alkali cations (see legend to Fig. 5). Secondly, proton flux exhibits very little dependence on H^+ concentration, which appears to contradict Eq. 8. These

"anomalies" have led to speculations that protons are transported through the bilayer by a mechanism that differs from that for other cations. This hypothesis is refuted by the data in Fig. 5, which show that there is no difference between H^+ and TEA^+ in the rate-limiting step of diffusion across the hydrophobic barrier. The anomalies of proton leak therefore reside in the terms f and P of Eq. 8.

The following observations appear to account for both anomalies of H^+ flux. Changing external pH or adding polyvalent cations (e.g., Mg^{2+} or spermine) to the medium profoundly affects TEA^+ leak, but has little effect on H^+ leak. This shows that the energy well for protons is shielded from the aqueous medium and therefore deeper than the energy well for cations. The observation that J_{H^+} is relatively unaffected by $[H^+]_o$ simply means that the proton energy well is nearly saturated with protons. Thus $C_1 > C_{10}$ for protons, contributing to a higher flux. We also observed that the activation enthalpy for H^+ flux is about 27 kJ/mol lower than that for TEA^+ ion. Together, these factors can readily account for the apparent anomalies in H^+ flux across mitochondrial and bilayer membranes: the hydronium ion equilibrates with an

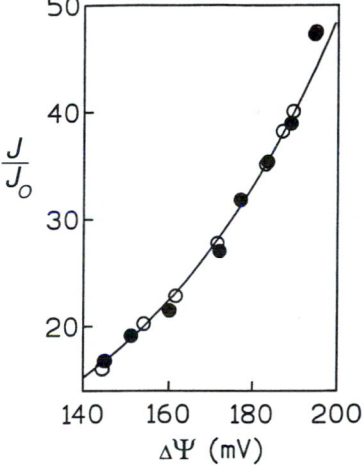

FIGURE 5. Fluxes of TEA^+ and H^+ in mitochondria exhibit exponential dependence on $\Delta\Psi$. J/J_o is plotted versus $\Delta\Psi$ over the range to which Eq. 8 applies. $J_o \equiv fPC_o$, where C_o is the aqueous concentration of TEA^+ and H^+, respectively, and f and P are defined in the text. J_o was obtained from the extrapolated intercept of a semilogarithmic plot, $\ln J$ versus $\mathscr{F}\Delta\Psi/RT$.

The first point to note from the figure is that the rate-limiting step of crossing the energy barrier is identical for TEA^+ and H^+, which probably crosses as hydronium ion. That is, the only difference between these ions is in the J_o term, which largely reflects events at the surface of the membrane. The second point is that this quantitative difference is very large indeed. The product, fP, is 2×10^{-10} cm/s for TEA^+ ion and 0.85×10^{-3} cm/s for H^+ ion. Thus, the apparent permeability constant for hydronium ions is 4.25×10^6 times greater than that for TEA^+, and 5.4×10^4 of this difference is accounted for by the fact that the activation enthalpy for proton flux is 27 kJ/mol less than that for TEA^+. This is largely a surface effect, reflecting extraction of the ions from their aqueous energy wells into the bilayer. The remainder is attributable to a 79-fold concentration of H^+ in its deeper energy well at the acylglycerol linkages. Thus, the apparent "pH" of the proton energy well is about 5.5, which is in reasonable agreement with other data, including the fact that the apparent pK_a values of fatty acids in the bilayer is 2–3 units higher than pK_a values in aqueous solution.

energy well located deeper in the membrane than the energy well for cations. The hydronium energy well is most probably located at the level of the phospholipid acylglycerol linkages. This energy well is not accessible to cations; it is shielded from the aqueous environment; and it is nearly saturated with H_3O^+.

VI. The Mitochondrial Proton Cycle—the Uncoupling Proteins

As we have seen, when inner membrane permeability to H^+ ions is increased, protons leak back into the matrix. If this leak is sufficiently high, electron transport is uncoupled from oxidative phosphorylation because redox energy is burned off as heat, and no energy is conserved to make ATP. This can readily be demonstrated in the laboratory, using uncouplers such as dinitrophenol or CCCP (as in Fig. 3). Protonophoretic uncouplers are weak acids that diffuse electroneutrally across the membrane, delivering a proton to the other side. By itself, this has no uncoupling effect because no charge movement has occurred. The anion, however, contains pi electrons, enabling charge delocalization in the protonophore structure and consequent membrane permeability to the anion. As diagrammed in Fig. 6, the anion and acid cycle across the membrane until the proton-motive force is dissipated.

Nature has designed proteins expressly for the purpose of uncoupling. The effect of uncoupling protein (UCP) is to short-circuit the insulating inner membrane, thereby converting energy to heat instead of to ATP. UCP1 plays a major role in providing thermogenesis to hibernating animals and to all mammalian newborns, including humans. UCP1 is inhibited by cytosolic ATP and is released from ATP inhibition by poorly understood mechanisms. Fatty acids are required for uncoupling by UCP1, and UCP1 acts by catalyzing the passive efflux of fatty acid anions from the matrix. The protonated fatty acid head-group then diffuses rapidly back into the matrix. Thus, UCP1 does not conduct protons, *per se;* rather, it enables fatty acids to behave as cycling protonophores, as shown in Fig. 7 (Garlid *et al.,* 1996a, 1998).

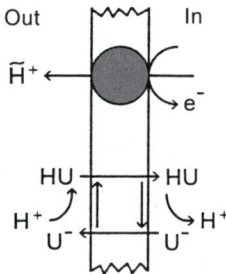

FIGURE 6. The mechanism of uncoupling by weak acid protonophores. When both the anion and acid can permeate the membrane, as is the case with all weak acid protonophores, their cycling leads to inward transport of H^+ ions which is coupled to outward transport of negative charge. This cycling draws current from the proton-motive batteries, diverting energy away from the ATP synthase and dissipating energy as heat. ATP synthesis is said to be **uncoupled** from respiration.

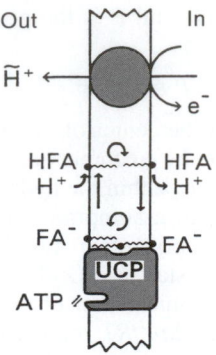

FIGURE 7. The mechanism of uncoupling by mitochondrial uncoupling protein. Uncoupling protein (UCP) contains a weak binding site for anions near the center of the membrane. This constitutes an energy well for anions and provides a low-resistance pathway for normally impermeant anions to cross the membrane. The physiological substrates of UCP are anions of free fatty acids (FAs). The FA head groups are located at the acylglycerol linkages of the phospholipid bilayer. When the FA is at the surface of UCP, the negative membrane potential drives the head group to the energy well at the center of the membrane. The FA then "flip-flops" and the head group is driven to the opposite surface. Here, it diffuses away from the protein and picks up a proton. The protonated FA freely diffuses across the membrane, during which it flip-flops again, and delivers a proton to the other side. Note that this uncoupling mechanism is qualitatively identical to that described in Fig. 4. Thus the role of UCP is to enable FAs to behave as cycling protonophores. Furthermore, FA cycling is highly regulated: UCP is inhibited from the cytosolic side by nucleotides, including ATP.

The human genome contains at least three uncoupling proteins designated UCP1, UCP2, and UCP3. UCP2 maps to regions of human chromosome 11 that have been linked to hyperinsulinemia and obesity, and it is hypothesized that UCP2 is the peripheral target for energy dissipation in the regulation of body weight. UCP2 is ubiquitously expressed in mammalian tissues, whereas UCP3 is expressed primarily in glycolytic skeletal muscle in humans, and may account for the thermogenic effect of thyroid hormone (Gong *et al.,* 1997). These aspects of this rapidly emerging area of research have been reviewed by Boss *et al.* (1998) and Jezek and Garlid (1998).

The transport functions and regulation of UCP2 and UCP3 have been characterized using recombinant proteins expressed in *E. coli* and reconstituted into liposomes for flux measurements. UCP2 and UCP3 both catalyze electrophoretic flux of protons and alkylsulfonates, and proton flux exhibits an obligatory requirement for fatty acids. Fatty acid–dependent proton transport by UCP2 and UCP3 is inhibited by purine nucleotides, but with considerably lower apparent affinities for nucleotides than those observed with UCP1 (Jaburek *et al.,* 1999). Based on these results, the properties of UCP2 and UCP3 are qualitatively identical to those of UCP1. Thus, they are functional uncoupling proteins, and their biophysical properties are consistent with a physiological role in energy dissipation. Nevertheless, additional investigation is required to determine their regulation *in vivo,* and their role in the regulation of body weight.

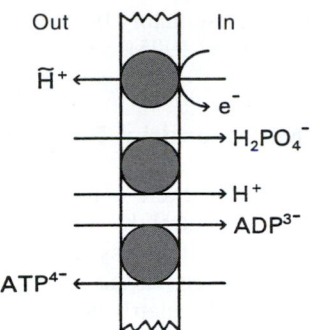

FIGURE 8. Mitochondrial transport of inorganic phosphate, ADP, and ATP. The phosphate carrier catalyzes electroneutral transport, and the ADP-ATP exchanger catalyzes electrophoretic transport. Uptake of P_i and ADP and expulsion of ATP use one electrogenically ejected proton.

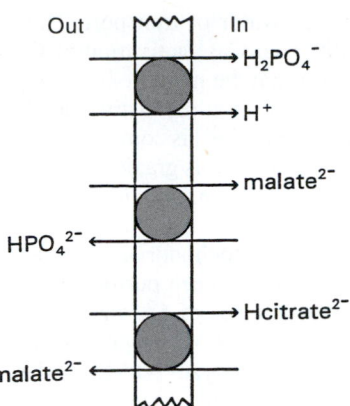

FIGURE 9. Mitochondrial transport of substrate anions. The dicarboxylic and tricarboxylic exchangers are electroneutral and coupled to the phosphate transporter by malate and citrate. No energy is expended during anion uptake by these mechanisms.

VII. The Anion Exchange Carriers

Mitochondria synthesize ATP in the matrix; consequently, the substrates for the ATP synthase, ADP and phosphate, must be imported across the inner membrane, and ATP must be exported for use by the cell. As shown in Fig. 8, nucleotides are exchanged on the ATP-ADP translocase (ANT) in a process involving outward movement of one negative charge. Although some of the proton-motive energy is utilized to export ATP, this is largely compensated by the fact that the F_1-F_0-ATPase can synthesize ATP at a lower phosphorylation free energy than that required by ATP-consuming processes of the cell. The phosphate carrier catalyzes electroneutral P_i-H^+ symport, or P_i-OH^- antiport, with the net result that it transports phosphoric acid.

The inner membrane also contains a variety of anion exchange carriers, which are designed to deliver substrates to the tricarboxylic acid cycle and which were first described by Chappell (1968) (see Fig. 9). The anion exchange carriers catalyze 1:1 exchange of anions; nevertheless, their distributions behave as if they were transported as fully protonated acids:

$$\frac{(A^-)_{in}}{(A^-)_{out}} = 10^{z\,\Delta pH} \qquad (9)$$

where z is the valence of the acid. This results from an arrangement of these exchange carriers in a cascade in which phosphate and malate are key intermediates. In addition to their normal substrates, the dicarboxylic acid exchanger also catalyzes malate-phosphate exchange, and the tricarboxylic acid exchanger also catalyzes malate-citrate exchange. In this way, both di- and tricarboxylic acids are linked to the phosphate carrier. Since the phosphate carrier transports fully protonated phosphate, the net result is that di- and tricarboxylic acids also enter as if they were fully protonated.

Many of the anion exchangers have been cloned and sequenced. They are all found to be 30–35 kDa proteins and are thought to function as homodimers. All those identified to date contain a tripartite structure and are considered to belong to a common gene superfamily.

VIII. The Mitochondrial Calcium Cycle

The mitochondrial Ca^{2+} cycle, diagrammed in Fig. 10, consists of three separate processes. Ca^{2+} is taken up by the Ca^{2+} channel at the expense of two ejected protons. Ca^{2+} is ejected by the Na^+-Ca^{2+} antiporter, which catalyzes an electrophoretic $3Na^+$-Ca^{2+} exchange, utilizing an additional ejected proton (Jung *et al.*, 1995). The three Na^+ ions taken up are then ejected by the electroneutral Na^+-H^+ antiporter.

The mitochondrial Ca^{2+} channel is characterized by high-capacity, low-affinity, electrophoretic Ca^{2+} uptake. In experiments with partially purified Ca^{2+} channels reconstituted into liposomes, the K_m for Ca^{2+} ranged between 7 and 20 μM, and the V_{max} was 130 μmol/mg protein per min, about 250-fold greater than that observed in intact mitochondria. La^{3+} is a competitive inhibitor of Ca^{2+} uptake, and ruthenium red is a noncompetitive inhibitor. In the reconstituted system Ca^{2+} transport exhibited hyperbolic dependence on $[Ca^{2+}]$, whereas in intact mitochondria, Ca^{2+} appears to

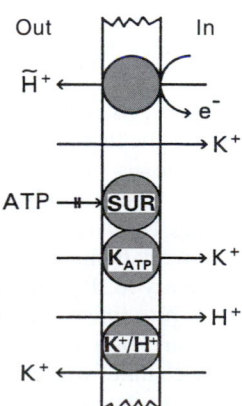

FIGURE 10. The mitochondrial Ca^{2+} cycle. Ca^{2+} enters the matrix via the electrophoretic Ca^{2+} channel and is ejected by the electrophoretic Na^+-Ca^{2+} antiporter, utilizing three ejected protons per Ca^{2+} ion taken up. The Na^+ is then expelled by the electroneutral Na^+-H^+ antiporter.

be an allosteric activator of transport. This raises the interesting possibility that Ca^{2+} activation of Ca^{2+} uniport in intact mitochondria may be mediated by a separate regulatory protein which is lost upon reconstitution.

Free mitochondrial $[Ca^{2+}]$ is comparable to cytosol $[Ca^{2+}]$ *in vivo,* in spite of the enormous gradient for electrophoretic Ca^{2+} uptake. This disequilibrium is maintained in heart mitochondria by an electrophoretic Na^+-Ca^{2+} antiporter (Jung *et al.,* 1995). The cardiac mitochondrial Na^+-Ca^{2+} antiporter is a 110-kDa protein that has been purified and reconstituted into liposomes. In the presence of K^+, the K_m for Ca^{2+} is 300 nM, and the K_m for Na^+ is 7 μM. In the absence of Ca^{2+}, the reconstituted antiporter catalyzes Na^+-Li^+ and Na^+-K^+ exchange; however, K^+-Ca^{2+} exchange was not observed (Li *et al.,* 1992). Liver mitochondria appear to have Na^+-independent as well as a Na^+-dependent Ca^{2+} efflux pathways, but these have not been well characterized.

The Na^+-H^+ antiporter has been identified as a 59-kDa inner membrane protein, and the purified protein is reconstitutively active (Garlid *et al.,* 1991). Unlike the plasma membrane Na^+-H^+ antiporters, the mitochondrial carrier appears to lack regulation, causing Na^+ to equilibrate with the ΔpH. Like the plasma membrane carriers, the mitochondrial version is competitively inhibited by Li^+ and unaffected by K^+.

Ca^{2+} is a second messenger, signalling a need to increase cellular work. Because increased work requires a higher rate of ATP production, this message must be relayed to the mitochondrial matrix. The physiological role of the mitochondrial Ca^{2+} cycle is to regulate matrix Ca^{2+} activity in response to signals from the cytosol. Intramitochondrial Ca^{2+} is required to activate the phosphorylase that converts pyruvate dehydrogenase to its active form, and α-ketoglutarate dehydrogenase is allosterically activated by matrix $[Ca^{2+}]$ in the physiological range (Denton and McCormack, 1990, Gunter *et al.,* 1994).

Recent studies on isolated hepatocytes strongly suggest that the mitochondrial Ca^{2+} cycle is designed to respond to *oscillations* in cytosolic $[Ca^{2+}]$. A static elevation in cytosolic $[Ca^{2+}]$ causes a transient spike in mitochondrial $[Ca^{2+}]$ and a transient increase in pyridine nucleotide reduction, associated with activation of the Ca^{2+}-sensitive mitochondrial dehydrogenases. The energetic effect decayed much more slowly than did matrix $[Ca^{2+}]$. IP_3-mediated oscillations in cytosolic $[Ca^{2+}]$, on the other hand, were matched by oscillations in mitochondrial $[Ca^{2+}]$ and a sustained activation of the mitochondrial dehydrogenases. Maintaining high activity is thought to be due to the fact that the effects of Ca^{2+} activation decline more slowly than $[Ca^{2+}]$ itself. From a metabolic point of view, these results indicate that mitochondria can discriminate between slow increases in cytosolic $[Ca^{2+}]$, associated with homeostatic mechanisms, and rapid oscillations secondary to IP_3 signaling. The kinetics of the Ca^{2+} channel and its proximity to the release sites on the endoplasmic reticulum are well suited to such discrimination. The high-capacity channel will be activated only when local $[Ca^{2+}]$ is high, due to its allosteric activation by Ca^{2+}. Ca^{2+} uptake in response to slower oscillations in cytosol $[Ca^{2+}]$ will rapidly be cleared by the efflux mechanism(s). Similarly, an electrophoretic Na^+-Ca^{2+} exchanger, such as exists in heart mitochondria, would drive rapid Ca^{2+} efflux

to permit the observed oscillatory behavior of mitochondrial $[Ca^{2+}]$ (Hajn'oczky *et al.,* 1995).

IX. The Mitochondrial Potassium Cycle

The mitochondrial K^+ cycle consists of electrophoretic K^+ influx and electroneutral K^+ efflux across the inner membrane (see Fig. 11). It is essential that mitochondria regulate net K^+ flux to zero in the steady state; otherwise, K^+ leak would cause the matrix to swell and eventually lyse. This regulation is provided by the K^+-H^+ antiporter, which ejects exactly the amount of K^+ that is taken in. The primary role of the K^+-H^+ antiporter is to provide **volume homeostasis** to mitochondria *in vivo* in order to maintain the vesicular integrity necessary for oxidative phosphorylation (Garlid, 1988).

Regulation of the K^+-H^+ antiporter is mediated by reversible binding of Mg^{2+} and H^+ to the K^+-H^+ antiporter on its matrix side. The activity of these ions decreases with uptake of K^+ salts, causing a graded, compensatory activation of K^+ efflux in response to increases in matrix volume. The K^+-H^+ antiporter has been identified as an 82-kDa inner membrane protein. It has been purified and reconstituted into proteoliposomes and shown to catalyze electroneutral K^+-H^+ antiport (Li *et al.,* 1990).

The mitochondrial K_{ATP} channel (mitoK_{ATP}) also plays an important role in volume homeostasis. When mitoK_{ATP} is open, the added K^+ conductance is thought to compensate for the lower driving force for K^+ influx (lower $\Delta\Psi$) in ischemia and in high ATP-consuming states of the cell. This is discussed in the next sections.

Although it possesses certain distinctive characteristics, mitoK_{ATP} exhibits striking similarities to plasma membrane K_{ATP} channels (pmK_{ATP})—for example, mitoK_{ATP} reacts with

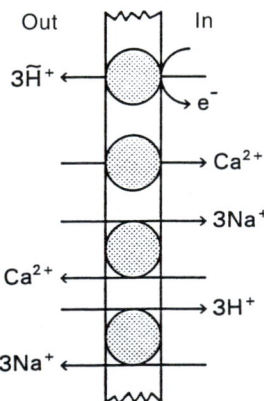

FIGURE 11. The mitochondrial K^+ cycle. Electrogenic proton ejection drives significant K^+ uptake by diffusive leak. In addition, the inner membrane contains a K_{ATP} channel, which is highly regulated by nucleotides, CoA esters, and pharmacological agents. The K_{ATP} channel consists of at least two subunits, a K^+-specific channel, and a sulfonylurea receptor (SUR). Net K^+ flux is regulated to zero in the steady state. Compensatory K^+ efflux is provided by the electroneutral K^+-H^+ antiporter, which is regulated by matrix Mg^{2+} and H^+ and is exquisitely sensitive to changes in matrix volume.

147

the same set of biochemical and pharmacological ligands, including adenine and guanine nucleotides, long-chain acyl-CoA esters, sulfonylureas, and K^+ channel openers. These similarities suggest that mitoK$_{ATP}$, like the pmK$_{ATP}$ of the pancreatic β-cell, is a heteromultimeric complex consisting of an inward rectifying K^+ channel (mitoKIR) and a sulfonylurea receptor (mitoSUR). Indeed, we have recently identified and purified these two proteins from the mitochondrial inner membrane. MitoSUR is a 63-kDa sulfonylurea-binding protein, and mitoKIR is a 55-kDa protein. Because of its strong resemblance to pmK$_{ATP}$, we infer that mitoK$_{ATP}$ belongs to the same gene family.

MitoK$_{ATP}$ is regulated by a rich variety of metabolic and pharmacological ligands. It is inhibited with high affinity by ATP, long-chain acyl-CoA esters, the antidiabetic sulfonylurea, glyburide, and 5-hydroxydecanoate. The ATP-inhibited channel is opened with high affinity by guanine nucleotides and K^+ channel openers such as cromakalim and diazoxide. There is indirect evidence that mitoK$_{ATP}$ is opened *in vivo* by phosphorylation.

The K^+ cycle will draw electrogenic H^+ current from the ETS, and the question arises whether opening mitoK$_{ATP}$ will significantly affect $\Delta\Psi$ or ΔpH across the inner membrane. To answer this question, we need an estimate of the magnitude of K^+ flux through mitoK$_{ATP}$ when fully open. The V_{max} of the K^+-H^+ antiporter is about 350 nmol/mg·min. It is clear that this value must be greater than the sum of K^+ fluxes through leak and mitoK$_{ATP}$ pathways. In agreement, estimates from studies on intact mitochondria indicate a maximum flux through mitoK$_{ATP}$ of about 100 nmol/mg·min. This is an exceedingly low rate, amounting to less than 5% of the maximum rate of proton pumping by heart mitochondria. Based on this estimate, opening mitoK$_{ATP}$ will not affect the membrane potential by more than 3 to 4 mV. Moreover, any perturbation of the pH gradient will largely be canceled by associated movement of phosphate and other anions on their electroneutral carriers. In this way, net K^+ flux is converted to net transport of K^+ salts, which will be accompanied by osmotically obligated water. Finally, net K^+ flux will not have a significant effect on matrix K^+ **concentration,** which is about 150 mM, *in vivo*. From these considerations, we may conclude that opening mitoK$_{ATP}$ will primarily affect matrix volume and that its effects on respiration, $\Delta\Psi$, and ΔpH will be negligible.

X. The Physiological Role of MitoK$_{ATP}$ in Heart

In view of the enormous driving force for K^+ uptake by diffusion, why do mitochondria possess a K^+ channel? A plausible role may be deduced from considering the situation in the working heart. When the heart goes from a resting state to a high-work state, ATP production and consumption may increase as much as eight-fold. This increased current through the ETS will cause $\Delta\Psi$ to drop, as shown in Fig. 5, and K^+ diffusion into the matrix will drop as an exponential function of $\Delta\Psi$, as shown in Fig. 3. Indeed, if $\Delta\Psi$ drops by 35 mV, diffusive K^+ influx will drop by 50%. If mitoK$_{ATP}$ does not open, matrix volume will contract until the K^+-H^+ antiporter senses the drop in volume, resulting in a lower

steady state volume in the high-work state. We have estimated the extent of this volume contraction in isolated mitochondria, and it amounts to 10–15% of matrix volume. Importantly, addition of a K^+ channel opener, such as diazoxide, reverses the matrix contraction caused by high phosphorylation rates. Thus, mitoK$_{ATP}$ is well-suited to maintain constant matrix volume when $\Delta\Psi$ falls.

When mitoK$_{ATP}$ is opened under these conditions, the additional conductance pathway compensates for the reduced driving force, thereby minimizing the matrix contraction that would otherwise occur during high ATP synthesis. Why is it important to prevent such a small contraction of the mitochondrial matrix? To answer this question, we return to the metabolic channeling by creatine kinase. We recall that an intact mitoCK assembly, involving association of mitoCK with itself and with the ATP-ADP translocase (ANT), is essential for the high-work state. These associations are strongly volume-dependent, referring of course to the volume of the intermembrane space (IMS). When the matrix contracts, the IMS will correspondingly expand. Moreover, expansion of the IMS will be greatly amplified, because its volume is normally very small. If IMS expansion is not prevented, mitoCK will dissociate during the high-work state, precisely when metabolic channeling through this complex is most needed. Thus, we hypothesize that matrix contraction, which would normally accompany high phosphorylation rates, must be prevented by opening mitoK$_{ATP}$ in order for metabolic channeling to proceed. The hypothesis predicts that increased work states in heart cannot proceed if mitoK$_{ATP}$ is blocked, and preliminary evidence supports this prediction (see Fig. 12). The signal to open mitoK$_{ATP}$ is not known, but it is assumed to derive from the signal leading to increased contraction rates (elevated cytosolic $[Ca^{2+}]$) and most likely involves phosphorylation of mitoK$_{ATP}$.

XI. MitoK$_{ATP}$ as the End Effector of Protection Against Ischemia-Reperfusion Injury

Ischemia is interruption of blood flow to a region of tissue, as in myocardial infarct or stroke. During ischemia, cytosolic $[Ca^{2+}]$ rises, and ATP falls due to lack of oxygen. Ischemic cells remain viable during ischemia and respire rapidly upon reperfusion; however, they cannot build up ΔG_P to levels needed for contraction, and the myocytes undergo contracture. Reactive oxygen species are formed at a high rate in the presence of elevated $[Ca^{2+}]$, causing oxidation of key thiol groups on mitochondrial proteins and damage to the plasma membrane. Eventually, the mitochondria become irreversibly damaged and the cell undergoes necrosis.

Many investigators have studied factors protecting tissues from ischemia-reperfusion injury, with most attention being focused on preconditioning—designating a treatment administered **before** the test ischemia. At least four distinct types of preconditioning have been identified (reviewed in Grover and Garlid, 1999): (1) **Ischemic preconditioning** denotes a short period of ischemia that protects the heart from a subsequent ischemic episode of more prolonged duration. (2) **Calcium preconditioning** refers to protection by

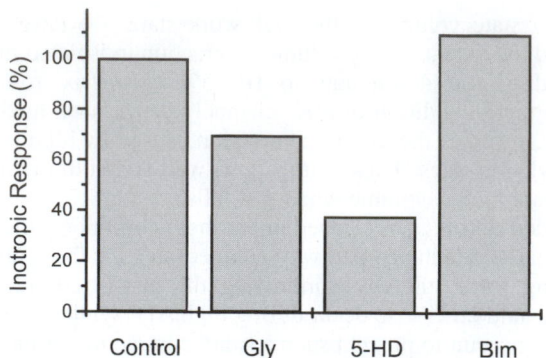

FIGURE 12. Requirement for open MitoK$_{ATP}$ in positive inotropy. Bar graph contains contractility data from 12 human atrial strips, 3 for each treatment. In each condition, isometric tension was stimulated by 10 μM dobutamine. **Control** (100%)—No other treatment. Developed tension increased 60% after superfusion of dobutamine, which is typical of this preparation. **Gly**—preparation was pretreated with 1 μM glyburide, a nonspecific blocker of K$_{ATP}$ channels. Glyburide caused the positive inotropic effect of dobutamine to drop by 30%. **5-HD**—preparation was pretreated with 300 μM 5-hydroxydecanoate, a specific blocker of the mitochondrial K$_{ATP}$ channel. 5-HD causes the positive inotropic effect of dobutamine to drop by 60%. **Bim**—preparation was pretreated with 0.1 μM bimakalim, a nonspecific opener of K$_{ATP}$ channels. Bimakalim had no significant effect on the positive inotropic effect of dobutamine, suggesting that mitoK$_{ATP}$ was already opened in response to the dobutamine (Puddu and Garlid, unpublished data).

a transient increase in intracellular [Ca^{2+}] prior to ischemia. (3) **Volatile anesthetic preconditioning** refers to protection following anesthesia with the fluothane class of anesthetics. (4) **K$^+$ channel opener preconditioning** refers to protection by drugs that open K$_{ATP}$ channels. In heart, each of these modes of preconditioning has been shown to improve cardiac function, reduce infarct size, and increase time to contracture. Protein kinase C has been shown to be a mediator of the first two types of preconditioning and is probably involved in the third as well.

By 1997, a consensus view had emerged from a wide variety of studies that the end effector of all modes of preconditioning is a K$_{ATP}$ channel, and this was generally assumed to be the plasma membrane K$_{ATP}$ channel. This created a conflict because of mounting experimental evidence against involvement of pmK$_{ATP}$ in the cardioprotective effects of K$_{ATP}$ openers (reviewed in Grover and Garlid, 1999). The suggestion was made that mitoK$_{ATP}$ may be the end effector of ischemic preconditioning, and moreover, a unique pharmacology was uncovered that could be used to test this hypothesis (Garlid and Garlid, 1996). Thus, mitoK$_{ATP}$ is unusually sensitive to diazoxide, whereas the cardiac pmK$_{ATP}$ is relatively insensitive to this K$^+$ channel opener. Furthermore, mitoK$_{ATP}$ is blocked by 5-hydroxydecanoate (5-HD), which has no effect on pmK$_{ATP}$ (Jaburek *et al.*, 1998).

The results of experiments in the perfused rat heart model of ischemia provided straightforward support for the mitoK$_{ATP}$ hypothesis (Garlid *et al.*, 1997). There was concentration-dependent cardioprotection in the low micromolar range of diazoxide. In agreement with its ineffectiveness in opening pmK$_{ATP}$, diazoxide did not cause shortening of action potential duration.

Moreover, it was shown that 5-HD blocks cardioprotection by K$_{ATP}$ openers without affecting pmK$_{ATP}$. These differential effects of 5-HD and diazoxide are fully consistent with the hypothesis that mitoK$_{ATP}$ is the receptor for cardioprotection by K$^+$ channel openers and for 5-HD blockade of pharmacological protection and ischemic preconditioning. The hypothesis that mitoK$_{ATP}$ is the end effector of ischemic protection is under active investigation and has received experimental support from several laboratories.

The mechanism of mitoK$_{ATP}$ cardioprotection is probably similar to the mechanism described for its role in the transition to the high-work state. Thus, ischemia causes ΔΨ to fall and the matrix will then contract, leading to disruption of the architecture of the IMS. MitoCK disruption, which is one of the earliest events to occur during ischemia (Kay *et al.*, 1997), will lead to accelerated hydrolysis of ATP by mitochondria and will prevent a concerted response to the reintroduction of oxygen. Opening mitoK$_{ATP}$, and maintenance of IMS volume at the expense of ATP hydrolysis, will prevent these events in the short term, thereby extending the duration of ischemia that the cell can survive.

Most organs are protected by ischemic preconditioning. Protection in brain was found to be mediated by K$_{ATP}$, and preconditioning was abolished by glyburide (Heurteaux *et al.*, 1995). It will be of interest to determine whether mitoK$_{ATP}$ is the end effector of ischemic protection in brain and other organs.

XII. The Mitochondrial Permeability Transition (MPT)

In agreement with postulate 3 of the chemiosmotic theory, isolated, respiring mitochondria normally have a very low permeability to ions and polar nonelectrolytes such as sucrose. However, when they are induced to take up a massive overload of Ca^{2+} through the Ca^{2+} channel, the permeability of the inner membrane to solutes with molecular weight below 1500 Da rapidly increases (Hunter and Haworth, 1979; for excellent reviews with different perspectives, see Bernardi, 1999, and Crompton, 1999). The conductance of the MPT has been estimated at 1000 pS, and opening of this large hole causes ions, including protons, to equilibrate across the membrane with consequent depolarization and uncoupling of oxidative phosphorylation.

MPT opening is prevented by Ca^{2+} chelation, Mg^{2+} and ADP, antioxidants and thiol reductants, pyridine nucleotide reductants, and the immune suppressor cyclosporin A. These agents can also reverse opening of MPT (cyclosporin A reversal requires the presence of Mg^{2+} or ADP) but only if they are added very early in the process. It is presumed that cyclosporin A binds intramitochondrial cyclophilins, which are necessary for MPT opening (Connern and Halestrap, 1994). The discovery of relatively high-affinity blockade of MPT by cyclosporin A (Crompton *et al.*, 1988) has led to an explosion of work on this process.

The primary trigger for Ca^{2+}-induced opening of MPT appears to be oxidation of one or more specific membrane protein thiols due to mitochondrially generated reactive oxygen species. Oxidation of a specific protein among many that are damaged in the presence of ROS would account for the temporary reversibility of MPT opening (Kowaltowski and Vercesi, 1999). Despite 20 years of intense investigation, MPT is still a collection of phenomenology, and no specific protein responsible for the activity has yet been identified. There is considerable, but not conclusive, evidence identifying this protein as the ANT (Hunter and Haworth, 1979; Halestrap and Davidson, 1990), and some workers have obtained evidence for participation of the outer membrane porin (voltage-dependent anion channel) as well (Crompton, 1999).

The pathophysiological role of MPT is even more uncertain. At least five different roles have been ascribed to this process, but it seems likely that only the first three are relevant *in vivo*. (1) MPT opening plays a major role in necrotic cell death, most notably triggering the end-stage of ischemia-reperfusion injury when mitochondria *in vivo* are progressively damaged by exposure to high levels of both Ca^{2+} and ROS (Kowaltowski and Vercesi, 1999). (2) MPT opening in individual mitochondria may signal onset of autophagy, an important factor in the normal turnover of mitochondria (Lemasters *et al.*, 1998). (3) The proteins of MPT are involved in assembly and release of proapoptotic proteins without MPT opening or outer membrane rupture (Crompton, 1999). (4) MPT opens, and the outer membrane ruptures and releases cytochrome *c* to trigger the apoptotic cascade (Susin *et al.*, 1998). (5) MPT opens and closes under physiological conditions, perhaps to function as a Ca^{2+} release mechanism (Ichas *et al.*, 1997).

The proposal that MPT opening and closing are normal physiological processes is based on interesting observations on isolated cells under highly nonphysiological conditions (Ichas *et al.*, 1997; Petronilli *et al.*, 1999). It would be easier to envision a role in normal physiology if MPT were identified with a specific protein rather than an oxidized state of the ANT, a protein essential for mitochondrial function. Another perspective arises from considerations of molecular abundance. It can be estimated that one open MPT per mitochondrion is sufficient to account for the transport phenomenology. Using the number of mitochondria per mg of mitochondrial protein, this works out to 10^{-14} mol MPT/mg. On the other hand, the ANT, one of the most abundant inner membrane proteins, amounts to as much as 10^{-9} mol/mg. Cyclophilin is also highly abundant, about 10^{-10} mol/mg (Connern and Halestrap, 1994). Thus, cyclophilin is present in a more than ten-thousand-fold excess over MPT, which raises doubts about a physiological role for a cyclophilin-regulated MPT. The excess of matrix cyclophilin strongly implies that it has other functions and, therefore, that cyclosporin A, whose pharmacology is largely unknown, has other effects. This is a largely undeveloped area of research that raises questions about the common assumption that cyclosporin A effects on intact cells are due solely to inhibition of MPT.

In summary, it is reasonable that ROS-induced oxidation of thiols on the translocator leads to modification of membrane proteins and development of membrane holes, with the ultimate result of necrotic cell death or autophagy of individual mitochondria. The proteins of MPT may play a role in apoptosis, but it is unlikely that MPT opening and outer membrane rupture is required for the apoptotic cascade (see next section). Thus, pore opening is largely a pathological event, and it seems unlikely that MPT opening and closing is a normal physiological process.

XIII. Mitochondrial Involvement in Apoptosis

In apoptosis, cellular components are broken down and sequestered in membrane-enclosed vesicles (apoptotic bodies), which are then taken up by other cells. This packaging of the products of cell death avoids the inflammatory response to release of cell contents, as occurs in necrosis. The central players in apoptosis are caspases, a large family of cysteine proteases. As might be expected of such an important process, their regulation is complex and involves a branching cascade of signals (Thornberry and Lazebnik, 1998). The **end effector** caspases, which cause the cellular disassembly, are activated by **initiator caspases,** which are themselves activated by specific pathways. Thus, caspase-8 is an initiator caspase that is activated by plasma membrane death receptors of the tumor necrosis factor receptor family (Ashkenazi and Dixit, 1998). Caspase-9 is an initiator caspase that is activated by a variety of cytotoxic agents, including mitochondrial cytochrome *c* (Liu *et al.*, 1996). Since the caspase cascade is cytosolic, and cytochrome *c* is localized in the intermembrane space, this requirement is taken to imply that cytochrome *c* is released across the outer membrane. Three other proapoptotic proteins have also been identified that are released from the intermembrane space of mitochondria: procaspase-3, procaspase-9, and apoptosis-inducing factor. Both pro- and anti-apoptotic members of the Bcl-2 oncogene family have also been found to be associated with mitochondria, including Bcl-2, Bid, and Bax (reviewed in Crompton, 1999).

There is considerable debate over the pathway by which cytochrome *c* crosses the outer membrane. One contingent favors the hypothesis that the MPT opens, and the mitochondrial matrix expands to an extent that ruptures the outer membrane, thereby releasing cytochrome *c* and other caspase-activating proteins (Susin *et al.*, 1998). This hypothesis appears to be supported by the finding that cyclosporin A blocks apoptosis in some systems. On the other hand, mitochondrial morphology appears normal during apoptosis *in vivo* (Wyllie *et al.*, 1980), and rupture of the outer membrane seems a crude step in such a highly regulated process. Moreover, ATP is required for apoptosis, suggesting that mitochondria are not permeabilized by damage in this process. Crompton (1999) suggests a novel mechanism in which the **proteins** of MPT are involved in cytochrome *c* release without opening MPT or rupturing the outer membrane. Thus, a more attractive model is that proapoptotic proteins induce a channel in the outer membrane that is large enough to pass cytochrome *c*. Evidence favoring this hypothesis has recently been presented by Shimizu *et al.* (1999).

One can envision MPT triggering apoptosis at the periphery of an ischemic infarct, where partial damage has occurred, initiating the necrotic pathway and opening MPT with eventual release of cytochrome *c*. If necrosis proceeds sufficiently slowly, and if mitochondrial damage is too severe to be handled by autophagy, these cells may proceed to death by necroapoptosis (Lemasters, 1999), a combination of the two mechanisms of cell death.

XIV. Summary

Mitochondria have their own unique physiology, which is directed toward maintenance of vesicular integrity and proper functioning in energy transduction. The setting in which this physiology takes place is explained by the elegant postulates of the chemiosmotic theory. Of particular importance is the very high electrical membrane potential required for ATP synthesis. Several transport systems have evolved that either protect against adverse consequences of high membrane potential or exploit this feature to benefit the organism. The electrical driving force causes significant inward leak of protons, which dissipates energy and reduces the efficiency of oxidative phosphorylation. On the other hand, the energy dissipation of proton back-flux has also been exploited for thermogenesis in hibernating animals and newborns. The high membrane potential also drives significant diffusive uptake of K^+ across the membrane, resulting in osmotic swelling of the matrix and threatening the vesicular integrity of the organelle. Excess swelling is prevented, and volume homeostasis is maintained by an electroneutral K^+-H^+ antiporter, which is finely regulated to maintain zero net K^+ flux in the face of fluctuations in inward K^+ leak. The high driving force for K^+ uptake has also been exploited to drive K^+ inward through the highly regulated K_{ATP} channel. This additional K^+ influx compensates in the heart for work-induced drops in $\Delta\Psi$, thereby maintaining matrix volume and preserving the architecture of the intermembrane space. Mitochondria have exploited the high membrane potential to drive Ca^{2+} rapidly, both inward, via the Ca^{2+} channel, and outward, via the Na^+-Ca^{2+} antiporter. The mitochondrial Ca^{2+} cycle thus allows oscillatory cytosolic Ca^{2+} signals to be transmitted to the matrix, where the high frequency of oscillations causes sustained activation of enzymes that synthesize NADH, a major substrate for the electron transport chain. In this way, the Ca^{2+} signal not only directs the cell to increase its ATP-consuming activities, but also directs the mitochondrion to increase its supply of reducing equivalents to meet the demands of increased ATP production.

The great resurgence of interest in mitochondrial physiology is due to the realization that mitochondria play a more central role in pathophysiology than previously thought. Thus, the apoptotic pathway is activated by cytochrome c and other proteins released from the intermembrane space. The new uncoupling proteins are thought to participate in the pathways that regulate body weight. The mitochondrial K_{ATP} channel is proposed as the end effector of a variety of preconditioning stimuli that protect cells from ischemia-reperfusion damage. Both mitoK$_{ATP}$ and mitoCK appear to be essential for the heart to work at maximum rates. It is to be hoped that the next decade will see substantial progress in each of these emerging areas of research.

The reader who wishes to explore ion transport and bioenergetics in greater detail is fortunate indeed in the availability of two excellent and readable texts by Nicholls and Ferguson (1992) and Gennis (1989). Both are highly recommended.

Acknowledgments

This work relies on major contributions from my colleagues and students who are listed as coauthors of our publications. These studies were funded in part by grants GM31086, GM55324, and DK56273 from the National Institutes of Health.

Bibliography

Ashkenazi, A., and Dixit, V. M. (1998). Death receptors: signaling and modulation. *Science* **281,** 1309–1312.

Balaban, R. S., Kantor, H. L., Katz, L. A., and Briggs, R. W. (1986). Relation between work and phosphate metabolite in the *in vivo* paced mammalian heart. *Science* **232,** 1121–1123.

Bernardi, P. (1999). Mitochondrial transport of cations: channels, exchangers and permeability transition. *Physiol. Rev.* **79,** 1127–1155.

Boss, O., Muzzin, P., and Giacobino, J.-P. (1998). The uncoupling proteins, a review. *Eur. J. Endocrinol.* **139,** 1–9.

Chappell, J. B. (1968). Systems used for the transport of substrates into mitochondria. *Brit. Med. Bull.* **24,** 150–157.

Connern, C. P., and Halestrap, A. P. (1994). Recruitment of mitochondrial cyclophylin to the mitochondrial inner membrane under conditions of oxidative stress that enhance the opening of an inner membrane Ca^{2+}-dependent pore. *Biochem. J.* **302,** 321–324.

Crompton, M., Ellinger, A., and Costi, A. (1988). Inhibition by cyclosporin A of a Ca^{2+}-dependent pore in heart mitochondria activated by inorganic phosphate and oxidative stress. *Biochem. J.* **255,** 357–360.

Crompton, M., Virji, S., and Ward, J. M. (1998). Cyclophilin-D binds strongly to complexes of the voltage-dependent anion channel and the adenine nucleotide translocase to form the permeability transition pore. *Eur. J. Biochem.* **258,** 729–735.

Crompton, M. (1999). The mitochondrial permeability transition pore and its role in cell death. *Biochem. J.* **341,** 233–249.

Deamer, D. W., and Nichols, J. W. (1989). Proton flux mechanisms in model and biological membranes. *J. Membr. Biol.* **107,** 91–103.

Denton, R. M., and McCormack, J. G. (1990). Ca^{2+} as a second messenger within mitochondria of the heart and other tissues. *Ann. Rev. Physiol.* **52,** 451–456.

Dzeja, P. P., Zeleznikar, R. J., and Goldberg, N. D. (1998). Adenylate kinase: kinetic behavior in intact cells indicates it is integral to multiple cellular processes. *Mol. Cell. Biochem.* **184,** 169–182.

Garlid, K. D. (1988). Mitochondrial volume control. *In* "Integration of Mitochondrial Function" (Lemasters, J. J., Hackenbrock, C. R., Thurman, R. G., and Westerhoff, H. V., Eds.). Plenum, New York, pp. 257–276.

Garlid, K. D., Semrad, C., and Zinchenko, V. (1992). Does redox slip contribute significantly to mitochondrial respiration? *In* "Modern Trends in Biothermokinetics" (Schuster, S., Rigoulet, M., Ouhabi, R., and Mazat, J.-P., Eds.). Plenum, New York, pp. 287–293.

Garlid, K. D., Beavis, A. D., and Ratkje, S. K. (1989). On the nature of ion leaks in energy-transducing membranes. *Biochim. Biophys. Acta* **976,** 109–120.

Garlid, K. D., Shariat-Madar, Z., Nath, S., and Jezek, P. (1991). Reconstitution and partial purification of the Na^+-selective Na^+/H^+ antiporter of beef heart mitochondria. *J. Biol. Chem.* **266,** 6518–6523.

Garlid, K. D., Orosz, D. E., Modriansky, M., Vassanelli, S., and Jezek, P. (1996a). On the mechanism of fatty acid-induced proton transport by mitochondrial uncoupling protein. *J. Biol. Chem.* **271,** 2615–2620.

Garlid, K. D., Paucek, P., Yarov-Yarovoy, V., Sun, X., and Schindler, P. A. (1996b). The mitochondrial K$_{ATP}$ channel as a receptor for potassium channel openers. *J. Biol. Chem.* **271,** 8796–8799.

Garlid, K. D. (1996). Cation transport in mitochondria—the potassium cycle. *Biochim. Biophys. Acta* **1275,** 123–126.

Garlid, K. D., Paucek, P., Yarov-Yarovoy, B., Murray, H. N. M., Darbenzio, R. B., D'Alonzo, A. J., Lodge, N. J., Smith, M. A., and Grover, G. J. (1997). Cardioprotective effect of diazoxide and its interaction with mitochondrial ATP-sensitive potassium channels: possible mechanism of cardioprotection. *Circ Res.* **81,** 1072–1082.

Garlid, K. D., Jaburek, M., and Jezek, P. (1998). The mechanism of proton transport mediated by mitochondrial uncoupling proteins. *FEBS Lett.* **438,** 10–14.

Gennis, R. B. (1989). "Biomembranes: Molecular Structure and Function," Springer-Verlag, New York.

Gnaiger, E., Steinlechner-Maran, R., Méndez, G., Eberl, T., and Margreiter, R. (1995). Control of mitochondrial and cellular respiration by oxygen. *J. Bioenerg. Biomembr.* **27,** 583–596.

Gong, D. W., He, Y., Karas, M., and Reitman, M. (1997). Uncoupling protein-3 is a mediator of thermogenesis regulated by thyroid hormone, β3-adrenergic agonists, and leptin. *J. Biol. Chem.* **272,** 24129–24132.

Grover, G. J., and Garlid, K. D. (1999). ATP-sensitive potassium channels: a review of their pharmacology in myocardial ischemia. *J. Mol. Cell. Cardiol.,* in press.

Gunter, T. E., Gunter, K. K., Sheu, S.-S., and Gavin, C. E. (1994). Mitochondrial calcium transport: physiological and pathological relevance. *Amer. J. Physiol.* **267,** C313–C339.

Hajn'oczky, G., Robb-Gaspers, L. D., Seitz, M. B., and Thomas, A. P. (1995). Decoding of cytosolic calcium oscillations in the mitochondria. *Cell* **82,** 415–424.

Halestrap, A. P., and Davidson, A. M. (1990). Inhibition of Ca^{2+}-induced large-amplitude swelling of liver and heart mitochondria by cyclosporin A is probably caused by the inhibitor binding to mitochondrial-matrix peptidyl-prolyl cis-trans isomerase and preventing it interacting with the adenine nucleotide translocase. *Biochem. J.* **268,** 153–160.

Heurteaux, C., Lauritzen, I., Widmann, C., and Lazdunski, M. (1995). Essential role of adenosine, adenosine A1 receptors, and ATP-sensitive K$^+$ channels in cerebral ischemic preconditioning. *Proc. Natl. Acad. Sci. USA* **92,** 4666–4670.

Hunter, D. R., and Haworth, R. A. (1979). The Ca^{2+} induced membrane transition in mitochondria. I. The protective mechanisms. *Arch. Biochem. Biophys.* **195,** 453–459.

Ichas, F., Jouaville, L. S., and Mazat, J. P. (1997). Mitochondria are excitable organelles capable of generating and conveying electrical and calcium signals. *Cell* **89,** 1145–1153.

Jaburek, M., Yarov-Yarovoy, V., Paucek, P., and Garlid, K. D. (1998). State-dependent inhibition of the mitochondrial K$_{ATP}$ channel by glyburide and 5-hydroxydecanoate. *J. Biol. Chem.* **273,** 13578–13582.

Jaburek, M., Varecha, M., Gimeno, R. E., Dembski, M., Jezek, P., Tartaglia, L. A., Zhang, M., Burn, P., and Garlid, K. D. (1999). Transport function and regulation of mitochondrial uncoupling proteins 2 and 3. *J. Biol. Chem.* **274,** 26003–26007.

Jezek, P., and Garlid, K. D. (1998). Mammalian mitochondrial uncoupling proteins. *Int. J. Biochem. Cell Biol.* **30,** 1163–1168.

Jung, D. W., Baysal, K., and Brierley, G. P. (1995). The sodium-calcium antiport of heart mitochondria is not electroneutral. *J. Biol. Chem.* **270,** 672–678.

Kay, L., Saks, V. A., and Rossi, A. (1997). Early alteration of the control of mitochondrial function in myocardial ischemia. *J. Mol. Cell. Cardiol.* **29,** 3399–3411.

Kowaltowski, A. J., and Vercesi, A. E. (1999). Mitochondrial damage induced by conditions of oxidative stress. *Free Radical Biol. & Med.* **26,** 463–471.

Lemasters, J. J. (1999). Necrapoptosis and the mitochondrial permeability transition: shared pathways to necrosis and apoptosis. *Am. J. Physiol.* **276,** G1–G6.

Lemasters, J. J., Nieminen, A. L., Qian, T., Trost, L. C., Elmore, S. P., Nishimura, Y., Crowe, R. A., Cascio, W. E., Bradham, C. A., Brenner, D. A., and Herman, B. (1998). The mitochondrial permeability transition in cell death: a common mechanism in necrosis, apoptosis and autophagy. *Biochim. Biophys. Acta* **1366,** 177–196.

Li, W., Shariat-Madar, Z., Powers, M., Sun, X., Lane, R. D., and Garlid, K. D. (1992). Reconstitution, identification, purification, and immunological characterization of the 110-kDa Na$^+$/Ca^{2+} antiporter from beef heart mitochondria. *J. Biol. Chem.* **267,** 17983–17989.

Li, X., Hegazy, M. G., Mahdi, F., Jezek, P., Lane, R. D., and Garlid, K. D. (1990). Purification of a reconstitutively active K$^+$/H$^+$ antiporter from rat liver mitochondria. *J. Biol. Chem.* **265,** 15316–15322.

Liu, X., Kim, C. N., Yang, J., Jemmerson, R., and Wang, X. (1996). Induction of apoptotic program in cell-free extracts: requirement for dATP and cytochrome *c*. *Cell* **86,** 147–157.

Mitchell, P. (1961). Coupling of phosphorylation to electron and hydrogen transfer by a chemiosmotic type of mechanism. *Nature* **191,** 144–148.

Mitchell, P. (1966). Chemiosmotic coupling in oxidative and photosynthetic phosphorylation. *Biol. Rev.* **41,** 445–502.

Mitchell, P., and Moyle, J. (1969). Estimation of membrane potential and pH difference across the cristae membrane of rat liver mitochondria. *Eur. J. Biochem.* **7,** 471–484.

Nicholls, D. G., and Ferguson, S. J. (1992). "Bioenergetics 2," Academic Press, London.

Petronilli, V., Miotto, G., Canton, M., Brini, M., Colonna, R., Bernardi, P., and Di Lisa, F. (1999). Transient and long-lasting openings of the mitochondrial permeability transition pore can be monitored directly in intact cells by changes in mitochondrial calcein fluorescence. *Biophys. J.* **76,** 725–734.

Saks, V. A., Khuchua, Z. A., Vasilyeva, E. V., Belikova, O. Y., and Kuznetsov, A. V. (1994). Metabolic compartmentation and substrate channelling in muscle cells. *Mol. Cell. Biochem.* **133/134,** 155–192.

Saks, V., Dos Santos, P., Gellerich, F. N., and Diolez, P. (1998). Quantitative studies of enzyme-substrate compartmentation, functional coupling and metabolic channelling in muscle cells. *Mol. Cell. Biochem.* **184,** 291–307.

Shimizu, S., Narita, M., and Tsujimoto, Y. (1999). Bcl-2 family proteins regulate the release of apoptogenic cytochrome *c* by the mitochondrial channel VDAC. *Nature* **399,** 483–487.

Susin, S. A., Zamzami, N., and Kroemer, G. (1998). Mitochondria as regulators of apoptosis: doubt no more. *Bioch. Biophys. Acta* **1366,** 151–165.

Thornberry, N. A., and Lazebnik, Y. (1998). Caspases: enemies within. *Science* **281,** 1312–1316.

van Deursen, J., Heerschap, A., Oerlemans, F., Ruitenbeek, W., Jap, P., ter Laak, H., and Wieringa, B. (1993). Skeletal muscles of mice deficient in muscle creatine kinase lack burst activity. *Cell* **74,** 621–631.

Vendelin, M., Kongas, O., and Saks, V. (1999). Regulation of mitochondrial respiration in heart cells analyzed by reaction-diffusion model of energy transfer. *Amer. J. Physiol.* In press.

Wan, B., Dounen, C., Duszynsky, J., Salama, G., Vary, T. C., and LaNoue, K. F. (1993). Effect of cardiac work on electrical potential gradient across mitochondrial membrane in perfused hearts. *Amer. J. Physiol.* **265,** H453–H460.

Williamson, J. R., Ford, G., Illingworth, J., and Safer, B. (1976). Coordination of citric acid cycle activity with electron transport flux. *Circ. Res.* **38,** I-39–I-51.

Wyllie, A. H., Kerr, J. F., and Currie, A. R. (1980). Cell death: the significance of apoptosis. *Intl. Review Cytol.* **68,** 251–306.

Yamazaki, R. K. (1975). Glucagon stimulation of mitochondrial respiration. *J. Biol. Chem.* **250,** 7924–7930.

Richard G. Sleight and Michael A. Lieberman

9

Signal Transduction

I. Introduction

Coordination of metabolic activities among cells, tissues, and organs is mediated by the action of extracellular signals. A variety of agents act as extracellular signals, including eicosanoids (such as prostaglandins), growth factors, hormones, neurotransmitters, and pheromones (sex attractants). These agents function by binding to specific receptors located at the cell surface. **Signal transduction** is the process whereby an external chemical signal evokes an intracellular metabolic change. Some agents such as acetylcholine and γ-aminobutyric acid (GABA) bind to receptors that also function as ion channels. Binding to these **ligand-gated channels** has a direct effect on ion flux. However, the majority of extracellular signals work indirectly by complex cascade systems that have the effect of amplification. This allows a small number of extracellular signaling molecules to influence the function of a large number of proteins to produce the desired physiological response. In this chapter, two mechanisms of signal transduction will be discussed: the generation of second messengers and receptor phosphorylation.

II. Second Messengers

While some signaling molecules, such as the steroid hormones, diffuse through the plasma membrane and bind to specific receptors present in the cytosol, most extracellular signals interact with target cells by binding to high affinity receptors located at the cell surface. Molecules that bind specifically to protein receptors are called **ligands.** Structural analogs of natural ligands that bind to receptors are called **agonists** or **antagonists.** Agonists mimic the effect of the natural ligands. Antagonists bind to receptors, but this binding does not lead to a biological response. Thus antagonists block the effects of natural ligands and agonists.

Binding of ligands (or agonists) to cell surface receptors causes the receptors to undergo a conformational change. In some cases, the altered conformation of the receptor initiates a series of events leading to the formation or release of an intracellular signal that alters the metabolism of the target cell. Although extracellular signaling molecules are rarely called *first messengers,* intracellular signaling molecules are generally called **second messengers.**

Second messengers may arise by two different mechanisms (Fig. 1). One mechanism involves receptor-mediated activation of an enzyme that catalyzes the production of the second messenger. For example, when glucagon binds to its receptor, adenylate cyclase is activated, catalyzing the synthesis of cyclic AMP (cAMP), a second messenger. Cyclic AMP is synthesized on the cytoplasmic face of the plasma membrane and is released into the cytosol where it signals for the activation or inhibition of intracellular enzymes.

A second mechanism for the production of intracellular signals is the opening or closing of ion channels in the plasma membrane. A good example of this is neurotransmitter action at neuromuscular junctions of skeletal muscle. Although in-depth coverage of muscle physiology is not given until Section VI of this volume, this system demonstrates how ions can act as second messengers. Acetylcholine (released from the motor nerve terminal) acts as an extracellular signal at neuromuscular junctions, and binds to a specific receptor in the postsynaptic membrane of skeletal muscle fibers. Binding causes a conformational change in the receptor ion channel complex that leads to an influx of cations (mostly Na^+) into the cell.[1] This influx results in depolarization of the plasma membrane, which triggers a propagating action potential. The action potential depolarization activates Ca^{2+} channels which allow Ca^{2+} influx into the cytoplasm. The increased free Ca^{2+} levels trigger muscle contraction.

A. G Proteins

In many instances, the effect of a ligand binding to its receptor is indirect. That is, the receptor itself does not produce the second messenger. **Guanine nucleotide-binding proteins (G proteins),** which are located at the inner leaflet of the plasma membrane, act as a link between receptors and

---

[1]This is an example of a ligand-gated ion channel.

153

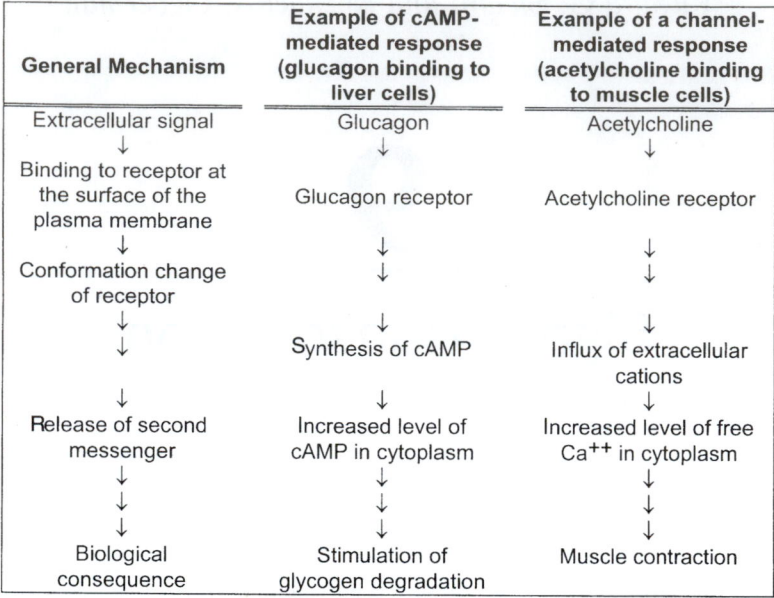

General Mechanism	Example of cAMP-mediated response (glucagon binding to liver cells)	Example of a channel-mediated response (acetylcholine binding to muscle cells)
Extracellular signal ↓	Glucagon ↓	Acetylcholine ↓
Binding to receptor at the surface of the plasma membrane ↓	Glucagon receptor ↓	Acetylcholine receptor ↓
Conformation change of receptor ↓	↓	↓
↓	Synthesis of cAMP ↓	Influx of extracellular cations ↓
Release of second messenger ↓	Increased level of cAMP in cytoplasm ↓	Increased level of free Ca⁺⁺ in cytoplasm ↓
↓	↓	↓
Biological consequence	Stimulation of glycogen degradation	Muscle contraction

FIGURE 1. Simplified overview of signal transduction and examples of two different mechanisms. A generalized mechanism for signal transduction is presented in the rightmost column. Some steps of cAMP-mediated and channel-mediated signal transduction processes are presented in the center and rightmost columns. Note that the arrows do not necessarily represent single-step processes.

the enzymes that generate second messengers (Table 1). These proteins are firmly bound to the membrane by hydrophobic interactions and are therefore considered to be integral (or intrinsic) proteins. In their unstimulated state, G proteins are heterotrimers, containing α, β, and γ subunits. Both a guanine nucleotide-binding site and GTPase activity are associated with the α subunit. At least 20 different G proteins exist.

The activation of G proteins by ligand-receptor complexes is diagrammed in Fig. 2. In the unstimulated state, there is no association between G protein and receptor, and the trimeric G protein has GDP bound to its α subunit. When a ligand (e.g., hormone or neurotransmitter) binds to the receptor, a conformational change occurs allowing the receptor and G protein to associate. Binding of the G protein to the receptor affects the guanine nucleotide binding site of the α subunit such that it releases GDP and binds GTP. The binding of GTP causes the G protein trimer to dissociate to a G_{α}-GTP complex and a $G_{\beta\gamma}$ dimer. The G_{α}-GTP complex and/or the $G_{\beta\gamma}$ dimer may activate or inhibit enzymes that will lead to the generation of second messengers. Because much more is known about the actions of the G_{α}-GTP complex as compared to that of the $G_{\beta\gamma}$ dimer, the role of the complex will be emphasized in this chapter. The action of the G_{α}-GTP complex is short lived, due to the GTPase activity of the α subunit. When the bound GTP is hydrolyzed to GDP, the G_{α} subunit becomes inactive and binds to the $G_{\beta\gamma}$ dimer, reforming the trimeric G protein.

Several effectors have been identified that regulate the duration of G_{α}-GTP complex activity. For example, members of the recently discovered regulator of the G protein signaling family (RGS family) interact with the G_{α}-GTP complex and stimulate GTPase activity. It has been postulated that each subset of the G_{α} family has its own corresponding RGS.

TABLE 1   Examples of Biological Processes Mediated by G-Proteins

Target Tissue	Effector	G-protein	Activity Modified	Major Response or Function
Adipose	Epinephrine	$G_s(\alpha_s\beta_\gamma)$	Adenylate cyclase	Activation of hormone-sensitive lipase
Blood vessels	Angiotensin	$G_q(\alpha_q\beta_\gamma)$	Phospholipase C	Vasoconstriction
Brain	Endorphins	G_o	Ca^{2+} and K^+ channels	Alter electrical activity of neurons
Eye	Light	G_t (transducin)	cGMP phosphodiesterase	Visual processing
Heart	Acetylcholine	$G_i(\alpha_i\beta_\gamma)$	K^+ channel	Decrease heart rate
Kidney	Antidiuretic hormone	G_s	Adenylate cyclase	Water retention
Liver	Glucagon	G_s	Adenylate cyclase	Glycogen degradation
Nose	Odorants	G_{olf}	Adenylate cyclase	Odor detection

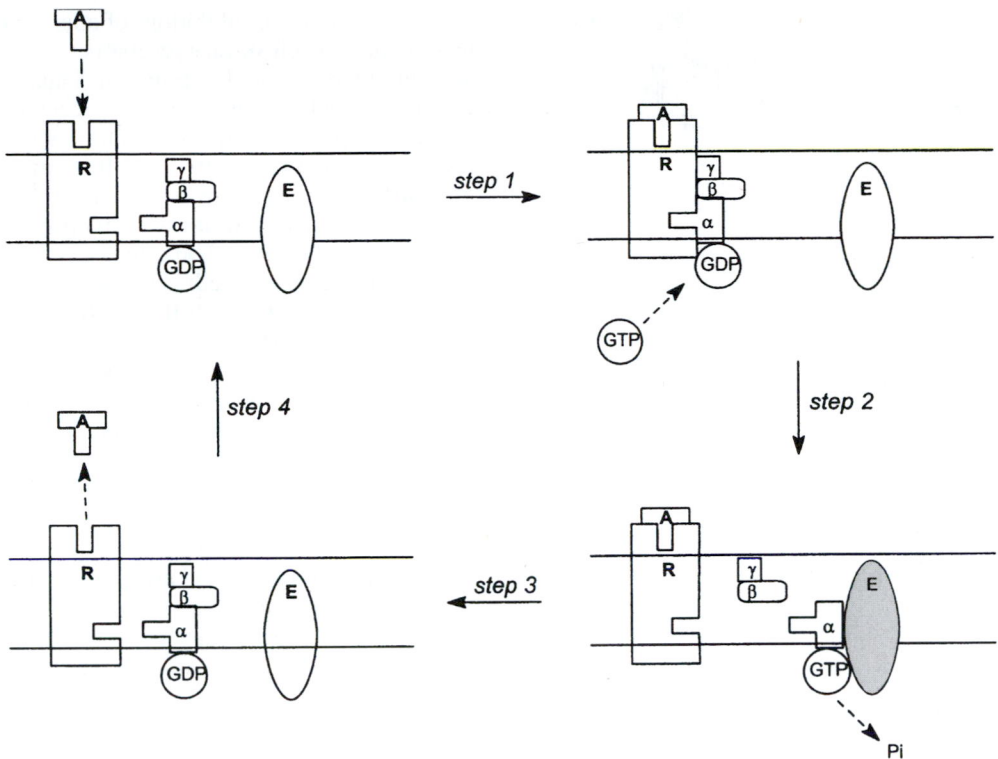

FIGURE 2. The role of G-proteins in signal transduction. In the resting state (top left), the G-protein exists as a trimer containing α, β, and γ subunits. The G_α subunit is bound to GDP. *Step 1:* First, the agonist (A) binds to the receptor (R). The activated receptor binds the G-protein causing it to lose its affinity for GDP and bind GTP. *Step 2:* Binding of GTP to the G_α subunit causes it to dissociate from both the receptor and the $\gamma\beta$ complex and bind to the target enzyme (E, e.g. PKA or PLC). Depending on the class of the G_α subunit (e.g., $G_{s\alpha}$ or $G_{i\alpha}$) the target enzyme is either activated or inhibited. *Step 3:* Hydrolysis of GTP to GDP by G_α subunit occurs. This causes the subunit to lose its affinity for the target enzyme and the trimeric G-protein complex reforms. This complex can reassociate with the receptor if the agonist is still bound. *Step 4:* Hormone dissociation from the receptor returns the system to the resting state. Note: The $G_{\beta\gamma}$ dimer can also bind to proteins and modulate their activities.

Many ion channels can be activated *in vitro* by specific G proteins in the absence of cytosol. Based on this finding, it has been postulated that G proteins may act directly on some ion channels and that no soluble second messenger is required for activation. A direct connection between G proteins and ion channels has yet to be established, thus it is possible that some other membrane components are interposed between the G protein and ion channels. Because the mechanisms of direct G protein regulation of ion channels remains obscure, this chapter focuses on the role of G proteins in the production of second messengers.

Surprisingly few inherited disorders have been associated with mutations in genes coding for G proteins. Examples of such disorders include Albright hereditary osteodystrophy, acromegaly, and McCune-Albright syndrome. All of these disorders appear to arise from mutations of G_s. Many more instances of mutations in G-protein-coupled receptors have been associated with specific diseases. For example, germ-line mutations in the rhodopsin, cone opsins, and V2 vasopressin receptors are responsible for retinitis pigmentosa, color blindness, and nephrogenic diabetes insipidis, respectively.

B. Cyclic AMP as a Second Messenger

Many extracellular signals work by altering the activity of **adenylate cyclase,** which catalyzes the formation of the second messenger cAMP from ATP (Fig. 3). Adenylate cyclase activity is located at the inner leaflet of the plasma membrane. The activity of adenylate cyclase is regulated by ligand-receptor complexes that act indirectly through two G proteins. $\mathbf{G_s}$ (α_s^* subunit) stimulates adenylate cyclase activity, while $\mathbf{G_i}$ (α_i^* subunit) inhibits the enzyme. These two G proteins have identical β and γ subunits, but different α subunits. The α_s and α_i subunits can be distinguished biochemically by the action of bacterial toxins. **Cholera toxin** enzymatically cleaves NAD^+ and attaches an ADP-ribose moiety to G_s. This **ADP-ribosylation** inhibits the GTPase activity of G_s, and thus allows G_s to remain active and permanently activate adenylate cyclase. **Pertussis toxin** permanently inactivates the G_i protein, thereby preventing inhibition of adenylate cyclase. This toxin works by ADP-ribosylation of G_i, which prevents GTP binding. Thus, GDP remains bound, preventing GTP activation of the α_i subunit, and G_i cannot inhibit adenylate cyclase.

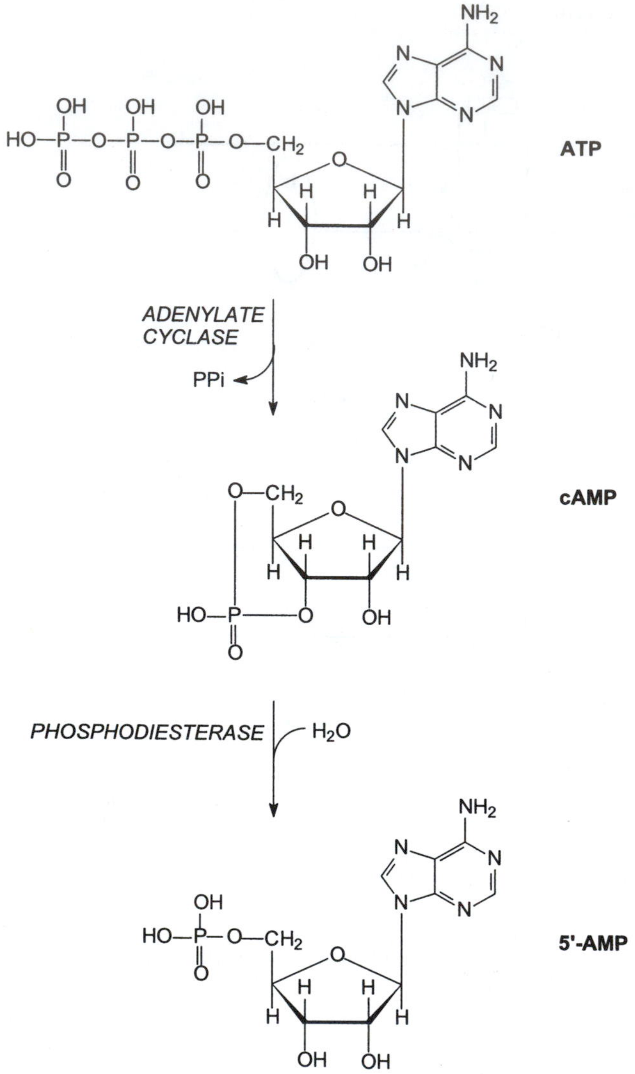

FIGURE 3.   Production and destruction of cAMP.

An increase in intracellular cAMP level causes different responses in different cell types (Table 2). For example, glucagon binds to cells in both liver and fat tissue. In both tissues, the hormone works through G proteins to activate adenylate cyclase and increase cAMP. In hepatocytes, this stimulates glycogen breakdown, while in adipocytes the breakdown of triacylglycerols (fat) is stimulated.

Many different extracellular signals may cause an increase in cAMP in a single cell type. The action of four different hormones on adipocytes provides a good example. Adrenocorticotropic hormone (ACTH), epinephrine, glucagon, and thyroid-stimulating hormone (TSH) all bind to their respective receptors at the surface of adipocytes. Each hormone causes an increase in cAMP followed by a stimulation of triacylglycerol breakdown. Although the four hormones bind to four different receptors, each receptor reacts with a common pool of G proteins and the G proteins, in turn, stimulate adenylate cyclase.

The activation (or inhibition) of adenylate cyclase by ligand binding to cell surface receptors only exists while the α subunit of the G protein is transient. Thus, adenylate cyclase activity is quickly shut down when receptor binding ceases. The cAMP produced by the action of adenylate cyclase has a short half-life in the cytoplasm. It is quickly hydrolyzed to 5´-AMP by the action of one or more **cAMP phosphodiesterases** (Fig. 3). **Caffeine** and **theophylline** are known inhibitors of cAMP phosphodiesterase.

An increase in intracellular cAMP concentration usually elicits a response through the **protein kinase A pathway** (Fig. 4). Cyclic AMP released into the cytoplasm binds to a cAMP-dependent protein kinase (protein kinase A or PKA). This kinase is composed of two catalytic subunits and two regulatory subunits. In its tetrameric form, the enzyme is inactive. When cAMP binds to the regulatory subunits, the tetramer dissociates into two active catalytic subunits and a dimer of regulatory subunits. Active protein kinase A catalyzes the transfer of a phosphoryl group from ATP to target proteins. The proteins are usually phosphorylated at serine or threonine residues. Depending on the cell type, different proteins are phosphorylated. This explains how a common pathway affects a variety of physiological responses.

Phosphorylation of proteins by PKA is the initial unique step in the process of translating receptor binding to biological response. This phosphorylation causes target enzyme activities to change, or structural proteins to alter their structure. For example, pyruvate kinase activity is regulated by reversible phosphorylation. When glucagon binds to receptors on hepatocytes, adenylate cyclase is activated, cAMP levels increase, PKA is activated, and pyruvate kinase is phosphorylated. The phosphorylated form of pyruvate kinase is inactive. This is one way that glucagon binding leads to an inhibition of hepatic glycolysis. The process is reversed when the hormone is no longer bound to the receptor. Dephosphorylation of pyruvate kinase and other enzymes phosphorylated by PKA occurs by the action of an enzyme called **protein phosphatase.** The complex regulation of protein phosphatase is under intense study.

Hundreds of protein kinases regulate cellular function, and perhaps a thousand or more proteins are targets for these kinases. Dephosphorylation of all of these targets is catalyzed by a set of four protein phosphatases (called PP-1, -2A, -2B, and -2C). Protein phosphatase-1 is responsible for dephosphorylation of many proteins, including muscle glycogen phosphorylase and glycogen synthase. This phosphatase is specifically inhibited by cellular protein inhibitors-1 and -2. The role of these proteins in the regulation of glycogen metabolism is described in the previous chapter. Like PP-1, PP-2A has a broad substrate specificity. The activity of PP-2A can be modulated *in vitro* by polyamine; however, it is not known if this occurs *in vivo*. The physiological roles of PP-2B (Ca^{2+} dependent) and PP-2C (Mg^{2+} dependent) are not completely understood.

Signal transduction results in **signal amplification.** A single receptor-ligand complex activates several G proteins, and each G_α protein activates (or inhibits) several adenylate cyclase molecules. Each molecule of adenylate cyclase pro-

TABLE 2   Examples of Hormone-Induced Responses Mediated by cAMP

Hormone	Target Tissue	Major Response
Adrenocorticotropic hormone (ACTH)	Adrenal cortex	Cortisol secretion
Epinephrine	Muscle, Liver	Glycogen breakdown
Epinephrine	Heart	Increase in heart rate
Epinephrine, ACTH, glucagon, thyroid-stimulating hormone	Adipose	Triacylglycerol breakdown
Follicle-stimulating hormone (FSH)	Ovarian follicle	Secretion of 17β-estradiol
Luteinizing hormone (LH)	Ovarian follicle	Ovulation, progesterone secretion, formation of the corpus luteum
Parathormone	Bone	Bone resorption
Thyrocalcitonin	Bone	Inhibits bone resorption
Thyroid-stimulating hormone	Thyroid	Thyroxin secretion
Vasopressin	Kidney	Water resorption

duces many cAMP molecules, that in turn activate numerous protein kinase A molecules. Every molecule of PKA then phosphorylates many target proteins and affects their activities or structure. A single binding event at the cell surface could alter the activity of thousands of intracellular molecules, leading to the biological response of the cell.

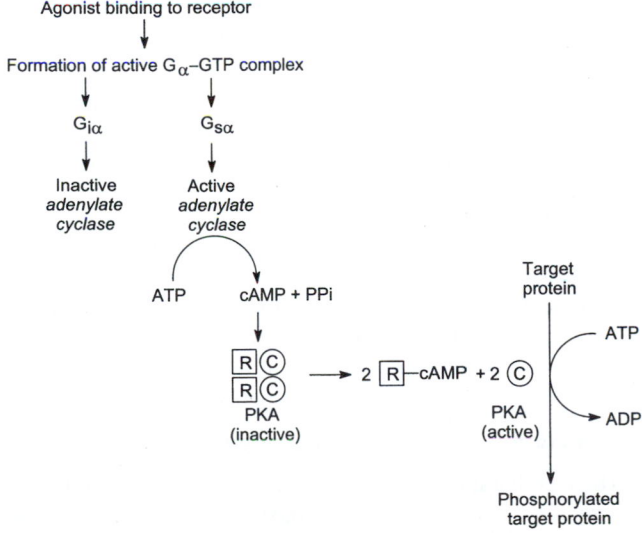

FIGURE 4.   The protein kinase A pathway. Agonist binding to receptors results in the activation of G_α as described in Fig. 2. If the receptor is specific for stimulatory G-proteins (G_s), then an activated $G_{s\alpha}$-cAMP complex is produced. This complex binds and activates adenylate cyclase, located at the inner leaflet of the plasma membrane. Active adenylate cyclase catalyzes the production of cAMP. cAMP-dependent protein kinase (PKA) normally exists in the cytosol as a tetramer of two regulatory and two inactive catalytic subunits. The catalytic subunits are released from the tetramer and become active when cAMP binds the regulatory subunits. Active PKA catalyzes the transfer of phosphate from ATP to target proteins. Depending on which target protein is phosphorylated, activity of the target protein may increase or decrease.

C. Cyclic GMP as a Second Messenger

Like cAMP, cGMP can act as a second messenger (Table 3). In many systems, the levels of cAMP and cGMP move in opposite directions in response to ligand-receptor interactions. In these systems, cAMP and cGMP often have opposing effects. Although signal transduction in the visual process occurs across membrane disks in rod cells of the retina, the process is remarkably similar to signal transduction across the plasma membrane, and provides a good example of cGMP signaling (Fig. 5). As described in detail in Chapter 48 on phototransduction, the sequence of events in the visual process begins with the absorption of a photon of light by rhodopsin. In this system, the photon may be considered the first messenger and rhodopsin the receptor. Rhodopsin is an integral membrane protein that spans across the membrane thickness of the membrane disks in rod cells. Photoactivation of rhodopsin results in a conformational change and allows it to bind a G protein called **transducin** (G_t). Like other G proteins, transducin contains α, β, and γ subunits. While bound to the rhodopsin, the α subunit exchanges GTP for GDP and is released from the complex. As diagrammed in Fig. 5, this subunit then binds to the inhibitory γ subunit of a heterotrimeric cyclic nucleotide phosphodiesterase (PDE) and removes the inhibitory subunit, thus acti-

TABLE 3   Examples of Biological Processes Mediated by cGMP

Target Tissue	Effector	Major Response
Eye	Light	Visual processing
Kidney	Atrial natriuretic factor (ANF)	Increased Na+ excretion
Intestine	*E. coli* endotoxin	Decreased water absorption
Heart	Nitric oxide	Decreased forcefulness of contractions

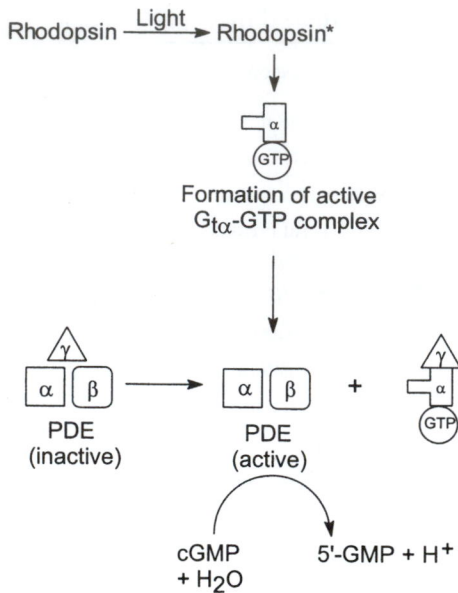

FIGURE 5. Signal transduction in the visual process as it occurs in rod cells. In a process similar to that presented in Fig. 2, photoactivated rhodopsin interacts with transducin (G_t). When the α subunit of transducin ($G_{t\alpha}$) binds to GTP, it dissociates from the β and γ subunits and associates with the inhibitory γ subunit of cyclic nucleotide phosphodiesterase (PDE). The active $\alpha\beta$ dimer of phosphodiesterase reduces cellular cGMP levels.

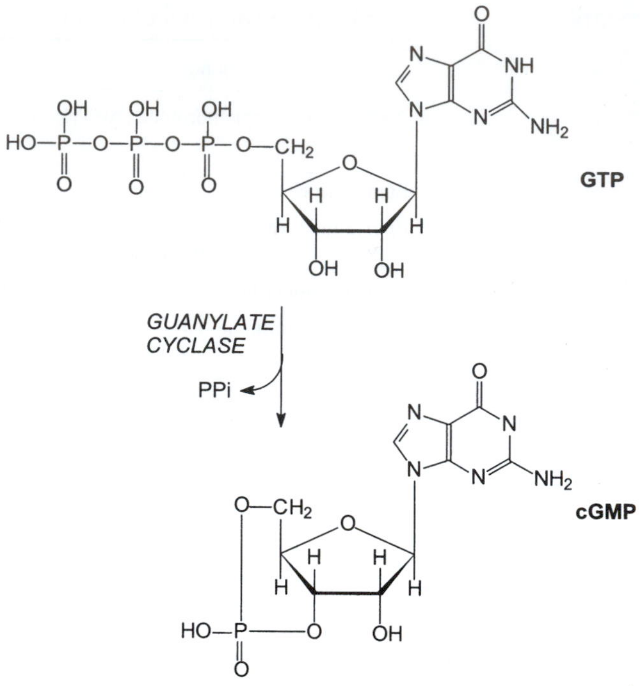

FIGURE 6. Production of cGMP by guanylate cyclase.

vating the PDE. The active PDE converts cGMP to 5'-GMP. The reduction of cytoplasmic cGMP results in the closing of plasma membrane Na^+ channels[2] and a hyperpolarization of the membrane. A series of events then occurs, including synaptic transmission between the photoreceptors and retinal bipolar cells, and between bipolar cells and optic neurons. Action potentials in optic nerve fibers transmit a visual signal to the brain.

In rod cells, a sustained high level of cGMP is the cytoplasmic signal that produces steady basal activity of Na^+ channels. In this example, the loss (lowering) of the second messenger (cGMP) by a phosphodiesterase leads to a physiological response, namely a visual signal. **Guanylate cyclase** (Fig. 6), the enzyme that catalyzes the production of cGMP from GTP, does not appear to be regulated in this process. In other systems, the activity of guanylate cyclase is regulated.

Two isozymes of guanylate cyclase are involved in signal transduction. One of the isozymes is bound to the plasma membrane and responds to receptor binding by increasing cGMP production. The resulting increase in cytoplasmic cGMP activates a **cGMP-dependent protein kinase (PKG)**. This kinase phosphorylates intracellular proteins that produce the observed biological effect. For example, when blood volume is increased, the heart releases **atrial natriuretic factor (ANF)**[3]. Binding of ANF

to the kidney causes an increase in PKG activity and the production of cGMP, which signals the kidney to increase excretion of Na^+. Loss of Na^+ causes a loss of water (diuresis) which leads to a decrease in blood volume. In addition, PKG phosphorylates L-type Ca^{2+} channels in vascular smooth muscle cells. The phosphorylated channels are inhibited, which causes a decrease in Ca^{2+} influx and leads to vasodilation.

A second isozyme of guanylate cyclase is present in the cytoplasm. This isozyme is activated by nitric oxide (NO). Nitric oxide can be produced by many cell types in response to environmental stimuli. It is also produced by the spontaneous breakdown of the common vasodilators, nitroglycerin and nitroprusside. In the heart and systemic circulation, activation of cytoplasmic guanylate cyclase by nitric oxide results in vasodilation and arteriolar relaxation.

D. Calcium as a Second Messenger

The fact that most intracellular Ca^{2+} is bound to membranes and myofilaments or sequestered into special compartments (organelles) explains the disparity between measurements of total cellular Ca^{2+} (~10^{-3} M) and free cytoplasmic Ca^{2+} (~10^{-7} M) concentrations. The low level of cytoplasmic Ca^{2+} is maintained by the action of several transporters that move Ca^{2+} either out of the cell or into organelles, such as mitochondria, endoplasmic reticulum (ER), or sarcoplasmic reticulum in muscle. Intracellular Ca^{2+} concentrations can be increased rapidly either by transient opening of plasma membrane Ca^{2+} channels or by release of Ca^{2+} sequestered in the sarcoplasmic reticulum (or ER). Both mechanisms are activated by binding of extracellular signals to plasma membrane receptors. Transient changes in intracellular free Ca^{2+} levels regulate diverse processes (Table 4).

---

[2]Thus, the Na^+ channels are open in the dark, giving rise to a steady depolarizing dark current.

[3]ANF is a peptide, and therefore is also known as atrial natriuretic peptide (ANP).

TABLE 4 Examples of Processes Regulated
by Intracellular Free Ca^{2+} Levels

Tissue	Key protein regulated	Process
Liver	Glycogen phosphorylase kinase	Glycogen degradation
Muscle	Troponin C	Muscle contraction
Brain	Tyrosine hydroxylase	Production of DOPA

Most of the effects of Ca^{2+} as a second messenger are explained by the action of **calmodulin.** Calmodulin is found in all cells, both as a free polypeptide molecule and as a component of some multisubunit enzyme complexes. A typical cell contains more than 10^7 molecules of calmodulin, constituting as much as 1% of total cell protein. Calmodulin is a single polypeptide chain of 148 amino acids and contains four high-affinity Ca^{2+} binding sites. Each Ca^{2+} binding site consists of a specialized helix-loop-helix structural motif called an EF hand.[4]

The binding of Ca^{2+} to calmodulin causes a large conformational change. In the case of free calmodulin, binding of Ca^{2+} causes it to bind to target proteins and either activate or inhibit them. For example, Ca^{2+}-ATPase in muscle sarcoplasmic reticulum is normally inactive. When cytoplasmic Ca^{2+} levels are high, Ca^{2+}-calmodulin complexes form, which bind and activate the Ca^{2+}-ATPase. Cytoplasmic Ca^{2+} is then pumped into the lumen of the sarcoplasmic reticulum. Some proteins have one or more molecules of calmodulin as subunits. Glycogen phosphorylase kinase is a 1.3-million-Da complex composed of 16 subunits ($\alpha_4\beta_4\gamma_4\delta_4$). The four δ subunits are calmodulin. The binding of Ca^{2+} to the calmodulin

[4]The term EF hand comes from the original studies used to illustrate calcium binding by this motif. These studies were performed using parvalbumin in which the helices labeled E and F are involved in calcium binding.

subunit of glycogen phosphorylase kinase changes the conformation of the complex and partially activates the kinase.

Calcium- and cAMP-regulated cellular activities often overlap. In some instances, a single protein can be regulated by both Ca^{2+} and cAMP. As described above, Ca^{2+} activates glycogen phosphorylase kinase. The same enzyme can also be activated by phosphorylation through the action of a PKA. In some cells, the levels of cAMP and free Ca^{2+} are interdependent. For example, cAMP-dependent kinases can regulate Ca^{2+} channels and pumps by phosphorylation, while calmodulin regulates enzymes responsible for cAMP production and degradation.

E. Second Messengers Generated by Lipid Hydrolysis

1. Phosphatidylinositol Cycle

Phospholipids containing inositol account for 2–8% of total cellular phospholipids. There are three major inositol-containing lipids, **phosphatidylinositol (PI), phosphatidylinositol-4-phosphate (PIP),** and **phosphatidylinositol-4,5-bisphosphate (PIP$_2$).** In response to external signals, cells may hydrolyze PIP$_2$ to form three distinct second messengers, **inositol-1,4,5-trisphosphate** ($I_{1,4,5}P_3$ or **IP$_3$**), **diacylglycerol (DAG),** and **arachidonic acid** (Figs. 7 and 8). IP$_3$ binds to a specific Ca^{2+} channel in the ER and signals for the release of Ca^{2+} into the cytoplasm. DAG activates a specific protein kinase called protein kinase C (PKC).

Arachidonic acid may be converted into a family of biologically active compounds known as the **eicosanoids** (e.g., **leukotrienes, thromboxanes,** and **prostaglandins**). The eicosanoids are local hormones (known as autacoids) because they are extremely short lived. They alter the activities of both the cells in which they are synthesized and adjoining cells. Eicosanoids have multiple effects including stimulation of the inflammatory response, regulation of blood flow, control of ion transport, and modulation of synaptic transmission.

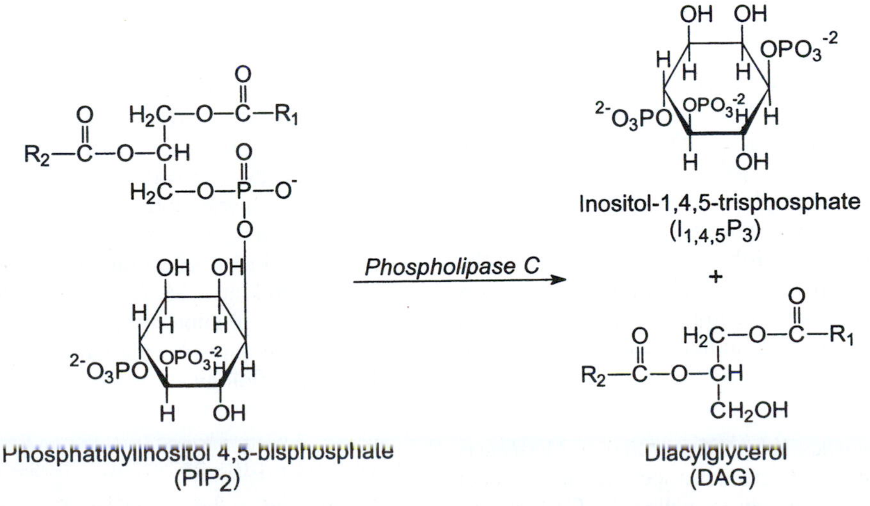

FIGURE 7. The hydrolysis of PIP$_2$ by phospholipase C produces two second messengers $I_{1,4,5}P_3$ and diacylglycerol.

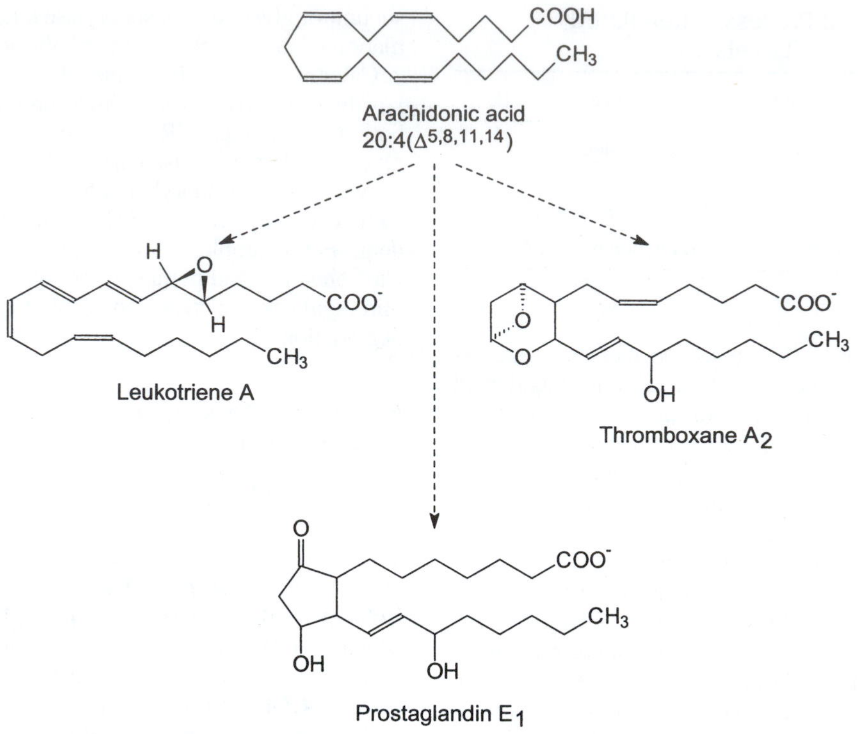

FIGURE 8.   Structure of arachidonic acid and some eicosanoids for which it is a precursor.

The enzyme that hydrolyzes PIP_2 to DAG and IP_3 is a phosphatidylinositol-specific **phospholipase C (PLC).** To date, twelve different **isoforms** of PLC have been identified, which are classified into four groups (α, β, γ, δ) based on structure. There appear to be two mechanisms for activating PLC. In some instances, the binding of an extracellular signaling molecule to a plasma membrane receptor causes the activation of a G protein called G_q. This G protein has a unique α subunit that activates $\beta 1$-phospholipase C (PLCβ1) when bound to GTP. Whether other isoforms of PLC are activated by other G-proteins is currently unknown.

A second method for PLC activation is **phosphorylation.** PLCγ can be activated directly by **tyrosine kinase** activity present in the receptor-ligand complex. A discussion of receptor kinase activity is presented in Section III of this chapter.

Diacylglycerol and IP_3 are recycled to PIP_2, as shown in Fig. 9. This constitutes the PI cycle. IP_3 is hydrolyzed to inositol by the action of specific phosphatases. Alternatively, IP_3 can be phosphorylated to form IP_4 ($I_{1,3,4,5}P_4$), or polyphosphoinositides containing five or six phosphates, and then hydrolyzed by phosphatases to inositol. These enzymes ensure that the lifetime of IP_3 in the cells is short. It has been suggested that the ability of Li^+ to inhibit complete removal of phosphate groups from phosphoinositides in brain neurons is correlated with its beneficial therapeutic effect in individuals suffering from manic depression.

a. IP$_3$ and Other Polyphosphoinositides as Second Messengers
As stated above, IP_3 binds and activates a specific Ca^{2+} channel in the ER. This channel is sometimes called the Ca^{2+} release channel, the IP_3-binding protein, or the ryanodine receptor. The

channel is a glycoprotein composed of four 260-kDa subunits. At least five different **isoforms** of the channel are believed to arise by **alternative splicing.** Although it is widely believed that most of the Ca^{2+} released by IP_3 binding comes from the ER, Ca^{2+} channels are also present in the nuclear membrane and *cis*-Golgi apparatus. The over one-hundred-fold increase in free cytoplasmic Ca^{2+} levels caused by IP_3 modulates the activities of several Ca^{2+}-sensitive proteins as described above.

Proteins that bind IP_4 and polyphosphoinositides containing five or six phosphates have been identified, but their function remains obscure. Recent evidence suggests that inositol "high" polyphosphates have a role in vesicular trafficking and the Ras signaling pathway.

b. Diacylglycerol as a Second Messenger   Diacylglycerol produced by the action of PLC binds to a family of Ca^{2+}-phospholipid-dependent protein kinases called protein kinase C (PKC). Binding of DAG to PKC in the presence of Ca^{2+} and phosphatidylserine results in activation of the kinase. PKC contains a catalytic domain and regulatory domain. Under some conditions, the regulatory domain is proteolytically cleaved, leaving a constitutively active protein, sometimes called protein kinase M. Active PKC uses ATP to phosphorylate serine or threonine residues of target proteins that are cell specific. These phosphorylations affect the proteins' activities and lead to a biological response. A partial list of signal transduction pathways involving PKC is presented in Table 5.

2. Sphingolipids as Second Messengers

Agonist-induced stimulation of sphingomyelinase results in the production of ceramide. Activation of sphingomyelinase

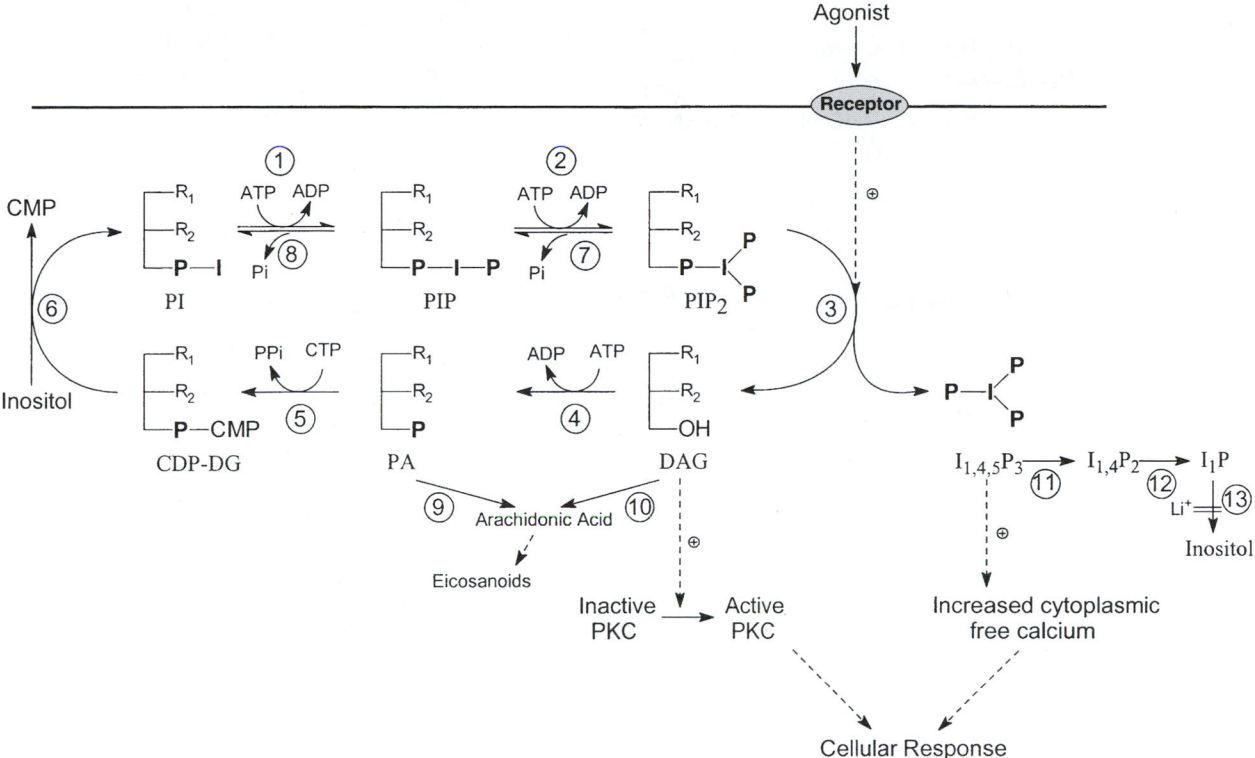

FIGURE 9. Overview of the PI cycle and activation of protein kinase C. The numbered enzymes are: 1. phosphatidylinositol kinase, 2. phosphatidylinositol-4-phosphate kinase, 3. phosphatidylinositol-specific phospholipase C, 4. diacylglycerol kinase, 5. phosphatidate cytidylyltransferase, 6. phosphatidylserine synthase, 7. PIP_2 5-phosphatase, 8. PIP 4-phosphatase, 9. phosphatidate phosphatase, 10. diacylglycerol lipase, 11. $I_{1,4,5}P_3$ 5-phosphatase, 12. $I_{1,4}P_2$ 4-phosphatase, and 13. I_1P phosphatase.

and accumulation of ceramide has been linked to several different molecules such as tumor necrosis factor, interleukin $\beta1$, nerve growth factor, interferon-γ, Fas, and chemotherapeutic agents. During the past few years, an increasing number of cell specific responses to increased intracellular ceramide levels have been identified, including inhibition of cell division, induction of apoptosis, modulation of prostaglandin secretion, and stimulation of interleukin-2 secretion. Several investigators have identified protein kinases and phosphoprotein phosphatases that are activated by ceramide. It has been hypothesized that through the actions of these enzymes, ceramides may play a role in "cross-talk" between the phosphatidylinositol and phospholipase D-mediated cycles.

Sphingosine and sphingosine 1-phosphate are potent regulators of a variety of proteins involved in signal transduction. Various growth factors increase the intracellular concentration of sphingosine 1-phosphate by activating sphingosine kinase. For example, sphingolipids appear to play a role in the stimulation of cell proliferation by platelet-derived growth factor (PDGF). Interestingly, they have also been implicated as endogenous mediators of apoptosis. This is a rapidly expanding area of research that holds great promise for our understanding of complex, coordinated signaling systems.

3. Phosphatidic acid as a second messenger

Early studies examining the PI cycle and production of DAG by PLC often suggested that more DAG was produced than could be accounted for by the breakdown of PIP_2. Analysis of the fatty acid composition of the DAG suggested that it might have come from PC. Recently, it has been determined that ligand-induced breakdown of PC is caused mainly by **phospholipase D (PLD)**. PLD hydrolyzes phospholipids such as PC, PI, and phosphatidylethanolamine (PE) to form phosphatidic acid (PA) and the free polar head group of the phospholipid substrate (Fig. 10). A unique feature of most PLDs is the ability to catalyze a transphosphatidylation reaction, utilizing short-chain alcohols as phosphatidyl group acceptors. For example, the enzyme can replace the choline moiety of PC with ethanol to produce phosphatidylethanol (Fig. 10).

TABLE 5 Examples of Physiological Events Mediated by Protein Kinase C

Agent	Tissue	Effect
Acetylcholine	Pancreas	Insulin release
Angiotensin	Adrenal cortex	Aldosterone release
Thrombin	Platelets	Serotonin release
Thyrotropin-releasing hormone	Pituitary	Prolactin secretion

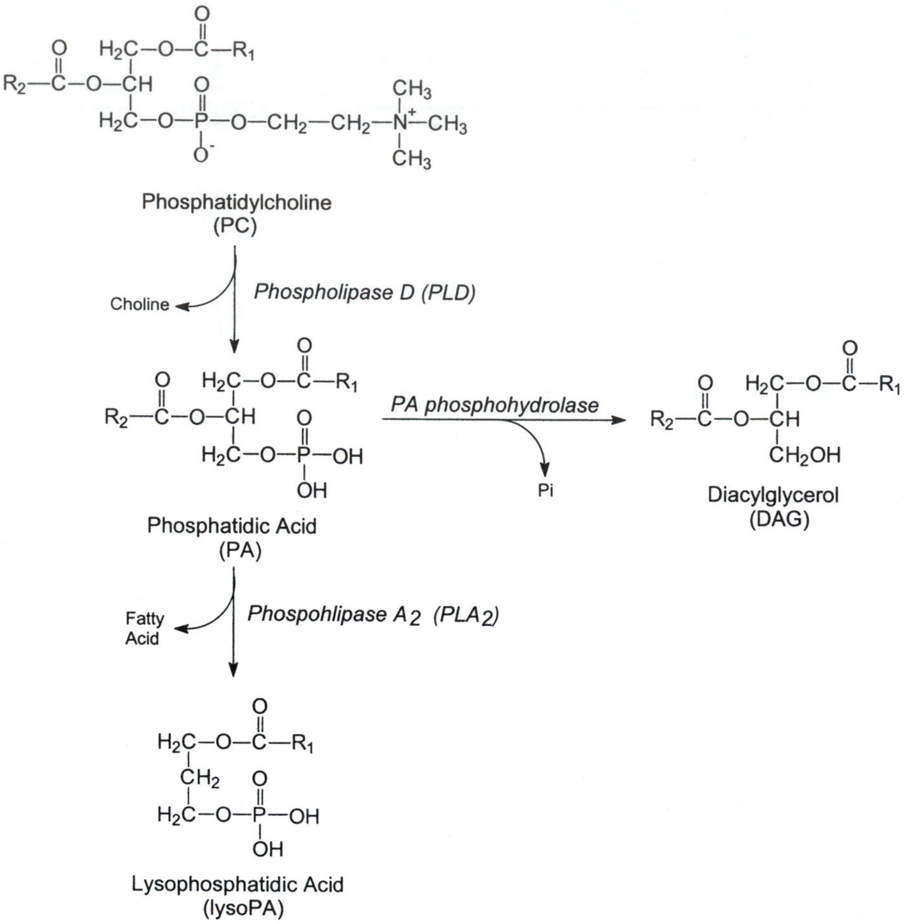

FIGURE 10.   Action of phospholipase D, phospholipase A2, and phosphatidate phosphohydrolase on glycerophospholipids. In this figure, phosphatidylcholine is used as an example of a glycerophospholipid.

A variety of external stimuli have been shown to activate PLD including hormones, growth factors, neurotransmitters, and cytokines. Although the agonists bind to receptors linked to heterotrimeric G proteins or to receptors with intrinsic tyrosine kinase activity, there is no direct evidence that either G proteins or receptor tyrosine kinases directly regulate PLD activity. Currently, regulation of PLD appears to occur via activation of PKC and/or the ARF and Rho families of small GTPases.

The phosphatidic acid produced by ligand-activated PLD has been implicated in a variety of cellular functions including granular secretion in neutrophils, actin polymerization in endothelial cells, stimulation of prostaglandin synthesis, and in the activation of PLCγ, c-Ras, and PIP-4-kinase. Phosphatidic acid is readily hydrolyzed to DAG and inorganic phosphate by highly active phosphatidate phosphohydrolases (Fig. 10) present in most cells. It has frequently been postulated that because PI-derived DAG has a short half-life, the function of PC-derived DAG may be to maintain long-term PKC activation. However, some recent reports have suggested that DAG produced by PC degradation may not be able to activate PKC. The other breakdown product of PA, lysoPA, has been linked to a variety of biological responses in several cell types. These responses include stress fiber formation, focal adhesion, and proliferation of epidermal cells. The biological responses to LysoPA are mediated by specific G-protein-coupled cell surface receptors.

It has been postulated that phosphatidic acid phosphohydrolases terminate the signaling of one class of bioactive lipids while at the same time generating products that are potent signaling molecules. Therefore, PA phosphohydrolases may act as switches between signaling pathways.

TABLE 6   Examples of Growth Factors with Receptors Having Cytoplasmic Tyrosine Kinase Activities

Colony stimulating factor 1
Epidermal growth factor
Insulin-like growth factor 1
Insulin
Nerve growth factor
Platelet-derived growth factor
Transforming growth factor α

III. Signaling by Receptor Phosphorylation

A family of receptors exists that function by directly phosphorylating target proteins. Binding of high-affinity ligands (e.g., insulin) to these receptors induces a conformational change that activates a **tyrosine kinase** activity located within a cytoplasmic domain. This kinase activity may result in **autophosphorylation** of the receptor and/or phosphorylation of intracellular target proteins.

The response of most **growth factor receptors** is mediated by their tyrosine kinase activity. Table 6 lists some of the receptors for growth factors known to be dependent on tyrosine kinase activity.

Activated receptor tyrosine kinases transmit information by two mechanisms: protein **phosphorylation** and **protein-protein binding.** The action of platelet-derived growth factor (PDGF) receptor requires both of these mechanisms. PDGF binds to a cell surface receptor that possesses cytoplasmic, ligand-activated, tyrosine kinase activity. Binding of the ligand to the receptor activates the tyrosine kinase activity. This kinase uses ATP to phosphorylate itself (autophosphorylation) and the γ isoform of phospholipase C. The autophosphorylated receptor and PLC_γ bind to each other and the phospholipase becomes activated. In addition to PLC_γ, the receptor binds to two other proteins, namely, phosphatidylinositol 3-kinase and **GTPase activating protein (GAP).** Both of these proteins are phosphorylated by the receptor's kinase. Phosphatidylinositol 4-kinase and phosphatidylinositol 4-phosphate 5-kinase may also bind to the receptor. Thus it appears that the receptor can assemble the enzymes required to produce PIP_2 and the enzyme required to hydrolyze PIP_2 to the second messengers IP_3 and DAG (Fig. 11).

The pathways through which tyrosine kinase receptors transmit signals to the nucleus, resulting in alterations in transcription, and stimulation of cell proliferation or differentiation are called the **MAP kinase** (mitogen activated protein kinase) pathways. At present, three MAP kinase pathways have been identified and named according to the MAP kinase that is activated: ERK (extracellular regulated kinase), JNK/SAPK, and p38 MAP. The best understood pathway, ERK, (Fig. 12) is described below.

Binding of a growth factor to a tyrosine kinase receptor results in autophosphorylation of critical tyrosine residues on the cytoplasmic portion of the receptor. Once the growth factor receptor has been phosphorylated, an adapter protein, GRB2 (growth factor receptor binding protein 2), binds to the phosphorylated tyrosine residues on the activated receptor. GRB2 contains a second domain, which allows it to bind to proteins containing a polyproline region. One such protein is a guanine nucleotide exchange protein called son of sevenless (SOS). Binding of SOS to GRB2 activates SOS which in turn accelerates the exchange of GTP for GDP on a small GTP-binding protein known as c-Ras. c-Ras is similar to the α-subunits of the heterotrimeric G -proteins in that it is a GTP binding protein, which is active in its GTP-bound form, inactive when bound to GDP, and contains an intrinsic GTPase activity. The GTPase activity of c-Ras is slow acting, but serves to limit the length of time that c-Ras is present in its GTP-bound active state.

When activated, membrane-bound c-Ras recruits and activates a serine/threonine kinase known as c-Raf. When acti-

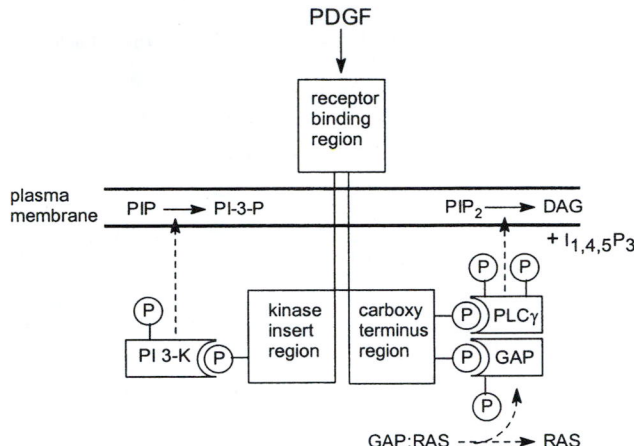

FIGURE 11. Interactions between the PDGF receptor and intracellular proteins after binding of PDGF to the receptor. Binding of PDGF to its receptor results in autophosphorylation of the receptor. This phosphorylation allows for the binding and activation of the γ isozyme of PI-specific phospholipase C (PLC_γ), GTPase activating protein (GAP), and PI 3-kinase.

vated, c-Raf phosphorylates and activates the dual specificity protein kinases MEK1 and MEK2. The activated MEKs then phosphorylate and activate a MAP kinase (e.g., ERK1 and ERK2). The activated MAP kinase then phosphorylates a group of target proteins, on either serine or threonine residues, which alter the activity of the target proteins. The main MAP kinase substrates are transcription factors (e.g., c-Myc, Elk-1, ATF2, and factors that control the expression of *c-fos* and *c-jun*), chromatin proteins (e.g., histone H3 and HMG-14) and Ser/Thr effector kinases (e.g., MaPKAP-K1, MNK, MSK, and RSK-B).

Historically, the MAP kinase pathways have been associated with growth factor receptors that display intrinsic tyrosine kinase activity; however, an increasing number of G-protein-coupled receptors have been found to activate the MAP kinase pathways. Recent data suggest that activation of MAP kinases by G-coupled receptors is preceded by $G_{\beta\gamma}$-mediated tyrosine phosphorylation of a docking protein called Shc. This has been demonstrated for G-protein-coupled receptors such as those for endothelin, thyrotropin releasing hormone, and thrombin. It appears that phosphorylated Shc can bind to GRB2 and initiate the ERK pathway. In certain cells, elevation of cAMP levels leads to inhibition of the MAP kinase pathway. The exact mechanism for the cross-talk between the two pathways remains unknown.

Recent analyses of signaling pathways, initiated by receptor tyrosine kinases at the plasma membrane, have suggested that signal-transduction pathways play a critical role in the regulation of protein and membrane trafficking. These findings suggest that within the next decade, we will identify many more cellular functions that are controlled by an interlinked group of control processes.

IV. Other Signaling Mechanisms

A. Two-Component Systems

To respond to environmental challenges, bacteria must adjust their structure, physiology, and behavior. The simplest signaling

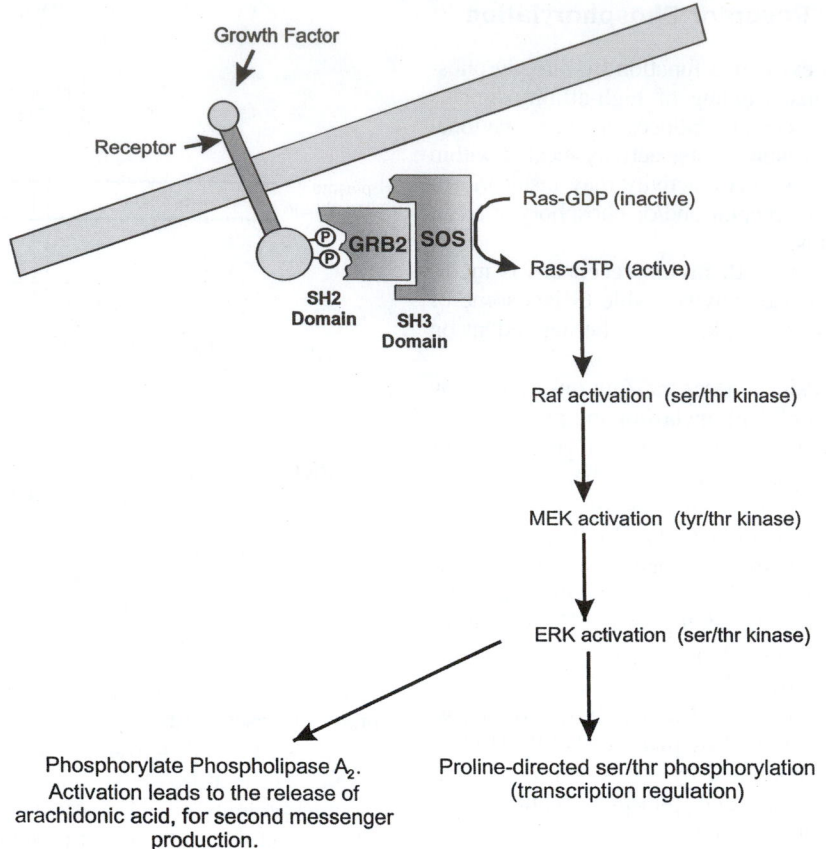

Growth Factor

Receptor

GRB2

SOS

SH2 Domain

SH3 Domain

Ras-GDP (inactive)

Ras-GTP (active)

Raf activation (ser/thr kinase)

MEK activation (tyr/thr kinase)

ERK activation (ser/thr kinase)

Phosphorylate Phospholipase A$_2$.
Activation leads to the release of
arachidonic acid, for second messenger
production.

Proline-directed ser/thr phosphorylation
(transcription regulation)

FIGURE 12. Schematic representation of the ERK mitogen activated protein kinase pathway. Raf is an example of a MAP kinase kinase kinase (MAPKKK). MEK1 and MEK2 are examples of MAP kinase kinases. ERK1 and ERK2 are examples of MAP kinases. MAP kinases can phosphorylate a wide range of substrates on serine and threonine residues within a consensus site where the phosphorylated residue is followed by a proline.

systems contain a sensor that monitors an environmental parameter and a response regulator that mediates an adaptive response. Sensors are transmembrane proteins located in the cytoplasmic membrane. Response regulators are cytoplasmic proteins that, when activated, can alter gene expression.

Communication between sensors and response regulators involves phosphorylation and dephosphorylation reactions. Sensors (also called sensory kinases) have an auto-kinase activity that catalyzes an ATP-dependent phosphorylation of a conserved histidine residue. The high-energy phosphohistidine acts as an intermediate for the subsequent transfer of phosphate to an aspartate residue present in the response regulator. Phosphorylation of the aspartate residue activates the response regulator. A phosphatase domain within response regulators autocatalyzes the removal of the phosphate from the aspartate residue. This causes inactivation of the response regulator and occurs with half-lives ranging from a few seconds to several minutes.

Two-component systems have been identified in more than 25 distinct types of signal transduction pathways, across a wide variety of prokaryotic genera. Many eukaryotic proteins contain domains with sequences similar to those found in prokaryotic sensors and response regulators, suggesting that similar signaling mechanisms exist in eukaryotes.

B. Protein Tyrosine Phosphatase Signaling

Transmembrane protein tyrosine phosphatases (PTPases) have been identified in many cell types. These phosphatases contain an extracellular domain of varying architecture connected by a transmembrane peptide to one or two intracellular PTPase domains. PTPases typically reverse signals generated by protein-tyrosine kinases. However, a few PTPases, such as SHP-2 and probably receptor protein tyrosine phosphatase (RPTP), play positive roles in signaling pathways. It has been estimated that the total number of PTPases encoded by the mammalian genome is between 100 and approximately 2000. To date, more than 80 PTPases have been cloned and characterized. Although a large number of potential mammalian targets of the PTPases are known, a relatively limited number of these are actively targeted by PTPases *in vivo*. Good examples of PTPase function can be found in mitogenic signaling pathways. For example, multiple PTPase species appear to effect the function of MAP kinases.

During the past few years, efforts to elucidate the functions and targets of PTPases have greatly increased. These efforts will undoubtedly continue to increase and reveal new pathways of signal transduction regulation and cross communication.

V. Summary

Signal transduction is the process whereby an external chemical signal elicits an intracellular metabolic change. The process begins with the binding of specific ligands to receptors located at the surface of the plasma membrane. The receptors respond to the binding of agonists in several different ways.

Most signaling appears to involve receptor activation of a GTP-binding protein (G protein). The activated G proteins interact with enzymes that produce second messenger molecules. For example, G protein activation of phosphatidylinositol specific phospholipase C results in the hydrolysis of PIP_2 to diacylglycerol and IP_3. Both of these act as second messengers transferring information from the plasma membrane to intracellular sites. IP_3 binds to receptors at the surface of the endoplasmic reticulum, which triggers the release of membrane bound Ca^{2+} into the cytoplasm. Diacylglycerols activate PKC, which catalyzes the ATP-dependent phosphorylation of intracellular proteins. Both Ca^{2+} release and protein phosphorylation affect the activities of intracellular enzymes. Changes in cellular enzyme activities lead directly to a physiological response by the cell. For example, thrombin activation of platelets is mediated by the hydrolysis of phosphoinositides.

Another signaling pathway regulated by the action of G proteins is the PKA pathway. In this pathway, binding of an agonist to a receptor leads to the activation of adenylate cyclase and thus the production of cAMP. Cyclic AMP is a second messenger that activates PKA (cyclic AMP-dependent protein kinase). PKA catalyzes the ATP-dependent phosphorylation of cell-specific target proteins. It is the change in activity of these target proteins that ultimately leads to the physiological response of cells bound to an agonist.

A second mechanism of signal transduction involves cell surface receptors that have tyrosine kinase activities associated with their cytoplasmic domains. Binding of agonist to these receptors directly activates their kinase domains. The kinase activity catalyzes the phosphorylation of intracellular proteins and often causes autophosphorylation. This phosphorylation may directly cause changes in enzyme activity or may indirectly affect activity by first promoting binding of proteins to the receptor.

Binding of extracellular signals to their receptors can result in the modulation of ion channels. In these cases, ions act as second messengers. This process is the least understood mechanism of signal transduction.

Bibliography

Barik, S. (1996). Protein phosphorylation and signal transduction. *Subcell. Biochem.* **26,** 115–64.

Brautigan, D. L. (1997). Phosphatases as partners in signaling. *Adv. Second Messenger Phosphoprotein Res.* **31,** 113–124.

Brindley, D. N., and Waggoner, D. W. (1996). Phosphatidate phosphohydrolase and signal transduction. *Chem. Phys. Lipids* **80,** 45–57.

Brown, A. M., Yatani, A., VanDongen, M. J., Kirsch, G. E., Codina, J., and Birnbaumer, L. (1990). Networking ionic channels by G proteins. *In* "G Proteins and Signal Transduction" (N. M. Nathanson, and T. K. Harden, Eds.), Society of General Physiologists Series, Vol. 45, pp. 1–9, Rockefeller University Press, New York.

Chernoff, J. (1999). Protein tyrosine phosphatases as negative regulators of mitogenic signaling. *J. Cell. Physiol.* **180,** 173–181.

Clapham, D. E., and Neer, E. J. (1997). G-protein beta gamma subunits. *Annu. Rev. Pharmacol. Toxicol.* **37,** 167–203.

Cobb, M. H. (1999). MAP kinase pathways. *Prog. Biophys. Molec. Biol.* **71,** 479–500.

Daniel, L. W., Sciorra, V. A., and Ghosh, S. (1999). Phospholipase D, tumor promoters, proliferation and prostaglandins. *Biochem. Biophys. Acta* **1439,** 265–276.

Denninger, J. W., and Marletta, M. A. (1999). Guanylate cyclase and the NO/cGMP signaling pathway. *Biochem. Biophys. Acta* **1411,** 334–350.

English, D., Cui, Y., and Siddiqui, R. A. (1996). Messenger functions of phosphatidic acid. *Chem. Phys. Lipids* **80,** 117–132.

Exton, J. H. (1999). Regulation of phospholipase D. *Biochem. Biophys. Acta* **1439,** 121–133.

Fukuda, M., and Mikoshiba, K. (1997). The function of inositol high polyphosphate binding proteins. *BioEssays*, 593–603.

Hamm, H. E. (1998). The many faces of G proteins. *J. Biol. Chem.* **273,** 669–672.

Hidaka, H., and Ishikawa, T. (1992). Molecular pharmacology of calmodulin pathways in the cell functions. *Cell Calcium* **12,** 465–472

Hoch, J. A., and Thomas, S. J. (eds.) (1995). "Two-Component Signal Transduction." ASM Press, Washington, D. C.

Kyriakis, J. M., and Avruch, J. (1996). Sounding the alarm: Protein kinase cascades activated by stress and inflammation. *J. Biol. Chem.* **271,** 24 313–24 316.

Levitzke, A. (1996). Targeting signal transduction for disease therapy. *Curr. Opin. Cell Biol.* **8,** 239–244.

Lopez-Ilasaca, M. (1998). Signaling from G-protein-coupled receptors to mitogen-activate protein (MAP)-kinase cascades. *Biochem. Pharmacol.* **56,** 269–277.

Luberto, C., and Hannun, Y. A. (1999). Sphingolipid metabolism in the regulation of bioactive molecules. *Lipids* **34,** s5–s11.

Martelli, A. M., Sang, N., Borgatti, P., Capitani, S., and Neri, L. M. (1999). Multiple biological responses activated by nuclear protein kinase C. *J. Cell. Biochem.* **74,** 499–521.

Missiaen, L., Parys, J. B., De Smedt, H., Sienaert, I., Bootman, M. D., and Casteels, R. (1996). Control of the Ca^{2+} release induced by myo-inositol trisphosphate and the implication in signal transduction. *Subcell. Biochem.* **26,** 59–95.

Spiegel, A. M. (ed.) (1998). "G Proteins, Receptors, and Disease." Humana Press, Totowa New Jersey.

Spiegel, S., Foster, D., and Kolesnick, R. (1996). Signal transduction through lipid second messengers. *Curr. Opin. Cell Biol.* **8,** 159–67.

Streoli, M. (1996) Protein tyrosine phosphatases in signaling. *Curr. Opin. Cell Biol.* **8,** 182–188.

John R. Dedman and Marcia A. Kaetzel

10

Calcium as an Intracellular Second Messenger: Mediation by Calcium-Binding Proteins

I. Introduction

Sidney Ringer provided the first report relating tissue and cellular function with Ca^{2+} in 1883. He demonstrated that Ca^{2+} was necessary for normal regular contractions of the isolated frog heart. Following this landmark study, Ca^{2+} became an essential component of physiological saline solutions. There have been numerous studies relating Ca^{2+} and cell functions including fertilization, development, differentiation, adhesion, growth, division, movement, contraction, and secretion. This evidence demonstrates a primary regulatory role for ionized Ca^{2+} in biological systems. Ca^{2+} has also been associated with a number of diseases, particularly those of the muscular and nervous systems, in which this ion plays an important role in contraction and neurotransmitter release.

Understanding the mechanism of Ca^{2+} action has required approaches and expertise from distinct fields. Ca^{2+} is unique compared with other second messengers, which are formed as metabolic intermediates, such as cyclic nucleotides, inositol phosphates, and diacylglycerol. Ca^{2+} is a divalent elemental metal and is not converted to any other form as a part of its cellular regulatory properties. Ca^{2+} is unlike other metal ions such as K^+ and Na^+, which are involved in membrane potentials and excitability, or Mg^{2+} and Zn^{2+}, which act as enzyme cofactors involved in the catalysis of metabolic intermediates. The fact that Ca^{2+} has been associated with a wide variety of cellular functions brings attention to the fact that the blocking of one Ca^{2+}-regulated function could very likely affect interdependent secondary and tertiary functions.

II. Determination of Ca^{2+} Involvement in Physiological Processes

The most direct approach to understanding the involvement of Ca^{2+} in a given physiological activity has been to follow the tradition of Ringer, that is, through the reduction of extracellular Ca^{2+}. For example, the reduction of $[Ca^{2+}]_o$ from 1 to 0.1 mM markedly alters cell growth, adhesion, secretion, and motility. A second approach has been to use various pharmacological agents. As discussed in other chapters, Ca^{2+} channel modulators have also proven useful in probing cellular systems. These distinct chemicals act by binding to the membrane Ca^{2+} channel, thereby blocking (Ca^{2+} channel agonist) the influx of extracellular Ca^{2+}. The antibiotic A23187 has been shown to cage divalent metal ions such as Ca^{2+} and to thereby act as an ionophore to facilitate their movement across biological membranes (Fig. 1). The A23187 ionophore binds Ca^{2+}, Mn^{2+}, and Mg^{2+} with respective affinities of 210:2:1. Under specified conditions, A23187 can provide evidence for a role for Ca^{2+} in a given biological response. The ionophore elevates intracellular

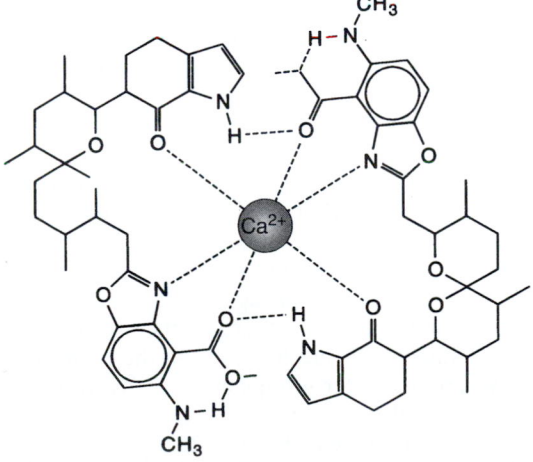

FIGURE 1. Molecular structure of A23187. The charged Ca^{2+} ions are extremely impermeable to the plasma membrane. Two molecules of ionophore A23187 form a molecular cage around the charged Ca^{2+} ion. The complex is membrane soluble and dissociates in the cytosol. Intracellular Ca^{2+} is then elevated.

Copyright © 2001 by Academic Press. All rights of reproduction in any form reserved.

Ca^{2+} and, for example, can induce the activation of lymphocytes, platelets, and sea urchin egg development.

III. Ca^{2+} as an Intracellular Signal

The total intracellular Ca^{2+} concentration has been estimated to be approximately 1 mM, a value similar to extracellular Ca^{2+} values. Early studies used model systems such as the skinned muscle fiber, in which the sarcolemma was physically removed. In such a system, the Ca^{2+} ion concentration can be regulated in the bathing solution, while contraction and glycogenolysis are monitored. This approach demonstrated that Ca^{2+} was active at micromolar levels, and that it was rapidly sequestered from the cytosol. Conclusions drawn from such studies indicated that contraction and metabolism are regulated at micromolecular Ca^{2+} levels. It was suspected that Ca^{2+} acted by moving from one cellular compartment to another, that is, into and out of the cytosol. Ca^{2+} channels and pumps were identified using isolated intact vesicles through which the uptake of $^{45}Ca^{2+}$ was monitored. Convincing evidence for the movement of Ca^{2+} into and out of cellular compartments using isotopic methods in intact tissues proved difficult, since the ion was in a dynamic state of flux between the various intracellular organelles, the cytoplasm, and the extracellular fluids.

During the past 20 years, however, several generations of Ca^{2+} indicators have been developed. Many coelenterates have the ability to glow by using Ca^{2+}-activated luminescent proteins. Aequorin was the first photoprotein to be used to measure intracellular free Ca^{2+}. Photoproteins can be used to measure free Ca^{2+} levels within the range of 0.1–10 μM. The use of aequorin, however, is limited since it is a relatively large molecule, with a molecular weight of approximately 20 000, and hence must be injected into cells. Early studies with aequorin required the use of large cells for microinjections. For example, the permeability of gap junctions to Ca^{2+} ions into large dipteran salivary glands was determined using aequorin luminescence with image intensification techniques. Under these conditions, nanograms of aequorin can be injected into a cell, which gives the cytosol the capacity to produce 10^4 to 10^8 photons per second. At these levels, the blue light luminescent output is readily measurable with sensitive photomultipliers.

More recently, a series of fluorescent indicator dyes that bind Ca^{2+} with high affinity and selectivity has been developed. The most commonly used is Fura-2. Since the free acids of these fluorescent indicators are not membrane permeable, the carboxyl acid groups are esterified during their chemical synthesis to make them permeable to the plasma membrane. Intact cells are loaded with the permeable form of the dye and the ester groups are then hydrolyzed in the cytoplasm by cellular esterases, thereby trapping the fluorescent probe within the cells. Fura-2 has been used in many biological systems. The use of Fura-2, in combination with high-resolution fluorescence video microscopy and computer-assisted image analysis systems, has provided valuable insight into understanding the intracellular Ca^{2+} signal (Berridge, 1990):

1. The resting intracellular ionized Ca^{2+} level is approximately 0.1 μM. Ca^{2+} ions not complexed to counterions or proteins are frequently referred to as being *free*. Cell stimulation by a variety of agents causes a transient rise in intracellular free Ca^{2+}; the increase is variable, lasting for a fraction of one second to minutes. The amplitude of the Ca^{2+} spike also varies from tissue to tissue.

2. The elevation in Ca^{2+} may be uniform throughout a cell or group of cells or highly localized to specific regions of individual cells.

3. In many cellular systems, the Ca^{2+} signal occurs as a wave, beginning at a discrete initiation site and then moving across the cell. Fertilization is a cellular process that displays this Ca^{2+} wave phenomenon. The Ca^{2+} influx is initiated at the point of sperm-egg contact, from where it spreads as a propagated wave toward the opposite pole. This Ca^{2+} signal initiates cellular reactions to prevent polyspermy (penetration of the ovum by more than one sperm).

4. In many cell systems, the intracellular level of Ca^{2+} oscillates. The frequency depends on several factors, including the cell type and cellular effectors such as hormones, neurotransmitters, growth factors, and cytokines. Most oscillations occur at a periodicity of 20–60 s when the cell is at "rest." When the cell is stimulated with a hormonal agonist, the frequency can increase to less than 5 s or result in a sustained elevation of up to 100 s. The oscillation frequency in Ca^{2+} may represent a periodic code, which can distinguish extracellular effector type and concentration.

IV. Creation of the Ca^{2+} Signal

Cells are able to maintain a resting level of Ca^{2+} of less than 0.1 μM in an environment where the extracellular Ca^{2+} is 1 mM or greater. This gradient is achieved because the plasma membrane is relatively impermeable to Ca^{2+} and also contains ATP-driven Ca^{2+} pumps and Na^+-Ca^{2+} exchangers. In addition, the endoplasmic reticulum sequesters cytosolic Ca^{2+}, also by ATP-driven pumps. When $[Ca^{2+}]_i$ increases to above 1 μM, mitochondria will internalize Ca^{2+}. Mitochondria that are heavily loaded with Ca^{2+} reflect a distressed cellular state: under such conditions, the mitochondria develop dense granules containing complexed Ca^{2+}. Collectively, these pump and exchanger systems maintain a resting $[Ca^{2+}]_i$ of less than 0.1 μM.

Initiation of the calcium signal is achieved from two primary sources, the extracellular fluid and sequestered internal stores. The plasma membrane can contain multiple species of Ca^{2+}-specific channels as determined by electro-physiological or pharmacological methods. In general, there are three channel types: voltage-dependent, ligand-gated, and mechanical (i.e., stretch-activated). The opening of these channels allows a rush of Ca^{2+} to enter the cytoplasm, producing a localized increase in intracellular Ca^{2+}. The mechanism of Ca^{2+} release from intracellular stores evaded elucidation for many decades. It was shown that inositol 1,4,5-trisphosphate (IP_3) causes Ca^{2+} mobilization from internal vesicular stores, primarily the endoplasmic reticulum (Berridge, 1993). IP_3 is a phospholipase C (PLC) hydrolysis product of phosphatidylinositol 4,5-bisphosphate (PIP_2), which is induced during cell stimulation. Recent evidence indicates that not all of the endoplasmic reticulum-sequestered Ca^{2+} is IP_3 sensitive, and release is prompted by cyclic ADP-ribose (Lee, 1993). A second product of PLC hydrolysis of PIP_2 is diacylglycerol (DAG).

Stimulation of the membrane phospholipase C is through receptor binding by agonists and is G protein regulated. GTPγS, a nonmetabolizable analogue of GTP, can artificially activate PLC production of IP_3 and DAG. G proteins also appear to regulate several processes, including the exchange of Ca^{2+} from IP_3-insensitive to IP_3-sensitive pools, the gating properties of plasma membrane Ca^{2+} channels, and the generation of cyclic AMP. The microinjection of GTPγS causes a marked increase in Ca^{2+} oscillation frequency and amplitude and stimulation of many cellular processes.

V. Mediation of Ca²⁺ Signal

Intracellular Ca^{2+}, like cyclic AMP, acts as a second messenger. Cyclic AMP is mediated by a limited number of receptor proteins, leading to the activation of specific protein kinases. There are, however, numerous intracellular Ca^{2+}-binding proteins; therefore, the intracellular Ca^{2+} signal has many possible bifurcations of action. A simple example is found in skeletal muscle, where elevated Ca^{2+} binds to troponin C to cause myofibrillar contraction at the expense of ATP. An independent, simultaneous Ca^{2+} pathway is mediated by a second Ca^{2+} receptor protein, calmodulin, which activates phosphorylase kinase. This activation initiates glycogenolysis, which leads to the regeneration of expended ATP (Fig. 2).

The most completely described family of intracellular Ca^{2+}-binding proteins to date is characterized by a protein structure known as the EF hand. The binding site is achieved through side-chain coordination of Ca^{2+} within a helix-loop-helix composed of precisely spaced amino acids (Fig. 3). This EF hand is found in a number of proteins as determined from the primary amino acid sequence and, in a few cases, has been confirmed by direct Ca^{2+} binding. The sequence data for the Ca^{2+}-dependent protease calpain suggest that it may have resulted from the fusion of a gene encoding four EF hand domains with a thiol protease gene (Ohno et al., 1984). Likewise, α-actinin may have resulted from the fusion of a gene encoding an actin-binding protein with a gene for a single EF hand domain (Noegel et al., 1987). In many cases the EF hand is nonfunctional. For example, the Ca^{2+} insensitivity of the muscle form of α-actinin is due to the imprecise positioning of the amino acids shown in Fig. 3. The identification of a number of putative Ca^{2+}-binding proteins is based on sequence similarity to members of the troponin C/calmodulin superfamily. For example, cell cycle yeast mutants, which carry the temperature-sensitive allele of *CDC31*, are blocked in spindle-pole body duplication (Baum et al., 1986). This gene has 42% sequence similarity with human calmodulin and may contain functional EF hands. A second cell division cycle gene, *CDC24*, is required for bud formation. Sequence analysis of *CDC24* indicates two potential Ca^{2+}-binding EF hands (Miyamoto et al., 1987).

Troponin C is present only in skeletal and cardiac muscle; calmodulin is present in all cells. The binding of four Ca^{2+} ions causes conformational changes with the formation of an active state. Ca^{2+}-bound calmodulin has very high affinity for its target proteins, which act as additional steps in mediating the original Ca^{2+} signal. The calmodulin target proteins include protein kinases, protein phosphatases, hydrolases, nitric oxide synthetase, and ion channels.

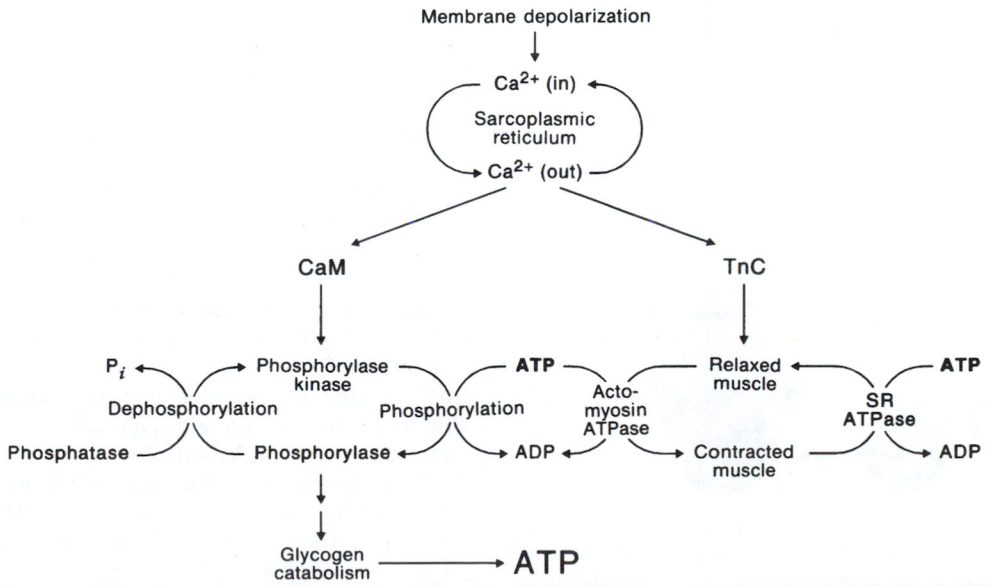

FIGURE 2. Independent pathways of Ca^{2+} regulation in skeletal muscle. Depolarization of the sarcolemma causes release of Ca^{2+} stored in the sarcoplasmic reticulum. This elevated Ca^{2+} level triggers contraction, an ATP-consuming process. The same signal Ca^{2+} also binds calmodulin (CaM), which initiates glycogenolysis, an ATP-producing process. This parallelism allows for metabolic coordination.

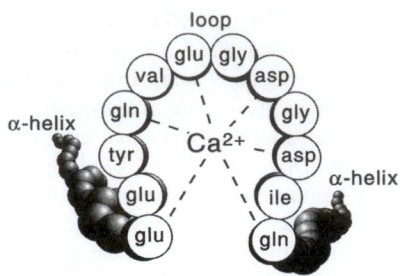

FIGURE 3. The EF hand Ca^{2+} binding pocket. Many intracellular Ca^{2+}-receptor proteins contain precisely positioned amino acids in a loop between highly structured α-helical coil. The Ca^{2+} ion is coordinated in this loop, which alters the overall structure of the protein and, in turn, its cellular activity.

Sequence analysis and site-directed mutagenesis studies have provided a general model for calmodulin regulation of its target proteins (Fig. 4). The target protein contains a pseudosubstrate attached to a flexible region of the polypeptide chain. This sequence acts as an endogenous inhibitor of the enzyme. Ca^{2+}-activated calmodulin clamps around an adjacent target site and physically displaces the

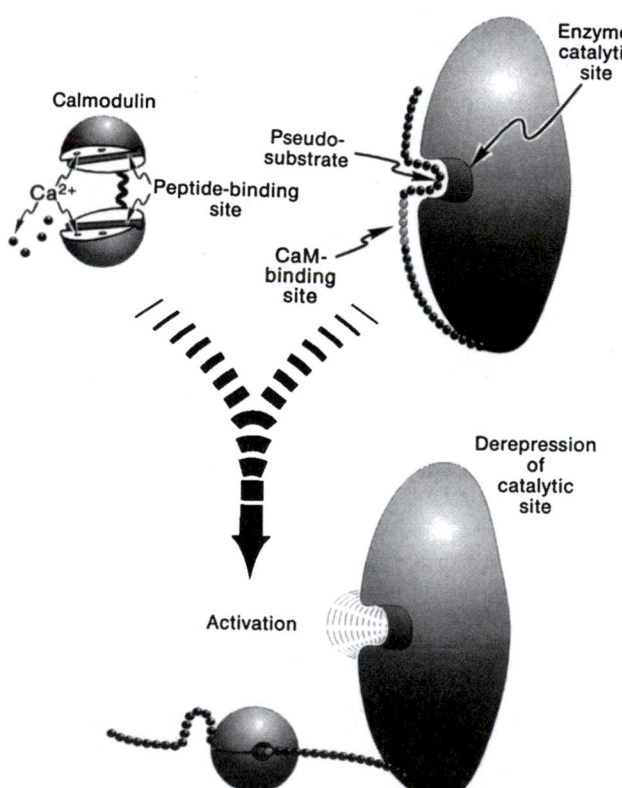

FIGURE 4. Mechanism of action of calmodulin regulation of target proteins. Calmodulin is molecularly structured as two opposing lobes containing Ca^{2+}-binding sites connected by a flexible peptide hinge. Ca^{2+} binding causes the formation of complementary grooves on each, which then allows binding to specific regulatory sites on target proteins. Ca^{2+}-dependent binding causes derepression of the enzyme active site by displacing an endogenous inhibitory pseudosubstrate.

pseudosubstrate from the active site, causing a derepression or activation of the enzyme. This process is reversed by the reduction of intracellular Ca^{2+} levels, which reduces the affinity of calmodulin for the target site and autoinhibition of the enzyme occurs. A smooth muscle contraction is a paradigm for calmodulin mediation of the Ca^{2+} signal (Fig. 5).

One approach to understanding the cellular role of a protein is to correlate subcellular localization with function. This information can be effectively obtained through the use of antibodies that are monospecific for the targeted protein. As shown in Fig. 6, when affinity-purified antibodies were used to localize calmodulin during mitosis, the spindle poles were stained most brightly. This result suggested that calmodulin is involved in mitotic function. Indeed, microtubules depolymerize in the presence of Ca^{2+}-bound calmodulin both *in vitro* and in microinjected cells (Dedman *et al.,* 1982). Many other cellular functions have been assigned to calmodulin through a combination of localization, biochemical, and genetic approaches (Davis, 1992).

VI. Ca^{2+}-Calmodulin Dependent Protein Kinase II

Many of the calmodulin regulated kinases, including phosphorylase kinase (Fig. 2) and myosin light chain kinase (Fig. 5), directly reflect the changes in free cytosolic Ca^{2+} through the phosphorylation of their respective specific protein substrates. Ca^{2+}-calmodulin dependent protein kinase II (CaM kinase II) is distinct in that it is multi-functional, having numerous substrates, and can develop many active states including Ca^{2+}-CaM autonomy (Braun and Schulman, 1999). The individual subunits of CaM kinase II assemble through their association domains to form a flower-like structure of 8–12 "petals" (Fig. 7). In low cytosolic free Ca^{2+}, the naive enzyme is inactive. When levels are elevated, free intracellular Ca^{2+} binds CaM which, in turn, binds to the regulatory domain of the CaM kinase II subunits and derepresses the active site. "Autophosphorylation" of adjacent CaM-bound subunits occurs at threonine-286 (pT286) during this activation period. Autophosphorylation renders the individual catalytic sites constitutively active and independent from Ca^{2+}-CaM regulation. A second "wave" of elevated Ca^{2+} allows additional sites to be activated by Ca^{2+}-CaM and provides a further opportunity for transphosphorylation of adjacent pT286. After a series of Ca^{2+} waves, the enzyme complex becomes increasingly autonomous from Ca^{2+}-CaM regulation. In addition, pT286 phosphorylation produces a 1000-fold increase in affinity for CaM and effectively "traps" CaM, making the Ca^{2+} mediator protein temporally unavailable for other target proteins. When Ca^{2+} levels are reduced, and calmodulin is released, the regulatory domain may be "capped" by phosphorylation of T305 and/or T306 in the CaM binding domain. Rebinding of CaM in the event of subsequent Ca^{2+} increases is prevented until pT305 and pT306 are dephosphorylated; pT286 autonomy is not affected. Autonomy and capping are reversed by protein phosphatases that are regulated by cAMP. The subunits are resensitized to Ca^{2+}-CaM following dephosphorylation of T305 and T306.

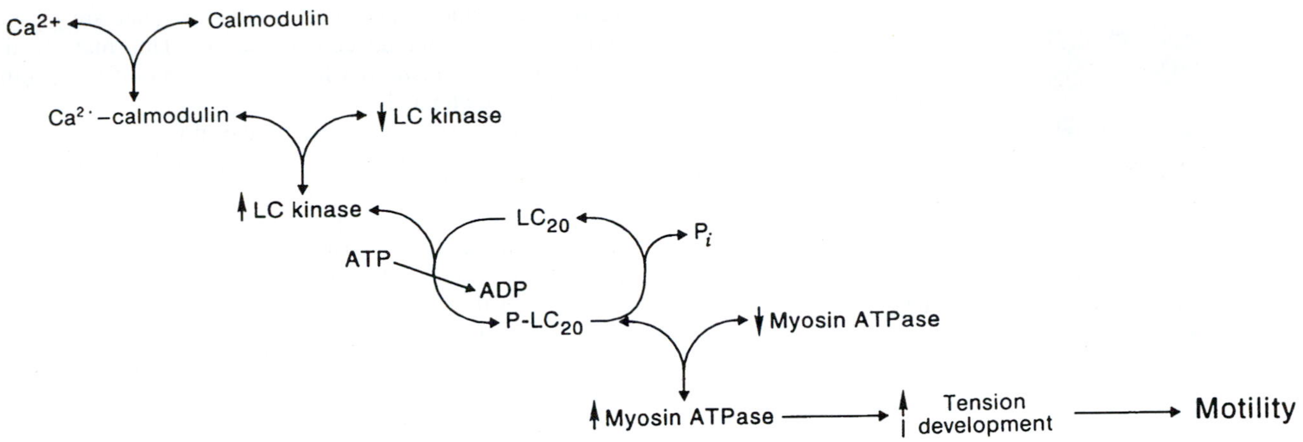

FIGURE 5. Calmodulin regulation of nonstriated muscle contraction. An increase in intracellular Ca^{2+} binds to calmodulin, causing its increased affinity for the target protein myosin light chain kinase (LC kinase). The activated kinase specifically phosphorylates one of the myosin light chains, which, in turn, increases the activity of myosin and its affinity for actin. Cross-bridge formation develops contraction. This process is reversed after cytosolic reduction in Ca^{2+}.

The metabolic sculpturing of CaM KII reflects the changes in free Ca^{2+} levels, the availability of CaM, autophosphorylation and protein phosphatase activities. These events allow CaM KII to develop numerous active states reflecting the physiologic state of the cell. Dynamic changes in second messenger levels, amplitude and frequency of Ca^{2+} oscillations, and cAMP levels, are decoded by CaM KII which, in turn, phosphorylates and modifies the activity of numerous enzymes including Ca^{2+}-channels, Ca^{2+}-pumps, synaptic proteins, and components of the nuclear envelope. In summary, CaM KII is a molecular microprocessor that integrates and stores cellular signals.

VII. Annexins: Calcium-Dependent Phospholipid-Binding Proteins

A number of laboratories with distinct experimental goals have identified a family of Ca^{2+}-dependent phospholipid-binding proteins. These proteins have the common property of binding (annexing) membranes (Crumpton and Dedman, 1990). Sequence data indicate that there are 10 unique mammalian annexins. Annexins are also present in the slime mold *Dictyostelium,* in the sponge, in the coelenterate *Hydra,* in the insect *Drosophila,* and in the mollusk *Aplysia,* as well as in higher plants.

The sequence organization of the family is highly conserved (Fig. 8). All except annexin VI are composed of a core of four repeated domains: annexin VI is composed of eight domains. Each domain is approximately 70 amino acids in length. The sequence conservation for each ranges between 40 and 60% when individual annexins are compared. The amino terminus of each protein is unique, suggesting that this region may confer functional differences to the proteins. This property has been confirmed with annexin II, in that the amino terminus binds a subunit, p11, and forms an actin-binding heterotetramer. Calcium-dependent phospholipid binding is a property of the four-domain core of each protein. *In vitro* functions of annexins—which include membrane binding and

fusion, ion channel activity, modulation of ion channel activity, inhibition of phospholipase A_2, and inhibition of blood coagulation—all require this property. The core does not contain the classic EF hand calcium-binding motif. Coordination of Ca^{2+} is accomplished through a unique, discontinuous binding loop. Chemical cross-linking data indicate that monomeric annexin in solution self-associates into trimers, which then form higher aggregates when bound to phospholipid vesicles. This observation is consistent with the diffraction pattern obtained from two-dimensional crystals.

Collectively, protein structural data, biophysical reconstitution studies, and subcellular localization results allowed development of the following model to describe the cellular function of the annexins. In the resting cell when free Ca^{2+} concentrations are low, the annexin exists as a soluble monomer. During cell stimulation, free Ca^{2+} concentrations subjacent to the cell membrane rise to micromolar levels (Llinas *et al.,* 1992). The annexin would bind to target proteins associated with a phospholipid surface, then organize into trimers, hexamers, and higher aggregates, ultimately forming an extended hexagonal array around the target protein (Andree *et al.,* 1992; Concha *et al.,* 1992). The immunofluorescent localization studies of annexins support this model. A sheet of annexin multimers lining the inner membrane leaflet would locally alter membrane properties such as fluidity and sequestration of specific phospholipids. Changes in membrane properties have been shown to modify specific membrane protein function (Sweet and Schroeder, 1988; Bennett, 1985). Such a submembranous scaffolding maintains and stabilizes the membrane. For example, dystrophin is localized to the inner surface of the sarcolemma in normal skeletal muscle. When absent, as in Duchenne muscular dystrophy, the plasma membrane is unstable and the fibers rapidly turn over (Koenig *et al.,* 1988). In addition, the Ca^{2+}-dependent lining of the membrane would sterically block the translocation of phospholipid-binding proteins such as protein kinase C and cellular phospholipases. The Ca^{2+}-dependent self-association on

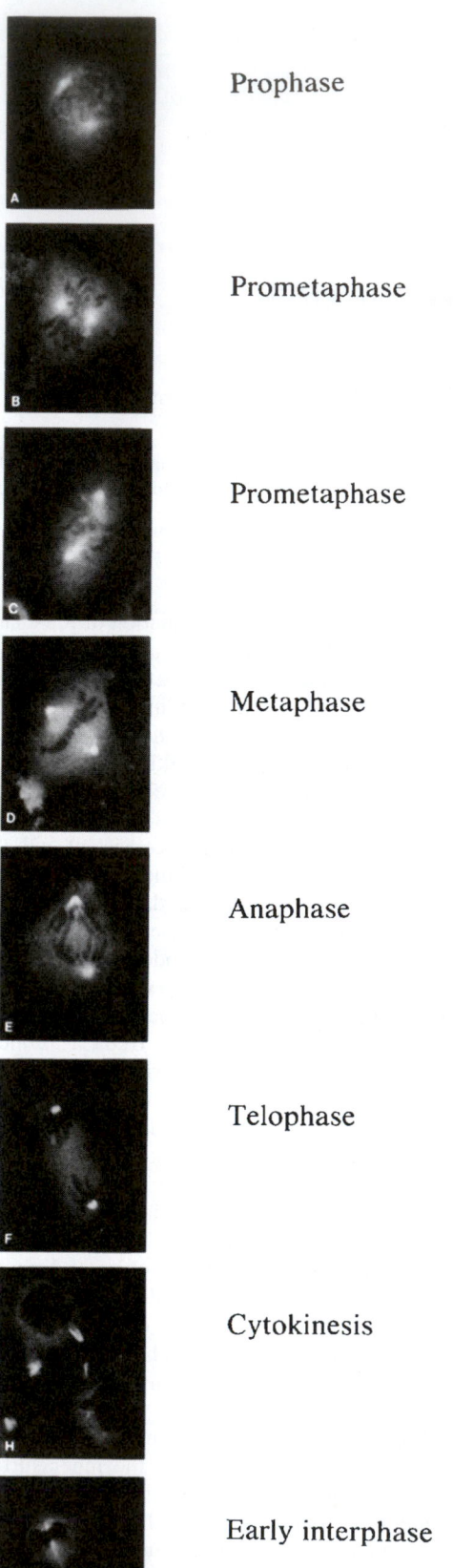

Prophase

Prometaphase

Prometaphase

Metaphase

Anaphase

Telophase

Cytokinesis

Early interphase

membrane surfaces represents a novel mechanism of second-messenger coupled cell regulation. This may be the regulatory mechanism by which annexins modify the gating activity of the channels.

Immunolocalization studies have proved valuable in providing insight into evaluating cellular function of the annexins. The individual annexins are associated with secretory granules, the endoplasmic/sarcoplasmic reticulum, actin bundles, and the plasma membrane. Annexin IV, for example, is expressed in many epithelia and is concentrated along the apical membrane, subjacent to the lumen of the organ (Fig. 9) (Kaetzel *et al.,*1994). This region is the cellular site of fluid secretion into the lumen. Recent studies indicate that this annexin regulates chloride-ion efflux, which produces an electrochemical gradient to draw sodium ions transcellularly across the epithelium. This salt causes water to follow because of hyperosmotic pressure. The luminal fluid is required for normal tissue function. Abnormal fluid secretion is involved in the pathologies of cholera and cystic fibrosis.

VIII. Protein Kinase C

The protein kinase C family is a third mediation pathway of intracellular Ca^{2+} action. One class of protein kinase C isozymes is activated by Ca^{2+}, which increases the affinity of the enzyme for phosphatidylserine. This ligand binding targets the translocation of protein kinase C to the plasma membrane. Once bound to the membrane surface, the enzyme can be further stimulated by diacylglycerol (DAG) (Fig. 10). This latter metabolite is a product of G-protein-activated phospholipase C hydrolysis of membrane PIP_2. Protein kinase C stimulates DNA synthesis and is the cellular mechanism by which many tumor promoters act (Nishizuka, 1992). A major widely distributed cellular substrate for protein kinase C is MARCKS, myristoylated alanine-rich C-kinase substrate (Aderem, 1992). This protein binds to the plasma membrane, calmodulin, and actin and has been associated with secretion, motility, vesicle trafficking, and transformation through the rearrangement of the actin cytoskeleton. Actin filament bundling by MARCKS is regulated by Ca^{2+}-dependent calmodulin binding, which is, in turn, regulated by protein kinase C phosphorylation. The kinase has numerous protein substrates in the cell; however, the precise mechanism of cellular regulation is not fully understood.

There are additional distinct intracellular Ca^{2+} binding proteins that have been well characterized in biochemical terms (see Smith *et al.,* 1990). Ca^{2+} has the responsibility of regulating a large number of unrelated cellular activities such as cell growth, secretion, motility, and transport. The ubiquitous Ca^{2+} signal is discriminated through the individual Ca^{2+}-mediator proteins (Fig. 11).

FIGURE 6. Subcellular localization of calmodulin in mitotic cells. Monospecific antibody and immunofluorescence identifies calmodulin to be concentrated at the poles of the spindle apparatus.

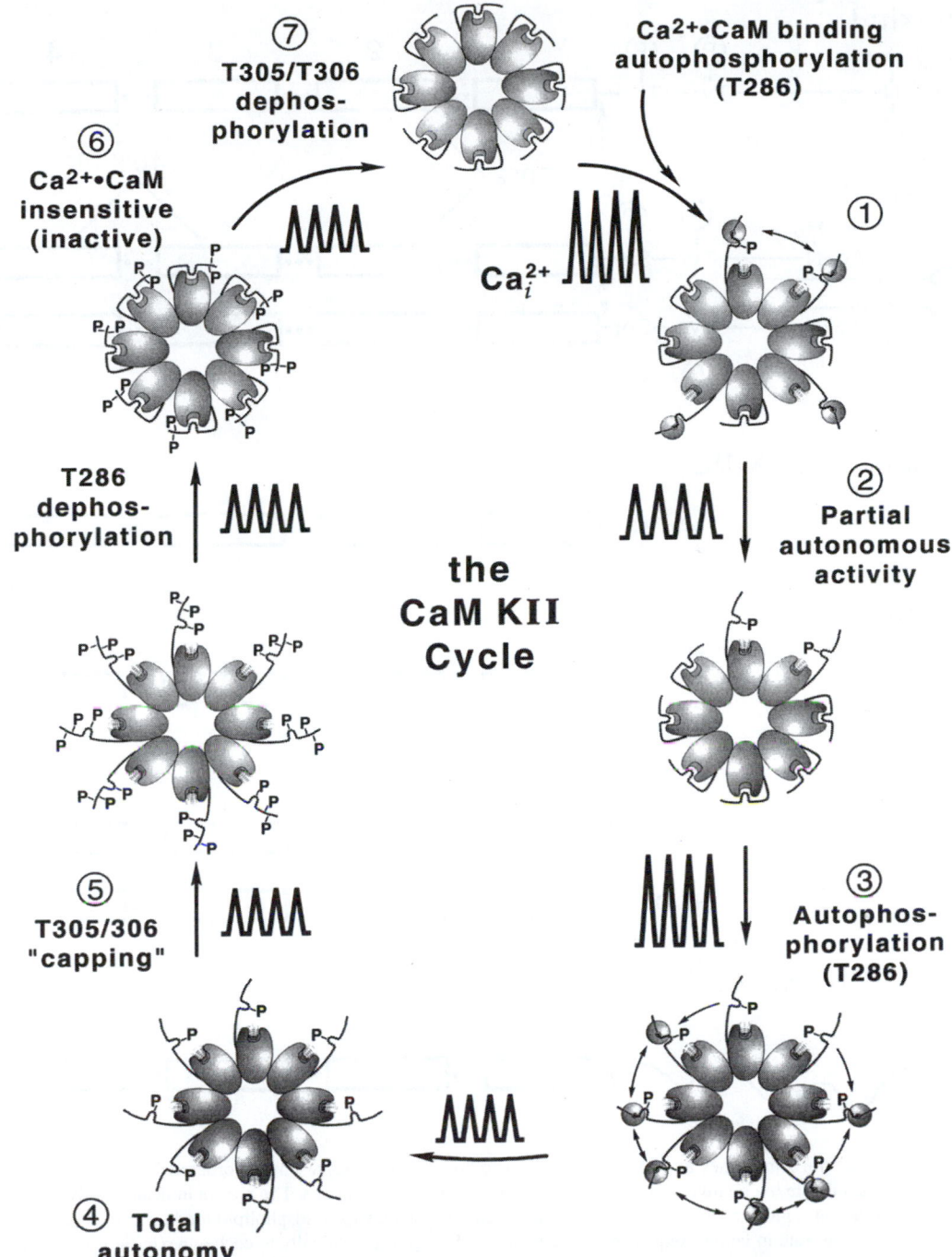

FIGURE 7. Ca^{2+}-CaM–dependent Kinase II. (1) The active site of each subunit is repressed by an autoregulatory domain. (2–5) Ca^{2+}-CaM binds to this domain and adjacent subunits are transphosphorylated at threonine-286 (T286). These autophosphorylated subunits are active and autonomous from Ca^{2+}-CaM regulation. (6) The regulatory domain can be further phosphorylated at threonine-305 and/or 306 (T305/T306), modifications which inhibit CaM binding. (7) Dephosphorylation of T286 renders the enzyme inactive and insensitive to Ca^{2+}-CaM. (1) Dephosphorylation of T305 and T306 returns the enzyme to its Ca^{2+}-CaM–dependent state.

IX. Current Perspectives

Scientific progress is not restricted by effort, but is dependent on insight and advances in technology. Technical progress in protein chemistry in the late 1960s and early 1970s led to the identification and characterization of the Ca^{2+} receptors, tro-
ponin C and calmodulin. Development of recombinant DNA technology in the mid-1970s provided precise knowledge of the molecular evolution and structural conservation of these proteins. In the early 1980s, Ca^{2+}-sensitive dyes and computer-assisted image processing allowed visualization of transient spatial changes in intracellular Ca^{2+}. Detailed information is

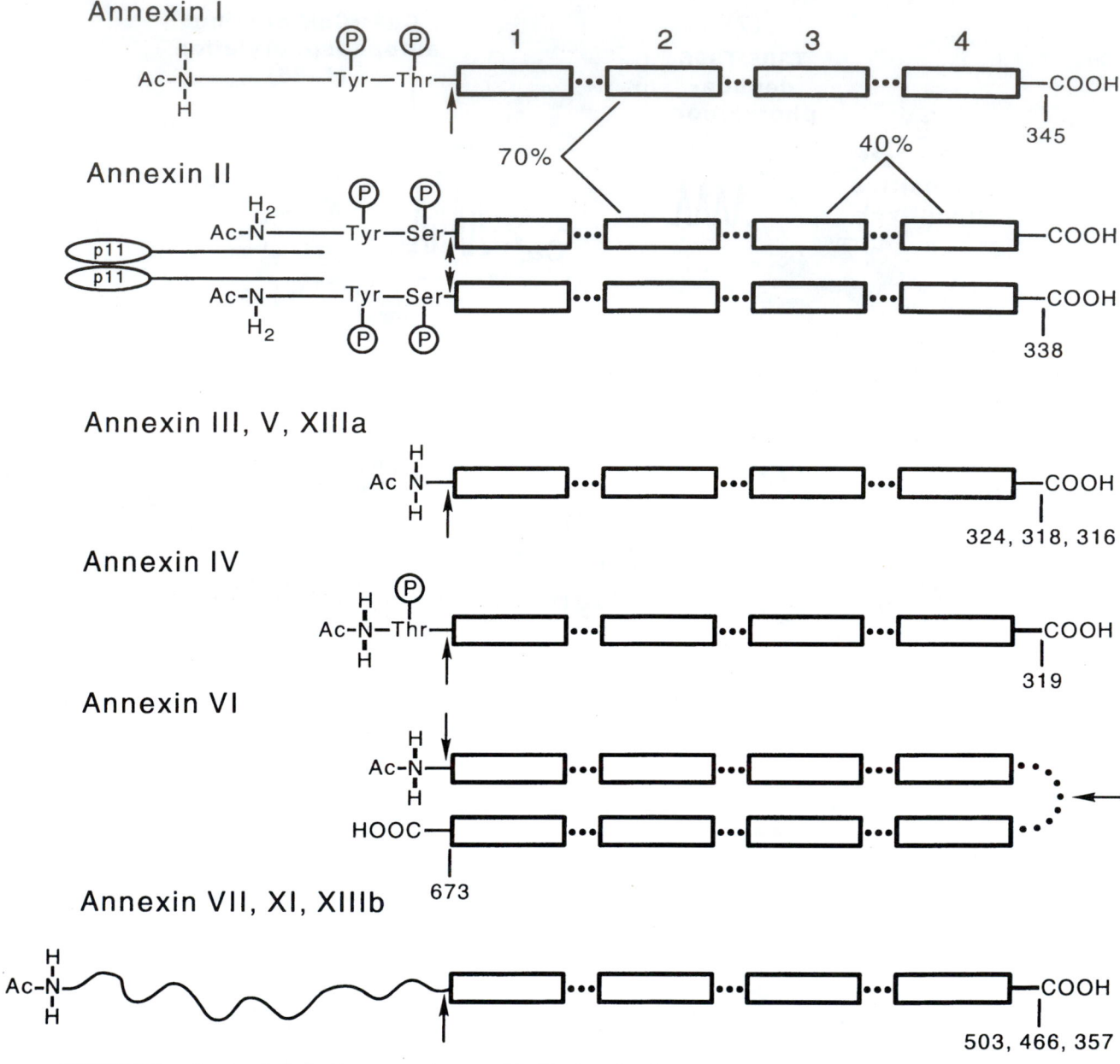

FIGURE 8. Structural similarities among the annexins. The boxed regions represent the ordered 70-amino-acid repeat domains. The numbers indicate the relative sequence similarities between repeat domains. The arrows indicate highly sensitive sites of proteolysis. The repeat regions form a highly structured, protease resistant, Ca^{2+}-phospholipid binding core. The amino terminal "tails" are highly variable in length, sequence, and structure and provide individually to each annexin family member.

being obtained on the cellular components that are essential for maintaining these levels, including Ca^{2+} channels, Na^{+}-Ca^{2+} exchangers, and Ca^{2+}-ATPases. The original observations of Hokin and Hokin (1953) concerning the turnover of membrane phosphatidylinositols were finally appreciated some 20 years later. New insights in second-messenger action revealed that the hydrolysis of PIP_2 to form IP_3 and DAG could coordinate, respectively, the release of intracellular Ca^{2+} from the endoplasmic reticulum and the activation of protein kinase C. Although the mechanisms involved in maintaining intracellular Ca^{2+} homeostasis are being defined, understanding the

events that couple stimuli to specific cellular responses has resisted biochemical definition. The elucidation of physiological regulation by Ca^{2+} through specific Ca^{2+} mediator proteins is not a trivial undertaking. Initial implications of function can be explored by cellular microinjection of inhibitor peptides, antibodies, and anti-sense RNA or by *in vitro* reconstitution. Information obtained from these inhibition experiments, however, is limited because it does not fully reflect the intact organism. For example, although our knowledge of the molecular aspects and regulatory properties of calmodulin *in vitro* is impressive, little is known of the functional role of calmodulin

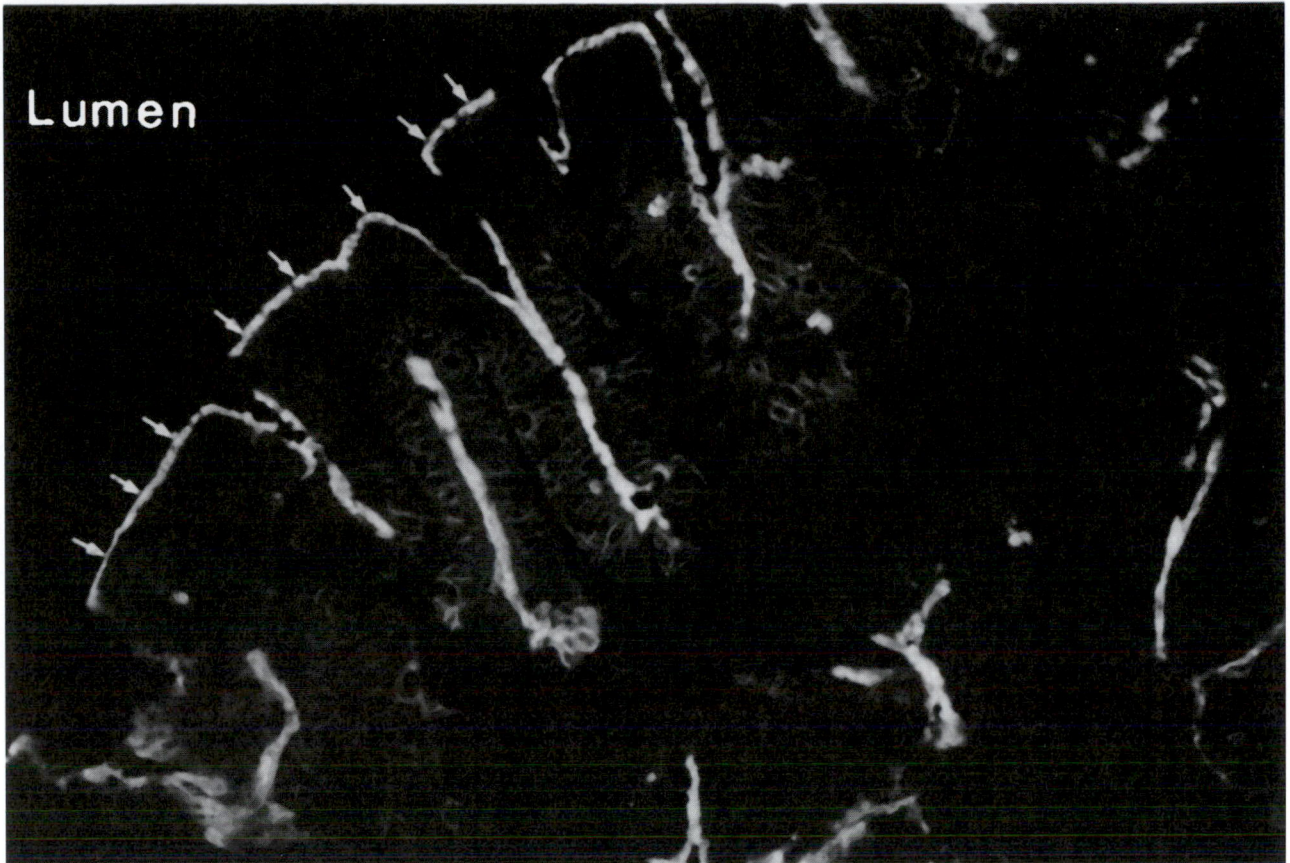

FIGURE 9.   Localization of annexin IV in the rat fallopian tube. Monospecific antibody identifies annexin IV as concentrated along the membrane bordering the lumen of the oviduct of the ciliated, columnar epithelia (arrows). This region is the cellular site of fluid secretion.

in the living cell or the whole animal, other than by inference. An attractive approach to obtaining this information is genetic manipulation. Fungi, such as yeast, are powerful models for the introduction and study of site-directed and temperature-sensitive mutations, and fundamental questions regarding the regulatory role Ca^{2+} in growth, division, and secretion have been addressed (see Davis, 1992). The usefulness of such hap-

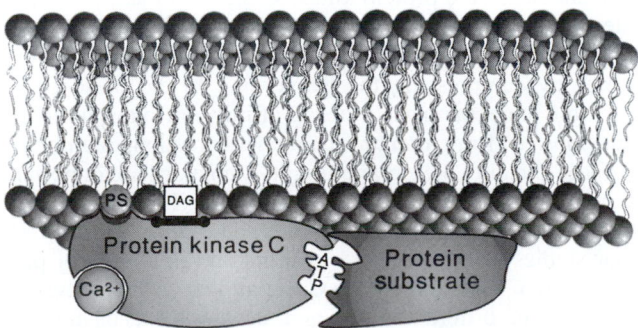

FIGURE 10.   Regulation of protein kinase C. Elevations in intracellular Ca^{2+} cause translocation of PKC to the plasma membrane where phosphatidylserine (PS) and diacylglycerol (DAG) further activate the enzyme.

loid systems, however, is limited to genes that are expressed in these organisms and to nonlethal mutations in these genes.

The cellular function of the diverse Ca^{2+}-binding proteins that are expressed in highly differentiated cells remains elusive. The respective pathways can be manipulated at many levels. Synthetic genes can be introduced into cultured cells and the resulting phenotypic changes monitored. In addition, specific genes can be deleted or "knocked out" in whole animals using homologous recombination of genes in mouse embryonic stem cells (See Frohmen and Martin, 1989).

Interpretation of targeted gene "knockout" studies is, however, complicated by the fact that the animal lacks the gene in its genome throughout its development. Phenotypic consequences within individual cell types resulting from gene disruptions may be due to defects that occur during embryonic development or inadequate communication with other defective cells types.

An alternative approach to understanding the precise role of proteins associated with the intracellular Ca^{2+} signal is to design and construct dominant-negative genes that, when expressed, neutralize the function of specific proteins. For example, peptides consisting of the calmodulin-binding site on target proteins (see Fig. 4) are extremely potent inhibitors of calmodulin. Wang *et al.* (1995, 1996) constructed a synthetic

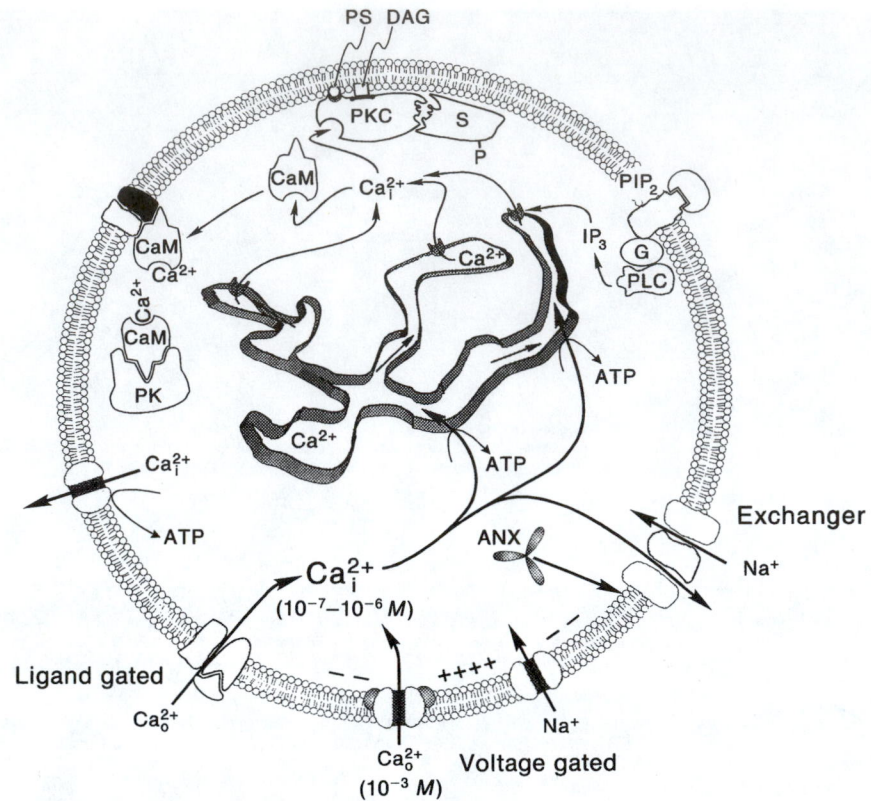

FIGURE 11.   Interplay of cellular components involved in creating and mediating the intracellular Ca^{2+} signal. Low intracellular Ca^{2+} levels are the result of pumps and ion exchangers. Ca^{2+} transients develop from influx through voltage and ligand gated channels on the plasmalemma and from IP_3-gated channels located in the endoplasmic/sarcoplasmic reticulum. IP_3 is generated from extracellular stimuli through a G protein (G)-regulated phospholipase C (PLC). The intracellular Ca^{2+} signal is mediated by three primary pathways: calmodulin (CaM), annexins (Anx), and protein kinase C (PKC).

gene that produced a string of these calmodulin-binding peptides. This peptide inhibitor was targeted to the nucleus of mouse-lung epithelial cells. By neutralizing the activity of nuclear calmodulin, it was shown that the cells did not synthesize DNA and the embryonic lung did not develop. Dominant-positive genes can also be designed to elucidate Ca^{2+}-regulated cellular systems. The amino acid sequence of the pseudosubstrate domain on calmodulin target proteins (see Fig. 4) can be altered to abolish inhibition of the catalytic site. This mutated enzyme becomes autonomous from Ca^{2+}-calmodulin regulation and remains active even during periods of low intracellular Ca^{2+}. Mayford *et al.* (1996) have shown, for example, that expression of an autonomously active CaM kinase II in mouse brain causes deficits in memory.

The understanding of Ca^{2+}-coupled stimulus responses has advanced through direct visualization of Ca^{2+} transients, the identification of Ca^{2+}-binding proteins, determination of their subcellular localization and their biochemical role through reconstitution of a cellular function, and genetic manipulation of intact cells. Although many candidates for Ca^{2+} mediator proteins are currently under investigation, it is anticipated that more Ca^{2+}-binding proteins will be identified as further research continues to solve the stimulus-response coupling puzzle.

X. Summary

Ca^{2+} is a ubiquitous intracellular regulator of cellular function. Levels of Ca^{2+} are controlled by a variety of channels, exchangers, and pumps found in the plasmalemma and internal membranes. Cell stimulation causes the intracellular Ca^{2+} to increase transiently. This second-messenger signal is then mediated by Ca^{2+}-binding proteins. There are three primary molecular mechanisms of transmitting the signal: calmodulin, annexins, and protein kinase C. Each of these pathways intersects and can be cross-regulatory. Ca^{2+}-calmodulin binds to specific sites on target proteins and activates enzymes by derepressing the active site. In the presence of Ca^{2+}, the annexins have a strong affinity for phospholipids. Annexins can form an interlocking network along membrane surfaces and alter membrane fluidity. Annexins are important in regulating membrane ion conductances. Protein kinase C is regulated by Ca^{2+}, phospholipid, and diacylglycerol. Cellular studies using specific activators and inhibitors of protein kinase C have shown this Ca^{2+} pathway to be involved in cell growth, differentiation, and development of tumors. Physiological, cellular, and molecular techniques are being used in combination to define the precise cellular roles of Ca^{2+}-binding proteins.

Bibliography

Aderem, A. (1992). The MARCKS brothers: A family of protein kinase C substrates. *Cell* **71**, 713–716.

Andree, H.A.M., Stuart, M.C.A., Hermens, W.T., Reutelingsperger, C.P.M., Hemker, H.C., Frederik, P.M., and Willems, G.M. (1992). Clustering of lipid-bound annexin V may explain its anticoagulant effect. *J. Biol. Chem.* **267**, 17907–17912.

Baum, P., Furlong, C., and Byers, B. (1986). Yeast gene required for spindle pole body duplication: homology of its product with Ca^{2+}-binding proteins. *Proc. Natl. Acad. Sci. USA* **83**, 5512–5516.

Bennett, V. (1985). The membrane skeleton of human erythrocytes and its implications for more complex cells. *Annu. Rev. Biochem.* **54**, 273–304.

Berridge, M.J. (1990). Calcium oscillations. *J. Biol. Chem.* **264**, 9583–9586.

Berridge, M.J. (1993). Inositol trisphosphate and calcium signaling. *Nature* **361**, 315–361.

Braun, A. P., and Schulman, H. (1999). Structural examination of autoregulation of multifunctional calcium/calmodulin-dependent Protein Kinase II. *J. Biol. Chem.* **274**, 26199–26208.

Concha, N.O., Head, J.F., Kaetzel, M.A., Dedman, J.R., and Seaton, B.A. (1992). Annexin V forms calcium-dependent trimeric units on phospholipid vesicles. *FEBS Lett.* **314**, 159–162.

Crumpton, M.J., and Dedman, J.R. (1990). Protein terminology tangle. *Nature* **345**, 212.

Davis, T.N. (1992). What's new with calcium? *Cell* **71**, 557–564.

Dedman, J.R., Welsh, M.J., Kaetzel, M.A., Pardue, R.L., and Brinkley, B.R. (1982). Localization of calmodulin in tissue culture cells. *In* "Calcium and Cell Function" (W.Y. Cheung, Ed.), Vol. 3, Academic Press, New York.

Frohman, M.A., and Martin, G.R. (1989). Cut, paste, and save: New approaches to altering specific genes in mice. *Cell* **56**, 145–147.

Heizmann, C.W., and Hunziker, W. (1991). Intracellular calcium-binding proteins: More sites than insights. *Trends Biochem. Sci.* **16**, 98–103.

Hodgkin, A.L., and Keynes, R.D. (1957). Movements of labeled calcium in squid giant axons. *J. Physiol.* **138**, 253–281.

Hokin, M.R., and Hokin, L.E. (1953). Enzyme secretion and the incorporation of ^{32}P into phospholipids of pancreas slices. *J. Biol. Chem.* **203**, 967–977.

Kaetzel, M.A., Hazarika, P., and Dedman, J.R. (1989). Differential tissue expression of three 35–kDa annexin calcium-dependent phospholipid-binding proteins. *J. Biol. Chem.* **264**, 14463–14470.

Kaetzel, M.A., Cahn, H.C., Dubinsky, W.P., Dedman, J.R., and Nelson, D.J. (1994). A role for annexin IV in epithelial cell function: Inhibition of calcium-activated chloride conductance. *J. Biol. Chem.* **269**, 5297–5302.

Koenig, M., Monaco, A.P., and Kunkel, L.M. (1988). The complete sequence of dystrophin predicts a rod-shaped cytoskeletal protein. *Cell* **53**, 219–228.

Llinas, R., Sugimori, M., and Silver, R.B. (1992). Microdomains of high calcium concentration in a presynaptic terminal. *Science* **256**, 677–679.

Lee, H.C. (1993). Potential of calcium- and caffeine-induced calcium release by cyclic ADP-ribrose. *J. Biol. Chem.* **268**, 293–299.

Mayford, M., Bach, M.E., Huang, Y.Y., Wang, L., Hawkins, R.D., and Kandel, E.R. (1966). Control of memory formation through related expression of a CaMKII transgene. *Science* **274**, 1678–1683.

Meers, P., Daleke, D., Hong, K., and Papahadjopoulos, D. (1991). Interactions of annexins with membrane phospholipids. *Biochem.* **30**, 2903–2908.

Miyamoto, S., Ohya, Y., Ohsumi, Y., and Anraku, Y. (1987). Nucleotide sequence of the CLS4 (CDC24) gene of *Saccharomyces cerevisiae*. *Gene* **54**, 125–132.

Nishizuka, Y. (1992). Intracellular signaling by hydrolysis of phospholipids and activation of protein kinase C. *Science* **258**, 607–614.

Noegel, A., Witke, W., and Schleicher, M. (1987). Calcium-sensitive nonmuscle α-actinin contains EF-hand structures and highly conserved regions. *FEBS Lett.* **221**, 391–396.

Ohno, S., Emori, Y., Imajoh, S., Kawaskai, H., Kisaragi, M., and Suzuki, K. (1984). Evolutionary origin of a calcium-dependent protease by fusion of genes for a thiol protease and a calcium-binding *Nature* **312**, 566–570.

Ringer, S. (1883). A further contribution regarding the influence of the different constituents of the blood of the contraction of the heart. *J. Physiol.* **4**, 29–43.

Smith, V.L., Kaetzel, M.A., and Dedman, J.R. (1990). Stimulus-response coupling: the search for intracellular calcium mediator proteins. *Cell Regul.* **1**, 165–172.

Sweet, W.D., and Schroeder, F. (1988). "Lipid Domains and the Relationship to Membrane Function." R. Liss, New York. p. 1742.

Wang, J., Campos, B, Jamieson, A. Kaetzel, M., and Dedman, J. (1995). Functional elimination of calmodulin with the nucleus by targeted expression of an inhibitor peptide. *J. Biol. Chem.* **270**, 30245–30248.

Wang, J., Moreira, K., Campos, B., Kaetzel, M., and Dedman, J. (1996). Targeted neutralization of calmodulin in the nucleus blocks DNA synthesis and cell cycle progression. *Biochim. Biophys. Acta* **1313**, 223–228.

Edward F. Nemeth

11

Regulation of Cellular Functions by Extracellular Calcium

I. Introduction

The signaling function of cytoplasmic Ca^{2+} has long been appreciated, and it is now well established that cytoplasmic Ca^{2+} acts to control a variety of cellular responses such as exocytotic secretion, cellular differentiation and proliferation, and muscular contraction (see Chapter 10). These actions of cytoplasmic Ca^{2+} are usually mediated by proteins such as *calmodulin* that bind Ca^{2+} with high affinity, commensurate with the relatively low concentrations of Ca^{2+} within the cytoplasm ($[Ca^{2+}]_i$; 100 nM). These high-affinity Ca^{2+}-binding proteins are essentially *intracellular Ca^{2+} receptors* that transduce changes in $[Ca^{2+}]_i$ into functional cellular responses.

Other important roles for Ca^{2+} are found outside the cell. The levels of extracellular Ca^{2+} in the blood and interstitial fluids modulate a number of vital cellular processes, including thrombosis, muscle and nerve excitability, and proper bone formation. Many of these events are mediated by proteins that bind extracellular Ca^{2+} in concentrations of 0.1 to 1 mM, such as proteins involved in cellular adhesion (Clark and Brugge, 1995). In this capacity, extracellular Ca^{2+} functions mostly in a permissive role and does not by itself elicit a functional cellular response, as cytoplasmic Ca^{2+} does.

An entirely different role of extracellular Ca^{2+} is performed in many of the cells that participate in systemic Ca^{2+} homeostasis, such as the cells in the parathyroid glands and kidney. These cells are capable of detecting very small changes in the concentration of extracellular Ca^{2+}, and in these cells extracellular Ca^{2+} by itself can profoundly influence cellular activity. It has been suspected for some time that extracellular Ca^{2+} might act in this messenger role, but the molecular mechanism explaining this action of extracellular Ca^{2+} has come to be understood only within the last decade. The novel molecular component of this extracellular Ca^{2+}-sensing mechanism is a **G-protein-coupled receptor** that uses extracel-

lular Ca^{2+} as its primary physiological ligand. Thus just as there are intracellular Ca^{2+} receptors responding to changes in $[Ca^{2+}]_i$, so too there are *extracellular Ca^{2+} receptors* present on the surface of some cells, which respond to small changes in the ambient level of Ca^{2+}. Not surprisingly, the data revealing this novel mechanism of signal transduction derive from studies of cells that are normally involved in maintaining Ca^{2+} homeostasis in the plasma and extracellular fluids, especially **parathyroid cells.** Therefore it is appropriate to briefly consider this homeostatic mechanism.

II. Systemic Calcium Homeostasis

Calcium is the most abundant mineral in the body and the average adult human body contains about 1 kg of calcium. Most of the calcium in the body is chemically complexed with phosphate or bound to proteins and deposited in bones and teeth. Only about 0.1% of total body calcium is in the plasma and extracellular fluids. In most mammals, the concentration of total calcium in the plasma is 2.4 mM, but only about half of this is free ionized calcium; the remaining calcium is bound to serum proteins and to various inorganic anions such as phosphate and citrate. It is the ionized form of calcium (Ca^{2+}) that is biologically active in the extracellular environment, and it is this form that is monitored by the "calciostat" maintaining **systemic Ca^{2+} homeostasis.** The principal factors in this mechanism are **parathyroid hormone** (PTH) and **1,25-dihydroxyvitamin D_3.** The former is an 84-amino-acid protein secreted by the parathyroid glands and the latter is a steroid-like hormone produced by the kidneys. PTH releases Ca^{2+} from bone, increases renal Ca^{2+} absorption, and decreases renal phosphate absorption. An additional action of PTH in the kidney is the stimulation of the synthesis of 1,25-dihydroxyvitamin D_3. 1,25-dihydroxyvitamin D_3 then acts in the intestine to increase absorption of dietary Ca^{2+}. All these actions tend to increase the level of Ca^{2+} in the circulation. Increased levels of plasma Ca^{2+} in turn act in a **negative feedback** capacity to depress secretion of PTH. There is a reciprocal relationship between the level of plasma Ca^{2+} and that of PTH, and this simple feedback loop is the primary mechanism that regulates systemic Ca^{2+} homeostasis in humans (Mundy, 1989; Brown, 1994; Figs. 1 and 2).

III. The Calcium Receptor

Because the inhibitory effect of extracellular Ca^{2+} on PTH secretion was readily observed using parathyroid cells *in vitro,* there was never much doubt that extracellular Ca^{2+} acts directly on these cells to regulate PTH secretion. The perplexing issue was how small changes in extracellular Ca^{2+} concentrations (0.05 to 0.1 mM) could cause profound changes in hormone secretion. It was not even known, for example, if extracellular Ca^{2+} acted at the cell surface or entered the parathyroid cell to affect PTH secretion.

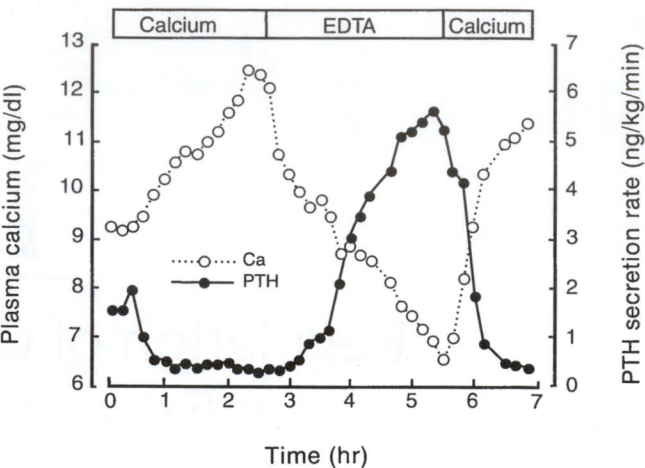

FIGURE 2. Reciprocal relationship between circulating levels of PTH and Ca^{2+}. The data are derived from studies in anesthetized calves where the plasma level of Ca^{2+} was varied either by infusing calcium chloride to raise plasma levels of Ca^{2+} or by infusing the divalent cation chelator EDTA to lower plasma levels of Ca^{2+} (Mayer and Hurst, 1978). Plasma levels of PTH rapidly rise or fall when plasma Ca^{2+} levels decrease or increase.

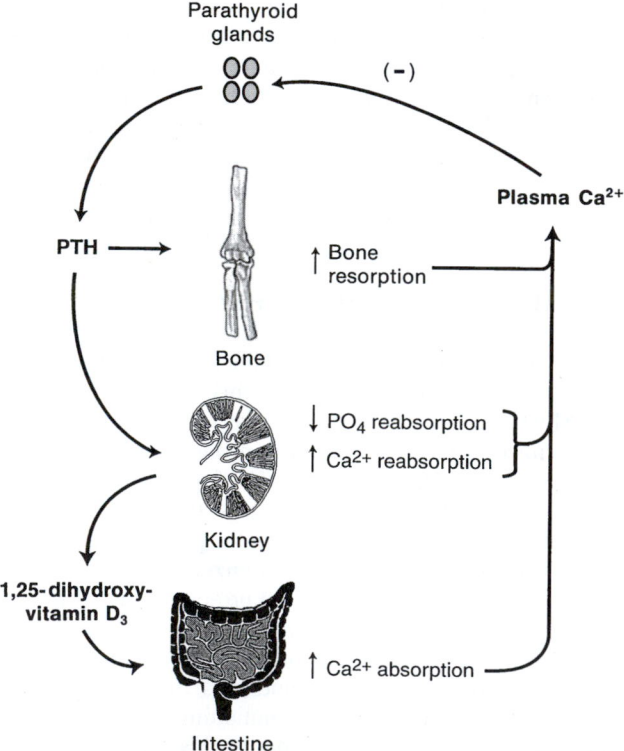

FIGURE 1. Regulation of systemic Ca^{2+} homeostasis in humans. The key player is PTH, which mobilizes Ca^{2+} from bone and conserves Ca^{2+} by the kidney. It also stimulates the formation of 1,25-dihydroxyvitamin D_3 in the kidney. 1,25-dihydroxyvitamin D_3 acts on the intestine to increase Ca^{2+} absorption from ingested food. All these actions tend to increase the plasma level of Ca^{2+}. Increases in the circulating levels of Ca^{2+} act on the parathyroid gland (there are four parathyroid glands in humans) to depress secretion of PTH. These mechanisms explain the reciprocal relationship between circulating levels of PTH and Ca^{2+}.

The results from a rather simple series of experiments in the mid-1980s provided the conceptual framework for understanding how certain cells might sense extracellular Ca^{2+}. It was proposed that parathyroid cells possess a cell surface Ca^{2+} receptor which enables them to detect and respond to small changes in the levels of extracellular Ca^{2+} (Nemeth and Scarpa, 1986). Key insights into the molecular aspects of cell surface receptors and the signaling mechanisms to which they couple, were also made around this time. The ability of extracellular Ca^{2+} to regulate similar mechanisms in parathyroid cells was taken as additional evidence for a cell surface Ca^{2+} receptor (Nemeth, 1990; Brown, 1991). Although the hypothesis of a cell surface receptor for an atom was greeted with skepticism, if not downright ridicule for many years, the issue was finally settled by the cloning in February of 1993 of a cDNA encoding the parathyroid cell Ca^{2+} receptor (Brown *et al.,* 1993). The Ca^{2+} receptor is structurally and functionally similar to many other cell surface receptors. The major difference is that the primary physiological ligand for the Ca^{2+} receptor is an inorganic ion rather than an organic molecule. As such, it is the first cell surface receptor for an inorganic ion to be structurally identified and functionally characterized.

A. Structure of the Ca^{2+} Receptor

The nucleotide sequence of the human parathyroid Ca^{2+} receptor encodes a protein of 1078 amino acids; it is 92% and 93% identical, respectively, with the rat kidney (1079 amino acids) and bovine parathyroid (1085 amino acids) Ca^{2+} receptors (Garrett *et al.,* 1995a). Hydropathy analysis of the deduced amino acid sequence reveals a membrane-spanning domain containing seven transmembrane regions

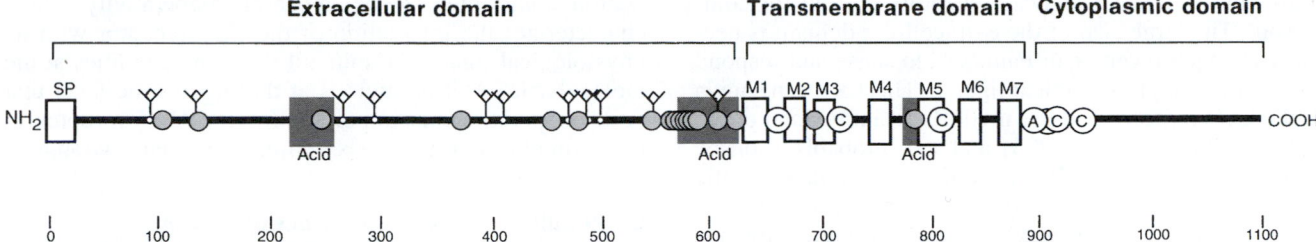

FIGURE 3. Schematic representation of the human Ca^{2+} receptor protein structure. The primary protein structure is composed of 1078 amino acids and is predicted to be composed of an NH_2-terminal extracellular domain ($\approx$620 amino acids), transmembrane domain ($\approx$250 amino acids) and an intracellular carboxy-terminal domain ($\approx$200 amino acids). The hydrophobic transmembrane-spanning regions are indicated by open blocks (M1 to M7). In the cytoplasmic domain and in the cytoplasmic loops linking some transmembrane-spanning domains are the five potential sites for protein kinase C (circled C). On the human receptor, there are two potential sites for protein kinase A phosphorylation (circled A) in the intracellular domain. There are two regions enriched in acidic amino acids in the extracellular domain and one in the extracellular loop linking M4 and M5 (shaded boxes). The locations of 17 cysteine residues, conserved between the Ca^{2+} receptor and metabotropic glutamate receptors, are indicated by shaded circles. Eleven potential sites of N-linked glycosylation are present in the extracellular domain (branches), and the signal peptide sequence (SP) is shown at the NH_2-terminus.

(Fig. 3) and places the Ca^{2+} receptor within the G-protein-coupled receptor superfamily.

Structural homologies based on deduced amino acid sequences have been used to classify the hundreds of different G-protein-coupled receptors into three large families (Kolakowski, 1994). Families A and B contain receptors for biogenic amines, autocoids, peptides, and proteins; Family C contains the Ca^{2+} receptor. Included in Family C are **metabotropic glutamate receptors** (mGluRs), γ-aminobutyric acid type B-receptors ($GABA_BRs$), and receptors in the vomeronasal organ (VNORs). The VNORs are thought to be receptors for pheromones although their functional significance in humans and New World primates is controversial. $GABA_BRs$ and mGluRs are neurotransmitter receptors expressed throughout the central nervous system and in some peripheral tissues receiving innervation. Depending on the algorithm used for sequence alignment, receptors within Family C are 15 to 20% identical and 25 to 40% homologous. The Ca^{2+} receptor shares the highest degree of structural similarity with the VNORs.

The Ca^{2+} receptor possesses 11 potential N-linked glycosylation sites located in the extracellular domain. Mannose and complex carbohydrates are the sugars attached to the Ca^{2+} receptor posttranslationally and these additions explain why the protein expressed at the cell surface has a molecular weight which is higher than that expected from its amino acid sequence. The glycosylation of the receptor is important for transporting the receptor through the Golgi apparatus to the cell surface (Ray *et al.*, 1998). In cells expressing mutated receptors which lack about half the glycosylation sites, the amount of receptors reaching the cell surface is greatly reduced. At least three glycosylation sites are necessary for any expression at the cell surface. Those receptors arriving at the cell surface, however, still respond to extracellular Ca^{2+} so glycosylation does not appear to greatly alter the functional activity of Ca^{2+} receptors.

The receptors in Family C contain numerous cysteine residues, most of which are present in the extracellular domain. The Ca^{2+} receptor and mGluRs are most similar in this respect, each containing 20 cysteines, 17 of which are in the extracellular domain and are precisely conserved. These cysteines probably determine the secondary (and perhaps tertiary) structure of these receptors, and their conservation suggests that the secondary structures of mGluRs and the Ca^{2+} receptor are similar.

Although once thought to be a feature only of tyrosine kinase receptors and immunoglobulin receptors, there is now extensive evidence showing that many, and perhaps all, G-protein-coupled receptors undergo dimerization (Hébert and Bouvier, 1998). The Ca^{2+} receptor also exists as a dimer and this is probably the major tertiary structure of the receptor at the cell surface (Bai, 1999). Certain cysteine residues are critical for dimerization of the Ca^{2+} receptor but it is still uncertain if dimerization is critically dependent on covalent disulfide-linkages between two receptor molecules or if hydrophobic interactions are more important for dimerization. It is also not known if Ca^{2+} receptors are inserted into the plasma membrane as dimers or if such oligomerization occurs subsequent to delivery at the cell surface. Of particular importance are the functional and pharmacological properties of the Ca^{2+} receptor that might differ between monomer and dimer states of the receptor.

B. Sensing of Ligand by the Ca^{2+} Receptor

Receptors in Family C are set apart from other G-protein-coupled receptors by possessing a very large extracellular domain and a relatively long C-terminal cytoplasmic domain which together make these receptors larger than other G-protein-coupled receptors (Fig. 3). The human Ca^{2+} receptor, for example, has an extracellular domain composed of 612 amino acids—this domain alone is larger than the receptors in Family A or B even when the entire amino acid sequence of these receptors is used for comparison. The large size of the extracellular domain is necessary for each receptor in Family C to sense and respond to its respective physiological ligand. This has been shown for the Ca^{2+} receptor and the mGluRs (Hammerland et al., 1999; Takahashi et al., 1993) and will probably soon be shown true for $GABA_BRs$ as well. (It may take a little longer to demonstrate a similar function for the extracellular domain of VNORs because the physiological ligands for these receptors are unknown.) The dominant if not exclusive role played by the extracellular domain is a feature which distinguishes Family C receptors from other G-protein-coupled receptors. Most receptors for peptides or proteins use the extracellular loops linking transmembrane helices together with the extracellular domain to bind their physiological ligand. Some smaller ligands, like biogenic amines, bind exclusively in the transmembrane domain of their cognate receptor (Beck-Sickinger, 1996). It is one of those curious features of nature that receptors sensing some of the smallest ligands use the largest domains to do so.

Although the extracellular domain of the Ca^{2+} receptor is critical for responding to extracellular Ca^{2+}, essentially nothing is known about how extracellular Ca^{2+} interacts with the receptor. It might even be misleading to speak of receptor "binding" since the apparent affinity is so low compared to other ligand-receptor interactions. So it is not surprising that the Ca^{2+} receptor does not possess any **"EF hand" motifs** that are characteristic of high-affinity calcium-binding proteins such as calmodulin (see Chapter 10). These latter proteins function as intracellular Ca^{2+} receptors and sense micromolar changes in the cytoplasmic Ca^{2+} concentration, whereas extracellular Ca^{2+} receptors sense millimolar changes in the extracellular Ca^{2+} concentration. There are regions of the Ca^{2+} receptor, however, that are enriched with acidic amino acids. Two are in the extracellular domain and one is on the extracellular loop linking transmembrane-spanning segments 4 and 5 (see Fig. 3). Two of these regions are similar to those found in low-affinity calcium-binding proteins such as **calreticulin.** All of these acidic amino acids are conserved among various species homologues of Ca^{2+} receptor proteins. These regions, enriched in negative charges, could conceivably play a role in the "binding" of extracellular Ca^{2+} by the receptor.

Unlike most other G-protein-coupled receptors, the Ca^{2+} receptor responds over a very narrow ligand concentration range, and this results in a very steep concentration-response curve (see Fig.6). Such curves are usually indicative of the phenomenon of "cooperativity" and predict that more than one Ca^{2+} interacts with the receptor to elicit a maximal response. Indeed, the Hill coefficient, which is an estimate of the stoichiometry of ligand-receptor interaction, yields a value of 3. Another feature of cooperativity is that the interaction of the first ligand increases the probability of interacting with the second ligand. In some cases, the interaction is an increased affinity for the ligand. For a G-protein-coupled receptor, the coupling affinity between the receptor and a G protein could also explain such cooperativity. Finally, it is quite possible that receptor dimerization contributes to the apparent cooperativity which characterizes the interaction of the Ca^{2+} receptor with its physiological ligand. Despite all these uncertainties at the molecular level, it is obvious at the organismic level that such cooperativity is important for the Ca^{2+} receptor to maintain plasma Ca^{2+} levels within a very narrow range.

C. Regulation of the Ca^{2+} Receptor Gene

The Ca^{2+} receptor gene, which is located on human chromosome 3q 13.3-21, consists of seven exons spanning at least 45 kbp. The human Ca^{2+} receptor promoter has no TATA or CAAT boxes and, in this respect, is similar to other G-protein-coupled receptor genes. In human parathyroid cells, an exon 5 to 6 alternative splice isoform has been identified and appears to be functionally equivalent to the major expression product (Garrett et al., 1995a).

Although the effects of altered plasma levels of Ca^{2+} or 1,25-dihydroxyvitamin D_3 on expression of the Ca^{2+} receptor gene have been studied in rats, a conclusive picture has yet to emerge. These are clearly important factors to assess because chronic states of hypocalcemia and/or lowered 1,25-dihydroxyvitamin D_3 (as in chronic renal failure) are believed to contribute importantly to parathyroid cell growth and the amount of PTH stored in the cells (Drüeke, 1995; Martin and Slatopolsky, 1994). Increased circulating levels of PTH accompanied by hyperplasia of the parathyroid glands characterize secondary hyperparathyroidism. Understanding the actions, if any, of the Ca^{2+} receptor that contribute to the development of secondary hyperparathyroidism clearly has important implications for disease management. Pharmacologically altering the activity of the Ca^{2+} receptor might provide a means of diminishing the severity of or preventing the secondary hyperparathyroidism that accompanies renal failure (Nemeth, 1996; Nemeth and Fox, 1999).

D. Diseases Caused by Mutations in the Ca^{2+} Receptor Gene

Familial benign hypocalciuric hypercalcemia (FBHH) is a rare autosomal dominant disorder characterized by mild hypercalcemia. This disease results from mutations in the Ca^{2+} receptor gene (Nemeth and Heath, 1995; Pearce and Brown, 1996). So far, more than 38 different mutations have been identified in kindreds with FBHH; nearly all are single amino acid substitutions, and most occur within the extracellular domain of the Ca^{2+} receptor. These mutations partially or totally inhibit the ability of extracellular Ca^{2+} to activate the Ca^{2+} receptor (Pollak et al., 1993). This genetic disorder therefore results from the expression of mutated Ca^{2+} receptors that make parathyroid cells less sensitive to regulation by extracellular Ca^{2+}. FBHH is the heterozygous manifestation of these Ca^{2+} receptor gene mutations. The homozygous condition, which is life-threatening, is called **neonatal severe hyperparathyroidism (NSHPT).** In the homozygous disease, the parathyroid gland is completely unresponsive to extracellular Ca^{2+}; severe hypercalcemia with greatly elevated levels of PTH ensue and total parathyroidectomy is usually necessary soon after birth.

The converse of FBHH and NSHPT is **autosomal dominant hypoparathyroidism (ADH).** In this disorder, mutations in the Ca^{2+} receptor gene produce a protein that is excessively sensitive to activation by extracellular Ca^{2+} (Pollak *et al.,* 1994; Bai *et al.,* 1996). Consequently, normal levels of plasma Ca^{2+} now depress PTH secretion to a much greater extent. Lower circulating levels of PTH will lead to lower levels of Ca^{2+}, which explains the hypocalcemia characteristic of ADH. The deviant Ca^{2+} receptor essentially lowers the **"set point"** for plasma Ca^{2+} concentration, so that systemic Ca^{2+} concentrations are maintained at a lower value. In this manner, the Ca^{2+} receptor can be considered the "thermostat" that monitors and maintains the levels of extracellular Ca^{2+}. In this particular capacity, the Ca^{2+} receptor serves as the body's "calciostat." Turning the calciostat up (as in ADH) or down (as in FBHH or NSHPT) lowers or raises the circulating levels of Ca^{2+}. In the aggregate, these molecular genetic studies identify the Ca^{2+} receptor as the key entity controlling systemic Ca^{2+} homeostasis.

E. Other Extracellular Ca^{2+} Receptors?

"Curiously, and despite much effort, no subtypes of the Ca^{2+} receptor have yet been identified, and Ca^{2+} receptors expressed in different tissues all seem to be products of a single gene." This sentence was present in the second edition of the *Sourcebook* and there is no reason to change it for the third. There was some transient excitement regarding research on this topic during the intervening period, but in one case the receptors turned out to be those for GABA (the $GABA_B$Rs mentioned above). The other case involved the discovery of the VNORs which, in the human, never quite become true receptors but rather appear only as expressed pseudogenes. Hence the view held by some that VNORs are not functional in humans—something to remember when considering perfumes which claim to attract the opposite sex.

Some reasons for supposing the existence of subtypes of the Ca^{2+} receptor arise from the sensitivity of different cells to extracellular Ca^{2+} and the pharmacological properties of these cells. **Keratinocytes** (skin cells), for example, respond to extracellular Ca^{2+} concentrations between 0.05 and 0.5 mM, far lower than circulating levels of Ca^{2+} (Bikle *et al.,* 1996). In contrast, osteoclasts (cells that resorb bone) respond to higher levels of extracellular Ca^{2+} (4 to 10 mM) (Zaidi *et al.,* 1993). Moreover, the pharmacology of extracellular Ca^{2+}-mediated responses in osteoclasts is very different from that in parathyroid cells. While these differences could reflect the existence of a new subtype of Ca^{2+} receptor, they can be explained by any number of alternative molecular events known to affect the function of G-protein-coupled receptors. These include the various conformational states the receptor can assume within the plasma membrane of a specific cell type, the kind of G-protein to which it couples, and the presence or absence of various accessory proteins in the membrane or cytoplasm (King and Wilson, 1999). It is

clear that the pharmacological properties of G-protein-coupled receptors can vary greatly depending upon the cellular environment in which they are expressed. Thus, a "systems" view will be necessary to fully understand the functional significance of receptors in the body.

While the issue of Ca^{2+} receptor subtypes will sort itself out in time, there is the possibility that other proteins, structurally unrelated to the Ca^{2+} receptor, might function as sensors for extracellular Ca^{2+}. For example, a 500 kDa glycoprotein, structurally related to **low density lipoprotein (LDL) receptors,** has been postulated to act as a Ca^{2+} receptor on parathyroid cells (Lundgren *et al.,* 1994). LDL receptors are known to have the ability to bind extracellular Ca^{2+} and some other divalent cations. However, this large protein has been difficult to express in heterologous systems, and it is not yet known if this protein functions in a manner akin to that of the Ca^{2+} receptor.

F. Pharmacology of the Ca^{2+} Receptor

The Ca^{2+} receptor, which is rather promiscuous, senses a variety of positively charged inorganic ions and organic compounds. Many divalent cations and nearly all the trivalent cations activate the Ca^{2+} receptor with a rank order of potency as follows: $La^{3+} > Gd^{3+} > Be^{2+} > Ca^{2+} = Ba^{2+} > Sr^{2+} > Mg^{2+}$. Of these, Mg^{2+} is the only cation likely to have any impact on Ca^{2+} receptor activity under physiological conditions. In addition to inorganic polycations, many organic polycations also activate the Ca^{2+} receptor; these include polyamines (spermine), aminoglycoside antibiotics (neomycin), polyamino acids (polylysine), and proteins (protamine). In general, there is some correlation between net positive charge of these compounds and their potency in activating the Ca^{2+} receptor. Nonetheless, there are exceptions to this rule that indicate that the Ca^{2+} receptor is sensing not just net positive charge. Like Ca^{2+}, all these polycations are believed to act largely, if not exclusively, in the extracellular domain of the Ca^{2+} receptor. It is conceivable that regions of the receptor containing a high density of acidic amino acids, and therefore a high density of negative charge, are involved in the binding of these inorganic and organic polycations.

The organic polycations are not useful tools for investigating the cellular physiology of parathyroid cells or other extracellular Ca^{2+}-sensing cells because they lack the potency and specificity necessary for many *in vitro* studies and all *in vivo* studies. A structurally distinct class of compounds, typified by NPS R-467, is used in superior pharmacological probes that overcome these problems. These **phenylalkylamine derivatives** carry only one positive charge at physiological pH and, at nanomolar concentrations, act selectively on the Ca^{2+} receptor to inhibit PTH secretion (Nemeth *et al.,* 1998). These compounds act in a stereoselective manner at the Ca^{2+} receptor, and the *R*-enantiomers are 10- to 100-fold more potent than the corresponding *S*-enantiomers. The mechanism of action of the phenylalkylamine compounds differs from those of the inorganic and organic polycations. Unlike the polycations, these phenylalkylamine compounds require extracellular Ca^{2+} for their activity. Thus these compounds fail to mobilize intracellular Ca^{2+} in the absence of extracellular Ca^{2+}, although a polycation such as neomycin is fully competent in doing so. These phenylalkylamines behave as **positive allosteric modulators** to increase the sensitivity of the Ca^{2+} receptor to activation by extracellular Ca^{2+}, thereby shifting the concentration-response curve to the left. These different mechanisms are also manifest in the concentration-response curves to polycations and to phenylalkylamines. The curve for polycations is extremely steep and apparently reflects cooperative binding of the ligand to the receptor, whereas that for phenylalkylamines is more conventional and suggests a 1:1 ligand:receptor complex. The phenylalkylamines bind in the transmembrane domain of the Ca^{2+} receptor (Hammerland *et al.*, 1999).

Compounds that mimic or potentiate the actions of extracellular Ca^{2+} at the Ca^{2+} receptor have been termed **calcimimetics**. As such, they behave as receptor agonists to mobilize intracellular Ca^{2+} and inhibit PTH secretion. Two types of calcimimetics can be distinguished: Type I, exemplified by the inorganic and organic polycations, which bind largely in the extracellular domain and mimic extracellular Ca^{2+}, and Type II, typified by the phenylalkylamines, which bind within the transmembrane domain of the Ca^{2+} receptor and potentiate the actions of extracellular Ca^{2+}.

Type II calcimimetics have been used to activate the Ca^{2+} receptor *in vivo* and, when administered to animals or humans, cause time- and dose-dependent decreases in plasma levels of PTH and Ca^{2+}. Circulating levels of PTH start to fall sooner than those of Ca^{2+}, as expected, since it is this decrease in PTH that leads to hypocalcemia (Fig. 4). These findings, together with those from molecular genetic studies, provide solid evidence that the Ca^{2+} receptor is the essential regulator of systemic Ca^{2+} homeostasis.

More recently, potent and selective small organic antagonists of the Ca^{2+} receptor have been developed. These compounds have been termed **calcilytics** and are the only substances known that antagonize the Ca^{2+} receptor. Calcilytic compounds stimulate secretion of PTH *in vitro* and *in vivo* (Fig. 5) and, together with the Type II calcimimetics, are pharmacological tools possessing the requisite potency and selectivity necessary to explore the role of the Ca^{2+} receptor in physiological processes beyond systemic Ca^{2+} homeostasis.

IV. Calcium Receptor-Dependent Regulation of Cellular Functions

There are a number of cells throughout the body whose activity is regulated by changes in the concentration of extracellular Ca^{2+}. For many of these cells, it is not certain that the effects of extracellular Ca^{2+} are mediated by a parathyroid-like Ca^{2+} receptor. However, there is clear evidence for the expression of this receptor in certain cells and some understanding of how this receptor regulates cellular responses. Discussed next are the three cell types that have furnished most of the information about how the Ca^{2+} receptor regulates cellular activity.

A. Parathyroid Cells

These are the classic cells long known to be responsive to small changes in the concentration of extracellular Ca^{2+}. For

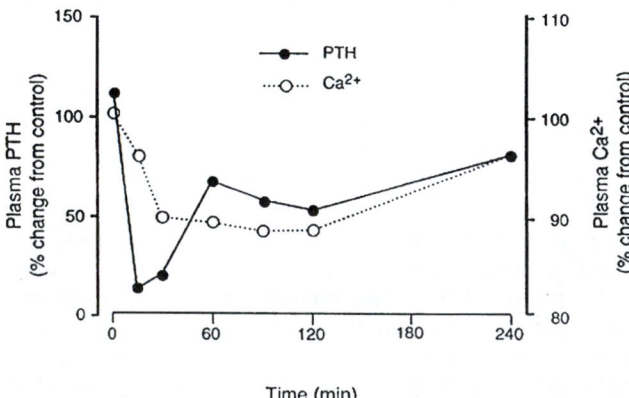

FIGURE 4. Activation of the Ca^{2+} receptor *in vivo* lowers plasma levels of PTH and Ca^{2+}. Normal rats were given 10 mg/kg NPS R-467 orally at time 0. There is a rapid fall in plasma levels of PTH followed by a decrease in plasma Ca^{2+} levels. Both parameters return to normal over several hours. Insert: The structure of NPS R-467, the R-enantiomer.

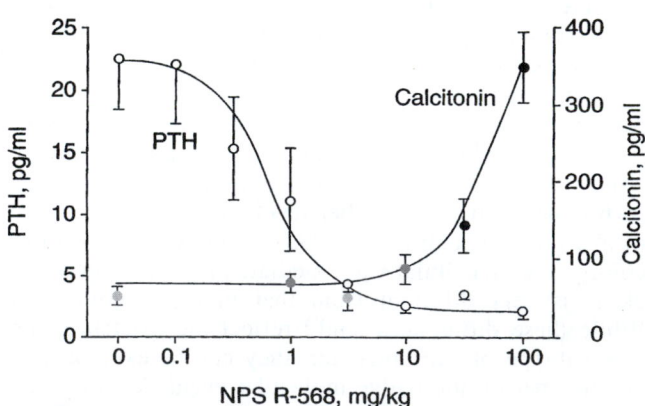

FIGURE 5. Preferential effects of a Type II calcimimetic compound on plasma levels of PTH. Orally administered NPS R-568 (the chloro-analog of NPS R-467) was about 40-fold more potent at depressing secretion of PTH than at stimulating secretion of calcitonin. (From Fox *et al.*, 1999, *Journal of Pharmacology and Experimental Therapeutics*, **290**: 480)

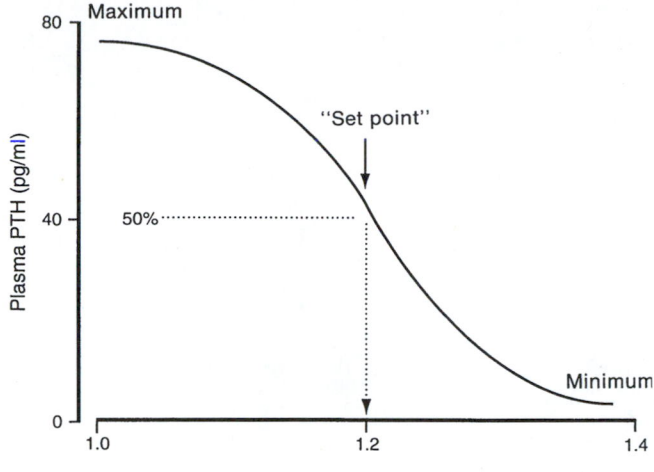

FIGURE 6. Small changes in the plasma levels of Ca^{2+} cause large changes in the plasma levels of PTH. In the normal adult human, plasma levels vary between 10 and 65 pg/ml and the plasma Ca^{2+} levels are maintained between 1.12 and 1.23 mM. The concentration of plasma Ca^{2+} causing a 50% decrease in the maximal levels of plasma PTH (achieved by lowering extracellular Ca^{2+} until no further increase occurs) is called the "set point" for extracellular Ca^{2+}. The Ca^{2+} receptor determines the set point and thereby functions as a "calciostat" to maintain systemic Ca^{2+} homeostasis.

these cells, extracellular Ca^{2+} is the primary physiological stimulus regulating secretion of PTH. The sensitivity of the parathyroid cell to the ambient Ca^{2+} concentration is impressive: minimal and maximal rates of PTH secretion are obtained over a concentration difference of only 0.5 mM (range of 1.0 to 1.5 mM). The concentration of serum Ca^{2+} that suppresses maximal PTH levels by 50% is called the **"set point"** for extracellular Ca^{2+} (Brown, 1983), and it is usually around the normal serum level of 1.2 mM (Fig. 6). Because of the steep concentration-response relationship between extracellular Ca^{2+} and PTH, very small changes in the concentration of extracellular Ca^{2+} cause large changes in the secretion of PTH (see Fig. 6).

The Ca^{2+} receptor on parathyroid cells is coupled to the activation of phospholipase C through a G protein (Fig. 7). The identity of this G protein is not yet certain, although it is believed to be either $G_{\alpha 11}$ or $G_{\alpha q}$. The G protein linking the Ca^{2+} receptor to phospholipase C is not inhibited by pertussis toxin. Activation of phospholipase C results in the rapid formation of IP_3, leading to the mobilization of Ca^{2+} from some nonmitochondrial intracellular store, presumably some component of the endoplasmic reticulum. There is additionally an influx of extracellular Ca^{2+} through **voltage-insensitive channels.** Thus when the concentration of extracellular Ca^{2+} is increased, there is a rapid and transient increase that is followed by a lower yet sustained increase in $[Ca^{2+}]_i$ (Fig. 8).

It is not at all clear how the influx of extracellular Ca^{2+} in parathyroid cells is regulated. Many Type I calcimimetics are just as efficacious as extracellular Ca^{2+} in evoking the mobilization of intracellular Ca^{2+}, yet they fail to promote

the influx of extracellular Ca^{2+} (see Fig. 8). Thus, activating the Ca^{2+} receptor is not sufficient to trigger influx of extracellular Ca^{2+}. Presumably, the Ca^{2+} receptor is not directly linked to an influx channel. The influx of extracellular Ca^{2+} might be regulated by an intracellular signal or by depletion of intracellular stores ("capacitive" Ca^{2+} influx; Petersen, 1996; see Chapter 55).

The Ca^{2+} receptor is also coupled to adenylate cyclase through a pertussis toxin-sensitive G_i-like protein. Type II calcimimetics potentiate the effects of extracellular Ca^{2+} on cyclic AMP levels, $[Ca^{2+}]_i$, and PTH secretion. These findings suggest that the Ca^{2+} receptor is the only molecular entity involved in sensing changes in the concentration of extracellular Ca^{2+}, and that the Ca^{2+} receptor can couple to two distinct transmembrane signaling systems in parathyroid cells.

Curiously, Ca^{2+} receptor-mediated increases in $[Ca^{2+}]_i$ are associated with an inhibition of PTH secretion (see Figs. 4 and 8). This is in contrast to most other secretory cells in which increases in $[Ca^{2+}]_i$ stimulate secretion (see Chapter 42) and it suggests that cytoplasmic Ca^{2+} may have an inhibitory effect on exocytotic secretion in parathyroid cells. At present, however, it is uncertain whether cytoplasmic Ca^{2+} acts as the principal intracellular signal regulating PTH secretion. Although increases in $[Ca^{2+}]_i$ in parathyroid cells almost invariably inhibit secretion, changes in PTH secretion can be dissociated from changes in $[Ca^{2+}]_i$, and this suggests that additional or alternative intracellular signals play an important role in regulating PTH secretion.

There is much evidence showing that increases in $[Ca^{2+}]_i$ arising from influx of extracellular Ca^{2+} are not importantly involved in the regulation of PTH secretion (Nemeth and Scarpa, 1987). Some of this evidence derives from the use of Type I calcimimetics that, as mentioned earlier, mobilize intracellular Ca^{2+} but do not allow influx of extracellular Ca^{2+}. Moreover, blocking the influx of extracellular Ca^{2+} (with channel blockers) does not affect the ability of extracellular Ca^{2+} to inhibit secretion of PTH. Nonetheless, most of these polycations inhibit secretion of PTH just as well as extracellular Ca^{2+}, which mobilizes intracellular Ca^{2+} and allows the influx of extracellular Ca^{2+} (see Fig. 8). These finding show that transient increases in $[Ca^{2+}]_i$ arising from mobilization of intracellular Ca^{2+} are associated with the inhibition of PTH secretion. Yet there is still no consensus that cytoplasmic Ca^{2+} plays the dominant role in regulating secretion of PTH.

Various other intracellular signaling mechanisms have been postulated to play a role in controlling PTH secretion, but no compelling molecular model has yet emerged. One signaling molecule that has been studied extensively is **protein kinase C** (PKC). When PKC is inhibited in parathyroid cells, or when cellular levels are depleted by downregulation of PKC activity, extracellular Ca^{2+} is still capable of regulating PTH secretion in a normal manner. However, PKC does play a role in modulating the sensitivity of the Ca^{2+} receptor to extracellular Ca^{2+}. PKC, probably by phosphorylation of the Ca^{2+} receptor, depresses the sensitivity of parathyroid cells to regulation by extracellular Ca^{2+}. Thus, PKC seems to act in a negative-feedback manner to dampen

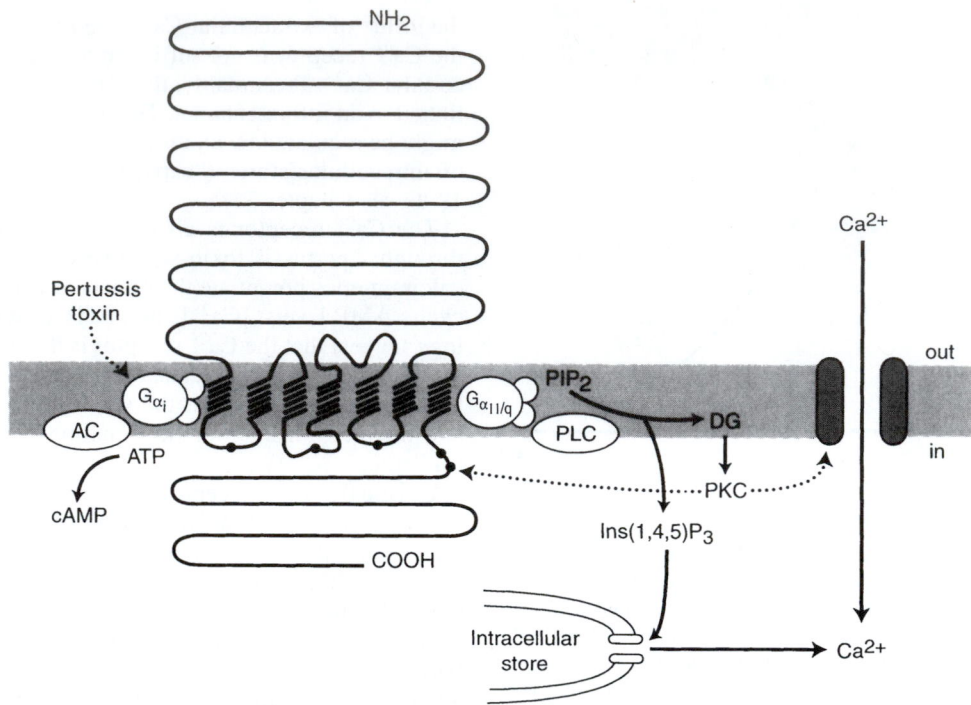

FIGURE 7. Schematic representation of some of the mechanisms known to be coupled to the Ca^{2+} receptor in parathyroid cells. The G proteins linking the Ca^{2+} receptor to adenylate cyclase (AC) and phospholipase C (PLC) are still uncertain, as discussed in the text. Activation of the Ca^{2+} receptor by extracellular Ca^{2+} results in the rapid formation of inositol 1,4,5-trisphosphate $[Ins(1,4,5,)P_3]$ and mobilization of intracellular Ca^{2+}. There is additionally an influx of extracellular Ca^{2+} through voltage-insensitive channels. Inhibitory pathways are shown by dotted lines. Pertussis toxin inhibits the G protein coupling the Ca^{2+} receptor to adenylate cyclase, but not that coupling the receptor to phospholipase C. Protein kinase C can phosphorylate the Ca^{2+} receptor at five potential sites (filled circles on cytoplasmic regions), and it also increases influx of extracellular Ca^{2+}.

transmembrane signaling mechanisms linked to the Ca^{2+} receptor (see Fig. 7; Racke and Nemeth, 1993a, b and 1994).

In addition to the rapid effects of extracellular Ca^{2+} on PTH secretion, there are other cellular responses apparently regulated by extracellular Ca^{2+} that become manifest over a day or two. Prolonged hypocalcemic conditions increase the synthesis of preproPTH, whereas hypercalcemic conditions decrease expression of the gene for PTH (Martin and Slatopolsky, 1994). Although it is reasonable to suppose that the Ca^{2+} receptor also mediates these long-term effects of extracellular Ca^{2+} on parathyroid cells, this additional function is not yet proven.

B. Parafollicular Cells

Scattered throughout the thyroid gland are parafollicular or C cells that secrete the hormone **calcitonin,** a 32-amino-acid peptide (Azria, 1989). The C cell, like the parathyroid cell, has long been known to respond to changes in the ambient level of Ca^{2+}, but unlike PTH secretion, calcitonin secretion is stimulated by elevated levels of extracellular Ca^{2+}. Calcitonin acts mostly on bone to inhibit **bone resorption** and therefore diminishes the flux of Ca^{2+} from bone into the general circulation. Calcitonin therefore lowers serum levels of Ca^{2+} and thus acts in opposition to PTH. In some species, calcitonin plays an important role in maintaining the serum level of Ca^{2+}. The physiological sig-

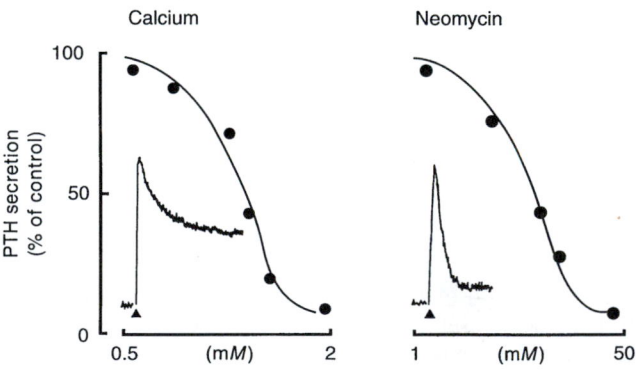

FIGURE 8. Different effects on Type I calcimimetics on $[Ca^{2+}]_i$, but not PTH secretion, in parathyroid cells. The traces show increases in $[Ca^{2+}]_i$ when the concentration of extracellular Ca^{2+} is increased from 0.5 to 2 mM (left panel) or when neomycin is added (50 mM; right panel). Extracellular Ca^{2+} causes a transient and sustained increase in $[Ca^{2+}]_i$, whereas neomycin causes only a transient increase. Nonetheless, both calcimimetics inhibit PTH secretion to the same degree.

nificance of calcitonin in systemic Ca^{2+} homeostasis in humans, however, is still a matter of debate. When administered in pharmacological doses, calcitonin can lower plasma Ca^{2+} levels, but physiological levels of this hormone do not figure prominently in the moment-to-moment regulation of systemic Ca^{2+} homeostasis in humans (McDermott and Kidd, 1987).

Most of our knowledge regarding the cellular physiology of C cells is based on studies using cells derived from rat or human **medullary thyroid carcinoma (MTC)** tumors. Because MTC cells might not express the normal C-cell phenotype, the accumulated data might not always accurately reflect the cellular physiology or normal parafollicular cells.

Increasing the concentration of extracellular Ca^{2+} evokes rapid increases in $[Ca^{2+}]_i$ in a number of different rat and human MTC cell lines (Raue and Sherübl, 1995). In contrast to PTH secretion, increases in $[Ca^{2+}]_i$ in C cells are associated with a stimulation of calcitonin secretion. The C cell, therefore, behaves like most other secretory cells, where cytoplasmic Ca^{2+} acts to stimulate exocytotic secretion. The C cell also differs from the parathyroid cell in the mechanisms used by extracellular Ca^{2+} to increase $[Ca^{2+}]_i$. C cells arise from the **neuroectoderm** and thus display a phenotype more characteristic of neurons. C cells express a number of *voltage-sensitive* ion channels, including **L-** and **T-type Ca^{2+} channels.** Increases in $[Ca^{2+}]_i$ and calcitonin secretion elicited by extracellular Ca^{2+} in MTC cells are blocked by inhibitors of voltage-sensitive Ca^{2+} channels such as **dihydropyridines.** Conversely, activators of voltage-sensitive Ca^{2+} channels increase $[Ca^{2+}]_i$ and stimulate calcitonin secretion. These compounds are without effect on $[Ca^{2+}]_i$ or PTH secretion in parathyroid cells (Muff *et al.*, 1988). It has been difficult to demonstrate mobilization of intracellular Ca^{2+} in C cells. It seems that influx of extracellular Ca^{2+} through voltage-sensitive, L-type Ca^{2+} channels accounts for most of the increase in $[Ca^{2+}]_i$ elicited by extracellular Ca^{2+}.

Unknown at present is how the Ca^{2+} receptor couples to these voltage-sensitive Ca^{2+} channels to permit influx of extracellular Ca^{2+}. It seems that extracellular Ca^{2+} acts initially on the Ca^{2+} receptor to allow influx of extracellular Ca^{2+}, because the human MTC cell line TT, which fails to respond to extracellular Ca^{2+}, does not express the Ca^{2+} receptor (Garrett *et al.*, 1995b). However, these TT cells also do not express certain subtypes of voltage-sensitive Ca^{2+} channels, and this confounds the interpretation. Perhaps the most compelling evidence so far derives from the use of selective Type II calcimimetics. Such compounds increase $[Ca^{2+}]_i$ and stimulate calcitonin secretion in various MTC cell lines, but not in TT cells, which lack the Ca^{2+} receptor.

The comparative pharmacology of the parathyroid and the parafollicular Ca^{2+} receptors is a textbook example showing the influence of cellular systems on receptor pharmacology. Despite identical receptors on both cell types there are a number of Type I calcimimetics (Mg^{2+}, for example) which fail to activate the Ca^{2+} receptor on MTC cells. Although these differential effects of Mg^{2+} might be peculiar to MTC cells, other studies performed *in*

vivo (therefore assessing Ca^{2+} receptor activity in authentic parafollicular cells) clearly show that the pharmacology of the Ca^{2+} receptor differs in these two cell types. Thus, Type II calcimimetic compounds are 10- to 40-fold more potent in suppressing plasma levels of PTH than they are in increasing those of calcitonin (Fox *et al.*, 1999). One possible explanation mentioned above, for these differences involves postreceptor mechanisms that are peculiar to the cell. Another possibility is that the density of receptors on parathyroid cells is much greater than that on parafollicular cells, and this possibly makes the parathyroid cells more sensitive to these compounds. In either case, the functional properties of the Ca^{2+} receptor are dependent on the cell in which it is expressed.

C. Renal Epithelial Cells

After the parathyroid gland, the Ca^{2+} receptor is most densely expressed in the kidney (Riccardi *et al.*, 1995). mRNA encoding the Ca^{2+} receptor can be detected along most of the nephron, but highest levels are expressed in the **thick ascending limb (TAL)** of the loop of Henle and in the **inner medullary collecting duct (IMCD).** In the TAL, where much of the Ca^{2+} (and Mg^{2+}) reabsorption from the nephron takes place, the Ca^{2+} receptor is located largely on the basolateral side of the tubule and is thus positioned to sense changes in the plasma concentration of Ca^{2+}. The Ca^{2+} receptor in the TAL couples to mechanisms that control the lumen-positive transepithelial voltage, which, in turn, drives Ca^{2+} (and Mg^{2+}) reabsorption by a paracellular pathway (Hebert, 1996). Thus, in hypercalcemic conditions, the TAL Ca^{2+} receptor on the basolateral surface of the epithelial cell is activated and Ca^{2+} reabsorption is diminished. Conversely, hypocalcemic conditions and consequent low Ca^{2+} activity will increase Ca^{2+} reabsorption in the TAL. In this segment of the nephron, the Ca^{2+} receptor acts to control the amount of Ca^{2+} that will be removed in the urine. One of the more convincing pieces of evidence supporting this role of the Ca^{2+} receptor in the TAL derives from molecular genetic studies. Recall that patients suffering from FBHH are hypocalciuric despite their hypercalcemia. These patients continue to reabsorb an inappropriate amount of Ca^{2+}. Ca^{2+} receptors in the TAL of patients with FBHH are less sensitive to extracellular Ca^{2+}. These mutant receptors do not perceive the hypercalcemia so Ca^{2+} reabsorption is not diminished as it normally would be.

In the IMCD, the Ca^{2+} receptor is localized mostly in vesicles within the epithelial cell and also on the luminal membrane. When luminal Ca^{2+} increases, there is a compensatory decrease in the amount of water reabsorption, and this effect is seemingly mediated by the Ca^{2+} receptor. The model proposed holds that activation of Ca^{2+} receptors on the luminal surface of collecting duct cells inhibits the exocytotic insertion of **aquaporin** (water) channels in the luminal membrane (Brown and Hebert, 1995). This action results in less reabsorption of water, thereby diluting the urine with respect to Ca^{2+}. This effect makes sense in the context of systemic Ca^{2+} homeostasis because if the plasma levels rise, then the kidney is called on to remove excess Ca^{2+} from the body. The increased levels of Ca^{2+} appearing in the urine must be diluted to avoid the formation of kidney stones.

D. Other Extracellular Ca^{2+}-Sensing Cells

The number of different cell types expressing the Ca^{2+} receptor continues to grow and one must consider the possibility that extracellular Ca^{2+} regulates a variety of bodily functions besides those involved in systemic Ca^{2+} homeostasis. Indeed, it now seems that many kinds of cells are sensitive to extracellular Ca^{2+}. Some caution in interpreting the physiological significance of these findings is warranted. Many of the studies claiming cellular sensitivity to extracellular Ca^{2+} used flawed experimental designs. Typically, the studies involved exposing the cell to a buffer which lacked divalent cations, followed by a rapid and large increase in the concentration of extracellular Ca^{2+}. Bathing cells in buffers lacking divalent cations tends to increase membrane permeability so that the subsequent addition of extracellular Ca^{2+} will increase $[Ca^{2+}]_i$ and this, in turn, will affect cellular responses. Another feature of many studies which limits their significance is the use of transformed cell lines, which are far easier to study than the authentic cell type from which they are derived. This convenience, however, is not without cost, and the phenotype of the transformed cell line often lacks one or more characteristics of the parental cell. Moreover, the cellular phenotype can change during culture. Yet even when more rigorous criteria are applied, a wide variety of cell types still emerges which express the Ca^{2+} receptor and seem to respond to extracellular Ca^{2+} under physiological and/or pathophysiological conditions (Table 1). It seems difficult to escape the conclusion that extracellular Ca^{2+} plays a general role as an extracellular signal affecting the activity of numerous cells throughout the body.

V. Summary

A biological phenomenon that is gaining recognition is the peculiar ability of extracellular Ca^{2+} to regulate the activity of certain cells in the body. Many of these cells, such as parathyroid cells, parafollicular cells in the thy-

TABLE 1   Extracellular Ca^{2+}-Sensing Cells

Cell Type	Function
Parathyroid	PTH secretion and synthesis; cellular proliferation
Parafollicular	Calcitonin secretion
Kidney	
TAL	Ca^{2+} and Mg^{2+} reabsorption
IMCD	Water transport
Pancreas	
Endocrine	
Exocrine	Unknown
Epithelial ducts	
G cells	Gastrin secretion
Osteoblasts	Proliferation

roid (C cells), and certain renal tubule cells, are involved in maintaining systemic Ca^{2+} homeostasis. These cell types, unlike most body cells, alter their activity in response to small, physiological changes in the concentration of Ca^{2+} in the plasma or the extracellular fluids. These "extracellular Ca^{2+}-sensing cells" express a Ca^{2+} receptor in the plasma membrane that enables them to detect and respond to small changes in the level of extracellular Ca^{2+}. The Ca^{2+} receptor is structurally and functionally similar to other plasma membrane receptors that sense changes in the concentration of extracellular ligands, and translate these ligand-receptor interactions into intracellular signals that alter cellular activity. The major difference is that the Ca^{2+} receptor binds to an inorganic ion rather than an organic molecule; the Ca^{2+} receptor is the first example of an "inorganic ion receptor."

Extracellular Ca^{2+} can thus act in a messenger capacity much like any other extracellular signal, such as a neurotransmitter or hormone. This newly discovered function of extracellular Ca^{2+} complements the well-known ability of cytoplasmic Ca^{2+} to act as an intracellular signal. Thus both extracellular Ca^{2+} and intracellular Ca^{2+} convey information and do so by interacting with Ca^{2+}-binding proteins. Intracellular Ca^{2+} receptors, such as calmodulin, bind Ca^{2+} with high affinity (nanomolar to low micromolar levels), whereas the extracellular Ca^{2+} receptor binds Ca^{2+} with much lower affinity (millimolar levels), consistent with the very different concentrations of Ca^{2+} in the cytoplasm and extracellular space.

The Ca^{2+} receptor is a G-protein-coupled receptor and is structurally homologous to metabotropic glutamate receptors. Both these receptor types possess an unusually large extracellular domain that binds the respective physiological ligand, extracellular Ca^{2+} or glutamate. Although there is no conclusive evidence at present for subtypes of the Ca^{2+} receptor, it seems likely that homologous yet distinct receptors will be discovered.

The Ca^{2+} receptor couples to a number of different transmembrane signaling mechanisms, depending on the particular cell type where it is expressed. In parathyroid cells, the Ca^{2+} receptor couples to phospholipase C and adenylate cyclase to increase $[Ca^{2+}]_i$ and decrease cAMP levels and secretion of PTH. The coupling to adenylate cyclase is mediated by a G_i-like protein, whereas that to phospholipase C is mediated either by G_{11} or G_q. In C cells, activation of the Ca^{2+} receptor allows the influx of extracellular Ca^{2+} through voltage-sensitive Ca^{2+} channels, leading to an increase in $[Ca^{2+}]_i$ and the secretion of calcitonin. In renal epithelial cells, entirely different intracellular signaling mechanisms are coupled to the Ca^{2+} receptor, and these intracellular signals regulate Ca^{2+} and Mg^{2+} reabsorption and water transport.

Both genetic and pharmacological studies show that the Ca^{2+} receptor is the essential molecular entity maintaining systemic Ca^{2+} homeostasis. Mutations in the Ca^{2+} receptor that decrease its sensitivity to extracellular Ca^{2+} result in hypercalcemic disorders, such as FBHH and NSHPT, and the latter is often fatal if not treated. Mutations in the Ca^{2+} receptor that increase its sensitivity to extracellular Ca^{2+} result in hypocalcemic disorders such as ADH. Moreover, calcimimetic compounds that activate the Ca^{2+} receptor cause decreases in circulating levels of PTH and Ca^{2+}. The last studies (Nemeth and Fox, 1999) also suggest that the Ca^{2+} receptor is a novel molecular target for drugs useful in treating a variety of bone and mineral disorders such as hyperparathyroidism and osteoporosis.

Most of what is known about how the Ca^{2+} receptor regulates cellular activity derives from the study of cells involved in systemic Ca^{2+} homeostasis. Yet it is clear that many other types of cells, which have little if any role in systemic Ca^{2+} homeostasis, also express the Ca^{2+} receptor. The function of the Ca^{2+} receptor in most of these cells is far from understood. The widespread distribution of the Ca^{2+} receptor, however, suggests that extracellular Ca^{2+} has an important messenger function throughout the body. These recent discoveries add a new dimension to physiology and show that extracellular Ca^{2+} (and perhaps other inorganic ions) must now be considered a variation on the theme of chemical communication in the body.

Bibliography

Azria, M. (1989). "The Calcitonins. Physiology and Pharmacology." Karger, New York.

Bai, M., Quinn, S., Trivedi, S., Kifor, O., Pearce, S. H. S., Pollak, M. R., Krapcho, K., Hebert, S. C., and Brown, E. M. (1996). Expression and characterization of inactivating and activating mutations in the human Ca_o^{2+}-sensing receptor. *J. Biol. Chem.* **271**, 19537–19545.

Bai, M. (1999). Structure and function of the extracellular calcium-sensing receptor (Review). *Int. J. Molec. Med.* **4**, 115–125.

Beck-Sickinger, A. G. (1996). Structural characterization and binding sites of G-protein-coupled receptors. *Drug Disc. Today* **1**, 502–513.

Bikle, D. D., Ratnam, A., Mauro, T., Harris, J., and Pillai, S. (1996). Changes in calcium responsiveness and handling during keratinocyte differentiation. Potential role of the calcium receptor. *J. Clin. Invest.* **97**, 1085–1093.

Brown, E. M. (1983). Four parameter model of the sigmoidal relationship between parathyroid hormone release and extracellular calcium concentration in normal and abnormal parathyroid tissue. *J. Clin. Endocrinol. Metab.* **56**, 572–581.

Brown, E. M. (1991). Extracellular Ca^{2+}-sensing, regulation of parathyroid cell function, and role of Ca^{2+} and other ions as extracellular (first) messengers. *Physiol. Rev.* **11**, 371–411.

Brown, E. M. (1994). Homeostatic mechanisms regulating extracellular and intracellular calcium metabolism. *In* "The Parathyroids" (J.P. Bilezikian, R. Marcus, and M.A. Levine, Eds.), Raven Press, New York, pp. 15–54.

Brown, E. M., and Hebert, S. C. (1995). A cloned Ca^{2+}-sensing receptor: A mediator of direct effects of extracellular Ca^{2+} on renal function? *J. Am. Soc. Nephrol.* **6**, 1530–1540.

Brown, E. M., Gamba, G., Riccardi, D., Lombardi, M., Butters, R., Kifor, O., Sun, A., Hediger, M. A., Lytton, J., and Hebert, S. C. (1993). Cloning and characterization of an extracellular Ca^{2+}-sensing receptor from bovine parathyroid. *Nature* **366**, 575–580.

Clark, E. A., and Brugge, J. S. (1995). Integrins and signal transduction pathways: The road taken. *Science* **268**, 233–239.

Drüeke, T. B. (1995). The pathogenesis of parathyroid gland hyperplasia in chronic renal failure. *Kidney Intl.* **48**, 259–272.

Fox, J., Lowe, S. H., Conklin, R. L., Petty, B. A., and Nemeth, E. F. (1999). Calcimimetic compound NPS R-568 stimulates calcitonin secretion but selectively targets parathyroid gland Ca^{2+} receptor in rats. *J. Pharm. and Exper. Therapeutics* **290**(2), 480–486.

Garrett, J. E., Capuano, I. V., Hammerland, L. G., Hung, B. C. P., Brown, E. M., Hebert, S. C., Nemeth, E. F., and Fuller, F. (1995a). Molecular cloning and functional expression of human parathyroid calcium receptor cDNAs. *J. Biol. Chem.* **270**, 12,919–12,925.

Garrett, J. E., Tamir, H., Kifor, O., Simin, R. T., Rogers, K. V., Mithal, A., Gage, R. F., and Brown, E. M. (1995b). Calcitonin-secreting cells of the thyroid express an extracellular calcium receptor gene. *Endocrinology* **136**, 5202–5211.

Hammerland, L. G., Krapcho, K. J., Garrett, J. E., Alasti, N., Hung, B. C. P., Simin, R. T., Levinthal, C., Nemeth, E. F., and Fuller, F. H. (1999). Domains determining ligand specificity for Ca^{2+} receptors. *Molec. Pharm.* **77**, 642–648.

Hebert, S. C. (1996). Extracellular calcium-sensing receptor: Implications for calcium and magnesium handling in the kidney. *Kidney Intl.* **50**, 2129–2139.

Hébert, T. E., and Bouvier, M. (1998). Structural and functional aspects of G protein-coupled receptors oligomerization. *Biochem. Cell Biol.* **76**, 1–11.

King, F. D., and Wilson, S. (1999). Recent advances in 7-transmembrane receptor research. *Curr. Opin. Drug Disc. Devel.* **2**(2), 83–95.

Kolakowski, L. F. (1994). GCRDb: A G-protein-coupled receptor database. *Recep. Channels* **2**, 1–7.

Lundgren, S., Hjälm, G., Hellman, P., Ek, B., Juhlin, C., Rastad, J., Klareskog, L., Åkerström, G., and Rask, L. (1994). A protein involved in calcium sensing of the human parathyroid and placental cytotrophoblast cells belongs to the LDL-receptor protein superfamily. *Exp. Cell Res.* **212**, 344–350.

Martin, K. J., and Slatopolsky, E. (1994). The parathyroids in renal disease. *In* "The Parathyroids" (J.P. Bilezikian, R. Marcus, and M.A. Levine, Eds.), Raven Press, New York, pp. 711–719.

Mayer, G. P., and Hurst, J. G. (1978). Sigmoidal relationship between parathyroid hormone secretion rate and plasma calcium concentration in calves. *Endocrinology* **102**, 1036–1042.

McDermott, M. T., and Kidd, G. S. (1987). The role of calcitonin in the development and treatment of osteoporosis. *Endocr. Rev.* **8**, 377–390.

Muff, R., Nemeth, E. F., Haller-brem, S., and Fischer, J. A. (1988). Regulation of hormone secretion and cytosolic Ca^{2+} by extracellular Ca^{2+} in parathyroid cells and C-cells: Role of voltage-sensitive Ca^{2+} channels. *Arch. Biochem. Biophys.* **265**, 128–135.

Mundy, G. R., (1989). "Calcium Homeostasis: Hypercalcemia and Hypocalcemia." Martin Dunitz, London.

Nemeth, E. F., and Scarpa, A. (1986). Cytosolic Ca^{2+} and the regulation of secretion in parathyroid cells. *FEBS Lett.* **203** (1), 15–19.

Nemeth, E. F., and Scarpa. A. (1987). Are changes in intracellular free calcium necessary for regulating secretion in parathyroid cells? *Ann. NY Acad. Sci.* **493**, 542–551.

Nemeth, E. F. (1990). Regulation of cytosolic calcium by extracellular divalent cations in C-cells and parathyroid cells. *Cell Calcium* **11**, 323–327.

Nemeth, E. F., and Heath, H., III. (1995). The calcium receptor and familial benign hypocalciuric hypercalcemia. *Curr. Opin. Endo. Diabetes* **2**, 556–561.

Nemeth, E. F. (1996). Calcium receptors as novel drug targets. *In* "Principles of Bone Biology" (J. P. Bilezikian, L. G. Raisz, and G. A. Rodan, Eds.), Academic Press, New York, pp. 1019–1035.

Nemeth, E. F., Steffey, M. E., Hammerland, L. G., Hung, B. C. P., Van Wagenen, B. C., DelMar, R. G., and Balandrin, M. F. (1998). Calcimimetic with potent and selective activity on the parathyroid calcium receptor. *Proc. Natl. Acad. Sci. USA* **95**, 4040–4045.

Nemeth, E. F., and Fox, J. (1999). Calcimimetic compounds: a direct approach to controlling plasma levels of parathyroid hormone in hyperparathyroidism. *TEM* **10**(2), 66–71.

Pearce, S. H., S., and Brown E. M., (1996). The genetic basis of endocrine disease. Disorders of calcium ion sensing. *J. Clin. Endo. Metab.* **81**, 2030–2035.

Petersen, C. C. H. (1996). Store operated calcium entry. *Semin. Neurosci.* **8**, 293–300.

Pollak, M. R., Brown, E. M., Chou. Y. H. W., Hebert, S. C., Marx, S. J., Steinmann, B., Levi, T., Seidman., C. E., and Seidman, J. G. (1993). Mutations in the human Ca^{2+}-sensing receptor gene cause familial hypocalciuric hypercalcemia and neonatal severe hyperparathyroidism. *Cell* **75**, 1297–1303.

Pollak, M. R., Brown, E. M., Estep, H. L., McLaine P. N., Kifor, O., Park J., Hebert, S. C., Seidman, C. E. and Seidman, J. G. (1994). Autosomal dominant hypercalcaemia caused by a Ca^{2+}-sensing receptor gene mutation. *Nat. Genet.* **8**, 303–307.

Racke, F. K., and Nemeth E. F. (1993a). Cytosolic calcium homeostasis in bovine parathyroid cells and its modulation by protein kinase. *C.J. Physiol.* **468.** 163–176.

Racke, F. K., and Nemeth, E. F. (1993b). Protein kinase C modulates hormone secretion regulated by extracellular polycations in bovine parathyroid cells. *J. Physiol.* **468,** 163–176.

Racke, F. K., and Nemeth E. F. (1994). Stimulus-secretion coupling in parathyroid cells deficient in protein kinase C activity. *Am. J. Physiol.* **267,** E429–E438.

Raue, F., and Scherubl, H. (1995). Extracellular calcium sensitivity and voltage-dependent calcium channels in C cells. *Endocr. Rev.* **16,** 752–764.

Ray, K., Clapp, P., Goldsmith, P. K., and Spiegel, A. M. (1998). Identification of the sites of N-linked glycosylation on the human calcium receptor and assessment of their role in cell surface expression and signal transduction. *J. Biol. Chem.* **273**(51), 34558–34567.

Riccardi, D., Park, J., Lee, W. -S., Gamba, G., Brown, E. M., and Hebert, S. C. (1995). Cloning and functional expression of a rat kidney extracellular calcium-sensing receptor. *Proc. Natl. Acad. Sci. USA* **92**, 131–135.

Takahashi, K., Tsuchida, K., Tanabe, Y., Masu, M., and Nakanishi, S. (1993). Role of the large extracellular domain of metabotropic glutamate receptors in agonist selectivity determination. *J. Biol. Chem.* **268**, 19341–19345.

Zaidi, M., Alam, A. S. M. T., Huang, C. L. -H., Pazianas. M., Bax, C. M. R., Bax. B. E., Moonga, B. S., Bevis, P. J. R., and Shankar, V. S. (1993). Extracellular Ca^{2+} sensing by the osteoclast. *Cell Calcium* **14**, 271–277.

Nelson D. Horseman and J. Wesley Pike

12

Cellular Responses to Hormones

I. Introduction

Hormones are secreted biochemical substances that affect the function of cells within an individual by binding to receptors and eliciting specific reactions from their target cells. Hormones affect cellular functions by altering rates of gene expression and metabolism, but are not themselves substrates for the processes that they regulate. Therefore, hormones are primarily information carriers. The first hormones were identified based on their secretion from specialized endocrine (internal secretion) glands. In recent years a large number of non-endocrine hormones have been discovered using advanced methods of cell culture, biochemical purification, and recombinant DNA technology. These *paracrine* (locally secreted) hormones have many novel features. Homeostasis and development are controlled through the actions of endocrine and paracrine hormones on target cells, which each express receptors for a limited array of hormones.

Because hormones do not directly participate in the processes they control, we conceive of them as initiating *signal transduction* protocols, that lead to the appropriate cellular outcomes. In this chapter we will cover examples of signal transduction mechanisms for two broad classes of hormones: the lipophilic hormones such as steroids, thyroxine, retinoids and vitamin D, which diffuse into cells and bind to intracellular receptors; and the hydrophilic peptide and amine hormones, which interact with receptors that are exposed on the surface of cells.

II. Actions of Lipophilic Hormones via Intracellular Receptors

A. Signal Transduction Through Intracellular Receptors: Basic Principles

The sex and adrenal steroids and the thyroid, retinoic acid, and vitamin D hormones are members of a chemically diverse set of endocrine signaling molecules that are produced and regulated in response to both internal and environmental cues. Acting on both distant and local tissue targets, they exert regulatory control over a myriad of specific cellular functions associated with virtually every vertebrate organ system. These actions impact, for example, cellular metabolism, reproductive function, nutrient homeostasis, and behavior. A common biological feature of each of these hormones is their capacity to regulate cellular differentiation. The mechanism through which lipophilic hormones dictate specific biological responses involves the ability of these hormones to enter the nucleus and modulate the expression of single genes and gene networks. This action is selective by virtue of the presence in cells of individual receptors for each of these hormones, and the presence of these receptors represents the primary determinant of tissue response to the cognate hormone. Nuclear receptors belong to a large gene family of transcription factors that recognize their respective endocrine signals and respond accordingly. In this section, we describe the progress over the past decade that has enhanced our understanding of how these signaling pathways modify gene expression.

In contrast to peptide hormones, growth factors, and cytokines, the lipophilic hormones are not limited in their ability to gain entry into the cell (Beato, 1989; Katzenellenbogen *et al.,* 1996). Due to their solubility in the membrane and their small size, they are believed to enter the cell by diffusion through the cell membrane; this energy-independent process apparently permits them to ultimately reach the cell nucleus where they associate though a high-affinity interaction with a specific nuclear receptor (Fig. 1). While the bulk of the members of the nuclear receptor family are believed to reside in the nucleus prior to ligand activation, the receptors for the glucocorticoids (GR) and the mineralocorticoids (MR) appear to be the single exceptions in that they are found in the cytoplasm (Evans, 1988; Beato *et al.,* 1995; O'Malley, 1990). In this specific case, interaction with glucocorticoid or mineralocorticoid ligand induces translocation of the receptors to the nucleus. The interaction of hormonal ligand with its receptor triggers a series of events which culminate in alterations in gene expression. Allosteric changes in the receptor induced by ligand lean to

Steroid, thyroid, retinoid, and 1,25(OH)$_2$ D$_3$ hormones

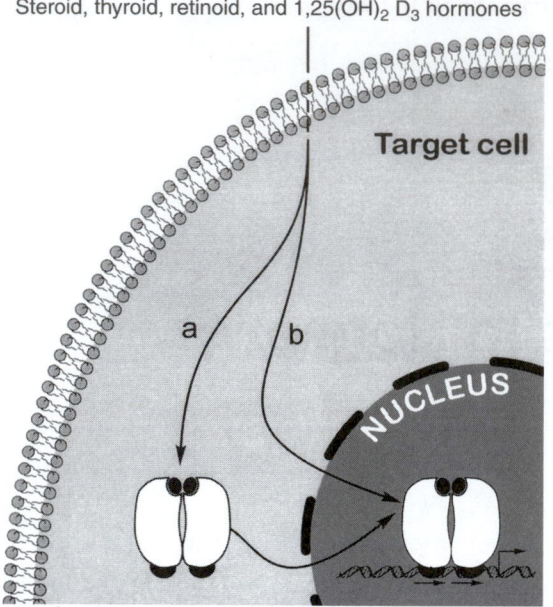

FIGURE 1. Model for the molecular mechanism of action of steroid, thyroid, retinoid, and vitamin D hormones. Hormonal ligand enters the cell by diffusion and interacts with its cognate receptor. Activation by the ligand leads to the interaction of receptor with responsive genes and the modulation of gene expression. Active receptors are comprised of monomers, homodimers, or heterodimers (see text). (a) Model for glucocorticoid and mineralocorticoid receptors, wherein the receptor is cytoplasmic in the absence of ligand. Upon ligand activation the receptor undergoes nuclear translocation and eventually binds to the regulatory region of a modulated gene. (b) Model for most of the nuclear receptors wherein the receptor is located in the nucleus and following ligand activation becomes bound to the regulatory region of hormone-responsive genes.

dissociation or associated proteins which function as inhibitors of receptor DNA-binding ability and transcriptional capacity, formation of functional protein units, binding of the units to specific DNA sequence elements located adjacent to hormone sensitive gene promoters, and activation or repression of transcription (Fig. 1). The capacity of a hormone to regulate transcription is dependent upon the presence of the receptor for that hormone in a cell, but the quantitative and qualitative nature of the response can be both gene-specific and cell-specific depending upon the presence and activity of additional cellular protein factors. Cellular mechanisms which serve to inactivate receptor signals within the nucleus are largely unknown, although the induction of enzymes by the hormone-activated receptor, that degrade the corresponding hormonal signal, is not uncommon.

B. The Nuclear Receptor Gene Family

1. Ligand-Activated Nuclear Receptors

The genes for all of the receptors which mediate the actions of traditional nonpeptide endocrine hormones have been cloned (Fig. 2). These include the receptor for the glucocorticoids (GR), mineralocorticoid receptor (MR), estrogen receptors (ERα, β), progesterone receptor (PR), androgen receptor (AR), thyroid hormone receptors (TRα, β), retinoic acid receptors (RARα, β, γ), 9-*cis*-retinoic acid receptors (RXRα, β, γ) and the vitamin D receptor (VDR) (Gronemeyer, 1993; Truss and Beato, 1993; Mangelsdorf and Evans, 1995; Mangelsdorf *et al.*, 1995; Beato *et al.*, 1995). Single genes encode certain of the receptors whereas other receptors are derived from multiple genes. In the latter case, these receptors may exhibit differences in tissue distribution and in gene regulatory activity as a result of slight but significant differences in their primary structure. During transcription, receptor genes can also undergo alternative splicing events that often lead to multiple mRNA transcripts that encode receptor gene products that exhibit different functional activities. The existence of multiple genes for receptors, alternative splicing events within single genes, and differential use of translation start sites within the mRNA transcripts result in a heterogeneous mix of receptor gene products, as well as the possibility for highly diverse responses to ligand within cells, depending upon the particular receptor gene product being expressed.

2. Orphan Receptors

The cloning of the receptors for known endocrine hormones and the observation that all belonged to a structurally related family of genes precipitated a search for additional gene members in vertebrate tissues and in nonvertebrate species. This search has resulted in the identification of numerous additional genes which are clearly members of the nuclear receptor gene family (Fig. 2) (O'Malley, 1990; Mangelsdorf and Evans, 1995). While none of these orphan receptors appear currently to mediate the actions of known endocrine hormones, several appear to facilitate the actions of local cellular factors such as 9-*cis*-retinoic acid and prostaglandin J2 (Forman *et al.*, 1995) as well as cellular metabolic intermediates such as farnesol, arachidonic acid, and perhaps certain cholesterol derivatives. Current efforts are focused on determining whether novel ligands indeed exist for these receptors, although activation pathways independent of ligand are clearly possible. Moreover, in addition to functioning as gene activators, there is evidence that certain of the orphans may play largely negative roles in transcription, acting not only to repress the transcriptional activity of specific genes, but also to function on other genes as repressors of the inducing actions of ligand-activated nuclear receptors. Novel biological roles for several of these orphan receptors have been defined.

3. Invertebrate Receptor Genes

A large number of genes from invertebrates have been identified as belonging to the steroid receptor gene family (Fig. 2) (Mangelsdorf *et al.*, 1995; Thummel, 1995). The genes include those from *Drosophila*, *Caenorhabditis elegans*, and other metazoan species. Perhaps the most interesting member of the nuclear receptor family found in *Drosophila melanogaster* is the receptor for ecdysone, the metamorphic or molting hormone of insects. This hormone has long been associated with its capacity to induce chromosomal puffing at specific sites on chromatin and to induce complex networks of genes. Interestingly, as will be described below, the

Genes	Species*	Ligand
GR	1	GLUCOCORTICOIDS
MR	1	MINERALOCORTICOIDS
PR	1	PROGESTERONE
AR	1	TESTOSTERONE
EcR	2	ECDYSONE
FXR	1	FARNESOIDS
LXR α,β	1	
VDR	1	$1,25(OH)_2D_3$
xONR	1	
MB67 α,β	1	
CeF11A1.3	2	
TR α,β	1	THYROID HORMONE
CeB0280.8	2	
CeC2B4.2	2	
COUP α,β,γ	1,2	
RXR $\alpha,\beta,\gamma,(\delta,\epsilon)$	1,2	9-cis-RA
TR2-11 α,β	1,2	
HNF-4	1,2	
TLL	1,2	
CeF21D12.4	2	
GCNF	1	
PPAR $\alpha,\beta,\gamma,\delta$?	1	EICOSANOIDS, PROSTAGLANDINS
RAR α,β,γ	1	RETINOIC ACID
CNR14	2	
E78	2	
Rev-Erb α,β	1	
E75A	2	
ROR α,β,γ	1	
DHR3	2	
NGFI-B α,β,γ	1,2	
FTZ-F1 α,β	1,2	
CeF11C1.6	2	
ERR α,β	1	
ER α,β	1	ESTROGEN
KNIRPS α,β,γ	2	
CeE2H1.6	2	
CEF43C1.4	2	
CeKO6A1.4	2	
CeODR7	2	
CeZK418.1	2	
CeF16H9.1	2	

FIGURE 2.   Members of the nuclear receptor superfamily of genes. The figure documents cloned nuclear receptor family members. Known ligands are indicated on the right.

ecdysone receptor functions as a heterodimer with a nuclear protein partner called *ultraspiracle*. *Ultraspiracle* represents the insect homolog of vertebrate retinoid X receptor (RXR) genes, which in turn function as permissive heterodimer partners with the thyroid receptor (TR), the retinoic acid receptor (RAR), the vitamin D receptor (VDR), and several others. The presence of nuclear receptors in insects suggests that this highly successful receptor family evolved prior to the divergence of invertebrates and vertebrates.

C. The Highly Conserved Structural Domains of the Nuclear Receptors

The nuclear receptors display a highly modular domain structure composed of a DNA-binding domain of 60 to 70 amino acids and an extended carboxy-terminal ligand-binding domain of 300 amino acids separated by a highly flexible hinge of 100 to 150 amino acids, as seen in Fig. 3 (Evans, 1988). Many of the individual functions of each domain are retained following their separation through enzymatic cleavage. The DNA-binding domain is highly conserved among members of the nuclear receptor superfamily and represents the hallmark of this transcription factor family. In addition to containing determinants of specific DNA binding, this domain also contains subregions which are involved in directing the protein to the nucleus following synthesis and stabilizing the protein's interaction with DNA through dimerization with a protein partner. The carboxy-terminal domain of the nuclear receptors, which is over 300 amino acids in length, exhibits subregions of homology across members of the family and contains hydrophobic residues that create a high affinity lipophilic pocket exquisitely selective for the receptor's cognate

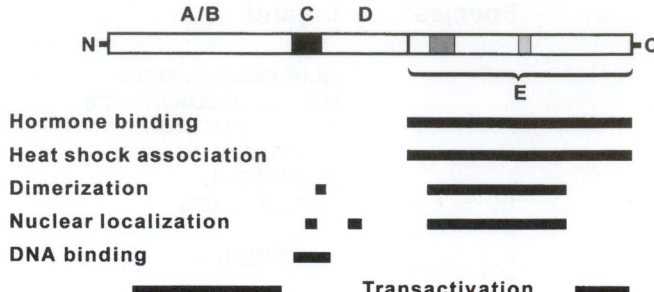

FIGURE 3.   Functional domain structure of the nuclear receptor superfamily. The nuclear receptors (NR) are separated into four regions designated A/B, C, D, and E. Regions of homology within the nuclear receptor family are shaded. Regions involved in known specific functions of the receptor are indicated below the diagram.

ligand. Like that of the DNA-binding domain, this region also exhibits additional functional activities, including a dimer interface for interaction with the receptor's protein partner, and additional protein/protein interfaces that are integrally involved in the receptor's ability to regulate transcription. Finally, receptors for estrogen, progesterone, androgens, and the glucocorticoids contain an extensive amino-terminal domain that exhibits additional functions associated with transcriptional activation. This complexity in structural organization of the nuclear receptor family implies diverse, highly flexible, and specific activities during the regulation of gene expression.

1. The DNA-Binding Domain of the Nuclear Receptors

The DNA-binding domain of the nuclear receptor family is comprised of two highly conserved zinc finger structures. These structures contain several invariant amino acids that include two sets of four cysteine residues that have been shown to coordinate two zinc ions (Berg, 1989). The two zinc ions hold the DNA-binding domain in an active conformation structurally; in their absence the finger motifs are nonfunctional and sensitive to proteolytic degradation. The three-dimensional structures of the DNA-binding domains of the receptors for the glucocorticoids, estrogen and 9-*cis*-retinoic acid have been established (Hard *et al.*, 1990; Schwabe *et al.*, 1990; Luisi *et al.*, 1991; Lee *et al.*, 1993; Schwabe *et al.*, 1993) In addition, the three-dimensional structure of the TR and RXR DNA-binding domains bound to DNA as heterodimers has also been determined. The observed structures of these regions of the nuclear receptors confirm and extend earlier predictions which were made based upon biochemical evaluations. The two zinc finger modules are known to serve very different functional roles during association with DNA (Mader *et al.*, 1989). The first functions to dictate the specificity of DNA interaction in the major groove on a core DNA sequence element, whereas the second functions to establish contacts with the receptor's protein partner. These contacts stabilize and strengthen the binding of the two proteins on the DNA sequence. Different regions located within the second zinc module, however, participate in this dimerization function depending upon the organization of the DNA-binding site and the orientation of the two receptor molecules which become bound to that site.

2. The Ligand-Binding Domain of the Nuclear Receptors

In contrast to the DNA-binding domain, the region of the receptor that is required for ligand binding extends to over several hundred amino acids. Since the ligand itself is rather small relative to this region, it is clear that the protein pocket which makes direct contact with the ligand is determined through proper folding of different components of the protein domain itself. Indeed, several regions of homology have been defined within this domain that exist among the nuclear receptor family. These regions of homology, however, participate not only in creating a ligand-binding pocket, but also serve several additional functional roles that involve protein/protein interactions; two such interactions are those that lead to homodimer or heterodimer formation with the receptor's protein partner, and those which mediate contact with the general transcriptional apparatus. Additional interactions include receptor associations with either inhibitor or repressor proteins that function to prevent receptor DNA-binding and transcriptional activity in the absence of ligand, or that facilitate transcriptional repression while bound to DNA in the presence of the activating ligand (Horlein *et al.*, 1995). Recent studies have led to the elucidation of the crystal structure of ligand binding domains of TR, RAR, and RXR (Bourguet *et al.*, 1995; Renaud *et al.*, 1995; Wagner *et al.*, 1995; Rastinejad *et al.*, 1995). Crystallization of the first two was achieved in the presence of the corresponding ligands whereas the structure of the RXR ligand-binding domain was determined in the absence of its cognate ligand. These studies have revealed the three-dimensional structure of the nuclear receptors to be that of an antiparallel α-helical sandwich comprised of as many as twelve helices. Importantly, the binding of hormone leads to a significant repositioning of several of these α-helices, including the most carboxy-terminal one that is known to be involved in mediating contacts with the general transcription apparatus. Helices 9 and 10 participate in the formation of receptor homo- or heterodimers. The ligand-binding pocket is comprised of extended clusters of residues that include helices 1, 3, 5, a β turn between 5 and 6, a loop between 6 and 7, 11 and 12, and the loop between 11 and 12. While the crystal structure of a complete receptor has yet to be determined, these studies with both the DNA-binding domain and the ligand-binding domain have provided significant insight into structure-function relationships which exist for this class of transcription factors.

3. Additional Nuclear Receptor Domains

Two additional domains are found within the nuclear receptor gene family. The first is a region common to all the receptors that serves to link the DNA-binding domain with the ligand-binding domain. It is not conserved within the receptor gene family. One characteristic of this region is flexibility; depending upon the orientation of the DNA sequence elements to which the individual receptors bind, the hinge region allows 180° swiveling of the two domains which surround it. It is likely that this region also subserves other as yet unknown functions. The second domain is a large amino-terminal region found predominantly in receptors for estrogen, progesterone, androgens, and the glucocorticoids. This domain contains additional regions that function to make direct or indirect protein-protein contacts with the general transcriptional apparatus. The three-dimensional structure of this region has not been determined, either independently or in association with the DNA-binding domain or other portions of the receptor molecule.

D. DNA-Binding Motifs within Nuclear Receptors

Nuclear receptors interact with unique sequences of DNA located within the promoters for hormone-regulated genes. These associations represent the first in a series of steps which result in transcriptional modulation. Research over the past decade on both naturally regulated genes as well as synthetically derived DNA sequences has revealed the nature of these elements which serve as selective binding sites for members of the nuclear receptor family. As seen in Fig. 4, three classes of DNA-binding sites have emerged which reflect the three modes of DNA-binding characteristic of members of this gene family—homodimeric DNA binding, heterodimeric binding, and monomeric binding (Umesono and Evans, 1989; Umesono et al., 1991; Perlmann et al., 1993). The core DNA-binding site within promoters is comprised of a hexanucleotide sequence. In the case of nuclear receptors which bind to DNA as monomers, this core sequence together with a relatively well-defined short stretch

of nucleotides located immediately 5', comprise the DNA-binding site. The sequence of the core as well as the adjacent nucleotides represent determinants of specificity. Numerous orphan receptors interact with DNA as monomers. The nuclear receptors which bind to DNA as dimers interact with core elements similar to those just described. In contrast, however, these core elements are repeated to provide the two similar and adjacent binding sites necessary to accommodate the association of a dimeric nuclear receptor. Two dissimilar motifs arise as a result of this repeating structure: one wherein the core sequence is repeated in a direct fashion, and the second in which the core sequence is inversely repeated in palindromic fashion. Those receptors which function as homodimers such as the GR, ER, PR, and AR, bind to palindromically repeated core sequences. In contrast, receptors which function as heterodimers in combination with a dissimilar nuclear receptor protein partner, such as TR, RAR, and VDR, bind to directly repeated motifs. The importance of the highly flexible hinge region is highlighted here. Clearly, dimeric protein binding to palindromic DNA sequences requires a symmetric interface between the two protein partners in a head-to-head configuration, whereas dimeric binding to directly repeated DNA sequences requires an asymmetric interface wherein the proteins are arranged head to tail. Research has revealed that the DNA-binding domain of the nuclear receptors configure in a head-to-head or head-to-tail arrangement depending upon the structure of the DNA binding site. The carboxy-terminal ligand-binding domains, in contrast, retain their symmetric head-to-head configurations irrespective of whether the DNA-binding site is palindromic on direct in orientation. These two possibilities are accomplished through the hinge region of the nuclear receptors, which allows for an independent swiveling of the DNA-binding domain relative to the ligand-binding domain. Finally, the individual sequences of the core elements determine the ability and therefore specificity of the interaction by homodimeric receptors such as ER, PR, and GR. In contrast, directly repeated core sequences are generally similar; specificity is determined not through a difference in the sequence, but rather through the number of nucleotides located between the two core half-sites. Thus, binding sites for the VDR, TR, and RAR are determined largely by nucleotide spacings of three, four, and five, respectively; additional arrangements have also been determined.

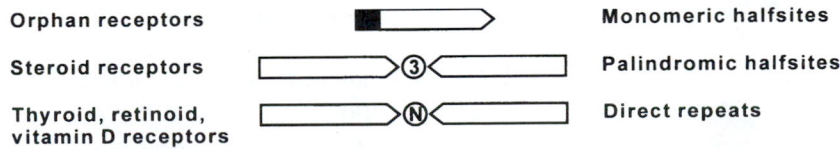

FIGURE 4. Organization of DNA sequence elements located within gene promoters that mediate the actions of the nuclear receptor family. A core element or monomeric binding site is indicated with an arrow. Receptors that bind to this core element as monomers contact additional DNA sequence located 5' to the core element. Receptors that bind as homodimers interact with two core elements arranged in a palindromic array. These core elements are separated by three base pairs and receptor specificity is determined by core element sequence. Receptors which bind as heterodimers interact with two core elements arranged in a directly repeated fashion. Receptor binding specificity is determined to a large extent in this motif through the number of nucleotides separating the two core elements.

E. Modes of Nuclear Receptor DNA Binding

As described above, nuclear receptor family members interact with DNA in three different ways: as monomers, homodimers, or heterodimers (Fig. 5) (Glass, 1994). The class of receptors that function as heterodimers, however, includes the TRs, RARs, VDR and several others that mediate the actions of intracellularly-produced ligands. Each of these receptors bind to specific hormone response elements in combination with the common nuclear receptor partner RXR. RXR appears to be a generally permissive partner; activation of the heterodimer is achieved through the ligand which binds the signaling partner (Glass, 1994; Kleiwer *et al.*, 1992; Yu *et al.*, 1991). The evolution of the heterodimer mechanism wherein a common protein plays a functional role in the activities of several apparently unrelated biochemical pathways is a repeated theme in cellular biology. The implication of such a mechanism is that competition for the permissive RXR protein pool must arise within cells that express more than one of the RXR-utilizing receptors. This may have profound biological consequences in that apparently unrelated endocrine signaling systems appear to converge and compete during the course of their actions to regulate gene expression.

F. The Role of the Ligand in Nuclear Receptor Activation

Lipophilic hormonal ligands initiate a series of events that culminate in the modulation of gene expression. These events are largely the result of conformational changes induced by the ligand upon association with its specific nuclear receptor. Considerable evidence exists for this conformation change upon ligand binding. Early studies demonstrated that

the sensitivity to endogenous proteases was reduced following association with ligand. The use of limited proteolytic digestion *in vitro* to define the effect of ligand on receptor continues to support the idea that ligand induces selective and specific conformational changes. Although a study of the effect of ligand on the three-dimensional structure of a single receptor has not yet been made, a comparison of the three-dimensional structures of unliganded RXR with liganded RAR and TR reveal a major repositioning of helical structures within the receptors in the liganded state.

1. Dissociation of Inhibitory Proteins

Receptors for the sex and adrenal steroids associate in the absence of ligand with a complement of inhibitory proteins that include hsp90 (Pratt *et al.*, 1988). These proteins are believed to dissociate as a result of ligand-induced changes in conformation that occur within the steroid receptors. This dissociation releases free receptor monomers which in turn associate with other monomers to form active homodimers which bind as high affinity complexes on DNA (Fig. 5). While receptors for T3, retinoic acid, 1,25-dihydroxyvitamin D, and other nonsteroidal ligands do not appear to interact with these inhibitory complexes, it is likely that inhibitor proteins perhaps unrelated to the heat-shock macromolecules exist. If so, a similar ligand-mediated mechanism might release these receptors as well. Since TR, RAR, and VDR form heterodimers with unliganded RXRs, however, an additional ligand-independent mechanism must be proposed when makes available RXR monomers for heterodimer formation. The dissociation of specific inhibitor proteins from TR and the RARs identified as nuclear receptor corepressor (NcoR) (Horlein *et al.*, 1995) have been demonstrated in response to cognate ligands. These inhibitor proteins, however, differ from the typical inhibitor proteins associated with the sex steroid receptors in that they serve a transcriptional repressive function. Considerable evidence exists for additional proteins that function to repress transcription. Disruption of these protein/protein interactions provides additional evidence for ligand-induced conformational changes.

2. Formation of Nuclear Receptor Dimers

Ligand-induced conformational changes also promote the formation of nuclear receptor dimers. Thus, binding of ligand to TR, VDR, and perhaps others leads to an increase in the affinity of the ligand-activated receptor for RXR (Fig. 5). These observations suggest an additional role for the hormone in facilitating the creation of functional receptor units.

3. Receptor Contact with the General Transcription Apparatus

Modulation of gene expression requires the interaction of a nuclear receptor with the general transcriptional apparatus. A final and perhaps anticipated role for the ligand appears to be its capacity to regulate this process. The mechanism whereby the nuclear receptors modulate the activity of the general transcriptional apparatus again involves additional modulatory proteins termed coactivators or corepressors

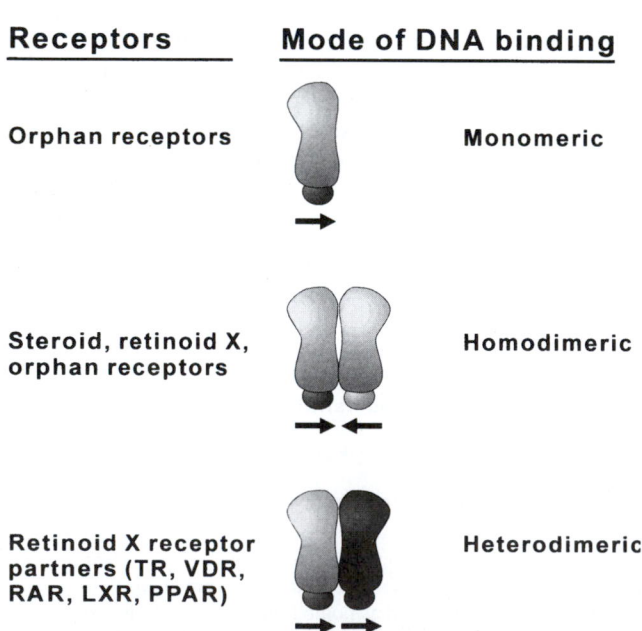

Receptors   Mode of DNA binding

Orphan receptors   **Monomeric**

Steroid, retinoid X, orphan receptors   **Homodimeric**

Retinoid X receptor partners (TR, VDR, RAR, LXR, PPAR)   **Heterodimeric**

FIGURE 5. Moles of DNA binding by nuclear receptors. Nuclear receptors interact with DNA sequence motifs as monomers, homodimers, and heterodimers.

(Danielian *et al.*, 1992). Thus, the conformational changes induced by the ligand lead not only to dissociation of inhibitors or repressors and to the formation of active receptor dimers, but also facilitate the association of receptors with comodulators that are directly able to contact the initiation complex of certain genes (Fig. 6). A variety of comodulators exist that mediate this contact. Most importantly, since these proteins are not expressed ubiquitously, but rather in tissue-specific patterns, they provide a means whereby cellular response can be graded or altered following induction of the general signaling pathway by ligand.

III. Cellular Actions of Protein and Amine Hormones via Plasma Membrane Receptors

A. Signal Transduction from the Plasma Membrane: Basic Principles

Because of the selective permeability of the plasma membrane, most hormones cannot traverse the membrane independently; therefore, they bind to integral membrane proteins that serve as sensors of the extracellular hormone concentration. Hormones that bind to membrane receptors include molecules as diverse as large multisubunit proteins, small peptides, and amino acid derivatives. As a general principle, binding of a hormone to its receptor causes the synthesis of one or more diffusible intracellular mediators, which have been termed *second messengers* or *intracellular signal transducers*. The synthesis of these signal transducers is initiated by hormone-induced changes in the conformation of the receptor, or its oligomerization with other membrane-associated proteins. Microbes and single-celled eukaryotes use mechanisms similar to hormonal signal transduction to integrate their responses to nutrients and other environmental cues; therefore, we can infer that hormone response systems evolved from nutrient transport and sensor mechanisms that originated early during evolution. Many of the features of vertebrate signal transduction are conserved in single-celled organisms such as yeast (Herskowitz, 1995), making these simple organisms valuable research tools in contemporary molecular endocrinology.

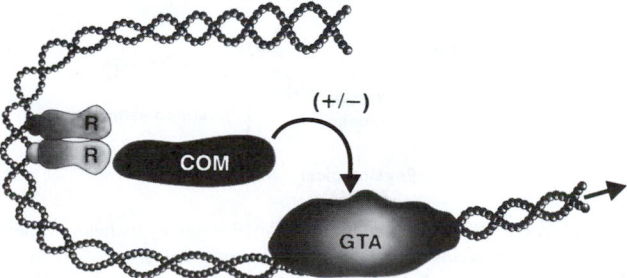

FIGURE 6. Nuclear receptor contact with the general transcriptional apparatus (GTA). Nuclear receptor dimers (or monomers) contact the GTA via comodulators. These transcription factors enable the nuclear receptor to stimulate (coactivator) or repress (corepressor) the basal or existing transcription of the gene.

The intracellular effectors that are controlled by hormonal signal transduction include proteins that reside in a variety of cell compartments. Some, such as the heterotrimeric GTP-binding proteins (G proteins) and various ion channels, are plasma membrane or organellar membrane proteins; others, including numerous protein kinases, are strictly cytosolic proteins; while still others, such as CREB (cAMP-response element binding protein) and Stat proteins (Signal transducer and activator of transcription) are either resident in the nucleus or are translocated to the nucleus after their activation. The variety of intracellular processes that are controlled by intracellular signal transduction pathways reflects the capacity of hormones to bring about integrated, adaptive cellular responses that promote either homeostasis or directed developmental changes.

Protein phosphorylation plays a central role in hormonal signal transduction. In the past two decades, it has become clear that protein phosphorylation is a ubiquitous regulatory process that affects all levels of cell signaling. The selectivity of signal transduction mechanisms is partly accounted for by the substrate specificities of the various protein kinases in signaling pathways. Each signal-transducing kinase is represented by multiple isoforms that can catalyze phosphorylation of different substrates.

Signal transduction molecules are highly compartmentalized within the cell. Transduction of signals from cell surface-bound receptors to diverse intracellular compartments results in the movement of information from the sensor to ultimate effectors. The means by which this information flow occurs are very poorly understood. What is clear is that most signal transduction pathways are initiated in multiprotein complexes (receptor signaling complexes), which include not only the receptor, but also a variety of coupling proteins, kinases, and substrates. The addressing of proteins to the receptor signaling complex, and then away from it to effector compartments, provides an important mechanism for conferring signaling selectivity. This addressing phenomenon is accounted for by various conserved protein/protein interaction domains and post-translational modifications. Over the past ten years the functions of some of these addressing sequences have been elucidated. These include the SH2 (Src Homology 2) and SH3 domains, which bind to phosphorylated tyrosines and proline-rich motifs, respectively, and carboxy-terminal prenylation signals, which direct the addition of lipid moieties that anchor signaling proteins in the plasma membrane (Pawson, 1995a, b; Zhang and Casey, 1996).

Transduction from the membrane to intracellular compartments occurs by two means. First, small, readily diffusible molecules that are referred to as second messengers carry information from one compartment to others. Examples of these second messengers include cAMP, cGMP, inositol 1, 4, 5-trisphosphate (IP$_3$), and calcium. After diffusing to their effector compartment, these second messengers bind to proteins (kinases, channels, etc.) and cause conformational alterations that increase or decrease their activation state. Secondly, a conformational change induced by protein phosphorylation can lead directly to *readdressing* so that a protein is forwarded to the compartment in which it exerts its effect, such as the nucleus.

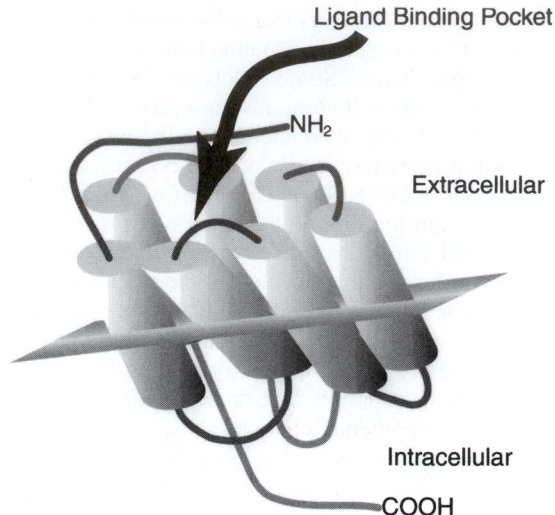

FIGURE 7. Diagrammatic structure of heptahelical receptors that activate G-protein-coupled signal transduction. The seven α-helices are designated by cylinders, and connecting loops by lines. The amino-terminus (NH_2) is in the extracellular compartment, and the carboxyl-terminus is in the intracellular compartment. Ligands interact with a pocket created by the articulation of the seven helices (either directly or indirectly, see text). Conformational changes transmitted by the displacement of the α-helices lead to activation of intracellular signaling.

These general principles of signal transduction are played out in a rich variety of specific mechanisms. To understand some of these signaling pathways we will focus primarily on those mechanisms that lead to altered control of gene expression. Other chapters in this volume reflect their focus on membrane permeability and transport mechanisms, and cellular metabolism. The reader should refer to them for more information on those topics.

B. Signaling via Heptahelical Membrane Receptors

1. Signaling by Activation of Trimeric GTP-binding Proteins

A large family of receptors share a structure that contains seven membrane-spanning α-helices (heptahelical receptors, Fig. 7), and all of these receptors interact with and activate GTP-binding proteins that consist of three subunits (α, β, γ). Extracellular ligands activate the heptahelical receptors by causing conformational changes to the α-helices and the intracellular carboxyl terminus of the receptors. The transmembrane α-helices of the receptors form a binding pocket that can be activated by multiple mechanisms. Small ligands can enter the pocket and bind directly, and large ligands and nonpeptides can activate the binding pocket by interacting with the amino-terminal portion of the molecule and the extracellular loops. The fact that ligands can activate these receptors through multiple binding mechanisms appears to account for the enormous variety of heptahelical receptors expressed from mammalian genomes (at least 1000 receptor genes in the human). Hormones that act through heptahelical receptors range from single amino acid derivatives such as epinephrine

to multisubunit pituitary hormones, such as Luteinizing Hormone and Follicle Stimulating Hormone (Fig. 8).

"Heterotrimeric G proteins" couple receptors to processes that result in the generation of second messengers such as cAMP (Fig. 8). Receptor signaling complexes for heptahelical receptors include the receptor-ligand pair, a trimeric G protein, and at least one effector, such as adenylyl cyclase. G proteins are anchored to the plasma membrane by virtue of a C-terminal prenylation signal in the γ subunit. The α subunit binds GTP and hydrolyzes it to GDP. The activation-inactivation cycle of the G proteins begins with the inactive trimeric protein in which GDP is bound to the α subunit and the protein is associated with the membrane through the γ subunit. The G protein complex is capable of binding with receptor proteins that are also in the membrane. Binding of ligand to its receptor induces a conformational shift and the G protein that is associated with the receptor displaces GDP and binds GTP. The binding of GTP causes the α subunit to be released from the complex, allowing it to interact with an effector molecule, such as adenylyl cyclase or phospholipase C (Spiegel *et al.*, 1995). Binding of the α subunit to these effector enzymes leads to synthesis of their respective second messengers, cAMP and IP_3. Inactivation of the G protein occurs when GTP is hydrolyzed to GDP. GDP-bound α subunit undergoes a conformational change that increases its affinity for the $\beta\gamma$ complex, completing the activation-inactivation cycle. The GTPase activity intrinsic to the α subunit has a slow rate constant, so α stays activated and dissociated from $\beta\gamma$ for an extended period of time, leading to amplification of the hormone signal by virtue of repeated rounds of second-messenger synthesis.

Signal diversity is generated by multiple G protein α subunit isoforms, which each couple to particular effector proteins (Spiegel *et al.*, 1995). Adenylyl cyclase is regulated by stimulatory ($G_{\alpha s}$) and inhibitory ($G_{\alpha i}$) subunits. $G_{\alpha q}$ activates phospholipase Cβ, leading to generation of IP_3. The diversity of α subunit isoforms has been realized only in re-

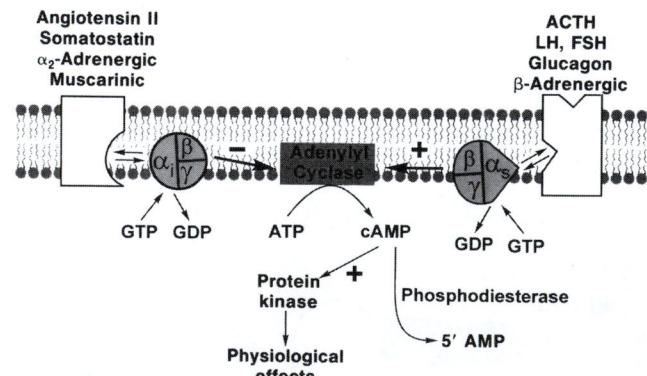

FIGURE 8. Bivalent regulation of cAMP synthesis by heterotrimeric GTP-binding proteins. Representative ligands are listed, which act through receptors coupled to either stimulatory (as) or inhibitory (at) G-protein subunits. Activation of G-protein function mediated by ligand-receptor interaction causes exchange of GTP for GDP, and dissociation of the a subunit from its bg partners. The a subunits interact with adenylyl cyclase to either accelerate, or inhibit cAMP synthesis. cAMP is degraded by cAMP phosphodiesterase to inactive AMP.

TABLE 1 The Functional Diversity of α Subunits among Mammalian Heterotrimeric GTP-Binding Process Proteins

α Subunit	Localization	Effectors
G_s	Ubiquitous	Adenylyl cyclasel Ca^{2+} channel
G_{olf}	Olfactory	Adenylyl cyclase
G_{t1}	Rod photoreceptors	cGMP phosphodiesterase
G_{t2}	Cone Photoreceptors	cGMP phosphodiesterase
G_{gust}	Taste cells	?
G_{i1}	Neural>others	Adenylyl cyclase K$^+$ channel
G_{i2}	Ubiquitous	Adenylyl cyclase K$^+$ channel
G_{i3}	Others>neural	Adenylyl cyclase K$^+$ channel
G_0	Neural, endocrine	Ca^{2+} channel
G_z	Neural, platelets	?
G_q	Ubiquitous	PLCβ
G_{11}	Ubiquitous	PLCβ
G_{14}	Liver, lung, kidney	PLCβ
$G_{15/16}$	Blood cells	PLCβ
G_{12}	Ubiquitous	?
G_{13}	Ubiquitous	?

cent years (Table 1) and the details of effector modulation by these isoforms is only partly understood. Signaling by G proteins to the nucleus converges on the transcription factor CREB (cAMP response element binding protein), which is a 341 amino acid protein that is activated by phosphorylation of serine 133 (Fig. 9). cAMP-dependent protein kinase (PKA) and calcium-calmodulin-dependent protein kinase II (CAM kinase II) each phosphorylate this site. CREB is able to integrate numerous G protein signals because it responds to changes in both cAMP and calcium, each of which is controlled by multiple G protein isoforms.

Activation of $G_{\alpha q}$ leads to calcium release from intracellular stores in response to the second messenger IP$_3$. IP$_3$ is a product of the breakdown of the membrane phospholipid phosphatidylinositol 4,5-bisphosphate (PIP$_2$) by phospholipase C, and diacylglycerol is generated simultaneously. IP$_3$ provokes calcium release from the endoplasmic reticulum and calcium activates numerous targets, including calmodulin and calcium-calmodulin-dependent proteins, and protein kinase C (PKC). PKC is localized to membrane compartments because it binds to diacylglycerol, as well as to calcium. Recent identification of multiple PKC isoforms has allowed a renewed focus on the downstream effectors that are regulated by PKC. It is unclear how PKC participates in nuclear gene regulation, but the presence of hormone-activated PKC in the nucleus suggests that it may be an important regulator of transcription or RNA processing (Goss *et al.*, 1994).

The α subunits of G proteins are the most heterogeneous in structure and function. However, the $\beta\gamma$ complex is also represented by multiple isoforms. The primary functions of the $\beta\gamma$ complexes are to bind the inactive α subunits and to localize the G protein complex to the membrane. However, it has recently been shown that the liberated $\beta\gamma$ complexes can directly activate the MAP kinase pathway (see Section III.C) by interacting with the Ras protein (Luttrell *et al.*, 1995).

2. Transcriptional Regulation by G Protein-Activated Pathways

Genes that are regulated by cAMP share a regulatory DNA sequence termed the cAMP response element (CRE) (Pestell and Jameson, 1995). The CRE consists of an octanucleotide sequence that is identical, or closely related, to the palindromic consensus 5'-TGACGTCA. The transcription factor that binds to the CRE is called CRE Binding protein (CREB). CREB is a member of a family of transcription factors termed basic-leucine zipper (bZIP) factors, which bind to related response elements. The bZIP proteins form homodimers and heterodimers by virtue of interactions between their leucine zipper motifs, and they interact with DNA through basic residues located in the C-terminal DNA binding domain (Fig. 9) (Meyer and Habener, 1993). CREB can form homodimers that activate transcription. It can also form heterodimers that either activate or inhibit transcription by interacting with isoforms of two closely related bZIP proteins, CREM (CRE modifier), and ATF-1 (activating transcription factor-1) (Habener, 1990). At least seven CREM isoforms are generated by alternative splicing, leading to a rich potential for selective modulation of transcriptional activity.

CREB is phosphorylated on a critical regulatory site (Ser 133) by cAMP-dependent protein kinase, calcium-calmodulin (CaM) kinase II and CaM kinase IV. Other phosphorylation sites in the molecule can downregulate CREB activity in response to a variety of kinases (Meyer and Habener, 1993).

The analysis of transcriptional activation by CREB led to the identification of a large (molecular weight 300 000) CREB binding protein (CBP) that has a high affinity for the phosphorylated CREB dimer. CBP enhances the transcriptional activation driven by CREB (Kwok *et al.*, 1994). CBP was the first identified representative of a wide variety of *coactivator* proteins that form complexes with hormone-inducible transcription factors. These factors enhance, or in some cases inhibit, the transcriptional activity of regulated genes by altering the interactions of upstream transcription factors with the basal transcription machinery. The mechanisms for most of these coactivator proteins are not yet understood. One of the mechanisms by which CBP increases transcriptional activity is by acetylating histones that are bound to the DNA, thereby increasing the ability of proteins to gain access to specific regions of genes (Ogryzko *et al.*, 1996).

CREB is a focal point of regulation by virtue of its ability to be phosphorylated by multiple hormone-regulated kinases, and to interact with other proteins such as CREM and CBP. The

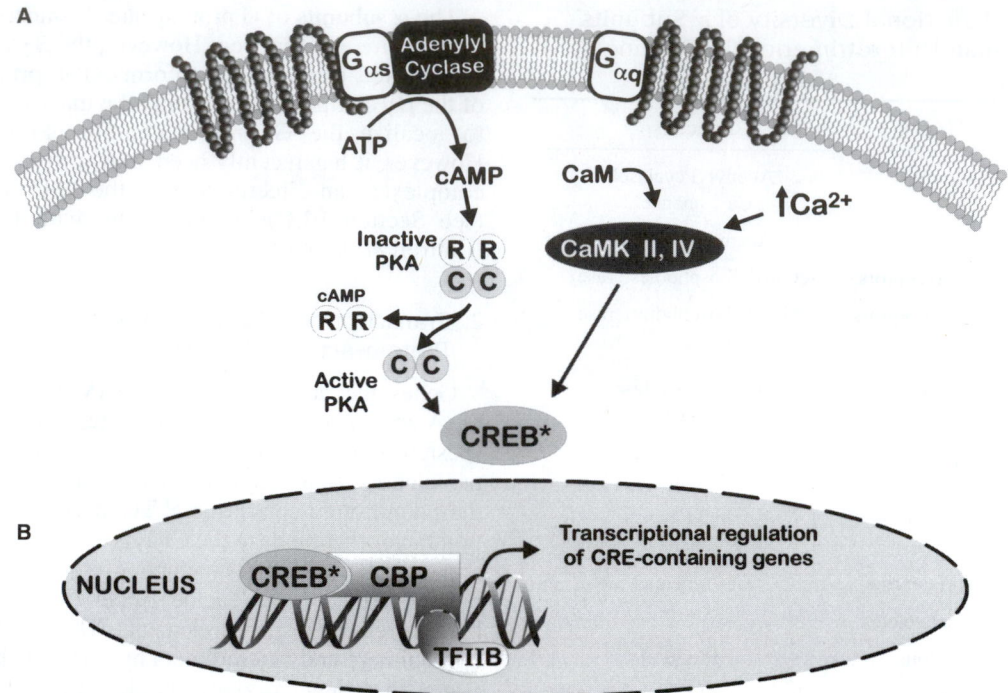

FIGURE 9. *(A)* Regulation of nuclear gene transcription by G-protein-mediated signaling. Both cAMP, generated in response to $G_{\alpha s}$ coupling, or calcium-calmodulin, generated in response to $G_{\alpha q}$, causes phosphorylation of the transcription factor cAMP response element binding protein (CREB*). Protein kinase A catalytic subunits (C) are released from the regulatory subunits (R) in response to cAMP binding to the R subunits. CREB interacts with DNA and other nuclear proteins, such as CBP and TFIIB both directly and indirectly. *(B)* The domain structure of CREB protein. CREB contains a large N-terminal domain that is involved in transcriptional activation and is regulated by phosphorylation on Ser 133, as well as other sites. The C-terminal DNA-binding domain includes both a region rich in basic amino acids and a leucine zipper motif, which is involved in homo- and hetero-dimerization.

commonness of CRE and CRE-like DNA sequences among inducibly regulated genes results in ensembles of coregulated genes in individual cells, as well as diverse cell-specific responses among cell types.

C. Signaling via Tyrosine Phosphorylation

Tyrosine kinase activity is a property that is shared among certain virus-encoded oncogenes and with proteins associated with normal cell signal transduction (Hunter, 1987). This discovery has driven a major advance in our understanding of cellular signal transduction for hormones and other molecules. Two types of hormone-regulated tyrosine kinase pathways are used by vertebrate cells. In some pathways hormones bind to receptors that have intrinsic tyrosine kinase activity in their intracellular domains, allowing direct activation of tyrosine phosphorylation. Examples of receptors that mediate this type of signaling include those for insulin, insulin-like growth factor I (IGF I), epidermal growth factor (EGF), platelet-derived growth factor (PDGF), and fibroblast growth factor (FGF). Other types of receptors, including those for growth hormone (GH), prolactin, erythropoietin, and a variety of hematopoietic cytokines have tyrosine kinases associated with them noncovalently, and the

kinase is activated indirectly following ligand binding. Tyrosine phosphorylation leads to association of SH2 domain-containing proteins in the receptor signaling complex (Fig. 10). Tyrosine phosphorylation accounts for only 1% of the protein-bound phosphate in a typical cell, and serine/threonine phosphorylation accounts for the remaining 99%. Phosphorylation on serine and threonine residues is used for many purposes that are not associated with signal transduction, such as creating a highly charged protein that can chelate cations. In contrast, all known tyrosine kinase reactions are directly associated with signal transduction events.

Tyrosine kinases operate in part by stimulating a cascade of serine/threonine kinase reactions that impact numerous intracellular targets. The prototypic kinase cascade is initiated when tyrosine phosphorylation of a receptor recruits growth factor receptor-binding protein-2 (Grb-2) and a guanine nucleotide exchange factor (GEF) to the receptor signaling complex. GEFs (of which there are several types) convert inactive GDP-bound Ras to active GTP-Ras by catalyzing the exchange of GDP for GTP. Ras-GTP is anchored to the membrane by a C-terminal lipid moiety and binds the serine/threonine kinase Raf-1. Raf-1 is the first of a series of serine/threonine kinases that phosphorylate, in stepwise fashion, other kinases, culminating in phosphorylation of a

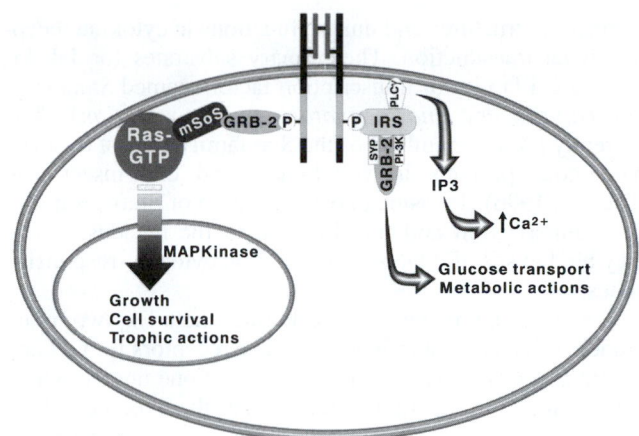

FIGURE 10. Insulin-regulated signaling pathways. The insulin receptor is a tetramer consisting of extracellular α subunits and transmembrane β subunits that are linked by disulfide bridges. The intracellular domains include a tyrosine kinase region (cross-hatched) and several phosphorylatable tyrosines. Phosphorylated tyrosine residues serve as docking sites for IRS proteins and GRB-2. Each of these is phosphorylated on tyrosine residues also, and recruits SH2 domain-containing effector proteins. The known effector proteins include the guanine nucleotide exchange factor mSoS, the SH2 phosphatase SYP, phosphatidylinositol-3' kinase (PI-3K), and phospholipase Cγ (PLCγ).

mitogen-activated protein kinase (MAP kinase) (Campbell *et al.*, 1995). MAP kinases are unusual bifunctional kinases in that they phosphorylate proteins simultaneously on both serine/threonine residues and on tyrosines. Each of the members of this kinase-activating cascade is represented by multiple isoforms that have different substrate specificities and intracellular localizations. A major goal in signal transduction research is to resolve the mechanisms that allow selectivity amid an array of closely related protein kinases and kinase substrates.

1. Hormone Receptors That Have Intrinsic Tyrosine Kinase Activity

The insulin and epidermal growth factor receptors are archetypes for receptors that have intrinsic tyrosine kinase activity. The signal transduction pathways of these receptors share many features. The insulin receptor (INS-R) is a heterotetrameric protein that contains two identical α subunits that are extracellular and bind directly to insulin, and two identical β subunits that traverse the plasma membrane. The β subunits are covalently linked to the α subunit, and to each other, by disulfide bridges (Fig. 10). The intracellular portion of the INS-R β subunit has a tyrosine kinase domain and several phosphorylatable tyrosine residues. Binding of insulin to the extracellular domain causes a conformational change that results in autophosphorylation of several tyrosine residues in the β subunit. The main SH2 domain protein that docks to the INS-R is called insulin receptor substrate, which is represented by two isoforms (IRS-1 and IRS-2).

The function of IRS proteins is to serve as a scaffold on which the INS-R signaling complex is constructed from nu-

merous effector proteins. IRS-proteins are large (MW 185 000) proteins that contain several potentially phosphorylatable tyrosines, at least eight of which are phosphorylated by the INS-R (Myers and White, 1996). Association of IRS proteins with the INS-R is mediated by a particular phosphotyrosine residue on the INS-R, but the IRS proteins do not have conventional SH2 domains, so the binding to INS-R is by some alternative, as yet unclear, phosphotyrosine-recognition motif. IRS proteins recruit several SH2 domain proteins that have specific effector activities (Myers and White, 1996). These include the regulatory p85 subunit of phosphotidylinositol-3' kinase (PI-3 kinase), SYP (src-homology phosphotyrosine phosphatase), c-src (and related src-family kinases) and Grb2. Grb2 serves as a docking protein for mSoS (mammalian homolog of drosophila *son-of-sevenless*). The mSoS protein represents one of the types of GEF, which in this case is employed by the INS-R to activate Ras and a kinase cascade that culminates in MAP kinase activation.

The most well-studied effector pathway for receptor tyrosine kinases has been the MAP kinase cascade. There are two main branches of the MAP kinase family: the ERKs (extracellular signal regulated kinases), which activate a ribosomal protein kinase, S6 kinase, and at least two nuclear transcription factors, Elk and Ets; and the JNKs (Jun N-terminus kinases), which phosphorylate the nuclear transcription factor c-Jun, and related proteins. These substrates are believed to represent the "business end" of the signaling pathway, at which specific changes in the activity of the translation and transcription machinery are imposed. Less well studied is another important effector pathway, the PI-3 kinase pathway. PI-3 kinase is capable of generating, in collaboration with phospholipases, unique second messengers that mediate some of the rapid metabolic effects of insulin, such as accelerated glucose uptake (Saltiel, 1996). Phospholipid products activated by PI-3 kinase bind to, and stimulate the activity of protein kinase B (PKB), which is related to cAMP-dependent protein kinase and protein kinase C. PKB promotes cell survival by preventing the activation of the apoptotic pathway (Frank *et al.*, 1997).

Epidermal growth factor receptor (EGF-R) also contains intrinsic tyrosine kinase activity in its intracellular domain. But unlike INS-R, the EGF-R is a single polypeptide chain that includes both an extracellular ligand-binding domain and an intracellular kinase domain. Binding of EGF to its receptors causes receptor dimerization, bringing the kinase domains of the receptor pair into close proximity where they transphosphorylate tyrosines in the intracellular domain. SH2 domain proteins associate with the receptor through the phosphotyrosine residues (Fig. 11). Some of the proteins that associate with the EGF-R are identical to those that associate with the INS-R, while others appear to be unique. For instance, whereas both EGF-R and INS-R recruit Grb-2 and PI-3 kinase, the IRS proteins that are important for insulin signaling do not appear to be involved in EGF signaling.

Receptor tyrosine kinases exert some of their most important actions through activating phospholipid turnover, and generating low molecular weight second messengers. The

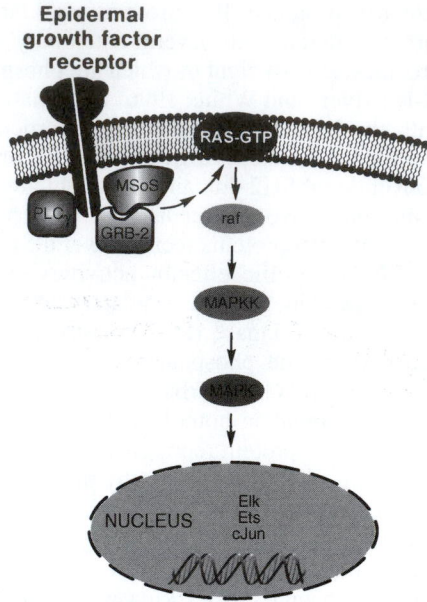

FIGURE 11. Epidermal growth factor (EGF) signal transduction to the nucleus. EGF binding to its receptor causes dimerization and activates the tyrosine kinase function of the intracellular domain. Binding of GRB-2 to the receptor results in activation of membrane-bound Ras through the action of mSoS. Ras recruits Raf-family ser/thr kinases, which phosphorylate one of several MAP kinase kinases (MAPKK), leading to phosphorylation and activation of one or more MAP kinases (ERK or JNK). MAP kinases phosphorylate several nuclear transcription factors, as well as cytoplasmic target proteins that are also important for gene expression.

SH2 domain protein phospholipase Cγ (PLCγ) is activated as a consequence of recruitment to the signaling complexes, and tyrosine phosphorylation in its regulatory domain. PLCγ catalyzes the release of IP$_3$, which generates calcium release from intracellular stores. In this regard the receptor tyrosine kinases converge with the G-protein coupled receptors, which generate IP$_3$ release by activating the PLCβ isoform. IP$_3$-stimulated calcium release may activate numerous effector proteins, such as PKC, calmodulin, CaM kinase II, CREB, and others. However, it is unclear which, if any, of these known calcium-sensitive proteins participate in physiological effects of the receptor tyrosine kinases.

2. The Jak-Stat Pathway and Cytokine Receptor Signaling

Growth hormone, prolactin, erythropoietin, interferons, and a variety of interleukins and hematopoietic cytokines bind to receptors that lack intrinsic tyrosine kinase activity and bind to a unique family of nonreceptor kinases called the *Janus kinases (Jak)*. The Jak proteins were named with the Roman gatekeeper god Janus in mind, but the discovery of new tyrosine kinases has become routine enough that many have interpreted Jak to mean "Just another kinase." However, although the receptor tyrosine kinases and src family kinases are closely related in structure and function, the Jaks have

distinctive structures and unique functions in cytokine receptor signal transduction. The primary substrates for Jak kinases are a family of transcription factors named *Stats (signal transducers and activators of transcription)*. The currently known members of the Stat family consist of seven paralogous proteins in vertebrates, and one insect Stat (Darnell, 1996). Tyrosine phosphorylation of Stats promotes their dimerization and translocation to the nucleus, where they bind to specific DNA sequences on cytokine-responsive genes.

The cytokine receptor superfamily includes two main branches that are distantly related. The receptors for GH and interleukin-6 (IL-6) are representatives of one family, which is large and diverse, and is referred to as the Type I cytokine receptor family; interferon (IFN) receptors comprise the other family, which is less diverse. Similar to the INS-R, IFN receptors are made up of covalently coupled subunits that contain extracellular and intracellular portions. Many of the details of Jak-Stat signaling were initially discovered for interferon signaling, and later found to be widely applicable to other hormones (Darnell, 1996).

Type I cytokine receptors are single transmembrane polypeptides that are characterized by a distinctive signature in the extracellular domain consisting of two pairs of cysteines and a tryptophan–serine–X–tryptophan–serine (WSXWS) motif. The ligands for the Type I cytokine receptors are collectively referred to as "helix bundle peptide hormones" and they share a similar 3-D structure that consists of four α-helices arranged as two antiparallel pairs. The basic features of signaling in the cytokine receptor superfamily are the following: binding of the hormone causes activation of a Jak-family kinase, which tyrosine-phosphorylates a latent Stat transcription factor. Signaling via Type I cytokine receptors is initiated by the formation of either homodimers (in the case of GH) or heterodimers (in the case of IL-6) (Horseman and Yu-Lee, 1994).

To initiate GH-R signaling, a single GH molecule draws together two GH-R molecules. The two binding sites on GH, referred to as sites 1 and 2, have completely different sequences, and the affinity of site 1 is much higher than that of site 2. Although the sequences of the ligand at site 1 and site 2 are different, there is a single binding surface on the GH-R so that the receptors are drawn together in a mirror image conformation (Fig. 12). Homodimerization of the GH-R brings the intracellular domains in close proximity, and receptor-bound Jak2 phosphorylates sites on both the receptor and on Jak2 itself. This creates docking sites for SH2 domain proteins. Stat proteins bind to the receptor via the SH2 domains in their N-terminus, and are then phosphorylated by Jak2 on a tyrosine in their C-terminus. Stat5 is the primary GH activated Stat protein, although Stat 1 and Stat3 are also activated. Activation of Jak2 by GH creates docking sites for SH2 domain proteins other than Stats. As with the insulin or EGF receptor, this results in the recruitment of Grb-2, Shc, mSos, and the activation of a MAP kinase cascade. GH-stimulated Stat proteins and MAP kinase substrates participate in transcriptional activation of genes that are involved in GH actions, including insulin-like growth factor 1, c-fos, and the pro-

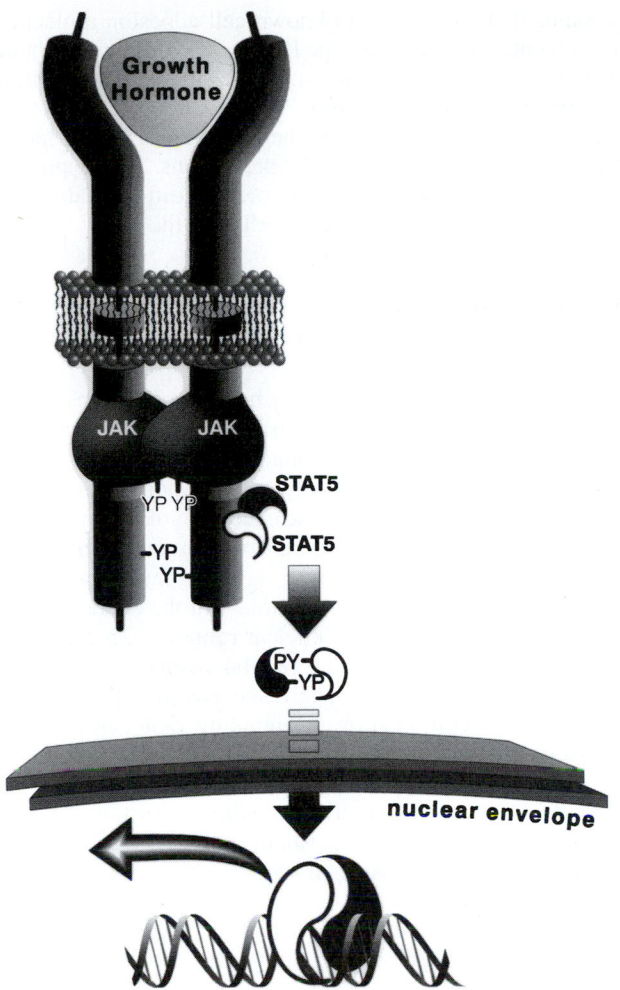

FIGURE 12. Growth hormone signal transduction through the Jak-Stat pathway. GH binding to its receptor draws together a receptor pair into a homodimer with which the tyrosine kinase Jak2 is associated. Jak2 phosphorylates tyrosine residues on itself, the GH receptor, and on Stat proteins, represented here by Stat5. Stat5 phosphorylation results in nuclear translocation and dimerization, followed by binding to sites on DNA that accelerate the transcription of GH-inducible genes.

tease inhibitor Spi 2.1 (Carter-Su *et al.*, 1996). Like GH, signal transduction for prolactin and erythropoietin is initiated by homodimerization of their receptors, and activation of Jak2.

IL-6 is representative of many cytokines that bind to Type I cytokine receptors that heterodimerize after ligand binding (Hibi *et al.*, 1996). The α subunit of the IL-6 receptor is the primary ligand binding moiety, but it has only a short intracellular domain and does not have any capability to transduce a signal to the cell's interior. The β subunit, which is also known as gp130, has no significant affinity for IL-6, but it associates with the liganded α subunit, and the long intracellular domain of gp130 binds and activates both Jak1 and Jak2 (Ihle *et al.*, 1995; Hibi *et al.*, 1996). The gp130 signaling subunit is shared by several other cytokine receptors, including those for leukemia inhibitory factor (LIF) and oncostatin M (OSM). Stat3 is the primary sub-

strate for the gp 130-Jak complex. Activation of the IL-6 receptor in the liver results in the transcription of several acute-phase inflammatory mediators that have Stat3 binding sites in their promoters.

Cytokine receptors induce both terminally differentiated cell functions and cell proliferation during hematopoiesis and organ development. The Stat proteins are connected to the induction of differentiated cell functions, but Jak activation is necessary for both proliferation and differentiation in response to cytokines. Bifurcation of the cytokine signaling pathways downstream of Jak activation leads to both generalized signaling pathways that regulate trophic functions such as growth and cell survival (i.e., MAP kinase, PKC, PKB), and to the differentiation-associated activation of Stat proteins (Horseman and Yu-Lee, 1994; Ihle *et al.*, 1995) (Fig. 13). It is not yet clear how the relative activity in each of these paths is established to allow cells

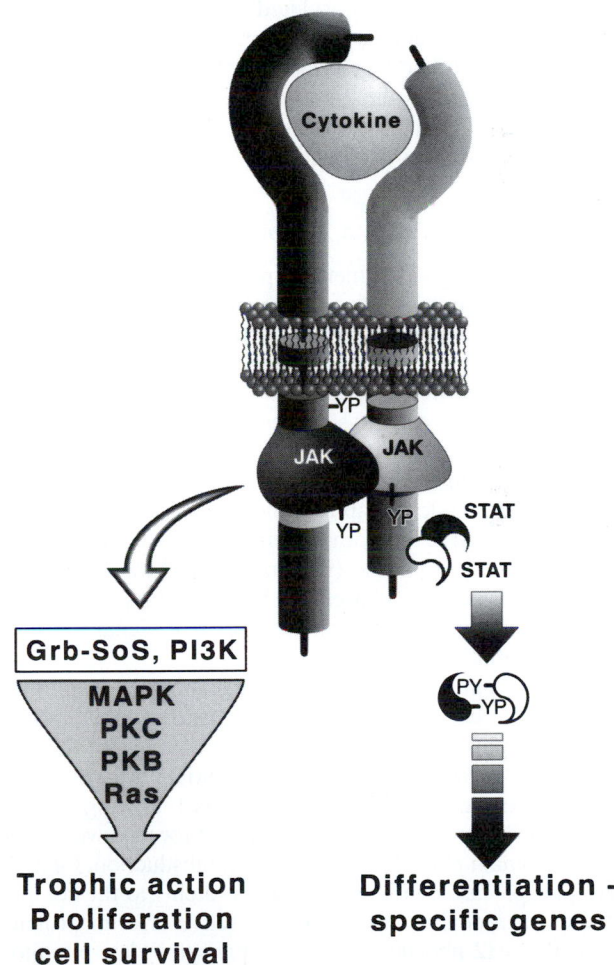

FIGURE 13. Cytokine receptors activate multiple signaling pathways. Binding to cytokine receptors activate Jak kinases, which in turn phosphorylate both Stat proteins and Grb-2 which recruits multiple SH2 domain proteins. Stat signaling is primarily associated with tissue-specific expression of differentiation-associated genes. Multiple effector molecules (Ras, MAP kinase, PKC, PKB, and others) are synthesized in response to cytokines and may contribute to generalized trophic actions, cell proliferation, and cell survival.

to proliferate or terminally differentiate in response to a cytokine signal.

D. Signaling via Hormone Receptors That Are Serine/Threonine Kinases

Transforming growth factor β (TGFβ), bone morphogenetic protein, activin, and inhibin are members of a family of hormones called *cystine knot proteins*, which are vital developmental regulators. The receptors for this family of polypeptide hormones were recently discovered to consist of an extracellular ligand-binding domain and an intracellular ser/thr kinase domain (Massagué, 1992; ten Dijke *et al.*, 1996). The kinase activity of the receptors is essential for signal transduction by these receptors. These ligands cause their receptors to aggregate into tetrameric complexes that phosphorylate sites on the receptors and effector proteins. A group of proteins called SMADs, named for related genes in *C. elegans* (small) and *drosophila* (mothers against decapentaplegic), are phosphorylated upon ligand binding, and translocate to the nucleus where they bind to DNA elements on hormone-regulated genes (Massagué, 1998). Although the nuclear targets of this family of receptors are poorly understood at this time, several genes that are regulated by TGFβ have been characterized.

E. Signaling via Protein Tyrosine Phosphatases

The pervasive involvement of protein phosphorylation in signal transduction is complemented by a variety of phosphatase mechanisms that are both positive and negative regulators of cell functions. Protein tyrosine phosphatases (PTPases), in particular, appear to include members that are highly specific signal transducing molecules. PTPases generally have highly specific "addressing domains" that target them to cell compartments, or to associations with other proteins. These include membrane anchors, cytoskeletal association motifs, SH2 domains, transmembrane helices, and fibronectin-like domains (Dixon, 1996; Tonks and Neel, 1996). Two types of PTPases, the SH2 domain phosphatases and the receptor-like PTPases, are positive signal transducers in mammalian cells.

The SH2 domain phosphatases, for example, Syp in mammals and the corkscrew gene product in drosophila, are recruited to phosphotyrosine residues in receptor signaling complexes. These phosphatases are essential for the activation of some, but not all, downstream events following receptor activation. Two mechanisms may account for the involvement of SH2 domain phosphatases in signal transduction. First, the SH2 phosphatase can recruit unique proteins to the complex which, when dephosphorylated, are active effector proteins. Secondly, SH2 phosphatases can dephosphorylate inhibitory isoforms of signaling proteins, thereby increasing the effectiveness of positive effector proteins.

The receptor-like PTPases have a single transmembrane domain with the catalytic region oriented to the inside of the cell, and a presumptive ligand-binding domain region outward. The CD45 protein is a T-cell surface antigen that encodes a receptor-like PTPase for which the ligand is unknown. Some of the receptor-like PTPases have extracellular domains that are similar to known cell-adhesion molecules. These contain fibronectin Type III repeats or immunoglobulin-like repeats that are known to associate with extracellular matrix factors (Dixon, 1996). Presumably, the activity of these PTPases is regulated by their binding to matrix factors and through direct cell-to-cell interactions. These proteins may, therefore, play important roles in differentiation and density-dependent regulation of cell growth.

IV. Integration of Signal Transduction Pathways in Health and Disease

Regulation of cell functions by hormones is necessary for both homeostasis and vectoral differentiation. The actions of endocrine hormones drive changes in cell functions in collaboration with a host of locally secreted hormones and cell matrix factors, providing for the execution of appropriate responses to constantly changing environmental conditions. The widespread use of common signaling reactions, such as phosphorylation-dephosphorylation, and the sharing of signaling proteins, as with the nuclear retinoid X-receptor and various kinases, assures that cellular responses are highly coordinated and efficient. The responses of cells are determined by the combination of signaling molecules that are activated at a given time, and by the kinetics of their activation. In the case of MAP kinases, transient activation appears to favor cell differentiation, whereas sustained activation favors proliferation (Marshall, 1995).

Human disease states occur when signal pathways are inappropriately activated or disrupted by mutations or infectious agents. A multitude of human diseases are associated with defects in hormone signaling. Diseases arise at every level of signaling; defects are found in the control of hormone production, synthesis, secretion, transport, and in the hormonal signal in target tissues. Defects occur at various levels within the signaling pathways from receptor proteins themselves through downstream effectors (Latchman, 1996).

Cholera toxin and pertussis (whooping cough) toxin cause disease by catalyzing ADP-ribosylation of $G_{\alpha s}$ and $G_{\alpha i}$, respectively. In the case of cholera toxin, the inappropriate cAMP synthesis in the gut leads to dysregulated water secretion, uncontrollable diarrhea, and death from dehydration. Many virus-encoded oncogenes, such as v-src and v-ras, are constitutively active forms of normal cellular signaling proteins. Each of the nuclear hormone receptor genes is represented by at least one genetic disease caused by mutation of the receptor. For example, testicular feminization results from mutations that inactivate the androgen receptor, and vitamin D-resistant rickets results from mutations in the DNA-binding region of the VDR (Hughes *et al.*, 1991). Elucidation of the details of signaling pathways will continue to lead to advances in diagnosis, prevention, and therapy for numerous diseases.

It is important to recognize that the phrase "signal transduction" is merely a metaphor for the processes that are initiated by hormones interacting with cells. There are many ways in which this metaphor is inadequate. For example, most everyday signals, such as a red light or a green light, can be interpreted unambiguously. In contrast, the meanings of hormone signals are always very ambiguous except in the context of a myriad of prior events and coregulatory factors. Cells continuously integrate multiple signals by coupling each of the signaling components to multiple receptors. Remarkably, these integrated, combinatorial regulatory systems produce all of the organism's normal physiological responses and disease states through a highly conserved repertoire of intracellular signaling molecules.

V. Summary

Hormones are regulatory molecules that are secreted from either circumscribed endocrine glands or cells of nonendocrine organs. Hormones activate signal transduction protocols that are highly conserved throughout eukaryotic organisms. Lipophilic hormones, such as steroids, thyroxine, retinoids, and vitamin D enter cells passively and bind to intracellular receptors. Many of these receptors, which are part of a large family of transcription factors, bind known hormones, and others, collectively referred to as orphan receptors, may bind unknown ligands, or function independently of ligand binding. The nuclear receptors are characterized by a conserved DNA binding domain flanked by gene activation and ligand-binding domains. Upon interaction with the cognate hormone, or a hormone analog, the nuclear receptor undergoes significant conformation changes that convert it to an active transcription factor. Hydrophilic protein and amine hormones interact with receptors that are integral membrane proteins. Ligand binding to membrane receptors induces both conformational changes and oligomerization of collaborating signal transduction proteins. A large family of membrane-bound receptors share a structure that includes seven membrane-spanning α helices (heptahelical receptors). These receptors interact with heterotrimeric GTP-binding proteins, which elicit changes in numerous effector pathways, such as cyclic nucleotide synthesis, phospholipid mobilization, and Ca^{2+} influx. Other membrane-associated receptors activate tyrosine phosphorylation of specialized signaling proteins. The protein-tyrosine kinase enzymatic activity may be either an intrinsic property of the intracellular domain of the hormone receptor, or a property of a receptor-associated protein. Tyrosine phosphorylation creates docking sites for effector proteins that include several families of kinases and latent transcription factors. Physiological integration of signal transduction is essential to normal physiology, and defects in signaling are a common feature of disease states.

Bibliography

Beato, M. (1989). Gene regulation by steroid hormones. *Cell* **56**, 335–344.

Beato, M., Herrliche, P., and Schutz, G. (1995). Steroid hormone receptors: Many actors in search of a plot. *Cell* **83**, 851–857.

Berg, J. M. (1989). DNA-binding specificity and steroid receptors. *Cell* **57**, 1065–1068.

Bourguet, W., Ruff, D., Chambon, P., Gronemeyer, H., and Moras, D. (1995). Crystal structure of the ligand-binding domain of the human nuclear receptor RXRα. *Nature* **375**, 377–382.

Campbell, J. S., Seger, R., Graves, J. D., Graves, L. M., Jensen, A. M., and Krebs, E. G. (1995). The MAP kinase cascade. *Recent Prog. Horm. Res.* **50**, 131–159.

Carter-Su, C., Schwartz, J., and Smit, L. S. (1996). Molecular mechanisms of growth hormone action. *Annu. Rev. Physiol.* **58**, 187–207.

Danielian, P. S., White, R., Lees, J. A., and Parker, M. G. (1992). Identification of a conserved region required for hormone-dependent transcriptional activation by steroid hormone receptors. *EMBO J.* **11**, 1025–1033.

Darnell, J. E., Jr. (1996). The JAK-STAT pathway: Summary of initial studies and recent advances. *Recent Prog. Horm. Res.* **51**, 391–404.

Dixon, J. E. (1996). Protein tyrosine phosphatases: their roles in signal transduction. *Recent Prog. Horm. Res.* **51**, 405–415.

Evans, R. M. (1988). The steroid and thyroid hormone receptor superfamily. *Science* **240**, 889–895.

Forman, B. M., Tontonoz, P., Chen, J., Brun, R. P., Spiegelman, B. M., and Evans, R. M. (1995). 15-Deoxy-delta^{12,14}-prostaglandin J$_2$ is a ligand for the adipocyte determination factor PPARγ. *Cell* **83**, 803–812.

Frank, T. F., Kaplan, D. R., and Cantley, L. C. (1997). P13K: Downstream AKTion blocks apoptosis. *Cell* **88**, 435–437.

Glass, C. (1994). Differential recognition of target genes by nuclear receptor monomers, dimers, and heterodimers. *Endocr. Rev.* **15**, 391–407.

Goss, V. L., Hocevar, B. A., Thomson, L. J., Stratton, C. A., Burns, D. J., and Fields, A. P. (1994). Identification of nuclear beta II protein kinase C as a mitotic lamin kinase. *J. Biol. Chem.* **269**, 19074–19080.

Gronemeyer, H. (1993). Transcriptional activation by nuclear receptors. *J. Receptor Res.* **13**, 667–691

Habener, J. F. (1990). Cyclic AMP response element binding proteins: A cornucopia of transcription factors. *Mol. Endocrinol.* **4**, 1087–1094.

Hard, T., Kellenbach, E., Boelens, R., Maler, B. A., Dahlman, K., Freedman, L. P., Carlstedt-Duke, J., Yamamoto K. R., Gustafsson, J. A., and Kaptein, R. (1990). Solution structure of the glucocorticoid receptor DNA-binding domain. *Science* **249**, 157–160.

Herskowitz, I. (1995). MAP kinase pathways in yeast: For mating and more. *Cell* **80**, 187–197.

Hibi, M., Nakajima, K., and Hirano, T. (1996). IL-6 cytokine family and signal transduction: A model of the cytokine system. *J. Mol. Med.* **74**, 1–12.

Horlein, A. J., Naar, A. M., Heinsel, T., Torchia, J., Gloss, B., Kurokawa, R., Ryan, A., Kamei, Y., Soderstrom, M., Glass, C. K., and Rosenfeld, M. G. (1995). Ligand-independent repression by the thyroid hormone receptor mediated by a nuclear receptor co-repressor. *Nature* **377**, 397–404.

Horseman, N. D. and Yu-Lee, L. - Y. (1994). Transcriptional regulation by the helix bundle peptides: Growth hormone, prolactin and hematopoietic cytokines. *Endocr. Rev.* **15**, 627–649.

Hughes, M. R., Malloy, P. J., O'Malley, B. W., Pike, J. W., and Feldman, D. (1991). Genetic defects of the 1,25-dihydroxyvitamin D receptor. *J. Receptor Res.* **11**, 699–716.

Hunter, T. (1987). A thousand and one protein kinases. *Cell* **50**, 823–829.

Ihle, J. N., Witthun, B. A., Quelle, F. W., Yamamoto, K., and Silvennoinen, O. (1995). Signaling through the hematopoietic cytokine receptors. *Annu. Rev. Immunol.* **13**, 369–398.

Katzenellenbogen, J. A., O'Malley, B. W., and Katzenellenbogen, B. S. (1996). Tripartite steroid hormone receptor pharmacology: Interaction with multiple effector sites as a basis for the cell- and promoter-specific action of these hormones. *Mol. Endocrinol.* **10**, 119–131.

Kleiwer, S. A., Umesono, K., Mangelsdorf, D. J., and Evans, R. M. (1992). Retinoid X receptor interacts with nuclear receptors in retinoic acid, thyroid, and vitamin D signaling. *Nature* **355**, 446–449.

Kwok, R. P. S., Lundblad, J. R., Chrivia, J. D., Richards, J. P., Bachinger, H. P., Brennan, R. G., Roberts, S. G. E., Green, M. R., and Goodman, R. H. (1994). Nuclear protein CBP is a coactivator for the transcription factor CREB. *Nature* **370,** 223–226.

Latchman, D.S. (1996). Transcription factor mutations and disease. *N. Engl. J. Med.* **334,** 28–33.

Lee, M. S., Kliewer, S. A., Provencal, J., Wright, P. E., and Evans, R. M. (1993). Structure of the retinoid X receptor, a DNA-binding domain: α helix required for homodimeric DNA binding. *Science* **260,** 1117–1121.

Luisi, B. F., Xu, W., Otwinowski, Z., Freedman, L. P., Yamamoto, K. R., and Sigler, P. B. (1991). Crystallographic analysis of the interaction of the glucocorticoid receptor with DNA. *Nature* **352,** 497–505.

Luttrell, L. M., van Biesen, T., Hawes, B. E., Koch, W. J., Touhara, K., and Lefkowitz, R. J. (1995). $G_{\beta\gamma}$ subunits mediate mitogen-activated protein kinase activation by the tyrosine kinase insulin-like growth factor 1 receptor. *J. Biol. Chem.* **270,** 16495–16498.

Mader, S., Kumar, V., deVereneuil, H., and Chambon, P. (1989). Three amino acids of the oestrogen receptor are essential to its ability to distinguish an oestrogen from a glucocorticoid-responsive receptor. *Nature* **338,** 271–274.

Mangelsdorf, D. J. and Evans, R. M. (1995). The RXR heterodimer and orphan receptors. *Cell* **83,** 841–850.

Mangelsdorf, D. J., Thummel, C., Beato, M., Herrliche, P., Schutz, G., Umesono, K., Blumberg, B., Kastner, P., Mark, M., Chambon, P., and Evans, R. M. (1995). The nuclear receptor superfamily: The second decade. *Cell* **83,** 835–839.

Marshall, C. J. (1995). Specificity of receptor tyrosine kinase signaling: Transient versus sustained extracellular signal-regulated kinase activation. *Cell* **80,** 179–185.

Massagué, J. (1992). Receptors for the TGFβ family. *Cell* **69,** 1067–1070.

Massagué, J. (1998). TGF-β signal transduction. *Annu. Rev. Biochem.* **67,** 753–791.

Meyer, T. E., and Habener, J. F. (1993). Cyclic adenosine 3', 5'-monophosphate response element-binding protein (CREB) and related transcription-activating deoxyribonucleic acid-binding proteins. *Endocr. Rev.* **14,** 269–290.

Myers, M. G., Jr., and White, M. F. (1996). Insulin signal transduction and the IRS proteins. *Annu. Rev. Pharmcol. Toxicol.* **36,** 615–658.

O'Malley, B. W. (1990). The steroid receptor superfamily: More excitement predicted for the future. *Mol. Endocrinol.* **4,** 363–369.

Ogryzko, V. V., Schiltz, R. L., Russanova, V., Howard, B. H., and Nakatani, Y. (1996) The transcriptional coactivators p300 and CBP are histone acetyltransferases. *Cell* **87,** 953–959.

Pawson, T. (1995a). Protein modules and signalling networks. *Nature* **373,** 573–580.

Pawson, T. (1995b). SH2 and SH3 domains in signal transduction. *Adv. Cancer Res.* **64,** 87–110.

Perlmann, T., Rangarajan, P. N., Umesono, K., and Evans, R. M. (1993). Determinants for selective RAR and TR recognition of direct repeat HREs. *Genes Dev.* **7,** 1411–1422.

Pestell, R. G. and Jameson, J. L. (1995). Transcriptional regulation of endocrine genes by second-messenger signaling pathways. *In*

"Molecular Endocrinology" (B. D. Weintraub, Ed.) pp. 59–76. Raven Press, New York.

Pratt, W., Jolly, D. J., Pratt, D. V., Hollenberg, S. M., Giguere, V., Cadepon, F. M., Schweizer-Groyer, G., Cartelli, M. G., Evans, R. M., and Baulieu, E. E. (1988). A region in the steroid-binding domain determines formation of the non-DNA-binding glucocorticoid receptor complex. *J. Biol. Chem.* **263,** 267–273.

Rastinejad, F., Perlmann, T., Evans, R. M., and Sigler, P. B. (1995). Structural determinants of nuclear receptor assembly on DNA direct repeats. *Nature* **375,** 203–211.

Renaud, .J-P., Natacha, R., Ruff, M., Vivat, V., Chambon, P., Gronemyer, H., and Moras, D. (1995). Crystal structure of the RARγ ligand-binding domain bound to all-*trans* retinoic acid. *Nature* **378,** 681–689.

Saltiel, A. R. (1996). Diverse signaling pathways in the cellular actions of insulin. *Am. J. Physiol.* **270,** E375–E385.

Schwabe, J. W. R., Chapman, L., Finch, J. T., and Rhodes, D. (1993). The crystal structure of the oestrogen receptor DNA-binding domain bound to DNA: how receptors discriminate between their response elements. *Cell* **75,** 567–578.

Schwabe, J. W. R., Neuhaus, D., and Rhodes, D. (1990). Solution structure of the DNA-binding domain of the oestrogen receptor. *Nature* **348,** 458–461.

Speigel, A. M., Shenker, A., Simonds, W. F., Weinstein, L. S. (1995). G-protein dysfunction in disease. *In* "Molecular Endocrinology" (B. D. Weintraub, Ed.) pp. 297–318. Raven Press, New York.

ten Dijke, P., Miyazono, K., and Helden, C. H. (1996). Signaling via hetero-oligomeric complexes of type I and type II serine/threonine kinase receptors. *Curr. Opinion Cell Biol.* **8,** 139–145.

Thummel, C. S. (1995). From embryogenesis to metamorphosis: The regulation and function of *Drosophila* nuclear receptor superfamily members. *Cell* **83,** 871–877.

Tonks, N. K., and Neel, B. G. (1996). From form to function: Signaling by protein tyrosine phosphatases. *Cell* **87,** 365–368.

Truss, M., and Beato, M. (1993). Steroid hormone receptors: Interaction with deoxyribonucleic acid and transcription factors. *Endocr. Rev.* **14,** 459–479.

Umesono, K., and Evans, R. M. (1989). Determinants of target gene specificity for steroid/thyroid hormone receptors. *Cell* **57,** 1139–1146.

Umesono, K., Murikami, K. K., Thompson, C. C., and Evans, R. M. (1991). Direct repeats as selective response elements for the thyroid hormone, retinoic acid, and vitamin D_3 receptors. *Cell* **65,** 1255–1266.

Wagner, R. L., Apriletti, J. W., McGrath, M. E., West, B. L., Baxter, J. D., and Fletterick, R. J. (1995). A structural role for hormone in the thyroid hormone receptor. *Nature* **378,** 690–697.

Yu, V., Delsert, C., Andersen, B., Holloway, J. M., Devary, O. V., Naar, A. M., Kim, S. Y., Boutin, J.-M., Glass, C. K., and Rosenfeld, M. G. (1991). RXR: A coregulator that enhances binding of retinoic acid, thyroid hormone, and vitamin D receptors to their cognate response elements. *Cell* **67,** 1251–1266.

Zhang, F. L., and Casey, P. J. (1996). Protein prenylation: Molecular mechanisms and functional consequences. *Annu. Rev. Biochem.* **65,** 241–269.

Section

II

Membrane Potential, Transport Physiology, Pumps, and Exchangers

Jeffrey C. Freedman and Nicholas Sperelakis

13

Diffusion and Permeability

I. Introduction

In order to understand the fundamentals of membrane transport, as well as the mechanisms for development of the resting electrical potential of cells, it is first necessary to consider the basic processes of **diffusion** and membrane **permeability**. Therefore, this chapter discusses fundamental principles that are utilized in subsequent chapters in this book.

Molecules of gases and liquids, and of all dissolved solutes, are continuously in motion. The velocities of individual molecules vary tremendously, as do their kinetic energies, in accordance with the Maxwell-Boltzmann distribution. Diffusion is the process whereby particles in a gas, liquid, or solid tend to intermingle due to their spontaneous motion caused by thermal agitation. Any diffusing substance tends to move from regions of higher concentration to regions of lower concentration, until the substance is uniformly distributed at equilibrium. At any given temperature above absolute zero, the molecules of a substance continue to move, even at equilibrium, but the net movement is zero. In the absence of **convection**, which refers to bulk flow of solvent containing solutes such as that caused by stirring, the movement of the molecules is by diffusion only. Diffusion occurs because of the random thermal motions of the molecules. Molecular collisions that transfer momentum are more probable in a region of higher concentration. Particles flow from a region of high concentration to one of low concentration whenever the concentration is greater behind than in front of any particle; that is, whenever a gradient of concentration exists, such a particle has a greater probability of being bumped from behind than from in front. The net result of these collisions imparts a forward movement to the center of mass of the particles, thus creating a flow of particles down the concentration gradient. Consequently, if a region of high solute concentration is adjacent to one of low solute concentration, separated by an imaginary plane, it is probable that more molecules per unit of time will be crossing the plane from the side of higher concentration to the side of lower concentration than in the opposite direction. Thus there are **fluxes**, or movements of molecules, in both directions (known as **unidirectional fluxes**), but the **net flux** is from the side of higher concentration to the side of lower concentration.

Now if the imaginary plane were replaced with a thin membrane permeable to the molecules, then the same situation would apply; the particles would diffuse from the side of higher concentration to the side of lower concentration across the membrane. In a living cell it is often assumed that the inside and outside solutions are relatively well stirred, and diffusion of most substances through the cell membrane is much slower than that through a free solution. Therefore our primary concern is about the diffusion of substances through the cell membrane, the rate-limiting step. For simplicity we assume that the solutions on either side are well stirred and that there are no concentration gradients within the bulk solutions on either side of the membrane (although in reality there probably are unstirred layers near the membrane). We confine ourselves here to the diffusion of small molecules or ions across membranes. For thin membranes, we first consider the case of diffusion without partitioning, as illustrated on the left side of Fig. 1, in which the diffusing molecule is freely able to enter the membrane. We will then introduce the effect of partitioning into the membrane, a process that involves a change of chemical potential upon

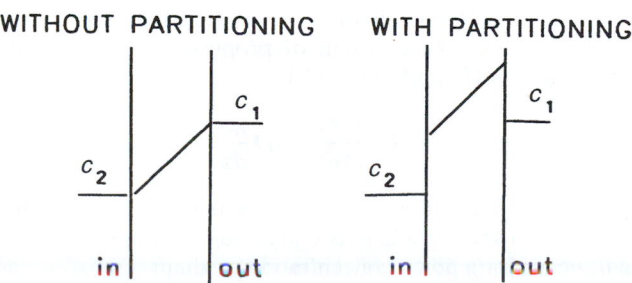

FIGURE 1. Concentration gradients across the membrane during diffusion with and without partitioning.

entrance into the membrane phase, as illustrated on the right side of Fig. 1. In either case, we disregard structural effects that were illustrated by Danielli (1943) as a series of potential energy barriers within the membrane.

II. Fick's Law of Diffusion

The fundamental law describing diffusion was discovered empirically by Adolph E. Fick (1829–1901), Demonstrator of Anatomy in Zürich, who in 1855 noted an analogy between diffusional transport and Fourier's law describing the flow of heat. Fick's law of diffusion was deduced theoretically in 1860 by James Clerk Maxwell (1831–1879) from the kinetic theory of gases, and it should be understood from the outset that the derivation of Fick's law makes the following assumptions: (1) the continuum hypothesis applies, that is, the molecular concentration is sufficiently high to ensure the validity of statistical laws; (2) the average duration of a collision is small compared to the average time between collisions so that all collisions are binary, a condition that would pertain to dilute solutions; (3) the distribution of molecular properties varies only slowly in time and space; (4) the particles move independently so that there is no correlation between the properties of any two particles; (5) quantum effects are negligible so that classical mechanics can be used to describe molecular collisions; (6) energy, momentum, and mass are conserved in every molecular collision; and (7) in liquids, the diffusing particles are much larger than the solvent molecules. These assumptions necessarily make the quantitative description of diffusion an approximation. A more accurate and complete description, especially for diffusion in liquids, depends upon knowledge of the structure of the medium in which diffusion occurs.

Consider two compartments separated by a membrane of thickness l, in which a substance diffuses from a cell (denoted as compartment 2 with volume V_2 and high concentration c_2) across the cell membrane into the medium (denoted as compartment 1 with volume V_1 and low concentration c_1). We assume that the temperature and pressure are the same in both compartments. For such a system, **Fick's first law of diffusion** states that the flux density (J), given by dN/Adt mol/(cm^2·s), or the number (dN) of moles that diffuse across a unit area (A, cm^2) of the membrane during an interval of time (dt, s), in a direction (x) perpendicular to the membrane, is directly proportional to the concentration gradient dc/dx (mol/cm^3/cm) across the membrane, the constant of proportionality being the **diffusion coefficient D** (cm^2/s).

$$J = \frac{dN_2}{Adt} = -D\frac{dc}{dx} \qquad (1)$$

The diffusion coefficient D, which is the flux density when the concentration gradient is unity, indicates how much flux will occur for a given concentration gradient.

For a thin membrane, when the solution is well stirred on either side, the **concentration gradient** in the steady state is equal to the difference in the concentration ($\Delta c = c_2 - c_1$) of the solute on both sides of the membrane divided by the thickness of the membrane (Δx), so that

$$J = -D\frac{\Delta c}{\Delta x} \qquad (2)$$

Assume that the concentration gradient across the membrane is linear over the thickness (l) of the membrane, which is a reasonable assumption for a thin membrane. Then $\Delta c/\Delta x = (c_2 - c_1)/l$, and the flux density ($J$) is

$$J = -D\frac{c_2 - c_1}{l} \qquad (3)$$

Note that when the concentration gradient is defined as the internal minus the external concentration, then the minus sign in Fick's law connotes that an efflux is negative and an influx is positive. Thus the net flux is equal to the concentration gradient times the diffusion coefficient. If c_1 is zero, then according to Fick's law, the flux density is linearly related to the concentration of the solute on the side from which the flux originates. Consequently, if one experimentally finds a linear dependence of flux on concentration, then the mechanism of transport is consistent with, but not proof of, simple diffusion. Finding a linear dependence of flux on concentration over a particular range of concentrations does not prove that diffusion is the mechanism of transport because the flux could deviate from linearity at a higher concentration.

To examine the rate of change of concentration (dc_2/dt), divide both sides of Eq. 1 by the intracellular volume (V_2) and multiply both sides by A, giving

$$\frac{dc_2}{dt} = \frac{dN_2}{V_2 dt} = -\frac{DA}{V_2}\frac{dc}{dx} = -\frac{DA}{V_2 l}\left(c_2 - c_1\right) \qquad (4)$$

Integration of Eq. 4 shows that the time course of diffusion, or the change in intracellular concentration of a diffusing substance as a function of time, is exponential (see the Appendix following this chapter). For two-dimensional diffusion, the time required for diffusion to become 50, 63, or 90% complete varies directly with the square of the distance, and inversely with the diffusion coefficient. Thus, diffusion is extremely fast over short distances (e.g., 10–1000 nm), but is exceedingly slow over long distances (e.g., 1 cm). For a small particle like K^+ or acetylcholine (ACh^+), with a D value of about 1×10^{-5} cm^2/s, the time required for 90% equilibration to be reached over a distance of 1 μm is about 1 ms. Table 1 shows how the time required changes as a function of distance. The chef knows that the time required for a meat roast to cook in the middle is dependent on the thickness of the roast, and the physiologist knows that there is a critical thickness (e.g., 0.5–1.0 mm) of a muscle bundle in an incubation bath that allows adequate diffusion of oxygen to the core of the strip, no matter how vigorously the bath is oxygenated.

III. Diffusion Coefficient

The diffusion coefficient (D) is a constant for a given substance and membrane under a given set of conditions. The dif-

TABLE 1 Calculations of Time Required for Diffusion to Become 90% Complete

Distance, λ	Time, t
100 Å	0.1 μs
0.1 μm	0.01 ms
1 μm	1 ms
10 μm	100 ms
100 μm	10 s
1 mm	16.7 min
1 cm	28 h

Note: Diffusion time varies with square of diffusion distance. These values are for a substance having a diffusion coefficient (D) of about 1×10^{-5} cm^2/s in free solution. The diffusion time is inversely proportional to D. Einstein's approximation equation is $\lambda = \sqrt{Dt}$, where λ is the mean displacement of the diffusing particles along the x-axis at any time (t).

fusion coefficient of substances in free solution is dependent upon molecular size (and shape, for large molecules). For ions, the smaller the unhydrated radius, the greater the charge density, which means that more water molecules are held in the hydration shells giving a larger hydrated radius; the larger water shell causes diffusion to be slower (see Chapter 1). The hydrated and unhydrated radii of some relevant ions are given in Table 2. Thus, although Na$^+$ has a lower atomic weight than K$^+$, Na$^+$ attracts and holds a larger hydration shell because of its greater charge density (same charge as K$^+$, but smaller unhydrated ion size), and therefore Na$^+$ diffuses in water considerably slower than K$^+$ (ratio of about 1:2).

The diffusion coefficient is inversely related to the resistance to free diffusion and is very much less in water than in a gas. **Graham's law** states that the diffusion coefficient of a gas is inversely proportional to the square root of the molecular weight, and is often given in the following form:

$$DM^{1/2} = \text{constant} \tag{5}$$

Graham's law was originally discovered by experiment, and was later derived theoretically by Maxwell. From the kinetic theory of gases, the diffusion coefficient is related to molecular parameters as follows:

$$D = \left(\frac{kT}{\pi}\right)^{3/2} \frac{2N_A}{\sigma^2 \, pM^{1/2}} \tag{6}$$

where k is Boltzmann's constant, T is the absolute temperature, N_A is Avogadro's number, σ is the radius of the particle, p is the pressure, and M is the molecular mass (see Cunningham and Williams, 1980).

The random molecular motions due to thermal energy are counteracted by intermolecular attractive forces, but in dilute gases these forces are small because the molecules are relatively far apart. The kinetic energy of gas molecules is given by $nMv^2/2$, where n is the number of moles, M is their molecular weight, and v is their mean velocity. At a given temperature, the average velocities of the molecules are inversely proportional to the square root of their masses. Because the velocity of the colliding molecules influences the speed of diffusion, the diffusion coefficient is also inversely proportional to the square root of the molecular weight, as stated in Graham's law.

The **Einstein–Stokes equation** applies to diffusion in liquids when spherical solute molecules are much larger than the solvent molecules, in which case the diffusion coefficient

TABLE 2 Physicochemical Properties of Selected Ions

Ion	Limiting equivalent conductivity (cm^2/S·eq)	Activity coefficient[a]	Crystallographic radius (Å)	Singly hydrated radius[b] (Å)	Hydrated volume[c] (Å^3)	Heat of hydration (kcal/g ion)	Number of water molecules[c]
Na$^+$	50.10	0.778	0.96	3.67	150	−115	(4–6)
K$^+$	73.50	0.770	1.33	4.05	(150)	−90	(3–6)
Cl$^-$	76.35	—	1.81	3.92	90	−59	(0–6)
Mg^{2+}	53.05	0.528	0.65	3.60	360	−501	12
Ca^{2+}	59.50	0.518	0.99	3.70	310	−428	10
Sr^{2+}	59.45	0.515	1.13	3.85	310	−381	10
Ba^{2+}	63.63	0.508	1.35	4.08	290	−347	9–10

Note: Values taken from R. A. Robinson and R. H. Stokes, *Electrolyte Solutions*, London, Butterworth & Co. (Publishers), Ltd., 1959, and from J. O. M. Bockris and B. E. Conway, *Modern Aspects of Electrochemistry*, London, Butterworth & Co. (Publishers), Ltd., 1954.

[a] Activity coefficient for each cation given for the Cl$^-$ salt at 0.1 M and at 25 °C.

[b] Singly hydrated radius equals crystallographic radius plus 2.72 Å (diameter of water molecule) for cations with crystal radii between 0.9 and 1.7 Å, and crystal radius plus 2.23 Å for anions. Values for singly hydrated radii taken from Mullins (1961).

[c] Values in parentheses are estimated.

of the solute was found by Einstein (1926) to have the following relationship to the viscosity (η) of the solvent and to the radius (r) of the solute particle:

$$D = \frac{RT}{6\pi\eta r N_A} \tag{7}$$

where R is the gas constant, T is the absolute temperature, and N_A is Avogadro's number. This equation is more applicable to the diffusion of solutes of colloidal size than to molecules that approximate the size of the solvent.

Temperature also affects the rate of diffusion. The relative increase in the diffusion coefficient when the temperature (T) is raised by 10 °C is known as Q_{10}. From the Einstein-Stokes equation,

$$\frac{D_{T+10}}{D_T} = \frac{T+10}{T}\frac{\eta_T}{\eta_{T+10}} \tag{8}$$

In dilute solutions for which the viscosity doesn't vary much over a range of 10 °C, Q_{10} should be the same for all molecules. Neglecting changes in η, Q_{10} is 1.03 between 25 °C and 35 °C. A value of Q_{10} much greater than 1 is often taken to indicate that some process other than diffusion is probably responsible for transport.

In liquids, the solute and solvent molecules are close together and within each others' spheres of mutual attraction. Hence, for any solute to move, it must first break away from its surrounding molecules. Diffusion in liquids consequently occurs in a discontinuous manner; a molecule is able to move only when it has acquired sufficient energy by collision to break away from its neighbors and push aside other molecules. Thus, the molecule diffuses in a series of jumps, each jump requiring a certain critical **activation energy**. When the attractive forces between molecules are weak, diffusion is rapid. Danielli (1943) derived from kinetic theory a relationship for D that includes the temperature coefficient (Q_{10}) raised to a power as follows:

$$DM^{1/2}Q_{10}^{(T+10)/10} = \text{constant} \tag{9a}$$

or,

$$D \propto \frac{1}{M^{1/2}Q_{10}^{(T+10)/10}} \tag{9b}$$

This relationship is applicable at constant temperature. The Q_{10} for diffusion of a substance in water depends on the activation energy necessary for a jump. If a large amount of energy is required (high Q_{10}), only a few molecules have the necessary energy to diffuse at any moment; raising the temperature increases the number of molecules with the required energy and thereby appreciably speeds up the net rate of diffusion. If only a small amount of activation energy is required (low Q_{10}), a greater fraction of the molecules possesses this minimal energy at any moment, and so raising the temperature has less of an effect on the net rate of diffusion. In either case, elevation of temperature increases the speed of the molecules and causes an increase in the rate of diffusion. When Q_{10} is close to 1.0, then Eqs. 9a and 9b reduce to Graham's law (Eq. 5). The Q_{10} for diffusion of Na$^+$ or K$^+$ in water is only about 1.22. Equations 9a and 9b also apply

when the permeability coefficient (κ), as defined in Section IV, is substituted for the diffusion coefficient (D) of the solute in the membrane.

The cell membrane constitutes a barrier to diffusion. In penetrating through the lipid bilayer matrix, a noncharged small solute molecule must move as follows: (1) detach itself from its surrounding solvent molecules and jump into the membrane phase; (2) move through the thickness of the membrane, perhaps by a series of small jumps over energy barriers; and (3) detach itself from the membrane environment and jump into the solvent phase again on the opposite side of the membrane. The permeability of the membrane to the solute depends greatly on the activation energy required for the molecule to jump into and out of the membrane. If the activation energy is high, the permeability is low and the Q_{10} is high; if the activation energy is low, then the permeability is great and the Q_{10} is low.

IV. Diffusion Across a Membrane with Partitioning

When a solute partitions into the membrane, or is excluded from it, the concentration gradient across the membrane phase itself differs from that calculated from the internal and external concentrations in the bulk solution. For example, suppose that a solute is at 10 mM in the internal solution and at 1 mM in the external solution; the concentration gradient would be $c_2 - c_1 = 10 - 1 = 9$ mM. Now further suppose that the solute dissolves into the membrane phase such that the concentration just within the membrane at the internal boundary is 10-fold greater than in the internal solution, and the same ten-fold partitioning occurs at the external boundary. With such partitioning, the concentrations in the membrane would be 100 mM at the internal boundary, and 10 mM at the external boundary, constituting a gradient of $100 - 10 = 90$ mM, or a tenfold higher concentration gradient within the membrane than the gradient based on the concentrations in the internal and external bulk solutions.

We can now apply Fick's first law to the case of diffusion across a membrane with partitioning of the solute into the membrane in a partition/diffusion model of transport. Letting the superscript m denote the concentrations in the membrane phase, we have from Eqs. 3 and 4,

$$J = -\frac{D}{l}\left(c_2^m - c_1^m\right) \tag{10}$$

$$\frac{dc_2}{dt} = -\frac{DA}{lV_2}\left(c_2^m - c_1^m\right) \tag{11}$$

The partition coefficient (β) relates the concentration just within the membrane at the boundary to the external or internal concentrations. In a homogeneous membrane, β is assumed equal at the internal and external boundaries.

$$\beta = \frac{c_1^m}{c_1} = \frac{c_2^m}{c_2} \tag{12}$$

Substituting the expressions for β into Eq. 11 results in the following transport equation:

$$\frac{dc_2}{dt} = -\frac{\beta D}{l}\frac{A}{V_2}\left(c_2 - c_1\right) \tag{13}$$

For a partition/diffusion process, the **permeability coefficient** (κ, cm/s) is defined as the product of the partition coefficient (β) and the diffusion coefficient (D) divided by the thickness (l) of the membrane:

$$\kappa = \frac{\beta D}{l} \tag{14}$$

Note that although diffusion coefficients apply to diffusion in free solution or in membranes, permeability coefficients apply only to membranes. The higher the diffusion coefficient for movement of a substance across a membrane, the higher the permeability coefficient.

The **permeability** (P, s^{-1}) itself is defined as the permeability coefficient (κ) times the area (A) of the membrane divided by the volume (V) of the cell:

$$P = \kappa\frac{A}{V} = \frac{\beta D}{l}\frac{A}{V} \tag{15}$$

so that from Eq. 13,

$$\frac{dc_2}{dt} = -P\left(c_2 - c_1\right) \tag{16}$$

Therefore the permeability (P) is the rate of change of concentration with time (dc_2/dt) for a concentration gradient of unity. In general, the permeability is the proportionality constant that relates the rate of change of concentration with time to the concentration gradient. Note that the unit of P in Eqs. 15 and 16 is s^{-1}, whereas the permeability coefficient (κ) defined in Eq. 15 has units of cm/s. **In electrophysiology, however, the symbol P is often also used to represent the permeability coefficient with units of cm/s.** We shall see that the permeability depends on the model chosen for the mechanism of membrane transport.

Now we can understand the basis for Overton's classical rule that the rate of diffusion of many substances across cell membranes correlates with the oil/water partition coefficient. The rate of permeation of a water-soluble substance is primarily determined by how easily the substance passes from water to lipid; its oil/water partition coefficient is a measure of this ease. The more a substance dissolves in the membrane, the greater its concentration gradient within the membrane, and the faster the rate of diffusion across the membrane. The classical data for permeation of nonelectrolytes including alcohols, amides, and various substituted ureas into algae, red blood cells, and other cells is a striking example of this correlation (see Lieb and Stein, 1986, for a particularly insightful discussion).

The diffusion coefficient across cell membranes for various substances is generally greater when the molecular size of the substance is small and when the lipid solubility is high; that is, small molecules of high lipid solubility (i.e., less polar and nonpolar molecules) penetrate most quickly through the membrane. Most nonpolar molecules pass directly through the lipid bilayer matrix of the membrane, that is, not through special sites (e.g., water-filled pores or channels). Small charged ions (e.g., Na$^+$, K$^+$, Cl$^-$, Ca^{2+}) apparently pass through water-filled channels, some of which

have a voltage-dependent gating mechanism; these channels can exhibit a high degree of selectivity for specific ions, with the selectivity orders not being solely based on the hydrated or unhydrated sizes of the ions. One type of ionophore, valinomycin, forms a hydrophobic cage around K$^+$. The K$^+$-ionophore complex is lipid-soluble and passes rapidly through the lipid bilayer matrix.

Transport proteins are normally present in the cell membrane for net uptake of nutrients such as glucose and amino acids. Some of these are for downhill transport only, that is, down the electrochemical gradient. One type of protein-mediated transport in cells (e.g., for glucose) is known as **facilitated diffusion**, a type of transport that contrasts with the so-called simple diffusion across the lipid bilayer matrix as described in the preceding paragraphs. Another type of mediated transport is known as **exchange diffusion**, in which one molecule of substance (or ion) inside the cell is exchanged for one molecule outside the cell, in which case there is no net movement. In either type of mediated transport, the rate of downhill movement of the substance across the membrane is enhanced by the transport protein. Mediated transport systems differ from simple diffusion in that they exhibit **saturation kinetics**; that is, the rate of transport increases with an increase in the substrate concentration up to a maximum, after which the rate of transport levels off due to a finite number of available sites. Other characteristics of mediated transport systems include **competitive inhibition**, in which two or more substances may compete at the same site for binding to the transporter, and **noncompetitive inhibition**, in which a nontransported substance may bind to the transporter at a site different from the transport site and thereby alter or prevent binding of the usual substrate. Mediated transport exhibits greater **specificity** than simple diffusion; for example the rate of mediated transport of D-glucose is much greater than L-glucose, a difference that could not occur with simple diffusion. Because mediated transport resembles an enzymatic process involving protein conformational changes, the temperature coefficient (Q_{10}) is also higher than for simple diffusion.

V. Electrodiffusion

Combining Eqs. 11, 12, and 14, the flux density (J) across the membrane for a partition/diffusion process is given by

$$J = -\kappa\left(c_2 - c_1\right) \tag{17}$$

Hence the net flux density of a nonelectrolyte across a membrane is equal to the permeability coefficient (κ) times the difference in concentration across the membrane. The permeability coefficient for K$^+$ (κ_K) across resting striated muscle membrane is about 1×10^{-6} cm/s. Some values for κ_K (represented by the symbol P_K in cm/s) are given in Table 3 for some cardiac tissues.

The **unidirectional influx** (J_i) and **efflux** (J_o) are defined by

$$J_i = -\kappa c_1 \tag{18a}$$

$$J_o = -\kappa c_2 \tag{18b}$$

TABLE 3 Summary of Internal K^+ and Na^+ Concentrations and Permeabilities for Some Selected Heart Tissues

Preparation	Resting potential (mV)	P_K (cm/s × 10^{-7})	P_{Na} (cm/s × 10^{-7})	P_{Na}/P_K	$[K^+]_i$ (mM)	$[Na^+]_i$ (mM)	Ref.
Chick embryonic, 19 days old	—	3.10	0.053	0.017	122	15	Carmeliet *et al.* (1976)
Rabbit papillary	−77.5	—	—	—	135 82.6[a]	32.7 5.7[a]	Lee and Fozzard (1975)
Rabbit papillary	−86	—	—	—	119[b]	—	Akiyama and Fozzard (1975)
Rabbit ventricular	−76	—	—	—	83.1[a]	—	Fozzard and Lee (1976)
Cow Purkinje	−75	1.66	—	—	—	—	Carmeliet and Verdonck (1977)
Sheep Purkinje	—	—	—	—	160	—	Carmeliet and Bosteels (1969)

Note: Unless otherwise specified, all ion concentrations are based on total tissue analyses, and all permeability coefficients are for tissues bathed in normal Ringer's solution.

[a] Measured with ion-selective microelectrodes.

[b] Calculated from the internal K^+ concentration and activity coefficient.

The **net flux** is defined as the difference between the efflux and influx. Thus,

$$J = J_o - J_i = -\kappa \left(c_2 - c_1 \right) \tag{19}$$

Equations 18a and 18b show that the influx of a substance is equal to its permeability coefficient times the external concentration (c_1), whereas the efflux is equal to the permeability coefficient times the internal concentration (c_2). Conversely, the permeability coefficient is equal to the ratio of flux to concentration; influx to c_1 or efflux to c_2.

Fick's law of diffusion (Eq. 1) applies only to uncharged molecules. If there is a net charge on the molecule, then the unidirectional and net fluxes are also determined by any electrical field that may exist across the membrane. The equation for the net flux is then in the general form

$$J = \kappa f(E_m) \left(c_1 - c_2 e^{E_m \mathscr{F}/RT} \right) \tag{20}$$

where $f(E_m)$ is some function of the **electrical potential difference** across the membrane, and $\mathscr{F}$, R, and T are the Faraday constant, the gas constant, and absolute temperature, respectively. Note the similarity of Eq. 20 to Eq. 17, except for the membrane potential terms. The electrical potential difference across the membrane (E_m) equals $\psi_2 - \psi_1$, where ψ_2 is the inside potential and ψ_1 is the outside potential. When $f(E_m)$ is equal to $E_m \mathscr{F}/RT$, this term is dimensionless ($E_m \mathscr{F}$ is the electrical energy, whereas RT is the thermal energy); $RT/\mathscr{F} = 0.026$ volts at 37 °C. Similarly, the term $e^{E_m \mathscr{F}/RT}$ is dimensionless; therefore, the units for flux are the same as those given in Eq. 17.

For a monovalent cation, **Einstein's law of diffusion** states that

$$D = ukT = uRT/\mathscr{F} \tag{21}$$

where u is the **electrophoretic mobility** of the ion through the membrane and has units of a velocity per unit driving voltage gradient (cm/s per V/cm), and k is the Boltzmann constant ($=R/\mathscr{F}$).

Substituting Eq. 21 into the definition of the permeability coefficient (Eq. 14) gives

$$\kappa = \beta u RT / l \mathscr{F} \tag{22}$$

where β is the partition coefficient (dimensionless) for the ion between the bulk solution and the edge of the membrane, and l is the membrane thickness (cm). Equation 22 indicates that the permeability coefficient of an ion is directly related to the electrophoretic mobility of the ion through the membrane.

The Nernst-Planck equation states that the total flux density, J_i (mol/cm^2·s), of the ith species of ion across the membrane is the sum of the flux density due to diffusion and the flux density due to the electrical potential:

$$J_i = J_{i\,(\text{diffusion})} + J_{i\,(\text{electric})} \tag{23}$$

The net outward flux due to diffusion, given by Fick's law for one-dimensional flow across the membrane (Eq. 1), is proportional to the concentration gradient, dc_i/dx:

$$J_{i\,(\text{diffusion})} = -D_i \frac{dc_i}{dx} \tag{24}$$

where D_i (cm^2/s) is the diffusion coefficient of the ith species.

Combining Fick's law (Eq. 24) and Einstein's law (Eq. 21) gives the Nernst law of diffusion,

$$J_{i\,(\text{diffusion})} = -\frac{u_i RT}{\mathscr{F}} \frac{dc_i}{dx} \tag{25}$$

The flux density due to the electrical potential gradient across the membrane is given by Planck's law:

$$J_{i\,(\text{electric})} = -u_i c_i z_i \frac{d\psi}{dx} \tag{26}$$

To understand Planck's law, consider a cubic element of volume dV (cm^3), with area A (cm^2) and length dx (cm), containing dn moles of ions, adjacent to the membrane, as shown in Fig. 2. Then the rate of flow of ions across the membrane (dn/dt, mol/s), is

$$\frac{dn}{dt} = \frac{dn}{dV}\frac{dV}{dt} = \frac{dn}{dV}\frac{A\,dx}{dt} \qquad (27)$$

Since the concentration c equals dn/dV, and the velocity v equals dx/dt, then the molar flux density (J, mol/cm^2·s) of ions is the product of the concentration and the velocity:

$$J\left(\text{mol/cm}^2\text{·s}\right) = \frac{dn}{A\,dt} = cv \qquad (28)$$

The velocity in turn is the product of the ionic mobility (u) and the electric field ($d\psi/dx$). Hence, the molar flux density is:

$$J\left(\text{mol/cm}^2\text{·s}\right) = -cu\frac{d\psi}{dx} \qquad (29)$$

Note that a positive flux is obtained with a negative electric field. The molar flux density is converted to a flux density of ionic equivalents by multiplying by the ionic valence (z, eq/mole):

$$J\left(\text{eq/cm}^2\text{·s}\right) = -zcu\frac{d\psi}{dx} \qquad (30)$$

Thus from Eqs. 23, 25, and 30, the total flux (J_i) of the ith species is given by the Nernst-Planck-Einstein electrodiffusion equation:

$$J_i = -\frac{u_i RT}{\mathscr{F}}\frac{dc_i}{dx} - u_i c_i z_i \frac{d\psi}{dx} \qquad (31)$$

The current density (I_i, C/cm^2·s), carried by the ith ion is the product of the flux density (mol/cm^2·s) and the Faraday (C/eq).

$$I_i = J_i \mathscr{F} = -u_i RT\frac{dc_i}{dx} - u_i c_i z_i \mathscr{F}\frac{d\psi}{dx} \qquad (32)$$

The foregoing discussion shows that the flux of an ion is proportional to the gradient of electrochemical potential ($\tilde{\mu}$) for that ion. The net flux is from the side of greater electrochemical potential to the side of lesser electrochemical potential.

$$J_i = -c_i u_i \frac{d\tilde{\mu}_i}{dx} \qquad (33)$$

The **electrochemical potential** ($\tilde{\mu}$) is a measure of the useful energy, and its units are in joules/mole (just as voltage is in joules/coulomb). Electrochemical potential is composed of a chemical part (μ_c) and an electrical part (μ_e), that is,

$$\tilde{\mu} = \mu_c + u_e \qquad (34)$$

The chemical part (μ_c) is given by

$$\mu_c = \mu_c^0 + RT\ln a \qquad (35a)$$

where μ_c^0 is the chemical potential at standard temperature and pressure, and a is the activity (activity coefficient, γ, times the concentration). The electrical part (μ_e) is given by

$$\mu_e = z\mathscr{F}\psi \qquad (35b)$$

In animal cells, since there is neither a significant temperature difference nor a substantial pressure difference across the cell membrane, the maximal electrochemical potential for Na$^+$ inside is given by

$$\tilde{\mu}_2^{\text{Na}} = \mu_{\text{Na}}^0 + RT\ln\left(\gamma\left[\text{Na}^+\right]_2\right) + z\mathscr{F}\psi_2 \qquad (36)$$

where ψ_2 is the internal potential. The difference in electrochemical potential between inside and outside ($\Delta\tilde{\mu} = \tilde{\mu}_2 - \tilde{\mu}_1$) then is

$$\Delta\tilde{\mu}_{\text{Na}} = RT\ln\frac{\left[\text{Na}^+\right]_2}{\left[\text{Na}^+\right]_1} + z\mathscr{F}\Delta\psi \qquad (37)$$

where $\Delta\psi = \psi_2 - \psi_1$ and is the same as E_m. The activity coefficients cancel out, assuming $\gamma_2 = \gamma_1$.

The electrical current (i, in amp) carried by Na$^+$ is equal to the flux (J', in mol/s) times $z\mathscr{F}$ or

$$i = J'z\mathscr{F} \qquad (38)$$

or

$$I = Jz\mathscr{F} \qquad (39)$$

where I is the current density in amp/cm^2, and J is the flux density in mol/cm^2·s).

VI. Ussing Flux Ratio Equation

As pointed out in Eqs. 18a and 18b, the unidirectional fluxes are determined also by the permeability coefficient. Ussing (1949) developed the so-called flux ratio equation, in which the ratio of influx to efflux is used (permeability cancels out). Specifically, for the ratio of Na$^+$ fluxes (by simple electrodiffusion), the following applies:

$$\frac{J_2^{\text{Na}}}{J_1^{\text{Na}}} = \frac{\left[\text{Na}^+\right]_1}{\left[\text{Na}^+\right]_2}e^{-E_m\mathscr{F}/RT} \qquad (40)$$

Thus for Na$^+$, the ratio of influx : efflux (passive) in a resting membrane (assuming a resting potential of −80 mV) is

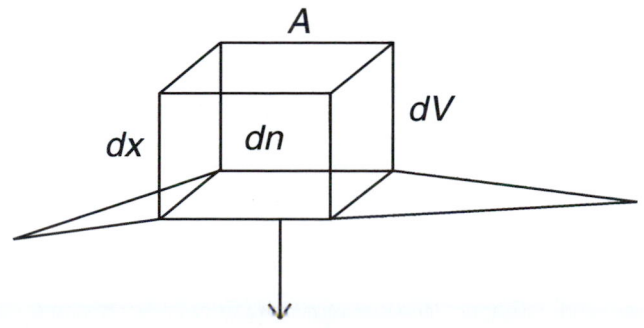

FIGURE 2. A cubic volume element (dV) of length dx and area A containing dn molecules passing across a membrane.

$$\frac{J_2^{Na}}{J_1^{Na}} = \frac{150\,mM}{15\,mM}\, e^{+80\,mV/26\,mV} = 217$$

Thus the passive influx of Na$^+$ should be 217 times greater than the passive efflux, because of the large electrochemical gradient directed inward.

The flux ratio for K$^+$ would be

$$\frac{J_2^{K}}{J_1^{K}} = \frac{4\,mM}{150\,mM}\, e^{+80\,mV/26\,mV} = 0.574$$

Thus the passive influx should be 0.574 times the passive efflux. The K$^+$ equilibrium potential (E_K) is only slightly greater (more negative by about 14 mV) than the resting potential, and so the passive flux ratio should be close to one.

A modified form of Eq. 40 can be obtained by substituting the ratio of ions with $e^{E_i\mathscr{F}/RT}$ (derived from the Nernst equation), giving

$$\frac{J_2^{i}}{J_1^{i}} = e^{-(E_m - E_i)\mathscr{F}/RT} \qquad (41)$$

where E_i is the **equilibrium potential** for the cation in question, and J_2^i and J_1^i are the inward and outward fluxes, respectively, for ion i. The newer sign convention refers inside solution to outside solution. Thus, for Na$^+$ we have

$$\frac{J_2^{Na}}{J_1^{Na}} = e^{-[-80\,mV - (+60\,mV)]/26\,mV} = 217$$

This value of 217 obtained for the flux ratio is identical to that obtained from Eq. 40. One advantage of Eq. 41 is that it is obvious at a glance that when $E_i = E_m$, the flux ratio is exactly 1.0, because $e^0 = 1$, where e is the base of the natural logarithm. Thus, if Cl$^-$ is passively distributed so that $E_{Cl} = E_m$, its flux ratio should be 1.0; that is, influx equals efflux, so there is no net flux. The larger the difference between E_i and E_m, that is, the farther the ion is from equilibrium, the greater the flux ratio.

Sometimes Eqs. 40 and 41 do not fit the experimental facts. The data are better fitted if the exponential term contains another factor (n), which is an empirical factor:

$$\frac{J_2^{i}}{J_1^{i}} = e^{-n(E_m - E_i)\mathscr{F}/RT} \qquad (42)$$

The best fit of the data is when n has a value of 2.5 to 4.0, depending on the membrane under investigation. One interpretation given to n is that if the length of the water-filled pore that the ion must traverse across the membrane is much longer than the ion diameter, as is likely, then for so-called **single-file diffusion,** n hits on the same side are required for the ion to complete its journey across the membrane. One could consider, for example, that there are three potential energy wells, or a chain of reactive sites, along the length of the pore, and the only way for the ion to escape the well is to receive a kinetic bump from an adjacent ion in the file. Complete permeation of an ion through the pore is more likely to happen if the ion is moving in the same direction as the majority of ions, that is, down the electrochemical gradient. Therefore, this factor (Eq. 42) makes the flux ratio much greater than would otherwise be predicted.

VII. Summary

This chapter described diffusion of uncharged particles and charged ions across membranes with and without partitioning, and provided the relevant equations that govern such diffusion. The relationship between the diffusion coefficient (D) and the permeability coefficient (P) was described, and the factors that determine these coefficients were discussed. The dependence of P on the mobility of an ion through the membrane under a voltage gradient was also discussed. Electrochemical potential was defined, and the interconversion between flux and current was developed. Finally, the Ussing flux ratio equation was presented, and examples of its significance were given. Relating to this, the concept of potential energy wells and barriers was presented, describing the movement of an ion through an ion channel. The exponential time course of diffusion is derived in this chapter's appendix.

Appendix: Exponential Time Course of Diffusion

Solutions of the diffusion equation generalized for three dimensions, in compartments of various geometries such as sheets, cylinders, and spheres may be found in treatises by Jacobs (1967), Crank (1956), and Jost (1960).

For a two-compartment system, denoted by subscripts 1 and 2, the total number of molecules N_T in the system is constant, $N_T = c_1 V_1 + c_2 V_2$, so that solving for c_1, we first obtain a relationship that will simplify the treatment of specific cases.

$$c_1 = \frac{N_T}{V_1} - c_2 \frac{V_2}{V_1} \tag{A-1}$$

As shown in the following two cases, the time course of diffusion is an exponential process.

Case 1. Efflux from a cell into a large volume V_1 of medium of constant chemical concentration c_1.

Mathematically, we assume that V_1 is so much greater than V_2 that c_1 remains constant (i.e., equal to N_T/V_1 in Eq. A-1). Hence, from Eq. 4,

$$\frac{dc_2}{(c_2 - c_1)} = -\frac{DA}{lV_2} dt \tag{A-2}$$

Integrating both sides,

$$\int_{c_2^0}^{c_2^t} \frac{dc_2}{(c_2 - c_1)} = -\frac{DA}{lV_2} \int_0^t dt \tag{A-3}$$

where c_2^0 is the internal concentration at $t = 0$ and c_2^t is at time t. Thus,

$$\ln \frac{c_2^t - c_1}{c_2^0 - c_1} = -\frac{DA}{lV_2} t \tag{A-4}$$

and

$$\frac{c_2^t - c_1}{c_2^0 - c_1} = e^{-DAt/lV_2} \tag{A-5}$$

Solving for c_2^t,

$$c_2^t = c_1 + \left(c_2^0 - c_1\right) e^{-DAt/lV_2} \tag{A-6}$$

Note that at $t = 0$, $c_2^t = c_2^0$, and as t approaches infinity, c_2^t approaches c_1, as expected.

Case 2. Influx from a medium of constant chemical composition c_1 into a cell of constant small volume V_2.

As before, according to Fick's first law with a linear concentration gradient, the influx dN_1/Adt is given by

$$\frac{dN_1}{Adt} = -D\frac{dc}{dx} = -D\frac{(c_2 - c_1)}{l} \tag{A-7}$$

From the conservation of mass, $dN_1 = -dN_2$. Dividing by V_2, and rearranging with c_1 constant, gives

$$\frac{dc_2}{\left(c_1 - c_2\right)} = -\frac{DA}{lV_2} dt \tag{A-8}$$

Again, integrating both sides,

$$\int_{c_2^0}^{c_2^t} \frac{dc_2}{c_1 - c_2} = -\frac{DA}{lV_2} \int_0^t dt \tag{A-9}$$

Thus,

$$\ln \frac{c_1 - c_2^t}{c_1 - 0} = -\frac{DA}{lV_2} t \tag{A-10}$$

and solving for c_2^t,

$$c_2^t = c_1 \left(1 - e^{-DAt/lV_2}\right) \tag{A-11}$$

Note that at $t = 0$, $c_2^t = 0$, and as t approaches infinity, c_2^t approaches c_1, again as expected.

The half-time of equilibration, $t_{1/2}$, is found by letting $c_2^t = c_1/2$,

$$\frac{c_1}{2} = c_1 \left(1 - e^{-kt_{1/2}}\right) \tag{A-12}$$

where $k = DA/lV_2$. Solving for $t_{1/2}$,

$$t_{1/2} = \frac{\ln 2}{k} = \frac{0.693}{k} \tag{A-13}$$

Note that the half-time for completion of diffusion does not depend on the concentration gradient, but merely on the diffusion coefficient, the area and the thickness of the membrane, and the volume of solution.

Bibliography

Akiyama, T., and Fozzard, H. A. (1975). Influence of potassium ions and osmolality on the resting membrane potential of rabbit ventricular papillary muscle with estimation of activity and the activity coefficient of internal potassium. *Cir. Res.* **37**, 621–629.

Bockris, J. O'M., and Conway, B. E. (1954). "Modern Aspects of Electrochemistry." Butterworths, London.

Carmeliet, E., and Bosteels, S. (1969). Coupling between Cl flux and Na or K flux in cardiac Purkinje fibers: influence of pH. *Arch. Int. Physiol. Biochim.* **77**, 57–72.

Carmeliet, E., and Verdonck. F. (1977). Reduction of potassium permeability by chloride substitution in cardiac cells. *J. Physiol. (London)* **265**, 193–206.

Carmeliet, E. E., Horres, C. R., Lieberman, M., and Vereecke, J. S. (1976). Developmental aspects of potassium flux and permeability of the embryonic chick heart. *J. Physiol. (London)* **254**, 673–692.

Crank, J. (1956). "Mathematics of Diffusion." Oxford University Press, New York.

Cunningham, R. E., and R. J. J. Williams. (1980). "Diffusion in Gases and Porous Media." Plenum Press, New York.

Danielli, J. (1943). The theory of penetration of a thin membrane: Appendix. *In* "The Permeability of Natural Membranes" (H. Davson and D. F. Danielli, Eds.), Cambridge University Press, London.

Davson, H. (1964). "A Textbook of General Physiology," 3rd ed. Little, Brown, and Company, Boston.

Einstein, A. (1926). "Investigations on the Theory of Brownian Movement." (R. Furth, ed., and A. D. Cowper, Transl.). Methuen, London; reprinted by Dover Publications, Inc., 1956.

Fozzard, H. A., and Lee, C. 0. (1976). Influence of changes in external potassium and chloride ions on membrane potential and intracellular potassium ion activity in rabbit ventricular muscle. *J. Physiol. (London)* **256**, 663–689.

Hodgkin, A. L., and Huxley, A. F. (1952). Currents carried by sodium and potassium ions through the membrane of the giant axon of Loligo. *J. Physiol. (London)* **116**, 449–472.

Jacobs, M. H. (1967). "Diffusion Processes." Springer-Verlag, New York.

Jost, W. (1960). "Diffusion in Solids, Liquids, Gases." Academic Press, NY, 1960.

Lee, C. O., and Fozzard, H. A. (1975). Activities of potassium and sodium ions in rabbit heart muscle. *J. Gen. Physiol.* **65**, 695–708.

Lieb, W. R., and Stein, W. (1986). Simple diffusion across the membrane bilayer. In: "Transport and Diffusion across Cell Membranes." Academic Press, New York, Chapter 2, pp. 69–112.

Mullins, L. J. (1961). The macromolecular properties of excitable membranes. *Ann. N. Y. Acad. Sci.* **94**, 390–404.

Robinson, R. A., and Stokes, R. H. (1959). "Electrolyte Solutions." Butterworths, London.

Sperelakis, N. (1979). Origin of the cardiac resting potential. *In* "Handbook of Physiology: Vol. 1, The Cardiovascular System" (R. M. Berne and N. Sperelakis, Eds.), pp. 187–267. Am. Physiological Society, Bethesda, MD.

Sperelakis, N. (1995). "Electrogenesis of Biopotentials," Kluwer Publishing Co., New York.

Ussing, H. H. (1949). Distinction by means of tracers between active transport and diffusion. The transfer of iodide across isolated frog skin. *Acta Physiol. Scand.* **19**, 43–56.

Nicholas Sperelakis

14

Origin of Resting Membrane Potentials

I. Introduction

The cell membrane exerts tight control over the electrical activity and the contractile machinery during the process of excitation-contraction (electromechanical) coupling. Some drugs and toxins exert primary or secondary effects on the electrical properties of the cell membrane and thereby exert effects, for example, on automaticity, arrhythmias, and force of contraction of the heart. Therefore, for an understanding of the mode of action of therapeutic drugs, toxic agents, neurotransmitters, hormones, and plasma electrolytes on the electrical activity of nerve and muscle, it is necessary to understand the electrical properties and behavior of the cell membrane at rest and during excitation. The first step in gaining such an understanding is to examine the electrical properties of nerve and muscle cells at rest, including the origin of the resting membrane potential (E_m). The resting E_m and action potential (AP) result from properties of the cell membrane and the ion distributions across it.

II. Passive Electrical Properties

A. Membrane Structure and Composition

As discussed in previous chapters, the cell membrane is composed of a bimolecular leaflet of phospholipid molecules (e.g., phosphatidylcholine, phosphatidylethanolamine) with protein molecules floating in the lipid bilayer. The nonpolar hydrophobic ends of the phospholipid molecules project toward the middle of the membrane, and the polar hydrophilic ends project toward the edges of the membrane bordering on the water phases (Fig. 1). This orientation is thermodynamically favorable. The lipid bilayer membrane is about 50–70 Å thick, and the phospholipid molecules are about the right

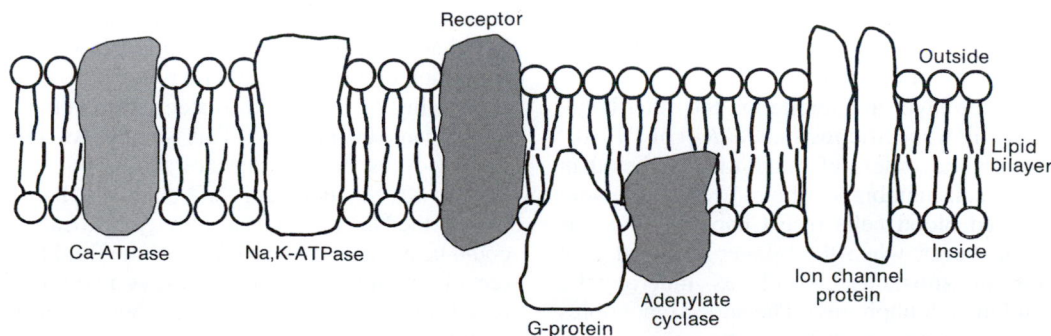

FIGURE 1. Diagrammatic illustration of cell membrane substructure showing the lipid bilayer. Nonpolar hydrophobic tail ends of the phospholipid molecules project toward the middle of the membrane, and polar hydrophilic heads border on the water phase at each side of the membrane. Lipid bilayer is about 50–70 Å thick. For simplicity, the cholesterol molecules are not shown. Large protein molecules protrude through entire membrane thickness or are inserted into one leaflet only, as depicted. These proteins include various enzymes associated with the cell membrane as well as membrane ionic channels. Membrane has fluidity so that the protein and lipid molecules can move around in the plane of the membrane and fluorescent probe molecules inserted into the hydrophobic region of the membrane have freedom to rotate.

length (30–40 Å) to stretch across half of the membrane thickness. Cholesterol molecules are in high concentration in the cell membrane (of animal cells) and are inserted between the phospholipid molecules, giving a phospholipid: cholesterol ratio of about 1.0. Some of the large protein molecules, called integral membrane proteins, protrude through the entire membrane thickness; for example, the Na^+,K^+-ATPase, Ca^{2+}-ATPase, and the various ion channel proteins protrude through both leaflets, whereas other proteins are inserted into one leaflet (inner or outer) only. These proteins "float" in the lipid bilayer matrix, and the membrane has fluidity (reciprocal of microviscosity), such that the protein molecules can move around laterally in the plane of the membrane. Some ion channels, for example, fast Na^+ channels of the node of Ranvier of myelinated neurons, are tethered in place to the cytoskeleton by anchoring proteins, such as ankyrin.

The outer surface of the cell membrane is lined with strands of mucopolysaccharides (the cell coat or glycocalyx) that endow the cell with immunochemical properties. The cell coat is highly charged negatively and therefore can bind cations, such as Ca^{2+}. Treatment with neuraminidase to remove sialic acid residues destroys the cell coat.

B. Membrane Capacitance and Resistivity

Lipid bilayer membranes made artificially have a specific membrane capacitance (C_m) of 0.4–1.0 µF/cm², which is close to the value for biologic membranes. The capacitance of natural cell membranes is due to this lipid bilayer matrix. A capacitor consists of two parallel plate conductors separated by a dielectric material of high resistance (e.g., oil). The factors that determine the value of the capacitance of a membrane are given in the equation

$$C_m = \frac{\epsilon A_m}{\delta} \frac{1}{4\pi k} \qquad (1)$$

where A_m is the membrane area (in cm²) and k is a constant (9.0 x 10^{11} cm/F). Calculation of membrane thickness (δ) from Eq. 1, assuming a measured membrane capacitance (C_m) of 0.7 µF/cm² and a dieletric constant (ϵ) of 5, gives 63 Å. Most oils have dielectric constants of 3–5. The more dipolar the material, the greater the dielectric constant (e.g., water, which is very dipolar, has a value of 81, compared with a value of 1.000 for a vacuum or air). Capacitors are discussed in more detail in Chapter 70.

The artificial lipid bilayer membrane, on the other hand, has an exceedingly high specific resistance (R_m) of 10^6–10^9 Ω·cm², which is several orders of magnitude higher than that of the biologic cell membrane (about 10^3 Ω·cm²), with the exception of red blood cells (see Chapter 23). R_m is greatly lowered, however, when the bilayer is doped with certain proteins or substances, such as macrocyclic polypeptide antibiotics (ionophores). The added ionophores may be of the ion-carrier type, such as valinomycin, or of the channel-former type, such as gramicidin. Therefore, the presence of proteins that span across the thickness of the cell membrane must account for the relatively low resistance (high conductance) of the cell membrane. These proteins include those associated with the voltage-dependent gated ion channels of the cell membrane. In summary, the

capacitance is due to the lipid bilayer matrix, and the conductance is due to proteins inserted in the lipid bilayer.

The dielectric property of the cell membrane is very good. For a resting E_m of 80 mV and a thickness of 60 Å, the voltage gradient sustained across the membrane is 133 000 V/cm. Thus, the cell membrane tolerates an enormous voltage gradient.

C. Membrane Fluidity

The electrical properties and the ion transport properties of the cell membrane are determined by the molecular composition of the membrane. The lipid bilayer matrix even influences the function of the membrane proteins; for example, the Na^+,K^+-ATPase activity is affected by the surrounding lipid. A high cholesterol content lowers the fluidity of the membrane. The polar portion of cholesterol lodges in the hydrophilic part of the membrane, and the nonpolar part of the planar cholesterol molecule is wedged between the fatty acid tails, thus restricting their motion and lowering fluidity. A high degree of unsaturation and branching of the tails of the phospholipid molecules raises the fluidity; phospholipids with unsaturated and branched-chain fatty acids cannot be packed tightly because of steric hindrance due to their greater rigidity. Chain length of the lipids also affects fluidity. Low temperature decreases membrane fluidity, as expected. Ca^{2+} and Mg^{2+} may diminish the charge repulsion between the phospholipid headgroups; this allows the bilayer molecules to pack together more closely, thereby constraining the motion of the tails and reducing fluidity. Each phospholipid tail occupies about 20–30 Å², and each headgroup about 60 Å² (Jain, 1972). Membrane fluidity changes occur in muscle development and in certain disease states such as cancer, muscular dystrophy (Duchenne type), and myotonic dystrophy.

The hydrophobic portion of local anesthetic molecules may interpose between the lipid molecules. This separates the acyl chain tails of the phospholipid molecules further, reducing the van der Waals forces of interaction between adjacent tails, and thus increasing the membrane fluidity. Local anesthetics depress the resting conductance of the membrane and the voltage-dependent changes in g_{Na}, g_K, and g_{Ca}. That is, the local anesthetics produce a nonselective depression of all ionic conductances of the membrane. At least part of this depression could be an indirect effect of the anesthetics on the fluidity of the lipid matrix. At the concentration of a local anesthetic required to completely block excitability, its estimated concentration in the lipid bilayer is more than 100 000/µm². This should be compared with a density of fast Na^+ channels of about 20–100/µm², and even less for K^+ channels and Ca^{2+} channels. Part of the depression of the Na^+,K^+-ATPase activity by local anesthetics also could be explained by an effect on the fluidity, although a direct effect on the protein enzyme is also possible. For additional information on fluidity, the reader is referred to Chapters 4 and 5.

D. Potential Profile across Membrane

The cell membrane has fixed negative charges on its outer and inner surfaces. The charges are presumably due to acidic phospholipids in the bilayer and to protein molecules either embedded in the membrane (islands floating in the lipid bilayer matrix) or tightly adsorbed to the surface of the membrane. Most proteins have an acid isoelectric point, so at a pH near 7.0 they possess a net negative charge. The charge at the outer surface of the cell membrane, with respect to the solution bathing the cell, is known as the **zeta potential.** This charge is responsible for the electrophoresis of cells in an electric field, the cells moving toward the anode (positive electrode) because unlike charges attract. This surface charge affects the true potential difference (PD) across the membrane, as shown in Fig. 2A. At each surface, the fixed charge

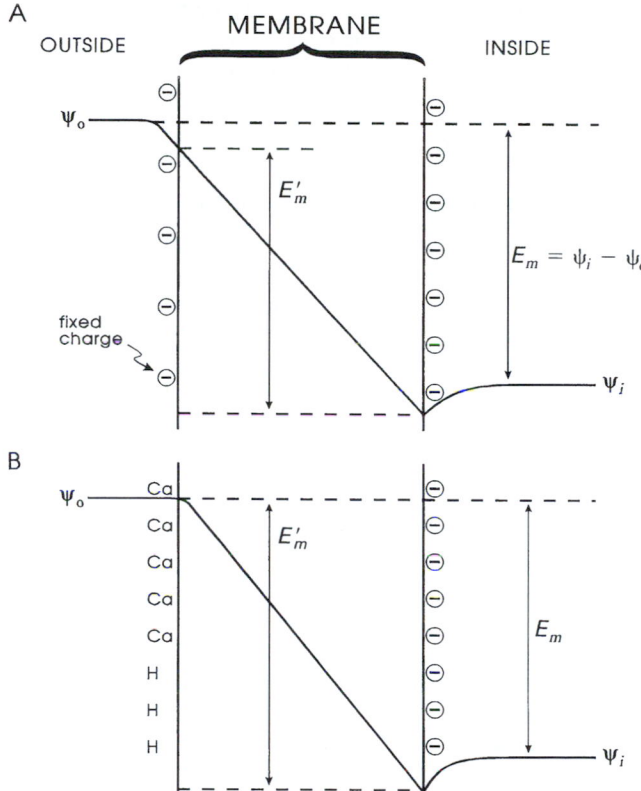

FIGURE 2. Potential profile across the cell membrane. (A) Because of fixed negative charges (at pH 7.4) at outer and inner surfaces of the membrane, there is a negative potential that extends from the edge of the membrane into the bathing solution on both sides of the membrane. This surface potential falls off exponentially with distance into the solution. Magnitude of the surface potential is a function of the charge density. Ψ_o is the electrical potential of the outside solution, Ψ_i is that of the inside solution, and membrane potential (E_m) is the difference ($\Psi_i - \Psi_o$). E_m is determined by the equilibrium potentials and relative conductances. Profile of the potential through the membrane is shown as linear (the constant-field assumption), although this need not be true for the present purpose. If the outer surface potential is exactly equal to that in the inner surface, then the true transmembrane potential (E'_m) is exactly equal to the (microelectrode) measured membrane potential (E_m). (B) If the outer surface potential is different from the inner potential, for example, by increasing the extracellular Ca^{2+} concentration or decreasing the pH to bind Ca^{2+} or H^+ to more of the negative charges, then the E'_m is greater than the measured E_m. Diminution of the inner surface charge decreases E'_m. The membrane ion channels are controlled by E'_m.

produces an electric field that extends a short distance into the solution and causes each surface of the membrane to be slightly more negative (by a few millivolts) than the extracellular and intracellular solutions. The potential theoretically recorded by an ideal tiny electrode, as the electrode is driven through the solution perpendicular to the membrane surface, should become negative as the electrode approaches within a few angstroms of the surface. The potential difference between the membrane surface and the solution declines exponentially as a function of distance from the surface. The length constant is a function of the ionic strength (or resistivity) of the solution: the lower the ionic strength, the greater the length constant. The magnitude of the PD depends on the density of the charge sites (number of chemical groups per unit of membrane area); the number of ionized charge groups is also affected by the pH and ionic strength.

The membrane potential (E_m) measured by an intracellular microelectrode is the potential of the inner solution (Ψ_i) minus the potential of the outer solution (Ψ_o); that is,

$$E_m = \Psi_i - \Psi_o \qquad (2)$$

The true PD across the membrane (E'_m), however, is really that PD directly across the membrane, as shown in Fig. 2A. If the surface charges at each surface of the membrane are equal, then $E'_m = E_m$. If the outer surface charge is decreased to zero by extra binding of protons or cations (such as Ca^{2+}), then the membrane becomes slightly hyperpolarized ($E'_m > E_m$), although this is not measurable by the intracellular microelectrode (Fig. 2B). Conversely, if the inner surface charge were neutralized and the outer surface charge restored, then the membrane would become slightly depolarized ($E'_m < E_m$). Again, this change is not measurable by the microelectrode, which can only measure the PD between the two solutions.

Because the membrane ionic conductances are controlled by the PD directly across the membrane (i.e., by E'_m and not by E_m), changes in the surface charges (e.g., by drugs, ionic strength, or pH) can lead to apparent shifts in the electrical threshold potential, mechanical threshold potential (the E_m value at which contraction of muscle just begins), activation curve, and inactivation curve (discussed in subsequent chapters). For example, elevation of extracellular Ca^{2+} concentration ($[Ca^{2+}]_o$) is known to raise the threshold potential (i.e., the critical depolarization required to reach electrical threshold), as expected from the small increase in E'_m that should occur.

III. Maintenance of Ion Distributions

A. Resting Potentials and Ion Distributions

The transmembrane potential in resting nerve and muscle cells varies with the type of cell. In myocardial cells, it is about −80 mV (Table 1). The resting E_m or maximum diastolic potential in cardiac Purkinje fibers is somewhat greater (about −90 mV), whereas that in the nodal cells of the heart is lower (about −60 mV). In nerve cells, the resting potential is about −70 mV, whereas in skeletal muscle fibers the value is close to −80 mV. In most smooth muscle cells, the resting

TABLE 1 Comparison of the Resting Potentials in
Different Types of Cells

Cell type	Resting potential (mV)
Neuron	−70
Skeletal muscle (mammalian)	−80
Skeletal Muscle (frog)	−90
Cardiac muscle (atrial and ventricular)	−80
Cardiac Purkinje fiber	−90
Atrioventricular nodal cell	−65
Sinoatrial nodal cell	−55
Smooth muscle cell	−55
Red blood cell (human)	−11

E_m is lower, about −55 mV. The values of the resting E_m for different types of cells are summarized in Table 1.

The ionic composition of the extracellular fluid bathing the muscle cells is similar to that of the blood stream. It is high in Na^+ (about 145 mM) and Cl^- (about 100 mM), but low in K^+ (about 4.5 mM). The Ca^{2+} concentration is about 2 mM, but about half is bound to serum proteins. In contrast, the intracellular fluid has a low concentration of Na^+ (about 15 mM or less) and Cl^- (about 6–8 mM), but a high concentration of K^+ (about 150–170 mM). These ion distributions are listed in Table 2. The free intracellular Ca^{2+} concentration ($[Ca^{2+}]_i$ is about 10^{-7} M or less, but during contraction it may rise as high as 10^{-5} M. The total intracellular Ca^{2+} is much higher (about 2 mM/kg), but most of this is bound to molecules such as proteins or is sequestered into compartments such as mitochondria and the sarcoplasmic reticulum (SR). Most of the intracellular K^+ is free, and it has a diffusion coefficient only slightly less than that of K^+ in free solution, consistent with the tortuosity of the diffusion paths around intracellular structures. Thus, under normal conditions, the cell maintains

TABLE 2 Summary of the Ion Distributions in Most
Types of Cells and the Equilibrium Potentials
Calculated from the Nernst Equation

Ion	Extracellular distribution (mM)	Intracellular distribution (mM)	Equilibrium potential (mV)
Na^+	145	15	+60
Cl^-	100	5^a	−80
K^+	4.5	150	−94
Ca^{2+}	1.8	0.0001	+130
H^+	0.0001	0.0002	−18

^{a}Assuming Cl^- is passively distributed and resting E_m is −80 mV. The extracellular H^+ concentration is given for pH 7; it would be 40 nM at pH 7.4.

an internal ion concentration markedly different from that in the medium bathing the cells, and it is these ion concentration differences that underlie the resting potential and excitability. The existence of the resting potential enables action potentials (APs) to be produced in those types of cells that have excitability. The ion distributions and related pumps and exchange reactions are depicted in Fig. 3.

Inhibition of the Na^+-K^+ pump (e.g., by cardiac glycosides such as digitalis) causes the ion concentration gradients to gradually run down or dissipate. The cells lose K^+ and gain Na^+, Cl^-, and water. Therefore, the K^+ and Na^+ equilibrium potentials (E_K and E_{Na}) become smaller and the cells become depolarized (see Section IV). The depolarization causes the cells to gain Cl^- (see Section III.C), and therefore also to gain water because of the resultant gain in osmotic strength, causing the cells to swell.

Although the topic of Na^+,K^+-ATPase, and the Na^+-K^+ pump is discussed in detail in Chapter 17, a brief description is given here.

B. Na+ and K+ Distribution and the Na+-K+ Pump

The intracellular ion concentrations are maintained differently from those in the extracellular fluid, by active ion transport mechanisms that expend metabolic energy to transport specific ions against their concentration or electrochemical gradients. These ion pumps are located in the cell membrane at the cell surface and probably also in the transverse tubular membrane of striated muscle cells. The major ion pump is the Na^+-K^+-linked pump, which pumps Na^+ out of the cell against its electrochemical gradient, while simultaneously pumping K^+ in against its electrochemical gradient (Fig. 3). The coupling between Na^+ and K^+ pumping is obligatory, since in zero $[K^+]_o$, the Na^+ can no longer be pumped out. That is, a coupling ratio of 3 Na^+:0 K^+ is not possible. The coupling ratio of Na^+ pumped out to K^+ pumped in is generally 3:2. The Na^+-K^+ pump is half-inhibited (K_i value) when $[K^+]_o$ is lowered to about 2 mM.

If the ratio were 3:3, the pump would be electrically neutral or nonelectrogenic. A PD across the membrane would not be produced directly, because the pump would pull in three positive charges (K^+) for every three positive charges (Na^+) it pushed out. When the ratio is 3:2, the pump is electrogenic and directly produces a PD that causes E_m to be greater (more negative) than it would be otherwise, solely on the basis of the ion concentration gradients and relative permeabilities or net diffusion potential (E_{diff}). A coupling ratio of 3 Na^+:1 K^+ would produce a greater electrogenic pump potential (V_p). Under normal steady-state conditions, the contribution of the Na^+-K^+ electrogenic pump potential to $E_m(\Delta V_p)$ is only a few millivolts in myocardial cells; the contribution is greater in smooth muscle cells (6–8 mV) and mammalian skeletal muscle fibers (12–16 mV) (see Section VII).

The driving mechanism for the Na^+-K^+ pump is a membrane ATPase, the Na^+,K^+-ATPase, which spans the membrane and requires both Na^+ and K^+ ions for activation. This enzyme requires Mg^{2+} for activity, and it is actually inhibited by Ca^{2+}. ATP, Mg^{2+}, and Na^+ are thus required at the inner

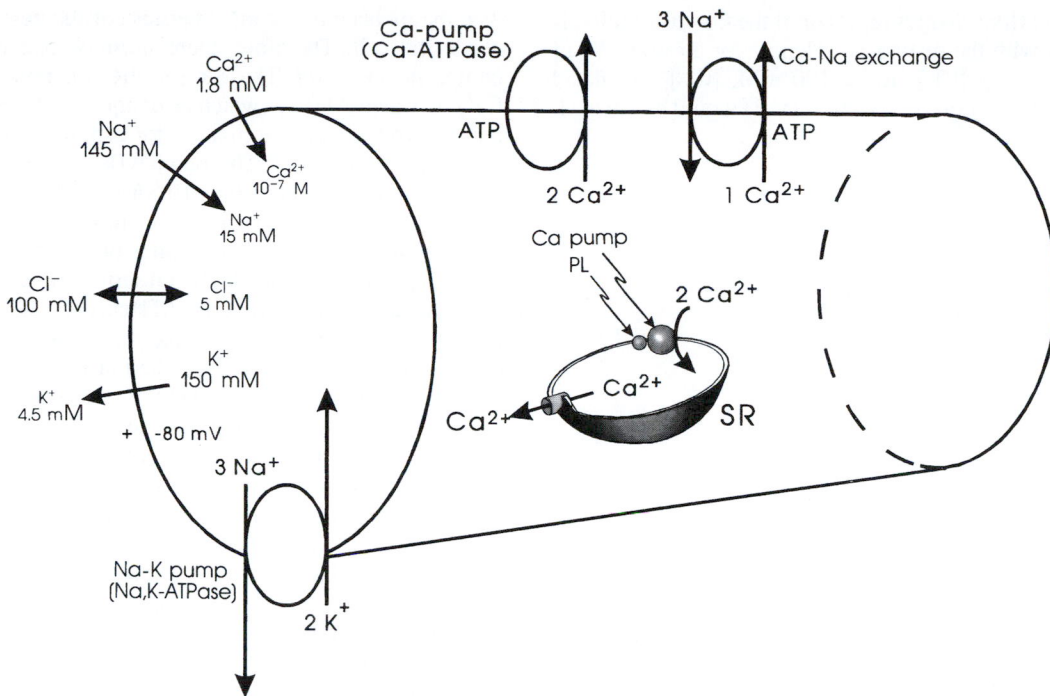

FIGURE 3. Intracellular and extracellular ion distributions in a myocardial cell or skeletal muscle fiber. Also given are polarity and magnitude of the resting potential. Arrows give direction of the net electrochemical gradient. Na⁺-K⁺ pump is located in the cell surface and the T-tubule membranes. A Ca²⁺-ATPase/Ca²⁺ pump, similar to that in the SR, is located in the cell membrane. A Ca²⁺-Na⁺ exchange carrier is located in the cell membrane.

surface of the membrane, and K^+ is required at the outer surface. A phosphorylated intermediate (aspartate residue) of the Na^+,K^+-ATPase occurs in the transport cycle, its phosphorylation being Na^+ dependent and its dephosphorylation being K^+ dependent (for references, see Sperelakis, 1979). The pump enzyme usually drives three Na^+ ions out and two K^+ ions in for each ATP molecule hydrolyzed. The Na^+,K^+-ATPase is specifically inhibited by the cardiac glycosides (digitalis drugs) acting on the outer surface. The pump enzyme is also inhibited by vanadate ion and by sulfhydryl (SH) reagents [such as N-ethylmaleimide (NEM), mercurial diuretics, and ethacrynic acid], thus indicating that the SH groups are crucial for activity.

Blockade of the Na^+-K^+ pump produces only a small immediate effect on the resting E_m: a small depolarization of about 2–16 mV, depending on cell type, representing the contribution of V_p to $E_m (\Delta V_p)$. Because excitability and generation of APs are almost unaffected at short times, excitability is independent of active ion transport. However, over many minutes, depending on the ratio of surface area to volume of the cell, the resting E_m slowly declines because of gradual dissipation of the ionic gradients. The progressive depolarization depresses the rate of rise of the AP, and hence the propagation velocity, and eventually all excitability is lost. Thus, a large resting potential and excitability, although not immediately dependent on the Na^+-K^+ pump, are ultimately dependent on it.

The rate of Na^+-K^+ pumping in excitable cells must change with the amount of electrical activity to maintain relatively constant intracellular ion concentrations. A higher frequency of APs results in a greater overall movement of ions down their electrochemical gradients, and these ions must be repumped to maintain the ion distributions. For example, the cells tend to gain Na^+, Cl^-, and Ca^{2+} and to lose K^+. The factors that control the rate of Na^+-K^+ pumping include $[Na^+]_i$ and $[K^+]_o$. In cells that have a large surface area to volume ratio (such as small-diameter nonmyelinated axons), $[Na^+]_i$ may increase by a relatively large percentage during a train of APs, and this would stimulate the pumping rate. Likewise, an accumulation of K^+ externally occurs and also stimulates the pump. As mentioned above, the K_m value for K^+, that is, the concentration for half-maximal rate, is about 2 mM. It has been shown that $[K^+]_o$ is significantly increased during the long AP plateau in cardiac muscle.

C. Cl⁻ Distribution

In many invertebrate and vertebrate nerve and muscle cells, Cl^- ion does not appear to be actively transported; that is, there is no Cl^- ion pump (or Cl^--ATPase). In such cases, Cl^- distributes itself passively (no energy used) in accordance with E_m. In such a case, E_{Cl} is equal to E_m in a resting cell. For example, in mammalian myocardial cells, Cl^- seems to be close to passive distribution, because $[Cl^-]_i$ is at, or only slightly above, the value predicted by the Nernst equation from the resting E_m (for references, see Sperelakis, 1979). When passively distributed, $[Cl^-]_i$ is low because the negative potential inside the cell (the resting potential) pushes out the negatively charged Cl^- ion (like charges repel) until the Cl^- distribution is at equilibrium with the resting E_m. Hence, for a resting E_m of -80 mV, and taking $[Cl^-]_o$ to be 100 mM, $[Cl^-]_i$ calculated from the Nernst equation (see Section IV.B) would be 4.9 mM.

$$E_{Cl} = E_m \qquad (3)$$

$$E_m = +61\,\text{mV} \, \log \frac{[Cl^-]_i}{[Cl^-]_o}$$

$$= -61\,\text{mV} \, \log \frac{[Cl^-]_o}{[Cl^-]_i}$$

$$\frac{-80\,\text{mV}}{-61\,\text{mV}} = \log \frac{100\,\text{mM}}{[Cl^-]_i}$$

$$\frac{100\,\text{mM}}{[Cl^-]_i} = \text{antilog} \, \frac{-80\,\text{mV}}{-61\,\text{mV}} = \text{antilog}\,1.31 = 20.5$$

$$[Cl^-]_i = \frac{100\,\text{mM}}{20.5} = 4.88\,\text{mM}$$

During the AP, the inside of the cell goes in a positive direction, and a net Cl^- influx (outward Cl^- current, I_{Cl}) will occur and thus increase $[Cl^-]_i$. The magnitude of the Cl^- influx depends on the Cl^- conductance (g_{Cl}) of the membrane:

$$I_{Cl} = g_{Cl}(E_m - E_{Cl}) \qquad (4)$$

This equation is discussed later in Section V. Thus, the average level of $[Cl^-]_i$ in excitable cells should depend on the frequency and duration of the AP, that is, on the mean E_m averaged over many AP cycles.

In smooth muscle, $[Cl^-]_i$ is often much higher (ca. 30 mM) than the value of 12.5 mM predicted from passive distribution:

$$E_{Cl} = E_m = -55\,\text{mV}$$

$$[Cl^-]_i = \frac{[Cl^-]_o}{\text{antilog} \, \dfrac{E_m}{-61\,\text{mV}}}$$

$$= \frac{100\,\text{mM}}{\text{antilog} \, \dfrac{-55\,\text{mV}}{-61\,\text{mV}}}$$

$$= \frac{100}{7.973}$$

$$= 12.5\,\text{mM}$$

The elevated $[Cl^-]_i$ in smooth muscle could be due to an exchange carrier (e.g., $Cl^- - HCO_3^-$ exchange) or to a cotransporter (e.g., $Na^+ - K^+ - Cl_2^-$ transport) (see Chapter 20).

D. Ca^{2+} Distribution

1. Need for Calcium Pumps

For the positively charged Ca^{2+} ion, there must be some mechanism for removing Ca^{2+} from the cytoplasm. Otherwise, the cell would continue to gain Ca^{2+} until there was no electrochemical gradient for net influx of Ca^{2+}. Ca^{2+} loading would occur until the free $[Ca^{2+}]_i$ in the cytoplasm was even greater

than that outside (ca. 2 mM) because of the negative potential inside the cell. Therefore, there must be one or more Ca^{2+} pumps in operation. The SR (or ER) membrane contains a Ca^{2+}-activated ATPase (which also requires Mg^{2+}) that actively pumps two Ca^{2+} ions from the cytoplasm into the SR lumen at the expense of one ATP. This pump ATPase is capable of pumping down the Ca^{2+} to less than 10^{-7} M. The Ca^{2+}-ATPase of the SR is regulated by an associated low-molecular-weight protein, **phospholamban**. Phospholamban is phosphorylated by cyclic AMP-dependent protein kinase (on serine 16)[1] and, when phosphorylated, stimulates the Ca^{2+}-ATPase and Ca^{2+} pumping (by a derepression process). The sequestration of Ca^{2+} by the SR is essential for muscle relaxation. The mitochondria also can actively take up Ca^{2+} almost to the same degree as the SR, but this Ca^{2+} pool probably does not play an important role in normal excitation-contraction coupling processes.

However, the resting Ca^{2+} influx and the extra Ca^{2+} influx that enter with each AP must be returned to the interstitial fluid. Two mechanisms have been proposed for this (for references, see Sperelakis, 1979): (1) a Ca^{2+}-ATPase, similar to that in the SR, is present in the sarcolemma, and (2) a Ca^{2+}-Na^+ exchange occurs across the cell membrane. It has been reported that there is a Ca^{2+}-ATPase in the sarcolemma of myocardial cells (Dhalla et al., 1977; Jones et al., 1980) and smooth muscle (Daniel et. al., 1977) that actively transports two Ca^{2+} outward against an electrochemical gradient, using one ATP in the process. Phospholamban is not associated with the sarcolemmal Ca^{2+}-ATPase.

2. Ca_i^{2+}-Na_o^+ Exchange Reaction

The Ca_i^{2+}-Na_o^+ exchange reaction exchanges one internal Ca^{2+} ion for 3 external Na^+ ions via a membrane carrier molecule (see Fig. 3). This reaction is facilitated by ATP, but ATP is not hydrolyzed (consumed) in this reaction. Instead, the energy for the transport of Ca^{2+} against its large electrochemical gradient comes from the Na^+ electrochemical gradient. That is, the uphill transport of Ca^{2+} is coupled to the downhill movement of Na^+. Effectively, the energy required for this Ca^{2+} movement is derived from the Na^+,K^+-ATPase. Thus, the Na^+-K^+ pump, which uses ATP to maintain the Na^+ electrochemical gradient, indirectly helps to maintain the Ca^{2+} electrochemical gradient. Hence, the inward Na^+ leak is greater than it would be otherwise. A complete discussion of Ca^{2+}-Na^+ exchange is given in Chapter 19.

The **energy cost** (ΔG_{Ca}, in joules/mole) for pumping out Ca^{2+} ion is directly proportional to its electrochemical gradient. These energetic equations are as follows (where ΔG is the change in free energy, z is the valance, and $\mathscr{F}$ is the Faraday constant):

$$\Delta G_{Ca} = z\mathscr{F}(E_m - E_{Ca}) \qquad (5)$$

The energy available from the Na^+ distribution is directly proportional to its electrochemical gradient.

$$\Delta G_{Na} = z\mathscr{F}(E_m - E_{Na}) \qquad (6)$$

---

[1] Phosphorylation of threonine-17 also occurs *in vivo* by the Ca-CAM-PK.

Depending on the exact values of $[Na^+]_i$ and $[Ca^{2+}]_i$ at rest in a muscle cell, the energetics would be about adequate for an exchange ratio of 3 Na^+:1 Ca^{2+}. An exchange ratio of 3:1 would produce a small depolarization due to a net inward flow of current (3 Na^+ in to 1 Ca^{2+} out) via this electrogenic Ca^{2+}_i-Na^+_o exchanger. That is, there is a net positive charge moving inward for every cycle of the exchanger. This net exchanger current can be measured in whole-cell voltage clamp studies when all ionic currents and Na^+-K^+ pump currents are blocked.

The exchange reaction depends on relative concentrations of Ca^{2+} and Na^+ on each side of the membrane and on relative affinities of the binding sites to Ca^{2+} and Na^+. Because of this Ca^{2+}_i-Na^+_o exchange reaction, whenever the cell gains Na^+, it will also gain Ca^{2+} because the Na^+ electrochemical gradient is reduced, and the exchange reaction becomes slowed. The Ca^{2+}_i-Na^+_o exchange process has been proposed

as the mechanism of the positive inotropic action (i.e., more forceful contraction) in the heart resulting from cardiac glycoside inhibition of the Na^+-K^+ pump.

In addition, when the membrane is depolarized during the AP plateau, the exchange carriers will exchange the ions in reverse, namely, internal Na^+ for external Ca^{2+}, and thus increase Ca^{2+} influx. The net effect of this mechanism is to elevate $[Ca^{2+}]_i$. Such reversed Ca^{2+}_o-Na^+_i exchange appears to be a significant source of Ca^{2+} for contraction in cardiac muscle of some species.

The ratio of free energy changes for Na^+ to Ca^{2+} was calculated for a coupling ratio of 3 Na^+:1 Ca^{2+} ($3\Delta G_{Na}/\Delta G_{Ca}$) and a $[Na^+]_i$ of 15 mM, and plotted as a function of membrane potential, for different $[Ca^{2+}]_i$ levels (Fig. 4A). This plot allows a simple assessment of how the directionality of the exchanger is affected by E_m, that is, forward mode versus reverse mode of operation, and how the reversal potential of

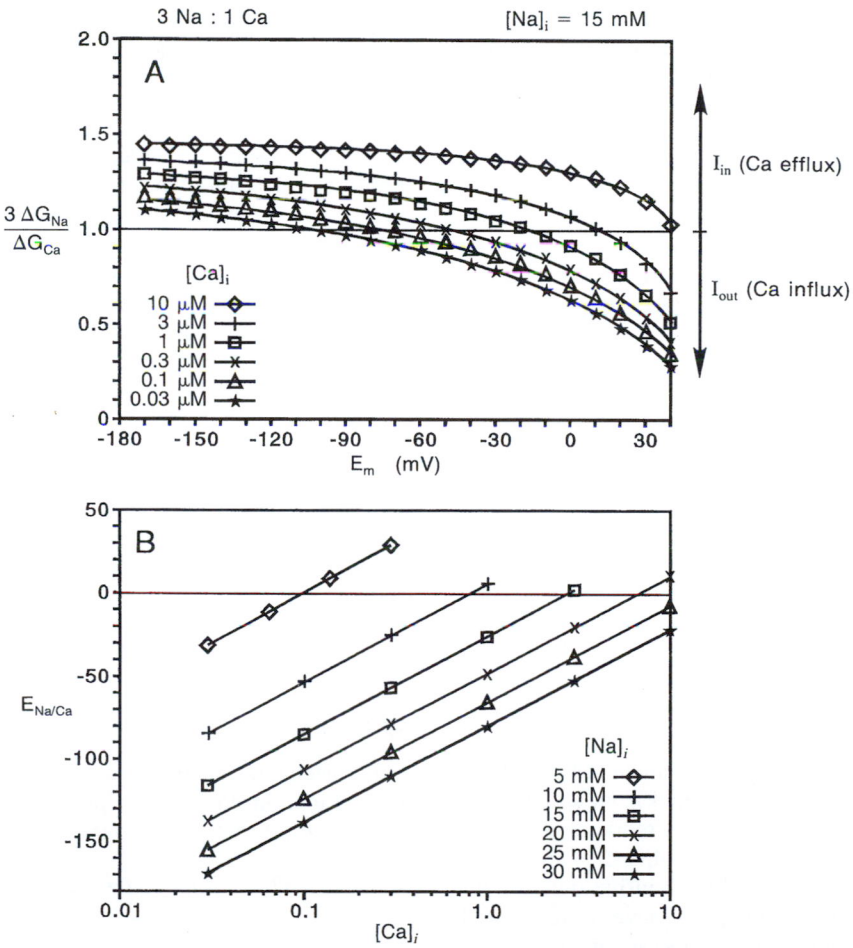

FIGURE 4. Electrogenic Na^+-Ca^{2+} exchange for $[Na^+]_i$ of 15 mM for a coupling ratio of 3 Na^+:1 Ca^{2+}. (A) Calculated ratio of the free energies for Na^+ versus Ca^{2+} ($3\Delta G_{Na}/\Delta G_{Ca}$) plotted as a function of membrane potentials (E_m) for six different intracellular Ca^{2+} concentrations ($[Ca^{2+}]_i$). At a ratio of 1.0, $3\Delta G_{Na} = -\Delta G_{Na}$ and the sum equals zero: $3\Delta G_{Na} + \Delta G_{Ca} = 0$. Therefore, the exchange would be at equilibrium at the E_m value at which each curve crosses the ratio of 1.0 line; that is, this gives the value of the exchanger equilibrium potential, E_{Na-Ca}. At ΔG ratios > 1.0, there is a net inward current carried by Na^+ ion, coupled with net Ca^{2+} efflux from the cell. This represents the **forward mode** of operation of the exchanger. At ΔG ratios < 1.0, there is a net outward current carried by Na^+ ion, coupled with net Ca^{2+} influx into the cell. This reflects the **reverse mode** of operation of the exchanger. (B) The **reversal potential** (E_{Na-Ca}) for the Na^+-Ca^{2+} exchanger is plotted on the ordinate as a function of $[Ca^{2+}]_i$ on the abscissa. The family of curves is for six different $[Na^+]_i$ levels.

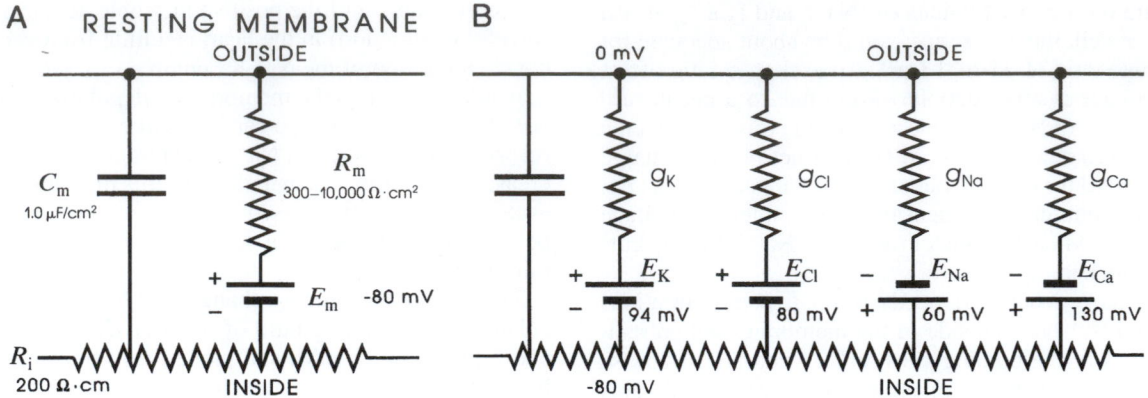

FIGURE 5. Electrical equivalent circuits for a cell membrane at rest. (A) Membrane as a parallel resistance/capacitance circuit, the membrane resistance (R_m) being in parallel with the membrane capacitance (C_m). Resting potential (E_m) is represented by an 80-mV battery in series with the membrane resistance, the negative pole facing inward. (B) Membrane resistance is divided into its four component parts, one for each of the four major ions of importance: K⁺, Cl⁻, Na⁺, and Ca²⁺. Resistances for these ions (R_K, R_{Cl}, R_{Na}, and R_{Ca}) are parallel to one another and represent totally separate and independent pathways for permeation of each ion through the resting membrane. These ion resistances are depicted as their reciprocals, namely, ion conductances (g_K, g_{Cl}, g_{Na}, g_{Ca}). Equilibrium potential for each ion (e.g., E_K), determined solely by the ion distribution in the steady state and calculated from the Nernst equation, is shown in series with the conductance path for that ion. Resting potential of −80 mV is determined by the equilibrium potentials and by the relative conductances.

the exchanger is shifted by [Ca²⁺]$_i$. When the ratio is 1.0, $3\Delta G_{Na} = -\Delta G_{Ca}$, and the sum equals zero: $3\Delta G_{Na} + -\Delta G_{Ca} = 0$. Therefore, the exchanger would be at equilibrium. For ΔG ratios > 1.0, there is a net inward current carried by Na⁺ ion, coupled with net Ca²⁺ efflux from the cell. This represents **forward mode** of operation of the exchanger. For ΔG ratios < 1.0, there is a net outward current carried by Na⁺ ion, coupled to a net Ca²⁺ influx into the cell. This represents **reverse mode** of operation of the exchanger.

The **equilibrium potential** or **reversal potential** for the Ca²⁺-Na⁺ exchanger (E_{Na-Ca}), for an exchange ratio of 3 Na⁺:1 Ca²⁺ is

$$E_{Na-Ca} = 3E_{Na} - 2E_{Ca} \tag{7}$$

where E_{Na} and E_{Ca} are the equilibrium potentials for Na⁺ and Ca²⁺, respectively, as calculated from the Nernst equation. Thus, the E_{Na-Ca} varies with the [Na⁺]$_i$ level and during changes in [Ca²⁺]$_i$ levels that occur with contraction. E_{Na-Ca} is less negative (more positive) when [Ca²⁺]$_i$ is elevated, and shifts to more negative potentials when [Na⁺]$_i$ is elevated (Fig. 4B).

Therefore, when a myocardial cell changes from the resting potential (ca. −80 mV) to the AP plateau (ca. +20 mV), simultaneous with [Ca²⁺]$_i$ being elevated from about 0.1 to 3 µM, the exchanger switches to reverse mode of operation, with Ca²⁺ influx. As stated previously, this Ca²⁺ influx can be a significant source of the total Ca²⁺ influx during excitation-contraction coupling. If the [Na⁺]$_i$ level were 25 or 30 mM, the exchanger would always operate in the reverse mode at the physiological E_m levels.

IV. Equilibrium Potentials

A. Equivalent Electrical Circuit

The electrical equivalent circuit of the cell membrane at rest is depicted in Fig. 5. In Fig. 5A, the membrane is indicated as

a parallel resistance (R_m) and capacitance (C_m). As discussed previously, the capacitance results from the lipid bilayer matrix of the membrane, and the resistance or conductance (G_m) results from the protein ion channels floating in the lipid bilayer and spanning it. The resting potential is depicted as a battery in series with R_m, with the negative pole facing inside the cell.

In Fig. 5B, the equivalent circuit for the resting membrane is shown further expanded, by breaking up the membrane conductance (G_m) into its four component parts, one for each ion of major importance electrophysiologically, namely K⁺, Cl⁻, Na⁺, and Ca²⁺. The respective conductances are g_K, g_{Cl}, g_{Na}, and g_{Ca}. The equilibrium potential (E) for each ion is placed in series with the conductance pathway for that ion, as depicted. The polarity of each battery is as shown; namely, the pole facing inwards is negative for K⁺ and Cl⁻ and positive for Na⁺ and Ca²⁺. These polarities are based on the directions of the concentration gradients and charge on the ions. The reasons and mechanisms are discussed as follows.

B. Nernst Equation

For each ionic species distributed unequally across the cell membrane, an equilibrium potential (E_i) or battery can be calculated for that ion from the Nernst equation (for 37 °C),

$$E_i = \frac{-RT}{z\mathscr{F}} \ln \frac{C_i}{C_o} \tag{8}$$

$$E_i = \frac{-2.303\,RT}{z\mathscr{F}} \log \frac{C_i}{C_o} \tag{8a}$$

$$E_i = \frac{-61\,\text{mV}}{z} \log \frac{C_i}{C_o} \tag{8b}$$

where C_i is the internal concentration of the ion, C_o is the extracellular concentration, R is the gas constant (8.3 J/mol·K), T is the absolute temperature in kelvins (K = 273 + °C), $\mathscr{F}$

is the Faraday constant (96 500 C/eq), and z is the valence (with sign). Thus, $z\mathscr{F}$ = C/mol. Taking the $RT/\mathscr{F}$ constants and the factor of 2.303 for conversion of natural log (ln) to log to the base of 10 ($\log_{10}$) gives[2]

$$\frac{-RT}{\mathscr{F}} \ln N = \frac{-2.303\, RT}{\mathscr{F}} \log N = -61\,\text{mV} \log N \quad (8c)$$

Therefore, the Nernst equation (Eq. 8) becomes

$$E_i = \frac{-2.303\, RT}{z\mathscr{F}} \log \frac{C_i}{C_o} \quad (8d)$$

$$E_i = \frac{-61\,\text{mV}}{z} \log \frac{C_i}{C_o} \quad (8e)$$

The −61 mV constant (2.303 $RT/\mathscr{F}$) becomes −59 mV at 22 °C. A derivation of the Nernst equation is given in the Appendix (part I) to this chapter. The Nernst equation gives the **potential difference (PD) (electrical force) that would exactly oppose the concentration gradient (diffusion force)**.

Only very small charge separation (Q, in coulombs) is required to build a very large PD.

$$E_m = \frac{Q}{C_m} \quad (9)$$

where C_m is the membrane capacitance. This is further discussed in the next section.

For the ion distributions given previously (see Table 2), the approximate equilibrium potentials are

$$E_{Na} = +60\,\text{mV}$$
$$E_{Ca} = +130\,\text{mV}$$
$$E_K = -93\,\text{mV}$$
$$E_{Cl} = -80\,\text{mV}$$

The sign of the equilibrium potential represents the inside of the cell with reference to the outside (see Fig. 5). Because Na^+ is higher outside (ca. 145 mM) than inside (ca. 15 mM), the positive pole of the Na^+ battery (E_{Na}) is inside the cell. The concentration gradient for Ca^{2+} is in the same direction as for Na^+ (1.8 mM $[Ca^{2+}]_o$ and about 1×10^{-7} M $[Ca^{2+}]_i$), and so the positive pole of E_{Ca} is inside. K^+ is higher inside (ca. 150 mM) than outside (ca. 4.5 mM), and so the negative pole is inside. Because Cl^- is higher outside (ca. 100 mM) than inside (ca. 5 mM), the negative pole is inside. Voltages are, by convention, given for the inside with respect to the outside.

C. Concentration Cell

In a concentration cell (essentially a two-compartment system separated by a membrane), the **side of higher concentration becomes negative for cations** (positive ions) and **positive for anions** (negative ions). Any ion whose equilibrium potential is different from the resting potential (e.g., −80 mV for a myocardial cell or skeletal muscle fiber) is off equilibrium and therefore must effectively be pumped at the expense

of energy. In many cell types, only Cl^- ion appears to be at or near equilibrium, whereas Na^+, Ca^{2+}, and K^+ are actively transported. Even H^+ ion is off equilibrium, E_H being closer to zero potential (see Table 2). If H^+ were passively distributed, the negative intracellular potential would pull in more H^+ ions, causing $[H^+]_i$ to increase, making the cell interior more acidic.

The mechanism for development of the **equilibrium potential** is depicted in Fig. 6. To show the development of an equilibrium potential, we can use an artificial membrane (e.g., one made of celloidin) to separate two solutions, that is, to form a concentration cell. This membrane contains negatively charged pores, which therefore allows cations (like K^+) to pass through, but prevents anions (like Cl^-) from passing. This is because like charges repel one another, and unlike charges attract one another. Therefore, in this particular membrane, K^+ is permeant and Cl^- is impermeant. If one side (side 1) contains a salt like KCl at a concentration higher (e.g., 0.10 M) than that in the other side (side 2) (e.g., 0.01 M), then a steady PD is very quickly built up across the membrane. As can be calculated from the Nernst equation,

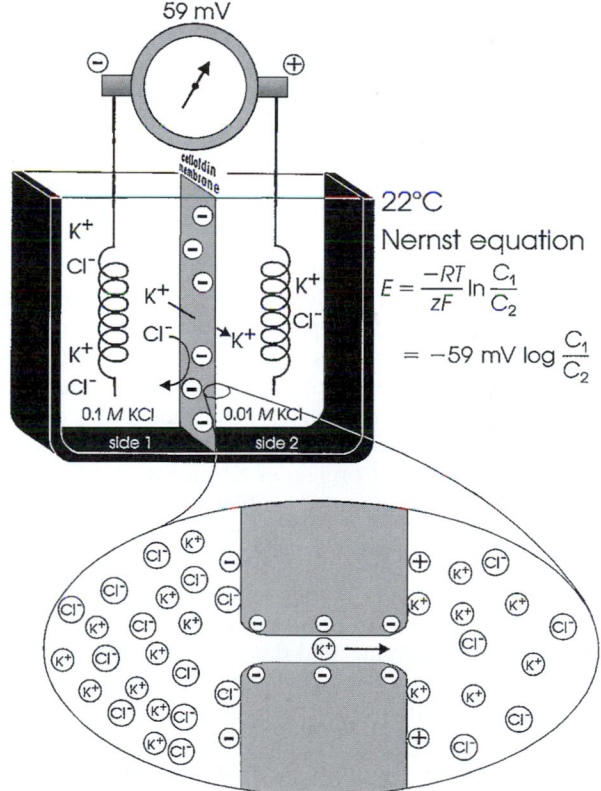

FIGURE 6. Upper diagram: Concentration cell diffusion potential developed across artificial membrane containing negatively-charged pores. The membrane is impermeable to Cl^- ions, but permeable to cations such as K^+. Concentration gradient for K^+ causes a potential to be generated, the side of higher K^+ concentration becoming negative. Lower diagram: Expanded diagram of water-filled pore in the membrane, showing the permeability to K^+ ions, but lack of penetration of Cl^- ions. Potential difference is generated by charge separation, a slight excess of K^+ ions being held close to the right-hand surface of the membrane; a slight excess of Cl^- ions is aggregated close to the left surface.

---

[2] ln (or $\log_e$, or $\log_{2.717}$) is the log to the base of e (2.717), and $\log_{10}$ is the log to the base of 10.

for a tenfold difference in concentration of the permeant monovalent cation (K$^+$), the PD would be −61 mV at 37 °C (or −59 mV at a room temperature of 22 °C).

This PD is between the two solutions and expressed across the membrane. Side 1 (side of highest K$^+$ concentration) becomes negative with respect to side 2. The PD is developed because of the tendency for diffusion (diffusion force) from high concentration to low concentration. This is based on the **random thermal motion** of the ions (particles), somewhat related to **Brownian motion** of larger particles. That is, the side of higher concentration has a **greater probability** of K$^+$ ions moving from side 1 to side 2 than in the reverse direction, based on the greater number of particles, all moving in random directions. Therefore, there will be a loss of positive charges (K$^+$ ions) from side 1 and a gain of positive charges in side 2.

Because negative charges (Cl$^-$ ion) cannot accompany the positive charges, as the membrane was made impermeable to anions, a **charge separation** is built up across the membrane. It can now be readily understood why side 1, the side of higher cation concentration, becomes negatively charged (due to loss of positive charges) and why side 2, the side of lower cation concentration, becomes positively charged (due to gain of positive charges). The charge separation is very tiny, and they stay plastered very close to the membrane. That is, for the example depicted in Fig. 6, side 1 will have the small excess of Cl$^-$ ions held very close to the membrane, and side 2 will have the small excess of K$^+$ ions held very close to the membrane. The force holding them there is called the electrostatic or Coulombic force, based on the attraction between unlike charges. This is related to Coulomb's law and the nature of capacitors discussed in Chapter 70.

In the two bulk solutions, the **law of electroneutrality** is upheld; that is, for every cation (K$^+$) there is a nearby anion (Cl$^-$). Thus, the **charge separation** occurs only directly across the membrane and is very tiny with respect to the total number of charges in the two solutions. In fact, after equilibrium is reached (within a few seconds), the most sensitive chemical analyses would fail to detect the very slight decrease in K$^+$ in side 1 or gain of K$^+$ on side 2.

Thus, the system comes to equilibrium quickly and with very little charge (K$^+$ and Cl$^-$) separation. That is, K$^+$ does not continue to have a net movement from side 1 to side 2 until the concentrations become equal. Why not? The answer is that the **small charge separation produces a large PD** across the membrane, and this PD is in such a polarity that it antagonizes further net movement of K$^+$ from side 1 to side 2. That is, the positive voltage that is developing on side 2 repels the positively charged K$^+$ ions because **like charges repel**. At equilibrium, these two forces become equal and opposite, and there is no further net movement of ions. Unidirectional fluxes of K$^+$ ions would still occur (because of their random thermal motion), but these would be equal and opposite and so no further **net flux** would occur.

In the example selected, KCl was used. However, any salt, such as NaCl, CaCl$_2$, or Na$_2$SO$_4$, could have been illustrated. If a divalent cation like Ca^{2+} were used, then from the Nernst equation, the same tenfold concentration gradient would develop a potential of only half, namely 30.5 mV (37 °C) or 29.5 mV (22 °C). The reason that this factor is half, rather than double as one might guess from the fact that the charge is double, is that the Nernst equation gives the PD that exactly opposes the diffusion force due to the concentration gradient, as stated in the paragraph above. Therefore, because the charge is double, **only half the voltage is necessary to effectively oppose the concentration force**. If the cation in question were trivalent, such as La^{3+}, then the 2.303 $RT/z\mathscr{F}$ factor would be one-third of 61 mV, or about 20.3 mV. It is for this reason that it is convenient to give the Nernst equation in the form shown in Eq. 8b, namely with the factor (at 37 °C) being −61 mV/z. This allows easy calculation of the equilibrium potential for an ion of any charge and sign (polarity).

That is, the sign and charge should be used, for example, +1 for K$^+$ or Na$^+$, +2 for Ca^{2+}, and −1 for Cl$^-$. When the ion in question is an anion like Cl$^-$, then −1/−1 gives a plus (+). Because [Cl$^-$]$_o$ > [Cl$^-$]$_i$, this concentration ratio can be inverted by changing the sign of the 2.30 $RT/z\mathscr{F}$ factor back to negative (−). That is, changing the sign of the factor in front of a log ratio simply inverts the ratio.

Finally, if the concentration cell depicted in Fig. 6 were made with a membrane that had positively charged pores, then everything would be reversed. The membrane would be permeable to Cl$^-$ and impermeable to K$^+$, and E_{Cl} would be +59 mV (at 22 °C) and +61 mV (at 37 °C). Again, the voltage is given for side 1 with respect to side 2. Thus, in dealing with an anion, the side of higher concentration becomes positive (due to a small loss of Cl$^-$) and the side of lower concentration becomes negative (due to a small gain of Cl$^-$). The separated charges again are plastered very close to the membrane: K$^+$ ions on side 1 and Cl$^-$ ions on side 2.

D. Activity Coefficient

Thus, an equilibrium potential can be calculated for any species of ion that is distributed unequally across a membrane. All that one needs to know are the concentrations in the two solutions and the charge (and temperature). Actually, we should use the **activity** (a) of the ion in question in the two solutions, instead of concentration. Thus the Nernst equation given in Eq. 8b becomes (for K$^+$, for example)

$$E_K = \frac{-61\,\text{mV}}{+1} \log \frac{a_K^i}{a_K^o}$$

The activity of an ion (in molar) can be obtained by multiplying the concentration of the ion (in molar) by the activity coefficient (γ) for the ion:

$$a = c \cdot \gamma \qquad (10)$$

In the biological case, the activity coefficients are relatively close (0.7 to 0.9) to 1.0 for Na$^+$, K$^+$, and Cl$^-$ in **both** the extracellular and the intracellular solutions. Therefore, in these cases, using the concentrations gives a good approximation. However, in the case of Ca^{2+}, the activity coefficient in the intracellular solution especially is substantially lower, and so this would affect the calculated value of E_{Ca}.

E. Nernst-Planck Equation

The basic Nernst equation has been modified in several ways for special situations. For example, in the concentration cell depicted in Fig. 6, a cell in which a single salt (both ions of same valence) is distributed across the membrane at two different concentrations, the PD developed across the membrane (E_m) can be calculated from the equation

$$E_m = \frac{U_c - U_a}{U_c + U_a} \frac{-61\,mV}{z} \log \frac{[salt]_1}{[salt]_2} \quad (11)$$

where $[salt]_1$ and $[salt]_2$ are the concentrations of the salt on side 1 and side 2, and U_c and U_a are the **mobilities** of the cations and anions, respectively, through the membrane. Thus, when the mobilities (or permeabilities) of the cation and anion are equal ($U_c = U_a$), E_m is zero, regardless of the equilibrium potentials. When the anion is impermeable ($U_a = 0$), the mobility fraction in Eq. 11 becomes 1.0, and the equation reduces to the simple Nernst equation. When the cation is impermeable ($U_c = 0$), the fraction becomes -1.0, and the same numerical value of E_m is produced, but of the opposite sign. Equation 11 can be used to calculate E_m for any combination of U_a and U_c. For example, if $U_a = 0.5\ U_c$, then for the problem illustrated in Fig. 6, E_m is about -20 mV (side 1 negative). Thus, the membrane potential is related to the relative mobilities.

F. Energy Wells

Ions do not just "fall" through a water-filled pore in the membrane (protein ion channel) down an electrochemical gradient. Instead, an ion may bind to several charged sites on its journey through the channel pore. The K^+ ion depicted in the bottom of Fig. 6, for example, is shown as binding to three negatively **charged sites** within the pore. These may be considered energy wells, and the ion must gain kinetic energy to become dislodged from this energy well to pass over the next **energy barrier** and into the next energy well. This energy comes from the ion being hit by another ion just entering the pore, producing a billiard ball effect. Some evidence for this model was presented in Chapter 13, which discussed the Ussing flux ratio equation.

From the measured value of the conductance of single ion channels (e.g., 20 pS, range of 10–300 pS), how many ions that pass through a single channel per second can be estimated. This number is about 6 000 000 ions/s. Therefore, the average transit time for a single ion to cross the membrane (50–70 Å thick) is about 0.17 µs.

G. Half-Cell Potentials

In the measurement of biological potentials, care must be taken not to introduce artifacts, such as reversible electrode half-cell potentials. See the Appendix (part B) to this chapter for a brief discussion of this topic.

V. Electrochemical Driving Forces and Membrane Ionic Currents

A. Electrochemical Driving Forces

The electrochemical driving force for each species of ion is the algebraic difference between its equilibrium potential, E_i, and the membrane potential, E_m. The total driving force is the sum of two forces: an electrical force (the negative potential in a cell at rest tends to pull in positively charged ions, because **unlike charges attract**) and a diffusion force (based on the concentration gradient) (Fig. 7); that is,

$$\text{driving force} = E_m - E_i \quad (12)$$

Thus, in a resting cell, the driving force for Na^+ is

$$(E_m - E_{Na}) = -80\,mV - (+60\,mV) = -140\,mV \quad (12a)$$

The negative sign means that the driving force is directed to bring about net movement of Na^+ inward. The driving force for Ca^{2+} is very large and is directed inward:

$$(E_m - E_{Ca}) = -80\,mV - (+129\,mV) = -209\,mV \quad (12b)$$

The driving force for K^+ is

$$(E_m - E_K) = -80\,mV - (-94\,mV) = +14\,mV \quad (12c)$$

Hence, the driving force for K^+ is small and directed outward. The driving force for Cl^- is nearly zero for a cell at

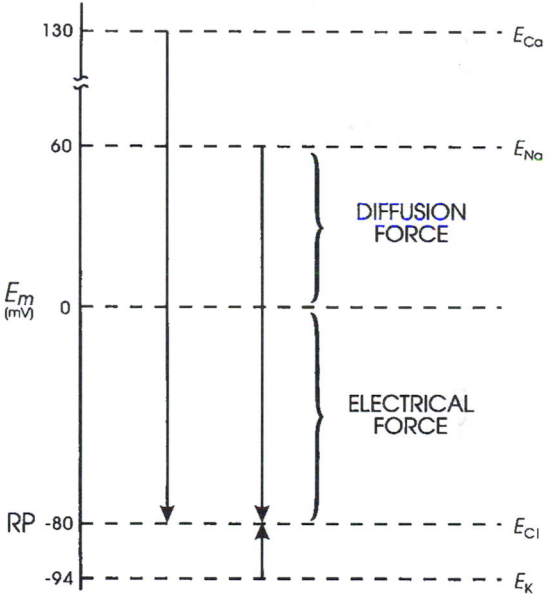

FIGURE 7. Representation of the electrochemical driving forces for Na^+, Ca^{2+}, K^+, and Cl^-. Equilibrium potentials for each ion (e.g., E_{Na}) are positioned vertically according to their magnitude and sign: they were calculated from the Nernst equation for a given set of extracellular and intracellular ion concentrations. Measured resting potential (RP) is assumed to be -80 mV. Electrochemical driving force for an ion is the difference between its equilibrium potential (E_i) and the membrane potential (E_m), that is ($E_m - E_i$). Thus, at rest, the driving force for Na^+ is the difference between E_{Na} and the resting E_m; if E_{Na} is $+60$ mV and resting E_m is -80 mV, the driving force is 140 mV; that is, the driving force is the algebraic sum of the diffusion force and the electrical force, and is represented by the length of the arrows in the diagram. Driving force for Ca^{2+} (about 210 mV) is even greater than that for Na^+, whereas that for K^+ is much less (about 14 mV). Direction of the arrows indicates the direction of the net electrochemical driving force, namely, the direction for K^+ is outward, whereas that for Na^+ and Ca^{2+} is inward. If Cl^- is passively distributed, then its distribution across the cell membrane can only be determined by the net membrane potential; for a cell sitting a long time at rest, $E_{Cl} = E_m$ and there is no net driving force.

rest in which Cl^- is passively distributed (e.g., neuron, myocardial cell, skeletal muscle fiber); that is,

$$(E_m - E_{Cl}) = -80\,mV - (-80\,mV) = 0 \qquad (13)$$

However, during the AP, **when E_m is changing, the driving force for Cl^- becomes large,** and there is a net driving force for inward Cl^- movement (Cl^- influx is an outward Cl^- current). Similarly, the driving force for K^+ outward movement increases during the AP, whereas those for Na^+ and Ca^{2+} decrease.

B. Membrane Ionic Currents

The net current for each ionic species (I_i) is equal to its **driving force** times its **conductance** (g_i, reciprocal of the resistance) through the membrane. This is essentially Ohm's law,

$$I = \frac{V}{R} = g \cdot V \qquad (14)$$

modified to reflect the fact that, in an electrolytic system, the total force tending to drive net movement of a charged particle must take into account both the electrical force and the concentration (or chemical) force. Thus, for the four ions, the net current can be expressed as

$$I_{Na} = g_{Na}(E_m - E_{Na}) \qquad (15)$$

$$I_{Ca} = g_{Ca}(E_m - E_{Ca}) \qquad (16)$$

$$I_{K} = g_{K}(E_m - E_{K}) \qquad (17)$$

$$I_{Cl} = g_{Cl}(E_m - E_{Cl}) \qquad (18)$$

In a resting cell, Cl^- and Ca^{2+} can be neglected, and the Na^+ current (inward) must be equal and opposite to the K^+ current (outward) to maintain a steady resting potential:

$$I_K = -I_{Na} \qquad (19)$$

$$g_K(E_m - E_K) = -g_{Na}(E_m - E_{Na}) \qquad (19a)$$

Thus, although in the resting membrane the driving force for Na^+ is much greater than that for K^+, g_K is much larger than g_{Na}, so the currents are equal. Hence, there is a continuous leakage of Na^+ inward and K^+ outward, even in a resting cell, and the system would run down if active pumping were blocked. Because the ratio of the Na^+ to K^+ driving forces ($-140\,mV/-14\,mV$) is 10, the ratio of conductances (g_{Na}/g_K) will be about 1:10. The fact that g_K is much greater than g_{Na} accounts for the resting potential being close to E_K and far from E_{Na}.

VI. Determination of Resting Potential and Net Diffusion Potential (E_{diff})

A. Determining Factors

For given ion distributions, which normally remain nearly constant under usual steady-state conditions, the resting potential is determined by the **relative membrane conductances** (g) or **permeabilities** (P) for Na^+ and K^+ ions. That is, the resting potential (of about $-80\,mV$ in cardiac muscle or skeletal muscle) is close to E_K (about $-94\,mV$) because $g_K \gg g_{Na}$ or $P_K \gg P_{Na}$. There is a direct proportionality between P and g at constant E_m and concentrations. From sim-

ple circuit analysis (using Ohm's law and Kirchhoff's laws), one can prove that the membrane potential will always be closer to the battery (equilibrium potential) having the lowest resistance (highest conductance) in series with it (see Figs. 5 and 7). In the resting membrane, this battery is E_K, whereas in the excited membrane it will be E_{Na} (or E_{Ca}), because there is a large increase in g_{Na} and/or g_{Ca} during the AP.

Any ion that is passively distributed cannot determine the resting potential; instead, the resting potential determines the distribution of that ion. Therefore, Cl^- is not considered for myocardial cells, skeletal muscle fibers, and neurons because it seems to be passively distributed. However, transient net movements of Cl^- across the membrane do influence E_m; for example, washout of Cl^- (in Cl^--free solution) produces a transient depolarization, and reintroduction of Cl^- produces a transient hyperpolarization. Cl^- movement is also involved in the production of inhibitory postsynaptic potentials (IPSPs) (see Chapter 41).

Because of its relatively low concentration, coupled with its relatively low resting conductance, the Ca^{2+} distribution has only a relatively small effect on the resting E_m and can be ignored.

B. Constant-Field Equation

A simplified, but most useful, version of the Goldman-Hodgkin-Katz constant-field equation can be given (for 37 °C):

$$E_m = -61\,mV\ \log \frac{[K^+]_i + \dfrac{P_{Na}}{P_K}[Na^+]_i}{[K^+]_o + \dfrac{P_{Na}}{P_K}[Na^+]_o} \qquad (20)$$

This equation shows that for a given ion distribution, the resting E_m is determined by the P_{Na}/P_K ratio, the **relative permeability** of the membrane to Na^+ and K^+. For myocardial cells and skeletal muscle fibers, the P_{Na}/P_K ratio is about 0.04, whereas for nodal cells of the heart and smooth muscle cells, this ratio is closer to 0.10 or 0.20.

Inspection of the constant-field equation shows that the numerator of the log term will be dominated by the $[K^+]_i$ term [since the $(P_{Na}/P_K)[Na^+]_i$ term will be very small], whereas the denominator will be affected by both the $[K^+]_o$ and $(P_{Na}/P_K)[Na^+]_o$ terms. This relationship thus accounts for the deviation of the E_m versus log $[K^+]_o$ curve from a straight line (having a slope of 61 mV/decade) in normal Ringer solution (Fig. 8). When $[K^+]_o$ is elevated ($[Na^+]_o$ being reduced by an equimolar amount), the denominator becomes more and more dominated by the $[K^+]_o$ term, and less and less by the $(P_{Na}/P_K)[Na^+]_o$ term. Therefore, in bathing solution containing high K^+, the constant-field equation approaches the simple Nernst equation for K^+, and E_m approaches E_K. As $[K^+]_o$ is raised stepwise, E_K becomes correspondingly reduced, because $[K^+]_i$ stays relatively constant; therefore, the membrane becomes more and more depolarized (see Fig. 8).

A more detailed discussion of the constant-field equation and its other variants is given in Section III of the Appendix to this chapter.

When $[K^+]_o$ is elevated (e.g., to 8 mM) in some types of cells, a hyperpolarization of up to about 10 mV may be

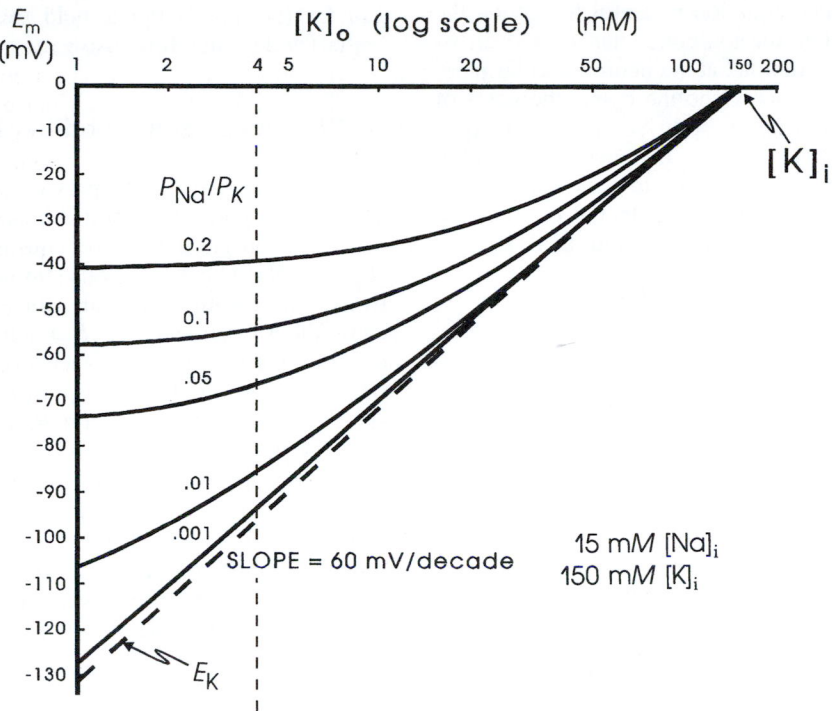

FIGURE 8. Theoretical curves calculated from the Goldman constant-field equation for resting potential (E_m) as a function of $[K^+]_o$. Family of curves is given for various P_{Na}/P_K ratios (0.001, 0.01, 0.05, 0.1, and 0.2). K^+ equilibrium potential (E_K) calculated from the Nernst equation (broken straight line). Curves calculated for a $[K^+]_i$ of 150 mM and a $[Na^+]_i$ of 15 mM. Calculations made holding $[K^+]_o + [Na^+]_o$ constant at 154 mM; that is, as $[K^+]_o$ was elevated, $[Na^+]_o$ was lowered by an equimolar amount. Change in P_K as a function of $[K^+]_o$ was not taken into account for these calculations. Point at which E_m is zero gives $[K^+]_i$. The potential reverses in sign when $[K^+]_o$ exceeds $[K^+]_i$.

produced. Such behavior is often observed in cells with a high P_{Na}/P_K ratio (due to low P_K) and therefore a low resting E_m, such as in young embryonic hearts. This hyperpolarization could be explained by several factors: (a) stimulation of the electrogenic Na^+ pump current (I_p), (b) an increase in P_K (and therefore g_K) due to $[K^+]_o$ effect on P_K, and (c) an increase in g_K (but not P_K) due to the concentration effect. A similar explanation may apply to the fall-over in the E_m versus log $[K^+]_o$ curve when $[K_o]$ is lowered to 1 mM and less, hence depolarizing the cells. This effect is prominent in rat skeletal muscle, for example (see Fig. 7 of Chapter 51).

C. Chord Conductance Equation

An alternative method of approximating the membrane resting potential (E_m) is the chord conductance equation. The word **chord** means a straight line connecting two points on a curve, and here specifically refers to the **average slope** of a nonlinear steady-state voltage-current curve, that is, a straight line from any point on the curve through the origin (zero applied current). (In contrast, slope conductance is the tangent at any point on the curve.) Thus,

$$E_m = \frac{g_K}{\Sigma g} E_K + \frac{g_{Na}}{\Sigma g} E_{Na} + \frac{g_{Cl}}{\Sigma g} E_{Cl} + \frac{g_{Ca}}{\Sigma g} E_{Ca} \quad (21)$$

where g_K, g_{Na}, g_{Cl}, and g_{Ca} are the membrane conductances for K^+, Na^+, Cl^-, and Ca^{2+}, respectively, and Σg is the total conductance (sum of all ionic partial conductances). The ratio of $g_K/\Sigma g$, for example, is the **relative or fractional conductance** for K^+.

The chord conductance equation can conveniently take into account all ions, including divalent cations, that are distributed unequally across the membrane. The ions important to membrane potentials (including action potentials, postsynaptic potentials, and receptor potentials) are K^+, Na^+, Cl^-, and Ca^{2+}. As discussed previously, Cl^- cannot help in determining the resting potential if it is passively distributed. Thus, Eq. 21 can be rewritten, omitting the Cl^- term, as

$$E_m = \frac{g_K}{\Sigma g} E_K + \frac{g_{Na}}{\Sigma g} E_{Na} + \frac{g_{Ca}}{\Sigma g} E_{Ca} \quad (21a)$$

For simplicity, we can ignore the Ca^{2+} term also, giving

$$E_m = \frac{g_K}{g_K + g_{Na}} E_K + \frac{g_{Na}}{g_K + g_{Na}} E_{Na} \quad (21b)$$

where Σg is now equal to $g_K + g_{Na}$.

The chord conductance equation can be derived simply from Ohm's law and from circuit analysis for the condition when net current is zero ($I_{Na} + I_K = 0$). (See part IV of the Appendix to this chapter for the derivation.) The equation holds true whenever the net current across the membrane is zero, as for the resting potential.

The chord conductance equation is useful for giving the membrane potential when the ion conductances and distributions are known. For example, at the neuromuscular junction, the neurotransmitter acetylcholine opens the gates of many ionic channels that allow both Na^+ and K^+ to pass through equally well (that is, $g_{Na} = g_K$). Hence, the potential that the postsynaptic membrane tends to seek when maximally activated (i.e., the equilibrium potential or so-called **reversal potential** for the end-plate potential, EPP) is

$$E_{EPP} = \frac{1}{2}(-94\,\text{mV}) + \frac{1}{2}(+60\,\text{mV})$$
$$= -17\,\text{mV} \tag{21c}$$

A disadvantage of the chord conductance equation is that it gives nearly a straight line for the E_m versus log $[K^+]_o$ plot (actually a slight bend in the opposite direction at low $[K^+]_o$). In contrast, the constant-field equation gives the complete bending of the curves (for different P_{Na}/P_K ratios) (see Section VI.B).

The chord conductance equation again illustrates the important fact that the g_K/g_{Na} ratio determines the resting potential. When $g_K \gg g_{Na}$, then E_m is close to E_K; conversely, when $g_{Na} \gg g_K$ (as during the spike part of the AP), E_m shifts to close to E_{Na} or to E_{Ca} (in the case of many types of smooth muscle cells).

The chord conductance equation can be rewritten, using resistances instead of conductances, and may then be called the **chord resistance equation,**

$$E_m = \frac{R_K}{R_K + R_{Na}} E_{Na} + \frac{R_{Na}}{R_K + R_{Na}} E_K \tag{21d}$$

where R_K and R_{Na} are the K^+ and Na^+ resistances, which are the reciprocals of the conductances ($R_K = 1/g_K$ and $R_{Na} = 1/g_{Na}$). Note that in this equation the positions of the two batteries are interchanged. This equation can be derived by simply substituting the two reciprocals given above into the chord conductance equation. It can also be derived by circuit analysis, as discussed in Section V of the Appendix to this chapter. This Appendix section also shows how simple circuit analysis can be used to determine what the resting potential should be, without using either the Goldman constant-field equation or the chord conductance equation.

D. Net Diffusion Potential, E_{diff}

In the presence of ouabain (short-term exposure only) to inhibit the Na^+-K^+ pump and V_p, the resting potential that remains reflects the net diffusion potential, E_{diff}. E_{diff} is determined by the ion concentration gradients for K^+ and Na^+ and by the relative permeability for K^+ and Na^+. When the Na^+-K^+ pump is operating, there is normally a small additional contribution of V_p to the resting E_m of about 2–16 mV, depending on cell type (discussed in following section).

Inhibition of the Na^+-K^+ pump for long periods will gradually run down the ion concentration gradients. The cells lose K^+ and gain Na^+, and therefore E_K and E_{Na} become smaller. The cells thus become depolarized (even if the relative permeabilities are unaffected), which causes them to gain Cl^- (because $[Cl]_i$ was held low by the large resting potential) and to therefore also gain water (cells swell).

VII. Electrogenic Sodium Pump Potentials

A brief summary of the previous principles is as follows: The Na^+-K^+ pump is responsible for maintaining the cation concentration gradients. The equilibrium potentials for K^+ (E_K) and Na^+ (E_{Na}) are about -94 mV and $+60$ mV, respectively. The resting potential value is usually near E_K, because the K^+ permeability (P_K) is much greater than P_{Na} in a resting membrane. The exact resting membrane potential (E_m) depends on the P_{Na}/P_K ratio, myocardial cells and skeletal muscle fibers having P_{Na}/P_K ratios of 0.01–0.05, whereas smooth muscle or nodal cells of the heart have a ratio closer to 0.10–0.15. In the various types of cells, the resting E_m has a smaller magnitude (i.e., is less negative) than E_K by 10–40 mV. If there were no **electrogenic pump potential** contribution to the resting potential (that is, as though the Na^+-K^+ pump was only indirectly responsible for the resting potential by its role in producing the ionic gradients), E_m would equal E_{diff}.

However, a direct contribution of the pump to the resting E_m can be demonstrated. For example, if the Na^+-K^+ pump is blocked by the addition of ouabain, there usually is an immediate depolarization of 2–16 mV, depending on the type of cell. Thus the direct contribution of the electrogenic Na^+-K^+ pump to the measured resting E_m is small under physiologic conditions (but very important).

However, under conditions in which the pump is stimulated to pump at a high rate (e.g., when $[Na^+]_i$ or $[K^+]_o$ is abnormally high) the direct electrogenic contribution of the pump to the resting potential can be much greater, and E_m can actually exceed E_K by as much as 20 mV or more. For example, if the ionic concentration gradients are allowed to run down (e.g., by storing the tissues in zero $[K^+]_o$ and at low temperatures for several hours), then after the tissues are allowed to restart pumping, the measured E_m can exceed the calculated E_K (e.g., by 10–20 mV) for a time (Fig. 9). The Na^+ loading of the cells is facilitated by placing them in cold low or zero $[K^+]_o$ solutions, because external K^+ is necessary for the Na^+-K^+-linked pump to operate; K_m of the Na^+,K^+-ATPase for K^+ is about 2 mM. After several hours in such a solution, the internal concentrations of Na^+, K^+, and Cl^- approach the concentrations in the bathing Ringer solution, and the resting potential is very low (<-30mV) (see Chapter 15). The cells are then transferred to a pumping solution, which is the appropriate Ringer solution containing normal K^+ and at normal temperature. Under such conditions, the pump turns over at a maximal rate, because the major control over pump rate is $[Na^+]_i$ and $[K^+]_o$. The low initial E_m also stimulates the pump rate, because the energy required to pump out Na^+ is less. The measured E_m of such Na^+ preloaded cells increases rapidly and more rapidly than E_K, as shown in Fig. 9. After this transient phase, however, a crossover of the two curves occurs, so that E_K again exceeds E_m, as in the physiologic condition. **Cardiac glycosides** prevent or reverse the transient hyperpolarization beyond E_K. The possibility that ionic conductance changes (e.g., an in-

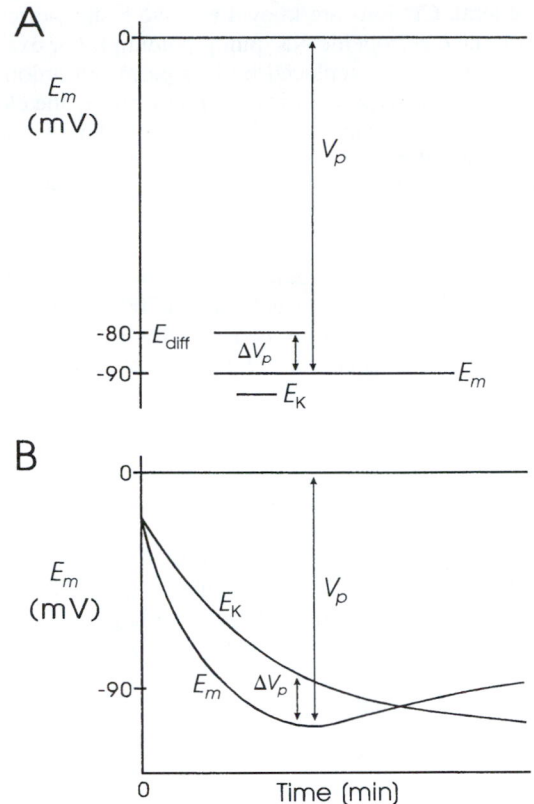

FIGURE 9. Diagrammatic representation of an electrogenic sodium pump potential. (A) Muscle cell in which the net ionic diffusion potential (E_{diff}, function of ion equilibrium potentials and relative conductances) is −80 mV, yet exhibits a measured membrane resting potential (E_m) that is greater. Difference between E_m and E_{diff} represents the contribution of the electrogenic pump to the resting potential. The electrogenic pump potential must be equal to V_p. The contribution of the electrogenic pump potential to the resting potential ($E_m − E_{diff}$) is equal to ΔV_p. (B) Cell that was run down (Na⁺ loaded, K⁺ depleted) over several hours by inhibition of Na⁺-K⁺ pumping, resulting in a low resting potential. Returning the muscle cell to a pumping solution allows the resting E_m to rebuild as a function of time. Buildup in E_m occurs faster than buildup in E_K, as illustrated.

crease in g_K or a decrease in g_{Na}) can account for the observed hyperpolarization, can be ruled out whenever E_m exceeds (is more negative than) E_K.

Rewarming cells previously cooled leads to the rapid restoration of the normal resting potential (within 10 min), whereas recovery of the intracellular Na⁺ and K⁺ concentrations is slower. During prolonged hypoxia, the resting potential of cardiac muscle decreases much less than E_K decreases (a difference of about 25 mV); the electrogenic pump attempts to hold the resting potential constant, despite dissipating ionic gradients.

Another method used to demonstrate that the pump is electrogenic is to inject Na⁺ ions into the cell through a micropipette. This procedure rapidly produces a small transient hyperpolarization, which is immediately abolished or prevented by ouabain. The **pump current** and the rate of Na⁺ extusion increase in proportion to the amount of Na⁺ injected. To prove that the pump is electrogenic, it must be demonstrated that the hyperpolarization produced in an intact muscle is not the result of enhanced pumping of an electroneutral pump. This could cause depletion of external K⁺ in a restricted diffusion space just outside the cell membrane, leading to a larger E_K and thereby to hyperpolarization. Depletion could occur if the Na⁺-K⁺ pump pumped in K⁺ faster than it could be replenished by diffusion from the bulk interstitial fluid.

The electrogenic Na⁺ pump is influenced by the membrane potential. From energetic considerations, depolarization should enhance the electrogenic Na⁺ pumping, whereas hyperpolarization should inhibit it. This is because depolarization reduces the electrochemical gradient (and hence the energy requirements) against which Na⁺ must be extruded whereas hyperpolarization increases the gradient. Thus there should be a distinct potential, more negative than E_K, at which Na⁺ pumping is prevented (e.g., a **pump equilibrium potential**). A value close to −140 mV was reported for cardiac cells and rat skeletal muscle fibers.

Any method used to increase membrane resistance increases the contribution of the pump to the resting potential (Fig. 10); that is, the electrogenic Na⁺ pump contribution must be augmented under conditions that increase membrane resistance. The contribution of the pump potential to the measured E_m is the difference in E_m when the pump is operating versus that immediately after the pump has been stopped by the addition of ouabain or zero $[K^+]_o$. Consequently, it appears as though the contribution from the electrogenic pump potential (ΔV_p) was in series with the net cationic diffusion potential (E_{diff}),

$$E_m = E_{diff} + R_m I_p = E_{diff} + \Delta V_p \qquad (22)$$

where I_p is the electrogenic component of the pump current, and E_{diff} is the E_m that would exist solely on the basis of the ionic gradients and relative permeabilities in the absence of an electrogenic pump potential (as calculated from the constant-field equation). Equation 22 states that E_m is the sum of E_{diff} and a voltage (IR) drop produced by the **electrogenic pump current** across R_m. The electrogenic pump potential (V_p) can be considered to be in parallel with E_{diff} (Fig. 10). Because the **density of pump sites** is more than a thousand-fold greater than that of Na⁺ and K⁺ channels in resting membrane, there is no relation between the pump pathway (the active flux path) and R_m (the passive flux paths); that is, the pump path and the passive conductance paths are in parallel. The pump potential should be considered the full potential between zero and the maximum negative pump potential (V_p) while the pump is pumping (see Fig. 9).

One possible equivalent circuit for an electrogenic Na⁺ pump that takes into account some of the known facts is given in Fig. 10. The pump pathway is in parallel with the resistance pathways. The pump resistance (R_p) is estimated to be about ninefold higher than R_m. If so, the pump resistance acts to minimize a short-circuit path to E_{diff} when the pump potential is low or zero (pump inhibited). The pump potential contribution to E_m (ΔV_p) is a function of membrane resistance (R_m); the higher the R_m (R_p constant), the more nearly

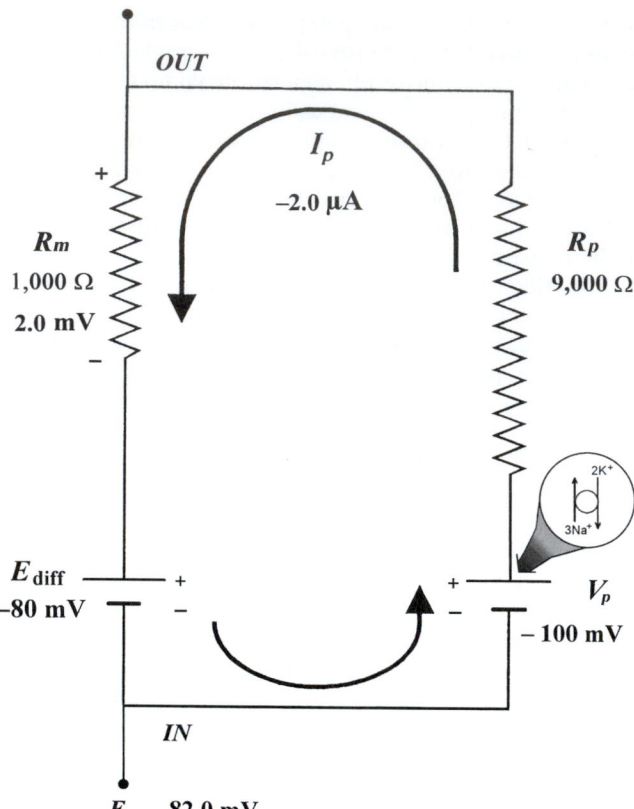

FIGURE 10. Hypothetical electrical equivalent circuit for electrogenic sodium pump. Model consists of a pump pathway in parallel with the membrane resistance (R_m) pathway. This model fits the evidence that the pump is independent of short-range membrane excitability and that the pump proteins and channel proteins are embedded in the lipid bilayer as parallel elements. Net diffusion potential (E_{diff}, determined by the ion equilibrium potentials and relative permeabilities) of −80 mV is depicted in series with R_m. Pump leg is assumed to consist of a battery in series with a fixed resistor (pump resistance, R_p) that does not change with changes in R_m and whose value is ninefold higher than R_m. Pump battery is charged up to some voltage (e.g., V_p of −100 mV) by a pump current generator. Net electrogenic pump current is developed by the pumping in of only 2 K$^+$ ions for every 3 Na$^+$ ions pumped out. For the values given in the figure (namely, R_m of 1000 Ω, E_{diff} of −80 mV, R_p of 9000 Ω, and V_p of −100 mV), it may be calculated by circuit analysis that the measured membrane potential (E_m) is −82.0 mV: that is, the direct electrogenic pump potential contribution to the resting potential is −2.0 mV. The calculated pump current (I_p) is −2.0 μA.

E_m approaches V_p. The pump battery is charged to some voltage by a pump current generator. If the pump is stopped by ouabain, V_p goes to zero. Using circuit analysis for the values of the parameters given in Fig. 10, E_m would be −82.0 mV, moderately close to E_{diff} (−80 mV) (Table 3). If R_m is raised twofold (to 2000 Ω), E_m would be −83.6 mV. Thus this circuit clearly gives a pump potential contribution to E_m that is dependent on R_m. The higher R_m is relative to R_p, the more E_m reflects V_p. If E_{diff} is made smaller (e.g., in smooth muscle cells having a higher P_{Na}/P_K ratio), then the relative contribution of the pump potential to E_m becomes greater (see Table 3).

In general, Cl$^-$ ions are known to have a short-circuiting effect on the electrogenic Na$^+$ pump potential. For example, if the external Cl$^-$ is replaced by less permeant anions, the magnitude of the hyperpolarization produced by the electrogenic Na$^+$ pump is substantially increased. This Cl$^-$ effect could be caused by the lowering of membrane resistance in the presence of Cl$^-$. The greater the R_m, the greater the contribution of the electrogenic pump potential to resting E_m (see Fig. 10 and Table 3).

The density of Na$^+$-K$^+$ pump sites, estimated by specific binding of [^{3}H]ouabain, is usually about 700–1000/μm^2. The turnover rate of the pump is generally estimated to be 20–100/s. The **pump current (I_p)** has been estimated as

$$I_p = \frac{\Delta V_p}{R_m} \tag{23}$$

where ΔV_p is the pump potential contribution. Values of about 20 pmol/(cm^2·s) were obtained. A density of 1000 sites/μm^2 (10^{11} sites/cm^2) times a turnover rate of 40/s gives 4 x 10^{12} turnovers/(cm^2·s). If 3 Na$^+$ are pumped with each turnover, this gives 12 × 10^{12} Na$^+$ ions/(cm^2·s); dividing by Avogadro's number (6.02 × 10^{23} ions/mol) yields 20 x 10^{-12} mol/(cm^2·s), which is the same value as the 20 pmol/(cm^2·s) measured. The net pump current would be less, depending on the amount of K$^+$ pumped in the opposite direction, that is, depending on the coupling ratio (e.g., 3 Na$^+$:2 K$^+$). Whenever the Na$^+$-K$^+$ pump is stimulated to turn over faster, for example, by increasing [Na$^+$]$_i$ or [K$^+$]$_o$, the electrogenic pump current is increased.

Ion flux (J) can be converted to current (I) by the relationship

$$I = J \cdot z \mathscr{F}$$

$$\frac{A}{cm^2} = \frac{mol}{s \cdot cm^2} \frac{C}{mol} \tag{24}$$

TABLE 3 Summary of Calculations of Resting Potential (E_m) for a Model Having an Electrogenic Pump Potential (V_p) in Parallel with Net Diffusion Potential (E_{diff}), as Depicted in Fig. 10

E_{diff} mV	R_m Ω	V_p mV	I_p μA	V_{Rp} mV	ΔV_p mV	RP mV
−80	1000	−100	−2.0	+18	−2.0	−82
−80	2000	−100	−1.82	−16.4	−3.64	−83.6
−80	1000	0	+8.0	−72	+8.0	−72
−50	1000	−100	−5.0	+45	−5.0	−55
−50	2000	−100	−4.55	+40.9	−9.1	−59.1
−50	1000	0	+5.0	−45	+5.0	−45

Pump resistance, R_p, was assumed to have constant value of 9000 Ω. R_m, membrane resistance; ΔV_p contribution of V_p to the measured E_m. E_m was calculated from the equation

$$E_m = \left(\frac{R_m}{R_m + R_p}\right) V_p + \left(\frac{R_p}{R_m + R_p}\right) E_{diff}$$

Thus, a flux of 20 pmol/(cm^2·s) is equal to approximately 2 μA/cm^2 (20×10^{-12} mol/(s·cm^2) $\times 0.965 \times 10^5$ C/mol). Since $\Delta V_p = I_p \times R_m$, if R_m were 1000 Ω·cm^2 and I_p were 2 μA/cm^2, the electrogenic pump contribution to E_m would be 2 mV ($E_m = E_{\text{diff}} + I_p R_m$).

Two K$^+$ ions are usually carried in for every 3 Na$^+$ ions moved out. Because the pump is **electrogenic**, that is, produces a net current (and hence potential) across the membrane, then the amount of K$^+$ pumped in must be less than the amount of Na$^+$ pumped out; for example, the Na$^+$/K$^+$ **coupling ratio** must be 3:2 (or 3:1). The coupling ratio cannot be 3:0, because of the well-known fact that external K$^+$ must be present for the pump to operate. The coupling ratio might be increased under some conditions, for example, when [Na$^+$]$_i$ is elevated. If the coupling ratio were to increase (e.g., to 3:1), the pump potential contribution would become larger, for a constant pumping rate.

The pump current may be stimulated by increasing the **turnover rate** of each pump site or by increasing the **number of pump sites**. In skeletal muscle, insulin has been reported to increase the number of Na$^+$-K$^+$ pump sites in the sarcolemma by increasing the rate of translocation from an internal pool, thereby increasing the pump current. β-Adrenergic agonists, like isoproterenol, stimulate the pump current by cyclic AMP/protein kinase A phosphorylation of the pump.

The electrogenic pump potential has **physiologic importance** in cells. Although small, the electrogenic pump potential contribution to the resting potential could have significant effects on the level of inactivation of the fast Na$^+$ channels, and hence on propagation velocity. Further, an electrogenic pump potential could act to delay depolarization under adverse conditions (e.g., ischemia and hypoxia) and would act to speed repolarization of the normal resting potential during recovery from the adverse conditions. It is crucial that the excitable cell maintain its normal resting potential as much as possible, because of the effect of small depolarizations on the AP rate of rise and conduction velocity and the complete loss of excitability with larger depolarizations. For example, the rate of firing of pacemaker nodal cells of the heart is affected significantly by very small potential changes.

In cells in which there are lower resting potentials (e.g., smooth muscle cells and cardiac nodal cells) (see Table 1), the electrogenic pump potential contribution can be larger (see Table 3). **Sinusoidal oscillations in the Na$^+$-K$^+$ pumping rate** could produce oscillations in E_m, which could exert important control over the spontaneous firing of the cell. The period of enhanced pumping hyperpolarizes the cell and suppresses automaticity, whereas slowing of the pump leads to depolarization and consequently to triggering of APs. Oscillation of the pump rate would be brought about by oscillating changes in [Na$^+$]$_i$. For example, the firing of several APs should raise [Na$^+$]$_i$ (nodal cells have a small volume to surface area ratio) and stimulate the electrogenic pump. The increased pumping rate, in turn, hyperpolarizes and suppresses firing, thus allowing [Na$^+$]$_i$ to decrease again, and removing the stimulation of the pump; the latter condition depolarizes and triggers spikes, and the cycle could be repeated. It was concluded that the electrogenic Na$^+$ pump

in rabbit sinoatrial nodal cells might be one factor that modulates the heart rate under physiologic conditions. When stimulated at a high rate, cardiac Purkinje fibers and nodal cells undergo a transient period of inhibition of automaticity after cessation of the stimulation, known as **overdrive suppression of automaticity.** Stimulation of the electrogenic pump due to elevation in [Na$^+$]$_i$ is the major cause of this phenomenon.

VIII. Summary

Most of the factors that determine or influence the resting E_m of cells were discussed in this chapter. The structural and chemical composition of the cell membrane was briefly examined and correlated with the resistive and capacitive properties of the membrane. The factors that determine the intracellular ion concentrations in cells were examined. These factors include the **Na$^+$-K$^+$-coupled pump,** the Ca^{2+}-Na$^+$ exchange reaction, and the sarcolemmal Ca^{2+} pump. The Na$^+$-K$^+$ pump enzyme, Na$^+$,K$^+$-ATPase, requires both Na$^+$ and K$^+$ for activity, and transports 3 Na$^+$ ions outward and usually 2 K$^+$ ions inward per ATP hydrolyzed. Cardiac glycosides are specific blockers of this transport ATPase. The Na$^+$-K$^+$ pump is not directly related to excitability, but only indirectly related by its role in maintaining the Na$^+$ and K$^+$ concentration gradients.

The carrier-mediated **Ca^{2+}-Na$^+$ exchange reaction** is driven by the Na$^+$ electrochemical gradient; that is, the energy for transporting out internal Ca^{2+} by this mechanism comes from the Na$^+$,K$^+$-ATPase. The Ca^{2+}-Na$^+$ exchange reaction exchanges one internal Ca^{2+} ion for three external Na$^+$ ions when working in the forward mode in cells at rest. During the AP depolarization, for example, in myocardial cells, the energetics cause the Ca^{2+}-Na$^+$ exchanger to operate in reverse mode, allowing Ca^{2+} influx.

The mechanism whereby the **ionic distributions** give rise to diffusion potentials was discussed, as were the factors that determine the magnitude and polarity of each ionic **equilibrium potential.** The equilibrium potential for any ion and the transmembrane potential determine the total **electrochemical driving force** for that ion, and the product of this driving force and membrane conductance for that ion determine the net ionic current or flux. The net ionic movement can be inward or outward across the membrane, depending on the direction of the electrochemical gradient.

The key factor that determines the resting E_m—in the absence of any electrogenic pump potential contributions—is the **relative permeability** of the various ions, particularly of K$^+$ and Na$^+$, that is, the P_{Na}/P_K ratio (or g_{Na}/g_K ratio), as calculated from the Goldman constant-field equation. The major physiologic ions that have some effect on the resting E_m or on the APs are K$^+$, Na$^+$, Ca^{2+}, and Cl$^-$. The Ca^{2+} electrochemical gradient has only a small direct effect on the resting E_m, although low external Ca^{2+} can affect the permeabilities and conductances for the other ions, such as Na$^+$ and K$^+$. Elevation of internal Ca^{2+} can increase the permeability to K$^+$ by activating Ca^{2+}-operated K$^+$-selective $I_{K(Ca)}$ channels.

Cl$^-$ is usually **passively distributed** according to the membrane potential, that is, not actively transported. However, there is some evidence indicating that $[Cl^-]_i$ may be about twice as high as that predicted from E_m in some cells like smooth muscle cells; if so, this would give an E_{Cl} value of about 18 mV less negative than the resting E_m. Before one can conclude that there is a Cl^- pump directed inward, however, the calculated E_{Cl} (concentrations corrected for activity coefficients) must be proven to be significantly more positive than the mean resting E_m of the cell averaged over time; for example, any spontaneous APs must be taken into account. If Cl^- is passively distributed, it cannot determine the resting E_m. However, transient net movements of Cl^- ions, for example, during the AP, can and do affect the E_m, particularly when g_{Cl} is high.

Elevation of $[K^+]_o$ to more than the normal concentration of about 4.5 mM decreases the K^+ equilibrium potential (E_K), as predicted from the Nernst equation ($[K^+]_i$ about constant), and depolarization is produced. Sometimes, however, some hyperpolarization is produced at a $[K^+]_o$ level between 5 and 9 mM. In addition, lowering $[K^+]_o$ to 0.1 mM often produces a prominent depolarization. These effects are usually explained on the basis that (1) P_K is lowered in low $[K^+]_o$ and elevated in higher $[K^+]_o$ and (2) an electrogenic Na^+-K^+ pump potential is inhibited at a low $[K^+]_o$ (K_m of about 2 mM).

Not only is the resting E_m the **potential energy storehouse** that is drawn upon for production and propagation of the APs, but because the membrane voltage-dependent cationic channels are inactivated with sustained depolarization, the rate of rise of the AP, and hence propagation velocity, is critically dependent on the level of the resting E_m. For example, a relatively small elevation of K^+ concentration in the blood has dire consequences for functioning of the heart.

The contribution of the Na^+-K^+ pump to the resting E_m depends on (1) the coupling ratio of Na^+ pumped out to K^+ pumped in, (2) the turnover rate of the pump, (3) the number of pumps, and (4) the magnitude of the membrane resistance. The **electrogenic pump potential** is in parallel to the net ionic diffusion potential (E_{diff}), determined by the ionic equilibrium potentials and by the relative permeabilities. The contribution of the electrogenic pump potential to the measured resting E_m of cells varies from 2 to 16 mV, depending on the type of cell. Thus, the immediate depolarization produced by complete Na^+-K^+ pump stoppage with cardiac glycosides is only a few millivolts in cells like myocardial cells. Of course, long-term pump inhibition produces a larger and larger depolarization as the ionic gradients are dissipated. The rate of Na^+-K^+ pumping, and hence the magnitude of the electrogenic pump contribution of E_m, is controlled primarily by $[Na^+]_i$ and by $[K^+]_o$. The electrogenic pump potential might be **physiologically important** to various tissues, particularly the heart, under certain conditions that tend to depolarize the cells, such as transient ischemia or hypoxia. In such cases, the actual depolarization produced may be less because of a relatively constant pump potential in parallel with a diminishing E_{diff}. The electrogenic pump potential may also affect **automaticity** of the nodal cells of the heart as well as other types of cells that exhibit automaticity.

Appendix

I. Derivation of Nernst Equation

The Nernst equation may be derived form the general equation for the free energy change (ΔG_c) resulting from both osmotic work and electrical work for transporting 1 mole of cation (c^+) across a membrane. Thus,

$$\Delta G_c = RT \ln \frac{[c^+]_i}{[c^+]_o} + z\mathscr{F}E_m \qquad (A\text{-}1)$$

where R is the gas constant; T is absolute temperature; $[c^+]_i$ and $[c^+]_o$ are the internal and external c^+ concentrations, respectively; z is the valence; $\mathscr{F}$ is the Faraday constant; and E_m is membrane potential. The first term on the right side of this equation, $RT \ln ([c^+]_i/[c^+]_o)$, gives the osmotic work for transporting a mole of particles across the membrane against a concentration gradient. The second term, $z\mathscr{F}E_m$, gives the electrical work for transporting 1 mole of charged particles across the membrane against an electrical gradient. The sum of these two terms then gives the total work required. At equilibrium, the change in free energy for moving one or only a few particles across the membrane must be zero ($\Delta G = 0$). Therefore,

$$0 = RT \ln \frac{[c^+]_i}{[c^+]_o} + z\mathscr{F}E_m \qquad (A\text{-}2)$$

and

$$z\mathscr{F}E_m = -RT \ln \frac{[c^+]_i}{[c^+]_o} \qquad (A\text{-}2a)$$

or

$$E_m = \left(\frac{-RT}{z\mathscr{F}}\right) \ln \frac{[c^+]_i}{[c^+]_o} \qquad (A\text{-}2b)$$

which is the Nernst equation.

Sine the Faraday constant ($\mathscr{F}$) is equal to the charge on an electron (e, in coulombs) times Avogadro's number (N_A, number of ions per mole), then

$$\frac{RT}{\mathscr{F}} = \frac{RT}{N_A Q_e} = \frac{kT}{Q_e} \qquad (A\text{-}3)$$

where k (the Boltzmann constant) is equal to the gas constant (R) divided by Avogadro's number (N_A), that is, the energy (in joules) of an ion per kelvin; and Q_e is the charge (in coulombs) on an electron (namely, 1.6×10^{-19} C/e$^-$).

II. Half-Cell Potentials

In measuring biological potentials, care must be taken not to introduce artifacts, such as half-cell potentials. This section will give a brief description of electrode half-cell potentials. For example, if two beakers containing NaCl at 0.1 and 0.01 M were joined by a salt bridge (agar-NaCl), and if a Ag-AgCl half-cell electrode were placed in each beaker, than a PD of 59 mV would be recorded between the two electrodes, because the potential of each half-cell, reversible to Cl$^-$ ions, would be different (Fig A1). In this example, the beaker containing the higher Cl$^-$ concentration would be **negative**, and the one with the lower Cl$^-$ concentration would be **positive**. The AgCl coat of the electrode immersed in the lower Cl$^-$ concentration would have the greater tendency to solubilize and ionize, leaving this electrode positive. Conversely, the AgCl coat of the electrode immersed in the higher Cl$^-$ concentration would have the lower tendency to solubilize and actually would tend to deposit more AgCl, stealing a positive charge from the wire and thus leaving that electrode negatively charged. A positive potential is applied to electroplate the Ag wires with AgCl by electrophoresing Cl$^-$ to the Ag wire, as shown in Fig A1.

Note that the resting potential recorded in biological cells by an intracellular microelectrode is not a function of the half-cell potentials (i.e., an artifact), because the solutions bathing the half-cells (e.g., Ag–AgCl wires or calomel half-cells) remain constant; that is, the half-cell potentials stay the same whether the microelectrode is inside or outside the cell. The two half-cell potentials are nearly equal in magnitude and so cancel each other. Any small amount of difference between the two half-cell potentials (e.g., a few millivolts) when the two electrodes are in the same Ringer's solution is arbitrarily called the zero potential (in practice, with the microelectrode in position, any small microelectrode tip potential, e.g., up to 5 mV, would be included in the zeroing procedure). The resting potential of the cell is added in series with half-cell potential, and thus the recording system gives the true transmembrane resting potential.

III. Constant-Field Equation

An important modification of the Nernst equation in common use for calculating the membrane potential, or for determining the P_{Na}/P_K ratio, is the Goldman-Hodgkin-Katz constant-field equation (Goldman, 1943; Hodgkin and Katz, 1949),

$$E_m = \frac{-RT}{\mathscr{F}} \ln \frac{P_K[K^+]_i + P_{Na}[Na^+]_i + P_{Cl}[Cl^-]_o}{P_K[K^+]_o + P_{Na}[Na^+]_o + P_{Cl}[Cl^-]_i} \qquad (A\text{-}4)$$

A

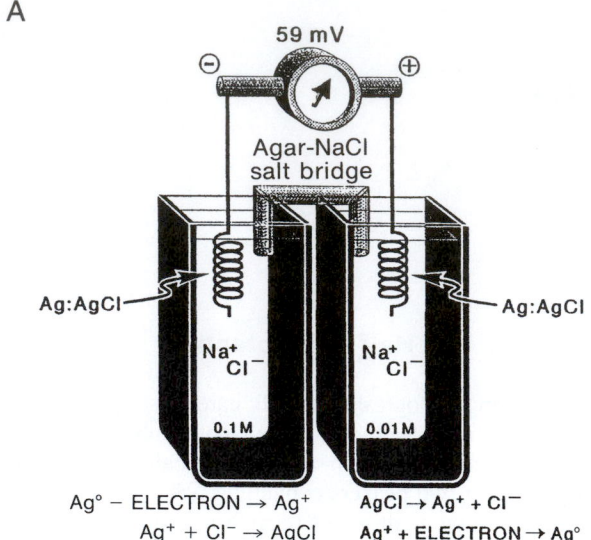

B

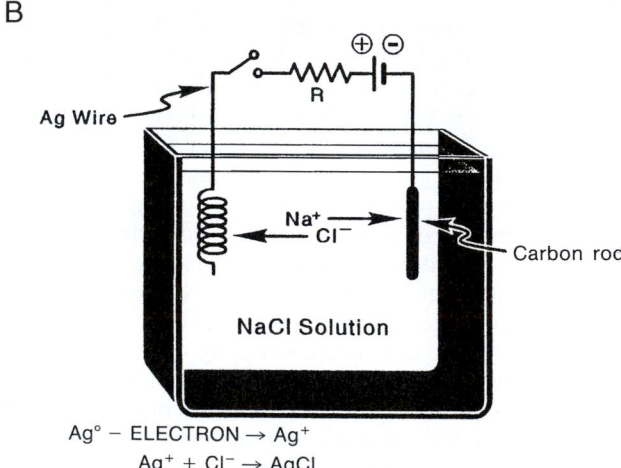

FIGURE A1. Half-cell electrode potentials: (A) Two Ag-AgCl half-cell electrodes are bathed in two different solutions containing Cl⁻ ion at different concentrations (0.1 and 0.01 M, in figure). The Ag-AgCl half-cells are reversible to Cl⁻, and therefore the half-cell potential depends on the Cl⁻ concentration in which the electrode is bathed. Electrode bathed in the highest Cl⁻ concentration has a more negative value than the electrode in lower Cl⁻ concentration. Thus the two half-cell potentials are not equal to one another, and their potentials do not cancel out, as is normally true. There is a net PD produced by the two unequal half-cell potentials, the electrode in the higher Cl⁻ concentration being negative and the electrode in the lower Cl⁻ concentration being positive. The two solutions are connected by an agar salt bridge to complete the circuit. (B) For electroplating a silver wire electrode with AgCl, the silver wire must be made positive so that Cl⁻ ions can be electrophoresed through the solution to react with silver atoms to plate AgCl.

where P_K, P_{Na}, and P_{Cl} are the membrane permeabilities for K^+, Na^+, and Cl^-, respectively. The $P_K [K^+]_i$ product, for example, is given in the units of a flux (mol/s per cm²), but the entire right-hand term (fraction) is dimensionless. Plugging in the numerical values for the constants and converting from natural logarithm, or ln, to logarithm to the base 10, or log (ln N = 2.3 log N) gives

$$E_m = -61 \,\text{mV} \log \frac{P_K[K^+]_i + P_{Na}[Na^+]_i + P_{Cl}[Cl^-]_o}{P_K[K^+]_o + P_{Na}[Na^+]_o + P_{Cl}[Cl^-]_i} \quad \text{(A-5)}$$

and E_m is expressed in mV. Dividing the right-hand term by P_K gives

$$E_m = -61 \,\text{mV} \log \frac{[K^+]_i + \dfrac{P_{Na}}{P_K}[Na^+]_i + \dfrac{P_{Cl}}{P_K}[Cl^-]_o}{[K^+]_o + \dfrac{P_{Na}}{P_K}[Na^+]_o + \dfrac{P_{Cl}}{P_K}[Cl^-]_i} \quad \text{(A-6)}$$

Again, the right-hand term (fraction) is dimensionless. Any ion that is passively distributed, that is, not actively pumped, however, cannot determine the resting potential, because the distribution of that ion must follow the resting potential. Therefore, when Cl⁻ is passively distributed, it is not considered because resting potential cannot be determined by Cl⁻. As a result Eq. A-6 can be reduced to

$$E_m = -61 \,\text{mV} \log \frac{[K^+]_i + \dfrac{P_{Na}}{P_K}[Na^+]_i}{[K^+]_o + \dfrac{P_{Na}}{P_K}[Na^+]_o} \quad \text{(A-7)}$$

Equation A-7 is one of the most useful forms of the constant field equation, because if E_m is measured, and if the internal and external ion concentrations are known, the P_{Na}/P_K ratio can be calculated. Thus for given ionic gradients, the resting potential is determined by the P_{Na}/P_K ratio (i.e., the relative permeabilities of the cell membrane to Na^+ and K^+) and not by the absolute permeabilities. For simplicity, Ca^{2+} is ignored here as a factor contributing to the resting potential. The relationship between the permeability coefficient for an ion (P_i) and the membrane conductance for that ion (g_i) is complex and involves several terms, including membrane potential.

Figure 8 of Chapter 14 gives the expected resting potential as a function of the P_{Na}/P_K ratio for a muscle cell (having a $[K^+]_i$ of 150 mM) bathed in normal Rigner's solution ($[K^+]_o$ of 4 mM. $[Na^+]_o$ of 150 mM), assuming an $[Na^+]_i$ value of 15 mM. As can be seen in the figure, a P_{Na}/P_K ratio of 0.1 gives a resting potential close to −60 mV, whereas a ratio of 0.01 gives a potential close to −85 mV. Some muscle cells, for example, smooth muscle and young embryonic myocardial cells, have a low resting potential (at a $[K^+]_o$ of 4 mM) of about −50 mV, presumably because of a high P_{Na}/P_K ratio of about 0.15 rather than smaller ionic gradients. The P_{Na}/P_K ratio can be high because of either a high P_{Na} or a low P_K, or both; in most cases the main reason appears to be a low P_K.

One advantage the constant-field equation has over the chord conductance equation is that it nicely accounts for the bend at low $[K^+]_o$ in the E_m versus log $[K^+]_o$ curves (see Fig. 8 in Chapter 14). As can be seen in Fig. 8, which presents theoretical curves calculated form Eq. A-7, the higher the P_{Na}/P_K ratio, the greater the deviation from a straight line as $[K^+]_o$ is lowered. As mentioned previously, from this equation, one can deduce that as $[K^+]_o$ is lowered and $[Na^+]_o$ is concomitantly elevated, the denominator of the right-hand term becomes more and more dominated by the Na^+ term, for any given P_{Na}/P_K ratio. Because the numerator is relatively fixed, E_m is more influenced by E_{Na} as $[K^+]_o$ is lowered more and more. Thus, this relationship accounts for the deviation from the straight line for E_K.

Although the $P_{Na}/P_K = 0.05$ curve is almost linear at high $[K^+]_o$ with a slope of 60 mV/decade, the membrane does not necessarily become "purely K^+-selective," as is often stated, because for these theoretical calculations, the P_{Na}/P_K ratio was held constant over the entire $[K^+]_o$ range. There is some evidence, however, that P_K itself increases as $[K^+]_o$ increases, but this effect was not taken into consideration in Fig. 8. g_K is a function of $[K^+]_o$, namely, $g_K \propto P_K [K^+]_o$. Finally, the increased bending for the higher P_{Na}/P_K ratios can again be seen from Eq. A-7; at a given $[K^+]_o$ (e.g., 4 mM) the denominator is more and more dominated by the Na^+ term as the P_{Na}/P_K ratio is increased more and more.

The order of selectivity of the resting membrane for the alkali metal ions generally is in the following sequence, from the highest permeability to the lowest: $K^+ > Rb^+ > Cs^+ > Na^+ > Li^+$. For example, the relative permeabilities (assigning $P_K = 1$) in squid giant axon (Baker *et al.,* 1968) are

$$P_K > P_{Rb} > P_{Cs} > P_{Na} > P_{Li}$$
$$1.0 \quad 0.69 \quad 0.19 \quad 0.17 \quad 0.12$$

(A-8)

In frog sartorius (Mullins, 1961) the values are

$$P_K > P_{Rb} > P_{Cs} > P_{Na}$$
$$1.0 \quad 0.54 \quad 0.11 \quad 0.04$$

(A-9)

The P_{Na}/P_{Ca} ratio in frog sartorius is about 3 (Mullins, 1961).

So far in our discussion, Ca^{2+} has been ignored. It can be demonstrated that Ca^{2+} has only a negligible effect on the resting potential, even if it has a permeability equal to, or 10 times greater than, Na^+. This is because of the relatively low extracellular and intracellular concentration of free Ca^{2+} ion compared with those of the K^+ and Na^+ ions. A modified version of the Goldman constant-field equation, which includes a Ca^{2+} term, is[3]

$$E_m = -60 \text{ mV} \log \frac{(B - A) + \sqrt{y}}{2\left(A - 4P_{Ca}[Ca^{2+}]_i\right)}$$

(A-10)

where

$$A = P_K[K^+]_i + P_{Na}[Na^+]_i$$
$$B = P_K[K^+]_o + P_{Na}[Na^+]_o$$
$$y = (B - A)^2 + 4\left(A + 4P_{Ca}[Ca^{2+}]_i\right)\left(B + 4P_{Ca}[Ca^{2+}]_o\right)$$

For simplification, the analogous Cl^- terms ($+P_{Cl}[Ca^{2+}]_o$ in the definition of A and $+P_{Cl}[Cl^-]_i$ term in B) have been omitted, assuming Cl^- to be passively distributed.

Calculations made from Eq. A-10 demonstrate some interesting points: (1) For the same permeabilities ($P_{Ca} = P_{Na}$), Ca^{2+} has much less effect on E_m than does Na^+, because of the lower Ca^{2+} concentrations and because of the square root function for the Ca^{2+} concentrations. For example, the depo-

larization produced by taking into account the Ca^{2+} ion is only +0.4 mV for a P_{Na}/P_K ratio of 0.1. (2) Even when P_{Ca} is set equal to 10 times P_{Na}, the effect of Ca^{2+} on the resting potential is still relatively small (e.g., +3.5 mV for a P_{Na}/P_K ratio of 0.01, and +8.0 mV for a P_{Na}/P_K ratio of 0.1). (3) The effect of Ca^{2+} is somewhat greater when the P_{Na}/P_K ratio is higher (as in cardiac nodal cells or smooth muscle). (4) The effect of taking into account Ca^{2+} is considerably less at high $[K^+]_o$ values.

Thus, these calculations support the view that Ca^{2+} can be virtually ignored in discussion of the ionic basis of the resting potential. This agrees with the well-known fact that a variation in $[Ca^{2+}]$ throughout a relatively wide range has a negligible effect on the resting potential (see, for example, Sperelakis, 1972). Further, these conclusions have implications about the relative importance of Na^+ versus Ca^{2+} background currents (inward) during genesis of the pacemaker potential (concomitant with the decrease in g_K and I_K). Finally, it should be emphasized that, for example, when $P_{Ca} = P_K$, g_{Ca} does not equal g_K, because of the concentration differences. To calculate g_K from a given P_K, one must use the appropriate equation that takes into account the concentrations and membrane potential. If $g_{Ca} = g_K$, then the membrane potential is halfway between E_K and E_{Ca}, and Ca^{2+} would have a much greater effect on the resting potential.

IV. Derivation of Chord Conductance Equation

Ohm's law states that the current (I) is equal to the voltage (E) either divided by the resistance (R) or multiplied by the conductance ($g = 1/R$)

$$I = \frac{E}{R} = gE$$

(A-11)

When dealing with solutions, the voltage or driving force must take into account both the concentration force and the electrical force. In this case, the ionic current (I_i) is a product of the conductance for a given ion times the total driving force on that ion ($E_m - E_i$) and may be expressed as

$$I_i = g_i(E_m - E_i)$$

(A-12)

In a resting cell membrane (stable resting potential), the total ionic current must be zero; otherwise, the membrane potential would change. Therefore, the K^+ current in the outward direction (I_K) must be equal and opposite to the Na^+ current (I_{Na}) entering the cell (neglecting Ca^{2+}, Cl^-, and minor ions) expressed as

$$I_K = -I_{Na}$$

(A-13)

Therefore,

$$I_K + I_{Na} = 0$$

(A-14)

Substituting the equations for ionic currents from Eq. A-12,

$$g_K(E_m - E_K) + g_{Na}(E_m - E_{Na}) = 0$$

(A-14)

Algebraic manipulations give

$$0 = g_K E_m - g_K E_K + g_{Na} E_m - g_{Na} E_{Na}$$
$$g_K E_m + g_{Na} E_m = g_K E_K - g_{Na} E_{Na}$$

(A-15)

$$E_m(g_K + g_{Na}) = g_K E_K + g_{Na} E_{Na}$$

---

[3]This equation was kindly provided by Professor D. E. Goldman.

Rearrangement gives

$$E_m = \frac{g_K}{g_K + g_{Na}} E_K + \frac{g_{Na}}{g_K + g_{Na}} E_{Na} \qquad \text{(A-16)}$$

Equation A-16 is the chord conductance equation. The ratios $g_K/(g_K + g_{Na})$ and $g_{Na}/(g_K + g_{Na})$ are the **fractional conductances** (relative) and are dimensionless.

If Cl^- were to be included, the same steps in the derivation would give the chord conductance equation containing a Cl^- term,

$$E_m = \frac{g_K}{\Sigma g} E_K + \frac{g_{Na}}{\Sigma g} E_{Na} + \frac{g_{Cl}}{\Sigma g} E_{Cl} \qquad \text{(A-17)}$$

$\Sigma g = g_K + g_{Na} + g_{Cl}$. However, if Cl^- is passively distributed (in equilibrium at the resting E_m), then Cl^- cannot be involved in determining the resting potential (although transient movements of Cl^- can affect the membrane potential when Cl^- is shifted off equilibrium during an AP or postsynaptic potential).

On the other hand, the Ca^{2+} ion is actively transported and is off equilibrium, so its conductance influences the resting potential. Thus, the chord conductance equation containing the Ca^{2+} term is

$$E_m = \frac{g_K}{\Sigma g} E_K + \frac{g_{Na}}{\Sigma g} E_{Na} + \frac{g_{Ca}}{\Sigma g} E_{Ca} \qquad \text{(A-18)}$$

$$\Sigma g = g_K + g_{Na} + g_{Ca}$$

The chord conductance equation, of course, can be written using resistances rather than conductances. For Eq. A-16 using only K^+ and Na^+ terms, substitution of $R = 1/g$ in the equation and algebraic manipulation gives the equation

$$E_m = \frac{R_K}{R_K + R_{Na}} E_{Na} + \frac{R_{Na}}{R_K + R_{Na}} E_K \qquad \text{(A-19)}$$

Note that the E_{Na} and E_K terms are interchanged from the chord conductance equation. This form of the equation might be termed the **chord resistance equation.**

The chord conductance equation applies only to those situations in which the net ionic current is zero, such as when the membrane is at rest. This equation is derived simply from Ohm's law, and one advantage it has over the constant-field equation is that it can more easily include divalent cations such as Ca^{2+}.

V. Circuit Analysis Applicable to Cell Membrane

Using Ohm's and Kirchhoff's laws and logic, it is possible to see why in the nerve or muscle cell, the K^+ battery dominates the resting potential, whereas the Na^+ or Ca^{2+} batteries, or both, dominate the peak of the AP. The circuit in Fig. A2 will be used to show that the battery having the lowest resistance in series with it is the battery that is the most expressed across the network. Before analyzing the circuit rigorously, we can consider three conditions and make some qualitative judgments: (1) If the left resistor (R_1) equals the right resistor (R_2) (regardless of their absolute values), the PD across the network is +150 V (upper terminal positive with respect to the

lower terminal), that is, halfway between both batteries because both should be equally expressed, (2) if R_2 is made infinite (e.g., open circuit in branch 2) and R_1 is finite, then the PD is exactly +100 V, since the right battery (E_2) cannot be expressed at all, and (3) if R_1 is much less than R_2, then the PD approaches +100 V because E_1 is dominant.

The circuit in Fig. A2 can also be analyzed quantitatively. For example, if $R_1 = 10\ \Omega$, and $R_2 = 990\ \Omega$, the exact PD may be reasoned from the following analysis. The current (I) has one magnitude; that is, it is constant throughout this simple closed circuit, but the current flows upward in branch 2 and downward in branch 1. This occurs because the right battery (E_2) is larger than E_1, and so the net driving force for the net current is in the direction as indicated in the figure. Therefore, the voltage drops produced across R_1 and R_2 are in opposite polarities, as shown in the figure. The voltage drop across R_1 adds to E_1 to make a greater PD across branch 1, like two batteries in series (+ −, + −). In contrast, the voltage drop across R_2 subtracts from E_2 to make a smaller PD, like two batteries back to back (− +, + −). Therefore, the following two equations can be written for the PD across branch 1 [$(PD)_1$] and across branch 2 [$(PD)_2$]:

$$(PD)_1 = E_1 - IR_1 \qquad \text{(A-20)}$$

$$(PD)_2 = E_2 - IR_2 \qquad \text{(A-21)}$$

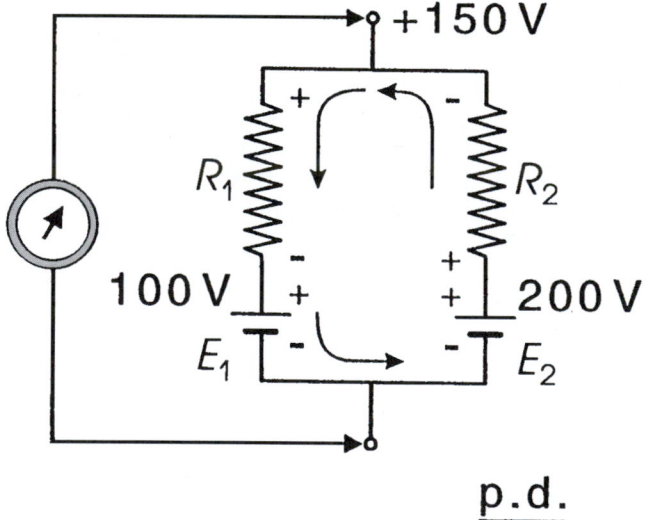

		p.d.
(1)	$R_1 = R_2$	+ 150 V
(2)	$R_2 = \infty$	+ 100 V
(3)	$R_1 = 10\ \Omega$ $R_2 = 990\ \Omega$	+ 101 V

FIGURE A2.   Circuit diagram of the circuit analysis applicable to the cell membrane, showing why the resting potential of a cell is determined by the relative permeabilities (or conductances). Battery having the lowest resistance in series with it is the battery most expressed across such a network.

To solve these equations, we must first calculate the current (I). The net driving force for I is equal to $E_2 - E_1$; hence from Ohm's law, the net current is equal to $(E_2 - E_1)$ divided by the total resistance ($R_1 + R_2$):

$$I = \frac{E_2 - E_1}{R_2 + R_1} \qquad \text{(A-22)}$$

$$I = \frac{200\,\text{V} - 100\,\text{V}}{990\,\Omega + 10\,\Omega}$$

$$= \frac{100\,\text{V}}{1000\,\Omega} = 0.1\,\text{A}$$

Now we can enter this value for I into Eq. A-20

$$(\text{PD})_1 = E_1 - IR_1$$

$$= 100\,\text{V} - (-0.1\,\text{A})(10\,\Omega)$$

$$= 100\,\text{V} - (-1\,\text{V})$$

$$= 100\,\text{V} + 1\,\text{V}$$

$$= 101\,\text{V}$$

The negative sign in the current is because the current in branch 1 produces a voltage drop that adds to E_1. Because the two branches are connected by zero resistances, that is, they are effectively the same points, $(\text{PD})_1$ must equal $(\text{PD})_2$:

$$(\text{PD})_1 = (\text{PD})_2 \qquad \text{(A-23)}$$

Therefore, we can also calculate the PD by substituting into Eq. A-21.

$$(\text{PD})_2 = E_2 - IR_2$$

$$= 200\,\text{V} - (+0.1\,\text{A})(990\,\Omega)$$

$$= 200\,\text{V} - (99\,\text{V})$$

$$= 101\,\text{V}$$

Thus, the two methods check.

To summarize, it has been quantitatively demonstrated that the battery with the lowest series resistance is the battery most expressed across this network. This analysis holds true regardless of the absolute values of the resistances or batteries or the polarity of each battery. Other methods of circuit analysis can be used to calculate the PD across such a network, but this method is one of the simplest.

The following chord resistance equation, analogous to the chord conductance equation, can also be derived and used to calculate the PD across the network:

$$\text{PD} = \frac{R_1}{R_1 + R_2} E_2 + \frac{R_2}{R_1 + R_2} E_1 \qquad \text{(A-24)}$$

$$= \frac{10\,\Omega}{10\,\Omega + 990\,\Omega} 200\,\text{V} + \frac{990\,\Omega}{10\,\Omega + 990\,\Omega} 100\,\text{V}$$

$$= \frac{10}{1000} 200\,\text{V} + \frac{990}{1000} 100\,\text{V}$$

$$= 2\,\text{V} + 99\,\text{V}$$

$$= 101\,\text{V}$$

This equation again emphasizes the point that it is the relative resistances that determine which battery is most expressed.

Bibliography

Baker, P. F. (1968). Nervous conduction: Some properties of the ion-selective channels which appear during the action potential. *Br. Med. Bull.* **24**, 179–182.

Carmeliet, E., and Vereecke, J. (1979). Electrogenesis of the action potential and automaticity. *In* "Handbook of Physiology" (R. M. Berne, and N. Sperelakis, Eds.) pp. 269–334. American Physiological Society, Bethesda, MD.

Cole, K. S. (1968). "Membranes, Ions and Impulses: A Chapter of Classical Biophysics." University of California, Berkeley.

Daniel, E. E., Kwan, C. Y., Matlib, M. A., Crankshaw, D., and Kidwai, A. (1977). Characterization and Ca^{2+}-accumulation by membrane fractions from myometrium and artery. *In* "Excitation-Contraction Coupling in Smooth Muscle" (R. Casteels, T. Godfraind, and J. C. Ruegg, Eds.), pp. 181–188. Elsevier-North-Holland, Amsterdam.

Dhalla, N. S., Ziegelhoffer, A., and Hazzow, J. A. (1977). Regulatory role of membrane systems in heart function. *Can. J. Physiol. Pharmacol.* **55**, 1211–1234.

Gadsby, D. C., and Nakao, M. (1989). Steady-state current-voltage relationship of the Na-K pump in guinea pig ventricular myocytes. *J. Gen. Physiol.* **94**, 511–537.

Glitsch, H. G. (1972). Activation of the electrogenic sodium pump in guinea-pig auricles by internal sodium ions. *J. Physiol. (London)* **220**, 565–582.

Goldman, D. E. (1943). Potential, impedance, and rectification in membranes. *J. Gen. Physiol.* **27**, 37–60.

Henn, F. A., and Sperelakis, N. (1968). Stimulative and protective action of Sr^{2+} and Ba^{2+} on (Na^+, K^+)-ATPase from cultured heart cells. *Biochim. Biophys. Acta* **163**, 415–417.

Hermsmeyer, K., and Sperelakis, N. (1970). Decrease in K^+ conductance and depolarization of frog cardiac muscle produced by Ba^{2+}. *Am. J. Physiol.* **219**, 1108–1114.

Hodgkin, A. L., and Katz, B. (1949). The effect on sodium ions in electrical activity of the giant axon of the squid. *J. Physiol. (London)* **108**, 37–77.

Irisawa, H. (1978). Comparative physiology of the cardiac pacemaker mechanism. *Physiol. Rev.* **58**, 461–498.

Jain, M. K. (1972). "The Bimolecular Lipid Membrane: A System." Van Nostrand, New York.

Jones, I. R., Maddock, S. W., and Besch, H. R., Jr. (1980). Unmasking effect of alamethicin on the (Na^+, K^+)-ATPase, beta-adrenergic receptor-coupled adenylate cyclase, and cAMP-dependent protein kinase activities of cardiac sarcolemmal vesicles. *J. Biol. Chem.* **255**, 9971–9980.

McDonald, T. F., and MacLeod, D. P. (1971). Maintenance of resting potential in anoxic guinea pig ventricular muscle: Electrogenic sodium pumping. *Science* **172**, 570–572.

Meech, R. W. (1972). Intracellular calcium injection causes increased potassium conductance in Aplysia nerve cells. *Comp. Biochem. Physiol.* **42A**, 493–499.

Mullins, L. J. (1961). The macromolecular properties of excitable membranes. *Ann. NY Acad. Sci.* **94**, 390–404.

New, W., and Trautwein, W. (1972). Inward membrane currents in mammalian myocardium. *Pflugers Arch.* **334**, 1–23.

Noble, D. (1975). "Initiation of the Heartbeat." Oxford University Press (Clarendon), London.

Noma, A., and Irisawa, H. (1974). Electrogenic sodium pump in rabbit sinoatrial node cell. *Pflugers Arch.* **351**, 177–182.

Pelleg, A. Vogel, S., Belardinelli, L., and Sperelakis, N. (1980). Overdrive suppression of automaticity in cultured chick myocardial cells. *Am. J. Physiol.* **238**, H24–H30.

Sperelakis, N. (1972). (Na^+, K^+)-ATPase activity of embryonic chick heart and skeletal muscles as a function of age. *Biochim. Biophys. Acta* **266**, 230–237.

Sperelakis, N. (1979). Origin of the cardiac resting potential. *In* "Handbook of Physiology, Vol. 1, The Cardiovascular System" (R. M. Berne, and N.

Sperelakis, Eds.), pp. 187–267. American Physiological Society, Bethesda, MD.

Sperelakis, N. (1980). Changes in membrane electrical properties during development of the heart. *In* "The Slow Inward Current and Cardiac Arrhythmias" (D. P. Zipes, J. C. Bailey, and V. Elharrar, Eds.), pp. 221–262. Martinus Nijhoff, The Hague.

Sperelakis, N. (1993). Origin of the resting membrane potential. *In* "Physiology" (N. Sperelakis and R. Banks, Eds.), pp. 29–48. Little, Brown, Boston.

Sperelakis, N. (1993). Basis of the resting potential. *In* "Physiology and Pathophysiology of the Heart," 3rd ed. Kluwer Academic Publishers, New York.

Sperelakis, N. (1995). "Electrogenesis of Biopotentials." Kluwer Academic Publishers, Boston.

Sperelakis, N., and Fabiato, A. (1995). Electrophysiology and excitation-contraction coupling in skeletal muscle. *In* "The Thorax: Vital Pump" (C. Roussos, Ed.), second edition, pp. 33–95. Marcel Dekker. New York.

Sperelakis, N., and Lehmkuhl, D. (1966). Ionic interconversion of pacemaker and nonpacemaker cultured chick heart cells. *J. Gen Physiol.* **49,** 867–895.

Sperelakis, N., Schneider, M., and Harris, E. J. (1967). Decreased K^+ conductance produced by Ba^{2+} in frog sartorius fibers. *J. Gen. Physiol.* **50,** 1565–1583.

Trautwein, W., and Kassebaum, D. G. (1961). On the mechanism of spontaneous impulse generation in the pacemaker of the heart. *J. Gen. Physiol.* **45,** 317–330.

Vassalle, M. (1970). Electrogenic suppression of automaticity in sheep and dog Purkinje fibers, *Circ. Res.* **27,** 361–377.

Nicholas Sperelakis

15

Gibbs-Donnan Equilibrium Potentials

I. Introduction

Because intracellular cytoplasm contains many colloids, including large nondiffusible polyvalent electrolytes, a **Donnan equilibrium** can be established across the cell membrane with an accompanying transmembrane **Gibbs-Donnan (G-D) potential.** The resting potential of most cells in the body, including nerve and muscle cells, however, is not due to a Donnan equilibrium, and the normal resting potential is not a Gibbs-Donnan potential, as is erroneously stated in some textbooks. **In the true Donnan equilibrium, all diffusible ions are in equilibrium across the membrane.** But many ions—like Na^+, K^+, Ca^{2+}, and H^+—in nerve and muscle cells are not in equilibrium; that is,

$$E_{Na} \neq E_m$$

$$E_K \neq E_m$$

$$E_{Ca} \neq E_m$$

and

$$E_H \neq E_m$$

On the other hand, Cl^- is at equilibrium (i.e., passively distributed) in many vertebrate cells; namely,

$$E_{Cl} = E_m$$

In addition, a large internal pressure and concomitant swelling of animal cells would occur if a Donnan equilibrium were allowed to become established. The action of two types of cation pumps keeps the Donnan osmotic pressure from developing and keeps certain cations out of equilibrium. Thus, a second important function of the Na^+-K^+ pump is the **regulation of cell volume.** The Na^+-K^+ pump actively pumps 3 Na^+ ions out (to 2 K^+ ions pumped in) with each cycle. The pump action decreases the osmotic pressure of the cytoplasm and prevents cell swelling. Inhibition of active ion transport by any means leads to osmotic swelling because of the establishment of the Donnan equilibrium.

Under such conditions, the cells gain Na^+, Cl^-, Ca^{2+}, and H_2O, and they lose K^+.

Thus, the Gibbs-Donnan potential is **passive;** that is, energy is not necessary to its establishment. In contrast, the resting potential is actively generated (indirectly or directly) by the action of the Na^+-K^+ pump. The Gibbs-Donnan potential is usually less than -20 mV, whereas the resting potential is -40 to -100 mV, depending on the cell type (and its ratio of P_{Na} to P_K).

II. Mechanism for Development of the Gibbs-Donnan Potential

In the **Gibbs-Donnan equilibrium,** a small membrane potential is established even though the biological membrane involved, or the artificial membrane used in a laboratory experiment, may be equally permeable to the small diffusible ions used. For the example illustrated in Fig. 1, where aqueous solutions of 0.1 M $(Na^+)_n$-proteinate^{n-} (side 1) and 0.1 M NaCl (side 2) are initially placed on the two sides of a two-compartment chamber separated by a membrane, and if $g_{Na} = g_{Cl}$ in this membrane, then at equilibrium, $E_{Na} = E_{Cl} = -18$ mV. The side containing the protein anion becomes negative with respect to the other side. Thus, because both diffusion potentials have the same polarity (as well as magnitude), a potential difference (PD) occurs across the membrane, even though conductances for Na^+ and Cl^- across the membrane may be equal. The osmotic pressure of the solution on side 1 containing the nonpermeant protein is greater than that on side 2.

In the G-D equilibrium, all permeant ions are in electrochemical equilibrium across the membrane, that is, they are passively distributed, and there is no net electrochemical driving force:

$$(E_m - E_{Na}) = 0$$

$$(E_m - E_{Cl}) = 0$$

243

A

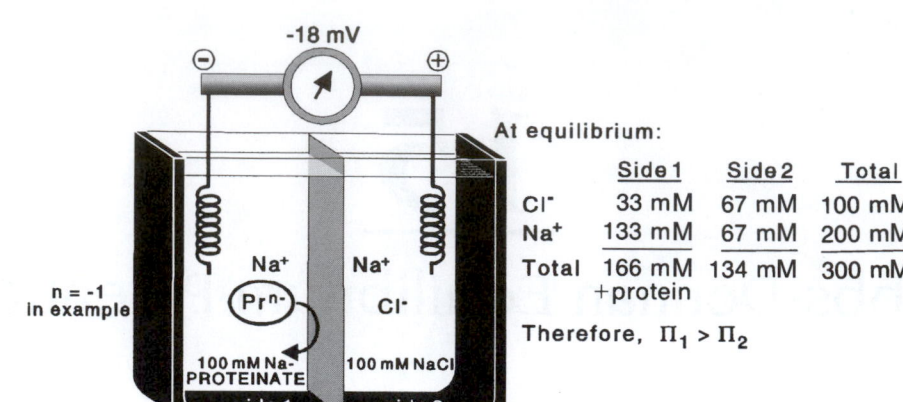

B

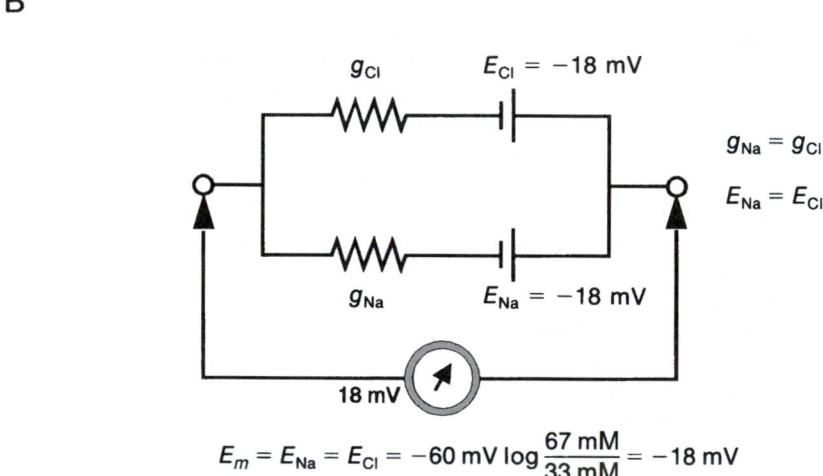

$$E_m = E_{Na} = E_{Cl} = -60 \text{ mV} \log \frac{67 \text{ mM}}{33 \text{ mM}} = -18 \text{ mV}$$

FIGURE 1. Gibbs-Donnan potential. (A) Gibbs-Donnan experiment. Diagram depicts the experimental arrangement for obtaining Gibbs-Donnan potential. A membrane freely permeable to all small ions, but impermeable to the large protein molecules, is used to separate two solutions, only one of which (side 1) contains protein. Side containing the protein becomes negative, with respect to the other side, by a small voltage (–18 mV in example). This membrane potential (E_m) does not depend on active ion transport or on selective permeability properties of the membrane, as normal cell resting potential does. The diffusible ions (Na^+ and Cl^- in example), however, become unequally distributed across the membrane, and it is their diffusion potentials ($E_{Na} = E_{Cl}$) that produce the Gibbs-Donnan potential. (B) Equivalent circuit for Gibbs-Donnan experiment (depicted in A) demonstrating that $E_m = E_{Na} = E_{Cl}$, the Na^+ and Cl^- batteries being of equal magnitude and of the same sign. Therefore, the relative conductances of the membrane to Na^+ and Cl^-, whether equal or not, are irrelevant to the potential (g_{Cl} and g_{Na} are conductances for Cl^- and Na^+, respectively).

A more complete explanation for the development of the Gibbs-Donnan potential follows. The Gibbs-Donnan potential (which is an equilibrium PD) does not depend on metabolic energy. Therefore, this discussion applies to a cell that either has no ATP for pumping ions against electrochemical gradients or has had its Na^+-K^+ pump completely blocked by either ouabain or another agent. The Gibbs-Donnan potential is passively produced by the concentration gradients for diffusible electrolytes (e.g., Na^+ and Cl^-) across a membrane. These ion gradients are caused by the presence of one or more large nondiffusible (with respect to the membrane) polyvalent electrolytes (e.g., **nega-tively charged proteins**) on one side of the membrane, as is present in all biological cells. In essence, the negatively charged protein molecules (at pH 7) inside the cell attract cations (e.g., Na^+ or K^+) and repel anions (e.g., Cl^-). Therefore, in the Gibbs-Donnan situation, the inside of the cell has a higher concentration of Na^+ (or K^+) and a lower concentration of Cl^- than has the solution bathing the cell. The equilibrium potentials for Na^+ (E_{Na}) and for Cl^- (E_{Cl}) are equal in magnitude and are of the same sign, thereby producing a PD across the membrane. The PD is negative on the inside (side containing the protein) and usually is about –20 mV or less.

III. Gibbs-Donnan Equilibrium

To quantitate the ion distributions produced at equilibrium and the PD developed, let us examine the artificial system shown in Fig. 1A. In this system, a chamber is separated into two compartments by a collodion membrane, which has small uncharged pores that allow Na^+ and Cl^- ions, but not large protein molecules, to diffuse through. A 100 mM solution of Na^+ proteinate is added to one side (compartment 1), and a 100 mM solution of NaCl to the other side (compartment 2). An electrode is positioned on each side so that the PD across the membrane can be recorded (37 °C). Let us assume that the Na^+ proteinate is completely ionized and, for simplicity, that the protein has a net negative charge of only one.

Thus there is, at the first instant, no diffusion force for Na^+, but there is diffusion force for Cl^-, because Cl^- is 100 mM in compartment 2, and 0 mM in compartment 1. Na^+ must accompany the diffusion of Cl^- from side 2 to side 1, because the **principle of electroneutrality** in the bulk solution cannot be violated (i.e, there must be an equal number of cations and anions). So one relation that must be true when the system comes to equilibrium is that

$$\left[Na^+\right]_2 = \left[Cl^-\right]_2 \tag{1}$$

In actuality, there is a small charge separation directly across the membrane to account for the PD; that is, side 2 of the membrane has a small excess of Na^+ ions, and side 1 has a small excess of Cl^- ions. Such a charge separation is very small, but is necessary to develop a PD across the membrane ($V = Q/C$), and is discussed in the preceding chapter on the resting potential.

The principle of electroneutrality also requires that the increase in Na^+ on side 1 must be exactly equal to the increase in Cl^- on side 1. Thus, the concentration difference of Na^+ that is built up at equilibrium must be exactly equal to the final concentration difference for Cl^-. This is because the large initial gradient for Cl^- is what drives the Na^+ to make its gradient. Therefore, it must also be true that

$$\frac{\left[Na^+\right]_1}{\left[Na^+\right]_2} = \frac{\left[Cl^-\right]_2}{\left[Cl^-\right]_1} \tag{2}$$

Cross-multiplying gives

$$\left[Na^+\right]_1\left[Cl^-\right]_1 = \left[Na^+\right]_2\left[Cl^-\right]_2 \tag{3}$$

Another way of considering this is that E_{Na} must equal E_{Cl}, and therefore, using the respective Nernst equations (see Chapter 14), we can write

$$E_{Na} = E_{Cl} \tag{4}$$

$$\frac{-61\,mV}{+1} \log\frac{\left[Na^+\right]_1}{\left[Na^+\right]_2} = \frac{-61\,mV}{-1} \log\frac{\left[Cl^-\right]_1}{\left[Cl^-\right]_2}$$

$$= \frac{-61\,mV}{+1} \log\frac{\left[Cl^-\right]_2}{\left[Cl^-\right]_1} \tag{5}$$

Dividing both sides by –61 mV and removing the log gives

$$\frac{\left[Na^+\right]_1}{\left[Na^+\right]_2} = \frac{\left[Cl^-\right]_2}{\left[Cl^-\right]_1} \tag{2}$$

Equation 3 indicates that, at equilibrium, the product of the diffusible ions on side 1 must be equal to the product of the diffusible ions on side 2. From the Nernst equation, the relationships

$$E_{Na} = \frac{-RT}{z\mathscr{F}} \ln\frac{\left[Na^+\right]_1}{\left[Na^+\right]_2} \tag{6}$$

$$= \frac{-61\,mV}{+1} \log\frac{\left[Na^+\right]_1}{\left[Na^+\right]_2} \tag{7}$$

and

$$E_{Cl} = \frac{-61\,mV}{-1} \log\frac{\left[Cl^-\right]_1}{\left[Cl^-\right]_2} \tag{8}$$

can be given because Cl^- is negative ($z = -1$), whereas Na^+ is positive ($z = +1$). Equation 8 is the same as (note that a negative sign in front of a log inverts the ratio)

$$E_{Cl} = -61\,mV \log\frac{\left[Cl^-\right]_2}{\left[Cl^-\right]_1} \tag{9}$$

because $[Na^+]_1/[Na^+]_2 = [Cl^-]_2/[Cl^-]_1$, as Eq. 2 indicates, and from Eqs. 7 and 9, it is clear that Eq. 4 holds true; that is, $E_{Na} = E_{Cl}$.

IV. Quantitation of the Gibbs-Donnan Potential

For quantitation, let us use x to indicate the amount (in mM) of Cl^- or Na^+ that shifted from side 2 to side 1 at equilibrium. Then the amount of Na^+ on side 2 is 100 mM – x (the original amount minus the amount lost); Cl^- on side 2 is also 100 mM – x, because $[Na^+]_2 = [Cl^-]_2$. The Na^+ on side 1 at equilibrium is 100 mM + x (the original amount plus the amount gained), and the Cl^- on side 1 is simply x. These parameters may be listed as follows:

$$\left[Na^+\right]_2 = 100\,mM - x$$
$$\left[Cl^-\right]_2 = 100\,mM - x$$
$$\left[Na^+\right]_1 = 100\,mM + x$$
$$\left[Cl^-\right]_1 = x$$

The value for x can be obtained by substituting these values into Eq. 3:

$$\left[Na^+\right]_1\left[Cl^-\right]_1 = \left[Na^+\right]_2\left[Cl^-\right]_2 \tag{3}$$

$$(100 + x)x = (100 - x)(100 - x)$$

$$100x + x^2 = 10\,000 - 200x + x^2$$

$$300x = 10\,000$$

$$x = 33.3$$

Thus, at equilibrium

$$\left[Cl^-\right]_1 = 33\,mM$$

$$\left[Na^+\right]_1 = (100 + 33) = 133\,mM$$

$$\left[Cl^-\right]_2 = (100 - 33) = 67\,mM$$

$$\left[Na^+\right]_2 = (100 - 33) = 67\,mM$$

These values are also given in Fig. 1A. Note that all the equations and conditions are obeyed. The Gibbs-Donnan potential produced then may be calculated by substituting into Eqs. 7 and 8:

$$E_{Na} = \frac{-61\,mV}{+1} \log \frac{133\,mM}{67\,mM} \qquad (10)$$

$$= -18\,mV$$

and

$$E_{Cl} = \frac{-61\,mV}{-1} \log \frac{33\,mM}{67\,mM} \qquad (11)$$

$$= -18\,mV$$

Hence, $E_{Na} = E_{Cl}$ (Eq. 4). That is, the two diffusion potentials are equal in magnitude and of the same sign. The PD across the membrane is −18 mV; side 1 containing the protein is negative. Therefore, relative permeability of the membrane to Na^+ and Cl^- is irrelevant. The equivalent circuit for this example at equilibrium is given in Fig. 1B.

V. Osmotic Considerations

We should note that, at equilibrium, the sum of Na^+ and Cl^- on side 1 (166 mM) is greater than that on side 2 (134 mM). In addition, there is 100 mM protein on side 1. Thus, the total osmotic concentration on side 1 is 266 mOsm (milliosmolar), compared to 134 mOsm on side 2. Therefore, there is a large osmotic gradient between the two sides. Water moves from side 2 to side 1 (i.e., water accompanies the net movement of Na^+ and Cl^-) until the hydrostatic pressure head buildup is sufficient to oppose further net movement of water. As expected, the biological cell swells when a Gibbs-Donnan equilibrium is allowed to develop following a blockade of active ion transport for long periods.

The total osmotic concentration, [osm], on each side, at equilibrium, may be summarized as follows:

$$[osm]_1 = \left[Na^+\right]_1 + \left[Cl^-\right]_1 + \left[protein\right]_1 \qquad (12)$$

$$[osm]_2 = \left[Na^+\right]_2 + \left[Cl^-\right]_2 \qquad (13)$$

Substitution gives

$$[osm]_1 = 133\,mM + 33\,mM + 100\,mM$$

$$= 266\,mM$$

$$[osm]_2 = 67\,mM + 67\,mM$$

$$= 134\,mM$$

The osmotic pressure (Π, in atm) of each solution is equal to the osmotic concentration in osmol/L (C) times the osmotic coefficient (i) times the gas constant (R, 0.082 L·atm/mol·K) times the absolute temperature (T, in K)

$$\Pi = iCRT \qquad (14)$$

where C is the number of osmoles per liter of solution. In the example depicted in Fig. 1, a hydrostatic pressure of 3.17 atm would need to be applied to side 1 to prevent this compartment from gaining water from side 2 (at 20 °C and assuming $i = 1.0$)

$$\Pi = i\Delta CRT \qquad (15)$$

$$= (1.0)(266\,mM - 134\,mM)\left(0.082\,\frac{L\cdot atm}{mol\cdot K}\right)(273 + 20)K$$

$$= \left(0.132\,\frac{mol}{L}\right)\left(0.082\,\frac{L\cdot atm}{mol\cdot K}\right)(293\,K)$$

$$= 3.17\,atm$$

The situation illustrated in Fig. 1 is actually more complex because the net water movement into side 1 acts to dilute the ion concentrations building up there, and therefore a true Gibbs-Donnan equilibrium can become established only if the net water movement is stopped, that is, by allowing an osmotic pressure gradient to develop by making side 1 a closed, or rigid, system. Otherwise, theoretically all of the water and NaCl eventually would move out of side 2.

The example of a G-D equilibrium in Fig. 1 could have been illustrated using another salt, such as KCl, instead of NaCl, or two or more salts.

The extra osmotic pressure in side 1 (or inside a cell) produced by the presence of the negatively charged proteins and other impermeant large charged molecules is known as the **colloid osmotic pressure (COP)**. The COP is also important for water movement across the capillary wall, which separates the blood plasma (containing impermeant proteins) and the interstitial fluid (ISF). At the arterial end of the capillary, the intracapillary hydrostatic blood pressure exceeds the COP, so water moves out of the capillary into ISF space; at the venous end, the COP exceeds the capillary hydrostatic pressure, so water moves into the capillary. In the midregion of the capillary, the two pressures are about equal, and there is no net water flow. Thus, there is a circulation of fluid distributed along the length of the capillary, and this idea is generally known as the **Starling hypothesis.**

VI. Summary

In summary, a Gibbs-Donnan equilibrium becomes established and a Gibbs-Donnan potential is developed across the cell membrane of cells under conditions in which metabolism and energy production have been inhibited or the Na^+-K^+ pump has been inhibited by digitalis. The G-D equilibrium occurs because of the large impermeant charged macromolecules, such as proteins, inside the cell. The G-D equilibrium does not require energy for its establishment; that is, it is passive. This contrasts with the normal resting potential of the cell, which requires active ion transport and use of metabolic energy to establish large ionic electrochemical gradients.

The G-D potential is usually less than −20 mV, whereas the resting potential is considerably greater. In the G-D equilibrium, all permeable ions are in equilibrium across the membrane, whereas this is not true for the normal resting potential. The equilibrium potentials for all permeant ions (e.g., E_K, E_{Cl}) are of equal magnitude and polarity. The G-D potential is developed even if the cell membrane had equal permeability or conductance for all small ions, whereas the normal resting potential requires different permeabilities for Na^+ and K^+, namely a low P_{Na}/P_K ratio. In the G-D equilibrium, the osmolarity of the cell becomes higher than the interstitial fluid bathing the cell, and so the cell tends to gain water and swell (unless prevented from doing so by a rigid cell wall, such as in plant cells).

When equilibrium is established, the product of the concentrations of the permeant ions inside the cell is equal to that outside the cell, and the concentrations of the anions and cations outside the cell must be equal (law of electroneutrality). The gains in cations and anions inside the cell also must be equal to each other. From these required conditions, the equation can be solved algebraically to give the final concentrations at equilibrium, and from this, the calculated potential difference across the membrane.

Bibliography

Davson, H. (1964). "A Textbook of General Physiology," 3rd ed. Little, Brown, Boston.

Sperelakis, N. (1979). Origin of the cardiac resting potential. *In* "Handbook of Physiology, Vol. 1, The Cardiovascular System" (R. M. Berne and N. Sperelakis, Eds.), pp. 187–267. American Physiological Society, Bethesda, MD.

Sperelakis, N. (1995). "Electrogenesis of Biopotentials." Kluwer Publishing Co., New York.

Steven M. Grassl

16

Mechanisms of Carrier-Mediated Transport: Facilitated Diffusion, Cotransport, and Countertransport

I. Introduction

The selective and regulated passage of ions and nonelectrolytes across the cell membrane is an essential component of cellular homeostasis. The maintenance of cell pH and volume and the accumulation of nutrients for protein synthesis and cell metabolism are physiological processes that depend on membrane transport for cells to thrive. Cell membranes are composed of phospholipids organized as a bilayer (5 nm) of two closely opposed leaflets separating the intracellular from the extracellular space. The hydrophobic properties of phospholipids make the cell membrane an impermeable barrier excluding the transfer of hydrophilic solutes that are either charged (anions and cations) or uncharged (nonelectrolytes). The selective passage of hydrophilic solutes across the hydrophobic barrier, a physiological property known as **membrane permeability**, is mediated by the presence of membrane transport proteins that span the phospholipid bilayer. Transport proteins may be functionally subdivided into channels, pumps, and carriers according to differences in the mechanism mediating ion and nonelectrolyte transport. This chapter describes the mechanisms of **carrier-mediated transport,** which include **facilitated diffusion**, **cotransport**, and **countertransport.**

II. Electrochemical Potential

Transport mechanisms may be distinguished thermodynamically according to their ability to mediate active or passive transport. **Active transport** is defined as movement of a solute from a region of low electrochemical potential on one side of the cell membrane to a region of higher electrochemical potential on the opposite side.

Passive transport is defined as movement of a solute from a region of high electrochemical potential on one side of the cell membrane to a region of lower electrochemical potential on the opposite side. The **electrochemical potential** of a solute is the partial molar free energy of the solute or the potential to do work when a difference in electrochemical potential exists across the cell membrane. The electrochemical potential of a solute on either side of the cell membrane is a function of the solute activity (or concentration in dilute solution), the solute charge and valence, and the electrical potential. The difference in electrochemical potential, therefore, reflects the magnitude of the difference in transmembrane solute concentration and the difference in transmembrane voltage factored by the charge and valence of the solute. Notably, for solutes without charge, such as nonelectrolytes, solute free energy is neither increased nor decreased by electrical potential, and only the chemical potential of the solute is considered. Thus, the electrochemical or chemical potential difference of a solute across the cell membrane may be considered a driving force acting on solute transport. In the absence of an electrochemical potential difference or a driving force for solute transport, transport mechanisms that are **passive** mediate equal solute transport in the forward and reverse direction across the membrane resulting in no net transport. For anions and cations, this would occur when the chemical and electrical driving forces acting on solute transport are equal and opposite in direction across the membrane such that the net driving force is zero. For nonelectrolytes, this would occur in the absence of a solute concentration gradient where transmembrane solute concentrations are equal. In both instances where no net transport occurs, the ion and nonelectrolyte are at electrochemical equilibrium with the driving forces acting on the transported solutes. Where an electrochemical potential difference for a solute exists across the cell membrane, the direction of net solute transport by a passive transport mechanism will depend on the direction and magnitude of the chemical and/or electrical driving forces acting on the solute. The chemical and electrical potential difference of a charged solute or ion may occur as opposing driving forces of unequal magnitude, with the direction of net solute transport determined by the direction of the larger driving force. In no instance would a passive transport mechanism mediate net transport of a charged solute in a direction across the cell membrane that opposed both the chemical and electrical driving forces acting on the ion. The same limitation holds for net transport of nonelectrolytes in a direction that opposes the chemical driving force acting on the solute.

Active transport mechanisms may be distinguished from passive transport mechanisms by the ability to generate and maintain an electrochemical or chemical potential difference for ions and nonelectrolytes across the cell membrane. This requires net transfer of ions or nonelectrolytes across the membrane in a direction that is opposed by the prevailing electrical gradient and/or chemical concentration gradients as driving forces acting on the transported solutes. To perform the work of moving solutes "uphill" against an electrical gradient and/or a chemical concentration gradient, active transport mechanisms require energy. The source of energy driving active transport is the hydrolysis of ATP. The direct or indirect coupling of active transport to ATP hydrolysis distinguishes **primary active transport** from **secondary active transport**. Primary active transport mechanisms such as the ion translocating ATPases or pumps are directly coupled to ATP hydrolysis and thermodynamically transduce the energy released upon ATP hydrolysis to the energy stored in the formation of an ion electrochemical potential difference. Secondary active transport mechanisms such as **cotransporters** and **countertransporters** are indirectly coupled to ATP hydrolysis and thermodynamically transduce the energy from one solute electrochemical potential difference to the energy stored in the formation of a second solute electrochemical potential difference. The indirect coupling of secondary active transport to ATP hydrolysis arises from the intermediate formation and ATP dependence of the solute electrochemical potential difference that drives secondary active transport.

III. Carrier-Mediated Transport Mechanisms

A. Facilitated Diffusion

Facilitated diffusion or **uniport** is the simplest form of carrier-mediated transport and results in the transfer of large hydrophilic molecules (sugars, amino acids, nucleotides, and organic acids and bases) across the cell membrane. Transport by facilitated diffusion is passive and reversible, with the direction of net transport into or out of the cell determined by the direction of the electrochemical potential difference of the transported solute. Net transport by facilitated diffusion may continue in either direction

until the solute is at equilibrium with the electrical and/or chemical driving forces acting on the solute. At equilibrium, facilitated diffusion of a solute occurs equally in both directions resulting in no net transport. The transport mechanism mediating facilitated diffusion may be modeled as shown in Fig. 1. The two main features of the transport mechanism are an association and dissociation of the transported solute with the transport protein, and a change in the conformation of the transport protein which makes the occupied or unoccupied site of solute interaction accessible from either side of the membrane (Fig. 1A). The transport mechanism may be considered in greater detail as a four-step process (Fig. 1B). First, solute S associates with the transport protein C facing side 1 (labeled "outside" in the figure) to form a solute-carrier complex SC. Second, the solute-carrier complex undergoes a conformational change that reorients the solute-carrier complex to face side 2 (labeled "inside" in the figure). Third, the solute dissociates with the transport protein on side 2. Fourth, the unoccupied carrier undergoes a second conformational change to reorient the solute association site to face side 1. Net solute transport from side 1 to side 2 occurs when an unoccupied solute association site is reoriented from side 2 to side 1. However, as the solute concentration on side 2 increases, net solute transport from side 1 to side 2 decreases because a greater proportion of the unoccupied carrier becomes associated with solute on

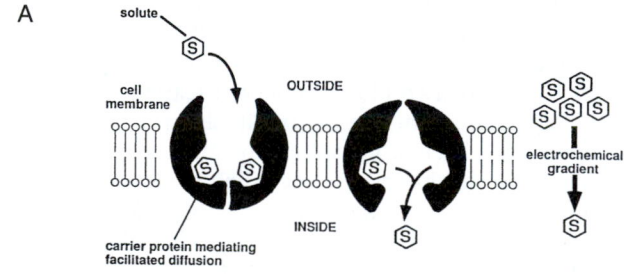

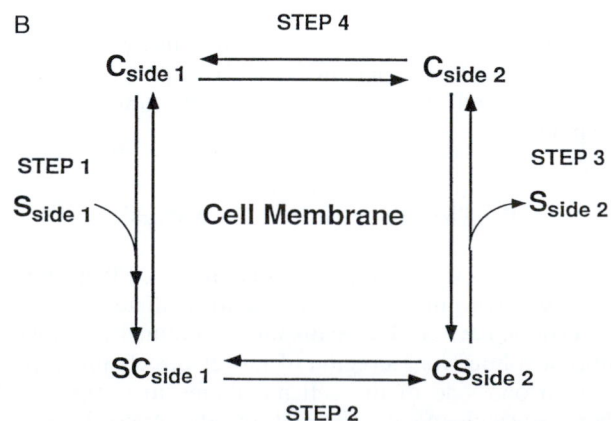

FIGURE 1.   Conceptual and kinetic model of facilitated diffusion.

side 2 and undergoes reorientation moving solute to side 1. When the solute concentrations on side 1 and side 2 are equal, no net solute transport will occur because the occupancy and reorientation of the carrier will be equal at both sides of the membrane. A principle feature of facilitated diffusion illustrated by the model is the interaction with solutes exclusively at one side of the membrane or the other but never at both sides simultaneously. This functional property of facilitated diffusion invokes the need for a conformational change in the protein, alternately orienting the solute association site from side to side across the membrane. Each step in the process is reversible and is linked to the preceding and succeeding step by rate constants defining the rate of solute association and dissociation on either side of the membrane and the rate of conformational change for transporter reorientation in either direction when associated with or without solute. In general, the rate of solute association and dissociation with the transporter occurs more rapidly than the rates of conformational change. Furthermore, the conformational change reorienting the sidedness of the transporter occurs at a faster rate for the solute-occupied transporter than for the solute-unoccupied transporter. The relative slowness of the rate

constant for the conformational change reorienting the solute-unoccupied transporter (step 4) makes it the rate-limiting step in the process of facilitated diffusion.

The **unidirectional rate** of solute transport across the cell membrane mediated by facilitated diffusion is dependent on the solute concentration at the side where transport originates. This is analogous to the substrate concentration dependence of the rate of product formation mediated by an enzyme. As shown in Fig. 2, at low solute concentrations the rate of solute transport increases linearly with solute concentration, and as solute concentration increases further, the rate of solute transport increases less and less as it approaches a maximum value at high solute concentrations (Fig. 2A). The relationship of the unidirectional solute transport rate to solute concentration shown in Fig. 2 is described mathematically by an equation analogous to the Michaelis-Menten equation from enzyme kinetics

$$V = \frac{[S]V_{max}}{K_m + [S]} \tag{1}$$

where V is determined experimentally as the initial rate of solute transport, [S] is the solute concentration, V_{max} is the maximal rate of solute transport, and K_m is a constant described by the solute concentration where V is one-half the V_{max}. The validity of using the Michaelis-Menten equation in a kinetic analysis of facilitated diffusion requires experimental conditions in which the rate of solute transport reflects solute transport only in one direction. Accordingly, initial rates of solute transport across the cell membrane are measured in the absence of a significant solute concentration on one side of the membrane. The Michaelis-Menten equation may be rearranged to obtain a linear relation as shown in Fig. 2B, where the kinetic parameters K_m and V_{max} are the x-intercept and 1/slope, respectively. The kinetic parameters of solute transport are functional properties that characterize different facilitated diffusion mechanisms as well as transport of different solutes by the same facilitated diffusion mechanism. The K_m value characterizes the affinity of solute association-dissociation with the transporter such that a lower or higher K_m value reflects a greater or lesser affinity respectively. The accuracy of the K_m value as a measure of solute affinity is further dependent on initial rate determinations performed at solute concentrations both above and below the K_m value. The V_{max} value is a measure of the number of transporters present in the membrane and the time required for a transporter to undergo one complete transport cycle or turnover. The maximum velocity of transport occurs when solute association and dissociation with the transporter are not rate-limiting steps in the transport process or when the transporter is saturated. Under these conditions the maximal velocity of solute transport is only as fast as the rate-limiting step in the transport process, which is the conformational change of the solute dissociated transporter.

The facilitated diffusion of a solute may be inhibited in the presence of other solutes that interact with, but are not necessarily transported by, the same transporter. The nature of interaction of the inhibitor with the transporter may be assessed by observing the effect of the inhibitor on the kinetic parameters characterizing the transport mechanism.

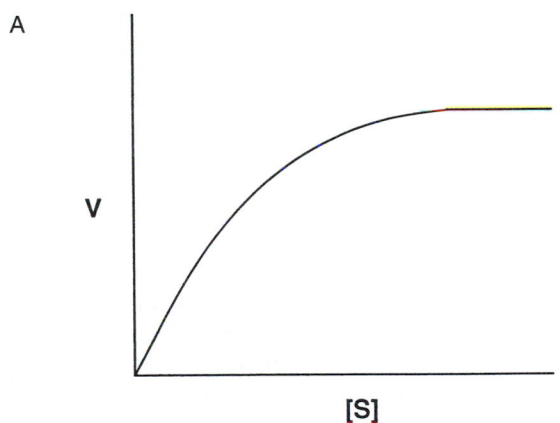

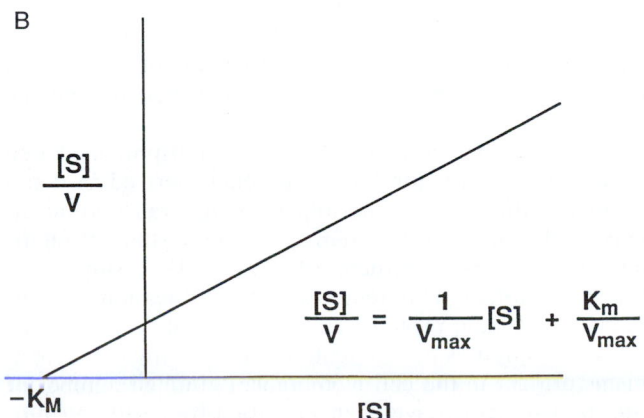

$$\frac{[S]}{V} = \frac{1}{V_{max}}[S] + \frac{K_m}{V_{max}}$$

FIGURE 2. Relationship of solute transport rate (V) to solute concentration [S].

An effect of the inhibitor to decrease the maximal velocity of solute transport, without an effect on the K_m value of the solute, characterizes **noncompetitive inhibition.** Noncompetitive inhibition is not reversed by increasing the concentration of the transported solute and therefore results from an interaction of the inhibitor with both the solute-associated and -dissociated transporter at a second site that does not interact with the transported solute. An effect of the inhibitor to increase the apparent K_m value of the solute, without an effect on the maximal velocity of transport, characterizes **competitive inhibition.** Competitive inhibition is reversed by increasing the concentration of the transported solute, and therefore results from an interaction of the inhibitor with only the solute-dissociated transporter at the same site that interacts with the transported solute.

The same facilitated diffusion mechanism may mediate the transport of multiple solutes that share common chemical and physical determinants such as negatively charged carboxyl groups or positively charged amino groups. The nature of these chemical determinants and their spatial position in the solute molecule permits recognition and interaction with the solute association site of the transport protein. The solute specificity of a facilitated diffusion mechanism is a functional property of the transporter characterized by solute-specific differences in the kinetic parameters of solute transport. A rank order of the relative affinity of the transporter for different solutes may be compiled by determination of the K_m values for multiple solutes transported by the same mechanism. The transporter is more highly specific for solutes with the lowest K_m values and less specific for solutes with the highest K_m values. However, solutes with higher affinity for the transporter do not necessarily have a greater maximal velocity of transport than solutes with a lower affinity. A comparison of the relative affinities and maximal velocities of solute transport determined experimentally further suggests which solutes are most likely to be transported in a physiological setting where multiple solutes are present at different concentrations.

The kinetic parameters characterizing solute transport by facilitated diffusion may differ depending on the direction of unidirectional solute transport. The K_m value as a measure of solute affinity for transport may be larger or smaller for solute transport mediated in one direction across the cell membrane when compared to the K_m value for solute transport in the reverse direction. However, the kinetic asymmetry of facilitated diffusion is subject to a thermodynamic limitation which requires the product of all the rate constants associated with the stepwise transport in the inward and outward direction be equal. Accordingly, a directional asymmetry in the K_m value of solute transport must be matched by a corresponding asymmetry in the maximal velocity of transport such that the ratio of V_{max} to K_m for transport in either direction is equal. Thus, a kinetic asymmetry of solute transport results from an increase or decrease of both kinetic parameters where an increased or decreased affinity for solute is matched by a corresponding decrease or increase in maximal velocity of transport, respectively.

In contrast to nonelectrolytes, the facilitated diffusion of charged solutes such as organic anions and cations is **electrogenic** and results in the net transfer of charge in the direction of net solute transport across the membrane. As a consequence of mediating net charge transfer across the membrane, the facilitated diffusion of charged solutes will increase or decrease the electrical potential difference across the membrane depending on the sign and magnitude of the solute charge and on the direction of net transport. As a further consequence of mediating net charge transfer, the rate and direction of electrogenic solute transport by facilitated diffusion is sensitive to membrane potential, which must be considered as an additional driving force acting on electrogenic solute transport. This results from the voltage sensitivity of at least one step in the transport process mediating the facilitated diffusion of charged solutes. A voltage-dependent increase or decrease in the kinetic parameters of electrogenic solute transport may indicate an effect of voltage on the affinity of the transporter for the charged solute (K_m) and/or on the conformational change reorienting the solute-associated or -dissociated transporter (V_{max}).

Facilitated diffusion mechanisms are present in the membranes of all cells and many different mechanisms exist in the same membrane of a single cell. The common features of these different facilitated diffusion mechanisms include (1) their presence as integral membrane proteins spanning the lipid bilayer; (2) a broad or narrow range of substrate specificity; (3) inhibition by physiological and nonphysiological solute analogs; (4) solute saturation conforming to Michaelis-Menten kinetics; and (5) evidence of an alternate side-to-side reorientation of the solute association site. Whereas solute transport via channels and facilitated diffusion are both mediated and passive, sharing the common functional properties of inhibition and saturation, other functional criteria distinguish these two forms of transport. In contrast to solute transport by facilitated diffusion, most channels mediate transport of inorganic anions or cations and have a solute specificity limited by the physical size of the ion. Channels are accessible for solute association from both sides of the membrane simultaneously, and mediate transport at a rate orders of magnitude faster than facilitated diffusion, which is rate-limited by the relatively slow conformational changes reorienting the sidedness of the carrier. In further contrast to solute transport mediated by facilitated diffusion, the rate-limiting step in ion transport mediated by channels is the association and dissociation of an ion with the channel, not the actual ion translocation through the channel.

The functional properties of solute transport mediated by facilitated diffusion have been characterized in detail for many different cell- and solute-specific transport mechanisms. A major question remaining in the study of facilitated diffusion is the structure-function relationship of the transport protein. This requires the identification of individual membrane proteins as the structural correlates of transport activities of different facilitated diffusion mechanisms present in the cell membrane. A limited number of transport proteins have been identified by a difficult biochemical process involving detergent solubilization of the cell membrane, affinity purification of the protein, and functional reconstitution of transport activity to prove its identity. The identification of many more transport proteins has been recently achieved using the techniques of molecular biology to clone the genes coding for these membrane proteins. The molecular cloning of a transporter gene begins with the preparation of mRNA from a tissue rich in a functionally well-characterized transporter activity. A high transporter activity suggests an abundance of transporter protein as well as transporter mRNA from which the protein is translated. The total mRNA is reverse transcribed to complementary DNA and individual cDNA molecules are inserted into plasmids or minichromosomes used to transform *Escherichia coli* bacterial cells. The transformed *E. coli* are grown to form a library of thousands of colonies or clones each reproducing a different cDNA plasmid. The library serves to renew and amplify a continuous source of cDNA plasmids which may be screened for possible hybridization with short DNA sequences thought to code for various regions of the transporter gene. To assess the function of the putative transporter gene, mRNA may be reverse transcribed from the plasmid cDNA of positively identified clones and injected into frog oocytes for translation or expression of the corresponding protein. The identity of the transporter gene is proven by the mRNA-dependent expression of a transport activity with functional properties similar to those characterizing transport activity in the tissue of origin. The transporter gene may also be identified without prescreening the library for hybridization by systematically preparing mRNA from the entire cDNA library and assessing functional expression of transport activity. The cDNA plasmid coding for the transporter gene may be sequenced to obtain the deduced amino acid sequence, or primary structure, and molecular weight of the transport protein. The deduced amino acid sequence of the transport protein may be further analyzed for the presence and concentration of hydrophobic amino acids at multiple sites along the peptide chain. Typically, 10–12 regions of relative hydrophobicity may be identified in the transporter amino acid sequence and are thought to represent the putative transmembrane domains of the transport protein. The location of the transmembrane domains further suggests the size and topological location of intracellular and extracellular peptide loops between the transmembrane domains as well as the amino and carboxy termini of the transport protein. A topological model depicting the putative secondary protein structure of a mammalian facilitated diffusion transporter for glucose is shown in Fig. 3. Further analysis of the amino acid sequence for the presence of asparagine in the extracellular peptide loops may suggest potential sites for N-linked

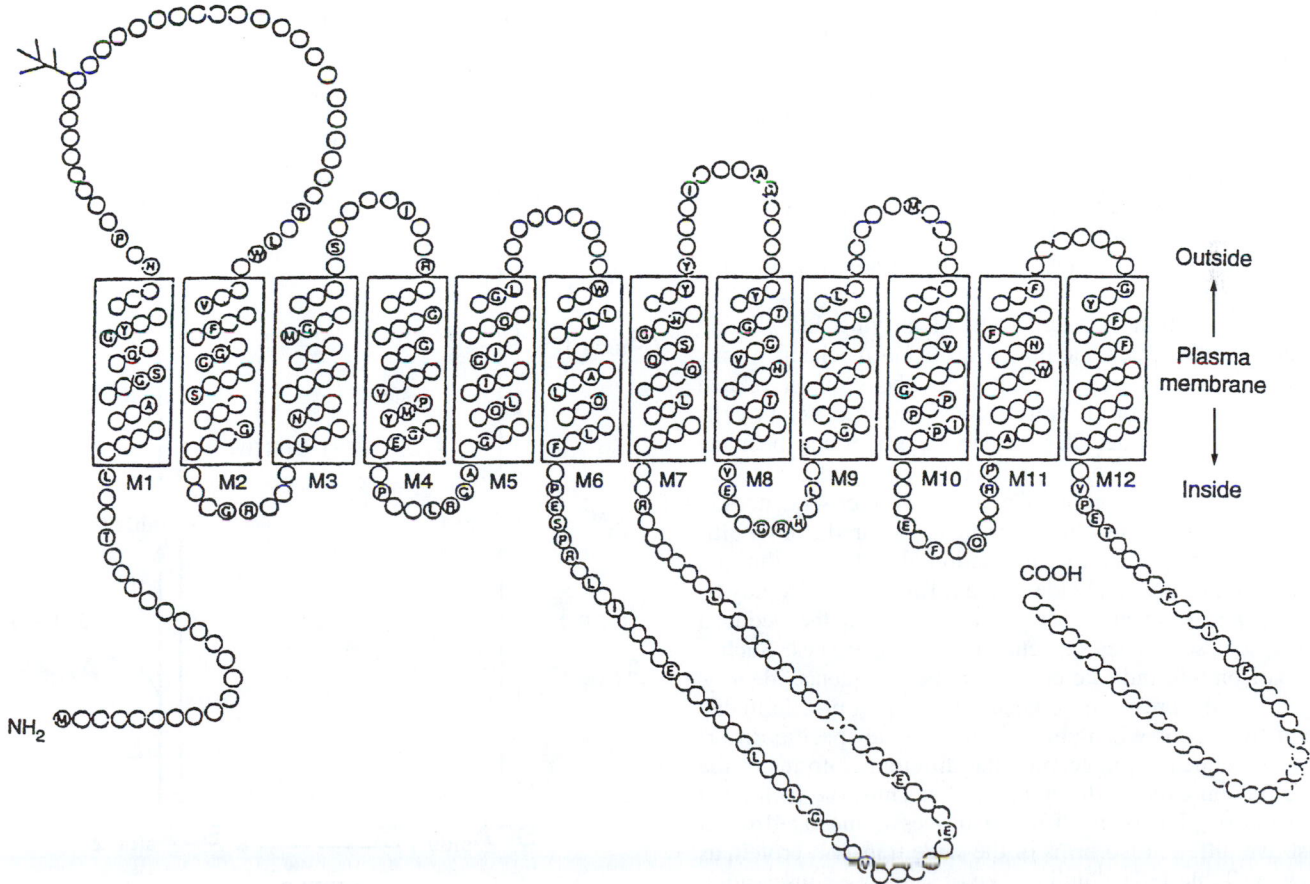

FIGURE 3. Consensus structure of the mammalian D-glucose carriers. (Reprinted with permission from Bell, G. I. (1991), Molecular defects in diabetes mellitus, *Diabetes* 40: 413–422. Copyright © 1991 American Diabetes Association.)

TABLE 1 D-Glucose Carrier Isoforms

Designation	Function
GLUT1	Basal uptake in placenta, brain, kidney, colon
GLUT2	Primarily Found in transport in liver cells
GLUT3	Basal uptake in the brain
GLUT4	Insulin-stimulated uptake in the skeletal muscle, cardiac muscle, and adipose tissue
GLUT5	Absorption in small intestines

glycosylation of the transport protein. Likewise, the presence of serine, threonine, and tyrosine in consensus phosphorylation sites in intracellular peptide loops suggests that the phosphorylation state of the transporter may regulate its activity.

The structure-function relationship of the protein mediating facilitated diffusion may be investigated upon identification of the primary structure or amino acid sequence of the protein. Mutant transport proteins may be engineered with single or multiple amino acid deletions, or with substitutions in the transmembrane domains, the intra- and extracellular peptide loops between transmembrane domains, and in the amino and carboxy termini. The functional properties of the engineered transport proteins may be assessed by heterologous expression in frog oocytes or cell cultures and compared to the native protein. Thus, the individual amino acids or amino acid sequences important for solute recognition by, and interaction with, the transport protein may be identified as well as those involved in the conformational change reorienting the solute-associated and -dissociated transporter. The mutational analysis of transporter function is limited by null mutations which result in the expression of nonfunctional transport proteins or by the absence of expression.

The information obtained from identifying and cloning the gene coding for a facilitated diffusion mechanism may be used to survey the distribution of the same or related transport proteins in different tissues and in different cells within the same tissue. Oligonucleotide probes may be made from short sequences of the transporter gene and used to detect the presence of complementary transporter sequences in cDNA from tissue libraries, in mRNA prepared from multiple tissues, or by *in situ* hybridization of mRNA in thin tissue slices. The positively identified mRNA or cDNA may be isolated and sequenced for comparison with the deduced amino acid sequences present in different tissues. Such a comparison will indicate either complete sequence identity, suggesting the presence of the same transport protein in different tissues, or will indicate minor tissue-specific differences in sequence, suggesting that different isoforms of the same protein exist in different tissues. Thus, the facilitated diffusion of glucose in different tissues is mediated by at least five different isoforms of the same transport protein as shown in Table 1. The high degree of sequence conservation among different isoforms of the same transport protein is consistent with a common solute specificity and mechanism

of solute translocation across the membrane. The minor tissue-specific differences in sequence distinguishing isoforms of the same protein presumably reflect a unique variation in the functional properties and/or regulation of the transporter serving the specialized physiology of the tissue.

B. Cotransport

Cotransport, or **symport,** is a form of secondary active transport that mediates net transfer of a solute across the cell membrane from a place of low solute electrochemical potential to a place of higher solute electrochemical potential. The source of energy driving secondary active transport of a solute against its electrochemical potential difference arises from a coupling to the transport of a second solute from a place of higher electrochemical potential to a place of lower electrochemical potential. The coupling of the driving solute to the driven solute results in the cotransport of both solutes in the same direction across the cell membrane. Cotransport mechanisms are reversible and will mediate net transport either into or out of the cell until the opposing electrochemical potential differences of the driven solute and the driving solute become equal across the cell membrane. The direction of net transport mediated by a cotransport mechanism may be determined by either solute, depending on which solute has the larger driving force and by the direction of the larger driving force across the membrane. The mechanism mediating cotransport may be modeled as a six-step process as shown in Fig. 4: (1) An

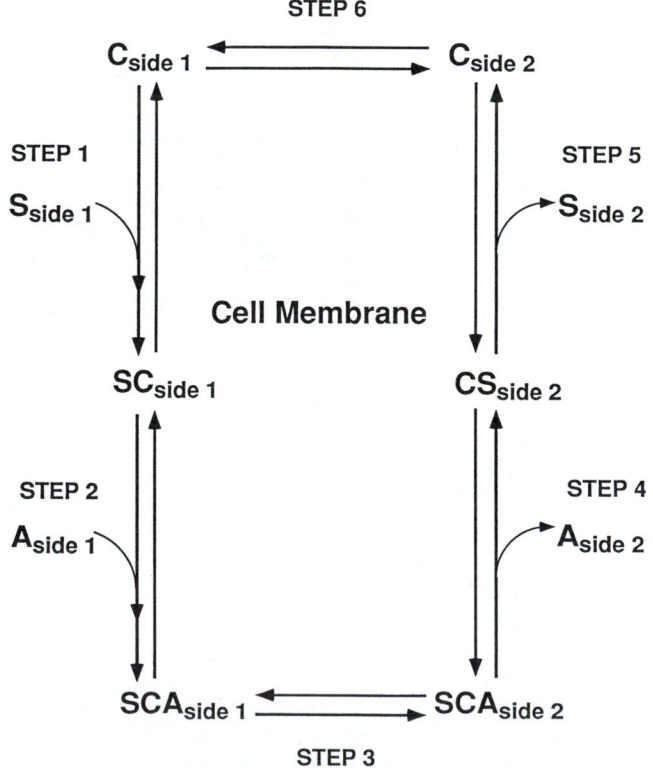

FIGURE 4. Kinetic model of cotransport.

association of solutes S and (2) A with the carrier facing side 1 to form the SCA solute-carrier complex, (3) a conformational change reorienting the SCA carrier complex to face side 2, (4) dissociation of A and (5) S with the carrier facing side 2, and (6) a second conformational change reorienting the solute-free carrier to face side 1. The cotransport mechanism may be further modeled in three different variations depending on the solute interaction with the carrier. Solute interaction with the carrier may be **random**, with either solute associating and dissociating first with the carrier, or solute association-dissociation may be **ordered**, in which either one of the solutes must associate or dissociate first with the carrier before the second solute may associate or dissociate. Note, as shown in Fig. 4 for an ordered interaction of solute with the carrier, that the second solute associating with the carrier at one side of the membrane is the first solute to dissociate with the carrier at the opposite side. Neither form of the partially associated carrier SC or AC may undergo a conformational change reorienting the sidedness of the partially associated carrier and therefore the cotransport mechanism mediates only the coupled transport of both solutes but does not mediate the uncoupled transport of either solute. Each step in the cotransport process is reversible and is linked to the preceding and succeeding step by rate constants defining the rate of solute association and dissociation on either side of the membrane and by the rate of conformational change for carrier reorientation in either direction when associated with or without both solutes. Similar to facilitated diffusion, solute association-dissociation with the carrier occurs faster than the conformational change reorienting the sidedness of the carrier and the rate-limiting step in the cotransport process is the conformational change reorienting the solute-dissociated carrier.

The unidirectional rate of solute transport across the cell membrane mediated by a cotransporter is dependent on the concentration of both solutes at the side where cotransport originates. When either solute concentration is constant, the cotransport rate of the second solute will approach a maximum value with increasing solute concentration and the saturation kinetics of the cotransport process may be analyzed using the Michaelis-Menten equation to describe the kinetic parameters K_m and V_{max} for each solute. The random or ordered association of cotransported solutes with the transporter may be determined from the mutual dependence of kinetic parameters on cotransported solute concentrations. As shown in Fig. 4, where the association of solutes S and A with the cotransporter at side 1 is ordered and S associates first, solute A will have the same maximal rate of transport (V_{max}) when determined at different concentrations of solute S. However, the K_m value for solute A will decrease when determined at increasing concentrations of solute S, indicating the initial association of S with the cotransporter effectively increases its affinity for association with solute A. In contrast, the V_{max} and K_m values characterizing transport of solute S, the first solute to associate with the cotransporter, will both vary when determined at different solute A concentrations. An increase in solute A concentration will increase the maximal rate of solute S through the cotransport mechanism and will also effectively

increase its affinity for solute S as reflected by a decreased K_m value. The effect of solute A on the maximal rate of solute S transport arises from its ordered association, as the second solute, to form the SCA solute-carrier complex, which may then transfer both solutes across the membrane. Thus, where a mutual solute concentration dependence of the kinetic parameters characterizing the transport of both solutes is determined for a cotransport mechanism, an ordered association-dissociation of solutes with the cotransporter may be indicated and a random association-dissociation of solutes may be excluded.

The cotransport of a solute may be inhibited by the presence of other solutes that interact with, but are not necessarily transported by, the same cotransport mechanism. Similar to facilitated diffusion, the competitive or noncompetitive nature of inhibitor interaction with the cotransport mechanism may be assessed by determining the effect of the inhibitor on the kinetic parameters characterizing transport of both solutes by the cotransport mechanism.

The same cotransport mechanism may mediate transport of multiple solutes that share common chemical and physical determinants. Similar to facilitated diffusion, the solute specificity of a cotransport mechanism may be characterized by determining the kinetic parameters of different solutes transported by the same cotransport mechanism. A comparison of the relative affinities and maximal velocities of solute transport determined experimentally will suggest the relative magnitudes of solute transport mediated by the cotransporter *in vivo*.

The cotransport mechanism modeled in Fig. 4 will be at a steady state and mediate equal transport of S and A in both directions across the membrane when the driving forces acting on the cotransported solutes are equal and opposite. This would occur when the transmembrane concentration difference or gradient of S in one direction is equal to the transmembrane concentration difference or gradient of A in the opposition direction. At the steady state, where no net solute transport occurs, the solute driving forces will be in thermodynamic equilibrium with each other and the transmembrane concentration ratios of S and A may be described by

$$\frac{S_1}{S_2} = \frac{A_2}{A_1} \qquad (2)$$

Thus, in the presence of a ten-fold concentration gradient of S, a cotransport mechanism coupling the transport of S and A across the membrane may generate up to, but not exceeding, a ten-fold concentration gradient of A. Because the cotransport mechanism effectively transduces the energy stored in the concentration gradient of S to the energy stored in the concentration gradient of A, it would be thermodynamically impossible for the cotransport mechanism to generate more than a ten-fold concentration gradient of A.

The cotransport process modeled in Fig. 4 describes the coupled transport of one molecule of S together with one molecule of A, or an S to A coupling ratio of 1:1. However, the stoichiometric coupling of solutes mediated by a cotransport mechanism may exceed unity, resulting in cotransport of unequal amounts of solute. This may be modeled kinetically as an extension of Fig. 4 to include an extra step of solute association on side 1 prior to carrier reorientation to face side 2 and an extra step of solute dissociation on side 2 prior to carrier reorientation to face side 1. The kinetic parameter (K_m) characterizing the solute affinity of a cotransport mechanism must account for each additional solute association or dissociation at mutually exclusive sites in the cotransport protein. The thermodynamic equilibrium of solute driving forces defining the steady state, where no net solute transport occurs, is profoundly affected by the stoichiometric coupling of solutes by the cotransport mechanism. Where the stoichiometric coupling is other than 1:1, the solute driving forces will be in thermodynamic equilibrium at transmembrane solute concentration ratios raised to the power of their respective coupling coefficient as shown.

$$\left(\frac{S_1}{S_2}\right)^s = \left(\frac{A_2}{A_1}\right)^a \qquad (3)$$

The coupling coefficients s and a are the number of S and A molecules that must associate with the carrier before a conformational change may occur reorienting the sidedness of the solute-carrier complex. The stoichiometric coupling of solutes by a cotransport mechanism is significant because in the presence of a ten-fold concentration gradient of S, a cotransport mechanism coupling the transport of two molecules of S to one molecule of A may generate a one-hundred-fold concentration gradient of A!

Similar to the facilitated diffusion of charged solutes, cotransport mechanisms may also be electrogenic and mediate net charge transfer across the membrane in the direction of net solute transport. In general, the electrogenicity of cotransport mechanisms arises from the transport of the inorganic cation sodium coupled to the transport of organic and inorganic anions and cations such as negatively charged amino acids, mono- and dicarboxylic acids, sulfate, and phosphate or positively charged amino acids and amines as well as nonelectrolytes such as sugars. In most instances, the coupled transport of charged or uncharged solutes with the transport of sodium results in the net transfer of positive charge in the direction of net solute transport into the cell. The positive electrogenicity of solute transport mediated by a sodium-coupled cotransport mechanism will tend to depolarize the cell membrane potential difference and results from at least one voltage-sensitive step in the process accounting for electrogenic cotransport. A voltage-dependent increase or decrease in the kinetic parameters characterizing an electrogenic cotransport mechanism may indicate an effect of voltage on the affinity of the cotransporter for either or both solutes (K_m) and/or on the conformational change reorienting the solute-associated or -dissociated transporter (V_{max}). However, because the rate-limiting step in the process accounting for cotransport is the conformational change reorienting the solute-dissociated transporter, a voltage-dependent increase in the rate of cotransport must result from the voltage sensitivity and acceleration of at least this step in the cotransport process. The voltage sensitivity of an electrogenic cotransport mechanism makes the voltage difference across the cell membrane a driving force acting on solute cotransport. Thus, the rate, magnitude, and direction of solute transport mediated by an electrogenic sodium cotransport mechanism is determined by two driving forces including the transmembrane voltage difference as well as the transmembrane sodium and solute concentration gradients. An electrogenic cotransport mechanism will be at steady state and mediate equal solute transport in both directions across the membrane when the electrical and chemical forces driving cotransport are at equilibrium across the membrane. For an uncharged solute S, and a charged solute A, the driving forces may be related by

$$RT \ln\left(\frac{S_1}{S_2}\right) = -RT \ln\left(\frac{A_2}{A_1}\right) + z\mathscr{F}\left(\Psi_2 - \Psi_1\right) \qquad (4)$$

where R is the universal gas constant, T is absolute temperature, z is the valence of A, $\mathscr{F}$ is the Faraday constant, and $(\Psi_2 - \Psi_1)$ is the transmembrane voltage difference. Thus, in the presence of a ten-fold sodium concentration gradient and a transmembrane voltage difference of 60 mV, a cotransport mechanism mediating 1:1 Na-to-solute transport may generate a one-hundred-fold uncharged solute gradient.

The functional properties of solute transport mediated by sodium-dependent cotransport mechanisms have been characterized in detail for many different cell- and solute-specific cotransporters. Like facilitated diffusion, a major question remaining in the study of sodium-dependent cotransport mechanisms is the structure-function relationship of the cotransport protein. Only recently have several sodium-dependent cotransport proteins been identified using a molecular biological approach to clone the genes coding for these membrane proteins. The primary structure of sodium-dependent cotransport proteins suggests topological features similar to membrane proteins mediating facilitated diffusion. These include the presence of 10–12 putative transmembrane domains, the location of extracellular N-linked glycosylation sites, and the location of intracellular phosphorylation sites. The structure-function relationship of the sodium-glucose cotransport protein SGLT1 has been investigated in mutagenesis studies of the cotransport protein. Intestinal glucose-galactose malabsorption is a genetic disorder resulting from multiple mutations localized in two putative transmembrane domains of SGLT1. Individual functional analysis of the different mutant SGLT1 proteins suggested the identity and location of specific amino acids that participate in the association of glucose and the inhibitor phloridzin with the cotransport protein as well as the amino acids involved in the conformational changes reorienting the solute-dissociated cotransporter. In addition to identifying the amino acids affecting the functional properties of the cotransport protein, mutagenesis studies of SGLT1 further suggest the identity and location of amino acids that participate in the membrane trafficking of cotransport proteins from its site of synthesis in the cell interior to the cell surface.

Three isoforms of the sodium-glucose cotransporter have been identified (SGLT1, SGLT2, SGLT3) and are distinguished by a high (SGLT1) or low (SGLT2, SGLT3) affinity for glucose, by differences in substrate specificity, and by tissue distribution in small intestine (SGLT1) and/or renal proximal tubule (SGLT1, SGLT2). The structural basis for isoform-specific differences in function may be assessed by functional studies of **chimeric proteins** composed of different peptide sequence combinations from different isoforms. Thus, the chimera composed of the amino terminus including the initial seven transmembrane domains of SGLT3 and the final five transmembrane domains including the carboxy terminus of SGLT1 is observed to have the substrate specificity of the SGLT1 isoform. Accordingly, this finding indicates that the topological location of the glucose association site in the sodium-glucose cotransport protein is in a region of the peptide sequence extending from the eighth transmembrane domain to the carboxy terminus.

C. Countertransport

Countertransport, or **antiport,** is a form of secondary active transport that, like cotransport, may also mediate net transfer of a solute across the cell membrane in a direction against the electrochemical potential gradient of the solute. The energy driving countertransport of a solute against its electrochemical potential difference arises from a coupling to the transport of a second solute moving across the cell membrane from a place of higher electrochemical potential to a place of lower electrochemical potential. In contrast to cotransport, the coupling of the driving solute to the driven solute in countertransport results in transport of solutes in the opposite direction across the cell membrane. Thus, countertransporters mediate the exchange of intracellular and extracellular solutes across the cell membrane. Countertransport mechanisms are reversible and will mediate net transport either into or out of the cell until the opposing electrochemical potential differences of the driven solute and the driving solute become equal across the cell membrane. The direction of net transport mediated by a countertransport mechanism may be determined by either solute, depending on which solute has the larger driving force and the direction of the larger driving force across the membrane. The mechanism mediating countertransport may be modeled as a six-step process as shown in Fig. 5. The principle features of the countertransport mechanism are (1, 4) a mutually exclusive association of either solute S or A with the carrier facing side 1 or side 2 to form the SC or AC carrier complex, (2, 5) a conformational change reorienting the sidedness of the solute-carrier complex, and, (3, 6) a dissociation of solute with carrier on the opposite side of the membrane. In contrast to the process of facilitated diffusion or cotransport, the solute-free carrier does not undergo a conformational change in countertransport and only the SC or AC carrier complex may undergo a conformational change reorienting the sided access of carrier to solute. It is the absence of a conformational change reorienting the sidedness of the solute-free carrier that explicitly distinguishes countertransport from facilitated diffusion and confers the ability to mediate net solute transport against an electrochemical potential difference. Countertransport mechanisms may mediate the exchange of the same solutes (**homoexchange**) or different solutes (**heteroexchange**) and will mediate net solute transport only when exchanging different solutes. Each step describing the process of countertransport is reversible and is linked to the preceding and succeeding step by rate constants defining the rate of solute association and dissociation on either side of the membrane and by the rate of conformational change reorienting the sidedness of the solute-associated carrier. The conformational change reorienting the sidedness of the solute-associated carrier occurs more slowly than solute association-dissociation with the carrier and is considered the rate-limiting step in the countertransport process.

The unidirectional rate of solute transport across the cell membrane mediated by a countertransport mechanism depends on the concentration of both solutes at opposite sides of the membrane. When either solute concentration is constant, the countertransport rate of the second solute will approach a maximum value with increasing solute concentration and the saturation kinetics of the countertransport process may be analyzed using the Michaelis-Menten equation to obtain the kinetic parameters K_m and V_{max} for each of the exchanged solutes. The solute concentration dependence of the rate of solute transport by countertransport may be different for transport occurring into or out of the cell. A kinetic asymmetry of solute transport mediated by countertransport results from a difference in the apparent affinity (K_m) for solute at either side of the membrane and not from a difference in the maximal rate of transport (V_{max}). This limitation in kinetic asymmetry is a unique property of countertransport arising from the inability of the solute-dissociated carrier to reorient across the membrane. While kinetic asymmetry is not a property of all countertransport mechanisms, the asymmetry in solute transport mediated by a countertransporter is observed to occur for all possible combinations of solutes exchanged by the countertransport mechanism.

The countertransport of a solute may be inhibited by the presence of other solutes that interact with, but are not necessarily transported by, the same countertransport mechanism.

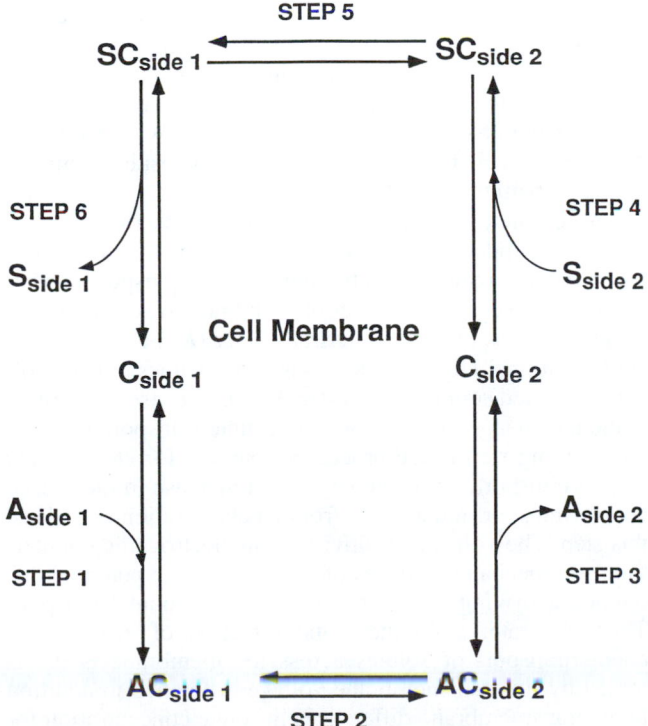

FIGURE 5.   Kinetic model of countertransport.

The competitive or noncompetitive nature of interaction of the inhibitor with the countertransport mechanism may be assessed by determining the effect of the inhibitor on the kinetic parameters characterizing transport of exchanged solutes by the countertransport mechanism. The same countertransport mechanism may mediate the exchange of multiple pairs of solutes that share common chemical and physical determinants. Countertransport mechanisms may be broadly categorized as either anion or cation exchange mechanisms such as the chloride-bicarbonate countertransporter present in kidney, intestine, and red blood cells or the ubiquitous pH-regulating sodium-proton countertransporter. The solute specificity of a countertransport mechanism may be characterized by determining the kinetic parameters of different solutes proven to be countertransported by the same mechanism. A comparison of the relative affinities and maximal velocities of solute transport determined experimentally will suggest the relative magnitudes of solute transport mediated by the countertransporter *in vivo*.

The countertransport mechanism modeled in Fig. 5 will be at a steady state and mediate equal transport of S and A from side 1 to side 2, and from side 2 to side 1, when the driving forces acting on solutes S and A are equal. For countertransport this will occur when the transmembrane concentration differences of S and A are equal and in the same direction across the cell membrane. At the steady state, where no net solute transport occurs, the solute driving forces will be in thermodynamic equilibrium with each other and the transmembrane concentration ratios of S and A resulting from countertransport may be described by

$$\frac{S_1}{S_2} = \frac{A_1}{A_2} \tag{5}$$

Thus, in the presence of a ten-fold concentration gradient of S, a countertransport mechanism coupling exchange of S for A may generate up to, but not exceeding, a ten-fold concentration gradient of A. Because the countertransport mechanism effectively transduces the energy stored in a ten-fold concentration gradient of S to the energy stored in a concentration gradient of A, it would be thermodynamically impossible for the countertransport mechanism to generate more than a ten-fold concentration gradient of A. The important difference distinguishing cotransport and countertransport at the steady state is the sidedness or direction of solute concentration gradients across the membrane. Cotransport is at steady state when solute concentration gradients are of equal magnitude but in opposite direction across the membrane, whereas countertransport is at steady state when solute concentration gradients are of equal magnitude and in the same direction across the membrane.

The countertransport process modeled in Fig. 5 describes the coupled exchange of one molecule of S for one molecule of A, or an S-to-A coupling ratio of 1:1. Whereas most countertransport mechanisms mediate the coupled exchange of single solutes, the stoichiometric coupling of solutes may exceed unity, resulting in unequal amounts of solute transported to opposite sides of the membrane. This may be modeled kinetically as an exten-

sion of Fig. 5 to include an extra step of solute association on side 1 prior to carrier reorientation to face side 2 and by an extra step of solute dissociation on side 2. The kinetic parameter (K_m) characterizing the solute affinity of a countertransport mechanism must account for each additional solute association-dissociation at mutually exclusive sites in the countertransport protein. Similar to cotransport, the thermodynamic equilibrium of solute driving forces coupled by the countertransport mechanism and defining the steady state is profoundly affected by the stoichiometric coupling of solutes. Where the stoichiometric coupling of solutes by a countertransport mechanism is other than 1:1, the solute driving force will be in thermodynamic equilibrium at transmembrane solute concentration ratios raised to the power of their respective coupling coefficient as shown.

$$\left(\frac{S_1}{S_2}\right)^s = \left(\frac{A_1}{A_2}\right)^a \tag{6}$$

Thus, in the presence of a ten-fold concentration gradient of S, a countertransport mechanism coupling the exchange of two molecules of S for one molecule of A may generate a one-hundred-fold concentration gradient of A in the same direction across the membrane!

Most countertransport mechanisms mediate the coupled exchange of single solutes with the same positive or negative charge and same valence and are therefore electroneutral and do not mediate net charge transfer across the membrane. However, when countertransport mechanisms mediate the coupled exchange of solutes of the same charge but different valence or mediate an unequal stoichiometric coupling of solutes of the same charge, a net charge transfer will occur across the cell membrane. Thus, the countertransport mechanism mediating sodium-calcium exchange across the cell membrane couples the influx of three sodium ions to the efflux of one calcium ion, resulting in a net transfer of positive charge into the cell. The positive electrogenicity of the sodium-calcium countertransport mechanism will tend to depolarize the cell membrane potential and results from the voltage sensitivity of at least one step in the process describing a complete sodium-calcium countertransport cycle. A voltage-dependent increase or decrease in the kinetic parameters characterizing an electrogenic countertransport mechanism would suggest an effect of membrane potential on the solute affinities of the countertransporter (K_m) and/or on the conformational change reorienting the sidedness of the solute-associated countertransporter (V_{max}). Because the conformational change reorienting the countertransporter is the rate-limiting step in the process accounting for electrogenic countertransport, a voltage-dependent increase in the rate of countertransport must result from an acceleration of at least this step. The voltage sensitivity of an electrogenic countertransport mechanism makes the cell membrane potential difference a driving force acting on solute countertransport. Thus, the rate, magnitude, and direction of electrogenic countertransport of solutes across the membrane is determined by the transmembrane voltage difference in addition to the transmembrane differences in solute concentration for both solutes coupled by the countertransport mechanism. An electrogenic countertransport mechanism will be at steady state, and mediate equal solute transport in both directions across the membrane, when the electrical and chemical forces driving countertransport are at equilibrium. For the 1:1 exchange of charged solutes S and A with valences z_S and z_A, respectively, the driving forces may be related by

$$RT \ln\left(\frac{S_1}{S_2}\right) = RT \ln\left(\frac{A_1}{A_2}\right) + \left(z_S - z_A\right) \mathscr{F} \left(\Psi_2 - \Psi_1\right) \quad (7)$$

The electrogenicity of a countertransport mechanism may arise from the exchange of solutes with the same charge and valence, but with an unequal stoichiometric coupling. In this instance the driving forces acting on electrogenic countertransport must be factored by the stoichiometric coupling coefficients s and a for solutes S and A, respectively, and may be related by

$$RT \ln\left(\frac{S_1}{S_2}\right)^s = RT \ln\left(\frac{A_1}{A_2}\right)^a + \left(sz_S - az_A\right) \mathscr{F} \left(\Psi_2 - \Psi_1\right) \quad (8)$$

Thus, in the presence of a ten-fold sodium concentration gradient and a transmembrane voltage difference of 60 mV, an electrogenic countertransport mechanism mediating the 3:1 exchange of sodium and calcium ions, respectively, may generate a maximum calcium concentration gradient of ten-thousand-fold across the cell membrane!

The functional properties of solute transport mediated by countertransport mechanisms have been characterized in detail for many different cell- and solute-specific countertransporters. Like facilitated diffusion and cotransport, the structure-function relation of membrane proteins mediating countertransport continues to be an important question under study. The techniques of modern molecular biology have resulted in the recent cloning of several genes coding for countertransport proteins. Analysis of the deduced amino acid sequences of countertransport proteins suggests topological features similar to membrane proteins mediating facilitated diffusion and cotransport, including the presence of 10–12 putative transmembrane domains, the location of extracellular N-linked glycosylation sites, and the location of intracellular phosphorylation sites. Recently, low-resolution electron density maps of the red blood cell chloride-bicarbonate exchange protein have been obtained from two-dimensional crystals of the countertransport protein studied by electron microscopy. The membrane domain of the countertransport protein exists as a dimeric structure with three electron-dense regions in each monomer. Within each monomer the configuration of electron densities suggests two closely apposed peptide structures which are spatially separated from a third structure by a peptide domain existing in two different conformations.

The structure-function relationship of the anion exchange protein AE1 has been investigated in mutagenesis studies of the countertransport protein. The functional significance of single amino acids or domains of multiple amino acids on the interaction with transport inhibitors at the extracellular side of AE1 has been studied to determine the nature and topological location of amino acids involved in substrate interaction and translocation. These mutagenesis studies involving both the deletion and substitution of amino acids indicate that several positively charged lysines at the surface of AE1 influence, and are necessary for, substrate recognition by the countertransport protein. The genes coding for three anion exchange proteins (AE1, AE2, AE3) have been identified and their gene products may be functionally distinguished by differences in inhibitor sensitivity and by the intracellular pH dependence of anion exchange activity. Multiple species-specific isoforms of the AE1 gene have been identified and the predominant sites of expression are the erythropoietic tissues and the acid-secreting cells of the renal cortical collecting duct. In contrast, AE2 gene expression occurs more widely in epithelial and nonepithelial tissues and AE3 gene expression is predominant in brain and heart. The putative transmembrane amino acid sequences of AE1, AE2, and AE3 are approximately 70% identical. The aligning portion of the N-terminal cytoplasmic domains of AE1 and AE2 are only 30% identical, with little or no corresponding sequence identity among AE1, AE2, or AE3 at the N-terminus. The observed amino acid sequence identity in the putative transmembrane domains of AE1, AE2, and AE3 is consistent with the common functional properties of the three anion exchange proteins. The divergent amino acid sequences noted in the N-terminal regions suggest possible cell- and tissue-specific differences in the regulation of anion exchange activity and/or in the intracellular sorting and membrane targeting of anion exchange proteins. Mutations in the human AE1 gene in the form of nucleotide deletions, insertions, and substitutions result in inheritable diseases of the red blood cell which manifest multiple clinical phenotypes ranging from mild to severe. The single amino acid substitution characterizing mutant AE1 in Band 3 Memphis has essentially no effect on red blood cell anion exchange activity. However, the nine-amino-acid deletion characterizing AE1 in Southeast Asian ovalocytosis (SAO) results in a heterozygous genetic disorder of red blood cells with only half the normal complement of functional anion exchangers. Interestingly, the red blood cell membrane of SAO patients is more resistant to plasmodial invasion, a property that may confer a selective survival advantage against malarial infection and thereby perpetuate this genetic disorder.

Bibliography

Alper, S. L. (1994). The band 3-related AE anion exchanger gene family. *Cell Physiol. Biochem.* **4**, 265–281.

Aronson, P. S. (1981). Identifying secondary active solute transport in epithelia. *Am. J. Physiol.* **240** (Renal Fluid Electrolyte Physiol. **9**), F1–F11.

Heinz, E. (1978). "Mechanics and Energetics of Biological Transport." Springer-Verlag, New York.

Heinz, E. (1981). "Electrical Potentials in Biological Membrane Transport." Springer-Verlag, New York.

Neame, K. D., and Richards, T. G. (1972). "Elementary Kinetics of Membrane Carrier Transport." Blackwell Scientific, Oxford.

Steel, A., and Hediger, M. (1998). The molecular physiology of sodium- and proton-coupled solute transporters. *News Physiol. Sci.* **13**, 123–131.

Stein, W. D. (1990). "Channels, Carriers and Pumps." Academic Press, New York.

David M. Balshaw, Lauren A. Millette, and Earl T. Wallick

17

Sodium Pump Function

I. Introduction

As primordial life forms were evolving, it became necessary to separate the internal milieu of developing cells from their varied, sometimes harsh, external surroundings. For this purpose, biological membranes developed (see Chapters 4 and 5). This compartmentalization enabled the evolution of other processes that allowed for functional specialization of cells. One such process that evolved was the active pumping of ions through the phospholipid membrane barrier. Early forms of biological pumps that evolved in prokaryotes were proton pumps that regulated intracellular pH (Gogarten *et al.*, 1989). As biological complexity increased, more elaborate **transport activities** evolved to improve the control of the organism over its environment. As early life forms became exposed to various ionic environments, a need to develop some means of regulating cellular volume and of maintaining cytosolic homeostasis arose. Since water follows the movement of ions as they are transported between compartments, it became imperative to develop mechanisms to control the flow of ions, such as Na^+ and K^+, through the plasma membrane. In eukaryotic cells, these processes are controlled by the **sodium pump.** The enzyme that fuels the sodium pump is the Na^+,K^+-ATPase.

Na^+,K^+-ATPase is a member of the P-type ATPase superfamily, which differs structurally and functionally from both the F-type ATPases (ATP-synthases present in prokaryotes, chloroplasts, and mitochondria) and V-type ATPases (e.g., the H^+ pump located in vacuolar membranes of eukaryotic cells). The mechanism of action of the P-type ATPases involves formation of a transient, covalently phosphorylated intermediate during their reaction cycle. Structurally, they include at least one catalytic subunit with a mass ranging from 70 to 130 kDa. There are P-type ATPases in both prokaryotes (e.g., copper and cadmium ion pumps) and eukaryotes. The eukaryotic P-type ATPases can be further subdivided into two groups. One group of eukaryotic P-type ATPases consists only of a single subunit, designated α, and includes the sarco-endoplasmic reticulum Ca^{2+}-ATPase (SERCA), the plasma membrane Ca^{2+}-ATPase, and the H^+-ATPase found in yeast and plants. Another grouping of the eukaryotic P-type ATPase family contains an additional subunit (β), and includes the gastric H^+,K^+-ATPase and the Na^+,K^+-ATPase. The α-subunit of the H^+,K^+-ATPase and the Na^+,K^+-ATPase show approximately 60% homology in amino acid sequence, and the structure of the genes encoding the two enzymes is very similar. The α-subunit of SERCA is approximately 30% homologous in amino acid sequence to the α-subunit of the Na^+,K^+-ATPase. Not surprisingly, the function and mechanism of action of this family of enzymes are closely related.

II. Na$^+$-K$^+$ Transport

Figure 1 shows the distribution of Na^+, K^+, and Cl^- across the plasma membrane of a typical mammalian cell with a resting membrane potential of -70 mV. There is little driving force for chloride ions, since their Nernst potential (see Chapter 1) is -69 mV, near the resting membrane potential. Potassium ions will tend to flow out of the cell, since their equilibrium potential (-91 mV) is more negative than the transmembrane potential: Conversely, Na^+ will have a very strong force driving it into the cell, since both the chemical and electrical gradients (equilibrium potential of $+64$ mV) favor Na^+ uptake. The pumping of Na^+ out of the cell and K^+ into the cell, and hence regulation of cytosolic ionic concentrations and cell volume, therefore requires sufficient energy (derived from the hydrolysis of ATP) to overcome these driving forces. This, by definition, is the **sodium pump activity.**

The nonequivalent transport of three Na^+ for two K^+ across the membrane maintains transmembrane gradients for these ions, and this became a convenient driving force for the secondary transport of metabolic substrates. In addition, the nonequivalent transport is electrogenic and leads to the generation of a transmembrane electrical potential, allowing cells to become excitable. The presence of the ionic gradients provided a mechanism for the regulation of many of the processes essential for the viability and stability of the cell. These include maintenance of intracellular pH via the activity of the Na^+-H^+ antiporter, contractility by means of K^+ activation of the actomyosin ATPase as well as Na^+-Ca^{2+} exchange, and

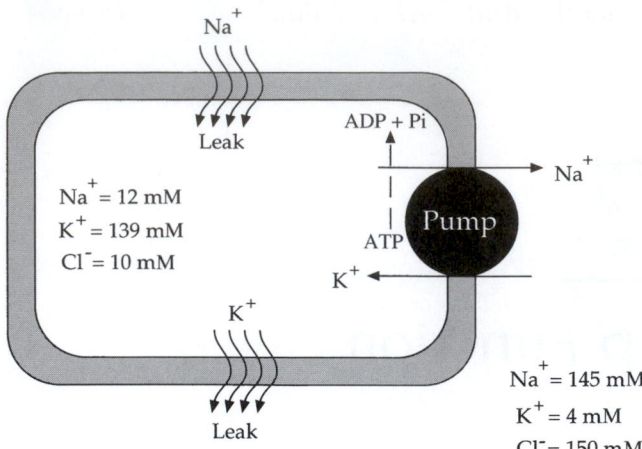

FIGURE 1.   Distribution of sodium, potassium, and chloride ions in a typical mammalian cell.

metabolism through symport of Na^+ with metabolites such as sugars, amino acids, and neurotransmitters.

The internal environment has a high concentration of K^+ and a low concentration of Na^+. Internal biological functions (such as the contractility of muscle cells) have, therefore, evolved to be activated by K^+ and inhibited by Na^+. If a cell is damaged or becomes starved for energy, the concentration of K^+ will decrease and the concentration of Na^+ will increase because of either an increased leak of the ions with their gradients or a decreased pumping against the gradient. This decrease in K^+ and increase in Na^+ leads to inhibition of energy-utilizing processes and allows other compensatory mechanisms to attempt to repair any damage.

III. Transport Mechanism

Early studies of the pump used intact systems such as giant squid axons and erythrocyte ghosts, enabling both the direct measurement of ion fluxes using radioactive tracers and indirect measurements using electrophysiological techniques. Initial studies found that the appearance of ^{22}Na in the bath solution following microinjection into the squid axon required that K^+ be present in the extracellular bath solution. This suggested that Na^+ efflux (transport out of the cell) was coupled to K^+ influx (transport into the cell). Experiments performed in parallel showed that, under physiological conditions, the efflux of ^{22}Na and influx of ^{42}K occurred at a coupling ratio of 3 Na^+:2 K^+:1 ATP. More recent studies have demonstrated that this ratio is not fixed, but can vary depending on the concentrations of the ions both within and outside the cell (reviewed in Lauger, 1991; Tepperman *et al.*, 1997).

Transport studies using erythrocytes allowed researchers to determine the ion selectivity of the sodium pump that intracellular activation of the pump explicitly requires. These sites have an intrinsic affinity[1] of 0.2 mM for Na^+. Potassium competes (but does not activate transport) with the Na^+ sites with an intrinsic affinity

of 9 mM (Garay-Garrahan, 1973). In the presence of normal concentrations of internal K^+ (see Fig. 1), the apparent affinity for Na^+ at the intracellular site is approximately 3 mM. The extracellular transport sites are much less specific for potassium since Rb^+, Cs^+, Li^+, Tl^+, NH_4^+, and even Na^+ can activate the ATPase (Bell *et al.*, 1977). The intrinsic affinities of the extracellular transport sites for K^+ and Na^+ are approximately 0.2 and 9–16 mM, respectively (Sachs, 1977). In the presence of normal concentrations of external Na^+ (see Fig. 1), the affinity for the extracellular K^+ is approximately 2 mM. Affinities for cations derived from studies of Na^+ and K^+ activation of hydrolysis of ATP in fragment enzyme preparations (Lindenmayer *et al.*, 1974) are in agreement with the studies of intact erythrocytes.

Studies with axons and erythrocytes also demonstrated that the activation of the pump is dependent on the presence of intracellular ATP and that, in particular, it is specific for MgATP. The products of the hydrolysis, ATP and P_i (inorganic phosphate), remain within the cell. The dependence of the pump activity on metabolically derived ATP was confirmed by the finding that metabolic inhibitors such as azide, cyanide, and dinitrophenol greatly reduce the pump activity.

Schatzmann (1953) reported that the sodium pump activity was specifically inhibited by cardiac glycosides, plant-derived steroids that have been used for 200 years to treat heart failure (Withering, 1785). Skou isolated an enzyme found in the plasma membrane whose hydrolysis of ATP was *activated by both Na^+ and K^+* (Skou, 1957). This is in contrast to the majority of cytosolic enzymes, which are activated by K^+ and inhibited by Na^+. This Na^+,K^+-ATPase is specifically inhibited by cardiac glycosides in the same concentration range that inhibits transport in intact cells, demonstrating that this enzyme is responsible for the sodium pump activity. For his work, Skou was awarded the 1997 Nobel Prize in Chemistry.

Studies using partially purified, fragmented membrane preparations of Na^+,K^+-ATPase found that, in the presence of Na^+, Mg^{2+}, and ATP-$\gamma^{32}P$, the γP was covalently transferred to the enzyme, forming an acid-stable acyl phosphate bond. This established the Na^+,K^+-ATPase as a member of the P-type family of ATPases. When K^+ is added to the enzyme that has been phosphorylated in the presence of Na^+, Mg^{2+}, and ATP, the phosphate group is rapidly cleaved and the enzyme is allowed to cycle, continuing to hydrolyze ATP (Fig. 2). Both N-ethylmaleimide and oligomycin decrease the sensitivity of the enzyme to K^+-induced dephosphorylation and increase the sensitivity of phosphorylated enzyme to ADP (rev. in Glynn, 1993). This led to the concept of two primary enzyme conformations, both of which can exist in either a phosphorylated or an unphosphorylated form. These are designated E_1, which has a high affinity for intracellular Na^+, Na_{in}^+, and E_2, which has a high affinity for extracellular K^+, (K_{ex}^+). The cycling of the enzyme between the E_1 forms, binding cations at the intracellular face, and E_2 forms, binding cations at the extracellular face, results in the transport of the ions through the membrane. This concept, illustrated in Fig. 2, is referred to as the **Post-Albers scheme** in recognition of the contribution of these two scientists.

In the E_1 conformation, the enzyme has high affinity for both intracellular Na^+ and MgATP. Following the binding of both of these ligands, the γ phosphate group of ATP is transferred to an aspartyl residue on the enzyme, forming a "high-

---

[1] Intrinsic affinity is the affinity of the ion in the absence of competing ions and is obtained by extrapolation to a concentration of zero competitor.

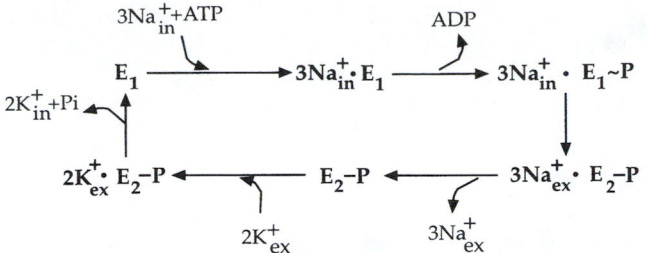

FIGURE 2. Post-Albers reaction mechanism for the sodium pump.

energy" phosphoenzyme. In this high-energy state, the phosphate group can be readily returned to ADP, resynthesizing ATP. In the forward cycle of the enzyme, the "high energy" residing in the phosphate bond is used to drive a change in the conformation of the enzyme, resulting in the exposure of Na^+ to the extracellular surface. This "low-energy" phosphoenzyme, with Na^+ bound, is now said to be in an E_2 conformation. The phosphate group in this state cannot be easily transferred back to ADP to form ATP. The E_2P conformation has low affinity for extracellular Na^+; therefore, it is released. This conformation has high affinity for extracellular K^+. After K^+ binds at the extracellular surface, the phosphatase activity of the enzyme is activated, the phosphate group is cleaved, and K^+ is translocated and released on the intracellular surface. The enzyme is once again in the E_1 conformation and capable of binding Na^+ and ATP, completing the cycle. Because of the cycling between these two conformations, the P-type ATPases are also referred to as E_1E_2-ATPases.

Additional evidence in support of the two-conformation hypothesis came from studies using controlled proteolysis. Jorgensen found that in the presence of Na^+, specific sequences for trypsin proteolysis were exposed, designated T2 and T3 (see Section IV and Fig. 4). In addition, a site for chymotrypsin, C3, was observed. In the presence of K^+, the T2 site is still cleaved, but an additional site, T1, is now exposed, while the T3 and C3 site are now hidden (Jorgensen, 1977). Fluorescence measurements of enzyme labeled at K501 with fluorescein isothiocyanate (FITC) also supported the hypothesis that the enzyme undergoes structural rearrangements dependent on ligand conditions. FITC labeling is competitive with ATP; however, mutation of the labeled residue did not affect the activity of the enzyme. It is likely, therefore, that the inhibition of ATP binding following FITC modification is due to an indirect, steric effect. Nonetheless, a change in the fluorescence properties of the FITC following the addition of Na^+ or K^+ indicates that the conformation of the active site is changed (Hegyvary and Jorgensen, 1981). These two types of studies have clearly demonstrated that the enzyme is undergoing structural rearrangement when exposed to different ionic conditions.

It is known that the Post-Albers scheme (see Fig. 2) is overly simplistic. Perhaps a more realistic view of the reaction mechanism is that it consists of a series of steps whereby the ions are bound, occluded, and released from the enzyme (Forbush, 1987a). The **occluded state** refers to a form of the enzyme where the ions are trapped within the membrane and are not accessible from either the extracellular or intracellular compartments. This becomes intuitively obvious if one considers the movement of a 2-Å molecule being moved through a 50- to 70-Å thick membrane against its concentration gradient. Figure 3 illustrates a modified version of the Post-Albers scheme that is more consistent with recent findings. First, in the E1 conformation, it has been proposed that the Na^+ ions bind in a multistep process, with the rate of binding for the first two sodium ions being so fast as to be indistinguishable, followed by the third Na^+ binding at a slower rate (Heyse et al., 1994). In the presence of Na^+, MgATP is bound with high affinity and the γ phosphate is transferred to the enzyme. This occurs with the simultaneous occlusion of either two or three Na^+ ions. The sodium ions are then translocated and released on the extracellular surface; it is believed that the release of sodium is an ordered process. One ion is released first, followed by the remaining two. The release of the sodium ions allows for the binding of the potassium ions, again in an ordered fashion, with positive cooperativity, such that the binding of the first potassium to the extracellular surface increases the affinity for the second (Tepperman et al., 1997). If the concentration of extracellular K^+ is low, it is possible that only one K^+ will bind and be transported. The binding of potassium results in a rapid dephosphorylation of the enzyme along with occlusion of the ions. In the absence of intracellular ATP, the rate of K^+ release to the intracellular surface is very slow. Following the binding of ATP the rate of deocclusion is greatly increased. Occlusion studies by Forbush (1987b) have demonstrated that the release of K^+ in the absence of ATP following transport is a multistep, ordered process in which one ion is released at a rapid rate while the second ion is released more slowly. This process is consistent with a flickering gate model, in which the exposure of the cation binding sites limits the release to a single ion at a time. The release of K^+ allows Na^+ to bind and increases the affinity of the enzyme for ATP, thus completing the enzyme cycle.

Figure 3 does not, however, take into account that internal K^+ binds to (competes with) the Na^+ transport sites, or that extracellular Na^+ is able to bind to (compete with) the external K^+ transport sites. Extracellular Na^+ has been implicated additionally as an allosteric effector of K^+ transport (Sachs, 1977). From the mechanism depicted in Fig. 3, it is clear that the coupling ratio does not have to be fixed at 3 Na^+:2 K^+ under all conditions. The study by Tepperman et al. demonstrated that the only models that fit their data were models in which translocation could take place when either one or two K^+ were bound to the enzyme (Tepperman et al., 1997).

IV. Na$^+$,K$^+$-ATPase Structure

The Na$^+$,K$^+$-ATPase is composed of two subunits (Fig. 4) in a one-to-one stoichiometry: the **alpha** (α) or catalytic subunit and the **beta** (β) or glycoprotein subunit (Lane et al., 1973). The α subunit contains all of the known binding sites for ligands that regulate pump function. The complete primary structure of the sheep α- and β-subunits has been determined by the cloning of cDNAs corresponding to the

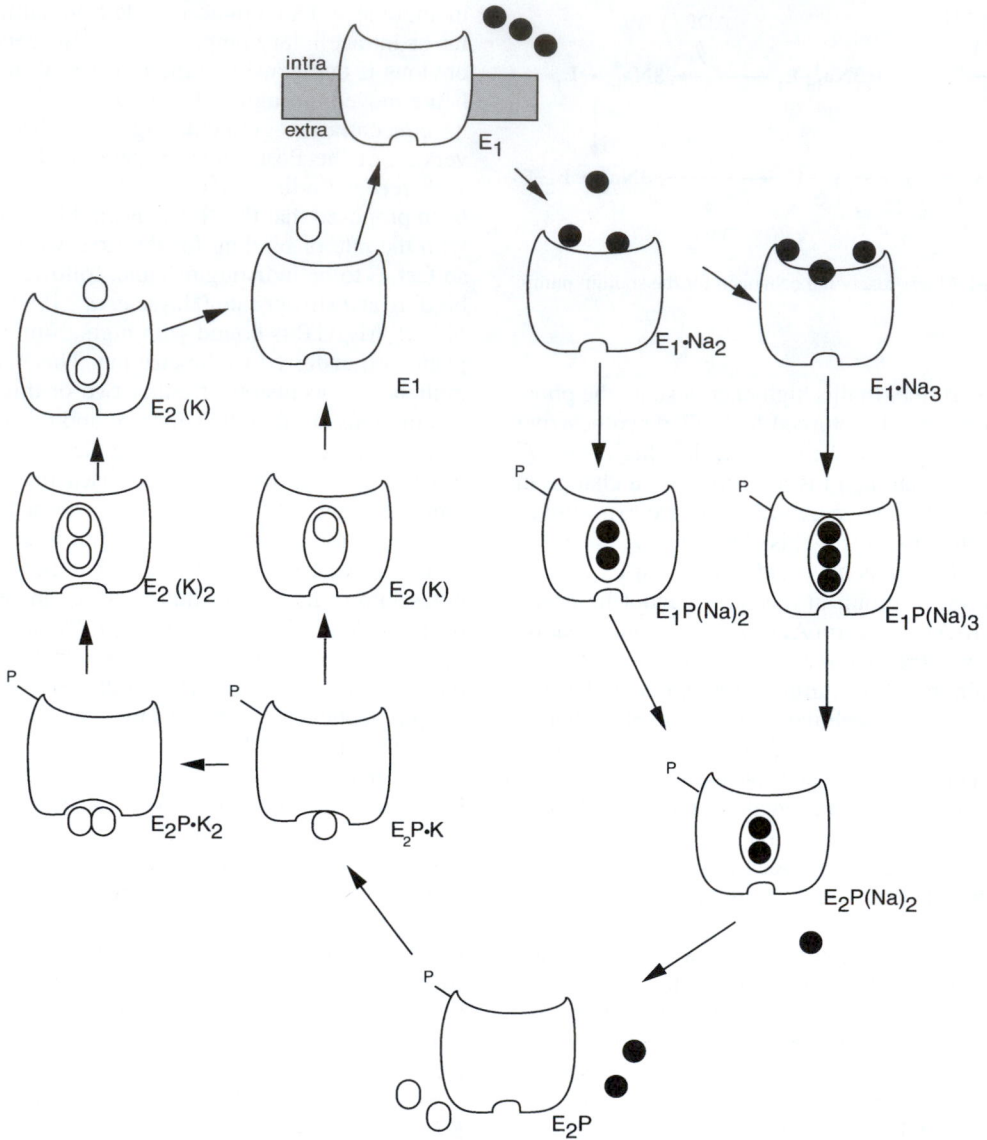

FIGURE 3. Modified Post-Albers scheme showing binding, occlusion, and release of cations (Na^+ and K^+) from both the intracellular and extracellular surfaces, and possible noncanonical flux modes. Enzyme forms that contain occluded cations are shown with the cations in parenthesis (i.e., $E_1P(Na)_3$). Enzyme forms with cations bound are shown with a dot (i.e., $E_1 \cdot Na_3$). Na^+ is designated by filled circles and K^+ by open circles. For simplicity, the binding and hydrolysis of ATP are omitted.

mRNA for both subunits (Shull *et al.*, 1985, 1986a). The sheep kidney α-subunit consists of a protein of M_r 112 117 (1016 amino acid residues) and the β-subunit consists of 302 amino acid residues with a core protein of M_r of 34 939 (Shull *et al.*, 1986b). The α1-subunit exists as multiple **isoforms,** and the primary sequences of these isoforms from a variety of species, including the α1, α2, α3, and α4 from the human and rat, have been determined (rev. in Lingrel *et al.*, 1990, and see Shamraj and Lingrel, 1994).

The distribution of the four α isoforms is tissue specific (demonstrated in Fig. 5) and is both hormonally and developmentally regulated. The **α1 isoform** has been found in all tissues and cell types examined and is considered to be the housekeeping form of the enzyme. The **α2 isoform** has been found primarily in skeletal muscle and brain, with smaller

amounts in the adult hearts of most species, including human. The **α3 isoform** has been found primarily in brain, but also in neonatal rat heart and adult human heart. The **α4 isoform** has been found only in rat and human testis, at an mRNA level comparable to the level of α1 in the brain. The sequence of the four rat isoforms shows 80–85% homology. The sequence of the α1 isoforms is also highly conserved between species. Based on hydropathy plots and analogy with the H^+,K^+-ATPase and the Ca^{2+}-ATPase, the α-subunit of Na^+,K^+-ATPase is predicted to have 10 membrane-spanning regions labeled H1 to H10, as shown in Fig. 4.

The β-subunit of sheep kidney contains three N-linked oligosaccharide sites and seven cyst(e)ine residues. The one free sulfhydryl, C44, is postulated to be in the membrane-spanning region. The other six residues participate in three

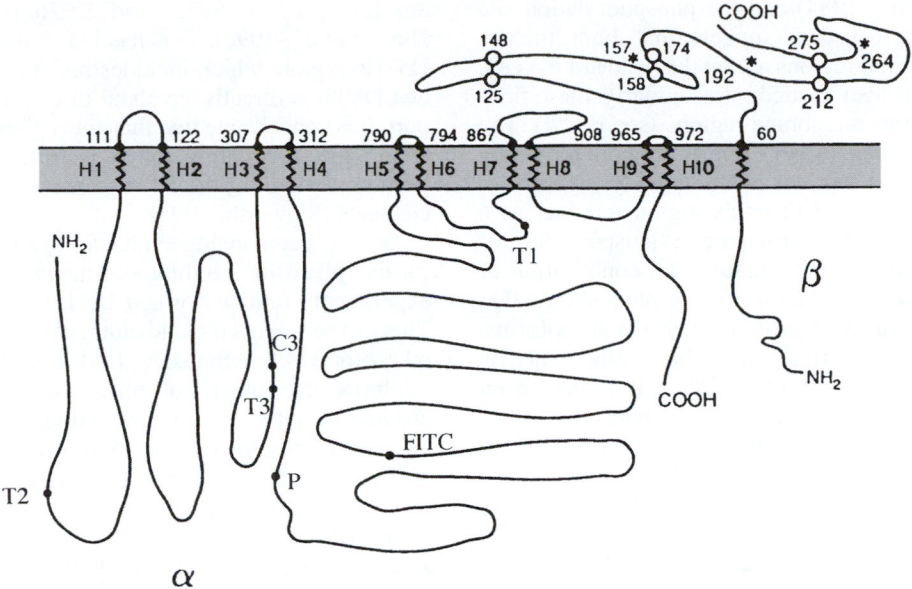

FIGURE 4. The predicted transmembrane model of the sheep α_1- and β_1-subunits of the Na$^+$,K$^+$-ATPase. The membrane-spanning helices are represented by H1–H10. Asterisks (*) identify the three glycosylation sites, and the open circles denote the position of the disulfide bridges on the β-subunit. The sites of tryptic digestion have been designated as T1 (R438), T2 (K30), T3 (R262), and chymotryptic digestion, C3 (L266). The FITC binding site is at residue K501 and the phosphorylation site (P) is at D369.

disulfide bonds (Kirley, 1989). Based on hydropathy plots, the β1-subunit possesses only one transmembrane region (see Fig. 4). Three isoforms of the β-subunit have been described (Shull *et al.,* 1988). The **β1 isoform** described previously is the major form expressed in virtually all tissues, with β2 occurring primarily in brain. Although the two mammalian isoforms exhibit only 35% homology, the location of the glycosylation sites, the conserved cysteine residues, and similarity of the hydropathy plots suggest that these two isoforms possess high structural homology. Interestingly, the β2 isoform appears to be identical to the adhesion molecule on glia (AMOG). A third isoform, β3, has only been observed during early development in *Xenopus laevis.*

The physiological function of the β-subunit is unknown. Expression of functional Na$^+$,K$^+$-ATPase activity in yeast, which do not contain a Na$^+$,K$^+$-ATPase, requires that both the α- and β-subunits be cotransfected (Horowitz *et al.,* 1990). This indicates that the β-subunit is required for the proper assembly of the α-subunit into the membrane and for the exit of this $\alpha\beta$ complex from the endoplasmic reticulum and proper insertion into the cell membrane (rev. by Lingrel *et al.,* 1990; McDonough *et al.,* 1990; Sweadner, 1989). The β-subunit of the enzyme also alters the susceptibility of the α-subunit to proteolytic enzymes, further showing its involvement in stabilizing the enzyme's structure. Furthermore, this subunit has been postulated to affect cation affinity, suggesting that it may undergo conformational changes associated with enzyme function (Chow and Forte, 1995). The β-subunit of Na$^+$,K$^+$-ATPase is believed to be in close proximity to the cardiac glycoside binding site on the α-subunit because it can be labeled with photoaffinity derivatives of digitoxin. The β-subunit of H$^+$,K$^+$-ATPase and Na$^+$,K$^+$-ATPase are highly homologous and, in fact, the β-subunit of the H$^+$,K$^+$-ATPase can replace the native Na$^+$,K$^+$-ATPase β-subunit and allow functional expression of α (Horisberger *et al.,* 1991).

The α-subunit of the transport ATPases have in common the phosphorylation of an aspartic acid residue (D369,[2] see Fig. 4) as a part of their enzymatic cycle. The large intracellular loop between H4 and H5 has been shown, by chemical modification and ATP analog labeling, to contain the ATP-binding site (in the region of K501 where FITC is covalently

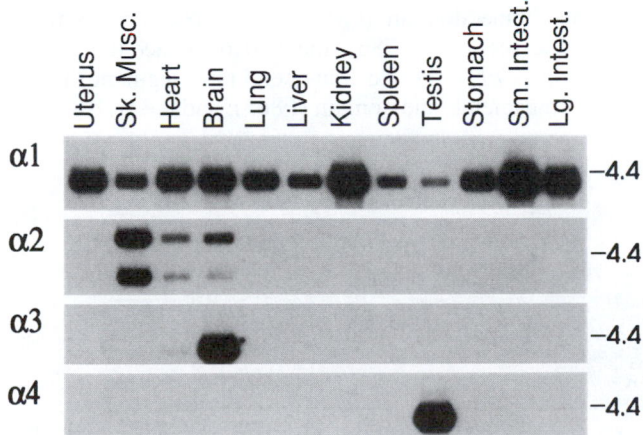

FIGURE 5. Northern blot analysis of the mRNA expression levels of the four α-subunit isoforms in various tissues of rats. (From Shamraj, O. I., and Lingrel, J. B. (1994). A putative fourth Na$^+$,K$^{(+)}$-ATPase alpha-subunit gene is expressed in testis. *Proc. Natl. Acad. Sci. USA* **91,** 12952–12956. Copyright 1994 National Academy of Sciences, U.S.A.)

---

[2] The numbering system used in this chapter is for the sheep α1 isoform.

attached; Kirley *et al.,* 1984) and the phosphorylation site (D369). Immunological data suggest that both the N-terminal and C-terminal regions are on the inside of the cell. Since the ATP site is also located intracellularly, these findings orient the 10 transmembrane regions (see Fig. 4). The N-terminal region contains a large number of both positively and negatively charged residues, including a lysine-rich sequence. It has been argued that this region may act as a cation-selective gate and/or participate in transport via salt-bridge formation. Studies investigating the contribution of the N-terminal domain to cation affinity have shown that this region, which varies greatly between the α isoforms, does reduce the apparent affinity for cations, albeit only by a factor of 2 to 3 (Horisberger *et al.,* 1991). It has also been suggested that the sequence between the phosphorylation site (D369) and the H4 transmembrane region may act as an energy-transduction region (Shull *et al.,* 1985).

Another area of research pertains to both the internal and external cation binding sites. It has been difficult to probe the internal Na$^+$ sites because of the difficulty of controlling internal Na$^+$ and the plethora of compensatory Na$^+$ transport mechanisms. Information about the interactions of external Na$^+$ and K$^+$ with the Na$^+$,K$^+$-ATPase has come from ^{86}Rb uptake in whole cells and [^{3}H]ouabain experiments in fragmented membrane preparations. These techniques have yielded apparent affinities for K$^+$ and Na$^+$ and detailed information about their interactions through the analysis of complex mechanistic models. The α3 isoform has a higher affinity for extracellular K$^+$ and lower affinity for intracellular Na$^+$ than do α1 and α2 (Munzer *et al.,* 1994). In addition, site-directed mutagenesis has been used in an effort to determine the amino acids involved in cation binding. Although the initial hypothesis was that negatively charged residues in the predicted transmembrane region would be involved in cation binding and transport, these studies have suggested that the binding site does not consist of a single residue, but rather is a **binding pocket** composed of a number of residues. The most significant effects have been observed by replacement of the Asp residue at position 804 in sheep α1, where every replacement that has been made resulted in a non-functional enzyme. At residue 808, the only surviving mutation is the conservative replacement of aspartate with glutamate. Other residues (shown in Fig. 6) that affect cation affinity, possibly through indirect mechanisms,

are E327, E779, S775, and D926 (summarized in Van Huysse *et al.,* 1996). This has led to the hypothesis that the H5–H6 region, which includes residues S775, E779, D804, and D808, is directly involved in cation binding and transport. It is conceivable that this entire domain may shift in the membrane, facilitating ion transport, similar to what has been postulated for the S4 segment of the voltage-gated ion channels (Sigworth, 1993).

The most convincing evidence regarding the nature of the cardiac glycoside binding site has arisen from the elegant experiments first performed by Price and Lingrel (1988). These experiments took advantage of the fact that the rodent α1 isoform is one thousand-fold less sensitive to glycoside inhibition than other isoforms. Expressing an "insensitive" enzyme in HeLa cells, whose endogenous enzyme is glycoside sensitive, allowed growth in media with a high concentration of ouabain, a water-soluble cardiac glycoside. If the cells do not express an insensitive transfected enzyme, they die in 0.2 μM ouabain. Expression, however, of mutant enzyme with decreased sensitivity to glycosides will allow the cells to live in 0.2 μM ouabain. The amino acid sequence of the H1–H2 region differs substantially between sheep and rat α1-subunits. Using chimeric proteins and site-directed mutagenesis, it was determined that the presence of the charged residues, R111 and D122, in the rat α1 isoform accounts for the decreased ouabain sensitivity observed for the rat (Price and Lingrel, 1988). Most species have uncharged residues in this position (e.g., sheep α1 has Q111 and N122), and these species are sensitive to cardiac glycosides. It has been hypothesized that, following the binding of the cardiac glycosides, these noncharged residues partition into the membrane, resulting in a conformation of the enzyme from which the rate of glycoside release is very slow, and thus the affinity very high.

In addition to these two residues, D121, also in the H1H2 extracellular loop, has been shown to affect the glycoside affinity. Mutagenesis of other regions has revealed that the glycoside binding pocket spans the majority of the extracellular and transmembrane topology (see Fig. 6) (summarized in Palasis *et al.,* 1996). These diverse regions include the H1 transmembrane domain (C104 and Y108), the H5 transmembrane segment (F786), the H5H6 extracellular "hairpin" loop (L793), the H6 transmembrane segment (T797), the H7 transmembrane domain (F863), and the H7H8 extra-

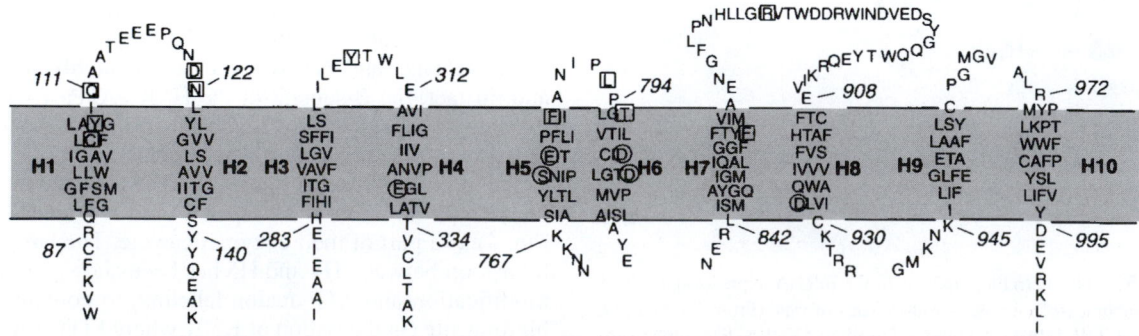

FIGURE 6.   The transmembrane amino acid residues of the sheep α1 sequence. The boxed residues represent amino acids implicated in affecting glycoside sensitivity, and encircled residues are those implicated in altering cation affinity.

cellular loop (R880) (see Fig. 6). It is interesting to note that the H5H6 region has been implicated in glycoside binding and also appears to be central in the cation-transport processes of the enzyme. Ouabain or K^+ is able to protect the enzyme from cleavage under extensive tryptic digestion at the borders of the H5H6 region, suggesting that this region is involved in the conformational transitions of the enzyme (Lutsenko and Kaplan, 1993).

V. Cardiac Glycosides

Plant extracts containing cardiac glycosides have been used for many years as medicines (for their therapeutic effect) and as poisons (for their toxic effect) by diverse groups that include ancient Egyptians, Romans, Chinese, and African tribesmen (Hoffman and Bigger, 1985). Withering (1785) noted the use of foxglove (*Digitalis purpurea*) extracts (which contain cardiac glycosides) in treatment of dropsy (pedal edema secondary to heart failure). For the next hundred years, digitalis (plant extracts of cardiac glycosides, containing primarily digoxin) was used, frequently at excessive doses, for the treatment of a range of diseases. By the early 20th century, digitalis was known to increase the force of myocardial contractility (**positive inotropic effect**) and thus to be useful in the treatment of congestive heart failure. The toxic effects of digitalis, however, are frequent and can be severe, even fatal. In spite of this, digoxin (the drug used clinically) is still one of the most widely used drugs, being the sixth most commonly prescribed drug in the United States in 1995 (Anon., 1996).

The mechanism of action now generally accepted is termed the **sodium pump lag hypothesis** and postulates that therapeutic concentrations of digoxin inhibit a moderate (30–40%) fraction of the membrane Na^+,K^+-ATPase. This, in turn, causes a slight increase in internal Na^+ that has two significant effects: (1) it inhibits the forward operation of the Na^+-Ca^{2+} exchanger (Chapter 19), slowing down the rate at which Ca^{2+} exits the cell (note that if Na^+ concentration is raised above approximately 20 mM, the Na^+-Ca^{2+} exchanger can be forced into reverse mode and drive Ca^{2+} influx); (2) it stimulates the 60–70% fraction of remaining uninhibited pumps, which tend to return the internal Na^+ concentration to homeostasis. Thus, therapeutic levels of digoxin produce transient increases in internal Na^+ in heart muscle, which lead to transient increases in internal Ca^{2+}, which in turn stimulate the release of a larger amount of Ca^{2+} from internal stores in the sarcoplasmic reticulum (Chapter 55). The released calcium interacts with contractile proteins to increase the force of contraction. Toxic concentrations of digoxin lead to extensive (>60%) inhibition of Na^+-K^+ transport, so that restoration of normal diastolic levels of Na^+ and Ca^{2+} is not possible. Digoxin also indirectly produces sympathetic and parasympathetic effects which can result in cardiac arrhythmias. The first step in both the therapeutic and toxic effects is clearly mediated via inhibition of Na^+,K^+-ATPase (see review by Thomas *et al.*, 1989).

Three factors determine the affinity of cardiac glycosides and their analogs for Na^+,K^+-ATPase: (1) the nature (and concentration) of the physiological ligands (Mg^{2+}, Na^+, ATP, P_i, and K^+) bound to the enzyme, (2) the structure of the drug, and (3) the structure of the receptor. In the following scheme, LRD represents the ternary complex of drug (D), receptor (R), and physiological ligand(s) (L). The dissociation constant for the rapid binding of the physiological ligand to receptor is represented by K_L, and the **association** and **dissociation rate** constants for binding of drug to receptor are represented by k_1 and k_{-1}, respectively. **The dissociation constant K_D is equal to k_{-1}/k_1.**

$$R + L \overset{K_L}{\leftrightarrow} LR + D \underset{k_1}{\overset{k_{-1}}{\rightleftharpoons}} LRD$$

For a given cardiac glycoside, the on-rate, and consequently the K_D, is very sensitive to the nature of the ligand(s) present in the medium. Inorganic phosphate, Mg^{2+}, Na^+, and ATP stimulate and K^+ inhibits the rate of binding, leading many to hypothesize that phosphorylation is necessary for ouabain binding. By varying the concentration of the ligand, the dissociation constant for the binding of the ligand to the enzyme can be obtained. The dissociation constant (K_L) for the Mg^{2+} site is 0.2 mM, and when saturated, increases the affinity for ouabain by a factor of 300 over that in presence of buffer alone. The stimulation by Na^+ ($K_L = 14$ mM) is competitively inhibited by K^+ ($K_L = 0.2$ mM). The similarity of these K_L values to those derived from transport studies suggests that these monovalent cation regulatory sites for cardiac glycosides are equivalent to the transport sites. In contrast to the on-rate, the off-rate of a specific cardiac glycoside from a specific isoform of the enzyme is not influenced to the same degree by the presence of physiological regulatory ligands. This could be the result of a **ligand-induced conformational change,** such that the enzyme-drug complex initially formed is transformed into a second, more stable complex whose rate of dissociation is independent of the regulatory ligands present.

Mutation of the phosphorylation site, D369N, produces an enzyme that cannot be phosphorylated, cannot hydrolyze ATP, and cannot translocate Na^+ and K^+. Expression of the sheep $\alpha 1$ mutant in NIH 3T3 cells, whose endogenous Na^+,K^+-ATPase is insensitive, produces a membrane form of the Na^+,K^+-ATPase that can still bind ouabain and can be used as a probe for the physiological ligands (Kuntzweiler *et al.*, 1995). The mutant enzyme still binds ouabain with high affinity, demonstrating that phosphorylation is not necessary for ouabain binding. The stimulatory effect of phosphate on the binding of ouabain, in the presence of Mg^{2+}, is absent in the mutant, showing that the effect of phosphorylation occurs at D369. The fact that ATP and MgATP still bind to the mutant enzyme but inhibit ouabain binding—in contrast to the wild type, in which they stimulate ouabain binding—demonstrates that the Mg-nucleotide site is still present. Three Na^+ and two K^+ still bind to the mutant enzyme with only slight changes in affinity and still affect ouabain binding (in the presence of Mg^{2+}). This indicates that Na^+ and K^+ can induce conformational changes in the enzyme that affect the affinity for ouabain independent of whether the enzyme is phosphorylated. This is consistent with the hypothesis that ouabain prefers to bind to the E_2 conformation. This study also demonstrated that there are distinct sites for Mg^{2+} and MgATP that can be simultaneously occupied.

The structure of the cardiac glycosides also affects their affinity. Classic structure-activity relationship studies of cardiac glycosides, reviewed by Thomas *et al.* (1989), concluded that three primary domains of the cardiac glycosides contribute to their biological activity. The first of these is a **steroid ring** system with a *cis* configuration of A/B and C/D rings, unique to the cardiac glycosides, as well as 3β- and 14β-hydroxyl groups (Fig. 7). The cardiac glycosides also require the presence of an unsaturated β-lactone on carbon 17. The second domain is a five-membered $\alpha\beta$ unsaturated **lactone ring** in the cardenolide family of the cardiac glycosides, and a six-membered diunsaturated lactone ring in the bufadienolides, a related family of compounds (found primarily in toads) with similar pharmacological action, but significantly higher affinity. Reduction of the double bond(s) in the lactone ring reduces the biological activity. The third domain is a **carbohydrate** moiety. Replacement of the 3β-hydroxyl group of the aglycone with a sugar residue increases affinity by dramatically slowing the rate of dissociation of the cardiac glycoside from the receptor. The presence of the rhamnose moiety on ouabain, for example, increases the affinity ten- to thirty-fold. Cardiac glycosides lacking a carbohydrate group are referred to as genins (i.e., digoxigenin is digoxin without the tridigitoxose moiety on C3, but with a 3β-hydroxyl group; see Fig 7).

VI. Summary

The Na⁺,K⁺-ATPase is one of the most vital proteins present in animal cells, as demonstrated by the remarkable degree of conservation across evolutionary lines and the fact that it has been found in every animal cell type studied. This enzyme places a high metabolic burden on the cell, using a substantial portion of the total cellular energy production in maintaining the osmotic balance and ionic gradients across the plasma membrane. This enzyme is crucial for the survival of animal cells, as demonstrated by the fact that transgenic knockout (either $\alpha 1$ or $\alpha 2$ isoform) mice do not survive (Lingrel, personal communication, 1997). In addition to its role in maintaining homeostasis, inhibition of the pump has significant therapeutic relevance by virtue of the secondary increase in Ca^{2+} concentration leading to an increase in cardiac contractility.

The pump activity is due to an integral membrane protein consisting of two subunits, α and β, and belonging to the P-type family of ATPases. This enzyme hydrolyzes ATP in the presence of Na^+ and K^+. The α-subunit has at least four isoforms and the β-subunit three, each of which is differentially expressed. The cation affinities for the activation of ion transport, ATP hydrolysis, and effects on glycoside binding agree well with each other, indicating that the same binding sites are involved in all three activities.

Although the enzyme has been studied in great detail over the past 40 years, there remain many unanswered questions that can be placed into three basic categories: (1) What is the three-dimensional structure of the enzyme? (2) What is the mechanism of cation transport? (3) What are the structural domains that make up ligand-binding sites?

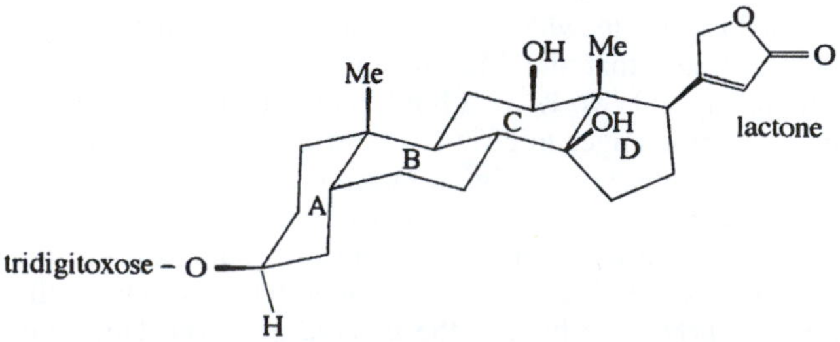

Bibliography

Anon. (1996). Top 200 drugs of 1995. *Pharmacy Times*, April, p. 29.

Bell, M. V., Tonduer, F., and Sargent, J. R. (1977). The activation of sodium-plus-potassium ion-dependent adenosine triphosphatase from marine teleost gills by univalent cations. *Biochem. J.* **163**, 185–187.

Chow, D. C., and Forte, J. G. (1995). Functional significance of the beta-subunit for heterodimeric P-type ATPases. *J. Exp. Biol.* **198**, 1–17.

Forbush, B. (1987a). Na⁺, K⁺, and Rb⁺ Movements in a single turnover of the Na/K pump. *Curr. Top. Membr. Transp.* **28**, 19–39.

Forbush, B. (1987b). Rapid release of ⁴²K or ⁸⁶Rb from two distinct transport sites on the Na,K-Pump in the presence of P_i or vanadate. *J. Biol. Chem.* **262**, 11116–11127.

Garay, R. P., and Garrahan, P. J. (1973). The interaction of sodium and potassium with the sodium pump in red cells. *J. Physiol. (London)* **231**, 297–325.

Glynn, I. M. (1993). All hands to the sodium pump. *J. Physiol.* **462**, 1–30.

Gogarten, J. P., Rausch, T., Bernasconi, P., Kibak, H., and Taiz, L. (1989). Molecular evolution of H⁺-ATPase I. Methanococcus and Solfolobus are monophyletic with respect to eukaryotes and eubacteria. *Z. Naturforsch.* **44**, 641–650.

Hegyvary, C., and Jorgensen, P. L. (1981). Conformational changes of renal sodium plus potassium ion-transport adenosine triphosphatase labeled with fluorescein. *J. Biol. Chem.* **256**, 6296–6303.

Heyse, S., Wuddell, I., Apell, H.-J., and Sturmer, W. (1994). Partial reactions of the Na,K-ATPase: determination of rate constants. *J. Gen. Physiol.* **104**, 197–240.

Hoffman, B. F., and Bigger, J. T. (1985). Digitalis and allied cardiac glycosides. *In* "Pharmacological Basis of Therapeutics" (A. G. Gilman, W. S. Goodman, T. W. Rall, and F. Murad, Eds.), pp. 716–747. Macmillan, New York.

Horisberger, J. D., Jaunin, P., Reuben, M. A., Lasater, L. S., Chow, D. C., Forte, J. G., Sachs, G., Rossier, B. C., and Geering, K. (1991). The H,K-ATPase beta-subunit can act as a surrogate for the beta-subunit of Na,K-pumps. *J. Biol. Chem.* **266**, 19131–19134.

Horowitz, B., Eakle, K. A., Scheiner-Bobis, G., Randolph, G. R., Chen, C. Y., Hitzeman, R. A., and Farley, R. A. (1990). Synthesis and assembly of functional mammalian Na,K-ATPase in yeast. *J. Biol. Chem.* **265**, 4189–4192.

Jorgensen, P. L. (1977). Purification and characterization of (Na⁺ + K⁺)-ATPase. VI. Differential tryptic modification of catalytic functions of the purified enzyme in presence of NaCl and KCl. *Biochim. Biophys. Acta* **466**, 97–108.

Kirley, T. L. (1989). Determination of three disulfide bonds and one free sulfhydryl in the β subunit of (Na,K)-ATPase *J. Biol. Chem.* **264**, 7185–7192.

Kirley, T. L., Wallick, E. T., and Lane, L. K. (1984). The amino acid sequence of the fluorescein isothiocyanate reactive site of lamb and rat

FIGURE 7. The structure of the most widely used cardiac glycoside, digoxin.

kidney Na⁺- and K⁺-dependent ATPase. *Biochem. Biophys. Res. Commun.* **125**, 767–773.

Kuntzweiler, T. A., Wallick, E. T., Johnson, C. L., and Lingrel, J. B. (1995). Amino acid replacement of D369 in the sheep $a1$ isoform eliminates ATP and phosphate stimulation of [H³]ouabain binding to the Na⁺,K⁺-ATPase without altering the cation binding properties of the enzyme. *J. Biol. Chem.* **270**, 16206–16212.

Lane, L. K., Copenhaver, G. H., Lindenmayer, G. E., and Schwartz, A. (1973). Purification and characterization of and [³H]ouabain binding to the transport adenosine triphosphatase from outer medulla of canine kidney. *J. Biol. Chem.* **248**, 7197–7200.

Lauger, P. (1991). "Electrogenic Ion Pumps," Vol. 5. Sinauer Associates, Sunderland, MA.

Lindenmayer, G. E., Schwartz, A., and Thompson, H. K., Jr. (1974). A kinetic description for sodium and potassium effects on (Na⁺ + K⁺)-adenosine triphosphatase: a model for a two-nonequivalent site potassium activation and an analysis of multiequivalent site models for sodium activation. *J. Physiol. (London)* **236**, 1–28.

Lingrel, J. B., Orlowski, J., Shull, M. M., and Price, E. M. (1990). Molecular genetics of Na,K-ATPase. *In* "Progress in Nucleic Acids Research and Molecular Biology," Vol. 38, pp. 37–89. Academic Press, London.

Lutsenko, S., and Kaplan, J. H. (1993). An essential role for the extracellular domain of the Na,K-ATPase β-subunit in cation occlusion. *Biochemistry* **32**, 6737–6743.

McDonough, A. A., Geering, K., and Farley, R. A. (1990). The sodium pump needs its β subunit. *FASEB J* **4**, 1598–1605.

Munzer, J. S., Daly, S. E., Jewell-Motz, E. A., Lingrel, J. B., and Blostein, R. (1994). Tissue- and isoform-specific kinetic behavior of the Na,K-ATPase. *J. Biol. Chem.* **269**, 16668–16676.

Palasis, M., Kuntzweiler, T. A., Arguello, J. M., and Lingrel, J. B. (1996). Ouabain interactions with the H5–H6 hairpin of the Na,K-ATPase reveal a possible inhibition mechanism via the cation binding domain. *J. Biol. Chem.* **271**, 14176–14182.

Price, E. M., and Lingrel, J. B. (1988). Structure-function relationships in the Na,K-ATPase α-subunit: site-directed mutagenesis of glutamine-111 to arginine and asparagine-122 to aspartic acid generates a ouabain-resistant enzyme. *Biochemistry* **27**, 8400–8408.

Sachs, J. R. (1977). Inhibition of the Na–K pump by external sodium. *J. Physiol. (London)* **264**, 449–470.

Schatzmann, H. J. (1953). Herzglykoside als Hemmstoffe für den activen Kalium und Natrium-transport durch die Erythrocytenmembran. *Helv. Physiol. Pharmacol. Acta* **11**, 346–354.

Shamraj, O. I., and Lingrel, J. B. (1994). A putative fourth Na⁺,K⁽⁺⁾-ATPase alpha-subunit gene is expressed in testis. *Proc. Natl. Acad. Sci. USA* **91**, 12952–12956.

Shull, G. E., Schwartz, A., and Lingrel, J. B. (1985). Amino-acid sequence of the catalytic subunit of the (Na⁺,K⁺)-ATPase deduced from a complementary DNA. *Nature* **316**, 691–695.

Shull, G. E., Greeb, J., and Lingrel, J. B. (1986a). Molecular cloning of three distinct forms of the Na,K-ATPase alpha subunit from rat brain. *Biochemistry* **25**, 8129–8132.

Shull, G. E., Lane, L. K., and Lingrel, J. B. (1986b). Amino-acid sequence of the β-subunit of the (Na⁺,K⁺)-ATPase deduced from a cDNA. *Nature* **321**, 429–431.

Shull, G. E., Young, R. M., Greeb, J., and Lingrel, J. B. (1988). Amino acid sequences of the α and β subunits of the Na,K-ATPase. *Prog. Clin. Biol. Res.* **268A**, 3–18.

Sigworth, F. J. (1993). Voltage gating of ion channels. *Quart. Rev. Biophys.* **27**, 1–40.

Skou, J. C. (1957). The influence of some cations on an adenosine triphosphatase from peripheral nerves. *Biochim. Biophys. Acta* **23**, 394–401.

Sweadner, K. J. (1989). Isozymes of the Na⁺,K⁺-ATPase. *Biochim. Biophys. Acta* **988**, 185–220.

Tepperman, K., Millette, L. A., Johnson, C. L., Jewell-Motz, E. A., Lingrel, J. B., and Wallick, E. T. (1997). Mutational analysis of glutamate 327 of Na⁺,K⁺-ATPase reveals stimulation of ⁸⁶Rb uptake by external K. *Am. J. Physiol.*, **273 (cell physiol. 42)**, C2065–C2079.

Thomas, R., Gray, P., and Andrews, J. (1989). Digitalis: its mode of action, receptor and structure-activity relationship. *In* "Advances in Drug Research" (B. Testa, Ed.), Vol. 19, pp. 311–562. Academic Press, New York.

Van Huysse, J. W., Kuntzweiler, T. A., and Lingrel, J. B. (1996). Critical effects on catalytic function produced by amino acid substitutions at Asp804 and Asp808 of the $\alpha1$ isoform of Na,K-ATPase. *FEBS Lett.* **389**, 179–185.

Withering, W. (1785). "An Account of the Foxglove and Some of Its Medicinal Uses: With Practical Remarks on Dropsy and Other Diseases" C. G. J. and J. Robinson, London.

Istvan Edes and Evangelia G. Kranias

18

Ca^{2+}-ATPases

I. Introduction

An important role of Ca^{2+} in muscle contraction was first indicated a century ago by Ringer (1883), who demonstrated that the frog's heart would not contract in the absence of extracellular Ca^{2+}. Since then, it has been shown that Ca^{2+} is a physiological regulator for the contractile proteins and several other enzymes and processes in muscle. This chapter will focus on the role of the various Ca^{2+}-ATPases in maintaining Ca^{2+} homeostasis in the cell, with special emphasis on the **sarcoplasmic reticular** (SR) Ca^{2+}-ATPase(s), which is the primary regulator of the Ca^{2+} levels and thus contractility in muscle.

During the cardiac action potential, Ca^{2+} enters the cell via Ca^{2+} channels, which also act as dihydropyridine receptors (Fig. 1). This Ca^{2+} can either activate the myofilaments directly or produce the release of additional Ca^{2+} from the SR. The SR Ca^{2+}-release channel in cardiac and skeletal muscle also acts as a ryanodine receptor and spans the gap between the transverse tubule and the SR ("foot" protein). Furthermore, it has been shown that the outer cell membrane Ca^{2+} channel is located close to the SR Ca^{2+} channel. Thus, the excitation-contraction coupling apparently involves the sarcolemmal Ca^{2+} channel and the SR Ca^{2+}-release channel with the Ca^{2+} current through the sarcolemmal channel being responsible for the initiation of Ca^{2+} release from the SR (Fig. 1). In skeletal muscle, the sarcolemmal membrane depolarization itself apparently is responsible for the induction of SR Ca^{2+} release. The relative importance of release from the SR in activation of the cardiac muscle contraction varies from preparation to preparation, but in the heart of mammals it usually accounts for 40–70% of the Ca^{2+} required (Bers, 1991).

The rising cytosolic Ca^{2+} concentration induces contraction through binding to troponin C, which activates a chain of conformational changes, allowing the thin and thick filaments to interact. Subsequently, Ca^{2+} is dissociated from troponin C and is rapidly removed from the cytosol by various systems, resulting in relaxation. At least three processes are responsible for the removal of Ca^{2+} to end contraction (Fig. 1): (a) the SR Ca^{2+} pump, which actively translocates

Ca^{2+} at the cost of ATP into the SR system; this is thought to be the most important process in mediating relaxation; (b) the Na$^+$-Ca^{2+} exchanger, which transports Ca^{2+} out of the cell during diastole; and (c) the sarcolemmal Ca^{2+}-ATPase, which also extrudes Ca^{2+} from the cell.

The SR is a tubular network, which serves as a sink for Ca^{2+} ions during relaxation and as a Ca^{2+} source during contraction. In cardiac muscle, about 60–70% of the intracellular Ca^{2+} released during systole is taken up by the SR (Bers, 1991), and the remaining amount is extruded from the cell by the Na$^+$-Ca^{2+} exchanger and the sarcolemmal Ca^{2+}-ATPase.

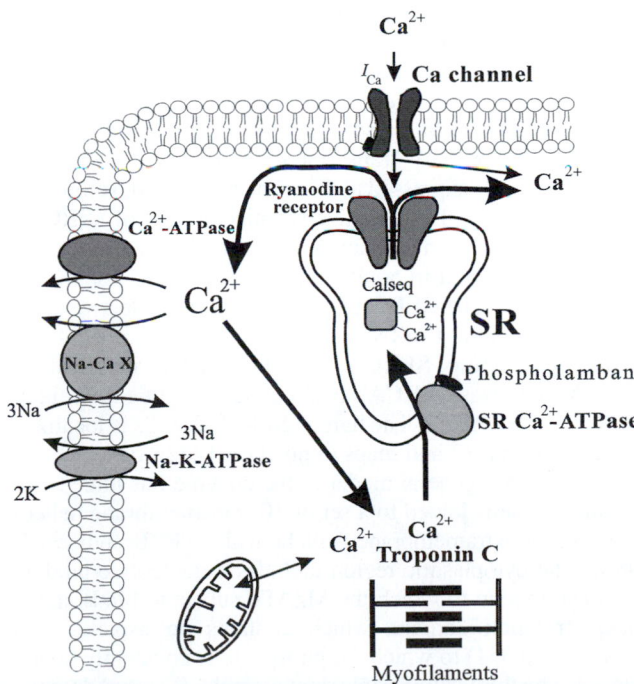

FIGURE 1. Schematic diagram of Ca^{2+} fluxes in cardiac cell. Na-CaX, Na$^+$-Ca^{2+} exchanger; Calseq., calsequestrin; I_{Ca}, slow inward Ca^{2+} current; SR, sarcoplasmic reticulum.

TABLE 1   Structure and Distribution of the
Sarcoplasmic Reticular Ca²⁺-ATPase (SERCA) Isoforms

Gene	Splice	Tissue
SERCA1[a]	a[b]	Adult fast skeletal muscle
SERCA1	b	Neonatal fast skeletal muscle
SERCA2	a	Cardiac/slow skeletal muscle
SERCA2	b	Smooth muscle/nonmuscle
SERCA3	?	Various tissues

[a]The SERCA numbers identify different gene products.
[b]The letters a and b indicate spliced isoforms.

The SR in both skeletal and cardiac muscles contains an acidic protein, calsequestrin (Fig.1), which binds 40–50 mol of Ca²⁺ per mol of protein. The binding and release of Ca²⁺ by calsequestrin is believed to be an integral step of excitation-contraction coupling, but the details of this process are still not fully understood. Mitochondria can also accumulate large amounts of Ca²⁺ under pathological conditions (ischemia, Ca²⁺ overload, etc.) (Bers, 1991).

II. Sarcoplasmic Reticular (SR) Ca²⁺-ATPase

A. Properties of SR Ca²⁺-ATPase

The major protein in the SR membrane is the Ca²⁺-ATPase (M_r 100 000), representing about 40% of the total protein in cardiac SR. The cardiac SR Ca²⁺-ATPase can create intraluminal Ca²⁺ concentrations of 5–10 mM. Recombinant DNA studies revealed that the **SR or endoplasmic reticulum (ER) Ca²⁺-ATPase family** (SERCA) is the product of at least three alternatively spliced genes, producing a minimum of five different proteins (Burk *et al.*, 1989) (Table 1). SERCA1 is expressed in fast skeletal muscle, and alternative splicing of the 3' end of the primary transcript gives rise to two mRNA forms, which are expressed at different stages of development (Brandl *et al.*, 1986). Alternatively, spliced forms of SERCA2 have been detected in cardiac muscle and slow skeletal muscle (SERCA2a) and in adult smooth muscle and nonmuscle tissues (SERCA2b). SERCA3 is expressed in a selective manner, with the highest mRNA levels in intestine, spleen, lung, uterus, and brain. SERCA2 is about 85% identical to SERCA1, whereas SERCA3 is about 75% identical to either SERCA1 or SERCA2. The human SERCA2 gene is localized on chromosome 12 and maps to position 12q23-q24.1.

The proposed general model of the enzyme has three cytoplasmic domains joined to a set of 10 transmembrane helices by a narrow extramembrane pentahelical stalk (Brandl *et al.*, 1986). The cytoplasmic region includes a nucleotide-binding site, or a domain to which the MgATP substrate binds, and a phosphorylation domain, which contains an aspartic acid residue (Asp 351) to which the phosphate is covalently bound (Fig. 2). The third cytoplasmic domain is the β-strand domain, whose function is still not fully understood. In skeletal muscle, 2 mol of Ca²⁺ is transported per mole of ATP hydrolyzed. In cardiac muscle, a similar stoichiometry is expected, but this

ratio has been generally found to be lower (0.4–1.0 mol Ca²⁺/mol ATP). Ca²⁺ has been shown to bind to a region involving several of the membrane-spanning α-helices (M4, M5, and M6) on the cytoplasmic side. The important amino acid residues constituting the Ca²⁺ binding sites are Glu 309 on the M4 transmembrane segment; Glu 771 on the M5 transmembrane segment; and Asn 796, Thr 799, and Asp 800 on the M6 transmembrane segment. It has been proposed that Glu 771 and Thr 799 are associated with the first Ca²⁺ binding site (site 1), while Glu 309 and Asn 796 are associated with the binding of the second Ca²⁺ ion (site 2). Asp 800 was suggested to donate ligands to both Ca²⁺ binding sites (MacLennan *et al.*, 1997; Andersen and Vilsen, 1998). During the Ca²⁺ transport cycle, the enzyme undergoes a transition from a high-affinity state to a low-affinity state for Ca²⁺, and the ions are translocated from the binding sites into the lumen of the SR (Fig. 2). This reaction pathway is characterized by the covalent phosphorylated Ca²⁺-ATPase form (E_1~P), when the energy of ATP is transferred to an acylphosphoprotein intermediate (Fig. 3). E_1~P rapidly becomes E_2–P when the energy contained originally in the acylphosphoprotein is transduced into the translocation of bound Ca²⁺ into the SR ("mario-nette" model, see Fig. 2). Subsequently, the acid-labile intermediate (E_2–P) decomposes to enzyme (E_2) and inorganic phosphate. In this model it is assumed that the phosphorylation of the enzyme at Asp 351 triggers a series of conformational changes in which the high affinity Ca²⁺ binding sites are disrupted, access to the

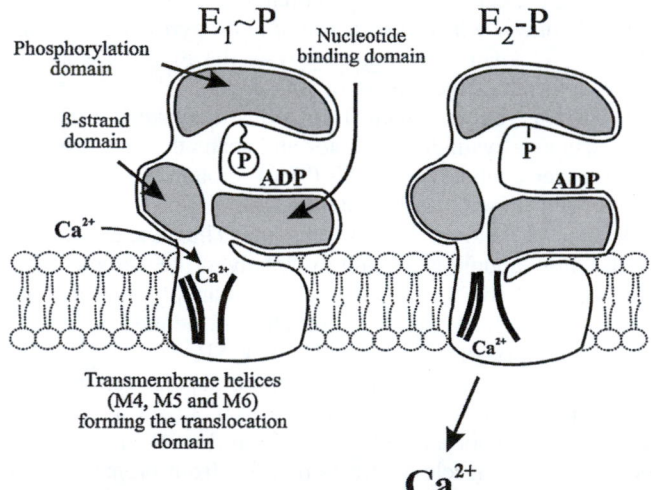

FIGURE 2.   Model illustrating Ca²⁺ translocation by SERCA-type Ca²⁺ pumps. In E_1~P conformation, Ca²⁺ binds to the high-affinity binding sites in the cytosol. The energy of the hydrolyzed ATP triggers a series of conformational changes and transforms the E_1~P intermediate to the E_2–P intermediate. These conformational changes are directly coupled to alterations in the orientation of the transmembrane regions, leading to Ca²⁺ release into the lumen of the sarcoplasmic reticulum.

sites by cytoplasmic Ca^{2+} is closed off, and access to the sites from the luminal surface is gained (E_2 conformation) (MacLennan *et al.*, 1992; MacLennan *et al.*, 1997).

The Ca^{2+}-free form of the enzyme exists in two different conformational states: one with low affinity for Ca^{2+} (E_2) and one with high affinity for Ca^{2+} (E_1) (Fig. 3). The conversion of E_2 to E_1 is proposed to be the rate-limiting step in the cycle. Thapsigargin (a plant sesquiterpene lactone) has been shown to interact specifically with the M3 transmembrane segment of the E_2 form of all members of the SR Ca^{2+}-ATPase family and to inhibit enzyme activity even at subnanomolar concentrations (Lytton *et al.*, 1991). The E_1 form of the enzyme has been stabilized and crystallized in the presence of lanthanide (La^{3+}) or Ca^{2+} ions (Dux *et al.*, 1985). On the other hand, vanadate ions in the absence of Ca^{2+} induced the formation of E_2-type crystals. E_1-type crystals consist of single chains of Ca^{2+}-ATPase molecules evenly spaced on the surface of the SR. E_2-type crystals consist of dimer chains of ATPase molecules forming an oblique surface lattice. The transition between E_1 and E_2 conformation may involve a shift in the monomer-oligomer equilibrium (Dux *et al.*, 1985). Recently, it has been shown that the cardiac SR Ca^{2+}-ATPase (SERCA2) can be phosphorylated by the Ca^{2+}/CAM–dependent protein kinase at Ser 38 (Toyofuku *et al.*, 1994a). However, the physiological role of this phosphorylation is still not fully understood.

B. Regulation of SR Ca²⁺-ATPase by Phospholamban

1. Structure of Phospholamban

In cardiac muscle, slow-twitch skeletal muscle, and smooth muscle, the SR contains the low-molecular-weight protein phospholamban, which can be phosphorylated by various protein kinases. The phosphorylation and dephosphorylation of phospholamban, which comprises 3–4% of the SR membrane protein, regulate the Ca^{2+}-ATPase activity in the SR

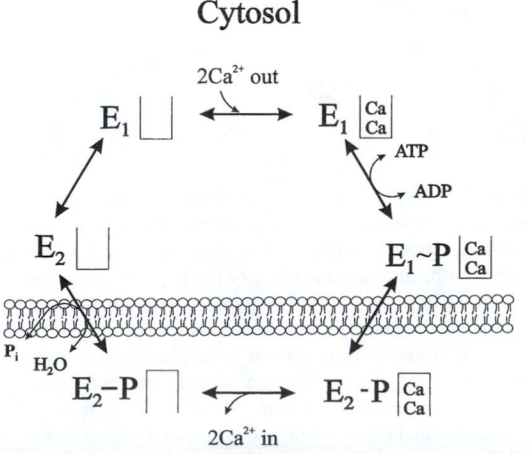

Cytosol

Lumen of SR

FIGURE 3. Reaction scheme of sarcoplasmic reticular Ca^{2+}-ATPase.

membrane. However, the exact stoichiometry of phospholamban to SR Ca^{2+}-ATPase is not known. In early studies, a stoichiometric relationship of 1 or 2 mol phospholamban per 1 mol of Ca^{2+}-ATPase was proposed for cardiac SR membranes (Colyer and Wang, 1991). In reconstituted systems, a molar ratio of 3:1 of phospholamban-Ca^{2+}-ATPase was necessary to obtain the maximal regulatory effects (Reddy *et al.*, 1995). Moreover, the "functional unit" of phospholamban is not currently clear. Experiments using electron paramagnetic resonance spectroscopy revealed that phospholamban is present primarily as pentamers in the SR membrane (Cornea *et al.*, 1996). In the dephosphorylated form, a substantial fraction of phospholamban monomers (20%) exists, but upon phosphorylation, phospholamban appears to form mainly pentamers, which is due to changes in the isoelectric point (from 10 to 6.7) of the protein (Simmerman and Jones, 1998).

The complete amino acid sequence of phospholamban has been determined for various tissues and species. There is currently no evidence for the existence of any isoforms for this protein and the phospholamban gene has been mapped to human chromosome 6 (Fujii *et al.*, 1991). The calculated molecular weight of phospholamban is 6080 (Fujii *et al.*, 1987), and the protein has been proposed to contain two major domains (Fig. 4): a hydrophilic domain (domain I) with two unique phosphorylatable sites (Ser 16 and Thr 17), and a hydrophobic C-terminal domain (domain II) anchored into the SR membrane. The hydrophylic domain (amino acids 1–30) has been further divided into two subdomains: domain Ia (amino acids 1–20) and Ib (amino acids 21–30). Domain Ia has a net positive charge in the dephosphorylated form and consists of an α-helix followed by a Pro residue at position 21 (stalk region). Domain Ib has been suggested to be relatively unstructured (Simmerman and Jones, 1998; Mortishire-Smith *et. al.*, 1995). The hydrophobic domain (amino acids 31–52) forms an α-helix in the SR membrane (Fig. 4).

Phospholamban migrates as a 24- to 28-kDa pentamer on SDS gels and dissociates into dimers and monomers upon boiling in SDS before electrophoresis. Spontaneous aggregation of phospholamban into pentamers was also observed upon expression of this protein in bacteria or in mammalian cells. Site-specific mutagenesis experiments identified Cys (Cys 36, Cys 41, and Cys 46), Leu (Leu 37, Leu 44, and Leu 51), and Ile (Ile 40 and Ile 47) residues in the hydrophobic transmembrane domain as essential amino acids for phospholamban pentamer formation (Fujii *et al.*, 1989; Simmerman *et al.*, 1996). The leucine and isoleucine amino acids are suggested to form five zippers in the membrane that stabilize the pentameric form of the protein with a central pore (Fig. 5), defined by the surface of the hydrophobic amino acids (Simmerman *et al.*, 1996). Based on this pentameric self-association of phospholamban, a channel function for this protein has been proposed (Wegener *et al.*, 1986).

Monoclonal antibodies raised against phospholamban stimulate SR Ca^{2+} uptake (Morris *et al.*, 1991). Furthermore, removal of phospholamban from the SR or uncoupling phospholamban from the Ca^{2+}-ATPase (using detergents, high-ionic-strength solutions, or polyanions such as heparin sulfate) markedly increases the affinity of the SR Ca^{2+} pump for Ca^{2+}. These findings suggest that the dephosphorylated form of phospholamban is an inhibitor of the SR Ca^{2+}-ATPase.

Cytosol

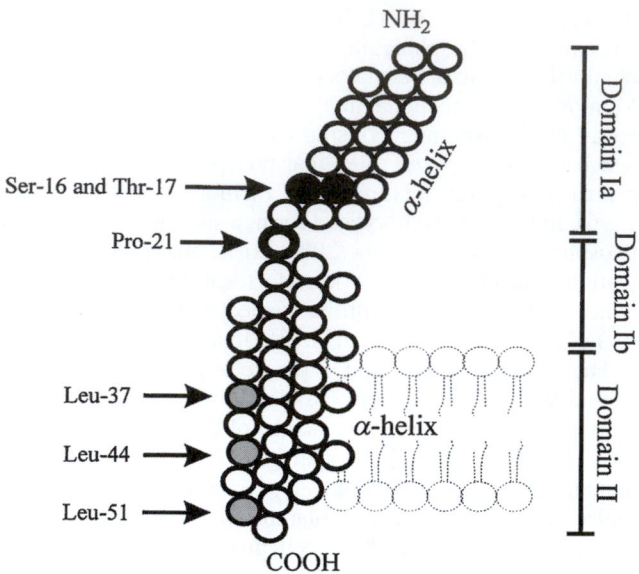

FIGURE 4. Molecular model of the structure of phospholamban. The cytoplasmic α-helix (domain Ia, residues 8–20) is interrupted by Pro 21 (heavy circle). Residues 22–32 (domain Ib) are relatively unstructured and may interconvert between transient conformations; residues 33–52 constitute the transmembrane domain II (α-helix). Ser 16 and Thr 17 (black circles) are the adjacent phosphorylation sites. The shaded circles indicate the leucines (Leu 37, Leu 44, and Leu 51), which are important for the phospholamban subunit interactions (pentamer formation).

The region of phospholamban interacting with the Ca^{2+}-ATPase may involve amino acids 2–18 (Morris *et al.,* 1991; Toyofuku *et al.,* 1994b). Based on these reports, the simplest model for the interaction between the phospholamban cytoplasmic domain and the SR Ca^{2+}-ATPase is one in which the highly positively charged region of phospholamban (residues 7–16) interacts directly with a negatively charged region on the surface of the Ca^{2+}-ATPase (Lys-Asp-Asp-Lys-Pro-Val 402) to modulate the inhibitory interactions between the two proteins (Fig. 6) (Toyofuku *et al.,* 1994c; Kimura *et al.,* 1997; MacLennan *et al.,* 1998). This association is disrupted by phosphorylation of Ser 16 or Thr 17 in phospholamban, because the positive charges of the phospholamban cytosolic domain are partially neutralized by the phosphate moiety in this vicinity. Phosphorylation of phospholamban by the cAMP-dependent protein kinase at Ser 16 is associated with local unwinding of the α-helix at position 12–16 resulting in conformational changes in the recognition unit of the protein (Mortishire-Smith *et al.,* 1995).

Interestingly, phospholamban peptides, corresponding to the hydrophobic membrane-spanning domain, also affect Ca^{2+}-ATPase activity by lowering its affinity for Ca^{2+} (Sasaki *et al.,* 1992; Kimura *et al.,* 1996). The importance of the membrane-spanning region of phospholamban in inhibiting SR Ca^{2+}-ATPase activity has been clarified recently by Kimura *et al.,* (1997). It was shown that substitution of the pentamer-stabilizing residues (Leu 37, Leu 44, Leu 51, Ile 40, and Ile 47) in the membrane-spanning region (domain II) by Ala resulted in monomeric mutants, which were more effective inhibitors of the SR Ca^{2+}-ATPase activity than wild-type phospholamban. These phospholamban monomeric mutants were called "supershifters" because they decreased the apparent affinity of SR Ca^{2+}-ATPase more effectively than wild-

This "depression hypothesis" has been confirmed by studies using purified Ca^{2+}-ATPase and purified or recombinant phospholamban in reconstituted systems. Inclusion of phospholamban resulted in inhibition of the SR Ca^{2+}-ATPase activity in reconstituted vesicles or cells (Kim *et al.,* 1990; Reddy *et al.,* 1995; Reddy *et al.,* 1996). Cyclic AMP phosphorylation of phospholamban reversed its inhibitory effect on the Ca^{2+} pump. Recently, the inhibitory role of phospholamban on SR and cardiac function has been directly confirmed using transgenic animal models. Overexpression of the protein (phospholamban-overexpressing mice) was associated with inhibition of SR Ca^{2+} transport, Ca^{2+} transient, and depression of basal left ventricular function (Kadambi *et al.,* 1996). On the other hand, partial (phospholamban-heterozygous mice) or complete ablation of the protein (phospholamban-deficient mice) in mouse models was associated with increases in SR Ca^{2+} transport and cardiac function (Luo *et al.,* 1994; Luo *et al.,* 1996). Actually, a close linear correlation between the levels of phospholamban and cardiac contractile parameters was observed, indicating that phospholamban is a prominent regulator of myocardial contractility. These findings suggest that changes in the level of this protein will result in parallel changes in SR function and cardiac contraction.

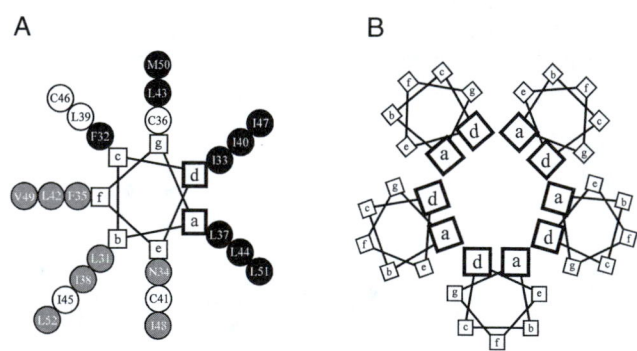

FIGURE 5. Heptad repeat model of the transmembrane domain of phospholamban monomer (A) and pentamer (B). (A) Residues 31–52 of monomeric phospholamban are configured as a 3.5 residues/360° turn helix with positions from a to g of the heptad repeat-squared. The Leu and Ile residues constituting zippers are localized to positions a and d, respectively. Darkly shaded circles represent mutations that enhance the inhibitory function of phospholamban by enhanced monomer formation (destabilization of pentamer structure). Lightly shaded circles represent mutations that reduce inhibitory function. These "loss-of-function" mutants are all located on the exterior face of each helix in a pentamer (positions b, e, and f). (B) The phospholamban pentamer model shows the interaction between monomers at positions a and d to form the leucine zipper (heavy squares). (Adapted from Simmerman *et al.,* (1996) and MacLennan *et al.,* (1998).)

type phospholamban. Thus, it was proposed that monomeric phospholamban is the active form, which is involved in the interaction with SR Ca²⁺-ATPase. Furthermore, the scanning alanine-mutagenesis studies (Kimura *et al.*, 1997) have identified the amino acid residues in the transmembrane domain of phospholamban (Leu 31, Asn 34, Phe 35, Ile 38, Leu 42, Ile 48, Val 49, and Leu 52), which are associated with loss of function. These amino acids are located on the exterior face of each helix in the pentameric assembly of phospholamban (opposite from the pentamer-stabilizing face) (see Fig. 5).

A schematic representation of interaction of phospholamban with SR Ca²⁺-ATPase is shown in Fig. 6. It has been proposed that the phospholamban monomer is the active species for interaction with the SR Ca²⁺-ATPase and the pentamers are regarded as functionally inactive forms of phospholamban (Kimura *et al.*, 1997; MacLennan *et al.*, 1998). Phosphorylation of phospholamban monomers promotes association into inactive pentamers. Thus, two important steps for SR Ca²⁺-ATPase inhibition have been suggested: (1) dissociation of monomeric phospholamban from dephosphorylated pentamers (K_{d1}); and (2) binding of phospholamban monomers to the SR Ca²⁺-ATPase (K_{d2}). These dissociation constants (K_{d1} and K_{d2}) will control both the concentration of phospholamban monomers and the concentration of units in which monomers are associated with the SR Ca²⁺-ATPase (Kimura *et al.*, 1997; and MacLennan *et al.*, 1998). There are at least two interaction sites between phospholamban and the SR Ca²⁺-ATPase (Fig. 6): one in the cytoplasmic domains of the two proteins and another one within the transmembrane sequences. The interaction between the hydrophobic membrane-spanning regions is associated with inhibition of the apparent affinity of SR Ca²⁺ ATPase for Ca²⁺ (K_{Ca}). The interaction between the cytosolic phospholamban domain Ia and the SR Ca²⁺-ATPase modulates the inhibitory interaction in the transmembrane region (domain II) through long-range coupling. Disruption of the cytosolic interactions (domain Ia) by phosphorylation of phospholamban or binding of a phospholamban antibody re-

sults in disruption of the inhibitory intramembrane interactions. However, resolution of the exact molecular mechanism by which phospholamban inhibits the SR Ca²⁺-ATPase Ca²⁺ affinity and the concomitant regulation of SR Ca²⁺ transport will have to await the development of a new methodology that allows detection of protein-protein interactions in a membrane environment.

2. In *Vitro* Studies on Regulation of SR Ca²⁺-ATPase

In the early 1970s, it was suggested that the effects of various catecholamines on cardiac function may be partly attributed to phosphorylation of the SR by cAMP-dependent protein kinase(s). It soon became clear that the substrate for the protein kinase (PK) was not the SR Ca²⁺-ATPase but phospholamban. Various other high- and low-molecular-weight SR proteins were also identified as minor substrates for cAMP-dependent PK, but only the changes in the phosphorylation of phospholamban were associated with functional alterations of the cardiac SR.

Cardiac SR membranes contain an endogenous cAMP-dependent PK and a Ca²⁺-CaM–dependent PK that have been shown to phosphorylate phospholamban independently of each other (Kranias,1985a). Phosphorylation by cAMP-dependent PK occurred on Ser 16, whereas Ca²⁺-CaM–dependent PK catalyzed exclusively the phosphorylation of Thr 17 (Simmerman *et al.*, 1986). Phosphorylation by either kinase was shown to result in stimulation of the SR Ca²⁺-ATPase activity and the initial rates of SR Ca²⁺ transport. Stimulation was associated with an increase in the apparent affinity of the SR Ca²⁺-ATPase for Ca²⁺ (K_{Ca}).

In vitro, phospholamban is phosphorylated by two additional PKs: PK-C and a cGMP-dependent PK. Protein kinase C (Ca²⁺/phospholipid–dependent PK) phosphorylated the protein at a site distinct from those phosphorylated by either cAMP-dependent PK or Ca²⁺-CaM–dependent PK (Movsesian *et al.*,1984). Phosphorylation stimulated the SR Ca²⁺-ATPase activity and it was postulated that this activity played a role in the action of agents known to stimulate phosphoinositide (PI) hydrolysis, since one product of PI hydrolysis, diacylglycerol, is an activator of PK-C. Cyclic GMP–dependent PK was shown to phosphorylate phospholamban on the same residue (Ser 16) as that phosphorylated by cAMP-dependent PK (Raeymakers *et al.*, 1988). This phosphorylation stimulated cardiac SR Ca²⁺ transport, similar to the effects of cAMP-dependent PK. Furthermore, the stimulatory effects on Ca²⁺ transport, mediated by cGMP-dependent phosphorylation of phospholamban, were also observed in smooth muscle, and this may be of particular interest because some vasodilators act by increasing cGMP levels in vascular smooth muscle.

The presence of endogenous PKs in cardiac SR necessitates the presence of phosphoprotein phosphatase(s) for the reversible regulation of the Ca²⁺ pump. Protein phosphatases have been generally classified into type 1 and type 2. Type 1 phosphatase is inhibited by nanomolar concentrations of the protein inhibitor-1 and inhibitor-2, whereas type 2 phosphatases are unaffected. In heart muscle, both types of phosphatase have been reported to be present and both can dephosphorylate phospholamban (Kranias and Di Salvo, 1986;

Cytosol

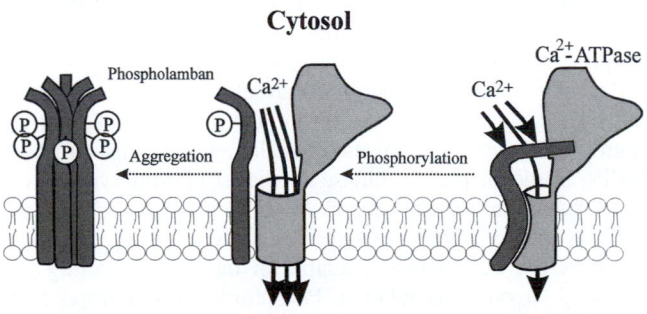

Lumen of SR

FIGURE 6. Model for regulation of SR Ca²⁺-ATPase by phosphorylated and nonphosphorylated phospholamban. Phosphorylation of phospholamban disrupts the interaction between the two proteins so that the inhibition of the Ca²⁺-ATPase is relieved. Note that both the cytosolic domain and the membrane-spanning region of phospholamban are involved in the phosphorylation-mediated conformational change to relieve the inhibition. Phosphorylation of phospholamban monomers promotes association into inactive phosphorylated pentamers.

MacDougall *et al.*, 1991). A type 1 protein phosphatase has been shown to be associated with cardiac SR membranes, and this activity could catalyze the dephosphorylation of both the cAMP-dependent PK and the Ca^{2+}-CaM–dependent PK phosphorylated sites (Ser 16 and Thr 17) on phospholamban (Steenaart *et al.*, 1992). Dephosphorylation was associated with a reduction in the stimulatory effects of PKs on the SR Ca^{2+} pump (Kranias, 1985b).

The SR phosphatase is similar to the skeletal muscle protein phosphatase I_G (PPI_G), which is composed of a catalytic (C) subunit and a G subunit (MacDougall *et al.*, 1991). The G subunit may become phosphorylated by cAMP-dependent PK, and this causes release of the C subunit from the SR vesicles or glycogen particles into the cytosol, rendering phospholamban in the phosphorylated state and thus capable of stimulating SR Ca^{2+} transport. *In vivo* studies have also shown that the phosphatase activity associated with cardiac SR membranes may be regulated by cAMP-dependent processes. β-Adrenergic stimulation of intact beating hearts was associated with inhibition of the SR phosphatase activity, and this inhibition correlated with increases in the phosphorylation status of inhibitor-1 (Neumann *et al.*,1991). Thus, regulation of the SR phosphatase activity may be one of the mechanisms by which cells achieve amplification of the cAMP-dependent processes.

3. In Vivo Studies on Regulation of SR Ca^{2+}-ATPase

The phosphorylation of SR proteins and their regulatory effects on the SR Ca^{2+}-ATPase activity have been studied in perfused hearts from various animal species whose ATP pool was labeled with [^{32}P]orthophosphate. Microsomal fractions enriched in SR were prepared from hearts freeze-clamped during stimulation with different agonists (catecholamines, forskolin, phosphodiesterase inhibitors, phorbol esters) and analyzed by gel electrophoresis and autoradiography for ^{32}P incorporation. β-adrenergic agonist (isoproterenol) stimulation of the perfused hearts produced an increase in ^{32}P incorporation into phospholamban (Kranias and Solaro, 1982; Lindemann *et al.*, 1983). The stimulation of ^{32}P incorporation into phospholamban was associated with an increased rate of Ca^{2+} uptake into SR membrane vesicles and an increased SR Ca^{2+}-ATPase activity (Lindemann *et al.*, 1983; Kranias *et al.*, 1985).

These biochemical changes were associated with increases in left ventricular functional parameters (contractility and relaxation). The *in vivo* phosphorylation of phospholamban was specific only for inotropic agents that increased the cAMP content of the myocardium (β-adrenergic agonists, forskolin, and phosphodiesterase inhibitors). On the other hand, positive inotropic interventions, which increased the intracellular Ca^{2+} level by cAMP-independent mechanisms (α-adrenergic agonists, ouabain, and elevated [Ca^{2+}]), failed to stimulate phospholamban phosphorylation and relaxation. Calmodulin inhibitors (fluphenazine) attenuated the isoproterenol-induced phosphorylation of phospholamban (Lindemann and Watanabe, 1985a), and it was shown that at steady-state isoproterenol exposure, phospholamban contains equimolar amounts of phosphoserine (pSer 16) and phosphothreonine (pThr 17). Phosphorylation of Ser 16 however, correlated most closely with changes in cardiac function in beating hearts (Talosi *et al.*, 1993). Based on these results and recent

findings in transgenic animals (Luo *et al.*, 1997) it is proposed that prevention of Ser 16 phosphorylation (Ser 16 → Ala mutation) results in attenuation of the β-adrenergic response in mammalian hearts, and that phosphorylation of Ser 16 is a prerequisite for Thr 17 phosphorylation.

The muscarinic agonist acetylcholine attenuated the increases in cAMP levels, phosphorylation of phospholamban, and the SR Ca^{2+}-ATPase activity produced either by β-adrenergic stimulation or by phosphodiesterase inhibition (using isobutylmethylxanthine) (Lindemann and Watanabe, 1985b). Protein kinase C and cGMP-dependent PK, which have been shown to phosphorylate phospholamban *in vitro*, failed to demonstrate similar effects in beating guinea pig hearts in response to stimuli that activate PK-C or elevate the cGMP levels (Edes and Kranias, 1990; Huggins *et al.*,1989). Thus, the physiological relevance of PK-C and PK-G in beating hearts is not clear at present.

The functional alterations in the SR Ca^{2+}-ATPase activity may explain, at least partly, the activating and relaxing effects of β-adrenergic agents in cardiac muscle (Figs. 7 and 8). The cAMP-dependent phosphorylation of phospholamban under either *in vitro* or *in vivo* conditions increases the rate of SR Ca^{2+} transport and SR Ca^{2+}-ATPase activity. Such an increase in Ca^{2+} transport is expected to contribute primarily to the relaxing effects of catecholamines (Fig. 7). An additional mechanism, which contributes to the increased phosphorylation of phospholamban upon β-adrenergic stimulation, is the phosphorylation of the phosphatase inhibitor protein by the stimulated cAMP-dependent kinase. This phosphorylation results in inactivation of protein phosphatase 1 and, thus, inhibition of dephosphorylation of phospholamban during the action of catecholamines (Fig. 7). The increased phosphorylation of phospholamban and the increased Ca^{2+} levels accumulated by the SR would lead to the availability of higher levels of Ca^{2+} to be subsequently released for binding to the contractile proteins (Fig. 7). The critical and prominent role of phospholamban in the mediation of β-adrenergic functional responses was also confirmed in transgenic animal studies. Cardiac myocytes or work-performing heart preparations from phospholamban-deficient mice exhibited largely attenuated responses to β-adrenergic agonist stimulation (Luo *et al.*, 1996; Wolska *et al.*, 1996), indicating that phospholamban is a key phosphoprotein in the heart's responses to β-adrenergic agonists.

Phosphorylation of other myocardial phosphoproteins has also been suggested to be involved in the mediation of positive inotropic and lusitropic effects of β-adrenergic agonists. Cyclic AMP–dependent protein kinase–mediated phosphorylation of the α_1-subunit of the Ca^{2+} channel (Fig. 8) is associated with an increase in the voltage-dependent Ca^{2+} current (I_{Ca}), which enhances the Ca^{2+} levels available in the cytosol during β-adrenergic agonist stimulation. Phosphorylation of troponin I has been shown to decrease the sensitivity of myofilaments for Ca^{2+} both in intact myocardium and skinned fibers (Kranias *et al.*, 1985). The desensitization of myofibrils is accompanied by an increased off-rate of Ca^{2+} from troponin C, which could contribute to faster relaxation (Fig. 8). In addition, phosphorylation of the SR Ca^{2+}-release channel (ryanodine receptor) by Ca^{2+}-CaM–dependent protein kinase may stimulate Ca^{2+} release from the SR vesicles and contribute to the elevation of intracellular Ca^{2+} levels during systole. Thus, the enhanced Ca^{2+} influx

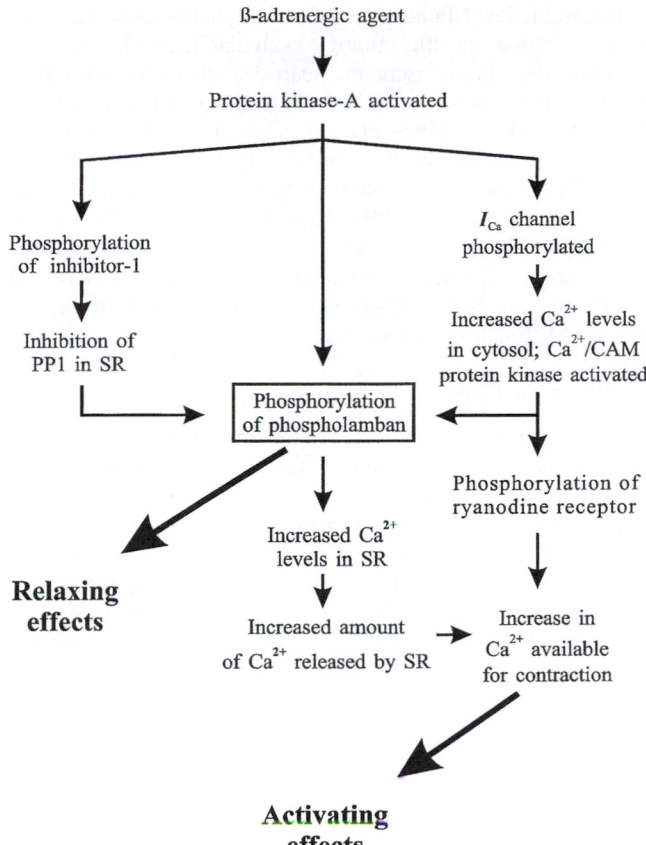

FIGURE 7. Schematic diagram of possible relaxing and activating effects of β-adrenergic agents in the heart. PP1, protein phosphatase 1.

of numerous investigations. It is assumed that in hypo- and hyperthyroid hearts the altered gene expression of the cardiac SR proteins, and hence the changes in the intracellular Ca^{2+} transients, is the most important determinant of the altered myocardial function. It was shown that the velocity of ATP-dependent Ca^{2+} transport and the Ca^{2+}-ATPase activity are specifically increased in SR vesicle preparations from hyperthyroid compared with euthyroid hearts (Beekman et al., 1989). Opposite changes were noted for hypothyroid animals compared with euthyroid ones (Beekman et al., 1989).

Examination of the steady-state mRNA levels of the cardiac SR Ca^{2+}-ATPase and the ryanodine receptor revealed a significant increase (140–190%) in hyperthyroid and a marked decline (40–50%) in hypothyroid animals (Arai et al., 1991). The changes in mRNA levels for the Ca^{2+}-ATPase in hypothyroid and hyperthyroid conditions also reflected changes in the protein amounts of the enzyme in these hearts (Kiss et al., 1994; Kiss et al., 1998). Interestingly, in the case of phospholamban, the regulator of the Ca^{2+}-ATPase, and calsequestrin, there was no coordinated regulation with respect to the Ca^{2+}-ATPase. In fact, both the relative mRNA level and the protein content of phospholamban were reported to decrease in hyperthyroid animals, whereas there was no change noted in the calsequestrin mRNA level upon L-thyroxine treatment. In hypothyroid hearts an opposite trend

across the sarcolemma, together with the increased Ca^{2+} levels to be released from the SR, may result in an elevation of the Ca^{2+} available for the contractile machinery, leading to an increase in the amplitude of contraction (Fig. 8).

C. SR Ca²⁺-ATPase in Cardiac Diseases

The complex regulation of the SR function clearly indicates that even small disturbances in SR Ca^{2+} handling may result in profound changes and deterioration of normal myocardial function. The fast removal of Ca^{2+} by the SR Ca^{2+}-ATPase during diastole and the subsequent rapid release through the SR Ca^{2+} channel (ryanodine receptor) at the beginning of contraction are prerequisites for normal diastolic and systolic function. We briefly outline in the next section the alterations in the SR Ca^{2+}-ATPase in the major cardiac diseases.

1. SR Ca²⁺-ATPase in Hyperthyroidism and Hypothyroidism

Thyroid hormones are important regulators of myocardial contractility and relaxation. Chronic increases in thyroid hormone levels lead to cardiac hypertrophy, with increases in the heart rate and cardiac output as well as left ventricular contractility and velocity of relaxation. On the other hand, opposite effects are associated with a hypothyroid condition. The mechanisms underlying these changes have been the subject

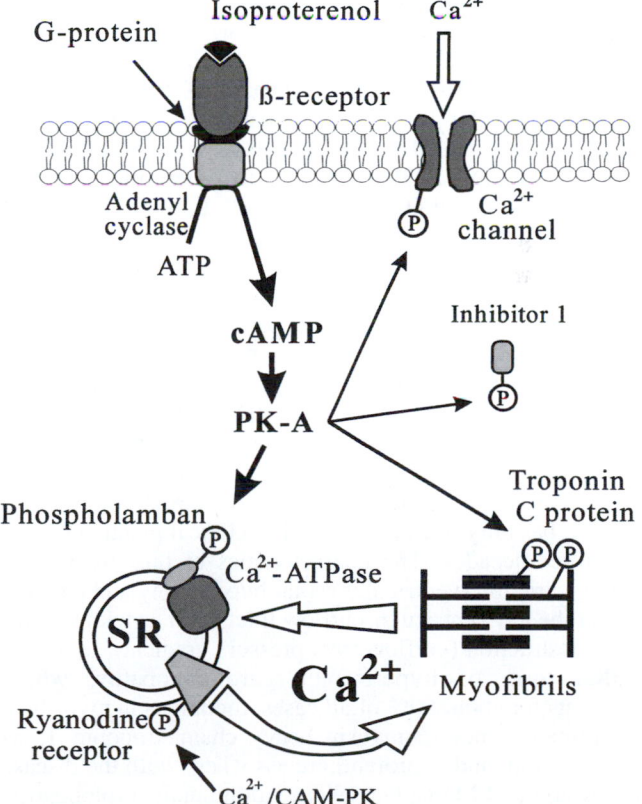

FIGURE 8. Effects of β-adrenergic agents on protein phosphorylation in cardiac cells. Increased intracellular cAMP levels activate the cAMP-dependent protein kinase(s), which phosphorylates various proteins (phospholamban, inhibitor-1, Ca^{2+}-channel and myofibrillar proteins) and increases the rates of SR Ca^{2+} uptake and release.

was noted since the protein amount of phospholamban was found to be increased as compared to the euthyroid or hyperthyroid animals (Kiss *et al.*, 1994; Kiss *et al.*, 1998). Consequently, the phospholamban-Ca^{2+}-ATPase protein ratio was highest in the hypothyroid animals, followed by euthyroid and hyperthyroid animals. These changes in the phospholamban-Ca^{2+}-ATPase ratio were associated with coordinate alterations in the SR Ca^{2+} uptake, affinity of the SERCA2 for Ca^{2+}, and myocardial function (Kiss *et al.*, 1994; Kimura *et al.*, 1994).

These changes indicate that the SR proteins responsible for Ca^{2+} uptake and release (Ca^{2+}-ATPase and ryanodine receptor) are coordinately regulated in hypothyroid and hyperthyroid hearts and provide a simple explanation for the altered Ca^{2+} release and reuptake capacity and hence the myocardial function under these conditions.

2. SR Ca^{2+}-ATPase in Cardiomyopathies

Dilated cardiomyopathy is a frequent form of cardiac muscle disease and is characterized by an impaired systolic function and dilatation of both ventricles (systolic pump failure). In various animal models of primary and secondary dilated cardiomyopathy, it was shown that both SR Ca^{2+} binding capacity and uptake were depressed because of the decreased activity and protein level of the SR Ca^{2+}-ATPase (Edes *et al.*, 1991). In some studies of human idiopathic dilated cardiomyopathy, decreases were noted for both SR Ca^{2+} uptake rates and Ca^{2+}-ATPase activity (Limas *et al.*, 1987; Unverferth *et al.*, 1988) as well as myocardial Ca^{2+} handling (Gwathmey *et al.*, 1987; Beuckelmann, *et al.*, 1992). Examination of the mRNA levels in left ventricular biopsies from patients with dilated cardiomyopathy revealed a significant decrease in mRNA content for the SR Ca^{2+}-ATPase relative to other mRNA forms (Mercadier *et al.*, 1990; Arai *et al.*, 1993). In contrast, other authors were unable to detect a decrease in SR Ca^{2+} uptake activity (Movsesian *et al.*, 1989) or the immunodetectable levels of the SR Ca^{2+}-ATPase protein (Schwinger *et al.*, 1995) in the left ventricular myocardium from patients with idiopathic dilated cardiomyopathy. Furthermore, the gating mechanism of the SR Ca^{2+}-release channel was recently reported to be abnormal in dilated cardiomyopathy (D'Agnolo *et al.*, 1992), and it was suggested that defective excitation-contraction coupling is involved in the pathogenesis of this disease.

Another type of cardiomyopathy, hypertrophic cardiomyopathy, has only been recognized in clinical practice for the last three decades. The characteristics of this disease are asymmetric interventricular septal hypertrophy and narrowing of the left ventricular outflow tract, with or without outflow obstruction (outflow tract pressure gradient). In the familial form of hypertrophic cardiomyopathy, which accounts for about 60% of all cases, mutations in myofibrillar protein genes (β-myosin heavy chain, troponin T, α-tropomyosin, and C protein) are associated with the disease (Schwartz and Mercadier 1996). Additionally, prolongation of the Ca^{2+} transient, abnormal Ca^{2+} handling, and a decline in SR Ca^{2+}-ATPase mRNA levels are reported to be characteristic for human hypertrophic cardiomyopathy (Gwathmey *et al.*, 1987; Mercadier *et al.*,1990), which may explain the diastolic function impairment in this disease.

In chronic heart failure due to hemodynamic overload, irrespective of the specific etiology (valvular heart disease, cardiomyopathy, chronic ischemic heart disease, or hypertension), a reduction was observed in both the number and the activity of the SR Ca^{2+} pump (Limas *et al.*, 1987; Unverferth *et al.*, 1988). Furthermore, a close correlation was obtained between the SR Ca^{2+}-ATPase mRNA or protein levels and the myocardial function (Mercadier *et al.*, 1990; Hasenfuss *et al.*, 1994). Interestingly, the Na^+-Ca^{2+} exchanger gene expression was reported to be increased in failing human hearts and it was hypothesized that the upregulation of this protein may compensate for the depressed SR function (Studer *et al.*, 1994).

3. SR Ca^{2+}-ATPase in Ischemia

A brief period of ischemia (10–20 min) induces reversible tissue damage in cardiac muscle, resulting in a "stunned" myocardium. This condition is characterized by regional contractile abnormalities (declines in both systolic and diastolic function) that persist for several hours despite the absence of necrosis. These hemodynamic changes are associated with a reduction in SR Ca^{2+} transport (Limbruno *et al.*, 1989; Krause *et al.*, 1989). The maximal activity of the SR Ca^{2+}-ATPase was found to be depressed, and the Ca^{2+} sensitivity of this enzyme was decreased (Krause *et al.*, 1989). Furthermore, a decrease in the coupling ratio (mol Ca^{2+}/mol ATP) was observed in the SR membranes isolated from the stunned myocardium, which was suggested to be the result of an increase in the Ca^{2+} permeability of the SR membrane. The SR Ca^{2+}-release process was also found to be impaired in the stunned myocardium due to a reduction of the number of ryanodine receptors (Zucchi *et al.*, 1994). These data suggest that complex modifications of the SR function occur in the stunned myocardium, which are at least partly responsible for the contractile impairment found in this condition.

In long-lasting myocardial ischemia, gradual declines in both SR Ca^{2+}-ATPase activity and Ca^{2+} uptake were found, which may be due to degradation of the SR Ca^{2+}-ATPase (Akiyama *et al.*, 1986; Schoutsen *et al.*, 1989). Ischemia was also shown to result in a gradual decrease in the phosphorylation status of phospholamban under both *in vitro* (Schoutsen *et al.*, 1989; Lamers *et al.*, 1986) and *in vivo* conditions (Bartel *et al.*, 1989), and this correlated with a decrease in the SR Ca^{2+}-ATPase activity. Thus, it has been postulated that the long-lasting ischemia-induced progressive inactivation of the SR Ca^{2+} pump not only is a consequence of the specific loss of enzyme activity, but may also be related to the altered characteristics of phospholamban (Schoutsen *et al.*, 1989). A combination of various pathogenic factors has been suggested to be responsible for the reduced SR function and the final tissue necrosis in the ischemic myocardium. These pathological factors include pH reduction (acidosis), activation of intracellular proteolytic enzymes, and increased generation of free radicals.

III. Other ATPases

The Ca^{2+} regulation in eukaryotic cells involves a complex mechanism that maintains a low background Ca^{2+} concentration (usually 0.1–0.2 μM) in the cell interior. Eukaryotic cells generally satisfy their Ca^{2+} demands by extracting Ca^{2+} from their own internal stores, but it is also evident that the long-term regulation of the Ca^{2+} gradient across the plasma membrane is a result of the concerted operation of the importing (Ca^{2+} channel) and exporting (SR Ca^{2+} pump; Na^+-Ca^{2+}exchanger and Ca^{2+} pump of the surface membrane) Ca^{2+} systems (see Section IIA). The plasma membrane Ca^{2+}-ATPase is a low-capacity system possessing a very high Ca^{2+} affinity, which enables the enzyme to interact with Ca^{2+} at low intracellular concentrations. Consequently, its function is continuous and presumably satisfies the fine tuning of Ca^{2+} homeostasis.

A. General Properties of Plasma Membrane Ca²⁺-ATPase(s)

The plasma membrane Ca^{2+}-ATPase (molecular mass of 140 kDa) general kinetic mechanism follows the pattern of the SR Ca^{2+}-ATPase. ATP phosphorylates an Asp residue to yield an acid-stable phosphorylated intermediate. The elementary steps of the cycle are probably similar in both SR and plasma membrane Ca^{2+}-ATPases (Schatzmann, 1989). The stoichiometry between transported Ca^{2+} and hydrolyzed ATP is only 1.0 for the plasma membrane Ca^{2+} pump. The administration of La^{3+} under various experimental conditions has been associated with an increase in the steady-state phosphoenzyme level of the plasma membrane Ca^{2+}-ATPase, and this increase possibly results from stabilization of the aspartyl phosphate (inhibition of hydrolysis of the phosphate group). The other classic inhibitor of Ca^{2+} pumps, vanadate, has been found to be a potent inhibitor of the plasma membrane Ca^{2+}-ATPase even at low concentrations (Bond and Hudgins, 1979).

Calmodulin stimulates the plasma membrane Ca^{2+}-ATPase by direct interaction with the enzyme. It has been shown that the stimulation results from a combined effect on the affinity for Ca^{2+} (K_m) and the maximal transport rate (V_{max}) (Carafoli, 1992). The calmodulin-binding domain of the Ca^{2+} pump has been suggested to function as a repressor of the enzymatic activity (autoinhibitory function), and calmodulin may relieve this inhibition (Carafoli, 1992). A common mechanism in the autoinhibition of plasma membrane Ca^{2+}-ATPase and phospholamban inhibition of SR Ca^{2+}-ATPase has been suggested. In both proteins, the interacting sites are amphiphilic and located in the cytoplasmic region. The interaction occurs with homologous regions in the SR and plasma membrane Ca^{2+}-ATPases close to the phosphorylation sites. In the absence of calmodulin, the plasma membrane Ca^{2+} pump can be activated by several other compounds. Polyunsaturated fatty acids and acidic phospholipids (phosphatidylinositol, phosphatidylinositol 4-phosphate, and phosphatidylinositol 4,5-diphosphate) have been reported to be good activators and, since they are present in the plasma membrane, they may be important regulators of the Ca^{2+}-ATPase under *in vivo* conditions (Carafoli, 1991). Phosphorylation of the enzyme by cAMP-dependent PK or PK-C has also been reported to stimulate plasma membrane Ca^{2+}-ATPase activity. The cAMP-dependent phosphorylation occurs C-terminally to the calmodulin-binding domain and the phosphorylation-mediated activation may likewise be significant *in vivo*. The PK-C phosphorylation occurs in the calmodulin-binding domain, inhibiting the binding of calmodulin to the plasma membrane Ca^{2+}-ATPase and lowering the

autoinhibitory potential of this domain (Hofmann *et al.*, 1994). Cyclic GMP–dependent PK has also been reported to stimulate the plasma membrane Ca^{2+} pump in vascular smooth muscle, but the Ca^{2+}-ATPase enzyme was not found to be the substrate for this kinase.

B. Primary Structure and Topography of Plasma Membrane Ca²⁺-ATPase(s)

The complete amino acid sequence of the plasma membrane Ca^{2+} pump has been deduced from rat and human cells (Shull and Greeb, 1988; Verma *et al.*, 1988). It appears that the plasma membrane Ca^{2+}-ATPase (PMCA) isoforms are encoded by a multigene family, and additional variability is produced by alternative RNA splicing of each gene transcript (Table 2). The regions important for the catalytic function and the transmembrane domains are highly conserved, with no observed diversity. The isoform diversity seems to alter primarily the regulatory characteristics of the enzyme and it can be regarded as an adaptation to tissue specificity.

The secondary structure of the plasma membrane Ca^{2+}-ATPase is similar to that of the SR Ca^{2+}-ATPase (Shull and Greeb, 1988). The enzyme contains 10 putative transmembrane helices, which are connected on the outside of the plasma membrane by short loops. Three primary domains (about 80% of the pump protein) protrude into the cytoplasm. The first domain corresponds to the transducing unit, which couples ATP hydrolysis to Ca^{2+} translocation. The second protruding domain contains the aspartyl phosphate site (phosphorylation domain). The C-terminal portion of this domain can also be labeled by ATP analogs and contains a "hinge" region that permits the movement of aspartyl phosphate and the ATP-binding site. The third C-terminal protruding domain contains the calmodulin-binding sequence and the phosphorylation sites for protein kinase C and cAMP-dependent protein kinase. The latter is not present in all isoforms. As in the SR Ca^{2+}-ATPase, selective mutations in the M4 and M6 transmembrane segments in the plasma membrane Ca^{2+}-ATPase have been associated with loss of its ability to form the ATP- and Ca^{2+}-dependent phosphorylated intermediate and to transport Ca^{2+} (Guerini *et al.*, 1998).

IV. Summary

The Ca^{2+} levels in muscle are primarily regulated by the sarcoplasmic reticulum (SR) network, which serves as a sink for Ca^{2+} ions during relaxation and as a Ca^{2+} source during contraction. In cardiac muscle, most of the intracellular Ca^{2+} released during systole is taken up by the SR through its Ca^{2+}-ATPase. This translocation of Ca^{2+} from the cytosol into the SR lumen uses ATP as the energy source, and is characterized by the formation of a phosphorylated intermediate (E_1~P) for the Ca^{2+}-ATPase.

In cardiac muscle, slow-twitch skeletal muscle, and smooth muscle, the Ca^{2+}-ATPase is regulated by a low-molecular-weight phosphoprotein called phospholamban. In its dephosphorylated form, phospholamban is an inhibitor of the Ca^{2+}-ATPase and phosphorylation relieves this inhibition. Phosphorylation of phospholamban occurs by cAMP-dependent, cGMP-dependent, Ca^{2+}-calmodulin–dependent,

TABLE 2 Distribution of the Human Plasma Membrane Ca^{2+}-ATPase (PMCA) Isoforms

Gene	Tissue distribution	Level of expression	Inhibition by calmodulin
PMCA1[a]	Ubiquitous	High	Medium sensitive
PMCA2	Restricted (brain high)	High	Highly sensitive
PMCA3	Restricted (brain low)	Low	N/A[b]
PMCA4	Ubiquitous	Medium	Medium sensitive

[a]PMCA numbers identify different gene products.
[b]N/A = data not available.

and Ca^{2+}-phospholipid–dependent protein kinases *in vitro*. However, *in vivo* studies have indicated that phospholamban is phosphorylated only by cAMP-dependent and Ca^{2+}-calmodulin–dependent protein kinases in intact beating hearts. A phospholamban phosphatase activity has been reported to be present in SR membranes, which can dephosphorylate this regulatory protein and reverse its stimulatory effects on the Ca^{2+}-ATPase.

Alterations in the SR Ca^{2+}-ATPase activity and its regulation by phospholamban have been shown to occur in cardiac diseases such as hypothyroidism, hyperthyroidism, hypertrophy, heart failure, and ischemia. In most instances, alterations in Ca^{2+}-ATPase activity correlated with alterations in its mRNA levels and ventricular function.

Another Ca^{2+}-ATPase, which is also important for maintaining Ca^{2+} homeostasis in muscle, is the plasma membrane Ca^{2+}-ATPase. This enzyme transports Ca^{2+} to the extracellular space and uses ATP as its energy source, similar to the SR Ca^{2+}-ATPase. The plasmalemmal Ca^{2+}-ATPase may be distinguished from the SR Ca^{2+}-ATPase primarily by its distinct sensitivity to La^{3+}, vanadate, and calmodulin.

The primary structure of the various Ca^{2+} pumps has been published, and there is a growing interest in further use of molecular biological approaches and specifically site-directed mutagenesis for these enzymes to obtain more information about their structural-functional relationships. The ultimate question is, what is the precise mechanism by which Ca^{2+} is transported across the ATPases? Site-directed mutagenesis studies and construction of molecular models have already given some information along these lines and hopefully will provide further data that will finally answer this question. Furthermore, in the absence of appropriate crystallographic data, a deeper understanding of the mechanisms that regulate the Ca^{2+}-ATPases under normal and pathological conditions may elucidate the structural-functional relationships in these enzymes and their role in maintaining Ca^{2+} homeostasis in the cell.

Acknowledgments

This study was supported by National Institutes of Health Grants HL-26057 and P40RR12358 and the Hungarian Academy of Sciences Grant OTKA T 020691.

Bibliography

Akiyama, K., Konno, N., Yanagishita, T., Tanno, F., and Katagiri, T. (1986). Ultrastructural changes in the sarcoplasmic reticulum in acute myocardial ischemia. *Jpn. Circ. J.* **50**, 829–838.

Andersen J. P., and Vilsen, B. (1998). Structure-function relationships of the calcium binding sites of the sarcoplasmic reticulum Ca^{2+}-ATPase. *Acta Physiol. Scan.* **163**, 45–54.

Arai, M., Otsu, K., MacLennan, D. H., Alpert, N. R., and Periasamy, M. (1991). Effect of thyroid hormone on the expression of mRNA encoding sarcoplasmic reticular proteins. *Circ. Res.* **69**, 266–276.

Arai, M., Alpert, N. R., MacLennan, D. H., Barton, P., and Periasamy, M. (1993). Alterations in sarcoplasmic reticulum gene expression in human heart failure. A possible mechanism for alterations in systolic and diastolic properties of the failing myocardium. *Circ. Res.* **72**, 463–469.

Bartel, S., Karczewski, P., and Krause, E-G. (1989). Phosphorylation of phospholamban and troponin I in the ischemic and reperfused heart: attenuation and restoration of isoprenaline responsiveness. *Biomed. Biochim. Acta* **48**, 108–113.

Beekman, R. I., Hardeveld, C., and Simonides, W. S. (1989). On the mechanism of the reduction by thyroid hormone of β-adrenergic relaxation rate stimulation in rat heart. *Biochem. J.* **259**, 229–236.

Bers, D. M. (1991). Ca regulation in cardiac muscle. *Med. Sci. Sports Exer.* **23**, 1157–1162.

Beuckelmann, D. J., Näbauer, Erdmann, E. (1992). Intracellular calcium handling in isolated ventricular myocytes from patients with terminal heart failure. *Circulation* **85**, 1046–1055.

Bond, G. H., and Hudgins, P. (1979). Kinetics of inhibition of Na^+, K^+-ATPase by Mg^{2+}, K^+ and vanadate. *Biochemistry* **18**, 325–331.

Brandl, C. J., Green, N. M., Korczak, B., and MacLennan, D. H. (1986). Two ATPase genes: homologies and mechanistic implications of deduced amino acid sequences. *Cell* **44**, 597–607.

Burk, S. E., Lytton, J., MacLennan, D. H., and Shull, G. E. (1989). cDNA cloning, functional expression and mRNA tissue distribution of a third organellar Ca^{2+} pump. *J. Biol. Chem.* **264**, 18 561–18 568.

Carafoli, E. (1991). The calcium pumping ATPase of the plasma membrane. *Annu. Rev. Physiol.* **53**, 531–547.

Carafoli, E. (1992). The Ca^{2+} pump of the plasma membrane. *J. Biol. Chem.* **267**, 2115–2118.

Colyer, J., and Wang, J. H. (1991). Dependence of cardiac sarcoplasmic reticulum calcium pump activity on the phosphorylation status of phospholamban. *J. Biol. Chem.* **266**, 17 486–17 493.

Cornea, R. L., Jones, L. R., Autry, J. M., and Thomas, D. D. (1997). Mutation and phosphorylation change the oligomeric structure of phospholamban in lipid bilayers. *Biochemistry* **36**, 2960–2967.

D'Agnolo, A., Luciani, G. B., Mazzucco, A., Gallucci, V., and Salviati, G. (1992). Contractile properties and Ca^{2+} release activity of the sarcoplasmic reticulum in dilated cardiomyopathy. *Circulation* **85**, 518–525.

Dux, L., Taylor, K. A., Tin-Beall, H. P., and Martonosi, A. (1985). Crystallization of the Ca²⁺-ATPase of sarcoplasmic reticulum by calcium and lanthanide ions. *J. Biol. Chem.* **260,** 11 730–11 743.

Edes, I., and Kranias, E. G. (1990). Phospholamban and troponin I are substrates for protein kinase C *in vitro* but not in intact beating guinea pig hearts. *Circ. Res.* **67,** 394–400.

Edes, I., Talosi, L., and Kranias, E. G. (1991). Sarcoplasmic reticulum function in normal heart and in cardiac disease. *Heart Failure* **6 (6)** 221–237.

Fujii, J., Ueno, A., Kitano, K., Tanaka, S., Kadoma, M., and Tada, M. (1987). Complete complementary DNA-derived amino acid sequence of canine cardiac phospholamban. *J. Clin. Invest.* **79,** 301–304.

Fujii, J., Maruyama, K., Tada, M., and MacLennan, D. H. (1989). Expression and site-specific mutagenesis of phospholamban. Studies of residues involved in phosphorylation and pentameric formation. *J. Biol. Chem.* **264,** 12 950–12 955.

Fujii, J., Zarain-Herzberg, A., Willard, H. F., Tada, M., and MacLennan, D. H. (1991). Structure of the rabbit phospholamban gene, cloning of the human cDNA, and assignment of the gene to human chromosome 6. *J. Biol. Chem.* **266,** 11 669–11 675.

Guerini, D., Garcia-Martin, E., Zecca, A., Guidi, F., Carafoli, and E. (1998). The calcium pump of the plasma membrane: membrane targeting, calcium binding sites, tissue-specific isoform expression. *Acta Physiol. Scand.* Suppl. **643,** 265–273.

Gwathmey, J. K., Copelas, L., MacKinnon, R., Schoen, F. J., Feldman, M. D., Grossman, W., and Morgan, J. P. (1987). Abnormal intracellular calcium handling in myocardium from patients with end-stage heart failure. *Circ. Res.* **61,** 70–76.

Hasenfuss, G., Reinecke, H., Studer, R., Meyer, M., Pieske, B., Holtz, J., Holubarsch, C., Posival, H., Just, H., Drexler, H. (1994). Relation between myocardial function and expression of sarcoplasmic reticulum Ca²⁺-ATPase in failing and nonfailing human myocardium. *Circ. Res.* **75,** 434–442.

Hofmann, F., Anagli, J., and Carafoli, E. (1994). Phosphorylation of the calmodulin binding domain of the plasma membrane Ca²⁺ pump by protein kinase C reduces its interaction with calmodulin and with its pump receptor site. *J. Biol. Chem.* **269,** 24 298–24 303.

Huggins, J. P., Cook, E. A., Pigott, J. R., Mattinsley, T. J., and England, P. J. (1989). Phospholamban is a good substrate for cGMP-dependent protein kinase *in vitro*, but not in intact cardiac or smooth muscle. *Biochem. J.* **260,** 829–835.

Kadambi, V. J., Ponniah, S., Harrer, J. M., Hoit, B. D., Dorn, G. W., Walsh, R. A., and Kranias, E. G. (1996). Cardiac-specific overexpression of phospholamban alters calcium kinetics and resultant cardiomyocyte mechanics in transgenic mice. *Clin. Invest.* **97,** 533–539.

Kim, H. W., Steenaart, N. A. E., Ferguson, D. G., and Kranias, E. G. (1990). Functional reconstitution of the cardiac sarcoplasmic reticulum Ca²⁺-ATPase with phospholamban in phospholipid vesicles. *J. Biol. Chem.* **265,** 1702–1709.

Kimura, Y., Otsu, K., Nishida, K., Kuzuya, T., and Tada, M. (1994). Thyroid hormone enhances Ca²⁺ pumping activity of the cardiac sarcoplasmic reticulum by increasing Ca²⁺ ATPase and decreasing phospholamban expression. *J. Mol. Cell. Cardiol.* **26,** 1145–1154.

Kimura, Y., Kurzydlowski, K., Tada, M., and MacLennan, D. H. (1996). Phospholamban regulates the Ca²⁺-ATPase through intramembrane interaction. *J. Biol. Chem.* **271,** 21 726–21 731.

Kimura, Y., Kurzydlowski, K., Tada, M., and MacLennan, D. H. (1997). Phospholamban inhibitory function is activated by depolymerization. *J. Biol. Chem.* **272,** 15 061–15 064.

Kiss, E., Jakab, G., Kranias, E. G., and Edes, I. (1994). Thyroid hormone-induced alterations in phospholamban protein expression. Regulatory effects on sarcoplasmic reticulum Ca²⁺ transport and myocardial relaxation. *Circ. Res.* **75,** 245–251.

Kiss, E., Brittsan, A. E., Edes, I., Grupp, I. L., Grupp, G., and Kranias, E. G. (1998). Thyroid hormone–induced alterations in phospholamban-deficient mouse hearts. *Circ. Res.* **83,** 608–613.

Kranias, E. G., and Solaro, R. J. (1982). Phosphorylation of troponin I and phospholamban during catecholamine stimulation of rabbit heart. *Nature* **298,** 182–184.

Kranias, E. G. (1985a). Regulation of Ca²⁺ transport by cyclic 3',5'-AMP-dependent and calcium-calmodulin–dependent phosphorylation of cardiac sarcoplasmic reticulum. *Biochim. Biophys. Acta* **844,** 193–199.

Kranias, E. G. (1985b). Regulation of calcium transport by protein phosphatase activity associated with cardiac sarcoplasmic reticulum. *J. Biol. Chem.* **260,** 11 006–11 010.

Kranias, E. G., Garvey, J. L., Srivastava, R. D., and Solaro, R. J. (1985). Phosphorylation and functional modifications of sarcoplasmic reticulum and myofibrils in isolated rabbit hearts stimulated with isoprenaline. *Biochem. J.* **226,** 113–121.

Kranias, E. G., and Di Salvo, J. (1986). A phospholamban protein phosphatase activity associated with cardiac sarcoplasmic reticulum. *J. Biol. Chem.* **261,** 10 029–10 032.

Krause, S. M., Jacobus, W. E., and Becker, L. C. (1989). Alterations in cardiac sarcoplasmic reticulum calcium transport in the postischemic "stunned" myocardium. *Circ. Res.* **65,** 526–530.

Lamers, J. M., De Jonge-Stinis, J. T., Hülsman, W. C., and Verdouw, P. D. (1986). Reduced *in vitro* ³² P-incorporation into phospholamban-like protein of sarcolemma due to myocardial ischaemia in anaesthetized pigs. *J. Mol. Cell. Cardiol.* **18,** 115–125.

Limas, C. J., Olivari, M. T., Goldenberg. I. F., Levine, T. B., Benditt, D. G., and Simon, A. (1987). Calcium uptake by cardiac sarcoplasmic reticulum in human dilated cardiomyopathy. *Cardiovasc. Res.* **21,** 601–605.

Limbruno, U., Zucchi, R., Ronca-Testoni, S., Galbani, P., Ronca, G., and Mariani, M. (1989). Sarcoplasmic reticulum function in the "stunned" myocardium. *J. Mol. Cell. Cardiol.* **21,** 1063–1072.

Lindemann, J. P., Jones, L. R., Hathaway, D. R., Henry, B. G., and Watanabe, A. M. (1983). β-Adrenergic stimulation of phospholamban phosphorylation and Ca²⁺-ATPase activity in guinea pig ventricles. *J. Biol. Chem.* **260,** 4516–4525.

Lindemann, J. P., and Watanabe, A. M. (1985a). Phosphorylation of phospholamban in intact myocardium. Role of Ca²⁺-calmodulin–dependent mechanisms. *J. Biol. Chem.* **260,** 4516–4525.

Lindemann, J. P., and Watanabe, A. M. (1985b). Muscarinic cholinergic inhibition of β-adrenergic stimulation of phospholamban phosphorylation and Ca²⁺ transport in guinea pig ventricles. *J. Biol. Chem.* **260,** 122–133.

Luo, W., Grupp, I. L., Harrer, J., Ponniah, S., Grupp, G., Duffy J. J., Doetschman, T., and Kranias, E. G. (1994). Targeted ablation of phospholamban gene is associated with markedly enhanced myocardial contractility and loss of β-adrenergic stimulation. *Circ. Res.* **75,** 401–409.

Luo, W., Wolska, B. M., Grupp, I. L., Harrer, J. M., Haghighi, K., Ferguson, D. G., Slack, J. P., Grupp, G., Doetschman, T., Solaro, R. J., and Kranias, E. G. (1996). Phospholamban gene dosage effect in the mammalian heart. *Circ. Res.* **78,** 839–847.

Luo, W., Chu, G., Sato, Y., Zhou, Z., Kadambi, V., and Kranias, E. G. (1997). Transgenic approaches to define the functional role of dual site phosphorylation of phospholamban. *J. Biol. Chem.* **273,** 4734–4739.

Lytton, J., Westlin, M., and Hanley, M. R. (1991). Thapsigargin inhibits the sarcoplasmic or endoplasmic reticulum Ca-ATPase family of calcium pump. *J. Biol. Chem.* **266,** 17 067–17 071.

MacDougall, L. K., Jones, J. R., and Cohen, P. (1991). Identification of the major protein phosphatases in mammalian cardiac muscle which dephosphorylate phospholamban. *Eur. J. Biochem.* **196,** 725–734.

MacLennan, D. H., Clarke, D. M., Loo, T. W., and Skerjanc, I. S. (1992). Site-directed mutagenesis of the Ca²⁺-ATPase of sarcoplasmic reticulum. *Acta Physiol. Scand.* **146,** 141–150.

MacLennan, D. H., Rice, W. J., and Green N. M. (1997). The mechanism of Ca²⁺ transport by sarco(endo)plasmic reticulum Ca²⁺-ATPases. *J. Biol. Chem.* **272,** 28 815–28 818.

MacLennan, D. H., Kimura, Y., and Toyofuku, T. (1998). Sites of regulatory interaction between Ca^{2+}-ATPases and phospholamban. *Ann. NY Acad. Sci.* **175,** 31–42.

Mercadier, J. J., Lompre, A. M., Duc, P., Boheler, K. R., Fraysse, J. B., Wisnewsky, P., Allen, P. D., Komajda, M., and Schwartz, K. (1990). Altered sarcoplasmic reticulum Ca^{2+}-ATPase gene expression in the human ventricle during end-stage heart failure. *J. Clin. Invest.* **85,** 305–309.

Morris, G. L., Cheng, H., Colyer, J., and Wang, J. H. (1991). Phospholamban regulation of cardiac sarcoplasmic reticulum (Ca^{2+}-Mg^{2+})-ATPase. Mechanism of regulation and site of monoclonal antibody interaction. *J. Biol. Chem.* **266,** 11 270–11 275.

Mortishire-Smith, R. J., Broughton, H., Garsky, V. M., Mayer, E. J., and Johnson, R. G. (1998). Structural studies on phospholamban and implications for regulation of the Ca^{2+}-ATPase. *Ann. NY Acad. Sci.* **853,** 63–78.

Movsesian, M. A., Nishikawa, M., and Adelstein, R. S. (1984). Phosphorylation of phospholamban by calcium-activated, phospholipid–dependent protein kinase. *J. Biol. Chem.* **259,** 8029–8032.

Movsesian, M. A., Bristow, M. R., and Krall, J. (1989). Ca^{2+} uptake by cardiac sarcoplasmic reticulum from patients with idiopathic dilated cardiomyopathy. *Circ. Res.* **65,** 1141–1144.

Neumann, J., Gupta, R. C., Schmitz, W., Scholz, H., Nairn, A. C., and Watanabe, A. M. (1991). Evidence for isoproterenol-induced phosphorylation of phosphatase inhibitor-1 in the intact heart. *Circ. Res.* **69,** 1450–1457.

Raeymakers, L., Hofmann, F., and Casteels, R. (1988). Cyclic GMP-dependent protein kinase phosphorylates phospholamban in isolated sarcoplasmic reticulum from cardiac and smooth muscle. *Biochem. J.* **252,** 269–273.

Reddy, L. G., Jones, L. R., Cala, S. E., O`Brian, J. J., Tatulian, S A., and Stokes, D. L. (1995). Functional reconstitution of recombinant phospholamban with rabbit skeletal Ca^{2+}-ATPase. *J. Biol. Chem.* **270,** 9390–9397.

Reddy, L. G., Jones, L. R., Pace, R. C., and Stokes, D. L. (1996). Purified, reconstituted cardiac Ca^{2+}-ATPase is regulated by phospholamban but not by direct phosphorylation with Ca^{2+}/calmodulin–dependent protein kinase. *J. Biol. Chem.* **271,** 14 964–14 970.

Ringer, S. A. (1883). A further contribution regarding the influence of different constituents of the blood on the contraction of the heart. *J. Physiol.* **4,** 29–42.

Sasaki, T., Inui, M., Kimura, Y., Kuzuya, T., and Tada, M. (1992). Molecular mechanism of regulation of Ca^{2+}-ATPase by phospholamban in cardiac sarcoplasmic reticulum. *J. Biol. Chem.* **267,** 1674–1679.

Schatzmann, H. J. (1989). The calcium pump of the surface membrane and of the sarcoplasmic reticulum. *Annu. Rev. Physiol.* **51,** 473–485.

Schoutsen, B., Blom. J. J., Verdouw, P. D., and Lamers, J. M. (1989). Calcium transport and phospholamban in sarcoplasmic reticulum of ischemic myocardium. *J. Mol. Cell. Cardiol.* **21,** 719–727.

Schwartz, K., and Mercadier, J-J. (1996). Molecular and cellular biology of heart failure. *Curr. Opinion Cardiol.* **11,** 227–236.

Schwinger, R. H. G., Böhm, M., Schmidt, U., Karczewski, P., Bavendiek, U., Flesch, M., Krause, E-G., and Erdmann, E. (1995). Unchanged protein levels of SERCA II and phospholamban but reduced Ca^{2+} uptake and Ca^{2+}-ATPase activity of cardiac sarcoplasmic reticulum from dilated cardiomyopathy patients compared with patients with nonfailing hearts. *Circulation* **92,** 3220–3228.

Shull, G. E., and Greeb, J. (1988). Molecular cloning of two isoforms of the plasma membrane Ca^{2+} transporting ATPase from rat brain. Structural and functional domains exhibit similarity to Na^+,K^+-and other cation transport ATPases. *J. Biol. Chem.* **263,** 8646–8657.

Simmerman, H. K. B., Collins, J. H., Theibert, J. L., Wegener, A. D., and Jones, L. R. (1986). Sequence analysis of phospholamban. Identification of phosphorylation sites and two major structural domains. *J. Biol. Chem.* **261,** 13 333–13 341.

Simmerman, H. K. B., Kobayashi, Y. M., Autry, J. M., and Jones, L. R. (1996). A leucine zipper stabilizes the pentameric membrane domain of phospholamban and forms a coiled-coil pore structure. *J. Biol. Chem.* **271,** 5941–5946.

Simmerman, H. K. B., and Jones, L. R. (1998) Phospholamban: protein structure, mechanism of action, and role in cardiac function. *Physiol. Rev.* **78,** 921–947.

Steenaart, N. A. E., Ganim, J. R., DiSalvo, J., and Kranias, E. G. (1992). The phospholamban phosphatase associated with cardiac sarcoplasmic reticulum is a type 1 enzyme. *Arch. Biochem. Biophys.* **293,** 17–24.

Studer, R., Reinecke, H., Bilger, J., Eschenhagen, T., Böhm, M., Hasenfuss, G., Just, H., Holtz, J., and Drexler, H. (1994). Gene expression of the cardiac Na^+-Ca^{2+} exchanger in end-stage human heart failure. *Circ. Res.* **75,** 443–453.

Talosi, L., Edes, I., and Kranias, E. G. (1993). Intracellular mechanisms mediating the reversal of β-adrenergic stimulation in intact beating hearts. *Am. J. Physiol.* **264,** H791–H797.

Toyofuku, T., Kurzydlowski, K., Narayanan, N., and MacLennan, D. H. (1994a). Identification of Ser^{38} as the site of cardiac sarcoplasmic reticulum Ca^{2+}-ATPase that is phosphorylated by Ca^{2+}/calmodulin–dependent protein kinase. *J. Biol. Chem.* **269,** 26 492–26 496.

Toyofuku, T., Kurzydlowski, K., Tada, M., and MacLennan, D. H. (1994b). Amino acids Glu^2 to Ile^{18} in the cytoplasmic domain of phospholamban are essential for functional association with the Ca^{2+}-ATPase of sarcoplasmic reticulum. *J. Biol. Chem.* **269,** 3088–3094.

Toyofuku, T., Kurzydlowski, K., Tada, M., and MacLennan, D. H. (1994c). Amino acids Lys-Asp-Asp-Lys-Pro-Val^{402} in the Ca^{2+}-ATPase of cardiac reticulum are critical for functional association with phospholamban. *J. Biol. Chem.* **269,** 22 929–22 932.

Unverferth, D. V., Lee, S. W., and Wallick, E. T. (1988). Human myocardial adenosine triphosphate activities in health and heart failure. *Am. Heart J.* **115,** 139–146.

Verma, A. K., Filoteo, A. G., Stanford, D. R., Wieben, E. D., Penniston, J. T., Strehler, E. E., Fischer, R., Heim, R., Vogel, G., Mathews, S., Strehler-Page, M. A., James, P., Vorherr, T., Krebs, J., and Carafoli, E. (1988). Complete primary structure of a human plasma membrane Ca^{2+} pump. *J. Biol. Chem.* **263,** 14 152–14 159.

Wegener, A. D., Simmerman, H. K. B., Liepnieks, J., and Jones, L. R. (1986). Proteolytic cleavage of phospholamban purified from canine cardiac sarcoplasmic reticulum. Generation of a low resolution model of phospholamban structure. *J. Biol. Chem.* **261,** 5154–5159.

Wolska, B. M., Stojanovic, M. O., Luo, W., Kranias, E. G., and Solaro, R. J. (1996). Effect of ablation of phospholamban on dynamics of cardiac myocyte contraction and intracellular Ca^{2+}. *Am. J. Physiol.* **271,** C391–C397.

Zucchi, R., Ronca-Testoni, S., Yu, G., Galbani, P., Ronca, G., and Mariani, M. (1994). Effect of ischemia and reperfusion on cardiac ryanodine receptors—sarcoplasmic reticulum Ca^{2+} channels. *Circ. Res.* **74,** 271–280.

John H.B. *Bridge*

19

Na+-Ca2+ Exchange Currents

I. Introduction

The existence of Na+-Ca2+ exchange was first postulated in 1964 by both Repke (1964) and Langer (1964) as a consequence of their studies on the contractility of heart muscle. Three years later Baker and colleagues (1967) provided the first report documenting Na+-Ca2+ exchange in giant squid axons. Shortly after this Reuter and Seitz (1968) presented the first complete study describing Na+-Ca2+ countertransport in heart. Based on studies of isotopic fluxes, these authors proposed that two Na+ ions were coupled to the extrusion of a single Ca2+ ion in a modified exchange diffusion process. Blaustein and Hodgkin (1969) then published the results of their studies on squid axons. They recognized that the distribution of free Ca2+ could not be predicted on simple electrochemical principles. However, cyanide (which was expected to block metabolic processes) failed to prevent the efflux of 45Ca, so that it seemed unlikely that a metabolic pump was involved in Ca2+ extrusion. However, this efflux of Ca2+ was (among other things) dependent upon external Na+. Blaustein and Hodgkin therefore concluded that, in unpoisoned axons, some or possibly all of the energy for extruding Ca2+ ions came from the inward movement of Na+ down its electrochemical gradient.

These early studies were seminal and provided impetus for an enormous number of subsequent investigations that have led not only to a study of Na+-Ca2+ exchange currents, but to the molecular cloning and elucidation of the structure of the exchanger molecule itself. This in turn introduced the possibility of studying the relationship between molecular structure and function.

This chapter provides a brief description of current knowledge of Na+-Ca2+ exchange currents. The ease with which heart cells can be patch-clamped, together with the presence of a vigorous exchange activity, doubtless explains the fact that most of our information on exchange current comes from this tissue. Na+-Ca2+ exchange currents have been measured in other cell types including the squid giant axon (Matsuoka *et al.*, 1997). In this chapter we will not deal with Na+-Ca2+ exchange in the vertebrate rod outer segment

(ROS). It is now known that this exchange process is different from that found in heart muscle. For example, under physiological conditions the exchange involves not only the Na+ and Ca2+ gradients but the K+ gradient as well. In addition, the structure of the ROS exchanger is quite different from the heart exchanger (Achilles *et al.*, 1991). There have been extensive studies of the currents associated with the ROS Na+-Ca2+ exchange. The reader interested in the electrogenicity of this exchange can refer to the brief review by Lagnado and McNaughton (1990) and the more recent and extensive review on Na+-Ca2+ exchange by Blaustein and Lederer (1999). Although measurements of exchange current are difficult in many tissues, it is now possible to express Na+-Ca2+ exchangers in frog oocytes. This together with the development of giant excised patches that can be voltage-clamped has made studies of the relationship between structure of the exchanger and function much easier.

II. Structure, Topology, and Distribution of the Na+-Ca2+ Exchanger

The cardiac Na+-Ca2+ exchanger was the first to be isolated and cloned (Philipson *et al.*, 1988; Nicoll *et al.*, 1990). This exchanger is now referred to as NCX1. The functional exchanger is known to be 980 amino acids long (this excludes a signal peptide of 32 amino acids) with nine transmembrane segments (Fig. 1) (Nicoll *et al.*, 1999). A large intracellular loop separates two sets of hydrophobic domains and it is these hydrophobic domains that contain the transmembrane segments. It is now known that the intracellular loop is not necessary for transport (Matsuoka *et al.*, 1993). However two regions—the $\alpha 1$ and $\alpha 2$ repeats, which are the sequences that span repeats 2 and 3 and 8 and 9—are similar and appear to be important in transport. There are also repeats in the large cytoplasmic loop. These are known as the $\beta 1$ and $\beta 2$ repeats. The function of these repeats is unknown. The cytoplasmic loop is, however, of great functional importance, particularly in the regulation of exchange (see Section IX.B).

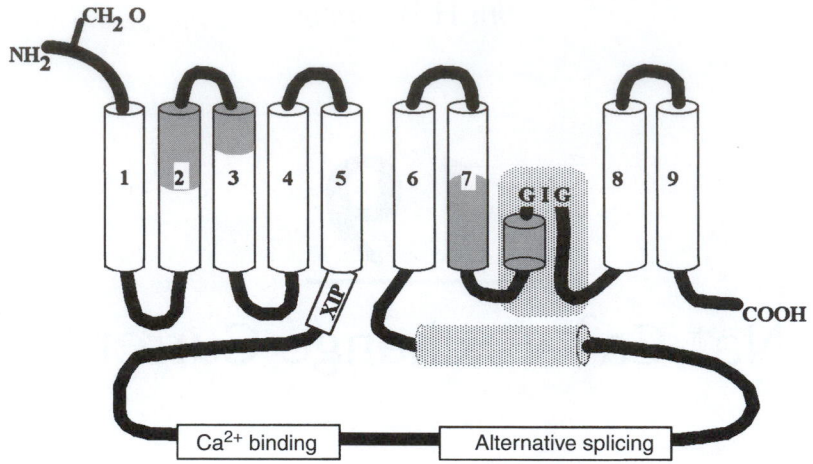

FIGURE 1. A proposed model of the Na^+-Ca^{2+} exchanger. Membrane-spanning segments are labeled 1 to 9. A large hydrophilic cytoplasmic domain contains a receptor that binds exchanger inhibitory peptide (XIP) and this site is thought to be associated with the phenomenon of Na^+-dependent inactivation. In addition, this cytoplasmic domain contains a secondary Ca^{2+} regulatory site and a region encoded by six small exons from which splice variants arise. For further details on the structure of the exchanger the reader is referred to Philipson and Nicoll (2000). (This figure was kindly provided by Dr. Kenneth Philipson.)

The topology of almost the entire exchanger has been mapped (Nicoll *et al.*, 1999). A couple of features of this topology are particularly interesting. The ninth hydrophobic segment has been proposed as a reentrant membrane loop that is similar to the pore-forming loop region of ion channels. The α repeats are currently modeled to be on the opposite side of the membrane (Fig. 1), which resembles that found in aquaporin water channels.

III. The Phylogeny of the Na^+-Ca^{2+} Exchanger

The number of proteins belonging to the exchange family is increasing. These proteins are defined by their similarity of structure including the presence of α repeats and multiple transmembrane segments. Recent analysis indicates four families of exchangers, which constitute a "superfamily." This is explained in some detail in the review by Philipson and Nicoll (2000). The first is the **NCX family.** These are all mammalian exchangers and include NCX1, the first exchanger to be isolated, as well as NCX2 and NCX3. It is on these exchangers, particularly NCX1, that most studies of currents that are generated by exchange have been performed. Other exchangers in the NCX family (of which there are eight) are found, for example, in the nematode *C. elegans* and also in the fruit fly *Drosophila*. The NCX family is found in a wide variety of mammalian species including human, dog, rabbit, cow, rat, and guinea pig as well as in various nonmammalian species.

A second family of exchangers is designated **NCKX** and consists of proteins similar to the Na^+-$(Ca^{2+}+K^+)$ exchanger initially cloned from retinal tissue. NCKX1 and NCKX2 are mammalian. An additional eight exchangers have been cloned from various invertebrates. These exchangers are structurally quite different from the NCX family and there have been considerable studies of the electrogenic properties of the NCKX ROS exchanger.

Two other families of proteins, one of which is bacterial and the other of which functions as a yeast Ca^{2+}-H^+ exchanger, are known to exist. However, discussion of these is outside the scope of this chapter since currents have not been measured in these exchangers.

IV. Isoforms of the Na^+-Ca^{2+} Exchanger

The large intracellular loop of NCX1 can be spliced in different ways to produce a variety of splice variants. A region of the loop (see Fig. 1) contains six small exons that are used in different combinations to produce tissue-specific splice variants which produce **isoforms** of the exchanger. The physiological significance of tissue-specific isoforms is unclear. Presumably they evolved to perform functions required only by those tissues in which they are found. Recent studies support this (Omelchenko *et al.*, 1998). These workers observed, for example, that Na^+ and Ca^{2+} regulated two splice variants of the exchanger found in *Drosophila* quite differently.

V. Energetics of Na^+-Ca^{2+} Exchange

Reuter and Seitz in their classic study proposed that two Na^+ ions exchanged for a single Ca^{2+} ion in a modified exchange diffusion process. Based upon this proposal, one would not expect steady-state exchange activity to produce a measurable electric current. However, the existence of exchange currents is a well-established fact, and one that can be appreciated by consideration of the energetic and stoichiometric properties of the exchange as they are currently understood. It is convenient (but not necessarily correct) to represent the transmembranous exchange reaction as a sequential (simultaneous) process,

$$nNa_i^+ + Ca_o^{2+} \leftrightarrow nNa_o^+ + Ca_i^{2+}$$

where n is the stoichiometric coefficient of the exchange reaction. If the forward and reverse reaction rates are equal, the exchange reaction is at equilibrium. Even if n is greater than 2, there can be no net charge movement and, hence, no electric current generated at equilibrium. An electric current can only be measured when the exchange is displaced from equilibrium. While the net reaction rate at equilibrium is zero, the unidirectional rates might be substantial (Axelsen and Bridge, 1985). As soon as the forward and reverse exchange rates differ from one another, net ion translocation takes place; provided that the stoichiometric coefficient is appropriate (i.e., n greater than 2) and no ions of opposite charge are cotransported, one can expect to measure an electric current as a consequence of exchange.

If electrochemical forces solely determine net movement of ions through the Na⁺-Ca²⁺ exchanger, classical thermodynamics may be used to calculate both the direction of exchange and the conditions under which we may expect equilibrium to occur. Having the capacity to do this is of enormous value when designing experiments to measure exchange currents. The electrochemical potential difference or driving force ($n\Delta\bar{\mu}$) producing exchange is the difference between n times the electrochemical potential difference or force producing sodium movement ($n\Delta\bar{\mu}_{Na}$) and calcium movement ($\Delta\bar{\mu}_{Ca}$). Driving force can be expressed in terms of membrane potential (E_m), Na⁺ equilibrium potential (E_{Na}), and Ca²⁺ equilibrium potential (E_{Ca}). Thus, we may write

$$\Delta\bar{\mu} = n\Delta\bar{\mu}_{Na} - \Delta\bar{\mu}_{Ca} \tag{1}$$

At equilibrium, $\Delta\bar{\mu} = 0$, so that

$$n\Delta\bar{\mu}_{Na} - \Delta\bar{\mu}_{Ca} = 0 \tag{2}$$

$$\Delta\bar{\mu}_{Na}(mV) = \frac{RT}{\mathscr{F}} \ln \frac{[Na^+]_o}{[Na^+]_i} + E_m = E_{Na} + E_m \tag{3}$$

$$\Delta\bar{\mu}_{Ca}(mV) = \frac{RT}{\mathscr{F}} \ln \frac{[Ca^{2+}]_o}{[Ca^{2+}]_i} + 2E_m = 2E_{Ca} + 2E_m \tag{4}$$

Substitution yields

$$nE_{Na} - 2E_{Ca} - (n-2)E_m = 0 \tag{5}$$

The exponential form of Eq. 5 is

$$\frac{[Ca^{2+}]_o}{[Ca^{2+}]_i} = \left(\frac{[Na^+]_o}{[Na^+]_i}\right)^n \exp\left(-(n-2)\frac{E_m\mathscr{F}}{RT}\right) \tag{6}$$

These are the equations that can be used to predict the equilibrium conditions of the exchanger. Before doing so, one needs to know the **stoichiometric coefficient** (n) of the exchanger. This issue has been the subject of a lengthy debate and numerous investigations. However, there now appears to be broad agreement that three Na⁺ ions are exchanged for a single Ca²⁺ ion in most mammalian systems studied (Bridge and Bassingthwaighte, 1983; Kimura et al., 1987; Bridge et al., 1990; Crespo et al., 1990). There is also good evidence that exchange stoichiometry is 3:1 in barnacle fibers (Rasgado-Flores and Blaustein, 1987).

An elegant (though somewhat indirect) demonstration of exchange stoichiometry was provided by Reeves and Hale (1984). Their study not only produced a value for the exchange stoichiometry, but provided an excellent example of the way that the foregoing energetic principles may guide experimental design. These authors took advantage of the fact that exchange equilibrium can be achieved simply by appropriate adjustment of the Na⁺ and Ca²⁺ electrochemical gradients. Bovine sarcolemmal vesicles containing Na⁺-Ca²⁺ exchanger were equilibrated with solutions of both Na⁺ and ⁴⁵Ca. Under these circumstances, the equilibrium may be described by Eq. 5. After treating the membrane with valinomycin in the presence of KCl, known membrane potentials were established that caused disequilibrium of the exchanger and either Ca²⁺ entry or exit. By adjusting the Na⁺ gradient, it was possible to precisely null the tendency of membrane potential to produce Ca²⁺ movement. Thus, if any of the quantities E_m, E_{Na}, and E_{Ca} are held constant, the relationship between the other two may be found. The point at which Ca²⁺ movement was nulled by Na⁺ gradient is given by

$$(n-2)E_m = nE_{Na} \tag{7}$$

By nulling Ca²⁺ movement over a range of membrane potentials, the value for n that these authors obtained was 2.97 ± 0.03, which is close to the currently accepted value of 3.0.

Equipped with a value for the stoichiometric coefficient of exchange, some useful parameters can be calculated. Heart muscle is used as an illustrative example to calculate expected reversal potentials for the exchange. It is assumed that a resting ventricular cell maintains intracellular Na⁺ and Ca²⁺ at concentrations of 10 mM and 100 nM respectively, and extracellular Na⁺ and Ca²⁺ at 140 mM and 2.0 mM. For $n = 3$, the reversal potential is given by (see Eq. 5)

$$E_{rev} = 3E_{Na} - 2E_{Ca} = -50\,mV \tag{8}$$

For the foregoing conditions E_{rev} may be calculated to be −50 mV.

VI. Methods and Problems Associated with the Measurement of Na⁺-Ca²⁺ Exchange Current

Several recent techniques have greatly facilitated the isolation of exchange current. First, the **whole-cell ruptured patch voltage clamp** in conjunction with intracellular dialysis (Hamill et al., 1981) has probably contributed most to the initial isolation of exchange currents. To record whole-cell current with the ruptured patch technique, a microelectrode or patch pipette is pressed onto the cell membrane and suction is applied, usually with a syringe or directly by the experimentalist. This often results in a high-resistance seal that forms between the pipette and the cell membrane. The resistance of this seal is ideally >10¹⁰ ohms and it is therefore referred to as a gigaseal. This is usually accompanied by the formation of an Ω structure as the membrane is pulled into the pipette (see Chapter 26). Continued suction ruptures the Ω, and this creates continuity between the cell interior and the pipette solution. By pulling the sealed pipette from the cell one may isolate an inside-out patch. It is also possible

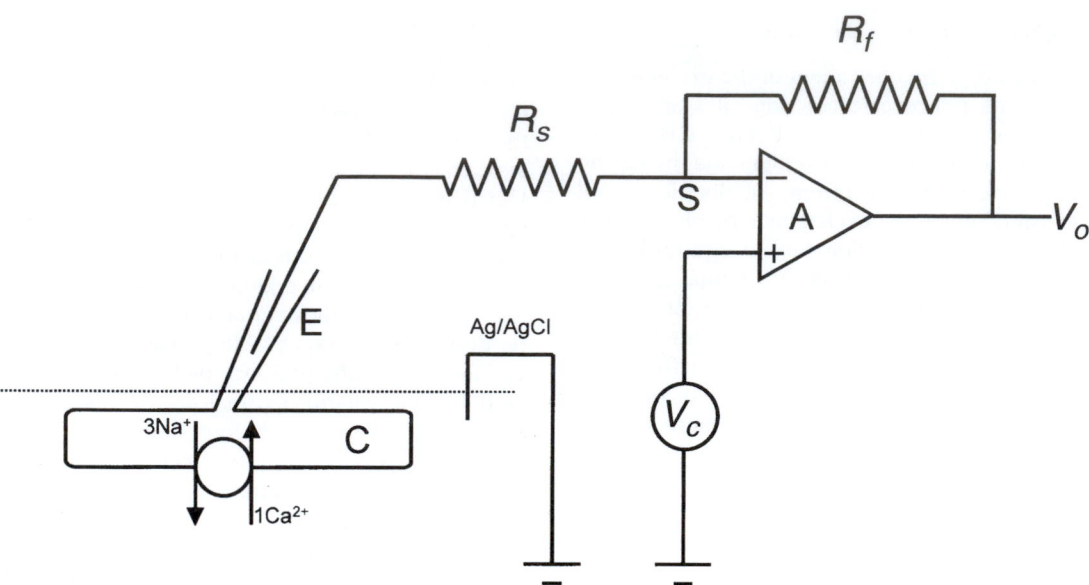

FIGURE 2. An arrangement suitable for measuring Na$^+$-Ca^{2+} exchange current in a ventricular heart cell. Under these circumstances the exchanger is producing a net outward current across the cell membrane by exchanging three internal Na$^+$ ions with a single external Ca^{2+} ion. This produces outward movement of one net charge. The voltage-clamp circuit consists of a high-gain amplifier A. Any deviations from the command potential V_c produced by the exchanger are compensated by feedback from the output through the feedback resistor R_f. The amplifier is connected to the cell through the electrode E, which is sealed to a ruptured patch. The output of this amplifier is proportional to the current flowing through the feedback resistor R_f and hence the cell C. The success of this method depends upon minimizing voltage drops across the series resistance R_s. The bath in which the cell is maintained is grounded by a Ag/AgCl electrode connected to a salt bridge (not shown).

with the methods illustrated in Chapter 26 to create an outside-in patch. Hilgemann (1989) has refined this technique so that very large membrane patches can be isolated from appropriately treated cell membranes.

A simplified arrangement for voltage-clamping heart cells and recording exchange current through a ruptured patch is depicted in Fig. 2. Essentially the same arrangement applies to the voltage-clamping of an isolated patch. The micropipette attached to the cell is connected via a (nonpolarizable) electrode to the front end of a high-gain amplifier known as a current-to-voltage converter. The most significant feature of this high amplifier is the feedback loop R_f and its branch point (the summing junction). In operational amplifiers (of which this current-to-voltage converter is an example), this branch point has special significance. Negative feedback through the feedback loop tends to cancel the input signal at S. A command potential V_c is applied to the noninverting input (+). A property of this amplifier is that as a result of negative feedback through R_f, current always flows so that the voltage at the summing junction is identical to the voltage V_c (the command potential). The potential difference between the summing junction and ground is therefore ideally equal to V_c. This potential difference therefore falls across the electrode resistance (and any other resistances in series with the membrane resistance R_m). Provided that the product of this series resistance (R_s) and the exchange current I_{Na-Ca} ($R_s x I_{Na-Ca}$) is small, most of the potential difference V_c occurs across the membrane. Feedback through R_f always compensates varying input at the summing junction, thus ensuring that the voltage across the membrane is held constant (i.e., is clamped).

Exchange current I_{Na-Ca} passing across the membrane produces a potential difference across the membrane. As this potential difference causes the membrane potential to depart from V_c, current (equal in magnitude to the exchange current) flows into the summing junction. In reality additional circuitry is required to reduce noise in the recording system while increasing its fidelity. This is dealt with in greater detail in Chapter 26 by Pun and Lecar.

The current flowing through a heart cell membrane, I_m, consists of two components: a capacity current and an ionic current I_i. The capacity current $I_C = C \, dV/dt$ and the membrane current is denoted I_m. Therefore

$$I_m = C \frac{dV}{dt} + I_i + I_{Na-Ca} \qquad (9)$$

The ionic current consists of ions flowing through ion channels as well as those translocated by ion exchange mechanisms (principally the Na$^+$-Ca^{2+} exchange and the Na$^+$ pump). For purposes of illustration, the exchange current has been separated from the ionic current in Eq. 9. Under voltage-clamp $dV/dt = 0$,

$$I_m = I_i + I_{Na-Ca} \qquad (10)$$

If one can remove or minimize I_i (which includes the Na$^+$ pump current and all contaminating channel currents) without affecting exchange current I_{Na-Ca}, one may measure a membrane current that is comprised mainly of I_{Na-Ca}.

How will the activity of an electrogenic Na$^+$-Ca^{2+} exchange affect membrane potential? A convenient equivalent circuit to explain this is depicted in Fig. 3. This simplified equivalent resting membrane circuit consists of a membrane capacitance that is in parallel with a battery E_m, which is largely responsible for producing the membrane potential. The membrane resistance R_m is modeled as the internal resistance of this battery. Under resting conditions the reversal potential for the Na$^+$-Ca^{2+} exchange in heart is approximately –40 mV. At –80 mV the exchange will produce a net inward current, which can be modeled as a current generator that is in parallel with the membrane battery and capacitance (Fig. 3). At the steady state, when the resting membrane potential is not changing, $C\, dV/dt = 0$. Thus the net membrane potential is composed of contributions from ionic currents (mainly through K$^+$ channels), exchange currents, and the Na$^+$ pump current (Eq. 10). The exchange will produce an IR drop across the membrane resistance so that the contribution of Na$^+$-Ca^{2+} current to the membrane potential ($V_{\text{Na-Ca}}$) at rest is given by

$$V_{\text{Na-Ca}} = R_m I_{\text{Na-Ca}} \qquad (11)$$

A large inward exchange current produced by elevated intracellular [Ca^{2+}] may be of the order 200 pA. If we assume that the resting exchange current is one-tenth of this value and given a resting R_m of 50 mΩ, the exchange will contribute about 1 mV to the resting membrane potential. In reality the value is likely to be less than this but cannot be specified with certainty because reliable measurements of resting current are not available.

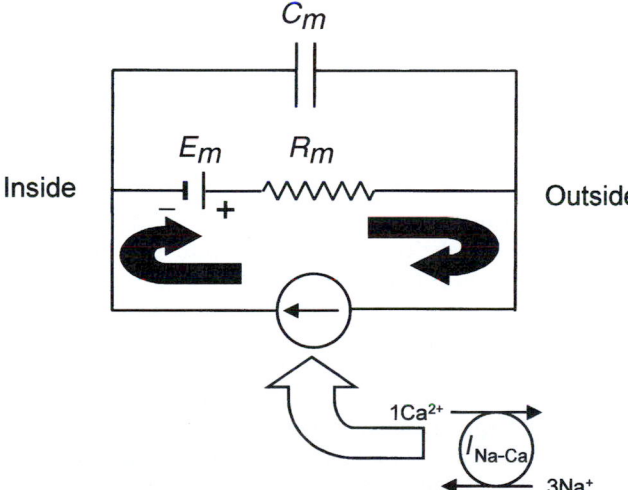

FIGURE 3. A simple equivalent circuit for a resting heart cell. The resting membrane potential E_m is typically about –80 mV. This resting membrane potential is modeled as a battery in series with an internal resistance of about 50 MΩ. The membrane capacitance is in series with the battery. The Na$^+$-Ca^{2+} exchanger may be considered to be a current generator in parallel with the battery and membrane resistance (symbolized by an open circle containing an arrow). At rest, with the exchanger steadily extruding Ca^{2+}, a small depolarizing current $I_{\text{Na-Ca}}$ causes a small voltage drop across the membrane resistance R_m. The solid arrows indicate the direction of current. The circle containing an arrow indicates a current generator (the Na$^+$-Ca^{2+} exchange). Thus for each cycle of the exchanger a single net charge is conveyed from the cell exterior to its interior.

The first observations of exchange current were not published until nearly 20 years after Reuter and Seitz originally reported Na$^+$-Ca^{2+} exchange activity in heart tissue. It is worth considering some of the problems associated with the measurement of exchange current that might account for this lengthy hiatus.

To stimulate exchange so that it can be measured, one can either change electrochemical gradients (and hence the exchange driving forces) by using a voltage-clamp step or one can abruptly change external ion composition with a rapid switching device. In the case of the voltage-clamp step, the experimentalist must have a complete understanding of what interfering currents will be activated, as well as some means of inhibiting them. If exchange is to be activated by changing ionic gradients, then these changes must be sufficiently rapid that they are not immediately dissipated by the exchange activity. It is difficult to rapidly change or control extracellular ionic composition in multicellular preparations because diffusion distances are large and diffusion may be hindered by a surface structure. In multicellular preparations, voltage-clamp is usually established with small microelectrodes, which are unsuitable for cell dialysis. It is, therefore, very difficult to control the intracellular ionic composition. This doubtless contributed to the difficulty of isolating exchange currents in multicellular preparations. Despite these difficulties a measurement of exchange currents has been reported in multicellular preparations (Horackova and Vassort, 1979).

Recently, it has proved feasible to use the patch-clamp technique in conjunction with wide-tipped microelectrodes (10-μm diameter) to obtain an isolated patch whose surface area is approximately 75 μm^2 (Fig. 2). A conventional isolated patch is of the order 2–3 μm^2. These **giant patches** are sufficiently large to permit the detection of exchange currents (Hilgemann, 1989). Their great virtue is that they permit relatively easy access to either side of the membrane. Thus, solution adjacent to the external surface of the sarcolemma can be changed by changing the pipette solution. The internal surface of the sarcolemma can be changed rapidly by changing the bathing solution. Clearly this method of voltage-clamp allows the experimentalist considerable control over the forces that drive exchange.

Several other techniques have recently become available that improve exchange current isolation. For example, forward exchange current in intact cells can be activated by abruptly elevating intracellular Ca^{2+}, which is then transported to the cell exterior. This has recently been accomplished in an elegant fashion by Niggli and Lederer (1991b). These workers used the compound DM-dinitrophen that is commonly referred to as *caged* Ca^{2+}. This can be introduced into the cell interior through a patch pipette. Upon appropriate irradiation with UV light, Ca^{2+} is released from the caged Ca^{2+} with extremely rapid kinetics. This abruptly elevates cytosolic Ca^{2+} and transiently stimulates forward Na$^+$-Ca^{2+} exchange as the released Ca^{2+} is pumped to the cell exterior.

As we have indicated, to activate exchange it is desirable to change external ionic composition (and the electrochemical gradients driving exchange) extremely rapidly in comparison with the time required for the exchange to dissipate these changes. It is now possible to change external solutions surrounding heart cells or isolated patches with a half-time of about 20 ms (this includes exchange of the unstirred layer). One method (Spitzer and Bridge, 1989) consists of placing a cell with attached microelectrodes in one of two adjacent microstreams of solution. The boundary separating these streams is abruptly moved across the cell so that it is placed in the adjacent stream. This rapid switching method has proved valuable in stimulating both inward and outward exchange currents (Chin *et al.*, 1993).

VII. Isolation of Na$^+$-Ca^{2+} Exchange Current

A. Whole-Cell Patch-Clamp Studies

The first clear evidence that an electric current was generated by Na$^+$-Ca^{2+} exchange was provided simultaneously by Kimura and coworkers (1986) and by Mechmann and Pott (1986). Given the probable stoichiometry of the exchange, electrophysiologists expected to be able to measure a current generated by the exchange. By using the whole-cell patch-clamp technique together with intracellular dialysis, Junko Kimura and her associates were able to isolate the Na$^+$-Ca^{2+} exchange current. These investigators voltage-clamped guinea pig ventricular myocytes with single microelectrodes. The pipette contained a dialyzing solution completely deficient in Na$^+$ and in which Ca^{2+} was buffered to a value of 73 nM with EGTA. The superfusing external solution contained no Ca^{2+}. Under these circumstances no exchange could take place. Very little change in current was observed when external Ca^{2+} was reapplied and subsequently removed. However, an outward current was generated after changing the pipette solution for one containing 30 mM Na$^+$ and then applying 1.0 mM external Ca^{2+}. Under these circumstances Ca^{2+} entered the cell in exchange for internal Na$^+$, which was extruded. This current (Fig. 4) is clearly attributable to the operation of electrogenic Na$^+$-Ca^{2+} exchange. This conclusion was strengthened by several additional observations. Upon removal of external Ca^{2+} the current was turned off. The current was reduced when lower concentrations of dialyzing Na$^+$ were used and enlarged by increasing external Ca^{2+}. Thus, the current exhibited the expected dependency on both Na$^+$ and Ca^{2+}. The current was blocked by La^{3+}, which is known to block Na$^+$-Ca^{2+} exchange. The current also exhibited **voltage dependence** (see Section X). One might have expected sustained current under these circumstances, but this was not observed (Fig. 4) and the current exhibited a tendency to decay. One plausible explanation for this is that excessive Ca^{2+} entry displaces protons when it is buffered by the dialyzing EGTA. The resulting acidification in the vicinity of the exchanger should inhibit exchange activity. An alternative possibility is that EGTA fails to buffer incoming Ca^{2+}, which accumulates in the vicinity of the exchanger. The resulting collapse of the Ca^{2+} gradient would also slow exchange.

Kimura and colleagues reported the measurement of an outward current corresponding to the entry of Ca^{2+} and extrusion of Na$^+$ from the cell. Mechmann and Pott (1986) demonstrated the existence of an inward exchange current corresponding to Ca^{2+} extrusion and Na$^+$ entry into a cardiac cell. In one experiment, these authors voltage-clamped spherical atrial cells

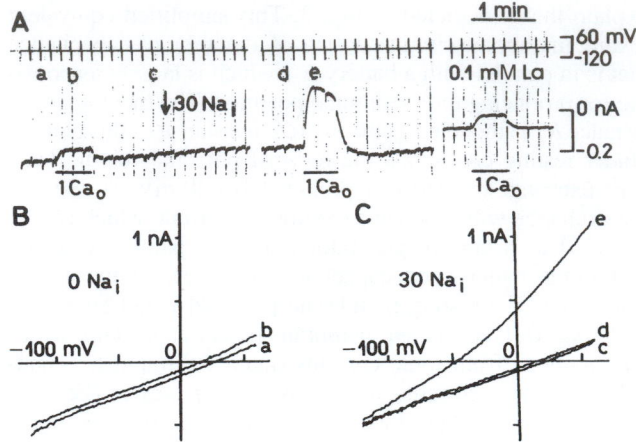

FIGURE 4. (A) Voltage-clamp record from a single myocyte showing voltage (upper trace) and current (lower trace). A ramp pulse from +60 mV to –120 mV was given from a holding potential of –30 mV every 10 s at a ramp speed of about 0.2 V/s. Current-voltage relationships were then constructed. An application of 1 mM Ca^{2+} in the absence of intracellular Na$^+$ (zero pipette Na$^+$) produced a small outward current. To exchange the pipette solution, a piece of tapered polyethylene tube was inserted into the pipette. The arrow indicates when the solution change was started. The holding current shifted outward slightly on loading Na$^+$. The middle current record shows 1 mM Ca^{2+} superfusion in the presence of 30 mM pipette Na$^+$. The Ca^{2+}-induced outward current appears to decay after reaching a peak. Current undershoot is seen after washing off Ca^{2+}. These phenomena may be due to Ca^{2+} accumulation immediately below the membrane of acid induced by binding of Ca^{2+} to EGTA. (B) *I-V* relations measured before (a) and during (b) 1mM Ca^{2+} superfusion in the absence of Na^+_i. (C) *I-V* curves before (c) and after (d) loading Na^+_i and during subsequent 1mM Ca$^{2+}_o$ application (e). (Reprinted from Mechmann and Pott (1986) with kind permission of the authors and *Nature,* copyright 1986, Macmillan Magazines Ltd.)

from an adult guinea pig. Transient inward current occurred spontaneously in this preparation. These transient inward currents seemed to depend on intracellular Ca^{2+} because they were abolished if the cell was dialyzed with 1.0 mM EGTA. Moreover, they could not be elicited in the presence of caffeine, which depletes the SR of Ca^{2+}. Presumably the transient release of SR Ca^{2+} activated these currents. The currents also showed a dependence on extracellular Na$^+$ and voltage as expected of a Na$^+$-Ca^{2+} exchange current. The more negative the voltage, the larger the inward current. This is reasonable and simply indicates that at more negative potentials the exchange rate is greater so that Ca^{2+} released from the SR is removed rapidly from the cell.

Shortly after these early demonstrations of exchange current, Hume and Uehara (1986a, b) were able to demonstrate Na$^+$-Ca^{2+} exchange currents in isolated frog atrial cells by using the whole-cell patch-clamp technique in combination with intracellular dialysis. It is fair to say that since this time measurement of exchange current has become both reliable and routine. An example of an inward exchange current measured in a ventricular cell is displayed in Fig. 5.

The development of the giant patch technique has also facilitated measurement of the exchange current produced by the squid exchanger NCX-SQ1 (He *et al.*, 1998) (see Section X).

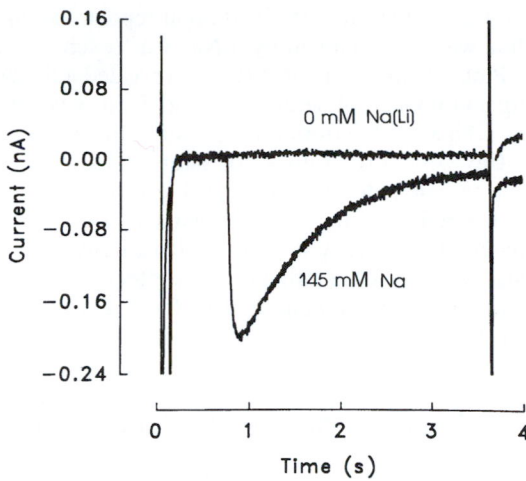

FIGURE 5. Transient inward Na⁺-Ca²⁺ exchange. To activate this current, a guinea pig ventricular cell was tetanized with voltage-clamp pulses in the presence of ryanodine and in the absence of external Na⁺. This resulted in an elevation of cytosolic Ca²⁺ and a sustained contraction (not shown). Abrupt application of 145 mM extracellular Na⁺ while the cell was held at –40mV produced mechanical relaxation and activated a transient inward Na⁺-Ca²⁺ exchange current. As intracellular Ca²⁺ declined, the current decayed. (Reprinted with permission of the *Annals of the New York Academy of Science*.)

B. Na⁺-Ca²⁺ Exchange Current Reversal Potential

It is desirable when studying any ionic current to be able to demonstrate a **reversal potential** for that current. The existence of reversal potentials is a thermodynamic necessity and provides one of the least ambiguous ways of identifying an ionic current. However, it is not always a straightforward matter to demonstrate reversal potentials. For example, if exchange current were extremely small over a large voltage range in the vicinity of the reversal potential, then the precise potential at which zero current occurred could be extremely difficult to specify. Moreover, subsarcolemmal spaces from which diffusion is restricted can produce changes in ionic concentration that confound measurement of the exchange reversal potential. For the Na⁺-Ca²⁺ exchange in heart the reversal potential is given by Eq. 8. For the conditions of their experiment, Kimura and coworkers (1986) (see Fig. 4) calculated that the reversal potential of exchange ought to be –131 mV. Their results indicate that exchange current becomes zero at a potential close to this value. Ehara and colleagues (1989) extended studies of the exchange reversal potential. With the fairly specific exchange inhibitor Ni, these authors were able to show that over a wide range of external Na⁺ values the measured exchange reversal potential conformed to theoretical expectation. Moreover, in view of the wide range over which the reversal potential remains constant, the stoichiometry is unlikely to vary.

C. Na⁺-Ca²⁺ Exchange Current (Transport) and Structure

Some information is now available on the relationship between molecular structure and transport function. It appears that exchange activity (and presumably exchange current) is extremely sensitive to mutations in the α repeat region (see Fig. 1). For example, a nonfunctional exchanger can be produced if Ser 110 is mutated to Ala. It appears therefore that the gene duplication event that produced the α repeat regions is of special functional importance. In general, exchange activity is not affected by mutations in the intracellular loop; however, mutations of transmembrane segment 2 dramatically affect ion selectivity. Normally Li⁺ is not transported by the exchanger. However, when Thr 103 is mutated to a Val, Li⁺-Ca²⁺ exchange can occur (Doering *et al.*, 1998). Mutations of this nature could prove extremely valuable as probes of the function of the exchanger, as in cardiac excitation contraction coupling.

VIII. Ionic Dependencies of Na⁺-Ca²⁺ Exchange Current

It is of considerable physiological significance that the Na⁺-Ca²⁺ exchange current is extremely sensitive to internal Na⁺. There are two aspects of this sensitivity to consider, one purely thermodynamic and the other kinetic. If we assume that a resting heart cell contains 100 nM free cytosolic Ca²⁺ and is bathed in 2.0 mM Ca²⁺ and 140 mM Na⁺, we can use Eq. 8 to calculate the way the exchange reversal potential varies with internal Na⁺. The results are displayed in Fig. 6. It is apparent that with 10 mM intracellular Na⁺, 140 mM extracellular Na⁺, and 100 nM intracellular free Ca²⁺, the reversal potential for exchange is –50 mV. Since the resting membrane potential is at least 30 mV negative to this, at rest the exchanger will extrude Ca²⁺. However, a modest increase in intracellular Na⁺ to 12 mM would change the reversal potential by 14 mV to –64 mV. Were the Na⁺ to accumulate to 15 mM, then the reversal potential would be –80 mV. Since the resting potential is likely to be close to this value the exchange would be close to, or at, equilibrium and would be incapable of extruding intracellular Ca²⁺. Further Na⁺ accumulation

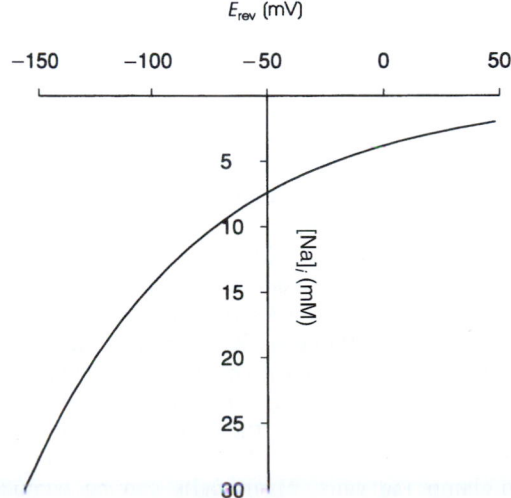

FIGURE 6. Calculated variation of exchange reversal potential with intracellular Na⁺. Intracellular Ca²⁺ was set at 100 nM and extracellular Ca²⁺ was 2.0 mM. Extracellular Na⁺ was 140 mM.

would cause the exchange to reverse and transmit Ca^{2+} into the cell. Na^+ accumulation will tend to bring the exchange closer to its equilibrium potential and, therefore, increase the likelihood of Ca^{2+} entry via the exchanger upon membrane depolarization. Cardiac glycosides, which, depending on dose, partially or completely block the sodium pump, tend to produce an accumulation of intracellular Na^+ (Sheu and Fozzard, 1982). Part of the basis of their inotropic effect resides in their shifting the reversal potential of the Na^+-Ca^{2+} exchange to more negative values. In extreme cases, this will prevent Ca^{2+} extrusion which, in the face of continued Ca^{2+} leak, will lead to Ca^{2+} accumulation.

The Na^+-Ca^{2+} exchange current also exhibits a physiologically important kinetic dependence upon internal Na^+. This issue has been investigated directly by Miura and Kimura (1989), who studied outward exchange current in guinea pig ventricular cells under voltage-clamp. Intracellular Ca^{2+} was buffered to 100 nM and intracellular Na^+ was varied by varying the pipette Na^+ concentration. External Na^+ was reduced to zero and a unidirectional exchange reaction was activated by applying 0.2 mM Ca^{2+} as rapidly as possible to the cell exterior. The results demonstrate that exchange current density exhibits a steep and somewhat sigmoid dependence on internal Na^+. The K_m for this dependency is approximately 20 mM and Hill plots reveal a Hill coefficient of 1.9. It is notable that an application of rate theory revealed that the voltage dependence of the exchange did not depend on intracellular Na^+ (see Section X). It is clear from these results that at least Ca^{2+} influx on the exchange will be extremely sensitive to fluctuations in intracellular Na^+ produced, for example, by increases in stimulation rate or cardiac glycosides. It is unfortunate that we have no information on the intracellular Na^+ dependence of inward exchange current which corresponds to Ca^{2+} efflux. However, studies with isotopic fluxes reveal that Na^+ and Ca^{2+} compete for transport on the exchanger. For example, Na^+ on the same side of the cardiac sarcolemmal membrane as Ca^{2+} inhibits Ca^{2+} flux through the exchanger with a K_i of about 15 mM (Reeves and Sutko, 1983). Finally, the net inward exchange current does depend on external Na^+. This has been investigated by Kimura and coworkers (1986), who showed that inward current exhibited a sigmoidal dependence on external Na^+ with a $K_{1/2}$ of 87.5 mM and a Hill coefficient of 2.9.

Net inward exchange current (i.e., Ca^{2+} extrusion in exchange for Na^+ entry) is also dependent on intracellular Ca^{2+}. The relationship between Ca^{2+} concentration and current has been measured in two different ways. Miura and Kimura (1989) used whole-cell patch-clamp on guinea pig myocytes to determine K_m (Ca^{2+}). They controlled intracellular Ca^{2+} by dialyzing the cell with various EGTA-buffered solutions and determined a K_m (Ca^{2+}) to be 0.6 μM. With a different approach, Barcenas-Ruiz and colleagues (1987) used guinea pig ventricular myocytes to measure exchange current and intracellular Ca^{2+} simultaneously. Intracellular Ca^{2+} was measured with the Ca^{2+}-sensitive indicator Fura–2, introduced through the dialyzing pipette used to voltage-clamp the cells. Heart cells can be tetanized or caused to go into contracture in the presence of ryanodine. Various clamp pulses up to 120 mV in amplitude (holding potential of –80 mV) produced slow and sustained in-

creases in Ca^{2+}. This Ca^{2+} declined upon repolarization and the decline was accompanied by a Na^+-Ca^{2+} exchange current tail. Plots of current against $[Ca^{2+}]_i$ revealed a linear relationship and no sign of saturation (Fig. 7). It is not possible to extract a K_m value from this data since the relationship between Ca^{2+} and current does not saturate. However, it is reasonable to conclude that the K_m value is well above the 0.6 μM obtained by Kimura and colleagues. Discrepancies of this nature are not easy to resolve. It is, however, worth noting that EGTA has the rather curious effect of increasing the affinity of the exchanger for Ca^{2+} (Trosper and Philipson, 1984).

IX. Regulation of Na^+-Ca^{2+} Exchange Current

The regulation of Na^+-Ca^{2+} exchange current has largely been studied in NCX1. We will consider four types of regulation. First, intracellular Ca^{2+} regulates the exchange current. The exchange current also exhibits a regulatory phenomenon known as **Na^+-dependent inactivation.** Thus the exchanger not only transports Na^+ and Ca^{2+}, but these ions are capable of regulating its function as well. Exchange current is also known to be regulated by phosphatidylinositol-4,5-bisphosphate (PIP_2), and finally, by phosphorylation. In addition, a great deal has been learned about regulatory function and its relationship to the structure of the exchanger.

A. Regulation by Na^+ and Ca^{2+}

There are two Ca^{2+} binding sites on the intracellular surface of the cell membrane: one is the primary transport site, and the other is a high-affinity Ca^{2+} site required for secondary ionic regulation. It is now apparent that Ca^{2+} influx in exchange for internal Na^+ cannot occur if intracellular Ca^{2+} is reduced below a certain critical level. The first description of this secondary regulation of Na^+-Ca^{2+} exchange by internal Ca^{2+} was provided by Baker and McNaughton (1976) in their study of squid axons. In their more recent studies of heart cell membrane, both Kimura and colleagues (1986) and Hilgemann (1990) have shown that when internal Ca^{2+} is sufficiently reduced, exchange current is inactivated. It appears that in dialyzed myocytes the K_D for the regulatory site is about 50 nM. However, in isolated patches, the K_D appears to be from 0.1 to 0.3 μM. If this is true, the regulatory sites in heart muscle would always be saturated and serve little function. If, on the other hand, the lower value obtained in dialyzed cells is correct, then Ca^{2+} regulation would acquire significance as a regulator of both Ca^{2+} influx and efflux on the exchanger. Why the regulatory site exists remains somewhat unclear.

Na^+-dependent inactivation of exchange current is a rather curious regulatory phenomenon first described by Hilgemann and coworkers (1992). Outward Na^+-Ca^{2+} exchange currents were activated in giant sarcolemmal patches by increasing Na^+ on the cytoplasmic side of the patch. Ca^{2+} was then transported from the pipette to the cytoplasmic side of the patch in exchange for Na^+. With 60 mM Na^+ on the pipette side of the patch, application of 60 mM

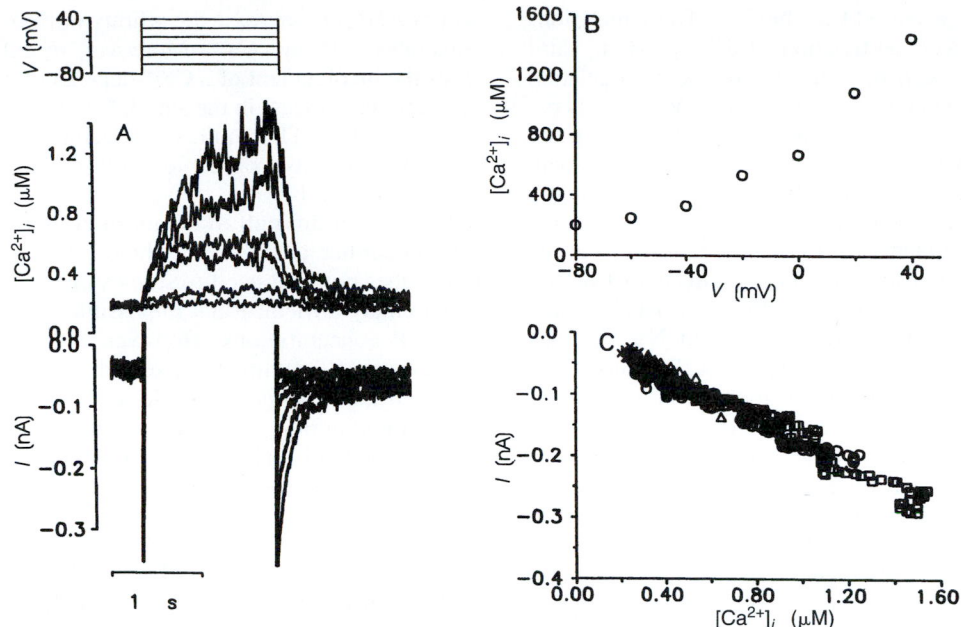

FIGURE 7. Changes in $[Ca^{2+}]_i$ and membrane current attributed to Na^+-Ca^{2+} exchange in a single guinea pig ventricular myocyte under voltage-clamp. Micropipette $[Na^+]$ was 7.5 mM. (A) Simultaneous recordings of $[Ca^{2+}]_i$ and membrane current. The holding potential was –80 mV, and depolarizing pulses lasting for 1.5 s were given from –60 mV to +40 mV. Outward currents during the depolarizing pulse are off-scale. (B) Voltage dependence of the change in $[Ca^{2+}]_i$. Values plotted are average $[Ca^{2+}]_i$ over the last 200 ms of the depolarizing pulse. (C) The relationship between $[Ca^{2+}]_i$ and membrane current after repolarization. The current, I, has been plotted as a function of $[Ca^{2+}]_i$ at 2 ms intervals during the first second after repolarization from +40 mV (circles), 0 mV (triangles), –20 mV (diamonds), and –40 mV (crosses). [Reprinted with permission from Bridge, J. H. B., Smolley, J. R., and Spitzer, K. W. (1990). The relationship between charge movements associated with I_{Ca} and I_{Na-Ca} in cardiac myocytes. *Science* **248**, 376–378. Copyright 1990 American Association for the Advancement of Science.]

Na⁺ on the cytoplasmic side of the patch activated outward current in the presence of a large Ca²⁺ gradient (at 0 mV membrane potential). This outward current declined with a time constant of approximately 1 s. This decay rate is far too low to represent an initial turnover of exchangers and it appears to be a true regulatory phenomenon. An example of this sort of behavior is displayed in Fig. 8. Here, application of 100 mM Na⁺ to the cytoplasmic surface activated a transient outward current.

B. Relationship between Ca²⁺ and Na⁺ Regulation and Structure

Advances in the study of the molecular biology of the Na⁺-Ca²⁺ exchanger have led to a study of the relationship between ionic regulation and the structure of the exchange molecule. Both Ca²⁺ secondary regulation and Na⁺-dependent inactivation can be removed if the cytoplasmic surface of an excised patch is treated with chymotrypsin (Hilgemann, 1990). From this it may be inferred that these regulatory properties are related in some way to a cytoplasmic domain of the exchange protein. Recent evidence suggests that the sarcolemmal Na⁺-Ca²⁺ exchanger is regulated by intracellular Ca²⁺ at a high-affinity binding site separate from the Ca²⁺ transport site. It was first suggested by Matsuoka and colleagues (1993) that the large hydrophilic loop of the exchanger was the binding site for regulatory Ca²⁺. Levitsky and colleagues (1994) have identified a high-affinity Ca²⁺-binding

region comprising amino acids 371–521. This is a large region of the cytoplasmic loop and Levitsky and colleagues have inferred that substantial secondary structure is necessary for high-affinity Ca²⁺ binding. Detailed functional studies to measure exchange currents on giant patches from oocytes expressing both wild-type and mutant exchangers have revealed additional information on the Ca²⁺ regulation of Na⁺-Ca²⁺ exchange. Mutation of certain Asp residues within two acidic

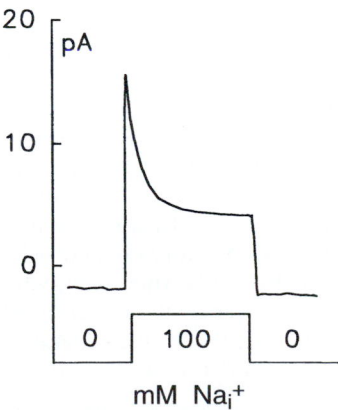

FIGURE 8. Secondary modulation of outward Na⁺-Ca²⁺ exchange current in giant excised inside-out patches (37 °C, 0 mV). Current transients activated with 100 mM Na⁺ (substituted for 100 mM Cs⁺). (This figure was kindly provided by Dr. Donald Hilgemann.)

segments markedly decreased Ca^{2+} binding. These mutations exhibited parallel effects on functional Ca^{2+} regulation. Until recently it has only been possible to study Ca^{2+} regulation when the exchanger operates in the reverse mode. This is because regulatory and transported Ca^{2+} are on opposite sides of the membrane so that the concentration of regulatory Ca^{2+} can be manipulated without affecting the concentration of transported Ca^{2+}. Matsuoka and coworkers (1995) have been able to exploit a class of mutants with low affinity for Ca^{2+} at the regulatory site to demonstrate regulatory effects of Ca^{2+} on forward mode exchange. Finally, the affinity of both wild and mutant Na^+-Ca^{2+} exchangers for transported Na^+ declines with declining regulatory Ca^{2+} concentration. This further indicates that Ca^{2+} regulation modifies transport properties and does not simply control the fraction of exchangers that are available in an active state.

Some information is now available on the part of the exchange molecule that is involved in Na^+-dependent inactivation (Matsuoka *et al.*, 1997). By studying mutations in the XIP region of the exchanger molecule, these authors have concluded that the endogenous XIP region is primarily involved in movement of the exchanger into and out of the Na^+-induced inactive state. The name XIP refers to exchanger inhibitory peptide, which has the same sequence as the XIP region and is a potent inhibitor of exchange activity.

C. PIP₂

Hilgemann has been able to show that outward exchange current in isolated giant sarcolemmal patches is stimulated by MgATP (Hilgemann, 1990). With this preparation it is possible to control MgATP on the cytoplasmic side of the cell membrane. This stimulation by MgATP appears to be a different process than that described in squid axon. For example, it does not appear to involve a protein-dependent kinase (Collins *et al.*, 1992). Hilgemann has also shown that brief treatment of membrane patches with chymotrypsin abolished modulation of exchange current by MgATP. However, recent evidence has been obtained for a mechanism that seems likely to explain regulation by ATP (Hilgemann and Ball, 1996). It now appears that phosphatidylinositol-4,5-bisphosphate (PIP_2) can strongly activate the cardiac Na^+-Ca^{2+} exchanger. Apparently ATP generates PIP_2 from phosphatidylinositol (PI). A number of observations suggest that the generation of PIP_2 can explain the regulation of Na^+-Ca^{2+} exchange by ATP: (1) PI-specific phospholipase C can abolish the action of ATP, which can in turn be restored by the addition of exogenous PI. (2) The effect of ATP can be reversed by a PIP_2-specific phospholipase C. However, the stimulatory effect of ATP can be mimicked by exogenous PIP_2. (3) Aluminum, which binds with high affinity to PIP_2, reverses the effect of ATP on the exchanger. It therefore appears that the generation of PIP_2 is an important regulator of Na^+-Ca^{2+} exchange activity in heart.

D. Phosphorylation

It has been known for some time that exchange activity in squid giant axons could be regulated by ATP. The main effect of ATP is to increase the affinity of the exchanger for its substrates Na^+ and Ca^{2+}. Moreover, recent evidence suggests the involvement of a Ca^{2+}-dependent protein kinase. It appears that, at least in the squid, Na^+-Ca^{2+} exchange that is stimulated by ATP requires intracellular Ca^{2+} and can be mimicked by hydrolyzable ATP analogs (DiPolo and Beauge, 1987a, b).

It has been difficult to demonstrate regulatory effects of ATP in cardiac tissues. The whole-cell isolated patch technique does not lend itself to studies of this nature because it is extremely difficult if not impossible to control intracellular ATP concentrations. However, recently Iwamoto and Shigekawa (1998) directly detected phosphorylation in cardiac and aortic smooth muscle using immunoprecipitation and autoradiography. It is also now known that exchanger from amphibian heart is down-regulated by isoproterenol via the protein kinase.

X. Current-Voltage Relationships and Voltage Dependence of Na^+-Ca^{2+} Exchange Current

The origin of the voltage dependence of Na^+-Ca^{2+} exchange has yet to be explained. However, current-voltage relationships obtained under a variety of ionic conditions can provide a great deal of information that forms a basis for discussing possible mechanisms of voltage dependence. Here we will try and show what can and cannot be concluded about voltage-dependent mechanisms from available current-voltage data. Before interpreting current-voltage relationships, the experimentalist should be confident that he or she is dealing with a pure current. Thus, considerable care in eliminating contaminating currents with appropriate inhibitors and an understanding of which currents (besides exchange current) are activated over the voltage range of interest is essential.

Most Na^+-Ca^{2+} exchange current-voltage relationships that have been measured so far do not show a region of negative slope. This is consistent with the idea that only a single rate-limiting charge translocation step exists in the reaction pathway. Were a second voltage-dependent step to exist in which charge moved in the opposite direction, then one might expect a region of slight negative slope. This is because the increasing voltages that stimulated forward exchange would necessarily begin to retard the movement of exchange in the opposite direction, with resulting decline in net transport and hence current. However, conditions under which a region of negative slope in an exchange current-voltage relationship have been detected recently. Using the giant patch technique, Hilgemann and coworkers (1991b) have measured outward exchange current-voltage relationships as a function of extracellular Ca^{2+} (Fig. 9). When extracellular Ca^{2+} is 0.1 mM, voltage dependence is diminished and at extreme depolarization the slope of the current-voltage relationship becomes discernibly negative. It is currently believed that most of the exchange voltage dependence is associated with Na^+ translocation. Hilgemann has suggested that the negative slope could be explained if Ca^{2+} passes through a small fraction of the membrane field before binding to the carrier. This would constitute a second

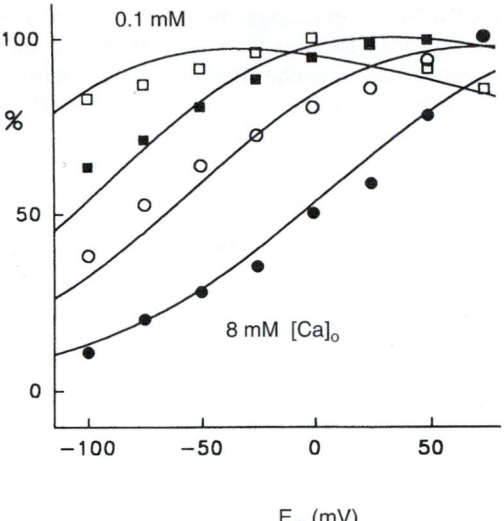

FIGURE 9. Ion and voltage dependencies of cardiac Na^+-Ca^{2+} exchange current conform to consecutive exchange model with voltage dependencies at extracellular Na^+ release and binding. Flattening and saturation of outward $I_{Na\text{-}Ca}$-voltage relations as extracellular $[Ca^{2+}]$ is lowered from 8 mM (closed circles) to 2 mM (open circles) to 0.4 mM (closed squares) to 0.1 mM (open squares). All results are normalized to the largest current occurring in the current-voltage relation. Note the negative slope with strong depolarization at 0.1 mM. (Reprinted with kind permission of the author and *Nature*.)

voltage-dependent step associated with Ca^{2+} translocation in the transport pathway.

A second property expected of current-voltage relationships for electrogenic exchange reactions is that they should saturate at extreme voltages. This is because reaction steps whose rate increases with voltage should become sufficiently rapid that they cease to be rate limiting for the entire reaction sequence. The reaction may then be rate limited by a voltage-independent step, at which point reaction rate will cease to be responsive to voltage. Many current-voltage relationships do show clear evidence of saturation. For example, the measurements of outward current depicted in Fig. 9 clearly exhibit saturation. Though not so obvious, measurements of inward current by Bridge and colleagues (1991) and by Miura and Kimura (1989) also show signs of saturation at extremely negative voltages. In contrast, recent results (from giant patches) under zero trans conditions suggest little saturation of inward exchange current at very negative potentials (Matsuoka and Hilgemann, 1992). The origin of these discrepancies is not clear, but may be related to whether or not net or unidirectional exchange is producing the inward current. Moreover, current-voltage relationships obtained in whole cells may be complicated by other partial reactions (e.g., Ca^{2+} dissociation from intracellular buffers) not present in the giant patch experiments.

The way that exchange current-voltage relationships depend on ionic conditions can also yield information about voltage-dependent steps in the reaction sequence that leads to exchange. With voltage-clamped giant patches under zero trans conditions—that is, Ca^{2+} in the pipette (extra-

cellular) side and Na^+ in the bath solution—that superfuse the cytoplasmic surface, outward current-voltage relationships showed a striking dependence on pipette (extracellular) Ca^{2+} (Fig 9). However, when the extracellular (pipette) Ca^{2+} is reduced to very low values, voltage dependence tends to disappear. An attractive explanation for this result is that as extracellular Ca^{2+} is reduced, a voltage-independent step or a step with very modest voltage dependence (presumably associated with Ca^{2+} translocation) starts to become rate limiting, with the result that exchange loses its voltage dependence. However, Ca^{2+} translocation is not independent of voltage in all species. In the squid Na^+-Ca^{2+} exchanger NCX-SQ1, charge movement accompanies Ca^{2+} translocation rather than Na^+ translocation, as has been shown rather elegantly by He and coworkers (1998). These authors examined giant patches from oocytes expressing either NCX1 or NCX1-SQ1. Na^+ and Ca^{2+} were respectively inside the pipette. This forced Na^+- and Ca^{2+}-binding sites to the intracellular surface. Rapid application of Na^+ or Ca^{2+} to the intracellular (outside) surface of the patch produces a transient half-reaction, which in turn was expected to produce a transient charge movement if the half reaction is electrogenic. It appears that in NCX1 transient current is produced by application of Na^+, but in the case of NCX1-SQ1 application of Ca^{2+} produces an electrogenic half-reaction. Thus it seems that in the squid, Ca^{2+} translocation is electrogenic (Fig. 10).

The dependence of outward current-voltage relationships on Na^+ is also interesting. Hilgemann and colleagues (1991b), using the giant patch technique, have measured outward current-voltage relationships when extracellular Na^+ was 100

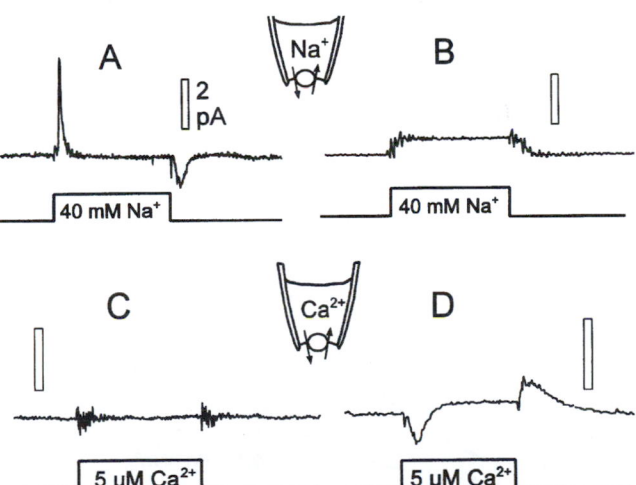

FIGURE 10. Identification of electrogenic reactions of NCX1 and NCX-SQ1 using concentration jumps. (A) Current transients recorded from an NCX1-expressing patch when 40 mM cytoplasmic Na^+ was applied and removed in the presence of 20 mM extracellular Na^+. (B) Typical lack of current transients recorded from an NCX-SQ1-expressing patch when 40 mM cytoplasmic Na^+ was applied and removed in the presence of 20 mM extracellular Na^+. (C) Inward NCX1 current activated when a solution with 5 μM free Ca^{2+} was applied as in A. (D) Inward NCX-SQ1 current activated when a solution of 5 μM free Ca^{2+} was applied as in B. (Modified from He *et al.*, (1998) and reprinted with kind permission from the *Journal of General Physiology*.)

and 400 mM (Fig. 11). At 400 mM there was a pronounced flattening of the current-voltage relationship. There are two possible (and not necessarily unrelated) explanations for this. If we assume that some step in the Na^+ translocation pathway is both rate limiting and voltage dependent, it is possible that as extracellular Na^+ increases, this step ceases to be rate limiting because it is accelerated. Were some other voltage-independent step to become rate limiting, one would expect a flattening of the current-voltage relationship as it became dominated by the voltage-independent step, as is in fact observed. Alternatively, if Na^+ binding to some external site were voltage dependent, one would expect voltage dependence of this binding to be lost as the site becomes saturated (Lagnado and McNaughton, 1990).

These results seem to suggest that somewhere in the Na^+ translocation pathway a voltage-dependent (charge translocation) step may be involved. In this regard the data obtained with giant patches, where ideal zero trans conditions are most likely to be achieved, are particularly compelling.

XI. Mechanism of Na^+-Ca^{2+} Exchange

A discussion of the evidence for and against various kinetic schemes to account for Na^+-Ca^{2+} exchange is beyond the scope of this chapter. The interested reader is referred to Khananshvili (1991). Here we will simply show the way in which measurement of exchange current can be used to infer mechanism. It should, however, be clearly understood that the measurement of isotopic fluxes and rapid mixing techniques provide indispensable tools to investigate mechanism but will not be discussed here.

For the Na^+-Ca^{2+} exchange, two different basic mechanisms for ion translocation can be considered: the **Ping-Pong or consecutive mechanism** and the **sequential or simultaneous mechanism**. Although it has been difficult to distinguish between these mechanisms, recent data using the patch-clamp technique provide strong evidence in favor of the Ping-Pong mechanism, diagrammed in its simplest form in Fig. 12. There is only one set of binding sites that binds either Ca^{2+} or Na^+ and the translocation of Na^+ and Ca^{2+} are separate events. For example, Ca^{2+} binds to the carrier E′ at the inner (cytoplasmic) surface and is translocated to the exterior, where it is released. The carrier E″ then can either bind Na^+ or Ca^{2+} at the external face, which is then transported to the internal surface where it is in turn released. A basic property of the Ping-Pong mechanism is that Na^+-Na^+ and Ca^{2+}-Ca^{2+} exchange are reversible partial reactions of the exchange. As such, it should be possible to isolate them (see Fig. 10). These partial reactions do not exist as elementary steps in the simultaneous mechanism.

Partial reactions of the exchanger have been observed (Hilgemann et al., 1991b). Using giant sarcolemmal patches, rapid application of Na^+ to the intracellular surface caused binding and translocation of intracellular Na^+ to the pipette side. This produced a transient current which could be blocked by exchange blockers including the exchanger inhibitory peptide (XIP) (Li et al., 1991). Similar currents were measured in giant oocyte patches in which the cloned exchanger had been expressed. Control oocytes did not exhibit this current. It appears that charge translocation is associated with Na^+ translocation and that the Na^+ translocation step can be isolated, consistent with the idea that a ping-pong mechanism is operative.

Further support for a consecutive model of exchange comes from work by Niggli and Lederer (1991a). These authors measured very small transient Ca^{2+} currents induced by the photo release of caged Ca^{2+} DM-dinitrophen. Insofar as these currents could be inhibited by known blockers of the Na^+-Ca^{2+} exchange, they appear to be associated with exchange. Since they are unaffected by Na^+ they are presumably associated with a partial reaction of the exchange. These authors have speculated that this current represents a charge move-

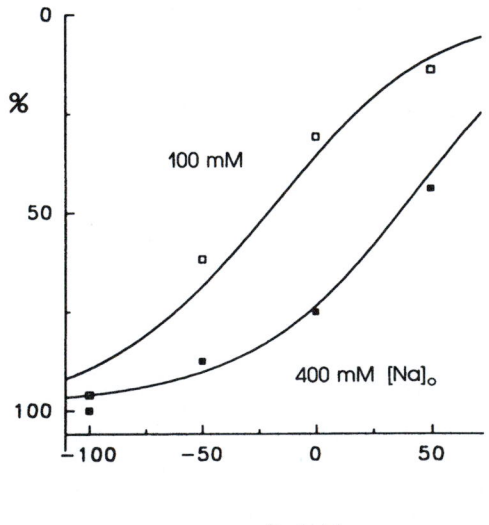

FIGURE 11. Flattening and saturation of inward I_{Na-Ca}-voltage relationships when extracellular $[Na^+]$ is increased from 100 mM to 400 mM. Na^+-MES was replaced by Ca^{2+}-MES in the pipette, as osmolarity for all extracellular solutions was equal. 20 μM free $[Ca^{2+}]$ on the cytoplasmic side. I_{Na-Ca} magnitudes in each current-voltage relationship were normalized to values at −100 mV.

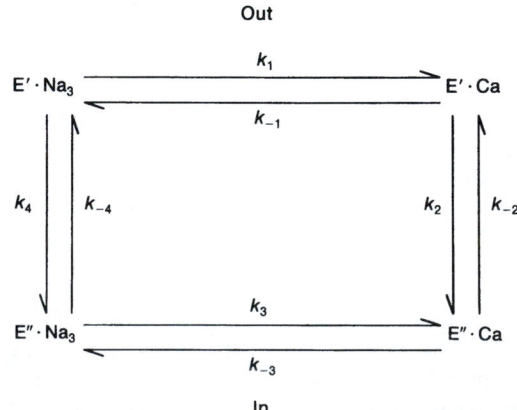

FIGURE 12. Diagram of the Ping-Pong mechanism of ion translocation.

ment associated with Ca^{2+} binding and the associated conformational change of the exchange molecule. This suggests that (as we have already seen) a voltage-dependent step might be associated in some way with the partial reactions leading to Ca^{2+} translocation. It is becoming clear that we as yet do not know exactly how many charge-translocating steps are actually in the exchange pathway. However, evidence is accruing that the mechanism is consecutive in nature.

Two additional pieces of evidence suggest that a consecutive mechanism could account for Na$^+$-Ca^{2+} exchange. First, Hilgemann and colleagues have demonstrated that the apparent affinity of one ion for the exchanger is a function of the concentration of the other (1991a). This is a requirement of a sequential reaction scheme. Moreover, Kimura has employed classic enzyme kinetics to outward exchange currents under assumed zero trans conditions. Her most recent results suggest that while it is difficult to discriminate between a simultaneous and consecutive reaction, the available evidence favors a consecutive reaction (Li and Kimura, 1991).

XII. Na$^+$-Ca^{2+} Exchange Currents during the Cardiac Action Potential

During every action potential, a Ca^{2+} current is activated and a modest quantity of Ca^{2+} enters the cell. As we shall discuss, it now seems likely that **reverse exchange** takes place, which also causes some Ca^{2+} entry. During a steady-state train of action potentials, this continual Ca^{2+} entry must in some way be compensated. It is therefore likely that forward exchange during the repolarizing phase of the action potential extrudes the Ca^{2+} entering during the initial phase of the action potential and thus maintains a beat-to-beat homeostasis. Direct evidence that Ca^{2+} entering the cell through Ca^{2+} channels can be extruded by the Na$^+$-Ca^{2+} exchange has been obtained by Bridge and colleagues (1990). Guinea pig ventricular cells treated with caffeine can be tetanized in the absence of extracellular Na$^+$ with a voltage-clamp pulse from –40 mV to +10 mV. During this clamp an inward Ca^{2+} current can be measured (Fig. 13). Although the SR is depleted by caffeine, the enlarged Ca^{2+} current can apparently produce sufficient Ca^{2+} entry to cause contractures. It should be emphasized that in these experiments the pipette contained no Na$^+$ so that in the absence of extracellular Na$^+$, Ca^{2+} entry by reverse exchange was unlikely. At the peak of the contracture, rapid application of extracellular Na$^+$ produced prompt mechanical relaxation and activated an inward transient current. This current was most likely due to forward Na$^+$-Ca^{2+} exchange because it could not be activated when intracellular Ca^{2+} was buffered with EGTA. If the Na$^+$-Ca^{2+} exchange current extruded all the entering Ca^{2+}, and if we further assume that three Na$^+$ ions exchange with a single Ca^{2+} ion, it follows that the integral of the exchange current is one-half that of the Ca^{2+} current. The relationship between the integrals of exchange current and Na$^+$-Ca^{2+} exchange current was best explained by assuming that three Na$^+$ ions exchange with a single Ca^{2+} ion and that the exchange extruded all the entering Ca^{2+}. It seems therefore that the Na$^+$-Ca^{2+} exchange does have the capacity to extrude all Ca^{2+} entering during the duty cycle.

It now seems likely that the activity of Na$^+$-Ca^{2+} exchange is profoundly modified by the cardiac action potential. It is also

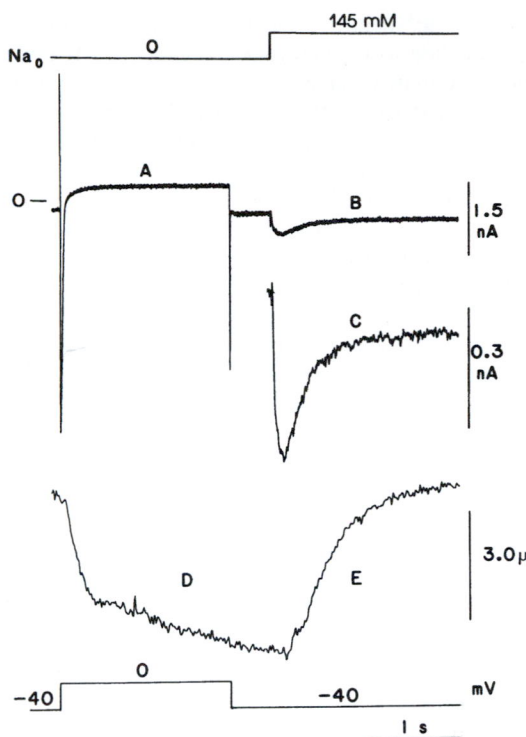

FIGURE 13. Two-second voltage-clamp pulses in the absence of Na$^0_+$ and presence of 10.0 mM caffeine cause contraction. Rapid application of Na^0+ 500 ms after repolarization causes relaxation. (A) I_{Ca} elicited by membrane depolarization. (B) The application of Na$^+$ produces putative transient inward I_{Na-Ca}. (C) This current is displayed on an expanded scale. (D) Contraction (cell shortening) activated by I_{Ca} recorded in A. (E) After repolarization to –40 mV, relaxation does not occur until 500 ms after the clamp pulse when Na$^+$ is suddenly applied. [Reprinted with permission from Barcenas-Ruiz, L., Beuckelmann, D. J. and Wier, W. G. (1987). Sodium-calcium exchange in heart: membrane currents and changes in (Ca^{2+})$_i$. *Science* **238,** 1720–1722. Copyright 1987 American Association for the Advancement of Science]

likely that the exchange current in part determines the duration of the cardiac action potential. The first comprehensive discussion of this topic was made by Mullins (1979). We will consider the behavior of the Na$^+$-Ca^{2+} exchange during the ventricular action potential. The guinea pig ventricular action potential is approximately 300 ms in duration. During the initial part of this action potential Ca^{2+} is released from the sarcoplasmic reticulum and this causes a rise of Ca^{2+} in the cytosol. Peak values for cytosolic free Ca^{2+} are probably 1–2 µM. The Ca^{2+} transient rises to a peak in approximately 50 ms, whereas the peak of the upstroke of the action potential occurs in approximately 2 ms. After the upstroke of the action potential, but before intracellular Ca^{2+} has risen appreciably, the membrane potential becomes positive to the exchange reversal potential, which is about –50 mV. Therefore, an outward exchange current is generated, and this will be accompanied by Ca^{2+} entry and Na$^+$ exit. This current has recently been implicated in the "triggering" of SR Ca^{2+} release (Leblanc and Hume, 1990). As the Ca^{2+} released from the SR begins to rise and the membrane begins to repolarize, the membrane potential will become negative to the reversal potential, the exchange current will reverse its direction, and Ca^{2+} will be extruded in exchange for Na$^+$. As the intracellular Ca^{2+} begins to decline, inward exchange current will also decline to resting values. The same principles presumably govern the behavior of the atrial cell. Earm and Noble (1990) have modeled the time course of Ca^{2+} current, Na^+-Ca^{2+} exchange current, and the Ca^{2+} transient during the rabbit atrial action potential (Fig. 14). The Na^+-Ca^{2+} exchange current is the main depolarizing current during the plateau. Moreover, the exchange activity required to maintain the late plateau is precisely sufficient to balance Ca^{2+} influx (Ca^{2+} current + Na^+-Ca^{2+} exchange) during the early part of the action potential. It should be appreciated that, regardless of the value of intracellular Ca^{2+}, membrane repolarization will tend to stimulate inward exchange current. On the other hand, the decline of the Ca^{2+} transient will tend to reduce exchange current. Therefore the relationship between the Ca^{2+} transient and membrane repolarization will largely determine the time course of inward exchange current and therefore the pattern of Ca^{2+} extrusion.

XIII. Na^+-Ca^{2+} Exchange Currents and Excitation-Contraction Coupling

Contractions occur in heart cells when Ca^{2+} released from intracellular stores known as the sarcoplasmic reticulum (SR) activates the contractile elements (for a more detailed account of excitation-contraction the reader is referred to Bers (1991)). This release is coupled to electrical excitation at the surface membrane and the whole process is often referred to as **excitation-**

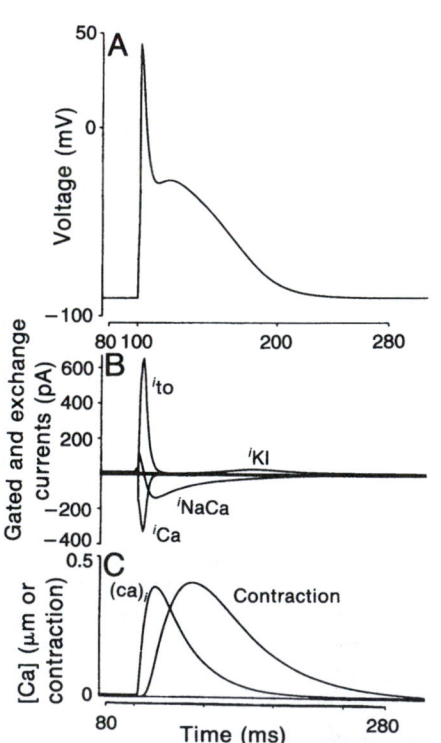

FIGURE 14.   The Earm-Noble model of the single rabbit atrial cell, based on the multicellular model of Hilgemann and Noble (1987). (A) Computed action potential. (B) Computed currents. (C) $[Ca^{2+}]_i$ and contraction (Earm and Noble, 1990). (Reprinted with kind permission of the author and the Royal Society.)

contraction coupling. The pioneering studies on skinned fibers by Alexander Fabiato lead to at least two fundamental findings (for a discussion of these see Stern and Lakatta (1992)). The first was that Ca^{2+} can be released from the SR when the concentration of Ca^{2+} in the vicinity of the SR is abruptly increased. This small increase in the concentration of Ca^{2+} leads to a much larger release of SR Ca^{2+}, so the system is one of inherently high gain. This process is usually referred to as **Ca^{2+}-induced Ca^{2+} release** or CICR. A second and important property of CICR is that, under normal circumstances, the release is graded with the size of the Ca^{2+} increase that induces the release. *A priori* such a system might be expected to be regenerative since the Ca^{2+} that is released from the SR ought to stimulate further release. A discussion of why SR Ca^{2+} release in heart is not normally regenerative is beyond the scope of this chapter, which nevertheless will discuss some of the evidence that suggests that Na^+-Ca^{2+} exchange currents are involved in CICR.

It is now known that large tetrameric Ca^{2+}-release channels (sometimes referred to as ryanodine receptors) are embedded in the sarcoplasmic reticulum (Saito *et al.*, 1988). These apparently respond to elevations of Ca^{2+} in their vicinity by gating the release of Ca^{2+} from the SR (Stern and Lakatta, 1992). Therefore a molecular basis for the early observations by Fabiato has been established. A central question for those studying excitation-contraction in heart is, What in intact cells produces the rise in intracellular Ca^{2+} that gates the SR release channel and triggers SR Ca^{2+} release?

It is now well established that the L-type Ca^{2+} current is principally involved in triggering SR Ca^{2+} release in mammalian ventricular cells (Lopez-Lopez *et al.*, 1994; London and Krueger, 1986; Beuckelmann and Wier, 1988; Cheng *et al.*, 1993). Because SR Ca^{2+} release is graded with the size of the increase in Ca^{2+} concentration that induces the release, it follows that if L-type Ca^{2+} channels trigger or induce the release of Ca^{2+}, the extent of SR Ca^{2+} release also should be graded with the size of the L-type Ca^{2+} current. Since the size of the Ca^{2+} current has a bell-shaped dependence on voltage, the finding that the rate or extent of SR Ca^{2+} release (or the magnitude of triggered contractions) is (under appropriate conditions) also bell-shaped lends strong support to the idea that the Ca^{2+} current is a trigger for SR Ca^{2+} release. For example, a detailed study by Beuckelmann and Wier (1988) has clearly established the bell-shaped relationship between triggered Ca^{2+} transients and voltage.

However, a number of studies in ventricular cells have revealed, under certain circumstances, a more complex relationship between the voltage dependence of triggered contractions and Ca^{2+} current (Nuss and Houser, 1992; Litwin *et al.*, 1998; Vornanen *et al.*, 1994). In particular, tension measurements did not follow a simple bell-shaped relationship with voltage but rather showed a sigmoid relationship. At positive potentials shortening did not decline steeply with voltage. Studies by Litwin and coworkers (1998) on guinea pig cells under voltage-clamp and dialyzed with various Na^+ solutions indicate that while the shape of the Ca^{2+} current/voltage relationship was independent of pipette Na^+, the triggered shortening-voltage relationship showed a striking dependence on pipette Na^+ (Fig. 15). At positive potentials the shortening-voltage relationship departed from a simple bell shape and the extent of this departure depended on the concentration of dialyzing Na^+.

If L-type Ca^{2+} current is blocked extremely rapidly, triggered contractions are reduced but not abolished (Levi *et al.*, 1996). Thus if Ca^{2+} currents are elicited under voltage-clamp it is possible to record what could be a triggered contraction or a triggered Ca^{2+} transient. It therefore seems likely that another process besides the Ca^{2+} current is involved, triggering the release of SR Ca^{2+}. The most likely process is in fact the Na$^+$-Ca^{2+} exchange. Recently Grantham and Cannell (1996) used voltage-clamp pulses shaped like an action potential to infer the magnitude and trajectory of Na$^+$-Ca^{2+} exchange currents during the initial part of an action potential. They found that the magnitude of the exchange current was somewhat less than 30% of the magnitude of the Ca^{2+} current that occurred during the initial part of the action potential (Fig 16). As the authors point out, this is not a negligible current and is consistent with the idea that at least some of the triggered SR Ca^{2+} release could be due to the activity of the Na$^+$-Ca^{2+} exchange. It is worth mentioning that if the Na$^+$-Ca^{2+} exchange is capable of contributing part of the trigger for SR Ca^{2+} release, then this component of the trigger will be extremely sensitive to intracellular Na$^+$. In this regard it has been proposed that the initial Na$^+$ current might provide enough Na$^+$ accumulation (provided the accumulation takes place in a restricted space) in the vicinity of the Na$^+$-Ca^{2+} exchangers to enhance triggering by reverse Na$^+$-Ca^{2+} exchange current (Leblanc and Hume, 1990; Lipp and Niggli, 1994). A difficulty with the idea that exchange can under physiological circumstances trigger significant Ca^{2+} release is that it is extremely small. Litwin and colleagues (1998) have suggested, for example, that the exchange may sum its effects with the Ca^{2+} current in a highly nonlinear way to augment triggering by Ca^{2+} current. It is possible that one function of the exchange is to increase the probability with which a Ca^{2+} current triggers exchange in a highly nonlinear system. It is therefore not yet clear what contribution Na$^+$-Ca^{2+} exchange currents make to triggering SR Ca^{2+} release under physiological circumstances. However, since the exchanger is regulated by intracellular Na$^+$, Ca^{2+}, and voltage, it seems

likely that its contribution to triggering (either direct or indirect) will be both complex and variable.

XIV. Summary

At first, study of the Na$^+$-Ca^{2+} exchange current was impeded because multicellular preparations did not permit adequate control of the driving forces producing exchange. However, methods for isolating single cells together with the patch-clamp technique made it possible to isolate exchange current and to provide reliable measurements of its properties. Now that the exchange molecule has been cloned, it is possible to express the mammalian exchanger in other cell types including frog oocytes. Currents may now be measured in these cell types with the giant

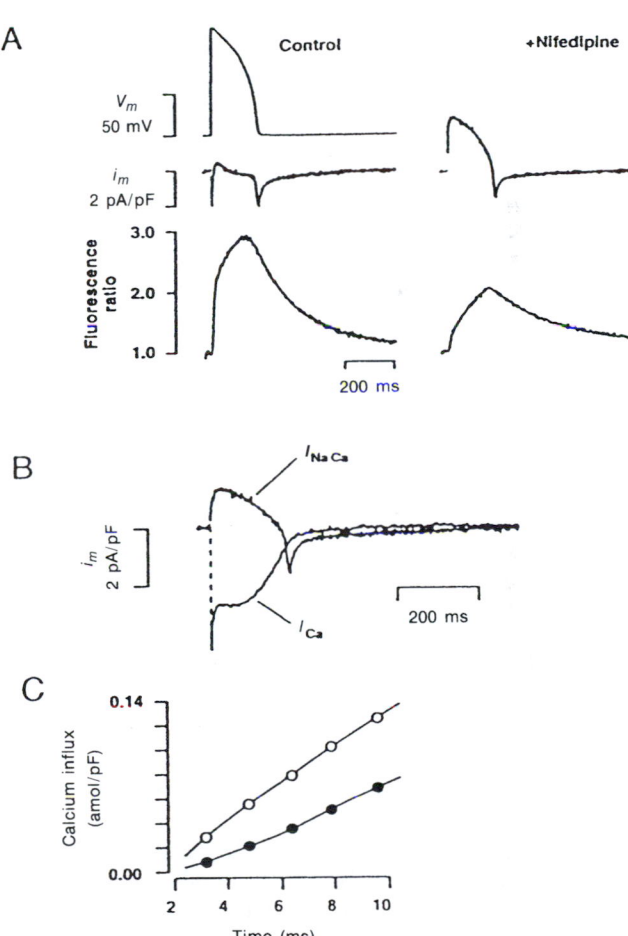

FIGURE 16. Membrane currents and Ca^{2+} influx in a myocyte with intact Na$^+$-Ca^{2+} exchange and SR release in response to an action potential (AP). The myocyte was at steady state, having been stimulated continuously at 0.2 Hz with a train of at least 20 APs. (A) Top, AP command; middle, membrane current; and bottom, [Ca^{2+}] transient time course in the absence of nifedipine or, on the right, immediately after blockade of I_{Ca} by 10 μmol/L nifedipine. (B) Time course of AP-evoked nifedipine-sensitive current (I_{Ca}) and calculated I_{Na-Ca}. (C) Cumulative Ca^{2+} influx via I_{Ca} (open circles) and the calculated I_{Na-Ca} (closed circles) at the beginning of the AP. Cell capacitance, 150 pF. Similar results were obtained in at least four other cells. (Reprinted with permission of *Circulation Research* from Grantham and Cannell (1996).)

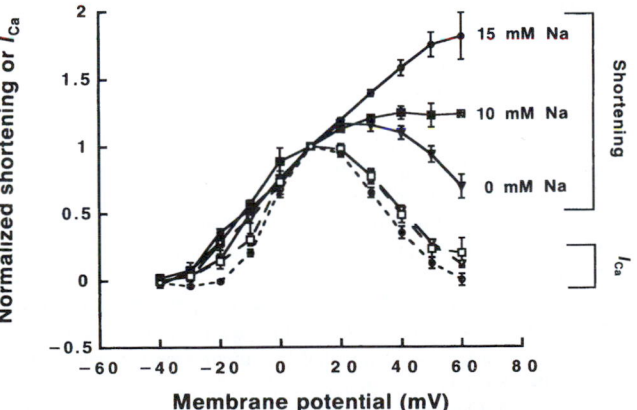

FIGURE 15. The relationship between voltage and I_{Ca} is bell-shaped regardless of dialyzing Na$^+$ concentration. However, the relationship between triggered shortening and voltage depends upon the concentration of dialyzing Na$^+$. When dialyzing Na$^+$ is nominally zero mM, the relationship between voltage and the extent of shortening approaches a bell shape. (Reprinted with permission of the *Proceeding of the New York Academy of Sciences*.)

patch technique or with the whole-cell patch-clamp method. These preparations and methods introduce the possibility of studying mutated forms of the exchanger with a view to understanding the relationship between structure and function.

Studies with isolated giant patches have resulted in significant advances in our understanding of both the regulation and mechanism of exchange activity. The phenomenon of Na^+-dependent inactivation of exchange was first identified in isolated patches and our understanding of secondary Ca^{2+} regulation as well as regulation by PIP_2 and phosphorylation have been significantly advanced by this technique. Advances in the molecular biology of the Na^+-Ca^{2+} exchanger—which include the elucidation of the primary structure of the exchange molecule together with the development of recombinant DNAs coding for the Na^+-Ca^{2+} exchanger, their expression in systems suitable for current measurement, and, finally, the production of various mutations—have already resulted in studies of the relationship between structure and function.

Measurements of both whole-cell currents as well as currents in giant patches have produced evidence in favor of the idea that a consecutive mechanism can explain exchange activity. Moreover, at least two laboratories have claimed to have isolated currents associated with partial reactions of the exchanger consistent with the idea that a consecutive (Ping-Pong) reaction scheme is operative. Preliminary results seem to suggest that much of the voltage dependence of the exchange resides in rate-limiting steps associated with Na^+ translocation in NCX1, although not with NCX-SQ1, in which voltage dependence appears to be associated with Ca^{2+} translocation. With the availability of the techniques of modern molecular biology it seems likely that we may expect considerable progress in the study of the relationship between the structure and function of the exchange molecule. Elegant measurements of whole-cell currents have resulted in a plausible model of the behavior of the Na^+-Ca^{2+} exchange during the cardiac cycle. The way that the Na^+-Ca^{2+} exchange contributes to excitation coupling is not understood but is an active area of research. The advent of peptide inhibitors like XIP together with methods for rapidly changing extracellular solutions suggest that we may soon gain more insight into the contribution of both inward and outward exchange currents to the events of excitation-contraction coupling. Most recently, Na^+-Ca^{2+} exchange currents have been proposed as a trigger for SR Ca^{2+} release under physiological conditions. Thus in the last 30 years the study of exchange currents has expanded enormously to provide not only insight into the way that the Na^+-Ca^{2+} exchange controls intracellular Ca^{2+} at the whole-cell level, but also into the details of the molecular mechanism of the exchange reaction itself.

Bibliography

Achilles, A., Friedel, U., Haase, W., Reilander, H., and Cook, N. (1991). Biochemical and molecular characterization of the sodium-calcium exchange from bovine rod photoreceprors. *Ann. N.Y. Acad. Sci.* **639**, 234–244.

Axelsen, P. H., and Bridge, J. H. B. (1985). Electrochemical ion gradients and the Na/Ca exchange stoichiometry. *J. Gen. Physiol.* **85**, 471–478.

Baker, P. F., Blaustein, M. P., Hodgkin, A. L., and Steinhardt, R. A. (1967). Effect of sodium concentration on calcium movements in giant axons of *loligo forbesi. J. Physiol. (Lond.).* **192**, 43–44.

Baker, P. F., and McNaughton, P. A. (1976). Kinetics and energetics of calcium efflux from intact squid giant axons. *J. Physiol.* **259**, 103–144.

Barcenas-Ruiz, L., Beuckelmann, D. J. and Wier, W. G. (1987). Sodium-calcium exchange in heart: membrane currents and changes in $(Ca^{2+})_i$. *Science* **238**, 1720–1722.

Bers, D. M. (1991). Excitation-contraction coupling and cardiac contractile force. Kluwer Academic Publishers, Dordrecht.

Beuckelmann, D. J., and Wier, W. G. (1988). Mechanism of release of calcium from sarcoplasmic reticulum of guinea pig cardiac cells. *J. Physiol. (Lond.)* **405**, 233–255.

Blaustein, M. P., and Hodgkin, A. L. (1969). The effect of cyanide on the efflux of calcium from squid axons. *J. Physiol.* **200**, 497–527.

Blaustein, M. P., and Lederer, W. J. (1999). Sodium/calcium exchange: its physiological implications. *Physiol. Rev.* **79**, 763–854.

Bridge, J. H. B., and Bassingthwaighte, J. B. (1983). Uphill sodium transport driven by an inward calcium gradient in heart muscle. *Science* **219**, 178–180.

Bridge, J. H. B., Smolley, J., Spitzer, K. W., and Chin, T. K. (1991). Voltage dependence of sodium-calcium exchange and the control of Ca extrusion in the heart. *Ann. N. Y. Acad. Sci.* **639**, 34–47.

Bridge, J. H. B., Smolley, J. R., and Spitzer, K. W. (1990). The relationship between charge movements associated with I_{Ca} and $I_{Na\text{-}Ca}$ in cardiac myocytes. *Science* **248**, 376–378.

Cheng, H., Lederer, W. J., and Cannell, M. B. (1993). Calcium sparks: elementary events underlying excitation-contraction coupling in heart muscle. *Science* **262**, 740–744.

Chin, T. K., Spitzer, K. W., Philipson, K. D., and Bridge, J. H. B. (1993). The effect of exchanger inhibitory peptide (XIP) on sodium-calcium exchange current in guinea pig ventricular cells. *Circ Res.* **72**, 497–503.

Collins, A., Somlyo, A. V., and Hilgemann, D. W. (1992). The giant cardiac membrane patch method: stimulation of outward Na^+-Ca^{2+} exchange current by MgATP. *J. Physiol.* **454**, 27–57.

Crespo, L. M., Grantham, C. J., and Cannell, M. B. (1990). Kinetics stoichiometry and role of the Na-Ca exchange mechanism in isolated cardiac myocytes. *Nature* **345**, 618–621.

DiPolo, R., and Beauge, L. (1987a). Characterization of the reverse Na-Ca exchange in squid axons and its modulation by Ca_i and ATP. *J .Gen. Physiol.* **90**, 505–525.

DiPolo, R., and Beauge, L. (1987b). In squid axons ATP modulates Na^+-Ca^{2+} exchange by a Ca^{2+}_i dependent phosphorylation. *Biochim. Biophys. Acta* **897**, 347–354.

Doering, A. E., Nicoll, D. A., Lu, Y., Lu, L., Weiss, J. N., and Philipson, K. D. (1998). Topology of a functionally important region of the cardiac Na^+/Ca^{2+} exchanger. *J. Biol. Chem.* **273**, 778–783.

Earm, Y. E., and Noble, D. (1990). A model of the single atrial cell: relation between calcium current and calcium release. *Pro. R. Soc. Lond. B.* **240**, 83–96.

Ehara, T., Matsuoka, S., and Noma, A. (1989). Measurement of reversal potential of Na^+-Ca^{2+} exchange current in single guinea-pig ventricular cells. *J. Physiol.* **410**, 227–249.

Grantham, C. J., and Cannell, M. B. (1996). Ca^{2+} influx during the cardiac action potential in guinea pig ventricular myocytes. *Circ. Res.* **79**, 194–200.

Hamill, O. P., Marty, E., Neher, E., Sakmann, B., and Sigworth, F. (1981). Improved patch-clamp techniques for high resolution current recording from cells and cell-free membrane patches. *Pflugers Arch.* **391**, 85–100.

He, Z., Tong, Q., Quednau, B. D., Philipson, K. D., and Hilgemann, D. W. (1998). Cloning, expression, and characterization of the squid Na^+-Ca^{2+} exchanger (NCX-SQ1). *J. Gen. Physiol.* **111**, 857–873.

Hilgemann, D. W. (1989). Giant excised cardiac sarcolemmal membrane patches: Sodium and sodium-calcium exchange currents. *Pflugers Arch.* 1–3.

Hilgemann, D. W. (1990). Regulation and deregulation of cardiac Na^+-Ca^{2+} exchange in giant excised sarcolemmal membrane patches. *Nature* **344**, 242–245.

Hilgemann, D. W., and Ball, R. (1996). Regulation of cardiac Na+,Ca2+ exchange and KATP potassium channels by PIP2. *Science* **273**, 956–959.

Hilgemann, D. W., Collins, A., Cash, D. P., and Nagel, G. A. (1991a). Cardiac Na+-Ca2+ exchange system in giant membrane patches. *Ann. N. Y. Acad. Sci.* **639**, 126–139.

Hilgemann, D. W., Matsuoka, S., Nagel, G. A., and Collins, A. (1992). Steady-state and dynamic properties of cardiac sodium-calcium exchange. Sodium-dependent inactivation. *J. Gen. Physiol.* **100**, 905–932.

Hilgemann, D. W., Nicoll, D. A., and Philipson, K. D. (1991b). Charge movement during Na+ translocation by native and cloned cardiac Na+/Ca2+ exchanger. *Nature* **352**, 715–718.

Horackova, M., and Vassort, G. (1979). Sodium-calcium exchange in regulation of cardiac contractility. *J. Gen. Physiol.* **73**, 403–424.

Hume, J. R., and Uehara, A. (1986a). "Creep currents" in single frog atrial cells may be generated by electrogenic Na-Ca exchange. *J. Gen. Physiol.* **87**, 857–884.

Hume, J. R., and Uehara, A. (1986b). Properties of "creep currents" in single frog atrial cells. *J. Gen. Physiol.* **87**, 833–855.

Iwamoto, T., and Shigekawa, M. (1998). Differential inhibition of Na+/Ca2+ exchanger isoforms by divalent cations and isothiourea derivative. *Am. J. Physiol.* **275**, C423–430.

Khananshvili, D. (1991). Mechanism of partial reactions in the cardiac Na-Ca exchange system. *Ann. N. Y. Acad. Sci.* **639**, 85–95.

Kimura, J., Miyamae, S., and Noma, A. (1987). Identification of sodium-calcium exchange currents in single ventricular cells of guinea-pig. *J. Physiol.* **384**, 199–222.

Kimura, J., Noma, A., and Irisawa, H. (1986). Na-Ca exchange current in mammalian heart cells. *Nature* **319**, 596–599.

Lagnado, L., and McNaughton, P. A. (1990). Electrogenic properties of the Na-Ca exchange. *J. Membr. Biol.* **113**, 177–191.

Langer, G. A. (1964). Kinetic studies of calcium distribution in ventricular muscle of the dog. *Circ. Res.* **15**, 393–405.

Leblanc, N., and Hume, J. R. (1990). Sodium current–induced release of calcium from cardiac sarcoplasmic reticulum [see comments]. *Science* **248**, 372–376.

Levi, A. J., Li, J., Spitzer, K. W., and Bridge, J. H. B. (1996). Effect on the Indo-1 transient of applying Ca2+ channel blocker for a single beat in voltage-clamped guinea-pig cardiac myocytes. *J. Physiol.* **494(3)**, 653–673.

Levitsky, D. O., Nicoll, D. A., and Philipson, K. D. (1994). Identification of the high affinity Ca-binding domain of the cardiac Na-Ca exchanger. *J. Biol. Chem.* **269**, 22 847–22 852.

Li, J., and Kimura, J. (1991). Translocation mechanism of cardiac Na-Ca exchange. *Ann. N. Y. Acad. Sci.* **639**, 48–60.

Lipp, P., and Niggli, E. (1994). Modulation of Ca2+ release in cultured neonatal rat cardiac myocytes: insight from subcellular release patterns revealed by confocal microscopy. *Circ Res.* **74**, 979–990.

Litwin, S. E., Li, J., and Bridge, J. H. (1998). Na-Ca exchange and the trigger for sarcoplasmic reticulum Ca release: studies in adult rabbit ventricular myocytes. *Biophys. J.* **75**, 359–371.

London, B., and Krueger, J. W. (1986). Contraction in voltage-clamped, internally perfused single heart cells. *J. Gen. Physiol.* **88**, 475–505.

Lopez-Lopez, J. R., Shacklock, P. S., Balke, C. W., and Wier, W. G. (1994). Local, stochastic release of Ca2+ in voltage-clamped rat heart cells: visualization with confocal microscopy. *J. Physiol. (Lond.)* **480**, 21–29.

Matsuoka, S., and Hilgemann, D. W. (1992). Steady-state and dynamic properties of cardiac sodium-calcium exchange. Ion and voltage dependencies of the transport cycle. *J. Gen. Physiol.* **100**, 963–1001.

Matsuoka, S., Nicoll, D. A., He, Z., and Philipson, K. D. (1997). Regulation of the cardiac Na-Ca exchanger by the endogenous XIP region. *J. Gen. Physiol.* **109**, 1–14.

Matsuoka, S., Nicoll, D. A., Hryshko, L. V., Levitsky, D. O., Weiss, J. N., and Philipson, K. D. (1995). Regulation of the cardiac Na+-Ca2+ exchanger by Ca2+: mutational analysis of the Ca2+-binding domain. *J. Gen. Physiol.* **105**, 403–420.

Matsuoka, S., Nicoll, D. A., Reilly, R. F., Hilgemann, D. W., and Philipson, K. D. (1993). Initial localization of regulatory regions of the cardiac sarcolemmal Na+-Ca2+ exchanger. *Proc. Natl. Acad. Sci. USA* **90**, 3870–3874.

Mechmann, S., and Pott, L. (1986). Indentification of Na-Ca exchange current in single cardiac myocytes. *Nature* **319**, 597–599.

Miura, Y., and Kimura, J. (1989). Sodium-calcium exchange current. *J. Gen. Physiol.* **93**, 1129–1145.

Mullins, L. J. (1979). The generation of electric currents in cardiac fibers by Na/Ca exchange. *Am. J. Physiol.: Cell Physiol.* **5(2)**, C103–C110.

Nicoll, D. A., Longoni, S., and Philipson, K. D. (1990). Molecular cloning and functional expression of the cardiac sarcolemmal Na+-Ca2+ exchanger. *Science* **250**, 562–565.

Nicoll, D. A., Ottolia, M., Lu, L., Lu, Y., and Philipson, K. D. (1999). A new topological model of the cardiac sarcolemmal Na+-Ca2+ exchanger. *J. Biol. Chem.* **274**, 910–917.

Niggli, E., and Lederer, W. J. (1991a). Molecular operations of the sodium-calcium exchanger revealed by conformation currents. *Nature* **349**, 621–624.

Niggli, E., and Lederer, W. J. (1991b). Photorelease of Ca2+ produces Na-Ca exchange currents and Na-Ca exchange "gating" currents. *Ann. N. Y. Acad. Sci.* **639**, 61–70.

Nuss, H. B., and Houser, S. R. (1992). Sodium-calcium exchange-mediated contractions in feline ventricular myocutes. *Am. J. Physiol.: Heart Circ. Physiol.* **263**, H1161–H1169.

Omelchenko, A., Dyck, C., Hnatowich, M., Buchko, J., Nicoll, D. A., Philipson, K. D., and Hryshko, L. V. (1998). Functional differences in ionic regulation between alternatively spliced isoforms of the Na+-Ca2+ exchanger from *Drosophila melanogaster*. *J. Gen. Physiol.* **111**, 691–702.

Philipson, D. D., and Nicoll, S. A. (2000). Sodium-calcium exchange: a molecular perspective. *Ann. Rev. Physiol.* in press.

Philipson, K. D., Longoni, S., and Ward, R. (1988). Purification of the cardiac Na+-Ca2+ exchange protein. *Biochim. Biophys. Acta* **945**, 298–306.

Rasgado-Flores, H., and Blaustein, M. P. (1987). Na/Ca exchange in barnacle muscle cells has a stoichiometry of 3 Na+/1 Ca2+. *Am. J. Physiol.: Cell Physiol.* **252(21)**, C499–C504.

Reeves, J. P., and Hale, C. C. (1984). The stoichiometry of the cardiac sodium-calcium exchange system. *J. Biol. Chem.* **259**, 7733–7739.

Reeves, J. P., and Sutko, J. L. (1983). Competitive interactions of sodium and calcium with the sodium-calcium exchange system of cardiac sarcolemmal vesicles. *J. Bio. Chem.* **258(5)**, 3178–3182.

Repke, K. (1964). Übersichten über den biochemischen Wirkungsmodus von Digitalis. *Klin. Wochenschr.* **41**, 157–165.

Reuter, H., and Seitz, N. (1968). The dependence of calcium efflux from cardiac muscle on temperature and external ion composition. *J. Physiol.* **195**, 451–470.

Saito, A., Inui, M., Radermacher, M., Frank, J., and Fleischer, S. (1988). Ultrastructure of the calcium release channel of sarcoplasmic reticulum. *J. Cell Biol.* **107**, 211–219.

Sheu, S. S., and Fozzard, H. A. (1982). Transmembrane Na+ and Ca2+ electrochemical gradients in cardiac muscle and their relationship to force development. *J. Gen. Physiol.* **80**, 325–351.

Spitzer, K. W., and Bridge, J. H. B. (1989). A simple device for rapidly exchanging solution surrounding a single cardiac cell. *Am. J. Physiol.: Cell Physiol.* **256**, C441–C447.

Stern, M. D., and Lakatta, E. G. (1992). Excitation-contraction coupling in the heart: the state of the question. *FASEB J.* **6**, 3092–3100.

Trosper, T. L., and Philipson, K. D. (1984). Stimulatory effect of calcium chelators on Na-Ca exchange in cardiac sarcolemmal vesicles. *Cell Calcium* **5**, 211–222.

Vornanen, M., Shepherd, N., and Isenberg, G. (1994). Tension-voltage relations of single myocytes reflect Ca release triggered by Na/Ca exchange at 35 degrees C but not 23 degrees C. *Am. J. Physiol.* **267**, C623–C632.

Francisco J. Alvarez-Leefmans

20

Intracellular Chloride Regulation

I. Introduction

The importance of chloride ions (Cl⁻) in cell physiology has not been fully recognized until recent years, in spite of the fact that Cl⁻, together with bicarbonate, is the most abundant free anion in living animal cells. More important, it is now well established that Cl⁻ plays vital roles in cell physiology and, contrary to what is often stated, is not distributed in thermodynamic equilibrium across most cells. We know now that Cl⁻ is actively transported and tightly regulated in virtually all cells and that the intracellular chloride concentration, $[Cl^-]_i$, of a cell at steady state is determined by the relative contribution of various anion transporting systems that include conductive Cl⁻ channels as well as several cotransporters and exchangers. During the last decade, the molecular structure and the working mechanisms of these specialized Cl⁻ transport systems have started to be elucidated. In spite of these advances and contrary to the overwhelming evidence gathered over the last two decades, most contemporary cell physiology and neuroscience textbooks still state that Cl⁻ ions are passively distributed across the plasma membrane of vertebrate cells. The section on Cl⁻ transport in some textbooks is often relegated to one paragraph, often in small print! Moreover, because of this erroneous view, Cl⁻ is usually omitted from the Goldman-Hodgkin-Katz equation for resting membrane potential.

What was the origin of this erroneous view that still has not been completely eradicated? It appears that this standpoint stemmed from two key experimental observations made in the 1950s and 1960s. First, work on frog skeletal muscle showed that Cl⁻ was "passively distributed" between the cytoplasm and the extracellular fluid (Hodgkin and Horowicz, 1959). Later on, it was found that the "passive distribution" for Cl⁻ across the muscle membrane is more apparent than real. It is true that the equilibrium potential for Cl⁻ (E_{Cl}) closely follows the value of the transmembrane potential (E_m), because the skeletal muscle plasma membrane has a relatively high permeability to Cl⁻ (P_{Cl}). For all practical purposes, Cl⁻ appears passively distributed across the sarcolemma. However, as we will see, the fact that $E_m = E_{Cl}$ does not preclude that Cl⁻ is actively transported in this or any other cell type. In fact, there is compelling evidence for an inward active transport of Cl⁻ in skeletal muscle, but the relatively high P_{Cl} masks any effects that the operation of this transport system may have on $[Cl^-]_i$ (Fig. 1). Second, early work on Cl⁻ transport in red blood cells confirmed the view that the Cl⁻ transmembrane distribution was thermodynamically passive and, in addition, showed that Cl⁻ crossed the membrane extremely rapidly. This latter finding (for a long time erroneously interpreted as being only the result of high passive P_{Cl}) made it quite likely that Cl⁻ could remain at thermodynamic equilibrium (see Chapter 23). These two observations were generalized and virtually all cells were thought to have a very high P_{Cl} and a thermodynamically passive Cl⁻ transmembrane distribution, implying that Cl⁻ was not actively transported. Moreover, to make things worse for the case of Cl⁻, cations claimed center stage during this period with attention focused on the Na⁺ and Ca²⁺ action potentials, the K⁺ channels, and the Na⁺-K⁺ pump. Thus, the belief that Cl⁻ played no major role in membrane physiology and the attention directed to the study of cations combined to cause research on Cl⁻ transport to be largely neglected for many years.

Research over the last couple of decades has led to a dramatic change in this simplistic view. First, we now know that most animal cells exhibit a nonequilibrium distribution of Cl⁻ across their plasma membranes. Some cells actively extrude Cl⁻, others actively accumulate it, but few cells ignore it. By virtue of being distributed out of electrochemical equilibrium, Cl⁻ does serve as a key player in a variety of cellular functions such as intracellular pH regulation (Chapter 22); cell volume regulation (Chapter 21); transepithelial salt transport; synaptic signaling (in both the depolarizing and hyperpolarizing directions); neuronal growth, migration, and targeting; membrane potential stabilization; regulation of transport systems; and K⁺ scavenging. In the present chapter some examples are presented to illustrate how Cl⁻ is involved in some of these processes.

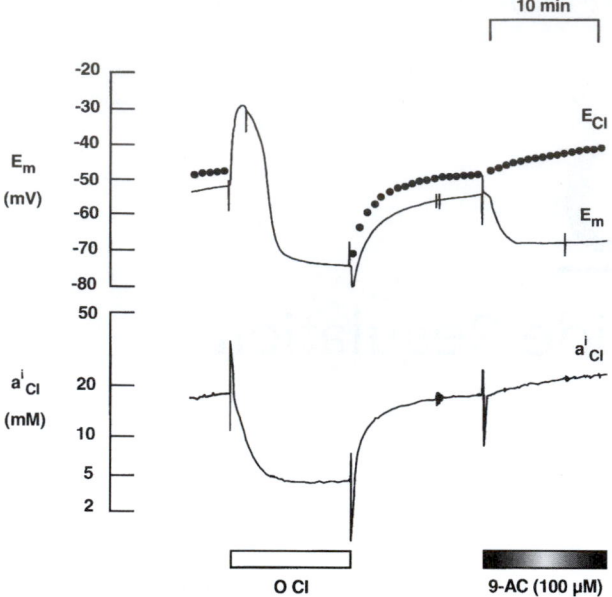

FIGURE 1. Active accumulation of Cl⁻ in rat skeletal muscle unmasked by reducing the resting Cl⁻ conductance. Membrane potential (E_m) and the intracellular Cl⁻ activity (a_{Cl}^i) were measured simultaneously in a rat lumbrical muscle during the removal and readdition of external Cl⁻ and the inhibition of the Cl⁻ conductance by application of 9-anthracene carboxylic acid (9-AC). The Cl⁻ equilibrium potential (E_{Cl}), calculated from the recorded a_{Cl}^i and a_{Cl}^o, using Eq. 1 has been plotted as filled circles above the recording of E_m. Note that E_{Cl} was slightly positive with respect to E_m, indicating that a_{Cl}^i was maintained just above equilibrium. Upon reduction of the Cl⁻ conductance, the difference between E_m and E_{Cl} increased significantly. The recording was made with a double-barreled microelectrode, one barrel sensing E_m and the other the a_{Cl}^i. The preparation was maintained in the nominal absence of CO_2 in solutions equilibrated with 100% O_2. The mechanism responsible for the uphill accumulation of Cl⁻ is a Na⁺-K⁺-Cl⁻ cotransporter. (Modified from Aickin, 1990.)

II. Passive and Nonpassive Cl⁻ Distribution across the Plasma Membrane

The Cl⁻ equilibrium potential, E_{Cl}, is defined by the Nernst equation:

$$E_{Cl} = \frac{RT}{\mathcal{F}} \ln \frac{a_{Cl}^i}{a_{Cl}^o} \qquad (1)$$

where a_{Cl}^i and a_{Cl}^o are the intracellular and extracellular Cl⁻ activities and R, T, and $\mathcal{F}$ have their usual thermodynamic meanings. If Cl⁻ is **passively distributed**, E_m will have the same value as E_{Cl} and therefore

$$E_m = \frac{RT}{\mathcal{F}} \ln \frac{a_{Cl}^i}{a_{Cl}^o} \qquad (2)$$

The predicted intracellular Cl⁻ activity at electrochemical equilibrium $(a_{Cl}^i)_{eq}$ can be obtained from Eq. 2:

$$\left(a_{Cl}^i\right)_{eq} = a_{Cl}^o \, e^{E_m \mathcal{F}/RT} \qquad (3)$$

Since the activity of an ion i (a_i) is equal to $\gamma_i \cdot [i]$, where γ_i is the activity coefficient and $[i]$ is the concentration of ion

i, it follows that if the extracellular and intracellular activity coefficients for Cl⁻ are the same, Eq. 3 becomes

$$\left[Cl^-\right]_i = \left[Cl^-\right]_o \, e^{E_m \mathcal{F}/RT} \qquad (4)$$

Equation 4 allows us to calculate the predicted $[Cl^-]_i$ at electrochemical equilibrium. This equation can also be derived from the Ussing flux ratio equation as explained in Chapter 13. At equilibrium Cl⁻ influx, J_{Cl}^i, and Cl⁻ efflux, J_{Cl}^o, will be equal and hence their ratio will be 1. Hence this equation predicts the value that $[Cl^-]_i$ should have if it was passively distributed across the plasma membrane.

If Cl⁻ is **nonpassively distributed** it means that it is actively transported, and this implies energy consumption. Active Cl⁻ transport (i.e., transport against Cl⁻ electrochemical equilibrium or uphill transport) requires demonstrating that $[Cl^-]_i$ (or a_{Cl}^i) is maintained at a level different from that predicted by Eq. 4; in other words, E_{Cl} and E_m will be different. In most cells, $[Cl^-]_i$ is maintained at a value different from that predicted from Eq. 4. However, as mentioned, equal values of E_m and E_{Cl} do not preclude the presence of active Cl⁻ transport, as is the case for skeletal muscle cells, which have a substantial electrodiffusional Cl⁻ permeability that "shunts" the effects of an inwardly directed active Cl⁻ transport system (Harris and Betz, 1987; Aickin, 1990).

In most cell types of both vertebrates and invertebrates, in which Cl⁻ is not in electrochemical equilibrium, the $[Cl^-]_i$ is maintained at a higher value than that predicted for a passive distribution across the plasma membrane. This implies the presence of transport mechanisms that actively accumulate Cl⁻. Examples include adult sensory neurons, sympathetic ganglion cells, epithelial cells, leukocytes, and both smooth and cardiac muscle cells (Fig. 2). In some other cells, particularly in cortical neurons, $[Cl^-]_i$ is maintained at a value lower than that predicted for a passive Cl⁻ distribution across the plasma membrane due to the presence of active Cl⁻ extrusion mechanisms.

III. Active Transport Mechanisms for Cl⁻

Intracellular Cl⁻ levels are determined by the relative contributions of the various Cl⁻ transport systems that may be present in the plasma membrane of a given cell type. A variety of systems capable of transporting Cl⁻ across the plasma membrane have been identified in vertebrate and invertebrate cells. They include various types of conductive Cl⁻ channels and a series of specialized carriers, all of which are secondary active transport systems. In animal cells there is no firm evidence for the existence of an active Cl⁻ pump, that is, a primary active transport mechanism directly energized by ATP. Chloride channels have been the subjects of other chapters of this book (Chapters 29 and 39) and therefore they will only be briefly mentioned here. The carrier protein molecules that transport Cl⁻ include (1) an electroneutral Cl⁻-HCO_3^- exchanger that plays a central role in intracellular pH regulation (see Chapters 22 and 23) and also serves as an uphill Cl⁻-accumulating system in some cases, such as cardiac and smooth muscle cells (Vaughan-Jones, 1986; Aickin, 1990), as shown in Fig. 2; and (2) the electroneutral cation-chloride

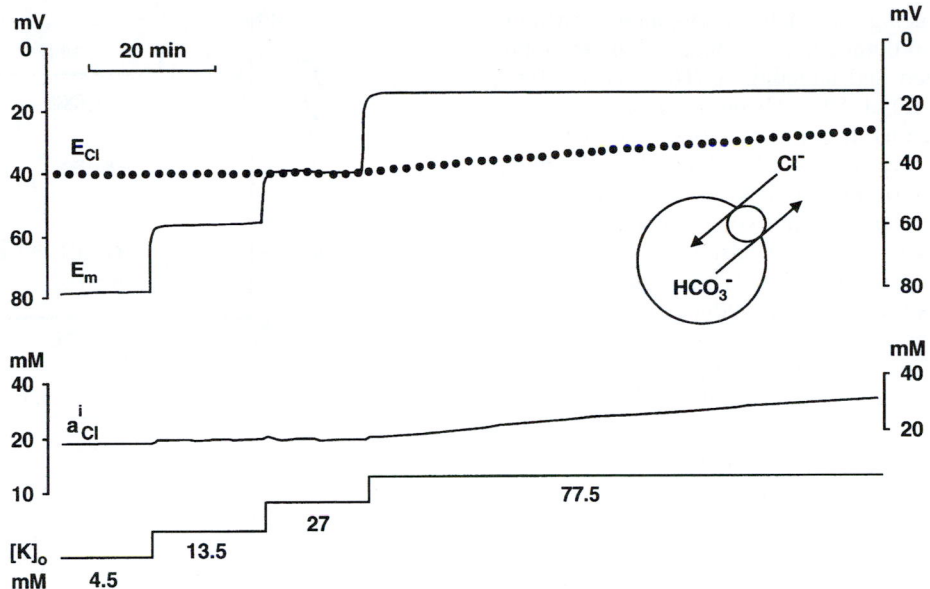

FIGURE 2. Nonpassive Cl⁻ distribution in mammalian heart muscle. Recordings were made with a double-barreled Cl⁻-selective microelectrode in a sheep cardiac Purkinje fiber. Modest increases in external $[K^+]_o$ caused changes in membrane potential (E_m) that were not followed by changes in the intracellular Cl⁻ activity (a^i_{Cl}). Hence, the latter was very much independent of E_m. At physiological E_m, (i.e., when $[K^+]_o = 4.5$ mM), a^i_{Cl} was about four times higher than predicted for a passive distribution. In these cells, the mechanism responsible for uphill accumulation of Cl⁻ is a Cl⁻-HCO⁻₃ exchanger (inset). (Modified from Vaughan-Jones, 1982.)

cotransporter family, which is the focus of this chapter. This family encompasses seven members: a Na⁺-Cl⁻ cotransporter (NCC), two Na⁺-K⁺-2Cl⁻ cotransporters (NKCC1 and NKCC2), and four Na⁺-independent K⁺-Cl⁻ cotransporters (KCC1, KCC2, KCC3, and KCC4).

All of these cation-Cl⁻ cotransporters, as well as the Cl⁻-HCO⁻₃ exchanger, are **secondary active transports.** In this kind of transport, the translocation of the substrate is coupled to the translocation of another substrate that uses the same carrier, whether in the opposite direction (countertransport, antiport, or exchanger) or in the same direction (cotransport or symport). The energy that is consumed in this type of transport process comes secondarily or indirectly from ATP. In other words, in these transporters, the flow of the substrate that is being translocated up its electrochemical gradient is coupled to the flow of a second substrate down its electrochemical gradient; the gradient of this second substrate is maintained by a primary active transport system, that is, one that consumes metabolic energy directly derived from the hydrolysis of ATP. For example, let us consider the Na⁺-Cl⁻ cotransporter (NCC). The uphill movement of Cl⁻ effected by this electroneutral cotransporter system is driven by the energy stored in the Na⁺ chemical potential gradient, which in turn is maintained by the ATP-requiring Na⁺-K⁺ pump. The Na⁺ that enters the cell is pumped back out of the cell via the Na⁺-K⁺ pump. Hence there is no net gain of intracellular Na⁺. In essence, energy is directly invested into a primary active transport mechanism (the Na⁺-K⁺ pump) that is responsible for extruding Na⁺ from the cell (in exchange for K⁺) and thereby maintaining a low and constant intracellular Na⁺ concentration. The Na⁺-Cl⁻ cotransport mechanism

can then convey the uphill movement of Cl⁻ energized by the downhill flow of Na⁺.

IV. Electroneutral Na⁺-K⁺-Cl⁻ Cotransporters

The Na⁺-K⁺-Cl⁻ cotransporters are a class of membrane proteins that mediate the coupled, electrically neutral movement of Na⁺, K⁺, and Cl⁻ ions across the membrane of many animal cells. Two distinct but highly homologous isoforms of the Na⁺-K⁺-Cl⁻ cotransporter (NKCC) proteins have been identified to date, by cDNA cloning, sequencing, and functional expression (Delpire *et al.*, 1994; Gamba *et al.*, 1994; Payne and Forbush, 1994; Payne *et al.*, 1995; Yerby *et al.*, 1997). One isoform (NKCC2) has at least three alternatively spliced variants and is found exclusively in the kidney. The other isoform (NKCC1) is found in nearly all cell types, whether epithelial or not. Under normal physiological conditions, the NKCC works as an active ("uphill") transport system for accumulation of Cl⁻ in cells. The NKCC maintains $[Cl^-]_i$ at levels above the predicted electrochemical equilibrium. The high $[Cl^-]_i$ is used in epithelial tissues to promote net salt transport and in certain nerve cells it makes possible the depolarizing action of transmitters that open Cl⁻ channels, such as GABA and glycine. The GABA-induced depolarizations are the basis for the phenomenon of presynaptic inhibition in the spinal cord (Alvarez-Leefmans *et al.*, 1988, 1998). Abolition of the depolarization generated by the outwardly directed Cl⁻ gradient blocks presynaptic inhibition. Moreover, during early neuronal development, GABA has a widespread depolarizing action (Cherubini *et al.*, 1991; Ben-Ari *et al.*, 1994) that seems crucial in promoting Ca²⁺ influx (Yuste and

Katz, 1991), influencing important developmental phenomena like neuronal proliferation, differentiation, and migration and neurite extension and targeting (LoTurco *et al.,* 1995; Obrietan and van den Pol, 1996; Owens *et al.,* 1996).

There is evidence that in some cell types the NKCC functions to offset osmotically induced cell shrinkage by mediating the net influx of ions and water (see Chapter 21). Whether it serves to maintain cell volume under euvolemic conditions is less clear. The NKCC also may play an important role in the cell cycle. Overexpression of the NKCC1 cotransporter gene has been shown to induce cell proliferation and phenotypic transformation in mouse fibroblasts (Panet *et al.,* 2000). NKCC has also been suggested to function as a K^+ uptake mechanism, thereby keeping $[K^+]_i$ at constant levels, or as an extracellular K^+ buffer, keeping $[K^+]_o$ constant and thereby preventing K^+ accumulation in the extracellular space (Haas and McManus, 1985; Walz, 1992; Payne, 1997; Wu *et al.,* 1998; Wong *et al.,* 1999; Alvarez-Leefmans *et al.,* 2001).

A. Basic Features of the Na^+-K^+-Cl^- Cotransporters

Four functional features define the NKCCs. (1) Ion translocation by the NKCC requires the simultaneous presence of all three ions (Na^+, K^+, and Cl^-) on the same side of the membrane. (2) The cotransport process is electroneutral, with a stoichiometry of $1\,Na^+:1\,K^+:2\,Cl^-$ in nearly all cases. (3) The transporter can work in forward and backward modes; that is, net transport may occur into or out of the cells, the magnitude and direction of this transport being determined by the sum of the chemical potential gradients of the transported ions (Lytle *et al.,* 1998). (4) Transport is inhibited by the so-called *loop diuretics,* derivatives of 5-sulfamoyl benzoic acid, which include furosemide, bumetanide, and benzmetanide (O'Grady *et al.,* 1987 and Fig. 7).

1. Absolute Requirement for All Three Ions on the Same Side of the Membrane

NKCC-mediated fluxes of all three ions require the presence of the other two ions on the side of the membrane from which the flux originates (termed the *cis*-side). This property has been definitely proved by measuring individual fluxes of Na^+, K^+, and Cl^- and looking at their interdependence. Perhaps the most stringent test for this coupling of ion fluxes has been carried out in internally dialyzed squid giant axons. This preparation has the unique advantage of permitting the simultaneous control of the composition of the intracellular as well as the extracellular milieu while measuring unidirectional fluxes. Using this preparation, Russell and his coworkers (cited in Russell, 2000) have demonstrated the absolute requirement for the *cis*-side presence of all three ions both for unidirectional influx and efflux. Each of the three co-ions was systematically removed while measuring the bumetanide-sensitive influx of the other two, using a double-label flux approach. This is illustrated in Fig. 3, which shows the effects on ^{36}Cl (open circles) and ^{24}Na (filled squares) influx when an axon was sequentially superfused with external solutions without K^+ (0 K^+), with K^+ (10 mM), and finally with K^+ (10 mM) plus bumetanide (10 μM) (Fig. 3A). Clearly the pres-

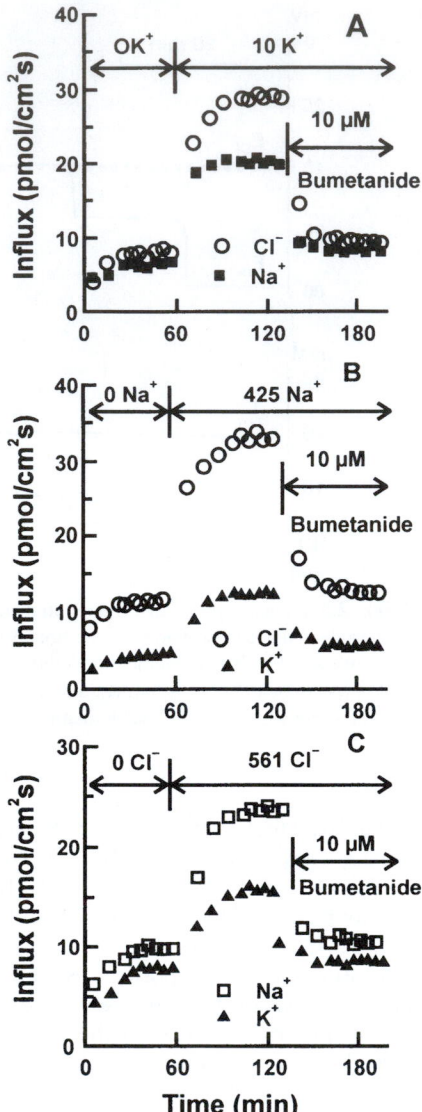

FIGURE 3. Dependence of cotransporter influxes of Na^+, K^+, and Cl^- on simultaneous presence of all three ions in external fluid. Experiments were carried out on internally dialyzed squid axons. Internal fluid in all cases contained the following (in mM): 400 K^+, 0 Na^+, 0 Cl^-, 8 Mg^{2+}, and 4 ATP. Ouabain and tetrodotoxin were present to block the Na^+-K^+ pump and Na^+ channels, respectively. (A) Effect of external K^+ on $^{24}Na^+$ and $^{36}Cl^-$ influxes. After a control period (0 K^+) in which the external K^+ was replaced with N-methyl-D-glucammonium (NMDG), 10 mM K^+ was applied, and influxes of both $^{24}Na^+$ (■) and $^{36}Cl^-$ (○) increased. Finally, bumetanide was added, and both influxes decreased to levels similar to those seen in the absence of external K^+. (B) Effect of external Na^+ on $^{42}K^+$ (▲) and $^{36}Cl^-$ (○) influxes. After a control period in which the external Na^+ was removed and replaced with NMDG (0 Na^+), 425 mM Na^+ was applied, leading to increases in both $^{42}K^+$ and $^{36}Cl^-$ influxes. Subsequent application of bumetanide caused influxes to decrease to values close to those observed in 0 Na^+. (C) Effect of external Cl^- on $^{42}K^+$ (▲) and $^{22}Na^+$ (□) influxes. In the control condition (0 Cl^-), gluconate substituted for Cl^-. Note that upon admission of the external solution containing 561 mM Cl^-, both $^{42}K^+$ and $^{24}N^+$ increased. Subsequent application of bumetanide caused influxes to decrease to levels near those observed in the absence of external Cl^-. (From Russell, 2000.)

ence of extracellular K^+ increased the influx of both Cl^- (by ~21 pmol·cm^{-2}·s^{-1}) and Na^+ (by ~14 pmol·cm^{-2}·s^{-1}), and these increases were reversed by bumetanide. Figure 3 also shows a similar protocol applied to examine the external Na^+ (Fig. 3B) and Cl^- (Fig. 3C) requirement. Note that in each case, the increase in the influx of the two co-ions caused by supplying the third one is reversed by bumetanide treatment.

The absolute requirement for the *cis*-side presence of Na^+, K^+, and Cl^- for the uphill accumulation of Cl^- mediated by NKCC was first demonstrated in vertebrate cells using different methodologies. Examples include Ehrlich ascites tumor cells (Geck *et al.*, 1980), Madin-Darby canine kidney cell line (McRoberts *et al.*, 1982); frog dorsal root ganglion (DRG) neurons (Alvarez-Leefmans *et al.*, 1988), and duck red cells (Haas *et al.*, 1982). In frog sensory neurons, measurements were made using Cl^--selective intracellular microelectrodes. Measurements of changes in intracellular Cl^- activity (a^i_{Cl}) with intracellular microelectrodes reflect **net** Cl^- fluxes occurring through the plasma membrane. Figure 4 shows an example of one experiment in which the cell body of a single neuron was penetrated with a double-barreled electrode (inset), one barrel containing a liquid anion exchanger (the Cl^--sensitive barrel) and the other a reference electrolyte to measure E_m. The advantages of this method are obvious: it is possible to measure simultaneously intracellular a^i_{Cl} and E_m. Moreover, in contrast with methods that measure total $[Cl^-]_i$, which include free-plus-bound or sequestered Cl^-, the microelectrodes give a measurement of free Cl^- in the cytosolic space (Alvarez-Leefmans *et al.*, 1990). The dependence on extracellular Na^+ and K^+ of the bumetanide-sensitive active accumulation of Cl^- was explored in these cells. To this end, the effects of Na^+- or K^+-free solutions on a^i_{Cl} were studied in the steady state and during the active reaccumulation from Cl^--free solutions. A typical result for the effects of external Na^+ removal on a^i_{Cl} is shown (Fig. 4). The basal a^i_{Cl} (upper trace) and E_m (lower trace) were recorded in the standard frog Ringer's solution (FR) for about 3 min after impalement. The initial E_m was ~ −50 mV and a^i_{Cl} was 25 mM. The latter value was twice the a^i_{Cl} predicted for an equilibrium distribution according to Eq. 3. Then, the cell was exposed for about 3.5 min to a solution in which Na^+ was eliminated (0 Na^+) and replaced mole by mole with the impermeant cation N-methyl-D-glucammonium, keeping a^o_{Cl} at the same level as in the FR solution. This resulted in an immediate decrease in a^i_{Cl} accompanied by a membrane hyperpolarization. The latter was due to the suppression of a resting electrodiffusional Na^+ permeability. Both E_m and a^i_{Cl} returned toward control levels upon readmission of FR solution to the bath. Observations like these indicate that external Na^+ is required to **maintain** internal chloride above electrochemical equilibrium in the resting state.

To explore if external Na^+ is required for the active reaccumulation of Cl^- after intracellular Cl^- depletion, external Cl^- was removed (0 Cl^-) and replaced mole by mole with the impermeant anion gluconate, keeping the other ions as in control FR. This also resulted in a decrease in Cl^-. The reaccumulation of Cl^- was then observed in the Na^+-free solution containing normal Cl^- (0 Na^+) and finally in standard FR. It was found that the rate of Cl^- reaccumulation in the Na^+-free solution was slowed to about 15% of its value in standard FR solution.

These observations indicate that **extracellular Na^+ is required for active reaccumulation of Cl^- after cell Cl^- depletion.** The concurrent changes in E_m and calculated values for E_{Cl} in the same experiment are shown in Fig. 5A, which illustrates more clearly the effects of external Na^+ on the distribution of Cl^- across the cell membrane. External Na^+ removal reduced the **basal** intracellular accumulation of Cl^-, as reflected by the reduction in the difference between E_m (gray circles) and E_{Cl} (filled circles). In addition, Na^+ removal after Cl^- depletion impeded the reaccumulation of Cl^- even though E_{Cl} was now more negative than E_m, a situation in which the direction of the Cl^- gradient was reversed, favoring Cl^- influx. Therefore, it can be concluded that external Na^+ is required to

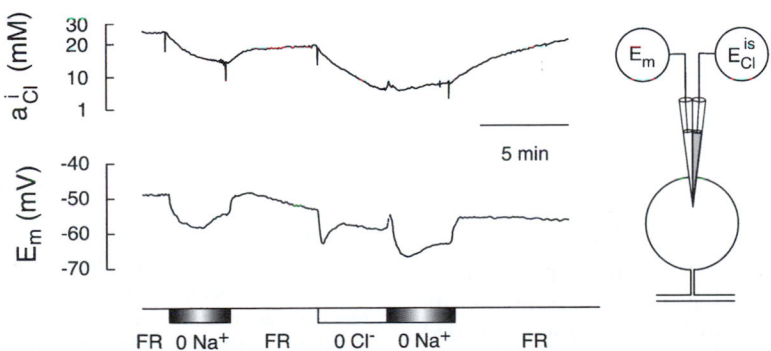

FIGURE 4. Effect of external Na^+ removal on intracellular Cl^- activity (a^i_{Cl}) and on membrane potential (E_m) in a frog dorsal root ganglion neuron. The recording was made with a double-barreled microelectrode, one barrel sensing E_m and the other a^i_{Cl}. Inset (right), diagram of a double-barreled electrode where E^{is}_{Cl} denotes the Cl^--selective microelectrode potential across the barrel filled with a liquid anion exchanger. When the microelectrode is inside the cell, E^{is}_{Cl} will be the sum of a potential proportional to a^i_{Cl} plus E_m. To obtain a signal proportional to a^i_{Cl}, the potential on the E_m barrel is subtracted from E^{is}_{Cl}, giving the differential signal $E_m - E^{is}_{Cl}$, which is proportional to a^i_{Cl} (top trace). During the times indicated by the bottom bars, the bathing solution was changed from frog Ringer (FR) to a Na^+-free solution (0 Na^+), in which the external Na^+ was replaced with N-methyl-D-glucammonium, and back to FR. Note that removal of external Na^+ led to a decrease in a^i_{Cl}. The cell was then exposed to Cl^--free solution (0 Cl^-), which also produced a decrease in a^i_{Cl}. Then the reaccumulation of internal Cl^- was monitored in the absence of external Na^+ (0 Na^+) and in FR, that is, in the presence of external Na^+. (Modified from Alvarez-Leefmans *et al.*, 1988.)

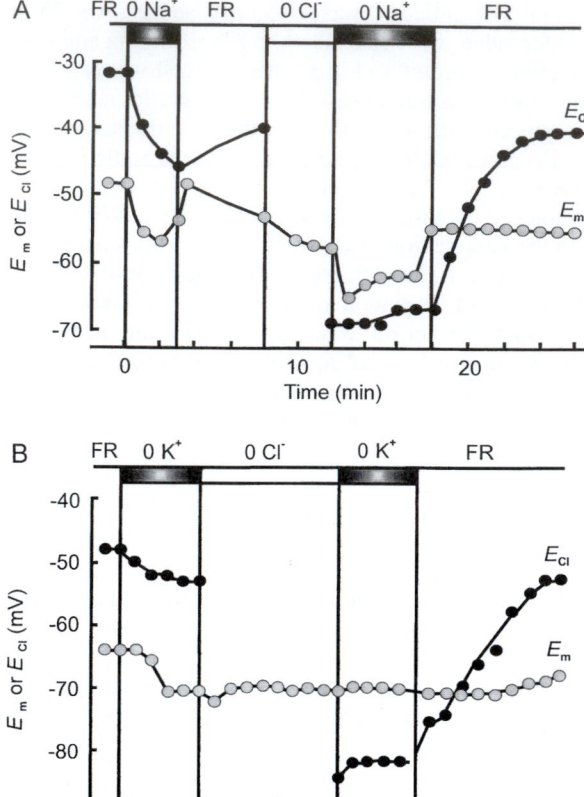

FIGURE 5. (A) Effects of external Na^+ on the intracellular accumulation of Cl^-. Membrane potential (E_m, gray dots) and the calculated Cl^- equilibrium potential (E_{Cl}, black dots) from measured a_{Cl}^i and a_{Cl}^o are plotted on the same time scale using data from the experiment illustrated in Fig. 4. (B) Effects of external K^+ on the intracellular accumulation of Cl^-. Symbols as in A. Data were obtained from an experiment similar to that shown in Fig. 4 but in which external K^+ was removed instead of external Na^+. (Modified from Alvarez-Leefmans *et al.*, 1988.)

accumulate Cl^- above its equilibrium distribution and for keeping it out of equilibrium in basal conditions.

The effects of extracellular K^+ on a_{Cl}^i were similar to those observed for Na^+. Figure 5B shows a plot of E_m and E_{Cl} from an experiment similar to that shown in Fig. 4 but in which external K^+ was removed during the periods indicated. Removal of K^+ was accompanied by an immediate reduction in a_{Cl}^i revealed in the graph by the changes in E_{Cl} toward more negative values. In the absence of extracellular K^+ the reaccumulation of Cl^- in standard FR solution was impeded. Summarizing, the steady-state a_{Cl}^i levels in DRG cells depend on the simultaneous presence of extracellular Na^+ and K^+. Similarly, the active reaccumulation of Cl^- after intracellular Cl^- depletion requires the simultaneous presence of Na^+ and K^+ in the extracellular medium.

2. Electroneutral Cotransport Process: Stoichiometry of 1 Na^+:1 K^+:2 Cl^-

Geck and coworkers (1980), working on Ehrlich ascites tumor cells, provided the first comprehensive evidence for

the existence of the electrically neutral cotransport of Na^+, K^+, and Cl^- ions, with a stoichiometry of 1Na^+:1K^+:2Cl^-. A solidly documented exception is the squid axon cotransporter which has a stoichiometry of 2Na^+:1K^+:3Cl^-, probably reflecting its distinct molecular structure (Russell, 2000). There is now general agreement that the NKCC is electrically silent, which means that the overall transport process of Na^+, K^+, and Cl^- via the cotransporter is not driven by the transmembrane voltage, nor does the transport process directly generate a membrane current that may affect the transmembrane voltage. As correctly pointed out by Russell (2000), the electrical silence of the NKCC is now well accepted by workers in the field, yet direct evidence for the assertion is surprisingly sparse. Electrical silence is often inferred from the apparent stoichiometry of the cotransport process, but arriving at this conclusion involves a circular argument.

Conclusive proof of electrical silence requires direct and simultaneous measurement of transmembrane potential and cotransporter-mediated ion fluxes. Only two reports in the literature fulfill these requirements, one for the squid axon NKCC cotransporter (Russell, 1984) and one for frog dorsal root ganglion cells (Alvarez-Leefmans *et al.*, 1988, 1998). The latter is, so far, the only existing direct evidence for the electroneutrality of Cl^- transport through NKCC in a vertebrate cell. The measurements were made with double-barreled Cl^--selective microelectrodes as already described (see Fig. 4).

Figure 6 shows an example of one experiment in which a frog DRG cell was penetrated with a double-barreled electrode

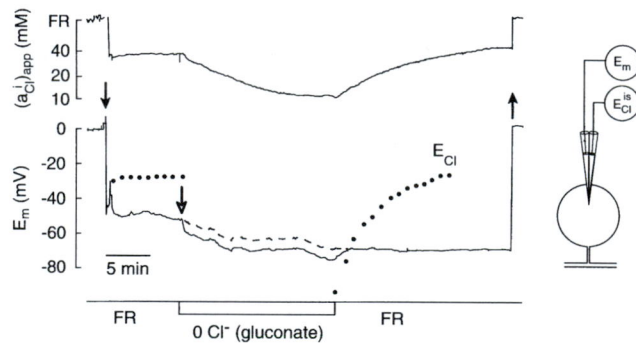

FIGURE 6. Electrically silent, NKCC-mediated, transmembrane movements of Cl^-. The intracellular Cl^- activity, a_{Cl}^i, was measured with a double-barreled Cl^--selective microelectrode in a frog DRG cell body. After obtaining steady-state readings, the effects of removing and re-adding external Cl^- on a_{Cl}^i and E_m were recorded. The upper trace shows the differential signal ($E_m - E_{Cl}^{is}$) which indicated $(a_{Cl}^i)_{app}$. The lower trace is the membrane potential (E_m). The dashed line above a segment of the E_m trace is the E_m value after subtraction of liquid junction potentials. At the periods indicated by the bottom marks the bathing solution was changed from standard frog Ringer's solution (FR) to a Cl^--free solution in which Cl^- was replaced by gluconate. The points superimposed on the E_m trace show the estimated E_{Cl} calculated from measured a_{Cl}^o and a_{Cl}^i. Note that "uphill" accumulation of Cl^- following Cl^- depletion occurred without changes in E_m. Filled arrows indicate electrode penetration and withdrawal from the cell. The open downward arrow indicates the time at which steady-state measurements were obtained. (Modified from Alvarez-Leefmans *et al.*, 1988.)

(inset). At the time indicated by the downward filled arrow, the cell was penetrated, and E_m stabilized at about −55 mV (indicated by clear arrow). The deflection in the differential signal (upper trace) at this time corresponded to an **apparent** intracellular Cl⁻ activity $(a_{Cl}^i)_{app}$ of about 40 mM, which, after correction for background intracellular interference produced by other native intracellular anions on microelectrode response (Alvarez-Leefmans, 1990), gave an a_{Cl}^i of 30 mM. If Cl⁻ was passively distributed, a_{Cl}^i as predicted from Eq. 3 should be only 10.2 mM. Clearly, **intracellular Cl⁻ was almost three times the value expected for an equilibrium distribution.** E_{Cl} was −27 mV, and the driving force for Cl⁻, $(E_m - E_{Cl})$, was 28 mV. The filled circles superimposed on the E_m trace (lower record in Fig. 6) correspond to the values of E_{Cl} calculated using Eq. 1. Removal of external Cl⁻ and its replacement with an impermeant anion (gluconate) led to a rapid decrease in $(a_{Cl}^i)_{app}$, reaching a final steady-state level of 10 mM in about 10 min. The recovery of intracellular Cl⁻ started immediately after external Cl⁻ was restored. Note that the net Cl⁻ movements occurred without changes in E_m, that is, they were **electroneutral**. The small rapid changes that can be observed in the E_m trace upon removal and readmission of external Cl⁻ were due to liquid junction potentials at the reference (bath) electrode. Subtraction of this liquid junction potential yielded the dashed line superimposed on the recorded E_m trace. If Cl⁻ movements occurred through an electrodiffusional pathway (i.e., channel-mediated) or through an electrogenic "carrier," E_m should change both upon Cl⁻ efflux and upon Cl⁻ influx. Clearly that was not the case. Since the DRG cell membrane has a relatively low Cl⁻ permeability (P_{Cl}), the movements of Cl⁻ occurred largely through a nonelectrodiffusional pathway, that is, an electroneutral carrier mechanism. As already mentioned, the circles superimposed on the E_m trace in Fig. 6 correspond to the calculated values of E_{Cl}. On exposure to Cl⁻-free Ringer's solution, the cell was depleted of Cl⁻ and, immediately after restoring the external Cl⁻, E_{Cl} was transiently **below** E_m. However, during most of the recovery period in Ringer's solution, the movement of Cl⁻ occurred against its electrochemical potential difference across the cell membrane. Since the Cl⁻ fluxes were bumetanide-sensitive and dependent on the simultaneous presence of external Na⁺ and K⁺, this is direct evidence for the presence of an "uphill" (i.e., active) process for accumulation of Cl⁻ in DRG cells through an electroneutral membrane transport mechanism (NKCC). The presence of NKCC protein in DRG cells has been recently corroborated using immunolabeling with monoclonal (Alvarez-Leefmans et al., 1998, 1999, 2001) and polyclonal antibodies (Sung et al., 2000). Moreover, in the latter study it was shown that in a mouse knock-out of NKCC1, the cotransporter disappears from DRG cells, rendering them incapable of uphill accumulation of Cl⁻.

Because the overall transport of the ions through the NKCC is an electroneutral process, it follows that during a single turnover of the cotransporter the sum of the cations (Na⁺ + K⁺) crossing the membrane will be equal to the sum of the Cl⁻ crossing the membrane. Therefore, the simplest stoichiometry is 1Na⁺:1K⁺:2Cl⁻, which happens to be the one that has been found in all vertebrate cells in which it has been determined (O'Grady et al., 1987; Russell, 2000).

3. Magnitude and Direction of Cotransport Process Determined by the Sum of the Chemical Potential Gradients of the Transported Ions

We have seen that the transport process carried out by NKCC is electrically neutral, which means that it neither affects nor is affected by the transmembrane potential. Consequently, the magnitude and direction of the driving force for net ion movement must be calculated from the chemical, rather than the electrochemical potential gradients of all the transported ions.

The NKCC is a secondary active transport mechanism and hence it is expected to be **reversible,** that is, to mediate ion fluxes into or out of the cell, the net direction of cotransport depending on the overall net free energy (ΔG) of the system (see Chapter 1). If ΔG is negative, it means that the direction of the cotransport will be inward, favoring uptake. If ΔG is zero, the system is at equilibrium. Since NKCC is electroneutral and has a 1 Na⁺:1 K⁺:2 Cl⁻ stoichiometry, ΔG is given by the following equation:

$$\Delta G = \Delta\mu_{Na,K,Cl} = \Delta\mu_{Na} + \Delta\mu_K + 2\Delta\mu_{Cl} \qquad (5)$$

In other words, the driving force for the electroneutral 1Na⁺, 1K⁺, 2Cl⁻ cotransport ($\Delta\mu_{Na,K,Cl}$) is the sum of the **chemical** potential differences of Na⁺ ($\Delta\mu_{Na}$) and K⁺ ($\Delta\mu_K$) plus twice the Cl⁻ potential difference ($\Delta\mu_{Cl}$), since two Cl⁻ ions are translocated per cycle.

From the definition of chemical potential for each of the ions involved, it follows that

$$\Delta G = \Delta\mu_{Na,K,Cl}$$

$$= RT \ln \frac{[Na^+]_i}{[Na^+]_o} + RT \ln \frac{[K^+]_i}{[K^+]_o} + 2RT \ln \frac{[Cl^-]_i}{[Cl^-]_o} \qquad (6)$$

Hence

$$\Delta G = \Delta\mu_{Na,K,Cl} = RT \ln \frac{[Na^+]_i [K^+]_i [Cl^-]_i^2}{[Na^+]_o [K^+]_o [Cl^-]_o^2} \qquad (7)$$

Using Eq. 7, Russell (2000) has calculated ΔG for several representative cell types and reached the conclusion that, under physiological conditions, ΔG is negative, therefore favoring net ion uptake. For vertebrate cells ΔG values range from −0.98 up to −8.86 kJ/mol. The system reaches equilibrium when $\Delta G = \Delta\mu_{Na,K,Cl} = 0$, that is, when the product of the concentrations of Na⁺ and K⁺ times the square of the Cl⁻ concentration on one side of the membrane is matched by the product of these ions in the *trans* side:

$$[Na^+]_i [K^+]_i [Cl^-]_i^2 = [Na^+]_o [K^+]_o [Cl^-]_o^2 \qquad (8)$$

Since the transporter is electroneutral, thermodynamic equilibrium will not be affected by E_m. Assuming for the sake of argument that Na⁺, K⁺, Cl⁻ cotransport is the only Cl⁻ transport system operating in the steady-state in a cell, that it has a 1 Na⁺:1 K⁺:2 Cl⁻ stoichiometry, and that there are no kinetic constraints, the expected $[Cl^-]_i$ when the cotransport reaches thermodynamic equilibrium is given by the following equation derived from Eq. 8:

$$[Cl^-]_i = \left([Na^+]_o [K^+]_o [Cl^-]_o^2 / [Na^+]_i [K^+]_i \right)^{1/2} \qquad (9)$$

For a "typical" mammalian cell (e.g., $[K^+]_i$ = 140 mM, $[K^+]_o$ = 4.5 mM, $[Na^+]_i$ = 15 mM, $[Na^+]_o$ = 140 mM, $[Cl^-]_o$ = 110 mM), Eq. 9 predicts that when the system attains thermodynamic equilibrium, $[Cl^-]_i$ would be ~60 mM. Most cells that are endowed with NKCC do not have such a high intracellular Cl^- level. Why is that? Are there parallel mechanisms that prevent the NKCC from achieving thermodynamic equilibrium? The answer is yes. The first issue that must be considered is that, besides the NKCC, other Cl^- transport systems such as Cl^- channels, the anion exchanger (Cl^--HCO_3^-), or a KCl cotransporter (KCC) could and in fact do coexist in the same cell with NKCC. The final $[Cl^-]_i$ will be determined by the coordinated functional interaction between the various transporter systems present in the cell. For instance, in skeletal muscle, active Cl^- uptake via NKCC is shunted by a large Cl^- conductance (Aickin, 1990; Wong et al., 1999). In this case, the $[Cl^-]_i$ will be determined by a balance between the net influx of Cl^- mediated by the NKCC and the net Cl^- efflux occurring through the electrodiffusional leak (see Fig. 1). Since the latter is larger, $[Cl^-]_i$ in skeletal muscle is low, and E_{Cl} is close to E_m (see Sections I and II). In addition to Cl^- channels, in most cells, NKCC coexists with KCC. Under physiological conditions the KCC mediates net efflux of K^+ and Cl^- from cells. Hence NKCC and KCC have opposing roles. This interaction between two cotransporters working in opposite directions is ideally suited for achieving tight regulation of an intracellular ion, in this case Cl^- (and K^+).

The functional interaction between KCC and NKCC has been studied recently in human embryonic kidney cells (HEK-293) overexpressing the "housekeeping" isoform of the KCC, KCC1 (Gillen and Forbush 1999). The results show that overexpression of KCC1 increases baseline endogenous NKCC activity and that a reduction of $[Cl^-]_i$ effected by KCC1, is at least partly responsible for this activation. This result is in agreement with previous observations showing that lowering $[Cl^-]_i$ activates NKCC, and increasing $[Cl^-]_i$ inhibits its activity. In other words, intracellular Cl^- serves the role of a negative feedback signal in controlling the activity of the NKCC. Ussing (1982) first suggested that the cotransporter might be activated by a fall of $[Cl^-]_i$. However, the definitive proof came from studies in the internally dialyzed squid axon, in which it was shown that elevation of $[Cl^-]_i$ inhibits bumetanide-sensitive unidirectional fluxes of all three cotransported ions in both the influx and efflux direction (Breitwieser et al., 1990, 1996; Russell, 2000). The question arises as to how changes in $[Cl^-]_i$ might exert a regulatory effect on the NKCC. Lytle and Forbush (1992) proposed that $[Cl^-]_i$ affected the degree of phosphorylation of the cotransporter and hence its activity, as had been originally suggested by Russell and coworkers (Altamirano et al., 1988). The current view is that a rise in $[Cl^-]_i$ promotes a reduction in the degree of phosphorylation of the NKCC protein, thereby inhibiting its activity. Whether this is a direct effect of Cl^- on the NKCC protein or is mediated by a protein kinase or phosphatase is still to be determined. However, most of the evidence supports the view that phosphorylation regulates the activity of the cotransporter (Haas et al., 1995; O'Donnell et al., 1995; Lytle, 1997). The residues that appear to be phosphorylated are serine and threonine (Lytle and Forbush, 1992). Interestingly, although the NKCC is a secondary active transport system, it has an absolute requirement for ATP. However, in contrast with primary active transport mechanisms, NKCC does not use ATP as a source of energy. The effects of ATP on NKCC are mediated via a protein phosphorylation/dephosphorylation mechanism (Russell, 2000).

4. Loop Diuretics Inhibit NKCC

Loop diuretics are organic anions derived from 5-sulfamoylbenzoic acid that inhibit NKCC. They are called "loop diuretics" because they inhibit Na^+ reabsorption by the thick ascending limb of the loop of Henle, by blocking the NKCC isoform located in the apical membrane of these cells (i.e., NKCC2). However, they also block NKCC1. The relative efficacy (IC_{50} values) and order of potency for blocking the NKCC was investigated in winter flounder intestine (O'Grady et al., 1987) and is shown in Fig. 7. Their order of potency to block NKCC was benzmetanide > bumetanide > piretanide > furosemide > amino-piretanide. Besides their clinical importance, these substances, particularly bumetanide, have played a crucial role in the identification of the NKCC proteins and their subsequent cloning (Xu et al., 1994). Moreover, they have provided information that has been used to interpret ion binding to the NKCC. Bumetanide is by far the most commonly used loop diuretic for research on NKCC. It reversibly inhibits the NKCC in a variety of preparations by a concentration-dependent mechanism, with a half inhibitory constant (IC_{50}) of ~1×10^{-7} M (Russell, 2000). NKCC2 is more sensitive to bumetanide than NKCC1 (Isenring et al., 1998). However, none of these compounds are specific toward the NKCC. Thus, used alone and/or in relatively high concentrations, not even bumetanide can provide positive proof that a given function is mediated by the NKCC. Other Cl^- transport systems that can be inhibited by loop diuretics include the Cl^--HCO_3^- exchanger, Ca^{2+}-activated and GABA-gated Cl^- channels, and KCC (for references see Alvarez-Leefmans, 1990 and Russell, 2000). Loop diuretics are anionic in the physiological range with very high lipid solubility and hence they do cross cell membranes (Alvarez-Leefmans et al., 1990). This raises the question of their site of action. It has been suggested but not proven that bumetanide binds exclusively to the external face of the NKCC (Haas, 1994), but whether it acts also from the inside remains to be determined.

B. Molecular Structure and Distribution of Na^+-K^+-Cl^- Cotransport Proteins

Presently, two distinct Na^+-K^+-Cl^- cotransporter isoforms have been identified by cDNA cloning and expression: NKCC1 (also referred to as BSC2 or CCC1) and NKCC2 (also known as BSC1 or CCC2). A cDNA encoding NKCC1 was first cloned from the shark rectal gland, a well-characterized secretory epithelium with high density of Na^+-K^+-Cl^- cotransporters in its basolateral membrane (Xu et al., 1994). Subsequently, cDNAs encoding NKCC1 have been cloned from other secretory epithelia, including

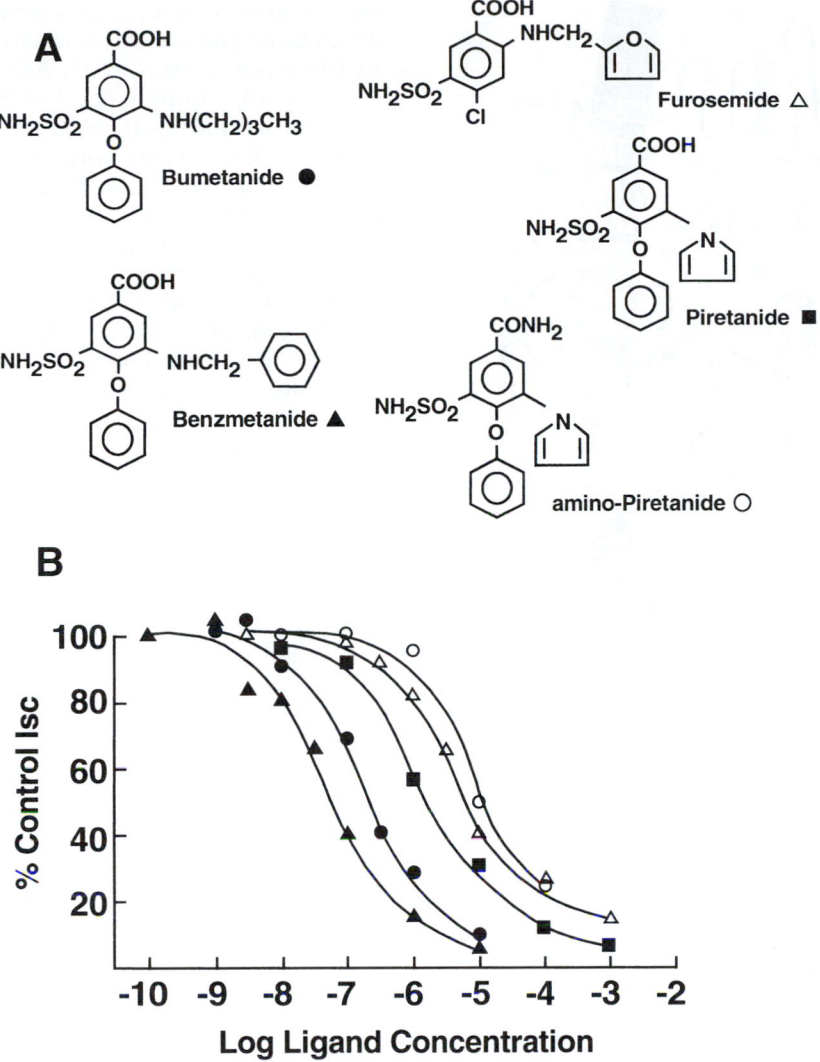

FIGURE 7. (A) Structures of several 5-sulfamoyl benzoic acid derivatives that affect NKCC in flounder intestine. (B) Dose-response curves for these derivatives on short-circuit current measured in flounder intestine. Order of potency is benzmetanide (5×10^{-8} M; $n = 4$), bumetanide (3×10^{-7} M; $n = 7$), piretanide (3×10^{-6} M; $n = 4$), furosemide (7×10^{-6} M; $n = 4$), amino-piretanide (1×10^{-5} M; $n = 4$). (Modified from O'Grady *et al.*, 1987.)

T84 cells, a human colonic epithelial cell line (Payne *et al.*, 1995), and cultured inner medullary collecting duct cells (mIMCD-3) from mouse kidney (Delpire *et al.*, 1994). Both of these mammalian NKCC1 proteins share a high degree of homology with the shark NKCC1 (71–74%) and even higher identity with each other (91%). More recently, a cDNA encoding the bovine aortic endothelial cell NKCC was also cloned and sequenced, and the predicted protein exhibits ~95% amino acid identity with the T84 cell cotransporter sequence (Yerby *et al.*, 1997).

NKCC1 comprises ~1200 amino acids and has a molecular weight of 130–132 kDa. The predicted secondary structure for NKCC1 from shark rectal gland shows 12 α-helical transmembrane (TM) domains and hydrophilic amino- and carboxy-terminal regions flanking a central hydrophobic domain (Fig. 8). Whereas the amino terminus displays considerable variability, the large carboxyl terminus is well con-

served among the different NKCC proteins. All identified Na$^+$-K$^+$-Cl$^-$ cotransporters are glycoproteins and possess consensus sites for N-linked glycosylation within a large hydrophilic (extracellular) loop between TM segments 7 and 8, as shown in Fig. 8 (Haas and Forbush, 1998). In fact, the protein in its physiological plasma membrane environment is glycosylated. It is believed that glycosylation enhances cell-surface expression of these proteins. In addition, NKCC1 has several phosphorylation sites within the predicted amino- and carboxy-terminal domains (labeled "P" in Fig. 8).

Northern blot analysis has revealed the presence of NKCC1 in a wide variety of tissues, including salivary gland, stomach, lung, trachea, and pancreas, all possessing secretory epithelia. This analysis as well as immunohistochemistry and *in situ* hybridization also demonstrated the presence of NKCC1 mRNA in nonepithelial tissues such

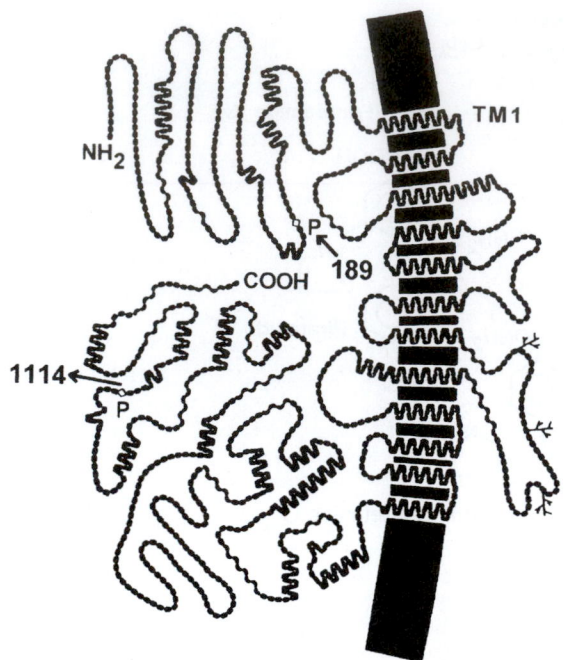

FIGURE 8. Model of the shark rectal gland Na^+-K^+-$2Cl^-$ cotransport protein (NKCC1), based on its cDNA sequence and hydropathy and secondary structural analysis. Each link symbolizes a single amino acid residue (1191 total). "NH_2" and "COOH" indicate amino- and carboxy-termini, respectively. Branched lines indicate potential glycosylation sites in the extracellular loop formed between putative transmembrane (TM) segments 7 and 8. Biochemically identified phosphorylated threonine (Thr) residues (positions 189 and 1114) are indicated by open diamonds with an adjacent "P." The proposed secondary structure of this protein with 12 TM helices and large intracellular amino- and carboxy-terminal domains is shared by other members of the cation-chloride cotransporter (CCC) superfamily that have been cloned and sequenced (see Fig. 13). (Modified from Haas and Forbush III, 1998.)

as heart, skeletal muscle, and vertebrate sensory neurons and axons, as well as in Schwann cells (Delpire *et al.,* 1994; Xu *et al.,* 1994; Payne *et al.,* 1995; Plotkin *et al.,* 1997; Sung *et al.,* 2000; Alvarez-Leefmans *et al.,* 1998, 2001).

Immunofluorescence studies using anticotransporter antibodies (Lytle *et al.,* 1995) have shown that in intact secretory epithelia, NKCC1 protein is localized in the basolateral membrane. The exception is the rat choroid plexus (CP) where NKCC1 is located in the apical membrane (Plotkin *et al.,* 1997). The cotransporters that are localized in the basolateral membrane mediate a net salt influx into the cells, and in doing so act in concert with apical Cl^- channels and basolateral K^+ channels to produce net salt and fluid secretion in these epithelia. The location of NKCC1 in the apical membrane of CP argues against its involvement in ion secretion into cerebrospinal fluid (see Section IV.D). It has been proposed that the CP cotransporter is constitutively active and it functions in series with Ba^{2+}-sensitive K^+ channels to reabsorb K^+ from cerebrospinal fluid to blood (Wu *et al.,* 1998). The location of NKCC in the paranodal region of the Schwann cells suggest a similar role in K^+ uptake from

the peri-axonal space, thereby preventing K^+ accumulation which could produce abnormal changes in axonal excitability (Alvarez-Leefmans, *et al.,* 2001).

The NKCC2 isoform has thus far been identified exclusively in the kidney (Gamba *et al.,* 1994; Mount *et al.,* 1998). NKCC2 shares only ~60% amino acid sequence identity with NKCC1. In fact, these two proteins are the products of independent genes. In the mouse, Delpire and his coworkers (1994) have localized the mNKCC1 gene to chromosome 18, whereas the mNKCC2 gene has been localized to chromosome 2. Studies in rabbit kidney revealed that there are at least three splice variants of NKCC2—A, B, and F (Payne and Forbush, 1994)—all of which are differentially distributed within the rabbit kidney: one form restricted to the cortex (variant B), one restricted to the medulla (variant F), and a third equally distributed between both the cortex and the medulla (variant A). It is now known that the three splice variants are likely to be a general feature of the mammalian kidney (Gamba *et al.,* 1994). Additional splice variants of mouse NKCC2 have been recently reported (Mount *et al.,* 1999a). The functional role of all these splice variants is not well understood but it is conceivable that they underlie heterogeneity of ion transport in the thick ascending limb of the loop of Henle (TALH).

NKCC2 is a protein smaller than NKCC1, with a deduced amino acid sequence of ~1100 residues and a molecular mass of 120–121 kDa, which is consistent with the size (150 kDa) of the glycosylated transporter. The difference between the two isoforms can be accounted for almost entirely by the additional 80 amino acids at the amino terminus of NKCC1. In fact, the amino terminus is the most divergent region of the entire protein both between the two isoforms and comparing the same isoform between species (Haas and Forbush, 1998).

NKCC2 is located at the apical membrane of the epithelial cells of the thick ascending limb of the loop of Henle (TALH), the main pharmacological target of the loop diuretics. NKCC2 plays a central role in transcellular absorption of Na^+ and Cl^- by the medullary and cortical TALH and plays a secondary role in the paracellular transport of Na^+, Ca^{2+}, and Mg^{2+} by this nephron segment (Mount *et al.,* 1998). NKCC2 is considered one of the major proteins involved in salt and fluid reabsorption in the kidney; 15–25% of filtered Na^+ is reabsorbed in the TALH, most of it via the NKCC2. Salt absorption by the TALH is crucial to countercurrent multiplication in the renal medulla and the excretion of concentrated urine. The TALH also functions in renal acid secretion, since NH_4^+ can compete with the K^+-binding site on the NKCC2 (Good, 1994). The NH_4^+ generated by the renal proximal tubule is thus reabsorbed from the tubule lumen in the TALH and excreted from the renal interstitium by more distal nephron segments. NKCC2 is also located in the apical membrane of the macula densa cells (Obermüller *et al.,* 1996; Mount *et al.,* 1998). The macula densa is a specialized group of tubular cells situated at the point of apposition of renal tubules with their parent glomeruli. These cells control two important physiological processes, tubuloglomerular feedback and tubular regulation of renin release. An increase in luminal Na⁺-Cl⁻ activity at the macula densa decreases glomerular filtration by inducing constriction of the afferent renal arteriole. Increases in luminal Na⁺-Cl⁻ activity at the macula densa also inhibit renin release from juxtaglomerular cells in the afferent arteriole. Thus NKCC2 plays a crucial role in the regulation of fluid volume and osmolarity.

C. A Kinetic Model of Na⁺-K⁺-2 Cl⁻ Cotransport

Normal operation of the vertebrate NKCC requires the binding to the cotransporter of all the ions involved (i.e., 1 Na⁺, 1 K⁺, and 2 Cl⁻) on the same side of the membrane before ion translocation occurs. The order of binding of the cotransported ions does not seem to occur at random. Evidence shows a preferred order of ion binding resulting from an allosteric effect such that the binding of one ion increases the apparent affinity of a binding site for the next ion. The best-tested model to date is one based on a systematic study of ion dependencies of self-exchange fluxes mediated by the cotransporter in duck and human red blood cells (Duhm, 1987; Haas, 1994; Lytle *et al.*, 1998). Two partial reactions of the cotransported cycle have been revealed: K⁺-K⁺ exchange in normal high-K⁺ cells and Na⁺-Na⁺ exchange in high-Na⁺ cells. The internal and external ion requirements of each mode can be explained by a reaction cycle based on ordered binding and glide symmetry. The latter means that the first ion to bind on one side of the "carrier" will be the first ion to be released on the other.

Figure 9 shows the cotransport model based on ordered ion binding and glide symmetry. Net cotransport influx would require that the ions bind to the cotransporter in the order shown. In the initial loading of the empty carrier at the outer face of the membrane, binding of a Na⁺ in the pocket (reaction 2) induces formation of a Cl⁻ site, and binding of that Cl⁻ (reaction 3) creates a site that can bind K⁺, and so forth. All four ions are occluded momentarily in a transitional state (E₄) without access to either side of the membrane. Reaction 5 represents a conformational change of the NKCC such that the binding sites are now accessible from the intracellular compartment, and reactions 6–9 represent the ordered release of ions. Reaction 10 leads to a conformational change that involves the reorientation of the totally unloaded cotransporter to an outwardly facing conformation. The model has been quantitatively tested using simulations based on flux data from duck red blood cells, HeLa cells, and epithelial cells (Benjamin and Johnson, 1997). A critique of the model showing its weaknesses and strengths has appeared recently (Russell, 2000).

D. Functions of the Na⁺-K⁺-2 Cl⁻ Cotransport

The net transport of Na⁺, K⁺, and Cl⁻ can serve a number of physiological functions depending on the cell type in question. Some of these functions have already been highlighted in Section IV. The central role that NKCC plays in net salt and fluid secretion across absorptive and secretory

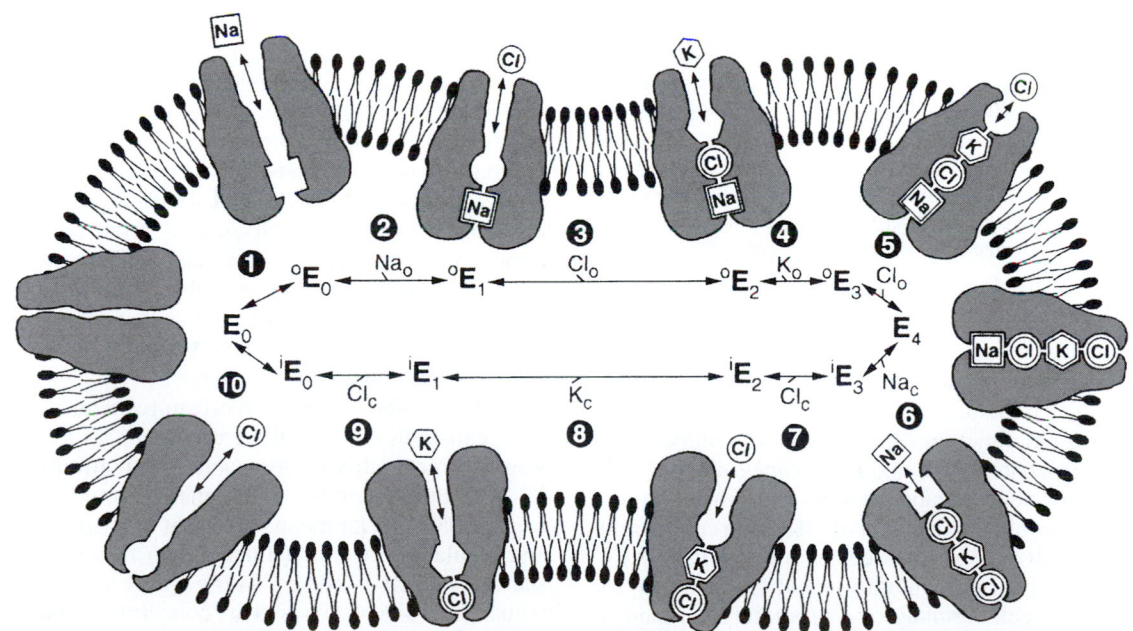

FIGURE 9. Hypothetical model of ordered ion binding with glide symmetry of Na⁺-K⁺-2Cl⁻ cotransport. Ten intermediate states of the transporter are depicted. The model assumes that cotransport is completely reversible and that only fully loaded (E₄) and completely unloaded (E₀) forms of cotransporter protein are capable of changing orientation from inward- to outward-facing (and vice versa). The idea is that when all sites are occupied (E₄), a major conformational change occurs, allowing bound ions to unload on the opposite side of the membrane in the order they were bound (first on, first off). When the carrier is completely unloaded (E₀), the empty form can also undergo a major conformational change, allowing it to begin reloading ions at the opposite side of the membrane, thus facilitating net cotransport. Numbers refer to order of binding, reorientation, and release for one complete influx cycle. (Modified from Lytle *et al.*, 1998.)

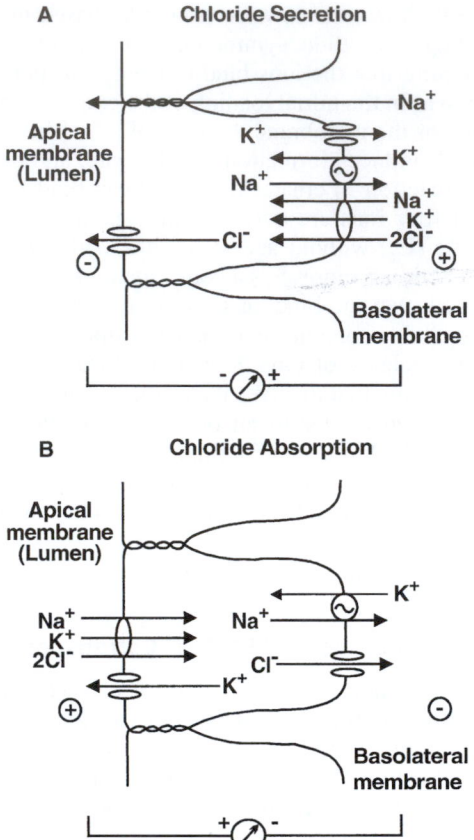

A   Chloride Secretion

Apical
membrane
(Lumen)

Basolateral
membrane

B   Chloride Absorption

Apical
membrane
(Lumen)

Basolateral
membrane

FIGURE 10.   Mechanism for active Cl^- absorption and secretion in Cl^- transporting epithelia. (A) Transport model for exocrine gland Cl^--secreting epithelium (e.g., salivary glands or trachea epithelium). Intracellular $[Cl^-]$ is maintained above electrochemical equilibrium by NKCC across the basolateral membrane. Na^+ influx is balanced by efflux via Na^+,K^+-ATPase; K^+ influxes are balanced by efflux via K^+ channels. Transepithelial secretion is induced by activation of apical-membrane Cl^- channels. (B) Model of Cl^- transport in thick ascending limb of Henle's loop, a Cl^--absorptive epithelium. Cl^- entry across the apical membrane is mediated by NKCC. Most basolateral Cl^- efflux proceeds through Cl^- channels. (Modified from Reuss, 1997.)

epithelia has already been mentioned. In cells of Cl^--absorptive (e.g., distal renal tubules) and Cl^--secretory (e.g., salivary glands) epithelia, NKCC serves as the major Cl^- entry pathway and functions in concert with three other transport proteins, namely Cl^- and K^+ channels and the Na^+-K^+ pump, to carry out net transepithelial movements of salt (Fig. 10). The possible involvement of NKCC in cell volume regulation has also been mentioned. In some cells, NKCC appears to be involved in regulatory volume increase in response to osmotic cell shrinkage, by promoting net gain of K^+, Cl^-, and osmotically obligated water (see Chapter 21 and Russel (2000) for a more detailed account).

Another role of NKCC is to maintain $[Cl^-]_i$ at higher than equilibrium values in vertebrate sensory spinal neurons, making possible the phenomenon of presynaptic inhibition. In vertebrate spinal cord, presynaptic inhibition is primarily due to depolarization of primary afferent nerve terminals by the neurotransmitter GABA (γ-aminobutyric acid), which leads

to a reduction in neurotransmitter output (Rudomin and Schmidt, 1999). The prevailing idea is that GABA released from interneurons depolarizes primary afferent fibers via axo-axonic synapses. These depolarizations inactivate Na^+ channels sufficiently to block action potential invasion into the primary afferent terminals, thereby inhibiting transmitter release. The depolarizing inward current results from opening of $GABA_A$-gated Cl^- channels, with the consequent efflux of Cl^- which transiently drives the transmembrane potential toward the equilibrium potential for Cl^-. This peculiar depolarizing action of an inhibitory transmitter is possible because the $[Cl^-]$ inside the nerve terminals of primary sensory neurons is higher than predicted from a passive distribution, thereby providing the necessary driving force for electrodiffusional Cl^- efflux. The nonpassive distribution of Cl^- is generated and maintained by NKCC (Alvarez-Leefmans et al., 1988, 1998; Sung et al., 2000). Moreover, due to their small volume/surface ratio, presynaptic nerve terminals are expected to be partially depleted of solutes (Cl^- and K^+) and water upon GABA action. It has been postulated that NKCC is capable of restoring ionic gradients and osmotic balance altered as a consequence of neurotransmitter action (Fig. 11). The Na^+-K^+-Cl^- cotransporter present in sensory neurons is similar to that characterized in the plasma membrane of a variety of non-neuronal cells, and has been recently identified with immunofluorescence techniques with anticotransporter antibodies (Plotkin et al., 1997; Alvarez-Leefmans et al., 1999, 2001; Sung et al., 2000).

It has been found that the expression of NKCC is developmentally regulated in postnatal rat brain (Plotkin et al., 1997; Clayton et al., 1998). Consistent with this finding, in embryonic neurons, GABA has a widespread depolarizing action (Ben-Ari, 1994; Cherubini et al., 1991). The latter seems crucial in promoting Ca^{2+} influx (Yuste et al., 1991; Obrietan and van den Pol, 1996), influencing important developmental events such as neuronal proliferation, differentiation, and migration and neurite extension and targeting (LoTurco et al., 1995; Obrietan and van den Pol, 1996). Again, the depolarizing effect of GABA results mainly from an efflux of Cl^- through $GABA_A$-gated anion channels. This outward Cl^- current is possible because the equilibrium potential for chloride (E_{Cl}) is kept at a more positive value than the transmembrane potential (E_m) by means of an inwardly directed NKCC that actively accumulates Cl^-.

NKCC exhibits some of the features required for an efficient extracellular K^+ buffer; that is, it has a strong net inwardly directed driving force and a high affinity for external K^+ (Payne et al., 1995). In fact, there is some evidence for its role as a K^+-uptake mechanism that contributes to buffering and regulation of extracellular K^+ in many tissues. For instance, NKCC1 is present in the apical membrane of mammalian choroid plexus and in concert with other transport systems coexisting in these polarized cells, plays a central role in the reabsorption of K^+ from the cerebrospinal fluid to the blood, thereby contributing to the buffering and regulation of brain interstitial $[K^+]$ (see Xu et al., 1998, and Fig. 12). NKCC is also prominently located in the paranodal region of Schwann cells of myelinated axons, a region prone to extracellular K^+ accumulation resulting from axonal impulse activity (Chiu, 1991). This suggests the possible involvement

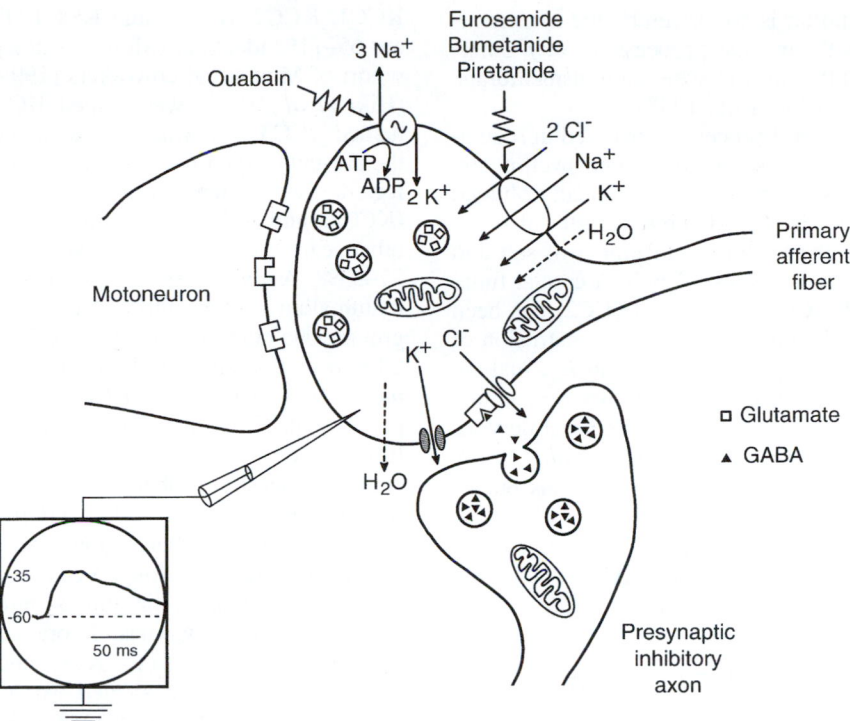

FIGURE 11. Schematic representation of a synapse made by a primary afferent on a motoneuron and the axo-axonic contact made by a GABAergic interneuron (presynaptic inhibitory axon). Release of GABA from the presynaptic inhibitory axon causes an efflux of Cl^- from the primary afferent ending, resulting in a depolarization that is "recorded" by a microelectrode inserted into the terminal (lower left). The depolarization activates K^+ channels with consequent K^+ efflux. The outwardly directed Cl^- gradient is generated and maintained by Na^+-K^+-$2Cl^-$ cotransport. Putative osmotic water fluxes accompanying GABA-induced ion movements are indicated with dashed arrows. (Reproduced from Alvarez-Leefmans *et al.*, 1998.)

of NKCC in K^+ uptake from the periaxonal space, thereby regulating the external $[K^+]$ and maintaining the axonal excitability in this critical location (Alvarez-Leefmans *et al.*, 2001).

V. Electroneutral K^+-Cl^- Cotransporters

The K^+-Cl^- cotransporter (KCC) is an integral membrane protein that mediates the obligatorily coupled, electroneutral movement of K^+ and Cl^- across the plasma membrane of many animal cells. As an electroneutral secondary active transport mechanism, the direction of the net movement of K^+ and Cl^- by the cotransporter is determined solely by the sum of the chemical potential gradients of the two ions (Lauf and Adragna, 1996). Under normal physiological conditions, the outwardly directed K^+ chemical potential gradient maintained by the Na^+-K^+ pump drives Cl^- uphill against its chemical potential gradient. Hence, under normal conditions, the KCC is an efflux pathway for K^+ and Cl^-. However, the cotransporter is bidirectional and can mediate net ion efflux or influx, depending upon the prevailing K^+ and Cl^- chemical potential gradients. Although the stoichiometry has actually never been determined, it is believed that the transport process involves one-for-one movement of K^+ with Cl^-. Whether it is $(n+1)K^+:(n+1)Cl^-$ remains to be determined. This means that the overall transport process of K^+

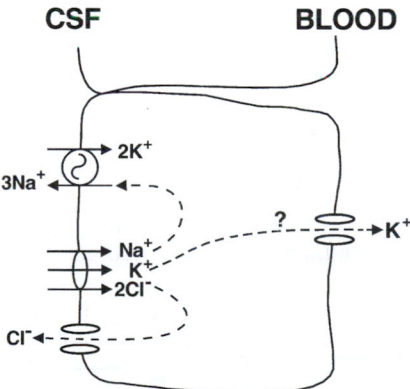

FIGURE 12. Model of Na^+-K^+-$2Cl^-$ cotransporter function in choroid plexus (CP) epithelial cells. Na^+-K^+-$2Cl^-$ cotransporter and Na^+,K^+-ATPase are localized in the apical membrane facing cerebrospinal fluid (CSF). K^+ is taken up into CP cells via the Na^+-K^+-$2Cl^-$ cotransporter, and it exits at a similar rate via K^+ channels in basolateral membrane (facing blood). It is proposed that the cotransporter plays a central role in the reabsorption of K^+ from CSF to the blood and in the buffering and regulation of brain interstitial K^+ concentration. (Modified from Wu *et al.*, 1998.)

and Cl⁻ via the cotransporter is not driven by the transmembrane voltage, nor does the transport process directly generate a membrane current that may change the transmembrane voltage (Brugnara *et al.*, 1989; Kaji, 1993).

In some cells, this transport process is involved in regulatory volume decrease in response to osmotic cell swelling, by promoting an efflux of K^+ and Cl^- and osmotically obliged water (see Chapter 21). The K^+-Cl^- cotransporter also appears to be involved in the vectorial movement of salt and water across certain epithelia (Reuss, 1997). A crucial function of an isoform of K^+-Cl^- cotransport (KCC2) has been proposed for neurons, whereby effecting active extrusion of Cl^-, it keeps E_{Cl} at more negative values than E_m, making possible the hyperpolarizing action of inhibitory neurotransmitters like GABA and glycine in the adult central nervous system (Alvarez-Leefmans, 1990; Rivera *et al.*, 1999). Another isoform, KCC3, is also present in neurons but its function still remains to be determined (Hiki *et al.*, 1999; Mount *et al.*, 1999b). Unlike KCC2, which is exclusively present in neurons (Williams *et al.*, 1999), KCC3 is also present in muscle, lung, heart, and kidney (Mount *et al.*, 1999b).

A. Basic Features of the K⁺-Cl⁻ Cotransporters

A key feature of KCCs is that the thiol-alkylating reagent N-ethylmaleimide (NEM) stimulates transport (Lauf *et al.*, 1992; Lauf and Adragna, 2000). No other membrane transporter is stimulated by NEM. The cotransporter can also be activated by osmotic cell swelling (Dunham and Ellory, 1981; Lauf *et al.*, 1992). Reducing $[Mg^{2+}]_i$ stimulates the cotransporter (Delpire and Lauf, 1991). This is probably due to a reduction in $[MgATP]_i$, the substrate for a kinase that inhibits KCC. Likewise, KCC is stimulated by the protein kinase inhibitor staurosporine (Bize and Dunham, 1994). Unlike either the NKCC or the NCC, the KCC can be inhibited by the disulfonic acid stilbenes such as DIDS (Delpire and Lauf, 1992). KCCs are also inhibited by loop diuretics, but, in general, the sensitivity for furosemide > bumetanide. In general, their sensitivity to loop diuretics is significantly lower than that of NKCCs. They can also be inhibited by dihydroindenyloxyalkanoic acid (DIOA). The transport process is independent of Na^+.

B. Molecular Structure, Distribution, and Functions of K⁺-Cl⁻ Cotransport Proteins

To date four isoforms of this member of the cation-coupled Cl^- cotransporter family have been cloned and sequenced:

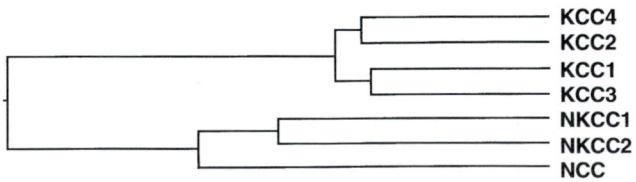

FIGURE 13. Phylogenetic relationships between mammalian cation-chloride cotransporters. A phylogenetic comparison of protein sequences indicates that the four K⁺-Cl⁻ cotransporters evolved separately from the Na⁺-dependent transporters, forming a distinct subfamily. (Modified from Mount *et al.*, 1999b.)

KCC1, KCC2, KCC3, and KCC4. The four KCC proteins are 65–71% identical (Mount *et al.*, 1999b). In the original report of Mount and coworkers (1999b), the isoform KCC3 (Hiki *et al.*, 1999) was named KCC4 and the latter was named KCC3. This was clarified in a note added in proof. In the present chapter the corrected nomenclature has been followed. Two splice variants of KCC3 have been cloned (KCC3a and KCC3b). The identity between the KCCs and other cation-chloride cotransporters is in the range of 27–33%. A phylogenetic tree (Fig. 13) indicates that the mammalian cation-chloride cotransporters fall into two groups, one composed of the NKCCs and the NCC, and the other encompassing the four KCCs. Direct sequence alignments show that the four KCCs form two subgroups, KCC1 paired with KCC3 and KCC4 paired with KCC2. The tree further shows that all these cotransporters evolved from a common ancestor, probably a primitive ion transport protein from a cyanobacterium (orf733) (Gillen *et al.*, 1996).

The seven mammalian cation-chloride cotransporters share a basic membrane topology. A central core of 12 TM domains is flanked by hydrophilic amino- and carboxyl-terminal domains that have a cytoplasmic orientation. The major structural difference between the KCCs and the NKCCs is the position of a large glycosylated extracellular loop, which is predicted to occur between TM5 and TM6 in the KCCs and between TM7 and TM8 in the NKCCs (Fig. 8) and the NCC. Homology is most marked in the TM domains, the intracellular loops, and the cytoplasmic carboxyl terminus (Mount *et al.*, 1999a, b).

KCC1 was the first isoform to be cloned (Gillen *et al.*, 1996). It is the isoform present in red blood cells and to date it is the one that has been best characterized from the functional perspective (Lauf *et al.*, 1992; Lauf and Adragna, 1996, 2000). It has a deglycosylated molecular mass of ~120 kDa. Besides erythroid cells, KCC1 is found widely expressed in mammalian tissues (brain, kidney, heart, lung, liver, stomach, colon, spleen, and placenta). This ubiquitous tissue distribution suggests that KCC1 may represent a "housekeeping" isoform responsible for cell volume regulation and maintenance, and the regulation of $[Cl^-]_i$. Heterologous expression of KCC1 in HEK-293 cells showed that it can be stimulated by cell swelling and N-ethylmaleimide, and inhibited by furosemide (K_i of ~ 40 μM) and by bumetanide (K_i of ~ 60 μM). Clearly the sensitivity to loop diuretics of KCC1 is different than that of NKCCs. The latter are more sensitive to bumetanide than furosemide, and the K_i for bumetanide is almost an order of magnitude lower. The gene for KCC1 is at chromosomal location 16q22.1.

KCC2 was originally described as a neuronal-specific isoform (Payne *et al.*, 1996). Shortly after its initial description KCC3 was cloned and shown also to be present in the brain. However, the detailed localization of KCC3 in the brain is still to be determined. KCC2 is present in cell bodies and dendrites of various neuronal types, but appears to be absent in axons and glial cells (Williams *et al.*, 1999). It has been postulated that KCC2 is the isoform that actively extrudes Cl^- from nerve cells, rendering E_{Cl} more negative than E_m, thereby setting an inwardly directed electrochemical gradient for Cl^- to enable the hyperpolarizing action of GABA- or glycine-gated Cl⁻ channels. Hence KCC2 seems to be crucial for the mechanism of postsynaptic inhibition in the adult brain. Moreover, KCC2 expression is developmentally regulated and this explains why GABA responses, which are depolarizing in embryonic nerve cells, become hyperpolarizing in the adult brain (Rivera *et al.*, 1999). However, KCC2 might not be the only candidate for such an important function (see below). KCC2 has a predicted molecular mass of ~124 kDa. Western blots of membranes isolated from rat brain show that the glycosylated molecular mass of KCC2 is ~140 kDa, and upon deglycosylation the protein migrates to near ~125 kDa, which is close to the predicted molecular mass (Williams *et al.*, 1999). KCC2 is also stimulated by NEM and inhibited by furosemide (K_i of ~ 25 μM) and bumetanide (K_i of ~ 55 μM). DIDS and dihydroindenyloxyalkanoic acid (DIOA) also inhibit it. The main functional difference from other isoforms is that KCC2 has been claimed not to be stimulated by osmotic swelling (Payne, 1997), although a critical assessment of this lack of volume sensitivity is still lacking. KCC2 also exhibits all the features required for an efficient $[K^+]_o$ buffer: (1) it is reversible, and the direction of net transport is sensitive to subtle changes in $[K^+]_o$; (2) the affinity of the cotransporter for external K^+ is compatible with the range of the observed $[K^+]_o$ (3–10 mM), allowing a high rate of cotransport activity. These requirements seem to be a feature of both KCC2 (K_m ~ 5.2 mM) and KCC3 (K_m ~ 10 mM). Hence it has been postulated that KCC2 could act as a $[K^+]_o$ buffer (Payne, 1997).

KCC3 was simultaneously discovered by three groups (Hiki *et al.*, 1999; Mount *et al.*, 1999b; Race *et al.*, 1999). It also has a deglycosylated molecular mass of ~120 to 128 kDa. It is expressed in endothelial cells, brain, heart, placenta, liver, lung, pancreas, skeletal muscle, and kidney. It is localized to human chromosome 15q13. KCC3 expressed in HEK cells was activated by cell swelling, by decreasing $[Mg^{2+}]_i$, by staurosporine, and prominently by NEM. It is inhibited by bumetanide (K_i ~ 100 μM), furosemide (K_i ~ 103 μM), and DIOA (K_i ~ 9.3 μM). Hence the best inhibitor of KCC3 is DIOA. The latter totally inhibits KCC3 at 50 μM. The function of KCC3 remains speculative. Given its reversibility and affinities for K^+ (K_m ~ 10 mM), it has been proposed that, like KCC2, it may function as a buffer of extracellular K^+ (Race *et al.*, 1999).

Mount and coworkers (1999b) cloned KCC4. It is prominently expressed in kidney and heart. The KCC4 gene is located on chromosome 5p15. KCC4 has been functionally expressed in *Xenopus* oocytes. Using this heterologous expression system, it has been possible to establish functional comparisons between KCC1 and KCC4. Cell swelling activates both of them, but KCC4 is one order of magnitude more sensitive than KCC1. The function of KCC4 is unknown.

C. Thermodynamics of K⁺-Cl⁻ Cotransport

Assuming that the K⁺-Cl⁻ cotransport is the sole system transporting Cl⁻, and that $[K^+]_i$ is held constant by the Na⁺-K⁺ pump, the driving force for the cotransport $\Delta\mu_{K,Cl}$ is the sum of the chemical potential differences of K^+ and Cl^- as expressed in the following equation:

$$\Delta\mu_{K,Cl} = \Delta\mu_K + \Delta\mu_{Cl} \tag{10}$$

It follows that

$$\Delta\mu_{K,Cl} = RT \ln \frac{[K^+]_i}{[K^+]_o} + RT \ln \frac{[Cl^-]_i}{[Cl^-]_o} \tag{11}$$

This system attains thermodynamic equilibrium when

$$\Delta\mu_{K,Cl} = 0 \tag{12}$$

and therefore, for a 1:1 stoichiometry,

$$\frac{[K^+]_o}{[K^+]_i} = \frac{[Cl^-]_i}{[Cl^-]_o} \tag{13}$$

and hence the $[Cl^-]_i$ at equilibrium will be

$$[Cl^-]_i = \frac{[Cl^-]_o [K^+]_o}{[K^+]_i} \tag{14}$$

The $[K^+]_o$ at which the flux reverses from outward to inward or vice versa (flux reversal point or FRP) is

$$FRP = [K^+]_o = \frac{[K^+]_i [Cl^-]_i}{[Cl^-]_o} \tag{15}$$

VI. Electroneutral Na⁺-Cl⁻ Cotransporter

The Na⁺-Cl⁻ cotransporter (NCC) was the first member of the cation-chloride cotransporter family to be identified (Gamba *et al.*, 1993). It is also known as CC3 or TSC. It was isolated from the epithelial cells of the urinary bladder of the winter flounder (*Pseudopleuronectes americanus*). Its mammalian homologue was isolated from rat renal cortex (Gamba *et al.*, 1994). Functional expression of this cotransporter in *Xenopus laevis* oocytes demonstrated the known characteristics of the cotransporter in native tissue: sensitivity to thiazide diuretics (metolazone > cyclothiazide > hydrochlorothiazide > chlorothiazide), insensitivity to amiloride and loop diuretics, K^+ independence, lower affinity for Cl^- than for Na^+, and a 1Na⁺:1Cl⁻ stoichiometry; hence, it is an electroneutral transporter. The molecular mass of the core (deglycosylated) protein is ~110 kDa. It has ~45% amino acid identity with NKCC1; 45–48% amino acid identity with NKCC2, and ~25% amino acid identity with KCC1 and KCC2.

In the rat, the cotransporter is almost exclusively expressed in kidney and to a lesser extent in osteoblasts. In the distal convoluted tubule (DCT) of the mammalian kidney, the primary apical membrane entry pathway for Na⁺ is through the thiazide-sensitive Na⁺-Cl⁻ cotransporter (TSC). The natriuretic effect of the thiazide diuretics underlies their therapeutic effect in edema states and hypertension. In addition, luminal thiazide in the DCT has a hypocalciuric effect, implicating TSC in the fine-tuning of Ca^{2+} excretion (Mount *et al.*, 1998).

VII. Cation-Chloride Cotransporters as Targets for Disease

Given their physiological importance, it is not surprising that altered activities of the cation-chloride cotransporters have been implicated in human disease. For instance, knockout mice in which the gene encoding for NKCC1 was disrupted show deafness and imbalance, the latter due to inner ear dysfunction (Delpire *et al.*, 1999; Flagella *et al.*, 1999). In the inner ear, NKCC1 is expressed at high levels on the basolateral membranes of marginal cells of the stria vascularis. K^+ secretion by marginal cells is responsible for the high concentration of K^+ in the endolymph and for the endocochlear potential, which is eliminated by perilymphatic or systemic application of bumetanide or furosemide. These observations and the ototoxicity of loop diuretics suggest that NKCC1 contributes to the high rates of K^+ uptake needed to maintain K^+ currents across the apical membrane of the marginal cell and that NKCC1 plays a critical role in hearing. Histological analysis of the inner ear of mice lacking NKCC1 revealed a collapse of the membranous labyrinth, consistent with a critical role for NKCC1 in transepithelial K^+ movements involved in generation of the K^+-reach endolymph and the endocochlear potential. These new knockout mice models provide a molecular explanation for the ototoxic effect of loop diuretics. In addition, these models will allow identification of the functions of NKCC1 in normal animals. For instance, male NKCC1-deficient mice have been found to be infertile because of defective spermatogenesis (Pace *et al.*, 2000). Moreover, dorsal root reflexes are increased in cutaneous neurogenic inflammation, possibly due to up-regulation of NKCC1 in primary afferent neurons (Willis, 1999).

Mutations in the NKCC2 gene have been identified in Barter's syndrome families. Barter's patients present with hypokalemic alkalosis and hypercalciuria. Recently identified genetic mutations in the human NCC have been linked to hypokalemic alkalosis, hypomagnesemia, and hypocalciuria, referred to as Gittleman's syndrome. The abnormalities of the syndrome can be explained by a loss of NCC function (Simon *et al.*, 1996).

Cultured chick cardiac cells possess a functional KCC. Activation of α_1-adrenoreceptor of isolated rat hearts stimulates a K^+ efflux pathway that is partially sensitive to loop diuretics. The cell swelling induced by cardiac ischemia also stimulates a significant efflux of K^+ that is mediated mainly via KCC. It has been proposed that this K^+ efflux plays an important role in the genesis of arrhythmias following myocardial ischemia (see Mount *et al.*, 1999b for references).

VIII. Summary

Intracellular chloride (Cl^-), together with bicarbonate (HCO_3^-), is the most abundant free anion in living cells. Most animal cells exhibit a nonequilibrium distribution of Cl^- across their plasma membranes. Some cells actively extrude Cl^-, others actively accumulate it, but few cells ignore it. By virtue of being distributed out of electrochemical equilibrium, Cl^- does serve as a key player in a variety of cellular functions such as intracellular pH regulation; cell volume regulation; transepithelial salt transport; synaptic signaling (in both the depolarizing and hyperpolarizing directions); neuronal growth, migration, and targeting; membrane potential stabilization; regulation of transport systems; and K^+ scavenging. The $[Cl^-]_i$ is determined by the interaction of various anion-transporting systems present in a cell, which include conductive Cl^- channels as well as several cotransporters and exchangers. The carrier protein molecules that transport Cl^- include an electroneutral Cl^--HCO_3^- exchanger that plays a central role in intracellular pH regulation and can also serve as an uphill Cl^--accumulating system in some cases (e.g., cardiac and smooth muscle), and the electroneutral cation-chloride cotransporter family. This family encompasses seven members: a Na^+-Cl^- cotransporter (NCC), two Na^+-K^+-Cl^- cotransporters (NKCC1 and NKCC2), and four Na^+-independent K^+-Cl^- cotransporters (KCC1, KCC2, KCC3, and KCC4). The cotransporter proteins share a common predicted membrane topology, with 12 putative transmembrane segments flanked by long hydrophilic amino- and carboxyl-terminal cytoplasmic domains. The molecular identification of these transporters has had a significant impact on the study of their function, regulation, and pathophysiology.

Acknowledgments

I am grateful to Drs. Peter Lauf and Gerardo Gamba for enriching discussions, Mr. Víctor Blanco for his critical reading of the manuscript, and Mr. José Rodolfo Fernández for helping with the illustrations. The author was partly supported by a Guggenheim fellowship and by a grant from the National Institute of Neurological Disorders and Stroke (NS-29227), both of which are gratefully acknowledged.

Bibliography

Aickin, C. C. (1990). Chloride transport across the sarcolemma of vertebrate smooth and skeletal muscle. *In* "Chloride Channels and Carriers in Nerve, Muscle and Glial Cells" (F. J. Alvarez-Leefmans, J. M. Russell, Eds.), pp. 209–249. Plenum Press, New York.

Altamirano, A. A., Breitwieser, G. E., and Russell, J. M. (1988). Vanadate and fluoride effects on Na-K-Cl cotransport in squid giant axon. *Am. J. Physiol.* **254,** C582–C586.

Alvarez-Leefmans, F. J. (1990). Intracellular Cl^- regulation and synaptic inhibition in vertebrate and invertebrate neurons. *In* "Chloride Channels and Carriers in Nerve, Muscle, and Glial Cells" (F. J. Alvarez-Leefmans, J. M. Russell, Eds.), pp. 109–158. Plenum Press, New York.

Alvarez-Leefmans, F. J., Gamiño, S. M., Giraldez, F., and Noguerón, I. (1988). Intracellular chloride regulation in amphibian dorsal root ganglion neurones studied with ion-selective microelectrodes. *J. Physiol.* **406,** 225–246.

Alvarez-Leefmans, F. J., Giraldez, F., and Russell, J. M. (1990). Methods for measuring chloride transport across nerve, muscle and glial cells. *In* "Chloride Channels and Carriers in Nerve, Muscle, and Glial Cells" (F. J. Alvarez-Leefmans, J. M. Russell, Eds.), pp. 3–66. Plenum Press, New York.

Alvarez-Leefmans, F. J., León-Oléa, M., Mendoza, J., Alvarez, F. J., Antón, B., and Garduño, R. (2001). Immunolocalization of the Na^+,K^+,Cl^- cotransporter in peripheral nervous tissue of vertebrates. *Neurosci.* (in press).

Alvarez-Leefmans, F. J., León-Oléa, M., and Alvarez, F. J. (1999). Distribution of NKCC-immunoreactivity in the fibers and cell bodies of spinal sensory neurons in the frog, cat and rat. *Society for Neuroscience Abstracts* **25(2),** 2208.

Alvarez-Leefmans, F. J., Nani, A., and Márquez, S. (1998). Chloride transport, osmotic balance, and presynaptic inhibition. *In* "Presynaptic Inhibition and Neural Control" (P. Rudomin, R. Romo, and L. M. Mendell, Eds.), pp. 50–79. Oxford UP, New York.

Ben-Ari, Y., Tseeb, V., Raggozzino, D., Khazipov, R., and Gaiarsa, J. L. (1994). γ-Aminobutyric acid (GABA): a fast excitatory transmitter which may regulate the development of hippocampal neurones in early postnatal life. *Prog. Brain Res.* **102,** 261–273.

Benjamin, B. A., and Johnson, E. A. (1997). A quantitative description of the Na,K,2Cl cotransporter and its conformity with experimental data. *Am. J. Physiol.* **273,** F473–F482.

Bize, I., and Dunham, P. B. (1994). Staurosporine, a protein kinase inhibitor, activates K-Cl cotransport in LK sheep erythrocytes. *Am. J. Physiol.* **266,** C759–C770.

Breitwieser, G. E., Altamirano, A. A., and Russell, J. M. (1990). Osmotic stimulation of Na⁺,K⁺,Cl⁻ cotransport in squid giant axons is [Cl⁻]ᵢ dependent. *Am. J. Physiol.* **258,** C749–C753.

Breitwieser, G. E., Altamirano, A. A., and Russell, J. M. (1996). Elevated [Cl⁻]ᵢ and [Na⁺]ᵢ inhibit Na⁺,K⁺,Cl⁻ cotransport by different mechanisms in squid giant axons. *J. Gen. Physiol.* **107,** 261–270.

Brugnara, C., Van Ha, T., and Tosteson, D. C. (1989). Role of chloride in potassium transport through a K-Cl cotransport system in human red blood cells. *Am. J. Physiol.* **256,** C994–C1003.

Cherubini, E., Gaiarsa, J. L., and Ben-Ari, Y. (1991). GABA: an excitatory transmitter in early postnatal life. *Trend Neurosci.* **14,** 515–519.

Chiu, S. Y. (1991). Functions and distribution of voltage-gated sodium and potassium channels in mammalian Schwann cells. *Glia* **4,** 541–558.

Clayton, G. H., Owens, G. C., Wolff, J., and Smith, R. L. (1998). Ontogeny of cation-Cl⁻ cotransporter expression in rat neocortex. *Developmental Brain Res.* **109,** 281–292.

Delpire, E., and Lauf, P. K. (1991). Magnesium and ATP dependence of K-Cl cotransport in low-K⁺ sheep red blood cells. *J. Physiol.* **441,** 219–231.

Delpire, E., and Lauf, P. K. (1992). Kinetics of DIDS inhibition of swelling-activated K-Cl cotransport in low K sheep erythrocytes. *J. Membr. Biol.* **126,** 89–96.

Delpire, E., Lu, J., England, R., Dull, C., and Thorne, T. (1999). Deafness and imbalance associated with inactivation of the secretory Na-K-2Cl co-transporter. *Nature Genetics* **22,** 192–195.

Delpire, E., Rauchman, M. I., Beier, D. R., Hebert, S. C., and Gullans, S. R. (1994). Molecular cloning and chromosome localization of a putative basolateral Na-K-2Cl cotransporter from mouse inner medullary collecting duct (mIMCD-3) cells. *J. Biol. Chem.* **269,** 25 677–25 683.

Duhm, J. (1987). Furosemide-sensitive K (Rb) transport in human erythrocytes: modes of operation, dependence on extracellular and intracellular Na, kinetics, pH dependency and the effect of cell volume and N-ethylmaleimide. *J. Membr. Biol.* **98,** 15–32.

Dunham, P .B., and Ellory, J. C. (1981). Passive potassium transport in low potassium sheep red cells: dependence upon cell volume and chloride. *J. Physiol. Lond.* **318,** 511–530.

Flagella, M., Clark, L. L., Miller, M. L., Erway, L. C., Giannella, R. A., Andringa, A., Gawenis, L. R., Kramer, J., Duffy, J. J., Doetschman, T., Lorenmz, J. N., Yamoah, E. N., Cardell, E. L., and Shull, G. E. (1999). Mice lacking the basolateral Na-K-2Cl cotransporter have impaired epithelial chloride secretion and are profoundly deaf. *J. Biol. Chem.* **247,** 26946–26955.

Gamba, G., Salzberg, S. N., Lombardi, M., Miyanoshita, A., Lytton, J., Hediger, M. A., Brenner, B. M., and Hebert, S. C. (1993). Primary structural and functional expression of a cDNA encoding the thiazide-sensitive, electroneutral sodium-chloride cotransporter. *Proc. Natl. Acad. Sci. USA* **90,** 2749–2753.

Gamba, G., Miyanoshita, A., Lombardi, M., Lytton, J., Lee, W-S, Hediger, M. A., and Hebert, S. C. (1994). Molecular cloning, primary structure, and characterization of two members of the mammalian electroneutral sodium-(potassium)-chloride cotransporter family expressed in kidney. *J. Biol. Chem.* **269,** 17 713–17 722.

Geck, P., Pietrzyk, C., Buckhardt, B-C, Pfeiffer, B., and Heinz, E. (1980). Electrically silent cotransport of Na, K, and Cl in Ehrlich cells. *Biochim. Biophys. Acta* **600,** 432–447.

Gillen, C. M., Brill, S., Payne, J. A., and Forbush III, B. (1996). Molecular cloning and functional expression of the K-Cl cotransporter from rabbit, rat and human. *J. Biol. Chem.* **271,** 16 237–16 244.

Gillen, C. M., and Forbush III, B. (1999). Functional interaction of the K-Cl cotransporter (KCC1) with the Na-K-Cl cotransporter in HEK-293 cells. *Am. J. Physiol.* **276,** C328–C336.

Good, D. W. (1994). Ammonium transport by the thick ascending limb of the Henle's loop of mammalian nephron. *Annu. Rev. Physiol.* **56,** 623–647.

Haas, M. (1994). The Na-K-Cl cotransporters. *Am. J. Physiol.* **267,** C869–C885.

Haas, M., and Forbush III, B. (1998). The Na-K-Cl cotransporters. *J. Bioenerg. Bioeng.* **30,** 161–172.

Haas, M., and McManus, T. J. (1985). Effect of norepinephrine on swelling-induced potassium transport in duck red cells. Evidence against a volume-regulatory decrease under physiological conditions. *J. Gen. Physiol.* **85,** 649–667.

Haas, M., McBrayer, D., and Lytle, C. (1995). [Cl⁻]ᵢ-dependent phosphorylation of the Na-K-Cl cotransport protein of dog tracheal epithelial cells. *J. Biol. Chem.* **270,** 28 955–28 961.

Haas, M., Schmidt III, W. F., and McManus, T. J. (1982). Catecholamine-stimulated ion transport in duck red cells. Gradient effects in electrically neutral (Na-K-2Cl) co-transport. *J. Gen. Physiol.* **80,** 125–147.

Harris, G. L., and Betz, W. J. (1987). Evidence for active chloride accumulation in normal and denervated rat lumbrica muscle. *J. Gen. Physiol.* **90,** 127–144.

Hiki, H., D'Andrea, R. J., Furze, J., Crawford, J., Woollatt, E., Sutherland, G. R., Vadas, M. A., and Gamble, J. R. (1999). Cloning, characterization, and chromosomal location of a novel human K⁺-Cl⁻ cotransporter. *J. Biol. Chem.* **274,** 10 661–10 667.

Hodgkin, A. L., and Horowicz, P. (1959). The influence of potassium and chloride ions on the membrane potential of single muscle fibers. *J. Physiol.* **148,** 127–160.

Isenring, P., Jacoby, S. C., Payne, J. A., and Forbush III, B. (1998). Comparison of the Na,K,Cl cotransporters. NKCC1, NKCC2, and the HEK-203 cell Na-K-Cl cotransporter. *J. Biol. Chem.* **273,** 11 295–11 301.

Kaji, D. M. (1993). Effect of membrane potential on K-Cl cotransport in human erythrocytes. *Am. J. Physiol.* **264,** C376–C382.

Lauf, P. K., and Adragna, N. C. (1996). A thermodynamic study of electroneutral K-Cl cotransport in pH- and volume-clamped low K sheep erythrocytes with normal and low internal magnesium. *J. Gen. Physiol.* **108,** 341–350.

Lauf, P. K., and Adragna, N. C. (2000). K-Cl cotransport: properties and molecular mechanism. *Cell Physiol. Biochem.* **10** (in press).

Lauf, P. K., Bauer, V., Adragna, N. C., Fujise, H., Zade-Oppen, A. M. M., Ryu, K. H., and Delpire, E. (1992). Erythrocyte K-Cl cotransport: properties and regulation. *Am. J. Physiol.* **263,** C917–C932.

LoTurco, J. J., Owens, D. F., Heath, M. J. S., Davis, M. B. E., and Kriegstein, A. R. (1995). GABA and glutamate depolarize cortical progenitor cells and inhibit DNA synthesis. *Neuron* **15,** 1287–1298.

Lytle, C. (1997). Activation of the avian erythrocyte Na-K-Cl cotransport protein by cell shrinkage, cAMP, fluoride, and calyculin-A involves phosphorylation at common sites. *J. Biol. Chem.* **272,** 15 069–15 077.

Lytle, C., and Forbush III, B. (1992). The Na-K-Cl cotransport protein of shark rectal gland. II. Regulation by direct phosphorylation. *J. Biol. Chem.* **267,** 25 438–25 443.

Lytle, C., McManus, T. J., and Haas, M. (1998). A model of Na-K-2Cl cotransport based on ordered ion binding and glide symmetry. *Am. J. Physiol.* **274**, C299–C309.

Lytle, C., Xu, J. C., Biemesderfer, D., and Forbush III, B. (1995). Distribution and diversity of Na-K-Cl cotransport proteins: a study with monoclonal antibodies. *Am. J. Physiol.* **269**, C1496–C1505.

McRoberts, J. A., Erlinger, S., Rindler, M .J., and Saier, M. H. (1982). Furosemide-sensitive salt transport in the Madin-Darby canine kidney cell line: evidence for the cotransport of Na^+,K^+ and Cl^-. *J. Biol. Chem.* **257**, 2260–2266.

Mount, D. B., Delpire, E., Gamba, G., Hall, A. E., Poch, E., Hoover, R., and Hebert, S. C. (1998). The electroneutral cation-chloride cotransporters. *J. Exp. Biol.* **201**, 2091–2102.

Mount, D. B., Baekgaard, A., Hall, A. E., Plata, C., Xu, J., Beier, D. R., Gamba, G., and Hebert, S. C. (1999a). Isoforms of the Na-K-2Cl cotransporter in murine TAL. I. Molecular characterization and intrarenal localization. *Am. J. Physiol.* **276**, F347–F358.

Mount, D. B., Mercado, A., Song, L., Xu, J., George, A. L., Delpire, E., and Gamba, G. (1999b). Cloning and characterization of KCC3 and KCC4, new members of the cation-chloride cotransporter gene family. *J. Biol. Chem.* **274**, 16 355–16 362.

Obermüller, N., Kunchaparty, S., Ellison, D. H., and Bachmann, S. (1996). Expression of the Na-K-2Cl cotransporter by macula densa and thick ascending limb cells of rat and rabbit nephron. *J. Clin. Invest.* **98**, 635–640.

Obrietan, K., and van den Pol, A. (1996). Growth cone calcium elevation by GABA. *J. Comp. Neurol.* **372**, 167–175.

O'Donnell, M. E., Martinez, A., and Sun, D. (1995). Endothelial Na-K-Cl cotransport regulation by tonicity and hormones: phosphorylation of cotransport protein. *Am. J. Physiol.* **269**, C1513–C1523.

O'Grady, S. M., Palfrey, H. C., and Field, M. (1987). Characteristics and functions of Na-K-Cl cotransport in epithelial tissues. *Am. J. Physiol.* **253**, C177–C192.

Owens, D. F., Boyce, L. H., Davis, M. B. E., and Kriegstein, A. R. (1996). Excitatory GABA responses in embryonic and neonatal cortical slices demonstrated by gramicidin perforated-patch recordings and calcium imaging. *J. Neurosci.* **16**, 6414–6423.

Pace, A. J., Lee, E., Athirakul, K., Coffman, T. M., O'Brien, D. A., and Koller, B. H. (2000). Failure of spermatogenesis in mouse lines deficient in the $Na^+-K^+-2Cl^-$ cotransporter. *J. Clin. Invest.* **105**, 441–450.

Panet, R., Marcus, M., and Atlan, H. (2000). Overexpression of the $Na^+/K^+/Cl^-$ cotransporter gene induces cell proliferation and phenotypic transformation in mouse fibroblasts. *J. Cell Physiol.* **182**, 109–118.

Payne, J. A. (1997). Functional characterization of the neuronal-specific K-Cl cotransporter: implications for $[K^+]_o$ regulation. *Am. J. Physiol.* **273**, C1516–C1525.

Payne, J. A., and Forbush III, B. (1994). Alternatively spliced isoforms of the putative Na-K-Cl cotransporter are differentially distributed within the rabbit kidney. *Proc. Natl. Acad. Sci. USA* **91**, 4544–4548.

Payne, J. A., Stevenson, T. J., and Donaldson, L. F. (1996). Molecular characterization of a putative K-Cl cotransporter in rat brain. A neuronal-specific isoform. *J. Biol. Chem.* **271**, 16 245–16 252.

Payne, J. A., Xu, J-C, Haas, M., Lytle, C. Y., Ward, D., and Forbush III, B. (1995). Primary structure, functional expression, and chromosomal localization of the bumetanide-sensitive Na-K-Cl cotransporter in human colon. *J. Biol. Chem.* **270**, 17 977–17 985.

Plotkin, M. D., Snyder, E. Y., Hebert, S. C., and Delpire, E. (1997). Expression of the Na-K-2Cl cotransporter is developmentally regulated in postnatal rat brains: a possible mechanism underlying GABA's excitatory role in immature brain. *J. Neurobiol.* **33**, 781–795.

Race, J. E., Makhlouf, F. N., Logue, P. J., Wilson, F. H., Dunham, P. B., and Holtzman, E. J. (1999). Molecular cloning and functional characterization of KCC3, a new KCl cotransporter. *Am. J. Physiol.* **277**, C1210–C1219.

Reuss, L. (1997). Epithelial transport. In "Handbook of Physiology," Section 14 (American Physiological Society), pp. 309–388. Oxford University Press, New York.

Rivera, C., Voipio, J., Payne, J. A., Ruusuvuori, E., Lahtinen, H., Lamsa, K., Pirvola, U., Saarma, M., and Kaila, K. (1999). The K^+/Cl^- co-transporter KCC2 renders GABA hyperpolarizing during neuronal maturation. *Nature* **397**, 251–255

Rudomin, P., and Schmidt, R. F. (1999). Presynaptic inhibition in the vertebrate spinal cord revisited. *Exp. Brain Res.* **129**, 1–37.

Russell, J. M. (1984). Chloride in the squid giant axon. *Curr. Top. Membr. Transp.* **22**, 177–193.

Russell, J. M. (2000). Sodium-potassium-chloride cotransport. *Physiol. Rev.* **80**, 211–276.

Simon, D. B., Nelson-Williams, C., Bia, M. J., Ellison, D., Karet, F. E., Molina, A. M., Vaara, I., Iwata, F., Cushener, H. M., Koolen, M., Gainza, F. J., Gitelman, H. J., and Lifton, R. P. (1996). Gitelman's variant of Bartter's syndrome, inherited hypokalaemic alkalosis, is caused by mutations in the thiazide-sensitive Na-Cl cotransporter. *Nature Genetics* **12**, 24–30.

Sung, K-W, Kirby, M., McDonald, M. P., Lovinger, D. M., and Delphire, E. (2000). Abnormal $GABA_A$ receptor-mediated currents in dorsal root ganglion neurons isolated from Na-K-2Cl cotransporter null mice. *J. Neurosci.* **15**, 7531-7538.

Ussing, H. H. (1982). Volume regulation of frog skin epithelium. *Acta Physiol. Scand.* **114**, 363–369.

Vaughan-Jones, R. D. (1982). Chloride activity and its control in skeletal and cardiac muscle. *Phil. Trans. R. Soc. Lond. B* **299**, 537–548.

Vaughan-Jones, R. D. (1986). An investigation of chloride-bicarbonate exchange in the sheep cardiac purkinje fibre. *J. Physiol.* **379**, 377–406.

Walz, W. (1992). Role of Na/K/Cl cotransport in astrocytes. *Can. J. Physiol. Pharmacol.* **70**, S260–S262.

Williams, J. R., Sharp, J. W., Kumari, V. G., Wilson, M., and Payne, J. A. (1999). The neuron-specific K-Cl cotransporter, KCC2. *J. Biol. Chem.* **274**, 12 656–12 664.

Willis, W. D. (1999). Dorsal root potentials and dorsal root reflexes: a double-edged sword. *Exp. Brain Res.* **124**, 395–421.

Wong, J. A., Fu, L., Schneider, E. G., and Thomason, D. B. (1999). Molecular and functional evidence for $Na^+-K^+-2Cl^-$ cotransporter expression in rat slow-twitch skeletal muscle. *Am. J. Physiol.* **277**, R154–R161.

Wu, Q., Delpire, E., Hebert, S. C., and Strange, K. (1998). Functional demonstration of $Na^+-K^+-2Cl^-$ cotransporter activity in isolated, polarized choroid plexus cells. *Am. J. Physiol.* **275**, C1565–C1572.

Xu, J. C., Lytle, C., Zhu, T., Payne, J. A., Benz, E., and Forbush III, B. (1994). Molecular cloning and functional expression of the bumetanide-sensitive Na-K-Cl cotransporter. *Proc. Natl. Acad. Sci. USA* **91**, 2201–2205.

Yerby, T. R., Vibat, C. R., Sun, D., Payne, J. A., and O'Donnell, M. E. (1997). Molecular characterization of the Na-K-Cl cotransporter of bovine aortic endothelial cells. *Am. J. Physiol.* **273**, C188–C197.

Yuste, R., and Katz, L. C. (1991). Control of postsynaptic Ca^{2+} influx in developing neocortex by excitatory and inhibitory neurotransmitters. *Neuron* **6**, 333–344.

Clive M. Baumgarten and Joseph J. Feher

21

Osmosis and Regulation of Cell Volume

I. Introduction

In whole blood, erythrocytes are biconcave disks about 7 μm in diameter and 2 μm thick. When diluted in a solution of 0.9% NaCl (w/v), erythrocytes retain this shape. When diluted with higher concentrations of salt, the erythrocytes shrink, appearing as spheres with spikes all over their surface. These cells are described as **crenated.** If erythrocytes are diluted with a markedly lower concentration of salt, the cells swell. They first become spherical and then, if the solution is sufficiently low in salt, the cells burst and release their contents. These simple observations give rise to the concept of tonicity. **Tonicity** is operationally defined as the ability of a solution to shrink or swell specified cells. Thus, an **isotonic** solution induces no volume change when placed in contact with the cells. The tonicity of the solution is equal to the tonicity of the cell's contents. **Hypertonic** solutions shrink cells, whereas **hypotonic** solutions increase cell volume. A solution that is isotonic for one type of cell may or may not be isotonic for others.

All animal cells shrink or swell on exposure to **anisotonic** solutions. Figure 1 shows the volume response of single isolated heart cells on exposure to hypertonic or hypotonic solutions. In hypotonic solution, myocytes swelled to more than 1.5 times their initial volume. The swelling was complete within 2 min, the new volume was stable, and volume returned to normal upon return of the isotonic solution. In hypertonic solution, cell volume decreased to about ~0.65 times normal, and the original volume was restored upon return of the isotonic solution.

The data in Fig. 1 show that volume changes in anisotonic media are very rapid. What is moving when the cells swell or shrink? What routes do these substances take? Are there homeostatic mechanisms that limit swelling and shrinking? If so, how are the compensatory mechanisms engaged? The answers to these questions are not yet complete. The purpose of this chapter is to provide the basis for understanding regulation of cell volume through the exchange of water and solutes across the plasma membrane.

II. Water Movement across Model Membranes

A. Definition of Osmosis

Osmosis refers to the movement of fluid across a membrane in response to differing concentrations of solutes on the two sides of the membrane. Osmosis has been used since antiquity to preserve foods by dehydration with salt or sugar. The removal of water from a tissue by salt was referred to as **imbibition.** This description comes from the notion that

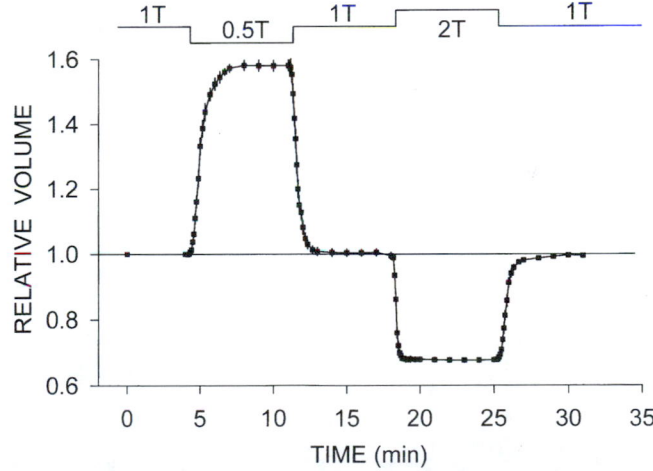

FIGURE 1. Response of isolated rabbit ventricular myocytes to osmotic stress. Cell volume was initially measured in isotonic solution (1T). Myocytes rapidly swelled 58% in a hypotonic solution with an osmolarity 0.5 times that of 1T and rapidly shrank 33% in a hypertonic solution with an osmolarity 2 times that of 1T. Cell volume was stable for the duration of perfusion with either hypotonic or hypertonic media and rapidly returned to its control value when 1T solution was readmitted. Volume was measured by digital video microscopy, and relative volume was calculated as volume_test/volume_1T. Solution osmolarity was adjusted by varying the concentration of mannitol. (From Suleymanian and Baumgarten (1996). Reproduced from *The Journal of General Physiology,* 1996, **107,** 503–514, by copyright permission of The Rockefeller University Press.)

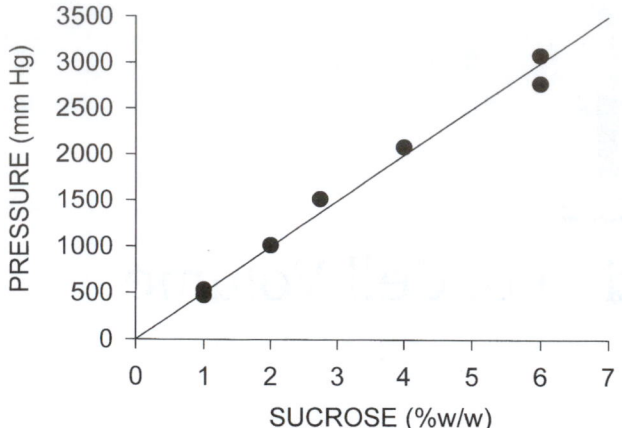

FIGURE 2. Plot of data from Pfeffer (1877) for the osmotic pressure of sucrose solutions. A copper ferrocyanide precipitation membrane was formed in the walls of an unglazed porcelain cup. The membrane separated a sucrose solution in the inner chamber from water in the outer chamber. The inner chamber was then attached to a manometer and sealed. The linear relation between the pressure measured with this device and the sucrose concentration were the experimental impetus for deriving van't Hoff's law.

sium ferrocyanide to form a copper ferrocyanide precipitation membrane on the surface of the vessel. He used this membrane to separate a sucrose solution inside the vessel from water outside and found a volume flow from the water side to the sucrose side. Pfeffer observed that the flow was proportional to the sucrose concentration. Further, a pressure applied inside the vessel produced a filtration flow proportional to the pressure. He found that a closed vessel containing a sucrose solution would develop a pressure proportional to the concentration of sucrose. He recognized this as an equilibrium state in which the pressure balanced the osmosis caused by the sucrose solution. Pfeffer's original data for the osmotic pressure of sucrose solutions are plotted in Fig. 2. He defined **osmotic pressure** as the hydrostatic pressure necessary to stop osmotic flow across a barrier (e.g., a membrane) that is impermeable to the solute. This concept is illustrated in Fig. 3. Osmotic pressure is a property intrinsic to the solution and is measured at equilibrium, when the pressure-driven flow exactly balances the osmotic-driven flow. By defining osmotic pressure in this way, we assign a positive value to an apparent reduction in pressure brought about by dissolving the solute. Thus, fluid movement occurs from the solution of low osmotic pressure (water) to the solution of high osmotic pressure, opposite in direction to the hydraulic flow of water from high to low hydrostatic pressure.

these solutes attracted water from material they touched. In 1748, J. A. Nollet used an animal bladder to separate chambers containing water and wine. He noted that the volume in the wine chamber increased, and, if this chamber was closed, a pressure developed. He named the phenomenon **osmosis** from the Greek $\omega\sigma\mu o\varsigma$, meaning thrust or impulse.

Pfeffer (1877) provided early quantitative observations on osmosis. He made an artificial membrane in the walls of an unglazed porcelain vessel by reacting copper salts with potas-

An ideal **semipermeable** membrane is required for determining osmotic pressure. These membranes are permeable to water but absolutely impermeable to solute. The concept of osmotic pressure differs from tonicity in that tonicity compares two solutions separated by a specific nonideal membrane. If the membrane is highly permeable to solute as well as to water, no water flow will occur, and therefore, the externally applied pressure required to stop osmosis is zero. This observation makes it plain that the **effective** osmotic pressure, which is

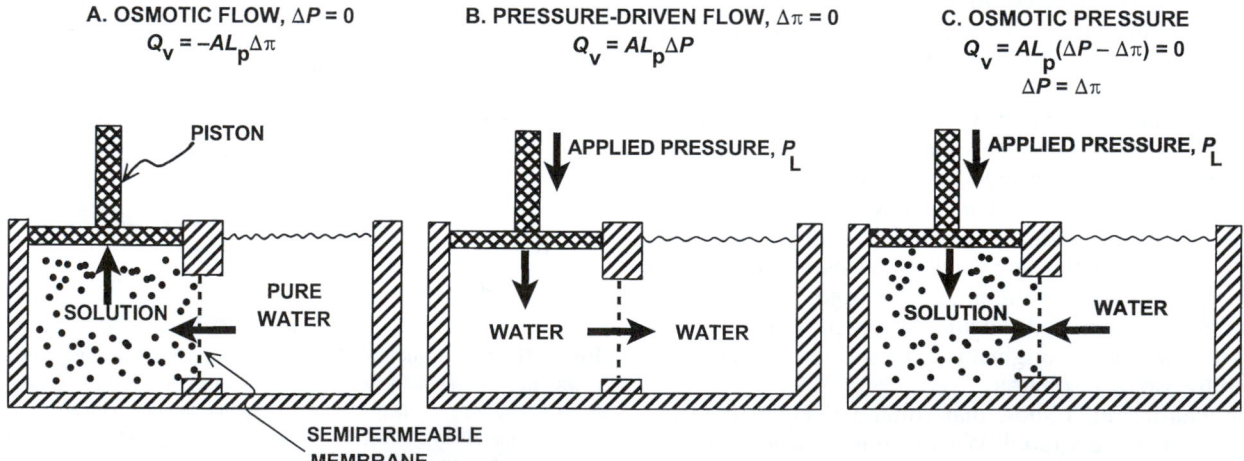

FIGURE 3. Equivalence of hydrostatic and osmotic pressures in driving fluid flow across a membrane. (A) An ideal, semipermeable membrane is freely permeable to water, but is impermeable to solute. When the membrane separates pure water on the right from solution on the left, water moves to the solution side. This water flow is **osmosis.** The flow, Q_v, in cm$^3 \cdot$s^{-1}, is linearly related to the difference in osmotic pressure, $\Delta\pi$, by the area of the membrane, A, and the hydraulic conductivity, L_p. Positive Q_v is taken as flow to the right. The flow causes expansion of the left compartment and movement of the piston (which is assumed to be weightless). (B) Application of a pressure, P_L, to the left compartment forces water out of this compartment, across the semipermeable membrane. The flow is linearly related to the pressure difference between the two compartments. (C) Application of a P_L so that $\Delta P = \Delta\pi$ results in no net flow across the membrane. The osmotic pressure of a solution is **defined** as the pressure necessary to stop water movement when the ideal, semipermeable membrane separates water from the solution.

TABLE 1 Units for Calculation of Osmotic Pressure

Pressure units	1 atm equivalent	Gas constant (R)	Solute osmolyte concentration, $\sum C_s$
atm	1	0.082 L·atm·mol^{-1}·K^{-1}	mol·L^{-1}
mm Hg	760	62.36 L·mm Hg·mol^{-1}·K^{-1}	mol·L^{-1}
Pa = N·m^{-2}	1.013×10^5	8.314 N·m·mol^{-1}·K^{-1}	mol·m^{-3} = mol·(1000 L)$^{-1}$
dyn·cm^{-2}	1.013×10^6	8.314×10^7 dyn·cm·mol^{-1}·K^{-1}	mol·cm^{-3}

Osmolarity (osmol·L^{-1}) is defined as the concentration of osmotically active particles, osmolytes, in mol·L^{-1}. Therefore, the units osmoles and moles cancel in the calculation of osmotic pressure.

measured with a real membrane, must be due to some interaction of the membrane with the solute because pressure depends on both the specific solute and the specific membrane.

B. van't Hoff's Law

From Pfeffer's data and thought experiments considering gases in equilibrium with water, van't Hoff (1887) argued that the osmotic pressure should be given by

$$\pi = RT \sum C_s \tag{1}$$

where π is the usual symbol for osmotic pressure, R is the gas constant, T is the temperature in kelvins, and C_s is the concentration of solute particles in solution. This equation is known as **van't Hoff's law.** Table 1 lists common units for osmotic pressure along with the values and units of R and C_s needed to make the calculation.

The concentration used in van't Hoff's law, $\sum C_s$, refers to the number of osmotically active particles that are formed upon dissolution of the solute. For example, organic compounds such as glucose ideally yield one particle, whereas strong salts such as NaCl or CaCl$_2$ ideally yield two (Na$^+$ and Cl$^-$) or three (Ca^{2+} and two Cl$^-$) particles. The **osmolarity** of a solution equals $\sum C_s$ and is expressed in osmol per liter to indicate that we are referring to the number of osmotically active particles, termed **osmolytes,** rather than the concentration of the solute. An alternative scale, **osmolality,** defines $\sum C_s$ per kilogram of solvent. Although the osmolal scale better describes the osmotic pressure in van't Hoff's equation, the osmolar scale is more generally used in physiological studies. As we shall see, van't Hoff's law is a limiting law that is true only for dilute solutions. In this limit of dilute solutions, both osmolal and osmolar concentration scales converge to the same results.

To illustrate the magnitude of osmotic pressure, ideal solutions of 10 mM glucose or 5 mM NaCl, which dissociates into 2 particles, both have an osmolarity of 10 mosmol·L^{-1} and an osmotic pressure at 37 °C of 0.082 L·atm·mol^{-1}·K^{-1} $\times$ 310 K $\times$ 0.01 mol·L^{-1} = 0.254 atm or 193 mm Hg. Thus, the osmotic pressure of even dilute solutions are large in comparison to normal hydrostatic pressures in physiological systems.

C. Thermodynamic Derivation of van't Hoff's Law

One of the conclusions of chemistry is that all spontaneous processes are accompanied by a **decrease in free energy.**

The total free energy of a solution can be divided among its components. This parceling out of the Gibbs free energy, G, is embodied in the concept of **chemical potential,**

$$\mu_i = \left(\frac{\partial G}{\partial n_i} \right)_{T, P, n_k} \tag{2}$$

where μ_i is the chemical potential of component i, n_i is the number of moles of component i, and $i \neq k$. The chemical potential of a component of a solution consists of three terms: a standard potential, which refers to the chemical energy involved in the formation of the material from standard states; a compositional term, which depends on the presence of other constituents; and a work term, encompassing other work required (per mole) to bring additional material into the solution. The work term in the chemical potential of water is $\int d(\overline{V}_w P)$, where $\overline{V}_w$ is the volume of water per mole and P is the pressure. The **electrochemical potential** of ions in solution requires the inclusion of an electrical work term, $\int z_i \mathscr{F} d\psi$, where z_i is the ion's valence, $\mathscr{F}$ is Faraday's constant, and ψ is voltage.

In the case of a solution separated from pure water by an ideal semipermeable membrane, water movement will occur when there is a difference in the chemical potential of water on the two sides of the membrane, such that water movement will result in a decrease in free energy. When the pressure applied to the solution is equal to the osmotic pressure, equilibrium is established, and the chemical potential of water is equal on both sides of the membrane; no net water movement occurs. This equality of chemical potential is written as

$$\mu^0 + \overline{V}_w P_L + RT \ln a_{w,L} = \mu^0 + \overline{V}_w P_R + RT \ln a_{w,R} \tag{3}$$

where the subscripts L and R refer to the left and right sides of the semipermeable membrane, μ^0 is the chemical potential of liquid water in its standard state (pure water at 1 atm pressure), and a_w is the activity of water. For an ideal solution, the activity of water can be replaced by its **mole fraction,** X_w

$$a_w = X_w = \frac{n_w}{n_w + n_s} \tag{4}$$

where n_w and n_s are the moles of water and solute, respectively. The balance of the chemical potential can be written as

$$\mu^0 + \overline{V}_w P_L + RT \ln X_{w,L} = \mu^0 + \overline{V}_w P_R + RT \ln X_{w,R} \tag{5}$$

Consider the situation in Fig. 3, where pure water is on the right side of the membrane and a solution is on the left. $X_{w,R}$ = 1.0, and thus, $\ln X_{w,R}$ = 0. Rearranging we find

$$\overline{V}_w \left(P_L - P_R \right) = -RT \ln X_{w,L} \tag{6}$$

The mole fractions of water and solute in a solution must sum to 1.0. This is expressed as

$$X_{w,L} + X_{s,L} = 1$$
$$\ln X_{w,L} = \ln \left(1 - X_{s,L} \right) \tag{7}$$

where $X_{s,L}$ is the mole fraction of solute in the solution on the left. In dilute solutions, $X_{s,L} \ll 1.0$, and thus, $\ln (1 - X_{s,L}) \approx -X_{s,L}$. Substitution of this approximation in Eq. 6 gives

$$P_L - P_R = \frac{RT}{\overline{V}_w} X_{s,L} \tag{8}$$

The left-hand side of Eq. 8 is just the osmotic pressure, π, which is equal to the extra pressure that must be applied to the solution on the left side in order to establish equality of the chemical potential of water on the two sides of the membrane. For physiological studies, it is convenient to express π in terms of concentration. From the definition of mole fraction and the assumption of dilute solutions ($n_s \ll n_w$), we get

$$\pi = \frac{RT}{\overline{V}_w} X_{s,L} = \frac{RT}{\overline{V}_w} \frac{n_s}{n_s + n_w} \approx \frac{RTn_s}{n_w \overline{V}_w}$$
$$= RT \frac{n_s}{V} = RTC_s \tag{9}$$

where $n_w \overline{V}_w \approx V$, the total volume of solution, and C_s is the concentration of impermeable solute on the solution side of the membrane. This last expression is the van't Hoff equation for the osmotic pressure, Eq. 1. The thermodynamic derivation entails two assumptions: (1) the solution is sufficiently dilute as to approach ideality, and (2) the solution is incompressible so that $\int d(\overline{V}_w P) = \overline{V}_w \Delta P$. It is important to recognize that Eq. 9 is not exact for physiological solutions. Rather, it is an approximation that is strictly true only for dilute ideal solutions.

The van't Hoff equation is based on thermodynamics, and as such, it tells us nothing about the rate of osmosis or the mechanism by which it occurs. Conceivably, the semipermeable membrane could be like a sieve that allows water to pass freely while blocking solute movement. Alternatively, solvent could dissolve in the membrane, whereas solute is insoluble. Both of these models would exhibit osmotic flow from the region of low osmotic pressure (pure water) to that of high osmotic pressure (impermeant solute solution). The mechanism by which osmosis occurs must be determined by methods of chemical kinetics, and must be determined for every membrane-solvent pair.

D. Other Colligative Properties of Solutions

The thermodynamic derivation given previously indicates that osmotic pressure (and osmotic flow) originates in the lowering of the chemical potential of water by the amount $\sim RTX_s$ when solute is dissolved. Several other properties of solutions also are a consequence of the lowered chemical potential of water because of dissolution of solutes. Together,

these are called the **colligative** properties (from the Latin, *ligare,* meaning to bind) and include **osmotic pressure, vapor pressure depression, boiling point elevation** and **freezing point depression.** Consider two open compartments enclosed in a chamber. One compartment contains pure water and the other a solution of a nonvolatile solute. The vapor pressure above a solution is defined as the partial pressure of water vapor in equilibrium with the solution. Since the vapor pressure of pure water is higher than that of the solution, water vapor above pure water will be at a higher pressure than that above the solution. As a result, water vapor will diffuse from the water side to the solution side. At the surface of the solution, water vapor will condense because the vapor pressure there will be higher than the equilibrium vapor pressure for the solution. Thus, water will move from the pure water to the solution side. In short, "osmosis" would occur through the "semipermeable membrane" represented by the surfaces of the two fluids and the intervening air. This illustrates the strong connection among the colligative properties of solutions. Laboratory osmometers typically use either vapor pressure depression or freezing point depression to determine the total solute concentration in an aqueous solution.

E. Osmotic Pressure of Nonideal Solutions

As discussed above, the van't Hoff equation is an approximation that adequately describes the osmotic pressure for dilute solutions. Its derivation requires the assumptions that the solutions are dilute and that the solutions are ideal. Here *ideal* means that Raoult's law (vapor pressure is proportional to mole fraction of solvent) is valid for the solution (Kiil, 1989; Hildebrand, 1955). Because the behavior of real solutions is not ideal, the van't Hoff equation must be modified to include a correction term, the osmotic coefficient (ϕ_s)

$$\pi = RT \sum \phi_s C_s \tag{10}$$

At physiological concentrations, the osmotic coefficients for NaCl and CaCl$_2$ are 0.93 and 0.85, respectively. This means the osmolarity of 150 mM NaCl is $0.93 \times 2 \times 150 = 279$ mosmol·L^{-1} and the osmolarity of 150 mM CaCl$_2$ is $0.85 \times 3 \times 150 = 382.5$ mosmol·L^{-1}. The osmotic coefficients for electrolytes vary with temperature, concentration, and the chemical nature of the electrolyte. For most electrolytes, $\phi < 1.0$ for dilute solutions, due to weak attraction of the ions. At higher concentrations, ϕ increases to exceed 1.0. Values for the osmotic coefficients for electrolytes can be found in Robinson and Stokes (1959) or can be calculated from the parameters tabulated by Pitzer and Mayorga (1973). These osmotic coefficients are corrections to van't Hoff's law due to interactions only for the particular solute. When more than one solute is present, interactions could occur that are not accounted for by the osmotic coefficients. Therefore, calculations of the osmotic pressure of a mixture of solutes, even when osmotic coefficients are used, are only approximations.

Nonelectrolytes and polyelectrolytes, especially proteins, also show marked departure from van't Hoff's law with increasing concentration. According to Eq. 10, the osmotic coefficient for a single solute can be calculated as

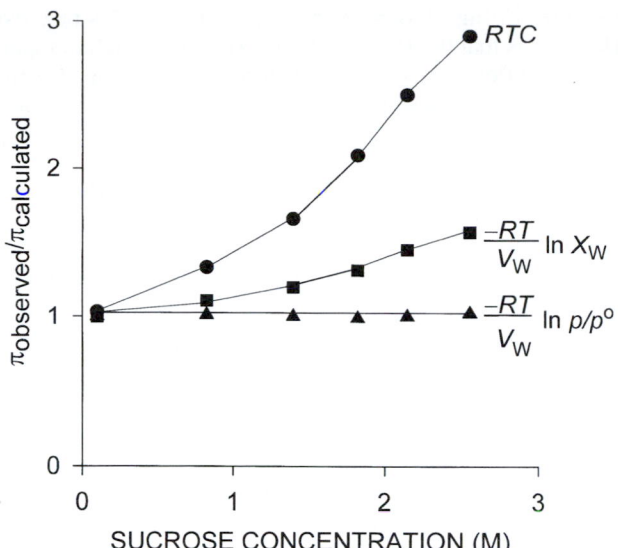

FIGURE 4. Osmotic coefficients as a function of sucrose concentration. Plotted are the molar osmotic coefficient, defined as ϕ_s in Eq. 11 (●), obtained by dividing the observed osmotic pressure by RTC; the rational osmotic coefficient, g, defined in Eq. 15 (■), obtained by dividing the observed osmotic pressure by $(-RT/\bar{V}_w) \ln X_w$; and the observed osmotic pressure divided by that predicted by vapor pressure measurements $(-RT/\bar{V}_w) \ln p/p^0$ according to Eq. 13 (▲). Deviation of the molar osmotic coefficient from 1.0 means that the van't Hoff law fails to adequately describe the osmotic pressure at high concentrations, but is accurate for dilute solutions. The van't Hoff law requires the assumption of dilute solution and ideal behavior. Deviation of the rational osmotic coefficient from 1.0 means that the solution is not ideal, as the equation requires this assumption. The nearly perfect agreement between the theoretical osmotic pressure predicted from vapor pressure measurements illustrates the connection between these two colligative properties. Data from Glasstone (1946).

$$\phi_s = \frac{\pi_{observed}}{\pi_{calculated}} = \frac{\pi_{observed}}{RTC_s} \tag{11}$$

The osmotic coefficient for sucrose is plotted against sucrose concentration in Fig. 4. The osmotic coefficient is nearly 1.0 in dilute solutions, but approaches 3 in saturated sucrose solutions. Thus, the van't Hoff equation successfully describes the osmotic pressure of dilute solutions, but fails at high solute concentrations. The failure of the van't Hoff equation for highly concentrated solutions is due to deviation of reality from the assumptions used to derive the equation, that solutions are dilute and ideal. The osmotic coefficient accounts for these deviations.

For high solute concentrations, we can calculate the osmotic pressure from the mole fraction of water without assuming a dilute solution by identifying $\pi = P_L - P_R$ in Eq. 6

$$\pi = -\frac{RT}{\bar{V}_w} \ln X_w \tag{12}$$

This equation still requires the assumption of ideal solution behavior: the activity of water is equal to its mole fraction. The expression for osmotic pressure without assuming a dilute solution or ideality is given by

$$\pi = -\frac{RT}{\bar{V}_w} \ln a_w = -\frac{RT}{\bar{V}_w} \ln \frac{p}{p^0} \tag{13}$$

where p and p^0 are the vapor pressures of the solution and pure water, respectively.

The rational osmotic coefficient, g, accounts for nonideal behavior and is defined as

$$\ln a_w = g \ln X_w \tag{14}$$

Then, from Eqs. 12–14, we find that

$$g = \frac{\pi_{observed}}{\pi_{calculated}} = \frac{\pi_{observed}}{-\frac{RT}{\bar{V}_w} \ln X_w} \tag{15}$$

The rational osmotic coefficient is closer to 1.0, but still deviates significantly at higher sucrose concentrations where solution behavior is further from ideal (Fig. 4). In contrast, the ratio of the observed osmotic pressure to the theoretical osmotic pressure calculated from vapor pressure measurements, according to Eq. 13, is very close to 1.0 throughout the entire concentration range. This shows the validity of Eq. 13 and the absolute correlation between vapor pressure depression and osmotic pressure as different measures of the same phenomenon, the lowering of the activity of solvent water by the dissolution of solute. Equation 12 does not adequately describe the variation of π with C_s because it requires ideal adherence to Raoult's law (vapor pressure is proportional to X_w); van't Hoff's limiting law further deviates from a linear relationship between π and C_s because it requires the additional approximation of dilute solutions. Despite these limitations in the high concentration domain, van't Hoff's law remains a good approximation for electrolyte solutions in the physiological range.

Because of their importance in physiological systems, the nonideality of the osmotic pressure of protein solutions requires special comment. Adair (1928) found that the observed osmotic pressure increased faster than the concentration in hemoglobin solutions, as shown in Fig. 5. Part of the osmotic pressure was due to the unequal distribution of ions across the semipermeable membrane caused by electric charge on the immobile protein molecules. This is the Gibbs-Donnan equilibrium, discussed in more depth later. The contribution of the Gibbs-Donnan distribution to osmotic pressure is small, however, and nearly all of the non-linearity between π and C_s is due to the protein itself. From the data obtained by Adair (1928), $\phi_{Hb} = 4.03$ at the concentration of hemoglobin within erythrocytes (34.4 g hemoglobin per 100 ml of solution).

The observed osmotic pressure of solutions of plasma proteins also increases more rapidly than concentration, but the degree of deviation from linearity is different for different proteins. Thus, serum albumin shows marked deviation, whereas γ-globulins are more nearly linear. The empirical fits to the concentration-dependence of osmotic pressure are given by Landis and Pappenheimer (1963) as

$$\pi_{albumin} = 2.8C + 0.18C^2 + 0.012C^3$$
$$\pi_{globulins} = 1.6C + 0.15C^2 + 0.006C^3 \tag{16}$$
$$\pi_{plasma\ proteins} = 2.1C + 0.16C^2 + 0.009C^3$$

In each of these three equations, the first term represents the limiting law of van't Hoff.

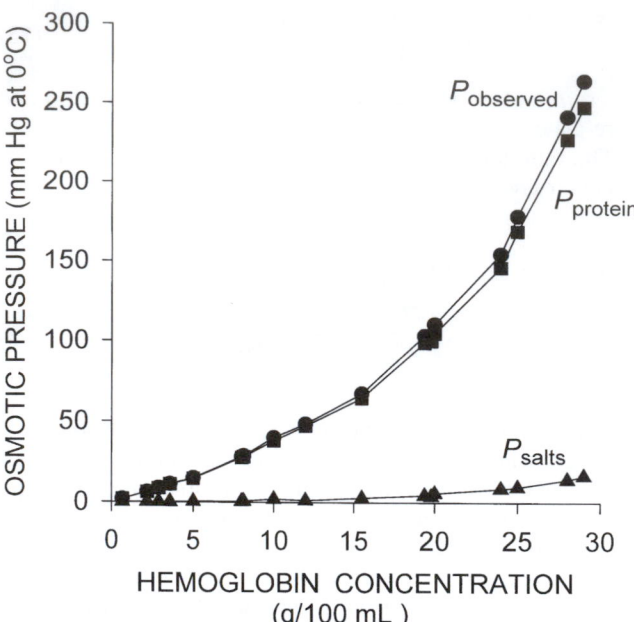

FIGURE 5. Dependence of the observed osmotic pressure on hemoglobin concentrations. $P_{observed}$ is the observed osmotic pressure (●). P_{salts} is the contribution of the salts to the osmotic pressure as calculated from the Gibbs-Donnan distribution and the van't Hoff equation (▲). Pprotein is the contribution of the protein itself to the observed osmotic pressure, calculated as $P_{observed} - P_{salts}$ (■). Data from Adair (1928).

The rather large ϕ_s for proteins and polymers is due in part to **excluded volume effects.** That is, proteins and polymers exclude solvent from a larger volume than inorganic ions. The lowering of the free energy of solvent water upon dissolution of solute, which gives rise to osmosis, can be calculated from the increase of entropy on mixing. This entropy of mixing depends on the volume occupied by the solute. From considerations of the excluded volume, it can be shown (Tanford, 1961) that the expected osmotic pressure is given as

$$\pi = RTC\left(1 + \frac{\overline{V}_s}{\overline{V}_w}C\right) \quad (17)$$

F. Equivalence of Osmotic and Hydrostatic Pressure

As mentioned earlier, Pfeffer originally observed a linear relationship between the flow rate and the concentration of solute. This is expressed as

$$J_v = -L_p\left(\pi_L - \pi_R\right) = -L_p \Delta\pi \quad (18)$$

where J_v is the volume flux in cm$^3 \cdot$s^{-1} per unit area of membrane, L_p is variously called the **filtration coefficient, hydraulic conductivity,** or **hydraulic permeability,** and $\Delta\pi$ is the osmotic pressure difference. A positive J_v in Eq. 18 represents flux from the left to the right compartment, and this is the order in which the osmotic pressure difference is taken. The minus sign before L_p indicates flux is from the region of low osmotic pressure to the region of high osmotic

pressure. In Fig. 3A, $\pi_L > \pi_R$, $\Delta\pi > 0$, and J_v is negative. This means that the flux is from the right to the left compartment. The flow across an extent of membrane is just the flux times the area exposed to the driving forces, expressed as

$$Q_v = -AL_p \Delta\pi \quad (19)$$

where A is the area of the membrane and Q_v is the flow in units of cm$^3 \cdot$s^{-1}.

In the absence of solute, the volume flow across Pfeffer's artificial membrane was also linearly related to the hydrostatic pressure

$$J_v = L_p\left(P_L - P_R\right) = L_p \Delta P$$
$$Q_v = AL_p \Delta P \quad (20)$$

In a study on collodion membranes, Meschia and Setnikar (1958) found that the proportionality constant for hydrostatic pressure-driven filtration was the same as the constant relating flow and osmotic pressure. That means that the L_p in Eq. 19 is the same as the L_p in Eq. 20. Thus, not only can the osmotic flow be nulled by opposing osmotic pressure with an equal but opposite hydrostatic pressure, but the equivalent proportionality implies that the mechanism of volume flow is also identical for osmotic and hydraulic flow. The equivalence of osmotic and hydrostatic pressures allows us to write

$$Q_v = AL_p\left[\left(P_L - P_R\right) - \left(\pi_L - \pi_R\right)\right]$$
$$= AL_p(\Delta P - \Delta\pi) \quad (21)$$

This equation describes the net flow that would be observed in the presence of both hydrostatic and osmotic pressure differences across a semipermeable membrane.

G. Reflection Coefficient

Equation 1, van't Hoff's law, describes the relation between osmotic pressure and concentration when a solution is separated from water by an ideal semipermeable membrane. Recall that a semipermeable membrane is defined as absolutely impermeable to the solute. Real membranes may not fit this ideal; they may be somewhat permeable to the solute. When membranes are permeable to the solute, the measured osmotic pressure is actually less than that predicted by van't Hoff's law. This phenomenon has led to a second membrane parameter, σ, the **reflection coefficient** which is defined as

$$\sigma = \frac{\pi_{observed}}{\pi_{theoretical}} = \frac{\pi_{observed}}{\phi_s RTC_s} \quad (22)$$

The reflection coefficient derives its name from the idea that all of the collisions of solute with a semipermeable membrane will result in the solute being reflected back into the solution. The reflection coefficient for an ideal membrane is 1.0. For a permeable solute, some fraction of the collisions with the membrane will result in permeation of the membrane, so that $\sigma < 1.0$, and the observed osmotic pressure will be less than that predicted by van't Hoff's law. The value of σ is not simply the fraction of collisions that penetrate the membrane. It involves discrimination by the membrane between solvent and solute. Thus, σ is a parameter that is different for every membrane-solute pair. A vapor-pressure osmometer or a freezing-point osmometer would still register the proper osmolarity of the solution, however (for nonvolatile solutes). A molecular interpretation of the origin of the reflection coefficient is given later.

Permeation of the solute should reduce osmotic flow along with osmotic pressure. In the presence of both hydrostatic pressure differences and concentration differences across a membrane, the resulting volume flow is given by

$$Q_v = AL_p \left[(P_L - P_R) - \left(\sum_i \sigma_i \pi_{i,L} - \sum_i \sigma_i \pi_{i,R} \right) \right] \quad (23)$$

where σ_i is the reflection coefficient of solute i, and $\pi_{i,L}$ and $\pi_{i,R}$ are the osmotic pressures of solute i on the left and right sides of the membrane, respectively. The π_i in this equation is that given by van't Hoff's law; its multiplication by the reflection coefficient, σ, gives the effective osmotic pressure. The consequence of a combination of hydrostatic and osmotic pressures on the flow across a membrane is shown in Fig. 6.

III. Mechanisms of Osmosis

The ultimate cause of osmosis is the reduction of the chemical potential of water in a solution. This thermodynamic statement and equations derived from it tell us nothing about the rate of osmosis or its mechanism. Several possible mechanisms have been investigated. As will be developed, classes of models can be distinguished by comparing the proportionality between applied force, which is either a pressure or concentration gradient, and water flow.

HYDROSTATIC AND OSMOTIC PRESSURE

$$Q_v = AL_p(\Delta P - \sigma \Delta \pi)$$

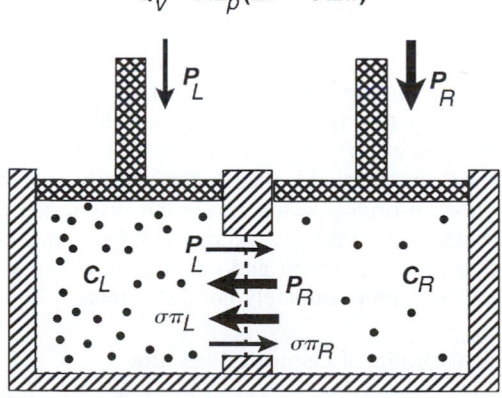

FIGURE 6. Net flow in the presence of osmotic and hydrostatic pressures. For a real membrane, the effective osmotic pressure on the left, $\sigma \pi_L$, causes flow toward the left, while the applied hydrostatic pressure, P_L, drives flow to the right. A similar situation occurs on the right. The **net** flow is driven by the balance of the forces, $\Delta P - \sigma \Delta \pi$, and is proportional to the area, A, and the hydraulic conductivity, L_p.

A. Microporous Membranes

1. Osmotic and Pressure-Driven Flow through Porous Membranes

The equivalence of L_p for osmotic and pressure-driven flow suggests a common mechanism. At least three different models have been proposed to explain flow across membranes: (1) hydrodynamic flow through a porous membrane; (2) diffusion through the membrane; and (3) nonhydrodynamic flow through narrow pores. As we shall see, it is likely that biological membranes are not modeled well by any one of these. Despite this, we shall consider these model membranes because investigators have relied heavily on them to clarify their thinking.

First, we consider a porous membrane as a model for understanding osmotic and hydrostatic pressure-driven flow and derive L_p, the proportionality constant relating pressure and flow. We assume the membrane is a flat, thin sheet of thickness δ. We imagine that the membrane is pierced by right-cylindrical pores of radius r, and the number of pores, N, per unit area is $n = N/A$. The membrane separates two compartments of water that are at different hydrostatic pressures. If we assume that the pores are large enough for laminar flow to occur, then the filtration flow will be given by the Poiseuille equation

$$q_v = \frac{\pi r^4}{8\eta\delta} \Delta P \quad (24)$$

where q_v is the flow per pore in $cm^3 \cdot s^{-1}$, η is the viscosity of the fluid, δ is the thickness of the membrane (equal to the length of the pore), and ΔP is the pressure difference across the pore. The π in this equation is the geometric ratio, 3.14..., and should not be confused with the symbol for osmotic pressure. Since the flow through N pores is just Nq_v, the observed macroscopic flux and flow are

$$J_v = \frac{Nq_v}{A} = nq_v = \frac{n\pi r^4}{8\eta\delta} \Delta P$$

$$Q_v = AJ_v = Nq_v = \frac{N\pi r^4}{8\eta\delta} \Delta P \quad (25)$$

Recall here that J_v is the flux, or flow per unit area of membrane, and Q_v is the flow in units of volume per unit time.

Equation 25 describes the steady-state flow of water across the membrane. Because at steady-state there is no buildup or depletion of water, there is no difference in the flow of water at any two points in the pore. Consequently, pressure changes linearly with distance through the pore, and the gradient of pressure, $\Delta P/\delta$, is constant.

A comparison of Eq. 25 with Eq. 20 indicates that the hydraulic conductivity is

$$L_p = \frac{n\pi r^4}{8\eta\delta} \quad (26)$$

Thus, L_p is a parameter determined by the viscosity of the fluid and by membrane characteristics including the number of pores per unit area, n, the pore radius, and the membrane thickness.

2. Diffusional Permeability of Porous Membranes: P_d

In the absence of a pressure gradient, solute and water cross a porous membrane by diffusion through the pores. If we assume that the membrane is impermeable at all other points, the permeability is given by Fick's first and second law of diffusion

$$j_s = -D\frac{\partial C}{\partial x}$$
$$\frac{\partial C}{\partial t} = D\frac{\partial^2 C}{\partial x^2} \quad (27)$$

where j_s is the flux of solute through one pore. The second expression describes the time dependence of the concentration profile over distance, x, within the pore. At steady-state, the concentration profile no longer changes. This means that $\partial C/\partial t = \partial^2 C/\partial x^2 = 0$; the concentration gradient is linear; and $\partial C/\partial x = (C_L - C_R)/(0-\delta) = -\Delta C/\delta$, where δ is the thickness of the membrane. Total solute flux across the entire membrane, J_s, is

$$J_s = \frac{n\pi r^2 D}{\delta}\Delta C \quad (28)$$

where n is the number of pores per unit area of membrane. According to this equation, the observed macroscopic flux of solute across a porous membrane is linearly related to the concentration difference by a coefficient that includes properties of the solute (diffusion coefficient) and the membrane (thickness, pore density, and pore cross-sectional area). The **permeability** of the membrane to solute, p_s, includes several parameters in Eq. 28 that are difficult to obtain experimentally; p_s relates solute flux, J_s, to the difference in concentration across the membrane

$$J_s = p_s\Delta C \quad (29)$$

From Eqs. 28 and 29, p_s is defined as

$$p_s = \frac{n\pi r^2 D}{\delta} \quad (30)$$

Isotopic water on one side of a porous membrane is distinguishable from ordinary water and may be viewed as a solute. Thus, water itself will obey these equations. This allows us to define the diffusional permeability of water, P_d, for a porous membrane

$$P_d = \frac{n\pi r^2 D_w}{\delta} \quad (31)$$

where D_w is the diffusion coefficient of water. The units of P_d are $cm\cdot s^{-1}$. Note that multiplication of a permeability by a concentration, as in Eq. 29, gives a flux with units of $mol\cdot cm^{-2}\cdot s^{-1}$.

3. Evidence for Pores: P_f/P_d Ratio

In the absence of a concentration gradient, pressure-driven water flow gives rise to a second permeability constant termed the **filtration permeability** or **osmotic permeability**, P_f, which has units of $cm\cdot s^{-1}$. Mauro (1957) realized that the proportionality constants relating pressure and con-

centration-gradient-driven water flow, P_f and P_d, provide evidence for the mechanism of transport. The ratio P_f/P_d should be 1.0 if water crosses by a dissolution-diffusion process. Mauro (1957) recognized that the flux of water in response to a pressure gradient could be partitioned into two components, diffusional and nondiffusional (e.g., bulk flow), and that the diffusional component of water flux, J_w, would obey the Nernst-Planck equation

$$J_w = -\frac{n\pi r^2 D_w}{RT}C_w\frac{d\mu_w}{dx} \quad (32)$$

where C_w is the concentration of water. In the case where only a hydrostatic pressure is applied, $d\mu_w = \overline{V}_w dP$, and $C_w\overline{V}_w = 1$. Assuming steady-state flows and a uniform membrane, $dP/dx = -\Delta P/\delta$, and Eq. 32 becomes

$$J_w = \frac{n\pi r^2 D_w}{RT\delta}\Delta P \quad (33)$$

This flow of water is in units of moles of water per second per cm^2 of membrane. It can be converted to units of volume per second per cm^2 (the units of J_v) by multiplying by the volume of water per mole, or $\overline{V}_w$:

$$J_v = \overline{V}_w J_w = \frac{\overline{V}_w n\pi r^2 D_w}{RT\delta}\Delta P \quad (34)$$

This equation relates the volume flux to the pressure difference across the membrane.

The total volume flux was earlier given as $J_v = L_p\Delta P$ (see Eq. 20). If diffusional flux is the only component of volume flux, Eqs. 20 and 34 may be combined to give

$$L_p = \frac{\overline{V}_w n\pi r^2 D_w}{RT\delta} \quad (35)$$

Part of this expression for L_p incorporates P_d. Insertion of Eq. 31 into Eq. 35 gives

$$P_f = \frac{L_p RT}{\overline{V}_w} = \frac{n\pi r^2 D_w}{\delta} = P_d \quad (36)$$

This definition of P_f converts L_p into a parameter having the same units as P_d, thereby allowing direct comparison of filtration and diffusional permeabilities. The equality of P_f and P_d obtained in Eq. 36 is dependent on the condition that the flow of water in response to a hydrostatic pressure difference is due only to diffusional processes. Thus, for a purely diffusional process, $P_f/P_d = 1$. In contrast, Mauro (1957) found that P_f/P_d was 727 in collodion membranes. That is to say, pressure-driven water movement was much greater than expected from a diffusional process. From this he concluded that pressure-driven and osmotic flow across these membranes was predominately nondiffusional.

4. Physical Origin of Osmotic Pressure

If a porous membrane separates a solution containing only impermeant solutes from pure water, we observe experimentally that water flows through the membrane from the pure water to the solution side. The flow is proportional to the osmotic pressure of the solution times L_p (see Eq. 20). The question is, what causes this water movement?

Because the membrane is impermeable to solute, solute cannot enter the pores, and the fluid in the pore is pure water. Consider a water molecule in the middle of the pore. How does the water "know" to move toward the solution side? It appears there are only two possible answers to this question. Either there is a concentration gradient of water within the pore, or there is a pressure gradient within the pore. These two possibilities are not mutually exclusive, but diffusion-driven water flow and pressure-driven water flow are often thought of as separate mechanisms. The dichotomy reflects the notion that water is an incompressible fluid. Water is not absolutely incompressible, however. The coefficient of compressibility is given as

$$\beta = -\frac{1}{V}\left(\frac{\partial V}{\partial P}\right)_T \tag{37}$$

and the value of β for water is 4.53×10^{-5} atm^{-1}. The equation for the volume of water is

$$V = V^0\left(1 - \beta P\right) \tag{38}$$

where V^0 is the volume at a standard temperature and pressure (1 atm), and P is the pressure in excess of 1 atm. The coefficient of compressibility is virtually constant in the range -500 to $+1000$ atm. Equation 38 indicates that application of a negative pressure of 2.5 atm would expand a pure water solution by 0.01%, which corresponds to a change in the concentration of water of about 5 mM. Looked at the other way, an expansion of water of only 0.01% would induce a negative pressure of 2.5 atm, equal to the osmotic pressure of a 0.1 molal solution at 37 °C.

Dainty (1965) proposed a model in which he considered the density of water immediately within the pore opening on the solution side. As solute molecules cannot enter the pore, Dainty reasoned that the concentration of water within the pore must be higher than in the solution. Because of this difference in concentration, he argued that water would diffuse into the solution side faster than water could diffuse into the pore. The resulting net movement of water toward the solution side would lower the density of water in the pore, thereby creating a reduced pressure. Bulk movement of water down its pressure gradient would follow.

This explanation of the origin of osmotic pressure supposes that the driving force is actually water diffusing down its concentration gradient. The data in Fig. 7 show, however, that the concentration of water cannot be the major determinant of the colligative properties of solutions. In Fig. 7A, the water concentration in solutions of sucrose and glucose are plotted against the concentration of the solute. The water concentration is indeed decreased by dissolving solute, but sucrose, being almost twice as large as glucose, displaces almost twice as much solvent. As shown in Fig. 7B, however, the colligative properties, represented here by the freezing point depression, depend **only** on the concentration of solute. Solutions with equal solute concentrations but different water concentrations have the same freezing point.

An alternative view of the physical origin of the osmotic pressure begins with the notion of pressure as a force divided by an area. The macroscopic concept of pressure relies on the averaging over time of the myriad of collisions that produce the pressure. By Newton's law,

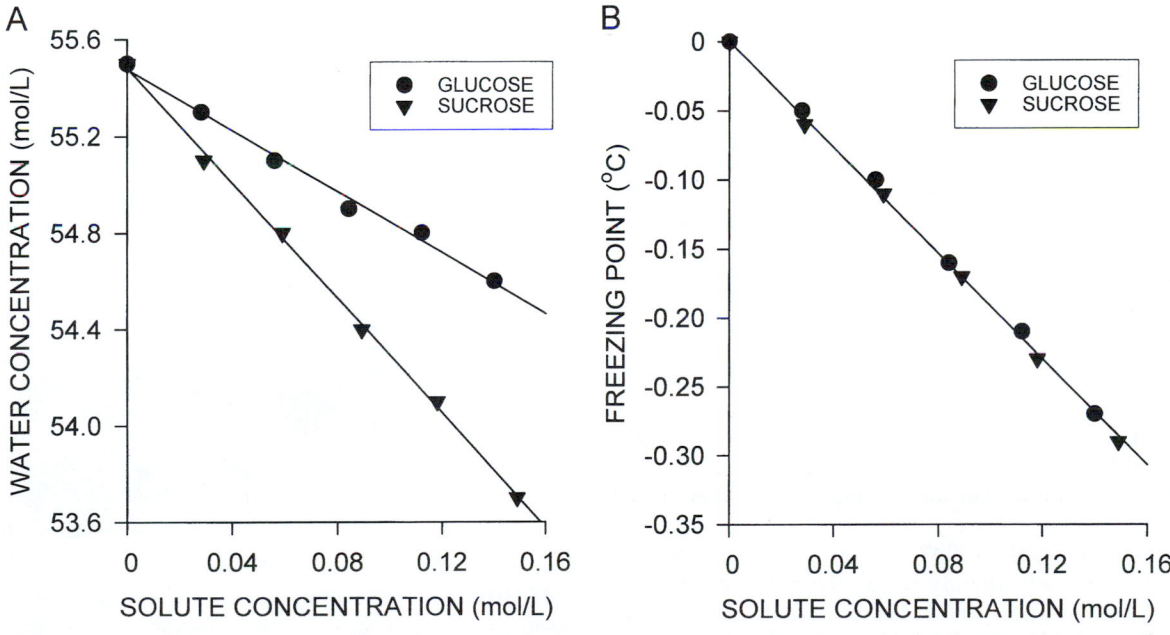

FIGURE 7. Effect of solute concentration on water concentration and freezing point depression in glucose and sucrose solutions. (A) Because solute displaces water, the water concentration decreases with increasing solute concentration. Sucrose is nearly twice the size of glucose. Consequently, there is less water in a sucrose solution having the same molarity as a glucose solution. (B) The freezing point depression, however, is dependent only on the solute concentration. It is the **mole fraction** of water that determines the colligative properties of solutions (the osmotic pressure, vapor pressure depression, boiling point elevation, and freezing point depression). (Data from *The Handbook of Chemistry and Physics*, Chemical Rubber Company, Cleveland, OH, 1965.)

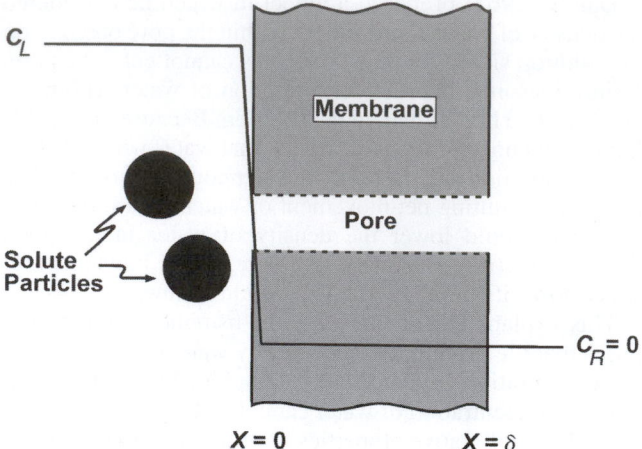

C_L

Membrane

Pore

Solute
Particles

$C_R = 0$

$X = 0$ $X = \delta$

FIGURE 8.   Concentration profile in the vicinity of a pore in a microporous membrane. The membrane is impermeable at all places except the pores, where water may penetrate but solute particles (solid circles) are too large to enter the pore. The concentration of solute in the bulk solution (C_L) must fall to zero upon entering the pore. The steep concentration gradient is accompanied by diffusion towards the pore that is balanced by reflection of solute by collision with the membrane.

force is the time derivative of momentum. An elastic collision of a solvent or solute molecule with the walls of the vessel results in a momentum change of $2mv$, where m is the mass of the molecule and v is its velocity, which contributes to the pressure against the vessel wall. At the entrance to the pore, however, solute molecules cannot transfer their momentum to the interior of the pore because they collide with the rim of the pore and are reflected back into the solution. Thus, the water molecules immediately inside the pore experience a momentum deficit that is equal to the component of pressure contributed by the solute molecules in the bulk phase.

Figure 8 shows the one-dimensional concentration profile of solute molecules near a pore opening. Because there is a steep solute gradient, there should be a diffusion of solute toward the pore opening. However, the actual steady-state flux of solute in this direction is zero because of the force exerted on the solute molecules by the membrane. The equation that describes the solute flux, J_s, in units of mol·cm^{-2}·s^{-1}, is

$$J_s = -D \frac{\partial C(x)}{\partial x} + \frac{D}{RT} f C(x) \quad (39)$$

where $C(x)$ is the concentration of solute at position x, D is the solute diffusion coefficient in units of cm^2·s^{-1}, f is the force per solute molecule, and R and T have their usual meanings.

Villars and Benedek (1974) derived an equation for the drop in pressure immediately inside the pore on the solution side by setting the flux in Eq. 39 to zero and analyzing the net force on a plug of volume near the pore. Under steady-state conditions with zero J_s through the pore, Eq. 39 gives

$$f C(x) = RT \frac{\partial C(x)}{\partial x} \quad (40)$$

where $f C(x)$ is the force per molecule times the number of molecules per unit volume, or the force per unit volume. Figure 9 shows a volume element near the opening of the pore on the solution side. We consider the forces acting on the element of fluid with an area A from a point x well within the bulk solution to a point $x + \Delta x$ just inside the pore. We assume that this element is in mechanical equilibrium; although it may be moving, it is not accelerated or decelerated. The forces acting on the volume are contact forces on the edges of the volume and additional forces acting on the solute molecules alone to counteract the diffusive flux. At mechanical equilibrium the sum of the forces must be zero. This is written as

$$F_s + F_c = 0 \quad (41)$$

where F_s is the total force acting on the solutes in the volume, and F_c is the net contact force due to the pressure from the adjacent volume elements. The net contact forces are the result of pressure acting over an area

$$F_c = AP(x) - AP(x + \Delta x) \quad (42)$$

The forces acting on the volume due to the solute particles is given by integrating Eq. 40

$$F_s = \int_x^{x+\Delta x} f C(x) \, dV \quad (43)$$

Inserting the volume element $dV = A \, dx$ and $f C(x) = RT \, \partial C(x)/\partial x$ from Eq. 40, we obtain

$$F_s = ART \int_x^{x+\Delta x} \frac{\partial C(x)}{\partial x} \, dx$$

$$= ART \left[C(x + \Delta x) - C(x) \right] \quad (44)$$

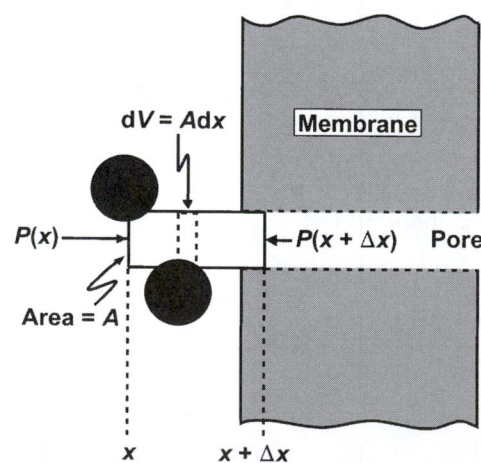

$dV = A dx$

Membrane

$P(x) \rightarrow$ $\leftarrow P(x + \Delta x)$ **Pore**

Area = A

x $x + \Delta x$

FIGURE 9.   Forces acting on a volume element immediately adjacent to a pore opening. The ideal, porous, semi-permeable membrane separates pure water on the right from solution of impermeant solute on the left. The volume element has an area, A, equal to the cross-sectional area of the pore. The pressure in the bulk phase at position x is $P(x)$. The pressure at position $x + \Delta x$, just within the pore, is $P(x + \Delta x)$. The net force on the element is the sum of the forces at both ends plus the forces acting only on the solute particles (solid circles) within the element.

Since $C(x + \Delta x) = 0$ because solute particles are not in the pore, this becomes

$$F_s = -ART\,C(x) \tag{45}$$

where the negative sign indicates that F_s is directed to the left. Inserting Eqs. 42 and 45 into Eq. 41, we have

$$AP(x) - AP(x + \Delta x) = ART\,C(x) \tag{46}$$

or

$$P(x + \Delta x) = P(x) - RTC_L \tag{47}$$

This last equation indicates that the pressure experienced by the volume of fluid immediately inside the pore is less than the bulk pressure, $P(x)$, by the amount RTC_L, where C_L is the concentration of impermeant solute in the solution on the left of the membrane. This analysis is consistent with the intuitive idea that water movement from the water side to the solution side of a semipermeable membrane must be due to a real force, which appears in this analysis to be due to the momentum deficit, and thus pressure deficit, within the pore on the solution side.

5. Physical Interpretation of the Reflection Coefficient, σ

The microporous semipermeable membrane presented previously distinguishes between solvent water and solute on the basis of pore and solute size. That is, the solute is too large to enter the pore and so cannot cross the membrane. If the pores were somewhat larger or the solute molecules smaller, the solute could enter the pore, but with a lower probability than water because of the tight fit. In this case, rather than the solute being absolutely impermeant, the membrane would allow its slow passage. How does this affect the situation? Let us suppose that solute molecules that hit the rim of the pore before entry are reflected back into the bulk solution. This is shown diagrammatically in Fig. 10, looking down the axis of the pore perpendicular to the surface of the membrane. The area of the pore that is accessible to solute is

$$A_s = \pi(r - a)^2 = \pi r^2\left(1 - \frac{a}{r}\right)^2 \tag{48}$$

where a is the radius of the solute molecule. Assuming that the radius of water molecules (0.75 Å) is negligible compared to the pore's radius, the ratio of areas available to solute and solvent water is

$$\frac{A_s}{A} = \frac{\pi r^2\left(1 - \dfrac{a}{r}\right)^2}{\pi r^2} = \left(1 - \frac{a}{r}\right)^2 \tag{49}$$

The fraction of collisions of solute molecules with the pore opening that are reflected back, compared with those of water, is approximated by the ratio of the area of the gray annulus in Fig. 10 to the cross-sectional area of the pore. This is identified with the reflection coefficient

$$\sigma = \frac{A - A_s}{A} = 1 - \frac{A_s}{A}$$

$$\sigma = 1 - \left(1 - \frac{a}{r}\right)^2 \tag{50}$$

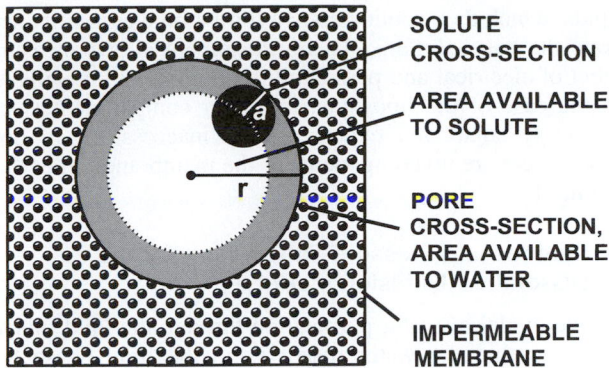

FIGURE 10. A physical interpretation of the reflection coefficient, σ, for a solute and a porous membrane. The view is down the pore in a direction perpendicular to the membrane. The pore is modeled as a right circular cylinder of radius r. The model assumes that any contact of the membrane with a solute particle of radius a (solid circle) will result in reflection of the particle back into the solution. The area available to solute is indicated in white. The total area of the pore, white plus gray annulus, is available for water movement.

According to this view, the concentration of solute immediately within the pore would not be zero, as in the case when the solute was impermeant, but instead would be $(1 - \sigma)C$, and the excluded concentration would be σC. Thus, the momentum deficit inside the pore would be due only to the excluded solute and would be equal to σRTC, which is equal to $\sigma\pi$.

In addition to entrance effects, the concentration profile within the pore will be influenced by the combination of diffusion through the pore, solvent drag due to movement of fluid in response to pressure gradients, and interaction of the solutes with the nonlinear velocity profile within the pore. Laminar flow through long pores is characterized by a parabolic velocity profile, with a motionless layer of fluid adjacent to the pore walls and most rapid flow in the center. Large solute molecules will span several layers of velocity, thereby distorting the velocity profile and changing the solute molecule's velocity. Various equations have been derived to relax the assumption of negligible water radius and to relate the effective filtration area of the solute to the geometric radius of the pore (Hobbie, 1978; Villars and Benedek, 1974; Renkin, 1954). Although models assuming hydrodynamic flow through pores have been useful, their applicability to osmotic flow across biological membranes remains an open question.

B. Lipid Bilayer Membranes: the Dissolution-Diffusion Model

There are two types of lipid bilayer membranes that are useful models of membranes: the lipid vesicle and the planar bilayer membrane. The lipid vesicle is a small spherical shell of lipid usually produced by sonicating a dispersion of lipid in water. Planar bilayer membranes consist of a thin film of phospholipids formed over a small hole in a partition between two aqueous compartments. The film is formed by "painting" the hole with a nonpolar solvent containing the lipids, which then spontaneously form a bilayer to fully separate the two solutions. This arrangement allows measurement of electrical and permeability properties of the bilayer. In the case of the liposome, the inner compartment is exceedingly small and experimentally inaccessible. In both cases, there are no components of the membranes other than the lipid.

1. Osmotic and Pressure-Driven Flow for the Dissolution-Diffusion Model: P_f

One model of water permeation through the lipid bilayer supposes that the water dissolves in the lipid phase and crosses the membrane by simple diffusion. In this model, the solution in contact with the membrane is one phase, while the hydrophobic core of the membrane is a second phase. Equilibrium of water in the solution phase with water in the membrane phase is described by equating the chemical potential of water in the two phases

$$
\mu^0_{w\,(\text{solution})} + RT \ln X_{w\,(\text{solution})} + P\overline{V}_{w\,(\text{solution})}
$$
$$
= \mu^0_{w\,(\text{membrane})} + RT \ln X_{w\,(\text{membrane})} + P\overline{V}_{w\,(\text{membrane})} \tag{51}
$$

where $X_{w\,(\text{solution})}$ and $X_{w\,(\text{membrane})}$ are the mole fractions of water in the solution in equilibrium with the membrane and in the membrane phase, respectively. The partition coefficient is defined as

$$
K_w = \frac{X_{w\,(\text{membrane})}}{X_{w\,(\text{solution})}}
$$
$$
= \exp \frac{\mu^0_{w\,(\text{solution})} - \mu^0_{w\,(\text{membrane})} + \Delta P\overline{V}_w}{RT} \tag{52}
$$

Consider the case where only osmotic pressure drives water flow and hydrostatic pressure across the membrane, ΔP, is 0. Since water generally partitions poorly into hydrocarbon solvents, we may assume that the mole fraction of water in the membrane phase is low. That is, the water concentration is dilute, and we may replace the mole fraction of water with its concentration

$$
C_{w\,(\text{membrane})} \simeq \frac{X_{w\,(\text{membrane})}}{\overline{V}_{\text{lipid}}} \tag{53}
$$

where $\overline{V}_{\text{lipid}}$ is the partial molar volume of lipid in the membrane. If the concentration of water immediately inside the membrane is in equilibrium with the solution in contact with the membrane, we combine Eqs. 52 and 53 to get

$$
C_{w\,(\text{membrane})} \simeq K_w \frac{X_{w\,(\text{solution})}}{\overline{V}_{\text{lipid}}} \tag{54}
$$

For dilute solutions, this is approximated by

$$
C_{w\,(\text{membrane})} \simeq K_w \frac{\left(1 - \overline{V}_w C_s\right)}{\overline{V}_{\text{lipid}}} \tag{55}
$$

where C_s is the solute concentration. The concentration of water immediately inside the left side of the membrane, $C_{w,L}$, is given by Eq. 55 where C_s is the concentration of sol-

ute in the solution on the left side of the membrane. A similar expression pertains to the concentration of water immediately inside the membrane on the right side ($C_{w,R}$). From the concentrations of water at both faces of the membrane, its diffusion across the membrane is given by Fick's law

$$
J_w = -D^m_w \frac{\left(C_{w,L} - C_{w,R}\right)}{(0 - \delta)} \tag{56}
$$

where D^m_w is the diffusion coefficient of water in the membrane phase. Substitution from Eq. 55 into Eq. 56 gives

$$
J_w = -\frac{D^m_w K_w \overline{V}_w \left(C_{s,L} - C_{s,R}\right)}{\overline{V}_{\text{lipid}} \delta}
$$
$$
= -\frac{D^m_w K_w \overline{V}_w \Delta C_s}{\overline{V}_{\text{lipid}} \delta} \tag{57}
$$

The flux of water, J_w, is in units of moles of water per second per cm^2 of membrane. It is converted to units of volume flow by multiplying by $\overline{V}_w$, as in Eq. 34, to obtain

$$
J_v = \overline{V}_w J_w = -\frac{D^m_w K_w \overline{V}^2_w \Delta C_s}{\overline{V}_{\text{lipid}} \delta}
$$
$$
= -\frac{D^m_w K_w \overline{V}^2_w}{\overline{V}_{\text{lipid}} RT\delta} RT\Delta C_s \tag{58}
$$

The last term on the right is the osmotic pressure difference, $\Delta\pi$. This equation relates the volume flux to the osmotic pressure difference when the mechanism of water flow is dissolution and diffusion. Comparison to the earlier description of osmotic flow by Eq. 18 allows us to identify L_p as

$$
L_p = \frac{D^m_w K_w \overline{V}^2_w}{\overline{V}_{\text{lipid}} RT\delta} \tag{59}
$$

From the definition, $P_f = L_p RT/\overline{V}_w$, we get the expression for P_f as

$$
P_f = \frac{D^m_w K_w \overline{V}_w}{\overline{V}_{\text{lipid}} \delta} \tag{60}
$$

Equation 58 was derived for an osmotic gradient ($\Delta C_s > 0$) in the absence of a hydrostatic gradient ($\Delta P = 0$). The expression relating volume flux and pressure when $\Delta C_s = 0$ can be derived by returning to Eq. 51 and setting the mole fractions of water on the two sides of the membrane equal, while the pressures differ. The result is that exactly the same L_p is derived for pressure-driven flow as for osmotic flow when the mechanism is by rapid dissolution of water followed by slow diffusion through the lipid membrane phase (Finkelstein, 1987).

2. Diffusional Water Permeability through Lipid Membranes: P_d

The permeability of lipid membranes to a diffusional water flux is expressed as

$$J_w = P_d \Delta C_w \tag{61}$$

where P_d is the diffusional permeability and ΔC_w is the difference in water concentration across the membrane. The overall permeation of the membrane by water is a consequence of three steps: dissolution into the membrane phase at the left interface; diffusion across the membrane phase; and reversal of dissolution at the right interface. If we assume, as we did in the derivation of P_f, that the rate-limiting step is diffusion through the membrane phase, then Eq. 61 may be written as

$$J_w = D_w^m \frac{\Delta C_{w \, (membrane)}}{\delta} \tag{62}$$

From Eq. 54, this is

$$J_w = \frac{D_w^m K_w}{\overline{V}_{lipid}} \frac{\Delta X_{w \, (solution)}}{\delta} \tag{63}$$

Because $\Delta X_w = \overline{V}_w \Delta C_w$, this becomes

$$J_w = \frac{D_w^m K_w \overline{V}_w}{\overline{V}_{lipid} \, \delta} \Delta C_w \tag{64}$$

P_d can be identified by comparing Eqs. 64 and 61

$$P_d = \frac{D_w^m K_w \overline{V}_w}{\overline{V}_{lipid} \, \delta} \tag{65}$$

3. P_f/P_d Ratio for the Dissolution–Diffusion Model

The expressions for P_f in Eq. 60 and P_d in Eq. 65 derived for a lipid membrane under identical assumptions (equilibrium at the interfaces with relatively slow diffusion across the membrane), indicate that the P_f/P_d ratio for diffusive flow of water across lipid membranes should be 1.0. Cass and Finkelstein (1967) measured the osmotic and diffusive permeability of planar lipid bilayers and found that, within experimental uncertainty, the ratio was indeed 1.0. The uncertainty arose mainly in the determination of P_d because of the presence of unstirred layers adjacent to the planar lipid bilayer. These unstirred layers are an additional diffusional barrier that affects the experimental determination of P_d much more than P_f. In the flux equations, permeability appears as a conductance relating a flux (J_w or J_s) to a driving force (ΔC_w or ΔC_s). Thus, the inverse of permeability is like a resistance. For a membrane in series with unstirred layers, the total resistance is the inverse of the observed permeability, which is the sum of the resistances offered by the individual barriers, the membrane and the unstirred layers. We write this as

$$\frac{1}{P_{d \, (obs)}} = \frac{1}{P_d} + \frac{1}{P_u} \tag{66}$$

From this equation, it is plain that if P_u, the combined permeability of the unstirred layers on both sides of the membrane, is less than infinity, the observed P_d will be less than the actual P_d of the membrane alone. Since diffusion of water through the unstirred layer is given by Fick's law, Eq. 66 may be rewritten as

$$\frac{1}{P_{d \, (obs)}} = \frac{1}{P_d} + \frac{\delta_L + \delta_R}{D_w} \tag{67}$$

where δ_L and δ_R are the equivalent unstirred layer thickness on the left and right sides of the membrane.

Unstirred layers also can affect measurement of the coefficients for pressure gradient-driven flow, P_f and L_p, but the error introduced is much less than for concentration gradient-driven flow, P_d (Barry and Diamond, 1984; Finkelstein, 1987), and was safely ignored in evaluating the P_f/P_d ratio (e.g., Cass and Finkelstein, 1967). The osmotic flow sweeps solute toward the membrane on the side with lower osmolarity and away from the membrane on the other side. As long as convection is faster than diffusion, this diminishes the transmembrane osmotic gradient, reducing J_v for the apparent $\Delta \pi$ and causing underestimation of P_f and L_p. The observed P_f, $P_{f \, (obs)}$, is given as

$$P_{f(obs)} = P_{f(membrane)} \exp\left(-J_v \delta/D_s\right) \tag{68}$$

where $P_{f \, (membrane)}$ is the true membrane parameter, δ is the unstirred layer thickness, J_v is the volume flux, and D_s is the diffusion coefficient of the osmolyte. The error can be minimized by determining P_f with a small $\Delta \pi$ so that J_v is small. Because P_f and L_p are proportional (Eq. 36), the errors are also proportional.

The value of P_d and P_f varies with the lipid composition of the membrane and ranges from 1×10^{-5} cm·s^{-1} to 5×10^{-3} cm·s^{-1} (Deamer and Bramhall, 1986). Cholesterol, which generally reduces the fluidity of lipid bilayers, reduces P_d and P_f progressively with increasing cholesterol content (Finkelstein and Cass, 1968). The **unidirectional** flux across a lipid membrane equals $P_d \times C_w$. Using a typical P_d of 1×10^{-3} cm·s^{-1} and a C_w of 55 mol·L^{-1}, the unidirectional flux across a membrane is 5.5×10^{-5} mol·cm^{-2}·s^{-1}. By comparison, the unidirectional flux of water across a distance $\delta = 5$ nm (approximately the thickness of the lipid bilayer) can be calculated as $(D_w \times C_w)/\delta$. Using 3×10^{-5} cm^2·s^{-1} for D_w, the unidirectional flux of water in water is 3.3 mol·cm^{-2}·s^{-1}. Thus, water flux through the membrane is about 60 000 times slower than that through water. Nevertheless, the water flux is still enormous. Taking 0.7 nm^2 as the average area of a typical phospholipid in the bilayer, the unidirectional water flux corresponds to about 2.2×10^5 water molecules passing each phospholipid molecule each second.

C. Flow through Narrow Pores: P_f / P_d Ratio

The equations derived earlier for P_f for a porous membrane required the assumption that the pores were large enough to allow laminar flow as described by the Poiseuille equation. Suppose that the pores are so narrow that water passes through the pores in single file. It is clear that laminar flow cannot occur here, and the Poiseuille equation does not apply. As shown in Fig. 11, the narrow pore restricts free diffusion in the pore because diffusion of one water molecule from one position in the pore to the next requires its neighbor to move away to provide a vacancy. In this way, diffusion within the restricted geometry of the pore becomes a collective property of all of the molecules in the pore. The likelihood that a tracer molecule will diffuse all the way

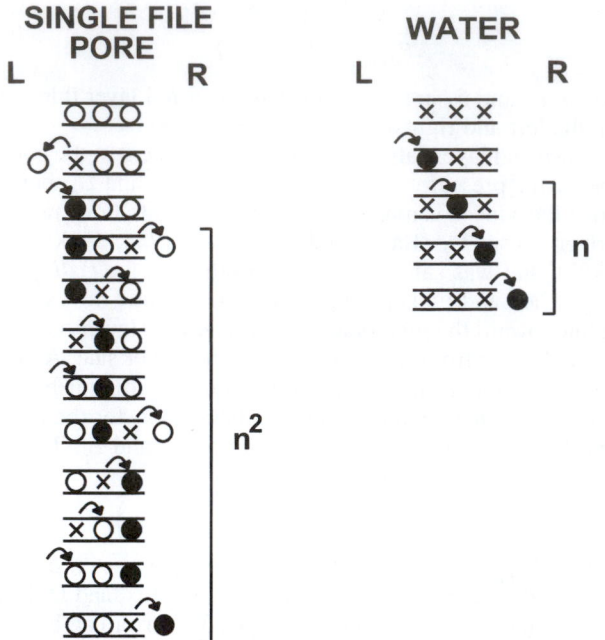

FIGURE 11. Jumping model of diffusion of water through a narrow pore. In the model of free water diffusion (right), tracer water (solid circles) makes successive jumps from one vacancy (×) to another. In order to move three places, it must make three jumps. In a narrow pore (left), tracer water cannot move to the next position in the pore unless a vacancy is present because unlabeled water (open circles) cannot move out of the way. Movement of tracer water from one site to the next in the pore requires a vacancy to diffuse all the way through the pore. Since the vacancy requires three steps to diffuse through the pore, tracer water diffusion through three steps in a narrow pore requires nine jumps. The result is that single-file diffusion becomes a collective property of the single file, and is slower than free water diffusion.

through the pore will depend on the number of water molecules in the pore, as movement of tracer water (solid circle) through the pore requires the movement of a vacancy (×) all the way through the pore.

Suppose that there are N water molecules in a pore. We may assume that the length of the pore, δ, is proportional to N, and that the water molecules reside in more or less specific positions separated by δ/N. In the free diffusion of liquid water in the bulk solution, according to Fick's first law of diffusion, the flux of water is proportional to $1/\delta$ or $1/N$. In the case of a single-file diffusion through a narrow pore, the diffusion of a vacancy through the water within the pore looks exactly like the diffusion of water through a series of vacancies in free diffusion. Thus, the flux of the vacancy also is proportional to $1/N$. The movement of tracer from one position to the next in the pore requires the diffusion of a vacancy all the way through the pore, so that the flux of tracer one step is proportional to $1/N$. In order to diffuse all the way through the pore, the tracer must make N such jumps. The unidirectional flux of tracer over N steps of equal flux is given by Stein (1976):

$$J_{0 \to N} = \frac{J_{ij}^N}{N J_{ij}^{N-1}} = \frac{1}{N} J_{ij} \qquad (69)$$

where $J_{0 \to N}$ is the unidirectional flux of tracer across N steps (all the way through the membrane) and J_{ij} is the unidirectional flux of tracer over one step. Since J_{ij} requires diffusion of a vacancy all the way through the pore, it is proportional to $1/N$. The unidirectional diffusional flux of tracer through a single-file pore is thus proportional to $1/N^2$, rather than $1/N$. Pressure-driven flow, on the other hand, remains inversely proportional to pore length. The theoretical analysis of both pressure-driven flow and diffusive flow through a single-file pore has led to the conclusion that the proportionality constants relating P_f to $1/N$ and P_d to $1/N^2$ are identical. The result, for a single-file pore, is

$$\frac{P_f}{P_d} = N \qquad (70)$$

where N is the number of water "binding sites" within the pore (Finkelstein, 1987). A more recent theoretical analysis suggests that this result is true near equilibrium, but the ratio of osmotic to diffusive permeability may exceed N for a membrane far from equilibrium (Hernandez and Fischbarg, 1992).

D. Mechanism of Water Transport across Lipid Bilayer Membranes

The previous discussion suggests that water transport across lipid bilayer membranes occurs by rapid dissolution at the membrane interface followed by diffusion across a hydrocarbon-like interior. Three observations strongly support this mechanism: (1) the P_f/P_d ratio, after correction for unstirred layers, appears to be close to 1.0 (Cass and Finkelstein, 1967; Andreoli and Troutman, 1971); (2) insertion of pore-forming antibiotics such as nystatin, amphotericin B or gramicidin A increases the P_f/P_d ratio to small values of N (between 3 and 5) (Holz and Finkelstein, 1970; Rosenberg and Finkelstein, 1978) and (3) the activation energy for water transport, typically 10–20 kcal·mol^{-1}, is larger than the activation energy for free water diffusion, ~4 kcal·mol^{-1}, which should apply if water moves through a pore (Solomon, 1972; Fettiplace and Haydon, 1980). However, experimental studies of water permeation across liposomes appear to give a different answer. The P_f of liposomes, measured by turbidimetric methods, is not affected by the chain length of the lipids or the degree of saturation (Carruthers and Melchior, 1983; Jansen and Blume, 1995), whereas water solubility in hydrocarbons is affected by chain length (Schatzberg, 1963). Further, the activation energy of P_f in liposomes is only 3.2 kcal·mol^{-1} (Carruthers and Melchior, 1983), similar to the activation energy for free water diffusion. The P_f of liposomes is discontinuous near phase transitions of the lipid and P_f/P_d ratios from 7 to 23 have been reported (Jansen and Blume, 1995). From these observations, it appears that liposomes do not behave as planar lipid bilayers, perhaps because of membrane defects induced by the marked curvature of the membrane in these small structures. The activation energy has been taken as diagnostic of whether water traverses the membrane through pores or by diffusion through the lipid itself (Finkelstein, 1987; Verkman, 1993). Low activation energies of water transport are associated with a high P_f or L_p in membranes containing water channels or pores, whereas high activation energies are associated with a low P_f or L_p, indicating diffusional water transport in membranes without pores or when pores are blocked by mercurials (Table 2).

In the dissolution-diffusion mechanism, water encounters a minimum of three sequential barriers: dissolution on one side of the membrane, transport across the hydrocarbon-like interior, and then removal from the membrane on the opposite side. Assuming rapid equilibration at the interfacial regions is equivalent to assuming that the barriers there are insignificant compared to the barrier of diffusion, but there is no *a priori* reason to make this assumption. The alternate view proposes that the rate-limiting step is lateral movement of the phospholipid head groups that creates a transient defect required for penetration of water into the interfacial region of the membrane (Trauble, 1971; Haines, 1994). Water transport through the hydrophobic core requires vacancies within the bilayer that form when hydrocarbon chains make *gauche-trans-gauche* kinks caused by the rotation of carbon-carbon bonds in the hydrocarbon tails. Kinks in the hydrocarbon tails propagate rapidly down acyl chains and provide sufficient space for water. Experimental support for this model comes from studies showing addition of cardiolipin to phosphatidylcholine liposomes decreases P_f without changing bilayer fluidity by stabilizing head group interactions (Shibata et al., 1994). These two models make very different assumptions, and the detailed mechanism of water permeation through lipid bilayers remains uncertain.

TABLE 2 Hydraulic Conductivity and Its Apparent Activation Energy

Tissue	L_p 10^{-10} L$\cdot$N$^{-1}\cdot$s^{-1}	E_a kcal$\cdot$mol^{-1}
Water channels		
RBC, human	18.0	3.9
RBC, beef	18.2	4.0
RBC, dog	23.0	4.3
Prox. tubule BLM-V, rabbit	21.9	2.5
Liposomes + AQP1	30.8	3.1
Water channels + mercurials		
RBC, human + PCMBS	1.3	11.6
Prox. tubule BLM-V, rabbit + Hg	4.4	8.2
Diffusional		
RBC, chicken	0.6	11.4
Intestinal brush border, rat	0.9	9.8
Ventricle, rabbit	1.2	10.5
Liposomes	1.9	16.0
PC bilayer	1.6	13.0
PC/chol bilayer	0.4	12.7
Water self-diffusion	—	4.2

Hydraulic conductivity, L_p, and activation energy, E_a, are characteristic of the mechanism of water transport. High L_p and low E_a are typical of membranes containing functioning water channels, whereas low L_p and high E_a indicate diffusional water transport. Note that 10^{-10} L$\cdot$N$^{-1}\cdot$s^{-1} [SI units] = $10^{-12}\cdot$cm$^3\cdot$dyn$^{-1}\cdot$s^{-1} [cgs units]. RBC, red blood cell; BLM-V, basolateral membrane vesicles; AQP1, aquaporin-1; PCMBS, *p*-chloromercuribenzene sulfonate; PC, phosphatidylcholine; chol, cholesterol. For references, see Suleymanian and Baumgarten (1996).

IV. Water Movement across Cell Membranes

In the prior sections, we considered osmotic- or pressure-driven flow and diffusive flow of water across membranes that were characterized as: (a) porous membranes with pores large enough to allow laminar flow; (b) lipid membranes with no pores, but allowing water permeation by a dissolution-diffusion mechanism; and (c) membranes containing narrow pores. We found expressions for the osmotic permeability, P_f, and the diffusive permeability, P_d, for each and found that the membranes could be distinguished in principle by the ratio P_f/P_d: large values of P_f/P_d indicate a porous membrane, $P_f/P_d = 1$ signifies a diffusive mechanism, and small values of $P_f/P_d > 1$ are characteristic of narrow pores. What are the permeabilities of real biological membranes, and what do these permeabilities tell us about the routes of water transport through membranes?

A. Rate of Water Exchange: Experimental Measure of P_d

Paganelli and Solomon (1957) measured the diffusional exchange of water across erythrocyte membranes by rapidly mixing a suspension of the cells with an isotonic buffer with added tracer ^{3}H$_2$O. The mixture was forced down a tube and samples of the extracellular water were obtained by filtration at various distances, corresponding to various times of exchange. Paganelli and Solomon found that the half-time for exchange of ^{3}H$_2$O was 4.2 ms at room temperature. This means that 90% of all of the water within an erythrocyte is exchanged with extracellular water every 14 ms. This is an extraordinarily rapid rate of exchange. Erythrocytes are sufficiently small that the presence of an unstirred layer within the cells does not appreciably affect the determination of P_d (i.e., $(\delta_L + \delta_R)/D_w << 1/P_d$; see Eq. 67).

Diffusional exchange of water across erythrocyte membranes also can be measured by nuclear magnetic resonance (NMR) spectroscopy. Relaxation of the nuclear spin states of hydrogen is much slower inside a cell than outside when a relatively impermeant paramagnetic ion like Mn^{2+} is added to the extracellular solution. This allows calculation of P_d because the relaxation of the spin states is then effectively limited by permeation through the membrane. The values of P_d for the erythrocyte determined by isotopic or NMR methods cluster around 4×10^{-3} cm$\cdot$s^{-1} (Solomon, 1989).

B. Rate of Osmotic Flow: Experimental Measure of P_f and L_p

According to Eq. 23, the hydraulic conductivity, L_p, can be determined experimentally as

$$L_p = \left(\frac{-Q_v}{A \sigma \Delta \pi} \right)_{\Delta P = 0} \quad (71)$$

The osmotic permeability, P_f, can then be calculated as $P_f = L_p RT/\bar{V}_w$ (Eq. 36). The experimentally determined values reflect the rate of water flow, Q_v, in the presence of a known osmotic pressure difference, $\Delta \pi$, produced by a solution with a known reflection coefficient, $\sigma \approx 1.0$, over a known surface area, A.

The rate of water movement into or out of erythrocytes in response to mixing with hypertonic or hypotonic media has been measured by using light scattering as an index of erythrocyte volume. The experiments are similar to those used to determine P_d; a suspension of erythrocytes is mixed with media of defined osmolality, and the mixture then flows down a tube passing through an observation cell. Light scattering is monitored at known distances down the tube, and cell volume changes are calculated from changes in light scattering. This and other methods give values for L_p that cluster around 1.8×10^{-11} cm$^3\cdot$dyn$^{-1}\cdot$s^{-1} (Solomon, 1989). Using $R = 8.314 \times 10^7$ dyn$\cdot$cm$\cdot$mol$^{-1}\cdot$K^{-1}, $\bar{V}_w = 18$ cm$^3\cdot$mol^{-1}, and $T = 298$ K, this average value of $\bar{L}_p$ corresponds to a P_f of about 2.5×10^{-2} cm$\cdot$s^{-1}.

C. Water Channels in Biological Membranes

The ratio of P_f/P_d for the erythrocyte membrane described earlier is about 5. Although there is some uncertainty in this ratio, it is clearly in excess of the value of 1.0 predicted for a diffusive mechanism. This suggests that there are pores in the erythrocyte membrane. The actual value of P_f/P_d for the pore alone cannot be obtained from just this information, however, because the erythrocyte membrane is actually a mosaic of lipid bilayer and pores; water moves through both in parallel, so the permeabilities of the components add. Experiments that block the pore suggest that the P_f/P_d ratio for the pore alone is about 10 and that 90% of water flux is via the pore, whereas 10% crosses the lipid bilayer (Macey, 1979; Finkelstein, 1987).

Two distinct means of inhibiting water transport provide additional evidence for proteinaceous pores. Mercurial sulfhydryl reagents, such as HgCl$_2$, p-chloromercuribenzoate (PCMB) and p-chloromercuribenzene sulfonate (PCMBS), decrease the erythrocyte P_f by a factor of 10 and P_d by less than a factor of 2, and the osmotic and diffusional permeabilities become equal (Macey and Farmer, 1970; Macey et al., 1972). Concurrently, the activation energy for permeation is increased from about 4 kcal$\cdot$mol^{-1} expected for water-filled pores to >10 kcal$\cdot$mol^{-1}, a value typical of artificial lipid bilayers (see Table 2). Mercurials inhibit water transport primarily by targeting protein SH groups because their effect is fully and rapidly reversed by cysteine. The second inhibitor is radiation. High doses of radiation inhibit water transport in both erythrocytes (van Hoek et al., 1992) and renal brush-border membrane vesicles (van Hoek et al., 1991). The characteristics of radiation inactivation suggest that the target is the size of a 30-kDa protein and are inconsistent with the entire membrane or transient defects serving as the major water pathway.

Although these studies and others, especially in epithelia (e.g., Verkman, 1993), made it clear that water pores or channels must exist, until recently their identity and characteristics remained mysterious. Are what we call "water channels" simply water moving through open ion channels or through an ion-exchanger or ion-cotransporter? Are water channels specific, admitting water but excluding ions? These questions have been answered with the cloning and expression of a family of water channels called aquaporins (Agre et al., 1993; King and Agre, 1996; Verkman et al., 1996; Heymann et al., 1998; Nielsen et al., 1999).

1. Aquaporins

Water channels are related to the membrane integral protein (MIP) family (20–40% homology). The first water channel identified was cloned from a human bone marrow library by Preston and Agre (1991) based on the sequence of a purified protein of uncertain function. Initially called CHIP28 (channel-forming integral protein, 28 kDa), this protein was later redesignated aquaporin-1 (AQP1), and MIP-1 is referred to as AQP0. Ten mammalian aquaporins are known, and more than 100 related proteins have been found in amphibians, Drosophila, plants, E. coli, and yeast, but not all of these conduct water (Heyman et al., 1998; Nielsen et al., 1999; Heymann and Engel, 1999).

The cloned AQP protein exhibits all of the characteristics of a water channel. AQP1 expressed in Xenopus oocytes induces up to 30-fold increases in P_f that is blocked by HgCl$_2$ (Preston et al., 1992). Reconstitution of purified AQP1 protein in liposomes verified that AQP1 itself, rather than modulation of an endogenous oocyte membrane protein, was responsible for mercurial-sensitive water permeation (Zeidel et al., 1992). Furthermore, incorporation of AQP1 into liposomes reduced the activation energy of P_f from 16.1 kcal$\cdot$mol^{-1}, characteristic of permeation through the bilayer, to 3.1 kcal$\cdot$mol^{-1}, characteristic of water passing through water-filled pores. Moreover, AQP1 appears to be active and highly selective for water without requiring regulatory subunits or cofactors (Preston et al., 1992; Zhang et al., 1993).

There are about 2×10^5 AQP1 molecules per erythrocyte (Zeidel et al., 1992). Taking the erythrocyte's membrane area as 1.35×10^{-6} cm^2 (Solomon, 1989), this corresponds to a density of about 1.5×10^{11} AQP1 cm^{-2} or 1500 AQP1 µm^{-2}. Despite this remarkable density of pores, the water permeability of a human erythrocyte is only ~10 to 50 times greater than that of a phosphatidylcholine/cholesterol bilayer (Fettiplace and Haydon, 1980). Thus, 30–150 aquaporin channels are needed to equal the permeability of 1 µm^2 of bilayer.

AQP1 contains 269 amino acid residues and was postulated to have 6 membrane-spanning domains based on an analysis of hydrophilicity (Preston and Agre, 1991) and selective proteolysis of protein loops that face the intra- or extracellular side (Preston et al., 1994). The topology for the AQP1 and the other members of the AQP family is shown in Fig. 12. Mercurial-inhibition, N-glycosylation, and PKA phosphorylation sites have been identified. There is an internal homology between the halves of AQP designated repeat-1 and -2. The greatest homology among AQPs is in the segments surrounding asparagine-proline-alanine (NPA) motifs in loops B and E. These loops are thought to be arranged anti-parallel and are postulated to dip back into the membrane to form an hourglass-shaped pore represented in the cartoon in Fig. 12 (Jung et al., 1994). It is hypothesized that four highly conserved polar residues, E17, N76, N192, E142, comprise a critical portion of the water permeation pathway (Heymann et al., 1998).

The biochemical behavior of solubilized erythrocyte AQP1 suggested that it is a noncovalently-linked tetrameric structure, and electron microscopy of negatively stained AQP1 confirmed this (Walz *et al.*, 1994). Coexpression of Hg-sensitive and -insensitive recombinant proteins indicates that each AQP monomer forms an independent functioning pore, however (Preston *et al.*, 1993). An exception to the tetrameric design of AQPs is AQP4. AQP4 forms large multimeric square arrays in the end-feet of astrocytes that surround capillaries (Rash *et al.*, 1998). X-ray diffraction patterns at 3.5- and 6-Å resolution have been obtained from two-dimensional crystalline AQP1 arrays (Jap and Li, 1995; Walz *et al.*, 1997). Electron density contour maps confirmed a tetrameric arrangement of monomers, and each

monomer appeared to have a central low density core, presumably the permeation pathway, surrounded by six tilted high density regions thought to represent membrane-spanning α-helicies.

Immunohistochemistry, western blot (protein determination), and northern and *in situ* hybridization and RNase protection assays (mRNA determinations) have identified broad but only partially overlapping distributions of AQP homologs in regions where water permeability is high (Hasegawa *et al.*, 1993; Nielsen *et al.*, 1993; Zhang *et al.*, 1993; Umenishi *et al.*, 1996; Nielsen *et al.*, 1999; Ma and Verkman, 1999). For example, AQP1 is located in erythrocytes, renal proximal

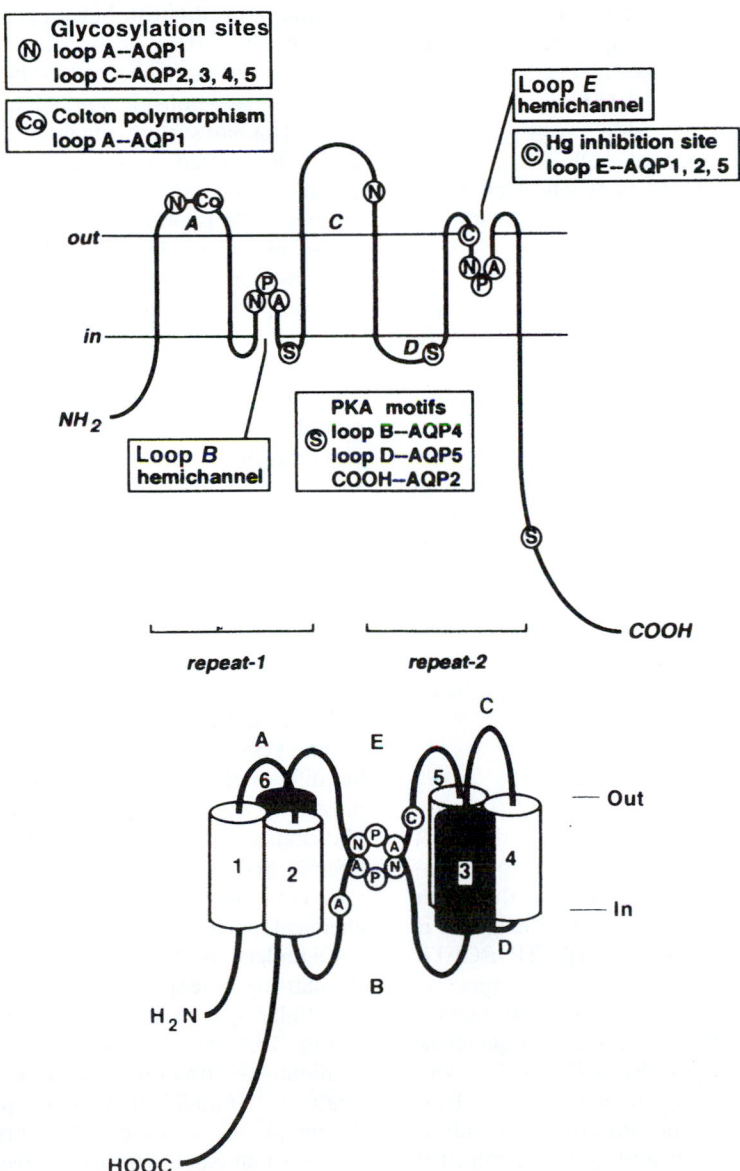

FIGURE 12. AQP family portrait. Cartoon illustrating the postulated topology of AQP as a structure with six membrane spanning α-helices (top). Cysteine (C) responsible for inhibition by mercurials and consensus PKA phosphorylation (S) and glycosylation (N) sites are also shown. Loops B and E are envisioned to dip into the membrane with the two NPA motifs forming the narrow neck of an hourglass-shaped, water-filled pore (bottom). Single letter codes are standard amino acid abbreviations. The position of the Colton polymorphic blood group antigen (Co) is noted. [Reproduced from King and Agre (1996) with permission, from the *Annual Review of Physiology*, volume 58, © 1996, by Annual Reviews, Inc., and from Jung *et al.*, (1994) with permission.]

tubules and the descending thin limb of the Loop of Henle (but not in the collecting duct, where water permeability is controlled by vasopressin), the choroid plexus, the iris, ciliary and lens epithelia and corneal endothelium of the eye, lung alveolar capillaries and epithelium, red splenic pulp (with erythrocyte precursors), colonic crypt epithelium, and nonfenestrated capillary and lymphatic endothelium in a number of organs including cardiac, skeletal, and smooth muscle. The exception with regard to broad distribution is AQP2. AQP2 underlies the vasopressin-regulated water permeation pathway of the renal collecting duct apical membrane that maintains water balance (Nielsen *et al.*, 1995; Nielsen *et al.*, 1999). Mutation of AQP2 is responsible for inherited nephrogenic diabetes insipidus, and targeting and expression defects may cause acquired forms of this disease. On the other hand, AQP2 expression is upregulated in pregnancy and congestive heart failure, states associated with enhanced water retention (Nielsen *et al.*, 1999).

Several cautions are warranted when drawing physiological interpretations from the localization of AQP. For example, AQP1 mRNA is higher in cardiac homogenate than in homogenate from any other organ (Umenishi *et al.*, 1996), but functionally, the story is different. In isolated myocytes, L_p is very low, and the activation energy is high, about 10 kcal·mol^{-1} (Suleymanian and Baumgarten, 1996) (see Table 2). Thus, AQPs do not significantly contribute to water transport across the myocyte membrane.

Organs comprise cells with diverse functions and often different requirements for water transport. Because of its localization in vascular tissue, AQP detected in homogenates should not be attributed to the principal cells of the organ without confirmation. Even careful immunohistochemistry and *in situ* hybridization may lead the physiologist astray. The problem is the high density of AQP necessary to significantly affect P_f. If, for example, the density of AQP1 in the erythrocyte membrane was 10 AQP1 μm^{-2} rather than 1500 AQP1 μm^{-2}, AQP1 would still be detected by modern techniques, but it would make a physiologically insignificant contribution to P_f.

2. Other Channels and Transporters

The identification of specific water channels does not exclude the possibility that water flux through ion channels or transporters significantly contributes to the water permeability of the membrane. One interesting example is the cystic fibrosis conductance regulator (CFTR). CFTR functions as a cAMP-regulated Cl$^-$ channel. Hasegawa *et al.* (1992) recently found that CFTR expressed in oocytes also acts as a water channel with an estimated single channel P_f comparable to that of AQP1. Both P_f and Cl$^-$ conductance were increased by cAMP. The effect of the CFTR channel on P_f is large but not unique. Pore-forming antibiotics such as gramicidin, nystatin, and amphotericin, are reported to induce a more modest P_f. Transporters also may contribute to P_f. Fischbarg *et al.* (1990) found that expression of Na$^+$-independent glucose transporters in oocytes increases water permeability, and Solomon *et al.* (1983) suggested that some of the water permeability of erythrocytes was contributed by the Cl$^-$-HCO$_3^-$ exchanger.

However, water flux mediated by these transporters does not account for the macroscopic properties of water transport.

V. Regulation of Cell Volume under Isosmotic Conditions

A. Gibbs-Donnan Equilibrium

Because ions exert an effective osmotic pressure, the distribution of ions affects water flow across the cell membrane and thus cell volume. The starting point for understanding these effects is the theoretical ideas of Gibbs that were first demonstrated experimentally by Donnan. The phenomenon is now called **Gibbs-Donnan equilibrium** or simply **Donnan equilibrium.** Macknight and Leaf (1977) elegantly describe the history of how these ideas were applied to cell volume regulation, and Overbeek (1956) provides a detailed derivation and considers nonideal solution behavior.

In a Donnan system, the membrane permits the movement of small charged solutes (e.g., K$^+$ and Cl$^-$) between two compartments but restricts the movement of large charged species such as proteins, which are usually polyvalent and negatively charged at intracellular pH. It is the inability of one (or more) charged species to distribute freely between the compartments that profoundly influences the distribution of the mobile ions and, consequently, water. For a cell, it is the cell membrane that restricts the movement of large ions. A membrane is not required to establish a Donnan equilibrium, however. All that is needed is a means of restricting one charged species to a single compartment. Donnan equilibria can arise in gels consisting of charged structural components (e.g., an ion-exchange column) because the charges on the gels are fixed in place. Donnan equilibria can also arise in cells after their membranes have been removed, leaving a cytoplasmic "gel" consisting of interconnected cellular proteins. A Donnan system represents a true equilibrium, although as we will see, an equilibrium is not attained under the usual biological conditions.

There are three important characteristics in a Donnan equilibrium system: (1) an unequal distribution of ions; (2) a potential difference between the compartments; and (3) an osmotic pressure. These characteristics can be understood from the application of a basic thermodynamic principle to the system. Species that can cross the membrane, including both mobile ions and water, will distribute themselves so that their electrochemical potential (μ_i) is the same in both compartments.

To understand how a Donnan equilibrium develops, imagine starting an experiment by "filling" the cell and the extracellular space with solutions of concentrations indicated in Fig. 13A. Since ions move down their electrochemical gradients, the first question is: what are the electrochemical gradients? Initially, there is no potential difference across the membrane because both internal and external solutions start with an equal number of positive and negative charges. There is no concentration gradient for K$^+$ because the amount of K$^+$ is the same on both sides of the membrane. Due to the presence of impermeant anions inside the cell, however, [Cl$^-$]$_o$ must be greater than [Cl$^-$]$_i$. Consequently, Cl$^-$ will enter the cell, moving down its electrochemical gradient, which is initially just the concentration gradient.

Inward movement of Cl^- makes the inside of the cell negative with respect to the outside. The developing inside negative potential has two consequences. First, it slows the rate of further influx of Cl^-. Second, the electrical gradient causes the accumulation of K^+ inside the cell. As K^+ moves down its electrochemical gradient in the direction set by the potential gradient, an opposing concentration gradient is created. Finally, an equilibrium is reached when the electrical and concentration gradients for both K^+ and Cl^- are equal in magnitude and opposite in direction. This satisfies the requirement that μ_K and μ_{Cl}, the electrochemical potential for K^+ and Cl^-, respectively, are the same in both compartments.

During the evolution of the equilibrium, the influx of K^+ and Cl^- are equal on a macroscopic scale and are said to be coupled by the requirement for macroscopic electroneutrality. This is simply a shorthand for the idea that any difference in rates of anion and cation transmembrane flux (a separation of charge) gives rise to a potential that equalizes the fluxes (e.g., if the influx of K^+ was greater than that of Cl^-, an inside positive potential would develop, slowing K^+ influx and accelerating Cl^- influx until the rates matched exactly). In the example in Fig. 13A, both $[K^+]_i$ and $[Cl^-]_i$ increased by ~78.5 mM. The total entry of K^+ and Cl^- cannot be precisely equal,

however, because establishing a potential difference implies that a separation of charge, albeit quite small, must have taken place. If this were a spherical cell with a radius of 20 μm, an inequality of the K^+ and Cl^- fluxes of less than 1 ion per 100 000 would cause a potential difference of 100 mV.[1]

The basis for Donnan equilibrium can be expressed in terms of the electrochemical potentials[2] of ions that cross the membrane, here K^+ and Cl^-. For ideal solutions, μ is given by

$$\mu_{Ko} = \mu^0 + RT \ln\left[K^+\right]_o + z_K \mathcal{F}\psi_o$$
$$\mu_{Ki} = \mu^0 + RT \ln\left[K^+\right]_i + z_K \mathcal{F}\psi_i \tag{72a}$$

and

$$\mu_{Clo} = \mu^0 + RT \ln\left[Cl^-\right]_o + z_{Cl} \mathcal{F}\psi_o$$
$$\mu_{Cli} = \mu^0 + RT \ln\left[Cl^-\right]_i + z_{Cl} \mathcal{F}\psi_i \tag{72b}$$

where the subscripts o and i represent the outside and inside of a cell, μ^0 is the chemical potential of the standard state, ψ is the potential of the compartment, z_i is the valence of species i, and R, T, and $\mathcal{F}$ have their usual meanings. At equilibrium, each permeant species distributes so that μ is

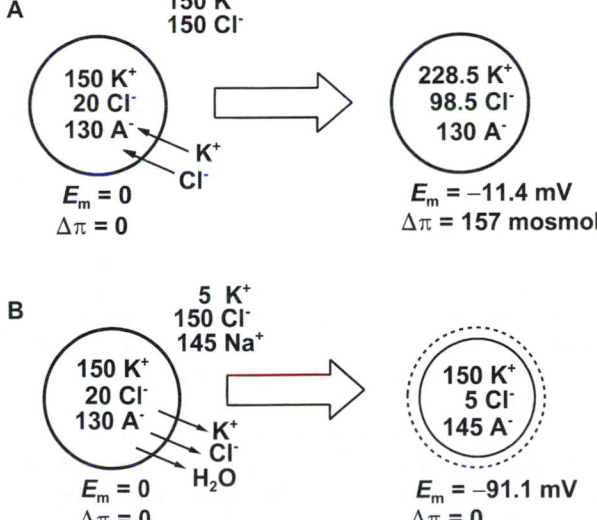

A

150 K^+
150 Cl^-

150 K^+
20 Cl^-
130 A^-

228.5 K^+
98.5 Cl^-
130 A^-

K^+
Cl^-

$E_m = 0$
$\Delta\pi = 0$

$E_m = -11.4$ mV
$\Delta\pi = 157$ mosmol/L

B

5 K^+
150 Cl^-
145 Na^+

150 K^+
20 Cl^-
130 A^-

150 K^+
5 Cl^-
145 A^-

K^+
Cl^-
H_2O

$E_m = 0$
$\Delta\pi = 0$

$E_m = -91.1$ mV
$\Delta\pi = 0$

FIGURE 13. Development of a Donnan equilibrium. (A) Cell bathed in 150 mM KCl is "filled" with 150 mM K^+, 20 mM Cl^-, and 130 mM A^-. Initially, there is no potential or osmotic gradient (left). K^+ and Cl^- enter the cell until $[K^+]_o \cdot [Cl^-]_o = [K^+]_i \cdot [Cl^-]_i$, and the ionic fluxes establish a potential and osmotic gradient (right). An equilibrium is achieved only if the membrane is rigid and cell volume is constant. With a biological membrane, however, water enters the cell, cell volume increases, and the $[K^+]_i \cdot [Cl^-]_i$ decreases. This causes further entry of solute and solvent, and the cycle repeats until the cell bursts. (B) Double-Donnan system. Part of the extracellular K^+ is replaced by Na^+, which is assumed to be impermeant. With impermeant ions on both sides of a membrane (left), equilibrium is attained (right) without an osmotic pressure gradient. In this case, K^+, Cl^-, and water leave the cell, until at equilibrium $[K^+]_o \cdot [Cl^-]_o = [K^+]_i \cdot [Cl^-]_i$. The amount of A^- in the cell is fixed, and thus, the decrease in cell volume increases $[A^-]_i$.

---

[1]The amount of charge, q, necessary to establish a potential difference, E_m, is related to the specific capacitance of the membrane, C_m, and its area, A, by

$$q = C_m A E_m$$

Assuming a specific membrane capacitance of 1 μF/cm², as is typical of most biological membranes, the amount of charge necessary to develop a potential of 100 mV in a spherical cell with a radius of 20 μm is

$$q = (1\times10^{-6}\ F/cm^2)[\ 4\pi(20\times10^{-4}\ cm)^2](0.1\ V)\left(\frac{C/V}{F}\right)$$

$$q = 5.0\times10^{-12}\ C$$

This can be converted to a change in the concentration, ΔC, of ions in the cell

$$\Delta C = (q/\mathcal{F})/\vartheta$$

where $\mathcal{F}$ is Faraday's constant, and ϑ is used for the volume of the cell here to distinguish it from voltage. For the same 20-μm-radius cell

$$\Delta C = (5.0\times10^{-12}\ C/96\ 500\ C/mol)/[\frac{4}{3}\pi(20\times10^{-4}\ cm)^3(\frac{1L}{1000\ cm^3})]$$

$$\Delta C = 1.5\times10^{-6}\ mol/L$$

Thus, a potential gradient of 100 mV would develop if ion influx were to increase $[K^+]_i$ by only 1.5 μM more than it increased $[Cl^-]_i$.

[2]Formally, as presented previously in Eq. 3 for the chemical potential of a neutral species, μ_i should be defined in terms of the mole fraction rather than the concentration of a component and an additional term, $P\overline{V}$, representing the pressure times the partial molar volume, should be added. For dilute solutions, the mole fraction of a solute closely approximates its concentration (see Eqs. 5 and 9). The $P\overline{V}$ terms are ignored in the derivation of the ion distribution for a Donnan equilibrium because to simplify we initially assume the membrane is rigid (see Fig. 13A), and then, we consider a situation without a pressure gradient where the $P\overline{V}$ terms cancel (see Fig. 13B). Derivations of Donnan equilibrium retaining the mole fraction and the pressure-volume terms can be found in Overbeek (1956) and Lakshminarayanaiah (1984).

identical inside and outside the cell. Equating the expressions for μ and simplifying gives

$$RT \ln \left[K^+ \right]_o + z_K \mathcal{F} \psi_o = RT \ln \left[K^+ \right]_i + z_K \mathcal{F} \psi_i$$

$$RT \ln \left[Cl^- \right]_o + z_{Cl} \mathcal{F} \psi_o = RT \ln \left[Cl^- \right]_i + z_{Cl} \mathcal{F} \psi_i \quad (73)$$

Membrane potential, E_m, is measured as $\psi_i - \psi_o$. Rearranging and substituting in for E_m and z gives

$$E_m = E_K = -\frac{RT}{\mathcal{F}} \ln \frac{\left[K^+ \right]_i}{\left[K^+ \right]_o}$$

$$E_m = E_{Cl} = -\frac{RT}{\mathcal{F}} \ln \frac{\left[Cl^- \right]_o}{\left[Cl^- \right]_i} \quad (74)$$

These expressions are the **Nernst equilibrium potentials** for K^+ (E_K) and Cl^- (E_{Cl}). Thus, when the mobile ions attain their equilibrium distribution in a Donnan system, the potential between the compartments, E_m, simultaneously equals both E_K and E_{Cl}. Equating the Nernst potentials and simplifying gives the ratio of intra and extracellular ions predicted by Donnan equilibrium

$$\frac{\left[K^+ \right]_i}{\left[K^+ \right]_o} = \frac{\left[Cl^- \right]_o}{\left[Cl^- \right]_i} \quad (75)$$

or expressed another way

$$\left[K^+ \right]_i \cdot \left[Cl^- \right]_i = \left[K^+ \right]_o \cdot \left[Cl^- \right]_o \quad (76)$$

In a Donnan equilibrium, the KCl product inside a cell equals the KCl product outside.

This simple rule implies that increasing extracellular K^+ (e.g., by replacing Na^+) will cause a cell obeying Donnan equilibrium to take up K^+ and Cl^- and swell. Ion movements in a number of tissues appear to follow Donnan equilibrium, at least for some conditions. Boyle and Conway (1941) made careful measurements of $[K^+]_i$, $[Cl^-]_i$, and cell water in frog sartorius muscles while varying extracellular KCl. For $[K^+]_o$ greater than 6 mM, the experimental ratio of the KCl products, $([K^+]_o \cdot [Cl^-]_o)/([K^+]_i \cdot [Cl^-]_i)$, was very nearly 1.0, as predicted by Eq. 76. The deviation at low $[K^+]_o$ occurs because Na^+ influx causes E_m to deviate from E_K. Accompanying changes in cell water in these experiments and in separate experiments in which K^+ replaced Na^+ also agreed with theory.

The expectations for a Donnan equilibrium are calculated for the system shown in Fig. 13A. When the intracellular and extracellular KCl products are equal, $E_m = E_K = E_{Cl} = -11.4$ mV. In addition, because the number of ions inside the cell is much greater than that outside, the Donnan system has established a significant osmotic gradient. Assuming ideal behavior, the osmolarity inside the cell, ~457 mosmol·L^{-1}, is about 1.5 times that outside, 300 mosmol·L^{-1}. The resulting osmotic pressure can be calculated as follows:

$$\Delta \pi = RT \Delta C \quad (77)$$

$$= (0.082 \text{ L·atm·K}^{-1} \cdot \text{mol}^{-1})(310 \text{ K})(0.457 - 0.300 \text{ mol·L}^{-1})$$

$$= 3.99 \text{ atm}$$

Because most cells are readily permeable to water, the osmotic pressure generated here would cause water to enter the cell. This influx of water dilutes the intracellular ion content, and the product $[K^+]_i \cdot [Cl^-]_i$ must fall below $[K^+]_o \cdot [Cl^-]_o$. As a result, more KCl would enter the cell, followed again by more water, in an endless cycle that would lead to destruction of the cell. That is to say, the simple Donnan system described in Fig. 13A fails to reach equilibrium when typical cell membrane properties are assumed.

B. Double-Donnan or Pump-Leak Hypothesis

How can we reconcile the failure of the simple Donnan equilibrium model (Fig. 13A) and the experimental studies demonstrating behavior consistent with Donnan equilibrium? The Donnan system can be stabilized in two ways. A hydrostatic pressure could be applied to balance the osmotic pressure and arrest transmembrane water movement. In view of the enormous pressures required, this is not a realistic solution for animal cells. Alternatively, equilibrium can be attained by restricting an ionic species to the extracellular compartment just as one is restricted to the intracellular compartment. This arrangement is referred to as a **double-Donnan** (Leaf, 1959) or **pump-leak system** (Tosteson and Hoffman, 1960) and is illustrated diagrammatically in Fig. 13B. Assume that both Na^+ and macromolecular anions, A^-, are restricted to the extracellular and intracellular compartments, respectively. Filling the cell with the same concentrations of K^+, Cl^- and A^- as before, we find now that $[K^+]_i \cdot [Cl^-]_i > [K^+]_o \cdot [Cl^-]_o$, and thus KCl must leave the cell to establish Donnan equilibrium. In the process, water follows the KCl, adjusting cell volume until the KCl, products are equal. A consequence of water flow is that the concentration of the impermeant intracellular ion exactly equals the concentration of the impermeant extracellular ion. At equilibrium, $E_m = E_K = E_{Cl} = -91.1$ mV. In contrast to the simple Donnan system, the osmotic pressure developed by intracellular macromolecules and their counterions (sometimes referred to as **colloid osmotic pressure**) is exactly balanced in the double-Donnan system by the osmotic pressure developed by ions restricted to the extracellular fluid and their counterions. As a consequence, there is no net osmotic pressure across the membrane, and the system is stable.

Another approach for considering the volume change expected in the system in Fig. 13B is to calculate the effective osmotic pressure of each compartment. We can suppose that the reflection coefficients, σ, are 1.0 for both A^- and Na^+ because they are impermeant, and that the reflection coefficients are 0 for both K^+ and Cl^- because they freely pass the membrane. Extracellular and intracellular osmotic pressures are given by

$$\pi = \sum_n \sigma_n C_n$$

$$\pi_o = 1 \cdot \left[Na^+ \right]_o + 0 \cdot \left[K^+ \right]_o + 0 \cdot \left[Cl^- \right]_o \quad (78)$$

$$\pi_i = 1 \cdot \left[A^- \right]_i + 0 \cdot \left[K^+ \right]_i + 0 \cdot \left[Cl^- \right]_i$$

Because only the impermeant species contribute to osmotic pressure, the only way for the cell to reach osmotic equilibrium, $\pi_o = \pi_i$, is to alter cell volume until $[Na^+]_o = [A^-]_i$. As a result, the ratio of $volume_{t=\infty}/volume_{t=0}$ must equal $[A^-]_i/[Na^+]_o$.

It is important to realize that an ion does not need to be impermeant to be **effectively** restricted to the extracellular space and thus counterbalance the osmotic pressure developed by intracellular macromolecules. Na^+ is permeant, but it adequately plays this role anyway. As long as the leak of Na^+ down its electrochemical gradient into the cell is matched by its transport back out, cell volume will remain stable. Consequently, the existence of a pump that actively extruded Na^+ against its concentration gradient was postulated to explain cell volume (Leaf, 1956), and subsequently the Na^+, K^+-ATPase, which extrudes 3 Na^+ while taking up 2 K^+ at the cost of ATP hydrolysis, was identified. Because energy is consumed to extrude Na^+, the cell is in a **steady state** rather than a true equilibrium.

C. Modulation of the Na⁺-K⁺ Pump

One implication of the double-Donnan or pump-leak model is that the Na^+-K^+ pump is ultimately responsible for cell volume regulation. Perturbations that alter passive Na^+ entry must lead to offsetting changes in the rate of Na^+ extrusion by the Na^+-K^+ pump, or cell volume will change. Because the K_m of the Na^+-K^+ pump, for intracellular Na^+ is close to the physiological $[Na^+]_i$, alterations in Na^+ influx automatically give rise to a compensatory modulation of Na^+ efflux. Nevertheless, metabolic or pharmacologic inhibition of the Na^+-K^+ pump should lead to a net gain of Na^+ and anions coupled by macroscopic electroneutrality and result in a swelling of the cell.

The effect of pump inhibition has been examined extensively (Macknight and Leaf, 1977; Macknight, 1988). The predicted cell swelling has been reported in many tissues, including brain slices, kidney slices, renal tubules, hepatocytes, and sheep erythrocytes, when the Na^+-K^+ pump is inhibited by cardiac glycosides (e.g., ouabain) or by depleting ATP. It is equally clear, however, that swelling after pump inhibition in renal cortex, liver slices, various muscle preparations, lymphocytes, and human erythrocytes is very slow or even absent, perhaps reflecting a low Na^+ permeability.

Several processes may affect the response of cell volume to Na^+-K^+ pump inhibition, and the outcome in some tissues depends on the experimental conditions. Rather than accumulating Cl^- with Na^+, cells might instead lose K^+ to satisfy macroscopic electroneutrality. An equivalent gain of Na^+ and loss of K^+ replaces one osmotically active particle with another, causing no change in cell volume. Closer consideration of this mechanism indicates that it can only be a holding action, however. The loss of intracellular K^+ eventually must lead to a reduction of the K^+ gradient and a less negative E_m. This will lead to accumulation of Cl^- with Na^+ and cell swelling. Nevertheless, a loss of K^+ must slow the swelling that otherwise would have occurred, and this would explain the absence of swelling in cells with appropriate Na^+ and K^+ permeabilities over the time course of experiments.

The possibility that a mechanism other than the Na^+-K^+ pump can extrude Na^+ has been considered. Although controversial, a cardiac glycoside-insensitive but metabolically dependent volume regulation mechanism in kidney that does not incorporate the Na^+-K^+ pump has been described (for review, see Macknight and Leaf, 1977). In addition, circulating erythrocytes from a number of carnivores, including dog, cat, bear, and ferret, lack a functioning Na^+-K^+ pump and must regulate their volume by a different mechanism (Sarkadi and Parker, 1991). In these cells, Na^+ efflux must depend on the gradient of other ions.

D. Isosmotic Volume Regulation

Although principles of the pump-leak or double-Donnan model are correct and still relevant to the regulation of cell volume, it has become apparent that neither the leak nor the pump is constant. Not only is the control of these fluxes more complex than originally envisioned, but a myriad of other transport processes also contribute to cell volume regulation. Stated simply, constant cell volume under isosmotic conditions implies an equality of intra- and extracellular osmolarity that is perpetuated by a continuous balance of the efflux and influx of osmolytes. The transport processes involved include, but are not limited to: (1) ion and organic osmolyte channels; (2) the Na^+-K^+ pump; (3) the Na^+-K^+-$2Cl^-$, K^+-Cl^-, and Na^+-Cl^- cotransporters, which transport the specified ions in one direction; (4) Na^+-dependent sugar and amino acid cotransport; (5) Na^+-Ca^{2+}; exchange, which exchanges 3 Na^+ for 1 Ca^{2+}; and (6) osmotically neutral exchangers that indirectly provide a net solute flux. For example, Na^+-H^+ exchange allows the cell to accumulate Na^+, but the H^+ removed is replaced by dissociation of H^+ from intracellular buffers. Cl^--HCO_3^- and Na^+-H^+ exchange can operate in parallel to mediate a net influx of Na^+ and Cl^- in exchange for H^+ and HCO_3^-, which are converted to CO_2 and H_2O by the action of carbonic anhydrase. It should be noted that CO_2 freely crosses the cell membrane ($\sigma = 0$) and does not directly contribute to solution tonicity. Also, the direct extrusion of water by these two parallel exchangers is negligible compared to the osmotic water gain caused by the accumulation of Na^+ and Cl^-.

Of the number of transporters that participate in volume regulation in any given type of cell, which are most important in regulating cell volume under isosmotic conditions? No simple answer can be offered. The importance of each process to cell volume regulation depends critically upon the tissue and species under consideration as well as the conditions. Moreover, transporters and channels are modulated by multiple signaling pathways, and they extensively interact by altering membrane potential or the concentration of the transported species. If the rates of ion transport are not correctly matched, cells will inappropriately shrink or swell. The precise maintenance of cell volume exemplifies the need for sensitive and complex regulatory mechanisms. Attempts to mathematically integrate the fluxes and study their interaction have been made based on the cell's requirement for macroscopic electroneutrality and osmotic equilibrium and the equations governing ion fluxes (Jakobsson, 1980). For erythrocytes, the nonideal behavior of hemoglobin has been added (Bookchin et al., 1989). Although these simplified models correctly predict a number of observations, they fail to explain others. In short, we remain a long way from a complete quantitative description of the processes underlying cell volume regulation. In the following sections, we will discuss three examples of isosmotic regulation of cell volume that illustrate some of the underlying principles.

1. Na⁺-K⁺-2Cl⁻ Cotransport in Heart

Recent findings indicate that the Na^+-K^+-$2Cl^-$ cotransporter plays a critical role in regulating cardiac myocyte cell volume under isosmotic conditions. As in other tissues, Na^+-K^+-$2Cl^-$ cotransport conveys osmolytes into cardiac cells under physiological conditions. Because **net** transmembrane fluxes control cell volume, a decreased osmolyte influx is equivalent to increased efflux. Therefore, inhibition of the Na^+-K^+-$2Cl^-$ cotransport by bumetanide, for example, favors a reduction of cell volume. Consistent with this idea, bumetanide decreases the volume of atrial and ventricular myocytes by about 10% in less than 5 min, and myocyte volume is stable at this new level (Drewnowska and Baumgarten, 1991). The Na^+-K^+-$2Cl^-$ cotransporter cannot operate without Na^+ and Cl^- in the extracellular fluid, and removing either ion renders bumetanide ineffective. These data imply that ion uptake by Na^+-K^+-$2Cl^-$ cotransport in the heart must be responsible for a significant osmolyte flux under isosmotic conditions and that other transport processes are incapable of fully compensating when this flux is removed. In contrast, myocyte volume was unchanged after inhibiting the Na^+-K^+ pump with 10 μM ouabain (Drewnowska and Baumgarten, 1991) or by cooling to 9 °C (Drewnowska et al., 1991) for 20 min. At least in the short term, cardiac cell volume in isosmotic solution is influenced more by Na^+-K^+-$2Cl^-$ cotransport than by the Na^+-K^+ pump.

Modulation of Na^+-K^+-$2Cl^-$ cotransport by intracellular messengers such as cGMP may provide a physiological means of modulating cell volume in heart tissue. Figure 14 shows the effects of elevating intracellular cGMP in three ways: (1) by adding 8-Br-cGMP, a membrane-permeant analog of cGMP; (2) by adding atrial natriuretic factor (ANF), a natriuretic, diuretic, and vasodilatory hormone released by the heart that elevates cGMP by activating guanylate cyclase; and (3) by adding sodium nitroprusside (SNP), a vasodilator, that also activates guanylate cyclase. In each case, cell volume decreased. Furthermore, blocking cGMP-specific phosphodiesterase with zaprinast (M&B22948) augmented the effect of ANF. Based on its sensitivity to bumetanide and the requirement for ions transported by Na^+-K^+-$2Cl^-$ cotransport, cGMP-dependent volume decreases were shown to be due to an inhibition of Na^+-K^+-$2Cl^-$ cotransport by cGMP (Clemo et al., 1992; Clemo and Baumgarten, 1995). Interestingly, **lowering** cGMP levels by inhibiting guanylate cyclase with LY83583 resulted in a small amount of cell swelling. Thus, changing cGMP from its physiological level in either direction altered cell volume. The mechanism and evidence for isosmotic regulation of cell volume in the heart are summarized in Fig. 15.

Does the same mechanism regulate cell volume under isosmotic conditions in other tissues? Perhaps it does in some cells, such as vascular endothelium, in which cGMP inhibits Na^+-K^+-$2Cl^-$ cotransport. In other cells, Na^+-K^+-$2Cl^-$ cotransport is stimulated by cAMP, cGMP, or a PKC-dependent pathway, and perhaps by a number of other signaling pathways

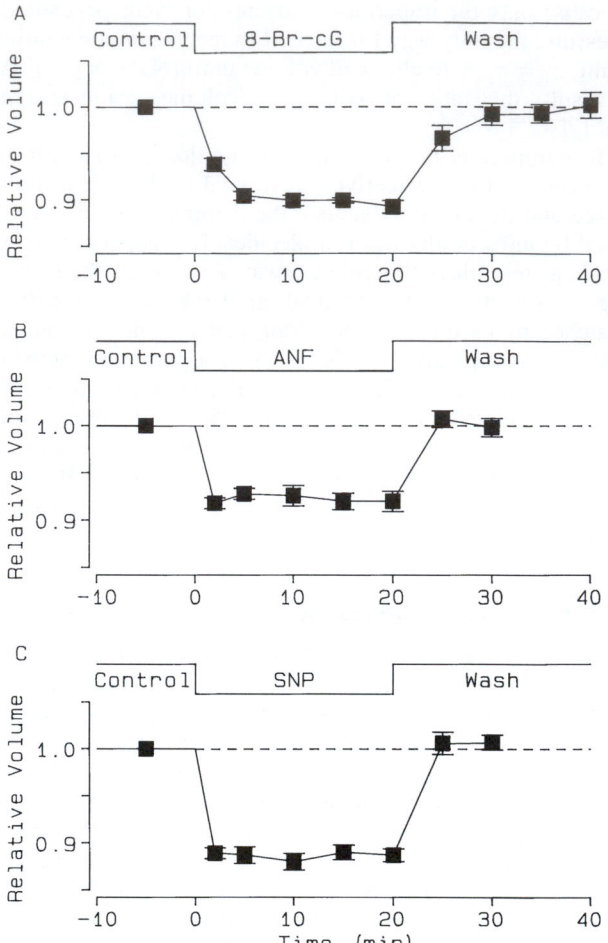

FIGURE 14. Changes in the volume of isolated ventricular myocytes during a 20-min exposure to: (A) 10 μM 8-Br-cGMP (8-Br-cG), a permeable cGMP analog; (B) 1 μM atrial natriuretic factor (ANF); and (C) 100 μM sodium nitroprusside (SNP). These agents reversibly decreased cell volume 11%, 8%, and 11%, respectively ($n = 5$ for each time point). The decreases in cell volume were caused by inhibition of Na^+-K^+-$2Cl^-$ cotransport by cGMP. Relative cell volume was measured and calculated as described in the legend of Fig. 1. (From Clemo et al. (1992). Reproduced from *The Journal of General Physiology,* 1992, **100,** 89–114, by copyright permission of The Rockefeller University Press.)

(Palfrey and O'Donnell, 1992; Palfrey, 1994). This diversity in the control of Na^+-K^+-$2Cl^-$ cotransport may be related to the variety of transporter isoforms that have been identified and cloned (Haas, 1994; Kaplan et al., 1996). Although the physiological significance of this diversity in the control of ion transport is not well understood, it must lead to diversity in the regulation of cell volume.

2. Hormones and Substrate Transport in Liver

Another interesting example of isosmotic volume regulation is found in hepatocytes. An impressive number of hormones induce either cell swelling or cell shrinkage at physiological concentrations, and these actions are related to their control of liver metabolism (Häussinger and Lang, 1991; Häussinger et al., 1994; Agius et al., 1994; Häussinger, 1998). Na^+-H^+ exchange, Na^+-K^+-$2Cl^-$ cotransport, and the Na^+-K^+ pump are

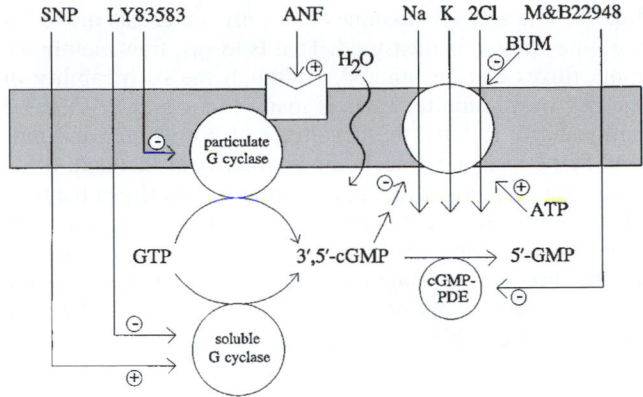

FIGURE 15. Schematic diagram of the action of atrial natriuretic factor (ANF) and cGMP on cardiac cell volume. Binding of ANF activates guanylate cyclase and increases intracellular cGMP levels. By one or more steps, cGMP inhibits Na^+-K^+-$2Cl^-$ cotransport. Reducing ion influx by this means is equivalent to increasing **net** ion efflux, and cell shrinkage ensues. LY83583 inhibits guanylate cyclase, thereby blocking the effect of ANF, and zaprinast (M&B22948) potentiates the effect of ANF by inhibiting cGMP-specific phosphodiesterase (PDE). Sodium nitroprusside (SNP) also increases cGMP levels, and bumetanide (BUM) directly inhibits the cotransporter; both cause cell shrinkage. (From Clemo *et al.* (1992) Reproduced from *The Journal of General Physiology*, 1992, **100**, 89–114, by copyright permission of The Rockefeller University Press.)

stimulated in rat liver cells by insulin. The net effect is that insulin increases $[K^+]_i$, $[Na^+]_i$ and $[Cl^-]_i$ and causes cells to swell by about 12%. This swelling is prevented by bumetanide, a blocker of Na^+-K^+-$2Cl^-$ cotransport, or by amiloride, a blocker of Na^+-H^+ exchange. In contrast, glucagon shrinks hepatocytes by about 14%. Instead of directly opposing the action of insulin, glucagon reduces cell volume by increasing K^+ and Cl^- efflux through ion channels. Other agents that swell hepatocytes include bradykinin and phenylephrine. Shrinking is initiated by adenosine, 5-HT, vasopressin, and cAMP.

Hepatocytes also swell as a result of Na^+-dependent amino acid cotransport in isosmotic media (Häussinger and Lang, 1991; Boyer *et al.*, 1992; Häussinger *et al.*, 1994; Häussinger, 1998). Exposure to amino acids that are accumulated with Na^+ (e.g., alanine, glutamine, glycine, hydroxyproline, phenylalanine, proline, and serine) causes cell swelling. These effects occur at amino acid levels found physiologically in the portal vein. For example, glutamine provokes up to a 10% swelling with a half-maximal effect at approximately 0.7 mM. Amino acid cotransport gives rise to an inward current, and Cl^- enters to maintain macroscopic electroneutrality. Instead of loading the cell, Na^+ is pumped out, mainly by the Na^+-K^+ pump, leaving an accumulation of K^+ and Cl^-. In contrast to these amino acids, substances not accumulated by the liver, such as glucose and leucine, do not affect hepatocyte volume. Cell swelling caused by Na^+-dependent amino acid cotransport has also been observed in intestine and renal proximal tubule.

3. Na^+-Ca^{2+} Exchange in Carnivore Erythrocytes

The Na^+-K^+ pump in erythrocytes from dogs, cats, ferrets, and bears ceases to function as cells mature. As a result, $[K^+]_i$

and $[Na^+]_i$ are similar to $[K^+]_o$ and $[Na^+]_o$. This poses a special problem for cell volume regulation. How can these cells offset the osmotic pressure generated by impermeant intracellular molecules and avoid swelling without a Na^+-K^+ pump to make Na^+ effectively impermeant? Carnivore erythrocytes solve this dilemma by extruding 3 Na^+ in exchange for 1 Ca^{2+} via the Na^+-Ca^{2+} exchanger (Parker, 1973; Sarkadi and Parker, 1991). In most cells, the electrochemical gradients for Na^+ and Ca^{2+} favor the efflux of Ca^{2+}. In these erythrocytes, however, the gradients favor Ca^{2+} entry because the Na^+ gradient is reduced, and Na^+-Ca^{2+} exchange operates in what is called the *reverse mode*. To stabilize their volume, carnivore erythrocytes also must have a means of maintaining the Ca^{2+} gradient (i.e., low $[Ca^{2+}]_i$). This is accomplished by an ATP-dependent Ca^{2+} pump in the plasma membrane. Thus, as in other cells, maintaining cell volume in the face of impermeant intracellular colloids requires the expenditure of energy in a pump-leak mechanism. In this case, ATP is expended by a plasma membrane Ca^{2+} pump rather than by the Na^+-K^+ pump.

VI. Regulation of Cell Volume under Anisosmotic Conditions

A. Osmometric Behavior of Cells

Because the permeability of most cell membranes to water is much greater than that to solutes, cells swell or shrink when placed in an environment that is hyposmotic or hyperosmotic, respectively. Water rapidly flows to equalize its chemical potential, μ_w, inside and outside the cell. The initial volume response is often close to that predicted for an ideal osmometer from van't Hoff's law. An example is shown in Fig. 16, which illustrates the response of rabbit ventricular myocytes to solutions with osmolarities ranging from 195 to 825 mosmol/L (0.60 to 2.55 times isotonic). Relative cell volume (V), calculated as $V_{test}/V_{isosmotic}$, is plotted against the inverse of relative osmolarity, $\pi_{isosmotic}/\pi_{test}$, and the data are fit to

$$V = (1 - V_b)(\pi_{isosmotic}/\pi_{test}) + V_b \qquad (79)$$

By definition, relative volume is 1.0 at a relative osmolarity of 1.0. Two conclusions can be reached from the data in Fig. 16. First, as expected from van't Hoff's law, the relationship between relative cell volume and the inverse of relative osmolarity is linear. Second, the intercept of the relationship on the volume axis, V_b, is 0.34, which is significantly different from 0. This is interpreted as meaning that a fraction of cell volume is **osmotically inactive,** that is, it apparently does not participate in the response to anisosmotic solutions.

Several arguments can be made to justify the observation of an osmotically inactive volume. The expectation from the simplest model is that **cell water** should vary in proportion to osmolarity. However, not all of cell volume is water. The volume of nonaqueous components such as small solutes and proteins, which represent 25–30% of the cell on a weight/weight basis, is unaffected by water movements. Even measurements of cell water show nonideal behavior, however (Macknight and Leaf, 1977; Solomon, 1989), so additional explanations are necessary. One suggestion is that a fraction of cell water is intimately associated with cell proteins or membranes and thereby is **bound** or **structured** and unavailable as solvent (e.g., LeNeveu *et al.*, 1976; Hinke, 1980). Although the state of water molecules adjacent to proteins and membranes must be different from that in the bulk phase of the cytoplasm, in light of NMR, intracellular ion activity, and other data, most investigators believe that virtually all water (~95%) is available as solvent (Shporer and Civan, 1977; Hladky and Rink, 1978). Another possibility is that the behavior of intracellular macromolecules is concentration-dependent. For example, the osmotic coefficient and charge on hemoglobin increase with its concentration as red cells shrink, and anions are drawn in to maintain electroneutrality. These phenomena are important in explaining water movement in red cells (Freedman and Hoffman, 1979), but their importance in other tissues remains uncertain. A third possibility is that intracellular compartments such as mitochondria, nuclei, endoplasmic reticulum, and sarcoplasmic reticulum of muscle cells may undergo volume changes that are not proportional to those of the whole cell. Differential responses to an osmotic challenge are expected because the plasmalemma and intracellular membranes possess distinct arrays of transporters and ion channels, and each sees a unique environment. Most methods for determining cell water or cell volume fail to distinguish between cytoplasmic and total cell water or volume (for a method that does distinguish these, see Reuss, 1985).

A crucial assumption made in determining osmotically inactive volume also may affect the value obtained for V_b in Eq. 79. The analysis assumes that only water has moved at the time volume is measured. That is to say, transmembrane ionic fluxes can be ignored. Although the permeability of the cell membrane to water is many times greater than the permeability to ions, the net fluxes of both water and ions start at the instant extracellular osmolarity is changed. If ion fluxes significantly affect intracellular osmolarity at the time of measurement, the extrapolated osmotically inactive volume will be imprecise. If in addition the ion fluxes depend on the direction or magnitude of the osmotic gradient, the plot of relative volume versus $\pi_{isosmotic}/\pi_{test}$ can become nonlinear (e.g., Grinstein *et al.*, 1984).

B. Compensatory Regulation of Cell Volume

Although an osmotic gradient initiates cell swelling or shrinkage, the initial volume response is not maintained in a wide variety of cells. Cell swelling activates compensatory processes that lead to an efflux of osmolytes and a reduction of cell volume. This is called a **regulatory volume decrease (RVD)**. Similarly, cell shrinking activates an influx of osmolytes in some cells, leading to a compensatory swelling referred to as a **regulatory volume increase (RVI)**. RVD and RVI nearly restore the original cell volume in some cells, are far less complete in others, and are absent in a few types of cells.

Regulatory volume effects are thought to be adaptive and were first identified in nucleated duck erythrocytes (for a review, see Kregenow, 1981). Examples of an RVD and RVI taken from work by Grinstein *et al.* (1983) are shown in Fig. 17. Exposure of human peripheral blood lymphocytes to a solution made hypotonic by 50% dilution with water leads to a rapid, 1.6-fold increase in cell volume (Fig. 17B). Then, over about 10 min, an RVD returned cell volume almost completely to its initial value. On switching back to isotonic solution, cell volume shrank to less than the control value and then was restored by an RVI. RVDs and RVIs may differ in magnitude, however. A much less complete RVI was observed when lymphocytes in isotonic solution were shrunk in media made hyperosmotic by adding of 300 mosmol·L^{-1} of NaCl, and the RVI was absent when the lymphocytes were challenged with 300 mosmol·L^{-1} of sucrose instead of NaCl (Fig. 17A). The RVD also could be eliminated, for example, by cooling lymphocytes to 4 °C (Grinstein *et al.*, 1984). Thus, regulation of cell volume following an osmotic challenge depends on the particulars of the perturbation as well as the cell under study (compare Figs. 1 and 17).

C. Transport Processes Responsible for RVD and RVI

How do cells gain or lose osmotic equivalents in anisosmotic media? The mechanisms underlying RVDs and RVIs have been extensively characterized in a variety of cell types and exhaustively reviewed (Hoffmann and Simonsen, 1989; Chamberlin and Strange, 1989; Grinstein and Foskett, 1990; Sarkadi and Parker, 1991; Häussinger and Lang, 1991; McCarty and O'Neil, 1992; Strange, 1994; Hoffmann and Dunham, 1995; Lang *et al.*, 1998; O'Neil, 1999). In general, cells undergo RVD or RVI by translocating Na$^+$, K$^+$, and Cl$^-$,

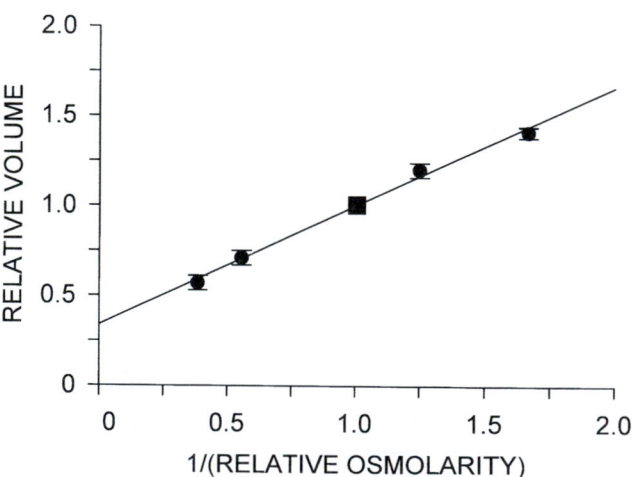

FIGURE 16. Relationship between relative cell volume and the inverse of relative osmolarity in cardiac ventricular myocytes. Data from 38 measurements of relative cell volume in anisosmotic solutions (0.6T, 0.8T, 1.8T, and 2.6T) were fit to a least squares regression line constrained to pass through (1, 1) (square) because relative cell volume is 1 in isotonic solution by definition. The extrapolated intercept on the relative volume axis, 0.34, represents the fraction of cell volume that is osmotically inactive (see Eq. 79). Volume measurements, calculation of relative cell volume, and means of adjusting osmolarity are as described for Fig. 1. (Reproduced from Drewnowska and Baumgarten (1991).)

and a variety of channels, exchangers, cotransporters, and pumps can participate. In some cases, organic osmolytes (e.g., taurine, betaine, sorbitol, and urea) are transported instead of, or in addition to, inorganic ions. Table 3 lists processes that are activated by altering the volume of various cells, and Fig. 18 illustrates some of the mechanisms diagrammatically. The compensatory mechanisms invoked vary with the tissue, species, and conditions under which osmotic stress is applied.

The primary mechanism of RVD in a number of cell types is activation of conductive pathways, and this process will be discussed in more detail. Cell swelling, acting directly or via a messenger, opens ion channels that allow increased efflux of K^+ and Cl^-, and H_2O follows. Although the openings of cation and anion channels are independent events, K^+ and Cl^- efflux are tightly coupled by the need to maintain macroscopic elec-

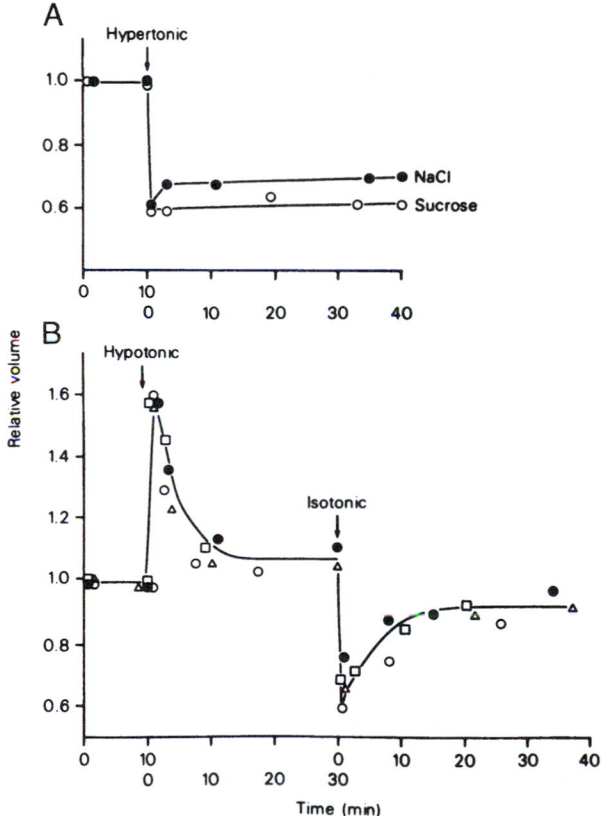

FIGURE 17. (A) Effect of hypertonic solution (2T) on relative cell volume of human peripheral blood mononuclear lymphocytes. Solution osmolarity was increased by adding 300 mosmol·L^{-1} of NaCl or sucrose. Cells shrank 40% in hypertonic solution, and a small regulatory volume increase (RVI) was observed in the NaCl solution. Points are representative of four experiments. (B) Response to hypotonic solution (0.5T). After swelling by nearly 60%, a regulatory volume decrease (RVD) virtually restored cell volume to its initial value in 10 min. On returning to isotonic solution, cell volume decreased below its control value, and an RVI lead to full recovery. Different symbols represent separate experiments selected from over 40. Cell volume was measured with a Coulter counter that determines volume from the change in electrical resistance of a column of solution as cells pass through an aperture. (From Grinstein *et al.* (1983). Reproduced from *The Journal of General Physiology*, 1983, **82**, 619–638, by copyright permission of The Rockefeller University Press.)

troneutrality. If only the anion or cation channel were to open, the resulting change in E_m would rapidly make ion efflux self-limiting and arrest volume regulation. The effect of this coupling of K^+ and Cl^- fluxes is illustrated in Fig. 19A, which shows RVDs in Ehrlich ascites cells (Hoffmann *et al.*, 1986; Hoffmann and Dunham, 1995). At time zero, cells were switched from 300 to 150 mosmol·L^{-1} media. Cells rapidly swelled to about 1.9 times their initial volume and then underwent an RVD that returned relative cell volume to about 1.3 within 5 min. When K^+ conductance was increased by pretreating cells with 0.5 µM gramicidin, a K^+ ionophore, the compensation by the RVD was more rapid and larger in magnitude, returning cell volume to nearly its control level in 2 min. When K^+ channels were blocked with 1 µM quinine, however, only a feeble RVD took place. These data argue that conductive K^+ efflux cannot keep up with Cl^- efflux in control cells and limits the rate of RVD (Hoffmann *et al.*, 1986). Thus cell swelling must increase Cl^- conductance more than K^+ conductance, leading to greater efflux of Cl^- than of K^+. Consistent with this idea, a depolarization is observed during the RVD. The activation of Cl^- conductance by swelling is only transient, however. The ability of gramicidin to induce RVD decays with time, as shown in Fig. 19B. In contrast to the high selectivity of Cl^--dependent cotransporters, RVD is also supported by Br^-, NO_3^- and SCN, suggesting the Cl^- channel responsible for RVD poorly discriminates among these anions. Furthermore, RVD can be suppressed by Cl^- channel blockers such as inacrinone (MK-196) and diphenylamine-2-carboxylate (DPC) but not by the cotransport inhibitors. Taken together, the data in Ehrlich ascites cells provide strong evidence that two independent channels are responsible for RVD instead of, for example, activation of a single K^+-Cl^- cotransporter. Organic osmolytes also permeate a class of swelling-activated Cl^- channels referred to as volume-sensitive organic osmolyte-anion channels (VSOAC) and are released by a number of cells to regulate volume (Jackson *et al.*, 1994) . VSOAC have a high permeability to taurine ($P_{taurine}/P_{Cl} = 0.75$) and glutamate ($P_{glutamate}/P_{Cl} = 0.20$) and other amino acids and small organic molecules are permeant (Bandarali and Roy, 1992).

Whereas many types of cells exhibit extensive RVDs, fewer cells exhibit robust RVIs. RVIs are usually due to an accumulation of Na^+, Cl^-, and in some cases, K^+, which occurs over minutes. The primary mechanisms underlying RVIs are acceleration of Na^+-K^+-$2Cl^-$ or Na^+-Cl^- cotransport and coupled Na^+-H^+ and Cl^--HCO_3^- exchange. Volume recovery can be blocked by application of appropriate inhibitors such as bumetanide for Na^+-K^+-$2Cl^-$ cotransport, amiloride for Na^+-H^+ exchange, and SITS (4-acetamido-4'-isothiocyano-stilbene-2,2'-disulfonic acid) for Cl^--HCO_3^- exchange. RVI in most cells is readily observed when simple salt solutions bathe the cells, but in renal cortical collecting duct and proximal tubules, butyrate, acetate, or other metabolizable fatty acids must be in the perfusate to support RVI. Substrate metabolism may provide H^+ and HCO_3^- to support Na^+-H^+ and Cl^--HCO_3^- exchange in these cells (for a review, see McCarty and O'Neil, 1992).

D. Organic Osmolytes

Adaptation to hyperosmolarity over a longer term also occurs in some cells and is mediated by an accumulation of

TABLE 3 Ionic Mechanisms of Regulatory Responses to Anisosmotic Solutions

Transport mechanism activated	Cell types
Cell swelling-induced regulatory volume decrease	
K^+ and Cl^- conductances	Frog urinary bladder
	Chinese hamster ovary cells
	Ehrlich ascites tumor cells
	Frog skin
	HeLa carcinoma cells
	Human platelets
	Human granulocytes
	Human lymphocytes
	Intestinal 407 cells
	Madin-Darby canine kidney (MDCK) cells
	Necturus enterocytes
	Necturus gallbladder
	Rabbit renal proximal convoluted tubule
	Rat hepatocytes
K^+-Cl^- cotransport	Avian, dog, fish, human, rabbit, and low-K^+ sheep erythrocytes
	Ehrlich ascites tumor cells (Ca^{2+} depleted)
	Necturus gallbladder
Coupled K^+-H^+ and Cl^--HCO_3^- exchange	*Amphiuma* erythrocytes
Na^+-Ca^{2+} exchange	Dog and ferret erythrocytes
Organic osmolyte efflux	Crustacean muscle and myocardium
	Ehrlich ascites tumor cells
	Elasmobranch and molluscan erythrocytes
Cell shrinkage-induced regulatory volume increase	
Na^+-K^+-$2Cl^-$ cotransport	Astrocytes
	C6 glioma cells
	Duck, fish, rat, and human erythrocytes
	Ehrlich ascites tumor cells
	Frog skin
	HeLa cells
	Rat kidney medullary thick ascending limb
	3T3 cells
Coupled Na^+-H^+ and Cl^--HCO_3^- exchange	Amphibian gallbladder
	Dog and amphibian erythrocytes
	Human lymphocytes
	Mouse medullary thick ascending limb
	Rabbit renal proximal straight tubule
	Ehrlich ascites tumor cells
Na^+-Cl^- cotransport	Ehrlich ascites tumor cells
	Necturus gallbladder
K^+ and Cl^- conductances, inhibited	Madin-Darby canine kidney (MDCK) cells
Organic osmolyte influx	Many animal and plant cells, bacteria and fungi

For references, see Yancey *et al.* (1982), Hoffmann and Simonsen (1989), Chamberlin and Strange (1989), Grinstein and Foskett (1990), Wolff and Balaban (1990), Sarkadi and Parker (1991), Häussinger and Lang (1991), McCarty and O'Neil (1992), Boyer *et al.* (1992), Strange (1994), Hoffmann and Dunham (1995), Lang *et al.* (1998), and O'Neil (1999).

organic osmolytes, including amino acids, polyols, and urea (Yancey *et al.*, 1982; Chamberlin and Strange, 1989; Wolff and Balaban, 1990; Garcia and Berg, 1991; Yancey, 1994; Berg, 1995). Three of these organic osmolytes, taurine, betaine, and inositol, are taken up by Na^+-dependent or Na^+- and

Cl^--dependent cotransporters. The taurine and betaine transporters have been cloned and belong to the same family as the Na^+- and Cl^--dependent norepinephrine, GABA, dopamine, serotonin, and proline transporters (Uchida *et al.*, 1992). The taurine transporter is found in a variety of cell types. The

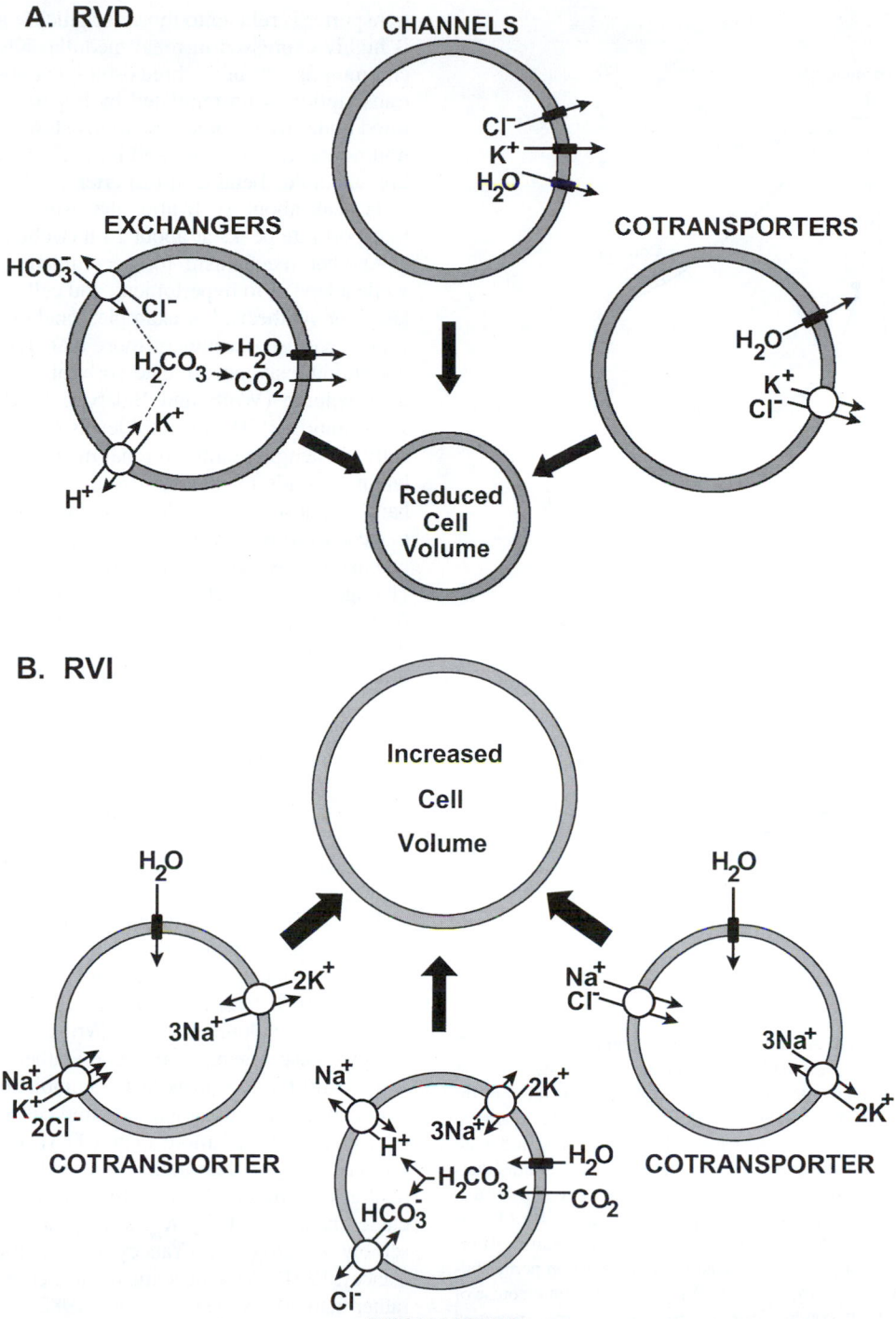

FIGURE 18. Schematic diagrams of the transport processes involved in (A) regulatory volume decrease (RVD) and (B) regulatory volume increase (RVI). Following cell swelling, a compensatory reduction in cell volume (RVD) may result from increased K^+ and Cl^- conductance (channels), activation of functionally coupled Cl^--HCO_3^- and K^+-H^+ exchange (exchangers), or activation of K^+-Cl^- cotransport (cotransporters). Following cell shrinking, a compensatory increase in cell volume (RVI) may result from activation of the Na^+-K^+-$2Cl^-$ or Na^+-Cl^- cotransporters (cotransporters) or activation of functionally coupled Cl^--HCO_3^-) and Na^+-H^+ exchange (exchangers). The Na^+-K^+ pump extrudes the Na^+ that enters during RVI, so that cells gain K^+ and Cl^-.

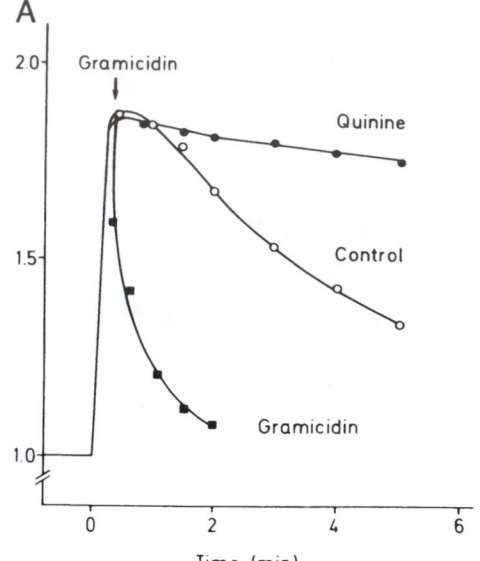

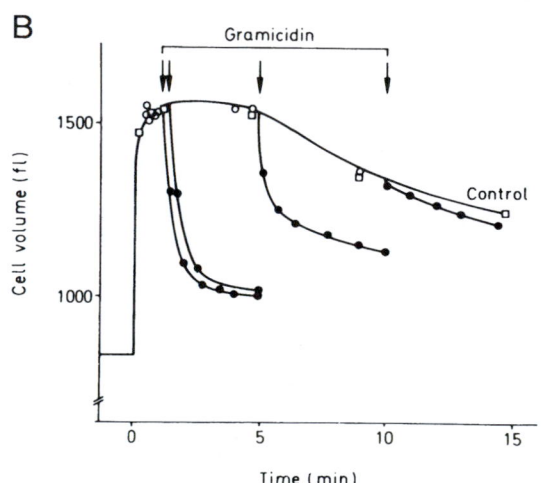

FIGURE 19. (A) Effects of quinidine (1 mM) and gramicidin (5 μM) on regulatory volume decrease (RVD) in Ehrlich ascites cells suspended in a Na$^+$-free choline medium with 1 mM Ca^{2+}. Swelling was induced by diluting the medium to 0.5T. Quinidine blocks Ca^{2+}-activated K$^+$ channels, and gramicidin increases cation conductance (K$^+$ conductance in the absence of Na$^+$). The data indicate that K$^+$ permeability is rate-limiting during the RVD. (B) The swelling-induced increase in Cl$^-$ conductance is transient. Same protocol as in as (A), except that quinidine (1 mM) was added to the hypotonic media to reduce K$^+$ conductance (control). At selected times, 0.5 μM gramicidin was added to increase cation permeability. The ability of gramicidin to induce an RVD follows the time-course of the swelling-induced Cl$^-$ conductance. Relative cell volumes measured with a Coulter counter, as described for Fig. 17. (Reproduced with permission from Hoffmann *et al.* (1986).)

distribution of mRNA levels is kidney > ileum > brain > liver > heart (Berg, 1995). This sequence for taurine transporter mRNA is surprising because heart has the highest intracellular taurine concentration among these tissues. The mRNA level for the betaine transporter is also highest in the kidney (Yamauchi *et al.*, 1992). A different family of transporters is responsible for Na$^+$-dependent inositol uptake. The inositol

transporter is related to those for glucose and nucleosides and is highly expressed in renal medulla (Kwon *et al.*, 1992). A common feature of all three osmolyte transporters is that gene transcription is up-regulated by hypertonic stress. Cells cultured under hypertonic conditions slowly increase their maximum rates of transport, and intracellular concentrations rise. For example, betaine transporter mRNA levels in MDCK cells peak about 16 h after the osmotic challenge, and the transport rate peaks at about 24 h (Uchida *et al.*, 1993).

Another mechanism for accumulating organic osmolytes while adapting to hypertonicity and cell shrinkage is to modulate their synthesis. For example, renal medullary cells, which experience osmolarities of more than 1200 mosmol·L^{-1} during antidiuresis, accumulate sorbitol and inositol to regulate their volume (Wolff and Balaban, 1990; Garcia and Berg, 1991; Sands, 1994). Sorbitol levels rise 12–24 h after osmotically challenging cultured renal medullary cells, and accumulation depends on the induction of aldose reductase by enhanced transcription. Aldose reductase converts exogenous glucose to sorbitol. Sorbitol is synthesized from stored glycogen in other tissues. This requires coordinated up-regulation of glycogen phosphorylase and hexosekinase to produce glucose 6-phosphate, an intermediate for sorbitol production, and the down-regulation of phosphofructokinase to prevent consumption of substrates for sorbitol production.

Inhibition of degradation is another mechanism for accumulating organic osmolytes when cells are challenged by hypertonicity. This is the main mechanism by which MDCK and medullary collecting duct accumulate glycerophosphocholine (GPC) (Berg, 1995). In response to hypertonicity or elevated urea, GPC-choline phosphodiesterase is inhibited. Under some conditions, synthesis of GPC by phospholipase also may be enhanced.

The accumulation of organic osmolytes is metabolically expensive, especially considering that many cells dump organic osmolytes to respond acutely to swelling (Chamberlin and Strange, 1989; Rasmusson *et al.*, 1993; Jackson *et al.*, 1994, Hoffmann and Dunham, 1995). Why does the cell expend extra energy to use organic compounds rather than inorganic ions? Apparently the reason is that accumulation of inorganic ions may perturb protein structure within cells. The Hofmeister series, first described more than 100 years ago, lists inorganic ions according to their ability to alter the solubility and conformation of proteins. Nonspecific effects of inorganic salts include changes in V_{max}, K_m, tertiary structure, and subunit assembly of enzymes (Yancey *et al.*, 1982; Somero, 1986; Yancey, 1994). Why are some organic compounds accumulated rather than others? Yancey *et al.* (1982) categorized osmolytes as nonperturbing (stabilizing) or perturbing (destabilizing). Elevated concentrations of nonperturbing osmolytes are compatible with normal enzyme function, whereas high concentrations of perturbing osmolytes are not. This distinction could be due either to direct interaction of osmolytes with enzymes or substrates or to effects on hydration, solubility, or charge interactions of proteins. Organic solutes that are nonperturbing generally are uncharged [e.g., trimethylamine N-oxide (TMAO), glycerol] or zwitterionic (e.g., betaine, taurine), although negatively charged octopine is used by some cells. In contrast, most perturbing organic osmolytes are positively charged (e.g., arginine, guanidinium). Neutral urea, however, perturbs proteins.

Finally, nonperturbing osmolytes can counteract the destabilizing effects of perturbing osmolytes. Several organisms accumulate perturbing osmolytes, such as urea, in fixed ratios with nonperturbing osmolytes.

E. Signaling Pathways Underlying RVD and RVI

How do cells detect alterations in their volume and activate the transport processes underlying compensatory volume regulation? Do the transporters themselves sense cell volume and respond by changing their activity? Alternatively, does water movement simply change the concentration of a critical regulatory substance? Answers to these questions are just beginning to emerge. It appears that the signaling pathways modulated by cell volume are as diverse as the transporters that respond.

1. Anisosmotic Media

An obvious candidate for signaling a change in cell volume is the composition of the anisosmotic media itself; that is, the ionic strength or concentrations of ions in the bathing media might initiate a regulatory volume response. Changing concentration or ionic strength must affect various ion transport processes and transmembrane ion fluxes to some degree. Nevertheless, these factors seem relatively unimportant in volume regulation because comparable responses are observed when nonelectrolytes such as mannitol or sucrose replace a large fraction of the electrolytes in the bathing media. RVDs also are initiated in isosmotic solutions after swelling caused by sugar or amino acid uptake. Thus it appears that regulation is initiated by the volume change itself rather than the composition of the bathing media.

2. Membrane Potential

Another possibility is that regulatory volume responses are initiated by dilution or concentration of intracellular K^+ via an effect on E_m. A two-fold change in volume without compensatory K^+ fluxes would alter $[K^+]_i$ and cause an 18 mV change in E_m in a cell that conforms to the Nernst equation. Fluxes through K^+ and Cl^- channels are voltage-dependent, reflecting both the electrochemical driving force and the voltage-dependent conductance. For many K^+ channels, the current-voltage relationship is highly non-linear. Furthermore, the Na^+-K^+ pump and Na^+-Ca^{2+} exchange are both voltage-dependent because they mediate a net movement of charge across the membrane. Despite this, it is unlikely that changes in E_m are the primary cause of RVDs or RVIs in most cells. Regulatory responses have been observed with changes in cell volume of <5% (Hoffmann and Simonsen, 1989), which would directly alter E_m by only ~1 mV. This is too small to have a significant effect. Nevertheless, it is clear that E_m can modulate RVDs due to activation of K^+ and Cl^- channels. In most instances, the increase in Cl^- conductance is greater than that for K^+ (Hoffmann and Simonsen, 1989; Grinstein and Foskett, 1990). This leads to a significant depolarization during the RVD due to a greater passive efflux of Cl^- than K^+, and the depolarization equalizes the anion and cation fluxes (due to macroscopic electroneutrality). Depolarization may have additional effects that support the RVD. The sensitivity of certain K^+ channels to intracellular Ca^{2+} (i.e., Ca^{2+}-activated K^+ channels) is increased by depolarization, and other K^+ channels are directly opened by depolarization.

3. Cytoskeleton

The cytoskeleton consists of three main elements: actin filaments (F-actin), which are microfilaments 5–7 nm in diameter that are double-stranded α-helical polymers of globular actin (G-actin); microtubules, which are hollow tubes 25 nm in diameter that are made from tubulin monomers arranged in 13 threads; and intermediate filaments (IF), which are 10-nm-diameter strands composed of tissue-specific proteins, such as keratin in epithelial cells and desmin in muscle cells (Bershadsky and Vasiliev, 1988; Luna and Hitt, 1992; Mills et al., 1994). In nonmuscle cells, the majority of actin filaments are associated with the cell membrane. F-actin is tied together and to integral membrane proteins by ankyrin, spectrin, MARKS (myristolated acid-rich C-kinase substrate), and other binding proteins to form a structural unit. This scaffolding undergoes constant reorganization as G-actin polymerizes and depolymerizes in response to various stimuli. The main role of microtubules in mature cells is thought to be the transport of vesicles within the cell using specific microtubule-associated proteins (MAPs) to attach to kinesin and dynein, which act as molecular motors. The functions of IF are not fully understood. Because IF binds to ankyrin and to desmosomal plaques, it is thought that these filaments have a structural role and link organelles to the membrane or cytoskeleton. Intermediate filaments are phosphorylated and dephosphorylated by protein kinases and phosphatases, and it is likely that their phosphorylation state regulates function.

Alterations in cell volume lead to deformation or reorganization of the cytoskeleton, and ideas of how these effects may be linked to cell volume regulation have been proposed (Chamberlin and Strange, 1989; Sachs, 1989; Sarkadi and Parker, 1991; Mills et al., 1994; Hoffmann and Dunham, 1995; Henson, 1999). One possibility is that the cytoskeleton might mechanically resist cell swelling, but the evidence for this is inconclusive.[3] Ion channels, exchangers, and cotransporters are structurally anchored in the membrane by specific components of the cytoskeleton (Luna and Hitt,

---

[3]For eukaryotic cells, it is generally assumed that the osmotic gradient across a membrane is negligible. This view arises because the thin bilayer membrane is too fragile to resist the substantial forces developed by even small differences in osmotic pressure. A tension of ~10 dyn·cm^{-1} is sufficient to rupture erythrocyte (Evans et al., 1976), protoplast (Wolfe et al., 1986), or lipid bilayer (Needham and Nunn, 1990) membranes. The relationship between tension on the membrane and transmembrane pressure is given by the law of Laplace. For a thin-walled spherical cell

$$P = 2T/r$$

where P is pressure, T is tension, and r is the cell's radius. If we assume a radius of 10 μm, a lytic tension is developed by a pressure of 2×10^4 dyn·cm^{-2}, which equals ~15 mm Hg. This is the osmotic pressure developed by only ~0.0008 osmol·L^{-1}. Hence, the membrane cannot support a sufficient hydrostatic pressure to offset an osmotic gradient.

Before accepting this conclusion, it is necessary to consider the effective radius of the cell in view of the geometry of the cytoskeleton (Jacobson, 1983). For example, there are ~10^5 copies of ankyrin per erythrocyte. If all these attach integral membrane proteins to the cytoskeleton and are evenly distributed, the membrane is strengthened by load-bearing cytoskeletal elements at ~40 nm intervals. Even if we assume an effective radius of 100 nm, lytic tension now requires a pressure of 2×10^6 dyn·cm^{-2}, which equals ~1500 mm Hg or nearly 2 atm. This is equivalent to the osmotic pressure generated by 0.078 osmol·L^{-1}.

1992), and their interaction may help regulate cell volume. Agents that disrupt the microfilaments, such as the cytochalasins, modify the regulatory responses to both hypotonic and hypertonic stress by inhibiting K^+ and Cl^- channels that promote RVD and stimulating Na^+-K^+-$2Cl^-$ cotransport responsible for RVI (Chamberlin and Strange, 1989; Mills *et al.*, 1994; Hoffmann and Dunham, 1995). In a few cases, cell volume under isosmotic conditions is affected (McCarty and O'Neil, 1992). Microtubules also may be involved in some cell lines. Colchicine, which prevents microtubule polymerization, decreases macrophage volume by 20% (Mills *et al.*, 1994). The shrinkage is inhibited by SITS, suggesting that either Cl^--HCO_3^- exchange or Cl^- channels are involved. Additionally, water and ion channels are inserted reversibly into the membrane in response to volume perturbations in a process that can be blocked by cytochalasin B (Lewis and de Moura, 1982). Furthermore, mechanical deformation of the membrane and supporting structures can modulate biochemical signaling systems such as the cAMP and protein kinase C cascades (Watson, 1991; Richter *et al.*, 1987). Emerging evidence suggests that protein kinases and phosphatases may be specifically localized by cytoskeletal binding proteins to sites adjacent to their target transport proteins. Moreover, the extracellular matrix is coupled to the cytoskeleton by integrins which regulate complex signaling cascades including focal adhesion kinase (FAK) and extracellular signal regulated kinases (ERK/MAPK) (Boudreau and Jones, 1999). These developments are important because it remains obscure how the cytoskeleton exerts its regulatory effect on ion transporters and whether or not constraints imposed by 3-dimensional tissues *in situ* affect cell volume.

Another role for the cytoskeleton involves widely distributed mechanosensitive and volume-sensitive ion channels (Sachs, 1989; Morris, 1990; Hu and Sachs, 1997; Vandenberg *et al.*, 1996; Wright and Rees, 1998). The probability of channel opening is increased by osmotic stretch or mechanical deformation. It has been proposed that mechanical forces detected by a cytoskeletal protein, spectrin, can directly affect cation channel gating, whereas the gating of volume-sensitive anion channels is regulated by their phosphorylation state. Most stretch-activated channels poorly discriminate between permeant species (e.g., the stretch-activated cation channel admits Na^+, K^+, and Ca^{2+}), but some are highly selective for K^+. There are also reports of mechanosensitive channels that are inactivated by stretch (Sachs, 1989; Morris, 1990). Mechanosensitive channels are potentially important in cell volume regulation as both sensors and effectors. The evidence that they are opened by osmotic stretch is convincing, and they carry substantial currents, which might directly or indirectly lead to volume regulation. For example, it has been argued that K^+-selective stretch-activated channels in molluscan heart cells pass sufficient current to alter $[K^+]_i$ by 1% in 1 s when the cell is voltage-clamped away from E_K (Brezden *et al.*, 1986), and that poorly selective cation stretch-activated channels in *Necturus* choroid epithelium raise Ca^{2+} sufficiently in 100 s to open Ca^{2+}-activated K^+ channels and initiate an RVD (Christensen, 1987). Such calculations must be regarded as estimates, however, because they do not reflect E_m under the relevant conditions. Nevertheless, Ca^{2+} entry via stretch-activated channels has been implicated in volume regulation in several types of cells (Foskett, 1994) and arguments favoring the idea that stretch-activated channels participate in cell volume regulation are accumulating (Sackin, 1994). Recently we found that Gd^{3+}, a blocker of nonselective stretch-activated cation channels, reduced swelling of both intact (i.e., unclamped) and voltage-clamped cardiac myocytes in hypoosmotic solutions, and that 9-anthracene carboxylic acid (9-AC), a blocker of stretch-activated anion channels, increased swelling in the same solutions (Suleymanian *et al.*, 1995; Clemo and Baumgarten, 1997; Clemo *et al.*, 1998, 1999). Opposite effects of the two blockers are expected because anions and cations travel in opposite directions under these conditions. In contrast, Gd^{3+} and 9-AC have negligible effects on myocyte volume in isosmotic solution when stretch-activated channels are expected to be closed. Interestingly, the same volume-sensitive cation and anion channels became persistently activated and modulated cell volume under isosmotic conditions in a canine model of congestive heart failure (Clemo *et al.*, 1998, 1999)

4. Calcium

A role for Ca^{2+} in cell volume regulation has been recognized for many years (Pierce and Politis, 1990; McCarty and O'Neil, 1992; Foskett, 1994; Hoffmann and Dunham, 1995). RVD is blocked by removing extracellular Ca^{2+} in *Amphiuma* erythrocytes, *Necturus* gallbladder, proximal convoluted and straight tubule, intestine 407, and osteosarcoma cells. Extracellular Ca^{2+} is not a requirement for RVD in lymphocytes and Ehrlich ascites cells, but RVD is more rapid in Ehrlich ascites cells when Ca^{2+} is present. In some tissues it appears that the Ca^{2+} involved in RVD enters, at least in part, through dihydropyridine-sensitive, L-type Ca^{2+} channels. Significant Ca^{2+} entry can also occur via nonselective stretch-activated cation channels. Block of RVD by lanthanides and disruption of the cytoskeleton, both of which affect stretch-activated cation channels, is consistent with this possibility. Instead of the entry of extracellular Ca^{2+}, release of Ca^{2+} from intracellular stores is critical for RVD in a variety of cells, including lymphocytes, Ehrlich ascites, intestine 407, and opossum kidney cells. Depletion of internal Ca^{2+} stores eliminates RVD in these cells; it is restored by extracellular Ca^{2+} and the Ca^{2+} ionophore, A23187. $[Ca^{2+}]_i$ has been shown to increase during swelling of *Amphiuma* erythrocytes using arsenazo III as a Ca^{2+} indicator. More recently, fluorescent Ca^{2+} indicators, quin-2 and fura-2, have been used to demonstrate increases in $[Ca^{2+}]_i$ that accompany cell swelling in urinary bladder, osteosarcoma, lymphoma, and proximal and straight convoluted tubule cells. On the other hand, $[Ca^{2+}]_i$ remains unchanged during volume changes in some cells, including human lymphocytes. More evidence for elevated Ca^{2+} comes from patch-clamp studies. In a number of tissues, recordings of single channel activity established that Ca^{2+}-activated K^+ channels open more frequently during RVD than under isotonic conditions. This suggests an increase in $[Ca^{2+}]_i$ because the probability that these channels open increases as $[Ca^{2+}]_i$ increases. Although studied in less detail, a decrease in $[Ca^{2+}]_i$ has been implicated in RVI in *Amphiuma* erythrocytes, Ehrlich ascites cells, and lymphocytes.

The mechanism by which Ca^{2+} modulates cell volume appears to vary (see Table 4). Ca^{2+} has direct effects on ion channels, but additional signaling mechanisms may be involved. For example, calmodulin inhibitors can block the increased K^+ conductance in lymphocytes, Ehrlich ascites cells, and *Necturus* gallbladder; the increased Cl^- conductance in Ehrlich ascites cells; and increased K^+-H^+ exchange in *Amphiuma* erythrocytes. In other cells, modulation of protein kinase C and leukotriene synthesis by Ca^{2+} have been proposed as signals in volume regulation.

5. Phosphorylation

The activities of many of the transporters discussed here are modified by phosphorylation (Parker, 1992; McCarty and O'Neil, 1992; Palfrey, 1994; Hoffmann and Dunham, 1995). This raises the possibility that cell volume alterations initiate an RVD or RVI by either increasing or decreasing the fraction of transporters in the phosphorylated state. Until recently, supporting data have been lacking. Over the last few years, however, strong evidence for this idea has come from studies in several tissues. We will discuss some of these data from erythrocytes in detail.

Pewitt *et al.* (1990) studied RVI in duck erythrocytes and determined that activation of Na^+-K^+-$2Cl^-$ cotransport on shrinking is caused by phosphorylation. Both cAMP-dependent and cAMP-independent protein kinase phosphorylate the Na^+-K^+-$2Cl^-$ cotransporter (or possibly a regulatory protein), but cAMP levels in duck erythrocytes are not affected by osmotic stress. Pewitt *et al.* (1990)

found that the protein kinase inhibitors K252a and H-9 prevent transporter activation on shrinking. Conversely, an inhibitor of serine and threonine protein-phosphatases, okadaic acid, which slows protein dephosphorylation, stimulates Na^+-K^+-$2Cl^-$ cotransport under isotonic conditions. These changes in the activity of the transporter with phosphorylation and with shrinking apparently result largely from a modulation of the number of functioning transporters, as detected by bumetanide binding, rather than from a modulation of their turnover rate. At about the same time, Jennings and al-Rohil (1990) and Jennings and Schulz (1991) developed evidence that K^+-Cl^- cotransport in rabbit erythrocytes, which is responsible for RVD, is activated by a dephosphorylation. They discovered that swelling inhibits a protein kinase distinct from protein kinases A and C. An RVD occurred only after a slow dephosphorylation, now identified as due to a Type 1 protein phosphatase (PP1) that is blocked by calyculin A (Starke and Jennings, 1993). Parker *et al.* (1991) obtained similar results in dog erythrocytes.

Parker *et al.* (1991) also recognized the important reciprocal coordination of K^+-Cl^- cotransport and Na^+-H^+ exchange by phosphorylation and dephosphorylation during both RVDs and RVIs in mammalian erythrocytes. This strategy is illustrated in Fig. 20 and can be summarized as follows: (1) Shrinking activates and swelling inhibits a protein kinase. (2) On shrinking, activated protein kinase rapidly phosphorylates regulatory sites associated with the K^+-Cl^- cotransporter and the Na^+-H^+ exchanger (or the Na^+-K^+-$2Cl^-$ cotransporter in duck erythrocytes). (3) Phosphorylation inhibits the K^+-Cl^- cotransporter, reducing osmolyte efflux, but stimulates the Na^+-H^+ exchanger or

TABLE 4 Intracellular Signaling Pathways for RVD and RVI

Signal	Effector	Cell type
Ca^{2+}	K^+ conductance	*Necturus* gallbladder
		Frog urinary bladder
		Ehrlich ascites cells
		Human lymphocytes
	Cl^- conductance	Ehrlich ascites cells
	Taurine efflux	Elasmobranch and molluscan erythrocytes
	Na^+-H^+ exchange	Human lymphocytes
	K^+-H^+ exchange	*Amphiuma* erythrocytes
Phosphorylation	Na^+-H^+ exchange	Human lymphocytes
	taurine efflux	Elasmobranch erythrocytes
	Na^+-K^+-$2Cl^-$ cotransport	Duck erythrocytes
	K^+-Cl^- cotransport	Duck, rabbit, and dog erythrocytes
Leukotrienes	K^+ conductance	Ehrlich ascites cells
	Cl^- conductance	Ehrlich ascites cells
cAMP	Na^+-H^+ exchange	Mouse thick ascending limb of Henle (mTALH) cells
	Cl^--HCO_3^- exchange	Mouse thick ascending limb of Henle (mTALH) cells
	Na^+-Cl^- cotransport	*Necturus* gallbladder
G proteins	Na^+-H^+ exchange	Barnacle skeletal muscle
Voltage	K^+ channels	Human lymphocytes

For references, see Chamberlin and Strange (1989), Sarkadi and Parker (1991), and McCarty and O'Neil (1992), Strange (1994), Hoffmann and Dunham (1995), Lang *et al.* (1998), O'Neil (1999), Häussinger and Schliess (1999).

Na^+-K^+-$2Cl^-$ cotransporter, stimulating osmolyte uptake and leading to an RVI. (4) Conversely, slow dephosphorylation on swelling stimulates K^+-Cl^- cotransport and inhibits Na^+-H^+ exchange or Na^+-K^+-$2Cl^-$ cotransport, leading to an RVD. Thus, the transporters underlying ion influx and efflux are regulated reciprocally by the activity of a protein kinase that reflects cell volume.

How does cell volume govern the activity of a protein kinase? Several possible detectors have been considered: cell shape or cytoskeletal deformation; the concentration of an impermeant intracellular cofactor such as Mg^{2+}; and a concept referred to as macromolecular crowding. The first possibility was already discussed, but experimental evidence suggests cell shape does not regulate phosphorylation of the relevant transport proteins, at least in erythrocytes. The last two possibilities will be considered next.

6. Mass Action Model

Increasing intracellular Mg^{2+} activates Na^+-H^+ exchange and inhibits K^+-Cl^- cotransport, and it has been suggested that Mg^{2+} might act by activating a kinase. Before accepting the idea that Mg^{2+} or another intracellular ion is the volume sensor, it is necessary to explain the steep dependence of ion transport on cell volume. Jennings and Schulz (1990) illustrated one possible answer for K^+-Cl^- cotransport with a theoretical mass action model. They assumed: (1) the volume sensor (e.g., Mg^{2+}) is an impermeant intracellular species; (2) the kinase and phosphatase are soluble enzymes; and (3) the sensor inhibits dephosphorylation. They then described the model in terms of three first-order Michaelis-Menten expressions. Taking into account that the concentrations of sensor, kinase, and phosphatase vary inversely with volume, they were able to reproduce the steep volume dependence of experimental K^+ influx data. As the authors emphasized, however, a good fit of the experimental data does not prove that the model is correct. Rather, it illustrates only that a simple dilution mechanism can give rise to a steep volume dependence of transport if dilution has different effects on the activity of enzymes that regulate the transporter (e.g., Fig. 20).

7. Macromolecular Crowding

The concept of **macromolecular crowding** comes from the idea that proteins do not behave ideally in solution at concentrations in the physiological range. We have already mentioned that the osmotic coefficient for hemoglobin and other proteins increases steeply with concentration, and that Freedman and Hoffman (1979) used this fact to explain water movement in red cells. Nonideal behavior is thought to be a more general phenomenon, however. Minton (1983, 1990, 1994) has argued that the kinetics and equilibria of enzymes (macromolecules) are markedly altered by the presence of inert macromolecules that occupy more than a few percent of the total solution volume. Just as one hemoglobin molecule affects another, macromolecules that are neither substrate nor product affect the behavior of their macromolecular neighbors in solution. This results because crowding reduces the solution volume accessible to a macromolecule by excluded-volume effects illustrated in Fig. 21. An excluded volume means that solution behavior is nonideal, and the chemical potential, μ_i, and activity, a_i, of a macromolecule is increased by crowding. Consequently, reaction rates are affected. On the other hand, small solutes (e.g., ions) are unaffected by the same concentration of macromolecules. Several examples are worth noting. Minton (1983) showed that the specific activity of glyceraldehyde-3-phosphate dehydrogenase decreased dramatically as the concentration of bovine serum albumin, β-lactoglobulin, polyethylene glycol (PEG), or ribonuclease in the reaction medium was increased. This was explained by suggesting that crowding favored the formation of tetramers of the enzyme that possess a lower catalytic specific activity than monomers. Similarly, the cohesion of complementary ends of λ DNA can be increased up to 2000-fold by albumin, Ficoll 70, or PEG (Zimmerman and Harrison, 1985), and the activity of T4 polynucleotide kinase is augmented by PEG (Harrison and Zimmerman, 1986). Protein concentrations within cells are sufficient to give significant excluded-volume effects (Zimmerman and Trach, 1991).

How does macromolecular crowding relate to cell volume regulation? Perhaps the activity of the kinase governing the

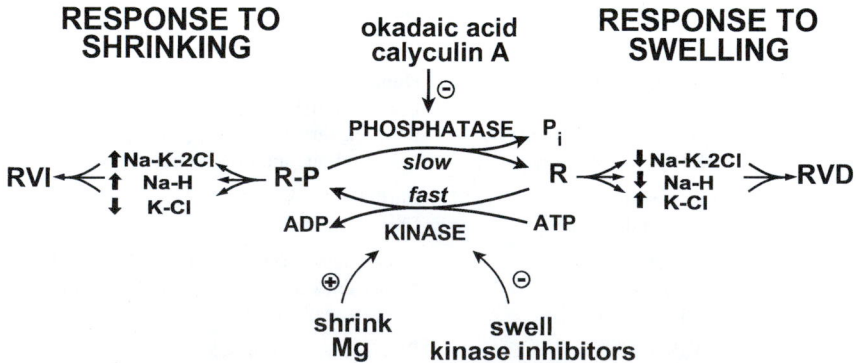

FIGURE 20. Schematic diagram of the mechanism of coordination of Na^+-K^+-$2Cl^-$ cotransport and Na^+-H^+ and Na^+-K^+ exchange during cell swelling and shrinking. R-P and R represent phosphorylated and dephosphorylated regulatory sites that modulate the activity of transporters leading to a regulatory volume decrease (RVD) or regulatory volume increase (RVI). Changes in cell volume affect protein kinase activity. Several interventions that inhibit or stimulate the protein kinase and phosphatase are indicated.

phosphorylation state of transporters decreases as macromolecular crowding is lessened during cell swelling. This would, for example, activate K+-Cl− cotransport and inactivate Na+-H+ exchange (see Fig. 20), and an RVD would ensue. Colclasure and Parker (1991, 1992) provided experimental support for this hypothesis (see also Sarkadi and Parker, 1991; Parker, 1992). They osmotically ruptured dog erythrocytes and then allowed the "ghosts" to reseal in a hypotonic medium. This gave resealed ghosts with about 1/4 normal volume but with a normal protein (hemoglobin or hemoglobin plus albumin) concentration. During osmotic challenge, K+-Cl− cotransport and Na+-H+ exchange were regulated at about the same protein concentration as in normal erythrocytes even though the volumes of the resealed ghosts were vastly different. This hypothesis is also consistent with studies on the kinetics of volume regulation by

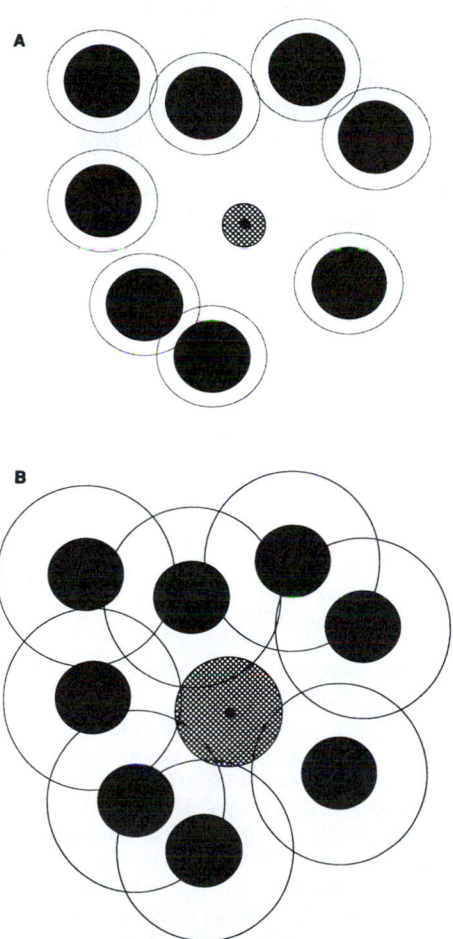

FIGURE 21. Diagram of macromolecular crowding. Background macromolecules (solid) in solution differentially affect the behavior of small molecules, such as ions (A) and other macromolecules (B). A test molecule (cross-hatched) cannot enter the excluded volume represented by a circle with a radius equal to the sum of the radii of the test molecule and the background macromolecule. The excluded volume is much greater for a test macromolecule (B) than for a test small molecule (A). Restricting the volume that a test molecule can enter increases its activity, a, and chemical potential, μ. Consequently, both reaction rates and equilibria are altered. (Reproduced with permission from Minton (1994). Copyright CRC Press, Boca Raton, Florida.)

Jennings and al-Rohil (1990), who concluded that the swelling-sensitive step was a decrease in the rate of phosphorylation rather than an increase in the rate of dephosphorylation.

The new insight that macromolecular crowding may be a mechanism for volume transduction is an interesting possibility. Minton *et al.* (1992) have presented a model quantitatively accounting for volume-dependent stimulation of ion fluxes on this basis. It is important to recognize that operation of the scheme depends on the components having appropriate sensitivity. In erythrocytes, for example, the activities of both the kinase and the phosphatase should be reduced by swelling. Consequently, as the rate of protein phosphorylation falls, so does the rate of protein dephosphorylation. Whether this leads to an increase or decrease in the fraction of transporters in the phosphorylated state must depend both on the relative effects of macromolecular crowding and the amount of substrate available for each enzyme. With injudicious choices for the parameters (e.g., pathological interventions), this mechanism might lead to an inappropriate RVI rather than an RVD.

VII. Summary

The study of mechanisms underlying osmosis and the regulation of cell volume under both isosmotic and anisosmotic conditions has been fruitful. We understand in substantial detail how water and ions cross the membrane. Where do we go from here? Many directions are possible, and as many details are missing as are known. For example, the identification and cloning of water channels raise several important questions. What is it about the protein structure that makes this a water channel? How does water interact with the channel? From a theoretical perspective, how is osmotic pressure sensed and how does osmosis occur through a channel structure that is different in important ways from the well-explored hydrodynamic models? From the perspective of regulation of ion transport, much remains to be understood about how cells sense swelling and shrinking and how a cell decides on its optimal volume. There are also unanswered questions concerning the regulation of volume regulatory ion transporters by cellular messengers, metabolic demands, and pathological states. In short, we can look forward to many more fruitful years of research on these topics.

Acknowledgments

We thank J. Maghirang for preparing numerous figures. Supported by National Institutes of Health grants HL-24847 and HL-46764, and Grants-in-Aid from the American Heart Association and its Virginia Affiliate.

Bibliography

Adair, G. S. (1928). A theory of partial osmotic pressures and membrane equilibrium with special reference to the application of Dalton's law to haemoglobin solutions in the presence of salts. *Proc. Roy. Soc. London* **120A,** 573–603.

Agius, L., Peak, M., Beresford, G., al-Habori, M., and Thomas, T. (1994). The role of ion content and cell volume in insulin action. *Biochem. Soc. Trans.* **22,** 518–522.

Agre, P., Preston, G. M., Smith, B. L., Jung, J. S., Raina, S., Moon, C., Guggino, W. B., and Nielsen, S. (1993). Aquaporin CHIP: the archetypal molecular water channel. *Am. J. Physiol.* **265,** F463–F476.

Andreoli, T. E., and Troutman, S. L. (1971). An analysis of unstirred layers in series with "tight" and "porous" lipid bilayer membranes. *J. Gen. Physiol.* **57,** 464–478.

Bandarali, U., and Roy, G. (1992). Anion channels for amino acids in MDCK cells. *Am. J. Physiol.* **263,** C1200–C1207.

Barry, P. H., and Diamond, J. M. (1984). Effects of unstirred layers on membrane phenomena. *Physiol. Rev.* **64,** 763–873.

Berg, M. B. (1995). Molecular basis of osmolyte regulation. *Am. J. Physiol.* **268,** F983–F996.

Bershadsky, A. D., and Vasiliev, J. M. (1988). "Cytoskeleton." Plenum Press, New York.

Bookchin, R. M., Ortiz, O. E., Freeman, C. J., and Lew, V. L. (1989). Predictions and tests of a new integrated reticulocyte model: implications for dehydration of sickle cells, *In* "The Red Cell: Seventh Annual Ann Arbor Conference." Alan R. Liss Publishers, New York, pp. 615–625.

Boudreau, N. J., and Jones, P. L. (1999). Extracellular matrix and integrin signalling: the shape of things to come. *Biochem. J.* **339,** 481–488.

Boyer, J. L., Graf, J., and Meier, P. J. (1992). Hepatic transport systems regulating pH$_i$, cell volume, and bile secretion. *Annu. Rev. Physiol.* **54,** 415–438.

Boyle, P. J., and Conway, E. J. (1941). Potassium accumulation in muscle and associated changes. *J. Physiol.* **100,** 1–63.

Brezden, B. L., Gardner, D. R., and Morris, C. E. (1986). A potassium-selective channel in isolated Lymnaea stagnalis heart muscle cells. *J. Exp. Biol.* **123,** 175–190.

Carruthers, A., and Melchior, D. L. (1983). Studies of the relationship between bilayer water permeability and bilayer physical state. *Biochemistry* **22,** 5797–5807.

Cass, A., and Finkelstein, A. (1967). Water permeability of thin lipid membranes. *J. Gen. Physiol.* **20,** 1765–1784.

Chamberlin, M. E., and Strange, K. (1989). Anisosmotic cell volume regulation: a comparative view. *Am. J. Physiol.* **257,** C159–C173.

Christensen, O. (1987). Mediation of cell volume regulation by Ca^{2+} influx through stretch-activated channels. *Nature* **330,** 66–68.

Clemo, H. F., and Baumgarten, C. M. (1995). cGMP and atrial natriuretic factor regulate cell volume of rabbit atrial myocytes. *Circ. Res.* **77,** 741–749.

Clemo, H. F., and Baumgarten, C. M. (1997). Swelling-activated Gd^{3+}-sensitive cation current and cell volume regulation in rabbit ventricular myocytes. *J. Gen. Physiol.* **110,** 297–312.

Clemo, H. F., Feher, J. J., and Baumgarten, C. M. (1992). Modulation of rabbit ventricular cell volume and Na$^+$/K$^+$/2Cl$^-$ cotransport by cGMP and atrial natriuretic factor. *J. Gen. Physiol.* **100,** 89–114.

Clemo, H. F., Stambler, B. S., and Baumgarten, C. M. (1998). Persistent activation of a swelling-activated cation current in ventricular myocytes from dogs with tachycardia-induced congestive heart failure. *Circ. Res.* **83,** 147–157.

Clemo, H. F., Stambler, B. S., and Baumgarten, C. M. (1999). Swelling-activated chloride current is persistently activated in ventricular myocytes from dogs with tachycardia-induced congestive heart failure. *Circ. Res.* **84,** 157–165.

Colclasure, C. G, and Parker, J. C. (1991). Cytosolic protein concentration is the primary volume signal in dog red cells. *J. Gen. Physiol.* **98,** 881–892.

Colclasure, C. G, and Parker, J. C. (1992). Cytosolic protein concentration is the primary volume signal for swelling-induced [K-Cl] cotransport in dog red cells. *J Gen Physiol* **100,** 1–10.

Dainty, J. (1965). Osmotic flow. *Symp. Soc. Exptl. Biol.* **19,** 75–85.

Deamer, D. D, and Bramhall, J. (1986). Permeability of lipid bilayers to water and ionic solutes. *Chem. Phys. Lipids* **40,** 167–188.

Drewnowska, K., and Baumgarten, C. M. (1991). Regulation of cellular volume in rabbit ventricular myocytes: bumetanide, chlorothiazide, and ouabain. *Am. J. Physiol.* **260,** C122–C131.

Drewnowska, K., Clemo, and H. F., Baumgarten, C. M. (1991). Prevention of myocardial intracellular edema induced by St. Thomas' Hospital cardioplegic solution. *J. Molec. Cell Cardiol.* **23,** 1215–1221.

Evans, E., Waugh, R., and Melnik, L. (1976). Elastic area compressibility modulus of red cell membrane. *Biophys. J.* **16,** 585–595

Fettiplace, R., and Haydon, D. A. (1980). Water permeability of lipid membranes. *Physiol. Rev.* **60,** 510–550.

Finkelstein, A. (1987). "Water Movement Through Lipid Bilayers, Pores, and Plasma Membranes. Theory and Reality." Wiley-Interscience, New York, NY.

Finkelstein, A., and Cass, A. (1968). Permeability and electrical properties of thin lipid membranes. *J. Gen. Physiol.* **52,** 145s–172s.

Fischbarg, J., Kuang, K., Vera, J. C., Arant, S., Silverstein, S. C., Loike, J., and Rosen, O. M. (1990). Glucose transporters serve as water channels. *Proc. Natl. Acad. Sci. USA* **87,** 3244–3247.

Foskett, J. K. (1994). The role of calcium in the control of volume-regulatory pathways. *In* "Cellular and Molecular Physiology of Cell Volume Regulation" (K. Strange, Ed.) CRC Press, Boca Raton, FL. pp. 259–277.

Freedman, J. C., and Hoffman, J. F. (1979). Ionic and osmotic equilibria of human red blood cells treated with nystatin. *J. Gen. Physiol.* **74,** 157–185.

Garcia, A., and Berg, M. B. (1991). Role of organic osmolytes in adaption of renal cells to high osmolarity. *J. Memb. Biol.* **119,** 1–13.

Glasstone, S. (1946). "Textbook of Physical Chemistry." Van Nostrand, Princeton, N. J.

Grinstein, S., and Foskett, J. K. (1990). Ionic mechanisms of cell volume regulation in leukocytes. *Annu. Rev. Physiol.* **52,** 399–414.

Grinstein, S., Clarke, C., A., and Rothstein, A. (1983). Activation of Na$^+$/H$^+$ exchange in lymphocytes by osmotically-induced volume changes and by cytoplasmic acidification. *J. Gen. Physiol.* **82,** 619–638.

Grinstein, S., Rothstein, A., Sarkadi, B., and Gelfand, E. W. (1984). Responses of lymphocytes to anisotonic media: volume-regulating behavior. *Am. J. Physiol.* **246,** C204–C215.

Haas, M. (1994). The Na-K-Cl cotransporters. *Am. J. Physiol.* **267,** C869–C885.

Haines, T. H. (1994). Water transport across biological membranes. *FEBS Lett.* **346,** 115–122.

Harrison, B., and Zimmerman, S. B. (1986). T4 polynucleotide kinase: macromolecular crowding increases the efficiency of reaction at DNA termini. *Anal. Biochem.* **158,** 307–315.

Hasegawa, H., Skach, W., Baker, W., Calayag, M. C., Lingappa, V., and Verkman, A. S. (1992). A multifunctional aqueous channel formed by CFTR. *Science* **258,** 1477–1479.

Hasegawa, H., Zhang, R., Dohrman, A., and Verkman, A. S. (1993). Tissue-specific expression of mRNA encoding rat kidney water channel CHIP28k by *in situ* hybridization. *Am. J. Physiol.* **264,** C237–C245.

Häussinger, D. (1998). Osmoregulation of liver cell function: signalling, osmolytes and cell heterogeneity. *Contrib. Nephrol.* **123,** 185–204.

Häussinger, D., and Lang, F. (1991). Cell volume in the regulation of hepatic function: a mechanism for metabolic control. *Biochim. Biophys. Acta.* **1071,** 331–350.

Häussinger, D., and Schliess, F. (1999). Osmotic induction of signaling cascades: role in regulation of cell function. *Biochem. Biophys. Res. Commun.* **255,** 551–555.

Häussinger, D., Lang, F., and Gerok, W. (1994). Regulation of cell function by the cellular hydration state. *Am. J. Physiol.* **267,** E343–E355.

Henson, J. H. (1999). Relationships between the actin cytoskeleton and cell volume regulation. *Microsc. Res. Tech.* **47**, 155–162.

Hernandez, J. A., and Fischbarg, J. (1992). Kinetic analysis of water transport through a single-file pore. *J. Gen. Physiol.* **99**, 645–662.

Heymann, J. B., and Engel, A. (1999). Aquaporins: phylogeny, structure, and physiology of water channels. *News Physiol. Sci.* **14**, 187–193.

Heymann, J. B., Agre P., and Engel, A. (1998). Progress on the structure and function of aquaporin 1. *J. Struct. Biol.* **121**, 191–206.

Hildebrand, J. H. (1955). Osmotic pressure. *Science* **121**, 116–119.

Hinke, J, A, M. (1980). Water and electrolyte content of the myofilament phase in the chemically skinned barnacle fiber. *J. Gen. Physiol.* **75**, 531–551.

Hladky, S. B., and Rink, T. J. (1978). Osmotic behaviour of human red blood cells: an interpretation in terms of negative intracellular fluid pressure. *J. Physiol.* **274**, 437–446.

Hobbie, R. K. (1978). "Intermediate Physics for Medicine and Biology." John Wiley, New York. pp. 126–132.

Hoffmann, E. K., and Dunham, P. B. (1995). Membrane mechanisms and intracellular signalling in cell volume regulation. *Int. Rev. Cytol.* **161**, 173–262.

Hoffmann, E. K., and Simonsen, L. O. (1989). Membrane mechanisms in volume and pH regulation in vertebrate cells. *Physiol. Rev.* **69**, 315–382.

Hoffmann, E. K., Lambert, I. H., and Simonsen, L. O. (1986). Separate, Ca^{2+}-activated K^+ and Cl^- transport pathways in Ehrlich ascites tumor cells. *J. Memb. Biol.* **91**, 227–244.

Holz, R., and Finkelstein, A. (1970). The water and nonelectrolyte permeability induced in thin lipid membranes by the polyene antibiotics nystatin and amphotericin B. *J. Gen. Physiol.* **56**, 125–145.

Hu, H., and Sachs, F. (1997). Stretch-activated ion channels in the heart. *J Mol. Cell. Cardiol.* **29**, 1511–1523.

Jackson, P. S., Morrison, R., and Strange, K. (1994). The volume-sensitive organic osmolyte-anion channel VSOAC is regulated by nonhydrolytic ATP binding. *Am. J. Physiol.* **267**, C1203–C1209.

Jacobson, B. S. (1983). Interaction of the plasma membrane with the cytoskeleton: an overview. *Tissue Cell* **15**, 829–852.

Jakobsson, E. (1980). Interactions of cell volume, membrane potential, and membrane transport parameters. *Am. J. Physiol.* **238**, C196–C206.

Jansen, M., and Blume, A. (1995). A comparative study of diffusive and osmotic water permeation across bilayers composed of phospholipids with different head groups and fatty acyl chains. *Biophys. J.* **68**, 997–1008.

Jap, B. K., and Li, H. (1995). Structure of the osmo-regulated H_2O-channel, AQP-CHIP, in projection at 3.5 Å resolution. *J. Mol. Biol.* **251**, 413–420.

Jennings, M. L., and al-Rohil, N. (1990). Kinetics of activation and inactivation of swelling-stimulated K^+/Cl^- transport. The volume sensitive parameter is the rate constant for inactivation. *J. Gen. Physiol.* **95**, 1021–1040.

Jennings, M. L., and Schulz, R. K. (1990). Swelling-activated KCl cotransport in rabbit red cells: flux is determined mainly by cell volume rather than shape. *Am. J. Physiol.* **259**, C960–C967.

Jennings, M. L., and Schulz, R. K. (1991). Okadaic acid inhibition of KCl cotransport. Evidence that protein dephosphorylation is necessary for activation of transport by either cell swelling or N-ethylmaleimide. *J. Gen. Physiol.* **97**, 799–817.

Jung, J. A., Preston, G. M., Smith, B. L., Guggino, W. B., and Agre, P. (1994). Molecular structure of the water channel through aquaporin CHIP: the hourglass model. *J. Biol. Chem.* **269**, 14648–14654.

Kaplan, M. R., Mount, D. B., and Delpire, E. (1996). Molecular mechanisms of NaCl cotransport. *Annu. Rev. Physiol.* **58**, 649–668.

Kiil, F. (1989). Molecular mechanisms of osmosis. *Am. J. Physiol.* **256**, R801–R808.

King, L. S., and Agre P. (1996). Pathophysiology of the aquaporin water channels. *Annu. Rev. Physiol.* **58**, 619–648.

Kregenow, F. M. (1981). Osmoregulatory salt transporting mechanisms: control of cell volume in anisotonic media. *Ann. Rev. Physiol.* **43**, 493–505.

Kwon, H. M., Yamauchi, A., Uchida, S., Preston, A. S., Garcia-Perez, A., Berg, M. B., and Handler, J. S. (1992). Cloning of a Na^+/myo-inositol cotransporter, a hypertonicity stress protein. *J. Biol. Chem.* **267**, 6229–6301.

Lakshminarayanaiah, N. (1984). "Equations of Membrane Biophysics." Academic Press, Orlando, FL, pp. 107–118.

Landis, E. M., and Pappenheimer, J. R. (1963). Exchange of substances through the capillary walls. *In* "Handbook of Physiology," Vol. 2, Sec. 2, American Physiological Society, Washington, DC, pp. 961–1034.

Lang, F., Busch, G. L., and Volkl, H. (1998). The diversity of volume regulatory mechanisms. *Cell. Physiol. Biochem.* **8**, 1–45.

Leaf, A. (1956). On the mechanism of fluid exchange of tissues *in vitro. Biochem. J.* **62**, 241–248.

Leaf, A. (1959). Maintenance of concentration gradients and regulation of cell volume. *Ann. NY Acad. Sci.* **72**, 396–404.

LeNeveu, D. M., Rand, R. P., and Parsegian, V. A. (1976). Measurement of forces between lecithin bilayers. *Nature* **259**, 601–603.

Lewis, S. A., and de Moura, J. L. (1982). Incorporation of cytoplasmic vesicles into apical membrane of mammalian urinary bladder epithelium. *Nature* **297**, 685–688.

Luna, E. J., and Hitt, A. L. (1992). Cytoskeleton-plasma membrane interactions. *Science* **258**, 955–964.

Ma, T., and Verkman, A. S. (1999). Aquaporin water channels in gastrointestinal physiology. *Am. J. Physiol.* **517**, 317–326.

Macknight, A. D. C. (1988). Principles of cell volume regulation. *Renal Physiol. Biochem.* **3–5**, 114–141.

Macknight, A. D. C., and Leaf, A. (1977). Regulation of cellular volume. *Physiol. Rev.* **57**, 510–573.

Macey, R. I. (1979). Transport of water and nonelectrolytes across red cell membranes. *In* "Membrane Transport in Biology," Vol. II Transport Across Single Biological Membranes. (G. Giebisch, Tosteson D. C., and H. H. Ussing, Eds.), Springer Verlag, Berlin, pp. 1–57.

Macey, R. I., and Farmer, R. E. L. (1970). Inhibition of water and solute permeability in human red cells. *Biochim. Biophys. Acta.* **211**, 104–106.

Macey, R. I., Karan, D. M., and Farmer, R. E. L. (1972). Properties of water channels in human red cells. *In* "Biomembranes," Vol. 3, Passive Permeability of Cell Membranes. (Kreuzer, F., and Slegers, J. F. G., Eds.), Plenum Press, New York. pp. 331–340.

Mauro, A. (1957). Nature of solvent transfer in osmosis. *Science* **126**, 252–253.

McCarty, N. A., and O'Neil, R. G. (1992). Calcium signaling in cell volume regulation. *Physiol. Rev.* **72**, 1037–1061.

Meschia, G., and Setnikar, I. (1958). Experimental study of osmosis through a collodion membrane. *J. Gen Physiol.* **42**, 429–444.

Mills, J. W., Schweibert, E. M., and Stanton, B. A. (1994). The cytoskeleton and cell volume regulation. *In* "Cellular and Molecular Physiology of Cell Volume Regulation." (K. Strange, Ed.), CRC Press, Boca Raton. pp. 241–258.

Minton, A. P. (1983). The effect of volume occupancy upon the thermodynamic activity of proteins: some biochemical consequences. *Molec. Cell. Biochem.* **55**, 119–140.

Minton, A. P. (1990). Holobiochemistry: the effect of local environment upon equilibria and rates of biochemical reactions. *Int. J. Biochem.* **22** 1063–1067.

Minton, A. P. (1994). Influence of macromolecular crowding on intracellular association reactions: possible role in volume regulation. *In* "Cellular and Molecular Physiology of Cell Volume Regulation," (K. Strange, Ed.), Academic Press, New York. pp. 181–190.

Minton, A. P., Colclasure, G. C., and Parker, J. C. (1992). Model for the role of macromolecular crowding in regulation of cellular volume. *Proc. Natl. Acad. Sci. USA* **89**, 10504–10506.

Morris, C. E. (1990). Mechanosensitive ion channels. *J. Memb. Biol.* **113**, 93–107.

Needham, D., Nunn, R. S. (1990). Elastic deformation and failure of lipid bilayer membranes containing cholesterol. *Biophys. J.* **58**, 997–1009.

Nielsen, S., Smith, B. L., Christensen, E. I., Knepper, M. A., and Agre, P. (1993). CHIP28 water channels are localized in constitutively water-permeable segments of the nephron. *J. Cell. Biol.* **120**, 371–383.

Nielsen, S., Chou, C. L., Marples, D., Christensen, E. I., Kishore, B. K., and Knepper, M. A. (1995). Vasopressin increases water permeability of kidney collecting duct by inducing translocation of aquaporin-CD water channels in rat kidney inner medulla. *Proc. Natl. Acad Sci. USA* **92**, 1013–1017.

Nielsen, S., Kwon, T. H., Christensen, B. M., Pomeneur, D., Frokiaer, J., and Marples, D. (1999). Physiology and pathophysiology of renal aquaporins. *J. Am. Soc. Nephrol.* **10**, 647–663

O'Neil, W. C. (1999). Physiological significance of volume regulatory transporters. *Am. J. Physiol.* **276**, C995–C1011.

Overbeek, J. T. G. (1956). The Donnan equilibrium. *Prog. Biophys. Biophys. Chem.* **3**, 57–84.

Paganelli, C. V., and Solomon, A. K. (1957). The rate of exchange of tritiated water across the human red cell membrane. *J. Gen. Physiol.* **41**, 259–277.

Palfrey, H. C. (1994). Protein phosphorylation control in the activity of volume-sensitive transport systems. *In* "Cellular and Molecular Physiology of Cell Volume Regulation," (K. Strange, Ed.), CRC Press, Boca Raton, FL. pp. 201–214.

Palfrey, H. C., and O'Donnell, M. E. (1992). Characteristics and regulation of the Na/K/2Cl cotransporter. *Cell. Physiol. Biochem.* **2**, 293–307.

Parker, J. C. (1973). Dog red blood cells. Adjustment of density *in vitro. J. Gen. Physiol.* **62**, 147–156.

Parker, J. C. (1992). Volume-activated cation transport in dog red cells: detection and transduction of the volume stimulus. *Comp. Biochem. Physiol.* **102A**, 615–618.

Parker, J. C., Colclasure, G. C, and McManus, T. J. (1991). Coordinated regulation of shrinkage-induced Na/H exchange and swelling-induced [K-Cl] cotransport in dog red cells. Further evidence from activation kinetics and phosphatase inhibition. *J. Gen. Physiol.* **98**, 869–880.

Pewitt, E. B., Hegde, R. S., Haas, M., and Palfrey, H. C. (1990). The regulation of Na/K/2Cl cotransport and bumetanide binding in avian erythrocytes by protein phosphorylation and dephosphorylation: effects of kinase inhibitors and okadaic acid. *J. Biol. Chem.* **265**, 20747–20756.

Pfeffer, W. (1877) "Osmotische Untersuchungen. Studien zur Zellmechanik." Wilhelm Engelmann, Leipzig. [Translated by G. R. Kepner and E. J. Tadelmann (1985) "Osmotic investigations. Studies on cell membranes." Van Nostrand Reinhold, New York.]

Pierce, S. K., and Politis, A. D. (1990). Ca^{2+}-activated cell volume recovery mechanisms. *Annu. Rev. Physiol.* **52**, 27–42.

Pitzer, K. S., and Mayorga, G. (1973). Thermodynamics of electrolytes. II. Activity and osmotic coefficients for strong electrolytes with one or both ions univalent. *J. Phys. Chem.* **77**, 2300–2308.

Preston, G. M., and Agre, P. (1991). Isolation of the cDNA for erythrocyte integral membrane protein of 28 kilodaltons member of an ancient channel family. *Proc. Natl. Acad. Sci. USA* **88**, 11110–11114.

Preston, G. M., Carroll, T. P., Guggino, W. B., and Agre, P. (1992). Appearance of water channels in *Xenopus* oocytes expressing red cell CHIP28 protein. *Science* **256**, 385–387.

Preston, G. M., Jung, J. S., Guggino, W. B., and Agre, P. (1993). The mercury-sensitive residue at cysteine-189 in the CHIP28 water channel. *J. Biol. Chem.* **268**, 17–20.

Preston, G. M., Jung, J. S., Guggino, W. B., and Agre, P. (1994). Membrane topology of aquaporin CHIP: analysis of functional epitope scanning mutants by vectorial proteolysis. *J. Biol. Chem.* **269**, 1668–1673.

Rash, J. E., Yasumura, T., Hudson, C. S., Agre, P., and Nielsen, S. (1998). Direct immunogold labeling of aquaporin-4 in square arrays of astrocyte and ependymocyte plasma membranes in rat brain and spinal chord. *Proc. Natl. Acad. Sci. USA* **95**, 11981–11986.

Rasmusson, R. L., Davis, D. G., and Lieberman, M. (1993). Amino acid loss during volume regulatory decrease in cultured chick heart cells. *Am. J. Physiol.* **264**, C136–C145.

Renkin, E. M. (1954). Filtration, diffusion, and molecular sieving through porous cellulose membranes. *J. Gen. Physiol.* **38**, 225–243.

Reuss, L. (1985). Changes in cell volume measured with an electrophysiological technique. *Proc. Natl. Acad. Sci. USA* **82**, 6014–6018.

Richter, E. A., Cleland, P. J. F., Rattigan, S., and Clark, M. G. (1987). Contraction-associated translocation of protein kinase C. *FEBS Lett.* **217**, 232–236.

Robinson, R. A., and Stokes, R. H. (1959). "Electrolyte Solutions." Butterworth, London. pp. 480–490.

Rosenberg, P. A., and Finkelstein, A. (1978). Water permeability of gramicidin A-treated lipid bilayer membranes. *J. Gen. Physiol.* **72**, 341-350.

Sachs, F. (1989). Ion channels as mechanical transducers. *In* "Cell Shape: Determinants, Regulation, and Regulatory Role." (W. D.Stein, and F. Bronner, Eds.), Academic Press, San Diego, CA. pp. 63–92.

Sackin, H. (1994). Stretch-activated ion channels. *In* "Cellular and Molecular Physiology of Cell Volume Regulation," (K. Strange, Ed.), CRC Press, Boca Raton, FL. pp. 215–240.

Sands, J. M. (1994). Regulation of intracellular polyols and sugars in response to osmotic stress. *In* "Cellular and Molecular Physiology of Cell Volume Regulation." (K. Strange, Ed.), CRC Press, Boca Raton. pp 133–144.

Sarkadi, B, and Parker, J. C. (1991). Activation of ion transport pathways by changes in cell volume. *Biochim. Biophys. Acta.* **1071**, 407–427.

Schatzberg, P. (1963). Solubilities of water in several normal alkanes from C_7 to C_{16}. *J. Phys. Chem.* **67**, 776–779.

Shibata, A., Ikawa, K., Shimooka, T., and Terada, H. (1994). Significant stabilization of the phosphatidylcholine bilayer structure by incorporation of small amounts of cardiolipin. *Biochim. Biophys. Acta.* **1192**, 71–78.

Shporer, M., and Civan, M. M. (1977). The state of water and alkali cations within the intracellular fluids: the contribution of NMR spectroscopy. *Curr. Top. Membr. Transport.* **9**, 1–69.

Solomon, A. K. (1972). Properties of water in red cells and synthetic membranes. *In* "Biomembranes," Vol. 3, Passive Permeability of Cell Membranes. (F. Kreuzer, and J. F. G. Slegers, Eds.), Plenum Press, New York. pp. 299–330.

Solomon, A. K. (1989). Water channels across the red blood cell and other biological membranes. *Methods in Enzymology* **173**, 192–222.

Solomon, A. K., Chasan, B., Dix, J. A., Lukacovic, M. F., Toon, M. R., and Verkman, A. S. (1983). The aqueous pore in the red cell membrane: band 3 as a channel for anions, cations, nonelectrolytes, and water. *Ann. NY Acad. Sci.* **414**, 97–124.

Somero, G. N. (1986). Protons, osmolytes, and fitness of internal milieu for protein function. *Am. J. Physiol.* **251**, R197–R213.

Starke, L. C., and Jennings, M. L. (1993). K-Cl cotransport in rabbit red cells: further evidence for regulation by protein phosphatase type 1. *Am. J. Physiol.* **264**, C118–C124.

Stein, W. D. (1976). An algorithm for writing down flux equations for carrier kinetics, and its application to co-transport. *J. Theor. Biol.* **62**, 467–478.

Strange, K. (1994). "Cellular and Molecular Physiology of Cell Volume Regulation." CRC Press, Boca Raton.

Suleymanian, M. A., and Baumgarten, C. M. (1996). Osmotic gradient-induced water permeation across the sarcolemma of rabbit ventricular myocytes. *J. Gen. Physiol.* **107**, 503–514.

Suleymanian, M. A., Clemo, H. F., Cohen, N. M., and Baumgarten, C. M. (1995). Stretch-activated channel blockers modulate cell volume in cardiac ventricular myocytes. *J. Mol. Cell. Cardiol.* **27**, 721–728.

Tanford, C. (1961). "Physical Chemistry of Macromolecules." John Wiley, New York.

Tosteson, D. C., and Hoffman, J. F. (1960). Regulation of cell volume by active cation transport in high and low potassium sheep red cells. *J. Gen. Physiol.* **44,** 169–194.

Trauble, H. (1971). The movement of molecules across lipid membranes: a molecular theory. *J. Memb. Biol.* **4,** 193–208.

Uchida, S., Kwon, H., Yamauchi, A., Preston, A., Marumo, F., and Handler, J. (1992). Molecular cloning of the cDNA for an MDCK cell Na$^+$ and Cl$^-$-dependent taurine transporter that is regulated by hypertonicity. *Proc. Natl. Acad. Sci. USA* **89,** 8230–8234.

Uchida, S., Kwon, H., Yamauchi, A., Preston, A., Kwon, H., and Handler, J. (1993). Medium tonicity regulates expression of the Na$^+$- and Cl$^-$-dependent betaine transporter in Madin-Darby canine kidney cells by increasing transcription of the transporter gene. *J. Clin. Invest.* **91,** 1604–1607.

Umenishi, F., Verkman, A. S., and Gropper, M. A. (1996). Quantitative analysis of aquaporin mRNA expression in rat tissues by RNase protection assay. *DNA Cell. Biol.* **15,** 475–480.

Vandenberg, J. I., Rees, S. A., Wright, A. R., and Powell, T. (1996) Cell swelling and ion transport pathways in cardiac myocytes. *Cardiovasc. Res.* **32,** 85–97.

van Hoek, A. N., Hom, M. L., Luthjens, L. H., de Jong, M. D., Dempster, J. A., and van Os, C. H. (1991). Functional unit of 30 kDa for proximal tubule water channel as revealed by radiation inactivation. *J. Biol. Chem.* **266,** 16633–16635.

van Hoek, A. N., Luthjens, L. H., Hom, M. L., van Os, C. H., and Dempster, J. A. (1992). A 30 kDa functional size for the erythrocyte water channel determined by in situ radiation inactivation. *Biochem. Biophys. Res. Comm.* **184,** 1331–1338.

van't Hoff, J. H. (1887). Die Rolle des osmotischen Druckes in der Analogie zwischen Lösungen und Gasen. *Z. Physik. Chemie.* **1,** 481–493 [Translated by G. L. Blackshear, (1979) *In* "Cell Membrane Permeability and Transport," (G. R. Kepner., Ed.) Dowden, Hutchinson and Ross, Stroudsburg, PA.]

Verkman, A. S. (1993). "Water Channels." R. G. Landis Co., Austin, TX.

Verkman, A. S., van Hoek, A. N., Ma, T., Frigeri, A., Skach, W. R., Mitra, A., Tamarappoo, B. K., and Farinas, J. (1996). Water transport across mammalian cell membranes. *Am. J. Physiol.* **270,** C11–C30.

Villars, F. M., and Benedek, G. B. (1974) "Physics with Illustrative Examples from Medicine and Biology. Statistical Physics," Vol. 2. Addison-Wesley, Reading, MA.

Walz, T., Smith, B. L., Agre, P., and Engel, A. (1994). The three-dimensional structure of human erythrocyte aquaporin CHIP. *EMBO J.* **13,** 2985–2993.

Walz, T., Hirai, T., Murata, K., Heymann, J. B., Mitsuoka, K., Fijiypshi, Y., Smith, B. L., Agre, P., and Engel, A. (1997). The three-dimensional structure of aquaporin-1. *Nature* **387,** 624–627.

Watson, P. A. (1991). Function follows form: generation of intracellular signals by cell deformation. *FASEB J.* **5,** 2013–2019.

Wolfe, J., Dowgert, M. F., and Steponkus, P. L. (1986). Mechanical study of the deformation and rupture of the plasma membranes of protoplasts during osmotic expansions. *J. Memb. Biol.* **93,** 63–74.

Wolff, S. D., and Balaban, R. S. (1990). Regulation of the predominant renal medullary organic solutes *in vivo. Annu. Rev. Physiol.* **52,** 727–746.

Wright, A. R., and Rees, S. A. (1998). Cardiac cell volume: crystal clear or murky waters? A comparison with other cell types. *Pharmacol. Ther.* **80,** 89–121.

Yamauchi, A., Uchida, S., Kwon, H., Preston, A., Robey, R., Garcia-Perez, A., Berg, M., and Handler, J. (1992). Cloning of a Na$^+$ and Cl$^-$ dependent betaine transporter that is regulated by hypertonicity. *J. Biol. Chem.* **267,** 649–652.

Yancey, P. H. (1994), Compatible and counteracting solutes. *In* "Cellular and Molecular Physiology of Cell Volume Regulation." (K. Strange, Ed.), CRC Press, Boca Raton, FL. pp 81–177.

Yancey, P. H., Clark, M. E., Hand, S. C., Bowlus, R. D., and Somero, G. N. (1982). Living with water stress: Evolution of osmolyte systems. *Science* **217,** 1214–1222.

Zeidel, M. L., Ambudkar, S. V., Smith, B. L., and Agre, P. (1992). Reconstitution of functional water channels in liposomes containing purified red cell CHIP28 protein. *Biochemistry* **31,** 7436–7440.

Zhang, R., Skach, W., Hasegawa, H., van Hoek, A. N., and Verkman, A. S. (1993). Cloning, functional analysis and cell localization of a kidney proximal tubule water transporter homologous to CHIP28. *J. Cell. Biol.* **120,** 359–369.

Zimmermann SB, and Harrison B. (1985). Macromolecular crowding accelerates the cohesion of DNA fragments with complementary termini. *Nucleic. Acids. Res.* **13,** 2241–2249.

Zimmerman, S. B., and Trach, S. (1991). Estimation of macromolecular concentrations and excluded volume effects for the cytoplasm of *Escherichia coli. Mol. Biol.* **222,** 599–620.

Robert W. Putnam

22

Intracellular pH Regulation

I. Introduction

Intracellular pH is an important aspect of the intracellular environment. Changes in intracellular pH can potentially affect virtually all cellular processes, including metabolism, membrane potential, cell growth, movement of substances across the surface membrane, state of polymerization of the cytoskeleton, and ability to contract in muscle cells. Changes of intracellular pH are also often one of the responses of cells to externally applied agents, including growth factors, hormones, and neurotransmitters. Further, many organelles, such as lysosomes, mitochondria, and endosomal vesicles, maintain an organellar pH that is different from the cytoplasmic pH (pH_i), and these pH differences have important functional consequences for those organelles. It is thus not surprising to find that cells have elaborated a variety of mechanisms that enable them to regulate their intracellular pH. In this chapter, we will discuss the pH level in the cytoplasm and various compartments of a cell, the variety of mechanisms available to a cell to regulate its pH_i, and the functional consequences of changes in pH_i.

II. pH and Buffering Power

The concept of pH was first introduced in 1909 by Sörenson and defined as $-\log[H^+]$. This term was a more convenient way to express the concentration of an ion that is present at very low concentrations. Incorporating the concept of activity, pH is defined as

$$pH = -\log\left(a_H\right) = -\log\left(\gamma_H[H^+]\right) \tag{1}$$

where a_H is the activity of H^+ and γ_H is the activity coefficient of H^+. At normal intracellular ionic strength, γ_H is about 0.83.

Protons tend to bind to macromolecules, and thus are usually present at very low concentrations in biological solutions. This property is the basis for **buffering power.** A variety of weak acids and bases can bind H^+ through reversible equilibrium binding reactions. Thus, a weak acid in solution obeys the equilibrium reaction

$$HA \rightleftharpoons H^+ + A^- \tag{2}$$

where HA is the weak acid (e.g., lactic acid) and A^- is the conjugate weak base (e.g., lactate). This equilibrium is described by an apparent equilibrium constant, K_a', as

$$K_a' = a_H \cdot \frac{[A^-]}{[HA]} \tag{3}$$

This equation is more familiar in its logarithmically transformed expression,

$$pH = pK_a' \log\frac{[A^-]}{[HA]} \tag{4}$$

where pK_a' is $-\log K_a'$. This equation, better known as the **Henderson-Hasselbalch equation,** describes the thermodynamic equilibrium that holds for a weak acid in a solution of constant pH. The Henderson-Hasselbalch equation is most commonly used in its specialized form for the total reaction of the hydration of CO_2 and the dissociation of the resulting carbonic acid into H^+ and bicarbonate as

$$pH = pK_a' + \log\frac{[HCO_3^-]}{\alpha \cdot P_{CO_2}} \tag{5}$$

where α is the solubility coefficient of CO_2 in a given solution and P_{CO_2} is the partial pressure of CO_2 in that solution. Two important facts can be deduced from the Henderson-Hasselbalch equation. First, in any weak acid solution, there will be a finite amount of both A^- and HA. For example, if HCO_3^- is added to a solution, CO_2 will be generated and thus be present. Conversely, if CO_2 is bubbled through a solution, HCO_3^- will be produced. Second, this equation can be used to calculate any of the variable parameters if the other three are known. For instance, if a solution is equilibrated with a gas of known P_{CO_2}, the $[HCO_3^-]$ in that solution can be calculated from Eq. 5 once the pH has reached a stable value (values for pK_a' are readily available).

On addition of H^+ (or OH^-) to a solution, the pH will change. However, if the solution contains weak acids (or bases), many of the added protons (or hydroxyl ions) will be bound up, thus minimizing the change in the concentration of free H^+ and thereby minimizing the change in pH. Since these substances minimize the change in pH upon addition of acid or base, weak acids and bases are referred to as *buffers*. The definition of the buffering power (β) of a solution is

$$\beta = \frac{d[B]}{dpH} \qquad (6)$$

where $d[B]$ is the amount of base added to the solution and dpH is the change in pH of the solution due to that base addition. The addition of acid to the solution is equivalent to a negative addition of base, $-d[B]$. The units of β are mM/pH unit.

An example will indicate the importance of buffering power to maintaining the pH of a solution. If 1 mM NaCl is added to a solution, $[Na^+]$ and $[Cl^-]$ increase by 1 mM (for simplicity, the effects of the nonideal activity coefficients will be ignored). However, the addition of 1 mM HCl to a solution that has a pH of 7.0 and a buffering power of 10 mM/pH unit will cause that $[Cl^-]$ to increase by 1 mM, but will cause the pH to decrease by only about 0.28 pH unit, to 6.72. Thus, of the added 1 mM of H^+, only 0.091 µM remain free, that is, only 1 out of every 11 000 added H ions remains free. The rest are bound to the weak acid buffers. If the buffers had not been present, the same addition of 1 mM HCl would have changed the solution pH by about 4 units to pH 3. This clearly demonstrates that the presence of buffers in a solution markedly blunts the effects of added acid or base.

A buffer can act as either a **closed buffer** or an **open buffer.** A closed buffer is one in which the total buffer concentration remains constant. Most of the commonly used laboratory buffers, such as Hepes or Tris, operate as closed buffers in solution. If a buffer is a weak acid ($HA \rightleftharpoons A^- + H^+$), then it operates as a closed buffer when the concentration of the total acid ($[A]_T = [HA] + [A^-]$) remains constant. Such a buffer can become protonated or deprotonated, but the total amount of buffer does not change. The following is a mathematical expression for the buffering power of a weak acid acting as a closed buffer:

$$\beta_{closed} = \frac{2.303[A]_T\, K_a'\, a_H}{\left(K_a' + a_H\right)^2} \qquad (7)$$

where K_a' is the apparent dissociation constant of the weak acid. Several conclusions can be derived from this equation. When pH is very high ($a_H \rightarrow 0$) or very low ($a_H \rightarrow \infty$), β_{closed} approaches 0. β_{closed} reaches a maximum when $a_H = K_a'$ (i.e., when pH = pK_a') and $\beta_{closed}^{max} = 0.58[A]_T$. The relationship between pH and β_{closed} for a theoretical closed buffer is shown in Fig. 1.

In contrast to the conditions for a closed buffer, if the protonated (or uncharged) form of a buffer remains constant (i.e., $[HA]$ = constant), then the buffer operates as an open buffer. The most common example of an open buffer in solution is CO_2/HCO_3^-. Such a solution contains HCO_3^- and is equilibrated with gaseous CO_2 (usually by bubbling). If acid is added to such a solution, H^+ combines with HCO_3^- and forms additional CO_2. Since the solution is in equilibrium with a

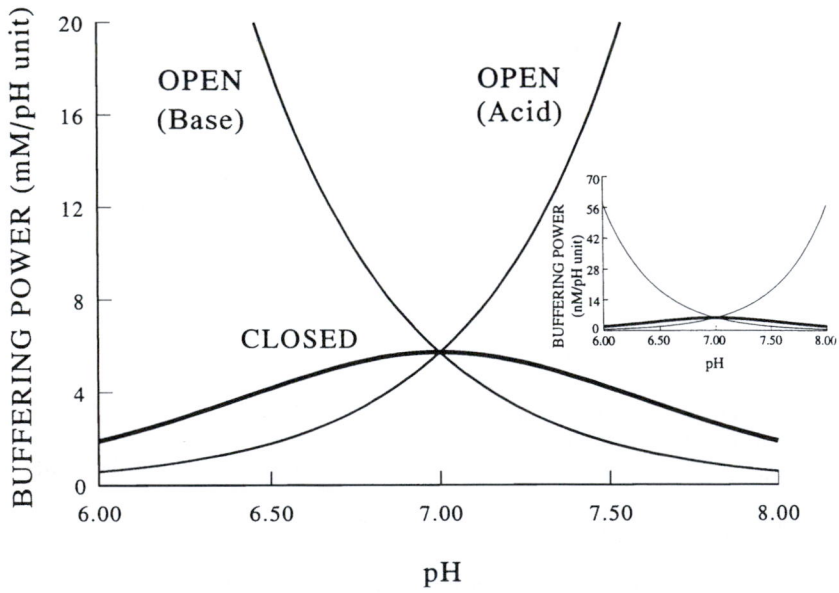

FIGURE 1.   The pH dependence of a buffer operating as a closed or open buffer. The buffering powers for both a weak acid buffer ($H^+ + A^- \rightleftharpoons HA$) and weak base buffer ($H^+ + B \rightleftharpoons BH^+$) are plotted. The buffers are assumed to have a pK_a' of 7.0 and to have the same buffering power at pH 7.0 whether operating as an open or closed buffer. Note that while closed buffers have maximal buffering power when pH = pK_a', the buffering power for an open buffer is higher at pH values higher than pK_a' (for weak acid buffers) or at pH values lower than pK_a' (for weak base buffers). Inset: the same plot with a different scale for the ordinate. Note how much higher the open buffering power is compared with the closed buffering power at pH values well above (weak acid) or below (weak base) the value of pK_a'.

fixed P_{CO_2}, the additional CO_2 diffuses from the solution and is removed. Thus, under these conditions, the amount of the uncharged form of the buffer (CO_2) remains constant, whereas the total buffer amount goes down because the [HCO_3^-] has decreased. The buffering power of an open buffer is given by

$$\beta_{open} = 2.303[A^-] \qquad (8)$$

where [A^-] = [HCO_3^-] for the open buffering power of CO_2/HCO_3^-. The relationship between pH and β_{open} for a theoretical open buffer is shown in Fig. 1.

Open buffers differ from closed buffers in two important respects. First, unlike closed buffers, the buffering power of open buffers is not maximal at their pK_a' values. In fact, open buffers become better buffers at the more extreme values of pH (Fig. 1). β_{open} becomes larger with alkalinization for weak acids and with acidification for weak bases. Second, open buffers have a much higher buffering power than closed buffers under similar conditions (Fig. 1). Note, however, that neither Eq. 7 nor Eq. 8 includes any indication of the nature of the buffer. Thus, all closed buffers are equally potent when the pH is at their pK_a' and all open buffers are equally potent when they have the same [A^-].

Since most uncharged substances are substantially more permeant through cell membranes than charged species, **almost all weak acids and bases can act as open buffers in cells.** For example, a weak acid that has a pK_a' of 4.0 acts like a closed buffer in solution, and since its pK_a' is 4.0, it would be a poor buffer at pH 7.0. However, inside a cell at pH 7.0, this weak acid acts like an open buffer. Any added H^+ will bind to the anionic form of the buffer. The uncharged buffer molecule formed will readily diffuse from the cell and be removed by the blood. Thus, although this weak acid is a poor buffer in the medium, it can contribute substantially to the buffering power inside a cell if present at sufficient concentration.

Complex solutions, like blood, contain multiple buffers. The **total buffering power** (β_{total}) of such solutions will be the sum of the various buffers; that is, buffers operate independently in solution. Thus,

$$\beta_{total} = \sum \beta_{closed} + \sum \beta_{open} \qquad (9)$$

Finally, it is inherent in the definition of buffering power (Eq. 6) that buffering power is a coefficient that can be used to convert a change of pH into a change in the amount of proton equivalents moved. This relationship has practical application. For instance, in the study of the regulation of intracellular pH, the activity of a transporter that moves H^+ across the surface membrane (such as the Na^+-H^+ exchanger, see Section VI) is determined by measuring the rate of change of pH_i (dpH_i/dt). The movement of H^+ on this transporter is accompanied by Na^+, whose movement is measured as a radioisotopic flux (amount of Na^+ influx per unit time). To compare the flux of Na^+ with the flux of H^+, the rate of change of pH_i needs to be converted to the amount of H^+ moved per unit time. This is accomplished by the following equation

$$J_H = \frac{dpH}{dt} \cdot \beta_{total} \qquad (10)$$

where J_H is the flux of protons and has units of mM H^+ per unit time. Using this equation, the flux of H^+, calculated from the measured rate of pH change, can be directly compared to the flux of another ion determined with radioisotopes.

III. Intracellular pH

Protons are just like any other cation, except for three distinguishing characteristics: (1) H ions are a dissociation product of water molecules ($H_2O \rightleftharpoons H^+ + OH^-$) and thus are always present in aqueous solutions, (2) H ions are present at very low concentrations in most solutions, and (3) H ions have much higher mobility than other cations. However, the equilibrium distribution and movement of H^+ across biological membranes are governed by the same principles that govern the movement of all other ions across biological membranes.

Originally, protons were assumed to be at equilibrium across biological membranes because of their very high mobility. Assuming an extracellular pH (pH_o) of 7.4, a V_m of –60 mV (inside negative), and assuming that H ions are passively distributed across the membrane (i.e., at equilibrium), pH_i would be 6.4 (calculated from the Nernst equation). However, at such an intracellular pH, metabolism and a variety of other cellular functions would be impaired. With the advent of modern reliable techniques for measuring intracellular pH, including **pH-sensitive glass microelectrodes** and **pH-sensitive fluorescent dyes** (see Appendix), it was shown that in the majority of cells pH_i was between 6.8 and 7.2, well above the calculated value for equilibrium pH. It is now clear that for most cells (with the notable exception of red blood cells), pH_i is considerably more alkaline than it would be if protons were at passive equilibrium across the cell membrane.

The question still remains how pH_i can be well above the equilibrium value, since H^+ should be highly permeant to most cell membranes. In fact, the **permeability of H^+** across biological membranes has been estimated to be between 10^{-4} and 10^{-2} cm/s, about four orders of magnitude higher than typical K^+ permeabilities. However, it is the conductance, and not the permeability, of H^+ that is crucial. **Conductance** is a measure of ion flux and is a function of both the permeability and the free concentration of an ion. H ions have low conductance across biological membranes despite their high permeability, because they are present in such low concentrations (10^{-7} M free concentration for H^+ versus 10^{-1} M free concentration for K^+). Thus, the acidifying influx of H ions (down their electrochemical gradient) will be small, and these H ions can easily be removed from the cell by membrane transport systems (see Section VI).

In summary, most cells have a cytoplasmic pH that is more alkaline than the value calculated assuming equilibrium of H^+ across the cell membrane, and pH_i for most cells is about 6.8–7.2.

IV. Organellar pH

Several intracellular organelles independently control their internal pH, which differs from the cytoplasmic pH (Fig. 2). These organelles include **mitochondria** and **acidic intracellular organelles.**

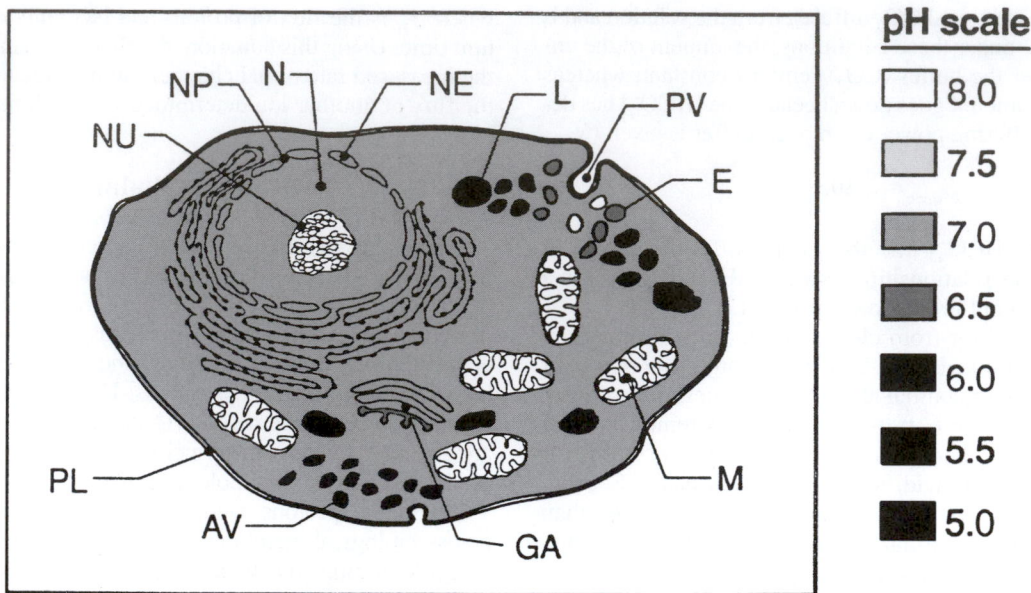

FIGURE 2.   A diagram of an idealized cell with the gray scale representing different values of pH (note calibration scale on the right). The cytoplasm has a fairly uniform pH of about 7.0, but note the alkaline mitochondria, the increasing acidification in the Golgi apparatus and in the endosomes, and the very acidic lysosomes. AV—acidic vesicles; E—endosomes; GA—Golgi apparatus; L—lysosomes; M—mitochondria; N—nucleus; NE—nuclear envelope; NP—nuclear pore; NU—nucleolus; PL—plasmalemma; PV—pinocytotic vesicle.

A. Mitochondria

One of the major roles of mitochondria is the production of ATP. A proton gradient across the inner mitochondrial membrane is required for the production of ATP. The electron transport chain translocates protons from the mitochondrion to the cytoplasm across the inner mitochondrial membrane. This proton extrusion creates an electrical and chemical gradient for proton influx into the mitochondrion. This H^+ influx occurs through a membrane-bound ATPase that produces ATP upon passive proton flux back into the mitochondrion. This is known as the **chemiosmotic hypothesis.**

The extrusion of H^+ to establish a proton gradient renders mitochondria alkaline relative to the cytoplasm by about 0.3–0.5 pH unit. Thus, intramitochondrial pH can be between 7.5 and 8.0. In addition, this pH gradient is maintained in the face of considerable acid loads, indicating that mitochondria may well be able to regulate their internal pH independently of cytoplasmic pH.

B. Acidic Intracellular Organelles

Organelles with a markedly acidic interior are those involved in either the endocytic pathway or the secretory pathway. The acidic organelles involved in endocytosis include **coated pits, endosomes** (i.e., prelysosomal endocytic vesicles), and **lysosomes.** Acidic vesicles in the secretory pathway include the **Golgi apparatus** (or at least part of it) and **storage granules** for amines (e.g., chromaffin granules involved in catecholamine secretion) and peptides (e.g. secretory granules in the endocrine pancreas). The pH is not known for all of these

compartments, but can be as low as 4.5–5.0 in lysosomes and 5.0–5.7 in endosomes and secretory granules. The pH in the Golgi apparatus seems to fall the farther the compartment is from the nucleus, with values of 6.2–6.4 in the medial Golgi and values as low as 5.9 in the trans Golgi network (TGN).

The acidic internal environment of these various organelles is believed to be necessary for their function. For example, the primary function of lysosomes is the biochemical degradation of macromolecules, and these organelles contain a large number of hydrolytic enzymes whose pH optima are about pH 5.0. This serves as protection for the cell. If the lysosomes should leak, the hydrolytic enzymes would be inactivated by the high cytoplasmic pH, thus preventing the indiscriminate degradation of important macromolecules. Receptor-mediated endocytosis involves the internalization of ligand-receptor complexes in endocytic vesicles, or endosomes. Acidification of these vesicles is essential for the dissociation of the ligand from the receptor within the endosome. Once the ligand has dissociated, the internalized receptor is recycled to the surface membrane, while the ligand is delivered to the lysosome. Finally, secretory granules (and perhaps part of the Golgi) serve to accumulate macromolecules to be secreted, and often mediate processing or modification of these substances. The maintenance of an acidic environment in these granules can be crucial to both of these functions. There is evidence that the large outward H^+ gradient is used to accumulate biogenic amines in amine secretory granules. This accumulation may be mediated by an H^+-amine exchanger in the granule membrane. These accumulated substances can be biochemically modified into their final form, and the enzymes responsible for these modifications often have low pH optima or depend on the availability of organic compounds that will accumulate only in acidic compartments. For all of these acidic compartments, then, it is clear that their proper functioning depends on the maintenance of a low internal pH.

C. Nucleus

The nucleus is separated from the cytoplasm by a double-membrane system that has large nuclear pores. Large macro-molecules (up to 5 kDa) readily permeate the nuclear pores and rapidly come to equilibrium between the nucleus and cy-toplasm. Given this high degree of permeability of the nu-clear membrane, it was believed that nuclear pH could not differ much from cytoplasmic pH. However, recent measure-ments clearly show that nuclear pH is from 0.1 to 0.5 pH unit more alkaline than cytoplasmic pH. These data imply that the nuclear pores do indeed represent a barrier to the free move-ment of H ions between the nucleus and cytoplasm.

V. Maintenance of a Steady-State pH$_i$

If a cell is maintaining a **steady-state pH$_i$** (i.e., the pH of the cytoplasm), the rate of acid loading must be equal to the rate of acid extrusion from the cell (in these terms it is ex-actly equivalent if an acid molecule moves in one direction

or a base molecule moves in the opposite direction) (Fig. 3). Several processes can contribute to acid loading of a cell, in-cluding metabolic production of acid, passive influx of H$^+$ across the cell membrane, leakage of H$^+$ from acidic inter-nal compartments, pumping of H$^+$ from alkaline internal compartments, and active influx of H$^+$ or active extrusion of base. Conversely, acid extrusion includes metabolic con-sumption of H$^+$, sequestration of H$^+$ in internal compart-ments, and active extrusion of acid or active influx of base.

A. Metabolic Production and Consumption of Acids

Several metabolic reactions involve the production of H$^+$ (Table 1). These processes include the production of CO_2, gly-colysis (through the generation of lactic and pyruvic acids), formation of creatine phosphate, ATP hydrolysis, lipolysis, triglyceride hydrolysis, the generation of superoxide, and the operation of the hexose monophosphate shunt. Obviously, when these reactions run in the opposite direction (e.g., creatine phosphate hydrolysis, ATP formation, or consumption of CO_2), there is a net consumption of H$^+$ and the cell will alkalinize.

When the cell is at steady state, these reactions must be bal-anced, and thus the levels of cellular metabolites such as ATP, creatine phosphate, CO_2, and lactate will also be at steady state. However, during transient periods, these metabolic reactions

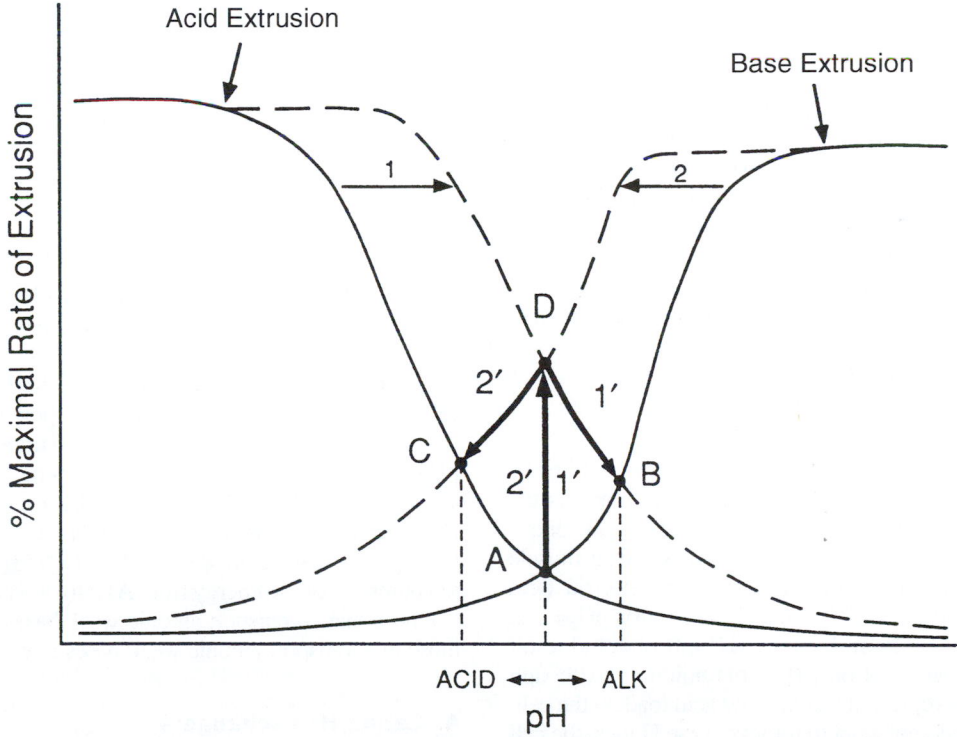

FIGURE 3. A model for the regulation of steady-state pH$_i$. In the cell, steady-state pH$_i$ is determined as the point where acid-extruding processes are balanced by base-extruding (or acid influx) pathways (point A). If acid-extruding processes are activated (arrow 1—represented as a shift to the right of the pH versus % maximal rate curve), as occurs when growth factors activate Na$^+$-H$^+$ exchange, the rate of acid extrusion exceeds the rate of base extrusion and the cell alkalinizes (arrows 1′). Eventually, the cell alkalin-izes to the point where acid and base extrusion are again equal and a new alkaline steady state pH$_i$ is reached (point B). In contrast, if base extrusion is activated (arrow 2—represented as a shift to the left of the pH versus % maximal rate curve), as occurs with an in-crease in the activity of Cl$^-$-HCO$_3^-$ exchange, the rate of base extrusion exceeds acid extrusion and the cell will acidify (arrows 2′). Eventually, the cell acidifies to the point where acid and base extrusion are again equal and a new acidic steady-state pH$_i$ is reached (point C). If acid and base extrusion are activated to the same extent, the steady state pH$_i$ will remain unchanged (point D).

TABLE 1 Examples of Various Metabolic Reactions That Involve the Generation or Consumption of H^+

Glycolysis[a]	Glucose + $2MgADP^- + 2P_i^{2-} \rightarrow 2(lactate)^-$ $+ 2MgATP^{2-}$ $2MgATP^{2-} \rightarrow 2MgADP^- + 2P_i^{2-} + 2H^+$
Glycogenolysis[b]	Glycogen + $3P_i^{2-} + 3MgADP^- + H^+ \rightarrow$ $3MgATP^{2-} + 2(lactate)^- + glycogen$
Creatine phosphate hydrolysis	H^+ + creatine phosphate^{2-} + $MgADP^-$ $\rightarrow MgATP^{2-}$ + creatine
Lipolysis	Triglyceride $\rightarrow 3(palmitate)^- + 3H^+$ $3(palmitate)^- + 3MgATP^{2-} + 3CoA^{4-} \rightarrow$ $3(palmitoyl\ CoA)^{4-} + 3AMP^{2-} + 6P_i^{2-}$ $+ 3H^+ + 3Mg^{2+}$
Superoxide formation	$NADPH + 2O_2 \rightarrow 2O_2^- + NADP^+ + H^+$
Hexose-monophosphate shunt	$G6P^- + 12NADP^+ + 6H_2O \rightarrow NADPH$ $+ P_i^- + 12H^+ + 6CO_2$

[a] At pH 7.2 and high [Mg^{2+}], $2H^+$ are always generated but the balance of H^+ produced by the two reactions varies with pH and Mg.

[b] Net H^+ production when hydrolysis of the 3MgATP is taken into account.

can result in net changes in metabolite concentrations and thus contribute markedly to changes in pH_i. For instance, during periods of ischemia, cellular lactate and CO_2 may accumulate while ATP will be hydrolyzed. All of these reactions can contribute to the observed cellular acidification. The hydrolysis of creatine phosphate during ischemia consumes H^+ and will blunt the cellular acidification. Another example of a transient effect of metabolism on cellular pH is the initial acidification seen upon activation of neutrophils with phorbol esters. In this case, stimulation of the production of superoxide and activation of the hexose monophosphate shunt results in increased production of H^+.

B. Passive Transmembrane Flux of H^+

Although the H^+ permeability of most biological membranes is quite high ($P_H \approx 10^{-3}$ cm/s), the actual H^+ flux across the membrane is quite low because of the low free H^+ concentration. For example, the putative H^+ flux across a frog muscle fiber can be calculated. Assuming a constant electric field across the membrane, $P_H = 10^{-3}$ cm/s, $V_m = -90$ mV, $pH_i = 7.2$, $pH_o = 7.35$, and $\beta_{total} = 26$ mM, passive H^+ influx would result in a cellular acidification of only 0.02 pH unit/hour. Although negligible, this does represent a continued acid load on the cell, and if mechanisms do not exist to remove these H ions, the cell will eventually acidify toward the equilibrium pH_i value.

H^+ currents associated with **proton channels** have been reported in several cells, including snail neurons, salamander oocytes, and a number of mammalian cells. All of these channels are activated by depolarization and the resulting H^+ currents alkalinize the cell. Other characteristics of these channels include inhibition by divalents (such as Zn^{2+} and Cd^{2+}), gating by both intracellular and extracellular pH, and

low single-channel conductance. These channels have also been shown to have a very high temperature dependence, an unusual feature for an ion channel. It has recently been proposed that these channels can be classified into four isoforms based on channel kinetic properties. These various isoforms include *n* channels (in snail neuron), *o* channels (in oocyte), *e* (for epithelial) channels (in rat alveolar epithelium and frog proximal tubule), and *p* (for phagocyte) channels (in macrophage, neutrophil, CHO, microglia, and myotube, among others). The significance of these channels is not known, but they may contribute to the maintenance of pH in a restricted submembrane space within the cell or contribute to pH_i regulation in cells undergoing prolonged depolarization, such as the prolonged depolarizing fertilization potential in oocytes.

C. Internal Compartments

If the pH values of intracellular compartments, such as mitochondria and lysosomes, are at steady state, then they should have no impact on cytoplasmic pH. However, under pathological conditions where mitochondria or acidic intracellular compartments are rendered leaky or mitochondria take up cytoplasmic Ca^{2+} in exchange for H^+, these compartments could influence pH_i. Under normal conditions, though, these compartments should contribute little to the maintenance of a steady-state pH_i, especially owing to their relatively small volume compared with total cell volume.

It is clear that, at the very least, cells face a continuous acid load from passive H^+ influx. In addition, under many conditions of metabolic stress, cells also experience a metabolic acid load. Thus, cells must possess active extrusion mechanisms to maintain a steady-state pH_i well above the equilibrium value for pH_i.

VI. Active Membrane Transport of Acids and Bases

Several integral proteins within the surface membrane of cells are specialized for the active transport of acids and bases across the membrane. Because of their importance to cellular pH regulation, these transport pathways have been extensively studied and can be divided into five classes: (1) those that move H^+ directly in exchange for another cation; (2) those that move HCO_3^-, or an associated species like CO_3^{2-}; (3) H^+-ATPases (proton pumps) that use energy from ATP hydrolysis to transport H^+; (4) those that cotransport anionic weak bases with Na^+; and (5) those that transport anionic weak bases in exchange for Cl^-.

A. Cation-H^+ Exchangers

The best characterized of the cation-H^+ exchangers is the **Na^+-H^+ exchanger (NHE)** (model 1 in Fig. 4). This exchanger responds to cellular acidification by extruding one H^+ in exchange for the influx of one Na^+ (1:1 stoichiometry means that the exchanger is **electroneutral**, i.e., it does not involve net charge movement). There are now known to be at least five isoforms of the Na^+-H^+ exchanger in mammalian cells. NHE-1 is found in virtually all cells, is inhibited by the loop diuretic

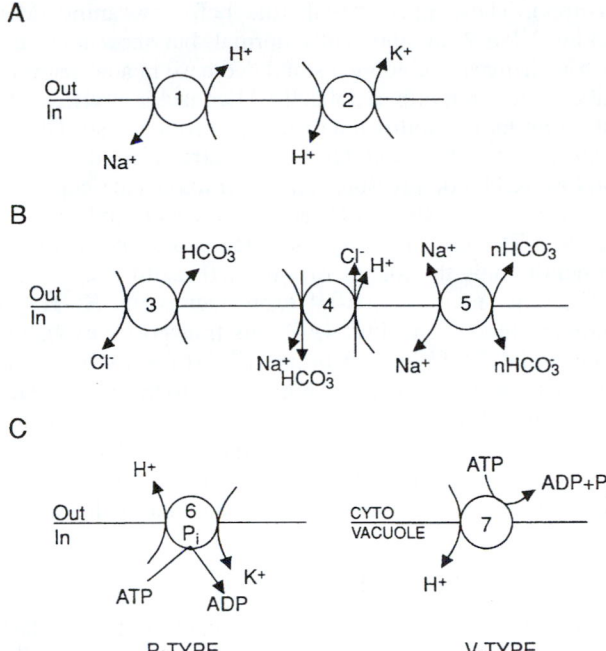

is largely expressed in the brain of rats, especially in neuronal cell bodies. Other NHE isoforms undoubtedly exist in mammalian cells. A sixth isoform, NHE-6, has been described in human mitochondria. In addition, a Cl-dependent Na$^+$-H$^+$ exchanger, described in colonic crypt cells, might represent yet another mammalian NHE isoform. Other NHE isoforms also exist in cells from lower vertebrates, invertebrates, plants, and bacteria. For instance, an unusual isoform, β-NHE, that is 50–75% homologous to NHE-1 and is activated by cAMP (unlike NHE-1) has been described in trout red blood cells. The significance of these different isoforms is still largely unclear, but given the variety of functions performed by NHE in cells, it is not surprising that multiple forms have arisen.

The basic structure of the five mammalian plasma membrane NHE isoforms is similar (Fig. 5A), containing two major domains, a transmembrane domain and a cytoplasmic domain.

FIGURE 4. Models of several different types of pH-regulating transporters. (A) Cation-H$^+$ exchangers. Included in this group are the alkalinizing Na$^+$-H$^+$ exchanger (1) and the acidifying K$^+$-H$^+$ exchanger (2). (B) HCO$_3^-$-dependent transporters. This group includes the Cl$^-$-HCO$_3^-$ exchanger (band 3 from red blood cells) (3). Another transporter in this group is the (Na$^+$+HCO$_3^-$)-Cl$^-$ exchanger (4). Shown here is merely one possible model for this exchanger. This transporter could involve the influx of two HCO$_3^-$ instead of the influx of one HCO$_3^-$ and the efflux of an H$^+$. Alternatively, one CO$_3^{-2}$ or two NaCO$_3^-$ could be transported in. These four different variants have the same effect on pH, but can be partially distinguished kinetically. The Na$^+$-HCO$_3^-$ cotransporter is shown as model 5. This transporter is electrogenic, mediating the movement of n HCO$_3^-$ for each Na$^+$, and is thus sensitive to membrane potential (V_m). The direction of cotransport depends on the value of V_m. (C) Two different types of H$^+$-ATPases. The P-type H$^+$-ATPase (6), typified by the electroneutral gastric H$^+$,K$^+$-ATPase, involves a phosphorylated intermediate. The V-type H$^+$-ATPase (7), typified by the vacuolar H$^+$-ATPase, does not produce a phosphorylated intermediate.

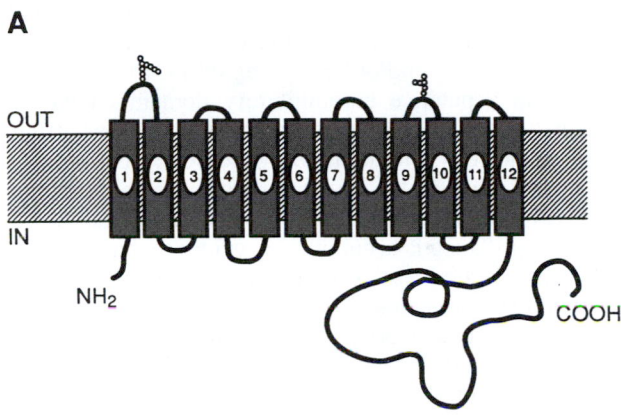

A

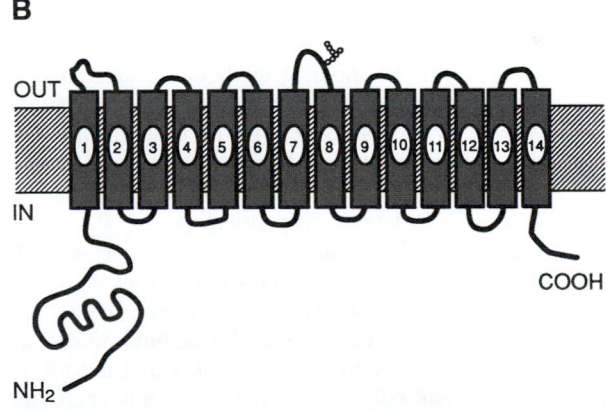

B

amiloride and its analogs, can be activated by a variety of agents, and has a molecular weight of 91 000. NHE-2, which has only 50% amino acid homology with NHE-1, is found predominantly in the gastrointestinal tract (GI tract) and kidney, is much less sensitive to inhibition by amiloride analogs than NHE-1, and has a molecular weight of about 91 000. NHE-3 has about 40% amino acid homology with NHE-1, has a molecular weight of 93 000, is expressed largely in the GI tract and kidney, and is not readily inhibited by amiloride. NHE-4 also has about 40% amino acid homology with NHE-1, is the smallest NHE isoform with a molecular weight of 81 000, is largely expressed in the GI tract (with some expression in the uterus, brain, and kidney), and is not very sensitive to inhibition by amiloride. NHE-5 has recently been characterized from humans and rats. This isoform is most similar to NHE-3 (62% amino acid homology), but has a larger molecular weight, 99 000, than any of the the other NHE isoforms. This isoform is not very sensitive to inhibition by amiloride analogs. NHE-5

FIGURE 5. Models of (A) the structure of the Na$^+$-H$^+$ exchanger (NHE-1) and (B) the Cl$^-$-HCO$_3^-$ exchanger (AE-1). The Na$^+$-H$^+$ exchanger is believed to have 12 membrane-spanning regions with a large cytoplasmic domain at the C-terminal end. This cytoplasmic domain contains several potential phosphorylation sites that are important to the regulation of the Na$^+$-H$^+$ exchanger. The N-terminal end is also cytoplasmic. Two potential glycosylation sites are shown. The Cl$^-$-HCO$_3^-$ exchanger is believed to have 14 membrane-spanning regions, and both terminal ends are cytoplasmic. The large cytoplasmic domain is on the N-terminal end in the Cl$^-$-HCO$_3^-$ exchanger. This cytoplasmic domain (in red blood cells at least) contains many binding sites, including those for the cytoskeletal element ankyrin, for hemoglobin, and for some glycolytic enzymes.

The transmembrane domain (N-terminal ~500 amino acids) is believed to have 12 membrane-spanning regions with an N-linked glycosylation site on the first external loop. This means that NHE is a glycoprotein. The cytoplasmic domain (C-terminal ~300 amino acids) represents a large cytoplasmic region with available serine phosphorylation sites. Both the N- and C-terminal ends of NHE are cytoplasmic. It is believed that the cytoplasmic C-terminal end is involved in the regulation of NHE while the membrane-spanning regions are responsible for Na^+ and H^+ transport. This is borne out by the expression of NHE mutants that lack the C-terminal end in fibroblasts that do not have NHE. In these cells, Na^+-H^+ exchange activity is seen, but can no longer be activated by growth factors.

A wide variety of substances have been shown to activate NHE, including hormones, neurotransmitters, growth factors, and the extracellular matrix. Many of these factors are believed to activate the NHE by phosphorylating serine residues on the C-terminal end. These factors may also affect the NHE through the binding of a regulatory protein with the exchanger and/or by binding of a Ca^{2+}-calmodulin complex to a putative autoinhibitory domain on the exchanger. Activation often involves an increase in the affinity of the exchanger's internal binding site for cytoplasmic H^+ (see arrow 1, Fig. 3). In some cases, activation of NHE may also involve interaction between the exchanger and the cell cytoskeleton. There is accumulating evidence that NHE isoforms may function best as homodimers. The functional significance of dimerization is unclear. Cell acidification also dramatically activates NHE. This activation is due to an internal allosteric H^+ binding site on NHE (distinct from the H^+ transport site) that increases exchange when occupied. Finally, decreased cell volume (cell shrinkage) also activates NHE. This activation does not involve phosphorylation of NHE and appears to be mediated by a unique activation pathway.

Many signaling pathways can activate NHE, depending on the stimulus. For example, growth factor activation of NHE is usually mediated by an elevation of intracellular Ca^{2+} and/or by an increased activity of protein kinase C. These pathways commonly involve phosphoinositide breakdown. Activation of NHE by cell shrinkage appears to involve ATP- and GTP-dependent pathways.

Much progress has been made recently to correlate NHE function with structure. As stated above, the transmembrane domain contains the Na^+ and H^+ transport sites. In addition, this domain contains the external amiloride binding site and the internal H^+ allosteric binding site, although the pK for this latter site is markedly acid in the absence of the C-terminal cytoplasmic domain. The C-terminal cytoplasmic domain contains several putative phosphorylation sites and is involved in the regulation of NHE. This region also contains the autoinhibitory domain, the calmodulin binding site, and a possible binding site for the putative cytoplasmic regulatory factor. The C-terminal domain may also contain sites able to interact with cytoskeletal proteins.

Transgenic mice now exist that lack NHE-1, NHE-2 and NHE-3. All 3 types of mice are viable, but mice lacking NHE-1 have slowed postnatal development and by 14 days exhibit neurologic symptoms (primarily gait problems and seizures). These mice usually die before weaning. Mice lacking NHE-2 are apparently normal, but show long-term gastric changes, including a small decrease in acid secretion and degeneration of parietal cells. Mice lacking NHE-3 also have long-term viability but exhibit gastrointestinal (GI) and renal defects. These defects cause diarrhea, mild acidosis, and lowered blood pressure. These symptoms are consistent with a role for NHE-3 in GI and renal HCO_3^- and fluid absorption. The use of transgenic mice should be helpful in further defining the role of various NHE isoforms.

The other major cation-H^+ exchanger is the **K^+-H^+ exchanger** (model 2 in Fig. 4A). This transporter exchanges intracellular K^+ for extracellular H^+ and results in cellular acidification. K^+-H^+ exchange has been found in nucleated red blood cells and mediates solute efflux during regulatory volume decrease (see Section VII.H). It has also been observed in retinal pigment epithelial cells, where it enables the cell to regulate pH_i in the face of an alkaline load.

B. HCO_3^--Dependent Transporters

These transporters are actually a superfamily of related transport proteins that affect pH_i and are characterized by their ability to transport HCO_3^- (or a related species like CO_3^{2-}) and by the ability of disulfonic stilbene derivatives to inhibit them. The three major types of HCO_3^--dependent transporters are Cl^--HCO_3^- exchange (band 3 from red blood cells, termed AE for anion exchanger), (Na^++HCO_3^-)-Cl^- exchange (often termed Na^+-dependent or Na^+-driven Cl^--HCO_3^- exchange), and electrogenic Na^+-HCO_3^- cotransport (termed NBC for Na^+ bicarbonate cotransport).

The operation of **Cl^--HCO_3^- exchange (AE)** (model 3 in Fig. 4B) has been studied most extensively in red blood cells. This electroneutral transporter involves a 1:1 exchange of Cl^- for HCO_3^- (although many other ions, such as SO_4^{2-}, can also be transported under specific conditions). As with the Na^+-H^+ exchanger, the Cl^--HCO_3^- exchanger has several isoforms, denoted AE 1, 2, and 3. These various isoforms have about 80–90% homology. AE 1 is the band 3 transporter from red blood cells and is the smallest isoform ($M_r = 115\,000$). This protein has as many as 14 membrane-spanning domains (helices) and a large N-terminal cytoplasmic domain that contains various binding sites, including one for ankyrin (Fig. 5B). AE 2 and 3 are larger ($M_r = 145\,000$–$165\,000$). AE 2 appears to be the housekeeping anion exchanger and is widely distributed. It is activated by cellular alkalinization and returns pH_i to normal by extruding base (HCO_3^-) and thereby reacidifying the cell. This exchanger may also possess an internal allosteric regulatory site that activates the exchanger at **alkaline** values of pH_i (see the base extrusion curve in Fig. 3). In addition to its role in the regulation of pH_i, Cl^--HCO_3^- exchange may also play a role in regulating intracellular Cl^-. The function of AE 3 is not yet clear but it has a far more restricted distribution than AE 2, being found only in the heart and the central nervous system.

The **(Na^++HCO_3^-)-Cl^- exchanger** (model 4 in Fig. 4B) was originally described as the pH-regulating transport system in invertebrate nerve and muscle preparations, but has since been found in a wide variety of cells. Because it transports Na^+, it operates in the opposite way than Na^+-independent Cl^--HCO_3^- exchange (i.e., the anion exchangers described above). Thus, the $(Na^++HCO_3^-)$-Cl^- exchanger exchanges one external Na^+ for one internal Cl^- and neutralizes the equivalent of two internal protons. As such, this exchanger is electroneutral and mediates the alkalinization of the cell in response to an acid load. The ability of this exchanger to neutralize two acid equivalents could be achieved by the influx of two HCO_3^-, the influx of one HCO_3^- in exchange for the efflux of one H^+, the influx of one CO_3^{2-}, or the influx of an ion pair ($NaCO_3^-$). In squid axon, this exchanger has been suggested to involve $NaCO_3^-$-Cl^- exchange, whereas in barnacle muscle fibers it must be another variant, suggesting that this exchanger also has at least two isoforms. The structure of this exchanger is not currently known.

The **Na^+-HCO_3^- cotransporter (NBC)** (model 5 in Fig. 4B) was originally identified in renal epithelial cells but has since been found in a number of other cell types as well. It mediates the movement of one Na^+ with one, two, or three HCO_3^-, depending on the cell type in which the transporter is located. This transporter differs from the other HCO_3^--dependent transporters in that it does not require Cl^- and, in the case where the stoichiometry is 1:2 or 1:3, it is electrogenic. In proximal tubule epithelial cells, NBC is involved in HCO_3^- reabsorption by mediating HCO_3^- efflux across the basolateral membrane. In other cells, the cotransporter is proposed to mediate HCO_3^- influx and contribute to the regulation of intracellular pH in the face of an acid load. The renal cotransporter (1:3 stoichiometry) has been suggested to involve the cotransport of one Na^+ with one HCO_3^- and one CO_3^{2-}.

The NBC and several of its variants have recently been cloned. The first mammalian NBC to be cloned is from rat kidney (rkNBC). This is a glycoprotein consisting of 1035 amino acids (M_r ~130 000). Its structure is similar to other membrane transport proteins, containing at least 10 membrane-spanning regions, a large extracellular loop between membrane-spanning segments 5 and 6, and large cytoplasmic domains (one each at the N- and C-terminal ends). These cytoplasmic domains are probably the site of cotransport regulation and contain putative phosphorylation sites for protein kinases A and C, casein kinase II, and tyrosine kinase. A human kidney clone (hkNBC), with 97% homology to rkNBC, has recently been described. These kidney NBCs are electrogenic, with a stoichiometry of one Na^+ transported for every three HCO_3^- ions.

An NBC from human heart (hhNBC) has recently been described. This clone (along with an identical one from human pancreas) is identical to rkNBC, except that hhNBC has a longer N-terminus (by 44 amino acids); hhNBC is also electrogenic, but with a stoichiometry of 1:2. A similar clone has been isolated from rat brain (rbNBC). This clone has longer C- and N-terminal ends, a stoichiometry of 1:2, and is predominantly localized to cortical neurons but not astrocytes. Finally, a novel clone has been isolated from vascular smooth muscle cells (also found in testis and spleen cells). This clone (NBC$_N$-1) has a very large N-terminus, is **electroneutral** (stoichiometry of 1:1), and is apparently not inhibited by DIDS.

These NBC clones are about 30–35% identical to AE transport proteins, indicating that these transporters are indeed part of a superfamily. It will be of interest to see the homology of the $(Na^++HCO_3^-)$-Cl^- exchanger with NBC and AE, once the former has been cloned.

C. H^+-ATPases (Proton Pumps)

There are at least three known varieties of **H^+-ATPases:** (1) F_0-F_1-type ATPase; (2) E_1-E_2 or P-type ATPase; and (3) vacuolar or V-type ATPase. The **F_0-F_1-type ATPase** is found in mitochondria (see Chapter 8). This ATPase has also been called the ATP synthase, since it functions to produce ATP when H^+ moves down its electrochemical gradient. F_0-F_1-type ATPase has a lollipop shape and a membrane-spanning region, F_0, that forms the putative H^+ pore through the membrane as well as an extrinsic head region, F_1, that contains the ATPase activity. The F_0 subunit is quite large, with 5 subunits and a molecular weight of about 380 000. The F_1 region has 4 subunits and a molecular weight of about 100 000. These two regions are connected by a stalk that contains several subunits, one of which confers sensitivity to oligomycin. The F_0-F_1-type ATPase can be inhibited by azide and N,N'-dicyclohexylcarbodiimide (DCCD) in addition to oligomycin.

The **P-type ATPases** (model 6 in Fig. 4C) are characterized by forming a phosphorylated intermediate upon ATP hydrolysis. The classic example of such an ATPase is the Na^+,K^+-ATPase. An example of a **P-type H^+-ATPase** is the H^+,K^+-ATPase, best characterized from the apical membrane of gastric glands, where it is responsible for acid secretion into the stomach. This ATPase has also been implicated in the acidification of the urine and reabsorption of K^+ by the kidney and in the establishment of an H^+ gradient across yeast plasma membrane. In contrast to the F_0-F_1-ATPase, the H^+,K^+-ATPase exchanges one K^+ for one H^+ and is thus electroneutral. The H^+,K^+-ATPase has a molecular weight of about 110 000 and has a structure similar to those of other membrane-bound ATPases with several membrane-spanning regions and a large cytoplasmic domain containing the ATP hydrolysis site. Inhibition by vanadate is characteristic for P-type ATPases.

The third type of H^+-ATPase is the vacuolar, or V-type, ATPase (model 7 in Fig. 4C) which is found in yeast and plant vacuoles as well as in several eukaryotic cells (e.g., kidney cells, osteoclasts, and macrophages) and organelles (e.g., endosomes, lysosomes, secretory granules, and Golgi apparatuses). **V-type ATPases** are more like the F_0-F_1-type ATPase than the P-type ATPase in that they do not form phosphorylated intermediates, are quite large (>400 kDa), assume a lollipop shape, and are electrogenic. The major function of V-type ATPases is the acidification of intracellular organelles (see Section IV.B), which is important for proper protein targeting and handling. These ATPases are characterized by their lack of sensitivity to vanadate and oligomycin and by their inhibition by N-ethylmaleimide (NEM), DCCD, and bafilomycin. While P- and F_0-F_1-type ATPases are found in both prokaryotes and eukaryotes, the V-type ATPases are only found in eukaryotes and therefore presumably evolved more recently.

D. Na^+-Organic Anion Cotransport

In a variety of organisms, renal proximal tubule cells have been shown to have **Na^+-organic anion cotransporters** (model 1 in Fig. 6). These cotransporters mediate the influx

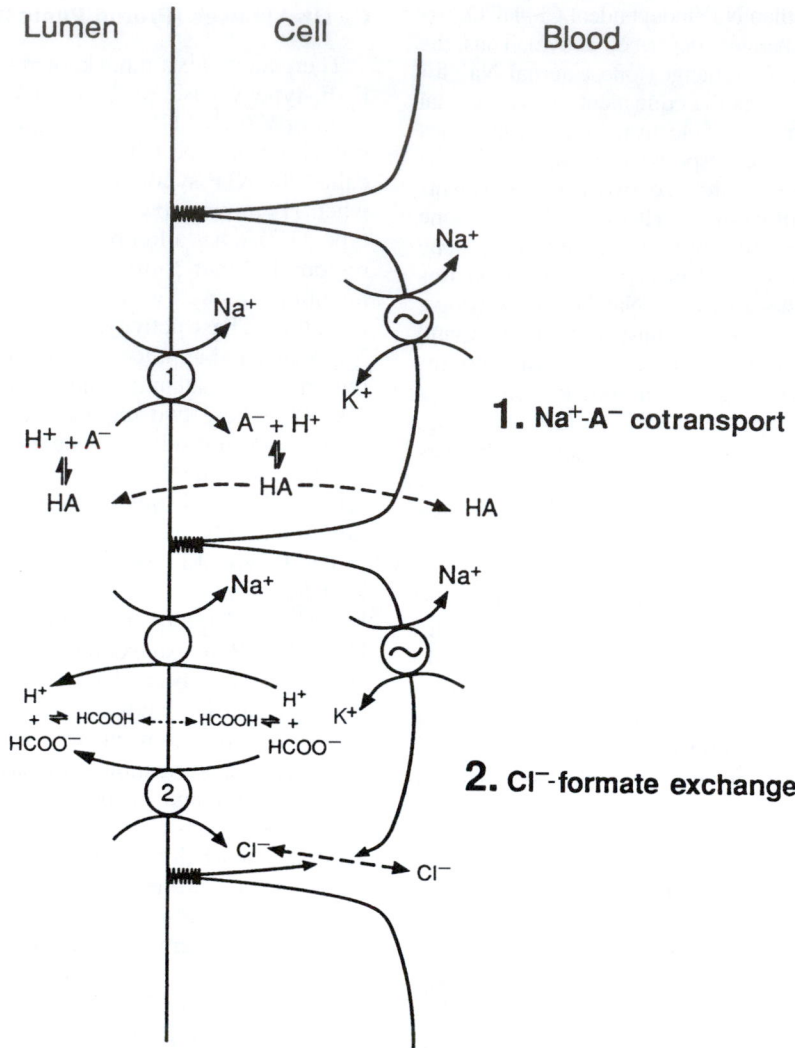

FIGURE 6. A model of anionic weak base fluxes in the renal proximal tubule mediated by either cotransport with Na^+ (1) or in exchange for Cl^- (2). An example of 1 is the Na^+-acetate cotransporter. The Cl^--formate exchanger is depicted by 2. These transporters are apparently involved in the luminal entry of Na^+ or Cl^- in NaCl-reabsorbing epithelia like the renal proximal tubule. Solid lines represent ion fluxes mediated by membrane transporters. Dotted lines represent passive diffusion of molecules across the membrane. Cl^- movements across the basolateral membrane are mediated by an ion-selective Cl^- channel.

of a Na^+ with an organic anionic weak base, such as lactate or acetate. Such cotransport would cause cellular alkalinization due to the entry of base. The anionic base would bind a proton upon entry and alkalinize the cell. These transport systems are likely to be part of a mechanism designed for the transepithelial movement of organic molecules and should be considered to **affect** pH_i rather than **regulate** it.

E. Chloride-Organic Anion Exchange

Another class of organic anion transporters has been described in renal proximal tubule cells. These transporters involve the exchange of organic anions for inorganic anions such as Cl^- or OH^-. For example, as part of the mechanism for NaCl reabsorption, an exchanger that mediates Cl^- influx (from the lumen) in exchange for formate efflux resides in the apical membrane of proximal tubule epithelia. This

Cl^--formate exchanger (model 2 in Fig. 6) can be inhibited by disulfonic stilbene derivatives and is functionally similar to the Cl^--HCO_3^- exchanger. However, the Cl^--formate exchanger is distinct from the Cl^--HCO_3^- exchanger. The operation of this transporter during NaCl reabsorption should result in epithelial cell acidification. Other transporters involve the movement of organic anions including urate and oxalate. When these anions are transported into the cell, they will result in cellular alkalinization.

VII. Cellular Functions Affected by Intracellular pH

A wide variety of cellular processes and properties are affected by intracellular pH, and perhaps in some way all cell functions are influenced by the level of pH_i. It is impossible to discuss all of the various effects, but many of the most important will be highlighted in the following sections (Fig. 7).

A. Cellular Metabolism

The fact that cellular metabolism can affect pH_i was discussed previously (see Section V.A). It has been appreciated for many years that the converse is also true; that is, changes of pH_i can affect cellular metabolism. Theoretically, because pH will affect the charge on ionizable groups in proteins, it would be anticipated that changes in pH_i could change the configuration of proteins and affect their activity. Such an effect of pH_i has been well documented for two key metabolic enzymes. Phosphofructokinase, a key glycolytic enzyme that converts fructose 6-phosphate (F6P) to fructose 1,6-diphosphate (FDP), has an exquisite pH sensitivity in the physiological range (6.5–7.5), its activity decreasing with a decrease of pH_i. The actual pH sensitivity is dependent on the cellular levels of F6P and 5'-AMP. Similarly, the conversion of phosphorylase (which catalyzes the metabolism of glycogen) from its inactive to active form is inhibited by a decrease in pH_i.

Two observations can be made from these findings. First, the pH dependence of enzyme activity is often affected by the concentrations of other factors, including substrates and other effectors. Thus, caution must be exercised in relating the *in vitro* pH profile of an enzyme to cellular conditions. Second, the general reaction of metabolic enzymes to a decrease in pH is a reduction in activity. This suggests that a decrease in pH_i could be used to prevent growth or to put a cell in a dormant state, as has been observed for many cells (see Section VII.G).

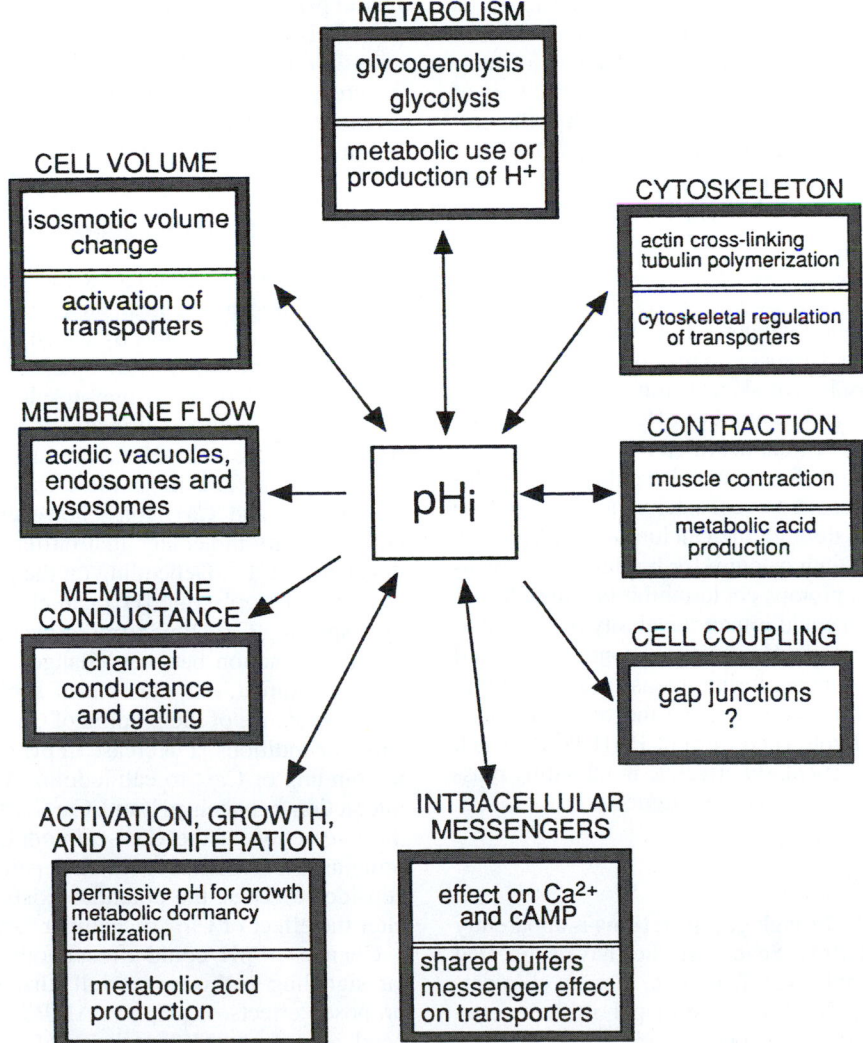

FIGURE 7. A summary of the cellular processes affected by intracellular pH. Each box represents a different cellular process. A two-headed arrow indicates that pH_i affects the process and that the process can also affect pH_i. Single-headed arrows suggest that pH affects that cellular function but that the cellular process probably does not have a major impact on pH_i. Within a box, the top half indicates the cellular processes affected by pH_i, while the bottom half indicates the mechanisms by which the process can affect pH_i.

B. Cytoskeleton

Changes in pH_i have been shown to affect the cytoskeleton, which can lead to changes in cell shape or motility. One example of such an effect on the cytoskeleton is the pH dependence of **actin filament** cross-linking to form gels. This cross-linking is mediated by actin-binding proteins whose ability to interact with actin is pH-dependent. In some cases, cell alkalinization increases the actin cross-linking to form a gel state in the cytoplasm or to form networks of microfilament bundles. In other cells, alkalinization reduces the cross-linking of actin filaments. The cell-specific responses of cytoskeletal cross-linking are probably due to different pH profiles of actin-binding proteins from different cells.

Changes in cellular pH can also affect the polymerization of cytoskeletal elements, such as **tubulin**. It has been shown that in some cells alkalinization can cause depolymerization of tubulin and disaggregation of microtubules within the cell. There are probably many other pH-dependent components to cytoskeletal assembly and function. It should be pointed out, however, that the conditions that lead to a change in pH_i are often accompanied by changes in intracellular calcium concentration and by phosphorylation of the cytoskeleton, and it is not always clear which is the predominant effector of cytoskeletal changes. Nevertheless, it is clear that changes in pH_i can play a major modulatory role in at least some alterations of the cytoskeleton.

C. Muscle Contraction

Intracellular acidification is known to reduce the ability of contractile cells to generate tension. This effect is particularly marked in cardiac muscle, but skeletal muscle also shows a reduced contractility at acidic values of pH_i. There are several possible ways by which cellular acidification could influence contractility: changes in surface channels could reduce cellular excitability, low pH could prevent calcium release from the sarcoplasmic reticulum through the calcium release channel, protons could compete with calcium for binding to the regulatory protein troponin, protons could inhibit the myofibrillar ATPase, or acidification could impair the ability of the cell to generate ATP. The effect of pH on muscle contraction could also be indirect. For example, during intense muscle activity inorganic phosphate (P_i) accumulates in the cell. A reduced pH will increase the diprotonated form of P_i ($H_2PO_4^-$), which has been shown to be particularly effective in inhibiting muscle force development during muscle fatigue.

D. Cell-Cell Coupling

The coupling of cells through **gap junctions** is apparently affected by intracellular pH. Several studies have suggested that a fall in pH_i uncouples gap junctions, thereby eliminating cell-cell coupling. The direct role of pH_i in the uncoupling process has been questioned, and it appears that changes in intracellular Ca^{2+} levels are responsible for uncoupling in some cells. The control of gap junction conductance may differ from cell to cell, but even in cells where Ca^{2+} is the primary regulator, changes in pH_i probably still have a modulatory effect on gap junction conductance.

E. Membrane Conductance

Ion-selective channels require the presence of charges within the channel proteins for proper ion conduction and channel gating. If these charges have pK values within the physiological range, these channels could well be affected by changes in pH. Indeed, the conductances of many channels are affected by changes in either pH_o or pH_i, including the tetrodotoxin-sensitive Na^+ channel, the delayed-rectifying and inward-rectifying K^+ channels, and Cl^- channels. Through the effect on conductance of membrane channels, changes in pH can affect the excitability of nerve and muscle cells and alter the membrane potential in all cells.

Recently, an interesting pH-sensitive K^+ channel has been described, the TASK channel (for TWIK-related acid-sensitive K^+ channel). The TASK channel exhibits rectification in asymmetric K^+ solutions that is consistent with the Goldman-Hodgkin-Katz equation, is noninactivating, is not voltage sensitive, and is highly sensitive to changes in **external** pH, with channel conductance falling with decreased pH_o. This type of leak channel is believed to be partly responsible for the resting membrane potential (especially in pancreas, placenta, and brain cells) and could explain the relationship between changes of pH_o and membrane potential in these cells.

F. Intracellular Messengers

Changes in pH_i can affect the levels of important intracellular signaling molecules, such as **Ca^{2+}** and **cAMP.** There are several possible ways by which pH can affect intracellular Ca^{2+}. An elevation of cytoplasmic H^+ can activate mitochondrial Ca^{2+}-H^+ exchange, resulting in a sequestering of H^+ within the mitochondria and an elevation of cytoplasmic $[Ca^{2+}]$. A decreased pH_i can reduce Ca^{2+} entry across the plasmalemma. The most direct interaction between cytoplasmic H^+ and Ca^{2+} ions, however, results from shared buffers. Many molecules that buffer H^+ will also bind, and thus buffer, Ca^{2+}. Depending on the relative affinities, an elevation of cytoplasmic $[H^+]$ can elevate intracellular $[Ca^{2+}]$ by displacing Ca^{2+} ions from intracellular buffer sites.

The interaction between changes of pH_i and $[Ca^{2+}]_i$ can also be indirect. An example of such an interaction is the pH-dependence of the binding of Ca^{2+} to calmodulin. Under certain conditions, a decrease in pH can be shown to reduce the binding of Ca^{2+} to calmodulin. Another example of the interaction between pH and Ca^{2+} is the pH-dependence of the interaction of the Ca^{2+}-calmodulin complex with other proteins, the direction of which depends on the protein being considered. Thus, the potential exists for changes of pH_i to alter the effect of Ca^{2+} on cellular function.

Changes of pH_i could affect another important intracellular signaling pathway as well, that involving cAMP. The proposed effects of pH_i on cAMP are based on the pH dependence of adenylyl cyclase (AC—the enzyme that synthesizes cAMP) and the cyclic nucleotide phosphodiesterase (PDE—the enzyme that hydrolyzes cAMP). In most cells, PDE apparently has a rather constant activity over the physiological range of pH (6.5–7.5). However, depending on the cell, an increase in pH can either markedly increase or decrease AC activity. Thus, alkalinization can result in either an increase or a decrease in cellular cAMP levels.

Given the pervasive effect of changes of pH_i on proteins, it is likely that pH_i effects on other signaling pathways, such as those mediated by cGMP or phosphoinositide metabolism, also exist. It should be noted, however, that the physiological significance of these pH_i effects on signaling pathways is often not clear, especially for cells that normally should see only small fluctuations of pH_i.

Some cells in the body function to sense changes in external pH, either in direct response to external acid (acid-sensing taste bud cells) or to elevated external CO_2 (glomus cells of the carotid body and central chemosensitive neurons). In these cells, extracellular acidification results in a **maintained** intracellular acidification, with pH recovery mechanisms being inhibited (most likely by decreased pH_o). It is believed that this decrease in pH_i inhibits K^+ channels, resulting in cell depolarization and increased generation of action potentials. Thus, in these chemosensitive cells, a decrease in pH_i apparently acts as an intracellular signal.

G. Cell Activation, Growth, and Proliferation

One of the most active areas of research on the role of intracellular pH in cell function has been the study of the role of changes of pH_i early in cell proliferation. These studies grew out of early observations that, shortly after fertilization of sea urchin eggs, egg pH increased markedly (roughly 0.4 pH unit), and this rise in pH was necessary for the initiation of growth by fertilization. These observations were followed by others on mammalian cells showing that a variety of growth-promoting agents, including epidermal growth factor (EGF), platelet-derived growth factor (PDGF), insulin, vasopressin, and serum albumin, similarly induced a cellular alkalinization (of about 0.1–0.2 pH unit) shortly after exposure. All of these alkalinizing effects are mediated by activation of Na^+-H^+ exchange. These growth-promoting agents activate the exchanger by activating cellular signaling pathways, which increase the affinity of the exchanger for internal H^+ ions and increase its activity, thereby alkalinizing the cell (see arrow 1 in Fig. 3).

It was initially hypothesized that the growth factor-induced increase in pH was part of a suite of early signals that are required for initiation of cell growth and proliferation. Cellular alkalinization was believed to contribute to the initiation of growth by activating key cellular enzymes that were then either direct effectors of growth (e.g., metabolic enzymes) or activators of other systems.

The direct signaling role of increases of pH_i in cell activation has been questioned. Changes in pH_i by themselves do not promote cell growth or division. Further, it has been shown that a number of cells have a higher pH_i in the presence of 5% CO_2 than in its absence. However, most of the initial experiments on the pH_i responses to growth factors had been done in the absence of CO_2. Upon repeating a number of these experiments under conditions more similar to physiological conditions (presence of 5% CO_2), the initial alkalinization upon exposure to stimulatory factors was not always seen, and in some cells the initial response was indeed an **acidification.** Thus, the current view of the role of pH_i in cell activation, growth, and

proliferation is that pH_i plays a permissive role. That is, cells will only grow and proliferate if pH_i is above a certain critical value, regardless of how that value is obtained. If a cell had a pH_i above the critical value before exposure to a stimulatory agent, no change in pH_i would be required for cell growth. In fact, the cell could acidify and still grow as long as its pH_i stays above the critical value.

In reality, the role of pH_i as a signal to initiate cell growth may depend on the cell and the activating agent. It is likely that a rapid and marked rise in pH_i is one of the critical early steps necessary for the initiation of growth in fertilized sea urchin eggs. Even more dramatic is the nearly 1-pH-unit increase of pH_i in *Artemia* (brine shrimp) embryos upon arousal from anaerobic dormancy by exposure to oxygen. Undoubtedly, this large rise in pH_i is crucial for the transition from metabolic dormancy in these organisms. On the other hand, in many mammalian cells, the rather modest increase in pH_i on exposure to an activating agent is probably of limited physiological significance, especially given the variability in the degree and direction of pH changes observed in different experimental conditions.

Finally, even under conditions where a change of pH_i is not observed in response to a stimulatory agent, pH-regulating transport systems can still be shown to be activated. It has been hypothesized that although a change of pH_i may not be crucial for the stimulation of cell growth, the initiation of this growth may confront the cell with an acid or alkaline load. In this regard, activation of the pH-regulating transport systems by growth-promoting agents could be viewed as preparatory, enabling the cell to better maintain a constant pH_i during a period of high metabolic activity.

H. Cell Volume Regulation

Most pH-regulating transporters move ions such as Na^+ and Cl^- in exchange for a proton equivalent (H^+ or HCO_3^-). Because the transported proton equivalents are buffered (H^+ by HCO_3^- and protein buffers and HCO_3^- by formation of CO_2), they are osmotically "invisible." For example, virtually all of the H^+ transported by the Na^+-H^+ exchanger derive from internal buffers and upon efflux from the cell are buffered by external buffers. Thus, the Na^+-H^+ exchanger mediates the net import of one osmotically active particle (Na^+), and this import of osmolytes will be accompanied by the influx of water and cell swelling. Therefore, the Na^+-H^+ exchanger, in addition to contributing to pH_i regulation, can mediate the regulation of cell volume.

Other pH-regulating transporters can similarly mediate cell volume changes. For example, the Cl^--HCO_3^- exchanger transports Cl^- into the cell. The HCO_3^- that leaves combines with an H^+ and is removed as CO_2. Thus, like Na^+-H^+ exchange, Cl^--HCO_3^- exchange contributes to cell swelling. In fact, these two exchangers often act in concert to result in the net influx of NaCl (and therefore water) into the cell. The Na^+-HCO_3^- and Na^+-anionic weak base cotransporters would be ideally suited to mediate net solute transfer and therefore cell volume change. Finally, the $(Na^+ + HCO_3^-)$-Cl^- exchanger should not contribute to cell volume regulation because it mediates the entry of one osmotically active ion (Na^+) for the efflux of another (Cl^-).

Cell volume can be rapidly changed by exposure to anisosmotic media, with hypertonic media causing cell shrinkage, and hypotonic media causing cell swelling. Many cells respond to shrinkage with a **regulatory volume increase (RVI)** that involves the net uptake of solutes, and therefore water, so that cells swell back toward the initial cell volume (Fig. 8). In several different types of cells, RVI has been shown to involve an activation of Na^+-H^+ exchange. This exchanger, often in association with the Cl^--HCO_3^- exchanger, results in NaCl influx and a regulatory volume increase. The mechanism by which cell shrinkage activates the Na^+-H^+ exchanger is not fully understood, but interestingly, unlike most other activation pathways, cell shrinkage apparently does not result in phosphorylation of the exchanger. A possible involvement of the cell cytoskeleton in activating the Na^+-H^+ exchanger upon cell shrinkage is currently being investigated.

Recently, it has been shown that shrinkage can also activate $(Na^+ + HCO_3^-)$-Cl^- exchange in some cells. This observation is interesting since, as stated previously, this exchanger does not mediate any net solute movement and thus should not directly contribute to cell volume regulation. This activation of $(Na^+ + HCO_3^-)$-Cl^- exchange by cell shrinkage suggests that pH-regulating transporters may be activated by shrinkage to alkalinize the cell regardless of whether they contribute to volume regulation. It is not clear what benefit a shrunken cell derives from becoming alkaline, but it may involve pH-dependent cytoskeletal rearrangements (see Section VII.B).

In response to swelling, most cells exhibit a **regulatory volume decrease (RVD)**. RVD involves the efflux of solutes accompanied by water and therefore cell shrinkage back toward the initial cell volume (Fig. 8). In at least one cell type, the nucleated red blood cell, RVD has been shown to be mediated by K^+-H^+ exchange (K^+ efflux and H^+ influx) in association with Cl^--HCO_3^- exchange (Cl^- efflux and HCO_3^- influx).

Cell volume can also be altered under isosmotic conditions by an imbalance of solute influx and efflux. For instance, during periods of active pH recovery from acidification, the Na^+ influx mediated by the Na^+-H^+ exchanger could result in cell swelling. Thus, changes in pH_i and the response to them could result in an alteration of cell volume.

It is clear that the regulation of intracellular volume and intracellular pH are highly linked in most cells. This linkage is due in part to the use of many of the same membrane transport systems for the regulation of cell pH and volume. In any given cell type, these transporters may respond predominantly either to changes in pH_i or to changes in cell volume.

I. Intracellular Membrane Flow

The intracellular flow of membranes is affected by changes of pH within acidic vacuolar compartments. In cells, these vacuolar compartments are often involved in the movement of membranes, membrane-bound proteins, and soluble proteins around the cell. In addition, components of the vacuolar system are involved in the synthesis, processing, and degradation of various proteins. This system includes the endoplasmic reticulum, the Golgi apparatus, lysosomes, and endosomes. Movement of materials through this system can be divided into the endocytic and exocytic pathways. The endocytic pathway is involved in the uptake of external macromolecules, the degradation or delivery to the cell of these macromolecules, and the down-regulation of surface proteins; it includes coated pits, endosomes, and lysosomes. The exocytic pathway delivers newly-synthesized proteins to a variety of sites, including the surface membrane or extracellular space. Many of the compartments within this vacuolar system are acidic (see Section IV. B), and the maintenance of an acidic interior is critical for the functioning of these compartments. This criticality has been shown by the marked disturbance of endocytic and exocytic pathways by a number of agents, such as chloroquine and ammonia, which alkalinize these compartments. In addition, inhibition of vacuolar H^+-ATPase results in alkalinization of the acidic compartments and can lead to inhibition of endocytosis and exocytosis. Thus the maintenance of a proper pH in acidic intracellular compartments is crucial for continued and proper intracellular membrane flow.

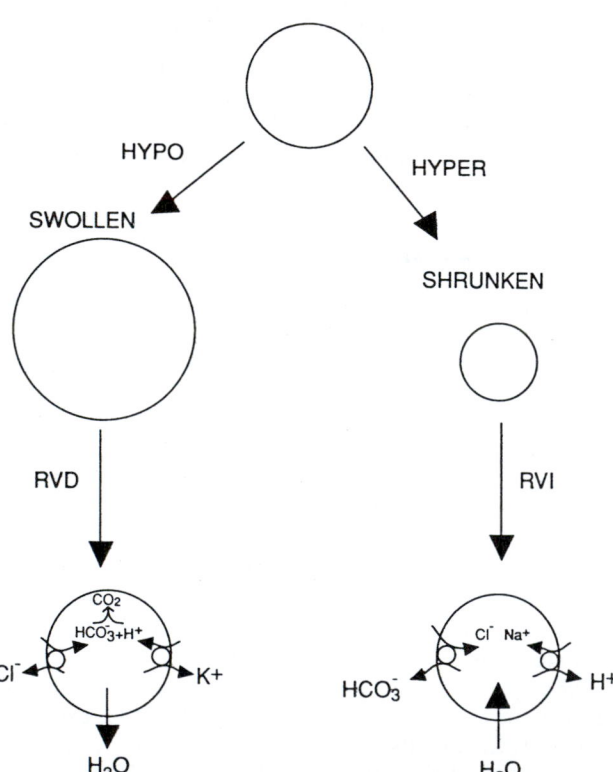

FIGURE 8. The role of pH-regulating transporters in cellular volume regulation. Upon cell swelling in hypotonic media, KCl has been shown to be removed from nucleated red blood cells by functionally coupled operation of the K^+-H^+ and the Cl^--HCO_3^- exchangers. The removal of KCl from the cell causes a loss of water from the cell and thus cell shrinkage in a process called regulatory volume decrease (RVD). Upon cell shrinkage in hypertonic media, NaCl enters the cell by parallel operation of Na^+-H^+ and Cl^--HCO_3^- exchangers. The NaCl entry results in water influx and cell swelling in a process called regulatory volume increase (RVI).

VIII. Summary

Virtually all of the H^+ ions within a cell are buffered by reversible binding to weak acids and bases, resulting in a low free H^+ ion activity. Therefore, the activity of free H^+ ions within the cytoplasm is usually expressed as cytoplasmic pH (pH_i), defined as $pH = -\log(a_H)$, which is a more convenient scale for molecules at low activities.

Cytoplasmic pH is an important aspect of the intracellular milieu and can affect nearly all aspects of cell function. In most cells, pH_i is maintained at a value of about 7.0, well alkaline with respect to the equilibrium pH_i, calculated on the assumption that H^+ ions are at equilibrium across the membrane. The fact that pH_i is alkaline to its equilibrium value creates a passive acidifying influx of H^+. In fact, most cells face a continuous acid load due not only to this acidifying influx but to metabolic acid production and leakage from internal acidic compartments as well. Such challenges to a stable pH_i can be blunted by cellular buffers, but the only way to fully regulate pH_i is through the activity of membrane-bound transporters. These transporters fall into five categories: (1) cation-H^+ exchangers, such as the alkalinizing Na^+-H^+ exchanger and the acidifying K^+-H^+ exchanger; (2) HCO_3^--dependent transporters, such as the (Na^++HCO_3^-)-Cl^- and the Cl^--HCO_3^- exchangers and the Na^+-HCO_3^- cotransporter; (3) H^+-ATPases or proton pumps; (4) Na^+-organic anion cotransporters; and (5) Cl^--organic anion exchangers.

Changes in pH_i can affect many cellular functions. Cell metabolism can be affected by changes in pH, predominantly because of pH-sensitive metabolic enzymes, such as phosphofructokinase. Changes of pH_i have also been shown to affect the cross-linking and polymerization of cytoskeletal elements such as actin and tubulin. The loss of the ability of muscle cells to generate tension (muscle fatigue) has been correlated with a decrease of pH_i. Cell pH is also believed to have a modulatory role in gap junctions and many ion-selective channels. Further, changes of pH_i, mediated by activation of the Na^+-H^+ exchanger, may serve as an intracellular signal for the promotion of cell growth and proliferation. It is significant that changes in pH_i can affect other intracellular signals, such as cellular Ca^{2+} and cAMP levels, suggesting a complex interaction among cellular signaling systems.

Many pH-regulating transporters move an osmotically active ion, such as Na^+ or Cl^-, in exchange for a buffered (and thus osmotically "invisible") ion like H^+ or HCO_3^- and thus mediate the net movement of solute into or out of the cell. This net movement of solute will be accompanied by a net water flow and result in a change in cell volume. Thus, many pH-regulating transporters, in addition to contributing to pH_i regulation, can mediate the regulation of cell volume.

The pH of certain organellar compartments can differ from the value of pH_i and these differences in pH are important for organellar function. For example, mitochondria maintain an internal pH of 7.5, about 0.5 unit more alkaline than pH_i. This pH gradient across the mitochondrial membrane is essential for the major function of mitochondria, the production of ATP. Further, several intracellular organelles in the vacuolar system (e.g., endosomes, lysosomes, and storage granules) maintain an internal pH of 5–6, well below pH_i. These organelles contribute to the movement of membranes, membrane-bound proteins, and soluble proteins around the cell and their acidic pH is essential for this function.

Given the importance of pH to so many cellular functions, it is not surprising that cells have elaborated highly regulated mechanisms to control pH_i.

Bibliography

Al-Awqati, Q. (1986). Proton-translocating ATPases. *Annu. Rev. Cell Biol.* **2,** 179–199.

Ammann, D., Lanter, F., Steiner, R. A., Schulthess, P., Shijo, Y., and Simon, W. (1981). Neutral carrier based hydrogen ion selective microelectrode for extra- and intracellular studies. *Anal. Chem.* **53,** 2267–2269.

Aronson, P. S., and Boron, W. F. (Eds.). (1986). "Na^+-H^+ Exchange, Intracellular pH, and Cell Function," Vol. 26 in "Current Topics in Membranes and Transport," Academic Press, New York.

Aronson, P. S. (1989). The renal proximal tubule: a model for diversity of anion exchangers and stilbene-sensitive anion transporters. *Annu. Rev. Physiol.* **51,** 419-441.

Aronson, P. S., Nee, J., and Suhm, M. A. (1982). Modifier role of internal H^+ in activating the Na^+-H^+ exchanger in renal microvillus membrane vesicles. *Nature* **299,** 161–163.

Attaphitaya, S., Park, K., and Melvin, J. E. (1999). Molecular cloning and functional expression of a rat Na^+/H^+ exchanger (NHE5) highly expressed in brain. *J. Biol. Chem.* **274,** 4383–4388.

Baird, N. R., Orlowski, H., Szabo, E. Z., Zaun, H. C., Schultheis, P. J., Menon, A. G., and Shull, G. E. (1999). Molecular cloning, genomic organization, and functional expression of Na^+/H^+ exchanger isoform 5 (NHE5) from human brain. *J. Biol. Chem.* **274,** 4377–4382.

Bell, S. M., Schreiner, C. M., Schultheis, P. J., Miller, M. L., Evans, R. L., Vorhees, C. V., Shull, G. E., and Scott, W. J. (1999). Targeted disruption of the murine Nhe1 locus induces ataxia, growth retardation, and seizures. *Am. J. Physiol.* **276,** C788–C795.

Bidani, A., and Brown, S. E. S. (1990). ATP-dependent pH_i recovery in lung macrophages: evidence for a plasma membrane H^+-ATPase. *Am. J. Physiol.* **259,** C586–C598.

Bock, G., and Marsh, J. (Eds.). (1988). "Proton Passage across Cell Membranes," Ciba Foundation Symposium 139, Wiley, New York.

Boron, W. F. (Ed.). (1986). Special topic: acid/base physiology. *Annu. Rev. Physiol.* **48,** 347–413.

Boron, W. F., and Boulpaep, E. L. (1983). Intracellular pH regulation in the renal proximal tubule of the salamander. Basolateral HCO_3^- transport. *J. Gen. Physiol.* **81,** 53–94.

Boron, W. F., and De Weer, P. (1976). Intracellular pH transients in squid giant axons caused by CO_2, NH_3, and metabolic inhibitors. *J. Gen. Physiol.* **67,** 91–112.

Busa, W. B., and Nuccitelli, R. (1984). Metabolic regulation via intracellular pH. *Am. J. Physiol.* **246,** R409–R438.

Cala, P. M. (1980). Volume regulation by *Amphiuma* red blood cells. The membrane potential and its implications regarding the nature of the ion-flux pathways. *J. Gen. Physiol.* **76,** 683–708.

Chamberlin, M. E., and Strange, K. (1989). Anisosmotic cell volume regulation: a comparative view. *Am. J. Physiol.* **257,** C159–C173.

Counillon, L., and Pouysségur, J. (1995). Structure-function studies and molecular regulation of the growth factor activatable sodium-hydrogen exchanger (NHE-1). *Cardiovascular Res.* **29,** 147–154.

DeCoursey, T. E. (1998). Four varieties of voltage-gated proton channels. *Frontiers in Biosci.* **3,** 477–482.

DeCoursey, T. E., and Cherny, V. V. (1994). Voltage-activated hydrogen ion currents. *J. Membr. Biol.* **141,** 1994.

Duprat, F., Lessage, F., Fink, M., Reyes, R., Heurteaux, C., and Lazdunski, M. (1997). TASK, a human background K^+ channel to sense external pH variations near physiological pH. *EMBO J.* **16**, 5464–5471.

Durham, J. H., and Hardy, M. A. (Eds.). (1989). "Bicarbonate, Chloride, and Proton Transport Systems," Vol. 574 in "Annals of the New York Academy of Sciences." New York Academy of Sciences, New York.

Edmonds, B. T., Murray, J., and Condeelis, J. (1995). pH regulation of the F-actin binding properties of *Dictyostelium* elongation factor 1α. *J. Biol. Chem.* **270**, 15222–15230.

Gevers, W. (1977). Generation of protons by metabolic processes in heart cells. *J. Molec. Cell. Cardiol.* **9**, 867–874.

Grinstein, S. (Ed.). (1988). "Na$^+$/H$^+$ Exchange." CRC Press, Boca Raton, FL.

Grinstein, S. (1996). Non-invasive measurement of the luminal pH of compartments of the secretory pathway. *The Physiologist* **39**, 144.

Grinstein, S., and Rothstein, A. (1986). Mechanisms of regulation of the Na$^+$/H$^+$ exchanger. *J. Membr. Biol.* **90**, 1–12.

Häussinger, D. (Ed.). (1988). "pH Homeostasis. Mechanisms and Control." Academic Press, New York.

Hille, B. (1992). "Ionic Channels of Excitable Membranes." Sinauer Associates, Sunderland, MA.

Hochachka, P. W., and Mommsen, T. P. (1983). Protons and anaerobiosis. *Science* **219**, 1391–1397.

Hoffmann, E. K., and Simonsen, L. O. (1989). Membrane mechanisms in volume and pH regulation in vertebrate cells. *Physiol. Rev.* **69**, 315–382.

Karniski, L. P., and Aronson, P. S. (1985). Chloride/formate exchange with formic acid recycling: a mechanism of active chloride transport across epithelial membranes. *Proc. Natl. Acad. Sci. USA* **82**, 6362–6365.

Kopito, R. R., and Lodish, H. F. (1985). Primary structure and transmembrane orientation of the murine anion exchange protein. *Nature* **316**, 234–238.

Kotyk, A., and Slavik, J. (1989). "Intracellular pH and Its Measurement." CRC Press, Boca Raton, FL.

Llopis, J., McCaffery, J. M., Miyawaki, A., Farquhar, M. G., and Tsien, R. Y. (1998). Measurement of cytosolic, mitochondrial, and Golgi pH in single living cells with green fluorescent proteins. *Proc. Natl. Acad. Sci. USA* **95**, 6803–6808.

Lowe, A. G., and Lambert, A. (1983). Chloride-bicarbonate exchange and related transport processes. *Biochim. Biophys. Acta* **694**, 353–374.

Madshus, I. H. (1988). Regulation of intracellular pH in eukaryotic cells. *Biochem. J.* **250**, 1–8.

Masuda, A., Oyamada, M., Nagaoka, T., Tateishi, N., and Takamatsu, T. (1998). Regulation of cytosol-nucleus pH gradients by K$^+$/H$^+$ exchange mechanism in the nuclear envelope of neonatal rat astrocytes. *Brain Res.* **807**, 70–77.

Murer, H., Hopfer, U., and Kinne, R. (1976). Sodium/proton antiport in brush-border membranes isolated from rat small intestine and kidney. *Biochem. J.* **154**, 597–604.

Noël, J., and Pouysségur, J. (1995). Hormonal regulation, pharmacology, and membrane sorting of vertebrate Na$^+$/H$^+$ exchanger isoforms. *Am. J. Physiol.* **268**, C283–C296.

Nosek, T. M., Fender, K. Y., and Godt, R. E. (1987). It is diprotonated inorganic phosphate that depresses force in skinned skeletal muscle fibers. *Science* **236**, 191–193.

Nuccitelli, R., and Deamer, D. W. (Eds.). (1981). "Intracellular pH: Its Measurement, Regulation, and Utilization in Cellular Function." A. R. Liss, New York.

Palokangas, H., Metsikkö, K., and Väänänen. (1994). Active vacuolar H$^+$ ATPase is required for both endocytic and exocytic processes during viral infection of BHK-21 cells. *J. Biol. Chem.* **269**, 17577–17585.

Pouysségur, J., Sardet, C., Franchi, A., L'Allemain, G., and Paris, S. (1984). A specific mutation abolishing Na$^+$/H$^+$ antiport activity in hamster fibroblasts precludes growth at neutral and acidic pH. *Proc. Natl. Acad. Sci. USA* **81**, 4833–4837.

Reusch, H. P., Lowe, J., and Ives, H. E. (1995). Osmotic activation of a Na$^+$-dependent Cl$^-$/HCO$_3^-$ exchanger. *Am. J. Physiol.* **268**, C147–C153.

Rink, T. J., Tsien, R. Y., and Pozzan, T. (1982). Cytoplasmic pH and free Mg^{2+} in lymphocytes. *J. Cell Biol.* **95**, 189–196.

Ritucci, N. A., Chambers-Kersh, L., Dean, J. B., and Putnam, R. W. (1998). Intracellular pH regulation in neurons from chemosensitive and nonchemosensitive areas of the medulla. *Am. J. Physiol.* **275**, R1152–R1163.

Ritucci, N. A., Dean, J. B., and Putnam, R. W. (1997). Intracellular pH response to hypercapnia in neurons from chemosensitive areas of the medulla. *Am. J. Physiol.* **273**, R433–R441.

Romero, M. F., and Boron, W. F. (1999). Electrogenic Na$^+$/HCO$_3^-$ cotransporters: cloning and physiology. *Annu. Rev. Physiol.* **61**, 699–723.

Roos, A., and Boron, W. F. (1980). The buffer value of weak acids and bases: origin of the concept, and first mathematical derivation and application to physico-chemical systems. The work of M. Koppel and K. Spiro (1914). *Respir. Physiol.* **40**, 1–32.

Roos, A., and Boron, W. F. (1981). Intracellular pH. *Physiol. Rev.* **61**, 296–434.

Sardet, C., Franchi, A., and Pouysségur, J. (1989). Molecular cloning, primary structure, and expression of the human growth factor-activatable Na$^+$/H$^+$ antiporter. *Cell* **56**, 271–280.

Sardet, C., Counillon, L., Franchi, A., and Pouysségur, A. (1990). Growth factors induce phosphorylation of the Na$^+$/H$^+$ antiporter, a glycoprotein of 110 kD. *Science* **247**, 723–726.

Schultheis, P. J., Clarke, L. L., Meneton, P., Harline, M., Boivin, G. P., Stemmermann, G., Duffy, J. J., Doetschman, T., Miller, M. L., and Shull, G. E. (1998). Targeted disruption of the murine Na$^+$/H$^+$ exchanger isoform 2 gene causes reduced viability of gastric parietal cells and loss of net acid secretion. *J. Clin. Invest.* **101**, 1243–1253.

Schultheis, P. J., Clarke, L. L., Meneton, P., Miller, M. L., Soleimani, M., Gawenis, L. R., Riddle, T. M., Duffy, J. J., Doetschman, T., Wang, T., Giebisch, G., Aronson, P. S., Lorenz, J. N., and Shull, G. E. (1998). Renal and intestinal absorptive defects in mice lacking the NHE3 Na$^+$/H$^+$ exchanger.

Seksek, O., Biwersi, J., and Verkman, A. S. (1995). Direct measurement of *trans*-Golgi pH in living cells and regulation by second messengers. *J. Biol. Chem.* **270**, 4967–4970.

Seksek, O., and Bolard, J. (1996). Nuclear pH gradient in mammalian cells revealed by laser microspectrofluorimetry. *J. Cell Sci.* **109**, 257–262.

Soleimani, M., and Aronson, P. S. (1989). Ionic mechanism of Na$^+$-HCO$_3^-$ cotransport in rabbit renal basolateral membrane vesicles. *J. Biol. Chem.* **264**, 18302–18308.

Thomas, J. A., Buchsbaum, R. N., Zimniak, A., and Racker, E. (1979). Intracellular pH measurements in Ehrlich ascites tumor cells utilizing spectroscopic probes generated *in situ*. *Biochemistry* **18**, 2210–2218.

Thomas, R. C. (1974). Intracellular pH of snail neurones measured with a new pH-sensitive glass micro-electrode. *J. Physiol. (London)* **238**, 159–180.

Thomas, R. C. (1978). "Ion-sensitive Intracellular Microelectrodes. How to Make and Use Them." Academic Press, New York.

Trivedi, B., and Danforth, W. H. (1966). Effect of pH on the kinetics of frog muscle phosphofructokinase. *J. Biol. Chem.* **241**, 4110–4111.

Wakabayashi, S., Ikeda, T., Iwamoto, T., Pouysségur, J., and Shigekawa, M. (1997). Calmodulin-binding autoinhibitory domain controls "pH-sensing" in the Na$^+$/H$^+$ exchanger NHE1 through sequence-specific interaction. *Biochemistry* **36**, 12854–12861.

Wakabayashi, S., Shigekawa, M., and Pouysségur, J. (1997). Molecular physiology of vertebrate Na$^+$/H$^+$ exchangers. *Physiol. Rev.* **77**, 51–74.

Yun, C. H. C., Tse, C. -M., Nath, S. K., Levine, S. A., Brant, S. R., and Donowitz, M. (1995). Mammalian Na$^+$/H$^+$ exchanger gene family: structure and function studies. *Am. J. Physiol.* **269**, G1–G11.

Appendix: Techniques for pH Measurement

The study of intracellular pH and its regulation was enabled by the development of the first reliable pH-sensitive microelectrode in 1974 by Roger Thomas, called the **Thomas recessed-tip microelectrode.** These electrodes are constructed from special glass that is conductive to H ions only. Small capillaries of this pH-sensitive glass are pulled into fine tip electrodes (tip diameter $\approx$ 1 μm) and these tips are sealed by heating. Such an electrode will respond with a Nernstian slope (about 59 mV/pH unit) to changes in pH. However, to reliably measure pH_i, one must assure that the pH-sensitive surfaces of this electrode are exposed to cytoplasm only. This is achieved by carefully lowering the sealed-tip pH electrode into a larger (about 2 μm) open-tip microelectrode (the shielding microelectrode) constructed from $AlSiO_4$ glass. The tip of the inner pH electrode is brought within a few micrometers of the open tip of the outer electrode (the distance BT, between tips in Fig. A-1). The pH electrode is heated, under internal pressure. The pH glass, which melts at a lower temperature than the $AlSiO_4$ glass, is pushed against the inner face of the shielding microelectrode and forms a high-resistance glass-glass seal (Fig. A-1). The inner electrode is removed above the seal and what remains is the tip of the pH electrode sealed near the tip of the shielding electrode (Fig. A-1). This recessed-tip electrode is filled with a buffered conducting solution and connected via a fine chlorided silver wire to an electrometer. The tip of the pH electrode below the seal (exposed length, EXPL in Fig. A-1) will respond to changes in the pH of the fluid trapped within the recess volume (striped area in Fig. A-1). When a cell is impaled with such an electrode, the fluid in the recess volume is replaced (by diffusion) with cytoplasm. A signal is generated across this electrode that is the sum of the actual membrane potential (V_m) and a Nernstian signal based on the pH difference between the electrode filling solution and the cytoplasm in the recess volume. To derive a signal that is directly proportional to the intracellular pH, a conventional open-tip KCl-filled microelectrode is placed within the same cell to measure V_m and the signals from the two electrodes are subtracted (Fig. A-2).

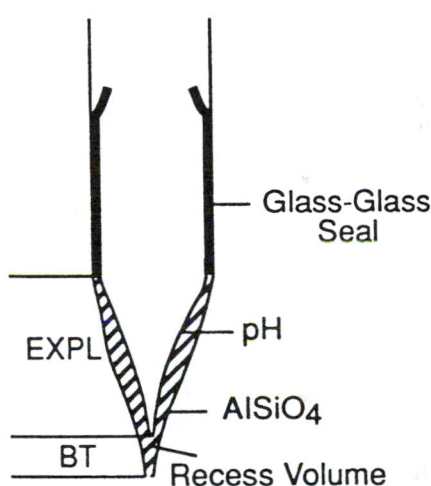

FIGURE A-1. A diagram of a pH-sensitive recessed-tip glass microelectrode showing the sealed tip of a microelectrode constructed from pH-sensitive glass fused within the tip of a larger shielding aluminosilicate electrode. BT—between-tips distance; EXPL—length of exposed pH-sensitive electrode below glass-glass seal.

The recessed-tip microelectrode gave some of the first continuous, reliable measurements of intracellular pH and initiated over two decades of intensive study of pH_i. However, this electrode has drawbacks. Because of its large size and the need to impale a cell with two electrodes, its use is restricted to fairly large cells. Further, the electrode has a high resistance (often greater than 100 GΩ) and must be used with a high input-impedance amplifier. Finally, because of the need for diffusional exchange between the recess volume and the cytoplasm, these electrodes have a slow response time (time constant no faster than 15 seconds).

A different type of pH microelectrode can be constructed using pH-sensitive resins (Fig. A-3). Like pH-sensitive glass, these resins are conductive to H ions only. A bit of this resin is placed in the tip of a conventional microelectrode and the electrodes are back-filled with a buffered conducting

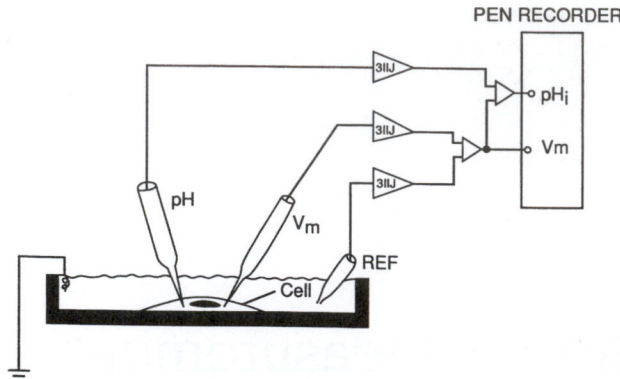

FIGURE A-2. Measurement of pH_i. Diagram of the apparatus used to measure intracellular pH with microelectrodes. The cell is impaled with a pH-sensitive and V_m-sensitive microelectrode. These electrodes are referenced to an open-tip flowing KCl reference electrode. Electrode voltages are amplified by Analog Devices 311J High Input Impedance Operational Amplifiers. The difference in electrode voltage between the V_m and reference electrodes is equal to membrane potential, and the difference in electrode voltage between the pH and V_m electrodes is a signal that is proportional to pH_i. These difference signals are plotted on a pen recorder.

medium (Fig. A-3). When a cell is impaled with these **pH-sensitive resin microelectrodes,** a signal is generated that is the sum of the V_m and a Nernstian signal due to the pH difference between the electrode and the cytoplasm. Thus a cell must be impaled with a V_m microelectrode when using the resin electrodes as well as the recessed electrodes. These pH-sensitive resin microelectrodes can penetrate much smaller cells because of their smaller tip diameters (usually less than 1 μm) and are faster than the recessed-tip electrodes. Further, pH-sensitive resin microelectrodes can readily be constructed in one barrel of a double-barreled elec-

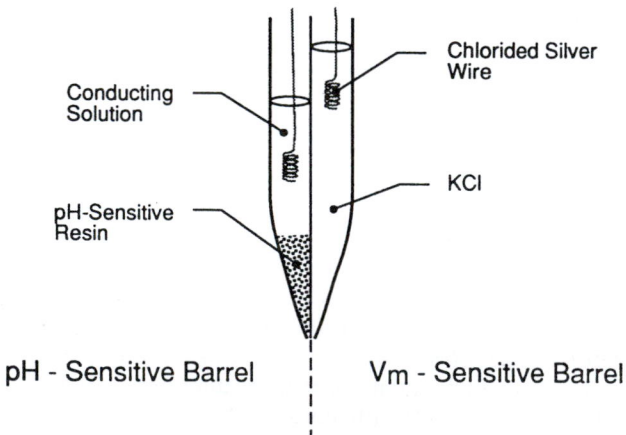

FIGURE A-3. A diagram of a double-barreled pH-sensitive resin microelectrode. The tip of one barrel of a double-barreled electrode is filled with a pH-sensitive resin and then back-filled with a conducting solution. This is the pH-sensitive barrel. The other barrel, the V_m-sensitive barrel, is filled with KCl. Each barrel is connected to an amplifier via a chlorided silver wire immersed in the conducting solution of that barrel.

trode (the other barrel filled with KCl to measure V_m). Thus, the pH-sensitive resin microelectrodes have supplanted the recessed-tip microelectrode for most studies of pH_i.

Another reliable method for measuring intracellular pH is based on the fact that a number of dyes bind H ions reversibly and this binding affects the fluorescence of these dyes. The most common **pH-sensitive dye** in use today is a derivative of fluorescein, biscarboxyethyl carboxyfluorescein (BCECF) (Fig. A-4). An excitation spectrum for this dye can be obtained by shining light of different wavelengths (between 300 and 600 nm) on a solution containing this dye and measuring the fluorescence emitted at 535 nm. When such an excitation spectrum is gathered for the dye at different values of solution pH, the resulting spectra can be superimposed (Fig. A-5). Such curves reveal that when the dye is excited at 440 nm, its fluorescence is the same regardless of pH (called the isoexcitation point), whereas the dye is maximally sensitive to pH when excited at 500 nm, fluorescence increasing as pH increases.

The charged sites on the dye are often rendered neutral by attaching acetoxymethyl ester groups (BCECF/AM). In this form, the dye readily permeates the cell. Once inside, cell esterases cleave the AM groups from the dye, once again creating the charged form, BCECF. This form is impermeant and remains trapped within the cytoplasm. The fluorescence of this dye within cells can be measured using a spectrofluorimeter.

The dye within cells can slowly leak out or photobleach during the course of an experiment. Either of these processes would result in a decreased fluorescence signal and appear as a decrease in pH_i. To prevent this artifact, the dye fluorescence is often collected at excitation wavelengths of both 440 and 500 nm and a fluorescence ratio (R_{fl}), Fl_{500}/Fl_{440}, calculated. In this ratio, Fl_{440} is a measure of the amount of dye present and thus serves as a normalizing factor. R_{fl} is proportional to intracellular pH and is not susceptible to leakage or photobleaching artifacts. The fluorescence ratio can be calibrated by exposing cells to a high extracellular $[K^+]$ (similar to intracellular $[K^+]$) and nigericin, a K^+-H^+-exchanging ionophore. Under these conditions it is assumed that the nigericin will equilibrate intracellular and extracellular pH. Upon exposing dye-loaded cells to a variety of extracellular pH values and measuring R_{fl}, a calibration curve can be constructed (Fig. A-6). The use of pH-sensitive fluorescent dyes to study pH_i has several

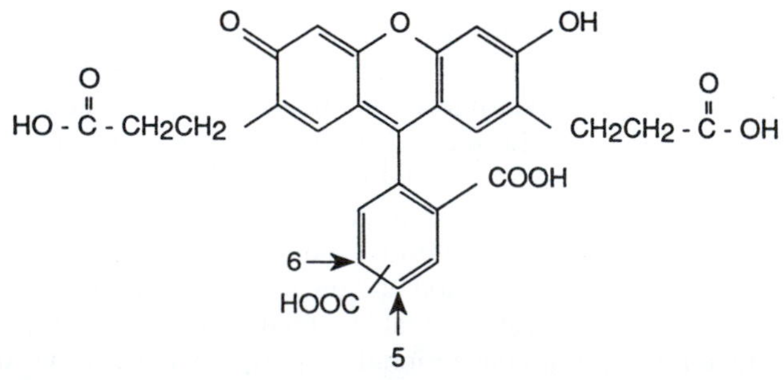

FIGURE A-4. A diagram of the pH-sensitive fluorescent probe BCECF.

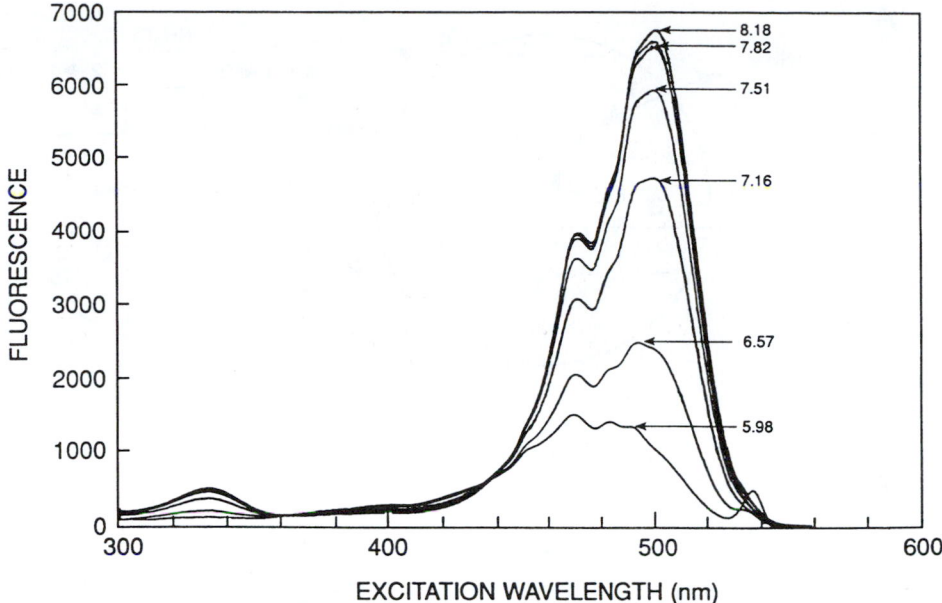

FIGURE A-5. Excitation spectra of BCECF: the fluorescent emission (at 535 nm) of BCECF when excited at wavelengths from 300–600 nm. Each curve represents the excitation spectrum of BCECF at a different pH (indicated to the right of each spectrum). The ordinate represents fluorescence in arbitrary units.

advantages over microelectrodes. This technique can be used on cells of any size, including preparations of intracellular organelles. Further, it is a relatively simple and reliable experimental technique. Finally, this technique can be used to visualize pH within single cells or parts of cells, using optical imaging techniques.

To study the ability of cells to actively extrude acid or base from the cell, techniques must be available to alter intracellular pH experimentally. In some large cells, this has been achieved directly by injecting acid into cells, by passing current through microelectrodes, or by internal dialysis through tubing threaded through the cytoplasm of the cell. More re-

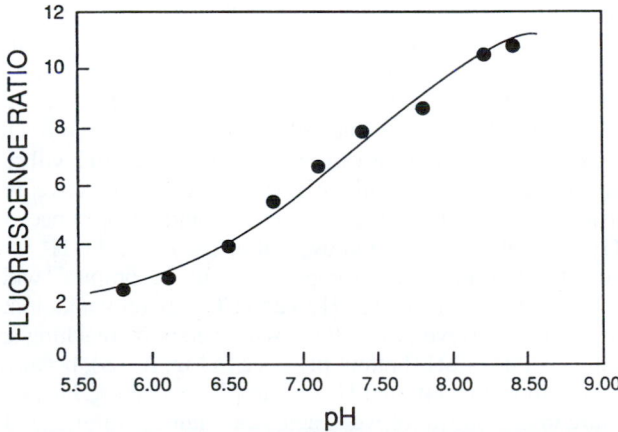

FIGURE A-6. A calibration curve for BCECF derived from data shown in Fig. A-5. The ratio of the emitted fluorescence at excitation wavelengths of 500 and 440 nm ($R_{fl} = Fl_{500}/Fl_{440}$) is plotted against the pH of the solution and a sigmoid titration curve is obtained.

cently, changes in pH_i have been accomplished by internal perfusion of cells using whole-cell patch-clamp electrodes. However, the most commonly used experimental method to modify pH_i is by external exposure of cells to weak acids or bases. One of the most popular of such techniques is the **NH_4Cl prepulse technique.** Cells are exposed to an external solution containing NH_4Cl. External NH_3, being uncharged, enters the cell (arrow 1 in Fig. A-7A) far more rapidly than external NH_4^+. In the cell, the NH_3 combines with an H ion to form NH_4^+ (arrow 2 in Fig. A-7A), thereby alkalinizing the cell. This alkalinization will continue until the internal and external concentrations of NH_3 are the same, at which time no more NH_3 will enter and the pH_i will reach a new steady-state value alkaline relative to the original pH_i. If NH_4^+ is unable to enter the cell, no further change in pH_i will occur until external NH_4Cl is removed, at which time cytoplasmic NH_4^+ will dissociate to NH_3 and H^+, and all the NH_3 will diffuse from the cell, returning pH_i back to its original value (top pH trace in Fig. A-7A). However, if NH_4^+ has some membrane permeability, it will enter the cell (arrow 3 in Fig. A-7A), largely driven by the negative internal membrane potential. The NH_4^+ that enters will dissociate into NH_3 and H^+. The newly formed NH_3 will diffuse from the cell, leaving H^+ in the cell. Thus, a shuttle is established whereby NH_4^+ enters the cell and NH_3 leaves the cell (arrows 3, 2, 1 and 4 in Fig. A-7A). For each cycle of the shuttle, an H ion is added to the cell and so the cell slowly acidifies in the maintained presence of external NH_4Cl (termed *plateau acidification*). Upon removal of external NH_4Cl, all the original NH_4^+ formed upon exposure will dissociate and regenerate NH_3 (which diffuses from the cell) and H^+. However, the cell will contain extra H^+ due to the operation of the shuttle and thus pH_i will undershoot its initial value, achieving a more acid pH_i than it

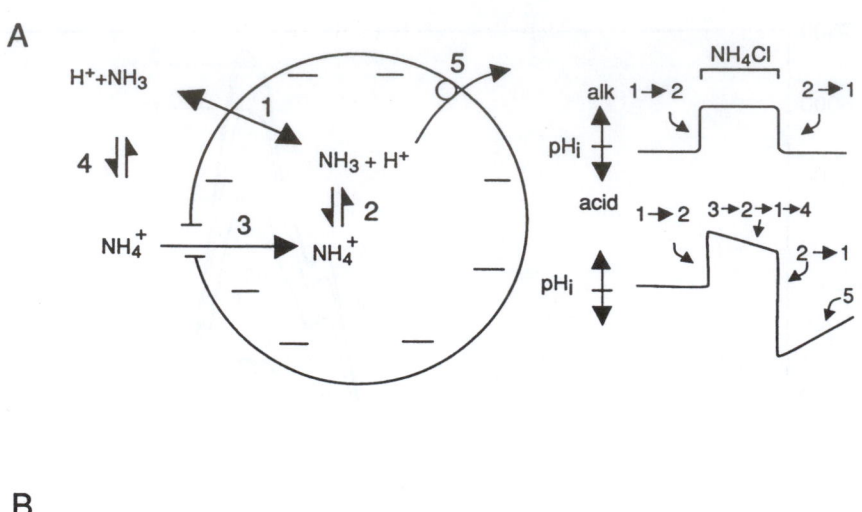

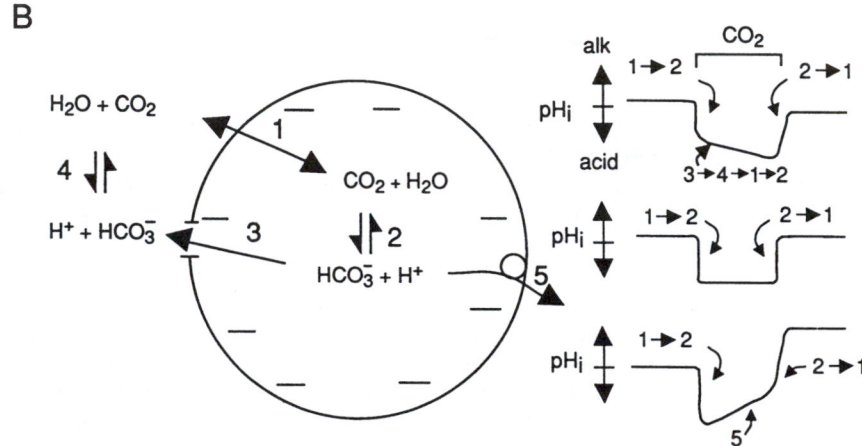

FIGURE A-7. (A) A diagram of the effects on intracellular pH of a transient exposure of a cell to external NH$_4$Cl. (B) A diagram of the effects on intracellular pH of a transient exposure of a cell to external CO$_2$ and HCO$_3^-$. The meaning of the various numbers and the pH traces to the right are given in the Appendix text.

had before the NH$_4$Cl exposure. The degree of this undershoot is dependent on the amount of NH$_4$Cl initially added, the membrane permeability to NH$_4^+$, and the duration of exposure to NH$_4$Cl. The net effect of an NH$_4$Cl prepulse is to acidify the cell, and it is entirely analogous to an injection of acid. If a cell possesses membrane transport systems for active H$^+$ extrusion (arrow 5 in Fig. A-7A), pH$_i$ will return toward its initial pH value, a process termed **pH recovery** (lower pH trace in Fig. A-7A).

The other common way to alter pH$_i$ is by exposure of cells to weak acids. The most common weak acid is carbonic acid, in the form of CO$_2$. This process is analogous to the NH$_4$Cl prepulse. Upon exposure to a solution containing CO$_2$ and HCO$_3^-$, CO$_2$ rapidly enters the cell (arrow 1 in Fig. A-7B), hydrates and dissociates to form internal HCO$_3^-$ and H$^+$ (arrow 2 in Fig. A-7B). The addition of H$^+$ internally acidifies the cell. The cell will continue to acidify until internal and external CO$_2$ are equal. At this point, if HCO$_3^-$ is impermeant, no further change in pH will occur, and pH will return to its initial value upon removal of extracellular CO$_2$ (middle

pH trace in Fig. A-7B). However, if HCO$_3^-$ can move across the membrane, a shuttle will be established. HCO$_3^-$ will leave the cell (arrow 3 in Fig. A-7B), mostly driven by the negative membrane potential, internal CO$_2$ will hydrate and dissociate, and more CO$_2$ will enter the cell. Thus, HCO$_3^-$ will leave the cell and CO$_2$ will enter the cell, adding an internal H$^+$ for every cycle of the shuttle (arrows 3, 4, 1 and 2 in Fig. A-7B). The cell will slowly acidify (top pH trace in Fig. A-7B). If, however, the cell possesses transmembrane H$^+$ extrusion mechanisms (arrow 5 in Fig. A-7B), pH will recover back toward the initial pH$_i$ value because of active extrusion of internal H$^+$ even in the maintained presence of CO$_2$. If external CO$_2$ is removed after recovery, the H$^+$ initially formed upon CO$_2$ exposure will recombine with HCO$_3^-$ and leave the cell as H$_2$O and CO$_2$, thereby alkalinizing the cell. However, the cell pH will overshoot, reaching an alkaline value of pH (bottom pH trace in Fig. A-7B), because of the removal of internal H$^+$ during pH recovery. Thus, exposure to or removal of weak acids can induce cellular acidification or alkalinization, respectively.

Jeffrey C. Freedman

23

Membrane Transport in Red Blood Cells

I. Introduction

Red blood cells are prototypical of more complicated cells and have long been a favorite object of study for cellular physiologists. According to Jacobs (1962, p. 1014), "The first serious osmotic study of an animal cell (the mammalian erythrocyte)" was conducted by H.J. Hamburger in 1895, at a time when the importance of the plasma membrane was not widely understood. In their landmark review of membrane permeability, Davson and Danielli (1943) recognized that the ion permeability of red blood cells, together with electrical impedance measurements, lipid extraction studies, and the birefringence of red-cell ghosts, are all consistent with the postulation of a bimolecular lipid membrane, as first proposed for red cells in 1925 by Gorter and Grendel (see Chapter 4). The structural, functional, metabolic, and transport properties of normal and abnormal red blood cells have been well described in detailed monographs (Henderson, 1928; Ponder, 1948; Whittam, 1964; Bishop and Surgenor, 1964; Harris and Kellermeyer, 1970; Surgenor, 1974; Yoshikawa and Rapoport, 1974; Ellory and Lew, 1977; Agre and Parker, 1989; Raess and Tunnicliff, 1990; Ohnishi and Ohnishi, 1994). Methods for studying red cells have also been summarized (Ellory and Young, 1982; Beutler, 1986; Shohet and Mohandas, 1988).

Current research reports indicate that many fundamental unsolved problems remain for further investigation. More than 8000 articles concerning red blood cells were published during the past 5 years alone, including some 325 reviews on the mechanisms of membrane transport, protein associations and genetic defects in the cytoskeleton, oxygen transport and rheology, hematopoiesis and cellular senescence, blood preservation and transfusion medicine, sickle cell anemia and other hemoglobinopathies, metabolic control and enzymopathies, and altered red cells in systemic and infectious diseases, such as hypertension and malaria. Indeed, Ponder's view (1948, p.1) that "there is scarcely a fundamental problem in General Physiology which does not have a relation, of one kind or another, to the problems which have arisen in connection with the erythrocyte" remains just as valid today as 50 years ago.

II. Membrane and Cytoskeleton

Mammalian red blood cells are highly differentiated for their primary function of oxygen and carbon dioxide transport, and from a structural point of view are the simplest of all eukaryotic cells. Devoid of mitochondria, endoplasmic reticulum, ribosomes, Golgi apparatus, and lysosomes, and lacking a nucleus, mammalian red blood cells are free of the complexities associated with intracellular organellar compartments, and thus have served as a classic model system for studying how ions, nutrients, and other solutes cross the plasma membrane. However, with electron probe microanalysis (Lew *et al.*, 1985) and NMR studies (Murphy *et al.*, 1987), the elevated Ca^{2+} in red cells from patients with sickle cell anemia was found sequestered in intracellular vesicles, which were also noted to occur in normal red cells. Observations of Ca^{2+}-sequestering vesicles in red cells are reminiscent of the ATP-dependent endocytosis that occurs in intact red cells and in isolated membranes in response to oxidants (Penniston *et al.*, 1979). Another type of compartmentalization proposed in red cells is a membrane pool estimated to contain some 500–600 molecules of ATP (Proverbio *et al.*, 1988). ^{32}P-labeling experiments in ghosts, and in inside-out vesicles (Mercer and Dunham, 1981), indicated that the membrane-bound glycolytic enzymes glyceraldehyde 3-phosphate dehydrogenase and phosphoglycerokinase form ATP, which comprises a membrane-bound pool that directly provides substrate for the Na^+,K^+-ATPase; the structural basis for this membrane pool of ATP is unknown.

Most mammalian red blood cells normally exhibit a biconcave discoidal shape. Under abnormal conditions red cells may be transformed into spiculated forms known as **echinocytes,** or into cup-shaped forms called **stomatocytes** (for reviews and scanning electron micrographs, see Bessis *et al.*, 1973; Bessis, 1974; and chapter 14 by Bull and Brailsford in Agre and Parker, 1989). Red cells from camels are an exception, being nonnucleated but with a biconvex ellipsoidal shape, resembling that of nucleated amphibian and avian red cells (for review, see chapter 3 by Ngai and

Lazarides in Agre and Parker, 1989). The red-cell membrane is one of the easiest to isolate (Dodge *et al.*, 1963), has been extensively studied, and is one of the best characterized of all cell membranes. The sidedness and orientation of proteins in the seven major bands found after electrophoresis of human red blood cell membranes in sodium dodecyl sulfate (SDS) polyacrylamide gels have been described by Steck (1974). Studies of membrane transport were greatly facilitated by optimizing conditions for "resealing" isolated membranes to make **resealed ghosts,** a technique that enabled manipulation of the ionic composition of the intracellular as well as the extracellular solutions (Hoffman *et al.*, 1960). Despite extensive studies of hemolysis, the mechanism of the formation and annealing of a hole or holes in the membrane sufficiently large to permit the passage of Na^+, K^+, metabolites, and hemoglobin is still not well understood (for review, see Hoffman, 1992). In the red-cell membrane, as in many other eukaryotic cell membranes, the phospholipids are distributed asymmetrically. The neutral phospholipids—phosphatidylcholine and sphingomyelin—are preferentially located in the outer hemileaflet of the lipid bilayer. The neutral phosphatidylethanolamine and the negatively charged phosphatidylinositol and phosphatidylserine are preferentially located in the inner hemileaflet, with the negative charges facing the cytoplasm. An aminophospholipid-specific translocase, or **flipase,** uses $Mg^{2+}ATP$ to catalyze the inward transport of phospholipids from the outer to the inner hemileaflet with the following specificity: phosphatidylserine > phosphatidylethanolamine > phosphatidylcholine. The flipase exhibits stereospecificity in acting only on L-isomers and is inhibited by vanadate and by oxidation of protein sulfhydryl groups. Elevated intracellular calcium, such as occurs during normal red-cell senescence as well as in red cells from patients with sickle cell anemia, causes loss of the normal **phospholipid asymmetry,** leading especially to excess phosphatidylserine in the outer hemileaflet. Under these conditions, the external membrane surface acquires procoagulant activity and promotes thrombosis (for review, see Diaz and Schroit, 1996).

The external surface of the red cell membrane is coated with adsorbed albumin and some plasma globulins. Because of the sialic acids on **glycophorin,** an integral membrane glycoprotein, the cell surface is negatively charged (for review, see chapter 27 by Seaman in Surgenor, 1974). The **cytoskeleton** is an organized polygonal fibrous network, about 60 nm thick containing spectrin and actin (for review, see chapter 1 by Gardner and Bennett in Agre and Parker, 1989; Bennett and Gilligan, 1993). Sides of the regular five- or six-sided polygons are formed by **spectrin,** a flexible filamentous protein 200 nm in length. Vertices of the polygons are formed by **β-actin** protofilaments 30–40 nm in length consisting of 12–14 actin monomers; both grooves of the actin filaments contain **tropomyosin,** which binds to **tropomodulin** at junctional complexes. At the mid-region of the spectrin filaments a junctional ternary complex formed by **ankyrin** and band 4.2 protein links spectrin to the membrane-spanning integral band 3 protein. At the ends of the spectrin filaments, an additional linkage site to band 3 and to **glycophorin C** may be provided by protein 4.1. Spectrin-

actin junctional complexes also contain the actin-bundling protein **dematin** (band 4.9) and the calmodulin-binding protein **adducin,** which functions to cap and to regulate the length of the actin filaments and which may be involved in Ca^{2+}-dependent alterations in cytoskeletal structure (Kuhlman *et al.*, 1996). Measurements of lateral translational diffusion showed that a fraction of band 3 protein is immobile because of its linkage with spectrin, whereas another fraction is capable of slow diffusion in the plane of the membrane (Golan and Veatch, 1980; for review, see chapter 13 by Golan in Agre and Parker, 1989). A small amount of **myosin** is also found in the red-cell cytoskeleton, but whether active tension is generated or regulated in mature red cells is not known. Presumably, the cytoskeleton is actively involved during **enucleation** in developing red cells and during **diapedesis,** or egress of red cells from the bone marrow to the peripheral circulation.

III. Intracellular Environment

The red-cell membrane and associated cytoskeleton enclose a viscous cytoplasmic solution of the oxygen-binding pigment **hemoglobin** at a concentration of 34 g/100 ml cells, corresponding to 5.2 mM, 7.3 millimolal, or 44 g Hb/100 g cell water. This concentration is near the threshold for gelation, but hemoglobin is one of the most soluble of all proteins and constitutes more than 98% of red-cell protein by mass. The hemoglobin $\alpha_2\beta_2$ tetramers have a molecular weight of 64 373 Da and are approximately spheroidal with dimensions of 65 Å × 55 Å × 50 Å. In the interior of red blood cells, hemoglobin is nearly close-packed, with the distance between the surfaces of neighboring proteins averaging only about 20 Å (Ponder, 1948). With the 86 carboxylates of aspartic and glutamic acids, and with the 98 basic amino and amine groups of arginine, lysine, and histidine per hemoglobin tetramer, the total cellular concentration of the titratable amino acids of human hemoglobin is around 0.9 M. Thus, hemoglobin contributes significantly, depending on intracellular pH, to a high intracellular ionic strength. Since the charged amino acids are located on the surface of hemoglobin (see Antonini and Brunori, 1971), and taking the radius of hemoglobin to be 28 Å, the average distance between charged sites is only about 7 Å. About 7600 water molecules per hemoglobin tetramer occupy the narrow interstices between the protein molecules. The hydration of hemoglobin in dilute solution is 0.2–0.3 g water/ g Hb (see Antonini and Brunori, 1971), representing about 15% of the intracellular water. When the tortuosity of the surface of soluble proteins is taken into account, as much as 30% of red-cell water could reside in the first monolayer around the protein surface.

To determine the **mean ionic activities** of the intracellular KCl and NaCl in this concentrated charged environment, studies were conducted in which the red-cell membrane was rendered permeable to cations by exposure of the cells to the channel-forming antibiotic nystatin, thus allowing K^+, Na^+, and Cl^- to reach **Gibbs-Donnan equilibrium**. In these experiments sufficient extracellular sucrose was added to prevent cell swelling by balancing the colloid osmotic pressure of hemoglobin and other impermeant cell solutes (see Chapter 15). The **mean ionic activity coefficient** of KCl and NaCl in the concentrated intracellular hemoglobin solution was found to be within 2% of that in the extracellular solution (Freedman and Hoffman, 1979a). Moreover, permeant nonelectrolytes also have equilibrium ratios of intracellular to extracellular concentrations within 10% of unity (Gary-Bobo, 1967). In view of the high intracellular ionic strength and the high volume fraction of cell water in direct contact with hemoglobin, it is both curious and remarkable that intracellular salts and nonelectrolytes appear to behave as if in dilute solution. Either the intracellular solution is indeed like a dilute solution, or alternatively, the expected effect of protein-solvent interactions in altering the activity of intracellular solutes is offset by the effect of interactions between proteins and the solutes themselves.

IV. Metabolism and Life Span

The red cell also has relatively simple metabolic pathways (Grimes, 1980; Beutler, 1986), at least in comparison with most other cells. Catalogs of red-cell enzymes list about 140 enzymes (see the chapter by Friedemann and Rapoport in Yoshikawa and Rapoport, 1974; and chapter 3 by Pennell in Surgenor, 1974). **Glycolysis** produces ATP and lactate from glucose, inorganic phosphate, and exogenous purine in the form of adenine, adenosine, or inosine. A mathematical model of red-cell glycolysis was proposed by Rapoport *et al.* (1974). The **pentose shunt** provides reducing equivalents in the form of glutathione, NADH, and NADPH, which, together with catalase, superoxide dismutase, glutathione peroxidase, glutathione reductase, and methemoglobin reductase, act to prevent the oxidation of protein sulfhydryl groups and of Fe^{2+} in hemoglobin. No tricarboxylic acid cycle, cytochrome system, or lipid catabolism or utilization are known to occur in red cells, although cholesterol and phospholipids do exchange with plasma lipids, and some fatty acids may be incorporated into membrane phospholipids (for review, see Shohet, 1976). Protein synthesis does not occur in mature red cells, and there is no DNA replication or transcription, and no RNA metabolism or gene action. Red cells in humans make up 40–45% of the blood volume, a fraction known as the **hematocrit**; thus, they are readily accessible, and are easily separable from leukocytes and platelets by centrifugation or filtration. In short-term experiments, red cells may be studied in simple buffered isotonic salt solutions (e.g., 145 mM NaCl, 5 mM KCl, 5 mM Hepes buffer, pH 7.4). For longer experiments, glucose is added to prevent the decline of ATP; for even longer term experiments, red cells survive *in vitro* at room temperature or at 37 °C for many days, albeit with morphological heterogeneity, in a chemically defined culture medium that includes vitamin cofactors, an exogenous purine for synthesis of ATP, and amino acids to support the synthesis of glutathione (Freedman, 1983). *In vitro*, red blood cells can be studied free from the uncontrollable variables of the intact organism.

Each of the 25 trillion red blood cells in normal adult humans lives for about 120 days. At a turnover rate of about

1%/day, some 250 billion new red cells are released from the bone marrow each day, a rate that corresponds to 3 million cells per second! This may seem like a lot, but 3 million red blood cells occupy less than a microliter of volume, since each biconcave discoidal cell occupies only 87 μm^3 and has a surface area of 133 μm^2, a diameter of 8 μm, and a thickness of 2.4 μm at the rim and 1.0 μm at the center. After 120 days, senescent human red cells bind **IgG autoantibody** and are then recognized by macrophages in the initial stage of **erythrophagocytosis**, a process that leads to the recycling of iron, amino acids, and other essential red-cell constituents. Considerable evidence indicates that the antigenic recognition sites are composed of clusters of an oxidatively denatured form of band 3 protein (for review, see chapter 9 by Low in Agre and Parker, 1989; Kay, 1991). The normal function of band 3 protein, also called **capnophorin** or **AE1 (anion exchange protein 1)**, is to mediate the obligate electroneutral exchange of Cl^- for HCO_3^- across the red-cell membrane during gas exchange in the pulmonary and systemic capillaries. The comparative biochemistry and physiology of red blood cells is instructive, with many differences in metabolic and membrane transport properties, as well as oxygen transport properties, known to occur among different animal species (for examples, see Willis, 1992). Whereas human red cells live for 120 days, the **life span** of dog red cells is 60 days, and that of mouse red cells is only 40 days. The life spans L of mammalian red blood cells correlate with body weight W according to the relation $L = 69W^{0.12}$, as illustrated in the log-log plot in Fig. 1, but the physiological basis for this striking correlation is not understood (for discussion, see Vácha and Znojil, 1981). The life span data suggest that red cells contain a biological clock that is obviously not directly determined or controlled in the mature cell by gene transcription or translation.

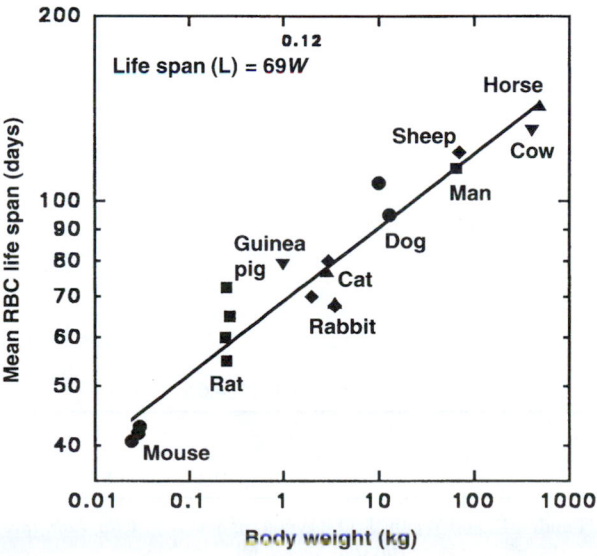

FIGURE 1. Mean life span (days) of mammalian red blood cells versus body weight (kg). (Data from Vásha and Znojil (1981) and various other sources.)

V. Membrane Transporters in Red Blood Cells

Among eight mammalian species (Table 1), the intracellular concentration of K^+ ranges from 135 mM in human red blood cells to only 8 mM in cat and dog red cells, whereas intracellular Na^+ varies from only 17 mM in humans to 142–162 mM in dog and cat; in contrast, the extracellular low K^+ and high Na^+ are relatively constant among species, as are intracellular and extracellular Cl^-. Cation transport in red blood cells was understood classically in terms of the **pump-leak theory** (Tosteson and Hoffman, 1960), which quantitatively accounted for the differing steady-state Na^+ and K^+ concentrations in red blood cells found in two genetic phenotypes of sheep designated as high K^+ (HK) and low K^+ (LK). High-K^+ sheep red cells were found to have a relatively high number of Na^+-K^+ pumps per cell and high Na^+-K^+ pump fluxes with relatively low ouabain-insensitive leakage fluxes. Anions such as Cl^-, HCO_3^-, and OH^- are passively distributed and appear to follow the **double-Donnan equilibrium**, as explained later (for reviews on red-cell transport, see chapter 3 by Passow in Bishop and Surgenor, 1964; chapter 15 by Sachs, Knauf, and Dunham in Surgenor, 1974; Ellory and Lew, 1977; Agre and Parker, 1989; Raess and Tunnicliff, 1990). The ionic composition of the intracellular and extracellular solutions may be depicted on bar graphs, as shown for human red blood cells in Fig. 2. In human red cells, the Na^+,K^+-ATPase (see Chapter 17) specifically selects the K^+ from the Na^+-rich medium, and pumps it against the K^+ concentration gradient (and electrochemical gradient) into the cytoplasmic solution. The same

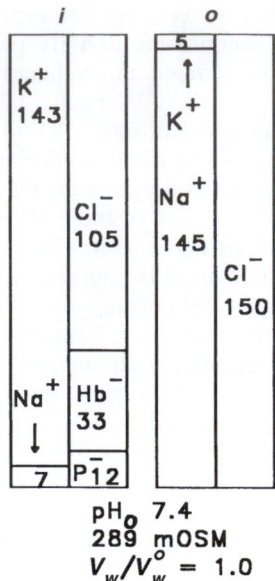

FIGURE 2. Ionic composition of the intracellular solution in human red blood cells and of the extracellular solution, designated by *i* and *o*, respectively. Organic phosphates are represented by P^-.

ionic pump specifically selects Na^+ from the K^+-rich cytoplasmic solution and extrudes it from the cell against the Na^+ concentration gradient. This coupled active transport of three internal Na^+ for two external K^+ uses metabolic energy obtained from the hydrolysis of ATP (for review, see Hoffman, 1986; chapters 6 and 16 by Mercer *et al.* and by Kaplan in Agre and Parker, 1989). In the steady state, at the same time that K^+ is actively accumulated in the cell, it is also continually leaking out of the cell at the same rate through parallel pathways down its concentration gradient. The same is true for Na^+ leaking into the cell. Steady-state distributions of K^+ and Na^+ are thus achieved by a balance between active pumping and passive leakage. If the Na^+-K^+ pump of human red blood cells is completely inhibited, the loss of K^+ and gain of Na^+ is so slow that the leakage fluxes would continue unabated for at least 30 days before Gibbs-Donnan equilibrium is approached (Fig. 3); however, the cells would hemolyze before reaching equilibrium because of the inability of the plasma membrane to withstand any significant osmotic pressure associated with the Gibbs-Donnan equilibrium, as discussed in Section VI.

A striking correlation between the properties of the Na^+,K^+-ATPase and cation fluxes in red blood cells constitutes impressive evidence that the Na^+,K^+-ATPase is the ion pump that mediates the fluxes of Na^+ and K^+ across the red-cell membrane: (1) Both the Na^+,K^+-ATPase and the control of cation transport are located in the membrane; (2) the Na^+,K^+-ATPase and cation transport are both stimulated specifically by intracellular ATP, by intracellular Na^+, and by extracellular K^+; (3) the cation concentrations required for half-maximal activation of the Na^+,K^+-ATPase and cation transport are the same; (4) cardiac glycosides, which are impermeant, specifically inhibit the Na^+,K^+-ATPase and cation transport without directly affecting other transporters or fluxes; (5) cardiac glycosides in-

TABLE I Concentrations (mM) of Potassium, Sodium, and Chloride in Mammalian Red Blood Cells and Plasma

Species	Intracellular			Extracellular		
	$[K^+]$	$[Na^+]$	$[Cl^-]$	$[K^+]$	$[Na^+]$	$[Cl^-]$
Man[a]	135	17	77	3.7	138	116
Baboon[b]	145	24	78	4.7	157	115
Rabbit[b]	142	22	80	5.5	150	110
Rat[b]	135	28	82	5.9	152	118
Horse[b]	140	16	85	5.2	152	108
Sheep[c]						
HK	124	13				
LK	17	119				
Dog[d]	8	162	80	4.6	165	123
Cat[b]	8	142	84	4.6	158	112
Mean[b]			81	4.9	153	115
SD			3	0.7	8	5

[a]Funder, J., and Wieth, J. O. (1966a,b). *Scand. J. Clin. Lab. Invest.* **18**, 151–166, and *Acta Physiol. Scand.* **68**, 234–245.

[b]Bernstein, R. E. (1954). *Science* **120**, 459–460.

[c]Dunham, P. B. (1992). *Comp. Biochem. Physiol.* **102A**, 625–630.

[d]Parker, J. C. (1973). *J. Gen. Physiol.* **61**, 146–157.

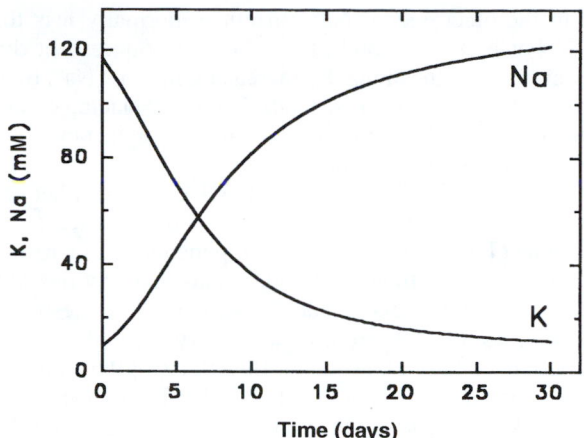

FIGURE 3. Loss of K$^+$ and gain of Na$^+$ when the Na$^+$-K$^+$ pump of human red blood cells is completely inhibited. (Theoretical plot by P. R. Pratap using the integrated model of Lew and Bookchin, 1986.)

hibit the Na$^+$,K$^+$-ATPase and cation transport half-maximally at the same concentration and with the same molecular specificity in that inhibition of both processes requires an unsaturated lactone group; (6) K$^+$ antagonizes the effect of cardiac glycosides on both the Na$^+$,K$^+$-ATPase and cation transport; and (7) the Na$^+$,K$^+$-ATPase is reversible, and net synthesis of ATP occurs when the Na$^+$ and K$^+$ concentration gradients are reversed. Moreover, the activity of the Na$^+$,K$^+$-ATPase parallels the magnitude of cation fluxes over a twenty-five thousand-fold range in a variety of tissues, and active transport of Na$^+$ and K$^+$ against their respective concentration gradients has been observed directly by measurement of net fluxes in resealed ghosts.

Läuger (1995) considered that the fluctuating energy barriers in ion pumps and exchangers may be similar to those in ion channels (see also DeFelice and Blakely, 1996). With this idea in mind, it is of considerable interest that **palytoxin**, isolated from a marine soft coral, reversibly increases the cation conductance of red cell membranes over the native Cl$^-$ conductance, and that the effects of the toxin are totally blocked by external ouabain and prevented by internal vanadate. In the presence of the toxin, the conductive cation selectivity becomes K$^+$ > Rb$^+$ > Cs$^+$ > Na$^+$ > Li$^+$, corresponding to Eisenman sequence IV (see Chapter 1). It appears that palytoxin "opens a 10 pS channel . . . at or near each pump site" (Tosteson *et al.*, 1991). Although this channel probably represents a perturbed form of the native cation permeation pathway in the Na$^+$,K$^+$-ATPase, further characterization might provide new insights as to how channels might function within pumps. A related problem in red cells concerns the relationship of DIDS-sensitive Cl$^-$ conductance to the DIDS-sensitive Cl$^-$-HCO$_3^-$ exchanger, as described in Section VII.

Whereas the Na$^+$-K$^+$ pump is by far the most widely distributed transporter that regulates the monovalent cation composition of cells, the red blood cells of carnivores (e.g., dogs, cats, ferrets, bears) contain high Na$^+$ and low K$^+$ (see Table 1). The immature nucleated red cells of dogs do contain the usual high K$^+$ and low Na$^+$ contents and Na$^+$-K$^+$ pumps, but during maturation primary active transport of

Na$^+$ and K$^+$ ceases. In mature dog red cells, intracellular [Na$^+$] is slightly less than extracellular [Na$^+$], whereas intracellular [K$^+$] is slightly greater than extracellular [K$^+$]. These cells, like human red cells, also contain a potent Ca^{2+}-ATPase in their membranes that maintains submicromolar concentrations of intracellular [Ca^{2+}] in the face of millimolar extracellular [Ca^{2+}] (for review, see chapter 17 by Vincenzi in Agre and Parker, 1989; see also Chapter 18). However, dog, ferret, and bear red cells, unlike human red cells, also contain a Na$^+$-Ca^{2+} countertransporter that couples the passive influx of Ca^{2+} to the efflux of Na$^+$, thus resulting in maintenance of a slight reduction in intracellular [Na$^+$] in the steady state (see Chapter 19). The slight excess of intracellular K$^+$ is consistent with an inside negative membrane potential.

In addition to the many structural and mechanistic studies on active pump fluxes mediated by the Na$^+$,K$^+$-ATPase and the Ca^{2+}-ATPase, much progress has also occurred in understanding the mechanisms of passive cation transport (for review, see chapter 18 by Parker and Dunham in Agre and Parker, 1989). Unlike muscle and nerve, red cells have a low resting permeability to both Na$^+$ and K$^+$, but a high conductive permeability to Cl$^-$ that is about 100 times greater than that to cations. Consequently, Cl$^-$ is in equilibrium with the resting membrane potential E_m,

$$E_m = \frac{RT}{\mathcal{F}} \ln \frac{[Cl^-]_i}{[Cl^-]_o}$$

where R is the gas constant, T is the absolute temperature, $\mathcal{F}$ is the Faraday constant, and $RT/\mathcal{F}$ is 25.5 mV at 23 °C. For the ratio of $[Cl^-]_i/[Cl^-]_o = 77/116 = 0.66$, the resting potential E_m is −11 mV, inside negative. In the steady state, ionic currents of K$^+$ and Na$^+$ both flow passively across the membrane down their respective concentration gradients. Assuming that the currents of monovalent ions are independent, that the membrane is symmetrical, that the transmembrane electric field is constant, and that ions first partition into the membrane and then diffuse across driven by concentration and electrical gradients, the theory of **electrodiffusion** gives the membrane potential as

$$E_m = -\frac{RT}{\mathcal{F}} \ln \frac{P_K[K^+]_i + P_{Na}[Na^+]_i + P_{Cl}[Cl^-]_o}{P_K[K^+]_o + P_{Na}[Na^+]_o + P_{Cl}[Cl^-]_i}$$

where P_K, P_{Na}, and P_{Cl} are the constant field permeabilities (s^{-1}). Since Cl$^-$ is at equilibrium, we may substitute Cl$_i$ = Cl$_o e^{E_m\mathcal{F}/RT}$ into the preceding Godman-Hodgkin-Katz equation. Rearranging and simplifying gives

$$E_m = -\frac{RT}{\mathcal{F}} \ln \frac{P_K[K^+]_i + P_{Na}[Na^+]_i}{P_K[K^+]_o + P_{Na}[Na^+]_o}$$

In other words, all permeant ions—cations and anions—have some relationship to the electrical potential E_m across the membrane. Solving this expression for the ratio P_K/P_{Na} gives

$$\frac{P_K}{P_{Na}} = \frac{[Na^+]_o e^{-\phi} - [Na^+]_i}{[K^+]_i - [K^+]_o e^{-\phi}}$$

where ϕ is the reduced potential $E_m\mathcal{F}/RT$. Using standard values for human red cells from Table 1 for $[K^+]_i$, $[K^+]_o$,

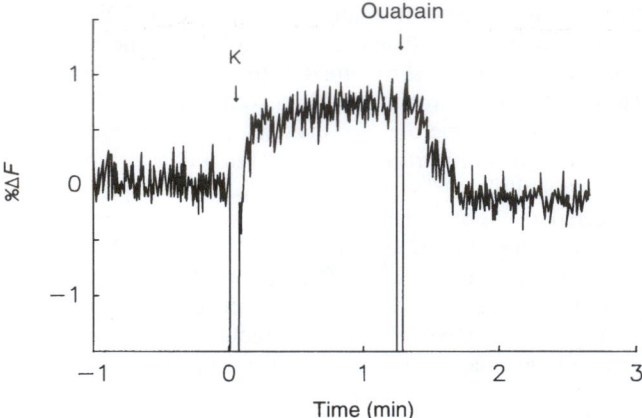

FIGURE 4. Electrogenic potential of the Na$^+$-K$^+$ pump in human red blood cells, as monitored with the oxonol dye WW781. Intracellular Cl$^-$ has been replaced with SO$_4^{2-}$, followed by treatment with DIDS. In Na$^+$-free medium, the pump is activated by addition of 10 mM K$^+$, and then inhibited with 33 millimolal ouabain, as indicated. An upward deflection indicates hyperpolarization. (Data of P. R. Pratap and J. C. Freedman.)

[Na$^+$]$_i$, [Na$^+$]$_o$, and ϕ yields $P_K/P_{Na} = 1.5$, much less than the corresponding ratio in nerve and muscle. Thus, the resting cation diffusion potential of human red blood cells is consistent with the membrane having a slightly greater constant-field permeability for K$^+$ than for Na$^+$. However, the measured ground permeabilities to Na$^+$ and K$^+$ appear to be nonselective and more consistent with a single-barrier model than with electrodiffusion (Zade-Oppen et al., 1988). Away from the steady state, such as following cell shrinkage by exogenously added extracellular sucrose, the gradient of the highly permeant Cl$^-$ will dominate and determine the membrane potential of human red blood cells until a new steady state is reached (see Bisognano et al., 1993). The electrogenic current of the Na$^+$,K$^+$-ATPase is one-third of the active Na$^+$ efflux; this current flows across the membrane electrical resistance, which for normal human red blood cells has been estimated at about 10^6 Ω cm^2 (for review, see Hoffman et al., 1980). The electrogenic potential

due to the electrogenic Na$^+$ current is normally less than 1 mV, but is demonstrable (Fig. 4) using fluorescent dyes after optimizing the signal by increasing internal Na$^+$, by increasing the membrane resistance through replacing cell Cl$^-$ with SO$_4^{2-}$, and by inhibiting the anion conductance with DIDS (Dissing and Hoffman, 1990).

As illustrated in Fig. 5, red cell passive fluxes, either conductive or electroneutral, are composed of at least nine phenomenologically separable components, a much more complex situation than assumed in classical theories (Van Slyke et al., 1923; Jacobs and Stewart, 1947; Tosteson and Hoffman, 1960). The pump-leak theory for red cells has been extended to model some of the effects of the additional cotransport pathways (Milanick and Hoffman, 1986; Lew and Bookchin, 1986). Pathways for passive electrolyte transport in human red blood cells include cotransporters for Na$^+$-K$^+$-2 Cl$^-$ and K$^+$-Cl$^-$ (for review, see Lauf et al., 1992; Hoffman and Dunham, 1995), countertransporters for Na$^+$(Li$^+$)-Na$^+$ (e.g., Duhm and Becker, 1977) and Na$^+$-H$^+$ (Escobales and Canessa, 1986), Ca^{2+}-activated K$^+$ channels (for review, see Schwarz and Passow, 1983), Cl$^-$-HCO$_3^-$ exchange mediated by capnophorin (AE1 or band 3 protein) (for review, see Knauf, 1979; Passow, 1986; chapter 19 by Gunn et al., in Agre and Parker, 1989; Jennings, 1992; Reithmeier, 1993), Cl$^-$ conductance (Freedman et al., 1988, 1994; Freedman and Novak, 1997), and HCl cotransport (Bisognano et al., 1993). Specific red-cell transporters also mediate the transmembrane movement of glucose, nucleosides, lactate and other organic anions, oxidized glutathione, choline, and amino acids (for review, see Lefevre, 1961; the chapters by Srivastava and by Martin in Ellory and Lew, 1977; Agre and Parker, 1989; Raess and Tunnicliff, 1990). Water flows through pores made of aquaporin (for review, see Agre, 1996).

Some of the most frequently used inhibitors in studies of red-cell membrane transport are listed in Table 2. In isotope flux studies, active transport of Na$^+$ and K$^+$ has often been equated with the "ouabain-sensitive fraction" of Na$^+$ efflux or K$^+$ influx. Similarly, Na$^+$-K$^+$-2 Cl$^-$ cotransport is operationally measured as the "ouabain-insensitive, bumetanide-sensitive" fraction of a Na$^+$ flux, whereas Na$^+$(Li$^+$)-Na$^+$ countertransport is the "ouabain- and bumetanide-insensitive,

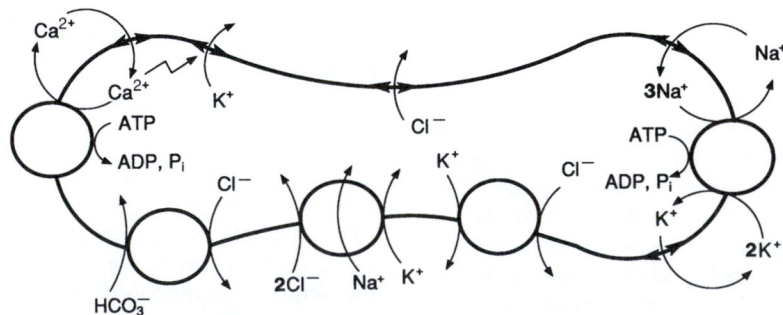

FIGURE 5. The principal membrane transport systems of human red blood cells. Starting with the Na$^+$-K$^+$ pump and the passive leakage pathways for Na$^+$ and K$^+$ on the right, and proceeding clockwise around the membrane, are K$^+$-Cl$^-$ cotransport, Na$^+$-K$^+$-2Cl$^-$ cotransport, Cl$^-$-HCO$_3^-$ exchange, the Ca^{2+} pump with a Ca^{2+} leakage pathway, the Ca^{2+}-activated K$^+$ channel, and Cl$^-$ conductance. The steady-state concentrations of K$^+$, Na$^+$, and Ca^{2+} represent a balance between active pumping and passive leakage. Cl$^-$ and HCO$_3^-$ are passively distributed at equilibrium with a membrane potential of −11 mV.

TABLE 2 Some Inhibitors of Membrane Transport in Red Blood Cells

Transporter	Inhibitor
Pumps	
Na$^+$,K$^+$-ATPase	Ouabain, vanadate
Ca^{2+}-ATPase	La^{3+}, vanadate, eosin
GSSG[a]	Fluoride
Cotransport	
Na$^+$-K$^+$-2Cl$^-$	Bumetanide, furosemide
K$^+$-Cl$^-$	DIOA[b]
	Calcyculin A, okadaic acid
H$^+$-Cl$^-$	DIDS[c]
Countertransport	
Cl$^-$-HCO$_3^-$	DIDS[c], DNDS[d]
	Phlorizin, phloretin
Na$^+$(Li$^+$)-Na$^+$	Phloretin
Na$^+$-H$^+$	Amiloride
Facilitated diffusion	
Glucose	Phlorizin, phloretin
	Cytochalasin B
Nucleosides	Nitrobenzylthioinosine
Lactate	PCMBS[e], DTNB[f]
Choline	NEM[g]
Channels	
Ca^{2+}(K$^+$)	Charybdotoxin

[a]Oxidized glutathione
[b][(Dihydroindenyl)oxy]alkanoic acid
[c]4,4′-Diisothiocyano-2,2′-disulfonic acid stilbene
[d]4,4′-Dinitro-2,2′-dinitrosulfonic acid stilbene
[e]para-Chloromercuribenzenesulfonate
[f]5,5′-Dithio-bis(2-nitrobenzoate)
[g]N-Ethylmaleimide

phloretin-sensitive" fraction. Fluoride and phloretin are relatively nonspecific and act on more than one transporter. Eosin is a relatively new inhibitor for the Ca^{2+}-ATPase. Some inhibitors bind reversibly to the transporter (e.g., DNDS to band 3), whereas others inhibit irreversibly by formation of covalent bonds (e.g., DIDS to band 3). Inhibitors may also act indirectly on associated enzymes that modulate transport. For example, the inhibition of a Type 1 protein phosphatase by calyculin A or okadaic acid prevents the activation of K$^+$-Cl$^-$ cotransport during cell volume regulation.

VI. Ionic and Osmotic Equilibrium and Cell Volume Regulation

Human red blood cells contain 66% water (100 × g water/g packed cells), as is easily determined by drying a weighed pellet of sedimented cells to constant weight in a vacuum oven, taking into account suitable corrections for trapped volume. According to the **Boyle-van't Hoff law**, which is derived from the Gibbs equation (see Dick, 1959), the **osmotic pressure** (π) of a solution containing N moles of solute dissolved in V_w liters of water is given by

$$\pi = \frac{\phi N R T}{V_w} = \phi c R T$$

where ϕ is the osmotic coefficient of the solute, R is the gas constant, T is the absolute temperature, and c $(= N/V_w)$ is the total concentration of dissolved solutes. The **osmolality** ϕc is the total osmolal concentration of all dissolved solutes (see Chapter 21). The **osmotic coefficient** ϕ is unity for ideal dilute solutions; deviations of ϕ from unity are due to solute-solvent interactions and other nonidealities. The osmotic pressure π_i of the intracellular solution is the sum of the osmotic pressures of each component of the mixture,

$$\pi_i = \sum \frac{\phi_j N_j RT}{V_w} = RT \sum \phi_j C_j$$

where c_j $(= N_J/V_w)$ is the concentration of the jth solute. Alternatively,

$$\pi_i = \frac{\Phi_i N_T RT}{V_w} = \Phi_i c_T RT$$

where Φ_i is the osmotic coefficient of the intracellular mixture, a parameter that equals the sum of the osmotic coefficients of the components of the mixture, weighted according to their respective mole fractions.

Water will cross the membrane until the intracellular osmotic pressure equals the extracellular osmotic pressure, or $\pi_i = \pi_o$, as expressed by

$$\sum \phi_{j,i} C_{j,i} = \sum \phi_{j,o} C_{j,o}$$

An **isosmotic** solution is defined as having an osmolality equal to that of normal plasma, or 289 milliosmolal; an **isotonic** solution will maintain the normal volume of cells incubated during physiological experiments. In isotonic solution, designated by the superscript ^{0},

$$\pi_i^0 = \frac{\Phi_i^0 N_T^0 RT}{V_w^0}$$

Dividing π_i by π_i^0, setting $\pi_i = \pi_o$ and $\pi_i^0 = \pi_o^0$, and rearranging yields the following expression for the equilibrium cell water content relative to that of cells in isotonic solution:

$$\frac{V_w}{V_w^0} = \frac{\left(\Phi_i N_T\right) \pi_o^0}{\left(\Phi_i^0 N_T^0\right) \pi_o}$$

This expression quantitates how much cell shrinkage will occur in **hypertonic** solution ($V_w/V_w^0 < 1$ when $\pi_o > \pi_o^0$), and how much cell swelling will occur in **hypotonic** solution ($V_w/V_w^0 > 1$ when $\pi_o > \pi_o^0$).

The osmotic coefficient of hemoglobin ϕ_{Hb} rises approximately with the square of hemoglobin concentration (Fig. 6B; see also Ross and Minton, 1977), resulting in significant deviations of red cells from ideal osmotic behavior (Dick, 1959; Freedman and Hoffman, 1979a). For an ideal osmometer, ϕ_{Hb} and the osmotic coefficients of other solutes would be unity. At a normal [Hb] of 7.3 millimolal, ϕ_{Hb} is 2.85; in contrast, the osmotic coefficients of KCl and NaCl

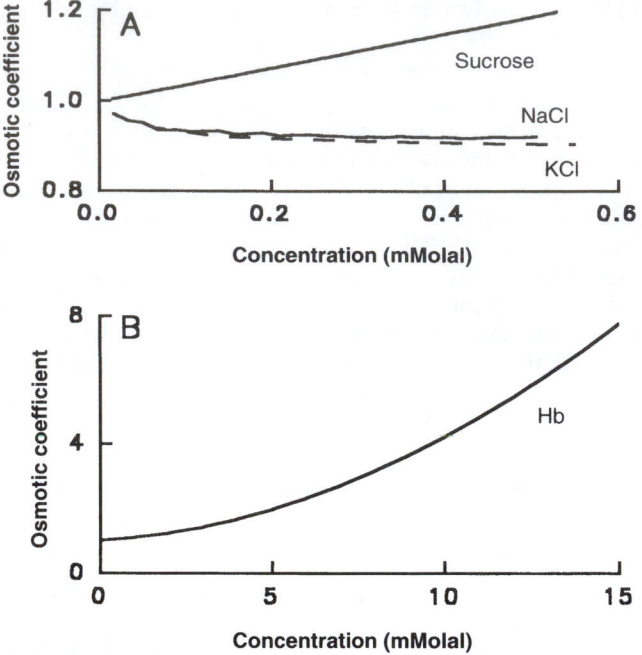

FIGURE 6. (A) Osmotic coefficients of NaCl, KCl, and sucrose. (B) Osmotic coefficient of hemoglobin. (Data for NaCl, KCl, and sucrose are from the "CRC Handbook of Chemistry and Physics," 58th Ed. (1978–79) (R. C. Weast and M. J. Astle, Eds.), p. D-261, and from R. A. Robinson and R. H. Stokes, "Electrolyte Solutions," 2nd Ed. (1959). Butterworths, London. Data for hemoglobin are from Adair, as adapted from Fig. 7 of Freedman and Hoffman, (1979), *J. Gen. Physiol.* **74**, p. 177, by copyright permission of The Rockefeller University Press, and is drawn according to $\phi_{Hb} = 1 + 0.0645[Hb] + 0.0258[Hb]^2$.)

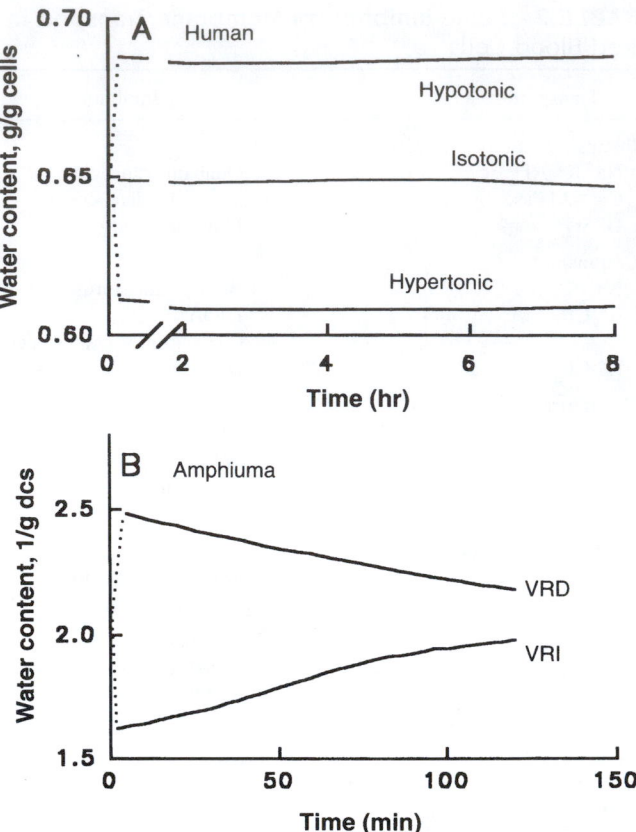

FIGURE 7. (A) Stable osmotic response of human red blood cells. Washed human red cells were incubated in hypotonic, isotonic, or hypertonic media, and the water contents were determined at selected times by drying packed cells to constant weight in a vacuum oven. (Data of J. C. Freedman and S. Fazio.) (B) Volume regulatory decrease, VRD (upper trace), or volume regulatory increase, VRI (lower trace), when *Amphiuma* red cells are incubated in hypotonic or hypertonic media, respectively. For red cells from *Amphiuma,* but not from humans, the red-cell water content spontaneously returns back toward its normal value. (Bottom panel adapted from Figs. 1 and 2 of P. M. Cala, *The Journal of General Physiology,* 1980, Vol. 76, pp. 688–689.)

decrease with increasing concentration and are 0.93 at physiological concentrations, whereas that of sucrose rises slightly with increasing concentration (Fig. 6A). Mature human red blood cells maintain stable volumes when placed in isotonic solution, and rapidly adjust their volumes to new stable levels that are maintained over time in hypotonic or hypertonic salt solutions (Fig. 7A). In contrast, human reticulocytes and red cells from many other species, including duck, dog, the salamander *Amphiuma,* trout, skate, rabbit, and a low-K^+ genetic variant of sheep, all possess the capability of altering their membrane permeability in response to an altered extracellular osmolarity in such a way as to alter their total intracellular solute concentration N_T and return their cell volumes back to normal (for review, see Hoffmann and Dunham, 1995; see also Chapter 21). Examples of the **volume regulatory decrease (VRD)** and **volume regulatory increase (VRI)** exhibited by *Amphiuma* red blood cells are shown in Fig. 7B. The swelling-activated response usually involves an increase in K^+-Cl^- cotransport, as in duck, sheep, and rabbit red cells, but may also involve formation of a channel related to band 3 that permits efflux of osmolytes such as taurine, as in red cells from skate and trout. An exception seems to be net KCl efflux resulting from activation of K^+-H^+ exchange coupled with Cl^--HCO_3^- exchange, as proposed for *Amphiuma* red cells. The shrinkage-activated response involves activation of Na^+-K^+-$2Cl^-$ cotransport in

duck red cells, Na^+H^+ countertransport in parallel with Cl^--HCO_3^- exchange in *Amphiuma* red cells, and Na^+-H^+ exchange in dog red cells. The mechanisms by which cells "sense" their altered volume prior to switching on a transport pathway, and "sense" their normal volume when turning off a transport pathway, are the subjects of current active investigation. Thus, the amount of water in red blood cells is determined by the total concentration (osmolality) of salts (cations and anions), and by proteins and organic metabolites, as well as by the nonideal effects expressed by the osmotic coefficients of hemoglobin and other intracellular solutes.

An understanding of the **Gibbs-Donnan equilibrium** is critically important for properly incubating red cells and for understanding their response to altered pH and osmolarity. Human red blood cells, for example, swell in acid and shrink in alkaline media (Fig. 8). Red cells swell and hemolyze on exposure to compounds such as nystatin that elevate

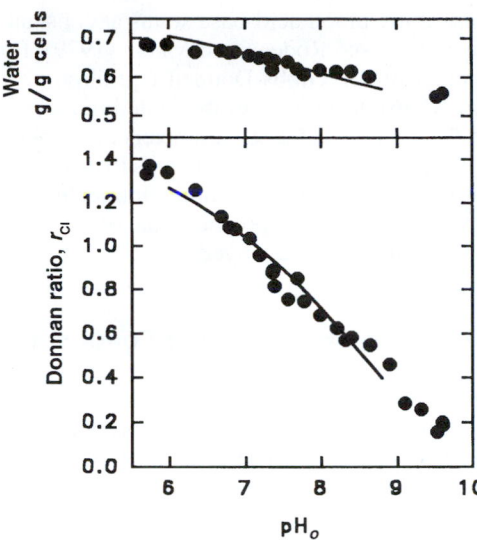

FIGURE 8. Comparison of measured (points) and predicted (lines) Donnan ratios in human red blood cells. The equilibrium Donnan ratios for Cl^- (bottom panel) and the cell water contents (top panel) were determined at varied extracellular pH. (Adapted from Fig. 3 of R. B. Gunn, M. Dalmark, D. C. Tosteson, and J. O. Wieth, (1973). *J. Gen. Physiol.* **61**, p. 193, by copyright permission of The Rockefeller University Press.) The solid lines are predictions of the program IONIC. (Adapted from Fig. 3 of J. C. Freedman and J. F. Hoffman, (1979) *J. Gen. Physiol.* **74**, 174, with copyright permission of The Rockefeller University Press.)

the normal low permeability of the membrane to cations, a phenomenon termed **colloid osmotic hemolysis** by Wilbrandt. As a human red blood cell swells, its membrane surface area remains constant as the biconcave disc gains water and converts to a spherical shape at the **critical hemolytic volume**, which is 1.65 times the original volume (for review, see Hoffman, 1992). The membrane, which has a surface tension of about 1 dyn/cm, but an area expansion modulus of ~450 dyn/cm, is unable to withstand more than about 4% stretching (see chapter 15 by Berk, Hochmuth, and Waugh in Agre and Parker, 1989), and consequently ruptures with further swelling. A Gibbs-Donnan equilibrium occurs whenever a membrane separates an aqueous salt solution from a salt solution that also contains impermeant charged electrolytes, such intracellular charged proteins and organic phosphates (see Chapter 15). For human red cells, the permeability of the membrane to Na^+ and K^+ is 100 times less than that to Cl^- and HCO_3^-, so that certain cellular responses can be understood by considering the cells to approximate a **double-Donnan** system. From the principle of bulk electroneutrality in the extracellular and intracellular solutions,

$$Cl_o^- = K_o^+ + Na_o^+$$

and

$$Cl_i^- = Na_i^+ + K_i^+ + z\,Hb$$

where z is the average net charge (μeq/mol Hb) on impermeant cell solutes including hemoglobin and organic phosphates. The **Donnan ratio** r_{Cl}, for Cl^- is

$$r_{Cl} = \frac{Cl_i^-}{Cl_o^-} = \frac{Na_i^+ + K_i^- + z\,Hb}{K_o^+ + Na_o^+} = e^{E_m\mathscr{F}/RT}$$

This expression illustrates most clearly that the double-Donnan equilibrium consists of one Donnan equilibrium being established by the slowly permeant ("functionally impermeant") cations, and the other by hemoglobin (Hb) and organic phosphates. The net charge on hemoglobin depends on intracellular pH as its constituent amino acids are titrated. The titration curve of hemoglobin is approximately given by

$$z = -m\left(pH_i - pI\right)$$

where m is the slope of the titration curve or **buffer capacity** ($m = -12$ eq/mol Hb_4/pH_o), and pI is the **isoelectric point** of the cell contents, or the pH_i where the net charge is zero (pI = 6.8 at 25 °C in human red cells). In ectothermic vertebrates, maintaining the constancy of the charge z on intracellular proteins may be more important for homeostasis than the constancy of extracellular pH (for review, see Reeves, 1977). If HCl is added to a suspension of red cells, the intracellular pH will decrease, hemoglobin will become more positively charged, and since Na^+ and K^+ are so much less permeant than Cl^-, the highly permeant anion will enter the cells to balance the increased positive charge on hemoglobin. Since the total intracellular solute concentration has now increased because of entry of Cl^-, the red cells swell (see Fig. 8). In contrast, if NaOH is added to a suspension of red cells, the intracellular pH increases, hemoglobin becomes more negatively charged and Cl^- leaves the cell, resulting in cell shrinkage (see Fig. 8). Model calculations show that at the isoelectric pH, the sum of the cation concentrations is greater in plasma than in the intracellular solution, in accordance with the double-Donnan concept; at the physiological pH of 7.4, however, cell shrinkage raises the intracellular cation concentration so that it nearly equals the plasma cation concentration, thus obscuring the double-Donnan effect. In human red cells, NO_3^-, I^-, SCN^-, and methane sulfonate ($CH_3SO_3^-$), but not methyl sulfate ($CH_3SO_4^-$), all bind to hemoglobin and alter its net charge, with a consequent small but significant change in cell volume in accordance with the Gibbs-Donnan equilibrium (Payne *et al.*, 1990). The Donnan equilibrium shifts in the expected manner on experimental alteration of intracellular phosphates (Duhm, 1971) and in abnormal human red cells containing HbC, which has an increased positive charge (Brugnara *et al.*, 1985).

Colloid osmotic hemolysis occurs when the red-cell membrane is exposed to pore-forming ionophores such as nystatin or gramicidin, or when the normal low permeability to cations increases on exposure to detergents or other chemicals. At equilibrium at the isoelectric pH, where the Donnan ratio equals unity and the membrane potential is zero, the intracellular concentrations of Na^+, K^+, and Cl^- would equal those of the extracellular solution. The osmotic pressure of hemoglobin would be unbalanced by any extracellular solute and would draw water into the cell, leading to swelling and hemolysis. Colloid osmotic hemolysis occurs at any pH when the membrane is permeant both to cations and to anions. The Na^+ pump, working against the leakage pathways and any significant cotransport pathways, establishes the steady-state intracellular cation concentrations, while Cl^- and HCO_3^- move to maintain electroneutrality.

In red blood cells, for given steady-state concentrations of K^+ and Na^+, as set by the balance of pumps and leaks, the equilibrium cellular electrolyte concentrations and water contents are determined by the following three principles:

1. Bulk electroneutrality
2. Osmotic equality of the intracellular and extracellular solutions
3. Equality of the Donnan ratios for all permeant ions of the same charge

The Donnan ratio for monovalent anions is equal to the inverse of the Donnan ratio for protons or any other highly permeant monovalent cations. For human red blood cells, the nonideal equations expressing these constraints, including the activity and osmotic coefficients of the salts and of hemoglobin, have been formulated into a computer program designated IONIC (Freedman and Hoffman, 1979a; Bisognano *et al.,* 1993; see also Raftos *et al.,* 1990). For a given composition of the extracellular medium, and certain cellular parameters, the program predicts the cell volume, the intracellular pH and ion concentrations, the membrane potential, and the charge on hemoglobin and organic phosphates. The predictions of the model generally agree with experimental results within a few percent; for example, the predicted and experimentally determined dependence of the Donnan ratios for Cl^- on pH are shown in Fig. 8. For a model describing blood acid-base status at constant or variable temperature, see Rodeau and Malan (1979). Beginning with a model of the Gibbs-Donnan equilibrium, Lew and Bookchin (1986) developed an integrated model of red-cell transport that includes kinetic parameters describing most of the known transport systems, and begins to describe the time course of the changes in cell volume, intracellular salt concentrations, and membrane potential that occur when the extracellular conditions are changed.

VII. Anion Exchange and Conductance

The transport of carbon dioxide from the peripheral tissues to the lungs is facilitated by the production of bicarbonate by red blood cells in a cyclic reaction scheme known as the **Jacobs-Stewart cycle.** In systemic tissue capillaries, carbon dioxide diffuses easily across the red cell membrane into the cells, whereupon **carbonic anhydrase** (CA) catalyzes its hydration to carbonic acid (Fig. 9). The acid then dissociates into a proton and bicarbonate. The proton is buffered by hemoglobin, which when protonated has a reduced affinity for oxygen, a phenomenon known as the **Bohr effect.** The bicarbonate passes across the red cell membrane into the venous blood by undergoing a rapid electroneutral exchange with Cl^-, a reaction known as the **chloride shift.** A small proportion of carbon dioxide forms a covalent **carbamino** compound with amino groups on hemoglobin. Production of HCO_3^- by red blood cells minimizes acidification of venous blood and increases its **carbon dioxide car-**

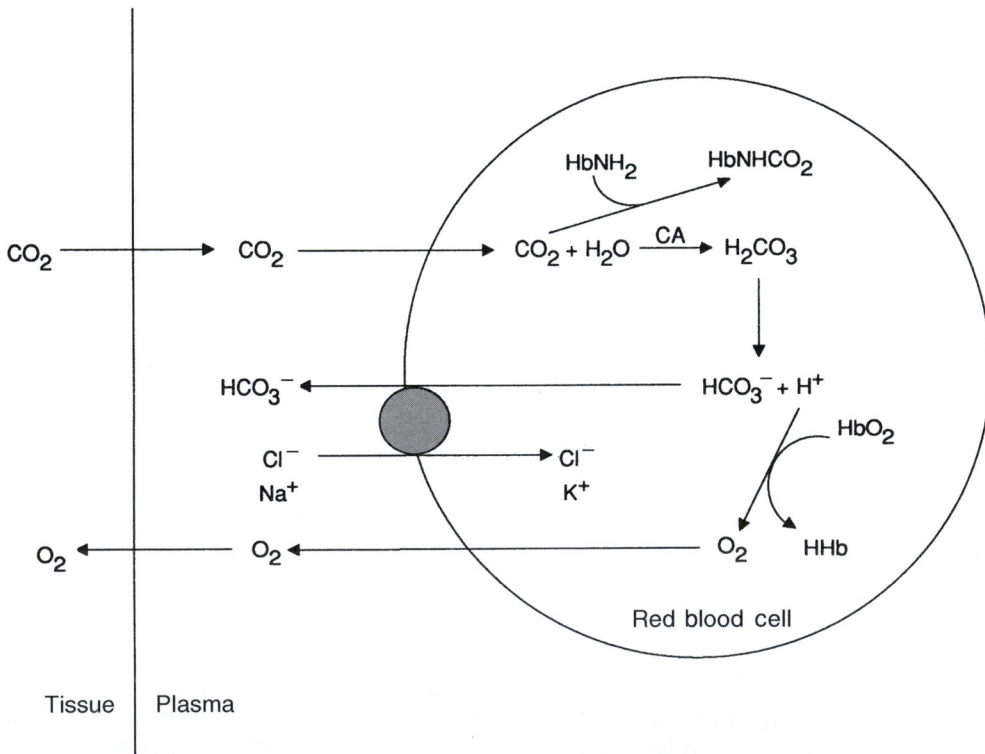

FIGURE 9.   Jacobs-Stewart cycle for red blood cells.

rying capacity. Red cell Cl^--HCO_3^- exchange is mediated by the membrane domain of **capnophorin**, a protein also denoted as anion exchanger AE1. This integral glycoprotein, having a molecular weight of 101 700 Da, makes up 25% of the mass of membrane protein in red blood cells (Steck, 1974) and is the predominant protein located in band 3 of SDS polyacrylamide gels.

DIDS is the most potent of a series of stilbene derivatives originally found by Cabantchik and Rothstein to inhibit red cell anion transport. DIDS inhibits 99.999% of Cl^--HCO_3^- exchange, but only 67% of net Cl^- efflux. Molecular biological experiments have identified the location of the DIDS binding site. The cDNA from mice and humans that codes for band 3 protein (AE1) was cloned and sequenced (for review, see Alper, 1991). Band 3 protein has also been functionally expressed in *Xenopus laevis* toad oocytes microinjected with mRNA prepared from human and mouse cDNA clones. Site-directed mutagenesis of the mouse protein indicates that one of the isothiocyanate (NCS) groups on DIDS binds covalently to Lys 558, which is located on the extracellular side of the 65-kDa chymotryptic N-terminal fragment. The stoichiometry of binding is one DIDS per capnophorin monomer (for review, see Passow *et al.*, 1992).

The membrane domain of capnophorin has been reconstituted with lipids and crystallized in two-dimensional arrays (Reithmeier, 1993; Wang *et al.*, 1994). At 20 Å resolution, reconstructed images indicate a dimeric structure with a cavity between the two monomers of band 3. ^{35}Cl nuclear magnetic resonance studies of capnophorin suggest an hourglass type of structure in which internal and external hemichannels lead to the transport site, which can exist in inward- and outward-facing conformations (Falke and Chan, 1986b). Cl^--HCO_3^- exchange follows Ping-Pong kinetics in which a conformational change occurs only after binding an anion alternately from the inward- and outward-facing conformations (for review, see Fröhlich and Gunn, 1986). DIDS lies in the outer hemichannel between the transport site and the extracellular medium, partially blocking the outward-facing transport site (Falke and Chan, 1986a). Structure-activity studies with a series of analogs of DIDS suggested a model of the DIDS binding site that included positively charged groups providing electrostatic stabilization of the sulfonates on DIDS, with adjacent hydrophobic and electron-donor centers. Experiments using the technique of fluorescence resonance energy transfer indicate that DIDS is located only 34–42 Å from sulfhydryl reagents bound to cysteine residues on the 40 000-Da aminoterminal cytoplasmic domain of capnophorin, a finding that is also consistent with DIDS residing in a cleft in the outer hemichannel. Moreover, substrate anions such as Cl^- traverse only 10–15% of the full transmembrane electric field when they move from the extracellular medium and bind to the outward-facing transport site (Jennings *et al.*, 1990). This observation is consistent with the hourglass model and indicates a low electrical resistance for the outer hemichannel, because if the resistance of the outer hemichannel were high, the transmembrane voltage would have dropped more than the 10–15% that was observed.

The classical view that the distribution of red-cell Cl^- is simply determined by the double-Donnan equilibrium was consistent with the observations that ^{38}Cl equilibrates within 1 s, which is 10^6 times faster than K^+, and that the Donnan ratios for Cl^-, HCO_3^-, and OH^- are all equal. The concept of rapid net movement of Cl^- across the red-cell membrane changed dramatically when experiments with gramicidin and valinomycin revealed that Cl^- conductance is about 10^4 times slower than electroneutral Cl^- exchange, yet still about 10^2 times greater than the cation conductance. Thus net transport of cations normally occurs over a time course of hours, net movement of Cl^- occurs over minutes, and exchange of Cl^- for HCO_3^- occurs in seconds.

In addition to electroneutral anion exchange, capnophorin is also believed to mediate the net flux of Cl^-, which could contribute to loss of salt and water during pathophysiological dehydration of the red cell, such as occurs during the formation of irreversibly sickled cells. Studies have revealed DIDS-sensitive and DIDS-insensitive components of the net Cl^- fluxes, and the DIDS-insensitive component increases with increasing extents of membrane hyperpolarization (Freedman *et al.*, 1988, 1994; Freedman and Novak, 1997).

Because it has not proven possible to use microelectrodes to drive currents or to clamp the voltage across the membrane of human red cells, ionophores have been used instead to increase the permeability to cations, thus permitting ionic currents to flow under the influence of diffusion potentials, which can then be measured indirectly. Consider the equivalent circuit shown in Fig. 10. The concentration gradients of K^+, Na^+, and Cl^- across the membrane are represented by batteries (E_K, E_{Na}, and E_{Cl}) in parallel with the membrane capacitance C_m. Each battery is in series with its respective ionic conductance (g_K, g_{Na}, and g_{Cl}). The electrical potential inside the cell is E_m, while that outside is at ground. Treatment of red cells with valinomycin or with gramicidin in Na^+-free medium, increases the normally low value of g_K above that of g_{Cl} and allows the flow of the ionic currents, i_K

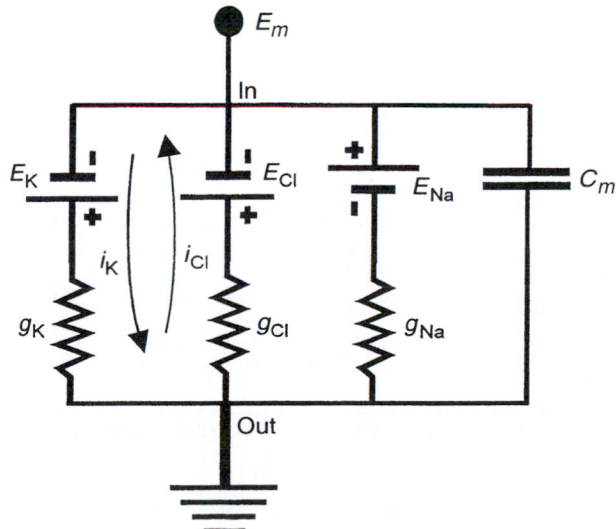

FIGURE 10. Equivalent electrical circuit for human red blood cells treated with valinomycin, or with gramicidin in Na^+-free medium. Arrows indicate the direction of positive current flow when the potassium conductance g_K is increased by the addition of valinomycin or of gramicidin in Na^+-free medium.

and i_{Cl}, around the loop (see Fig. 10, arrows). At low $[K^+]_o$ the net efflux of K^+ corresponds to the outward current i_K, which discharges the potassium battery. The magnitude of i_K depends on $[K^+]_o$, which is easily varied. The efflux of K^+ hyperpolarizes the membrane potential from its resting value near -11 mV to a value of some -60 mV, thus driving a net efflux of Cl^-, corresponding to an inward current i_{Cl} that charges the chloride battery. The K^+ and Cl^- currents are opposite in sign and approximately equal in magnitude; a small disparity of less than 10% is accounted for by a proton current. Whereas the total net current across the membrane must be zero in the absence of an external circuit, the individual ionic currents induced by the addition of ionophores can be either measured directly or inferred from the rate of cell shrinkage. Fluorescent potentiometric indicators that monitor the voltages continuously show that the voltage remains at a steady value after the addition of valinomycin or gramicidin. Steady levels of dye fluorescence are seen with the oxacarbocyanine dye, diO-C_6(3) (Hoffman and Laris, 1974), or with the thiadicarbocyanine, diS-C_3(5) (Fig. 11), or indodicarbocyanine, diI-C_3(5), dyes (Freedman and Hoffman, 1979b; Freedman and Novak, 1989). Human red blood cells treated with valinomycin or with gramicidin thus behave as if voltage-clamped by their intrinsic ion concentration batteries, E_K and E_{Cl}, instead of by an external circuit. This alternative voltage-clamp experiment yields current-voltage curves characterizing the conductance to Cl^-.

The change in voltage upon addition of ionophores has been estimated from the change in the extracellular pH of unbuffered DIDS-treated red-cell suspensions (Macey *et al.*, 1978) using the proton ionophore FCCP, a method that is also useful for calibrating fluorescent potentiometric indicators (Freedman and Novak, 1989; Bifano *et al.*, 1984). The inward-rectifying current-voltage curve found for the DIDS-insensitive fraction of Cl^- net transport (Fig. 12) deviates markedly from the nearly linear outward-rectifying curve predicted by the constant field theory for the inward concentration gradient of Cl^-. The inward-rectifying current-voltage curve is also inconsistent with certain single-occupancy multiple-barrier models, but instead is consistent either with a single barrier located near the center of the transmembrane electric field, or with a voltage-gated mechanism according to which half the channels are open at -27 mV, and with an equivalent gating charge of -1.2 ± 0.3 (Freedman and Novak, 1997). Further experiments are needed to clarify the relationship of DIDS-insensitive Cl^- conductance inferred for red cells voltage-clamped with ionophores to the anion-selective channels recorded with the patch-clamp technique (Schwarz *et al.*, 1989), and to those seen after fusing vesicles from red-cell suspensions into planar lipid bilayers (Freedman and Miller, 1984). It is also important to know if the DIDS-sensitive fraction of Cl^- conductance is indeed mediated by the band 3 exchanger, in which case further studies could attempt to discover a unified kinetic scheme for exchange and conductance mediated by the same transporter.

VIII. Cytotoxic Calcium Cascade

Despite the importance of Ca^{2+} as a trigger and modulator in a variety of cell activities (see Chapter 1), too much intracellular Ca^{2+} is harmful to cells in general and to red blood cells in particular. Elevated intracellular Ca^{2+} stimulates a cytotoxic cascade of pathophysiological and biochemical changes. Some of these events are depicted in Fig. 13. Subjecting human red blood cells to shear stress increases the passive influx of Ca^{2+} as seen in suspension (Larsen *et al.*, 1981), or during flow through synthetic microlattices (Brody

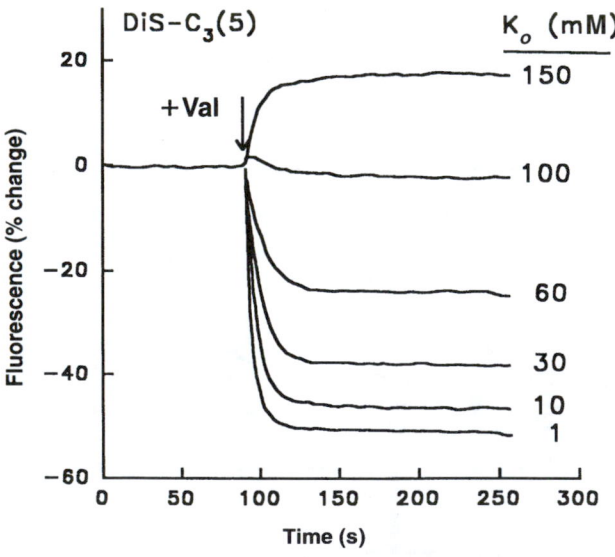

FIGURE 11. Steady levels of fluorescence of diS-C_3(5) following the addition of valinomycin to suspensions of human red blood cells in x mM KCl (as indicated), $150 - x$ mM NaCl, and 5 mM Hepes buffer (pH 7.4 at 23 °C). Downward deflections indicate hyperpolarization, whereas upward deflections indicate depolarization. (Adapted from Fig. 2 of Freedman, J. C., and Novak, T. S. (1989). *Methods Enzymol.* **172,** 108.)

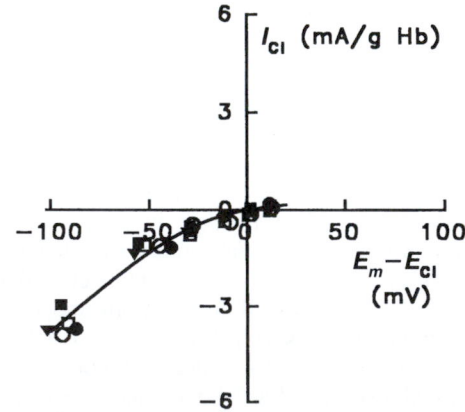

FIGURE 12. Current-voltage curve for DIDS-insensitive Cl^- conductance of human red blood cells. The solid line represents the best fit either of a single-barrier model, or of a mechanism involving voltage gating of ionic channels. (Adapted from Fig. 5 of Freedman, J. C., and Novak, T. S. (1997). *J. Gen. Physiol.* **107,** 207, with copyright permission of The Rockefeller University Press.)

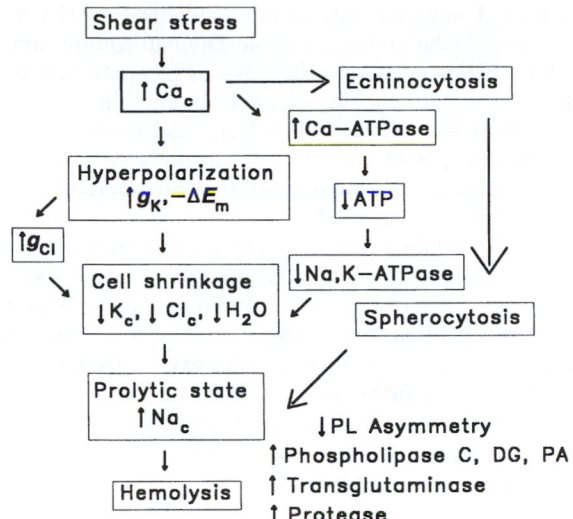

FIGURE 13. The cytotoxic calcium cascade of human red blood cells. The scheme illustrates some of the effects of elevated intracellular Ca^{2+}. (Adapted from Fig. 3 of Freedman, J. C. *et al.,* (1988). "Cell Physiology of Blood." p.222, with copyright permission of The Rockefeller University Press.)

et al., 1995). When intracellular Ca^{2+} rises to about 3 μM, the K^+ permeability dramatically increases, a phenomenon first described by Gardos and known as the **Gardos effect.** The increased K^+ permeability is due to the opening of a Ca^{2+}-activated K^+ channel, resulting in loss of intracellular K^+ and Cl^-, accompanied by cell shrinkage (for review, see Schwarz and Passow, 1983). Elevated $[Ca^{2+}]_c$ also stimulates the Ca^{2+}-ATPase, causing depletion of ATP within minutes (Crespo *et al.,* 1987), which in turn inhibits the Na^+,K^+-ATPase. In addition to causing substrate depletion, Ca^{2+} also inhibits the Na^+,K^+-ATPase with a potency that depends on a regulatory factor (for review, see Yingst, 1988). At higher levels of intracellular Ca^{2+}, the smooth biconcave discoidal form of human red cells converts first to an echinocytic form with spicules protruding from the membrane, and then to a spherocytic form with release of microvesicles. ATP depletion is prevented by inhibiting the ATPases with vanadate, but echinocytosis still develops (Freedman *et al.,* 1988). The echinocytogenic effect of Ca^{2+} is a striking example of the profound influence that Ca^{2+} can exert on cell morphology. At still higher levels, Ca^{2+} activates a transglutaminase that cross-links cytoskeletal proteins and reduces **cell deformability**. Ca^{2+} also activates degradative proteases and a phospholipase C that converts membrane phosphatidylcholine (lecithin) to diacylglycerol.

The various effects of elevated intracellular Ca^{2+} are induced at widely differing concentrations. Addition of A23187 stimulates loss of KCl at 0.1 μM free external Ca^{2+}, and the membrane hyperpolarizes within 2 s (Freedman and Novak, 1983). The threshold for Ca^{2+}-induced echinocytosis (assessed by the morphologic index) is 2 orders of magnitude greater, at 10 μM $[Ca^{2+}]_f$. It is possible to set external Ca^{2+} at a level sufficient to induce hyperpolarization, but below the threshold for echinocytosis, thus showing that the membrane potential does not directly influence cell shape

(Bifano *et al.,* 1984). Under these conditions, the cells are able to crenate because simply increasing extracellular Ca^{2+} produces echinocytes. Under similar conditions, the Ca^{2+}-induced decrease in cell deformability, as assessed by a decreased elongation index during flow at set shear stress (measured in an **Ektacytometer**), occurs at 1 mM Ca^{2+}. The separate apparent potencies of Ca^{2+}, spanning 5 orders of magnitude, for inducing hyperpolarization, echinocytosis, and reduced deformability are consistent with the idea that Ca^{2+} acts at independent sites to initiate these elements of the cytotoxic calcium cascade (Freedman *et al.,* 1988).

The relationships of the activation by Ca^{2+} of a transglutaminase, a phospholipase C, and a protease to the loss of phospholipid asymmetry (see Lin *et al.,* 1994), and to the changes in permeability, cell volume, and morphology are complex. Lipid alterations, including phosphoinositide hydrolysis and phosphatidic acid production, occur at 1–10 μM Ca^{2+}. Whereas μM Ca^{2+} dramatically affects the viscosity of complexes of spectrin, actin, and band 4.1 (Fowler and Taylor, 1980), the relative contributions of lipid and cytoskeletal alterations to the Ca^{2+}-induced shape changes are still unclear.

The final consequence of increased $[Ca^{2+}]_c$ *in vitro,* after prolonged cell shrinkage due to loss of KCl, is the development of **prolytic** cells leaky to Na that then undergo colloid osmotic hemolysis. Experiments suggest that cell shrinkage, rather than Ca^{2+} itself, results in the gain of Na^+ (Crespo *et al.,* 1987). The observation that comparable gains of Na^+ occur in cells simply shrunken by sucrose, as upon exposure to valinomycin or to Ca^{2+} plus calcium ionophore A23187 at low external K^+, suggests that Ca^{2+} is not essential in causing elevated $[Na^+]_c$. Other flux studies and assessment of relative cell volume distributions indicate that a subpopulation of cells becomes highly permeable to Na^+ after shrinkage, induced either osmotically or by Ca^{2+}-induced loss of KCl. An apparent steady-state level of elevated $[Na]_c$ in the population of cells is attained because a fraction of the cells become prolytic while another fraction hemolyzes. An interesting question is what determines which cells enter the prolytic state.

A role of the cytotoxic calcium cascade in the formation of dehydrated irreversibly sickle cells (ISCs) is indicated by the high levels of intracellular Ca^{2+} sequestered in sickle cells, and by the ability of charybdotoxin and clotrimazole, two blockers of Ca^{2+}-K^+ channels, to inhibit the formation of ISCs (Ohnishi *et al.,* 1989; Brugnara *et al.,* 1996). It should be noted that K^+-Cl^- cotransport has very low activity in normal mature red cells, but has elevated activity in immature reticulocytes and in sickle cells, and is also activated by acidosis (Brugnara *et al.,* 1989). The relative contributions of K^+-Cl^- cotransport and net loss of K^+ and Cl^- after activation of Ca^{2+}-K^+ channels is an important question pertinent to the development of pharmacological interventions designed to ameliorate the painful vaso-occlusive crises suffered by patients with sickle cell disease.

IX. Summary

The availability, structural and metabolic simplicity, and ease of manipulating the intracellular as well as the extracellular solutions of red blood cells have made them a favorite subject for the study of membrane transport. The plasma membrane and underlying cytoskeleton have been extensively characterized. A flipase that transports phospholipids from the outer to the inner hemileaflet contributes to phospholipid asymmetry in the membrane. The high concentration of hemoglobin, nearly close-packed and hydrated with a large volume fraction of intracellular water, contributes to a high intracellular ionic strength. Curiously, the intracellular activities of KCl, NaCl, and nonelectrolytes appear to be nearly ideal. Red-cell metabolism supplies ATP as substrate for the Na^+,K^+-ATPase and Ca^{2+}-ATPase ionic pumps and prevents the oxidation of Fe^{2+} in hemoglobin. The life spans of red blood cells from various mammalian species correlate with body mass, suggesting the existence of a biological clock that is obviously independent of gene transcription or translation in the mature nonnucleated cells.

The red blood cells of all species normally maintain a submicromolar intracellular concentration of Ca^{2+}, as set by the Ca^{2+}-ATPase pumping against a small passive inward leakage. The red blood cells of most species have high steady-state concentrations of K^+ but low Na^+, as set by a balance between pumping by the Na^+,K^+-ATPase and parallel leaks. An exception is the red cells of carnivores, which, unlike other red cells, have high intracellular Na^+ and low intracellular K^+. These cells lack a Na^+,K^+-ATPase, but do have a Na^+-Ca^{2+} exchanger that couples the efflux of Na^+ to the influx of Ca^{2+}. Whereas human red blood cells are unable to regulate their volume in anisotonic media, the red cells of many other species use the gradients of K^+ and Na^+ and activate their K^+-Cl^- and Na^+-K^+-$2Cl^-$ cotransporters to return their volume to normal in hypotonic and hypertonic media, respectively. Alternatively, in red cells of trout and skate, an efflux of the osmolyte taurine is used in cell volume regulation.

In human red blood cells, the conductive permeability to Cl^- is 100 times greater than that to K^+ or Na^+. Consequently, the membrane potential is consistent with the Cl^- equilibrium potential of -11 mV, and with a constant field permeability ratio P_K/P_{Na}, of 1.5. The electrogenic potential of the Na^+,K^+-ATPase is normally less than 1 mV. Away from the steady state, Cl^- will dominate the membrane potential. Treatment of human red blood cells with valinomycin, or with gramicidin in Na^+-free media, results in clamping of the voltage by the intrinsic ion concentration batteries. Analysis of individual ionic currents flowing under the influence of diffusion potentials yields current-voltage curves that characterize the conductance to Cl^-. An important problem for future study is to understand the "channel-like" properties of exchangers and pumps.

According to the Boyle-van't Hoff law, the total intracellular and extracellular osmolalities are equal at osmotic equilibrium. Significant deviations from osmotic ideality are due to the osmotic coefficient of Hb rising with the square of Hb concentration. For given cation contents of human red cells, the double-Donnan equilibrium describes the passive distribution of Cl^- and water across the red-cell membrane and accounts quantitatively for the effects of pH on cell swelling and shrinkage, as well as for colloid osmotic

hemolysis. Computer models are available for (1) red cell glycolysis, (2) the nonideal double-Donnan equilibrium, (3) acid-base balance, and (4) the kinetics of membrane transport. These models begin to enable prediction of the time course of changes in the cell volume, the intracellular pH, the membrane potential, and the intracellular salt concentrations when the extracellular conditions are altered.

When intracellular Ca^{2+} rises too much for the Ca^{2+}-ATPase to maintain the steady state, a cytotoxic calcium cascade leads to depletion of ATP, inhibition of the Na^+,K^+-ATPase, activation of Ca^{2+}-K^+ channels, cellular dehydration, loss of phospholipid asymmetry, echinocytosis, membrane vesiculation, activation of phospholipase C, transglutaminase, and protease, and hemolysis. Blockers of Ca^{2+}-K^+ channels reduce the formation of dehydrated, irreversibly sickled cells both *in vitro* and *in vivo* and are currently being tested as a new treatment for sickle cell disease.

Acknowledgment

The author thanks Dr. Philip B. Dunham of the Biology Department, Syracuse University, for reading this chapter.

Bibliography

Agre, P. (1996). Pathophysiology of the aquaporin water channels. *Annu. Rev. Physiol.* **58**, 619–648.

Agre, P., and Parker, J. C. (Eds.). (1989). "Red Blood Cell Membranes: Structure, Function, Clinical Implications." Marcel Dekker, New York.

Alper, S. L. (1991). The band 3–related anion exchanger (AE) gene family. *Annu. Rev. Physiol.* **53**, 549–564.

Antonini, E., and Brunori, M. (1971). "Hemoglobin and Myoglobin in Their Reaction with Ligands." North-Holland, Amsterdam.

Bennett, V., and Gilligan, D. M. (1993). The spectrin-based membrane skeleton and micron-scale organization of the plasma membrane. *Annu. Rev. Cell Biol.* **9**, 27–66.

Bessis, M. (1974). "Corpuscles. Atlas of Red Blood Cell Shapes." Springer-Verlag, New York.

Bessis, M., Weed, R. I., and Lebiond, P. F. (Eds.). (1973). "Red Cell Shape. Physiology, Pathology, Ultrastructure." Springer-Verlag, New York.

Beutler, E. (1986). "Red Cell Metabolism, Vol. 16. Methods in Hematology." Churchill Livingstone, New York.

Bifano, E. M., Novak, T. S., and Freedman. J. C. (1984). The relationship between the shape and the membrane potential of human red blood cells. *J. Membr. Biol.* **82**, 1–13.

Bishop, C., and Surgenor, D. M. (1964). "The Red Blood Cell. A Comprehensive Treatise." Academic Press, New York.

Bisognano, J. D., Dix, J. A., Pratap, P. R., Novak, T. S., and Freedman, J. C. (1993). Proton (or hydroxide) fluxes and the biphasic osmotic response of human red blood cells. *J. Gen. Physiol.* **102**, 99–123.

Brody, J. P., Han, Y., Austin, R. H., and Bitensky, M. (1995). Deformation and flow of red blood cells in a synthetic lattice: evidence for an active cytoskeleton. *Biophys. J.* **68**, 2224–2232.

Brugnara, C., Kopin, A. S., Bunn, H. F., and Tosteson, D. C. (1985). Regulation of cation content and cell volume in hemoglobin erythrocytes from patients with homozygous hemoglobin C disease. *J. Clin. Invest.* **75**, 1608–1617.

Brugnara, C., Ha, T. V., and Tosteson, D. C. (1989). Acid pH induces formation of dense cells in sickle erythrocytes. *Blood* **232**, 487–495.

Brugnara, C., Gee, B., Armsby, C. C., Kurth. S., Sakamoto, M., Rifai, N., Alper, S. L., and Platt, O. S. (1996). Therapy with oral clotrimazole induces inhibition of the Gardos channel and reduction of erythrocyte dehydration in patients with sickle cell disease. *J. Clin. Invest.* **97,** 1227–1234.

Crespo, L. M., Novak, T. S., and Freedman. J. C. (1987). Calcium, cell shrinkage, and prolytic state of human red blood cells. *Am. J. Physiol.* **252** (*Cell Physiol.* **21**), C138–C152.

Davson, H., and Danielli, J. F. (1943). "The Permeability of Natural Membranes." Cambridge University Press, Cambridge, UK.

DeFelice, L. J., and Blakely, R. D. (1996). Pore models for transporters? *Biophys. J.* **70,** 579–580.

Diaz, C., and Schroit, A. J. (1996). Role of translocases in the generation of phosphatidylserine asymmetry. *J. Membr. Biol.* **151,** 1–9.

Dick, D. A. T. (1959). Osmotic properties of living cells. *Int. Rev. Cytol.* **8,** 387–448.

Dissing, S., and Hoffman, J. F. (1990). Anion-coupled Na efflux mediated by the human red blood cell Na/K pump. *J. Gen. Physiol.* **96,** 167–193.

Dodge, J. T., Mitchell, C., and Hanahan, D. J. (1963). The preparation and chemical characteristics of hemoglobin-free ghosts of human erythrocytes. *Arch. Biochem. Biophys.* **100,** 119–130.

Duhm, J. (1971). Effects of 2,3–diphosphoglycerate and other organic phosphate compounds on oxygen affinity and intracellular pH of human erythrocytes. *Pflügers Arch.* **326,** 341–356.

Duhm, J., and Becker, B. F. (1977). Studies on the lithium transport across the red cell membrane. IV. Interindividual variations in the Na^+-dependent Li^+ countertransport system of human erythrocytes. *Pflügers Arch.* **370,** 211–219.

Ellory, J. C., and Lew, V. L. (1977). "Membrane Transport in Red Cells." Academic Press, New York

Ellory, J. C., and Young, J. D. (1982). "Red Cell Membranes—A Methodological Approach." Biological Techniques Series. Academic Press, New York.

Escobales, N., and Canessa, M. (1986). Amiloride-sensitive Na^+ transport in human red cells: evidence for a Na/H exchange system. *J. Membr. Biol.* **90,** 21–28.

Falke, J. J., and Chan, S. I. (1986a). Molecular mechanism of band 3 inhibitors. I. Transport site inhibitors. *Biochemistry* **25,** 7888–7894.

Falke, J. J., and Chan, S. I. (1986b). Molecular mechanism of band 3 inhibitors. 2. Channel blockers. *Biochemistry* **25,** 7895–7898.

Fowler, V., and Taylor, D. L. (1980). Spectrin plus band 4.1 cross-link actin. Regulation by micromolar calcium. *J. Cell Biol.* **85,** 361–376.

Freedman, J. C. (1983). Partial requirements for *in vitro* survival of human red blood cells. *J. Membr. Biol.* **75,** 225–231.

Freedman, J. C., and Hoffman, J. F. (1979a). Ionic and osmotic equilibria of human red blood cells treated with nystatin. *J. Gen. Physiol.* **74,** 157–185.

Freedman, J. C., and Hoffman, J. F. (1979b). The relation between dicarbocyanine dye fluorescence and the membrane potential of human red blood cells set at varying Donnan equilibria. *J. Gen. Physiol.* **174,** 187–212.

Freedman, J. C., and Miller, C. (1984). Membrane vesicles from human red blood cells in planar lipid bilayers. *Ann. NY Acad. Sci.* **435,** 541–544.

Freedman, J. C., and Novak, T. S. (1983). Membrane potentials associated with Ca-induced K conductance in human red blood cells. Studies with a fluorescent oxonol dye, WW781. *J. Membr. Biol.* **72,** 59–74.

Freedman, J. C., and Novak, T. S. (1989). Optical measurements of membrane potential in cells, organelles, and vesicles. *Methods Enzymol.* **172,** 102–122.

Freedman, J. C.., and Novak, T. S. (1997). Electrodiffusion, barrier, and gating analysis of DIDS-insensitive chloride conductance in human red blood cells treated with valinomycin or gramicidin. *J. Gen. Physiol.* **109,** 201–216.

Freedman, J. C., Biffno, E. M., Crespo, L. M., Pratap, P. R., Wallenga, R., Bailey, R. E., Zuk, S., and Novak, T. S. (1988). Membrane potential and the cytotoxic Ca cascade of human red blood cells. *In* "Cell Physiology of Blood" (R. B. Gunn and J. C. Parker, Eds.), pp.

218–231. Society of General Physiologists Series, The Rockefeller University Press, New York.

Freedman, J. C., Novak, T. S., Bisognano, J. D., and Pratap, P. R. (1994). Voltage dependence of DIDS-insensitive chloride conductance in human red blood cells treated with valinomycin or gramicidin. *J. Gen. Physiol.* **104,** 961–983.

Fröhlich, O., and Gunn, R. B. (1986). Erythrocyte anion transport: the kinetics of a single-site obligatory exchange system. *Biochim. Biophys. Acta* **864,** 169–194.

Gary-Bobo, C. M. (1967). Nonsolvent water in human erythrocytes and hemoglobin solutions. *J. Gen. Physiol.* **50,** 2547–2564.

Golan, D. E., and Veatch, W. (1980). Lateral mobility of band 3 in the human erythrocyte membrane studied by fluorescence photobleaching recovery: evidence for control by cytoskeletal interactions. *Proc. Natl. Acad. Sci. USA* **77,** 2537–2541.

Grimes, A. C. (1980). "Human Red Cell Metabolism." Blackwell, Oxford.

Harris, J. W., and Kellermeyer, R. W. (1970). "The Red Cell. Production, Metabolism, Destruction: Normal and Abnormal," rev. ed. Harvard University Press, Cambridge, MA.

Henderson, L. J. (1928). "Blood. A Study in General Physiology." Yale University Press, New Haven, CT.

Hoffman, J. F. (1986). Active transport of Na^+ and K^+, by red blood cells. *In* "Physiology of Membrane Disorders," 2nd ed. (T. E. , Andreoli, J. F. Hoffman. D. F. Fanestil, and S. G. Schultz, Eds.), pp. 221–231. Plenum Press, New York.

Hoffman, J. F. (1992). On red blood cells, hemolysis and resealed ghosts. *In* "The Use of Resealed Erythrocytes as Carriers and Bioreactors" (M. Magnani and J. R. DeLoach, Eds.). *Adv. Exper. Biol. Med.,* **326,** 1–15. Plenum, New York.

Hoffman, J. F., and Laris, P. C. (1974). Determination of membrane potential in human and *Amphiuma* red blood cells by means of a fluorescent probe. *J. Physiol. (London)* **239,** 519–552.

Hoffman, J. F., Kaplan, J. H., Callahan, T. J., and Freedman, J. C. (1980). Electrical resistance of the red cell membrane and the relation between net anion transport and the anion exchange mechanism. *Ann. NY Acad. Sci.* **341,** 357–360.

Hoffmann, E. K., and Dunham, P. B. (1995). Membrane mechanisms and intracellular signalling in cell volume regulation. *Int. Rev. Cytol.* **161,** 173–262.

Jacobs, M. H. (1962). Early osmotic history of the plasma membrane. *In* "Symposium on the Plasma Membrane, New York Heart Association, Inc." *Circulation* **26,** 1013–1021.

Jacobs, M. H., and Stewart, D. R. (1947). Osmotic properties of the erythrocyte. XII. Ionic and osmotic equilibria with a complex external solution. *J. Cell Comp. Physiol.* **30,** 79–103.

Jennings, M. L. (1992). Inorganic anion transport. *In* "The Structure of Biological Membranes" (P. Yeagle, Ed.), pp. 781–832. CRC Press, Boca Raton, FL.

Jennings, M. L., Schulz, R. K., and Allen, M. (1990). Effects of membrane potential on electrically silent transport. Potential-independent translocation and asymmetric potential-dependent substrate binding to the red blood cell anion exchange protein. *J. Gen. Physiol.* **96,** 991–1012.

Kay, M. M. B. (1991). *Drosophila* to bacteriophage to erythrocyte: the erythrocyte as a model for molecular and membrane aging of terminally differentiated cells. *Gerontology* **37,** 5–32.

Knauf, P. A. (1979). Erythrocyte anion exchange and the band 3 protein: transport kinetics and molecular structure. *Curr. Top. Membr. Transp.* **12,** 249–363.

Kuhlman, P. A., Hughes, C. A., Bennett, V., and Fowler, V. M. (1996). A new function for adducin. Calcium/calmodulin-regulated capping of the barbed ends of actin filaments. *J. Biol. Chem.* **271,** 7986–7991.

Larsen, R. L., Katz, S., Roufogalis, B. D., and Brooks, D. E. (1981). Physiological shear stresses enhance the Ca^{2+} permeability of human erythrocytes. *Nature* **294,** 667–668.

Lauf, P. K., Bauer, J., Adragna, N. C., Fujise, H., Zade-Oppen, A. M. M., Ryu, K. H., and Delpire, E. (1992). Erythrocyte K-Cl cotransport: properties and regulation. *Am. J. Physiol.* **263** (*Cell Physiol.* **32**), C917–C932.

Läuger, P. (1995). Conformational transitions of ionic channels. *In* "Single Channel Recording," 2nd ed. (B. Sakmann and E. Neher, Eds.), ch. 22, pp. 651–662. Plenum, New York.

LeFevre, P. G. (1961). Sugar transport in the red blood cell: structure-activity relationships in substrates and antagonists. *Pharm. Rev.* **13**, 39–70.

Lew, V. L., and Bookchin, R. M. (1986). Volume, pH and ion content regulation in human red cells: analysis of transient behavior with an integrated model. *J. Membr. Biol.* **92**, 57–74.

Lew, V. L., Hockaday, A., Sipulveda, M.-I., Somlyo, A. P., Somlyo, A. V., Ortiz, O. E., and Bookchin, R. M. (1985). Compartmentalization of sickle-cell calcium in endocytic inside-out vesicles. *Nature* **315**, 586–589.

Lin, S., Yang, E., and Huestis, W. H. (1994). Relationship of phospholipid distribution to shape change in Ca^{2+}-crenated and recovered human erythrocytes. *Biochemistry* **33**, 7337–7344.

Macey, R. I., Adorante, J. S., and Orme, F. W. (1978). Erythrocyte membrane potentials determined by hydrogen ion distribution. *Biochim. Biophys. Acta* **512**, 284–295.

Mercer, R. W., and Dunham. P. B. (1981). Membrane-bound ATP fuels the Na/K pump. *J. Gen. Physiol.* **78**, 547–568.

Milanick, M., and Hoffman, J. F. (1986). Ion transport and volume regulation in red blood cells. *Ann. NY Acad. Sci.* **488**, 174–186.

Murphy, E., Berkowitz, L. R., Orringer, E., Levy, L., Gabel, S. A., and London, R. E. (1987). Cytosolic free calcium levels in sickle red blood cells. *Blood* **69**, 1469–1474.

Ohnishi, S. T., and Ohnishi, T. (1994). "Membrane Abnormalities in Sickle Cell Disease and Other Red Blood Cell Disorders." CRC Press, Boca Raton, FL.

Ohnishi, S. T., Katagi, H., and Katagi, C. (1989). Inhibition of the *in vitro* formation of dense cells and of irreversibly sickled cells by charybdotoxin, a specific inhibitor of calcium-activated potassium efflux. *Biochem. Biophys. Acta* **1010**, 199–203.

Passow, H. (1986). Molecular aspects of band 3 protein-mediated anion transport across the red blood cell membrane. *Rev. Physiol. Biochem. Pharmacol.* **103**, 61–203.

Passow, H., Wood, P. G., Lepke, S., Müller, H., and Sovak, M. (1992). Exploration of the functional significance of the stilbene disulfonate binding site in mouse band 3 by site-directed mutagenesis. *Biophys. J.* **62**, 98–100.

Payne, J. A., Lytle, C., and McManus, T. J. (1990). Foreign anion substitution for chloride in human red blood cells: effect on ionic and osmotic equilibria. *Am. J. Physiol.* **259** (*Cell Physiol.* **28**), C819–C827.

Penniston, J. T., Vaughan, L., and Nakamura, M. (1979). Endocytosis in erythrocytes and ghosts: occurrence at 0 °C after ATP preincubation. *Arch. Biochem. Biophys.* **198**, 339–348.

Ponder, E. (1948). "Hemolysis and Related Phenomena." Grune and Stratton, New York.

Proverbio, F., Shoemaker, D. G., and Hoffman, J. F. (1988). Functional consequences of the membrane pool of ATP associated with the human red blood cell Na/K pump. *In* "The Na$^+$,K$^+$-Pump, Part A: Molecular Aspects" (J. C. Skou, J. G. Norby, A. B. Maunsbach, and M. Esmann, Eds.). Prog. Clin. Biol. Res. **268A**, 561–567, Alan R. Liss, New York.

Raess, B. U., and Tunnicliff, G. (Eds.). (1990). "The Red Cell Membrane: A Model for Solute Transport." Humana Press, Clifton, NJ.

Raftos, J. E., Bulliman, B. T., and Kuchel, P. W. (1990). Evaluation of an electrochemical model of erythrocyte pH buffering using

^{31}P nuclear magnetic resonance data. *J. Gen. Physiol.* **95**, 1185–1204.

Rapoport, T. A., Heinrich, R., Jacobasch, G., and Rapoport, S. (1974). A linear steady-state treatment of enzymatic chains. A mathematical model of glycolysis of human erythrocytes. *Eur. J. Biochem.* **42**, 107–120.

Reeves, R. B. (1977). The interaction of body temperature and acid-base balance in ectothermic vertebrates. *Annu. Rev. Physiol.* **39**, 559–586.

Reithmeier, R. A. F. (1993). The erythrocyte anion transporter (band 3). *Curr. Opin. Struct. Biol.* **3**, 515–523.

Rodeau, J., and Malan, A. (1979). A two-compartment model of blood acid-base state at constant or variable temperature. *Resp. Physiol.* **36**, 5–30.

Ross, P. D., and Minton, A. P. (1977). Analysis of non-ideal behavior in concentrated hemoglobin solutions. *J. Mol. Biol.* **112**, 437–452.

Salhany, J. (Ed.). (1989). "Erythrocyte Band 3 Protein." CRC Press, Boca Raton, FL.

Schwarz, W., and Passow, H. (1983). Ca^{2+}-activated K$^+$ channels in erythrocytes and excitable cells. *Annu. Rev. Physiol.* **45**, 359–374.

Schwarz, W., Grygorczyk, R., and Hof, D. (1999). Recording single-channel currents from human red cells. *Methods Enzymol.* **173**, 112–121.

Shohet, S. B. (1976). Mechanisms of red cell membrane lipid renewal. *In* "Membranes and Disease" (L. Bolis, J. F. Hoffman, and A. Leaf, Eds.), pp. 61–74. Raven Press, New York.

Shohet, S. B., and Mohandas, N. (Eds.). (1988). "Red Cell Membranes, Vol. 19, Methods in Hematology." Churchill Livingstone, New York.

Steck, T. L. (1974). The organization of proteins in the human red blood cell membrane. A review. *J. Cell Biol.* **62**, 1–19.

Surgenor, D. MacN. (Ed.). (1974). "The Red Blood Cell," 2nd ed., Vols. I and II. Academic Press, New York.

Tosteson, D. C., and Hoffman, J. F. (1960). Regulation of cell volume by active cation transport in high and low potassium sheep red cells. *J. Gen. Physiol.* **44**, 169–194.

Tosteson, M. T., Halperin, J. A., Kishi, Y., and Tosteson, D. C. (1991). Palytoxin induces an increase in the cation conductance of red cells. *J. Gen. Physiol.* **98**, 969–985.

Vácha, J., and Znojil, V. (1981). The allometric dependence of the life span of erythrocytes on body weight in mammals. *Comp. Biochem. Physiol.* **69A**, 357–362.

Van Slyke, D. D., Wu, H., and McLean, F. C. (1923). Studies of gas and electrolyte equilibria in the blood. V. Factors controlling the electrolyte and water distribution in the blood. *J. Biol. Chem.* **56**, 765–849.

Wang, D. N., Sarabia, V. E., Reithmeier, R. A. F., and Kühlbrandt, W. (1994). Three-dimensional map of the dimeric membrane domain of the human erythrocyte anion exchanger, band 3. *EMBO J.* **13**, 3230–3235.

Whittam, R. (1964). "Transport and diffusion in red blood cells." Williams and Wilkins, Baltimore.

Willis, J. S. (1992). Symposium on diversity of membrane cation transport in vertebrate red blood cells. An overview. *Comp. Biochem. Physiol.* **102A**, 595–596.

Yingst, D. R. (1988). Modulation of the Na,K-ATPase by Ca and intracellular proteins. *Annu. Rev. Physiol.* **50**, 291–303.

Yoshikawa, H., and Rapoport, S. M. (1974). "Cellular and Molecular Biology of Erythrocytes." University Park Press, Baltimore.

Zade-Oppen, A. M., Adragna, N. C, and Tosteson, D. C. (1988). Effects of pH, potential, chloride and furosemide on passive Na$^+$ and K$^+$ effluxes from human red blood cells. *J. Membr. Biol.* **102**, 217–225.

SECTION

III

Membrane Excitability and Ion Channels

SECTION

III

Membrane Excitability and Ion Channels

Nicholas Sperelakis

24

Cable Properties and Propagation of Action Potentials

I. Introduction

The resting potential (RP) of cells enables the electrogenesis of **action potentials** (APs) and excitability. In this chapter, we examine the mechanism for propagation of the APs and excitability from one part of a neuron or muscle to a distal part.

It is imperative that the body be able to transmit a signal from one point to another very rapidly. The only way that this can be accomplished is by an electrical mechanism. Blood flow and diffusion and signaling molecules are much too slow to allow rapid signaling. In contrast, electricity flows very quickly, at the speed of light (3×10^8 m/s) in a copper wire or about one-ninth the speed of light in a water solution (like the composition of the body). Therefore, the body makes use of electricity for rapid signaling in the nervous system, skeletal muscle, heart, and smooth muscles. Propagation velocity is about 120 m/s in our fastest nerve fibers, about 6 m/s skeletal muscle, about 0.5 m/s in heart, and about 0.05 m/s in smooth muscle (Table 1).

One example of the need for very fast communication or signaling is the process of walking. Very rapid signals must travel from the motor cortex of the brain, down to the lower spinal cord region, and out the motor axons to the skeletal muscles of the lower extremities (Fig. 1). In this process, the signal crosses one or more **synapses,** which are regions in which one neuron ends and the next one begins, and in which a special chemical neurotransmitter signal is involved (Fig. 2). At the termination of each branch of a motor nerve axon on the skeletal muscle fiber, there is another synapse, known as the **neuromuscular junction** or **motor end plate** (see Fig. 1). The signal crosses the neuromuscular junction and gives rise to an AP in the muscle fiber that propagates in both directions from the motor end plate. The muscle AP elicits contraction. Receptors in the muscles (e.g., stretch receptors) transmit information (in the form of propagating APs) back into the central nervous system (CNS) Thus, in walking, there is a continuous rapid flow of information and instructions to the muscles in both directions: out of the CNS and into the CNS. Therefore, even a relatively simple

skeletal activity such as walking would not be possible without a very rapid signaling system. To illustrate, in various demyelinating diseases (e.g., caused by some viruses, heavy metals, autoimmune reactions), loss of the myelin sheath around the myelinated nerve fibers causes propagation to become slowed and impaired in the affected nerve fibers, with associated uncoordination and partial paralysis.

II. Frequency-Modulated Signals

Because propagating all-or-none APs are all very similar to each other (in shape, duration, amplitude, rate of rise, and propagation velocity), to make the signal stronger or weaker, the body increases or decreases, respectively, the frequency of the APs. That is, the body uses a **frequency-modulated (FM) system**, rather than an amplitude-modulated (AM) system (see

TABLE 1 Conduction Velocity as a Function of Fiber Diameter in Nerve Axons and Muscle Fibers

Fiber type	Fiber diameter (µm)	Propagation velocity (m/s)	Velocity/ diameter (m/s/µm)
Myelinated axons	20	120	6.0
	12	70	5.8
	5	30	6.0
Nonmyelinated axons	1.5	2.0	1.3
	1.0	1.3	1.3
Squid giant axons (20 °C)	500	25	0.05
Skeletal muscle fibers	50	6	0.12
Cardiac muscle fibers	15	0.5	0.03
Smooth muscle fibers	5	0.05	0.01

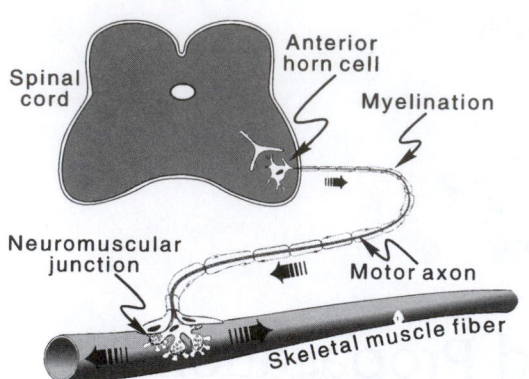

FIGURE 1. Schematic diagram of a motor axon, with the cell body (soma) in the anterior horn of the spinal cord and its terminal branches ending on skeletal muscle fibers (only one muscle fiber depicted) to form the neuromuscular junctions (motor end plates). Each motor axon with its attached skeletal muscle fibers is known as the motor unit. The motor axons are of large diameter (e.g., 20 μm) and are myelinated, and therefore propagate at fast velocities (e.g., 120 m/s). As a consequence of the chemical synaptic transmission process at the neuromuscular junction, an AP is initiated in the muscle fiber and propagates in both directions, bringing about contraction.

Fig. 2). It is a digital system composed of on-off identical signals. At each synapse, the signal becomes graded in amplitude rather than all-or-none: the greater the amplitude and duration of the local postsynaptic potential, the higher the frequency of APs triggered. The same is true of the local graded receptor potential generated at some sensory organs/receptors. Stronger signals translate into a higher frequency of impulses, and weaker signals correspond to a lower frequency of impulses.

III. Cable Properties

A. Biological Fiber as a Cable

An electrical cable consists of two parallel conductors separated by insulation material, for example, two copper wires

separated by rubber. Usually one of the conductors is arranged as a tubular sleeve surrounding a central solid rod (wire), as depicted in Fig. 3A. The equivalent electrical circuit for a cable is shown in Fig. 3B: two parallel conductors (wires) separated by a transverse resistance (R) shown distributed along the length of the cable. The resistance of the conductors is so small compared with the transverse insulation resistance that it is assumed to be zero. In the case of the biological cable (a long narrow nerve fiber or skeletal muscle fiber), one parallel conductor is the inside fluid (cytoplasm) and the second parallel conductor is the outside fluid surrounding (bathing) the cell (the interstitial fluid). Because the conductivity of biological fluid is much less (i.e., much higher resistance) than that of copper wire, and because the cross-sectional area of the cell is so small, the inside longitudinal resistance is high and cannot be ignored (Fig. 3C). The outside longitudinal resistance is relatively small, as compared with the inside, because of the larger volume (cross-sectional area) of fluid available to carry the outside current, and therefore is assumed to be negligible.

In addition, there is a stray capacitance distributed along the length of the cable (Fig. 3D), because a capacitance occurs when two parallel conductors ("plates") are separated by a high-resistance dielectric material. The dielectric constant of materials is related to vacuum, which is assigned a value of 1.0000; air has a value very close to vacuum, oils have a value of 3–6, and that of pure water is 81. The biological membrane, which has a matrix of phospholipid molecules, has a dielectric constant of about 5, typical of oils. The higher the dielectric constant, the higher the capacitance; the closer the parallel plates, the higher the capacitance. Because the biological membrane is so thin (approx. 70 Å or 7 nm), its capacitance is relatively high: all cell membranes have a membrane capacitance (C_m) of about 1.0 μF/cm^2 (where F stands for farads). Capacitors and the dielectric constant are discussed further in Chapter 70.

B. Length Constant

In the electric cable depicted in Fig. 3A and B, a voltage (or signal) applied at one end would be transmitted to a dis-

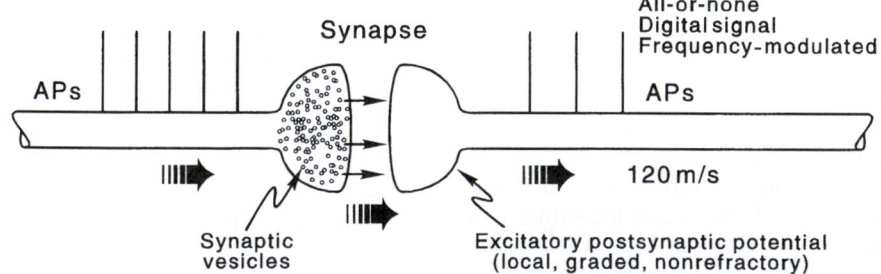

FIGURE 2. Diagram of a chemical excitatory synapse between two nerve fibers. An AP in the presynaptic fiber (left) brings about the release of the neurotransmitter at its nerve terminal. The transmitter molecules rapidly diffuse the short distance (e.g., 0.1 μm) across the synaptic cleft, bind to receptor sites on the postsynaptic membrane, and open the associated nonselective ion channels (Na$^+$,K$^+$ mixed conductance) complexed to the receptor (i.e., these are ligand-gated ion channels). The associated synaptic current depolarizes the postsynaptic membrane, producing the excitatory postsynaptic potential (EPSP); this depolarization spreads passively into the adjacent conductile membrane (excitable) because of cable properties, thereby triggering one or more APs in the postsynaptic axon. The EPSPs are local, graded in amplitude, and nonrefractory whereas the APs are all-or-none (maximal), refractory, and propagated actively. Thus, the AM synaptic process gives rise to an FM or digital signal. The strength of a biological response (e.g., contraction, secretion, sensation) is a function of the frequency of the signal. That is, higher AP frequency corresponds to stronger contraction or stronger sensation (in the sensory nervous system).

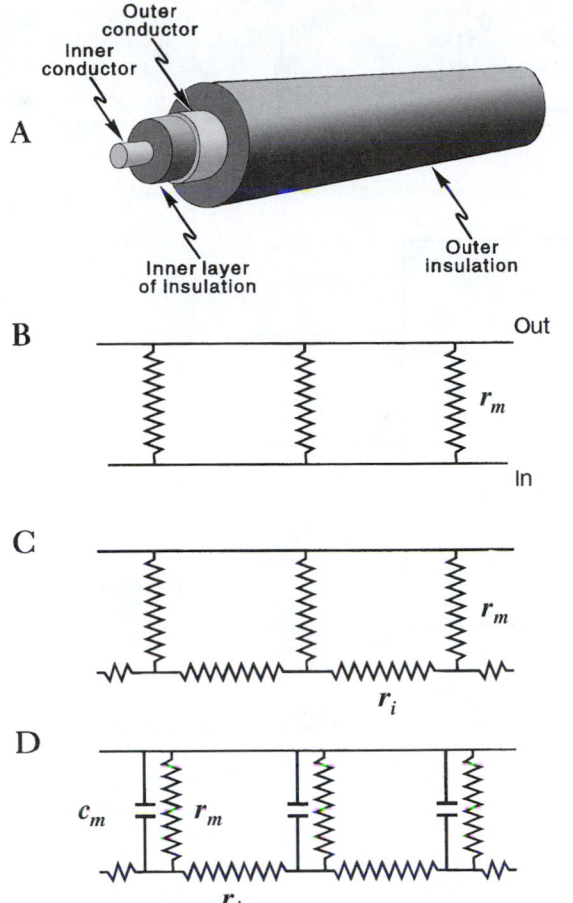

$$V_x = V_o \frac{1}{e} \tag{2a}$$

$$= V_o \frac{1}{2.718}$$

$$= 0.37 V_o$$

The mathematical solution to Eq. 1 is

$$V_x = \text{antiln}\left(\ln V_o - \frac{x}{\lambda}\right) \tag{2b}$$

$$V_x = \text{antilog}\frac{2.303 \log V_o - (x/\lambda)}{2.303} \tag{2c}$$

Hence, the distance at which the voltage decays to 37% of the initial value gives the length constant, λ. In nerve fibers and skeletal muscle fibers, λ has a value of only about 1–3 mm.

Therefore, the relatively short λ, compared with the length of the neuron (e.g., over 1.0 m for a lumbar anterior horn cell motor neuron) or skeletal muscle fiber, means that a signal applied at one end (or midpoint) would fall off very quickly with distance along the fiber (Fig. 5). If the length constant were 1.0 mm, then at 4 mm, the signal would become negligible. Hence, the electrical signal cannot be conducted passively in the biological cable, because it would decrement and disappear over relatively short distances. The AP (signal) is amplified to a constant value at each point (or each node) in the membrane, as discussed in Chapter 25. That is, conduction is active, not passive, with

FIGURE 3. A coaxial cable and associated electrical equivalent circuit (A, B), and the electrical equivalent circuit for a biological cable (C, D). (A) The coaxial cable consists of an inner conductor (e.g., copper wire) and an outer concentric conductor separated by a layer of insulation (e.g., rubber). (B) The equivalent circuit for a coaxial cable. The inner and outer conductors are depicted as having nearly zero resistances, and the transmembrane insulation resistance is shown distributed along the length of the cable. (C) In the biological cable, the inner conductor is the cytoplasm (axoplasm or myoplasm), which is not of negligible resistivity, and therefore is depicted as r_i distributed along the length of the fiber. The transverse insulation resistance is the cell membrane resistance (r_m). (D) Addition of the capacitance elements to the biological cable (c_m), which arise due to the lipid bilayer matrix of the cell membrane.

tant end with little or no **decrement** (diminution or attenuation), and the so-called **length constant** would be very long or nearly infinite. In the biological cable (Fig. 3C and D), however, a signal applied at one end rapidly falls off (decays) in amplitude as a function of distance, with a relatively short length constant (λ). This decay in voltage is exponential (Fig. 4A). An exponential process gives a straight line on a semilogarithmic plot (log V versus distance) (Fig. 4B). In a cable, the relationship between the voltage at any distance (*x*) from the applied voltage (V_o) is

$$V_x = V_o e^{-x/\lambda} \tag{1}$$

Thus, when $x = \lambda$

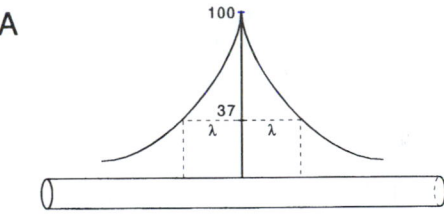

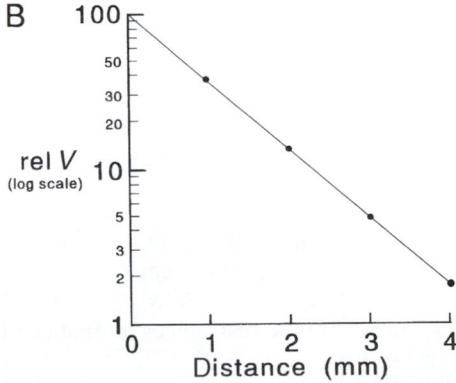

FIGURE 4. Length constant (A) of the biological cable (fiber). (A) The exponential decay of voltage on both sides of an applied current (voltage) as a function of distance is diagrammed. The distance at which the voltage falls to 1/e, or 36.7% of initial voltage (at x = 0), gives the λ value. (B) When the voltage is plotted on a log scale against distance, a straight line is obtained.

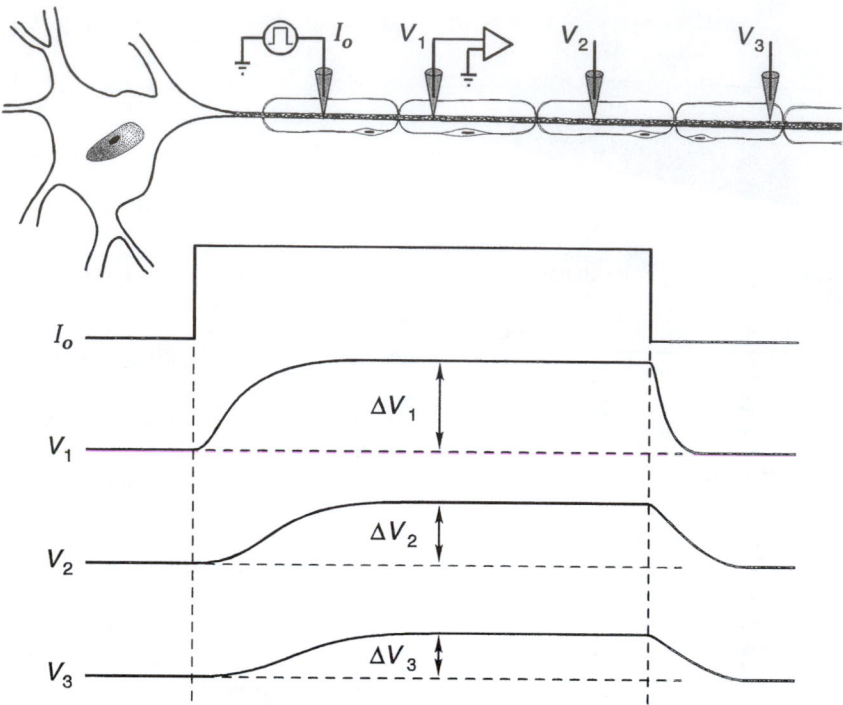

FIGURE 5.   Diagram of a motor axon and how the voltage signal would decay with distance if the neuron were only a passive cable (nonexcitable). The voltage traces illustrate the voltage signals that would be simultaneously recorded at three different points along the axon (V_1, V_2, and V_3) from the site of injection of a rectangular current pulse (I_o). As depicted, the amplitude of the steady-state voltage pulse rapidly falls off with distance, because the length constant (λ) of the biological cable is short (e.g., 1 mm) compared with the length of the axon (e.g., 1000 mm). Therefore, an active response of the cell membrane at each point is required to faithfully propagate the signal. The voltage recorded at point V_1 (ΔV_1) also illustrates that the membrane potential changes in an exponential manner, both at the beginning of an applied rectangular current pulse and at the end. This exponential change results from the capacitance of the cell membrane, the time constant (τ) being a product of the resistance and capacitance ($R_m C_m$). The time it takes for the voltage to decay to $1/e$ (36.7%) of the initial (maximal) value at the end of the pulse, or to build up to 63.3% ($1 - 1/e$) of the maximal value at the beginning of the pulse, gives the τ value.

energy being put into the signal at each point to prevent any decay of the signal.

The parameters that determine the length constant of a cable are the square root of the ratio of the transverse resistance (r_m) to the sum of the inside (r_i) and outside (r_o) longitudinal resistances,

$$\lambda = \sqrt{\frac{r_m}{r_i + r_o}} \qquad (3)$$

$$cm = \sqrt{\frac{\Omega \cdot cm}{\dfrac{\Omega}{cm} + \dfrac{\Omega}{cm}}}$$

where r_m, r_i, and r_o are the resistances normalized for a unit length (1 cm) of fiber.

For surface fibers of a nerve or a muscle bundle bathed in a large volume conductor, r_o is negligibly small, and Eq. 3 reduces to

$$\lambda = \sqrt{\frac{r_m}{r_i}} \qquad (4a)$$

$$\lambda = \sqrt{\frac{R_m}{R_i} \frac{a}{2}} \qquad (4b)$$

$$cm = \sqrt{\frac{\Omega \cdot cm^2 \cdot cm}{\Omega \cdot cm}}$$

where a is the fiber radius, and R_m and R_i are the membrane resistance and longitudinal cytoplasmic resistance, respectively, normalized for both length (1 cm) and cell diameter. Thus, the greater the membrane resistance and the smaller the internal longitudinal resistance (larger cell diameter), the greater the λ value. We will see later that the propagation velocity is a function of λ; that is, larger diameter fibers propagate faster. We will also see that myelination increases the effective membrane resistance (R_m) and lowers the effective capacitance, thereby increasing propagation velocity.

C. Time Constant

Because of the large capacitance of the cell membrane (C_m) the membrane potential (E_m) cannot change instantaneously upon application of a step current pulse. Instead, E_m

changes in an exponential (negative) manner (see Fig. 5) on both the charge and the discharge. The membrane time constant (τ_m) is a product of the resistance (R_m) and capacitance of the membrane and can be expressed as

$$\tau_m = r_m c_m \qquad (5a)$$

$$= R_m C_m \qquad (5b)$$

$$s = \Omega \cdot F$$

where Ω is the resistance in ohms, and F is the capacitance in farads.

The discharge of the membrane (parallel RC network) is given by

$$V_t = V_{max} e^{-t/\tau} \qquad (6)$$

where V_t is the voltage at any time t (at the site of current injection), and V_{max} is the final maximum voltage attained during the pulse. When $t = \tau$,

$$V_t = V_{max} \frac{1}{e} \qquad (7a)$$

$$= V_{max} \frac{1}{2.718}$$

$$= 0.37 V_{max}$$

Hence, the time at which the voltage decays to 37% of the initial (maximal) value gives the **time constant**, τ (see Fig. 5). In nerve fibers and skeletal muscle fibers, τ_m has a value of about 1.0 ms.

When the membrane is charging, there is a similar exponential (negative) process, with the identical time constant (see Fig. 5). The corresponding relationship is given by

$$V_t = V_{max} \left(1 - e^{-t/\tau} \right) \qquad (7b)$$

The time it takes for the voltage to build up to 63% of its final value gives the time constant. When $t = \tau$,

$$V_t = \left(1 - \frac{1}{e} \right) V_{max} \qquad (7c)$$

$$= (1 - 0.37) V_{max}$$

$$= 0.63 V_{max}$$

Thus, the time constant can be measured on the build-up of the pulse (time to reach 63% of the final voltage) or on the decay (time to reach 37% of the initial voltage).

D. Input Resistance

The **input resistance** (R_{in}) of a muscle or nerve fiber is essentially the resistance that an intracellular microelectrode "looks" into when a small current (DC pulse) is injected. Thus, R_{in} is determined by the change in membrane potential (V_o or $V_{x=0}$) at steady state produced at the site of injection ($x = 0$) by the injection of a known (measured) amount of current (I_o or $I_{x=0}$), based on Ohm's law.

$$R_{in} = \frac{V_o}{I_o} \qquad (8)$$

If the microelectrode is near the middle of a very long fiber (or "infinite" cable), then R_{in} is related to the other cable parameters (e.g., to the internal longitudinal resistance r_i and the DC length constant λ_{DC}):

$$R_{in} = 0.5 \, r_i \, \lambda_{DC} \qquad (9)$$

$$\Omega = \frac{\Omega}{cm} \cdot cm$$

The factor of 0.5 is present because current flows in both directions from the microelectrode. If the microelectrode is at one end of the long fiber, then the factor of 0.5 is removed. Equation 9 indicates that the R_{in} is greater when r_i and/or λ_{DC} are greater.

Combining Eqs. 8 and 9 gives

$$\frac{V_o}{I_o} = 0.5 \, r_i \, \lambda_{DC} \qquad (10a)$$

or

$$V_o = I_o \, 0.5 \, r_i \, \lambda_{DC} \qquad (10b)$$

This equation indicates that the change in membrane potential at $x = 0$ is equal to the applied current times the input resistance.

To obtain the change in membrane potential at any point x along the fiber cable (V_x), then an exponential term ($e^{-x/\lambda}$) must be added to account for the exponential decay of voltage over distance:

$$V_x = I_o \, 0.5 \, r_i \, \lambda_{DC} \, e^{-x/\lambda} \qquad (11)$$

This equation is identical to Eq. 1, because the $I_o 0.5 r_i \lambda_{DC}$ term is equal to V_o (Eq. 10b):

$$V_x = V_o \, e^{-x/\lambda} \qquad (1)$$

Because $\lambda_{DC} = \sqrt{r_m / r_i}$ (Eq. 4a), Eq. 9 can be given as

$$R_{in} = 0.5 \, r_i \sqrt{\frac{r_m}{r_i}} = 0.5 \sqrt{r_m r_i} \qquad (12)$$

and Eq. 11 can also be given as

$$V_x = I_o \, 0.5 \sqrt{r_m r_i} \, e^{-x/\lambda} \qquad (13)$$

Thus, input resistance is proportional to $\sqrt{r_m r_i}$. This is logical because the higher the longitudinal resistance of the fiber's axoplasm (or myoplasm), which is a function of the resistivity of the axoplasm and fiber diameter (see Section XI.A–C of Chapter 70), and the higher the membrane resistance, the higher the input resistance should be. The input resistance can be calculated from the electrical equivalent circuit for a cable.

The input impedance can also be given in terms of R_m and R_i, the specific resistance of the cell membrane and resistivity of the axoplasm, respectively (see Section XI.A, B of Chapter 70). Since $R_m = 2\pi a r_m$ and $R_i = \pi a^2 r_i$, then Eq. 12 can be converted to

$$R_{in} = 0.5 \sqrt{\frac{R_m}{2\pi a} \frac{R_i}{\pi a^2}} \qquad (14a)$$

$$= 0.5 \sqrt{\frac{R_m R_i}{2 \pi^2 a^3}} \qquad (14b)$$

Thus, the input resistance is directly proportional to the square root of the membrane-specific resistance and the resistivity of the axoplasm, and inversely proportional to the square root of the fiber radius raised to the third power.

Calculation of the **input impedance** (Z_{in}) is more complicated because it also depends on the membrane capacitance (C_m) and on the frequency (f) of the alternating current (AC) used. According to Katz (1966), the input impedance of a long fiber (current injected at one end) is given by

$$Z_{in} = \sqrt{\frac{R_m R_i}{2\pi^2 a^3 \sqrt{1 + 4\pi^2 f^2 R_m^2 C_m^2}}} \qquad (15)$$

$$\Omega = \sqrt{\frac{(\Omega \cdot cm^2)(\Omega \cdot cm)}{cm^3 \sqrt{(s^{-2})(s^2)}}}$$

A discussion of impedance is given in Section VIII of Chapter 70 and a discussion of the AC length constant (λ_{AC}) is given in Appendix 2 to this chapter.

E. Local Potentials

In contrast to the active propagation of APs, synaptic potentials and sensory receptor potentials are not actively propagated. Such potentials decay exponentially (from their source of initiation) along the fiber cable, as described previously. Therefore, postsynaptic potentials and receptor potentials are **local potentials.** When local potentials are in the depolarizing direction, they can give rise to APs, which are propagated; when hyperpolarizing, they act to inhibit production of APs. These local potentials are similar to the local excitatory response (see Chapter 25), in that both are confined to a local region; however, the electrogenesis of the two is different. As stated before, the neuromuscular junction is an excitatory type of chemical synapse and produces **excitatory postsynaptic potentials** (EPSPs), known here as **end-plate potentials** (EPPs).

Most synaptic potentials are graded, that is, they can add on to one another, both in time and in space (temporal summation and spatial summation), to produce larger responses. Larger synaptic potentials exert a greater stimulatory or inhibitory effect on the production of APs.

IV. Conduction of Action Potentials

A. Local-Circuit Currents

The generation of APs is described in Chapter 25. This section examines the mechanism for their rapid propagation (conduction). **Propagation** occurs by means of the local-circuit currents that accompany the propagating APs, as depicted in Fig. 6. Such currents exist because, when two points are at a different potential (voltage) in a conducting medium, current (I) will flow between the two points, as governed by **Ohm's law** ($I = V/R$).

At the peak of the AP in one region of the fiber, the inside of the membrane at that region becomes positive with respect to the outside. The inside is also positive with respect to the inside cytoplasm at a region downstream from the active region. Therefore, current flows through the cytoplasm from the active region (current source) to the adjacent inac-

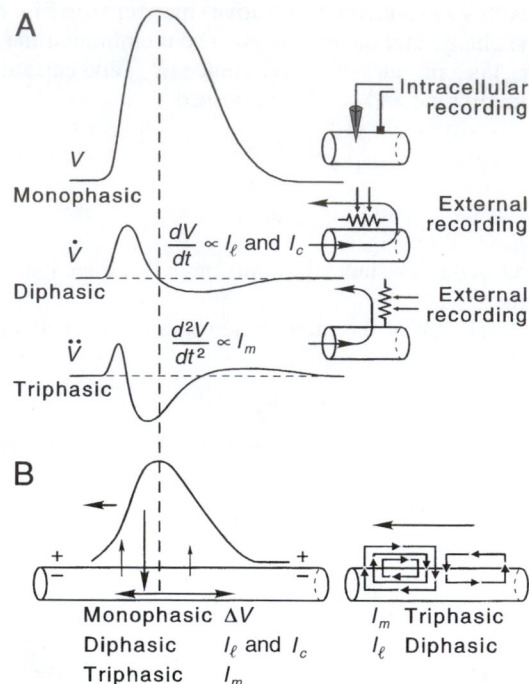

FIGURE 6. (A) Schematic representation of the first ($\dot{V}$) and second ($\ddot{V}$) time derivatives of the AP spike and the longitudinal and radial currents associated with the propagating spike. The first derivative (dV/dt or $\dot{V}$) is proportional to the capacitive current ($I_c = C_m\, dV/dt$) and to the longitudinal (axial) current and is **diphasic,** having an intense forward phase and a less intense backward phase, as depicted in the diagram at the lower right. The second derivative (d^2V/dt^2 or $\ddot{V}$) is proportional to the radial transmembrane current (I_m), and is **triphasic,** having a moderately intense initial outward phase, then a very intense inward phase (the "current sink"), followed by a least intense second outward phase, as depicted in the lower diagrams. (B) The arrows at the lower left also depict the three phases of the membrane current and the two phases of the axial current. dV/dt can be recorded externally by a pair of closely spaced (relative to the spike wavelength) electrodes arranged parallel to the fiber axis, as illustrated. d^2V/dt^2 can be recorded by a pair of electrodes arranged perpendicular to the fiber axis, as depicted. The vertical dashed line indicates that when the slope of the spike goes to zero at the peak of the spike, dV/dt is zero; dV/dt is maximum at about the middle of the rising phase of the spike.

tive region, then out of the fiber across the cell membrane, then through the interstitial fluid back to the active region (current "sink"), and finally through the membrane of the active region. This completes the closed loop for the current.

The outward current through the membrane of the inactive region produces an IR voltage drop (Ohm's law), positive inside to negative outside. This acts to depolarize this region, because the polarity of the voltage drop is opposite to that of the resting potential (negative inside, positive outside). When the depolarization exceeds the threshold potential, an AP is triggered. Thus, the inactive region becomes converted to an active region. This process is repeated in each segment of fiber, thus resulting in movement (propagation) of the impulse sequentially down the fiber.

If we examine a propagating AP in the middle region of a fiber (Fig. 6), we see that there is also a small backflow of current internally, coupled with a corresponding small forward flow externally, associated with the repolarizing phase of the AP. Thus, as the AP propagates down the fiber, from right to left, there is a simultaneous double flow of local-circuit current: clockwise flow associated with the rising phase of the AP, and counterclockwise flow associated with the repolarizing phase of the AP.

The internal longitudinal current, sweeping past a transverse plane of the fiber, has two phases, first forward (right to left), and then reverse (left to right). The external longitudinal current also has two phases: first left to right and then right to left.

The transverse membrane current, flowing outward in a plane perpendicular to the membrane, has three consecutive phases: first outward (still passive membrane), then inward (active membrane), and finally outward again (still active membrane).

It is the local-circuit current flow that enables the electrocardiogram (ECG), electromyogram (EMG), electroencephalogram (EEG), and electroretinogram (ERG) to be recorded from the body surface over the tissue of interest (heart, skeletal muscle, brain, and eye). The internal longitudinal current is confined to the cytoplasm of the fiber, but the external current can use whatever conducting fluid is available (e.g., the entire torso volume conductor) because of the principle that parallel resistors give a lower total resistance (or current takes the path of least resistance). Thus, this external local-circuit current causes the skin to be at different potentials, and these differences can be recorded (as the ECG, etc.).

B. Propagation Velocity Determinants

The factors that determine active velocity of propagation (θ) include (1) fiber diameter, (2) length constant (λ), (3) time constant (τ_m), (4) local-circuit current intensity, (5) threshold potential, and (6) temperature. Some of these factors are interrelated, such as fiber diameter and length constant (since λ is proportional to the square root of the radius), and length constant and time constant (since both have a dependence on R_m). Propagation velocity is directly proportional to length constant and inversely proportional to time constant as

$$\theta \propto \frac{\lambda}{\tau_m} \qquad (16a)$$

By substituting Eq. 4a for λ and Eq. 5a for τ_m, we have

$$\theta \propto \frac{1}{C_m} \sqrt{\frac{a}{R_m R_i}} \qquad (16b)$$

Thus, propagation velocity is directly proportional to the square root of fiber diameter or radius (a) and inversely proportional to membrane capacitance (C_m). The larger the fiber diameter, the lower the absolute longitudinal resistance of the intracellular cytoplasm (principle of resistors in parallel), and therefore the greater the amount of local-circuit current flowing longitudinally and the greater the length constant. For example, it is well known that the larger the diameter of nerve fibers, the faster they propagate (Table 1).

Equation 16b shows that if C_m can be reduced (by myelination), then θ should increase in proportion. This is discussed in Section IV.C.

In addition, θ depends on the intensity of the local-circuit current, and hence on the rate of rise of the AP. The greater the AP rate of rise (max dV/dt), the greater the longitudinal current and the transmembrane capacitive current (I_c). Therefore, all other factors being constant, faster rising APs propagate faster. The AP rate of rise depends on the density of the fast Na^+ channels that carry inward current, on C_m, and on temperature. Max dV/dt decreases with increased C_m, with cooling, and with partial depolarization (due to the h_∞ versus E_m relationship discussed in Chapter 25). Cooling slows the rate of all chemical reactions, particularly those with a high Q_{10} (activation energy), such as the ion conductance changes in activated membrane.

Finally, the threshold potential (V_{th}) affects propagation velocity. If the threshold were to be shifted to a more positive voltage (more depolarized), then it would take longer for a given point in the membrane to reach threshold (and explode) during propagation of an AP from upstream. A greater critical depolarization (difference between RP and V_{th}) would be required to bring the membrane to threshold. Therefore, propagation velocity would be slowed.

As stated earlier, some of these factors are interrelated, and some actually exert opposing effects.

The foregoing discussion applies to nonmyelinated nerve axons and skeletal muscle fibers. An electron micrograph of small bundles of nonmyelinated nerve fibers enveloped by Schwann cells is shown in Fig. 7. In myelinated nerve fibers, propagation velocity is greatly increased by the myelin sheath, as discussed in Section IV.C.

C. Saltatory Conduction

The nerve cable has been vastly improved by the evolutionary development of **myelination** in vertebrates. An electron micrograph of a myelinated nerve fiber is shown in Fig. 8. The myelin sheath improves the cable by increasing the effective R_m by about one-hundred-fold and decreasing the effective C_m by about one-hundred-fold. This increases the length constant, λ, and tends to decrease the time constant τ_m. Although τ_m tends to increase with the increase in effective R_m (due to the myelin sheath), the decrease in C_m counteracts this effect, thus acting to hold τ_m almost constant. Thus, Eqs. 16a and 16b predict that propagation velocity should increase with myelination.

One consequence of myelination, therefore, is that propagation velocity is greatly increased (Table 1). Another consequence is that the energy cost of signaling is greatly decreased, because passive ion leaks are limited and active current losses are restricted to the small nodes of Ranvier, which are spaced relatively far apart. An electron micrograph of a node of Ranvier is shown in Fig. 9. At each node, the length of exposed (naked) cell membrane is only a few micrometers. The internodal distance is about 0.5–2.0 mm (depending on fiber diameter), and the width (length) of each node is only about 0.5–3 μm. The node forms an annulus around the entire perimeter of the fiber. Therefore the degree of energy-requiring active ion transport (Na^+-K^+ and Ca^{2+}) required to maintain the steady-state ion

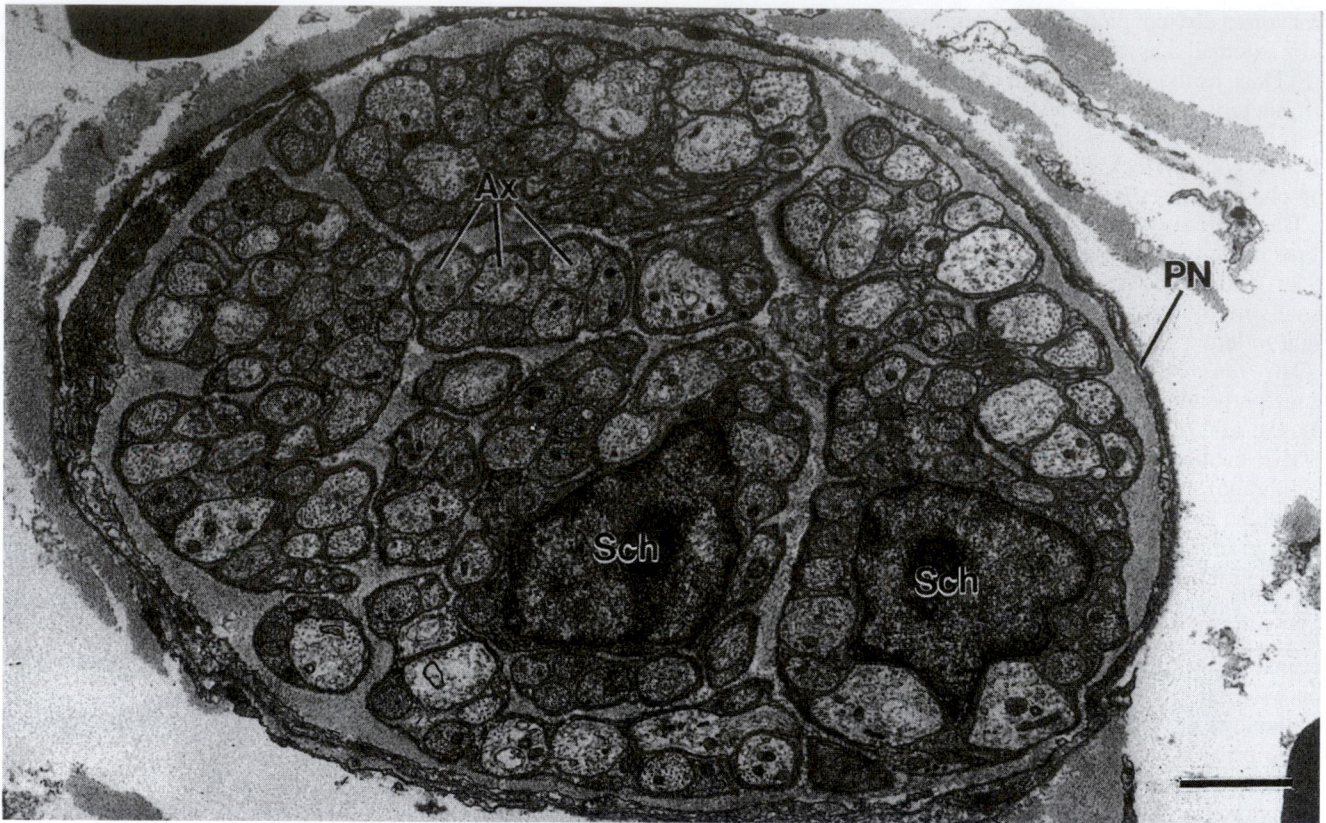

FIGURE 7. Electron micrograph of an autonomic nerve of mouse heart cut in transverse section. The unmyelinated axons (Ax) are arranged in small groups (bundles) that are engulfed by the cytoplasm of a Schwann cell. That is, the nonmyelinated axons are embedded in, and surrounded by, a Schwann cell. Two Schwann cell nuclei (Sch) can be seen. The axon bundles are separated and surrounded by a collagenous matrix. The entire nerve is surrounded by a perineurial sheath (PN). The scale bar at lower right equals 2 μm. (Micrograph courtesy of Dr. Mike Forbes, University of Virginia.)

distributions and to hold the system in a state of high potential energy is greatly reduced. For example, the amount of Na^+ gained and K^+ lost per impulse is reduced as a result of myelination. The rate of oxidative metabolism in myelinated fibers reflects this lowered energy requirement.

The myelin sheath is produced by the wrapping of the Schwann cell repeatedly around the nerve fiber in a spiral, forming 20–200 wrappings, depending on axon diameter. That is, the larger axons have a thicker myelin sheath. For the purpose of our discussion, we will assume an average of 100 wrappings. The myelin sheath covers the nerve axon as a coat sleeve and is interrupted at each node. The cytoplasm of the Schwann cell in the region of the myelin sheath is nearly completely extruded during its formation, so that the sheath consists essentially of 100 cell membranes stacked in series. Because resistors in series are added to calculate the total resistance, the effective transmembrane resistance is increased one-hundred-fold, but because the total capacitance of capacitors in series is calculated like resistors in parallel, the effective capacitance is reduced one-hundred-fold. Since λ is directly proportional to $\sqrt{R_m}$, λ is increased accordingly. As described in the section on the factors influencing propagation velocity, increasing λ and lowering C_m increase θ.

Myelinated nerves usually have an optimal amount of myelin, which is an amount such that the ratio of diameter of axon cylinder (naked axon) to total fiber (including myelin sheath) is about 0.6–0.8. Assuming a maximal total diameter feasible (the body must pack many circuits within a limited space, e.g., sciatic nerve bundle), a greater fraction of myelin, by infringing on the diameter of the axis cylinder, would raise r_i too high, causing θ to decrease. Thus, there are two opposing factors in determining the degree of optimal thickness of the myelin sheath: the more the myelin, the greater the decrease in C_m and the increase in effective R_m, but the smaller the diameter of the axis cylinder and therefore the higher the r_i.

In **saltatory conduction** (Latin *saltare*, to jump), the impulse jumps from one node to the next. The internodal membrane does not fire an AP. This is due to two reasons: (1) the internodal membrane is much less excitable (e.g., much fewer fast Na^+ channels) and (2) the depolarization of the neuron cell membrane at the internodal region is only about 1/100th of that at the node. The latter occurs because the IR voltage drop across the internodal cell membrane is only 1/100th of that across the entire series resistance network (neuron cell membrane plus 100 layers of Schwann cell membrane) (Kirchhoff's laws dealing with voltage drops across resistors in series). Even though the internal potential in the internodal region swings positive (e.g., to +30 mV) when the adjacent nodes fire, the potential at the

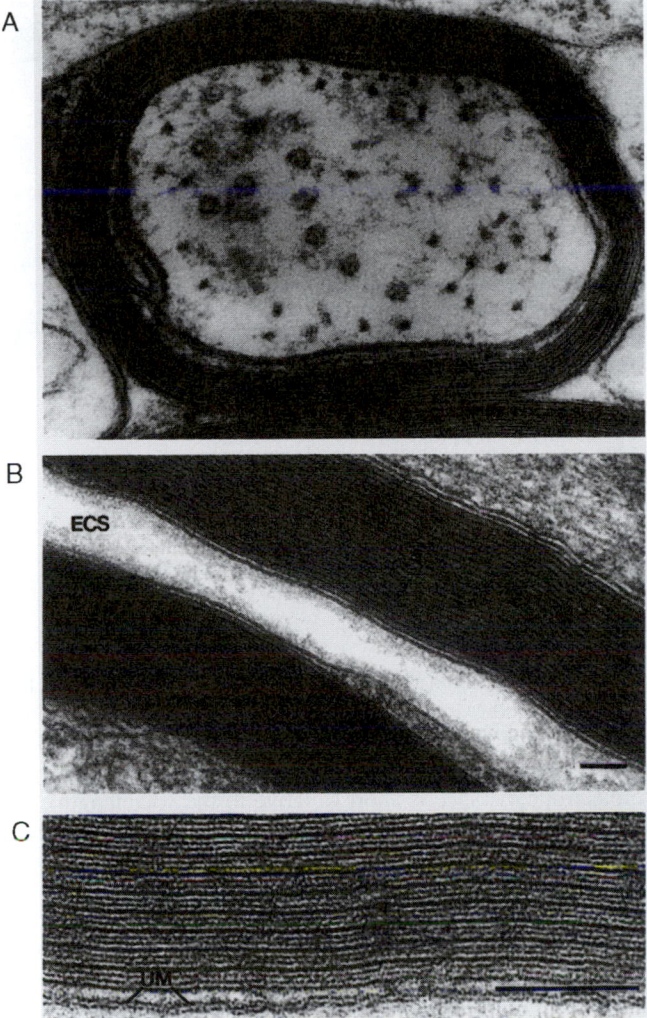

FIGURE 8. Myelin sheath of myelinated nerve fibers. (A) Myelinated nerve fiber in spinal cord of rat cut in cross-section. There are about seven wrappings of the Schwann cell around the axon, thus giving about 14 membranes in series with the cell membrane of the axon (axolemma). The internal **mesaxon** is visible at the left side (lower) of the axon, and represents the beginning of the spiraling of the Schwann cell membranes (cytoplasm squeezed out) to form the myelin sheath surrounding the axon. On the upper left side, a process of an oligodendrocyte abuts on the myelin sheath. The axon shown is closely plastered up against the myelin sheath of a neighboring axon (bottom of figure). **Microtubules** and **neurofilaments** are visible in the cytoplasm of the axon. Magnification of 168000×. (From J. Rhodin, "An Atlas of Histology," Oxford University Press, New York, 1975.) (B) Higher magnification of two adjacent nerves fibers (in phrenic nerve of rat) cut in longitudinal section to illustrate an axon with a thicker myelin sheath (approx. 23 wrappings = 46 membranes). Note that between the thicker (dense) lines are two thinner lines, which can be seen more clearly in C. The extracellular space (ECS) between the two axons is labeled. The calibration bar at the lower right equals 0.1 μm. (C) Still higher magnification of a portion of one of the myelin sheaths shown in the middle panel to better illustrate the periodicity of the myelin membranes. The dense lines alternate with a pair of thin lines. The space in between the pair of thin lines is extracellular space. The dense lines represent the intracellular (cytoplasmic) region of the Schwann cell process, with the cytoplasm extruded, thus allowing the inner (cytoplasmic face) leaflets of the cell membrane of the Schwann cell to come into close contact. The outer

outer surface of the internodal membrane also swings nearly as positive (e.g., to +29 mV). Therefore, the depolarization of the neuronal membrane at the internode is only about 1 mV, which is well below threshold, and so the internodal membrane does not fire. The potential that controls the membrane conductances (activates the voltage-dependent ion channels) is the pd (potential difference) directly across the membrane and not the absolute potential on either side.

As stated previously, the internodal membrane has only a few fast Na^+ channels, whereas the nodal membrane has a very high density. Since the cell membrane is fluid and proteins can diffuse (float) laterally in the lipid bilayer matrix, what keeps the fast Na^+ channel proteins confined (at high density) in the nodal region? It appears that there are special anchoring proteins (e.g., ankyrin) that anchor the ion channel proteins to the cytoskeletal framework, thus preventing their lateral movement into the internodal membrane.

The effect of myelin and saltatory propagation is to make propagation much faster. For example, a 20-μm-diameter myelinated nerve fiber conducts even faster than a 1000-μm (1-mm)-diameter nonmyelinated nerve fiber (e.g., the giant axon in squid, lobster, earthworm): 120 m/s versus 25–50 m/s. Thus, invertebrates, to achieve fast conduction in some essential circuits, must resort to giant neurons, resulting in a lower r_i and hence fast conduction. Because of space/size limitations, only a few critical neurons can be made giant in diameter. In vertebrates, on the other hand, a large fraction of the nerve fibers in the peripheral nerves is myelinated for fast propagation.

We saw earlier that, in nonmyelinated axons and skeletal muscle fibers, θ should vary with the square root of the cell diameter or radius ($a^{0.5}$). In myelinated axons, θ varies with the first power of the cell radius (a^1) because θ varies with λ^2 as indicated by

$$\theta \propto \frac{\lambda^2}{\tau_m} = \frac{a}{2 R_i C_m} \qquad (17)$$

The dependence of conduction velocity on diameter of myelinated and nonmyelinated fibers is summarized in Table 1.

D. Wavelength of the Impulse

We can calculate the wavelength of the AP, which is the length of the axon simultaneously undergoing some portion of the AP. The wavelength is equal to propagation velocity

(extracellular face) leaflets of the Schwann cell membrane are seen as the thin lines. Thus, each cell membrane has the appearance of a double line, representing the hydrophilic surfaces of the cell membrane; the hydrophobic region of the membrane is the clear region between the dense line and the contiguous thin line on either side. UM = unit cell membrane. The scale bar on the lower right represents 0.1 μm. (Micrographs B and C courtesy of Dr. Mike Forbes, University of Virginia.)

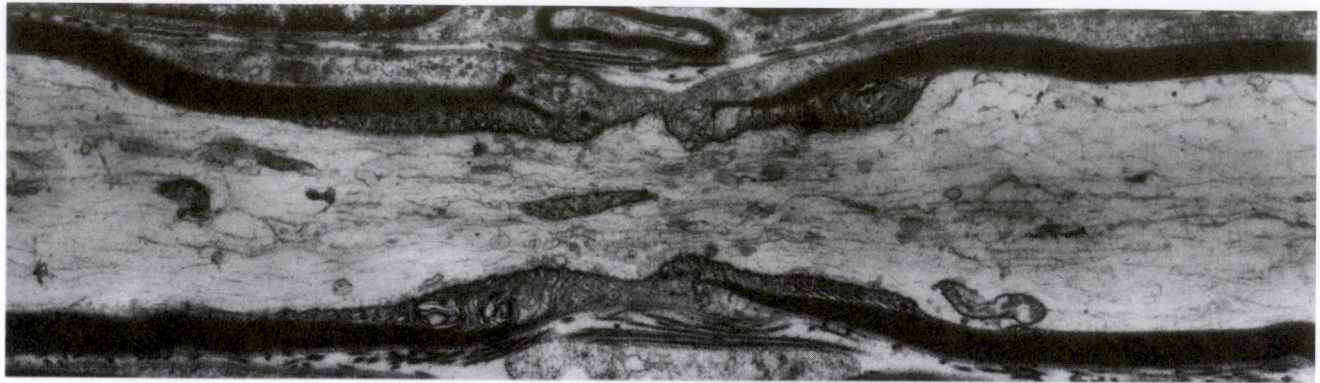

FIGURE 9. Electron micrograph of one node of Ranvier in a single myelinated nerve axon of rat sciatic nerve cut in longitudinal section. As can be seen, the dark myelin bordering the axon on each side is interrupted near the middle of the micrograph, leaving the neuronal cell membrane nude of myelin at the node region. However, cytoplasmic processes of the Schwann cells cover the nodal membrane. Collagenous fibrils of the endoneurium are visible at the region of the node, peripheral to the Schwann cell cytoplasm. The myelin tapers and thins as it approaches the node, and there is a frayed appearance caused by the successive laminae of the myelin sheath terminating as cytoplasmic swellings. The axoplasm contains neurofilaments and microtubules. Magnification of 9000×. (From J. Rhodin, "An Atlas of Histology," Oxford University Press, New York, 1975.)

(θ) times the duration of the AP (APD$_{100}$); thus,

$$\text{wavelength} = \theta \times \text{APD}_{100} \qquad (18)$$

$$\text{cm} = (\text{cm/s}) \times \text{s}$$

$$\text{distance} = \text{velocity} \times \text{time}$$

Note the similarity of this relationship to that for the wavelength of an electromagnetic radiation: wavelength = velocity of light/frequency of the radiation. The reciprocal of frequency is the period (duration of one cycle).

The wavelength in a large myelinated nerve axon is about 12 cm: 120 m/s × 1.0 ms. In a skeletal muscle fiber, it is about 1.8 cm (6 m/s × 3.0 ms). In a smooth muscle bundle, the wavelength is only about 1.5 mm (5 cm/s × 30 ms).

V. External Recording of Action Potentials

A. Monophasic, Diphasic, and Triphasic Recording

As discussed previously, local-circuit currents accompany the propagating AP in each fiber. The intracellular and extracellular longitudinal currents are diphasic; that is, initially they travel in the forward direction intracellularly and then in the reverse direction. The forward direction current is intense (high current density) and the reverse direction current is weak (low current density).

The transmembrane radial currents are **triphasic;** that is, the first phase is outward (moderate intensity), the second phase is inward (high intensity), and the third phase is outward (low intensity). The first phase (outward) gives rise to the passive exponential foot of the AP, and is due to the passive cable spread of voltage and current. The second phase (inward) corresponds to the large inward fast Na$^+$ current, which occurs during the later portion of the rising phase and peak of the AP. The third phase (outward) corresponds to the net outward current (K$^+$), which occurs during the repolarizing phase.

These longitudinal and radial currents can be recorded by suitably placed external electrodes. The extracellular longitudinal currents can be recorded by two electrodes (bipolar) placed close together along the length of the fiber. If the interelectrode distance is short (relative to the wavelength), an approximate first (time) derivative of the AP is obtained (see Figs. 6 and 10C).

The extracellular radial currents can be recorded by two electrodes placed close together in a plane perpendicular to the fiber axis. This gives an approximation of the second (time) derivative of the AP (see Fig. 6).

The internal axial currents are confined to the cytoplasm, whereas the external longitudinal currents can use the entire interstitial fluid space of the nerve bundle or muscle or even the entire torso (so-called volume conductor), since current takes the path of least resistance (resistors in parallel). As mentioned in Section IV.A, this allows the recording of the ECG from the body surface and the EMG from the skin overlaying an activated skeletal muscle. The ECG and EMG consist essentially of diphasic potentials, reflecting the external longitudinal currents during propagation of APs.

When the two external electrodes are placed far apart (with respect to the wavelength) along a nerve or muscle fiber, the diphasic recording has two phases that are about equal (Fig. 10A). The proximal electrode records the wave of negativity (associated with the propagating AP) first, and then returns to isopotential. When the wave reaches the second electrode, the wave is recorded by it in reversed polarity (because current flow through the voltmeter is reversed). If the AP were now prevented from reaching the second (distal) electrode by crushing this region of the fiber or elevating [K$^+$]$_o$ to depolarize it, then a monophasic recording would be obtained (Fig. 10B). This monophasic recording would most resemble the true AP recorded by a microelectrode impaled into a fiber to record the transmembrane potential, but would be much smaller in amplitude.

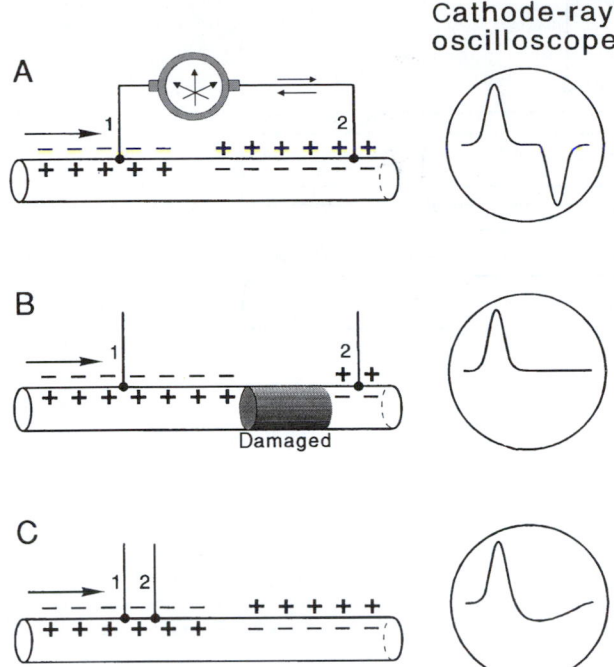

Cathode-ray oscilloscope

FIGURE 10. Diagram of the waveforms that would be recorded externally during propagation of an AP in a single fiber. (A) When the two electrodes are far apart (relative to the AP wavelength), a diphasic recording is obtained, with the two phases being symmetrical and separated by an isopotential segment. The two phases are due to the current flow through the voltmeter recorder being first in one direction and then in the opposite direction. (B) If the fiber between these two electrodes is damaged (e.g., by crushing) or depolarized (by elevated $[K^+]_o$), so the AP cannot sweep past the second electrode (2), then this second phase is prevented and the recording is monophasic. (C) If the two electrodes depicted in part A are brought progressively closer, then the isopotential segment would shorten and disappear. If electrode 2 is brought very close to electrode 1, so that the interelectrode distance is short relative to the wavelength, then the second phase is smaller than the first phase, and the record resembles the first derivative of the true AP.

B. Compound Action Potential

When one records the APs externally, the records are graded and not all-or-none, as in the case of the true APs recorded intracellularly from single fibers (see Chapter 25). That is, the signal recorded becomes larger and larger, up to a maximum amplitude as the intensity of stimulation is increased. This is the so-called **compound action potential.** It is graded because, as a greater and greater fraction of the fibers is activated, the external longitudinal currents associated with the all-or-none AP in each fiber cut across the recording electrodes, thereby producing a larger signal. The AP in each individual fiber is always all-or-none. The amplitude of the signal is determined by the resistance between the electrodes multiplied by the amount of current flowing through this resistance ($V = IR$). The recording of compound action potentials is diagrammed in Fig. 11.

The compound action potentials can be demonstrated by recording the EMG from a human subject when one elec-

trode is placed on the skin of the ventral forearm and the other (reference) electrode on the wrist of the same arm. Then, as the subject voluntarily produces stronger and stronger contraction to flex the hand, the electrical signals picked up become greater and greater in amplitude and frequency. The amplitude becomes larger because more muscle fibers are simultaneously activated. This is known as **fiber recruitment.** The frequency also increases, because the motor nerves fire at a higher frequency, causing the muscle fibers to fire at a higher frequency and thus producing a more powerful tetanic contraction.

VI. Summary

Although the biological cable (i.e., nerve fiber or skeletal muscle fiber) is the best possible, it is a relatively poor cable with a short length constant and relatively long time constant. Therefore, for faithful and rapid signal transmission over long distances, energy must be put into the system at each point along the way. The system evolved is that of AP generation, which are all-or-none signals of constant amplitude and constant propagation velocity, in addition to having refractory periods and sharp thresholds. This is a frequency-modulated system, in which increasing strength of sensation or motor response follows from an increase in frequency of the AP signals.

AP propagation occurs by means of local-circuit currents. The transmembrane current has three phases: outward, inward, and outward. The internal and external longitudinal currents have two phases: forward and backward (for internal) or backward and forward (for external). The external currents use the path of least resistance, enabling electrograms (e.g., ECG, EMG) to be recorded from the body surface. The compound AP is graded in amplitude, reflecting the summation of the external currents generated from each fiber that is activated; that is, the more fibers simultaneously activated, the greater the amplitude of the electrogram signal.

Propagation velocity is faster the larger the diameter of the fiber, the longer the length constant, and the lower its time constant and capacitance. The myelin sheath evolved by vertebrates enables much faster propagation velocity and at a lower energy cost. Myelination raises the effective membrane resistance and lowers the effective capacitance, and excitability occurs only at the short nodes of Ranvier that periodically interrupt the myelin sheath. Therefore, the AP signal jumps from node to node in a saltatory pattern of conduction.

Bibliography

Cole, K. S. (1968). "Membranes, Ions and Impulses: A Chapter of Classical Biophysics." University of California Press, Berkeley.

Davis, L., Jr., and Lorente de No, R. (1947). Contribution to the mathematical theory of the electrotonus. *Stud. Rockefeller Inst. Med. Res.* **131,** 442–496.

Hodgkin, A. L., and Rushton, W. A. H. (1946). The electrical constants of a crustacean nerve. *Proc. R. Soc. Lond. (Biol.)* **133,** 444–479.

Jack, J. J. B., Noble, D., and Tsien, R. W. (1975). "Electric Current Flow in Excitable Cells." Clarendon Press, Oxford.

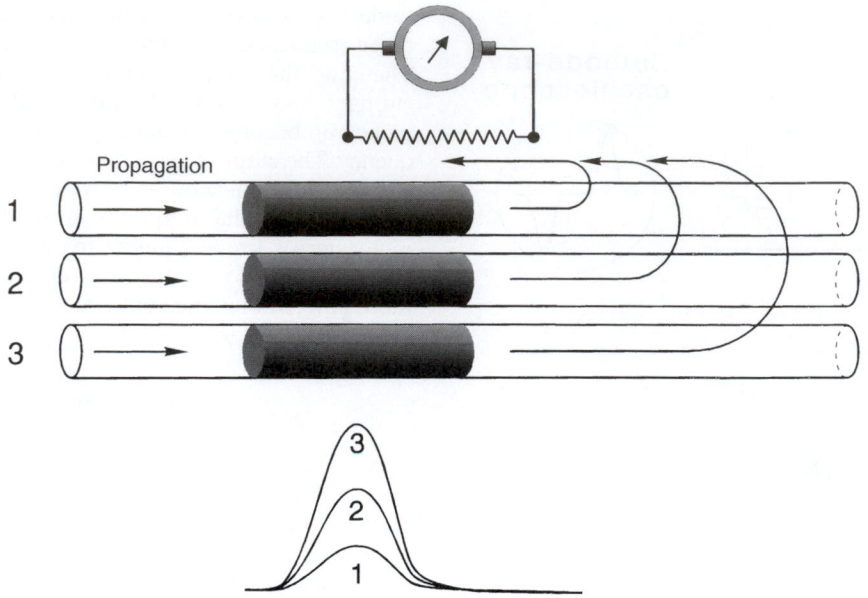

FIGURE 11.   Diagram of a compound action potential in an isolated nerve trunk, such as the frog sciatic nerve, recorded externally by a pair of longitudinal electrodes. The voltmeter (oscilloscope) records the voltage (*IR*) drop across the resistance (fluid) between the two electrodes. If only fiber 1 is activated, the current passing between the electrodes is small, and the voltage recorded is small. If fiber 2 is simultaneously activated with fiber 1, then the amount of current is doubled, and the voltage is doubled. When all three fibers are simultaneously activated, the current is tripled, and the voltage is tripled. Therefore, the externally recorded compound action potential is graded because it reflects the electrical activity of numerous fibers, each of which produces an all-or-none (nongraded) AP.

Katz, B. (1966). "Nerve, Muscle, and Synapse." McGraw-Hill, New York.

Lakshminarayanaiah, N. (1984). "Equations of Membrane Biophysics." Academic Press, Orlando, FL.

Rail, W. (1977). *In* "Handbook of Physiology" (J. M. Brookhart and V. B. Mounteastle, Eds.), Vol. 1, pp. 39–97. American Physiological Society, Bethesda, MD.

Sperelakis, N. (1979). Origin of the cardiac resting potential. *In* "Handbook of Physiology" (R. M. Berne and N. Sperelakis, Eds.), Vol. 1, pp. 187–267. American Physiological Society, Bethesda, MD.

Sperelakis, N. (1992). Cable properties and propagation mechanisms. *In* "Physiology" (N. Sperelakis and R. O. Banks, Eds.), pp. 83–97. Little, Brown, Boston.

Sperelakis, N., and Fabiato, A. (1985). Electrophysiology and excitation-contraction coupling in skeletal muscle. *In* "The Thorax: Vital Pump" (C. Roussous and P. Macklem, Eds.), pp. 45–113. Marcel Dekker, New York.

Sperelakis, N., and Mann, J. E., Jr. (1977). Evaluation of electric field changes in the cleft between excitable cells. *J. Theor. Biol.* **64,** 71–96.

Taylor, R. E. (1963). Cable theory. *In* "Physical Techniques in Biological Research" (W. L. Nastuk, Ed.), Vol. 6, pp. 219–262. Academic Press, New York.

Appendix 1: Propagation in Cardiac Muscle and Smooth Muscles

1. Background

Chapter 24 discusses propagation in cells that are long cables, such as nerve fibers and skeletal muscle fibers. Propagation is more complex in tissues composed of assemblies of short cells, such as cardiac muscle and visceral smooth muscles. These short cells may be considered to be short or truncated cables. In such a truncated cable cell, its true length constant λ is much longer than the cell's length. Therefore, there is relatively little voltage fall-off (decay) over the length of each cell. It follows that the entire cell undergoes an action potential (AP) nearly simultaneously. Yet propagation velocity θ is much slower in cardiac muscle (ca. 0.5 m/s) and smooth muscles (ca. 0.05 m/s) than in skeletal muscle (ca. 5 m/s) or nonmyelinated nerve fibers (ca. 2 m/s). Part of the reason for the slower θ concerns fiber diameter (i.e., diameter of cardiac muscle fiber is about 15 μm compared to about 60 μm for skeletal muscle fibers). Equation 16b in Chapter 24 indicates that in a simple cable, θ is a function of $\sqrt{a}$ (where a is the fiber radius). Thus, the theoretical θ for cardiac muscle (θ_c) should be

$$\theta_c = \sqrt{\frac{15\,\mu m}{60\,\mu m}} \times \theta_{sk}$$

$$= 0.5 \times 5 \text{ m/s}$$

$$= 2.5 \text{ m/s}$$

where θ_{sk} is the propagation velocity for skeletal muscle, and it is assumed that the other parameters given in Eq. 16b in Chapter 24 are the same in the two types of muscle.

This theoretical velocity of 2.5 m/s is substantially greater than the actual 0.5 m/s. Therefore, at least one other factor must determine velocity in cardiac muscle, and that is the high resistance of the junctional membranes (the intercalated disc membranes in the case of cardiac muscle). It is still controversial as to exactly how high the junctional resistance is. When gap junctions are present, the gap junction channels span the intercellular junction, and so serve to lower the cell-to-cell resistance (see Chapter 31). However, the hearts of the lower vertebrates (e.g., amphibians, reptiles) either have no gap junctions or they are very sparse and tiny. In addition, even parts of mammalian hearts and some visceral smooth muscles (e.g., longitudinal muscle layer of intestine) do not appear to contain gap junctions. If so, another mechanism may be involved for cell-to-cell

propagation in those muscles in which there is a virtual absence of gap junctions. One of the mechanisms proposed is known as the **electric field model** (Sperelakis *et al.*, 1989). This model is discussed in this appendix.

First, however, let us return to the problem that propagation in cardiac muscle is about five times slower than it should be if the gap junction channels served to interconnect thoroughly (electrically) the cytoplasm of two contiguous cells lying end to end. We also stated previously that the short cardiac muscle cell should show no voltage decay, and so the entire length of the cell should undergo the AP nearly simultaneously (i.e., there is nearly infinite propagation velocity within each cell). Yet the overall propagation velocity in the tissue is relatively slow. Clearly, a large time delay must occur at each cell junction. In fact, most of the propagation time is consumed at the cell junctions. This has been demonstrated experimentally. Propagation in cardiac muscle has been shown to be actually **discontinuous** or **saltatory** in nature (Spach *et al.*, 1981; Rudy and Quan, 1987a, b; Sperelakis *et al.*, 1989). Therefore, the presence of a large number of gap junction channels is insufficient to reduce the junctional resistance enough to allow a chain of cells to behave as a simple cable. This relatively high junctional resistance has relevance to heart block and fibrillation, but this is outside the scope of this chapter.

A similar analysis can be done for visceral smooth muscle. The theoretical θ for visceral smooth muscle (θ_s) should be

$$\theta_s = \sqrt{\frac{5\,\mu m}{60\,\mu m}} \times \theta_{sk}$$

$$= 0.289 \times 5 \text{ m/s}$$

$$= 1.44 \text{ m/s}$$

This theoretical velocity of 1.44 m/s is much greater than the actual velocity of 0.05 m/s. Thus, propagation velocity in smooth muscle is only about 3.5% of what it should be based on fiber diameter.

The lower maximum rate of rise of the AP (+ max dV/dt) in cardiac muscle (ca. 200 V/s) and smooth muscle (ca. 5 V/s), as compared with skeletal muscle (ca. 600 V/s), is another factor that contributes to the lower than expected θ in cardiac muscle and smooth muscle. In addition, the higher the extracellular resistance, which depends on the tightness of packing of the fibers in the muscle, the slower the velocity (e.g., see Jack *et al.*, 1975).

II. Experimental Facts

Some of the key experimental facts relevant to the transmission of excitation from one cell to the next in cardiac muscle and visceral smooth muscle can be summarized as follows. These tissues can be enzymatically separated into their individual cells, and the individual single cells are viable and functional. Gap junctions are absent in the hearts of lower vertebrates and in some regions of mammalian hearts, as well as in some visceral smooth muscles, as stated previously. The length constant λ of cardiac muscle and smooth muscle tissues, when measured properly, is relatively short, that is, less than 0.5 mm (i.e., not much more than about one cell length). The input resistance (R_{in}) of cardiac muscle and smooth muscle, when measured properly, is relatively high, about 5–40 MΩ. The short λ and high R_{in} suggest that the cells are not profusely connected by low-resistance pathways (e.g., by gap junction channels). Thus, even when gap junctions are present, the junctional membranes constitute a substantial barrier to current flow from one cell to the next.

As stated in Section I, the true λ of individual cells is much greater than the cell length, so that there is almost no voltage decay in a single cell and that the entire cell fires an AP nearly simultaneously. Therefore, most of the propagation time is consumed at the cell junctions, and propagation in these tissues is a discontinuous process, in contrast to a continuous process for skeletal muscle fibers and nonmyelinated nerve fibers. Propagation velocity in cardiac muscle and smooth muscle is slower than what can be accounted for by the smaller fiber diameter and lower $+$max dV/dt. The cell-to-cell transmission process is quite labile, somewhat like that in synaptic transmission.

The reader is referred to several review-type articles by Sperelakis and colleagues for additional details and evidence for some of the statements made here. Considerable data have been published concerning the degree of spread of electrotonic current between neighboring cells in cardiac muscle and visceral and vascular smooth muscles. Only one example is presented here for frog cardiac (ventricular) muscle. In these experiments, electrode 1 was used to record voltage and to inject current (using a bridge circuit), whereas electrode 2 recorded voltage only. As illustrated in Fig. A-1, using a pair of microelectrodes whose tips were spaced 11 μm apart, in some double impalements there was no electrotonic current spread between the two electrodes (e.g., Fig. A-1A), whereas in other impalements there was substantial spread of current (e.g., Fig. A-1B). In one unusual case (Fig. A-1C–D), the double impalement first showed good spread (i.e., interaction) between the electrodes (C), but then due to muscle contraction, one electrode left that cell and impaled a neighboring cell having a normal resting potential (right portion of Fig. A-1C); now there was no significant interaction between the two neighboring cells (D). In the impalements in which there was substantial interaction between the electrodes (e.g., Fig. A-1B, C), it was proposed that the two electrodes had impaled the same cell (e.g., both electrodes recorded low resting potentials, perhaps due to damage caused by the two electrodes impaling one cell). In the impalements in which there was little or no interaction (e.g., Fig. A-1A, D), it was proposed that the electrodes had impaled neighboring cells. This interpretation is most clear in Fig. A-1D, in which the two electrodes

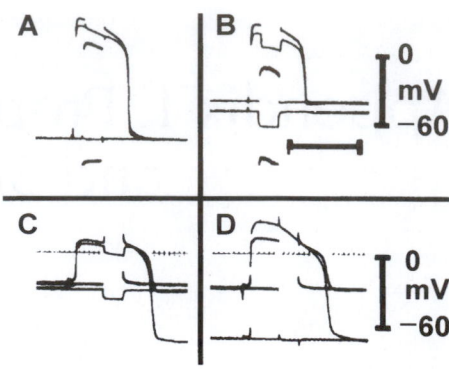

FIGURE A-1. Three typical experiments (A, B, and C–D) measuring the spread of electrotonic current between two closely spaced intracellular microelectrodes in intact frog ventricular trabeculae at rest and during the plateau of the action potential. The interelectrode distance was 11 μm. Rectangular hyperpolarizing current pulses (approx. 150 ms in duration) were applied. Two successive sweeps of the oscilloscope are superimposed in each panel. (A) Interaction at rest and during the plateau was nearly 0%. Capacitive transients are only seen on the trace from electrode 2, whereas a large maintained hyperpolarization occurred at electrode 1. (B) In another impalement in which the cell was injured by the two electrodes, the resting potential was low and the degree of interaction was high. (C–D) In another impalement in which the cell was damaged, the resting potential was low, and the degree of interaction was high (electrode 1 deflection not shown because of bridge imbalance). The contraction accompanying the action potential caused one of the electrodes to become dislodged from that cell and penetrate into a neighboring cell that had a normal resting potential; the degree of interaction then became nearly zero. (Reproduced with permission from Tarr, M., and Sperelakis, N. (1964). *Am. J. Physiol.* **207,** 691–700.)

recorded markedly different resting potentials. Thus, there appears to be little or no spread of current between neighboring cells in frog cardiac muscle. If so, then propagation of excitation must occur by some other means.

III. Electric Field Model

A. Electric Field Effect

Sperelakis and colleagues developed an electric field hypothesis for propagation of APs in cardiac muscle for situations in which there were no functioning gap junction channels. A computer simulation model for cell-to-cell propagation in cardiac muscle was developed and progressively improved since the mid-1970s (Sperelakis and Mann, 1977; Sperelakis *et al.,* 1985; Sperelakis, 1987; Picone *et al.,* 1991). The model allows electrical transmission to occur between adjacent excitable cells by means of the electric field effect in the very narrow junctional cleft between the contiguous cells (Sperelakis and Mann, 1977). This electric field model does not require low-resistance channels (gap junctions) between cells. The major requirements of the model are that the pre- and postjunctional membranes (pre-JM and post-JM) be ordinary excitable membranes, and that these membranes be very closely apposed to one another (i.e., the junctional cleft be very narrow, about 10 nm). When the pre-JM fires, the cleft between the cells becomes negative with respect to ground (the interstitial fluid surrounding the cells), and this negative cleft potential (about –40 mV) acts to depolarize the post-JM by an equal amount (namely, 40 mV) and brings it to threshold. (The inner surface of the post-JM remains at nearly constant potential with respect to ground.) This, in turn, brings the surface membrane of the postjunctional cell to threshold.

Figure A-2A illustrates propagation of an AP along a chain of 10 cells by the electric field effect. In this computer simulation of cardiac muscle, propagation of overshooting APs occurred at a constant velocity of 32 cm/s, and the maximum rate of rise of the AP averaged 209 V/s. As can be seen, the upstroke of each AP exhibited a **break** or **step,** reflecting the junctional transmission process.

In the model, a plot of propagation time as a function of distance along a chain of cells has a staircase shape, indicating that almost all propagation time is consumed at the cell junctions and that excitation of each cell is virtually instantaneous (Picone *et al.,* 1991; Sperelakis *et al.,* 1991) (see Fig.

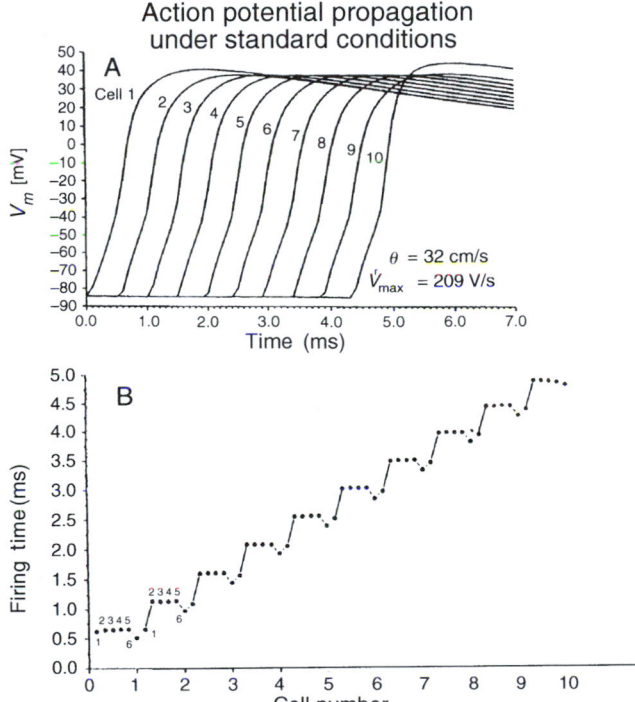

FIGURE A-2. Electric field model for propagation in cardiac muscle using a computer simulation. (A) Successful propagation of an AP at constant velocity along a chain of 10 cells under standard conditions. Propagation velocity was 32 cm/s; max *dV/dt* was 209 V/s. Note the small step, or prepotential, on the AP upstroke. (B) Plot illustrating that junctional delays occupy most of the propagation time. Firing time for an AP to spread along the chain of 10 cells under standard conditions (corresponding to part A) is plotted against distance. Within each cell, the units are numbered 1 through 6 from left to right, 1 corresponding to the post-JM (input), 2–5 the surface membrane, and 6 the pre-JM (output). The AP travels along the surface membrane at a high velocity. At the junctions between cells, there is a significant conduction delay as the AP "jumps" across the cleft. This clearly demonstrates the discontinuous nature of AP propagation. (Adapted with permission from Figs. 2 and 3 of Sperelakis, N., Ortiz-Zuazaga, H., and Picone, J. B. (1991). *Innov. Tech. Biol. Med.* **12,** 404-414.)

A-2B). In this figure, it can also be seen that the prejunctional membrane fires a fraction of a millisecond **before** the surface membrane of the same cell, as required by the electric field model. Propagation was found to be strongly dependent on radial cleft resistance (R_{jc}) and the junctional membrane properties. There was an optimal R_{jc} for maximum propagation velocity under any given conditions. This model is consistent with many experimental facts about propagation in cardiac muscle, and provides an alternative mechanism for AP propagation that does not require low-resistance pathways to transfer excitation directly between adjacent cells. The electric field model can also account for the fact that propagation in cardiac muscle is actually discontinuous or saltatory in nature (Spach *et al.,* 1981; Rudy and Quan, 1987a, b).

B. High Density of Fast Na$^+$ Channels at Intercalated Discs

For this mechanism to work efficiently, there is a requirement for the prejunctional membrane to fire an AP a fraction of a millisecond before the contiguous surface membrane (Sperelakis and Mann, 1977). That is, the prejunctional membrane should be more excitable than the surface membrane, which would cause it to reach threshold first. It was suggested that this situation would be achieved if there were a greater density of fast Na$^+$ channels in the junctional membranes (intercalated discs) than in the contiguous surface membranes (Sperelakis and Mann, 1977). This would make the intercalated discs more excitable and give them a lower threshold than the surface membrane. To examine this possibility, a polyclonal antibody raised against fast Na$^+$ channels (from rat brain) was used to immunolocalize the fast Na$^+$ channels in rat atrial and ventricular tissues. In immunofluorescence examination, intense labeling was observed associated with the intercalated discs of both atrial and ventricular cells (Ferguson, D., Sperelakis, N., and Angelides, K. J., unpublished observations) (Fig. A-3). This fluorescence was more intense than that of the cell surface membrane. These findings are in agreement with results reported by Cohen and Levitt (1993).

Therefore, it is likely that, not only are fast Na$^+$ channels present at the intercalated disc membranes, they are present in a higher concentration (density) than in the surface cell membrane. A higher density of fast Na$^+$ channels in the junctional membrane would cause it to have a lower threshold than the surface membrane, and so it would reach threshold and discharge first. Thus, not only are the intercalated discs composed of excitable membranes, but they also may have greater excitability than the contiguous surface membranes. This would be analogous to the initial segment of the axon of the anterior horn neuron having a lower threshold, and therefore discharging before the soma or proximal dendrites (where the excitatory synapses are actually located). Consistent with these findings, it was reported that K$^+$ channels are also localized at the intercalated discs (Mays *et al.,* 1995). Therefore, the electric field model is a plausible mechanism for cell-to-cell transmission of excitability in cardiac muscle and in visceral smooth muscle. Electric field effects may also occur between closely spaced contiguous neurons in the central nervous system.

FIGURE A-3. Immunofluorescence localization of Na⁺ channels in adult rat cardiac muscle (ventricle) using polyclonal antibody to the fast Na⁺ channels of rat brain. The antibody is most densely localized at the cell junctions (intercalated discs), but also stained the surface cell membrane. These results suggest that the intercalated disks are highly excitable membranes, even more excitable than the surface cell membrane, consistent with the requirement of the electric field model for propagation between cells not connected by low-resistance tunnels (gap junction connexons). (From Ferguson, D., Sperelakis, N., and Angelides, K. J., unpublished observations.)

IV. Electronic Model for Simulation of Propagation

Sperelakis and colleagues (1990) constructed an electronic model to simulate an excitable membrane, and this model was used as a new circuit model of propagation. Four such circuits were successfully made to interact with one another through capacitive coupling to simulate propagation over four cells. Adjustment of one parameter either slightly depolarized and caused repetitive spontaneous APs, or hyperpolarized slightly and depressed excitability, causing partial block (e.g., 2:1, 3:1, 4:1) or complete block of propagation from cell to cell at the cell junctions. When four such cells were connected head to tail in a closed loop, reentry of excitation occurred and could keep going for many seconds before dying out (due to failure at one of the labile junctions). Several configurations of external networks were used in this model to study possible electrical field coupling in cell chains.

In a subsequent study (Ge *et al.*, 1993), 12 such units were arranged to model two adjacent cells: 4 units for each surface membrane and 1 for each junctional membrane. The first unit was stimulated to threshold, and AP propagation spread over the two cells was recorded. The time delay at each junction between two adjacent cells was measured as a function of R_{jc}, the radial shunt resistance at the cell junction. The experimental results showed that the time delay was about 1–2 ms when R_{jc} values were changed

from infinite down to about 100 kΩ; the time delay increased at lower R_{jc} values, and propagation was blocked when R_{jc} was below 10 kΩ. Excitation of cell 1 caused hyperpolarization of the post-JM (unit 7) and depolarization of the other units in cell 2, due to current flow prior to triggering of the AP in cell 2. Raising the effective coupling resistance (R_c) between the two cells increased the conduction delay. As expected, the junctional delay increased when C_j was lowered. Stimulating at the pre-JM (unit 6) resulted in bidirectional propagation. Making the pre-JM inexcitable did not prevent excitation of cell 2, reflecting the presence of longitudinal current flow. However, the post-JM (unit 7) only fired an AP when unit 6 was active, reflecting the electrical field effect across the junction. Thus, in this model, both local-circuit current and the electrical field effect play roles in the transfer of excitation.

Records from an electronic model of two myocardial cells with a cell junction between them are illustrated in Fig. A-4. In this example, the first cell was set to be spontaneously active and fire a spontaneous cardiac-like AP. All four surface-membrane units in cell 1 (U2, 3, 4, 5) fired an AP nearly simultaneously. However, there was a short delay (e.g., ca. 1 ms) in the firing of the unit (U6) representing the prejunctional membrane (at the intercalated disc). Firing of U6 led to the firing of all units (U7, 8, 9, 10, 11, 12) of cell 2 after a junctional delay of about 2 ms. As can be seen, the firing of the prejunctional membrane (U6) drove inward hyperpolarizing current through the postjunctional membrane (U7) and outward depolarizing current through the other units (U8, 9, 10, 11, 12) of cell 2.

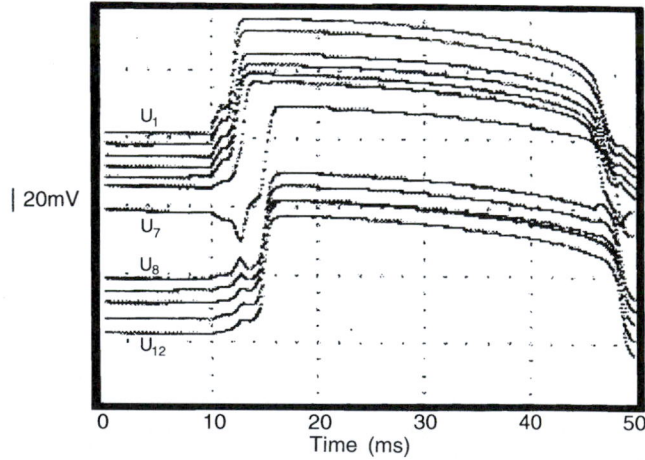

FIGURE A-4. Action potentials produced in an electronic model of two heart cells with a cell junction between them. There were six excitable-circuit units in each cell: four representing the surface cell membrane (U2, 3, 4, 5 and U8, 9, 10, 11) and one for each junctional membrane (U1, 6 and U7, 12). The parameters in the first cell (cell 1) were adjusted so that it fired spontaneously in response to pacemaker potential depolarization. Note the nearly simultaneous firing of all units in cell 1, and after a slight junctional delay time, the nearly simultaneous firing of all units of cell 2. (Reproduced with permission from Fig. 5A of Ge, J., Sperelakis, N., and Ortiz-Zuazaga, H. (1993). *Innov. Tech. Biol. Med.* **14,** 404–420.)

Bibliography

Cohen, S. A., and Levitt, L. K. (1993). Partial characterization of the rHl sodium channel protein from rat heart using subtype-specific antibodies. *Circ. Res.* **73**, 735–742.

Cole, W. C., Picone, J. B., and Sperelakis, N. (1988). Gap junction uncoupling and discontinuous propagation in the heart: a comparison of experimental data with computer simulations. *Biophys. J.* **53**, 809–818.

Ge, J., Sperelakis, N., and Ortiz-Zuazaga, H. (1993). Simulation of action potential propagation with electronic circuits. *Innov. Tech. Biol. Med.* **14**, 404–420.

Jack, J. J. B., Noble, D., and Tsien, R. W. (1975). "Electric Current Flow in Excitable Cells." Oxford University Press, Oxford.

Mann, J. E., Jr., Sperelakis, N., and Ruffner, J. A. (1981). Alterations in sodium channel gate kinetics of the Hodgkin-Huxley equations on an electric field model for interaction between excitable cells. *IEEE Trans. Biomed. Eng.* **28**, 655–661.

Mays, D. J., Foose, J. M., Philipson, L. H., and Tamkun, M. M. (1995). Localization of the Kv1.5K$^+$ channel protein in explanted cardiac tissue. *J. Clin. Invest.* **96**, 282–292.

Picone, J. B., Cole, W. C., and Sperelakis, N. (1989). Discontinuous conduction in cardiac muscle. *In* "Cell Interactions and Gap Junctions" (N. Sperelakis and W. C. Cole, Eds.), Vol. II, pp. 143–154. CRC Press, Boca Raton, FL.

Picone, J. B., Sperelakis, N., and Mann, J. E., Jr. (1991). Expanded model of the electric field hypothesis for propagation in cardiac muscle. *Math. Comp. Mod.* **15**, 17–35.

Rudy, Y., and Quan, W. L. (1987a). A model study of the effects of the discrete cellular structure on electrical propagation in cardiac tissue. *Circ. Res.* **61**, 815–823.

Rudy, Y., and Quan, W.-L. (1987b). Effects of the discrete cellular structure on electrical propagation in cardiac tissue. *In* "Activation, Metabolism, and Perfusion of the Heart—Simulation and Experimental Models" (S. Sideman and R. Beyar, Eds.), pp. 61–76. Martinus Nijhoff, Dordrecht.

Spach, M. S., Miller, W. T., III, Geselowitz, D. B., Barr, R. C., Kootsey, J. M., and Johnson, E. A. (1981). The discontinuous nature of propagation in normal canine cardiac muscle. Evidence for recurrent discontinuity of intracellular resistance that affects the membrane currents. *Circ. Res.* **48**, 39–54.

Sperelakis, N. (1987). Electrical field model for electric interactions between myocardial cells. *In* "Activation, Metabolism, and Perfusion of the Heart—Simulation and Experimental Models" (S. Sideman and R. Beyar, Eds.), pp. 77–113. Martinus Nijhoff, Dordrecht.

Sperelakis, N., and Mann, J. E., Jr. (1977). Evaluation of electric field changes in the cleft between excitable cells. *J. Theor. Biol.* **64**, 71–96.

Sperelakis, N., and Picone, J. (1986). Cable analysis in cardiac muscle and smooth muscle bundles. *Innov. Tech. Biol. Med.* **7**, 433–457.

Sperelakis, N., Marschall, R., and Mann, J. E. (1983). Propagation down a chain of excitable cells by electric field interactions in the junctional clefts: effect of variation in extracellular resistances, including a "sucrose gap" simulation. *IEEE Trans. Biomed. Eng.* **30**, 658–664.

Sperelakis, N., LoBrocco, B., Mann, J. E., Jr., and Marschall, R. (1985). Potassium accumulation in intercellular junctions combined with electric field interactions for propagation in cardiac muscle. *Innov. Tech. Biol. Med.* **6(1)**, 24–43.

Sperelakis, N., Picone, J. B., and Mann, J. E., Jr.(1989). Electric field model for electric interactions between cells: an alternative mechanism for cell-to-cell propagation. *In* "Cell Interactions and Gap Junctions" (N. Sperelakis and W. C. Cole, Eds.), Vol. II, pp. 191–208. CRC Press, Boca Raton, FL.

Sperelakis, N., Rollins, C., and Bryant, S. H. (1990). An electronic analog simulation for cardiac arrhythmias and reentry. *J. Cardiovasc. Electrophysiol.* **1**, 294–302.

Sperelakis, N., Ortiz-Zuazaga, H., and Picone, J. B. (1991). Fast conduction in the electric field model for propagation in cardiac muscle. *Innov. Tech. Biol. Med.* **12(4)**, 404–414.

Tarr, M., and Sperelakis, N. (1964). Weak electronic interaction between contiguous cardiac cells. *Am. J. Physiol.* **207**, 691–700.

Richard D. Veenstra

Appendix 2: Derivation of the Cable Equation and the AC Length Constant

I. Membrane Properties

The biological membrane has both resistive and capacitive elements associated with it due to the ion transporters and channels that permit the translocation of electrical charge across the membrane lipid bilayer that separates the cytoplasm from the extracellular fluid. Fig. A-5 illustrates an axon and the equivalent electrical circuit representation in a manner analogous to Fig. 3 in Chapter 24. For a uniform linear cable of infinite length ($>5\lambda$), the voltage (V_x) at distance x from the point source of the applied voltage (V_o) decays exponentially according to the expression $V_x = V_o e^{-x/\lambda}$ where λ is the length constant for the axon (see Eq. 1 in Chapter

24). Since the equivalent membrane circuit at any point also has a capacitive element (c_m) in parallel with the membrane resistance (r_m), the voltage at point x will decay following the termination of a long-duration applied voltage ($>5\tau$) in proportion to the value of $r_m c_m$ or time constant for the membrane. Another important concept that is not intuitively obvious from the linear cable diagram is that r_m and the internal resistance (r_i) are not constants, but depend inversely on the radius of the axon. This is because, in simplest terms, two equivalent resistors in parallel yield half the resistance of their initial value (like two traffic lanes can carry twice as many cars as one, all else being equal). The membrane resistance varies according to the surface area of the membrane while the extracellular (r_o) and cytoplasmic resistances vary according to the volume of the extracellular fluid or cytoplasm. For a simple right cylinder as illustrated in Fig. A-5, the circumference of the axonal membrane is equivalent to $2\pi a$ and the cross-sectional area is equivalent to πa^2, where a is the radius of the axon. It follows that for a cylindrical axon, the surface to volume ratio (S_V) is proportional to $2/a$. Resistivity (R) is classically defined as the resistance of a 1 cm^2 area of membrane. For these reasons the extracellular and internal resistivities, R_o and R_i, have units of $\Omega \cdot$cm. Since r_m varies inversely with membrane surface area, $r_m = R/$area and the membrane resistivity has units of $\Omega \cdot$cm^2. Every centimeter of membrane also has a specific membrane capacitance, C_m, equivalent to 1 μF/cm^2 for biological lipid bilayers. Hence, C_m is expressed in units of F/cm^2 since the capacity increases in direct proportion to the area. For a right cylindrical cable, the values of the resistance and capacitive elements at any point x are related to their specific resistivities and capacitance by the following expressions:

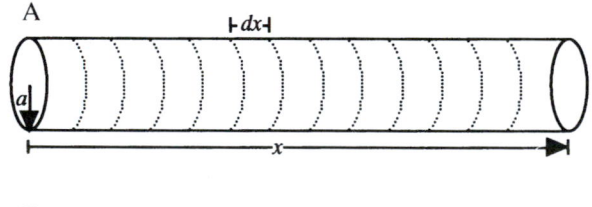

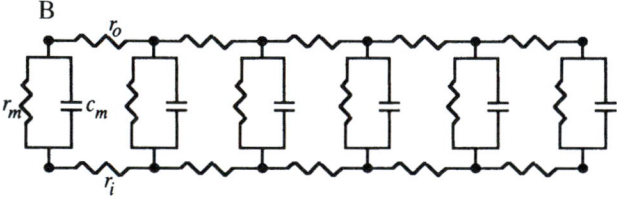

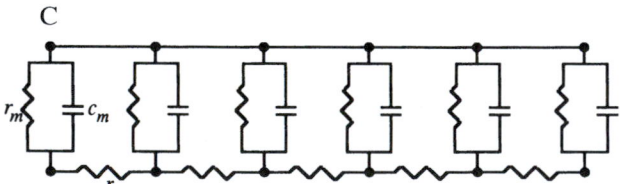

FIGURE A-5. Diagram of a linear cable. (A) Cylindrical model of a linear cable of length x and radius a. In order to derive equations for axial and transverse current flow, the cable is evaluated in small increments of distance (dx) along its entire length. (B) Equivalent electrical circuit for the cable illustrated in A with an axial internal core resistance, r_i, an external resistance, r_o, and a parallel resistor-capacitor, ($r_m c_m$) circuit which represents a biological membrane. (C) The equivalent circuit for the cable assuming there is no external voltage gradient (i.e., $r_o = 0$, no resistance to external current flow).

$$r_m \,(\text{in } \Omega \cdot \text{cm}) = R_m / 2\pi a \qquad (A\text{-}1)$$

$$r_i \,(\text{in } \Omega/\text{cm}) = R_i / \pi a^2 \qquad (A\text{-}2)$$

$$r_o \,(\text{in } \Omega/\text{cm}) = R_o / \text{area}_o \qquad (A\text{-}3)$$

and

$$c_m \,(\text{in } \mu\text{F/cm}) = 2\pi a C_m \qquad (A\text{-}4)$$

The actual value for any length dx of the cable can be determined by r_m/dx, $r_i \, dx$, $r_o \, dx$, and $c_m \, dx$, assuming that the radius a remains constant. The volume of the extracellular space is usually considered to be infinite, so the value of R_o

is determined only by the ionic composition of the extracellular fluid. This latter assumption is not true when restricted extracellular spaces are encountered (e.g., tight junctions or tortuous extracellular pathways in bundles of packed fibers). In this case, significant barriers to ionic diffusion are encountered in the extracellular space and the geometry of the intercellular space must be considered. Commonly, the external resistance is assumed to be negligible and r_o is omitted from further consideration. For now, we will consider the first case when $r_o = 0$ (Fig. A-5C).

II. Linear Cable Theory

A. Passive Cable Properties

From Ohm's law, we know that $V = IR$, so in order for a voltage to develop across the membrane (V_m) there must be current flowing across the membrane. For a point source of constant current (DC = direct current) injection at the center ($x = 0$) of an infinitely long cable, V_m will decay uniformly along the length of the cable in both directions. At each point along the cable, the membrane current, i_m, will generate a membrane voltage, V_m, equivalent to $i_m r_m$, or $i_m = V_m/r_m$. Since each segment of membrane has a capacitive element, there must also be a capacitive current (i_c) component equivalent to $c_m\, dV_m/dt$. That is, the total membrane current (I_m) is the sum of the ionic current (I_i) through the resistance and the capacitive current (I_c) through the capacitance: $I_m = I_i + I_c$. Hence,

$$I_m = \left(V_m/r_m\right) + c_m\, dV_m/dt \tag{A-5}$$

However, V_m cannot be constant along the entire length of the cable, because the internal axial current flow, i_i, would have to remain constant along the entire cable in order for this to occur; that is, there must no leakage of internal axial current across the membrane along the length of the cable. Simply put,

$$I_m = -i_i/dx \tag{A-6}$$

where dx is an infinitesimally small distance along the length of the cable. It follows that the change in V_m over the distance dx is $dV_m/dx = -i_i r_i$. Combining this expression with Eqs. A-5 and A-6 gives

$$I_m = \left(1/r_i\right) d^2V_m/dx^2 \tag{A-7a}$$

$$= \left(V_m/r_m\right) + c_m\, dV_m/dt \tag{A-7b}$$

This is referred to as the linear **cable equation.** Multiplying both sides of the equation by r_m yields the following expression:

$$\left(r_m/r_i\right) d^2V_m/dx^2 = V_m + \left(c_m r_m\right) dV_m/dt \tag{A-8}$$

From this version of the cable equation, it is apparent that the value of V_m at the point $x + dx$ (dV_m/dx) is directly proportional to the square root of r_m/r_i. The value of $\sqrt{r_m/r_i}$ is referred to as the **length constant** of the cable:

$$\lambda = \sqrt{r_m/r_i} \tag{A-9a}$$

These expressions refer to the passive spread of voltage along the length of a linear cable. It should now become ap-

parent that an infinite length of cable is considered to be $> 5\lambda$ since $V_x/V_o < 1\%$ when $x \geq 5\lambda$. Other conditions will alter the value of λ. For example, when $r_o > 0$,

$$\lambda = \sqrt{\frac{r_m}{r_i + r_o}} \tag{A-9b}$$

Also, cardiac muscle fibers, unlike axons or skeletal muscle fibers, do not possess a continuous cytoplasm. Instead, each cardiac cell is coupled to its neighbor by gap junctions that add to the internal resistance of the muscle fiber. Under these conditions,

$$\lambda = \sqrt{\frac{r_m}{r_i + r_j + r_o}} \tag{A-9c}$$

It is also apparent that the change in voltage with respect to time at the point $x + dx$ is directly proportional to ($c_m r_m$), which is defined as the membrane **time constant:**

$$\tau_m = c_m r_m \tag{A-10}$$

For the same reasons given for considering an infinite length of cable to be $> 5\lambda$, an infinitely long duration current pulse is considered to be $> 5\tau$.

B. Active Cable Properties: Action Potential Propagation

The previous conditions apply to the resting muscle fiber or nerve axon, provided that the fiber or axon is several λ long and its geometry approximates a linear cable (e.g., no branch points for axial current flow). However, conduction of the action potential involves the spread of current flow along a passive membrane cable segment from an advancing active membrane segment. The transition from a passive to an active membrane response (i.e., the initiation of an action potential) requires that a certain threshold voltage level be achieved in order for the segment of membrane undergoing a voltage change in response to passive current spread (as defined by the cable equation in Section II.A) to generate an active response. We will see later that the excitation threshold is not actually a voltage value; rather, it is equivalent to the amount of charge required to depolarize the membrane and produce an active current response. During the action potential, the active portion of the membrane becomes a source of depolarizing current which contributes to the distal spread of the voltage deflection. r_m, or $1/g_m$, is actually the sum of several ionic components resulting from the presence of specific proteins (ion channels and pumps) that permit the selective movement of ions across the membrane. These discrete resistive membrane elements are associated with the activity of ion channels (e.g., Na^+, K^+, Ca^{2+}, and Cl^- channels) and pumps (e.g., Na^+,K^+-ATPase, Na^+-Ca^{2+} exchanger) associated with the cell membrane. Since each ion possesses a net charge (valence), the translocation of an ion across the membrane produces a current (I_i). The activities of the protein channels and transporters are not constant, but are regulated by a variety of mechanisms including membrane voltage (V_m). Hence, at different points in time (dt), r_m (and R_m) is not constant, but rather is a function of voltage and time, $r_m(V,t)$. So a voltage deflection dV_m alters r_m, which further affects V_m. This constitutes a feedback loop where r_m and V_m are dependent variables (see Chapter 24).

In a passive cable, any voltage deflection caused by the decrement in axial current flow in that particular segment of membrane is initially opposed by existing membrane currents, which counteract the voltage deflection and maintain the resting (passive) state. This is a **negative feedback** mechanism and is predominant at all times except during an action potential. When the voltage deflection exceeds the **threshold** for excitation, certain ion channels are activated, which leads to the generation of more ionic current and a greater voltage deflection in the same direction as the initiating response. Hence, the voltage response of the membrane becomes additive to the initial passive dV_m and a **positive feedback loop** is transiently produced. If allowed to persist, a positive feedback loop leads to prolonged excitation, which prevents the return to the resting state and the generation of any further action potentials.[1] Therefore, the positive feedback loop, in order to be effective in generating an action potential, must also be terminated once that action potential is initiated.

The threshold for excitation is the critical value of V_m (V_{th}) that must be achieved in order for the active membrane response to commence. However, as observed from the equivalent circuit and cable equation, there are both resistive and capacitive elements to the membrane current. Current is also equivalent to the derivative of charge flowing per unit time (e.g., coulomb per second). Hence, the resistive and capacitive current components of i_m can be expressed in terms of dQ/dt, where Q is the charge in coulombs. From Faraday's law applied to our segment of membrane,

$$Q_m = C_m V_m \tag{A-11}$$

So the threshold for excitation is actually the amount of charge (Q_{th}) that must flow across the membrane to achieve the required membrane voltage (V_{th}) to initiate the action potential. Once initiated, the action potential propagates down the length of the cable at a rapid rate. The **conduction velocity** (θ) is the distance traveled per unit time, or dx/dt. Returning to our cable equation (Eq. A-8) and solving for dx/dt, we obtain

$$\lambda^2 (1/\theta^2)\, d^2V_m/dt^2 = V_m + \tau_m\, dV_m/dt \tag{A-12}$$

It is now evident that the rate at which the action potential propagates along the length of passive cable elements depends upon the length and time constants of the membrane or

$$\theta^2 \propto \lambda/\tau \tag{A-13}$$

Now we have successfully related the cellular membrane properties (r_m, r_i, c_m, etc.) to the passive cable properties (λ, τ), and subsequently to the active cable properties (θ) involved in the conduction of an action potential.

To summarize, the cable equation can be expressed in several equivalent forms:

---

[1] One example of overamplification and positive feedback is the annoying hum of the microphone heard over the audio speaker system of any soundstage production.

$$I_m = (V_m/r_m) + c_m\, dV_m/dt \tag{A-5}$$

$$I_m = (1/r_i)\, d^2V_m/dx^2 \tag{A-7a}$$

$$(r_m/r_i)\, d^2V_m/dx^2 = V_m + (c_m r_m)\, dV_m/dt \tag{A-8}$$

$$\lambda^2 (1/\theta^2)\, d^2V_m/dt^2 = V_m + \tau_m\, dV_m/dt \tag{A-12}$$

Another useful derivation of the cable equation provides insight into how cell size, or more specifically cell surface/volume ratio (S_V), influences propagation (Joyner *et al.*, 1983):

$$\frac{r_m}{R_i + S_v}\frac{1}{\theta^2}\, d^2V_m/dt^2 = V_m + c_m r_m\, dV_m/dt \tag{A-14}$$

Several additional derivations are possible which depend on the boundary conditions applied to the cable such as length, extracellular isopotentiality ($r_o = 0$), voltage uniformity (i.e., voltage-clamp), and uniformity of geometry (e.g., fiber radius a or branch points).

C. Finite Spatial and Temporal Responses

1. Uniform Cable

Let us consider further our linear cable of uniform diameter $2a$, length x, and external resistance $r_o = 0$. Throughout the passive cable analysis described above we have only considered the case where the voltage displacement along the length of this cable is produced by a constant current (DC) source, I_0, located at the exact center ($x = 0$) of the cable. We have shown that $\lambda^2 = r_m/r_i$ or

$$\lambda^2 = \frac{(R_m/2\pi a)}{(R_i/\pi a^2)} \tag{A-15a}$$

$$= \frac{aR_m}{2R_i} \tag{A-15b}$$

or

$$\lambda = \sqrt{\frac{R_m}{R_i}\frac{a}{2}} \tag{A-15c}$$

and that $\tau_m = c_m r_m$ (Eq. A-10) or

$$\tau_m = (R_m/2\pi a)(2\pi a C_m) \tag{A-16a}$$

$$= R_m C_m \tag{A-16b}$$

This implies that λ^2 (and consequently θ^2) $\propto (a/2)$, whereas τ_m is independent of fiber radius and membrane area (provided R_m is constant). Under steady-state conditions $dV_m/dt = 0$, so the cable equation becomes

$$0 = \lambda^2\, d^2V_m/dx^2 - \tau_m\, dV_m/dt - V_m \tag{A-17a}$$

which further reduces to

$$0 = \lambda^2\, d^2V_m/dx^2 - V_m \tag{A-17b}$$

To solve this differential equation, let us define $X = x/\lambda$ (Taylor, 1963; Jack *et al.*, 1983). It follows that $dX/dx = 1/\lambda$, so $\lambda^2 =$

$(dx/dX)^2$ and the equation becomes $0 = d^2V_m/dX^2 - V_m$. The solution to dX/dx is e^X and $1/(dX/dx) = e^{-X}$. The solution to our differential equation therefore becomes

$$V_m = V_o e^{-X} \qquad (A\text{-}18a)$$

or

$$V_m = V_o e^{-x/\lambda} \qquad (A\text{-}18b)$$

where V_o = the value of V_m at $x = 0$. As $x \to 0$, $V_m \to V_o$ and spatial uniformity ($V_m = V_o$) of membrane voltage is approached. This is identical to Eq. 1 in Chapter 24.

Another consideration is the time required for the membrane to decay from its steady-state voltage once a current stimulus is terminated. Again,

$$0 = \lambda^2 \, d^2V_m/dx^2 - \tau_m \, dV_m/dt - V_m \qquad (A\text{-}17a)$$

except in this case $d^2V_m/dx^2 = 0$, since we are observing the same point x at different times during the current injection and $dV_m/dt \neq 0$. Now our equation reduces to

$$0 = \tau_m \, dV_m/dt + V_m \qquad (A\text{-}19)$$

which is again a differential equation. Let us define $T = 1/\tau_m$, and it follows that $dT/dt = 1/\tau_m$ (Taylor, 1963; Jack *et al.*, 1983). Our differential equation takes the form

$$0 = dV_m/dT + V_m \qquad (A\text{-}20)$$

which has the solution

$$V_m = V_o e^{-t/\tau} \qquad (A\text{-}21)$$

where V_o = the value of V_m when $t = 0$. This is identical to Eq. 6 in Chapter 24. In conclusion, for a uniform fiber, V_m decays over distance and time by a first-order exponential decay process.

2. Nonuniform Cable

These solutions to the differential equations apply only to the specified conditions described above. For a nonuniform cable (but r_o still $= 0$), the generalized cable equation $0 = d^2V_m/dX^2 - dV_m/T - V_m$ will have different complex solutions. The solutions to several different boundary conditions are derived in Rall (1977). *In situ*, the condition of voltage uniformity does not apply to a muscle fiber or nerve axon due to a large r_i. Hence, for a constant current injection (I_o) at $X = 0$ and time T, the cable equation has the general solution of (Hodgkin and Rushton, 1946)

$$V_m = \left(\frac{r_i I_0 \lambda}{4}\right) \left\{ e^{-X}\left[1 - \mathrm{erf}\left(\frac{X}{2\sqrt{T}} - \sqrt{T}\right)\right] \right.$$
$$\left. - e^X \left[1 - \mathrm{erf}\left(\frac{X}{2\sqrt{T}} + \sqrt{T}\right)\right] \right\} \qquad (A\text{-}22)$$

For $X = x$ and $t = \infty$, this equation reduces to

$$V_m = V_o e^{-x/\lambda} \qquad (A\text{-}18b)$$

which is the same as the solution derived for the decay in V_m as a function of distance. However, the special solution for V_m as a function of time at $x = 0$ has the solution

$$V_m = V_o \,\mathrm{erf}\sqrt{t/\tau} \qquad (A\text{-}23)$$

which differs from the previous derivation of the time-dependent voltage response for a spatially uniform cable. Since erf (1) = 0.84, when $t = \tau$, $V_m = 0.84V_o$ for the nonuniform fiber or axon instead of $0.63V_o$ in the case of the uniform cable (Eq. 24, Jack *et al.*, 1983). This means that V_m will rise and fall faster in a long fiber with a point source of current injection than in a short cable where spatial voltage uniformity is approximated ($x/\lambda \to 1$) or a long fiber made short by injecting current uniformly over distance x. Hence, under non–voltage-clamp conditions, the 84% time provides the measure of $r_m c_m$, not the 63% time as indicated in Eq. A-21 (Jack *et al.*, 1983).

D. Input Resistance

The magnitude of the steady-state membrane voltage deflection produced by a DC current source will be determined by Ohm's law, $V_o = I_o R_{in}$ where the **input resistance** (R_{in}) of the cable represents the effective (net) resistance of the preparation. The input resistance is dependent on the geometry of the preparation and is not a direct measure of r_m unless the preparation is under the conditions of spatial voltage uniformity and leakage around the current and/or voltage electrodes is negligible. R_{in} is best defined as the resistance of the preparation resulting from the internal resistance and membrane resistance over the length of the fiber. For intracellularly injected current, all current must flow across the internal resistance and will proportion itself between axial and membrane current flow according to the length constant as previously derived. Since half of the injected current will flow in either direction down the long axis of the cable,

$$V_o = r_i I_o \lambda/2 \qquad (A\text{-}24)$$

where $R_{in} = r_i \lambda$, or

$$R_{in} = \sqrt{r_m r_i}/2 \qquad (A\text{-}25a)$$

$$= \frac{1}{2}\sqrt{\frac{R_m R_i}{2\pi^2 a^3}} \qquad (A\text{-}25b)$$

Hence, the input resistance of a fiber is proportional to $\sqrt{1/a^3}$ (inversely proportional to the square root of the volume). With these derivations it should be apparent how the length constant, time constant, voltage, conduction velocity, and input resistance of a one-dimensional fiber or axon are related to the parameters of radius, membrane resistance, axial resistance, and current. Other derivations are based on different boundary conditions (e.g., Rall, 1977).

E. Alternating Current

When an **alternating current (AC)** is applied to an RC circuit, the solution is more complex. Capacitive current, I_c, is $C_m \, dV/dt$, and for a sine wave, $dV/dt = 0$ when the current is at its maximum (I_p). Simply stated, the voltage across the capacitor (V_c) lags behind the current across the resistor by 90°. The current across the resistor is now a sine wave and takes the form of $V_p \sin(\omega t)/r_m$ where V_p = peak voltage of the sine wave. Since $dV_p \sin(\omega t)/dt = V_p\omega \cos(\omega t) = V_p\omega \sin(\omega t - 90°)$, the current equation (Eq. A-5) for our circuit becomes

$$I_p = V_p \sin(\omega t)/r_m + c_m V_p \omega \sin(\omega t - 90°) \quad \text{(A-26)}$$

where $\omega = 2\pi f$ and f = frequency of the sine wave (in Hz). Ohm's law states that $V = IR$, so it follows that $1/\omega C$ has units of resistance (Ω). Let us define the capacitive **reactance,**

$$X_c = -1/\omega C \quad \text{(A-27)}$$

Let us further define **impedance,**

$$Z = R + jX \quad \text{(A-28)}$$

where R is the purely resistive component, X is the reactance of the circuit, and $j = \sqrt{-1}$. Actually, the membrane should always be thought of as having an impedance. In the DC case, $X_c = 0$ because $\omega = 0$ (since there is no frequency component) and the membrane exhibits only a resistance. Since the axial and external resistances in our linear cable do not contain a capacitive component, R_i and R_o are unaffected. Recall that the steady-state solution for voltage as a function of distance x is

$$V_m = V_o e^{-x/\lambda} \quad \text{(A-18b)}$$

where $V_o = (r_i I_o \lambda)/2$, and $\lambda = \sqrt{r_m/r_i}$. We can rewrite Eq. A-18b by substituting for V_o and λ in the above expression, thus obtaining

$$V_m = \left(I_o/2\right)\sqrt{r_m r_i}\, e^{-x/\sqrt{r_m/r_i}} \quad \text{(A-29a)}$$

Substituting z_m for r_m in the above expression results in the following expression:

$$V_m = \left(I_o/2\right)\sqrt{z_m r_i}\, e^{-x/\sqrt{z_m/r_i}} \quad \text{(A-29b)}$$

The **membrane impedance,** z_m, is a complex number consisting of **real** (resistance) and **imaginary** (reactance) components. So the above expression should be further subdivided into real and complex numbers. Let us define a **propagation constant,** γ, which is a complex number composed of an **attenuation factor,** α, and a **phase constant,** β, such that

$$\gamma = \alpha + j\beta \quad \text{(A-30)}$$

The expression for the peak voltage of our circuit is

$$V_p = \left(i_p/2\right)\sqrt{|z_m| r_i}\, e^{-\gamma x} \quad \text{(A-31)}$$

where $|z_m|$ is the magnitude of the impedance, $\sqrt{R^2 + X^2}$ (Eisenberg and Johnson, 1970; see also Chapter 70). It follows from Eq. A-31 that $\lambda_{AC} = 1/\gamma$. It is easier to solve for the above expression by assuming the frequency of the injected current is high and $z_m \approx jX_c = j/\omega c_m$. This occurs when $f > 1/(2\pi r_m c_m)$. $X_c = r_m$ when $f = 1/(2\pi r_m c_m) = f_b$, where f_b is the **cut-off** or **corner frequency** for the circuit. Equation A-31 now becomes

$$V_p = \left(i_p/2\right)\sqrt{jr_i/\omega c_m}\, e^{-\gamma x} \quad \text{(A-32)}$$

and $\gamma = 1/\lambda = 1/\sqrt{z_m/r_i} = 1/\sqrt{j/\omega c_m r_i} = \alpha + j\beta$. Solving for α and β, we obtain

$$\alpha = 1/\sqrt{2/\omega c_m r_i} = \sqrt{\omega c_m r_i/2} \quad \text{(A-33)}$$

and

$$j\beta = j\sqrt{\omega c_m r_i/2} \quad \text{(A-34)}$$

Since $\lambda_{AC} = 1/\gamma$, it follows that $\lambda_{AC} = 1/\alpha = \sqrt{2/\omega c_m r_i}$ or

$$\lambda_{AC} = \sqrt{2/2\pi f c_m r_i} = \sqrt{1/\pi f c_m r_i} \quad \text{(A-35)}$$

When $f < f_\beta$, $r_m > X_m$, and $z_m \approx r_m$, so $\lambda = \sqrt{r_m/r_i}$, which is equivalent to the DC case. The net effect of the frequency dependence of λ_{AC} is that the length constant becomes shorter at higher frequencies (e.g., instantaneous steps in membrane voltage).

There is also an analogous expression for the **input impedance** of the cable that is derived from the previous expression for the **input resistance:**

$$R_{in} = \sqrt{r_m r_i}\,/2 = \frac{1}{2}\sqrt{\frac{R_m R_i}{2\pi^2 a^3}} \quad \text{(A-25a, b)}$$

The resistance and capacitor in series form a low-pass filter as we have already observed in Eqs. A-21 and A-23. When $f < 1/(2\pi RC)$, the voltage output (V_{out}) relative to the voltage input (V_{in}) (in response to an AC source) is unity. The ratio of V_{out}/V_{in} is defined as the **gain** of the circuit and is equivalent to

$$V_{out}/V_{in} = 1/\sqrt{\omega^2 R_m^2 C_m^2 + 1} = 1/\sqrt{4\pi^2 f^2 R_m^2 C_m^2 + 1} \quad \text{(A-36)}$$

since the magnitude of the output voltage is determined by the expression

$$V_{out} = (V_{in}|X_c|/Z_m) \quad \text{(A-37)}$$

The equivalent expression for the input impedance of the cable becomes

$$Z_{in} = \frac{1}{2}\sqrt{\frac{R_m R_i}{2\pi^2 a^3\left(4\pi^2 f^2 R_m^2 C_m^2 + 1\right)}} \quad \text{(A-38)}$$

This equation was originally derived by Falk and Fatt (1964) in their analysis of frog sartorius muscle fiber linear cable properties. The expression is again divided by two since the source of current injection was the center of the cable and current will flow in both directions along the long axis of the fiber.

Bibliography

Eisenberg, R. S., and Johnson, E. A. (1970). Three-dimensional electrical field problems in physiology. *Prog. Biophys. Molec. Biol.* **20,** 5–65.

Falk, G., and Fatt, P. (1964). Linear electrical properties of striated muscle fibres observed with intracellular electrodes. *Proc. R. Soc. Lond. B* **160,** 69–123.

Hodgkin, A. L., and Rushton, W. A. H. (1946). The electrical constants of a crustacean nerve fibre. *Proc. R. Soc. Lond. B.* **133,** 444–479

Jack, J. J. B., Noble, D., and Tsien, R. W. (1983). "Electric Current Flow in Excitable Cells." Clarendon Press, Oxford.

Joyner, R. W., Picone, J., Veenstra, R., and Rawling, D. (1983). Propagation through electrically coupled cells. Effects of regional changes in membrane properties. *Circ. Res.* **53,** 526–534.

Rall, W. (1977). Core conductor theory and cable properties of neurons. *In* "Handbook of Physiology" (J. M. Brookhart and V. B. Mountcastle, Eds.), Vol. 1, pp. 39–97. American Physiological Society, Bethesda, MD.

Taylor, R. E. (1963). Cable theory. *In* "Physical Techniques in Biological Research" (W. L. Nastuk, Ed.), Vol. 6B, pp. 219–262, Academic Press, New York.

Nicholas Sperelakis

25

Electrogenesis of Membrane Excitability

I. Introduction

Excitability is an intrinsic membrane property that allows a cell to generate an electrical signal or action potential (AP) in response to stimuli of sufficient magnitude. The elongated nerve axon serves to transmit information in the form of APs over long distances. The AP mechanism is required to propagate a uniform signal in a nondecremental manner. In muscle cells, the AP serves to spread excitation over the entire cell surface and is involved in triggering cell contraction.

The energy source for the generation of the AP is stored in the excitable cell itself. An initial depolarization is produced by a stimulus and triggers the intrinsic AP mechanism. The immediate source of energy for the AP comes from the transmembrane ionic gradients for K^+ and Na^+, which act like a battery. The K^+ ion concentration gradient is mainly responsible for the generation of the resting potential, which causes an excess of negative charge to build up on the inner surface of the membrane. Upon depolarization to threshold, the Na^+ ion electrochemical driving force, which is directed inward, causes a large and rapid inward Na^+ current that generates the AP upstroke. Over a longer time frame, the Na^+-K^+ pump is responsible for the generation of the Na^+ and K^+ ionic gradients and for their maintenance and restoration after repetitive AP activity. The Na^+-K^+ pump derives chemical energy from the hydrolysis of ATP. The bases of the resting potential and active ion transport have been discussed in earlier chapters.

Important technological improvements have led to significant advances in the understanding of the basis of membrane excitability. In the early 1900s, several theories were proposed to explain the mechanism that produces the AP. Julius Bernstein (1902, 1912) proposed that the excitable cell membrane, at rest, was selectively permeable to K^+ ions (producing the resting potential), and that during excitation the membrane became permeable to all ions (producing the AP). About the same time, Overton (1902) had demonstrated that Na^+ ions were essential for excitability.

By the 1940s, improvements had been achieved in electronic instrumentation, especially in the high-input impedance amplifiers necessary to record bioelectric phenomena using tiny intracellular microelectrodes. In addition, biophysicists began to study the squid giant axon (500–1000 μm in diameter), which permitted insertion of relatively large intracellular electrodes, yielding the first measurements of the true transmembrane potential. The transmembrane potential is recorded as the difference between an intracellular and an extracellular electrode (Fig. 1). The findings from the squid giant axon were successfully applied to the smaller diameter (1–20 μm) neurons found in the vertebrate nervous system (Fig. 2).

II. Action Potential Characteristics

Action potentials in a given fiber (e.g., myelinated nerve fiber) have the properties of being all-or-none, having a sharp threshold, and having a refractory period. All impulses look alike, being very similar in shape, amplitude, and duration. Thus, they constitute a digital system. Strength of the signal is conveyed by changing the frequency of the impulses, as discussed in the preceding chapter on propagation. The all-or-none property means that the single signal is not graded in amplitude (i.e., not an amplitude-modulated system) and that the signal is either zero ("off" or "no") or maximum ("on" or "yes"). These characteristics of the AP are discussed in the next sections.

A. Local-Circuit Currents

During propagation of an AP down a nerve fiber, current flow accompanies the propagating change in membrane voltage. This current is called the **local-circuit current**, and has both longitudinal and radial (transverse) components that make a complete circuit (Fig. 3A, B). Propagation and local-circuit currents have been thoroughly discussed in the preced-

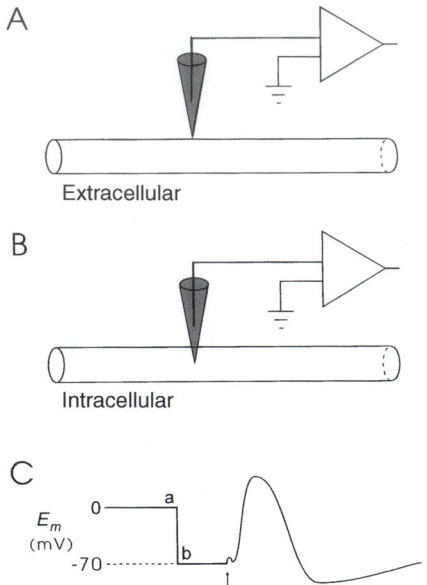

FIGURE 1. Recording the transmembrane potential of a squid giant nerve axon. The transmembrane potential is measured as the potential difference between the intracellular and extracellular electrodes. (A) The microelectrode is outside the axon, and measures 0 mV (segment a in C). (B) As the electrode is advanced and crosses the membrane, the resting potential of −70 mV is measured (segment b in C). The cell membrane makes a tight seal around the glass electrode. (C) The membrane potential (E_m) is depicted. Stimulation (at arrow) elicits an action potential.

ing chapter. For our purpose here, we focus on the longitudinal component. The intracellular and the extracellular (external) longitudinal current are exactly equal in amplitude, but flow in opposite directions (Fig. 3A, B). The intracellular current is, of course, confined to the cross-sectional area of the nerve fiber (neuroplasm), whereas the extracellular current can use the entire extracellular volume (so-called **volume conductor**). If a single nerve fiber or nerve bundle is mounted in air, then the extracellular action current is confined to the surface film of fluid adhering to the single fiber or the interstitial fluid space between fibers in the bundle.

To illustrate the principles involved, let us record externally with a pair of electrodes from a single nerve fiber (e.g., a

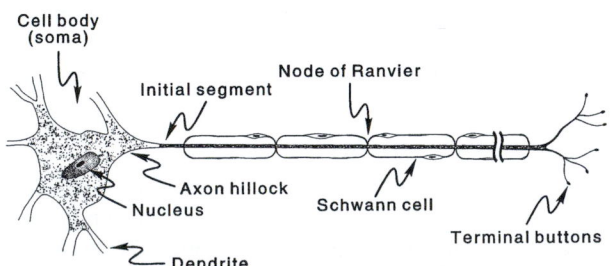

FIGURE 2. Diagram of a motor neuron with a myelinated axon. The major structural elements are diagrammed, including the cell body, initial segment, Schwann cell, and node of Ranvier.

squid giant axon) bathed in air (Fig. 3C). As the AP moves from left to right past the pair of recording electrodes, the first electrode becomes negative with respect to the second electrode. Therefore, the voltmeter swings in one direction (an upward deflection is defined as negative in extracellular recording) due to the potential difference (pd). When the wavefront moves to a position between the two electrodes, the voltmeter will record almost zero voltage. When the AP reaches the second electrode, the voltmeter will swing in the opposite direction, because the second electrode is now negative with respect to the first. Thus, the electrodes will record a biphasic change in voltage as the AP sweeps past. Thus, when recording externally, the AP is often described as a wave of negativity sweeping down the fiber; when recording internally, the AP is a wave of positivity.

The reason for the wave of negativity in external recording is that the fluid between the two electrodes constitutes a resistance (R), and longitudinal current (I) flowing through this fluid resistance produces an IR voltage drop (Ohm's law: $V = IR$). As the wavefront approaches the first electrode, the external local-circuit current is from right to left and produces an IR drop, positive (second electrode) to negative (first electrode). Conversely, when the AP passes beyond the first electrode, the action current is now reversed in direction and produces an IR drop, positive (first electrode) to negative (second electrode). This produces the biphasic voltage recording.

Propagation velocity is a function of axon diameter and myelination, such that the larger the diameter, the greater the velocity. **Myelinated axons** conduct much faster than **nonmyelinated axons**. The factors that determine propagation velocity and the mechanism of saltatory conduction are discussed in the preceding chapter.

B. Threshold and All-or-None Property

Nerve membrane responses near the site of application of brief current pulses vary depending on the magnitude and direction of the pulses (Fig. 4). Anodal (inward) currents produce hyperpolarization, and cathodal (outward) currents produce depolarization. Hyperpolarizing and subthreshold depolarizing responses are graded in magnitude according to the stimulus current. However, as can be seen, a somewhat higher intensity outward current produces a depolarizing response with a different waveform and a longer lasting duration. This condition is referred to as the **local excitatory state** (Fig. 4A, B). It occurs when a small area of the membrane near the stimulus electrode comes close to threshold, but it does not generate an AP. This local membrane activity is not propagated and decays with distance along the axon. A slightly stronger stimulus is, however, sufficient to bring the membrane potential of a large enough membrane area to the threshold potential, thereby initiating an all-or-none AP. At the threshold potential, there is a greater amount of inward (depolarizing) than outward (repolarizing) current, and the membrane will continue to depolarize.

In regard to the applied stimulus, outward current depolarizes the membrane (IR drop across the resting membrane is positive inside, negative outside), whereas inward current hyperpolarizes (IR drop is negative inside).

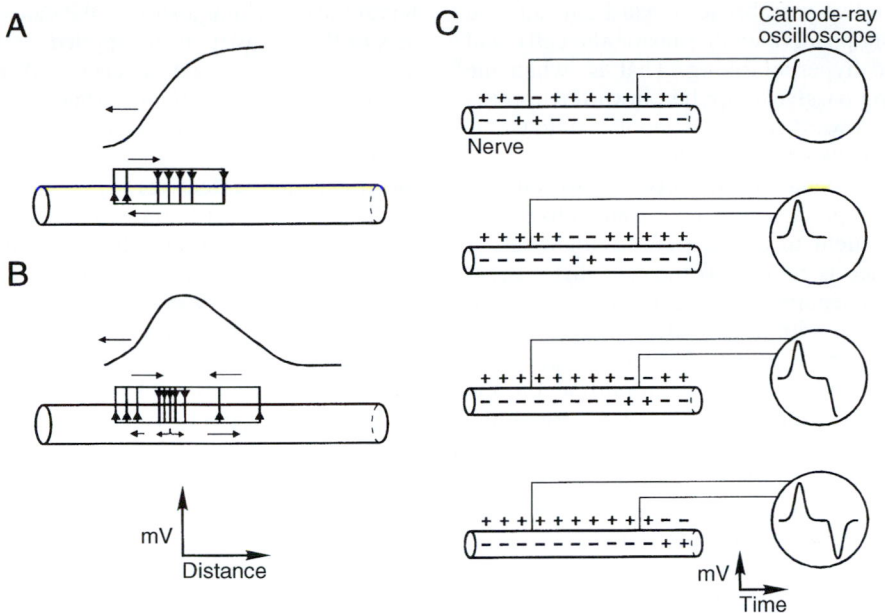

FIGURE 3. Diagram of the local-circuit current in a nerve fiber during an AP. (A) The current during the rising phase. (B) The current during the entire AP. The AP is depicted as propagating from right to left. (C) The actual registration of the signal is depicted for an AP propagating from left to right past a pair of electrodes spaced relatively far apart.

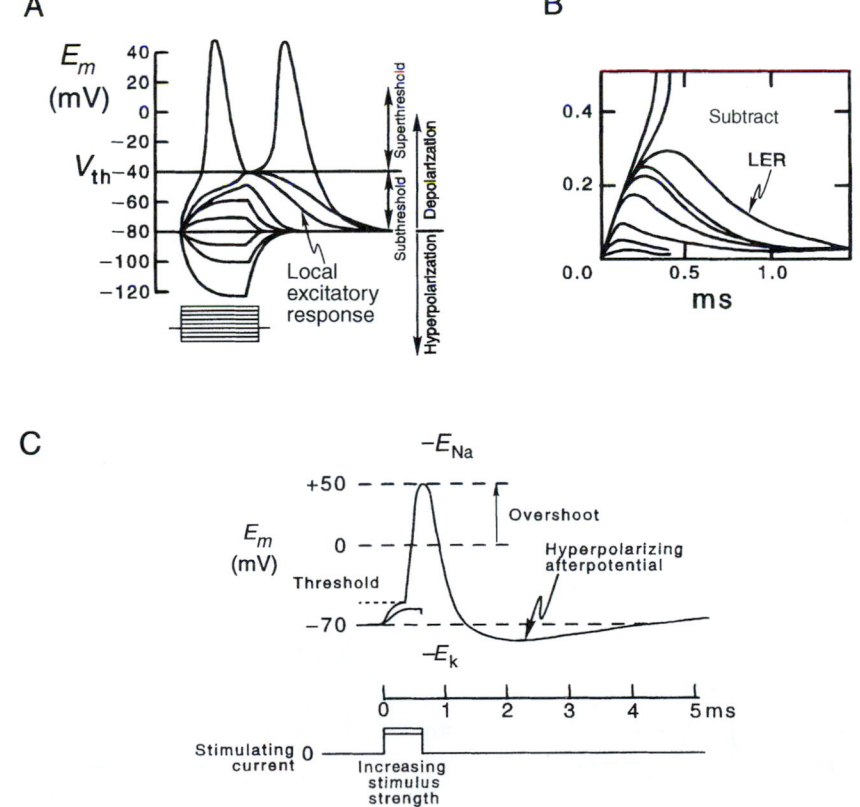

FIGURE 4. Diagrammatic sketches depicting initiation of the nerve impulse by membrane depolarization. (A) Membrane responses to depolarizing and hyperpolarizing current pulses are shown. The nonlinear local excitatory response (LER) occurs just below the threshold for the all-or-none AP. (B) Illustration of the LER obtained by subtraction of the membrane responses to hyperpolarizing current pulses (passive only) from the responses to depolarizing pulses (passive plus active). (C) A nerve AP is depicted, illustrating the sharp threshold, overshoot, and hyperpolarizing afterpotential.

In contrast, for the active membrane, inward currents are depolarizing (bringing positive charge inside the cell), and outward currents are hyperpolarizing. That is, when the membrane is behaving passively, applied inward current is hyperpolarizing and outward current is depolarizing; this is equivalent to a circuit external to a battery. In contrast, when the membrane is behaving actively, inward current generated is depolarizing, and outward current is hyperpolarizing; this is equivalent to an internal circuit within a battery. This difference is related to the fact that current flows from positive to negative in the external circuit, and from negative to positive within the battery itself.

A subthreshold depolarization is defined as one that does not reach threshold to elicit an AP. It is proportional to the applied stimulus, and it decrements with distance along the nerve axon cable. If a stimulus current is of sufficient magnitude (i.e., enough positive charges are transferred into the cell), then the resulting membrane depolarization reaches a critical value, called the **threshold potential**, at which an AP is initiated (see Fig. 4).

The AP parameters, including overshoot, duration, and rate of rise, are characteristic for each type of excitable cell. For example, the duration of the AP of the squid giant axon is about 1 ms, whereas the cardiac AP lasts for several hundred milliseconds. These differences in the APs subserve the functions performed by the difference excitable tissues. The overshoot and a hyperpolarizing afterpotential (following the spike) are illustrated in Fig. 4C.

C. Refractoriness

Once an AP is initiated, a finite and characteristic time must elapse before a second AP can be generated. This time interval is called the **refractory period**, and its value depends on the type of excitable cell. Cells with long-duration APs (e.g., myocardial cells) have long refractory periods; cells with brief APs (e.g., neurons) have short refractory periods. That is, the refractory periods are proportional to the AP duration.

Two types of refractory periods are usually defined: an **absolute refractory period** and a **relative refractory period** (Fig. 5). The absolute refractory period denotes the in-terval during which a second AP cannot be elicited, regardless of the intensity of the applied stimulus. During the relative refractory period, a second AP may be elicited, provided that a greater than normal stimulus is applied. The second AP often is subnormal in amplitude and rate of rise. Therefore, the physiologically important refractory period is the **functional (effective) refractory period**, which is defined by the highest frequency of APs that the excitable cell (e.g., neuron) can propagate. For example, if a myelinated nerve axon can propagate impulses up to 1000/s, then the functional refractory period is 1.0 ms. The triggering of a second impulse at a given point (or node) is limited by the amount of action current available from an active point (or node) upstream (unlike an electronic stimulator). Therefore, the functional refractory period encompasses all of the absolute refractory period and part of the relative refractory period.

The absolute refractory period extends from when threshold (V_{th}) is reached at the initial portion of the rising phase of the AP to when repolarization has reached a level of about −50 mV. During further repolarization beyond −50 mV (e.g., to −70 mV), a larger and larger fraction of the fast Na^+ channels recovers from inactivation, and so more fast channels are again available to be reactivated to produce another AP. This is the period that inscribes the relative refractory period. The greater the degree of repolarization (toward the resting potential), the larger the subsequent AP.

In addition to voltage, time is a factor in the recovery of the ion channels. Therefore, the absolute and relative refractory periods persist briefly beyond the theoretical voltages, and the relative refractory period actually slightly exceeds the AP duration.

Membrane excitability is greatly altered during the refractory periods. Excitability is zero during the absolute refractory period, and is depressed during the relative refractory period, becoming less and less depressed as the membrane repolarizes back to the resting potential (Fig. 6).

The afterpotentials that many cells exhibit also affect membrane excitability: hyperpolarizing afterpotentials depress excitability (greater critical depolarization required to reach V_{th}) and depolarizing afterpotentials enhance excitability. The latter produces a supernormal period of ex-

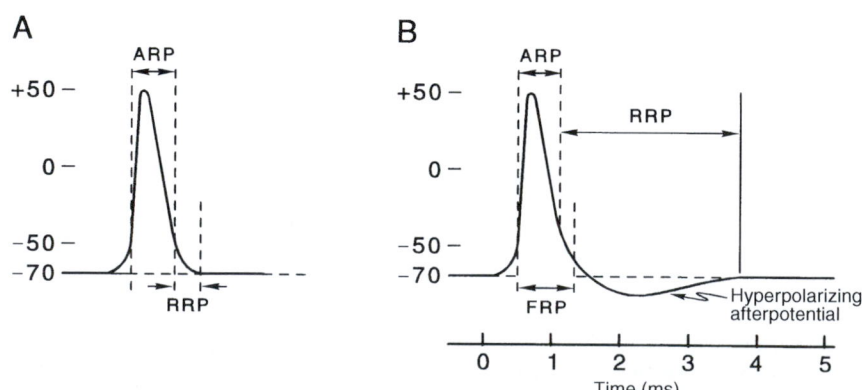

FIGURE 5. Refractory periods of nerve action potentials, both without (A) and with (B) a hyperpolarizing afterpotential. The absolute (ARP), relative (RRP), and functional (FRP) refractory periods are labeled.

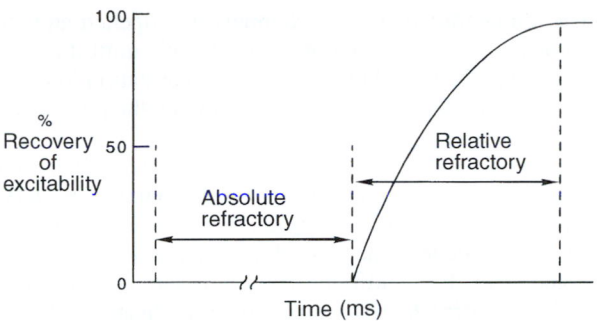

FIGURE 6. The time course of recovery of nerve excitability during the relative refractory period. As shown, during the absolute refractory period, an AP cannot be elicited, and there is a progressive increase in excitability during the relative refractory period.

citability, and the former blends into and extends the relative refractory period. Thus, for example, since neurons usually exhibit hyperpolarizing afterpotentials, the functional refractory period described previously actually includes part of the afterpotential.

Propagation velocity is slowed during the relative refractory period, achieving the normal value at the end of the relative refractory period. Propagation velocity is slightly faster than normal during the supernormal period of excitability.

D. Strength-Duration Curve

Whether the threshold potential is reached depends on the amount of charge transferred across the membrane. Figure 7 shows that the total charge transfer across the membrane required to produce excitation is approximately constant (since $Q = IT$). It is an approximate rectangular hyperbola ($xy = k$) over the sharply bending region of the curve. The strength-duration (S-D) curve can be derived from the equation for the exponential charge of the membrane capacitance.

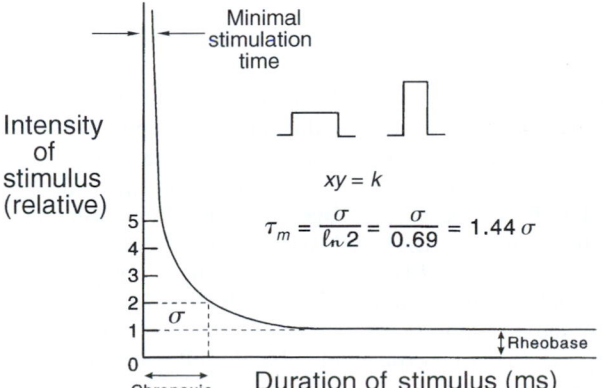

FIGURE 7. Strength-duration curve for AP initiation in excitable membranes. The intensity of rectangular stimulating pulses is plotted against their duration for stimuli that are just sufficient to elicit an AP. The rheobase current and chronaxie (σ) are indicated.

The S-D curve deals only with the stimulus parameters (i.e., strength and duration of the applied current pulses) necessary to bring the membrane to threshold. It shows that the greater the duration of the applied pulse, the smaller the current intensity required to just excite the fiber. The asymptote parallel to the x-axis is the **rheobase**, which is the lowest intensity of current capable of producing excitation, even when the current is applied for infinite time (practically, >10 ms for myelinated nerve fibers). The asymptote parallel to the y-axis is the **minimal stimulation time**, which is the shortest duration of stimulation capable of producing excitation, even when huge currents are applied.

The rheobase is useless when comparing the excitability of one nerve with another because only the relative current intensity is meaningful. Furthermore, it is difficult to measure the stimulation time of a current with the intensity of the rheobase because it is an asymptote. Thus, a graphic measurement is made of the time during which a stimulus of double the rheobasic strength must act in order to reach threshold. This time is the **chronaxie**. Chronaxie values tend to remain constant regardless of geometry of the stimulating electrodes. The shorter the chronaxie, the more excitable the fiber. The chronaxie value for normal myelinated nerve fibers is about 0.7 ms. Some nerve pathologies in humans can be detected early by changes in their chronaxies.

Measurement of chronaxie in the laboratory is also valuable because it provides an easy method for measuring the value of the membrane time constant τ_m (see Chapter 24 on cable properties and Chapter 70 for a review of electricity). In brief, the relationship between chronaxie (σ) and time constant (τ_m) is

$$\tau_m = \frac{\sigma}{\ln 2} = \frac{\sigma}{0.69} = 1.44\sigma \qquad (1)$$

Thus, τ_m is 1.44 times the value of σ. Therefore, σ is analogous to a half-time for a first-order reaction, whose rate constant is the reciprocal of the τ_m ($k = 1/\tau_m$).

The S-D curve indicates that current pulses of very short duration (e.g., < 0.1 ms) are less effective for stimulation. Thus, sinusoidal alternating current (AC) at frequencies above 10 000 Hz is less capable of stimulation. Another way to view this is that, because the membrane impedance decreases greatly at high frequencies (since the cell membrane is a parallel RC network), the potential difference that can be produced across the membrane by current flow across it (*IR* or *IX* drops) is very small. Hence, AC of very high frequency has less tendency to electrocute, and the energy of such currents can be dissipated as heat in body tissues, and thus may be used in diathermy for therapeutic warming of injured tissues.

E. Accommodation

Accommodation, in electrophysiology, refers to the loss of sensitivity of a cell to an applied stimulus. Sensory organs exhibit the property of accommodation, as do many neurons and other excitable membranes, such as skeletal muscle fibers. When a rectangular (square-wave) current pulse is used to depolarize a quiescent motor neuron (step depolarization) from the resting potential (e.g., −70 mV) to the threshold potential (V_{th}) or beyond, the neuron quickly responds with an all-or-none AP. However, if the applied pulse is ramp-shaped (triangular), the neuron may or may not respond, even if the normal V_{th} is exceeded, depending on the slope of the ramp. If the slope of the ramp is steep, the neuron will respond, but at a higher V_{th} level (more **critical depolarization** is required). If the slope is shallow, the neuron will fail to fire an AP, regardless of the level to which it is depolarized. This is accommodation. That is, when the membrane is depolarized gradually, the stimulus is ineffective in producing an AP response (Fig. 8A).

This is not true of pacemaker cells (Fig. 8B). **Automatic (spontaneously discharging) cells** do not exhibit accommodation of their membrane to stimuli capable of evoking AP responses (e.g., nodal pacemaker cells of the heart and some sensory neurons). Such cells will discharge an AP no matter how gradually the membrane is brought to the V_{th} point. In fact, the pacemaker potential in cardiac nodal cells is of the ramp type, producing depolarization to V_{th} over a period of about 200–800 ms.

The explanation for this phenomenon of accommodation is as follows. As the membrane is slowly depolarized toward V_{th}, the positive feedback cycle between E_m, Na⁺ conductance (g_{Na}), and I_{Na} begins to operate (beginning at about 80% of the critical depolarization). Therefore, some of the fast Na⁺ channels are turned on (activated), only to inactivate spontaneously (I-gates close) within 1–2 ms. If a critical number (**critical mass**) of fast Na⁺ channels are not activated simultaneously, then the positive feedback cycle does not become explosive, and a regenerative AP is not produced. That is, the slow depolarization does not allow a critical number of fast Na⁺ channels to be simultaneously in the

open conducting state. The channels that opened and spontaneously inactivated cannot be reactivated until they return to the resting state, which requires repolarization to near the resting potential. Hence, they are lost from the pool of available channels.

Added to this is the fact that K⁺ channels (delayed rectifier type) will open during the slow depolarization, thus increasing K⁺ conductance (g_K). Whenever g_K is increased, excitability becomes depressed, because it tends to repolarize and keep the membrane from depolarizing (to produce an AP); lowering R_m also lowers the effectiveness of the depolarizing current.

Therefore, accommodation to low-slope stimuli occurs for two reasons: (1) spontaneous inactivation of fast Na⁺ channels that have been activated, and therefore lack of a simultaneously open critical mass and (2) increase in g_K, which depresses excitability.

The lack of accommodation in automatic cells (e.g., SA nodal cells of the heart) then may be due to (1) less spontaneous ion channel inactivation and (2) less g_K increase during the applied ramp stimulus or natural ramp pacemaker potential. Both of these conditions apparently apply to cardiac nodal cells. The inward current responsible for the rising phase of the AP is not a fast Na⁺ current, but rather a slow Ca²⁺ current, which inactivates very slowly, and the kinetics of the turn-on of the delayed rectifier K⁺ current is also very slow.

As stated previously, accommodation also occurs in sensory organs. For example, some stretch receptors accommodate to a sustained stretch. When the stretch is first applied, there is a burst of APs. But the bursting frequency of discharge gradually slows down and then stops, even though the stretch is maintained.

F. Anodal-Break Excitation

Excitation occurs on the **make** (the beginning) of a square-wave depolarizing stimulus. If the applied stimulus duration is very long (relative to the AP duration), repetitive firing of APs will occur if the membrane is of the nonaccommodating type. If the membrane is of the accommodating type, then only the initial AP is produced, because accommodation occurs.

If the cathode (negative) and anode (positive) electrodes are placed directly on an isolated single nerve axon (e.g., squid giant axon), then an AP will be triggered at the cathode region on the make of the square-wave stimulus. This happens because, as depicted in Fig. 9A, depolarization occurs under the cathode, whereas hyperpolarization occurs under the anode. However, something unexpected occurs under the anode on the **break** of the stimulating pulse; namely, an AP is triggered from this hyperpolarized region of the axon (Fig. 9B). The explanation for this is that ion channel changes occur during the hyperpolarization, such that the excitability of that membrane is transiently increased (lower V_{th} point) immediately following cessation of the applied pulse.

The increase in excitability is due to two factors: (1) there is an increase in h_∞ during the hyperpolarization, reflecting that almost 100% of the fast Na⁺ channels have their I-gates open, and hence are capable of conducting (open state) when the

A

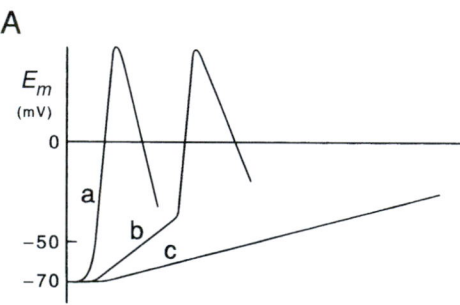

B

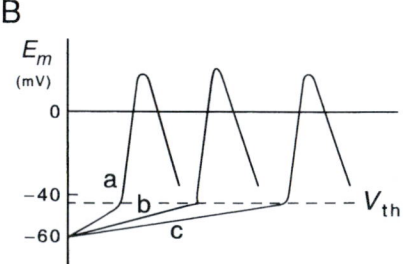

FIGURE 8. (A) The process of accommodation in response to a ramp stimulus in a motor neuron and (B) contrasted with the lack of accommodation in a nodal pacemaker cell from the heart. In A, as the slope of the ramp stimulus is decreased (from a to c), the action potential is delayed (b), and then fails completely (c). V_{th}, threshold potential.

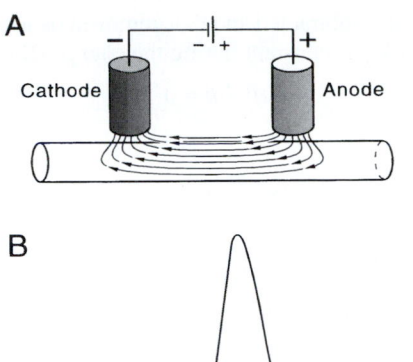

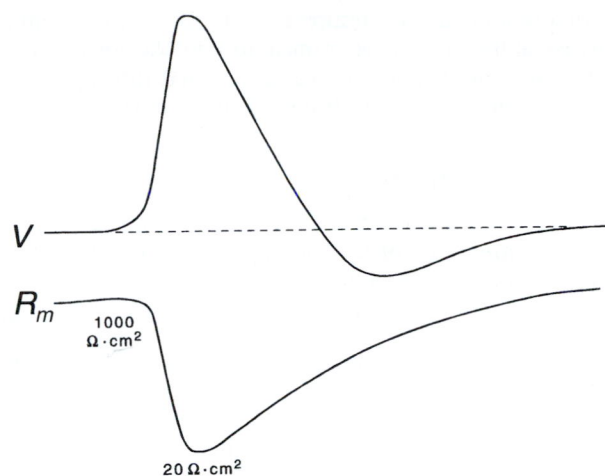

FIGURE 9. Current flow under the anode and cathode during extracellular stimulation (A) and anodal-break stimulation of an action potential (B). (A) Excitation occurs under the cathode on the make of a current pulse and under the anode at the break of the rectangular pulse. (B) An intracellularly applied hyperpolarizing current pulse produces an anodal-break response. RP, resting potential.

FIGURE 10. Time course for the decrease in membrane resistance (R_m) to the passage of current measured during the action potential (V) in a nerve axon. R_m falls from about 1000 $\Omega \cdot cm^2$ at rest to about 20 $\Omega \cdot cm^2$ near the peak of the AP. (Adapted from Cole, K. S., and Curtis, H. J. (1939). *J. Gen. Physiol.* **22**, 649–687.)

membrane is depolarized on removal of the hyperpolarizing pulse ($g_{Na} = \max g_{Na} \, m^3h$); and (2) there is a decrease in n_∞ during the hyperpolarization, hence decreasing g_K ($g_K = \max g_K \, n^4$). These equations are discussed in the following section. The changes in the h and n parameters persist for a short period after termination of the applied pulse, and hence increase membrane excitability during this brief period and trigger an AP. This is called **anodal-break excitation** and **postanodal enhancement of excitability** (due to the postanodal depolarization).

In contrast, under the cathode, after termination of the applied pulse, an opposite change occurs in E_m and excitability. The membrane is hyperpolarized transiently and excitability is depressed. This is known as **postcathodal hyperpolarization** and **postcathodal depression of excitability.** In Hodgkin-Huxley terms, this phenomenon also is due to two factors: (1) decrease in h_∞ during the depolarization, reflecting a smaller fraction of fast Na^+ channels having their I-gates open and hence incapable of conducting and thereby decreasing g_{Na}; and (2) increase in n_∞, hence increasing g_K. The increased g_K and decreased g_{Na} produce hyperpolarization (refer to Chapter 4), and thereby depress excitability.

III. Electrogenesis of Action Potential

Early experimentation in electrophysiology focused on determining the mechanism for the generation of the AP. One important finding by Cole and Curtis (1939) was that during the AP, the membrane resistance (but not the capacitance) changed dramatically (Fig. 10). The large reduction in membrane resistance during the AP supported the hypothesis that the AP resulted from a large increase in the ionic permeability of the membrane.

To determine which ionic species might be involved in generating the AP, subsequent experimentation was directed

toward varying the concentrations of the different ions bathing the axon. Figure 11 shows a classic experiment in which the concentration of Na^+ ions bathing the squid axon was altered. It was found that the overshoot and the rate of rise of the AP were proportional to $[Na^+]_o$. This result was

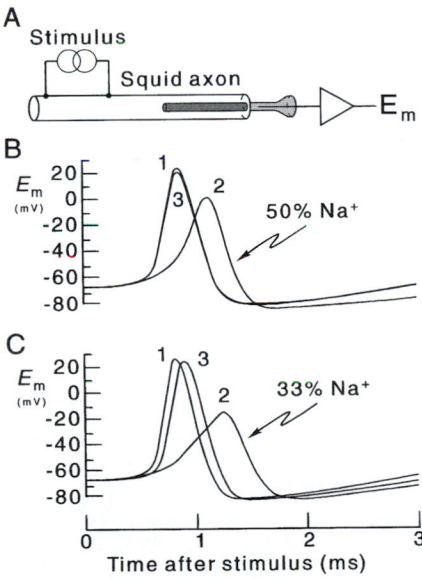

FIGURE 11. Na^+ dependence of the nerve action potential. (A) Diagram depicts the method used to record the transmembrane AP (E_m) from a squid axon. (B and C) The voltage traces show the reduction in AP amplitude as the Na^+ concentration of the solution bathing the axon was reduced to 50% (B) or to 33% (C) of normal. APs were recorded with 100% of the normal Na^+ concentration (traces labeled 1), after reduction of the Na^+ concentration (traces 2), and after return to normal Na^+ concentration (traces 3). (Adapted from Hodgkin, A. L., and Katz, B. (1949). *J. Physiol.* (*London*) **108**, 37–77.)

the first indirect demonstration that the AP resulted from an increase in the membrane permeability to Na^+ ions. Several years later, this hypothesis was confirmed directly by the voltage-clamp method, as discussed in the next section.

A. Voltage-Clamp Method

The membrane current (I_m) that generates the AP is composed of **ionic current** (I_i) and **capacitive current** (I_c) according to the relationship

$$I_m = I_i + I_c \qquad (2)$$

The flow of ionic currents across their respective resistive membrane pathways (or channels) causes a change in the membrane potential (from Ohm's law: $V = IR$). The change in membrane voltage causes a capacitive current to flow as

$$I_c = C_m \, dV/dt \qquad (3)$$

where dV/dt is the rate of change of the AP. Substituting Eq. 3 into Eq. 2 yields

$$I_m = I_i + C_m \, dV/dt \qquad (4)$$

Since the membrane potential during an AP is constantly changing, it would be difficult to separate the contributions of these interacting ionic and capacitive components. In addition, the total ionic current is composed of multiple individual currents carried by specific ions.

To analyze and separate the membrane currents into their capacitive and ionic components, a revolutionary method, called **voltage clamping**, was introduced in the early 1950s by Cole and by Hodgkin and Huxley. A diagram of the voltage-clamp method is shown in Fig. 12. During a voltage-clamp experiment, the membrane potential is held constant (clamped) by a negative-feedback amplifier, and the amount of current that is necessary to perform this task is recorded. Since the membrane potential (V_m) is held constant, the capacitive current is equal to zero. However, at the very beginning of the depolarizing clamp pulse, there is a large transient capacitive current that occurs, and again at the "off" of the pulse; these can

be electronically subtracted and thus removed by a special procedure. Since V_m is constant during the clamp pulse,

$$dV/dt = 0$$

and

$$I_c = 0$$

Therefore,

$$I_m = I_i$$

The voltage-clamp experiment gives the magnitude and time course of the ionic currents at a given clamp potential. By clamping the membrane to many different potentials, information about the flow of ionic currents and the underlying conductance changes during the AP is obtained.

B. Voltage-Clamp Analysis

In the voltage-clamp experiments, the various ion currents (such as Na^+, Ca^{2+}, or K^+ currents) can be isolated from the total ionic current and analyzed individually. For example, in the squid axon experiments, the total ionic current consists of an early inward current followed by a delayed outward current (Fig. 13). By varying $[Na^+]_o$, the early inward current can be shown to be carried by Na^+ ions. Similarly, by changing $[K^+]_o$, the delayed outward current can be shown to be carried by K^+ ions. The Na^+ and K^+ currents can also be separated by blocking their pathways through the membrane. Na^+ channels can be blocked with **tetrodotoxin** (TTX), derived from the ovaries of Japanese puffer fish, and K^+ channels can be blocked by several inorganic ions and organic compounds, including tetraethylammonium ions (TEA$^+$) and 4-aminopyridine (4-AP). The current remaining can then be

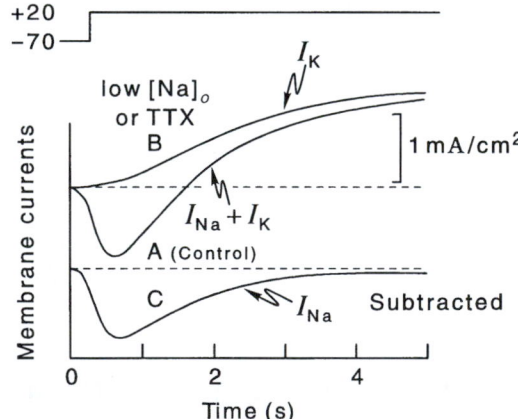

FIGURE 13. The ionic currents that flow when a squid giant axon in seawater is clamped from its resting potential (-70 mV) to a transmembrane potential of $+20$ mV. Trace A shows the net inward Na^+ current (I_{Na}) and outward K^+ current (I_K) in normal medium. Trace B shows the net ionic current when the axon is placed in artificial seawater with most of the Na^+ replaced by choline$^+$ (an impermeant cation), so that the intracellular and extracellular Na^+ concentrations are equal. This current is due to K^+ only. TTX also can be used to block I_{Na}. Trace C shows the difference between curves A and B, which represents I_{Na}. (Redrawn from Hodgkin, A. L., and Huxley, A. F. (1952a). *J. Physiol.* (*London*) **116**, 449.)

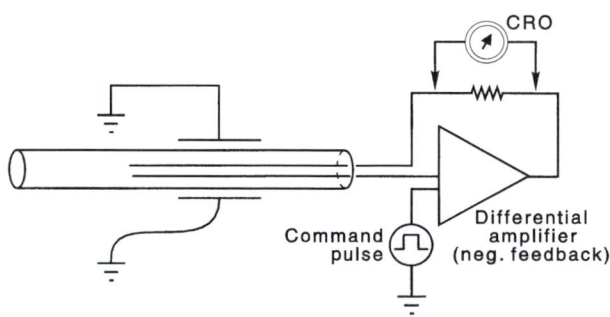

FIGURE 12. The voltage-clamp method used in a squid giant axon. The two wires inserted into the axon are used to measure membrane potential (V) and to pass current (I). The high-gain negative-feedback amplifier compares the command pulse with the membrane potential, and outputs the amount of current necessary to hold the membrane potential constant (clamped). The magnitude of the feedback current can be measured as the IR voltage drop across a resistor and displayed on a cathode-ray oscilloscope (CRO).

subtracted from the total ionic current to reveal the time course for the current that was blocked.

The **current-voltage relationship** is obtained from measurements of the peak inward Na$^+$ current and peak outward K$^+$ current during a series of voltage clamp steps (Fig. 14). Depolarizing voltage steps to just above the resting potential first produce a small outward current. In this voltage region, the membrane behaves in an ohmic fashion. With greater depolarization, the inward I_{Na} is activated and the *I-V* relationship displays a negative slope or **negative resistance** region. At potentials above the peak current of the *I-V* curve, a positive slope is seen, and the current magnitude decreases as E_{Na} is approached, and actually becomes outward at voltages above E_{Na}. The voltage at which the current reverses in direction is the **reversal potential**. The reason that the current diminishes and reverses at potentials approaching E_{Na} and beyond is that the net electrochemical driving force for Na$^+$ ions first becomes smaller and smaller and then outwardly directed, whereas the conductance for Na$^+$ ions remains constant and high over this entire voltage range, as indicated by

$$I_{Na} = g_{Na}(E_m - E_{Na}) \tag{5}$$

The outward K$^+$ current activates above -20 mV and increases with depolarization as

$$I_K = g_K(E_m - E_K) \tag{6}$$

The voltage-clamp experiments have revealed the most fundamental property of the ionic conductances of excitable membrane: namely, that the conductances are both voltage dependent and time dependent (Fig. 15). Both g_{Na} and g_K activate with depolarization, but with different time courses (Figs. 15 and 16). With time, g_{Na} spontaneously turns off, or inactivates. That is, the I_{Na} currents shuts off within 1–2 ms.

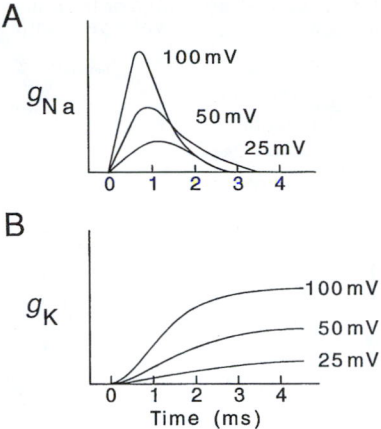

FIGURE 15. Voltage dependence and time dependence of the changes in Na$^+$ conductance (g_{Na}) and K$^+$ conductance (g_K) during voltage-clamp of the squid giant axon. The numbers refer to the magnitude of depolarization (in mV) from the resting potential. (A) g_{Na} turns on rapidly and then spontaneously declines over a brief time period. (B) g_K turns on more slowly and is sustained in amplitude during the entire clamp pulse. (Redrawn from Hodgkin, 1964).

A number of biological toxins that act on specific ion channels have been discovered. For example, TTX mentioned previously has a very high affinity for the fast Na$^+$ channel of nerve and some types of muscle cells. It binds to the fast Na$^+$ channel and blocks the passage of Na$^+$ ions through the channel. Several different types of toxin, including **batrachotoxin** (BTX), inhibit the inactivation process of the Na$^+$ channel, so that the Na$^+$ currents are greatly prolonged once activated. Such toxins have proven to be valuable tools in analyzing voltage-clamp currents and understanding ion channel function. The ion channel toxins are discussed in great detail in a Chapter 37.

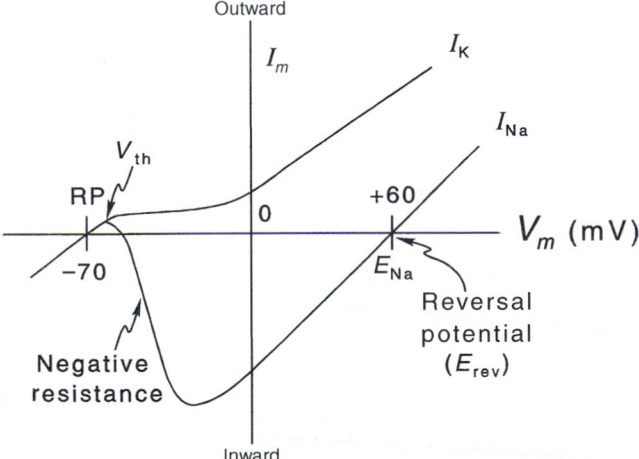

FIGURE 14. Current-voltage relationship for the peak early inward current and delayed outward current obtained from a squid axon under voltage-clamp. The inward current is carried by Na$^+$ (I_{Na}), the outward current by K$^+$ (I_K). The reversal potential for I_{Na} is the voltage at which the current changes from inward to outward. The region of the I_{Na} curve that has a negative slope is known as negative resistance. RP, resting potential; V_{th}, threshold; E_{Na}, Na$^+$ equilibrium potential; I_m, total membrane current; V_m, membrane potential. (Redrawn from Hodgkin, 1964.)

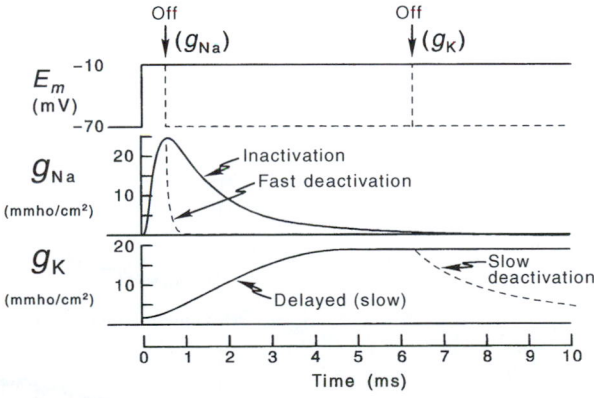

FIGURE 16. Na$^+$ conductance (g_{Na}) and K$^+$ conductance (g_K) changes are shown in response to voltage-clamp of the squid giant axon from the resting potential to a membrane potential (E_m) of -10 mV. g_{Na} inactivates after a short time, even when the voltage-clamp is maintained; in contrast, g_K remains elevated until the clamp is released. When the clamp pulse is terminated early, the g_{Na} increase is quickly turned off (deactivated); the turn-off of g_K is considerably slower. (Redrawn from Hodgkin, A. L. (1958). *Proc. R. Soc. London* (Biol.) **B148**, 1.)

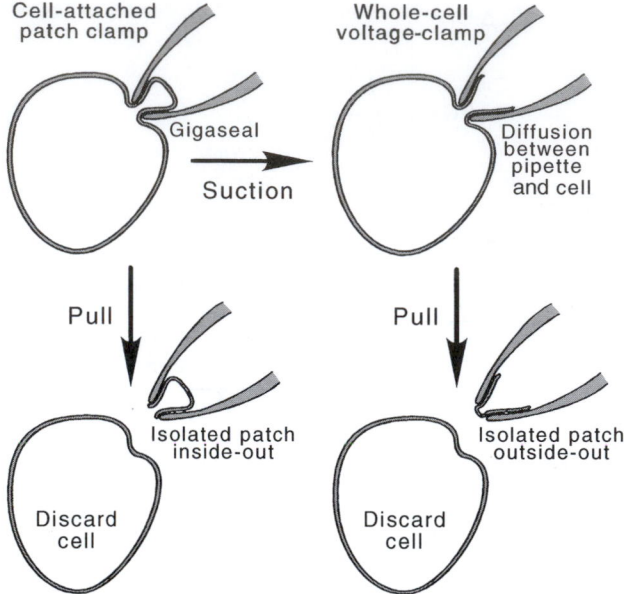

FIGURE 17. The whole-cell voltage-clamp technique in isolated single cells to record the macroscopic current (e.g., I_{Na}, I_{Ca}, or I_K)from the entire cell membrane. (Upper left) Cell-attached patch mode, produced by applying light suction to the patch pipette to produce a gigaseal (10^9 Ω). This mode is used to record the microscopic currents from only one channel or from a few channels. (Upper right) Whole-cell clamp mode produced by applying strong suction to the patch pipette to blow out the membrane patch and allowing the lumen of the pipette to be continuous with the lumen of the cell. This mode is used to record the macroscopic currents from the entire complement of ion channels in the cell membrane. (Bottom) Preparation of isolated membrane patches. (Reprinted with permission from Hamill, O. P., and Sakmann, B. (1981). *Nature* **294**, 462–464.)

C. Whole-Cell Voltage-Clamp

A new electrophysiological method, the whole-cell voltage-clamp technique, has enabled researchers to examine the basis of excitability at the single cell level, for example, for myocardial cells or smooth muscle cells. The method allows the recording of the macroscopic currents that flow through the assembly of ion channels in the cell membrane. The details of this method are given in the following chapter. In brief, to record the whole-cell current, a small-tipped glass pipette, known as a patch pipette, is pressed against the cell membrane, and negative pressure is applied to the interior of the pipette to draw a small patch of membrane into its tip. A high-resistance seal (e.g., 10^{10} Ω) spontaneously forms (Fig. 17). Then, strong suction is applied to the patch pipette to blow out the membrane patch and allow the lumen of the pipette to be continuous with the lumen of the cell. This technique allows recording of the whole-cell current, and a complete voltage-clamp analysis can be done for small cells, as described previously for giant axons. This technique also allows some control over the intracellular content of the cell; for example, a substance can be introduced into the cell by diffusion from the patch pipette solution.

Whole-cell voltage-clamp records obtained from isolated single uterine smooth muscle cells (18-day pregnant rat) are illustrated in Fig. 18. The condition in Fig. 18 is arranged so that outward K^+ currents are blocked (by high Cs^+ concentration in the patch pipette), and any inward currents carried by Ca^{2+} or Na^+ ions can be recorded. As shown in Fig. 18, there are two inward currents: an initial fast current and a later slow current. The initial fast current is carried by Na^+ ion and this Na^+ current is blocked by TTX; the later slow current is carried by Ca^{2+} ions. The presence of functional fast Na^+ channels is unusual for most smooth muscles, and

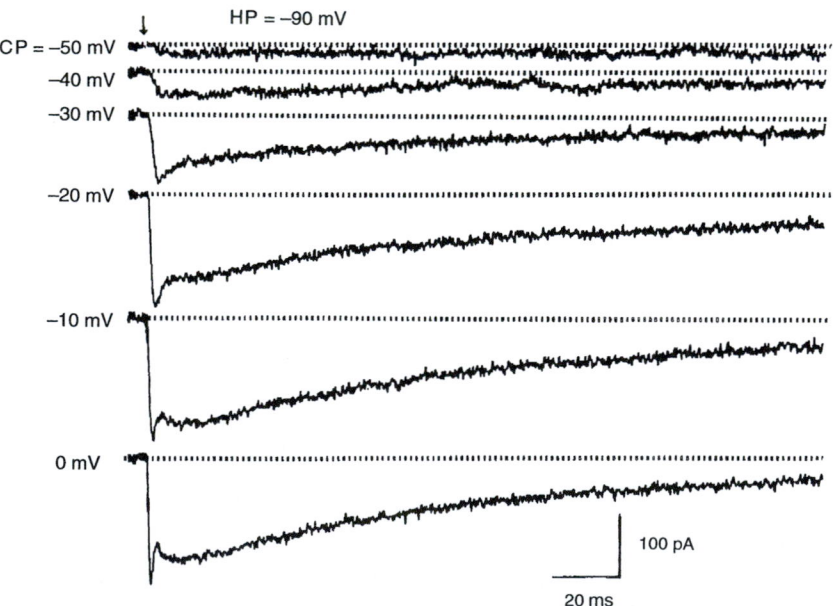

FIGURE 18. Two types of inward current recorded from isolated single myometrial cells from pregnant rat using whole-cell voltage-clamp with a patch pipette technique. Depolarizing potential steps (−50 to 0 mV) were applied from a holding potential (HP) of −90 mV. The arrow indicates the beginning of a voltage step that continues to the end of the trace. The bath contained K^+-free solution with 150 mM Na^+ and 2 mM Ca^{2+}. The pipette solution contained high Cs^+. Cell capacitance was 80 pF. CP, command potential. (Reproduced with permission from Ohya, Y., and Sperelakis, N. (1989). *Am. J. Physiol.* **257**, C408–C412.)

it was discovered that such channels developed during pregnancy and reached their maximum close to term.

The information gained from such whole-cell voltage-clamp experiments, performed on a variety of excitable cell types, has greatly increased our knowledge of the properties and regulation of the many types of ionic currents. The macroscopic or whole-cell current (I) is the product of the number of functional channels in the cell membrane (N) times the probability that the "average" channel is open during the clamp step (p_o) times the single-channel current (i), giving

$$I = iNp_o \qquad (7)$$

The single-channel current (in pA), open probability (p_o), and single-channel conductance (γ) are obtained from patch-clamp experiments, in which the activity of a single channel (or a few channels) can be monitored in a small patch of membrane (e.g., 1–10 μm^2) that is electrically isolated from the rest of the cell membrane (cell-attached patch) or excised

(isolated patch). Opening and closing of the gates of the single channel reflect conformational changes in the channel protein. This sophisticated technique, which earned the Nobel Prize in physiology and medicine in 1991 for E. Neher and B. Sakmann, allows the biophysical study of the behavior of a single protein molecule and thus represents electrophysiology at the molecular level. Although this patch-clamp technique and analysis is discussed in detail in the following chapter, a brief illustration of actual records is given here (Fig. 19).

D. Overview of Action Potential Generation

The increase in g_{Na} and resulting increase in inward I_{Na} cause the regenerative depolarization of the AP. The depolarization is limited by the approach of the membrane potential toward E_{Na} and by the Na inactivation process. As the membrane is depolarized, both g_K and the driving force for I_K increase, and the outward I_K repolarizes the membrane.

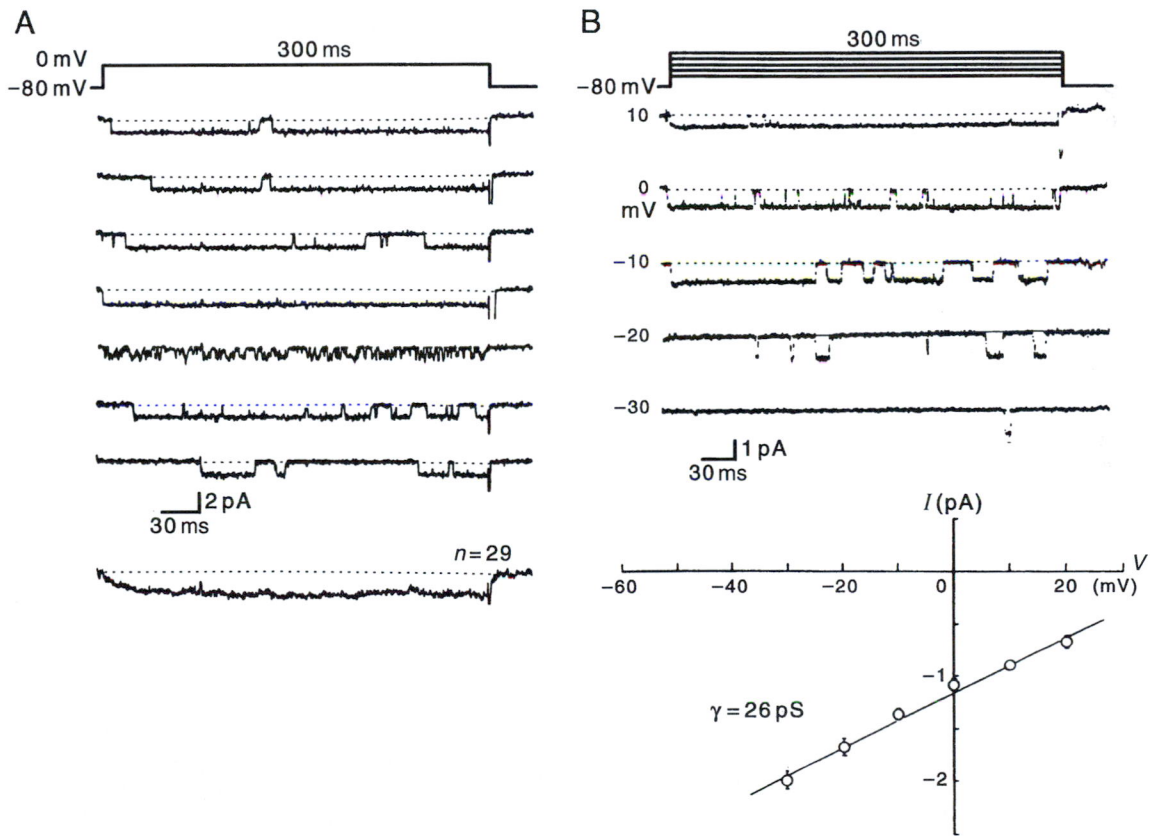

FIGURE 19. Single-channel currents recorded from a slow (L-type) Ca^{2+} channel, using the cell-attached patch-clamp technique, in a single cardiomyocyte isolated from a young (3-day-old) embryonic chick heart (ventricular cell). These Ca^{2+} channels in young embryonic/fetal chick and rat heart cells exhibit a high incidence of long openings. (A) Inward currents were evoked by seven depolarizing voltage step pulses (300-ms duration; 0.5 Hz) to 0 mV from a holding potential (HP) of −80 mV. Pulse protocol is given at the top. The ensemble-averaged current from 29 such pulses is given at the bottom. (B) Similar recordings in another cell illustrating the inward currents recorded at five different command step potentials (−30, −20, −10, 0, and +10 mV) from a HP of −80 mV. Again note the presence of numerous long openings. Also note that the amplitude of the unitary current became smaller at the higher (more positive) command potentials. The current amplitudes are plotted below in the current-voltage curve (each point being the mean ± SE of 5–10 experiments). The data points were fitted by a straight line giving a slope conductance of 26 pS for the single-channel conductance. (Modified from Tohse, N., and Sperelakis, N. (1990). *Am. J. Physiol.* **259**, H639–H642; and Tohse, N., and Sperelakis, N. (1991). *Circ. Res.* **69**, 325–331.)

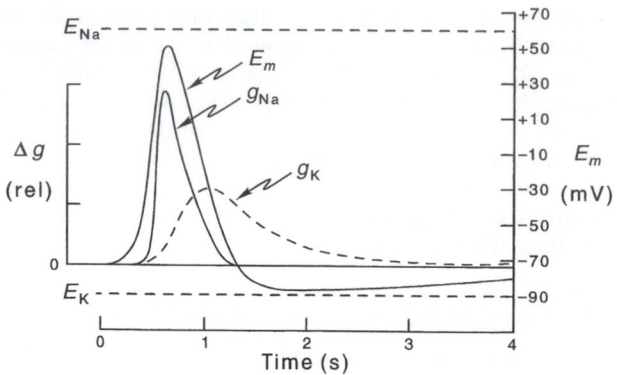

FIGURE 20. The relative conductance for Na$^+$ (g_{Na}) and K$^+$ (g_K) during an action potential in a nerve fiber. The rising phase of the AP is caused by an increase in g_{Na}. The falling phase of the AP is due to the rise of g_K (delayed rectification) and to the decrease in g_{Na} (Na$^+$ inactivation). The hyperpolarizing afterpotential is explained by the fact that g_K remains elevated for a short time following repolarization, tending to hold the membrane potential (E_m) near the K$^+$ equilibrium potential (E_K). E_{Na}, Na$^+$ equilibrium potential; Δg, change in conductance. (Redrawn from Hodgkin, A. L., and Huxley, A. F. (1952). *J. Physiol.* **117**, 500–544.)

The increase in g_K is **self-limiting**; that is, the increase in g_K produces repolarization, which, in turn, shuts off the increase in g_K.

The slow kinetics of the turn-off of g_K result in a transient hyperpolarization, the hyperpolarizing afterpotential. During the hyperpolarizing afterpotential, the membrane potential is brought closer to E_K than at rest. The membrane conductance changes that occur during the AP are shown in Fig. 20.

The time course for the ionic currents during the nerve AP is shown in Fig. 21. The total ionic current (I_i) is sepa-

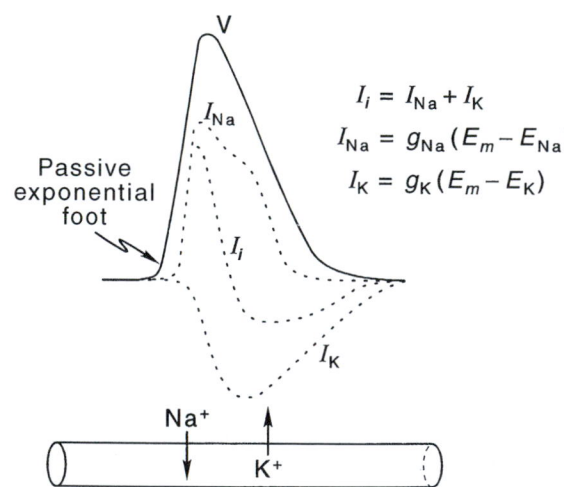

$$I_i = I_{Na} + I_K$$
$$I_{Na} = g_{Na}(E_m - E_{Na})$$
$$I_K = g_K(E_m - E_K)$$

FIGURE 21. Ionic currents that flow during the nerve action potential. The total current (I_i) is separated into an inward Na$^+$ current (I_{Na}) and an outward K$^+$ current (I_K). I_i is the algebraic sum of I_{Na} and I_K. The appropriate equations for I_i, I_{Na}, and I_K are given. Also depicted is the fact that a net inward Na$^+$ flux occurs during the rising phase of the AP and a net K$^+$ efflux occurs during the repolarizing phase. (Redrawn from Hodgkin and Huxley, 1952a.)

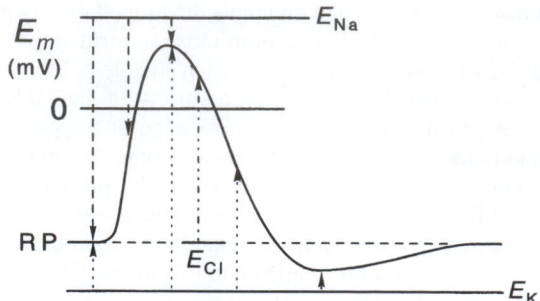

FIGURE 22. Driving forces for Na$^+$ and K$^+$ currents during the action potential. The total electrochemical driving force is equal to membrane potential (E_m) minus the equilibrium potential for the ion (E_i): ($E_m - E_i$). As depicted, the driving force for the Na$^+$ current decreases during the AP, whereas that for K$^+$ increases. Even when Cl$^-$ is passively distributed, a net driving force for Cl$^-$ influx (outward Cl$^-$ current) occurs during the AP.

rated into its two major components, I_{Na} and I_K. Since I_K is slower to activate than I_{Na}, the inward I_{Na} predominates initially, giving rise to the upstroke of the AP. Later, I_K dominates, causing a net outward current that repolarizes the membrane, that is, is partly responsible for the downstroke of the AP.

As stated in Eqs. 5 and 6, the specific ionic currents are a product of the membrane conductance for the ionic species and the electrochemical driving force exerted on the ion. Thus, the driving forces on Na$^+$ and K$^+$ ions continually change during the time course of the AP, as diagrammed in Fig. 22. At the resting potential, there is a large driving force for Na$^+$ to flow into the cell, since ($E_m - E_{Na}$) is large. Conversely, at the peak of the AP, ($E_m - E_{Na}$) is at its lowest value, and the driving force for Na$^+$ entry is small. In contrast, the driving force for K$^+$ efflux is largest at the peak of the AP, when ($E_m - E_K$) is maximal. It is important to remember that ionic current flow depends on both conductance and driving force. There is no net current if only one, but not the other, is present.

The biological elements of the excitable membrane may be represented in terms of an electrical equivalent circuit model, as shown in Fig. 23. In the circuit model, currents flow through the individual conductance pathways for each type of ion. The conductances for Na$^+$ and K$^+$ ions are variable and depend on the transmembrane potential and time. Batteries (positive pole directed inwardly for Na$^+$ ions and outwardly for K$^+$ ions) provide the driving forces for current flow. A passive leak conductance for Cl$^-$ ions is also included in the model. If the correct values for the elements are incorporated and varied over time, the model circuit will generate an AP.

E. Fast Na$^+$ Channel Activation

During an AP, the increase in g_{Na} is related to the membrane potential E_m in a **positive-feedback** fashion, or "vicious cycle" (Fig. 24A). That is, a small depolarization leads to an increase in g_{Na}, which allows a larger inward I_{Na}, which causes further depolarization. This greater depolarization produces a greater increase in g_{Na}. This positive feedback process is "explosive,"

Outside

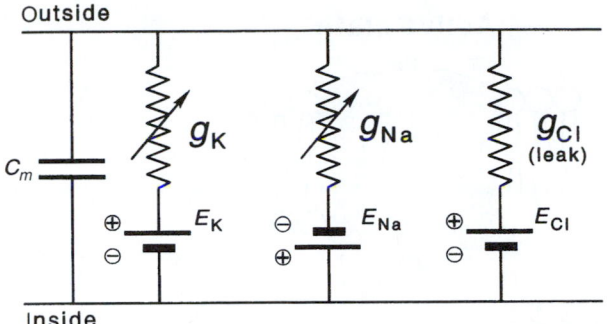

Inside

FIGURE 23. Hodgkin-Huxley electrical equivalent circuit for the squid giant nerve axon. The K$^+$ conductance (g_K) is in series with K$^+$ equilibrium potential (E_K) and g_{Na} is in series with E_{Na}. The arrows indicate that g_{Na} and g_K vary with voltage and time. The low conductance for Cl$^-$ (g_{Cl}) was termed the **leak conductance**. C_m, membrane capacitance. (Redrawn with permission from Hodgkin, A. L. (1964). "The Conduction of the Nervous Impulse." Charles C Thomas, Springfield, IL.)

with a sharp trigger point (threshold) resulting from the exponential (positive) relationship between g_{Na} and E_m (Fig. 24B). It is this positive-feedback relationship that accounts for the negative resistance (slope) in the current-voltage curve (Fig. 14). As Fig. 24B shows, g_{Na} reaches a maximum (saturates) at positive potentials that gives maximal activation of the population of fast Na$^+$ channels.

The fast Na$^+$ channels (and the slow Ca^{2+} channels) have a double gating mechanism: an inactivation gate (I-gate) and an activation gate (A-gate) (Fig. 25). For a channel to be conducting, both the A-gate and the I-gate must be open; if either one is closed, the channel is nonconducting. The A-gate is located somewhere near the middle of the channel; it is not at the outer surface because even TTX does not prevent the movement of this gate, and it is not at the inner surface because proteases perfused internally do not affect it. The A-gate is closed at the resting E_m and opens rapidly on depolarization; in contrast, the I-gate is open at the resting E_m and closes slowly on depolarization. The time course of the changes in the m and h variables during a depolarizing clamp step is schematized in Fig. 26.

In the Hodgkin-Huxley (1952) analysis, the opening of the A-gate requires simultaneous occupation of three negatively charged sites by three positively charged (m^+) particles. The activation variable, m, is the probability of one site being occupied, and m^3 is the probability that all three sites are occupied; therefore,

$$g_{Na} = \overline{g}_{Na} m^3 h \tag{8}$$

where h is the inactivation variable and $\overline{g}_{Na}$ is the maximum conductance.

Hodgkin and Huxley found that, in squid giant axon, there was an e-fold increase in g_{Na} per 4–6 mV of depolarization. To convert to ln from $\log_{10}$, the 2.303 $RT/z\mathscr{F}$ factor of 58 mV (at 20 °C) becomes 58 mV/2.303, or 25.2 mV. Therefore, they concluded that the voltage-sensitive activation gate must have about 4 to 6 unitary charges (25/6 to 25/4).

F. Gating Current

Since the gating of the ionic channels is voltage dependent, a part of the channel protein contains a charged group or dipole that can sense the electric field across the membrane and move in response to a change in transmembrane voltage. When the gating region moves, it causes a shift in the overall conformation of the channel, which allows it to conduct ions. A gating current (I_g) that corresponds to the movement of the charged m^+ particles (or rotation of an equivalent dipole) has been measured. The gating current is very small in intensity and is measured by subtracting the linear capacitive current (from a hyperpolarizing clamp step) from the total capacitive current (linear plus nonlinear) that occurs with a depolarizing clamp step beyond threshold. The outward I_g precedes the inward I_{Na}. Tetrodotoxin does not block I_g, although it does block I_{Na}. Thus, the gating cur-

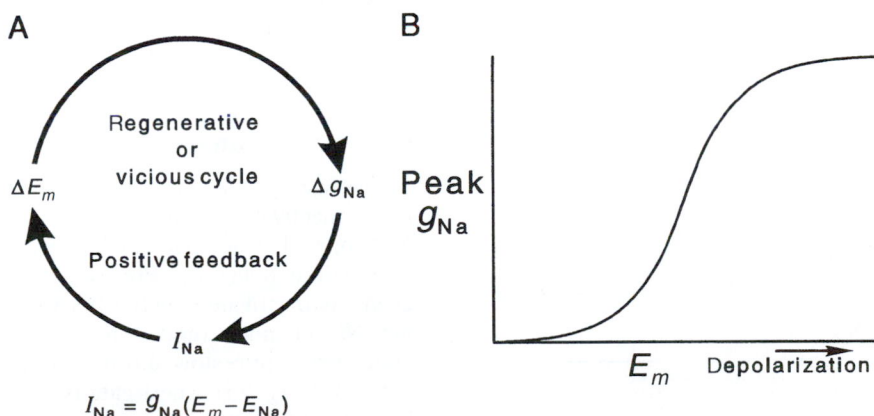

$$I_{Na} = g_{Na}(E_m - E_{Na})$$

FIGURE 24. The positive-feedback relationship between Na$^+$ conductance (g_{Na}) and membrane potential E_m, leading to the all-or-none action potential. (A) The increase in g_{Na} allows an increase in the inward Na$^+$ current, $I_{Na} = g_{Na} (E_m - E_{Na})$, which is depolarizing and so triggers a further increase in g_{Na}. This explosive feedback cycle is caused by the voltage dependency of the gated fast Na$^+$ channels. (B) Plot of g_{Na} versus depolarization, showing the initial exponential (positive) increase in g_{Na} as a function of voltage, followed by saturation with greater depolarization, thus giving a sigmoidal relationship. (Adapted from Hodgkin, A. L. (1964). "The Conduction of the Nerve Impulse." Charles C. Thomas, Springfield, IL.)

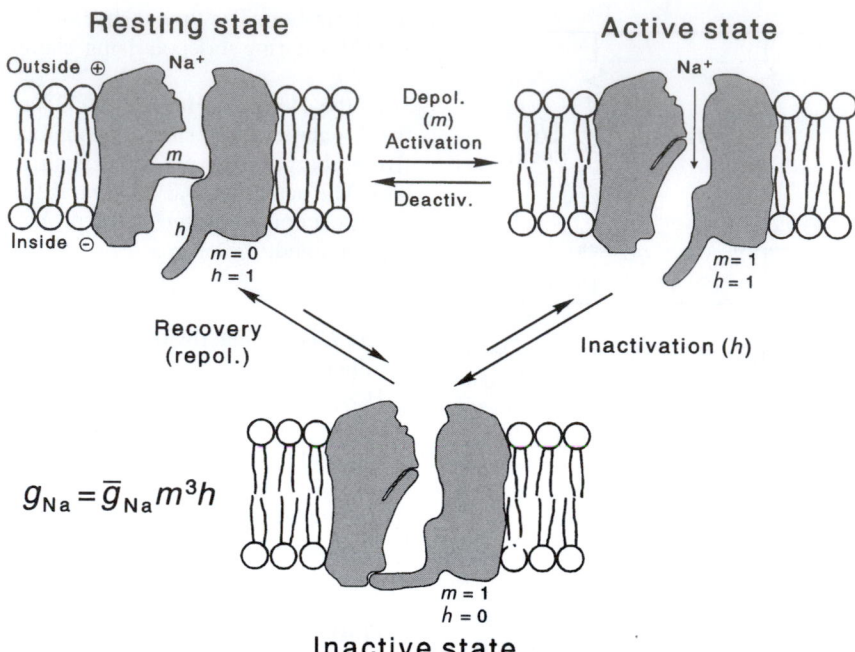

Resting state

Outside ⊕   Na⁺

Inside ⊖

$m = 0$
$h = 1$

Depol.
(*m*)
Activation

Deactiv.

Active state

Na⁺

$m = 1$
$h = 1$

Recovery
(repol.)

Inactivation (*h*)

$$g_{Na} = \bar{g}_{Na} m^3 h$$

$m = 1$
$h = 0$

Inactive state

FIGURE 25. The three hypothetical states of the fast Na⁺ channel, based on the Hodgkin-Huxley model. In the resting state, the activation gate (A or *m*) is closed and the inactivation gate (I or *h*) is open: $m = 0$, $h = 1$. Depolarization to the threshold or beyond activates the channel to the active state, the A-gate opening rapidly and the I-gate still being open: $m = 1$, $h = 1$. The activated channel spontaneously inactivates to the inactive state because of delayed closure of the I-gate: $m = 1$, $h = 0$. The recovery process on repolarization returns the channel from the inactive state to the resting state, thus making the channel again available for reactivation. Na⁺ is depicted as being bound to the outer mouth of the channel and poised for entry down its electrochemical gradient when both gates are open (active state of channel). The reaction between resting state and the active states is reversible, whereas the other reactions may be less reversible. The Ca²⁺ slow channels pass through similar states.

rent is a nonlinear outward capacitive current (not an ionic current) obtained during depolarizing clamps that reflects movement of the A-gates from the closed to the open configuration. The linear capacitive current results mainly from the lipid bilayer matrix (see Chapter 70), whereas the nonlinear capacitive current arises from the charge movement associated with the A-gates of the protein channels.

Specific charged residues are located along the primary sequence of amino acids that comprise the channel polypeptide,

and they can "sense" the transmembrane electric field and move in response to changes in the electric field. The movement of these voltage-sensing or gating residues initiates a conformational change in the channel structure, which then permits ions to flow through the central pore region of the alpha subunit. The movement of the gating charge produces the small, but measurable, gating current. Gating currents have been recorded from several types of ion channels and provide information concerning the steps leading to channel opening.

G. Na⁺ Inactivation

The fast I_{Na} lasts only for 1–2 ms because of the spontaneous inactivation of the fast Na⁺ channels. That is, the fast Na⁺ channels inactivate quickly, even if the membrane were to remain depolarized (Fig. 26). (In contrast, the slow Ca²⁺ channels inactivate slowly.) Inactivation is produced in the fast Na⁺ channels (and in the slow Ca²⁺ channels) by the voltage-sensitive slow closing of the inactivation gate (I-gate) (see Fig. 25). The I-gate is located near the inner surface of the membrane, as evidenced by the fact that addition of proteolytic enzyme to the inside of a perfused giant axon chops off the I-gate and eliminates inactivation (protease added outside does not have this effect). The I-gate is presumably charged positively to allow it to move with changes in the membrane potential. During depolarization, the inside of the membrane becomes less negative or more positive,

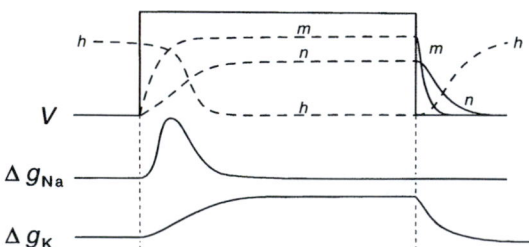

h

m

n

m

h

n

V

Δg_{Na}

Δg_K

FIGURE 26. Sketch of the time course of the changes in the Hodgkin-Huxley *m*, *n*, and *h* factors during a depolarizing voltage-clamp step. The lower traces give the resulting time course for the changes in Na⁺ and K⁺ conductances during the step. Because g_{Na} is proportional to the product of the *m* and *h* variables, it returns to about the original level within approximately 2 ms after the clamp step is applied.

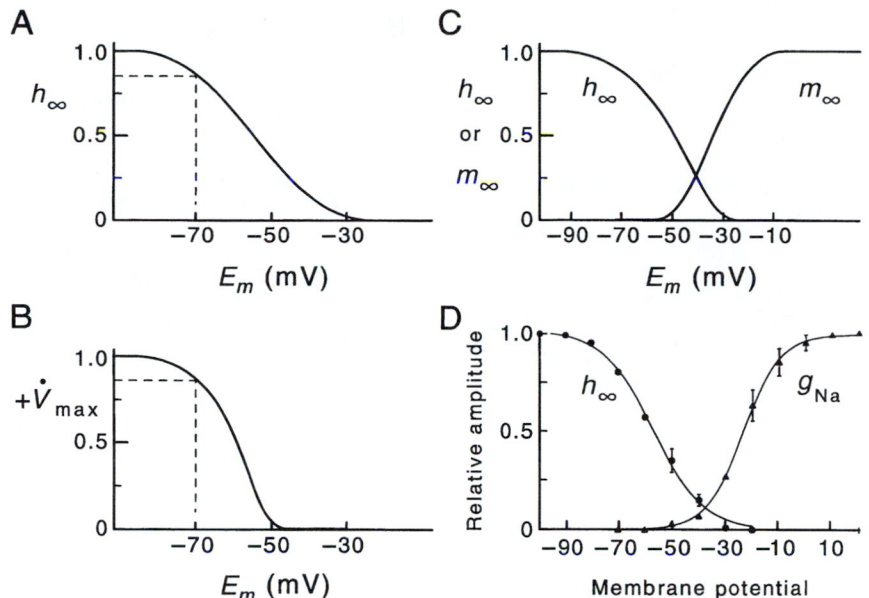

FIGURE 27. Voltage inactivation of the fast Na^+ channels as a function of the membrane potential (E_m). (A) h_∞ plotted against E_m, where h_∞ is the inactivation factor of Hodgkin-Huxley. The h_∞ represents h at infinite time or steady state. This graph illustrates that fast Na^+ channels begin to inactivate at about −75 mV, and nearly complete inactivation occurs at about −30 mV. (B) Maximal rate of rise of the AP (max dV/dt) as a function of resting E_m. Max dV/dt is a measure of the inward current intensity, which is dependent on the number of channels available for activation. Therefore, max dV/dt decreases as h_∞ decreases. (C) Plot of both steady-state inactivation (h_∞) and activation (m_∞) against E_m to illustrate the overlap of the two curves, depicting the window current region. (D) Steady-state activation and inactivation curves for the fast Na^+ current [$I_{Na(f)}$] recorded using the whole-cell voltage-clamp technique from uterine smooth muscle cells isolated from late pregnant (18-day) rats. For the inactivation curve, conditioning pulses of various amplitudes were applied for 2 s before a test pulse to 0 mV. The activation curve was obtained by measuring peak amplitude of I_{Na} elicited by various command potentials from a HP of −90 mV. The two curves were obtained by fitting the data to Boltzmann distributions. Each point represents the mean ± SE. Note the overlap between the activation and inactivation curves, resulting in a steady-state window current between about −50 and −30 mV. (Data from Y. Inoue and N. Sperelakis, unpublished observations.)

and this causes the I-gate to close. At the normal resting potential, the I-gate is open, but the A-gate is closed.

The voltage dependency of inactivation is given by the h_∞ versus E_m curve (Fig. 27). The inactivation variable h (probability function) varies between 0 and 1.0, perhaps reflecting occupation of a negatively charged site by a positively charged h particle; h_∞ is the value of h at infinite time (>10 ms) or at steady state. When $h = 1.0$, the I-gates of all of the fast Na^+ channels are in the open configuration; conversely, when $h = 0$, all the I-gates are closed. Since g_{Na} at any time is equal to the maximal value ($\bar{g}_{Na}$) times $m^3 h$ (see Eq. 8), when $h = 0$, $g_{Na} = 0$.

When $h = 1.0$, $g_{Na} = \bar{g}_{Na}$ (if $m = 1.0$). At the normal resting potential, h_∞ is nearly 1.0 and it diminishes with depolarization, becoming zero at about −30 mV. The maximal rate of rise of the AP (max dV/dt) is directly proportional to the net inward I_{Na}, which is directly proportional to g_{Na}; the decrease in h_∞ is the cause of decrease in g_{Na}. At about −30 mV, $h_\infty = 0$ and there is complete inactivation of the fast Na^+ channels. Therefore, depolarization by any means (e.g., elevated $[K^+]_o$ or applied depolarizing current pulses) decreases g_{Na} and excitability disappears at about −50 mV (Fig. 27B). The AP disappears at −50 mV (rather than −30 mV) because of a minimum current density requirement for a regenerative and propagating response.

A plot of the inactivation curve along with the activation curve is given in Fig. 27C to illustrate the overlap in these curves, giving rise to a so-called **window current** (or steady-state inward current) over a certain voltage region. The presence of the window current means that, over the voltage range of about −50 to −30 mV, there is a steady inward Na^+ current passing through about 10–20% of the fast Na^+ channels that are always open (on a rotating population basis).

The slow Ca^{2+} channels (Fig. 28B) behave much the same way as the fast Na^+ channels (Fig. 28A) with respect to activation and inactivation, with one main difference being the voltage range over which the slow channels operate. For example, inactivation occurs between −50 and 0 mV for the slow channels, compared with between −100 and −30 mV for the fast Na^+ channels. Another major difference is that slow channels inactivate much more slowly than the fast channels; that is, they have a long inactivation time constant (τ_{inact}). In myocardial cells, the h variable for the slow channel is referred to as the f variable, and the m variable as the d variable and the exponent is 2.0:

$$g_{Ca} = \bar{g}_{Ca}\, d^2 f \tag{9}$$

The amino acid sequences of Na^+ and Ca^{2+} channels are known, as well as their putative tertiary structure (Fig. 29).

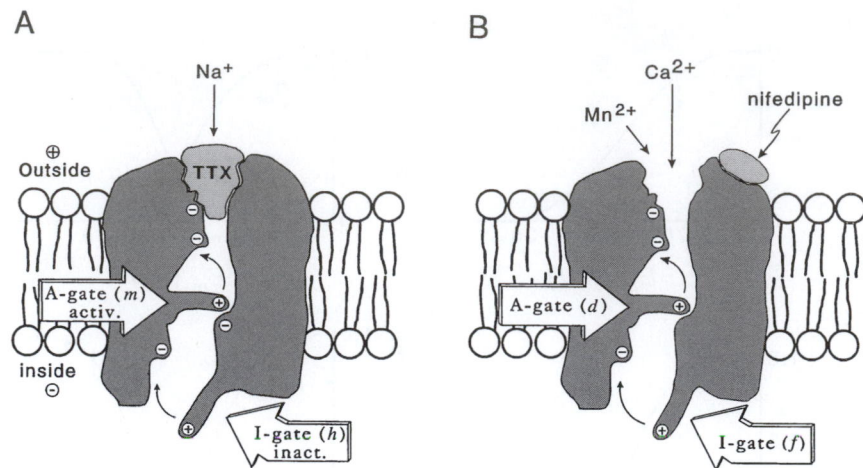

FIGURE 28. (A) Illustration of a fast Na$^+$ channel and (B) a slow Ca^{2+} channel. The ionic channels are proteins that float in the lipid bilayer matrix of the cell membrane. These voltage-dependent channels have two gates, as depicted: the activation gate (A or m) and the inactivation gate (I or h). Conformational changes in the protein could serve as the gates. The I-gate is located near the inner surface of the membrane because proteolytic enzymes added internally destroy this gate. The A-gate is located somewhere in the middle of the channel; TTX does not prevent movement of this gate. TTX binds to the outer mouth of the channel and plugs it physically, as depicted. The gates are presumably charged positively, so that on depolarization—inside going less negative (i.e., in a positive direction)—the gates are electrostatically repelled outward. The A-gate opens relatively quickly (e.g., 0.25 ms), whereas the I-gate closes relatively slowly (e.g., inactivation time constant of 1–2 ms). Therefore, for a short period, both gates are open and the channel is in the conducting mode with Na$^+$ entering the cell down its electrochemical gradient. However, when the I-gate closes (after about 2 ms), the channel again becomes nonconducting. During recovery (reactivation) upon repolarization, the gates return back to their original resting positions. The slow Ca^{2+} channels (B) behave similarly to the fast channels (A), except that their gates apparently move more slowly on a population basis. The slow-channel conductance activates, inactivates, and recovers more slowly. The slow-channel gates operate over a different voltage range than the fast channels (i.e., less negative, more depolarized). TTX does not block the slow channels. Drugs such as nifedipine block the slow channels (but not the fast Na$^+$ channels) by binding to the channel and somehow inhibiting it. In addition, the voltage inactivation curve of the Ca^{2+} slow channels is shifted to the right, so that inactivation begins at about −45 mV and is not complete until about −5 mV. The slow channels also have a higher activation (threshold) potential of about −35 mV (compared with about −55 mV for the fast Na$^+$ channel). The activation gate variable is known as d, and the inactivation gate variable as f.

Na$^+$ and Ca^{2+} channels consist of several subunits, one of which contains the water-filled pore through which the ions pass. Ion channels have one or more sites that can be phosphorylated, and phosphorylation alters their behavior. For example, in cardiac muscle, the slow Ca^{2+} channel activity is increased by adrenaline, a hormone that increases the cyclic AMP level and leads to Ca^{2+} channel phosphorylation (see Chapter 33). The molecular structure of ion channels is discussed in more detail in a later chapter.

H. Recovery

Any Na$^+$ or Ca^{2+} channel that has been activated and then spontaneously inactivated must go through a recovery process before it can return to the resting state from which it can be reactivated (see Fig. 25). The recovery process is dependent on voltage and time. The membrane must be repolarized beyond about −50 mV before the recovery process can begin (i.e., traveling up the h_∞ versus E_m curve). At any given E_m, time is necessary for the recovery process to occur, namely, the time required for the charged A-gates and I-gates to move back to their resting configuration (A-gate closed, I-gate open) with the electric field. The recovery process is rapid for fast Na$^+$ channels (e.g., 1–10 ms) and less rapid for the slow Ca^{2+} channels. The recovery process of the slow channels is slowed by organic calcium antagonist drugs.

I. K$^+$ Activation

The K$^+$ channel (outward-going delayed rectifier) is generally believed to have only an A-gate, because it does not inactivate. This gate is thought to be located near the inner surface of the membrane, because TEA$^+$ blocks the K$^+$ channel more readily from the inner surface. The block is use dependent or frequency dependent: as the A-gate opens, the TEA$^+$ molecule can bind in the channel behind the gate. The A-gate is believed to be positively charged, and depolarization (inside going positive) opens the gate. In the Hodgkin-Huxley analysis of squid giant axon, the A-gate opens when four positively charged n^+ particles simultaneously occupy four negatively charged sites. If n is the probability that one site is occupied, then n^4 is the probability that all four sites are occupied; therefore,

$$g_K = \bar{g}_K\, n^4 \qquad (10)$$

The power to which n is raised varies in different tissues.

J. Model for Activation and Inactivation of Na$^+$ and K$^+$ Channels

As stated previously, in the Hodgkin-Huxley analysis, the Na$^+$ conductance g_{Na} is controlled by three charged activating m particles and one blocking h particle, which move with changes in the electric field across the membrane to ei-

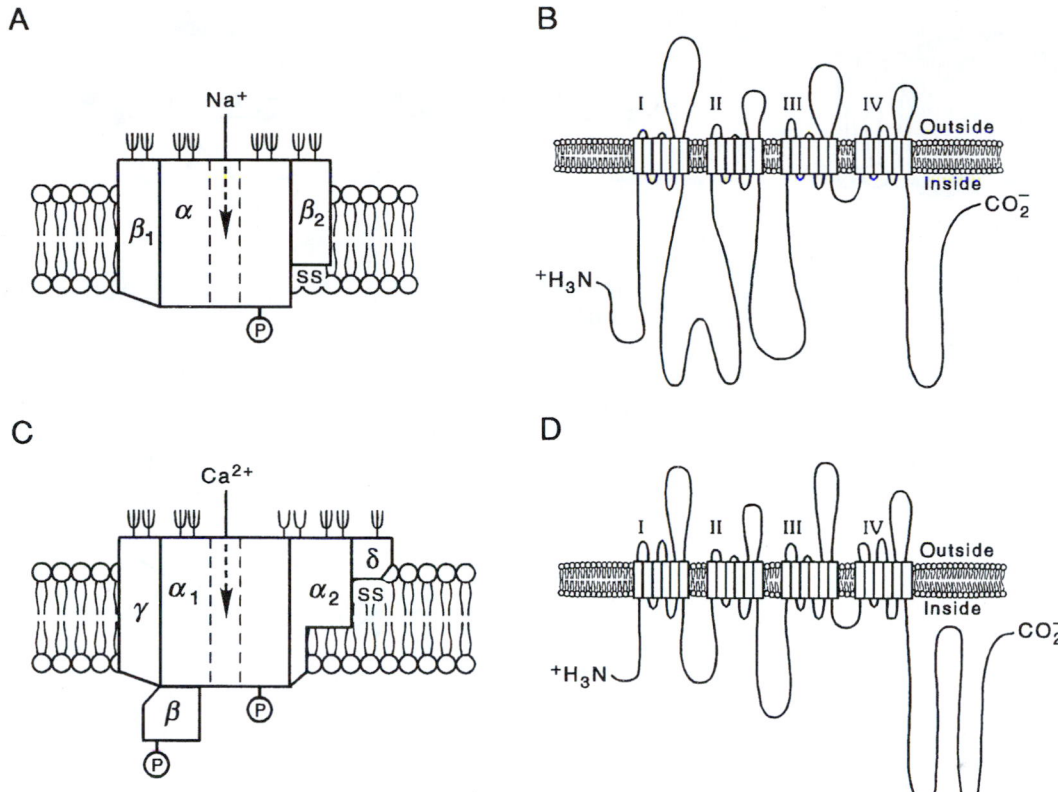

FIGURE 29. Models of the fast Na^+ channel and the slow Ca^{2+} channel proteins based on structural data. (A) The Na^+ channel has multiple protein subunits, labeled α, β_1, and β_2. The α subunit is the one that contains the water-filled pore through which Na^+ passes. One site that can be phosphorylated by cyclic AMP–dependent protein kinase (PK) is present on the α subunit. The β_2 subunit has a disulfide bond (SS). (B) The structure of the central pore-forming α subunit for the Na^+ channel. There are four homologous intramembrane polypeptide repeat domains, connected by intracellular polypeptide segments or loops. The four domains are arranged into a circular structure within the plane of the membrane to form a channel. Each domain consists of six units that span the membrane, as depicted. Both the carboxy (COO$^-$) and amino (NH$_3^+$) termini are intracellular. (C) The Ca^{2+} channel also is composed of several subunits: α_1, α_2, β, γ, and δ. The α_1 subunit contains the pathway by which Ca^{2+} ions traverse the channel. Two sites can be phosphorylated by the cAMP-PK, one on the α_1 subunit and the other on the β subunit. As depicted, the β and δ subunits do not span the lipid bilayer, and the δ subunit possesses a disulfide bond. Ribosylation sites are present on the outer surfaces of the subunits. (D) The polypeptide structure of the α_1 subunit of the slow Ca^{2+} channel is similar to that for the fast Na^+ channel, with four repeat membrane-spanning domains connected by intracellular and extracellular loops. (Reprinted with permission from Catterall, W. (1988). *Science* **242**, 50–61.)

ther occupy or unoccupy certain sites on the Na^+ channel protein. If m is the probability of one favorable site being occupied, then $m \times m \times m$ or m^3 is the probability that all three sites are occupied simultaneously. The activation A-gate cannot be fully opened unless all three sites are occupied. A simplistic variant is to consider that the A-gate of the Na^+ channel actually consists of three separate subgates or subdoors, as illustrated in Fig. 30. For a Na^+ ion to pass through the channel, all three subdoors must be open.

The value of h is the probability of one favorable site being occupied that opens the I-gate, and this site is already occupied at the normal resting potential (I-gate open), and becomes unoccupied with depolarization. Therefore,

$$g_{Na} = \bar{g}_{Na} m^3 h \qquad (8)$$

where $\bar{g}_{Na}$ is the maximum g_{Na} possible (primarily a function of density of channels), m is the activation variable, and h is the inactivation variable.

The K^+ conductance g_K is controlled by four charged activating n particles, which move with depolarization to occupy four sites on the K^+ channel protein. If n is the probability of one favorable site being occupied, then $n \times n \times n \times n$ or n^4 is the probability that all four sites will be occupied simultaneously. There is no inactivation variable for the K^+ channel, because g_K does not exhibit a major decrease during a short depolarizing voltage-clamp pulse (e.g., 20 ms). Therefore,

$$g_K = \bar{g}_K n^4 \qquad (10)$$

where $\bar{g}_K$ is the maximum g_K and n is the activation variable.

Based on the preceding analysis, Hodgkin and Huxley (1952a, b, c) then gave the differential equations that govern the values of n, m, and h using a pair of rate constants, α and β, for the forward reaction and reverse reaction, respectively. The subscripts for the α and β rate constants identify which variable they pertain to (i.e., n, m, or h). The α and β rate constants depend only on membrane potential (at constant

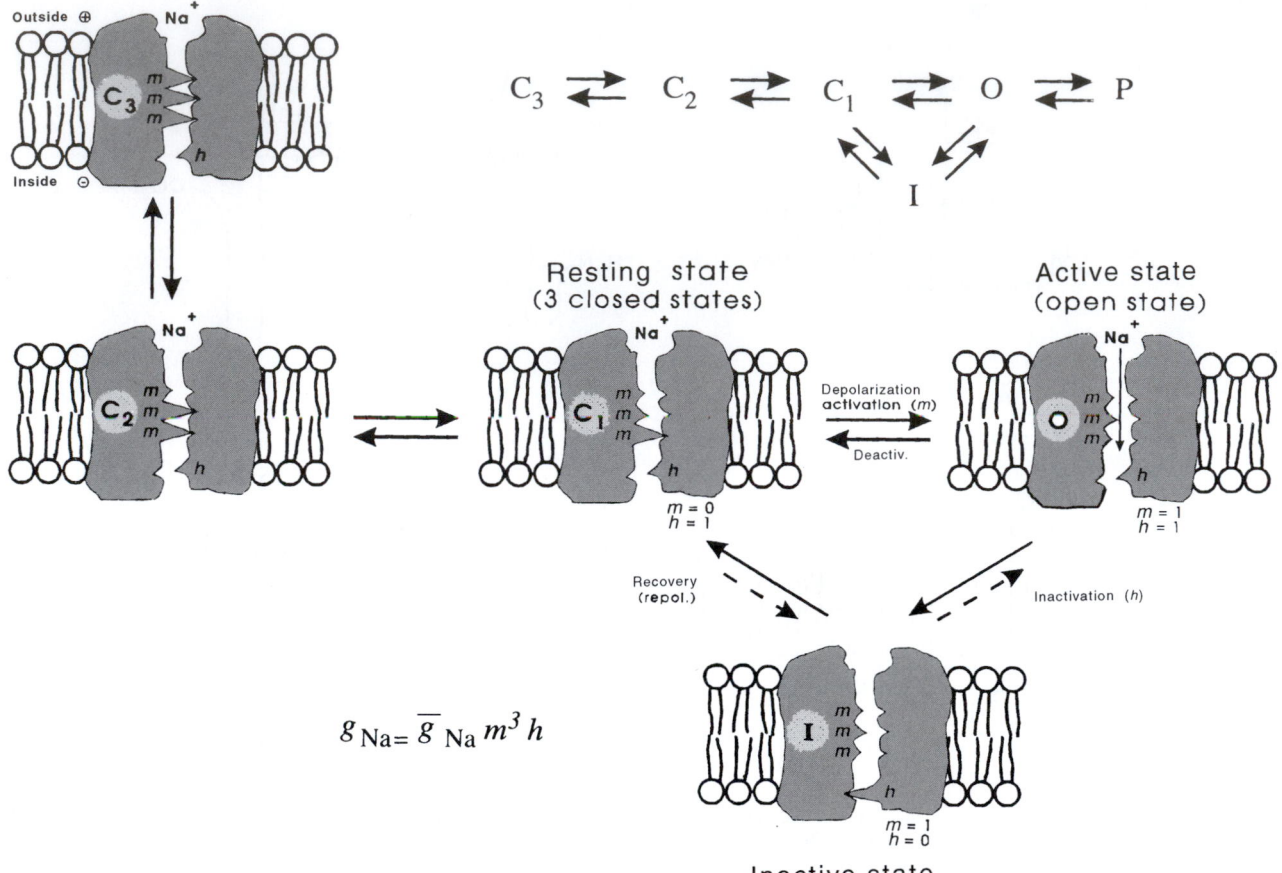

$$g_{Na} = \overline{g}_{Na}\, m^3 h$$

FIGURE 30. Diagram depicting the modified Hodgkin-Huxley view of the three states of the fast Na$^+$ channel. As shown, there is evidence for the existence of three closed states of the channel (C$_3$, C$_2$, and C$_1$). The activation gate (A-gate) is depicted as consisting of three parts, each one being controlled by one m^+ particle. For the A-gate to be open to allow Na$^+$ ions to pass, all three subgates must be open, and therefore all m sites must be occupied. The C$_3$ closed state is depicted as having all three subgates closed, the C$_2$ closed state with two subgates closed (one open), and the C$_1$ closed state with one subgate closed (two open). For the open state (O), of course, all three subgates must be open. P, plugged channel; I, inactive.

temperature and [Ca^{2+}]$_o$). A schematic model is given in Fig. 31, and the corresponding equations are

$$\frac{dn}{dt} = \alpha_n(1 - n) - \beta_n n \qquad (11)$$

$$\frac{dm}{dt} = \alpha_m(1 - m) - \beta_m m \qquad (12)$$

$$\frac{dh}{dt} = \alpha_h(1 - h) - \beta_h h \qquad (13)$$

where n is the fraction of sites occupied, and therefore $(1 - n)$ is the fraction of sites unoccupied. Therefore,

$$\text{rate of occupation} = \alpha_n(1 - n) \qquad (14)$$

$$\text{rate of unoccupation} = \beta_n n \qquad (15)$$

$$\begin{aligned}\text{rate of change of } n = \ &\text{rate of occupation}\\ &- \text{rate of unoccupation}\end{aligned} \qquad (16)$$

or

$$\frac{dn}{dt} = \alpha_n(1 - n) - \beta_n n \qquad (11)$$

The same analysis can be applied to the m and h factors. The effect of making the inside of the fiber more positive (i.e., depolarizing) is to increase α_n, α_m, and β_h and to decrease β_n, β_m, and α_h. One must remember that the h particles occupy the favorable site at the resting potential (I-gate open), and this site becomes unoccupied with depolarization (I-gate closes). This is opposite to the situation for the m and n particles (and A-gate).

In voltage-clamp experiments, at a given clamp step, the membrane potential is fixed and the differential equations (Eqs. 11–13) lead to exponential expressions for n, m, and h, and the K$^+$ and Na$^+$ conductances can then be calculated. When this was done, there was a good fit of the calculated curves with the experimental data points for the changes in g_{Na} and g_K with time at various clamp steps (see Fig. 15).

K. Mechanisms of Repolarization

The AP is terminated primarily by the turn-on of g_K, the **delayed rectification**. It is called the delayed rectifier be-

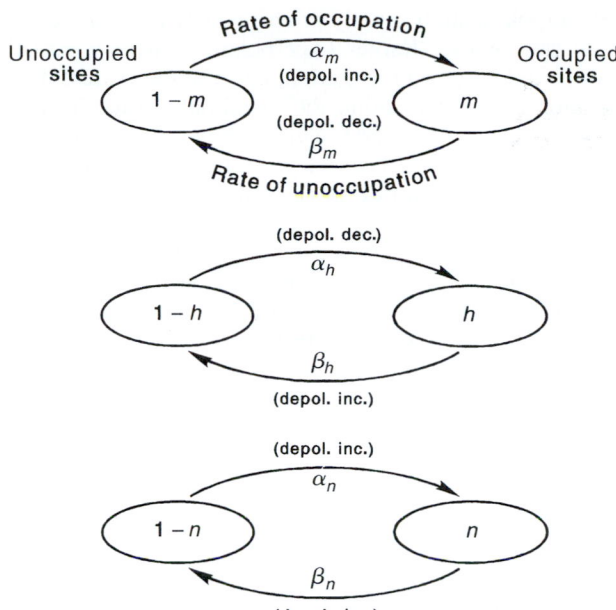

FIGURE 31. Schematic diagram based on the Hodgkin-Huxley model for activation and inactivation of Na$^+$ channels and K$^+$ channels. α and β are the rate constants for occupation and unoccupation, respectively, of sites on the channel protein by the m^+, h^+, and n^+ particles. See text for further explanation.

cause its turn-on is slower and delayed with respect to the turn-on of the fast Na$^+$ channels. The increase in g_K acts to bring E_m toward E_K (about −90 mV), since the membrane potential at any time is determined mainly by the ratio of g_{Na}/g_K (see Chapter 14). This type of g_K channel is activated by depolarization and turned off by repolarization. Therefore, this g_K channel is self-limiting, in that it turns itself off as the membrane is repolarized by its action.

In addition to the g_K turn-on, there is also some turn-off of g_{Na}, which contributes to repolarization. Two reasons for g_{Na} turn-off are (1) spontaneous inactivation of fast Na$^+$ channels that had been activated, that is, closing of their I-gate (inactivation τ of 1–3 ms); and (2) a reversible shifting of activated channels directly back to the resting state because of the rapid repolarization occurring due to the g_K mechanism. Theoretically, it would be possible to have an AP that would repolarize (but slowly) even if there were no g_K mechanism, because the g_{Na} channels would spontaneously inactivate, and so the g_{Na}/g_K ratio and E_m would slowly be restored to their original resting values. Turn-on of g_K acts to sharpen repolarization of the AP and allows higher frequency of impulses.

L. Skeletal Muscle Repolarization

The two mechanisms for sharp repolarization of the AP discussed previously for neurons also apply to skeletal muscle. But there is an important third factor involved in repolarization of the skeletal muscle AP, the Cl$^-$ current. The Cl$^-$ permeability (P_{Cl}) and conductance (g_{Cl}) are very high in skeletal muscle. In fact, P_{Cl} of the surface membrane is much higher than P_K, the P_{Cl}/P_K ratio being about 3–7. However, as discussed in Chapter 14, the Cl$^-$ ion is passively distributed, or nearly so, and thus cannot determine the resting potential under steady-state conditions. However, net Cl$^-$ movements inward (hyperpolarizing) or outward (depolarizing) can and do affect E_m transiently until reequilibration occurs. There is no net electrochemical driving force for Cl$^-$ current (I_{Cl}) at the resting potential, since

$$E_m = E_{Cl} \tag{17a}$$

and

$$(E_m - E_{Cl}) = 0 \tag{17b}$$

However, during AP depolarization, there is a larger and larger driving force for outward I_{Cl} (i.e., Cl$^-$ influx), since

$$I_{Cl} = g_{Cl}(E_m - E_{Cl}) \tag{18}$$

where E_{Cl} is the Cl$^-$ equilibrium potential calculated from the Nernst equation (see Fig. 22). In other words, the large electric field that was keeping Cl$^-$ out (i.e., $[Cl^-]_i \ll [Cl^-]_o$) is diminishing during the AP, and so Cl$^-$ ions enter the fiber. This Cl$^-$ entry is hyperpolarizing, and therefore tends to repolarize the membrane more quickly than would otherwise occur. That is, repolarization of the AP is sharpened by the Cl$^-$ mechanism. The higher the g_{Cl}, the greater this effect.

To illustrate some of the above points, if skeletal muscle fibers are placed into Cl$^-$-free Ringer's solution (e.g., methanesulfonate substitution), depolarization and spontaneous APs and twitches occur for a few minutes until most or all of the $[Cl^-]_i$ is washed out. After equilibration, the resting E_m returns to the original value (ca. −90 mV for frog muscle), clearly indicating that Cl$^-$ does not determine the resting potential and that net Cl$^-$ efflux produces depolarization. Readdition of Cl$^-$ to the bath produces a rapid large hyperpolarization (e.g., to −120 mV) due to net Cl$^-$ influx, and E_m then slowly returns to the original value (−90 mV) as Cl$^-$ reequilibrates (i.e., redistributes itself passively). These same effects would occur in cardiac muscle, smooth muscle, and nerve, but to a lesser extent, because P_{Cl} is much lower in these tissues.

M. Types of K$^+$ Channels

The cell membrane of some cells (e.g., myocardial cells) has at least five separate K$^+$ channels. As discussed previously, one type of voltage-dependent K$^+$ channel is the usual K$^+$ channel found in many types of excitable membranes. This channel slowly opens (increasing total g_K) on depolarization and is the so-called **delayed rectifier** [i_K or $I_{K(del)}$]. This channel allows K$^+$ to pass readily outward down the usual electrochemical gradient for K$^+$, and so is also known as the **outward-going rectifier.** This delayed rectifier channel in myocardial cells turns on much more slowly than in nerve, skeletal muscle, or smooth muscle, and therefore accounts for the long duration of the AP. The activation of this channel produces the increase in total g_K that terminates the cardiac AP plateau (so-called phase 3 repolarization) (see Chapter 52).

A second type, known as the I_{K1} channel, allows K$^+$ ion to pass more readily inward than outward, the so-called inward-going rectifier or anomalous rectification.[1] This kinetically fast channel is responsible for the rapid decrease in K^+ conductance on depolarization (and increase in conductance with repolarization), helps to set the resting potential, and helps bring about the terminal repolarization of the cardiac AP (phase 3). The I_{K1} channel has been found to appear at a certain stage of development of the heart.

A third type of K^+ channel is activated by elevation of $[Ca^{2+}]_i$ and is therefore known as the Ca^{2+}-activated K^+ channel or $I_{K(Ca)}$. With Ca^{2+} influx and internal release of Ca^{2+} during the AP and contraction, these channels are activated and help the $I_{K(del)}$ channels to repolarize the AP, that is, to bring the membrane back to the resting potential. The presence of this type of K^+ channel has been reported for many types of excitable cells. In actuality, two subtypes of $I_{K(Ca)}$ channel have been found: one that has a high conductance of about 400 pS (big or maxi-K channel) and one that has a lower conductance of about 120 pS (small or mini-K channel).

A fourth type of K^+ channel present in some types of cells, such as myocardial cells, is kinetically fast (compared to $I_{K(del)}$) and provides a rapid outward K^+ current that produces a small amount of initial repolarization, known as phase 1 repolarization in cardiac cells. This occurs immediately following the rapidly rising spike portion of the AP and is known as the transient outward current (I_{to}) in myocardial cells and I_A in neurons. There is some evidence that Cl^- current may contribute to I_{to} (Cl^- influx provides an outward I_{Cl}, which is repolarizing in direction).

A fifth type of K^+ channel is sensitive to ATP, the K_{ATP} channel, and provides a current known as $I_{K(ATP)}$. This channel is regulated by ATP, such that in normal myocardial or smooth muscle cells, this K^+ channel is inhibited (masked or silent). However, in ischemic or hypoxic conditions, when the ATP level is lowered, the $I_{K(ATP)}$ channels become unmasked and provide a large outward I_K that prematurely shortens the cardiac AP. This channel provides a protection mechanism for the heart; namely, the ischemic region of the heart develops very abbreviated APs, and hence contraction is greatly depressed. This effect acts to conserve ATP in the afflicted cells, enabling full recovery if the blood flow returns to normal after a short time period.

IV. Effect of Resting Potential on Action Potential

Any agent that affects the resting potential has important repercussions on the AP. Depolarization reduces the rate of rise of the AP, and thereby also slows its velocity of propagation. A slow spread of excitation throughout the nerve or muscle will interfere with its ability to act efficiently. This effect is progressive as a function of the degree of depolarization. If nerve, skeletal muscle fibers, or cardiac muscle cells are depolarized to about −50 mV by any means, then the rate of rise goes to zero and all excitability is lost.

---

[1]It is anomalous in the sense that this channel turns off with depolarization and turns on with hyperpolarization.

Hyperpolarization usually produces only a small increase in the rate of rise. Larger hyperpolarization may actually slow the velocity of propagation, because the critical depolarization required to bring the membrane to its threshold potential is increased, and it can cause propagation block.

The explanation for the effect of resting E_m (or takeoff potential) on maximum rate of rise (max dV/dt) of the AP is based on the sigmoidal h_∞ versus E_m curve (see Fig. 27A). The I (h)-gates are open in a resting membrane and close with depolarization. As discussed previously, h, the inactivation variable for the fast Na^+ conductance, is a probability factor that deals with the open ($h = 1.0$) versus closed ($h = 0$) positions of the inactivation gate of the channel (see Fig. 25). At the resting potential of −80 mV, h_∞ is 0.9–1.0 and diminishes with depolarization, becoming nearly zero at about −30 mV.

The resting potential also affects the duration of the AP. With polarizing current, depolarization lengthens the AP, whereas hyperpolarization shortens it. In contrast, when elevated $[K^+]_o$ levels are used to depolarize the cells, the AP is shortened. One important determinant of the AP duration is g_K. Agents or conditions that increase g_K, such as elevation of $[K^+]_o$, tend to shorten the duration. In contrast, agents that decrease g_K or slow its activation, such as Ba^{2+} ion or TEA^+, tend to lengthen the AP duration. Because of anomalous rectification (i.e., a decrease in g_K with depolarization and an increase with hyperpolarization), depolarization by current prolongs the AP and hyperpolarization shortens it.

Other factors are also important in determining the AP duration. For example, agents that slow the closing of the I-gates of the fast Na^+ channels, such as veratridine, prolong the AP. High rates of activity generally shorten AP duration, for example, by an increase in $[Ca^{2+}]_i$ (resulting from an increase in $[Na^+]_i$) producing an increase in g_K (the $g_{K(Ca)}$ channel).

V. Electrogenesis of Afterpotentials

The APs of nerve and muscle cells usually consist of two components: an initial spike followed by an early afterpotential (Fig. 32A, B). The early afterpotentials may be of two types: depolarizing or hyperpolarizing. In addition, late afterpotentials, both depolarizing or hyperpolarizing, may appear following a brief train of spikes (Fig. 32C, D). The electrogenesis of the early and late afterpotentials is different. The early afterpotentials are due to a conductance change, whereas the late afterpotentials may be due to K^+ accumulation or depletion in restricted diffusion spaces and to electrogenic pump stimulation.

A. Early Depolarizing Afterpotentials

The AP spike in skeletal muscle fibers is immediately followed by a prominent depolarizing afterpotential (also called a negative afterpotential, based on the old terminology used in external recording) (Fig. 32A). The early depolarizing afterpotential of frog skeletal fibers is about 25 mV (immediately after the spike component), and gradually decays to the resting potential within 10–20 ms. It results from

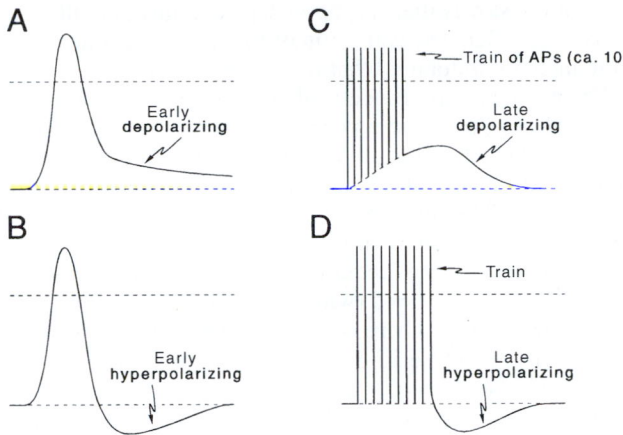

FIGURE 32. Examples of the different types of afterpotentials. (A) Early depolarizing (negative) afterpotential recorded after a single AP in a skeletal muscle fiber. (B) Early hyperpolarizing (positive) afterpotential recorded after a single AP in a nerve fiber. (C) Late depolarizing afterpotential recorded after a train (e.g., 10) of APs in a skeletal muscle fiber. (D) Late hyperpolarizing afterpotential recorded following a train of APs in a nerve terminal.

the fact that the **delayed rectifier K⁺ channel** that opens during depolarization to terminate the spike is less selective for K⁺ (ca. 30:1, K⁺:Na⁺) than is the K⁺ channel in the resting membrane (ca. 100:1, K⁺:Na⁺). Therefore, from the constant-field equation (see Chapter 14), one can predict that the membrane should be partly depolarized when the membrane is dominated by this delayed rectifier K⁺ conductance. Thus, the early depolarizing afterpotential is apparently due to the persistence of, and slow decay of, this less selective K⁺ conductance.

B. Early Hyperpolarizing Afterpotentials

Neurons, pacemaker heart cells, and vascular smooth muscle cells often exhibit early hyperpolarizing (positive) afterpotentials (Fig. 32B). These are due to the **delayed rectifier K⁺ conductance** increase (which terminates the spike) persisting after the spike, thereby bringing E_m closer to E_K. The maximum amplitude possible for the afterpotential is the difference between E_K and the normal resting potential. The time course of this afterpotential is determined by the decay of the K⁺ conductance increase.

C. Late Depolarizing Afterpotentials

The late depolarizing afterpotential results from **accumulation of K⁺ ions in the transverse (T) tubules** (Fig. 32C). During the AP depolarization and turn-on of the g_K (delayed rectifier), there is a large driving force for K⁺ efflux from the myoplasm coupled with a large K⁺ conductance, resulting in a large outward K⁺ current ($I_K = g_K (E_m - E_K)$) across all surfaces of the fiber, namely, the surface sarcolemma and T-tubule walls. The K⁺ efflux at the fiber surface membrane can rapidly diffuse away and mix with the relatively large in-

terstitial fluid (ISF) volume, whereas the K⁺ efflux into the T-tubules is trapped in this **restricted diffusion space.** The resulting high $[K^+]_{TT}$ decreases E_K across the T-tubule membrane and thereby depolarizes this membrane. Because of cable properties, part of this depolarization is transmitted to the surface sarcolemma and is recorded by an intracellular microelectrode. The K⁺ accumulation in the T-tubules can only be dissipated relatively slowly by diffusion out of the mouth of the T-tubules and by active pumping back into the myoplasm across the T-tubule wall. Thus, the decay of the late afterpotential will be a function of these two processes.

The amplitude and duration of the late depolarizing afterpotential of frog skeletal fibers are functions of the number of spikes in the train and their frequency. That is, the greater the spike activity, the greater the amplitude and duration of the late depolarizing afterpotential. If the train consists of 20 spikes at a frequency of 50/s, a typical value for the amplitude of the afterpotential is about 20 mV. When the diameter of the T-tubules is increased by placing the fibers in hypertonic solutions, the amplitude of the afterpotential decreases, as expected, because of the greater dilution of the K⁺ ions accumulating in the T-tubule lumen. When the **T-tubular system is disrupted** and disconnected from the surface membrane by the glycerol osmotic shock method, the late depolarizing afterpotential disappears (whereas the early depolarizing afterpotential persists).

A slow relaxation of a component of the K⁺ conductance increase may contribute to the late afterpotential.

D. Late Hyperpolarizing Afterpotentials

Some cells, such as nonmyelinated neurons, exhibit late hyperpolarizing afterpotentials following a train of spikes (Fig. 32D). These hyperpolarizing afterpotentials are due to the **Na⁺-K⁺ pump,** because inhibition of the pump by any means (such as ouabain) abolishes them. Two mechanisms have been proposed for the hyperpolarization. (1) Hyperpolarization occurs because there is an increased **electrogenic Na⁺ pump potential (V$_p$),** stimulated both by an increase in the $[Na^+]_i$ (since these neurons are small in diameter, and hence have a large surface area/volume ratio) and by an increase in $[K^+]_o$ (since these axons are surrounded by Schwann cells and hence have a narrow intercellular cleft and restricted diffusion space). (2) Hyperpolarization occurs due to an increased E_K caused by K⁺ depletion in the intercellular cleft because of the stimulated Na⁺-K⁺ pump overpumping the K⁺ back in. It is generally believed that the first mechanism is the most probable.

E. Importance of Afterpotentials

All afterpotentials have physiological importance because they alter the excitability and propagation velocity of the cell. A depolarizing afterpotential should enhance excitability (lower threshold), and a hyperpolarizing afterpotential should depress excitability to a subsequent AP. This is because the critical depolarization required to reach the threshold potential would be decreased or increased, respectively. A large late depolarizing afterpotential, such as that due to K⁺ accumulation in the T-tubules, can trigger repetitive APs under certain pathological conditions.

The effect of afterpotentials on velocity of propagation is complex because there are two opposing factors: (1) the change in critical depolarization required and (2) the change in maximal rate of rise of the AP, which is a function of the takeoff potential (h_∞ versus E_m curve). For example, during a depolarizing afterpotential in skeletal muscle fibers, the critical depolarization required is decreased, but the maximal rate of rise of the AP is also decreased. Therefore, these two factors exert opposing effects.

VI. Summary

Membrane excitability is a fundamental property of nerve and muscle cells (skeletal, cardiac, and smooth), as well as certain other cell types, such as some endocrine cells. An excitable cell is one that, in response to certain environmental stimuli (electrical, chemical, or mechanical), generates an all-or-none electrical signal or AP. The AP is sometimes called an "impulse," for example, a nerve impulse. The AP is triggered by a depolarization of the membrane, which is produced by the applied stimulus. The depolarization initiates an increase in the membrane permeability to Na$^+$ ions, which then flow into the cell, causing a transient reversal in the membrane potential. A slower increase in the permeability of the membrane to K$^+$ ions contributes to the repolarization of the membrane, in addition to the spontaneous inactivation of the Na$^+$ channels. Some cells display the property of automaticity; that is, they produce APs spontaneously without any externally applied stimulus.

In some excitable cell types (such as cardiac), a slow inward Ca^{2+} current contributes to the long plateau of the AP, and in others (such as smooth muscle) the Ca^{2+} current itself generates the upstroke. The membrane currents that contribute to the AP can be studied by the voltage-clamp method, which allows isolation and characterization of each membrane current as a function of membrane potential and time. Ionic currents flow across the membrane by means of numerous ion-specific protein channels. Each channel molecule has a region that senses the transmembrane potential and acts as a gate to open or close the channel to ion passage through its central pore.

Some types of channels (i.e., Na$^+$ channels) have a second gating system that closes (or inactivates) the channel during a maintained depolarization. The pattern of minute currents that flow through individual voltage-dependent ion channels can be studied using the patch-clamp method. Single-channel current measurement and structural information have given a greater understanding of the molecular basis of membrane excitability.

The APs in vertebrate nerve fibers consist of a spike followed by a hyperpolarizing afterpotential. A large fast inward Na$^+$ current, passing through fast Na$^+$ channels, is responsible for electrogenesis of the spike, which rises rapidly (~1000 V/s). Subsequently, a small inward Ca^{2+} current, passing through kinetically slower channels, may be involved in excitation-secretion coupling at the nerve terminals. The nerve cell membrane has voltage-dependent K$^+$ channels which allow K$^+$ ions to pass readily outward down the electrochemical gradient for K$^+$. This channel population is responsible for repolarization. This K$^+$ channel, which opens more slowly than g_{Na} upon depolarization, is called the delayed rectifier. The activation of this channel produces the large increase in total g_K that terminates the AP.

The nerve AP amplitude is about 110 mV, from a resting potential of −70 mV to a peak overshoot potential of about +40 mV. The duration of the AP (at 50% repolarization) ranges between 0.5 and 1 ms, depending on the species and temperature. The threshold potential (V_{th}) for triggering of the fast Na$^+$ channels is about −55 mV; a critical depolarization of about 15 mV is required to reach V_{th}. The turn-on of the fast Na$^+$ is very rapid (within 0.2 ms), and E_m is brought rapidly toward E_{Na}. There is an explosive (positive exponential initially) increase in g_{Na} caused by a positive-feedback relationship between g_{Na} and E_m.

In certain nerve cells, as well as muscle cells, as E_m depolarizes it crosses V_{th} (about −35 mV) for the slow Ca channels. Turn-on of the slow Ca^{2+} conductance (g_{Ca}) and I_{Ca} is slow and tends to bring E_m toward E_{Ca}. The peak I_{Ca} is considerably smaller than the peak fast I_{Na}. From the voltage-clamp current versus voltage curves, the maximum inward fast current and slow current occur at an E_m of about −20 and +10 mV, respectively. The currents decrease at more depolarized E_m levels because of the diminution in electrochemical driving force as the membrane is further depolarized, even though the conductance remains high. At the reversal potential (E_{rev}) for the current, the electrochemical driving force goes to zero and reverses direction with greater depolarization.

In some smooth muscle cells, the slow Ca^{2+} current itself is sufficient to depolarize the membrane and generate a regenerative slowly rising AP in the absence of the fast Na$^+$ current.

Bibliography

Bernstein, J. (1902). Untersuchengen zür Thermodynamik der bioelektrischen Ströme. *Pflügers Arch.* **92**, 521.

Bernstein, J. (1912). "Elektrobiologie." Braunschweig, Vieweg.

Catterall, W. A. (1988). Structure and function of voltage-sensitive ion channels. *Science* **242**, 50–61.

Cole, K. S. (1949). Dynamic electrical characteristics of the squid axon membrane. *Archs. Sci. Physiol.* **3**, 253–258.

Cole, K. S., and Curtis, H. J. (1939). Electric impedance of the squid giant axon during activity. *J. Gen. Physiol.* **22**, 649–687.

Hamill, O. P., Marty, A., Neher, E., Sakmann, B., and Sigworth, F. J. (1981). Improved patch-clamp technique for high resolution current recording from cells and cell-free membrane patches. *Pflügers Arch.* **391**, 85–100.

Hamill, O. P., and Sakmann, B. (1981). Multiple conductance states of single acetylcholine receptor channels in embryonic muscle cells. *Nature* **294**, 462–464.

Hille, B. (1984). "Ionic Channels of Excitable Membrane." Sinauer Associates, Sunderland, MA.

Hodgkin, A. L., (1958). Ionic movements and electrical activity in giant nerve fibres. *Proc. Roy. Soc. B.* **148**, 1.

Hodgkin, A. L., (1964). "The Conduction of the Nervous Impulse." Liverpool University Press, Liverpool, UK.

Hodgkin, A. L., and Huxley, A. F. (1952a). Currents carried by sodium and potassium ions through the membrane of the giant axon of *Loligo. J. Physiol. (London)* **116**, 449–472.

Hodgkin, A. L., and Huxley, A. F., (1952b). The components of membrane conductance in the giant axon of *Loligo. J. Physiol. (London)* **116**, 473–496.

Hodgkin, A. L., and Huxley, A. F., (1952c). The dual effect of membrane potential on sodium conductance in the giant axon of *Loligo*. *J. Physiol. (London)* **116,** 497–506.

Hodgkin, A. L., and Huxley, A. F., (1952d). A quantitative description of membrane current and its application to conduction and excitation in nerve. *J. Physiol. (London)* **117,** 500–544.

Hodgkin, A. L., and Katz, B. (1949). The effect of sodium ions on the electrical activity of the giant axon of the squid. *J. Physiol. (London)* **108,** 37–77.

Katz, B. (1966). "Nerve, Muscle and Synapse." McGraw-Hill, New York.

Ohya, Y., and Sperelakis, N. (1989). Fast Na^+ and slow Ca^{2+} channels in single uterine muscle cells from pregnant rat. *Am. J. Physiol. Cell* **257,** C408–C412.

Overton, E. (1902). Beiträge zur allgemeinen Muskel- und Nervenphysiologie. *Pflügers Arch.* **92,** 346.

Sperelakis, N. (1988). Electrical properties of cells at rest and maintenance of the ion distributions. *In* "Physiology and Pathophysiology of the Heart" (N. Sperelakis, Ed.), pp. 59–82. Kluwer, New York.

Sperelakis, N., and Fabiato, A. (1985). Electrophysiology and excitation-contraction coupling in skeletal muscle. *In* "The Thorax: Vital Pump" (C. H. Roussos and P. Macklem, Eds.), pp. 45–113. Marcel Dekker, New York.

Sperelakis, N., and Josephson, I. (1992). Basis of membrane excitability. *In* "Essentials of Physiology" (N. Sperelakis and R. O. Banks, Eds.), pp. 49–97. Little, Brown, Boston.

Tohse, N., and Sperelakis, N. (1990). Long-lasting openings of single slow (L-type) Ca^{2+} channels in chick embryonic heart cells. *Am. J. Physiol.* **259,** H639–H642.

Tohse, N., and Sperelakis, N. (1991). cGMP inhibits the activity of single calcium channels in embryonic chick heart cells. *Circ. Res.* **69,** 325–331.

Raymund Y. K. Pun and Harold Lecar

26

Patch-Clamp Techniques and Analysis

I. Introduction

Since the initial publication of this chapter the number of articles dealing with the technique of patch-clamping has increased several fold. Numerous books, reviews, and chapters have been written on this topic. It is not possible to cover all aspects of the technique because of the vastness of the literature. Any omission of relevant papers is unavoidable. We attempt to make this chapter as general as possible to provide the reader with a flavor of the technique. We have rearranged the contents, reduced the description of the variety of patch-clamp techniques, and included a brief description of whole-cell current analysis. We hope that this addition will provide a basis for the comparison of results obtained between whole-cell data versus single-channel measurements.

The **patch-clamp technique** revolutionized electrophysiology by revealing the activity of individual molecular ion channels involved in electrical signaling in excitable cells. Just as the development of the voltage-clamp 50 years ago led to the understanding of nerve excitation in terms of specific transmembrane ionic currents, the development of the patch-clamp or **gigaseal technique** provided the impetus for analyzing excitation phenomena at the molecular level in terms of the behavior of specific molecular ion channels.

This chapter and the several that follow cover the use of the patch-clamp technique in studies of both macroscopic ionic currents and single ion channels. The chapter first focuses on the technique and its application in studies of single channels or unitary conductances, and then covers the technique of whole-cell "tight-seal" voltage-clamp, an offshoot of the patch-clamp technique that is now the most widely used method of examining membrane excitability.

The ionic basis of the action potential was established in the early 1950s by Hodgkin and Huxley (1952a, b, c) using voltage-clamp on the squid giant axon. As described in Chapter 25, they demonstrated that depolarization of the membrane elicits an inward transient Na^+ current, followed by an outward K^+ current. The Na^+ current is activated rapidly, initially following a sigmoidal time course, reaching a peak within 0.5–1 ms, then declining more slowly (2–5 ms; measurement was made at 6.3 °C). The K^+ current also follows a sigmoidal time course, but with a longer delay, and is more or less sustained if the membrane remains depolarized.

From these and later studies (radioactive isotope measurements), it was inferred that these currents reflect the movement of ions across the membrane via separate ion-specific pathways, which are activated by changes in the transmembrane electric field.

Experiments with neurotoxins showed that ion flux could be blocked by binding to sites of such low density that the rate at which ions flow across a single conducting site was high (10^6–10^7 ions/s). Such high unitary transport rates can only be explained by localized pores that must gate open and closed to regulate the intensity of ion flux. The random gating of individual ion channels at equilibrium produces large current fluctuations, both as electrical noise in intact cells and as quantized single-channel jumps in small patches of membrane. Pharmacological dissection with site-specific channel-blockers further enables investigators to isolate currents from different species of channels, each with its own characteristic ion selectivities and gating kinetics.

Thus, even before the isolation and cloning of channel proteins, ion channels were identified as intrinsic membrane proteins, operating as discrete independent conducting pathways. Channel gating gives rise to the transient changes in membrane potential that directly trigger nerve excitation as well as the ionic messengers that trigger muscle contraction and secretory processes. In general, an influx of cations (or efflux of anions) will depolarize the membrane potential and increase the generation of action potentials or firing rate, whereas an efflux of cations (or influx of anions) will hyperpolarize and decrease excitability. The Na^+ and K^+ channels that underlie action potential generation are **voltage-gated channels**, because the rates of opening and closing of these channels are sharply dependent on changes in the membrane potential. Another class of channels, known as **ligand-gated channels**, is activated through neurotransmitter or hormone binding. A prototype of this class of channel is the nicotinic acetylcholine (ACh) channel present at the skeletal neuromuscular junction, which was the first ligand-gated channel to be studied thoroughly (see Chapter 40). Many other types of voltage-gated channels and ligand-gated channels have now been discovered and characterized (discussed in Chapters 27, 40 and 47).

Before the invention of the patch-clamp technique, studies using lipid bilayers doped with various channel-forming peptides and proteins showed how certain reconstituted ion channels open in discrete steps or jumps (Ehrenstein and Lecar, 1977). The implication was that channels undergo random transitions between discrete states of different conductance, and that the biasing of transition rates between the conductance states by stimuli such as changes in the applied electric field leads to macroscopic excitability phenomena. Direct confirmation of the existence of channels with discrete states of conductance in intact cell membranes came with the advent of the patch-clamp.

II. Patch-Clamp or Gigaseal Technique

To measure the minute ionic currents carried by an individual protein channel, one must be able to voltage-clamp an electrically isolated patch of membrane containing only a few channels, hence the term "patch-clamp." The currents recorded from the patch must also have a background noise level sufficiently low for the single-channel jumps to be detectable. Neher and Sakmann (1976) first measured single-channel currents in a cell membrane by pressing a blunt-tipped pipette against the surface of a denervated muscle fiber (freed from overlying cells and connective tissue by collagenase treatment), and feeding the signal through a low-noise head stage current-to-voltage converter. A schematic diagram of current measurement through an isolated membrane patch is shown in Fig. 1. The equivalent circuit of this arrangement is shown in Fig. 2A. Detailed analysis of the circuit and the effects of the various components (e.g., shunt resistance, cell resistance) on the minimization of background noise are described by Hamill and coworkers (1981), Fenwick and colleagues (1982), and Barry and Lynch (1991). Several recent books provide a wealth of information on all aspects of the patch-clamp tech-

nique (Sakmann and Neher, 1995; Rudy and Iverson, 1992; Ashley, 1995; Boulton *et al.*, 1995).

A. Noise in Patch-Clamp

The main consideration in the design of a patch-clamp experiment is to minimize the background electrical noise that interferes with and obscures the single-channel current fluctuations. The minimum detectable current-jump amplitude in the presence of interfering electrical noise depends not only on the intensity of the noise but also on the duration of the jump. Shorter duration pulses and short-duration flickering channels require broader receiver bandwidth in order to be resolved with some fidelity. The background noise generated in the membrane, the patch pipette, and the recording electronics is spread out in frequency. Hence, design of the patch-clamp requires some compromise between minimizing the bandwidth for noise reduction and increasing it for fidelity of recording. Patch-clamp amplifiers achieve the optimum compromise by employing a first-stage current-to-voltage converter, in which gain and effective noise are minimized at the expense of slow response, and by having a second-stage frequency booster, which reconstructs the sharp channel jumps.

Background noise is generated in all the conductance pathways in parallel with the channels. Thus, particular attention must be paid to the thermal noise generated across the pipette-to-membrane seal and across the pipette wall. In early patch-clamp experiments, the current noise generated across the pipette-to-membrane seal, whose rms amplitude is inversely

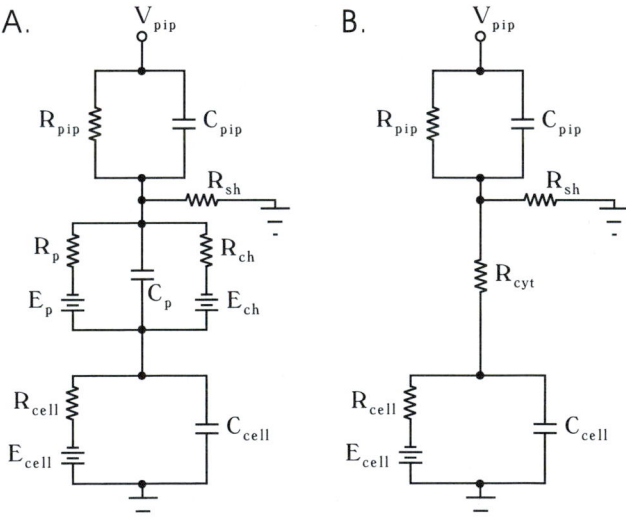

FIGURE 2. Equivalent circuit diagrams for (A) cell-attached patch-clamp recording and (B) whole-cell tight-seal recording (See Section IV.A). V_{pip}, R_{pip}, and C_{pip} in the equivalent circuit diagram denote the voltage, resistance, and capacitance of the pipette, respectively; R_{sh} denotes the shunt of seal resistance; E_p, R_p, and C_p denote the driving potential, resistance, and capacitance of the patch of membrane, respectively; E_{cell}, R_{cell}, and C_{cell} denote the membrane potential, resistance, and capacitance of the cell, respectively; and R_{cyt} is the resistance of the cytoplasm after establishing whole-cell tight-seal recording. The series or access resistance is the sum of R_{pip} and R_{cyt}. The series resistance and the current flowing across this resistor determine the voltage drop across the pipette.

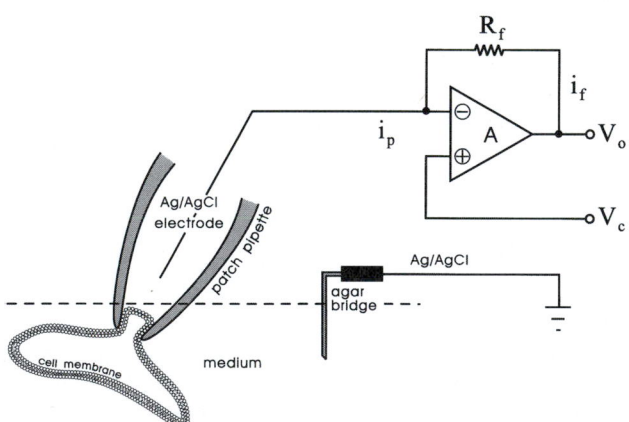

FIGURE 1. Schematic diagram depicting cell-attached patch-clamp recording. A Ag/AgCl wire immersed in the pipette solution acts as the recording electrode. The ground is another Ag/AgCl electrode that interfaced the bath with an agar bridge (1–3% agar in 0.9% saline). The signal input is into amplifier A. Under the voltage-clamp condition, the voltage to the input ($\ominus$ junction of A) is maintained at the same potential as V_c, the command potential, such that $V_o = -V_c$. Under this condition, $i_f R_f = -i_p R_f$, where i_f is the feedback current, i_p the pipette current, and R_f the feedback resistance.

proportional to the square root of seal resistance, was the limiting factor in patch-clamp sensitivity. The multi-gigaohm seal reduces this seal noise to a minimal level, leaving the noise generated across the pipette walls by several different mechanisms as the main source. Current flowing through the layer of electrolyte clinging to the pipette wall is transmitted across the pipette-wall capacitance, which acts as a high-pass filter. The pipette noise is suppressed by making the electrode of highly resistive glass and/or coating the walls with a hydrophobic sealant. The use of quartz glass electrodes (Levis and Rae, 1993) coupled with a cooled capacitor-feedback headstage (Axon Instruments, Inc.) can further reduce the electrical noise level. The detailed analysis of the sources of electrical noise operating at different frequencies and the principles underlying design of a patch-clamp are discussed by Hamill and coworkers (1981). When all the sources of background noise are taken into account, a patch with a 100 GΩ seal and 2 kHz recording bandwidth allows the observation of 1-ms channel jumps of amplitude as low as 0.05 pA. This gives some indication of the smallest unit of conductance observable by patch-clamp at present (order of 1 pS).

The detection of single-channel jumps is regarded as the *sine qua non* of transport by channels—sites of intense flux—as opposed to carriers or other transport pathways, for which flux rate is limited by the rate at which ion-bearing sites can be translocated across the membrane. Thus detection of single-channel currents constitutes a sufficient condition for identifying channel transport but not a necessary one. There may be membrane-spanning pathways that transport too slowly to yield observable single-channel jumps. In such cases, the unitary conductances might be inferred from macroscopic membrane noise analysis, as discussed in Section IV.D.

B. Variations of the Patch-Clamp Technique

Variations in the technique also proved to be successful in determining channel properties (Hamill *et al.*, 1981). The method used in studying the behavior of single channels residing within the electrically isolated patch of membrane described earlier is known as the **cell-attached patch**, since the cell is firmly attached to the recording pipette (Fig. 3, left). A cell-attached patch is so well isolated from the rest of the membrane that it can be used to distinguish between direct and indirect second-messenger activation of postsynaptic channels. In addition, the relatively large size of the patch pipette facilitates solution access to the tip, allowing rapid changes in pressure and solution. Since the suction on a tightly sealed patch can be controlled with some stability, the pressure patch-clamp is virtually the only method for studying stretch-sensitive ion channels, which gate in response to changes in membrane tension and are probably of importance in various mechanosensing processes. Internal perfusion of the patch pipette allows for rapid change of solution on either side of an excised membrane patch so that changes in ligand-gated channel opening can be observed in response to step changes in neurotransmitter concentration.

Examination of ligand-gated channels can also be accomplished using the outside-out configuration (discussed later). One drawback of the cell-attached configuration is the inability to determine accurately the cell's membrane potential, which has an effect on the patch potential. To determine the actual potential across the membrane patch, one can measure the cell's potential with another recording pipette and then set the voltage of the patch accordingly, or alternatively, depolarize the cell with high-K$^+$ solution to "zero" the membrane potential before setting the membrane patch potential. A less accurate means of determining the patch potential is either to use a potential relative to that of the cell (e.g., +20 mV to rest) or to estimate the membrane potential before imposing a voltage across the patch.

When a gigaohm seal is formed, there is a strong bond between the glass surface of the patch pipette and the membrane. The membrane patch is stuck to the pipette and can be excised from the cell by a variety of methods, as illustrated in Fig. 3. If the pipette is withdrawn and pulled away from the cell, the piece of membrane (estimated area of 1–10 μm^2) sealed against the pipette is torn away from the cell. Provided this action does not alter seal resistance drastically, channel activities can still be recorded. Occasionally, a membrane vesicle (instead of a simple planar patch) is formed at the tip following excision from the cell, but the vesicle can be broken by passing the pipette rapidly across the fluid-air interface. This variation is known as the isolated

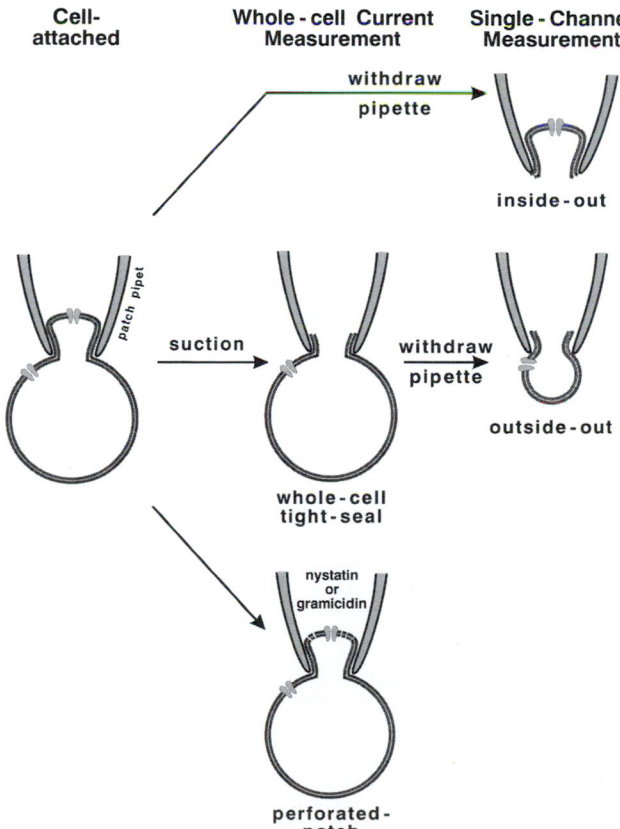

FIGURE 3. Various configurations of patch-clamp (single-channel current) and whole-cell tight-seal (macroscopic current) recordings. For a perforated patch, the antibiotic used could be nystatin, amphotericin, or gramicidin. See text for description.

inside-out patch because the cytoplasmic side of the membrane is now facing the outside or bath solution (Fig. 3, top right).

The excised patch permits the experimenter to control the voltage across the membrane patch, and thereby to effect the voltage-clamp of single-channel currents over a wide voltage range (±200 mV). This configuration can be used to investigate the effects of externally applied second messengers on the behavior of the channel. Occasionally, there is a loss or decrease in channel activity with time. This is probably because regulatory components that are not tightly bound to the channel protein can diffuse away from the membrane and become lost. It may be possible to sustain channel activity for a longer duration by addition of appropriate agents such as BAY-K 8644 to the bath medium of an inside-out patch as reported by Ohya and Sperelakis (1989).

Another variation of the excised-patch technique is the **outside-out patch**, in which the external surface of the membrane faces the external bath medium (Fig. 3, middle right). This configuration is comparable to the normal whole-cell tight-seal recording mode (see Section IV.A), with the exception that only a small patch of membrane is being examined. The outside-out patch is obtained by withdrawing the pipette after the establishment of the whole-cell configuration, extruding a tube of membrane, which eventually pinches off to form a membrane patch with the outside surface facing the bath. The outside-out mode of recording is most useful for studying ligand-gated channels, because it is possible to vary the concentration of a ligand in the bath and determine its effect on channel activity. With a cell-free patch a known specific ligand-activated channel can be used as a bioassay for detecting neurotransmitter in the neighborhood

of a synapse, providing a novel solution to the venerable problem of transmitter identification.

III. Single-Channel Analysis

Single-channel recording reveals the discrete steps of ionic current that occur when individual ion channels form nonconducting (closed) and conducting (open) states. Under steady-state conditions, channels fluctuate randomly between different conformations over time periods typically on the order of milliseconds, but ranging from 10^{-5} s to several seconds for different channels. At the single-channel level, the kinetics of the gating process is characterized by statistical distributions of inter-event times, which can be related to the rates of state-to-state transition. Single-channel data can provide a detailed description of the kinetic behavior of a single protein. When single-channel data are summed, as shown in Fig. 4, one obtains an ensemble current that reflects the current flowing across the entire cell surface. The figure illustrates that though the macroscopic current can be the same in magnitude and time course, the behavior of the channel in each of the cases shown is very different, from single opening per stimulus (Fig. 4A), to reopening of the channel during the stimulus (Fig. 4B), to multiple openings during the stimulus. The behavior of the channel depends on the kinetics scheme described in the figure legend.

The random gating transitions lead to macroscopic excitability phenomena, because the rates of transition between states can be regulated by specific stimuli, such as the change in transmembrane electric field or the binding of ligands. Appropriate stimuli acting on the channel structure are

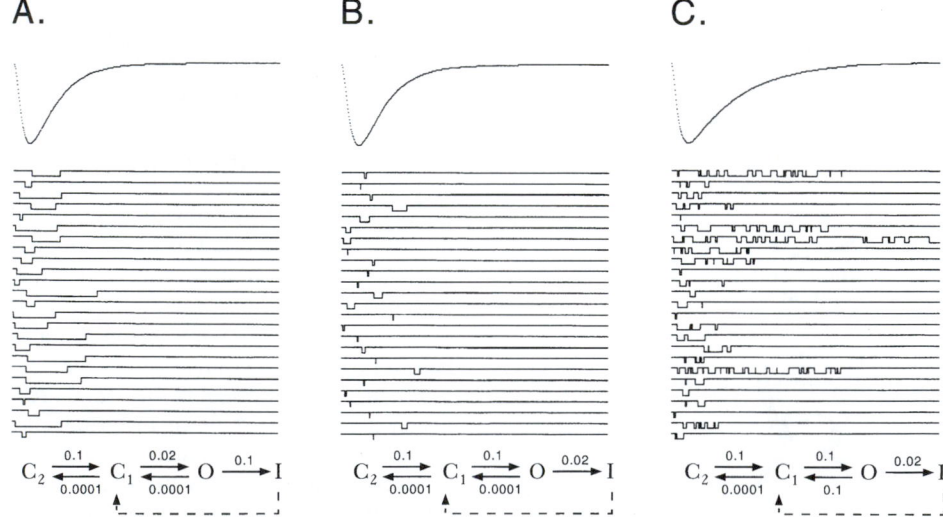

FIGURE 4. Computer-simulated single-channel records and ensemble current. Simulated single Na^+-channel records were generated by a computer program based on a $C \rightarrow C \rightarrow O \rightarrow I$ kinetic scheme. The rates used to calculate channel openings are shown at the bottom of the figures. Macroscopic currents are obtained by summation of more than 40 000 single-channel records. (A) Slow activation and fast inactivation. (B) Based on the Hodgkin-Huxley model (fast activation and slow inactivation). (C) A very slow inactivation rate. The macroscopic currents from schemes A and B are identical, except that about a quarter of the number of traces is required to generate the summed current for scheme B. Scheme C has comparable macroscopic current with a slightly slower inactivation. Note that the single-channel behavior for each case is very different. The dashed line indicates a possible connection between the inactivated (I) and closed (C) state. (Figure provided by Dr. Shirley Bryant.)

able to alter the free energy difference between conformational states, and thereby bias the rates of transition between them. The transition rates are highly temperature sensitive, suggesting that the various states are separated by relatively high free energy barriers. The physical features of the channel structure that allow sensitive coupling to different physical stimuli thus act as sensors for transducing the various types of stimuli into electrical signals.

A. Families of Gated Channels

Patch-clamp experiments have revealed families of channels obeying various modes of gating transduction:

1. *Voltage-sensitive gating.* The Na^+, K^+, and Ca^{2+} channels of nerve and muscle have very steeply voltage dependent opening probabilities. These channels have highly charged subunits that must be displaced by the transmembrane electric field in order to effect channel opening. The steepness, an e-fold change in open-state probability per 3–6 mV, suggests that the equivalent of four to six elementary (univalent) charges moved across the membrane to bring about channel opening.
2. *Ligand-sensitive gating.* Postsynaptic channels and other chemosensitive channels have extracellular receptors that bind transmitter substances. Transmitter binding alters the free energies of the open and closed states to facilitate opening while the transmitter is bound to its receptor site. Fast synaptic channels, such as the ACh-sensitive channel of the neuromuscular junction and the GABA channel of inhibitory synapses, have two ligand-binding sites leading to some cooperative action of the ligands in activating the channel.
3. *Internal messenger-activated gating.* These are channels sensitive to the binding of intracellular ligands, such as Ca^{2+} and cyclic nucleotides. These channels are generally gated as a steep function of messenger concentration, suggesting that they have multiple binding sites that must be occupied for maximum rate of opening.
4. *G-protein gating.* These are channels whose gating is modified by membrane-bound G-proteins, which in turn are coupled to separate receptor molecules that release them when activated by the appropriate ligand. The G-proteins diffuse laterally within the membrane and activate the channels they encounter (see Chapter 34).
5. *Stretch-activated gating.* An ubiquitous class of channels can be activated by mechanical tension applied to the membrane. In specific stretch transducer cells, these channels can be coupled to elastic elements of the cytoplasmic structure (see Chapters 36 and 44).

B. Random Nature of Channel Gating

For all of the different gating mechanisms, channel activation is always observed as a random switching between discrete conformational states of the channel protein. The stimulus does not deterministically drive the channel to switch conformations in a fixed time, or to change conductance gradually. Rather, stimuli that induce gating act to bias

the rates of an ongoing thermally driven random process. This provides the rationale for modeling channel gating as a stochastic process, in which a system having a fixed number of discrete states passes from one state to another via first-order kinetic transitions.

If the channels reside in any specific metastable state long enough on the average that the transitions are not affected by prior history or the manner of entry into the state, the interstate transitions will be memoryless. Such a **Markov process** can be described by a set of equations relating the occupancy probabilities of the states to the transition rates between every pair of states.

The single-channel currents do not represent all the transitions among the states but only those that yield a measurable electric current jump. Some of the transitions are electrically silent, although their effects are implicit on the statistics of the channel transitions. A process in which only some of the underlying transitions are directly observable is called an **aggregated Markov process.**

Generally, a gated ion channel is a system with multiple open and closed states represented by a Markov process with N states, and hence described by a set of N probabilities P_i, which obey the set of first-order equations

$$dP_i / dt = \sum k_{ij} P_j - \sum k_{ji} P_i = \sum Q_{ij} P_j \qquad (1)$$

Here, the k_{ij} are the transition rates from the jth to the ith state. The transition rates of a particular model can be gathered in an array that is the $N \times N$ matrix Q. The transitions can be partitioned into groups as observable transitions between an open or closed state or as unobservable transitions between two closed states or two open states. Thus the system can be partitioned into two classes of states and the matrix Q can be partitioned to show explicitly how the system moves between the two classes of states. These partitioned matrices can be used to predict the distributions of open and closed dwell times, which are multiexponential functions representing sojourns in either of the two collections of states. If all the Q_{ij} were nonzero, so that every state was connected to every other state, the system would be too complex to analyze. However, most of the kinetic schemes that might be suggested on physical or aesthetic grounds do not involve transitions between every pair of states.

To describe observable quantities, we divide the states into two classes, open and closed. The matrix of the transition probabilities can be thought of schematically as

$$Q = \begin{bmatrix} \textbf{open–open} & \textbf{open–closed} \\ \textbf{closed–open} & \textbf{closed–closed} \end{bmatrix}$$

Here, the diagonal submatrices represent unobservable transitions that occur during a sojourn in a complex of states having the same conductance, whereas the off-diagonal submatrices represent the observable transitions by which a channel opens or closes.

Figure 5 shows examples of kinetic schemes that have been proposed for different gated channels. Each of the schemes is illustrated by a diagram indicating the states and the allowed transitions between them. The matrix of transition rates shows all the independent rate constants needed to specify a model and the dependence of the rates on external

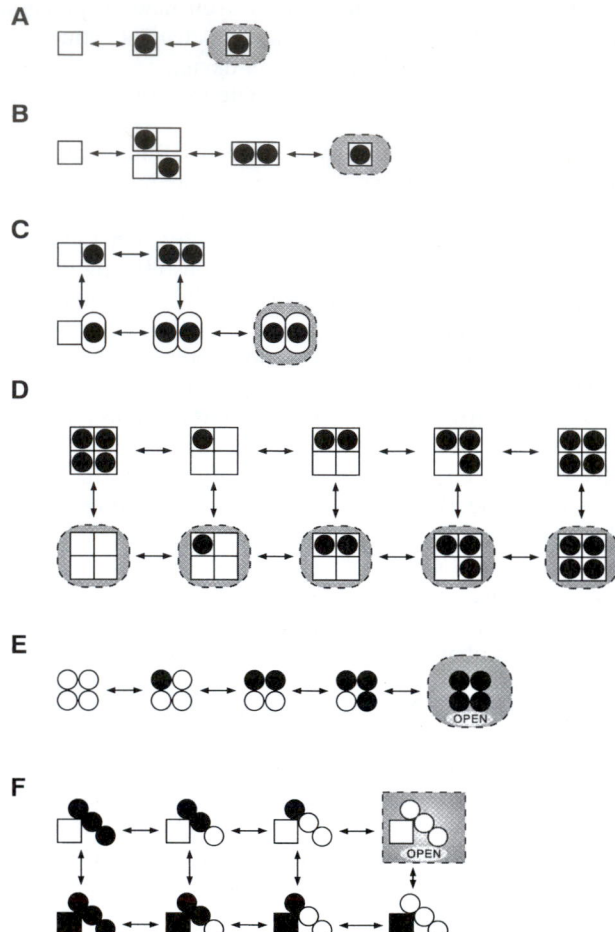

FIGURE 5. Kinetic diagrams for various models of gating. (A–D) Schemes of increasing complexity for ligand-activated gating. Subunits of the channel protein are shown under transitions to activated states on ligand binding, with further activation leading to channel opening. Open states are indicated by shading. (E and F) Kinetic diagrams for voltage-gated channels, which would follow Hodgkin-Huxley kinetics, literally interpreted. The K⁺ channel is opened when four activation subunits are simultaneously in the open position. The state of the Na⁺ channel is determined by the joint activation of three subunits and a separate inactivation subunit.

parameters such as membrane potential or ligand concentration. Colquhoun and Hawkes (1981, 1982) show how the observable statistical properties of single-channel records are related to the underlying kinetics model.

Three schemes are shown that might describe agonist-activated channels. The first is a hypothetical scheme (Fig. 5A, first proposed by del Castillo and Katz) in which the binding of an agonist by a closed channel leads to an activated closed state capable of undergoing transitions to the open state. This simple three-state chain fits much of the data on postsynaptic activation.

The major family of postsynaptic channels is known to have two agonist-binding subunits per channel, so that one

can hypothesize various ways in which the binding of two agonists leads to channel opening. Two two-agonist schemes are shown in Fig. 5B and C. In B, first one then the other agonist molecule binds before their concerted action leads to channel opening. In C, each bound agonist allows the binding subunit to undergo a conformational change, and the opening results from the sequential effect of both bindings. These more accurate models lead to chains of greater length. Current analysis on chemically activated channels attempts to use single-channel data to distinguish between alternatives of this sort.

A more complex scheme, which has been applied to explain the Ca^{2+}-activated K^+ channel and glutamate-activated channels, involves allosteric interactions. A ligand-activated channel has ligand-binding subunits, and each binding to a subunit stabilizes open conformations that can exist with any number of bound ligands. This scheme, shown in Fig. 5D, forms a network of connected states for which there are multiple open states as well as closed states.

As examples of schemes representing voltage gating, we can look at the Hodgkin-Huxley kinetic schemes for the K^+ and Na^+ channels of nerve taken literally. The K^+ channel is represented by four identical and independent voltage-activated subunits. The Na^+ channel is represented by three independent activation subunits and an independent autonomous inactivating subunit. These lead to the kinetic diagrams shown in Fig. 5E and F, a chain of five states for the hypothetical K^+ channel and a ladder network of eight states for the Na^+ channel, which has the three states: closed, open, and inactivated.

More exact modeling that must account for the enormous amount of voltage-clamp and gating-current data would lead to more complex schemes. Structural data suggest that the Na^+ channels have four charged subunits, which are too large and too close to each other to move independently. There is evidence that an inactivation gate exists as a separate unit but that its closing depends on the state of the activation gates. Schemes incorporating all the known information about Na^+ channel gating will undoubtedly be complex and somewhat different from the phenomenological Hodgkin-Huxley scheme.

The dynamics of the channel protein in any scheme are summarized by the time-varying vector of the state: probabilities

$$\boldsymbol{p}(t) = [p_1, p_2, \ldots p_N] \qquad (2)$$

In classical kinetic measurements, we observe the entry or exit of the system into the complex of open states in response to some perturbation. The time course of such a relaxation follows a sum of exponentials,

$$\boldsymbol{p}(t) = \boldsymbol{p}(\infty) + \boldsymbol{p}(0) \sum A_k \exp(\mu_k t) \qquad (3)$$

where the μ_k are the eigenvalues or relaxation rates for a particular kinetic scheme. A system of N states will show relaxations composed of $N - 1$ exponentials. These exponential relaxation times are functions of the rates that connect the substates of the system. Most kinetic schemes will have far more than just $N - 1$ rates. There are at least $2(N - 1)$ rates, and there can be as many as $N(N - 1)/2$. Classical kinetic measurements, such as the measurement of current transients by voltage-clamp, provide considerable information for specifying the transition scheme of channel gating, but there is not enough information to resolve ambiguities fully in comparing competing schemes. Single-channel measurements, by recording more than just the behavior of the ensemble of channels, can extract additional information from observations of the detailed fluctuations of the stochastic process itself.

1. Dwell-Time Distributions

In most of these schemes, there is a single major open state, but the channel may wander around among its closed configurations before it gets into the open state. The extent of this wandering determines what the distribution of closed dwell times might look like. We can think of this as representative of some free energy profile that governs the rates of transition. If there is a single dominant barrier much higher than thermal energy, the open-closed transition is a Poisson process, which has a single exponential closed dwell-time distribution. If there are several dominant barriers so that there are a small number of stable closed states, then the transitions among them are governed by a finite-state Markov process, and the distribution of dwell times is a sum of exponentials.

For a finite-state Markov process, the two observable conditions are represented by two dwell-time distributions, $F_c(t)$, the distribution of closed intervals, and $F_o(t)$, the distribution of open intervals. These are given by

$$F_c(t) = \sum A_{c,i} \exp(R_{c,i} t) \tag{4}$$

$$F_o(t) = \sum A_{o,i} \exp(R_{o,i} t) \tag{5}$$

where the parameters $A_{c,i}$, $A_{o,i}$, $R_{c,i}$, and $R_{o,i}$ are functions of the rate constants in the Markov model of the gating process (Colquhoun and Hawkes, 1981). Figure 6A shows a channel undergoing a sequence of random transitions among a set of closed states leading to the observable pattern of current jumps depicted at the top of the figure. The underlying sojourn in the set of closed states leads to the multiexponential closed time distribution. In general, the distribution of dwell times can be a complicated function of the rates in the stochastic matrix. Only for relatively simple kinetic models can the distributions $F_c(t)$ and $F_o(t)$ actually be written in terms of the rates, Q_{ij}. For a complex of N_c closed states, the dwell-time distribution yields $2N_c - 1$ parameters, N_c time constants, and $N_c - 1$ amplitudes. Thus, for a model with N_o open states and N_c closed states, the two dwell-time distributions give $2(N - 1)$ relations among the rates.

We have assumed that a channel has a small number of closed states through which it passes, so that the dwell-time distribution is characteristic of a Markov process of reasonably low order and can be fit to a small number of exponentials. Another possibility for the molecular dynamics of excursion through the closed configurations is that the channel takes an enormous number of small steps through an extensive array of closed states that are separated by small free energy barriers. This would lead to a diffusion-like process for the sojourn in the complex of closed states, which is shown schematically in Fig. 6B. Such a process

A

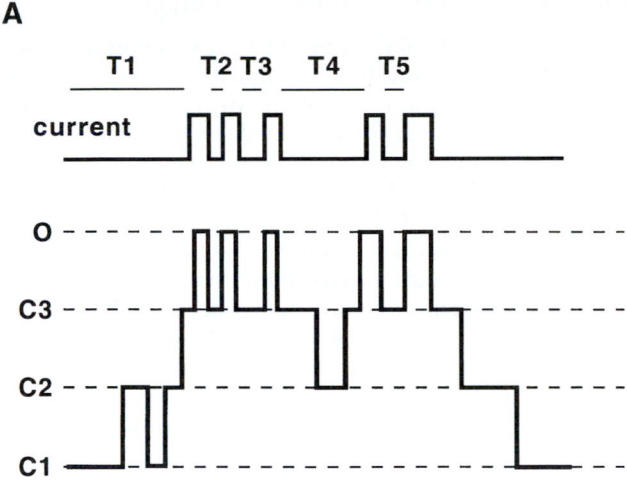

B

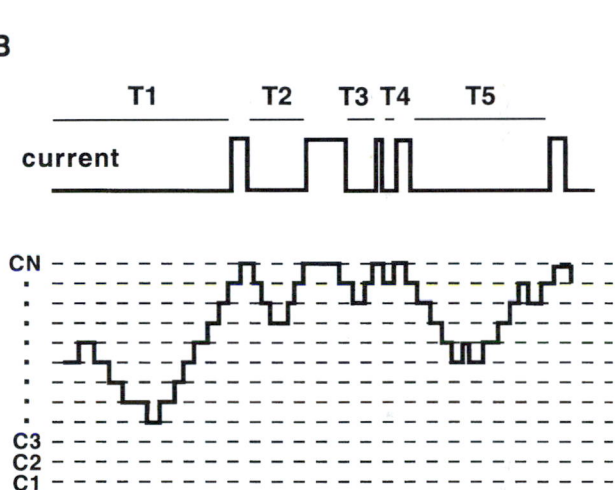

FIGURE 6. Dwell-time distributions. (A) Finite-state Markov process showing multiexponential dwell-time distribution. (B) Diffusive kinetics showing power-law distribution.

would lead to a distribution of dwell times that asymptotically obeys a power law varying between $t^{-3/2}$ and $t^{-1/2}$ (Millhauser et al., 1988).

The long tail distribution expresses the tendency of a diffusion variable to wander away from the channel opening goal and get lost for a time in its vast state space leading to a relatively higher density of long shut intervals. An even higher level of complexity has sometimes been suggested for protein conformational transitions: a fractal-like free energy topography, which would lead to chaotic wandering through the energy space and a closed time distribution that falls off as t^{-1} (Leibovitch et al., 1987). There has been some suggestion that the dwell-time distributions of certain channels fit power laws. However, most single-channel data fit well with multiexponential distributions, allowing interpretation in terms of a finite-state Markov process.

2. Correlations and Two-Time Distributions

When the distributions of open and closed time are multiexponential, they indicate a multiplicity of closed or open states. In such cases, observations of the correlations between successive open or closed times can yield information about the different routes by which the states of the system interconvert. The Markov assumption limits the extent of correlation since a Markov process starting from a known state has a future evolution completely independent of what happened before entry into the initial state. However, some correlation between successive occupancies can be seen and used to discriminate between kinetic schemes.

Consider an example in which correlation analysis can shed light on the gating mechanism. Say an agonist-activated channel has two distinct open lifetimes. Knowing that there are two agonist receptors, we might attribute the two lifetimes to two different schemes: (1) single-agonist bindings lead to short-lived openings and two-agonist binding leads to more long-lived openings, or (2) two agonists are needed to reach the open state but there are two different open states, one short lived and one long lived. The alternative schemes are shown in Fig. 7.

We can see that although these two schemes might lead to the same two-exponential distribution of open times, they will lead to different bursting patterns or different correlations between the lifetimes of successive openings. In Fig. 7A, if we consider experiments at low agonist concentration, so that C_1-C_2 transitions are infrequent compared to C_1-O_1 or C_2-O_2, then C_1-O_1 and C_2-O_2 transitions will have the opportunity to repeat, these repeated transitions leading to bursting and to correlations between successive openings.

The alternative model (Fig. 7B) does not lead to correlations; every closing of either open state leads back to state C_2 and during the sojourn in C_2, the channel loses memory of which open state it last came from (Markov definition). A study of correlations can distinguish between these two alternative models.

One measure of **correlation** is the correlation coefficient for successive lifetimes.

$$K(T_1, T_2) = \frac{\langle T_1 T_2 \rangle - \langle T_1 \rangle \langle T_2 \rangle}{(\langle T_1^2 \rangle - \langle T_1 \rangle^2)(\langle T_2^2 \rangle - \langle T_2 \rangle^2)} \tag{6}$$

For the two schemes discussed, scheme A gives $K = 0.3$–0.5 depending on the C_1-C_2 interconversion rate. For scheme B, K must be equal to zero.

All the kinetic information obtainable from stationary patch-clamp data is contained in two-dimensional dwell-time histograms (Fredkin and Rice, 1986). A two-dimensional dwell-time density is defined as $P_2(T_1, T_2)$ equals the probability in a particular condition (e.g., membrane potential) that a channel is in the conducting state I from $T = 0$ to $T = T_1$ and is in conductance J from $T = T_1$ to $T = T_1 + T_2$.

Thus, if the two-time distributions can be measured, they can be compared to a model in which the component exponential of a multiexponential distribution can be fit. However, the amount of information available may generally be considerably less than the amount needed to specify the kinetic model fully. For a kinetic model of N states, the

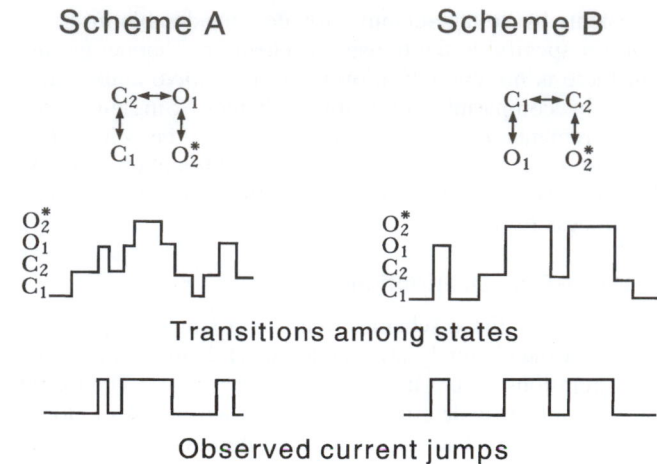

FIGURE 7. Two kinetic schemes that can be distinguished by measuring correlations between successive open-state lifetimes. Both schemes would give a two-exponential open-state lifetime distribution, but (A) the successive openings are correlated, where the two open states are connected, and (B) are uncorrelated, where they are kinetically disjoint.

system is described by as many as $N(N - 1)$ rates. The number of rates determinable from the patch-clamp data is

$$H = 2 \sum_{i}^{N} \sum_{j}^{N} N_i N_j \tag{7}$$

Thus, there are $G = N(N - 1) - H$ rates that cannot be determined from stationary patch-clamp data and must be fixed by hypothesis or from other experimental data such as nonstationary voltage-clamp. For channels with many closed states and a few open states, it can be seen that H is in fact small compared to $N(N - 1)$.

C. Nonstationary Analysis

Nonstationary single-channel analysis is analogous to voltage-clamp. The experimenter prepares the channel system in a specific set of states by controlling a rate-determining parameter, such as membrane potential. Channels are then suddenly perturbed by a stimulus, such as a step depolarization of potential, which changes some of the transition rates, causing the system of channels to relax to a new equilibrium. The time course of the random transitions during the course of equilibration is what is observed. By repeating pulses, one obtains the statistics of ensemble distributions of outcomes. Voltage-gated channels, which inactivate rapidly, or agonist-gated channels, which desensitize quickly, do not give steady-state responses and are better studied by nonstationary methods.

One example of the kind of information obtainable is the measurement of first-latency distributions (the distribution of times between application of the stimulus and the first opening of the channel) (Horn and Vandenberg, 1984). Such distributions can be analyzed to obtain the sequence of kinetic steps in a chain of closed states leading to the initial channel opening. So, for example, in the Hodgkin-Huxley K+ channel scheme, a chain of four closed states precedes opening. A hyperpolarizing initial voltage maximizes the population in the leftmost closed states. The probability of arriving at the open state in a particular time interval is equal to the probability of a particular dwell-time for the sojourn in the collection of closed states. Such an experiment is indicated schematically in Fig. 8A. The form of the distribution times to first opening can be predicted from the kinetic model by solving the equations of the modified scheme in which no return is allowed from the open state. The first latency distribution is analogous to the closed-state dwell-time distribution but gives information specifically about the sequence of closed states through which the activating channel passes.

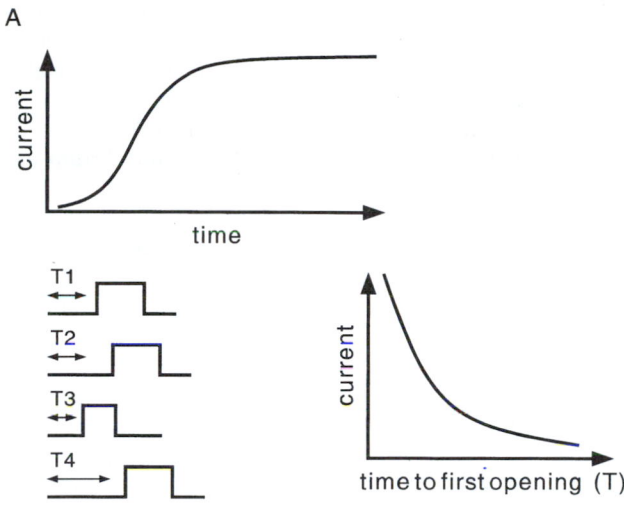

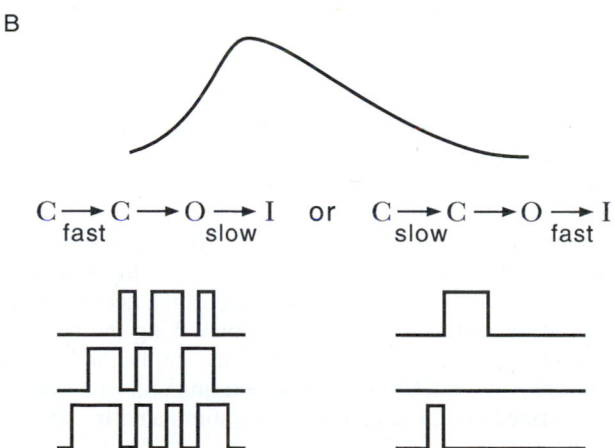

FIGURE 8. Nonstationary properties of channel lifetime statistics. (A) Distribution of first-latency times. Different kinetic schemes will lead to different excursions through closed states on the way to first opening following a step in membrane potential. (B) Distribution of number of openings per inactivating event. Different kinetic models of the opening and inactivating of a voltage-gated channel that give the same macroscopic transient will lead to different predictions for transient bursts of channel opening following a step change in potential.

Another example of a nonstationary determination is the measurement of the number of times that a channel opens before inactivation (Aldrich *et al.*, 1983; Vandenberg and Horn, 1984). Such a measurement can resolve ambiguities in the macroscopic description of inactivation. In a voltage-clamp experiment on a Na^+ channel, the transient conduction increase represents the product of a fast activation process and an independent, slow, voltage-dependent inactivation process. However, the actual shape of the transient is not really sensitive to the choice of activation and inactivation rates and a very similar fit can be obtained with a slow activation and fast inactivation.

At the single-channel level, however, the two situations differ, as illustrated in Fig. 8B (see also Fig. 4). In the first case nearly all channels open on depolarization and then individually slowly fluctuate closed. In the second picture, only a fraction of the channels opens at any time and the apparent slow abatement comes about because new channels are still entering the open population. Single-channel voltage-clamp experiments show the actual number of openings that an individual channel can undergo before it inactivates and thus make an unambiguous fit to the ratio of activation to inactivation rates.

D. Model Discrimination

In single-channel analysis, one takes advantage of the ability to observe the actual stochastic process that underlies channel gating. Because even conceptually simple pictures of the molecular conformation changes lead to rather complex kinetic schemes, the single-channel statistics do not yield sufficient information to specify the scheme completely. What they do provide, as illustrated earlier, is accurate criteria for distinguishing between candidate schemes. The detailed picture of channel gating must also be guided by insights derived from the emerging knowledge of channel molecular structure.

IV. Whole-Cell Currents

A. Whole-Cell Tight-Seal Voltage-Clamp

The tight-seal patch-clamp technique was invented for investigating the behavior of a single channel in an isolated patch of membrane. A modification of the technique, in which the patch is punctured, provides the basis for a novel voltage-clamp technique for studying currents that flow across the entire membrane of a cell. When more suction is applied to the pipette after sealing onto the membrane, the membrane patch underneath the tip can break, and the pipette solution is now continuous with the inside of the cell. This creates a low–electrical resistance pathway between the pipette and cell, making it possible to control the voltage of the cell. This variation is known as the **whole-cell tight-seal voltage-clamp** (see Fig. 3; the term "tight seal" is used here to distinguish it from the whole-cell "loose-patch" voltage-clamp technique described by Almers *et al.*, 1987).

Figure 2B shows the equivalent circuit for the combination of resting cell membrane and micropipette electrode. To see why the patch technique is so useful, we can compare the recording situation to recording with a conventional microelectrode. For conventional recording, when a microelectrode impales a small cell, typical values are $R_{pip} = 100$ MΩ. The voltage drop across the series resistance is comparable to or greater than that across the cell membrane and the shunt current provides a significant leakage current. Thus the membrane potential can only be controlled by a feedback system using either a second microelectrode for monitoring membrane potential or some sort of intermittent switching. These techniques are complicated, are limited in the response time for changing voltage, and generate considerable electrical noise.

In contrast the "patch" junction gives a pipette resistance of 4–10 MΩ and a seal resistance of $\geq$ 10 GΩ, so that the voltage drop across the series resistance is small compared to that across the cell membrane and the leakage current is relatively small. In consequence, a patched cell can be voltage-clamped with the same type of current-to-voltage converter used in patch-clamp.

If a step of voltage is placed across the patch pipette, the voltage rise time can be seen to be (assuming $R_{cell} \gg R_{pip}$)

$$\tau = R_{pip} C_{cell} \tag{8}$$

For a small cell of $R_{pip} = 5$ MΩ and $C_{cell} = 5$ pF, $\tau = 25$ μs. The membrane potential, compared to the command potential, is

$$V_m/V_c = R_{cell}/(R_{cell} + R_{pip}) \tag{9}$$

with $R_{cell} \approx 10$ GΩ, $V_m/V_c = 0.99$. Thus, a small cell can be voltage-clamped without any compensation for series resistance or membrane capacitance. In general, R_{cell} scales inversely with cell area and C_m scales proportional to cell area, so that both of these factors will be worse when cell size is increased, limiting the applicability to cells less than ~20 μm in diameter. Patch-clamp circuitry uses capacitance cancellation and series compensation circuits to counteract these limitations. Limitations and design of the whole-cell patch-clamp are discussed in Marty and Neher (1983), Lecar and Smith (1985), and Strickholm (1995a, b).

Another major focus of patch-clamp analysis is its combination with molecular biology of identified membrane proteins. A number of expression systems have proven amenable to electrophysiology: *Xenopus* oocytes, mammalian cells, insect cells, and yeast. Perhaps the most popular system is the *Xenopus* oocyte, an enormous cell that can be microinjected with RNA to express a great variety of ion channels (Smart and Krishek, 1995). The oocyte is too large to be effectively whole-cell clamped using a patch electrode, and has been studied more by microelectrode techniques. However, current measurements can be done on macropatches of oocyte membrane—2–5 μm patches excised with hard alumino silicate pipettes (Stuhmer *et al.*, 1991).

B. Perforated-Patch Method

The same low-resistance access that permits whole-cell voltage-clamp also has the often unwanted effect of allowing the cytoplasmic solution to interdiffuse with the much larger volume of the patch pipette. This often leads to the loss or rundown of membrane and cytoplasmic constituents that are vital for maintaining channel activity. The perforated-patch method (see Fig. 3; Horn and Marty, 1988; Waltz, 1995) was originally invented to prevent the loss of response due to rundown. This method involves the use of the antibiotics nystatin or amphotericin, which form ion-permeable pores in the membrane without the loss of large molecules from the cytoplasm. The perforated patch creates an electrically low-resistance pathway between the pipette and the inside of the cell. The channels created by the antibiotics are more permeable to monovalent cations than monovalent anions, and are thought to be impermeable to divalents. Because of this ion selectivity, it is necessary to include SO_4^{2-} ions in the pipette solution to counteract the Donnan equilibrium potential that arises when only Cl^- is present in the pipette solution. More recent use of gramicidin (Kyrozis and Reichling, 1995), which has induced pores that are selectively permeable to monovalent cations and negligible to chloride ions, appears to ameliorate the problem of changing intracellular free Cl^- concentrations. Avoiding the change of chloride equilibrium may be critical when studying ligand-gated or voltage-gated chloride channels.

The tip of the pipette is first filled with antibiotic-free recording solution, then back-filled with the antibiotic-containing recording solution. Following the formation of a high seal resistance, time is allowed for the antibiotic to diffuse down to the tip and form pores in the patch membrane. The time interval (usually between 5–15 min) depends on how much antibiotic-free solution is in the tip and the concentration of the antibiotic used, parameters that must be determined experimentally for each cell type. The series resistance under this mode of recording is usually high (> 20 MΩ), making adequate control of the voltage in voltage-jump studies difficult. A shift in the current-voltage relation curve to the right following a reduction of current amplitudes (e.g., when studying drug antagonism) and a very sudden or sharp increase in the rising limb of the current-voltage relation curve are indicative of this problem. A slow time-to-peak current is also associated with high series resistance. The perforated-patch method is, therefore, more appropriate for studying regulation of small-amplitude currents or ligand-gated currents.

An important area of application of the perforated patch method is the extension of patch-clamping to brain-slice preparations, allowing individual neurons in a network to be analyzed by voltage-clamp without disturbing the network (Plant *et al.*, 1995). Slice preparations are available from many areas of the CNS, so that neurons and glia can be characterized *in situ*, and as they are stimulated by afferent synaptic inputs.

C. Voltage-Clamp Analysis

Voltage-clamp analysis is the primary means of characterizing the electrical behavior of the ion-conducting systems within the membrane. In a voltage-clamp experiment, ion current transients are recorded in response to step changes of membrane potential. When membrane potential is stepped, an ensemble of voltage-gated ion channels will undergo transient changes in conductance as the number of activated channels changes in time. The current-voltage relation of a membrane with gated channels is thus dynamic, characterized by a relaxing voltage-dependent conductance. The current flowing in response to a voltage step from an initial V_o to a command voltage V_c obeys

$$I(V_c, V_o, t) = G(V_c, V_o, t)\, I(V_c - E_r) \tag{10}$$

Here I is the instantaneous current-voltage relation of an open channel with reversal potential. E_r is determined by the ion selectivity of the particular channel and the prevailing ion concentrations. G is the instantaneous value of conductance whose transient behavior reflects the change in channel-opening probability.

The dynamic behavior of G is determined by the kinetics of transition of a channel between its various allowed states. In terms of the stochastic schemes of Eq. 10, the observed G represents the arrival of channels into their open states. The delay in the activation of G represents the sequence of transitions of the channels among various closed states before the actual openings. The kinetic model for explaining the voltage-clamp transients corresponds to the transient analysis of the mean conductance from the stochastic system. The transient changes are characterized by sums of exponentials with characteristic relaxation times. This corresponds to the sum of exponentials description of the open- and closed-state dwell-times of the stochastic analysis.

The classic kinetic models of channel activation and inactivation, such as the Hodgkin-Huxley model, were derived from voltage-clamp data and are equivalent to special cases of the general stochastic models in the earlier discussion. Often the aim of voltage-clamp experiments is to provide a quicker characterization of a particular species of channel rather than a full description of the often complicated channel-gating dynamics. One convenient way of doing this is to clamp the cell to a linearly increasing voltage signal (ramp clamp) and then obtain directly an approximate dynamic current-voltage curve. Depending on the placement of the voltage-dependent conductance region vis-à-vis the reversal potential for the channel, a voltage-gated channel will exhibit either a rectifying current-voltage relation or a distinct negative resistance region. When several species of channels coexist, one attempts to isolate the response of particular channels using specific channel-blocking drugs or by changing the ionic composition of the solutions (chemical dissection). A review of the practical considerations of analysis of whole-cell patch-clamp recordings is given by Sontheimer (1995).

D. Membrane Noise Analysis

In whole-cell recording on small cells, the dominant electrical noise observed at frequencies less than 1 kHz is the noise generated by the random opening and closing of ion channels. Although the individual gating transitions may not be resolvable in such noise records, the unit conductances and some of the kinetic properties of the channels can be obtained from analysis of the noise records.

The unit conductance can be inferred from a measurement of the fluctuation noise power. For a membrane with N in-

dependently gating channels, each with unit current i and probability p of being in the open state, the average membrane current will be $I = Npi$ and the variance of the current (or the average fluctuating power in a unit resistor) is

$$\overline{I^2} - \overline{I}^2 = Np(1 - p)i^2 \tag{11}$$

To infer the unit current, i, one measures the ratio of variance to mean,

$$\text{variance/mean} = (1 - p)i \tag{12}$$

which approaches the unit current as p is made small.

Kinetic properties of the gating can be obtained by analysis of the frequency spectrum of the noise. For a Markov process, the spectrum by channel fluctuating noise is given by a characteristic spectrum, which is the sum of Lorenzian spectra,

$$P(\omega) = \sum_j [A_j / (1 + \omega^2 \tau_j)] \tag{13}$$

where the parameters A_j and τ_j are related to the transition probabilities of the kinetic scheme.

Fluctuation analysis is particularly useful for determining the properties of channels that either cannot be studied by patch-clamp or have conductances too small to be resolved by the single-channel method.

E. Capacitance Measurements

Whole-cell voltage-clamp can be used to monitor changes in membrane capacitance during exocytosis or secretion. Secretion of hormones, enzymes, and transmitters occurs when messenger-containing cytoplasmic vesicles fuse with the plasmalemmal membrane and disgorge their contents. In the process, the vesicular membranes add to the surface area of the cell, leading to an increase in the cell membrane capacitance. The capacitance changes caused by the fusion of individual vesicles can be monitored to provide a detailed picture of the quantal events underlying secretion (see Penner and Neher, 1989; Lindau and Neher, 1988; Neher and Marty, 1982).

Neher and Marty (1982) first observed the unitary capacitance changes in bovine adrenal chromaffin cells. They injected a sine wave of known amplitude into a cell under voltage-clamp. The voltage and current signals are then fed into a lock-in amplifier, which measures a phase angle and separates the current response into its resistive and capacitive components. The phase angle is determined by changing the capacitance compensation knob of the amplifier and adjusting the angle until the resistive component output does not alter (Neher and Marty, 1982). The capacitance is then determined from an angle orthogonal to the measured phase angle. An alternative approach is to find the phase angle by changing a resistance in series with the recording ground (Fidler and Fernandez, 1989; note that the phase angle determined by this approach is different from that described by Neher and Marty). A simpler means of determining the phase angle has been proposed (Zierler, 1992). Other approaches to measure membrane capacitance include the use of two sinusoidal frequencies (Donnelly, 1994; Rohlicek and Schmid, 1994) and theoretical calculations (Niu et al., 1995). Capacitance change as a means to measure exocytosis or secretion has been used extensively to study the molecular mechanisms of release in various preparations, including mast cells, pancreatic β cells, nerve terminals, neurons, and neuroendocrine cells.

Although it has been suggested that the observed change in membrane capacitance may include a component of gating currents (Horrigan and Bookman, 1994), other studies have found a good correlation between capacitance increase and the exocytotic events. In mast cells from the beige mouse, capacitance increase can be measured only after the observed fusion of vesicles (Zimmerberg et al., 1987). In adrenal chromaffin cells and mast cells, the increase in amperometric signal (amperometry is an electrochemical method that is sensitive to monoamines (dopamine, norepinephrine, and 5-hydroxytryptamine) correlates with capacitance increase (Chow et al., 1992; Alvarez de Toledo et al., 1993; Haller et al., 1998). In spite of its high sensitivity and accuracy in detecting secretion, amperometry and the less-invasive optical method have become the methods of choice for studying secretion. This is because of the ease of doing amperometry (Wightman et al., 1991, see Koh and Hille, 1999) and the improvements in membrane fluorescent dyes (Cochilla et al., 1999; Murphy, 1999). With the rapid advances in molecular techniques, the ability to generate mutant or variant animals and cells devoid of or with modified vesicular proteins will undoubtedly lead to the unraveling of the molecular mechanisms underlying the once elusive exocytotic process.

V. Summary

The patch-clamp was originally devised to measure single-channel currents in an isolated patch of cell membrane. However, because patch electrodes can make tight stable molecular bonds with membrane surfaces, the patch-clamp technique can be exploited in a remarkable number of ways to study ion currents at both the single-channel and whole-cell level (Sigworth, 1986). Variations of the patch-clamp technique (cell-attached, inside-out, and outside-out) have unique advantages for studying different types of ion-channel gating at the single-channel level. Whole-cell voltage-clamp and its variation, the perforated-patch method, are now probably the most commonly used methods to characterize ionic currents in excitable cells. An example of the versatility of patch methods can be seen in a recent symposium on olfactory transduction in which many different cell preparations previously inaccessible to electrophysiology were studied by all the different variants of the patch-clamp (Spielman and Brand, 1995).

The different variations or modes of patch recordings further permit the experimenter to alter the environment and examine the gating and behavior of a single-channel protein. Under the cell-attached mode it is possible to evaluate the behavior of a channel protein in its more or less native environment. Under the inside-out mode, since the inner membrane is made to face the surrounding medium, one can study whether the behavior of the protein is altered by addition of substances, for example, second messengers, or changing of the ionic conditions, for example, al-

tering the Ca^{2+} concentrations to determine the gating effects of Ca^{2+} on the Ca^{2+}-dependent K^+ channel. Using the outside-out mode one can examine the detailed gating kinetics of agonist-activated channels with varying agonist concentrations.

The whole-cell tight-seal voltage-clamp allows experimenters to study a whole world of cells that were previously not accessible to recording with conventional sharp-tipped electrodes. Moreover, the ability to include substances in the pipette solutions allows one to manipulate the cell's interior environment and study the modulation of ionic currents. Finally, a variation of the whole-cell recording technique, the perforated-patch recording technique, allows for noninvasive access to the cell interior under which ions can permeate the patch but large molecules do not leak out of the cytoplasm. Thus, there appears to be a patch-clamp technique for almost any measurement situation, a remarkable consequence of the fact that clean glass surfaces can bond to cell membranes.

Bibliography

Aldrich, R. W., Corey, D. P., and Stevens, C. F. (1983). A reinterpretation of mammalian sodium channel gating based on single channel recording. *Nature (London)* **260,** 436–411.

Almers, W., Stanfield, P. R., and Stuhmer, W. (1987). Lateral distribution of sodium and potassium channels in frog skeletal muscle: measurements with a patch clamp technique. *J. Physiol. (London)* **336,** 261–284.

Alvarez de Toledo, G., Fernandez-Chacon, R., and Fernandez, J. M. (1993). Release of secretory products during transient vesicle fusion. *Nature (London)* **363,** 554–558.

Ashley, R. H. (Ed.) (1995). "Ion Channels: A Practical Approach." Oxford University Press, New York.

Barry, P. H., and Lynch, J. W. (1991). Liquid junction potentials and small cell effects in patch-clamp analysis. *J. Membr. Biol.* **121,** 101–117.

Boulton, A. A., Baker, G. B., and Walz, W. (1995). Neuromethods. *In* "Patch-Clamp Applications and Protocols," Vol. 26. Humana Press, Totowa, NJ.

Chow, R. H., Von Ruden, L., and Neher, E. (1992). Delay in vesicle fusion revealed by electrochemical monitoring of single secretory events in adrenal chromaffin cells. *Nature (London)* **356,** 60–63.

Cochilla, A. J., Angleson, J. K., and Betz, W. J. (1999). Monitoring secretory membrane with FM1-43 fluorescence. *Annu. Rev. Neurosci.* **22,** 1–10.

Colquhoun, D., and Hawkes, A. G. (1981). On the stochastic properties of single ion channels. *Proc. R. Soc. London B* **211,** 205–235.

Colquhoun, D., and Hawkes, A. G. (1982). On the stochastic properties of bursts of single ion channel openings and clusters of bursts. *Proc. R. Soc. London B* **300,** 1–59.

Donnelly, D. F. (1994). A novel method for rapid measurement of membrane resistance, capacitance, and access resistance. *Biophys. J.* **66,** 873–877.

Ehrenstein, G., and Lecar, H. (1977). Electrically gated ionic channels in lipid bilayers. *Q. Rev. Biophys.* **10,** 1–34.

Fenwick, E. M., Marty, A., and Neher, E. (1982). A patch-clamp study of bovine chromaffin cells and of their sensitivity to acetylcholine. *J. Physiol. (London)* **331,** 577–597.

Fidler, N., and Fernandez, J. M. (1989). Phase tracking: an improved phase detection technique for cell membrane capacitance measurements. *Biophys. J.* **56,** 1153–1162.

Fredkin, D., and Rice, J. A. (1986). On aggregated Markov chains. *J. Appl. Prob.* **23,** 208–214.

Haller, M., Heinemann, C., Chow, R. H., Heidelberger, R., and Neher, E. (1998). Comparison of secretory responses as measured by membrane capacitance and by amperometry. *Biophys. J.* **74**, 2100–2113.

Hamill, O. P., Marty, A., Neher, E., Sakmann, B., and Sigworth, F. J. (1981). Improved patch-clamp techniques for high-resolution current recording from cells and cell-free membrane patches. *Pflügers Arch.* **391**, 85–100.

Hodgkin, A. L., and Huxley, A. F. (1952a). Currents carried by sodium and potassium ions through the membrane of the giant axon of *Loligo. J. Physiol. (London)* **116**, 449–472.

Hodgkin, A. L., and Huxley, A. F. (1952b). The components of membrane conductance in the giant axon of *Loligo. J. Physiol. (London)* **116**, 473–496.

Hodgkin, A. L., and Huxley, A. F. (1952c). A quantitative description of membrane current and its application to conduction and excitation in nerve. *J. Physiol. (London)* **117**, 500–514.

Horn, R., and Marty, A. (1988). Muscarinic activation of ionic currents measured by a new whole cell recording method. *J. Gen. Physiol.* **92**, 145–159.

Horn, R., and Vandenberg, C. A. (1984). Statistical properties of single sodium channels. *J. Gen. Physiol.* **84**, 505–534.

Horrigan, F. T., and Bookman, R. J. (1994). Releasable pools and the kinetics of exocytosis in adrenal chromaffin cells. *Neuron* **13**, 1119–1129.

Koh, D.-S., and Hille, B. (1999). Rapid fabrication of plastic-insulated carbon-fiber electrodes for micro-amperometry. *J. Neurosci. Method.* **88**, 83–91.

Kyrozis, A., and Reichling, D. B. (1995). Perforated-patch recording with gramicidin avoids artifactual changes in intracellular chloride concentration. *J. Neurosci. Methods* **57**, 27–35.

Lecar, H., and Smith, Jr., T. G. (1985). Voltage clamping small cells. *In* "Voltage-Clamping and Patch Clamping with Microelectrodes" (T. G. Smith, Jr., H. Lecar, S. J. Redman, and P. W. Gage, Eds.), American Physiological Society, pp. 231–256. Bethesda, MD.

Leibovitch, L. S., Fishbarg, J., Koniarek, J. P., Todorova, I., and Wang, M. (1987). Fractal model of ion-channel kinetics. *Biochim. Biophys. Acta* **896**, 173–180.

Levis, R. A., and Rae, J. L. (1993). The use of quartz patch pipettes for low noise single channel recording. *Biophys. J.* **65**, 1666–1677.

Lindau, M., and Neher, E. (1988). Patch-clamp techniques for time-resolved capacitance measurements in single cells. *Pflügers Arch.* **411**, 137–146.

Marty, A., and Neher, E. (1983). Tight-seal whole-cell recording. *In* "Single-Channel Recording" (B. Sakmann, and E. Neher, Eds.), Plenum Press, pp. 107-122. New York and London.

Millhauser, G. L., Saltpeter, E. E., and Oswald, R. E. (1988). Diffusion models of ion-channel gating and the origin of power-law distributions from single-channel recording. *Proc. Natl. Acad. Sci. USA* **85**, 1503–1507.

Murphy, V. N. (1999). Optical detection of synaptic vesicle exocytosis and endocytosis. *Curr. Opin. Neurobiol.* **9**, 314–320.

Neher, E., and Marty, A. (1982). Discrete changes of cell membrane capacitance observed under conditions of enhanced secretion in bovine adrenal chromaffin cells. *Proc. Natl. Acad. Sci. USA* **79**, 6712–6716.

Neher, E., and Sakmann, B. (1976). Single-channel currents recorded from membrane of denervated frog muscle fibres. *Nature (London)* **260**, 779–802.

Niu, A., Zhou, N., Xie, R., and Pun, R. Y. K. (1995). Computational method to determine capacitance changes in secretory cells. *FASEB J. Abstracts* **9(2)**, A674 #3910.

Ohya, Y. and Sperelakis, N. (1989). Modulation of single slow (L-type) calcium channels by intracellular ATP in vascular smooth muscle cells. *Pflügers Arch.*, **414**, 257–264.

Penner, R., and Neher, E. (1989). The patch-clamp technique in the study of secretion. *Trends in Neurosci.* **12**, 159–163.

Plant, T. D., Eilers, J., and Konnerth, A. (1995). Patch-clamp technique in brain slices. *In* "Neuromethods," Vol. 26: "Patch Clamp Applications and Protocols" (A. A. Boulton, G. B. Baker, and W. Waltz, Eds.) Humana Press, Totowa, NJ.

Rohlicek, V., and Schmid, A. (1994). Dual-frequency method for synchronous measurement of cell capacitance, membrane conductance and access-resistance on single cells. *Pflügers Arch.* **428**, 30–38.

Rudy, B., and Iverson, L. E. (1992). Ion channels. *In* "Methods in Enzymology," Vol. 207. Academic Press, San Diego.

Sakmann, B., and Neher, E. (1995). "Single-Channel Recording," 2nd ed. Plenum Press, New York.

Sigworth, F. J. (1986). The patch clamp is more useful than anyone had expected. *Fed. Proc.* **45**, 2673–2677.

Smart, T. G., and Krishek, B. J. (1995). *Xenopus* oocyte microinjection and ion-channel expression. *In* "Neuromethods," Vol. 26: "Patch Clamp Applications and Protocols" (A. A. Boulton, G. B. Baker, and W. Waltz, Eds.) Humana Press, Totowa, NJ, pp. 259–305.

Sontheimer, H. (1995). Whole cell patch clamp recordings. *In* "Neuromethods," Vol. 26: "Patch-Clamp Applications and Protocols," (A. A. Boulton, G. B. Baker, and W. Waltz, Eds.) Humana Press, Totowa, NJ, pp. 37–73.

Spielman, A.I., and Brand, J.G., Eds., (1995). "Experimental Cell Biology of Taste and Olfaction: Current Techniques and Protocols." pp. 299–379. CRC Press, Boca Raton, FL.

Strickholm, A. (1995a). A supercharger for single electrode voltage and current clamping. *J. Neurosci. Methods* **61**, 47–52.

Strickholm, A. (1995b). A single electrode voltage, current- and patch-clamp amplifier with complete stable series resistance compensation. *J. Neurosci. Methods* **61**, 53–66.

Stuhmer, W., Conti, F., Stocker, M., Pongs, O., and Heinemann, S. H. (1991). Gating currents of inactivating and non-inactivating potassium channels expressed in *Xenopus* oocytes. *Pflügers Arch.* **418**, 423–429.

Vandenberg, C. A., and Horn, R. (1984). Inactivation viewed through single sodium channels. *J. Gen. Physiol.* **84**, 505–534.

Waltz, W. (1995). Perforated patch-clamp technique. *In* "Neuromethods," Vol. 26: "Patch Clamp Applications and Protocols" (A. A. Boulton, G. B. Baker, and W. Waltz, Eds.) Humana Press, Totowa, NJ, pp. 155–171.

Wightman, R. M., Jankowski, J. A., Kennedy, R. T., Kawagoe, K. T., Schroeder, T. J., Leszczyszyn, D. J., Near, J. A., Diliberto, E. J., and Viveros, O. H. (1991). *Proc. Natl. Acad. Sci. USA*, **88**, 10 754–10 758.

Zierler, K. (1992). Simplified method for setting the phase angle for use in capacitance measurements in studies of exocytosis. *Biophys. J.* **63**, 854–856.

Zimmerberg, J., Curran, M., Cohen, F. S., and Brodwick, M. (1987). Simultaneous electrical and optical measurements show that membrane fusion precedes secretory granule swelling during exocytosis of beige mouse mast cells. *Proc. Natl. Acad. Sci. USA* **84**, 1585–1589.

Simon Rock Levinson and William A. Sather

27

Structure and Mechanism of Voltage-Gated Ion Channels

I. Introduction: How Is Ion Channel Structure Studied?

Previous chapters have described how the flow of ions across cell membranes is the basis for electrical excitability and signaling in the nervous system. These flows are controlled by a special class of macromolecules known as **ion channels** that form gated pores in the cell membrane. The structure of ion channels and the current thinking about how these structures form transmembrane pores that gate to open and closed states in response to changes in membrane voltage are discussed in this chapter. Subsequent chapters will describe other important aspects of ion channels, such as interactions with drugs and toxins, modulation by intracellular messengers, and specific ion channel types found in intracellular organelles and involved in synaptic transmission.

What are the questions that should be addressed in this chapter? Most fundamentally, we wish to know how these macromolecular structures give rise to the important functional properties of ion channels, namely, the formation of transmembrane pores, the selective transport of specific ions, and the ability to open and close the pore; that is, how do ion channels function as molecular mechanisms? As we will see, these considerations also lead into a brief consideration of the structural diversity among ion channels and the origin and possible purposes of such diversity.

Cellular ion channels are basically proteins. In the earliest structural studies, biochemical methods were employed to purify channel molecules from excitable tissues. This approach has been essential to the current state of knowledge of ion channel structure. However, ion channels have certain physicochemical properties that have limited the amount of information obtained solely through biochemical characterization. This bottleneck has been broken recently through the application of recombinant DNA methodologies, which have uncovered the primary structures of numerous ion channel types, while providing a powerful means to study higher order structure and function. However, this latter approach

has its own limitations, and ultimately we will need to "see" more directly the actual structure of the channels to understand more fully their molecular mechanisms of action. In this regard, it is encouraging that significant progress has been made recently in elucidating the structure of purified channels using x-ray crystallographic techniques.

II. Biochemistry of Ion Channels: Purification and Characterization of Voltage-Gated Channels

Before the advent of recombinant DNA methods, ion channels were extensively studied by first purifying channels from an excitable tissue and then characterizing the purified molecules by a combination of chemical and physical techniques. While this biochemical approach has fallen somewhat out of use in favor of DNA cloning techniques, it seems certain to reemerge as a necessary adjunct to structural studies that now seem feasible (see Section H). Thus a brief introduction to channel biochemistry seems appropriate here. The first voltage-gated channel to be purified was the Na^+ channel from the electric eel *Electrophorus electricus*, and the account of how this was achieved will serve to illustrate the basic rationale for the biochemical approach (see Miller *et al.,* 1983).

The basic procedure for purification of membrane-associated proteins (Fig. 1) is as follows. First, an enriched fraction of membranes is usually prepared by disrupting the tissue and its cells mechanically and subsequently separating the insoluble fraction containing membranes and connective tissue components from the soluble fraction consisting of cytoplasmic protein. Although relatively nonspecific, this step also often gives substantial enrichment of ion channel preparations. Next, the membranes are solubilized through the use of detergents. This step disperses the membranes into minute droplets of lipid, protein, and detergent known as micelles. This is often the most problematic step

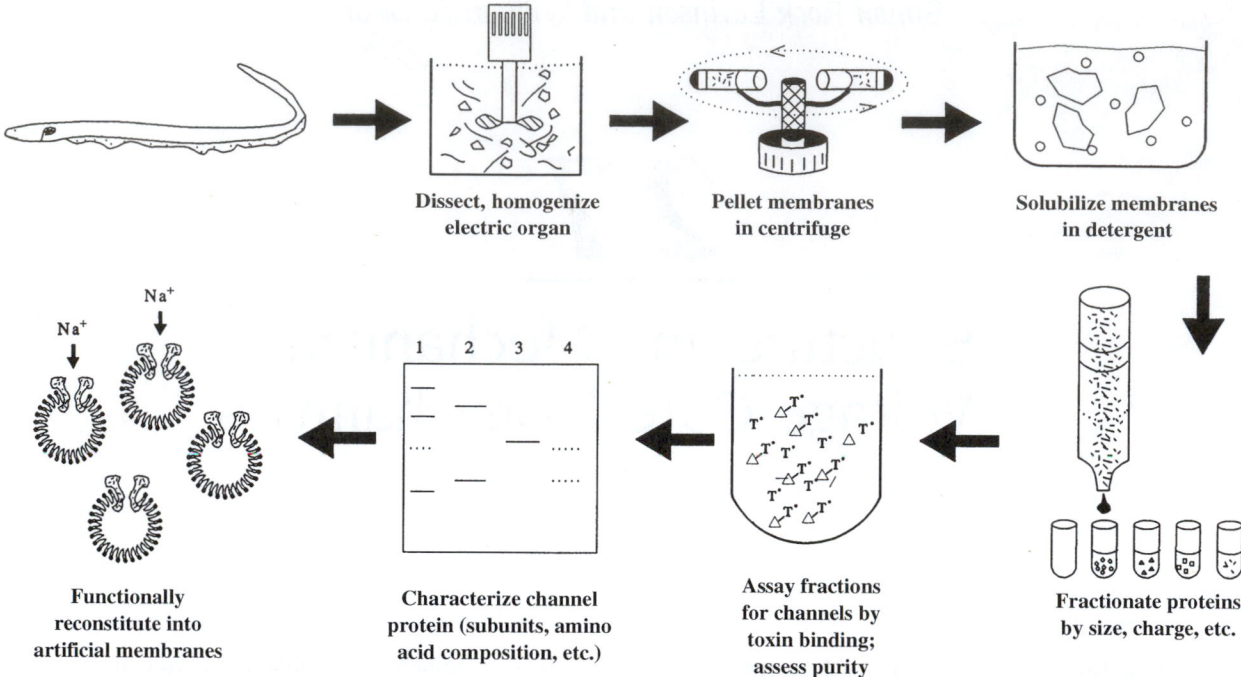

FIGURE 1. Scheme used for the biochemical isolation of Na$^+$ channels from electric tissue of the eel *Electrophorus electricus*. Purification protocols devised for other ion channels utilize similar general principles but differ in details.

of a purification, because the micellar fraction must be dispersed well enough so that no more than one protein molecule is present in each micelle. Too much detergent often results in irreversible denaturation of channels, hence preventing their identification during the purification process. Thus, the solubilization process must be optimized, and often many different detergents at various concentrations are tested under a plethora of conditions before the right combination is found. Finally, the dispersed micelles are fractionated based on their physical properties or affinities for certain compounds until a pure sample of channel protein is isolated.

A. Electric Fish Are a Rich Source of Ion Channels

The first requirement for purification of any protein is to have a rich source. Actually, this can be a serious difficulty, since ion channels are present in very small amounts in most excitable tissues. The rarity of channels might at first seem paradoxical in view of the ubiquity of excitability phenomena in nearly all cell membranes. However, the reader should recall that only minute numbers of channels are needed to mediate the flow of ions required to cause the changes in transmembrane voltage during electrical excitation (Table 1).

Fortunately, there exist some highly specialized "freaks" of nature that use large ionic currents for purposes other than signaling and stimulus transduction. These are the strongly electric fish, which are capable of generating powerful electrical discharges to stun prey and defend themselves. The electric organs of these animals contain large numbers of ion channels that are present at high density on the surface of the excitable cells of the electric organ (see Chapter 60). As a re-

sult, the first two ion channels to be purified were the acetylcholine receptor channel of the electric ray *Torpedo* and the Na$^+$ channel of the electric eel *Electrophorus*.

B. Toxins and Drugs as Markers for Ion Channels during Purification

Having a rich source of channels does not guarantee a successful isolation of channels. One must also have a very specific and sensitive assay for the presence of solubilized chan-

TABLE 1 Sodium Channel Densities in Selected Excitable Tissues

Tissue	Surface density channels/μm^2	Tissue density μg channel/g tissue)
Mammalian		
Vagus nerve (unmyelinated)	110	28
Node of Ranvier	2100	—
Skeletal muscle (various)	2066–557	6–14
Other animals		
Squid giant axon	166–533	~1
Frog sartorius muscle	280	6
Electric eel electroplax (excitable surface)	550	38
Garfish olfactory nerve	35	96
Lobster walking leg nerve	90	24

nels to follow the process of fractionation. Unfortunately, unlike enzymes or other chemically active proteins, the activity of ion channels is to transport ions across membranes. Since this barrier is disrupted during the first stage of purification, the compartments essential to channel function no longer exist, and hence the electrically detectable flow of ions through channels no longer occurs. Instead, one may use the binding of powerful neurotoxins to channels to assess the degree of purity attained during fractionation.

For Na^+ channels, there exist highly selective inhibitors of sodium transport known as guanidinium toxins. The most commonly used of these are tetrodotoxin (TTX), obtained from certain species of toxic puffer fish, and saxitoxin (STX), a product of certain dinoflagellates that "bloom" during oceanic "red tides". Either toxin may be radioactively labeled, and these substances very specifically interact with Na^+ channels from a wide variety of sources with high affinity. To assess purity, one may thus compare the number of toxin-binding sites in a fraction with the amount of protein present.

C. Isolation of Channel Molecules Based on Fractionation Procedures

With a good source of channels, efficient solubilization procedures, and a specific and sensitive assay, one may proceed to attempts at purification. For Na^+ channels, fractionation based on charge, followed by size fractionation of the mixed channel/lipid/detergent micelle, has resulted in highly purified preparations of Na^+ channels. Alternatively, in some protocols, channel proteins have been specifically purified using affinity chromatography, in which channels are selectively retained by a column to which channel-binding substances, such as drugs, toxins, or antibodies, are attached.

Skeletal muscle Ca^{2+} channels have also been extensively purified and characterized using much the same approaches as those described for Na^+ channels. In this case, it has been the L-type channels that were isolated, using radiolabeled drugs of the dihydropyridine class to tag channels irreversibly in the intact muscle membrane. This allowed the investigators to isolate the channel protein by following the fractionation of the bound radiolabeled drug (Sharp *et al.*, 1987).

D. Information about Channel Structure from Channel Purification

There is much to be learned about channel structure from the physicochemical characteristics of the purified molecule. Some of this information forms the basis for separating the molecule from others in membrane extracts (such as size or charge differences), as described previously, whereas other measurements may be done on the purified material itself. Thus, for Na^+ channels, it was found that the channel is very negatively charged. Furthermore, in detergent solution the channel appears quite large; in micellar form, its size is on the order of several million molecular weight. Both characteristics are important clues regarding the chemical composition of Na^+ channels.

The subunit composition of a channel is an especially important characteristic to be determined from the purified ma-

terial. For a number of enzymes, separate functions have been attributed to different subunits; hence breaking down the structural entities represents an important basis for further structure-function correlations. In brief, one usually determines the subunit composition of the purified material by analyzing it using denaturing electrophoresis, in which the intersubunit bonds are broken and each subunit separated according to size in an electric field. The most commonly used technique is sodium dodecyl sulfate–polyacrylamide gel electrophoresis (SDS-PAGE), in which each polypeptide species appears as a separate band on staining the polyacrylamide sieve used to separate polypeptides. Unfortunately, this is often more difficult than it appears, since (1) one is often not sure whether different bands on a gel are contaminants or true subunits and (2) the stoichiometry of subunits is difficult to establish because of variations in the intrinsic staining intensity among various polypeptides. As a result, considerable effort is expended by various researchers to determine (and argue about!) subunit composition using a wide variety of approaches. For the voltage-gated Na^+ channel, a single large polypeptide apparently accounts for pore formation, ion selectivity, and gating. This protein appears on SDS-PAGE as a broad band at high molecular weight.

Given the large size of the eel electric organ, enough protein can be purified to allow study of its chemical composition. Not too surprisingly, Na^+ channels have an elevated proportion of hydrophobic amino acids compared to soluble globular proteins, likely reflecting domains of the channel that insert into the lipid membrane. However, the channel molecule appears to be much more hydrophobic than can be accounted for from its amino acid composition. Thus, further chemical analysis has found that a large number of lipid molecules are bound to the eel protein. Finally, it has been found that Na^+ channels are modified by the presence of extensive domains of carbohydrate. This glycosylation makes up about 30% of the molecule (by weight).

Na^+ channels have also been purified from mammalian brain and muscle tissue. Like the eel channel, they consist primarily of heavily glycosylated large polypeptides. However, unlike the electric organ channel, they are found in association with one or several smaller polypeptides that have been thought to be channel subunits (Table 2). The large polypeptide common to Na^+ channels from all sources is referred to as alpha (α) while smaller subunits have been designated as beta (muscle) or $\beta 1$ and $\beta 2$ (brain). At present, the role of these accessory polypeptides in channel function is not certain, although recent evidence suggests that they may aid in channel expression and modulate channel gating (Isom *et al.*, 1992).

For the skeletal muscle L-type Ca^{2+} channel, it was found that although a single large polypeptide ($\alpha 1$) was affinity labeled by the dihydropyridine (DHP) marker, at least four other polypeptides—$\alpha 2, \beta$, gamma (γ), delta (δ)—copurified with the DHP label. As with the Na^+ channel, $\alpha 1$ seems to have the mechanisms required for ion channel operation. However, there is evidence that the other subunits are important accessory proteins that may be involved in the key role of this channel as the voltage sensor in excitation-contraction coupling, which is described in Chapter 42.

TABLE 2 Subunit Composition of Voltage-Gated Cation Channels

Channel/tissue	Subunit	Molecular weight[a]
Na+		
Eel electric organ	α	260 (208)
Rat brain	α	260 (220)
	$\beta 1$	36 (23)
	$\beta 2$	33
Rat skeletal muscle	α	260 (209)
	β	36
Chick heart	α	235
Ca2+		
L-type, rabbit skeletal muscle	$\alpha 1$	170 (212)
	$\alpha 2$	143 (125)[b]
	β	54 (58)
	γ	30 (25)
	δ	24–27 (27)[b]
K+ channel		
Drosophila Shaker A	—	65–85 (70)
Rat drk1	—	130 (95)

[a]First set of values is the apparent molecular weight determined biochemically (SDS-PAGE); weights in parentheses were obtained from cloned DNA sequences.

[b]The biochemically observed calcium channel $\alpha 2$ and δ subunits are derived from proteolysis of a full-length $\alpha 2$ translation product during biosynthesis.

Also shown in Table 2 are data for voltage-gated K+ channels. The characterization of K+ channel subunits was achieved primarily through molecular genetic means, as discussed in Section III.F.

E. Reconstitution of Purified Proteins Confirms Their Identity as Channels

An important criterion for purification was the reconstitution of the isolated material into artificial membranes and the demonstration that voltage-gated, sodium-selective, toxin-inhibitable channels were formed (see Miller, 1986). Thus, these experiments showed that purified toxin-binding material could be reinserted into small artificial lipid vesicles (liposomes) or planar lipid films (lipid bilayers) where they retained their ability to transport sodium selectively in response to voltage changes. These reconstituted channels had the same ability to respond to different pharmacological compounds, such as toxins and anesthetics, as the channels in natural tissue (rev. in Catterall, 1992). Similar experiments have demonstrated that purified DHP receptors can form functional Ca2+ channels in bilayers (Catterall, 1988).

F. Limitations of Biochemical Characterization of Channels

The biochemical approach to channel purification thus gave important information about the size, chemical compo-

sition, and polypeptide makeup of Na+ and L-type Ca2+ channels. Classically, the next steps would be to sequence the purified material and attempt to obtain a high-resolution structure from crystallographic approaches. However, although these techniques have worked well with a small number of soluble globular proteins, for several reasons they have been relatively unsuccessful with membrane proteins in general and voltage-gated channels in particular. First, sequencing a large polypeptide requires that it be enzymatically or chemically cleaved into overlapping smaller fragments amenable to Edman degradation techniques (which can only sequence from 25 to 50 residues at a time). In such an approach, each small fragment must be separately purified from the fragmented preparation. For the large polypeptides of Na+ and Ca2+ channels, the number of fragments generated by such cleavages is just too great for all of them to be purified separately. Furthermore, many of the interesting membrane-spanning segments do not fragment or isolate well because of their high degree of hydrophobicity ("fears water"). Finally, the amphipathic (i.e., polarized hydrophilic/hydrophobic) nature of membrane proteins prevents their crystallization; instead, in the absence of lipids or detergents, such purified preparations usually form amorphous aggregates or precipitates with little intrinsic order. Hence, the elaboration of detailed structure of biochemically purified channels has been hampered by the unfavorable physical properties of membrane proteins.

III. Channel Structure Investigation through Manipulation of DNA Sequences Encoding Channel Polypeptides

Fortunately, the limitations of classic biochemistry may be partly overcome by the application of techniques that allow one to identify and sequence DNA that encodes channel peptides. The next three subsections describe how the powerful methods of molecular biology are being applied to questions of interest to the cell physiologist. These approaches have given important insights into the structure of channels and how they work. This aspect of channel biology is considered in Sections III.A and III.B. However, molecular biology has also shown us that voltage-gated channels exist in a rich diversity of forms even within a single organism. This aspect of channel biology is described in Section III.C. First, a brief description of how recombinant DNA methods are used in such studies is provided for the novice reader (Fig. 2).

A. Primary Structure of Ion Channels Determined Using Recombinant DNA Technology

One starts with the same tissues shown to be enriched in channel protein because these are also probably enriched in messenger RNAs (mRNAs) encoding these channels. Using the retroviral enzyme reverse transcriptase, one may then make **complementary DNA copies** (cDNAs) of the mRNAs. Using enzymatic "scissors" (called restriction enzymes), one may then insert each cDNA obtained into a circular DNA **plasmid**. These plasmids have the ability to be replicated along with other genetic material in bacteria, and

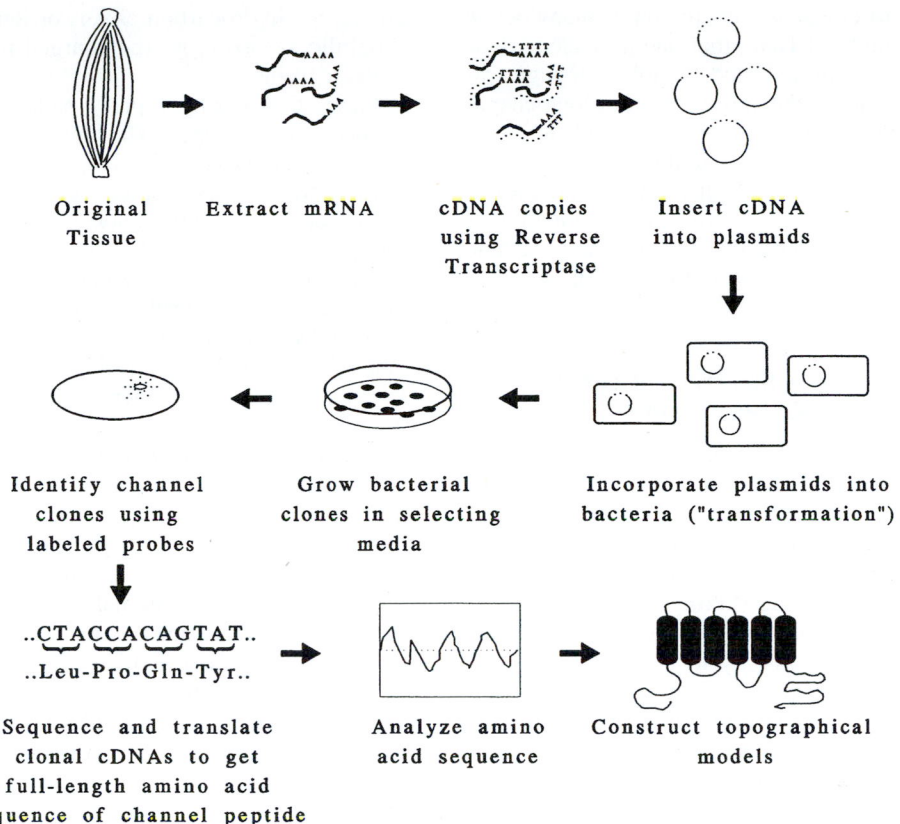

Original Tissue → **Extract mRNA** → **cDNA copies using Reverse Transcriptase** → **Insert cDNA into plasmids**

Identify channel clones using labeled probes ← **Grow bacterial clones in selecting media** ← **Incorporate plasmids into bacteria ("transformation")**

..CTACCACAGTAT..
..Leu-Pro-Gln-Tyr..

Sequence and translate clonal cDNAs to get full-length amino acid sequence of channel peptide → **Analyze amino acid sequence** → **Construct topographical models**

FIGURE 2. General strategy used to determine ion channel structure using recombinant DNA approaches.

also they encode an enzyme that will destroy certain antibiotics that would otherwise kill their bacterial hosts. The recombinant plasmids are next introduced into a bacterial host so that each bacterium will contain no more than a single plasmid with its unique cDNA insert. The bacteria are then diluted and spread over a plate of agar containing an antibiotic. Thus, bacteria without a plasmid are killed, whereas plasmid-containing bacteria can survive and replicate to form visible colony plaques. Since each such colony arises from a single original plasmid-transformed cell, they are known as **clones**. Each agar plate may contain many thousands of such clones, each clone having many cells with plasmids containing the identical cDNA insert.

The problem is to identify which of the many colonies has a particular cDNA that encodes for the channel protein of interest, among all of the cloned cDNAs that encode the myriad proteins of the original tissue. The severity of this problem can be appreciated from the fact that even highly enriched tissues such as eel electric organ may have at most only 1 message out of 10 000 or so that encodes a Na^+ channel. Hence, there must be a rapid method for screening the colonies to identify the channel cDNA clones. This is where information from biochemical purification proves to be vital.

Despite the near impossibility of totally sequencing large Na^+ or Ca^{2+} channel polypeptides, from a proteolytic digest one can readily purify a homogeneous preparation of a particular fragment. This fragment may then be partially sequenced

using standard Edman degradation techniques. In practice, one only needs a few (perhaps six or so) residues of sequence to be reasonably sure that a unique segment of the protein is encoded. From this short amino acid segment, one can then synthesize (by automated means) a short DNA fragment, called an **oligonucleotide,** that encodes the segment. Further, this synthetic DNA may be radiolabeled, and thus becomes a binding probe to identify cDNA inserts. Identification is done by placing a piece of special paper briefly on top of the agar plate, and lifting off a few bacteria from each colony plaque onto the paper. The paper is then placed in a special solution that lyses the bacteria but binds the released DNA immediately on the surface. The radiolabeled synthetic probe may then be washed over the paper; it will bind by complementary interactions to the antisense strand of clonal plasmids encoding the channel protein containing the original sequence. After unbound probe is washed away, the paper is overlaid in the dark with x-ray film. Channel clones are thus identified as dark spots (usually only a few) on the film, and may be traced back to the plaque on the original agar plate. This plaque is then lifted from the plate with a toothpick and grown in a nutrient broth to produce many bacteria. This technique provides enough cells from which the plasmids are purified by simple chemical means, and the cDNA insert may be readily sequenced. Once the cDNA sequence is known, the genetic code is applied to derive the primary amino acid sequence of the channel protein.

This description of clonal sequencing omits many details. For example, the length of DNA that one may clone or sequence at one time is limited, and complex strategies are often required to sequence the entire length of long channel polypeptides. In addition, ion channels that are heteromeric complexes will require that each subunit polypeptide be separately sequenced. Despite these apparent complexities, an organized group of researchers using automated equipment can screen hundreds of thousands of clones and sequence thousands of bases of interesting cDNA inserts in a relatively short time.

B. Analysis of the Primary Structure of Large Polypeptide Voltage-Gated Channels

The electroplax Na$^+$ channel was the first voltage-gated channel to be sequenced (Noda *et al.*, 1984). As suggested by the biochemistry, it was found to be a long polypeptide of 1820 amino acids (Fig. 3). Subsequent cloning of mammalian channels showed them to be similar in size and sequence (see Numa and Noda, 1986). What can one learn about channel structure from such primary sequence data? Typically, one subjects the primary sequence data to several forms of analysis to derive a model of the appearance of the higher order structure. The first striking characteristic revealed in the Na$^+$ channel sequence was the presence of four repeats (of about 200 amino acids each) that were highly (but not completely) homologous to one another. Furthermore, these internal repeats were found to have great potential for forming membrane-spanning domains of the channel. From such analyses, the model shown later in Fig. 6 was obtained.

How are such analyses performed? Basically the analyses occur in two or three separate steps. First, one analyzes the sequence of the peptide for regions of high hydrophobicity, on the expectation that membrane-spanning regions should have a high proportion of hydrophobic amino acids. To do this, hydropathy values that reflect its relative hydrophobicity are assigned to each amino acid residue. The assignment of such values is based largely on indirect physical measurements of the hydrophobicity, such as the ability of a given residue to partition from an aqueous solution to an oil phase. Such values may or may not reflect the chemical behavior of a residue in a long polypeptide chain, where chemical properties may be further influenced by neighboring residues. As a result, a number of such hydropathy tables have been developed by various investigators who have significant differences in the way they assign hydropathy weights to a given residue. One of the most popular tables is that of Kyte and Doolittle (1982) shown in Fig. 4. In most cases, the assignments are qualitatively obvious, especially for expected hydrophobic groups (for ex-

ample, the hydrocarbon chains or leucine or valine) or hydrophilic groups (e.g., the charged residues such as glutamate or lysine).

Figure 5 shows hydropathy plots for the Na$^+$ channel sequences given in Fig. 3. The plot is obtained by averaging the hydropathy values of a fixed segment (perhaps 7–15 residues). Starting with the first residues at the N-terminal end, this average is computed and the value is plotted versus the number of the residue in the middle of the segment. This averaging "frame" is then shifted by one residue toward the carboxyl end, the new hydropathy average is recomputed (i.e., the change due to the loss of the N-terminal side residue and the gain of the next C-terminal side residue in the sequence), and the value is plotted by incrementing the residue number by one. This **frame-shift** average is thus continued until the end of the protein is reached. The purpose of using the averaging frame is to smooth out sharp variations in hydropathy, so that a pattern may be more readily seen.

As can be seen in Fig. 5, there is a repeated pattern of hydropathy for Na$^+$ channels that reflects the conserved homology of the four internal repeat domains I–IV. Within each domain, there are five or six peaks of hydrophobicity that are each about 20 amino acids long. This is significant because such a length is considered optimal to span the hydrophobic interior of the lipid bilayer. However, hydrophobicity alone is not sufficient to predict a membrane-spanning region; instead, the residues must assume a favorable secondary structure, most commonly assumed to be an α-helix. In the helical conformation the hydrophobic side chains project radially away from the axis of the helix and presumably into the surrounding lipid. Such a configuration prevents the very hydrophilic peptide backbone of the protein from exposure to the hydrophobic lipid environment and thus represents a favorable low free energy arrangement of the protein-lipid interface.

How are such secondary structures predicted from primary sequence? This too is a highly empirical process (Chou and Fasman, 1978). Basically, investigators have analyzed in detail a number of proteins for which there are high-resolution x-ray structures available at the amino acid level. By cataloging the frequencies with which given amino acids or short combinations of residues are found in helices, β-sheets, or turns, a set of empirical rules has been developed to predict the secondary structures that a given sequence of amino acids will most probably form (see Fig. 5). Although the general accuracy of such predictions has been debated, when combined with hydropathy information the method is apparently rather accurate for the prediction of membrane-spanning domains in membrane proteins. This is probably because of the severe thermodynamic constraints confronting polypeptides needing to cross a boundary of high hydrophobicity while interfacing with a highly

---

FIGURE 3. Amino acid sequences of three Na$^+$ channels obtained by cDNA cloning techniques. Rat I and II were obtained from rat brain cDNA libraries, while the eel Na$^+$ channel is that expressed in the electric organ of *Electrophorus electricus*. Boxed residues show homology among the three channels, while dashed lines designate absent segments. Labeled brackets underneath the sequences refer to parts of the highly conserved putative transmembrane repeats (see text and Figs. 6 and 7 for details). (From Numa and Noda, 1986.)

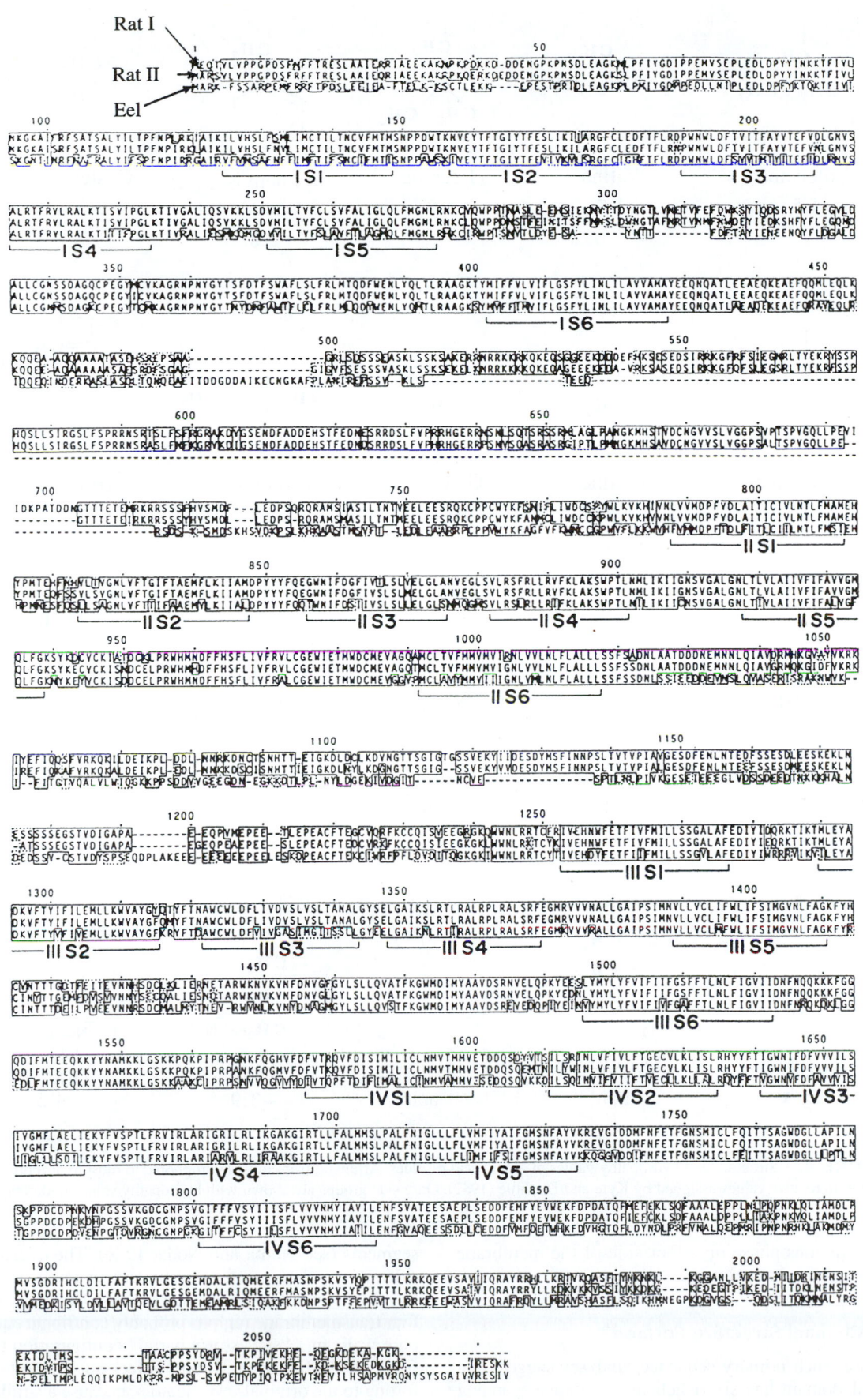

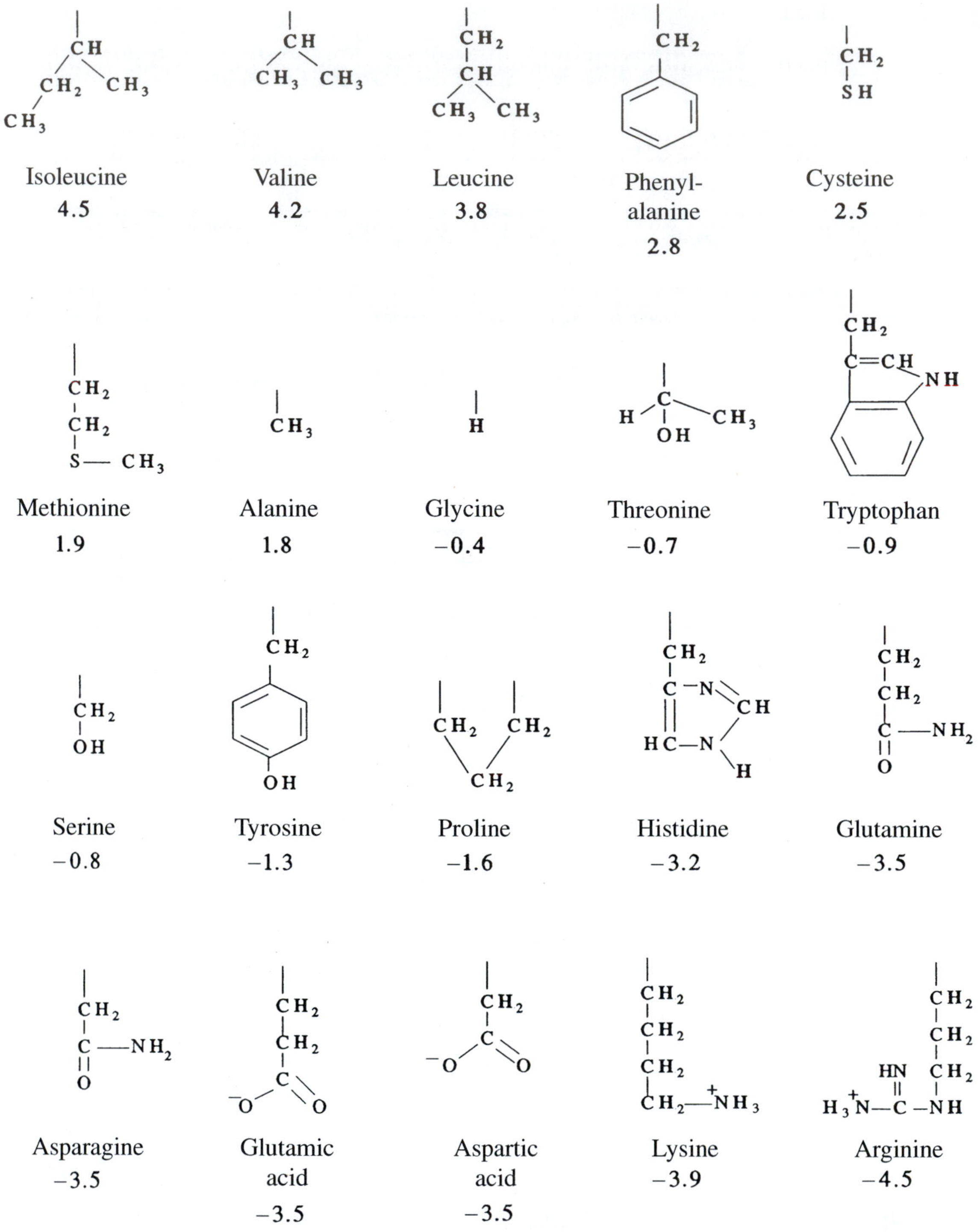

FIGURE 4. Structure and hydropathy values of amino acid residues. Arranged from most hydrophobic to most hydrophilic according to relative values assigned by Kyte and Doolittle (1982). Only side groups are shown with hydropathy values underneath.

hydrophilic aqueous phase on either side of the membrane and in the aqueous channel pore.

C. How Is Channel Structure Formed?

In any case, such primary sequence analyses suggest that each internal domain has six α-helical membrane-spanning segments (see Numa and Noda, 1986). The question to consider next is how such an arrangement might form a transmembrane pore for the passage of ions. The answer is that the four transmembrane regions probably contribute equally to the pore walls in a "staves of a barrel" configuration (Fig. 6; see Guy and Seetharamulu, 1986). In fact, such a model was comforting to the original investigators, because a similar arrange-

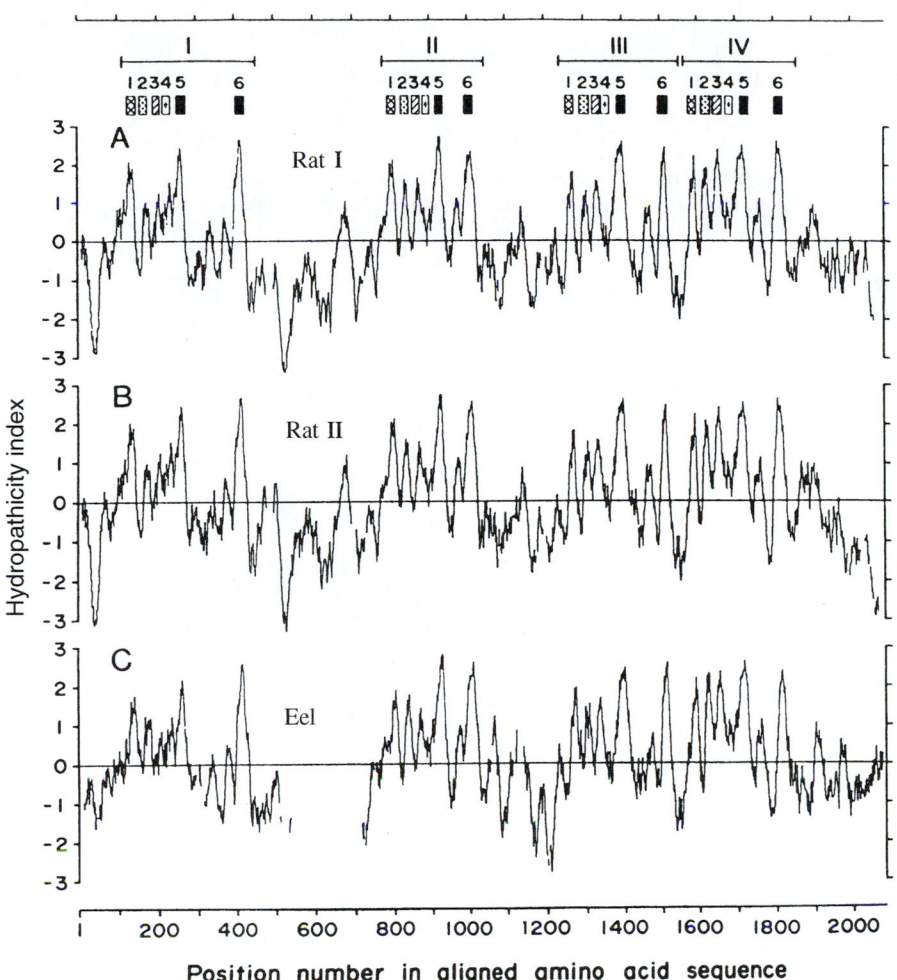

FIGURE 5.   Hydropathy plots for the three Na⁺ channel sequences of Fig. 3. Brackets and numbered patterns at top refer to the locations of homologous repeats and transmembrane segments (see text and Fig. 7). (From Numa and Noda, 1986.)

ment had previously been shown to exist among the separate subunits of the synaptic acetylcholine-gated channel, the first ion channel to be purified and sequenced via cDNA cloning.

D. Sequence Homology among the Transmembrane Domains of Ion Channels

Analysis of the DHP-sensitive Ca^{2+} channel $\alpha 1$ peptide revealed that it also has four internal homologous repeats (Fig. 7; see Tanabe *et al.,* 1987). Furthermore, these repeats (but not the intervening loops) were found to be highly homologous to the Na⁺ channel repeats. The "staves of a barrel" configuration in which either homologous internal repeats (for Na⁺ and L-type Ca^{2+} channels) or highly homologous subunits (as in the case of transmitter-gated and K⁺ channel–see Section III.G) form a pore seems to be a wide-spread architecture among ion channels in general.

E. How Are Topographical Predictions for Channel Structure Tested?

The modeling approach based on primary structure is highly uncertain and needs to be empirically confirmed for

each channel. One of the most popular ways this has been done is through the use of site-directed antibodies raised to synthetic peptides encoding a short stretch (i.e., 10–20 residues) of the channel of which one wishes to know the topographical orientation. These antibodies may be applied to channels in their native membranes or intact cells, and the relative sidedness of the epitope may be determined from the side of the membrane to which the antibody is bound. The approach has been used for Na⁺ channels in several laboratories, and the topographical model shown in Fig. 6 has largely been confirmed (see Catterall, 1992). Similar approaches can be used in which impermeant group-reactive chemical reagents are applied to either intact cells or liposomes with oriented channels, followed by some method to identify the site of interaction on the protein (e.g., sequencing of proteolytic fragments). Alternatively, a synthetic, but highly antigenic, short epitope may be recombinantly inserted into the channel sequence and the mutant expressed in cultured cells. In this case, the same antibody may be used in all experiments to determine on which side of the membrane the epitope tag is located as a function of its position in the native channel sequence.

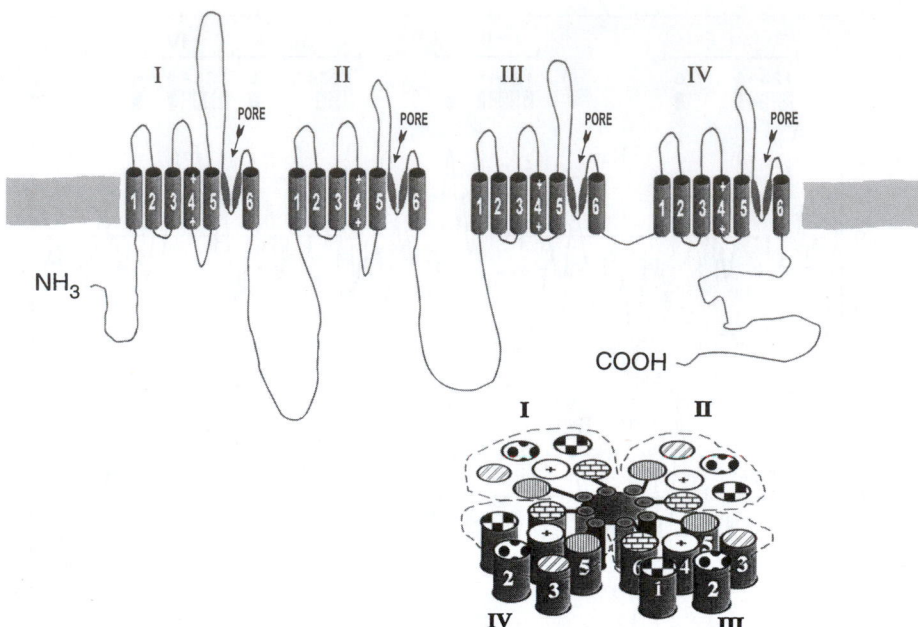

FIGURE 6.   Model of Na⁺ channel structure based on analysis of primary sequence data. The top drawing shows linear placement of four homologous domains, loops, and termini (outer surface of membrane is at the top). Possible pore and selectivity regions are labeled and shown as the angled segments between transmembrane segments 5 and 6. The drawing on the bottom right shows a hypothetical arrangement of homology repeats in the "staves of a barrel" configuration to form a pore (placement of transmembrane segments 1–6 is somewhat arbitrary).

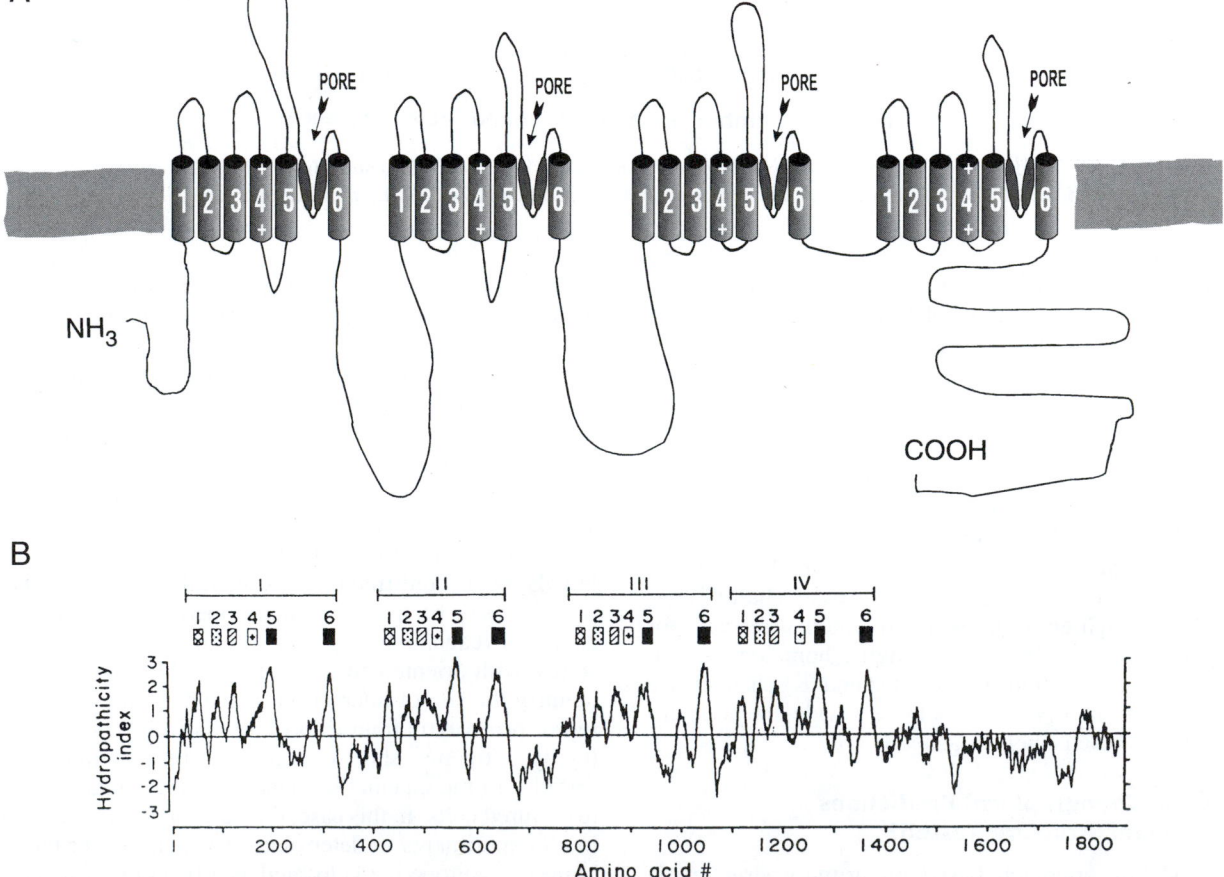

FIGURE 7.   Model and hydropathy plot (bottom) of L-type skeletal muscle Ca²⁺ channel α-1 subunit. Compare to Figs. 5 and 6.

F. The Genetic Approach Used to Identify Nucleic Acid Clones Coding for K⁺ Channels

K⁺ channels are extremely important in excitation and transduction phenomena. Partly because of the way these channels were cloned and partly because their structure is a simplified version of the long Na⁺ and Ca²⁺ channels, they are very valuable systems with which to study the general structures and mechanisms of the voltage-gated ion channel class. Such channels were not originally cloned via the usual biochemical purification–oligonucleotide probe approach described previously because there were few good chemical labels like tetrodotoxin or DHP.

A highly creative, yet complex, approach was taken independently by several laboratories that used behavioral mutants of the fruit fly *Drosophila* (e.g., see Papazian *et al.,* 1987). In brief, mutant strains that had uncontrolled shaking when exposed to ether were identified. Electrophysiology then established that the muscles and certain neurons of such flies had missing or altered K⁺ currents of a specific type known as A-type currents that, like sodium currents, inactivate on depolarization. The locus of the genetic defect causing this behavior could be seen as an actual physical defect (translocation) in the polytene salivary chromosomes of these so-called *Shaker* mutants. Since extensive libraries of cloned identified chromosomal DNA fragments of *Drosophila* have long been available (called genomic clones), the investigators were able to obtain the DNA fragment of normal flies on which the mutant defect occurred. This allowed them to sequence such fragments, and by comparison with *Shaker* mutant sequences and by the identification of homologous cDNA fragments (derived as described earlier), the sequence of the A-type K⁺ channel could be inferred. Since the *Shaker* mutation allowed a completely genetic elucidation of the channel primary amino acid sequence, the A-current it manifests has become known as the *Shaker*-type K⁺ channel.

G. K⁺ Channels Are Homologous to a Single Internal Repeat of Na⁺ Channels

An initially surprising finding when the *Shaker* A-type channel sequence was analyzed was that the length of the sequence was much shorter than that of the other voltage-gated channels, but contained a single domain that was reminiscent of (but strictly speaking not very homologous to) the internal repeat domains of the Na⁺ or Ca²⁺ channels (Fig. 8; see Pongs, 1992). An obvious hypothesis for channel structure for the *Shaker* channel was that four molecules of this sequence (i.e., subunits) combined to form the "staves of a barrel" structure postulated for the long-channel polypeptides. This tetrameric assembly model has recently received strong support from several elegant experiments (MacKinnon, 1991).

IV. Molecular Mechanisms of Channel Function: How Does One Investigate Them?

We now discuss the methods that can be used to determine which parts of an elucidated channel sequence are involved in the basic channel properties of pore formation, ion selectivity, and voltage-dependent gating. Here, too, the

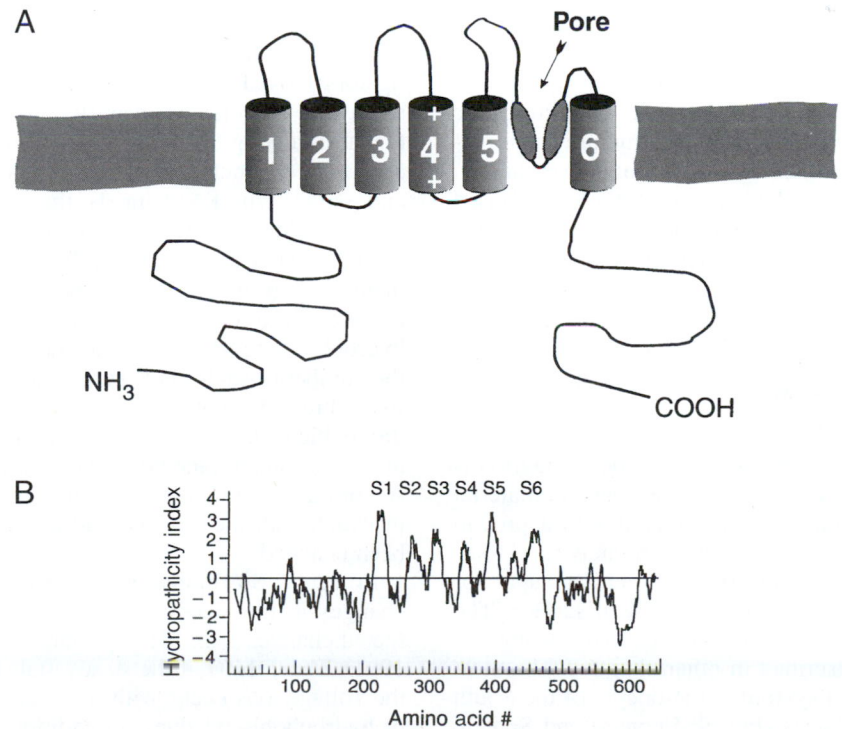

FIGURE 8. (A) Model and (B) hydropathy plot of *Shaker* B-type K⁺ channel. Compare to Figs. 5, 6, and 7.

methods of recombinant DNA technology have proved to be of great use in that they allow the investigator to change the nucleic acid sequence encoding a channel in the manner desired. Using such mutagenesis techniques, the basic idea is to alter the amino acid sequence via deletion, addition, or change in the primary sequence in specific locations and then to test how such changes affect the function of the mutant channels. The hope is that by obtaining many such structure-function correlations, one might not only infer the part of the sequence that underlies, for example, channel activation, but also infer how this structure works as a mechanism.

A. Choosing Interesting Sites and Segments as Targets for Mutagenesis

How does one decide which of the many amino acids in a channel polypeptide to change? This is where the modeling approach described previously is so important, for it provides clues or testable hypotheses that focus mutagenesis attempts on reasonable guesses. For example, one might assume that a highly conserved sequence in channels from phylogenetically distant animals might reflect a functionally essential part of the molecule that cannot withstand evolutionary alteration without being deleterious. Another strategy might be to compare the regions of divergence in the sequence of the various voltage-gated channels in the hope of finding candidate regions involved in different functions among channels. For example, the difference in ion selectivity among Na^+, K^+, and Ca^{2+} channels must lie somewhere in the variations in their amino acid sequence. Both approaches have had a certain degree of success, although "homology" is a relative term, and in reality large areas of channels are highly variable, even among isotypes of channels from the same organism (see Section V). This makes it difficult to guess which of these many differences relates to specific functions.

Another approach is to take advantage of certain genetic disorders in which mutations in channel structure occur naturally with functional consequences. A number of human diseases and animal models have been shown to be caused by channel defects. The sequence of such channels can give important clues into structure-function relationships in channels, while directing artificial mutagenesis experiments (see Chapter 39).

B. Use of Expression Systems in Mutagenesis Studies

Another essential part of mutagenesis experimentation is the ability to introduce artificially altered genetic material into a cell that is capable of translating it into a protein molecule, processing the molecule appropriately (i.e., folding the polypeptide, assembling subunits, adding sugars or lipids), and inserting the protein into the cell surface. The protein then may be functionally characterized using the biophysical methods described in other chapters. The most popular such **expression system** is the oocyte of the South African frog *Xenopus laevis* (Fig. 9; Leonard and Snutch, 1991). In this system one prepares mRNA from clones of the mutant channel and injects this message into the large (1–2 mm in diameter) oocytes with a micropipette. After waiting a few days for channel synthesis and membrane insertion to occur in sufficient numbers, one observes the expressed ionic currents via voltage-clamp methods, using large microelectrodes to impale the cell. Alternatively, one may also use a very large diameter electrode to patch-clamp a large area of membrane; such macropatch methods give better voltage control than the impaled electrode technique while allowing macroscopic current characterization. Finally, conventional single-channel analysis may be done in all the usual modes using standard patch pipettes. Thus, the oocyte allows the investigator to compare both the macroscopic and single-channel properties of the mutants with the normal wild-type channel.

Although other expression systems in which the altered genetic material may be introduced into cells transiently via viruses or integrated permanently into the host genome (stable transfection) have been developed more recently, the use of such systems is beyond the scope of this chapter. Suffice it to say that the frog oocyte is currently the most popular and productive expression system with which to study structure-function relationships.

Because of its great potential in elucidating channel mechanisms, mutagenesis is being used by a large number of investigators. As a result, both new and altered concepts of channel function appear frequently in the scientific literature. In the brief account that follows, we can only summarize some of the fairly firm basics of gating, pore formation, and ion selectivity.

C. Domains Involved in Voltage-Dependent Activation

The ability to respond to changes in transmembrane potential is a hallmark of voltage-gated channels. Thus investigators looked for common motifs in channel sequences suggestive of voltage sensor domains that might target mutagenesis among the large number of residues in cloned voltage-gated channels. In comparing Na^+, DHP-sensitive Ca^{2+}, and *Shaker*-type K^+ channels, the fourth predicted helix in the repeat/subunit motif was seen to display a highly conserved sequence pattern (see Figs. 6, 7, and 8). This segment, known as S4, consists of a repeated triad of a positively charged residue (Arg or Lys), followed by two highly hydrophobic residues (such as Val, Leu, Ile; Fig 10). While the number of such triads in the various S4 segments ranges from three to eight, its structure of both charged and hydrophobic residues suggests an element capable of responding to transmembrane voltage changes (see Catterall, 1992). Accordingly, a number of experiments have now been done in which both the charged and hydrophobic elements have been changed.

In brief, S4 mutagenesis usually causes significant changes in the steady-state activation behavior of the affected channel (Fig. 11; Papazian *et al.*, 1991; Lopez *et al.*, 1991). Thus, shifts of the steady-state activation curve along the voltage axis occur with most changes to either charged or hydrophobic residues. In addition, changes in activation slope are frequently seen.

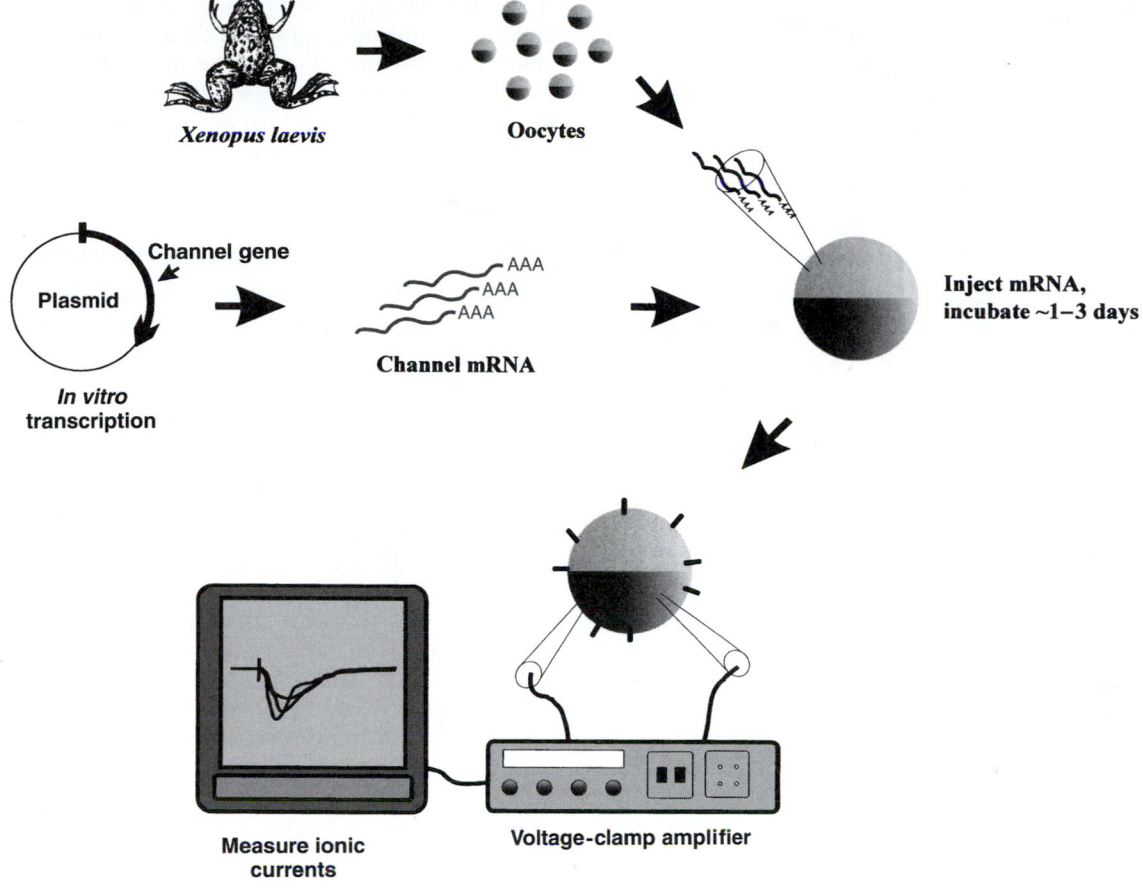

FIGURE 9.   *Xenopus* oocyte expression system. For details, see text.

The question is whether the observed changes can be explained in terms of a mechanism by which S4 participates in gating (instead of the general hypothesis that S4 is some part of a gating structure). Unfortunately, to date the results obtained from mutagenesis do not discriminate very well among the various possible roles for S4 in gating, for example, as a gating sensor, a part of the transduction linkage to a gating element, or simply an unrelated part of channel structure. Thus, the magnitude or direction of changes in activation behavior does not systematically correlate with changes to S4 charge or sequence. In addition, others have questioned whether the charged groups in the S4 segments can completely account for all the gating charge (see Chapter 20) that moves when channels activate (see Sigworth, 1995). In general, the mutagenesis approach is inherently limited in its ability to distinguish between alterations that directly affect a mechanistically important part of the molecule and those that indirectly cause functional effects through structural changes that propagate from mechanistically unrelated domains. This problem arises because protein structures are notoriously sensitive to long-range effects of such alterations, generically known as allosteric ("other site") effects.

	+	+	+	+	+	+	+	+
Electric eel NaCh IV-S4	L F R	V I R	L A R	I A R	V L R	L I R	A A K	G I R
Rat brain NaCh IV-S4	L F R	V I R	L A R	I G R	I L R	L I K	G A K	G I R
Rat skeletal muscle NaCh IV-S4	L F R	V I R	L A R	I G R	V L R	L I R	G A K	G I R
Drosophila NaCh S4	L L R	V V R	V F R	I G R	I L R	L I K	A A K	G I R
Rabbit skel. muscle CaCh IV-S4	S S A	F F R	L F R	V M R	L I K	L L S	R A E	G V R
Mouse brain KCh S4	I L R	V I R	L V R	V F R	I F K	L S R	H S K	
Drosophila Shaker S4	I L R	V I R	L V R	V F R	I F K	I S R	H S K	

FIGURE 10.   Comparison of S4 segments from different voltage-gated channels. Note the repeating triadic motif of positive charge (R, Arg; K, Lys) followed by two hydrophobic residues. Hatched areas show conservation of charge among channels; all amino acids designated by single letter codes.

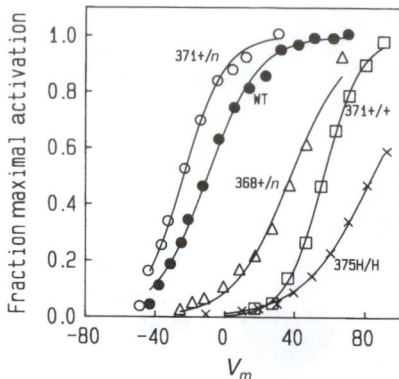

FIGURE 11. Effect of S4 mutations on *Shaker* K$^+$ channel activation expressed in oocytes. Shown are steady-state activation curves for wild-type (WT) and various single amino acid changes. +/n, change from a positively charged residue to a neutral one; +/+, a change of positively charged residue for an other charged residue; H/H, replace one hydrophobic residue with another hydrophobic residue. (Data redrawn from Papazian *et al.*, 1991, and Lopez *et al.*, 1991.)

D. Limitations of Mutagenesis in the Study of Channel Mechanisms

The naive reader can understand the problem from a crude analogy in which many of us as children tried to understand the workings of a mechanical watch by a similar approach using simple tools. The child understands the function of the mechanism only by the movement of the hands. However, as we all know, a mechanical watch is quite a delicate machine. Thus, even the act of carelessly opening the case may cause the mechanism to cease working. Are we to conclude that the case is therefore part of the mechanism? Given access to the mechanism itself, the situation is highly complex, since the movement of many parts is required for the ultimate movement of the hands. For example, the result of stopping the second hand by interfering with a given gear tells one very little of the role that gear plays in how the watch operates.

For some time it has been known that the interactions of chemical substances or structural alterations with a site on one side of a protein can dramatically affect the function of another site at some distance, for example, on the other side of the protein. This concern and the potentially complex nature of channel-gating mechanisms should motivate the responsible practitioner of mutagenesis to display caution in interpreting results in mechanistic terms.

How, then, in the face of such limitations does one make progress in elucidating the mechanisms of channel function? In fact, the approach is philosophically identical to the way one attempts to prove any hypothesis, namely, to establish a circumstantial case by as many independent tests as possible. In the example of S4 and mutagenesis, for example, one might wish to demonstrate that mutations to any other part of the molecule have no effect on voltage-dependent activation as a sufficient (but not necessary) test. Unfortunately, mutations to other parts of channel sequence can have equally profound effects as S4 alterations (for example, to the pore-forming region described previously; see

Yool and Schwarz, 1991). Thus, while these other mutations do not disprove a role for S4 as a sensor (or any other role), they do suggest, as in the mechanical watch, that a number of segments of the sequence are either directly involved in a complex mechanism of voltage gating or indirectly affect a gating process through allosteric/structural influences.

E. The Mechanism of Channel Inactivation

Genetic engineering studies of the inactivation process currently provide a better example of what one may learn mechanistically from the mutagenesis approach. Some of the most important such studies have involved the *Shaker* A-type channel. Here early workers found important clues to the segment involved in the inactivation process by comparing naturally occurring variants of the channel (**isotypes**; see Section V). Thus, Fig. 12 shows the pattern of four such isotypes, called A, B, C, and D (Timpe *et al.*, 1988). *Shaker* B, C, and D were identified from further testing of cDNA clones using probes constructed from the original *Shaker* A sequence as described earlier. When expressed in oocytes, these isotypes displayed very different kinetics of inactivation. Analysis of the regions of homology and differences among the isotypes shown in Fig. 12 indicates that fast inactivation properties correlate most strongly with the nature of the N-terminal segment (thought to be cytoplasmic; see Fig. 8).

Further evidence for the role of this segment in inactivation was obtained by constructing channel chimeras in which the *Shaker* A clone was modified by replacing its N-terminal segment with those from other isotypes. In each case the mutant channel assumed the inactivation properties of the isotype donating the N-terminal segment, despite the fact that the rest of the molecule had the sequence of the fast-inactivating *Shaker* isotype (Aldrich *et al.*, 1990).

Finally, investigators mutagenized specific regions of the fast-inactivating *Shaker* B segment to localize those residues involved in the inactivation process (Hoshi *et al.*, 1990). Both by deleting small segments and by replacing others it was found that the crucial part of the molecule was a stretch of residues from the positions 6–83 (i.e., the N-terminal tail of the molecule seen in Fig. 8). In this segment there were distant effects of the alterations. Thus, mutations to the residues in the 23–83 residue region tended to produce changes in inactivation kinetics, whereas changes to those in the N-terminal side (6–22) of the segment tended to destroy inactivation completely (see Fig. 14).

What might be the mechanism of *Shaker* inactivation? Much earlier Armstrong and coworkers had reported that internal perfusion of proteases in the squid destroyed the ability of Na$^+$ channels to inactivate. On the basis of this and other experiments, Armstrong proposed that the inactivation mechanism is a cytoplasmic ball attached to a peptide tether that swings into an open channel to block it from the inside (Fig. 13; Armstrong and Bezanilla, 1977). In this hypothesis the ball blocks the channel by binding to a receptor in the inside of the pore that is accessible only when

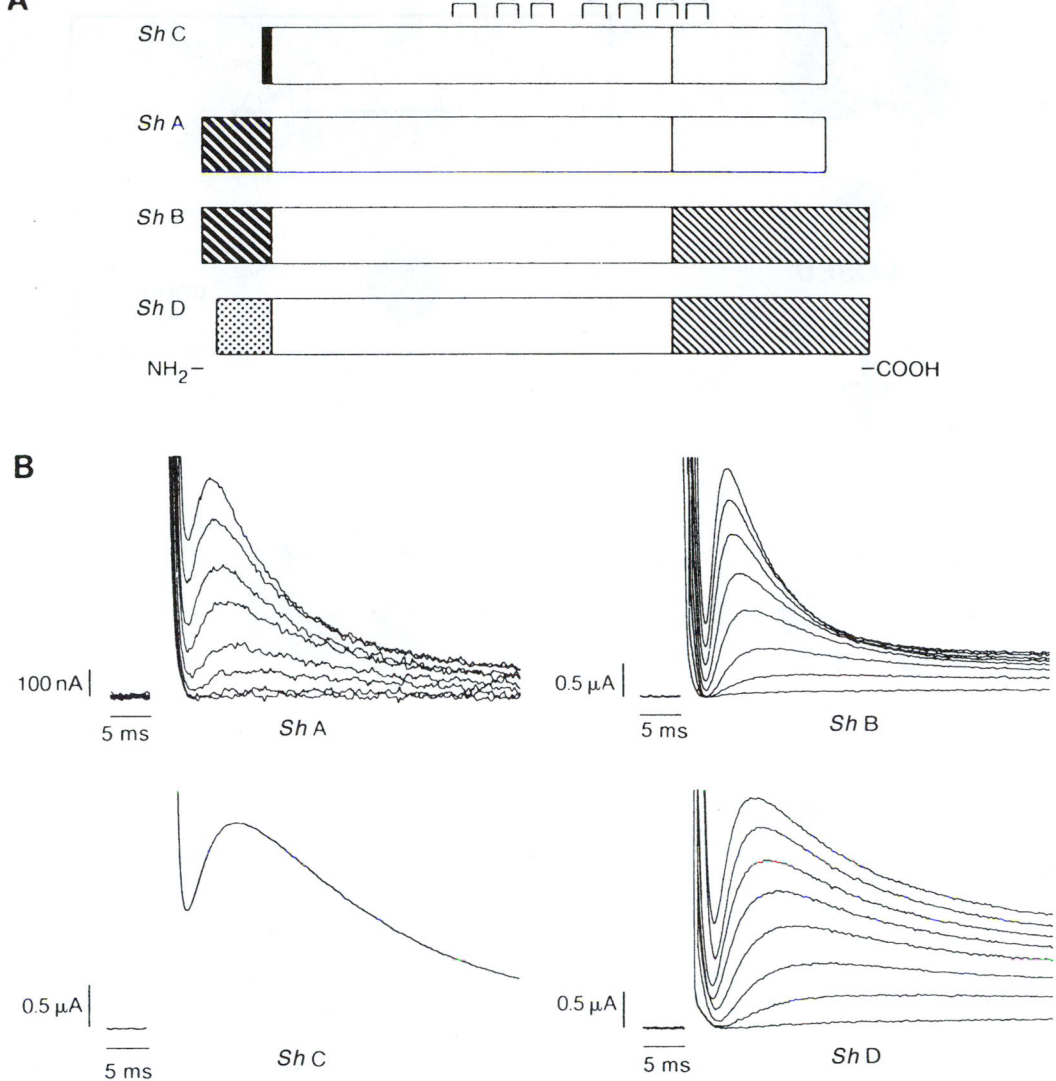

FIGURE 12. Gating kinetics of alternatively spliced *Shaker* isoforms expressed in oocytes. (A) Comparison of sequences, with patterned regions representing alternatively spliced domains that are different among isoforms (other unpatterned regions are identical in sequence). (B) Both current traces obtained from oocyte voltage-clamp show currents resulting from each isoform. (From Timpe *et al.* 1988. Copyright 1988 by Cell Press.)

the channel has opened (activated). The reader may recall the evidence that inactivation is coupled to activation, and hence may be intrinsically voltage independent. Thus, the ball-and-chain model also explains these properties since the cytoplasmic ball would experience little of the transmembrane field but could swing into the channel only after it had activated and revealed its receptor (hence the coupling properties).

In the case of the *Shaker* channel, Aldrich and colleagues proposed that the N-terminal segment that they identified was just such a mechanism, since changes to the length of the 23–83 segment affected kinetics (as might be expected from changing the length of the chain), while changes to the N-terminal segment (such as alterations of charge) that eliminated inactivation were postulated to prevent the ball from binding to its receptor in the activated channel.

How might such a model be further tested? Here the investigators performed a particularly elegant experiment (Fig. 14; Zagotta *et al.*, 1990). First they constructed a mutant *Shaker* channel in which inactivation had been eliminated by a deletion of its own N-terminal segment. Next, a synthetic peptide that had the exact sequence of the deleted region was prepared. The mutant channel was then expressed in oocytes, where both excised macropatch and single-channel observations could be made. The investigators found that when the peptide was applied to the solution bathing the inside surface of the membrane, inactivation was restored to these mutant channels. Furthermore, the kinetics of the restored inactivation depended on the concentration of the peptide in the bath, and the inactivation could be reversed by washing away the peptide-containing solution. These observations are thus consistent with a direct blocking interaction of the peptide with a site in the open channel,

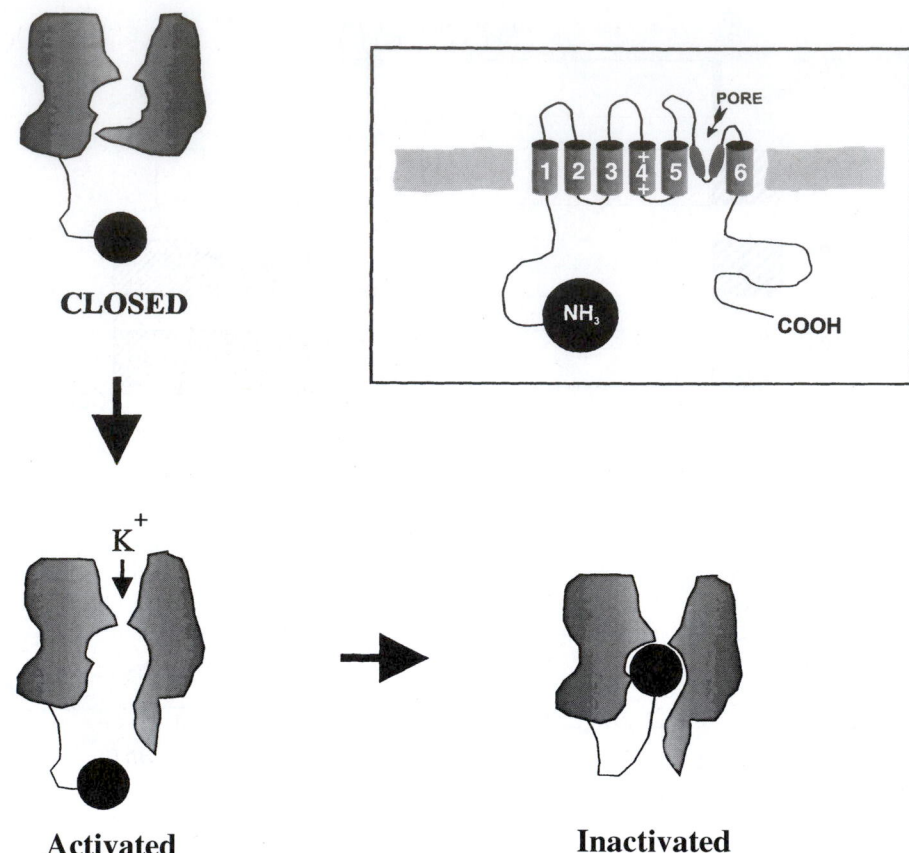

FIGURE 13. Ball-and-chain model of inactivation. Not shown are the three other balls that the other subunits of the channel will contribute.

and constitute strong evidence for the ball-and-chain mechanism of inactivation in A-type K+ channels.

F. Other Gating-Related Domains in Voltage-Sensitive Channels

Analysis of the homology of the various cloned Na+ channels revealed that the longest segment of conservation lay in the postulated internal loop between repeats III and IV, even among animals of some phylogenetic distance (see Fig. 3). This high degree of conservation suggested a functionally important role for the segment. This idea was initially tested by constructing mutant Na+ channels in which the channel was expressed in oocytes in two pieces, with the genetic "cut" within this segment (Stuhmer *et al.*, 1989). Interestingly, such artificial dimeric channels were assembled and expressed in the oocyte and displayed relatively normal activation and conductance; however, such channels did not inactivate. Further, site-directed antibodies raised to a synthetic peptide encoding this segment were found to slow or eliminate inactivation when they were applied to the cytoplasmic surface of cells expressing normally inactivating channels (Vassilev *et al.*, 1989). Finally, mutagenesis directed to this segment produced channels with altered inactivation (Catterall, 1992).

However, the sequence of this segment has little resemblance to the N-terminal ball-and-chain *Shaker* domain,

and it is additionally tethered at both ends. Nonetheless, several lines of evidence now suggest that this domain operates analogously to the *Shaker* ball-and-chain in producing Na+ channel inactivation. Most notably, via mutagenesis to the III–IV loop, one can produce "inactivationless" sodium channels; in such mutants, inactivation can be restored with the internal application of the same synthetic *Shaker* peptide that restored inactivation to non-inactivating potassium channels in the experiments described above.

For DHP-sensitive Ca^{2+} channels, an isotype of the originally cloned skeletal muscle channel has been cloned from the heart. In the parent tissues these channels share some characteristics, but differ greatly in the speed by which their calcium currents activate (heart channel currents being much more rapid in responding to voltage). Although the isotypes share considerable homology, there is still enough difference in the sequences to make it difficult to guess which residues might be responsible for activation kinetics. This problem was elegantly addressed by Beam, Tanabe, and coworkers, who made chimeras of the two channels (Fig. 15; Tanabe *et al.*, 1991). To focus the mutagenesis it was reasoned that only the predicted transmembrane repeats should be involved in voltage-gated activation. Thus, the chimeras swapped the homologous repeats among constructs. As shown in Fig. 15, the kinetics of activation correlate almost completely with the kinetics of the donor of the

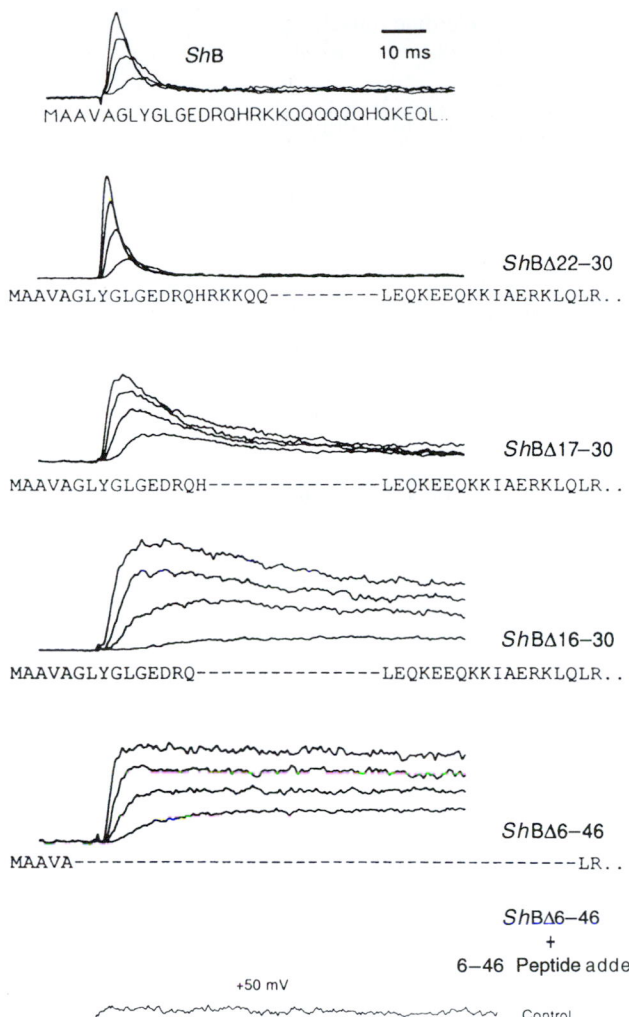

FIGURE 14. Identification of a putative ball and chain in *Shaker* B channels expressed in oocytes. Panels 2–5 from the top show the effect of deleting increasing lengths of a segment near the N terminus of the *Shaker* protein. Note that the deletion of residues 6–46 results in a noninactivating K⁺ current. The bottom panel shows that inactivation properties of this latter mutant may be restored by application of the synthetic 6-46 peptide to the inner channel surface in a dose-dependent fashion. (Reorganized from Hoshi *et al.*, 1990, and Zagotta *et al.*, 1990.)

first (I) repeat. This finding suggests that despite their approximate homology, the repeats in a given channel contribute differently to gating properties. This is perhaps expected from the observation that the S4 regions among these domains in different channels have somewhat different structures and numbers of charges (e.g., see Fig. 10). Having identified the domain responsible, these investigators have been able to narrow their search to shorter segments, and ultimately will identify the residues involved in this activation phenomenon (Nakai *et al.*, 1994).

G. Pore Formation and Ion Selectivity

Pores are the defining structures of all ion channels. As such, they provide both an aqueous pathway for ions to traverse the membrane as well as a means for discriminating among ions for access to this path. Studies of the mechanisms of pores provide an especially vivid example of the integration of theoretical modeling, mutagenesis, and molecular structural approaches.

Even before the advent of structural studies, Hille hypothesized that ion selectivity and pore formation were performed by the same parts of the channel structure. Experimental evidence also indicated that charged or polar oxygen groups lined the narrowest region of the pore, where cation recognition occurs (see Hille, 1992). This latter finding was a sensible one considering the chemistry of cations in solution. In aqueous solution, metal cations such K⁺, Na⁺, and Ca²⁺ are stabilized by electrostatic interaction with the electronegative oxygen atoms of water molecules. In the narrowest region of the pore, water molecules must be removed in order for the channel to recognize the preferred cation; lining this narrow region with oxygen-bearing groups provides a chemically favorable **selectivity filter.**

Based on this early work, investigators sought short protein segments that might line the aqueous pore with oxygen atoms. A candidate sequence was initially identified using sophisticated analysis of primary sequence information (Guy and Seetharamulu, 1986). This sequence is 20–25 amino acids long and forms part of the loop connecting the fifth and sixth predicted transmembrane helices in the voltage-gated cation channels (see Figs. 6, 7, and 8). Tests of this putative pore-lining loop's (P-loop) role in pore formation were soon forthcoming.

One type of test was to determine whether site-directed mutations in the P-loop sequence affected the action of known pore-blocking agents on heterologously expressed mutant channels. Among the agents tested in this way were the K⁺ channel-blockers tetraethylammonium and charybdotoxin and the Na⁺ channel-blocker tetrodotoxin (MacKinnon and Yellen, 1990; MacKinnon *et al.*, 1990; Terlau *et al.*, 1991). Block by the externally effective agents tetrodotoxin or charybdotoxin was specifically affected by mutations at the extracellular end of the P-loop. Tetraethylammonium block of K⁺ channels, which occurs by blocker binding at either end of the narrow pore, was altered by mutations at both ends of the P-loop. Results such as these confirmed that the P-loop does indeed form at least part of the pore in voltage-gated ion channels. Does the P-loop also form the ion selectivity filter?

The answer to this question also comes from site-directed mutagenesis studies. In *Shaker*-type K⁺ channels, point mutations introduced into the P-loop profoundly altered the channel's preference for K⁺ over Rb⁺ or NH₄⁺ as shown in Fig.16 (Yool and Schwarz, 1991). However, nonconducting channels were produced whenever mutations were introduced into one stretch of eight amino acids in K⁺ channel P-loops. This so-called **K⁺ channel signature sequence** is highly conserved in K⁺ channels found in bacteria to humans, and it was therefore suspected of forming the narrow selectivity filter in K⁺ channels.

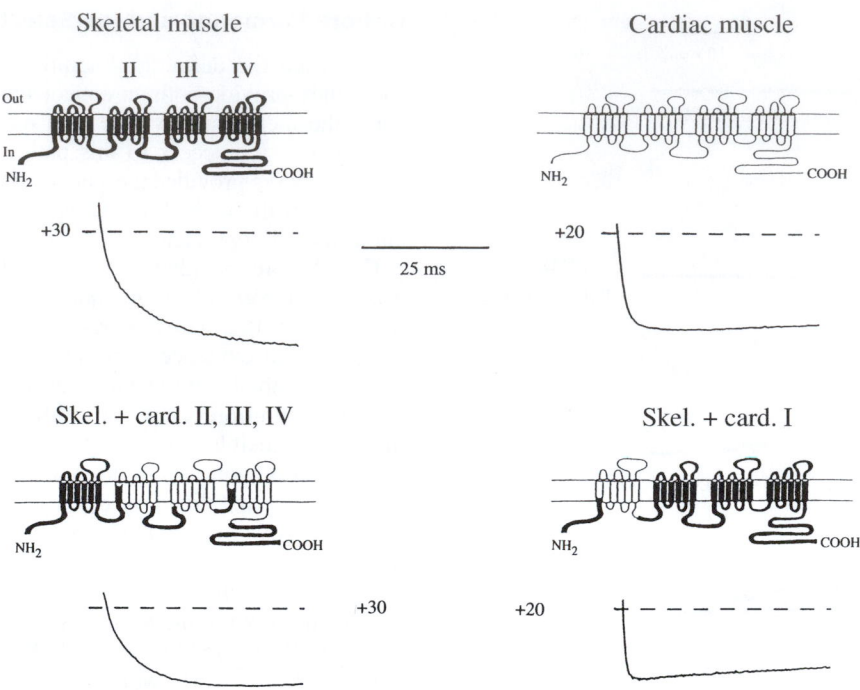

FIGURE 15. Identification of regions determining activation kinetics in L-type Ca^{2+} channels from skeletal muscle and heart. In these experiments hybrids (chimeras) of skeletal muscle and cardiac channels were made in various combinations to localize the domains responsible for the differences in Ca^{2+} current activation kinetics in heart and skeletal muscle. cDNA encoding each construct was injected in cultured muscle cells (myotubes) from mutant mice unable to synthesize active L-type channels. Thus these mutant cells would express only currents resulting from the genetically engineered DNA introduced into them. The top two panels show the current kinetics resulting from DNA encoding the wild-type skeletal muscle and cardiac channels. The bottom panels show two chimeras in which current kinetics correlate with the donor of repeat domain I. Most of the other possible combinations of channels were also made, with the same correlation of kinetics with domain I. (Redrawn from Tanabe *et al.* 1991.)

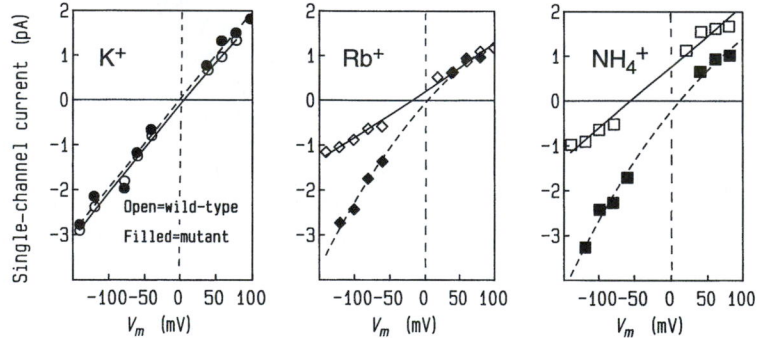

FIGURE 16. Alteration of *Shaker* B K^+ channel ion selectivity caused by a mutation in the putative pore-forming region. In these experiments, Phe 433 was mutated to Ser. The oocyte system used to express both wild-type and mutant channels is described in the text. Single-channel currents were measured using patch-clamp methods in which the ion indicated in each panel was applied to the external surface of the channel with K^+ on the inside in all cases. In these biionic conditions, inward (negative) currents will be carried mainly by the test ion, while outward (positive) currents will be mostly K^+ ion fluxes. As can be seen from the increased inward currents, permeability of the channel to Rb and NH_4 ions was greatly enhanced in the mutants, while K^+ was unaffected. (Data redrawn from Yool and Schwarz, 1991.)

Site-directed mutations in the P-loops of Na^+ channels and Ca^{2+} channels also affect ion selectivity (Heinemann *et al.*, 1992; Yang *et al.*, 1993). In Na^+ channels, a locus of four residues has been identified that confers selectivity for Na^+ over Ca^{2+} (Fig. 17). Each of the four P-loops of a Na^+ channel contributes one residue to the locus, which includes two carboxylate-bearing residues (Asp, Glu) and two noncarboxylate residues (Lys, Ala). Substitution of Glu residues for the Lys and Ala residues dramatically reduces Na^+ selectivity, and allows significant amounts of Ca^{2+} to permeate this mutant Na^+ channel. The analogous locus in Ca^{2+} channels is comprised of four Glu residues. Functional analysis of the effects of mutations in this locus has shown that these four Glu residues form the core of the selectivity filter in Ca^{2+} channels (Ellinor *et al.*, 1995).

	Motif I					Motif II					Motif III					Motif IV			
	-2	-1	0	1	2	-2	-1	0	1	2	-2	-1	0	1	2	-2	-1	0	1	2
CaCh	T	M	E	G	W	T	G	E	D	W	T	f	E	G	W	T	G	E	A	W
NaCh	T	Q	D	x	W	C	G	E	-	W	T	F	K	G	W	T	S	A	G	W

FIGURE 17. Alignment of Ca^{2+} and Na^+ channel pore sequences. Position 0 in the alignments (shaded) marks the critical residues in the selectivity filter (EEEE locus in Ca^{2+} channels; DEKA locus in Na^+ channels). High-voltage–activated calcium channels are identical to one another in these pore sequences, except at the ñ1 position in motif III: the "f" indicates that this position is occupied by either Phe (F) or Gly (G). In sodium channels, the residue at the +1 position in motif I (x) varies depending on channel isoform. Na^+ channel sensitivity to tetrodotoxin depends partly on the residue present at this position.

As described above, oxygen atoms have long been thought to line the selectivity filter in voltage-gated, cation-selective channels. The presence of multiple carboxylate groups in the selectivity filters of Ca^{2+} and Na^+ channels is consonant with the earlier work, although, in principle, the carboxylates may either project into the pore or away from it. In fact, Ca^{2+} channel carboxylate groups apparently project into the aqueous pore, and thereby provide the electronegative oxygen atoms needed to stabilize Ca^{2+} there. Evidence supporting this picture of selectivity filter structure derives from the fact that protonation of these carboxylate groups blocks current through Ca^{2+} channels; mutational replacement of the carboxylates by nonprotonatable groups predictably alters proton block. Orientation of selectivity filter side-chains has also been investigated using a technique called the "substituted cysteine accessibility method." In this method, a sulfhydryl-containing Cys residue is substituted for the wild-type residue, and a sulfhydryl-modifying agent is bath-applied to the mutant channel. If the side-chain of the substituted Cys projects into the pore, where it may be accessible to the sulfhydryl-modifying agent, then the modifying agent will attach itself to the Cys sulfhydryl and obstruct ion flow. Interruption of channel current thus reports that a particular side-chain is oriented into the aqueous pore, whereas absence of block indicates that the side-chain faces away from the pore. Application of this method to Na^+ and Ca^{2+} channels indicates that the side-chains in the carboxylate-containing loci of these channels all project into the pore (Chiamvimonvat *et al.*, 1996). The extreme narrowness of K^+ channel pores has prevented this technique from being effectively employed with these channels. However, the selectivity filter of K^+ channels may be structurally distinct from that of Ca^{2+} channels and perhaps Na^+ channels.

H. Crystal Structure of a Bacterial K$^+$ Channel Pore Region

Results of these site-directed mutagenesis studies have been strikingly confirmed by the first x ray crystallographic structure obtained for a member of the family of voltage-gated ion channels (Doyle *et al.*, 1998). This structure was solved for a bacterial K^+ channel, which though not in fact voltage-gated, is nevertheless a member of the family of voltage-gated ion channels: most significantly, it possesses the hallmark of all K^+ selective channels, the K^+ channel signature sequence. The crystal structure reveals that the selectivity filter is formed by part of the P-loop, with two K^+ ions stabilized therein, probably by main-chain carbonyl oxygen atoms (Fig. 18). That uncharged carbonyl oxygens apparently line the selectivity filter in the bacterial K^+ channel actually agrees with predictions, based on mutagenesis studies, for other K^+ channels.

Why might K^+ channels have evolved a selectivity structure that employs neutral, main-chain carbonyl oxygen atoms, whereas their evolutionary descendants, Ca^{2+} channels, apparently line their selectivity filter with negatively

Extracellular

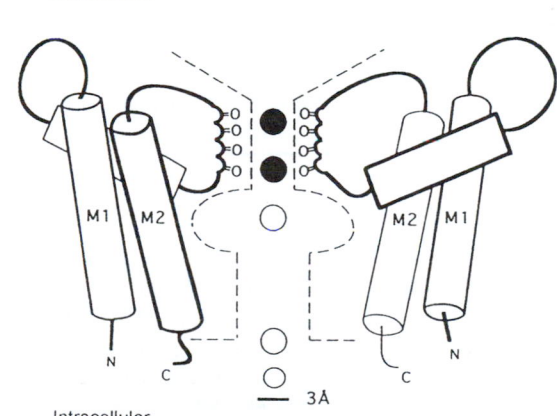

Intracellular

FIGURE 18. Cartoon illustration of the bacterial K^+ channel structure. Only two of the four subunits of the channel are shown. Each subunit has a P-loop and two helices (M1, M2), similar to the structure of inwardly rectifying K^+ channels. The segment of the P-loop that lines the ion permeation pathway projects carbonyl oxygen atoms into the pore. A pair of K^+ ions is shown stabilized in the selectivity filter by the rings of carbonyl oxygens. (Adapted with permission from Choe, S., and Robinson, R. (1998). An ingenious filter: the structural basis for ion channel selectivity. *Neuron* **20**: 821–823. Copyright 1998 Cell Press.)

charged, side-chain carboxylate oxygens? Perhaps the negative charge of carboxylate oxygens is necessary in selecting for Ca^{2+}, a charge-dense divalent cation, whereas uncharged carbonyl oxygen atoms are ideal in selecting for certain monovalent cations, such as K^+. How then to account for Na^+ channel selectivity structure? Na^+ channels have evolved from Ca^{2+} channels, and the Na^+ channel selectivity filter apparently includes two side-chain carboxylate oxygens. K^+ and Na^+ have the same ionic charge but Na^+ has a higher charge density owing to its smaller ionic radius. As in the case for Ca^{2+}, the relatively high charge density of Na^+ may require carboxylate oxygens for selectivity, though only two rather than the four present in the Ca^{2+} channel selectivity filter.

V. Isoforms of Voltage-Gated Channels as Part of a Large Superfamily

One of the most striking results of recombinant DNA analysis of channel clones has been the discovery of numerous channel *isoforms* expressed within the same organism and even within the same cell. Naturally, some degree of homology among channels of different organisms was expected, and electrophysiological recordings showed that channels could be functionally diverse at the cellular level. Even a structural similarity among Na^+, Ca^{2+}, and K^+ channels had been predicted by some investigators. However, the number and diversity of channels within a single organism revealed by molecular biology were rather unexpected. Channels of a functionally similar class (usually based on ion selectivity) are now called *isotypes* or **isoforms** by analogy to the phenomenon of isoenzymes.

We have previously discussed the *Shaker* isotypes. *Shaker*-derived sequences have also been used to construct probes that have discovered other families of K^+ channels in flies,

mammals, and other species (Salkoff *et al.*, 1992). These families are so extensive that they challenge investigators to develop appropriate nomenclature for their classification (Table 3; Chandy and Gutman, 1995). In any case, although they are all to some degree homologous with the original *Shaker* family, they have much higher degrees of sequence homology among members of their class. Functionally, they run the gamut from inactivating A-type channels to noninactivating channels of the classic delayed rectifier sort. Interestingly, there is usually a much higher degree of sequence homology among channels of the same class from different animals than there is between channels of separate classes in the same animal. For example, *Shaker* channels from *Drosophila* are much more homologous to *Shaker* channels of mouse brain than they are to the *Drosophila Shab*-type channels.

Recently several other subfamilies of K^+ channels have been discovered in mammalian and invertebrate genomes. Two of these, the *erg* and KQT subfamilies, are important in mammalian cardiac function. Mutants of these channels have been associated with serious inherited heart and nervous system disorders in humans, in particular the long QT syndrome discussed in other chapters in this book. In any case, in mammals the number of K^+ channel isoforms is already quite large: currently there are 12 known potassium channel subfamilies totaling over 100 isoforms, with more being discovered at frequent intervals (see Coetzee *et al.*, 1999).

Considerable diversity is also found among Na^+ and Ca^{2+} channels (Agnew and Trimmer, 1989; Hofmann *et al.*, 1994). For the main channel-forming α subunits (see Figs. 6 and 7), there are thought to be at least ten Na^+ channel and nine Ca^{2+} channel genes in the mammalian genome. Some of these isoforms are expressed very selectively in certain tissues, such as brain, or in only certain cells types or at very specific stages of development.

The term "isoform" is rather nonspecific in that it does not convey a sense of relative relatedness among channels. An evolving system of classification refers to the set of voltage-gated channels as a **superfamily**, while the selectivity types (i.e., Na^+, Ca^{2+}, or K^+ channels) form families. Because of the impressive degree of diversity found for K^+ channels (see Table 3), investigators have formally proposed a uniform system of nomenclature in which the closely related isoforms such as *Shaker* are referred to as subfamilies. A separate nomenclature for isoforms of Ca^{2+} channel subunits has also been developed. Further consideration of channel classification will undoubtedly be useful in understanding the process of channel evolution.

A. How Do Isoforms Arise?

In general, two separate mechanisms that create such diversity have been identified. First, there is the expected situation in which distinctly separate genes encode different channels in the same beast. Such is largely the situation for mammalian Na^+ channel and Ca^{2+} channel isoforms. However, a second mechanism has been found, called **alternative splicing**, in which a given segment of a channel polypeptide may be encoded in several different stretches of

TABLE 3 Properties of Major Potassium Channel Subfamilies in Fruit Flies and Mammals

Drosophila gene	Homologous mammalian genes (no.)	Properties
Shaker	Kv1.1–1.7 (7)	Extensively spliced in fly; inactivation fast to slow
Shab	Kv2.1, 2.2 (2)	Limited splicing in fly; intermediate inactivation kinetics
Shaw	Kv3.1–3.4 (4)	Limited splicing in fly and mammal;slow inactivation, delayed rectifier characteristics
Shal	Kv4.1–4.3 (3)	Limited splicing in fly; slow inactivation, delayed rectifier characteristics
eag	h-erg (3)	Limited splicing in fly and mammals; slow-activation delayed rectifier

genomic DNA known as **exons**. This is the case for *Shaker* in *Drosophila*; thus, in Fig.12, the various *Shaker* isoforms are generated by mixing and matching the transcripts of different exons. During transcription some sort of regulatory mechanism exists that decides which one of the alternative coding RNAs for a given segment will be incorporated in the final mRNA. The other undesired homologous segments are then excised from the transcript and the desired RNA segments are spliced together. Currently, there is evidence for the operation of both mechanisms in vertebrates and invertebrates. However, most mammalian systems studied to date appear to rely more heavily on separate genes than alternative splicing, whereas in the fruit fly both mechanisms appear common.

Finally, for all multisubunit channels there is the possibility of diversification through various combinations of each set of subunit isotypes. The best-studied examples are the heteromeric formation of channels by different K^+ channel subunit isotypes (Salkoff *et al.*, 1992). For example, if the same cell synthesized *Shaker* A and *Shaker* B channel subunits, one might find heteromeric complexes of A and B types as well as the homomeric types. In fact, experiments with the oocyte expression system into which both types of mRNA have been injected show that such heteromeric assembly can occur, yielding functional channels with characteristics different from homomultimers of either channel type. Thus, it is theoretically possible that distinct channel properties may be generated by heteromeric assembly of K^+ channel subunits. Whether this occurs in nature is currently being investigated.

On the other hand, oocyte experiments show that heteromeric association among subunits from different families (for example, *Shaker* and *Shal*) does not occur. Apparently different families have unique structural domains that allow only homomeric assembly among members of their family. This discovery has led to a recent set of chimeric experiments in which domains among a fly *Shaker* channel and a distantly related mammalian delayed rectifier channel (*drk1*) were exchanged in an attempt to identify the segments specifying assembly specificity (Li *et al.*, 1992). Such a segment was found on the N-terminal part of the peptide. Thus, substitution of the *Shaker* segment into the *drk1* cDNA produced a channel *drk1* subunit that was able to coassemble with native *Shaker*.

B. Why Are Channels So Diverse in a Given Organism?

One obvious reason for the presence of so many channel isotypes in the same animal is that each isotype has different functional properties that are appropriate to the function of its host cell. For example, high-frequency, repetitive firing requires short-duration action potentials that could be created by rapidly gating Na^+ channels and fast-inactivating A-type K^+ channels. In addition, isotypes may be differentially sensitive to intracellular modulators (see Chapter 28) that allow cellular mechanisms to alter the properties of specific channels selectively in response to physiological effectors. On the other hand, many of the discovered isotypes have no significant functional differences among

them. In this case it is possible that it is not the differences in the channels themselves that are important, but rather the way the isotypes are expressed. Thus, in a number of cases it has been shown that expression of one isotype over another occurs developmentally or in response to a physiological stimulus (Ribera and Spitzer, 1992). Such specific expression is thought to be controlled by genetic elements outside the coding region for the channel, known as regulatory elements. These stretches of DNA respond specifically to one of many possible soluble factors by increasing (in the case of promoters and enhancers) or inhibiting (silencers) transcription of the channel gene (see Maue *et al.*, 1990). Separate genes, then, potentially allow the cell to express selectively a given channel in response to a given stimulus or at a specific time during development. Finally, there is the possibility that a given isotype sequence encodes information used by the cellular machinery to localize or cluster it in very discrete locations, such as the nodes of Ranvier in myelinated nerve fibers. (Dugandzija-Novakovic *et al.*, 1995), neuronal cell bodies, or dendrites (Maletic-Savatic *et al.*, 1994). The study of the diversity of channels and their biological purpose will continue to be a highly active area of cell biology for some time.

VI. Future Directions

The current molecular biological approaches will no doubt be productive for some time. However, it is likely that better understanding of channel mechanism will require more direct and precise resolution of the higher order structure of channel molecules. The problems inherent in a crystallographic approach have been discussed. However, the problem is currently being solved by recombinantly synthesizing smaller segments of channel proteins that may be amenable to crystal formation. To be useful, these small segments must retain a significant part of the structure they have in the original protein (a serious concern). Alternatively, small segments may be structurally studied using advanced spectroscopic techniques such as magnetic resonance. Finally, some important information has been obtained by direct imaging of purified channels using electron microscopy, and modern instruments that promise greatly improved resolution at the molecular level are being developed. Finally, the study of the genetic regulation of channel expression will be expanded through the further study of regulatory domains in the DNA. This will undoubtedly increase our appreciation of ion channels as important elements in the metabolism of the entire organism.

VII. Summary

The structure, mechanism, and expression of ion channels are currently intense areas of research interest. The first advances in these areas were made by the application of biochemical techniques to the problems of channel purification. Such studies have told us the size of intact channels and their subunit composition, and have yielded clues regarding functionally important domains (such as pores, gates, modulation sites, and nonprotein modifications). For Na$^+$ and Ca^{2+} channels, attention has been focused on a large polypeptide (so-called α peptide) found in each purified preparation. These α proteins apparently have all the molecular apparatus required for channel operation, while smaller associated subunits may play other roles in channel modulation, synthesis, or cellular localization. Purified material has also been important in the construction of probes to identify clones of channel-encoding DNA fragments.

Recent work has applied recombinant DNA technology to determine the primary amino acid sequence of channels through cloning. These sequences have revealed widespread sequence homology among the superfamily of voltage-gated ion channels, for example, between Na$^+$ and Ca^{2+} channel α polypeptides. Furthermore, the large α proteins have four domains of homology within their sequences, suggesting that pore formation occurs via a common "staves of a barrel" architecture. K$^+$ channel sequences, determined via a genetic approach, are relatively shorter, and they partially correspond to a single internal repeat of the larger α peptides. Hence, K$^+$ channels are thought to consist of tetramers with each subunit contributing to the pore wall. Other analyses of amino acid sequences have given clues regarding the mechanistically and structurally important domains of ion channels. One analysis was based on thermodynamic (hydropathy) considerations to identify possible membrane-spanning domains. In addition, the sequences were subjected to an empirical analysis that predicts regions of secondary structure, such as helices and sheets. From such predictions detailed models of channel structure and function have been developed. These models have been tested using immunological and mutagenesis approaches, and candidate domains for ion selectivity, pore wall formation, gating, and modulation have been identified. However, detailed knowledge of the molecular mechanisms of channel function is still limited.

Finally, molecular studies have revealed an astonishing diversity of ion channels at all levels of organization. In particular, the genome of most organisms can express multiple channel isotypes, sometimes coexisting within the same living cell. Much remains to be learned about channel function and the role of channel isotypes in the biology of both simple and complex organisms.

Bibliography

Agnew, W. S., and Trimmer, J. (1989). Molecular diversity of voltage-sensitive sodium channels. *Annu. Rev. Physiol.* **51**, 401–418.

Akabas, M. H., Stauffer, D. A., Xu, M., and Karlin, A. (1992). Acetylcholine receptor channel structure probed in cysteine-substitution mutants. *Science* **258**, 307–310.

Aldrich, R.W., Hoshi, T., and Zagotta, W. N. (1990). Differences in gating among amino terminal variants of *Shaker* potassium channels. *Cold Spring Harbor Symp. Quant. Biol.* **55**, 19–27.

Armstrong, C. M., and Bezanilla, F. (1977). Inactivation of the sodium channel. II. Gating current experiments. *J. Gen. Physiol.* **70**, 567–590.

Barchi, R. L. (1983). Protein components of the purified sodium channel from rat skeletal muscle sarcolemma. *J. Neurochem.* **40**, 1377–1385.

Bennett, E., Urcan, M. S., Tinkle, S. S., Koszowski, A. G., and Levinson, S. R. (1997). Contribution of sialic acid to the voltage-dependence of sodium channel gating: a possible electrostatic mechanism. *J. Gen. Physiol.* **109**, 327–343.

Catterall, W. A. (1988). Molecular properties of dihydropyridine sensitive calcium channels in skeletal muscle. *J. Biol. Chem.* **263**, 3535–3538.

Catterall, W. A. (1992). Cellular and molecular biology of voltage-gated sodium channels. *Physiol. Rev.* **72 Suppl.**, S15–S48.

Chandy, K. G., and Gutman, G. A. (1995). Voltage-gated K$^+$ channel genes. *In* "CRC Handbook of Receptors and Channels," ed. R. A. North, pp.1–71. CRC Press, Boca Raton, FL.

Chiamvimonvat, N., Perez-Garcia, M. T., Ranjan, R., Marban, E., and Tomaselli, G. F. (1996). Depth asymmetries of the pore-lining segments of the Na$^+$ channel revealed by cysteine mutagenesis. *Neuron* **16**; 1037–1047.

Chou, P. Y., and Fasman, G. D. (1978). Empirical predictions of protein conformation. *Annu. Rev. Biochem.* **47**, 251–276.

Coetzee, W. A., Amarillo, Y., Chiu, J., Chow, A., Lau, D., McCormack, T., Moreno, H., Nadal, M. S., Ozaita, A., Pountney, D., Saganich, M., Vega-Saenz de Miera, E., and Rudy, B. (1999). Molecular diversity of K$^+$ channels. *Ann. NY Acad. Sci.* **868**, 233–285.

Doyle, D. A., Cabral, J. M., Pfuetzner, R. A., Kuo, A., Gulbis, J. M., Cohen, S. L., Chair, B. T., and MacKinnon, R. (1998). The structure of the potassium channel: molecular basis of K$^+$ conduction and selectivity. *Science* **280**, 69–77.

Dugandzija-Novakovic, S., Koszowski, A. G., Levinson, S. R., and Shrager, P. (1995). Clustering of K$^+$ channels and node of Ranvier formation in remyelinating axons. *J. Neuroscience* **15**, 492–503.

Ellinor, P. T., Yang, J., Sather, W. A., Zhang, J-F, and Tsien, R. W. (1995) Ca^{2+} channel selectivity at a single locus for high affinity Ca^{2+} interactions. *Neuron* **15**, 1121–1132.

Guy, H. R., and Seetharamulu, P. (1986). Molecular model of the action potential sodium channel. *Proc. Natl. Acad. Sci. USA* **83**, 508–512.

Heinemann, S. H., Terlau, H., Stuhmer, W., Imoto, K. and Numa, S. (1992). Calcium channel characteristics conferred on the sodium channel by single mutations. *Nature* **356**, 441–443.

Hille, B. (1992). "Ionic Channels of Excitable Membranes." Sinauer Associates, Sunderland, MA.

Hofmann, F., Biel, M., and Flockerzi, V. (1994). Molecular basis for Ca^{2+} channel diversity. *Ann. Rev. Neurosci.* **17**, 399–418.

Hoshi, T., Zagotta, W. N., and Aldrich, R. W. (1990). Biophysical and molecular mechanisms of *Shaker* potassium channel inactivation. *Science* **250**, 533–538.

Isom, L. L., De Jongh, K. S., Patton, D. E., Reber, B .F. X., Offord, J., Charbonneau, H., Walsh, K., Goldin, A. L., and Catterall, W. A. (1992). Primary structure and functional expression of the β_1 subunit of the rat brain sodium channel. *Science* **256**, 839–842.

Kyte, J., and Doolittle, R. F. (1982). A simple method for displaying the hydropathic character of a protein. *J. Mol. Biol.* **157**, 105–132.

Leonard, J., and Snutch, T. P. (1991). The expression of neurotransmitter receptors and ion channels in *Xenopus* oocytes. *In* "Molecular Neurobiology: A Practical Approach" (D. M Glover and B. D. Hanes, Eds.), pp. 161–182. Oxford University Press, New York.

Levinson, S. R., Thornhill, W. B., Duch, D. S., Recio-Pinto, E., and Urban, B. W. (1990). The role of nonprotein domains in the function and synthesis of voltage-gated sodium channels. *In* "Ion Channels," Vol. 2 (Toshio Narahashi, Ed.) Vol 2, pp. 33–64, Plenum Press, New York.

Li, M., Jan, Y. N., and Jan, L. Y. (1992). Specification of subunit assembly by the hydrophilic amino-terminal domain of the *Shaker* potassium channel. *Science* **257**, 1225–1230.

Lopez, G. A., Jan, Y. N., and Jan, L. Y. (1991). Hydrophobic substitution mutations in the S4 sequence alter voltage-dependent gating in *Shaker* K$^+$ channels. *Neuron* **7**, 327–336.

MacKinnon, R., Heginbotham, L., and Abramson, T. (1990). Mapping the receptor site for charybdotoxin, a pore-blocking potassium channel inhibitor. *Neuron* **5**, 767–771.

MacKinnon, R., and Yellen, G. (1990). Mutations affecting TEA blockade and ion permeation in voltage-activated K+ channels. *Science* **250**, 276–279.

MacKinnon, R. (1991). Determination of the subunit stoichiometry of a voltage-activated potassium channel. *Nature (London)* **350**, 232–235.

Maletic-Savatic, M., Lenn, N. J., and Trimmer, J. S. (1994). Differential spatiotemporal expression of K+ channel polypeptides in rat hippocampal neurons developing *in situ* and *in vitro*. *J. Neurosci.* **15**, 3840–3851.

Maue, R. A., Kraner, S. D., Goodman, R. H., and Mandel, G. (1990). Neuron-specific expression of the rat brain type II sodium channel gene is directed by upstream regulatory elements. *Neuron* **4**, 22–231.

Miller, C. (Ed.) (1986). "Ion Channel Reconstitution." Plenum, New York.

Miller, J. A., Agnew, W. S., and Levinson, S. R. (1983). Principal glycopepptide of the tetrodotoxin/saxitoxin binding protein from *Electrophorus electricus*: isolation and partial physical and chemical characterization. *Biochemistry* **22**, 462–470.

Nakai, J., Adams, B. A., Imoto, K., and Beam, K. G. (1994). Critical roles of the S3-segment and S3–S4 linker of repeat I in activation of L-type calcium channels. *Proc. Natl. Acad. Sci. USA* **91**, 1014–1018.

Noda, M., Shimizu, S., Tanabe, T., Takai, T., Kayano, T., Ikeda, T., Takahashi, H., Nakayama, H., Kanaoka, Y., Minamino, N., Kangawa, K., Matsuo, H., Raftery, M. A., Hirose, T., Inayama, S., Hayashida, H., Miyata, T., and Numa, S. (1984). Primary structure of *Electrophorus electricus* sodium channel deduced from cDNA sequence. *Nature* **312**, 121–127.

Numa, S., and Noda, M. (1986). Molecular structure of sodium channels. *Ann. NY Acad. Sci.* **479**, 338–355.

Papazian, D. M., Timpe, L. C., Jan, Y. N., and Jan, L. Y. (1987). Cloning of genomic and complementary DNA from *Shaker*, a putative potassium channel gene from *Drosophila*. *Science* **237**, 749–753.

Papazian, D. M., Timpe, L. C., Jan, Y. N., and Jan, L. Y. (1991). Alteration of voltage-dependence of *Shaker* potassium channel by mutations in the S4 sequence. *Nature* **349**, 305–10.

Pongs, O. (1992). Molecular biology of voltage-dependent potassium channels. *Physiol. Revs.* **72 Suppl.**, S69–S88.

Recio-Pinto, E., Duch, D. S., Urban, B. W., Thornhill, W. B., and Levinson, S. R. (1990). Neuraminidase treatment modifies the function of eel sodium channels reconstituted in planar lipid bilayers. *Neuron* **5**, 675–684.

Ribera, A. B., and Spitzer, N. C. (1992). Developmental regulation of potassium channels and the impact on neuronal differentiation. *In*

"Ion Channels," Vol. 3, pp. 1–38 (T. Narahashi, Ed.), Plenum Press, New York.

Salkoff, L., Baker, K., Butler, A., Covarrubias, M., Pak, M. D., and Wei, A. (1992). An essential set of K+ channels conserved in flies, mice, and humans. *TINS* **15**, 161–166.

Sharp, A. H., Imagawa, T., Leung, A. T., and Campbell, K. P. (1987). Identification and characterization of the dihydropyridine-binding subunit of the skeletal muscle dihydropyridine receptor. *J. Biol. Chem.* **262**, 12 309–12 315.

Sigworth, F. J. (1995). Charge movement in the sodium channel. *J. Gen. Physiol.* **106**, 1047–1051.

Stuhmer, W., Conti, F., Suzuki, H., Wang, X., Noda, M., Yahagi, N., Kubo, H., and Numa, S. (1989). Structural parts involved in the activation and inactivation of sodium channels. *Nature* **339**, 597–603.

Tanabe, T., Adams, B. A., Numa, S., and Beam, K. G. (1991). Repeat I of the dihydropyridine receptor is critical in determining sodium channel activation kinetics. *Nature* **352**, 800–803.

Tanabe, T., Takeshima, H., Mikami, A., Flockerzi, V., Takahashi, H., Kangawa, K., Kojima, M., Matsuo, H., Hirose, T., and Numa, S. (1987). Primary structure of the receptor for calcium channel blockers from skeletal muscle. *Nature* **328**, 313–318.

Terlau, H., Heinemann, S. H., Stuhmer, W., Pusch, M., Conti, F., Imoto, K., and Numa, S. (1991). Mapping the site of block by tetrodotoxin and saxitoxin of sodium channel II. *FEBS Lett.* **293**, 93–96.

Thornhill, W. B., Wu, M. B., Wu, X., Morgan, P. T., and Margiotta, J. F. (1996). Expression of Kv1.1 delayed rectifier in *lec* mutant Chinese hamster ovary cells reveals a role for sialidation in channel function. *J. Biol. Chem.* **271**, 19 093–19 098.

Timpe, L. C., Jan, Y. N. and Jan, L. Y. (1988). Four cDNA clones from the *Shaker* locus of *Drosophila* induce kinetically distinct A-type potassium currents in *Xenopus* oocytes. *Neuron* **1**, 659–667.

Trimmer, J. S. (1991). Immunological identification and characterization of a delayed rectifier K+ channel polypeptide in rat brain. *Proc. Natl. Acad. Sci. USA* **88**, 10 764–10 768

Vassilev, P., Scheuer, T. and Catterall, W. A. (1989). Inhibition of inactivation of single sodium channels by a site-directed antibody. *Proc. Natl. Acad Sci USA* **86**, 8147–8151

Yang, J., Ellinor, P. T., Sather, W. A., Zhang, J-F, and Tsien, R. W. (1993). Molecular determinants of Ca2+ selectivity and ion permeation in L-type Ca2+ channels. *Nature* **366**, 158–161

Yool, A. J. and Schwarz, T. L. (1991). Alteration of ionic selectivity of a K+ channel by mutation of the H5 region. *Nature* **349**, 700–704

Zagotta, W. N., Hoshi, T. and Aldrich, R. W. (1990). Restoration of inactivation in mutants of *Shaker* potassium channels by a peptide derived from ShB [see comments]. *Science* **250**, 568–571.

Michael M. Behbehani

28

Biology of Neurons

I. Introduction

The general functions of a neuron are to integrate chemical (and, in the case of electrical synapses, electrical) afferent signals and convert the result to action potentials, and to change the action potential into chemical release of neurotransmitters at the nerve terminals. To accomplish these tasks, the neuron must be able to (1) synthesize receptors for all the transmitters it responds to, and move these receptors to appropriate locations. (2) Maintain ion gradients by operating a variety of ion pumps and exchangers. (3) Synthesize the neurotransmitters it releases and all the enzymes necessary for their synthesis. (4) Transport the transmitters from the cell body to the synaptic terminals, and (5) transport macromolecules and membrane segments to and from the synaptic terminals and the cell body. A close look at the functions listed above show that the neuron must manufacture distinct types of molecules. Since the majority of macromolecules that are utilized by the neuron are proteins, the cell has developed complex machinery for manufacturing specific proteins and their transport to specific sites within the neuron.

II. Ultrastructure

A nerve cell consists of a cell body (the **soma**), a **dendritic tree**, an **axon**, and **synaptic terminals**. Although all neurons have the same types of components, there are significant differences in the morphology of nerve cells. Based on the number of processes that originate from the soma, the neurons have been classified as unipolar, bipolar or multipolar (Fig. 1). The **unipolar** cells have only a single process that originates from the cell body. Unipolar cells are found in invertebrates and contain an axon that originates from a specialized dendrite instead of the soma. The sensory neurons in the vertebrates are examples of **bipolar** cells. These neurons have two processes originating from the soma: one process, the axon, connects the soma to the central nervous system; the second process, the sensory nerve, ends in specialized sensory receptors. The majority of cells within the

central nervous system are **multipolar**. The axons of these neurons initiate from a specialized region of the soma, called the **axon hillock**, and end in multiple branches containing the synaptic terminals.

Nerve cells have a large nucleus and a nucleolus. The nucleus is enclosed by a specific type of membrane called the **nuclear envelope**. This envelope contains pores called **nuclear pores**. These pores are open to the cytosolic compartment of the cell. The nucleus of the cell is in continuity with the **endoplasmic reticulum** (ER), which is a major organelle of the cell (Fig. 2). The DNA of the neuron is located in the nucleus and is transcribed there to make RNA for the synthesis of a large variety of proteins and macromolecules. The mRNA that encodes each macromolecule is released from the nucleus, moves through the nuclear pores, combines with **ribosomes** to form **polysomes** (Alberts *et al.*, 1989). In general, three types of proteins are synthesized by the mRNAs released from the nucleus.

One class of proteins is the **cytosolic proteins**. These proteins are soluble proteins and their ribosomes remain in the cytosol (the cytoplasm minus organelles). Among the major proteins in this class are those that form the **cytoskeleton** of the cell, including neurofilaments, tubulines, actins, actin-associated proteins, and enzymes that catalyze reactions within the cell.

The second class of proteins is the **nuclear proteins** and **mitochondrial proteins**. The ribosomes for these proteins are attached to the ER and translocate into the ER lumen as they are synthesized. In this process, a branch made of carbohydrate rich in mannose may be added to each protein to form an N-linked glycosyl chain. The attachment of the ribosomes of these proteins to the ER produces the roughness of the ER and for this reason, these classes of the ER are called the **rough endoplasmic reticulum** (RER). The ribosomal RNA in the RER stain with several basic dyes including cresyl violet, toluidine blue, and methylene blue. Because of this histological property, the RER is also called the **Nissl substance**.

The third class of proteins that is encoded by nuclear mRNA is proteins that are destined to become constituents

479

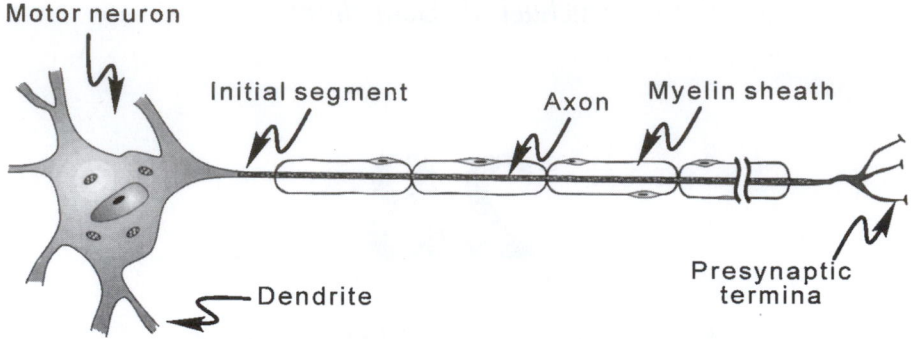

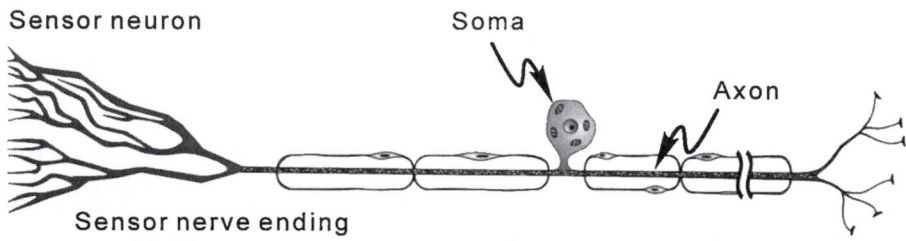

FIGURE 1. Morphological characteristics of different types of neurons. (A) A typical unipolar neuron. This type of cell consists of a large soma, a large myelinated axon, and synaptic terminals. The initial segment is the site where the APs are generated. (B) A bipolar sensory neuron such as a dorsal root ganglion cell. This type of neuron is a bipolar cell with one process located in the periphery and the second branch projecting into the central nervous system. The terminals of the peripheral branch can be sensory receptors, for example nociceptors, or can innervate a variety of other types of sensory receptors. (C) The type of multipolar cell found in invertebrates. The axons of these cells originate from dendrites that are connected to the soma. (Adapted from Hall, Z. W. (1992). "An Introduction to Molecular Neurobiology." Sinauer Associates, Sunderland, MA.)

of the cell membrane and membranes of the organelles within the neuron, and secretory proteins that are released at the nerve terminals. These proteins include neuropeptides that act as neurotransmitters or neuromodulators and growth factors. The polysomes for these proteins attach themselves to the cytoplasmic side of the RER.

Proteins that are synthesized in the RER and are destined to be released enter a tightly packed stack of intercellular membrane called the **Golgi apparatus**. During their passage through this region the proteins go through the posttranslational modification process. The majority of neuropeptides that act as neurotransmitters are modified in this process. For example, the N-terminal Glu of neurotensin is changed to pyroglutamate and the C-terminal amino acid of substance P is amiditated. Following posttranslational modification, secretory proteins are inserted into **vesicles** that bud off the Golgi apparatus. Some proteins are released into the cytoplasm. Other proteins are released after the vesicles have reached the cell membrane by exocytosis. This type of release is called **constitutive secretion**. If the proteins are to be released at the axon terminals, the vesicles are transported by microtubules to the axon terminals and are stored. When the terminal is depolarized, calcium concentration is increased, and the vesicles are released. This type of release is called **regulated secretion**. Neurons contain two types of vesicles that are released through the regulated release process. The **dense-core vesicles** contain neuropeptides and **small synaptic vesicles** contain the classical transmitters such as monoamines and acetylcholine.

III. Neuronal Cytoskeleton

The neuronal cytoskeleton is essential for growth of neurons during development, for arborization of dendritic branches, for maintenance of the shape of the neuron, and for transfer of macromolecules from the cell body to the terminals and from the terminals to the cell body (Mitchison and Kirschner, 1988). Considering the complexity of neurons, each of the above processes is highly regulated and involves synthesis and polymerization of a variety of proteins.

The cytoskeleton consists of filaments that cross-link to form a tight meshwork. This network retains its shape even in the absence of the cell membrane. For example, in electrophysiological recording from squid axon, the cytoplasm is sometimes extruded from the axon. The extruded cytoplasm retains its cylindrical shape and many of its functions as long as ATP can be supplied (Vale *et al.*, 1985, Schnapp *et al.*, 1986).

The cytoskeleton contains three major proteins: microfilaments, neurofilaments, and microtubules. These proteins are essential for the axoplasmic flow that is essential for the survival of the nerve cell.

A. Microfilaments

Microfilaments are formed from **actin**, a 43-kDa globular protein. It is about 7 nm in diameter and is composed

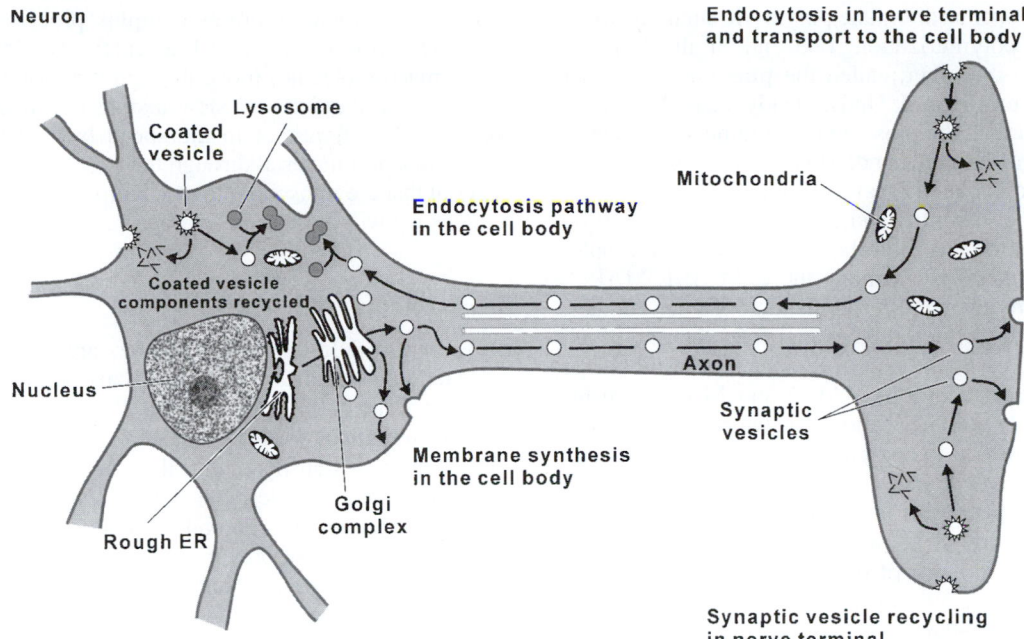

Neuron

Endocytosis in nerve terminal
and transport to the cell body

Lysosome

Coated
vesicle

Mitochondria

Endocytosis pathway
in the cell body

Coated vesicle
components recycled

Axon

Nucleus

Synaptic
vesicles

Membrane synthesis
in the cell body

Golgi
complex

Rough ER

Synaptic vesicle recycling
in nerve terminal

FIGURE 2. The cell nucleus and structures involved in transportation of macromolecules that are synthesized within the nucleus. The figure shows the nucleus, nucleolus, rough endoplasmic reticulum (ER), lysogomes, ribosomes, Golgi complex, and mitochondria. The transport of macromolecules is denoted by arrows. Macromolecules that are synthesized by the nucleus are transferred to the endoplasmic reticulum and then to the Golgi complex. Macromolecules that are needed in the axon terminals are transported by anterograde and slow axoplasmic flow. Proteins and macromolecules are transported from the terminals to the cell body by retrograde axoplasmic flow. (Adapted from Hall, Z. W. (1992). "An Introduction to Molecular Neurobiology." Sinauer Associates, Sunderland, MA.)

of two strands (each 3–5 nm in diameter) of polymerized globular (G) actin monomers arranged in a helix. These strands are asymmetric and each monomer has a pointed tip and a barbed end. Because the globular monomers are asymmetric, they polymerize tip to tail and form polar structures. Some axonal microfilaments are oriented longitudinally. The microfilaments are attached to the plasma membrane through associated proteins linked to actin including spectrin, ankyrin, vinculin, and talin. The principal anchoring protein in neurons and other cells of the body is **fodrin** (or neural spectrin), which is transported down the axon at the same velocity as actin. Microfilaments are also able to interact with proteins in the extracellular matrix (e.g., laminin and fibronectin) through their association with a family of membrane-spanning proteins called **integrins**.

B. Neurofilaments

Neurofilaments are the most abundant fibrillar components in axons and are the "bones" of the cytoskeleton. These filaments are about 10 nm in diameter and are related to a family of proteins, which includes vimentin, glial fibrillary acidic protein, and desmin. Neurofilaments are composed of three polypeptide subunits: NF-H (112 kDa), NF-M (120 kDa), and NG L (68 kDa). Neurofilaments are built with fibers that twist around each other to produce coils of increasing thickness. The thinnest units are monomers that form coiled-coil heterodimers. These dimers form a tetrameric complex that becomes the **protofilament**. Two

protofilaments become a **protofibril**, and four protofibrils are helically twisted to form the 10-nm neurofilament (Bershadsky and Vasiliev, 1988; Frixione, 2000).

C. Microtubules

Microtubules are long polar polymers about 25 nm in diameter. They are usually constructed of 13 linearly arranged α-tubulin and β-tubulin dimers called **protofilaments**. Each filament is about 5 nm in diameter. The monomeric subunit of microtubules is **tubulin**. Each monomer binds two guanosine triphosphate (GTP) molecules, or one GTP and one guanosine diphosphate (GDP) molecule. In the axon, they are oriented longitudinally with polarity always in the same direction. This arrangement is presumably important for the directional specificities of the two forms of fast axonal transport. Although axonal or dendritic microtubules can be as long as 0.1 mm, they usually do not extend the full length of the axon or dendrite, and are not continuous with microtubules in the cell body.

Both **monomeric** and **polymeric** components of tubulin and actin are present in the axon. The self-assembly of tubulin and actin requires triphosphates, and depends on the concentration of each component. At a concentration called the **critical concentration**, the rate of subunit added to the polymeric component is the same as the rate at which the polymeric component is disassembled. If the concentration of monomeric subunits is larger than its critical concentration, the monomers assemble into polymeric units. However, if the concentration of monomeric subunits is less than the critical concentration, the polymeric subunits disassemble. During polymerization, one end of the filament grows faster than the other, called the **plus end**; the other end is called the **minus end**. Under steady state, the net addition of monomers at the plus end is the same as the rate of dissociation from the minus end. This process can occur simultaneously and is called **treadmilling**.

The microtubules' stability and the rates of polymerization and depolymerization are controlled by several microtubule-associated proteins (MAP). Among these, MAP 1, MAP 2, and tau protein are involved in stabilization of microtubules. The binding of tau to microtubules increases the rate of association and decreases the rate of dissociation of tubulin at the growing end. In addition, MAP 2 and MAP 3 are involved in the rate of polymerization. The tau protein has been the focus of attention in the past decade because of its possible role in Alzheimer's disease. This degenerative disease produces total loss of short-term and eventually long-term memory. The brains of Alzheimer's patients contain extracellular deposits that are made of beta amyloid protein Aβ that form amyloid plaques. Within the neurons the amyloid deposits form neurofibrillary tangles. These lesions contain paired helical filaments (PHF). More recent study has established that tau is the major protein that forms the PHFs.

IV. Axoplasmic Flow

Macromolecules that are needed for the function of the nerve cell are synthesized in the cell body. In order for the nerve terminals to have access to these macromolecules, they have to be transported from the cell body to the terminals. Typically, cell bodies and nerve terminals are at considerable distances from each other. For example, the cell body of a spinal motor neuron that innervates muscles around the ankle in a 180-cm tall man is more than 1 m away from its terminals. The separation between cell body

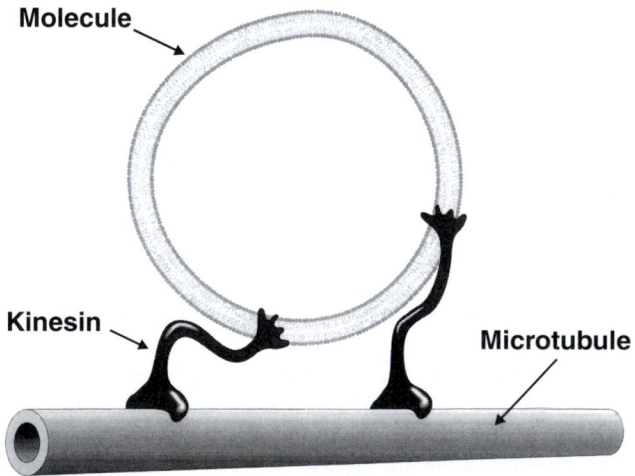

FIGURE 3. A proposed mechanism by which the motor molecule of the anterograde axoplasmic flow, kinesin, transports molecules. Molecules that are transported by this system bind to kinesin molecules. Kinesins from cross-bridges with microtubules at the plus end translocate on the microtubules toward the minus end. (Adapted from Hirokawa *et. al.*, 1989.)

and nerve terminals requires complex processes that have specialized components (see Fig. 2) (Brady, 1991).

The macromolecules move through the axon by two types of processes, the fast and slow axoplasmic flow. There are considerable differences in the speed by which these processes transport material through the axon and in the types of material that are transported by each process. In the following sections we will discuss these processes.

A. Fast Axoplasmic Flow

Synaptic vesicles or their precursors are transported to the nerve terminals, in most cases, by fast axoplasmic flow, which moves at a rate of 200–400 mm/day. Particles move in a stop-and-go intermittent or saltatory fashion (Terasaki *et al.*, 1995). Fast anterograde transport can still occur in axons that are severed from their cell bodies, indicating that this type of transport is not dependent on the cell body. In addition, fast axoplasmic flow is independent of protein synthesis. In contrast, this process is critically dependent on *oxidative metabolism*.

There are two types of fast axoplasmic flow processes: fast anterograde axonal transport (forward-moving, i.e., moving from cell body to the terminals), and fast retrograde axonal transport (backward moving, i.e., moving from the nerve terminals to the cell body).

1. Fast Anterograde Axonal Transport

Fast anterograde transport is involved in transport of tubulovesicular structures, synaptic vesicles, membrane-associated proteins, neuropeptides, neurotransmitters, and associated enzymes. Fast anterograde transport in the axon is based on microtubules that provide a stationary track on which specific organelles move in a saltatory fashion. A major component of these microtubules is a motor molecule called **kinesin**. Kinesin is a rod-shaped tetrameric ATPase consisting of two α subunits and two β subunits. Each of the α subunits has a molecular weight of about 115–130 kDa, and consists of 340 amino acids and an N-terminal force-generating domain. Each of the light chains (β subunit) has a molecular weight of 62–70 kDa (Bloom *et al.*, 1988; Yang *et al.*, 1989). The holoenzyme $\alpha_2\beta_2$ has a molecular weight of about 270 000. High-resolution electron microscopy analysis indicates that kinesin is an 80-nm rod with two globular heads approximately 10 nm in diameter. The two heavy chains are arranged in parallel, with a kink in the middle of each chain. The light chains form a fan-shaped tail (Hirokawa *et al.*, 1989). The ATP-binding site is located in the head region and is also the site of binding to the microtubules. The tail region contains the binding site to the membrane surface (Cry *et al.*, 1991). There is evidence that kinesin is anchored to the membrane by a protein called **kinectin** (Kumar *et al.*, 1995).

Kinesins form the cross-bridges between the moving membranous organelles, which have the appearance of little feet walking along the microtubules (Fig. 3). It has been postulated that the kinesin molecule, which is analogous to myosin in structure, interacts with other proteins such as actin or dynein to form cross-bridges (see Chapters 56 and 58). It has been proposed that as kinesin hydrolyzes a molecule of ATP, it undergoes a series of conformational changes. When the kinesin is attached to a filament, one of the configurational changes introduces strain into the proteins. When this strain is relieved, it causes the kinesin to move. Studies of the motility of kinesin indicate that during each cycle of ATP hydrolysis, kinesin moves a distance approximately equal to its length toward the next binding site on the filament. Recently studies have shown that the force generated by a single kinesin molecule is approximately 5 pN.

2. Fast Retrograde Axonal Transport

Materials are transported from nerve terminals to the cell body either for degradation or for restoration and reuse. The rate of fast retrograde transport is 50–100 mm/day. Before being transported, these materials are packaged in large membrane-bound organelles that are part of the lysosomal system. The materials that are transported by the fast retrograde transport are prelysosomal vesicles, multivesicular bodies, multilamellar bodies, growth factors, and recycled proteins. The retrograde transport also utilizes microtubules. However, the motor molecule for fast retrograde transport is a form of **dynein** (Wang *et al.,* 1995), which also is a microtubule-associated ATPase (MAP-lC). Dynein is approximately 40 nm long and is a two-headed structure with two complex polypeptides. It has two heavy chains and several light chains (Vallee and Bloom, 1991). Dynein is distributed in punctate fashion in neuronal tissue (Hirokawa, *et al.,* 1990), in association with membrane-bound organelles (Brady, 1991).

B. Slow Axoplasmic Flow

Slow axoplasmic flow is used for transport of cytoskeletal elements and soluble proteins. This type of transport is more complex than fast anterograde (Terada *et al.,* 1996). It consists of at least two kinetic components. The slow component travels at a rate of 0.1–2.5 mm/day, and carries the subunits of neurofilaments and the tubulin subunits of the microtubule. The faster component of slow axoplasmic flow travels at a rate of 2–6 mm/day and transports actin, clathrin and associated proteins, spectrin, glycolytic enzymes, and calmodulin. Approximately 2–4% of the protein transported by this process is actin (MW 43 000), which polymerizes to form microfilaments. Other proteins that are transported by this process are calmodulin and neural myosin or a myosin-like protein, clathrin (180 000-MW protein involved in coating of synaptic vesicles for recycling).

V. Regulatory Mechanisms for Axonal Transport

The factors that regulate axonal transport and allow the delivery of transported macromolecules to their appropriate location are not well understood. In the case of synaptic vesicles, phosphorylation of **synapsin-I** is an important mechanism that may be involved in the vesicles being transported by the fast axoplasmic flow to localize at the synaptic terminals. When it is dephosphorylated, synapsin I binds tightly to synaptic vesicles; upon phosphorylation, it is released from the synaptic vesicles. Phosphorylated synapsin-I has no effect on axonal transport; however, when synapsin becomes dephosphorylated, it inhibits movement of organelles that are bound to the microfilaments (McGuinness *et al.,* 1989). It has been proposed that dephosphorylated synapsin binds to the passing vesicles and confines them in the presynaptic terminals. It has also been suggested that phosphorylation and dephosphorylation are general mechanisms that allow localization of macromolecules to their specialized locations in the neuron.

VI. Summary

The neuron consists of four major components: the soma, the dendritic tree, the axon, and axon terminals. The DNA of the neuron is located and transcribed in the nucleus to make RNA for the synthesis of a large variety of proteins and macromolecules. One class of proteins, the cytosolic proteins, are soluble and their ribosomes remain in the cytosol. Among the major proteins in this class are those that form the cytoskeleton of the cell including neurofilaments, tubulins, actins , actin-associated proteins, and enzymes that catalyze reactions within the cell. The second class of proteins are nuclear proteins and mitochondrial proteins. The ribosomes for these proteins are attached to the ER and translocate into the ER lumen as they are synthesized. The third class of proteins that are encoded by nuclear mRNA are (1) proteins that are destined to become constituents of the cell membrane and membranes of the organelles within the neuron, and (2) secretory proteins that are released at the nerve terminals. These proteins include neuropeptides that act as neurotransmitters or neuromodulators and growth factors.

The cytoskeleton of neurons is essential for their growth during development and for arborization of dendritic branches. The cytoskeleton consists of filaments that crosslink to form a tight meshwork that retains its shape even in the absence of the cell membrane. The cytoskeleton contains three major proteins: microfilaments, neurofilaments, and microtubules. These proteins are essential for the axoplasmic flow which is necessary for the survival of the nerve cell.

Microfilaments are formed from actin. They are attached to the plasma membrane through associated proteins linked to actin including spectrin, ankyrin, vinculin, and talin. The principal anchoring protein in neurons and other cells of the body is fodrin (or neural spectrin). Neurofilaments are the most abundant fibrillar components in axons and are the "bones" of the cytoskeleton. These filaments, about 10 nm in diameter, are related to a family of proteins that includes vimentin, glial fibrillary acidic protein, and desmin. Microtubules are long polar polymers about 25 nm in diameter. In the axon, they are oriented longitudinally with polarity always in the same direction. This arrangement is presumably important for the directional specificities of the two forms of fast axonal transport.

Macromolecules that are needed for the function of the nerve cell are synthesized in the cell body and transported to the other regions of the cell using three types of transport mechanisms: fast anterograde axonal transport, fast retrograde axonal transport, and slow axoplasmic flow.

Fast axoplasmic flow that moves at a rate of 200–400 mm/day is involved in the transport of synaptic vesicles or their precursors. Fast axoplasmic flow is independent of protein synthesis, but is critically dependent on oxidative metabolism. Fast anterograde transport is involved in transport of tubulovesicular structures, synaptic vesicles, membrane-associated proteins, neuropeptides, neurotransmitters, and associated enzymes. Fast anterograde transport in the axon is based on microtubules that provide a stationary track on which specific organelles move in a saltatory fashion. A major component of these microtubules is a motor molecule called kinesin.

The fast retrograde mechanism transports macromolecules at a rate of 50–100 mm/day. The major macromolecules that are transported by fast retrograde transport are prelysosomal vesicles, multivesicular bodies, multilamellar bodies, growth factors, and recycled proteins. Retrograde transport also utilizes microtubules; however, the motor molecule for fast retrograde transport is a form of dynein.

Slow axoplasmic flow is used for transport of cytoskeletal elements and soluble proteins. This type of transport is more complex than fast anterograde. It consists of at least two kinetic components. (1) The slow component travels at a rate of 0.1–2.5 mm/day and carries the subunits of neurofilaments and the tubulin subunits of the microtubule. (2) The faster component of slow axoplasmic flow travels at a rate of 2–6 mm/day and transports actin, clathrin and associated proteins, spectrin, glycolytic enzymes, and calmodulin. Approximately 2–4% of the protein transported by this process is actin, which polymerizes to form microfilaments. Other proteins that are transported by this process are calmodulin and neural myosin or a myosin-like protein, clathrin.

The factors that regulate axonal transport and allow the delivery of transported macromolecules to their appropriate location are not well understood. In the case of synaptic vesicles, phosphorylation of synapsin-I is an important mechanism that may be involved in the vesicles being transported by the fast axoplasmic flow to localize at the synaptic terminals.

Bibliography

Alberts, B., Bray, D., Lewis, J., Raff, M., Roberts, K., and Watson, J. D. (1989). "Molecular Biology of the Cell." second edition. Garland, New York.

Bershadsky, A. D., and Vasiliev, J. M. (1998). "Cytoskeleton." Plenum, New York.

Bloom, G. S., Wagner, M. C., Pfister, K. K., and Brady, S. T. (1988). Native structure and physical properties of bovine brain kinesin and identification of the ATP-binding subunit polypeptide. *Biochemistry* **27,** 3409–3416.

Bomsel, M., Parton, R., Kuznetsov, S. A., Schroer, T. A., Gruenberg, J. (1990). Microtubule- and motor-dependent fusion *in vitro* between apical and basolateral endocytic vesicles from MDCK cells., *Cell* **62,** 719–731.

Brady, S. T. (1991). Molecular motors in the nervous system. *Neuron* **7,** 521–533.

Dabora S. L., Sheetz, M. P. (1988). Cultured cell extracts support organelle movement on microtubules *in vitro*. *Cell Motil. Cytoskel.* **10,** 482–495.

Frixione, E. (2000). Recurring views on the structure and function of the cytoskeleton: a 300-year epic. *Cell Motil. Cytoskeleton* **46,** 73–94.

Gelfand, V. I., and Bershadsky, A. D. (1991). Microtubule dynamics: mechanism, regulation, and function. *Annu. Rev. Cell Biol.* **7,** 93–116.

Gilbert, S. P., Allen, R. D., Sloboda, R. D. (1985). Translocation of vesicles from squid axoplasm on flagellar microtubules. *Nature* **315,** 245–248.

Hirokawa, N., Bloom, G. S., and Vallee, R. B. (1985). Cytoskeletal architecture and immunocytochemical localization of microtubule-associated proteins in regions of axons associated with rapid axonal transport: the beta,beta'-iminodipropionitrile-intoxicated axon as a model system. *J. Cell Biol.* **101,** 227–239.

Hirokawa, N., Funakoshi, T., Sato-Harada, R., and Kanai, Y. (1996). Selective stabilization of tau in axons and microtubule-associated protein 2C in cell bodies and dendrites contributes to polarized localization of cytoskeletal proteins in mature neurons. *J. Cell Biol.* **132,** 667–679.

Hirokawa, N. (1993). Mechanism of axonal transport. Identification of new molecular motors and regulations of transports. *Neurosci. Res.* **18,** 1–9.

Hirokawa, N., Pfister, K. K., Yorifuji, H., Wagner, M. C., Brady, S. T., Bloom, G. S. (1989). Submolecular domains of bovine brain kinesin identified by electron microscopy and monoclonal antibody decoration. *Cell* **56,** 867–878.

Jacob, J. M., and O'Donoghue, D. L. (1995). Direct measurement of fast axonal transport rates in corticospinal axons of the adult rat. *Neurosci. Lett.* **197,** 17–20.

Kumar, J., Yu, H., and Sheetz, M. P. (1995). Kinectin, an essential anchor for kinesin-driven vesicle motility. *Science* **267,** 1834–1837.

Li, L., Chin, L. S., Shupliakov, O., Brodin, L., Sihra, T. S., Hvalby, O., Jensen, V., Zheng, D., McNamara, J. O., and Greengard, P. (1995). Impairment of synaptic vesicle clustering and of synaptic transmission, and increased seizure propensity, in synapsin I–deficient mice. *Proc. Natl. Acad. Sci. USA* **92,** 9235–9239.

McLean, W. G., Kanje, M., and Remgard, P. (1993). An *in vitro* system for the study of slow axonal transport. *Brain Res.* **613,** 295–299.

Mitchison, T., and Kirschner, M. (1998). Cytoskeletal dynamics and nerve growth. *Neuron.* **1,** 761–772.

Rodionov, V. I., Gyoeva, F. K., Tanaka, E., Bershadsky, A. D., Vasiliev, J. M., and Gelfand, V. I. (1993). Microtubule-dependent control of cell shape and pseudopodial activity is inhibited by the antibody to kinesin motor domain. *J. Cell Biol.* **123,** 1811–1820.

Schnapp, B. J., Vale, R. D., Sheetz, M. P., and Reese, T. S., (1986). Microtubules and the mechanism of directed organelle movement. *Ann. NY Acad. Sci.* **466,** 909–918.

Sheetz, M. P. (1999). Motor and cargo interactions. *Eur. J. Biochem.* **262,** 19–25.

Takenaka, T., Kawakami, Hori, H., Hashimoto, Y., Hiruma, H., and Kusakabe, T. (1998). Axoplasmic transport and its signal transduction mechanism. *Jpn J. Physiol.* **48,** 413–420.

Terada, S., Nakata, T., Peterson, A. C., Hirokawa, N. (1996). Visualization of slow axonal transport *in vivo*. *Science* **273,** 784–788.

Terasaki, M., Schmidek, A., Galbraith, J. A., Gallant, P. E., and Reese, T. S. (1995). Transport of cytoskeletal elements in the squid giant axon. *Proc. Natl. Acad. Sci. USA* **92,** 11 500–11 503.

Vale, R. D. (1987). Intracellular transport using microtubule-based motors. *Annu. Rev. Cell Biol.* **3,** 347–378.

Vale, R. D., Reese, T. S., and Sheetz, M. P. (1985). Identification of a novel force-generating protein, kinesin, involved in microtubule-based motility. *Cell* **42,** 39–50.

Vale, R. D., Schnapp, B. J., Reese, T. S., and Sheetz, M. P. (1985). Movement of organelles along filaments dissociated from the axoplasm of the squid giant axon. *Cell* **40,** 449–454.

Vallee, R. B., and Bloom, G. S. (1991). Mechanisms of fast and slow axonal transport. *Annu. Rev. Neurosci.* **14,** 59–92.

Viancour, T. A., and Kreiter, N. A. (1993). Vesicular fast axonal transport rates in young and old rat axons. *Brain Res.* **628,** 209–217.

Wang, C., Asai, D. J., and Robinson, K. R. (1995). Retrograde but not anterograde bead movement in intact axons requires dynein. *J. Neurobiol.* **27,** 216–226.

Yang, J. T., Laymon, R. A., and Goldstein, L. S. (1989). A three-domain structure of kinesin heavy chain revealed by DNA sequence and microtubule binding analyses. *Cell* **56,** 879–889.

Bernd Nilius and Guy Droogmans

29

Ion Channels in Nonexcitable Cells

I. Introduction

For a long time, electrophysiological methods were focused on cells that generate action potentials by activating voltage-dependent ion channels such as Na^+, K^+, and Ca^{2+} channels. These cells—neurons, nerve fibers, cardiac cells, and skeletal muscle fibers—were conventionally named **excitable cells**. Small cells, like most blood cells, epithelial cells, or extremely flat cells such as endothelial cells, which do not evoke regenerative action potentials, were not accessible by the hitherto applied microelectrode techniques. They only became the object of intensive electrophysiological research after the advent of the patch clamp technique developed by the 1991 Nobel prize winners Erwin Neher and Bert Sakmann. The long-standing dogma that excitable cells express voltage-gated ion channels and nonexcitable cells lack these channels, is challenged by the discovery of almost all kinds of channels in cells that do not generate action potentials under physiological conditions. Nonexcitable cells possess such a tremendous variety of functionally important ion channels that they are "exciting" in the best sense of the word.

This chapter cannot cover all types of ion channels found in nonexcitable cells, but will focus on a selection of ion channels in epithelial cells, especially those that function as secreting or electrolyte- and fluid-reabsorbing cells, endothelium, tumor cells, and endocrine cells. In the last part of the chapter we will discuss some physiological functions of ion channels coupled to other transporting proteins.

II. Types of Ion Channels in Nonexcitable Cells

A. Amiloride-Sensitive Na^+ Channels

Various epithelial cells in the proximal tubules, distal colon, sweat ducts, and trachea reabsorb sodium, a function coupled to channels which are approximately 20 times more permeable for Na^+ than K^+. This channel, called the **epithelial sodium channel (ENaC)**, is expressed at the luminal (apical, mucosal) side of these tissues and allows passive Na^+ entry. The confinement of Na^+ channels to this side of the cell is an essential feature of Na^+-reabsorbing epithelia and seems to occur via binding of a channel subunit to the apical cytoskeleton, possibly to the cytoskeletal protein ankyrin. A Na^+,K^+-ATPase at the abluminal side (interstitial, serosal, basolateral, or blood side) pumps the Na^+ absorbed at the luminal side out of the cell. K^+-ions, accumulated by the Na^+,K^+-ATPase, recycle via basolateral K^+ channels. Thus, the Na^+ channels provide a vectorial transport of Na^+ through these cells (see also Fig. 17).

The trademark of the Na^+ channel in sodium-absorbing cells is the block by amiloride, which plugs the open channel with high affinity (K_i between 0.1 and 0.3 μM). This Na^+ pathway is characterized by a small single-channel conductance of approximately 5 pS. The conductance is Na^+-dependent, and saturates with increasing extracellular Na^+ concentrations (K_m between 20 and 75 mM). Increased cytosolic Na^+ also blocks the channel and exerts a feedback inhibition that prevents an excessive increase in $[Na^+]_i$, e.g., if the Na^+-K^+ pump is blocked. Also a decrease in intracellular pH (pH_i) and an increase in intracellular Ca^{2+} concentration, $[Ca^{2+}]_i$ block the channel. This Ca^{2+}-induced block is decreased if the intracellular H^+ concentration is increased, consistent with a competition between Ca^{2+} and protons to bind to a cytoplasmic site.

Another functionally important hallmark of this channel is its activation by hormones, such as ADH (antidiuretic hormone or vasopressin) and aldosterone, which stimulate Na^+ entry in Na^+-reabsorbing epithelial cells. ADH stimulation enhances adenylate cyclase activity and elevates intracellular cAMP (cyclic-3′,5′-adenosine monophosphate). This hormone action increases the number of functional channels in the membrane either by inserting Na^+-channels in the plasma lemma from a pool of channel-bearing vesicles originating from the endoplasmic reticulum, or by activating "sleeping channels" present in the plasma lemma. The enhanced intracellular cAMP may activate these channels by stimulating a cAMP-dependent protein kinase (protein kinase A, PKA) that catalyses phosphorylation of the pore-forming channel or some of its subunits. PKA may also affect regulatory proteins which facilitate insertion of Na⁺ channels into the plasma membrane. Conversely, protein kinase C (PKC) inhibits the channel.

Other hormones, such as atrial natriuretic factor (ANF), mechanical forces (stretch), changes in osmolality of the basolateral fluid, PKA, PKC, tyrosine kinases, G proteins, and leukotrienes also modulate ENaC. ANF down-regulates EnaC, probably via a decrease in aldosterone secretion. As discussed before, ADH modulates the number of available channels as well as the channel open probability. The effects of G proteins may be mediated by phospholipase A_2 and thus via arachidonic acid and its metabolites (leukotrienes).

Importantly, ENaC is also down-regulated via a direct protein-protein interaction with the cystic fibrosis (CF) transmembrane conductance regulator (CFTR, see Section II.F.4 or Chapter 36). This effect is absent for the ΔF508 mutant, which does not mature and is retained in the Golgi stack. Patients with CF therefore show an up-regulated activity of ENaC which is pathophysiologically important. Protein-protein interaction possibly occurs between the cytoplasmic domains and is modulated by actin.

The molecular biology of this channel turned out to be very complex. A major breakthrough was the identification in 1994 of three subunits of the renal ENaC, that is, αrENaC, βrENaC, and γrENaC. The α subunit, which forms the channel pore, has a molecular mass of 79 kDa, as predicted by its nucleotide sequence, whereas the translation product migrates at 92 kDa depending on glycosylation. Expression of the α subunit in oocytes induced functional channels, but larger expression levels and currents were observed if the β (72 kDa) and γ (75 kDa) subunits were coexpressed. ENaC is probably a heterotrimeric or even pentameric structure, each subunit consisting of two membrane-spanning regions connected by a cysteine-rich long extracellular loop. The homology between the subunits is approximately 35%. The structure of the channel is depicted in Fig. 1. Interestingly, the three rENaC subunits share a very high degree of homology with the **degenerins** encoded by the genes *mec-4, mec-6,* and *mec-10* from the nematode *Caenorhabditis elegans,* which are involved in mechanosensation and possibly function as mechanosensitive channels. The mec-4 subunit (from the gene *mec-4*) is not a functional mechanosensitive channel but becomes so after coassembly with the other subunits.

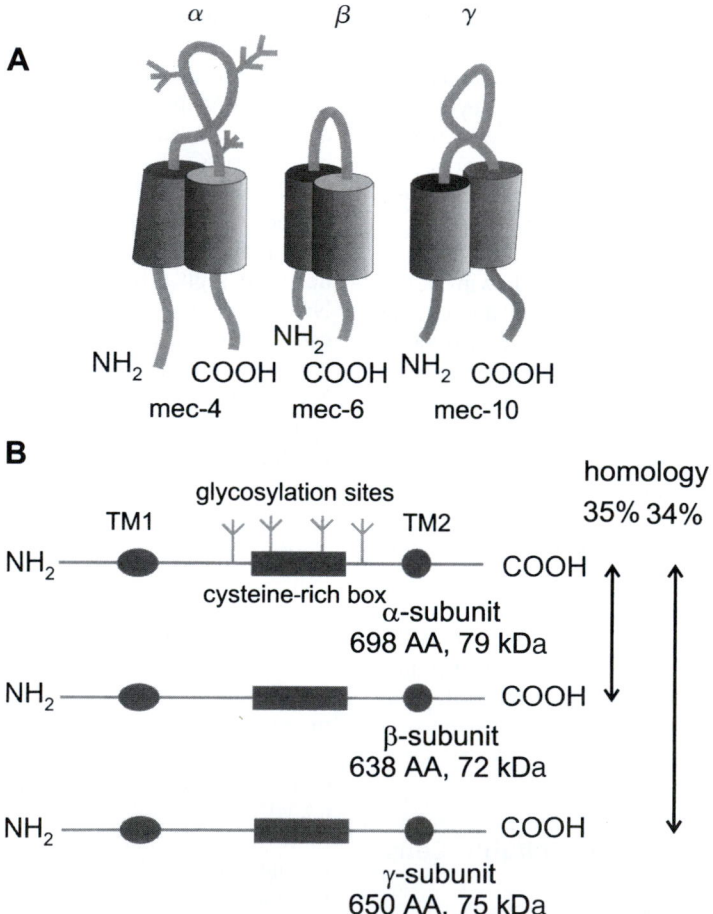

FIGURE 1.   Structure of the amiloride-sensitive Na⁺ channel. (A) The channel is a heterotrimer consisting of three subunits with two membrane-spanning helices and a long extracellular loop. The channel might be structurally related to the *mec-4, 6,* and *10* encoded mechanosensitive channels from *C. elegans*. (B) Linear models of the three subunits showing the conserved cysteine box, glycolysation sites, and the degree of homology. (Reprinted by permission from *Nature* 342: 762–766 (1989), copyright 1989 Macmillan Magazines Ltd.)

This feature thus resembles the normal function of the heterotrimeric rENaC. On the other hand, the amiloride-sensitive Na^+ channel also shows mechanosensitive features.

The amiloride-sensitive Na^+ channel is encoded by human chromosome 7 in a region that flanks the cystic fibrosis gene locus (see Section II.F.4). Importantly, a gain of function of ENaC is observed in an inherited form of hypertension, Liddle's syndrome. Truncated C-termini of the β and γ subunits are found in patients with this syndrome, that is characterized by hypertension, pseudohypoaldosteronism, hyperkalemia, dehydration, and metabolic acidosis. The disease is autosomal dominant and is possibly due to insertion of a stop codon at Arg 564 in the subunits. These truncations do not alter channel activity, but increase the number of channels in the apical membrane. Another important aspect is a missense mutation in the β and γ subunit of Liddle's patients located in a conserved domain region (PPPxY) in the cytoplasmic COOH terminus. This region is present in all subunits that bind to WW-domains, that is, protein modules of 38–40 amino acids involved in protein-protein interactions. For ENaC, the binding partner might be Nedd-4, an ubiquitin-protein ligase involved in internalization and degradation. This defect might also explain the increased number of channels in Liddle's disease. A similar mechanism for the formation of signaling complexes by the clustering of trp-channels with PDZ-proteins, such as InaD, will be discussed in Section II.E. Interestingly, the mutation in the PPPxY region also increases channel activity because the mutated ENaC channels lack the auto-inhibition induced by intracellular Na^+ in wild-type channels.

B. K^+ Channels

Most, if not all nonexcitable cells, such as secretory epithelial cells, liver cells, and endothelial cells, express different kinds of K^+ channels. These channels belong to at least six subfamilies: Kv1 (shaker Kv1.1–1.7), Kv2 (shab Kv2.1–2.2), Kv3 (shaw Kv3.1–3.4), Kv4 (Kv4.1–4.4), ether-a-go-go (eag or HERG), and the subfamily of both voltage- and Ca^{2+}-dependent K^+ channels. In the latter family, big conductance Ca^{2+}-activated K^+ channels with a conductance between 180 and 250 pS (BK_{Ca} channels or **slo**) can be distinguished from intermediate conductance channels (conductance between 20 and 80 pS, IK_{Ca} channels) and small conductance Ca^{2+}-activated K^+ channels (SK_{Ca}, conductance between 4 and 14 pS).

The blueprint of all these channels is similar: they contain 6 TMs with TM4 positively charged, but the N- and C-termini differ enormously. Probably, these channels form all tetramers. The pore region between TM5 and TM6 is characterized by a very conserved GYG motif. At least four β subunits are known for the Kv families, Kvβ1.1– 1.3, Kvβ2.1. These subunits modulate channel kinetics mainly by binding to the C and N terminus. Unlike these subunits, slo type channels have a functionally important β subunit with two transmembrane helices (see next section).

1. Voltage-Dependent Ca^{2+}-Activated K^+ Channels

We will focus mainly on Ca^{2+}-activated K^+ channels, because the release of intracellular Ca^{2+} is associated in many nonexcitable cells with activation of Ca^{2+}-dependent K^+ channels and a concomitant hyperpolarization, which increases the inwardly directed driving force for Ca^{2+}. The existence of high-conductance maxi K^+ channels was first described in salivary and lacrimal acinar cells. The high-conductance channel is highly selective for K^+, and its activation is both Ca^{2+}- and voltage-dependent. Increased $[Ca^{2+}]_i$ and depolarization enhance the open probability of the channel. Depolarization decreases the apparent dissociation constant of Ca^{2+} for activation of the channel. This negative feedback of membrane hyperpolarization on the channel provides an ideal tool for fine tuning of channel activity by membrane potential and $[Ca^{2+}]_i$. Channel activity is also affected by phosphorylation via cAMP-dependent protein kinases. Any rise in $[Ca^{2+}]_i$ induced by agonists or mechanical responses via stretch-activated channels (SACs, see Section II.D) will activate Ca^{2+}-activated K^+ channels. These channels also play a significant role by modulating the driving force for Ca^{2+} influx in nonexcitable cells stimulated by agonists which release intracellular Ca^{2+} and open an influx pathway for Ca^{2+} (see Section III.C). They are also involved in the volume regulation of cells (**regulatory volume decrease**, RVD, see Section III.A) by initiating a loss of intracellular K^+ and a concomitant loss of cell water resulting in a decrease of cell volume.

BK_{Ca} has been cloned from the *Drosophila* slo-gene. The mammalian voltage- and Ca^{2+}-dependent **maxi-K^+ channel** (**Kv slo family**) contains six membrane-spanning hydrophobic regions S1–S6, S4 being responsible for its voltage dependence. The linker between segment S5–S6 forms the pore region (P-region) which is highly conserved for K^+ channels. A large protein domain which probably includes 4 helix structures (7 to 10) is appended to the C-terminus and confers to the channel its Ca^{2+}-dependent gating. High-conductance K^+ channels have a subunit architecture consisting of a channel forming α subunit and a regulatory β subunit. The N-terminus is probably located extracellularly and is bound to the cell membrane by a scaffolding domain called TM0 (see Fig. 2). Not all cells that express the α subunit of BK_{Ca} coexpress the β subunit (e.g., endothelial cells). A functional indication for the absence of the β subunit is the lack of effect of the BK_{Ca} opener DHS-I, which is only effective if the β subunit is present. BK_{Ca} channels function in the absence of the β subunit. The sensitizing effect of the β subunit on BK_{Ca} consists of a leftward shift of the open channel probability (see Fig. 2). Pharmacologically, BK_{Ca} is selectively blocked by the scorpion venom **charybdotoxin** (CTX, IC_{50} approximately 50 nM) and the even more selective **iberiotoxin** (ITX). It is also sensitive to **tetraethylammoniumchloride** (TEA, IC_{50} approximately 1 mM, 100 % block at 10 mM), d-tubocurarine, and quinine. Mg^{2+} voltage-dependently blocks the channel from the outside. BK_{Ca} channels can be activated by the benzimidazolone compounds NS004 and NS1619 (in the μM range), which also induce an increase in the open probability by shifting their voltage dependence toward more negative potentials.

Intermediate IK_{Ca} are inwardly rectifying and have a conductance between 20 and 80 pS in symmetrical K^+, and 15 pS at physiological extracellular K^+. These channels were first functionally described as **Gardos channels** in red blood cells. They also form a subfamily from which the human and the mouse IK1 are very well characterized. Charybdotoxin,

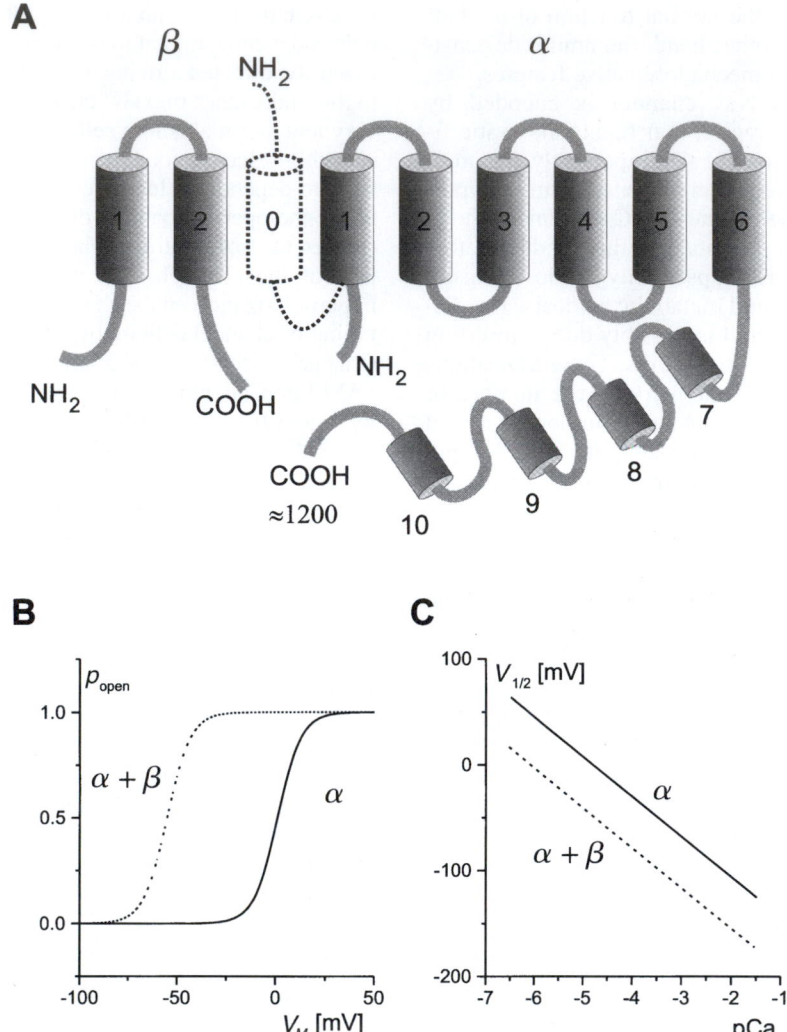

FIGURE 2. Structure of Ca^{2+} activated K^+ channels (BK_{Ca}, slo) and functional effects of coexpression of the β-subunit (A) BK_{Ca} channels consist of 6 TM. Likely, an N-terminus helix (TM0) forms a scaffolding domain for binding of the β-subunit which has only 1 TM. N- and C-termini of both subunits are intracellular. The BK_{Ca} α-subunit has—probably cytoplasmic—helix regions, which are denoted 7–10. (B) The open probability of the channel-forming α-subunit of BK_{Ca} is voltage- and Ca^{2+} dependent. Expression of the β-subunit enhances the opening probability of BK_{Ca} at normal membrane potentials of nonexcitable cells. (C) An increase in $[Ca^{2+}]_i$ (pCa) shifts the potential for half-maximal activation ($V_{1/2}$, panel C) towards more negative potentials. Expression of the β-subunit shifts the $V_{1/2}$-Ca relation toward more negative potentials.

clotrimazole, quinine, and tetrabutylammonium-chloride (TBA) are efficient blockers of these channels.

Small conductance K^+-channels with a conductance between 4 and 14 pS have also been observed in many nonexcitable cells. At least five members of this family are known. The overall structure is similar, that is six TMs, TM4 positively charged, N- and C-terminus cytoplasmic. Interestingly, these channels lack voltage dependence. SK channels are both apamin-sensitive and insensitive. Most of them show pronounced inward rectification and are blocked by extracellular TBA, apamin, and d-tubocurarine.

2. Inwardly Rectifying K^+ Channels

Inwardly rectifying K^+ (IRK) channels are present in some but not all nonexcitable cells (e.g., they are absent in

juxtaglomerular epitheloid cells, several types of endothelial cells, and mast cells). IRKs are responsible for stabilization of the resting potential near the K^+ equilibrium potential, E_K, and are extremely important for K^+ movements during fluid secretion and reabsorption. The molecular structure of inwardly rectifying K^+ channels was disclosed in 1993. All channels belong to a superfamily, Kir, characterized by a basic structure of 2 TM with the in-between pore-region which resembles that of voltage-activated K^+ channels (see Fig. 3). Five different members with a length between 320 and 500 amino acid residues can be functionally distinguished:

1. The constitutively open Kir2.1–2.3.
2. The G-protein–regulated Kir3.1–3.4 (also known as GIRK1-4).

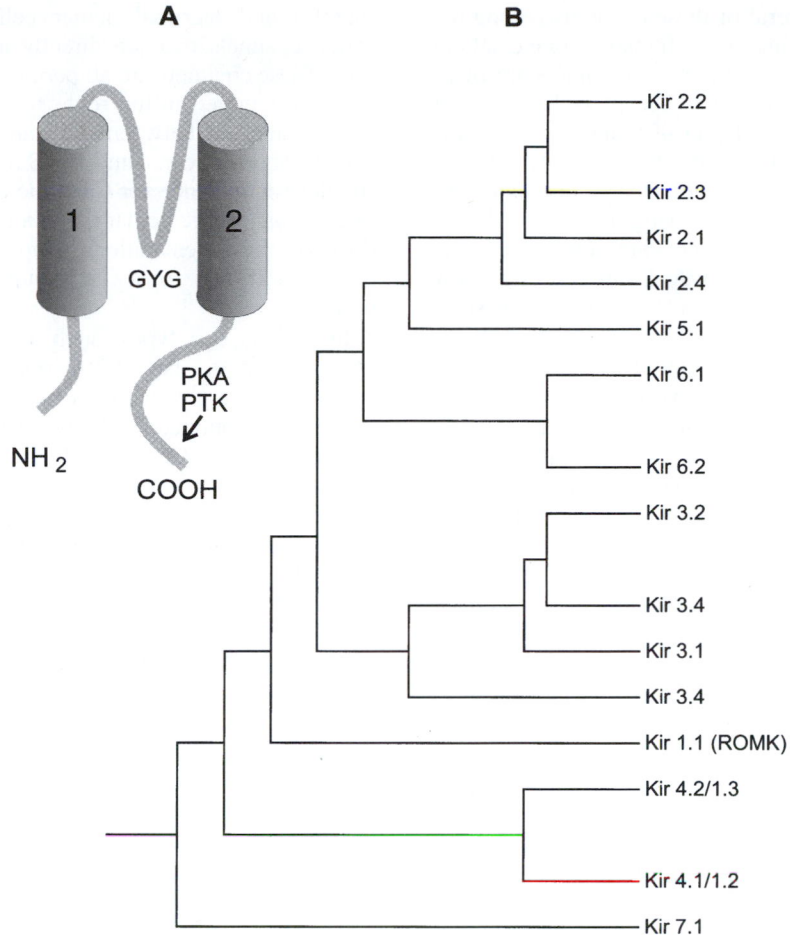

FIGURE 3. The phylogenetic tree of the 15 identified Kir channel genes. (A) Kir channels are members of a two-TM family with a highly conserved GYG motif in the K⁺ pore region. (B) The members of the Kir family that have been identified.

3. The ATP-sensitive Kir6.1 and 6.2. SUR1 and 2, two members of the ABC transporter family, confer to these channels their ATP sensitivity.
4. The ATP-dependent Kir4.1, 5.1, and 1.1. Kir1.1, also known as ROMK (renal outer medulla K⁺ channel), plays an important role in electrolyte secretion and reabsorption in the kidney, and is involved in the pathogenesis of Bartter's syndrome (discussed later).
5. The poorly rectifying Kir7.1, expressed in epithelial cells of lung, kidney, testis, plexus choroidae, and hippocampal neurons. Its single channel conductance is K⁺-independent and does not obey the square-root-dependence on extracellular K⁺ concentration, which is characteristic for all other Kir members.

The strongly inwardly rectifying Kir2 channel is constitutively open and therefore important for the stabilization of the membrane resting potential. Its two transmembrane regions are highly conserved. The channel consists of approximately 420 AA and a highly conserved TIGYG-H5 motif in the pore region. The C-terminus contains a phosphorylation motif, RRESEI, in which the serine seems to be important for channel regulation. Inward currents through these channels can be measured at potentials negative to E_K, but outward currents are almost absent. Inward rectification depends on intracellular Mg^{2+} being pushed as a plug in the open channel by depolarization. Block by intracellular polyamines such as spermine and spermidine is functionally important. Mg^{2+} and spermine, spermidine, and putrescine block appear to be connected to D172. The conductance of the channel at high extracellular K⁺ concentrations (150 mM $[K^+]_e$) is between 25 and 30 pS, and declines with the square root of $[K^+]_e$. Agonists, such as angiotensin II, vasopressin, and histamine, block the inward rectifier in some cell types, e.g., endothelial cells, mast cells, enterocytes, and juxtaglomerular cells. This block can be mimicked by GTPγS and is probably G-protein mediated. Block of the inward rectifier depolarizes these cells and decreases the driving force for Ca^{2+} influx, e.g., via Ca^{2+}-release-activated channels (CRAC) or other Ca^{2+}-permeable nonselective channels.

C. Nonselective Cation Channels

Nonselective cation channels (NSC) form a flourishing family of ion channels, most of them being only functionally characterized. Neither of these channels discriminates between Na^+ and K^+. Several of these channels belong to a family of channels consisting of two transmembrane helices, which also includes ENaC. Some NSC channels are members of the purinergic ligand-gated receptor-channel complexes (the P_{2X} family, see Chapter 40), another two-transmembrane helix family). Others are mechanically activated (mscL, mechanosensitive channel, large conductance), gated by protons (ASIC, acid sensing ion channels), or mammalian analogues of the *C. elegans* degenerin family (MDEG, mammalian degenerins). A discussion of all NSC channels described in nonexcitable cells is outside the scope of this chapter.

Many nonexcitable cells (epithelial cells, fibroblasts, adipocytes, etc.) express nonselective cation channels that are either activated by binding of intracellular Ca^{2+} (NSC_{Ca}) or by cyclic nucleotides (NSC_{CN}) such as cAMP or cGMP (cyclic-3′,5′-guanosine monophosphate). These channels do not discriminate between monovalent cations, and are impermeable to anions. Most of them are Ca^{2+}-permeable and therefore involved in Ca^{2+} signaling. Various members of the exploding family of trp-related channel proteins appear to NSCs.

NSC_{Ca} was first described in epithelial cells. This classical nonselective channel has a single-channel conductance between 25 and 35 pS. In many cell types (e.g., thyroid follicular cells, pancreatic cells, lacrimal gland cells, insuloma tumor cells) activation of NSC_{Ca} upon hormonal stimulation (e.g., acetylcholine or cholecystokinin in exocrine pancreatic cells) is mediated by an increase in $[Ca^{2+}]_i$. Since this channel is also permeable to Ca^{2+} (e.g., in neutrophils and in endothelial cells), it provides an entry route for Ca^{2+} into the cell, that exerts positive feedback on its activation. Their molecular structure is not known, although members of the trp family that have Ca^{2+}-calmodulin binding sites in their primary structure (see Section II.E) are possible candidates.

Multiple activation mechanisms of NSC have been described in nonexcitable cells. Cell shrinking activates NSCs in renal cortical collecting duct cells. In other kidney epithelial cells, NSCs are either inhibited or activated by stretch. Ca^{2+}-permeable NSCs are probably involved in apoptosis in insulin-secreting cells, since hydrogen peroxide and NAD (NSC_{NAD}) activate them and induce Ca^{2+} overload. NSCs gated by protons have been also described (NSC_H).

The physiological role of NSC channels in nonexcitable cells might be at least twofold. Not only do they provide an entry pathway for Ca^{2+}, but their activation also causes depolarization, which may activate or deactivate voltage-dependent ion channels. Ca^{2+} entry exerts a positive feedback on its own entry, whereas depolarization reduces the electrochemical Ca^{2+} gradient and provides a negative feedback for Ca^{2+}-entry via other non–voltage-gated channels (see Section III.C).

Two families of NSC channels will be discussed in more detail: the receptor-operated NSC and the cyclic-nucleotide- and voltage-activated NSC.

1. Receptor-Operated NSC (ROC)

In addition to Ca^{2+}-activated NSCs, many excitable cells (smooth muscle, cardiac cells, neurons) and nonexcitable (parotid and lacrimal acinar cells) express nonselective cation channels that are directly activated by extracellular ATP. These channels are all permeable to Ca^{2+} and provide a receptor-operated influx route for Ca^{2+}. The single-channel conductance is between 20 and 30 pS for monovalent cations, and between 4 and 10 pS for Ca^{2+}-ions. In spite of its smaller conductance for Ca^{2+}, the channel has in some cells (e.g., in pancreatic acinar cells) a higher selectivity for Ca^{2+} than for monovalent cations, supposedly due to the existence of a high affinity for a Ca^{2+}-binding site in the conducting pore.

In various cell types, such as epithelial cells, lymphocytes, and platelets, ATP is released by exocytosis, then binds to purinergic receptors, and activates NSCs with a single-channel conductance between 10 and 40 pS and a higher permeability for Ca^{2+} than for monovalent cations. These channels desensitize quickly after the binding of the agonist. P_{2X} receptors are distinct from P_{2Y} receptors, which belong to the G-protein-coupled serpent receptor family (GCRP). Seven P_{2X} receptors, sharing 35–48% identity, have been identified by molecular cloning. The previously described lytic ATP receptor, which forms large pores in macrophages and plays a role in cell death, has now been identified as P_{2X}. The length of these receptors is between 379 and 595 amino acid (AA) residues. All seven members have two transmembrane-spanning domains (TM), the first begins about 30 AA residues from the N-terminus, the second near residue 320. The loop between the two transmembrane domains contains 67 AA abundant in glycine, that are conserved in all seven members. A cysteine-rich domain and glycosylation suggest topological similarities with ENaC.

2. Cyclic Nucleotide and Voltage-Activated NSC, HAC, or HCN

Receptor activation by an agonist is in many nonexcitable cells often coupled to a G-protein. Activation of this G-protein may either directly activate the channel, or stimulate an effector system (e.g., adenylate cyclase, AC) that produces an intracellular messenger (e.g., cAMP) which gates the channel or activates kinases that phosphorylate the channel protein.

Various nonexcitable cells express cyclic nucleotide-activated nonselective cation channels, which belong to a large family of ion channels first described in cones, rods, and olfactory cells. Both N- and C-termini of these channel proteins are cytoplasmic with the binding site for cyclic nucleotides located in the C-terminus. These channels are directly gated by binding of cAMP or cGMP at the cyclic nucleotide binding domain (CNBD). An unusual feature is a sequence motif in TM4 that has so far only been found in voltage-gated Na^+, K^+, and Ca^{2+} channels. This motif consists of four to ten positively charged arginine or lysine residues at every third position separated by two hydrophobic amino acids. Phototransduction and signaling in olfactory cells depends on the modulation of such nonselective cation channels by cGMP and cAMP respectively. The channels have a conductance of approximately 20 pS for monovalent cations in the absence of Ca^{2+} and Mg^{2+}. In pineal gland cells, intracellular cGMP (but not cAMP) opens NSC with a single-channel conductance of 15 to 25 pS (for details see Chapter 40).

Four new members of this family, voltage dependent channels gated by hyperpolarizing voltages (hyperpolarization activated cation channels, **HAC**, or hyperpolarization activated cation channels, nucleotide-sensitive **HCN**) with a length between 780 and more than 900 AA have been cloned recently. These channels consist of 6 TM and are modulated by binding of cyclic nucleotides on a CNBD. TM4 is highly positively charged and contains at least 10 arg and lys residues. These channels have a high overall similarity with the cone cyclic nucleotide channels (CNG) (Chapter 47) and also with eag K^+ channels (see Chapter 27). They activate slowly at hyperpolarizing potentials and are permeable for monovalent but not for divalent cations, Na^+ being more permeable than K^+ (P_{Na}:$P_K \approx 0.15$–0.3). Their single channel conductance is probably very small. The role of these channels in nonexcitable cells is not yet clear. Figure 4 gives an overview of the members of this NSC family.

D. Mechanosensitive Cation Channels

Mechanoelectrical properties of hair cells and sensory mechanoreceptors provide an intriguing example of mechanosensing ion channels (see Chapters 44, 45, 46). Ion channels acting as mechanosensors also occur in nonsensory, nonexcitable cells. They are activated by stretch (stretch-activated channels, SACs) or by shear stress (shear-stress-activated channels, SSCs), and have been found in kidney cells, mesangial cells, tumor cells (such as neuroblastoma cells), astrocytes, endothelial cells, osteoblasts, blood cells, and oocytes, for example.

Some SACs are selectively permeable for anions, others form nonselective cation channels, while still others are selectively permeable for K^+. SSCs, which have not yet been studied at the single-channel level, either form entry pathways selective for K^+, or a pathway that is nonselective for cations with generally a higher permeability for Ca^{2+} than for monovalent cations. Reversible block by micromolar gadolinium (Gd^{3+}) and lanthanum (La^{3+}) is a common trademark for all mechanosensitive channels. Most of these channels in nonexcitable cells are also blocked by amiloride.

Because of their large conductance of about 130 pS, Cl^- selective SACs are probably different from volume-activated Cl^- channels. K^+-selective SACs have a low single-channel-conductance of approximately 7 pS in hepatocytes, but 30 to 50 pS in epithelial cells. Nonselective cation SACs in hepatocytes have a single-channel conductance of 16 pS. In the basolateral and apical membrane of renal tubule cells, and in endothelial cells, these channels have a conductance of approximately 16–37 pS for Na^+ and K^+ and 15 pS for Ca^{2+}. Nonselective SACs with a conductance of 55 and 116 pS were described in gall bladder cells. All nonselective SACs appear to be permeable for Ca^{2+}.

Endothelial cells are an intriguing example of mechanosensing cells in which mechano-sensitive ion channels are of paramount functional importance. These cells are subjected to flow-related forces such as biaxial-tensile stress (stretch perpendicular to their surface) and shear stress (a tangential force that is generated by friction between blood and the endothelial surface). Many processes in endothelial cells are linked to changes in blood flow, e.g., secretion of prostacyclin (PGI_2), endothelium derived relaxing factor (nitric oxide NO, EDRF), expression of tissue plasminogen activator (tPA), plasminogen-activator inhibitor (PAI-1), several adhesion molecules, endothelin, monocyte-chemoattractant protein (MCP-1), NO-synthase, activation of early response genes and small G-proteins, cytoskeletal rearrangement, cell cycle entry, long term responses, such as adaptive changes in cytoskeleton, vessel-remodeling, and others.

Mechanically gated ion channels, SACs (stretch-activated channels) or SSCs (shear-stress-activated channels, Figs. 5 and 6), are possible candidates for the functional link between physical forces and biological responses. A typical nonselective cation-permeable SAC in endothelium has been described. This channel is permeable for Ca^{2+} and has a single-channel-conductance of 19 pS in isotonic Ca^{2+} solution, and 56 and 40 pS in isotonic K^+ and Na^+ solutions. The channel is about six times more permeable for calcium than for sodium (permeation ratio P_{Ca}/P_{Na} between 1.2 and 8.4). Modulation of stretch-activated ion channels seems to depend on changes in the cytoskeleton which project membrane distortions from a large area of the plasma membrane to the site of a single-ion-channel (Fig. 6B).

Two types of shear-stress-activated channels (SSCs) have been described in endothelial cells. A K^+-selective

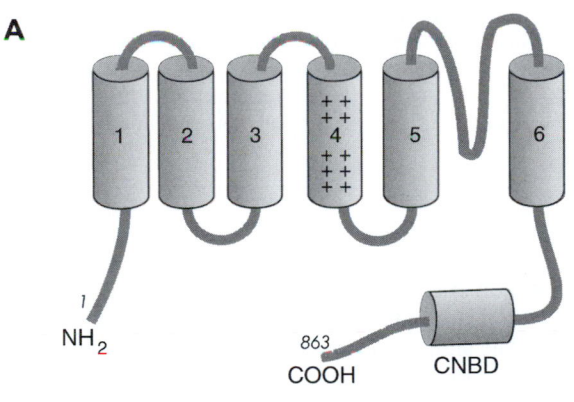

A

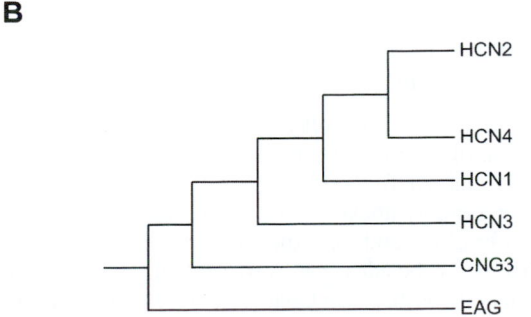

B

FIGURE 4. Structure of the hyperpolarization-activated and nucleotide sensitive cation selective channels (HAC). Structure of a typical HAC channel with at least ten positive charges in TM4 and a nucleotide binding domain, CNBD, and phylogenetic tree of the HAC family. The photoreceptor NSC, CNG3, and the K^+ channel EAG are shown for comparison.

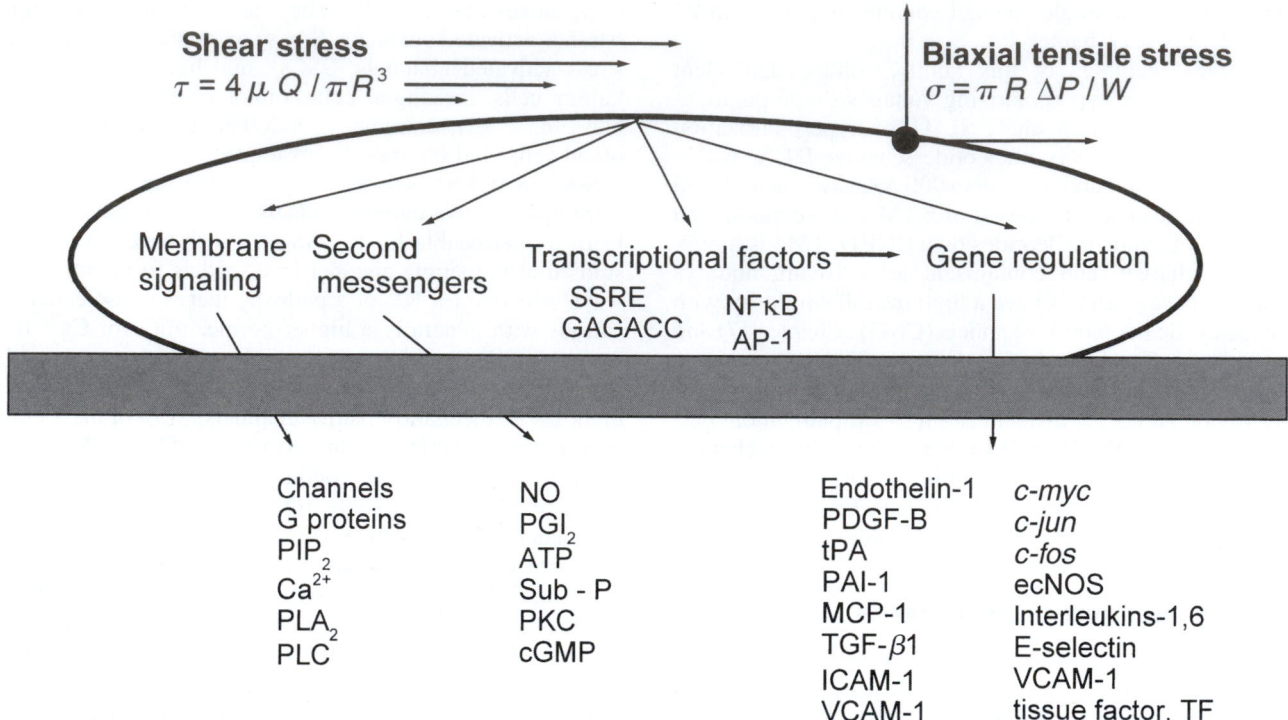

FIGURE 5. Mechanical activation of endothelial cells and related cell responses. The cell can be activated by shear stress, τ, which depends on blood viscosity, μ, flow, Q, and vessel radius, R. Biaxial tensile stress, σ, depend on radius, R, transmural pressure, P, and wall thickness, W. Cells respond immediately (in the seconds and minutes range) by channel activation, Ca^{2+} release, activation of phospholipase A and C, PIP_2 breakdown, synthesis and release of NO, PGI_2, release of different messengers as ATP, substance P. They react in the range of minutes to a few hours by modulation of the expression of endothelin (ET-1), transcriptional factors ($NF_\kappa B$, AP-1, nuclear factors, sensors are probably shear-stress response elements in the gene-promotor region, SSRE with the typical GAGACC-sequence), early response genes (*c-myc, c-fos, c-jun*), platelet-derived growth factor (PDGF-B), tissue-plasminogen activator (tPA), plasminogen-activator inhibitor (PAI-1), monocyte-chemoattractant protein (MAP-1), endothelial cell NO synthase (ecNOS), tissue factor TF for initiation of blood clotting, several cell adhesion proteins (ICAM, VCAM), interleukins, and selectins.

SSC acting as a mechanosensor that induces hyperpolarization in response to shear stress in the physiological range between 0.5 and 25 dyn/cm^2. This channel differs from the classical inwardly rectifying K$^+$ channels (Kir2.0 family) but shares some similarities with ROMK channels (Kir1.0 family). Opening of this channel apparently induces non–EDRF-dependent vasodilatation via electrical coupling of endothelial cells to smooth muscle cells. Hyperpolarization of endothelial cells would also spread to neighboring smooth muscle cells, causing relaxation. As discussed previously, opening of K$^+$ channels induces hyperpolarization and therefore an increased driving force for Ca^{2+} entry into the endothelial cell. Such a mechanism would increase [Ca^{2+}]$_i$ and trigger the synthesis and release of vasoactive compounds. Sensitivity of K$^+$ and Cl$^-$ channels to mechanical activation seems to be a more widespread phenomenon, and also includes Ca^{2+}-activated SK$_{Ca}$, IK$_{Ca}$, and BK$_{Ca}$ channels. Other transmembrane currents activated by shear stress appear to be associated with the opening of nonselective ion channels which are more permeable for Ca^{2+} than for Na$^+$ or Cs$^+$.

The first revolutionary step from the molecular viewpoint was the cloning of mechanically gated channels from the en-

velope of the bacteria *Escherichia coli*. This channel probably has a multimeric structure consisting of four to five 15 kDa subunits of 136 amino acids and four transmembrane spanning regions. The amino acid sequence of these channels shows no similarity with other putative eukaryotic mechanosensitive channels identified in the nematode *Caenorhabditis elegans*.

E. Ca^{2+}-Entry and Store-Operated Ca^{2+} Channels, The trp Family

Intracellular Ca^{2+} ions play a fundamental role in linking information of receptor stimulation at the plasma membrane to various distinct functions in nonexcitable cells, like enzyme secretion, synthesis of various compounds, control of cell proliferation, and cell differentiation. Different Ca^{2+}-entry pathways besides the well-studied voltage-operated Ca^{2+}-channels (**VOCs**, see Chapter 27) have been described. Receptor-operated Ca^{2+} channels (**ROCs**) mediate Ca^{2+}-influx as a consequence of agonist binding to its membrane receptor (e.g., ATP to P$_{2x}$ receptors, see above). Other Ca^{2+}-channels are indirectly activated by a second messenger via a G-protein or tyrosine-kinase and are often termed second-messenger-operated Ca^{2+} channels (**SMOCs**). The initial rise

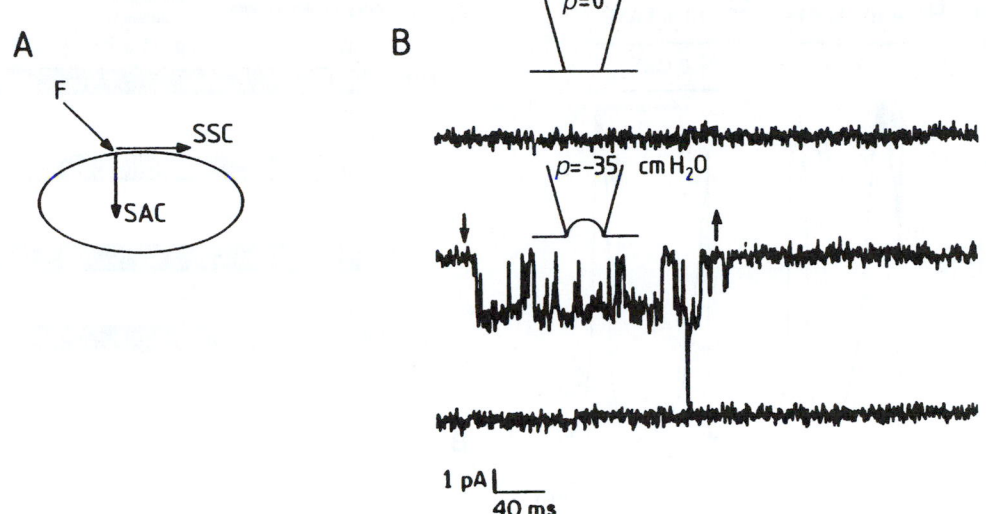

FIGURE 6. Mechanical-activated ion channels in endothelial cells. (A) A cell is challenged by a mechanical force F that generates at the cell surface a force perpendicular to the surface (stretch that opens stretch-activated channels, SACs) and tangential to the surface (shear stress that opens shear-stress-activated channels, SSCs). (B) Example of a SAC in endothelial cells. Opening of a nonselective cation channel in an inside-out patch by application of suction pulses (between the arrows) of 1.3 to 2.7 dyn/cm^2 through the pipette. The pipette contains 150 mM Na$^+$, the bath solution 150 mM K$^+$. Under this asymmetrical condition, single-channel-conductance is approximately 40 pS. (Modified from Lansman *et al.*, 1987. Reprinted by permission from *Nature* 325: 811–813 (1987), copyright 1987 Macmillan Magazines, Ltd.)

in [Ca^{2+}]$_i$ upon receptor stimulation originates in most non-excitable cells from Ins(1,4,5)P$_3$ mediated Ca^{2+} release from intracellular stores. However, a sustained Ca^{2+} influx giving rise to a plateau-like intracellular Ca^{2+} signal and depending on extracellular Ca^{2+} concentration is required for the long-lasting effects of agonists on cell function (Fig. 7A–C). This Ca^{2+} influx pathway could also be involved in the refilling of intracellular Ca^{2+} stores upon termination of the stimulus. Apparently, several pathways for Ca^{2+} entry exist, that is, pathways controlled by the filling degree of the intracellular Ca^{2+} stores (Ca^{2+}-release-activated Ca^{2+} channels, **CRAC**), and nonselective cation channels permeable for Ca^{2+} and directly or indirectly activated by agonists (Fig. 7D).

Various names in addition to CRAC have been used for Ca^{2+} entry channels coupled to the filling state of intracellular Ca^{2+} stores, such as depletion-operated channels (**DAC**), capacitative Ca^{2+}-entry channels (**CCE**), and store-operated Ca^{2+}-channels (**SOC**). A tremendous effort has been made in the last three years to identify and isolate mammalian genes that may encode for these channels. These efforts started mainly from the cloning of gene homologues of the *Drosophila melanogaster* trp channels (trp stands for transient receptor potential, a spontaneous mutation in *Drosophila*). Two *Drosophila* genes have been identified, the first (*trp*) encodes for a highly Ca^{2+}-selective channel protein (trp) that might be responsible for store-operated Ca^{2+}-influx, the other one (*trpl*, trp-like) encodes for a less Ca^{2+}-selective channel, trpl. Trp is activated by all known maneuvers that release intracellular Ca^{2+} but also directly by polyunsaturated fatty acids (PuFA), including DAG and arachidonic acid derivatives. Trpl is insensitive to store depletion by thapsigargin but is also gated by the PuFAs mentioned. Both proteins have a

putative channel structure with 6 membrane-spanning (TM) helices, a conserved pore region between TM5 and TM6, and three ankyrin-like motifs in the N-terminus which could bind to the ankyrin-binding motif of other proteins (e.g., the Ins(1,4,5)P$_3$ receptor). The TM4 domain is, in contrast with voltage-gated Ca^{2+} channels, not positively charged. The carboxy-terminal domain of trp contains a lys-pro motif, which is repeated twenty-seven times, and a highly charged sequence of asp-lys-asp-lys-lys-glu-(arg/gly)-asp, which occurs nine times. These sequences, which might be responsible for sensing intraluminal-Ca^{2+}, are absent in trpl. The fact that the dependence of trpl activation on store filling is less pronounced than that of trp activation, might be related to these differences. Both proteins also have several calmodulin-binding regions, which may mediate the Ca^{2+}-dependent modulation of store-related Ca^{2+} entry. The length of trp proteins varies considerably between 800 and 1200 AA, but the N-terminal ankyrin binding motifs and the TM regions are highly conserved.

Mammalian trp-related proteins are currently believed to form CCE channels gated by receptor-mediated stimulation of phospholipase Cβ (PLCβ) which induces Ca^{2+} release from intracellular stores in the endoplasmic reticulum. PLCβ, which is activated by heteromeric G proteins from the α_q family, catalyzes the hydrolysis of phosphatidyl 4,5-bisphosphate (PIP$_2$) to produce 1,2-diacylglycerol and inositol 1,4,5-trisphosphate (Ins(1,4,5)P$_3$). Both of these products are able to activate members of the trp family. This expanding family contains presently at least three members from humans, six members from mice, four from rats, seven bovine members, and also nonmammalian members, such as *Caenorhabditis elegans*, *Calliphora vinina*, *Xenopus laevis*, and *Loligo forbes* (squid) (see Fig. 8).

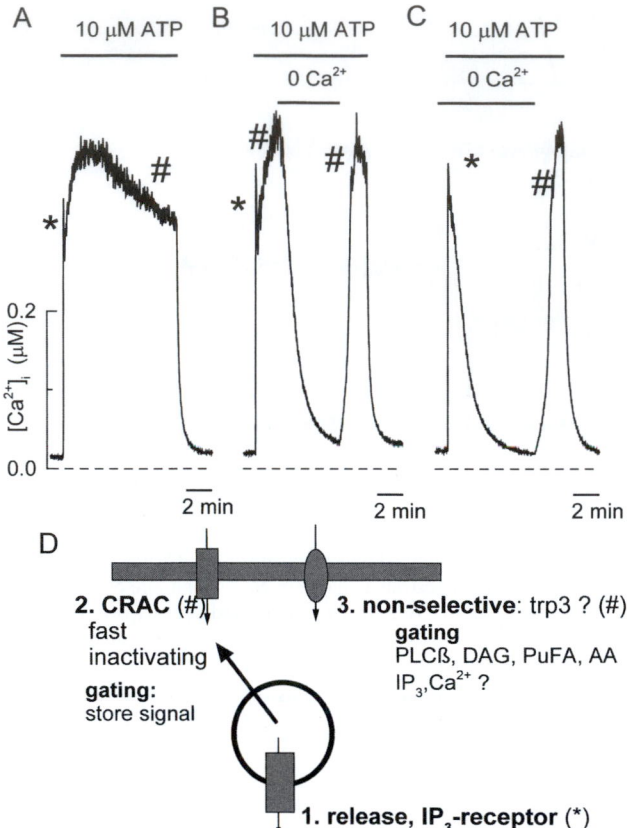

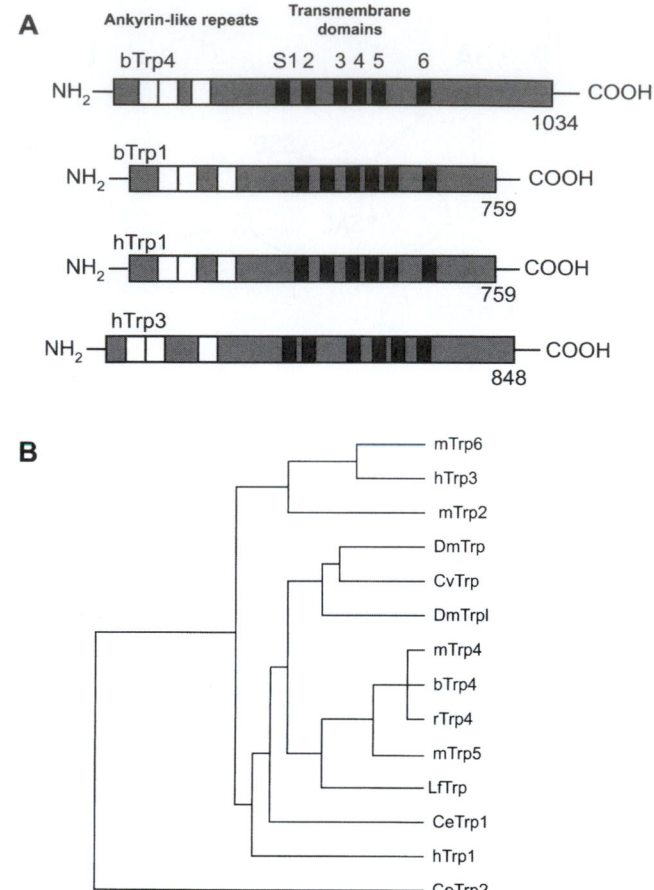

FIGURE 7. Ca^{2+} entry in nonexcitable cells. (A) Activation of a Ca^{2+} transient in a vascular endothelial cell by the vasoactive agonist ATP. A short transient peak is followed by a large, long-lasting, plateau-like Ca^{2+} signal. (B) If $[Ca^{2+}]_e$ is washed out during agonist stimulation, $[Ca^{2+}]_i$ decreases, indicating that an influx of Ca^{2+} is activated during the plateau phase. The Ca^{2+} signal reappears after adding $[Ca^{2+}]_e$. (C) The agonist induces only the first initial peak if applied in the absence of $[Ca^{2+}]_e$. Reapplication of Ca^{2+} in the presence of the agonist still induces elevation of $[Ca^{2+}]_i$. The initial peak (*) is due to release of intracellular Ca^{2+} from $Ins(1,4,5)P_3$ sensitive store. The plateau (#) depends on $[Ca^{2+}]_e$ and is mainly Ca^{2+} influx. (D) Mechanism of $[Ca^{2+}]_i$ increase. At least two different entry pathways are suggested. CRAC is activated by release of Ca^{2+} from intracellular Ca^{2+} stores (1, *). This channel seems to inactivate rapidly in the presence of high $[Ca^{2+}]_i$. Another pathway is a nonselective cation channel, possibly the PLCβ-dependently gated trp3 channel which is activated by polyunsaturated fatty acids, such as DAG and arachidonic acid derivatives. $[Ca^{2+}]_i$ is probably involved, but a direct activation by $Ins(1,4,5)P_3$ cannot be completely excluded (see text for detailed discussion).

FIGURE 8. Structure and phylogenetic tree of mammalian trp-encoded proteins. (A) trp1, 3, and 4, three members of the trp family, consist of six transmembrane spanning regions, and a putative pore region between S5 and S6. Possible interaction sites with the cytoskeleton are the three ankyrin-like motifs in the N-terminus. (B) The trp family (m: mouse, h: human, Dm: *Drosophila melanogaster,* Cv: fly *Calliphora vicina,* b: bovine, r: rat, Lf: squid *Loligo forbes,* Ce: *Caenorhabditis elegans*). This tree is far from complete, e.g., eight bovine members are now known, the *Xenopus laevis* trp is cloned, etc.

$Ins(1,4,5)P_3$-induced Ca^{2+} release and CRAC activation are of general interest because they are intimately linked by luminal calcium (Ca^{2+} inside the store organelles) to the regulation of many cell functions such as intracellular protein traffic, secretion of mature processed proteins from the endoplasmic reticulum (ER), retention of the partially processed proteins, and control of cell proliferation. The first evidence for a Ca^{2+}-entry mechanism dependent on store depletion in nonexcitable cells came from experiments in which Ca^{2+} stores were emptied by application of an agonist in Ca^{2+}-free external solutions in the presence of thapsigar-

gin or BHQ to selectively block the SERCa Ca^{2+}-pumps responsible for refilling the stores. Under these conditions, reapplication of extracellular Ca^{2+} in the absence of an agonist induces an increase in $[Ca^{2+}]_i$ due to Ca^{2+} entry from the extracellular space via channels opened by store depletion. An ionic current through highly Ca^{2+}-selective low conductance channels associated with Ca^{2+} release from intracellular $Ins(1,4,5)P_3$-sensitive Ca^{2+} stores has been first described in mast cells and T-lymphocytes.

Ca^{2+} influx following the discharge of Ca^{2+} stores seems in other cells to be mediated by a less Ca^{2+}-selective pathway which may even be independent of store-depletion. The signal transduction mechanism linking changes in intraluminal Ca^{2+} to the opening of plasma membrane Ca^{2+} channels is not yet understood. A direct physical contact between $Ins(1,4,5)P_3$ receptors in the endoplasmic reticulum and the putative plasma membrane channel as well as various second messengers, such as a low molecular weight Ca^{2+} influx factor (CIF, 500 Da), an unidentified tyrosine kinase or phosphatase, the arachidonic acid metabolite 5,6-epoxyeicosatrienoic acid (5,6-EET), cGMP, DAG or metabolized arachidonic acid derivatives, have been proposed to explain this cross talk between Ca^{2+} stores and membrane channels. It has also been suggested that $Ins(1,3,4,5)P_4$, or $Ins(1,4,5)P_3$, or a decrease of $[Ca^{2+}]_i$ in a restricted region of the subplasmalemmal space may directly gate the entry channels. A direct coupling between trp channels (trp3) and parts of the $Ins(1,4,5)P_3$-receptor may interact to activate gating.

Not all members of the putative trp channel family have been analyzed at the functional level. So far four sub-families have emerged:

1. The short trp3 and trp6 with approximately 800 AA form store-independent non-selective cation channels probably activated via the DAG pathway.
2. The longer trp4 and trp5 are Ca^{2+} selective and store-operated.
3. Trp1 is probably nonselective but store-operated.
4. The trp2 subfamily is still uncertain (see Fig. 6).

Interestingly, some other channels, including the capsaicin receptor (VR1) which forms cation selective channels, the Ca^{2+}-permeable ECaC channel (discussed later), and growth-factor-regulated cation channels (GRC), show some structural similarities with the trp family.

The permeation mechanism through trp channels is also not yet resolved. The endogenous CRAC channels are highly selective for Ca^{2+} with a tiny single-channel conductance of approximately 30 fS in the presence of extracellular Ca^{2+}, but 36–40 pS for monovalent cations in the absence of extracellular Ca^{2+}. trp3 and trp6 have a conductance between 25 and 66 pS and are not very Ca^{2+}-selective.

It is not clear how these channel proteins are organized, but the extreme functional heterogeneity observed in expression experiments points to a multiheteromeric structure. Proper heteromerization may also depend on unidentified auxiliary proteins. It is known, for example, that the *Drosophila* protein InaD (from the mutant "inactivation no afterpotential") contains five PDZ domains and is physically linked to the rhodopsin PLC (NorpA), a protein kinase C (InaC), and the $G_q\alpha$-protein photoactivated by rhodopsin. This whole structure forms a signaling complex in which trp and trpl are integrated. A similar protein hINal (human InaD like) has been cloned in humans, but is twice as long as the *Drosophila* InaD (1525 versus 675 AA).

F. Cl⁻ Channels

Cl⁻ channels, probably present in every animal cell, are involved in various cell functions such as regulation of cell volume, fluid and electrolyte transepithelial transport, regulation of intracellular pH in cooperation with a variety of transporters, such as the band III bicarbonate-chloride exchanger, acidification of intracellular organelles, and more general in electrogenesis and setting the membrane potential in nonexcitable cells. Because Cl⁻ is often passively distributed, Cl⁻ transport has a low energy cost and is able to drag water and other electrolytes such as Na^+. Some Cl⁻

channels also form pathways for larger molecules and osmolytes.

1. The ClC Family of Voltage-Dependent Cl⁻ Channels

The ClC family is a large family of Cl⁻ channels which are completely different structurally from previously cloned Cl⁻ channels, such as ligand-gated Cl⁻ channels (GABA- and glycine receptor) and the cAMP-activated cystic fibrosis transmembrane conductance regulator (CFTR, see Section II.F.4). Nine ClC genes have been identified in mammals, including humans. Similar genes are present in plant cells, yeast, and bacteria. The structure of these channels is not completely known, the most likely localization of the 13 hydrophobic stretches is shown in Fig. 9: most of them are transmembrane domains, but TM4 and TM13 are believed to be respectively extracellular and intracellular. TM9–12 span the plasma membrane three to five times. The C-terminal region comprises two cystathine beta-synthase (CBS) domains (TM13 is CBS2). These domains comprise sorting signals containing aromatic amino acids (FVTID), which are important for ClC sorting. The location of the pore region is still uncertain.

ClC-2 is ubiquitously expressed; ClC-Ka and ClC-Kb are kidney specific; ClC-3, ClC-4, and ClC-5 show a broad expression spectrum and include nonexcitable cells such as endothelium, liver, and kidney. Most likely, ClC-6 and ClC-7 are intracellular anion channels. ClC-1 is exclusively expressed in skeletal muscle and will not be further discussed here. It may turn out that most members of the ClC-family are functionally important Cl⁻ channels in nonexcitable cells.

ClC-2, which seems to be a housekeeping Cl⁻ channel, is composed of 907 amino acids. It is activated by hyperpolarization, shows inward rectification, and is more permeable for iodide than chloride. The single-channel-conductance is between 3–5 pS. It can also be activated by cell swelling and might therefore be a volume-sensor in nonexcitable cells. The N-terminus of this channel probably comprises a "ball" structure ("essential" region) which is tethered to the channel by a polypeptide "chain" and normally closes the channel. Mutations in the essential region induce constitutively open channels. One of the intriguing hypotheses predicts that cell swelling removes the "ball" closure of the channel and thus activates a Cl⁻ flux. These channels are certainly different from the so-called "volume regulated anion channels" (VRAC) that will be discussed later. Of the known functional ClC channels, ClC-2 is the only one expressed in secretory epithelia. Its apical location in small intestine, airway, and renal epithelia suggests a role in fluid and electrolyte transport.

The kidney specific ClC channels (ClC-Ka,b in humans, ClCK-1,2 in rats) are responsible for Cl⁻ reabsorption in Henle's loop, the thick ascending limb (TAL), and the distal tubules. They are important for concentrating the urine. ClC-Ka knockout mice show massive diuresis (nephrogenic diabetes insipidus). Mutants of ClC-Kb at protein 687 AA (encoded by a gene on chromosome 1p36) lead to Bartter's syndrome III. This is an autosomal recessive disease, which is characterized by hypokalemic acidosis, low blood pressure,

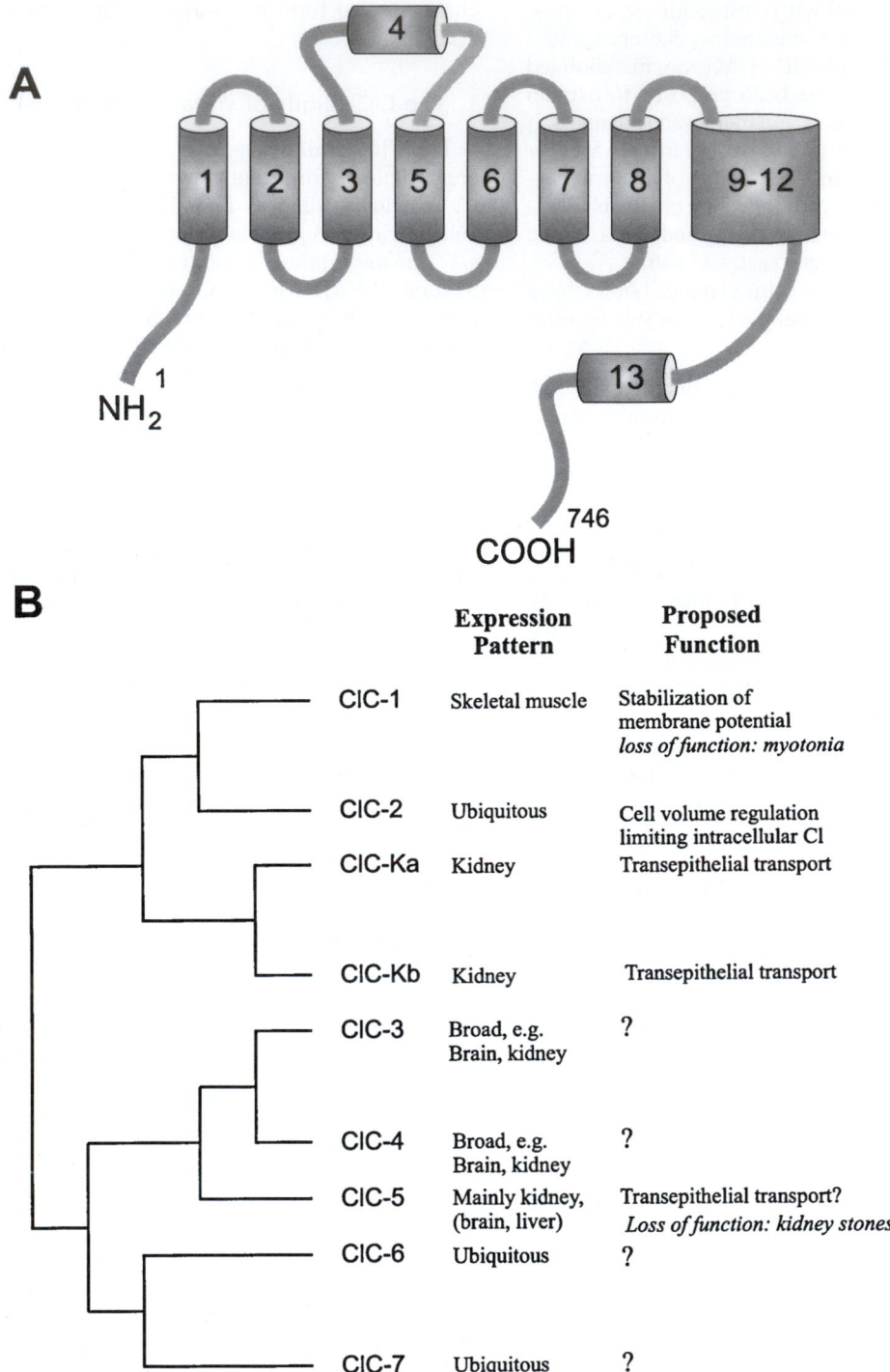

FIGURE 9. Structure and phylogenetic tree of mammalian ClC channels. (A) Structure of ClC-5, a member of the ClC family which is expressed in kidney epithelial cells. (B) Phylogenetic tree of the nine mammalian ClC members including the expression pattern. For detailed discussion see text.

renal salt wasting, hyperaldosteronism, and hypercalciuria. Mutations in the $Na^+,K^+,2Cl^-$ transporter (NKCC) and ROMK channels (see Section II.B.2) induce similar defects. Under normal conditions, Cl^- is taken up into the cells by the NKCC together with K^+, which leaves the cell via luminal ROMK channels. Loss of function of ClC-Kb at the ablumi-

nal, basolateral membrane will block Cl^- reabsorption (see also Fig. 19B).

ClC-3 has recently been discussed as a plasmalemmal, volume-activated channel. ClC-5, a 746 AA 83 kDa protein, is a strongly outwardly rectifying channel expressed in the apical membrane of epithelial cells, in proximal tubules of the kidney, and colocalizes with v-type H⁺ pumps. It is therefore most likely an electrical shunt for the H⁺-ATPase which supports acidification of early endosomes. This mechanism is important for endocytosis, sorting, and degradation of internalized proteins. Mutations of the ClC-5 gene localized on chromosome Xp11.22 induce a defect in the acidification of endosomes, which promotes protein accumulation in tubules and kidney stone formation. This disease, known as Dent's disease, is characterized by low molecular weight proteinuria, hypercalciuria, nephrocalcinosis, and nephrolithiasis. The third member of this group, ClC-4, is functionally not yet understood.

The function of the ubiquitously expressed putative ClC-6 and ClC-7 Cl⁻ channels is unclear. They probably form intracellular anion channels.

2. Ca²⁺-Activated Cl⁻ Channels

Cl⁻ channels activated by an increase in $[Ca^{2+}]_i$ seem to be ubiquitously distributed among nonexcitable cells and are abundant in secretory and reabsorbing epithelial cells, endothelium, intestine, pituitary cells, and hepatic cells. The first evidence for the existence of a Ca²⁺-activated Cl⁻ channel (**CaCC**) was obtained in rat lacrimal glands. CaCCs are involved in fluid and electrolyte transport in epithelial cells, in the maintenance of pH balance, in osmo- and volume regulation, in setting the driving force for Ca²⁺ influx, and may be directly linked to vectorial transport via a "push and pull" mechanism (discussed later). The conductance of these channels is very small and ranges from 1 to 5 pS. The whole cell current is nevertheless substantial because the number of channels is huge (e.g., 5000 to 20 000 in acinar cells, compared to approximately 150 "maxi"-K⁺ channels). Currents through CaCC, $I_{Cl,Ca}$, activate slowly at positive potentials and decay rapidly at negative potentials. They are strongly outwardly rectifying, but instantaneous I-V curves obtained from tail current do not show rectification. A typical example of $I_{Cl,Ca}$ is shown in Fig. 10. The K_D for Ca²⁺ activation strongly depends on membrane potential and ranges from 400–600 nM at –100 mV to 50–80 nM at +80 mV, pointing to high affinity binding sites for Ca²⁺. Steady-state and kinetic behavior of this current can be described by a model which assumes activation of the channel by two identical, independent, sequential Ca²⁺ binding steps preceding a final Ca²⁺-independent transition from the closed to the open state of the channel

$$\text{closed} \underset{\beta_1}{\overset{\alpha_1 \cdot Ca}{\rightleftharpoons}} \text{closed}^{Ca} \underset{\beta_1}{\overset{\alpha_1 \cdot Ca}{\rightleftharpoons}} \text{closed}^{Ca}_{Ca} \underset{\beta_2}{\overset{\alpha_2}{\rightleftharpoons}} \text{open}$$

The putative binding site for Ca²⁺ is approximately 10–15% within the membrane electric field from the cytoplasmic site.

The molecular nature of this channel is not yet resolved. Recently, a number of related membrane proteins have been cloned, including the endothelial adhesion protein Lu-ECAM, the bovine bCLICA1, murine mCLCA1, and human hCLCA1, 2 and 3 proteins which are believed to represent putative Cl⁻ channels. Currents, showing some similarity with $I_{Cl,Ca}$ can be observed in HEK cells expressing these proteins. Proteins of this family are characterized by a pre-

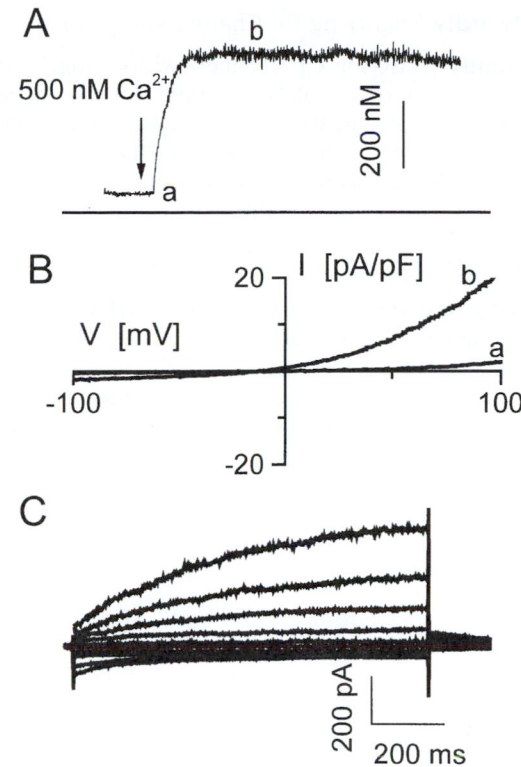

FIGURE 10. Kinetic properties of the endothelial Ca²⁺-activated Cl⁻ current. (A) Time course of the changes in intracellular Ca²⁺ after breaking into the cell with a pipette solution buffered at 500 nM Ca²⁺. The arrow indicates the point at which the membrane in the patch was disrupted. (B) Membrane currents measured during linear voltage ramps from –150 to +100 mV applied before the rise of $[Ca^{2+}]_i$ (a) and after a constant concentration has been reached (b). The activated current is strongly outwardly rectifying. (C) Current traces recorded during voltage steps at 500 nM $[Ca^{2+}]_i$. Voltage steps were applied from a holding potential of –50 mV. Steps ranged from –100 to +100 mV in increments of 20 mV. Note the slow activation at positive potentials and the fast deactivation at negative potentials and after returning to the holding potential.

cursor of approximately 130 kDa of between 900 and 940 AA residues. This precursor is cleaved to form heterodimers of approximately 90 and 35 kDa. The most likely topology is 5 TM with an extracellular glycosylated N-terminus that contains a number of cysteine residues that are conserved among all members, and an intracellular C-terminus. The above-mentioned cleavage site is located in the intracellular loop between TM3 and TM4. The proteins comprise several consensus sites for PKC phosphorylation.

A Ca²⁺ activated Cl⁻ channel with properties clearly different from the above-described CaCC can be activated by a Ca²⁺ calmodulin-dependent protein-kinase II, CaMKII. Interestingly, this channel is inhibited by Ins(3,4,5,6)P₄, an endogenous inositol-tetraphosphate that also inhibits Ca²⁺-stimulated Cl⁻ secretion. The effect seems to be specific for that particular isoform, since Ins(1,4,5,6)P₄, Ins(1,3,4,5)P₄, Ins(1,3,4,6)P₄, and Ins(1,3,4,5,6)P₅ [the latter being the immediate precursor of Ins(3,4,5,6)P₄] are all ineffective.

3. Outwardly Rectifying Cl⁻ Channels (ORCC)

Outwardly rectifying Cl⁻ channels with a single-channel conductance of approximately 50 pS are abundant in epithelial cells, especially in the apical membranes of secretory cells, and are defective in cystic fibrosis. The molecular nature of this channel as well as its mechanism of activation are not clear. ORCC are possibly activated by ATP.

4. cAMP-activated Cl⁻ channels (CFTR)

cAMP-activated Cl⁻ channels are expressed in most epithelial cells (airway epithelial cells, rectal gland cells, pancreatic duct epithelial cells, carcinoma cells). This channel is encoded by the *cftr* gene (cystic fibrosis transmembrane conductance regulator), a member of the ABC (ATP-binding cassette) transporter family. *cftr* was the first cloned epithelial Cl⁻ channel. The *cftr* gene is localized on chromosome 7q31 and has 27

exons from which several splice variants exist. The channel is a 169 kDa protein with 1480 AA residues. It has the typical TNRNT structure, consisting of two transmembrane regions (N-terminus T1, C-terminus T2) with each 6 TM, two nucleotide binding folds (N1 and N2). Both domains have as conserved nucleotide-binding domains Walker A (GxxGxG)-, LSGGQ-, and Walker B (Rx₇hhhhD)-motifs, where x refers to any amino acid, h to a hydrophobic residue. N1 and N2 have p21-ras folds, which points to a structural homology of N1 and N2 with the catalytic sites of G proteins. CFTR is further characterized by the ABC family unique regulatory R-domain (R, approximately between AA residues 610–835, probably exon 13). N1-R-N2 are located in a close proximity (Fig. 11A). The intracellular link between TM2 and TM3 represents 30 AA residues encoded by exon 5 that are spliced out in the cardiac isoform of CFTR. This protein forms a Cl⁻ channel with a single-channel conductance between 6–8 pS and a pore diam-

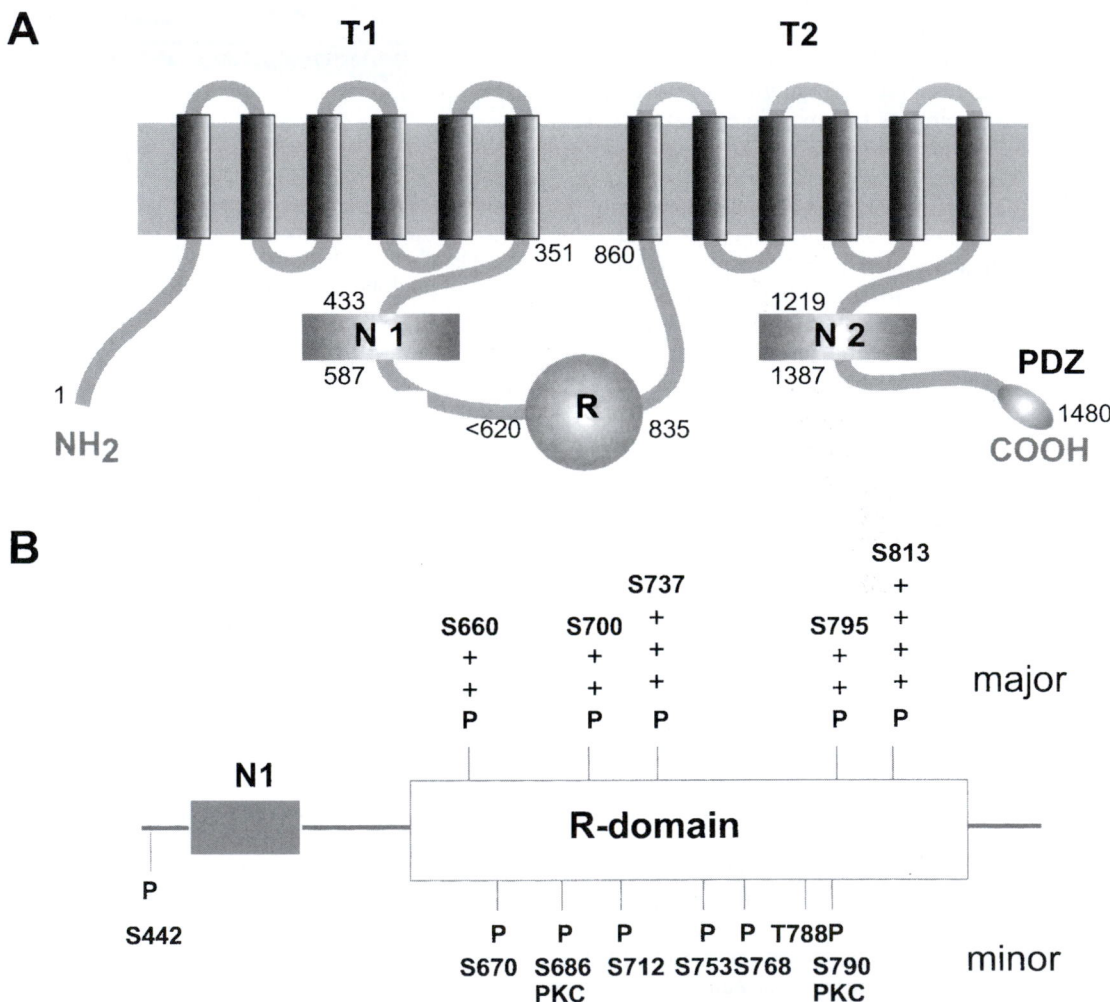

FIGURE 11. Structure of the cAMP-activated Cl⁻ channels encoded by the *cftr* gene. (A) Structure of the CFTR channel. N1,2 are the nucleotide binding domains which are in close proximity to the catalytic domain R. T1,2 indicate the membrane spanning domains with 6 TM each. The thick linker between TM2 and TM3 represents a stretch encoded by exon 5 which is absent in the heart isoform. The last 4 AA represent a PDZ domain-binding motif. (B) Structure of the R-domain indicating the PKA phosphorylation sites necessary for CFTR gating. Sites which induce major activation are indicated above the R domain, sites with less functional impact are drawn below the domain. Two PKC sites are also present in the R-domain.

eter of approximately 5.3 Å. The permeability sequence for anions of $Br^- > Cl^- > I^- > F^-$ is consistent with a weak field strength binding site. I^- blocks the multi-ion pore which is also permeable for HCO_3^-. The channel is activated by a cAMP-mediated stimulation of PKA and in the intestine also by a cGMP-mediated stimulation of PKG. The R-domain has 10 PKA phosphorylation sites, between S660 and S813. In human tissue, nine sites lie in a monobasic RxS sequence. In all other species these sites present a dibasic motif RRxS. Two serines belong to PKC consensus sites (S686 and S790, Fig. 11B). Gating only proceeds after several phosphorylation steps in the R-domain, which then allows ATP binding to the nucleotide-binding fold N_2. This step precedes channel opening. Gating is then further regulated by a cycle of ATP hydrolysis at N1 and N2 when R is highly phosphorylated. Dephosphorylation by protein phosphatases returns the channel into the closed form (for gating see Fig. 12).

CFTR, which is much more than a Cl^- channel, regulates other ion channels and transporters, indicating that the name **CFT-regulator** was chosen with much foresight. These regulatory properties might even be more important than its function as a Cl^- channel. The epithelial Na^+ channel ENaC is inhibited by CFTR. Its activity is up-regulated when CFTR is not expressed in the plasmalemma. ORCC is normally activated by CFTR and is defective in cystic fibrosis. CaCC and the volume-regulated anion channel VRAC (discussed later) are down-regulated in cells that coexpress CFTR. Basolateral ROMK

channels are activated by CFTR. The KvLQT-1 channel that has recently been detected in epithelial cells probably is a complex of IsK (minK) and eag channels (Chapter 27), and seems to be activated by CFTR and is therefore down-regulated in the absence of CFTR. CFTR also regulates water channels (aquaporins) and several transporters such as the Na^+-H^+ exchanger and probably the members of the AE (anion exchanger) family. Importantly, CFTR is involved in the cAMP-stimulated release of intracellular ATP, but the exact mechanism is unknown. This CFTR-mediated ATP release is probably responsible for the functional coupling between CFTR and ORCC, since the latter is modulated by ATP agonists.

Some proteins which are important for exocytosis either inhibit (e.g., syntaxin 1A) or stimulate CFTR (MUNC 18). CFTR also affects intracellular processes, such as the traffic of exocytotic vesicles and the acidification of the trans-Golgi network, which is related to sialylation of secretory proteins. The accumulation of asialo moieties can create a proinflammatory environment attracting neutrophils and enhancing the binding of *Pseudomonas aeruginosa* and other bacteria.

Potential mechanisms of interaction with other proteins might involve a direct protein-protein interaction via the PDZ-binding domain located at the C-terminus of CFTR. The last for AA residues 1477–1480 (DTRL) provide the PDZ-binding motif, which seems essential for at least the functional interaction with the Na^+-H^+ exchanger (NHE) regulator protein.

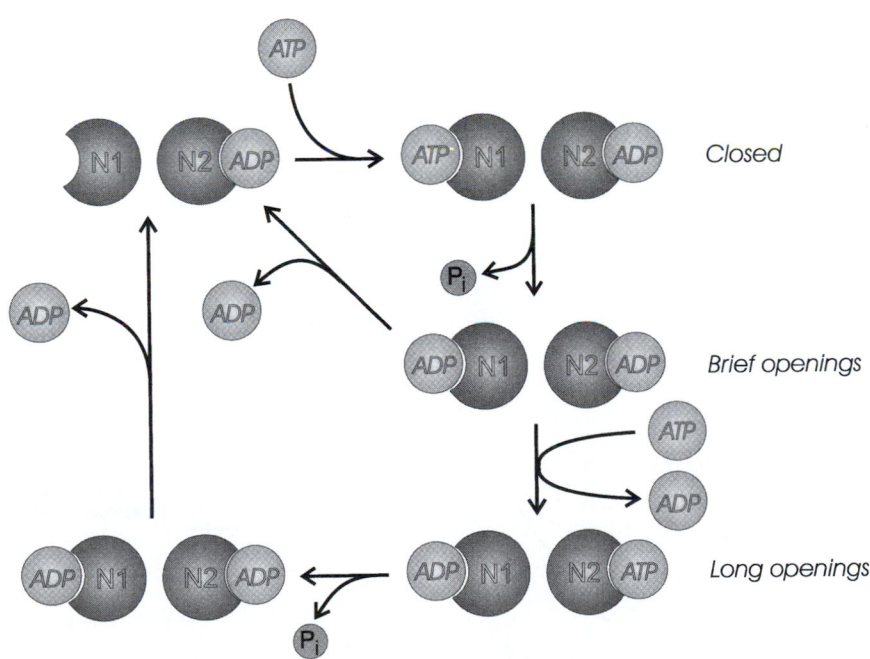

R strongly phosphorylated

FIGURE 12. Gating scheme of the cAMP-activated CFTR Cl^- channel. Shown are only the N- and C-terminus nucleotide-binding folds, N1 and N2. If R is highly phosphorylated, ATP binding and hydrolysis at both sites gate the channel. If N1 binds ATP and N2 binds ADP, the channel is still closed. Hydrolysis of ATP at N1 induces short channel openings. Openings are stabilized when ADP at N2 is exchanged by ATP. Hydrolysis and ADP unbinding at both sites close the channel. The channel is also closed when phosphorylation of R is terminated by phosphatases.

Various CFTR mutants lead to loss-of-functions which are phenotypically coupled to mucus abnormalities, defects in electrolyte transport, salty sweat, ileus, cilia dyskinesia in the airway epithelium, dilated bronchi, bronchiectasis, biliary cirrhosis, gallstones, pancreas insufficiency, severe malnutrition, and impairment of HCO_3^- transport. The most important mutant types are arranged in classes. Class 1 comprises defective translation mutants that all constitute truncated forms, and stop mutations in N1. Class 2 contains proteins which are incompletely glycosylated and cannot maturate. They are retained in the endoplasmic reticulum. The most important mutant, a deletion of phenylalanine at position 508 in N1, ΔF508, accounts for approximately 80% of the CF patients. Class 3 mutations interfere with ATP binding and hydrolysis and thus influence gating of the CFTR channel: Class 4 mutants represent mutations in the pore, causing functional channel defects due to changes in permeation. Typical mutations of this kind are R117H, R334W, and R234P. Class 5 summarizes mutations that do not cause CF, but CBAVD, the congenital bilateral absence of vas deferens. The exceptional not yet understood defects in regulator functions of CFTR are comprised in class 6.

5. Volume-Regulated Anion Channels (VRAC)

Anion channels that open in response to an increase in cell volume are abundantly expressed in most excitable and nonexcitable cells. The current through this volume-regulated anion channel (VRAC) is mainly carried by Cl^-, but the channel is also permeable for amino acids and organic osmolytes. The efflux of Cl^- and organic osmolytes through VRAC is functionally important for the regulation of cell volume. In addition, VRAC also contributes to the control of membrane potential and the driving force for Ca^{2+} influx, vectorial transport of solutes, and intracellular pH homeostasis. More speculatively, VRAC could assist in regulating cell growth and the transition from proliferation to differentiation. VRAC can also be activated by mechanical forces (shear stress, biaxial tensile stress) which induce changes in cell shape and possibly fold and unfold the plasma membrane. This function might be especially important for cells that sense mechanical forces, such as endothelial cells.

Currents through VRAC are outwardly rectifying, slowly inactivate at potentials positive to +60 mV, and show a voltage-dependent recovery from inactivation (Fig. 13). The permeability sequence of VRAC for anions is $SCN^- > I^- > NO_3^- > Br^- > Cl^- > HCO_3^- > F^-$ > gluconate > glycine > taurine > lactate > aspartate, glutamate. The sequence for amino acids is consistent with that for the rates at which these amino acids are lost from cells during hypo-osmotic swelling. The pore diameter estimated from the fit of the relative permeabilities of these anions to their Stoke's diameter was approximately 11 Å. The single-channel conductance is approximately 40–50 pS at positive potentials and 10–20 pS at negative potentials. Rectification is therefore a property of the open channel.

It is not known how mechanostimulation or cell swelling affect gating of VRAC. The annexin II-p11 complex, which is part of the cortical cytoskeleton and is involved in the formation of caveolae, seems to be important for activation of VRAC in

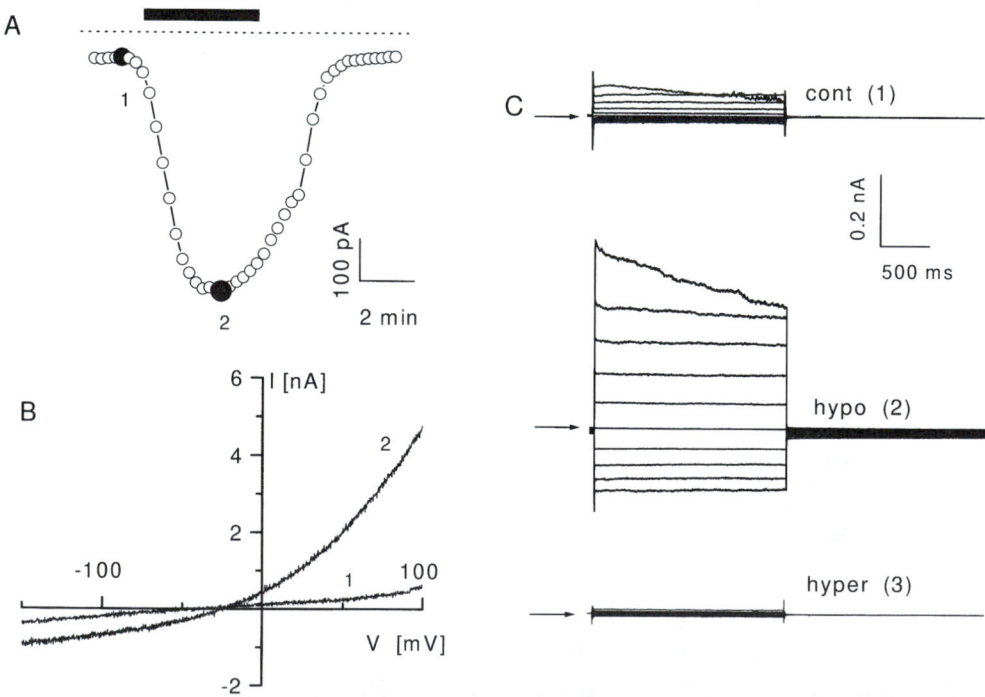

FIGURE 13. Activation of VRAC by hypotonic solutions. (A) Time course of the activation and deactivation of currents through VRAC (measured at −80 mV) in an endothelial cell from cultured pulmonary artery. The solid bar indicates the application of a solution which, with a 25% lower osmolality than the control solution, induces cell swelling. (B) I-V curves obtained at the time-points 1,2 indicated in (A). (C) Current traces under isotonic, hypotonic, or hypertonic conditions in response to a step-voltage protocol. Holding potential is 0 mV, 2-s steps range from −80 to +100 mV and are 20 mV spaced.

endothelial cells. The major protein in caveolae is the 178 AA caveolin-1 which is linked to many other signaling proteins such as tyrosine kinase, small GTPase (e.g., RhoA), Ca^{2+} pump proteins of the exocytotic machinery, and in endothelial cells NO synthase. The current density of VRAC correlates with the caveolin-1 content in various cell types. In cells with little caveolin-1, such as FRT and CaCo2 cells, VRAC has a very low current density. VRAC can also be activated under isovolumetric conditions. A decrease in intracellular ionic strength directly activates VRAC under conditions of constant osmolality and Cl^- concentration, a possible interpretation being an influence of ionic strength on protein tyrosine kinases (PTK). Intracellular dialysis with GTPγS also activates VRAC without any increase in cell volume, an effect probably mediated via small G-proteins. A possible candidate is the GTP-hydrolyzing RhoA-GTPase, since VRAC is down-regulated in cells pretreated with the *Clostridium limosum* exo-enzyme C3, which inactivates p21RhoA, and is enhanced by CNF1 (cytotoxic necrosis factor), a bacterial toxin that constitutively activates Rho GTPases. Inhibition of the RhoA-associated protein kinase (ROCK), a possible downstream target of p21 RhoA, with the specific blocker Y-27632, also down- regulates VRAC.

A tyrosine phosphorylation step downstream from the Rho-signaling cascade may be a critical event in the swelling-induced activation of the channel, since inhibitors of protein tyrosine kinases inhibit VRAC, whereas the protein tyrosine phosphatase (PTP) inhibitors Na_3VO_4 and dephosphatin potentiate it. It has recently been shown that the tyrosine kinase p56lck mediates activation of the swelling-induced Cl^- currents in lymphocytes. A scheme summarizing the possible steps involved in VRAC-gating is shown in Fig. 14.

The molecular nature of this elusive VRAC channel is still a matter of debate. P-glycoprotein, the product of the multidrug resistance 1 gene which functions as a drug transporter and possibly as an anion channel, has been proposed as a putative candidate for VRAC, but now even its role as a regulator of VRAC is questioned. A second candidate for VRAC was pI_{Cln}, where n refers to a putative extracellular nucleotide blocking site. Since the primary amino acid sequence of pI_{Cln} lacks transmembrane helices, it has been proposed that the pI_{Cln} channel is a homodimer consisting of four β strands which form an eight-stranded, antiparallel β-barrel transmembrane pore similar to that of porins. The most compelling evidence for pI_{Cln} being a plasma membrane channel was the presence of a GXGXG motif, being an essential component of the extracellular nucleotide binding site, which was believed to be essential for the nucleotide sensitivity of the VRAC current. It was reported that mutations of this site (G54A;G56A;G58A) prevent current inhibition by extracellular nucleotides, but these data could not be reproduced, which undermine the idea that pI_{Cln} is VRAC. Recent evidence has also shown that pI_{Cln} is a cytosolic rather than membrane protein that probably binds spliceosomes and regulates protein import into the nucleus. ClC-3, a membrane protein of the ClC family, is currently considered as a likely candidate for VRAC. Within the same family, ClC-2 is obviously volume-sensitive but its biophysical and pharmacological properties are completely different from those of VRAC. Also another candidate, the 72-amino-acid intrinsic membrane protein phospholemman with a single membrane-spanning domain is obviously not VRAC.

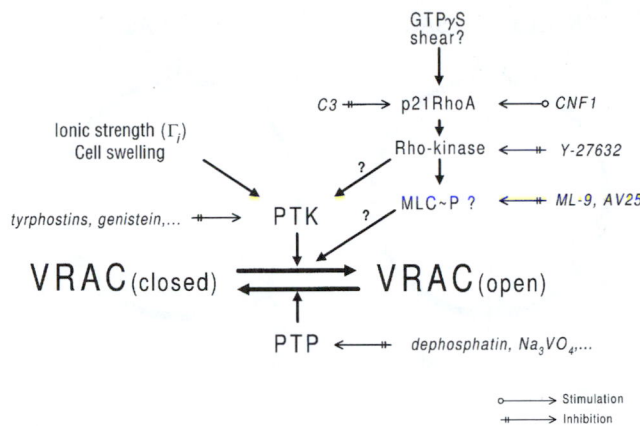

FIGURE 14. Gating Model of VRAC channels. Changes in intracellular ionic strength, Γ_i (likely due to water influx into the cell) or cell swelling activate VRAC probably via tyrosine protein kinase(s), PTK. VRAC can be blocked by diverse PTK inhibitors and can be potentiated—when activated—by inhibitors of protein tyrosine phosphatases (PTP). A second pathway might be involved during isovolumetric activation of VRAC. GTPγS and probably also shear stress (in endothelial cells) can activate a small G-protein, RhoA, which responds to downstream targets such as Rho-kinases. This pathway has been elucidated by use of specific activators or inhibitors of RhoA (CNF1, C3, respectively) and Rho-kinase (Y-27632). Downstream targets of Rho-kinase might involve myosin light chains. Rho-kinase inhibits MLC phosphatases, resulting in an increased level of phosphorylated MLC. This might be a step in VRAC activation. Inhibitors of myosin light chain kinase (MLCK), the specific peptide AV25, and ML-9 also inhibit VRAC.

III. Functional Role of Ion Channels in Nonexcitable Cells

A. Ion Channels and Cell Volume Regulation

Most if not all nonexcitable cells regulate their volume by activating either a **regulatory volume increase (RVI)** in shrunken cells or a **regulatory volume decrease (RVD)** in swollen cells. RVI is associated with an accumulation of $[Na^+]_i$ and $[Cl^-]_i$ and a concomitant entry of water to restore the original cell volume. RVD is induced by a cellular loss of K^+, Cl^-, and water.

Cell swelling opens stretch-activated channels (SACs), either K^+-selective or non-selective cation channels. K^+-selective SACs with a single-channel conductance between 30 to 50 pS are activated by stretch between 0 and -50 cm H_2O in most cells (e.g., in the basolateral renal proximal tubule), and induce a loss of intracellular K^+ which is necessary for RVD. Nonselective cation SACs are permeable for Ca^{2+} and provide a pathway for Ca^{2+} influx, which increases $[Ca^{2+}]_i$ and opens Ca^{2+}-activated K^+ channels. K^+ efflux associated with Cl^- efflux induces RVD. Quinine and quinidine block these K^+ channels and inhibit RVD. Both mechanisms for RVD mediated by SACs are schematically depicted in Fig. 15.

It is important to emphasize that RVD depends on coactivation of channels carrying ions of opposite charge and with a different equilibrium potential. This is illustrated in Fig. 16 for an epithelial cell from toad distal nephron in which cell swelling coactivates K^+ and Cl^- currents. Under isotonic conditions, cells are very tight and only a small current, I_{iso}, is

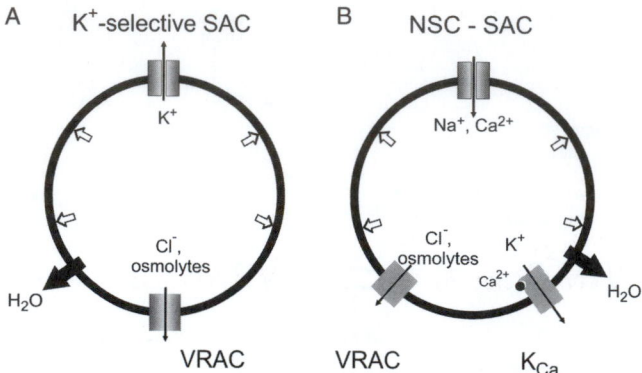

FIGURE 15. Involvement of ion channels in volume regulation. (A) Swelling of an epithelial cell activates K^+-selective stretch-activated ion channels (SACs) and possibly also Cl^- and osmolyte permeable volume-regulated anion channels (VRAC). Both events decrease the osmolality of the cell followed by outflow of water and decrease of the cell volume. (B) Cell swelling activates a Ca^{2+}-permeable nonselective SAC and VRAC. Ca^{2+} influx activates K^+ channels, likely BK_{Ca}, IK_{Ca}, SK_{Ca}. Both events induce loss of osmolytes and decrease the cell volume.

present (Fig. 16A). Swelling in hypotonic solution activates a large current (I_{hypo}) which represents the sum of a Cl^- and a K^+ current. The Cl^- component can be inhibited by the Cl^- channel blocker NPPB (a nitro-phenylpropylamino-benzoic acid derivative), the K^+ component by quinine (Fig.16B). The current in the presence of either blocker represents the current through the other channel, which reverses at its corresponding equilibrium potential. The cell depolarizes during swelling to the zero-current potential between E_K and E_{Cl}. There is no net current flow at this potential, but equal inward Cl^- and outward K^+ currents, and thus a Cl^- and K^+ osmolyte efflux. Block of either current component shifts the membrane potential towards the equilibrium potential of the other ion and thus prevents net efflux of both ions, which impedes RVD (Figs. 16C and D). It is only the coactivation of both currents that can induce RVD. Since VRAC is permeable for non-ionic osmolytes it also provides a voltage-independent pathway for efflux of osmolytes and RVD.

RVD and RVI do not only depend on activation of ion channels. The functional cooperation between volume-sen-

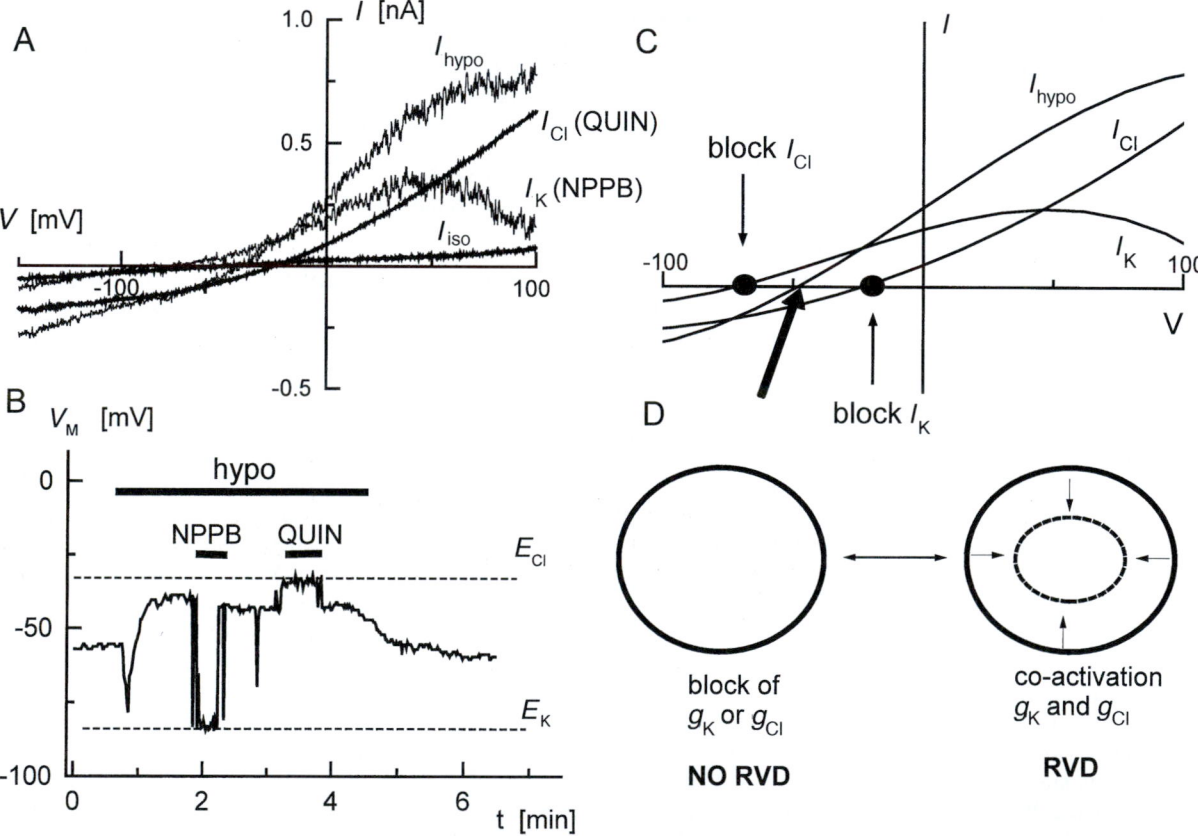

FIGURE 16. Coactivation of Cl^- and K^+ channels is necessary for regulatory-volume decrease (RVD). (A) A distal nephron epithelial cell (A6 cell from toad) is challenged with a 40% hypotonic solution. Swelling induces a large current (see change from I_{iso} to I_{hypo}). This current can be partially blocked by quinine (500 μM). The remaining current is a Cl^- current (I_{Cl}, QUIN, reversal potential close to $E_{Cl} = -15$ mV). The other component of the current can be blocked by NPPB (100 μM) and is a K^+ current (I_K, NPPB, reversal potential close to $E_K = -85$ mV). (B) Changes in membrane potential during cell swelling and after application of NPPB and quinine (for further explanation see text). (C) Schematic presentation of the swellings-activated currents. Block of one component will shift the membrane potential toward the reversal potential of the remaining component. Zero-current represents (thick arrow) inward Cl^- and outward K^+ currents, thus efflux of both osmolytes. (D) Block of one conductance will block RVD because of shifting the membrane potential to the reversal potential of the remaining component and thus inhibition of osmolyte efflux. Only **co-activation** of two currents can induce RVD.

sitive ion channels, cotransporters, and antiporters (e.g., Na^+-H^+ and Cl^--HCO_3^- antiporters, Na^+-K^+-$2Cl^-$ and K^+-Cl^- cotransporters) in the regulation of cell volume will be discussed in Chapter 21.

B. Role of Ion Channels in Vectorial Transport

Many functions of nonexcitable cells are associated with vectorial secretion and reabsorption. These functions are widespread in the body and only a few examples will be discussed here.

Na^+ reabsorption is a well-understood mechanism for vectorial transport that occurs in renal cells, intestine, colon, frog skin, and various other cell types. Ion channels and transporters are anchored by the cytoskeleton to strategically important sites in these cells. The ENaC channel is confined to the outer, apical, or luminal face, whereas Na^+-K^+ pumps and K^+ channels are localized at the inner, serosal, basolateral, or blood side (Fig. 17). The apical Na^+ channels are always open and provide an influx pathway for Na^+ into the cell. The Na^+ ions are extruded at the basolateral side by the Na^+,K^+-ATPase, which also transports K^+ into the cell that flows along its electrochemical gradient through a variety of K^+ channels in the basolateral membrane. This concerted action of ion channels and the Na^+-K^+ pump generates a pathway for Na^+ across the epithelial cell.

Cl^- secretion represents an example of the functional importance of Cl^- channels in controlling vectorial transport (Fig. 18). Acinar cells in secretory glands transport Cl^- against its electrochemical gradient from the interstitial side (blood side, basolateral side) into the cell mainly via the Na^+-K^+-$2Cl^-$ cotransporter (NKCC) fueled by the Na^+-gradient generated by the Na^+-K^+ pump. Na^+ and K^+ recycle via the Na^+-K^+ pump and basolateral K^+ channels. Cl^- secretion can be activated by

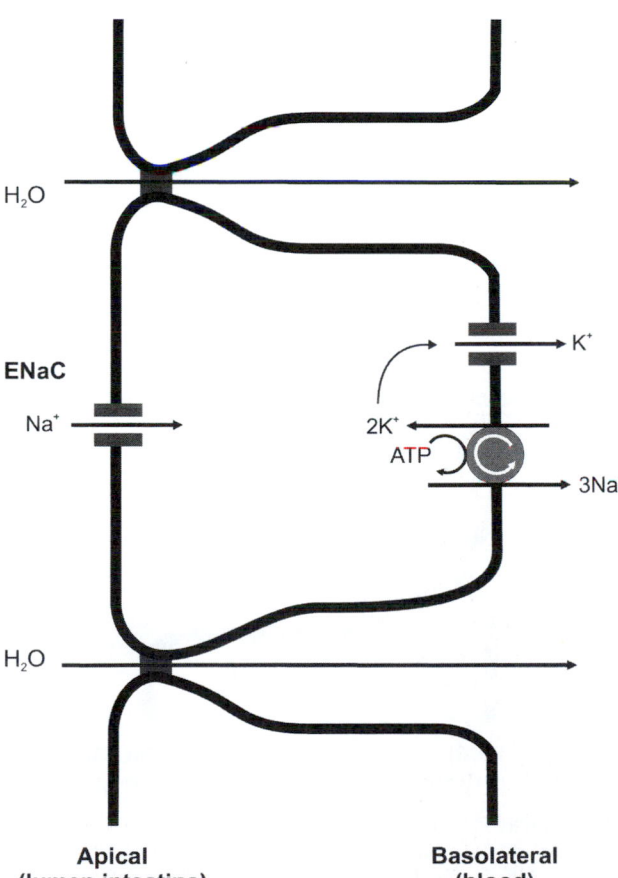

FIGURE 17. Na^+ reabsorption. In Na^+-reabsorbing epithelial cells, amiloride-sensitive Na^+ channels (ENaC) are located at the outer, mucosal, apical face. Na^+-K^+ pumps are mainly localized at the inner, serosal, basolateral face. This topology creates a Na^+ sink. K^+ recycles via pump and basolateral K^+ channels, e.g., ROMK.

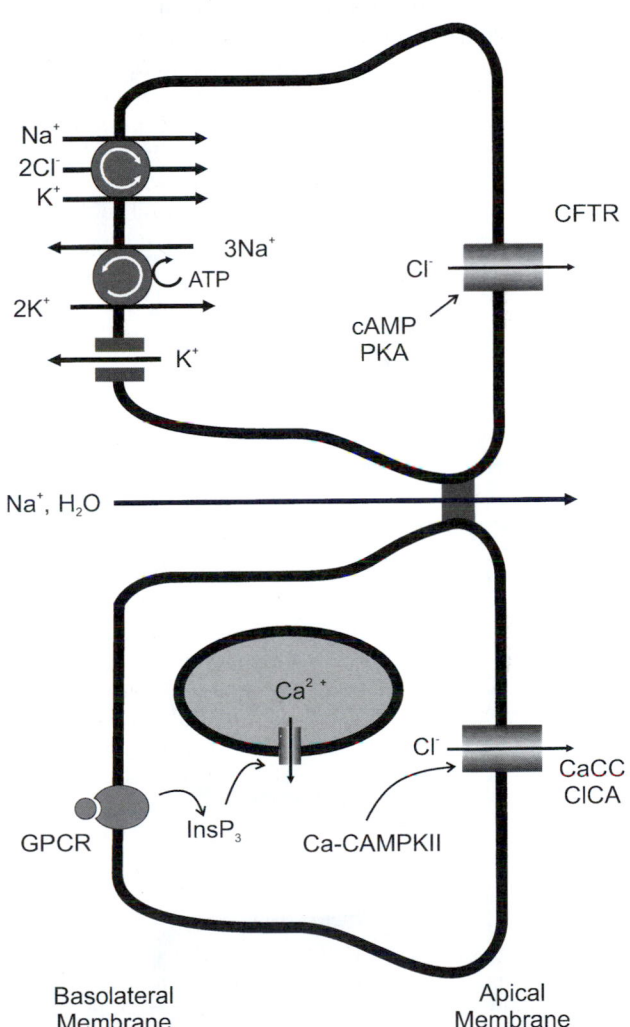

FIGURE 18. Transcellular transport of Cl^- in secretory epithelial cells–Transcellular Cl^- transport results from the asymmetric distribution of Na^+-K^+-$2Cl^-$ cotransporters, mainly located at the blood or interstitial site (abluminal), and cAMP-gated (CFTR) or Ca^{2+}-gated (CaCC, CLCA) Cl^- channels at the luminal side. Cl^- entry proceeds via the abluminal Na^+-K^+-$2Cl^-$ cotransporter, luminal secretion either via cAMP-activated CFTR channels stimulated by PKA (top) or via Ca^{2+}-activated Cl^- channels (bottom). The elevation in $[Ca^{2+}]_i$ occurs via release from intracellular stores due to activation of G-protein-coupled receptors (GPCR). Cl^- movement is passively followed by water flux through a paracellular pathway (for explanation see text).

two mechanisms: one is cAMP-dependent, the other Ca^{2+}-dependent. CFTR channels at the apical side of epithelial cells are stimulated by cAMP-mediated phosphorylation. Opening of these cAMP-activated Cl^- channels induces a transcellular transport of Cl^-. Defective CFTR channels in patients with cystic fibrosis, a secretory malfunction (see Section II.F.4), cannot be activated even if the secretory stimulus still generates the same amount of cAMP. Therefore the secretion and the coupled transport of water and Na^+ no longer occur in these patients. The opposite effect occurs during overproduction of cAMP in the presence of the Gs-protein activating cholera toxin. Secretory Cl^- channels are maximally activated, causing a massive secretion of fluid toward the luminal side.

The other mechanism that activates Cl^- secretion is initiated by the binding of agonists, e.g., acetylcholine or cholecystokinin, to their respective G-protein coupled receptors. Activation of these receptors releases Ca^{2+} from intracellular

$Ins(1,4,5)P_3$ sensitive stores and increases $[Ca^{2+}]_i$, which opens Cl^- channels at the luminal side (CaCC, CLCA, see Section II.F.3). These channels drain Cl^- into the lumen of the secretory duct, increasing the abluminal Cl^- gradient and activating NKCC. The inwardly transported Na^+ and K^+ recycle via the Na^+-K^+ pump and presumably via Ca^{2+}-activated K^+ channels in both luminal and abluminal membranes. The vectorial transport of Cl^- would be accompanied by a passive movement of Na^+ and H_2O through a paracellular pathway (see Fig. 18).

Approximately 50% of the filtered Cl^- is reabsorbed in the proximal tubules. An important component of the **Cl^- reabsorption** is passive and paracellular, and driven by the electrochemical gradient (Fig. 19A). In addition, Cl^--formate (and also Cl^--oxalate) exchangers have been identified that transport Cl^- uphill through the apical membrane. The recycling of formate (oxalate) occurs by an H^+-coupled transport in parallel with the Na^+-H^+ exchanger. Cl^- exits at the basolateral

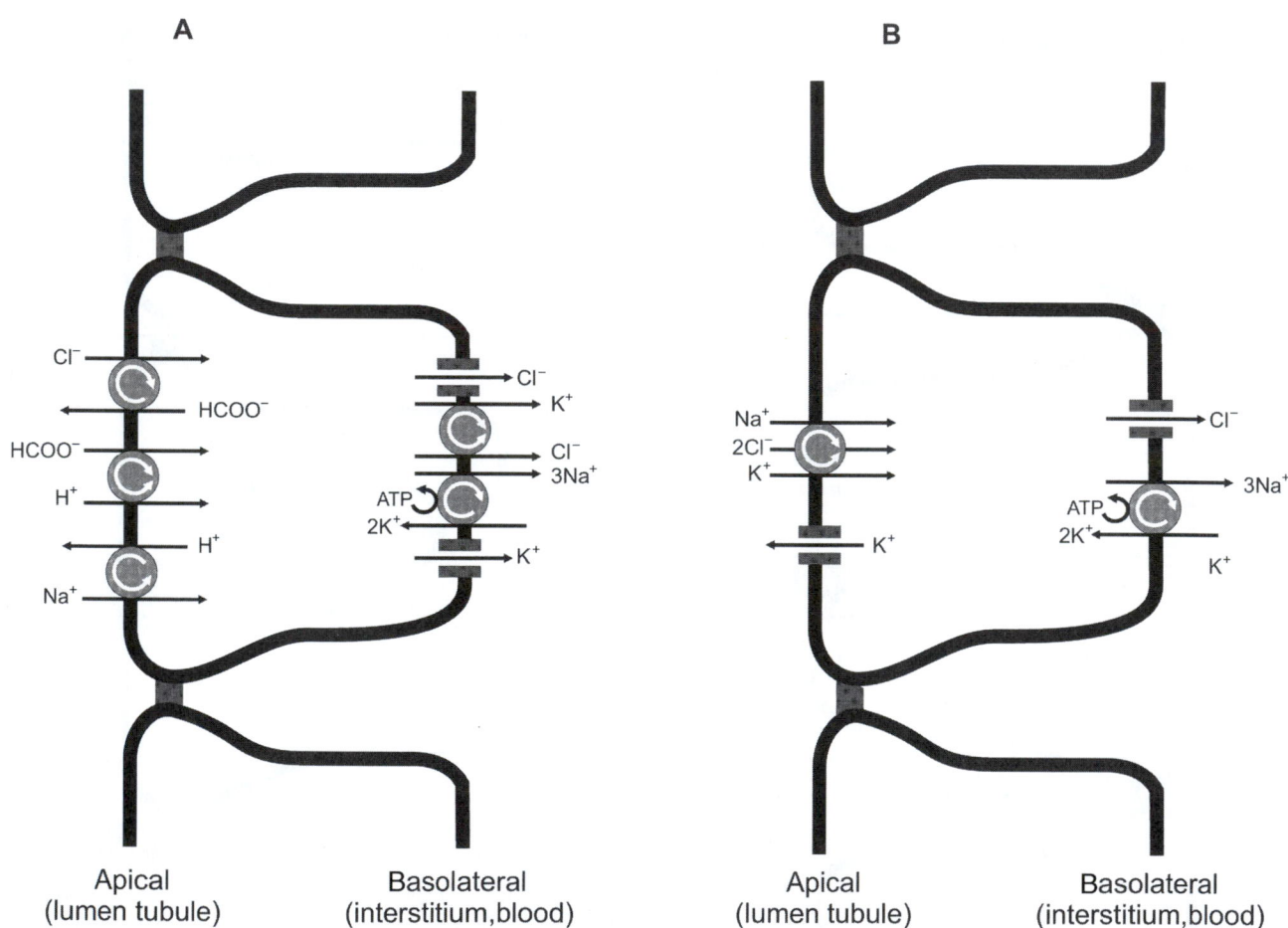

A	B
Apical (lumen tubule) Basolateral (interstitium, blood)	Apical (lumen tubule) Basolateral (interstitium, blood)

FIGURE 19. Cl^- reabsorption in the kidney. (A) Proximal tubule. Cl^- is reabsorbed from the luminal side facing the tubule fluid via exchange with a formate transporter ($HCOO^-$) and oxalate transporter. Formate (or oxalate, not shown) is transported into the cell via a H^+-$HCOO^-$ cotransporter. H^+ recycles in exchange for Na^+ transported by the Na^+-H^+ exchanger, NHE. The transcellular transport of Cl^- at the basolateral (abluminal) side proceeds via passive transport through Cl^- channels, most likely ClC-Ka and b. In addition, Cl^- leaves the cells via a K^+-Cl^- cotransporter. Na^+ is also secreted through the basolateral side via cooperative action of NHE and the Na^+,K^+-ATPase. K^+ recycles via pump and K^+ channel. (B) Distal tubule. (TAL: thick ascending limb; DCT: distal convolute tubules). At the luminal side Cl^- is cotransported with Na^+ and K^+ via the NKCC transporter. K^+ leaves the cells via ROMK1, a member of the Kir family. At the basolateral side, Cl^- exits the cells via the ClC-Kb channel. Na^+ passages through the cells via NKCC and the Na^+, K^+-ATPase. ROMK1, NKCC, and ClC-Kb are targets in Bartter's syndrome.

membrane via channels (possibly ClC-Ka,b) and the K^+-Cl^- cotransporter. Na^+ which enters the cell at the apical side via the Na^+-H^+ transporter exits through the basolateral membrane via the Na^+-K^+ pump, and K^+ recycles via basolateral K^+ channels (see Fig. 19). Therefore, a net Cl^- uptake is driven by the combined action of various exchangers, cotransporters, and pumps. As already mentioned, Cl^- secretion in the distal tubule (TAL, thick ascending limb, and DCT distal convolute tubules) is different: Cl^- enters the cell via NKCC, K^+ recycles through ROMK. The abluminal transport of Cl^- and Na^+ occurs via ClC-Kb channels and via NKCC and the basolateral Na^+-K^+ pump respectively. Defects in NKCC, ROMK, and ClC-Kb induce hypokalemic acidosis, low blood pressure, and renal salt-wasting, but also hypercalciuria by a mechanism not yet understood (see Fig. 19B).

The total blood Ca^{2+} includes a fraction bound to plasma proteins (albumin) and an unbound fraction which is ultra-filtrated. The ultrafiltrated fraction contains Ca^{2+} complexed by anions and free Ca^{2+}. Importantly, part of the free Ca^{2+} is reabsorbed in the kidney and the intestine. Vitamin D and PTH (parathyroid hormone) activate **Ca^{2+} reabsorption**. Transepithelial uptake of Ca^{2+} seems to be a three-step process consisting of Ca^{2+} entry into the cell down its electrochemical gradient, binding to calbindin, a 28-kDa Ca^{2+}-binding protein, and active extrusion out of the cell via a basolateral plasma membrane Ca^{2+}-pump (PMCa). Part of the active Ca^{2+} reuptake may also occur by a Na^+-Ca^{2+} ex-changer (Fig. 20). Recently a channel called EcaC has been cloned which is closely related structurally to the trp channel family and may be responsible for Ca^{2+} entry at the apical membrane.

Channel co-operation, as a nice example of the cooperation between ion channels in the organization of vectorial transport, has been proposed by E. Neher (Fig. 21). A rise in $[Ca^{2+}]_i$ in a stimulated pancreatic acinar cell opens Ca^{2+}-activated Cl^- channels at the luminal side, leading to a Cl^- efflux down its electrochemical gradient. At the abluminal side this rise in $[Ca^{2+}]_i$ opens Ca^{2+}-activated nonselective cation channels and Ca^{2+}-activated Cl^- channels, which depolarizes the abluminal membrane and reverses the electrochemical gradient for Cl^-, leading to a Cl^- influx. Because

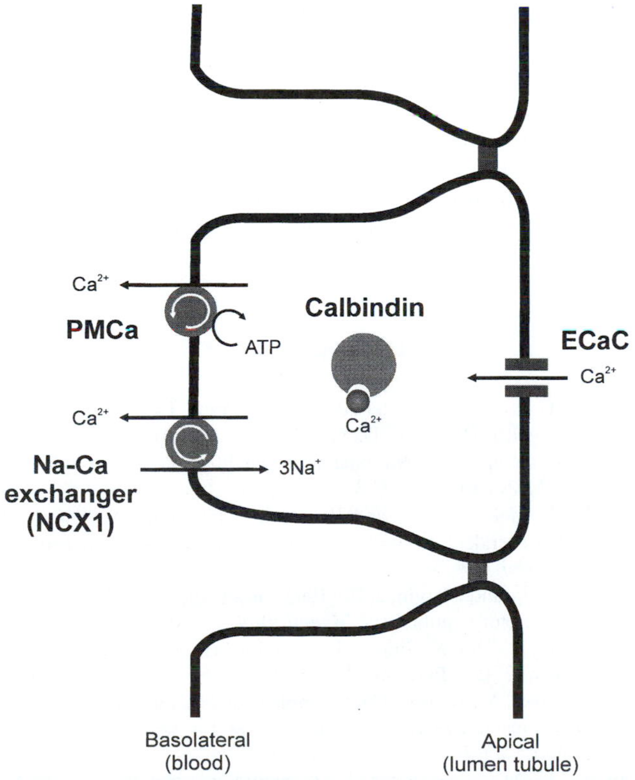

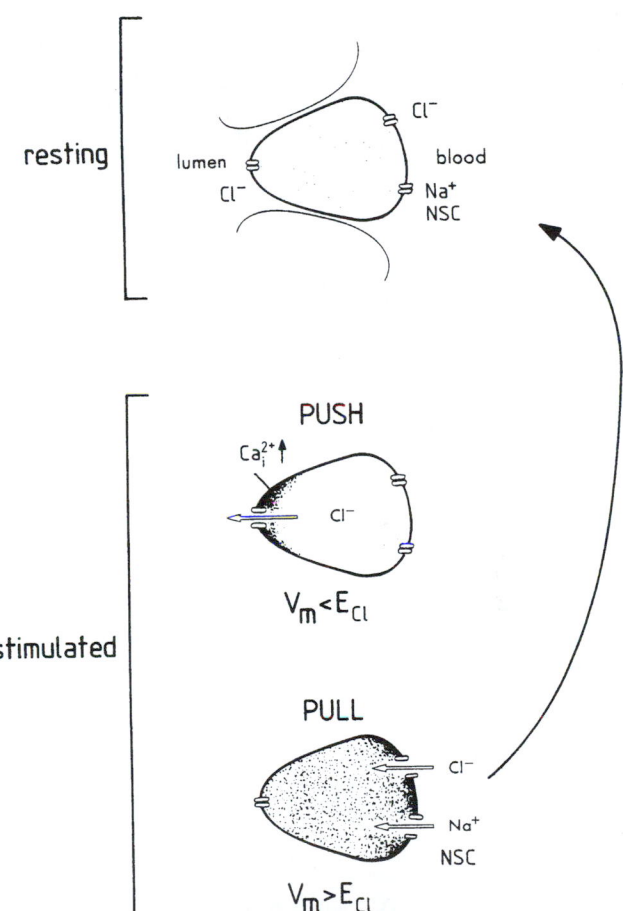

FIGURE 21. The push-pull hypothesis of vectorial transport. In a resting secretory pancreas acinus cell Cl^- channels and Ca^{2+} dependent NSCs are closed. Stimulation by a secretagogue induces first a luminal increase in $[Ca^{2+}]_i$ that activates Cl^- channels and induces a Cl^- efflux via CaCC, because the luminal membrane potential is more negative than the Cl^- equilibrium potential. The abliminal channels are still closed. A Ca^{2+} wave reaches the abluminal site with a delay. $[Ca^{2+}]_i$ is now higher at the abluminal side. Opening of Ca^{2+}-dependent NSC and Cl^- channels at the abluminal side depolarizes the membrane to a value more positive than the Cl^- equilibrium potential, resulting in Cl^- influx. The luminal channels are closed again. Sequestration of Ca^{2+} leads again to the resting state. [Modified after Kasai and Augustine (1990), with permission from *Nature* 348: 735–738 (1990), copyright 1990 Macmillan Magazines Ltd.]

FIGURE 20. Reabsorption of Ca^{2+} in kidney and intestine. Ca^{2+} enters the epithelial cell passively via the recently cloned ECaC channel, binds to calbindin, and is extruded at the abluminal side via the PMCa Ca^{2+} pump and probably also via the Na^+-Ca^{2+} exchanger.

Ca^{2+}-waves originate at the luminal side and reach the abluminal side with some delay, these events occur sequentially, that is, Cl^- is first "pushed" into the lumen and then "pulled" into the cell from the abluminal side (**push-pull model**).

C. Role of Ion Channels in Ca^{2+}-Signaling

Many cellular functions modulated by agonists are triggered by changes in intracellular Ca^{2+}, consisting of an initial transient Ca^{2+} rise generated by Ca^{2+} release from $Ins(1,4,5)P_3$-sensitive Ca^{2+} stores followed by a more sustained Ca^{2+}-plateau coupled to Ca^{2+} influx. It is mainly the latter increase that is essential to maintain, for example, the secretion or synthesis of various messengers. An important example is the Ca^{2+}-dependent synthesis of endothelium-derived relaxing factor EDRF. This factor, identical with the simple chemical molecule nitric oxide (NO), acts as an activator of the soluble guanylate cyclase that increases the intracellular cGMP concentration in neighboring cells. EDRF-NO synthesis and release are triggered during a long lasting elevation of $[Ca^{2+}]_i$ via a Ca^{2+}-calmodulin-NADPH-dependent NO synthase that catalyses oxidation of the N-guanidine terminus of L-arginine, and releases the easily diffusable NO. This sustained Ca^{2+} influx occurs via Ca^{2+}-selective channels (DACs, SOCs, CRACs) or via Ca^{2+}-permeable nonselective ion channels. Activation of Ca^{2+}-activated NSCs creates positive feedback, since the increase of $[Ca^{2+}]_i$ via these channels would further activate them. Another feedback is the opening of Ca^{2+}-activated K^+ channels, which induces hyperpolarization and increases the driving force for Ca^{2+} influx.

The renin secretion from juxtaglomerular epitheloid cells is another exciting example of cooperation between changes in $[Ca^{2+}]_i$ and ion channels. Exocytosis of renin in these cells is enhanced by swelling, but inhibited by an increase of $[Ca^{2+}]_i$. Increased $[Ca^{2+}]_i$ activates Ca^{2+}-dependent Cl^- channels and depolarizes these cells by shifting the membrane potential to the Cl^- equilibrium potential, which in turn promotes K^+ efflux through K^+ channels. This coactivation of K^+ and Cl^- channels decreases the osmolarity of the cell, which causes it to shrink and reduces renin secretion. Stimulation of these cells, for example, with angiotensin II (AT II), increases $[Ca^{2+}]_i$, which exerts negative feedback on the renin secretion.

IV. Summary

Ion channels in nonexcitable cells have been less extensively studied at the functional level than the channels involved in fast signaling processes via voltage-operated ion channels. Also, the molecular biological information about these channels is rather restricted. Nevertheless, these channels appear to be involved in the control of biologically important functions.

We have discussed several of these channels in this chapter:

- Na^+-selective, amiloride-blocked channels (ENaC) which regulate Na^+ absorption, and Na^+ secretion in epithelial cells

- Nonselective cation channels (NSC) that provide a pathway for Ca^{2+} entry and cause depolarization. They can be activated by an increase in $[Ca^{2+}]_i$ or by the binding of cyclic nucleotides. Recently, a voltage- and cyclic-nucleotide-gated family of hyperpolarization-activated channels (HAC) has been discovered. Other nonselective cation channels can be activated by extracellular agonists indirectly via G-protein dependent mechanisms or directly by extracellular messengers, such as ATP (P_{2X} family), or by intracellular cyclic nucleotides (e.g., cGMP, cAMP)

- K^+ channels, especially those activated by intracellular Ca^{2+} (voltage-dependent small-, intermediate-, and big-conductance channels), which modulate cellular signals caused by various agonists, and inwardly-rectifying K^+ channels

- Cl^- channels, which are mainly involved in secretory functions. They are activated by an increase in $[Ca^{2+}]_i$ (CaCCs or the recently identified CACL family), by cAMP via PKA-dependent protein kinase, or by changes in cell volume. Some of them are voltage-dependent (ClC)

- Most nonexcitable cells have developed a Ca^{2+} entry mechanism dependent on intracellular Ca^{2+} stores (SOCs, DACs, CRAC) and are responsible for long-lasting Ca^{2+} signals. Putative molecular candidates may belong to the recently described trp-channel family

- Various mechanically activated ion channels (K^+ channels, nonselective cation channels, Cl^- channels) involved in, for example, volume-regulation of many cells, and in shear-stress mediated release of EDRF in endothelial cells

We have also addressed the functional role of the ion channels in volume regulation and fine-tuning of the driving force for Ca^{2+} influx, and their involvement in vectorial transport, which is intimately linked to their localization at strategic sites of the cells and usually requires the cooperation of several channel types.

Bibliography

Abriel, H., Loffing, J., Rebhun, J. F., Pratt, J. H., Schild, L., Horisberger, J. D., Rotin, D., and Staub, O. (1999). Defective regulation of the epithelial Na^+ channel by Nedd4 in Liddle's syndrome. *J. Clin. Invest.* **103,** 667–673.

Arreola, J., Melvin, J. E., and Begenisich, T. (1996). Activation of calcium-dependent chloride channels in rat parotid acinar cells. *J. Gen. Physiol.* **108,** 35–47.

Begenisich, T., and Melvin, J. E. (1998). Regulation of chloride channels in secretory epithelia. *J. Membr. Biol.* **163,** 77–85.

Birnbaumer, L., Zhu, X., Jiang, M. S., Boulay, G., Peyton, M., Vannier, B., Brown, D., Platano, D., Sadeghi, H., Stefani, E., and Birnbaumer, M. (1996). On the molecular basis and regulation of cellular capacitative calcium entry: Roles for Trp proteins. *Proc. Natl. Acad. Sci. USA* **93,** 15195–15202.

Canessa, C. M., Horisberger, J. D., Schild, L., and Rossier, B. C. (1995). Expression cloning of the epithelial sodium channel. *Kidney Int.* **48,** 950–955.

Davies, P. F. (1995). Flow-mediated endothelial mechanotransduction. *Physiol. Rev.* **75,** 519–560.

Davies, P. F., and Tripathi, S. C. (1993). Mechanical stress mechanisms and the cell. An endothelial paradigm. *Circ. Res.* **72**, 239–245.

Foskett, J. K. (1998). ClC and CFTR chloride channel gating. *Annu. Rev. Physiol.* **60**, 689–717.

Gadsby, D. C., and Nairn, A. C. (1999). Control of CFTR channel gating by phosphorylation and nucleotide hydrolysis. *Physiol. Rev.* **79**, S77–S107.

Gandhi, R., Elble, R. C., Gruber, A. D., Schreur, K. D., Ji, H. L., Fuller, C. M., and Pauli, B. U. (1998). Molecular and functional characterization of a calcium-sensitive chloride channel from mouse lung. *J. Biol. Chem.* **273**, 32096–32101.

George, A. L., Jr. (1998). Chloride channels and endocytosis: ClC-5 makes a dent. *Proc. Natl. Acad. Sci. USA* **95**, 7843–7845.

Gruber, A. D., Elble, R. C., Ji, H. L., Schreur, K. D., Fuller, C. M., and Pauli, B. U. (1998). Genomic cloning, molecular characterization, and functional analysis of human CLCA1, the first human member of the family of Ca^{2+}-activated Cl^- channel proteins. *Genomics* **54**, 200–214.

Gruber, A. D., and Pauli, B. U. (1999). Molecular cloning and biochemical characterization of a truncated, secreted member of the human family of Ca^{2+}-activated Cl^- channels. *Biochim. Biophys. Acta.* **1444**, 418–423.

Gruber, A. D., Schreur, K. D., Ji, H. L., Fuller, C. M., and Pauli, B. U. (1999). Molecular cloning and transmembrane structure of hCLCA2 from human lung, trachea, and mammary gland. *Am. J. Physiol.* **45**, C1261–C1270.

Hamill, O. P., and McBride, D. W., Jr. (1994). The cloning of a mechano-gated membrane ion channel. *Trends Neurosci.* **17**, 439–443.

Hille, B. (1992). "Ionic Channels of Excitable Membranes." Sinauer Associates, Sunderland, MA.

Hoenderop, J. G. J., van der Kemp, A., Hartog, A., van de Graaf, S. F. J., van Os, C. H., Willems, P., and Bindels, R. J. M. (1999). Molecular identification of the apical Ca^{2+} channel in 1,25-dihydroxyvitamin D-3–responsive epithelia. *J. Biol. Chem.* **274**, 8375–8378.

Hoffmann, E. K., and Dunham, P. B. (1995). Membrane mechanisms and intracellular signalling in cell volume regulation. *Int. Rev. Cytol.* **161**, 173–262.

Hofmann, T., Obukhov, A. G., Schaefer, M., Harteneck, C., Gudermann, T., and Schultz, G. (1999). Direct activation of human TRPC6 and TRPC3 channels by diacylglycerol. *Nature* **397**, 259–263.

Jan, L. Y., and Jan, Y. N. (1997). Cloned potassium channels from eukaryotes and prokaryotes. *Annu. Rev. Neurosci.* **20**, 91–123.

Jensen, B. S., Strobaek, D., Christophersen, P., Jorgensen, T. D., Hansen, C., Silahtaroglu, A., Olesen, S. P., and Ahring, P. K. (1998). Characterization of the cloned human intermediate-conductance Ca^{2+}-activated K^+ channel. *Am. J. Physiol.* **275**, C848–C856.

Jentsch, T. J., Friedrich, T., Schriever, A., and Yamada, H. (1999). The CLC chloride channel family. *Pflugers Arch. Eur. J. Physiol.* **437**, 783–795.

Kasai, H., and Augustine, G. J. (1990). Cytosolic Ca^{2+} gradients triggering unidirectional fluid secretion from exocrine pancreas. *Nature* **348**, 735–738.

Kellenberger, S., Gautschi, I., Rossier, B. C., and Schild, L. (1998). Mutations causing Liddle syndrome reduce sodium-dependent downregulation of the epithelial sodium channel in the *Xenopus* oocyte expression system. *J. Clin. Invest.* **101**, 2741–2750.

Kirk, K., and Strange, K. (1998). Functional properties and physiological roles of organic solute channels. *Annu. Rev. Physiol.* **60**, 719–739.

Kunzelmann, K., and Schreiber, R. (1999). CFTR, a regulator of channels. *J. Membr. Biol.* **168**, 1–8.

Kurtz, A. (1990). Do calcium-activated chloride channels control renin secretion? *News Physiol. Sci.* **5**, 43–46.

Lansman, J. B., Hallam, T. J., and Rink, T.J. (1987). Single stretch-activated ion channels in vascular endothelial cells as mechanotransducers? *Nature* **235**, 811–813.

Ludwig, A., Zong, X., Jeglitsch, M., Hofmann, F., and Biel, M. (1998). A family of hyperpolarization-activated mammalian cation channels. *Nature* **393**, 587–591.

Montell, C. (1998). TRP trapped in fly signaling web. *Curr. Opin. Neurobiol.* **8**, 389–397.

Nilius, B., Eggermont, J., Voets, T., Buyse, G., Manolopoulos, V., and Droogmans, G. (1997). Properties of volume-regulated anion channels in mammalian cells. *Progress Biophysics and Molec. Biol.* **68**, 69–119.

Nilius, B., Sehrer, J., De Smet, P., Van Driessche, W., and Droogmans, G. (1995). Volume regulation in a toad epithelial cell line: Role of coactivation of K^+ and Cl^- channels. *J. Physiol. (London)* **487**, 367–378.

Nilius, B., Viana, F., and Droogmans, G. (1997). Ion channels in vascular endothelium. *Annu. Rev. Physiol.* **59**, 145–170.

North, R. A. (1996). Families of ion channels with two hydrophobic segments. *Curr. Opin. Biol.* **8**, 474–483.

Okada, Y. (1997). Volume expansion-sensing outward-rectifier Cl^- channel: Fresh start to the molecular identity and volume sensor. *Am. J. Physiol.* **273**, C755–89.

Oleson, S.-O., Clapham, D. E., and Davies, P. F. (1988). Haemodynamic shear stress activates a K^+ current in vascular endothelial cells. *Nature* **331**, 168–170.

Parekh, A. B., and Penner, R. (1997). Store depletion and calcium influx. *Physiol. Rev.* **77**, 901–930.

Rossier, B. C., Canessa, C. M., Schild, L., and Horisberger, J. D. (1994). Epithelial sodium channels. *Curr. Opin. Nephrol. Hypertens.* **3**, 487–496.

Schwiebert, E. M., Benos, D. J., Egan, M. E., Stutts, M. J., and Guggino, W. B. (1999). CFTR is a conductance regulator as well as a chloride channel. *Physiol. Rev.* **79**, S145–S166.

Sheppard, D. N., and Welsh, M. J. (1999). Structure and function of the CFTR chloride channel. *Physiol. Rev.* **79**, S23–S45.

Strange, K., Emma, F., and Jackson, P. S. (1996). Cellular and molecular physiology of volume-sensitive anion channels. *Am. J. Physiol.* **270**, C711–C730.

Tsien, R. W., and Tsien, R. Y. (1990). Calcium channels, stores, and oscillations. *Annu. Rev. Cell Biol.* **6**, 715–760.

Zhu, X., and Birnbaumer, L. (1998). Calcium Channels Formed by Mammalian Trp Homologues. *News Physiol. Sci.* **13**, 211–217.

A. Liévano and A. Darszon

30

Ion Channels in Sperm

I. Introduction

Fertilization is one of the most important biological events. It allows not only the generation of a new individual, but an opportunity for genetic recombination as well as an increased probability of species preservation over time. Complex signaling events between gametes are required for their final encounter and fusion, which starts a developmental program. The spermatozoa, a motile cell playing one of the key roles in this process, is very specialized. It lacks the machinery for protein or nucleic acid synthesis and has only a nucleus, mitochondria, a flagellum, an acrosomal vesicle, and a centriole pair. Figure 1A depicts sperm from the sea urchin (upper drawing) and from the mouse (lower drawing).

Sperm, like other types of cells, have ion channels that participate in fundamental responses to the outer layer of the egg that are required for fertilization. It has been shown that the flow of ions through the plasma membrane of sperm, particularly Ca^{2+}, participates crucially in the events leading to fertilization (Schackmann, 1989; Florman et al., 1998; Darszon et al., 1999). Indeed, sperm quickly respond to components of the outer layer of the egg with changes in their plasma membrane ion permeability. During its life span, a sperm must respond to different stimuli, thereby changing behavior until fusion is achieved with the egg to create a zygote.

II. Sperm Responses to Egg Components

A. Sea Urchin Sperm

Sea urchins are among the most useful and better known experimental models in fertilization because they undergo external fertilization in a simple medium (seawater) and each male can spawn about 10^{10} cells, making it possible to have large amounts of biological material for biochemical and biophysical manipulations. Furthermore, the cells respond to environmental stimuli rapidly (in seconds), synchronously, and in a compulsory order.

Sea urchin sperm are metabolically arrested in semen. When spawned into seawater, their respiration and motility quickly activate, and chemotaxis to egg components may help them to find the egg (rev. in Garbers, 1989; Darszon et al., 1999). Contact with the **egg jelly** induces dramatic morphophysiological changes in sperm through a complex process called the **acrosome reaction (AR)**, which is necessary for fertilization. The most conspicuous morphological change in this reaction is the exocytosis of the **acrosomal vesicle** and the extension of the **acrosomal tubule** (Fig. 1B, upper panel). The AR leads to the release of hydrolytic enzymes present in

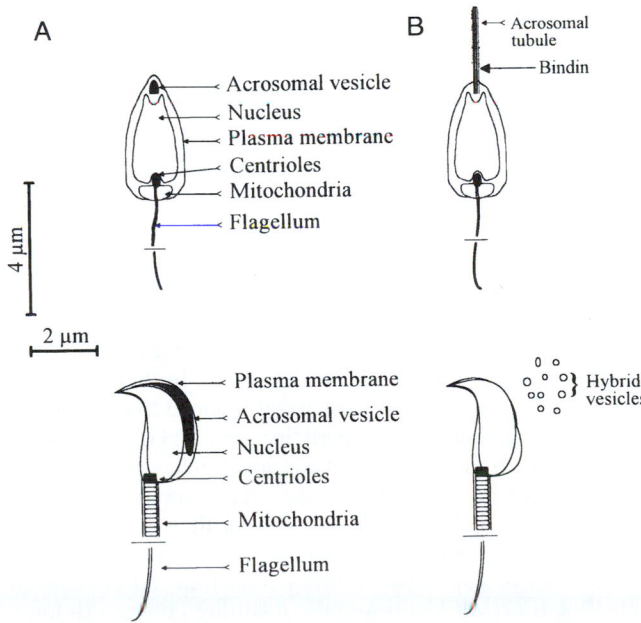

FIGURE 1. Schematic diagram of a sea urchin (upper half) and mouse (lower half) sperm (A) before and (B) after acrosome reaction. The sperm head contains all the cell organelles except the flagellum.

509

the acrosome that are required for sperm penetration of the egg's external layers, and to the exposure of new membrane surfaces specialized for fusion with the egg plasma membrane. When sperm undergo the AR, a protein called **bindin** (see Fig. 1B) is exposed and interacts species specifically with components in the egg. The AR is triggered by plasma membrane ion permeability changes that lead to modifications in membrane potential (E_m), intracellular pH (pH_i), and intracellular Ca^{2+} concentration ($[Ca^{2+}]_i$).

1. Responses to Egg Peptides

Small peptides contained in the egg jelly surrounding the sea urchin egg profoundly influence sperm physiology. For instance, pM concentrations of **speract,** a decapeptide isolated from *S. purpuratus* egg jelly, and **resact,** a similar peptide isolated from *Arbacia punctulata*, stimulate sperm phospholipid metabolism and respiration in sperm suspended in acidified seawater. At nM concentrations these peptides stimulate $^{22}Na^+$ and $^{45}Ca^{2+}$ uptake, H^+ and K^+ efflux, and increases in cGMP, cAMP, $[Ca^{2+}]_i$, and pH_i (Garbers, 1989; Schackmann, 1989; Ward and Kopf, 1993). They also regulate sperm motility (Ward *et al.*, 1985; Cook *et al.*, 1994) and appear to enhance fertilization (Suzuki and Yoshino, 1992). Chemotaxis has been demonstrated only in *A. punctulata*, where nM concentrations of resact attract sperm in a Ca^{2+}-dependent manner (Ward *et al.*, 1985).

Cross-linking experiments indicate that speract binds to a 77-kDa sperm plasma membrane protein in the flagella of *S. purpuratus* sperm; this protein was purified, sequenced, and cloned. It has been proposed that this receptor protein modulates a membrane guanylyl cyclase (rev. in Garbers, 1989). In *A. punctulata*, resact binds to a membrane-bound guanylyl cyclase, transiently activating it and changing its phosphorylation state (Trimmer and Vacquier, 1986). The phosphorylation state of guanylyl cyclase is pH_i- and $[Na^+]_o$-dependent (Trimmer and Vacquier, 1986; Garbers, 1989). The resact receptor is the first cloned and sequenced member of a family of guanylyl cyclases that are surface receptors participating in a new signal transduction pathway (Drewett and Garbers, 1994).

Speract triggers a transient hyperpolarization in *S. purpuratus* sperm flagella, in flagellar membranes, and in sperm, probably activating cGMP-modulated K^+ channels (rev. in Darszon *et al.*, 1999). This hyperpolarization stimulates a voltage-dependent Na^+-H^+ exchange in swollen (Babcock *et al.*, 1992; Reynaud *et al.*, 1993; Cook and Babcock, 1993a) and nonswollen sperm (Lee, 1984a, b; Schackmann and Chock, 1986), even though its Na^+:H^+ stoichiometry was estimated to be 1:1 (Lee, 1984a; Lee and Garbers, 1986). In flagellar membranes, GTPγS stimulates the speract-induced hyperpolarization, suggesting the possible participation of a G protein (Lee, 1988). Though G_i, G_s, and several low–molecular weight G proteins have been detected in sea urchin sperm (Ward and Kopf, 1993; Cuéllar-Mata *et al.*, 1995), their role in sperm physiology has not been established.

The relationship between the speract-induced increase in cGMP and cAMP levels and the resulting changes in ionic permeability of the sperm holds the key to understanding how sea urchin gametes manage to meet. Patch-clamping sea urchin sperm is difficult due to their tiny size (see Fig.

5C; Guerrero *et al.*, 1987); thus, to facilitate clamping, they must be swollen in diluted seawater (see Fig. 5D; Babcock *et al.*, 1992). Swollen sperm are spherical. (ca. 4 μm diameter) and retain their E_m, pH_i, and $[Ca^{2+}]_i$ regulation (Fig. 2A). Picomolar concentrations of speract activate a K^+-selective channel in swollen sperm, as indicated by patch-clamp experiments (Babcock *et al.*, 1992). It is not known how the increase in [cGMP] opens the TEA^+-insensitive K^+-selective channels that hyperpolarize sperm. Higher speract concentrations (>25 pM) transiently hyperpolarize the cells near the K^+ equilibrium potential. (E_K) and immediately repolarize them towards the resting potential. (E_r; Fig. 2A, upper left panel). As mentioned earlier, the hyperpolarization activates Na^+-H^+ exchange (Lee and Garbers, 1986; Gonzalez-Martinez *et al.*, 1992; Reynaud *et al.*, 1993). The increase in pH_i inhibits guanylyl cyclase (Trimmer and Vacquier, 1986) and stimulates adenylyl cyclase (Cook and Babcock, 1993a, b), which is also activated by a membrane potential hyperpolarization (Beltrán *et al.*, 1996) and increases in $[Ca^{2+}]_i$ (Garbers, 1989). The decrease in [cGMP] would diminish K^+ permeability (Cook and Babcock, 1993a) and contribute to repolarizing sperm.

In addition, nM concentrations of speract increase $[Ca^{2+}]_i$ and induce a Ca^{2+}-dependent depolarization beyond E_r in swollen sperm (see Fig. 2A; Babcock *et al.*, 1992; Reynaud *et al.*, 1993; Cook and Babcock, 1993a). These changes are inhibited by Ca^{2+}-channel–blockers like Co^{2+}, Ni^{2+}, and Zn^{2+} (Reynaud *et al.*, 1993; Cook and Babcock, 1993b). These Ca^{2+}-permeable channels allow Mn^{2+} to pass through and are regulated by cAMP (Cook and Babcock, 1993b). In normal sperm the hyperpolarization is small and fast and the depolarizing phase (1) is only partly diminished in the absence of external Ca^{2+}, (2) depends on external Na^+, (3) is poorly sensitive to Ca^{2+} channel-blockers, and (4) is partially blocked by high concentrations of TEA^+ and Ba^{2+} (Labarca *et al.*, 1996, 1997). Two (or more) ion channels with distinct selectivity and pharmacology might contribute to the depolarization triggered by speract in normal sea urchin sperm: a cAMP- and/or pH_i-regulated Ca^{2+} channel (Babcock *et al.*, 1992; Cook and Babcock, 1993b) and a cAMP-regulated K^+ channel that allows Na^+ flux into sperm (Labarca *et al.*, 1996). A cAMP-modulated K^+ channel whose open probability increases at hyperpolarizing potentials has been detected in flagellar membranes incorporated into planar lipid bilayers. Because this channel has a P_{K^+}/P_{Na^+} of 5, its opening would depolarize sperm. This channel would be activated by the hyperpolarization and the increase in [cAMP] induced by speract, and could explain part of the Na^+-dependence of the speract-induced repolarization (Labarca *et al.*, 1996). This channel could be involved in the modulation of motility (Darszon *et al.*, 1999). A cAMP-regulated mildly selective K^+ channel whose properties resemble those of the channel just described in planar bilayers has been cloned from sea urchin testis and shown to be present in sperm flagella (Gauss *et al.*, 1998; see Section III.A).

2. Acrosome Reaction

Contact of sperm with a glycoprotein-fucose sulfate polymer complex contained in the egg jelly, called **factor (FSG)**, triggers the AR (Trimmer and Vacquier, 1986; Keller and

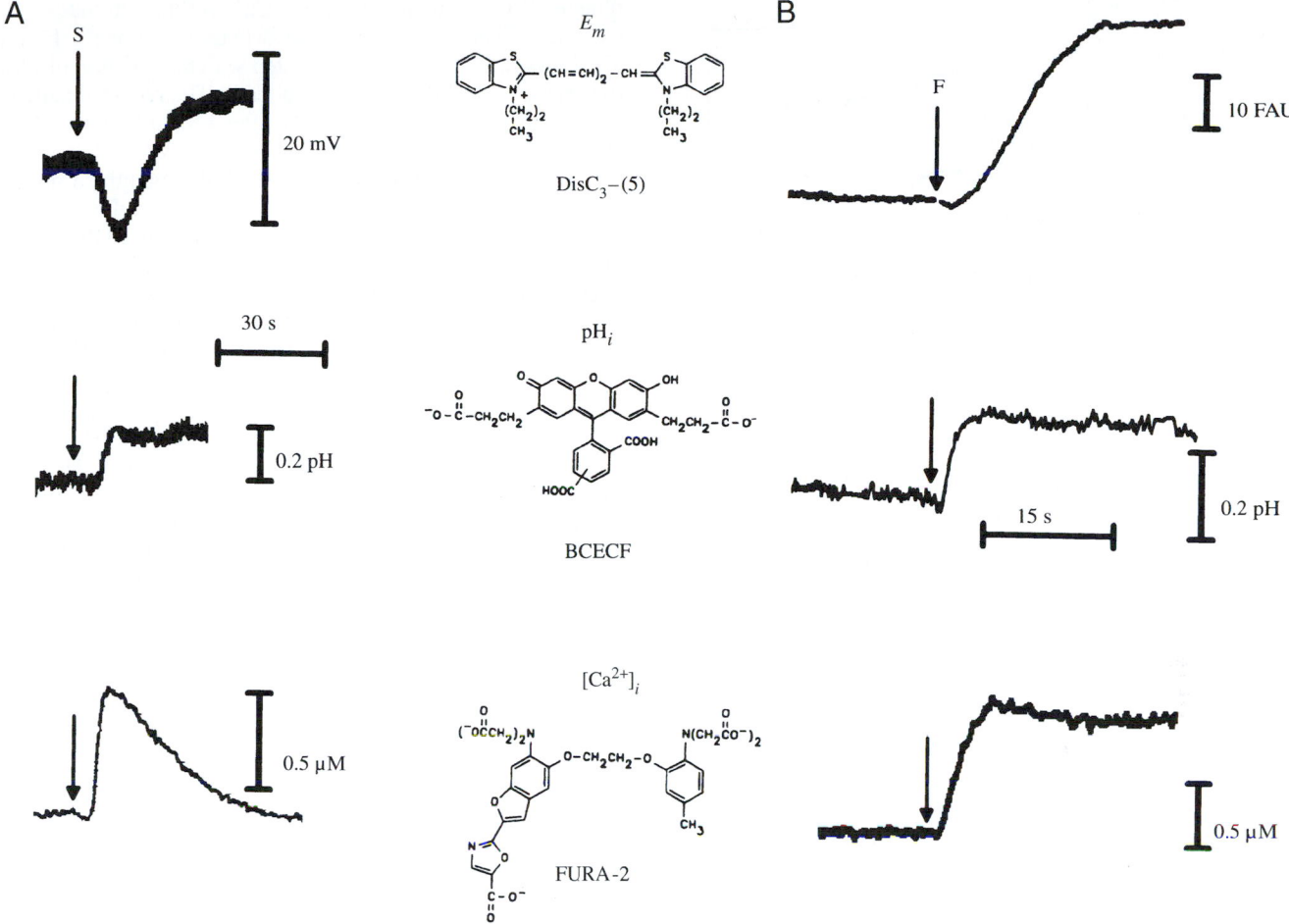

FIGURE 2. Membrane potential (E_m), intracellular pH (pH$_i$), and intracellular Ca^{2+} concentration ([Ca^{2+}]$_i$) changes. (A) *S. purpuratus* swollen sperm exposed to speract. (B) Normal sperm exposed to FSG. In all records, the arrow indicates 100 nM speract (S) or FSG (F) addition. An upward deflection indicates an increase in the measured parameter. The fluorescent probe used is indicated in the center of the figure: a cyanine dye DisC$_3$-(5) for E_m, BCECF for pH$_i$, and FURA-2 for [Ca^{2+}]$_i$. Sperm were swollen in tenfold diluted artificial seawater containing 20 mM MgCl$_2$ (DASW) (A). For pH$_i$ and [Ca^{2+}]$_i$ measurements, cells were loaded overnight with the permeant FURA-2-AM or BCECF-AM dyes at 4 °C in 0 Ca artificial seawater, pH 7.0. For E_m determinations, the cells were pre-equilibrated with 500 nM of the E_m-sensitive dye DisC$_3$-(5) during 2–3 min. (Records kindly provided by Marco González-Martínez, Enrique Reynaud, and Lucia de De La Torre.)

Vacquier, 1994). This reaction involves acrosomal vesicle exocytosis, which exposes material required for sperm-egg binding. These events lead to the extension of the acrosomal tubule, the latter being surrounded by the membrane destined to fuse with the egg (Ward and Kopf, 1993; Darszon *et al.*, 1999).

The AR requires external Ca^{2+} and Na$^+$ in seawater at pH 8.0. Exposure of sperm to FSG induces, within seconds, Na$^+$ and Ca^{2+} entry and H$^+$ and K$^+$ efflux (Schackmann, 1989; Ward and Kopf, 1993). These ion fluxes result in interrelated changes in E_m (González-Martínez and Darszon, 1987; Darszon *et al.*, 1999), [Ca^{2+}]$_i$ (Schackmann, 1989; Guerrero and Darszon, 1989a), and pH$_i$ (Lee *et al.*, 1983; Guerrero and Darszon, 1989b) (Fig. 2B). The FSG also raises cAMP levels, protein kinase A activity, turnover of inositol trisphosphate (InsP$_3$) and phospholipase D activity. It is not clear how these changes relate to the FSG-induced permeability changes (Garbers, 1989; Ward and Kopf, 1993).

When exposed to FSG, *L. pictus* sperm first transiently hyperpolarize and then depolarize (Fig. 3A). This hyperpolarization is K$^+$-dependent, probably being mediated by K$^+$ channels (González-Martínez and Darszon, 1987), and activates a Na$^+$-H$^+$ exchange that leads to an increase in pH$_i$ (Schackmann, 1989). It is not known if speract and FSG modulate the same Na$^+$-H$^+$ exchange. Antagonists to Ca^{2+} channels (verapamil and dihydropyridines) and K$^+$ channels (TEA$^+$) inhibit Ca^{2+} uptake and the AR in *S. purpuratus* sperm, indicating their mandatory participation in this process (Schackmann, 1989; Darszon *et al.*, 1994). Figure 3B shows the changes in E_m (upper trace), [Ca^{2+}]$_i$ (middle trace), and pH$_i$ (lower trace) in *L. pictus* sperm suspended in 0 K$^+$ seawater associated with a valinomycin-induced hyperpolarization; the latter is followed by a K$^+$-induced depolarization about 60 s after the hyperpolarization. The first arrow indicates the addition of valinomycin to increase the K$^+$ permeability of the membrane and to bring the E_m close to E_K. At

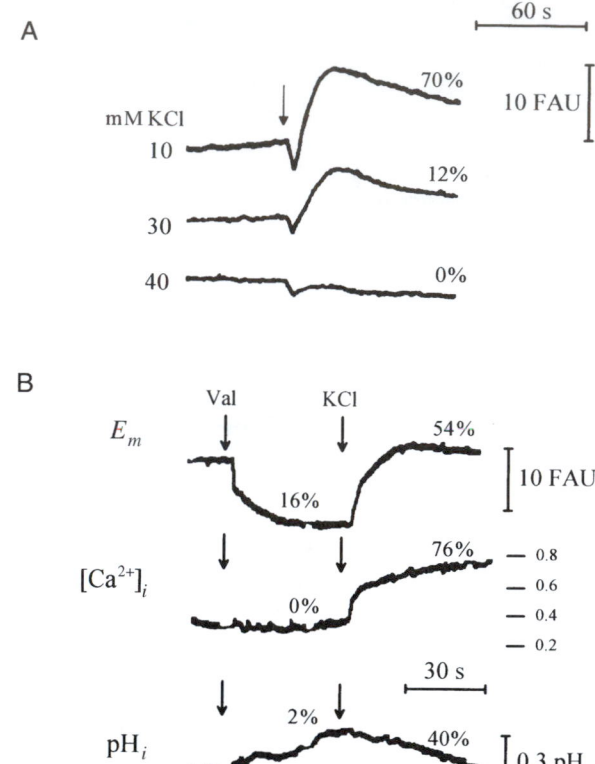

FIGURE 3. FSG-induced changes in E_m in normal sea urchin sperm. Percent numbers indicate AR; FAU indicates fluorescence arbitrary units. (A) *L. pictus* sperm's FSG-induced E_m changes at the indicated KCl concentrations (numbers on the left of each record; modified from González-Martínez and Darszon, 1987). The upper trace shows a control in normal artificial seawater (ASW). (B) Effect of valinomycin (Val; 2 μM) and subsequent KCl addition on E_m (upper trace), $[Ca^{2+}]_i$ (middle trace), and pH_i (lower trace) in *L. pictus* sperm incubated in 0 K ASW. Intracellular pH was measured with the fluorescent probe DMCF, and $[Ca^{2+}]_i$ with QUIN-2. Cells were loaded overnight with 10 μM of the permeant form of the dyes at 4 °C in 0 Ca ASW, pH 7.0, and E_m was measured as in Fig. 2. (Modified from González-Martínez *et al.*, 1992.)

this point, only a small percentage of sperm have reacted (at most 16%). However, the hyperpolarization has clearly increased pH_i which at about 60 s reaches the value attained during a normal AR. Addition of KCl to the medium at this time (second arrow) immediately depolarizes and increases $[Ca^{2+}]_i$ and the percentage of reacted sperm (>40%). A valinomycin-induced depolarization in high-KCl seawater is unable by itself to trigger the AR. Such experiments indicate that it is possible to artificially manipulate E_m and pH_i to induce an increase in $[Ca^{2+}]_i$, that the induction of the AR requires coordination between the increase in $[Ca^{2+}]_i$ and pH_i, and that sea urchin sperm have voltage-dependent Ca^{2+} channels (González-Martínez *et al.*, 1992).

Measuring $[Ca^{2+}]_i$ in sea urchin sperm has provided evidence of the participation of two different Ca^{2+} channels in the AR (Guerrero and Darszon, 1989a, b). Figure 4A depicts the changes in $[Ca^{2+}]_i$ associated with the FSG-induced AR. After the addition of FSG, $[Ca^{2+}]_i$ levels increase ten- to twenty-fold and remain high. In fact, once the AR has occurred, even

though $[Ca^{2+}]_i$ remains constant, Ca^{2+} influx continues, and Ca^{2+} accumulates in the mitochondria until cell death. Figure 4B shows $[Ca^{2+}]_i$ changes associated with the addition of FSG to sperm in seawater with Mn^{2+} added.[1] The record illustrates a biphasic behavior: immediately after FSG addition, a Ca^{2+} channel opens transiently; this channel is very selective for Ca^{2+} and does not allow Mn^{2+} influx. Subsequently, a second type of channel opens that is less selective than the first one, and allows the influx of Mn^{2+} that quenches the FURA-2 fluorescence. Thus, the first type of Ca^{2+} channel is a channel that opens by a still unknown mechanism based upon receptor occupancy. The first type of channel is blocked by verapamil and dihydropyridines (DHPs), and it shows inactivation. The other type is not blocked by these compounds, does not inactivate, and allows Mn^{2+} to permeate. This second type of channel is blocked by conditions that inhibit the increase in pH_i and the AR, but still support a transient increase in intracellular Ca^{2+}. Thus, the second channel is modulated by pH_i. The opening of the first type of channel is required for the opening of the second; blocking of the first inhibits Ca^{2+} uptake through the second and blocks the AR. How the two types of channels are coupled is still a mystery; however, both are important to fully achieve AR (Darszon and González-Martínez, unpublished). The first Ca^{2+} channel could be inactivated at the normal sperm resting potential (around −45 mV; González-Martínez and Darszon, 1987), like the low-threshold T-type channels

[1]This divalent cation has a forty-fold higher affinity for FURA-2 than Ca^{2+} and quenches its fluorescence.

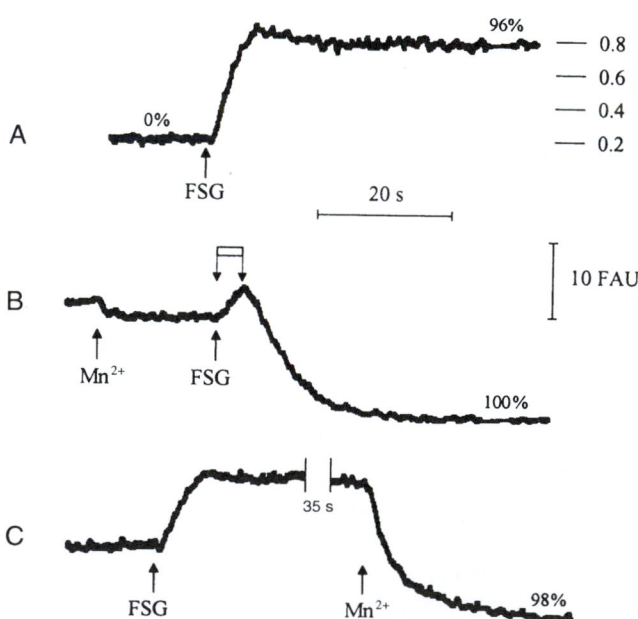

FIGURE 4. FURA-2 detection of Mn^{2+} influx through Ca^{2+} channels during the *S. purpuratus* sperm FSG-induced AR. (A) Control record showing the profile of $[Ca^{2+}]_i$ change induced by the addition of FSG. Mn^{2+} (3 mM) was added (B) before or (C) after FSG. Mn^{2+} influx is indicated by a decrease in fluorescence. Numbers on the right of (A) indicate the fraction of calcium-bound FURA-2. The percentage of FSG-induced AR is shown at the end of each record. (Modified from Guerrero and Darszon, 1989b.)

(Hille, 1992). The FSG-induced transient hyperpolarization could remove this inactivation and subsequently trigger the AR. It is known that rat and mouse spermatogenic cells express T-type Ca^{2+} channels in their membrane (Hagiwara and Kawa, 1984; Liévano et al. 1996; Arnoulta et al., 1996; see Section II.B.2), and that the mouse AR is blocked by µM concentrations of dihydropyridines and Ni^{2+} (Florman et al., 1992), matching the T-type Ca^{2+} channel's sensitivity to those compounds (Hille, 1992; Liévano et al., 1994, 1996; Arnoulta et al., 1996, 1999).

An approximately 210-kDa plasma membrane protein is the best candidate for the FSG sperm receptor. It is not known how the activated receptor triggers the AR, but it could directly regulate ion channels. This protein has species-specific affinity for egg jelly, and some monoclonal antibodies to it induce the AR (Moy et al., 1996). These antibodies bind to a narrow plasma membrane collar over the acrosome and along the entire flagellum (Trimmer and Vacquier, 1986). Recently this glycoprotein receptor, composed of 1450 residues and approximately 50% carbohydrate, has been cloned (Moy et al., 1996). It has 17 potential sites for N-linked, and 12 for O-linked, glycosylation, and only the extreme C-terminal appears to have a putative transmembrane region, strongly suggesting that this protein is not an ion channel. The amino-terminal portion has an EGF domain and two contiguous carbohydrate recognition domains with significant relatedness to those of the human macrophage mannose receptor (important in the activation of the classical complement pathway). The receptor contains a novel module (700 residues) that shares extensive homology with the human polycystic kidney disease protein (PKD1) (Moy et al., 1996). Autosomal dominant polycystic kidney disease is one of the most frequent human genetic diseases and is caused by mutations in PKD1 and PKD2. The mechanism by which these two proteins function is unknown, though it has been proposed that they modulate ion fluxes (Sanford et al., 1997).

B. Mammalian Sperm

Mammalian spermatozoa must undergo changes after leaving the testis to become competent for fertilization. These changes occur in the male reproductive tract (epididymal maturation) and in the female reproductive tract (capacitation and the AR). During maturation the sperm plasma membrane is modified with respect to surface charge, reactivity and distribution of its components, addition of epididymal-secreted proteins, and ionic permeability (rev. in Florman and Babcock, 1991).

1. Capacitation

Capacitation is essential for successful gamete interaction in mammals (Chang, 1951 Austin, 1951). It regulates the efficiency of acrosome exocytosis in sperm and coordinates it with egg contact to ensure fertilization (Baldi et al., 1996; Visconti and Kopf, 1998). It is worth stressing that unlike somatic secretory cells, which have many secretory vesicles, sperm only have a single large vesicle, the acrosome. Gamete interaction action and fusion requires exocytosis of this vesicle (Yanagimachi, 1994). Preventing premature AR

is vital for fertilization since sperm that have undergone this reaction too early have difficulties reaching the egg and fusing with it (Florman et al., 1998).

Mammalian sperm capacitation can also be accomplished in vitro by incubating ejaculated sperm in defined medium (Yanagimachi, 1994). Three key components are required for this process in mouse sperm: Ca^{2+}, $NaHCO_3$, and serum albumin (Visconti and Kopf, 1998). In addition to the changes in composition and distribution of plasma membrane lipids that accompany in vitro capacitation in mouse sperm, a time-dependent and cAMP-regulated increase in the protein tyrosine phosphorylation occurs. The serum albumin requirement has been related to its ability to remove cholesterol from the membrane. The decrease in cholesterol content is thought to alter membrane architecture, which results in changes, amongst them an elevation of cAMP levels in sperm (Visconti and Kopf, 1998). Furthermore, during capacitation, $[Ca^{2+}]_i$ and pH_i increase and membrane potential becomes more negative. These changes are important to endow sperm with the capacity to undergo the AR (Florman et al., 1998).

The pH_i increase that occurs during capacitation mainly involves a Na^+, Cl^-, and HCO_3^-–dependent mechanism (Zeng et al., 1996) and has been related to sperm cholesterol content (Cross and Razy-Faulkner, 1997). As discussed earlier, pH_i may influence sperm Ca^{2+} permeability, therefore, an acidic pH_i may contribute to maintain membrane potential and low $[Ca^{2+}]_i$, thus avoiding premature AR (Babcock and Pfeiffer, 1987; Darszon et al., 1999).

A K^+ permeability increase accompanies capacitation and significantly contributes to the hyperpolarization (~ -30 mV) that occurs during this process (Zeng et al., 1995; Arnoult et al., 1999). If mammalian sperm AC is sensitive to membrane potential, like the one from sea urchin sperm (Beltrán et al., 1996), this hyperpolarization could stimulate it. Increases in [cAMP] would activate PKA and result in protein phosphorylation. Voltage-dependent channels would also be affected by this membrane potential change, particularly the T-type Ca^{2+} channels likely to be present in sperm (see Section II. B.2) (Liévano et al., 1996; Arnoult et al., 1996a).

It is interesting that 30 mM external K^+ can inhibit capacitation in mouse sperm (Arnoult et al., 1999). This indicates that K^+ channels play a role in this event. K^+-selective channels blocked by TEA^+ have been recorded in bilayers containing rat sperm plasma membranes (Chan et al., 1997). On the other hand, voltage-dependent K^+-selective and TEA^+-sensitive currents were observed in rat spermatogenic cells. These currents were insensitive to external Ca^{2+} and decreased during spermatogenesis (Hagiwara and Kawa, 1984). Recently, transcripts for Kir5.1, a subunit of inwardly rectifying K^+ channels, were detected in rat testis. Antibodies against an external epitope of this subunit allowed its detection in spermatogenic cells and in mature sperm. Kir5.1 is unable by itself to form functional channels but can do so when coexpressed with Kir4.1 (Salvatore et al., 1999). Little is known about the regulation of K^+ channels in spermatogenic cells or in sperm.

2. Acrosome Reaction

The **zona pellucida (ZP)**, a thick extracellular glycoprotein coat surrounding the egg, is the main mediator of the sperm AR in mammals. **ZP3** is the murine ZP *sulfated glycoprotein* that displays the sperm-binding and AR-inducing activity of unfertilized eggs. Specific receptors for ZP3 on the plasma membrane of an acrosome-intact sperm overlying the acrosome must mediate its binding and induction of the AR. This Ca^{2+}-dependent reaction involves the fusion of the plasma and outer acrosomal membranes of the sperm head (see Fig. 1B, lower drawing). It exposes the sperm's inner acrosomal membrane, whose surface contains proteases and/or glycosidases thought to allow its penetration through the zona pellucida to reach the egg plasma membrane (Bleil, 1991; Ward and Kopf, 1993).

Identification of the sperm **surface receptor for ZP3** has been attempted by several laboratories. Among the candidate proteins are the following: in mouse sperm, a β–1,4 galactosyl transferase (Gong *et al.*, 1995), a hexokinase (Leyton and Saling, 1989), and the lectin sp56 (Bookbinder *et al.*, 1995); in guinea pig sperm, a hyaluronidase (Gmachl and Kreil, 1993). Trypsin-like proteins (Boettger-Tong *et al.*, 1993) and spermadhesins (Gao and Garbers, 1998) have been also proposed as receptors. The physiological relevance of many of these candidates is under active debate (McLeskey *et al.*, 1998; Darszon *et al.*, 1999). Multiple concerted and cooperative interactions between ZP3 and the sperm surface, possibly involving receptor aggregation, may be needed to achieve the signal transduction events that result in the AR (Leyton and Saling, 1989; Ward and Kopf, 1993). How these receptors convey information to initiate signal transduction in mammalian sperm is not known.

The ZP-induced AR is inhibited by pertussis toxin (PTX), a specific inactivator of the G_i class of heterotrimeric G proteins, in mouse, bovine and human sperm. Among the multiple species of G proteins found in mouse sperm, apparently G_{i1} and G_{i2} are preferentially activated by ZP (Ward *et al.*, 1994). It was reported that mouse sperm β-1,4 galactosyl transferase interacts with a G protein (Gong *et al.*, 1995). There are still many questions that need to be solved: How does the ZP3-activated receptor turn these G_i proteins on, which activities do they regulate, and are there ion channels among them?

A rise in $[Ca^{2+}]_i$ is an essential step in the ZP3 signaling path leading to the AR. External Ca^{2+} is required for the physiological AR, and ZP induces $[Ca^{2+}]_i$ and pH_i increases that precede exocytosis in single sperm loaded with fluorescent dyes as ion indicators. These changes are somehow mediated by G_i proteins since both are inhibited by PTX (Florman *et al.*, 1992). It turns out that the PTX-sensitive step in the ZP-induced AR is the pH_i increase (Arnoult *et al.*, 1996b).

Evidence has been provided for the presence of voltage-dependent Ca^{2+} **channels** in the plasma membrane of mammalian sperm (Babcock and Pfeifer, 1987; Florman *et al.*, 1992). Micromolar concentrations of DHPs are needed to block the AR and the increase in $[Ca^{2+}]_i$. Because typical L-type Ca^{2+} channels are blocked by submicromolar concentrations of these blockers, doubts emerged as to the nature of this voltage-dependent Ca^{2+} channel (Florman *et al.*, 1998; Darszon *et al.*, 1999). Depolarizing conditions, at elevated pH_i, that open Ca^{2+} channels sensitive to DHPs bypass the inhibition of the ZP3-induced exocytosis produced by PTX.

It is not known how pH_i and $[Ca^{2+}]_i$ are intimately related and finely tuned in mammalian sperm. The activation of voltage-dependent Ca^{2+} **channels** is a required step in the ZP3 signal transduction pathway (Florman *et al.*, 1992).

It has been suggested that, like in the sea urchin sperm (Guerrero and Darszon, 1989b), at least two different Ca^{2+} channels are present in mammalian sperm (Florman, 1994). Two phases of the ZP3-induced increase in $[Ca^{2+}]_i$ have been resolved using ion-selective fluorescent probes.

First, transient elevation of $[Ca^{2+}]_i$ occurs within 40–50 ms to values of approximately 10 μM and subsequently relaxes to resting values within the next 200 ms (Arnoult *et al.*, 1999). The characteristics of this transient in terms of its kinetics of activation and inactivation and pharmacology are consistent with the properties of low voltage–activated (LVA), T-type voltage-sensitive Ca^{2+} channels (Arnoult *et al.*, 1999). T-type Ca^{2+} currents are the only voltage-dependent Ca^{2+} currents expressed during the later stages of rodent spermatogenesis (Hagiwara and Kawa, 1984; Liévano *et al.*, 1996; Santi *et al.*, 1996; Arnoult *et al.*, 1996a). Like the LVA T-type channels from spermatogenic cells, both the AR and the transient increase in $[Ca^{2+}]_i$ are inhibited by micromolar concentrations of DHPs, pimozide, and Ni^{2+} (Arnoult *et al.*, 1996a, b). Therefore, it is likely that the fast transient increase in sperm $[Ca^{2+}]_i$ is mediated by a ZP3-dependent activation of T-type Ca^{2+} channels (Florman *et al.*, 1998; Darszon *et al.*, 1999).

How does ZP depolarize sperm to open Ca^{2+} channels? ZP or ZP3 have been reported to induce a 30-mV depolarization in bovine or mouse sperm (Arnoult *et al.*, 1996b); however, this membrane potential change seems too slow to activate T-type Ca^{2+} channels. Two newly cloned channels from testis mRNA, if present in mature sperm, could be candidates to accomplish a ZP3-induced depolarization: mSlo3, a channel with extensive sequence similarity to large-conductance K^+ channels activated by Ca^{2+} and voltage (Schreiber *et al.*, 1998), and a homologue of sea urchin sperm SPIH called hHCN4, which is a member of the pacemaker channel family and thus activated by hyperpolarization and regulated directly by CN (Seifert *et al.*, 1999) (see Section III for more details).

Second, the fast transitory response is followed by a much slower elevation in $[Ca^{2+}]_i$ which remains high while ZP is present. This slow response requires several seconds to several minutes to develop and AR occurs only after the high, sustained $[Ca^{2+}]_i$ is reached (Arnoult *et al.*, 1996a, b). The kinetic characteristics of the slow sustained elevation in $[Ca^{2+}]_i$ are incompatible with the properties of T-type Ca^{2+} channels (Bean and McDonough, 1998); therefore, at least another pathway for Ca^{2+} is necessarily involved in triggering the AR. Antagonists of LVA Ca^{2+} channels added before ZP3 also inhibit the sustained elevation in $[Ca^{2+}]_i$ (Arnoult *et al.*, 1996a). These results indicate that the transient increase in $[Ca^{2+}]_i$ due to the ZP-induced activation of T-type channels is necessary to open a second Ca^{2+} pathway that keeps $[Ca^{2+}]_i$ elevated to allow the AR (Florman *et al.*, 1998; Darszon *et al.*, 1999).

ZP3 signaling may achieve the sustained elevation of $[Ca^{2+}]_i$, releasing Ca^{2+} from an InsP$_3$-sensitive intracellular store (Walensky and Snyder, 1995) by a mechanism requiring prior Ca^{2+} influx through a T-type Ca^{2+} channel. Consistent with this possibility are the following findings: there are several isoforms of phospholipase C in sperm (Srivastava *et al.*, 1982; Vanha-Perttula and Kasurinen, 1989) and ZP3 stimulates $InsP_3$ production in these cells (Tomes *et al.*, 1996); $InsP_3$ receptors are present in acrosomal membranes (Walensky and Snyder, 1995; Trevino *et al.*, 1998); in digitonin-permeabilized mouse sperm exogenous $InsP_3$ causes $^{45}Ca^{2+}$ efflux from a nonmitochondrial pool (Walensky and Snyder, 1995); and agents that are expected to promote Ca^{2+} release from smooth endoplasmic reticulum pools in nonpermeabilized sperm promote elevations in $[Ca^{2+}]_i$ and the AR (Meizel and Turner, 1993; Blackmore, 1993). Nevertheless, how T-type Ca^{2+} channels, Ca^{2+} release from intracellular stores, and the AR are coordinated is yet not known. Recently, it was reported that a store-operated Ca^{2+} channel is present in the plasma membrane of mouse spermatogenic cells, as well as in immotile testicular sperm. This channel could be responsible for the sustained $[Ca^{2+}]_i$ elevation necessary for the AR (Santi *et al.*, 1998).

Ca^{2+}-permeable channels that are activated by the depletion of intracellular Ca^{2+} stores (SOCs) have been detected in many cell types. SOCs replenish the $[Ca^{2+}]$ in intracellular stores when it is are depleted during repetitive cycles of Ca^{2+} release and may also participate in signaling. Several types of SOCs have been described, such as the highly Ca^{2+}-selective ICRAC pathway, a number of distinct cation-selective channels, and certain mammalian homologs of the *Drosophila trp* gene (rev. in Parekh and Penner, 1997; Barritt, 1999). At least one member of the *trp* gene family is expressed in mammalian spermatogenic cells (Wissenbach *et al.*, 1998). Further work is needed to identify the protein components of this influx mechanism in sperm.

Progesterone and other progestins can induce PTX- and DHP-insensitive large extracellular Ca^{2+}-dependent increases in $[Ca^{2+}]_i$ which result in the AR in mammalian sperm (Thomas and Meizel, 1989; Meizel, 1997; Blackmore, 1999). Considering that this steroid activates phospholipase C (Thomas and Meizel, 1989), it could also lead to depletion of sperm Ca^{2+} stores and activation of a SOC. This would be in agreement with observations that progesterone and ZP3 initially activate different transduction systems (Tesarik *et al.*, 1993; Murase and Roldan, 1996) and yet act cooperatively to cause the AR (Roldan *et al.*, 1994). Progesterone metabolites have been shown to enhance the interaction of γ-aminobutyric acid (GABA) with the GABA receptor in central nervous system neurons. The **GABA receptor** is a multisubunit protein containing a Cl^- channel that has been detected in boar and ram sperm. It has been proposed that the fast progestin-induced human sperm responses may involve steroid interaction with a sperm steroid receptor-Cl^-channel complex similar to the $GABA_A$-Cl^- channel complex (Wistrom and Meizel, 1993). The first single-channel recordings made on mouse sperm have indicated the presence of cation channels and a Cl^- channel sensitive to niflumic acid, a blocker of anion channels, which in addition inhibits the mouse AR (Espinosa *et al.*, 1998).

III. Sperm Ion Channels

In spite of the explosion of knowledge on ion channels generated by the patch-clamp technique in the last few years, the properties of sperm ion channels are still relatively unknown. The reason for this is their small size and complex geometry (Fig. 1), which has hindered the use of conventional electrophysiological strategies in the characterization of the channels involved in sperm physiology. As mentioned before, one of the main advantages of working with sea urchin sperm is the large quantity of biological material available. This allows the isolation and characterization of different plasma membrane fractions, which can be reassembled to study sperm ion channels in model systems by different reconstitution strategies (rev. in Darszon *et al.*, 1994, 1996).

A. Sea Urchin Sperm Ion Channels

Single **K^+ channels** were first recorded in bilayers made at the tip of patch-clamp pipettes from monolayers generated from a mixture of lipid vesicles and isolated sperm flagellar membranes. Three types of K^+ channels were identified with conductances of 22, 46, and 82 pS (Fig. 5A). Two of them are blocked by TEA^+, which inhibits the AR (Liévano *et al.*, 1985). Although with great difficulty, single channels were recorded directly from sea urchin sperm heads using the patch-clamp technique. Single-channel events of 40, 60, and 180 pS were detected, and one of the channels observed was a K^+ channel (Fig. 5C; Guerrero *et al.*, 1987). As mentioned earlier, swelling *S. purpuratus* sperm significantly improved the success rate of patch formation and allowed the detection of a 2–5 pS K^+ channel that is activated by speract. Swollen sperm have opened new possibilities of directly studying the ion channels modulated by egg components and their regulation (Fig. 5D; Babcock *et al.*, 1992).

Two types of **Ca^{2+} channels** have been detected in *S. purpuratus* sea urchin sperm by fusing isolated plasma membranes into planar lipid bilayers (Fig. 5B): (1) a voltage-dependent channel of 50 pS conductance observed in 10 mM Ca^{2+}, and (2) a high-conductance channel having a main conducting state of 172 pS in 50 mM $CaCl_2$ and several subconductance states (Fig. 5B; Liévano *et al.*, 1990). This channel is strongly voltage dependent, showing a single main-conducting state, with rare closing events at voltages more positive than −25 mV, and displaying several subconductance states of lesser conductance at more negative potentials. The main-state conductance size sequence is $Ba^{2+} > Sr^{2+} > Ca^{2+}$, as in many other Ca^{2+} channels (Bean, 1989). The channel discriminates poorly between divalent and monovalent cations ($P_{Ca^{2+}}/P_{Na^+} = 5.9$), and is also permeable to Mg^{2+} when it is added to the *cis* side ($P_{Ca^{2+}}/P_{Mg^{2+}} = 2.8$). In contrast, addition of Mg^{2+} to the *trans* side blocks the channel in a voltage-independent manner (Liévano *et al.*, 1990). Two findings suggest the possible participation of the high-conductance Ca^{2+} channel in the sperm AR. Both Cd^{2+} and Co^{2+} block the channel at concentrations similar to those required to inhibit the AR and the Ca^{2+} uptake induced by egg jelly (Liévano *et al.*, 1990). In addition, Mg^{2+} blocks the high-conductance Ca^{2+} channel only when present in the *trans*

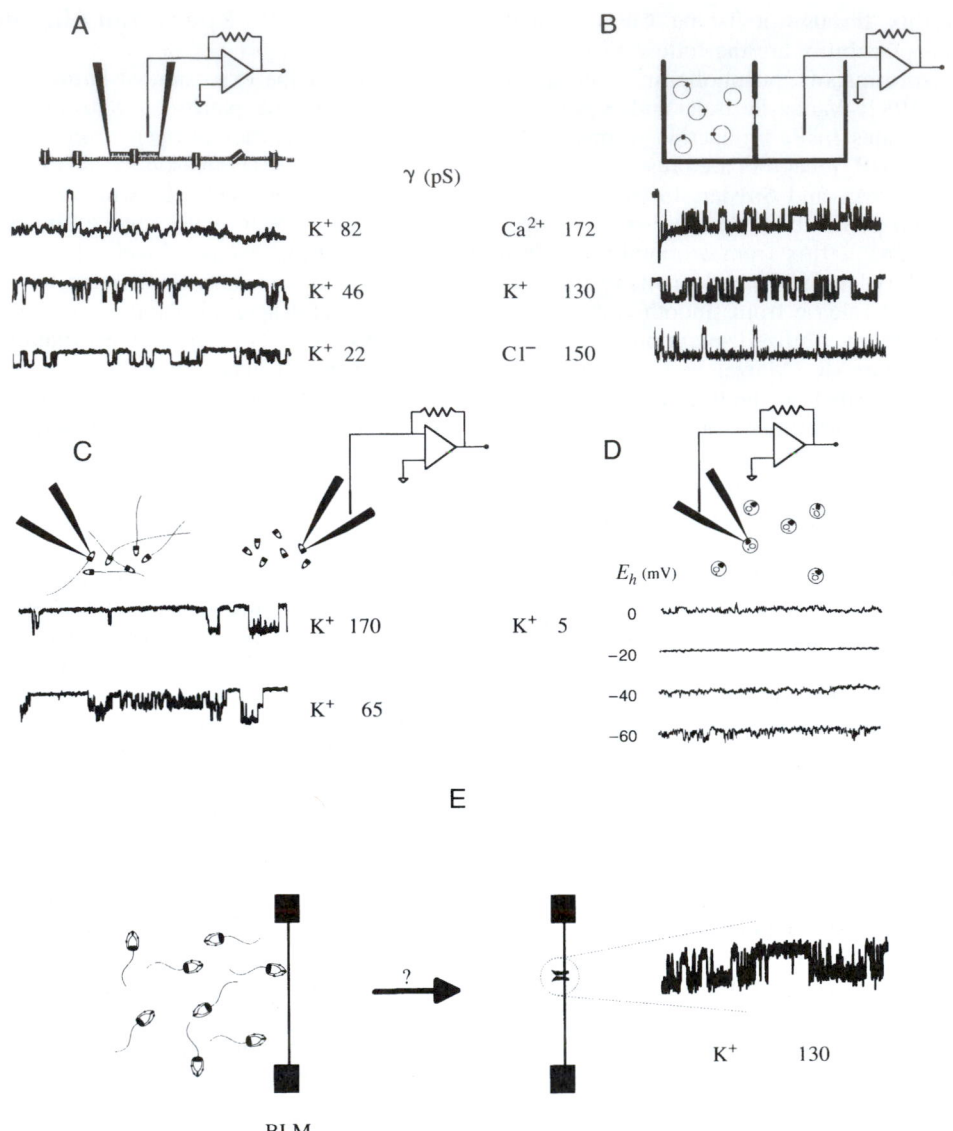

FIGURE 5. Strategies used for ion channel characterization in sperm cells. The channels detected with each technique are shown, indicating the main ion transported and their single-channel conductance value. (A) Bilayers at the tip of a patch-clamp pipette. (B) Black lipid membranes (BLM) with fused sperm plasma membrane vesicles. (C) On-cell patch-clamp recordings in sperm (left) and on deflagellated heads (right). (D) On-cell patch-clamp recordings in osmotically swollen sperm. (E) Direct ion channel transfer from cells to the BLM; see Section III.B (Modified from Darszon *et al.*, 1994.)

side. This could indicate that Mg^{2+} in seawater may modulate the influx of Ca^{2+} into sperm, either from the outside or after entering the cell. Verapamil and nisoldipine do not significantly modify the kinetics nor the conductance of the high-conductance channel seen in planar bilayers. Therefore, the high-conductance Ca^{2+} channel could be the second type of channel that participates in the AR, allows Mn^{2+} influx, and is pH_i-modulated.

Little is known about **anion channels** in these cells. The fusion of sperm plasma membranes into lipid bilayers allowed identification of a 150-pS anion channel (see Fig. 5B). This anion channel was enriched from detergent-solubilized sperm plasma membranes using a wheat germ agglutinin–Sepharose column. Vesicles formed from this

preparation were fused into black lipid membranes (BLM), yielding single channel anion-selective activity with similar properties to those found in the sperm membranes. The anion selectivity sequence found was $NO_3^- > CNS^- > Br^- > Cl^-$. This anion channel has a high open probability at the holding potentials tested, is partially blocked by DIDS, and often displays substates. DIDS blocks the AR in *S. purpuratus* sea urchin sperm by a still unknown mechanism. These results suggest that this Cl^- channel could be involved in the events that lead to the AR, or in determining the resting potential of sperm, which modulates this reaction (Morales *et al.*, 1994).

Ion channels are subject to multiple forms of regulation (Hille, 1992), and **cyclic nucleotides** can directly modulate them (Yau, 1994). Spermatozoa respond to the outer layer of eggs with changes in the second messenger levels ($[Ca^{2+}]_i$, cyclic nucleotides, IP_3) (Schackmann, 1989; Garbers, 1989; Ward and Kopf 1993). Recently a K^+-selective channel derived from sea urchin plasma membranes, upwardly regulated by cAMP, has been detected in planar bilayers. Its single-channel conductance is 103 pS in 100 mM KC1. The channel has a low open probability and is weakly voltage dependent. Addition of cAMP on the *cis* side (the side of membrane addition) up-regulates channel activity in a dose-dependent ($K_D = 200 \mu M$) and reversible fashion, increasing the open probability. Millimolar concentrations of Ba^{2+} or TEA on the *trans* side blocked the channel in a voltage-dependent fashion. The channel exhibited a low $P_{K^+}/P_{Na^+} \approx 5$, indicating a sizable permeability to Na^+; therefore, its opening would depolarize sperm. Hence, this channel could contribute to the depolarizing phase of the response to speract, and perhaps even during the AR as cAMP levels increase (Labarca *et al.*, 1996).

A cAMP-regulated K^+ channel with similar characteristics as the one described above in planar bilayers has been cloned from sea urchin testis and functionally expressed in HEK 293 cells (Gauss *et al.*, 1998). The cDNA encodes a 767-amino-acid polypeptide ($M_r \sim 88$ kDa) named SPIH with significant sequence similarity to cyclic nucleotide-gated (CNG) and ether-a-gogo (EAG/HERG) channels. The selectivity filter contains the GYG residues typical of K^+ channels; however, a conserved threonine on the N-terminal side of the GYG sequence is replaced by cysteine, and the aspartate immediately following GYG is replaced by lysine. The channel is only about four times more permeant to K^+ than to Na^+ and is much more sensitive to cAMP than to cGMP. Antibodies to this channel revealed its presence in sperm flagella (Gauss *et al.*, 1998). Because this channel is activated both by cyclic nucleotides and hyperpolarizing potentials, it is a member of a growing family of channels named **HCN channels** (Clapham, 1998). The channels of this new family are important in shaping the autonomous rhythmic activity of single neurons and the periodicity of network oscillations.

B. Ion Channels of Mammalian Sperm

Fewer reports have been published on single-channel activity from mammalian sperm. For example, it was shown that addition of a partially purified 110-kDa human sperm plasma membrane protein to a preformed lipid bilayer resulted in incorporation of cationic channels having a single-channel conductance of 130 pS (0.1 M NaCl). Apparently the channel was formed from functionally aggregated triplets (Young *et al.*, 1988). Another study described the detection of two types of Ca^{2+} **channels** in tip-dip bilayers[2] formed from liposomes containing boar sperm plasma membrane. The single-channel conductances detected were between 10–20 pS and 50–60 pS, and the smaller channels were partially blocked by unknown concentrations of ni-

trendipine and verapamil, and completely blocked by 0.5 mM La^{3+} (Cox and Peterson, 1989). A nonselective cation channel was also reported, both from cauda epididymal or ejaculated boar sperm plasma membranes incorporated into planar lipid bilayers. Both monovalent and divalent cations permeate through the channel. The channel displays voltage-independent kinetics and is blocked by high concentrations of verapamil or nitrendipine and by ruthenium red (Cox *et al.*, 1991).

CNG channels are heterooligomeric complexes formed from at least two subunits (α and β). The α subunit displays the channel activity, while β alone is not functionally active. Coexpression of α and β subunits yields channel species with different properties, when compared to homo-oligomeric channels (rev. in Kaupp, 1995). The first subunit cloned from bovine testis was α (Weyand *et al.*, 1994). It has 78% amino acid sequence homology to CNG channels in chicken photoreceptors and contains the cyclic nucleotide–binding site, pore sequence, transmembrane segments, and S4 voltage sensor motif characteristic of the CNG channel family. The single-channel conductance of the channel expressed in *Xenopus* oocytesis is 20 pS. It selects poorly between Na^+ and K^+, is blocked by Mg^{2+}, is permeable to Ca^{2+}, and has a much higher affinity for cGMP (>100 fold) than for cAMP. Small cGMP-induced currents associated with single-channel transitions of <10 pS were detected in vesicles thought to be sperm cytoplasmic droplets. Inside-out patches from human and bovine sperm responded to cGMP with similar small currents (Weyand *et al.*, 1994).

One short and several long less abundant transcripts of CNG channel β subunits have been identified in bovine testis (Wiesner *et al.*, 1998). Immunodetection revealed that the α subunit is present along the entire sperm flagellum, whereas the short β subunit is only found in the principal piece of the flagellum. Various combinations of αs and βs have different permeability to Ca^{2+}. Since the βs are distinctly localized in the flagellum, Ca^{2+} microdomains may exist, and be the basis for flagellar bending control (Wiesner *et al.*, 1998).

A new member of the family of channels activated by hyperpolarization and **CN, HCN** channels, was recently cloned and characterized (hHCN4). This channel is present in human thalamus, heart, and testis. Heterologously expressed hHCN4 channels in HEK cells have unusually slow kinetics of activation and inactivation. The channel is activated at negative potentials with the half-maximal activation ($V_{1/2}$) at –72 mV. cAMP displaces $V_{1/2}$ to more positive values by 11 mV. It is interesting that sperm may possess the appropriate potpourri of channels, as neurons do, to generate rhythmic activity that could modulate their flagellar beating (Seifert *et al.*, 1999). If the ZP3-induced increase in cAMP is fast enough (Ward and Kopf, 1993), this channel could even participate in the depolarization required to open T-type Ca^{2+} channels during the AR.

Recently another channel was cloned from mouse testis (Schreiber *et al.*, 1998). This channel, named **mSlo3**, is similar to Slo1, the large conductance K^+ channel activated by Ca^{2+} and voltage. In contrast to Slo1, mSlo3 is refractory to Ca^{2+}, but is activated by relatively large depolarizations and is

---

[2]Dip-tip bilayers are formed at the tip of patch-clamp electrodes by apposition of two monolayers; see Fig. 5A.

A

Ca²⁺ 380 pS

B

Ca²⁺ 9 pS

C

C⁺ 103 pS

D

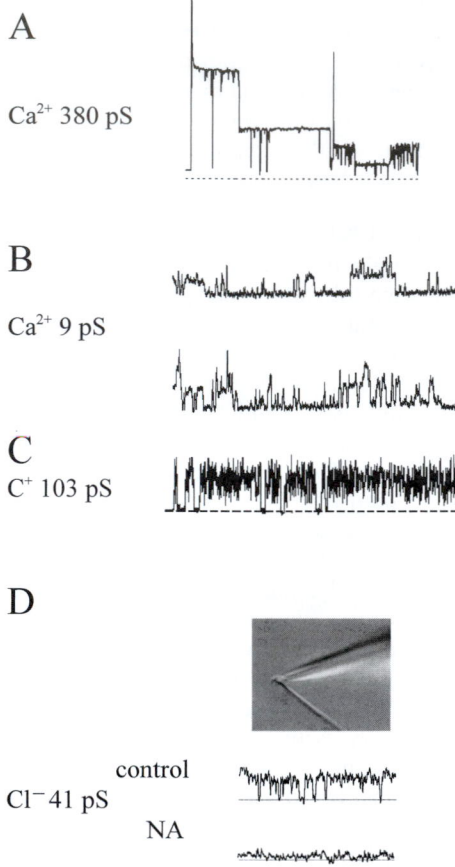

control

Cl⁻ 41 pS

NA

FIGURE 6. Planar bilayer (A, B, C) and patch-clamp (D) single-channel recordings of mammalian sperm ion channels. (A) A large-conductance Ca^{2+} channel from mouse sperm. (B) A small conductance Ca^{2+} channel from boar sperm plasma membrane. (C) A nonselective cationic channel from mouse sperm plasma membranes. (D) A Cl^- channel sensitive to niflumic acid (NA, 20 µM) recorded from the head of mouse sperm. (Modified from: (A) Beltran et al., 1994, (B) Tiwari-Woodroff et al., 1995; (C) Labarca et al., 1995; Espinosa et al., 1998.)

sensitive to pH. When expressed in a heterologous system, the channel's open probability at +80 mV changes from less than 1% at pH 7.0 to as much as 100% at pH 8.0. Since mSlo3 is poorly selective for K^+ over Na^+ ($P_{K^+}/P_{Na^+} = 5$), its opening would depolarize sperm. Possibly the ZP-induced increase in pH_i could activate mSlo3, if its voltage dependence in sperm is significantly displaced towards less positive potentials than in the heterologous system (half-activation voltage +70 mV). *In situ* hybridization revealed that the mSlo3 message is present in the seminiferous tubules, predominantly in maturing spermatocytes and in the later stages of spermatogenesis (Schreiber et al., 1998).

It has also been possible to incorporate ion channels into lipid bilayers directly from sea urchin and mouse spermatozoa (see Fig. 5E; Beltrán et al., 1994). The high-conductance Ca^{2+} channel, several K^+ channels, and a smaller voltage-dependent 10-pS Ca^{2+} channel (resembling the one from boar sperm described by Tiwari-Woodruff et al., 1995; Fig. 6B) have been recorded with this strategy. This approach

opens new avenues to explore cell-cell interactions, such as sperm-egg fusion, at the single-channel level. More recently, several channels were detected by fusing mouse sperm plasma membranes to planar bilayers: (1) an 80-pS anion channel with similar characteristics to the one found in sea urchin sperm plasma membranes, (2) a cation channel ($P_{Na^+}/P_{K^+} = 2.5$) with two modes of gating, and (3) a high-conductance Ca^{2+} channel (Fig. 6A; Labarca et al., 1995). This high-conductance Ca^{2+} channel resembles the one from *S. purpuratus* (Liévano et al., 1990). It is attractive to consider that this Ca^{2+} channel may be important in sperm physiology, since it is present in such diverse species (Beltrán et al., 1994). This latter channel was blocked by micromolar concentrations of ruthenium red, which inhibits the AR in sea urchin sperm (Labarca et al., 1995).

As mentioned earlier, sperm are differentiated terminal cells lacking the machinery for protein synthesis. Thus, during spermatogenesis, all ion channels required for cell function must be synthesized. Little is known about the biogenesis of ion channels during spermatogenesis, or their role in sperm differentiation. There is only one report, which found that rat spermatogenic cells have only low-threshold, inactivating Ca^{2+} currents that increase their density during rat spermatogenesis. This result could suggest a possible role for Ca^{2+} channels in sperm differentiation (Hagiwara and Kawa, 1984). The genotypic and phenotypic expression of Ca^{2+} channels is being studied in mouse pachytene spermatocytes (PS) and round (RS) and condensing spermatids (CS). These are the cell types in the last phases of spermatogenesis, before the loss of the cytoplasmic vesicle which leads to the formation of mature sperm (Kretzer and Kerr, 1988). Low-threshold T-type Ca^{2+} channels are the only detectable Ca^{2+} channels in mouse PS and RS (Fig. 7A; Liévano et al., 1996; Santi et al., 1996; Arnoult et al., 1996b, 1999). These T-type Ca^{2+} currents from spermatogenic cells are blocked by nifedipine, Ni^{2+} (Fig. 7B), and amiloride at concentrations that inhibit the mouse sperm AR and the uptake of Ca^{2+} that triggers it (Liévano et al., 1996; Arnoult et al., 1996a; Santi et al., 1996). As discussed in Section II.B.2, these findings are consistent with the involvement of a T-type Ca^{2+} channel in inducing this reaction.

RNAs isolated from purified fractions of PS, RS, and CS were explored for the presence of transcripts of the Ca^{2+} channel α_1 subunit, the one containing both the pore and the voltage sensor of voltage-dependent Ca^{2+} channels (Stea et al., 1995). Reverse transcription coupled to polymerase chain reaction (RT-PCR) with oligonucleotides specific for α_{1A}, α_{1B}, α_{1C}, α_{1D}, and α_{1E} showed that only α_{1E}, and to a much lesser extent α_{1A}, transcripts are present in PS, RS, and CS (Fig. 7C; Liévano et al., 1996). Recent studies using similar strategies revealed that transcripts for α_{1C} (Benoff, 1998; Espinosa et al, 1999), G, and H (Espinosa et al., 1999) are also present in spermatogenic cells. Immunodetection with specific antibodies has revealed the presence of α_{1A}, C, and E in mature sperm (Benoff, 1998; Westenbroek and Babcock, 1999). Correlating channel gene expression and function in spermatogenic cells will allow for a better understanding of how these ion channels participate in fertilization.

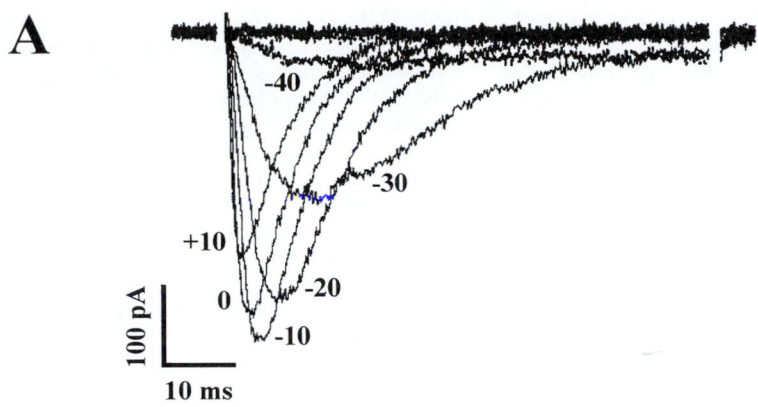

A

-40

-30

+10

0

-20

-10

100 pA

10 ms

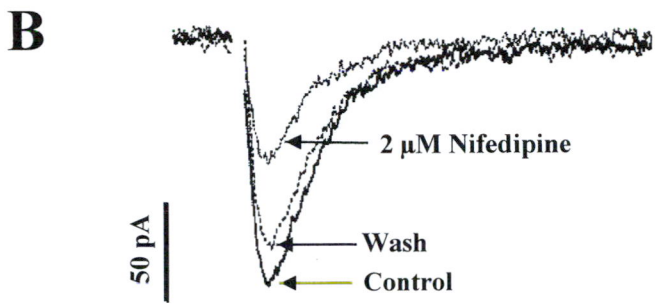

B

2 µM Nifedipine

Wash

Control

50 pA

C

α1 subunit gene	mRNA in spermatogenic cells	Reference
A	+	1
B	–	1
C	+	2,3
D	–	1
E	+	1
G	+	2
H	+	2
I	ND	—
S	ND	—

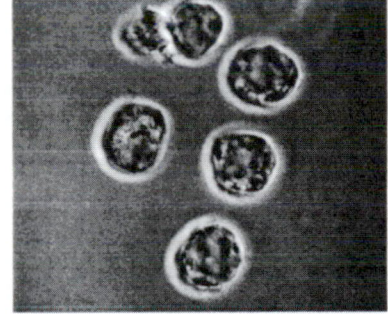

FIGURE 7. Low-threshold Ca^{2+} channels are expressed in mouse spermatogenic cells. (A) T-type channel records obtained from pachytene spermatocytes elicited by membrane depolarizations from −70 to +10-mV in 10 mV steps from a holding potential (E_h) of −80 mV. Numbers beside the records correspond to the value in mV of the applied test pulse. (B) Spermatogenic T-type channel is reversibly blocked by dihydropyridines. Superimposed records show the control, the effect of 2 µM nifedipine, and the recovery after washing nifedipine. (C) Expression of various α1 transcripts of voltage-dependent Ca^{2+} channels in spermatogenic cells detected by RT-PCR as described in the references. A + indicates that mRNA is present, a − indicates that mRNA is not present, and ND indicates that the presence of mRNA is not determined. References: 1, Liévano *et al.,* (1996); 2, Espinosa *et al.,* (1999); 3, Benoff (1998). The right side shows a phase-contrast image of a mixture of spermatogenic cells, mainly pachytene spermatocytes and round spermatids.

IV. Summary

Cell communication involves molecular mechanisms that are at the forefront of research in biology today, since they play a key role in determining the behavior of organisms. Successful gamete interactions requiring cell signaling and the crucial participation of ion channels determine the propagation of life.

Ionic fluxes play a fundamental role in activation of respiration and motility, chemotaxis, the sperm acrosome reaction (AR), and therefore in fertilization. Indeed, sperm are excitable cells that quickly respond to components from the outer layer of the egg, the jelly, with fast changes in their plasma membrane permeability. Model membranes formed from sperm components and patch-clamp techniques in whole cells have been used to detect, for the first time, the activity of single channels in the plasma membrane of sea urchin and mouse sperm. These techniques are now being applied to mammalian sperm, and together with studies of E_m, $[Ca^{2+}]_i$, and pH_i in whole sperm, have established the presence of K^+, Ca^{2+}, and Cl^- channels in this specialized cell and are helping to unravel their participation in chemotaxis and in the AR.

Sperm are very tiny cells (head diameter ca. 2–3 μm), which has hindered the characterization of their electrophysiological properties that would shed light on the molecular mechanisms leading to their fascinating, egg-induced behavioral changes. Sea urchin sperm can be swollen in diluted seawater, maintaining the regulation of their $[Ca^{2+}]_i$, pH_i and E_m. Swollen sperm can be much more easily patch-clamped than normal sperm, thus providing new avenues to study ionic channels and their regulation by egg factors and second messengers. Certain strategies developed for the sea urchin are being applied to mammalian sperm. There is presently background information about some of the ion channels present in these cells. Future study will determine the molecular mechanisms that regulate these channels in the cell. An alternative to identifying and functionally studying ion channels in sperm is to look for their expression in spermatogenic cells, combining molecular biological strategies and electrophysiology. Hopefully, this will allow a deeper understanding of the finely orchestrated events that lead to sperm activation, induction of the acrosome reaction, and in the end to the generation of a new individual.

Acknowledgments

This work was supported by grants from CONACyT, DGAPA-UNAM, and the Howard Hughes Medical Institute. The authors thank Carmen Beltrán for critically reading the manuscript, and José Luis de la Vega-Beltrán and Shirley Ainsworth for their technical support.

Bibliography

Arnoult, C., Cardullo, R. A., Lemos, J. R., and Florman, H. M. (1996a). Activation of mouse sperm T-type Ca^{2+} channels by adhesion to the egg zona pellucida. *Proc. Natl. Acad. Sci. USA* **93**, 13 004–13 009.

Arnoult, C., Kazam, I. G., Visconti, P. E., Kopf, G. S., Villaz, M., and Florman, H. M. (1999). Control of the low voltage-activated calcium channel of mouse sperm by egg ZP3 and by membrane hyperpolarization during capacitation. *Proc. Natl. Acad. Sci.* **96**, 6757–6762.

Arnoult, C., Zeng, Y., and Florman, H. (1996b). ZP3-dependent activation of sperm cation channels regulates acrosomal secretion during mammalian fertilization. *J. Cell Biol.* **134**, 637–645.

Austin, C. R. (1951). Observations on the penetration of the sperm into the mammalian egg. *Aust. J. Sci. Res. B* **4**, 581–596.

Babcock, D. F., and Pfeiffer, D. R. (1987). Independent elevation of cytosolic $[Ca^{2+}]$ and pH of mammalian sperm by voltage-dependent and pH-sensitive mechanisms. *J. Biol. Chem.* **262**, 15 041–15 047.

Babcock, D. F., Bosma, M. M., Battaglia, D. E., and Darszon, A. (1992). Early persistent activation of sperm K^+ channels by the egg peptide speract. *Proc. Natl. Acad. Sci. USA.* **89**, 6001–6005.

Baldi, B., Luconi, M., Bonaccorsi, L., Krausz, C., and Forti, G. (1996). Human sperm activation during capacitation and acrosome reaction: role of calcium, protein phosphorylation and lipid remodelling pathways. *Front. Biosci.* **1**, d189–205.

Barritt, G. J. (1999). Receptor-activated Ca^{2+} inflow in animal cells: a variety of pathways tailored to meet different intracellular Ca^{2+} signalling requirements. *Biochem. J.* **337**, 153–169.

Bean, B. P. (1989). Classes of calcium channels in vertebrate membranes. *Ann. Rev. Physiol.* **51**, 367–389.

Bean, B. P., and McDonough, S. I. (1998). Two for T. *Neuron* **20**, 825–828.

Beltrán, C., Darszon, A., Labarca, P., and Liévano, A. (1994). A high-conductance multistate Ca^{2+} channel found in sea urchin and mouse spermatozoa. *FEBS Lett.* **338**, 23–26.

Beltrán, C., Zapata, O., and Darszon, A. (1996). Membrane potential regulates sea urchin sperm adenylyl cyclase. *Biochem.* **35**, 7591–7598.

Benoff, S. (1998). Voltage-dependent calcium channels in mammalian spermatozoa. *Front. Biosci.* **3**, d1220–1240.

Blackmore, P. F. (1993). Thapsigargin elevates and potentiates the ability of progesterone to increase intracellular free calcium in human sperm: possible role of perinuclear calcium. *Cell Calcium* **14**, 53–60.

Blackmore, P. F. (1999). Extragenomic actions of progesterone in human sperm and progesterone metabolites in human platelets. *Steroids* **64**, 149–156.

Bleil, J. D. (1991). Sperm receptors of mammalian eggs. In "Elements of Mammalian Fertilization" Vol. 1 (P. M. Wassarman (Ed)), pp. 133–152. CRC Press, Boca Raton, FL.

Boettger-Tong, H., Aarons, D., Biegler, B., Lee, T., and Poirier G. R. (1992). Competition between zonae pellucidae and a proteinase inhibitor for sperm binding. *Biol Reprod.* **47**, 716–722.

Bookbinder, L. H., Cheng, A., and Bleil, J. D. (1995). Tissue- and species-specific expression of sp56, a mouse sperm fertilization protein. *Science* **269**, 86–89.

Chan, H. C., Zhou, T. S., Fu, W. O., Wang, W. P., Shi, L., and Wong, P. Y. D. (1997). Cation and anion channels in rat and human spermatozoa. *Biochim. Biophys. Acta.* **1323**, 117–129.

Chang, M. C. (1951). Fertilizing capacity of spermatozoa deposited into the fallopian tubes. *Nature* **168**, 697–698.

Clapham, D. E. (1998). Not so funny anymore: pacing channels are cloned. *Neuron* **21**, 5–7.

Cook, S. P., and Babcock, D. F. (1993a). Selective modulation by cGMP of the K^+ channel activated by speract. *J. Biol. Chem.* **268**, 22 402–22 407.

Cook, S. P,. and Babcock, D. F. (1993b). Activation of Ca^{2+} permeability by cAMP is coordinated through the pH_i increase induced by speract. *J. Biol. Chem.* **268**, 22 408–22 413.

Cook, S. P., Brokaw C. J., Muller, C. H., and Babcock, D. F. (1994). Sperm chemotaxis: egg peptides control cytosolic calcium to regulate flagellar response. *Dev. Biol.* **165**, 10–19.

Cox, T. and Peterson, R. N. (1989). Identification of calcium conducting channels in isolated boar sperm plasma membranes. *Biochem. Biophys. Res. Commun.* **161**, 162–168.

Cox, T., Campbell, P., and Peterson, R. N. (1991). Ion channels in boar sperm plasma membranes: characterization of a cation selective channel. *Mol. Reprod. Dev.* **30**, 135–147.

Cross, N. L., and Razy-Faulkner, P. (1997). Control of human sperm intracellular pH by cholesterol and its relationship to the response of the acrosome to progesterone. *Bio. Repro.* **56**, 1169–1174.

Cuéllar-Mata, P., Martínez-Cadena, G., Castellano, L. E., Aldana-Velóz, G., Novoa-Martínez, G., Vargas, I., Darszon, A., and García-Soto, J. (1995). Multiple G-binding proteins in sea urchin sperm: evidence for Gs and small G-proteins. *Develop. Growth and Diff.* **37**, 173–181.

Darszon, A., Labarca, P., Beltrán, C., García-Soto, J., and Liévano, A. (1994). Sea urchin sperm: an ion channel reconstitution study case. *Methods: A Companion to Methods Enzymol.* **6**, 37–50.

Darszon, S., Labarca, P., Nishigaki, T., and Espinosa, F. (1999). Ion channels in sperm physiology. *Physiol. Rev.* **79**, 481–510.

Darszon, A., Liévano, A., and Beltrán, C. (1996). Ion channels: key elements in gamete signaling. *Curr. Top. Devel. Biol.* **34**, 117–167.

Drewett, J. G., and Garbers, D. L. (1994). The family of guanylyl cyclase receptors and their ligands. *Endocrine Rev.* **15**, 135–162.

Espinosa, F., De La Vega-Beltrán, J. L., López-González, I., Delgado, R., Labarca, P., and Darszon, A. (1998). Mouse sperm patch-clamp recordings reveal single Cl⁻ channels sensitive to niflumic acid, a blocker of the sperm acrosome reaction. *FEBS Lett.* **426**, 47–51.

Espinosa, F., Lopéz-González, I., Serrano, C. J., Gasque, G., De La Vega-Beltrán, J. L., Trevino, C. L., and Darszon, A. (1999). Anion channel blockers differentially affect T-type Ca^{2+} currents of mouse spermatogenic cells, $\alpha1E$ currents expressed in *Xenopus* oocytes and the sperm acrosome reaction. *Devel. Gen.* **25**, 103–114.

Florman, H. M. (1994). Sequential, focal, and global, elevations of sperm intracellular Ca^{2+} are initiated by the zona pellucida during acrosomal exocytosis. *Dev. Biol.* **165**, 152–164.

Florman, H. M., Arnoult, C., Kazam, I. G., Li, C., and O'Toole, C. M. B. (1998). A perspective on the control of mammalian fertilization by egg-activated ion channels in sperm: a tale of two channels. *Biol. Reprod.* **59**, 12–16.

Florman, H. M., and Babcock, D. F. (1991). Progress toward understanding the molecular basis of capacitation. *In* "Elements of Mammalian Fertilization. I. Basic Concepts" (P. M. Wassarman, Ed.), pp. 105–132. CRC Press, Boca Raton, FL.

Florman, H., Corron, M. E., Kim, T. D-H. and Babcock, D. F, (1992). Activation of voltage-dependent calcium channels of mammalian sperm is required for zona pellucida-induced acrosomal exocytosis. *Dev. Biol.* **152**, 304–314.

Foltz, K. R., Partin, J. S., and Lennarz, W. J. (1993). Sea urchin egg receptor for sperm: sequence similarity of binding domain and hsp70. *Science* **259**, 1421–1425.

Gao, Z., and Garbers, D. L. (1998). Species diversity in the structure of zonadhesin, a sperm-specific membrane protein containing multiple cell adhesion molecule-like domains. *J. Biol. Chem.* **273**, 3415–3421.

Garbers, D. L., (1989). Molecular basis of fertilization. *Annu. Rev. Biochem.* **58**, 719–742.

Gauss, R., Seifert, R., and Kaupp, U. B. (1998). Molecular identification of a hyperpolarization-activated channel in sea urchin sperm. *Nature* **393**, 583–587.

Gong, X., Dubois, D. H., Miller, D. J., Shur, B. D. (1995). Activation of a G protein complex by aggregation of b-1,4-galactosyltransferase on the surface of sperm. *Science* **269**, 1718–1721.

Gmachl, M., and Kreil, G. (1993). Bee venom hyaluronidase to a membrane protein of mammalian sperm. *Proc. Natl. Acad. Sci. USA.* **90**, 3569–3573.

González-Martínez, M. T., and Darszon, A. (1987). A fast transient hyperpolarization occurs during the sea urchin sperm acrosome reaction induced by egg jelly. *FEBS Lett.* **218**, 247–250.

González-Martínez, M. T., Guerrero, A., Morales, E., de De La Torre, L., and Darszon, A. (1992). A depolarization can trigger Ca^{2+} uptake and the acrosome reaction when preceded by a hyperpolarization in *L. pictus* sea urchin sperm. *Dev. Biol.* **150**, 193–202.

Guerrero, A., and Darszon, A. (1989a). Egg jelly triggers a calcium influx which inactivates and is inhibited by calmodulin antagonists in the sea urchin sperm. *Biochim. Biophys. Acta* **980**, 109–116.

Guerrero, A., and Darszon, A. (1989b). Evidence for the activation of two different Ca channels during the egg jelly-induced acrosome reaction of sea urchin sperm. *J. Biol. Chem.* **264**, 19 593–19 599.

Guerrero, A., Sánchez, J. A., and Darszon, A. (1987). Single-channel activity in sea urchin sperm revealed by the patch-clamp technique. *FEBS Lett.* **220**, 295–298.

Hagiwara, N., and Kawa, K. (1984). Calcium and potassium currents in spermatogenic cells dissociated from rat seminiferous tubules. *J Physiol. (London)* **356**, 135–149.

Hille, B. (1992). "Ion Channels of Excitable Membranes," 2nd Ed. Sinauer Assoc. Inc. Pub., Sunderland, MA.

Kaupp, B. (1995). Family of cyclic nucleotide gated ion channels. *Curr. Op. Neurobiol.* **5**, 434–442.

Keller, S. T., and Vacquier, V. D. (1994). The isolation of acrosome reaction-inducing glycoproteins from sea urchin egg jelly. *Dev. Biol.* **162**, 304–312.

Kretzer, D. M., and Kerr, J. B. (1988). The cytology of the testis. *In* "The Physiology of Reproduction," Vol 1 (E. Knobil and J. D. Neill, Eds.), pp. 837–932. Raven Press, New York.

Labarca, P., Santi, C., Zapata, O., Beltrán, C., Liévano, A., Sandoval, Y., and Darszon, A. (1997). Possible participation of a cAMP regulated K^+ channel from the sea urchin sperm in the speract response. *In* "From Ion Channels to Cell-to-Cell Conversations" (R. Latorre, Ed.), pp. 147–168. Plenum Publishing Corporation, New York and London.

Labarca, P., Santi, C., Zapata, O., Morales, E., Beltrán, C., Liévano, A., and Darszon, A. (1996). A cAMP regulated K^+-selective channel from the sea urchin sperm plasma membrane. *Dev. Biol.* **174**, 271–280.

Labarca, P., Zapata, O., Beltrán, C., and Darszon, A. (1995). Ion channels from the mouse sperm plasma membrane in planar lipid bilayers. *Zygote* **3**, 199–206.

Lee, H. C. (1984a). Sodium and proton transport in flagella isolated from sea urchin spermatozoa. *J. Biol. Chem.* **259**, 4957–4963.

Lee, H. C. (1984b). A membrane potential-sensitive Na^+-H^+ exchange system in flagella isolated from sea urchin spermatozoa. *J Biol. Chem.* **259**, 15 315–15 319.

Lee, H. C. (1988). Internal GTP stimulates the speract receptor mediated voltage change in sea urchin spermatozoa membrane vesicles. *Dev. Biol.* **126**, 91–97.

Lee, H. C., and Garbers, D. L. (1986). Modulation of the voltage sensitive Na^+/H^+ exchange in sea urchin spermatozoa through membrane potential changes induced by the egg peptide speract. *J. Biol. Chem.* **261**, 16 026–16 032.

Lee, H. C., Johnson, C., and Epel, D. (1983). Changes in internal pH associated with the initiation of motility and the acrosome reaction of sea urchin sperm. *Dev. Biol.* **95**, 31–45.

Leyton, L., and Saling, P. (1989). 95 kDa sperm proteins bind ZP3 and serve as tyrosine kinase substrates in response to zona binding. *Cell* **57**, 1123–1130.

Liévano, A., Bolden, A., and Horn, R. (1994). Calcium channels in excitable cells: divergent genotypic and phenotypic expression of α_1-subunits. *Am. J. Physiol.* **267**, C411–C424.

Liévano, A., Sanchez, J., and Darszon, A. (1985). Single channel activity of bilayers derived from sea urchin sperm plasma membranes at the tip of a patch-clamp electrode. *Dev. Biol.* **112**, 235–295.

Liévano, A., Santi, C., Serrano, C. J., Trevino, C. L., Bellvé, A. R., Hernádez-Cruz, A., and Darszon, A. (1996). T-type Ca^{2+} channels and a_{1E} expression in spermatogenic cells, and their possible relevance to the sperm acrosome reaction. *FEBS Lett.* **388**, 150–154.

Liévano, A., Vega Saenz de Miera, E. C., and Darszon, A. (1990). Ca^{2+} channels from the sea urchin sperm plasma membrane *J. Gen. Physiol.* **95**, 273–296.

McLeskey, S. B., Dowds, C., Carballada, R., White, R. R., and Saling, P. M. (1998). Molecules involved in mammalian sperm-egg interaction. *Int. Rev. Cytol.* **177**, 57–113.

Meizel, S. (1997). Amino acid neurotransmitter receptor/chloride channels of mammalian sperm and the acrosome reaction. *Biol. Reprod.* **56**, 569–574.

Meizel, S., and Turner, K. O. (1993). Initiation of the human sperm acrosome reaction by thapsigargin. *J. Exp. Zool.* **267**, 350–355.

Morales, E., de De la Torre, L., Moy, G., Vacquier, V. D., and Darszon, A. (1994). Anion channels in the sea urchin sperm plasma membrane. *Mol. Reprod. Dev.* **36**, 174–182.

Moy, G., Mendoza, L. M., Schulz, J. R., Swanson, W. J., Glabe, C. G., and Vacquier, V. D. (1996). The sea urchin sperm receptor for egg jelly is a modular protein with extensive homology to the human polycystic kidney disease protein, PKD1. *J. Cell. Biol.* **133**, 809–817.

Murase, T., and Roldan, E. R. S. (1996). Progesterone and the zona pellucida activate different transducing pathways in the sequence of events leading to diacylglycerol generation during mouse sperm acrosomal exocytosis. *Biochem. J.* **320**, 1017–1023.

Parekh, A. B., and Penner, R. (1997). Store depletion and calcium influx. *Physiol. Rev.* **77**, 901–930.

Reynaud, E., de De la Torre, L., Zapata, O., Liévano, A., and Darszon, A. (1993). Ionic bases of the membrane potential changes induced by speract in swollen sea urchin sperm. *FEBS Lett.* **329**, 210–214.

Roldan, E. R. S., Murase, T., and Shi, Q.-X. (1994). Exocytosis in spermatozoa in response to progesterone and zona pellucida. *Science* **266**, 1578–1581.

Salvatore, L., D'Adamo, M. C., Polishchuk, R., Salmona, M., and Pessia, M. (1999). Localization and age-dependent expression of the inward rectifier K$^+$ channel subunit Kir5.1 in a mammalian reproductive system. *FEBS Lett.* **449**, 146–152.

Sanford, R., Sgotto, B., Aparicio, S., Brenner, S., Vaudin, M., Wilson, R. K., Chissoe, S., Pepin, K., Chothia, C., Hughes, J., and Harris, P. (1997). Comparative analysis of the polycystic kidney disease 1 (PKD1) gene reveals an integral membrane glycoprotein with multiple evolutionary conserved domains. *Hum. Mol. Gen.* **6**, 1483–1489.

Santi, C. M., Darszon, A., and Hernández, A. (1996). A dihydropyridine-sensitive T-type Ca^{2+} current is the main Ca^{2+} current carrier in mouse primary spermatocytes. *Am. J. Physiol.* **271**, C1583–C1593.

Santi, C. M., Santos, T., Hernández, A., and Darszon, A. (1998). Properties of a novel pH-dependent Ca^{2+} permeation pathway present in male germ cells with possible roles in spermatogenesis and mouse sperm function. *J. Gen. Physiol.* **112**, 33–53.

Schackmann, R. W. (1989) Ionic regulation of the sea urchin sperm acrosome reaction and stimulation by egg-derived peptides. *In:* "The Cell Biology of Fertilization" (H. Schatten and G. Schatten, Eds.) Academic Press, San Diego, CA, pp. 3–28.

Schackmann, R. W., and Chock, P. B. (1986). Alteration of intracellular [Ca^{2+}] in sea urchin sperm by the egg peptide speract. *J. Biol. Chem.* **261**, 8719–8728.

Schreiber, M., Wei, A., Yuan, A., Gaut, J., Saito, M., and Salkoff, L. (1998). Slo3, a novel pH-sensitive K$^+$ channel from mammalian spermatocytes. *J. Biol. Chem.* **273**, 3509–3516.

Seifert, R., Scholten, A., Gauss, R., Micheva, A., Lichter, P., and Kaupp, U. B. (1999). Molecular characterization of a slowly gating human hyperpolarization-activated channel predominantly expressed in thalamus, heart and testis. *Proc. Natl. Acad Sci. USA.* **96**, 9391–9396.

Srivastava, P. N., Brewer, J. M., and White, R. A. (1982). Hydrolysis of *p*-nitrophenylphosphorylcholine by alkaline phosphatase and phospholipase C from rabbit sperm-acrosome. *Biochem. Biophys. Res. Commun.* **108**, 1120–1125.

Stea, A., Wah Soong, T., and Snutch, T. P. (1995). Voltage-gated calcium channels. *In* "Handbook of Receptors and Channels. Ligand and Voltage-Gated Ion Channels." (A. North, Ed.), pp. 112–150. CRC Press, Boca Raton, FL.

Suzuki, N., and Yoshino, K. (1992). The relationship between amino acid sequences of sperm-activating peptides and the taxonomy of equinoids. *Comp. Biochem. Physiol.* **102B**, 679–690.

Tesarik, J., Carreras, A., and Mendoza, C. (1993). Differential sensitivity of progesterone- and zona pellucida–induced acrosome reactions to pertussis toxin. *Mol. Reprod. Dev.* **34**, 183–189.

Thomas, P., and Meizel, S. (1989). Phosphatidyl inositol 4,5-bisphosphate hydrolysis in human sperm stimulated with follicular fluid or progesterone is dependent upon Ca^{2+} influx. *Biochem. J.* **264**, 539–546.

Tiwari-Woodruff, S. K., and Cox, T. (1995). Boar sperm plasma membrane Ca^{2+}-selective channels in planar bilayers. *Am. J. Physiol.* **268**, C1284–C1294.

Tomes, C. N., McMaster, C. R., and Saling, P. M. (1996). Activation of mouse sperm phosphatidylinositol-4,5 bisphosphate-phospholipase C by zona pellucida is modulated by tyrosine phosphorylation. *Mol. Reprod. Dev.* **43**, 196–204.

Trevino, C. L., Santi, C. M., Beltrán, C., Hernández-Cruz, A., Darszon, A., and Lomeli, H. (1998). Localisation of 1P$_3$ and ryanodine receptors during mouse spermatogenesis: possible functional implications. *Zygote* **6**, 159–172.

Trimmer, J. S., and Vacquier, V. D. (1986). Activation of sea urchin gametes. *Ann. Rev. Cell Biol.* **2**, 1–26.

Vanha-Perttula, T., and Kasurinen, J. (1989). Purification and characterization of phosphatidylinositol-specific phospholipase C from bovine spermatozoa. *Int. J. Biochem.* **21**, 997–1007.

Visconti, P. E., and Kopf, G. S. (1998). Regulation of protein phosphorylation during sperm capacitation. *Biol. Reprod.* **59**, 1–6.

Walensky, L. D., and Snyder, S. H. (1995). Inositol 1,4,5-trisphosphate receptors selectively localized to the acrosomes of mammalian sperm. *J. Cell Biol.* **130**, 857–869.

Ward, G. E., Brokaw, C. J., Garbers, D. L., and Vacquier, V. D. (1985). Chemotaxis of *Arbacia punctulata* spermatozoa to resact, a peptide from the egg jelly layer. *J. Cell. Biol.* **101**, 2324–2329.

Ward, G. R., and Kopf, G. (1993). Molecular events mediating sperm activation. *Devel. Biol.* **158**, 9–34.

Ward, G. R., Storey, B. T., and Kopf, G. (1994). Selective activation of G$_{i1}$ and G$_{i2}$ in mouse sperm by the zona pellucida, the eggs extracellular matrix. *J. Biol. Chem.* **269**, 13 254–13 258.

Westenbroek, R. E., and Babcock, D. F. (1999). Discrete regional distributions suggest diverse functional roles of calcium channel α_1 subunits in sperm. *Dev. Biol.* **207**, 457–469.

Weyand, I., Godde, M., Frings, S., Welner, J., Muller, F., Altenhofen, W., Hatt, H., and Kaupp, B. (1994). Cloning and functional expression of a cyclic-nucleotide-gated channel from mammalian sperm. *Nature (London)* **368**, 859–863.

Wiesner, B., Weiner, J., Middendorff, K., Hagen, V., Kaupp, U. B., and Weyand, I. (1998). Cyclic nucleotide-gated channels on the flagellum control Ca^{2+} entry into sperm. *J. Cell Biol.* **142**, 473–484.

Wissenbach, U., Schroth, G., Phillipp, S., and Flockerzi, V. (1998). Structure and mRNA expression of a bovine trp homologue related to mammalian trp2 transcripts. *FEBS Lett.* **42**, 61–66.

Wistrom, C. A., and Meizel, S. (1993) Evidence supporting involvement of a unique human steroid receptor/Cl$^-$ channel complex in the progesterone-initiated acrosome reaction. *Dev. Biol.* **159**, 679–690.

Yanagimachi, R. (1994). Mammalian fertilization. *In* "The Physiology of Reproduction", (E. Knobil and J. D. Neil, Eds.) pp. 189–317. Raven Press, New York.

Yau, K-W. (1994). Cyclic nucleotide–gated channels: an expanding new family of ion channels. *Proc. Natl. Acad. Sci. USA* **91**, 3481–3483.

Young, G. P. H., Koide, S. S., Goldstein, M., and Young, J. D. E. (1988). Isolation and partial characterization of an ion channel protein from human sperm membranes. *Arch. Biochem. Biophys.* **262**, 491–500.

Zeng, Y., Clark, E. N., and Florman, H. M. (1995). Sperm membrane potential: hyperpolarization during capacitation regulates zona pellucida–dependent acrosomal secretion. *Dev. Biol.* **171**, 554–563.

Zeng, Y., Oberdorf, J. A., and Florman, H. M. (1996). pH regulation in mouse sperm: identification of Na$^+$-, Cl$^-$-, and HCO3$^-$-dependent and arylaminobenzoate-dependent regulatory mechanisms and characterization of their roles in sperm capacitation. *Dev. Biol.* **173**; 510–520.

William J. Larsen and Richard D. Veenstra

31

Biology of Gap Junctions

I. Introduction

In the late 1950s and early 1960s, physiologists who had been poking fine glass current-injecting and current-recording electrodes into neighboring cells within a variety of tissues made an interesting discovery. They found that while the injection of current into a cell caused a predictable shift in its nonjunctional membrane potential, it also caused a similar shift in the nonjunctional membrane potential of immediately adjacent cells (Fig. 1).

One interpretation of this finding was that the ions emanating from the injection electrode were able to flow freely from the injected cell to the adjacent cell and did so in preference to pathways leading to the extracellular medium or to the intercellular space. Moreover, several studies demonstrated that fluorescent dyes were selectively transferred from cell to cell when injected into the cytoplasm through glass injection pipettes. Many observations such as these led to the hypothesis that some cells were coupled (electrically or with respect to dye transfer) by **permeable cell junctions**.

Additionally, in the neuronal tissues in which this phenomenon had first been witnessed, it was found that action potentials generated in a presynaptic element could be passed to a postsynaptic element much faster than would occur if the presynaptic and postsynaptic elements were connected by chemical synapses (Fig. 2). Ironically, this evidence for the presence of permeable cell junctions that could serve as **electrical synapses** in the nervous system came to light soon after the common acceptance of Otto Loewi's findings that supported the idea that neuronal synapses were probably chemical in nature and not electrical as traditionally believed.

II. Advantages of Electrical Synapses in Excitable Cells

The utility of electrical synapses in excitable cells is apparent. They may pass action potentials or subthreshold electrical activity more rapidly from one neuronal element

to another than can their chemical counterparts. This advantage is especially obvious in neuronal pathways that serve as **escape mechanisms**, such as those in the tail muscles of crayfish and lobsters and those in the pectoral fins of fishes. In these cases, the rapidity of the animal's response to imminent danger has selective value. On the other hand, such permeable junctions connecting smooth muscle cells of the uterus, or cardiac muscle cells of the heart wall, provide a mechanism for the systematic cell-to-cell spread of depolarization that is required for **coordinated and effective contractile activity**.

III. Ubiquitous Membrane Permeable Junctions

Numerous studies published in the 1960s and 1970s supported the idea that virtually all cells in normal tissues (even

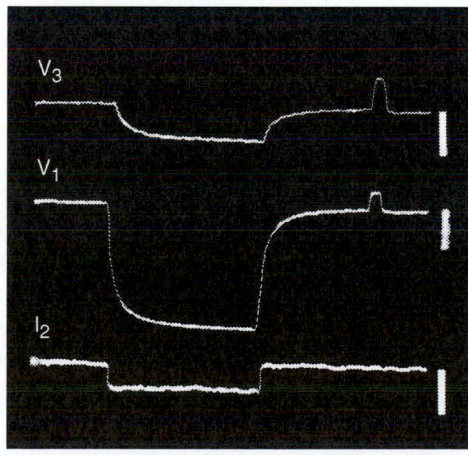

FIGURE 1. The injection of current into one cultured WI-26 cell (I2) within a monolayer produces a shift in potential within the injected cell (V1) and in a cell two or three cells removed from the injected cell (V3), indicating that ions may move freely between contacting cells. [From Furshpan, E. J., and Potter, D. D. (1968). Low-resistance junctions between cells in embryos and tissue culture. *Curr. Top. Dev. Biol.* **3,** 95, with permission.]

those in inexcitable tissues) were coupled by permeable cell junctions. Based on studies with native biological molecules and with tracers of different sizes, it was also suggested that the pores within vertebrate cell-to-cell junctions were approximately 1.2 nm in diameter and that molecules larger than about 1 kDa were excluded. Thus, it was suggested that the biological molecules capable of freely moving from cell to cell in normal tissues would include a wide range of ions and small metabolites including sugars, nucleotides, and nucleosides, and possible signalling molecules such as cyclic adenosine monophosphate (cAMP).

Since the pores initially appeared to select molecules only with respect to size, it was suggested that one function of permeable cell junctions was to buffer the concentrations of small metabolites throughout the tissue. Direct evidence for the cell contact-mediated exchange of metabolites, for example, was provided by metabolic cooperation experiments in which it was shown that a normal wild-type cell, able to incorporate thymidine into DNA, could transfer DNA precursor molecules (probably the nucleoside triphosphate form of thymidine) to mutant thymidine kinase-deficient cells only if the wild-type and mutant cell were in contact. In these "kiss-of-life" experiments, the mutant cells were then able to incorporate this nucleoside into DNA required for continued proliferation and survival.

IV. Structural Candidates for the Permeable Cell Junction

As electron microscopes first came into common use in the late 1950s, several different kinds of cell-cell junctions were discovered. The **tight junction**, or **occludens-type junction**, was characterized by the fusion of membranes of adjacent cells and so held some appeal as a potential permeable cell junction. **Septate junctions** were also good candidates for the permeable cell junction since they were characterized by cell-to-cell bridges or septa covering a large region of cell-cell apposition in many of the tissues shown to be well coupled by electrophysiological or dye-tracing techniques. The basic argument against their function as permeable cell junctions, however, was that their distribution among tissues and organisms was significantly more limited than was the coupling phenomenon. The septate junction, for example, could not be identified in coupled vertebrate cells.

V. Ultrastructural Characterization of Gap Junctions and Correlations with Cell Coupling

In a pioneering study, Revel and Karnovsky infiltrated the intercellular spaces of heart and liver with an electron-opaque dye (lanthanum hydroxide) and then examined these preparations in the electron microscope. They found regions where the membranes of adjacent cells were apposed to one another across a uniform lanthanum-infiltrated intercellular space about 2–4 nm in width (Fig. 3A). In addition, in cross section, they observed small unstained structures bridging the stain-filled gap between adjacent cells in these regions, and in en face views these bridges were packed hexagonally within the intercellular space. Ultimately, freeze-fracture studies pro-

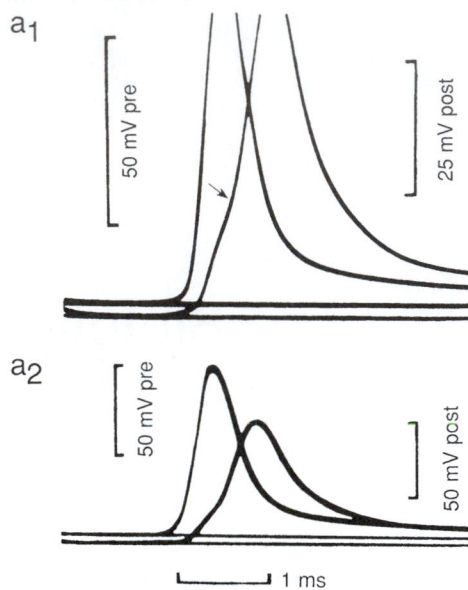

FIGURE 2.   The upper trace is an AP in the presynaptic element of a crayfish septate axon; the lower trace shows the subsequent AP in the postsynaptic element. The short lag of the latter following the former is typical of electrical synapses. The top and bottom traces were recorded at the same synapse at different amplifications. [From Furshpan, E. J., and Potter, D. D. (1959). Transmission at the giant motor synapses of the crayfish. *J. Physiol.* **145,** 289, with permission.]

vided evidence that these intercellular bridges were in continuity with structures that spanned the lipid bilayers of both cells and thereby could theoretically provide the structural foundation for the cell-to-cell conduit implied by earlier electrophysiological studies.

It was also very important that this cell-cell junction, which Revel and Karnovsky called the **gap junction**, was found in many tissues which were ion- or dye-coupled; including both inexcitable and excitable cells. While well-coupled inexcitable cells such as liver hepatocytes possessed large numbers of gap junctions, it was especially satisfying to find that these structures were prominent features of the cell-cell contact regions of excitable cells such as cardiac myocytes, and were prominent in neuronal systems that had been shown to possess electrical synapses. In addition, in the metabolic cooperation experiments described above, DNA precursor molecules were transferred between test cells only if they were capable of forming gap junctions in regions of cell-cell contact. Moreover, a large body of ultrastructural evidence indicated that gap junctions were components of virtually all tissues in multicellular organisms of the animal kingdom.

VI. Molecular and Structural Studies of Gap Junction Proteins

Application of biochemical and molecular approaches to the study of gap junctions proved to be more perplexing

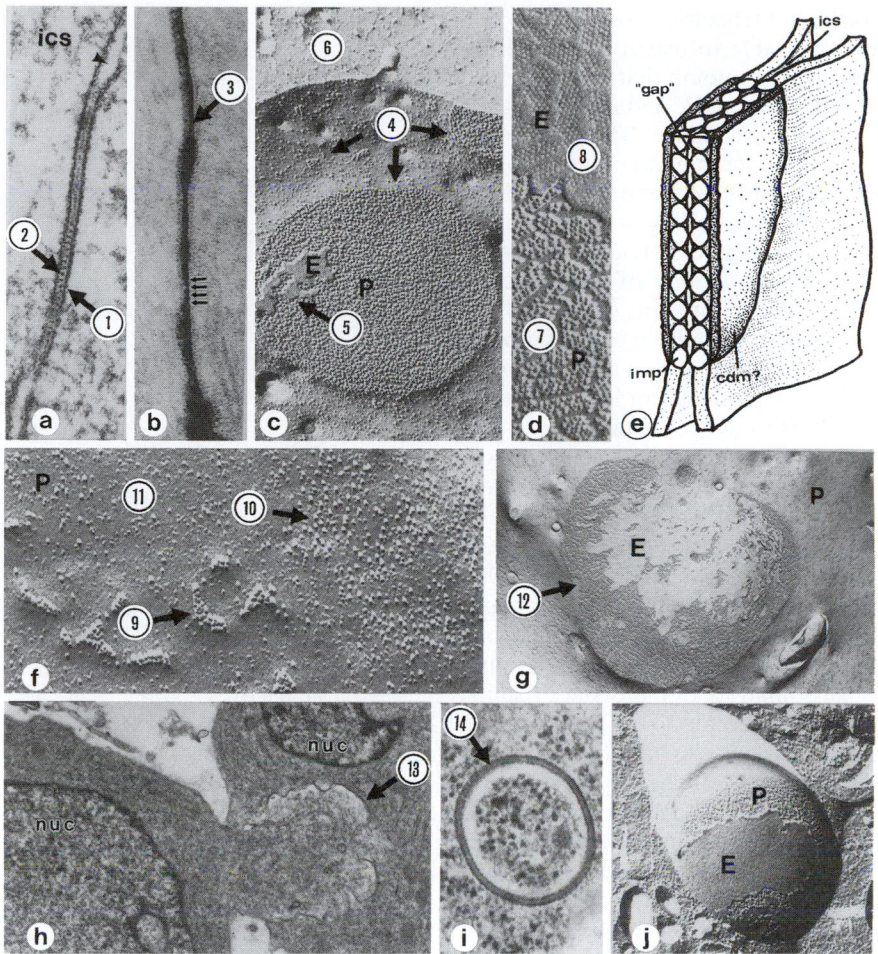

FIGURE 3. (a) Thin section of typical gap junction. The entire width of both apposed membranes and the intercellular space is about 18 nm. Dense material is often associated with cytoplasmic surface of gap junction (1). Some gap junctions are characterized by stained periodicites evident at level of "gap" (2). Magnification, 157,500×. [From Larsen, W. J., Skowron-Lomneth, C., and Carron, C. (1988). Gap junction modulation: Possible role in tumor cell behavior. *In* "Biochemical Mechanisms and Regulation of Intercellular Communication" (H. A. Milman and E. Elmore, Eds.), Vol. 14, p. 151. Princeton Scientific Publ. Co., Princeton, NJ, with permission.] (b) Lanthanum infiltrated gap junction. The lanthanum infiltrated intercellular "gap" (3) is about 2–4 nm. Magnification, 212,400× [Reproduced from *The Journal of Cell Biology*, 1975, vol. 67, p. 801, by copyright permission of the Rockefeller University Press.] (c) Freeze-fracture image of a gap junction. Each junctional membrane contains a set of particles and a set of pits. In vertebrate tissues, the particles adhere to the protoplasmic leaflet of the lipid bilayer (protoplasmic fracture face, P), whereas the pits remain associated with the extracellular leaflet (extracellular fracture face, E). Gap junction particles aggregate within particle-poor zones (4). The fracture may jump between the two membranes, resulting in adhesion of bits of the E-fracture face from the membrane of the adjacent cell (5) on the P-fracture face or vice versa. Occasionally, the fracture plane leaves the membrane and cuts into the cytoplasm (6). Magnification, 64,800×. [From Larsen, W. J. (1977). Structural diversity of gap junctions: A review. *Tissue Cell* **9,** 373, with permission.] (d) Enlargement of P- and E-fracture faces showing details of gap junction particles (7) and corresponding pits (8). Magnification, 139,000×. From Larsen, W. J. (1977). Structural diversity of gap junctions: A review. *Tissue Cell* **9,** 373, with permission.] (e) Simple model of gap junction showing particle-particle (imp) contact within the "gap" at the level of the intercellular space (ics). Cytoplasmic dense material (cdm) is particularily apparent in gap junctions in excitable tissues. [From Larsen, W. J. (1977). Structural diversity of gap junctions: A review. *Tissue Cell*, **9,** with permission.] (f) Gap junctions form by the aggregation (9) of single 11-nm particles (10) in particle-poor regions of the membrane (11). Magnification, 222,300×. [From Larsen, W. J., and Risinger, M. A. (1985). The dynamic life histories of intercellular membrane junctions. *In* "Modern Cell Biology" (B. Satir, Ed.), Vol. 4, p. 151. John Wiley & Sons, New York. Copyright 1985 John Wiley & Sons. Reprinted by permission of Wiley-Liss, Inc., a subsidiary of John Wiley & Sons, Inc.] (g) Very large gap junctions (12) containing thousands of gap junction particles, may represent terminal gap junction gaps. Magnification, 21,600× [From Larsen, W. J. (1977). Structural diversity of gap junctions: A review. *Tissue Cell* **9,** 373, with permission.] (h) The large gap junction caps may invaginate into the cell (13) through an endocytotic mechanism. Magnification, 14,400×. [From Larsen, W. J., and Tung, H. (1978). Origin and fate of gap junction vesicles in rabbit granulosa cells. *Tissue Cell* **10,** 585, with permission.] (i) Bimembranous gap junction vesicles (14) may pinch off from the invaginating junction to fuse with lysosomes before undergoing degradation. Magnification, 52,200×. [From Larsen, W. J., and Tung, H. (1978). Origin and fate of gap junction vesicles in rabbit granulosa cells. *Tissue Cell* **10,** 585, with permission.] (j) Freeze-fracture planes within the cytoplasm may reveal the P-face particles (P) and E-face pits (E) within cytoplasmic gap junction vesicles. Magnification, 38,700× [From Larsen, W. J., and Tung, H. (1978). Origin and fate of gap junction vesicles in rabbit granulosa cells. *Tissue Cell* **10,** 585, with permission.]

than originally envisioned, but tenacious investigators in a number of laboratories were able to overcome initial obstacles. It is now known that proteins that make up gap junctions in vertebrate tissues are basic and contain significant stretches of hydrophobic sequence. This is not surprising given the fact that these are integral membrane proteins. For these reasons, they proved difficult to isolate and purify. Extraction of tissues with boiling sodium dodecyl sulfate (SDS) proved to be the only reliable method for obtaining morphologically pure fractions of gap junction membranes for many years, but the proteins extracted from these fractions ran at variable molecular weights on polyacrylamide gels. It was controversial as to whether some of these bands were breakdown products, or other proteins associated with gap junctions in the cell membrane.

It was not until gentler, nondetergent extraction procedures were developed in the early 1980s that a consistent band on polyacrylamide gels with a predicted molecular weight of 26 kDa could be routinely isolated from rodent livers. Once this native protein was obtained and purified, it was possible to produce a polyclonal antibody. This antibody was used to screen an expression cDNA library to isolate and sequence its gene. This first sequence was published in 1986. Hydropathy analysis of the deduced amino acid sequence of this cDNA revealed a 32-kDa protein which could be interpreted as possessing four potential transmembrane regions, two extracellular loops, and an internal loop and amino and carboxyl termini that extended into the cytoplasm (Fig. 4). The mapping of these proteins with antibodies directed against specific sequences, and their cutting with specific proteases have largely confirmed this initial interpretation of the relationships of gap junction protein segments to the membrane. It was also postulated that six gap junction protein molecules called **connexins** constitute each gap junction intramembrane particle or **connexon**, thus providing a molecular basis for models of the permeable membrane junction deduced from earlier physiological and structural studies (Figs. 4 and 5). Indeed, more recent electron cryomicroscopy studies of frozen, rehydrated 2D crystals of a recombinant C-terminal truncated form of Cx43 has revealed even more detail regarding the relationship of connexins to the connexon, and the relationship of their specific transmembrane domains to the plasma membrane. These studies show that each connexin subunit contributes a transmembrane alpha helix that lines the aqueous pore, and an additional alpha helix adjacent to the surrounding membrane lipids. In addition, the images reveal that the apposing connexons that form each channel are staggered by about 30 degrees.

VII. Two Large Families of Gap Junction Proteins

Early structural studies, particularly those utilizing the freeze-fracture technique, demonstrated structural diversity among gap junctions distributed throughout the animal kingdom and within a variety of different tissues. Variability in some of these structural qualities was shown to depend upon

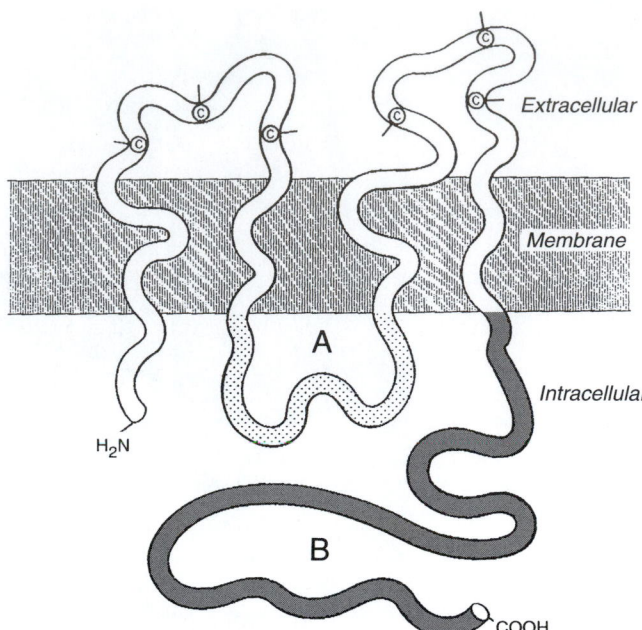

FIGURE 4. Structure and topology of a connexin. Both amino and carboxyl termini (B) are located in the cytoplasm along with an intermediate loop (A). Every connexin contains four membrane-spanning regions and two extracellular loops containing highly conserved cyteine residues (C). [From Beyer, E. C., Paul, D. L., and Goodenough, D. L. (1990). Connexin family of gap junction proteins. *J. Membr. Biol.* **116**, 187, with permission.]

the tissue in which the gap junction was located, or the phylum or species in which the gap junction was identified. While relationships between detailed gap junction structure and composition are not completely understood, it is now known that gap junctions in different tissues and in different phyla and species may be formed by different protein isoforms within two large families of gap junction proteins. The gap junctions in vertebrate tissues are formed by members of the **connexin family** of gap junction proteins. The gap junctions of invertebrate tissues appear to be formed by members of an analogous **innexin family** of gap junction proteins.

Members of the connexin family of gap junction proteins range in molecular weight from 26 kDa to 57 kDa (Fig. 6). Each connexin is named for its deduced molecular weight in kilodaltons so that the 26-kDa connexin in humans, for example, is called human connexin 26 (human Cx26), while the 30.3-kDa connexin in mice is called mouse connexin 30.3 (mouse Cx30.3). Sequence analysis supports the possibility that two major phylogenetic subfamilies of connexins diverged from each other between 1.3 and 1.9 billion years ago; one group ranging in size from about 26 kDa to 32 kDA and the other from about 33 kDa to 57 kDa. About a dozen different connexins have been identified in rodents; another dozen or so related isoforms have been described in other species, and the list is growing.

Typically, it appears that only one or a few connexin species may comprise the gap junctions within any given tissue. For example, Cx26 and Cx32 constitute gap junctions

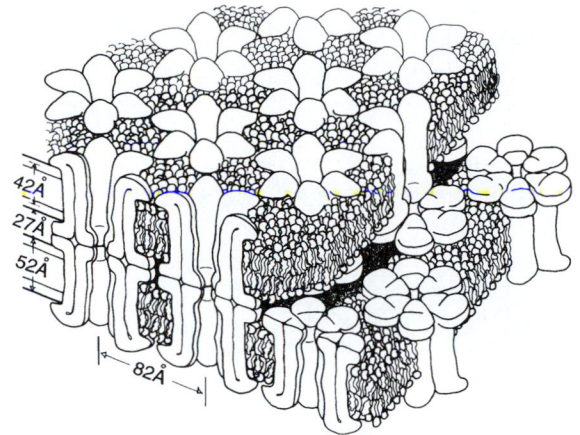

FIGURE 5. This diagram of a region of a gap junction isolated from mouse liver is based on electron microscope images and X-ray diffraction data. Each gap junction particle or connexon is composed of six gap junction protein molecules called connexins. Connexons in apposed membranes meet within the intercellular space. [Reproduced from *The Journal of Cell Biology*, 1977, **74**, p. 449, by copyright permission of the The Rockefeller University Press.]

of mammalian liver and have been shown to coexist within the same gap junction aggregates. On the other hand, the Cx32 and Cx43 documented in thyroid epithelium, appear to be segregated into separate gap junctions formed in different regions of the lateral cell membrane. Likewise, the distribution of three different connexins (Cx40, Cx43, and Cx45) within the human heart appears to be nonrandom. For example, Cx45 is the first connexin to appear in the embryonic mouse heart. Ultimately, however, its expression is restricted to the region of the SA node, while Cx40 is localized in the AV node and in myocytes of the atrium and conductive myocardium. In contrast, Cx43 expression occurs within the crista terminalis and other cardiomyocytes of the heart. As many as six different connexin genes (Cx30.1, Cx31, Cx31.1, Cx40, Cx43, and Cx45) are transcribed and translated in mouse embryos as early as the eight-cell stage. These studies imply differential functions for gap junctions composed of different connexins and consequently stimulate the following question: what part of the connexin molecule encodes functional specificity and how do these specificities differ from connexin to connexin?

It is known that the amino terminus, portions of the third transmembrane helix, cystine-containing regions of the extracellular loops, and serine residues of the carboxyl terminus are well conserved (Figs. 4 and 6). However, the cytoplasmic loops and cytoplasmic carboxyl termini vary significantly in length and amino acid sequence from one connexin to another. Moreover, heterotypic gap junctions can be formed only with connexons made of particular connexins, suggesting that sequence specificity within the extracellular domains may be relevant to gap junction assembly. Recent studies, therefore, have focused on the potential functional relevance of specific differences in amino acid sequences within the various domains of the connexin proteins (see Sections IX–XIV).

In invertebrates, proteins of the innexin family are able to form gap junctions based on ultrastructural criteria. In addi-

tion, studies with a *Xenopus* oocyte expression system have shown that some of the innexin proteins are able to form permeable cell-to-cell channels that may be voltage- or pH-sensitive. As many as 24 innexin genes have been identified in the nematode, *Caenorhabditis elegans*, while 6 innexin genes have been described in *Drosophila*. While the innexins bear no sequence homology to the connexin family of proteins, the structural similarities of innexins and connexins are dramatically similar. For example, like the connexins, the innexin proteins have four transmembrane domains, two extracellular loops, and cytoplasmic N- and C-termini. Moreover, the functions of the innexin proteins in invertebrates parallel functions of the connexins in vertebrate tissues. For example, the innexin encoded by the *shak*-B(neural) locus forms electrical synapses of the *Drosophila* nervous system, while the innexin encoded by the *eat*-5 locus forms permeable junctions responsible for synchronization of pharyngeal musculature in *Drosophila*.

VIII. Channels within Gap Junctions

Following the identification of the gap junction as the site of cell-to-cell transfer of ions and small hydrophilic molecules, a hypothesis developed that the pathway for this exchange consisted of an array of parallel aqueous channels. The central aqueous pore resides in each IMP particle in the freeze-fracture images or hexagonal bridge in the negatively stained micrographs of a gap junction plaque. An aqueous pore 16 nm in length with a channel diameter of approximately 14 Å should have a unitary conductance of 10^{-10} Ω (= 100 pS). Such a relatively high conductance should produce a detectable electrical signal in support of the channel

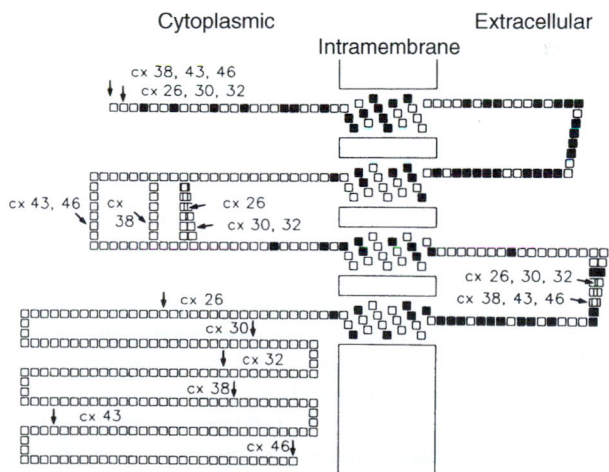

FIGURE 6. This diagram depicts the predicted structure of rat and *Xenopus* connexins. Each square represents an amino acid. The diagonal arrows mark points within the cytoplasmic and extracellular loops where the connexins differ in length. The vertical arrows indicate the amino and carboxyl termini for each connexin. Closed squares represent amino acids that are the same in all connexins. [From Bennett, M. V. L., Barrio, L. C., Bargiello, T. A., Spray, D. C., Hertzberg, E., and Saez, J. C. (1991). Gap junctions: New tools, new answers, new questions. *Neuron* **6**, 305, with permission. Copyright 1991 Cell Press.]

hypothesis for gap junction permeability. Small stepwise changes in the electrical coupling between paired *Xenopus* embryonic cells provided evidence for the existence of discrete coupling elements, but the low input resistance owing to the large cell size limited the resolution of the electrical signals. Adaptation of the two-cell voltage clamp to cell pairs of high input resistance achieved precise determination of unitary gap junction channel conductance. Concomitant current records from both voltage-clamped cells always appear as mirror images since each cell resides on opposite sides of the junction (Fig. 7). This **equal amplitude and opposite polarity** is the criterion used to define junctional current signals in the dual whole cell recording configuration. Channel conductance values of 120 and 160 pS were close to the original predicted value and provided supportive evidence for the existence of relatively large nonselective aqueous channels. Conductance values range from 30 to 300 pS for different gap junction channels under essentially physiological salt conditions. For an aqueous diffusion limited pore, the channel conductance will increase in direct proportion to $2\pi r^2$. For a long pore of 16 nm in length, a 300-pS gap junction channel requires a 26-Å diameter right cylindrical pore. It follows that higher conductance gap junction channels should exhibit less ionic selectivity and a higher molecular permeability than channels with smaller conductance values.

IX. Evidence for Charge Selectivity

The apparent permeability to a variety of water-soluble (hydrophilic) molecules (ions, cAMP, sugars, ATP, nucleosides, etc.) suggests the presence of a large-diameter aqueous pore with minimal selectivity based on electrical charge (equivalent valence at physiological pH). Several permeability studies on mammalian cell and invertebrate gap junctions were instrumental in assigning the commonly accepted molecular permeability limit of 1 kDa or diameter < 14 Å. However, two of these same studies used tagged fluorescent tracers with different valences and demonstrated that the molecular permeability limit was lower for molecules with higher negativity. Selective permeability of molecules > 600 Da and approximately 10 Å in diameter is not surprising since the size of the permeant molecule is approaching the estimated diameter of the pore. Placing any charged surfaces of the pore and the permeable molecule in close proximity to each other would enhance any electrostatic attractive or repulsive forces that might be present. This suggests that the pore of the gap junction channel contains fixed electronegative sites within the pore which reduce the permeability of large negatively charged molecules relative to their neutral or less negative counterparts.

Evidence for selectivity at the ionic level did not exist until the patch clamp methodology was adapted to the recording of gap junction channel currents. In two cellular preparations, the junctional membranes of the earthworm septate axon and rat lacrimal gland cells were found to be less permeable to Cl⁻ relative to K⁺ by ratios of 0.52 and 0.69, respectively. This modest selectivity among ions with nearly identical aqueous mobilities and diameters of less than 4 Å

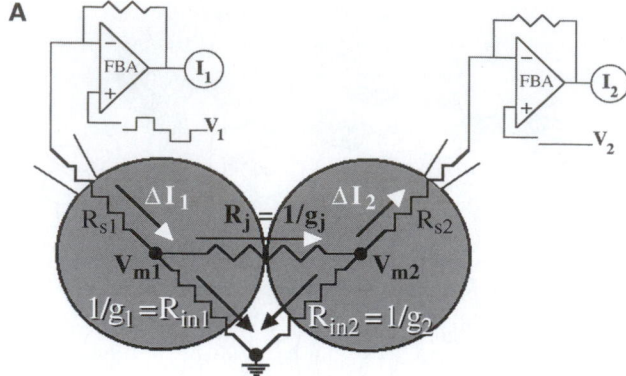

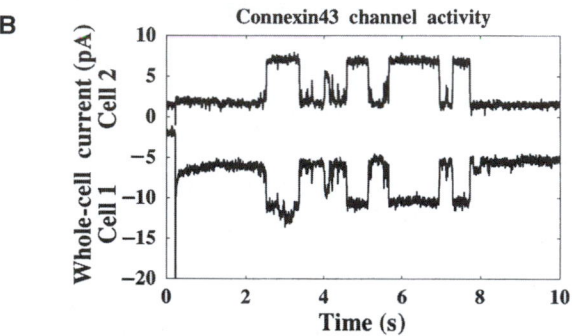

FIGURE 7. (A) Dual whole-cell recording configuration. Equivalent resistive circuit for the whole-cell patch-clamp recording configuration from two coupled cells. The voltage command potentials (V_1 and V_2) are applied via each negative feedback patch-clamp amplifier (FBA) through whole-cell patch electrodes. The difference between the command potentials and the cell membrane potentials (V_{m1} and V_{m2}) is minimized by reducing the series resistance (R_{s1} and R_{s2}) to less than 5% of the cell input resistance (R_{in1} and R_{in2}) and junctional resistance (R_j). Accurate junctional current signals are achieved when a majority of the applied current flows across the junction (white arrowheads) instead of the cell membrane (black arrowheads). This condition is readily obtained by using high input resistance cells ($R_{in} > 1$ GΩ). Conductances (g) are the reciprocal of the corresponding resistance element. (B) The whole-cell currents for each amplifier (I_1 and I_2) appear as simultaneous signals of opposite sign. These are unaltered whole-cell currents obtained from a Cx43-transfected N2A cell pair digitized at 1 kHz after low-pass filtering at 100 Hz. The transjunctional voltage was −30 mV.

suggests that there are weak interactions between the permeant ion and the wall of the pore. In a simple diffusion-limited pore, ions interact with water molecules but not with other ions. Recent observations from four different homotypic connexin hemichannels or gap junction channels report K⁺ to Cl⁻ permeability ratios of approximately 10:1. Selectivity among monovalent cations followed their relative aqueous mobility sequence. These observations suggest a common theme to the structure of the selectivity filter within the connexin channel pore. This proposed selectivity filter readily permits the passage of hydrated cations while restricting the simultaneous passage of similar aqueous anions by ten-fold. These observations provide direct evidence for cation-anion interactions within the gap junction pore that contradict aqueous diffusion theory. The observed permeability to atomic and large organic cations is consistent with a pore diameter of 11–13 Å for mammalian connexin channels.

The ability to transfer these electrical and chemical signals from cell to cell is important to tissue function. The most direct comparison of connexin-specific molecular permeability differences comes from the use of structurally similar molecules of varying size and valence. Two fluorescein derivatives with varying valences but essentially constant physical dimensions ($\approx$10 Å), 2',7'-dichlorofluorescein (diCl-F), and 6-carboxyfluorescein (6-CF), were readily permeable through rat Cx43 channels. In contrast, both dyes were less permeable through chicken Cx43, human Cx37, and rat Cx40 gap junctions. Chicken Cx45 gap junctions were permeable to diCl-F, but not 6-CF. This indicates that the more anionic 6-CF is restricted in its junctional permeability relative to diCl-F in a connexin-specific manner. Since Cx43 has a lower channel conductance than Cx40 and Cx37, no correlation between the maximum conductance and molecular permeability of the connexin channels exists. This lack of correlation between charge selectivity and channel conductance suggests that pore size and conductance are not directly related. Alternatively, recent channel theory proposes that the pore with the smaller diameter and same electrostatic (fixed structural) charge will exhibit the higher channel conductance due to the increased charge density. This updated charged membrane theory can explain the conductance and selectivity properties of known ion channels, including gap junctions.

X. Channel Properties of Different Connexins

The functional expression of different connexins, either by mRNA injection into *Xenopus* oocytes or stable expression in communication-deficient cell lines derived from mammalian tumors, has yielded tantalizing results. Single-channel unitary conductance values (the ease with which current flows) are connexin-specific. Channel conductance values vary by ten-fold, from 30 to 300 picosiemens (pS = $1/10^{12}$ Ω), with a majority of the channel conductance values in the 90 to 180 pS range. The reporting of channel conductance became more complicated due to the presence of multiple conductance states for several of the connexin channels. The ability of some connexins to form heterotypic (two connexons of different connexin composition) or heteromeric (two different connexins within the same connexon) gap junction channels adds another level of complexity to the function of channel conductance. The physiological relevance of these conductance differences is poorly understood, but one recent demonstration indicates that connexin composition modulates gap junction channel molecular permeability limits. Rat hepatocyte gap junctions contain almost exclusively Cx32 while mouse hepatocyte gap junctions are composed of a mixture of Cx26 and Cx32. Isolated mouse liver gap junctions exhibited a lower size limit to uncharged sugar molecules and cAMP than rat liver gap junctions, consistent with a smaller pore size, and they exhibited possible charge selectivity, contributed by Cx26 to otherwise Cx32 containing gap junctions.

The regulation of gap junction conductance by intracellular pH, intracellular calcium, protein kinases, and transjunc-

tional (cell-to-cell) voltage are also connexin-specific. The most significant differences in the amino acid sequences of the connexins occur within the cytoplasmic loop and carboxyl-terminal domains. These two cytoplasmic domains provide the basis for distinct conductance and regulatory properties upon the assembled multimeric connexin channels. Since a specific gene encodes for each connexin, tissue- and developmental-specific patterns of expression exist. Limited comparisons of the channel conductance and regulatory properties of the connexin channels to gap junctions in native cell types demonstrate the presence of at least two distinct connexin channel types. The presence of multiple connexin channels can explain the properties of the gap junctions found in many cell types. In one study, the expression of three embryonic chicken heart connexins, Cx42, Cx43, and Cx45, demonstrated that each connexin exhibited unique conductance properties, which coincided with the observance of multiple conductance states in cultured cardiac myocytes. Furthermore, mixing these three connexins in different proportions modeled the response to transjunctional voltage during different stages of cardiac development. Thus, the observed decline in Cx45 expression could significantly explain the decrease in the transjunctional voltage sensitivity of junctional conductance in the developing chicken heart. Developmental and hormonal changes in connexin expression occur in other tissues, suggesting a possible role for this mechanism in the modulation of gap junction communication in a variety of tissues.

XI. Gating by Ions and Second Messengers

Perhaps of greater physiological relevance than transjunctional voltage is the modulation of junctional communication by highly buffered intracellular cations (protons and calcium) and organic second messengers such as cAMP, diacylglycerol, and inositoltrisphosphate. Whether cations directly bind to a regulatory site (or sites) on the connexin or act through intermediate accessory proteins is unknown. This issue remains an important topic for investigation since it has been observed that interventions that lower intracellular pH also increase intracellular free calcium. Conversely, there is direct evidence that several of the connexins are substrates for phosphorylation by protein kinases such as PKA and PKC. Protein kinase-dependent phosphorylation often increases or decreases junctional communication, depending on the connexins expressed in each tissue. Physiological correlates at the channel level are rare, owing to the difficulty of observing channel activity under distinct ionic or phosphorylation conditions. Generally, cAMP increases junctional conductance and PKC has the opposite effect, at least in Cx43-containing cells. PKA and PKC may both phosphorylate the same serine residue (S233) of Cx32. In contrast, tyrosine phosphorylation by pp60[v-src] produces a rapid and reversible uncoupling of Cx43 gap junctions but has no effect on Cx32 gap junctions. Cx43 contains specific v-src SH2- and SH3-binding domains near a tyrosine phosphorylation site (Y265) by pp60[v-src]. Cx32 lacks these binding and tyrosine phosphorylation sites. There is increasing evidence that the diversity in the cytoplasmic loop and carboxyl-tail domains of different connexins confers different conductance and regulatory properties on the connexin-specific gap junctions in response to protein kinases and intracellular ions.

XII. Regulation of Functions of Connexin-Based Gap Junctions at Multiple Levels

As suggested above, a likely reason for differences in the sequence of the carboxyl tail of connexins may be in the regulation of gap junction-mediated tissue-specific or cell-specific biological activities. Indeed, gap junction regulatory mechanisms described so far appear to fall into one of two general categories. On the one hand, rapid changes in cell coupling that occur within seconds or minutes may reflect regulation of gating of preexisting pores. On the other hand, long term changes which occur within minutes or a few hours to days are likely to involve the modulation of synthesis, assembly, or degradation of connexins (Fig. 8). Existence of this latter scheme of modulation is supported by studies that show that the half-lives of several different connexins are relatively short, ranging from 1 to 3 hours, and by more recent direct videomicrography of living cells expressing green fluorescent protein tagged connexins. In these latter studies, connexins have been observed as they concentrate within the developing Golgi; as they move to the surface to form plaques in regions of cell-cell contact; and as large fragments of the connexin-positive plaques are internalized as apparent vesicles (Fig. 8). Moreover, a variety of studies support the idea that gap junctions may be regulated at virtually any stage of their formation, maturation, and degradation.

For example, the possible effect of cAMP on transcription of connexin mRNA and its ultimate translation is supported by studies which show that cAMP-mediated increases in coupling or in junction formation in some cells can be blocked by inhibitors of mRNA or protein synthesis. The potential for control at the level of transcription of connexins is also implied by the characterization of their 5' flanking sequences. For example, the 5' flanking sequence of the myometrial Cx43 gene has been shown to possess several consensus activator protein-1 (AP-1) binding sites as well as half-palindromic estrogen response elements. Since the AP-1 proteins, Fos and Jun, are expressed in response to increased estrogen, transcription of Cx43 may be indirectly up-regulated through its AP-1, *cis*-acting elements. In vascular smooth muscle, the transcription of Cx43 may be up-regulated by mechanical load induced by stretch, presumably via an up-regulation of *c-fos,* which may act in turn on AP-1 binding sites or upon other sites in the 5' flanking region of the Cx43 gene. Promoter analysis of other connexin genes have shown that their transcription may also be regulated by transcription factor binding sites. For example, two GC boxes, two GT boxes, a TTAAAA box, a YY1-like binding site, and a mammary gland factor binding site have been identified in the proximal promoter region of the human Cx26 gene. Likewise AP-1, AP-2, SP-1, TRE, and p53 transcription factor binding sites have been identified in the 5' flanking sequence of the mouse Cx40 gene while Cx-B2, and P1 and P2 promoters are implicated in the transcrip-

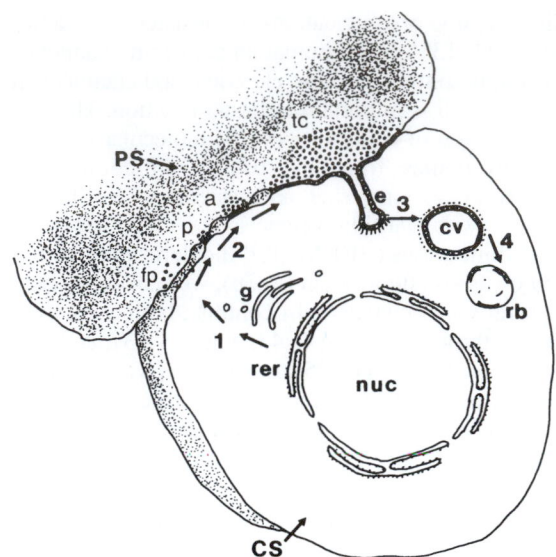

FIGURE 8. Scheme depicting (1) the synthesis of connexins within rough endoplasmic reticulum (rer), packaging in the Golgi (g), and transport of connexons from the cytoplasm (CS) to the plasma membrane; (2) their assembly in particle-poor regions of the membrane called formation plaques (fp) into small primary plaques (p), functional gap junction aggregates (a), and then into very large terminal caps (tc); (3) their endocytosis, which is facilitated by clathrin and an actin-mediated contractile mechanism; and (4) their degradation, following fusion with lysosomes resulting in the formation of residual bodies (rb).

tional regulation of Cx32.

Yet another level of control of gap junction function is revealed by studies that show the up-regulation of coupling or gap junction formation in the absence of transcription or translation, perhaps at a post-translational level of control. The direct application of dibutyryl cAMP (dbcAMP) to mammary tumor cells in culture, for example, increases the number of gap junctions as well as cell-cell coupling in the absence of an increase in Cx43 mRNA or protein. Likewise, connexin 43 trafficking within several cultured cell lines appears to be regulated by post-translational mechanisms. Specifically, the trafficking of Cx43 from the cytoplasm to the plasma membrane, and its assembly into functional gap junction aggregates may require phosphorylation of two serine phosphorylation consensus sites in the carboxyl tail of this connexin. For example, two cell lines were studied that could not form gap junctions or become coupled, but were able to synthesize significant amounts of Cx43, which was stored within vesicles in the cytoplasm. The Cx43 in these cells was phosphorylated at only one of its serine residues. When one of these lines was transfected with a gene for a cell adhesion molecule, L-CAM, the Cx43 in the transfected line was phosphorylated again, translocated to the plasma membrane and assembled into gap junctions. Further evidence for the phosphorylation-dependent trafficking and assembly of Cx43 in cardiomyocytes has been obtained in experiments with monensin, a reagent that inhibits translocation of proteins between Golgi cisternae. These studies suggest that phosphorylation of Cx43 to the mature form occurs in a compartment of the cell between the Golgi apparatus and the plasma membrane. Similar results are obtained in studies of effects of brefeldin A, which blocks both Cx43 phosphorylation and trafficking to the plasma membrane in a mammary tumor cell line. Finally, it has also been suggested that the process of connexon clustering that gives rise to functional gap junction plaques may be facilitated by the action of microfilaments. Thus the increase in Cx43-positive gap junctions and coupling in the absence of Cx43 mRNA synthesis following treatment of cultured prostatic cells with forskolin may be explained by effects of cAMP on post-translational processing of Cx43, namely their assembly into functional plaques and their translocation from the cytoplasm to the membrane.

Numbers of functional gap junctions within the plasma membrane may also be affected by regulation of their removal and degradation. For example, the treatment of some cells with phorbol esters such as 12-o-tetradecanoylphorbol-13-acetate (TPA) may stimulate the removal and degradation of gap junctions through phosphorylation of connexins by protein kinase C. Similarly, it is thought that the phosphorylation of mitogen-activated protein (MAP) kinase serine phosphorylation sequences in cultured rat liver cells treated with epidermal growth factor, may mediate the loss of functional gap junctions from the cell membrane. In another example, it has been shown that massive endocytosis of gap junctions in granulosa cells of the ovary is stimulated by high titers of follicle stimulating hormone; that endocytosis is facilitated by clathrin and microfilaments; and that the internalized gap junction vesicles fuse with lysosomes prior to degradation. (Fig. 8). Alternatively, it has been suggested that gap junction plaques and their connexins may be disposed of via a ubiquitin-mediated proteasomal proteolytic pathway. This mechanism is supported by studies showing that connexins accumulate in cells treated with proteasomal inhibitors and by immunogold studies that demonstrate the association of ubiquitin with gap junction plaques (Fig. 9).

In summary, it appears that the amount of functional gap junction membrane at the cell surface may be controlled at virtually any point in the gap junction's life history; namely through regulation of (1) transcription of connexin mRNA, (2) translation of connexin protein, and then through post-translational modifications, which affect the (3) assembly of functional plaques, (4) gating of junctional channels, and (5) degradation of functional gap junctions.

XIII. Specific Biological Functions of Gap Junctions

Early speculations regarding the general functional significance of gap junctions, such as the electrical coupling of some neurons and the buffering of metabolites within a tissue mass, have not been seriously challenged for the past three decades (see Sections I, II, and IX). Several recent studies, however, have begun to establish even more specific functional relationships between gap junctions and tissue activities and behaviors, most notably with respect to the regulation of smooth muscle contraction and ligand-mediated secretory activity.

Perhaps one of the best studied cases is that of the myometrial connexin, Cx43. It is known, for example, that the

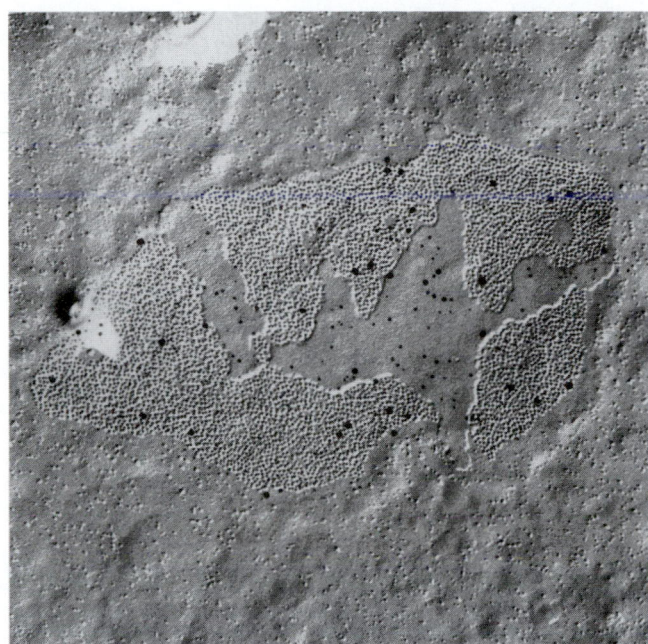

FIGURE 9. Image of gap junction in cultured HeLa cell transfected with Cx40 and Cx43. The replica was immunostained with antibodies to Cx40, Cx43, and ubiquitin. The 6-nm gold particles indicate Cx40, the 10-nm gold particles stain for Cx43, and the 15-nm gold particles indicate epitopes for ubiquitin.

myometrium of the nonpregnant uterus in a variety of species is virtually devoid of gap junctions and expresses very little, if any myometrial Cx43 mRNA or protein. In contrast, however, Cx43 mRNA is rapidly synthesized and translated and then gap junctions appear coincident with rising estrogen and declining progesterone titres just prior to parturition. Experimental manipulations of these hormones also predictably induce or inhibit the appearance of gap junctions in experimental models of preterm labor or tocolysis respectively (effective labor coincides with their induction while myometrial quiescence parallels their inhibition). For example, removal of the ovaries or the injection of RU-486 during the latter part of pregnancy effectively increases the estrogen : progesterone ratio, the expression of Cx43 mRNA and protein, the assembly of myometrial gap junctions, and concomitant preterm labor. Conversely, the injection of progesterone several days prior to term inhibits the formation of gap junctions and onset of labor. Thus, it is likely that transcription and translation of Cx43 are required for gap junction formation and coordinated myometrial contractions in humans. Moreover, the evidence for the presence of AP-1 binding sites in 5' flanking sequences of the myometrial Cx43 gene and the estrogen-mediated expression of Jun and Fos described above (see Section XII) is consistent with this notion. The function of myometrial gap junctions and contractility may also be regulated at the level of post-translational processing. For example, high levels of myometrial Cx43 appear a full day prior to delivery in rats (Fig. 10) but most of the Cx43 in these cells is located within perinuclear vesicles which also stain with an antibody against a Golgi-associated protein (Fig. 10B). Six to twelve hours prior to onset of delivery, however, Cx43 immunopositive plaques are observed in the plasma membrane (Fig. 10C). Conversely, cessation of delivery is accompanied by the loss of gap junctions from the cell surface by a process of endocytosis (Fig. 10E).

Gap junctions have also been implicated in the process of secretion and other ligand-mediated functions in many different kinds of cells. For example, gap junctions are well-developed in differentiated endocrine and exocrine cells and are typically sparse or absent in proliferating stem cells. Moreover, the induction of secretory activity itself appears to be correlated with increased numbers of gap junctions in luteal cells, adrenal cortical cells, pancreatic exocrine cells, mammary alveolar cells, and thyroid cells. Studies of osteoblasts have demonstrated a close correlation between the

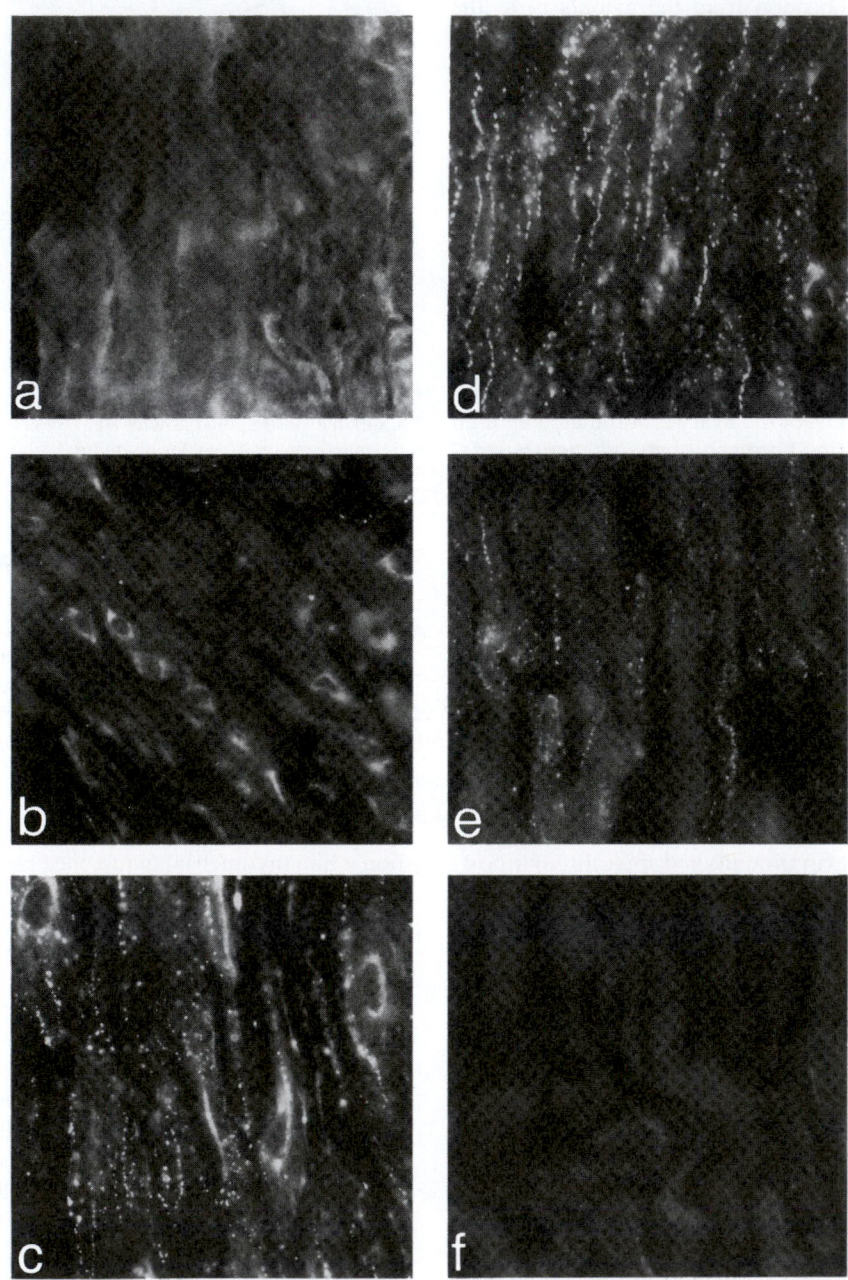

FIGURE 10.   Immunofluorescent staining of myometrium sections incubated with site-specific antibodies to the carboxyl terminus of Cx43 from pregnant rat sacrificed on day 19 (a) or day 21 (b), at midnight on day 21 (c), during delivery on day 22 (d), 6 h after delivery (e), or 24 h after delivery (f). Magnification, 500×. [From Hendrix, E. M., Mao, S. J. T., Everson, W., and Larsen, W. J. (1992). Myometrical connexin 43 trafficking and gap junction assembly at term and in preterm labor. *Mol. Reprod. Dev.* **33,** 27, with permission. Copyright 1992 Wiley-Liss, a division of John Wiley & Sons, Inc.]

effects of hormones on the generation of a second messenger, formation of gap junctions, and cell function. In addition, gap junctions appear to be more abundant in cultured confluent osteoblasts than proliferating cells and osteoblast cell coupling in culture and cAMP production are enhanced by the application of parathyroid hormone (PTH). Conversely, an analog of PTH which binds to PTH receptors but attenuates cAMP accumulation, results in a decrease in cell coupling. Finally, transfection of osteoblasts with antisense Cx43 mRNA results in significant concomitant reductions in coupling and in cAMP synthesis in response to PTH.

It has been suggested that the development of gap junctions may enhance the sensitivity of hormonally responsive cells to their specific ligand as a consequence of the transfer of cAMP through gap junctions. Thus cells not receiving direct stimulation by binding of the secretogogue to its receptor may also respond to the ligand-mediated production of cAMP in neighboring cells which possess the appropriate receptor. Such an hypothesis has been invoked to explain the maintenance of meiotic arrest of mammalian primary oocytes. These germ cells initiate the first meiotic division during embryonic life but are then almost immediately arrested at the first meiotic prophase stage. Typically, a primary oocyte does not resume meiotic maturation until it responds to an ovulatory surge of gonadotropic hormones following puberty. However, since meiosis spontaneously resumes when the oocyte is removed from the follicle and cultured in medium-lacking hormones, it has been postulated that the follicular environment is inhibitory to resumption of meiotic maturation. Meiotic arrest may be maintained by follicle cell-generated cAMP which enters the oocyte through a well-developed network of gap junctions. Conversely, it has been suggested that meiotic resumption may be signalled by the LH-mediated disruption of the gap junction pathway within the cumulus mass (following an ovulatory surge of gonadotropins), thus preventing cAMP manufactured within the follicle cells from entering the oocyte. Other hypotheses of both negative and positive regulation of meiotic resumption have also been proposed.

It has recently been proposed that gap junction connexins (specifically Cx43) may function in a role somewhat different than those described above; namely in protein trafficking. It has been shown, for example, that the second PDZ domain (of three) of the tight junction-associated protein ZO-1 binds to the extreme C-terminal end of the Cx43 molecule typically resulting in cell membrane colocalization of Cx43 and ZO-1 in cells that coexpress these proteins. In support of the possibility that Cx43 is required for shuttling of ZO-1 to the cell membrane, it was found that both ZO-1 and Cx43 fail to accumulate at the cell membrane in Cx43 knock-out astrocytes.

XIV. Gap Junctions in Human Disease and in Murine Models of Human Disease

A. Carcinogenesis

In the early 1970s, a hypothesis was formulated stating that the regulated and coordinated growth of cells in normal tissues requires the presence of permeable (communicating) cell junctions, while cells incapable of forming communicating junctions exhibit uncontrolled growth like the uncontrolled growth observed in cancer cells. Studies that compared the presence or absence of coupling or gap junctions in normal cells and in several tumors and cancer cell lines initially supported this hypothesis. For example, coupling or gap junctions were found to be deficient in certain hepatoma cells and in L-cell derivatives such as the clone-1D cell line in contrast to their normal counterparts. Indeed, more recent studies support these early observations. For example, an extensive study has examined the role of gap junctions in tumor progression in rat liver following initiation by diethylnitrosamine (DEN) and promotion by either phenobarbitol (PB) or 2,3-dichloro-7,8-dibenzo-p-dioxin (TCDD). Transcripts and protein levels of Cx26, Cx32, and Cx43 were measured and reductions in Cx26- and Cx32-positive gap junctions were observed in all resulting neoplasms. However, it was also demonstrated that these decreases were not always associated with reductions in specific m-RNA transcripts for these connexins, suggesting that the loss of junctions could result from the modulation of transcription or translation or from post-translational modification affecting assembly or degradation (see Section XII). Consistent with these observations, the transfection of HeLa cells with cDNA encoding Cx26, but not Cx40 or Cx43, results in inhibition of tumor formation in nude mice, suggesting that Cx26 gap junctions play a pivotal role in growth control. Similarly, the transfection of communication-deficient hepatoma cells (SKHep1) with cDNA encoding Cx32 results in slowing of the growth rate of tumors in nude mice, compared with tumors arising from communication-deficient parental SKHep1 cells.

In conflict with these findings, however, other highly malignant tumor cells have been shown to be well coupled or to possess large numbers of gap junctions. These include Novikoff hepatoma cells, human SW-13 adrenal cortical carcinoma cells, and murine B16 melanoma cells. Moreover, other studies have analyzed gap junctions during tumor progression in skin tumors and in hepatocarcinoma and have found that while gap junctions or coupling may disappear during the transformation of normal cells by cancer-causing agents (such as tumor promoters), their loss seems to follow, rather than lead the changes which transform these normal cells to tumor cells. In response to results such as these, a modified **defective communication** hypothesis was proposed to include cancer cells which might not necessarily exhibit an absence of, or obvious structural defect in the gap junction itself, but which might be defective with respect to some other part of the gap junction communication mechanism, for example, lack of ability of the cell to synthesize the postulated regulatory message, or defects in the receptor which translates the cell's activity. Other modifications of the hypothesis have also been advanced, proposing, for example, that some cancer cells (for example, cultured human SW13 cells and diethylstilbestrol-induced tumors of the proximal tubule in the Syrian hamster) are not able to maintain a steady-state gap junction level through a regulated balance of synthesis or degradation of gap junctions. Under some circumstances, these cells possess normal or even excessive gap junction complements and are mitotically quiescent, while under other conditions the cells lose most or all of their junctions, resulting in cancerous growth. A temporal relationship between gap junction formation and degradation, and the rate of proliferation is also exhibited by normal liver cells in partially hepatectomized rats. Normal rat liver cells remaining after partial hepatectomy initially lose their gap junctions and proliferate rapidly. By 40 hours after partial hepatectomy, the liver cells regain their gap junctions coincident with a reduction in the rate of DNA synthesis and cell division.

Another variation of the hypothesis is based upon a series of experiments that demonstrate that normal cells may have the ability to suppress tumor cell behavior in some normal-tumor cell cocultures as long as the tumor cells can also form gap junctions with the cocultured normal cells. Other tumor cells may not be able to form gap junctions with normal cells but may be able to form gap junctions between themselves. In these tumor cells, contact with normal cells does not suppress the transformed phenotype of the tumor cell. However, the ability of certain tumor cells to make gap junctions with one another may result in more efficient killing by antitumor agents by allowing the cell-to-cell movement of lethal metabolites formed in some, but not all of the tumor cells within the interconnected network. This phenomenon is called the **bystander effect**.

In conclusion, it is unlikely that the progression of all cancers is dependent upon the activity of gap junctions. It is also unlikely that mechanisms of gap junction-mediated growth control will be identical in all cells or tumors where the activity of gap junctions can be demonstrated to play a pivotal role.

B. Charcot-Marie-Tooth Disease and Mutations of Cx32

Spontaneous mutations of Cx32 result in a common peripheral neuropathy in humans, Charcot-Marie-Tooth disease, which affects 1 of every 2500 individuals. While this demyelinating disease may arise as a consequence of mutations of a peripheral myelin protein PMP22 (on chromosome 17) or of the myelin constituent Po (on chromosome 1), numerous families have been identified with an X-linked dominant form of the disease characterized by as many as 150 different mutations of Cx32 (CMTX). Mutations may involve single-base substitutions, formation of a premature stop codon, frame-shift, or elimination of an amino acid residue. They may occur in amino or carboxyl termini, cytoplasmic or extracellular loops, or in transmembrane regions of the connexin. The causal relationship between these diverse mutations of Cx32 and demyelination of peripheral nerves is obviously varied since the mutations are quite diverse and may or may not affect channel properties. In some cases, it seems possible that mutations may directly affect the function of Cx32 gap junctions in peripheral myelin by interfering with their possible function as ATP-sensitive hemichannels, their function in the exchange of nutrients between the perinuclear region of the Schwann cell and the Schmidt-Lantermann incisures and paranodal processes at the Node of Ranvier, or their possible function in signaling between internodes. In other cases, it is thought that the mutations may indirectly affect Cx32 function through effects on Cx32 synthesis and trafficking within the cell. Targeted knockout of connexin 32 in mice also results in changes in liver enzyme activities and in glucose mobilization, but most

interestingly, these Cx32-deficient mice have an increased susceptibility to hepatocarcinogenesis. Otherwise, they are vital and fertile.

C. Heart Function and Mutations of Cx43 and Cx40

While no specific heart defects related to mutations of Cx43 have been observed in humans, a series of studies with animal models of ischemia support a role for Cx43 gap junctions in recovery from injuries to the myocardium. For example, the number of gap junction plaques within the intercalated disc has been shown to decrease in the region of the border of the healing infarct. In addition, the spatial distribution of gap junctions is also abnormal in these areas. These observations along with studies that show the critical role of Cx43 in conduction (particularly in the ventricle) have led to suggestions that alterations in Cx43 levels and distributions may play an important role in development of infarct-mediated arrhythmias. Targeted knockouts of Cx43 and mice overexpressing Cx43 are characterized by abnormal development of the pulmonary (right ventricular) outflow tract resulting in their death immediately following birth. Enlargement of the right ventricle with thinning of the wall, attenuation of the ductus arteriosus, and formation of one or two abnormal pouches at the base of the pulmonary outflow tract were also observed. It has been suggested that these defects of conotruncal development may result from disruption of the cardiac neural crest. In addition, Cx43 knockout mice of both sexes have small gonads, resulting in part from deficiencies in germ cell development. Since these mice die shortly after birth, the gonads were cultured to determine their capacity for folliculogenesis. It was found that follicles in these ovaries develop only to the primary follicle stage. Targeted knockout of Cx40 results in defects consistent with its localization in the AV node and in conductive myocardium. Electrocardiography of Cx40 knockout animals was characterized by reduced conduction velocities and abnormalities characteristic of first-degree atrioventricular block with associated bundle branch block.

D. Cataract Formation and Mutations of Cx50

Mutations of connexin 50 have recently been linked to inherited congenital cataract in human lens. Since lens fibers are linked by extensive gap junctions composed of Cx50 (and Cx46), it is thought that the maintenance of junctional communication between lens fibers is necessary for the maintenance of normal transparency of the lens. Targeted knockout of Cx50 in mice leads to development of zonular pulverulent cataracts of the lens but also results in micropthalmia. These results thus implicate Cx50 not only in the maintenance of lens transparency but also in the growth of the eye.

E. Deafness and Mutations of Cx26

A variety of mutations of Cx26 including missense mutations and mutations resulting in frameshifts and formation of premature stop codons have been linked to both autosomal recessive and autosomal dominant hearing loss in humans. It has been suggested that these mutations affect the formation of gap junctions in the fibrocytes in the spiral ligament that underlies the stria vascularis of the cochlea leading to disruptions of ion and endolymph circulation. Targeted knockout of Cx26 results in embryonic lethality presumably as a consequence of placental dysfunction.

F. Fertility and Targeted Knockout of Cx37

While no human syndrome resulting from mutations of Cx37 is known, the targeted knockout of Cx37 in mice has been shown to result in the disruption of folliculogenesis. Follicles develop only to early antral stages, the oocytes fail to acquire meiotic competence, and ovulation does not occur.

XV. Summary

A structure serving as an electrical synapse in excitable tissues and as permeable cell-to-cell ion channels in inexcitable cells is the gap junction. Although the function of these structures as electrical synapses in neurons and cardiac muscle is generally understood, new studies relating the activity of specific connexins (and innexins) in particular conduction pathways (for example in the heart) may begin to specify the functions of gap junctions composed of different connexins. Developmental studies now seem to hold particular promise, while studies in the field of cancer cell biology continue to make progress, including analysis of the relationship between connexin expression, processing, assembly, gating, and degradation and transformation to malignancy. In addition, current studies of the properties of gap junction channels and of the cellular regulation of gap junction gating, formation, and turnover in relation to normal cell physiology may yield more specific functional information. Connexin-specific gap junction channels exhibit differences in electrical conductance and molecular permeability that are not directly correlated and dispute the contention that all gap junctions are large aqueous pores. Fortunately, DNAs encoding several connexins have been cloned and significant advances have been made in the correlation of sequence and function. Moreover, the application of this knowledge to the development of transgenic animals exhibiting gain or loss of function of specific connexins continue to lead to more specific insights into the functional significance of particular connexins and innexins.

Bibliography

Alcolea, S., Theveniau-Ruissy, M., Jarry-Gnichard, T., Maries, I., Tzouanacou, E., Chavin, J. P., Briand, J. P., Moorman, A. F., Lamers, W. H., and Gros, D. B. (1999). Downregulation of connexin45 gene products during mouse heart development. *Circ. Res.* **84**,1365–1379.

Bai, S., Schoenfeld, A., Pietrangelo, A., and Burk, R. D. (1995). Basal promoter of the rat connexin 32 gene: Identification and characterization of an essential element and its DNA-binding protein. *Mol. Cell. Biol.* **15**, 1439–1445.

Bennett, M. V. L., and Verselis, V. K. (1992). Biophysics of gap junctions. *Sem. Cell Biol.* **3**, 29–47.

Bennett, M. V. L., Barrio, L. C., Bargiello, T. A., Spray, D. C., Hertzberg, E., and Saez, J. C. (1991). Gap junctions: New tools, new answers, new questions. *Neuron* **6**, 305–320.

Bennett, M. V. L., Zheng, X., and Sogin, M. L. (1994). The connexins and their family tree. *In* "Molecular Evolution of Physiological Processes," pps. 223–233. Rockefeller University Press, New York.

Bevans, C. G., Kordel, M., Rhee, S. K., and Harris, A. L. (1998). Isoform composition of connexin channels determines selectivity among second messengers and uncharged molecules. *J. Biol. Chem.* **272**, 2808–2816.

Beyer, E. C., Paul, D. L., and Goodenough, D. L. (1990). Connexin family of gap junction proteins. *J. Membrane Biol.* **116**, 187–194.

Brink, P. R., and Dewey, M. M. (1980). Evidence for fixed charge in the nexus. *Nature* **285**, 101–102.

Brink, P. R., and Fan, S.-F. (1989). Patch clamp recordings from membranes which contain gap junction channels. *Biophys. J.* **56**, 579–593.

Budunova, I. V., Carbajal, S., Viaje, A., and Slaga, T. J. (1996). Connexin expression in epidermal cell lines from SENCAR mouse skin tumors. *Mol. Carcinog.* **15**, 190–201.

Chiba, H., Sawada, N., Oyamada, M., Kojima, T., Nomura, S., Ishii, S., and Mori, M. (1993). Relationship between the expression of the gap junction protein and osteoblast phenotype in a human osteoblastic cell line during cell proliferation. *Cell Struct. Func.* **19**, 419–426.

Coppen, S. R., Kodama, I., Boyett, M. R., Dobrzynski, H., Takagishi, Y., Honjo, H., Yeh, H. I., and Severs, N. J. (1999). Connexin45, a major connexin of the rabbit sinoatrial node, is co-expressed with connexin43 in a restricted zone at the nodal-crista terminalis border. *J. Histochem. Cytochem.* **47**, 907–918.

Cowan, D. B., Lye, S. J., and Langille, B. L. (1998). Regulation of vascular connexin43 gene expression by mechanical loads. *Circ. Res.* **82**, 786–793.

Curtin, K. D., Zhang, Z., and Wyman, R. J. (1999). *Drosophila* has several genes for gap junction proteins. *Gene* **232**, 191–201.

Darrow, B. J., Laing, J. G., Lampe, P. D., Saffitz, J. E., and Beyer, E. C. (1995). Expression of multiple connexins in cultured neonatal rat ventricular myocytes. *Circ. Res.* **76**, 381–387.

Donahue, H. J., McLeod, K. J., Rubin, C. T., Andersen, J., Grine, E. A., Hertzberg, E. L., and Brink, P. R. (1995). Cell-to-cell communication in osteoblastic networks: Cell line-dependent hormonal regulation of gap junction function. *J. Bone Min. Res.* **10**, 881–889.

Ewart, J. L., Cohen, M. F., Meyer, R. A., Huang, G. Y., Wessels, A., Gourdie, R. G., Chin, A. J., Park, S. M., Lazatin, B. O., Villabon, S., and Lo, C. W. (1997). Heart and neural tube defects in transgensis mice overexpressing the Cx43 gap junction gene. *Development* **124**, 1281–1292.

Filson, A. J., Azarnia, R., Beyer, E. C., Loewenstein, W. R., and Brugge, J. A. (1990). Tyrosine phosphorylation of a gap junction protein correlated with inhibition of cell-to-cell communication. *Cell Growth Diff.* **1**, 661–668.

Fishman, G. I., Moreno, A. P., Spray, D. C., and Leinwand, L. A. (1991). Functional analysis of human cardiac gap junction channel mutants. *Proc. Natl. Acad. Sci., USA* **88**, 3525–3529.

Flagg-Newton, J., Simpson, I., and Loewenstein, W. R. (1979). Permeability of the cell-to-cell membrane channels in mammalian cell junction. *Science* **205**, 404–407.

Forge, A., Becker, D., Casalotti, S., Edwards, J., Evans, W. H., Lench, N., and Souter, M. (1999). Gap junctions and connexin expression in the inner ear. *Novartis Found. Symp.* **219**, 134–150.

Furshpan, E. J., and Potter, D. D. (1959). Transmission at the giant motor synapses of the crayfish. *J. Physiol.* **145**, 289 325.

Furshpan, E. J., and Potter, D. D. (1968). Low-resistance junctions between cells in embryos and tissue culture. *Curr. Top. Dev. Biol.* **3**, 95–127.

Giepmans, B. N., and Moolenaar, W. H. (1998). The gap junction protein connexin43 interacts with the second PDZ domain of the zona occludens-1 protein. *Curr. Biol.* **8**, 931–934.

Goldberg, G. S., Martyn, K. D., and Lau, A. F. (1994). A connexin 43 antisense vector reduces the ability of normal cells to inhibit the foci formation of transformed cells. *Mol. Cytogen.* **11**, 106–114.

Goodenough, D. A., Goliger, J. A., and Paul, D. (1996). Connexins, connexons, and intercellular communication. *Ann. Rev. Biochem.* **65**, 475–502.

Goodenough, D. A., Simon A. M., and Paul, D. L. (1999). Gap junctional intercellular communication in the mouse ovarian follicle. *Novartis Found. Symp.* **219**, 226–235.

Haeflinger, J.-A., Bruzzone, R., Jenkins, N. A., Gilbert, D. J., Copeland, N. G., and Paul, D. L. (1992). Four novel members of the connexin family of gap junction proteins. *J. Biol. Chem.* **267**, 2057–2064.

Hendrix, E. M., Mao, S. J. T., Everson, W., and Larsen, W. J., (1992). Myometrial connexin 43 trafficking and gap junction assembly at term and in preterm labor. *Mol. Reprod. Develop.* **33**, 27–38.

Hendrix, E. M., Myatt, L., Sellers, S., Russell, P. T., and Larsen, W. J. (1995). Steroid hormone regulation of rat myometrial gap junction formation: Effects on Cx43 levels and trafficking. *Biol. Reprod.* **53**, 547–560.

Hille, B. (1992). "Ionic Channels of Excitable Membranes." 2nd Ed. Sinauer Assoc., Sunderland, MA.

Hoh, J. H., John, S. A., and Revel, J.-P. (1991) Molecular cloning and characterization of a new member of the gap junction gene family, connexin-31. *J. Biol. Chem.* **266**, 6524–6531.

Huelser, D. F., Rehkopf, B., and Traub., O. (1997). Dispersed and aggregated gap junction channels I identified by immunogold labeling of freeze-fractured membranes. *Exp. Cell Res.* **233**, 240–251.

Jordan, K., Solan, J. L., Dominguez, M., Sia, M., Hnad, A., Lampe, P. D., and Laird, D. W. (1999). Trafficking, assembly, and function of a connexin 43-green fluorescent protein chimera in live mammalian cells. *Mol. Biol. Cell* **10**, 2033–2050.

Juneja, S. C., Barr, K. J., Enders, G. C., and Kidder, G. M. (1999). Defects in the germ line and gonads of mice lacking connexin43. *Biol. Reprod.* **60**, 1263–1270.

Kamibayashi, Y., Oyamada, Y., Mori, M., and Oyamada, M. (1995). Aberrant expression of gap junction proteins (connexins) is associated with tumor progression during multistage mouse skin carcinogenesis. *Carcinogen.* **16**, 1287–1297.

Kanemitsu, M. Y., Loo, L. W. M., Simon, S., Lau, A. F., and Eckhart, W. (1997). Tyrosine phosphorylation of connexin 43 by v-src is mediated by SH2 and SH3 domain interactions. *J. Biol. Chem.* **272**, 22 824–22 831.

Kanter, H. L., Saffitz, J. E., and Beyer, E. C. (1992). Cardiac myocytes express multiple gap junction proteins. *Circ. Res.* **70**, 438–444.

Kenne, K., Fransson-Steen, R., Honkasalo, S., and Warngard, L. (1994). Two inhibitors of gap junction intercellular communication, TPA and endosulfan: Different effects on phosphorylation of connexin 43 in the rat liver epithelial cell line, IAR 20. *Carcinogen.* **15**, 1161–1165.

Khan-Dawood, F. S., Yang, J., and Dawood, M. Y. (1996). Expression of gap junction protein connexin-43 in the human and baboon corpus luteum. *J. Clin. Endocrinol. Metab.* **81**, 835–842.

Kiang, D. T., Jin, N., Tu, Z. J., and Lin, H. H. (1997). Upstream genomic sequence of the human connexin26 gene. *Gene.* **199**, 165–171.

Kirchhoff, S., Nelles, E., Hagendorff, A., Kruger O., Traub, O., and Willecke, K. (1998). Reduced cardiac conduction velocity and predisposition to arrythmias in connexin40-deficient mice. *Curr. Biol.* **8**, 299–302.

Laird, D. W., Puranam, K. L, and Revel, J.-P. (1991). Turnover and phosphorylation dynamics of connexin43 gap junction protein in cultured cardiac myocytes. *Biochem J.* **273**, 67–72.

Laird, D. W., Castillo, M., and Kasprzak, L. (1995). Gap junction turnover, intracellular trafficking and phosphorylation of con-

nexin43 in brefeldin A-treated rat mammary tumor cells. *J. Cell. Biol.* **131**, 1193–1203.

Landesman, Y., White, T. W., Starich T. A., Shaw, J. E., Goodenough, D. A., and Paul, D. L. (1999). Innexin-3 forms connexin-like intercellular channels. *J. Cell Science* **112**, 2391–2396.

Larsen, W. J., Wert, S. E., and Brunner, G. D. (1987) Differential modulation of follicle cell gap junction populations at ovulation. *Dev. Biol.* **122**, 61–71.

Lo, C. W., Cohen, M. F., Huang, G. Y., Lazatin, B. O., Patel, N., Sullivan, R., Pauken C., and Park, S. M. (1997). Cx43 gap junction gene expression and gap junctional communication in mouse neural crest cells. *Dev. Genet.* **20**, 119–132.

Loewenstein, W. R., Kanno, Y., and Socolar, S. J. (1978). Quantum leaps of conductance during formation of membrane channels at cell-cell junction. *Nature* **274**, 133–136.

Loewenstein, W. R. (1985). Regulation of cell-to-cell communication by phosphorylation. *Biochem Soc. Symp.* **50**, 43–58.

Loewenstein, W. R. (1987). The cell-to-cell channel of gap junctions. *Cell* **48**, 725–726.

Loo, L. W., Berestecky, J. M., Kanemitsu, M. Y. and Lau, A. F. (1995). pp60src-mediated phosphorylation of connexin 43, a gap junction protein. *J. Biol. Chem.* **270**, 12751–12761.

Maestrini, E., Korge, B. P., Ocana-Sierra, J., Calzolari, E., Cambiaghi, S., Scudder, P. M., Hovnanian, A., Monaco, A. P., and Munro, C. S. (1999). A missense mutation in connexin26, D66H, causes mutilating keratoderma with sensorineural deafness (Vohwinkel's syndrome) in three unrelated families. *Hum. Mol. Genet.* **8**, 1237–1243.

Makowski, L., Caspar, D. L. D., Phillips, W. C., and Goodenough, D. A. (1977). Gap junction structures. II. Analysis of the x-ray diffraction data. *J. Cell Biol.* **74**, 629–645.

Mesnil, M., Krutovskikh, V., Piccoli, C., Elfgang, C., Traub, O., Willicke, K., and Yamasaki, H. (1995). Negative growth control of HeLa cells by connexin genes: connexin specificity. *Cancer Res.* **55**, 629–639.

Moennikes, O., Buchmann, A., Ott, T., Willecke, K., and Schwarz, M. (1999). The effect of connexin32 null mutation on hepatocarcinogenesis in different mouse strains. *Carcinogenesis* **20**, 1379–1382.

Musil, L. S., and Goodenough, D. A. (1990). Gap junctional intercellular communication and the regulation of connexin expression and function. *Curr Opin. Cell Biol.* **2**, 875–880.

Neuhaus, I. M., Bone, L., Wang, S., Ionascescu, V., and Werner, R.. (1996). The human connexin32 gene is transcribed from two tissue-specific promoters. *Biosci. Rep.* **16**, 239–248.

Neveu, M. J., Hully, J. R., Babcock, K. L., Hertzberg, E. L., Nicholson, B. J., Paul, D. L., and Pitot, H. C. (1994). Multiple mechanisms are responsible for altered expression of gap junction genes during oncogenesis in rat liver. *J. Cell Sci.* **107**, 83–95.

Neyton, J., and Trautmann, A. (1985). Single-channel currents of an intercellular junction. *Nature* **317**, 331–335.

Nonner, W., and Eisenberg, B. (1998). Ion permeation and glutamate residues linked by Poisson-Nernst-Planck Theory in L-type calcium channels. *Biophys. J.* **75**, 1287–1305.

Orsino, A., Taylor, C. V., and Lye, S. J. (1996). Connexin-26 and connexin-43 are differentially expressed and regulated in the rat myometrium throughout late pregnancy and the onset of labor. *Endocrinol.* **137**, 1545–1553.

Pal, J. D., Berthoud, V. M., Beyer, E. C., Mackay, D., Shiels, A., and Ebihara, L. (1999). Molecular mechanism underlying a Cx50-linked congenital cataract. *Am. J. Physiol.* **276**, c1443–c1446.

Peters, N. S. 1996. New insights into myocardial arrythmogenesis: Distribution of gap-junctional coupling in normal ischaemic and hypertrophied human hearts. *Clin. Sci.* **90**, 447–452.

Piersanti, M., and Lye, S. J. (1995). Increase in messenger ribonucleic acid encoding the myometrial gap junction protein, connexin-43, requires protein synthesis and is associated with increased expression of the activator protein-1, c-fos. *Endocrinol.* **136**, 3571–3578.

Reaume, A. G., de Sousa, P. A., Kulkarni, S., Langille, B. L., Zhu, D., Davies, T. C., Juneja, S. C., Kidder, G. M., and Rossant, J. (1995). Cardiac malformation in neonatal mice lacking connexin43. *Science* **267**, 1831–1834.

Revel, J.-P., and Karnovsky, M. J. (1967). Hexagonal array of subunits in intercellular junctions in the mouse heart and liver. *J Cell Biol.* **38**, C7–C12.

Revel, J.-P., Hoh, J. H., John, S. A., Laird, D. W., Puranam, K., and Yancey, S. B. (1992). Aspects of gap junction structure and assembly. *Sem. Cell Biol.* **3**, 21–28.

Saez, J., Nairn, A. C., Czernik, A. J., Spray, D. C., Hertzberg, E. L., Greengard, P., and Bennett, M. V. L. (1990). Phosphorylation of connexin32, a hepatocyte gap-junction protein, by cAMP-dependent protein kinase, protein kinase C and Ca^{2+}/calmodulin-dependent protein kinase II. *Eur. J. Biochem.* **192**, 263-273.

Seul, K. H., Tadros, P. N., and Beyer, E. C. (1997). Mouse connexin40: Gene structure and promoter analysis. *Genomics* **46**, 120–126.

Simon, A. M., Goodenough, D. A., and Paul, D. L. (1998). Mice lacking connexin40 have cardiac conduction abnormalities characteristic of atriventricular block and bundle branch block. *Curr. Biol.* **8**, 295–298.

Simpson, I., Rose, B., and Loewenstein, W. R. (1977). Size limit of molecules permeating the junctional membrane channels. *Science* **195**, 294-296.

Spray, D. C., and Dermiietzel, R. (1995). X-linked dominant Charcot-Marie-Tooth disease and other potential gap junction diseases of the nervous system. *TINS* **18**, 256–262.

Swenson, K. I., Piwnica-Worms, H., McNamee, H., and Paul, D. L. (1990). Tyrosine phosphorylation of the gap junction protein connexin43 is required for the pp60v-src-induced inhibition of communication. *Cell Reg.* **1**, 989–1002.

Tu, Z. J., and Kiang, D. T. (1998). Mapping and characterization of the basal promoter of the human connexin26 gene. *Biochim. Biophys. Acta.* **1441**, 169–181.

Unger, V. M., Kumar N. M., Gilula, N. B., and Yeager, M. (1999). Three-dimensional structure of a recombinant gap junction membrane channel. *Science.* **283**, 1176–1180.

Veenstra, R. D. (2000). Ion permeation through connexin gap junction channels: Effects on conductance and selectivity. *In* "Current Topics in Membranes," (C. Perrachia, Ed.), Vol. 49, pp. 95–129.

Veenstra, R. D., and DeHaan, R. L. (1986). Measurement of single channel currents from cardiac gap junctions. *Science* **233**, 972–974.

Veenstra, R. D., Wang, H.-Z., Beblo, D. A., Chilton, M. G., Harris, A. L., Beyer, E. C., and Brink, P. R. (1995). Selectivity of connexin-specific gap junctions does not correlate with channel conductance. *Circ. Res.* **77**, 1156–1165.

Veenstra, R. D., Wang, H.-Z., Westphale, E. M., and Beyer, E. C. (1992). Multiple connexins confer distinct regulatory and conductance properties of gap junctions in developing heart. *Circ. Res.* **71**, 1277–1283.

Waldo, K. L., Lo, C. W., and Kirby, M. L. (1999). Connexin43 expression reflects neural crest patterns during cardiovascular development. *Dev. Biol.* **208**, 307–323.

Warn-Cramer, B. J., Lampe, P. D., Kurata, W. E., Kanemitsu, M. Y., Loo, L. W., Eckert, W., and Lau, A. F. (1996). Characterization of the mitogen-activated protein kinase phosphorylation sites on the connexin-43 gap junction protein. *J. Biol. Chem.* **271**, 3779–3786.

White, T. W., Goodenough, D. A., and Paul, P. L. (1998). Targeted ablation of connexin 50 in mice results in micropthalmia and zonular pulverulent cataracts. *J. Cell Biol.* **143**, 815–825.

Wohlburg, H., and Rohlmann, A. (1995). Structure-function relationships in gap junctions [review]. *Int. Rev. Cytol.* **157**, 315–373.

Yamaguchi, D. T., Huang, J. T., and Ma, D. (1995). Regulation of gap junction intercellular communication by pH in MC3T3-E1 osteoblastic cells. *J. Bone Min. Res.* **10**, 1891–1899.

Yamasaki, H. (1991). Aberrant expression and function of gap junctions during carcinogenesis. *Env. Health Prosp.* **93**, 191–197.

Yu, W., Dahl, G., and Werner, R. (1994). The connexin43 gene is responsive to oestrogen. *Proc. Roy. Soc. (London) Series B: Biol. Sci.* **255**, 125–132.

Louis J. DeFelice and Michele Mazzanti

32

Biophysics of the Nuclear Envelope

I. Introduction

Eukaryotic cells contain a nucleus, a spherical body that encloses characteristic organelles. These organelles include a thin nuclear envelope, one or more nucleoli, chromatin (the chromosomes, DNA attached to proteins, primarily histones), a proteinaceous nuclear lamina (which contacts the inner nuclear membrane, chromosomes, and the nuclear RNA), irregular granules of chromatin material, and linin (fine threads that associate with the chromatin granules). Amorphous nucleoplasm occupies the rest of the nuclear volume.

The nuclear envelope transports macromolecules, and the transport selectivity changes during the cell cycle and throughout development. Most textbooks describe the envelope as forming no barrier to small ions such as Ca^{2+}, Na^+, K^+, or Cl^-. In this view, the nucleus is a sievelike structure containing large aqueous pores. However, certain data suggest that the nuclear envelope can restrict the movement of intermediate-sized molecules, including peptides, amino acids, and sugars. Moreover, although the data are controversial, particular experiments support the idea that inorganic ions like Ca^{2+} can accumulate in, or be excluded from, the nucleus. Furthermore, the nucleus has an electrical potential, and it can swell or shrink in response to osmotic forces. One theory is that such phenomena derive from selective absorption by the nucleoplasm. This chapter explores another view, that the nuclear envelope selectively transports small ions and has electrical properties similar to the semipermeable plasma membrane.

The nuclear pore governs the movement of nucleic acids and large-molecular-weight proteins into and out of the nucleus: enzymes enter the nucleus to polymerize nucleic acids, RNA leaves for translation in the cytosol, and DNA-binding proteins penetrate for transcription and gene regulation. Selection is precise, and the nucleus eliminates from its domain the majority of proteins, many of which are smaller than the proteins it admits. During singular events, such as fertilization or viral infection, or under laboratory manipulations, such as gene transfection, the nuclear pore exhibits transitory

selection. This chapter explores the possibility that ion selection accounts for the electric and osmotic properties of the nucleus. The movements of ions across the nuclear envelope may act as regulatory counter currents for the transport of charged macromolecules. This assertion derives from the biophysical characteristics of nuclei and the presence of ion-selective channels in the nuclear envelope. Ion channels in the nuclear envelope have properties similar to ion channels in the plasma membrane. We further suggest that some channels observed in the envelope are part of the nuclear pore complex. This proposition introduces a paradox: how could a pore that transports large molecules through a large opening also select small ions? This chapter addresses the implications of this question and offers an explanation of its consequences.

II. Permeability of the Nuclear Envelope

Two mechanisms may explain the separation of molecules by the nuclear envelope. One mechanism involves specific solubility in the nucleoplasm (or exclusion from the cytoplasm). Separation of solutes between adjacent phases falls under the heading of the **Donnan equilibrium**. The Donnan equilibrium relies on the differential absorption of solutes into immiscible phases. If charged molecules partition into adjacent compartments, an electric potential difference results. If the nucleoplasm and the cytoplasm form such domains, and if they connect through open pores, the Donnan equilibrium could explain the nuclear resting potential.

Another mechanism involves a semipermeable membrane that separates the nucleoplasm from the cytoplasm. This mechanism, which characterizes cell plasma membranes, comes under the heading of the **Nernst-Planck** regime. Whereas one lipid bilayer forms the plasma membrane that surrounds the cell, two such membranes form the envelope that surrounds the nucleus. However, the double-membrane structure simplifies because **nuclear pores** span the envelope and act as channels between nucleoplasm and cytoplasm.

An early model of the double-membrane-spanning nuclear pore depicts the pore as two parallel rings, one on the

539

cytoplasmic side and the other on the nucleoplasmic side. Each ring consists of eight subunits that surround a central pore. We refer to this structure as the **nuclear pore complex**. The phrase nuclear pore complex applies to the entire structure, not merely to the opening in the center. The opening in the center (the nuclear pore) contains an electron-dense **central granule** that connects the two rings. The size and shape of the central pore help determine its selectivity to macromolecules. For example, RNA molecules elongate into cylindrical rods (diameter ~100 Å) as they enter the pore, and certain proteins larger than 100 Å in diameter cannot pass through the pore. However, the size of the pore does not entirely explain its selectivity, because 170 Å diameter gold particles coated with **nucleoplasmin** can enter the nucleus. Without nucleoplasmin, or after treatment with trypsin to remove nucleoplasmin, the large gold particles cannot enter. Therefore it is thought that certain proteins contain a signal that permits nuclear transport. This signal is specific, because a single amino acid substitution can abolish, for example, the transport of specific tumor antigens into the nucleus. Furthermore, a single base substitution can dramatically reduce the rate of tRNA transport. Some substances apparently block transport merely by getting in the way, for example, wheat germ agglutinin (WGA) accumulates on the cytoplasmic face of the nucleus and inhibits the uptake of nucleoplasmin. Such blocking cannot explain all the data, because WGA does not influence the rate of dextran entry. Thus nuclear transport is complex and highly varied and is unlikely to be explained by one mechanism.

The high rates of tRNA translocation in mammalian cell nuclei (10^9 molecules/min) exclude simple diffusion as the transport mechanism. Some experiments suggest a carrier mechanism for macromolecules, augmented by the observation that RNA transport requires ATP. Although diffusion accounts for only a small fraction of macromolecular transport, smaller molecules are translocated at rates below diffusion. For example, inulin (5.5 kDa) enters the nucleus at one-fifth its diffusion rate in water. Another example is sucrose, which passes through the nuclear envelope more readily than it does through the plasma membrane, but at one-third the rate predicted by diffusion. Furthermore, nuclei may even restrict small inorganic anions and cations. Evidence for this comes from *Drosophila* salivary gland cells, whose nuclei maintain an electrical potential of −15 mV with respect to the cytoplasm. Electrical potentials exist across other nuclear envelopes, although not always of the same sign or amplitude. Finally, Na^+ and K^+ ions have different concentrations in the nucleoplasm than in the cytoplasm, and nuclear Ca^{2+} levels change during cell differentiation and fertilization. These data are controversial, and not everyone agrees that ion gradients and electrical potentials exist across the nuclear envelope. Even if they do, it is unclear whether the nucleus restricts ions via the Donnan equilibrium or whether the nuclear envelope acts as a semipermeable membrane. These mechanisms are not exclusive; however, if some selection of small ions does occur via a semipermeable membrane, how can we reconcile this phenomenon with the passage of large macromolecules through the same openings? The next section offers a possible explanation.

III. Structure of the Nuclear Envelope

Two lipid bilayers surround the nucleus, and in this regard, the nucleus resembles a mitochondrion. In contrast, a single lipid bilayer surrounds other internal organelles, such as the Golgi apparatus, storage vesicles, and the endoplasmic reticulum. The double membrane structure of isolated nuclei remains intact *in vitro* and, in many ways, isolated nuclei may be treated as cells maintained in culture.

A. Isolation of Nuclei

Nuclei from invertebrate cells are easily dissected from starfish oocytes in the germinal vesicle stage, when the nuclei are large (30 μm in diameter). After the dissection, the nuclei can be kept for hours at cold temperatures (5 °C) in an intracellular-like solution (200 mM K_2SO_4, 20 mM NaCl, 10 mM Hepes, 10 mM EGTA, and sucrose to 1000 milliosmol). Nuclei are tolerant to other physiological solutions, such as extracellular media and seawater. **Mammalian cell nuclei** (10 μm in diameter) are readily obtained from mouse oocytes 12 h after superovulation. Upon fertilization, pronuclei or early embryo nuclei are collected 12 h after mating. Prior to removing the nucleus, the oocytes, zygotes, or embryos should be placed in an intracellular-like solution (120 mM KCl, 2 mM $MgCl_2$, 1.1 mM EGTA, 0.1 mM $CaCl_2$, 5 mM glucose, 10 mM Hepes, pH 7.4) so that when the nucleus comes out of the cell it experiences a normal ionic environment. Mammalian nuclei can also be isolated from differentiated tissue, such as liver cells, by shearing the tissue and centrifuging the homogenate. This results in a pellet of pure nuclei, which can then be re-suspended into a physiological solution. These techniques result in four categories of nuclei that we have used for comparative studies: germinal vesicles, pronuclei, two-cell embryo nuclei, and adult liver cell nuclei. It is unclear whether nuclei preserve their characteristics in culture conditions; however, zygote enucleation results in pronuclei that have the same resting potential as *in vivo*. Mouse liver nuclei removed by centrifugation have zero resting potential, but it is uncertain whether this condition comes from the isolation procedure.

B. Electron Microscopy of Nuclei

The procedures for obtaining the photographs and electron micrographs of nuclei become important when evaluating nuclear pore density. The dissected starfish nuclei were fixed in 1% glutaraldehyde, postfixed in 1% osmium tetroxide, dehydrated in alcohol, and dried in CO_2. The nuclei were then coated with gold for scanning electron microscopy. The mouse nuclei were fixed in 1% glutaraldehyde and 1% tannic acid, post-fixed in OsO_4, and dehydrated in alcohol, and 400–500 Å sections were stained with uranyl acetate and lead citrate. These dehydration and fixation procedures may introduce shrinkage and distortion of the nuclear envelope, causing an apparent density of nuclear pores that exceeds the *in vitro* density.

Figure 1 is an electron micrograph of three mouse oocytes packed together. Only a segment of each oocyte appears. The single darkly stained line, designated as **plasma membrane** (pm), represents the bilayer that surrounds the oocyte. The plasma membrane encompasses the **cytoplasm** (cy), which includes internal organelles such as mitochondria (mt) and the endoplasmic reticulum (er). The term **cytosol** (cy) refers to the cytoplasmic volume, excluding the organelles. The designation (cy) will stand either for the cytoplasm or the cytosol, depending on the context. In Fig. 2, parallel lines represent the lipid bilayer membrane. Figure 2 depicts the extracellular volume (ex), the cytosol (cy), the single-membrane-enclosed endoplasmic reticulum (er), and the double-membrane-enclosed nucleus (nu). Openings that penetrate the double membrane connect the nucleoplasm with the cytoplasm. The membranes that encircle the nucleus enclose a volume topologically equivalent to the lumen of the endoplasmic reticulum. This restricted volume has a particular name, the **cistern** (cs). In effect, cisterns are specialized organelles bound by one membrane. The membrane that defines the cistern is further categorized into the **outer membrane** and the **inner membrane**. Because the inner and outer membranes are close together, they are given the cumulative name, **nuclear envelope** (ne).

Figure 3 indicates that cistern membranes and endoplasmic reticulum membranes may join one another. However, cisterns may exist without any connection to the endoplasmic reticulum. The anatomy of the nucleus and its associations to cell structures varies with the developmental stage or the specialization of the cell. At some stages, the endoplasmic reticulum is virtually absent. Even in cells that have a well-developed endoplasmic reticulum, large regions of the envelope do not connect with the endoplasmic reticulum. Where they do connect, cisterns adjoin the endoplasmic reticulum via the outer membrane. The endoplasmic reticulum is either smooth or rough, depending on the absence or presence of ribosomes (dots in Fig. 3). The outer membrane of the nuclear envelope is also smooth or rough. The ribosomes on the endoplasmic reticulum are evident in the photograph in Fig. 1; if there are ribosomes on the outer membrane, they are not obvious.

The openings between the cisterns depicted in Figs. 2 and 3 are more than mere breaches in the nuclear envelope. They contain the nuclear pore complex. The nuclear pore complex is sometimes referred to simply as the nuclear pore (np). The reader must be careful, because "nuclear pore" also refers to the large central opening in the nuclear pore complex.

IV. Structure of the Nuclear Pore

The nuclear pore complex has a molecular weight of about 100 MDa, and it is comprised of nearly 100 gene products. The complex consists of eight symmetrical units that form a large central opening. We note the following difference between pore-forming proteins in the nuclear pore and pore-forming proteins in the plasma membrane: ion channels, gap junction proteins, receptor subunits, and neurotransmitter transporters span a single bilayer. These we call transmembrane (TM) proteins. Some nuclear-pore proteins also span a

single bilayer, but with a much different relationship to the flow of material. To illustrate, consider the structure of **gap junctions**. This example has special relevance to our topic because nuclear pore proteins and gap junction proteins both form openings through abutting membranes. Figure 4 illustrates membrane-spanning proteins of the nuclear pore complex, which insert at the division of the inner and outer membrane. Eight units (for simplicity, only two appear in Fig. 4), each composed of many proteins, form the central opening (center arrow, top drawing in Fig. 4). In addition, eight lateral openings exist (side arrows in Fig. 4 indicate two lateral openings). In gap junctions, six pairs of proteins (only one pair appears in Fig. 4, bottom drawing) define the opening. The distinction may seem artificial, because gap-junction proteins (connexins), as well as the subunits of ion channels and receptors, form parallel structures. However, the nuclear envelope has a distinct structure-function relationship between the TM proteins and transport. As far as we know, the functional connections made within the nuclear pore do not include cistern-to-cistern communication, but only cytoplasm-to-nucleoplasm communication.

Figure 5 illustrates the three-dimensional structure in more detail. The dark regions in the Fig. 5 reconstruction are the TM domains of the eight repeated units. These units also define eight additional pathways (as indicated in Fig. 4 by the lateral arrows). These pathways surround the central opening. We refer to this structure as eight-plus-one configuration, in analogy with the axoneme nine-plus-two nomenclature. Surprisingly, in a minority of cases more than eight units can form a nuclear pore complex. The significance of these higher order structures is unknown. Figure 6 defines the proposed relationship between the nuclear pore (np), the outer membrane (om), and the inner membrane (im) of the nuclear envelope (ne). The inner and outer membranes form the connecting loops that define openings between cisterns. The crosshatched regions in the drawing indicate the lateral openings that exist within the complex. Figure 6 also shows a feature mentioned earlier—the nuclear granule (or nuclear plug), illustrated as spool-shaped, that resides in the central opening. Recent work indicates that the nuclear plug moves within the pore to regulate transport in response to cistern Ca^{2+}. It has been suggested that the nuclear plug may include newly synthesized ribosomes.

The nuclear pore complex therefore, consisting of a large number of proteins, some of which are embedded in lipid bilayers, forms an eight-plus-one assembly that spans the double membrane. Some nuclei are virtually covered with nuclear pores. The high density of pores raises several interesting questions. What is the physical chemistry of the lipid/protein ratio in dense arrays? How is it possible to patch-clamp such arrays? In nuclei with a high density of pores, the relative amount of outer lipid membrane diminishes, and connections to the endoplasmic reticulum become sparse. Pore density varies between 1 and 120 per μm^2, depending on the specialization of the cell, and local pore densities may differ within the same nucleus.

Figure 7 illustrates the patch-clamp technique applied to isolated nuclei. In this technique, a glass electrode isolates a

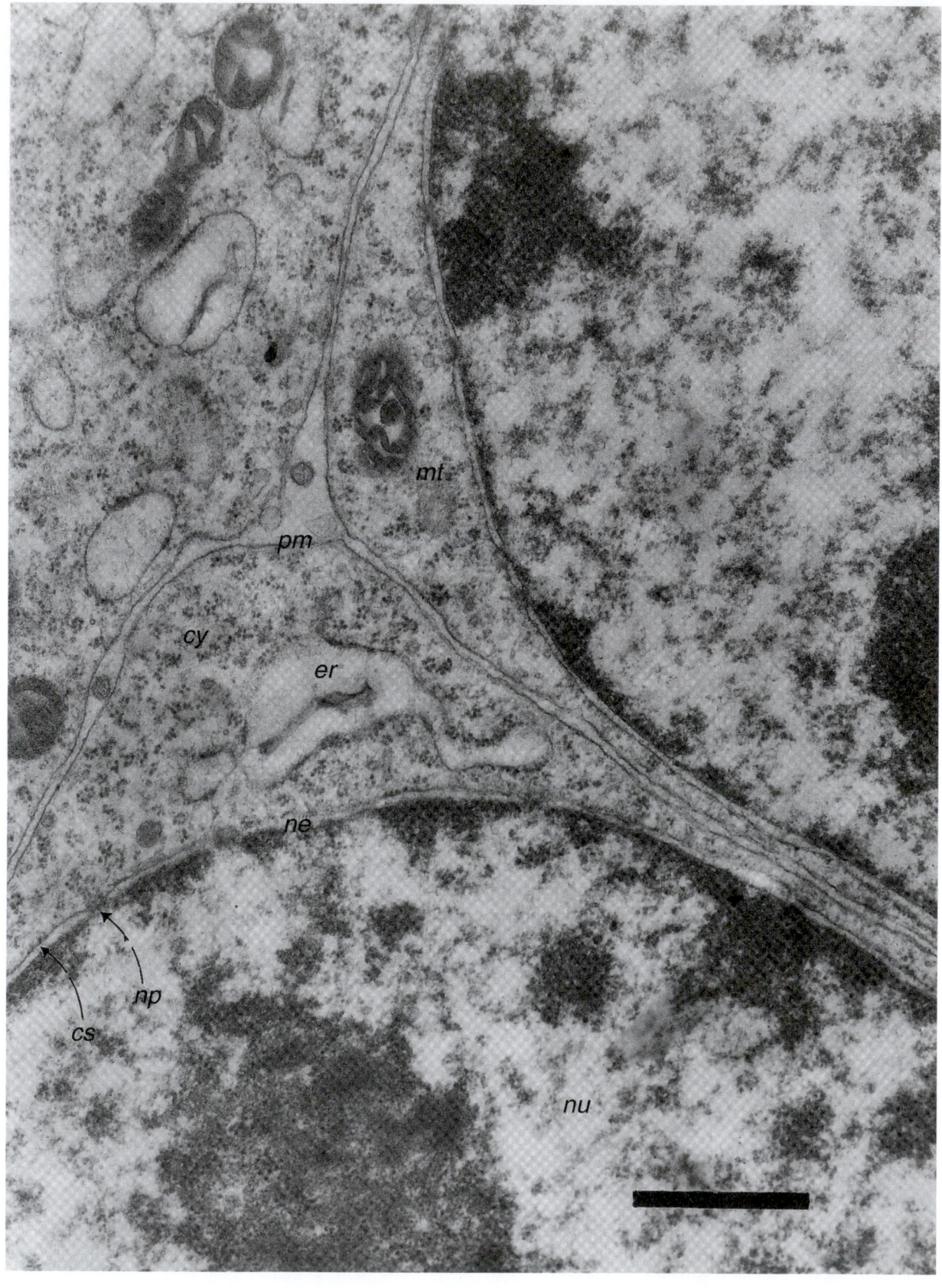

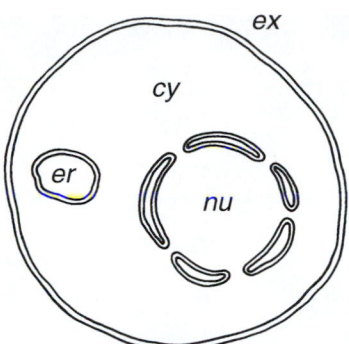

FIGURE 2. Cell spaces. The diagram defines the volumes within and around a cell: the extracellular space (ex); the cytosol (cy); the endoplasmic reticulum (er); and the nucleus (nu). The double lines represent lipid bilayers.

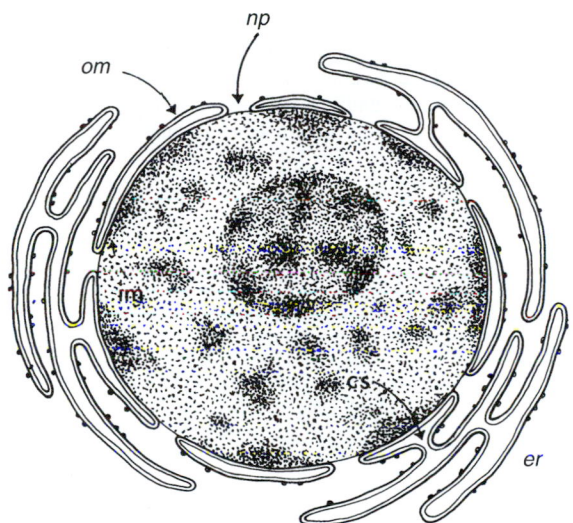

FIGURE 3. The nuclear envelope and the endoplasmic reticulum. This illustration shows the relationship between the outer membranes (om), the inner membranes (im), and the endoplasmic reticulum (er). The inner and outer membranes and the nuclear pore complexes (np) define the nuclear envelope. The central egg-shaped structure inside the nucleus represents the nucleolus. The dots along the outer membrane and on the endoplasmic reticulum represent ribosomes. The cistern (cs) of the nuclear envelope and the lumen of the endoplasmic reticulum form a continuous compartment within the cell.

patch of membrane on the outer surface of the nucleus. The applied voltage V (shown as a battery in the diagram) appears at the electrode tip, and the current meter (I) records the flow of ions through the patch. An isolated patch is 1–5 μm^2 in area. The voltage across the envelope is the difference between the nuclear resting potential and the pipette potential. This voltage also appears across the nuclear pores in the patch if the bath and the pipette solution are at ground potential. The possibility exists that channels in the outer and inner membranes connect the pipette solution to the nucleoplasm. This hypothetical pathway through the cistern would be in parallel with the nuclear pore. Let us now consider an actual patch-clamp experiment on a nucleus with a high density of pores.

Figure 8 shows a scanning electron micrograph of a starfish oocyte at the germinal vesicle stage. The nucleus is ~30 μm in diameter, larger than many specialized cells. The inset shows an isolated area on the outer membrane about 1 μm in diameter. The diameter of a nuclear pore complex is ~1000 Å (0.1 μm), which implies 100 nuclear pore complexes side-by-side in 1.0 μm^2 of membrane. Each doughnut-shaped object in the expanded field in Fig. 8 represents one nuclear pore complex; thus the pores cover the surface. Thus even if shrinkage has occurred during the fixation procedure, it seems likely that a nucleus-attached patch on a nucleus in this preparation contains dozens of pores.

V. Electrophysiology of the Nucleus

If nuclei like the one shown in Fig. 8 are dispersed in a dish, they can be manipulated and prepared for electrophysiology as if they were cells. As mentioned above, some nuclei have resting potentials. For example, in a mouse zygote, the *in situ* pronuclei have a resting potential of −10 mV with respect to the cytoplasm. If we eliminate the resting potential of the zygote plasma membrane (see Section II, Chapter 14), no immediate change occurs in the nuclear potential. We expect this, because the nuclear potential appears across the envelope; altering a voltage in series should not affect it. For excised pronuclei in an intracellular-like solution (120 mM K$^+$), the nuclear potential has the same value it had in the cell. Lowering K$^+$ concentration *in vitro* to 60 mM K$^+$ changes the nuclear potential to −20 mV. Table 1 summarizes experiments on *in vivo* and

FIGURE 1. Structure of the nucleus. Transmission electron micrograph of three mouse oocytes at the germinal vesicle stage. The plasma membrane (pm) surrounds the cytoplasmic compartment (cy), and the nuclear envelope (ne) surrounds the nuclear compartment (nu) within the cytoplasm. The micrograph demonstrates two additional membrane-bound compartments in the cytoplasm, mitochondria (mt) and the endoplasmic reticulum (er). The endoplasmic reticulum has most likely swollen during the fixation procedure. The nuclear envelope consists of a double-membrane structure interrupted by nuclear pores (np). The double membranes, designated as the inner membranes and the outer membranes (see Fig. 3), define a separate cytoplasmic compartment called the cistern (cs). The cistern may be continuous with the endoplasmic reticulum (see Fig. 3). The nuclei contain darkly stained material adjacent to the inner membrane, called nuclear lamina, and they contain the diffuse darkly stained material called chromatin. Samples were dehydrated with alcohol to propylene oxide, infiltrated and embedded in epon, thin sectioned, and stained in uranyl acetate and lead citrate. The bar length indicates one micron, and the bar width indicates 1000 angstroms. (From Mazzanti *et al.*, *J. Memb. Biol.* **121**:189–198 (1991).)

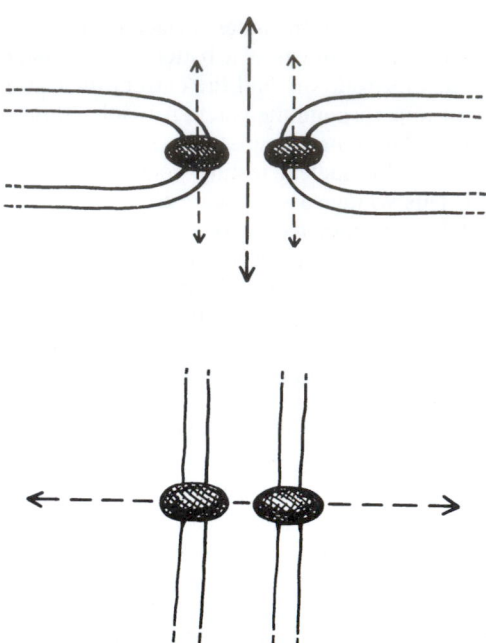

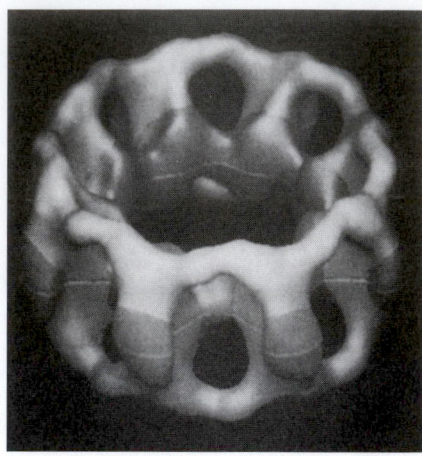

FIGURE 5.   Structure of the nuclear pore. A three-dimensional reconstruction based on an electron density map of the 100-MDa structure that forms the nuclear pore complex. The entire structure has an external diameter of about 1300 Å. The model proposes eight lateral channels that surround a large central pore. The central pore normally contains a proteinaceous plug that occupies most of the opening (see Fig. 6). The central plug is absent in Fig. 5. The reconstruction is based on electron microscopy and image analysis of detergent-released nuclear pore complexes from macronuclei of mature *Xenopus laevis* oocytes. [From Hinshaw *et al.*, Cell **69**:1133–1141 (1992). Copyright 1992 by Cell Press.]

FIGURE 4.   Nuclear pores versus gap junctions. A comparison between the structure and function of the nuclear pore (top) and the gap junction (bottom). The nuclear pore complex contains eight membrane-spanning subunits (two are shown here). Each subunit is a multiprotein structure that repeats eight times. Material flows through a central pore (the center arrow in the top figure) defined by the surrounding subunits. In addition, lateral nucleocytoplasmic pathways are located within each subunit (see Fig. 5). The shorter arrows in the top drawing illustrate the lateral pathways. The complete nuclear pore complex contains eight lateral pathways surrounding the central pore. Gap junctions have a total of six pairs of abutting membrane-spanning proteins. One such pair is illustrated in Fig. 4 (bottom). Apart from a different number of repeating subunits, gap junctions also have a different structural relationship to function when compared to nuclear pore complexes.

in vitro nuclei under different ionic conditions. The data suggest that the nuclear envelope selects K^+ ions. Different conditions may lead to other potentials. Frog oocyte and adult liver cell nuclei have no apparent resting potential, and starfish nuclei can maintain a positive potential with respect to the bath.

A. Measurement of Resting Potential

To detect the nuclear resting membrane potentials in intact cells requires a conventional glass microelectrode of high resistance (70 to 90 MΩ when filled with 3 M KCl). In such experiments, we penetrate both the outer and inner membranes. Thus the nuclear potential means the voltage across the entire double-membrane nuclear envelope. Ion channels exist in both the outer and inner membranes. For example, Cl^- channels have been reported on the surface of nuclei, as well as ligand-gated channels. However, no measurements exist from isolated cisterns, and we do not know the electrical potential of the cistern with respect to the cytoplasm or nucleoplasm. For *in situ* experiments on the nucleus, cells that are maintained in physiological solutions

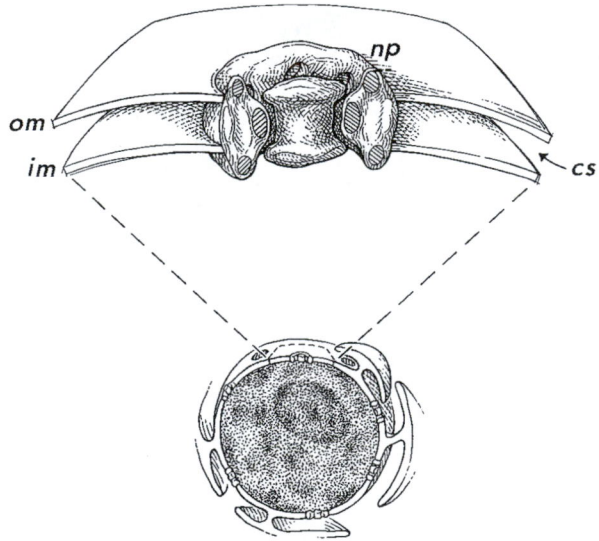

FIGURE 6.   The nuclear pore and the nuclear envelope. Illustration of the position of the nuclear pore (np) within the double-membrane structure that envelopes the nucleus. The two planar membranes illustrated in the diagram represent the outer membrane (om) and the inner membrane (im). The interconnecting subunits that make up the supramolecular nuclear pore complex insert into the lipid bilayer at the border between the inner and outer membranes. These subunits form eight channels that surround the main opening of the pore (see Fig. 5). The main opening contains a proteinaceous granule or "plug" (not shown in Fig. 5), which we have pictured as a nearly space filling, spool shaped object. Thus two kinds of pathways connect the cytoplasm and the nucleoplasm: the middle pore that contains the plug, and the eight encircling channels that riddle the proteins that form the central pore.

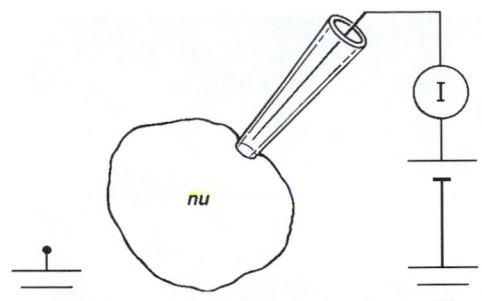

FIGURE 7. Patch-Clamp Technique. An illustration of a nucleus-attached patch-clamp experiment. The glass pipette contains a physiological solution connected by a Ag/AgCl electrode to a current meter (I) and a battery. The battery supplies a voltage, V_p, to the pipette that ideally appears at the outer membrane of the nucleus. The nucleus bathes in a physiological cytoplasm-like solution connected to ground. The potential across the nuclear envelope is the resting potential of the nucleus minus V_p.

(133 mM NaCl, 2 mM KCl, 1.5 mM $CaCl_2$, 0.5 mM $MgCl_2$, 10 mM Hepes, 5 mM dextrose, pH 7.4) have normal resting potentials. To eliminate the cell resting potential, we bathe in a solution that mimics the cytoplasm (120 mM KCl, 2 mM $MgCl_2$, 1.1 mM EGTA, 0.1 mM $CaCl_2$, 10 mM Hepes, 5 mM dextrose, pH 7.4).

Among the technical difficulties associated with measuring the nuclear resting potential, electrode tip potential gives the most problems. The structures that compose the nuclear envelope, the inner and outer membranes and the lamina may stick to the microelectrode and increase resistance and tip potential. To guard against such errors, check that the tip potential (the electrode voltage with the tip in the bath solution) returns to the baseline upon removal. This does not prove that the observed potential comes from the nuclear envelope, but it is an essential first step. Similar technical difficulties related to the value of nuclear potentials arise in measuring plasma membrane potentials.

B. Measurement of Single Channels from Nuclei

In mouse liver cell nuclei, about 50% of the attempts to patch succeed. Seals up to 50 GΩ are obtained using pipettes having an external tip diameter of about 1 μm and a resistance between 5 and 10 MΩ. In the single-channel experiments, 120 mM K^+ (cytoplasmic-like) solutions may be used to fill the electrode and to bathe the nucleus. To investigate the ionic selectivity of the nuclear envelope, NaCl may replace KCl. Single-channel currents can be recorded with a standard current-to-voltage converter (List EPC-7). Positive pressure in the pipette (~2 cm H_2O) keeps the electrode solution from mixing with the bath solution, flattens and cleans the nuclear surface, and helps form the seal. For unknown reasons, a negative potential (~V_p = −40 mV) applied to the electrode tip before touching the nucleus facilitates seal formation. Such negative potentials help form seals on most cells. After touching, a slight increase in electrode resistance can be noticed: the pressure can be reversed until a tight seal occurs, and then the suction can be released.

A patch-clamp experiment on the nucleus results in single-channel currents similar to those observed in the plasma membrane of most cells. Nuclear envelope channels were first observed in mouse oocytes. Figure 9 shows the results of a patch-clamp experiment on a starfish oocyte nucleus (as in Fig. 8). In Fig. 9, V_p is the voltage applied to the pipette (the battery in the external circuit, Fig. 7). The channel conductance in this experiment was about 100 pS. Not only long-lasting openings, but also numerous fast openings occur. These are too brief to resolve in this record. The downward deflections in Fig. 9 indicate positive electric current flowing into the nucleus. The downward deflection could also result from negative ions moving out of the nucleus.

Mouse oocyte nuclei have a lower pore density than starfish germinal vesicles; nevertheless, they display similar ion channel activity. The nuclear pore density depends on the tissue and on the stage of development of the cell, as we have stated. The freeze-fracture image shown in Fig. 10A is from an oocyte germinal vesicle, and that in Fig. 10B is from an adult liver cell nucleus. The pore density is 3–10 per μm^2 in Fig. 10A and 14–18 per μm^2 in Fig. 10B. Recordings from such surfaces show channels similar to those shown in Fig. 9. Figure 11 summarizes results from four different kinds of nuclei at various stages of cellular development: the mouse germinal vesicle, the pronucleus, a two-cell embryo nucleus, and a differentiated adult cell. In Fig. 11, the current is upward because the applied potential is negative. We may conclude from experiments on starfish oocyte nuclei and mammalian cell nuclei that channels in the nuclear envelope are a ubiquitous phenomenon. In addition, Fig. 11 shows for embryonic preparations (A, B, C) only one or two conductance levels, while in the adult preparation there are consistently more levels, as illustrated in the histograms to the right of the figure. Combining morphological and functional data, we have hypothesized a relationship between nuclear pores and ion channels. Specifically, the channels could represent an ionic pathway through the nuclear pore complex itself.

VI. Osmotic Effects in the Nucleus

Figure 12 shows that the nucleus undergoes changes in response to osmotic forces. However, these changes are not the simple osmotic effects expected from solution chemistry. In this experiment, the ionophore nystatin was used to make the plasma membrane permeable to monovalent cations. The extracellular bath and the cytoplasm are in electrical communication through nystatin channels. Adjusting bath osmolarity changes nuclear volume. Such permutations of the volume are difficult to see and even more difficult to quantify. Some responses occur in the way we would expect from osmotic changes: in hypertonic solutions the nucleus shrinks, and in hypotonic solutions the nucleus swells. However, too great an osmotic shock produces different effects, some of which are irreversible: if the nucleus shrinks in external 480 mM KCl, it subsequently swells irreversibly to the configuration shown in

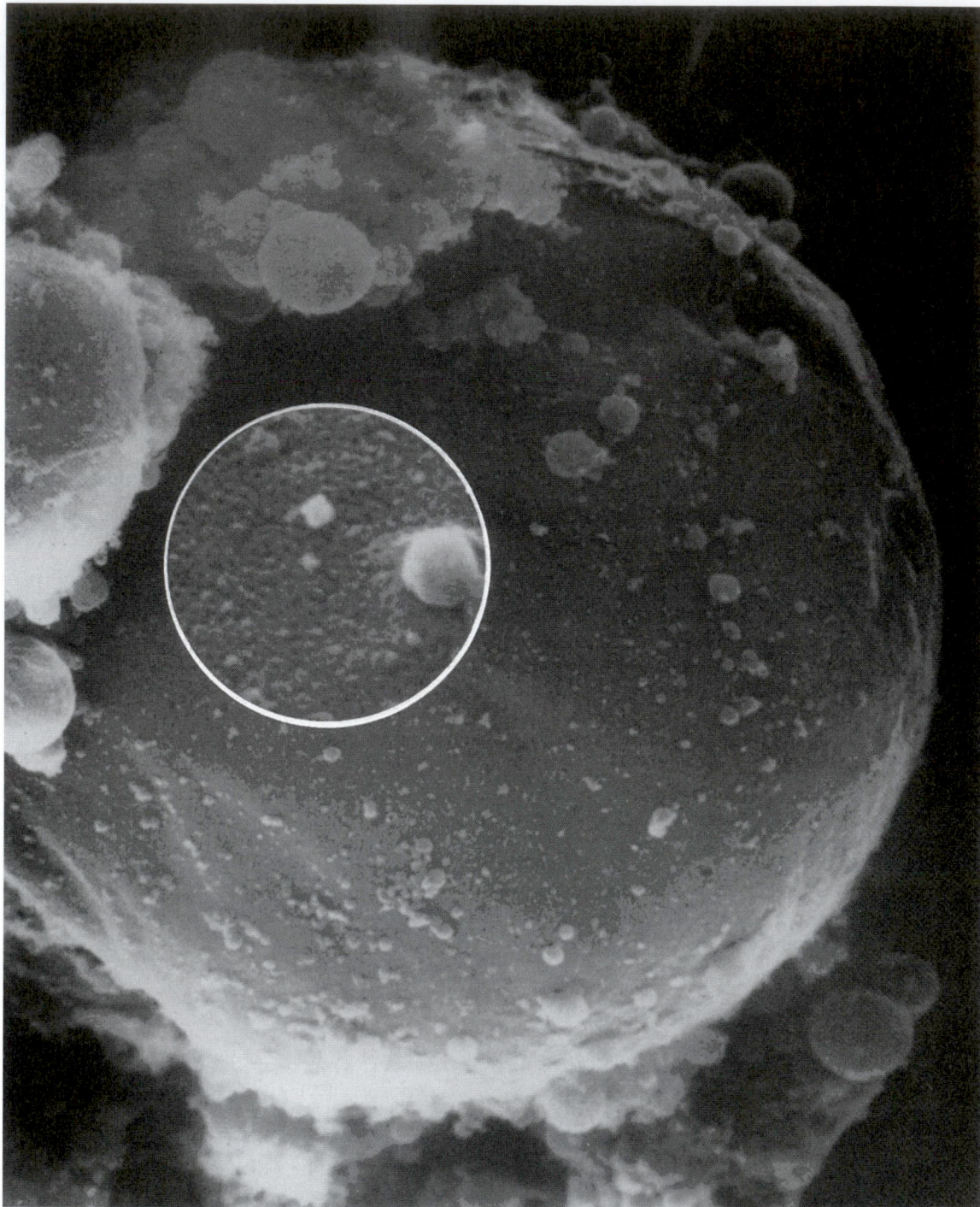

FIGURE 8. Scanning electron micrograph of a nucleus from a starfish oocyte. The figure shows a scanning electron micrograph of the surface of a germinal-vesicle-stage nucleus removed from a starfish oocyte (*Mathasterias glacialis*) by manual dissection. The diameter of the nucleus is approximately 30 μm. The diameter of the inset, which gives a magnified view of the outer membrane, is approximately 1 μm. Large areas of the external face of the nucleus appear relatively free of extraneous material. The whitish, globular matter is cellular debris stuck to the surface after dissection. The exterior surface of the dissected nucleus appears studded with numerous doughnut-shaped particles that we have interpreted as nuclear pore complexes (average density 60–110 per μm^2). This anatomy makes it impossible to patch the starfish nuclear envelope without encountering numerous NPC. (Courtesy of Luigia Santella, Stazione Zoologica, Naples, Italy.)

TABLE I   Results of Experiments on Nuclei Subjected to Different Ionic Conditions

(1) 2 K/133 Na		(2) 120 K/0 Na		(3) 120 K/0 Na	(4) 60 K/60 Na
cy	nu	cy	nu	nu	nu
−23	−11	0	−10	−10	−28
−18	−12	2.5	−9.5	−7	−18
−28	−9	3	−9	−9	−23
−25	−11	1	−11	−12	−25

Note: The ratios of K to Na (in mM) refer to extracellular (bath) concentrations. In columns (1) and (2) (zygote pronuclei *in situ*), the voltages (in mV) are as follows. cy = the potential of the cytoplasm with respect to the extracellular solution nu = the potential of the nucleoplasm with respect to the cytoplasm. In columns (3) and (4) (pronuclei *in vitro*), the voltages are the nuclear potential with respect to the bath.

Fig. 12D. The swelling that occurs in 480 mM KCl is due in part to internal structural changes in the chromatin. Distended nuclei remain stable for up to 30 min. Returning the nucleus to 120 mM KCl, after swelling with 480 mM KCl, causes the nucleus to disrupt. With $V_p = 0$ and 120 mM KCl in the patch and the bath, channels in the patch carry small inward currents. After perfusion with 480 mM KCl and waiting 5 min for the nucleus to swell, the direction of the current changes from inward to outward, and the amplitude increases. In 120 mM KCl, the conductance of the channels is 230 pS, as in the pronucleus. In 480 mM KCl, the conductance is 510 pS, and their reversal potential shifts negative by 33 mV. The size and direction of the potential shift indicate that the channel is K^+-selective. The existence of K^+-selective channels would explain the nuclear resting potential discussed above. More importantly, these results imply that ion channels in the patch span the double membrane, because the current pathway must join the pipette with the bath. It is conceivable that channels in the patch reside in the outer membrane, and that other channels in the inner membrane complete the circuit. Because the only structure known to connect cytoplasm to nucleoplasm is the nuclear pore, and because the patch contains many pores, it is likely that some structure within the nuclear pore complex itself forms the observed channels.

VII. Electrical and Diffusional Forces across the Nuclear Envelope

Solution chemistry is insufficient to explain the complex phases that make up nucleoplasm and cytoplasm. Nevertheless, the data suggest that the nuclear envelope can sustain electric and diffusion gradients similar to those that exist across the plasma membrane. In particular, channels recorded from the nuclear surface respond to transenvelope voltages, and the nucleus changes shape and volume with osmolarity. Thus we consider the hypothesis that the nuclear envelope acts as a barrier to the diffusion of small ions. Such gradients

could play a major role in the transport of macromolecules. We will now apply to the nucleus a well-known approach from cellular physiology.

The flux of any substrate through a boundary that sustains diffusional and electrical forces is

$$-\Phi = ukT\frac{dn}{dx} + zenu\frac{dV}{dx} \tag{1}$$

The minus sign comes about because flux occurs down a negative gradient (i.e., downhill). The symbol Φ represents the flux of the substrate, that is, the number of molecules that move per unit area per unit time. We use the symbol $-\!o\!-$

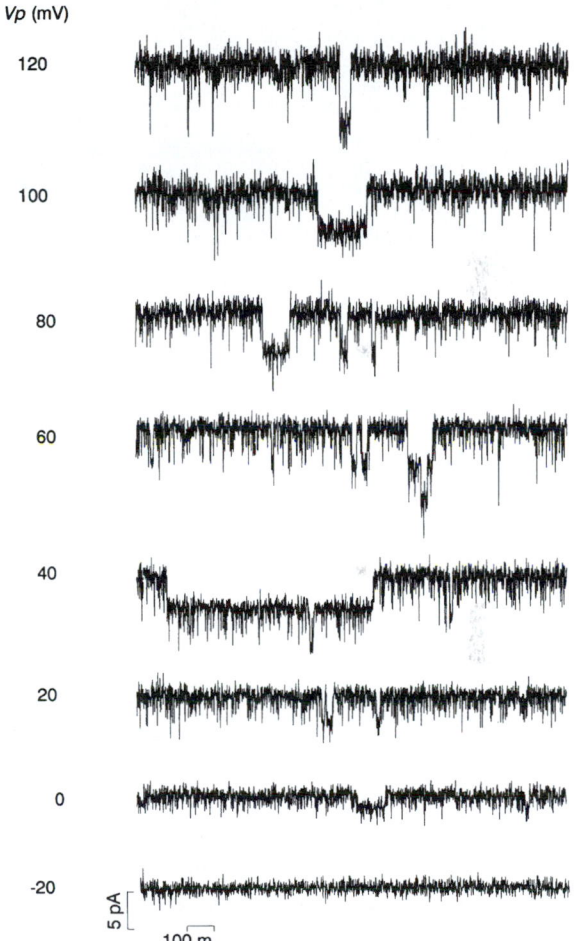

FIGURE 9. Nucleus-attached patch recordings from starfish germinal vesicle. The figure shows sample currents recorded from the surface of a germinal-vesicle-stage nucleus isolated from a starfish oocyte similar to Fig. 8. Channel openings appear as downward deflections, which signify positive charges flowing into the nucleus. Two categories of openings appear: brief inward currents barely resolved on this time scale, and longer openings that last several hundred milliseconds. In some traces (e.g., at 60 mV), simultaneous openings occur giving two current levels of the same amplitude. The voltages (V_p) at the side of each trace indicate the pipette potential with respect to the bath. The nucleus-attached electrode contains 200 mM KSO_4, 20 NaCl, 200 sucrose, 10 EGTA, 10 Hepes (pH 7.0) and the bath solution is comprised of natural seawater at room temperature. (From Santella *et al.*, 5th Intl. Conf. Cell Biol., Madrid (1992).)

A
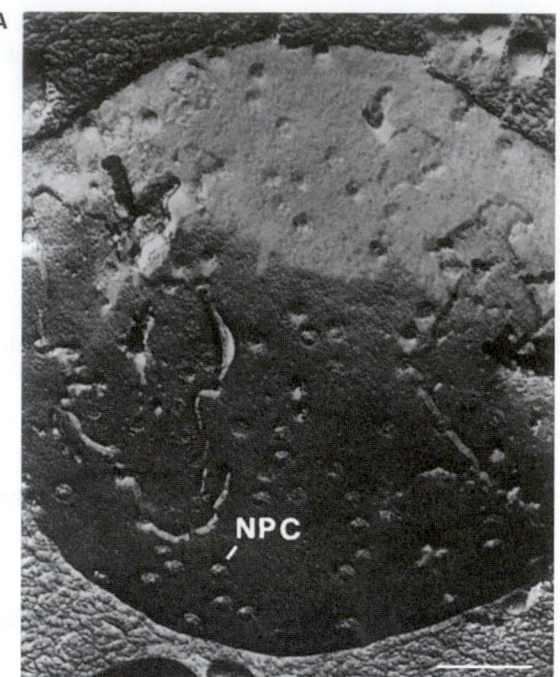

B

FIGURE 10. Nuclear envelope freeze fracture from mouse oocyte and adult mouse liver cells. The figures show scanning electron micrographs of a fracture through the outer membrane of the nuclear envelope of (A) mouse oocyte nucleus and (B) adult mouse liver cell nucleus. The doughnut shaped objects represent the nuclear pore complexes (NPC). Note the different pore density in the two preparations. The bar length indicates 1 micron. (Modified from Innocenti and Mazzanti, *J. Memb. Biol.* **131**:137–142 (1993).)

to stand for "has the units of." Thus $\Phi \multimap$ #/cm²·s, $n(x)$ is the substrate concentration, and $n \multimap$ #/cm³ at position x, $V(x)$ is the voltage (mV) at position x, u is the mobility of the substrate through the barrier, and $u \multimap$ (velocity/force), k is the Boltzmann constant, T is the absolute temperature in kelvins, z is the valence, and e is the electronic charge. These latter quantities will appear as the ratio

$$a = ze/kT = z\mathscr{F}/RT = z/25 \text{ mV}^{-1} \text{ at 23 °C}$$

where $\mathscr{F}$ is the Faraday constant and R is the gas constant. By considering the convention in Fig. 2 and integrating the flux equation between the nucleoplasm (nu) and the cytosol (cy), the expression for the flux becomes

$$-\Phi = ukT \frac{n_{nu} e^{-aV_{nu}} - n_{cy} e^{-aV_{cy}}}{\int e^{aV(x)} dx} \quad (2)$$

This assumes that u is constant and flux is in a steady state (no buildup or rundown of substrate; thus Φ is constant). Note that $D = ukT$, where D is the diffusion constant of the substrate. In the derivation of Eq. 2, no assumption was made about the spatial distribution of the voltage or concentration gradients, $V(x)$ or $n(x)$ except that they are constant. Only the values of n and V far from the envelope come into play. An unevaluated integral appears in the denominator of Eq. 2. To integrate this term, $V(x)$ must be known, the shape of the voltage gradient across the barrier; however, $V(x)$ is never known exactly and, in any case, it varies during transport. The term $\int e^{aV(x)} dx$ is therefore in-

tractable and is usually a free parameter. The flux equation (in its differential form, Eq. 1, or in its integral form, Eq. 2) is called the **Nernst-Planck** equation. If we assume $V(x)$ to be a linear function of x, the Nernst-Planck equation leads to the **Goldman equation**. Rather than assume a particular $V(x)$, we leave the integral undetermined and replace it by the symbol L

$$L = \int e^{aV(x)} dx \quad (3)$$

where $L \multimap$ length (cm) and $ukT/L = D/L \multimap$ velocity (cm/s); we call this ratio the permeability p, where $p \multimap$ (cm/s). Thus we can write for the flux of any substrate

$$-\Phi = p \left(n_{nu} e^{-aV_{nu}} - n_{cy} e^{-aV_{cy}} \right) \quad (4)$$

where n_{nu} is the concentration of the substrate in the nucleus and n_{cy} is its concentration in the cytoplasm. For a further discussion see ch. 2 in DeFelice (1981) and DeFelice (1997).

How would a theory used for cell physiology apply to the transport of macromolecules across the nuclear envelope? To answer this, we overlook the double membrane and consider the nucleus as a cell within a cell (Fig. 13). This ignores the anatomy of the cisterns and replaces the double membrane with a single boundary penetrated by nuclear pores. Let us first consider a carrier model. For specificity, suppose that K⁺ ions cannot get through the envelope unless they combine with the nuclear pore complex

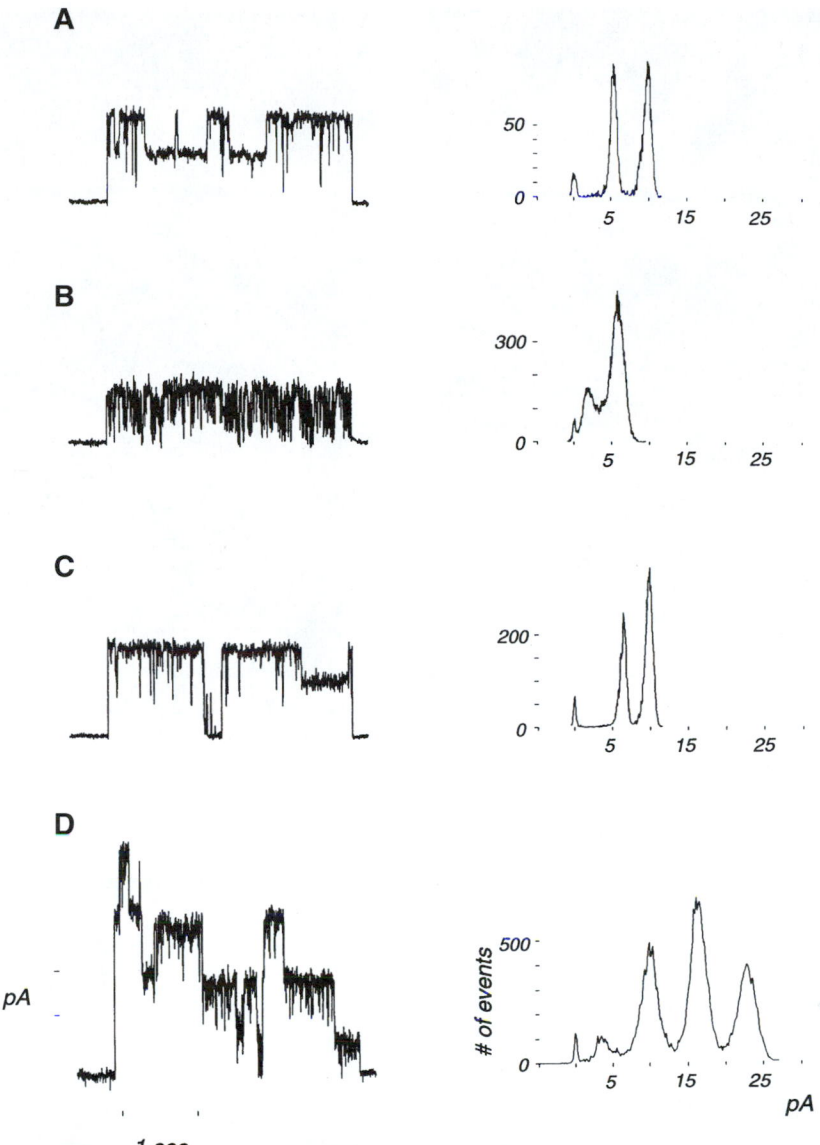

FIGURE 11. Nucleus-attached patch recordings from mouse nuclei during development. The figure shows large-conductance ion channels that appear on the surface of (A) germinal-vesicle-stage nucleus from a mouse oocyte, (B) pronucleus from a mouse zygote, (C) nucleus from two-cell mouse embryo, and (D) nucleus from an adult mouse liver cell. Amplitude probability histograms appear to the right of the representative trace. The traces and histograms result from stepping the pipette voltage from 0 mV to −25 mV. In this experiment, the bath and the nucleus-attached patch pipette contained an intracellular-like solution (120 mM KCl, 2 mM MgCl$_2$, 1.1 mM EGTA, 0.1 mM CaCl$_2$, 5 mM glucose, 10 mM Hepes, pH 7.4) at room temperature. (From Mazzanti *et al., J. Memb. Biol.* **121**:189–198 (1991).)

(np), and eight ions must interact before transport can occur. We select eight because of the geometry of the complex. Thus we expect eight (or a multiple of eight) interactions between the complex and the ions. Under this assumption, we write

$$8K^+ + P \rightleftharpoons K_8P$$

$$K_{kp} = \frac{[K]^8 [P]}{[K_8P]} \tag{5}$$

where P stands for the nuclear pore complex and K_{kp} is the equilibrium constant for the reaction. Next consider a macromolecule, M, that undergoes a similar interaction with P, but one molecule at a time

$$M + P \rightleftharpoons MP$$

$$K_{mp} = \frac{[M][P]}{[MP]} \tag{6}$$

K_{mp} is the equilibrium constant for the reaction. Assume that K$_8$P and MP can move across the envelope, but that K$^+$, M,

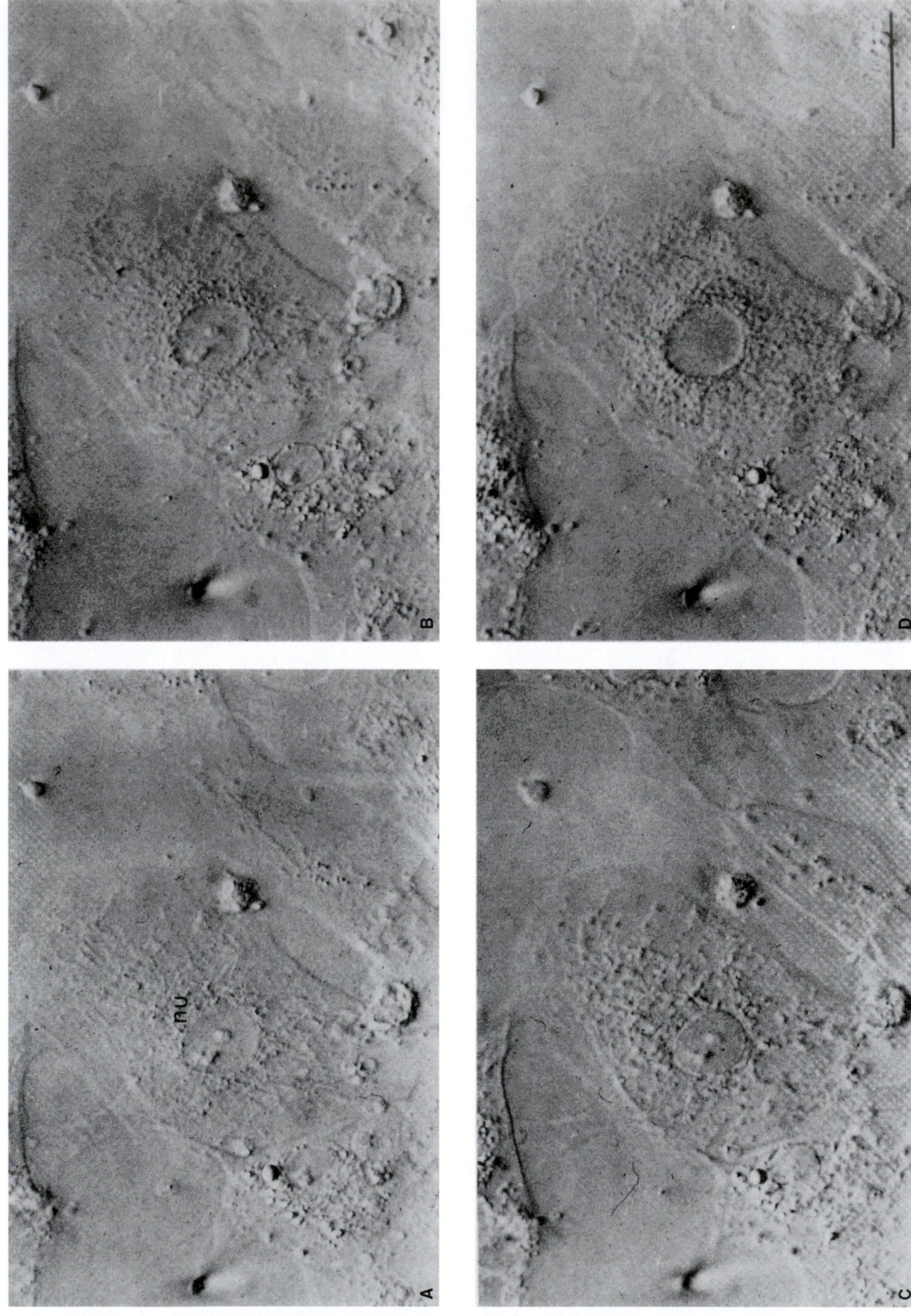

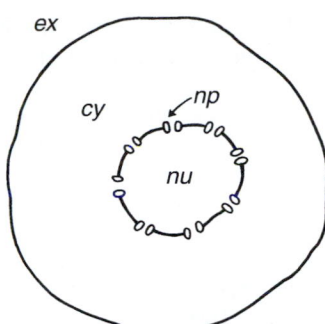

FIGURE 13. The nucleus as a cell within a cell. This diagram suppresses the double-membrane structure of the nuclear envelope, and it regards the nucleus (nu) as a membrane-bound organelle. In this model, only the nuclear pores (np) connect the nucleoplasm (nu) to the cytoplasm (cy). The nuclear envelope is, therefore, a semipermeable membrane analogous to the cell plasma membrane that separates the cytoplasm from the extracellular space (ex).

and P by themselves cannot. Now consider two fluxes: the flux of K_8P bound to the pore

$$-\Phi_{kp} = p_{kp}\left([K_8P]_{nu}\, e^{-aV_{nu}} - [K_8P]_{cy}\, e^{-aV_{cy}}\right) \quad (7)$$

and the flux of MP bound to the pore

$$-\Phi_{mp} = p_{mp}\left([MP]_{nu}\, e^{-aV_{nu}} - [MP]_{cy}\, e^{-aV_{cy}}\right) \quad (8)$$

where p_{kp} is the permeability of K_8P, and p_{mp} is the permeability of MP through the envelope. By hypothesis, let the transport of macromolecules and the transport couple to one another. We assume this is a property of the nuclear pore. Then the steady-state condition reduces to

$$\Phi_{kp} + \Phi_{mp} = 0 \quad (9)$$

This development treats the nuclear pore complex as a carrier and indicates that the concentration of the carrier does not change: if a certain number of these carriers in the kp state move to the left, then an equivalent number in the mp state move to the right. We may solve these equations for Φ_{mp}, the flux of macromolecules, in terms of the concentrations of the substrates, [K] and [M], and the concentration of the pores, [P], using

$$\begin{aligned}[K_8P] &= [K]^8 [P] \big/ K_{kp} \\ [MP] &= [M][P] \big/ K_{mp}\end{aligned} \quad (10)$$

The values for K_{kp}, K_{mp}, as well as the concentrations of the substrates, may differ on each side of the pore. The fixed

number of pores constrains the equations. Thus, we may write the flux of the macromolecules, Φ_{mp}, in terms of $[K]_{cy}$, $[K]_{nu}$, $[M]_{cy}$, and $[M]_{nu}$.

Now consider a channel model. Assume the nuclear pore complex consists of eight channels for ions and one channel for charged macromolecules, and no that other pathways exist between cytoplasm and nucleoplasm. Then the fluxes of ions and macromolecules are coupled through their electric charge without a separate assumption about the carrier nature of the pore. We may write the flux equations

$$-\Phi_k = p_k\left([K]_{nu}\, e^{-aV_{nu}} - [K]_{cy}\, e^{-aV_{cy}}\right) \quad (11)$$

$$-\Phi_m = p_m\left([M]_{nu}\, e^{-aV_{nu}} - [M]_{cy}\, e^{-aV_{cy}}\right) \quad (12)$$

We have dropped the p in the subscripts kp, etc., to distinguish the channel model from the carrier model. In the channel model, the steady-state condition reduces to this

$$8\Phi_k + z_m\Phi_m = 0 \quad (13)$$

where z_m is the valence of the macromolecule. Equation 13 leads to an expression for the flux of the macromolecules, Φ_m, in terms of the substrate concentrations on either side of the envelope. The flux of any charged molecule, including a macromolecule, is an electrical current. In general, the relationship between flux, Φ —o— #/cm²·s, and current, I —o— pA/cm², is

$$I = zen\Phi \quad (14)$$

The parameter z in this expression is the valence on the molecule whose concentration is n, and e is the electronic charge ($e = 1.6 \times 10^{-19}$ coulombs). For example, if ze were the charge of the K^+ ion, then $ze = +1e$, and $z_m e$ would represent the charge on the macromolecule under consideration. Thus we may use Eq. 14 to write an expression for the K^+ current and for the macromolecular current, and these two currents would be constrained by Eq. 13.

The carrier model or the channel model amounts to a balance of fluxes and currents. For example, if the charged macromolecule being transported is RNA, then we could write from the above channel model

$$I_K + I_{RNA} = 0 \quad (15)$$

A similar equation would hold for charged proteins. It may seem unusual to consider the movement of RNA or proteins as a current. However, we are already familiar with this concept in gel electrophoresis, in which nucleotides and proteins move under an electric field. Equation 15 hypothesizes that a comparable phenomenon to gel electrophoresis occurs at the nuclear envelope. Assume a rate of 10^9 RNA molecules/min, one electronic

FIGURE 12. The nucleus as an osmometer. Nuclei (*nu*) from 3T3 fibroblasts under four osmotic conditions. In all cases, the bath solution contains nystatin, a monovalent cation ionophore, at a concentration of 50–100 ug/ml. The nystatin pores form holes in the plasma membrane that connect the cytoplasm to the bath. (A) 120 mM KCl bath solution (isosmotic), (B) 240 mM KCl (hyperosmotic), (C) 60 mM KCl (hypo-osmotic), and (D) 480 mM KCl. In (A, B, and C) the effects are reversible, but in (D) the effects are irreversible. (Unpublished data.)

unit charge/base, and 10^4 bases/molecule. Then I_{RNA} would be the order of 10 nA for the entire nucleus and 1 pA for an individual nuclear pore. The point we wish to emphasize is that transport of a charged macromolecule must have a counter current. If electrical coupling between macromolecules and ions does occur, the ionic current and the voltage across the nuclear envelope could regulate transport of nucleic acids and proteins into and out of the nucleus.

VIII. Modulation of Ionic Nuclear Permeability by ATP

Patch-clamp experiments on isolated nuclei suggest that the channel activity recorded on the nuclear envelope surface represent a direct nucleocytoplasmic communication pathway. As previously stated, several paradoxes arise from this hypothesis. Results obtained from photobleaching techniques indicate that 40 kDa molecules are the upper limit for free diffusion through the pore. In terms of diameter this would mean 9 nm particles. A channel that is 9 nm in diameter, 80 nm in length, and contains a solution of 100 $\Omega \cdot$cm, has a conductance of about 1 nS. This value is significantly greater than the lowest channel conductance levels obtained using the patch-clamp recording technique, especially if we consider that the surface isolated by the patch pipette contains many nuclear pores. Pores of 1 nS conductance would shunt the patch and make the observation of smaller conductances impossible. Thus if the observed channels are in the outer membrane, the nuclear pores must be closed. On the other hand, if the channels are within the nuclear pore, their conductance is too low. Is it possible that, under the experimental conditions of the patch clamp recordings, the pores are closed? As mentioned previously, the nuclear pore contains a plug (Fig. 6) that may move under altered Ca conditions. Nuclear isolation procedures are traumatic: cytoplasmic components are lost and nucleo cytoplasmic cytoskeletal structures are disrupted. Patch-clamp recordings from nuclei inside the cell help overcome these shortcomings. Figure 14 shows single-channel events recorded *in situ* from *Xenopus* oocyte nuclei (developmental stage one). Adding 1 mM ATP to the pipette solution (a physiological concentration of ATP typical for cytoplasm) transforms sporadic channel openings into a macroscopic current with characteristic kinetics. Figure 15 shows the results of such experiments under three different conditions. The presence of ATP in the recording pipette is essential to record the macroscopic current illustrated in (Fig. 15B). In the absence of the nucleotide, either *in situ* or after nuclear isolation, individual channel openings are observed, as illustrated in Fig. 15C and E. However, the ensemble averages of (C) or (E) have the same kinetics as (B). Furthermore, whole-cell nucleus recordings have resulted in current traces similar to Fig. 15B, although the amplitude of the current is much larger.

We can estimate the upper limit of the current that should flow if all the pores in the patch were open. Freeze-fracture measurements on first-stage *Xenopus* oocytes indicate 5–8 pores per μm^2. A patch area of 2 μm^2 would contain 10–16 pores. Assuming the maximum expected value of 1 nS per pore, −25 mV would generate a patch

current between 250 and 375 pA. In experiments like that of Fig. 15B, the average peak current at −25 mV is 318 ± 34 pA (mean ± SD; $n = 22$). These results show that the channels observed in isolated nuclei are similar to the channels observed *in situ*, that the channel pathways responsible for the currents are the same with or without ATP, and that ATP increases the open probability. Thus, the experimental conditions of *in vitro* patch-clamp recordings apparently do not alter the properties of individual channels, although the open probability may vary. These data, and data from whole-nucleus preparations, support the hypothesis that channels recorded from the nuclear envelope represent a direct nucleocytoplasmic communication pathway.

IX. Cytoskeletal Interaction with the Nuclear Ionic Flux

Recent experiments have tested the hypothesis that most of the current flows through the nuclear pore. The first approach was to combine single-channel recording techniques on isolated rat liver nuclei with Atomic Force Microscopy (AFM). The sequence was as follows: after forming a seal on an isolated nucleus, the patch was stimulated with voltage steps to observe channel activity. The pipette was then withdrawn, detaching a patch from the nuclear envelope. It was then possible to transfer the patch on a rigid support and visualize its surface using AFM. In this way we showed that membrane patches with ion channel activity contained several nuclear pores (Fig. 16).

Unfortunately this procedure was unable to demonstrate how the number of channels and nuclear pores are related. Using a different approach, Tonini and colleagues have made a comparative study using neonatal and adult mouse liver nuclei. Using freeze-fracture techniques, Tonini showed that pore density in two preparations is substantially the same, however, electrophysiology experiments show that the number of channels having the same conductance (300 pS) are three- to fourfold less in adult preparation. Biochemical analysis of the cytoskeletal nuclear components, actin and myosin, also show marked differences, with more structural proteins present in adult nuclei. The possibility exists that the acto-myosin complex located in the nuclear pore represents an engine for gating a molecular iris formed by the pore subunits. The nuclear pore as a diaphragm is supported by other experiments: ATP and Ca^{2+} ions are nuclear envelope modulators and they are implicated in the actin-myosin interactions. AFM demonstrates that addition of ATP to the nuclear membrane modifies the pore structure. Finally, in the comparison between neonatal and adult nuclei, one way to increase the number of channels in adult preparations is to add ATP to the external solution. Thus the nuclear pore may act like a diaphragm with cycles of contraction and relaxation depending on the ATP or Ca^{2+} concentration. In this view, the cytoskeleton would then be an active part of the regulatory mechanism for nuclear transport. Adult preparations have no more than a few active channels per patch. In nucleus-attached experiments, with patches containing two current

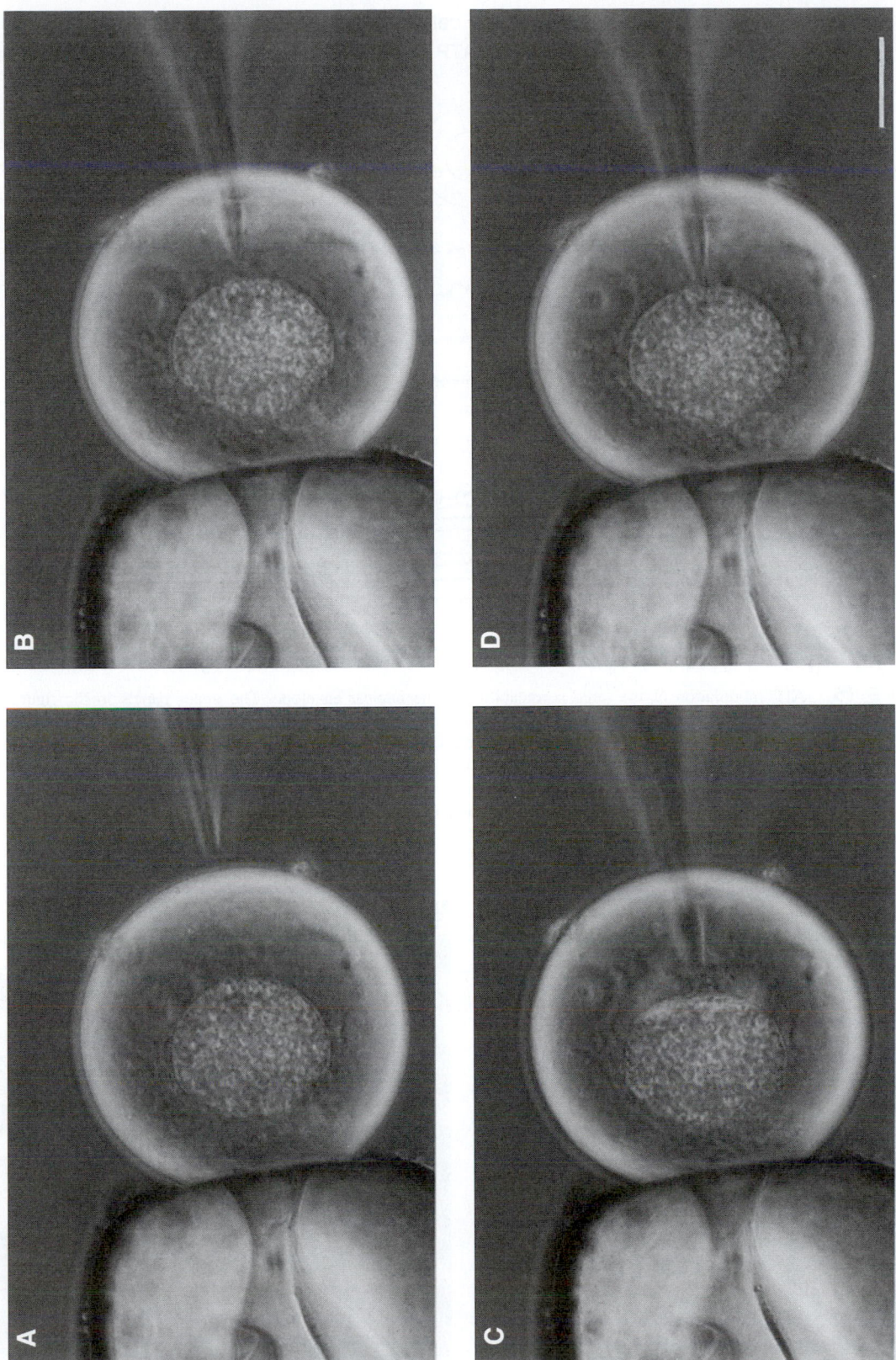

FIGURE 14. *In-situ* patch-clamp experiment on a *Xenopus* oocyte. Enzymatically isolated first-stage oocytes were transferred into experimental chamber. The oocytes were held in position by a holding pipette (A) via light suction. After perforating the plasma membrane (B) the patch-pipette (right) was pushed gently against the nuclear surface. (D) Before touching the nuclear envelope, positive pressure was applied to the patch-pipette solution. (C) The clear shadow in the oocyte cytoplasm is due to solution outflow, which was used to keep the electrode tip clean and to remove cytoplasmic material from the nuclear surface. (From Mazzanti *et al.*, *FASEB J.* **8**:231–236 (1994).)

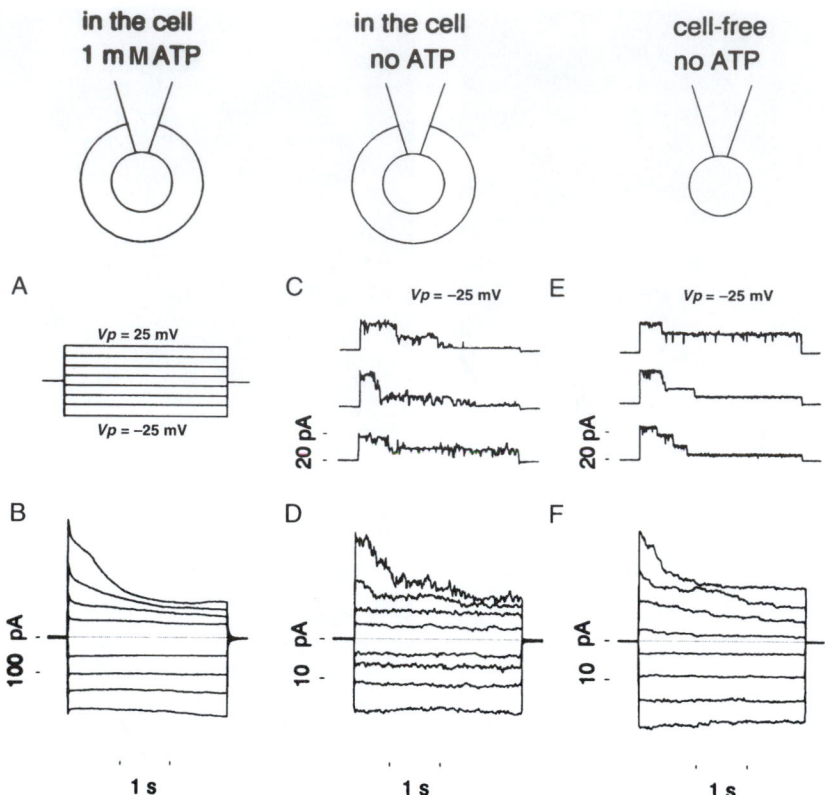

FIGURE 15. ATP modulation of the ionic permeability of the nuclear envelope. The upper panels in the figure show schematic drawings of the patch-clamp configurations corresponding to the experimental traces plotted below. (A) Voltage protocol used in the experiments. (B) Current traces at eight different test voltages from a nucleus-attached experiment in an intact oocyte. The pipette contained 1 mM ATP. (C) Nucleus-attached single-channel traces with −25 mV in the pipette. The patch electrode did not contain ATP. (D) Average current of 20 single-channel recordings obtained at each voltage step. In (E) and (F) experiments were performed after manual isolation of nuclei. The patch pipette contains no ATP. (E) depicts single-channel activity at −25 mV. (F) shows average currents at different test potentials. (From Mazzanti *et al., FASEB J.* **8**:231–236 (1994).)

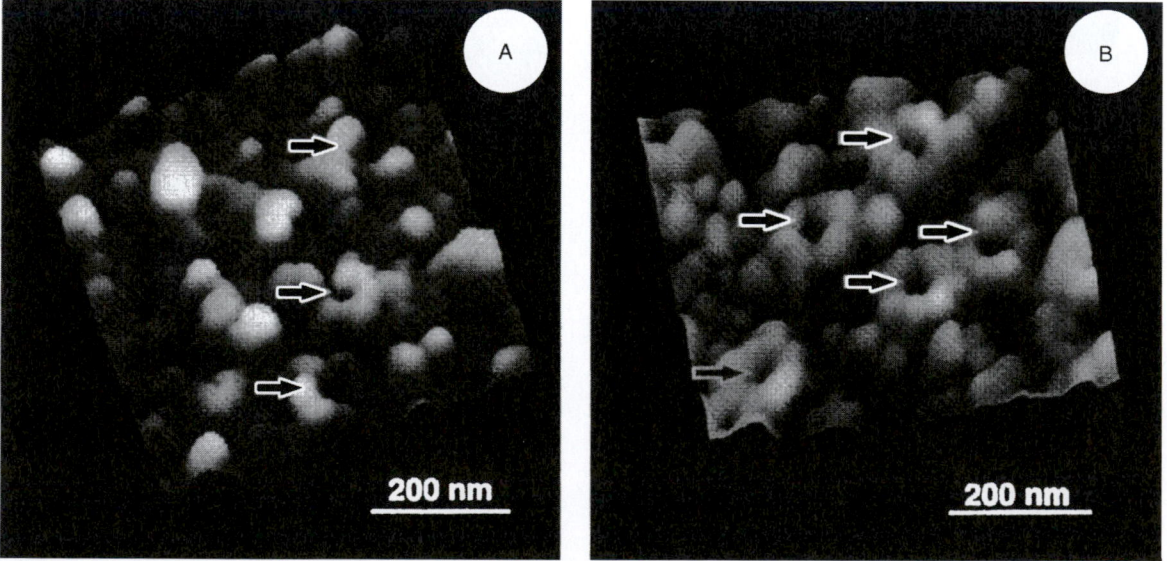

FIGURE 16. Atomic Force Microscopy view of nuclear envelope surface. Comparison between a nuclear membrane patch that was transferred from a patch pipette to the supporting substrate (a) and a nuclear membrane that was manually prepared (b). Nuclear pore complexes (indicated by the solid arrows) are clearly preserved by the manual preparation and resolved by AFM. In the transferred patch, the structures are less well preserved, probably due to the transfer process. The arrows indicate NPCs. Both the manually prepared nuclear envelopes and the transferred membrane patch shown in this figure are from *Xenopus laevis* oocyte nuclei. (From Danker *et al., Cell Biol. Int.,* **21**,11:747–757 (1998).)

levels corresponding to 300 pS conductance, the actin filament disrupter, cytochalasin, was slowly perfused around the nucleus. After 2–5 seconds, the number of single-channel current levels increased to 10 before loosing the seal (Fig. 17). Taking into account that the number of nuclear pore structures per patch is between 8 and 15 in this preparation, this experiment suggests that the acto-myosin complex gates the nuclear pore.

X. Summary

The intricacy of the nuclear pore complex, which contains perhaps as many as 100 proteins and possibly multiple nucleocytoplasmic pathways, indicates that the central pore is not simply a watery hole. The complex is rather a coupled transporter for ions and macromolecules. This conclusion seems plausible because the nucleus can partition ions, the nuclear envelope can maintain a resting potential, osmotic forces are manifest, and ion-selective channels exist in patches that contain dozens of nuclear pores.

How can the same structure that transports macromolecules be selective to small ions? One viewpoint would be that the observed channels lie not in the pore but in the inner and outer membranes. Nuclear pores would then be the sites for macromolecular transport, and channels in cisterns would be the sites of ion transport. Channels do exist in the cistern membranes, but we hypothesize that they also exist within the nuclear pore complex itself. The lateral openings shown in Fig. 5 may be the site of these channels. This hypothesis is attractive for several reasons, for example, macromolecular transport requires ATP, and *in situ* patch-clamp experiments indicate that ATP opens channels. If proteins and nucleic acids move as ions, ATP would regulate macromolecular transport

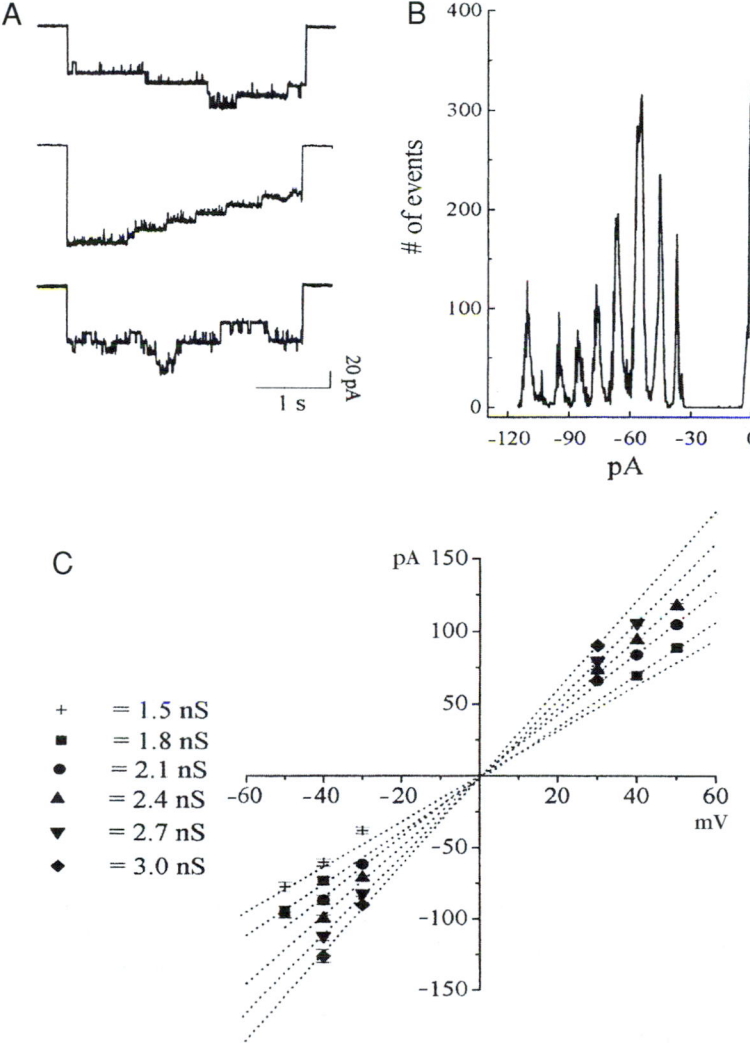

FIGURE 17. Cytochalasin effect on nuclear permeability. Cytocalasin modulates the nuclear pore. The panel on the top presents 3 different examples of single-channel current recordings obtained in steady-state conditions with −50 mV holding potential in the patch pipette. Before the interruption of the experimental traces (two parallel bars), the current has 1 or 2 levels. After a variable time of exposure to 10 μM cytocalasin, the number of levels increases to eight in this example before loosing the seal. The bottom panel shows amplitude histograms before (left) and after (right) cytocalasin treatment. (From Tonini *et al.*, *FASEB J.* **13**:1395–1403).

through channel openings. It is thus crucial to test whether the nuclear pore complex and the ion channels are linked to the same structure.

The majority of experiments indicate that the ion channels in the envelope select K^+ ions; however, Cl^- channels, ligand-gated ion channels, and Ca^{2+} pumps are also present. Channels in the outer and inner membranes do not preclude channels in the nuclear pore complex. Furthermore, the Donnan equilibrium and the semipermeable membrane may coexist at the nuclear envelope, as they do in the plasma membrane. Attributing the electrical potential of the nucleus to **either** a Donnan equilibrium **or** a selective membrane reflects early controversies concerning the plasma membrane. Curiously, a similar debate exists in another area of cell biology, the release of hormones and neurotransmitters from secretory vesicles. Secretory vesicles contain a gel matrix that restricts the movement of transmitters even after membrane fusion has occurred. In such cases, mechanisms that rely on the matrix, and mechanisms that rely on the membrane that surrounds the matrix, are not mutually exclusive.

It is interesting to speculate whether the structure of the nuclear pore complex resembles other macromolecular assemblies. One similarity regards the central granule (or plug), a protein within a protein motif that is common to other macromolecules. Criss Hartzell (Emory University) has noticed that the eight-plus-one structure of the nuclear pore complex is reminiscent of chaperones, which have a seven-plus-one structure, and we have already pointed to the axoneme with its nine-plus-one structure. Mu-Ming Poo (UC Berkeley) has asked whether parts of the assembled nuclear pore complex might not exist as independent units in the endoplasmic reticulum, to be reassembled into the complete nuclear pore complex. K^+-selective channels and protein-conducting channels reside in the reticulum and bear some similarity to the ionic and macromolecular pathways that we have discussed. The possibility that channel proteins and transporters in endoplasmic reticulum are components of the nuclear pore complex would have broad implications for pore assembly during development. These speculations may stimulate new experiments to help resolve the structural basis of nuclear pore function.

Acknowledgments

We wish to thank B.J. Duke for helping with experiments and preparing the figures. Unpublished work presented here was done at Stazione Zoologica in Naples, Italy with Luigia Santella and Brian Dale. NATO-CGR 190025 to BD and LJD supported this work.

Bibliography

Agutter, P. S. (1991). Role of the cytoskeleton in nucleocytoplasmic RNA and protein distributions. *Biochem. Soc. Trans.* **19**, 1094–1098.

Aidley, D. J. (1989). "The Physiology of the Excitable Cell," Third edition. Cambridge University Press, Cambridge, UK.

Akey, C. W. (1989) Interaction and structure of the nuclear pore complex revealed by cryoelectron microscopy. *J. Cell Biol.* **109**, 955–970.

Alberts, et al. (1989). "Molecular Biology of the Cell," 2nd ed., Ch. 8 Garland Publishing, New York.

Blobel, G., and Potter, V. R. (1966) Nuclei from rat liver: Isolation method that combines purity with high yield. *Science* **154**, 1662–1665.

Bonner, W. M. (1975). Protein migration into nuclei: I. Frog oocyte nuclei *in vivo* accumulate microinjected histones, allow entry to small proteins, and exclude large proteins. *J. Cell Biol.* **64**, 421–430

Bonner, W. M. (1975). Protein migration into nuclei: II. Frog oocyte nuclei accumulate a class of microinjected oocyte nuclear proteins and exclude a class of microinjected oocyte cytoplasmic proteins. *J. Cell Biol.* **64**, 431–437.

Bustamente, J. O. (1992). Nuclear ion channels in cardiac myocytes. *Pflugers Arch.* **421**, 473–485.

Bustamente, J. O. (1992). Nuclear electrophysiology. *J. Memb. Biol.* **138**, 105–112.

Carmo-Fonesca, M., and Hurt, E. C. (1991). Across the nuclear pores with the help of nucleoporins. *Chromosoma* **101**, 199–205.

Carter, K. C., Bowman, D., Carrington, W., Fargarty, K., McNeil, J. A., Fay, F. S., and Lawrence, J. B. (1993). A 3-D view of precursor messenger RNA metabolism within the mammalian nucleus. *Science* **259**, 1330–1335.

Century, T. J., Fenichel, I. R., and Horowitz, S. B. (1970). The concentration of water, Na and K ions in the nucleus and cytoplasm of amphibian oocytes. *J. Cell Sci.* **7**, 5–13.

Dale, B., DeFelice, L. J., Kyozuka, K., Santella, L., and Tosti, E. (1994). Voltage clamp of the nuclear envelope. *Proc. Royal Soc.* **B 225**, 119–124.

Danker, T., Mazzanti, M., Tonini, R., Rakowska, A., and Oberleithner, H. (1998). Using atomic force microscopy to investigate patch-clamped nuclear membrane. *Cell Biol. Int.* **21**, 747–757.

Dargemont, C., and Kuhn, L. C. (1992). Export of mRNA from micro-injected nuclei of *Xenopus laevis* oocytes. *J. Cell Biol.* **118**, 1–9.

DeFelice, L. J. (1981). "Introduction to Membrane Noise." Plenum Press, N.Y.

DeFelice, L. J. (1997). "Electrical Properties of Cells: Patch Clamp for Biologists." Plenum Press, N.Y.

Dingwall, C. (1990). Plugging the nuclear pore. *Nature* **346**, 512–514.

Dworetzky, S. I. and Feldherr, C. M. (1988). Translocation of RNA-coated gold particles through the nuclear pores of oocytes. *J. Cell. Biol.* **106**, 575–584.

Feldherr, C. M., and Akin, D. (1990). EM visualization of nucleocyto-plasmic transport process. *Elec. Micro. Rev.* **3**, 73–86.

Finkelstein, A. (1987). "Water Movement through Lipid Bilayers, Pores, and Plasma Membranes." John Wiley and Sons, New York.

Finlay, D. R., Newmeyer, D. D., Price, T. M., and Forbes, D. J. (1987). Inhibition of *in vitro* nuclear transport by a lectin that binds to nuclear pores. *J. Cell Biol.* **104**, 189–200.

Forbes, D. J. (1992). Structure and function of the nuclear pore complex. *Annu. Rev. Cell Biol.* **8**, 495–527.

Goldberg, M. W., Blow, J. J., and Allen, T. D. (1992). The use of field emission in-lens scanning electron microscopy to study the steps of assembly of the nuclear envelope. *J. Struct. Biol.* **108**, 257–268.

Hernandez-Cruz, A., Sala, F., and Conner, J. A. (1991). Stimulus-induced nuclear Ca signals in Fura-2 loaded amphibian neurons. *Ann. N.Y. Acad. Sci.* **635**, 416–420.

Hille, B. (1992). "Ionic Channels of Excitable Membranes." Second edition. Sinauer Associates, Sunderland, MA.

Hinshaw, J. E., Carracher, B. O., and Milligan, R. A. (1992). Architecture and design of the nuclear pore complex. *Cell* **69**, 1133–1141.

Horowitz, S. B., and Moore, L. C. (1974). The nuclear permeability intracellular distribution, and diffusion in Inulin in the amphibian oocyte. *J. Cell Biol.* **60**, 405–415.

Innocenti, B., and Mazzanti, M. (1993). Identification of a nucleocyto-plasmic ionic pathway by osmotic shock in isolated mouse liver nuclei. *J. Memb. Biol.* **131**, 137–142.

Jarnik, M., and Aebi, U. (1991). Toward a more complete 3-D structure of the nuclear pore complex. *J. Struct. Biol.* **110**, 883–894.

Lauger, P. (1991). "Electrogenic Ion Pumps." Second edition. Sinauer Associates, Sunderland, MA.

Loewenstein, W. R., and Kanno, Y. (1963). Some electrical properties of a nuclear membrane examined with a microelectrode. *J. Gen. Physiol.* **46**, 1123–1140.

Matzke, A. J. M., and Matzke, M. A., (1991). The electrical properties of the nuclear envelope, and their possible role in the regulation of eukaryotic gene expression. *Bioelectrochem. Bioenerg.* **25**, 357–370.

Maul, G. G. (1977). The nuclear and cytoplasmic pore complex: Structure, dynamics, distribution, and evolution. *Int. Rev. Cytol.* **6** (*Suppl.*), 75–186.

Mazzanti, M., DeFelice, L. J., Cohen, J., and Malter, H. (1990). Ion channels in the nuclear envelope. *Nature* **343**, 764–767.

Mazzanti, M., DeFelice, L. J., and Smith, E. F. (1991). Ion channels in murine nuclei during early development and in fully differentiated adult cells. *J. Memb. Biol.* **121**, 189–198.

Mazzanti, M., Innocenti, B., and Rigatelli, M. (1994). ATP dependent ionic permeability of nuclear envelope in *in-situ xenopus* oocyte nuclei. *FASEB J.* **8**, 231–236.

Moore, M. S., and Blobel, G. (1992). The two steps of nuclear import, targeting to the nuclear envelope and translocation through the nuclear pore, require different cytosolic factors. *Cell* **69**, 939–950.

Newport, J. W. and Forbes, D. J. (1987). The nucleus: Structure, function and dynamics. *Ann. Rev. Biochem.* **56**, 535–65.

Nicotera, P., McConkey, D. J., Jones, D. P., and Orrenius, S. (1989). ATP stimulates Ca uptake and increases the free Ca concentration in isolated rat liver nuclei. *PNAS* **86**, 453–457.

Nigg, E. A., Baeuerle, P. A., and Luhrmann, R. (1991). Nuclear import-export: In search of signals and mechanisms. *Cell* **66**, 15–22.

Overbeek, J. T. G. (1956). The Donnan Equilibrium. *Prog. Biophys. Mol. Biol.* **6**, 57–84.

Paine, P. L., Johnson, M. E., Lao, Y. T., Tluczek, L. J., and Miller, D. S. (1992). The oocyte nucleus isolated in oil retains *in vivo* structure and functions. *BioTechniques* **13**, 238–246.

Paine, P. L., Horowitz, S. B. (1980). The movement of material between nucleus and cytoplasm. *Cell Biol.* **4**, 299–338.

Perez-Terzic, C., Pyle, J., Jaconi, M., Stehno-Bittle, L., and Clapham, D. E. (1996). Conformational states of the nuclear pore complex induced by depletion of nuclear Ca stores. *Science* **273**, 1875–1877.

Richardson, W. D., Mills, A. D., Dilworth, S. M., Laskey, R. A., and Forbes D. J. (1992). Structure and function of the nuclear pore complex. *Annu. Rev. Cell Biol.* **8**, 495–572.

Santella, L. (1996). The cell nucleas: an Eldorado to future Ca research? *J. Membr. Biol.* **153**, 83–92.

Simon, S. M., and Blobel, G. (1991). A protein-conducting channel in the endoplasmic reticulum. *Cell* **65**, 371–380.

Stewart, M., Whytock, S., and Moir, R. D. (1991). Nuclear envelope dynamics and nucleocytoplasmic transport. *J. Cell Sci. Supp.* **14**, 79–82.

Stochaj, U., and Silver, P. (1992). Nucleoplasmic traffic of proteins. *Eur. J. Cell Biol.* **59**, 1–11.

Stricker, S. A., Centonze, V. E., Paddock, S. W., and Schatten, G. (1992). Confocal microscopy of fertilization-induced Ca dynamics in sea urchin eggs. *Dev. Biol.* **149**, 370–380.

Stehno-Bittel, L., Perez-Terzic, C., and Clapham, D. E. (1995). Diffusion across the nuclear envelope inhibited by depletion of the nuclear Ca store. *Science* **270**, 1835–1838.

Tabares, L. M., Mazzanti, M., and Clapham, D. E. (1991). Cl channels in the nuclear envelope. *J. Memb. Biol.* **123**, 49–54.

Tonini, R., Grohovaz, F., LaPorta, C. A. M., and Mazzanti, M. (1999). Gating mechanism of the nuclear pore complex channel in isolated neonatal and adult mouse. *FASEB J.* **13**, 1395–1403.

Unwin, P. N. T., and Milligan, R. A. (1982). A large particle associated with the perimeter of the nuclear pore complex. *J. Cell Biol.* **93**, 63–75.

Wagner, P., Kunz, J., Koller, A., and Hall, M. N. (1990). Active transport of proteins into the nucleus. *FEBS Lett.* **275**, 1–5.

Waybill, M. M., Yelamarty, R. V., Zhang, Y. L., Scaduto, R. C., Jr., LaNoue, K. F., Hsu, C. J., Smith, B. C., Tillotson, D. L., Yu, F. T., and Cheung, J. Y. (1991). Nuclear Ca transients in cultured rat hepatocytes. *Am. J. Physiol.* **261**, E49–E57.

Wente, S. R., Rout, M. P., and Blobel, G. (1992). A new family of yeast nuclear pore complex proteins. *J. Cell Biol.* **119**, 705–723.

Williams, D. A., Becker, P., and Fay, F. S. (1987). Regional changes in Ca underlying contraction of single smooth muscle cells. *Science* **235**, 1644–1648.

Yarmola, E. G., Zarudnaya, M. I., and Lazurkin, Y. S. (1985). Osmotic pressure of DNA solutions and effective diameter of the double helix. *J. Bio. Structure and Dynamics* **2**, 981–993.

Zasloff, M. (1983). tRNA transport from the nucleus in a eukaryotic cell: Carrier-mediated translocation process. *Proc. Natl. Acad. Sci. USA* **80**, 6436–6440

Nicholas Sperelakis and Gordon M. Wahler

33

Regulation of Ion Channels by Phosphorylation

I. Introduction

Considerable attention has been given during the past 25 years to phosphorylation of ion channels as a means whereby the activity of the channels can be regulated or modulated. There is evidence for such regulation or modulation of function of Ca^{2+}, K^+, Na^+, and Cl^- channels by phosphorylation, and biochemical evidence shows that one or a few sites on the channel proteins can be phosphorylated by various protein kinases. Most physiological evidence for such changes in ion channel function is based on the L-type Ca^{2+} channels of nerve, skeletal muscle, cardiac muscle, and vascular smooth muscle (VSM) and on the K^+ channels (delayed-rectifier type) of cardiac muscle and nerve. This chapter focuses primarily on the slow L-type Ca^{2+} channel of cardiac muscle in order to illustrate the important principles that are involved.

The voltage-dependent L-type Ca^{2+} channels in the myocardial cell membrane are the major pathway by which Ca^{2+} ions enter the cell during excitation for initiation and regulation of the force of contraction of cardiac muscle. The L-type Ca^{2+} channels have some special properties, including functional dependence on metabolic energy, selective blockade by acidosis, and regulation by the intracellular cyclic nucleotide levels. Because of the special properties of these slow channels, Ca^{2+} influx into the myocardial cell can be controlled by both extrinsic factors (such as autonomic nerve stimulation or circulating hormones) and intrinsic factors (such as intracellular pH or adenosine triphosphate (ATP) levels).

In myocardial cells, the Ca^{2+} influx that occurs during each cardiac cycle is regulated by cyclic nucleotides. This regulation is presumably mediated by phosphorylation(s) of the L-type Ca^{2+} channel protein. Phosphorylation of the Ca^{2+} channels (or of an associated regulatory protein) by cAMP-dependent protein kinase (PK-A) (Fig. 1) (1) increases the number of L-type Ca^{2+} channels available for voltage activation during the action potential (AP), (2) increases the probability of channel opening, and (3) increases channel mean open time. The greater density of open Ca^{2+} channels increases the inward Ca^{2+} current (I_{Ca}) during the

AP. The resulting increase in Ca^{2+} influx increases the force of contraction of the heart. In contrast to the response of the channel to phosphorylation by PK-A, phosphorylation of the channels or an associated regulatory protein by cGMP-dependent PK (PK-G) depresses the activity of the L-type Ca^{2+} channels (Wahler and Sperelakis, 1985; Wahler *et al.,* 1990).

II. Types of Ca^+ Channels

Four or five different subtypes of voltage-dependent Ca^{2+} channels have been described for nerve and muscle cells. The first three, found in sensory ganglion neurons (rat dorsal root ganglion) of the spinal cord, were called **L-type** (i.e., long-lasting or kinetically slow), **T-type** (i.e., transient or kinetically fast), and **N-type** (i.e., neither L-type nor T-type) Ca^{2+} channels (Nowycky *et al.,* 1985). More recently, another subtype was found initially in Purkinje neurons of the cerebellum, and hence called **P-type** (Llinas *et al.,* 1992). The **P-type** channel has since been identified at the nerve terminals of neuromuscular junctions of both vertebrates and invertebrates. Muscle fibers, in general, apparently possess only the L-type and T-type channels, with the T-type being very sparse or absent in some types of muscles. That is, in muscle cells, the major inward Ca^{2+} current (involved in excitation-contraction coupling) is the current through the L-type Ca^{2+} channels. This is true of skeletal muscle, cardiac muscle, and smooth muscles.

The N-type Ca^{2+} channel has a single-channel conductance (γ) of about 14–18 pS, which is between that of the T-type (8–12 pS) and L-type (18–26 pS). The N-type channel has been localized only to neurons so far. The P-type Ca^{2+} channel is high threshold (like the L-type), having an activation voltage of −45 to −35 mV, but is not blocked by the L-type channel blockers. The values reported for single-channel conductance (γ) in various cells range from 9–20 pS.

Table 1 summarizes the major differences between the L-type and T-type Ca^{2+} channels. As indicated, the kinetics of

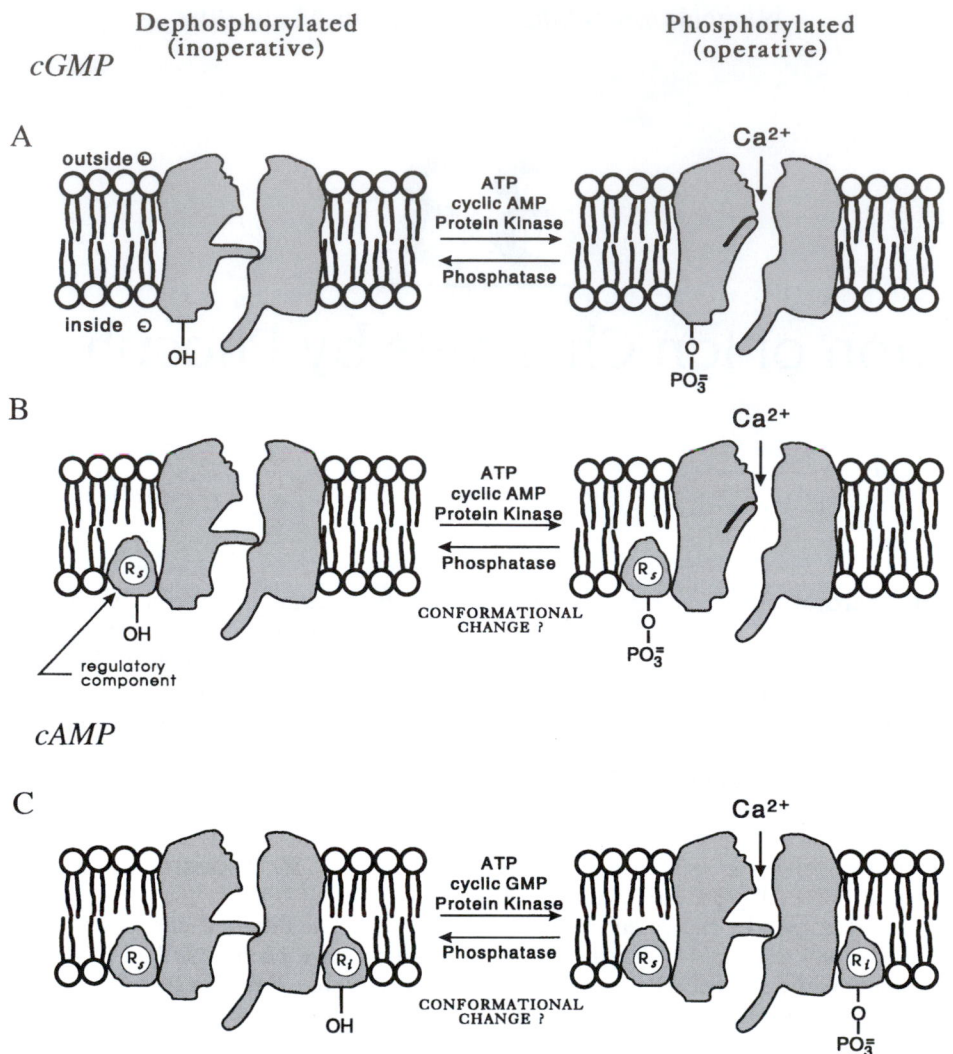

FIGURE 1.   Schematic model for an L-type Ca^{2+} channel in myocardial cell membrane in two hypothetical forms: dephosphorylated (or electrically silent) form (left diagrams) and phosphorylated form (right diagrams). The two gates associated with the channel are an activation gate and an inactivation gate. The phosphorylation hypothesis states that a protein constituent of the slow channel itself (A) or a regulatory protein associated with the slow channel (B) must be phosphorylated for the channel to be in a state available for voltage activation. Phosphorylation of a serine or threonine residue occurs by PK-A in the presence of ATP. Phosphorylation may produce a conformational change that effectively allows the channel gates to operate. The L-type channel (or an associated regulatory protein) may also be phosphorylated by PK-G (C), thus mediating the inhibitory effects of cGMP on the Ca^{2+} channel. (Adapted from Sperelakis and Schneider, 1976.)

activation and inactivation are slower for the L-type. That is, the L-type current $I_{Ca(L)}$ turns on (activates) more slowly and turns off (inactivates) more slowly. In addition, the voltage range over which these channels operate is different, the threshold potential and inactivation potential being higher (more positive or less negative) for the L-type Ca^{2+} channels. Therefore, the L-type channels are high threshold and the T-type channels are low threshold. The single-channel conductance is greater for the L-type Ca^{2+} channel: 18–26 pS versus 8–12 pS. The L-type Ca^{2+} channels are regulated by cyclic nucleotides and phosphorylation, whereas the fast T-type Ca^{2+} channels are not. Finally, the L-type Ca^{2+} channels are blocked by Ca^{2+} channel antagonist drugs (such as verapamil, diltiazem, and nifedipine) and opened by Ca^{2+}

agonist drugs (such as Bay K-8644, a dihydropyridine that is chemically very close to nifedipine), whereas the T-type Ca^{2+} channels are not. In some respects, the T-type Ca^{2+} channels behave like fast Na^+ channels, except that the T-type channels are Ca^{2+} selective (rather than Na^+ selective) and are not blocked by tetrodotoxin (TTX).

Table 2 summarizes the blocking or opening action of several drugs and toxins on the various subtypes of Ca^{2+} channels. As indicated, the L-type Ca^{2+} channel is blocked by the prototype Ca^{2+} antagonist drugs (verapamil, diltiazem, and nifedipine) and opened by the dihydropyridine Bay K-8644. The T-type channels are relatively selectively blocked by tetramethrine and by low concentrations of Ni^{2+} (e.g., 30 μM). Higher concentrations of Ni^{2+} block the L-

TABLE 1 Summary of Major Differences between the L-Type and T-Type Ca^{2+} Channels

Properties	L-type (Slow)	T-type (Fast)
Duration of current	Long-lasting (sustained)	Transient
Inactivation kinetics	Slower	Faster
Activation kinetics	Slower	Faster
Threshold	High (ca. -35 mV)	Low (ca. -60 mV)
Half-inactivation potential	ca. -20 mV	ca. -50 mV
Single-channel conductance	High (18–26 pS)	Low (8–12 pS)
Regulated by cAMP and cGMP	Yes	No
Regulated by phosphorylation	Yes	No
Blocked by Ca^{2+} antagonist drugs	Yes	No (slight)
Opened by Ca^{2+} agonist drugs	Yes	No
Permeation by Me^{2+}	Ba > Ca	Ba $\cong$ Ca
Inactivation by [Ca]$_i$	Yes	Slight (?)
Recordings in isolated patches	Runs down	Relatively stable

type channels as well. Possible blockers of the F-type channel (see below) are not known.

In addition to the Ca^{2+} channel subtypes described previously, a new subtype was discovered in 18-day-old fetal rat ventricular (heart) cells (Tohse *et al.*, 1992). A substantial fraction (e.g., 30%) of the total I_{Ca} remained in the presence of a high concentration (3 μM) of nifedipine (nifedipine-resistant I_{Ca}), and it was not blocked by diltiazem (another L-type channel blocker) or ω-conotoxin (N-type channel blocker). This novel Ca^{2+} current had a half-inactivation potential about 20 mV more negative than that of the L-type Ca^{2+} current. It was called **F-type** (fetal-type) Ca^{2+} current ($I_{Ca(F)}$). The single-channel conductance (γ) is not known.

Another difference discovered during the embryonic or fetal period, first observed in chick and later in rat, is that the L-type Ca^{2+} channels exhibit an unusually high incidence of very long openings, as observed in single-channel recordings (cell-attached patch) (Tohse and Sperelakis, 1990; Masuda *et al.*, 1995). The incidence of long openings diminishes during development and approaches the adult channel behavior. The adult behavior primarily consists of bursting patterns (rapid openings and closings of short durations, or flickering).

III. Cyclic AMP Stimulation of L-type Ca^{2+} Channels

A. Cyclic AMP

It has long been known that cyclic AMP (cAMP) modulates the functioning of the L-type Ca^{2+} channels (Shigenobu and Sperelakis, 1972; Tsien *et al.*, 1972; Reuter and Scholz,

TABLE 2 Summary of Drugs or Toxins that Block or Open the Various Subtypes of Ca^{2+} Channels

Channel type	Blockers	Openers
L-type	Verapamil, diltiazem, nifedipine	Bay K-8644
T-type	Tetramethrine, Ni^{2+} (30 μM)	—
N-type	ω-Conotoxin	—
P-type	ω-Agatoxin-IVA[a] Polyamine (FTX)[b]	—
F-type	—	—

[a]ω-Agatoxin-IVA, a polypeptide (5202 Da) from funnel-web spider venom (*Agenopsis aperta*), blocks P-type Ca^{2+} channels of rat Purkinje neurons with a K_D of about 2 nM.

[b]A smaller (ca. 234 Da) polyamine (FTX) extracted from venom of this spider also blocks the P-type channels (Mintz *et al.*, 1992). FTX block is antagonized by Ba^{2+} ion. P-type Ca^{2+} channels are involved in presynaptic Ca^{2+} influx and associated neurotransmitter release.

1977). β-adrenergic agonists, after binding to their specific receptors, lead to rapid stimulation of adenylate cyclase with resultant elevation of cAMP levels. Other neurotransmitters, hormones, and so on that stimulate cAMP production (e.g., histamine) similarly enhance I_{Ca}. Drugs that inhibit phosphodiesterase (the enzyme that hydrolyzes cyclic nucleotides) also cause an elevation of cAMP and increase of I_{Ca} (Fischmeister and Hartzell, 1991; Mubagwa et al., 1993; Kawamura and Wahler, 1994).

Additional evidence for the regulatory role of cAMP in heart cells includes the following: (1) Forskolin and the GTP analog GPP(NH)P, which directly activate adenylate cyclase, both induce Ca^{2+}-dependent slow APs (Josephson and Sperelakis, 1978; Wahler and Sperelakis, 1986) and enhance I_{Ca} (Hescheler et al., 1986). (2) cAMP injection into ventricular muscle cells induces Ca^{2+}-dependent slow APs in the injected cells within seconds (Vogel and Sperelakis, 1981; Li and Sperelakis, 1983; Bkaily and Sperelakis, 1985) or enhances I_{Ca} in isolated single cardiac cells (Irisawa and Kokubun, 1983). (3) A photochemical activation method for suddenly increasing the intracellular cAMP level enhances I_{Ca} in bullfrog atrial cells (Nargeot et al., 1983). (4) Single-channel analysis suggests that cAMP increases the number of functional L-type Ca^{2+} channels available and/or the probability of opening of a given channel (Cachelin et al., 1983; Trautwein and Hofmann, 1983; Bean et al., 1984). The β-adrenergic agonist isoproterenol increases the mean open time of single Ca^{2+} channels and decreases the intervals between bursts; the conductance of the single channel is not increased (Reuter et al., 1982). Therefore, the increase in the Ca^{2+} current produced by isoproterenol could be produced by the observed increase in mean open time of each channel and the probability of opening, as well as by an increase in the number of available channels.

There are gender differences in the cardiovascular responses to β-adrenergic stimulation. Thus, the β-adrenergic-mediated increases in the Ca^{2+} current (Vizgirda et al., 1997) and in cell shortening (Vizgirda et al., unpublished observations) of rat ventricular myocytes have been found to be approximately two-fold greater in males than females.

B. Phosphorylation Hypothesis

Because of the relationship between cAMP and the number of available L-type Ca^{2+} channels and because of the dependence of the functioning of these channels on metabolic energy, it was postulated that the L-type Ca^{2+} channel must be phosphorylated for it to become available for voltage activation (Shigenobu and Sperelakis, 1972; Tsien et al., 1972). The elevation of cAMP by a positive inotropic agent activates PK-A, which phosphorylates a variety of proteins in the presence of ATP. One protein that is phosphorylated is the L-type Ca^{2+} channel protein itself or a contiguous regulatory type of protein (see Fig. 1). Agents that elevate cAMP increase the fraction of the channels that are in the phosphorylated form and hence readily available for voltage activation. Phosphorylation could make the Ca^{2+} channel available for activation by a conformational change that allowed the activation gate to be opened upon depolarization (or increased the pore diameter). cAMP-mediated phosphor-

ylation of the Ca^{2+} channel may also act to increase I_{Ca} by reducing the normal blockade of the channel by Mg^{2+} (i.e., by altering the affinity of binding sites in the channel pore for Mg^{2+}) (Yamaoka and Seyama, 1998).

Cyclic AMP–dependent phosphorylation also modulates the activity of other channel types. For example, in the heart alone, other channels regulated by cAMP-dependent phosphorylation include the CFTR (cystic fibrosis transmembrane-conductance regulator) chloride channel (rev. in Gadsby and Nairn 1999) and the delayed-rectifier channel ($I_{K(del)}$) (e.g., Walsh and Kass, 1988).

Whatever the precise subcellular mechanism for the effect phosphorylation has on channel activity, in the phosphorylation model, the phosphorylated form of the L-type Ca^{2+} channel is the active (operational) form, and the dephosphorylated form is the inactive (inoperative) form. The dephosphorylated channels are virtually silent electrically (i.e., their opening probability approaches zero). Phosphorylation increases the probability of channel opening with depolarization. An equilibrium would exist between the phosphorylated and dephosphorylated forms of the channel under a given set of conditions. For example, fluoride ions (< 1 mM) increase the force of contraction of the heart and potentiate the slow Ca^{2+}-dependent APs and Ca^{2+} influx (I_{Ca}) without increasing the level of cAMP. Fluoride may act by inhibiting the phosphatase, which dephosphorylates the channel protein, thus prolonging the life span of the phosphorylated channel. This suggests that the rate of dephosphorylation, in addition to the rate of phosphorylation (see below), may also be an important determinant of the amplitude of I_{Ca}.

Based on the rapid decay of the response to microinjected cAMP (Fig. 2, top), the mean life span of a phosphorylated channel is probably only a few seconds at most, and it is possible that the channels are phosphorylated and dephosphorylated with every cardiac cycle. Agents that affect or regulate the phosphatase that dephosphorylates the channel would affect the life span of the phosphorylated channel. Thus, channel stimulation can be produced either by increasing the rate of phosphorylation (by PK-A activation) or by decreasing the rate of dephosphorylation (inhibition of the phosphatase).

C. Protein Kinase A

Intracellular injection of the catalytic subunit of PK-A induces and increases the slow Ca^{2+}-dependent APs and potentiates I_{Ca} (Osterreider et al., 1982; Bkaily and Sperelakis, 1984). Injection of an inhibitor (protein) of the PK-A into heart cells inhibits the spontaneous slow Ca^{2+}-dependent APs and I_{Ca} (Bkaily and Sperelakis, 1984; Kameyama et al., 1986). These results verify that the regulatory effect of cAMP is exerted by means of PK-A and phosphorylation.

Consistent with the phosphorylation hypothesis, L-type Ca^{2+} channel activity disappears within 90 s in isolated membrane inside-out patches (Reuter, 1983), but can be restored (in neurons) by applying the catalytic subunit of PK-A together with MgATP (Armstrong and Eckert, 1987). This is consistent with the washing away of regulatory components of the Ca^{2+} channels or of the enzymes necessary to phosphorylate the channel. Even in whole-cell voltage-

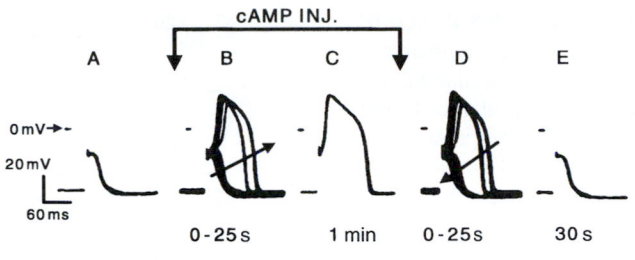

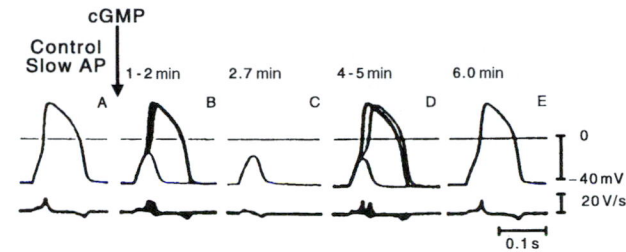

FIGURE 2.   Effects of intracellular injections of cyclic nucleotides. Upper row: Induction of Ca^{2+}-dependent slow APs in guinea pig papillary muscle by intracellular pressure injection of cAMP. The muscle was depolarized in 22 mM K_o to voltage-inactivate fast Na^+ channels. (A) Small graded response (stimulation rate 30/min). (B) Superimposed records showing the gradual appearance of slow APs upon cAMP injection over a 25-s period. (C) Presence of stable slow APs after injection for 1 min. (D) Gradual spontaneous depression of slow APs over a period of 25 s after the injection is stopped. (E) Complete decay of slow APs 30 s after cessation of cAMP injection. All records are from one impaled cell. (From Li and Sperelakis, 1983.) Lower row: Transient abolition of Ca^{2+}-dependent slow APs by pressure injection of cGMP. (A) Control slow AP. (B and C) 1–2 min after the onset of cGMP injection (10-s duration), the slow APs were depressed and then abolished. (D and E) At 4–6 min, the slow APs recovered spontaneously to control levels. All records from the same cell. (From Wahler and Sperelakis, 1985.)

clamp, there is a progressive run-down of the Ca^{2+} current, which is slowed or partially reversed by conditions that enhance PK-A phosphorylation.

D. Phosphatases

In ventricular cells, phosphatases have been shown to inhibit I_{Ca} (e.g., Kameyama *et al.,* 1986; Hescheler *et al.,* 1987; duBell *et al.,* 1996), consistent with the phosphorylation hypothesis. Additionally, phosphatase inhibitors, such as okadaic acid and microcystin, cause large increases in I_{Ca} (Hescheler *et al.,* 1988; Frace and Hartzell, 1993). There is some disagreement about the relative effectiveness of specific phosphatases (particularly type 1 versus type 2A) in reducing I_{Ca}, and whether basal I_{Ca} is significantly inhibited by phosphatases, or if only the cAMP-stimulated I_{Ca} (e.g., by a β-adrenergic agonist) is affected. There is overall agreement that dephosphorylation of the Ca^{2+} channel by phosphatase is an additional important regulatory step in determining the amplitude of I_{Ca}.

In addition to cardiac muscles, phosphatases have been shown to decrease the Ca^{2+} current in other cell types, such as neurons (Chad and Eckert, 1986) and vascular smooth muscle cells (e.g., Schumann *et al.,* 1997). Other types of channels have also been shown to be modulated by phosphatases. For example, the CFTR Cl^- channel is inactivated by phosphatase 2B (calcineurin) (Fischer *et al.,* 1998).

Type 1 protein phosphatase activity is inhibited by (at least) two low-molecular weight proteins, known as protein phosphatase inhibitor-1 (PPI-1) and protein phosphatase inhibitor-2 (PPI-2). PPI-1 is present in ventricular myocytes, and it may be an important additional component of the modulation of the phosphorylation/dephosphorylation cycle of the Ca^{2+} channel. The activity of PPI-1 is enhanced with phosphorylation by PK-A (Ahmad *et al.,* 1989; Gupta *et al.,* 1993, 1996). Thus, PK-A not only phosphorylates the Ca^{2+} channel, it also decreases the rate of channel dephosphorylation. Both actions stimulate channel activity.

IV. Cyclic GMP Inhibition of the Ca^{2+} Current

A. Cyclic GMP

The physiological role played by cyclic GMP (cGMP) on cardiac function is still controversial. A number of years ago, it was proposed that cGMP plays a role antagonistic to that of cAMP, namely, that there was a Yin-Yang relationship between cAMP and cGMP (Goldberg *et al.,* 1975). Since that time, considerable evidence has accumulated supporting this hypothesis. For example, 8-Br-cGMP (10^{-4} M) shortens the AP duration in rat atria (accompanied by a negative inotropic effect), and it was suggested that cGMP might decrease the Ca^{2+} conductance (Nawrath, 1977). Both acetylcholine (ACh), which increases cGMP levels, and 8-Br-cGMP, a more permeable analog of cGMP, reduce the upstroke velocity and duration of the Ca^{2+}-dependent slow AP in guinea pig atria (Kohlhardt and Haap, 1978). The abbreviation of AP duration also occurs following injection of cGMP into isolated guinea pig cardiomyocytes (Trautwein *et al.,* 1982).

Superfusion of isolated ventricular muscle with 8-Br-cGMP abolishes the Ca^{2+}-dependent slow APs and accompanying contractions (Wahler and Sperelakis, 1985). A similar inhibition by cGMP was shown for the slow APs of atrial muscle and Purkinje fibers (Mehegan *et al.*, 1985). Intracellular pressure injection of cGMP into ventricular cells was found to transiently depress or abolish slow APs more quickly (e.g., 1–2 min) (Wahler and Sperelakis, 1985) (Fig. 2, bottom). It was also demonstrated that 8-Br-cGMP inhibits the basal I_{Ca} (i.e., unstimulated by cAMP) in voltage-clamped ventricular myocytes (Wahler *et al.*, 1990; Haddad *et al.*, 1995) (Figs. 3 and 4).

Cyclic GMP inhibition of Ca^{2+} channel activity of embryonic chick heart was also demonstrated at the single-channel level (Tohse and Sperelakis, 1991) (Fig. 5). Cyclic GMP did not change the unit amplitude and slope conductance of the Ca^{2+} channel, but prolonged the closed times and shortened the open times.

The Ca^{2+} slow channels of young (3-day-old) embryonic chick heart cells often exhibit long-lasting openings (e.g., 300 ms) under normal conditions, especially at more positive command potentials (Tohse and Sperelakis, 1990, 1991). Long-lasting openings were much less frequently observed in 17-day-old embryonic cells. That is, the Ca^{2+} channels in early development naturally possess some mode 2 behavior, which is normally produced by Ca^{2+} channel agonists such as the dihydropyridine BayK-8644. Addition of 8-Br-cGMP to the bath of cells (3-day) exhibiting long openings completely inhibited Ca^{2+} channel activity (see Fig. 5). Long openings were also observed in fetal (12-day) rat ventricular cardiomyocytes (Masuda *et al.*, 1995).

In whole-cell voltage-clamp experiments on single ventricular cardiomyocytes from 17-day embryonic chicks, the

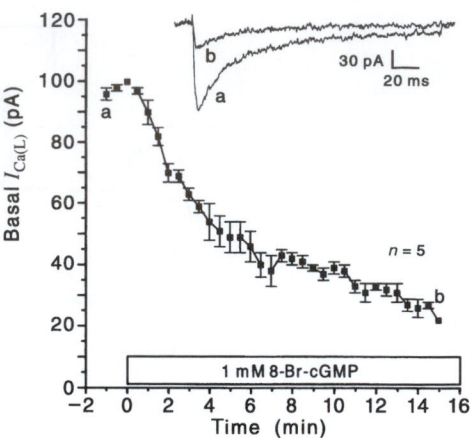

FIGURE 4. Time course of the inhibition of the basal $I_{Ca(L)}$ by 8-Br-cGMP (1 mM) in 17-day-old embryonic chick heart cells. Data points plotted are the mean ± standard error. The upper two traces show the original current recordings of $I_{Ca(L)}$ taken at the time points shown by the corresponding letters in the graph. $I_{Ca(L)}$ was elicited by 200-ms depolarizing pulses to +10 mV from a holding potential of −45 mV. Experiments conducted at room temperature. (From Haddad *et al.*, 1995.)

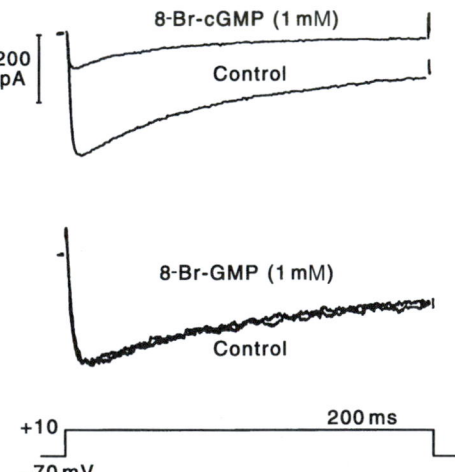

FIGURE 3. Effect of 8-Br-cGMP on the basal Ca^{2+} current in two cultured embryonic chick ventricular myocytes. Upper traces: Currents elicited by depolarizing pulses from −70 mV to +10 mV in the control bath solution and after a 10-min superfusion with a solution containing 1 mM 8-Br-cGMP. Note the large inhibition of $I_{Ca(L)}$. Lower traces: Currents elicited by depolarizing pulses in the control bath solution and after a 10-min superfusion with a solution containing 1 mM 8-Br-GMP, the noncyclic (and inactive) analog of 8-Br-cGMP. (Reproduced from Wahler *et al.*, 1990.)

stimulation of $I_{Ca(L)}$ produced by 8-Br-cAMP added to the bath could be completely reversed by the addition of 8-Br-cGMP. Similar results were obtained in experiments on early neonatal (2-day) rat ventricular myocytes; namely, 8-Br-cGMP antagonized the stimulation of $I_{Ca(L)}$ produced by 8-Br-cAMP (Fig. 6). (Masuda and Sperelakis, unpublished.) Therefore, the ratio of cAMP/cGMP apparently determines the degree of stimulation of $I_{Ca(L)}$, and even the basal I_{Ca} is inhibited by cGMP in some circumstances (see Section IV.B).

B. Mechanisms for the cGMP Inhibition of I_{Ca}

As noted above, cGMP regulates the functioning of the myocardial Ca^{2+} slow channels in a manner that is antagonistic to that of cAMP (Fig. 7). It is possible that the Ca^{2+} channel protein has a second site that can be phosphorylated by the cGMP-dependent protein kinase (PK-G) and that, when phosphorylated, inhibits the Ca^{2+} channel. Another possibility is that there is a second type of regulatory protein that is inhibitory when phosphorylated (see Fig. 1).

Another mechanism has been proposed for cGMP inhibition of I_{Ca}. This mechanism involves cGMP depression of the cAMP level, thereby leading to an **antiadrenergic effect**. Intracellular application of cGMP inhibited the I_{Ca} of frog ventricular myocytes only after the cAMP levels had been increased; there was no effect of cGMP on the basal I_{Ca} (Hartzell and Fischmeister, 1986; Fischmeister and Hartzell, 1987). It was concluded that cGMP inhibited I_{Ca} by activating a cGMP-stimulated isoform of phosphodiesterase (PDE II), resulting in increased degradation of cAMP. The cGMP-stimulated PDE mechanism for inhibition of I_{Ca} also occurs in human atrial cells (Rivet-Bastide *et al.*, 1997).

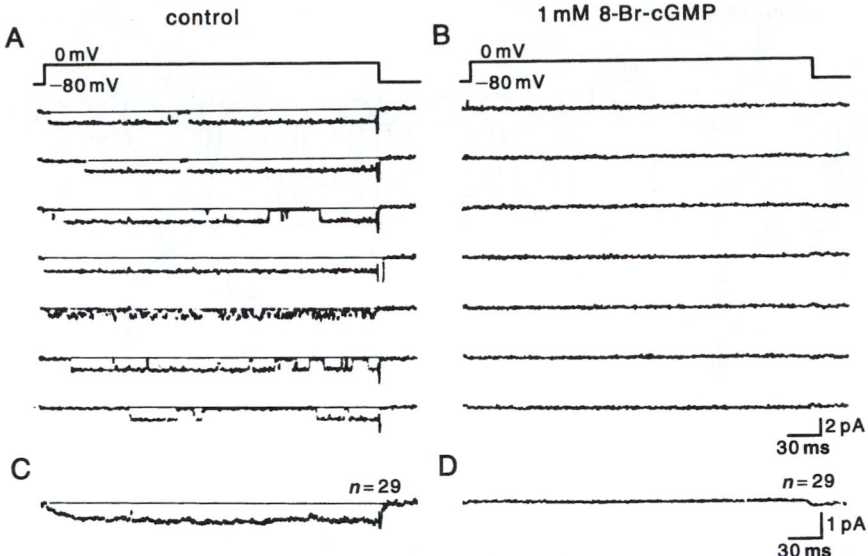

FIGURE 5. Current recordings from a cell-attached patch showing the effect of 8-Br-cGMP on the Ca^{2+} channel activity in a single myocardial cell isolated from a 3-day-old embryonic chick heart. Single-channel currents were evoked by depolarizing voltage pulses to 0 mV from a holding potential of −80 mV, at a duration of 300 ms and repetition rate of 0.5 Hz. (A and B) Examples of original current recordings from the same patch, before (A) and after (B) superfusion with 1.0 mM 8-Br-cGMP. (C and D) Ensemble-averaged currents calculated from the current recordings ($n = 29$). (Data from Tohse and Sperelakis, 1991.)

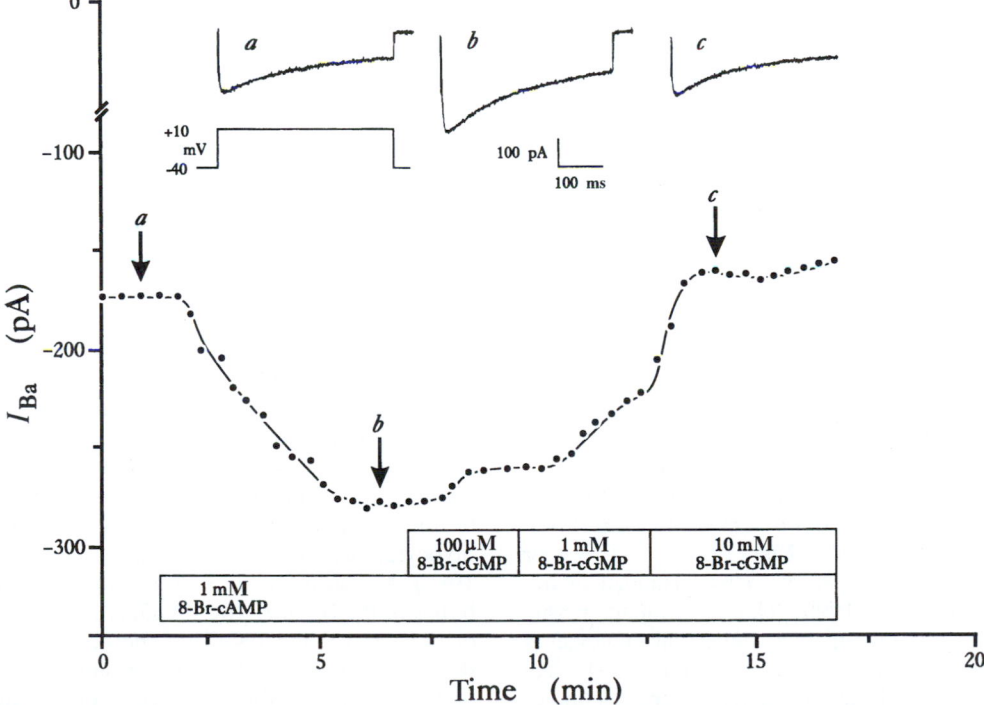

FIGURE 6. Antagonism of the stimulating effect of 8-Br-cAMP on $I_{Ca(L)}$ by 8-Br-cGMP in a single young neonatal rat ventricular myocyte. Upper tracings show three original current recordings of I_{Ca} corresponding to the three time points labeled in the lower graph. $I_{Ca(L)}$ was elicited by 300-ms depolarizing pulses to +10 mV from a holding potential of −40 mV. Ba^{2+} (20 mM) was used as the charge carrier. Experiments conducted at room temperature of 25 °C. (From Masuda and Sperelakis, unpublished.)

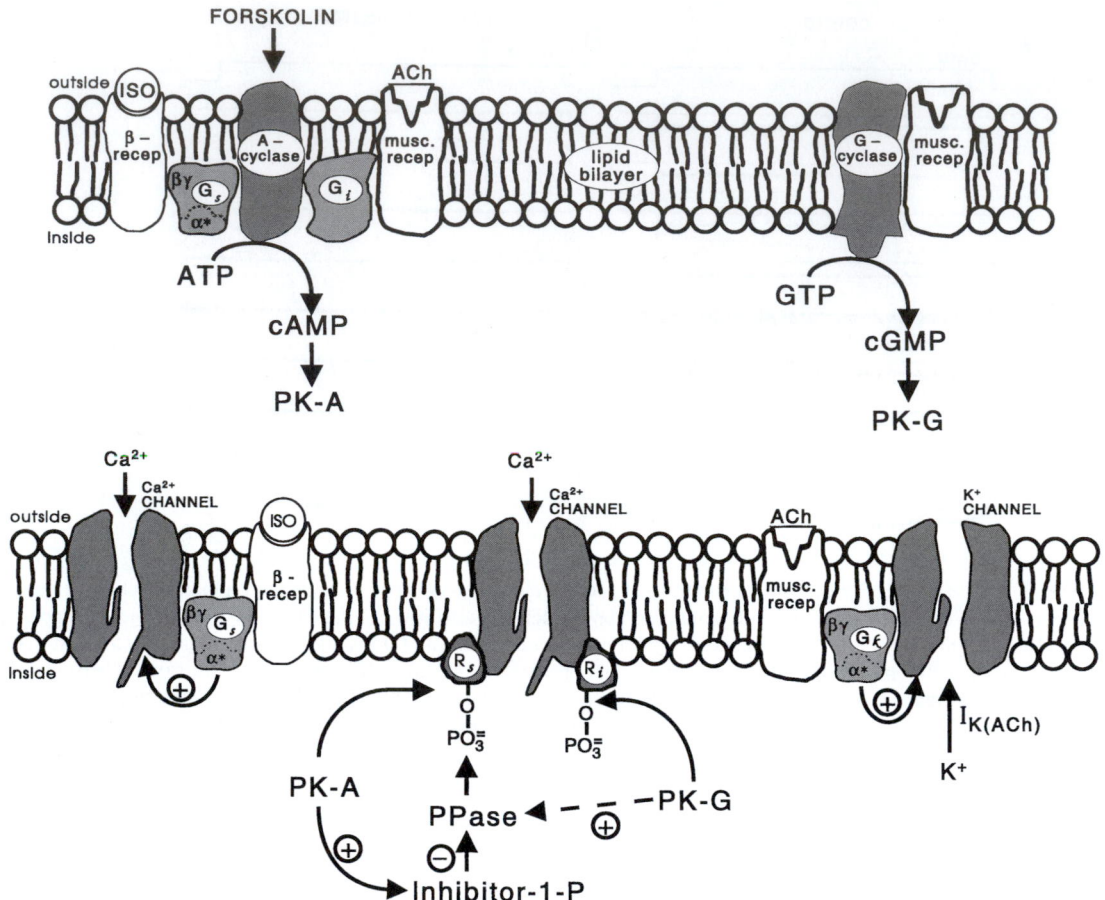

FIGURE 7.   Diagrammatic summary of the regulation of the L-type Ca^{2+} channels in the myocardial cell membrane and the mechanisms of action of some inotropic agents. The β-adrenergic agonists act via their receptor on a GTP-binding protein (G_s) to stimulate adenylate cyclase and cAMP production. The voltage-dependent myocardial Ca^{2+} channels are stimulated by cAMP, presumably because the channel (or an associated regulatory protein) must be phosphorylated for it to be in a form that is available for voltage activation. cGMP-dependent phosphorylation also regulates the Ca^{2+} channel in a manner antagonistic to cAMP, namely, producing inhibition. Thus, the muscarinic receptor activated by ACh can produce inhibition of Ca^{2+} influx by at least the four mechanisms depicted: (1) reversal of adenylate cyclase stimulation produced by β-agonists (or H_2 agonists); (2) stimulation of guanylate cyclase and production of cGMP; (3) activation of a K^+ channel ($I_{K(ACh)}$), which produces an outward K^+ current that terminates the AP earlier, and thereby voltage-deactivates the Ca^{2+} channels earlier, thus, indirectly reducing I_{Ca}; and (4) stimulation of a phosphatase (PPase), which may involve cGMP and PK-G. Mechanism (3) may be absent in ventricular myocardial cells.

However, this mechanism (cGMP stimulation of PDE) for cGMP inhibition of I_{Ca} does not appear to be present in mammalian ventricular myocytes. Thus, for example, a blocker of PDE II (EHNA) has no effect on basal or β-stimulated I_{Ca} in mammalian ventricular myocytes (Rivet-Bastide *et al.*, 1997). In mammalian and avian cardiomyocytes, the inhibition of I_{Ca} by cGMP is clearly not mediated by PDE II, but rather by PK-G (Levi *et al.*, 1989; Mery *et al.*, 1991, Wahler and Dollinger, 1995; Wahler *et al.*, 1990; Haddad *et al.*, 1995; Sumii and Sperelakis, 1995). Thus, the antiadrenergic effect of cGMP on I_{Ca} in mammalian ventricular myocytes is mimicked by direct intracellular application of PK-G (Levi *et al.*, 1989; Mery *et al.*, 1991), and the antiadrenergic action of cGMP is blocked by a selective inhibitor of PK-G (Wahler and Dollinger, 1995). Additionally, in some preparations (e.g., embryonic chick, neonatal rat), when PK-G is added to the patch pipette for diffusion into the cell during whole-cell

voltage-clamp, **basal** I_{Ca} is inhibited markedly and rapidly, with maximum inhibition reached in about 3–5 min (Figs. 8 and 9). Data from 17-day-old chick cardiomyocytes are illustrated in Fig. 8. A summary of the effect of PK-G is shown in Fig. 8B. Note that inhibition of basal I_{Ca} began about 80 s after breaking into the cell. Similar effects of PK-G infusion were observed in early neonatal rat ventricular myocytes, as illustrated in Fig. 9 (Sumii and Sperelakis, 1995). As can be seen, there is a rapid and prominent inhibition of the basal I_{Ca} by PK-G. Addition of H-8 (a blocker of both PK-A and PK-G) to the bath often causes a rapid restoration of I_{Ca} to about the original basal level. Addition of 1 mM 8-Br-cAMP can produce only a small stimulation of I_{Ca} in the continued presence of PK-G (see Fig. 8). Therefore, these findings indicate that the inhibitory effects of cGMP on I_{Ca} in mammalian and avian ventricular myocytes are mediated by activation of PK-G and resultant

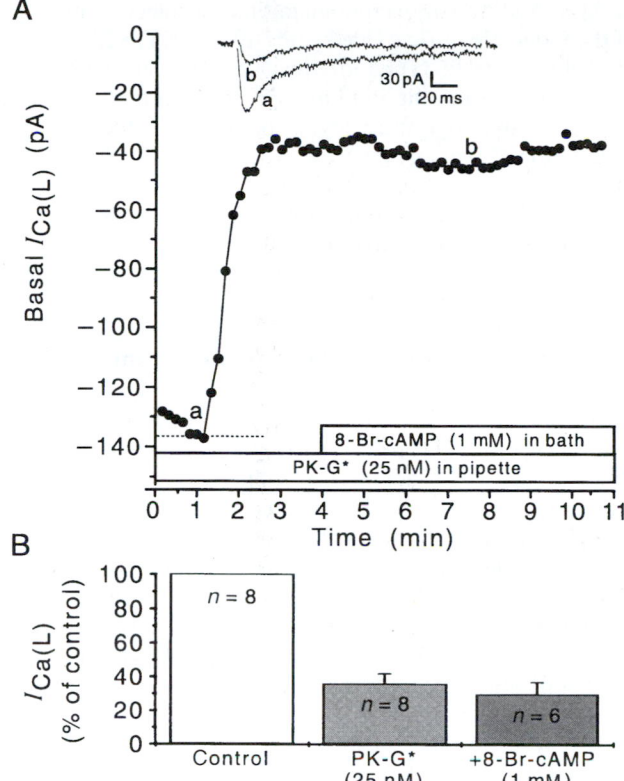

FIGURE 8. Inhibition of basal I_{Ca} of 17-day-old embryonic chick cardiomyocytes by PK-G. (A) PK-G (25 nM) was present in the patch pipette for diffusion into the cell during whole-cell voltage-clamp. Inhibition of basal I_{Ca} began within 70 s after breaking into the cell and reached maximum at about 2.5 min. Addition of 1 mM 8-Br-cAMP into the bath failed to reverse the inhibition produced by the PK-G. The two current traces illustrated at the top correspond to the time points labeled a and b in the graph. (B) Bar graph summary of the inhibition of basal $I_{Ca(L)}$ by PK-G in eight cells, and the lack of reversal by 8-Br-cAMP in six cells. Experiments were done at room temperature. (From Haddad *et al.*, 1995.)

phosphorylation, and that in some instances even the basal I_{Ca} is inhibited. Because 8-Br-cGMP is a potent activator of PK-G and does not stimulate cAMP hydrolysis, cGMP-induced inhibition of the basal activity of the Ca²⁺ channels (not prestimulated by cAMP) is likely to also be mediated by PK-G. In addition, the 8-Br-cGMP inhibition of slow APs in mammalian cardiac muscle occurs without a decrease in cAMP levels (Thakkar *et al.*, 1989), further indicating that the inhibition of I_{Ca} in mammalian ventricular muscle is independent of cGMP-mediated decreases in cAMP. A single protein of approximately 47 kDa has been found to be phosphorylated in guinea pig sarcolemmal preparations treated with 8-Br-cGMP (Cuppoletti *et al.*, 1988). Thus, this substrate may be a possible mediator of cGMP regulation of Ca²⁺ channels of the heart.

There are several isoforms of phosphodiesterase. Some isoforms preferentially degrade cAMP, whereas others hydrolyze both cAMP and cGMP relatively equally. The inhibitory action of exogenous cGMP on I_{Ca} can be mimicked

by reducing cGMP degradation through inhibition of a phosphodiesterase isoform that preferentially hydrolyzes cGMP (PDE V), rather than cAMP. Zaprinast, a selective inhibitor of PDE V, has an antiadrenergic effect on I_{Ca} similar to exogenous cGMP. That is, zaprinast inhibited the I_{Ca} of cardiac myocytes following stimulation by isoproterenol or forskolin. It did not significantly alter basal (unstimulated) I_{Ca} in standard whole-cell recording in which the intracellular Ca²⁺ concentration is unphysiologically low (Ziolo *et al.*, 1998). The effect of zaprinast was blocked by KT5823, an inhibitor of PK-G. In addition, zaprinast inhibited basal I_{Ca} when the perforated-patch technique was used (which maintains a more physiological intracellular environment) or when whole-cell recording was used with free Ca²⁺ concentration maintained intracellularly at a more physiological level (pCa = 7) (Ziolo *et al.*, 1998). These results indicate that inhibiting **endogenous** cGMP hydrolysis decreases both the basal and cAMP-stimulated I_{Ca} via a mechanism that involves PK-G under physiological conditions and at physiological cGMP concentrations.

C. Effects of Nitric Oxide on I_{Ca}

Nitric oxide (NO) is known to be an important regulator of diverse cellular functions in various tissues. Recent evidence suggests that NO may play an important role in regulating myocardial contractility. Many of the effects of NO are mediated through stimulation of guanylate cyclase activity and enhanced cGMP production. As noted above, numerous studies have shown that cGMP inhibits I_{Ca} in cardiac myocytes; thus, the depression of contractility by NO may be due to a cGMP-mediated inhibition of I_{Ca}.

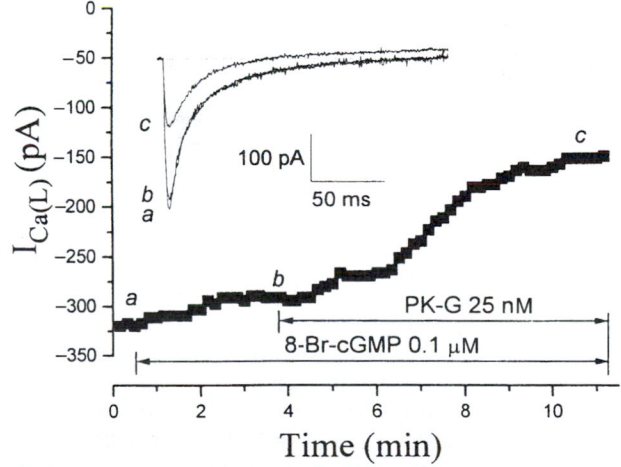

FIGURE. 9. Inhibition of basal I_{Ca} by PK-G (25 nM) in a ventricular myocyte from an early (4-day) neonatal rat heart. Time course of the effect of low doses of 8-Br-cGMP (0.1 μM) and PK-G (25 nM) on basal (not stimulated) $I_{Ca(L)}$. 8-Br-cGMP and PK-G were applied using the perfusion patch-pipette technique. As shown, when 8-Br-cGMP was applied in advance of the PK-G, it had only little effect, whereas subsequent addition of PK-G produced rapid inhibition (inset). Selected current traces of $I_{Ca(L)}$ (a, b, and c) at points denoted on the time-course curve. (Reproduced with permission from Sumii and Sperelakis, 1995.)

Several studies have used NO donors to examine the effect of NO on I_{Ca}. The NO donor SIN-1 (which releases NO in solution) has been shown to have an antiadrenergic effect on I_{Ca} in both frog and mammalian ventricular myocytes (Mery *et al.*, 1993; Wahler and Dollinger, 1995), similar to the effects of cGMP. In mammalian ventricular myocytes, the SIN-1 inhibition was blocked by a selective inhibitor of PK-G. In addition, in cells in which the response to β-agonists was relatively small, low doses of SIN-1 actually caused a stimulation of I_{Ca}, rather than an inhibition. A similar enhancement of β-stimulated I_{Ca} by low doses of pipette cGMP had been reported previously (Ono and Trautwein, 1991). Thus, while NO and cGMP generally have an inhibitory effect on I_{Ca}, in some instances they can enhance the response to β-stimulation of I_{Ca}. Although in the virtual absence of intracellular (pipette) Ca^{2+}, SIN-1 was reported to have no effect on ventricular I_{Ca}, when a physiological concentration of Ca^{2+} (pCa = 6.85) was present in the pipette solution, SIN-1 reduced basal I_{Ca} (Matsumoto, 1997). Thus, it is clear that cGMP can also inhibit basal I_{Ca}, at least in some preparations under physiological intracellular conditions.

D. Pathophysiological Effects of Nitric Oxide/cGMP

The enzyme that synthesizes NO from L-arginine exists in two basic isoforms, **inducible nitric oxide synthase** (iNOS or type II) and **constitutive nitric oxide synthase** (cNOS). The constitutive isoforms are further subdivided into two separate isoforms (types I and III). The constitutive isoform of NOS produces relatively small amounts of NO in a Ca-dependent fashion (Xie and Nathan, 1994). In cardiac myocytes, the constitutive isoform is normally present and functional (Schulz *et al.*, 1992). In contrast, iNOS (which produces much more NO than the constitutive isoform and for longer periods of time) is not normally present in most types of cells, including cardiac myocytes. However, it is expressed in cardiac myocytes under pathological conditions that involve exposure of the cells to cytokines. Thus, it may be that NO produced by cNOS mediates the physiological effects of NO, whereas the much higher levels of NO produced by iNOS may mediate the pathophysiological actions of NO in a number of pathological conditions of the heart. For example, preexposure of guinea pig ventricular myocytes to IL-1β for several hours has an antiadrenergic effect on I_{Ca} that is L-arginine-dependent and NO-mediated (Rozanski and Witt, 1994).

Transplanted hearts are exposed to a variety of cytokines during rejection and have an increased NO production and depressed contractility. Myocytes isolated from rejecting hearts exhibit parallel increases in NO (and cGMP) production (due to the expression of the inducible form of nitric oxide synthase) (Yang *et al.*, 1994; Ziolo *et al.*, 1998), and both a reduced basal contraction and inotropic response to β-adrenergic stimulation (Pyo and Wahler, 1995; Ziolo *et al.*, 1998). The reduced contractile function is due to an NO/cGMP-mediated inhibition of I_{Ca}. Thus, I_{Ca} is reduced in myocytes from rejecting transplanted rat hearts (allografts) compared to myocytes from nonrejecting transplanted control hearts (isografts). This reduction in I_{Ca} in the allograft myocytes is dependent on L-arginine (the precursor

of NO). Additionally, aminoguanidine (a selective inhibitor of the inducible nitric oxide synthase) and KT5823 (a selective inhibitor of PKG) rapidly reversed the elevated NO and cGMP production, the inhibition of I_{Ca}, and the contractile depression in allograft myocytes, but had no effect on I_{Ca} or contraction of isograft myocytes (Ziolo *et. al.*, 1996, 1998). In human transplant patients, there is expression of iNOS in rejecting hearts that correlated with increased cGMP levels and depressed contractility (Lewis *et al.*, 1996; Paulus *et al.*, 1997).

V. Inhibition by Muscarinic Agonists

The parasympathetic transmitter acetylcholine (ACh) exerts a negative inotropic effect on ventricular myocardium prestimulated by β-adrenergic agonists. Activation of the muscarinic receptor by ACh exerts an inhibitory effect on adenylate cyclase and cAMP levels via the G_i (inhibitory) coupling protein, reversing the stimulation of adenylate cyclase produced by G_s coupling protein due to, for example, activation of the β-adrenoceptor (see Fig. 7). This effect of ACh is accompanied by a reduction in PK-A activity and β-adrenergic-stimulated protein phosphorylation (George *et al.*, 1991). Thus, ACh may depress Ca^{2+} influx and contraction by reversing the cAMP elevation produced by various agonists. For example, ACh depresses the I_{Ca} of cultured chick ventricular cells that had been prestimulated by isoproterenol (Josephson and Sperelakis, 1978). ACh also reverses the electrophysiological effects of direct adenylate cyclase stimulation by forskolin (Wahler and Sperelakis, 1985). In addition to the antiadrenergic action of ACh on I_{Ca} due to the effects on G_i, muscarinic inhibition of the hyperpolarization-activated current (I_h) in nodal cells is mediated by the same G_i pathway (Accili *et al.*, 1998).

In ventricular cells, ACh inhibition of the L-type Ca^{2+} channels may also be due partly to elevation of cGMP levels. Muscarinic agonists are known to elevate cGMP (George *et al.*, 1970), and this elevation of cGMP would be expected to contribute to the inhibition of I_{Ca}. In some instances, the muscarinic inhibition of I_{Ca} has been attributed to NO release and cGMP elevation (e.g., Belhassen *et al.*, 1996; Han *et al.*, 1998). However, the involvement of NO in ACh inhibition of I_{Ca} is controversial (e.g., see Mery *et al.*, 1997).

It has been proposed that muscarinic agonists may also act to inhibit the Ca^{2+} channel by stimulation of one or more phosphatases, resulting in dephosphorylation of the channel (Ahmad *et al.*, 1989; Herzig *et al.*, 1995) (see Fig. 7). This mechanism, which may involve cGMP, would have the effect of decreasing the fraction of channels in the phosphorylated form, and therefore decreasing the Ca^{2+} influx. In this mechanism, the rate of phosphorylation is unaffected, but the rate of dephosphorylation is increased. In support of the hypothesis that the antiadrenergic effects of ACh are mediated in part by phosphatase activation, certain phosphatase inhibitors (such as okadaic acid) reduced the antiadrenergic effect of ACh (Herzig *et al.*, 1995). Thus, ACh may reduce Ca^{2+} channel activity by a number of different mechanisms. Additionally, ACh can indirectly reduce I_{Ca} in some cardiac cells due to activation of an ACh-activated K^+ channel, thereby shortening the AP and, thus, shutting off I_{Ca} earlier.

VI. Protein Kinase C and Calmodulin Protein Kinase

Protein kinase C (PK-C) is apparently involved in regulation of the myocardial slow Ca^{2+} channels. Angiotensin II and a high concentration of the α-adrenergic agonist phenylephrine cause a positive inotropic effect in cardiac muscle. These agonists stimulate the phosphatidylinositol cycle and generation of inositol trisphosphate (IP_3) and diacyl glycerol (DAG). IP_3 acts as a second messenger to release stored Ca^{2+} from the sarcoplasmic reticulum (SR). DAG and Ca^{2+} together activate PK-C, which phosphorylates a number of proteins. Phorbol esters (direct activators of PK-C) stimulate I_{Ca} in rat and chick hearts, but have been reported not to stimulate I_{Ca} in guinea pig heart. However, both phorbol esters and α-adrenergic agonists stimulate I_{Ca} in adult guinea pig ventricular myocytes when intact cells (i.e., not dialyzed) have been used (Woo and Lee, 1999).

Inhibitors of calmodulin (e.g., calmidazolium) inhibit the Ca^{2+}-dependent slow APs of heart cells, and subsequent injection of calmodulin reverses the inhibition. Apparently, maximal activation of the Ca^{2+} channels requires two separate phosphorylation steps (calmodulin-dependent and cAMP-dependent). These phosphorylation sites may be on the same protein or on two separate proteins.

VII. Na^+, K^+, and I_f Channels

The previous discussion dealt primarily with regulation of slow L-type Ca^{2+} channels in cardiac muscle. Table 3 summarizes the effects of the cyclic nucleotides on $I_{Ca(L)}$ in cardiac muscle, vascular smooth muscle (VSM), and skeletal muscle. As indicated previously, in myocardial cells, cAMP and cGMP generally act in an antagonistic manner, with cAMP stimulating and cGMP inhibiting I_{Ca}. In contrast, in VSM cells, cAMP and cGMP act in the same direction, both inhibiting $I_{Ca(L)}$. Similarly, in skeletal muscle fibers, both cyclic nucleotides act in the same direction, but here both cAMP and cGMP stimulate $I_{Ca(L)}$. In uterine smooth muscle, neither nucleotide has any effect on $I_{Ca(L)}$ (Table 4). Therefore, these tis-

TABLE 4 Listing of Some Channels Modulated by cAMP and PK-A Phosphorylation

Channel type	Tissue	Action
$I_{Ca(L)}$	Heart, nerve, skeletal muscle	Stimulated
$I_{Ca(L)}$	Vascular smooth muscle	Inhibited
$I_{Ca(L)}$	Uterine smooth muscle	No effect
$I_{K(del)}$	Heart	Stimulated
I_f	Heart	Stimulated
$I_{Na(f)}$	Heart	No effect[a]

[a]Variable effects have been reported for $I_{Na(f)}$, including slight inhibition, slight stimulation, and shifting of the voltage dependency.

sues provide examples of a variety of effects concerning the regulation exerted by cyclic nucleotides.

cAMP also has effects on other types of ion channels (see Table 4). For example, the following channels of heart, in addition to L-type Ca^{2+} channels, are stimulated by cAMP: (1) delayed-rectifier K^+ channels (Trautwein et al., 1982; Yazawa and Kameyama, 1990); (2) hyperpolarization-activated Na^+-K^+ I_f channels (DiFrancesco and Tromba, 1988); and (3) β-adrenergic-activated Cl^- channels (Ehara and Ishihara, 1990). The fast Na^+ channel has been reported to be slightly inhibited by cAMP (Ono et al., 1989), but others have reported no effect or slight stimulation (Ono et al., 1993).

The cAMP stimulation of $I_{K(del)}$ (the delayed-rectifier K^+ current) in myocardial cells, coupled with its stimulation of $I_{Ca(L)}$, serves to shorten the duration of the cardiac AP, whereas heart rate and force of contraction are increased by agents that elevate cAMP. For example, β-adrenergic agonists, such as isoproterenol or epinephrine, raise the cAMP level, thereby increasing automaticity and force of contraction.

In addition, cGMP was reported to stimulate the delayed-rectifier K^+ current (Ono and Trautwein, 1991) and the Cl^- current activated by β-adrenergic agonists (Tareen et al., 1991), perhaps due to elevation of cAMP levels caused by cGMP-mediated inhibition of a PDE. PK-G also stimulates the activity of Ca-activated K^+ channels in a variety of cell types. This effect of PK-G is inhibited by the phosphatase inhibitors okadaic acid and microcystin (Zhou et al., 1996; White et al., 1993). The catalytic subunit of the phosphatase PP-2A mimicked the effect of PK-G. Thus, it has been concluded that the activation of Ca-activated K channels by PK-G is mediated, at least in part, by dephosphorylation of the channels.

VIII. Summary

The slow L-type Ca^{2+} channels of the heart are stimulated by cAMP. Elevation of cAMP produces a very rapid increase in the number of Ca^{2+} channels available for voltage activation during excitation. The probability of a Ca^{2+} channel opening and the mean open time are increased.

TABLE 3 Summary of Effects of Cyclic Nucleotides on L-type Ca^{2+} Channels in Cardiac Muscle, Vascular Smooth Muscle, and Skeletal Muscle

Cyclic nucleotide	Cardiac muscle	Vascular smooth muscle	Skeletal muscle[a]
cAMP	Stimulation	Inhibition	Stimulation
cGMP	Inhibition[b]	Inhibition	Stimulation

[a]From bullfrog (From Kokate et al., 1993).
[b]Under certain conditions, cGMP can enhance I_{Ca} in cardiac muscle (see text).

Therefore, any agent that increases the cAMP level of the myocardial cell will tend to potentiate I_{Ca}, Ca^{2+} influx, and contraction. This action of cAMP is mediated by PK-A and phosphorylation of the slow Ca^{2+} channel protein or an associated regulatory protein (stimulatory type).

The myocardial L-type Ca^{2+} channels are also regulated by cGMP, in a manner that is opposite to that of cAMP. In mammalian hearts, the inhibitory effect of cGMP is mediated by PK-G and phosphorylation of a protein, for example, a regulatory protein (inhibitory type) associated with the Ca^{2+} channel. In addition, cGMP acts to stimulate a phosphatase that dephosphorylates the Ca^{2+} channel.

PK-C and calmodulin-PK may also play roles in regulation of myocardial Ca^{2+} channels, possibly mediated by phosphorylation of some regulatory type of protein. However, the effect is probably not as important as regulation of I_{Ca} by PK-A and PK-G.

Thus, the L-type Ca^{2+} channel is apparently a complex structure, including perhaps several associated regulatory proteins, which can be regulated by a number of factors intrinsic and extrinsic to the cell (see Fig. 7). In general, it appears that ion channels form functional complexes with their associated regulatory proteins in the cell membrane. Kinases and phosphatases may be physically associated with the channel protein, because, for example, Ca^{2+}-dependent K^+ channels incorporated into bilayers continue to be modulated by phosphorylation and dephosphorylation (Levitan, 1993).

cAMP and cGMP also have effects on the L-type Ca^{2+} channels in cells other than cardiac muscle, including neurons, smooth muscle, and skeletal muscle (see Table 4). In cardiac muscle, the two cyclic nucleotides generally have opposing effects, cAMP stimulating and cGMP inhibiting I_{Ca}. In some smooth muscles (e.g., vascular), both cyclic nucleotides act in the same direction, namely, both inhibit $I_{Ca(L)}$. In skeletal muscle, both cAMP and cGMP act in the same direction on $I_{Ca(L)}$, that is, to stimulate (see Table 3).

The cyclic nucleotides and resulting phosphorylation may also modulate the activity of several other types of ion channels, including K^+ channels (delayed-rectifier type), Cl^- channels (β-agonist activated type), fast Na^+ channels, and I_f pacemaker channels (see Table 4).

Bibliography

Accili, E. A., Redaelli, G., and DiFrancesco, D. (1998). Two distinct pathways of muscarinic current responses in rabbit sino-atrial node myocytes. *Pflügers Arch.* **437,** 164–167.

Ahmad, Z., Green, F. J., Subuhi, H. S., and Watanabe, A. M. (1989). Autonomic regulation of type 1 protein phosphatase in cardiac muscle. *J. Biol. Chem.* **264,** 3859–3863.

Armstrong, D., and Eckert, R. (1987). Voltage-activated calcium channels that must be phosphorylated to respond to membrane depolarization. *Proc. Natl. Acad. Sci. USA* **84,** 2518–2522.

Bean, B. P., Nowycky, M. C., and Tsien, R. W. (1984). Beta-adrenergic modulation of calcium channels in frog ventricular heart cells. *Nature* **307,** 371–375.

Belhassen, L., Kelly, R. A., Smith, T. W., and Balligand, J. L. (1996). Nitric oxide synthase (NOS3) and contractile responsiveness to adrenergic and cholinergic agonists in the heart. *J. Clin. Invest.* **97,** 1908–1915.

Bkaily, G., and Sperelakis, N. (1984). Injection of protein kinase inhibitor into cultured heart cells blocks calcium slow channels. *Am. J. Physiol. Heart Circ. Physiol.* **246,** H630–H634.

Bkaily, G., and Sperelakis, N. (1985). Injection of cyclic GMP into heart cells blocks the Ca^{2+} slow channels. *Am. J. Physiol. Heart Circ. Physiol.* **248,** H745–H749.

Bkaily, G., and Sperelakis, N. (1986). Calmodulin is required for a full activation of the calcium slow channels in heart cells. *J. Cyclic Nucleotide Protein Phosphorylation Res.* **11,** 25–34.

Bruckner, R., and Scholz, H. (1984). Effects of alpha-adrenoceptor stimulation with phenylephrine in the presence of propranolol on force of contraction, slow inward current and cyclic AMP content in the bovine heart. *Br. J. Pharmacol.* **82,** 223–232.

Cachelin, A. B., dePeyer, J. E., Kokubun, S., and Reuter, H. (1983). Ca^{2+} channel modulation by 8-bromo-cyclic AMP in cultured heart cells. *Nature* **304,** 462–464.

Chad, J. E., and Eckert, R. J. (1986). An enzymatic mechanism for calcium current inactivation in dialysed Helix neurones. *J. Physiol. (London)* **378,** 31–51.

Cuppoletti, J., Thakkar, J., Sperelakis, N., and Wahler, G. (1988). Cardiac sarcolemmal substrate of the cGMP-dependent protein kinase. *Membr. Biochem.* **7,** 135–142.

DiFrancesco, D., and Tromba, C. (1988). Muscarinic control of the hyperpolarization-activated current (I_f) in rabbit sino-atrial node myocytes. *J. Physiol. (London)* **405,** 493–510.

Dosemeci, A., Dhalla, R. S., Cohen, N. M., Lederer, W. J., and Rogers, T. B. (1988). Phorbol ester increases calcium current and stimulated the effects of angiotensin II on cultured neonatal rat heart myocytes. *Circ. Res.* **62,** 347.

duBell, W. H., Lederer, W. L., and Rogers, T. B. (1996). Dynamic modulation of excitation-contraction coupling by protein phophatases in rat ventricular myocytes. *J. Physiol.* **493,** 793–800.

Ehara, T., and Ishihara, K. (1990). Anion channels activated by adrenaline in cardiac myocytes. *Nature* **347,** 284–286.

Fischer, H., Illek, B., and Machen, T. E. (1998). Regulation of CFTR by protein phosphatase 2B and protein kinase C. *Pflügers Arch.* **436,** 175–181.

Fischmeister, R., and Hartzell, H. C. (1987). Cyclic guanosine 3',5'-monophosphate regulates the calcium current in single cells from frog ventricle. *J. Physiol. (London)* **387,** 455–472.

Fischmeister, R., and Hartzell, H. C. (1991). Cyclic AMP phosphodiesterases and Ca^{2+} current regulation in cardiac cells. *Life Sci.* **48,** 2365–2376.

Frace, A. M., and Hartzell, H. C. (1993). Opposite effects of phosphatase inhibitors on L-type calcium and delayed rectifier currents in frog cardiac myocytes. *J. Physiol.* **472,** 305–326

Gadsby, D. C., and Narin, A.C. (1999). Control of CFTR channel gating by phosphorylation and nucleotide hydrolysis. *Physiol. Rev.* **79,** Suppl. 1 S77–S107

George, E. E., Romano, F. D., and Dobson, J. G., Jr. (1991). Adenosine and acetylcholine reduce isoproterenol-induced protein phosphorylation of rat myocytes. *J. Molec. Cell. Cardiol.* **23,** 749–764.

George, W. J., Polson, J. B., O'Toole, A. G., and Goldberg, N. D. (1970). Elevation of guanosine 3', 5'-cyclic phosphate in rat heart after perfusion with acetylcholine. *Proc. Natl. Acad. Sci. USA* **66,** 398–403.

Goldberg, N. D., Haddox, M. K., Nicol, S. E., Glass, D. B., Sanford, C. H., Kuehl, F. A., Jr., and Estensen, R. (1975). Biological regulation through opposing influences of cyclic GMP and cyclic AMP: the Yin Yang hypothesis. *Adv. Cyclic Nucleotides* **5,** 307–330.

Gupta, R. C., Neumann, J., and Watanabe, A. M. (1993). Comparison of adenosine and muscarinic receptor-mediated effects on protein phosphatase inhibitor-1 activity in the heart. *J. Pharmacol. Exp. Therap.* **266,** 16–22.

Gupta, R. C., Neumann, J., Watanabe, A. M., Lesch, M., and Sabbah, H. N. (1996). Evidence for presence and hormonal regulation of protein phosphatase inhibitor-1 in ventricular cardiomyocyte. *Am. J. Physiol.* **270,** H1159–H1164.

Haddad, G. E., Sperelakis, N., and Bkaily, G. (1995). Regulation of calcium channel by cyclic GMP-dependent protein kinase in chick heart cells. *Mol. Cell. Biochem.* **148,** 89–94.

Han, X., Kobzik, L., Severson, D., and Shimoni, Y. (1998). Characteristics of nitric oxide–mediated cholinergic modulation of calcium current in rabbit sino-atrial node. *J. Physiol.* **509,** 741–754.

Hartzell, H. C., and Fischmeister, R. (1986). Opposite effects of cyclic GMP and cyclic AMP on Ca^{2+} current in single heart cells. *Nature* **323,** 273–275.

Herzig, S., Meier, A., Pfeiffer, M., and Neumann, J. (1995). Stimulation of protein phosphatases as a mechanism of the muscarinic-receptor-mediated inhibition of cardiac L-type Ca^{2+} channels. *Pflügers Arch.* **429,** 531–538

Hescheler, J., Kameyama, M., and Trautwein, W. (1986). On the mechanism of muscarinic inhibition of the cardiac Ca current. *Pflügers Arch.* **407,** 182–189.

Hescheler, J., Kameyama, M., Trautwein, W., Mieskes, G., and Soling, H. D. (1987). Regulation of the cardiac calcium channel by protein phosphatases. *Eur. J. Biochem.* **165,** 261–266.

Hescheler, J., Mieskes, G., Ruegg, J. C., Takai, A., and Trautwein, W. (1988). Effects of a protein phosphatase inhibitor, okadaic acid, on membrane currents of isolated guinea-pig cardiac myocytes. *Pflügers Arch.* **412,** 248–252.

Irisawa, H., and Kokubun, S. (1983). Modulation of intracellular ATP and cyclic AMP of the slow inward current in isolated single ventricular cells of the guinea-pig. *J. Physiol.* **338,** 321–327.

Josephson, I., and Sperelakis, N. (1978). 5'Guanylimidophosphate stimulation of slow Ca^{2+} current in myocardial cells. *J. Mol. Cell. Cardiol.* **10,** 1157–1166.

Kameyama, M., Hoffman, F., and Trautwein, W. (1986). On the mechanism of β-adrenergic regulation of the Ca^{2+} channel in the guinea pig heart. *Pflügers Arch.* **405,** 285–293.

Kawamura, A., and Wahler, G. M. (1994) Perforated-patch recording does not enhance the effect of 3-isobutyl-1-methylxanthine on cardiac calcium current. *Am. J. Physiol.* **266,** C1619–C1627.

Kohlhardt, M., and Haap, K. (1978). 8-Bromo-guanosine-3', 5'-monophosphate mimics the effect of acetylcholine on slow response action potential and contractile force in mammalian atrial myocardium. *J. Mol. Cell. Cardiol.* **10,** 573–578.

Levi, R. C., Alloatti, G., and Fischmeister, R. (1989). Cyclic GMP regulates the Ca-channel current in guinea pig ventricular myocytes. *Pflügers Arch.* **413,** 685–687.

Levitan, I. B. (1993). A kinase-phosphatase regulatory complex associated with a Ca^{2+}-dependent K$^+$ channel. *Neurosci. Facts* **4,** 12.

Lewis, N. P., Tsao, P. S., Rickenbacher, P. R., Xue, C., Johns, R. A., Haywood, G. A, von der Leyen, H., Trindade, P. T., Cooke, J. P., Hunt, S. A., Billingham, M. E., Valantine, H. A., and Fowler, M. B. (1996). Induction of nitric oxide synthase in the human cardiac allograft is associated with contractile dysfunction of the left ventricle. *Circulation* **93,** 720–729.

Li, T., and Sperelakis, N. (1983). Stimulation of slow action potentials in guinea pig papillary muscle cells by intracellular injection of cAMP, Gpp(NH)p, and cholera toxin. *Circ. Res.* **52,** 111–117.

Llinas, R., Sugimori, D., Hillman, E., and Cherskey, B. (1992). Distribution and functional significance of the P-type, voltage-dependent Ca^{2+} channels in the mammalian central nervous system. *Trends Neurosci.* **15,** 351–355.

Masuda, H., Sumii, K., and Sperelakis, N. (1995). Long openings of calcium channels in fetal rat ventricular cardiomyocytes. *Pflügers Arch.* **429,** 595–597.

Matsumoto, S. (1997). Effect of molsidomine on basal Ca^{++} current in rat cardiac cells. *Life Sciences* **60,** 383–390.

Mehegan, J. P., Muir, W. W., Unverferth, D. V., Fertel, R. H., and McGuirck, S. M. (1985). Electrophysiological effects of cyclic

GMP on canine cardiac Purkinje fibers. *J. Cardiovasc. Pharmacol.* **7,** 30–35.

Mery, P.-F., Abi-Gerges, N., Vandecasteele, G., Jurvicius, J., Eschenhagen, T., and Fischmeister, R. (1997). Muscarinic regulation of the L-type calcium current in isolated cardiac myocytes. *Life Sciences* **60,** 1113–1120.

Mery, P. F., Lohmann, S. M., Walter, U., and Fischmeister, R. (1991). Ca^{2+} current is regulated by cyclic GMP-dependent protein kinase in mammalian cardiac myocytes. *Proc. Natl. Acad. Sci. USA* **88,** 1197–1201.

Mery, P. F., Pavoine, C., Belhassen, L., Pecker, F., and Fischmeister, R. (1993). Nitric oxide regulates cardiac Ca^{2+}. Involvement of cGMP-inhibited and cGMP–stimulated phosphodiesterases through guanylyl cyclase activation. *J. Biol. Chem.* **268,** 26 286–26 295.

Mintz, I. M., Venema, V. J., Swiderek, K. M., Lee, T. D., Bean, B. P., and Adams, M. E. (1992). P-type calcium channels blocked by the spider toxin omega-Aga-IVA. *Nature* **355,** 827–829.

Mubagwa, K., Shirayama, T., Moreau, M., and Pappano, A. J. (1993). Effects of PDE inhibitors and carbachol on the L-type Ca current in guinea pig ventricular myocytes. *Am. J. Physiol.* **264,** H1353–H1363.

Nargeot, J., Nerbonne, J. M., Engels, J., and Lester, H. A. (1983). Time course of the increase in the myocardial slow inward current after a photochemically generated concentration jump of intracellular cAMP. *Proc. Natl. Acad. Sci. USA* **80,** 2395–2399.

Nawrath, H. (1977). Does cyclic GMP mediate the negative inotropic effect of acetylcholine in the heart? *Nature* **267,** 72–74.

Nowycky, M. C., Fox, A. P., and Tsien, R. W. (1985). Three types of neuronal calcium channels with different calcium agonist sensitivity. *Nature* **316,** 440–443.

Ono, K., and Trautwein, W. (1991). Potentiation by cyclic GMP of β-adrenergic effect on Ca^{2+} current in guinea-pig ventricular cell. *J. Physiol.* **443,** 387–404.

Ono, K., Fozzard, H. A., and Hanck, D. A. (1993). Mechanism of cAMP-dependent modulation of cardiac sodium channel current kinetics. *Circ. Res.* **72,** 807–815.

Ono, K., Kiyosue, T., and Arita, M. (1989). Isoproterenol, DBcAMP, and forskolin inhibit cardiac sodium current. *Am. J. Physiol.* **256,** C1131–C1137.

Osterreider, W., Brum, G., Hescheler, J., Trautwein, W., Flockerzi, V., and Hofmann, F. (1982) Injection of subunits of cyclic AMP-dependent protein kinase into cardiac myocytes modulates Ca^{2+} current. *Nature* **298,** 576–578.

Paulus, W. J., Kastner, S., Pujadas, P., Shah, A. M., Drexler, H., and Vandeheyden, M. (1997). Left ventricular contractile effects of inducible nitric oxide synthase in the human allograft. *Circulation* **96,** 3336–3342.

Pyo, R., and Wahler, G. M. (1995). Ventricular myocytes isolated from rejecting cardiac allografts exhibit a reduced β-adrenergic contractile response. *J. Molec. Cellular Cardiol.* **27,** 773–776.

Reuter, H. (1983). Calcium channel modulation by neurotransmitters, enzymes, and drugs. *Nature* **301,** 569–574.

Reuter, H., and Scholz, H. (1977). The regulation of calcium conductance of cardiac muscle by adrenaline. *J. Physiol.* **264,** 49–62.

Reuter, H., Stevens, C.-F., Tsien, R. W., and Yellen, G. (1982). Properties of single calcium channels in cardiac cell culture. *Nature* **297,** 501–504.

Rivet-Bastide, M., Vandecasteele, G., Hatem, S., Verde, I., Benardeau, A., Mercadier, J-J., and Fischmeister, R. (1997). cGMP-stimulated cyclic nucleotide phosphodiesterase regulates the basal calcium current in human atrial myocytes. *J. Clin. Invest.* **99,** 2710–2718.

Rozanski, G. J., and Witt, R. C. (1994). IL-1 inhibits β-adrenergic control of cardiac calcium current: role of L-arginine/nitric oxide pathway. *Am. J. Physiol.* **267,** H1753–H1758.

Schulz, R., Nava, E., and Moncada, S. (1992). Induction and potential biological relevance of a Ca^{2+}-independent nitric oxide synthase in the myocardium. *Br. J. Pharmacol.* **105,** 575–580.

Schumann, K. Romanin, C., Baumgartner, W., and Groschner, K. (1997). Intracellular Ca^{2+} inhibits smooth muscle L-type Ca^{2+} channels by activation of protein phosphatase type 2B and by direct interaction with the channel. *J. Gen Physiol.* **110**, 503–513.

Shigenobu, K., and Sperelakis, N. (1972). Ca^{2+} current channels induced by catecholamines in chick embryonic hearts whose fast Na$^+$ channels are blocked by tetrodotoxin or elevated K$^+$. *Circ. Res.* **31**, 932–952.

Sperelakis, N., and Schneider, J. A. (1976). A metabolic control mechanism for calcium ion influx that may protect the ventricular myocardial cell. *Am. J. Cardiol.* **37**, 1079–1085.

Sumii, K., and Sperelakis, N. (1995). Cyclic GMP–dependent protein kinase regulation of the L-type calcium current in neonatal rat ventricular myocytes. *Circ. Res.* **77**, 803–812.

Tareen, F. M., Ono, K., Noma, A., and Ehara, T. (1991). β-adrenergic and muscarinic regulation of the chloride current in guinea-pig ventricular cells. *J. Physiol.* **440**, 225–241.

Thakkar, J., Tang, S. B., Sperelakis, N., and Wahler, G. M. (1989). Inhibition of cardiac slow action potentials by 8-bromo-cyclic GMP occurs independent of changes in cyclic AMP levels. *Canad. J. Physiol. Pharmacol.* **66**, 1092–1095.

Tohse, N., and Sperelakis, N. (1990). Long-lasting openings of single slow (L-type) Ca^{2+} channels in chick embryonic heart cells. *Am. J. Physiol.* **259**, H639–H642.

Tohse, N., and Sperelakis, N. (1991). Cyclic GMP inhibits the activity of single calcium channels in embryonic chick heart cells. *Circ. Res.* **69**, 325–331.

Tohse, N., Kameyama, M., Sakiguchi, K., Shearman, M. S., and Kanno, M. (1990). Protein kinase C activation enhances the delayed rectifier K$^+$ current in guinea-pig heart cells. *J. Mol. Cell. Cardiol.* **22**, 725–734.

Tohse, N., Meszaros, J., and Sperelakis, N. (1992). Developmental changes in long-opening behavior of L-type Ca^{2+} (slow) channels in embryonic chick heart cells. *Circ. Res.* **71**, 376–384.

Trautwein, W., and Hoffman, F. (1983). Activation of calcium current by injection of cAMP and catalytic subunit of cAMP-dependent protein kinase. *Proc. Int. Union Physiol. Sci.* **15**, 75–83.

Trautwein, W., Taniguchi, J., and Noma, A. (1982). The effect of intracellular cyclic nucleotides and calcium on the action potential and acetylcholine response of isolated cardiac cells. *Pflügers Arch.* **392**, 307–314.

Tsien, R. W., Giles, W., and Greengard, P. (1972). Cyclic AMP mediates the action of adrenaline on the action potential plateau of cardiac Purkinje fibers. *Nature* **240**, 181–183.

Vizgirda, V. M., Schwertz, D. W., Ziolo, M. T., and Wahler, G. M., (1997). Sex differences in beta-adrenergic stimulation of the cardiac calcium current. *Circulation* **96**, 1993, (Abstract).

Vogel, S., and Sperelakis, N. (1981). Induction of slow action potentials by microiontophoresis of cyclic AMP into heart cells. *J. Mol. Cell. Cardiol.* **13**, 51–64.

Vogel, S., Sperelakis, N., Josephson, I. J., and Brooker, G. (1977). Fluoride stimulation of slow Ca^{2+} current in cardiac muscle. *J. Mol. Cell. Cardiol.* **9**, 461–475.

Wahler, G. M., and Dollinger, S. J. (1995). Nitric oxide donor SIN-1 inhibits mammalian cardiac calcium current through cGMP-dependent protein kinase. *Am. J. Physiol.* **268**, C45–C54.

Wahler, G. M., and Sperelakis, N. (1985). Intracellular injection of cyclic GMP depresses cardiac slow action potentials. *J. Cyclic Nucleotide Protein Phosphorylation Res.* **10**, 83–95.

Wahler, G. M., and Sperelakis, N., (1986). Cholinergic attenuation of the electrophysiological effects of forskolin. *J. Cyclic Nucleotide Protein Phosphorylation Res.* **11**, 1–10.

Wahler, G. M., Rusch, N. J., and Sperelakis, N. (1990). 8-Bromo-cyclic GMP inhibits the calcium channel current in embryonic chick ventricular myocytes. *Can. J. Physiol. Pharmacol.* **68**, 531–534.

Walsh, K. B., and Kass, R. S. (1988) Regulation of a heart potassium channel by protein kinase A and C. *Science* **242**, 67–69.

White, R. E., Lee, A. B., Shcherbatko, A. D., Lincoln, T. M., Schonbrunn, A., and Armstrong, D. L. (1993). Potassium channel stimulation by natriuretic peptides through cGMP-dependent dephosphorylation. *Nature* **361**, 263–266.

Woo, S. H., and Lee, C. O. (1999). Role of PKC in the effects of α$_1$-adrenergic stimulation on Ca^{2+} transients, contraction and Ca^{2+} current in guinea-pig ventricular myocytes. *Pflügers Arch.* **437**, 335–344.

Xie, Q., and Nathan, C. (1994). The high-output nitric oxide pathway: role and regulation. *J. Leukocyte Biol.* **56**, 572–582.

Yamaoka, K., and Seyama, I. (1998). Phosphorylation modulates L-type Ca channels in frog ventricular myocytes by changes in sensitivity to Mg^{2+} block. *Pflügers Arch.* **435**, 329–337.

Yang, X., Chowdhury, N., Cai, B., Brett, J., Marboe, C., Sciacca, R., Michler, R., and Cannon, P. (1994). Induction of myocardial nitric oxide synthase by cardiac allograft rejection. *J. Clin. Invest.* **94**, 714–721.

Yazawa, K., and Kameyama, M. (1990). Mechanism of receptor-mediated modulation of the delayed outward potassium current in guinea-pig ventricular myocytes. *J. Physiol.* **421**, 135–150.

Zhou, X. B., Ruth, P., Schlossmann, J., Hofmann, F., and Korth, M. (1996). Protein phosphatase 2A is essential for the activation of Ca^{2+}-activated K$^+$ currents by cGMP-dependent protein kinase in tracheal smooth muscle and chinese hamster ovary cells. *J. Biol. Chem.* **271**, 19 760–19 767.

Ziolo, M. T., and Wahler, G. M. (1998). Cyclic GMP inhibition of basal calcium current in mammalian cardiac myocytes requires physiological levels of intracellular calcium. *Biophys. J.* **74**, A157 (Abstract).

Ziolo, M. T., Dollinger, S. J., Roycroft, K. E., and Wahler, G. M. (1996). Nitric oxide-mediated depression of calcium currents in myocytes isolated from rejecting transplanted rat hearts. *Circulation* **94**, 1307 (Abstract).

Ziolo, M. T., Dollinger, S. J., and Wahler, G. M. (1998). Myocytes isolated from rejecting transplanted rat hearts exhibit reduced basal shortening which is reversible by aminoguanidine. *J. Mol. Cell. Cardiol.* **30**, 1009–1017.

Atsushi Inanobe and Yoshihisa Kurachi

34

Direct Regulation of Ion Channels by G Proteins

I. Introduction

Numerous extracellular stimuli, such as neurotransmitters, hormones, and autacoids as well as light and odors, regulate cellular responses via membrane receptors (Nicoll *et al.,* 1990; Kurachi, 1995; Wickman and Clapham, 1995; Jan and Jan, 1997). **Heterotrimeric GTP-binding proteins** (G proteins) directly couple hundreds of different types of membrane receptors which each possess seven transmembrane regions for a relatively small number of different **effectors:** adenylyl cyclase, phospholipase Cβ, cyclic GMP phosphodiesterase, phosphatidylinositol 3-kinase, and ion channels. Many ion channels are under the control of G proteins directly or indirectly. In this chapter, we will deal with the regulation of ion channels by the direct action of G proteins. The first example of this type of channel regulation was acetylcholine activation of the muscarinic K$^+$ channel in the heart (Kurachi, 1995; Yamada *et al.,* 1998), which is responsible for deceleration of the heartbeat upon stimulation of vagal nerves (Loewi, 1921). This K$^+$ channel also underlies the formation of the slow inhibitory postsynaptic potential in the central nervous system. It is also known that presynaptic Ca^{2+} channels are directly inhibited by G proteins (Dolphin, 1998). In addition, a number of other K$^+$ and Ca^{2+} channels, Na$^+$ channels, and Cl$^-$ channels have been suggested to be regulated by G proteins in a membrane-delimited manner in a variety of tissues, including cardiac, neuronal, and endocrine cells (Wickman and Clapham, 1995). Therefore, the direct G protein regulation of ion channels is an important cell signaling system.

II. The G Protein Cyclic Reaction Mediates Receptor-to-Channel Signal Transmission

The heterotrimeric G proteins consist of α, β, and γ subunits G$_\alpha$, G$_\beta$, and G$_\gamma$, respectively (Gilman, 1987). Up to now, at least 16 G$_\alpha$, 5 G$_\beta$, and 12 G$_\gamma$ genes have been identified. G$_\alpha$ has a binding site for guanine nucleotides (GTP and GDP) and

GTPase activity. G$_\alpha$ has been classified into four groups (G$_{\alpha s}$, G$_{\alpha i}$, G$_{\alpha q}$, and G$_{\alpha 12}$) based upon sequence similarity, coupling effectors, and sensitivity to the toxins of *Vibrio cholerae* (CTX) and *Bordetella pertussis* (PTX). Under physiological conditions G$_\beta$ and G$_\gamma$ tightly associate to form a dimer (G$_{\beta\gamma}$). Different G$_\beta$ and G$_\gamma$ could then form various combinations of the G$_{\beta\gamma}$ complex, although some preferences in the possible combinations exist: G$_{\gamma 2}$ can assemble with G$_{\beta 1}$ and G$_{\beta 2}$, but G$_{\gamma 1}$ forms dimers only with G$_{\beta 1}$, not with G$_{\beta 2}$ (Spring and Neer, 1994). However, the physiological implications of the possible heterogeneity of G$_{\beta\gamma}$ have remained unclear.

G$_\alpha$ exists predominantly in the inactive GDP-bound state (G$_\alpha$-GDP) in the absence of agonists (Fig. 1). G$_\alpha$-GDP has a high affinity for G$_{\beta\gamma}$, thereby forming a heterotrimer. Receptor stimulation substantially increases the rate of dissociation of GDP from G$_\alpha$, which results in marked acceleration of the GDP-GTP exchange reaction. Formation of GTP-bound G$_\alpha$ (G$_\alpha$-GTP) leads to dissociation of G$_{\beta\gamma}$ from G$_\alpha$-GTP. Either component of the G protein (G$_\alpha$-GTP or G$_{\beta\gamma}$) is now available to transduce signals to the downstream effectors. G$_\alpha$ has intrinsic GTP hydrolysis activity: its typical k_{cat} value is 1–5 min^{-1}. G$_\alpha$, therefore, hydrolyzes the bound GTP to GDP, thereby returning to the GDP-bound state (G$_\alpha$-GDP), which reassociates with G$_{\beta\gamma}$. This reaction terminates effector regulation. In the continuous presence of agonists, the heterotrimeric G protein restarts the cycle by interacting with an agonist-bound receptor.

III. Electrophysiological Evidence for K$_G$ Channel Activation Mediated by G Proteins

On stimulation of vagal nerves, acetylcholine (ACh) released from synaptic terminals decelerates the heartbeat and decreases atrioventricular conduction (Loewi, 1921; Loewi and Navaratil, 1926). It was shown that ACh causes hyperpolarization of the membrane in frog heart (Del Castillo and Katz, 1955), which accompanies an increase of K$^+$ efflux across the cell membrane

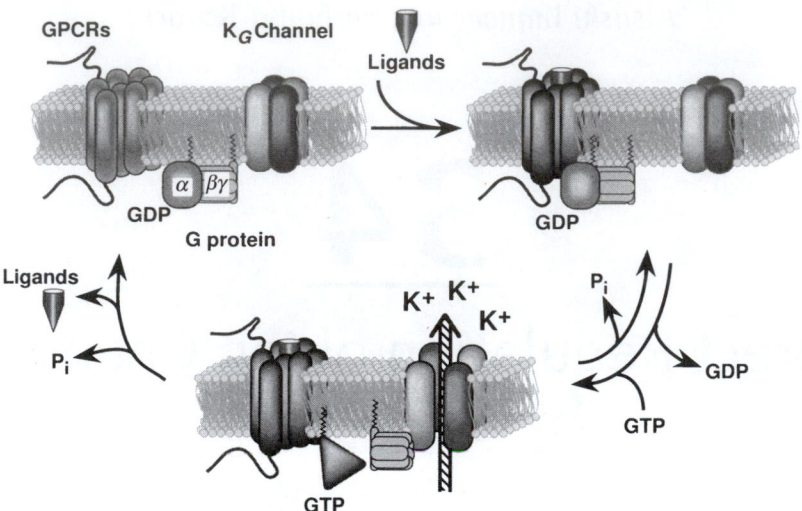

FIGURE 1. Schematic representation of the G protein cycle involved in the activation of the G-protein-gated inwardly rectifying K$^+$ channel (Gilman, 1987). GPCR, G-protein coupled receptor; G protein, trimeric GTP-binding protein; K$_G$ channel, G-protein gated inwardly rectifying K$^+$ channel.

(Hutter and Trautwein, 1955). It was proposed that ACh evokes activation of a specific population of K$^+$ channels, which were named **muscarinic** K$^+$ (K$_{ACh}$) channels (Trautwein and Dudel, 1958; Noma and Trautwein, 1978; Osterrieder et al., 1981; Sackman et al., 1983). The K$_{ACh}$ channel was the first ion channel shown to be directly regulated by G proteins. When atrial myocytes were treated with pertussis toxin (PTX), the activation of K$_{ACh}$ channels by the m$_2$-muscarinic and A$_1$-adenosine receptors was inhibited, indicating the involvement of G proteins in the receptor-dependent regulation of the channels (Pfaffinger et al., 1985; Breitwiester and Szabo, 1985; Kurachi et al., 1986a, b, and c). The membrane-delimited nature of the system was clearly indicated by cell-free inside-out membrane patch experiments (Fig. 2). In inside-out membrane patches from atrial myocytes, GTP (in the presence of extracellular agonists) (Kurachi et al., 1986a, b, and c; Ito et al., 1991) and GTPγS, a nonhydrolyzable analog of GTP (even in the absence of agonists) (Kurachi et al., 1986b, c; Ito et al., 1991), applied to the intracellular surface could activate the K$_{ACh}$ channel. These experiments led to the proposal that the channel was directly activated by G proteins. Furthermore, it was found that G$_{\beta\gamma}$ and not G$_\alpha$ was the subunit to activate the K$_{ACh}$ channel (Logothetis et al., 1987; Kurachi et al., 1989; Ito et al., 1992; Yamada et al., 1993, 1994; Wickman et al., 1994). Not only in atrial myocytes but also in neurons and endocrine cells, G-protein-gated inwardly rectifying K$^+$ (K$_G$) channels have been found to be regulated by G$_{\beta\gamma}$.

IV. Direct Coupling of K$_G$ Channel Subunits to G$_{\beta\gamma}$

The mammalian inward rectifier K$^+$ channel family (Kir) contains such diverse ion channels as the classical inward rectifier K$^+$ channel (Kir2.0), the ATP-sensitive K$^+$

channel (Kir6.0), and the G-protein-regulated K$^+$ channels (Kir3.0). At present, it is thought that each mammalian K$_G$ channel is made up of four Kir3.0 subunits. Kir3.1/GIRK1/KGA was the first K$_G$ channel subunit to be isolated from rat atrium (Kubo et al., 1993; Dascal et al., 1993). From a mouse brain cDNA library, two additional homologs of Kir3.1 were isolated and designated Kir3.2/GIRK2 and Kir3.3/GIRK3 (Lesage et al., 1994, 1995). Another homolog (Kir3.4/GIRK4/CIR) was isolated from rat atrial cDNA library and it was shown that the cardiac K$_{ACh}$ channel is a heteromultimer composed of Kir3.1 and Kir3.4 in atrial cell membranes (Krapivinsky et al., 1995). The Kir3.2 and Kir3.4 subunits can form functional homomeric K$_G$ channels in heterologous expression systems (Lesage et al., 1995; Kofuji et al., 1995; Duprat et al., 1995; Krapivinsky et al., 1995; Tucker et al., 1996; Velimirovic et al., 1996) and in native tissues (Corey and Clapham, 1998; Inanobe et al., 1999), while the homomeric K$_G$ channel composed of Kir3.1 is not functional (Krapivinsky et al., 1995; Kennedy et al., 1996; Hedin et al., 1996). Because Kir3.5/XIR, a Xenopus homolog of Kir3.4, is sometimes expressed in oocytes, a functional K$_G$ channel could occasionally be measured in Xenopus oocytes injected with Kir3.1 mRNA alone (Hedin et al., 1996).

Reuveny and colleagues (1994) presented evidence that Kir3.1, when coexpressed with G$_{\beta\gamma}$ in Xenopus oocytes, exhibited constitutive channel activity that was not affected by intracellular GTP, but inhibited by the G$_{\beta\gamma}$-binding site of β-adrenergic receptor kinase 1 (Fig. 3). An amino acid sequence of this G$_{\beta\gamma}$-binding site shows a certain level of similarity (~26%) with that of the C-terminus of Kir3.1 (between positions 318 and 455). They also found that truncation of the C-terminus of Kir3.1 at Leu at position 403, but not at Pro at position 462, resulted in loss of

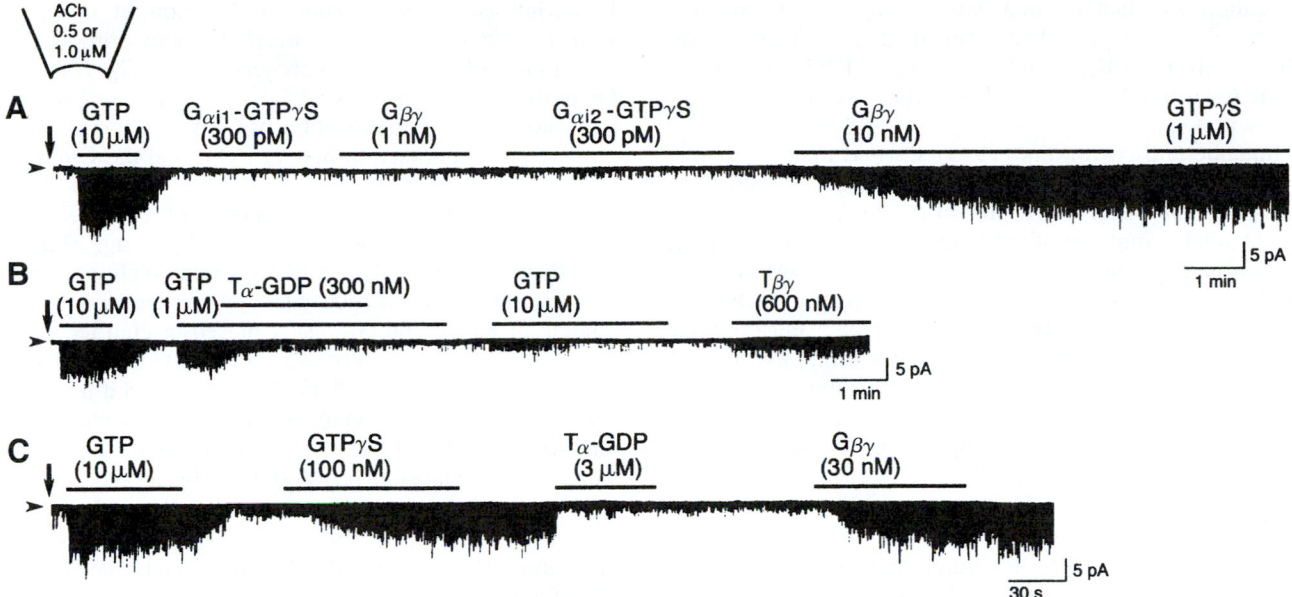

FIGURE 2. G-protein-regulation of K_{ACh} channels examined in inside-out patches of atrial cell membrane. (A) K_{ACh} channels were fully activated by $G_{\beta\gamma}$, but not by GTPγS-bound either $G_{\alpha i1}$ or $G_{\alpha i2}$. Holding potential was –60 mV. (B) K_{ACh} channel activity was inhibited by the GDP-bound transducin α subunit (T_α) and reactivated by the transducin βγ subunit ($T_{\beta\gamma}$). (C) GTPγS-evoked K_{ACh} channel activity was also inhibited by T_α-GDP and the channel could be reactivated by $G_{\beta\gamma}$. In these examples, activity of K_{ACh} channels is represented by downwards deflections of the membrane patch current record. (Reproduced from Kurachi (1995) and Yamada *et. al.* (1994), with permission.)

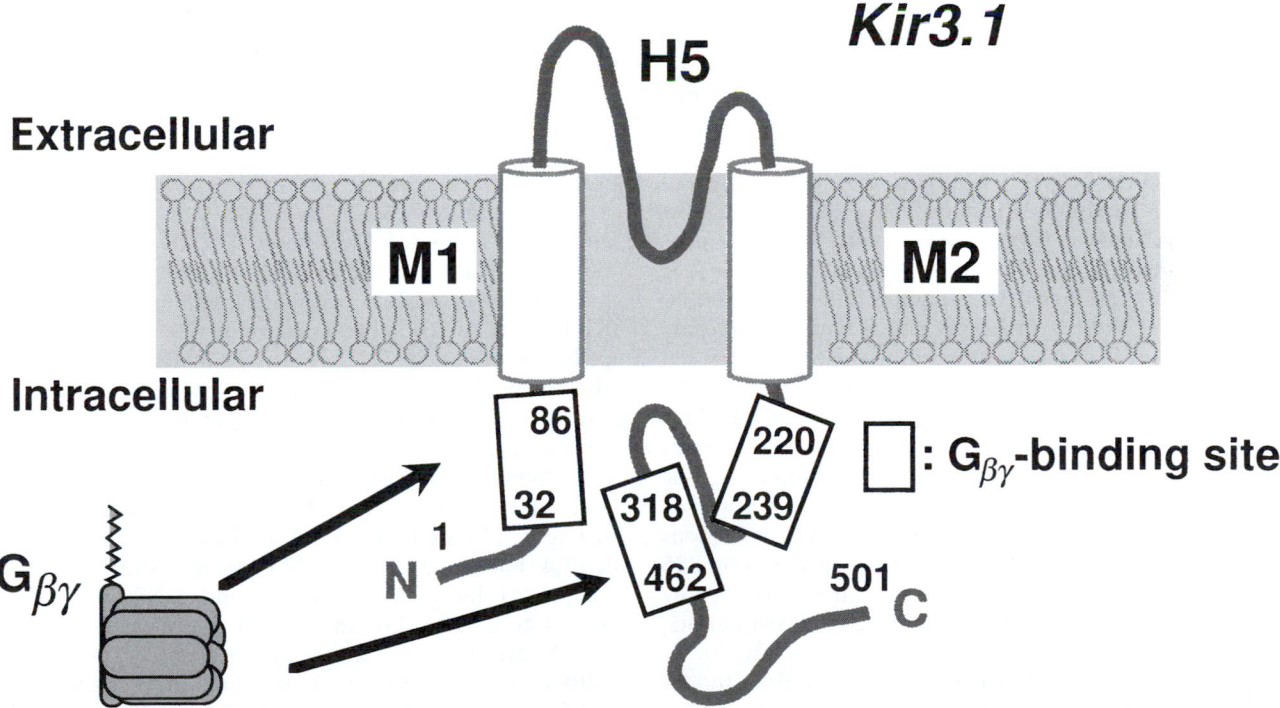

FIGURE 3. Schematic representation of the $G_{\beta\gamma}$-binding sites on the Kir3.1 subunit. The $G_{\beta\gamma}$-binding site in the N-terminal domain of Kir3.1 was reported to locate at a position between 32 and 86 (Slesinger *et al.,* 1995) and at a position between 34 and 86 (Huang *et al.,* 1997). In the C-terminus, two segments have been identified to bind to $G_{\beta\gamma}$. The proximal portion is the position between 220 and 239, which is in the vicinity of the domain corresponding to that of Kir3.4, exhibited the strongest $G_{\beta\gamma}$-binding activity (see Fig. 4; Krapivinsky *et al.,* 1998). The distal domain between 318 and 455 is the homologous region of the $G_{\beta\gamma}$-binding site of β-adrenergic receptor kinase 1 (Reuveny *et al.,* 1994; Huang *et al.,* 1995; Slesinger *et al.,* 1995; Kunkel and Peralta, 1995). This portion is separated into two fragments corresponding to amino acids 318 and 374 and amino acids 390 and 462 (Huang *et al.,* 1997). The former segment, whose homolog is found in the other Kir3.x subunits, seems to correspond to $G_{\beta\gamma}$ binding and the latter, which is unique to Kir3.1, enhances the affinity of the former.

functional K_G channels in *Xenopus* oocytes even when co-expressed with $G_{\beta1\gamma2}$. These results suggest that $G_{\beta\gamma}$ directly activates a K_G channel containing Kir3.1 and that it is the C-terminus of Kir3.1 that participates in its association with $G_{\beta\gamma}$.

Inanobe and coworkers (1995) showed that $G_{\beta\gamma}$ bound to the whole C-terminal domain of Kir3.1 (between positions 180 and 501). Using fusion proteins containing different deleted mutants of the C-terminal domain of Kir3.1, Huang and colleagues (1995) narrowed down the $G_{\beta\gamma}$-binding region in the C-terminus of Kir3.1 to a 190-amino-acid stretch (between positions 273 and 462; see Fig. 3). $G_{\beta\gamma}$ interacted with the fusion protein with ~1:1 stoichiometry with a calculated equilibrium binding constant (K_d) of ~0.5 μM. They further found that the $G_{\beta\gamma}$-binding domain is composed of two separate segments between positions 318 and 374 and between positions 390 and 462. The latter segment did not show significant $G_{\beta\gamma}$-binding activity by itself but enhanced the $G_{\beta\gamma}$-binding activity of the former segment. The segment between positions 390 and 462 contains a short amino acid sequence similar to the glutamine-X-X-glutamate-arginine (Q-X-X-E-R) motif that in adenynyl cyclase 2 is considered to be critical for regulation of the enzyme by $G_{\beta\gamma}$ (Chen *et al.*, 1995).

Interestingly, $G_{\beta\gamma}$ also bound to a segment in the N-terminus of Kir3.1 (see Fig. 3; Huang *et al.*, 1995; Kunkel and Peralta, 1995). The $G_{\beta\gamma}$ binding to the N-terminus also occurred with 1:1 stoichiometry but exhibited ~ 10 times lower affinity than that of C-terminal fusion proteins. They also found that the fusion proteins of the N- and C-terminal domains bind together and synergistically enhance the $G_{\beta\gamma}$-binding activity of each other.

The interaction between $G_{\beta\gamma}$ and the cytoplasmic domains of Kir3.1 may thus underlie the $G_{\beta\gamma}$-induced activation of K_G channels. Huang and coworkers (1995) constructed synthetic peptides possessing the partial amino acid sequence of the predicted $G_{\beta\gamma}$-binding domains of the N- and C-termini. These synthetic peptides inhibited not only the binding of $G_{\beta\gamma}$ to the corresponding fusion proteins, but suppressed the Kir3.1 K_G channel currents activated by $G_{\beta1\gamma2}$. Slesinger and colleagues (1995) expressed chimeras of Kir2.1/IRK1 and Kir3.1 in *Xenopus* oocytes and examined the response of the chimeric channels to $G_{\beta1\gamma2}$. The $G_{\beta1\gamma2}$-induced increase in channel activity was observed only when the chimeras contained the N- (between positions 32 and 86) and/or the C-terminal (between positions 325 and 501) domains of Kir3.1. A similar result was reported by using chimeras of Kir3.1 and Kir2.2 (Kunkel and Peralta, 1995).

Yan and Gautam (1996) showed that G_β binds to the N-terminus of Kir3.1 using the yeast two-hybrid system. Different types of G_β interacted with the N-terminal domain of Kir3.1 with different affinities. An N-terminal fragment of 100 amino acids of G_β interacted with the N-terminal domain of Kir3.1 as effectively as the whole G_β. This N-terminal domain of G_β includes the region responsible for the interaction between G_β and G_α, according to the analysis of the crystal structure (Wall *et al.*, 1995;

Lambright *et al.*, 1996). Thus, G_α-GDP might prevent $G_{\beta\gamma}$ from interacting with K_G channels by competing with the N-terminus of Kir3.1 for the N-terminus of G_β. Binding of G_β to the C-terminus of Kir3.1 was not clearly detected in this study. Other domains of G_β or G_γ might, therefore, participate in the interaction between $G_{\beta\gamma}$ and the C-terminus of Kir3.1.

Homomeric K_G channels composed of either Kir3.2 or Kir3.4 are known to be activated by $G_{\beta\gamma}$, suggesting that Kir3.0 subunits other than Kir3.1 have the ability to interact with $G_{\beta\gamma}$. Doupnik and colleagues (1996) examined the direct binding of Kir3.4 and $G_{\beta\gamma}$ using biosensor chip technology. They prepared the fusion protein of the C-terminus of Kir3.4 (amino acids 186–419) with GST, immobilized it on the sensor chip through GST, and measured the surface plasmon resonance with different concentrations of mobile $G_{\beta\gamma}$. The K_d value was calculated to be ~800 nM. Huang and coworkers (1997) demonstrated that the proximal portions of both N- and C-termini of Kir3.2, Kir3.3, and Kir3.4, which correspond to the segments between residues 34 and 86 and between 318 and 374 of Kir3.1, bind $G_{\beta\gamma}$ (see Fig. 3). However, the C-terminal domains of Kir3.2, Kir3.3, and Kir3.4 exhibited slightly lower affinity for $G_{\beta\gamma}$ binding than that of Kir3.1, probably due to the lack of a region corresponding to the distal C-terminus of Kir3.1 (between positions 390 and 462). They also found that the C-terminal domain of Kir3.1 interacts with the N-terminus of Kir3.4 and thereby synergistically enhances the $G_{\beta\gamma}$-binding activity. Therefore, the high-affinity $G_{\beta\gamma}$-binding site in the Kir3.1/Kir3.4 heteromeric channel might be formed through the interaction of the C-terminus of Kir3.1 subunit with the N-terminus of the Kir3.1 and/or Kir3.4 subunits. This interaction may at least in part underlie the higher channel activity yielded by coexpression of Kir3.1 and Kir3.4 than their homomeric channels.

Full-length K_G (K_{ACh}) channels (composed of Kir3.1 and Kir3.4) isolated by subunit-specific antibodies from bovine atrium exhibited $G_{\beta\gamma}$-binding activity (Krapivinsky *et al.*, 1995) with a K_d value of 55 nM. The Kir3.1 and Kir3.4 subunits individually expressed in Sf9 cells also bound $G_{\beta\gamma}$ with K_d values of 125 and 50 nM, respectively. Krapivinsky and coworkers (1998) demonstrated that the association of $G_{\beta\gamma}$ and the full-length K_G channel is attenuated in the presence of synthetic peptides corresponding to partial C-terminal sequences of Kir3.1 and Kir3.4 (see Fig. 3; Fig. 4). The interaction was strongly inhibited by Kir3.1 amino acids 364–383 and Kir3.4 amino acids 209–225, which show sequences unique to individual subunits, and weakly by Kir3.1 amino acids 220–239 and Kir3.4 amino acids 226–245, which are highly conserved between both subunits. Although these findings might not exclude that the region of Kir3.4 amino acids 209–225 forms the $G_{\beta\gamma}$-binding site, it was also shown that a corresponding peptide from the G protein–insensitive inward-rectifier Kir2.1 also attenuated the interaction of $G_{\beta\gamma}$ and K_G channels. Therefore, the definitive binding sites of $G_{\beta\gamma}$ to the K_G channel subunits have not yet been identified. These biochemical examinations, nevertheless, clearly reflect some aspects of the molecular events underlying the activation

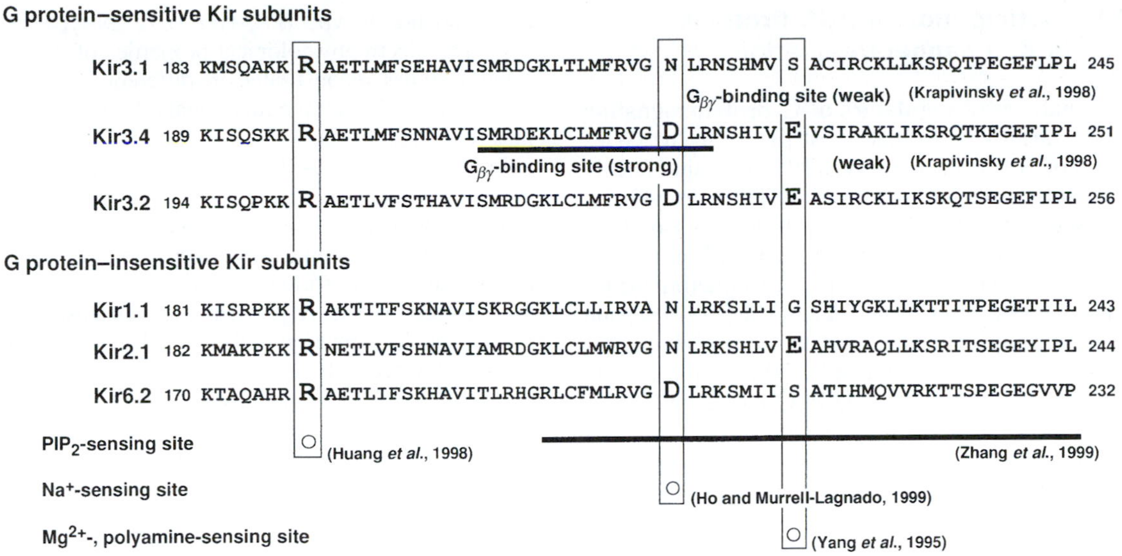

G protein–sensitive Kir subunits

Kir3.1 183 KMSQAKK **R** AETLMFSEHAVISMRDGKLTLMFRVG **N** LRNSHMV **S** ACIRCKLLKSRQTPEGEFLPL 245
$G_{\beta\gamma}$-binding site (weak)   (Krapivinsky et al., 1998)

Kir3.4 189 KISQSKK **R** AETLMFSNNAVISMRDEKLCLMFRVG **D** LRNSHIV **E** VSIRAKLIKSRQTKEGEFIPL 251
$G_{\beta\gamma}$-binding site (strong)   (weak)   (Krapivinsky et al., 1998)

Kir3.2 194 KISQPKK **R** AETLVFSTHAVISMRDGKLCLMFRVG **D** LRNSHIV **E** ASIRCKLIKSKQTSEGEFIPL 256

G protein–insensitive Kir subunits

Kir1.1 181 KISRPKK **R** AKTITFSKNAVISKRGGKLCLLIRVA **N** LRKSLLI **G** SHIYGKLLKTTITPEGETIIL 243

Kir2.1 182 KMAKPKK **R** NETLVFSHNAVIAMRDGKLCLMWRVG **N** LRKSHLV **E** AHVRAQLLKSRITSEGEYIPL 244

Kir6.2 170 KTAQAHR **R** AETLIFSKHAVITLRHGRLCFMLRVG **D** LRKSMII **S** ATIHMQVVRKTTSPEGEGVVP 232

PIP$_2$-sensing site   (Huang et al., 1998)   (Zhang et al., 1999)

Na$^+$-sensing site   (Ho and Murrell-Lagnado, 1999)

Mg^{2+}-, polyamine-sensing site   (Yang et al., 1995)

FIGURE 4. Alignment of amino acid sequences of the proximal C-terminal region of the inward-rectifier Kir subunits. The underlined sequences correspond to the synthetic peptides that attenuate the interaction of $G_{\beta\gamma}$ and either Kir3.1 or Kir3.4 (Krapivinsky et al., 1998). The critical residues to respond to PIP$_2$ (Huang et al., 1998; Zhang et al., 1999) and Na$^+$ ions (Ho and Murrell-Lagnado, 1999) as well as Mg^{2+} ions and polyamine (Yang et al., 1995) have been identified in this hydrophilic region. Bold letters represent these responding amino acid residues and the corresponding residues are boxed.

of the K_G channel by $G_{\beta\gamma}$. The multiple contacts between K_G channel subunits and $G_{\beta\gamma}$ seem to play an important functional role in K_G channel activation.

V. Modulation of K_G Channel Activity by PIP$_2$ and Na$^+$ Ions

K_G channels can be activated by cytosolic ATP and Na$^+$ (Sui et al., 1996). The former is considered to maintain the concentration of phosphatidylinositol 4,5-bisphosphate (PIP$_2$) close to the intracellular surface of the plasma membrane (Sui et al., 1998). The application of anti-PIP$_2$ antibody to the cytoplasmic side of inside-out patches attenuated the activities of constitutively active Kir channels (Kir1.1 and Kir2.1) as well as receptor-stimulated K_G channels (Kir3.1/3.4 and Kir3.2) which are constitutively inactive. PIP$_2$ added to the patches recovered the channel activity that had been inhibited by the antibody (Huang et al., 1998). Huang and colleagues (1998) also demonstrated that the C-terminus of Kir1.1 has PIP$_2$-binding activity with an affinity higher than that of Kir3.1. This may reflect the different character of the constitutively active Kir and inactive Kir channels. A requirement for PIP$_2$ in Kir channel activation was also found in the ATP-sensitive Kir channels in native cardiac myocytes (Hilgemann and Ball, 1996) and in the reconstitution system using Kir6.2 and SUR1 (Shyng and Nichols, 1998; Baukrowitz et al., 1998). Therefore, the interaction of PIP$_2$ and Kir channels might be one of the determinants for the constitutive activity of Kir channels (Huang et al., 1998). The hydrophilic proximal C-terminal region of Kir, especially at the Arg residue (position 190 in Kir3.1) that is conserved among all Kir subunits, has been identified to confer

PIP$_2$-binding activity (see Fig. 4). Zhang and colleagues (1999) also reported that Kir3.4 has a PIP$_2$-binding site between positions 214 and 252. When the amino residues were replaced by those of Kir2.1, the mutated channels exhibited higher affinity for PIP$_2$ than the Kir3.4 homomeric channel (see Fig. 4). The run-down K_G channel could be reactivated by PIP$_2$ but not by $G_{\beta\gamma}$. It was also observed that K_G channel activity was reduced by anti-PIP$_2$ antibody, which was not effectively activated by $G_{\beta\gamma}$. In contrast to these observations of reconstituted Kir channel activity, one report has presented the ineffectiveness of PIP$_2$ on the native K_{ACh} channel in rat atrial myocytes (Kim and Bang, 1999). Thus, although it might be possible that the $G_{\beta\gamma}$-induced K_G channel activation is due to $G_{\beta\gamma}$ increasing affinity of the channel for PIP$_2$ (Huang et al., 1998; Zhang et al., 1999), this remains rather controversial at present.

The recombinant Kir3.1/Kir3.4 channels in *Xenopus* oocytes and the native atrial K_{ACh} channel can be activated by internal Na$^+$ with an EC$_{50}$ of ~40 mM (Sui et al., 1996). Na$^+$ seems to activate not only heteromeric Kir3.1/Kir3.4 and Kir3.1/Kir3.2 channels, but also the homomeric Kir3.2 channel (Ho and Murrell-Lagnado, 1999). Because the Hill coefficients are close to 2 for the heteromers and close to 4 for the homomeric channel, the Na$^+$-sensitive site on the K_G channel is thought to be located on either of the Kir3.2 or Kir3.4 subunits, but not on Kir3.1. Ho and Murrell-Lagnado (1999) identified the Na$^+$ sensor of the K_G channel at the Asp 226 in the proximal C-terminal region of Kir3.2 (see Fig. 4). The corresponding amino acid is also Asp in Kir3.4, but Asn in Kir3.1. Thus, although $G_{\beta\gamma}$ generated by receptor stimulation is the predominant physiological trigger for activation of K_G channels, intracellular PIP$_2$ and Na$^+$ may also play roles in the modulation of K_G channel activity.

VI. Participation of RGS Proteins in K_G Channel Regulation

A growing number of **regulators of G protein signaling** (RGS) proteins are found to bind most tightly to the GDP-AlF$_4^-$ complex of G_α via the conserved RGS domain (Tesmer *et al.*, 1997) and accelerate the intrinsic GTP hydrolysis activity of G_α (see Fig. 1) (Berman *et al.*, 1996; Watson *et al.*, 1996). Originally, RGS proteins were identified as antagonists of G protein signaling: SST2 in the yeast pheromone response (Chan and Otte, 1982; Dietzel and Kurjan, 1987) and EGL-10 in the egg-laying behavior in *Caenorhabditis elegans* (Koelle and Horvitz, 1996), for example. Other well-known G protein signals such as $G_{\alpha q}$-mediated PLC activation (Huang *et al.*, 1997; Yan *et al.*, 1997), $G_{\alpha i}$-linked inhibition of adenylyl cyclase (Huang *et al.*, 1996) and $G_{\alpha q}$- and $G_{\alpha i}$-regulated MAP kinase activation (Druey *et al.*, 1996; Ingi *et al.*, 1998; Zhang *et al.*, 1999) also have been shown to be attenuated in the presence of overexpressed RGS proteins. The native K_{ACh} channel is known to exhibit rapid turn-on and turn-off responses to ACh (Kurachi, 1995; Yamada *et al.*, 1998), whereas the deactivation time constant of reconstituted K_{ACh} channel and m$_2$-muscarinic receptor in *Xenopus* oocytes is about 40-fold slower (Doupnik *et al.*, 1996). When RGS proteins were coexpressed with the heteromeric Kir3.1/Kir3.2 channel and $G_{i/o}$-coupled receptor in *Xenopus* oocytes or mammalian cell lines, both turn-on and turn-off rates of K_G channel to receptor stimulation were accelerated (Doupnik *et al.*, 1997; Saitoh *et al.*, 1997, 1999; Herlitze *et al.*, 1999). Because the RGS proteins behave as GTPase-activating proteins and thus reduce the quantity of GTP-bound G_α, their acceleration of the turn-off response could be due to the accelerated sequestration of $G_{\beta\gamma}$ by the increase of GDP-bound G_α. The acceleration of the turn-on rate by RGS proteins is not clearly understood (Chung *et al.*, 1998). The modulation of G protein signaling by RGS proteins seems to be an important factor in the regulation of the K_G channels in myocytes and neurons; its precise effect must now be examined.

VII. G-Protein-Inhibition of Calcium Channels

Calcium ions (Ca^{2+}) flowing through voltage-gated Ca^{2+} channels are a trigger for the rapid release of neurotransmitters, contraction, secretion of hormones, and initiation and propagation of action potentials in both nonexcitable and excitable cells (Hille, 1992; Dolphin, 1995). Ca^{2+} channel activity is controlled by a variety of cellular signaling elements such as protein kinase A^{pkA}-and C^{pkC}-dependent phosphorylation, second messengers including cGMP and arachidonic acids, and G proteins. These modulatory effects vary depending on the type of Ca^{2+} channels; for example, β-adrenergic stimulation enhances dihydropyridine-sensitive high-voltage-activated L-type Ca^{2+} channels in cardiac myocytes, while various neurotransmitters attenuate the activity of ω-conotoxin GVIA-sensitive N-type Ca^{2+} channels in dorsal root ganglion neurons. Voltage-dependent Ca^{2+} channels are composed of α_1, β, α_2/δ, and γ subunits. The α_1 subunit forms the pore structure. So far, 10 isoforms of Ca^{2+} channel α_1 subunit have been isolated: T-type (α_{1G}, α_{1H}, α_{1I}), L-type (α_{1S}, α_{1C}, α_{1D},

α_{1F}), and non-L-type (P/Q-type, α_{1A}; N-type, α_{1B}; R-type, α_{1E}). The electrophysiological properties of these isolated α_1 subunit clones reconstituted in different expression systems correspond well with the functional differences of the subfamilies (Dolphin, 1995).

In cardiac myocytes, β_1-adrenergic receptor stimulation obviously enhances the activity of L-type Ca^{2+} channels. This action is mediated by the successive reactions from receptor to the channel via second messengers: an agonist-occupied β_1-receptor dissociates trimeric G$_s$ to G$_{\alpha s}$-GTP and G$_{\beta\gamma}$, leading to the activation of adenylyl cyclase, an increase of intracellular cAMP, and the activation of PKA-mediated phosphorylation of the α_1 subunit of an L-type Ca^{2+} channel, which results in an increase of channel activity (Osterrieder *et al.*, 1982). In neurons, inhibitory neurotransmitters can attenuate the activity of the voltage-dependent Ca^{2+} channels at presynaptic nerve terminus (Bean, 1989; Hille, 1992; Takahashi and Momiyama, 1993; Wheeler *et al.*, 1994; Dolphin, 1995). These neurotransmitter receptors, which couple to G$_{i/o}$, include D$_2$-dopamine, α_2-adrenergic, GABA$_B$, 5-HT$_{1A}$, m$_2$-muscarinic, neuropeptide Y, A$_1$-purinergic, μ- and γ-opiate receptors. The actions of these receptors are prevented by PTX treatment, indicating that G proteins link the receptors and the Ca^{2+} channels. Heterogenous signaling systems are involved in neurotransmitter-mediated modulation of neuronal Ca^{2+} channel current (Hille, 1992). In some cases, G proteins may directly interact with the neuronal Ca^{2+} channels and inhibit them, while in other cases the modulation of the channel activity involves phosphorylation of the α_1 subunit (Stea *et al.*, 1995). In cases where G proteins are thought to inhibit the Ca^{2+} channel directly, the characteristic alterations of the Ca^{2+} current by such agonists as noradrenaline and adenosine includes slowing the activation kinetics and shifting the voltage dependency of activation gating in the positive direction (Kasai and Aosaki, 1989; Bean, 1989). The effect could also be interpreted as evidence that the agonists confer a voltage- and time-dependent inhibitory gating mechanism on the Ca^{2+} channel (Kasai, 1992).

VIII. G Protein $\beta\gamma$ Subunits Inhibit Neuronal Ca^{2+} Channels

Ikeda (1996) and Herlitze *et al.* (1996) provided convincing evidence to indicate that G protein-inhibition of neuronal Ca^{2+} channels is mediated by G$_{\beta\gamma}$, although G$_\alpha$-GTP had been proposed to mediate the response (Hescheler *et al.*, 1987; Kleuss *et al.*, 1991). Ikeda (1996) measured the N-type Ca^{2+} current in rat superior cervical ganglion (SCG) neurons that had been injected with various combinations of the plasmids of G protein subunits. In normal SCG cells noradrenaline (NA) slows voltage-dependent Ca^{2+} channel activation. Co-injection of G$_{\beta 1}$ and G$_{\gamma 2}$ plasmids, but not either alone, resulted in a Ca^{2+} current whose kinetic properties resembled those in the presence of NA. Furthermore, the wild-type G$_{\alpha o}$ plasmid did not change the effect of NA on the Ca^{2+} current, but the mutant G$_{\alpha o}$, which has higher affinity to GTP than GDP, did, probably due to the sequestration of G$_{\beta\gamma}$ generated by NA stimulation by GDP-bound G$_{\alpha o}$, but not by GTP-bound G$_{\alpha o}$ Herlitze and colleagues (1996) examined the effect of G$_{\beta\gamma}$ on the activity of the P/Q-type Ca^{2+} channel reconstituted with α_{1A}, β_{1b}, and

α_2/δ subunits in tsA-201 cells. Coexpression of $G_{\beta 2 \gamma 3}$ caused the alteration of the P/Q-type Ca^{2+} channel current in a way similar to application of GTPγS: both caused a depolarizing shift of the activation curve and decreased the steepness of voltage-dependent activation of the channels. In the presence of $G_{\beta \gamma}$, an agonist did not further affect the Ca^{2+} currents. Their results strongly indicate that the $G_{\beta \gamma}$ subunit is the functional arm for inhibition of both N- and P/Q-type Ca^{2+} channels by receptor stimulations.

IX. Direct Interaction of Voltage-Gated Ca^{2+} Channels and $G_{\beta \gamma}$

In the heterologous reconstitution system, the Ca^{2+} channel β subunit markedly affects the transmembrane Ca^{2+} current passing through the α_1 subunit (Mori et al., 1991; Castellano et al., 1993; Stea et al., 1994): The association of the β subunit with the α_1 subunit increases the current amplitude, shifts the voltage dependence of activation gating to hyperpolarized potentials, and causes voltage-dependent inactivation. The β subunit directly binds to a cytoplasmic link between repeat I and II of the α_1 subunit which is called AID (the α_1 subunit interaction domain) (Pragnell et al., 1994). The site that links the β subunit to the α_1 subunit has been mapped to a small region composed of 50 amino acid residues which is conserved among all β subunits (De Waard et al., 1994). The effects of the β subunit on the Ca^{2+} current are the opposite of those associated with G protein modulation of the channels (Bean, 1989). The G-protein-modulation of neuronal Ca^{2+} channels might be mediated through the interaction of $G_{\beta \gamma}$ and the β subunit.

The GABA$_B$ agonist (−)-baclofen inhibits voltage-dependent Ca^{2+} channel activity in cultured rat dorsal root ganglion neurons. Campbell and coworkers (1995) attempted to deplete β subunits from these cells by injecting an antisense oligonucleotide that reduced the Ca^{2+} current amplitude, but significantly enhanced the (−)-baclofen-induced inhibition of the channel activity. This observation indicates that G-protein-mediated inhibition of Ca^{2+} channels is prevented by the β subunit. Roche and colleagues (1995) found that the Ca^{2+} currents in Xenopus oocytes expressing either α_{1A} (P/Q-type) or α_{1B} (N-type) subunits, but not α_{1C} (L-type) subunits, are enhanced by the injection of GDPβS. GDPβS binds to G_α and blocks the exchange reaction of GDP for GTP (Gilman, 1987). The increase of Ca^{2+} current by GDPβS, therefore, could be considered to be due to the reduction of free $G_{\beta \gamma}$ from basally active G proteins in the oocytes. Because this GDPβS-induced enhancement of α_1 channel activity was prevented by the coexpression of the channel β_3 subunit, they speculated that the β subunit occupies the $G_{\beta \gamma}$-binding site(s) of the α_1 subunit. An antagonistic effect of the β subunit on μ-opioid receptor-induced inhibition of α_{1A} Ca^{2+} channel has also been shown in Xenopus oocytes by Bourinet and group (1996). These observations strongly support the hypothesis that the Ca^{2+} channel β subunit and $G_{\beta \gamma}$ bind to the pore-forming α_1 subunit in a competitive manner and/or that the β subunit induces conformational changes in the α_1 subunit that interfere with the interaction between $G_{\beta \gamma}$ and α_1 subunit.

Zamponi and colleagues (1997) and De Waard and colleagues (1997) examined the localization of the $G_{\beta \gamma}$-binding sites in the cytoplasmic region of α_1 subunits. Both found that the repeat I–II linker regions of α_{1A}, α_{1B}, and α_{1E} N-type Ca^{2+} channels specifically interact with $G_{\beta \gamma}$, while the corresponding regions in α_{1S} and α_{1C} L-type Ca^{2+} channels do not. Furthermore, the AID of G-protein-sensitive Ca^{2+} channel α_1 subunits (α_{1A}, α_{1B}, and α_{1E}) contains the putative $G_{\beta \gamma}$-binding motif (Q-X-X-E-R) (Chen et al., 1995) (Fig. 5A). De Waard and colleagues (1997) found that the amino acid residues

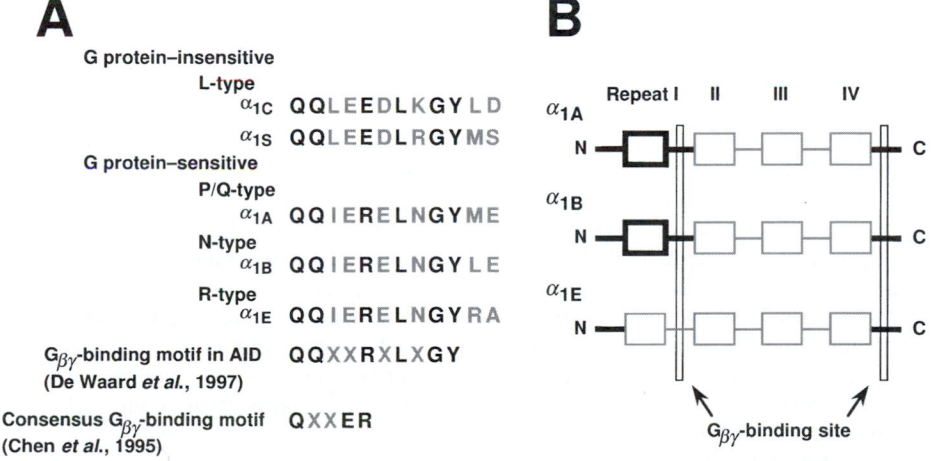

FIGURE 5. G-protein-modulation of the voltage-dependent Ca^{2+} channels. (A) Alignment of amino acid sequences of the repeat I–II linker of the α_1 subunit of voltage-dependent Ca^{2+} channels. De Waard and coworkers (1997) reported the critical amino acids (Q-Q-X-X-R-X-L-X-G-Y) in the AID for the association of $G_{\beta \gamma}$. The region contains the $G_{\beta \gamma}$-binding motif identified in adenylyl cyclase 2 (Chen et al., 1995). (B) Schematic representation of the α_1 subunits. In addition to the repeat I–II linker, the cytoplasmic C-termini of α_{1A}, α_{1B}, and α_{1E} subunits also exhibit $G_{\beta \gamma}$-binding activity (Qin et al., 1997). Both regions are boxed. The regions represented with black lines were found to confer aspects of G-protein-sensitivity to the voltage-dependent Ca^{2+} channels (Zhang et al., 1997; Page et al., 1998; Cantí et al., 1999).

critical for the association of $G_{\beta\gamma}$ and the α_1 subunit locate to the N-terminal part of the AID and that the essential motif in the AID site for binding to $G_{\beta\gamma}$ is Q Q X X R X L X G Y (Fig. 5A). They showed that the substitution of Arg (R) at position 5 (R5) in the AID of the α_{1A} subunit to Glu (E), which corresponds to this sequence in the G-protein-insensitive α_{1C} subunit, resulted in the loss of the G-protein-responsiveness of the α_{1A} subunit (De Waard et al., 1997). However, not only the AID but also the downstream sequence in the repeat I–II linker of the α_1 subunit was shown to mediate the interaction between the α_1 subunit and $G_{\beta\gamma}$ (Zamponi et al., 1997; De Waard et al., 1997). The repeat I–II linker is also the target of PKC-mediated phosphorylation (Stea et al., 1995). Zamponi and coworkers (1997) showed that the interaction of $G_{\beta\gamma}$ and the downstream portion of the AID site is attenuated by PKC-mediated phosphorylation, suggesting that two different signals—the direct modulation by G proteins and the PKC-mediated phosphorylation—converge at the repeat I–II linker to modulate the Ca^{2+} channel activity. These findings indicate that the association of $G_{\beta\gamma}$ with the repeat I–II linker is critical for the G-protein-inhibition of Ca^{2+} channel activity.

Although Zamponi and colleagues (1997) and De Waard and colleagues (1997) showed the importance of the AID in the G protein modulation of α_{1A} Ca^{2+} channel activity, this finding remains controversial. Zhang and coworkers (1997) reported that the replacement of R5 in the AID of the α_{1B} subunit with E, which corresponds to that of G-protein-insensitive α_1 subunits (α_{1S} and α_{1C}), and the exchange of the repeat I–II linker between α_{1A} and α_{1B} subunits do not significantly affect the G-protein sensitivity of the subunit (Fig. 5A, B). They also showed that the substitution of both repeat I and the C-terminal cytoplasmic tail of the α_{1A} subunit with those of α_{1B} alters the kinetics and the G-protein-responsiveness in a similar way to α_{1B} subunit and vice versa. Thus, not only the linker between repeat I and II, but also both repeat I and the C-terminal tail of the α_1 subunit, seem to confer the G-protein-responsiveness on the α_1 subunit. However, even when both regions of the α_{1B} subunit were substituted with those of the G-protein-insensitive α_{1C} subunit, the chimera channel could still respond to G-protein-stimulation (Zhang et al., 1997). This observation indicates the possibility that the other domains are also involved in the G-protein–mediated inhibition of the Ca^{2+} channels. In addition to the $G_{\beta\gamma}$-binding site in the repeat I–II linker, Qin and group (1997) found that $G_{\beta\gamma}$ also binds to the cytoplasmic domain (the position between 2036 and 2074 of α_{1E}) of C-terminal tails of G-protein-sensitive α_1 subunits (α_{1A}, α_{1B}, and α_{1E}) (Fig. 5B). The β subunit of the Ca^{2+} channel also associates with this $G_{\beta\gamma}$-binding region of the α_1 subunit in a competitive manner. The functional implication of the interaction of $G_{\beta\gamma}$ and the C-terminal tails of α_1 subunits in the G-protein-modulation of the channel activity was examined in chimera in which the C-terminal tail of α_{1E} was replaced with that of α_{1C} (EEEEc) and the substitution of the repeat I–II linker of α_{1E} with that of α_{1C} (EcEEE) (Qin et al., 1997) (see Fig. 5B). The chimera EEEEc lost the G-protein responsiveness, but the chimera EcEEE did not. Furthermore, they showed that the deletion mutant of α_{1E} at position 2035 in the cytoplasmic C-terminus expresses Ca^{2+} channel activity without the G-protein responsiveness, but that the G-protein-mediated inhibition remains in a mu-

tant deleted at position 2063. Therefore, the cytoplasmic domain in the C-terminus of the α_1 subunit also seems to participate in the G-protein-mediated modulation of Ca^{2+} channel activity. In addition to these observations, recent studies by Page and colleagues (1998) and Cantí and colleagues (1999) reported that the intracellular N-terminus of α_{1B} and α_{1E} subunits is the major determinant in the G protein modulation of the Ca^{2+} channel activity and that the repeat I–II linker is ineffective (see Fig. 5B). Therefore, $G_{\beta\gamma}$ appears to associate with α_1 subunits through multiple contacts which are located throughout the entire α_1 subunit, including the intracellular N- and C-terminus, transmembrane domain I, and the linker region of the repeat I–II. The multiple contacts of these molecules should correspond to the G-protein–induced modulation of Ca^{2+} channel activity with the slowing of activation kinetics, the shift of activation gating in the positive direction, and agonist-induced voltage-dependent inhibitory gating in the presynaptic terminus.

X. Conclusion

Progress in the last decade has revealed definite components in the receptor-operated membrane-delimited modulation of both inward-rectifier K^+ and voltage-dependent Ca^{2+} channels. Ligand-occupied PTX-sensitive receptors produce $G_{\beta\gamma}$ dissociated from G_α and the direct interaction of $G_{\beta\gamma}$ with the channel subunits leads the K_G channel to activate and the Ca^{2+} channel to alter its kinetics. In spite of this understanding of the molecular basis of G-protein-regulation, many questions remain: how the intramolecular elements regulate the channel activity, how $G_{\beta\gamma}$ interacts with and evokes the structural changes of the channels, and how the conformational change evokes the G-protein modulation of the ion channel activity. RGS proteins affect the activity of G proteins, so RGS proteins might act as accessory elements in G-protein-regulated systems—not only K_G channels, but also Ca^{2+} channels (Bünemann and Hosey, 1999). Each member of the RGS protein family shares the RGS domain, whereas the individual members contain pleiotropical protein motifs, like the PDZ (PSD-95, disc-large, and ZO-1) domain in RGS12, the PH (pleckstrin homology) domain in p113RhoGEF, and the GRK (G protein-coupled receptor kinases) domain in GRK2 (Hepler, 1999). Therefore, these RGS proteins might temporally and spatially influence the G-protein-dependent modulation of ion channel activity in myocytes and neurons. The question of the selectivity of signal transduction also remains; both K^+ and Ca^{2+} channels are under the control of many receptors even in a single cell. Thus, we need to further clarify the regulatory mechanism of ion channels at the molecular as well as the cellular levels.

Bibliography

Baukrowitz, T., Schulte, U., Oliver, D., Herlitze, S., Krauter, T., Tucker, S. J., Ruppersberg, J. P., and Fakler, B. (1998). PIP$_2$ and PIP as determinants for ATP inhibition of K_{ATP} channels. Science **282**, 1141–1144.

Bean, B. P. (1989). Neurotransmitter inhibition of neuronal calcium currents by changes in channel voltage dependence. *Nature* **340**, 153–156.

Berman, D. M., Wilkie, T. M., and Gilman, A. G. (1996). GAIP and RGS4 are GTPase-activating proteins for the G_i subfamily of G protein α subunits. *Cell* **86**, 445–452.

Breitwieser, G., and Szabo, G. (1985). Uncoupling of cardiac muscarinic and β-adrenergic receptors from ion channels by a guanine nucleotide analogue. *Nature* **317**, 538–540.

Bourinet, E., Soong, T. W., Stea, A., and Snutch, T. P. (1996). Determinants of the G-protein-dependent opioid modulation of neuronal calcium channels. *Proc. Natl. Acad. Sci. USA* **93**, 1486–1491.

Bünemann, M., and Hosey, M. M. (1999). Regulators of G protein signaling (RGS) proteins constitutively activate $G_{\beta\gamma}$-gated potassium channels. *J. Biol. Chem.* **273**, 31 186–31 190.

Campbell, V., Berrow, N. S., Fitzgerald, E. M., Brickley, K., and Dolphin, A. C. (1995). Inhibition of the interaction of G protein G_o with calcium channels by the calcium channel β-subunit in rat neurones. *J. Physiol.* **485**, 365–372.

Cantí, C., Page, K. M., Stephens, G. J., and Dolphin, A. (1999). Identification of residues in the N-terminus of α_{1B} critical for inhibition of the voltage-dependent calcium channel by $G_{\beta\gamma}$. *J. Neurosci.* **19**, 6855–6864.

Castellano, A., Wei, X., Birnbaumer, L., and Perez-Reyes, E. (1993). Cloning and expression of a neuronal calcium channel β subunit. *J. Biol. Chem.* **268**, 12 359–12 366.

Chan, R. K., and Otte, C. A. (1982). Isolation and genetic analysis of *Saccharomyces cerevisiae* mutants supersensitive to G1 arrest by a factor and alpha factor pheromones. *Mol. Cell. Biol.* **2**, 11–20.

Chen, J., Devivo, M., Dingus, J., Harry, A., Li, J., Sui, J., Carty, D. J., Bland, J. L., Exton, J. H., Stoffel, R. H., Inglese, J., Lefkowitz, R. J., Logothetis, D. E., Hidebrandt, J. D., and Inyengar, R. (1995). A region of adenylyl cyclase 2 critical for regulation by G protein $\beta\gamma$ subunits. *Science* **268**, 1166–1169.

Chung, H.-H., Yu, M., Jan, Y. N., and Jan, L. Y. (1998). Evidence that the nucleotide exchange and hydrolysis cycle of G proteins causes acute desensitization of G-protein gated inward rectifier K$^+$ channels. *Proc. Natl. Acad. Sci. USA* **95**, 11 727–11 732.

Corey, S., and Clapham, D. E. (1998). Identification of native atrial G-protein-regulated inwardly rectifying K$^+$ (GIRK4) channel homomultimers. *J. Biol. Chem.* **273**, 27 499–27 504.

Dascal, N., Schreibmayer, W., Lim, N. F., Wang, W., Chavkin, C., DiMagno, L., Labarca, C., Kieffer, B. L., Graveriaxu-Ruff, C., Trollinger, D., Lester, H. A., and Davidson, N. (1993). Atrial G protein-activated K$^+$ channel: expression cloning and molecular properties. *Proc. Natl. Acad. Sci. USA* **90**, 10 235–10 239.

Del Castillo, J., and Katz, B. (1955). Production of membrane potential changes in the frog's heart by inhibitory nerve impulses. *Nature* **175**, 1035.

De Waard, M., Pragnell, M., and Campbell, K. P. (1994). Ca^{2+} channel regulation by a conserved β subunit domain. *Neuron* **13**, 495–503.

De Waard, M., Liu, H., Walker, D., Scott, V. E. S., Gurnett, C. A., and Campbell, K. P. (1997). Direct binding of G-protein $\beta\gamma$ complex to voltage-dependent calcium channels. *Nature* **385**, 446–450.

Dietzel, C., and Kurjan, J. (1987). Pheromonal regulation and sequence of the *Saccharomyces cerevisiae* SST2 gene: a model for desensitization to pheromone. *Mol. Cell. Biol.* **7**, 4169–4177.

Dolphin, A. C. (1995). Voltage-dependent calcium channels and their modulation by neurotransmitters and G proteins. *Exp. Physiol.* **80**, 1–36.

Dolphin, A. C. (1998). Mechanism of modulation of voltage-dependent calcium channels by G proteins. *J. Physiol.* **506**, 3–11.

Doupnik, C. A., Dessauer, C. W., Slepak, V. Z., Gilman, A. G., Davidson, N., and Lester, H. A. (1996). Time resolved kinetics of direct $G_{\beta1\gamma2}$ interactions with the carboxyl terminus of Kir3.4 inward rectifier K$^+$ channel subunits. *Neuropharmacol.* **35**, 923–931.

Doupnik, C. A., Davidson, N., Lester, H. A., and Kofuji, P. (1997). RGS proteins reconstitute the rapid gating kinetics of $G_{\beta\gamma}$-activating inwardly rectifying K$^+$ currents. *Proc. Natl. Acad. Sci. USA* **94**, 10 461–10 466.

Druey, K. M., Blumer, K. J., Kang, V. H., and Kehrl, J. H. (1996). Inhibition of G-protein-mediated MAP kinase activation by a new mammalian gene family. *Nature* **379**, 742–746.

Duprat, F., Lesage, F., Guillemare, E., Fink, M., Hugnot J.-P., Bigay, J., Lazdunski, M., Romey, G., and Barhanin, J. (1995). Heterologous multimeric assembly is essential for K$^+$ channel activity of neuronal and cardiac G-protein-activated inward rectifiers. *Biochem. Biophys. Res. Commun.* **212**, 657–663.

Gilman, A. G. (1987). G proteins: transducers of receptor-generated signals. *Annu. Rev. Biochem.* **56**, 615–649.

Hedin, K, E., Lim, N, F., and Clapham, D. E. (1996). Cloning of a *Xenopus laevis* inwardly rectifying K$^+$ channel subunit that permits GIRK1 expression of I_{KACh} currents in oocytes. *Neuron* **16**, 423–329.

Hepler, J. R. (1999). Emerging roles for RGS proteins in cell signalling. *Trends Pharmacol. Sci.* **20**, 376–382.

Herlitze, S., Garcia, D. E., Mackie, K., Hille, B., Scheuer, T., and Catterall, W. A. (1996). Modulation of Ca^{2+} channels by G-protein $\beta\gamma$ subunits. *Nature* **380**, 258–262.

Herlitze, S., Ruppersberg, J. P., and Mark, M. D. (1999). New roles of RGS2, 5 and 8 on the ratio-dependent modulation of recombinant GIRK channels expressed in *Xenopus* oocytes. *J. Physiol.* **517**, 341–352.

Hescheler, J., Rosenthal, W., Trautwein, W., and Schultz, G. (1987). The GTP-binding protein, Go, regulates neuronal calcium channels. *Nature* **325**, 445–447.

Hilgemann, D. W., and Ball, R. (1996). Regulation of cardiac Na$^+$, Ca^{2+} exchange and KATP potassium channels by PIP$_2$. *Science* **273**, 956–959.

Hille, B. (1992). G protein-coupled mechanisms and nervous signaling. *Neuron* **9**, 187–195.

Ho, I. H., and Murrell-Lagnado, R. D. (1999). Molecular determinants for sodium-dependent activation of G protein-gated K$^+$ channels. *J. Biol. Chem.* **274**, 8639–8648.

Huang, C., Hepler, J. R., Gilman, A. G., and Mumby, S. M. (1997). Attenuation of G_i and G_q-mediated signaling by expression of RGS4 or GAIP in mammalian cells. *Proc. Natl. Acad. Sci. USA* **94**, 6159–6163.

Huang, C.-L., Slesinger, P. A., Casey, P. J., Jan, N. Y., and Jan, L. Y. (1995). Evidence that direct binding of $G_{\beta\gamma}$ to the GIRK1 G-protein–gated inwardly rectifying K$^+$ channel is important for channel activation. *Neuron* **15**, 1133–1143.

Huang, C.-L., Jan, Y. N., and Jan, L. Y. (1997). Binding of the G protein $\beta\gamma$ subunit to multiple regions of G protein-gated inward-rectifying K$^+$ channels. *FEBS Lett.* **405**, 291–298.

Huang, C.-L, Feng, S., and Hilgemann, D. W. (1998). Direct activation of inward rectifier potassium channels by PIP$_2$ and its stabilization by $G_{\beta\gamma}$. *Nature* **391**, 803–806.

Hutter, O. R., and Trautwein, W. (1955). Vagal and sympathetic effects on the pacemaker fibers in the sinus venosus of the heart. *J. Gen. Physiol.* **39**, 715–733.

Ikeda, S. R. (1996). Voltage-dependent modulation of N-type calcium channels by G protein $\beta\gamma$ subunits. *Nature* **380**, 255–258.

Inanobe, A., Morishige, K.-I., Takahashi, N., Ito, H., Yamada, M., Takumi, T., Nishina, H., Takahashi, K., Kanaho, Y., Katada, T., and Kurachi, Y. (1995). $G_{\beta\gamma}$ directly binds to the carboxyl terminus of the G protein-gated muscarinic K$^+$ channel, GIRK1. *Biochem. Biophys. Res. Commun.* **217**, 1022–1028.

Inanobe, A., Yoshimoto, Y., Horio, Y., Morishige, K.-I., Hibino, H., Matsumoto, S., Tokunaga, Y., Maeda, T., Hata, Y., Takai, Y., and Kurachi, Y. (1999). Characterization of G-protein-gated K$^+$ channels composed of Kir3.2 subunits in dopaminergic neurons of the substantia nigra. *J. Neurosci.* **19**, 1006–1017.

Ingi, T., Krumins, A. M., Chidiac, P., Brothers, G. M., Chung, S., Snow, B. E., Barnes, C. A., Lanahan, A. A., Siderovski, D. P., Ross, E. M., Gilman, A. G., and Worley, P. F. (1998). Dynamic regulation of RGS2 suggests a novel mechanism in G-protein signaling and neuronal plasticity. *J. Neurosci.* **18**, 7178–7188.

Ito, H., Sugimoto, T., Kobayashi, I., Takahashi, K., Katada, T., Ui, M., and Kurachi, Y. (1991). On the mechanism of basal and agonist-induced activation of the G protein-gated muscarinic K+ channel in atrial myocytes of guinea pig heart. *J. Gen. Physiol.* **98**, 517–533.

Ito, H., Tung, R. T., Sugimoto, T., Kobayashi, I., Takahashi, K., Katada, T., Ui, M., and Kurachi, Y. (1992). On the mechanism of G protein $\beta\gamma$ subunit activation of the muscarinic K+ channel in guinea pig atrial cell membrane: comparison with the ATP-sensitive K+ channel. *J. Gen. Physiol.* **99**, 961–983.

Jan, L. Y., and Jan, Y. N. (1997). Receptor-regulated ion channels. *Curr. Opin. Cell Biol.* **9**, 155–160.

Kasai, H., and Aosaki, T. (1989). Modulation of Ca-channel current by an adenosine analog mediated by a GTP-binding protein in chick sensory neurons. *Pflügers Arch.* **414**, 145–149.

Kasai, H. (1992). Voltage- and time-dependent inhibition of neuronal calcium channels by a GTP-binding protein in a mammalian cell line. *J. Physiol.* **448**, 189–209.

Kennedy, M. E., Nemec, J., and Clapham, D. E. (1996). Localization and interaction of epitope-tagged GIRK1 and CIR inward rectifier K+ channel subunits. *Neuropharmacol.* **35**, 831–839.

Kim, D., and Bang, H. (1999). Modulation of rat atrial G protein-coupled K+ channel function by phospholipids. *J. Physiol.* **517**, 59–74.

Kleuss, C., Hescheler, J., Ewel, C., Rosenthal, W., Schultz, C., and Wittig, B. (1991). Assignment of G-protein subtypes to specific receptors inducing inhibition of calcium currents. *Nature* **353**, 43–48.

Koelle, M. R., and Horvitz, H. R. (1996). EGL-10 regulates G protein signaling in the *C. elegans* nervous system and shares a conserved domain with many mammalian proteins. *Cell* **84**, 115–125.

Kofuji, P., Davidson N., and Lester, H. A. (1995). Evidence that neuronal G-protein-gated inwardly rectifying K+ channels are activated by $G_{\beta\gamma}$ subunits and function as heteromultimers. *Proc. Natl. Acad. Sci. USA* **92**, 6542–6546.

Krapivinsky, G., Gordon, E. A., Wickman, K., Velimirović, B., Krapivinsky, L., and Clapham, D. E. (1995). The G-protein-gated atrial K+ channel I_{KACh} is a heteromultimer of two inwardly rectifying K+-channel proteins. *Nature* **374**, 135–141.

Krapivinsky, G., Krapivinsky, L., Wickman, K., and Clapham, D. E. (1995). $G_{\beta\gamma}$ binds directly to the G protein-gated K+ channel, I_{KACh}. *J. Biol. Chem.* **270**, 29 059–29 062.

Krapivinsky, G., Kennedy, M. E., Nemec, J., Medina, I., Krapivinsky, L., and Clapham, D. E. (1998). $G_{\beta\gamma}$ binding to GIRK4 subunit is critical for G protein-gated K+ channel activation. *J. Biol. Chem.* **273**, 16 946–16 952.

Kubo, Y., Reuveny, E., Slesinger, P. A., Jan, Y. N., and Jan, L. Y. (1993). Primary structure and functional expression of a rat G-protein-coupled muscarinic potassium channel. *Nature* **364**, 802–806.

Kunkel, M. T., and Peralta, E. G. (1995). Identification of domains conferring G protein regulation on inward rectifier potassium channels. *Cell* **83**, 443–449.

Kurachi, Y., Nakajima, T., and Sugimoto, T. (1986a). Acetylcholine activation of K+ channels in cell-free membrane of atrial cells. *Am. J. Physiol.* **251**, H681–H684.

Kurachi, Y., Nakajima, T., and Sugimoto, T. (1986b). On the mechanism of activation of muscarinic K+ channels by adenosine in isolated atrial cells: involvement of GTP-binding proteins. *Pflügers Arch.* **407**, 264–274.

Kurachi, Y., Nakajima, T., and Sugimoto, T. (1986c). Role of intracellular Mg^{2+} in the activation of muscarinic K+ channel in cardiac atrial cell membrane. *Pflügers Arch.* **407**, 572–574.

Kurachi, Y., Ito, Y., Sugimoto, T., Katada, T., and Ui, M. (1989). Activation of atrial muscarinic K+ channels by low concentrations of $\beta\gamma$ subunits of rat brain G protein. *Pflügers Arch.* **413**, 325–327.

Kurachi, Y. (1995). G protein regulation of cardiac muscarinic potassium channel. *Am. J. Physiol.* **269**, C821–C830.

Lambright, D. G., Sondek, J., Bohm, A., Skiba, N. P., Hamm, H. E., and Sigler, P. B. (1996). The 2.0 Å crystal structure of a heteromeric G protein. *Nature* **379**, 311–319.

Lesage, F., Duprat, F., Fink, M., Guillemare, E., Coppola, T., Lazdunski, M., and Hugnot, J. P. (1994). Cloning provides evidence for a family of inward rectifier and G-protein coupled K+ channels in brain. *FEBS Lett.* **353**, 37–42.

Lesage, F., Guillemare, E., Fink, M., Duprat, F., Meurteaux, C., Fosset, M., Romey, G., Barhanin, J., and Lazdunski, M. (1995). Molecular properties of neuronal G-protein-activated inwardly rectifying K+ channels. *J. Biol. Chem.* **270**, 28 660–28 667.

Loewi, O. (1921). Über humorale Übertragbarkeit der Herznervenwirkung. *Pflügers Arch.* **189**, 239–242.

Loewi, O., and Navaratil, E. (1926). Übertragbarkeit der Herznevenwirking, X. Mitteilung. Über das Schicksal des Vagusstoffs. *Pflügers Arch.* **214**, 678–688.

Logothetis, D. E., Kurachi, Y., Galper, Y., Neer, E. J., and Clapham, D. E. (1987). The $\beta\gamma$ subunits of GTP-binding proteins activate the muscarinic K+ channel in heart. *Nature* **325**, 321–326.

Mori, Y., Friedrich, T., Kin, M.-S., Mikami, A., Nakai, J., Ruth, P., Bosse, E., Hofmann, F., Flockerzi, V., Furiichi, T., Mikoshiba, K, Imoto, K., Tanabe, T., and Numa, S. (1991). Primary structure and functional expression from complementary DNA of a brain calcium channel. *Nature* **350**, 398–402.

Nicoll, R. A., Malenka, R. C., and Kauer, J. A. (1990). Functional comparison of neurotransmitter receptor subtypes in mammalian central nervous system. *Physiol. Rev.* **70**, 513–565.

Noma, A., and Trautwein, W. (1978). Relaxation of the ACh-induced potassium current in the rabbit sinoatrial node cell. *Pflügers Arch.* **377**, 193–200.

Osterrieder, W., Yang, Q. F., and Trautwein, W. (1981). The time course of the muscarinic response to ionophoretic acetylcholine application to the S-A node of the rabbit heart. *Pflügers Arch.* **389**, 283–291.

Osterrieder, W., Hescheler, B. J., Trautwein, W., Flockerzi, V., and Hofmann, F. (1982). Injection of subunits of cyclic AMP-dependent protein kinase into cardiac myocytes modulates Ca^{2+} current. *Nature* **298**, 576–578.

Page, K. M., Cantí, C., Stephens, G. J., Berrow, N. S., and Dolphin, A. C. (1998). Identification of the amino terminus of neuronal Ca^{2+} channel α_1 subunits α_{1B} and α_{1E} as an essential determinant of G protein modulation. *J. Neurosci.* **18**, 4815–4824.

Pfaffinger, P. J., Martin, J. M., Hunter, D. D., Nathanson, N. M., and Hille, B. (1985). GTP-binding proteins couple cardiac muscarinic receptors to a K channel. *Nature* **317**, 536–538.

Pragnell, M., De Waard, M., Mori, Y., Tanabe, T., Snutch, T. P., and Campbell, K. P. (1994). Calcium channel β-subunit binds to a conserved motif in the I–II cytoplasmic linker of the α_1-subunit. *Nature* **368**, 67–70.

Qin, N., Platano, D., Okese, R., Stefani, E., and Birnbaumer, L. (1997). Direct interaction of G$\beta\gamma$ with a C-terminal G$\beta\gamma$-binding domain of the Ca^{2+} channel α_1 subunit is responsible for channel inhibition by G protein-coupled receptors. *Proc. Natl. Acad. Sci. USA* **94**, 8866–8871.

Reuveny, E., Slesinger, P. A., Inglese, J., Molales, J. M., Iñiguez-Lluhi, J. A., Lefkowitz, R. J., Bourne, H. R., Jan, Y. N., and Jan, L. Y. (1994). Activation of the cloned muscarinic potassium channel by G protein $\beta\gamma$ subunits. *Nature* **390**, 525–529.

Roche, J. P., Anantharam, V., and Treistman, S. N. (1995). Abolition of G protein inhibition of α_{1A} and α_{1B} calcium channels by co-expression of the β_3 subunit. *FEBS Lett.* **371**, 43–46.

Sackman, B., Noma, A., and Trautwein, W. (1983). Acetylcholine activation of single muscarinic K$^+$ channels in isolated pacemaker cells of the mammalian heart. *Nature* **303,** 250–253.

Saitoh, O., Kubo, Y., Miyatani, Y., Asano, T., and Nakata, H. (1997). RGS8 accelerates G-protein-mediated modulation of K$^+$ currents. *Nature* **390,** 525–529.

Saitoh, O., Kubo, Y., Odagiri, M., Ichikawa, M., Yamagata, K., and Sekine, T. (1999). RGS7 and RGS8 differentially accelerate G protein-mediated modulation of K$^+$ currents. *J. Biol. Chem.* **274,** 9899–9904.

Schultz, G., Rosenthal, W., Hescheler, J., and Trautwein, W. (1990). Role of G proteins in calcium channel modulation. *Annu. Rev. Physiol.* **52,** 275–292.

Shyng, S.-L., and Nichols, C. G. (1998). Membrane phospholipid control of nucleotide sensitivity of K$_{ATP}$ channels. *Science* **282,** 1138–1141.

Slesinger, P. A., Reuveny, E., Jan, Y. N., and Jan, L. Y. (1995). Identification of structural elements involved in G protein gating of the GIRK1 potassium channel. *Neuron* **15,** 1145–1156.

Spring, D. J., and Neer, E. J. (1994). A 14-amino acid region of the G protein γ subunit is sufficient to confer selectivity of γ binding to the β subunit. *J. Biol. Chem.* **269,** 22 882–22 886.

Stea, A., Tomlinson, W. J., Soong, T. W., Bourinet, E., Dubel, S. J., Vincent, S. R., and Snutch, T. P. (1994). Localization and functional properties of a rat brain α_{1A} calcium channel reflect similarities to neuronal Q- and P-type channels. *Proc. Natl. Acad. Sci. USA* **91,** 10 576–10 580.

Stea, A., Soong, T. W., and Snutch, T. P. (1995). Determinants of PKC-dependent modulation of a family of neuronal calcium channels. *Neuron* **15,** 929–940.

Sui, J. L., Chan, K. W., and Logothetis, D. E. (1996). Na$^+$ activation of the muscarinic K$^+$ channel by a G-protein-independent mechanism. *J. Gen. Physiol.* **108,** 381–391.

Sui, J. L., Petit Jacques, J., and Logothetis, D. E. (1998). Activation of the atrial K$_{ACh}$ channel by the $\beta\gamma$ subunits of G proteins or intracellular Na$^+$ ions depends on the presence of phosphatidylinositol phosphates. *Proc. Natl. Acad. Sci. USA* **95,** 1307–1312.

Takahashi, T., and Momiyama, A. (1993). Different types of calcium channels mediate central synaptic transmission. *Nature* **366,** 156–158.

Tesmer, J. J. G., Berman, D. M., Gilman, A. G., and Sprang, S. R. (1997). Structure of RGS bound to AlF$_4^-$-activated G$_{i\alpha1}$: stabilization of the transition state for GTP hydrolysis. *Cell* **89,** 251–261.

Trautwein, W., and Dudel, J. (1958). Zum Mechanismus der Membranwirkung des Acetylcholines as der Herzmusklefaster. *Pflügers Arch.* **266,** 324–334.

Tucker, S. J., Pessia, M., Moorhouse, A. J., Gribble, F., Ashcroft, F. M., Maylie, J., and Adelman, J. P. (1996). Heteromeric channel formation and Ca^{2+}-free media reduce the toxic effect of the *weaver* K$_{ir}$ 3.2 allele. *FEBS Lett.* **390,** 253–257.

Velimirovic, B. M., Gordon, E. A., Lim, A. F., Navarro, B., and Clapham, D. E. (1996). The K$^+$ channel inward rectifier subunits form a channel similar to neuronal G protein-gated K$^+$ channel. *FEBS Lett.* **379,** 31–37.

Wall, M. A., Coleman, D. E., Lee, E., Iñiguez-Lluhi, J. A., Posner, B. A., Gilman, A. G., and Sprang, S. R. (1995). The structure of the G protein heterotrimer G$_{i\alpha1}\beta_1\gamma_2$. *Cell* **83,** 1047–1058.

Watson, N., Linder, M. E., Druey, K. M., Kehrl, J. H., and Blumer, K. J. (1996). RGS family members: GTPase-activating proteins for heteromeric G-protein α-subunits. *Nature* **383,** 172–175.

Wheeler, D. B., Randall, A., and Tsien, R. W. (1994). Roles of N-type and Q-type Ca^{2+} channels in supporting hippocampal synaptic transmission. *Science* **264,** 107–111.

Wickman, K., and Clapham, D. E. (1995). Ion channel regulation by G proteins. *Physiol. Rev.* **75,** 865–885.

Wickman, K., Iñiguez-Lluhi, J., Davenport, P., Taussig, R. A., Krapivinsky, G. B., Linder, M. E., Gilman, A., and Clapham, D. E. (1994). Recombinant G$_{\beta\gamma}$ activates the muscarinic-gated atrial potassium channel I_{KACh}. *Nature* **368,** 255–257.

Yamada, M., Ho, Y.-K., Lee, R. H., Kontani, K., Takahashi, K., Katada, T., and Kurachi, Y. (1994). Muscarinic K$^+$ channels are activated by $\beta\gamma$ subunits and inhibited by the GDP-bound form of α subunit of transducin. *Biochem. Biophy. Res. Commun.* **200,** 1484–1490.

Yamada, M., Inanobe, A., and Kurachi, Y. (1998). G protein regulation of potassium ion channels. *Pharmacol. Rev.* **50,** 723–757.

Yamada, M., Jahangir, A., Hosoya, Y., Inanobe, A., Katada, T., and Kurachi, Y. (1993). G$_K$* and brain G$_{\beta\gamma}$ activate muscarinic K$^+$ channel through the same mechanism: intracellular substance-mediated mechanism of inward rectification. *J. Biol. Chem.* **268,** 24 551–24 554.

Yan, K., and Gautam, N. (1996). A domain on the G protein β subunit interacts with both adenylyl cyclase 2 and the muscarinic atrial potassium channel. *J. Biol. Chem.* **271,** 17 597–17 600.

Yan, Y., Chi, P. P., and Bourne, H. R. (1997). RGS4 inhibits G$_q$-mediated activation of nitrogen-activated protein kinase and phophoinositide synthesis. *J. Biol. Chem.* **272,** 11924–11927.

Yang, J., Jan, Y. N., and Jan, L. Y. (1995). Control of rectification and permeation by residues in two distinct domains in an inwardly rectifying K$^+$ channel. *Neuron* **14,** 1047–1054.

Zamponi, G. W., Bourinet, E., Nelson, D., Nargeot, J., and Snutch, T. P. (1997). Crosstalk between G proteins and protein kinase C mediated by the calcium channel α_1 subunit. *Nature* **385,** 442–446.

Zhang, H., He, C., Yan, X, Mirshahi, T., and Logothetis, D. E. (1999). Activation of inwardly rectifying K$^+$ channels by distinct PtdIns(4,5)P$_2$ interactions. *Nature Cell Biol.* **1,** 183–188.

Zhang, J.-F., Ellinor, T. P., Aldrich, R. W., and Tsien, R. W. (1997). Multiple structural elements in voltage-dependent Ca^{2+} channels support their inhibition by G proteins. *Neuron* **17,** 991–1003.

Zhang, Y., Neo, S. Y., Han, J., Yaw, L. P., and Lin, S.-C. (1999). RGS16 attenuates G$_{\alpha q}$-dependent p38 mitogen-activated protein kinase activation by platelet-activating factor. *J. Biol. Chem.* **274,** 2851–2857.

Noritsugu Tohse, Hisashi Yokoshiki, and Nicholas Sperelakis

35

Developmental Changes in Ion Channels

I. Introduction

Cellular functions and tissue structures change dramatically during development. Ion channels are responsible for cellular signaling and maintenance of the intracellular environment. For example, the Ca^{2+} channels allow Ca^{2+} influx into the cell, which acts as a second messenger that affects several structures: activation of enzymes, activation of some ion channels, and activation of the contractile proteins. The ion channels change during both the embryonic/fetal period and the neonatal period. These developmental changes include changes in the types, number, and kinetic properties of the ion channels. The developmental changes of ion channels are clearly observed in excitable cells (i.e., cardiomyocytes, skeletal muscle fibers, neurons) because their resting potential and action potential are progressively greatly altered during the developmental stages. For example, resting potential increases in amplitude during development, and large changes occur in the action potential rate of rise, overshoot, and duration. In general, the rate of rise increases markedly, the overshoot increases, and the duration decreases during development. This chapter focuses primarily on the ion channels of cardiomyocytes, skeletal muscle fibers, and neurons, where most is known about the developmental changes.

II. Cardiomyocytes

A. Resting Potential

In the early embryonic period, the heart is tubular in shape. This heart tube has no distinction yet between atria and ventricles. In the middle embryonic period, the heart tube twists and the cardiac loop is constructed. In this period, the ventricular portion becomes distinguished from the atrial portion.

The electrophysiological properties are also altered during development. The **resting potential** (RP) of the ventricular cells in the early embryonic/fetal period is low (e.g.,

−40 to −50 mV), and there is a gradual hyperpolarization during development. Finally, in the late embryonic period, the RP becomes nearly the level of adult cells (around −80 mV). During the hyperpolarization of the RP, a decrease in the permeability ratio for Na^+ and K^+ (P_{Na}/P_K ratio) has been observed (Sperelakis and Shigenobu, 1972; Sperelakis and Haddad, 1995). However, the developmental changes in the RP cannot be accounted for by changes in the intracellular ion concentrations because $[K^+]_i$ is already high in the early embryonic period (Sperelakis and Shigenobu, 1972; Sperelakis and Haddad, 1995). Although the Na^+-K^+ pump specific activity was found to be low in the early embryonic period (Sperelakis and Lee, 1971), the Na^+-K^+ pump is sufficient to maintain a high $[K^+]_i$ and low $[Na^+]_i$ because of the less leaky membrane (i.e., high in resistance) (Sperelakis and Shigenobu, 1972). Therefore, the developmental change in the RP is due to changes in membrane permeability (conductance) of the ions.

In the early embryonic period, the low RP of the ventricular portion is not stable, but exhibits a spontaneous depolarization, the **pacemaker potential** (phase 4 diastolic depolarization). The maximum diastolic potential increases (hyperpolarized) and the slope of the pacemaker potential progressively decreases during embryonic development. When the RP has attained the adult level in the late embryonic period, the pacemaker potential disappears. Thus, automaticity of the ventricular cells is lost by the middle embryonic period. Possible factors in the loss of automaticity are the decrease in the P_{Na}/P_K ratio and the resultant hyperpolarization. These factors are closely related to the increase in the inward-rectifier K^+ current ($I_{K(IR)}$) and the loss of the hyperpolarization-activated inward current (I_h or I_f) (see Sections II.E and II.G).

B. Action Potential

The **action potentials** (APs) get larger and rise faster during embryonic development. These changes are caused by the hyperpolarization of the RP in the middle embryonic period and by an increase in overshoot to about +30 mV. The maximal rate of depolarization (max dV/dt) progressively increases during development, from about 20 to about 200 V/s in the late embryonic stage, which is about the adult level. However, the time course of the increase in max dV/dt is not parallel to the increase in RP. The increase in RP precedes the increase in max dV/dt by several days. Therefore, this increase in max dV/dt is not simply due to the hyperpolarization, but is produced by a much greater number (density) of tetrodotoxin (TTX)-sensitive fast Na+ channels (see Section II.C).

The duration of the AP (e.g., at 50% repolarization, APD_{50}) hardly changes in the chick during development. The same is true of the guinea pig heart as well as many other mammalian species. However, in human atrial cells, the APD_{50} is significantly shortened (e.g., about 20%). The rat also shows a marked decrease in APD_{50} beginning in the late fetal period and extending through the first three weeks of the neonatal period, after which adultlike brief APs are attained (Kojima et al., 1990). Several factors contribute to this marked abbreviation of the AP in the rat, including increase in the transient outward current (I_{to}) and loss of the sustained component of the fast Na+ current (See Section II.C).

C. Na+ Channels

Slow Ca^{2+} channel current makes a major contribution to the upstroke of the AP in the early embryonic period. The TTX-sensitive fast Na+ current in ventricular cells increases markedly during development by a factor of about 10 in chick, rat, and murine hearts (Fujii et al., 1988; Sada et al., 1988, 1995; Conforti et al., 1993; Davies et al., 1996). Saxitoxin (STX, a specific blocker of the fast Na+ channels) binding reveals a marked increase in density of the fast Na+ channel protein during development of embryonic chick hearts (Renaud et al., 1981). This increase in number (density) of fast Na+ channels accounts for the large increase in max dV/dt of the AP that occurs during development. Therefore, the contribution of fast Na+ channels to the AP is progressively increased during development.

The heart greatly enlarges in size during development. Thus, the excitation wave must travel over longer distances in the larger hearts during the late embryonic and adult periods. A fast prolongation velocity of excitation is required to allow a synchronized contraction of ventricle, that is, to allow the heart to serve as an effective pump. The increase in max dV/dt during development would contribute to the required increase in propagation velocity. Another factor involved in propagation velocity is that the cell size (i.e., diameter) becomes much greater. It is well known that propagation velocity is a function of the square root of cell diameter (see Chapter 24).

The TTX sensitivity of the fast Na+ channels in avian cardiomyocytes (in the nanomolar range) is about 1000-fold greater than that for adult mammalian hearts (such as guinea pig), which are in the micromolar range (Sada et al., unpublished observation). This finding is in agreement with previous reports of the high sensitivity of embryonic chick hearts to TTX (Iijima and Pappano, 1979; Marcus and Fozzard, 1981; Fuji et al., 1988). The fast Na+ channels are completely blocked by 10 μM TTX in fetal rat cardiomyocytes (Conforti et al, 1993), and by 30 μM TTX in adult rat cardiomyocytes (Brown et al., 1981). It is not clear whether the high TTX sensitivity of chick embryonic hearts is due to a different isoform of the channel.

The TTX-sensitive fast Na+ current has a slow inactivating or sustained component. This component is small, but gives a relatively larger contribution in the embryonic period than in adult (Conforti et al., 1993). Sustained Na+ current, which is blocked by TTX, is observed in the early embryonic period of chicks. Reopening of some of the fast Na+ channels during a long depolarizing clamp step is one explanation for the small sustained component (Josephson and Sperelakis, 1989). The sustained component may reflect the **window current** produced by a balance between the activating (m) gate and the inactivating (h) gate (Sada et al., 1995) (see Chapter 25). In rat heart cells, the fast Na+ current has a slow inactivating component, and the time constant of the slow inactivating component decreases in neonatal cells compared to fetal cells (Conforti et al., 1993). Although the slow component of the Na+ current is small, inward current produced by the slow component helps to maintain the longer duration of the AP plateau in the fetal period. TTX, which does not affect ion channels other than the fast Na+ channels, shortens the AP duration in rat fetal cardiomyocytes (Fig. 1) (Conforti et al., 1993). A key factor that may contribute to the shortening of the AP duration during development of rat heart is the loss of the slow component of the Na+ current. However, in adult hearts, it appears that the sustained component of the Na+ current persists in the Purkinje fiber, because the AP plateau is substantially shortened by TTX. Another factor responsible is an increase in I_{to} carried primarily by K+ (see Section II.F).

D. Ca2+ Channels

Changes in the slow (L-type) Ca^{2+} current also occur during development of the heart. However, the direction of the change is opposite in avian versus mammalian hearts (Fig. 2). In rat hearts, the L-type Ca^{2+} current increases during development (Masuda et al., 1995), whereas in chick hearts it actually decreases (Tohse et al., 1992b). In chick early embryonic period, the current density of the L-type channels is 8 μA/cm², which is comparable to that in other adult animals (about 10 μA/cm²). The current density decreases during development to about 5 μA/cm² in the late embryonic period (Fig. 2B). However, the current density of the L-type Ca^{2+} channels of rat cardiomyocytes increases through the middle fetal, late fetal, and neonatal period (Masuda et al., 1995) (Fig. 2A). In mouse, the current density of L-type Ca^{2+} channels increases during the fetal period (Davies et al., 1996). Another investigation demonstrated that the current density in the neonatal period is actually larger than that in adult rat (Cohen and Lederer, 1988). That is, in development of rat heart, the current density increases, followed by a decrease. In contrast, in rabbit (Osaka and Joyner, 1991) and guinea pig (Kato et al., 1996) cardiomyocytes, the current density in the neonatal period is smaller than that in adults. Thus, the

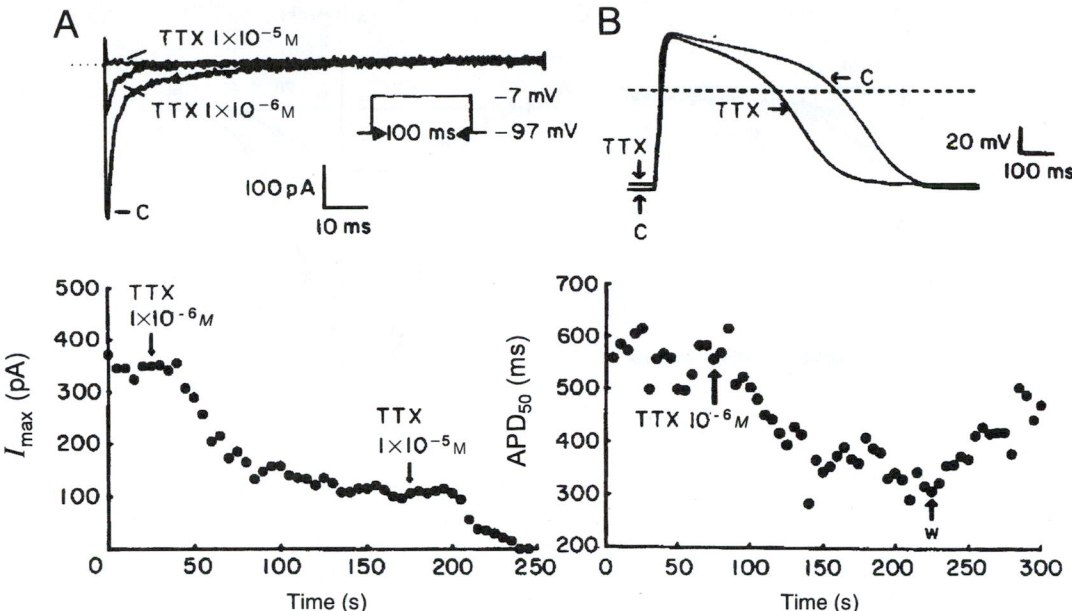

FIGURE 1. Effect of TTX on the Na⁺ current and the AP configuration recorded from a fetal rat cardiomyocyte. (A) Top panel: Superimposed current traces showing the Na⁺ current recorded before (C, Control) and after a 90-s exposure to 1 and 10 μM TTX. Holding potential (HP) was −97 mV, and the test potential was −7 mV. Bottom panel: Time course of the TTX effect. Steady-state responses were attained at about 1–2 min. (B) Effect of TTX on the AP configuration of a fetal cell recorded in current-clamp mode. Top panel: Superimposed traces (averaged from 10 consecutive records) showing APs before (C, Control) and after a 90-s exposure to 1 μM TTX. Bottom panel: Time course of the change in APD_{50} produced by TTX. Arrows indicate points of introduction and washout (w) of TTX. (Reproduced with permission from Conforti *et al.*, 1993.)

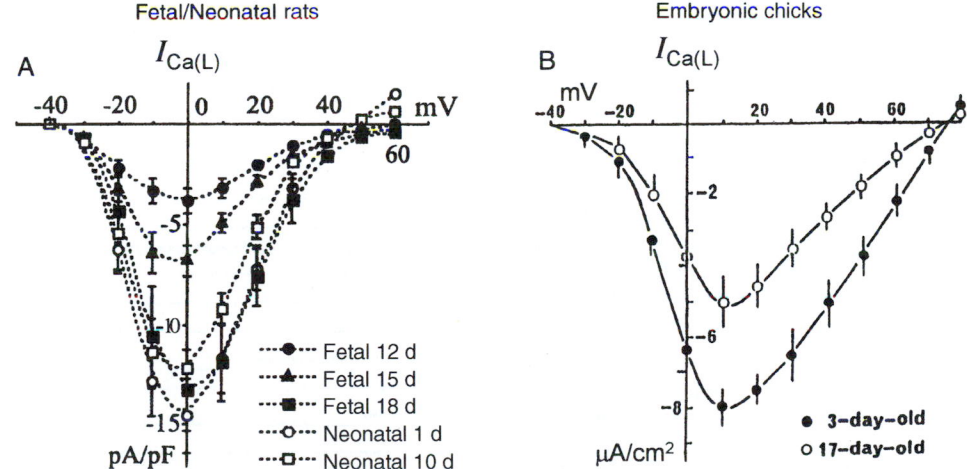

FIGURE 2. Developmental changes of $I_{Ca(L)}$ in (A) fetal/neonatal rats and (B) embryonic chicks. (A) Ba²⁺ currents through L-type Ca²⁺ channels ($I_{Ba(L)}$) were elicited by depolarizing steps from a HP of −40 mV (22 °C). Current-voltage curves (normalized as current density (in pA/pF)(mean ± SE)) are shown for the different developmental stages (from day 12 fetal to day 10 neonatal). (B) Changes in the density of $I_{Ca(L)}$ in isolated embryonic chick heart cells; 1.8 mM $[Ca^{2+}]_o$, 35 °C. (Panel A modified with permission from Masuda *et al.*, 1995. Long openings of calcium channels in fetal rat ventricular cardiomyocytes. *Pflügers Arch.* **429**, 595–597, copyright Springer-Verlag. Panel B: Tohse *et al.*, 1992b. Developmental changes in long-opening behavior of L-type Ca²⁺ channels in embryonic chick heart cells. *Circ. Res.* **71**, 376–384. Reproduced with permission. *Circulation Research*. Copyright 1992 American Heart Association.)

changes in the L-type Ca²⁺ channel density that occur during development are complex and vary from one species to another.

Other types of Ca²⁺ channels are also observed during development. In chick embryonic heart cells, it has been reported that the T-type channel is dominant in the early em-

bryonic period, but that the L-type current is dominant in the late embryonic period (Kawano and DeHaan, 1991). However, other reports indicate that the L-type current is also dominant in the early embryonic period (Tohse *et al.*, 1992a, b). In rat fetal cardiomyocytes, it was found that a substantial fraction of the total Ca²⁺ current is resistant to

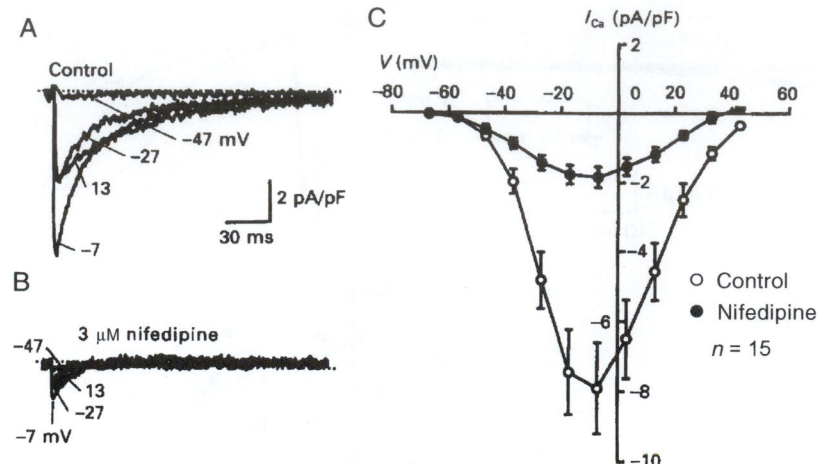

FIGURE 3.   Presence of fetal-type Ca^{2+} channels in fetal (18-day) rat cardiomyocytes. (A) Currents elicited by 300-ms (only 150 ms shown) depolarizing pulses to −47, −27, −7, and 13 mV from a HP of −87 mV. (B) In the presence of 3 μM nifedipine, a significant inward current remained at each potential. (C) Current–voltage relationship; data points given as mean ± SE. Nifedipine did not completely block the Ca^{2+} current, indicating the presence of a nifedipine-resistant Ca^{2+} current. (Reproduced with permission from Tohse *et al.*, 1992a.)

nifedipine (a relatively selective blocker of L-type Ca^{2+} channels) and other Ca^{2+} channel–blockers (Fig. 3) (Tohse *et al.*, 1992a). This nifedipine-resistant current is not blocked by an N-type Ca^{2+} channel–blocker, ω-conotoxin, and is only partially inhibited by 30 μM Ni^{2+}, which is a blocker of T-type Ca^{2+} channels. Therefore, this channel is called a **fetal-type** (F-type) Ca^{2+} channel. The F-type Ca^{2+} current is absent in adult heart cells. That is, in the fetal period, the total Ca^{2+} current has two main components: the L-type current, which is blocked by nifedipine, and the F-type current, which is not blocked by nifedipine.

In chick embryonic heart, unit conductance of the L-type Ca^{2+} channel is 26 pS (using 50 mM Ba^{2+} in the pipette), which is comparable to that in adult heart cells (Tohse and Sperelakis, 1990). The single-channel activity of the L-type Ca^{2+} channel in the embryonic cells was completely blocked by nifedipine.

The kinetics of opening of the L-type Ca^{2+} channels in embryonic heart cells is different from that in adult heart cells. Long-lasting openings of the channels occur relatively frequently, in addition to the more usual brief bursting openings observed in adult heart cells (Fig. 4). These long-lasting openings are similar to the mode 2 openings produced by Ca^{2+} agonists, such as the dihydropyridine Bay-K-8644. For example, Fig. 4 shows long openings that persist over the entire duration of the clamp pulse (i.e., 300 ms); the long openings are sometimes punctuated by brief

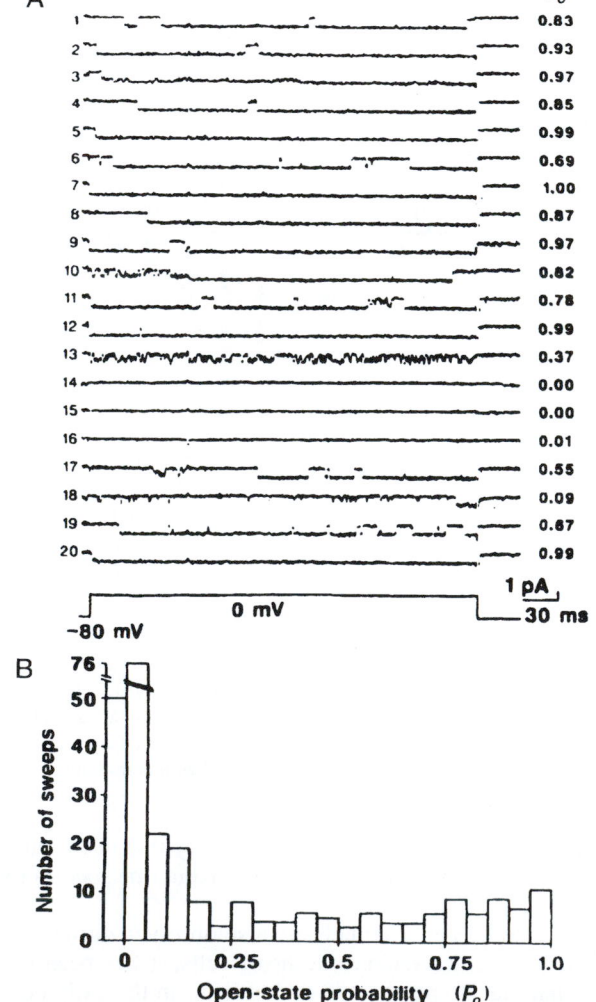

FIGURE 4.   Presence of long openings of the slow (L-type) Ca^{2+} channels in young embryonic (3-day) chick heart cell. (A) Single-channel activity elicited by consecutive command pulses to 0 mV (from a HP of −80 mV) every 2 s. Sweep-to-sweep variations of the probability of the channel opening (P_o) are given in the right-hand column. (B) A histogram of P_o data from nine cells (30 sweeps each). Note that many sweeps showed long openings and high P_o. (Reproduced with permission from Tohse and Sperelakis, 1990.)

closures. As can be seen, in many sweeps, the open probability (P_o) is close to 1.00. The long-lasting openings gradually disappear during development (Tohse *et al.*, 1992b; Masuda *et al.*, 1995).

E. Inward-Rectifier K⁺ Channels

It has been demonstrated that the inward-rectifier K⁺ current ($I_{K(IR)}$) of ventricular cells increases markedly during development in embryonic chick (Josephson and Sperelakis, 1990), fetal/neonatal rat hearts (Masuda and Sperelakis, 1993), and rabbit ventricular cells (Huynh *et al.*, 1992). The increase in $I_{K(IR)}$ is likely to be a major factor responsible for the increase in the RP (hyperpolarization) that occurs during development, concomitant with a decrease in membrane resistivity and in the membrane time constant ($\tau_m = R_m C_m$). The increase in $I_{K(IR)}$ channels can also account for the decrease in the P_{Na}/P_K ratio that occurs during development (Sperelakis and Shigenobu, 1972); namely, it results in an increase in the K⁺ permeability (P_K). Similar change in the P_{Na}/P_K ratio during development is shown in the skeletal muscle fibers (see Section III.A).

The increase in $I_{K(IR)}$ during development may be due to two factors: (1) an increase in the number of channel molecules and (2) an increase in single-channel conductance (Masuda and Sperelakis, 1993). The single-channel conductance in young fetal rat is much less than that in old fetus and neonate (Fig. 5). However, the mean open time of the channels is longer in young fetal cells than in old fetal and neonatal cells. These observations suggest that the structure (i.e., a different isoform) of the inward-rectifier K⁺ channel changes dramatically during development. On the other hand, a later study (Xie *et al.*, 1997)

suggests that the small conductance events in young fetal heart are sublevels of the large conductance channels in old fetal heart.

From neonate to adult, a further small increase is observed in rabbit ventricular cells (Huynh *et al.*, 1992), whereas some gradual decrease (due to an increase in cell membrane capacitance) is evident in rat ventricular cells (Xie *et al.*, 1997).

F. Voltage-Gated K+ Channels

Transient outward current (I_{to}) density, mainly carried by K⁺ ion, has been reported to increase during development. A substantial amount of I_{to} has been observed in early embryonic chick heart cells (Satoh and Sperelakis, 1990). In neonatal rat ventricular cells, the density of I_{to} was reported to increase (Kilborn and Fedida, 1990; Wahler *et al.*, 1994). In mouse (Wang and Duff, 1997), rabbit (Sanchez-Chapula *et al.*, 1994), and canine ventricular cells (Jeck and Boyden, 1992), a postnatal increase of I_{to} was also observed. This increase in I_{to} contributes to the abbreviation of the AP that occurs in the neonatal period.

Shal (Kv4) K⁺ channel (Kv4.1, Kv4.2 and Kv4.3) exhibits I_{to}-type K⁺ currents in several types of cells. On the other hand, I_{to} in rat heart is thought to be composed of Kv4.3 and Kv4.2 channels. In human (and probably canine) heart, the Kv4.3 (not Kv4.2) channel plays a major role in I_{to}. On the other hand, significant contributions by the Kvl.4 channel are proposed in endocardial I_{to} of ferret (and probably rat) ventricle as well as rabbit atrium. The Kvl.4 channel is characterized by its slower recovery kinetics. Consistent with these findings, the developmental changes in rat I_{to} would result from a consequence of the isoform switch from Kvl.4 to Kv4.2/Kv4.3 (Xu *et al.*, 1996; Shimoni *et al*, 1997). That is,

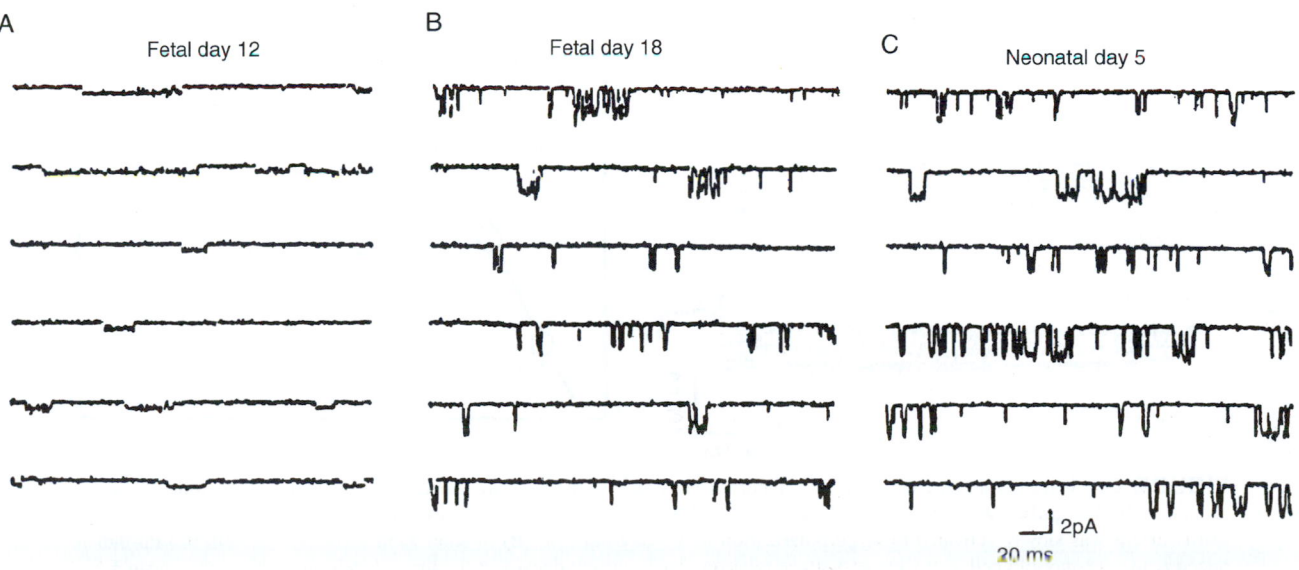

FIGURE 5. Developmental increase in $I_{K(IR)}$ in maturing rats. Single-channel activities illustrated for (A) 12-day fetal, (B) 18-day fetal, and (C) 5-day neonatal ventricular myocytes. Single-channel activities were recorded at −80 mV. (Reproduced with permission from Masuda and Sperelakis, 1993.)

the relative abundance of Kv1.4 mRNA seen in neonatal rat is attenuated during development, whereas Kv4.2/Kv4.3 predominates in adult. Thyroid hormone (T3) induces the molecular switch of I_{to}, and is thought to play an important role in postnatal development of I_{to} (Shimoni et al., 1997). In addition, basic fibroblast growth factor (bFGF), which has tyrosine kinase activity, may also promote the expression of I_{to} during development (Guo et al., 1995).

As I_{to} in neonatal rat ventricular cells is absent or very small, an ultrarapidly activating delayed-rectifier K⁺ current (I_{Kur}) predominates I_{to} only during the early neonatal period (Guo et al., 1997b). In cultured neonatal rat ventricular cells, IGF-I promotes the expression of I_{Kur} as well as Kv1.5 protein (Guo et al., 1997a). I_{Kur} in ventricular cells decreases during postnatal development, which is associated with a reduction of the Kv1.5 expression. However, no substantial change in Kv1.5 protein is also reported from neonate to adult in the same preparation (Xu et al., 1996).

In mouse fetal ventricular cells, a rapidly activating delayed-rectifier K⁺ current (I_{Kr}) is the dominant component of delayed-rectifier I_K, whereas a slowly activating delayed-rectifier K⁺ current (I_{Ks}) is lacking or very small (Davies et al., 1996; Wang et al., 1996). In early neonate (day one to three), I_{Ks} becomes dominant. However, both components disappear in adult mouse ventricular cells. Similarly, I_{Kr} in

rat ventricular cells functions during the fetal period, but is negligible in adults. On the other hand, the density of I_K (the sum of I_{Kr} and I_{Ks}) in guinea pig ventricles is smaller in fetal cells than in neonatal and adult cells, suggesting a developmental increase in I_K (Kato et al., 1996). No substantial changes in the kinetics and voltage dependency of I_K are observed during development.

G. Hyperpolarization-Activated Inward Current

The hyperpolarization-activated inward current (I_h or I_f), which is mainly carried by Na⁺ and K⁺ ions, is observed in early chick embryonic cardiomyocytes (Satoh and Sperelakis, 1993). This current progressively decreases during development and essentially disappears in the late embryonic period (Fig. 6). In rabbit sinoatrial node, a slope conductance of I_h in neonatal cells is larger than that in the adult cells (Accili et al., 1997).

The I_h is called the **pacemaker current** in adult cardiomyocytes. In Purkinje fibers, I_h plays a key role in pacemaker depolarization during the diastolic phase. In sinoatrial node cells, the contribution of I_h to pacemaker potential is still controversial (Irisawa et al., 1993) for two reasons: the time course of activation of I_h is too slow to account for the high frequency of the pacemaker, and the threshold potential

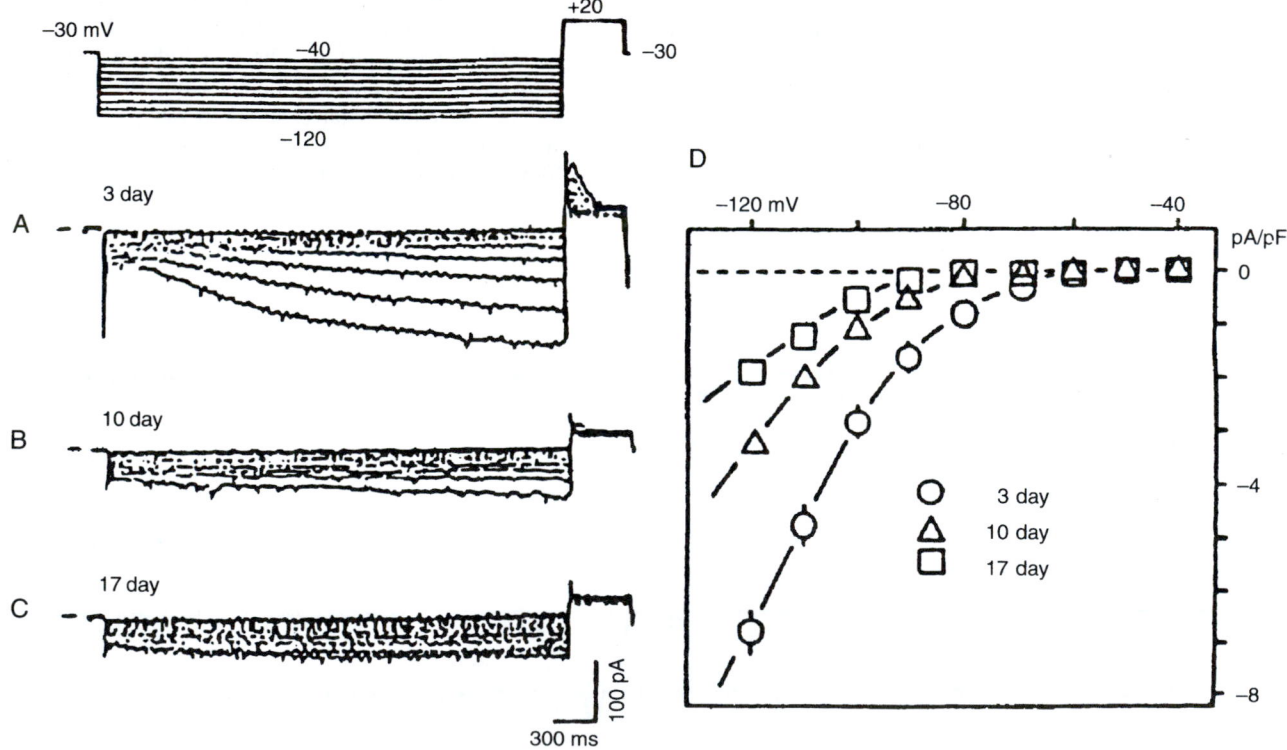

FIGURE 6. Developmental changes of the hyperpolarization-activated inward current (I_h) in young embryonic chick ventricular cells. Test pulses were applied between −40 and −120 mV, in 10-mV increments, from a HP of −30 mV. (A) A large inward current was slowly activated by hyperpolarization in a 3-day-old cell. (B) Smaller I_h in a 10-day-old cell. (C) Further reduced I_h in a 17-day-old cell. (D) Current-voltage relations for I_h current density at the three developmental stages (mean ± SE). (Reproduced from Eur. J. Pharmacol, **240,** H. Satoh and N. Sperelakis, Hyperpolarization-activated inward current embryonic chick cardiac myocytes: Developmental changes and modulation by isoproterenol and carbachol, pp. 283–290, Copyright 1993. With permission from Elsevier Science.)

for activation of I_h (close to −70 mV) is beyond the maximum diastolic potential (−60 to −70 mV) for the nodal cells. In chick embryonic cardiomyocytes, although the time course of decrease in I_h parallels the disappearance of the pacemaker potential, the contribution of I_h to the pacemaking may still be small (Satoh and Sperelakis, 1993).

H. Excitation-Contraction Coupling

Changes in the excitation-contraction coupling process also occur during development of the heart. In particular, the source of Ca^{2+} for producing contraction is altered during development (Fig. 7A) (Nakanishi et al., 1988). In fetal heart cells, the role of the sarcoplasmic reticulum (SR) is minimal, so that most of the Ca^{2+} required for contraction is derived from Ca^{2+} influx through the voltage-dependent Ca^{2+} channels (L-type and T-type) (i.e., originates from the extracellular space). In neonatal heart cells, the SR matures and plays a main role as the source of Ca^{2+} for contraction. Therefore, the Ca^{2+}-induced Ca^{2+} release from the SR compartment (Fabiato and Fabiato, 1978) becomes the more important system for contraction. That is, in adult heart cells, most of the Ca^{2+} for contraction comes from the internal SR stores. However, Ca^{2+} influx through the sarcolemma remains the determining factor for contractile force, because the Ca^{2+} influx controls the amount of Ca^{2+} released.

The spanning protein (foot protein) couples the L-type Ca^{2+} channel in the T-tubule wall membrane to the Ca^{2+}-release channel in the SR membrane. This protein has been reported to be absent in the immature fetal heart (Fig. 7B)

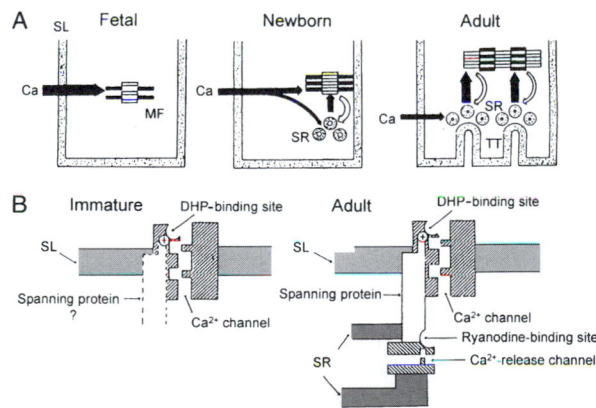

FIGURE 7. Schematic diagrams showing developmental changes in the excitation-contraction (E-C) coupling. (A) In the early fetus (left), Ca^{2+} entering the cell across the sarcolemma (SL) directly activates the myofilaments (MF). In the newborn (middle), Ca^{2+} influx across the SL induces some Ca^{2+} release from the sarcoplasmic reticulum (SR), which then contributes to the activation of the MFs. In the adult (right), the Ca^{2+} released from the SR is greater and more important. TT, T-tubule. (B) Model of E-C coupling mechanisms in immature and adult heart cells. In the immature cells (left), the protein linking the Ca^{2+} channel in the SL with the Ca^{2+}-release channel in the SR is absent. In the adult cells (right), the spanning protein is present, allowing Ca^{2+} release from the SR. The Ca^{2+} channel in the SL contains the dihydropyridine (DHP)-binding site. The sarcolemmal sensor responsible for intramembrane charge movement is shown as a plus (+) sign. (Reproduced in modified form, with permission, from (panel A) Nakanishi et al., 1988, and (panel B) Cohen and Lederer, 1988.)

(Cohen and Lederer, 1988). If so, then the release of Ca^{2+} is prevented from any SR that may be present. Thus, Ca^{2+} influx would be the main source of Ca^{2+} for contraction in the immature heart.

III. Skeletal Muscle Fibers

A. Resting Potential and Action Potential

In skeletal muscle fibers, the RP also increases during development and differentiation from the individual myoblast stage to the multinucleated myotube stage to the mature fiber. (The myoblasts fuse end to end to form the myotubes, which get progressively larger in length and diameter.) For example, the RP of cultured chick embryonic skeletal myocytes is low in the immature stage (short mononucleated myoblasts) and is dramatically hyperpolarized during differentiation (to long myotubes or mature fibers) (Fig. 8A) (Fischbach et al., 1971; Spector and Prives, 1977). The mature fibers in older cultures consist of multinucleated myotubes with cross-striations (i.e., aligned myofibrils). In the immature myoblasts, the RP generally is about −40 mV. However, the RP of the maturing myotubes is about −60 mV, which is approximately equal to that of adult skeletal muscles bathed in the same culture medium (Fig. 8A). A similar change in the RP occurs in rat skeletal myoblasts/myotubes during development (Ritchie and Fambrough, 1975). A progressive decrease in the P_{Na}/P_K ratio, which sets the RP closer to the equilibrium potential for K^+ (E_K), accounts for the developmental changes in the RP (Fig. 8B). The family of curves shown in Fig. 8B is similar to that reported in developing chick heart (Sperelakis and Shigenobu, 1972).

In human skeletal muscle (from biopsy), the RP increases (hyperpolarization) during culture (Iannaccone et al, 1987). Fetal myocytes exhibit RPs of about −35 mV in the early period of culture, and then hyperpolarize to about −50 mV in later culture. Therefore, the RP of human myocytes increases during development in vitro.

The AP configuration of skeletal muscle fibers also changes during development and differentiation. For example, in one study of chick embryonic myotubes, prolonged APs that persist for more than 500 ms, with prolonged contractions, were exhibited on day 5 in culture (Spector and Prives, 1977). By culture day 7, the myotubes exhibited primarily brief APs (less than 10 ms) and brief twitch contractions. (On day 5, small regions of the myotubes sometimes displayed brief APs with localized twitches in those regions.) Thus, the brief AP becomes more dominant during development that proceeds in culture.

A second study on cultured chick skeletal myoblasts/myotubes is depicted in Fig. 9. As can be seen, at the early stage, the AP is small and does not overshoot (the zero membrane potential level)(Fig. 9A, upper record). In the myotubes formed by fusion of myoblasts after a few more days in culture (days 7 and 11), the amplitude and maximum rate of the spikelike APs increased markedly during development while in culture (Fig. 9A, middle and lower records, and Fig. 9B). A plot of max dV/dt versus the number of days in culture is given in Fig. 9B. Since the APs were blocked by

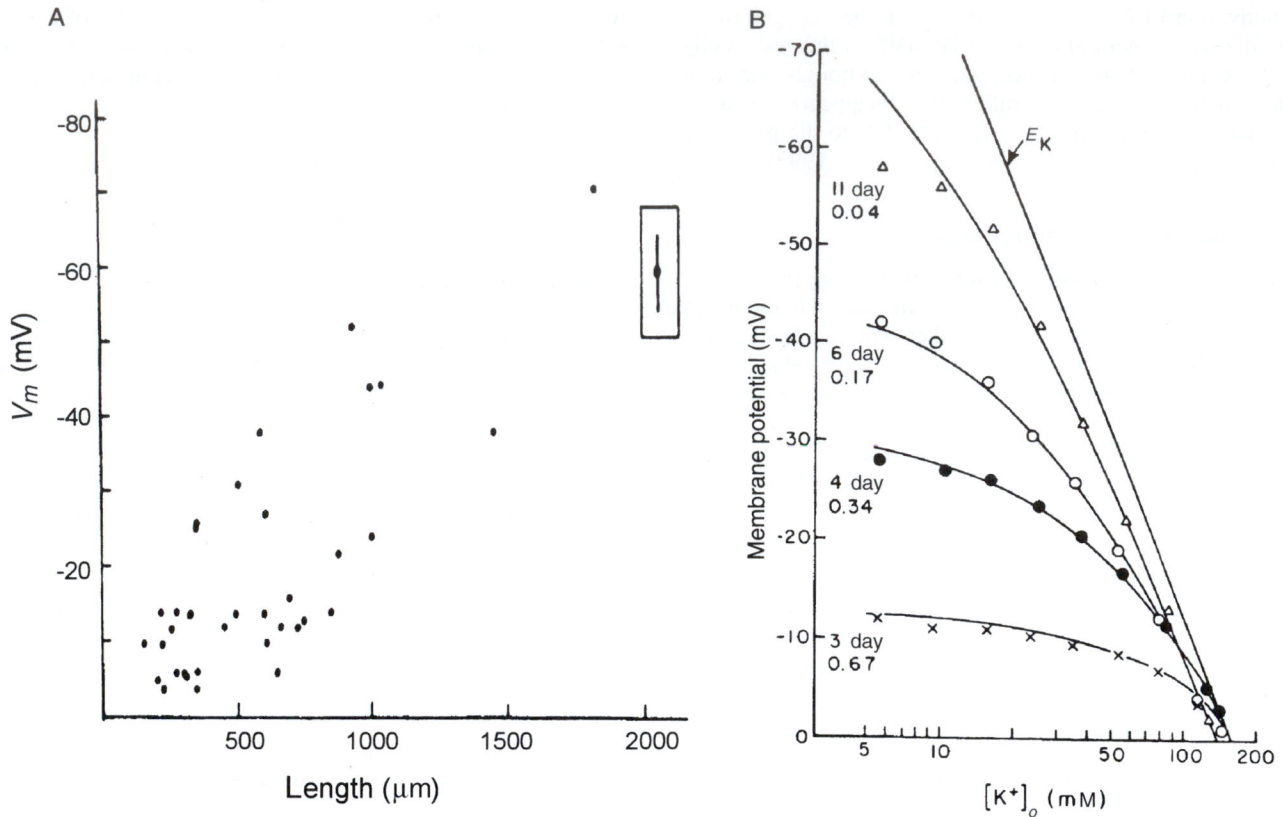

FIGURE 8. Developmental changes of resting potential in the skeletal myoblasts/myotubes from (A) embryonic chick and (B) fetal rat. (A) The relation between the RP and length of the chick skeletal myoblasts/myotubes in culture. The filled circle enclosed within the box is the mean (± 2SE) resting potential of 20 myotubes. (B) The relationship between the RP and external K^+ ion concentration ($[K+]_0$) for rat skeletal myotubes of different periods in culture. The Nernst equilibrium potential for K^+ is indicated by the straight line labeled E_K. The solid line for each myotube is the theoretical curve predicted by the Goldman equation for the P_{Na}/P_K ratios given at the left and the extrapolated $[K^+]_i$ value. [Reproduced with permission from (panel A) Fischbach *et al.*, 1971. *J. Cell. Physiol.* **78**, 289–300. Copyright 1971 John Wiley & Sons, Inc. Reprinted by permission of Wiley-Liss, Inc., a subsidiary of John Wiley & Sons, Inc., and (panel B) Ritchie and Fambrough, 1975. Reproduced from *The Journal of General Physiology* by copyright permission of The Rockefeller University Press.]

TTX in almost all myotubes tested (Fig. 9A, right side, and Fig. 9B), differentiation of the spikelike APs is due to progressively increased intensity of inward current through fast Na^+ channels (Kano and Yamamoto, 1977). That is, an increase in the number (density) of fast Na^+ channels allows a regenerative AP to be produced, and the maximum rate of rise to become faster and faster.

In myocytes of a marine tunicate (ascidian), AP duration abbreviates during the embryonic period (Greaves *et al.*, 1996). There is a progressive decrease in AP duration from the middle embryonic to the late embryonic period, and a corresponding increase in the rate of rise and fall of the AP. Spontaneous firing of APs is observed in most ascidian myocytes in the middle embryonic period. The automaticity progressively disappears during development.

B. Inward-Rectifier K^+ Channels

The RP of maturing myotubes gets closer to E_K because the P_{Na}/P_K ratio is gradually decreased during development (see Fig. 8B). These characteristics are also observed in

chick embryonic cardiomyocytes during development (Sperelakis and Shigenobu, 1972) and are produced by marked expression of the inward-rectifier K^+ channels in the surface membrane of the myocytes. Therefore, it seems likely that the hyperpolarization of the RP of skeletal myocytes in culture is produced by a similar developmental change of inward-rectifier K^+ channels.

The $I_{K(IR)}$ is present in skeletal muscle cells from early embryonic amphibian (Linsdell and Moody, 1995). Although there is a brief period (approximately 4 h) during which its density decreases, the overall trend is an increase during development.

In ascidian myocytes, $I_{K(IR)}$ exhibited dramatic changes during development (Fig. 10C) (Greaves *et al.*, 1996). In the ascidian, the inward-rectifier K^+ current is gained after fertilization of the egg. (The same change also occurs in other species.) When gastrulation ends (at 16 h after fertilization), the current density suddenly decreases from 4 to 0.5 pA/pF. After the tailbud stage (22 h after fertilization), the current density progressively increases again and reaches a value of 5 pA/pF before hatching. Because $I_{K(IR)}$ is one of the most

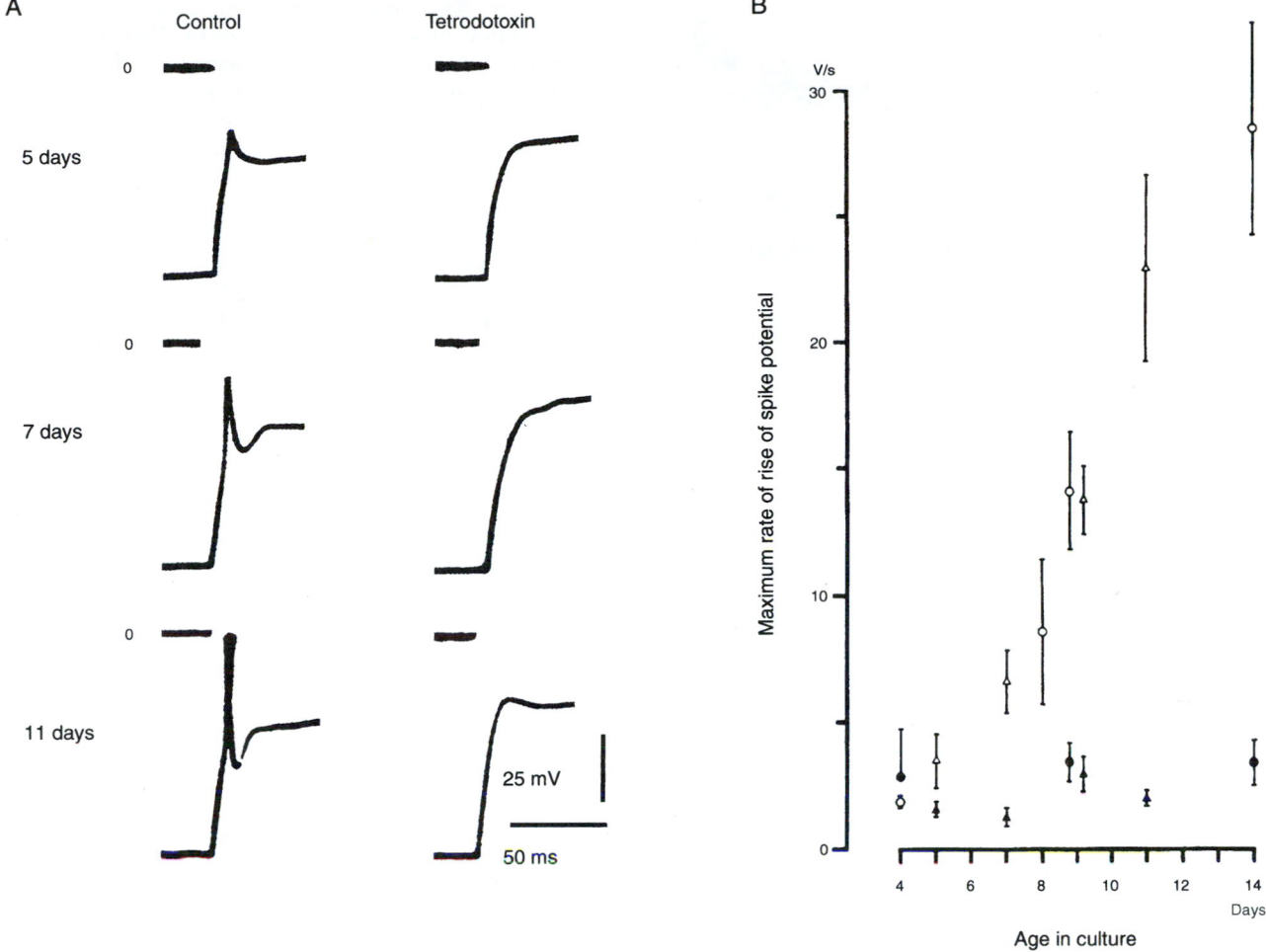

FIGURE 9. Spikelike APs and their maximum rate of rise in maturing skeletal myotubes from chick embryo. (A) Responses of three myotubes at different ages in culture: 5 days (upper), 7 days (middle), and 11 days (lower). Each pair of records is taken from the same myotube before (left) and after (right) application of TTX (10^{-7} M). Depolarizing current pulses were applied after the membrane potential was hyperpolarized to a standard level of −80 mV. The zero potential level is indicated. (B) Maximum rate of rise of spike potentials as a function of the age in culture (mean ± SE). Circles and triangles represent two different batches of cultures; filled symbol indicates presence of TTX. (From Kano and Yamamoto, 1977. Development of spike potentials in skeletal muscle cells differentiated *in vitro* from chick embryo. *J. Cell. Physiol.* **90**, 439–444. Copyright © 1977 *Journal of Cellular Physiology.* Reprinted by permission of Wiley-Liss, Inc., a subsidiary of John Wiley & Sons, Inc.)

important resting conductances (which stabilizes and helps to set the RP), this transient decrease and subsequent increase in the current density parallels the generation of spontaneous APs.

C. Ca²⁺ Channels and Na⁺ Channels

High voltage–activated Ca^{2+} channels have also been observed in ascidian embryo (Fig. 10A) (Greaves *et al.*, 1996). These Ca^{2+} channels exhibit inactivation, and may be an N-type Ca^{2+} channel because the current is blocked by conotoxin. The channels increase after the neurula stage (16 h after fertilization). After the tailbud stage, sustained Ca^{2+} channel activity (probably L-type) begins to increase, and dominates at the time of hatching. Low voltage–activating, rapidly inactivating Ca^{2+} channels (T-type) are detected in

about 50% of the cells at each stage of development. The contribution of the T-type channels to total Ca^{2+} influx is relatively small in comparison with the L-type and N-type channels.

In early embryonic amphibian skeletal myocytes, substantial Ca^{2+} currents and Na⁺ current appeared almost at the same time (after 10–12 h in culture) (Lindsdell and Moody, 1995). These channel currents continued to increase steadily over an observation period of 10–28 h in culture.

D. Delayed-Rectifier K⁺ Channels

The delayed-rectifier K⁺ current (I_K) in skeletal myocytes of early embryonic amphibian progressively increases during development in culture (Lindsdell and Moody, 1995). A similar increase in I_K occurs in ascidian myocytes after the

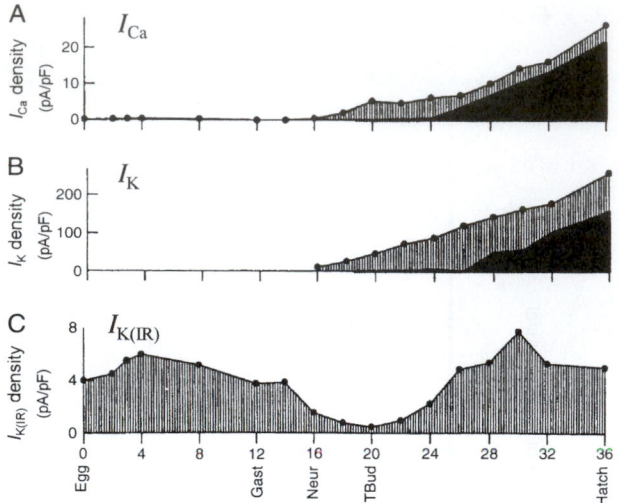

FIGURE 10. Development of Ca^{2+} and K^+ currents in skeletal muscle of a marine tunicate (ascidian). The plots start at fertilization (0 h). (A) Total Ca^{2+} current density (filled circles), with inactivating (hatched) and sustained (solid) components. (B) Total K^+ current density (filled circles), with voltage-dependent (hatched) and Ca^{2+}-dependent (solid) components. (C) Inward-rectifier K^+ current density. Gast, Neur, and Tbud indicate the stages of gastrula, neurula, and tailbud, respectively. (Modified with permission from Greaves *et al.,* 1996.)

neurula stage (16 h postfertilization) (Fig. l0B) (Greaves *et al.,* 1996). However, $I_{K(Ca)}$ progressively increases after 26 h postfertilization. The increase in $I_{K(Ca)}$ closely parallels the increase in sustained L-type Ca^{2+} current (Fig. 10A, B). These developmental changes in K^+ channels contribute to the abbreviation of AP duration during the late developmental period.

E. Acetylcholine Receptor/Channel

The nicotinic acetylcholine receptor/channel (nAChR) is essential to transmission at the neuromuscular junction (see Chapter 41). During development, the nAChR channels in embryonic muscles are converted to adult-type nAChR channels around the time of birth (Mishina *et al.,* 1986). The fetal nAChR channel is composed of α-, β-, γ-, and δ-subunits, and in the adult channel, the γ-subunit is substituted by an ϵ-subunit. In functional characteristics, the fetal channel exhibits a low conductance and long openings, compared with those of the adult channel. This conversion of the nAChR channel may be related to innervation of the muscles that occurs during development, because the fetal type of nAChR channel has also been observed in denervated muscles and the extrajunctional nAChR channel of fetal muscles.

F. Regulation of Expression of Ion Channels

Recently, it has been well analyzed that several humoral factors regulate expression of ion channels during development. Activin, a member of the TGFβ family, has proved to be a particularly potent inducing agent to form

different mesodermal cell types from animal cap cells. *In vitro* induction of animal cap *Xenopus* cells under culturing with activin triggers a whole cascade of developmental events, resulting in the differentiation of skeletal muscle (Currie and Moody, 1999) The developmental pattern of ion channel expression in muscle induced *in vitro* by activin is close to that of normal muscle. First, the same currents (I_K, $I_{K(IR)}$, I_{Na}, I_A, I_{Ca}) are expressed over a similar time course during differentiation. The sequence in which the currents are expressed is also maintained, with I_K and $I_{K(IR)}$ being expressed first, followed by I_{Na}, I_A, and I_{Ca} at slightly later stages. This study indicates that activin is very important for development of electrical activity in skeletal muscle.

IV. Neurons

A. Action Potential

The ionic dependence of the neuronal AP is altered during the early stages of embryonic development (Spitzer and Baccaglini, 1976; Spitzer *et al.,* 1994). Initially, the AP exhibits a prominent Ca^{2+} dependence (i.e., Ca^{2+}-dependent AP) and its duration is prolonged. Later in development, the AP duration becomes brief, and most of the inward current during the depolarizing phase is carried by Na^+ (i.e., Na^+-dependent AP). The Na^+-dependent APs continue until maturation of the neuron. For example, in embryonic amphibian neurons *in vivo*, the AP is prolonged and the rate of rise is slow at a relatively early stage (Fig. 11A). Removal of Na^+ ion does not affect the AP configuration at this stage, whereas it is almost abolished by Co^{2+} ion, an inorganic blocker of voltage-dependent Ca^{2+} channels. At the late em-

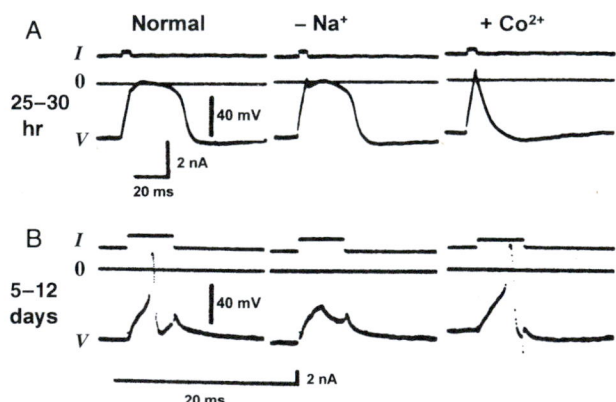

FIGURE 11. Developmental changes in APs of embryonic amphibian neurons at two different stages: (A) 25–30 h and (B) 5–12 days after fertilization of the egg. Depolarizing current (*I*) is applied to evoke the AP (*V*). The zero potential is shown by a solid line (0). (A) The AP is of long duration, and the rate of rise is slow at this early stage. This AP is little affected by removal of Na^+, but is abolished by Co^{2+}. (B) The AP at this late stage is brief and its amplitude is large. This AP is blocked by removal of Na^+ and is unaffected by Co^{2+}. (Modified from *Brain Res.,* **107**, N. C. Spitzer and P. I. Baccaglini, Development of the action potential in embryo amphibian neurons *in vivo*, pp. 610–616, Copyright 1976, with permission from Elsevier Science.)

bryonic stage, the AP becomes greatly abbreviated and loses the shoulder on its falling phase (Fig. 11B). This AP is completely blocked by removal of Na+ ion (or by TTX), but is unaffected by Co^{2+} (Spitzer and Baccaglini, 1976).

In rat spinal motoneurons at embryonic day 16, the AP upstroke may be dependent upon Na+ ion, although maximum rate of rise of AP is slow (about 20 V/s) (Gao and Ziskind-Conhaim, 1998). Duration of the AP at day 16 is prolonged (about 10 ms). At postnatal day 3, the AP upstroke is very fast (about 70 V/s) and duration of the AP is very short (2–3 ms).

B. Ca^{2+} Transient

Spontaneous transient elevation of intracellular Ca^{2+} is observed in developing neurons (Holliday and Spitzer, 1990). The spontaneous Ca^{2+} transient (recorded by use of fluorescent dyes, e.g., FURA-2, fluo-3, indo-1, etc.) is exclusively dependent on Ca^{2+} influx, because it is abolished either by removal of extracellular Ca^{2+} or by agents that block Ca^{2+} channels (Holliday and Spitzer, 1990; Spitzer, 1994). Therefore, the observed intracellular Ca^{2+} transient is due to Ca^{2+} influx from the extracellular space.

Two classes of spontaneous Ca^{2+} transients have been detected: rapid events, termed Ca^{2+} **spikes**, and slow events, termed Ca^{2+} **waves** (Gu et al., 1994). The incidence of these Ca^{2+} transients changes during development in culture. Ca^{2+}

spikes in the cell body (soma) are triggered by spontaneous APs and are rapidly propagated to the growth cone. Ca^{2+} spikes also use the intracellular Ca^{2+} store, because depletion of the store with caffeine substantially reduces their amplitude. Ca^{2+} spikes may be required for the normal appearance of transmitter GABA, since blocking of Ca^{2+} spikes by a Ca^{2+} channel–blocker prevents the acquisition of GABA immunoreactivity. The normal developmental increase in the activation kinetics of K+ currents is also prevented by the blocking of Ca^{2+} spikes.

Ca^{2+} waves occur often (about 10/h) in the growth cone, and they are not generally propagated to the soma. The Ca^{2+} waves in the soma occur at a lower frequency (about 2/h). Therefore, the Ca^{2+} waves are local and occur independently in separate growth cones of the same neuron. Because there seems to be some relation between external Ca^{2+} and the length of the neurite, Ca^{2+} waves in growth cones are likely to regulate neurite extension (Gu et al, 1994; Spitzer, 1994; Spitzer et al., 1994).

C. Voltage-Gated Ion Channels

In mature excitable cell membranes, the major inward currents consist of two ions, Na+ and Ca^{2+}, which are carried through voltage-gated Na+ channels and Ca^{2+} channels, respectively. Na+ and Ca^{2+} channels exhibit two patterns of development (Gottmann et al., 1988; O'Dowd et al., 1988). In

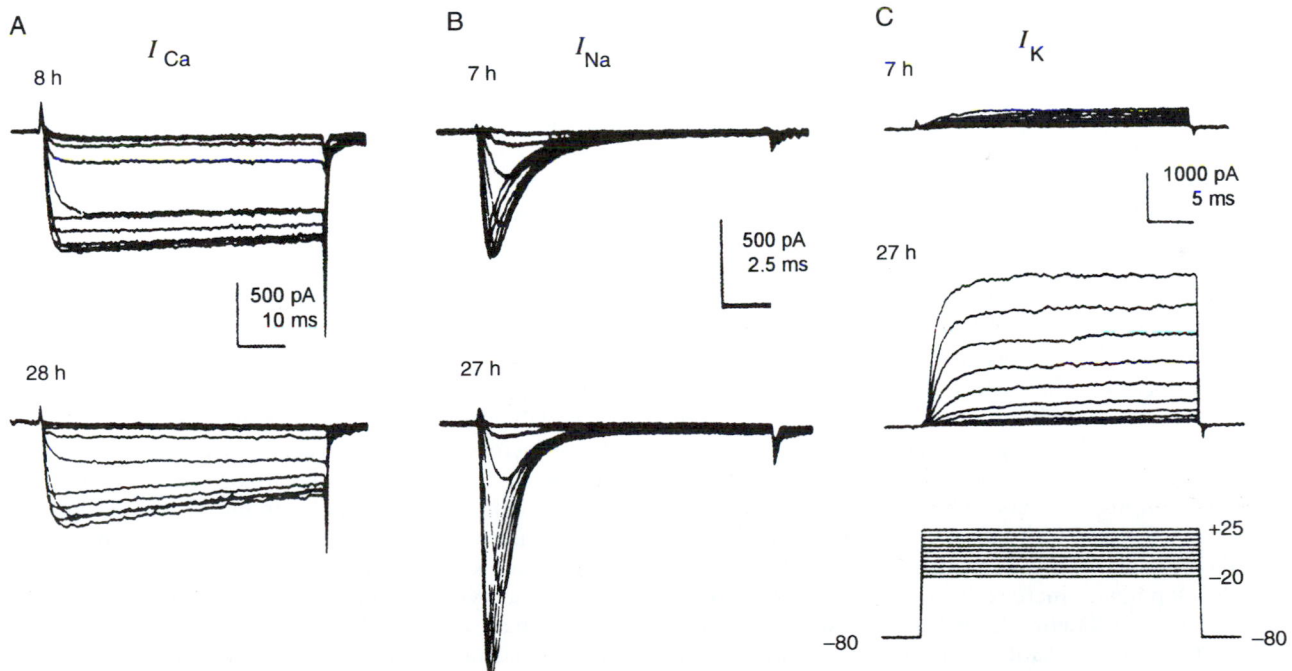

FIGURE 12. Developmental changes in ionic currents of cultured maturing neurons of embryonic amphibian. Records from cells in the early stage (7-8 h in culture) (upper traces) and the late stage (27–28 h) (lower traces). Records in A and B were obtained in presence of Cs+ (internal solution) and TEA (external solution) to block outward K+ currents. (A) Ca^{2+} currents (I_{Ca}) recorded in Na+-free solution. The amplitude of I_{Ca} was nearly equal at early and late stages. (B) Na+ current (I_{Na}) recorded in presence of Co^{2+} to block I_{Ca}. I_{Na} was smaller at the early stage than at the late stage. (C) K+ currents (I_{K}) recorded in the presence of Co^{2+} and TTX to block the inward currents (I_{Ca} and I_{Na}). I_{K} is small at the early stage, and gets larger in amplitude and faster in rate of activation at the late stage. Pulse protocols for parts A, B, and C are given in the bottom of panel C. [Modified with permission from O'Dowd et al., Development of voltage-dependent calcium, sodium and potassium currents in *Xenopus* spinal neurons. *J. Neurosci.* **8,** 796–800 (1988), Copyright 1988 by the Society for Neuroscience.]

the first pattern, Ca^{2+} channels appear earlier and develop faster than Na^+ channels. This explains, at least in part, the conversion of the Ca^{2+}-dependent AP to the Na^+-dependent one discussed earlier.

In the second pattern, Ca^{2+} channels and Na^+ channels become expressed almost at the same time. For example, in cultured embryonic amphibian neurons, Ca^{2+} currents are large even at the early stage of development, and the peak current density does not change from the early stage to the late, stage (Fig. 12A). Na^+ currents are also present, but the current density is small at the early stage. The peak density of the Na^+ current approximately doubles between the early and late stages of development in culture (Fig. 12B).

In ascidian embryos, Na^+ channels in neural cells dramatically change during the early developmental stage (Takahashi and Okamura, 1998). In unfertilized eggs, a small amount of Na^+ current (type A) is already observed. However, the type A Na^+ current is almost abolished after fertilization. After fertilization, fast-activating and inactivating Na^+ current (type C) with slow-inactivating Na^+ current (type B) is markedly increased. However, in blastomere, which develops into neurons, the type B current is abolished in a moment. Finally, the type C current, corresponding to neural-specific *TuNa I* gene, increases further and becomes responsible for the neural Na^+ spike.

As described earlier, the Ca^{2+} channel allows Ca^{2+} influx, and is responsible for the spontaneous Ca^{2+} transient in developing neurons. However, regulation of Ca^{2+} influx during development is not due to change in the density of the Ca^{2+} channels, because the degree of expression of Ca^{2+} channels may not change. The ratio of the K^+ current to the Ca^{2+} current (I_K/I_{Ca}) affects the configuration of the APs, because the inward I_{Ca} is deactivated earlier when the outward I_K is augmented, resulting in a decrease in the net inward current. In other words, any outward (K^+) current subtracts from the inward (Ca^{2+}) current, to give a lower net inward current. In addition, the change in the amplitude and the kinetics of I_K that occurs during development greatly affects the AP configuration and apparent ionic dependence (Barish, 1986; Lockery and Spitzer, 1992). Therefore, the I_K/I_{Ca} ratio determines whether Ca^{2+}-independent APs are exhibited, as well as influences the Ca^{2+} influx.

It is well known that Ca^{2+} channels are classified by their threshold voltage (low-voltage activated, LVA; high-voltage activated, HVA) or specific blockers (T-, L-, and N-type). In rat cortical neurons, the major Ca^{2+} channel subtype shifts from the LVA (L-type) Ca^{2+} current to the HVA Ca^{2+} current during the postnatal period (Tarasenko *et al.*, 1998).

The developmental increase in K^+ currents and change in their activation kinetics have been investigated in amphibian neurons in intact (isolated) spinal cord (Desarmenien *et al.*, 1993) and in dissociated cell culture (Fig. 12C)(Barish, 1986; O'Dowd *et al.*, 1988). The general rule regarding the order of appearance of ionic conductances in developing neurons is that K^+ currents appear first, before functional expression of the voltage-dependent inward currents (Spitzer *et al.*, 1994). However, the types of K^+ currents regulating the AP duration and the order of their maturation vary among different neuronal cells at early stages of development. In some neurons, delayed-

rectifier K^+ current precedes the appearance of the transient outward K^+ current (I_A or I_{to}). In contrast, in other neurons, the inactivating K^+ current (I_A) precedes expression of the delayed-rectifier K^+ current (Bader *et al.*, 1985; Aguayo, 1989; Beck *et al.*, 1992).

The Ca^{2+}-activated K^+ channels play a pivotal role for afterhyperpolarization and regulation of repetitive firing in neurons. In rat spinal motoneurons, the Ca^{2+}-activated K^+ current is markedly increased from the embryonic period to the postnatal period (Gao and Ziskin-Conhaim, 1998). This current increase shortens the duration of the action potential and produces the afterhyperpolarization. In addition, the Ca^{2+}-activated K^+ current reduces the frequency of action potential repetitive firing after birth.

D. Ligand-Gated Channels

During development, activity of the N-methyl-D-aspartate (NMDA) receptor channels decreases, and neuronal differentiation is supported by Ca^{2+} influx through the receptor channels. In some cases, NMDA receptors are expressed transiently during early stages of development, restricting an increase in intracellular Ca^{2+} at the early stages (Garthwaite *et al.*, 1987).

Glutamate receptors gated by kinate/AMPA (α-amino-3-hydroxymethyl-4-isoxazolepropionic acid) allow Ca^{2+} influx and elevation of intracellular Ca^{2+}. The Ca^{2+} permeability of glutamate receptors is governed by their subunit composition, as well as by single amino acid substitutions generated by RNA editing; both factors may be developmentally regulated (Hume *et al*, 1991). It has recently been reported that AMPA receptors express in the early stage of differentiation of *Xenopus* spinal neurons. Ca^{2+} influx to regulate the neural differentiation mainly flows through AMPA receptors soon after neurite initiation and before expression of NMDA receptors (Gleason and Spitzer, 1998).

V. Summary

Action potentials and resting potentials in excitable cells, such as cardiomyocytes, skeletal muscle fibers, and neurons, are greatly altered during development. In general, the rate of rise increases, the overshoot increases, the duration decreases, and the RP is hyperpolarized. These electrophysiological alterations are mainly produced by developmental changes in ion channels, that is, by changes in the types, number, and kinetic properties of the ion channels.

In cardiomyocytes, the density of inward-rectifier K^+ currents increases during development. This increase may result from changes in the single-channel conductance, as well as in the number and open probability of the channel, because the conductance of early fetal myocytes is much smaller than those of neonatal myocytes. The hyperpolarization of the RP during development can be accounted for by the increase in the density of $I_{K(IR)}$ and the resultant decrease in the P_{Na}/P_K ratio. In contrast, the hyperpolarization-activated inward current (I_h), which may affect automaticity, is dominant in early embryonic chick cardiomyocytes and disappears during development.

The fast Na^+ current also increases markedly during development. That is, there are few or no functional fast Na^+ channels present at the earliest stages, and the density of these channels increases progressively during development. The fast Na^+ current is responsible for the increase in the AP rate of rise (independent of the hyperpolarization of RP). The max dV/dt increases dramatically during development, for example, in chick heart, from 20 to 200 V/s.

The fast Na^+ current in embryonic/fetal hearts has a significant sustained (i.e., slow-inactivating or steady-state) component. The sustained component of the embryonic/fetal Na^+ current decreases during development, and this decrease contributes, at least in part, to the abbreviation of the AP duration.

Development of Ca^{2+} channels seems more complex. The density of total Ca^{2+} current in chick cardiomyocytes decreases during the developmental period from fetal to neonate. In rat and mouse, however, it increases from the fetal to the neonatal period, followed by a substantial decrease in the adult. In rabbit and guinea pig cardiomyocytes, the current density in the neonatal period is smaller than that in the adult. The total Ca^{2+} current is composed of currents through several different types of channels: L-type, T-type, and F-type. (1) The proportion of the T-type Ca^{2+} channel current in immature cells is generally more than that in mature cells, and it may actually disappear in adults. That is, the L-type Ca^{2+} channel current becomes more dominant in mature cells. (2) A nifedipine-resistant F-type Ca^{2+} channel current is also present in early fetal cardiomyocytes of rats. (3) Long-lasting openings of the L-type Ca^{2+} channels are relatively frequently observed in embryonic chick and fetal rat cardiomyocytes, but are quite unusual in adult cells.

Ca^{2+} influx through the Ca^{2+} channels is especially important for the excitation-contraction coupling process of fetal cardiomyocytes. This is because the SR function and/or the essential component for Ca^{2+} release from the SR is immature, and so Ca^{2+} influx from the extracellular space is the main source of Ca^{2+} for contraction.

The density of the transient outward current (I_{to}) increases during development. An increase in the density of the delayed-rectifier K^+ current (I_K) also occurs during early development, and there is no further change during the postnatal period. The change of these voltage-gated outward currents (i.e., and I_{to} and I_K) helps to abbreviate the AP duration during development.

In skeletal muscle fibers, there is an overall trend toward an increase in the density of $I_{K(IR)}$ during the early developmental period. However, there is a transient period during which the current density decreases. This transient period parallels the increased incidence of spontaneous firings, presumably due to a less stable RP. The major voltage-gated inward currents, Na^+ and Ca^{2+} currents, are already present in the early embryonic skeletal myocytes in culture, and they increase in intensity during development. These changes contribute to the increases in the AP rate of rise, overshoot, and propagation velocity. Prolonged APs and brief APs can be elicited in chick embryonic skeletal myotubes, depending on the stage of development. The prolonged APs are associated with long-lasting contractions of the myotubes. The brief APs become more dominant during development in

culture, and are associated with myotube twitches. The nicotinic acetylcholine receptor/channel, which is essential to transmission at the neuromuscular junction, is converted from fetal-type to adult-type, and this conversion may be related to innervation of the muscles that occurs during development. Activin, a member of the TGFβ family, is important for development of ion channels in skeletal muscles. Activin stimulates the expression of the same ion channels (I_K $I_{K(IR)}$, I_{Na}, I_A, I_{Ca}) in cultured skeletal myoblasts. The sequence in which currents are expressed is also maintained (I_K and $I_{K(IR)}$ are expressed first).

In neuronal cells, the ionic dependence of the AP is altered from being Ca^{2+} dependent (prolonged AP duration) to Na^+ dependent (brief AP duration) during development of amphibian embryo, although Na^+ current is already observed in early neural blastomere of ascidian embryos. The patterns of ion channel development vary among different types of neuronal cells, with faster development of Ca^{2+} channels in some cells. Another important factor that determines the ionic dependence of the AP is the developmental increase in I_K. Therefore, the I_K/I_{Ca} ratio is the major determinant of the conversion of the AP configuration, and influences Ca^{2+} influx during development. Two types of Ca^{2+} transients, Ca^{2+} spikes and Ca^{2+} waves, are present in developing neurons. Ca^{2+} spikes in cell body (soma) are generated by spontaneous APs and propagate to the growth cone. The incidence of these Ca^{2+} transients changes in developing neurons in culture. The activity of ligand-gated Ca^{2+}-permeable channels (such as NMDA receptor/channels and kinate/AMPA-gated receptor channels) are also altered during development. Therefore, the Ca^{2+} influx through the voltage-gated (LVA or HVA channels) and the ligand-gated Ca^{2+} channels, and the subsequent effects on intracellular Ca^{2+}, may affect the structural changes of developing neurons and help the establishment of the neuronal network.

As described earlier, ion channels exhibit dramatic changes during development in their type, structure, function, and distribution. These dynamic alterations are controlled by expression of genes coding ion channels, and may be essential to cellular growth and differentiation by affecting intracellular Ca^{2+} concentration and excitability. Thus, molecular biology, combined with electrophysiology, has enabled large advances in our understanding of cell function.

Bibliography

Accili, E. A., Robinson, R. B., and DiFrancesco, D. (1997). Properties and modulation of I_f in newborn versus adult cardiac SA node. *Am. J. Physiol.* **272**, H1549–H1552.

Aguayo, A. G. (1989). Post-Natal development of K^+ currents studied in isolated rat pineal cells. *J. Physiol. (London)* **414**, 238–300.

Bader, C. R., Bertrand, D., and Dupin, E. (1985). Voltage-dependent potassium currents in developing neurons from quail mesencephalic neural crest. *J. Physiol. (London)* **366**, 129–151.

Barish, M. E. (1986). Differentiation of voltage-gated potassium current and modulation of excitability in cultured amphibian spinal neurons. *J. Physiol. (London)* **375**, 229–250.

Beck, H., Ficker, E., and Heinemann, U. (1992). Properties of two voltage-activated potassium currents in acutely isolated juvenile rat dentate gyrus granule cells. *J. Neurophysiol.* **68**, 2086–2099.

Brown, A. M., Lee, K. S., and Powell, T. (1981). Sodium current in single rat heart muscle cells. *J. Physiol. (London)* **318,** 479–500.

Cohen, N. M., and Lederer, W. J. (1988). Changes in the calcium current of rat heart ventricular myocytes during development. *J. Physiol. (London)* **406,** 115–146.

Conforti, L., Tohse, N., and Sperelakis, N. (1993). Tetrodotoxin-sensitive sodium current in rat fetal ventricular myocytes—contribution to the plateau phase of action potential. *J. Mol. Cell. Cardiol.* **25,** 159–173.

Currie, D. A., and Moody, W. J. (1999). Time course of ion channel development in *Xenopus* muscle induced *in vitro* by activin. *Dev. Biol.* **209,** 40–51.

Davies, M. P., An, R. H., Doevendans, P., Kubalak, K. R., Chien, K. R., and Kass, R. S. (1996). Developmental changes in ionic channel activity in the embryonic murine heart. *Circ. Res.* **78,** 15–25.

Desarmenien, M. G., Clendening, B., and Spitzer, N. C. (1993). *In vivo* development of voltage-dependent ionic currents in embryonic *Xenopus* spinal neurons. *J. Neurosci.* **13,** 1275–2581.

Fabiato, A., and Fabiato, F. (1978). Calcium-induced release of calcium from the sarcoplasmic reticulum of skinned cells from adult human, dog, cat, rabbit, rat, and frog hearts and from fetal and newborn rat ventricles. *Ann. NY Acad. Sci.* **307,** 491–522.

Fischbach, G. D., Nameroff, M., and Nelson, P. G. (1971). Electrical properties of chick skeletal muscle fibers developing in cell culture. *J. Cell. Physiol.* **78,** 289–300.

Fujii, S., Ayer, R. K., Jr., and DeHaan, R. L. (1988). Development of the fast sodium current in early embryonic chick heart cells. *J. Membr. Biol.* **101,** 209–223.

Gao, B.-X., and Ziskind-Conhaim, L. (1998). Development of ionic currents underlying changes in action potential waveforms in rat spinal motoneurons. *J. Neurophysiol.* **80,** 3047–3061.

Garthwaite, G., Yamini, B., Jr., and Garthwaite, J. (1987). Selective loss of Purkinje and granule cell responsiveness to N-methyl-D-aspartate in rat cerebellum during development. *Brain Res.* **433,** 288–292.

Gleason, E. L., and Spitzer, N. C. (1998). AMPA and NMDA receptors expressed by differentiating *Xenopus* spinal neurons. *J. Neurophysiol.* **79,** 2986–2998.

Gottmann, K., Dietzel, I., Lux, H. D., Huck, S., and Rohrer, H. (1988). Development of inward currents in chick sensory and autonomic neuronal precursor cells in culture. *J. Neurosci.* **8,** 3722-3732.

Greaves, A. A., Davis, A. K., Dallman, J. E., and Moody, W. J. (1996). Co-ordinated modulation of Ca^{2+} and K^+ currents during ascidian muscle development. *J. Physiol. (London)* **497,** 39–52.

Gu, X., Olson, E. C., and Spitzer, N. C. (1994). Spontaneous neuronal calcium spikes and waves during early differentiation. *J. Neurosci.* **14,** 6325–6335.

Guo, W., Kamiya, K., and Toyama, J. (1995). bFGF promotes functional expressions of transient outward currents in cultured neonatal rat ventricular cells. *Pflügers Arch.* **430,** 1015–1017.

Holliday, J., and Spitzer, N. C. (1990). Spontaneous calcium influx and its roles in differentiation of spinal neurons in culture. *Dev. Biol.* **141,** 13–23.

Hume, R. I., Dingledine, R., and Heinemann, S. F. (1991). Identification of a site in glutamate receptor subunits that controls calcium permeability. *Science* **253,** 1028–1031.

Huynh, T. V., Chen, F., Wetzel, G. T., Friedman, W. F., and Klitzner, T. S. (1992). Developmental changes in membrane Ca^{2+} and K^+ currents in fetal, neonatal, and adult rabbit ventricular myocytes. *Circ. Res.* **70,** 508–515.

Iannaccone, S. T., Li, K. X., and Sperelakis, N. (1987). Transmembrane electrical characteristics of cultured human skeletal muscle cells. *J. Cell. Physiol.* **133,** 409–413.

Ijima, T., and Pappano, A. J. (1979). Ontogenetics increase of the maximal rate of rise of the chick embryonic heart action potential: relationship to voltage, time and tetrodotoxin. *Circ. Res.* **44,** 358–367.

Irisawa, H., Brown, H. F., and Giles, W. (1993). Cardiac pacemaking in the sinoatrial node. *Physiol. Rev.* **73,** 197–227.

Jeck, C. D., and Boyden, P. A. (1992). Age-related appearance of outward currents may contribute to developmental differences in ventricular depolarization. *Circ. Res.* **71,** 1390–1403.

Josephson, I. R., and Sperelakis, N. (1989). Tetrodotoxin differentially blocks peak and steady-state sodium channel currents in early embryonic chick ventricular myocytes. *Pflügers Arch.* **414,** 354–359.

Josephson, I. R., and Sperelakis, N. (1990). Developmental increases in the inwardly rectifying K^+ current of embryonic chick ventricular myocytes. *Biochim. Biopys. Acta* **1052,** 123–127.

Kano, M., and Yamamoto, M. (1977). Development of spike potentials in skeletal muscle cells differentiated *in vitro* from chick embryo. *J. Cell. Physiol.* **90,** 439–444.

Kato, Y., Masumiya, H., Agata, N., Tanaka, H., and Shigenobu, K. (1996). Developmental changes in action potential and membrane currents in fetal, neonatal and adult guinea pig ventricular myocytes. *J. Mol. Cell. Cardiol.* **28,** 1515–1522.

Kawano, S., and DeHaan, R. L. (1991). Developmental changes in the calcium currents in embryonic chick ventricular myocytes. *J. Membr. Biol.* **120,** 17–28.

Kilborn, M. J., and Fedida, D. (1990). A study of the developmental changes in outward currents of rat ventricular myocytes. *J. Physiol. (London)* **430,** 37–60.

Kojima, M., Sada, H., and Sperelakis, N. (1990). Developmental changes in beta-adrenergic and cholinergic interactions on calcium-dependent slow action potentials in rat ventricular muscles. *Br. J. Pharmacol.* **99,** 327–333.

Linsdell, P., and Moody, W. (1995). Electrical activity and calcium influx regulate ion channel development in embryonic *Xenopus* skeletal muscle. *J. Neurosci.* **15,** 4507–4514.

Lockery, S. R., and Spitzer, N. C. (1992). Reconstruction of action potential development from whole-cell currents of differentiating spinal neurons. *J. Neurosci.* **12,** 2268–2287.

Marcus, N. C., and Fozzard, H. (1981). Tetrodotoxin sensitivity in the developing and adult chick heart. *J. Mol. Cell. Cardiol.* **13,** 335–340.

Masuda, H., and Sperelakis, N. (1993). Inwardly rectifying potassium current in rat fetal and neonatal ventricular cardiomyocytes. *Am. J. Physiol.* **265,** H1107–H1111.

Masuda, H., Sumii, K., and Sperelakis, N. (1995). Long openings of calcium channels in fetal rat ventricular cardiomyocytes. *Pflügers Arch.* **429,** 595–597.

Mays, D. J., Foose, J. M., Philipson, L. H., and Tamkun, M. M. (1995). Localization of the Kv1.5 K^+ channel protein in explanted cardiac tissue. *J. Clin. Invest.* **96,** 282–292.

Mishina, M., Takai, T., Imoto, K., Noda, T., Takahashi, S., Numa, S., Mathfessel, C., and Sakmann, B. (1986). Molecular distinction between fetal and adult forms of muscle acetylcholine receptor. *Nature* **321,** 406–411.

Nakanishi, T., Sebuchi, M., and Takao, A. (1988). Development of the myocardial contractile system. *Experientia* **44,** 936–944.

O'Dowd, D. K., Ribera, A. B., and Spitzer, N. C. (1988). Development of voltage-dependent calcium, sodium and potassium currents in *Xenopus* spinal neurons. *J. Neurosci.* **8,** 792–805.

Osaka, T., and Joyner, R. W. (1991). Developmental changes in calcium currents of rabbit ventricular cells. *Circ. Res.* **68,** 788–796.

Po, S., Snyders., D. J., Tamkun, M. M., and Bennett, P. B. (1993). Heteromultimeric assembly of human potassium channels. *Circ. Res.* **72,** 1326–1336.

Renaud, J. F., Romey, G., Lombet, A., and Lazdunski, M. (1981). Differentiation of the Na^+ channel in embryonic heart cells: interaction of the channel with neurotoxin. *Proc. Natl. Acad. Sci. USA* **78,** 5348–5352.

Ritchie, A. K., and Fambrough, D. M. (1975). Electrophysiological properties of the membrane and acetylcholine receptor in developing rat and chick myotubes. *J. Gen. Physiol.* **66,** 327–355.

Sada, H., Ban, T., Fujita, T., Ebina, Y., and Sperelakis, N. (1995). Developmental change in fast Na+ channel properties in embryonic chick ventricular heart cells. *Can. J. Physiol. Pharmacol.* **73**, 1475–1484.

Sada, H., Kojima, M., and Sperelakis, N. (1988). Fast inward current properties of voltage-clamped ventricular cells of embryonic chick heart. *Am. J. Physiol.* **255**, H540–H553.

Satoh, H., and Sperelakis, N. (1990). Identification of and developmental changes in transient outward current in embryonic chick cardiomyocytes. *J. Dev. Physiol.* **20**, 149–154.

Satoh, H., and Sperelakis, N. (1993). Hyperpolarization-activated inward current in embryonic chick cardiac myocytes: developmental changes and modulation by isoproterenol and carbachol. *Eur. J. Pharmacol.* **240**, 283–290.

Sanchez-Chapula, J., Elizalde, A., Navarro-Polanco, R., and Barajas, H. (1994). Differences in outward currents between neonatal and adult rabbit ventricular cells. *Am. J. Physiol.* **266**, H1184–H1194.

Shimoni, Y., Fiset, C., Clark, R. B., Dixon, J. E., McKinnon, D., and Giles, W. R. (1997). Thyroid hormone regulates postnatal expression of transient K+ channel isoforms in rat ventricle. *J. Physiol. (Lond.)* **500**, 65–73.

Spector, I., and Prives, J. M. (1977). Development of electrophysiological and biochemical membrane properties during differentiation of embryonic skeletal muscle in culture. *Proc. Natl. Acad. Sci. USA* **74**, 5166–5170.

Sperelakis, N., and Haddad, G. E. (1995). Developmental changes in membrane electrical properties of the heart. *In* "Physiology and Pathology of the Heart," 3rd. ed. (N. Sperelakis, Ed.), pp. 669–700. Kluwer Academic Publishers, New York.

Sperelakis, N., and Lee, E. C. (1971). Characterization of (Na+, K+)-ATPase isolated from embryonic chick hearts and cultured chick heart cells. *Biochim. Biophys. Acta* **233**, 562–579.

Sperelakis, N., and Shigenobu, K. (1972). Changes in membrane properties of chick embryonic hearts during development. *J. Gen. Physiol.* **60**, 430–453.

Spitzer, N. C. (1994). Spontaneous Ca2+ spikes and waves in embryonic neurons: signaling systems for differentiation. *Trends Neurosci.* **17**, 115–118.

Spitzer, N. C., and Baccaglini, P. I. (1976). Development of the action potential in embryo amphibian neurons *in vivo. Brain Res.* **107**, 610–616.

Spitzer, N. C., Gu, X., and Olson, E. (1994). Action potentials, calcium transients and the control of differentiation of excitable cells. *Curr. Opin. Neurobiol.* **4**, 70–77.

Takahashi, K., and Okamura, Y. (1998). Ion channels and early development of neural cells. *Physiological Rev.* **78**, 307–337.

Tarasenko, A. N., Isaev, D. S., Eremin, A. V., and Kostyuk, P. G. (1998). Developmental changes in the expression of low-voltage–activated Ca2+ channels in rat visual cortical neurons. *J. Physiol. (London)* **509**, 385–394.

Tohse, N., and Sperelakis, N. (1990). Long-lasting openings of single slow (L-type) Ca2+ channels in chick embryonic heart cells. *Am. J. Physiol.* **259**, H639–H642.

Tohse, N., Masuda, H., and Sperelakis, N. (1992a). Novel isoform of Ca2+ channel in rat fetal cardiomyocytes. *J. Physiol. (London)* **451**, 295–306.

Tohse, N., Maszaros, J., and Sperelakis, N. (1992b). Developmental changes in long-opening behavior of L-type Ca2+ channels in embryonic chick heart cells. *Circ. Res.* **71**, 376–384.

Wahler, G. M., Dallinger, S. J., Smith, J. M., and Flemal, K. L. (1994). Time course of postnatal changes in rat heart action potential and in transient outward current is different. *Am. J. Physiol.* **267**, H1157–H1166.

Wang, L., and Duff, H. J. (1997). Developmental changes in transient outward current in mouse ventricle. *Circ. Res.* **81**, 120–127.

Xie, L.-H., Takano, M., and Noma, A. (1997). Development of inwardly rectifying K+ channel family in rat ventricular myocytes. *Am. J. Physiol.* **272**, H1741-H1750.

Xu, H., Dixon, J. E., Barry, D. M., Trimmer, J. S., Merlie, J. P., McKinnon, D., and Nerbonne, J. M. (1996). Developmental analysis reveals mismatches in the expression of K+ channel α subunits and voltage-gated K+ channel currents in rat ventricular myocytes. *J. Gen. Physiol.* **108**, 405–419.

Akikazu Fujita, Hiroshi Hibino, and Yoshihisa Kurachi

36

Regulation of Ion Channels by Membrane Proteins and Cytoskeleton

I. Introduction

Ion channels are integral membrane proteins that mediate ion permeation through cellular membranes. As movement of charged ions creates electrical currents, ion channels are the molecular structures responsible for the electrical properties of a cell. The probability of channel opening can be regulated by several gating mechanisms including voltage changes, ligand binding, or covalent modifications. In addition to conventional regulators of ion channel gating, it has become apparent that the cytoskeleton itself, or proteins associated with the cellular cytoskeleton, may also provide a major regulator of ion channel behavior.

The **cytoskeleton** forms fibrillar network structures throughout the cytosol, including the microenvironment surrounding ion channel proteins within the plasma membrane. The cytoskeleton is made of microfilaments, microtubules, intermediate filaments, and associated proteins (Gallo-Payet and Payet, 1995). Microfilaments are composed of G-actin monomers, which assemble into actin polymers (F-actin). Actin polymers link into three-dimensional frameworks, which interact with myosin filaments, or form cortical networks at the cell periphery. There is a dynamic equilibrium between F-actin (polymeric forms of actin) and G-actin monomers (Fig. 1A). The status of actin and the state of myofilament organization is set by diverse actin-binding proteins, including profilin (which binds G-actin), gelsolin (which caps F-actin), and filamin and α-actinin (which cross-link microfilaments). Microtubules are composed of α- and β-tubulin heterodimers that form tubular filaments. Polymerization, stabilization, and modulation of microtubule function are regulated by microtubule-associated proteins. The cytoskeleton is essential not only in the maintenance of cell shape and motility, but also in the distribution, stability, and function of integral membrane proteins (Bennett and Gilligan, 1993; Hitt and Luna, 1994).

Interactions between various ion channel proteins and cytoskeletal structures are numerous, and have been implicated in mediating spatial sorting of surface channel proteins, and in regulating of channel activity (Cantiello, 1995; Gomperts, 1996; Smith and Benos, 1996). This chapter provides a synopsis of the effects of cytoskeleton on ion channels.

II. Domain-Dependent Distribution of Ion Channels by Cytoskeleton-Associated and Cytoskeleton Proteins

A critical role for the cytoskeleton and associated proteins lies in the governance of ion channel distribution within specialized regions of plasma membranes. Such regulated distribution of ion channels at the cell surface is necessary for proper intra- and intercellular signaling, in particular within and between excitable cells, such as neurons and epithelial cells (Kennedy, 1993; Sheng, 1996). In the nervous system, electrical signaling is driven by the synchronized function of ion channels, which are typically localized at specific locations, such as the neuromuscular junction, node of Ranvier, or postsynaptic sites. Beyond the nervous system, association of cytoskeletal proteins, such as ankyrin and spectrin, with ion channels and ion transporters, including the Cl-HCO$_3$ exchanger and the α-subunit of the Na-K pump, has been reported in erythrocytes or epithelial cells (Cantiello, 1995; Smith and Benos, 1996). More recently, various K^+ channels including Kir1.1 and Kir4.1 have been found to be localized at the apical and basolateral sides of renal tubular epithelial cells, respectively (Lee and Hebert, 1995; Ito *et al.*, 1996). This section provides an overview of the interaction between cytoskeletal proteins and ion channels responsible for domain-dependent ion channel distribution (Table 1).

A. Rapsyn and Clustering of Nicotinic Acetylcholine Receptors at the Neuromuscular Junction

It is well established that domain-specific ion channel clustering of **nicotinic acetylcholine receptors** (nAChRs)

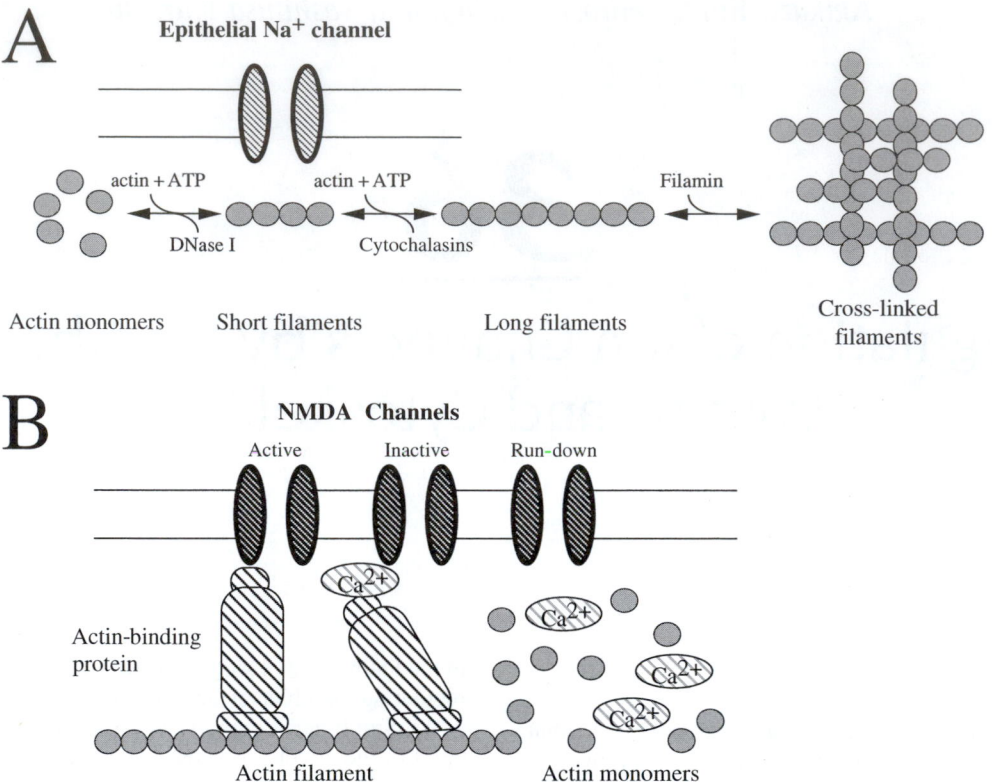

FIGURE 1. (A) A simplified scheme of dynamic interactions among various conformations of actin. In the presence of MgATP, monomeric G-actin nucleates into short actin filaments (e.g., tetramers) that anneal to produce long actin filaments. DNase I inhibits polymerization by binding with G-actin. Cytochalasins favor the transition from long to short filaments. Cross-linking proteins, such as filamin, induce the gelation of actin filaments. The length of actin filaments appears to be a determinant of channel activation. For example, an increase in the availability of short actin filaments enhances the probability for renal epithelial Na$^+$ channels to be open. (Modified from Cantiello, 1995; see also Gallo-Payet and Payet, 1995.) (B) Proposed model of actin-dependent regulation of NMDA channels. This model (Rosenmund and Westbrook, 1993) includes an actin-binding protein that dissociates from the NMDA channel in a Ca^{2+}-dependent manner, leading the channel to inactivate. Calcium-dependent actin filament depolymerization results in removal of the actin-binding protein and channel run-down.

at the neuromuscular junction is essential for synaptic efficacy during neurotransmission. At the neuromuscular junction (NMJ), nAChRs are localized at the motor end plate opposite to the presynaptic nerve terminal. In this specialized membrane domain, nAChRs are tightly clustered at a density of $10\,000/\mu m^2$, in contrast to the thousand-fold lower density of nAChRs found just outside of the motor end plate (Froehner, 1993). The spatially localized aggregation of nACh receptor/channel complexes ensures a rapid and robust response to released acetylcholine molecules, while a change in the cluster density of nAChRs strongly impacts postsynaptic response (Gomperts, 1996). The 43-kDa protein **rapsyn,** which is associated with the inner face of the postsynaptic membrane, clusters and localizes nAChRs and links them to the subsynaptic cytoskeleton (Fig. 2; Phillips *et al.,* 1991; Froehner, 1993). Mutational studies on rapsyn have identified the binding domains for nAChRs, as well as for the cytoskeleton, within the primary structure of rapsyn. The targeting of nAChR and rapsyn to the NMJ is further orchestrated by the nerve-derived factor, agrin (see Section III.A; Wallace, 1992).

Rapsyn can also cluster **dystroglycan,** a member of the **dystrophin-glycoprotein complex** (DGC) found along the sarcolemma of skeletal muscle and associated with the cy-

toskeleton, extracellular matrix (laminin), and with synapse-specific proteins, utrophin and β_2-syntrophin (Fig. 2; Apel and Merlie, 1995). By binding to DGC, rapsyn may, in turn, localize nAChR clusters to the synapse. Targeted disruption of the rapsyn gene abolishes nAChRs clustering in the NMJ, directly implicating rapsyn as essential for the immobilization of nAChRs in the postsynaptic membrane (Gautam *et al.,* 1995). The importance of protein association with DGC is further underscored by the demonstration that several muscular dystrophies, which exhibit improper function of the NMJ, progressive muscle wasting, and weakness, are caused by genetic errors within members of DGC, which results in loss of the cytoskeleton-extracellular matrix linkage (Campbell, 1995).

B. Ankyrin and Na$^+$ Channels in the Node of Ranvier

Segregation of ion channels to specialized regions of neurons is crucial for the propagation of action potentials (APs). Studies of the distribution of molecules and their lateral mobility (diffusion) in the membranes of neurons indicate that Na$^+$ channels are freely mobile on the neuronal cell body, but are immobile at the axon hillock, presynaptic terminal, and at focal points along the axon (Thompson *et al.,* 1988).

TABLE 1   Ion Channel Distribution Directed by Cytoskeleton-Associated Proteins

Channel	Location	Cytoskeleton-associated protein
Na$^+$ Channel	Node of Ranvier	Ankyrin[1]
Glycine receptor	Postsynaptic membrane	Gephyrin[2]
GAB$_A$ receptor	Postsynaptic membrane	Gephyrin[3] GABARAP[4]
GABA$_C$ receptor	Postsynaptic membrane	MAP1B[5]
NMDA receptor	Postsynaptic density	PSD-95[6], PSD-93/chapsyn 110[7] SAP97[8], SAP102[9], S-SCAM[10], CIPP[11] Yotiao[12], spectrin (fodrin)[13], α_2-actinin[14], NF-L[15], tubulin[16]
Kainate receptor	Postsynaptic membrane	PSD-95[17], SAP102[17], SAP97[17]
AMPA receptor	Postsynaptic membrane	GRIP[18], ABP[19], PICK1[20], NSF[21], Narp[22]
mGlu receptor	Postsynaptic membrane	Homers (Vesl)[23]
Shaker type K$^+$ channel (Kv1.4)	Various neural microdomain	PSD-95[24], PSD-93/chapsyn110[7,24], S-SCAM[10]
Inwardly rectifying K$^+$ channels		
Kir2.3	Forebrain	PSD-95[25]
Kir4.1	Retinal Müller cells, glial cells	SAP97[26]
Nicotinic Ach receptor	Neuromuscular junction	Rapsyn[27]
CFTR	Renal tubular epithelial cells	EBP50[28]
ENaC	Rental tubular epithelial cells Distal colon epithelial cells Lung epithelial cells	Nedd4[29]

Note: See the text for further details and abbreviations.

References: (1) Srinivasan *et al.*, 1988; (2) Kuhse *et al.*, 1995; (3) Craig *et al.*, 1996; (4) Wang *et al.*, 1999; (5) Hanley *et al.*, 1999; (6) Kennedy, 1993; Kornau *et al.*, 1995; (7) Kim *et al.*, 1996; (8) Niethammer *et al.*, 1996; (9) Müller *et al.*, 1996; (10) Hirano *et al.*, 1998; (11) Kurschner *et al.*, 1998; (12) Lin *et al.*, 1998; (13) Wechsler and Teichberg, 1998; (14) Wyszynski *et al.*, 1997; (15) Ehlers *et al.*, 1996; (16) van Rossum *et al.*, 1999; (17) Garcia *et al.*, 1998; (18) Dong *et al.*, 1997; (19) Srivastava *et al.*, 1998; (20) Xia *et al.*, 1999; (21) Nishimune *et al.*, 1998; Osten *et al.*, 1998; Song *et al.*, 1998; (22) O'Brien *et al.*, 1999; (23) Brakeman *et al.*, 1997; Kato *et al.*, 1997; Xiao *et al.*, 1998; (24) Kim *et al.*, 1995; Sheng, 1996; (25) Choen *et al.*, 1996; (26) Horio *et al.*, 1997; (27) Phillips *et al.*, 1991; Froehner, 1993; (28) Short *et al.*, 1998; (29) Staub *et al.*, 1996.

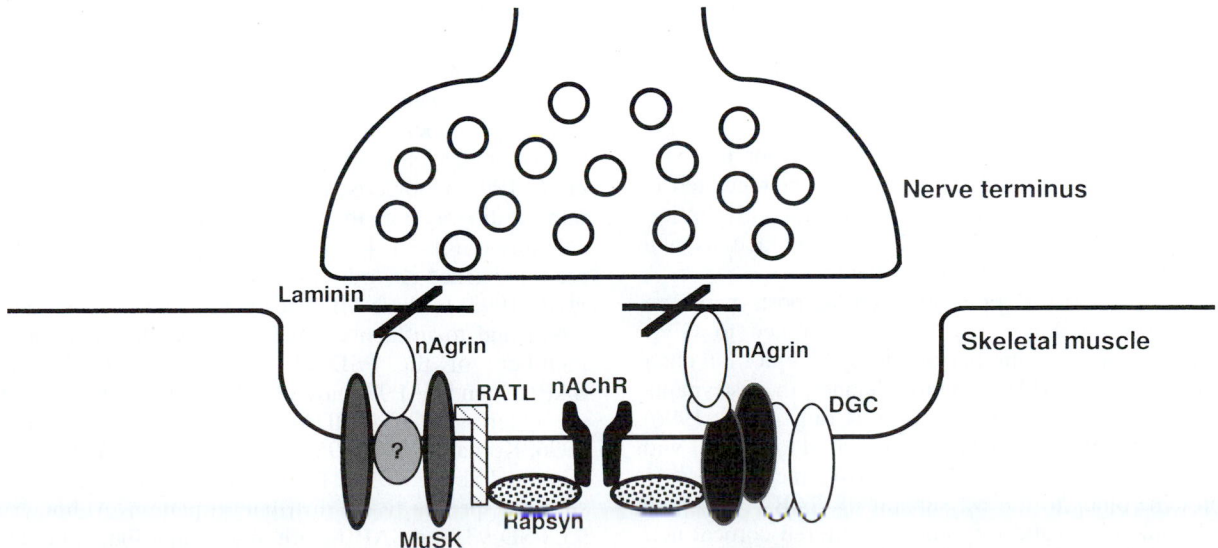

FIGURE 2.   Formation of the neuromuscular junction and associated postsynaptic proteins. nAgrin, neural agrin isoform; mAgrin, agrin isoform derived by muscle; DGC: dystrophin-glycoprotein complex; RATL, rapsyn-associated transmembrane linker.

In particular, a high density of voltage-gated Na$^+$ channels is found at the node of Ranvier, a specialized membrane domain in myelinated axons essential for fast propagation of nerve impulses. Such domain-dependent distribution of Na$^+$ channels has been ascribed to an interaction of channel proteins with a specific isoform of **ankyrin,** a cytoskeletal linker protein (Srinivasan *et al.,* 1988). Initially, ankyrins and Na$^+$ channels distribute diffusely on the premyelinated axolemma. Clustering of neurofascin and NrCAM, which are the members of the CAM (for cell adhesion molecule) family, is the first event prior to the expression of MAG (for myelin-associated glycoprotein) from Schwann cells and concentration of ankyrins and Na$^+$ channels at the node. The clusters of ankyrin and Na$^+$ channels localized at the edge of Schwann cell process then start to fuse to each other to form the mature membrane of the Ranvier node (Lambert *et al.,* 1997). The multivalent properties of ankyrin (i.e., a capacity to bind not only to membranes but also to ion channel proteins and adhesion molecules) play an important role in the coordinated recruitment and targeted distribution of Na$^+$ channels to the node of Ranvier, and thereby in AP conduction (Davis *et al.,* 1996).

C. Gephyrin and Postsynaptic Ion Channel/Receptor Mosaic

The molecular mechanisms underlying the postsynaptic ion channel/receptor mosaic within central nervous system (CNS) neurons have been related, at least in part, to cytoskeleton binding proteins. In particular, clustering of **inhibitory glycine receptors** (GlyR) at postsynaptic membranes, underneath glycine-releasing nerve terminals, has been ascribed to the association of these pentameric receptor/ion channel complexes with **gephyrin,** a 93-kDa tubulin-binding protein (Kuhse *et al.,* 1995). Specifically, a gephyrin-binding domain has been identified in the cytoplasmic loop of the β-subunit of GlyR, between the third and fourth transmembrane segment (Meyer *et al.,* 1995). Antisense inhibition of the expression of gephyrin prevents GlyR accumulation at the specialized postsynaptic membrane sites (Kirsch *et al.,* 1993). Furthermore, depolymerization of microtubules by demecolcine in cultured rat spinal cord neurons reduces the percentage of cells with postsynaptic gephyrin clusters and disperses postsynaptic GlyR clusters (Kirsch and Betz, 1995). This supports the notion that postsynaptic localization of GlyR is regulated by the gephyrin-mediated anchoring of receptor polypeptides to the subsynaptic cytoskeleton through interactions with microtubules and possibly microfilaments.

The essential role for gephyrin in setting postsynaptic receptor/channel topology may not be restricted to GlyR. Gephyrin also anchors another inhibitory receptor/ion channel complex, the GABA$_A$ receptor/channel, to postsynaptic membranes in hippocampal neurons (Craig *et al.,* 1996). Other anchoring proteins have been reported to interact with GABA$_A$ receptors. GABARAP (GABA$_A$-receptor–associated protein) binds to the γ2 subunit of GABA$_A$ receptors and colocalizes with the receptors in cultured cortical neurons (Wang *et al.,* 1999). GABARAP possesses amino acid sequence similarity to that of the light chain-3 of micro-

tubule-associated proteins 1A and 1B (MAP1A and MAP1B) and also a putative tubulin-binding motif in its N-terminal region. The interactions among GABA$_A$ receptors, GABARAP, and tubulin suggest a mechanism for the targeting and clustering of GABA$_A$ receptors. Another anchoring protein, MAP1B specifically interacts with the ρ subunits of the GABA$_C$ receptor but not with GABA$_A$ receptor subunits. GABA$_C$ receptors and MAP1B colocalize at postsynaptic sites on axonal terminals of bipolar cells in the rat retina (Hanley *et al.,* 1999). This specific interaction of MAP1B with GABA$_C$, but not with GABA$_A$, receptors may allow these receptor subtypes, which have distinct physiological and pharmacological properties, to be differentially localized at inhibitory synapses.

D. Postsynaptic Density Proteins and Glutamate Receptor/Channels

1. Distribution of Glutamate Receptor/Channels and Anchoring Proteins

Glutamate receptors, including the N-methyl-D-aspartate (NMDA), α-amino-3-hydroxy-5-methyl-4-isoxazole propionic acid (AMPA), kainate, and metabotropic glutamate (mGlu) receptors, are highly concentrated at postsynaptic sites at excitatory synapses in the CNS. These ionotropic neurotransmitter-gated receptor/channels are actually embedded in the **postsynaptic density** (PSD), an electron-dense thickening that represents a fibrous specialization of the submembrane cytoskeleton at the postsynaptic membrane. The four types of glutamate receptors differ in their subsynaptic distribution: mGlu receptors are located at the periphery, whereas NMDA and AMPA receptor/channels are in the central region of the PSD (Nusser *et al.,* 1994). This suggests that distinct mechanisms underlie the subsynaptic distribution of each glutamate receptor.

2. NMDA Receptor/Channels

Several proteins that contain PDZ domains have been reported to interact closely with the subunits of NMDA receptor/channels and to influence their locations (Tables 1 and 2). The postsynaptic density protein, PSD-95 (also known as the synapse-associated protein 90 kDa, or SAP90) is a cytoskeleton-associated protein of ~95 kDa abundant in the postsynaptic synaptosomal fraction (Kennedy, 1993; Kornau *et al.,* 1995; Gomperts, 1996). PSD-95 has been reported to interact directly with the C-terminal domains of five different subunits (NR1-3/1-4 and NR2A/2B/2C/2D; NR1 subtypes are represented according to the nomenclature of Hollmann *et al.,* 1993) of the NMDA receptor/channels (Kornau *et al.,* 1995) and to influence their clustered distributions. Other members of the PSD-95 family, PSD-93/chapsyn110, SAP102, and SAP97, have also been shown to interact with the C-terminal of NMDA receptor/channels (Kim *et al.,* 1996, Kornau *et al.,* 1995, Müller *et al.,* 1996, Niethammer, Kim, and Sheng, 1996). Each member of the PSD-95 family shows a specific tissue distribution pattern. Although PSD-95, PSD-93, and SAP102 are mainly distributed to the PSD, SAP97 exhibits presynaptic and axonal distributions and also is expressed in the epithelial cells of small intestine and

TABLE 2 Interacting Receptor/Channels and Motifs of Anchoring Proteins

Anchoring protein	Receptor/Channel	Motif	Interacting domain	Reference
MAGUK				
PSD-95/SAP90	NR1-3[a]	–STVV	PDZ-1, -2	1
	NR1-4[a]	–STVV	PDZ-1, -2	
	NR2A	–ESDV	PDZ-1, -2	
	NR2B	–ESDV	PDZ-1, -2	
	NR2C	–ESEV	PDZ-1, -2	
	NR2D	–ESEV	PDZ-1, -2	
	GluR6	–ETMA	PDZ-1, -2	2
	KA2	–PGGP–	SH3 and GK	2
		–PPRP–	SH3 and GK	
	Kv1.4	–ISSL	PDZ-1, -2	3
	Kir2.1	–ESEI	PDZ-1, -2	4
	Kir2.3	–ESRI	PDZ-1, -2	4
	Kir4.1	–ISNV	(ND)	5
PSD-93	NR2A	–ESDV	PDZ-2 (ND for PDZ-1)	6
/chapsyn110	NR2B	–ESDV	PDX-2 (ND for PDZ-1)	6
	Kv1.4	–ISSL	PDZ-1,-2	3
SAP102	NR2B	–ESDV	PDZ-1, -2, -3	7
	GluR6	–ETMA	PDZ-1	2
	KA2	–PGGP–	SH3 and GK	2
		–PPRP–	SH3 and GK	
SAP97/hdlg	NR2A	–ESDV	PDZ-1, -2	6
	NR2B	–ESDV	PDZ-1, -2	6
	GluR6	–ETMA	PDZ-1	2
	Kv1.4	–ISSL	PDZ-2 (ND for PDZ-1)	3
	Kir4.1	–ISNV	(ND)	5
S-SCAM	NR2A	–ESDV	PDZ-5	8
	Kv1.4	–ISSL	PDZ-5	8
Other PDZ-containing proteins				
GRIP	GluR2	–SVKI	PDZ-4, -5	9
	GluR3	–SVKI	PDZ-4, -5	
ABP	GluR2	–SVKI	PDZ-3, -5, -6	10
	GluR3	–SVKI	PDZ-3, -5, -6	
CIPP	NR2A	–ESDV	PDZ-3 (ND for PDZ-2)	11
	NR2B	–ESDV	PDZ-2, -3	
	NR2C	–ESEV	PDZ-3	
	NR2D	–ESEV	PDZ-3 (ND for PDZ-2)	
	Kir4.1	–ISNV	PDZ-2	11
	Kir4.2	–QSNV	PDZ-2, -3	11
PICK1	GluR2	–SVKI	PDZ	12
	GluR3	–SVKI	PDZ	
	GluR4C	–SIKI	PDZ	
	(not bind to GluR1	–ATGL		
	GluR4)	–SDLP		
EBP50	CFTR	–DTRL	PDZ-1, -2	13
EVH domain–containing proteins				
Homer	mGluR1α	–PPSPFR–	EVH domain	14
	mGluR5	–PPSFFR–	EVH domain	
	IP3R	–PPVPFR–	EVH domain	
Other anchoring proteins				
NSF	GluR2	–KRMKVAKNPQ–	All domains	15
	GluR4C	–KRMKVAKSAQ–	(N-terminal domain,	
	(not bind to		ATP domain 1 and	
	GluR1	–KRMKGFCLIP–	ATP domain 2)	
	GlueR3	–KRMKLTKNTQ–		
	GluR4)	–KRMKLTFSEA–		

TABLE 2   Interacting Receptor/Channels and Motifs of Anchoring Proteins (*Continued*)

Yotiao	NR1-1[a]	(C-terminal exon cassette C1 region)	970–1241aa Region of Yotiao	16
Spectrin (Fodrin)	NR1-1a[a] NR2A NR2B	(C-terminal domain) (C-terminal domain) (middle region of cytosolic C-terminal domain: 1086–1326 aa)	(ND)	17
α_2-Actinin	NR1 NR2B (not bind to NR2A)	(C-terminal exon cassette C0 region)	Rod domain of α_2-actinin	18

[a]NR1 subtypes are represented according to the nomenclature of Hoomann *et al.,* 1993. ND, not determined.
    References: (1) Kornau *et al.,* 1995; Kurschner *et al.,* 1998; (2) Garcia *et al.,* 1998; (3) Kim *et al.,* 1995; Sheng *et al.,* 1996; (4) Cohen *et al.,* 1996; (5) Horio *et al.,* 1997; (6) Kim *et al.,* 1995; Hiethammer *et al.,* 1996; (7) Müller *et al.,* 1996; (8) Hirano *et al.,* 1998; (9) Dong *et al.,* 1997; (10) Srivastava *et al.,* 1998; (11) Kurschner *et al.,* 1998; (12) Xia *et al.,* 1999; (13) Short *et al.,* 1998; (14) Tu *et al.,* 1998; (15) Nishimune *et al.,* 1998; Osten *et al.,* 1998; Song *et al.,* 1998; (16) Lin *et al.,* 1998; (17) Wechsler and Teichberg, 1998; (18) Wyszynski *et al.,* 1997.

choroid plexus and localized on their basolateral membrane (Müller *et al.,* 1995). In rat brain, PSD-93 was coimmunoprecipitated with PSD-95 and showed a somatodendritic expression pattern that overlapped partly with PSD-95. These data suggest that PSD-93, but not SAP97, and PSD-95 may interact at postsynaptic sites to form a multimeric scaffold for the clustering of NMDA receptors (Kim *et al.,* 1996). Another multivalent PDZ domain–containing protein, CIPP (channel-interacting PDZ-domain protein) was shown to bind to all four NR2 subunits (NR2A/2B/2C/2D) and Kir4.1 (Kurschner *et al.,* 1998). CIPP is expressed at particularly high levels in the cerebellum, inferior colliculus, vestibular nucleus, facial nucleus, and thalamus. Deep cerebellar nuclei, superior colliculus, dorsal transition zone, and brainstem, as well as the glomerular and mitral cell layers of the olfactory bulb, also contain considerable amounts of CIPP mRNA. Within the cerebellum, CIPP mRNA is detected in the Purkinje cell layer and the granule cell layer, but the subcellular localization of CIPP protein has not yet been determined.

These data also suggest that the specific interaction of NMDA receptor/channels with various proteins containing PDZ domains is important for determining the localization of the receptors to the PSD. In order to ascertain whether this is true, the genes encoding NR2 subunits that lack their C-terminal domain were expressed in mice (Mori *et al.,* 1998, Sprengel *et al.,* 1998). Specific phenotypes similar to those observed when the gene encoding the entire subunit is deleted were observed for both mutants. Sprengel and colleagues (1998) suggested that a disturbed recruitment of the signal-transducing machinery is the main cause for the phenotypes. On the other hand, Mori and coworkers (1998) visualized that the mutated NR2B subunits are not clustered and distribute to not only the PSD but also nonsynaptic regions. These data further support a role for PDZ domain–containing proteins in the maintenance of receptor clusters and the anchoring of these clusters in the subsynaptic scaffolds.

Anchoring proteins without PDZ domains were also reported to bind to NMDA receptor/channels and cluster them (Table 2). **Yotiao** interacts with the middle region of

the C-terminal cytoplasmic domain of NR1-1 (Lin *et al.,* 1998). Yotiao is localized at the NMJ as well as at neuronal synapses in brain. The immunostaining pattern of anti-Yotiao antibody in the NMJ was similar to that of the intermediate filament, desmin. It was also shown that Yotiao is colocalized with NR1 immunoreactivity in somatodendritic regions of pyramidal neuron in rat cerebral cortex. These data suggest that Yotiao may be a NR1-binding protein, which is also potentially involved in cytoskeletal attachment of NMDA receptor/channels at the PSD.

3. Kainate Receptor/Channels

Two subunits of kainate receptor, GluR6 and KA2, have been reported to interact with PSD-95 and SAP102 (see Table 2; Garcia *et al.,* 1998). Both PSD-95 and SAP102 are members of the membrane-associated guanylate-kinase (MAGUK) superfamily which is characterized by the existence of Src 3 homology (SH3) and guanylate kinase–like (GK) domains in the C-terminal region in addition to the presence of PDZ domains (Fig. 3; Gomperts, 1996; Sheng, 1996). Similar to NMDA receptors, GluR6 clustering is mediated by its interaction with the PDZ domain of PSD-95. In contrast, the clustering of KA2 is caused by its interaction with the SH3 and GK domains of PSD-95 (see Section II.G.1). These data suggest that NMDA and kainate receptor/channels may cluster at postsynaptic sites by interacting with PSD-95 and SAP102.

4. AMPA Receptors

The AMPA receptor/channels interact specifically with GRIP (glutamate receptor–interacting protein; Dong *et al.,* 1997) and its analog, ABP (AMPA receptor–binding protein; Srivastava *et al.,* 1998), and PICK1 (protein interacting with C kinase 1) (Table 2; Xia *et al.,* 1999). Each of these proteins, which contain one or more PDZ domains, interacts with C-termini of GluR2 and GluR3 subunits of AMPA receptor/channels and may be important in receptor targeting. **NSF** (N-ethylmaleimide-sensitive fusion protein) has been reported to interact directly and selectively with the intracellular C-terminal domain of the GluR2 and GluR4c subunits of

AMPA receptor/channels (Nishimune *et al.*, 1998, Osten *et al.*, 1998, Song *et al.*, 1998). NSF is a hexameric ATPase originally identified as a factor necessary for Golgi membrane fusion events. NSF is currently thought to be involved in the regulation of fusion of most intracellular membranes (Block *et al.*, 1988). At the presynaptic terminal, NSF forms a large complex with α-SNAP, synaptobrevin (VAMP), SNAP-25, and syntaxin, which all play critical roles in the regulation of neurotransmitter release (Sollner *et al.*, 1993; Rothman, 1994; Sudhof, 1995; Hay and Scheller, 1997). Recent studies have suggested that NSF is also involved in postsynaptic membrane fusion events (Hu *et al.*, 1998; Lledo *et al.*, 1998). Immunogold electron microscopy and biochemical fractionation studies have shown that NSF is localized at the dendrites of hippocampal neurons (Osten *et al.*, 1998) and enriched in the PSD (Song *et al.*, 1998). Moreover, intracellular perfusion of neurons with synthetic peptide that competes with the interaction of NSF and AMPA receptor/channel subunits rapidly decreases the amplitude of miniature excitatory postsynaptic currents (Song *et al.*, 1998). These results suggest that NSF may play an essential role not only in presynaptic neurotransmitter release, but also in the function of postsynaptic AMPA receptor/channels.

Narp (neuronal activity–regulated pentraxin) is a secreted protein encoded by an immediate-early gene (IEG) regulated by synaptic activity in brain (see Table 1; O'Brien *et al.*, 1999). Narp has been shown to be selectively enriched at excitatory synapses on neurons in hippocampus as well as spinal cord, and supposed to be an extracellular factor involved in the organization of excitatory synapses.

Narp binds to GluR1-3 subunits of AMPA receptor/channels. In cells cotransfected with Narp and AMPA receptor subunits, Narp proteins multimerize to form large surface clusters, which coaggregate with AMPA receptor/channels subunits. These data suggest that Narp is an extracellular aggregating factor for AMPA receptor/channels at excitatory synapses.

5. mGlu Receptors

Homer 1a (also called Vesl for VASP/Ena-related gene upregulated during seizure and LTP) binds to group 1 mGlu receptors (which are G protein–coupled receptors) (see Table 2; Brakeman *et al.*, 1997; Kato *et al.*, 1997). More recently, new members (Homer 1b, 1c, 2, and 3) of the Homer family have been identified and shown to bind to mGlu receptors (Xiao *et al.*, 1998). All members of the Homer family contain a domain that is homologous to the enabled/VASP (vasodilator-stimulated phosphoprotein) homology (EVH1) domain (Tu *et al.*, 1998; see Section II.G.2). In contrast to Homer 1a, all new members are constitutively expressed (3′ untranslated region sequences of them don't encode AUUUA repeats, which are included in Homer 1a and implicated in destabilizing mRNAs of IEGs) and possess a C-terminal coiled-coil (CC) domain (Fig. 3) that mediates self-multimerization (Table 3). CC-Homers forming natural complexes that cross-link mGlu receptors are enriched at the PSD. Homer 1a, which lacks a CC domain in its C-terminal region, does not multimerize and blocks the association of mGlu receptors with CC-Homer complexes. These data suggest a model in which the expression of Homer 1a induced by high neural activity competes

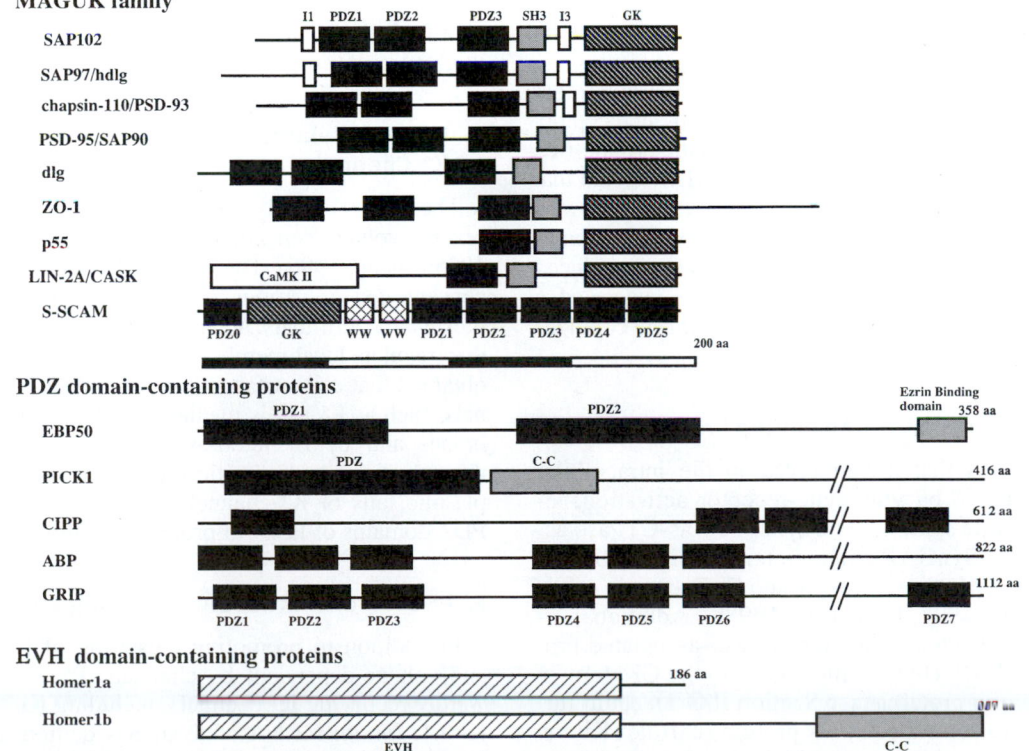

FIGURE 3. The structure of anchoring proteins. I1, insert 1; I3, insert 3; GK, guanylate kinase–like domain; CaMK II, Ca²⁺/calmodulin-dependent protein kinase II; C-C, coiled-coil domain; WW: WW domain; EVH, EVH domain.

TABLE 3 Protein-Protein Interactions of PDZ Domain–Containing
Proteins Other Than Receptor/Channels

PDZ-containing protein	Binding molecule	Interacting domain	Reference
PSD-95/SAP90	GKAP/SAPAP/DAP	GK	1
	Neuroligin	PDZ-3	2
	SynGAP	PDZ-1, -2, -3	3
	CRIPT	PDZ-3	4
	BEGAIN	GK	5
	MAPIA	GK	6
	nNOS	PDZ-2	7
PSD-93/chapsyn10	MAPIA	GK	6
SAP97/hdlg	MAPIA	GK	6
S-SCAM	GKAP/SAPAP/DAP	GK	8
	Neuroligin	PDZ-1	8
GRIP	ABP	ND[a]	9
ABP	GRIP	PDZ-4, -5, -6	9
	ABP[a]	PDZ-4, -5, -6	9
CIPP	Neurexin	PDZ-3	10
	Neuroligin	PDZ-2	10
CASK	Neurexin	PDZ (CASK has only one PDZ domain)	11
nNOS	CAPON	PDZ (nNOS has only one PDZ domain)	12

[a]ND, not determined. ABP can homomultimerize by itself.
References: (1) Takeuchi *et al.*, 1997; Kim *et al.*, 1997; Satoh *et al.*, 1997; (2) Irie *et al.*, 1997; (3) Kim *et al.*, 1998b; (4) Niethanmmer *et al.*, 1998; (5) Deguchi *et al.*, 1998; (6) Brenman *et al.*, 1998; (7) Brenmen *et al.*, 1996; (8) Hirano *et al.*, 1998; (9) Srivastava *et al.*, 1998; (10) Kurschner *et al.*, 1998; (11) Hata *et al.*, 1996; (12) Jaffrey *et al.*, 1998.

with constitutively expressed CC-Homers to modify synaptic mGlu receptor properties.

More recently, Tu and colleagues (1998) have shown that Homer proteins form a physical tether linking mGlu receptors with the IP3 receptors, both of which possess the proline-rich **Homer ligand motif** (PPXXFr; see Section II.G.2), and that these receptors are coimmunoprecipitated from brain with Homer. Expression of Homer 1a, the IEG form of Homer members, could decrease and delay mGlu receptor–induced intracellular Ca^{2+} release. These studies have suggested a novel mechanism in Ca^{2+} signaling in which an IEG could directly modify a specific synaptic function.

6. Scaffolding Functions of Anchoring Proteins

Various proteins that are involved in the intracellular signaling stimulated by glutamate-receptor activation, including nitric oxide synthase and synaptic Ras-GTPase activating protein (SynGAP), also interact with PDZ domain–containing proteins (Figs. 4 and 5; Brenman *et al.*, 1996; Chen *et al.*, 1998; Kim *et al.*, 1998b). Additional organizer proteins, such as guanylate kinase–associated protein (GKAP/SAPAP/ DAP), which binds to the GK domain of PDZ-containing proteins (see Section II.G.1), could further contribute to form a complex protein scaffold (Kim *et al.*, 1997). It is possible that the PDZ domain–containing proteins are, therefore, not only involved in the maintenance of the receptor constellation at synapses but might

also facilitate efficient signaling by keeping key enzymes in close proximity.

E. PSD-95-Related Proteins and Voltage-Gated K⁺ Channels

The PSD-95 family proteins interact also with subunits of several voltage-dependent K^+ (Kv) channels (Kim *et al.*, 1995; Sheng, 1996). The voltage-dependent K^+ channels are concentrated at various neuronal microdomains, including presynaptic terminals, nodes of Ranvier, and dendrites, where they regulate local membrane excitability. Evidence has been obtained that cell-surface clustering of *Shaker*-type K^+ channels, such as Kv1.4, is mediated by PSD-95 family of membrane- and cytoskeleton-associated proteins. This occurs through direct and specific binding of the C-terminal cytoplasmic tails of K^+-channel subunits to first and/or second PDZ domains of PSD-95 protein (see Table 2).

F. PSD-95 and Inwardly Rectifying K+ Channels

In addition to promoting clustering of NMDA receptors and voltage-dependent K^+ channels, PSD-95 also binds inwardly rectifying K^+ channels, including Kir2.1, Kir2.3, and Kir4.1. *In situ* mRNA analysis has demonstrated that both Kir2.3 and PSD-95 are specifically enriched and colocalized in granule cells of the dentate gyrus region of the hippocampus (Cohen *et al.*, 1996). PSD-95 and Kir2.3 were actually

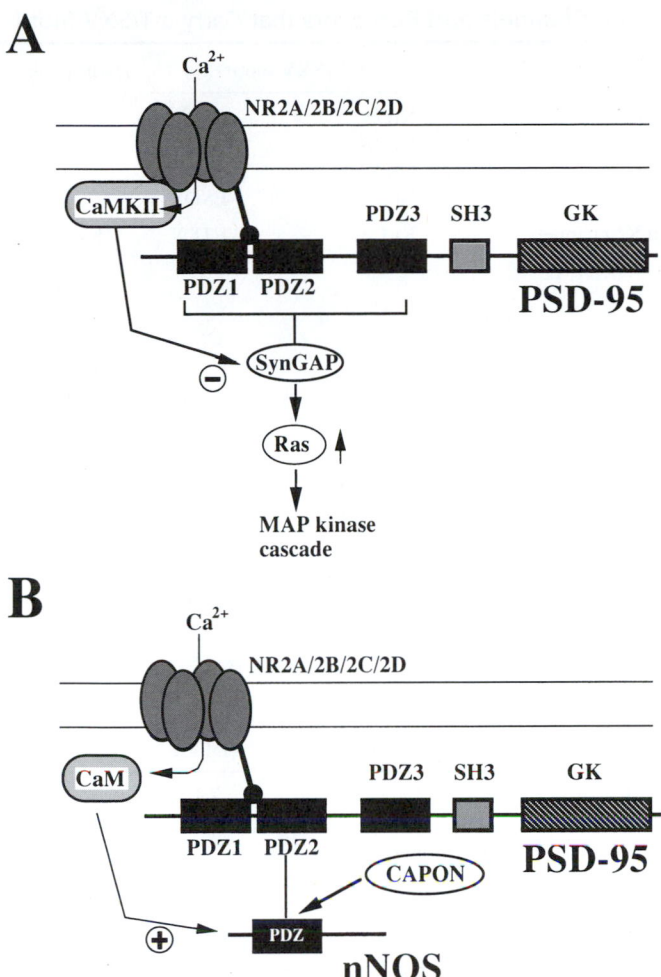

FIGURE 4. Scaffolding functions of SDP-95/SAP90. (A) Scaffolding of MAP kinase signaling. Active SynGAP at postsynaptic densities keeps the steady-state level of active Ras low near the synapse because of its rapid catalyzing hydrolysis of Ras-GTP to Ras-GDP. Activation of NMDA receptors produces an influx of Ca^{2+} that activates CaMKII at the postsynaptic density. CaMKII then phosphorylates and inactivates SynGAP, releasing the brake on the accumulation of active Ras-GTP and leading to increased activation of the MAP kinase cascade. In this manner, activation of the NMDA receptor may potentiate the action of any signal that leads to formation of Ras-GTP. Such potentiation would constitute yet another form of coincidence detection by the NMDA-type glutamate receptor. (B) Scaffolding of nNOS signalling. NMDA receptors are coupled to nNOS through a PSD-95. These interactions are mediated by PDZ domains. In this complex, nNOS is situated close to NMDA receptor–modulated calcium influx. Binding of CAPON results in a reduction of NMDA receptor/PSD-95/nNOS complexes, leading to decreased access to NMDA receptor–gated Ca^{2+} influx and a catalytically inactive enzyme.

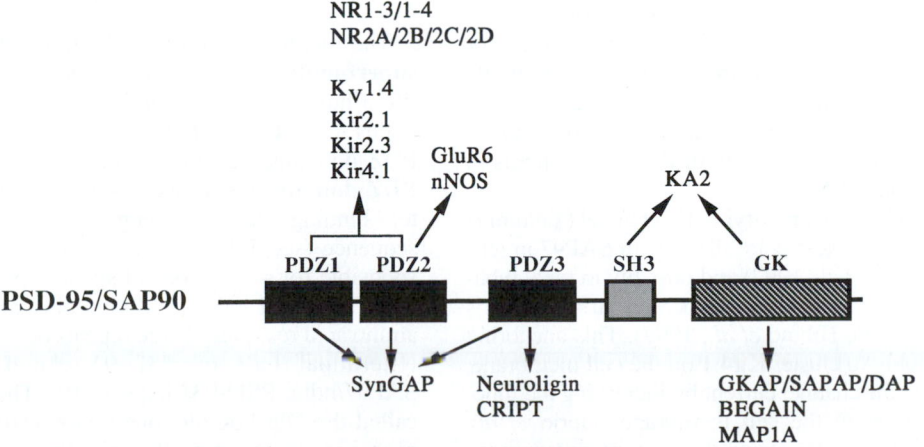

FIGURE 5. Domain-dependent protein-protein interactions of PSD-95/SAP90.

TABLE 4   Ion Channels and Receptors that Carry a T/SXV Motif

	E-T/SXV motif		Hydrophobic T/SXV motif	
NMDA receptors	NR2A	–ESDV		
	NR2B	–ESDV		
	NR2C	–ESDV		
	NR2D-2	–ESDV		
Voltage-gated K⁺ channels	Kv1.4	–ETDV	Kv1.1	–LTDV
	Kv1.5	–ETDV	Kv1.2	–LTDV
			Kv1.3	–FTDV
			Kv1.6	–LTEV
			Kv3.2b	–PSIL
			Kv3.3b	–PSIL
			Kv4.1	–ISSL
			Kv4.2	–VSAL
			Kv4.3	–VSAL
Inwardly rectifying K⁺ channels	IRK1/Kir2.1	–ESEI		
	IRK2/Kir2.2	–ESEI		
	IRK3/Kir2.3	–ESRI		
	GIRK2C/			
	Kir3.2c	–ESKV		
K⁺ channels with two-pore domains			cTBAK-1 (TASK-1)	–RSSV
G-protein-coupled receptors	β1 Receptor	–ETVV	5HT2A	–VSCV
			5HT2C	–ISSV
	MAS oncogene	–ETVV	VIP	–VSLV

S: possible phosphorylation site by PKA

Various kinds of channels and receptors can bind to the PDZ domains of PSD/SAP family proteins via a T/SXV motif in their C-termini.

coimmunoprecipitated from hippocampal synaptic membranes (Cohen *et al.*, 1996). These data suggest that Kir2.3 and PSD-95 bind to form a protein complex in CNS neurons.

The **neuronal G protein–gated K⁺ (K$_G$) channels** are heterotetramers of Kir3.1/GIRK1 and Kir3.2/GIRK2. In substantia nigra (SN), however, K$_G$ channels are composed of only Kir3.2 subunits. An immunological study has shown that Kir3.2 is localized specifically at the postsynaptic membrane on the dendrites of dopaminergic neurons (Inanobe *et al.*, 1999). Biochemical studies show that at least some of the K$_G$ channels in SN are composed of the splicing variants Kir3.2a/GIRK2a and Kir3.2c/GIRK2c. Kir3.2c, but not Kir3.2a, possesses the PDZ domain–interacting motif (–ESKV) in the C-terminal domain. The heterologously expressed K$_G$ channels composed of Kir3.2a and Kir3.2c or Kir3.2a alone are activated by G-protein-stimulation, while expression of Kir3.2c alone is not. These data suggest that the Kir3.2 splicing variants play distinct roles in the control of function and localization of some of the KG channels in dopaminergic neurons of SN.

Kir4.1, a glial cell inwardly rectifying K⁺ channel (Takumi *et al.*, 1995), is a channel protein colocalized with SAP97 in retinal Müller cells (Horio *et al.*, 1997) and possibly in renal tubular epithelium and the marginal cells in cochlear stria vascularis (Table 1; Ito *et al.*, 1996; Hibino *et al.*, 1997). This anchoring protein seems not only to cluster Kir4.1 on the cell membrane, but also to stimulate the channel current by increasing the functional channel number in the cell membrane (Horio *et al.*, 1997). In addition, it was also found that a **cardiac two-pore K⁺ channel** (cTBAK) possesses the C-terminal domain interacting with PSD-95 family proteins (Table 4; Kim *et al.*, 1998a). Therefore, many other K⁺ channels might also be under control of this family of anchoring proteins in various organs.

Based on the multiple protein-protein interactions, and the observation that PSD-95 and related proteins form oligomers, a scaffolding role for PSD-95 in organizing signaling cascades at the PSD has been proposed (Cohen *et al.*, 1996; Gomperts, 1996).

G. Protein-Protein Interactions of PDZ, SH3, GK, and EVH Domains

1. PDZ, SH3, and GK Domains

The **membrane-associated guanylate kinase (MAGUK) superfamily** of proteins is characterized by the existence of Src 3 homology (SH3) and guanylate kinase–like (GK) domains in the C-terminal region in addition to the presence of PDZ domains (see Fig. 3; Gomperts, 1996; Sheng, 1996). **PDZ domains** are viewed as a modular arrangement of protein-binding sites that recognizes a short consensus peptide sequence (see Tables 2 and 4) of large proteins and are responsible for some types of specific protein-protein association (Sheng, 1996). PDZ domains are composed of ~90 amino acid residues. Three repeats of the domains exist in the N-terminal half of PSD-95-related proteins (PSD-95, SAP97/hdlg, PSD-93/chapsyn110). The PDZ domain is also called the Dlg homologous region (DHR) or the GLGF repeat, because most of the initially identified PDZ domains contain Gly-Leu-Gly-Phe in the sequences. The C-terminal regions of the NR2 subunits of the NMDA receptor, *Shaker*-type K^+ channels, and Kir channels possess four highly conserved amino acids (–(E)S/TXV– motif) that are specifically recognized by the PDZ domains in PSD-95-related proteins (see Tables 2 and 4; Doyle *et al.*, 1996; Gomperts, 1996; Sheng, 1996). X-ray crystallography has revealed that the third PDZ domain of PSD-95 is composed of six β-sheets and two α-helices. The C-terminal peptide (–TDV and –QTSV) binds to the groove between the second β-sheet and the second α-helix and a carboxylate-binding loop provided by GLGF (Cabral *et al.*, 1996; Doyle *et al.*, 1996).

The ability of PDZ domains to function as independent modules for protein-protein interaction suggests that PDZ domain–containing polypeptides may be widely involved in the organization of proteins at specialized membrane domains (Kim *et al.*, 1995; Sheng, 1996). While *Shaker*, Kir, and NR2 proteins do not cluster when expressed alone, co-expression of them with PSD-95, SAP97, or PSD-93 results in the coclustering of channel and anchoring proteins (Kim *et al.*, 1995, 1996; Sheng, 1996; Horio *et al.*, 1997). This emphasizes the importance of PSD-95 proteins in directing the distribution of NMDA, voltage-dependent K^+, and Kir channels. The possible mechanisms of membrane protein clustering by PSD-95 have been hypothesized as follows.

It has been proposed that PSD-95 forms a dimer, using a head-to-head linkage that is mediated by disulfide bonds between the conserved N-terminal regions (Hsueh *et al.*, 1997). However, recently it was shown that the third and fifth cysteines from the N-terminal of PSD-95 are palmitoylated, which is indispensable for its clustering and targeting (Craven *et al.*, 1999). The third PDZ domain of PSD-95 interacts with neuroligin, a neuronal cell adhesion molecule (Irie *et al.*, 1997). These findings suggest that PSD-95 assembles receptors and channels, and then fixes them at the specialized membrane domain through its interaction with cell adhesion molecules.

Recent studies have revealed that although it has no guanylate kinase activity (Kistner *et al.*, 1995), the GK domain of the PSD-95/SAP90 family also acts as an anchoring domain to GKAP/SAPAP/DAP (Takeuchi *et al.*, 1997; Kim *et al.*, 1997; Satoh *et al.*, 1997), MAP1A (Brenman *et al.*, 1998), BEGAIN (Deguchi *et al.*, 1998), and the KA2 subunit of kainate receptor/channels (Garcia *et al.*, 1998) (the KA2 subunit also binds to the SH3 domain of PSD-95/SAP90, SAP102, as described in Section III.D.3) (see Fig. 5). GKAP/SAPAP/DAP also binds to a synaptic protein named S-SCAM (Hirano *et al.*, 1998).

S-SCAM is a MAGUK with an inverse structure composed of a GK domain at the N-terminus followed by two WW and five PDZ domains (see Fig. 3 and Table 3; Hirano *et al.*, 1998). The PDZ-5 domain of S-SCAM binds to NMDA receptors and the PDZ-1 binds to neuroligins (Hirano *et al.*, 1998). S-SCAM and PSD-95 are likely to concomitantly provide synaptic scaffolds for receptors and cell adhesion molecules.

2. EVH Domain

As described in Section II.D.5, Homer and its homologs bind to the group 1 mGlu receptors. The N-terminal region of

Homers, which is required for binding to mGlu receptors, was originally thought to possess homology to PDZ domains (Brakeman *et al.*, 1997). Others, however, pointed out that the sequence of Homer exhibits close homology to the members of the EVH family (enabled/VASP homology; Gertler *et al.*, 1996, Kato *et al.*, 1997, Ponting and Phillips, 1997). The proteins that contain the EVH1 domain include *Drosophila* enabled (termed Mena in mouse; Gertler *et al.*, 1996), vasodilator-stimulated phosphoprotein (VASP; Haffner *et al.*, 1995), and the Wiscott-Aldrich syndrome protein (WASP; Symons *et al.*, 1996). The putative EVH domain in Homer is highly conserved in the Homer-related family members (Xiao *et al.*, 1998). The EVH proteins, VASP and Mena, bind a proline-rich sequence, E/DFPPPPXD/E (Niebuhr *et al.*, 1997). This motif is present in ActA protein, which plays a role in a dramatic reorganization of the actin-based cytoskeleton in *Listeria* monocytogenes. The motif is also found in zyxin and vinculin, which are eukaryotic analogs of ActA. Group 1 mGlu receptors and inositol triphosphate (IP3) receptors possess proline-rich amino acids (–PPXXFr– motif) (see Table 2; Tu *et al.*, 1998). Actually, Homer members may form a physical tether linking mGlu receptors with the IP3 receptors, because IP3 receptors have been coimmunoprecipitated as a complex with Homer and mGlu receptors from brain, as described in Section II.D.5 (Tu *et al.*, 1998).

H. Microfilament Cross-Linking Proteins and Ion Channel Expression in Postsynaptic Domains

Other cytoskeleton-related proteins, such as α_2-actinin, are also involved in the regulation of NMDA receptor localization within the postsynaptic domain. The integrity of postsynaptic actin filaments mediated by α_2-actinin is, indeed, important for NMDA receptor function. A specific biochemical association between α_2-actinin and the cytoplasmic tail of the NMDA receptor subunits NR1 and NR2B has also been demonstrated (see Table 1). It has, therefore, been suggested that the actin-binding protein α_2-actinin could serve as an anchor protein to mediate NMDA receptor association with the postsynaptic actin cytoskeleton (Wyszynski *et al.*, 1997, 1998). Thus, the interaction of α_2-actinin with NMDA receptor proteins could be involved in the plasticity of excitatory synapses. In addition, it has been reported that overexpression of a dominant negative truncated α_2-actinin construct, which competes with the interaction of endogenous α_2-actinin with NMDA receptor/channels and actin, influences the run-down properties of the NMDA receptor/channels (Zhang *et al.*, 1998). α_2-Actinin could, therefore, not only act as a microfilament cross-linking protein but also as a modulator of the NMDA receptor/channels.

Brain spectrin, **fodrin**, a microfilament cross-linking protein, is a protein that links membrane proteins to the actin cytoskeleton. It has been shown to interact directly with the NR1-1a, NR2A, and NR2B subunits of the NMDA receptor but not with the GluR1 subunit of the AMPA receptor (see Tables 1 and 2; Wechsler and Teichberg, 1998). Fodrin binds to NR2B at sites distinct from those of α_2-actinin and members of the PSD-95/SAP90 family (see Table 2). Biochemical and electron microscopy data show that fodrin and NMDA receptor/channels colocalize at PSD. The fodrin-NR2B interactions are inhibited by Ca^{2+} at ~100 μM but not by Ca^{2+}/calmodulin (CaM) or by Ca^{2+}/CaM-dependent protein kinase II–mediated NR2B phosphorylation. The fodrin–NR1-1a interaction is unaffected by Ca^{2+} even at 1 mM but inhibited by CaM and also protein kinase A– and C–mediated phosphorylation of NR1-1a. The interaction of the NMDA receptor/channels with fodrin is extensively regulated by synaptic plasticity–related factors (Ca^{2+} and protein phosphorylation). This highly regulated interaction may underlie the morphological changes of neuronal dendrites that occur on increase in synaptic activity and plasticity.

The dissolution of microfilaments within dendritic spines by the depolymerizing agent latrunculin A shifted a significant portion of NMDA receptor clusters from synaptic to nonsynaptic locations (Allison et al., 1998). The clusters of AMPA receptors appear to be dependent on the microfilament integrity in pyramidal neurons but not in GABA neurons. The localization of a receptor subtype could, thus, be modulated differentially according to its microenvironment within different populations of neurons.

I. Neurofilaments and NMDA Receptors in Postsynaptic Membranes

A recent study has shown that NMDA receptors can bind not only to cytoskeleton-associated proteins, but also to the cytoskeleton itself. **Neurofilaments** (NFs) are intermediate filaments often considered to play a role in specifying the axonal diameter of large myelinated axons (Lee and Cleveland, 1994). Although less abundant and less organized, NFs are also present in neuronal dendrites (Peters et al., 1991; Benson et al., 1996). Ehlers and colleagues (1998) have identified within the yeast two-hybrid system the direct interaction of 68-kDa NF subunit NF-L (see Table 1) with NR1-1 but not NR1-2. NF-L shows a structure composed of head, rod, and tail domains. The interaction of NF-L with NR1-1 occurs between the C-terminus of NR1-1 and the rod domain of NF-L, a domain that interacts with other neurofilament subunits, NF-M and NF-H (Heins et al., 1993), and with other cytoskeletal proteins, such as microtubule-associated protein-2 (MAP2) and β-spectrin (Flynn et al., 1987; Frappier et al., 1991, 1992). Furthermore, it was found that NR1 and NF-L coexist in the dendrites of cultured hippocampal neurons (Ehlers et al., 1998). As described in the last section, the C-termini of NR1 and NR2B can also interact with α_2-actinin, which binds to actin. Because NF-L binds to β-spectrin, one of the actin-binding proteins (Frappier et al., 1991, 1992), it is postulated that a complex of NR1, NR2B, β-spectrin, NF-L, and actin is formed at the postsynaptic sites. Thus, the cellular distribution and functional properties of NMDA receptors may be regulated by the cytoskeleton-associated and cytoskeletal proteins.

J. Microtubule and Microtubule-Associated Proteins Interacting with NMDA Receptors in Postsynaptic Membranes

MAP1A (microtubule-associated protein 1A) has been reported to interact with PSD-93 (see Table 3; Brenman et al.,

1998). Immunological studies have shown that PSD-93 and MAP1A are colocalized in cell bodies and dendritic processes of cerebellar Purkinje neurons and that PSD-93 is concentrated along dendritic microtubules and at the PSD of the neurons (Brenman et al., 1998). In addition, the microtubule-associated protein, CRIPT (cysteine-rich interactor of PDZ3) binds to the third PDZ domain of PSD-95 (see Fig. 5 and Table 3; Niethammer et al., 1998). These data suggest that MAP1A and CRIPT are implicated in tethering PSD-93 and PSD-95, respectively, and NMDA receptor/channels to cytoskeletal tubulin. van Rossum and coworkers (1999) have shown the direct physical interaction between tubulin and the cytoplasmic C-terminal domains of the NR1 and NR2B subunits of NMDA receptor/channels (see Table 1). Although PSD-95 and CRIPT have been reported to be localized at the PSD, whether the microtubules are actually present within spines has not yet been elucidated (Harris and Kater, 1994; van Rossum and Hanisch, 1999). Further studies are needed to identify the effect of microtubules on the localization and function of NMDA receptor/channels.

In summary, evidence has been obtained to indicate that cytoskeleton-associated proteins (including rapsyn, ankyrin, gephyrin, PSD-95, α_2-actinin) and the cytoskeleton itself (NF-L, tubulin) direct and maintain the domain-dependent distribution of several ion channels (such as nAChRs and Na^+ and K^+ channels, as well as glycine, GABA, NMDA, kainate, and AMPA receptor/channel complexes and mGlu receptors) within the plasma membrane (see Tables 1 and 2).

III. Role of Phosphorylation in Cytoskeletal Protein–Directed Clustering of Ion Channels

Targeting of ion channels to discrete plasma membrane domains is a dynamic process and believed to be commonly regulated by additional enzymatic processes. Most commonly, evidence for protein phosphorylation has been obtained.

A. Tyrosine Kinase Activity and Clustering of Nicotinic Acetylcholine Receptors

During development, nAChRs are clustered to postsynaptic muscular junctions (NMJs) in response to release of the nerve-derived factor agrin (see Fig. 2, Wallace, 1992). Neural agrin-mediated signal transduction is mediated by receptor tyrosine kinase (Glass et al., 1996).

Tyrosine phosphorylation of nAChR β-subunit precedes aggregation of nAChRs. Inhibitors of tyrosine kinases abrogate agrin-induced aggregation of nAChRs (Wallace et al., 1991). Although it has not been demonstrated directly that phosphorylation of the nAChR β-subunit is required for aggregation of nAChRs, tyrosine kinases participate in the signaling that is activated by neural agrin (Wallace 1994; Meier et al., 1995; Ferns et al., 1996). Targeted disruption of the gene encoding MuSK, a receptor tyrosine kinase selectively localized at the postsynaptic muscle surface, disrupts neuromuscular synapse formation (DeChiara et al., 1996). Moreover, muscle cells taken from these animals are incapable of aggregating nAChRs in response to neural agrin (Glass et al., 1996). These results show that nAChR clustering requires not only rapsyn, as described in Section II.A, but also agrin and MuSK, which is required for neural agrin-mediated signaling. However, MuSK, is not, by itself, the agrin receptor, because MuSK and neural agrin don't bind to each other (Glass *et al.*, 1996). The receptor of neural agrin and the mechanism of phosphorylation of MuSK induced by neural agrin has not yet been found.

In cultured myotubes from rapsyn-deficient mice, although neural agrin still induces rapid phosphorylation of MuSK, the nAChR β-subunit does not become tyrosine phosphorylated, and nAChRs do not form clusters (Elizabeth *et al.*, 1997). This suggests that rapsyn serves as an adapter molecule to link MuSK activation to further downstream events that are important for the localization of molecules to the postsynaptic apparatus. The interaction between rapsyn and MuSK does not occur directly (Apel *et al.*, 1997). This result suggests that a linker molecule, RATL, confers binding (see Fig. 2).

B. Protein Kinase A Activity and Synaptic Channel Density

It has been recently indicated that phosphorylation by protein kinase A (PKA) is important for the interaction between an inwardly rectifying K$^+$ channel, Kir2.3/IRK3, and a cytoskeletal protein, PSD-95. A serine residue (Ser 440), located within the C-terminal tail of Kir2.3, is critical not only for interaction with PSD-95, but also serves as a substrate for phosphorylation by PKA. Actually, stimulation of PKA in intact cells causes rapid dissociation of Kir2.3 from PSD-95. Therefore phosphorylation and dephosphorylation of this serine residue in Kir channels may regulate the dynamic interaction between K$^+$ channels and the cytoskeleton (Cohen *et al.*, 1996).

A similar PKA phosphorylation site exists in the C-terminal tail of cTBAK-1 (TASK-1) that can interact with PSD-95 (see Table 4; Duprat *et al.*, 1997; Kim *et al.*, 1998). Thus, PKA-dependent modulation of the interaction between PSD-95 and the K$^+$ channel might be involved in regulation of channel function more widely than currently recognized.

Anchoring of PKA also appears important in the regulation of synaptic function. Specifically, it has been shown that anchoring of PKA by A-kinase–anchoring proteins (AKAPs) is required for the modulation of AMPA/kainate receptor/channels (Rosenmund *et al.*, 1994). Intracellular perfusion of hippocampal neurons with peptides derived from the conserved kinase-binding region of AKAPs prevents PKA-mediated regulation of AMPA/kainate currents as well as fast excitatory synaptic currents. Thus, positioning of kinases by anchoring proteins near their substrates, including ion channel complexes, may be essential in the regulation of the electrical properties of a cellular membrane.

IV. Regulation of Ion Channel Function by Cytoskeletal Proteins

As described earlier, the cytoskeleton forms a fibrillar network structure throughout the cytosol. Ion channels and other ion transport molecules within the plasma membrane

are surrounded by cytoskeletal strands, in particular, actin filaments (Ruknudin *et al.*, 1991; Horber *et al.*, 1995). Actin accounts for more than 20% of total cell proteins. It has been shown that actin and actin-binding proteins couple to several ion channels and ion transport molecules (Cantiello, 1995). This is of importance in epithelia and neurons, where maintenance of ion channels and transporters within specific membrane domains is vital for their normal function (Smith and Benos, 1996).

In addition to structural interactions, there is growing evidence for functional interactions between ion channels and the adjacent actin microfilament network in epithelia, nervous tissues, and heart (Cantiello and Prat, 1996).

A. Epithelial Ion Channel Function

1. Regulation of Epithelial Na$^+$ Channels (ENaC) by the Cytoskeleton

It has been established that epithelial Na$^+$ channels (ENaC) are linked to cytoskeleton structures, including microfilament cross-linking proteins such as ankyrin and spectrin (Smith and Benos, 1996). Using specific antibodies, the Na$^+$ channels were found to be colocalized at the apical membrane with actin and apically associated isoforms of ankyrin and spectrin (Smith *et al.*, 1991; Cantiello and Prat, 1996). A proline-rich region in the ENaC channel mediates its binding to the SH3 region of α-spectrin, which in turn maintains the polarized distribution of the channel to the apical membrane of renal epithelia (Rotin *et al.*, 1994). While such interaction may serve to determine the spatial distribution of ENaC channels, colocalization of actin filaments with ENaCs has also been related to the functional regulation of Na$^+$ channel activity (Cantiello, 1995). Agents that depolymerize actin filaments, such as cytochalasin D (see Fig. 1A), enhanced the open probability of a 9-pS ENaC channel in renal epithelial cells (Cantiello *et al.*, 1991). In contrast, DNase I, which stabilizes the pool of monomeric actin (see Fig. 1A), lacked such effect. The length of the actin filaments appears to be a determinant of channel activation. Addition of short actin filaments to excised membrane patches enhanced the probability for ENaC channels to be open, yet whenever actin was added after being polymerized to achieve predominantly long filaments, no Na$^+$-channel activation was observed (Cantiello *et al.*, 1991; Cantiello, 1995). This suggests that short actin filaments, but neither G-actin nor long actin filaments, are responsible for channel activation (Cantiello and Prat, 1996).

The actin-dependent regulation of channel activity may participate in the stretch-dependent activation of the renal ENaC (Awayda *et al.*, 1995). Under basal conditions, stabilized actin filaments may contribute to maintaining ENaC channels in the closed state, whereas actin depolymerization, either by stretch or a hormone (e.g., vasopressin), may result in channel activation by affecting the membrane environment or by interacting with other membrane-cytoskeleton proteins associated with the channel (Cantiello, 1995; Cantiello and Prat, 1996; Smith and Benos, 1996). The effect of actin on ENaC channels is modulated by phosphorylation through PKA (Prat *et al.*, 1993).

The **amiloride-sensitive renal epithelial Na$^+$ channels** are composed of three subunits of α-, β-, and γ-ENaC (Cannessa

TABLE 5   Effects of Actin Cytoskeleton on Selected Epithelial Channels

Channel	Tissue	Effect
9-pS Na⁺ channel	Renal epithelial (A6) cells	Activation by short actin filaments
CFTR	Transfected adenocarcinoma cell line	Activation by long actin filaments
33-pS Cl⁻ channel	Proximal tubular cell line	Activation by long actin filaments
30-pS K⁺ channel	Principal cells in cortical collecting duct	Inactivation by actin filament disrupters

et al., 1994). Mutations in β- and γ-subunits of ENaC are responsible for Liddle's syndrome (Shimkets *et al.*, 1994; Hansson *et al.*, 1995). These mutations cause deletion or dysfunction of PY motifs (PPPXY) of the subunits, resulting in loss of interaction of the ENaC with Nedd4 (neural precursor cells expressed developmentally down-regulated) protein through their WW domain (Staub *et al.*, 1996). Because Nedd4 facilitates down-regulation of the ENaC channel in normal conditions (Fig. 6A) and has no detectable effect on its single-channel properties, the Liddle mutation (Fig. 6B) causes an increase of the number of ENaC channels in the membrane, resulting in its high activity (Goulet *et al.*, 1998).

Nedd4 is a multimodular ubiquitin protein ligase (E3) composed of a Ca^{2+}/lipid-binding (C2) domain, three or four WW domains, and a C-terminal ubiquitin protein ligase hect domain (Staub *et al.*, 1996). Recently, Plant and colleagues (1997) showed that the C2 domain of Nedd4 is implicated in its Ca^{2+}-dependent localization in epithelial cells. Nedd4, endogenously expressed in MDCK cells, is redistributed from the cytosolic to the apical and lateral membrane domains. The C2 domain of Nedd4, expressed as a GST fusion protein, is sufficient to bind cellular membranes in a Ca^{2+}-dependent manner. The Nedd4 lacking its C2 domain and stably expressed in MDCK cells fails to mediate the Ca^{2+}-induced plasma membrane localization seen in wild-type Nedd4. These data indicate that the C2 domain of Nedd4 is involved in localizing the protein primarily to the apical region of polarized cells in response to Ca^{2+}.

ENaC is located at the apical membrane of epithelial cells (such as those in the distal nephron, distal colon, and lung epithelia; Staub *et al.*, 1997) and interacted with Nedd4 through its WW domains. ENaC was shown to be ubiquitinated *in vivo* although it is not yet known whether Nedd4 directly ubiquitinated (Staub *et al.*, 1997). Ubiquitination of cellular proteins usually serves to tag them for rapid degradation (Ciechanover, 1994; Jentsh and Schlenker, 1995). One possible model is that elevation of intracellular Ca^{2+} may target Nedd4 to the apical membrane where ENaC is located. This would allow the Nedd4 WW domains to associate with the channel and then the Nedd4 hect domain to ubiquitinate ENaC. Indeed, it was reported that amiloride-sensitive Na⁺-channel activity is inhibited by elevated intracellular Ca^{2+} levels.

2. Regulation of Epithelial Cl⁻ Channels by the Cytoskeleton

Chloride (Cl⁻) channels play an important role in fluid movement across epithelia. The actin cytoskeleton appears to regulate the behavior of the cystic fibrosis transmembrane regulator (CFTR), a low-conductance Cl⁻ channel predomi-

nantly expressed in the apical membrane of epithelia (Cantiello and Prat, 1996; Smith and Benos, 1996). Severing of the endogenous actin-cytoskeleton by cytochalasin D, or direct addition of exogenous actin, induces activation of Cl⁻ current in adenocarcinoma cells transfected with the human CFTR gene (Prat *et al.*, 1995). Thus, increase in the availability of short filaments (see Fig. 1A) by either disruption of preexisting filaments or *de novo* formation of new ones activates CFTR. In contrast, decrease in the number of short

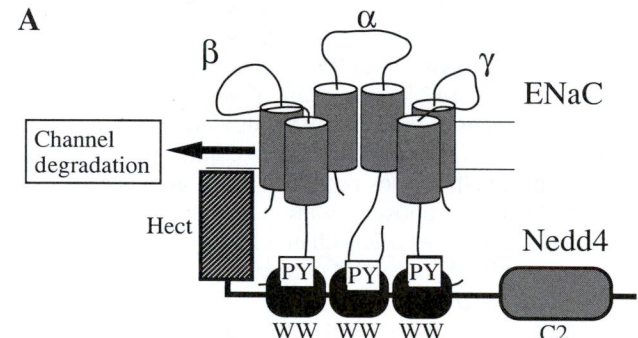

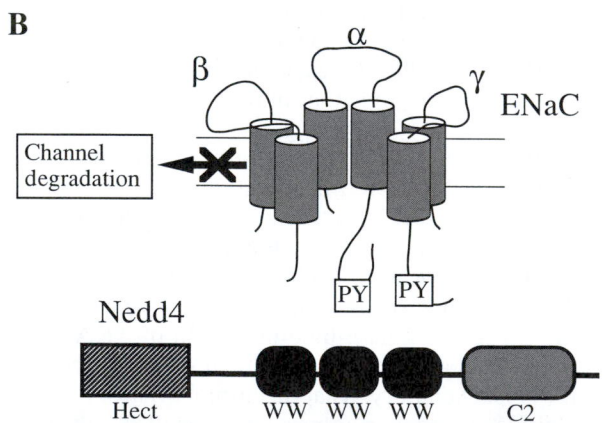

FIGURE 6.   A hypothetical model for ENaC-Nedd4 interaction in normal (A) and Liddle's syndrome (B). The three WW domains of Nedd4 bind to the PY motifs in the C-terminal of α-, β-, and γ-EnaC, thereby bringing the ubiquitin ligase (hect) domain in close proximity to EnaC. This allows the ubiquitination and subsequent degradation of the channel by proteosomes, resulting in decrease in channel numbers. In Liddle's syndrome, deletions/mutations within the PY motifs of β-ENaC (or γ-ENaC) lead to abrogation of Nedd4-WW binding, resulting in a lack of the putative suppressive effect of Nedd4, which may explain the increase in channel activity associated with the Liddle's phenotype. The role of the C2 domain is not known, but it may be associated with channel mobilization from the apical membrane.

actin filaments by preventing actin polymerization using DNase I, for example, or by bundling filaments with filamin, inhibits CFTR-associated channel activity (Cantiello and Prat, 1996).

In this regard, the actin-dependent regulation of CFTR activity appears similar to that of epithelial Na^+ channels (see Section IV.A.1; Cantiello, 1995), since long filamentous actin maintains the channels in the closed state while severing of actin into short filaments activates channels (Cantiello, 1995; Cantiello and Prat, 1996; Smith and Benos, 1996). Comparison of amino acid sequences of CFTR with known actin-binding proteins, such as severin and filamin, revealed putative actin-binding domain(s) within the nucleotide-binding folds of CFTR (Prat et al., 1995).

It should be pointed out, however, that the effect of the F-actin network–dependent regulation of Cl^- channels may not be uniform. For example, opening of a different Cl^- conductance, such as the 33-pS Cl^- channel present in renal proximal tubule epithelia or the channel responsible for Cl^- conductance in bronchial epithelia, is actually inhibited by cytochalasin D (Suzuki et al., 1993; Hug et al., 1995). The signaling pathway of the regulatory cell volume decrease, which includes disruption of F-actin, has been suggested to mediate the activation of a 305-pS Cl^- channel during cell swelling in renal collecting duct cells (Schwiebert et al., 1994). Regardless of the outcome on channel activity resulting from the modification in actin microfilament structure, these experimental data point toward a functional interaction between the actin cytoskeleton and epithelial Cl^- channels.

3. Regulation of Epithelial K^+ Channels by the Cytoskeleton

The membrane cytoskeleton is involved in the modulation of the low-conductance K^+ channel present in the apical membrane of the cortical collecting duct (Wang et al., 1994). This K^+ channel is inactivated by application of known disrupters of actin filaments, such as cytochalasins. Phalloidin, which stabilizes actin filaments, prevents cytochalasin-induced K^+ channel inactivation. Based on such findings, it has been proposed that the actin cytoskeleton is critically involved in the interaction between epithelial K^+ channel proteins and the lipid phase of the cell membrane (Wang et al., 1994).

Taken together, these findings indicate that the activity of various epithelial Na^+, K^+, and Cl^- channels is modulated by the agents acting on cytoskeletal structures (Table 5). This, in turn, suggests a role for the submembrane cytoskeleton in the regulation of ion channel function in epithelial tissues.

B. Neuronal Ion Channel Function

In addition to compartmenting and anchoring integral membrane proteins, the neuronal cytoskeleton has also been suggested to modulate neuronal excitability and synaptic plasticity through regulation of ion channel function (Fukuda et al., 1981; Rosenmund and Westbrook, 1993). The initial observation was that cytoskeleton breakdown decreases AP upstroke in dorsal root ganglion neurons and axons, apparently through inhibition of Na^+ and Ca^{2+} channels (Matsumoto and Sakai, 1979; Fukuda et al., 1981). Thereafter, cytoskeletal breakdown in neuronal tissue was also shown to affect membrane excitability through regulation of Ca^{2+} and NMDA channels (Johnson and Byerly, 1993; Rosemund and Westbrook, 1993). Here we summarize the studies that relate to the cytoskeleton-dependent regulation of ion channel activity.

1. Regulation of Neuronal Ca^{2+} Channels by the Cytoskeleton

It was first shown that the cytoskeletal disrupter, colchicine, causes a reduction of the upstroke velocity of APs in cultured neurons. From these data, it was inferred that Ca^{2+} channels interact with microtubules in neurons (Fukuda et al., 1981). The metabolic dependence on and inactivation by intracellular Ca^{2+} of Ca^{2+} channels were found to be mediated by an allosteric interaction between channel proteins and the cytoskeleton (Johnson and Byerly, 1993). Cytoskeletal disruption (by colchicine and cytochalasin B) prevents ATP from preserving Ca^{2+} channel activity, whereas cytoskeletal stabilizers (taxol and phalloidin) reduce both dependence of the channel activity on ATP and inactivation by intracellular Ca^{2+}. An allosteric interaction between the cytoskeleton and Ca^{2+} might trigger a conformational change in the cytoskeleton, which rapidly closes the adjacent Ca^{2+} channels (Johnson and Byerly, 1993). Cytoskeletal stabilizers would reduce Ca^{2+}-induced channel inactivation by restricting the Ca^{2+}-dependent conformational change in the cytoskeleton. Thus, it is proposed that Ca-dependent inactivation of Ca^{2+} current in neurons may be related to cytoskeleton integrity (Johnson and Byerly, 1993).

2. Regulation of Neuronal NMDA Channels by the Cytoskeleton

F-actin is a major component of the cytoskeleton in postsynaptic densities and dendritic spines. It is under dynamic regulation of both Ca^{2+}, which rapidly induces depolymerization, and adenosine triphosphate (ATP), which promotes repolymerization. Actin depolymerization influences NMDA channel activity in whole-cell recordings of cultured hippocampal neurons (Rosenmund and Westbrook, 1993). Specifically, the ATP- and Ca^{2+}-dependent **run-down** of NMDA channels (a parameter used to probe channel regulation) was prevented when actin depolymerization was blocked by phalloidin. This agent binds to F-actin and shifts the equilibrium between F-actin and actin monomers (G-actin) toward the polymerized state. Cytochalasins, which enhance actin-ATP hydrolysis (Sampath and Pollard, 1991), induced NMDA channel run-down, whereas taxol or colchicine, which stabilize or disrupt microtubule assembly, had no effect (Rosenmund and Westbrook, 1993). These results were interpreted to suggest that Ca^{2+} and ATP can influence NMDA channel activity by altering the state of actin polymerization (Rosenmund and Westbrook, 1993; see Fig. 1B). Thus, actin dynamics may contribute to calcium-dependent postsynaptic events, such as long-term depression.

C. Cardiac Ion Channel Function

In addition to epithelial and neuronal ion channels, indications were obtained for a functional interaction between the cytoskeleton and Na^+, Ca^{2+}, as well as ATP-sensitive K^+ channels expressed in cardiac myocytes. These data are based primarily on the ability of agents known to affect the cytoskeleton to modulate ion channel activity.

1. Cytoskeleton Modulates Gating of Voltage-Dependent Cardiac Na^+ Channels

Agents that interfere with actin polymerization, such as cytochalasin D, reduce whole-cell peak Na^+ current and slow the current decay in ventricular cardiac myocytes (Undrovinas *et al.*, 1995). Application of cytochalasin on the cytoplasmic side of inside-out patches results in reduction of peak open probability, accompanied with long bursts of Na^+ channel openings. These results were interpreted to indicate that cytochalasin D, through effects on the cytoskeleton, induces cardiac Na^+ channels to enter a mode characterized by a lower peak open probability but a greater persistent activity, as if the inactivation rate were slowed.

2. Cytoskeleton Disrupters Regulate L-Type Cardiac Ca^{2+} Channels

Initially, it was observed that cardiac excitability can be modulated by agents that target microtubules, such as tubulin (Lampidis *et al.*, 1992). Colchicine, which dissociates microtubules into tubulin, and taxol, which stabilizes microtubules, strongly influence the kinetics of L-type Ca^{2+} channels in intact cardiac cells (Galli and DeFelice, 1994). Colchicine increases the probability of Ca^{2+} channels in the closed state, whereas taxol increases the open probability of Ca^{2+} channels. Moreover, taxol lengthens the mean open time of Ca^{2+} channels. Neither taxol nor colchicine affects the number of Ca^{2+} channels (Galli and DeFelice, 1994). Several interpretations were proposed for these findings, including a direct interaction of tubulin with Ca^{2+} channels, or alternatively an action of taxol and/or colchicine through the buffering ability of the cytoskeleton to regulate the effective concentrations of inactivating ions near the mouths of channels (Galli and DeFelice, 1994). This relates to the concept that the dynamics of current-induced inactivation are dictated by restricted and heterogeneous spaces near the membrane, as well as by a transient local buffering within cells. In this regard, the structure of the cytoskeleton surrounding the mouths of channels could contribute to both compartmentalization and buffering. Thus, alterations in the structure of the cytoskeleton within the Ca^{2+} channel's microenvironment could participate in the phenomenon of channel inactivation, and thereby in the regulation of cell excitability (Galli and DeFelice, 1994).

3. Actin Filaments Regulate Cardiac ATP-Sensitive K^+ Channel Activity

The defining property of ATP-sensitive K^+ (K_{ATP}) channels is their inhibition by intracellular ATP, whereby these channels are viewed as a link between the metabolic state and electrical excitability of a cardiac cell (Terzic *et al.*,

1995). Opening of K_{ATP} channels in the myocardium is sensitive to the mechanical distortion of the membrane (Van Wagoner, 1993), suggesting that the integrity of the microenvironment surrounding K_{ATP} channels may play a role in modulating channel activity. Indeed, cytoskeletal disrupters, DNase I (Fig. 7), and cytochalasin B (but not antimicrotubule agents), have been found to antagonize the ATP-induced inhibition of cardiac K_{ATP} channels; they produced an apparent decrease in the sensitivity of K_{ATP} channels toward ATP-induced inhibition, which was partially restored by addition of purified actin subunits (Terzic and Kurachi, 1996). Taken together, these findings may fulfill the established criteria for a disrupter of actin microfilaments to regulate a specific ion channel (Cantiello, 1995) and support the notion that DNase I acts on actin filaments to modulate K_{ATP} channel activity (Terzic and Kurachi, 1996).

The subsarcolemmal actin microfilament network may be of importance in governing not only the ATP-dependent gating of the channel, but also the sulfonylurea-dependent K_{ATP} channel regulation. In addition to ATP, a major pharmacological property of K_{ATP} channels is their sensitivity to sulfonylurea drugs, which are considered among the more specific K_{ATP} channel ligands to inhibit channel activity. DNase I, when applied to the internal surface of excised membrane patches, also impaired the action of sulfonylurea drugs on myocardial K_{ATP} channel activity (Brady *et al.*, 1996). Specifically, this high-affinity actin-sequestering protein, which depolymerizes actin filaments, decreased the apparent sensitivity of K_{ATP} channels to inhibition by a prototype sulfonylurea, glyburide. The effect of DNase appeared mediated through binding to actin molecules since cytoskeletal strands are present in excised-membrane patches, whereas other known targets of DNase are absent (Ruknudin *et al.*, 1991; Horber *et al.*, 1995). Denatured DNase I could not antagonize glyburide-induced K_{ATP} channel inhibition, which is consistent with the notion that it is the native structure of the protein that is essential for DNase I to form 1:1 molar complexes with G-actin and prevent actin filament formation (Kabsch *et al.*, 1990). Co-incubation of DNase I with excess purified actin, which forms 1:1 molar complexes with DNase, prevented DNase action on K_{ATP} channels, suggesting that unoccupied binding sites for actin binding on the DNase molecule are important for the modulation of K_{ATP} channel regulation (Terzic and Kurachi, 1996; Brady *et al.*, 1996).

Further evidence for a functional linkage of K_{ATP} channels to the actin cytoskeleton was obtained from the observation that phalloidin, an actin filament stabilizing agent, could maintain channel activity and partially restore run-down channel activity (Furukawa *et al.*, 1996). It was thus proposed that for **fully activated** channels, long and polymerized F-actin filaments are required. **Partially run-down** channels were associated with short actin filaments capped by actin-binding proteins. **Completely run-down** channels were related to depolymerized G-actin (Furukawa *et al.*, 1996). Taken together, these results could be interpreted to indicate that cardiac K_{ATP} channels can be regulated by the assembly and disassembly of the actin cytoskeleton network (Terzic and Kurachi, 1996; Brady *et al.*, 1996; Furukawa *et al.*, 1996).

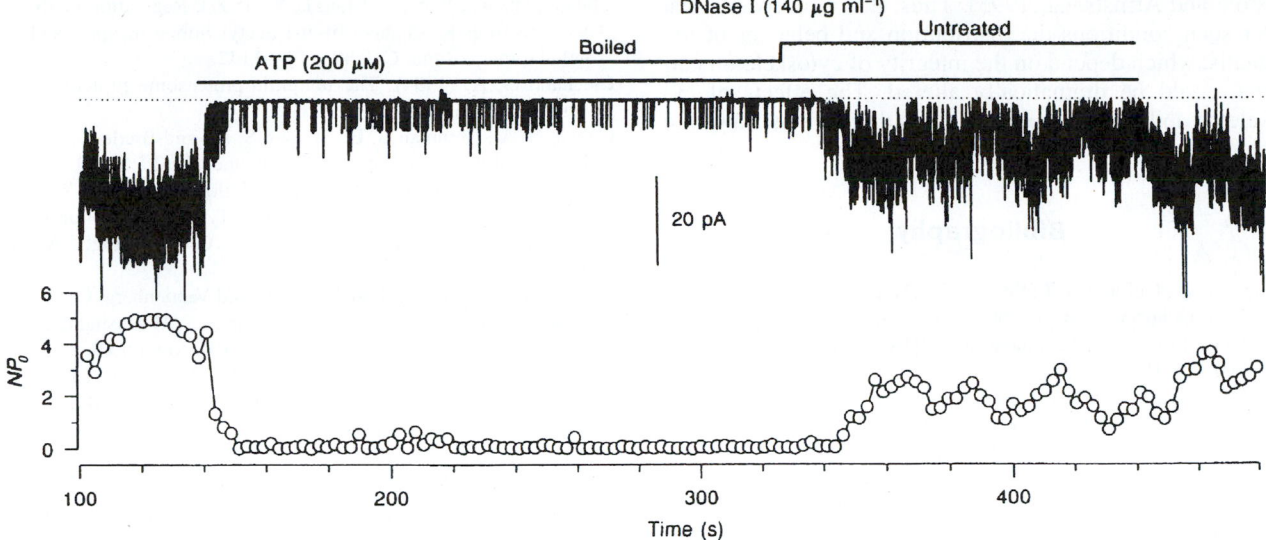

FIGURE 7. The actin microfilament disrupter, DNase I, enhances K_{ATP} channel opening. Untreated DNase I antagonized ATP-induced K_{ATP} channel inhibition. By contrast, DNase that has been denatured by boiling had no effect. Upper trace: Original trace record from an inside-out patch excised from a guinea pig ventricular cardiac cell. Lower trace: Channel open probability calculated over 2.5-s intervals. (From Terzic and Kurachi, 1996.)

V. Mechanosensitive Gating of Ion Channels and Cytoskeleton

The interaction of protein channels with the cytoskeleton has also been suggested in the gating of mechanosensitive and transduction channels. This is the case, for example, in certain specialized hair cells (Sachs, 1988). Also, in skeletal muscle, it has been shown that an absence of normal dystrophin (a spectrinlike component of the cortical cytoskeleton) is associated with altered mechanosensitive gating (Morris, 1995).

Gating of stretch-activated channels is thought to rely on forces between the cytoskeleton and the attached membrane channels. Although the biochemical basis of this interaction is uncertain, disruption of actin by cytochalasin alters behavior of mechanosensitive channels (Sachs, 1986). In the case of vertebrate hair cells, adaptation of the transduction current involves a Ca^{2+}- and actin-dependent mechanism. Ca^{2+} influx is believed to activate a molecular motor that maintains gating spring tension by moving along the actin core of the stereocilia (Hudspeth, 1989).

VI. Summary of Cytoskeleton Effects on Ion Channels

The cytoskeleton regulates ion channel function through integrated interactions with cytoskeleton-associated proteins, as well as dynamic regulation of its own state. Two main roles of the cytoskeleton in the regulation of ion channel function have been recognized: (1) targeted distribution of ion channel proteins within specialized domains of plasma membranes, and (2) modulation of ion channel activity.

It is now established that cytoskeleton-associated proteins, such as rapsyn, ankyrin, gephyrin, PSD-95, and α_2-actinin, target the distribution of Na^+ and K^+ channels,

nAChRs, and glycine, GABA, and NMDA receptor/channel complexes to specialized membrane domains including the postsynaptic membranes and the nodes of Ranvier (see Table 1). Cytoskeleton-dependent targeting and maintenance of ion channels at discrete plasma membrane sites, in turn, are regulated by catalytic processes, including protein phosphorylation by tyrosine kinases and PKA. Thus, structural interactions between the cytoskeleton, cytoskeleton-associated proteins, and channel/receptor subunits determine the highly specialized distribution of ion channel proteins within certain domains of plasma membranes. Such domain-dependent distribution and anchoring of ion channels is required for proper intra- and intercellular signaling.

In addition to structural interactions, functional interactions between ion channel and colocalized subplasmalemmal cytoskeletal networks have been described in epithelia, neurons, and cardiac myocytes. Apparently, the open probability of epithelial Na^+, K^+, and Cl^- channels is regulated by the length of actin filament networks (see Table 5). The cytoskeleton has also been suggested to modulate neuronal excitability and synaptic plasticity through modulation of ion channel activity, in particular through regulation of the open probability and kinetics of Ca^{2+} and NMDA channels. In the heart, more recently, the activities of Na^+, Ca^{2+}, and K_{ATP} channels have been shown to depend on the integrity of the cytoskeleton. Therefore, based on the current understanding of the relationship between the cytoskeleton and ion channels, it has become apparent that modulation of the cytoskeleton and associated proteins may represent an important means of regulating the physiology of ion channels, and thereby cellular functions, including signaling and excitability.

Moreover, disturbances of the cytoskeleton or associated proteins can occur under disease conditions, including muscular dystrophies (Campbell, 1995), as well as under pathophysiological conditions, such as ischemia and hypoxia

(Ganote and Armstrong, 1993). Thus, it is conceivable that under such conditions the distribution and behavior of ion channels, which depend on the integrity of cytoskeleton networks, could be dramatically altered. The effects of cytoskeleton and ion channel function in disease states await to be elucidated.

Bibliography

Allison, D. W., Gelfand, V. I., Spector, I., Craig, A. M. (1998). Role of actin in anchoring postsynaptic receptors in cultured hippocampal neurons: differential attachment of NMDA versus AMPA receptors. *J. Neurosci.* **18,** 2423–2436.

Apel, E. D., and Merlie, J. P. (1995). Assembly of the postsynaptic apparatus. *Curr. Opin. Neurobiol.* **5,** 62–67.

Awayda, M. S., Ismailov, I. I., Berdiev, B. K., and Benos, D. J. (1995). A cloned renal epithelial Na+ channel proteins display stretch activation in planar lipid bilayers. *Am. J. Physiol.* **268,** C1450–C1459.

Bennett, V., and Gilligan, D. M. (1993). The spectrin-based membrane skeleton and micron-scale organization of the plasma membrane. *Annu. Rev. Cell Biol.* **9,** 27–66.

Benson, D. L., Mandell, J. W., Shaw, G., and Banker, G. (1996). Compartmentation of alpha-internexin and neurofilament triplet proteins in cultured hippocampal neurons. *J. Neurocytol.* **25,** 181–196.

Brady, P. A., Alekseev, A. E. A., Aleksandrova, L. A., Gomez, L. A., and Terzic, A. (1996). A disrupter of actin microfilaments impairs sulfonylurea-inhibitory gating of cardiac K$_{ATP}$ channels. *Am. J. Physiol.* **271,** H2710–H2716.

Brakeman, P. R., Lanahan, A. A., O'Brien, R., Roche, K., Barnes, C. A., Huganir, R. L., and Worley, P. F. (1997). Homer: a protein that selectively binds metabotropic glutamate receptors. *Nature* **386,** 284–287.

Brenman, J. E., Chao, D. S., Gee, S. H., McGee, A. W., Craven, S. E., Santillano, D. R., Wu, Z., Haung, F., Xia, H., Peters, M. F., Froehner, S. C., and Bredt, D. S. (1996). Interaction of nitric oxide synthase with the postsynaptic density protein PSD-95 and α1-syntrophin mediated by PDZ domains. *Cell* **84,** 757–767.

Brenman, J. E., Topinka, J. R., Cooper, E. C., McGee, A. W., Rosen, J. T., Milroy, T., Ralston, H. J., and Bredt, D. S. (1998). Localization of postsynaptic density-93 to dendritic microtubules and interaction with microtubule-associated protein 1A. *J. Neurosci.* **18,** 8805–8813.

Cabral, J. H. M., Petosa, C., Sutcliffe, M. J., Raza, S., Byron, O., Poy, F., Marfatia, S. M., Chisti, A. H., and Liddigton, R. C. (1996). Crystal structure of a PDZ domain. *Nature* **382,** 649–652.

Campbell, K. P. (1995). Three muscular dystrophies: loss of cytoskeleton-extracellular matrix linkage. *Cell* **80,** 675–679.

Cannessa, C. M., Horisberger, J.-D., and Rossier, B. C. (1993). Epithelial sodium channel related to proteins involved in neurodegeneration. *Nature* **361,** 467–470.

Cannessa, C. M., Schild, L., Gary, B., Thorens, B., Gautschi, I., Horisberger, J. D., and Rossier B. C. (1994). Amiloride-sensitive epithelial Na+ channel is made of three homologous subunits. *Nature* **367,** 463–467.

Cantiello, H. F. (1995). Role of the actin cytoskeleton on epithelial Na+ channel regulation. *Kidney Int.* **48,** 970–984.

Cantiello, H. F., and Prat, A. G. (1996). Role of actin filament organization in ion channel activity and cell volume regulation. *Curr. Top. Membr.* **43,** 373–396.

Cantiello, H. F., Stow, J. L., Prat, A. G., and Ausiello, D. A. (1991). Actin filaments regulate epithelial Na+ channel activity. *Am. J. Physiol.* **261,** C882–C888.

Chen, H. J., Rojas-Soto, M., Oguni, A., and Kennedy, M. B. (1998). A synaptic Ras-GTPase activating protein (p135 SynGAP) inhibited by CaM kinase II. *Neuron* **20,** 895–904.

Chuang, H., Jan, Y. N., and Jan, L. Y. (1997). Regulation of IRK3 inward rectifier K+ channel by m1 acetylcholine receptor and intracellular magnesium. *Cell* **89,** 1121–1132.

Ciechanover, A. (1994). The ubiquitin-proteasome proteolytic pathway. *Cell* **79,** 13–21.

Cohen, N. A., Brenman, J. E., Snyder, S. H., and Bredt, D. S. (1996). Binding of the inward rectifier K+ channel Kir 2.3 to PSD-95 is regulated by protein kinase A phosphorylation. *Neuron* **17,** 759–767.

Colledge, M., and Froehner, S. C. (1997). Tyrosine phosphorylation of nicotinic acetylcholine receptor mediates Grb2 binding. *J. Neurosci.* **17,** 5038–5045.

Coulter, K. L., Perier, F., Radele, C. M., and Vandenberg, C. A. (1995). Identification and molecular localization of a pH-sensing domain for the inward rectifier potassium channel HIR. *Neuron* **15,** 1157–1168.

Craig, A. M., Banker, G., Chang, W., McGrath, M. E., and Serpinskaya, A. S. (1996). Clustering of gephyrin at GABAergic but not glutamatergic synapses in cultured rat hippocampal neurons. *J. Neurosci.* **16,** 3166–3177.

Craven, S. E., El-Husseini, A. E., and Bredt, D. S. (1999). Synaptic targeting of the postsynaptic density protein PSD-95 mediated by lipid and protein motifs. *Neuron* **22,** 497–509.

Davis, J. Q., Lambert, S., and Bennett, V. (1996). Molecular composition of the node of Ranvier—identification of ankyrin-binding cell adhesion molecules neurofascin and NrCAM at nodal axon segments. *J. Cell Biol.* **135,** 1355–1367.

DeChiara, T. M., Bowen, D. C., Valenzuela, D. M., Simmons, M. V., Poueymirou, W. T., Thomas, S., Kinetz, E., Compton, D. L., Rojas, E., Park, J. S., Smith, C., DiStefano, P. S., Glass, D. J., Burden, S. J., and Yancopoulos, G. D. (1996). The receptor tyrosine kinase MuSK is required for neuromuscular junction formation *in vivo. Cell* **85,** 501–512.

Deguchi, M., Hata, Y., Takeuchi, M., Ide, N., Hirano, K., Yao, I., Irie, M., Toyoda, A., and Takai, Y. (1998). BEGAIN (brain-enriched guanylate kinase–associated protein), a novel PSD-95/SAP90-binding protein. *J. Biol. Chem.* **273,** 26 269–26 272.

Doi, T., Fakler, B., Schultz, J. H., Schulte, U., Brandle, U., Weidemann, S., Zenner, H. P., Lang, J., and Ruppersberg, J. P. (1996). Extracellular K+ and intracellular pH allosterically regulate renal Kir1.1 channels. *J. Biol. Chem.* **271,** 17 261–17 266.

Dong, H., O'Brien, R. J., Fung, E. T., Lanhan, A. A., Worley, P. F., and Huganir, R. L. (1997). GRIP: a synaptic PDZ domain–containing protein that interacts with AMPA receptors. *Nature* **386,** 279–284.

Doyle, D. A., Lee, A., Lewis, J., Kim, E., Sheng, M., and MacKinnon, R. (1996). Crystal structures of a complexed and peptide-free membrane domain—molecular basis of peptide recognition by PDZ. *Cell* **85,** 1067–1076.

Duprat, F., Lesage, F., Fink, M., Reyes, R., Heurteaux, C., and Lazdunski, M. (1997). TASK, a human background K+ channel to sense external pH variations near physiological pH. *EMBO J.* **16,** 5464–5471.

Ehlers, M. D., Fung, E. T., O'Brien, R. J., and Huganir, R. L. (1998). Splice variant–specific interaction of the NMDA receptor subunit NR1 with neuronal intermediate filaments. *J. Neurosci.* **18,** 720–730.

Ehlers, M. D., Mammen, A. L., Lau, L. F., and Huganir, R. L. (1996). Synaptic targeting of glutamate receptors. *Curr. Opin. Cell Biol.* **8,** 484–489.

Elizabeth, D. A., Glass, D. J., Moscoso, L. M., Yancopoulos, G. D., and Sanes, J. R. (1997). Rapsyn is required for MuSK signaling and recruits synaptic components to a MuSK-containing scaffold. *Neuron* **18,** 623–635.

Falker, B., Brändle, U., Glowatzki, E., Weldemann, S., Zenner, H. P., and Ruppersberg, J. P. (1994). Strong voltage dependent inward rectification of inward rectifier K+ channels is caused by intracellular spermine. *Cell* **80,** 149–154.

Ferns, M., Deiner, M., and Hall, Z. W. (1996). Agrin-induced acetylcholine receptor clustering in mammalian muscle requires tyrosine phosphorylation. *J. Cell Biol.* **132**, 937–944.

Ficker, E., Taglialatela, M., Wible, B. A., Henley, C. M., and Brown, A. M. (1994). Spermine and spermidine as gating molecules for inward rectifier K+ channels. *Science* **266**, 1068–1072.

Frappier, T., Stetzkowski-Marden, F., and Pradel, L. A. (1991). Interaction domains of neurofilament light chain and brain spectrin. *Biochem. J.* **275**, 521–527.

Frappier, T., Derancourt, J., Pradel, L. A. (1992). Actin and neurofilament binding domain of brain spectrin beta subunit. *Eur. J. Biochem.* **205**, 85–91.

Froehner, S. C. (1993). Regulation of ion channel distribution at synapses. *Annu. Rev. Neurosci.* **16**, 347–368.

Fukuda, J., Kameyama, M., and Yamaguchi, K. (1981). Breakdown of cytoskeletal filaments selectively reduces Na and Ca spikes in cultured mammal neurons. *Nature* **294**, 82–85.

Furukawa, T., Yamane, Y., Terai, Y., Katayama, Y., and Hiraoka, M. (1996). Functional linkage of the cardiac ATP-sensitive K+ channel to the actin cytoskeleton. *Pflügers Arch.* **431**, 504–512.

Galli, A., and DeFelice, L. J. (1994). Inactivation of L-type Ca channels in embryonic chick ventricle cells: dependence on the cytoskeletal agents colchicine and taxol. *Biophys. J.* **67**, 2296–2304.

Gallo-Payet, N., and Payet, M. D. (1995). Excitation-secretion coupling. *In* "Cell Physiology" (N. Sperelakis, Ed.), pp. 465–482. Academic Press, San Diego.

Ganote, C., and Armstrong, S. (1993). Ischaemia and the myocyte cytoskeleton: review and speculation. *Cardiovasc. Res.* **27**, 1387–1403.

Garcia, E., Mehta, S., Blair, L. A. C., Wells, D. G., Shang, J., Fukushima, T., Fallon, J. R., Garner, C. C., and Maeshall, J. (1998). SAP90 binds and clusters kainate receptors causing incomplete desensitization. *Neuron* **21**, 727–739.

Gautam, M., Noakes, P. G., Mudd, J., Nichol, M., Chu, G. C., Sanes, J. R., and Merlie, J. P. (1995). Failure of postsynaptic specialization to develop at neuromuscular junctions of rapsyn-deficient mice. *Nature* **377**, 195–196.

Gertler, F. B., Niebuhr, K., Reinhard, M., Wehland, J., and Soriano, P. (1996). Mena, a relative of VASP and *Drosophila* enabled, is implicated in the control of microfilament dynamics. *Cell* **87**, 227–239.

Gillespie, S. K., Balasusramanian, S., Fung, E. T., and Huganir, R. L. (1996). Rapsyn clusters and activates the synapse-specific receptor tyrosine kinase MuSK. *Neuron* **16**, 953–962.

Glass, D. J., Bowen, D. C., Stitt, T. N., Radziejewski, C., Bruno, J., Ryan, T. E., Gies, D. R., Shah, S., Mattsson, K., Burden, S. J., DiStefano, P. S., Valenzuela, D. M., Dechiara, T. M., and Yancopoulos, G. D. (1996). Agrin acts via a MuSK receptor complex. *Cell* **85**, 513–523.

Gomperts, S. N. (1996). Clustering membrane proteins: it's all coming together with PSD-95/SAP90 protein family. *Cell* **84**, 659–662.

Goulet, C. C., Volk, K. A., Adams, C. M., Prince, L. S., Stokes, J. B. and Snyder, P. M. (1998). Inhibition of the epithelial Na+ channel by interaction of Nedd4 with a PY motif deleted in Liddle's syndrome. *J. Biol. Chem.* **273**, 30 012–30 017.

Graig, A. M., Banker, G., Chang, W., McGrath, M. E., and Serpinskaya, A. S. (1996). Clustering of gephyrin at GABAergic but not glutamatergic synapses in cultured rat hippocampal neurons. *J. Neurosci.* **16**, 3166–3177.

Haffner, C., Jarchau, T., Reinhard, M., Hoppe, J., Lohmann, S. M., and Walter, U. (1995). Molecular cloning, structural analysis and functional expression of the proline-rich focal adhesion and microfilament-associated protein, VASP. *EMBO J.* **14**, 19–27.

Hanley, J. G., Koulen, P., Bedford, F., Gordon-Weeks, P. R., and Moss, S. J. (1999). The protein MAP-1B links GABAc receptors to the cytoskeleton at retinal synapses. *Nature* **397**, 66–69.

Hansson, J. H., Nelson-Williams, C., Suzuki, H., Schild, L., Shimkets, R., Lu, Y., Canessa, C., Iwasaki, T., Rossier, B., and Lifton, R. P. (1995). Hypertension caused by a truncated epithelial sodium channel subunit: genetic heterogeneity of Liddle syndrome. *Nature Genet.* **11**, 76–82.

Harris, K. M., and Kater, S. B. (1994). Dendritic spines: cellular specializations imparting both stability and flexibility to synaptic function. *Annu. Rev. Neurosci.* **17**, 341–371.

Hata, Y., Butz, S., and Sudhof, T. C. (1996). CASK: a novel dlg/PSD95 homolog with an N-terminal calmodulin-dependent protein kinase domain identified by interaction with neurexins. *J. Neurosci.* **16**, 2488–2494.

Hay, J. C., and Scheller, R. H. (1997). SNAREs and NSF in targeted membrane fusion. *Curr. Opin. Cell Biol.* **9**, 505–512.

Hibino, H., Horio, Y., Inanobe, A., Doi, K., Ito, M., Yamada, M., Gotow, T., Uchiyama, Y., Kawamura, M., Kubo, T., and Kurachi, Y. (1997). An ATP-dependent inwardly rectifying potassium channel, K_{AB}-2 (Kir4.1), in cochlear stria vascularis of inner ear: its specific subcellular localization and correlation with the formation of endocochlear potential. *J. Neurosci.* **17**, 4711–4721.

Hille, B. (1992). Potassium channels and chloride channels. *In* "Ionic Channels of Excitable Membrane" (B. Hille, Ed.), pp. 115–139. Sinauer, Sunderland.

Hirano, K., Hata, Y., Ide, N., Takeuchi, M., Irie, M., Yao, I., Deguchi, M., Toyoda, A., Sudhof, T. C., and Takai, Y. (1998). A novel multiple PDZ domain–containing molecule interacting with N-methyl-D-aspartate receptors and neuronal cell adhesion proteins. *J. Biol. Chem.* **273**, 21 105–21 110.

Hitt, A. L., and Luna, E. A. (1994). Membrane interactions with the actin cytoskeleton. *Curr. Opin. Cell Biol.* **6**, 120–130.

Hollmann, M., Boulter, J., Maron, C., Beasley, L., Sullivan, J., Pecht, G., and Heinemann, S. (1993). Zinc potentiates agonist-induced currents at certain splice variants of the NMDA receptor. *Neuron* **10**, 943–954.

Horber, J. K. H., Mosbacher, J., Haberele, W., Ruppersberg, J. P., and Sackmann, B. (1995). A look at membrane patches with a scanning force microscope. *Biophys. J.* **68**, 1687–1693.

Horio, Y., Hibino, H., Inanobe, A., Yamada, M., Ishii, M., Tada, Y., Sato, E., Hata, Y., Takai, Y., and Kurachi, Y. (1997). Clustering and enhanced activity of an inwardly rectifying potassium channel, Kir4.1, by an anchoring protein, PSD-95/SAP90. *J. Biol. Chem.* **272**, 12 885–12 888.

Hsueh, Y.-P., Kim, E., and Sheng, M. (1997). Disulfide-linked head-to-head multimerization in the mechanism of ion channel clustering by PSD-95. *Neuron* **18**, 803–814.

Hu, B. R., Park, M., Martone, M. E., Fischer, W. H., Ellisman, M. H., and Zivin, J. A. (1998). Assembly of proteins to postsynaptic densities after transient cerebral ischemia. *J. Neurosci.* **18**, 625–633.

Hudspeth, A. J. (1989). How the ear's works work. *Nature* **341**, 397–401.

Hug, T., Koslowsky, T., Ecke, T., Greger, R., and Kunzelmann, K. (1995). Actin-dependent activation of ion conductances in bronchial epithelial cells. *Pflügers Arch.* **429**, 682–690.

Inagaki, N., Gonoi, T., Clement, J. P., Namba, N., Inazawa, J., Gonzalez, G., Aguilar-Bryan, L., Seino, S., and Bryan, J. (1995). Reconstitution of I_{KATP}: an inward rectifier subunit plus the sulfonylurea receptor. *Science* **270**, 1166–1170.

Inagaki, N., Gonoi, T., Clement, J. P., Wang, C. Z., Aguilar-Bryan, L., Bryan, J., and Seino, S. (1996). A family of sulfonylurea receptors determines the pharmacological properties of ATP-sensitive K+ channels. *Neuron* **16**, 1011–1017.

Inanobe, A., Yoshimoto, Y., Horio, Y., Morishige, K., Hibino, H., Matsumoto, S., Tokunaga, Y., Maeda, T., Hata, Y., Takai, Y., and Kurachi, Y. (1999). Characterization of G-protein–gated K+ channels composed of Kir3.2 subunits in dopaminergic neurons of the substantia nigra. *J. Neurosci.* **19**, 1006–1017.

Irie, M., Hata, Y., Takeuchi, M., Ichtchenko, K., Toyoda, A., Hirano, K, Takai, Y., Rosahl, T. W., and Sudhof, T. C. (1997). Binding of neuroligins to PSD-95. *Science* **277**, 1511–1515.

Isomoto, S., Kondo, C., and Kurachi, Y. (1997). Inwardly rectifying potassium channels: their molecular heterogeneity and function. *Jap. J. Physiol.* **47**, 11–39.

Isomoto, S., Kondo, C., Yamada, M., Matsumoto, S., Higashiguchi, O., Horio, Y., Matsuzawa, Y., and Kurachi, Y. (1996). A novel sulfonylurea receptor forms with BIR (Kir 6.2) a smooth muscle type ATP-sensitive K$^+$ channel. *J. Biol. Chem.* **271**, 24 321–24 324.

Ito, M., Inanobe, A., Horio, Y., Hibino, H., Isomoto, S., Ito, H., Mori, K., Tonosaki, A., Tomoike, H., and Kurachi, Y. (1996). Immuno-localization of an inwardly rectifying K$^+$ channel, K$_{AB}$-2 (Kir4.1), in the basolateral membrane of renal distal tubular epithelia. *FEBS Lett.* **388**, 11–15.

Jaffery, S. R., Snowman, A. M., Eliasson, M. J. L., Cohen, N. A., and Snyder, S. H. (1998). CAPON: a protein associated with neuronal nitric oxide synthase that regulates its interactions with PSD95. *Neuron* **20**, 115–124.

Jentsch, S., and Schlenker, S. (1995). Selective protein degradation: a journey's end within the proteasome. *Cell* **82**, 881–884.

Johnson, B. D., and Byerly, L. (1993). A cytoskeletal mechanism for Ca^{2+} channel metabolic dependence and inactivation by intracellular Ca^{2+}. *Neuron* **10**, 797–804.

Kabsch, W., Mannherz, D., Suck, D., Pai, E. F., and Holmes, K. C. (1990). Atomic structure of the actin: DNase I complex. *Nature* **347**, 37–44.

Kato, A., Ozawa, F., Saitoh, Y., Hirai, K., and Inokuchi, K. (1997). *vesl*, a gene encoding VASP/Ena family related protein, is upregulated during seizure, long-term potentiation and synaptogenesis. *FEBS Lett.* **412**, 183–189.

Kennedy, M. B. (1993). The postsynaptic density. *Curr. Opin. Neurobiol.* **3**, 732–737.

Kim, E., Cho, K.-O., Rothschild, A., and Sheng, M. (1996). Hetero-multimerization and NMDA receptor-clustering activity of chapsyn-110, a member of the PSD-95 family of proteins. *Neuron* **17**, 103–113.

Kim, D., Fujita, A., Horio, Y., and Kurachi, Y. (1998a). Cloning and functional expression of a novel cardiac two pore background K$^+$ channel (cTBAK-1). *Circ. Res.* **82**, 513–518.

Kim, E., Niethammer, M., Rothschild, A., Jan, Y. N., and Sheng, M. (1995). Clustering of *Shaker*-type K$^+$ channels by interaction with a family of membrane-associated guanylate kinases. *Nature* **378**, 85–88.

Kim, E., Naisbitt, S., Hsueh, Y.-P., Rao, A., Rothschild, A., Craig, A. M., and Sheng, M. (1997). GKAP, a novel synaptic protein that interacts with the guanylate kinase-like domain of PSD-95/SAP90 family of channel clustering molecules. *J. Cell Biol.* **136**, 669–678.

Kim, J. H., Liao, D., Lau, L.-F., and Huganir, R. L. (1998b). SynGAP: a synaptic RasGAP that associates with the PSD-95/SAP90 protein family. *Neuron* **20**, 683–691.

Kirsch, J., and Betz, H. (1995). The postsynaptic localization of the glycine receptor–associated protein gephyrin is regulated by the cytoskeleton. *J. Neurosci.* **15**, 4148–4156.

Kirsch, J., Wolters, I., Triller, A., and Betz, H. (1993). Gephyrin antisense oligonucleotides prevent glycine receptor clustering in spinal neurons. *Nature* **366**, 745–748.

Kornau, H.-C., Schenker, L. T., Kennedy, M. B., and Seeburg, P. H. (1995). Domain interaction between NMDA receptor subunits and the postsynaptic density protein PSD-95. *Science* **269**, 1737–1740.

Krapivinsky, G. B., Ackerman, M., Gordon, E., Krapivinsky, L., and Clapham, D. E. (1994). Molecular characterization of a swelling-induced chloride conductance regulatory protein, pI$_{Cln}$. *Cell* **76**, 439–448.

Kuhse, J., Betz, H., and Kirsch, J. (1995). The inhibitory glycine receptor: architecture, synaptic localization and molecular pathology of a postsynaptic ion-channel complex. *Curr. Opin. Neurobiol.* **5**, 318–323.

Kurachi, Y. (1996). Muscarinic and purinergic regulation of cardiac K$^+$ channels. *In* "Molecular Physiology and Pathophysiology of Cardiac Ion Channels and Transporters" (M. Morad, S. Ebashi, W. Trautwein, and Y. Kurachi, Eds.), pp. 177–186. Kluwar, Dordrecht.

Kurschner, C., Mermelstein, P. G., Holden, W. T., and Surmeier, D. J. (1998). CIPP, a novel multivalent PDZ domain, selectively interacts with Kir4.0 family members, NMDA receptor subunits, neurexins, and neuroligins. *Mol. Cell. Neurosci.* **11**, 161–172.

Lambert, S., Davis, J. Q., and Bennett, V. (1997). Morphogenesis of the node of Ranvier: co-clusters of ankyrin and ankyrin-binding integral proteins define early developmental intermediates. *J. Neurosci.* **17**, 7025–7036.

Lampidis, T. J., Kolonias, D., Savara, J. N., and Rubin, R. (1992). Cardiostimulatory and antiarrhythmic activity of tubulin-binding agents. *Proc. Natl. Acad. Sci. USA* **86**, 1256–1260.

Lee, M. K., and Cleveland, D. W. (1994). Neurofilament function and dysfunction: involvement in axonal growth and neuronal disease. *Curr. Opin. Cell Biol.* **6**, 34–40.

Lee, W.-S., and Hebert, S. C. (1995). ROMK inwardly rectifying ATP-sensitive K$^+$ channel. I. Expression in rat distal nephron segments. *Am. J. Physiol.* **268**, F1124–F1131.

Lin, J. W., Wyszynski, M., Madhavan, R., Sealock, R., Kim, J. U., and Sheng, M. (1998). Yotiao, a novel protein of neuromuscular junction and brain that interacts with specific splice variants of NMDA receptor subunit NR1. *J. Neurosci.* **18**, 2017–2027.

Lledo, P .M., Zhang, X., Sudhof, T. C., Malenka, R. C., and Nicoll, R. A. (1998). Postsynaptic membrane fusion and long-term potentiation. *Science* **279**, 399–403.

Lopatin, A. N., Makhina, E. N., and Nichols, C. G. (1994). Potassium channel block by cytoplasmic polyamines as the mechanism of intrinsic rectification. *Nature* **372**, 366–369.

Matsumoto, G., and Sakai, H. (1979). Microtubules inside the plasma membrane of squid giant axons and their possible physiological function. *J. Membr. Biol.* **50**, 1–14.

Meier, T., Perez, G. M., and Wallace, B. G. (1995). Immobilization of nicotinic acetylcholine receptors in mouse C2 myotubes by agrin-induced protein tyrosine phosphorylation. *J. Cell. Biol.* **131**, 441–451.

Meyer, G., Kirsch, J., Betz, H., and Langosch, D. (1995). Identification of a gephyrin binding motif on the glycine receptor beta subunit. *Neuron* **15**, 563–572.

Mori, H., Manabe, T., Watanabe, M., Satoh, Y., Suzuki, N., Toki, S., Nakamura, K., Yagi, T., Kushiya, E., Takahashi, T., Inoue, Y., Sakimura, K., and Mishina, M. (1998). Role of the carboxy-terminal region of the GluRÉ√2 subunit in synaptic localization of the NMDA receptor channel. *Neuron* **21**, 571–580.

Morris, C. E. (1995). Stretch-sensitive ion channels. *In* "Cell Physiology" (N. Sperelakis, Ed.), pp. 483–489. Academic Press, San Diego.

Müller, B. M., Kistner, U., Kindler, S., Chung, W. J., Kuhlendahl, S., Fenster, S. D., Lau, L.-F., Veh, R. W., Huganir, R. L., Gundelfinger, E. D., and Garner, C. C. (1996). SAP102, a novel postsynaptic protein that interacts with NMDA receptor complexes *in vivo*. *Neuron* **17**, 255–265.

Müller, B. M., Kistner, U., Veh, R. W., Chung, W. J., Cases-Langhoff, C., Becker, B., Gundelfinger, E. D., and Garner, C. C. (1995). Molecular characterization and spatial distribution of SAP97, a novel presynaptic protein homologous to SAP90 and the *Drosophila* disc-large tumor suppressor protein. *J. Neurosci.* **15**, 2354–2366.

Niebuhr, K., Ebel, F., Frank, R., Reinhard, M., Domann, E., Carl, U. D., Walter, U., Gertler, F. B., Wehland, J., and Chakraborty, T. (1997). A novel proline-rich motif present in ActA of *Listeria* monocytogenes and cytoskeletal proteins is the ligand for the EVH1 domain, a protein module present in the Ena/VASP family. *EMBO J.* **16**, 5433–5444.

Niethammer, M, Kim, E., and Sheng, M. (1996). Interaction between the C-terminus of NMDA receptor subunits and multiple of PSD-95 family of membrane-associated guanylate kinases. *J. Neurosci.* **16,** 2157–2163.

Niethammer, M., Valtschanoff, J. G., Kapoor, T. M., Allison, D. W., Weinberg, R. J., Craig, A. M., and Sheng, M. (1998). CRIPT, a novel postsynaptic protein that binds to the third PDZ domain of PSD-95/SAP90. *Neuron* **20,** 693–707.

Nishimune, A., Isaac, J. T. R., Molnar, E., Noel, J., Nash, S. R., Tagaya, M., Collingridge, G. L., Nakanishi, S., and Henley, J. M. (1998). NSF binding to GluR2 regulates synaptic transmission. *Neuron* **21,** 87–97.

Nusser, Z., Mulvihill, E., Streit, P., and Somogyi, P. (1994). Subsynaptic segregation of metabotropic and ionotropic glutamate receptors as revealed by immunogold localization. *Neurosci.* **61,** 421–427.

O'Brien, R. J., Xu, D., Petralia, R. S., Steward, O., Huganir, R. L., and Worley, P. (1999). Synaptic clustering of AMPA receptors by the extracellular immediate-early gene product Narp. *Neuron* **23,** 309–323.

Osten, P., Srivastava, S., Inman, G. J., Vilim, F. S., Khatri, L., Lee, L. M., States, B. A., Einheber, S., Milner, T. A., Hanson, P. I., and Ziff, E. B. (1998). The AMPA receptor GluR2 C-terminus can mediate a reversible, ATP-dependent interaction with NSF and α- and β-SNAPs. *Neuron* **21,** 99–110.

Papazian, D. M., Timpe, L. C., and Jan, Y. N., Jan, L. Y. (1991). Alteration of voltage-dependence of *Shaker* potassium channel by mutations in the S4 sequence. *Nature* **349,** 305–310.

Paulmichl, M., Li, Y., Wickman, K., Acherman, M., Peralta, E., and Clapham, D. (1992). New mammalian chloride channel identified by expression cloning. *Nature* **356,** 238–241.

Peters, A., Palay, S. L., and Webster, H. D. (1991). The fine structure of the nervous system: the neurons and supporting cells. Oxford University Press, New York.

Phillips, W. D., Kopta, C., Blount, P., Gardner, P. D., Steinbach, J. H., and Merlie, J. P. (1991). ACh receptor–rich membrane domains organized in fibroblasts by recombinant 43-kilodalton protein. *Science* **251,** 568–570.

Ponting, C. P., and Phillips, C. (1997). Identification of Homer as a homologue of the Wiskott-Aldrich syndrome protein suggests a receptor-binding function for WH1 domains. *J. Mol. Med.* **75,** 769–771.

Prat, A. G., Bertorello, A. M., Ausiello, D. A., and Cantiello, H. F. (1993). Activation of epithelial Na⁺ channels by protein kinase A requires actin filaments. *Am. J. Physiol.* **265,** C224–C233.

Prat, A. G., Xiao, Y.-F., Ausiello, D. A., and Cantiello, H. F. (1995). c-AMP-independent regulation of CFTR by the actin cytoskeleton. *Am. J. Physiol.* **268,** C1522–C1561.

Rosenmund, C., Carr, D. W., Bergeson, S. E., Nilaver, G., Scott, J. D., and Westbrook, G. L. (1994). Anchoring of protein kinase A is required for modulation of AMPA/kainate receptors on hippocampal neurons. *Nature* **368,** 853–856.

Rosenmund, C., and Westbrook, G. L. (1993). Calcium-induced actin depolymerization reduces NMDA channel activity. *Neuron* **10,** 805–814.

Rothman, J. E. (1994). Mechanisms of intracellular protein transport. *Nature* **372,** 55–63

Rotin, D., Bar-Sagi, D., O'Brodovich, H., Merilainen, J., Lehto, V. P., Canessa, C. M., Rossier, B. C., and Downey, G. P. (1994). An SH3 binding region in the epithelial Na⁺ channel (αrENaC) mediates its localization at the apical membrane. *EMBO J.* **13,** 4440–4450.

Ruegg, M. A., and Bixby, J. L. (1998). Agrin orchestrates synaptic differentiation at the vertebrate neuromuscular junction. *Trends Neurosci.* **21(1),** 22–27.

Ruknudin, A., Song, M. J., and Sachs, F. (1991). The ultrastructure of patch-clamped membranes: a study using high-voltage electron microscopy. *J. Cell Biol.* **112,** 125–134.

Sachs, F. (1986). Biophysics of mechanoreception. *Membr. Bio. Chem.* **6,** 173–195.

Sachs, F. (1988). Mechanical transduction in biological system. *Crit. Rev. Biomed. Engineer.* **16,** 141–169.

Sampath, S., and Pollard, T. D. (1991). Effects of cytochalasin, phalloidin and pH on the elongation of actin filaments. *Biochemistry* **30,** 1973–1980.

Satoh, K., Yatani, H., Senda, T., Kohu, K., Nakamura, T., Okumura, N., Matsumine, A., Kobayashi, S., Toyoshima, K., and Akiyama, T. (1997). DAP-1, a novel protein that interacts with the guanylate kinase–like domains of hDLG and PSD-95. *Genes Cells* **2,** 415–424.

Schwiebert, E. M., Mills, J. W., and Stanton, B. A. (1994). Actin-based cytoskeleton regulates a chloride channel and cell volume in a renal cortical collecting duct cell line. *J. Biol. Chem.* **269,** 7081–7089.

Sheng, M. (1996). PDZs and receptor/channel clustering: rounding up the latest suspects. *Neuron* **17,** 575–578.

Sheng, M. (1997). Glutamate receptors put in their place. *Nature* **386,** 221–223.

Shimkets, R. A., Warnock, D. G., Bositis, C. M., Nelson-Williams, C., Hansson, J. H., Schambelan, M., Gill, J. R. Jr., Ulick, S., Milora, R. V., Findling, J. W., Canessa, C. M., Rossier, B. C., and Lifton, R. P. (1994). Liddle's syndrome: heritable human hypertension caused by mutation in the β subunit of the epithelial sodium channel. *Cell* **79,** 407–414.

Short, D. B., Trotter, K. W., Reczek, D., Kreda, S. M., Bretscher, A., Boucher, R. C., Stutts, M. J., and Milgram, S. L. (1998). An apical PDZ protein anchors the cystic fibrosis transmembrane conductance regulator to the cytoskeleton. *J. Biol. Chem.* **273,** 19 797–19 801.

Smith, P. R., and Benos, D. J. (1996). Regulation of epithelial ion channel activity by the membrane-cytoskeleton. *Curr. Top. Membr.* **43,** 345–372.

Smith, P. R., Saccomani, G., Joe. E.-H., Angelides, K. J., and Benos, D. J. (1991). Amiloride-sensitive sodium channel is linked to the cytoskeleton in renal epithelial cells. *Proc. Natl. Acad. Sci. USA* **88,** 6971–6975.

Sollner, T., Bennett, M. K., Whiteheart, S. W., Scheller, R. H., and Rothman, J. E. (1993). A protein assembly-disassembly pathway *in vitro* that may correspond to sequential steps of synaptic vesicle docking, activation, and fusion. *Cell* **75,** 409–418.

Song, I., Kamboj, S., Xia, J., Dong, H., Liao, D., and Huganir, R. L. (1998). Interaction of the N-ethylmaleimide-sensitive factor with AMPA receptors. *Neuron* **21,** 393–400.

Sprengel, R., Suchanek, B., Amico, C., Brusa, R., Burnashev, N., Rozov, A., Hvalby, O., Jensen, V., Paulsen, O., Andersen, P., Kim, J. J., Thompson, R. F., Sun, W., Webster, L. C., Grant, S. G. N., Eilers, J., Konnerth, A., Li, J., McNamara, J. O., and Seeburg, P. H. (1998). Importance of the intracellular domain of NR2 subunits for NMDA receptor function *in vivo*. *Cell* **92,** 279–289.

Srinivasan, Y., Elmer, L., Davis, J., Bennett, V., and Angelides, K. (1988). Ankyrin and spectrin associate with voltage-dependent sodium channels in brain. *Nature* **333,** 177–180.

Srivastava, S., Osten, P., Vilim, F. S., Khatri, L., Inman, G., States, B., Daly, C., DeSouza, S., Abagyan, R., Valtschanoff, J. G., Weinberg, R. J., and Ziff, E. B. (1998). Novel anchorage of GluR2/3 to the postsynaptic density by the AMPA receptor–binding protein ABP. *Neuron* **21,** 581–591.

Staub, O., Dho, S., Henry, P. C., Correa, J., Ishikawa, T., McGlade, J., and Rotin, D. (1996). WW domains of Nedd4 bind to the proline-rich PY motifs in the epithelial Na⁺ channel deleted in Liddle's syndrome. *EMBO J.* **15,** 2371–2380.

Staub, O., Gautschi, I., Ishikawa, T., Breitschopf, K., Ciechanover, A., Schild, L., and Rotin, D. (1997a). Regulation of stability and function of the epithelial Na⁺ channel (ENaC) by ubiquitination. *EMBO J.* **16,** 6325–6336.

Staub, O., Yeger, H., Plant, P .J., Kim, H., Ernst, S. A., and Rotin, D. (1997b). Immunolocalization of the ubiquitin-protein ligase Nedd4 in tissues expressing the epithelial Na⁺ channel (ENaC). *Am. J. Physiol.* **272,** C1871–C1880.

Sudhof, T. C. (1995). The synaptic vesicle cycle: a cascade of protein-protein interactions. *Nature* **375,** 645–653.

Surprenant, A., Buell, G., and North, R. A. (1995). P2X receptors bring new structure to ligand-gated ion channels. *Trends Neurosci.* **18,** 224–229.

Suzuki, M., Miyazaki, K., Ikeda, M., Kawaguchi, Y., and Sakai, O. (1993). F-actin network may regulate a Cl⁻ channel in proximal tubule cells. *J. Membr. Biol.* **134,** 31–39.

Symons, M., Derry, J. M., Karlak, B., Jiang, S., Lemahieu, V., McCormick, F., Francke, U., and Abo, A. (1996). Wiskott-Aldrich syndrome protein, a novel effector for the GTPase CDC42Hs, is implicated in actin polymerization. *Cell* **84,** 723–734.

Takeuchi, M., Hata, Y., Hirano, K., Toyoda, A., Irie, M., and Takai, Y. (1997). SAPAPs: a family of PSD-95/SAP-90-associated proteins localized at the postsynaptic density. *J. Biol. Chem.* **272,** 11 943–11 951.

Takumi, T., Ishii, T., Horio, Y., Morishige, K.-I., Takahashi, N., Yamada, M., Yamashita, T., Kiyama, H., Sohmiya, K., Nakanishi, S., and Kurachi, Y. (1995). A novel ATP-dependent inward rectifier potassium channel expressed predominantly in glial cells. *J. Biol. Chem.* **270,** 16 339–16 346.

Terzic, A., and Kurachi, Y. (1996). Actin microfilament disrupters enhance K_{ATP} channel opening in patches from guinea-pig cardiomyocytes. *J. Physiol.* **492,** 395–404.

Terzic, A., and Kurachi, Y. (1998). Cytoskeleton effects on ion channels. *In* "Cell Physiology Source Book" (N. Sperelakis, Ed.), pp. 532–543. Academic Press, San Diego.

Terzic, A., Jahangir, A., and Kurachi, Y. (1995). Cardiac ATP-sensitive K⁺ channels: regulation by intracellular nucleotides and K⁺ channel–opening drugs. *Am. J. Physiol.* **269,** C525–C545.

Thompson, C L., Angelides, K. J., Velazquez, J. L. and Barnes, E. M. (1988). Distribution, mobility, and function of benzodiazepine receptors on primary cultures of vertebrate neurons. *Adv. Biochem. Psychopharmacol.* **45,** 135–149.

Tu, J. C., Xiao, B., Yuan, J. P., Lanahan, A. A., Leoffert, K., Li, M., Linden, D. J., and Worley, P. F. (1998). Homer binds a novel proline-rich motif and links group 1 metabotropic glutamate receptors with IP3 receptors. *Neuron* **21,** 717–726.

Undrovinas, A. I., Shander, G. S., and Makielski, J. C. (1995). Cytoskeleton modulates gating of voltage-dependent sodium channel in heart. *Am. J. Physiol.* **269,** H203–H214.

van Rossum, D., and Hanisch, U.-K. (1999). Cytoskeletal dynamics in dendritic spines: direct modulation by glutamate receptors? *Trends Neurosci.* **22,** 290–295.

van Rossum, D., Kuhse, J., and Betz, H. (1999). Dynamic interaction between soluble tublin and C-terminal domains of N-methyl-D-aspartate receptor subunits. *J. Neurochem.* **72,** 962–973.

Van Wagoner, D. R. (1993). Mechanosensitive gating of atrial ATP-sensitive potassium channels. *Circ. Res.* **72,** 973–983.

Waldmann, R., Champigny, G., Bassilana, F., Heurteaux, C., and Lazdunski, M. (1997). A proton-gated cation channel involved in acid-sensing. *Nature* **386,** 173–177.

Wallace, B. G. (1992). Mechanism of agrin-induced acetylcholine receptor aggregation. *J. Neurobiol.* **23,** 592–604.

Wallace, B. G. (1994). Staurosporine inhibits agrin-induced acetylcholine receptor phosphorylation and aggregation. *J. Cell Biol.* **125,** 661–668.

Wallace, B. G., Qu, Z., and Huganir, R. L. (1991). Agrin induces phosphorylation of the nicotinic acetylcholine receptor. *Neuron* **6,** 869–878.

Wang, H., Bedford, F. K., Brandon, N. J., Moss, S. J., and Olsen, R. W. (1999). GABA$_A$-receptor–associated protein links GABA$_A$ receptors and the cytoskeleton. *Nature* **397,** 69–72.

Wang, W.-H., Cassola, A., and Giebisch, G. (1994). Involvement of actin cytoskeleton in modulation of apical K channel activity in rat collecting duct. *Am. J. Physiol.* **267,** F592–F598.

Wechsler, A., and Teichberg, V. I. (1998). Brain spectrin binding to the NMDA receptor is regulated by phosphorylation, calcium and calmodulin. *EMBO J.* **17,** 3931–3939.

Wischmeyer, E., and Karschin, A. (1996). Receptor stimulation causes slow inhibition of IRK1 inwardly rectifying K⁺ channels by direct protein kinase A–mediated phosphorylation. *Proc. Natl. Acad. Sci. USA* **93,** 5819–5823.

Wyszynski, M., Lin, J., Rao, A., Nigh, E., Beggs, A. H., Craig, A. M., and Morgan, S. (1997). Competitive binding of α-actinin and calmodulin to the NMDA receptor. *Nature* **385,** 439–442.

Wyszynski, M., Kharazia, V., Shanghvi, R., Rao, A., Beggs, A. H., Craig, A. M., Weinberg, R., and Sheng, M. (1998). Differential regional expression and ultrastructual localization of α-actinin-2, a putative NMDA receptor–anchoring protein, in rat brain. *J. Neurosci.* **18,** 1383–1392.

Yamada, M., and Kurachi, Y. (1995). Spermine gates inward-rectifying muscarinic but not ATP-sensitive K⁺ channels in rabbit atrial myocytes. *J. Biol. Chem.* **270,** 9289–9294.

Yao, I., Hata, Y., Ide, N., Hirano, K., Deguchi, M., Nishioka, H., Mizoguchi, A., and Takai, Y. (1999). MAGUIN, a novel neuronal membrane-associated guanylate kinase–interacting protein. *J. Biol. Chem.* **274,** 11 889–11 896.

Xia, J., Zhang, X., Staudinger, J., and Huganir, R. L. (1999). Clustering of AMPA receptors by the synaptic PDZ domain–containing protein PICK1. *Neuron* **22,** 179–187.

Xiao, B., Tu, J. C., Petralia, R. S., Yuan, J. P., Doan, A., Breder, C. D., Ruggiero, A., Lanahan A. A., Wenthold, R. J., Worley, P. F. (1998). Homer regulates the association of group 1 metabotropic glutamate receptors with multivalent complexes of Homer-related, synaptic proteins. *Neuron* **21,** 707–716.

Zhang, S., Ehlers, M. D., Bernhardt, J. P., Su, C.-T., and Huganir, R. L. (1998). Calmodulin mediates calcium-dependent inactivation of N-Methyl-D-Aspartate receptors. *Neuron* **21,** 443–453.

SECTION
IV

Ion Channels as Targets for Toxins, Drugs, and Genetic Diseases

SECTION

IV

Ion Channels as Targets
for Toxins, Drugs, and
Genetic Disease

Kenneth M. Blumenthal

37

Ion Channels as Targets for Toxins

I. Introduction

Since the 1970s, a wide variety of neurotoxins have become potent tools in the armamentarium of the biochemist, biophysicist, physiologist, or pharmacologist interested in identifying, purifying, and characterizing voltage-sensitive ion channels of excitable membranes and in understanding the molecular details of their structure and function. The purpose of this chapter is to provide a brief description of some of these toxins, focusing on their chemical natures, similarities and differences, target macromolecules, and effects on transmembrane ionic fluxes. In this chapter, the ion channels are used as a framework, and in those cases where multiple-toxin binding sites have been demonstrated with a single channel, the relationships among these sites are discussed. A number of excellent reviews on this subject have been published recently (Catterall, 1995; Goldstein, 1996; Fozzard and Hanck, 1996; Armstrong and Hille, 1998).

The toxins are diverse both chemically and functionally. They include an ever-expanding array of polypeptides derived from marine invertebrates and terrestrial arthropods, alkaloids of diverse structures, heterocyclic compounds of the tetrodotoxin family, and synthetic insecticides. Functionally, these molecules are equally diverse. Toxins have been characterized that shift the voltage dependence of activation, delay channel inactivation, or block ion fluxes; some have multiple actions. Thus, in addition to being useful probes for purification of channel constituents, they have contributed significantly to our understanding of how these various processes are coupled to one another.

II. Voltage-Sensitive Sodium Channels

The neuronal sodium channel consists of three polypeptides having molecular masses of 260 (α), 36 (β_1), and 33 kDa (β_2) (Tamkun et al., 1984; Hartshorne and Catterall, 1984); whether the smaller subunits are present in skeletal muscle or cardiac channels remains unresolved. All of the

toxin binding sites characterized to date appear to be associated with the α-subunit, reconstitution of which into planar lipid bilayers restores many of the activities of the native channel (Tamkun et al., 1984). Microinjection of mRNA encoding this subunit into Xenopus oocytes results in expression of sodium channels that are functionally similar, though not identical, to those seen in nature. The α-subunit is organized into four repeated structurally homologous domains, each containing 300–400 amino acid residues. Each domain is predicted to include six transmembrane helices (designated S1–S6). There is also a seventh region (SS1–SS2), whose length varies among the four domains, that lies at least partially within the membrane. Recent data are consistent with the conclusion that these SS1–SS2 sequences contribute to the inner lining of the conducting pore and also provide its outer vestibule. A variety of mutually incompatible models for the secondary structure of this region exist. Whatever its structure, the binding sites for both tetrodotoxin/saxitoxin and the α-scorpion toxins have been associated with this region by both biochemical and mutagenic analyses, although very recent experiments point to a binding site for scorpion and sea anemone toxins within the S3-S4 extracellular linker of domain IV (Rogers et al., 1996, Benzinger et al., 1998). Figure 1 depicts the putative transmembrane organization of the α-subunit of this channel; as described in subsequent sections, this organization is shared by known K$^+$ channels and the α_1-subunit of the Ca^{2+} channel as well.

A series of elegant analyses, carried out mainly by Catterall, have clearly established the presence of four independent classes of neurotoxin binding sites on the rat brain Na$^+$ channel (Catterall, 1977; also, see summary, Table 1). These include sites (1) for the classic channel inhibitors **tetrodotoxin** and **saxitoxin,** (2) for activating alkaloids, such as **batrachotoxin** and **veratridine,** (3) for **polypeptide (α) toxins** from scorpion and sea anemone venoms, which delay channel inactivation, and (4) for a distinct set of **scorpion toxins (β-toxins)** that alter the voltage dependence of channel inactivation. Each of these four classes will be discussed.

625

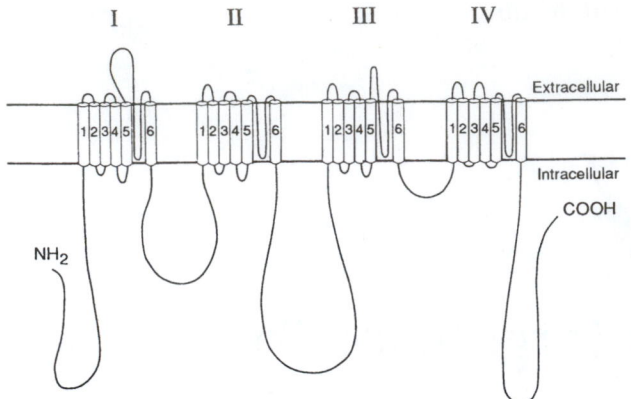

FIGURE 1. Proposed structure of the voltage-sensitive sodium channel. The channel consists of four homologous domains (I–IV), each consisting of six transmembrane segments (S1–S6). Cytoplasmic loop III–IV has been implicated in both activation and inactivation of the channel, while residues lying between S5 and S6 are involved in both ionic selectivity and binding of specific channel blockers such as TTX. (From McClatchey *et al.* (1992) *Cell* **68**, 769–774).

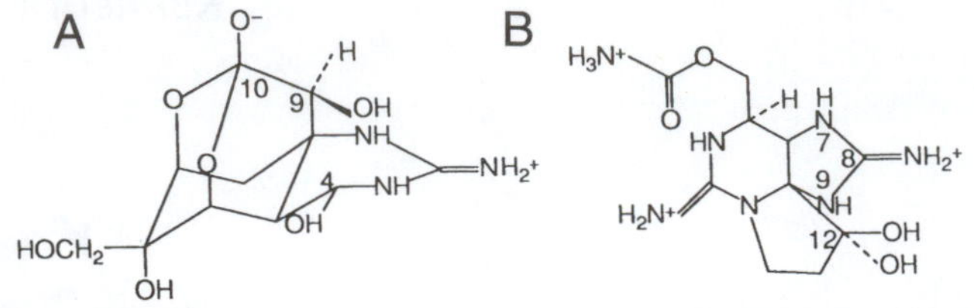

A. Site 1: Tetrodotoxin, Saxitoxin, and μ-Conotoxin

The presence of a paralytic toxin in newts, pufferfish, and other organisms has been known for well over 50 years, and the structure of this compound (Fig. 2), called tetrodotoxin, was first described in 1964. Narahashi, Moore, and collaborators demonstrated that the toxicity of tetrodotoxin (TTX) was attributable to its ability to block the increase in sodium conductance associated with the rising phase of the action potential in frog muscle. Because TTX and the structurally related, dinoflagellate-derived saxitoxin (STX) bind to the channel with high affinity (1–5 nM for the neuronal isoform), a property that is essentially unaltered by detergent solubilization, their discovery provided biochemists with an invaluable probe for purification of the channel protein.

FIGURE 2. The structures of (A) tetrodotoxin and (B) saxitoxin. For TTX, the guanidinium group and the C4, C9, and C10 hydroxyls are important for activity, while in saxitoxin, the C12 hydroxyl and the 7, 8, 9 guanidinium group are essential.

Because of the availability of TTX and STX, purification, biochemical characterization, and molecular cloning of voltage-dependent sodium channels from a variety of sources has been accomplished over the past ten years.

Tetrodotoxin and saxitoxin are heterocyclic guanidinium compounds that bind reversibly and with high affinity (K_d = 1–5 nM) to a site accessible from the external face of neuronal and muscle Na^+ channels. This K_d agrees well with that obtained from analysis of the effects of these toxins on Na^+ currents in these tissues. Binding of TTX and STX to cardiac Na^+ channels is of much lower affinity, with K_d values on the order of 1 μM. The cardiac channels are thus said to be TTX-resistant. Analyses of channels by site-directed mutagenesis (Noda *et al.*, 1989; Satin *et al.*, 1992) have implicated Cys 374 and Arg 377 of the cardiac channel as being important for their decreased TTX sensitivity.

Binding of TTX and STX blocks the Na^+ current, and was originally thought to involve interaction of toxin with the ionized form of a channel carboxyl group, which Hille proposed was associated with the **selectivity filter** of the channel. However, the notion that the carboxylate(s) important in TTX binding was also a part of the selectivity filter was abandoned when it was shown that chemical modification of channel carboxylates, which abolished binding, had no effect

TABLE 1 Toxin Effectors of Voltage-Sensitive Sodium Channels

Toxin	Binding affinity	Functional effect
Tetrodotoxin	1–5 nM (Nerve)	Blockade
	1–10 μM (Heart)	Blockade
Saxitoxin	1 nM (Nerve)	Blockade
	100 nM (Heart)	Blockade
Alkaloids		
Batrachotoxin	0.25 μM	Persistent activation
Grayanotoxin	> 1 mM	Persistent activation
Veratridine	50 μM	Persistent activation
Polypeptides		
α-Scorpion Toxins	1–2 nM	Delayed inactivation
β-Scorpion Toxins	0.5 nM	Shifts voltage dependence of activation
Anemone Toxins	10–1000 nM	Delayed inactivation
Dinoflagellate toxins		
Brevetoxins	μM range	Shifts voltage dependence of activation
Ciguatoxin	0.6 ng/ml	Induces depolarization
Goniopora	30–50 nM	Delayed inactivation

on ion selectivity. Subsequent structure-function analysis of TTX and STX, using both natural and synthetic analogs (Mosher, 1986; Shimizu, 1986), clearly delineated the importance of the guanidinium group of TTX (and the corresponding 7, 8, 9 guanidinium of the bifunctional STX) for activity. In addition, the C4, C9, and C10 hydroxyl groups of TTX and the C12 hydroxyl of STX are required. It is assumed that these hydroxyl groups form hydrogen bonds to as yet unknown acceptor sites on the channel. Recent data suggest that the TTX binding site is located near the external face of the channel and may lie outside of the transmembrane field.

In the absence of direct information on the three-dimensional structure of the channel, precise localization of the TTX binding site is impossible. While current models for the binding of both toxins still emphasize the importance of fixed negative charges at or near the channel mouth, it is clear that both molecules make multiple contacts to channel residues lying within the linker connecting the putative S5 and S6 helices. Site-directed mutagenesis of both Na$^+$ channels and K$^+$ channels indicates that this region is important both for ion selectivity and binding of channel-specific toxins. Thus, replacement of Arg 377 of the TTX-resistant cardiac (RH$_1$) channel by asparagine results in a small increase in TTX-sensitivity (Stuhmer et al., 1989a), whereas the mutation of Cys 374 to tyrosine renders the channel approximately 700-fold more sensitive to TTX (Satin et al., 1992). This latter result raises the possibility that the positively charged guanidinium groups of TTX and STX are actually bound not by ionic interactions but rather by interactions with the π-clouds contributed by a cluster of electron-rich aromatic residues located near the channel mouth. Analogous models have been proposed to account for acetylcholine binding to the nicotinic acetylcholine receptor, and tetraethylammonium binding to a variety of potassium channels. However, it is clear that interaction with aromatic residues is not the **sole** binding determinant, since mutagenesis has also identified a carboxyl group, Glu 387, essential for high-affinity TTX binding in the brain R$_{II}$ channel (Noda et al, 1989). Like Cys 374 and Arg 377, this residue is also located in the SS2 region. Thus, whatever the nature of the specific interactions involved in TTX binding, important determinants are clearly found in the SS1–SS2 region. A very recent development has utilized the ability of TTX to protect mutagenically created channel cysteine residues from reaction with sulfhydryl-specific reagents to develop topographic maps of the channel pore.

The corresponding S5–S6 linker regions from domains III and IV are also important determinants of ionic selectivity, as shown by the recent finding that the mutations Lys 1422 to glutamate and Ala 1714 to glutamate give the R$_{II}$ Na$^+$ channel an ionic selectivity similar to that of a Ca^{2+} channel. Interestingly, mutagenesis has also been used to show that the S5–S6 region of *shaker* K$^+$ channels is directly involved in both ion transit and interaction of inhibitors such as tetraethylammonium (TEA) and charybdotoxin. Tyr 379 of the K$^+$ channel RBK-1 has been identified as an important determinant of tetraethylammonium (TEA) binding (MacKinnon and Yellen, 1990), while the introduction of a tyrosine at position 449 of the *shaker* H4 channel greatly in-

creases its sensitivity to TEA. Thus, a leitmotif stressing the importance of aromatic residues in the binding of cationic ligands to ion channels is emerging (Dougherty, 1996).

A structurally unrelated toxin, designated **μ-conotoxin** or **geographutoxin,** is a competitive inhibitor of STX binding to muscle, but not nerve, Na$^+$ channels (K_d < 100 nM in muscle versus > 1 μM in brain; Cruz et al., 1985; Ohizumi et al., 1986). μ-Conotoxin is a 22-residue, hydroxyproline-containing peptide, tightly cross-linked by three disulfide bonds. A model for its three-dimensional structure, based on two-dimensional NMR measurements and simulated annealing protocols, has been proposed. The overall structure (Fig. 3) includes a series of tight turns in the N-terminal region and a short stretch of right-handed helix, while the core of the peptide encompasses what the authors refer to as a disulfide cage. A key feature of the structure is that the cationic side chains of arginines 1, 13, and 19, and lysines 8 and 16 all project into solution, away from the core of the molecule as defined by the disulfide bonds. Analysis of synthetic variants of μ-conotoxin has shown that Arg 13 is essential for activity: even conservative replacement by lysine reduces potency by about tenfold. In contrast, Arg 19 can be replaced by lysine with essentially no change in activity, while substitution at Arg 1 yields an intermediate result (Cruz et al., 1989). It is interesting that binding of both TTX/STX and the μ-conotoxins is at least partially dependent on guanidinium functions, despite the disparate chemistries of these molecules.

Unlike TTX and STX, the binding of μ-conotoxin is dependent on membrane potential, with K_d decreasing *e*-fold for every 34 mV depolarization (Cruz et al., 1989). TTX/STX-resistant (i.e., cardiac) channels are likewise resistant to μ-conotoxins. Very recently, a mutated form of μ-conotoxin has been used as a monitor of distance between the voltage sensor and channel pore (French et al., 1996).

B. Site 2: Batrachotoxin, Grayanotoxin, Veratridine, and Related Compounds

The existence of low molecular weight substances having dramatic effects on the gating properties of voltage-dependent Na$^+$ channels has been known for over 20 years. The most commonly used of these alkaloids are **batrachotoxin, grayanotoxin, and veratridine.** Despite their different biological origins (respectively, skin secretions of the frog *P. aurotaenia, Lilaceae,* and *Ericaceae* species) and dose-response curves (IC$_{50}$s are approximately 0.5, 10, and 25 μM for batrachotoxin, grayanotoxin, and veratridine), these molecules have very similar effects on channel function. They all appear to interact at a common binding site, which is allosterically coupled to both the TTX binding site described in the previous section and the α-scorpion toxin site to be discussed later (Catterall, 1977). Binding of any of the alkaloids to this site causes persistent activation of the channel, and leads to membrane depolarization. In addition, alkaloid-treated channels display a more relaxed ionic selectivity than do their naive counterparts. Thus, while the permeability order of untreated Na$^+$ channels is Na$^+$: guanidinium : K$^+$: Rb$^+$ = 1 : 0.13 : 0.09 : 0.01, veratridine increases the relative permeability for guanidinium, K$^+$, and Rb$^+$ by factors of 2.7,

4.5, and 10, respectively. Qualitatively similar results are obtained with the other alkaloids of this class and with combinations of alkaloid and polypeptide toxins from scorpion or sea anemone venoms.

The steroidal alkaloid batrachotoxin (BTX) was purified and characterized by Witkop, Daly, and their coworkers. In collaboration with Albuquerque, the effects of BTX have been measured in a variety of tissues including axons, brain slices and isolated synaptosomes, neuromuscular junctions, heart papillary muscle, and cardiac Purkinje fibers. Batrachotoxin binding has been measured directly in rat brain synaptosomes using a tritiated derivative, and the measured density of binding sites correlates well with that for TTX and STX. Binding of BTX to this site is competitively inhibited by veratridine with a K_i very similar to the $K_{0.5}$ for its activation of the channel. Because BTX acts from either side of the membrane, it had been speculated that this binding site might lie within the lipid bilayer, and a very recent study has demonstrated that a photoactivatable derivative of batrachotoxin specifically labels a site within the S6 transmembrane segment of domain I of the rat brain channel (Trainer et al., 1996). As discussed later, fluorescent derivatives of BTX have also been important for measuring distances between the various toxin binding sites by resonance

energy transfer experiments. Because the effects of BTX on membrane potential are dependent on the presence of extracellular Na^+ and are blocked by treatment with TTX, they are attributed to the toxin's ability to cause voltage-dependent Na^+ channels to activate at the resting potential, and to persist in the activated state. BTX also causes increased permeability to larger cations, consistent with allosteric coupling between the BTX binding site and the selectivity filter of the Na^+ channel.

Other alkaloids that appear to interact with the BTX site with similar functional effects include veratridine, grayanotoxin (GTX), and aconitine. While structurally distinct from one another, these highly hydrophobic compounds do have certain common features, such as the bridged ring systems and esterified aromatic moieties found in BTX, veratridine, and aconitine (Fig. 4). In the case of BTX, both the bridge and the esterified group are essential for activity.

For all of these toxins, the available data strongly suggest that interaction of a single toxin molecule per channel is sufficient for activation. The known biological effects of these toxins are ascribed to their ability to interact with the voltage-dependent Na^+ channel, although only BTX binding has been measured directly. Catterall (1977) proposed that the alkaloids bind preferentially to the activated conformation

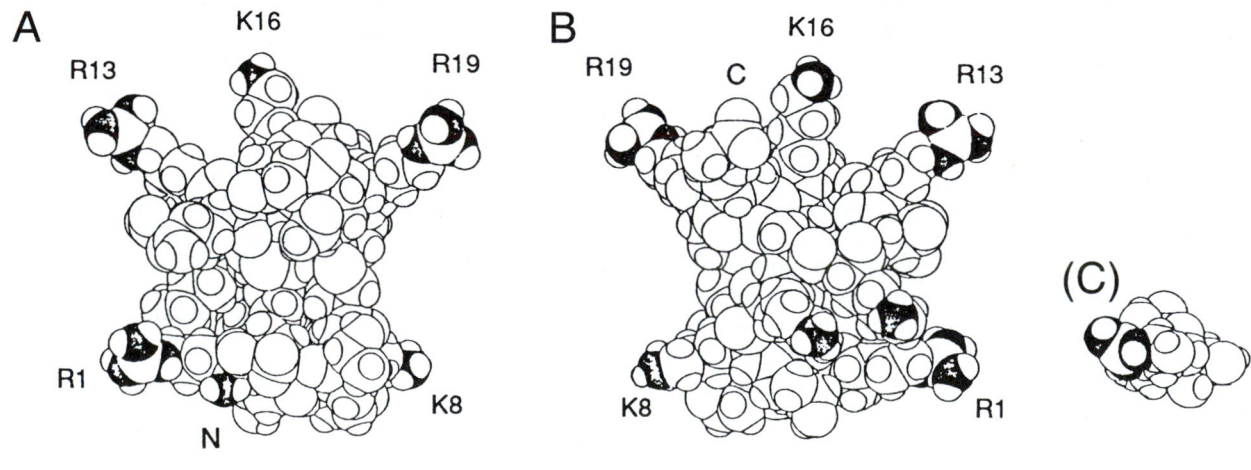

Variant	Sequence
GIIIA	RDCCTPPKKC KDRQCKPQRC CA
GIIIB	RDCCTPPRKC KDRRCKPMKC CA
GIIIC	RDCCTPPKKC KDRRCKPLKC CA
Consensus	RDCCTPP+KC KDROCKPX+C CA

FIGURE 3. μ-Conotoxin structures. Top: A space-filling model of the three-dimensional structure of μ-GIIIA, as deduced from two-dimensional nuclear magnetic resonance data. Black spheres indicate sidechain nitrogens of Arg, Lys, and the N-terminal group. A and B depict different faces of the molecule, while the structure of tetrodotoxin is shown in scale in C for comparison. (From Lancelin et al. (1991) Biochemistry **30**, 6908–6916.) Bottom: Amino acid sequences of three μ-conotoxin variants are shown in the single letter code with P designating hydroxyproline. Disulfide bonds link Cys 3 to Cys 15, Cys 4 to Cys 20, and Cys 10 to Cys 21. In the consensus sequence + = cationic residue; X = ambiguous character.

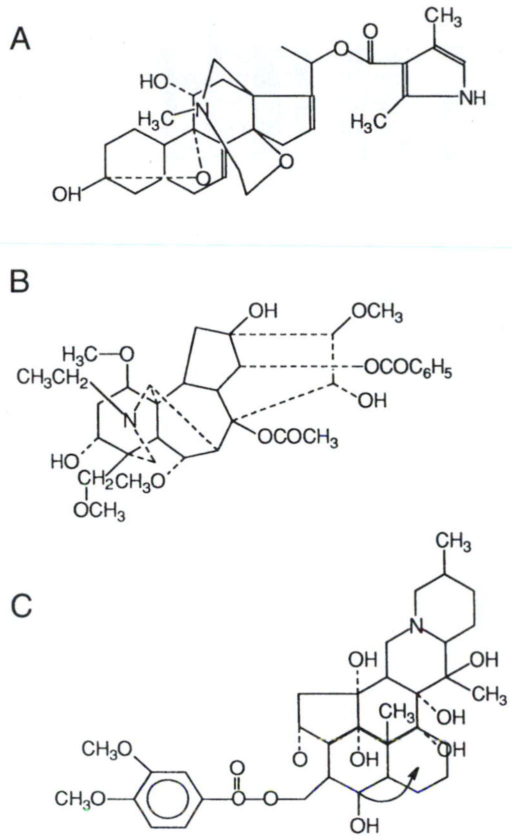

A

B

C

FIGURE 4. Structures of some major alkaloid agonists of the voltage-sensitive sodium channel. (A) Batrachotoxin, (B) Aconitine, (C) Veratridine.

of the Na^+ channel, shifting a pre-existing equilibrium between active and inactive states toward the former. Thus, toxin action would best be approximated by the allosteric model of Monod, Wyman, and Changeaux, in which the binding energy for a given toxin is expressed as a shift in the voltage dependence for channel activation. As will be discussed later, conformational changes in the channel induced by binding of toxins have been directly demonstrated, thus supporting this model.

C. Sites 3 and 4: Scorpion and Sea Anemone Toxins

An effect of scorpion venom on frog Na^+ channels was first demonstrated almost 30 years ago, and individual toxins from *Androctonus* venom were purified and sequenced beginning in the mid-1970s. Subsequently, toxins active on the sodium channel have been purified from the venoms of *Androctonus, Leiurus, Centruroides, Tityus,* and *Buthus* species. While these toxins exhibit certain common structural features, the pharmacology of the *Androctonus* and *Leiurus* toxins is quite distinct from that of the corresponding polypeptides from *Centruroides* and *Tityus*. The former (α toxin) group delays or abolishes channel inactivation and thus delays repolarization of the membrane through binding to neurotoxin site 3, while the latter (β-toxin) group shifts the voltage dependence

of activation to more negative potentials as a result of binding to site 4 (in the Catterall nomenclature), resulting in spontaneous activity. Binding of the α-toxins is dependent on membrane potential, with K_d values in the nanomolar range at the resting potential. In contrast, β-toxin binding is potential-independent and generally of slightly lesser affinity. The pharmacologic distinction between the α- and β-toxins should be emphasized, in view of the close relationship among these toxins at the level of covalent and tertiary structure.

The primary structures of representative scorpion toxins are depicted in Fig. 5. All scorpion toxins characterized to date are cationic polypeptides of molecular mass approximately 7000 Da that contain 4 disulfide bonds whose integrity is essential for activity; in addition, they display a significant degree of sequence homology, with approximately 30% of all positions being conserved among the known α-toxins. The β-toxins, exemplified by *Centruroides* toxin 3, are generally similar to those of the α group, but display less extensive homology. It is, however, possible that much of this sequence conservation within a group is essential simply for allowing proper folding of these toxins, similar to results obtained recently with the conotoxin family, but until residues important for interaction with the channel binding site have been definitively identified, this question will remain unresolved. Nonetheless, three-dimensional structures of representatives of each group show a good deal of similarity in the peptide backbones. With respect to structure-activity correlations, it has been shown that alkylation of Lys 56 of AaH-I is correlated with its inactivation, suggesting the importance of a cationic group at this position in the α-scorpion toxins, and that modification of Arg 2 and Arg 60 likewise results in inactivation. While other residues have been implicated in activity by treatment of one or another of the scorpion toxins with group-selective reagents, the results are somewhat fragmentary, and no overall patterns have as yet emerged. While at first glance this paucity of information might appear surprising, given the small size of the toxins and their ready availability, it must be remembered that prior to the advent of molecular biological approaches, the most accessible techniques (e.g., chemical modification using group-specific reagents) for identifying essential residues in proteins were developed to probe substrate binding to enzymes and their catalytic mechanisms. In general, these methods do not adapt well to proteins lacking superreactive residues, such as the toxins now under consideration. The ability to mutagenize peptide neurotoxins has in the past few years led to more extensive knowledge of the relationship between their structure and function, as exemplified by analysis of anemone toxins described in a later section.

The three-dimensional structures of a number of scorpion toxins are now known, including members of both the α- and β-classes (Fig. 6). As might be anticipated from the highly cross-linked nature of these proteins, both possess a dense core of secondary structure connected by two of the four disulfides; this core contains three strands of antiparallel β-sheet (residues 1–4, 37–41, and 46–50) and a short strand of α-helix (residues 23–32). In addition to several loops that protrude from this core region, a prominent feature of both structures is a relatively nonpolar surface region, which contains a number of conserved aromatic residues. Because chemical modification of sites external to this surface region

Representative α-toxins

```
AaH-I     KRDGYIVYPN  NCVYHCVPP-  -CDGLCKKNG  GSSGSSCFLV
AaH-II    VKDGYIVDDV  NCTYFCGRNA  YCNEECTKLK  GESGY-CQWA
Lqq-v     LKDGYIVDDK  NCTFFCGRNA  YCNDECKKKG  GESGY-CQWA
BoT-1     GRDAYIAQPE  NCVYECAQNS  YCNDLCTKNG  ATSGY-CQWL
BoT-XI    VKDGYIVDDV  NCTYFCGRNA  YCHEECTKLK  GESGY-CQWA

Consensus N+DGYIUOXX  NCßΦXCXOOO  YCOOXCOKXO  GOSGO-CXUU


AaH-I     -PSGLACWC-  KDLPDNVPIK  DTSRKCT
AaH-II    SPYGNACYCY  K-LPDHVRTK  GPGR-CH
LqQ-V     SPYGNACWCY  K-LPDRVSIK  EKGR-CN
BoT-I     GKYGNACWC-  KDLPDNVPIR  IPGK-CHF
BoT-XI    SPYGNACYCY  K-LPDHVRTK  GSPGRCH

Consensus OXOGXACΦCY  KDLPDOVOß+  OXOO+CO
```

Representative β-toxins

```
CsE-I     KEGYLVEKTG  CKKTCYKLGE  NDFC  RECKW  KHIGGSYGYC
TsS-G     KEGYLMDHEG  CKLSCFIRPS  -GYC  RECGI  KK--GSSGYC

Consensus KEGYLU-+OG  CKXOCΦXXXO  XOΦCORECXU  K+XXGSOGYC


CsE-I     YGFGCYCEGL  PDSTQTWPLP  -NKCT
TsS-G     AWPACYCYGL  PNWVKVWDRA  TNKC

Consensus XXXOCYCXGL  POXßOßWXXX  XNKC
```

FIGURE 5. Primary structures of representative α- and β-scorpion toxins. Sequences are shown in the single letter code, and disulfide bonds link Cys 12 to Cys 63, Cys 16 to Cys 36, Cys 22 to Cys 46, and Cys 26 to Cys 48 in both groups. Numbering is based on the sequence of AaH-II. Highlighted positions are either invariant or occupied by functionally equivalent residues within the α or β subgroup. In the consensus sequence, the coding is: + = cationic residue; O = polar residue; X = character undefined; U = nonpolar residue; Φ = aromatic residue; β = branched side chain. AaH, *Androctonus australis* Hector; Lqq, *Leiurus quinquestriatus quinquestriatus;* BoT, *Buthus occitanus Tunetanus;* CsE, *Centruroides sculpturatus* Ewing; TsS, *Tityus serrulatus.*

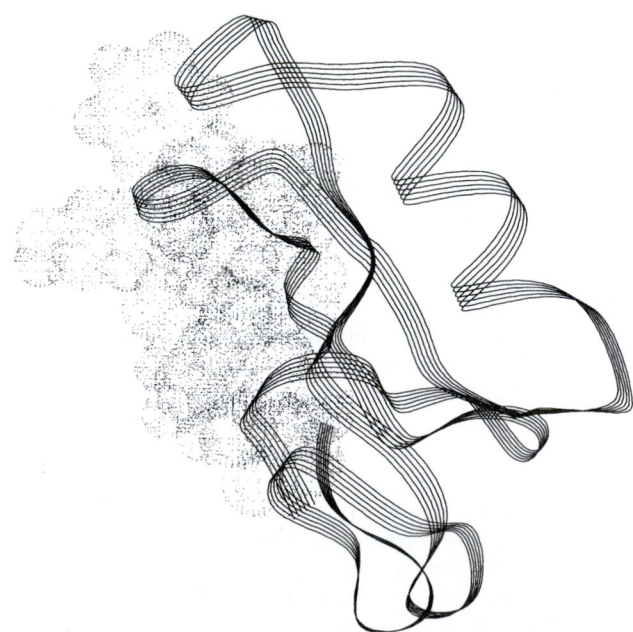

FIGURE 6. Three-dimensional structure of *Centruroides* toxin variant 3. Structural coordinates were obtained from the Brookhaven Protein Database, with permission. Although variant 3 is a β-toxin, the overall folding patterns of the α- and β-toxins are highly similar (Fontecilla-Camps *et al.* (1988) *Proc. Natl. Acad. Sci. USA* **85,** 7443-7447). The backbone of the molecule is depicted as a ribbon, and the hydrophobic surface proposed to be important in toxin-receptor interactions is shown as a dot surface. All side chains have been omitted for the sake of clarity.

```
AsI     GAPCKCKSDG PNTRGNSMSG TIWV--FGCP SGWNNCEGRA --HGYCCKQ
AsII    GVPCLCDSDG PSVRGNTLSG IIWL--AGCP SGWHNCKKHG PTIGWCCKQ
AsV     GVPCLCDSDG PSVRGNTLSG ILWL--AGCP SGWHNCKKHK PTIGWCCK
ApA     GVSCLCDSDG PSVRGNTLSG TLWLYPSGCP SGWHNCKAHG PTIGWCCKQ
ApB     GVPCLCDSDG PRPRGNTLSG ILWFYPSGCP SGWHNCKAHG PNIGWCCKK
ShI     -AACKCDDEG PDIRTAPLTG --TVDLGSCN AGWEKCASYY TIIADCCRKKK
RpII    -ASCKCDDDG PDVRSATGTG --TVDFWNCN EGWEKCTAVY TPVASCCRKK

Consensus  XUXCXCOODG POXROOOUOG ßUXUXXXOCX OGWOOCOOXX XXXOXCC+O
```

FIGURE 7. Primary structures of representative sea anemone toxins. Sequences are shown in the single letter code, with the numbering based on the sequence of anthopleurin A. Disulfide bonds link Cys 4 to Cys 46, Cys 6 to Cys 36, and Cys 29 to Cys 47. Highlighted positions are either invariant or contain functionally conservative replacements. In the consensus sequence, the coding is + = cationic residue; O = polar residue; X = character undefined; U = nonpolar residue; Φ = aromatic residue; β = branched side chain. As, *Anemonia sulcata;* Ap, *Anthopleura xanthogrammica;* Sh, *Stichodactyla helianthus;* Rp, *Radianthus paumotensis.*

is, in general, not deleterious to toxin function, and because Lys 56, referred to previously, lies within it, Fontecilla-Camps and coworkers first suggested that this region was essential for receptor binding. Interestingly, this surface is fairly well conserved in both the α- and β-toxins, despite their distinct pharmacologies and nonoverlapping binding sites. Indeed, all of the structural features discussed here are common amongst scorpion toxins regardless of binding site or species specificity.

In addition to the α-scorpion toxins, site 3 also binds a variety of polypeptide toxins derived from marine invertebrates: the best characterized of these are a diverse family of sea anemone toxins derived from *Anemonia, Anthopleura, Stichodactyla,* and *Radianthus* species (Catterall and Beress, 1978). Under appropriate conditions, certain of these polypeptides can function as cardiac stimulants. While also basic and rich in disulfides, the anemone toxins differ from those discussed previously in being somewhat smaller (46–50 residues versus 60–65 residues) and having three, rather than four, disulfides. Moreover, despite the wealth of structural information now available, no sequence homology has been found between neurotoxins from scorpions and anemones. Representative anemone toxin sequences are shown in Fig. 7.

Within the past few years, the solution structures of four anemone toxins have been solved (Fig. 8). These structures, all having a core of twisted, four-stranded, antiparallel β-pleated sheet (residues 2–4, 18–23, 31–34, and 42–47 in anthopleurin A numbering), demonstrate unequivocally that anemone and scorpion toxins are structurally unrelated. Nonetheless, anemone and α-scorpion toxins have very similar pharmacologies and display mutually competitive binding to the Na+ channel, suggesting that at least a limited degree of relatedness may yet remain to be found. It may be significant that all anemone toxins contain a large loop region (approximately residues 9–20) whose solution structure is very poorly defined. At least two of the amino acid residues known to be important for anemone toxin binding are found within this loop. In addition, biophysical analyses have provided evidence for the exposure of a number of hydrophobic residues on the surface of homologous anemone toxins. In anthopleurin B, the functional roles played by a subset of these residues have been analyzed using site-directed mutagenesis

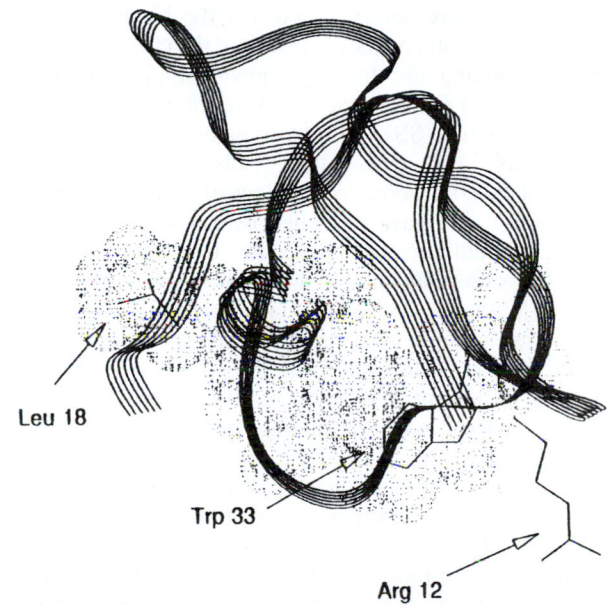

Leu 18

Trp 33

Arg 12

FIGURE 8. Three-dimensional structure of anthopleurin-B, a representative anemone toxin. The structure shown represents a model developed in the author's laboratory (Khera *et al.,* 1995) based on the coordinates of the homologous toxin from *Stichodactyla helianthus* (Fogh *et al.* (1990) *J. Biol. Chem.* **265,** 13016). The overall folding patterns for anemone toxins of known structure are similar. The model is also consistent with the recently determined solution structure of ApB (Monks *et al.,* (1995) *Structure* **3,** 791). The backbone is shown in ribbon format, and the side chains of residues which have been shown to be important determinants of binding affinity are annotated. The van der Waals surface depicts a region of partially exposed hydrophobic residues, including the required Leu 18 and Trp 33, which may be the functional equivalent of the hydrophobic surface involved in scorpion toxin binding to the sodium channel.

of a synthetic gene capable of encoding the toxin polypeptide. These studies have clearly demonstrated that both Leu 18 and Trp 33 play important roles in regulating high-affinity binding of this toxin to both neuronal and cardiac sodium channels, and further show that two other exposed hydrophobic sites, Ile

43 and Trp 45, are external to this binding epitope. Thus, the former pair of residues may act as the functional equivalent of the hydrophobic surface implicated in scorpion toxin binding (Dias-Kadambi *et al.*, 1996a,b). Indeed, charge-neutralizing mutations of all the ionizable groups in this toxin reveal that the contribution to affinity made by hydrophobic contacts is far more significant than any involving electrostatic interactions. Surprisingly, even Arg 14, which is conserved in all known anemone toxins, is revealed by mutagenesis to play at most a minor role in defining binding affinity (Khera and Blumenthal, 1994).

While sea anemone and α-scorpion toxins bind to the channel in a membrane-potential-dependent manner and their binding sites overlap at least in part, β-scorpion toxins bind to a demonstrably distinct site on the same protein and their binding is potential-independent. All of the known polypeptide toxin binding sites are structurally distinct from those for the alkaloid activators and guanidinium inhibitors. Affinity labeling experiments, using derivatives of either TTX or *Leiurus* toxin, have demonstrated that the binding sites for both toxins are found on the α-subunit of the channel. Like the TTX/STX binding site discussed previously, a photoactivatable derivative of *Leiurus* toxin labels residues between the S5 and S6 helices (i.e., SS1–SS2) of domains I and IV of the rat brain channel. In some cells, the β_1-subunit of the channel is also photolabeled, leading to the suggestion that the scorpion toxin binding site may be located near a subunit interface. Labeling of both subunits is specific, being blocked in the presence of excess unlabeled ligand, or by membrane depolarization. While these experiments demonstrate convincingly that scorpion toxin is capable of labeling the channel in the SS1–SS2 region of domains I and IV, they do not prove that interaction at this site is responsible for the physiologic consequences of scorpion toxin binding. Interestingly, recent genetic studies have highlighted the role of other extracellular sequences in modulating channel inactivation kinetics. The dominantly heritable disorder paramyotonia congenita, which displays an electrophysiologic phenotype identical to that of an anemone or scorpion toxin-treated, wild-type channel, has been mapped to a set of missense mutations in the S3–S4 linker region of domain IV of the muscle sodium channel (Cannon and Corey, 1993; Chahine *et al.*, 1994; Yang *et al.*, 1994). Biochemical analyses have demonstrated that Glu 1613, located in this region of the neuronal channel (Rogers *et al.*, 1996), is an important determinant of *Leiurus* toxin binding, while Benzinger *et al.* (1998) have shown that the analogous Asp 1612 of the cardiac channel interacts electrostatically with Lys 37 of the anemone toxin anthopleurin B.

D. Other Toxins Specific for the Na$^+$ Channel

A variety of nonproteinaceous substances of diverse (and in some cases, unknown) structure have been shown to interact with the Na$^+$ channel at sites distinct from those discussed previously. These include lipid-soluble dinoflagellate toxins (brevetoxins, ciguatoxin, and maitotoxin), the *Goniopora* toxins, and the pyrethroid insecticides. The brevetoxins induce the Na$^+$ channel to open at membrane potentials of –80 to –160 mV, and also abolish rapid inactivation; these effects lead to both generation of repetitive ac-

tion potentials and ultimately membrane depolarization. Both effects are blocked by TTX. Analyses of brevetoxin binding suggest the existence of a fifth ligand site on the channel, since brevetoxin binding fails to displace toxins bound to sites 1–4. The physiologic effects of ciguatoxin are very similar to those of the brevetoxins, while the mechanism of maitotoxin toxicity is unresolved at this time. *Goniopora* toxins, while not completely characterized, appear to be protein in nature, with molecular weights between 12 000 and 19 000 being reported from different preparations. The mode of action of at least one of these proteins appears to be similar to that of sea anemone toxins, but there is evidence that multiple *Goniopora* toxins with distinct molecular targets exist. Finally, the pyrethroids delay inactivation via interaction with a site distinct from that for polypeptide toxins having the same effect. Physiologic effects of pyrethroids are antagonized by TTX.

E. Relationships among Sites 1-4

The best data available from direct binding studies support the notion that each channel molecule contains single sites for each class of toxins. The next obvious question relates to identification of these sites and characterization of interactions among them. Angelides and coworkers have estimated the distances between the various toxin sites using resonance energy transfer between fluorescent derivatives of the bound ligands. In addition to confirming that the α and β sites are unique and separated by about 22 Å, these studies indicate that the TTX site is some 33 Å removed from that for β-toxins and the BTX site lies about 37 Å distant from that of α-toxins. Analogous experiments have also been used to verify that the sites involved are conformationally linked, with distances between a given pair of ligands changing upon binding of a third.

Because multiple subtypes of the Na$^+$ channel have been characterized, it is natural to wonder whether different toxins are directed against specific channel isoforms, either in a tissue-specific manner or not. As discussed below, this may be the case for K$^+$ channels. However, to date no convincing evidence has been presented that a given (e.g.) scorpion toxin distinguishes among Na$^+$ channel subtypes within an individual tissue, although it is clear that toxins do differ substantially in their affinities for channels that are expressed in a tissue-specific manner (Khera *et al.*, 1995). Obvious examples of this phenomenon would include the specificity of μ-conotoxins for muscle Na$^+$ channels, and the tenfold affinity preference displayed by some sea anemone toxins for cardiac, as compared to neuronal, Na$^+$ channels.

III. Voltage-Activated and Ca^{2+}-Activated Potassium Channels

Our knowledge of the structure, function, and diversity of potassium channels has increased dramatically within the past five years, with the application of combined molecular genetic, electrophysiologic, and crystallographic approaches (Miller, 1991). As will be seen, despite the enormous power of the genetic approach, the ability of polypeptide toxins specific for this family of channels to distinguish among subtypes is proving an invaluable tool. In addition, both polypeptides and nonproteinaceous blockers, such as tetraethylammonium ions and 4-aminopyridines, have proven to be valuable tools for development of structural models of the K^+ channel pore. These models are likely to apply to certain aspects of the homologous Na^+ and Ca^{2+} channels as well. In the case of K^+ channels, toxins from both arthropods and snakes have provided a wealth of pharmacologically significant data (summarized in Table 2), and very recently a new class of polypeptide K^+ channel blockers has been purified from sea anemone cnidocytes.

While voltage-dependent Na^+ channels exist in a relatively small number of molecular forms, the diversity seen in K^+ channels is vastly greater: some are ligand-coupled, while others are activated by either Ca^{2+} or depolarization. A large number of channels seem to exist even within this last category. In *Drosophila*, K^+ channels are encoded by genetic loci designated as *shaker, shal, shab, shaw,* and *eag,* and mammalian counterparts for each of these channel types have been identified. *Shaker* apparently encodes channels of the rapidly inactivating (or A) type, and transcripts from this locus display alternative splicing. It seems likely that this gives rise to channel forms that differ with respect to virtually all important properties: pharmacology, voltage-dependence, and unit conductance (Stuhmer *et al.,* 1989b). Multiple transcripts from the *shab* locus have also been observed, and the gene products have many of the properties associated with the delayed rectifier channel. The *shal* and *shaw* genes seem to encode A-type and delayed rectifier channels, respectively.

In addition to this multiplicity of voltage-gated K^+ channels, there exists an array of channels gated by Ca^{2+} as well. Because the unit conductances within this group vary quite widely, it is generally divided into high-conductance (BK or maxi-K, 130–300 pS) and small conductance (SK, 10–50 pS) subtypes. The amount of structural information presently available on these channels is significantly less than for the A-type channels, and because low-stringency hybridization screening using known K^+ channel probes has failed to identify members of this group, it is inferred that significant differences exist at the primary structure level despite some toxin cross-reactivity. Toxin binding may thus provide a means for purification of (e.g.) BK channels, and partial sequence analysis would then allow design of oligonucleotide probes, accessing additional structural information using classic molecular biological approaches.

The channels encoded by *shaker* and related genes have many structural features in common with the Na^+ channels discussed above, and these proteins display a limited degree of homology at the amino acid level. In both cases, the channel "subunit" contains six putative transmembrane helices with the N-terminus of the first being cytoplasmic. Helix S4, which contains a series of cationic groups separated by two hydrophobic residues, is postulated to serve as the voltage sensor, perhaps with some participation by charged residues in S2 and S3. As with the Na^+ channel, sequences joining the helices designated S5 and S6 are thought to provide the inner lining of the pore, and a great deal of evidence that these residues also provide the binding sites for channel-specific toxins has been accumulated in the past few years. The most obvious structural difference between these channels is that whereas the Na^+ channel α-subunit has a molecular mass of about 260 kDa and contains some 1800 amino acid residues grouped into 4 homologous domains, the K^+ channel monomer is approximately one-quarter this size. While the monomeric unit of the K^+ channel is thus small relative to its Na^+ and Ca^{2+} counterparts, coexpression of mRNAs encoding distinct channel isoforms in *Xenopus* oocytes results in the appearance of functional channels having intermediate properties, strongly suggesting that the physiologically relevant form of this channel is a noncovalent tetramer. The possibility of functional diversity arising from both alternative splicing and formation of hetero-oligomeric channels thus arises.

In 1998, MacKinnon's lab solved the crystal structure of the *Streptomyces lividans* K^+ channel (Doyle *et al.,* 1998). While this protein is only about 30% the size of the eukaryotic voltage-gated K^+ channels, it displays significant homology to (e.g.) the *shaker* and other eukaryotic voltage-gated channels, particularly within the putative pore-forming regions. As a result, the *S. lividans* channel represents an attractive starting point for modeling the three-dimensional structures of the eukaryotic pore. In the bacterial channel, this region contains a 10-residue α-helix (the pore helix) whose C-terminus is oriented toward the center of the pore joined to the C-terminal transmembrane helix by a 20-residue linker that lacks regular secondary structure. The helices flanking the pore region are proposed to be analogous to the S5 and S6 helices of eukaryotic K^+ channels.

Most of the known K^+-channel-specific toxins have been purified from scorpion, honeybee, and snake venoms. The first such agent to be characterized was purified by Miller from *Leiurus* venom, and given the colorful, if nondescriptive, name **charybdotoxin** (Miller *et al.,* 1985). Homologs of this polypeptide have been found in venoms from *Leiurus* (scyllatoxin and agitoxin), *Androctonus* (kaliotoxin), *Buthus*

TABLE 2 Polypeptide Blockers of K^+ Channels

Toxin	Channel type	K_d
Charybdotoxin	Ca^{2+}-activated K^+	3.5 nM
	Voltage-sensitive K^+	140 pM
Iberiotoxin	Ca^{2+}-activated K^+ (BK)	250 pM
Kaliotoxin	Ca^{2+}-activated K^+ (BK)	20 nM
Noxiustoxin	Ca^{2+}-activated K^+ (BK)	>300 nM
Scyllatoxin	Ca^{2+}-activated K^+ (SK)	80 pM
Apamin	SK	10–400 pM
Dendrotoxins	Voltage-sensitive K^+	100 pM

```
ChTX         QFTNVSCTTS KECWSVCQRL HNTSRGK-CMN KKCRCYS
ScTX         AFCNL- RMCQLSCRSL GLL--GK-CIG DKCECVKH
KaTX         GVEINVKCSGS PQCLKPCKDA GM-RFGK-CMN RKCHCTP
IbTX         QFTDVDCSVS KECWSVCKDL FGVDRGK-CMG KKCRCYQ
NxTX         TIINVKCTSP KQCSKPCKEL YGSSAGAKCMN GKCKCYNN


Consensus    XXßOUOCOXX XXCXXXCOOU XXXOXGO CUO OKCOCXX



Apamin       CNCKAPETAL CARRCQQH
```

```
αDtx         ZPRRKLCILH RNPGRCYDKI PAFYYNQKKK QCERFDWSGC GGNSNRFKTI EECRRTCIG
Dtx₁         QPLRKLCILH RNPGRCYQKI PARYYNQKKK QCEGFTWSGC GGNSNRFKTI EECRRTCIRl
Kalicludine  INKDCLLP MDVGRCRASH PRYYYNSSSK RCEKFIYGGC RGNANNFHTL EECKEKVCGl
```

FIGURE 9.   Primary structures of polypeptide toxins specific for K⁺ channels. Sequences are depicted in the single letter code, with disulfide bonds linking Cys 7 to Cys 28, Cys 13 to Cys 33, and Cys 17 to Cys 35 in charybdotoxin. Highlighted residues are invariant within a class. In the consensus sequence, the coding is + = cationic residue; O = polar residue; X = character undefined; U = nonpolar residue; Φ = aromatic residue; β = branched side chain. ChTX, charybdotoxin; ScTX, scyllatoxin; KaTX, kaliotoxin; IbTX, iberiotoxin; NxTX, noxiustoxin; αDTx, α-dendrotoxin; Dtx₁, dendrotoxin I; kalicludine, K⁺ channel blocker from *Anemonia sulcata*.

(iberiotoxin), and *Centruroides* (noxiustoxin) species. In addition, the honeybee peptides apamin and mast cell degranulating peptide, and a homologous family of snake (*Dendroaspis*) dendrotoxins have been shown to have K⁺-channel-blocking activity. Analogous to the case for sodium channels, the potassium channel toxins from scorpion venom comprise an homologous group in which essentially the only **absolutely** conserved residues are involved in disulfide bond formation (Fig. 9). Thus, one would reasonably expect these toxins to fold into a generally similar overall structure, but to differ in detail. It is therefore not surprising to realize that at least some of the K⁺ channel toxins are able to distinguish among channel subtypes. The honeybee toxins and dendrotoxins display no homology to any of the scorpion polypeptides.

Charybdotoxin (ChTX) was first identified in 1985 as a polypeptide from *Leiurus* venom capable of inhibiting Ca²⁺-activated K⁺ channels from skeletal muscle. Amino acid sequence analysis of the purified toxin revealed a primary structure completely distinct from that of other known scorpion toxins. The three-dimensional structure of ChTX was solved by NMR spectroscopy in 1991 (Fig. 10; Bontems *et al.*, 1991a,b; 1992). Its major features include a small, triple-stranded β-sheet encompassing residues 1–2, 25–29, and 32–36 linked by disulfide bonds (Cys 13–Cys 33 and Cys 17–Cys 35) to an α-helix including residues 10–19. Surprisingly, however, the overall fold of this molecule displays significant similarities to both the *Centruroides* and

Androctonus Na⁺-channel-specific toxins for which crystal structures had been obtained during the 1980s. It is important to recall that comparison of the sequences of the three toxins shows that only the positions of the disulfide bonds are conserved. However, in three dimensions, additional common features of these toxins are revealed, including

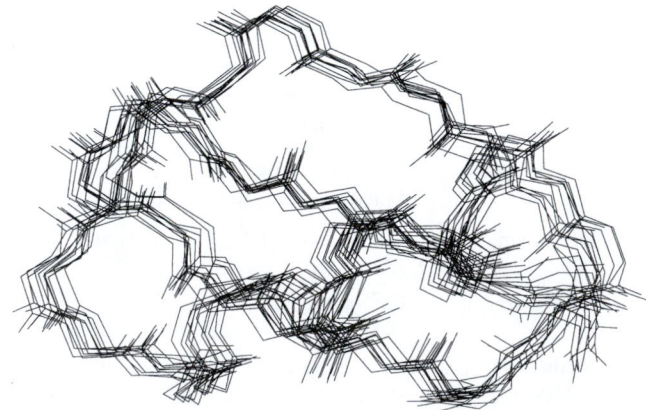

FIGURE 10.   Three-dimensional structure of charybdotoxin as deduced from NMR experiments. Positions of the backbone atoms of each of the 12 contributing structures in the PDB file pdb2crn.ent (available at www.rcsb.org) are depicted. Original data contributing to this structure were reported in Bontems *et al.* (1991a).

three disulfide bonds occurring in similar positions and orientations, the triple-stranded β-pleated sheet (residues 1–4, 37–41, and 46–50 of *Centruroides* toxin versus residues 1–2, 25–29, and 32–36 of charybdotoxin), the single strand of α-helix (residues 23–32 versus 10–19) and a significant proportion of extended structure (Fig. 11). The differences in comparing the structures of these long and short neurotoxins are most pronounced in loop regions connecting the elements of defined secondary structure. Thus, it would appear that the scorpion neurotoxins as a group share a common structural motif, apparent only at the level of three-dimensional structure, regardless of the nature of their target channel. The apparent similarities in channel folding thus seem to be mirrored in the structures of their "ligands." Interestingly, this motif appears to be similar to one observed in insect defensins, small peptides produced in response to injury that are thought to play a role in protecting insects against bacterial infections (Bontems *et al.*, 1992). It seems counterintuitive that the scorpion toxins, despite their vastly different pharmacologies, display conservation of tertiary structure, while anemone and scorpion α-toxins, which show very similar pharmacologies, do not. Perhaps this is simply indicative of the fact that our knowledge of the molecular interactions important for productive binding of these toxins to their target sites remains incomplete (however, see the following). However, it should be noted that the very recently characterized K$^+$-channel-blockers from anemones (**kalicludines** and **kaliseptines,** e.g.), which are functional homologs of charybdotoxin and dendrotoxin, appear to share neither this structural motif, sequence homology, nor even disulfide connectivities.

Because ChTX possesses a net charge of +5, it was predicted that ionic interactions with the channel would be important in its binding. This prediction is supported by the observations that (1) ChTX binding affinity is decreased 100-fold in a medium of increased ionic strength (300 mM salt); this reduction in binding is due to a decrease in toxin association and (2) chemical modification of the channel with the carboxylate-specific reagent trimethyloxonium (TMO) ion greatly reduces ChTX binding. The structure of the *S. lividans* channel reveals that the binding site for these cationic toxins is rich in acidic amino acid residues. Affinity also depends on membrane potential, with ChTX preferring the open conformation by approximately seven-fold. More recent studies with charybdotoxin mutants are consistent with hydrophobic interactions between channel sites and Met 29 and Tyr 36 of ChTX as also being important for binding (MacKinnon and Miller, 1989).

Although charybdotoxin was discovered because of its ability to block BK channels, evidence has since accumulated that its specificity is less well-defined. It now seems clear that voltage-dependent K$^+$ channels in frog ganglia, and SK channels from *Aplysia* among others, are also blocked by ChTX. However, because it affords us the ability to precisely map ChTX-channel interactions, the most interesting new target for ChTX is the *shaker* channel. To test the importance of ionic interactions in ChTX binding to this target, acidic residues located in extracellular sequences of the channel were altered by site-directed mutagenesis, and the kinetics of ChTX blocking measured. This strategy iden-

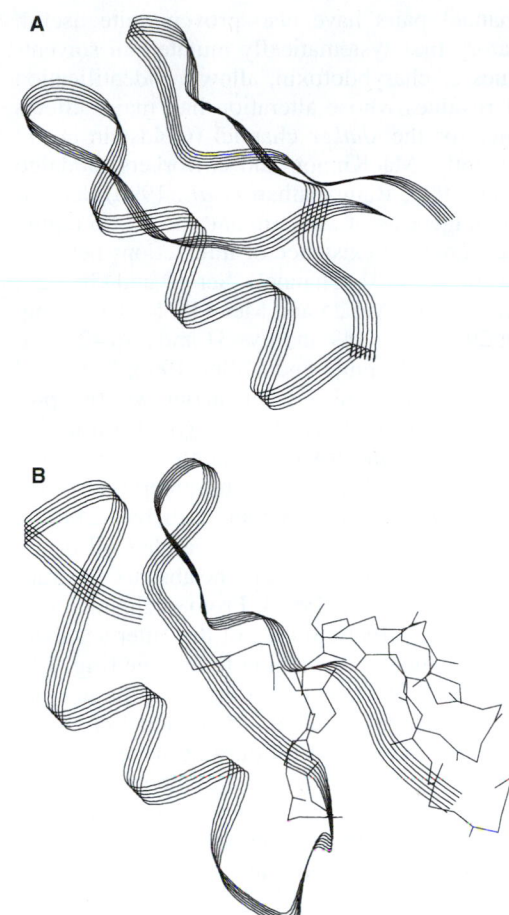

FIGURE 11. Comparison of the three-dimensional structures of (A) *Centruroides* variant 3 and (B) charybdotoxin. Backbone structures are shown as ribbons, representing the regions of highest similarity between the two polypeptides: α-helices 10–20 (charybdotoxin) and 22–32 (variant 3), and β-strands 1–2, 25–29, and 32–36 (charybdotoxin) and 1–4, 37–42, and 44–48 (variant 3). Images were created from the PDB files pdb2sn3 and pdb2crd, respectively, using InsightII for graphics display. Structural coordinates are reported in Bontems *et al.* (1991), and may be found in the PDB files cited at www.rcsb.org.

tified Glu 422, located in the SS1 region, as important for binding: substitution by aspartic acid had no effect while conversion to glutamine or lysine decreased blocking activity by 3- and 12-fold, respectively. Very recent data have identified another point mutation (Phe 425 to Gly) whose major consequence is to render the *shaker* channel approximately 2000 times more sensitive to ChTX. From both the nature of the substitution made and the magnitude of the change observed one might reasonably ascribe this increase to a major conformational change in the toxin binding site.

More recently, application of charybdotoxin and other homologous peptide blockers of potassium channels, notably agitoxin-2, as molecular calipers capable of providing topographical information on the vestibule of the potassium channel, has provided exciting results. For the sake of clarity, description of these results will be restricted to the charybdotoxin/agitoxin:*shaker* system, although analyses of other toxin:channel pairs have also proven quite useful. Miller's laboratory first systematically mutated all solvent-exposed residues of charybdotoxin, allowing identification of a subset of residues whose alteration had major effects upon its affinity for the *shaker* channel (Goldstein *et al.*, 1994). Subsequently, MacKinnon and coworkers (Hidalgo and MacKinnon, 1995; Ranganathan *et al.*, 1996) demonstrated by co-mutagenesis of agitoxin and the S5-S6 region of the *shaker* channel the existence of interactions between Arg 24 (toxin) and Asp 431 (channel), Ser 10 and Gly 425, Lys 27 and Thr 449, and Arg 25 and Met 448. Miller's group found that Met 29 and Thr 449 and Lys 31 and Asp 427 also form interacting pairs (Naranjo and Miller, 1996; Naini and Miller, 1996). Interestingly, this last interaction was first predicted based on previous models of the charybdotoxin:*shaker* complex, knowledge of the toxin's solution structure, and considerations of symmetry in the tetrameric channel. Importantly, docking and mutagenic analyses based on the crystal structure of the bacterial channel validate all of the predictions of interacting pairs made in the absence of structural data (MacKinnon *et al.*, 1998). From such studies as these, a low-resolution structural map of the outer vestibule of the eukaryotic channel is beginning to emerge (Fig. 12). Since nonpeptide pore blockers that function at the internal mouth of the channel are available, and channel chimeras and mutagenesis have identified residues at or near the cytoplasmic ends of S5 and S6 as essential to their binding, an analogous approach to mapping its structure may also prove fruitful. That the high-affinity polypeptide K$^+$ channel blockers from sea anemone venom are structurally distinct

from the charybdotoxin family makes them yet another potentially fruitful probe of pore three-dimensional structure (Tudor *et al.*, 1996).

The effects of ChTX on *shaker* homologs cloned from rat brain (RCK channels) have also been examined. These experiments indicate a correlation between toxin affinity and the presence of negative charges in the S3-S4 loop. While this correlation remains to be verified experimentally, involvement of the S3-S4 loop in charybdotoxin binding would be of special interest in view of the demonstrated involvement of this region of the Na$^+$ channel in inactivation (see page 332).

Because ChTX displays high-affinity interactions with so many different K$^+$ channels, the search for more specific probes has continued and a number of groups have purified and characterized new toxins whose targeting may be more stringent. **Iberiotoxin** (IbTX), isolated from *Buthus* venom on the basis of its ability to inhibit ChTX binding, is 70% homologous to ChTX but significantly less basic. IbTX inhibits ChTX binding to smooth muscle noncompetitively, consistent with its having a distinct binding site, and data available to date are consistent with IbTX recognizing only Ca^{2+}-activated channels in a variety of tissues (Galvez *et al.*, 1990). Studies with chemically-synthesized chimeras of the two toxins suggest that C-terminal sequences of each toxin dictate channel specificity to a large degree: a construct containing the N-terminal 19 residues of ChTX and the C-terminal 18 from IbTX has IbTX-like properties, and vice versa.

Two additional scorpion toxins have been shown to target the BK-type channel, at least preferentially. **Kaliotoxin,** recently purified and characterized from *Androctonus* venom, is approximately 50% homologous to ChTX, IbTX, and **Noxiustoxin** and has been shown to suppress the whole-cell Ca^{2+}-activated K$^+$ current in mollusk and rabbit sympathetic nerve. The affected channel is insensitive to both apamin and dendrotoxins, and is blocked by ChTX and TEA (Crest *et al.*, 1992). Noxiustoxin (NxTX), purified from *Centruroides* venom, is 50% identical to ChTX, but due to its relatively low affinity in many systems, and its apparent interaction with a multiplicity of channel types, it seems unlikely that NxTX will be widely useful for channel purification and/or discrimination studies. NxTX appears to bind most tightly to Jurkat cells and to the ChTX-sensitive channel from brain. In contrast, its affinity for Ca^{2+}-activated channels lies in the 0.1–1.0 μM range.

Two distinct, nonhomologous toxins directed at the SK channel are now known: **apamin** and **scyllatoxin;** the latter is 25% identical to ChTX, although most of this relatedness derives from alignment of the cysteine residues. Apamin is an 18-residue polypeptide containing two disulfide bonds; it has an α-helical core comprising residues 9–15 flanked by regions rich in β-turns. Analysis of synthetic analogs of apamin highlights the importance of Arginines 13 and 14 for activity. Electrophysiologic analyses and ion flux experiments indicate that apamin is directed against the SK channel in guinea pig hepatocytes, murine neuroblastoma cells, and cultured rat muscle cells. Iodinated apamin has been used to demonstrate the existence of high affinity (K_d = 10–400 pM) receptors in rat brain, with affinity increasing with [K$^+$]. The chemical na-

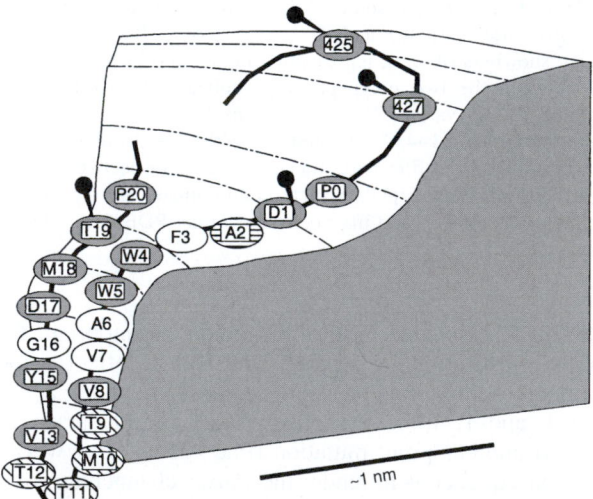

FIGURE 12. Proposed topography of the external pore and vestibule of the *shaker* K$^+$ channel. A hypothetical single *shaker* subunit, a 90° sector of the outer pore region and vestibule, is displayed. Residue numbers given are referenced to the P-region (P0-P20) or to *shaker* numbering. Push pins indicate residues whose positions relative to the pore axis were established by mapping with pore-blocking polypeptides. Shaded residues are predicted to project into the aqueous pore or vestibule, while unshaded residues are proposed to project away from the aqueous phase. No predictions have been made regarding side-chain orientations for those residues having diagonal striping. Reproduced from Lu and Miller (1995) *Science* **268,** 304, with permission.

ture of the apamin-binding protein has been studied by radiation inactivation and chemical cross-linking, but the results have been inconclusive, with multiple polypeptides being labeled.

Scyllatoxin seems to be directed against the same target as apamin, despite the fact that the two polypeptides display no sequence homology. In all systems analyzed to date, scyllatoxin competes with apamin for binding and/or exhibits the same functional effects as apamin. Although the latter toxin appears to bind more tightly (K_d is 15 pM for apamin and 80 pM for scyllatoxin), both proteins display very high affinities (Auguste *et al.*, 1992).

While the positions of defined helix and sheet structures are conserved in charybdotoxin and scyllatoxin (with the exception of the first β-strand, which is absent in the latter polypeptide), recent results indicate that these toxins, which act on different K^+ channel subtypes, are structurally distinct. In addition to lacking strand β-1, scyllatoxin also lacks many of the residues known to form an important hydrophobic cluster in charybdotoxin. Sequence homology suggests that this hydrophobic cluster is also maintained in noxius- and iberiotoxins. In addition, scyllatoxin can be distinguished from the others based upon positioning of cationic side chains. Three positively charged residues (Arg 25, Arg 34, and Lys 31) of charybdotoxin are replaced by leucine, aspartic acid, and glutamic acid, respectively, in scyllatoxin. Two of these residues, Arg 25 and Arg 34, have been shown to be important for interaction of charybdotoxin with the rat muscle Ca^{2+}-activated K^+ channel: mutagenesis to glutamine causes a large increase in the rate of toxin dissociation. Thus, scyllatoxin, which targets the SK channel, lacks these cationic sites but contains alternative ones that appear to be important for channel interaction. Modification of Arg 18 and His 38 of scyllatoxin appears to abolish its binding to rat brain channels. One possible interpretation of these data is that the charybdotoxin subgroup utilizes mainly positive residues in the β-sheet in its primary binding interaction, while scyllatoxin uses cationic sites in both the β-sheet and the α-helix.

Dendrotoxins comprise yet another family of toxins which target the K^+ channel and at present are the only snake (*Dendroaspis* species) toxins known to do so. Dendrotoxins (Dtx; four homologs known) are grossly similar to the toxins discussed previously, being highly cationic molecules containing 57–59 amino acid residues cross-linked by three disulfide bonds. While there is no apparent sequence homology between the Dtx's and the other toxins described previously, the Dtx's are clearly related to the small subunit of β-bungarotoxin and to bovine pancreatic trypsin inhibitor (BPTI). Very recently, the crystal structure of α-Dtx has been solved at 2 Å resolution, showing the similarity between Dtx and BPTI. Although the authors failed to discuss this structure in the context of those of other polypeptide neurotoxins, examination of the relevant structures does not reveal any obvious relationships. The similarity between Dtx and BPTI is even more curious in view of the fact that the latter protein binds to an internal site of the BK channel from skeletal muscle and that this binding is associated with the generation of subconducting states.

Dendrotoxins were first identified by their ability to stimulate neurotransmitter release at both central and peripheral neurons. Available electrophysiologic data now indicate that α-Dtx recognizes a variety of voltage-dependent K^+ channels. The Dtx sensitivity of RCK channels expressed in *Xenopus* oocytes has been analyzed, showing that rapidly inactivating channels are less sensitive than slowly inactivating ones, although all four types were responsive. To date, no data demonstrating an effect of the Dtx's on Ca^{2+}-activated currents have been published. Based on immunologic cross-reactivity, the α-Dtx receptor in mammalian brain is likely related to proteins encoded by the *shaker* locus, and mutations in the S5-S6 linker of RBK1 have been shown to strongly influence Dtx binding to this *shaker* homolog.

IV. Voltage-Dependent Calcium Channels

Voltage-dependent Ca^{2+} channels are a good deal more complex than those discussed in Section III, containing 5 subunits ranging in size from approximately 30 kDa to 170 kDa (Catterall, 1991). Nonetheless, certain structural features of the Ca^{2+} channel are strikingly similar to those of the Na^+ and K^+ channels. Most importantly, the α_1 subunit of the calcium channel is 29% identical to the α subunit of the rat brain R_{II} Na^+ channel, and the two proteins have functionally related but chemically distinct residues at about the same frequency. Application of the usual predictive structural algorithms to the α_1 subunit suggest the protein to be organized into four homologous structural domains, each containing the now-familiar six transmembrane helices. In addition, the sequence of the S4 helix suggests that, like its counterpart in Na^+ and K^+ channels, this region functions as the voltage sensor. Heterologous expression experiments have shown that the α_1 subunit is essential to the properties of dihydropyridine-sensitive Ca^{2+} channels.

Knowledge of the toxicology of Ca^{2+} channels is not extensive, at least as regards polypeptide effectors. There are no known scorpion, anemone, or snake toxins directed at this channel molecule. As of this writing, the only known peptide blockers of the Ca^{2+} channel are the omega conotoxins and a family of peptides from funnel web spiders that are structurally unrelated to the conotoxins; both groups prevent transmitter release at presynaptic terminals (Venema *et al.*, 1992). At least seven **ω-conotoxins** have been described; their common features include their cationic nature, the presence of three disulfide bonds, amidated C-termini, and a large number of hydroxylated amino acids, including in some cases hydroxyproline (Fig. 13). Two basic classes of these conotoxins are known, and homology across these classes amounts to about 40%. However, it should again be emphasized that this value may be misleading in view of the fact that approximately half the identical residues are involved in disulfide bonds. At least two classes of spider toxins, distinguishable by size and amino acid sequences, are now known and have been designated either as **agatoxins** (not to be confused with agitoxin, discussed previously) or curtatoxins. The shorter group (μ-toxins) contains 36–38 residues and is cross-linked by four disulfides; in addition, these μ toxins are distinguished from the polypeptides discussed previously in being generally less cationic. A second group of spider toxins (omega toxins) contains 65–75 residues and as many as six disulfides, and is more cationic

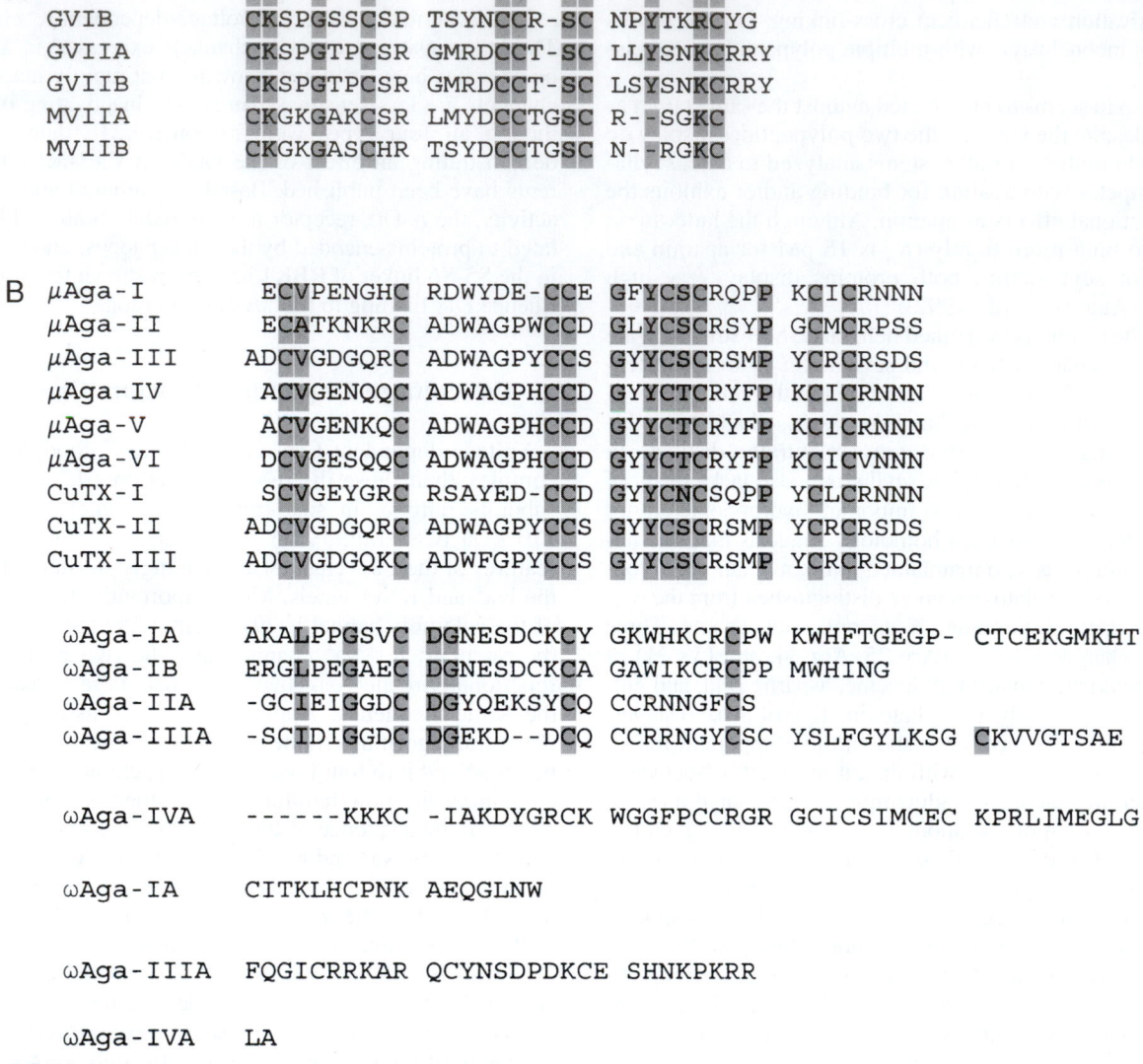

A

```
GVIA     CKSPGSSCSP  TSYNCCR-SC  NPYTKRCY
GVIB     CKSPGSSCSP  TSYNCCR-SC  NPYTKRCYG
GVIIA    CKSPGTPCSR  GMRDCCT-SC  LLYSNKCRRY
GVIIB    CKSPGTPCSR  GMRDCCT-SC  LSYSNKCRRY
MVIIA    CKGKGAKCSR  LMYDCCTGSC  R--SGKC
MVIIB    CKGKGASCHR  TSYDCCTGSC  N--RGKC
```

B

```
μAga-I     ECVPENGHC  RDWYDE-CCE  GFYCSCRQPP  KCICRNNN
μAga-II    ECATKNKRC  ADWAGPWCCD  GLYCSCRSYP  GCMCRPSS
μAga-III   ADCVGDGQRC  ADWAGPYCCS  GYYCSCRSMP  YCRCRSDS
μAga-IV    ACVGENQQC  ADWAGPHCCD  GYYCTCRYFP  KCICRNNN
μAga-V     ACVGENKQC  ADWAGPHCCD  GYYCTCRYFP  KCICRNNN
μAga-VI    DCVGESQQC  ADWAGPHCCD  GYYCTCRYFP  KCICVNNN
CuTX-I     SCVGEYGRC  RSAYED-CCD  GYYCNCSQPP  YCLCRNNN
CuTX-II    ADCVGDGQRC  ADWAGPYCCS  GYYCSCRSMP  YCRCRSDS
CuTX-III   ADCVGDGQKC  ADWFGPYCCS  GYYCSCRSMP  YCRCRSDS

ωAga-IA    AKALPPGSVC  DGNESDCKCY  GKWHKCRCPW  KWHFTGEGP-  CTCEKGMKHT
ωAga-IB    ERGLPEGAEC  DGNESDCKCA  GAWIKCRCPP  MWHING
ωAga-IIA   -GCIEIGGDC  DGYQEKSYCQ  CCRNNGFCS
ωAga-IIIA  -SCIDIGGDC  DGEKD--DCQ  CCRRNGYCSC  YSLFGYLKSG  CKVVGTSAE

ωAga-IVA   ------KKKC  -IAKDYGRCK  WGGFPCCRGR  GCICSIMCEC  KPRLIMEGLG

ωAga-IA    CITKLHCPNK  AEQGLNW

ωAga-IIIA  FQGICRRKAR  QCYNSDPDKCE  SHNKPKRR

ωAga-IVA   LA
```

FIGURE 13. (A) Primary structures of omega-conotoxin variants. The sequences are shown in the single letter code, with invariant residues or functionally conservative residues highlighted. Disulfide bonds link Cys 1 to Cys 16, Cys 15 to Cys 27, and Cys8 to Cys20. (B) Primary structures of representative spider toxins specific for Ca^{2+} channels. Highlighted positions are invariant within a group or occupied by functionally conservative replacements. Where complete sequences have not been determined, N-terminal sequences are shown. Aga, agatoxins; CuTX, curtatoxins.

than the shorter group. Sequences of representative spider toxins are also shown in Fig. 13, emphasizing their lack of relatedness to the conotoxins at the level of primary structure.

Omega conotoxins irreversibly block stimulus-evoked transmitter release in frog neuromuscular preparations; because the muscle retains the ability to respond to a direct stimulus, a presynaptic target for the toxin is indicated. That this target is the Ca^{2+} channel has since been confirmed in chick dorsal root ganglia (DRG). The effect of this toxin seems to be irreversible: in experiments in which DRG was treated with 100 nM toxin, then washed for 2 hours, no recovery of Ca^{2+} activity was observed.

Toxins isolated from funnel web spider venom are in general less well-characterized than the conotoxins as regards

their target channel. The unfractionated venom has been shown to block a dihydropyridine and ωCgTX-insensitive Ca^{2+} current detectable in *Xenopus* oocytes following microinjection of rat brain mRNA. However, because these spider venoms are known to contain both peptide and nonpeptide toxins, the chemical nature of the active substance in these experiments is unknown. More recent analyses using purified ω-agatoxins support the conclusion that these peptides target the Ca^{2+} channel. ω-Aga-IIA has been shown to inhibit binding of CgTX GVIA to chick brain membranes, while ω-Aga-IA and -IB do not, suggesting that the spider toxins may be able to discriminate among channel subtypes. Additional evidence consistent with this hypothesis arises from analysis of the effects of Aga-IA, -IIA, and -IIIA on

Ca^{2+} entry into depolarized chicken synaptosomes. In this system, the latter two polypeptides block entry, with 50% effective doses in the nanomolar range, while ω-Aga-IA is ineffective at 1 μM. Very recently, a new ω-agatoxin, ω-Aga-IVA, has been characterized and found to target P-type Ca^{2+} channels in synaptosomes; these channels are distinct from the N-type, and are resistant to both dihydropyridine and ω-CgTX blockade. Based on its sequence, ω-Aga-IVA may represent yet another class of spider toxins.

Because many distinct Ca^{2+} channel subtypes have been characterized electrophysiologically, it is natural to ask whether all are equally susceptible to the ω-conotoxins. Clearly, from the experiments described above, muscle channels are unaffected. However, even amongst neuronal channels, at least three subtypes are distinguishable by a variety of criteria, including gating kinetics, unit conductance, and antagonist pharmacology: Nowycky and coworkers, who studied them in chick DRG, designated these channels as T, N, and L-types. The L-type channel is distinguished by its sensitivity to dihydropyridines (DHPs), whereas T-types are not affected by these agents. ω-Conotoxins block T-type channels only weakly, whereas inhibition of neuronal L-type channels is essentially complete and irreversible in almost all cases; the irreversibility precludes an accurate determination of K_d.

Binding of ω-conotoxins to a variety of tissues has been analyzed, with K_d values in the subnanomolar range commonly reported. However, the validity of these values may fairly be questioned, given the irreversible nature of the toxin's effects. What is most clear is the lack of competition between the conotoxins and any class of organic channel blockers: dihydropyridine, benzothiazepine, and phenylalkylamine binding is unaffected in the presence of high concentrations of any of these drugs. Reported densities of ω-conotoxin binding sites are in the range of picomoles per milligram membrane protein in a variety of vertebrate neuronal preparations. Binding analyses also suggest that channels have either a DHP or a conotoxin site, although since L$_n$ channels can be blocked by both agents, exceptions to this rule must exist. In interpreting the literature on ω-conotoxin binding, it must be remembered that distinct forms of this toxin are made by different *Conus* species, and these toxins do not invariably have the same specificity.

ω-Conotoxins have been cross-linked to putative receptors in rat and chick brain. The variation in the molecular weight of this complex upon reduction suggests that the ω-conotoxin receptor is associated with the α_2-δ subunit of the Ca^{2+} channel. However, given the ability of the ω-conotoxins to block this channel, and the fact that the α_1 subunit is integral to ion permeation, it is reasonable to postulate interactions of the toxins with α_1 as well.

V. Other Toxins and Channels

A. α-Conotoxins and the Nicotinic Acetylcholine Receptor-Associated Channel

Although not targeted to ion channels, the α-conotoxins will be briefly discussed in the interest of covering all of the conotoxin classes. The α-conotoxins are the smallest of the conotoxin peptides known, comprising 13–15 amino acid residues, of which four are half-cystines involved in disulfide bonds. All of the known α-conotoxins are cationic, with formal charges of +1.5 to +3.5 at neutral pH. Sequences of the α-conotoxins are shown in Fig. 14. Conservative changes are tolerated at most positions without significant loss of activity, so that relatively little information on structure-function relationships can be derived from synthetic analogs. It is clear that the presence of two disulfides in such a short peptide imposes relatively severe constraints upon its three-dimensional structure. The proposed structure for α-conotoxin, based upon NMR measurements, includes β-turns at residues 5–8 and 9–12.

α-Conotoxins are directed at the nicotinic acetylcholine receptor channel, and mechanistically act like the snake toxins exemplified by α-bungarotoxin. In fact, the most widely accepted model for α-conotoxin structure places several of its side-chain functional groups into positions very similar to those seen in the active site of bungarotoxin. The postsynaptic site of action of α-conotoxin is consistent with its observed ability to compete with other nAChR antagonists like *d*-tubocurarine and α-BgTX. On this basis, α-conotoxin is, strictly speaking, not truly directed against the ion channel, and will not be discussed further.

B. Chlorotoxin and Chloride Channels

The toxicology of anion-selective channels is not nearly as well understood as that of the cation channels described previously, and therefore represents a fertile ground for future enquiry. The recent literature is replete with reports of chloride channels in a variety of tissues that are activated by diverse stimuli including membrane potential, cyclic AMP, volume, pH, etc. (Cuppoletti *et al.*, 1993; Pusch *et al.*, 1995a, b; Malinowska *et al.*, 1995). These channels, which are all members of the family designated as ClC, are formed by proteins of M_r 90–100 kDa, which most likely function as dimers (Fahlke *et al.*, 1997a). Hydropathy-based models predict that both their N- and C-termini are located in the cytoplasm, and are consistent with the presence of twelve transmembrane domains, which are modeled as α-helices. It should be noted, however, that these models are clearly still in an evolutionary state, with domains moving into or out of the membrane as recently as 1998. Structure-function analysis of ClC channels is still at a rather early stage, although N-terminal cytoplasmic sequences essential for channel inactivation have been identified by deletion scanning mutagenesis (Grunder *et al.*, 1992). This region is hypothesized to physically occlude the internal opening of the channel in a ball-and-chain-type inactivation mechanism, perhaps

G1	ECCNPACGRH YSC
G1A	ECCNPACGRH YSCGK
GII	ECCHPACGKH FSC
MI	GRCCHPACGKN YSC

FIGURE 14. Primary structures of α-conotoxins. The sequences are shown in the single letter code, with highlighted residues either invariant or allowing only functionally conservative substitution.

docking to sequences within the D7-D8 cytoplasmic linker (Jordt and Jentsch, 1997). More recently, mutagenesis coupled with chemical modification has implicated residues 240–260 and Cys 546 of human ClC-1 as lying in the outer vestibule of the channel pore (Fahlke *et al.*, 1997b; Kurz *et al.*, 1999). Clearly, high-affinity toxins analogous to TTX or charybdotoxin would be an invaluable tool in helping to limit the target size of such analyses. However, to date only a single peptide blocker of chloride channels has been identified. This polypeptide, isolated from *Leiurus* venom and designated as **chlorotoxin,** contains 36 amino acid residues cross-linked by four disulfide bonds, and displays no sequence homology to toxins from the same venom that target Na^+ or K^+ channels (DeBin *et al.*, 1993). It is, however, highly homologous to the scorpion insectotoxins derived from *Buthus eupeus* venom. Interestingly, while chlorotoxin blocks enterocyte chloride conductances with an affinity in the low micromolar range, it does so only when applied to the interior aspect of the channel; thus, its biological role is unlikely to involve chloride channel blockade. The homologous insectotoxins have been suggested to target the insect post-synaptic glutamate receptor, although supporting data for this proposal have not been presented, and the effect of chlorotoxin on this receptor is unknown.

The solution structure of chlorotoxin was recently solved by Lippens *et al.* (1995) using multidimensional NMR. Not surprisingly, the averaged structure, which contains a small three-stranded β-sheet packed against a single α-helix, is very similar to that of the insectotoxin. More unexpected is the observation that the overall fold of the molecule is rather similar to that of charybdotoxin, despite the lack of significant sequence homology between the two proteins. While structure-function relationships have not been addressed experimentally for either the insectotoxins or chlorotoxin, our recent ability to express active toxin in bacteria opens up this area for further exploitation.

VI. Summary

Neurotoxins have proven to be invaluable tools in the purification of many of the best-characterized cation channels, and in addition have contributed significantly to our understanding of how these macromolecules function in excitable tissues. As should be clear from the foregoing outline, this collection of molecules displays a great deal of diversity in both a chemical and a functional sense. Although many of the compounds discussed in the preceding sections are cationic, additional unifying structural themes are conspicuous by their absence. Even within the subgroup of polypeptide toxins, the most well-studied groups (i.e., anemone, scorpion, snail, and snake toxins) fail to display any intergroup relatedness at the level of primary structure. Indeed, this is rather surprising in view of the ability of (e.g.) anemone and α-scorpion toxins to compete for a common receptor. While all of the polypeptides do share a cationic character, and are rich in disulfide bonds, it remains to be proven that the former property is intimately related to their activity; the latter property is almost certainly a consequence of the necessity for stabilizing small peptides in an extracel-

lular environment. Perhaps the most striking observation is the relatedness between charybdotoxin and the Na^+ channel-directed scorpion toxins at the level of their three-dimensional structures. This clearly raises the possibility of common binding mechanisms having evolved.

Given the vast diversity of the venoms from which each of these toxins is derived, and the growing array of ion channels purified and/or channel activities characterized, it is likely that new and interesting toxins will continue to emerge in the future, and that their characterization will add significantly to our knowledge of channels and receptors in excitable tissues. Two areas of research that might prove especially fruitful involve using existing (or newly discovered) toxins to probe for specific channel isoforms in different tissues, and using behavioral assays to identify new venom components likely to have neuronal activities. This latter approach has already been applied to components of *Conus geographus* venom with very intriguing results, among them being the discovery of the so-called "sleeper" and "King Kong" peptides. It is likely that this surface has as yet only been scratched, and that future studies will unveil additional toxins that interact with new, and perhaps presently unknown, target sites.

Bibliography

Armstrong, C. M., and Hille, B. (1998). Voltage-gated ion channels and electrical excitability. *Neuron* **20,** 371–380.

Auguste, P., Hugues, M., Mourre, C., Moinier, D., Tartar, A., and Lazdunski, M. (1992). Scyllatoxin, a blocker of Ca-activated K-channels: structure-function relationships and brain localization of the binding sites. *Biochemistry* **31,** 648–654.

Benzinger, G. R., Kyle, J. W., Blumenthal, K. M., and Hanck, D. A. (1998). A specific interaction between the cardiac sodium channel and site-3 toxin anthopleurin B. *J. Biol. Chem.* **273,** 80–84.

Bontems, F., Roumestand, C., Boyot, P., Gilquin, B., Doljansky, Y., Menez, A., and Toma, F., (1991a). Three-dimensional structure of natural charybdotoxin in aqueous solution by^{1}H-NMR. Charybdotoxin possesses a structural motif found in other scorpion toxins. *Eur. J. Biochem.* **196,** 19–28.

Bontems, F., Roumestand, C., Gilquin, B., Menez, A., and Toma, F. (1991). Refined structure of charybdotoxin: common motifs in scorpion toxins and defensins. *Science* **254,** 1521–1523.

Bontems, F., Gilquin, B., Roumestand, C., Menez, A. and Toma, F. (1992). Analysis of side-chain organization on a refined model of charybdotoxin: structural and functional implications. *Biochemistry* **31,** 7757–7764.

Cannon, S. C., and Corey, D. P. (1993). Loss of Na-channel inactivation by anemone toxin mimics the myotonic state in HPP. *J. Physiol.* **466,** 501.

Catterall, W. A. (1977). Activation of the action potential Na-ionophore by neurotoxins: an allosteric model. *J. Biol. Chem.* **252,** 8669–8676.

Catterall, W. A. (1991). Functional subunit structure of voltage-gated calcium channels. *Science* **253,** 1499–1500.

Catterall, W. A. (1995). Structure and function of voltage-gated ion channels. *Annu. Rev. Biochem.* **64,** 493–531.

Catterall, W. A., and Beress, L. (1978). Sea anemone toxin and scorpion toxin share a common receptor site associated with the action potential sodium ionophore. *J. Biol. Chem.* **253,** 7393–7396.

Chahine, M., George, A. L. Jr., Zhou, M., Ji, S., Sun, W., Barchi, R. L., and Horn, R. (1994). Na channel mutation in paramyotonia congenita uncouple inactivation from activation. *Neuron* **12,** 281.

Crest, M., Jacquet, G., Gola, M., Zerrouk, H., Benslimane, A., Rochat, H., Mansuelle, P., and Martin-Eauclaire, M.-F. (1992). Kaliotoxin, a novel peptidyl inhibitor of neuronal BK-type Ca-activated K-channels characterized from *Androctonus mauretanicus mauretanicus* venom. *J. Biol. Chem.* **267,** 1640–1647.

Cruz, L. J., Gray, W. R., Olivera, B. M., Zeikus, R. D., Kerr, L., Yoshikami, D. and Moczydlowski, E. (1985). *Conus geographus* toxins that discriminate between neuronal and muscle sodium channels. *J. Biol. Chem.* **260,** 9280–9288.

Cruz, L. J., Kupryszewski, G., LeCheminant, G. W., Gray, W. R., Olivera, B. M., and Rivier, J. (1989). μ-Conotoxin GIIIA, a peptide ligand for muscle sodium channels: chemical synthesis, radiolabeling and receptor characterization. *Biochemistry* **28,** 3437–3442.

Cuppoletti, J., Baker, A. M., and Malinowska, D. H. (1993). Chloride channels of the gastric parietal cell that are active at low pH. *Am. J. Physiol.* **264,** C1609.

DeBin, J. A., Maggio, J. E., and Strichartz, G. R. (1993). Purification and characterization of chlorotoxin, a chloride channel ligand from the venom of the scorpion. *Am. J. Physiol.* **264,** C361–C369.

Dias-Kadambi, B. L., Combs, K. A., Drum, C. L., Hanck, D. A., and Blumenthal, K. M. (1996a). The role of exposed tryptophan residues in the activity of the cardiotonic polypeptide anthopleurin B. *J. Biol. Chem.* **271,** 23828–23835.

Dias-Kadambi, B. L., Drum, C. L., Hanck, D. A., and Blumenthal, K. M. (1996b). Leucine-18, a hydrophobic residue essential for high affinity binding of anthopleurin-B to the voltage sensitive sodium channel. *J. Biol. Chem.* **271,** 9422–9429.

Dougherty, D. A. (1996). Cation-π interactions in chemistry and biology: a new view of benzene, Phe, Tyr and Trp. *Science* **271,** 163–168.

Doyle, D. A., Cabral, J. M., Pfuetzner, R. A., Kuo, A., Gulbis, J. M., Cohen, S. L., Chait, B. T., and MacKinnon, R. (1998). The structure of the potassium channel: molecular basis of K-conduction and selectivity *Science* **280,** 69–77.

Fahlke C., Knittle, T., Gurnett, C. A., Campbell, K. P., and George, A. L. (1997a). Subunit stoichiometry of human muscle chloride channels. *J. Gen. Physiol.* **109,** 93–104.

Fahlke C., Yu, H. T., Beck, C. L., Rhodes, T. H., and George A. L. Jr. (1997b). Pore-forming segments in voltage-gated chloride channels. *Nature* **390,** 529–532.

Fogh, R. H., Kem, W. R., and Norton, R. S. (1990). Solution structure of neurotoxin 1 from the sea anemone *Stichodactyla helianthus*. *J. Biol. Chem.* **265,** 13 016–13 028.

Fozzard, H. A. and Hanck, D. A. (1996). Structure and function of voltage-dependent sodium channels: comparison of brain II and cardiac isoforms. *Physiol. Rev.* **76,** in press.

French, R. J., Sochaczewski, E. P., Zamponi, G. W., Becker, S., Kularatna, A. S., and Horn, R. (1996). Interactions between a pore-blocking peptide and the voltage sensor of the sodium channel: an electrostatic approach to channel geometry. *Neuron* **16,** 407–413.

Galvez, A., Gimenez-Gallego, G., Reuben, J. P., Roy-Contancin, L., Feigenbaum, P., Kaczorowski, G. J., and Garcia, M. L. (1990). Purification and characterization of a unique, potent, peptidyl probe for the high conductance calcium-activated potassium channel from venom of the scorpion *Buthus tamulus*. *J. Biol. Chem.* **265,** 11083–11090.

Goldstein, S. A. N. (1996). A structural vignette common to voltage sensors and conduction pores: canaliculi. *Neuron* **16,** 717–722.

Goldstein, S. A. N., Pheasant, D. J., and Miller, C. (1994). The charybdotoxin receptor of a *shaker* K⁺ channel: peptide and channel residues mediating molecular recognition. *Neuron* **12,** 1377–1388

Grunder, S., Thiemann, A., Pusch, M., and Jentsch, T. J. (1992). Regions involved in the opening of ClC-2 chloride channel by voltage and cell volume. *Nature* **360,** 759–762.

Hartshorne, R. P. and Catterall, W. A. (1984). The sodium channel from rat brain: purification and subunit composition. *J. Biol. Chem.* **259,** 1667–1675.

Hidalgo, P., and MacKinnon, R. (1995). Revealing the architecture of a K⁺ channel pore through mutant cycles with a peptide inhibitor. *Science* **268,** 307–310.

Hoshi, T., Zagotta, W. N., and Aldrich, R. W. (1990). "Biophysical and molecular mechanisms of *shaker* channel inactivation. *Science* **250,** 533–538.

Jordt, S.-E., and Jentsch, T. J. (1997). Molecular dissection of gating in the ClC-2 chloride channel. *EMBO J.* **16,** 1582–1592.

Kerr, L. M., and Yoshikami, D. (1984). A venom peptide with a novel presynaptic blocking action. *Nature* **308,** 282–284.

Khera, P. K., and Blumenthal, K. M. (1994). Role of the cationic residues Argenine-14 and Lysine-48 in the function of the cardiotonic polypeptide anthopleurin B. *J. Biol. Chem.* **269,** 921–926.

Khera, P. K., Benzinger, G. R., Lipkind, G., Drum, C. L., Hanck, D. A., and Blumenthal, K. M. (1995). Multiple cationic residues of anthopleuruin B that determine high affinity and channel isoform discrimination," *Biochemistry* **34,** 8533–8541.

Kurz, L. L., Klink, H., Jakob, I., Kuchenbecker, M., Benz, S., Lehmann-Horn, F., and Rudel, R. (1999). Identification of three cysteines as targets for the Zn²⁺-blockade of the human skeletal muscle chloride channel. *J. Biol. Chem.* **273,** 11687–11692.

Lippens, G., Najib, J., Wodak, S. J., and Tartar, A. (1995). NMR sequential assignments and solution structure of chlorotoxin, a small scorpion toxin that blocks chloride channels. *Biochemistry* **34,** 13–21.

MacKinnon, R., and Miller, C. (1989). Mutant potassium channels with altered binding of charybdotoxin, a pore-blocking inhibitor. *Science* **245,** 1382–1385.

MacKinnon, R., and Yellen, G. (1990). Mutations affecting TEA blockade and ion permeation in voltage-activated K channels. *Science* **250,** 276–279.

MacKinnon, R., Cohen, S. L., Kuo, A., Lee, A., and Chait, B. T. (1998). Structural conservation in prokaryotic and eukaryotic potassium channels. *Science* **280,** 106–109.

Malinowska, D. H., Kupert, E. Y., Bahinski, A., Sherry, A. M., and Cuppoletti, J. (1995). Cloning, functional expression and characterization of a PKA-activated gastric Cl– channel. *Am. J. Physiol.* **268,** C191–200.

Miller, C. (1991). 1990: Annus mirabilis of potassium channels. *Science* **252,** 1092–1096.

Miller, C., Moczydlowski, E., Latorre, R., and Phillips, M. (1985). Charybdotoxin, a protein inhibitor of single Ca-activated K channels from mammalian skeletal muscle. *Nature* **313,** 316–318.

Monks, S. A., Pallaghy, P. K., Scanlon, M. J., and Norton, R. S. (1995). Solution structure of the cardiostimulant polypeptide Anthopleurin-B. *Structure* **3,** 791–803.

Mosher, H. S. (1986). The chemistry of tetrodotoxin. *Ann. N.Y. Acad. Sci.* **479,** 32–43.

Naini, A. A., and Miller, C. (1996). A symmetry-driven search for electrostatic interaction partners in charybdotoxin and a voltage-gated K⁺ channel. *Biochemistry* **35,** 6181–6187.

Naranjo, D., and Miller, C. (1996). A strongly interacting pair of residues on the contact surface of charybdotoxin and a *shaker* K⁺ channel. *Neuron* **16,** 123–130.

Noda, M., Suzuki, H., Numa, S., and Stuhmer, W. (1989). A single point mutation confers tetrodotoxin and saxitoxin insensitivity on the sodium channel II. *FEBS Lett.* **259,** 213–216.

Ohizumi, Y., Nakamura, H., Kobayashi, J., and Catterall, W. A. (1986). Specific inhibition of saxitoxin binding to skeletal muscle sodium channels by geographutoxin II, a polypeptide channel blocker. *J. Biol. Chem.* **261,** 6149–6152.

Pusch, M., Steinmeyer, K., Koch, M. C., and Jentsch, T. J. (1995a). Mutations in dominant human myotonia congenita drastically alter the voltage dependence of the ClC-1 chloride channel. *Neuron* **15,** 1455–1463.

Pusch, M., Ludewig, U., Rehfeldt, A., and Jentsch, T. J. (1995b). Gating of the voltage-dependent chloride channel ClC-0 by the permeant anion. *Nature* **373,** 527–531.

Ranganathan, R., Lewis, J. H., and MacKinnon, R. (1996). Spatial localization of the K$^+$ channel selectivity filter by mutant cycle-based structure analysis. *Neuron* **16,** 131–139.

Rogers, J. C., Qu, Y., Tanada, T. N., Scheuer, T., and Catterall, W. A. (1996). Molecular determinants of high affinity binding of α-scorpion toxin and sea anemone toxin in the S3-S4 extracellular loop in domain IV of the Na-channel α subunit. *J. Biol. Chem.* **271,** 15950–15962.

Satin, J., Kyle, J. W., Chen, M., Bell, P., Cribbs, L. L., Fozzard, H. A., and Rogart, R. B. (1992). A mutant of TTX-resistant cardiac sodium channels with TTX-sensitive properties. *Science* **256,** 1202–1205.

Shimizu, Y. (1986). Chemistry and biochemistry of saxitoxin analogues and tetrodotoxin. *Ann. N.Y. Acad. Sci.* **479,** 24–31.

Stuhmer, W., Conti, F., Suzuki, H., Wang, X., Noda, M., Yahagi, N., Kubo, H., and Numa, S. (1989a). Structural parts involved in activation and inactivation of the sodium channel. *Nature* **339,** 597–603.

Stuhmer, W., Ruppersberg, J. P., Schroter, K. H., Sakmann, B., Stocker, M., Giese, K. P., Perschke, A., Baumann, A., and Pongs, O. (1989b). Molecular basis of functional diversity of voltage-gated potassium channels in mammalian brain. *EMBO J.* **8,** 3235–3244.

Tamkun, M. M., Talvenheimo, J. A., and Catterall, W. A. (1984). The sodium channel from rat brain: reconstitution of neurotoxin-activated ion flux and scorpion toxin binding from purified components. *J. Biol. Chem.* **259,** 1676–1687.

Trainer, V. L., Brown, G. B. and Catterall, W. A. (1996). Site of covalent labeling by a photoreactive batrachotoxin derivative near transmembrane segment IS6 of the sodium channel α subunit. *J. Biol. Chem.* **271,** 11261–11267.

Tudor, J. E., Pallaghy, P. K., Pennington, M. W., and Norton, R. S. (1996). Solution structure of ShK toxin, a novel potassium channel inhibitor from a sea anemoně. *Nature Stuct. Biol.* **3,** 317–320.

Venema, V. J., Swiderek, K. M., Lee, T. D., Hathaway, G. M., and Adams, M. E. (1992). Antagonism of synaptosomal calcium channels by subtypes of omega agatoxins. *J. Biol. Chem.* **267,** 2610–2615.

Yang, N., Ji, S., Zhou, M., Ptácek, L. J., Barchi, R. L., Horn, R., and George, A. L. Jr. (1994). Na channel mutations in paramyotonia congenita exhibit similar biophysical phenotypes *in vitro*. *Proc. Natl. Acad. Sci. USA* **91,** 12785.

I. *Rivolta, H. Abriel, and Robert S. Kass*

38

Ion Channels as Targets for Drugs

Many drugs exert their therapeutic effects by acting on the ion channels of various tissues. As illustrations, we will concentrate on only a few of the ion channels of the cardiac muscle. Electrical activity in the heart is generated by the summation of currents through multiple ion channels (Kass, 1995). The ventricular action potential is characterized by a long-lasting plateau period in which a balance is maintained between small inwardly and outwardly directed exchange and ion channel currents, and small changes in this balance can have severe functional consequences. During the past 10 years, a wealth of information has evolved characterizing the molecular structures that form the protein pathways for ion conduction in heart and other excitable cells. With the emerging molecular and structural information, true molecular pharmacological approaches to disease management are evolving with the concept of rational drug design being brought to fruition in the novel approaches applied to pharmacological intervention based on genetic defects in ion channel expression. This chapter will focus on recent progress in the development of a molecular pharmacological approach to targeting two types of cardiac ion channels: L-type calcium (Ca^{2+}) and voltage-gated sodium (Na^+) channels.

I. Calcium Channels

At least six types of voltage-gated calcium channels (N-, L-, P-, Q-, R-, and T-type) have been identified based on their pharmacological and/or biophysical properties (Tsien *et al.,* 1991; Birnbaumer *et al.,* 1994). In the heart, both T- and L-type channels contribute to cardiac electrophysiology, but L-type channels, the targets of organic calcium channel modulators and substrates for cAMP-dependent protein kinase A and protein kinase C, are unique in their importance to the maintenance of calcium homeostasis because they can be under pharmacological and/or neurohormonal control (Gutierrez *et al.,* 1994; Zhao *et al.,* 1994; Ma *et al.,* 1992; Tsien *et al.,* 1991; Sculptoreanu *et al.,* 1993a, b; Catterall *et al.,* 1991). Skeletal muscle L-type calcium channels are heteromultimeric proteins consisting of α_1-, β_2-, and α_2/δ-subunits (Hille, 1992; Catterall *et al.,* 1988; Gutierrez *et al.,* 1991) and it is likely that L-type calcium channels in other tissues are also multi-subunit proteins (Takahashi *et al.,* 1987; Schneider and Hofmann, 1988; Takahashi and Catterall, 1987; Ahlijanian *et al.,* 1990). Figure 1 shows a comparison of putative subunit structure of skeletal and cardiac calcium channels as well as potential phosphorylation sites.

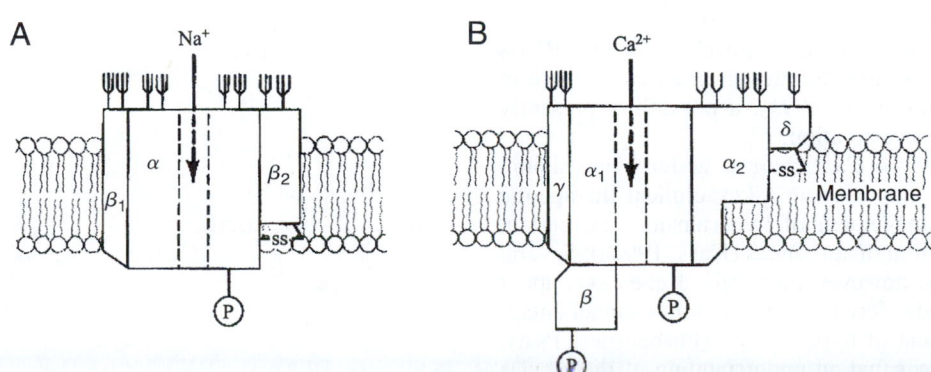

FIGURE 1. Block diagram illustrating subunit structures of voltage-gated sodium channels *(a)* and L-type calcium channels *(b)*. Sodium channels consist of a central pore-forming subunit (α) and two auxiliary β-subunits. L-type calcium channels also contain a pore-forming α-subunit and β-subunits, but also $\alpha_2\delta$- and γ-subunits (Catterall, 1988).

Copyright © 2001 by Academic Press. All rights of reproduction in any form reserved.

Functional roles of the auxiliary subunits have been studied by several groups with particular focus on the role of the β_2-subunit (Ruth *et al.*, 1989; Singer *et al.*, 1991; Gao *et al.*, 1997). When both α- and β-subunits of the L-type Ca^{2+} channel are coexpressed, the kinetics of I_{Ca} are more nearly normal than when the α-subunit is expressed alone (Lacerda *et al.*, 1991; Lory *et al.*, 1993; Nishimura *et al.*, 1993; Varadi *et al.*, 1991) and expression levels are enhanced Neely *et al.*, 1993). One role of the β-subunit is thus thought to involve the targeting of the α_1-subunit to the surface membrane. Similarly, α_2/δ-subunits increase gating charge and ionic current in recombinant channel activity (Bangalore *et al.*, 1996). The roles of auxiliary subunits in modulating native channel function continues to be an important area of investigation that will be important in unraveling drug and neuro-hormonal modulation of L-type calcium channels Hosey *et al.*, 1996).

L-type calcium channels inactivate in a voltage- and calcium-dependent manner (Yue *et al.*, 1990; Kass and Sanguinetti, 1984) and are the targets of the most extensively developed calcium channel pharmacology (Kass, 1994a). Perhaps best studied is the marked modulation of L-type calcium channels by the β-adrenergic (β-AR) signaling cascade. Action potentials in the heart are sensitive to catecholamines. Exposure of isolated tissue to norepinephrine increases pacemaker activity (Tsien, 1977), increases the height and affects the duration of the ventricular action potential plateau (Reuter and Scholz, 1977), and increases the strength of contraction, all of which have been shown to be due, at least in part, to modulation of L-type channels by β-AR stimulation (Cachelin *et al.*, 1983; McDonald *et al.*, 1994; Morad and Trautwein, 1968; Morad and Rolett, 1972; Reuter and Scholz, 1977; Reuter, 1983).

Calcium entry via L-type channels contributes the major source of calcium ions to load intracellular stores for subsequent release and activation of contractile proteins. The process linking calcium entry via sarcolemmal channels and release of calcium from the sarcoplasmic reticulum (SR) is thought to rely on calcium-induced calcium release from the SR due to rapid accumulation of calcium ions in a restricted space surrounding L-type calcium channels and ryanodine receptors (RYRs), the intracellular calcium channels of the SR (Marks, 1997; Oh *et al.*, 1997; Yamazawa *et al.*, 1997; Cannell *et al.*, 1994). This work does not rule out the interesting possibility that L-type calcium channels and RYRs are coupled with molecular bridges in heart as they are in skeletal muscle (Chu *et al.*, 1991), a possibility presently under experimental investigation.

Modulation of L-type Ca^{2+} channel activity by calcium channel antagonists has become a key clinical therapeutic approach to the management of hypertension and certain types of cardiac rhythm disturbances (Kass, 1994a). Several more recent reports, however, have raised questions about the effectiveness and safety of some calcium channel antagonists in the treatment of hypertension (Furberg and Psaty, 1995), making it clear that an understanding of the precise molecular targets of this important drug family is crucial to improved therapeutic efficacy.

The drugs that have received the most attention belong to three distinct chemical classes: (1) phenylalkylamines (ver-apamil, D-600); (2) benzothiazepines ((+) *cis* diltiazem); and the 1,4-dihydropyridines (PN 200-110, nitrendipine, nifedipine, nisoldipine) (Kass, 1994b) (Fig. 2). These drugs bind to distinct but allosterically coupled sites on the channel protein (Glossmann and Striessnig, 1990). The α_1-subunit of the L-type calcium channel contains the binding sites for the three major classes of calcium channel modulators described above (Catterall and Striessnig, 1992), and, when expressed in heterologous expression systems, is sufficient to encode channels with most of the biophysical and pharmacological properties of intact native channels (Hofmann *et al.*, 1993; Welling *et al.*, 1993a; Mori *et al.*, 1991; Mikami *et al.*, 1989) (see Fig. 1). Studies of the molecular site(s) and mechanisms of action of calcium channel blockers have focused on the biochemical and biophysical role of the α_1-subunit and its relationship to modulation of native L-type channels.

Site-directed mutagenesis studies have revealed specific residues that form the binding domains for all three classes of drugs. Emerging from these studies is the consensus view that domains III and IV of the α_1-subunit are crucial to modulation of L-type Ca^{2+} channels, and that it is very likely that multiple residues on the α_1-subunit interact in an allosteric manner to cause voltage-dependent modulation of channel gating (Ito *et al.*, 1997; Grabner *et al.*, 1996; Peterson *et al.*, 1996, 1997; Schuster *et al.*, 1996; Tang *et al.*, 1993). The most provocative model for the actions of these drugs has been developed by Catterall and his colleagues, who have proposed a domain interface model (Fig. 3) for the mode of action of dihydropyridine derivatives (Hockerman *et al.*, 1997). Most importantly, in this model, allosteric interactions between bound drug and two domains of the α_1-subunit cause conformational changes in the channel, which modify channel gating and control the entry of divalent ions.

Most DHP derivatives previously studied in detail are neutral compounds at physiological pH (Rodenkirchen *et al.*, 1982) and it is well established that neutral DHPs such as nitrendipine and nisoldipine inhibit L-type Ca^{2+} channels in a voltage-dependent manner, being approximately 100-fold

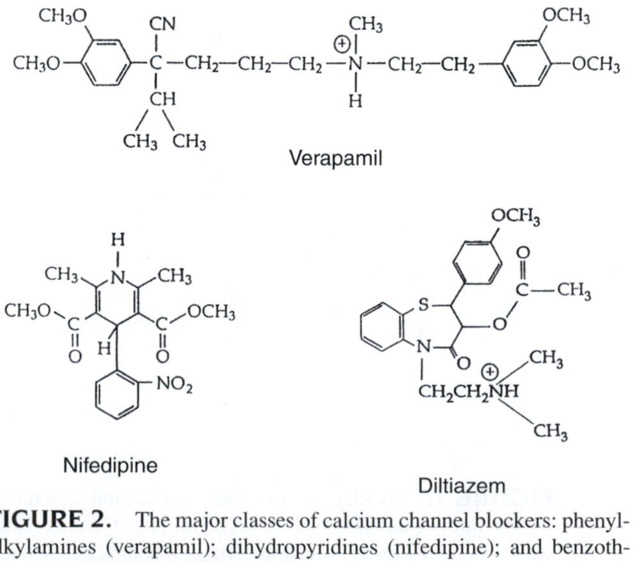

FIGURE 2. The major classes of calcium channel blockers: phenyl-alkylamines (verapamil); dihydropyridines (nifedipine); and benzothiazepines (diltiazem) (Hille, 1992).

IIIS6   IVS6

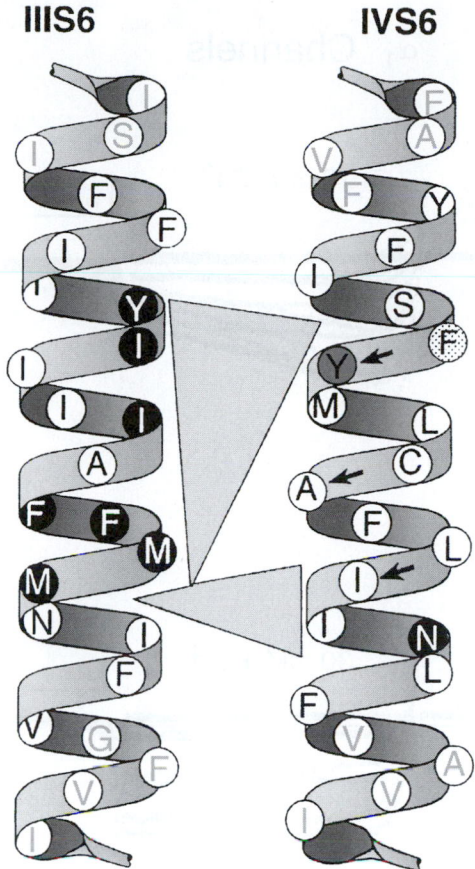

FIGURE 3. Domain interface model for dihydropyridine (DHP) binding in the α-subunit of the L-type calcium channel. Indicated are key amino acids in domain IIIS6 and domain IVS6 that when mutated change dihydropyridine by factors greater than five-fold. Black letters inside shaded circles represent amino acids that when mutated have significant but less than five-fold reduction in DHP binding. A schematized DHP ligand is illustrated by the triangle contacting the putative key binding residues. It is proposed that channel regulation is a consequence of allosteric interactions between the drug molecule and the two helices IIIS6 and IVS6. (From Catterall *et al.*, 1997.)

more potent at inhibiting channel activity at depolarized voltages. Subsequent repolarization to negative holding potentials readily reverses voltage-enhanced channel inhibition (Sanguinetti and Kass, 1984). Addition of a charged headgroup appears to change this relationship between channel modulation and membrane potential: channel inhibition by charged DHPs is not relieved upon repolarization (Kass, 1994a,b). The voltage dependence of neutral dihydropyridine channel modulation has been interpreted within the general framework of the modulated receptor hypothesis which has been very successful in predicting differences in the activity of charged and neutral Class Ib local anesthetics (see Section II). For dihydropyridines, binding affinity to inactivated channels is high; binding affinity to channels in the resting state is low (Sanguinetti and Kass, 1984), consistent with this hypothesis.

The interrelationship between the activity of DHPs and divalent ions is well established (Glossmann *et al.*, 1987). High-affinity dihydropyridine labeling of L-type channels in brain, cardiac, or smooth muscle membranes depends on the presence of divalent ions (Glossmann and Striessnig, 1990), as does the

binding of phenylalkylamines and DHPs to the purified DHP receptor (Flockerzi *et al.*,). High-affinity Ca^{2+} binding to glutamate residues in the putative pore-lining regions (SS1–SS2) of repeats I–IV in the calcium channel $\alpha_{1C\text{-}a}$-subunit is also a critical determinant of ion selectivity and permeation in $\alpha_{1C\text{-}a}$ Ca^{2+} channels (Yang *et al.*, 1993). Subsequent studies showed that high-affinity Ca^{2+} binding to the glutamate residues in the SS2 segments of domains III and IV also stabilizes the DHP receptor site in a high-affinity state (Hockerman *et al.*, 1995). Thus, for the first time, two independent studies demonstrated that high-affinity DHP-binding depends on Ca^{2+} coordination by glutamate residues within the L-channel pore. Conversely, these results imply that DHP binding modulates the binding of Ca^{2+} to these glutamates, which, in turn, will control Ca^{2+} movement through the channel. In combination with recent studies identifying the IVS6 residues critical to PAA and DHP-binding (Hockerman *et al.*, 1995), these data raise the possibility that the glutamate residues and the IVS6-DHP-binding site may be within close proximity to each other. Therefore, a positively charged headgroup added to DHP derivatives may promote ionic or allosteric interactions with the negative charges of the pore-lining glutamate residues in addition to the binding of the DHP moiety to its high-affinity site.

L-type calcium channels encoded by the smooth muscle splice variant of the α ($\alpha_{1C\text{-}b}$)-subunit respond to DHP derivatives differently than channel encoded by the cardiac splice variant ($\alpha_{1C\text{-}a}$) (Welling *et al.*, 1993b). In particular, it was found that channels encoded by $\alpha_{1C\text{-}b}$ cDNA were more sensitive to the DHP antagonist nisoldipine than were channels encoded by the $\alpha_{1C\text{-}a}$ splice variant (Fig. 4). Subsequent work has shown that this distinct pharmacological profile is due to alternative splicing of the IS6 segments of the 1C gene in cardiac and smooth muscle (Welling *et al.*, 1997). In addition, Welling and colleagues confirmed that $\alpha_{1C\text{-}b}$ is expressed in smooth muscle, but not cardiac, cells, linking the molecular pharmacology to the physiology of the tissues. These experimental results are important because they show that regions of the 1C subunit distinct from the DHP-binding domain influence drug activity, most likely through allosteric interactions. These results for DHP antagonists have been confirmed by others (Hu and Marban, 1998; Zuhlke *et al.*, 1998); however, differences in DHP agonist modulation of smooth muscle versus cardiac muscle L-type calcium channels has not yet been systematically addressed.

T-type calcium channels have voltage-dependent kinetics, ion permeability, and pharmacological properties that distinguish them from L-type calcium channels. They are resistant to block by dihydropyridines, inactivate in a voltage-dependent manner, and, most importantly, activate at voltages much more negative than L-type calcium channels (Bean, 1989). Because of the voltage dependence of T-type channels and the relatively small size of T-channel current in the ventricle (Bean, 1985), direct activation of contractile proteins and/or calcium-induced release of calcium from the SR is not likely (Balke *et al.*, 1993; Cannell *et al.*, 1995). T-type channel activity has been suggested to contribute to pacemaker activity in nodal cells (Irisawa and Hagiwara, 1991), which may be particularly important in the Purkinje fibers of the ventricle. The cloning of the neuronal T-type channel has opened the possibility of determining the molecular basis for these differences (Perez-Reyes *et al.*, 1998).

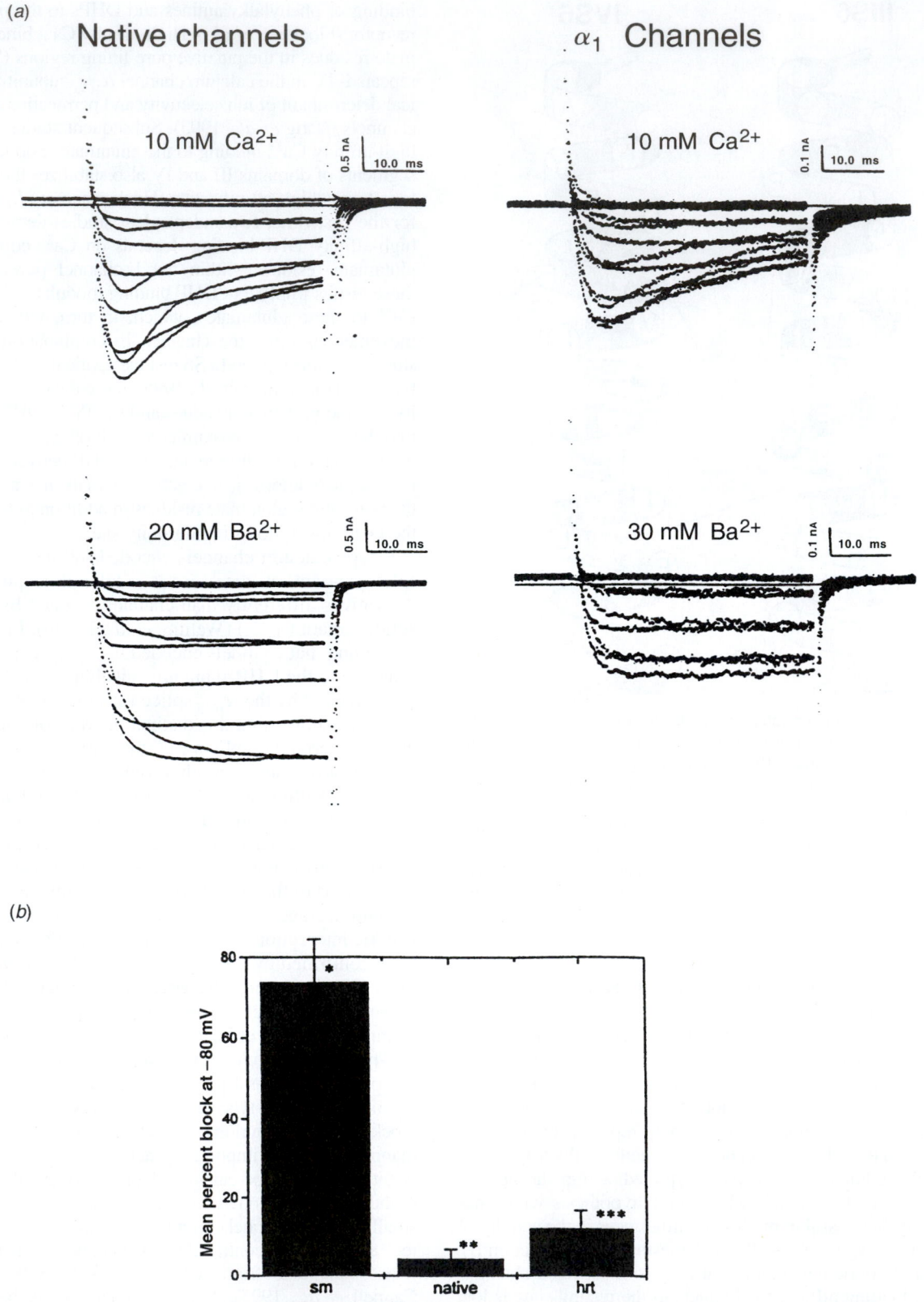

FIGURE 4. L-type calcium channels expressed in Chinese hamster ovary (CHO) cells, showing the greater sensitivity of smooth muscle (sm) than heart (hrt) channel variants to dihydropyridine block. (*a*) Native (left) and recombinant (right) L-type calcium channel activity when calcium (upper row) and barium (lower row) carries the charge. (*b*) Tonic block of recombinant channel activity by the DHP nisoldipine for $\alpha_{1C\text{-}a}$ (hrt) and $\alpha_{1C\text{-}b}$ (sm) α subunits. (From Welling, *et al.*)

II. Sodium (Na⁺) Channels

Voltage-gated Na$^+$ channels are also integral membrane proteins (Catterall, 1993) that not only control the movement of Na$^+$ and underlie the spread of excitation in ventricular and atrial muscle cells and in the Purkinje fiber network throughout the heart (Kass, 1994c), but also can contribute so-called window inward current, which prolongs action potential duration (Attwell et al., 1979). In most tissues the voltage-gated Na$^+$ channel is a heterotrimeric protein consisting of α (33 kDa)-, β_1 (36 kDa)-, and/or β_2 (33-kDa)-subunits (Catterall, 1993; Catterall and Epstein, 1992; Catterall et al., 1992), but only the α-subunit is needed for expression of recombinant channels, particularly for heart channels (Stuhmer et al., 1989a; Noda et al., 1989; Suzuki et al., 1988).

In heart, Na$^+$ channel activity underlies the rapid spread of electrical impulses in specialized conducting tissue of the atria and ventricle (Purkinje fibers) as well as the working muscle of both atrial and ventricular tissue (Kass, 1994d). Consequently, cell activity can be indirectly read from the electrocardiogram: the QRS interval reflects conduction time through the ventricle and hence the number of available Na$^+$ channels.

Prolongation of the QRS interval occurs when Na$^+$ channel activity is reduced by drugs or cellular conditions. Although the role of Na$^+$ channel currents in impulse propagation is well known, Na$^+$ channel activity can also contribute to the duration of the ventricular action potential. The action potential is shortened in the presence of TTX (a Na$^+$ channel–blocking toxin) or after sodium removal (Hiraoka et al., 1986; Colatsky, 1982) because a very small fraction of channels fail to enter the absorbing inactivated state of the channel and thus create what has been referred to as a **window current** through TTX-sensitive Na$^+$ channels. Most importantly, inherited mutations of the human Na$^+$ channel can affect this property of the channel and promote reopening of a larger fraction of channels after inactivation. This produces enhanced inward current that prolongs the action potential in forms of the long-QT syndrome linked to the gene that encodes the Na$^+$ channel α-subunit (Clancy and Rudy, 1999).

The selective pore of Na$^+$ channels is regulated by voltage-sensitive channel gates. According to Hodgkin and Huxley (Hodgkin and Huxley, 1952), at resting membrane potentials, Na$^+$ channels are in a resting state where the pore is closed by activation (m) gates. Membrane depolarization induces conformational changes (gating) that open m-gates, resulting in conduction of Na$^+$ ions through the pore (permeation) and Na$^+$ current. Continued depolarization triggers closure of an inactivation (h) gate, occlusion of the channel pore, and termination of the Na$^+$ current. Membrane repolarization returns the channel to the resting state by shutting m-gates and opening the h-gate. Na$^+$ channel gating is more complex than the fundamental workings detailed in this model (rev. in Patlak, 1991). However, the addition of a single-state, slow inactivation provides a satisfactory backdrop for considering binding and gating interactions relevant to local anesthetic (LA) action.

Local anesthetic drugs, such as lidocaine and mexilitine, block Na$^+$ channels in a voltage dependent manner. Investigation of the molecular basis for this action has provided valuable information about the structure and function of the Na$^+$ channel with a particular emphasis on channel gating. LA modulation of Na$^+$ currents and underlying channel gating is commonly investigated using electrophysiological studies; both single-channel and macroscopic current.

Our present understanding of LA action is founded on a number of key experimental observations that lead to hypotheses that account for tonic and phasic inhibition. Permanently charged, hydrophilic analogs of LAs (e.g., QX-314) are poorly membrane permeable and only block Na$^+$ currents when delivered intracellularly (Strichartz, 1977). In addition, QX-314 binding requires open channels, traverses 60% of the membrane electric field from the cytoplasmic channel mouth to reach its receptor, and does not prevent channel gating. Na$^+$ currents recover rapidly from phasic inhibition induced by hydrophobic LAs compared with permanently charged analogs, suggesting that uncharged molecules leak out from closed channels. Reducing external pH slows recovery from phasic inhibition by tertiary amine LAs, suggesting that hydrogen ions can pass the Na$^+$ channel selectivity filter and ionize the LA molecule (Hille, 1977). LAs enhance Na$^+$ channel inactivation such that voltage-dependent recovery from inactivation is shifted leftward, hyperpolarization relieves inhibition, and recovery from inactivation is slow (Schwarz et al., 1977). These results lead to the proposal of an intrapore LA receptor just inside the Na$^+$ channel selectivity filter involving hydrophobic and hydrophilic access routes (Hille, 1977).

Hille (1977) proposed the **modulated receptor hypothesis** in which a single LA receptor lies within the pore, between the selectivity filter and channel gates. Receptor occupation leads to cessation of ion flow and promotion of inactivation. Uncharged molecules gain access through a hydrophobic, transmembrane pathway. Charged molecules approach the receptor by an aqueous, hydrophilic pathway through the open cytoplasmic mouth (channel gates open) and the pore (see Fig. 1). Also, receptor affinity is modulated by channel state, where open and inactivated states bind drug avidly and resting states do not. Alternatively, Starmer and Courtney (1986) proposed that state-dependent affinity of an intrapore receptor arises from a changeless receptor that is guarded by channel gates, a **guarded receptor**. Here, receptor access is regulated by channel gates, where open channels provide greater receptor access. Such state-dependent accessibility manifests as affinity modulation but with greater mechanistic parsimony.

A. Mutation-Specific Drug Actions

Voltage-dependent block of Na$^+$ channel currents by antiarrhythmic drugs is a consequence of distinct interactions with different states of the voltage-gated Na$^+$ channel. According to the modulated receptor theory of drug/channel interactions, state-dependent activity of these drugs is a consequence of low-affinity interactions with channels in the rested, or closed, state and higher affinity interactions with channels in open and/or inactivated states that are evoked by depolarization (Hille, 1977; Hondeghem and Katzung, 1977). Two drugs, lidocaine and flecainide, differ in their modes of action in that lidocaine interacts preferentially with inactivated channels, and drug block is not necessarily dependent on channel openings (Bean et al., 1983; Ragsdale et al., 1996). In contrast, flecainide requires channels to open, and does not require channels to enter the inactivated state to promote block (Anno and Hondeghem, 1990; Ragsdale *et al.*, 1996).

The molecular determinants of inactivation of the voltage-gated Na$^+$ channel have been well studied and have been shown to be interrelated to the molecular determinants of local anesthetic modulation of sodium channels (Ragsdale and Avoli, 1998; Kambouris *et al.*, 1998). Multiple groups have provided molecular and biophysical evidence that fast inactivation of this channel occurs by the binding of an intracellular inactivation gate, the loop connecting homologous domains III and IV of the Na$^+$ channel, to regions around the inner mouth of the Na$^+$ channel pore through hydrophobic interactions (Stuhmer *et al.*, 1989b; Vassilev *et al.*, 1988; Rohl *et al.*, 1999; Chahine *et al.*, 1997). The docking site of the inactivation gate has been less clearly identified, but the short intracellular loops connecting the S4 and S5 segments in each domain of the channel (McPhee *et al.*, 1998) have been revealed as interaction sites, and two residues in the S6 segment of domain IV (F1764 and Y1774 in rat brain IIA channel) transmembrane segment IVS6 have been found to be key in modulating inactivation (McPhee *et al.*, 1995). Importantly, overlapping residues on IVS6 (F1764 and Y1771), when mutated, decrease the affinity for lidocaine and, to a lesser extent, flecainide (Ragsdale *et al.*, 1996) (Fig. 5).

That lidocaine and other local anesthetics stabilize the inactivation state of the channel is evidenced by the hyperpolarizing shift in the voltage dependence of inactivation and by the voltage range that potentiates its activity (Ragsdale *et al.*, 1996). Furthermore, mutation of the hydrophobic IFM motif of the III–IV intracellular loop into QQQ removes fast inactivation of brain IIA (West *et al.*, 1992) and cardiac Na$^+$ channels (Bennett *et al.*, 1995), and strongly decreases lidocaine TB and UDB (Bennett *et al.*, 1995). However, recent data suggest that transitions to (or conformations of) the active state may play a more important role in the actions of lidocaine block than previously considered (Vedantham and Cannon, 1999; Scheuer, 1999).

Multiple mutations of *SCN5A*, the gene that encodes the human heart Na$^+$ channel α-subunit, have been discovered and linked to two inherited cardiac arrhythmias: the long QT syndrome and the Brugada's syndrome. The fact that these mutations cause functional changes in expressed channel activity has created the unique opportunity to develop specific molecular therapeutic approaches to disease management based on specific functional changes in the channel proteins encoded by mutant genes (Keating and Sanguinetti, 1996).

Pharmacological analysis of mutant channels expressed heterologously has provided evidence that Na$^+$ channel blockers that preferentially interact with the inactivated state of the channel (Hille, 1977; Hondeghem and Katzung, 1977) block in a targeted manner the maintained current conducted by some mutant LQT (ΔKPQ mutant) channels (An *et al.*, 1996; Compton *et al.*, 1996; Dumaine *et al.*, 1996; Priori *et al.*, 1996; Wang *et al.*, 1997; Dumaine and Kirsch, 1998), shorten action potential duration in cellular studies (Shimizu and Antzelevitch, 1997; Schwartz *et al.*, 1995), and in preliminary studies correct QT prolongation in patients

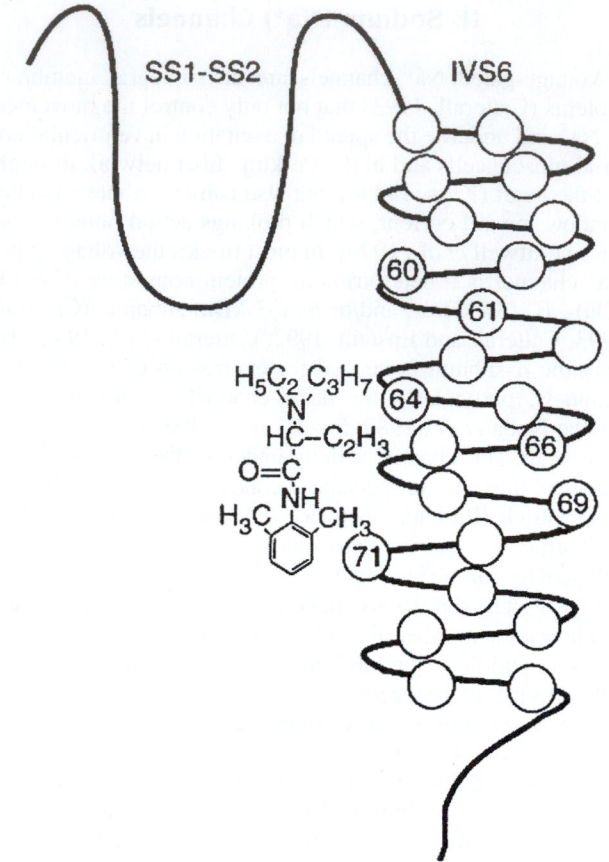

FIGURE 5. Schematic diagram of the putative interaction between lidocaine and key residues on domain IV transmembrane segment S6 of the rat brain Na channel. (Reprinted with permission from *Science*. Copyright 1994 American Association for the Advancement of Science.)

(Schwartz *et al.*, 1995; Rosero *et al.*, 1997). Figure 6B illustrates the maintained current carried by LQT-3 mutant (ΔKPQ) channels and preferential block of this current by lidocaine.

More recently, flecainide, which preferentially blocks open but not inactivated channels (Ragsdale *et al.*, 1996), has also been shown to be effective in inhibiting persistent Na$^+$ channel current in ΔKPQ mutant channels (Nagatomo *et al.*, 1997) and in shortening and normalizing QT intervals in patients carrying this gene mutation (Windle *et al.*, 1999). However other LQT mutations, such as the D1790G mutation of the Na$^+$ channel α-subunit (An *et al.*, 1998; Benhorin *et al.*, 1998) do not promote sustained inward currents and are not likely to be targeted by lidocaine.

Interestingly, clinical data suggest that the open channel blocker flecainide might be effective in controlling arrhythmias caused by the D1790G mutation (Benhorin *et al.*, 1999). Preliminary experimental data suggest that flecainide might be a reasonable candidate drug to selectively target D1790G mutant channels in carriers of this LQT-3 mutation due to unique effects of this mutation on flecainide interactions with the Na$^+$ channel (Abriel *et al.*, 2000). In order to extrapolate to possible modes of action in patients, estimates of plasma drug concentrations are useful.

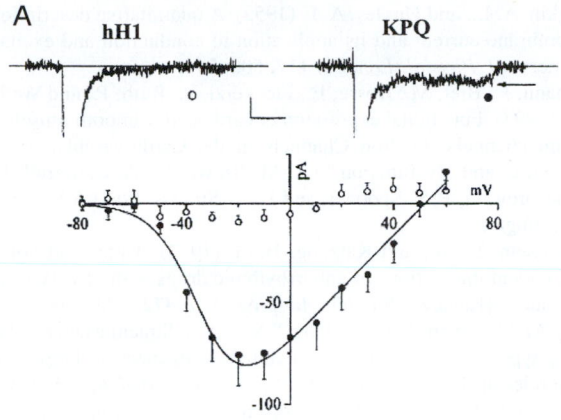

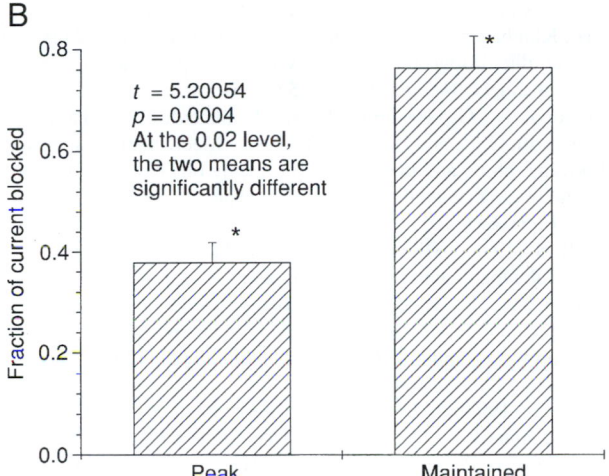

FIGURE 6. (*a*) The ΔKPQ deletion mutation of the cytoplasmic linker between domains III and IV of the human heart Na$^+$ channel α-subunit promotes maintained sodium channel current that is not present in wild-type a (hH1) channels. (*b*) Summary of experiments in which the fraction of peak inward Na$^+$ channel current (left bar) was compared with the fraction of late or maintained Na$^+$ channel current blocked (right bar) by the same concentration of lidocaine (100 μM). The late current is more than twice as sensitive to lidocaine as peak current. (From An *et al.*, 1996.)

These studies represent the first approach to selectively target ion channels that have been encoded by disease-causing mutant genes. Despite the fact that both Brugada's syndrome and long-QT syndrome are relatively rare diseases, the targeted therapy approach can serve as a model for more general cases that are likely to emerge, such as the broad category of acquired forms of cardiac arrhythmias.

Bibliography

Abriel, H., Wehrens, X. H. T., Benhorin, J., and Kass, R. S. (2000). Molecular pharmacology of the D1790G LQT-3 mutation: biophysical determinants of drug- and mutation-specific therapeutics. *Biophys. J.*, in press.

Ahlijanian, M. K., Westenbroek, R. E., and Catterall, W. A. (1990). Subunit structure and localization of dihydropyridine-sensitive calcium channels in mammalian brain, spinal cord, and retina. *Neuron* **4**, 819–832.

An, R. H., Bangalore, R., Rosero, S. Z., and Kass, R. S. (1996). Lidocaine block of LQT-3 mutant human Na$^+$ channels. *Circ. Res.* **79**, 103–108.

An, R. H., Wang, X. L., Kerem, B., Benhorin, J., Medina, A., Goldmit, M., and Kass, R. S. (1998). Novel LQT-3 mutation affects Na$^+$ channel activity through interactions between alpha- and beta1-subunits. *Circ. Res.* **83**, 141–146.

Anno, T., and Hondeghem, L. M. (1990). Interactions of flecainide with guinea pig cardiac sodium channels. Importance of activation unblocking to the voltage dependence of recovery. *Circ. Res.* **66**, 789–803.

Attwell, D., Cohen, I., Eisner, D., Ohba, M., and Ojeda, C. (1979). The steady state TTX-sensitive ("window") sodium current in cardiac Purkinje fibres. *Pflügers Arch.* **379**, 137–142.

Balke, C. W., Rose, W. C., O'Rourke, B., Mejia-Alvarez, R., Backx, P., and Marban, E. (1993). Biophysics and physiology of cardiac calcium channels. *Circ.* **87 Suppl. 7**, VII49–VII53.

Bangalore, R., Mehrke, G., Gingrich, K., Hofmann, F., and Kass, R. S. (1996). Influence of the L-type Ca-channel α$_2$/δ subunit on ionic and gating current in transiently-transfected HEK 293 cells. *Am. J. Physiol.* **270**, H1521–H1528.

Bean, B. P. (1985). Two kinds of calcium channels in canine atrial cells. Differences in kinetics, selectivity, and pharmacology. *J. Gen. Physiol.* **86**, 1–30.

Bean, B. P. (1989). Classes of calcium channels in vertebrate cells. *Ann. Rev. Physiol.* **51**, 367–384.

Bean, B. P., Cohen, C. J., and Tsien, R. W. (1983). Lidocaine block of cardiac sodium channels. *J. Gen. Physiol.* **81**, 613–642.

Benhorin, J., Goldmit, M., MacCluer, J., Blangero, J., Goffen, R., Leibovitch, A., Rahat, A., Wang, Q., Medina, A., Towbin, J., and Kerem, B. (1998). Identification of a new SCN5A mutation associated with the long QT syndrome. *Human Mutations* **12**, 72.

Benhorin, J., Medina, A., Taub, Y., Windman, M., Goldmit, M., Kerem, B., and Kass, R. S. (1999). Effects of flecainide in patients with a new SCN5A mutation: mutation-specific therapy for Long QT Syndrome? *Circulation*, in press.

Bennett, P. B., Valenzuela, C., Chen, L. Q., and Kallen, R. G. (1995). On the molecular nature of the lidocaine receptor of cardiac Na$^+$ channels. Modification of block by alterations in the alpha-subunit III–IV interdomain. *Circ. Res.* **77**, 584–592.

Birnbaumer, L., Campbell, K. P., Catterall, W. A., Harpold, M. M., Hofmann, F., Horne, W. A., Mori, Y., Schwartz, A., Snutch, T. P., and Tanabe, T. (1994). The naming of voltage-gated calcium channels. *Neuron* **13**, 505–506.

Cachelin, A. B., de Peyer, J. E., Kokubun, S., and Reuter, H. (1983). Ca channel modulation by 8-bromocyclic AMP in cultured heart cells. *Nature* **304**, 462–464.

Cannell, M. B., Cheng, H., and Lederer, W. J. (1994). Spatial non-uniformities in [Ca^{2+}]$_i$ during excitation-contraction coupling in cardiac myocytes. *Biophys. J.* **67**, 1942–1956.

Cannell, M. B., Cheng, H., and Lederer, W. J. (1995). The control of calcium release in heart muscle. *Science* **268**, 1045–1049.

Catterall, W., and Epstein, P. N. (1992). Ion channels. [Review]. *Diabetologia* **35 Suppl. 2**, S23–S33.

Catterall, W. A. (1988). Structure and function of voltage-sensitive ion channels. *Science* **242**, 50–61.

Catterall, W. A. (1993). Structure and function of voltage-gated ion channels. [Review]. *Trds Neurosci.* **16**, 500–506.

Catterall, W. A., Scheuer, T., Thomsen, W., and Rossie, S. (1991). Structure and modulation of voltage-gated ion channels. [Review]. *Ann. N. Y. Acad. Sci.* **625**, 174–180.

Catterall, W. A., Seagar, M. J., and Takahashi, M. (1988). Molecular properties of dihydropyridine-sensitive calcium channels in skeletal muscle. *J. Biol. Chem.* **263**, 3535–3538.

Catterall, W. A., and Striessnig, J. (1992). Receptor sites for Ca channel antagonists. *TIPS* **13,** 256–262.

Catterall, W. A., Trainer, V., and Baden, D. G. (1992). Molecular properties of the sodium channel: a receptor for multiple neurotoxins. [Review]. *Bull. Soc. Pathol. Exot.* **85,** 481–485.

Chahine, M., Deschenes, I., Trottier, E., Chen, L. Q., and Kallen, R. G. (1997). Restoration of fast inactivation in an inactivation-defective human heart sodium channel by the cysteine modifying reagent benzyl-MTS: analysis of IFM-ICM mutation. *Biochem. Biophys. Res. Commun.* **233,** 606–610.

Chu, A., Fill, M., Entman, M. L., and Stefani, E. (1991). Different Ca^{2+} sensitivity of the ryanodine-sensitive Ca^{2+} release channels of cardiac and skeletal muscle sarcoplasmic reticulum. *Biophys. J.* **59,** 102a.

Clancy, C. E., and Rudy, Y. (1999). Linking a genetic defect to its cellular phenotype in a cardiac arrhythmia. *Nature* **400,** 566–569.

Colatsky, T. J. (1982). Mechanisms of action of lidocaine and quinidine on action potential duration in rabbit Purkinje fibers. An effect on steady state sodium currents? *Circ. Res.* **50,** 17–27.

Compton, S. J., Lux, R. L., Ramsey, M. R., Strelich, K. R., Sanguinetti, M. C., Green, L. S., Keating, M. T., and Mason, J. W. (1996). Genetically defined therapy of inherited long-QT syndrome—correction of abnormal repolarization by potassium. *Circulation* **94,** 1018–1022.

Dumaine, R., and Kirsch, G. E. (1998). Mechanism of lidocaine block of late current in long Q-T mutant Na$^+$ channels. *Am. J. Physiol.* **274,** H477–H487.

Dumaine, R., Wang, Q., Keating, M. T., Hartmann, H. A., Schwartz, P. J., Brown, A. M., and Kirsch, G. E. (1996). Multiple mechanisms of Na$^+$ channel linked long-QT syndrome. *Circ. Res.* **78,** 916–924.

Flockerzi, V., Oeken, H. J., Hofman, F., Pelzer, D., Cavalie, A., and Trautwein, W. (1986). Purified dihydropyridine-binding site from skeletal muscle T-tubules is a functional calcium channel. *Nature* **323,** 66–68.

Furberg, C. D., and Psaty, B. M. (1995). Should dihydropyridines be used as first-line drugs in the treatments of hypertension—the con side. *Arch. Intern. Med.* **155,** 2157–2161.

Gao, T., Puri, T. S., Gerhardstein, B. L., Chien, A. J., Green, R. D., and Hosey, M. M. (1997). Identification and subcellular localization of the subunits of L-type calcium channels and adenylyl cyclase in cardiac myocytes. *J. Biolog. Chem.* **272,** 19 401–19 407.

Glossmann, H., and Striessnig, J. (1990). Molecular properties of calcium channels. *Rev. Physiol. Biochem. Pharmacol.* **114,** 1–105

Glossmann, H., Striessnig, J., Ferry, D. R., Goll, A., Moosburger, K., and Schirmer, M. (1987). Interaction between calcium channel ligands and calcium channels. *Circ. Res.* **61,** 130–36

Grabner, M., Wang, Z., Hering, S., Striessnig, J., and Glossmann, H. (1996). Transfer of 1,4-dihydropyridine sensitivity from L-type to class A (BI) calcium channels. *Neuron* **16,** 207–218.

Gutierrez, L. M., Brawley, R. M., and Hosey, M. M. (1991). Dihydropyridine-sensitive calcium channels from skeletal muscle. I. Roles of subunits in channel activity. *J. Biolog. Chem.* **266,** 16 387–16 394.

Gutierrez, L. M., Zhao, X. L., and Hosey, M. M. (1994). Protein kinase C–mediated regulation of L-type Ca channels from skeletal muscle requires phosphorylation of the alpha 1 subunit. *Biochem. Biophys. Res. Communic.* **202,** 857–865.

Hille, B. (1977). Local anesthetics: hydrophilic and hydrophobic pathways for the drug-receptor reaction. *J. Gen. Physiol.* **69,** 497–515.

Hille, B. (1992). "Ionic Channels of Excitable Membranes," 2 ed. Sinauer, Sunderland, MA.

Hiraoka, M., Sawada, K., and Kawano, S. (1986). Effect of quinidine on plateau currents of guinea-pig ventricular myocytes. *J. Molec. Cell. Cardio.* **18,** 1097–1106.

Hockerman, G. H., Peterson, B. Z., Johnson, B. D., and Catterall, W. A. (1997). Molecular determinants of drug binding and action on L-type calcium channels. [Review]. *Annu. Rev. Pharmacol. Toxicol.* **37,** 361–396.

Hodgkin, A. L., and Huxley, A. F. (1952). A quantitative description of membrane current and its application to conduction and excitation in nerve. *J. Physiol.(London)* **117,** 500–544.

Hofmann, F., Biel, M., Bosse, E., Flockerzi, V., Ruth, P., and Welling, A. (1993). Functional expression of cardiac and smooth muscle calcium channels. In "Ion Channels in the Cardiovascular System: Function and Dysfunction." (A. M. Brown, W. A. Catterrall, G. J. Kaczorowski, P. S. Spooner, and H. C. Strauss, Eds.) AAAS Press, Washington.

Hondeghem, L. M., and Katzung, B. G. (1977). Time- and voltage-dependent interactions of antiarrhythmic drugs with cardiac sodium channels. [Review]. *Biochim. Biophys. Acta* **472,** 373–398.

Hosey, M. M., Chien, A. J., and Puri, T. S. (1996). Structure and regulation of L-type calcium channels — a current assessment of the properties and roles of channel subunits. *Trds. Cardiovasc. Med.* **6,** 265–273.

Hu, H., and Marban, E. (1998). Isoform-specific inhibition of L-type calcium channels by dihydropyridines is independent of isoform-specific gating properties. *Mol. Pharmacol.* **53,** 902–907.

Irisawa, H., and Hagiwara, N. (1991). Ionic current in sinoatrial node cells. *J. Cardiovasc. Electrophys.* **2,** 531–540.

Ito, H., Klugbauer, N., and Hofmann, F. (1997). Transfer of the high-affinity dihydropyridine sensitivity from L-type to non–L-type calcium channels. *Mol. Pharmacol.* **52,** 735–740.

Kambouris, N. G., Hastings, L. A., Stepanovic, S., Marban, E., Tomaselli, G. F., and Balser, J. R. (1998). Mechanistic link between lidocaine block and inactivation probed by outer pore mutations in the rat micro1 skeletal muscle sodium channel. *J. Physiol. (London)* **512 (Pt 3),** 693–705.

Kass, R. S. (1994a). Dihydropyridine modulation of cardiovascular L-type calcium channels: molecular and cellular pharmacology. *In* "Ion Channels in the Cardiovascular System: Function and Dysfunction." (P. M. Spooner, A. M. Brown, W. A. Catterall, G. J. Kaczorowski, and H. C. Strauss, Eds.), pp. 425–440. Futura Publishing Co., Armonk, NY.

Kass, R. S. (1994b). Molecular pharmacology of cardiac L-type calcium channels. *In* "Handbook of Membrane Channels: Molecular and Cellular Physiology" C. Peracchia, pp. 187–198. Academic Press, Orlando, FL.

Kass, R. S. (1994c). Ionic basis of electrical activity in the heart. *In* "Physiology and Pathophysiology of the Heart," (N. Sperelakis, Ed.) Kluwer Academic, Norwell, MA.

Kass, R. S. (1994d). Ionic basis of electrical activity in the heart. *In* "Physiology and Pathophysiology of the Heart," (N. Sperelakis, Ed.) Kluwer Academic, Norwell, MA.

Kass, R. S. and Sanguinetti, M. C. (1984). Calcium channel inactivation in the cardiac Purkinje fiber. Evidence for voltage- and calcium-mediated mechanisms. *J. Gen. Physiol.* **84,** 705–726.

Kass, R. S. (1995). Ionic basis of electrical activity in the heart. *In* "Physiology and Pathophysiology of the Heart," (N. Sperelakis, Ed.) pp. 77–90. Kluwer Academic, Norwell, MA .

Keating, M. T., and Sanguinetti, M. C. (1996). Pathophysiology of ion channel mutations. *Curr. Opin. Gen. Devel.* **6,** 326–333.

Lacerda, A. E., Kim, H. S., Ruth, P., Perez-Reyes, E., Flockerzi, V., Hofmann, F., Birnbaumer, L., and Brown, A. M. (1991). Normalization of current kinetics by interaction between the $\alpha 1$ and β subunits of the skeletal muscle dihydropyridine-sensitive Ca^{2+} channel. *Nature* **352,** 527.

Lory, P., Varadi, G., Slish, D. F., Varadi, M., and Schwartz, A. (1993). Characterization of beta subunit modulation of a rabbit cardiac L-type Ca^{2+} channel alpha1 subunit as expressed in mouse L cells. *FEBS Lett.* **315(2),** 167–172.

Ma, J., Gutierrez, L. M., Hosey, M. M., and Rios, E. (1992). Dihydropyridine-sensitive skeletal muscle Ca channels in polarized planar bilayers. 3. Effects of phosphorylation by protein kinase C. *Biophys. J.* **63,** 639–647.

Marks, A. R. (1997). Intracellular calcium-release channels: regulators of cell life and death. [Rev.] *Am. J. Physiol.* **272,** H597–H605.

McDonald, T. F., Pelzer, S., Trautwein, W., and Pelzer, D. J. (1994). Regulation and modulation of calcium channels in cardiac, skeletal, and smooth muscle cells. [Rev.] *Physiol. Rev.* **74**, 365–507.

McPhee, J. C., Ragsdale, D. S., Scheuer, T., and Catterall, W. A. (1995). A critical role for transmembrane segment IVS6 of the sodium channel alpha subunit in fast inactivation. *J. Biolog. Chem.* **270**, 12 025–12 034.

McPhee, J. C., Ragsdale, D. S., Scheuer, T., and Catterall, W. A. (1998). A critical role for the S4–S5 intracellular loop in domain IV of the sodium channel alpha-subunit in fast inactivation. *J. Biolog. Chem.* **273**, 1121–1129.

Mikami, A., Imoto, K., Tanabe, T., Niidome, T., Mori, Y., Takeshima, H., Narumiya, S., and Numa, S. (1989). Primary structure and functional expression of the cardiac dihydropyridine-sensitive calcium channel. *Nature* **340**, 230–233.

Morad, M. and Rolett, E. L. (1972). Relaxing effects of catecholamines on mammalian heart. *J. Physiol.* **224**, 537–558.

Morad, M., and Trautwein, W. (1968). The effect of the duration of the action potential on contractions in mammalian heart tissue. *Pflügers Arch.* **299**, 66–82.

Mori, Y., Friedrich, T., Kim, M.-S., Mikami, A., Nakai, J., Ruth, P., Bosse, E., Hofmann, F., Flockerzi, V., Furuichi, T., Nikoshiba, K., Imoto, K., Tanabe, T., and Numa, S. (1991). Primary structure and functional expression from complementary DNA of a brain calcium channel. *Nature* **350**, 398–402.

Nagatomo, T., Fan, Z., Ye, B., January, C. T., and Makielski, J. C. (1997). Effects of flecainide on the long QT sodium channel syndrome. *Circulation* **96**, 677.

Neely, A., Wei, X., Olcese, R., Birnbaumer, L., and Stefani, E. (1993). Potentiation by the beta subunit of the ratio of the ionic current to the charge movement in the cardiac calcium channel. *Science* **262**, 575–578.

Nishimura, S., Takeshima, H., Hofmann, F., Flockerzi, V., and Imoto, K. (1993). Requirement of the calcium channel beta subunit for functional conformation. *FEBS Lett.* **324**, 283–286.

Noda, M., Suzuki, H., Numa, S., and Stuhmer, W. (1989). A single point mutation confers tetrodotoxin and saxitoxin insensitivity on the sodium channel II. *FEBS Lett.* **259**, 213–216.

Oh, S. T., Yedidag, E., Conklin, J. L., Martin, M., and Bielefeldt, K. (1997). Calcium release from intracellular stores and excitation-contraction coupling in intestinal smooth muscle. *J. Surg. Res.* **71**, 79–86.

Patlak, J. (1991). Molecular kinetics of voltage-dependent Na+ channels. [Rev.] *Physiol. Rev.* **71**, 1047–1080.

Perez-Reyes, E., Cribbs, L. L., Daud, A., Lacerda, A. E., Barclay, J., Williamson, M. P., Fox, M., Rees, M., and Lee, J. H. (1998). Molecular characterization of a neuronal low-voltage–activated T-type calcium channel. *Nature* **391**, 896–900.

Peterson, B. Z., Johnson, B. D., Hockerman, G. H., Acheson, M., Scheuer, T., and Catterall, W. A. (1997). Analysis of the dihydropyridine receptor site of L-type calcium channels by alanine-scanning mutagenesis. *J. Biolog. Chem.* **272**, 18 752–18 758.

Peterson, B. Z., Tanada, T. N., and Catterall, W. A. (1996). Molecular determinants of high affinity dihydropyridine binding in L-type calcium channels. *J. Biolog. Chem.* **271**, 5293–5296.

Priori, S. G., Napolitano, C., Cantu, F., Brown, A. M., and Schwartz, P. J. (1996). Differential response to Na+ channel blockade, beta-adrenergic stimulation and rapid pacing in a cellular model mimicking the SCN5A and HERG defects present in the long-QT syndrome. *Circ. Res.* **78**, 1009–1015.

Ragsdale, D. S., and Avoli, M. (1998). Sodium channels as molecular targets for antiepileptic drugs. *Brain Res. Rev.* **26**, 16–28.

Ragsdale, D. S., McPhee, J. C., Scheuer, T., and Catterall, W. A. (1996). Common molecular determinants of local anesthetic, antiarrhythmic, and anticonvulsant block of voltage-gated Na+ channels. *Proc. Natl. Acad. Sci. USA* **93**, 9270–9275.

Reuter, H. (1983). Calcium channel modulation by neurotransmitters, enzymes and drugs. *Nature* **301**, 569–574.

Reuter, H., and Scholz, H. (1977). The regulation of Ca conductance of cardiac muscle by adrenaline. *J. Physiol.* **264**, 49–62.

Rodenkirchen, R., Bayer, R., and Mannhold, R. (1982). Specific and non-specific Ca antagonists: a structure-activity analysis of cardiodepressive drugs. *Progr. Pharmacol.* **5**, 9–23.

Rohl, C. A., Boeckman, F. A., Baker, C., Scheuer, T., Catterall, W. A., and Klevit, R. E. (1999). Solution structure of the sodium channel inactivation gate. *Biochem.* **38**, 855–861.

Rosero, S. Z., Zareba, W., Robinson, J. L., and Moss, A. (1997). Gene-specific therapy for long QT syndrome: QT shortening with lidocaine and tocainide in patients with mutation of the sodium channel gene. *Ann. Noninvasive Electrocardiol.* **2**, 274–278.

Ruth, P., Rohrkasten, A., Biel, M., Bosse, E., Regulla, S., Meyer, H. E., Flockerzi, V., and Hofmann, F. (1989). Primary structure of the β subunit of the DHP-sensitive calcium channel from skeletal muscle. *Science* **245**, 1115–1118.

Sanguinetti, M. C., and Kass, R. S. (1984). Voltage-dependent block of calcium channel current in the calf cardiac Purkinje fiber by dihydropyridine calcium channel antagonists. *Circ. Res.* **55**, 336–348.

Scheuer, T. (1999). Commentary: a revised view of local anesthetic action: what channel state is really stabilized? *J. Gen. Physiol.* **113**, 3–6.

Schneider, T., and Hofmann, F. (1988). The bovine cardiac receptor for calcium channel blockers is a 195-kDa protein. *Eur. J. Biochem.* **174**, 369–375.

Schuster, A., Lacinova, L., Klugbauer, N., Ito, H., Birnbaumer, L., and Hormann, F. (1996). The IVS6 segment of the L-type calcium channel is critical for the action of dihydropyridines and phenylalkylamines. *EMBO J.* **15**, 2365–2370.

Schwartz, P. J., Priori, S. G., Locati, E. H., Napolitano, C., Cantu, F., Towbin, J. A., Keating, M. T., Hammoude, H., Brown, A. M., Chen, L. S. K., and Colatsky, T. J. (1995). Long QT syndrome patients with mutations of the SCN5A and HERG genes have differential responses to Na+ channel blockade and to increases in heart rate: implications for gene-specific therapy. *Circulation* **92**, 3381–3386.

Schwarz, W., Palade, P. T., and Hille, B. (1977). Local anesthetics. Effect of pH on use-dependent block of sodium channels in frog muscle. *Biophys. J.* **20**, 343–368.

Sculptoreanu, A., Rotman, E., Takahashi, M., Scheuer, T., and Catterall, W. A. (1993a). Voltage-dependent potentiation of the activity of cardiac L-type calcium channel alpha 1 subunits due to phosphorylation by cAMP-dependent protein kinase. *Proc. Natl. Acad. Sci. USA* **90**, 10 135–10 139.

Sculptoreanu, A., Scheuer, T., and Catterall, W. A. (1993b). Voltage-dependent potentiation of L-type Ca2+ channels due to phosphorylation by cAMP-dependent protein kinase. *Nature* **364**, 240–243.

Shimizu, W., and Antzelevitch, C. (1997). Sodium channel block with mexiletine is effective in reducing dispersion of repolarization and preventing *torsade des pointes* in LQT2 and LQT3 models of the long-QT syndrome. *Circulation* **96**, 2038–2047.

Singer, D., Biel, M., Lotan, I., Flockerzi, V., Hofmann, F., and Dascal, N. (1991). The roles of the subunits in the function of the calcium channel. *Science* **253**, 1553–1557.

Starmer, C. F., and Courtney, K. R. (1986). Modeling ion channel blockade at guarded binding sites: application to tertiary drugs. *Am. J. Physiol.* **251**, H848–H856.

Strichartz, G. (1977). Effects of tertiary local anesthetics and their quaternary derivatives on sodium channels of nerve membranes. [Proceedings]. *Biophys. J.* **18**, 353–354.

Stuhmer, W., Conti, F., Suzuki, H., Wang, X., Noda, M., Yahagi, N., Kubo, H., and Numa, S. (1989b). Structural parts involved in activation and inactivation of the sodium channel. *Nature* **339**, 597–603.

Stuhmer, W., Conti, F., Suzuki, H., Wang, X., Noda, M., Yahagi, N., Kubo, H., and Numa, S. (1989a). Structural parts involved in activation and inactivation of the sodium channel. *Nature* **339**, 597–603.

Suzuki, H., Beckh, S., Kubo, H., Yahagi, N., Ishida, H., Kayano, T., Noda, M., and Numa, S. (1988). Functional expression of cloned cDNA encoding sodium channel III. *FEBS Lett.* **228**, 195–200.

Takahashi, M., and Catterall, W. A. (1987). Dihydropyridine-sensitive calcium channels in cardiac and skeletal muscle membranes: studies with antibodies against the alpha subunits. *Biochemistry* **26**, 5518–5526.

Takahashi, M., Seagar, M. J., Jones, J. F., Reber, B. F. X., and Catterall, W. (1987). Subunit structure of dihydropyridine-sensitive calcium channels from skeletal muscle. *Proc. Natl. Acad. Sci. USA* **84**, 5478–5482.

Tang, S., Yatani, A., Bahinski, A., Mori, Y., and Schwartz, A. (1993). Molecular localization of regions in the L-type calcium channel critical for dihydropyridine action. *Neuron* **11**, 1013–1021.

Triggle, D. J. (1994). Molecular pharmacology of voltage-gated calcium channels. [Rev.]. *Ann. NY Acad. Sci.* **747**, 267–281.

Tsien, R. W. (1977). Effects of epinephrine on the pacemaker potassium current of cardiac Purkinje fibers. *J. Gen. Physiol.* **64**, 293–319.

Tsien, R. W., Ellinor, P. T., and Horne, W. A. (1991). Molecular diversity of voltage-dependent Ca^{2+} channels. [Rev.]. *TIPS* **12**, 349–354.

Varadi, G., Lory, P., Schultz, D., Varadi, M., and Schwartz, A. (1991). Acceleration of activation and inactivation by the β subunit of the skeletal muscle calcium channel. *Nature* **352**, 159–162.

Vassilev, P. M., Scheuer, T., and Catterall, W. A. (1988). Identification of an intracellular peptide segment involved in sodium channel inactivation. *Science* **241**, 1658–1661.

Vedantham, V., and Cannon, S. C. (1999). The position of the fast-inactivation gate during lidocaine block of voltage-gated Na^+ channels. *J. Gen. Physiol* **113**, 7–16.

Wang, D. W., Yazawa, K., Makita, N., George, A. L., and Bennett, P. B. (1997). Pharmacological targeting of long QT mutant sodium channels. *J. Clin. Invest.* **99**, 1714–1720.

Welling, A., Kwan, Y. W., Bosse, E., Flockerzi, V., Hofmann, F., and Kass, R. S. (1993a). Subunit-dependent modulation of recombinant L-type calcium channels: molecular basis for dihydropyridine tissue selectivity. *Circ. Res.* **73**, 974–980.

Welling, A., Kwan, Y. W., Bosse, E., Flockerzi, V., Hofmann, F., and Kass, R. S. (1993b). Subunit-dependent modulation of recombinant L-type calcium channels: molecular basis for dihydropyridine tissue selectivity. *Circ. Res.* **73**, 974–980.

Welling, A., Ludwig, A., Zimmer, S., Klugbauer, N., Flockerzi, V., and Hofmann, F. (1997). Alternatively spliced IS6 segments of the alpha 1C gene determine the tissue-specific dihydropyridine sensitivity of cardiac and vascular smooth muscle L-type Ca^{2+} channels. *Circ. Res.* **81**, 526–532.

West, J. W., Patton, D. E., Scheuer, T., Wang, Y., Goldin, A. L., and Catterall, W. A. (1992). A cluster of hydrophobic amino acid residues required for fast $Na(+)$-channel inactivation. *Proc. Natl. Acad. Sci. USA* **89**, 10 910–10 914.

Windle, J. R., Geletka, R. C., Moss, A. J., and Atkins, D. L. (1999). Normalization of ventricular repolarization with flecainide in patients with the LQT3 form (SCN5A) of Long QT Syndrome. *Circulation*, in press.

Yamazawa, T., Takeshima, H., Shimuta, M., and Iino, M. (1997). A region of the ryanodine receptor critical for excitation-contraction coupling in skeletal muscle. *J. Biolog. Chem.* **272**, 8161–8164.

Yang, J., Ellinor, P. T., Sather, W. A., Zhang, J. F., and Tsien, R. W. (1993). Molecular determinates of Ca^{2+} selectivity and ion permeation in L-type Ca^{2+} channels. *Nature* **366**, 158–161.

Yue, D. T., Backx, P. H., and Imredy, J. P. (1990). Calcium-sensitive inactivation in the gating of single calcium channels. *Science* **21**, 1735–1738.

Zhao, X. L., Gutierrez, L. M., Chang, C. F., and Hosey, M. M. (1994). The alpha 1-subunit of skeletal muscle L-type Ca channels is the key target for regulation by A-kinase and protein phosphatase-1C. *Biochem. Biophys. Res. Commun.* **198**, 166–173.

Zuhlke, R. D., Bouron, A., Soldatov, N. M., and Reuter, H. (1998). Ca^{2+} channel sensitivity towards the blocker isradipine is affected by alternative splicing of the human alpha1C subunit gene. *FEBS Lett.* **427**, 220–224.

Shirley H. Bryant and James Maylie

39

Ion Channels as Targets for Disease

I. Introduction

Ion channels are involved in many critical cellular processes as varied as the transmission of impulses in excitable tissues including skeletal muscle to transport across epithelial tissue. Altered function of membrane ion channels then would be expected to have serious consequences for the survival of the organism. Ion channels can be affected in a variety of ways by disease, both directly and indirectly, thereby serving as targets for diseases. Direct action on the channel protein structure occurs as a result of genetic mutations within the gene coding a channel subunit. Many of the genetic mutations that markedly alter the channel function result from a single base pair change affecting only a single amino acid. We recognize these as the hereditary diseases whose targets are ion channels. Indirect action includes any of the following: (1) abnormalities in regulatory mechanisms such as phosphorylation required for normal channel function, (2) presence of any number of kinds of toxic materials that block channels or prevent their synthesis, and (3) development of autoimmune disease directed at a channel protein or its regulators, and (4) altered gene expression of channel subunits.

Naturally occurring channel diseases have been reported within the general families of Cl^-, Na^+, Ca^{2+}, K^+, and transmitter-activated channels. With the recent introduction of rapid cloning methods and improvements in functional analysis of channels, especially the patch-clamp methodology, the rate of identification of dysfunctional channels causing disease has mushroomed. The diseases themselves, now being referred to as **channelopathies,** are becoming too numerous to discuss with any depth. It is our task in this chapter to make the reader aware of pertinent examples of disease for each of these general families of channels. The chapter is organized by channel type (i.e., Cl^-, Na^+, Ca^{2+}, K^+, and neurotransmitter gated) rather than by disease. Some disease entities (e.g., trinucleotide repeats in myotonia) may be caused by mutations that affect transcription or translation. Such diseases are discussed under each of the channel type headings. For a more detailed discussion of the molecular basis of ion channel diseases, consult *Ion Channels and Disease* (Ashcroft, 1999).

II. Ion Channel Diseases

A. Mutations of Ion Channel Genes

Genetic mutations can affect channel function in several ways: (1) a mutation in a gene coding a channel subunit giving rise to altered channel function, (2) altered gene expression of channel subunits affecting the number of channels incorporated into the membrane, and (3) mutation within a regulatory subunit affecting regulation of channel function.

Identification of genes involved in ion channel diseases involves two principal search strategies: (1) **functional** or **expression cloning** and (2) **positional cloning.** In expression cloning the gene is isolated on the basis of information about the expressed protein product, for example, its amino acid sequence or antibody reactivity, or its function (e.g., receptor/ligand reactivity or ion channel characteristics). cDNA libraries are screened using different types of probes including antibodies and oligonucleotides. The polymerase chain reaction (PCR) technique can be employed in several ways including the amplification of cDNA using oligonucleotides derived from the protein sequence. Functional cloning was used to characterize the genes for the CLC-1 Cl^- channel and the SkM-1 Na^+ channel discussed later. In contrast, positional cloning is the process of isolating the gene starting from information about its genetic or physical location in the genome. Often little is known about the function of the product. The work involved is enormous since it relies on the method of **chromosome walking** and identification of expressed sequences. In spite of the difficulties, 19 disease genes were identified in this way between 1989 and 1993 (Ballabio, 1993), including myotonic dystrophy and cystic fibrosis, which are discussed in this chapter.

B. Autoimmune Disease and Pleiotropic Effects

Ion channel diseases can be due to the production of antibodies directed at the ion channel protein. Three examples in which autoimmune disease affects channels are discussed. Two of these are **myasthenia gravis** and the **Lambert-Eaton syndrome.** These conditions lead to bouts of skeletal muscle weakness because neuromuscular transmission is compromised. The third example is a voltage-gated K^+ channel, which is the target of an autoimmune disease called **neuromyotonia.**

Mutations on one gene may influence the expression or functions of channels coded by different genes; these are referred to as **pleiotropic effects.** In **myotonic dystrophy** or **Steinert's disease** (see Section II.D), the aberrant expression of an apamin-sensitive small-conductance calcium-activated K^+ (SK) channel leads to skeletal muscle hyperexcitability and **myotonia.** The mechanisms for pleiotropic effects are not well understood.

C. Diseases, Linkage, Candidate Genes, and Candidate Channels

Common methods are evolving for the study of familial diseases. Two basic strategies are used: **positional cloning** and the **candidate gene.** The first approach, positional cloning, is used to identify the genetic locus even when there is no knowledge of the biophysical or biochemical abnormalities underlying the disease. This is done by examining affected individuals for genetic linkage to polymorphic markers located throughout the genome. In this way a region of the gene is found that is linked to the disease. Linkage is established when there is a low rate of recombination between the marker locus and the disease locus. The gene can be used to express the protein it encodes and then determine its function and how the disease mutations disturb this function. The cystic fibrosis transmembrane-regulator channel was discovered in this way.

The second approach, or candidate gene strategy, requires knowledge of functional abnormalities brought about by the disease. Genes responsible for maintaining the normal function of the process affected by the disease are then considered to be candidate genes for the disease. Genetic linkage to the disease is then sought, and if one is found the candidate gene is screened for mutations. These mutations can be incorporated into the wild-type DNA and expressed in heterologous systems. Then one can examine the biophysical and biochemical behavior and determine if the abnormality could account for the disease. The new rapid cloning techniques are largely responsible for the increased pool of known channel genes having known functions. This makes it easier to identify candidate channel genes; thus the candidate gene approach has been very successful for the study of the channelopathies. Most of the new channelopathies including the CLC-5 mutations causing kidney stones, the episodic ataxias, and the long-QT-interval diseases were discovered using the candidate gene approach.

From the small number of channelopathies now reported, we can begin to make certain generalizations with respect to excitable cells. In excitable cells channels often assume both excitatory and stabilizing roles. Therefore blocking the stabilizing Cl^- channel (e.g., the low–Cl^- channel myotonias) is tantamount to blocking the inactivation of the excitatory Na^+ channels (e.g., the Na^+ channel myotonias). Similarly, in the human heart the long-QT syndrome is linked either to K^+ channels that have a decreased P_{open} or to Na^+ channels that have an increased P_{open}. In the central nervous system the phenomenon of episodic ataxia is linked both to K^+ channels that have a decreased P_{open} (episodic ataxia type 1) or to Ca^{2+} channels that have an increased P_{open} (episodic ataxia type 2).

Sometimes a linkage to a mutant gene is present but the expressed channels containing the mutation do not behave abnormally. This is the case for hypokalemic periodic paralysis (HoPP) discussed in Section V.B.2. The linkage to a mutation in the gene coding for a skeletal muscle Ca^{2+} channel is clear, but heterologously expressed channels do not display behavior that would explain the disease. In another example, some mutant Cl^- channels linked to myotonia in patients when expressed in heterologous systems have normal behavior. One explanation for discrepancies of this type is that important factors necessary for phenotypic expression may be absent in the expressions system, be it a cultured mammalian cell or a *Xenopus* oocyte. The missing factors could include regulatory paths or chemical modifications (e.g., disulfide linkages) necessary for proper folding of the channel protein (George, 1995).

D. Phenomenon of Myotonia

Myotonia is a clinical sign of certain diseases that target skeletal muscle membrane ion channels. It is a relatively rare phenomenon, usually not life threatening, yet it has held the interest of scientists and physicians for more than a century after it was first described by Dr. J. Thomsen in 1875 (see Rudel and Lehmann-Horn, 1985, for early references). Myotonia is defined as an abnormal contraction of skeletal muscles following an evoked contraction. In a typical example, a patient with myotonia is asked to grip an object or hand for a few seconds and then asked to release his grip. In spite of the fact that the patient is no longer willing a contraction, i.e., causing nerve action potentials to excite the muscle, the contraction continues for several more seconds. The myotonic muscle fibers are hyperexcitable (i.e., have low threshold potential) and tend to fire repetitive action potentials on their own. The **afterdischarge** of action potentials leading to the **aftercontraction** has been shown to result from a slowly decaying depolarization of the transverse tubules by accumulated K^+ in their lumen (Adrian and Bryant, 1974). Myotonic fibers fire repetitively for many seconds to steady depolarization, whereas normal fibers fire a short burst for about one-tenth of a second. This characteristic produces a diagnostically distinct electromyographic recording, which sounds like a dive-bomber when played through a sound system.

Myotonic disease occurs as a result of ion channel dysfunction in the skeletal muscle excitable membranes, surface and tubular. Hereditary myotonias occur indirectly as a result of mutations in enzymes regulating the ion channels, or directly through mutations of genes coding for ion channels. Although normal muscle fiber excitability requires Na^+, K^+, and Cl^- channels, to date, only errors in Na^+ and Cl^- channels have been associated with myotonia. K^+ channels play an important role in action potential generation, so it is surprising that there are no natural K^+ channel myotonias.

The most frequently encountered myotonic disease is human **myotonia dystrophy** (DM). The mutation responsible for this autosomal dominant disease (actually, the most common inherited neuromuscular disorder) is an expanded CTG repeat within the 3^1 untranslated region of a gene encoding a protein with homology to serine-threonine protein kinases, DM protein kinase (DM kinase). The CTG repeat does not produce a defective ion channel and the original proposition was that the disease results from a reduction or elevation of steady-state levels of DM kinase. However, transgenic DM kinase–null mice do not replicate the symptoms of DM, suggesting that changes in DM kinase alone do not induce DM (Jansen *et al.*, 1996; Reddy *et al.*, 1996). Regardless, the hallmark feature of muscle biopsies obtained from DM patients is that they contain higher levels of SK channels than normal muscle (Renaud *et al.*, 1986). Additionally, application of the bee venom peptide toxin apamin, a blocker of SK channels, dramatically reduces muscle myotonia (Behrens *et al.*, 1994). Thus the aberrant expression of a K^+ channel gives rise to hyperexcitability and myotonia in DM muscle. Therefore the underlying genetic and molecular mechanisms in DM ultimately exert effects on the expression of an ion channel in muscle. However, the disease is generalized to many organ systems and the underlying molecular mechanisms may be different in other tissue.

Less frequently occurring, but better understood, are several nondystrophic muscle diseases due to mutations in the skeletal muscle membrane ion channels. These can be classified into (1) **Cl^- channel type** and (2) **Na^+ channel type**. The membrane properties characteristic of these two types are illustrated in Fig. 1.

1. Cl^- Channel Type

The main Cl^- channel–type myotonias include human **dominant myotonia congenita** (DMC, Thomsen's disease), human **recessive myotonia congenita** (RMC; also called recessive generalized myotonia), and, among animals, the dominant myotonia of goats (which we also consider a DMC) and the recessive myotonia **adr mouse**. The basic physiological problem in these myotonias is a low membrane Cl^- conductance (g_{Cl}) due to dysfunctional CLC-1 Cl^- channels. Normally, the mammalian CLC-1 Cl^- channels provide the major stabilizing steady conductance (i.e., g_{Cl}) necessary for normal excitability. When g_{Cl} falls below 25% of normal, the skeletal muscle membrane becomes hyperexcitable and myotonia results. Cl^- channel myotonia can be induced by agents that specifically block the CLC-1 channel such as anthracene-9-carboxylic acid (9-AC), and simulated by computer models of action potentials in which g_{Cl} is lowered (Bryant and Morales-Aguilera, 1971; Furman and Barchi, 1978; Adrian and Marshall, 1976). See Fig. 1 A–C and G.

2. Na^+ Channel Type

The main Na^+ channel myotonias, which are all dominant, include human and equine **hyperkalemic periodic paralysis,** human **paramyotonia congenita,** and human **Na^+ channel myotonia.** The common physiological factor in this group is a small fraction of Na^+ channels that inactivate more slowly and/or show abnormal reopenings compared to the wild type. The abnormally increased Na^+ currents result in hyperexcitability because a smaller stimulus can initiate an action potential and can then cause reexcitation with myotonic repetitive firing. Na^+ channel myotonia can be induced by agents that block Na^+ channel inactivation like the sea anemone toxin, ATX-II. Like the low-g_{Cl} myotonia, Na^+ channel myotonia has been simulated using computer models (Cannon *et al.*, 1993). See Fig. 1D–F and H.

III. Cl^- Channels

A. General Comments

Cl^- channels and nonspecific anion channels have only recently gained attention, yet the number of distinctly different types of Cl^- channels may in time exceed that of other channels (Vaughan and French, 1989; Pusch and Jentsch, 1994). Unfortunately, many of the newly described channels have not yet been assigned physiological functions. The functions already identified include osmotic pressure and stretch sensitivity, Cl^- concentration regulation, and membrane potential stabilization. With respect to the latter function, the channels can furnish a constant stabilizing membrane conductance in skeletal muscle, a transient agonist-activated stabilizing conductance in neurons (known as inhibition), and a hormonally activated stabilizing conductance in heart fibers. Based on structural similarity and function we can identify many classes of Cl^- channels. Examples of genetic diseases are given for three: (1) the CFTR channel important for normal airway function whose dysfunction produces cystic fibrosis, (2) the CLC-1 channel present mainly in skeletal muscle membranes that controls the excitability of muscle fibers and whose dysfunction leads to forms of myotonia, and (3) the CLC-5 channel of the kidney whose dysfunction leads to formation of kidney stones.

B. Cystic Fibrosis Transmembrane Regulator

Cystic fibrosis (CF) is an autosomal recessive disease that affects about 1 in 2000 Caucasians. The recessive gene is present in about 5% of that population, and it is the most common inherited lethal disease. It is caused by mutations (more than 100 have been identified) in the gene for the **cystic fibrosis transmembrane regulator,** which is a Cl^- channel known as CFTR. Figure 2A shows the primary structure of CFTR and the site of the major (70%) mutation, $\Delta 508f$, a deletion of a single phenylalanine at position 508. The primary sequence suggests 12 transmembrane helices, two hydrophilic nucleotide-binding sites, and a large cytosolic regulatory (or R) domain, which has many potential phosphorylation sites. This channel is regulated by phosphorylation and by nucleoside triphosphates. These agents act at the proposed regulatory areas.

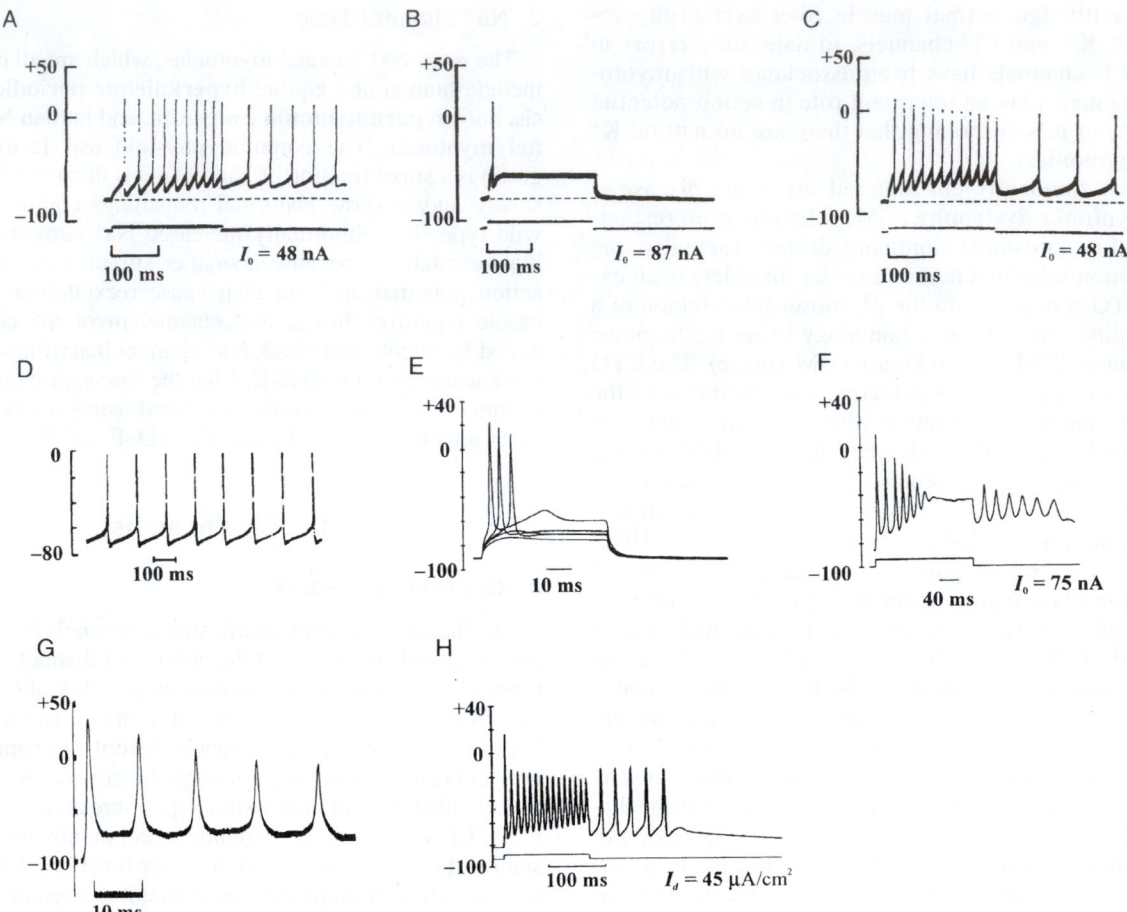

FIGURE 1. Myotonia due to Cl$^-$ and Na$^+$ channels. Tracings A through F are actual microelectrode recordings of membrane potentials from single mammalian skeletal muscle fibers. Except for trace D, in which the fiber was spontaneously active, the fibers were stimulated by long square-current pulses passed through a second microelectrode inserted into the same fiber close to the voltage-recording microelectrode. A lower trace representing this depolarizing current is shown in several of the traces and the value I_0 is the intensity of this pulse in nanoamperes. (A–C): Low g_{Cl} myotonia. (A) From a myotonic mutant goat fiber. Note the repetitive action potential and an afterdischarge when the current is turned off. (B) From a normal goat fiber for control. Note that a larger current than in trace A produced no myotonia. (C) Impermeant sulfate ion replaced Cl$^-$ ion bathing a normal fiber, thus blocking g_{Cl}. Myotonic responses are induced identical to those of the mutant fiber in trace A. Other g_{Cl} blockers such as 9 AC at 50 μM would give the same result. These goat fibers were studied at 38 °C (Adrian and Bryant, 1974). D–F Na$^+$ channel myotonia. (D) From a human patient with the PC mutation (Lehmann-Horn *et al.*, 1981). PC patients have normal g_{Cl} yet the fibers are myotonic. Cooling this fiber from 37 to 30 °C caused the depolarization, the small action potentials, and myotonia. (E and F) From normal rat fibers (Cannon *et al*; 1993). In trace E are five superimposed control measurements with 50-ms current pulses ranging from 90 to 175 nA (threshold at 115 nA); there is no myotonia. In trace F the fiber was exposed to 10 μM ATX-II, the sea anemone toxin that blocks Na$^+$ channel inactivation, and stimulated with a long pulse. Note the induction of myotonia, the baseline depolarization, and an afterdischarge. G–H Computer simulations of myotonia. (G) Simulates low-g_{Cl} myotonia. Shown are the first five surface membrane potentials calculated for propagating the action potential model, which has a simulated T-tubular system that accumulates K$^+$. These action potentials are from a myotonic train during a long depolarizing current pulse. After the pulse was turned off there was a short afterdischarge. The model was made myotonic by lowering the g_{Cl} term in a Hodgkin-Huxley mathematical model of the action potential adapted to skeletal muscle. The control calculation (not shown) with normal g_{Cl} gave only one action potential for a large stimulating current. (From Adrian and Marshall, 1976.) (H) Computer simulation of Na$^+$-channel myotonia (Cannon *et al.*, 1993). The mathematical model was similar to that used for the low-g_{Cl} myotonia except in this case the Na$^+$ channel currents were modified by adding 2% of the slowly inactivating HYPP mutant-type Na$^+$ currents to the normal Na$^+$ currents. Trace H shares many details with trace F from the toxin model of Na$^+$-channel myotonia, for example, the tendency for the average potential to move toward a depolarizing plateau during the pulse, followed by an afterdischarge after cessation of the current pulse. Both low-g_{Cl} and Na$^+$ myotonias require an intact T-tubular system for an afterdischarge. Detubulated real or simulated myotonic fibers remain myotonic but lack the afterdischarge.

Normal operation is assumed to be that the channel opens when the sites are occupied. Levels of cAMP determined by physiological needs of the cell would then regulate normal function. When mutant CFTR is present in the pulmonary epithelia it does not respond to regulation by cAMP and remains closed, not allowing Cl$^-$ ions to enter the cell, which prevents normal secretion in airway passages. This leads to accumulation of a thick mucus, which blocks the passages and leads to death in 90% of the patients before reaching adulthood. There are additional

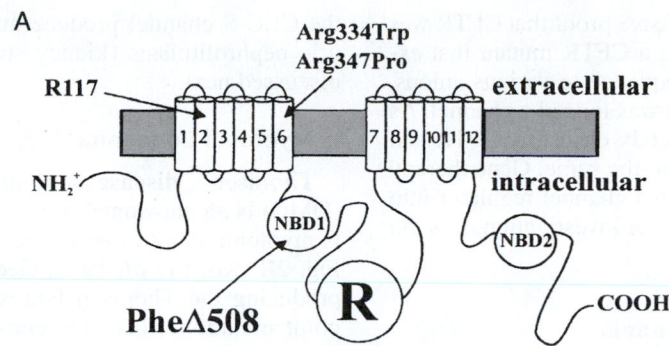

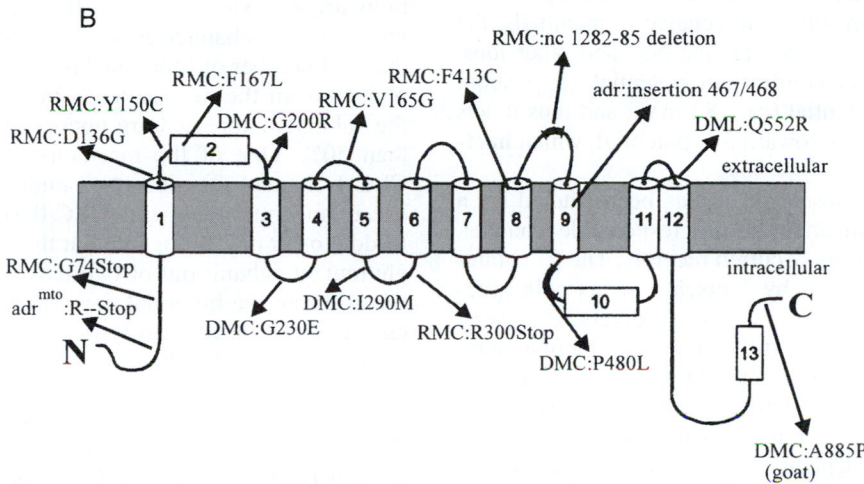

FIGURE 2. Cl^- channels. The primary structures as deduced from hydropathy plots and other information are given for the following Cl^- channels: (A) The CFTR channel located in pulmonary and other epithelia. The lower arrow points to NBD1 (nucleotide-binding domain 1), the location of the principal mutation PheΔ508, a deletion of phenylalanine at position 508, which prevents the channel from being regulated by cAMP and leads to 70% of the cases of severe CF. M6 is indicated as the site of two mutations at which arginine (334 and 347) is substituted by tryptophan and proline, respectively. M2 is the site of a similar mutation at position 117. These latter three mutations decrease the Cl^- conductance of the channel and lead to mild CF. (B) The CLC-1 channel of mammalian skeletal muscle produces the large stabilizing g_{Cl} necessary for normal excitability. Shown are some of the known mutations with the theoretical structural alterations of the CLC-1 channel leading to the following phenotypes exhibiting low–Cl^- conductance myotonia: DMC, RMC, DML (dominant myotonia levior), adr, and adr^{mto} (recessive myotonia in the myotonic mouse). All of the diseases indicated occur in human patients except for those conditions labeled as adr mouse or myotonic goat. Following each disease designation is the mutation given in standard single-letter amino acid codes and the residue position. For example, DMC: G230E is interpreted as dominant myotonia congenita in which at residue 230 a glutamate (E) is substituted for the wild-type glycine (G). The adr-mouse syndrome is due to the insertion of a transposon into the gene, which prevents expression of the channel. Shown here is the position (between 467 and 468) on the protein corresponding to the region on the gene affected by the transposon. This is for illustration only since the protein is never completely expressed. For more complete information on mutations of the CLCN1 gene that encodes for the human CLC-1 channel, see Lehmann-Horn and Rüdel (1996). (Redrawn and modified from Lehmann-Horn and Rüdel, 1996, using data from their Table 2.)

problems with secretion by the epithelia of the pancreas, gut, salivary glands, and other organs. One diagnostic feature of CF is a high NaCl content of sweat. A mutant CFTR can account for most of the pathological signs of the disease but some additional factors may also be involved (rev. in Warner, 1992). It has been recently argued that in spite of the fact that CFTR is a channel it may not be the actual Cl^- channel that furnishes the ion transport. CFTR may indeed act as a regulatory protein, which controls this channel as originally thought.

The discovery that an abnormal Cl^- channel was ultimately at fault came from resting potential measurements in epithelial cells from airways or sweat glands that always showed the CF cells to be hyperpolarized. Perfusing the luminal side of the cells with a Cl^--free (sulfate) medium increased the potential of normal cells as predicted, but had no effect on the diseased cells; in other words, Cl^- channels were not conducting properly. The major breakthrough occurred when molecular biologists sequenced the gene responsible for CF. The initial analysis of the proposed product suggested that it was a regulatory enzyme; hence, it received the name cystic fibrosis transmembrane regulatory factor or CFTR. Patch-clamp studies later showed that CFTR codes for a nonrectifying cAMP-activated channel of about 8 pS. The $\Delta508f$ mutation prevents the channel from responding to regulation by cAMP, although the defective

CFTR protein is expressed. Impressive proof that CFTR was a channel came from constructing a CFTR mutant that exhibited an altered permeability sequence to various anions, suggesting that the CFTR product was indeed a channel. As mentioned earlier, it is not completely clear that CFTR and the responsible channel are one and the same. Gene therapy by transfecting genes for the normal channel regulator into diseased pulmonary epithelia is under investigation. It is too early to evaluate these results.

C. Skeletal Muscle CLC-1 Channels

Skeletal muscle fiber membranes in their resting or unstimulated state have a relatively high g_{Cl}. This amounts to about 80–90% of the resting conductance in mammals, the remainder being K^+ and less than 1% Na^+ and other ions. The resting g_{Cl} has an equilibrium potential (E_{Cl}) about equal to the resting potential (ca. −82 mV), and thus it acts to stabilize the membrane toward this potential, which in effect prevents myotonia.

The g_{Cl} of skeletal muscle fibers has been studied for a long time, but identification of the responsible channel, known as CLC-1, has only occurred recently. The CLC family of Cl^- channels cloned by Jentsch and his colleagues (1995) presently includes about a dozen diverse members, and the number appears to be still increasing. The first member cloned was CLC-0 from electrocytes of the marine electric fish *Torpedo,* where it stabilizes the membrane on one side of the electric cells, thus allowing the cells to act as electric batteries (see Chapter 60). Because these electrocytes were derived embryologically from skeletal muscle, polypeptides from short sequences of the CLC-0 channel were used to probe a rat skeletal muscle cDNA library, which resulted in cloning of the CLC-1 channel (Steinmeyer *et al.,* 1991b). Development and innervation control the CLC-1 channel, whose primary structure is shown in Fig. 2B. The channel is not seen in embryonic muscle or in cultured muscle cells, or following denervation of the adult fiber. The corresponding mRNA was seen only in trace amounts in other tissues.

Monoclonal antibodies to CLC-1 react only with sarcolemmal membrane fragments, suggesting a purely sarcolemmal location of this channel (Gurnett *et al.,* 1995). This is in contrast to earlier functional studies, which placed mammalian g_{Cl} in both the T-tubular and surface membranes (e.g., Dulhunty, 1979). Possibly, there is a T-tubular isoform that does not bind the antibody, or the channels are not accessible in the T-tubular fragment. The location problem awaits further resolution.

The CLC Cl^- channel family has continued to expand with the reports of new members. At present this family, in addition to CLC-0 and CLC-1, consists of CLC-2, CLC-3, and CLC-4, which are ubiquitous mammalian channels of little understood function, possibly involved in cell volume regulation; CLC-K1 and CLC-K2, which are involved in reabsorption in the mammalian kidney; and CLC-S, an additional kidney channel. So far, only two members of the CLC family are associated with genetic disease. Mutations of the CLC-1 channel account for several classical hereditary myotonias in humans, mice, and goats; mutations in the CLC-S channel produce human hereditary hypercalciuric nephrolithiasis (kidney stones). These conditions are discussed next.

l. Myotonia Congenita

Thomsen's disease or **dominant myotonia congenita** (DMC) is an autosomal dominant condition in which there is myotonia due to loss of the stabilizing g_{Cl}. As shown in Fig. 2B (Koch *et al.* 1992; George *et al.* 1993), the defect producing the Thomsen family form of DMC is a single point mutation where the conserved glycine (G) (residue 230) is replaced by glutamic acid (E), shown on the figure as DMC:G230E. Some of the other known dominant mutations are also shown in Fig. 2B. The fact that a dominant mutation in a channel gene can cause a large reduction in the total number of functional normal channels requires explanation. In the heterozygous individual where only half the subunits expressed are abnormal, the g_{Cl} is much less than 50%. One of the mechanisms proposed is that the CLC-1 channel functions as a multimer, probably a dimer based on homology with CLC-0 (Ludewig *et al.* 1996; Middleton *et al.* 1996), and that the presence of even a single mutant subunit out of two makes the channel dysfunctional. From the binomial theorem about one-fourth of the channels will have wild-type subunits and the remaining three-fourths of the channels will have at least one mutant subunit and therefore be inactive. Experimental evidence for this possibility is that coexpression of a DMC mutant CLC-1 with the wild-type CLC-1 produces a reduction in g_{Cl} that is greater than one-fourth of that predicted for homomeric wild-type channels expressed by themselves (Steinmeyer *et al.* 1994). It is also of interest that the gating process of CLC-0 and CLC-1 is fairly unique in that these channels utilize the permeating chloride ion itself as the gating particle.

2. Recessive Myotonia Congenita

A large population of nondystrophic myotonic patients was studied, and two separate diseases corresponding to their dominant or recessive inheritance were distinguished (Becker, 1973). The smaller group (30%) were the autosomal dominant DMC patients who showed less myotonia on the average and whose physiology we discussed previously. The larger group (70%) were the autosomal recessive RMC patients who were generally more severely myotonic. RMC is also called **recessive generalized myotonia.** In RMC there is a reduced g_{Cl} due to a variety of mutations, some of which are shown in Fig. 2B. Additionally, some dysfunctional Na^+ channels may contribute to the myotonia in certain patients. This Na^+ channel abnormality may be a pleiotropic effect consequent to the CLC-1 mutation.

3. Myotonic Mouse

A mouse that had difficulties in righting itself when placed on its back was discovered at the Jackson laboratories (Mehrke *et al.* 1988). This condition was due to a recessive mutation, and at first was thought to be a neurological condition affecting the righting mechanism and thus received the name of **arrested development of righting response** or adr mouse. Later it became clear that the major problem was the myotonia of the skeletal muscles, and thus the term **myotonic mouse** is also used. The repetitive firing and other defects in excitability were shown to be due to a specific loss of g_{Cl}. The heterozygous animals that had nearly normal g_{Cl} behaved normally, which indicates that the normal allele in the heterozygous mouse is capable of coding for and expressing a normal density of Cl^- channels. Jentsch and colleagues (Steinmeyer *et al.* 1991a, b) have shown that the CLC-1 channel (see Fig. 2B) is also the major skeletal muscle Cl^- channel in the mouse. They further showed that the mutation was caused by a **transposon** (i.e., a nonsense code) inserted into this gene at a region that would otherwise code for an essential part of the channel protein. Actually, the defect is so severe that no mRNA for this channel was detected in the myotonic mouse. In addition to the transposon defect of the adr mouse, other missense and nonsense mutations have been reported in the mouse CLC-1 that lead to myotonic conditions. One of these is the adr^{mto} mouse.

4. Myotonic Goat

A dominant mutation in goats in which myotonia is the only clinical sign has been known for more than a century (see Bryant, 1979). These animals (so-called stiff, falling, or nervous goats) are still being raised for their interesting behavior on sudden movement, which often causes bizarre stiffness and sometimes falling. Based on human experience it is believed that the muscle effects are not painful. Electrophysiological studies on biopsied muscle fibers from these animals advanced our knowledge of the biophysics of myotonia, and in particular led to the discovery of myotonia due to low g_{Cl} discussed earlier. Also shown was the necessity of K^+ accumulation in the T-tubular system for maintaining an afterdischarge (Adrian and Bryant, 1974). The myotonia can be simulated by blocking g_{Cl} with 9-AC or other Cl^- channel blockers (Bryant and Morales-Aguilera, 1971), and by computer modeling of action potentials when the Cl^- conductance term, modeled as a leak conductance, is made low (Adrian and Marshall, 1976).

The mutation responsible for CLC-1 dysfunction in myotonic goat is the replacement of a highly conserved alanine by proline in the carboxy-terminus region, A885P. The functional effect of this mutation, determined in a human CLC-1 construct, is to shift the curve relating open probability (P_{open}) to membrane potential to the right by 45 mV. This is illustrated in Fig. 3. These results imply that at the physiological resting potential of −82 mV, if only mutant channels were present, the Cl^- conductance would be roughly 25% of a membrane having wild-type channels; this is similar to the reduction of Cl^- conductance recorded in native myotonic goat membrane. An exact comparison with the older studies would be difficult, since these data were usually averaged from mixed genetic populations having both mutant and wild-type alleles. The pattern of shifting the conductance curves toward more depolarized voltages also occurs in four dominant human myotonias (I290M, R317C, P480L, and Q552R), and may represent a biophysical phenotype for dominant Cl^- channel myotonic disease.

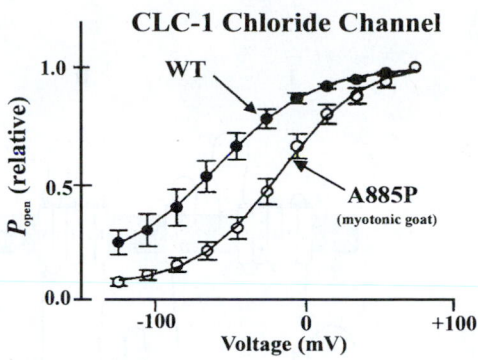

FIGURE 3. Myotonic goat CLC-1 Cl^- channel function. The relative open probabilities at the different membrane voltages were determined with a two-electrode voltage-clamp from *Xenopus* oocyte expressing wild-type (WT) or mutant (A885P) CLC-1 channel currents. The curves were fitted to single Boltzmann functions. The principal change observed is the right shift of the center potential, $V_{1/2}$, on the mutant channel by 45 mV from the wild-type control. This kind of shift is typical of shifts observed in several dominant human myotonias involving the CLC-1 channel. At the resting potential of −80 mV the mutant channels would have a lower probability of being open, and this would account for the low g_{Cl} and myotonia in the affected goat (Redrawn from Beck *et al*, 1996.)

D. Kidney CLC-5 Channel: Kidney Stone Diseases

The second member of the CLC family of Cl^- channels to be associated with genetic disease is CLC-5. This channel is strongly **outwardly rectifying,** unlike CLC-1, which is an **inward rectifier.** Kidney stones affect about 12% of males and 5% of females in the Western world, and in 45% of these patients, the disease is inherited. Dent's disease and two other types of kidney stone disease have been linked to chromosome Xpll.22. A microdeletion in a Dent's disease patient allowed the identification of a candidate gene (CLCN5) that codes for the renal Cl^- channel, CLC-5. Many types of mutations of this gene were then detected; these errors include nonsense, missense, donor splice mutations, and various deletions (Lloyd *et al.,* 1996).

A speculative explanation for the pathological effects of the mutant CLC-5 is that the protein leakage through the glomerulus is taken up by endocytosis in the proximal tubules, and subsequently degraded in a lysosomal compartment. The essential endocytotic uptake is postulated to require acidification of the endosomes, and the CLC-5 channel provides the Cl^- transport necessary to maintain the low pH in this compartment. Because the mutant CLC-5 does not transport Cl^-, the endosomal pH rises, endocytosis is impaired, protein accumulates in the tubule, and kidney stone formation is promoted.

IV. Na$^+$ Channels

A. General Comments

Voltage-gated Na$^+$ channels play an essential role in excitable cells where they are responsible for the rapid rise in the action potential upstroke and for the rapid conduction. Voltage-gated Na$^+$ channels belong to the general family of

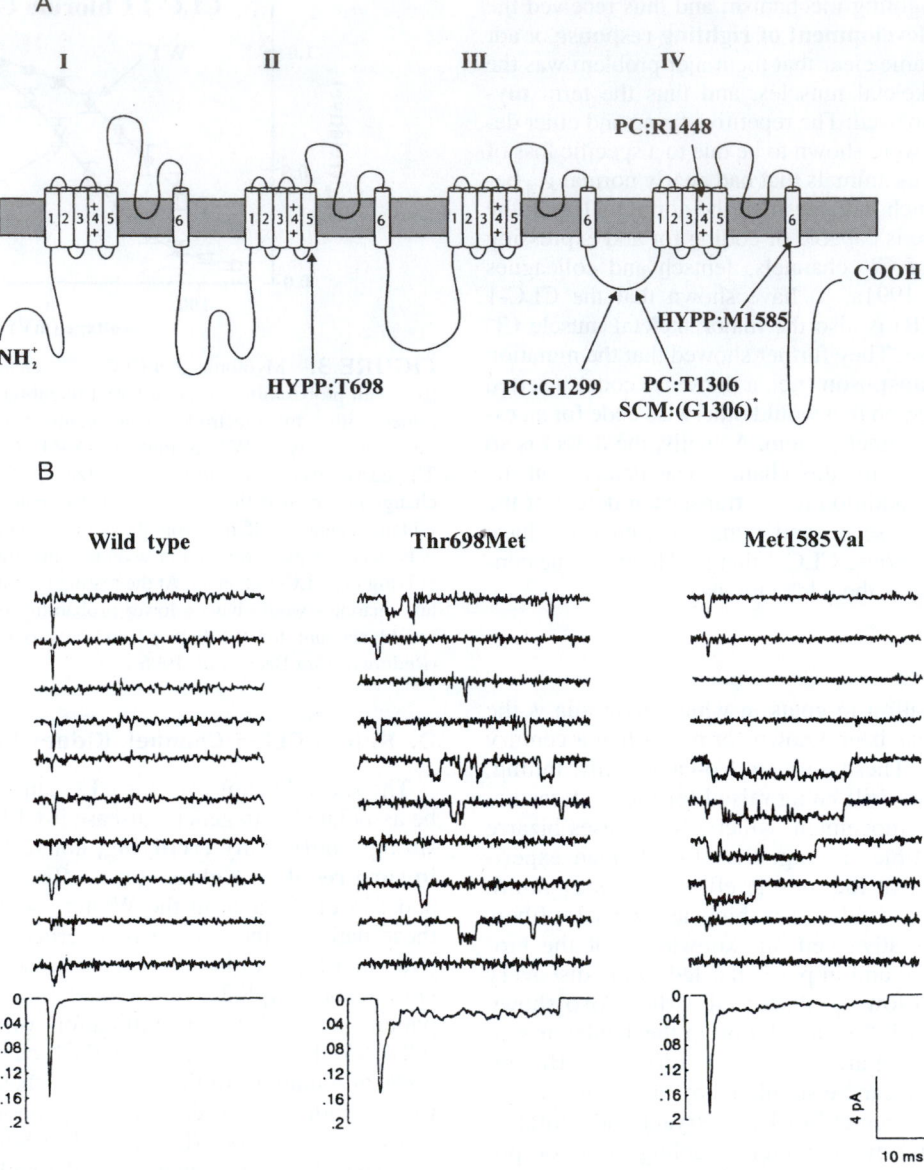

FIGURE 4. Skeletal muscle Na$^+$ channel. (A) Primary structure of the α-subunit of the SkM-1 skeletal muscle Na$^+$ channel coded by the gene SCN4A. Mutations are indicated that lead to PC (paramyotonia congenita), HYPP (hyperkalemic periodic paralysis), and SCM (Na$^+$ channel myotonia). *The amino acid position numbers for PC and HYPP follow those of Cannon and Strittmatter (1993); however, for the mutations at the conserved glycine causing SCM, the numbering system of Lerche *et al.*, (1993) was used. Unfortunately, the position numbers overlap here although the real amino acid locations are different. (B) Single Na$^+$ channel currents recorded in HEK293 cells after expressing the wild type and two genetically engineered constructs, Thr6698Met and Met1585Val, which correspond to natural mutations indicated in part A that cause HYPP. Ten single-channel current traces are shown above, and the averaged ensemble currents are shown below. Note that the wild-type channel shows brief single openings and a transient ensemble current. In contrast, the mutant Na$^+$ channels display long openings and reopenings giving averaged Na$^+$ currents with long noninactivating tails. These prolonged depolarizing inward currents lead to the myotonic behavior and, if sufficiently large, to the block of excitation and a flaccid paralysis. The HYPP mutant channels, unlike the PC channels, show a sensitivity to K$^+$ ions that further decreases their ability to inactivate. The PC channels, on the other hand, are sensitive to lowering the temperature, which has the same effect.

voltage-gated cation channels, which includes Ca^{2+} and K$^+$ channels. The primary α-subunit forming the Na$^+$ channel (Fig. 4A) consists of about 2000 amino acids. There are four repeats, each containing six putative transmembrane helices. Amino acid analysis and site-directed mutagenesis studies have determined that the fourth transmembrane segment in each repeat, called S4, acts as the voltage sensor. The cytoplasmic loop connecting repeats III and IV is necessary for inactivation, and the region between S5 and S6 in each repeat, the P region, contributes to formation of the hydrated, ion-conducting pore. Ca^{2+} channels share a similar molecular design. The K^+ channel, however, departs from the scheme in that genes encode for only a single repeat and the channel is assembled from four identical or differing α-subunit.

Well-described disease entities involving voltage-gated Na^+ channel mutations include those of the human *SCN4A* gene on chromosome 17q, which codes for skeletal muscle Na^+ channels. These are discussed as a group next.

B. Skeletal Muscle SkM-l and SkM-2 Channels

Multiple subtypes of Na^+ channels in skeletal muscle have been identified based on sensitivity to toxins and to antibodies, and these subtypes may be located preferentially in the T-tubules or surface membrane. Two Na^+ channels specific to skeletal muscle, known as SkM-1 (George *et al.*, 1992) and SkM-2 (Kallen *et al.*, 1990), have been cloned and sequenced. These channels are sometimes designated $\mu 1$ and $\mu 2$, respectively. Skeletal muscle Na^+ channels are heterodimers of a 260-kDa α-subunit and a 38-kDa β-subunit (Kraner *et al.*, 1985). Type SkM-1 is expressed in both innervated and denervated adult muscle and is blocked by nanomolar concentrations of tetrodotoxin (TTX) and μ-conotoxin. The TTX-insensitive SkM-2 is not found in innervated adult mammalian muscle, but appears within hours following denervation, reaching a maximum at 48 h and then decreasing. Type SkM-2 is expressed in early development and disappears as type SkM-1 increases; thus SkM-1 is the only adult form. Type SkM-2 is also seen *in vitro* in skeletal muscle culture (absence of innervation) where it accounts for about 30% of the functioning channels; its sequence is essentially the same as the normal TTX-insensitive Na^+ channel of heart (Rogart *et al.*, 1989).

l. Periodic Paralysis and Paramyotonia

In a number of inherited diseases, periodic bouts of paralysis occur. Some of these conditions are associated with either hyperkalemia or hypokalemia, and myotonia may also be present. Recent studies have brought considerable light to understanding many of these conditions. Several genetic forms of myotonia and periodic paralysis may be due to failure of a small fraction of the skeletal muscle voltage-gated Na^+ channels to inactivate normally (Rudel and Lehmann-Horn, 1985). **Hyperkalemic periodic paralysis** (HYPP) and **paramyotonia congenita** (PC) are related because they are due to autosomal dominant mutations of the α-subunit of the SkM-1 Na^+ channel located in the surface membrane and T-tubular membrane of mammalian skeletal muscle fibers. These channels are normally responsible for conducting the skeletal muscle action potential (AP) throughout the tubular membranes, which is necessary for excitation-contraction coupling. Figure 4A shows the locations of several PC and HYPP mutations of the α-subunit of the Na^+ channel (Fontaine *et al.*, 1990). In Fig. 4B, the lack of effect on activation and the profound effect on inactivation are clearly seen in mutant constructs having point mutations in these regions (residue numbers

differ slightly from Fig. 4A due to species). Note in the single-channel records the longer openings and repeated reopenings of the mutant constructs. This accounts for the longer persistence of the averaged mutant sodium currents. Another way of stating this is to say that the mutant Na^+ channels exhibit a decreased tendency to inactivate.

HYPP is an autosomal dominant mutation in which recurrent episodes of weakness occur with small elevations in the serum K^+ concentration. In this particular form of the disease, there is marked depolarization of skeletal muscle fibers during an attack, which makes the fiber inexcitable and leads to a flaccid paralysis of the muscle. Diseased biopsied fibers studied under voltage-clamp conditions yield TTX-sensitive Na^+ currents that do not completely inactivate; this has been shown to be the mechanism for depolarization (Lehmann-Horn *et al.*, 1987). Single-channel recordings confirming this finding (Cannon *et al.* 1991) show that single Na^+ channels recorded from HYPP biopsied fibers have prolonged open times and tend to open repetitively (see Fig. 4A, B). In high extracellular K^+ concentrations (around 10 mM compared with a normal of 4–5 mM), a small fraction (5–10%) of the mutant channels do not inactivate and this produces a constant open probability of between 0.02 and 0.05, yielding a steady depolarizing current. This current is large enough to account for the depolarization that inactivates the unaffected Na^+ channels, resulting in action potential failure (Cannon *et al.*, 1993).

To demonstrate that lack of Na^+ channel inactivation is the possible cause of some forms of myotonia and periodic paralysis, two approaches have been taken. In the first approach, an *in vitro* model was created in rat muscle exposed to a polypeptide toxin (ATX-II) obtained from a sea anemone. The toxin at 10 μm produced a noninactivating open probability at −10 mV of approximately 0.02, similar to that reported for myotubes from HYPP patients, and myotonia was apparent (Cannon and Corey, 1993). In the second approach, a computer simulation was developed based on modifications of the Hodgkin-Huxley equations adapted to skeletal muscle fibers. This approach is similar to that discussed for computer simulation of low-g_{Cl} myotonia. The simulated fiber, like the low-Cl^- model, required a T-tubular compartment to act as a diffusion-limited space in which activity-induced K^+ can accumulate. The computed APs for increasing degrees of incomplete inactivation of Na^+ channels effectively simulated normal, myotonic, and paralytic muscle. The simulated APs with altered Na^+ inactivation compared favorably with the abnormal APs produced pharmacologically with ATX-II (Cannon *et al.*, 1993).

PC, an autosomal dominant condition, has been also referred to as **paradoxical myotonia** because the signs of myotonia **increase** with use of the skeletal muscles, rather than diminish, and there is an increase in myotonia and a dramatic increase in the response of a muscle to percussion with local cooling. At least *in vitro*, low-Cl^- conductance myotonias are quite different since the signs are diminished with cooling below 27 °C (Furman and Barchi, 1978). If one replaces the sensitivity to cooling with the worsening of symptoms with elevated serum K^+ levels, there is a physiological resemblance

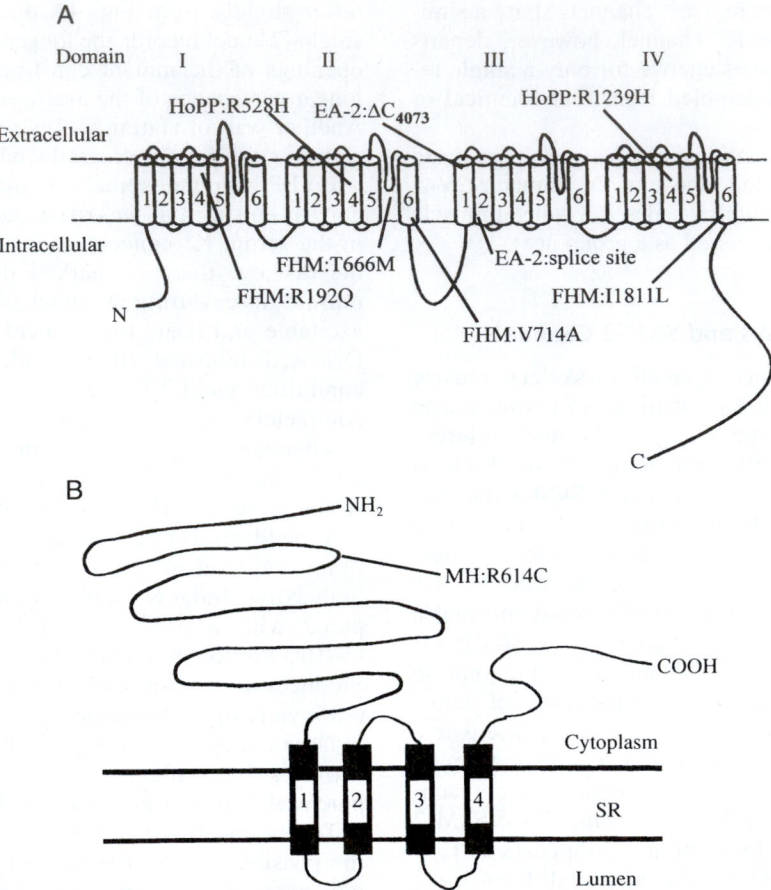

FIGURE 5. Ca^{2+} channels. (A) The primary structure of the a1-subunit of the human skeletal muscle type-1 voltage-sensor/channel (i.e., the DHP receptor) of the T-tubular membrane, and the neuronal P/Q-type Ca^{2+} channel are shown here superimposed on the same illustration. In skeletal muscle, two mutations linked to human hypokalemic periodic paralysis (HoPP) are indicated. Both mutations occur in segment 4 of domains II and IV, which is a region believed to be involved in gating of the channel. Using the same schematic, the primary structure is superimposed on the α1-subunit of the neuronal P/Q-type Ca^{2+} channel. The mutations causing familial hemiplegic migraine (FHM) and EA-2 are indicated. Note that the mutations causing FHM are of the missense type, resulting in single amino acid substitutions, whereas the two mutations for EA-2 both cause premature stops. (Redrawn and modified from Ophoff *et al.*, 1996.) (B) The primary structure of the human ryanodine receptor indicating the main mutation responsible for malignant hyperthermia (MH). The mutation at the conserved Arg 614 is a substitution for Cys in the cytoplasmic region of the molecule. In porcine MH, Arg 615 is substituted in the same way. One should note that except for the four membrane-spanning segments Ml through M4 (indicated here as simple 1 through 4), 90% of this molecule (5037 amino acids) is cytoplasmic. The functioning Ca^{2+} release channel consists of four of these units (i.e., a homotetramer) in the SR membrane. Activation to the T-tubular DHP receptor causes activation of the Ca^{2+} release channel in the process of excitation-contraction coupling. One of many speculations as to how this is accomplished is that a third molecule, triadin, interacts with the connector between domains II and III of the DHP receptor (shown in A), giving a mechanical coupling between the two systems.

---

between PC and HYPP. Both of these conditions have myotonia with episodic weakness and the similarities appear to be explained by the findings that both diseases are due to different mutations but on the same skeletal muscle Na⁺ channel gene (Ptacek *et al.*, 1993; McClatchey *et al.*, 1992).

2. Na⁺ Channel Myotonia

Some mutations in the SCN4A gene for the SkM-1 channel can produce in patients a relatively pure form of myotonia known as human **Na⁺ channel myotonia** (SCM) (Lerche *et al.*, 1993). This syndrome is devoid of dystrophy and weakness, and does not resemble PC or HYPP, al-

though the latter pair is also due to mutations in the same gene (see Fig. 4A). Glycine at position 1306 is completely conserved in all Na⁺ channel α-subunits in all species. This is an important region for control of fast Na⁺ channel inactivation and it has even been proposed that the two glycines at 1306 and 1307 may act as a hinge to block the channel pore (Kellenberger *et al.*, 1997). In families with SCM, the glycine at 1306 is substituted by glutamine, valine, or alanine. The glutamine substitution produces the most severe myotonia. None of the mutations in the patients affects inactivation as much as the substitution for phenylalanine at 1311, but the degree is sufficient to account for the myotonia. Patch-clamp studies of the SCM channel show a slowing of the rate of fast inactivation and an increased frequency of late channel openings. These parameters lead to an increase in the ratio of late current to peak current from control of 0.4% to 6%. The simulation experiments using ATX-II anemone toxin (Cannon and Corey, 1993) with computer modeling showed that myotonia occurred when only 2% of Na⁺ channels fail to inactivate, which corresponds to a late over-peak ratio of 5%. It should be clear that a larger late depolarizing current could result in restimulation and repetitive firing, that is, myotonia. Thus, SCM and its severity can be explained by the length and charge of the group substituted for Gly 1306.

C. Altered Neuronal Na⁺ Channel Function in Febrile Epilepsy

Febrile seizures are the most common seizure disorder, affecting approximately 3% of all children under six years of age. Some of these children develop ongoing epilepsy with afebrile seizures later in life. Linkage studies in one large family mapped this disorder to chromosome region 19q13.1 and a mutation in the voltage-gated Na⁺ channel β_1 subunit gene (SCN1B) was identified (Wallace *et al.*, 1998). The mutation changes a conserved cysteine residue and co-expression of mutant β_1 subunits with a brain Na⁺ channel α subunit in *Xenopus* oocytes shows that the mutation reduces the rate of Na⁺ channel inactivation. The complex inheritance pattern of this disorder raises the possibility of involvement of other Na⁺ channel subunit genes in febrile seizures and generalized epilepsies.

D. Cardiac Na⁺ Channel: Long-QT-Interval Disease

Like neuronal and skeletal muscle action potentials, the Na⁺ current is responsible for the initial upstroke of the action potential in cardiac ventricular and Purkinje cells. The Na⁺ channels quickly inactivate but the action potential remains depolarized for its duration due to the rectification of the inwardly rectifying K⁺ current and the activation of Ca²⁺ currents. The action potential is repolarized by the Ca²⁺-dependent inactivation of the Ca²⁺ current and activation of the delayed-rectifier K⁺ currents, I_{Kr} and I_{Ks}. The cardiac Na⁺ channel is coded by the SCN5A gene, and mutations in this gene produce aberrant Na⁺ channels in which fast inactivation is delayed (Roden *et al.*, 1995). The end result is a cardiac action potential that has a prolonged repolarization, which is seen as a distinctive long QT interval on the electrocardiogram. Associated with the repolarization defect is a serious arrhythmia leading to syncope (fainting), seizures, and sudden death from a distinctive ventricular tachycardia known as **torsades de pointes.** The long-QT syndrome is also genetically linked to sites other than the Na⁺ channel gene (Keating and Sanguinetti, 1996). These sites include mutations in the genes that encode I_{Kr} and I_{Ks}. (See Section VI.)

E. Non–Voltage-Gated Na⁺ Channels in Disorders Involving Epithelial Transport

Non–voltage-gated Na⁺ channels such as the epithelial Na⁺ channel (ENaC) control transepithelial Na⁺ reabsorption in the distal nephron, distal colon, and the airways. ENaC is a heteromere of three different but homologous subunits (α, β, and γ). Mutations within any of these subunits result in either loss or gain of function (rev. in Hummler and Horisberger, 1999). Loss of ENaC function results in **pseudohypoaldosteronism type 1,** which is characterized by urinary loss of Na⁺ and reduced K⁺ excretion. Severe forms of this syndrome are inherited as an autosomal recessive trait that may result in lethal episodes of hyponatremia, hypotension, and hyperkalemia. Certain mutations in the β- and γ-subunits result in a gain of ENaC function, **Liddle's syndrome,** giving rise to hypertension due to Na⁺ retention. Metabolic alkalosis and hypokalemia are often associated with this disorder.

V. Ca²⁺ Channels

A. General Comments

There are several types of Ca²⁺ channels, both agonist activated and voltage controlled. The primary structure of the α subunit of the L-type Ca²⁺ channel of skeletal muscle is shown in Fig. 5A. This channel is blocked by dihydropyridines (DHPs), so it is also known as the skeletal muscle DHP receptor; in addition it is the T-tubule voltage sensor. (For a review of the important voltage-gated Ca²⁺ channels, see Tsien *et al.*, 1991.) The Ca²⁺ release channel of the sarcoplasmic reticulum (Fig. 5B) can be reviewed from the references given in MacLennan and Phillips (1992).

B. Dihydropyridine Receptor

1. Muscular Dysgenesis

Mice that are homozygous for the autosomal recessive dysgenic gene mutation (mdg/mdg) lack excitation-contraction (EC) coupling and the slow Ca²⁺ current. These animals are therefore incapable of muscular movement and die at birth, presumably from respiratory paralysis. The physiological problem is related to lack of function of the DHP receptor. The mutant dysgenic animals have a five-fold decrease in DHP binding in skeletal muscle and a low level of mRNA encoding the DHP receptor; monoclonal antibodies detect no DPH receptors. On the other hand, cardiac DHP binding is normal, illustrating the fact that the L-type Ca²⁺ channels (or DHP receptors) are distinctly different in these two tissues. Muscular dysgenesis appears to have been the first documented example of a genetic disease in vertebrates that is produced by a defect in a structural gene coding for an ion channel (Tanabe *et al.*, 1988).

Although a human counterpart has not been reported for this mutation, basic studies with the dysgenic mouse model have helped confirm the role of the DHP receptor as both a voltage sensor and channel in the T-tubular membrane. A critical experiment was performed in which both EC coupling and slow Ca²⁺ current were restored in the cultured dysgenic muscle cells by microinjection of an expression plasmid (pCAC6) carrying cDNA for the missing DHP receptor (Tanabe *et al.*, 1988). The restoration of both voltage sensing and ion channel functions with the single cDNA confirmed the dual role of this protein in the T-tubule. These mice have been used extensively to examine the coupling between the DHP receptor and the sarcoplasmic reticulum (SR) Ca^{2+} release channel and to show conclusively the difference between cardiac- and skeletal-like EC coupling (Beam and Franzini-Armstrong, 1997).

2. Hypokalemic Periodic Paralysis

Hypokalemic periodic paralysis (HoPP) is a disease of skeletal muscle in which episodes of weakness occur in association with lowered serum K^+ levels. It is an autosomal dominant trait, but in one-third of the cases it appears spontaneously.

A typical attack of weakness induced by low serum K^+ can begin a few hours after a high carbohydrate meal, often when asleep, and can last for 4–24 h. Other precipitating factors include rest after exercise and situations where K^+ moves into muscle cells. For years this disease was thought to be involved somehow with abnormal pump (Na^+,K^+-ATPase) or cation channel function and the syndrome can be reasonably modeled in rats that have their skeletal muscle K^+ contents altered through insulin and/or low K^+ diet. It came as a surprise then when linkages were discovered to the gene that encodes the skeletal muscle L-type Ca^{2+} channel. Two mutations were described and are shown in Fig. 5A. In each mutation there is a replacement of a highly conserved arginine by histidine in the S4 segment, a region believed to be involved in channel gating. One replacement occurs in the D4 domain at Arg 1239 (Ptacek *et al.*, 1994), and the other occurs in the D2 domain at Arg 528 (Jurkat-Rott *et al.*, 1994). Recent studies of these mutations expressed in heterologous systems have thus far been disappointing in that they have not revealed abnormal functions of this channel that can explain HoPP. (For further discussion see Lehmann-Horn and Rudel, 1996.)

C. Ca^{2+} Release Channel

1. Malignant Hyperthermia

Malignant hyperthermia (MH) is a rare autosomal dominant condition in humans that predisposes these individuals to react to anesthesia with muscle rigidity, hypermetabolism, and high fever; the muscles become highly stimulated metabolically, with a consequent rise in body temperature (MacLennan and Phillips, 1992). A popular form of anesthesia involving use of halothane in conjunction with a depolarizing neuromuscular blocker, succinylcholine, can trigger MH. If not treated immediately these patients may die within a few minutes from ventricular fibrillation or within hours to days from neurological or renal complications. Although MH is rare, it is serious enough to be fatal in apparently healthy individuals undergoing anesthesia. MH is also seen in animals where it is often triggered by heat stress. MH-susceptible pigs have been recognized for many years (Lucke *et al.*, 1979) and genetic and physiological studies of the MH pigs have hastened our understanding of the comparable condition in human patients.

The mechanism of MH is a mutation in the skeletal muscle SR Ca^{2+} release channel (Fig. 5B). The specific mutations that account for the dysfunction of this channel are different in human and pigs. The Ca^{2+} release channel is a key element of the EC coupling process. The AP invades the T-tubular membrane causing movement of gating charge in the DHP-sensitive Ca^{2+} channel/voltage sensor, which in turn activates the SR Ca^{2+} release channel to release Ca^{2+} into the myoplasma to initiate contraction. The exact means by which the voltage sensor of the T-tubular membrane couples with the release channel of the SR is not yet fully understood. In addition, Ca^{2+} pumps located in the SR reaccumulate the released Ca^{2+}. These events and a scheme to explain the pathophysiology of MH are depicted in Fig. 6A and B.

The Ca^{2+} release channel is opened specifically by nanomolar concentrations and blocked by micromolar concentrations of the plant alkaloid ryanodine. Thus, this channel is also referred to as the **ryanodine receptor.** Using this high-affinity ligand, the ryanodine receptor was purified and several full-length cDNAs from different species were subsequently cloned. Two different genes encode the channel: RYR1 on chromosome 19 codes for the skeletal muscle channel, and RYR2 on chromosome 1 codes for the heart and brain channel. In pigs the cDNA sequences for RYR1 between normal and MH pigs predict a single amino acid difference, a cysteine for an arginine in the N-terminus (Fig. 5B). In human studies two mutations of RYR1 are associated with MH, but some MH patients have no linkage with this gene; further, mutations of this gene may cause unrelated forms of muscle disease. Human MH may therefore involve other gene mutations and interactions (see Gillard *et al.*, 1992). These major isoforms of the ryanodine receptor are the only known cellular binding sites for the ligand. The protein is a large homotetrameric complex constructed from 565-kDa subunits. The channel region is probably located in the membrane-spanning segments, which are 20% of the subunit from the C-terminal. The remainder of each subunit is cytoplasmic and the four subunits together form a cytoplasmic structure that make up the feet proteins of the triadic junction (Block *et al.*, 1988). This cytoplasmic region also contains the sites for coupling to the DHP receptor.

All of the signs of MH as seen in Fig. 6A and B can be explained by a defect in the regulation of intracellular Ca^{2+}. Sustained Ca^{2+} levels in the fiber would cause contracture, increased glycolytic and anaerobic metabolism with depletion of ATP, glucose, and oxygen, and overproduction of CO_2, lactic acid, and heat. A specific antagonist for the MH reaction is dantrolene Na^+, which is capable of blocking release of intracellular Ca^{2+} from the SR.

2. Central Core Disease

Another autosomal dominant disease of the RYR1 gene is **central core disease** (CCD). It is characterized by hypotonia and proximal weakness first seen in infants, but appears to be nonprogressive. The severity of the disease is variable within families. To date, three mutations causing the disease have been identified among affected families. The substitution of histidine for a conserved arginine (R2434H), which is absent in the general population, was identified in a large Canadian family. Two other missense mutations, I404M and R163C, have also been identified. One might presume that these channel defects should cause EC coupling malfunction and explain the observed pathology; however, this is still an open question (George, 1995).

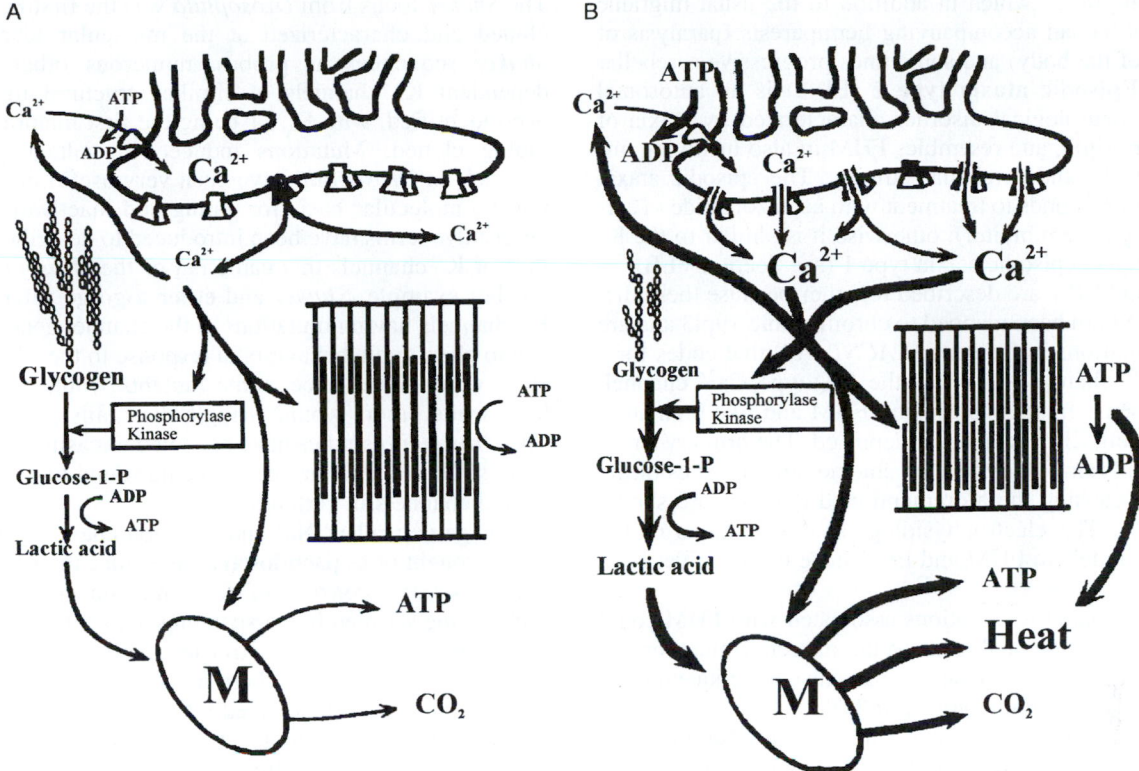

FIGURE 6. Malignant hyperthermia. (A) Normal physiological mechanism of EC coupling. The Ca^{2+} release channel (shown here as square elements) is a key element of the process known as excitation-contraction coupling. (B) MH muscle contraction cycle. Mutant Ca^{2+} release channels can explain the basic mechanism of MH. The mutations produce a release channel that is more sensitive to opening stimuli and fails to close rapidly, thus leading to abnormally high Ca^{2+} levels. Sustained Ca^{2+} levels in the fiber then would cause contracture, increased glycolytic and anaerobic metabolism with depletion of ATP, glucose, and oxygen, resulting in overproduction of CO_2, lactic acid, and heat. M indicates a mitochondrion. (Modified from MacLennan and Phillips, 1992; copyright © 1992 AAAS.)

D. Neuronal Ca^{2+} Channels

The neuronal Ca^{2+} channels are like the L-type skeletal muscle channels discussed earlier. They are multimeric entities consisting of one each of α_1-, α_2-, δ-, and β-subunits. The central α_1-subunit has the voltage sensor and the conducting pore. The neuronal channels discussed here differ from the skeletal muscle L-type channel in lacking the DHP receptor on the α_1-subunit. Two different classes of neuronal Ca^{2+} channel diseases are considered. The first, Lambert-Eaton syndrome, is due to autoimmune block of Ca^{2+} channels in motor nerve terminals leading to skeletal muscle weakness. The second class is a pair of neurological diseases related by the fact that they are caused by different types of mutations in the same gene that codes for a central nervous system Ca^{2+} channel. The types of mutations in the same channel determine the resultant form of the disease.

l. Lambert-Eaton Syndrome

In **Lambert-Eaton syndrome** autoimmune-generated antibodies reduce the number of voltage-activated Ca^{2+} channels in the presynaptic nerve terminal. This decreases the amount of Ca^{2+} entering the nerve terminal with each nerve impulse. Because Ca^{2+} entry is necessary for the release of transmitter release at the presynaptic terminal, the amount of acetylcholine delivered to the postsynaptic AChR channels will be reduced causing neuromuscular transmission failure and muscle weakness. There are many subtypes of neuronal Ca^{2+} channels; however, the type involved in Lambert-Eaton syndrome is specifically labeled by the marine snail toxin, ω-conotoxin (see Penn *et al.*, 1993). Lambert-Eaton syndrome has a presynaptic origin that is to be contrasted with myasthenia gravis (another form of neuromuscular weakness, discussed in Section VII.B), which has a postsynaptic origin. Patients with Lambert-Eaton syndrome often have an accompanying carcinoma and it has been asserted that the antibodies to the cancer cells also recognize the Ca^{2+} channel.

2. Hemiplegic Migraine and Episodic Ataxia Type 2

Migraine is a common neurological disorder characterized by recurrent severe headaches accompanied by phonophobia, photophobia, nausea, and vomiting. It occurs in about 24% of females and 12% of males, and varies greatly with regard to frequency, duration, and severity. **Familial hemiplegic migraine** (FHM3) is a rare autosomal dominant form of migraine, which in addition to the usual migraine symptoms has an accompanying hemiparesis (paralysis of one side of the body) and sometimes progressive cerebellar atrophy. **Episodic ataxia type 2** (EP-2) is an autosomal dominant neurological disorder characterized by ataxia of cerebellar origin, and resembles FHM in also involving migraine attacks and cerebellar atrophy. The episodic ataxia component responds to treatment with acetazolamide (a carbonic anhydrase inhibitor); otherwise it is similar to the K^+ channel defect episodic ataxia type 1 (see Section VI.B).

FHM and EP-2 are described together because these diseases have both been mapped to chromosome 19pl3 and are due to mutations in the gene (*CACNL1A4*) that codes for a brain Ca^{2+} channel known as the P/Q-type Ca^{2+} channel. Four missense mutations causing FHM and two premature stops causing EP-2 have been identified. The primary structure of the brain P/Q-type Ca^{2+} channel and the sites of the mutations causing the aforementioned diseases are shown in Fig. 5A. The electrophysiological dysfunctions of the mutant channels in FHM and EP-2 have not been yet been identified.

The four missense mutations associated with FHM could produce a situation analogous to the missense mutations of the skeletal muscle Na^+ channel that cause hyperkalemic periodic paralysis, paramyotonia, and Na^+ channel myotonia (see Section IV). In these latter conditions, the Na^+ channels have impaired inactivation, and by analogy in FHM the Ca^{2+} channels would exhibit increased activity. The cellular mechanism proposed is that increased Ca^{2+} activity would promote the process known as **spreading cortical depression** (Ophoff *et al.,* 1986), which would in turn facilitate initiation of a migraine attack.

On the other hand, the mutations occurring in EP-2 are classed as premature stops (i.e., frame shifts and splice-site mutations), and these would lead to truncated α-subunits consisting only of repeats I and II and part of III. Thus, the EP-2 channels would not be expected to function and probably would be present in lower density, the end result being decreased Ca^{2+} channel function in EP-2 with consequent dysfunction of the cerebellum, which would account for the ataxia.

VI. K^+ Channels

A. General Comments

K^+ channels serve many important functions in excitable tissues such as action potential repolarization and stabilization of the resting membrane potential. Therefore, it is not surprising that dysfunction of K^+ channels might be expected in neurological or cardiac disorders related to electrical activity. In fact, the very first mutation described in a mammalian K^+ channel causes a human neurological condition, episodic ataxia type 1, EA-1. Subsequently, the congenital long QT syndrome and inherited idiopathic epilepsy were linked to several different K^+ channels.

From a historical perspective, the defective K^+ channel in EA-1, $K_V1.1$, is a member of the *Shaker* K^+ channel family. The finding that $K_V1.1$ missense mutations underlie EA-1 presents analogies to the *Shaker* disorder in *Drosophila*.

The *Shaker* locus from *Drosophila* was the first K^+ channel cloned and characterized at the molecular level. Using *Shaker* sequences as probes, numerous other voltage-dependent K^+ channels of similar structural motif have been identified, with $K_V1.1$ being the first mammalian homolog cloned. Mutations induced in voltage-gated K^+ channels in *Drosophila* have been very useful in determining the molecular basis for gating and inactivation. Many descriptive terms have been introduced to describe the families of K^+ channels that came out of the *Drosophila* studies. For example, **Shaker** and **ether a-go-go** refer to those K^+ channels having mutations in the channel gene that give rise to phenotypic behaviors in response to the challenge of ether vapor. In essence, some mutants would shake their legs, while others would wriggle their tails provocatively. The *Drosophila* studies have become increasingly useful as more structural similarities are being found to *Drosophila*-related channels in vertebrates.

Voltage-dependent Na^+ and Ca^{2+} ion channels are sometimes thought of as **pseudotetramers** since their main functioning subunit consists of four interconnected domains, each having six membrane-spanning segments (Fig. 7A, B). K^+ channels, in contrast, are true tetramers with each subunit coded by a single gene. There are six membrane-spanning segments for the voltage-gated K^+ channels and only two in the case of non–voltage-gated inward-rectifying K^+ channels. The three main functional types are voltage gated, Ca^{2+} activated, and inward rectifying. Many have been cloned and they consist of homotetramers or heterotetramers within the same class. Thus, there is the possibility for mixing subunits that would allow for the evolution of designer channels to match particular different functional circumstances. There is a systematic classification of mammalian K^+ channels, and for greater detail one should consult Gutman and Chandy (1993).

B. Episodic Ataxia Type 1

Hereditary episodic ataxia type 1 (also called myokymia) is an autosomal dominant disease characterized by periodic attacks of motor imbalance, uncoordination, and involuntary tremor. These attacks are precipitated by emotional or physical stress. The condition is caused by point mutations in the $K_V1.1$ (or *KCNA1*) gene (located on chromosome 12p) of the *Shaker* K^+ subfamily, which codes for a delayed outward-rectifying K^+ channel present in neurons (Browne *et al.,* 1994). The predicted primary structure of the $K_V1.1$ channel and the known mutations causing EA-1 are shown in Fig. 7A. The $K_V1.1$ K^+ channel displays the C-type inactivation (conformational change at the mouth of the pore) and N-type inactivation is conferred by an associated β-subunit, $K_V\beta1$. A common defect shared by many of the mutant channels thus far examined is a reduction in current amplitude for homomeric EA-1 channels when expressed in *Xenopus* oocytes. Some of the mutations shifted the $V_{1/2}$ for activation in the depolarizing direction accelerated the rate of channel closure (Adelman *et al.,* 1995; Zerr *et al.,* 1998). Because heterozygous individuals show neurological signs and no homozygous individuals have yet been found, the many other types of K^+ channels that might be present are

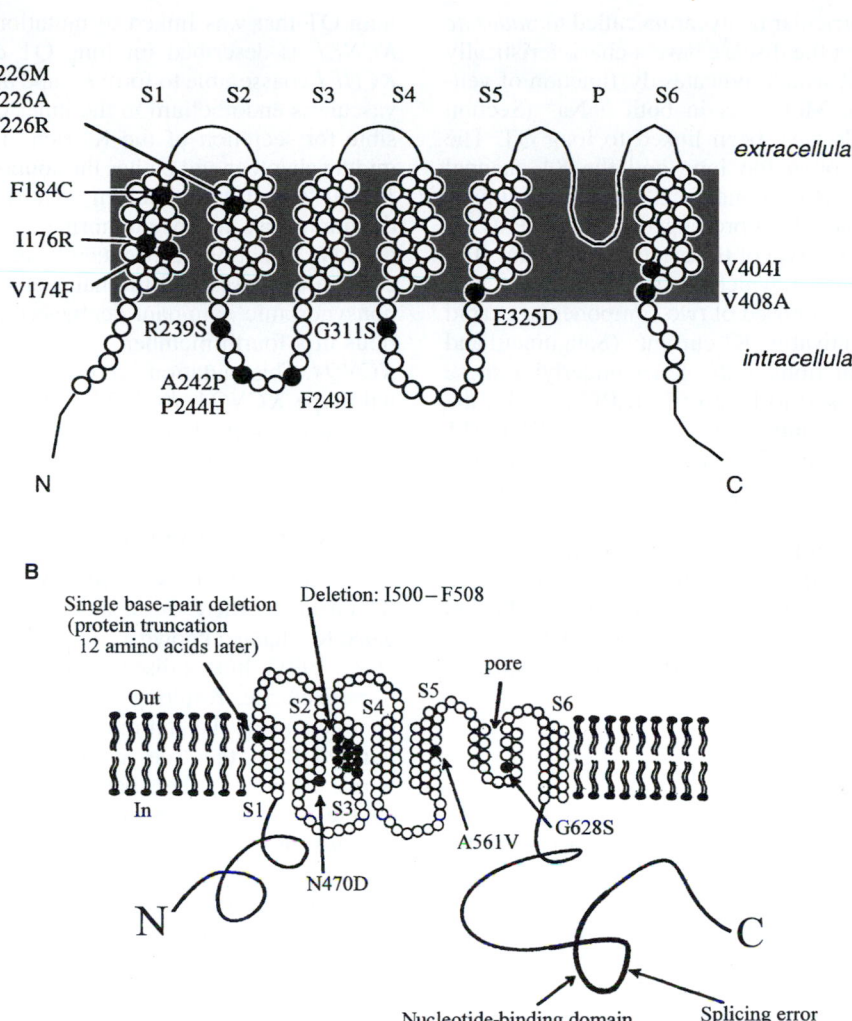

FIGURE 7. K+ channels. (A) Structure of the $K_V1.1$ K+ channel (*KCNA1* gene), which is implicated in the human neurological condition EA-1. Point mutations from 14 families having this condition are indicated. (B) Structure of the HERG (human ether a-go-go related gene) inward-rectifying K+ channel of the human heart, which is implicated in the K+ channel type of the long QT syndrome. Several types of mutations found in this condition are indicated. There are three point missense mutations, N470D, A561V, and G628S. There is also a deletion between I500 and F508, a protein truncation due to a single base-pair deletion in S1, and a splicing error in the carboxy-terminus. The errors in S1 and S3 are unlikely to form functional channels, whereas the point mutations might do so. (Redrawn and used with permission of Roden *et al.*, 1995.)

unable to take over the function of the $K_V1.1$ channel. This speaks against a current notion that a large number of different types of K+ channels have evolved to prevent just such a dependence. Coexpression studies show that the mutant subunits, in addition to forming homotetramers characteristic of the $K_V1.1$ channel, mix with those from the wild-type allele and assemble to form a variety of functionally defective heterotetramers *in vivo* (Zerr *et al.*, 1998).

The cellular mechanism suggested by these biophysical findings is that nerve cells having mutant channels would not effectively repolarize after each AP due to the impaired delayed rectifier function of the mutant channels. Though an animal model of EA-1 has not yet been developed, mice deficient of $K_V1.1$ may provide useful clues for the cellular role of $K_V1.1$. Homozygous $K_V1.1$ knockout mice, while not replicating the symptoms of EA-1, do have frequent generalized epileptic seizures (Smart *et al.*, 1998). Physiological studies show that deletion of $K_V1.1$ gives rise to tonic inhibitory tone on cerebellum Purkinje neurons and enhances excitability of the basket cells in the cerebellum by selectively increasing the likelihood of action potential propagation past axonal branch points (Zhang *et al.*, 1999). It is therefore likely that missense mutations of $K_V1.1$ linked to EA-1 will similarly affect inhibitory tone in the cerebellum, resulting in an imbalance between inhibitory and excitatory input and thereby destabilizing motor control under stress or exercise.

C. Long QT Interval Disease

Inherited **long QT syndrome** is a cardiac disorder characterized by fainting (syncope), seizures, and sudden death due to a particular ventricular tachycardia called *torsades de pointes*. Patients having the disease have a characteristically prolonged QT interval, which indicates dysfunction of ventricular repolarization. Mutations in both a Na$^+$ (Section IV.D) and K$^+$ channels have been linked to long QT. The Na$^+$ channel remains open too long and the K$^+$ channel opens less efficiently. Both conditions produce a similar physiological effect, namely, a prolonged cardiac AP as reflected in the long QT interval of the electrocardiogram. The delayed-rectifier K$^+$ current, which is a key regulator of cardiac repolarization, is composed of two components: a rapid (I_{Kr}) and a slow (I_{Ks}) activating K$^+$ current (Sanguinetti and Jurkiewicz, 1990). Mutations in the genes underlying these channels have been linked to long QT. HERG encodes the primary α-subunit for I_{Kr} and minK (*KCNE1*) and KVLQT1 (*KCNQ1*) encode subunits that coassemble to form I_{Ks}. Mutations in HERG and KVLQT1 are the most common cause of long QT. The primary structure of the HERG channel and some of the identified mutations causing long QT syndrome are shown in Fig. 7B. Though not all mutations in HERG or KVLQT1 have the same effect on K$^+$ channel function, the functional consequence of each is delayed cardiac repolarization with increased risk for *torsades de pointes* and sudden death.

D. Inherited Epilepsy of Newborns

Benign neonatal familial convulsions (BNFC) is a subtype of idiopathic generalized epilepsy that accounts for ~40% of epilepsy up to the age of 40. A search for the gene underlying this form of epilepsy converged with channel specialists seeking to identify the molecular mechanism underlying cholinergic modulation of excitability in the central nervous system. The acetylcholine sensitive M-current, which is a key regulator of neuronal excitability, has gained prominence as a potential contributor to normal cognition, dementia, and epilepsy. Thus when linkage studies identified mutations in genes encoding two new K$^+$ channel subunits, *KCNQ2* and *KCNQ3*, as responsible for BNFC (Biervert *et al.*, 1998; Charlier *et al.*, 1998; Singh *et al.*, 1998), speculation developed that these subunits might underlie the M-current. *KCNQ2* and *KCNQ3* belong to the *KCNQ* family that includes *KCNQ1*, one of the genes affected in long QT syndrome. *KCNQ2* and *KCNQ3* are expressed in overlapping patterns in the brain, and their coexpression in *Xenopus* oocytes yielded currents with gating and pharmacological properties similar to the neuronal M-current, which supported the idea that *KCNQ2* and *KCNQ3* do indeed underlie the M-current. To date the known BNFC mutations in *KCNQ2* and *KNCQ3* cause a reduction in current amplitude. BNFC thus results from a loss of function of the M-current in agreement with experimental data showing that muscarinic agonists, which suppress the M-current, cause acute status epilepticus and induce a chronic seizure.

E. Deafness

Two forms of deafness are related to mutations in different K$^+$ channels. **Jervell and Lange-Nielsen syndrome** is a recessive disorder characterized by bilateral deafness and

long QT that was linked to mutations in either *KCNQ1* or *KCNE1* as described for long QT disorders. *KCNQ1* and *KCNE1* coassemble to form I_{Ks} and are expressed in the stria vascularis endothelium in the inner ear. The stria is responsible for secretion of the K$^+$-rich endolymph that fills the middle chamber and bathes the sound-receptive hair cells. In Jervell and Lange-Nielsen syndrome, secretion of endolymph does not occur normally, presumably due to reduced I_{Ks} function. Consequently, the middle chamber of the cochlea collapses and the hair cells degenerate. A form of nonsyndromic dominant deafness has been linked to mutations in a fourth member of the *KCNQ* K$^+$ channel family, *KCNQ4*. This mutation exerts a dominant negative effect on wild-type *KCNQ4*. *KCNQ4* is expressed in the outer hair cells, which serve to mechanically amplify sound vibrations within the cochlea.

F. Acquired Neuromyotonia

In addition to the two human diseases due to mutations of K$^+$ channels, there is an autoimmune disease involving a gated K$^+$ channel known as **acquired neuromyotonia** (Sinha *et al.*, 1991). In this disease the peripheral motor nerves are hyperexcitable, resulting in repetitive firing and abnormal contractions of skeletal muscle. This must be clearly distinguished from the true myotonia discussed previously, in which the muscle fibers but not the nerve fibers are hyperexcitable. The two syndromes can be distinguished by administration of a neuromuscular blocker such as tubocurarine and, with direct muscle stimulation, the true myotonic fibers remain hyperexcitable, whereas in the neural type of myotonia the muscle fibers have normal excitability.

VII. Neurotransmitter-Gated Channels

A. General Comments

Several types of channels are gated open by neurotransmitters. The major ones can be classed by the neurotransmitter effective at their receptors. These include acetylcholine (ACh), glutamate, 5-HT, adenosine triphosphate (ATP), gamma-aminobutyric acid (GABA), and glycine. The selectivity of the channels can be for nonspecific cations, such as the ACh nicotinic channel, the K$^+$-selective ATP channels, or the Cl-selective glycine or GABA channels. This class of channels usually has five subunits, with one of the units occurring more than once. Only a few mutations among this class of channels have been described for human disease, although the channels have been targets of autoimmune disease. Our disease examples include a clinically relevant one involving the skeletal muscle nicotinic receptor channel, and a rare but instructive one involving the strychnine receptor glycine channel.

B. Nicotinic Acetylcholine Receptor: Myasthenia Gravis

In the human disease **myasthenia gravis,** the number of nicotinic acetylcholine receptors (AChRs) at the postsynaptic side of the neuromuscular junction is decreased, resulting in smaller postsynaptic currents and the tendency to block neuromuscular transmission. As a result patients experience weakness of skeletal muscles. The AChR is a membrane-bound glycoprotein that exists as a pentamer. Mammals have two main isoforms: the mature or innervated form having two α- and one each of β-, δ-, and ϵ-subunits, and the immature or denervated form in which a γ-subunit replaces the ϵ-subunit. The α-subunit has the ACh receptor and all five subunits are necessary for ligand-gated channel function during neuromuscular transmission. When ACh is released from the nerve terminal it diffuses to the AChR, and when two molecules occupy each α-receptor, the AChR channel can open, causing both Na^+ and K^+ currents to flow and resulting in depolarization of the postsynaptic membrane and excitation of the muscle fiber. Most of the myasthenia gravis syndromes are apparently due to a T-lymphocyte–dependent serum auto-antibody against AChR (see Penn *et al.*, 1993). Experi-mentally produced monoclonal antibodies have been shown to cause the channel to make kinetic transitions, leading to desensitization rather than activation. Myasthenia gravis may also occur very rarely due to heredity, reportedly by a mutation in the ϵ-subunit.

C. Glycine Receptor: Hyperekplexia

Glycine-activated Cl^- channels are found in many central neurons, where they function to produce inhibition. Activation of these channels causes an increase in g_{Cl}, which tends to clamp the membrane potential to the Cl^- equilibrium potential, which is often in the hyperpolarizing direction. The effects of the excitatory depolarizing channels are thus reduced causing inhibition. The mammalian glycine-activated channel is a heterooligomeric Cl^- channel that is inhibited by the plant toxin strychnine. The adult receptor is a pentamer consisting of three 48-kD α-subunits and two 58-kDa β-subunits. The α-subunit of the channel protein has the strychnine binding site and behaves as a channel when expressed in *Xenopus* oocytes. There are several subtypes of this channel and a large degree of sequence homology to the GABA, glutamate, and ACh ligand-gated ion channels.

In mammals small doses of strychnine cause highly exaggerated startle reactions to unexpected acoustic or mechanical stimulation. A response pattern resembling strychnine poisoning is seen in patients afflicted with the autosomal dominant disorder known as **hereditary hyperekplexia** (HHE) or familial startle disease (STHE) (Shiang *et al.*, 1993). Mutations in the α-subunit of the glycine-activated Cl^- channels have been shown to be the cause. The proposed structure of the α-subunit seen in Fig. 8 shows position 271 between domains M2 and M3, where a positively charged arginine is substituted by a neutral leucine or glutamine in the major mutations. These mutations lie at the presumed mouth of the channel, where altering the charge has a significant effect on channel function.

The fact that three α-subunits are required in the functional adult receptor helps explain the dominance of the inheritance. If the channel is rendered dysfunctional if only one mutant subunit is present, then even if there is only one mutant allele (the other producing normal subunits), the het-

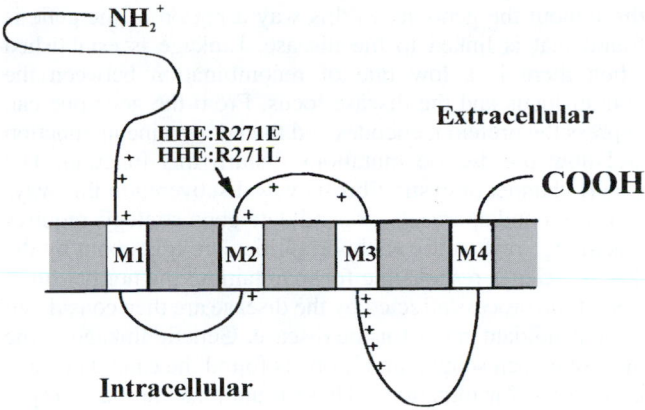

FIGURE 8. Glycine receptor Cl^- channel. The α-subunit of the glycine-activated, strychnine-blocked Cl^- channel (or GLRA1 channel) present in the mammalian nervous system. The functional pentameric channel is assembled from three of these α-subunits and two β-subunits (not shown). The arrow points to position 271 near the presumed mouth of the assembled channel where autosomal dominant mutations cause a charged lysine to be replaced by either an uncharged leucine or glutamine. The result is a blocked inhibitory channel leading to the nervous hyperexcitability called hereditary hyperekplexia (HHE).

erozygous individual would have only one-eighth or 12.5% of the normal functioning channels. This theory was discussed earlier for dominant MC.

VIII. Summary

Ion channels are often targets of disease, either directly through a mutation in the gene coding for some critical part of the channel protein structure, or indirectly through auto-immune disease or by defective coding of a critical ion channel regulator. We have seen examples of each case. When ion channel function is abnormal, we may get malfunction of excitable cells as occurs in the myotonias, periodic paralyses and episodic ataxias, or problems with ion transport as occurs in cystic fibrosis and malignant hyperthermia. No region of the channel protein appears to be immune to the effects of mutation. Effects are seen on gating, selectivity, and unit conductance. Often a change in a single amino acid accounts for the altered function of a channel or of its regulator. Ion channel diseases are often called channelopathies.

Modern electrophysiological methods, including the patch-clamp method, have made the monitoring of channel function very direct and convenient. Combining the new electrophysiology with the new molecular biology has been successful in offering us a staggering amount of information on normal and diseased channels in a remarkably short period of time.

Common methods are evolving for the study of familial diseases. Two basic strategies are used, positional cloning and the candidate gene. The first approach, positional cloning, is used to identify the genetic locus even when there is no knowledge of the biophysical or biochemical abnormalities underlying the disease by examining affected individuals for genetic linkage to polymorphic markers located throughout the genome. In this way a region of the gene is found that is linked to the disease. Linkage is established when there is a low rate of recombination between the marker locus and the disease locus. From the gene one can express the protein it encodes and then determine its function and how the disease mutations disturb this function. The CFTR channel of cystic fibrosis was discovered in this way.

The second approach, or candidate gene strategy, requires knowledge of functional abnormalities brought about by the disease. Genes responsible for maintaining the normal function of the process affected by the disease are then considered to be candidate genes for the disease. Genetic linkage to the disease is then sought, and if one is found the candidate gene is screened for mutations. These mutations can be incorporated into the wild-type DNA and expressed in heterologous systems. Then one can examine the biophysical and biochemical behavior and determine if the abnormality could account for the disease. The new rapid cloning techniques are largely responsible for the increased pool of known channel genes having known functions. This makes it easier to identify candidate channel genes; thus the candidate gene approach has been very successful for the study of the channelopathies. Most of the new channelopathies including the CLC5 mutations causing kidney stones, the episodic ataxias, and the long QT interval diseases were discovered using the candidate gene approach. With the increase in the number of channelopathies now described we can begin to make certain generalizations. We have seen that if more than one channel type is involved in control of cellular excitability, then either decreasing the P_{open} of the stabilizing channel or increasing the P_{open} of the depolarizing channel will lead to hyperexcitability of the excitable cell. Examples to date are reduced Cl^- channel function or decreased Na^+ channel inactivation in the myotonias, aberrant K^+ channel block expression or decreased Na^+ channel inactivation in the long QT syndromes, and reduced K^+ channel function or increased activation of Ca^{2+} channels in the two types of episodic ataxias.

Bibliography

Adelman, J. P., Bond, C. T., Pessia, M., and Maylie, J. (1995). Episodic ataxia results from voltage-dependent K^+ channels with altered functions. *Neuron* **15**, 1445–1454.

Adrian, R. H., and Bryant, S. H. (1974). On the repetitive discharge in myotonic muscle fibres. *J. Physiol.* **240**, 505–515.

Adrian, R. H., and Marshall, M. W. (1976). Action potentials reconstructed in normal and myotonic muscle fibres. *J. Physiol.* (*London*) **258**, 125–143.

Ashcroft, F. M. (1999). "Ion Channels and Disease." Academic Press, San Diego, CA.

Ballabio, A. (1993). The rise and fall of positional cloning? *Nature Genet.* **3**, 277–279.

Beam, K. G., and Franzini-Armstrong, C. (1991). Functional and structural approaches to the study of excitation-contraction coupling. *Methods Cell Biol.* **52**, 283–306.

Beck, C. L., Fahlke, C., and George, A. L., Jr. (1996). Molecular basis for decreased muscle chloride conductance in the myotonic goat. *Proc. Natl. Acad. Sci. USA* **93**, 11248-11252

Becker, P. E. (1973). Generalized non-dystrophic myotonia: the dominant (Thomsen) type and the recently identified recessive type. *New Devel. Electromyography Clin. Neurophysiol.* **1**, 407–412.

Behrens, M. I., Jalil, P., Serani, A., Vergara, F., and Alvarez, O. (1994). Possible role of apamin-sensitive K^+ channels in myotonic dystrophy. *Muscle Nerve* **17**, 1264–1270.

Biervert, C., Schroeder, B. C., Kubisch, C., Berkovic, S. F., Propping, P., Jentsch, T. J., and Steinlein, O. K. (1998). A potassium channel mutation in neonatal human epilepsy. *Science.* **279**, 403–406.

Block, B. A., Imagawa, T., Campbell, K. P. and Franzini-Armstrong, C. (1988). Structural evidence for direct interaction between the molecular components of the transverse tubule/sarcoplasmic reticulum junction in skeletal muscle. *J. Cell Biol.* **107**, 2587–2600.

Browne, D. L., Gancher, S. T., Nutt, J. G., Brunt, E. R. P., Smith, E. A., Kramer, P., and Litt, M. (1994). Episodic ataxia/myokymia syndrome is associated with point mutations in the human potassium channel gene, KCNA1. *Nature Genet.* **8**, 136–140.

Bryant, S. H. (1979). Myotonia in the goat. *Ann. NY Acad. Sci.* **317**, 314–325.

Bryant, S. H., and Morales-Aguilera, A. (1971). Chloride conductance of normal and myotonic goat fibres and the action of monocarboxylic aromatic acids. *J. Physiol.* (*London*) **219**, 367–383.

Bryant, S. H., Mambrini, M., and Entrikin, R. K. (1987). Chloride and potassium conductances are decreased in skeletal muscle fibers from the (mto) myotonic mouse. *Neuroscience* **465**, 17.

Cannon, S. C., and Corey, D. P. (1993). Loss of Na^+ channel inactivation by anemone toxin (ATX II) mimics the myotonic state in hyperkalaemic periodic paralysis. *J. Physiol.* (*London*) **466**, 501–520.

Cannon, S. C., and Strittmatter, S. M. (1993). Functional expression of sodium channel mutations identified in families with periodic paralysis. *Neuron* **10**, 317–326.

Cannon, S. C., Brown, R. H., Jr., and Corey, D. P. (1991). A sodium channel defect in hyperkalemic periodic paralysis: potassium-induced failure of inactivation. *Neuron* **6**, 619–626.

Cannon, S. C., Brown, R. H., Jr., and Corey, D. P. (1993). Theoretical reconstruction of myotonia and paralysis caused by incomplete inactivation of sodium channels. *Biophys. J.* **65**, 270–288.

Charlier, C., Singh, N. A., Ryan, S. G., Lewis, T. B., Reus, B. E., Leach, R. J., and Leppert, M. (1998). A pore mutation in a novel KQT-like potassium channel gene in an idiopathic epilepsy family *Nature Genet.* **18**, 53–55.

Dulhunty, A. F. (1979). Distribution of potassium and chloride permeability over the surface and T-tubule membranes of mammalian skeletal muscle. *J. Membr. Biol.* **45**, 293–310.

Fontaine, B., Khurana, T. S., Hoffman, E. P., Bruns, G. A. P., Haines, J. L., Trofatter, J. A., Hanson, M. P., Rich, J., McFarlane, H., Yasek, D. M., Romano, D., Gusella, J. F., and Brown, R. H. (1990). Hyperkalemic periodic paralysis and the adult muscle sodium channel alpha-subunit gene. *Science* **250**, 1000–1002.

Furman, R. E., and Barchi, R. L. (1978). The pathophysiology of myotonia produced by aromatic carboxylic acids. *Ann. Neurol.* **4**, 357–365.

George, A. L., Jr. (1995). Molecular genetics of ion channel diseases. *Kidney Int.* **48**, 1180–1190.

George, A. L., Kamisarof, J., Kallen, R. G., and Barchi, R. L. (1992). Primary structure of adult human skeletal muscle voltage-dependent Na^+ channel. *Ann. Neurol.* **31**, 131–137.

George, A. L., Crackower, M. A., Abdalla, J. A., Hudson, A. J., and Ebers, G. C. (1993). Molecular basis of Thomsen's disease (autosomal dominant myotonia congenita). *Nature Genet.* **3**, 305–310.

Gillard, E. F., Otsu, K., Fujii, J., Duff, C., DeLeon, S., Khanna, V. K., Britt, B. A., Warton, R. G., and MacLennan, D. H. (1992). Polymorphisms and deduced amino acid substitutions in the coding sequence of the ryanodine receptor (RYR1) gene in individuals with malignant hyperthermia. *Genomics* **13**, 1247–1254.

Gurnett, C. A., Kahl, S. D., Anderson, R. D., and Campbell, K. P. (1995). Absence of the skeletal muscle sarcolemma chloride channel CLC-1 in myotonic mice. *J. Biol. Chem.* **270**, 9035–9038.

Gutman, G. A., and Chandy, K. G. (1993). Nomenclature of mammalian voltage-dependent potassium channel genes. *Neurosciences* **5,** 101–106.

Hummler, E., and Horisberger, J. D. (1999). Genetic disorders of membrane transport. V. The epithelial sodium channel and its implication in human diseases. *Am. J. Physiol.* **276,** G567–G571.

Jansen, G. P., Groenen, P. J., Bachner, D., Jap, P. H., Coerwinkel, M., Oerlman, F., van den Broek, W., Gohlsch, B., Pette, D., Plomp, J. J., Molenaar, P. C., Nederhoff, M. G., van Echteld, C. J., Dekker, M., Berns, A., Hameister, H., and Wieringa, B. (1996). Abnormal myotonic dystrophy protein kinase levels produce only mild myopathy in mice. *Nature Genet.* **13,** 316–324.

Jentsch, T. J., Gunther, W., Pusch, M., and Schwappach, B. (1995). Properties of voltage-gated chloride channels of the CLC gene family. *J. Physiol.* (*London*) **482, Suppl. P,** 19S–25S.

Jurkat-Rott, K., Lehmann-Horn, F., Elbaz, A., Heine, R., Gregg, R. G., Hogan, K., Powers, P. A., Lapie, P., Vale-Santos, J. E., Weissenbach, J., and Fontaine, B. (1994). A calcium channel mutation causing hypokalemic periodic paralysis. *Human Mol. Genet.* **3,** 1415–1419.

Kallen, R. G., Sheng, Z.-H., Yang, J., Chen, L., Rogart, R. B., and Barchi, R. L. (1990). Primary structure and expression of a sodium channel characteristic of denervated and immature rat skeletal muscle. *Neuron* **4,** 233–242.

Keating, M. T., and Sanguinetti, M. C. (1996). Molecular insights into cardiovascular disease. *Science* **272,** 681–685.

Kellenberger, S., West, J. W., Catterall, W. A., and Scheuer, T. (1997). Molecular analysis of potential hinge residues in the inactivation gate of brain type IIA Na$^+$ channels. *J. Gen. Physiol.* **109,** 607–617.

Koch, M. C., Steinmeyer, K., Lorenz, C., Ricker, K., Wolf, F., Otto, M., Zoll, B., Lehmann-Horn, F., Grzeschik, K.-H., and Jentsch, T. J. (1992). The skeletal muscle chloride channel in dominant and recessive human myotonia. *Science* **257,** 797–800.

Kraner, S. D., Tanaka, J. C., and Barchi, R. L. (1985). Purification and functional reconstitution of the voltage-sensitive sodium channel from rabbit T-tubular membranes. *J. Biol. Chem.* **25,** 6341–6347.

Lehmann-Horn, F., and Rudel, R. (1996). Molecular pathophysiology of voltage-gated ion channels. *Rev. Physiol. Biochem. Pharmacol.* **128,** 195–268.

Lehmann-Horn, F., Rudel, R., Dengler, R., Lorkovic, H., Haass, A., and Ricker, K. (1981). Membrane defects in paramyotonia congenita with and without myotonia in a warm environment. *Muscle Nerve* **4,** 496–506.

Lehmann-Horn, F., Kuther, G., Ricker, K., Grafe, P., Ballanyi, K., and Rudel, R. (1987). Adynamia episodica hereditaria with myotonia: a non-inactivating sodium current and the effect of extracellular pH. *Muscle Nerve* **10,** 363–374.

Lerche, H., Heine, R., Pika, U., George, A. L., Jr., Mitrovic, N., Browatzki, M., Weiss, T., Rivet-Bastide, M., Franek, C., Lomonaco, M., Ricker, K., and Lehmann-Horn, F. (1993). Human sodium channel myotonia: slowed channel inactivation due to substitutions for a glycine within the III-IV linker. *J. Physiol.* (*London*) **470,** 13–22.

Lloyd, S. E., Pearce, S. H. S., Fisher, S. E., Steinmeyer, K., Schwappach, B., Scheinman, S. J., Harding, B., Bolino, A., Devoto, M., Goodyer, P., Rigden, S. P. A., Wrong, O., Jentsch, T. J., Craig, I. W., and Thakker, R. V. (1996). A common molecular basis for three inherited kidney stone diseases. *Nature* **379,** 445–449.

Lucke, J. N., Hall, G. M., and Lister, D. (1979). Malignant hyperthermia in the pig and the role of stress. *Ann. NY Acad. Sci.* **317,** 326–337.

Ludewig, U., Pusch, M., and Jentsch, T. J. (1996). Two physically distinct pores in the dimeric CLC-0 chloride channel. *Nature* **383,** 340–343.

MacLennan, D. H., and Phillips, M. S. (1992). Malignant hyperthermia. *Science* **256,** 789–794.

McClatchey, A. I., Van den Bergh, P., Pericak-Vance, M. A., Raskind, W., Verellen, C., McKenna-Yasek, D., Rao, K., Haines, J. L., Bird, T., Brown, R. H., and Gusella, J. F. (1992). Temperature-sensitive mutations in the III–IV cytoplasmic loop region of the skeletal muscle sodium channel gene in paramyotonia congenita. *Cell* **68,** 769–774.

Mehrke, G., Brinkmeier, H., and Jockusch, H. (1988). The myotonic mouse mutant ADR: electrophysiology of the muscle fiber. *Muscle Nerve* **11,** 440–446.

Middleton, R. E., Pheasant, D. J., and Miller, C. (1996). Homodimeric architecture of a CLC-type chloride ion channel. *Nature* **383,** 337–340.

Ophoff, R. A., Terwindt, G. M., Vergouwe, M. N., van Eijk, R., Oefner, P. J., Hoffman, S. M., Lamerdin, J. E., Mohrenweiser, H. W., Bulman, D. E., Ferrari, M., Haan, J., Lindhout, D., van Ommen, G. J., Hofker, M. H., Ferrari, M. D., and Frants, R. R. (1996). Familial hemiplegic migraine and episodic ataxia type-2 are caused by mutations in the Ca^{2+} channel gene CACNL1A4. *Cell* **87,** 543–552

Penn, A. S., Richman, D. P., Ruff, R. L., and Lennon, V. A. (Eds.). (1993). Myasthenia gravis and related disorders. *Ann. NY Acad. Sci.* **681,** 1–622

Ptacek, L. J., Gouw, L., Kwiecinski, H., McManis, P., Mendell, J. R., Barohn, R. J., George, A. L., Barchi, R. L., Robertson, M., and Leppert, M. F. (1993). Sodium channel mutations in paramyotonia congenita and hyperkalemic periodic paralysis. *Ann. Neurol.* **33,** 300–307.

Ptacek, L. J., Tawil, R., Griggs, R. C., Engel, A. G., Layzer, R. B., Kwiecinski, H., McManis, P. G., Santiago, L., Moore, M., Fouad, G., Bradley, P., and Leppert, M. F. (1994). Dihydropyridine receptor mutations cause hypokalemic periodic paralysis. *Cell* **77,** 863–868.

Pusch, M., and Jentsch, T. J. (1994). Molecular physiology of voltage-gated chloride channels. *Physiol. Rev.* **74,** 813–828.

Reddy, S., Smith, D. B., Rich, M. M., Leferovich, J. M., Reilly, P., Davis, B. M., Tran, K., Rayburn, H., Bronson, R., Cros, D., Balice-Gordon, R. J., and Housman, D. (1996). Mice lacking the myotonic dystrophy protein kinase develop a late onset progressive myopathy. *Nat. Genet.* **13,** 325–335.

Renaud, J. F., Desnuelle, C., Schmid-Antomarchi, H., Hugues, M., Serratrice, G., and Lazdunski, M. (1986). Expression of apamin receptor in muscles of patients with myotonic muscular dystrophy. *Nature.* **319,** 678–680.

Roden, D. M., George, A. L., Jr., and Bennett, P. B. (1995). Recent advances in understanding the molecular mechanisms of the long QT syndrome. *J. Cardiovasc. Electrophysiol.* **6,** 1023–1031.

Rogart, R. B., Cribbs, L. L., Muglia, L. K., Kephart, D. D., and Kaiser, M. W. (1989). Molecular cloning of a putative tetrodotoxin-resistant rat heart Na$^+$ channel isoform. *Proc. Natl. Acad. Sci. USA* **86,** 8170–8174.

Rudel, R., and Lehmann-Horn, F. (1985). Membrane changes in cells from myotonia patients. *Physiol. Rev.* **65,** 310–356.

Sanguinetti, M. C., and Jurkiewicz, N. K. (1990). Two components of cardiac delayed rectifier K$^+$ current. Differential sensitivity to block by class III antiarrhythmic agents. *J. Gen. Physiol.* **96,** 195–215.

Shiang, R., Ryan, S. G., Zhu, Y. Z., Hahn, A. F., O'Connell, P., and Wasmuth, J. J. (1993). Mutations in the α-subunit of the inhibitory glycine receptor cause the dominant neurological disorder, hyperekplexia. *Nature Genet.* **5,** 351–358.

Singh, N. A., Charlier, C., Stauffer, D., DuPont, B. R., Leach, R. J., Melis, R., Ronen, G. M., Bjerre, I., Quattlebaum, T., Murphy, J. V., McHarg, M. L., Gagnon, D., Rosales, T. O., Peiffer, A., Anderson, V. E., and Leppert, M. (1998). A novel potassium channel gene, KCNQ2, is mutated in an inherited epilepsy of newborns. *Nature Genet.* **18,** 25–29.

Sinha, S., Newson-Davies, J., Mills, K., Byrne, N., Lang, B., and Vincent, A. (1991). Autoimmune aetiology for acquired neuromyotonia (Isaac's Syndrome). *Lancet* **338,** 75–77.

Smart, S. L., Lopantsev, V., Zhang, C. L., Robbins, C. A., Wang, H., Chiu, S. Y., Schwartzkroin, P. A., Messing, A., and Tempel, B. L., (1998). Deletion of the K(V)1.1 potassium channel causes epilepsy in mice. *Neuron.* **20,** 809–19.

Smith, P. L., Baukrowitz, T., and Yellin, G. (1996). The inward rectification mechanism of the HERG cardiac channel. *Nature* **379,** 833–836.

Steinmeyer, K., Lorenz, C., Pusch, M., Koch, M. C., and Jentsch, T J. (1994). Multimeric structure of CLC-1 chloride channel revealed by mutations in dominant myotonia congenita (Thomsen). *EMBO J.* **13,** 737–743.

Steinmeyer, K., Klocke, R., Ortland, C., Gronemeier, M., Jockusch, H., Grunder, S., and Jentsch, T. (1991a). Inactivation of muscle chloride channel by transposon insertion in myotonic mice. *Nature* **354,** 304–308.

Steinmeyer, K., Ortland, C., and Jentsch, T. (1991b). Primary structure and functional expression of a developmentally regulated skeletal muscle chloride channel. *Nature* **354,** 301–304.

Tanabe, T., Beam, K. G., Powell, J. A., and Numa, S. (1988). Restoration of excitation-contraction coupling and slow calcium current in dysgenic muscle by dihydropyridine receptor complementary DNA. *Nature* **336,** 134–139.

Tsien, R. W., Ellinor, P. T., and Horne, W. A. (1991). Molecular diversity of voltage-dependent calcium channels. *Trends Pharmacol. Sci.* **12,** 349–354.

Vaughan, P. C., and French, A. S. (1989). Non–ligand activated chloride channels of skeletal muscle and epithelia. *Prog. Biophys. Mol. Biol.* **54,** 59–79.

Wallace, R. H., Wang, D. W., Singh, R., Scheffer, I. E., George, A. L., Jr., Phillips, H. A., Saar, K., Reis, A., Johnson, E. W., Sutherland, G. R., Berkovic, S. F., and Mulley, J. C. (1998). Febrile seizures and generalized epilepsy associated with a mutation in the Na^+-channel $\beta1$ subunit gene *SCN1B*. *Nature Genet.* **19,** 366–70.

Warner, J. O. (1992). Immunology of cystic fibrosis. *Br. Med. Bull.* **48,** 893–911.

Zerr, P., Adelman, J. P., and Maylie, J. (1998). Episodic ataxia mutations in Kv1.1 alter potassium channel function by dominant negative effects or haploinsufficiency. *J. Neurosci.* **18,** 2842–2848.

Zhang, C. L., Messing, A., and Chiu, S. Y. (1999). Specific alteration of spontaneous GABAergic inhibition in cerebellar Purkinje cells in mice lacking the potassium channel Kv1.1. *J. Neurosci.* **19,** 2852–2864.

SECTION V

Synaptic Transmission and Sensory Transduction

Gary L. Westbrook

40

Ligand-Gated Ion Channels

I. Introduction

Ion channels are fundamental signaling molecules in virtually all cells. Although channels are present on intracellular membranes of organelles, our current understanding of ion channel function originated with studies of ion channels in the plasmalemma of excitable cells. In the nervous system, voltage-gated ion channels mediate action potentials (APs) and trigger transmitter release, whereas ligand-gated channels are responsible for chemical signaling mediated by classical fast-acting neurotransmitters. Neurotransmitters also trigger slower synaptic responses that are mediated by G-protein-coupled receptors, as discussed in Chapter 41.

Until the 1980s, knowledge about the molecular properties of ion channel proteins was largely inferred from physiological and biophysical studies (reviewed in Hille, 1992). In retrospect, many of the predictions from these biophysical studies have been confirmed in dramatic fashion with the elucidation of the amino acid sequence of many voltage- and ligand-gated ion channels by molecular cloning. However, the increasing knowledge of the molecular structure has also revealed a number of unexpected findings and has begun to permit sophisticated correlations of protein structure with function (see, e.g., Miller, 1989; Montal, 1990; Unwin, 1993). These advances have led to an explosive increase in our understanding of the molecular operation of this important class of membrane proteins. These interesting molecules have captured the interest of many outstanding scientists whose studies of ion channels have led to several Nobel Prizes (see, e.g., Neher, 1992; Sakmann, 1992).

Studies of ion channels were pioneered by studies of the voltage-gated ion channels in squid axon. These studies established that a conductance change was associated with the AP, and that this conductance change could be attributed to selective increases in the membrane permeability to Na^+ and K^+ ions, with the energy for the process derived from the transmembrane ion gradients created by the Na^+,K^+-ATPase. Both the Na^+ and K^+ conductances increased with membrane depolarization. These findings suggested that there were discrete pores in the membrane that accounted for the Na^+ and K^+ conductance. However it was not until the 1970s that the existence of ion channels was confirmed, first by measurements of the statistical properties of currents through populations of ion channels, a technique called **fluctuation** or **noise analysis.** Later, this was refined by the introduction of **patch-clamp recording,** which allowed measurement of current through single ion channels (for review, see Neher, 1992; Sakmann, 1992). Patch-clamp recording revolutionized the study of ion channels, because it allowed detailed biophysical studies of the electrical activity of single molecules. Finally, protein purification and subsequent molecular cloning have confirmed that ion channels are a large and heterogeneous family of membrane proteins. Although the ligand-gated ion channels vary in their structural features, they share a common basic structure in that they are composed of multiple subunits arranged around a central water-filled pore. Each subunit is a polypeptide encoded by a separate gene. Generally, there are a large number of possible subunit combinations; thus the particular subunits expressed in a certain class of neurons can result in channels with distinct functional characteristics.

This chapter is divided into three topic areas. In the first, the categories of ligand-gated ion channels are briefly reviewed. The basic physiological properties and general molecular structure of the ligand-gated channels are then discussed using examples from the muscle nicotinic acetylcholine receptor (AChR) and the N-methyl-D-aspartate (NMDA)-type glutamate receptor. In the third section, the features of neuronal ligand-gated channels that are responsible for fast synaptic transmission are reviewed with an emphasis on the distinctive characteristics of each class of channel. Channels gated by acetylcholine (ACh), γ-aminobutyric acid (GABA), glycine, and glutamate are included as representative examples. Related topics on ion channels are covered extensively in other chapters in Sections III through V of this book. The reader is also referred to a large number of excellent monographs and reviews listed in the references for further details and original citations in this rapidly advancing field.

II. Classes of Ligand-Gated Ion Channels

Ligands that activate ion channels in nerve cell membranes can be divided into two major categories, neurotransmitters and intracellular ligands, as listed in Table 1. The best studied are neurotransmitters that are involved in fast chemical synaptic transmission. These include ACh, which is the transmitter at the vertebrate neuromuscular junction and in autonomic ganglia, and the amino acids, L-glutamate and GABA, which mediate the majority of fast excitatory and inhibitory synaptic transmission, respectively, in the vertebrate central nervous system (Betz, 1990). Each of these ligands also activates receptors that are coupled to second messengers via GTP-binding (G) proteins. G-protein-coupled receptors generally mediate slower and neuromodulatory transmembrane signaling (reviewed in Nicoll *et al.,* 1990). Other neurotransmitters that activate ligand-gated ion channels in various cell types include serotonin, glycine, histamine, and adenosine triphosphate (ATP). For example, the $5HT_3$ receptor activated by serotonin is a ligand-gated ion channel (reviewed in Julius, 1991). ATP is released from nerve terminals in several pathways and activates an ATP-gated channel in some spinal neurons that process incoming sensory information (reviewed in Bean, 1992; Burnstock, 1999).

In addition to neurotransmitter-gated ion channels, increasing attention has focused on ligand-gated ion channels that are activated by intracellular ligands. These ligands include Ca^{2+}, cyclic nucleotides, ATP, and inositol trisphosphate (IP_3). Intracellular ligands can increase (Ca^{2+}, cAMP, IP_3) or decrease (cGMP, ATP) activity of the associated channel. These channels play important roles in cell function. For example, Ca^{2+}-dependent potassium, chloride, and nonspecific cation channels modify the excitability of the nerve cell membrane and thus can alter the AP and release of transmitter from nerve terminals. The cyclic nucleotide (cAMP,

cGMP)-gated channels mediate sensory transduction in the visual and olfactory system. Light results in the closing of cGMP channels in vertebrate photoreceptors; odors trigger the opening of cAMP-gated channels in olfactory receptor neurons. See Chapters 47 and 49 for more discussion of channels gated by cyclic nucleotides.

III. Basic Physiological Features

Modern patch-clamp techniques allow investigators to define the three fundamental properties of any ion channel—**conductance, selective permeability,** and **gating**—in molecular terms. In addition to these fundamental properties, **modulation, channel block,** and **desensitization** are processes that affect the activity of ligand-gated ion channels, and are important in shaping the impact of channel activity on membrane excitability. The biophysical basis of these properties will be discussed first, and then the structural features of ligand-gated channels that control these properties will be considered. For a more detailed discussion, see Hille (1992).

Conductance reflects the flux of charged ions through the channel, and is measured in picoSiemens (pS). Most ligand-gated ion channels have a conductance of 5–50 pS, which corresponds to the movement of more than 1×10^6 ions per second through the channel pore. These high flux rates initially suggested that ligand-gated receptors contained a water-filled pore, because this rate is much higher than that predicted for other transport mechanisms, such as pumps or exchangers. The size of the single-channel conductance (γ) is characteristic for a given channel, although many ligand-gated channels also have subconductance levels (states). An example of a glutamate-activated channel in a hippocampal neuron with several subconductance states is shown in Fig. 1. Note that the conductance of the open channel is usually near 50 pS, but drops occasionally to 8, 35, 40, and 45 pS. The conductance of a channel increases as the permeant ion concentration is raised, but eventually saturates at concentrations well above physiological levels. This behavior can be described by Michaelis-Menten kinetics, suggesting that ions bind to sites within the pore rather than simply obeying the laws of free diffusion. Most permeant ions have a low K_m of approximately 100 mM (K_m is the concentration at which the conductance is half-maximal, as defined by the Michaelis-Menten equation); thus the ions bind for only a microsecond or so before continuing through the pore.

The current carried by the opening of a single-channel is measured in picoamperes (pA) with the single-channel current given by

$$i = \gamma (V - V_{eq}) \qquad (1)$$

where V is the membrane potential and V_{eq} is the equilibrium potential, at which no net current is measured. This equation is simply Ohm's law, where $(V - V_{eq})$ is the electrochemical driving force for ions that can pass through the channel. For example, if the extracellular and cytoplasmic concentrations of permeant ions are equal, then V_{eq} is 0 mV. Thus, as the membrane potential changes, the size of the sin-

TABLE 1   Ligand-Gated Ion Channels

Ligand	Receptor	Ion selectivity[a]
Neurotransmitters		
Acetylcholine	Muscle, neuronal AChRs	NS
Glutamate	AMPA, kainate, NMDA	NS
GABA	$GABA_A$, $GABA_C$	Cl
Glycine	GlyR	Cl
Serotonin	$5HT_3$ receptor	NS
ATP	ATP receptor	NS
Intracellular ligands		
Calcium	Calcium-dependent channels	K, Cl, NS
Cyclic nucleotides	cGMP and cAMP receptors	NS
ATP	ATP-dependent channel	K
IP_3	Calcium release channel	Ca

[a]Abbreviations for ion selectivity are NS (nonselective cation permeability, some with calcium permeability); Cl (permeable to chloride); K (permeable to potassium); and Ca (permeable to calcium).

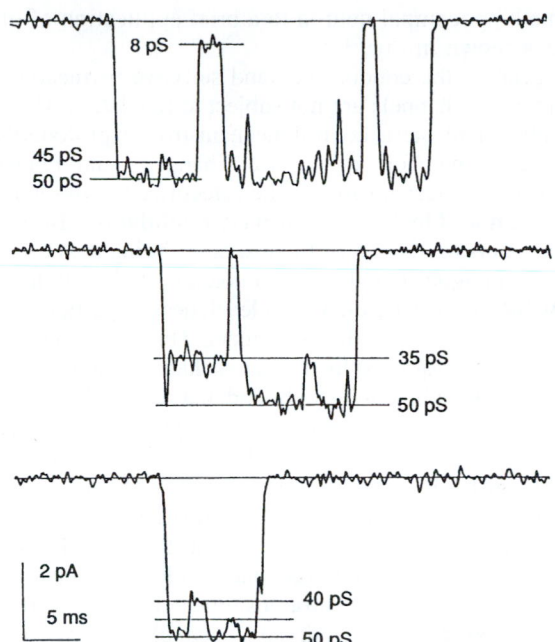

FIGURE 1. Ligand-gated channels can open to multiple conductance levels. Examples of NMDA-type glutamate channel recording in an outside-out membrane patch obtained from a cultured hippocampal neuron. Downward deflections indicate opening of the channel in the presence of ligand (in this case, 20 µM NMDA). The channel usually opens to the 50 pS level, but may switch to lower levels of 8, 35, 40 or 45 pS before closing to the baseline level. [Modified with permission from Jahr and Stevens (1987). *Nature* 325, 522–525. Copyright 1987 by Macmillan Magazines Limited.]

gle-channel current also changes; a plot of this relationship is called a **current-voltage** or ***I-V* plot.** For channels such as the muscle AChR, that are permeable to Na^+ and K^+, the reversal potential is 0 mV and the *I-V* curve is nearly linear. However some channels deviate from linear behavior and show inward or outward rectification (that is, the channel passes current in one direction better than in the other), analogous to the behavior of some voltage-gated K^+ channels.

Channels are not equally permeable to all ions, that is, they exhibit **selective permeability.** As listed in Table 1, most ligand-gated ion channels are permeable to either monovalent cations or anions. The selective permeability of channels is partly due to the physical size of the pore. For example, the muscle AChR is permeable to cations with diameters up to about 6.5 Å. In general, cationic ligand-gated ion channels, are less selective than voltage-gated ion channels, which are highly selective for Na^+, K^+, or Ca^{2+}, but more selective than gap junction channels, which are permeable to small molecular weight molecules such as cyclic nucleotides as well as to ions. Because anions and cations are approximately the same size, it is obvious that the pore dimensions cannot totally account for selective permeability. As discussed later, positively and negatively charged amino acid residues in the entrance to the channel and within the pore provide a means for this discrimination. Selective permeability is extremely impor-

tant in considering the function of an ion channel. For example, channels that are permeable to Na^+ and K^+ do not significantly alter the ion concentrations on either side of the membrane, but instead provide an electrical signal that depolarizes the neuron and brings the membrane potential closer to the threshold for AP generation. However, channels with significant Ca^{2+} permeability can transiently increase the cytoplasmic Ca^{2+} concentration, and thus act as a biochemical signal. Thus the relative Ca^{2+} permeability of ligand-gated channels such as the NMDA receptor and some nicotinic receptors is an important aspect of their role in neuronal function. The reader is referred to the monograph by Hille (1992) for a full discussion of channel permeability and its measurement.

Channel **gating** refers to the conformational change in the ion channel that is triggered by ligand binding. The conformational change results in a rapid switch between conducting (open) and nonconducting (closed) states of the channel. The gating behavior of channels has many parallels with the allosteric behavior of enzymes, with the binding of ligand providing the free energy necessary to maintain the channel in the open conformation. Gating between the open and closed configuration is extremely rapid (approximately 10 µs), and thus is beyond the resolution of standard recording methods. Each channel opening usually lasts a few milliseconds, but can, on some occasions, last up to several hundred milliseconds. Binding of more than one agonist molecule is usually necessary to open a channel. The binding of multiple agonist molecules creates a sigmoidal dose-response relationship, characteristic of the cooperative binding of substrates to enzymes. The steep activation created by sigmoidal activation kinetics prevents the channel from opening in the presence of a low concentration of agonist. This may be quite important in preventing desensitization of receptors at synapses. In general, the higher the binding affinity of the agonist, the longer the channel will remain open. This is because the channel can remain open (or reopen) until the agonist dissociates.

The gating of ligand-gated ion channels can be described by multistate kinetic diagrams as shown in Fig. 2A. Such state diagrams have been developed for many ligand-gated ion channels, based on the analysis of open and closed time distributions obtained in single channel recording. For example, the presence of two exponentials in an open time histogram implies the existence of two open states. The minimal kinetic model for a ligand-gated channel usually has at least four states, including closed and unbound, closed and bound, open and bound, and nonconducting (desensitized). For the case shown in Fig. 2A and B, note that the channel opens only when two agonist molecules are bound. For fast-acting neurotransmitters, the concentration of transmitter in the synaptic cleft is typically very high for a brief period. At high concentrations of agonist, the receptors quickly reach the fully liganded but closed state, this transition is determined by the product of the binding rate, r_b and the concentration of agonist. Thus the rising and falling phase of the synaptic response largely reflect the unbinding rate r_u, the channel opening and closing rates (β and α) and, in some cases, the rates in and out of the desensitized state, R_d. Classical pharmacological terms can be interpreted in terms of these kinetic schemes (see, e.g., Colquhoun, 1998). Agonists open the channel, while antagonists bind, but apparently do not cause a sufficient conformational change in the channel protein to open the channel. The fraction of time the bound channel spends in the open state is called the **open probability,** and may differ between agonists—an agonist that activates the channel with higher probability therefore has a higher efficacy in the nomenclature of classical receptor pharmacology.

Earlier studies of the activity of ligand-gated channels were generally limited to equilibrium measurements due to the limits in the speed of application of agonists to cells or cell-free patches. Such equilibrium measurements are difficult to interpret in terms of state diagrams such as that shown in Fig. 2. In addition, the duration of fast-acting transmitters such as glutamate is generally brief, suggesting that the synapse itself operates under nonequilibrium conditions. However, rapid solution exchange techniques now allow agonists to be applied within several hundred microseconds to membrane patches, usually in the "outside-out" configuration to allow agonist in the bath to access the extracellular face of the receptor. Many investigators have used these methods to analyze the gating of ligand-gated channels. Such approaches also appear to more closely resemble conditions at synapses and are particularly useful in evaluating the role of channel modulation, channel block, and desensitization on synaptic responses. An example of the response of a membrane patch

from a hippocampal neuron to a brief application of glutamate is shown in Fig. 3.

In general, the conductance and selective permeability of ligand-gated channels are not subject to regulation. However a number of allosteric control mechanisms can profoundly affect the gating (for review, see Changeux and Edelstein, 1998). These mechanisms can be categorized as **desensitization, channel block,** and **allosteric modulation. Desensitization** refers to the loss of the response during the continued presence of agonist, and was first noted in studies of the muscle AChR. At the single-channel level, desensitization reflects a nonconducting state of the channel. This phenomenon has now been seen for most ligand-gated channels, and can be described by including an extra bound, but nonconducting, state in the kinetic scheme (see Fig. 2A). Depending on the channel, the nonconducting state may be accessible from either the closed state (A_2R) or the open state (A_2R^*). Desensitization is thought to represent a distinct conformation of the receptor, but this has yet to be demonstrated at the molecular level. Very rapid desensitization may play a role in terminating the neuronal response to neurotransmitters at some synapses, whereas slower desensitization may actually prolong synaptic responses as channels reenter open states after transiting through desensitized states (see, e.g., Jones and Westbrook, 1996).

Ligand-gated channels can also be plugged by ions or drugs, a phenomenon referred to as **channel block.** If the blocking particle is charged, such as a large ion, the block will be influenced by the voltage across the membrane. Thus the reduction in channel activity will be more pronounced at some membrane potentials. The voltage-dependent block of NMDA-type glutamate channels by extracellular Mg^{2+} is an important example of this phenomenon (see Ascher and Nowak, 1987). At the single channel level, channel block is usually seen as rapid interruptions ("flickers") of the open state as illustrated in Fig. 4A and B. A characteristic feature of most forms of channel block is that the blocker is only effective when the channel is open (see, e.g., Pascual and Karlin, 1998). Thus the binding site for the blocker is thought to be within the pore of the channel, or at least at a site that is only available in the open conformation of the channel. Channel block is the mechanism of action for a number of drugs, including psychoactive compounds such as phencyclidine (PCP).

Channel gating is also subject to **modulation** by either covalent modifications such as phosphorylation, or via the noncovalent binding of modulators to the channel (e.g., Swope *et al.,* 1999; Changeux *et al.,* 1990). Most ligand-gated channels undergo phosphorylation of intracellular portions of the channel protein; however, the resulting effect on channel function is dependent on the specific channel and the type of kinase involved. Phosphate groups are highly charged, and thus can modify interactions in local regions. For example, phosphorylation of the AChR enhances desensitization, and may also be important in clustering of AChR molecules at sites of innervation. The list of noncovalent modulators is long and includes Ca^{2+}, toxins, drugs, and some endogenous substances such as protons and steroid hormones. These modulators can usually act in a noncompetitive manner, that is, they do not directly interfere with

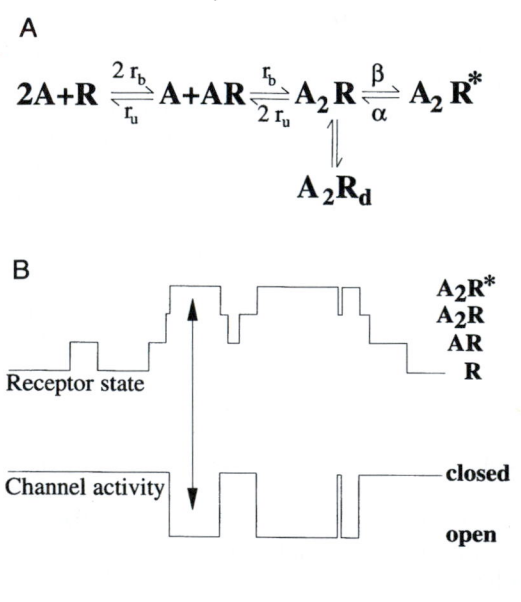

FIGURE 2. Kinetics of ligand-gated channels. (A) Example of a four-state kinetic scheme used to describe the behavior of ligand-gated ion channels. In this example, two agonist molecules (A) must bind before the channel can open (A_2R^*) or enter a nonconducting desensitized state (A_2R_d). The rate constants for agonist binding and unbinding are designated r_b and r_u, and the rate constants for channel opening and closing are designated as β and α. (B) Relationship of the state of the receptor to the open and closing of the channel. The channel opens only when the receptor is in the A_2R^* state as indicated by the arrows.

Hippocampal Neuron

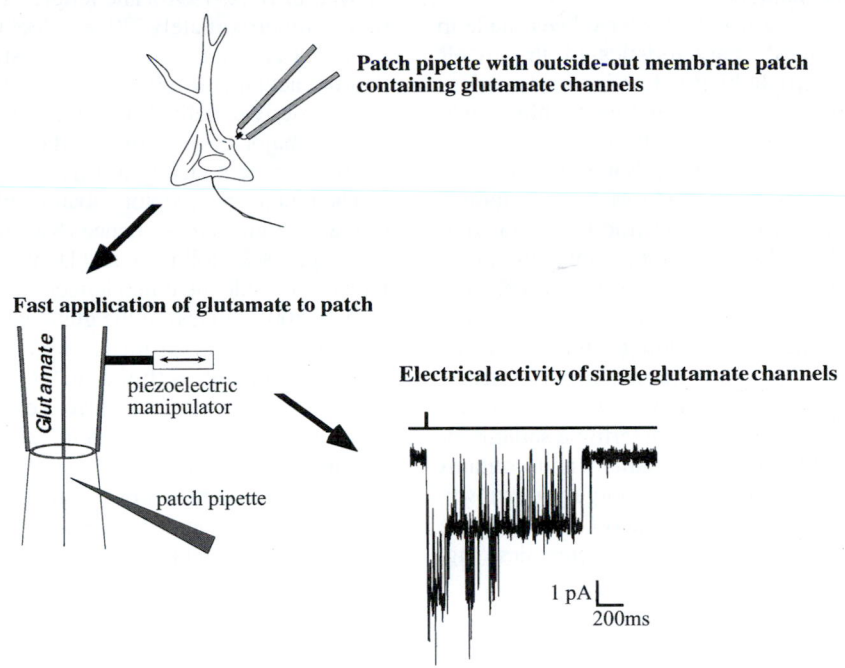

FIGURE 3. Use of rapid application methods to study kinetics of ligand-gated channels. Use of the patch-clamp technique allows small membrane patches to be removed from a cell such as a hippocampal neuron with the extracellular side of the membrane facing the bath (outside-out configuration). The tip of the pipette is then placed in the interface of two rapidly moving streams of solution. A piezoelectric manipulator moves the solutions a few microns to change the solution flow over the patch. Such methods allow solution exchange in less than 1 ms. Thus the response to short applications of ligand, similar as is thought to occur at synapses, can be recorded as openings of single channels. Note, in the response at the right, the opening of single glutamate channels of the NMDA type greatly outlasts the brief application of agonist. The duration of the glutamate application is indicated by the bar above the channel activity.

ligand binding. Most of these reagents affect gating by altering the rate of channel opening, or the time spent in the open state. For example, glycine binds to the NMDA channel, and results in an increased probability of opening (Fig. 4C and D). This is a dramatic example of allosteric regulation because the channel will not open unless both glycine and the transmitter (glutamate) are bound (for review, see McBain and Mayer, 1994). A less profound, but equally important, example is the upregulation of $GABA_A$ receptor activity by the benzodiazepines (Olsen *et al.*, 1991).

IV. Molecular Structure

The AChR from skeletal muscle is the prototypic ligand-gated ion channel, and its physiological properties have been extensively studied during the past 35 years (Changeux *et al.*, 1990; Karlin, 1991). The AChR also led the way in determining the molecular structure of ligand-gated channels. This analysis took advantage of the high density of AChRs in the electric organ of the ray (*Torpedo californica*), and the high-affinity snake toxin, α-bungarotoxin, that binds to muscle-type AChRs. Protein pu-

rification studies revealed an approximately 250-kDa complex with two α-bungarotoxin binding sites per complex, and separation on denaturing gels revealed polypeptides of 40, 50, 60, and 65 kDa designated as the α, β, γ, and δ subunits. Because of the molecular weight of the complex, and the binding of toxin molecules to the α subunit, a pentameric structure with two α subunits and single copies of β, γ, and δ was proposed. Using the partial amino acid sequence of the purified subunits, the cDNAs encoding AChRs in the *Torpedo* electric organ were cloned and sequenced in the early 1980s. Incorporation of the purified protein complex in lipid bilayers or expression of AChR mRNA in *Xenopus* oocytes demonstrated channels activated by acetylcholine, confirming the identity of the molecule. Consistent with the proposed combination of subunits in intact receptors, the expression of AChRs in oocytes was much greater when all four subunit mRNAs were included. The α subunit was essential in order to obtain responses to acetylcholine. The $\alpha_2\beta\gamma\delta$ pentamer occurs in embryonic muscle, but later in development the γ subunit is replaced by an ϵ subunit. This substitution results in an increase in the single-channel conductance and a decrease in the time the channel is open. A much greater heterogeneity of receptor subunits appears to be involved in the expression of ligand-gated channels at central synapses (Sections V–VII of this chapter).

During the past 20 years, great strides have been made in understanding the structural characteristics of the AChR channel. Based on the hydrophobicity of amino acid residues in the primary sequence, a transmembrane topology was proposed that included four putative transmembrane domains. A large N-terminal domain and a short C-terminal domain extend from the extracellular side of the membrane. This configuration predicts two cytoplasmic loops—a short one between M1 and M2 and a longer loop containing phosphorylation sites between M3 and M4. For the AChR, several residues in the N-terminal domain have been shown to be involved in forming the ACh binding site. The actual binding pocket appears to lie at the interface between the N-terminal domain of an α subunit and that of an adjacent subunit. The N-terminal domain may also determine subunit interactions during assembly of the pentamer, and contains carbohydrate residues whose function is largely unknown. The structures of the neuronal nicotinic and GABA/glycine receptor subunits have the same general characteristics (Fig.

5A), and have therefore been grouped together as a gene superfamily, perhaps suggesting a common ancestral gene (reviewed in Betz, 1990).The length of the transmembrane domains (approximately 20 residues) is consistent with an α-helical configuration, although structural studies suggest that the actual configuration of the transmembrane domains is not strictly α-helical. The M2 domains of the five subunits make a major contribution to the lining of the channel pore as shown schematically in Fig. 5B.

The glutamate receptor subunits differ substantially in their primary amino acid sequence (Nakanishi, 1992, Sommer and Seeburg, 1992; Hollmann and Heinemann, 1994). Notable differences include the much longer N-terminal domain (approximately 500 residues) of the glutamate subunits (Fig. 5A). The glutamate subunit transmembrane topology was initially modeled by homology with the AChR receptor family with both the N- and C-terminal domains as extracellular. However, subsequent analysis has revealed marked differences between the transmembrane topology of the AChR family and the glutamate receptor family. For example, the second hydrophobic domain of the glutamate subunits (M2 in the AChR subunits) appears to be a reentrant loop that enters and exits the mem-

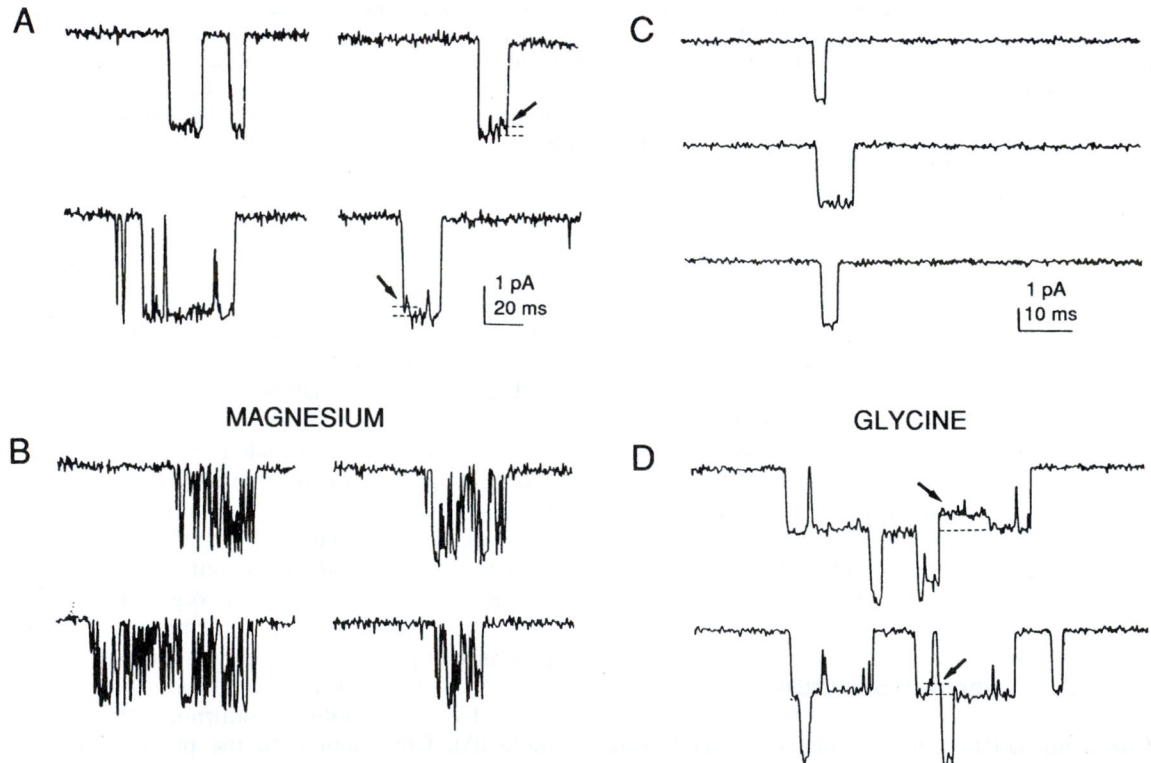

FIGURE 4. NMDA-type glutamate channels illustrate two common mechanisms of channel regulation: channel block and allosteric modulation. (A and B) Channels activated by NMDA in a membrane patch from a hippocampal neuron in the absence of extracellular Mg^{2+} remain open for several milliseconds before closing (A). However in the presence of Mg^{2+} (B), the openings are interrupted by brief closures due to plugging of the pore by Mg^{2+} ions. Downward deflections are due to channel opening. (C and D) The frequency of NMDA channel opening is increased in the presence of extracellular glycine (D) compared to glycine-free solutions (C). The stepwise increase in current in part D indicates that at least two channels were present in the membrane patch. Because glycine is now known to be required for NMDA channel opening, the occasional openings in part C probably reflect contamination of the solution with low levels of glycine. [Modified with permission from Ascher, R. and Nowak, L. (1987). *Trends Neurosci.* 10, 284–288.]

A

AChR family

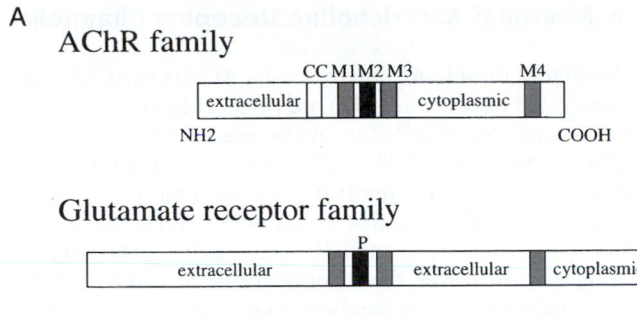

Glutamate receptor family

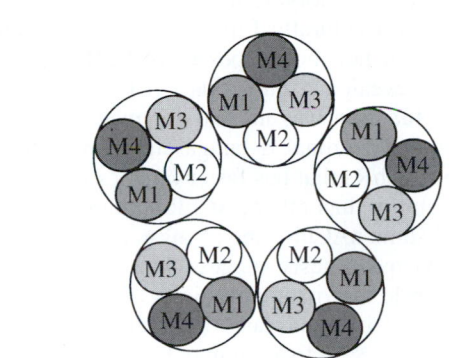

B

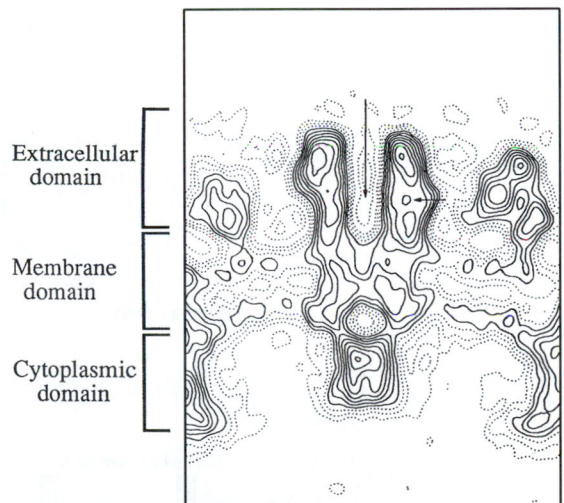

FIGURE 5. Schematic topology of ligand-gated channels. (A) The proposed membrane arrangement of polypeptide subunits for the AChR and glutamate receptors, diagrammed from the amino-terminal (NH2) to carboxyl-terminal (COOH) ends. The hydrophobic segments M1–M4 span the membrane with a long cytoplasmic loop between M3 and M4. The region marked CC in the AChR subunits contains cysteines that form disulfide bridges and is involved in ligand binding. Note the much larger extracellular domain of the glutamate receptors. (B) Cross-section through the membrane showing the proposed distribution of 5 subunits around a central pore of the AChR. The M2 domain lines the pore.

brane from the cytoplasmic side without fully traversing the membrane. This "P" loop shows striking homology to the P loop of voltage-gated channels that line the channel pore. A recently discovered prokaryotic glutamate receptor may provide the missing link between the voltage-gated potassium channels and glutamate receptors (Chen *et al.*, 1999). Secondly, the region between the third and fourth hydrophobic domains is extracellular in glutamate subunits whereas it is cytoplasmic in the AChR subunits. Finally the C-terminus of the glutamate subunits is cytoplasmic. These unique features provide important clues to the location of the transmitter binding site(s), the behavior of the channel pore and sites of interaction with other proteins that anchor and/or localize glutamate receptor subunits at synapses (see Section VII of this chapter).

The analysis of the primary amino acid sequence gives limited information about the three dimensional structure of the channel. Recently, crystals of AChRs from the *Torpedo* electric organ have been analyzed with the electron microscope (Unwin, 1993; Miyazawa *et al.*, 1999). The resolution of these studies provides a good representation of the overall shape and features of the channel (Fig. 6). Several features are striking,

including the large component of the receptor that protrudes out of the membrane into the extracellular space. In addition, the recent higher-resolution structure (4.6 Å, Miyazawa *et al.*, 1999) shows that the entrance to the channel (vestibule) is rather long and connected to the putative ligand binding sites by tunnels lined by twisted β-sheet strands of the contributing subunits. The vestibule is 25 Å in diameter, and ion flux may be influenced by electrostatic interactions with residues in the wall of the pore. The gate of the channel is formed by a narrow bridge across the transmembrane segment with narrow openings between the α-helical structure of the channel lining segments. It is also interesting that a cytoplasmic protein remains attached to the base of the channel protein. The attached protein is likely to be the 43-kDa protein that copurifies with AChRs, and is involved in clustering of receptors. Similar proteins linking ligand-gated ion channels to the cytoskeleton may be involved in localizing, anchoring, or influencing the gating of other channels, including the glycine receptor and the NMDA receptor (see, e.g., Sheng, 1996, 1997).

The evidence for M2 as the pore lining comes from several approaches. For example, a 23-amino-acid peptide fragment from the M2 region of the *Torpedo* δ AChR subunit forms a cation channel when expressed in lipid bilayers. More direct evidence comes from site-directed mutagenesis of conserved residues in this putative membrane-spanning segment. It has been suggested that the M2 region forms an α-helix with 3.6 residues per turn. However, studies using a technique known as scanning cysteine mutagenesis (see Karlin and Akabas, 1995, 1998) suggest that some portions of M2 are in a β-sheet configuration with the gate located near the intracellular extent of M2. This method has great

FIGURE 6. Cross-section of AChR obtained from crystallized postsynaptic membrane of *Torpedo* electric organ at 9-Å resolution. Parts of several receptors are shown, with a complete profile of one receptor along the axis of the pore. The dark lines indicate the channel protein, with arrows identifying the pore and a presumptive area ca. 30 Å above the membrane that may be the site of ACh binding. The pore narrows in the membrane-spanning region. The square contour lines at the cytoplasmic face of the receptor may be the 43-kDa protein that is involved in receptor clustering. [Modified with permission from Unwin (1993). Copyright 1993 Cell Press.]

application in structure-function studies of ion channels because substitution of other amino acid residues by cysteines usually does not affect channel properties. The cysteine residue can then form disulfide bonds with applied reagents, allowing investigators to systematically map residues that are exposed in particular conformations of the molecule, such as during channel opening.

Rings of negatively charged residues (glutamate or aspartate) at the ends of M2 for the cationic AChR and glutamate receptor subunits are thought to attract permeant cations, because mutation of these residues to neutral or positively charged amino acids results in a decrease in the single-channel conductance. Consistent with this hypothesis, glycine and GABA subunits have positively charged residues (lysine or arginine) near the ends of the M2 region that may attract permeant anions (Fig. 7B). The M2 region was assumed to be membrane-spanning because the amino acid residues are hydrophobic, that is, they have no charged sidechains (except of course for the charged residues at the ends of M2). Several small uncharged residues near the middle of the bilayer appear to line the channel lumen, as they are involved in the binding of some open channel blockers to the AChR. Mutations of the conserved leucine residue (position 12 in Fig. 7B) also cause a profound change in the response to ACh in the neuronal AChR subunit α_7, and is involved in the binding of the noncompetitive antagonist chlorpromazine (Section V of this chapter).

V. Neuronal Acetylcholine Receptor Channels

Similar to muscle AChRs, neuronal AChRs (nAChRs) are ligand-gated ion channels that are permeable to Na^+ and K^+, and in some cases Ca^{2+} (for review, see Luetje et al., 1990; McGehee and Role, 1995). ACh has long been known to activate channels mediating excitatory synaptic transmission in autonomic ganglia, and at the recurrent collateral synapse of spinal motoneurons onto Renshaw cells (inhibitory interneurons). However, the functional role of nAChRs in the central nervous system has been more difficult to establish despite the well-known effect of nicotine (via cigarette smoke) on human behavior (Dani and Heinemann, 1996). nAChRs differ from prototypic muscle AChRs both pharmacologically and structurally. First, α-bungarotoxin blocks muscle AChRs, but not most types of nAChRs. However, binding studies revealed α-bungarotoxin binding sites on neurons whose function was unclear for many years. It is now clear that the neuronal α-bungarotoxin receptor is a subtype of the nAChRs that has faster kinetics and a higher calcium permeability than other nAChR subtypes. The Ca^{2+} permeability of some nAChRs and changes in channel properties during synaptogenesis suggest that various nAChR subtypes play a role in synaptic plasticity and development. The activity of nAChRs may also be modulated by phosphorylation. The kinase activation may be due to the effects

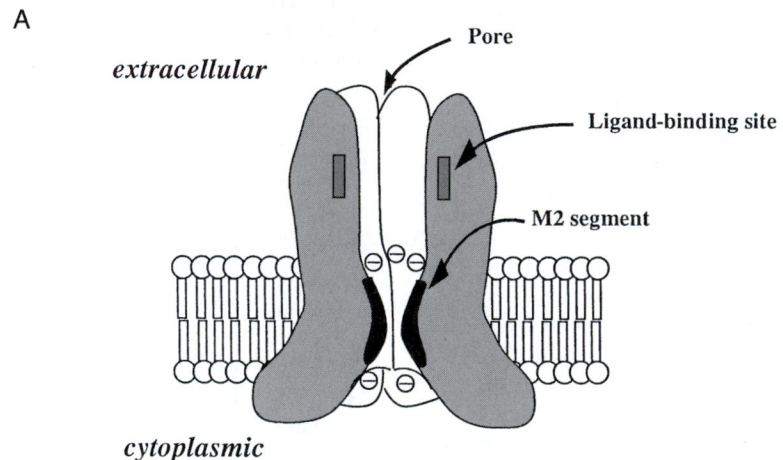

FIGURE 7. The M2 segment lines the channel pore. (A) Schematic of a multiple subunit AChR channel, as suggested by electron micrographs (See Fig. 6) and physiological data. The membrane-spanning region of the pore is lined by the M2 segment of each subunit, with negatively-charged residues at either end of M2. (B) Amino acid sequence of the α1 subunit of the AChR that forms a cation channel is compared with the α1 subunit of the GABA_A receptor that forms an anion channel. The charged residues are in bold, and conserved small uncharged residues are highlighted by gray bars. Note that the negatively-charged residues (glutamate, Glu) in the AChR subunit are replaced by positively charged residues (arginine, Arg) in the GABA_A subunit. These differences may account respectively for the cation and anion selectivity.

of neuropeptides such as vasoactive intestinal peptide and substance P that are coreleased with acetylcholine at some synapses.

Perhaps even more than for other ligand-gated channels, the cloning and characterization of subunit genes has greatly accelerated studies of both the structure and function of nAChRs (Steinbach, 1989; Role, 1992). The structure of nAChRs is comprised of a pentamer of subunits assembled around a central pore. But unlike the muscle AChRs, the nAChR pentamer can consist of only one or two classes of subunits, aptly named α and β. The subunits with cysteines at residues 192 and 193 (involved in binding of ACh to the α subunit of muscle AChRs) are designated as α; subunits lacking these cysteines are called β. To date, the genes for 8 α (α_{2-9}) and 3 β (β_{2-4}) subunits have been cloned from rat and chick brain cDNA libraries. The stoichiometry of each channel molecule appears to be 2 α and 3 β subunits, at least for channels comprised of α_4 and β_2 subunits. Expression of single subunits and combination of subunits in *Xenopus* oocytes (by injection of subunit mRNA) has begun to narrow the possible combinations of subunits in functional channels. For example, α_{2-4} do not express as homo-oligomeric (e.g., five α_2 subunits) channels, whereas α_7, α_8, and α_9 do. On the other hand, α_{5-6} do not express in oocytes even in combination with any of the current β subunits, although it is of course possible (even likely) that these subunits contribute to functional receptors *in vivo*. Likewise, combinations of subunits differ in their single-channel properties, ligand affinity, and sensitivity to antagonists such as neuronal bungarotoxin. Homo-oligomeric α_7 channels produce a rapidly desensitizing response to ACh. This subunit is being used to probe the structural substrates of channel properties such as desensitization, ion flux, and selectivity. Homo-oligomeric α_7, α_8, and α_9 channels are also sensitive to α-bungarotoxin, suggesting that α_7 (which is the predominant subunit of these three subunits in brain) is the α-bungarotoxin receptor. Recent studies suggest that presynaptic α-bungarotoxin receptors, particularly of the α_7 type with their high calcium permeability, can enhance the release of glutamate and other transmitters. The recently cloned α_9 subunit is unique due to its limited expression in the outer hair cells of the cochlea as well as its mixed muscarinic/nicotinic pharmacology.

Although many central neurons have responses to exogenous application of nicotinic agonists, the distribution of nAChR subunit genes revealed by *in situ* hybridization is wider than might have been predicted based on physiological studies. However, there is a reasonable correlation between pathways with functional nAChRs (e.g., the cholinergic input to the hippocampus and medial habenula from the medial septum and diagonal band), the presence of nicotinic binding sites, and the expression of nAChR subunits. Recent studies of nAChRs in different neuronal cell types has revealed a spectrum of channel properties and pharmacological characteristics, suggesting that individual neurons may express several combinations of nAChR subunits. Thus the simplest scenario—that each cell type expresses only one combination of nAChR subunits—is certainly incorrect. For example, *in situ* hybridization studies of neurons in the medial habenular nucleus have revealed two α and three β subunit genes.

However $\alpha_3\beta_4$ and $\alpha_4\beta_2$ appear to be common heteromers in the brain and autonomic ganglia, respectively. As for many classes of ligand-gated channels, the role of specific nAChR subunits is currently being investigated using subunit knockout mice. Interest in the nAChRs has been further fueled by recent studies indicating that mutations in the α_4 subunit cause a genetic form of human epilepsy.

VI. γ-Aminobutyric Acid and Glycine Receptor Channels

Synaptic inhibition in the central nervous system is mediated by channels gated by GABA and glycine. Virtually all neurons have GABA channels, while the distribution of glycine channels is restricted primarily to the brainstem and spinal cord. These two channels are selectively permeable to anions, principally Cl⁻ under physiological conditions. GABA-gated Cl⁻ channels are designated as GABA_A receptors to distinguish them from the G-protein-coupled GABA_B receptor (reviewed by Bormann, 1988). GABA_A channels are often localized on proximal dendrites of central neuron, but also are expressed on axon initial segments and distal dendrites. Because the Cl⁻ equilibrium potential in many neurons is more negative than the resting potential, the opening of GABA_A channels hyperpolarizes the cell, and thus reduces excitability. In addition to the hyperpolarization, the opening of large numbers of GABA_A channels lowers the resistance of the membrane, and effectively "shunts" excitation traveling down the dendrite from excitatory synapses on more distal dendritic branches. However, in some neurons, particularly during early development, the Cl⁻ equilibrium is more positive than the resting potential resulting in excitatory GABA_A responses. Such excitatory responses may act as a development signal in some cases.

The behavior of single GABA_A and glycine channels can be described by a kinetic scheme similar to that of the AChR, with the binding of two agonist molecules required for channel opening (see Macdonald and Twyman, 1992). Three conductance levels are apparent. These are approximately 19, 30, and 45 pS when the Cl⁻ concentration is equal on both sides of the membrane. The 30-pS level is the most frequently observed level for GABA_A channels, and 45 pS is generally more common for glycine channels. Analysis of the openings and closings of single GABA_A channels suggests that the channel may open briefly following the binding of a single GABA_A molecule, and into two longer-lived open states from the doubly liganded configuration. Receptors may close and reenter the longer-lived open states before the agonist dissociates, so-called **bursts** where short closings interrupt a series of openings. These bursts last tens of milliseconds. Desensitization of GABA_A channels results in long closed intervals that are grouped with bursts into **clusters** lasting up to several hundred milliseconds. These clusters are important in determining the duration of inhibitory postsynaptic potentials at some synapses (see Jones and Westbrook, 1996). By measuring the permeability of anions of different diameters, the pore size has been estimated to be 5.6 Å. GABA_A channels also become less active during prolonged recording. This process has been called *rundown,* and appears, at least in part, to be due to channel dephosphorylation, and may also be sensitive to intracellular Ca^{2+}. These are themes that are common to several ligand-gated ion channels.

The drugs that act on $GABA_A$ and glycine channels comprise a fascinating assortment of clinically important compounds (Olsen *et al.*, 1991). Because these channels are so important in synaptic inhibition, either enhancement or reduction in their activity can lead to profound changes in brain function, including amnesia (increased $GABA_A$ activity) or seizures (decreased $GABA_A$ activity). Antagonists for these receptor channels include strychnine, which blocks glycine receptors; bicuculline, which inhibits $GABA_A$ channels; and picrotoxin, which inhibits both channel types. The $GABA_A$ channel is also the target of sedative-hypnotic drugs, such as the benzodiazepines and barbiturates. Benzodiazepines (BDZ) increase the probability of channel opening, whereas barbiturates appear to act by prolonging long channel openings (bursts). The pharmacology of benzodiazepine modulation of the $GABA_A$ receptor is particularly interesting, because compounds can either enhance channel opening (BDZ agonists), reduce channel opening (BDZ inverse agonists), or block the effects of BDZ agonists (BDZ antagonists). $GABA_A$ receptor activity is also modulated by alcohol, volatile anesthetics such as isoflurane, and some steroid anesthetics (or their endogenous equivalents, the neurosteroids).

Using benzodiazepines and strychnine as selective ligands, $GABA_A$ and glycine receptors were purified as several polypeptides, each with molecular weights of approximately 50–60 kDa (reviewed in Betz *et al.*, 1999; DeLorey and Olsen, 1992). The solubilized receptor complex had a molecular weight of approximately 250 kDa, suggesting that, as for the AChR, multiple (four or five) subunits constitute a receptor. Molecular cloning has now identified a series of receptor subunits for both the $GABA_A$ and glycine receptor. The glycine subunits include the strychnine-binding subunit (α) and a β subunit, with a proposed stoichiometry of $\alpha_3\beta_2$. A larger number of $GABA_A$ subunits (seventeen) have been identified, including six α, three β, three γ, δ, ϵ, π, and two ρ (see Schofield *et al.*, 1990; Wisden and Seeburg, 1992; Macdonald and Olsen, 1994). The GABA and benzodiazepine binding sites appear to reside at the interface between an α subunit and a β or γ subunit (usually γ_2), respectively. Interestingly, the α_6 subunit has a low affinity for BDZ agonists, but still can bind BDZ inverse agonists or antagonists. This observation may explain the occurrence of benzodiazepine-insensitive $GABA_A$ receptors in some neurons. As for the nAChRs and glutamate receptors, the large number of $GABA_A$ subunits provides a formidable challenge in determining which combinations are the predominant ones in neurons. The combinations of subunits that form functional receptors is just beginning to be explored, but it is clear that different combinations can alter receptor properties. Subunit expression also varies during development and with neuronal cell type. Molecular studies are also beginning to reveal anchoring and regulatory proteins that interact with glycine and $GABA_A$ receptors such as gephyrin (reviewed in Fallon, 2000) and GABARAP (Wang *et al.*, 1999).

There is now increasing evidence for a distinct GABA-mediated chloride channel, termed the $GABA_C$ receptor, in some retinal neurons. This receptor can be activated by several conformationally restricted GABA analogs, including *cis*-4-aminocrotonic acid. The $GABA_C$ receptor is bicuculline-insensitive, weakly antagonized by picrotoxin, and not modulated by BDZs, barbiturates, or neurosteroids. These channels show distinct gating properties and conductance compared to $GABA_A$ receptors. Because the $GABA_C$ receptor shares some features with the ρ_1 subunit, that is highly expressed in retina, ρ_1 is likely to be one of the $GABA_C$ receptor subunits.

VII. Glutamate Receptor Channels

Although it was known from the early 1950s that L-glutamate was a neuroexcitant, it was not until the 1980s that the role of glutamate-gated ion channels in central synaptic transmission was widely accepted (see Collingridge and Lester, 1989; Westbrook and Jahr, 1989). It is now clear that glutamate receptors mediate a substantial portion of fast excitatory transmission in the brain and spinal cord through the simultaneous activation of two types of ion channels colocalized at excitatory synapses. Characterization of this family of ligand-gated ion channels was initially based on selective activation by the exogenous amino acid ligands: NMDA, kainate, and quisqualate. Because quisqualate also activates a G-protein-coupled receptor, AMPA (α-amino-3-hydroxy-5-methyl-4-isoxazole proprionic acid) has replaced quisqualate as one of the prototypic ligands. It soon became apparent that NMDA receptors were distinct from kainate and AMPA receptors, but for some time it was debated whether kainate and AMPA (i.e., non-NMDA) responses were due to the same or different receptors. It is now clear that AMPA and kainate receptors are distinct molecular entities (see below), although the agonist kainate can activate both classes of receptors.

Initial studies of native AMPA/kainate receptor channels in neurons demonstrated many of the same features as AChRs. In hippocampal pyramidal neurons, AMPA channel activation leads to brief openings (1–5 ms) of monovalent cation channels that show little or no voltage dependence. However, in some neurons, particularly interneurons, AMPA channels have a higher calcium permeability and are inwardly rectifying. AMPA receptors rapidly desensitize when gated by glutamate, and both agonist dissociation (unbinding) and desensitization (binding, but nonconducting) can contribute to the decay of the synaptic current. The behavior of AMPA channels seems well suited to the fast relay of information that characterizes the fast component of excitatory postsynaptic potentials in the brain.

The NMDA channel has been intensively investigated by neuroscientists in recent years, and provides perhaps the best example of the linkage between fundamental properties of ion channels, and the electrical activity of single cells and complex behavioral phenomena (for review see Bekkers and Stevens, 1990; McBain and Mayer, 1994). This linkage is primarily based on three features of NMDA channels: (1) voltage-dependent block of the open channel by extracellular Mg^{2+}; (2) a relatively high permeability to Ca^{2+}; and (3) slow channel kinetics resulting in long-lasting channel activity (see, e.g., Jahr and Lester, 1992). Because Mg^{2+} is positively charged, these ions sense the membrane electric field and are drawn into the open channel at negative membrane potentials. Mg^{2+} binds in the channel pore and impedes the flow of permeant ions, even though the ligand (glutamate) remains bound. However, as the cell membrane depolarizes during synaptic activity, Mg^{2+} falls out of the channel and permeant cations flow through the channel. Thus ion flux through the channel is voltage-dependent. This mechanism differs from other voltage-dependent ion channels such as the sodium or calcium channel, where voltage-dependence results from an intrinsic conformational change in the channel protein. However, the end result is the same; the NMDA channel becomes more effective with depolarization, and thus acts as a positive feedback on synaptic activation.

Also important to the cellular function of NMDA receptors is ion permeability. Unlike many AMPA channels, Ca^{2+} contributes a substantial fraction of the flux through NMDA channels. Based on measurements of the equilibrium (reversal) potential in different ionic solutions, the permeability ratio of Ca^{2+} to Na^+ has been estimated to be approximately 10. However, the rate of flux of Ca^{2+} through the NMDA channel is slower than Na^+ because Ca^{2+} transiently binds with higher affinity to sites within the channel (**energy wells**). Thus most of the current through open NMDA channels is due to Na^+. Nonetheless, the approximately 5–10% of the current that is carried by Ca^{2+} concentration is sufficient to act as a biochemical signal for such processes as the induction of long-term potentiation in the hippocampus, an experimental model of associative learning. Excessive activation of glutamate receptors may also cause neuronal damage, presumably through elevations of intracellular Ca^{2+} concentration with subsequent activation of proteases and free radical formation. This mechanism, called **excitotoxicity,** has been implicated in brain damage due to prolonged seizures, following strokes, and may also contribute to loss of neurons in several degenerative neurological diseases.

Because no high-affinity ligands were available for purification of glutamate channel proteins, the structure of this class of ligand-gated channels defied analysis for a number of years. Thus molecular biologists were forced to use the somewhat laborious task of expression cloning to isolate cDNA clones for glutamate receptor subunits. The first AMPA receptor subunit (GluR1) was isolated by this technique in 1989, and the first NMDA receptor clone (NMDAR1) was also isolated by expression cloning in 1991. As would be predicted from analysis of other ligand-gated ion channels, the proposed structure of glutamate channels involves multisubunit complexes surrounding a central pore. Using the initial AMPA and NMDA clones to screen for homologous sequences, approximately 17 related glutamate receptor subunits have now been isolated from rat and mouse cDNA libraries. Surprisingly, the sequence of glutamate channel subunits is not highly homologous to AChR, GABA, or glycine channels. In particular, glutamate channels have a very large extracellular domain constituting approximately 50% of the protein. Likewise, as discussed in Section IV of this chapter, the topology of glutamate subunits is very different from that of the AChR family of receptor channels.

The two major extracellular domains of the glutamate subunits, the N-terminus and the loop between the third and fourth hydrophobic domains, are homologous to a group of bacterial proteins that bind amino acids. This homology is also shared by the N-terminus of the single polypeptide metabotropic glutamate receptors. This led to a proposed model for the glutamate binding site based on the crystal structure of the periplasmic binding proteins (O'Hara et al., 1993). Using chimeric receptors composed of domains from two glutamate subunits with distinct pharmacology, the role of these two domains in binding of glutamate was confirmed (Stern-Bach et al., 1994). The analysis of crystals of the extracellular ligand binding domains of a glutamate receptor subunit has provided a more detailed understanding of this homology (Armstrong et al., 1998).

On the basis of sequence homology and the characteristics of the expressed receptors, the subunits can be grouped into three categories: four AMPA subunits (GluR1-4 or A-D), five high-affinity kainate subunits (GluR5-7, KA-1, KA-2) and six NMDA subunits (NMDA1, NMDA2A-D, NR3). Other related glutamate subunits (delta 1, delta 2) have been isolated, but their function remains unclear. The delta subunit is particularly intriguing because a mutation in this orphan subunit is responsible for the neurological phenotype of the *Lurcher* mouse. AMPA receptors have a low affinity for glutamate and kainate, compared to the so-called high-affinity kainate subunits. Some of the subunits can combine to form functional homo-oligomeric receptors, and this has revealed some interesting and curious phenomena. For example, in the M2 region of GluR1-4 (which forms at least part of the channel pore), the genes all code for a glutamine residue at one location. The RNA for GluR2, 5, and 6 is edited, however, in such a way that the expressed protein has an arginine residue at this site. This switch has a profound effect on the behavior of the channel. The current evoked by homo-oligomeric expression of unedited subunits shows marked inward rectification and an increased permeability to Ca^{2+}. The edited versions have a linear current-voltage relationship and are permeable only to monovalent cations (Na^+ and K^+). Because the synaptic response mediated by AMPA receptors in hippocampal pyramidal cells has a linear current-voltage relationship, it seems likely that most AMPA receptors in these neurons contain one or more edited copies of GluR2. However, in interneurons in cortex and hippocampus as well as in dorsal horn neurons of the spinal cord, AMPA receptors lacking edited GluR2 subunits show inward rectification and increased permeability to Ca^{2+}. Physiological studies have demonstrated a more restricted distribution of high-affinity kainate responses, including dorsal root ganglion neurons and both presynaptic and postsynaptic localization on some central neurons (for review, Chittajallu et al., 1999). Expression studies and *in situ* immunohistochemistry suggest that combinations of GluR5 or 6 with KA-2 may account for at least some of the high-affinity kainate responses.

Glutamate channels, particularly NMDA channels, are regulated both by phosphorylation and by a variety of allosteric mechanisms. The most dramatic is the action of glycine as a **coagonist** that is required for the opening of NMDA channels, and extracellular Mg^{2+} that blocks NMDA channels (Fig. 4). It appears that two molecules of glutamate and two molecules of glycine are required to activate an NMDA channel. Because it now appears that glutamate channels are tetramers, it is thought that each subunit probably contributes to agonist binding. Although AMPA channels do not require a coagonist, given the homology between AMPA and NMDA channels, it may be that four molecules of glutamate are also necessary to activate AMPA channels (e.g. see Rosenmund *et al.*, 1998). Other regulator factors that have been shown to affect NMDA channel activity include extracellular protons, Zn^{2+}, polyamines, redox potential, and intracellular Ca^{2+}. NMDA receptors can be regulated by protein kinase A and C and tyrosine kinases, whereas AMPA channel activity is upregulated by cAMP-dependent protein kinase. The impact of these regulatory mechanisms on synaptic function is an important area for future investigation.

As is the case for all synaptic receptor channels, AChR and glutamate channels are clustered at synaptic sites on muscle and neurons. For AchRs on muscle, the clustering-anchoring involves a complex cascade involving factors released from the neuron as well as a number of protein-protein interactions in the postsynaptic membrane, including the 43-kDa protein rapsyn (for review, see Froehner, 1993). Recent studies have revealed a number of candidate proteins in central neurons that are involved in the clustering and localization of glycine and glutamate receptor channels. Glutamate channels are imbedded in the postsynaptic density that contains receptors, regulatory proteins, and cytoskeletal proteins. Biochemical studies suggest that calmodulin and cytoskeletal proteins can interact with C-terminal domains of several NMDA receptor subunits at domains that are involved in the regulation of NMDA receptor desensitization, suggesting an intriguing link between dynamic regulation of channel activity (e.g. by compartmentalized regulatory proteins) and structural features such as channel anchoring and clustering. A family of cytoskeletal proteins with so-called PDZ domains appears to be involved in interactions with AMPA and NMDA subunits (for review, see Sheng, 1996; Ehlers *et al.*, 1996). Studies of these interactions may be important in synapse development as well as in the dynamic regulation of activity of ligand-gated channels at synapses.

VIII. Summary

Ligand-gated ion channels are membrane proteins that are fundamental signaling molecules in neurons. These molecules are localized in the plasmalemma and on intracellular organelles and can be gated by both intracellular and extracellular ligands. The neurotransmitter-gated ion channels discussed in this chapter mediate fast excitation and inhibition in the nervous system, and they have now been well characterized by physiological and molecular studies. Studies using the technique of voltage and patch-clamp recording have examined the three basic features of an ion channel: gating, conductance, and selective permeability. In general, ligand-gated ion channels can be described by kinetic models involving binding of two or more ligand molecules that induce a conformational change in the protein. As a result, a central, water-filled pore opens and conducts ions at very high rates of up to 10^7 ions per second.

Channel activity is terminated both when the channel closes or when it enters a nonconducting (desensitized) state. The ACh and glutamate receptors are cation channels, whereas GABA and glycine receptors are anion channels. Many of the genes coding for these channels have been cloned. The primary structure of the ACh, GABA and glycine channel subunits, deduced from the primary amino acid sequence, is highly homologous. Thus, this group of cDNAs constitutes a gene **superfamily** that probably evolved from a common ancestral gene. Glutamate channels comprise a separate but related family of channel subunits. The ligand-binding domain of glutamate channels shares homology with bacterial amino acid binding proteins, whereas the pore shares homology with the pore of voltage-dependent ion channels.

Each of the neurotransmitter-gated channels appears to be comprised of a hetero-oligomeric complex, which is comprised of four or five closely related subunits surrounding a central ion pore. For each receptor type there are a number of possible subunit combinations that can form functional channels. AChR subunits have four transmembrane domains with the second transmembrane domain lining the pore. Charged amino acid residues near the mouth of the pore are important in determining the selectivity and conductance of the individual channel types. The glutamate subunits have three transmembrane domains and a reentrant loop domain that forms a portion of the pore. A number of important drugs act by binding to the walls of the channel pore and obstructing the flow of ions through the open channel.

Molecular studies of these subunits have revealed more and more detailed structural features of the channel proteins that determine their gating behavior, conductance, and selective ion permeability. Ligand-gated ion channels are also highly regulated by allosteric and covalent modifications. Allosteric regulation constitutes an important mechanism for drug action in the brain. Recent progress in the study of ligand-gated ion channel function and structure has answered many long-standing questions, but has also raised many new and interesting issues concerning the function of this important class of membrane proteins.

Bibliography

Armstrong, N., Sun, Y., Chen G. Q., and Gouaux, E. (1998). Structure of a glutamate-receptor ligand-binding core in complex with kainate. *Nature* **395**, 913–917.

Ascher, P., and Nowak, L. (1987). Electrophysiological studies of NMDA receptors. *Trends Neurosci.* **10**, 284-288.

Bean, B. P. (1992). Pharmacology and electrophysiology of ATP-activated ion channels. *Trends Pharmacol. Sci.* **13**, 87–90.

Bekkers, J. M., and Stevens, C. F. (1990). Computational implications of NMDA receptor channels. *Cold Spring Harbor Symp. Quant. Biol.* **55**, 131–135.

Betz, H. (1990). Ligand-gated ion channels in the brain: the amino acid receptor superfamily. *Neuron* **5**, 383-392.

Betz, H., Kuhse, J., Schmieden, V., Laube, B., Kirsch, J., and Harvey, R. J. (1999). Structure and functions of inhibitory and excitatory glycine receptors. *Ann. NY Acad. Sci.* **868**, 667–676.

Bormann, J. (1988). Electrophysiology of GABA$_A$ and GABA$_B$ receptor subtypes. *Trends Neurosci.* **11**, 112–116.

Burnstock, G. (1999). Current status of purinergic signalling in the nervous system. *Prog. Brain Res.* **120**, 3–10.

Changeux, J.-P., and Edelstein, S. J. (1998). Allosteric receptors after 30 years. *Neuron* **21**, 959–980.

Changeux, J. P., Benoit, P., Bessis, A., Cartaud, J., Devillers-Thiery, A., Fontaine, B., Galzi, J. L., Klarsfeld, A., Laufer, R., Mulle, C., Nghiem, H. O., Osterlund, M., Piette, J., and Rehav, F. (1990). The acetylcholine receptor: functional architecture and regulation. *Adv. Second Messenger Phosphoprotein Res.* **24**, 15–19.

Chen, G. Q., Cui, C., Mayer, M. L., and Gouaux, E. (1999). Functional characterization of a potassium-selective prokaryotic glutamate receptor. *Nature* **402**, 817–821.

Chittajallu, R., Braithwaite, S. P., Clarke, V. R., and Henley, J. M. (1999). Kainate receptors: subunits, synaptic localization and function. *Trends Pharmacol. Sci.* **20**, 26–35.

Collingridge, G. L., and Lester, R. A. (1989). Excitatory amino acid receptors in the vertebrate central nervous system. *Pharmacol. Rev.* **41**, 143–210.

Colquhoun, D. (1998). Binding, gating, affinity and efficacy: the interpretation of structure-activity relationships for agonists and of the effects of mutating receptors. *Brit. J. Pharmacol.* **125**, 923–947.

Dani, J. A., and Heinemann, S. (1996). Molecular and cellular aspects of nicotine abuse. *Neuron* **16**, 905–908.

DeLorey, T. M., and Olsen, R. W. (1992). γ-Aminobutyric acid$_A$ receptor structure and function. *J. Biol. Chem.* **267**, 16747–16750.

Ehlers, M. D., Mammen, A. L., Lau, L. F., and Huganir, R. L. (1996). Synaptic targeting of glutamate receptors. *Curr. Opin. Cell. Biol.* **8**, 484–489.

Fallon, J. R. (2000). Building inhibitory synapses: exchange factors getting into the act? *Nat. Neurosci.* **3**, 5–6.

Froehner, S. (1993). Regulation of ion channel distribution at synapses. *Annu. Rev. Neurosci.* **16**, 347–368.

Hille, B. (1992). "Ionic Channels of Excitable Membranes." Second edition, Sinauer Associates Inc., Sunderland, MA.

Hollmann, M., and Heinemann, S. (1994). Cloned glutamate receptors. *Annu. Rev. Neurosci.* **17**, 31–108.

Jahr, C. E., and Lester, R. A. J. (1992). Synaptic excitation mediated by glutamate-gated ion channels. *Curr. Opin. Neurobiol.* **2**, 270–274.

Jones, M. V. and Westbrook, G. L. (1996). The impact of receptor desensitization on fast synaptic transmission. *Trends Neurosci.* **19**, 96–101.

Julius, D. (1991). Molecular biology of serotonin receptors. *Annu. Rev. Neurosci.* **14**, 335–360.

Karlin, A. (1991). Explorations of the nicotinic acetylcholine receptor. *Harvey Lectures* **85**, 71–107.

Karlin, A., and Akabas, M. H. (1995). Toward a structural basis for the function of nicotinic acetylcholine receptors and their cousins. *Neuron.* **15**, 1231–1244.

Karlin, A. and Akabas, M. H. (1998). Substituted-cysteine accessibility method. *Methods Enzymol.* **293**, 123–145.

Luetje, C. W., Patrick, J., and Seguela, P. (1990). Nicotine receptors in the mammalian brain. *FASEB J.* **4**, 2753–2760.

Macdonald, R. L., and Olsen, R. W. (1994). GABA$_A$ receptor channels. *Annu. Rev. Neurosci.* **17**, 569–602.

Macdonald, R. L., and Twyman, R. E. (1992). Kinetic properties and regulation of GABA$_A$ receptor channels. *Ion Channels* **3**, 315–343.

McBain, C., and Mayer, M. (1994). N-methyl-D-aspartic acid receptor structure and function. *Physiol. Rev.* **74**, 723–759.

McGehee, D. S., and Role, L. W. (1995). Physiological diversity of nicotinic acetylcholine receptors expressed by vertebrate neurons. *Annu. Rev. Physiol.* **57**, 521–546.

Miller, C. (1989). Genetic manipulation of ion channels: a new approach to structure and mechanism. *Neuron* **2**, 1195–1205.

Miyazawa, A., Fujiyoshi, Y., Stowell, M., and Unwin, N. (1999). Nicotinic acetylcholine receptor at 4.6 Å resolution: transverse tunnels in the channel wall. *J. Mol. Biol.* **288**, 765–786.

Montal, M. (1990). Molecular anatomy and molecular design of channel proteins. *FASEB J.* **4**, 2623–2635.

Nakanishi, S. (1992). Molecular diversity of glutamate receptors and implications for brain function. *Science* **258**, 597–603.

Neher, E. (1992). Ion channels for communication between and within cells. *Science* **256**, 498–502.

Nicoll, R. A., Malenka, R. C., and Kauer, J. A. (1990). Functional comparison of neurotransmitter receptor subtypes in mammalian central nervous system. *Physiol. Rev.* **70**, 513–565.

O'Hara, P. J., Sheppard, P. O., Thogersen, H., Venezia, D., Haldeman, B. A., McGrane, V., Houamed, K. M., Thomsen, C., Gilbert, T. L., and Mulvihill, E. R. (1993). The ligand-binding domain in metabotropic glutamate receptors is related to bacterial periplasmic binding proteins. *Neuron* **11**, 41–52.

Olsen, R. W., Sapp, D. M., Bureau, M. H., Turner, D. M., and Kokka, N. (1991). Allosteric actions of central nervous system depressants including anesthetics on subtypes of the inhibitory gamma-aminobutyric acid$_A$ receptor-chloride channel complex. *Ann. NY Acad. Sci.* **625**, 145–154.

Pascual, J. M., and Karlin, A. (1998). Delimiting the binding site for quaternary ammonium lidocaine derivatives in the acetylcholine receptor channel. *J. Gen. Physiol.* **112**, 611–621.

Role, L. W. (1992). Diversity in primary structure and function of neuronal nicotinic acetylcholine receptor channels. *Curr. Opin. Neurobiol.* **2**, 254–262.

Rosenmund, C., Stern-Bach, Y., and Stevens, C. F. (1998). The tetrameric structure of a glutamate receptor channel. *Science* **280**, 1596–1599.

Sakmann, B. (1992). Nobel Lecture. Elementary steps in synaptic transmission revealed by currents through single ion channels. *Neuron* **8**, 613–629.

Schofield, P. R., Shivers, B. D., and Seeburg, P. H. (1990). The role of receptor subtype diversity in the CNS. *Trends Neurosci.* **13**, 8–11.

Sheng, M. (1996). PDZs and receptor/channel clustering: rounding up the latest suspects. *Neuron* **17**, 575–578.

Sheng, M. (1997). Excitatory synapses. Glutamate receptors put in their place. *Nature* **386**, 221–223.

Sommer, B., and Seeburg, P. H. (1992). Glutamate receptor channels: novel properties and new clones. *Trends Pharmacol. Sci.* **13**, 291–296.

Steinbach, J. H. (1989). Structural and functional diversity in vertebrate skeletal muscle nicotinic acetylcholine receptors. *Annu. Rev. Physiol.* **51**, 353–365.

Stern-Back, Y., Bettler, B., Hartley, M., Sheppard, P. O., O'Hara, P. J., and Heineman, S. F. (1994). Agonist selectivity of glutamate receptors is specified by two domains structurally related to bacterial amino acid–binding proteins. *Neuron* **13**, 1345–1357.

Swope, S. L., Moss, S. J., Raymond, L. A., and Huganir, R. L. (1999). Regulation of ligand-gated ion channels by protein phosphorylation. *Adv. Second Messenger Phosphoprotein Res.* **33**, 49–78.

Unwin, N. (1993). Neurotransmitter action: opening of ligand-gated ion channels. *Neuron* **10**, Suppl., 31–41.

Wang, H., Bedford, F. K., Brandon, N. J., Moss, S. J., and Olsen, R. W. (1999). GABA(A)-receptor-associated protein links GABA(A) receptors and the cytoskeleton. *Nature* **397**, 69–72.

Westbrook, G. L., and Jahr, C. E. (1989). Glutamate receptors in excitatory neurotransmission. *Sem. Neurosci.* **1**, 103–114.

Wisden, W., and Seeburg, P. H. (1992). GABA$_A$ receptor channels: from subunits to functional entities. *Curr. Opin. Neurobiol.* **2**, 263–269.

Janusz B. Suszkiw

41

Synaptic Transmission

I. Introduction

The function of nerve cells is to receive, process, and transmit information. Intercellular transfer of signals among neurons or from neurons to effector cells is accomplished at specialized intercellular junctions called **synapses** and is by and large chemically mediated. A characteristic feature of a chemical synapses is the presence of a 20–100 nm wide extracellular space, the **synaptic cleft**, which separates the **presynaptic** (transmitting) and **postsynaptic** (receiving) elements of the synapse. This physical discontinuity prevents efficient transfer of current between the cells and necessitates intervention of a diffusible chemical transmitter substance. The fundamental aspect of chemically mediated transmission is the transduction of voltage signal into the release of neurotransmitter from the presynaptic neuron and transduction of transmitter binding to a specific receptor into voltage change in the postsynaptic cell.

The basic principles of chemical neurotransmission have been elaborated in a series of seminal experiments conducted on the vertebrate skeletal neuromuscular junction by B. Katz and his collaborators in the 1950s and 1960s (Katz, 1966). At about the same time, the pioneering work by J. C. Eccles and his coworkers on spinal motor neurons showed that the general principles of chemical transmission established at the vertebrate neuromuscular junction also apply to central synapses (Eccles, 1964). Since then, new insights into the mechanism of synaptic transmission have been gained on the molecular level through the application of contemporary techniques of molecular biology and electrophysiology.

This chapter focuses on chemical synaptic transmission; however, the reader should be cognizant of another form of signal transfer, referred to as **electrical transmission**. Electrical transmission is mediated by the **gap junctions** (Bennett, 1997). Gap junctions consist of channel aggregates that bridge the closely apposed cell membranes and thus provide electrical continuity between the communicating cells. Unlike the chemical synapses, gap junctions and the electrical coupling they mediate are not unique to the nervous system but are also found in many other tissues.

Gap junctions and electrotonic transmission are discussed in greater detail elsewhere in this volume.

II. Structure and Function of Chemical Synapses: An Overview

Based on their morphofunctional characteristics, synaptic junctions may be classified as directed, fast-acting or nondirected, slow-acting synapses. Although the configurations of synaptic junctions can vary considerably, all share common features (Fig. 1). The presynaptic element, usually an axon terminal bouton or axon varicosity, is characterized by the presence of membrane-bound spherical organelles, the **synaptic vesicles**, which store the transmitter prior to its release. A subset of synaptic vesicles clusters near the presynaptic **active zone**, a structural specialization of the presynaptic membrane for synaptic vesicle docking and exocytosis. The juxtaposed postsynaptic membrane contains specific receptors for the transmitter released from the presynaptic terminal. A thickening of postsynaptic membrane, the **postsynaptic density**, is frequently observed and thought to play a role in localizing the neurotransmitter receptors at the synaptic membrane. The width of the synaptic cleft at highly directed (point-to-point), fast-acting synapses typically varies from about 20–50 nm but can be considerably wider at nondirected, slow-acting synapses. The synaptic cleft contains proteinaceous material, including various cell adhesion molecules (Hall and Sanes, 1993), thought to serve as an adherent for stabilizing the synapse. The junction is enveloped by glial elements, which insulate the synapse as well as perform other functions such as, for example, transmitter removal from the synaptic cleft. For a detailed discussion of synaptic ultrastructure, the student may wish to consult the articles in Pappas and Purpura (1972) and a more recent review by Burns and Augustine (1995).

Transmission at chemical synapses is **unidirectional** from the presynaptic to postsynaptic element. The presynaptic terminals are specialized for transmitter biosynthesis,

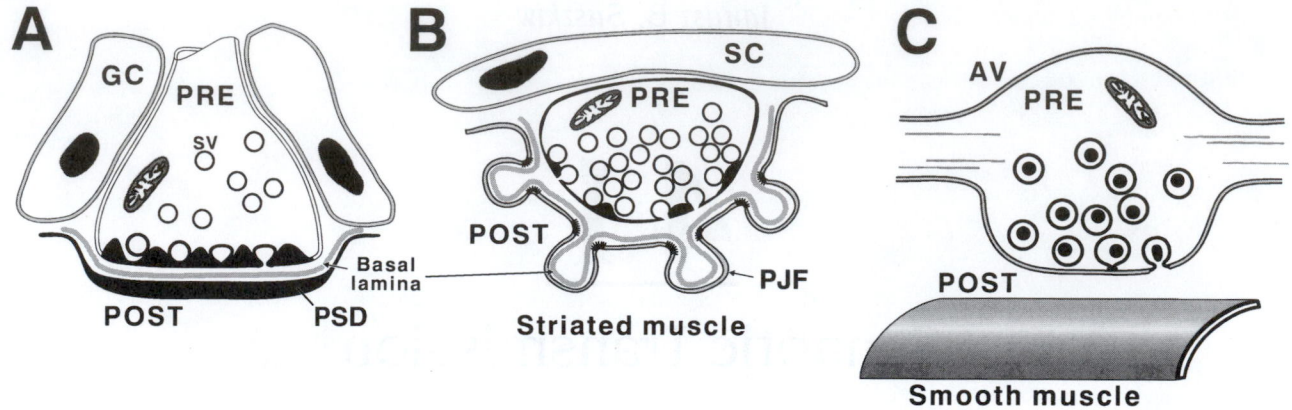

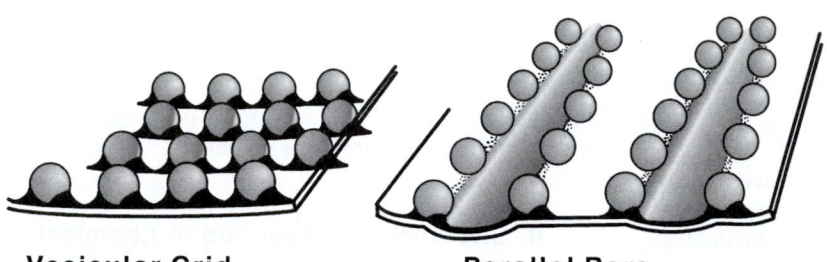

FIGURE 1.   Schematic representations of neuro-neuronal and neuro-effector synaptic junctions. (A) A directed neuro-neu-ronal synapse subserving fast synaptic transmission in the CNS. The presynaptic terminal bouton contains mitochondria (M), clear-cored, about 40 nm in diameter, synaptic vesicles (SV), a subset of which is seen docked at the presynaptic vesicular grid. The receptors are localized in the juxtaposed thickening of postsynaptic membrane, the postsynaptic density (PSD). The presy-naptic vesicular grid together with the juxtaposed PSD define the active zone of the synapse. The synaptic cleft is typically about 20 nm wide and contains dense material, with filamentous structures seen sometimes to span the cleft from pre-to-postsynaptic membrane. Glial cells (GC) cap and insulate the synapse. (B) The vertebrate skeletal neuromuscular junction (NMJ), a directed synapse subserving fast transmission from motor neuron to striated muscle. These large terminals are filled with many clear-cored, acetylcholine-containing synaptic vesicles. A few dense-cored vesicles can be also seen. A subset of clear-cored synaptic vesicles aggregate at the presynaptic active zone, which is defined by the presence of the presynaptic density. The presynaptic density appears to have the configuration of a bar defined by parallel rows of intramembrane particles, thought to represent the calcium channels, with two rows of vesicles attached on either side of the bar (bottom). The receptors for acetylcholine (AchR) are localized in the crests of the postjunctional folds (PJF) directly opposite the presynaptic active zones. The synaptic cleft at the vertebrate NMJ is typically 50 nm wide. The basal lamina within the synaptic cleft contains AChE, an enzyme that degrades ACh released into the synaptic cleft. The junction is insulated by Schwann cell (SC) processes. (C) A nondirected, slow-acting synapse between a presynaptic axon varicosity (AV) of a sympathetic neuron and smooth muscle. The varicosity contains large (>60 nm) dense-core, norepinephrine-containing vesicles. This synapse is characterized by the absence of distinct active zone structures and a variable but usually wide (>100 nm) separation between the presynaptic and postsynaptic elements of the junction.

packaging of transmitter into the synaptic vesicles, and vesicular transmitter exocytosis. The transmitter receptors in the postsynaptic membrane transduce transmitter binding into ionic current which generates the **postsynaptic potential (PSP)**. Postsynaptic potentials at fast synapses are generated by activation of **ionophoric,** or channel-forming, receptors. The synaptic potentials are typically fast in onset and last for only few milliseconds. The generation of the postsynaptic potentials is, however, not instantaneous but rather registers 0.3–0.5 ms after the arrival of the action potential at the presynaptic terminal. This **synaptic delay** is a characteristic feature of chemical synapses and reflects, for the most part, the time required for the molecular events associated with

the transmitter release. At nondirected, slow synapses, chan-nels are not directly transmitter-gated but rather are coupled to the receptor via G protein and second messenger systems. These receptors are referred to as **metabotropic.** The synap-tic potential generated by activation of the metabotropic re-ceptors is slower in onset and longer lasting.

The process of chemical neurotransmission involves (1) synthesis and vesicular uptake of neurotransmitter in presy-naptic terminal, (2) exocytotic release of vesicular transmit-ter into the synaptic cleft, (3) diffusion and binding of trans-mitter to postsynaptic receptors and generation of synaptic potential, and (4) termination of synaptic activity. As indi-cated above, transmitters are synthesized in the nerve ending of the presynaptic neuron and are concentrated and stored in synaptic vesicles. The intravesicular packet or **quantum** of transmitter that is stored in the vesicles is released by Ca^{2+}-dependent exocytosis. Exocytosis is triggered when an action potential (AP) in the presynaptic neurons invades and depolarizes the nerve terminal, thereby opening the presynaptic voltage-gated Ca^{2+} channels and allowing influx of Ca^{2+} into the terminal. The resulting transient rise in $[Ca^{2+}]_i$ triggers fusion of synaptic vesicles with the plasmalemma and transmitter exocytosis. The requirement for Ca^{2+} influx is absolute. In absence of Ca^{2+} the presynaptic AP will fail to trigger release.

Following its release into the synaptic cleft, transmitter combines with specific receptors in the postsynaptic membrane, causing a change in its permeability to specific ions. Change in membrane permeability to ions gives rise to synaptic current which, depending on ions involved, depolarizes or hyperpolarizes the postsynaptic membrane (Fig. 2). The depolarizing potential is called an **excitatory postsynaptic potential (EPSP)** because it tends to bring the cell membrane potential toward the threshold for an AP. EPSPs are associated with a change in the postsynaptic membrane permeability to Na^+ and K^+ ions and a net influx of positive charges (inward current carried by Na^+). The hyperpolarizing potential is called an **inhibitory postsynaptic potential (IPSP)**, because it tends to move or hold the membrane potential away from the threshold, thus decreasing the likelihood of an AP being fired. The IPSPs elicited by activation of ionophoric receptors at fast synapses are associated with an increase in the postsynaptic membrane permeability to Cl^-. PSPs associated with activation of metabotropic receptors usually involve change in membrane permeability to K^+ ions.

Transmitter release normally terminates within less than 1 ms. As the presynaptic AP decays and the terminal repolarizes back toward the resting potential, the depolarization-activated calcium channels reclose, Ca^{2+} influx ceases, and $[Ca^{2+}]_i$ is rapidly lowered to the prestimulus level. Transmitter action on the receptors in the postsynaptic membrane is terminated by diffusion and enzymatic degradation into an ineffective substance or clearance from the synaptic cleft by reuptake into the nerve terminals and/or glial cells. Reuptake and subsequent catabolism is the primary route of transmitter inactivation for most transmitters. The exception is acetylcholine, in which case inactivation is primarily by means of extracellular degradation of ACh to inactive acetate and choline.

III. Neurotransmission

A. Neurotransmitters and Neurotransmitter Receptors

Several criteria define a chemical substance as a neurotransmitter. (1) The biosynthetic enzymes for the synthesis of the substance must be present in the identified presynaptic neuron to catalyze the synthesis of transmitter in the nerve terminals. (2) The substance must be released by stimulation of the presynaptic neuron in a Ca^{2+}-dependent

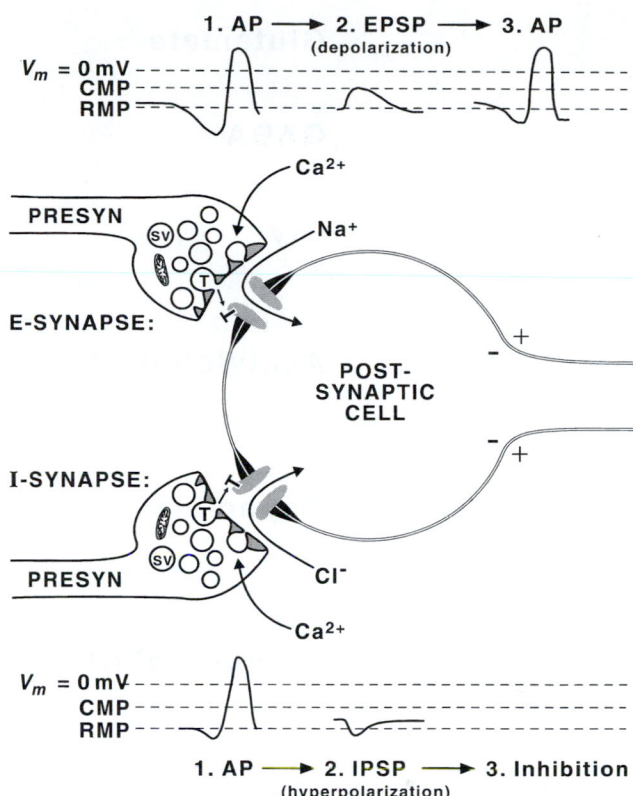

FIGURE 2. Fast excitatory versus inhibitory synaptic transmission. At an excitatory (E) synapse, the presynaptic action potential releases transmitters that activate cationic channels, resulting in an inward synaptic current carried by Na^+ ions and depolarizing potential, the excitatory postsynaptic potential (EPSP). EPSPs increase the likelihood of an action potential (AP) being fired by the postsynaptic cell. At an inhibitory (I) synapse, the transmitter activates receptor-gated chloride channels and influx of Cl^- ions, resulting in hyperpolarizing potential, the inhibitory postsynaptic potential (IPSP). The likelihood of an AP is diminished by activation of I-synapses, because IPSPs hold or move the membrane potential of the postsynaptic cell away from the threshold. RMP, resting membrane potential; CMP, critical membrane potential (threshold) at which an AP is triggered.

manner. (3) Application of the substance to the postsynaptic cell must mimic the actions of the neurally released substance. (4) A specific mechanism for inactivation of the transmitter substance, such as a selective uptake system in the presynaptic terminals or the presence of degradative enzyme(s), must be demonstrated at the synapse investigated.

Amino acid glutamate is the major excitatory neurotransmitter and γ-aminobutyric acid (GABA) and glycine serve as inhibitory transmitters in the CNS. Acetylcholine (ACh), dopamine (DA), norepinephrine (NE), epinephrine, serotonin (5-HT), histamine, and ATP are all recognized as neurotransmitters or neurotransmitter candidates. Acetylcholine and norepinephrine are the established transmitters in the peripheral nervous system. In addition there is good evidence that in the vascular smooth muscle, ATP is co-released and acts as a cotransmitter with NE. Neurons may also co-store and co-release various peptide hormones. These may act as cotransmitters or

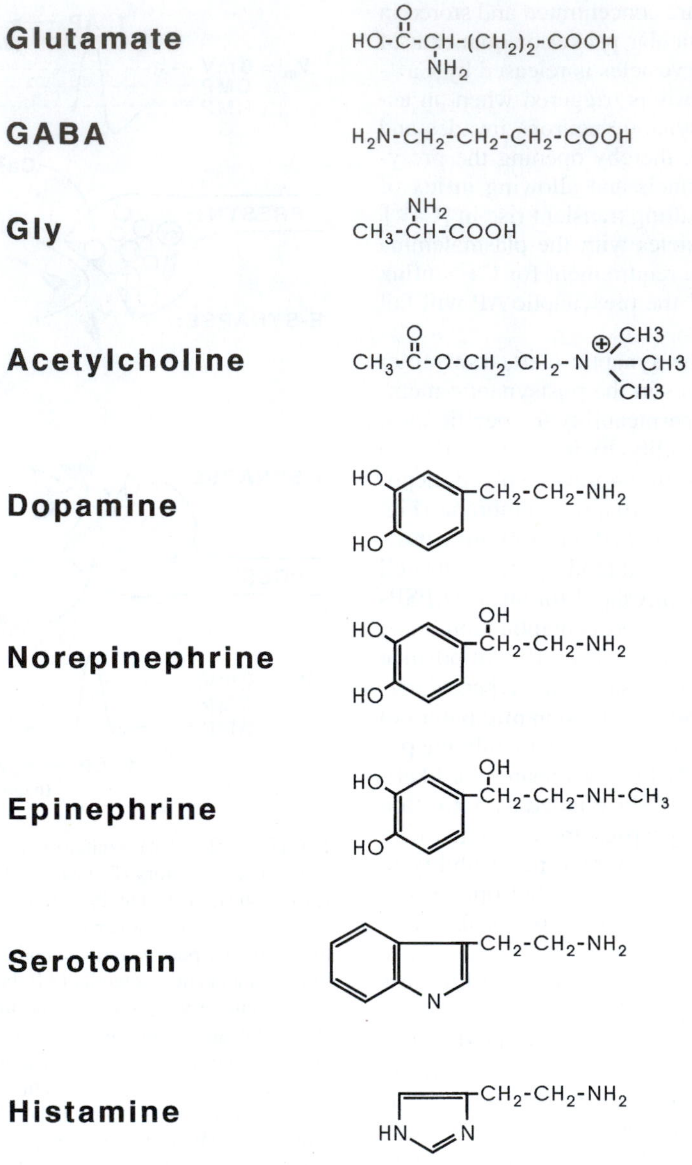

Glutamate $HO-\overset{\overset{O}{\|}}{C}-\underset{\underset{NH_2}{|}}{CH}-(CH_2)_2-COOH$

GABA $H_2N-CH_2-CH_2-CH_2-COOH$

Gly $CH_3-\underset{\overset{|}{NH_2}}{CH}-COOH$

Acetylcholine $CH_3-\overset{\overset{O}{\|}}{C}-O-CH_2-CH_2-\overset{\oplus}{N}\overset{CH3}{\underset{CH_3}{-CH3}}$

Dopamine

Norepinephrine

Epinephrine

Serotonin

Histamine

FIGURE 3.   Structural formulas of common neurotransmitters.

TABLE 1   Common Neurotransmitters and Major Receptor Types

Transmission type	Neurotransmitter	Receptor[a]
Amino acidergic	Glutamic acid	**Kainate, AMPA, NMDA**, mGluR
	Gamma-aminobutyric acid (GABA)	$GABA_A$, $GABA_B$, $GABA_C$
	Glycine	**Gly-R**
Cholinergic	Acetylcholine	**nAChR**, mAChR
Aminergic	Dopamine	D_1, D_2
	Norepinephrine	α_1, α_2, β_1, β_2
	Serotonin	$5HT_1$, $5HT_2$, **$5HT_3$**
	Histamine	H_1, H_2, H_3
Purinergic	ATP, adenosine	P_1, **P_{2x}**, P_{2y}

[a]Ionophoric or channel-forming receptors are indicated in bold. Regular lettering indicates metabotropic or G-protein-coupled receptors.

modulate the synaptic actions of conventional transmitters. Common low-molecular-weight transmitter substances and their chemical structures are shown in Fig. 3.

Neurotransmitter receptors (Table 1) are integral membrane proteins containing the transmitter recognition site and the transducer site. Ionophoric or channel-forming receptors include the muscle and neural nicotinic acetylcholine receptors; three pharmacologically distinguishable glutamate receptors that are selectively activated by analogs of glutamate—kainate, α-amino-3-hydroxy-5-methylisoxazole-4-propionic acid (AMPA), or N-methyl-D-aspartic acid (NMDA); the $5HT_3$ serotonin receptor; and the P_{2x} purinoceptor for ATP. These relatively nonspecific cationic channels permit passage of Na^+, K^+, and in some cases Ca^{2+} ions, but exclude anions. The anionic (Cl^-) channel-forming receptors include the $GABA_A$ and $GABA_C$ receptors and the glycine receptors (Gly-R).

Metabotropic receptors transduce the transmitter binding into a physiological response via G-protein/second messenger–coupled mechanisms. The synaptic actions of monoamines, dopamine, norepinephrine, epinephrine, serotonin, and histamine are exerted via the G-protein/second messenger–coupled mechanisms systems. The G-protein-coupled receptors also include the muscarinic ACh receptors, metabotropic glutamate receptor (mGluR), $GABA_B$, and several receptors for adenosine and ATP. Ligand-gated and G-protein-coupled channels are discussed elsewhere in this volume.

B. Biosynthesis, Storage, and Inactivation of Neurotransmitters

Conventional, low–molecular-weight neurotransmitters are synthesized locally in the nerve terminals and packaged into synaptic vesicles (SV) prior to release. The local synthesis of transmitter in the nerve terminals assures that transmitter is available for refilling the vesicles after they have released their contents into the synaptic cleft. The vesicular uptake of transmitters is mediated by specific transporters and is driven by an electrochemical gradient generated by electrogenic proton pump in the vesicular membrane (McMahon and Nicholls, 1991). Following release, the synaptic neurotransmitter action is terminated by diffusion, reuptake into presynaptic terminals or glial cells, and degradation. Reuptake of transmitters from the extracellular space is energetically coupled to a transmembrane Na^+ gradient and is mediated by specific high affinity transporters that are distinct from the vesicular transporters. Two

families of plasmalemmal transporters have been identified. The carriers for excitatory amino acids require the presence of extracellular Na^+, whereas the carriers for biogenic amines, GABA, and glycine require both Na^+ and Cl^- for uptake (Fig. 4). (Amara and Azzaro, 1993; Sonders and Amara, 1996).

The key enzyme(s) involved in transmitter synthesis are selectively expressed in the neurons and define their neurotransmitter phenotype. Choline acetyltransferase (ChAT) is the marker enzyme for cholinergic neurons where it catalyzes the synthesis of acetylcholine (ACh) from acetyl coenzyme A (AcCoA) and choline (Ch):

$$AcCoA + Ch \xrightarrow{ChAT} ACh$$

Acetyl coenzyme A derives from the tricarboxylic acid (TCA) cycle whereas choline is transported into nerve terminals from the extracellular medium by the sodium-dependent, high-affinity carrier system that is localized in the presynaptic terminals. Transport of choline and synthesis of ACh are tightly coupled and both the rate of choline uptake and ACh synthesis increase during activity, ensuring adequate supply of the transmitter. The vesicular ACh transporter (VAChT) mediates uptake of ACh from the cytosol into synaptic vesicles. The plasmalemmal Ch uptake is selectively inhibited by a chemical hemicholinium-3 (HC-3) whereas vesamicol blocks vesicular uptake of ACh. Either of these drugs will cause depletion of ACh and failure of cholinergic transmission. Following its release, the synaptic action of ACh is terminated by diffusion and acetylcholinesterase (AChE)-catalyzed hydrolysis to inactive acetate and choline. Much of the choline is recaptured by the nerve terminals and reutilized in new ACh synthesis.

Catecholaminergic neurons and their synaptic connections comprise dopaminergic, noradrenergic (norepinephrine), and adrenergic (epinephrine) systems. The rate-limiting step in catecholamine biosynthesis is the tyrosine hydroxylase (TH)-catalyzed hydroxylation of tyrosine to L-dihydroxyphenylalanine (L-DOPA), which is then decarboxylated by a nonspecific aromatic L-amino acid decarboxylase (L-AADC) to dopamine. Dopamine is taken up into synaptic vesicles and serves as a transmitter at dopaminergic synapses. In noradrenergic nerve endings, dopamine is further converted to norepinephrine by intravesicularly localized dopamine-β-hydroxylase (DβH). Methylation of norepinephrine to epinephrine at adrenergic synapses is catalyzed in cytosol by phenylethanolamine-N-methyltransferase (PNMT) in a reaction requiring S-adenosylmethionine as methyl donor:

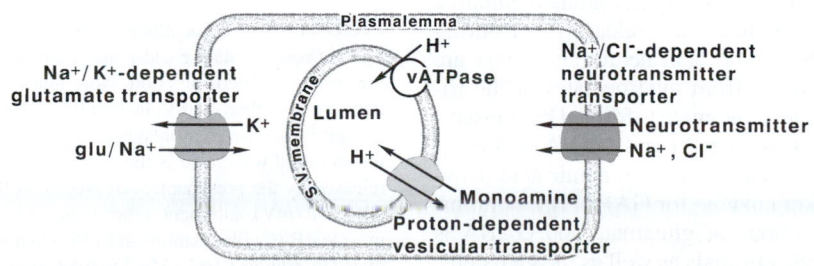

FIGURE 4. Neurotransmitter transport systems in vesicular and plasmalemmal membranes. (Based on Amara and Arriza, 1993.)

$$\text{tyrosine} \xrightarrow{\text{TH}} \text{L-DOPA}$$

$$\text{L-DOPA} \xrightarrow{\text{L-AADC}} \text{dopamine}$$

$$\text{dopamine} \xrightarrow{\text{D}\beta\text{H}} \text{norepinephrine}$$

$$\text{norepinephrine} \xrightarrow{\text{PNMT}} \text{epinephrine}$$

Vesicular monoamine transporters (VMATs) mediate the uptake of catecholamines into synaptic vesicles. Reserpine inhibits the VMATs and thereby depletes vesicular stores of catecholamines. Synaptic actions of catecholamines are terminated by high-affinity neuronal reuptake followed by intracellular degradation or reuptake into synaptic vesicles. The plasmalemmal catecholamine transporters are distinct from VMATs. Plasmalemmal reuptake is insensitive to reserpine but is blocked by psychoactive drugs such as cocaine and amphetamine, which thereby increase extracellular levels of catecholamines. The major catabolic pathway for catecholamines involves the mitochondrial monoamine oxidase (MAO)–catalyzed oxidative deamination to aldehydes followed by their rapid conversion by hydrogenases and reductases to corresponding acids and alcohols.

Serotonergic neurons utilize serotonin as the transmitter. Serotonin, an indoleamine, is formed by tryptophan hydroxylase–mediated hydroxylation of tryptophan to 5-hydroxytryptophan (5-HTP), followed by L-AADC-mediated decarboxylation of 5-HTP to 5-hydroxytryptamine (5-HT, serotonin):

$$\text{tryptophan} \xrightarrow{\text{Trp–hydroxylase}} \text{5-HTP} \xrightarrow{\text{L–AADC}} \text{5-HT}$$

Uptake of serotonin into synaptic vesicles appears to be mediated by the same or a closely related VMAT that mediates vesicular transport of catecholamines. The synaptic action of serotonin is terminated primarily by neuronal sodium-dependent reuptake, mediated by a transporter that is homologous to the plasmalemmal transporters for catecholamines. Following reuptake, serotonin is degraded by MAO-catalyzed oxidative deamination to 5-hydroxyindole acid aldehyde and 5-hydroxyindole acetic acid.

Glutamic acid, γ-aminobutyric acid (GABA), and glycine are the major amino acid neurotransmitters in the CNS. Because glutamic acid and glycine are common constituents of amino acid pools found in all cells, their use as neurotransmitters at glutamatergic and glycinergic synapses implies subcompartmentation within the respective nerve terminals. This subcompartment in all likelihood corresponds to synaptic vesicles (SV), which presumably selectively accumulate glutamic acid and glycine from the cytosol and release them during synaptic transmission. The amino acid neurotransmitters are synthesized in nerve terminals from intermediates of the tricarboxylic acid (TCA) cycle. Glutamate is formed by transamination of α-ketoglutaric acid (α-KA) and GABA is formed from glutamate in a reaction catalyzed by glutamic acid dehydrogenase (GAD), a marker enzyme for GABAergic neurons. Inactivation of synaptic actions of glutamate and GABA is through reuptake into nerve terminals as well as the glial cells, where both amino acids are converted to glutamine, which is

in turn exported to the nerve terminals and reutilized in the formation of glutamate:

$$\alpha\text{-KA}_{\text{(TCA)}} \xleftrightarrow{\text{transaminase}} \text{glutamate} \xleftarrow{\text{glutaminase}} \text{glutamine}$$

$$\text{glutamate} \xrightarrow{\text{GAD}} \text{GABA}$$

C. Transmitter Release

Transmitter release is evoked by arrival of an action potential at the presynaptic terminal. Depolarization of the nerve terminal membrane activates (opens) the plasmalemmal voltage-gated Ca^{2+} channels, allowing influx of extracellular Ca^{2+} into the terminal. The resultant transient increase of Ca^{2+} concentration near the plasmalemmal release sites triggers a cascade of biochemical events that culminate in synaptic vesicle-plasmalemma fusion and exocytosis of transmitter into the synaptic cleft. The process terminates upon nerve terminal repolarization when the Ca^{2+} channels deactivate (close), Ca^{2+} entry ceases, and intracellular Ca^{2+} returns to prestimulus levels.

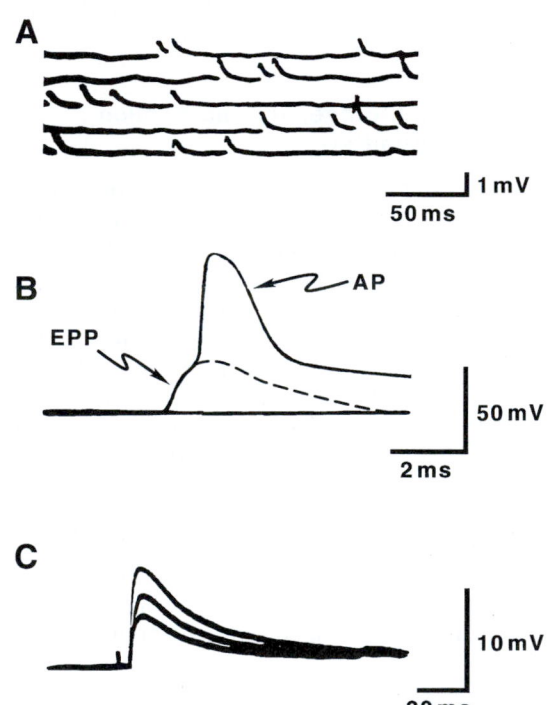

FIGURE 5. Intracellular recordings from the frog end plate. (A) Spontaneous, miniature end-plate potentials (MEPPs) recorded from resting (not stimulated) junction. Note that these small depolarizing potentials are less than 1 mV in amplitude and occur randomly. (B) Postsynaptic response to a presynaptic action potential. The initial hump on the recorded waveform is the end-plate potential (EPP) elicited by ACh released by the presynaptic AP. Note that the EPP is a large depolarization (> 40 mV), sufficient to bring the end plate to threshold and trigger a muscle AP. (C) Fluctuations in EPPs when transmitter output has been reduced by adding 10 mM Mg^{2+} to the bathing medium. (Adapted from Fatt and Katz, 1952, and del Castillo and Katz, 1954.)

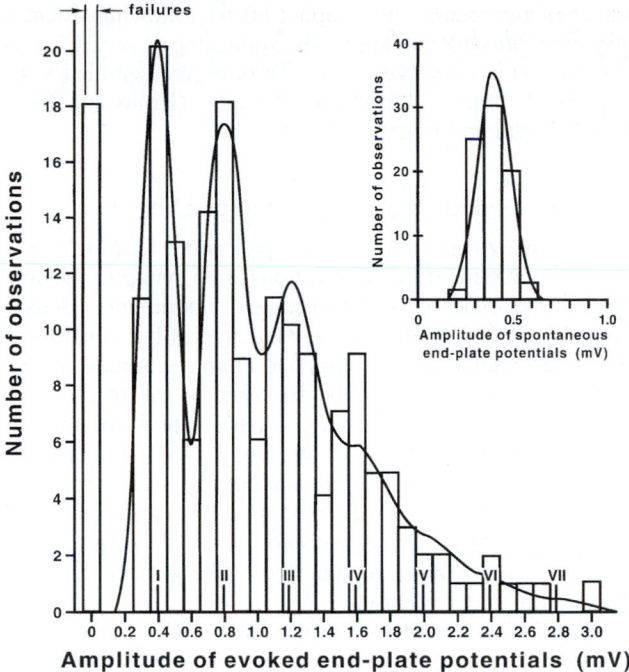

FIGURE 6. Distribution of EPP amplitudes recorded from mammalian end plate under conditions of reduced transmitter release in high (12.5 mM) Mg^{2+}. The inset shows a histogram of spontaneous potentials (MEPPs) recorded from resting junction. Note that EPP amplitudes group around multiples of mean MEPPs amplitude, and the number of experimentally observed failures (0 quanta released) and single, double, triple, or more quantal responses, fit the theoretical distribution (solid curve) calculated from the Poisson equation. (From Boyd & Martin, 1956.)

1. Quantal-Vesicular Hypothesis of Transmitter Release

The evolution of the current understanding of the mechanism of transmitter release began with the formulation of the **quantal-vesicular hypothesis** of transmitter release by Bernard Katz and his collaborators (Fatt and Katz, 1952; delCastillo and Katz, 1954). Katz and colleagues observed that in the resting neuromuscular junction, that is, in the absence of stimulation, nerve terminals spontaneously release ACh, giving rise to small depolarizations that occur at random intervals and average about 0.5 mV in amplitude (Fig. 5A). These small potentials behaved in all respects as miniature replicas of the end-plate potential (EPP) evoked by presynaptic APs (Fig. 5B) and therefore were called miniature end-plate potentials (MEPPs). The crucial insight into the relationship between MEPPs and EPPs was provided by the observation that when evoked release was reduced in low-Ca^{2+}/high-Mg^{2+} solutions, the size of EPPs fluctuated in a random manner (Fig. 5C) such that the EPP amplitudes appeared to be made up of integral multiples of the average MEPP amplitude and could be described by Poisson distribution (Fig. 6). Katz concluded that transmitter release is a stochastic process, consisting of random release of multimolecular packets or quanta of ACh each producing a unit response (MEPPs), and that the EPP is a summation of

many quantal units released nearly synchronously by a presynaptic AP.

For detailed discussion of the statistics of transmitter release, the interested reader is referred to Martin (1977). For the present purpose, it suffices to state that transmitter release can be described by a simple statistical expression:

$$m = nP$$

The parameter m is the average number of quanta released per presynaptic impulse when a large number of trials are performed and is called the **quantal content** of the EPP. The parameter n represents the number of quanta immediately available for release and most likely corresponds to either the population of synaptic vesicles associated with the presynaptic active zones or the number of release sites. P is the probability of any single quantum being released and primarily reflects the probability of a productive Ca^{2+}-dependent vesicle fusion with the plasmalemma as a function of Ca^{2+} concentration at the release sites. Reducing the availability of Ca^{2+} to enter terminals would reduce P, thus accounting for a reduction of transmitter release in low-Ca^{2+}/high-Mg^{2+} media. Conversely, increasing the concentration of Ca^{2+} at or near the release sites would tend to enhance the probability of exocytosis, that is, facilitate transmitter release.

2. Essential Role of Ca^{2+} in Depolarization-Release Coupling

The voltage-gated calcium channels in the presynaptic plasma membrane couple membrane depolarization to transmitter exocytosis. The essential role of Ca^{2+} influx in transmitter release was demonstrated by Katz and Miledi

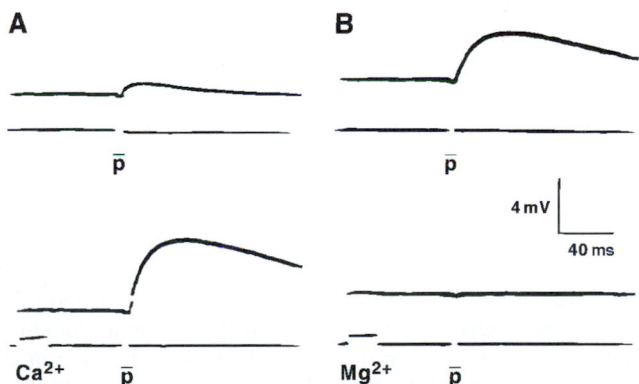

FIGURE 7. Calcium is required for transmitter release evoked by depolarization of nerve terminals. In the absence of Ca^{2+}, a depolarizing pulse (p) applied to a presynaptic nerve fails to evoke transmitter release and the postsynaptic response is nearly absent (A, top). Application of a pulse of Ca^{2+} to the neuromuscular junction shortly before applying stimulus (p) to the presynaptic nerve enables transmitter release as shown by the appearance of a postsynaptic response (A, bottom). The normal postsynaptic response in Ca^{2+}-containing media (B, top) is blocked by applying a pulse of high Mg^{2+} prior to the depolarizing pulse (p) (B, bottom). The failure to elicit transmitter release in the presence of high Mg^{2+} is due to block of Ca^{2+} influx into the terminals. (Adapted from Katz and Miledi, 1967.)

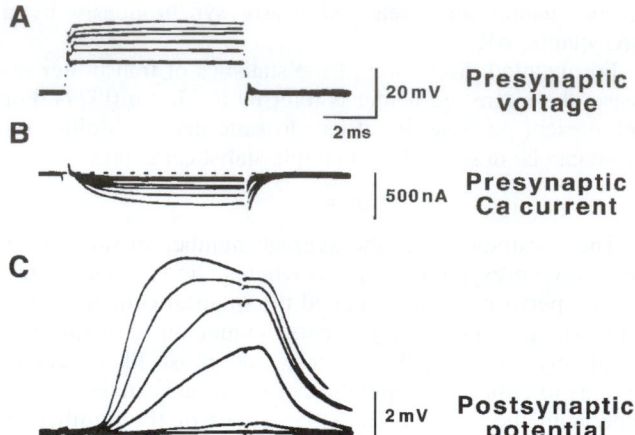

FIGURE 8. Experiments in the squid giant synapse illustrating the relationship between magnitude of presynaptic Ca^{2+} current and transmitter release monitored by recording postsynaptic potentials. Graded depolarizations of presynaptic terminals (A) evoke graded inward Ca^{2+} currents (B) that correlate with graded postsynaptic potentials (C), which reflect the amount of transmitter released. (Adapted from Llinas, 1977.)

(1967), who showed that depolarization of presynaptic terminals failed to evoke transmitter release when Ca^{2+} was absent or its entry into nerve terminals was prevented by high Mg^{2+} concentration in the extracellular medium (Fig. 7). In subsequent experiments carried out at the squid giant synapse, where it is possible to make intracellular recordings from both pre- and postsynaptic cells simultaneously, Miledi (1973) showed that injection of Ca^{2+} into presynaptic terminals elicited transmitter release, thus providing direct evidence that a rise in intracellular Ca^{2+} alone is sufficient for activation of the release process. Furthermore, using voltage-clamping to control the membrane potential of the presynaptic terminals at the squid giant synapse, Llinas (1977) showed that the quantity of synaptic transmitter released, monitored as the size of the postsynaptic potential, is related to the size of the presynaptic calcium current, which in turn depends on the extent of nerve terminal depolarization (Fig. 8).

The presynaptic calcium channels that couple membrane depolarization to transmitter release at fast synapses appear to be strategically localized near the release sites (Robitaille *et al.*, 1990). Opening of these channels results in domains of calcium entry, giving rise to localized subplasmalemmal calcium concentrations that may reach 0.1 mM or higher very rapidly and for a very brief duration (Smith and Augustine, 1988). These subplasmalemmal calcium transients represent more than a thousand-fold increase of $[Ca^{2+}]$ relative to the resting cytosolic calcium level of about 0.1 µM and provide a powerful trigger for focal synaptic vesicle exocytosis. The relationship between intracellular calcium concentration and the rate of exocytosis investigated in the terminals of goldfish retinal bipolar neurons indicates half-saturation at about 200 µM and cooperative interactions involving at least four calcium ions in activation of synaptic vesicle exocytosis (Heidelberger *et al.*, 1994). Trans-

mitter release occurs within about 60 µs following calcium entry into presynaptic terminals, indicating that Ca^{2+}-triggered exocytosis involves activation of a preassembled vesicle-plasmalemma docking/fusion complex (Bruns and Jahn, 1995; Sabatini and Regehr, 1996).

3. Exocytosis and Recycling of Synaptic Vesicles

The active zones in the presynaptic nerve terminals provide plasmalemmal specializations for synaptic vesicle docking and exocytosis. In ingenious experiments, Heuser and coworkers (1979) employed a specially constructed quick-freeze apparatus that enabled them to capture images of vesicle exocytosis at the active zone of the frog neuromuscular junction (Fig. 9). The comparison of the number of exocytotic pores captured in the quick-freeze experiment correlated with the estimated number of quanta released under similar conditions of stimulation, providing the most direct, structural evidence that exocytosis of synaptic vesicles at the active zone region is the likely mechanism of quantal transmitter release.

Synaptic vesicles within the terminal are distributed among the so-called readily available and reserve pools. The **readily available pool** is thought to correspond to the vesicles docked at the active zone and immediately available for release, whereas the **reserve pool** comprises the vesicles distributed within the terminal at some distance from the active zone. It is evident that in order to maintain the availability of quanta for release, exocytosis must be accompanied by mobilization of new vesicles from the reserve pool within the terminal to the plasmalemmal release sites. Mobilization of vesicles is thought to involve alteration in vesicle cytoskeleton interactions that are regulated by phospho-dephosphorylation of synaptic vesicle–associated proteins, synapsins. Dephosphorylated synapsin seems to stabilize vesicle-cytoskeletal interactions and inhibit vesicle mobilization, whereas phosphorylation of synapsin by the Ca^{2+}-calmodulin–dependent protein kinase II is thought to promote vesicle mobilization (Llinas *et al.*, 1991; Greengard *et al.*, 1993).

The mechanism of vesicle docking and initiation of exocytosis by Ca^{2+} has not been yet completely worked out; however, remarkable progress has been made in this direction during the past few years. Studies initially conducted in model systems such as mast cells or chromaffin cells and recently extended to synaptic preparations indicate that secretion is accompanied by a stepwise increase in cell capacitance, consistent with fusion of secretory granules with the cell membrane. Electrical measurements further suggest that the first event in exocytosis may be the formation of a pore that connects vesicle lumen with the extracellular space and may provide a channel for the release of soluble contents into the synaptic cleft and/or promote collapse and complete fusion of vesicle with the plasma membrane (Lindau and Almers, 1995; Matthews, 1996).

Beginning with the identification and cloning of synaptic proteins in the early 1990s (rev. in Südhof, 1995), extraordinarily rapid progress has been made during the last few years in defining the molecular machinery of exocytosis. Current evidence indicates that the synaptic exocytotic apparatus makes use of constitutive membrane fusion ma-

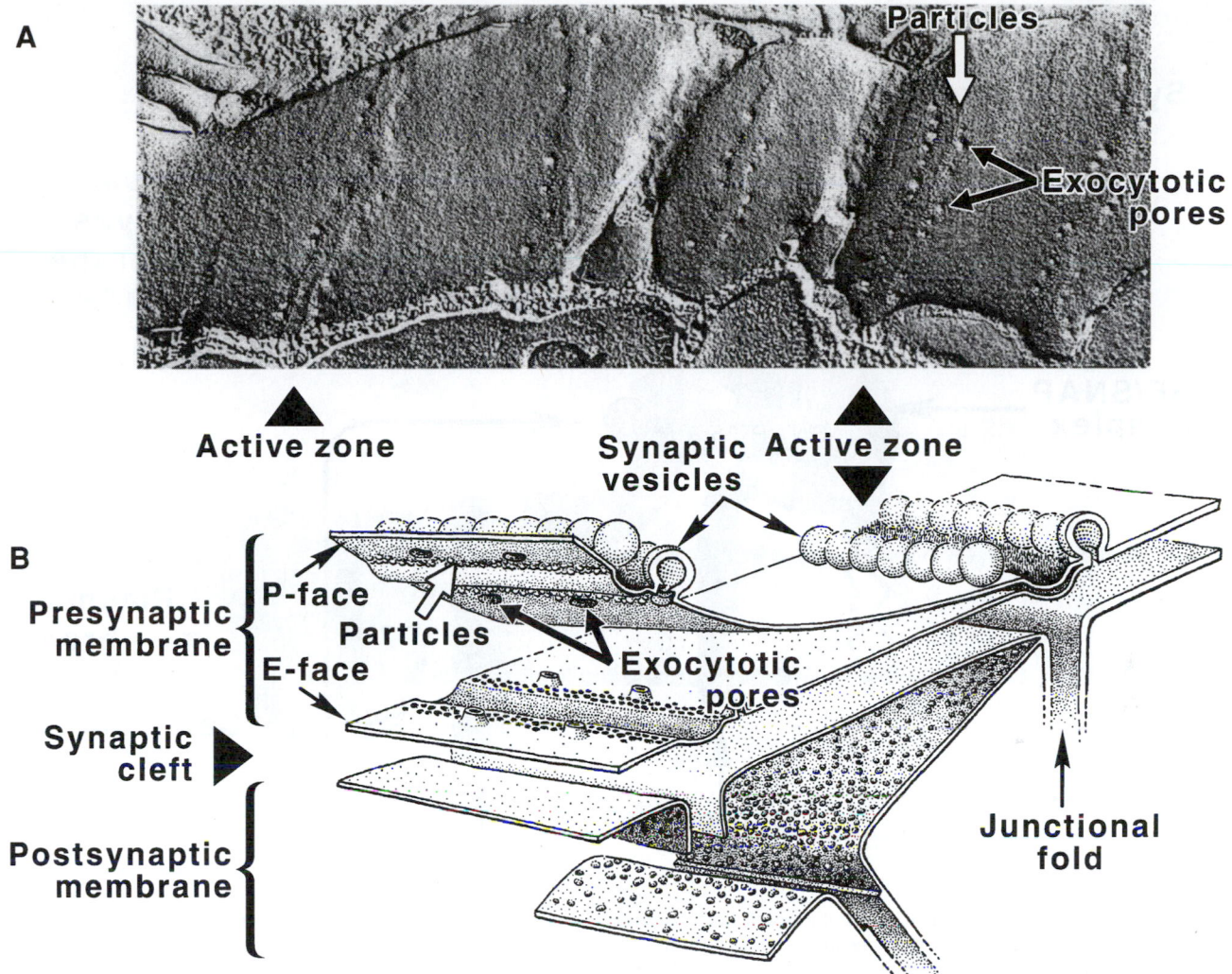

FIGURE 9.   Synaptic vesicle exocytosis captured by a rapid-freezing technique. (A) P-face of freeze-fractured replica of a stimulated and rapidly frozen nerve terminal showing images of synaptic vesicles caught in the process of exocytosis at the active zones. The junction was activated in presence of 1 mM 4-aminopyridine to enhance the level of transmitter release per single presynaptic impulse. (Adapted from Heuser *et al. J. Cell Biol.* 81: 275–300, 1979.) (B) Three-dimensional representation of pre- and postsynaptic membranes in relation to the freeze-fracture image above. The freeze-fracture splits membrane bilayers into protoplasmic (P-face) and extracellular (E-face) leaflets of the bilayer. The rows of intramembrane particles seen in P- and E-faces are thought to be the presynaptic Ca^{2+} channels. The openings in the P-face and their craterlike continuations in the E-face are the exocytotic pores formed when synaptic vesicles fuse with the plasmalemma along the parallel bars of the active zone. (Adapted from Heuser et al. *J. Cell Biol.* 81: 275–300, 1979.)

chinery that has been placed under control of a calcium sensor. The elements of the constitutive docking-fusion apparatus at synapses are the synaptic vesicle membrane-associated protein (VAMP) synaptobrevin and the plasma membrane proteins syntaxin and SNAP-25 (25-kDa, synaptosome-associated protein). The VAMP, syntaxin, and SNAP-25 belong to a class of highly conserved membrane-targeting proteins that serve as SNAP (soluble NSF attachment protein) receptors and hence are referred to as SNAREs (Söllner et al., 1993a, b). The SNAREs have been shown to associate into a stable, 7-S core complex that links the apposed vesicle and plasma membranes through formation of parallel bundles of four interacting α-helices,

with syntaxin and synaptobrevin each contributing one helix and SNAP-25 contributing two helices (Sutton *et al.*, 1998). The evidence for the crucial role of SNARE complexes in exocytosis has been provided by the observation that selective proteolysis of either VAMP, syntaxin, or SNAP-25 by clostridial neurotoxins prevents formation of the complex and inhibits transmitter release (Niemann *et al.*, 1994; Hayashi *et al.*, 1994). Biochemical studies indicate that the core complex is primed by ATP-dependent, SNAP-assisted binding of NSF (N-maleimide–sensitive factor). NSF is an ATPase whose activation is thought to play an important role in fusion by inducing a conformational change in syntaxin and disassembly of the SNARE

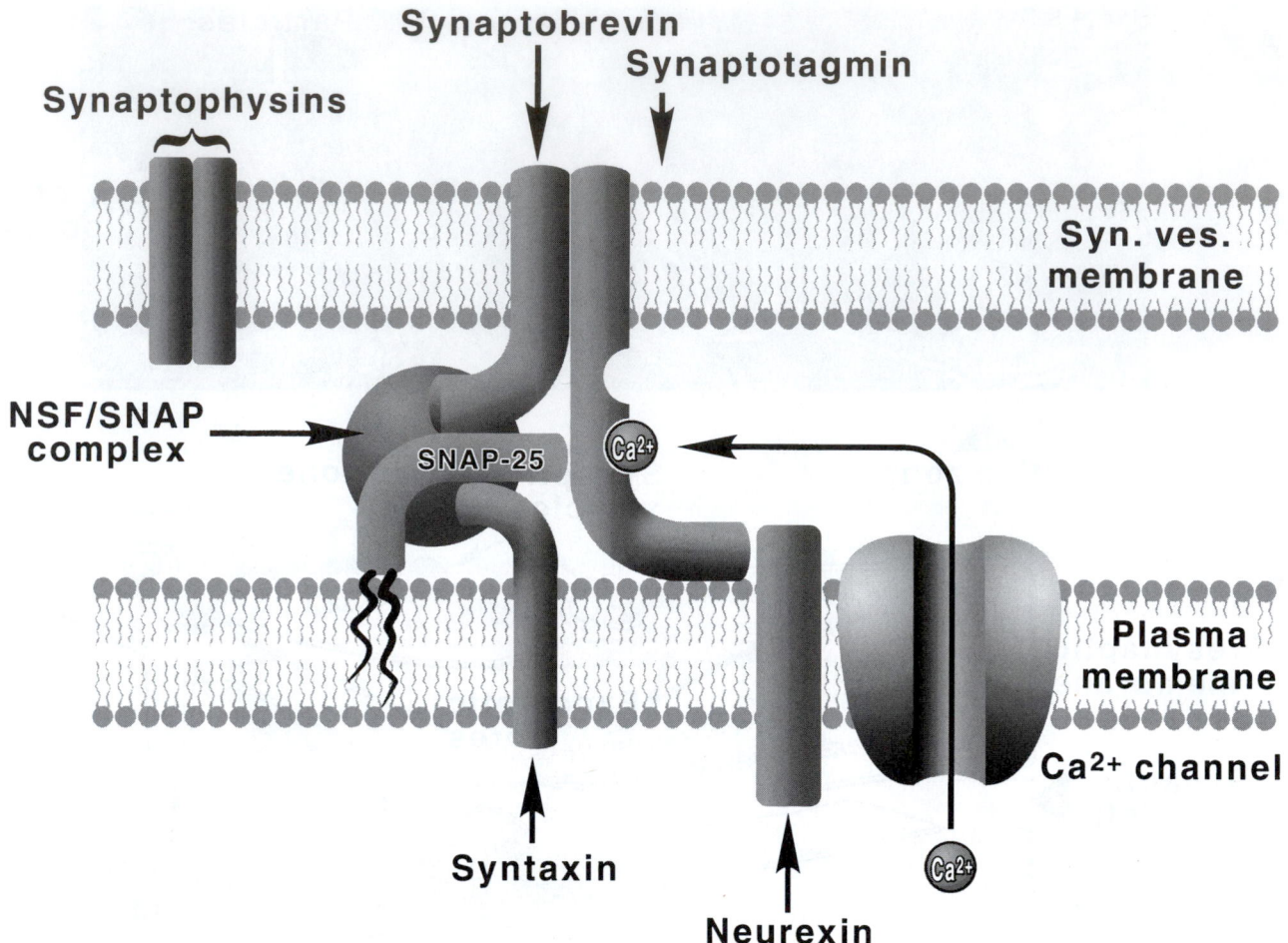

FIGURE 10.   A simplified model of a vesicle docking-fusion complex. The model incorporates interactions among the proteins of synaptic vesicles (synaptobrevins and synaptotagmins), plasma membrane proteins (syntaxins, SNAP-25, neurexin), and the soluble elements of constitutive fusion machinery (NSF, SNAP). Synaptobrevin, syntaxin, and SNAP-25 form a stable, 7-S core complex to which SNAP and NSF (an ATPase) bind to form a 20-S complex. Activation of NSF ATPase leads to disassembly of the complex, perhaps assisting in the final step of membrane fusion. The complex is thought to form in close proximity to the plasmalemmal calcium channels, with synaptotagmin serving as the key Ca^{2+} sensor. The plasma membrane protein neurexin interacts with synaptotagmin and may modulate its interactions with the core complex.

complex (Whiteheart *et al.*, 1994; Hanson *et al.*, 1997). The synaptic vesicle membrane protein synaptotagmin 1, whose cytosolic domains contain two Ca^{2+}-binding C2 motifs homologous to the C2 regulatory domain of protein kinase C, is believed to serve as the Ca^{2+} -sensor of exocytosis (Brose *et al.*, 1992; Shao *et al.*, 1977). Biochemical experiments indicate that Ca^{2+} activation of synaptotagmin is associated with simultaneous binding of its C2 domains to the membrane phospholipids and the SNARE complex. The kinetics of synaptotagmin–Ca^{2+}–core complex interactions occur on a microsecond time scale consistent with its postulated function as Ca^{2+}-trigger receptor of exocytosis (Davis *et al.*, 1999). Another protein that has been implicated in exocytosis is neurexin (Petrenko *et al.*, 1991), a plasmalemmal protein that provides a target for α-latrotoxin, a component of black widow spider venom that induces massive, Ca^{2+}-independent transmitter exocytosis. Through its capacity to associate with synaptotagmin, neurexin could somehow modulate interactions between

synaptotagmin and SNAREs. A simplified model of a vesicle docking-fusion complex is illustrated in Fig. 10. Several other synaptic vesicle– and plasma membrane--associated proteins have been identified that are likely to have either accessory or regulatory functions in vesicle exocytosis/endocytosis (Südhof, 1995); however, their precise function and mechanism of action still remain by and large unresolved.

Following the release of transmitter, exocytosis may be terminated in at least two ways. One is simply through reclosure of the pore and fission of vesicles at the active zone. These postexocytotic vesicles may then refill with locally synthesized transmitter and, by virtue of being already positioned close to or at the active zone, release the newly formed transmitter in preference to the reserve pool. This could explain the well-documented phenomenon of the preferential release of newly synthesized transmitter. A more generally accepted idea is that, following fusion and exocytosis, vesicle membrane collapses into the presynaptic plasmalemma to be retrieved in a series of transformations starting with clathrin-mediated endocytosis followed by endosomal fusion-budding and reformation of functional vesicles (Fig. 11). It is likely that both types, the rapid exo-endocytosis as well as the slower, clathrin-mediated endocytosis occur, but the extent to which one predominates over the other depends on the rate of presynaptic stimulation and possibly other factors. The evidence for local recycling of synaptic vesicles in the nerve terminals has been provided by the observation that vesicles become labeled with high–molecularweight markers such as horseradish peroxidase (HRP) or dextrans when motor nerve terminals are stimulated with these markers in the extracellular medium (Ceccarelli *et al.,* 1973). More recent evidence for vesicle recycling has been provided by tracking the movement of fluorescentlabeled synaptic vesicles during stimulation of motor nerve terminals at the frog neuromuscular junction (Betz and Bewick, 1992) and in other synaptic preparations (Ryan *et al.,* 1993; Lagnado *et al.,* 1996).

Vesicles undergoing local recycling within the nerve terminals are progressively degraded and replaced by vesicles that are formed *de novo* in the cell body and transported to the nerve terminals by fast axoplasmic transport. The half-life of vesicles has been estimated at 7–14 days. Evidently, synaptic vesicles can undergo numerous cycles of transmitter release-reloading before being replaced with new ones.

D. Generation of Postsynaptic Potentials at Fast Synapses

The nature of postsynaptic responses is determined by the type of receptor-gated ionic conductances that are activated in the postsynaptic membrane. At excitatory synapses, the transmitters activate receptor-gated channels that conduct cations, principally Na^+ and K^+. The net current through the synaptic channels is inward and carried by Na^+ ions, causing a depolarization of the postsynaptic membrane, or EPSP. The EPSPs generated at the skeletal neuromuscular junction are called end-plate potentials (EPPs) and those at other peripheral synapses, for example at nerve–smooth muscle junctions, are frequently referred to as excitatory junctional potentials, or EJPs. At inhibitory synapses, transmitters activate receptor-gated channels that conduct Cl^- ions. Influx of chloride ions through the synaptic channels tends to increase the negativity of the cell interior and hyperpolarizes the postsynaptic membrane. The hyperpolarizing postsynaptic potentials are called inhibitory postsynaptic potentials (IPSPs) because they tend to move the membrane potential away from the threshold.

1. Synaptic Current and Synaptic Equilibrium Potential

The synaptic current (i_S) flowing through a single transmitter-activated channel is determined by the channel

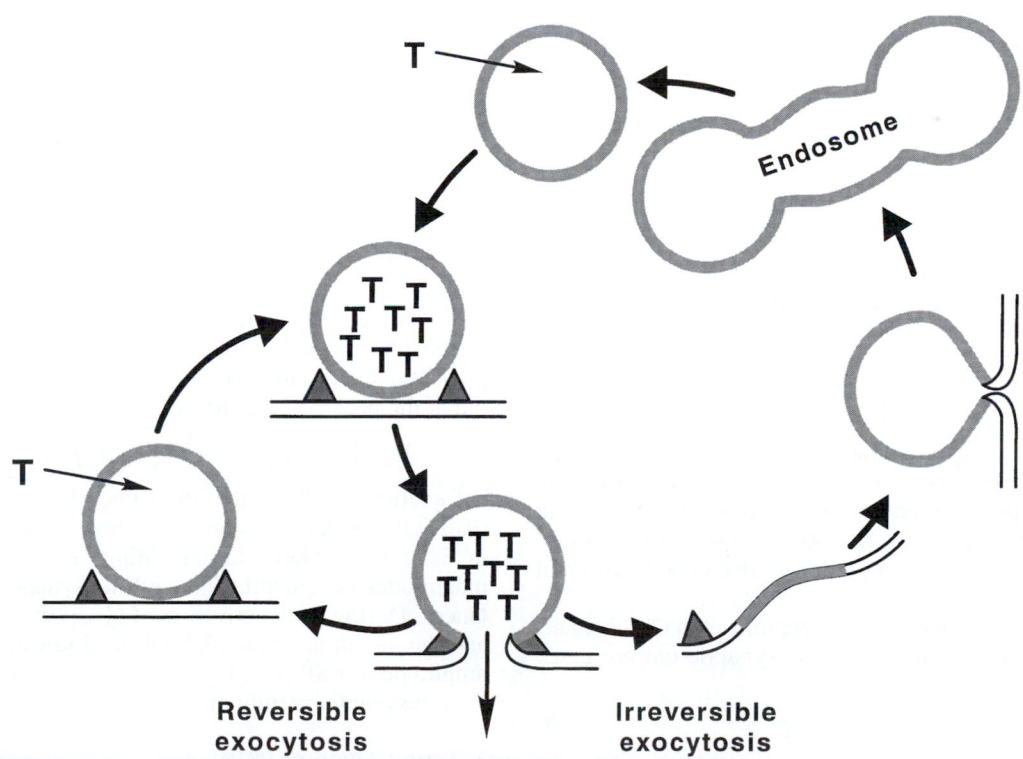

Reversible exocytosis

Irreversible exocytosis

FIGURE 11. Synaptic vesicle cycling in nerve endings. Following vesicle docking and fusion, synaptic vesicles can recycle through a short or long pathway. The short pathway (left) involves transient fusion between vesicle and plasma membrane, followed by fission and recharging of the emptied vesicle with new transmitter (T). In the long pathway (right), vesicle membrane collapses into the plasmalemma to be eventually retrieved and reformed via a sequence of steps, then reloaded with new transmitter.

conductance (γ_S) and the electrochemical driving force ($V_m - E_S$) acting on the ions moving through the channel:

$$i_S = \gamma_S (V_m - E_S) \tag{1}$$

where V_m and E_S are membrane potential and the synaptic equilibrium potential, respectively.

In resting synapse, most of the receptor-gated channels are closed and the conductance of the postsynaptic membrane to ions is very low. When transmitter is released by presynaptic impulse, it binds to its receptors in the postsynaptic membrane and opens the associated channels for a short, 1–2 ms duration. The resultant total synaptic current, I_S, is a sum of currents through all opened channels:

$$I_S = n\gamma_S (V_m - E_S) = g_S (V_m - E_S) \tag{2}$$

where n is the number of active channels and g_S is the total synaptic conductance ($n\gamma_S$).

At excitatory synapses, the transmitter-activated channels permit the passage of both Na^+ and K^+ ions with about equal ease and the excitatory postsynaptic current $I_{S(E)}$ is the sum of Na^+ and K^+ currents:

$$I_{S(E)} = I_{Na} + I_K = g_{Na}(V_m - E_{Na}) + g_K(V_m - E_K) \tag{3}$$

where g_{Na} and g_K are Na^+ and K^+ ion conductances, and E_{Na} and E_K are the Na^+ and K^+ equilibrium potentials, respectively. The direction and magnitude of ion flux is determined by the electrochemical driving forces ($V_m - E_i$). Because g_{Na} ($V_m - E_{Na}$) > $g_K(V_m - E_K)$, that is, $I_{Na} > I_K$, the net synaptic current is always inward and carried by Na^+ ions, resulting in membrane depolarization. As the membrane potential (V_m) becomes depolarized, the term ($V_m - E_{Na}$) decreases and ($V_m - E_K$) increases until equilibrium is reached, where the inward Na^+ current is exactly equal to the outward K^+ current,

$$I_{Na} = -I_K \tag{4}$$

or

$$g_{Na}(V_m - E_{Na}) = -g_K(V_m - E_K) \tag{5}$$

The membrane potential V_m at which this occurs is the synaptic equilibrium potential (E_S). By solving for V_m, it is seen that E_S is a weighted average of sodium and potassium equilibrium potentials:

$$E_S = \left(g_{Na}/(g_{Na} + g_K)\right)E_{Na} + \left(g_K/(g_K + g_{Na})\right)E_K \tag{6}$$

E_S is the limiting potential to which the postsynaptic membrane can be depolarized during the transmitter action. Any further depolarization beyond this point would result in $I_K > I_{Na}$ and reversal of net synaptic current from inward to outward direction. Therefore, the E_S is also called a **reversal potential** (E_r).

It is evident that activation of receptor-gated Cl^- channels at inhibitory synapses results in the synaptic current given by

$$I_{S(I)} = g_S (V_m - E_S) \tag{7}$$

where the equilibrium potential E_S is the chloride equilibrium potential, E_{Cl}. This is the limiting potential to which a synaptic membrane can be hyperpolarized during the action of transmitter at inhibitory synapse.

2. Relationship Between Synaptic Currents and Postsynaptic Potentials

The relationship between the synaptic current and postsynaptic potential can be analyzed in terms of an equivalent electrical circuit consisting of parallel synaptic and nonsynaptic branches, as is illustrated for an excitatory synapse in Fig. 12. The synaptic branch represents the synaptic receptor-gated conductance (g_S) in series with the synaptic battery of the synaptic equilibrium potential E_S. The nonsynaptic branch consists of membrane capacitance C_m and leakage channels (g_m) in series with the battery of the resting membrane potential, E_m.

During the synaptic action of transmitter at the E-synapses, the synaptic current (I_S) flows inward through the synaptic branch and outward through the parallel capacitive and resistive elements of the nonsynaptic branch as I_C and I_m. The direction of current flow at the I-synapses is a mirror image of that at excitatory synapses.

$$I_{S(E)} = -(I_C + I_m) \tag{8a}$$

and

$$-I_{S(I)} = -(I_C + I_m) \tag{8b}$$

where

$$I_C = C_m \, dV_m / dt \tag{9}$$

and

$$I_m = g_m(V_m - E_m) \tag{10}$$

At the onset of synaptic action most of the synaptic current flows through the capacitive branch because the outward driving force ($V_m - E_m$) on current flow through the nonsynaptic channels (g_m) is small. Once the membrane capacitance is discharged (depolarization) or charged (hyperpolarization) to its final value, all synaptic current exits through the leakage channels (g_m). Thus at the peak of synaptic activation, $I_C = 0$ and

$$I_{S(E)} = -I_m \tag{11a}$$

$$-I_{S(I)} = I_m \tag{11b}$$

at E- and I-synapses, respectively.

Substituting Eqs. 2 and 10 into Eq. 11 and solving for V_m yields the expression for the postsynaptic membrane potential at the peak of synaptic activation:

$$V_m = \left(g_S/(g_S + g_m)\right)E_S + \left(g_m/(g_S + g_m)\right)E_m \tag{12}$$

Equation 12 shows that the value of membrane potential (V_m) at the peak of synaptic activation is a weighted average of E_S and E_m, where the weighting factors are the relative magnitudes of synaptic (g_S) and nonsynaptic (g_m) conductances. During peak activation of synaptic channels, $g_S > g_m$ and the membrane potential will tend toward the E_S but the amplitude of PSP (i.e., $V_m - E_m$) will be influenced by g_m of the nonsynaptic membrane.

3. Time Course of PSPs

The synaptic potentials are electrotonic potentials: they decay passively as a function of time and distance. The time course of the rising phase of the synaptic potential is deter-

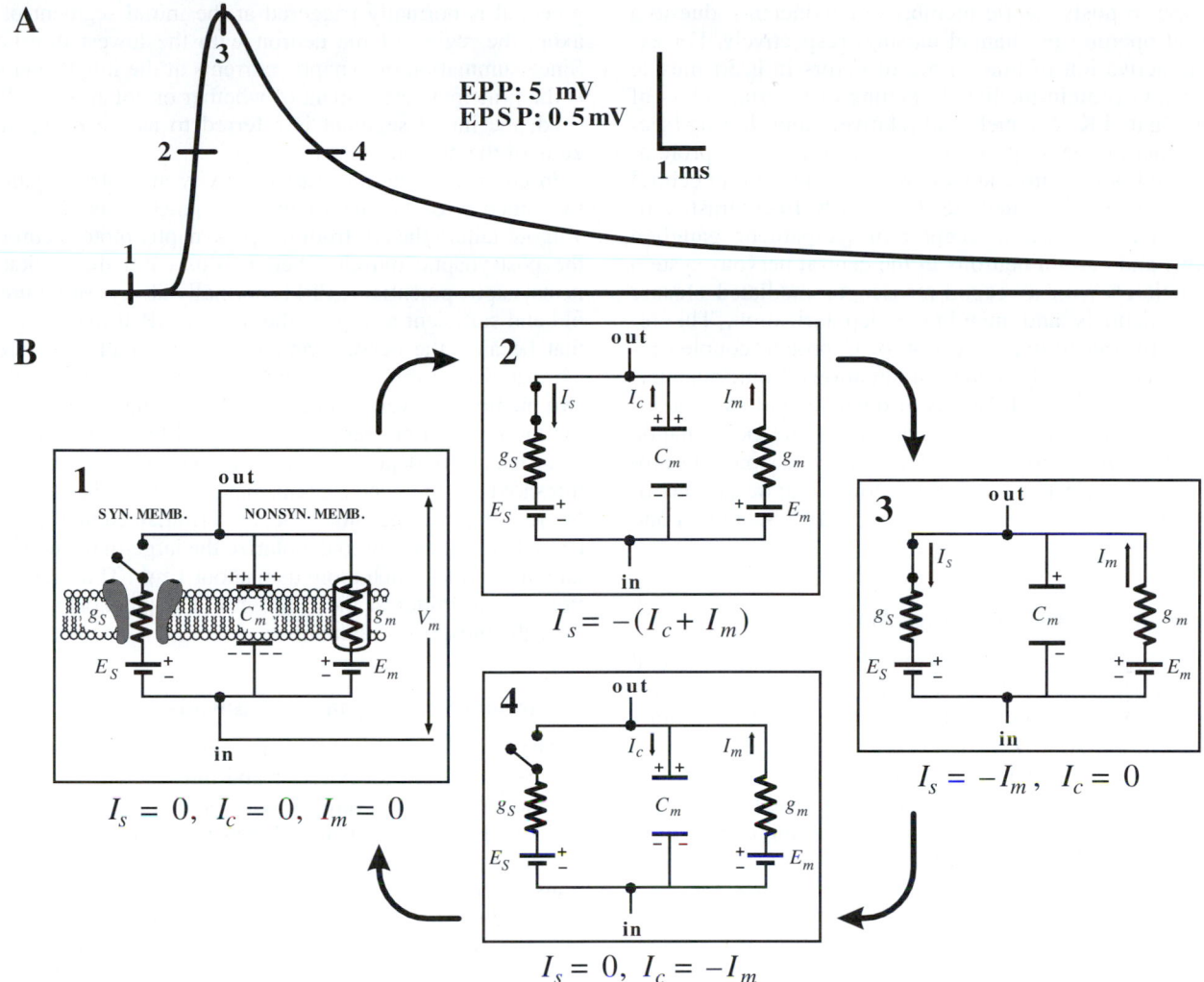

FIGURE 12. Equivalent electrical circuit description of excitatory synaptic potential. Four phases of the postsynaptic potential (A) and the corresponding equivalent circuit representations (B) are illustrated. (1). At rest, the transmitter-gated synaptic channels are closed (circuit open; no current flow; $dV/dt = 0$). (2) Onset of synaptic action (active phase). Synaptic channels are opened by transmitter binding to the receptor (the synaptic switch is closed) and I_s flows inward through the synaptic branch (g_s) and outward as I_c and I_m through the nonsynaptic branch. (3) At the peak of synaptic action, capacitance C_m has discharged to its final value, $I_c = 0$ and $I_s = -I_m$ ($dV/dt = 0$). (4) Passive decay phase. Synaptic channels have reclosed (synaptic switch open) and $I_s = 0$. The current through the nonsynaptic membrane consists of outward I_m and inward I_c, which recharges the membrane capacitance and repolarizes the membrane back to the prestimulus membrane potential. (Based on Kandel et al. "Principles of Neural Science" (2000). McGraw-Hill, New York. Reproduced with permission of The McGraw-Hill Companies.)

mined by both active and passive properties of the membrane. As the synaptic channels spontaneously reclose, the PSP decays in an exponential fashion. The decay of the PSP is purely a passive process whose time course is a function of the membrane time constant, τ. The membrane time constant is the time required for an electrotonic potential to decay to $1/e$ or 37% of its peak value. The time constant of neurons is in the range of 1–20 ms. The amplitude of the synaptic potentials decreases exponentially from the maximum recorded focally at the synapse as a function of the membrane length constant, γ. The length constant is the distance at which electrotonic potentials decay to $1/e$ or 37% of their amplitude at the point of origin. The length constant of dendrites is typically in the range 0.1 to 1 mm.

E. Slow Synaptic Transmission Mediated by G-Protein-Coupled Receptors

In contrast to fast synaptic potentials mediated by directly transmitter-gated channels, slow synaptic responses are mediated by a distinct family of proteins that transduce the transmitter binding into cellular responses through activation of GTP-binding regulatory proteins or G proteins. Binding of a transmitter to a specific receptor transforms the associated G protein into an active form that modulates activity of ionic channels at some distance from the receptor either through direct interaction with the channel and/or through second messenger systems. The slow synaptic transmission may involve either increase or decrease in postsynaptic membrane conductance due to a channel opening or channel closure, respectively. For example, activation of muscarinic receptors in heart muscle causes a G-protein-mediated opening of a certain class of voltage-gated K^+ channels and relatively long-lasting (seconds) membrane hyperpolarization. Similar, G-protein-modulated potassium channels K(G) are present in central neurons, where they mediate slow IPSPs. In contrast, activation of a muscarinic receptor in sympathetic ganglion neurons and certain neurons in the central nervous system is associated with a second-messenger–mediated closure of K^+ channels and membrane depolarization. The responses to transmitter activation of G-protein-coupled receptors need not be limited to modulation of ionic channels but also involve modifications of other regulatory proteins resulting in long-term modifications of the postsynaptic cell's physiology. For a more detailed discussion of G proteins and second messengers in slow synaptic actions of neurotransmitters, the reader is referred to Schulman and Hyman (1999).

F. Synaptic Integration versus Amplification

Central neurons can receive from several dozens to several thousand excitatory (E) and inhibitory (I) synaptic connections converging on the target neuron from a variety of other neurons in the brain. Whether or not a neuron discharges a propagated AP is determined by the number of E-synapses and I-synapses active at any one time. It should be noted that even in absence of IPSPs, activation of a single excitatory synapse would be insufficient to discharge an AP, because individual PSPs generated at central synapses are very small, usually in the range of 0.5–2 mV in amplitude. Therefore, summation of several EPSPs is usually necessary to bring the membrane of the postsynaptic neuron to the threshold potential.

The postsynaptic neurons integrate the synaptic potentials by adding the EPSPs and subtracting the IPSPs from the membrane potential at any instant of time. Algebraic summation of two or more topographically separated synapses that are activated nearly simultaneously is called **spatial summation**. The effectiveness of spatial summation depends on the membrane space constant which, it will be recalled, is the distance at which electrotonic potentials decay to 37% of their amplitude at the point of origin. Clearly, synaptic inputs separated by a distance smaller than the space constant can sum together more effectively than those that are separated by distances larger than the space constant. When a presynaptic neuron fires at a rate such that the interval between successive presynaptic APs is less than the duration of the PSP, each succeeding PSP adds to its predecessor. This process is referred to as **temporal summation**. Effectiveness of temporal summation depends on the membrane time constant (i.e., the time required for an electrotonic potential to decay to 37% of its peak value). The larger the time constant, the longer is the duration of the PSP and thus the greater the opportunity for summation of successive PSPs to occur. A propagated action potential will be triggered only if the net current is of sufficient magnitude to depolarize the neuronal membrane to the threshold. An action

potential is normally triggered at the initial segment of an axon, the region of the neuron with the lowest threshold. Since summation of synaptic currents at the initial segment is the principal determinant of whether or not an AP will be fired, the initial segment is referred to as the **integrative zone** of the neuron.

In contrast to the integrative activity at central synapses, the function of the neuromuscular junction is to transfer without failure the AP from the presynaptic motor neuron to the postsynaptic muscle fiber. In this case, the excitatory postsynaptic potential (EPP) is normally always suprathreshold and sufficient to trigger the muscle AP. It may be noted that because the nerve terminal is very small in diameter compared with the muscle fiber it innervates, even if these two membranes were contiguous, the current generated during the invasion of presynaptic terminal by an AP would be insufficient to depolarize the postsynaptic membrane to threshold due to impedance mismatch between the two membranes; that is, the small nerve terminal cannot provide enough action current to depolarize the large-diameter skeletal muscle fiber much more than about 1 mV. Thus, the function of transmitter at the neuromuscular junction is to amplify the presynaptic signal.

G. Modulation of Synaptic Transmission

The efficacy of signal transmission at chemical synapses can be modulated by extrinsic and intrinsic factors, including the pattern of ongoing activity as well as history of previous activity. The efficacy of transmission may be altered by mechanisms that affect the dynamics of presynaptic transmitter release and/or modify the postsynaptic receptor-mediated events. This modification may be short lasting or may persist for some time. Thus, synaptic modulation provides for fine-tuning of ongoing synaptic activity as well as for longer lasting changes that are likely to play an important role in learning processes.

1. Depression

Depression or fatigue of synaptic transmission refers to progressive reduction in the amplitudes of postsynaptic potentials in the course of prolonged, relatively high-frequency activation of presynaptic neurons, reflecting progressive depletion of releasable transmitter stores. Recovery from fatigue may take from minutes to hours.

2. Facilitation

Facilitation is a frequency–dependent increase in the amplitude of postsynaptic potentials evoked by closely spaced presynaptic action potentials. The increase in the amplitudes of succeeding PSPs reflects the progressively larger amount of transmitter released with each nerve impulse in the course of stimulation. The mechanism is thought to involve build-up of ionized Ca^{2+} within the terminals. That is, when presynaptic neuron is stimulated at certain frequency, the diffusion and clearance of Ca^{2+} from the release sites begins to lag, and the residual Ca^{2+} adds to the Ca^{2+} transient evoked by the next arriving impulse. Build-up of Ca^{2+} increases the probability (P) of transmitter quanta being released. Facilitation is a relatively short-lived process that lasts for a few seconds.

3. Post-Tetanic Potentiation

When the nerve is activated with relatively prolonged and/or high-frequency (tetanic) bursts of impulses, the quantity of transmitter is increased upon subsequent stimulation, even after a relatively long intervening rest period. This phenomenon is known as post-tetanic potentiation. In contrast to facilitation, post-tetanic potentiation may last for minutes and sometimes hours, suggesting a long-term modification of presynaptic function secondary to increase in cytosolic Ca^{2+} levels.

4. Long-Term Potentiation

Long-term potentiation (LTP) refers to a long-lasting increase in EPSP following tetanic stimulation in the presynaptic neurons. It is distinguished from post-tetanic potentiation (PTP) in that the latter is a strictly presynaptic phenomenon, whereas induction and expression of LTP involves both postsynaptic and presynaptic elements. LTP was first described (Bliss and Lomo, 1973) and analyzed most extensively in the hippocampus. In the CA1 region, the LTP involves a special glutamate receptor subtype, the NMDA receptor. The NMDA receptor-gated channel is normally blocked by Mg^{2+}, but can be activated by glutamate when the postsynaptic neuron is sufficiently depolarized so that the Mg^{2+} blockade of the channel is relieved. The induction of LTP appears to be associated with influx of Ca^{2+} and activation of Ca^{2+}-dependent protein kinases CaM kinase II and protein kinase C, and possibly other protein kinases in the postsynaptic dendritic spine, which leads to increased efficacy of synaptic transmission that may last for days to weeks. It is thought that while induction of LTP involves the postsynaptic events, the maintenance of LTP may be associated with long-term increase in the probability of transmitter release, a presynaptic event.

H. Presynaptic Receptors and Transmitter Release

Receptors for neurotransmitters are not confined to postsynaptic sites, but are also found on presynaptic nerve terminals. The modulation of transmitter release by presynaptic receptors that respond to transmitter released by another neuron is referred to as **heterosynaptic modulation**. Heterosynaptic modulation may involve either inhibition or facilitation of transmitter release. In addition, certain presynaptic autoreceptors recognize the cell's own neurotransmitter. In this case, the neuron's transmitter may modulate its own release by interacting with these receptors. This is called **automodulation**. For example, many cholinergic neuron terminals possess muscarinic autoreceptors, and ACh released from these terminals acts on the autoreceptors to inhibit its own release. Although the physiological role of presynaptic receptors has been a subject of debate, it is evident that they provide a potential mechanism for fine-tuning of transmitter release.

IV. Summary

The transmission of synaptic signals is mediated by chemical neurotransmitter substances. Neurotransmitters are synthesized in presynaptic terminals and stored in synaptic vesicles. Transmitter release is evoked by presynaptic APs, which activate influx of Ca^{2+} into terminals and trigger a Ca^{2+}-dependent exocytosis of transmitter from synaptic vesicles into the synaptic cleft. Once released, neurotransmitters activate specific receptor-gated channels in the postsynaptic cell and elicit a transient change in the membrane permeability to cations or anions. Fast synaptic transmission is mediated by ionophoric receptors. Slow synaptic transmission is mediated by G-protein-coupled receptors. Excitatory postsynaptic potentials (EPSPs) are associated with transmitter-induced increase in Na^+ and K^+ conductance of the synaptic membrane, resulting in net entry of positive charge carried by Na^+ and membrane depolarization. Inhibitory postsynaptic potentials (IPSPs) are associated with transmitter-activated influx of Cl^- and membrane hyperpolarization. The EPSPs at the skeletal neuromuscular junction are called end-plate potentials (EPPs). In a healthy neuromuscular junction, the EPPs are always large enough to depolarize the muscle membrane to threshold and trigger muscle APs. The EPSPs generated at any single neuro-neuronal synapse are usually too small to depolarize the postsynaptic neuron to threshold. Synaptic signals converging onto a neuron are normally integrated through summation of EPSPs and IPSPs, and an AP is triggered only when the resultant membrane potential reaches or exceeds the threshold. Chemical synaptic transmission is subject to modulation by intrinsic and extrinsic factors, including frequency and pattern of AP firing, which can either facilitate or depress the transmission across any given synapse.

Bibliography

Amara, S. G., and Arriza J. L. (1993). Neurotransmitter transporters: three distinct gene families. *Curr. Opin. Neurobiol.* **3,** 337–344.

Bennett, M. V. L. (1997). Gap junctions as electrical synapses. *J. Neurocyt.* **26,** 349–366.

Bennett, M. K., and Scheller, R. H. (1993). The molecular machinery for secretion is conserved from yeast to neurons. *Proc. Natl. Acad. Sci. USA* **90,** 2559–2563.

Betz, W. J., and Bewick, G. S. (1992). Optical analysis of synaptic vesicle recycling at the frog neuromuscular junction. *Science* **255,** 200–203.

Bliss, T. V. P., and Lomo, T. (1973). Long-lasting potentiation of synaptic transmission in the dentate area of the anaesthetized rabbit following stimulation of the perforant path. *J. Physiol.* **232,** 331–356.

Boyd, I. A., and Martin, A. R. (1956). The end-plate potential in mammalian muscle. *J. Physiol.* **132,** 30–38.

Brose, N., Petrenko, A. G., Südhof, T. C., and Jahn, R. (1992). Synaptotagmin: a calcium sensor on the synaptic vesicle surface. *Science* **256,** 1021–1025.

Bruns, D., and Jahn, R. (1995). Real-time measurement of transmitter release from single synaptic vesicles. *Nature* **377,** 62–65.

Burns, M. E., and Augustine, G. J. (1995). Synaptic structure and function: dynamic organization yields architectural precision. *Cell* **83,** 187–194.

Ceccarelli, B., Hurlbut, W. P., and Mauro, A. (1973). Turnover of transmitter and synaptic vesicles at the frog neuromuscular junction. *J. Cell Biol.* **54**, 30–38.

Davis, A. F., Bai, J., Fasshauer, D., Wolowick, M. J., Lewis, J. L., and Chapman, E. R. (1999). Kinetics of synaptotagmin responses to Ca^{2+} and assembly with the core SNARE complex onto membranes. *Neuron* **24**, 363–376.

delCastillo, J., and Katz, B. (1954). Quantal components of the endplate potential. *J. Physiol.* **124**, 560–573.

Dreyer, F., Peper, K., Akert, K., Sandri, C., and Moor, H. (1973). Ultrastructure of the "active zone" in the frog neuromuscular junction. *Brain Res.* **62**, 373–380.

Eccles, J. C. (1964). "The Physiology of Synapses." Academic Press, Inc, New York.

Fatt, P., and Katz, B. (1952). Spontaneous subthreshold activity at motor nerve endings. *J. Physiol.* **117**, 109–128.

Greengard, P., Valtorta, F., Czernik, A. J., and Benfenati, F. (1993). Synaptic vesicle phosphoproteins and regulation of synaptic function. *Science* **259**, 780–784.

Hall, Z. W., and Sanes, J. R. (1993). Synaptic structure and development: the neuromuscular junction. *Neuron* **72(suppl.),** 99–121.

Hanson, P. I., Roth, R., Morisaki, H., Jahn, R., and Heuser, J. E. (1997). Structure and conformational changes in NSF and its membrane receptor complexes visualized by quick-freeze/deep etch electron microscopy. *Cell* **90**, 523–535

Hayashi, T., McMahon, H., Yamasaki, S., Binz, T., Hata, Y., Südhof, T. C., and Niemann, H. (1994). Synaptic vesicle fusion complex: action of clostridial neurotoxins on assembly. *EMBO J.* **13**, 5051–5061.

Heidelberger, R., Heinemann, C., Neher, E., and Matthews, G. (1994). Calcium dependence of the rate of exocytosis in a synaptic terminal. *Nature* **371**, 513–515.

Heuser, J. E., Reese, T. S., Dennis, M. J., Jan, Y., Jan, L., and Evans, L. (1979). Synaptic vesicle exocytosis captured by quick freezing and correlated with quantal transmitter release. *J. Cell Biol.* **81**, 275–300.

Kandel, E. R., and Siegelbaum, S. A. (2000). Signalling at the nervemuscle synapse: directly gated transmission. ch. 11, pp. 187–206. *In* "Principles of Neural Science." E. R. Kandel, J. H. Schware, and T. M. Jessell (Eds.). McGraw-Hill, New York.

Katz, B. (1966). "Nerve, Muscle, and Synapse." McGraw-Hill Book Co., Inc.

Katz, B., and Miledi, R. (1967). The timing of calcium action during neuromuscular transmission. *J. Physiol.* **189**, 535–544.

Kuttler, S. W., and Nicholls, J. G. (1977). "From Neuron to Brain: A Cellular Approach to the Function of the Nervous System." Sinauer Associates, Inc., Sunderland, MA.

Lagnado, L., Gomis, A., and Job, C. (1996). Continuous vesicle cycling in the synaptic terminal of retinal bipolar cells. *Neuron* **17**, 957–967.

Lindau, M., and Almers, W. (1995). Structure and function of fusion pores in exocytosis and ectoplasmic membrane fusion. *Curr. Opin. Cell Biol.* **7**, 509–517.

Llinas, R. R. (1977). Calcium and transmitter release in squid synapse. *In* "Approaches to the Cell Biology of Neurons" (W. M. Cowan and J. A. Ferrendelli, Eds.) *Society for Neuroscience Symposia*, **II**, pp. 139–169. Society for Neuroscience, Bethesda.

Llinas, R. R., Gruner, J. A., Sugimori, M., McGuinness, T. L., and Greengard, P. (1991). Regulation by synapsin I and Ca^{2+}-calmodulin–dependent protein kinase II of transmitter release in squid giant synapse. *J. Physiol.* **436**, 257–282.

Martin, R. A. (1977). Junctional transmission II. Presynaptic mechanisms. In "Handbook of Physiology. The Nervous System," Vol. 1., ch. 10, pp. 329–355. American Physiological Society, Bethesda, MD.

McMahon, H. T., and Nicholls, D. G. (1991). The bioenergetics of neurotransmitter release. *Biochim. Biophys. Acta* **1059**, 243–264.

Matthews, G. (1996). Synaptic vesicle exocytosis and endocytosis: capacitance measurements. *Curr. Opin. Neurobiol.* 6, 358–364.

Maycox, P. R., Hell, J. W,. and Jahn, R. (1990). Amino acid neurotransmission: spotlight on synaptic vesicles, Trends Neurosci. 13, 83–87.

Miledi, R. (1973). Transmitter release induced by injection of calcium ions into nerve terminals. *Proc. Royal Soc.* **183**, 421–425.

Niemann, H., Blasi, J., and Jahn, R. (1994). Clostridial neurotoxins: new tools for dissecting exocytosis. *Trends Cell Biol.* **4**, 179–185.

Pappas, G. D., and Purpura, D. P. (Eds.). (1972). "Structure and Function of Synapses." Raven Press Publishers, New York.

Petrenko, A. G., Perin, M. S., Davletov, B. A., Ushkaryov, Y. A., Geppert, M., and Südhof, T. C. (1991). Binding of synaptotagmin to the a-latrotoxin receptor. *Nature* **353**, 65–68.

Robitaille, R., Adler, E. M., and Charlton, M. P. (1990). Strategic location of calcium channels at transmitter release sites of frog neuromuscular junction. *Neuron* **5**, 773–779.

Ryan, T. A., Reuters, H., Wendland, B., Schweizer, F. E., and Smith, S. J. (1993). The kinetics of synaptic vesicle recycling measured at single presynaptic boutons. *Neuron* **11**, 713–724.

Sabatini, B., and Regehr, W. G. (1996). Timing of neurotransmission at fast synapses in the mammalian brain. *Nature* **384**, 170–172.

Schulman, H., and Hyman, S. E. (1999). Intracellular signaling. In "Fundamental Neuroscience" (M. J. Zigmond, F. E. Bloom, S. C. Landis, J. L. Roberst, and L. R. Squire, Eds.), pp. 269–316 Academic Press, San Diego.

Shao, X., Davletov, B. A., Sutton, R. B., Südhof, T. C., and Rizo, J. (1997). Bipartite Ca binding motif in C2 domains of synaptotagmin and protein kinase C. *Science* **273**, 248–251.

Smith, S. J., and Augustine, G. J. (1988). Calcium ions, active zones and synaptic transmitter release. *Trends in Neurosci.* **11**, 458–464.

Söllner, T., Bennett, M., Whiteheart, S., Scheller, R., and Rothman, J. (1993a). A protein assembly-disassembly pathway in vitro that may correspond to sequential steps of synaptic vesicle docking, activation and fusion. *Cell* **75**, 409–418.

Söllner, T., Whiteheart, S. W., Brunner, M., Erdjument-Bromage, H., Geromanos, S., Tempst, P., and Rothman, J. E. (1993b). SNAP receptors implicated in vesicle targeting and fusion. *Nature* **362**, 318–324.

Sonders, M. S., and Amara, S. G. (1996) Channels in transporters. *Curr. Opin. Neurobiol.* **6**, 294–302.

Spray, D. C., Scemes, E., and Rozental, R. (1999). Cell-cell communication via gap junctions. In "Fundamental Neuroscience" (M. J. Zigmond, F. E. Bloom. S. C. Landis, J. L. Roberts, and L. R. Squire, Eds.), pp. 317–343, Academic Press, San Diego.

Südhof, T. C.(1995). The synaptic vesicle cycle: a cascade of proteinprotein interactions. *Nature*, **375**, 645–653.

Sutton, R. B., Fasshauer, D., Jahn, R., and Brunger, A. T. (1998). Crystal structure of a SNARE complex involved in synaptic exocytosis at 2.4 Å resolution. *Nature* **395**, 347–353.

Whiteheart, S. W., Rossnagel, K., Buhrow, S. A., Brunner, M., Jaenicke, R., and Rothman, J. E. (1994). N-ethylmaleimide–sensitive fusion protein: a trimeric ATPase whose hydrolysis of ATP is required for membrane fusion. *J. Cell. Biol.* **126**, 945–954.

Nicole Gallo-Payet and Marcel Daniel Payet

42

Excitation-Secretion Coupling

I. Introduction

The process of excitation-secretion coupling is completely different depending on the peptide/amine or lipid nature of the secretory products. The secretory process for peptide and amine molecules begins with the synthesis, modification, and sorting of the molecules to be secreted. Synthesis occurs in the rough endoplasmic reticulum (RER) and sorting occurs in the Golgi complex. The secretory molecules are packaged in secretory granules or vesicles, which are then transported to the cell periphery before they are released in the extracellular space by fusion with the plasma membrane. This complex process is named **exocytosis** or **reverse pinocytosis** and can be operated either by a constitutive or a regulated mechanism. Constitutive secretion is unregulated and closely follows the rate of synthesis of the secretory products. This form of secretion occurs in many cell types, including lymphocytes, hepatocytes, and pancreatic β cells. In regulated secretion, fusion of the secretory granules with the plasma membrane is initiated by a specific signal (ligand-receptor coupling), triggered by an increase in cytosolic calcium concentration ($[Ca^{2+}]_i$). In steroid-secreting cells (adrenal cortex, ovary, testis), the process of synthesis begins with cholesterol stored in lipid droplets followed by subsequent steps occurring in mitochondria and smooth endoplasmic reticulum. It is generally assumed that steroids are free to diffuse throughout the aqueous cytoplasm and lipid phase of the plasma membrane. Secretory vesicles are not present and secretion and/or release of steroids is tightly coupled to steroid synthesis.

Although the process of synthesis differs, stimulation of secretion of peptide hormones, neurotransmitters, and steroids involves similar cascades of molecular events. After binding to their specific receptors, the stimuli activate second messenger production, several cascades of phosphorylation/dephosphorylation of intracellular proteins, cytoskeleton reorganization, synthesis of new products, and

release of secretory products. Cytoskeleton and Ca^{2+} ion are certainly the most important players involved in this excitation-secretion coupling. However, while disruption of the actin network is necessary to trigger fusion of secretory vesicles with the cell membrane, a well-preserved organization seems important for steroid release.

II. Cellular Components Involved in Excitation-Secretion Coupling

A. Interaction of Cytoskeletal Structures with Transmembrane Signaling Molecules

The cytoskeletal elements are described in Chapter 6. Therefore the brief descriptions given here on microfilaments and microtubules are aimed at understanding the mechanisms involved in excitation-secretion coupling (Schmidt and Hall, 1998). In the living cell, actin filaments, F-actin (consisting of two staggered, parallel rows of monomers, G-actin, noncovalently bound and twisted into a helix), interact with several proteins, such as vinculin, α-actinin, villin, and fodrin, which cross-link and bundle microfilaments into a well-organized three-dimensional network. They can also cross-link myosin, forming a contractile network, or form a very dense network at the cell periphery, the cell cortex (Fig. 1A and Fig. 2A). This dynamic organization of microfilaments depends on a large and diverse group of actin-binding proteins, including profilin and caldesmon (which bind G-actin), and gelsolin and scinderin (which cap and sever F-actin). On the other hand, polymerization, stabilization, and modulation of microtubules depend on several microtubule-associated proteins (MAPs) that adorn the tubulin-containing core of the tubules. The complete cell cytoskeletal network includes interaction between microfilaments, microtubules, and intermediate filaments (see Figs. 1 and 2 and Chapter 6).

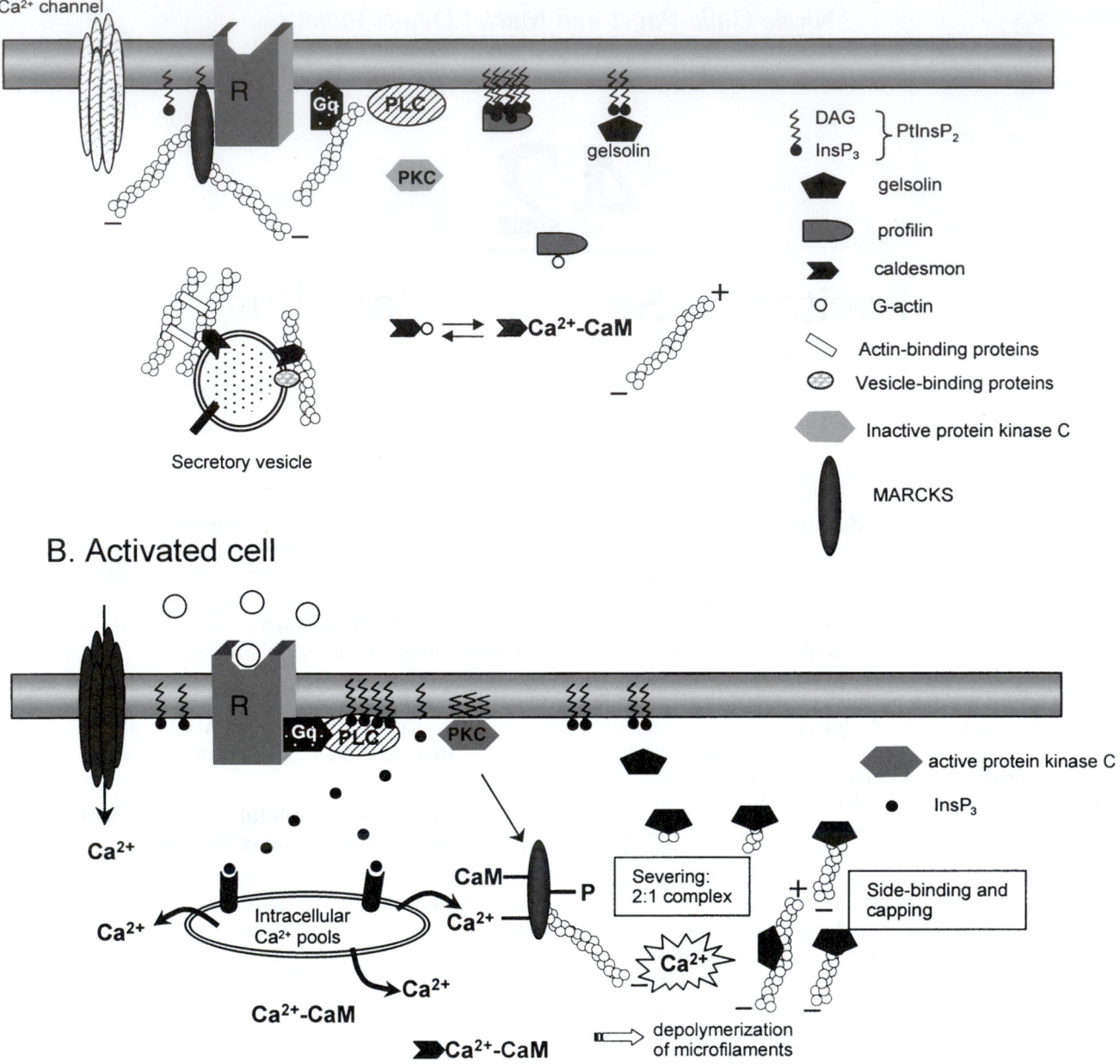

FIGURE 1. Interaction of cytoskeletal structures with transmembrane signaling. In resting cells (A), receptors, G proteins, phospholipase C (or adenylyl cyclase), and phosphoinositides are not linked together, but are associated with microfilaments or actin-associated proteins. For example, profilin interacts with high affinity with eight molecules of PtdInsP2, protecting them from hydrolysis by phospholipase C; gelsolin is also associated with $PtInsP_2$; the protein caldesmon is associated with both actin monomers and calmodulin; MARCKS (myristoylated, alanine-rich C kinase substrate) is a specific protein kinase C substrate associated with the cytoplasmic face of the membrane under resting conditions. In its nonphosphorylated form, MARCKS cross-links actin, favoring a rigid actin meshwork at the membrane level. In activated cells (B), ligand binding to its receptor activates phospholipase C, which hydrolyzes $PtdInsP_2$ and leads to $InsP_3$ and diacylglycerol (DAG). The rise in $[Ca^{2+}]_i$ (due to the $InsP_3$ binding to intracellular pools of Ca^{2+}) induces F-actin depolymerization. Moreover, calmodulin-caldesmon complex binds to Ca^{2+}, gelsolin binds, severs, and caps actin filaments into 2:1 complexes to the barbed ends. Activated protein kinase C (now at the membrane) phosphorylates MARCKS, which is released from the membrane. All these modifications decrease cell rigidity, making the actin network more plastic.

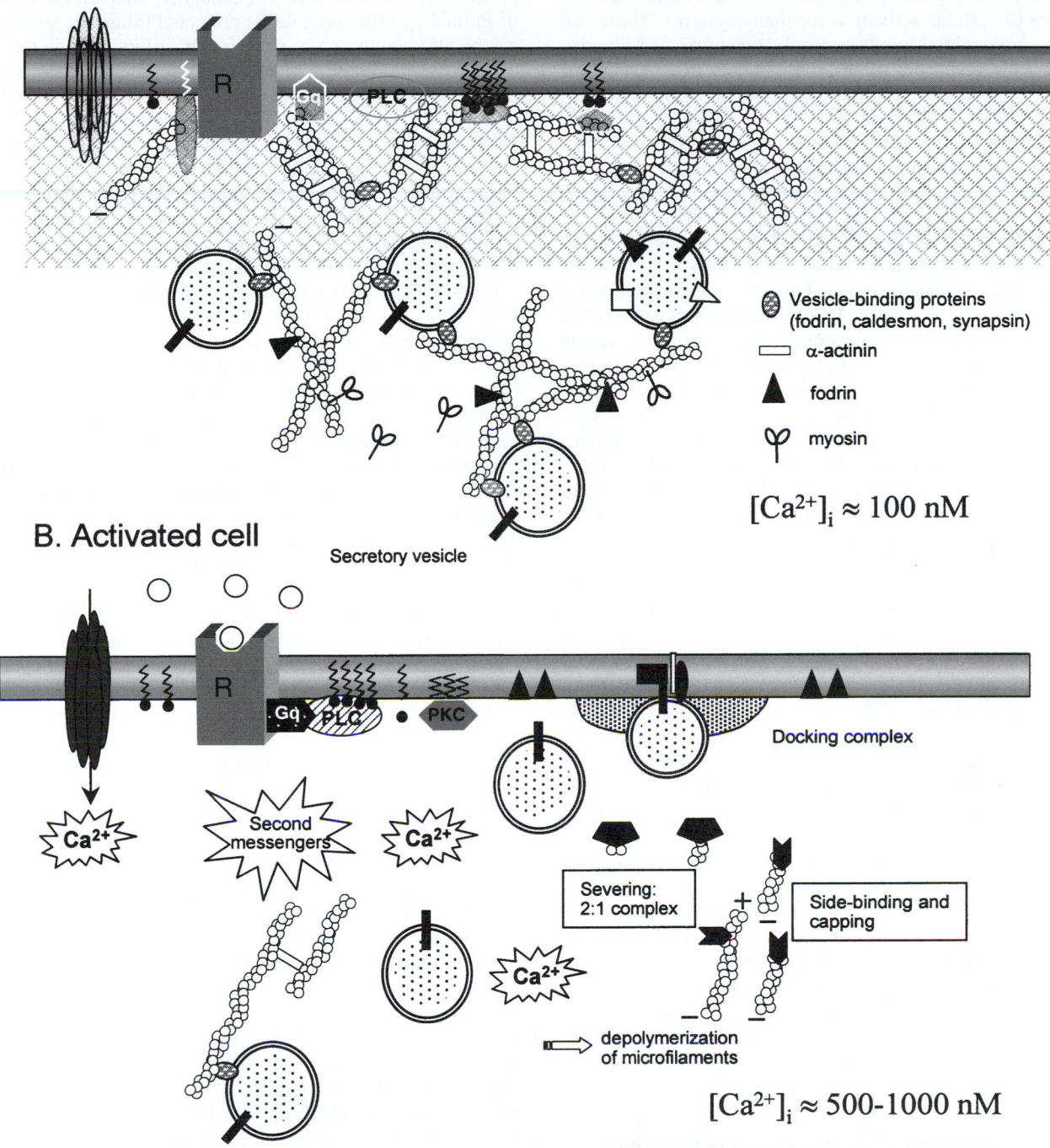

FIGURE 2. Dynamic changes in cytoskeleton during exocytosis. In resting cells ($[Ca^{2+}]_i \approx 100$ nM) (A), microfilaments interact with several proteins (including fodrin and α-actinin) that cross-link and bundle microfilaments into a well-organized three-dimensional network. In vesicle-secretory cells, microfilaments can also cross-link myosin, forming a contractile network, or form a very dense network at the cell periphery, the cell cortex. Activation of cells by appropriate stimulation (B) is associated with a reduction in the cell cortex rigidity, due to characteristic changes in cytoskeletal properties from "gel" (solid) to "sol" (liquid) states. These changes are primarily under the control of Ca^{2+}, which cooperates with phospholipid messengers to induce microfilament disruption. An increase in intracellular calcium ($[Ca^{2+}]_i$ (≈ 0.5–1 μM) induces several cellular modifications. In addition to capping and severing of the actin microfilaments (see Fig. 1), there is dissociation of actin from actin-associated proteins, such as fodrin, patching of fodrin along the plane of the plasma membrane (docking site); and dissociation of caldesmon from secretory vesicles. These events result in a decrease in viscosity, which favors movement of granules toward the plasma membrane releasing sites. Actin-myosin interactions could facilitate granule displacement in cytosol.

Association of cell surface molecules with the cytoskeleton is widely believed to be one of the earliest consequences of cellular activation for many systems. These cell surface molecules include not only receptors, but also **integrins,** the receptors of the components of extracellular matrix. Appropriate stimulation of receptors and integrins has immediate and profound effects on the organization and activity of the microfilaments (Aplin *et al.,* 1998). Several studies have shown that microfilament disruption (with cytochalasins) or microtubule disruption (with colchicine or vinblastine) increased exocytosis (i.e., peptide or amine secretion), although steroid hormone secretion is decreased or abolished. In other words, these observations indicate that cytoskeleton is a barrier in the former, but a requirement for the latter. In both cases, microfilaments are active players in the process of excitation-secretion coupling. Several studies indicate that receptors (either G-protein-coupled or tyrosine kinase types); α subunits of heterotrimeric G-proteins (α_i, α_s, α_q); mono-meric G proteins (Rho, Rac, Cdc42); second-messenger-activating enzymes (adenylyl cyclase, phospholipase C (PLC)), Ca^{2+}, K^+, or Cl^- channels (see Chapter 36); and proteins up or downstream from second messenger production (phosphatases, protein kinase C (PKC), mitogenic associated protein kinase (MAPK)) are associated with microfilaments. For example, anchoring of PKC-ϵ to F-actin is required for glutamate release in nerve endings of neuronal cells (Janmey, 1998). In general, these associations occur mainly via acting-binding proteins, which bind effectors or phosphoinositides through specific domains, called plecktrin homology domains, (PH domains). PH domains are essential for the membrane recruitment of several proteins that contain them and are frequently accompanied by other motifs (Src homology domains SH2 and SH3, and proline-rich, Dbl homology (DH), or GTP exchange domains), indicating that the recruitment of the PH domain protein would nucleate sites on the membrane for protein complex assembly (Inglese *et al.,* 1995; Janmey, 1998; Martin, 1998). In addition, close functional association between G proteins and microtubules has also been extensively described during the past five years (Popova *et al.,* 1997). Such observations indicate that the cytoskeleton operates as a matrix improving the efficiency of the signal transduction cascade and that actin-binding proteins are in large part responsible for these cytoskeleton-membrane receptor interactions.

B. Actin-Binding Proteins Important in Signaling

The physical properties of actin networks depend on the length of microfilaments and the architecture of the three-dimensional network formed by interaction between actin and several actin-binding proteins. The aim of this section is to point attention to the F-actin cross-linking proteins which may be regulated by Ca^{2+} and thus involved in the process of excitation-secretion coupling (Schmidt and Hall, 1998).

Profilin was the first actin-binding protein described in signaling. Profilin also contains a phosphatidylinositol bisphosphate (PtdInsP$_2$) binding site, and binding of PtdInsP$_2$ to this site triggers the dissociation of profilin from actin, promoting actin polymerization (Fig. 1A). PtdInsP$_2$ molecules bind to profilin with a stoichiometry of about 8:1. This 8:1 association of PtdInsP$_2$:profilin complexes protects PtdInsP$_2$ from cleavage by phospholipase Cγ. In a similar fashion, also in resting conditions, PLCγ may also bind profilin. However, when PLCγ becomes tyrosine phosphorylated (by appropriate ligand activation), its affinity for PtdInsP$_2$ increases to a level where profilin protection can be overcome, resulting in PtdInsP$_2$ hydrolysis. Hydrolysis of one or two of the eight PtdInsP$_2$ molecules bound to each profilin leads to a rapid decrease in PtdInsP$_2$ affinity and release of the remaining profilin-bound PtdInsP$_2$ molecules (Fig. 1B). Thus, the binding of profilin to PtdInsP$_2$ not only liberates polymerization-competent G-actin, but also affects hydrolysis of PtdInsP$_2$ by PLCγ (Forscher, 1989; Janmey, 1998).

Gelsolin is a Ca^{2+}-dependent F-actin severing molecule which, along with villin, fragmin, adseverin, and scinderin, is able to sever actin filaments. These proteins bind to actin filaments and bend and cleave them in a Ca^{2+}-dependent manner and afterwards cap the barbed filament ends. Severing actin filaments promotes cell cortex transition from a "gel" state to a "sol" state upon addition of Ca^{2+}. Gelsolin contains two spatially separate binding sites: a G-actin Ca^{2+}-sensitive site in the C-terminal domain and a PtdIns-sensitive site closer to the N-terminal portion. In resting cells ($[Ca^{2+}]_i \approx 100$ nM), the bulk of gelsolin is cytoplasmic and a small proportion is PtdInsP$_2$-associated. Gelsolin has little affinity for actin under these conditions and is in an actin-free state (Fig. 1A). When the cell is activated, there is a transient rise in $[Ca^{2+}]_i$ due to InsP$_3$ action or Ca^{2+} influx that then activates gelsolin, causing a 200-fold increase in its affinity for F-actin. This results in rapid filament side-binding, followed by severing and capping of any free barbed filament ends, inducing a dramatic disruption of existing actin network structure in the vicinity of Ca^{2+} elevation (Fig. 1B). The activities of these severing proteins are inhibited by binding to PtdInsP$_2$.

In the absence of PtdIns turnover, much of the PtdInsP$_2$ is likely to be tightly associated with profilin and thus unavailable to gelsolin. If gelsolin is activated under these conditions (for example by an increase in $[Ca^{2+}]_i$, independent of the PtdIns turnover), severing of actin networks is observed, but without subsequent actin reassembly. This leads to two possible modes of gelsolin activation: (1) Calcium influx produced by activation of PtdInsP-independent pathways (i.e., voltage- or agonist-gated Ca^{2+} channels) leads to actin severing and capping only. (2) In contrast, activation of these same Ca^{2+} channels concomitant with PtdInsP turnover results in severing and capping followed by actin polymerization, that is, actin remodelling (Forscher, 1989; Janmey, 1998).

In summary, association of actin-binding proteins with phosphoinositides causes opposite effects to those observed upon association with Ca^{2+}. In the former, such associations (profilin:PtInsP$_2$) promotes actin polymerization, favoring the cortical actin "gel" near the plasma membrane, while increased Ca^{2+} concentration favors association of actin-binding proteins (profilin, gelsolin) with Ca^{2+}, inducing depolymerization and thus "sol" state of actin filaments.

Caldesmon is a calmodulin-dependent actin-binding protein that, at low Ca^{2+} concentrations (100 nM), binds and cross-links actin monomers, inhibiting actin polymerization. Under these conditions, caldesmon interacts reversibly with secretory granules. At greater concentrations of Ca^{2+} (μM ranges), Ca^{2+}-calmodulin complex binds to caldesmon and reverses this inhibition. The flip-flop regulation of caldesmon may be important for secretory vesicle function during the changes in intracellular Ca^{2+} levels observed upon stimulation (Figs. 1 and 2).

MARCKS (myristoylated, alanine-rich C kinase substrate) is a specific protein kinase C substrate that is targeted to the membrane by its amino-terminal binding domain. In resting cells, MARCKS associates with the cytoplasmic face of the membrane. In its nonphosphorylated form, MARCKS cross-links actin, favoring a rigid actin meshwork at the membrane level (Fig. 1A). Activated protein kinase C phosphorylates MARCKS, which remains associated with actin filaments, but can no longer cross-link actin fibers, making the actin network more plastic (Fig. 1B). In addition, an increase in intracellular Ca^{2+} concentration promotes binding of calmodulin to MARCKS, inhibiting its actin cross-linking activity, again resulting in a less rigid actin meshwork. Thus, PKC induces a local destabilization of the actin skeleton through the phosphorylation of MARCKS. MARCKS is phosphorylated when synaptosomes are depolarized, suggesting a role in secretion (Arbuzova *et al.*, 1998).

Several newly identified proteins, such as the focal adhesion molecule (p125^{FAK}), paxillin, and the small GTP-binding protein Rho are also closely implicated in actin polymerization after hormonal stimulation. Tyrosine phosphorylation of p125^{FAK} and paxillin and their association with cytoskeleton and $\beta\gamma$ subunits of G proteins have been recently identified as early events in the action of several growth factors and G-protein-coupled receptors (such as angiotensin II and vasopressin). However, to date, their role has been ascribed to regulating cell adhesion, motility, or proliferation rather than secretion (Inglese *et al.*, 1995; Tapon and Hall, 1997; Hall, 1998).

C. Interaction between Microtubules and Microtubule-Associated Proteins in Signaling

As mentioned earlier, polymerization, stabilization, and plasticity of microtubules depend on several microtubule-associated proteins (MAPs), which differentially cross-link microtubules (Matus, 1988). Recent studies have shown that some MAPs could be implicated in the secretory process (Gundersen and Cook, 1999). MAPs are substrates for several protein kinases, including a Ca^{2+}-calmodulin-dependent kinase, a cAMP-dependent kinase, tyrosine kinases, and protein kinase C. Both cAMP (via protein kinase A) and Ca^{2+} (via Ca^{2+}-calmodulin kinase) lead to phosphorylation of MAP-2. Moreover, MAP-1 and MAP-2 are responsible for binding secretory granules to microtubules, either in cells from the anterior pituitary gland, the β cells of the endocrine pancreas or in synaptic vesicles of neurons. These results strengthen the probable role of MAPs in secretory processes.

D. Actin-Binding, Docking and Fusion Proteins of the Secretory Granules

Granule vesicles for peptides or amine products not only allow secretory tissues to store large amounts of secretory products in a relatively small volume, but also protect this material from intracellular degradation while providing a very efficient means for transporting and releasing fixed quantities of secretory material.

1. Actin-binding Proteins Associated with Secretory Granules

Induction of exocytosis is associated with a loss in the cell cortex rigidity, due to characteristic changes in cytoskeletal properties from "gel" to "sol" states. These changes are primarily under the control of Ca^{2+}, which cooperates with phospholipid messengers to induce microfilament disruption (Fig. 1B and 2B) (Tchakarov *et al.*, 1998). Moreover, several pieces of evidence indicate that microfilaments, rather than microtubules or intermediate filaments, are responsible for the mechanical changes initiated by Ca^{2+}. Chromaffin cells of the adrenal medulla and mast cells from the immune system synthesize and, along with neuronal synaptic vesicles, store and secrete large amounts of neurotransmitters or neuropeptides. These cells contain numerous electron-dense secretory granules, which discharge their contents into the extracellular space by exocytosis. The subplasmalemmal area is characterized by the presence of a highly organized cytoskeletal network. F-actin seems to be exclusively localized in this area and, together with specific actin-binding proteins, forms a dense viscoelastic gel.

Fodrin, vinculin, α-actinin, and caldesmon, four actin-binding proteins, as well as gelsolin and scinderin, two actin-severing proteins, are found in the plasmalemmal region. Moreover, fodrin, caldesmon, and α-actinin binding sites also exist on secretory granule membranes, indicating that actin filaments can also link to secretory granules (Fig. 2A). Chromaffin granules can be entrapped in this subplasmalemmal lattice, and thus the cytoskeleton acts as a barrier preventing exocytosis (Burgoyne, 1995; Burgoyne and Morgan, 1998a).

Synapsin I, a phosphoprotein and a substrate for protein kinase A and calmodulin-dependent protein kinase II (CaM kinase II), is associated with synaptic vesicles. Synapsin I also binds to spectrin and actin microfilaments, and may serve as an anchor between synaptic vesicles and the cytoskeleton. The affinity of synapsin I for synaptic vesicles is decreased by phosphorylation, and neurotransmitter release is preceded by a reversible phosphorylation of synapsin I. Therefore, synapsin I phosphorylation results in the release of synaptic vesicles from their anchorage sites on the cytoskeleton, thus allowing the vesicles to move to the active exocytosis zones.

2. Docking and Fusion Proteins

In addition to the actin-binding proteins, several proteins from the secretory vesicle membrane, the plasma membrane, and the bulk cytosol interact together to form the core complex (or a **docking complex**) with docking, priming, and fusioning properties (or **budding properties**). The most important of these properties are described next (Südhof, 1995; Fernandez-Chacon and Südhof, 1999).

The vesicular proteins synaptobrevins (or vesicle-associated membrane proteins, VAMPS) and the plasma membrane proteins syntaxin and synaptosome-associated protein (SNAP-25, M_r 25 kDa) form the three complex membrane proteins (called the core complex) essential to the regulated exocytosis machinery in neurons and neuroendocrine and endocrine cell types. Moreover, the potential involvement of the cytosolic proteins, N-ethylmalimide-sensitive fusion protein (NSF) and the soluble NSF attachment proteins (SNAPs) (α, β, γ isoforms), was demonstrated by the discovery that all these proteins interact to form a complex named SNARE (for SNAP receptor) (Fig. 3). Based on *in vitro* studies, a model for neurotransmitter release was proposed, in which synaptic vesicles become docked at the plasma membrane by a specific pairing of VAMP (the vesicle membrane, v-SNARE) with syntaxin 1 and SNAP-25 (the target membrane, t-SNARE). The proposed sequence for docking and priming is described hereafter (Südhof, 1995; Aroeti *et al.*, 1998; Burgoyne and Morgan, 1998a). Before and/or during docking, syntaxin is bound to Munc 18 and synaptophysin to synaptobrevin. These two complexes, syntaxin-Munc18 and synaptophysin-synaptobrevin, must dissociate in order for the core complex to be formed. Formation of this complex is considered to be the first step of vesicle priming. Syntaxin and SNAP-25 (t-SNARE) can then bind tightly together to form a high-affinity site for synaptobrevin (v-SNARE) located on the vesicle membrane; the core complex has a stoichiometry of

1:1:1. The trimeric core complex serves as a receptor (SNARE) for the soluble SNAPs (not related to SNAP-25). NSF will only interact with SNAPs (α, β) bounded on the trimeric complex. The NSF-SNAP receptor forms a multi-subunit particle that sediments at 20S; it may form the core of a generalized apparatus catalyzing bilayer fusion (Söllner *et al.*, 1993). NSF is a trimeric protein that cross-links multiple core complexes into a network. The core complex is then disrupted by enzymatic activity of NSF under ATP hydrolysis. Botulinum A and tetanus toxins are toxin proteases able to digest synaptobrevin, SNAP-25, and syntaxin. When entering in the nerve terminal, they irreversibly inhibit exocytosis. However, the number of docked granules is not decreased by the toxins, indicating that the primary function of the core complex is fusion and not docking. Hydrolysis of ATP by NSF is followed by ATP-independent steps (see later) sensitive to temperature, H^+, and Ca^{2+}. The last step is Ca^{2+}-sensitive and likely involves a Ca^{2+} sensor at the site of exocytosis.

Synaptotagmins (Syt) are membrane glycoproteins found in brain secretory vesicles of which eight forms have been cloned. One of these, synaptotagmin I (Syt I), plays a pivotal role in the Ca^{2+}-triggered neurotransmitter release as a Ca^{2+} sensor (Südhof and Rizo, 1996; Goda and Südhof, 1997; Burgoyne and Morgan, 1998b). The functional implication of multiple synaptotagmins is unknown. Syt I binds Ca^{2+} co-operatively and undergoes a Ca^{2+}-dependent conformational change; the coefficient of cooperativity (4) is similar to that observed for Ca^{2+}-triggered release. Syt I also binds phospholipids as a function of Ca^{2+} with high affinity (half max-

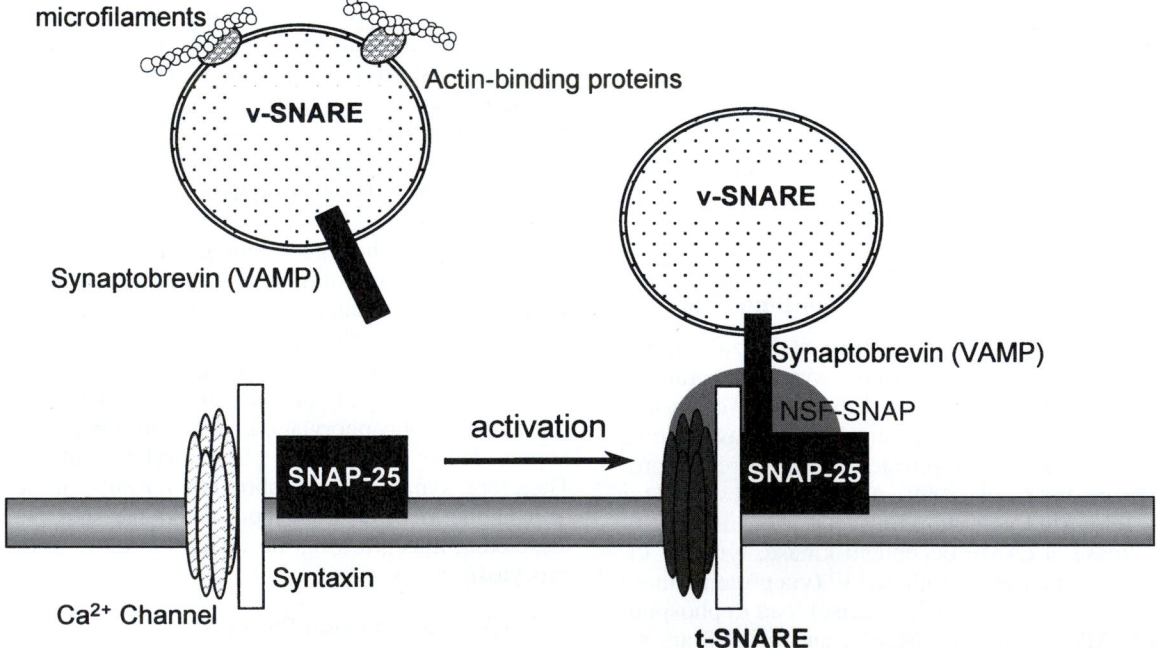

FIGURE 3. SNARE, the receptor involved in docking and fusion. The 20S particle that forms the core complex contains several interacting proteins. v-SNARE, related to synaptobrevin (VAMP) and located on the vesicle, binds to t-SNARE, related to syntaxin and SNAP-25 and located on the plasma membrane, to form the SNARE or SNAP receptor. SNAPs and NSF can then bind to the receptor to complete the core complex. Numerous SNARE-related proteins, each specific for a single kind of vesicle or target membrane, ensure vesicle-to-target specificity. Binding of synaptobrevin relies on the inhibition of the Ca^{2+} channel.

imal binding, 5–6 μM). Syntaxin, one of the proteins of the core complex, in addition to its role in the docking and fusion process of the vesicle, controls the entry of Ca^{2+} by the voltage-dependent Ca^{2+} channels (L- and N-type). Indeed, binding of syntaxin to the Ca^{2+} channel prevents the opening of the channel by membrane depolarization. When the vesicle docks to the membrane by binding to syntaxin and SNAP-25, the inhibition of syntaxin on the channel is relieved, allowing voltage-dependent opening of the channel (Fig. 3) (Geppert and Südhof, 1998).

In addition to phospholipids and syntaxin, Syt I binds to neuroxins, a family of neuronal cell surface proteins, and to AP-2, a protein complex involved in synaptic vesicle endocytosis. In addition to a Ca^{2+} sensitivity similar to Ca^{2+}-triggered release, the role of Syt I as a Ca^{2+} sensor has been illustrated in several ways. Knockout mice for Syt I have impaired Ca^{2+}-triggered transmitter release, but release can still be obtained by Ca^{2+}-independent agents such as hypertonic sucrose or the excitatory neurotoxin α-latrotoxin (the receptor for α-latrotoxin belongs to the neuroxin family). Synaptotagmin *Drosophila* mutants show a severe but incomplete block of neurotransmission with an altered Ca^{2+}-dependence in some mutants. Injection of synaptotagmin peptide in squid nerve terminal inhibits release and vesicles accumulate possibly by competing for a common effector. Synaptotagmins are also able to interact with non-neuronal syntaxin, indicating that they can play a role in a variety of cell types from endocrine and immune systems.

Synaptophysin and the related protein synaptogyrin are major integral membrane proteins of small presynaptic vesicles. The presence of phosphorylation sites for tyrosine kinase of the Src family could indicate that phosphorylation modulates protein activity. The primary structure was deduced from the cDNA sequence, leading to the proposition that synaptophysin and synaptogyrin span the membrane four times with N- and C-terminals located in the cytoplasm. Ubiquitous isoforms of these two proteins are co-expressed in all cells and could represent invariant components of trafficking organelles.

The annexin family includes several Ca^{2+}- and phospholipid-binding proteins with conserved structure. Two of these are thought to be involved in the mechanism of exocytosis. Synexin (annexin VII) is a calcium-binding protein (47 kDa) with four transmembrane domains. This protein demonstrates a voltage-dependent Ca^{2+} channel activity. Synexin is able to induce the aggregation of chromaffin vesicles in the presence of Ca^{2+}. A model for the synexin-driven Ca^{2+}-dependent membrane fusion has been proposed in which synexin monomers polymerize as the concentration of Ca^{2+} increases. The polymerized synexin forms a hydrophobic bridge between the two membranes. Annexin II (calpactin) is located on the cytoplasmic side of the plasma membrane in chromaffin cells. Sites of phosphorylation for protein kinase C and cAMP- and calmodulin-dependent protein kinases have been localized within the N-terminal domain; phosphorylation of these sites could inactivate the protein. A possible role for annexin II could be to link the granule with the plasma membrane (docking) and/or to induce fusion (Burgoyne, 1991).

Exocytotic fusion mechanisms also involve PtdIns transfer protein, PtdIns 4-kinase (PI4K), and PI5K to promote formation of $PtIns4,5P_2$ on the vesicle membrane (Martin, 1998).

III. Cellular and Molecular Events in Chromaffin, Mast Cells, and Neuronal Synaptic Vesicles

Exocytosis is an all-or-none phenomenon in which Ca^{2+} plays a pivotal role. Ca^{2+} is required for second messenger activity, for the control of cytoskeletal dynamics, and for the vesicle-plasma membrane fusion process. In neurons, neuroendocrine cells, and some endocrine cells, the electrical activity of the cell, the action potential, leads to the opening of voltage-dependent Ca^{2+} channels with a subsequent increase in cytosolic Ca^{2+}. Neurotransmitter release at synaptic and neuromuscular junctions, peptide hormone secretion, and catecholamine release by the adrenal medulla all belong to this class. In nonexcitable cells, the triggering Ca^{2+} signal is provided by release of Ca^{2+} from intracellular stores after appropriate stimulation. Depletion in Ca^{2+} from these intracellular pools activates an influx of Ca^{2+}, which is responsible for the sustained increase in $[Ca^{2+}]_i$ observed in many cell types. Moreover, this Ca^{2+} influx provides Ca^{2+} ions for the replenishment of internal stores.

A. Dynamic Changes in the Cytoskeletal Networks Are Required for Exocytosis

Secretion is a process that requires (1) the movement of secretory vesicles toward the plasma membrane, (2) the fusion of vesicles with the plasma membrane, and (3) subsequent release of secretory contents in the cell exterior.

As explained before, in resting conditions, actin filaments (F-actin) are preferentially localized in the cell surface and act as a barrier to the secretory granules, impeding their contact with the plasma membrane (Trifaró *et al.*, 1992). Stimulation of chromaffin or mast cells as well as neuronal synaptic vesicles produces disassembly of the actin network and removal of the barrier (Tchakarov *et al.*, 1998) (Fig. 2). Several observations support this concept: (1) Direct evidence for an actin barrier has come from the use of drugs that affect actin assembly and disassembly. Cytochalasin B and DNase I prevent actin assembly and drive the system toward net disassembly and increased secretion in permeabilized chromaffin cells. (2) Studies using fluorescent rhodamine-labeled phalloidin (a drug that stabilizes actin filaments *in vitro* and stops actin disassembly on stimulation) and actin antibodies have shown, in resting cells, a strong cortical fluorescent ring of filamentous actin. Cholinergic receptor stimulation produces a fragmentation of the fluorescent ring, leaving cell cortical areas devoid of fluorescence. These changes are accompanied by a decrease in F-actin associated with a concomitant increase in G-actin (Tchakarov *et al.*, 1998) (Fig. 4). (3) Results from the use of toxins also support the concept of a cortical actin barrier. Botulinum C2 toxin, which ADP-ribosylates actin and inhibits actin polymerization, enhances secretion in PC12 cells. In contrast, tetanus and botulinum A toxins, which

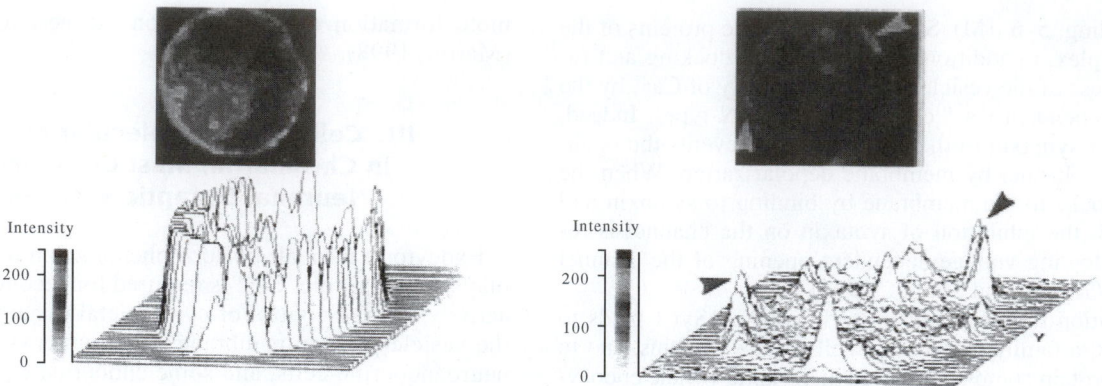

FIGURE 4. Rhodamine-phalloidin fluorescence of chromaffin cell cytoskeleton. Cultured chromaffin cells were incubated for 40 s in Locke's solution in the absence (A) or in the presence (B) of 10 μM nicotine. Cells were fixed and processed for fluorescence studies. Resting chromaffin cells showed a bright and continuous cortical fluorescent ring (A), while nicotinic-receptor-stimulated cells showed disruption in the cortical fluorescent ring. Some fluorescent patches are shown by arrowheads (B). Three-dimensional image analysis of the same patterns is shown below each cell. In control cells, there is a uniform cortical fluorescence intensity pattern, whereas in stimulated cells, the cortical fluorescent intensity pattern shows irregularities such as valleys and peaks. These peaks correspond to the patches observed in cells (Adapted with permission from Tchakarov *et al.* (1998)).

block actin disassembly, inhibit exocytosis upon cholinergic stimulation.

Several actin-binding proteins present in the cell cortex undergo changes upon stimulation. In a resting cell, fodrin is localized in the cell cortex. On stimulation, it rearranges into patches beneath the plasma membrane. This redistribution could be related to the clearing of exocytotic sites at the plasma membrane. Scinderin is a cytosolic protein that shortens actin filament length when Ca^{2+} is present in the medium. Stimulation induces both redistribution of scinderin from cytosol to cell cortex and F-actin disassembly, which precedes exocytosis. Thus, stimulation-induced redistribution of scinderin and F-actin disassembly produce subplasmalemmal areas of decreased cytoplasmic viscosity and high secretory vesicle mobility. All these processes require the presence of Ca^{2+} in the extracellular medium. Therefore, only secretagogues that induce Ca^{2+} entry are able to produce these effects (Fig. 2B).

In isolated chromaffin cells, stimulation with nicotinic agonists can result in secretion of about 30% of the total catecholamine. Electron microscopic observations show that a small number of granules lie within the exclusion zone of the cell cortex. This demonstrates the importance of changes in cortical actin to allow movement of the bulk of the granules involved in a full secretory response. Nevertheless, low levels of exocytosis, due to these granules in the cortical exclusion zone, could occur without generalized changes in cortical actin. Thus, in a physiological situation, where relatively few exocytotic events occur per stimulus, changes in cortical actin may not be necessary for the initial wave of exocytosis, but are required for the movement, into the cortical exclusion zone, of granules ready for the next stimulus. A similar picture has emerged from studies on the nerve terminal cytoskeletal phosphoprotein, synapsin I, which is believed to cross-link synaptic vesicles and release them following depolarization and phosphorylation of the synapsin I.

B. Physical Events Associated with the Fusion of Vesicles to Plasma Membrane

In regulated exocytosis, fusion of the secretory granules with the plasma membrane is triggered by an appropriate signal. The fusion process, following the triggering signal, can be very fast, such as in mammalian nerve terminals, with delays of less than 0.2 ms between the action potential and exocytosis. In some cells, however, delays of 0.2 s (chromaffin cells) to 50 s (mast cells) are observed. This delay is thought to be caused by the time required for production of second messengers possibly involved in exocytosis and removal of the cytoskeletal barrier that immobilizes the vesicles. However, the physical interactions between the granule membrane and the plasma membrane remain similar whether the exocytotic delay is fast or slow. Accordingly, the following fusion events will be described along general lines based on a sequence proposed by Almers (1990).

1. Capacitance Jump

Each time a secretory granule fuses with the plasma membrane, the total capacitance of the cell increases by a value proportional to the surface area of the new membrane added to the existing cell membrane. Assuming that biological membranes have a constant specific capacitance of about 1 $\mu F/cm^2$ allows a simple calculation of the granule size. Upon fusion of a single vesicle, the capacitance value increases abruptly to a new stable value as the membrane of the vesicle and the plasma membrane become continuous and the vesicle lumen opens into the extracellular space. Figure 5 illustrates the equivalent circuitry of a resting cell (Fig. 5A) and that of a cell undergoing exocytosis (Fig. 5B). During degranulation, several granules fuse with the plasma membrane, which produces a typical staircase recording (Fig. 5C). Degranulation in three different cell types are presented in Fig. 6: human neutrophil (Fig. 6A), guinea pig eosinophil

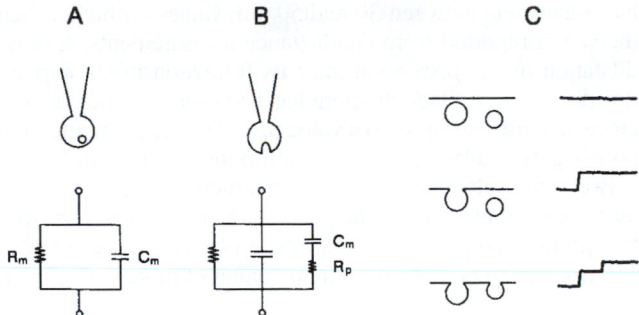

FIGURE 5. Capacitance measurement and equivalent electrical circuit. (A) The cell is patch-clamped in the whole-cell configuration. The cell is at rest and an unfused granule is shown near the plasma membrane. The equivalent circuit is represented by membrane resistance R_m in parallel with membrane capacitance C_m. The series resistance due mainly to the micropipette has been omitted for clarity. (B) Once the granule has fused with the plasma membrane, its capacitance C_v, proportional to the surface of membrane added, is added to the total capacitance of the cell. R_p is the fusion pore resistance, which will rapidly decrease as the pore dilates. (C) The fusion of a granule with the plasma membrane increases the capacitance by step. Granules of identical sizes induce the same increases in capacitance.

(Fig. 6B), and horse eosinophil (Fig. 6C). Note that the amplitude of the individual step capacitance is lower in human neutrophil than in horse eosinophil, reflecting the different sizes of the vesicles. A capacitance step amplitude histogram is built by measuring the step height of a large number of individual events. An example of this is illustrated in Fig. 6D for neutrophils. The capacitance step amplitudes range from 1 to 6 fF (1 to 6×10^{-15} F), with a greater number of events having an amplitude of 2 fF. Assuming a spherical shape, the diameter of the granule can be calculated from the step change in capacitance, δC_m in fF, by the relation

$$\delta C_m = \pi d^2 \qquad (1)$$

The frequency distribution of the sizes of the granules obtained from the capacitance step amplitude histogram is represented in Fig. 6E for guinea pig eosinophils. The distribution is fitted (smooth line) by the sum of two Gaussian curves with means of 520 and 590 nm (Lindau and Gomperts, 1991). This size distribution fits in well with the morphometric data obtained by direct microscopic observation of the secretory vesicles. Capacitance step measurements have been achieved in a variety of cell types, including adrenal chromaffin cells, mast cells, pancreatic acinar cells, neutrophils, eosinophils, basophils, pituitary lactotrophs, and nerve terminals derived from the posterior pituitary. Capacitance step values generally range between 1 and 30 fF, corresponding to granule diameters of 0.2 to 1 μm.

Giant vesicles with diameters ranging from 1 to 5 μm are found in mast cells from a strain of genetically defective beige mice (strain C57BL/6J-bg^j/bg^j). In these mice, mast cells and other granulocytes are unable to limit the size of their secretory vesicles; mast cells contain 10 to 40

giant vesicles that can easily be observed under photonic microscopy. These cells thus provide an ideal material for exocytosis studies and, for this reason, have been extensively used. The capacitance method offers the possibility to study degranulation in real time online; time analysis of the secretory process reveals that the granules fuse sequentially, one by one, with the plasma membrane. However, in mast cells, capacitance step analysis demonstrates the presence of step values greater than 60 fF, which could not be produced by the fusion of a single vesicle. A detailed analysis of the capacitance step histogram reveals a multimodal distribution of granule size, which indicates that the larger granules could be formed by the fusion of two to five single granules with each other.

2. Capacitance Flickering

The pattern of the staircase increase in capacitance during degranulation demonstrates that each step builds upon the previous one, indicating that the fusion event is irreversible. However, closer observation of capacitance jumps in mast cells reveals the existence of "on" and "off" steps. The "on" step is produced by the opening of a small-diameter pore, called a **fusion pore,** which adds the surface of the vesicle to that of the cell. Once opened, the fusion pore can close quickly and reopen (Fig. 7). Rapid oscillation between open and closed states gives the appearance of a flickering of the capacitance (Almers, 1990). Size distribution of capacitance steps during the flickering period shows that large steps are absent and that all steps remain in the range of values expected for single vesicle fusion events. The size of the fusion pore is proportional to its conductance. An unexpected result is that the reclosing of the pore occurs, not only in small-diameter pores (low conductance), but also in larger-sized

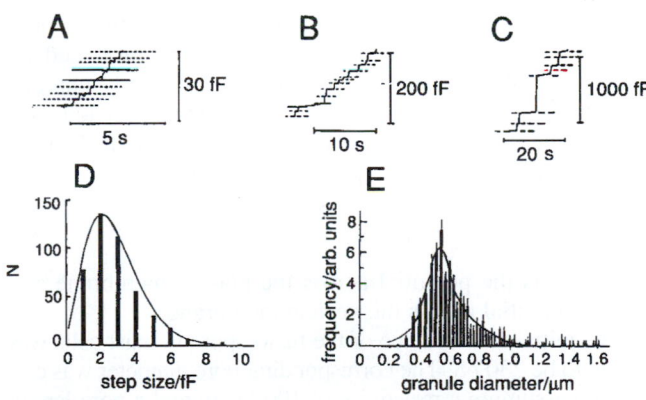

FIGURE 6. Analysis of step capacitance. Capacitance recordings obtained from different cells: (A) human neutrophils, (B) guinea pig eosinophils, (C) horse eosinophils. Note the staircase appearance of the recordings, which reflects the sequential fusion of individual granules and the size of the step capacitances, which are proportional to the size of the granules. (D) Frequency distribution of step sizes for human neutrophils. (E) Distribution of granule size for guinea pig eosinophils derived from capacitance measurement. (Adapted with permission from Lindau and Gomperts (1991)).

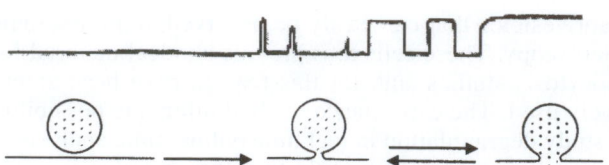

FIGURE 7. Flickering of the membrane capacitance. Fusion of the granule is reversible and can oscillate between an unfused state and a fused state. Each time the granule fuses with the plasma membrane, the capacitance of the cell increases by a step. If the fusion pore closes, the capacitance recovers its initial value; incomplete recovery indicates very brief closures exceeding the speed of the recording system. Eventually, a stable fused state can be reached.

pores having a conductance of several nS. Capacitance flickering is rather frequent in mast cells, but very few have been observed in eosinophils. This raises the question of the undetectable presence of flickering during the fusion of all vesicles before the irreversible fused state is attained. Indeed, the initial phase of fusion is a very fast process, which cannot be faithfully recorded by the speed of the recording techniques used. The earliest steps as well as short-lived flickering are certainly missed.

3. Fusion Pore

Freeze-fracture images of exocytosis in mast cells and neutrophils reveal the presence of narrow pores formed between the granules and the plasma membrane. The size of the pore increases as fusion progresses, allowing the release of vesicle contents. The presence of pores joining two secretory vesicles is also observed (Lindau and Gompertz, 1991). Electrically, the first opening of the fusion pore generates a brief transient current from which several parameters can be deduced.

a. Size of Fusion Pore. The conductance of the fusion pore can be calculated from the current transient produced by movement of the charges between two differently charged membranes, that is, the plasma membrane and vesicle membrane. The initial value of the current, I_0, is related to the initial pore conductance, g_0, by the relationship

$$g_0 = I_0 / \left(E_c - E_v \right) \qquad (2)$$

where E_c is the potential across the plasma membrane and E_v the potential across the vesicle membrane.

The initial conductance of the fusion pore in mast cells was found to be 230 pS. The corresponding pore diameter was calculated assuming a resistivity of 100 $\Omega \cdot$cm and a pore length of 15 nm. The abrupt increase in conductance corresponds to the all-or-none opening of a pore having an inner diameter of less than 2 nm. For comparison purposes, gap junction channels have conductances that vary from 80 to 240 pS and a diameter of approximately 2 nm. It can thus be proposed that the fusion pore is a large protein spanning across two membrane thicknesses, and having a structure and function resembling that of a channel. Data obtained from electron microscopic observations reveal that the smallest pores that can be observed

have diameters between 30 and 50 nm, values far higher than the values reported from conductance measurements. A rapid dilatation of the pore soon after its formation might explain this discrepancy. Once the pore has been formed, the conductance increases abruptly to a value near 250 pS. Thereafter, the pore begins to dilate, possibly by infiltration of lipid molecules between the subunits of the protein structure, followed by an increase in conductance. The pore conductance increases from 500 pS to 3000 pS in about 25 ms; a plateau is reached after 150 ms, corresponding to a pore diameter of more than 16 nm.

b. Does the Fusion Pore Leak? Once established, the diameter of the fusion pore is similar to that of the gap junction, between 1.5 and 2 nm. Since gap junctions allow the intercellular passage of molecules weighing up to 1.9 kDa, the question can be asked as to whether granule contents can leak through the fusion pore. In guinea pig eosinophils, the irreversible fusion of the granule with the plasma membrane is occasionally preceded by a long-lived fusion pore having a conductance of 70–250 pS. When granules were loaded with the fluorescent dye quinacrine, no release of the dye in the extracellular medium could be observed during the life of the fusion pore. Only after the fusion pore had completely dilated, and the vesicle had reached its irreversible state of fusion, was the dye released. These results indicate that the fusion pore is too narrow for the release of granule contents.

However, computations based on the diffusion of a small molecule such as histamine predict that a granule with a diameter of 0.8 mm should release its contents rapidly through the opening of the fusion pore. Recently, experimental proof was provided by Neher and collaborators using bovine chromaffin cells (Chow *et al.,* 1992). Secretion was measured by voltametry, while cells were studied by voltage-clamp. The cells were stimulated by depolarizing the membrane from a holding potential of −60 mV to a step potential of +10 mV for 25 ms to activate the Ca^{2+} channels. The amperometric signals, which represent the detection of the released catecholamine molecules by the carbon electrode (potential of 800 mV), were transient with a fast or slow rising phase and variable amplitudes. A histogram of integrals of current transient amplitude obtained on several cells showed that the mean charge transfer had a value of 0.76 pC, which is equivalent to the release of 2.36×10^6 molecules of catecholamine. Sometimes, larger events were detected, presumably corresponding to multigranular exocytosis. One interesting and surprising feature was that the majority of the fast-rising events were preceded by a small "foot" or "pedestal," as illustrated in Fig. 8A. The mean duration of the foot was 8.26 ms and the mean charge 34 fC, equivalent to 1.05×10^5 molecules (Fig. 8B). The foot was interpreted as reflecting a slow leakage of catecholamine molecules through the fusion pore formed during the early step of the fusion process. A second important result of this study was the discovery and quantification of a long latency period between the end of the stimulus and catecholamine release. The majority of the secretory events occurred 5 to 100 ms after the end of the electrical stimulus, which is rather long when compared to nerve endings, where the delay is about 1 ms. A complex cascade of intracellular events triggered by

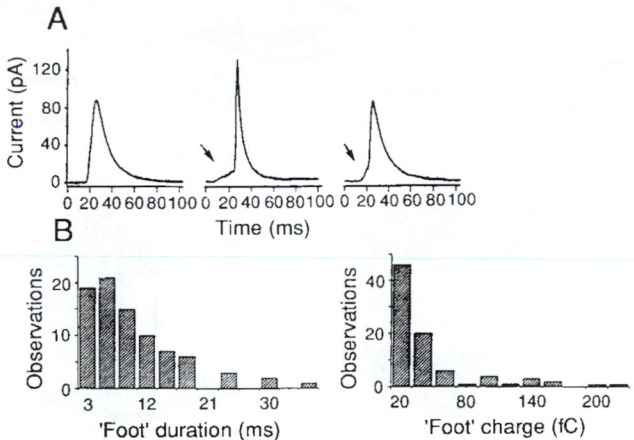

FIGURE 8. Amperometric current recorded in chromaffin cells. (A) The amperometric signal recorded with the voltametry method shows a fast rising phase followed by a slower decrease. Occasionally, the rising phase is preceded by a pedestal or foot. The foot signal is thought to be due to the leak of catecholamine by the fusion pore before its complete dilatation. (B) Histogram of the foot signal duration (left panel) and histogram of the charge of the foot signal (right panel), with a mean of 34 fC corresponding to 1.05×10^5 molecules. (Adapted with permission from Chow *et al.*, (1992). *Nature* **356**, 60–63.) (Copyright 1992 Macmillan Magazines Limited.)

the increase of cytosolic Ca^{2+} concentration could be responsible for this long latency.

4. Membrane Tension as a Driving Force for Fusion

As previously described, flickering is characterized by the opening and closing of the fusion pore. In some cases, it has been shown that after a period of flickering, the capacitance of the plasma membrane declines to a value lower than its initial value (Monck *et al.*, 1990). Figure 9A shows that this decrease in capacitance is paralleled by an increase in the conductance of the fusion pore (Fig. 9B),

thus establishing a relationship between the dilatation of the pore and decreased plasma membrane capacitance. Once the flickering stops, the conductance recovers its initial value but the whole-cell capacitance remains lower. These results are interpreted as reflecting a decrease of the plasma membrane surface due to a net transfer of material to the granule membrane. The difference between the "on" and "off" steps (found in one-half of the transient fusions with values between −2 to −4 fF) is proportional to the duration of contact between the plasma membrane and the secretory vesicle (Fig. 9C). The rate of cell surface area reduction is 0.16 $\mu m^2 \cdot s^{-1}$. The transfer of membrane is facilitated by the fact that the membrane of the secretory vesicle is under tension. Upon fusion, a movement of phospholipid molecules occurs from the plasma membrane to the granule membrane. A possible mechanism for generating tension in the granule membrane is osmotic swelling. However, fusion can proceed in isotonic, hypotonic, or hypertonic solutions with no change in the kinetics of capacitance increase.

5. Fusion Steps

Several lines of evidence favor the hypothesis that a pore-forming protein could be involved in the fusion process. The abrupt opening of the fusion pore with an initial conductance of 250 pS, similar to the conductance of the gap junction channel, and the occurrence of rapid flickering are the strongest arguments. Prior to their fusion, secretory granules are docked to the plasma membrane. In synaptic nerve endings, docking is localized to a restricted region and, upon stimulation, yields localized secretions. Docked granules have been observed in a variety of systems, including chromaffin cells. In these cells, localized secretion was also reported, depending on the applied stimulus.

C. Control of Exocytosis

Exocytosis occurs in a variety of electrically excitable and nonexcitable systems in response to receptor activation. The

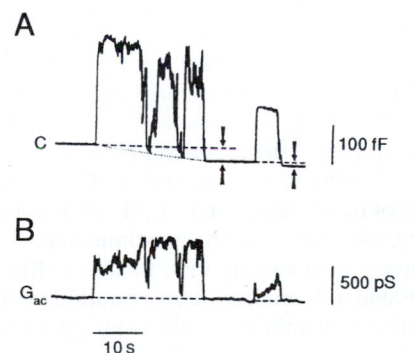

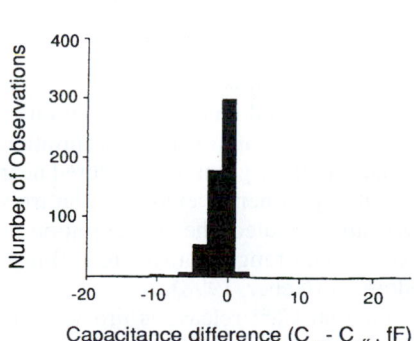

FIGURE 9. Decrease of total membrane capacitance after fusion. Capacitance (A) and conductance (B) measurements during transient fusion of a giant secretory granule from beige mouse mast cell. Capacitance and conductance increase on fusion. The fusion pore closes twice with a decrease in conductance and capacitance. Note that the conductance returns to its initial level but that the capacitance of the cell is lower than its initial value. This indicates a reduction of the surface of the plasma membrane due to a leak of lipid molecules toward the granule membrane. (C) Histogram showing the size distribution of the capacitance difference measured after transient fusion in mast cells. (Reproduced with permission from Monck *et al.* (1990)).

regulatory pathways that couple stimulation and secretion vary widely among cell types. In the following section, analysis of the factors controlling exocytosis mainly focuses on two well-studied systems: the chromaffin cell from the adrenal medulla, which belongs to the class of excitable cells, and the mast cell from the immune system, a nonexcitable cell.

1. Effectors of Exocytosis

a. Calcium Signaling and Sources of Calcium. In many cells, Ca^{2+} is the key signal for triggering exocytosis. In chromaffin cells, the resting Ca^{2+} concentration has a value ranging between 50 and 100 nM. Chromaffin cells can be stimulated by two different classes of acetylcholine receptors. The nicotinic receptor has a channel-like structure consisting of five transmembrane subunits. The binding of two acetylcholine molecules opens the channel, which leads to a large net influx of Na^+ ions. This influx causes a membrane depolarization, which can activate voltage-dependent Ca^{2+} channels. The nicotinic receptor channel is also permeable to K^+ ions and, to a lesser extent, Ca^{2+} ions. The muscarinic receptor belongs to the family of the seven-span transmembrane domain receptor proteins, which utilizes the G protein cascade pathway as its signal-transducing mechanism. Stimulation of chromaffin cells with the cholinergic agonist nicotine induces a rapid rise in $[Ca^{2+}]_i$ (measured with Ca^{2+}-sensitive fluorescent dye) up to 1 μM; this $[Ca^{2+}]_i$ increase is followed by F-actin disassembly (Fig. 4) and exocytosis. Muscarinic stimulation of chromaffin cells also increases $[Ca^{2+}]_i$ but does not induce secretion.

In a Ca^{2+}-free external medium, secretion is abolished regardless of the stimulus. This observation reinforces the fact that Ca^{2+} influx from the external medium is crucial for secretion in chromaffin cells. Several voltage-dependent Ca^{2+} channels have been described in chromaffin cells, namely: (a) the L-type dihydropyridine-sensitive channels, (b) the N-type ω-conotoxin-sensitive dihydropyridine-insensitive Ca^{2+} channels, and (c) the dihydropyridine-sensitive facilitation Ca^{2+} channels. Electrical depolarization, K^+ depolarization, and nicotinic stimulation open the Ca^{2+} channels, allowing an immediate influx of Ca^{2+} ions from the external medium. The video-imaging technique allows the recording of $[Ca^{2+}]_i$ with good spatial definition. As shown in Fig. 10 (right panel), the increase in $[Ca^{2+}]_i$ is restricted to the immediate vicinity of the plasma membrane after activation of Ca^{2+} channels. The requirement of high Ca^{2+} concentration for exocytosis implies that secretory granules are stored near the Ca^{2+} channels. Recent experimental evidence confirms that when Ca^{2+} channels are activated, the concentration of Ca^{2+} at the secretory sites could range from 10 to 100 μM, creating a Ca^{2+} microdomain (Neher, 1998).

In chromaffin cells, internal Ca^{2+} release is provided by two different pools: the $InsP_3$-sensitive pool and the Ca^{2+}-induced Ca^{2+} release pool (CICR). The activation of the PtdIns-specific phospholipase C by the G_q-protein-coupled receptor or by cytosolic Ca^{2+} elevation induces the hydrolysis of $PtdInsP_2$, thus generating two messengers: $InsP_3$ and diacylglycerol (DAG) (Figs. 1 and 2). The binding of $InsP_3$ to specific sites on the endoplasmic reticulum induces Ca^{2+}

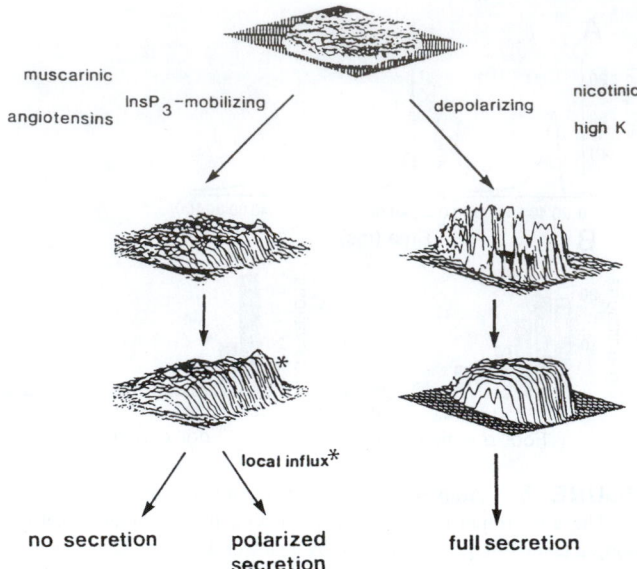

FIGURE 10. Ca^{2+} signal in a chromaffin cell in response to various agonists. The distribution of $[Ca^{2+}]_i$ is obtained by video-imaging. The figure illustrates the fact that level and distribution of $[Ca^{2+}]_i$ are important for full secretion. The right portion of the figure shows that a depolarization of the plasma membrane, induced by high $[K^+]_o$ or nicotinic agonist, opens voltage-activated Ca^{2+} channels with an immediate and high $[Ca^{2+}]_i$ increase at the periphery of the cell. Afterwards, $[Ca^{2+}]_i$ increases in the entire cell due to release of Ca^{2+} from internal stores; a full secretion is obtained. The left portion of the figure shows that release of Ca^{2+} from the IP_3-sensitive pool is not sufficient to induce secretion, although some localized secretion can be obtained in the region of the cell where $[Ca^{2+}]_i$ increase was the highest. (Adapted with permission from Burgoyne, (1991)).

release. $InsP_3$ receptors were also found in secretory granules of endocrine and neuroendocrine cells. Their activation by $InsP_3$ induces the release of Ca^{2+} ions from the granules and provides a localized increase of $[Ca^{2+}]_i$. Ca^{2+} ion alone or caffeine can trigger the release of Ca^{2+} from the second pool (CICR). The muscarinic receptor induces the release of Ca^{2+} from the $InsP_3$-sensitive store without Ca^{2+} channel activation. The observed Ca^{2+} increase is low and sometimes localized to one pole of the cell (Fig. 10, left panel). Despite this $[Ca^{2+}]_i$ increase, there is little or no stimulation of secretion at the site where $[Ca^{2+}]_i$ has increased. A more uniform $[Ca^{2+}]_i$ increase can be generated by release from the Ca^{2+}-induced Ca^{2+} release pool by caffeine or from the $InsP_3$ pool by introduction of GTP-γS into the cell. However exocytosis still remains unstimulated. These results conclusively demonstrate that, in chromaffin cells, only a considerable rise in Ca^{2+} concentration in the proximity of the plasma membrane is able to trigger exocytosis. This highly localized $[Ca^{2+}]_i$ increase can only be achieved via the Ca^{2+} channels, and not by a small increase due to release from internal stores.

Mast cells are able to synthesize, store, and secrete histamine. However, Ca^{2+} alone is not able to trigger histamine secretion. There is a requirement for a guanine nucleotide together with Ca^{2+} for exocytosis. Mast cells are stimulated by antigenic binding on cell surface IgE receptors or by compound 48/80. A sudden rise in Ca^{2+} concentration (released from $InsP_3$-sensitive pool) is observed, reaching a level of up to several micromolar. The signal is transient and $[Ca^{2+}]_i$ declines within several seconds to its original baseline value. Degranulation begins soon after this transient response. Two important features should be noted concerning $[Ca^{2+}]_i$ and secretion in mast cells: (a) the $[Ca^{2+}]_i$ transient is not dependent on the presence of Ca^{2+} in the external medium, and (b) simultaneous recording of capacitance and membrane conductance reveals that conductance is constant throughout the duration of exocytosis. These results indicate that Ca^{2+} entry through Ca^{2+} channels is not a required signal for exocytosis, but rather its release from internal stores. A Ca^{2+} current, called I_{CRAC} (calcium-release activated calcium), was described in mast cells (Hoth and Penner, 1992); I_{CRAC} is activated by the depletion of intracellular Ca^{2+} pools. The role of I_{CRAC} in nonexcitable cells could be to maintain a high $[Ca^{2+}]_i$ and to replenish empty Ca^{2+} stores after stimulation. A possible direct role in exocytosis has not yet been shown.

One conclusion is that Ca^{2+} release from internal stores is sufficient for exocytosis in mast cells. However, intracellular application of $InsP_3$ induces a transient $[Ca^{2+}]_i$ increase but does not trigger secretion, emphasizing that the signal is more complex. Application of the nonhydrolyzable GTP analog GTP-γS (100 μM) inside a mast cell (via a patch pipette) induces a transient $[Ca^{2+}]_i$ increase, after a delay of 10–20 s, followed by an increase in capacitance. GTP-γS activates G proteins, including the G_q protein coupled to PLC, with a resulting production of $InsP_3$ and DAG. Ca^{2+} is released from the $InsP_3$-sensitive pool by the same mechanism as with the direct application of $InsP_3$ described above. However, in this case, exocytosis is stimulated, leading to the conclusion that GTP-γS activates a G protein that plays a crucial role in exocytosis.

b. Guanine Nucleotides. As seen earlier, Ca^{2+} and a guanine nucleotide are necessary and sufficient effectors to ensure secretion in mast cells. The stable analog GTP-γS is the commonly used guanine nucleotide, although any ligand that binds to G protein is able to stimulate secretion. In chromaffin cells, two types of observations have been recorded: (1) GTP-γS increases the Ca^{2+} sensitivity of secretion in permeabilized cells, and (2) GTP analogs stimulate secretion in a Ca^{2+}-independent manner (Morgan and Burgoyne, 1990). GTP analogs do not enhance the secretion induced by high Ca^{2+} concentration (10 μM), which indicates that the two stimuli (GTP and Ca^{2+}) act on the same exocytotic pathway.

Evidence for the involvement of a GTP-binding protein in secretion has been obtained from a variety of cells: neutrophils; platelets; parathyroid; pituitary lactotroph, gonadotroph, and melanotroph; insulin-secreting cells; and pancreatic acinar cells. Various effects have also been observed in these cells: GTP or GTP analogs behave as effectors able to trigger a Ca^{2+}-independent secretion, or as modulators that increase Ca^{2+} sensitivity, or more surprisingly as inhibitors that block the exocytotic pathway at a late stage. The exact nature of the G proteins is not yet

known, although some analogy can be made with GTP proteins involved in vesicular traffic (Rothman and Orci, 1992).

The small Ras-like GTPases are involved in the formation, transport, and fusion of vesicles. In yeast, the small Ras-related GTP-binding protein SEC4 is required for targeting and/or fusion of vesicles to the plasma membrane. Rab3A from the rab family of proteins has been located in the membrane of neurosecretory vesicles. More than 30 Rabs have been identified. They have a regulatory role in secretion. In its GTP-bound form, Rab3A inhibits secretion, possibly by stabilizing a fusion-incompatible conformation. Under an appropriate signal, GTP is hydrolyzed to GDP with the help of GAP (a GTPase-activating protein); this exchange of GTP for GDP releases the inhibition. The fusion can then proceed to subsequent steps. Two accessory proteins are involved in the cycle: rabphilin-3A, which binds to the GTP-Rab3A form, and GDI (a GDP-dissociation inhibitor), which binds to the GDP-Rab3A form. Rabphilin is phosphorylated by many kinases and contains binding sites for Ca^{2+} and phospholipids. In endocrine cells, Nac2 binds GTP-Rab3A; this protein has a high similarity to the Rab3A-binding domain of rabphilin, but lacks the Ca^{2+} binding sites and phospholipids. GDI binds to Rab3A in its GDP form and removes it from the membrane. After dissociation of the GDI-GDP-Rab3A complex, GTP-Rab3A can bind again to another vesicle. Thus rabphilin-3A and GDI control the Rab3A cycle but do not directly participate in the fusion (Südhof, 1995; Geppert and Südhof, 1998)).

2. Modulators of Exocytosis

a. ATP. Permeabilized cells rapidly lose their secretory response to agonists unless ATP is present in the medium. However, the sequence of exocytosis can be separated in an early phase that requires MgATP to proceed and a late phase that is MgATP-independent. ATP could act at various levels of the secretory response, acting as a substrate for protein phosphorylation or modulating various kinases involved in secretion. The last ATP-requiring steps in exocytosis have been identified; they involve ATPase, NSF, and the formation of $PtdInsP_2$. NSF forms a large (20S) complex with the attachment proteins (SNAPs) and SNAP receptors (SNARES), which has been proposed to be the fusion particle (Fig. 3). The hydrolysis of ATP by NSF is thought to produce energy for the fusion (priming step). The formation of $PtdInsP_2$ by phosphatidylinositol 4-phosphatase-5-kinase and a phosphatidylinositol transfer protein requires the presence of ATP. $PtdInsP_2$ is thought to be involved in the interaction between cytoskeleton and secretory granules (docking step). Docking and priming of secretory granules are thus ATP-dependent. During vesicle transport through the Golgi, ATP hydrolysis by NSF occurs after docking, but in granule secretion no such evidence exists.

b. Cyclic Nucleotides. In chromaffin cells, cAMP concentration increases after cholinergic stimulation, leading to the hypothesis that cAMP could have a role in exocytosis. In pancreatic β cells, the cAMP-dependent protein kinase A enhances secretion. However, contradictory results have been reported on the effects of high and low cAMP concentrations on secretion modulation, that is, inhibition and potentiation, respectively. Nicotinic-induced secretion in chromaffin cells is inhibited by high cGMP concentration, whereas low concentrations potentiate the secretory response.

c. Calmodulin. A calmodulin-binding protein of 65 kDa, called 65-CMBP or p65, has been found in several secretory vesicles, such as synaptic, neurohypophyseal, chromaffin, platelets, and pancreatic islet granules. This protein also binds to phospholipids with high affinity. A possible role of calmodulin in exocytosis was investigated by blocking its action. The calmodulin antagonist calmidazolium inhibits secretion from intact chromaffin cells. Antibodies raised against calmodulin inhibit the Ca^{2+}-dependent binding of calmodulin to vesicle membrane by acting on the docking and/or fusion steps of exocytosis. However, some reports indicate that the less specific calmodulin inhibitor trifluoperazine (TFP) has no effect on permeabilized cells. At present, it appears that calmodulin is not essential for Ca^{2+}-dependent exocytosis, although interaction between calmodulin and cytoskeletal proteins should nevertheless be considered.

d. Protein Kinase C. A role for PKC in secretion could be inferred from the fact that any agonist that stimulates secretion also induces activation of PKC. PKC has a modulating role in Ca^{2+} affinity, by decreasing Ca^{2+} requirements for exocytosis. Inhibition of PKC by staurosporine or through down-regulation partially inhibits secretion. However, in gonadotrophs, PKC-stimulated luteinizing hormone exocytosis is independent of Ca^{2+}. Obviously, the role of PKC in exocytosis is not yet completely understood. It is possible that PKC is not essential for Ca^{2+}-dependent exocytosis (modulator), but that a second Ca^{2+}-independent pathway could coexist in which PKC acts as an effector of secretion. Indeed, it has been recently shown in chromaffin cells that PKC acts at a late stage in exocytosis before the final Ca^{2+}-sensitive step. The role of PKC would be to increase the size of the readily releasable pool (RRP) of secretory granules by speeding their maturation after they dock with the plasma membrane. Moreover, direct stimulation of PKC by the phorbol ester PMA (phorbol 1,2-myristate-1,3-acetate) leads to a disruption of the actin network near the plasma membrane, which increases the number of docked granules. Several proteins involved in the regulation of the cytoskeleton are substrates for PKC: annexin I, annexin II, and MARCKS (see Fig. 1).

e. Phospholipase A2 and Arachidonic Acid. A possible fusogen role of arachidonic acid was inferred from the fact that *in vitro,* granules aggregated by synexin and Ca^{2+} fuse together if arachidonic acid is added to the medium. As previously mentioned for PKC activation, arachidonic acid is produced each time secretion is stimulated. Inhibition of phospholipase A2 and arachidonic acid metabolism inhibits secretion of chromaffin cells due to a blockage of Ca^{2+} entry.

3. Secretory Granule Pools

In bovine chromaffin cells, increase in $[Ca^{2+}]_i$ triggers secretion at various rates: ultrafast secretion (time constant < 0.5 s),

fast secretion (time constant 3 s) and slow secretion (time constant 10 to 30 s). Nevertheless there is a weak correlation between the rate of secretion and $[Ca^{2+}]_i$ levels, the three types of responses are mainly observed for 10 to 50 μM, above 80 μM, and around 170 μM $[Ca^{2+}]_i$, respectively. The time constant of the increase in membrane capacitance can be considered as a measure of the hormone released from a particular store. This indicates that the secretory granules are in various states of releasability. The ultrafast response comes from vesicles docked to the plasma membrane and immediately available for release by an increase in $[Ca^{2+}]_i$; they belong to the immediately releasable pool (IRP). A second pool of granules are located near the membrane in a nearly releasable pool (NRP) and can be released within seconds of a rise in $[Ca^{2+}]_i$; the fast response originates from this pool. Finally, the slow response originates from vesicles from a depot store in the bulk cytoplasm (Neher and Zucker, 1993). Movement of granules from the NRP to the IRP is Ca^{2+}-dependent and could involve calpactin in the docking process. However, not all the docked granules are available for the last steps, opening of the granule induced by Ca^{2+}. The fastest response, the exocytotic burst, mobilizes only one-tenth of the docked granules, which indicates that after docking, granules undergo a maturation process in the IRP. Figure 11 describes a possible sequence for exocytosis in endocrine cells. PKC and ATP hydrolysis are involved in the docking and priming steps. Thereafter, three steps have been proposed based on their sensitivity to temperature, blockage by acidification, and Ca^{2+} ion (three or four ions) triggering; PKC is thought to play a role in this sequence. The pivotal role of ATP in secretion has been outlined; however, a large pool of granules can be released in the absence of ATP by an increase in $[Ca^{2+}]_i$. This indicates first that ATP hydrolysis by NSF is an early step in the priming process, and second that if the energy of hydrolysis powers fusion, it remains stored in the core complex until $[Ca^{2+}]_i$ increases (Parsons *et al.,* 1995; Gillis *et al.,* 1996).

IV. Hormone Release in Endocrine Cells

A. Polypeptide and Thyroid Hormones

Most of the events described previously could be applied for most of the endocrine secretory cells.

1. Secretion of Insulin

Insulin secretion by the pancreatic β cells provides an excellent example of a cellular activity that requires direction. Insulin is packaged in secretory vesicles, which have to migrate to the plasma membrane and fuse with it to release the entrapped insulin. Both microscopic and biochemical studies have shown that secretory granules are linked to microtubules, which direct attached vesicles to the cell surface. However, a cortical band of fine microfilaments is consistently observed in β cells. Alteration of this cell web by cytochalasin B is associated with an enhancement of glucose-induced secretion of insulin by isolated islets. This

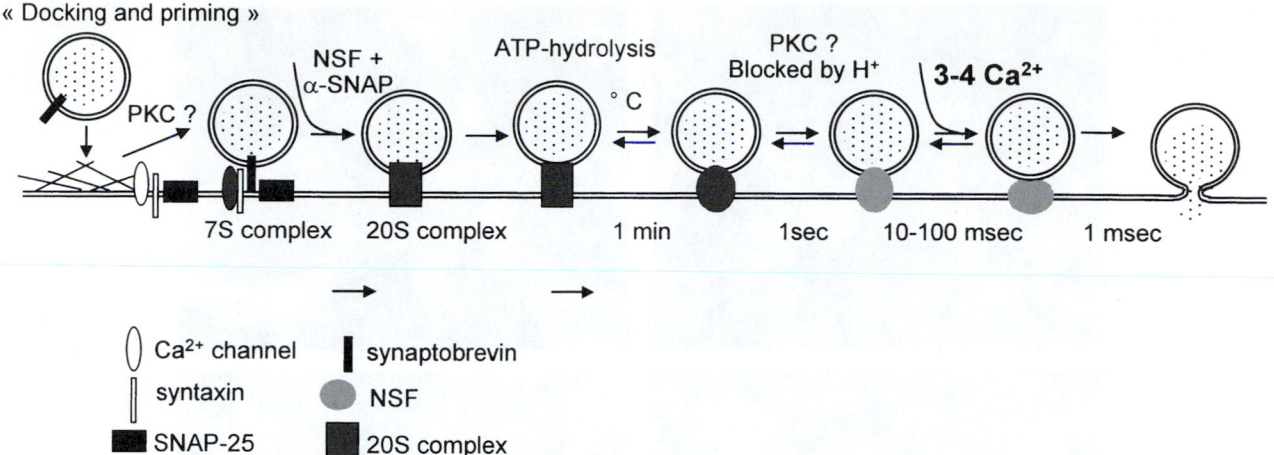

FIGURE 11. Docking and fusion of secretory granules in endocrine cells. Docking of vesicles involves the proteins synaptobrevin and syntaxin/SNAP-25. In the first step of docking, protein kinase C (PKC) could have a role in the disruption of the actin barrier. ATP hydrolyzed by NSF is thought to produce energy for fusion. Several steps have been identified after the ATP-dependent step based on their temperature, H^+, and Ca^{2+} dependence. PKC could be involved before the final Ca^{2+}-dependent step (Adapted with permission from Parsons *et al.* (1995) and Gillis *et al.* (1996)).

microfilamentous web plays an important role in the exocytosis of insulin secretory granules by controlling access to the cell membrane via a mechanism probably similar to that previously described for chromaffin cells. Ca^{2+} appears to initiate the cascade of events by which microtubules facilitate the displacement of granules toward the cell membrane. Glucose metabolism increases intracellular concentration of ATP, which closes the ATP-sensitive K^+ channels, consequently inducing cell depolarization and Ca^{2+} influx, while cAMP modifies the intracellular distribution of Ca^{2+} by increasing the cytosolic pool at the expense of Ca^{2+} bound to intracellular organelles. Protein kinase C also appears to be involved in the secretion of insulin.

2. Secretion of Pituitary Hormones

Anterior pituitary cells, in their diversity and heterogeneity, provide a rich source of models for secretory function. Secretion of the pituitary hormones is controlled by both specific hypothalamic-releasing peptides and neurotransmitters. Upon binding to their receptors, these agonists activate both Ca^{2+} influx by different types of Ca^{2+} channels (voltage-activated channels, ligand-activated channels, and second messenger-activated channels) and second messenger production, thus resulting in hormone secretion. As a second messenger, cAMP appears to be as potent as $InsP_3$ and PKC in interacting with Ca^{2+} to trigger secretion. As for insulin secretion, cytoskeletal structures are tightly associated with hormone release. Purified secretory granule membranes cosediment with microtubules, MAPs being involved in this association. This suggests that microtubules facilitate the movement of secretory granules from the Golgi apparatus to the plasma membrane, by providing tracks along which the granules can move. The granule membrane can then dissociate from the microtubules and fuse with the cell membrane, followed by exocytosis and release of the hormone into the circulation. Moreover, actin and microtubules are cross-linked by MAPs, forming three-dimensional networks. This cross-linking activity can be inhibited if MAPs are heavily phosphorylated. These observations suggest that MAPs might play an important role in the binding of secretory granules to tubulin and actin. Binding of actin to secretory granules suggests a role for actin in the final steps of exocytosis, as described previously.

B. Stimulation of Steroid Synthesis and Secretion

While disruption of the actin network is necessary to trigger fusion of secretory vesicles with the cell membrane, a well-preserved organization seems important for steroid release. Several studies have shown that microfilament disruption (with cytochalasins) or microtubule disruption (with colchicine or vinblastine) decreases or blocks second messenger production and steroid secretion, either from the adrenal cortex or gonads. In contrast with peptide hormones, steroids are not packaged in vesicles. Steroid hormones are synthesized from cholesterol contained in lipid droplets. The process of steroidogenesis begins in the mitochondria, where cholesterol is converted to pregnenolone, from which others steroids are synthesized. Stimuli induce rapid increase in the production and secretion of steroids, without cytoplasmic storage. More recent biochemical analysis and fluorescence studies indicate that ligand-receptor interaction induces a rapid polymerization of actin, demonstrated by a rise in the proportion of F-actin over G-actin and by an increased interaction of F-actin with the membrane (Fig. 12A and B versus C and D). Moreover, $G_{q/11}$ protein localization overlaps F-actin distribution, and cell activation is accompanied by a rapid translocation

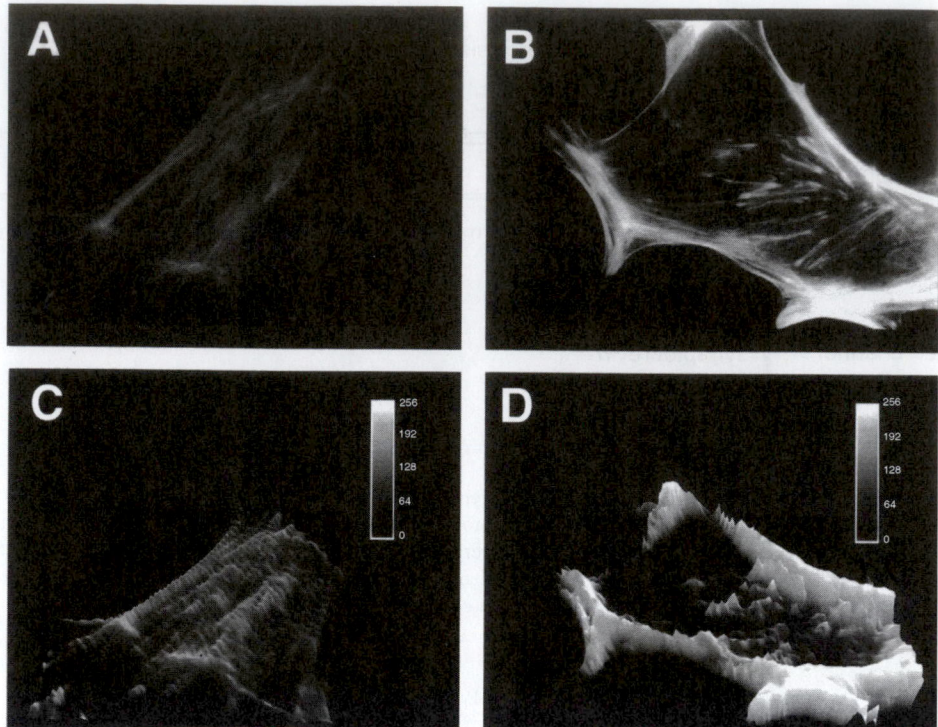

FIGURE 12.   Effect of Ang II on immunofluorescence labeling of actin in rat glomerulosa cells. Rat glomerulosa cells were cultured for three days on plastic coverslips and then incubated without (A) or with (B) 100 nM angiotensin II for 1 min. After formaldehyde fixation and permeabilization with 0.1% Triton X-100, cells were processed for immunofluorescence labeling of actin using rhodamine-phalloidin. C and D represents the same images as A and B after computer analysis. (Adapted from Côté *et al.* (1997)). Bars represent 13 μm.

of $G_{q/11}$ and F-actin from the cytosol to the membrane, an association essential in promoting phospholipase C activation (Côté *et al.,* 1997).

It has been shown that microfilaments are strictly required for both spontaneous and stimuli-induced corticosteroid secretions, including those utilizing the cAMP-dependent pathways (ACTH, serotonin, and vasoactive intestinal peptide), as well as those utilizing the phosphoinositide pathway (angiotensin II and acetylcholine). They are involved in a common and probably late step of steroidogenesis (translocation of the DOC from RER to mitochondria). In contrast, microtubules seem involved in an early step in the mechanism of hormone action, probably at the level of their interaction with the α-subunit of the G protein. Microtubules and microfilaments are closely associated with the plasma membrane. That colchicine and vinblastine stimulate basal steroidogenesis may be explained by the fact that most of the tubulin is linked to cholesterol, in lipid droplets. Thus, by acting on tubulin, antimicrotubular drugs could release cholesterol, causing an increase in basal steroid secretion (Feuillolley and Vaudry, 1996). Moreover, in adrenocortical cells, findings suggest that intermediate filaments could facilitate or increase the transport of cholesterol to mitochondria in response to ACTH and cAMP. Figure 13 summarizes the role of the three types of cytoskeletal fibers in the steroidogenic effects of the main stimuli of corticosterone and aldosterone secretions (Feuillolley and Vaudry, 1996).

V. Summary

The first step in the secretory process for peptide hormones and neurotransmitters involves synthesis, modification, and sorting of the molecules to be secreted. These secretory molecules are packaged in secretory granules or vesicles, which are then transported to the cell periphery, where they are released in the extracellular space by fusion with the plasma membrane. This complex process is named *exocytosis.* Exocytosis is an all-or-none phenomenon in which Ca^{2+} plays a pivotal role.

Interaction of cytoskeletal structures with transmembrane signaling is an important feature of excitation-secretion coupling. Actin-binding proteins are in large part responsible for cytoskeleton-receptor interactions. In resting cells, more than 95% of the $PtdInsP_2$ may be complexed with profilin. This interaction promotes actin polymerization. Following receptor activation, phospholipase C activity increases to a level where profilin protection can be overcome, resulting in $PtdInsP_2$ hydrolysis. In resting cells, the bulk of gelsolin is cytosolic. When the cell is activated, there is a transient rise in $[Ca^{2+}]_i$ due to $InsP_3$ action and/or Ca^{2+} influx, which activates gelsolin, causing a 200-fold increase in its affinity for actin. This results in rapid filament side-binding, followed by severing and capping, inducing a dramatic disruption of existing actin network structure in the vicinity of $[Ca^{2+}]_i$ elevation.

Dynamic changes in the cytoskeletal network are required for exocytosis. The subplasmalemmal area of secretory cells is characterized by the presence of a highly organized cy-

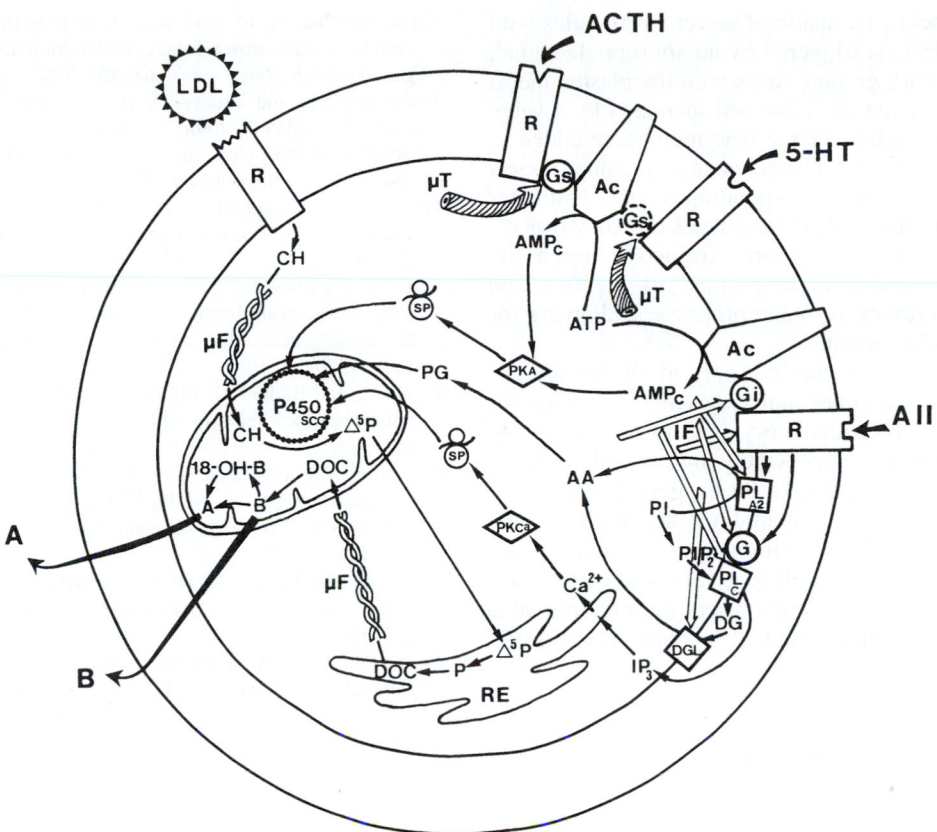

FIGURE 13. Cytoskeleton implication in adrenal corticosteroidogenesis. Microtubules (μT) are implicated in the action of ACTH and serotonin (5-HT), probably at a level of G_s protein. Microfilaments (μF) control the transfer of steroid precursors (cholesterol from lipid droplets or membrane low-density lipoproteins to mitochondria, and from mitochondria to endoplasmic reticulum). Intermediate filaments (IF) appear to be implicated in the action of angiotensin II (AII), between hormone-receptor coupling and second messenger production. A, aldosterone; AA, arachidonic acid; AC, adenylyl cyclase; cAMP, cyclic adenosine monophosphate; ATP, adenosine triphosphate; B, corticosterone; DAG, diacylglycerol; DOC, 11-deoxycorticosterone; 18-OH-B, 18-hydroxycorticosterone; P, progesterone; P450scc, P450scc cytochrome; Δ^5P, pregnenolone; PG, prostaglandins; PI, PtdIns; PIP_2, $PtdInsP_2$; PKA, protein kinase A; PKCa, protein kinase Ca^{2+}-dependent; PLA_2, phospholipase A_2; PLC, phospholipase C; PS, protein synthesis. (Reproduced with permission from Feuilloley *et al.* (1989)).

toskeletal network where F-actin, together with specific actin-binding proteins, forms a dense viscoelastic gel. Some actin-associated proteins, such as fodrin, caldesmon, gelsolin, and scinderin exist on secretory granule membranes, linking actin microfilaments to secretory granules. Caldesmon and synapsin I are other proteins that are associated with synaptic vesicles. Synapsin I binds to spectrin and actin microfilaments, and may serve as an anchor between synaptic vesicles and the cytoskeleton. Synapsin I phosphorylation results in the release of synaptic vesicles from their anchoring site on the cytoskeleton, allowing the vesicles to move to the active synaptic zones.

Secretion is a process that requires (1) the movement of secretory vesicles toward the plasma membrane, (2) the fusion of vesicles with the plasma membrane, and (3) subsequent release of secretory contents into the cell exterior. The process of secretion is mediated by contractile elements either associated with secretory vesicles or present elsewhere in the cell. As microfilaments (F-actin) are preferentially localized in the cortical surface of the chromaffin

cell, F-actin may act as a barrier to the secretory granules, impeding their contact with the plasma membrane. Upon stimulation, fodrin rearranges into patches beneath the plasma membrane. Such a redistribution could be related to the clearing of exocytotic sites at the level of the plasma membrane. Scinderin is a cytosolic protein that shortens actin filament length when Ca^{2+} is present in the medium. Stimulation induces both redistribution of scinderin from the cytosol to the cell cortex and F-actin disassembly, which precedes exocytosis. Docking of granules on the plasma membrane is an important step in exocytosis. Several proteins are involved in this process: synaptobrevin, syntaxin, and SNAP-25, which form the SNAP receptor (SNARE). Priming of the granules is also a pivotal step in exocytosis. N-ethylmalimide-sensitive fusion protein (NSF) primes the granule through ATP hydrolysis; once primed, an increase in $[Ca^{2+}]_i$ triggers the release. All these processes require the presence of Ca^{2+} in the extracellular medium. Therefore, only secretagogues that induce Ca^{2+} entry are able to produce these effects.

In regulated exocytosis, fusion of secretory granules with the plasma membrane is triggered by an appropriate signal. Every time a secretory granule fuses with the plasma membrane, the total capacitance of the cell increases by a value proportional to the surface of the new membrane added to the existing cell membrane. Capacitance step values generally range from 1 to 30 fF, corresponding to granule diameters of 0.2 to 1 μm. Freeze-fracture images of exocytosis reveal the presence of narrow pores formed between the granules and the plasma membrane called *fusion pores*. The size of the pore increases as fusion progresses, allowing for the release of vesicle contents.

Control of exocytosis occurs not only in electrically excitable cells, but also in nonexcitable systems in response to receptor activation. The regulatory pathways that couple stimulation and secretion vary widely among cell types. In many cells, Ca^{2+} is the key signal for triggering exocytosis. Ca^{2+} influx from the external medium is crucial for secretion, but internal Ca^{2+} release from internal stores also plays a pivotal role, depending on cell type. In some cases, Ca^{2+} alone is not able to trigger secretion, but the presence of a guanine nucleotide together with Ca^{2+} is necessary and sufficient for exocytosis.

Acknowledgments

The authors thank Dr. Mylène Côté for critical review of the revised version and fruitful discussions and Dusan Chorvat, Jr., from the International Laser Center, Bratislava (Slovakia), for image analysis.

Bibliography

Almers, W. (1990). *Exocytosis. Annu. Rev. Physiol.* **52**, 607–624.

Aplin, A., Howe, A., Alahari, S., and Juliano, R. (1998). Signal transduction and signal modulation by cell adhesion receptors: the role of integrins, cadherins, immunoglobulin-cell adhesion molecules, and selectins. *Pharmacol. Rev.* **50**, 197–263.

Arbuzova, A., Murray, D., and McLaughlin, S. (1998). MARCKS, membranes, and calmodulin: kinetics of their interaction. *Biochim. Biophys. Acta* **1376**, 369–379.

Aroeti, B., Okhrimenko, H., Reich, V., and Orzech, E. (1998). Polarized trafficking of plasma membrane proteins: emerging roles for coats, SNAREs, GTPases and their link to the cytoskeleton. *Biochim. Biophys. Acta* **1376**, 57–90.

Burgoyne, R. D. (1991). Control of exocytosis in adrenal chromaffin cells. *Biochim. Biophys. Acta* **1071**, 174–202.

Burgoyne, R. (1995). Mechanisms of catecholamine secretion from adrenal chromaffin cells. *J. Physiol. Pharmacol.* **46**, 273–283.

Burgoyne, R., and Morgan, A. (1998a). Analysis of regulated exocytosis in adrenal chromaffin cells: insights into NSF/SNAP/SNARE function. *Bioessays* **20**, 328–335.

Burgoyne, R., and Morgan, A. (1998b). Calcium sensors in regulated exocytosis. *Cell Calcium* **24**, 367–376.

Chow, R. H., Von Rüden, L., and Neher, E. (1992). Delay in vesicle fusion revealed by electrochemical monitoring of single secretory events in adrenal chromaffin cells. *Nature* **356**, 60–63.

Côté, M., Payet, M.-D., Dufour, M.-N., Guillon, G., and Gallo-Payet, N. (1997). Association of the G protein α_q/α_{11}-subunit with cytoskeleton in adrenal glomerulosa cells: role in receptor-effector coupling. *Endocrinology* **138**, 3299–3307.

Fernandez-Chacon, R., and Südhof, T. (1999). Genetics of synaptic vesicle function: toward the complete functional anatomy of an organelle. *Annu. Rev. Physiol.* **61**, 753–776.

Feuillolley, M., and Vaudry, H. (1996). Role of the cytoskeleton in adrenocortical cells. *Endocr. Rev.* **17**, 269–288.

Forscher, P. (1989). Calcium and polyphosphoinositide control of cytoskeletal dynamics. *Trends Neurosci.* **12**, 468–474.

Geppert, M., and Südhof, T. (1998). RAB3 and synaptotagmin: the yin and yang of synaptic membrane fusion. *Annu. Rev. Neurosci.* **21**, 75–95.

Gillis, K. D., Mössner, R., and Neher, E. (1996). Protein kinase C enhances exocytosis from chromaffin cells by increasing the size of the readily releasable pool of secretory granules. *Neuron* **16**, 1209–1220.

Goda, Y., and Südhof, T. (1997). Calcium regulation of neurotransmitter release: reliably unreliable? *Curr. Opin. Cell. Biol.* **9**, 513–518.

Gundersen, G., and Cook, T. (1999). Microtubules and signal transduction. *Curr. Opin. Cell. Biol.* **11**, 81–94.

Hall, A., (1998). Rho GTPases and the actin cytoskeleton. *Science* **279**, 509–514.

Hoth, M., and Penner, R. (1992). Depletion of intracellular calcium stores activates a calcium current in mast cells. *Nature* **355**, 353–356.

Inglese, J., Koch, W., Touhara, K., and Lefkowitz, R. (1995). G beta gamma interactions with PH domains and Ras-MAPK signaling pathways. *Trends Biochem. Sci.* **20**, 151–156.

Janmey, P. (1998). The cytoskeleton and cell signaling: component localization and mechanical coupling. *Physiol. Rev.* **78**, 763–781.

Lindau, M., and Gomperts, B. D. (1991). Techniques and concepts in exocytosis: focus on mast cells. *Biochim. Biophys. Acta* **1071**, 429–471.

Martin, T. (1998). Phosphoinositides as signaling molecules. *Annu. Rev. Cell Dev. Biol.* **14**, 231–264.

Matus, A. (1988). Microtubule-associated proteins: their potential role in determining neuronal morphology. *Annu. Rev. Neurosci.* **11**, 29–44.

Monck, J. R., Alvarez de Toledo, G., and Fernandez, J. M. (1990). Tension in secretory granule membranes causes extensive membrane transfer through the exocytotic fusion pore. *Proc. Natl. Acad. Sci. USA* **87**, 7804–7808.

Morgan, D., and Burgoyne, R. D. (1990). Stimulation of Ca^{2+}-independent catecholamine secretion from digitonin-permeabilized bovine adrenal chromaffin cells by guanine nucleotide analogues. Relationship to arachidonate release. *Biochem. J.* **269**, 521–526.

Neher, E., and Zucker, R. S. (1993). Multiple calcium-dependent processes related to secretion in bovine chromaffin cells. *Neuron* **10**, 21–30.

Neher, E., (1998). Vesicle pools and Ca^{2+} microdomains: new tools for understanding their roles in neurotransmitter release. *Neuron* **20**, 389–399.

Parsons, T. D., Coorssen, J. R., Horstmann, H., and Almers, W. (1995). Docked granules, the exocytic burst, and the need for ATP hydrolysis in endocrine cells. *Neuron* **15**, 1085–1096.

Popova, J. S., Garrison, J. C., Rhee, S. G., and Rasenik, M. M. (1997). Tubulin, Gq, and phosphatidylinositol 4,5-bisphosphate interact to regulate phospholipase $C\beta_1$ signalling. *J. Biol. Chem.* **272**, 6760–6765.

Rothman, J. E., and Orci, L. (1992). Molecular dissection of the secretory pathway. *Nature* **355**, 409–415.

Schmidt, A., and Hall, M. (1998). Signaling to the actin cytoskeleton. *Annu. Rev. Cell. Dev. Biol.* **14**, 305–338.

Söllner, T., Whiteheart, S. W., Brunner, M., Erdjument-Bromage, H., Geromanos, S., Tempst, P., and Rothman, J. (1993). SNAP receptors implicated in vesicle targeting and fusion. *Nature* **362**, 318–323.

Südhof, T.C. (1995). The synaptic vesicle cycle: a cascade of protein-protein interactions. *Nature* **375**, 645–653.

Südhof, T., and Rizo, J. (1996). Synaptotagmins: C2-domain proteins that regulate membrane traffic. *Neuron* **17**, 379–388.

Tapon, N., and Hall, A. (1997). Rho, Rac and Cdc42 GTPases regulate the organization of the actin cytoskeleton. *Curr. Opin. Cell Biol.* **9,** 86–92.

Tchakarov, L., Zhang, L., Rose, S., Tang, R., and Trifaro, J. (1998). Light and electron microscopic study of changes in the organization of the cortical actin cytoskeleton during chromaffin cell secretion. *J. Histochem. Cytochem.* **46,** 193–203.

Trifaró, J.-M., Vitale, M. L., and Rodriguez Del Castillo, A. (1992). Cytoskeleton and molecular mechanisms in neurotransmitter release by neurosecretory cells. *Eur. J. Pharmacol.* **225,** 83–104.

Stan Misler

43

Stimulus-Response Coupling in Metabolic Sensor Cells

I. Introduction

The term **sensory receptor** is usually applied to a limited group of cells, located near the surface of the organism, designed to transduce stimuli from the external environment (e.g., photons of light or air or water-borne molecules or pressure disturbances) into the release of a chemical transmitter, which ultimately signals neurons that project into the central nervous system. However, to maintain a "steady state," organisms must also detect changes in their internal environment. To do this, specialized **internal receptors** must sense changes (a) in the levels of circulating metabolic fuels (e.g., nutrient metabolites and O_2); (b) in the levels of metabolic wastes (e.g., CO_2); (c) in the ionic composition of the extracellular lymph-like fluid which bathes the cells; and (d) in the internal or lumenal tension in hollow tubular organs (e.g., blood vessels and gut).

Currently, among the **internal receptors,** those sensitive to metabolic changes are better understood than those responding to changes in ionic composition or lumenal tension. In this chapter, we shall examine, in some detail, sensory transduction by five types of **metabolic sensor cells.**

1. *Pancreatic islet cells:* These cells sense changes in the levels of several circulating metabolites, especially glucose and amino acids, and respond with secretion of the hormone insulin and glucagon. Insulin promotes the uptake of glucose and its storage as glycogen in muscle and liver and maintains fat stores in adipocytes, while glucagon mobilizes fat and glycogen stores.
2. *Cardiac and skeletal myocytes and some neurons:* These cells curtail their excitability during hypoxia and substrate deprivation, thereby saving themselves from severe hypometabolic injury.
3. *Chemoreceptor cells of carotid body:* These cells sense changes in the O_2 and CO_2 content of blood, and synapse onto sensory nerve fibers which project into the central nervous system (CNS) and help shape respiratory drive.
4. *Vascular smooth muscle cells:* These cells sense local changes in O_2 tension, and respond by regulating their tone, thereby locally controlling blood flow.
5. *Hypoxia-secreting renal interstitial cells and liver Ito cells:* In response to hypoxia, these cells secrete erythropoetin, the hormone critical for maturation and release from bone marrow of oxygen-carrying erythrocytes.

The first four types of metabolic sensor cells share a common property; each makes use of K^+ channels specially attuned to metabolic changes to transduce stimuli. Some also use the metabolic intermediates to modulate more distal processes in their scheme of **stimulus-response coupling.** The fifth cell makes use of what might be called **stimulus-synthesis coupling.** Familiarity with different patch-clamp recording configurations and basic concepts of excitation-secretion coupling in nerve cells is assumed. (See Chapters 26 and 42).

II. Stimulus-Secretion Coupling in the Pancreatic Islet Cells

The pancreas consists of islands of endocrine tissue, islets of Langerhans that regulate metabolite uptake, storage and release by liver, adipocytes, and skeletal muscle, surrounded by a sea of exocrine tissue, acini and their ducts that secrete alkaline digestive juices into the duodenum. Each islet contains about 1000 cells. Roughly 65% are insulin-producing β cells, triggered by an increase in serum glucose. These are the largest of the islet cells and have distinct crystals in their secretory granules. Roughly 20% are glucagon-producing α cells, largely triggered by a decrease in serum glucose. The remaining 15% are δ or PP cells secreting somatostatin or pancreatic polypeptide. After dispersal of the islet into single cells, α cells can be separated from β cells on the basis of their smaller size and decreased NAD^- autofluorescence. Though isolated islet cells secrete in response to appropriate stimuli, they secrete best when they are in clusters containing representatives of each cell type. Beta cells are electrically coupled. Alpha, beta, and delta cells all "talk" to each other via paracrine interactions; glucagon enhances stimulus-induced insulin secretion, whereas somatostatin inhibits it. Insulin, glucagon, and somatostatin all inhibit glucagon secretion.

The study of how a rise in plasma glucose leads to rapid secretion of insulin by β cells has provided fertile ground for the interaction of a wide range of disciplines in cell physiology. Since the mid 1970s, it has been widely appreciated that the release of insulin, which is stored in secretory granules, is dependent on extracellular Ca^{2+} concentration $[Ca^{2+}]_o$ (the **calcium hypothesis**), requires oxidative metabolism of glucose or other metabolite fuels (the **fuel hypothesis**) and follows the onset of complex electrical activity in the β cells (the **depolarization-secretion coupling hypothesis**). These features were first established using techniques of islet perifusion and radioimmunoassay of secreted insulin (see Fig. 1A). In these experiments, islets are mounted on a filter and exposed to a continuous flow of solution able to trap secreted insulin. Timed aliquots of solution are collected and the insulin content is measured by radioimmunoassay, that is, the ability of the secreted insulin to displace radioactively labeled insulin bound to an anti-insulin antibody. Note that insulin secretion induced by raising glucose from preprandial levels of 2–3 mM to levels above 5 mM occurs in two phases, a transient "first phase" and a sustained "second phase." *In vivo*, the "first phase" may saturate hepatic insulin receptors, as it is often not readily detected in assays of peripheral insulin levels. Note that the "second phase" of release is reversibly reduced by (1) transient reduction of the Ca^{2+} concentration of the perifusing solution or (2) addition of an inhibitor of glucose metabolism, in this case, sodium azide (NaN_3), an inhibitor of the mitochondrial respiration (see Fig. 1A). In parallel experiments it was found that islets bathed in preprandial levels of glucose rapidly released insulin in response to exposure to high $[K^+]$, provided adequate extracellular Ca^{2+} was present.

When the **fuel hypothesis** was extensively tested, insulin release was found to be (1) stimulated by amino acids, glycolytic intermediates, and ketone bodies shown to be metabolized by β cells but (2) inhibited by substances that specifically block uptake and metabolism of these substrates or by general inhibitors of mitochondrial function. The **calcium hypothesis** and the **depolarization-secretion coupling hypothesis** were tested with intracellular recording from islet cells in intact and perifused islets. Beta cells were found to maintain resting potentials of about –60 mV at fasting glucose concentrations. The resting potential varies with $[K^+]_o$ according to the Nernst equation, hence suggesting that it is largely due to K^+ permeability (P_K). Within several minutes of exposure to "secretagogue" concentrations of fuel metabolites, these cells depolarize, coincident with a decline in resting membrane conductance, and shortly thereafter exhibit bursts of action potentials whose overshoots are a function of $[Ca^{2+}]_o$. In cells loaded with a Ca^{2+} indicator, "spikes" of cytosolic Ca^{2+} coincide with bursts of action potentials (see Fig. 1B).

From these observations emerged an early critical unifying hypothesis: *fuel metabolism $\rightarrow$ reduced P_K $\rightarrow$ electrical activity $\rightarrow$ Ca^{2+} entry $\rightarrow$ insulin release*. It is worth mentioning

that this scheme also accommodates the actions of a class of hypoglycemic agents, called sulfonylureas. Though not related to fuel metabolites, and themselves not metabolized, sulfonylureas potentiate the action of glucose even when glucose metabolism is modestly inhibited. They reduce resting P_K, depolarize the cell, set off action potentials, and stimulate insulin secretion even at preprandial levels of glucose. As we shall discuss in short order, these substances have been found to block directly the K^+ conductance that is physiologically gated by cell metabolism.

Critical clues to the validity and the molecular details of this hypothesis have been provided by (1) patch clamp recording plus molecular biology of channels (cloning and site-directed mutagenesis) and (2) single-cell, real-time monitoring of exocytosis (or fusion of the granule membrane with the plasma membrane). In patch-clamped β cells, the workings of ATP-inhibited K^+ channels, whose closure underlies cell depolarization, have been analyzed in great molecular detail, in large part due to the identification of this molecule as the β cell's major sulfonylurea receptor. In patch-clamped β cells, depolarization-induced exocytosis can be measured electrically as an increase in membrane capacitance (C_m), while agonist- (e.g., sulfonylurea-) induced release can be monitored electrochemically as release of packets of native insulin or the false transmitter serotonin previously loaded into insulin granules (see Fig. 1C). Using either measure, exocytosis is found to be graded with the amount of Ca^{2+} entry induced by depolarization.

(Here it is worth recalling that exocytosis involves simultaneous fusion of the membrane of the secretory granule with the plasma membrane as well as the release of the granule content into the extracellular space. The membrane fusion event, an increase in plasma membrane surface area, is usually detectable as an increase in plasma membrane capacitance. Membrane capacitance is estimated from the ability of the membrane circuit to "phase shift," or displace in time, a sine wave of voltage excitation. The release of a packet of transmitter (or hormone) can be detected "biologically," by a receptive patch of membrane, or "electrochemically," at the surface of an electrode held at a potential sufficient to oxidize or reduce a component of the granule content. In either case, a brief spike of current is produced. When dealing with cells with large secretory granules, one can be more adequately convinced that exocytosis has occurred when it is possible to demonstrate the following points: (1) the jumps in C_m thought to represent single granule fusion events have amplitudes predictable from the surface area of the granule; (2) the single electrochemical spike (or "amperometric event") coincides in time with jumps in C_m and represents the release of at least many tens to hundreds of thousands of molecules that are either native to the granule or else preloaded as a marker. In the case of β cells, the preloaded marker is serotonin. It is worth remembering that capacitance and amperometry are complementary techniques; capacitance monitors net increase in total membrane surface area (exocytosis minus endocytosis), whereas amperometry often permits direct counting of quantal release, albeit from a fraction of the cell surface.)

On the basis of these and other experiments, over the past decade a more detailed "consensus hypothesis" has emerged

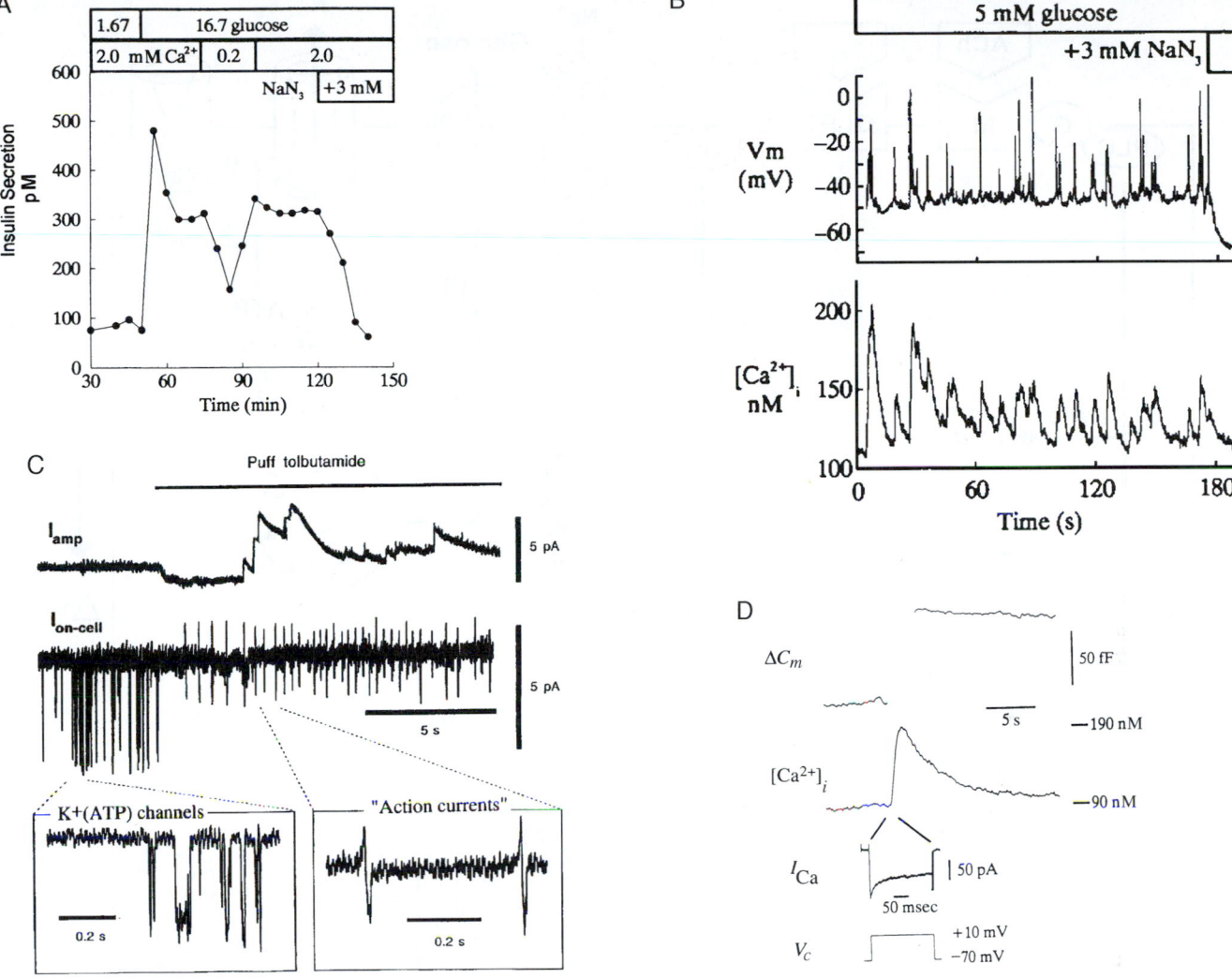

FIGURE 1. Overview of basic evidence supporting "fuel hypothesis," "calcium hypothesis," and "depolarization-secretion coupling hypothesis" of glucose-induced insulin release from pancreatic β cells. (A) Insulin secretion, measured by radioimmunoassay, from isolated intact human pancreatic islets of Langerhans. Note the rapid onset of biphasic insulin secretion on increasing glucose from 1.67 to 16.7 mM, the interruption of the plateau phase by reduction in extracellular [Ca^{2+}] from 0.2 to 2.0 mM, and the termination of secretion by the addition of sodium azide (NaN_3), an inhibitor of electron transport by mitochondrial cytochromes. Leucine, glyceraldehyde, and acetoacetate produce similar patterns of secretion in the presence of Ca^{2+}. (B) Simultaneous recording of electrical activity and cytosolic Ca^{2+} transients from single cryopreserved/thawed human β cells loaded with Ca^{2+} indicator FURA-2. Note that in the presence of 5 mM extracellular glucose, the cell fires individual or short trains of action potentials and that even single action potentials are sufficient to induce a Ca^{2+} transient. Subsequent addition of NaN_3 abruptly hyperpolarizes the cell to ~ –70 mV and blocks the appearance of Ca^{2+} transients. (Unpublished experiments by D. Barnett and S. Misler.) (C) Single cell exocytosis detected by amperometry and capacitance. Simultaneous cell-attached patch recording and amperometry from small clamp of islet cells preloaded with serotonin. The amperometric electrode was positioned at the confluence of cells. In 2 mM glucose, the $I_{on\text{-}cell}$ trace shows single-channel openings, which will later be identified as those of ATP-sensitive K^+ channels ($K^+(ATP)$). After addition of tolbutamide, single channels close, and "action currents" develop. After several impulses have fired, amperometric events are noted on the I_{amp} trace. Those events, slower than those observed when recording for single isolated cells, probably reflect simultaneous exocytotic discharge at several sites, each distant from the electrode. (D) Simultaneous-monitoring membrane capacitance and cytosolic Ca^{2+} from a voltage-clamped β cell. Note that a 200-ms depolarization from –70 to +10 mV increases global cytosolic [Ca^{2+}] by an estimated 100 μM and membrane capacitance by 75 fF. This corresponds to the fusion of 30–60 insulin granules, given that the C_m increase per granule is estimated at 1.25–2.5 fF.

for glucose-induced insulin secretion from β cells (see Fig. 2, right hand side). Uptake of glucose (predominantly via a Glut-2 type transporter) provides glucose entry into glycolysis (via glucokinase, a low affinity hexokinase), and the mitochondrial Krebs cycle results in the generation of ATP and

consumption of ADP (Step 1). This leads to the closure of ATP-inhibited K^+ channels, here abbreviated as $K^+(ATP)$ (Step 2). Against a background of tonic activity of nonselective cation channels that pass Na^+ inward. $K^+(ATP)$ channel closure results in the depolarization of the β-cell membrane

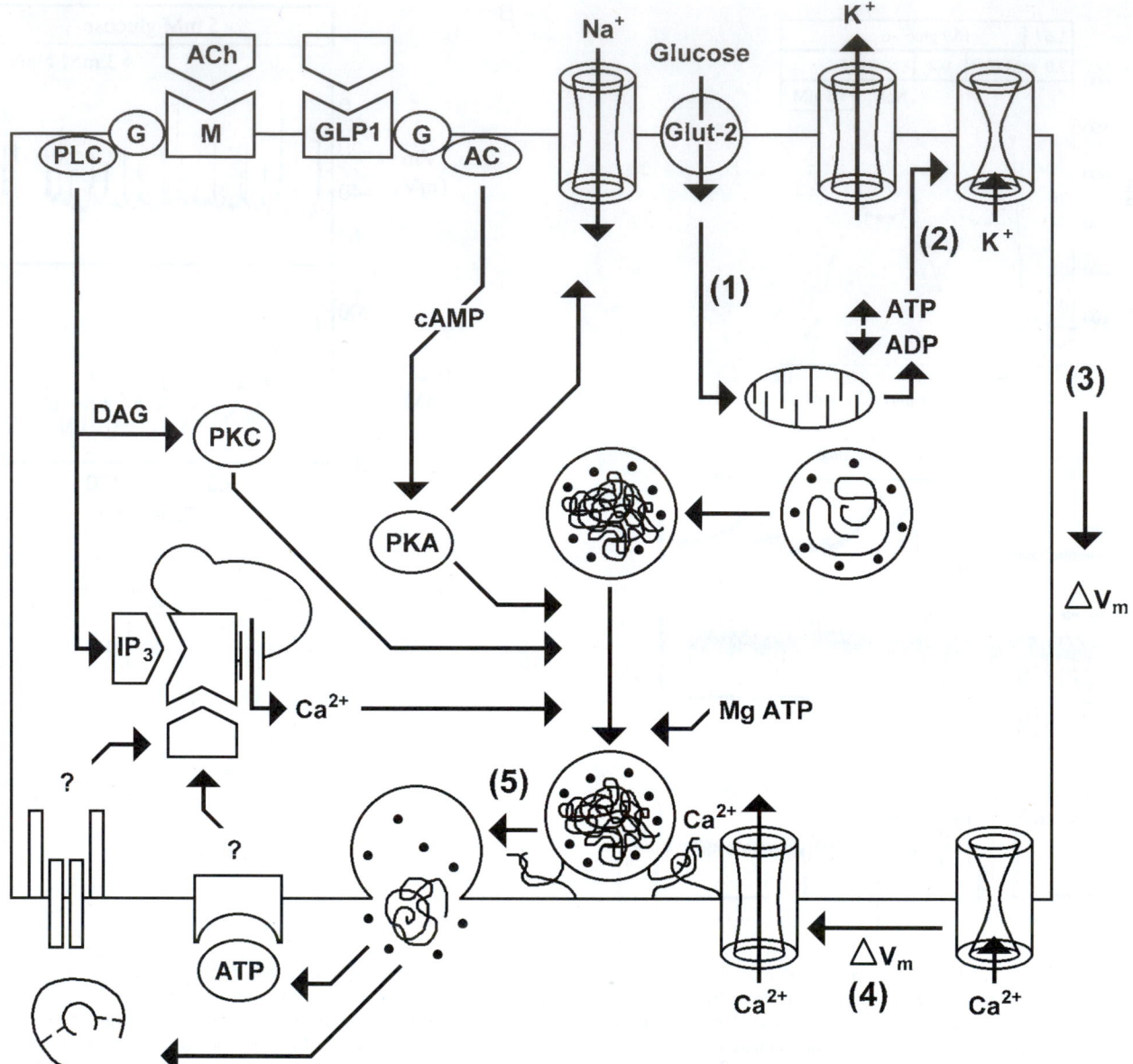

FIGURE 2. Overview of stimulus-secretion coupling in the pancreatic β cell. The right-hand side shows (1–5) a "consensus scheme" of stimulus-secretion coupling: glucose transport and phosphorylation (1), $\rightarrow$ to intermediate metabolism, ATP production/ADP consumption and $K^+(ATP)$ channel closure (2), $\rightarrow$ membrane depolarization (3), $\rightarrow$ opening of voltage-dependent Ca^{2+} channels (4), and finally Ca^{2+} binding to granule docking complex and granule fusion (5). Non selective cation channels that form the counterweight to $K^+(ATP)$ channels at rest are shown to the left of the Glut-2 transporter. Further to the left are the G-protein–linked receptor cascades for muscarinic, cholinergic agonists, and the incretin GLP-1. The latter two "priming" compounds produce second messengers capable of releasing Ca^{2+} from intracellular stores and/or activating protein kinases to modulate channel activity and granule mobilization. Lastly, the contents of the secretory granule, namely insulin and ATP, may "feed back" on the β cell via cell surface receptors and complex signaling cascades.

(Step 3). When V_m reaches the threshold for activating voltage-dependent Ca^{2+} and Na^+ channels, electrical activity and voltage-dependent Ca^{2+} entry begins (Step 4). The resultant rise in cytosolic Ca^{2+} triggers Ca^{2+}-dependent fusion of insulin granules with the plasma membrane (Step 5). Steps 4 and 5 are reminiscent of the process of depolarization-secretion coupling, well studied at nerve terminals and other electrically excitable endocrine cells (adrenal chromaffin cells, pituitary lactotropes and melanotropes, and terminals of the

supraoptic nucleus cells), with subtle differences that shall be discussed later.

However, this scheme does not answer some rather old questions. (1) How do enhancers of cAMP serve to restore stimulus-secretion coupling in metabolically and electrically active but nonsecreting β cells? (2) How, aside from possibly depolarizing the β cell, do anticipatory or modulatory factors such as gut hormones and acetylcholine work? The anticipatory-modulatory pathways of receptor-mediated stimulus-secretion coupling, as well as possible autocrine effects of insulin and ATP released by exocytosis, are outlined on the left of Fig. 3 and also will be discussed later. In fact, it may be possible that under very extreme conditions, modulatory pathways may result in some secretion independent of an inciting Ca^{2+} current or a detectable rise in global cytosolic Ca^{2+}. This might occur through the combination of a very localized rise in cytosolic Ca^{2+} with a vastly increased readiness of the secretory apparatus. However, it should never be forgotten that the "sui generis" of the β cell remains metabolic regulation of cell electrical activity via $K^+(ATP)$ channels.

A. Role of ATP-Sensitive K^+ Channels in Coupling Metabolism to β-Cell Depolarization

How do fuel metabolites produce β-cell depolarization? Early patch-clamp experiments, using the cell-attached mode of recording, revealed the presence of a very interesting K^+ channel that was open at fasting levels of extracellular glucose (2–3 mM), but rapidly closed on raising ambient glucose concentrations to those encountered after a meal (5–7 mM). Usually, channel closure is followed closely by the onset of electrical activity. This channel is open at the cell's resting potential, and its activity is not dependent on membrane voltage. However, the channel is inward rectifying (i.e., single channels pass inward flow of potassium better than outward). These channels have many features predictable from the **fuel metabolism** hypothesis: addition to the bath of any metabolized fuel closes the channel, yet the channel is reopened subsequently by the addition to the bath of inhibitors of metabolite transport or oxidation. The time course and relative change in channel activity seen in a cell-attached patch during a variety of metabolic maneuvers closely parallels the changes in whole-cell resting conductance (G_m), monitored as the change in membrane potential in response to a given test pulse of current applied (see Fig. 3A, B). The latter strongly suggested that the activity of this channel, which tends to move V_m to a level near the potassium equilibrium potential, E_K, critically controls the cell's resting potential. Since this channel population is the predominant type open at rest, its closure *en masse* would be needed to make the K^+ conductance of the cell comparable to that of other conductance pathways operative at rest. The latter are nonselective cation conductance channels and anion conductance channels with estimated equilibrium potentials of 0 or ~ -30 mV respectively. This would permit V_m to depolarize to the threshold for activating cell electrical activity (~ -45 mV). Another hint of the importance of this type of channel was that it is closed by oral hypoglycemic agents.

How is this channel gated? In the cell-attached patch-recording configuration, in which the exterior of the channel was isolated from bath glucose by a very tight pipette-to-membrane seal, it is safe to assume that the channel is gated by one or more metabolic intermediates. *A priori,* cell metabolism might affect a host of metabolic intermediates, ranging from high-energy phosphate compounds (e.g., ATP and ADP), to redox equivalents, to cytosolic H^+ or Ca^{2+}. Early experiments (see Fig. 3C) showed that after excision of the patch, in an inside-out excised patch configuration,

the metabolite-regulated K^+ channel was avidly gated by ATP at its inner surface, hence the name $K^+(ATP)$ channel. Free concentrations of ATP, or its nonhydrolyzable analogs (e.g. AMP-PNP) as low as 10 μM reduce channel activity by half even in the absence of Mg. This suggests that ATP closes this channel by directly "gating" it rather than "phosphorylating" it. In parallel with this, in whole-cell patch experiments where ATP dialyzes out of the cell (due to the absence of ATP in the pipette), resting G_m increases, and the cell hyperpolarizes. Interestingly, ATP inhibition of channel activity in the excised (inside-out) patch can be partially reversed by addition of MgADP to the bath. Hence, changes in the relative concentrations of ATP and ADP, which usually change in a reciprocal fashion as metabolism is altered, should be an important factor in physiological channel gating. In most cells, ATP concentrations change little with metabolism, but percentage-wise, ADP levels change more dramatically. Hence the channel may have been misnamed. To be sure, the activity of the $K^+(ATP)$ channel is also altered by other potential intermediates whose cytosolic concentrations change with metabolic stimulation (e.g., NADH/NAD ratio, H^+ and Ca^{2+}), but these effects are much smaller and require changes beyond the physiological range.

A puzzling question regarding the function of this channel is how its low K_d value (μM) for ATP-induced channel closure in the excised membrane patch can be reconciled with the millimolar-range concentration of ATP in the cytosol. Several factors mitigate this discrepancy. (1) *Complex effects of ADP on channel activity.* ADP competition for an ATP binding site, combined with allosteric action of ADP at an independent binding site, adjusts the half-maximal effective concentration (K_d) of ATP, measured in the presence of ADP, closer to the mM range. (2) *Compartmentation of the ATP and steep ATP gradients from source to sink.* It is likely that the overwhelming majority of ATP is of mitochondrial origin: mitochondrial inhibition totally prevents glucose-induced channel closure, while ketone bodies, amino acids, and their deamination products (e.g., leucine and ketoisovalerate), which directly enter mitochondria, stimulate channel closure even in the face of inhibition of glycolysis. In addition, given the random localization of these channels over the surface of cultured cells, it is likely that the channels are interspersed among ATPases. Hence, steep cytoplasmic gradients of ATP and ADP, due to *en passant* consumption of ATP and concomitant local production of ADP by neighboring plasma membrane ATPases, should make the ATP/ADP available to the channel different from the average cytosolic concentrations actually measured. (3) *Modulation of gating by other metabolic intermediates* such as long chain fatty acids. (4) *Artifactual enhancement of channel sensitivity to ATP in the excised patch by progressive washout of intrinsic channel inhibitors* such as anionic lipids (phosphoinositol phosphates), from the exposed membrane. Polyanionic molecules might specifically compete with ATP for binding to the channel or else contribute to a negative surface charge at the inner surface of the plasma membrane, thereby electrostatically repelling soluble anions.

Two features of the metabolite-regulated $K^+(ATP)$ channel proved critical to its cloning and structure-function analysis. (1) The channel functions as an inward rectifier K^+

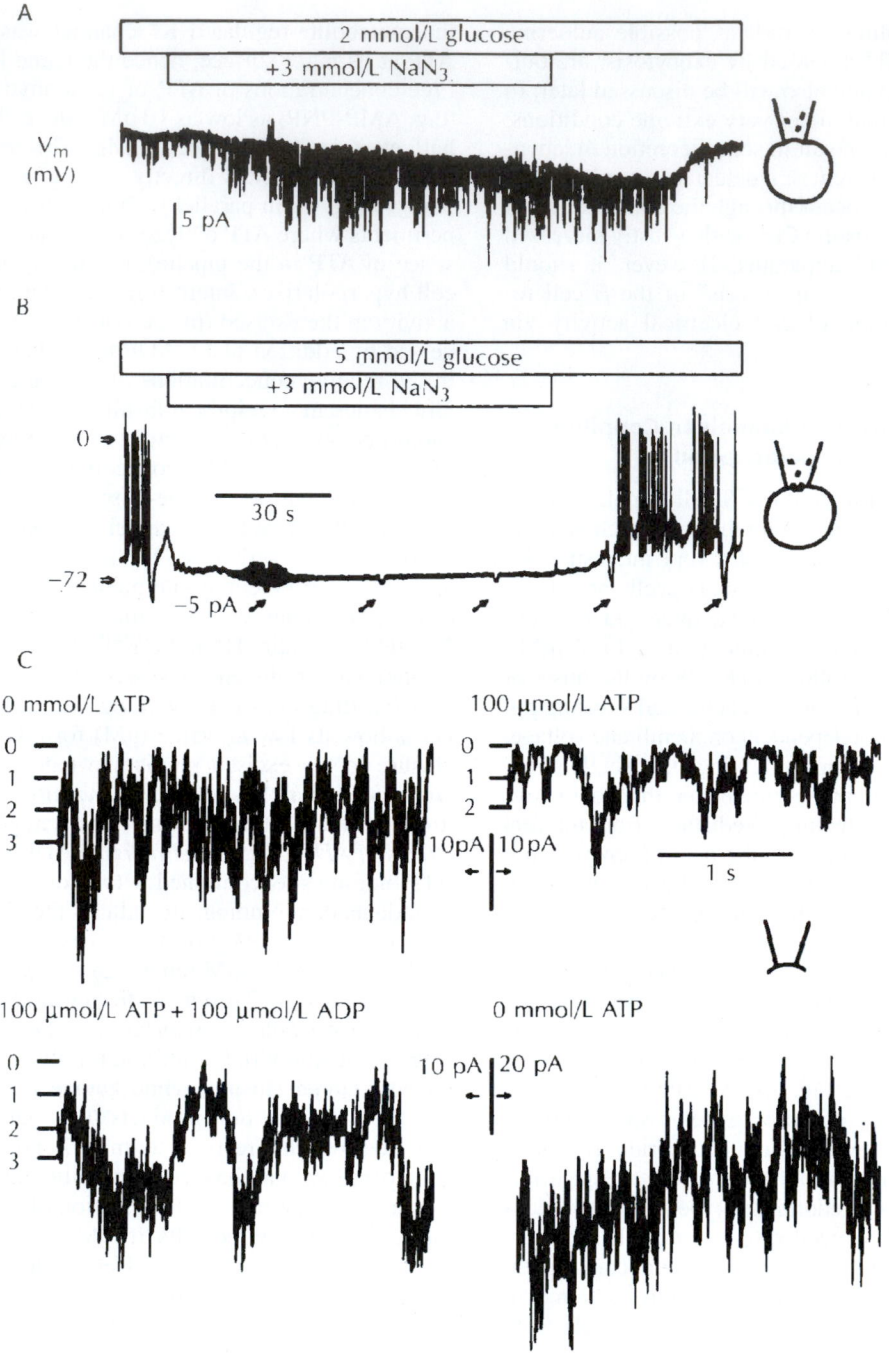

FIGURE 3. ATP-sensitive K+ channels and electrical activity in canine pancreatic islet β cells. (A) A cell-attached patch was formed with the pore-forming antibiotic nystatin in the pipette. Shortly thereafter, ion-channel activity was recorded before, during, and after application of sodium azide (NaN$_3$, 3 mM) to the low-glucose bath. (B) Electrical activity recorded 20 minutes later. By that time, the patch of membrane encompassed by the pipette had been permeabilized by insertion of nystatin channels, thereby permitting recording of membrane current or voltage from the whole cells. Reapplication of NaN$_3$ results in membrane hyperpolarization, abolition of electrical activity, and increased membrane conductance (signified by the reductions in transient hyperpolarizations accompanying –5 pA currents applied at arrows). V$_m$ denotes cell membrane potential. (C) Nucleotide gating of K+(ATP) channel activity in the excised patch. Addition of 100 micromolar ATP (100 µM) reduces channel activity by three-fold (C, upper right panel); further addition of ADP restores channel activity (C, lower left panel), whereas washout of ATP and ADP results in overshoot of channel activity (C, lower right panel) (D.M. Pressel and S. Misler, unpublished data).

channel and hence could be cloned as a member of that family. (2) The channel is the major locus of action of hypoglycemic (glucose-lowering) sulfonylureas and could be cloned as a sulfonylurea receptor (SUR). The sulfonylureas (e.g., **tolbutamide** and **glyburide** (glibenclamide)) enhance glucose-induced insulin release by β cells and are useful in the treatment of non–insulin-dependent diabetes mellitus, in which β cells produce significant quantities of insulin, but release it out of synchrony with the appearance of the glucose load. At concentrations equal to free plasma levels that enhance glucose-induced electrical activity and insulin secretion in intact animals, the **sulfonylureas** close K$^+$(ATP) channels in cell-attached or excised patches alike, while leaving voltage and calcium gated K$^+$ channels unaffected. Curiously, though K$^+$(ATP) channels in myocytes show similar conductance, kinetics, and ATP sensitivity, they have greatly reduced sulfonylurea sensitivity.

To reconstitute channel activity in a naive cell (e.g., *Xenopus* oocyte or clonal cell) it was necessary to transfect or inject the cell with cDNA or mRNA coding for (1) a standard inward rectifier type subunit (Kir6.2), comprised of two membrane-spanning domains separated by an H loop dipping into the bilayer, and (2) the SUR from β cells (SUR1) (see Fig. 4A, B). SUR 1 is a complex protein with multiple transmembrane domains (hydropathy plot analysis suggests 17!). Structurally it contains two motifs resembling those of the ATP-binding cassettes of members of the ABC transporter superfamily, the most famous of which are the cystic fibrosis gene product, CFTR, and the multidrug resistance p-glycoprotein. SUR1 has two distinct cytoplasmic loops (or facettes) that bind nucleotide di- and tri-phosphates in the presence of Mg (NBFs). The channel itself is likely to be an octamer consisting of an inner ring of four Kir6.2 subunits (the confluence of the four H loops forming the ion selective pore) and an outer ring of four SUR1 subunits. The two types of subunits are needed to chaperone each other into the membrane as well as to provide full channel gating; without SUR1, Kir6.2 is not expressed. Clinically, mutations in SUR1 have been identified in persistent hyperinsulinemic, hypoglycemia of infancy, a disease where K$^+$(ATP) channel activity is not detected even after metabolic inhibition and β cells are persistently depolarized and secreting. Point mutations to NBF result in channels that are closed by ATP in excised patches, but are not reopened by MgADP, while more extensive mutations result in the absence of channel activity under any test circumstance. In contrast, a C-terminal truncated mutant of Kir6.2 (Kir6.2ΔC26) forms K$^+$ channels gated by ATP in the absence of Mg. These channels have lower affinity for ATP (K_d ~10 μM in the absence of Mg) than the native or Kir6.2/SUR1 cloned channel (K_d ~100 μM) but show similar need for ATP cooperativity in channel gating ($n = 4$). However, these channels lack both inhibition of activity by sulfonylureas and enhancement of activity by MgADP. Hence, while Kir6.2 might be considered the ATP gating subunit, SUR1 with its NFBs is the ATP-hypersensitizing subunit as well as the subunit endowing the channel with special sensitivity to drugs and MgADP. A schematic model of K$^+$(ATP) channel gating is presented in Fig. 4C. SUR might have a natural ligand, a cytoplasmic protein called en-

dosulfine, recently demonstrated to displace sulfonylureas from SUR binding sites.

Analysis of mutant channels has also elucidated the origin of a number of previously puzzling effects. One example is the "refreshment effect" of MgATP. That is, rapid wash-in and wash-out of MgATP, even in concentrations of 10–100 μM, can restore channel activity that otherwise would run down in time in its absence (see Fig. 3C). This effect can now be partially explained by the finding that small concentrations of MgATP as well as MgADP activate the channel by binding to a "Walker motif" of the NFB in native forms of SUR1 but not in those SURs whose NBFs have been mutated. Likewise, channel modulation by Mg-guanidine nucleotides also occurs at these sites, suggesting that there is no need to invoke G proteins in GTP or GDP modulation of channel gating (though G-protein-coupled receptors may independently modulate channel activity). Both effects require hydrolyzable nucleotides and Mg, suggesting that some phosphorylation reaction may be involved. A second example is the ability of certain vasodilator drugs, such as diazoxide, to open K$^+$(ATP) channels, hyperpolarize the β cell and inhibit glucose-induced insulin secretion. These drugs also bind to the "Walker motif." A third example is the differential sulfonylurea sensitivity of K$^+$(ATP) channels in β cells as compared with those in cardiac myocytes. While both channels have similar Kir6.2 subunits, their SURs differ. Replacing transmembrane domains 13–16 of cardiac SUR (SUR2) with the corresponding region of SUR by engineering of a chimeric SUR restores sulfonylurea sensitivity.

Is this finely tuned coupling of cell metabolism to K$^+$(ATP) channel activity used as a sensory mechanism by the two other major glucose receptor cells of the body, namely the **satiety center** of the hypothalamus and the peripheral sweet **taste receptors?** Recently, K$^+$(ATP) channels have been sought in glucose-receptive (GR) neurons of the ventromedial nucleus (VMN) whose activity is associated with suppression of food intake. In these neurons, reduction of bath glucose or addition of the glycolytic inhibitor mannoheptulose results in hyperpolarization and cessation of electrical activity accompanied by an increase in resting G_m; oppositely, tolbutamide produces excitation. In cell-attached patch recordings, cessation of electrical activity is associated with increased activity of a K$^+$-selective channel, which is voltage-insensitive. After excision of the membrane patch, the latter channel is inhibited by cytoplasmic ATP. However, this K$^+$ channel is quite distinct from the K$^+$(ATP) channel in β cells: under identical excised patch recording conditions, the K_i for ATP inhibition is 2.3 mM (rather than 20 μM in β cells) and the single-channel conductance is 135 pS (rather than 55–65 pS in β cells). One very interesting feature of this channel is that it is opened, and the cell is hyperpolarized by leptin, a protein released by adipocytes, or fat-storing cells, in proportion to their triglyceride storage. Leptin also opens K$^+$(ATP) channels in β cells and appears to do this in a manner dependent on activation of PI-3 kinase. Does leptin serve as a negative feedback hormone, that is signaling adequate energy stores by downregulating both appetite-induced caloric intake and insulin-dependent depot storage? One indication of this

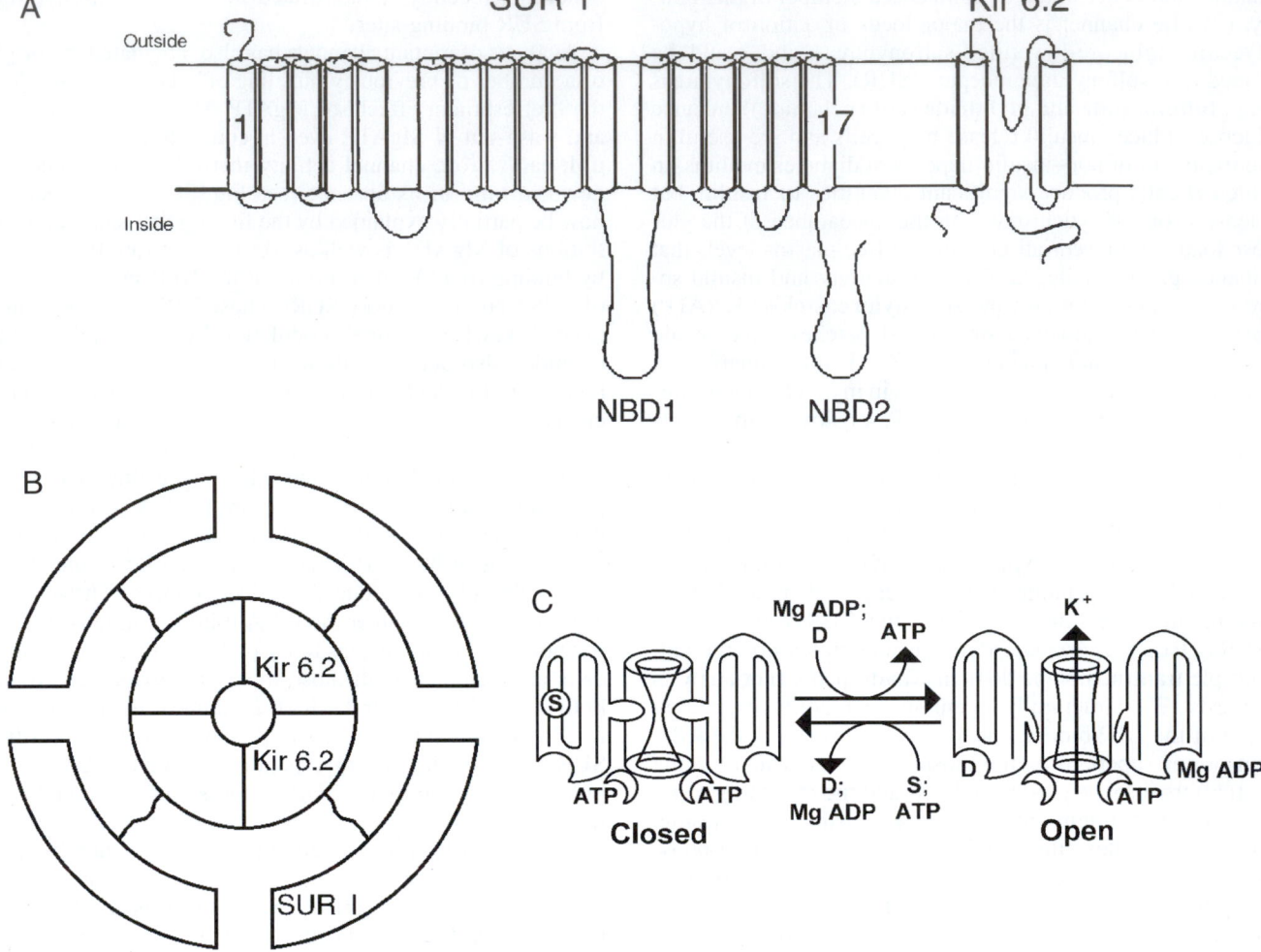

FIGURE 4. Illustrations depicting K$^+$(ATP) channel structure and function. (A) Transmembrane orientations of SUR1 and Kir6.2 subunits based on structural homology, hydropathy, and site-directed mutagenesis. (B) Proposed octamer array of SUR1 and Kir6.2 subunits. (C) Simplified reaction scheme in channel gating.

comes from obese, hyperphagic, hyperinsulinemic, and hyperglycemic db/db rats, where leptin has little influence on the firing rate of VMN neurons.

In peripheral taste receptors, glucose chemoreception may occur through G proteins and a second messenger cascade. In these cells, application of sucrose or saccharin reduces the amplitude of a voltage-dependent outward K$^+$ current. This effect can be mimicked by application of membrane-permeable analogs of cAMP. In excised patches, K$^+$ channels with similar features are closed by exposure of the patch to ATP and cAMP-dependent protein kinase, but not by exposure to ATP alone, suggesting that this K$^+$ channel is "gated" by phosphorylation.

B. Metabolic Regulation of Depolarization-Secretion Coupling?

Recall that closure of K$^+$(ATP) channels only depolarizes the cell. Calcium must enter the cells and insulin granules must be ready to be released. Does metabolism alter the "calcium trigger" or maintain the "readily releasable pool" of insulin granules?

1. The "Ca trigger"

When β cells, exposed to sufficient glucose, a cAMP enhancer, and near physiological temperature, are depolarized for many tens to hundreds of milliseconds, they exocytose in a Ca^{2+}-entry–dependent manner. Under these conditions, several studies have shown a ~3rd power dependence of exocytosis, as measured by amperometry or capacitance, on the total Ca^{2+} entering the cell (i.e., the integral of the Ca^{2+} current). Though most of the Ca^{2+} current is carried by L-type Ca^{2+} channels, in a subset of human β cells, where they are present in sufficient density, lower threshold T-type Ca^{2+} channels also support exocytosis. We might expect changes in cell metabolism to affect the gating of Ca^{2+} channels via phosphorylation or lipidation. To date, data supporting the hypothesis that enhanced glucose enhances Ca^{2+} channel activation or slows inactivation have been inconsistent. Another paradigm, however, which illustrates other aspects of depolarization-secretion coupling may provide some other clues.

Beta cells in intact islets secrete pulses of insulin that are time-linked to "bursts" of electrical activity. Longer "bursts" evoke larger pulses of insulin. Contrary to the situation at most nerve terminals and despite quite sizeable Ca^{2+} currents, it is very rare to see an isolated action potential, or a brief depolarization (10–50 ms in duration), provoke quantal release (here measured by amperometry). However, when the depolarization is extended out to many tens to several hundreds of milliseconds, release routinely commences in the course of the depolarization and often continues for several seconds after Ca^{2+} entry ceases, while cytosolic Ca^{2+} is falling (see Fig. 5). These data suggest that Ca^{2+} might have to "jump through a few hoops" before triggering secretion and that it might set up a "chain reaction" that outlasts the peak Ca^{2+} concentration (Fig. 6).

Several possibilities for this "slow-to-start, slow-to-stop" pattern of secretion are currently under investigation. *First,* with large-sized granules that are not clustered in any regular array, Ca^{2+} channels and granule release sites may not be all that closely co-localized. To be sure, the interspersed cytosol has many Ca^{2+}-buffering molecules. Hence, if the site where Ca^{2+} triggers exocytosis were, on average, a distance of 1–2 granule radii (say 200–300 nm) from the inner mouth of the Ca^{2+} channel site, local cytoplasmic buffers would need to be saturated before Ca^{2+} could appear at the release site. This would cause a slight delay in the onset of exocytosis. At this distance, during continuous Ca^{2+} entry there might be good spatial equilibration of Ca^{2+} so that the Ca^{2+}

concentration at the release site might closely follow the average Ca^{2+} concentration of the cell. This is in no way counterintuitive because it is well appreciated that Ca^{2+} and cytosolic Ca^{2+} concentrations as low as 1.5 µM evoke respectable rates of release (see Fig. 7B). Conversely, once Ca^{2+} no longer enters the cell, the buffer might serve as a continuing source for Ca^{2+} and support a "secretagogue" level of free cytosolic Ca^{2+} for some time. This feature has been examined and modeled extensively in adrenal chromaffin cells but has only been explored in a cursory manner in β cells.

Second, Ca^{2+} entry might be causing regenerative release of Ca^{2+} from some intracellular stores, thus providing a sort of propagating Ca^{2+} wave, while the increase in cytosolic Ca^{2+} might be stimulating Ca^{2+} uptake by other stores for later use. To date, while it is well appreciated that Ca^{2+} can be released from ER stores by the generation of IP_3, blockade of Ca^{2+} release from ER stores has yielded only small changes in depolarization-induced rises in cytosolic Ca^{2+} or exocytosis. However, the picture may be more complex in other situations. For example, when GLP-1 is applied to voltage-clamped β cells in the presence of 20 mM Ca^{2+}, "spontaneous" spikes of cytosolic Ca^{2+} occur between depolarizations. To be sure, there are multiple stores of intracellular Ca^{2+}, aside from the IP_3-releasable ones in ER. For example, mitochondria- and hormone-containing granules

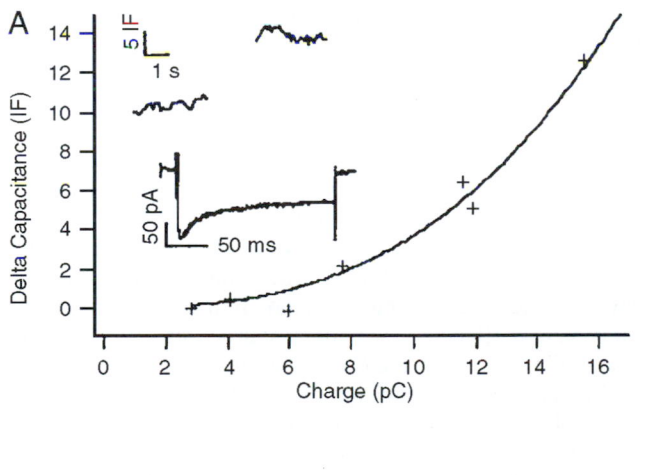

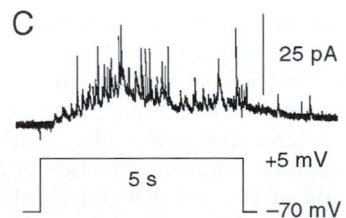

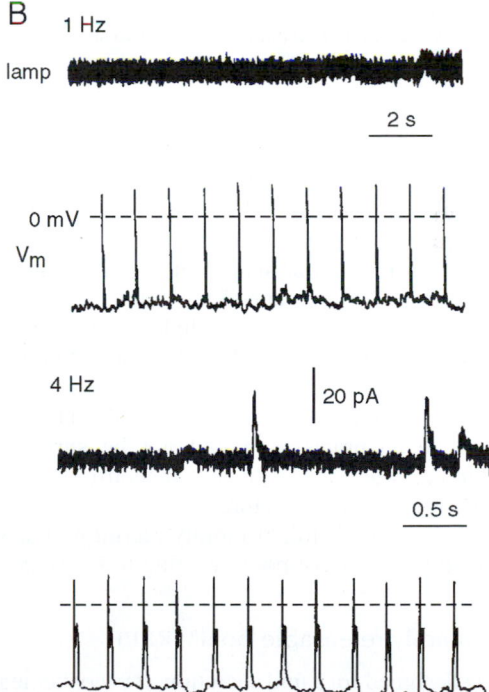

FIGURE 5. Aspects of Ca^{2+} dependence of exocytosis in β cells (A) Relationship of capacitance increase to total Ca^{2+} entry (integral of Ca^{2+} current Q_{Ca}) during β-cell depolarization. Ca^{2+} entry was altered by changing the membrane potential to which the cell was voltage-clamped for the 200-ms test stimulus (unpublished data by D. Barnett and S. Misler). (B) Time course of amperometric events during trains of action potentials stimulated at 1 Hz (upper panel) and 4 Hz (lower panel) Note that none of the 11 action potentials occurring at 1 Hz stimulation results in quantal release, while at 4 Hz stimulation quantal release begins after several action potentials. (C) Time course of amperometric events recorded during continuous depolarization to +10 mV, where Ca^{2+} entry is maximal. Note that the first release event is seen ~120 ms after start of voltage-clamp pulse, while release events continue, albeit at a low frequency, for nearly 3 s after depolarization is completed (unpublished data by Z. Zhou and S. Misler).

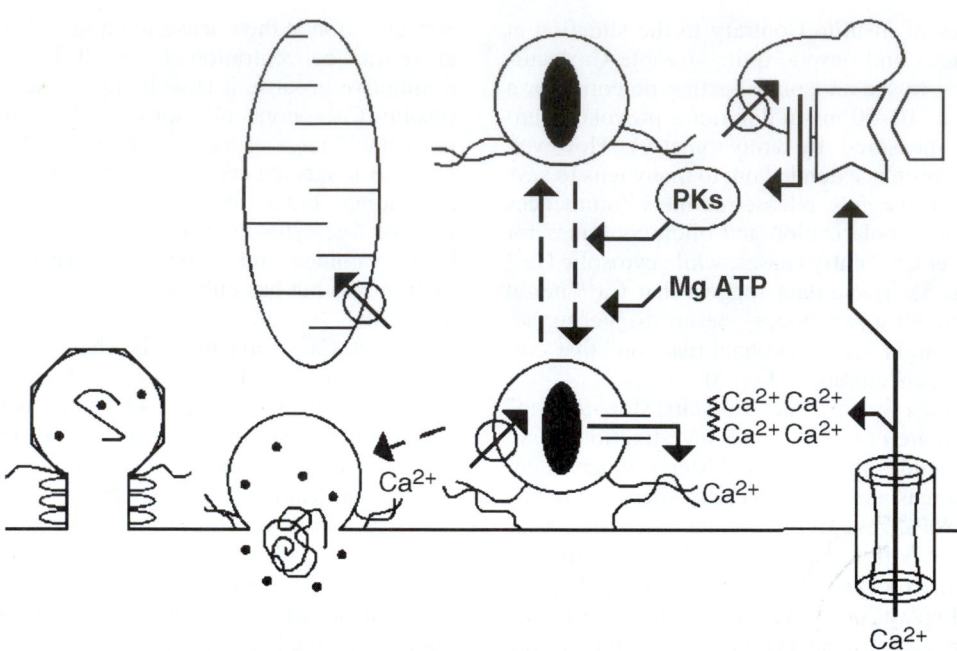

FIGURE 6. Calcium metabolism in β cells: potential sites for modulation of depolarization-secretion coupling. After calcium enters the cytoplasm, it is likely that much of it rapidly binds to either fixed or mobile buffers, the latter perhaps resembling the tetra-clawed EGTA. Some of the bound Ca^{2+} unbinds to diffuse further. Diffusing Ca^{2+} may then bind to one of the following: (1) a Ca^{2+}-binding protein (e.g. synaptotagmin) that is part of the docking complex that anchors a granule to the membrane and thereby directly triggers exocytosis, (2) a protein kinase, whose activity enhances the movement of granules along "motorized" tracks, (3) a binding site on Ca^{2+}-gated channels in Ca^{2+} storage organelles, including the secretory granule itself, which thereby triggers Ca^{2+}-induced Ca^{2+} release. Calcium uptake into organelles (granules, mitochondria, and ER) is probably slower (on the time scale of hundreds of milliseconds to seconds). After fusion with the plasma membrane, granule membrane may be retrieved piecemeal by a local "pinch off" mechanism, by using dynamin, or as part of a large endocytic "slurp."

might be expected to release as well as "mop-up" Ca^{2+}, but little is known about their release and uptake rates during various patterns of electrical stimulation or exposure to hormones and transmitters. Some recent evidence suggests that depletion of the Ca^{2+} content of the acidic compartment of cytoplasmic organelles, which includes secretory granules, reduces depolarization-induced Ca^{2+} release. In a metabolically stimulated cell where the Krebs cycle intermediates, NAD-derived redox equivalents and local ATP concentrations may be changing, it would not be surprising that changes in cell Ca^{2+} metabolism in response to Ca^{2+} influx might be a very dynamic feature.

Third, entering Ca^{2+} might rapidly recruit granules into a small, "readily releasable pool" waiting to be triggered.

2. The "readily releasable pool" (RRP)

In cells designed for rapid secretion, it is convenient to assume that vesicles are lined up against the plasma membrane and ready to fuse as soon as the appropriate level of second messenger is achieved in the cytoplasm. Morphologically, in β cells, this does not appear to be the case (see Fig. 6). When viewed under the electron microscope, very few of the ~10,000 granules in the β cell are within one granule diameter of the membrane. Physiologically, experiments with single cells reveal that secretion is actually quite "wimpy"

unless "revved up" by other maneuvers; with repeated depolarizations, exocytosis "poops out" and may take many tens of seconds to recover. This suggests that the RRP of available granules is actually quite small and that maintaining it, under high output demand, is a non trivial matter. However, in β cells, two ideal sources of "physiological revving" of secretion are (1) activation of protein kinases (A and C) and (2) the provision of ATP by enhanced substrate metabolism. **Protein kinases** are well known to facilitate the attachment of a variety of granules to tracks (microtubules, actin filaments) that bring them in close proximity to the membrane. In β cells, during feeding, insulin secretion is "primed" even before serum glucose rises, by release of two substance that activate protein kinases, acetylcholine from vagal fibers and incretins (e.g., glucagon-like peptide (GLP-1) from enterochromaffin cells. Activation of muscarinic ACh receptors results in activation of phospholipase C and production of IP_3 and diacylglycerol. Activation of incretin receptors results in activation of protein kinase A pathway and enhanced cytosolic cAMP levels. These ACh and GLP-1 pathways might enhance distal processes in excitation-secretion coupling, along with increasing the rate of β-cell depolarization in response to glucose, by affecting ATP-sensitive K^+ channels and/or the nonselective cation channels. Hence both background electrical activity and insulin release- and metabolite-induced electrical activity and insulin release could be enhanced. Changes in **cytosolic MgATP** would be expected to alter the fueling of vesicle transport processes as well as the molecular motors (e.g., NSF) that untwine paired attachment proteins (SNAREs) on granules and the target membrane. The latter permits SNAREs to subsequently intertwine with their cohorts on the complementary membrane, thereby accomplishing vesicle docking. In β cells, glucose is known to enhance insulin secretion evoked by elevated $[K^+]_o$ even when $K^+(ATP)$ channels are already closed, while metabolic inhibition can block insulin secretion, even as cytosolic $[Ca^{2+}]$ rises into the micromolar range.

Experiments monitoring exocytosis from single β cells support the motion that the RRP is small, easily depletable, and requires constant replenishment by processes that are cAMP, MgATP and temperature dependent. Three approaches have been used. In the first approach, the cell is rapidly depolarized, in the presence versus absence of a modulator, until there is no further increase in capacitance. An estimate of the total number of granules exocytosed provides a measure of RRP, while the time to recovery of full response provides a measure of the refilling rate. Then, a rough estimate of size and dynamics of vesicle pools, in the presence versus absence of a modulatory factor, is obtained by examining the time course of release after introduction of a given concentration of Ca^{2+} into cell. A final approach is to measure the capacitance increase, in the presence versus absence of the modulator, in response to instantaneous "uncaging," by flash photolysis, of Ca^{2+} bound to a chelator introduced into the cytosol, or "uncaging" a fraction of the modulator, now bound to a chelator, in the presence of a fixed level of cytosolic Ca^{2+}. Sample experiments demonstrating that stimulators of cytosolic cAMP enhance initial Ca^{2+}-entry-dependent, depolarization-evoked release, as well as maintain release with repeated depolarization, all with little attendant change in Ca^{2+} entry are shown in Fig. 7A. Sample experiments demonstrating that cAMP enhances both the initial maximum rate of exocytosis and the total exocytosis in cells dialyzed against a pipette with fixed concentrations of Ca^{2+} are shown in Fig. 7B. More recently, similar experiments using all three approaches have been performed at varying cytosolic levels of MgATP. ATP appears to modulate RRP as well.

Lastly, to be sure, anticipation of future high rates of release requires activity-stimulated, energy-dependent insulin synthesis, followed by processing in specialized membrane compartments. With prolonged continuous stimulation, β cells undergo several phases of insulin release; after an initial spurt of release lasting several minutes, insulin secretion wanes and then, beginning 10–15 min later, rises to a plateau that is sustained for hours. Also, islets rechallenged with glucose several hours after an initial brief bout of secretion display greater "peak" insulin release on the second round of stimulation and do this with newly synthesized insulin. Thus islets display a crude but effective form of "memory" for previous stimulation. As activity-stimulated protein synthesis can be triggered by changes in cell Ca^{2+}, cAMP, and ATP levels, their relative contributions to synthesis of competent new insulin granules is receiving intense scrutiny.

C. The Paradox of Stimulus-Secretion Coupling In the Glucagon-Secreting α Cell

Alpha cells secrete glucagon at preprandial levels of glucose (2–3 mM); secretion is further enhanced by amino acids such as arginine and depressed by higher levels of glucose. Curiously, under some conditions, the glucagon secretion is

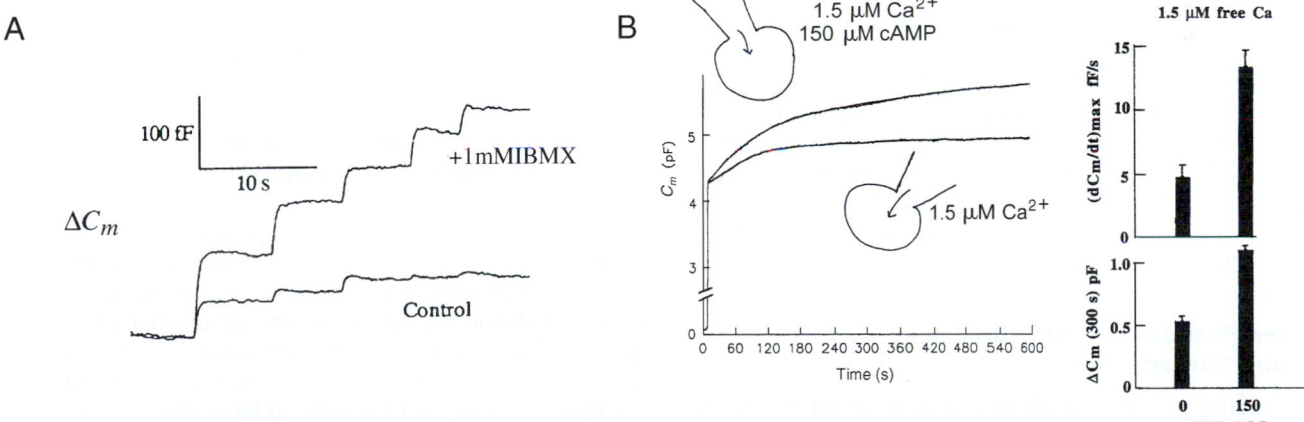

FIGURE 7. Enhancement of cytosolic cAMP appears to increase the pool of readily releasable granules. Test by repetitive stimulation. (A) Note that in the "control," repetitive depolarization produces rapid rundown of capacitance response so that by the fourth depolarization there is no evidence of further net exocytosis. In contrast, after addition of membrane permeable cAMP or a phosphodiesterase inhibitor, isobutylmethylxanthine (IBMX), not only is the response to initial depolarization enhanced, but response to 4th depolarization is nearly as large as the response to the 1st depolarization in the "control". In this cell, total Ca^{2+} entry over the entire series of depolarizations differed by less than 15% across the two runs. Perforated patch recording from rat β cell. Test by cell dialysis. (B) Cells with similar baseline capacitance (4–6 pF at break-in to the cell at time 0) were dialyzed against pipette solutions containing 1.5 μM Ca^{2+} versus 1.5 μM Ca^{2+} + 150 μM cAMP. Note that, in the presence of cAMP, not only was the maximum rate of rise of capacitance, abbreviated (dC_m/dt) max, on average nearly 2.5-fold greater, but the total capacitance increase over the entire 6 min of recording was on average nearly 2-fold greater.

stimulated by tolbutamide and inhibited by diazoxide, both at concentrations that alter insulin secretion from β cells. So, do α cells have K$^+$(ATP) channels and, if so, do these channels contribute to α-cell function?

In the presence of 0–3 mM glucose, α cells generate spontaneous Na$_o^+$- and Ca$_o^{2+}$-dependent action potentials from their resting potential of ~−60 mV. They hyperpolarize on their exposure to increased glucose (5–20 mM); they depolarize and show enhanced spike frequency on exposure to 10 mM arginine. Under voltage clamp control, depolarization of the α-cell results in Ca^{2+}-entry-dependent increases in cytosolic Ca^{2+} and increases in membrane capacitance, the latter enhanced by agents that increase cytosolic cAMP (e.g. β2 adrenergic agonists).

Single-channel and whole-cell patch-clamp recording from α cells reveals inward rectifier K$^+$ currents that have similar single-channel conductance and kinetics to those in β cells as well as similar K_d for inhibition by ATP and activation by MgADP and PIP$_2$. In situ hybridization identifies Kir6.2 and SUR1 subunits of characteristic of β-cell K$^+$(ATP) channels on glucagon-bearing cells; in fact, the densities of these subunits are even higher than those in β cells.

With apparently functional K$^+$(ATP) channels in place, the α cell's inability to respond to glucose or metabolic inhibition is probably related to their extraordinarily high cytosolic [ATP] and [ATP]/[ADP] ratio that remains virtually unchanged after increases in extracellular glucose. (α cells are much less efficient transporters and metabolizers of glucose than are β cells.) In contrast, the α cell's response to arginine may be related to the depolarizing effect of the electrogenic transport of this cationic amino acid coupled to the low threshold for excitability of this cell type (i.e., low resting membrane conductance and abundance of low-voltage–activated (T-type) Ca^{2+} channels). (Arginine also depolarizes β cells without closing K$^+$(ATP) channels, although it usually does not enhance electrical activity unless suprathreshold concentrations of glucose are also present.) However, the crucial element of physiological regulation of the α cell, the inhibition of its electrical activity and secretion by glucose, remain unexplained on the basis of K$^+$(ATP) channel activity. It is possible that the inhibitory effect of glucose results largely from a paracrine interaction within the islet. α cells display a robust GABA activated Cl$^-$ current, which, when stimulated, can abolish arginine-enhanced electrical activity. β cells are known to secrete GABA along with insulin. Hence secretagogue-induced β-cell exocytosis might be the "suppressor" of α-cell activity.

D. Implications for Pathophysiology and Therapeutics

Diabetes mellitus, a disease characterized by hyperglycemia and dysregulation of metabolite use and culminating in multiorgan system damage, comes in two general varieties. The insulin-dependent, ketosis-prone variety (IDDM) is usually an autoimmune disease of rapid onset, characterized by destruction of β cells as a result of massive assault by antibodies and cytokines. In contrast, the noninsulin-dependent, nonketosis-prone variety (NIDDM) is a spectrum of diseases characterized by increasing insulin resistance coupled with poorly timed, inadequate insulin secretion. One of its earliest indicators of this condition is the loss of the transient "first phase" of glucose-induced insulin secretion, thought to be "primed" by acetylcholine released by the vagus nerve and GLP-1 released by enterochrommaffin cells. Accompanying this is a "compensatory" increase in insulin release during the prolonged "second phase." Unfortunately, this contributes to downregulation or "tonic tune out" of insulin receptors. Possible defects in NIDDM have been proposed for every link along the β-cell's stimulus-secretion coupling cascade (i.e., from reduced affinity of glucose transporters to reduced efficiency of intracellular handling of Ca^{2+}). A recently explored model for NIDDM features insulin resistance of the β cell with chronically decreased release of Ca^{2+} from intracellular (ER) stores. No doubt, the single cell approaches outlined above will be increasingly brought to bear to dissect out cellular mechanisms underlying NIDDM.

Those approaches may be expected to spur the development of more "physiologically targeted" pharmacotherapy for NIDDM, including carefully timed, pre-prandial administration of (1) GLP-1 and other gut-secreted incretins and (2) highly selective sulfonylurea-type agents of shorter onset and duration of action than those currently marketed. To be sure, two realizations, (1) that α cells contain sulfonylurea-inhibited K$^+$(ATP) channels possibly of identical structure, and (2) that constant activation of α cells coupled with waning of β-cell function, may contribute to the progressive sulfonylurea insensitivity, are now promoting synthesis of hypoglycemic drugs more selectively targeted at β-cell function.

It is worth noting that sulfonylureas also modify the gating of some Cl conductance channels in β cells as well as the activity of Na$^+$-K$^+$ ATPase and the efficacy of depolarization-secretion coupling in a manner similar to the effects of protein kinase C enhancement. As mentioned above, the NBFs of SUR resemble those of CFTR. Spare CFTR subunits appear to interact with a variety of Cl channels and even with the epithelial Na$^+$ channel (EnaC) in epithelial cells. Do spare SUR subunits likewise interact with integral membrane or membrane associated proteins in β cells to modulate a variety of events in stimulus-secretion coupling?

III. Metabolic Sensing as Protection from Hypometabolic Injury

The hyperpolarization and increased K$^+$(ATP) channel activity displayed by β cells exposed to metabolic inhibitors that reduce cytosolic [ATP]/[ADP] suggests that a safeguard against hypoxic damage of an excitable cell might be hypoxia-induced opening of K$^+$(ATP) channels. This mechanism appears to apply to skeletal and cardiac myocytes and to some neurons. As in β cells, in these cells metabolically sensitive channels are very often K$^+$(ATP) channels consisting of Kir and SUR subunits, though often the SUR subunit (SUR2) has much lower affinity for some sulphonylureas than does SUR1. In addition, in skeletal myocytes, which produce lactate under hypoxic conditions, the K$^+$(ATP) channel displays pH$_i$ sensitivity that is the reverse of K$^+$(ATP) channels in β cells; in these myocytes, intracellular acidosis vigorously opens K$^+$(ATP) channels even at fixed [ATP]/[ADP]. Recent evidence suggests that in highly active regions of the brain, those cell types with higher density of K⁺(ATP) channels show most rapid excitability block in response to hypoxia and survive it best. However, what may be an adaptive feature to individual cells may have disastrous consequences for the entire organism. In skeletal muscle, a drop in tension development by some motor units results, via spinal reflex, in recruitment of other less active motor units. More extended fatigue due to inexcitability may preserve enough ATP to maintain sarcolemmal integrity and avert leak of myoplasmic contents such as myoglobin. (The latter condition, known as rhabdomyolysis, is the chief cause of extended incapacity and systemic illnesses, such as with acute renal failure, seen after intense bouts of exercise.) However, in contrast, in cardiac ventricle, block of excitability or excitation contraction coupling in one part of the electromechanical syncitium may predispose to an arrhythmia or severely dyskinetic contraction, while in neurons, excitability block can result in a depressed level of central consciousness and even coma.

As an example of hypoxia-induced excitation block, let us examine block of excitation-contraction coupling in the cardiac ventricular myocyte. Historically, this phenomenon was key to the original discovery of the K⁺(ATP) current. In these cells, deprivation of substrate or oxygen, or inhibition of substrate metabolism, was found to result in shortening of Ca^{2+}_o-dependent plateau phase of the action potential with attendant abbreviation or block of generation of contractile force. This was found to have occurred prior to detectable changes in membrane potential, or voltage-dependent Na⁺ and Ca²⁺ currents, and it was accompanied by major augmentation of resting K⁺ conductance. Critically, plateau phase shortening and reduced contraction were reversed by intracellular injection of MgATP. As shown in Fig. 8, the key to this phenomenon is very basic. The plateau phase of the ventricular action poten-

tial represents a delicate moment-to-moment balance between two opposing tendencies. These are (i) the "depolarizing tendency" of the voltage-dependent, but slowly inactivating Ca²⁺ conductance and (ii) the "repolarizing tendency" of the sum of a variety of K⁺ currents that are either slowly activating or slowly emerging from "inward rectification" (i.e., polyamine or Mg block). Increases in background K⁺ current provided by opening of K⁺(ATP) channel tip the balance in favor of repolarization, as soon as the huge inward Na⁺ current, underlying the rapid upstroke of the action potential, wanes. However, shortening of the broadly propagating action potential may reduce the refractory period and promote local impulse re-entry or rebound excitation. In addition, the period of reoxygenation may enhance arrhythmogenicity by promoting large transient inward currents (I_{ti}) which result in spontaneous depolarizations. Hence, it would appear to be better for the ventricle to anticipate rather than react to metabolic deprivation. Physiologically, this is exactly what happens: ventricular hypoxia results in dilation of coronary arterioles leading to greater supply of oxygen and metabolites (see Section V.A). However, as often happens in critical physiological processes, there is yet another factor at work. Short bouts of hypoxia, with opening of K⁺(ATP) channels, may actually "pre-condition" myocytes and ameliorate some of the effects of subsequent, more prolonged bouts. In that case, does prevention of hypoxia-induced opening of K⁺(ATP) channels, as might occur in the presence of a sulfonylurea, predispose to hypoxia-induced cell injury? Controversy still surrounds this issue.

It is worth mentioning other mechanisms that might contribute to ischemia-induced contractile failure. These include (i) altered $[Ca^{2+}]_i$ "homeostasis," (ii) changes in the binding affinity of the contractile apparatus for Ca²⁺ induced by accumulation of acid equivalents and (iii) activation of other K⁺ channels, such as muscarinic and G-protein gated K⁺ channels, by metabolic intermediates generated during hypoxia.

IV. Stimulus-Secretion Coupling in Carotid Chemoreceptor Cells

Neural control of respiratory drive is critical in maintaining physiologic levels of plasma O₂, CO₂ and H⁺. In the rhythmic, neuromuscular process of breathing, trains of action potentials, generated by pacemaker cells in the CNS, trigger motor neurons and activate periodic contractions of the inspiratory muscles (diaphragm and intercostal muscles). This action expands the chest, thereby sucking air of high-O₂, low-CO₂ content into the lungs. Gas exchange consists of net CO₂ diffusion from capillary to adjacent lung air spaces (alveoli) and net O₂ diffusion from alveoli to the capillary. Relaxation of inspiratory muscles, sometimes combined with the contraction of expiratory muscles, expels the air of high CO₂, low O₂ content from the lungs. A major stimulus for altering the pattern of respiratory drive is a drop in the partial pressure of O₂ dissolved in plasma (i.e., plasma pO_2); a secondary stimulus is a rise in plasma pCO_2 or a fall in pH. The carotid body is the organ that senses changes in plasma pO_2, pCO_2, and pH, and mediates changes in CNS respiratory drive. Located at the bifurcation of the carotid artery, a branch of the aorta, the carotid body consists of a central core of **chemoreceptor (or glomus) cells**;

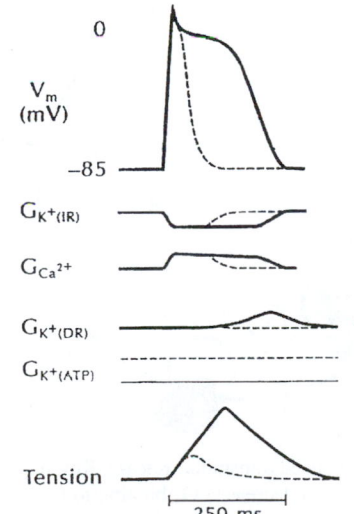

FIGURE 8. K⁺(ATP) channels and cardiac excitability. A steady-state enhancement of K⁺(ATP) conductance ($G_{K^+(ATP)}$), resulting from metabolic inhibition, shortens the time courses of the ventricular action potential and twitch tension and alters the time courses and magnitudes of the underlying ionic conductances. Solid lines indicate control; dashed lines indicate effects of metabolic inhibition.

these originate from the neural crest. Glomus cells synapse on dendritic endings of sensory nerve fibers that comprise the carotid sinus nerve travelling into the CNS. The glomus cell core is surrounded by more superficial glial, or sustentacular, cells, as well as by a dense network of highly porous capillaries. The glomus cell is the site of transduction of changes in plasma pO_2, pCO_2, and pH into changes in electrical activity of the afferent carotid sinus nerve. Decreases in pO_2, as well increases in pCO_2, or decreases in pH, lead to an increase in release of dopamine and probably to an increase in as-yet-unidentified transmitters from glomus cells, as well as to an increased action potential frequency in the dopamine-sensitive carotid sinus nerve. This contributes to increased respiratory drive.

A. Role of O_2-Sensitive K^+ Channels in Transduction of the Hypoxic Stimulus

Two divergent views have emerged concerning chemotransduction of hypoxia by glomus cells (see Fig. 9A). The first theory proposed that hypoxia induces cell depolarization, followed by opening of voltage-gated Ca^{2+} channels, Ca^{2+} influx, and synchronized exocytotic release of dopamine. This might give rise to sufficiently large excitatory post-synaptic potentials in the dendritic regions of single sinus nerve fibers to trigger propagating action potential. An alternative view proposed that hypoxia induces quantal release of transmitter in a Ca^{2+}-dependent manner that is independent of voltage-gated Ca^{2+} entry but dependent on the discharge of Ca^{2+} from intracellular stores. The latter should increase asynchronous quantal release of transmitter, generating a rise in the frequency of miniature excitatory post-synaptic potentials, and should enhance ongoing

electrical activity in the low-threshold carotid sinus nerve fiber. These contrasting viewpoints have arisen from data generated with different preparations from different species.

There are several lines of evidence supporting the classical scheme for "depolarization-secretion coupling" in glomus cells. First, glomi are electrically excitable sensory cells and contain HVA Ca^{2+} currents, delayed rectifier K^+ currents, and in some cases voltage-dependent Na^+ currents. Second, in some isolated cell preparations, glomus cells respond to hypoxia by depolarizing and generating action potentials. (However, in other preparations, glomus cells fire action potentials only in response to current injection or release of the cell from sustained hyperpolarization. These cells show no change in passive electrical activity in response to hypoxia.) Third, in glomus cells that fire in response to hypoxia, depolarization increases both cytosolic Ca^{2+}, measured with intracellular dyes, and dopamine release, measured by amperometry. In these cells, depolarization-secretion coupling is reduced by exposure to blockers of the HVA-type Ca^{2+} channels or by reduction in extracellular Ca^{2+}.

A very exciting development in chemoreceptor physiology consistent with the depolarization-secretion coupling scheme outlined above is the discovery that, in whole-cell recordings, lowering ambient pO_2 selectively and reversibly reduces the outward K^+ current flowing through delayed rectifier $K^+(DR)$ channels of the glomus cells. Reducing pO_2 from 160 mm Hg to 90 mm Hg reduces peak $K^+(DR)$ current by ~30%. This effect is not dependent on the concentrations of ATP (0–3 mM) or Ca^{2+} (<1 nM–0.5 mM) in the pipette. In outside-out patches of membrane, reduced ambient pO_2 re-

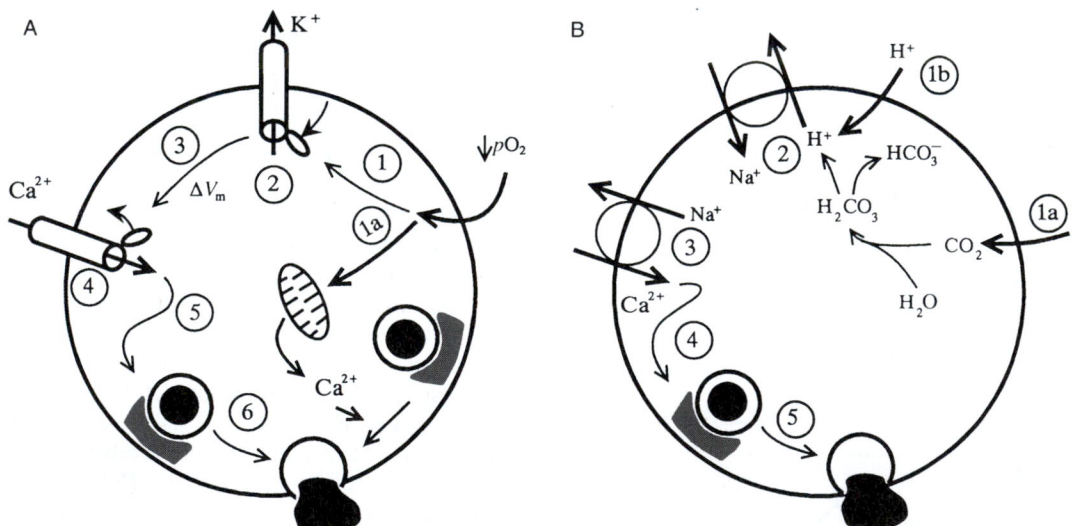

FIGURE 9.   Stimulus-secretion coupling in carotid chemoreceptor cells. (A) O_2-induced transmitter release. Key steps in classical scheme of depolarization-secretion coupling are outlined. Decreased pO_2 results in decreased O_2 binding to the intracellular receptor coupling to a specific voltage-dependent K^+ channel (1). This results in the closure of the channel (2), cell depolarization (3), increased probability of the opening of HVA Ca^{2+} channels (4), increased calcium entry (5), and the enhanced rate of fusion of dopamine-containing vesicles with the plasma membrane (6). Alternatively (see path beginning with 1a), hypoxia-induced rundown of the mitochondrial proton gradient might result in reduced ATP generation, slow release of Ca^{2+} from mitochondria or other stores, and ultimately a slow rise in spontaneous quantal release. (B) CO_2 and H-induced transmitter release. Increased pCO_2 or H entry (1b) results in decreased cytosolic pH and stimulation of the Na^+-H exchanger (2). The resultant influx in Na^+ stimulates the Ca^{2+}-Na^+ exchanger (3). This, in turn, increases Ca^{2+} influx (4) and Ca^{2+}-dependent exocytosis (5).

versibly decreases the probability of opening (P_o) of a 20-pS K$^+$(DR) type channel (see Fig. 10). The first reports suggested that *in vitro* this channel was most responsive to changes in O$_2$ over the range of pO_2 values between 110 and 150 mm Hg, in contrast to the intact carotid sinus nerve that actually fires optimally at pO_2s <70 mm Hg. However, the O$_2$ sensitivity of the channel can be shifted into a more physiological range by the addition of a membrane-permeant analog of cAMP, and recent data has shown K$^+$(DR)-type channel activity pO_2s ranging from 20 to 150 mm Hg. In whole-cell, current-clamp recordings, the effect of a reduction in pO_2 is (i) an increase in the rate of cell depolarization on release from maintained hyperpolarization, (ii) an increase in the frequency of spike activity in the resultant short train impulses, and (iii) an increase in AP overshoot. Hence, in glomus cells with some intrinsic spontaneous electrical pacemaker activity, O$_2$-dependent changes in K$^+$ current could alter the frequency of action potentials and large Ca^{2+} transients, thereby increasing [Ca^{2+}]$_i$-dependent transmitter release onto the afferent nerve. In glomus cells with little automaticity that maintain resting potentials of between −50 and −40 mV, closure of K$^+$(DR)-type channels could still cause steady-state depolarization "generator potential" and trigger electrical activity de novo. More extensive perforated-patch recordings will be needed to determine the range of electrical activity patterns exhibited by these cells.

These data suggest that O$_2$ maintains the activity of a delayed rectifier type K$^+$ channel through a novel gating mechanism. Several possibilities for the molecular mechanism of O$_2$ transduction and its relationship to channel gating have been suggested, but there is no definitive evidence for any of them. The first is that a heme protein, analogous to a subunit

of the O$_2$-carrying protein hemoglobin, is attached to the channel, or to a functional subunit of the channel. In this way a change in configuration of the protein, on losing O$_2$, would alter channel gating. The second is the presence of an oxidase which, on reduction of cytosolic pO_2, produces less hydrogen peroxide (H$_2$O$_2$), consequently altering the concentration of redox intermediates and thereby of channel conformation.

The alternative view of chemotransduction of hypoxia in glomus cells is that the rise in intracellular Ca^{2+} necessary for dopamine secretion is due to Ca^{2+} release from intracellular stores and that voltage-activated currents play only a secondary role, perhaps allowing replenishment of intracellular Ca^{2+} stores. This hypothesis has arisen from data generated from both isolated glomus cells and *in situ* carotid cell bodies. Mitochondrial poisons, such as cyanide, produce a condition known as **histotoxic hypoxia,** which mimics true hypoxia in stimulating carotid body nerve activity and respiratory drive. These poisons produce increases in intracellular Ca^{2+} that are unaffected by pharmacologic maneuvers designed to abolish or enhance the action potential, hence suggesting that increases in cytosolic Ca^{2+} arise from intracellular stores. Hypoxia and mitochondrial poisons produce a rise of cell NAD(P)H. Additionally, studies from *in situ* carotid bodies have shown (i) that action potentials induced in carotid sinus nerve by hypoxia are not blocked by drugs that block outward K$^+$ currents and (ii) that whole cell glomus membrane resistance is not changed in response to hypoxia. A proposed mechanism whereby hypoxia induces intracellular Ca^{2+} release is that low O$_2$ decreases mitochondrial efficiency, perhaps slowing electron transfer in the respiratory chain, and reducing the proton gradient across the mitochondrial inner membrane. A consequence of this could be the release of Ca^{2+} from mitochondria or decreased ATP pro-

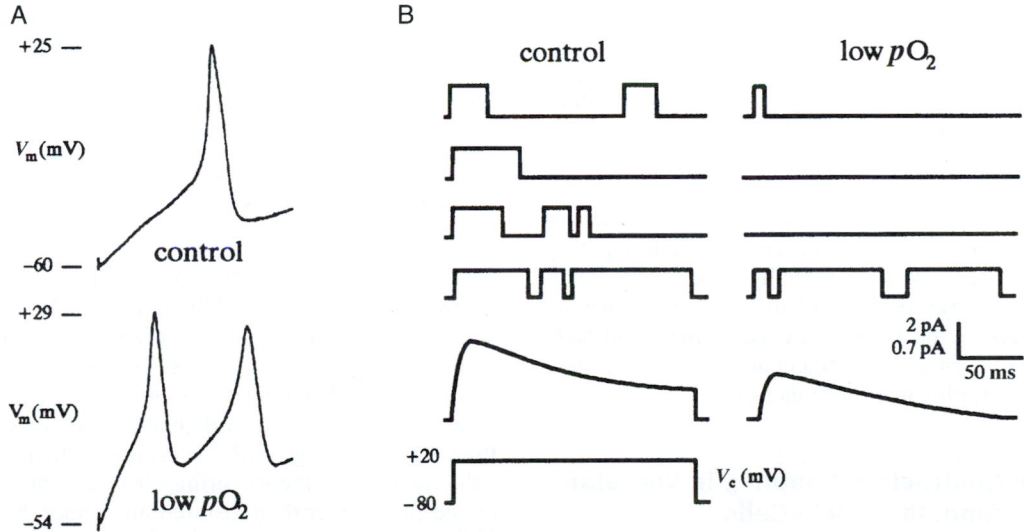

FIGURE 10. Origin of low-pO_2-induced enhancement of electrical activity in carotid glomus chemoreceptors. (A) Reduction in ambient pO_2 results in increased frequency of spontaneous AP activity, partly due to reduced rate of repolarization of the AP. Hence, any background depolarizing current might be more effective in raising the membrane potential toward threshold for firing a spike. (Reproduced from *The Journal of General Physiology,* 1989, Vol. 93, p. 979, Fig. 14. By permission of the Rockefeller University Press.) (B) Reduction in ambient pO_2 reduces the probability (P_o) that a K$^+$(DR) channel will open on depolarizing the cell from −80 to +20 mV. Lowest traces in each column represent the average of many individual channel current traces. (Idealized traces adapted from Ganfornina and Lopez Barneo, 1991.)

duction, resulting in slow Ca^{2+} release from other intracellular stores (e.g., the ER). For this mitochondrial-based hypothesis to work, it might be necessary that O_2-trapping properties of mitochondrial cytochrome oxidase and Ca^{2+}-storage properties of mitochondria in glomus cells differ significantly from their counterparts in other tissues.

B. Transduction of Increased pCO_2 and Decreased Plasma pH

Glomus cells rapidly equilibrate pH_i and pH_o. Hence, it is probable that with acidic stimuli the cell is actually sensing a fall in pH_i. But how does an increase in $[H^+]_i$ result in Ca^{2+}_o-dependent transmitter release, especially as H^+_i-induced release is insensitive to blockers of HVA Ca^{2+} channels, and HVA Ca^{2+} channels are often inhibited by reduction in pH_i? The proposed link between increased $[H^+]_i$ and Ca^{2+} entry in promoting dopamine release is that increased $[H^+]_i$ activates a Na^+_o-H^+_i exchanger which, in turn, elevates Na^+ and recruits a Na^+-Ca^{2+} exchanger. The activity of the latter exchanger is often augmented by cell depolarization; this may, in fact, occur because a drop in pH_i reduces K^+(DR) type K^+ current much as a drop in pO_2 does. An increase in background "spontaneous" quantal release of dopamine, which is not phase-linked to action potential activity, could increase steady-state depolarization of carotid nerve fibers and, perhaps, augment on-going impulse activity (see Fig. 9B). In considering this scheme, it should be cautioned that thus far, good evidence exists only for activation of an Na^+-H^+ exchanger.

To be sure, respiratory drive varies significantly according to age and physiologic status. Human fetuses do not exhibit regular respiratory movements *in utero,* whereas newborn infants often display periodic breathing different from breathing patterns seen in older children and adults. Disease states produce abnormal breathing patterns such as Cheyne-Stokes breathing, the diamond-shaped changes in respiratory excursions followed by a period of non-breathing (apnea), seen in stroke, congestive heart failure, and chronic hypoxia. In infants, prolonged apnea despite persistent hypoxia or hypercarbia can culminate in sudden infant death syndrome (SIDS). Recent data suggest that carotid body glomus cells from animals raised under hypoxic conditions show blunted electrophysiologic responses to hypoxia compared to cells from healthy animals. Disorders of breathing such as apnea and SIDS are seen in higher frequency in infants at increased risk for chronic hypoxia due to conditions such as prematurity or second-hand cigarette smoke exposure. Studies on the mechanism of changes in carotid body chemoreception in normal development and in various disease states should be a fruitful area for further inquiry.

V. Stimulus-Contraction Coupling in Vascular Smooth Muscle Cells

A. Hypoxic Vasodilation of Coronary and Mesenteric Vessels

Smooth muscle cells of resistance arterioles are very sophisticated metabolic sensors. In fact, they encompass a complete vasodilator reflex system in a single cell. In re-

sponse to a drop in ambient pO_2, these cells reduce their tension generation. This results in vessel relaxation and the local redistribution of O_2 supply to O_2-consuming tissue. As in the case of the cardiac myocyte, a key to the explanation of this phenomenon is the increase in K^+ conductance and membrane hyperpolarization, which precedes the fall in tension.

It is well appreciated that resistance vessels in the coronary and mesenteric (gut) circulation display resting tone. Their myocytes maintain a resting potential of -40 to -50 mV and have an abundance of K^+(ATP) and L-type (HVA) Ca^{2+} channels. In fact, in smooth muscle of mesenteric artery, the K_d value (μM) for ATP-induced closure of K^+(ATP) channels is roughly double that in cardiac myocytes, in principle making K^+(ATP) channels of these myocytes more sensitive to hypoxia than the K^+(ATP) channels in the heart. Under these circumstances, small changes in resting potential would be expected to affect resting tone. A small depolarization (of 5–10 mV) caused by closure of K^+(ATP) channels should result in increased opening of HVA Ca^{2+} channels and subsequent vasoconstriction (see Fig. 11A). In contrast, a small hyperpolarization of 5–10 mV, affected by addition of a K^+(ATP) channel opener or a dihydropyridine Ca^{2+} channel antagonist, should reduce resting tone and result in vasodilation. These predictions have been borne out experimentally. A modest drop in microenvironment pO_2, insufficient to affect cardiac excitation-contraction coupling, is sufficient to reduce cytosolic ATP in vascular smooth muscle cells, open K^+(ATP) channels, and cause vasodilation. (In bulk cardiac muscle *in situ,* preferential opening of smooth muscle K^+(ATP) channels might be further augmented by the release of adenosine by active cardiac tissue; extracellular adenosine has been shown to activate K^+(ATP) channels via a G-protein–dependent mechanism.) Hence preferential hypoxia-induced arteriolar dilation, with its attendant increases in local O_2 delivery, may spare cardiac myocytes the risk of hypoxia-induced alterations in excitation-contraction coupling. This scheme might work for the regulation of local blood supply to the brain and gut as well as to the heart.

B. Hypoxia-Induced Vasoconstriction of Pulmonary Vessels

In contrast to coronary, mesenteric, and cerebral vessels, pulmonary artery and its smooth muscle constrict in response to a drop in pO_2. This hypoxic pulmonary vasoconstriction (HPV) constitutes an adaptive response in the lung bed because it ensures that areas of the lung that are poorly oxygenated will receive less blood flow; the extra blood flow is "redirected" towards better oxygenated areas to optimize gas exchange. HPV is critical in fetal pulmonary development when the maturing air spaces are filled with secreted fluid rather than inspired air. Under these conditions, HPV maintains the relatively high pulmonary vascular resistance that shunts venous return around the low flow pulmonary bed, through the ductus arteriosus of the cardiac septum, and into the left heart. With inflation of the newborn's lungs to air containing substantially higher pO_2, O_2-induced pulmonary vasodilation occurs, thereby promoting blood flow through the pulmonary circulation. In the adult,

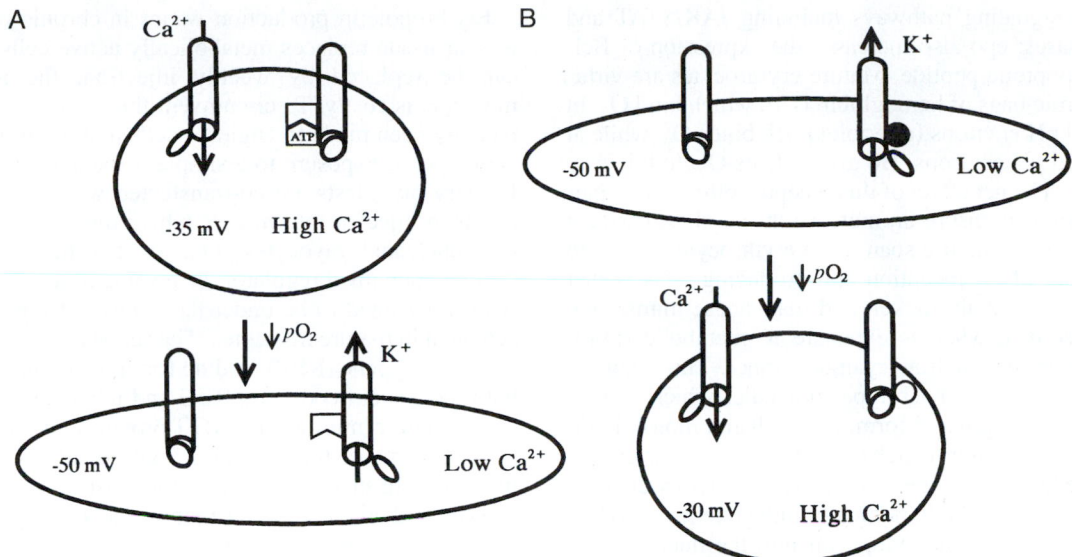

FIGURE 11. Hypoxia alters stimulus-contraction coupling in vascular myocytes. (A) Model of hypoxia-induced vasodilation in cardiac, cerebral, and mesenteric vessels. In normoxic conditions, $K^+(ATP)$ channels are largely closed, but voltage-dependent Ca^{2+} channels are open, $V_m \sim -35$ mV, and cytosolic Ca^{2+} is low. With a decrease in pO_2, cytosolic ATP levels drop, resulting in the opening of $K^+(ATP)$ channels, repolarization to ~ -50 mV, closure of Ca^{2+} channels, and myocyte relaxation. (B) Model of hypoxia-induced vasoconstriction in pulmonary arteries. In normoxic conditions, $V_m \sim -50$ mV, O_2-sensitive K^+ channels are largely open, and voltage-dependent Ca^{2+} channels are largely closed. Hence, cytosolic Ca^{2+} levels are low, and there is little resting tension. With a decrease in pO_2, O_2-sensitive K^+ channels close, resulting in depolarization, opening of Ca^{2+} channels, Ca^{2+} entry, and myocyte contraction.

HPV is useful in maintaining moment-to-moment matching of local ventilation to perfusion; this reduces the risk of hypoxia that can occur when a portion of the lung is poorly inflated. However, when the areas of local poor ventilation are widespread, or with chronic hypoxia (such as at high altitudes), chronic HPV is accompanied by smooth muscle proliferation (i.e., "work hypertrophy of muscle"). The end result is the development of increased resistance and pressure in the total pulmonary vascular bed. This "pulmonary artery hypertension" imposes an increased "afterload" on the right ventricle, thus becoming a stimulus for its hypertrophy.

What cellular mechanisms support HPV? Pulmonary artery myocytes maintain a low resting tension, a V_m of ~ -40 mV and a resting cytosolic $[Ca^{2+}]$ of < 100 nM. They respond to progressive hypoxia (e.g., a slow fall in pO_2 from 150 to 15 mm Hg) with a 15-mV depolarization, which is not affected by changes in $[Ca^{2+}]_o$, followed by Ca^{2+}-dependent increases in both cytosolic Ca^{2+} and tension. Given that V_m in these cells show a Nernstian relationship to $[K^+]_o$, these results suggest that membrane depolarization is due to a large decrease in Ca^{2+}_o-independent resting membrane conductance (e.g., GK). However, depolarization ultimately results in a small increase in G_{Ca}. This is sufficient to provide entry to trigger contraction. Tension is maintained so long as Ca^{2+} entry through voltage-gated Ca^{2+} channels exceeds Ca^{2+} efflux via the Na^+-Ca^{2+} exchanger. Whole cell voltage clamp experiments have provided good evidence that a voltage-activated, delayed rectifier type K^+ current seen at V_m values positive to -50 mV and, hence, open at rest, is significantly inhibited by hypoxia. Depolarization of a few mV would be sufficient to open HVA Ca^{2+} channels and result in vasoconstriction (see Fig 11B). Hence, as in the carotid chemoreceptor, hypoxia-induced reduction in the activity of $K^+(DR)$ type channels in a cell with a very low background G_m, appears to be responsible for depolarization and stimulus-response coupling.

An important paradox, which remains to be explained, is how pulmonary artery myocytes manage to "hide" $K^+(ATP)$ channels, known to be present in the sarcolemma, during hypoxia. In these cells, $K^+(ATP)$ channel openers abort high $[K^+]_o$-induced tension increase, while sulfonylureas enhance tension generation, as they do in resistance vessels, yet hypoxia does not open these channels. Are mechanisms similar to those in pancreatic α cells at play here?

VI. Coupling of Oxygen Sensing to Red Cell Production by Erythropoietin-Secreting Cells

The ultimate metabolic sensors are the erythropoietin-(epo-) secreting cells of the renal interstitium and liver. These modified fibroblasts respond slowly to hypoxia by increasing their synthesis and constitutive secretion of epo, which, in turn serves as a maturation factor for erythrocyte precursors in bone marrow. Epo binds to a membrane receptor, which serves as a scaffold for the activation of numerous growth and division signaling pathways including JAK/STAT and ras/MAP kinases; epo also increases the expression of Bcl-Xl, an anti-apoptotic peptide. Mature erythrocytes are virtually cytoplasmic bags of hemoglobin (Hb) which bind O_2. In high pO_2 and pH environs (arterioles), Hb binds O_2, while at lower pO_2 and pH environs, Hb gives off its O_2. In this way Hb ferries O_2. The net effect of this receptor-effector cell pair is that hypoxia can ensure an increase in blood O_2 carriage for up to 120 days, the life span of an erythrocyte. A clue to the mechanism of O_2-reception epo-producing cells is that their exposure to cobalt, nickel, and manganese mimics the effects of hypoxia, whereas exposure to metabolic (mitochondrial) inhibitors such as cyanide cannot. This suggests that the O_2 sensor is an Hb-type molecule, which can be locked into a deoxygenated form, rather than a molecule involved in red-ox transfer, which uses O_2 as the final acceptor. The epo gene has been cloned, and its cis-regulatory element has been found to bind a hypoxically inducible factor (HIF) that can transfer from the cytoplasm into the nucleus. (see Fig. 12A). Epo regulation is an example of a widespread system for gene induction by O_2; this system includes pathways for induction vascular growth factors and changing isoforms of glycolytic enzymes in metabolically active cells.

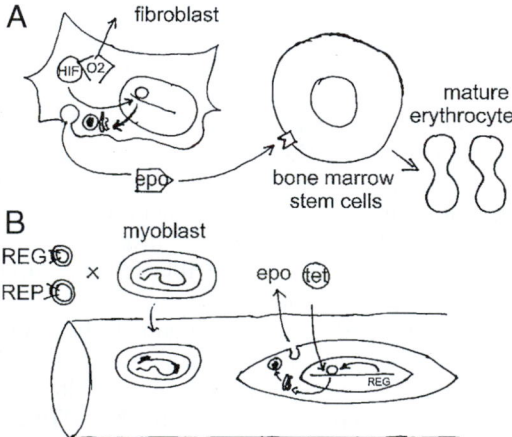

FIGURE 12. Native and bioengineered stimulus-synthesis coupling in erythropoietin production. (A) Stimulus-synthesis coupling in renal interstitial fibroblasts and hepatic Ito cells likely consists of triggering of epo gene transcription by a hypoxically induced factor (HIF), which is freed to enter the nucleus after O_2 dissociates from a heme pigment receptor. Newly synthesized epo is packaged and constitutively released. The kidneys are an excellent primary site for an O_2 sensor because under control conditions they are perfused with ~20% of the cardiac output. Circulating epo binds to erythroblasts (erythrocyte stem cells) in bone marrow and triggers their proliferation and maturation to enucleated erythrocytes that circulate as hemoglobin-packed, O_2 carrying sacs for an average of 120 days. (B) Genetically engineered stimulus-synthesis coupling for epo production in skeletal myocytes. Myoblasts are co-transfected with (i) a regulator gene (REG) under the control of mature muscle cells and (ii) a reporter gene (REP) coding for epo and under control of a tetracycline-sensitive product of the regulator gene. These co-transfected myoblasts are injected into muscle where they fuse with multinucleated myocytes. Under the control of exogenously added tetracycline (tet), myocytes are induced to synthesize and secrete epo.

Erythropoietin production wanes in chronic renal failure, as scar tissue replaces metabolically active cells. While epo can be replaced by weekly injection, the recombinant molecule is costly. To circumvent this, significant effort has recently been made to engineer cells that might tonically secrete epo on exposure to a simple, gene-activating stimulus. Primary myoblasts are co-transfected with two retroviruses and then injected into muscle where they fuse with mature, multinucleated myocytes. One of the transfecting retroviruses contains a regulator, or reverse transactivation, gene (REG) designed to be under the control of a promoter only activated in mature myocytes. The second retrovirus contains the reporter gene (REP), coding for epo, but under control of both the protein coded by REG and tetracycline (tet). Once in the mature muscle, the REG protein and tet control can tonically activate the epo gene. Significantly, when mice are injected with these engineered myoblasts, epo secretion can be switched on and off over months, depending on the availability of tet in the drinking water.

VII. Acknowledgments

Original data traces shown here were obtained in our lab through the support of NIH grant DK37380. I thank Todd Owyoung for preparing Figures 2 and 6, David Bryant for preparing Figures 9–11 and Drs. David Pressel and David Barnett for sharing insights and delicate turns of phrase during the preparation of a prior version of this chapter.

This chapter is dedicated to the memory of Golda Hazak, wit and sage of Kiryat Yam.

Bibliography

Ashcroft, F. M., and Grimble, F. M. (1999). K-ATP channels and insulin secretion; their role in health and disease. *Diabetologia* **42**, 9039–9196.

Ashcroft, F. M., and Rorsman, P. (1989). Electrophysiology of the pancreatic Beta cell. *Prog Biophys Mol Biol* **54**, 87–143.

Bohl, D., Naffakh, N., and Heard, J. M. (1997). Long-term control of erythropoietin secretion by doxycycline in mice transplanted with engineered primary myoblasts. *Nature Medicine* **3**, 299–304.

Bokvist, K., Olsen, H. L., Hoy, M., Gotfredsen, C. F., Holmes, W. F., Buschard, K., Rorsman, P., and Gromada, J. (1999). Characterisation of sulphonylurea and ATP-regulated K^+ channels in rat pancreatic A-cells. *Pflugers Arch—Eur J Physiol* **438**, 428–436.

Ganfornina, M. D., and Lopez-Barneo, J. (1991). Single K channels in membrane patches of arterial chemoreceptor cells are modulated by O_2 tension. *Proc Natl Acad Sci USA* **88**, 2927–2930.

Gonzalez, C., Almaraz, L., Obeso, A., and Rigual, R. (1992). Oxygen and acid chemoreception in the carotid body chemoreceptors. *TINS* **15**, 146–157.

Gromada, J., Holst, J. J., and Rorsman, P. (1998). Cellular regulation of islet hormone secretion by the incretin hormone glucagon-like peptide 1. *Pflugers Arch—Eur J Physiol* **435**, 583–594.

Harvey, J., McKay, N. G., Walker, K. S., Van der Kay, J., Downes, C. P., and Ashford, M. L. J. (2000). Essential role of ohosphoinositide 3-kinase in leptin induced K-ATP channel activation. *J Biol Chem* **275**, 4660–4669.

Inagaki, N., Gonoi, T., Clement IV, J. P., Namba, N., Inazawa, J., Gonzalez, G., Aguilar-Bryan, L., Seino, S., and Bryan, J. (1995). Reconstitution of I-K^+(ATP): an inward rectifier subunit plus the sulfonylurea receptor. *Science* **270**, 1166–1169.

Johnson, J. D., and Chang, J. P. (2000). Function- and agonist-specific Ca^{2+} signalling: the requirement for and mechanism of spatial and temporal complexity in Ca^{2+} signals. *Biochem Cell Biol* (in press).

Klingauf, J., and Neher, E. (1997). Modeling buffered Ca diffusion near the membrane: implications for secretion in neuroendocrine calls. *Biophysical J* **72**, 674–690.

Misler, S., Barnett, D. W., Pressel, D. M., and Gillis, K. D. (1992). Electrophysiology of stimulus-secretion coupling in human beta-cells. *Diabetes* **42**, 1220–1227.

Nichols, C. G., and Lederer, W. J. (1991). Adenosine triphosphate-sensitive potassium channels in the cardiovascular system. *Am J Physiol* **261**, H1675–H1686.

Nichols, C. G., Shyng, S.-L., Nestorowicz, A., Glaser, B., Clement J. P., Gonzalez, G., Aguilar-Bryan, L., Permutt, M. A., and Bryan, J. (1996). Adenosine diphosphate as an intracellular regulator of insulin secretion. *Science* **272**, 1785–1787.

Ratcliffe, P. J., Ebert, B. L., Ferguson, D. J. P., Firth, J. D., Gleadle, J. M., Maxwell, P .H. and Pugh, C. W., (1995). Regulation of the erythropoietin gene. *Nephrol Dial Transplant* **10**, 18–27.

Rorsman, P., Ashcroft, F. M., Berggren, P. O. (1991). Regulation of glucagon release from pancreatic A-cells. *Biochemical Pharmacology* **41**, 1783–1790.

Seino, S., Inagaki, N., Namba, N., and Gonoi, T. (1996). Molecular biology of the beta-cell ATP-sensitive K channel. *Diabetes Rev* **4**, 177–190.

Weir, E. K., and Archer, S. L. (1995). The mechanism of acute hypoxic pulmonary vasoconstriction: the tale of two channels. *FASEB J* **9**, 183–189.

Wollheim, C. B., Lang, J., and Regazzi, R. (1996). The exocytotic process of insulin secretion and its regulation by Ca and G-proteins. *Diabetes Rev* **4**, 276–297.

Zawar, C., Neumcke, B. (2000). Differential activation of ATP-sensitive potassium channels during energy depletion in Ca$_1$ pyramidal cells and interneurones of rat hippocampus. *Pflugers Arch—Eur J Physiol* **439**, 256–262.

Zhou, Z., and Misler, S. (1996). Amperometric detection of quantal secretion from patch-clamped rat pancreatic beta-cells. *J Biol Chem* **271**, 270–277.

Catherine E. Morris

44

Mechanosensitive Ion Channels
in Eukaryotic Cells

I. Introduction

The idea of mechanically gated channels has a long history and there is abundant evidence that mechanosensory specialist cells use **mechanosensitive** (MS) ion channels for mechanotransduction (French, 1992; Hudspeth,1997). Although several mechanosensory systems have been intensively studied by electrophysiologists and molecular biologists (e.g., hair cells, crayfish stretch receptors, and, for molecular work, *C. elegans*. mechanosensory neurons), nature has uncooperatively placed their mechanotransducing channels in structures that defy single-channel recording. Moreover, it is abundantly evident from cell biological, biophysical, and molecular work on hair cells and on a nematode model that specialized mechanotransduction involves a supramolecular assembly wherein a MS channel is connected fairly rigidly to intracellular and/or extracellular filamentous proteins (Tavernarakis and Driscoll, 1997). It is not likely to become possible to preserve the necessary cytoarchitecture while making single-channel recordings from these mechanotransducers.

It is somewhat ironic, therefore, that single-channel recordings from nonspecialized cells—bacteria, fungi, and plant and animal cells—reveal a plethora of MS ion channels. The irony is compounded because, from most of these cases, compelling evidence that the MS channels act as physiological mechanosensors is lacking. It is easy to point to physiological tasks requiring mechanosensors in bacteria, fungi, and plant and animal cells, but it has been frustratingly difficult to prove that these tasks are mediated by MS channels.

How far to extrapolate from single MS channel data to putative physiological function is a matter of controversy (Morris and Horn 1991; Gustin *et al.*, 1991); various workers offer substantially different perspectives, as is evident in recent reviews (Martinac, 1993; Morris, 1992; Sackin, 1995; Hamill and McBride, 1995; Sachs and Morris, 1998). For the completely nonselective MS channels in prokaryotes, a compelling body of evidence shows that the channels act as do-

or-die osmotic safety valves that open only if osmotic lysis is imminent (Batiza *et al.*, 1999). For eukaryotes, Hamill and McBride (1996, 1997) have provided exhaustive accounts covering the pharmacology and caveats relevant to MS channel recordings.

II. MS Channel Breakthroughs

MS channel breakthroughs have occurred on three fronts over the last half decade: electrophysiologic, genetic, and molecular. Although this chapter focuses on eukaryotic MS channels that can be studied by single-channel recording (Fig. 1), it will allude briefly to the genetic and molecular work. Hair cell mechanotransduction is described in this book by Marcus (see Chapter 46).

Breakthroughs in the molecular biology of MS channels center on *E. coli*, *C. elegans.*, and mammals. For the nematode *C. elegans.*, a large body of genetics by M. Chalfie and colleagues laid the groundwork for the subsequent molecular biology; the genetic approach continues to be crucial (rev. in Bargmann, 1994; Tavernarakis and Driscoll, 1997). From each preparation, a MS channel has been cloned, albeit with important caveats: for *E. coli*, it is unknown if the channel is mechanosensitive *in vivo*. For the *C. elegans*. case, the missing piece is just the opposite—the cloning strategy was based on *in vivo* mechanoreception, but functional expression of the putative channel in a heterologous system remains to be accomplished. Nevertheless, both are major findings, made all the more intriguing because they generate radically different pictures of channel mechanosensitivity. For mammals, an already cloned channel, TREK-1, was reexamined in light of its many functional similarities to the molluscan S-channel (Vandorpe and Morris, 1992), one of which is susceptibility to stretch activation (Patel *et al.*, 1998). TREK-1 has been shown to be stretch sensitive as, subsequently, were some other family members (Maingret *et*

A

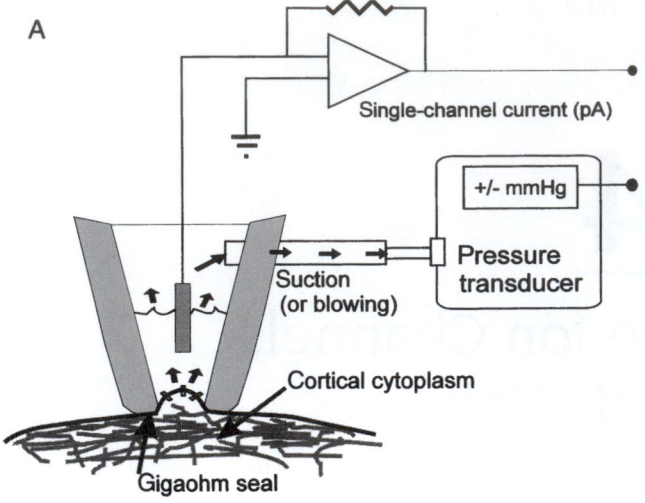

B

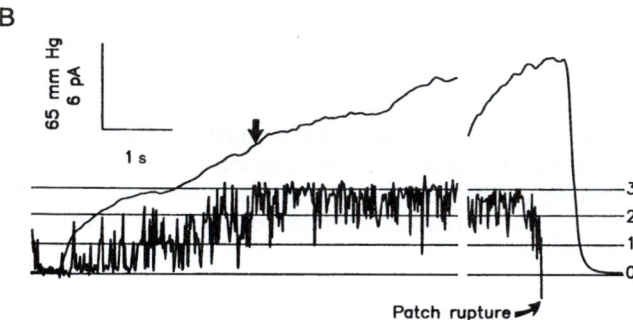

FIGURE 1. Recording MS channels using pipette pressure to create membrane tension. (A) The most commonly used single-channel recording configuration for MS channels. The fire-polished tip is ~1 μm in diameter. A glass-to-membrane gigaohm seal forms several μm from the tip; the underlying cytoplasm sustains some damage, both as a result of seal formation and from subsequent stimulation by suction or blowing. The patch can be excised to form an inside-out patch without abolishing the mechanosensitivity of MS channels. (B) A recording of SA channel events from an *Aplysia* neuron. Suction is increased progressively, activating a maximum of three channels, whose open probability saturates near −100 mm Hg, as indicated by the arrow on the pressure trace (see Vandorpe *et al.,* 1994).

al., 1999). But in contrast to the *E. coli* and *C. elegans* cases, there is yet no evidence that either molluscan or mammalian TREK-like channels are mechanotransducers.

The *E. coli* MS channel is the simplest of channels: a single small protein (~15 kDa) oligomerizes, probably as a pentamer, to form a high-conductance ion pore (1–3 nS) (see Batiza *et al.,* 1999). Though this channel, MscL, can exhibit its mechanosensitive behavior in an artificial lipid bilayer, it generally activates only at near-lytic tensions. Initially, the physiological role for the channel (osmotic safety valve) was uncertain because *E. coli* strains with the MscL gene knocked out had no "osmotic phenotype," but as indicated above, this difficulty has been overcome. Though MscL has no known eukaryotic homologs, it is biophysically fascinating beyond the bacterial context. A crystal structure (Chang *et al.,* 1998) allows for the testing of many hypotheses about the basis for mechanosensitiv-

ity: what aspects of channel structure determine the tension threshold for activation? Do open channel conformations occupy more membrane area than closed channels? What is the role of interactions between the channel protein and bilayer lipid?

In the case of the putative *C. elegans* mechanosensory channel, genetic analysis of mutant individuals is pivotal. Although electrophysiology has not been possible, analysis of stereotyped touch-sensitivity behavior (yes, whole animal behavior!) and of leaky channel-induced cell death in mutant worms has proven astonishingly effective in generating a detailed physical model (Fig. 2). The model features a pore (a heteromer of degenerin-like proteins, each with two hydrophobic membrane-spanning domains, as well as a chain of residues that may contribute to an aqueous cation-selective pore). A filamentous cytoskeletal element (stiffened microtubules) plus a linker molecule (a stomatin-like globular protein) and collagen in the extracellular matrix are all necessary for functional MS machinery in the sensory neurons. Given the elaborateness of the machinery, it is perhaps not surprising that heterologous expression of the degenerins—the putative channel proteins—has not yielded functional expression of a conductance. It may be impossible to gate the channel mechanically minus its auxiliary bells and whistles. There was, therefore, real excitement when it was realized that nematode MS channel degenerins are related by homology to mammalian epithelial Na$^+$ channels. Evidence for *in situ* mechanosensitivity in the latter channels, however, has not been forthcoming (Awayda and Subramanyam, 1998). Nevertheless, a compelling picture is emerging for the degenerins as *C. elegans* MS channels: they appear to be part of a supramolecular entity broadly comparable to what is envisaged for the mechanotransducer channel in vertebrate hair cells. In hair cells there is cell biological, biophysical, and (for the myosins) genetic evidence that a cation channel (identity unknown), an extracellular tip link (protein unknown but visible by electronmicroscopy and perturbable by low-calcium solutions), and actomyosin-linked elements contribute to the MS channel machinery.

The TREK-like channels are two pore–domain K$^+$ channels—dimers presumably of a monomer containing two K$^+$-selective pore-forming regions and four transmembrane segments. Expressed in mammalian cell lines, the recombinant channel is responsive, like *E. coli* MS channels in liposomes, to both membrane stretch and bilayer-distorting amphipaths (Patel *et al.,* 1998). A pH (intracellular)-sensitive region in the C-terminus of TREK 1 is also critically involved in mechanosensitivity and arachidonic acid activation (Maingret *et al.,* 1999). TRAAK, also mechanosensitive, is a variant found only in neuronal tissue and, significantly, is activated by diverse lipid-soluble anaesthetics (Patel *et al.,* 1999). This exciting class of channels should be fascinating to watch in the coming years, especially once channel crystals are analyzed. Hopefully, some of its members will prove to be physiological mechanotransducers in some situations.

The electrophysiological milestone was a (molecularly unidentified) channel whose MS activity has been followed in osmosensory neurons from the single-channel level to

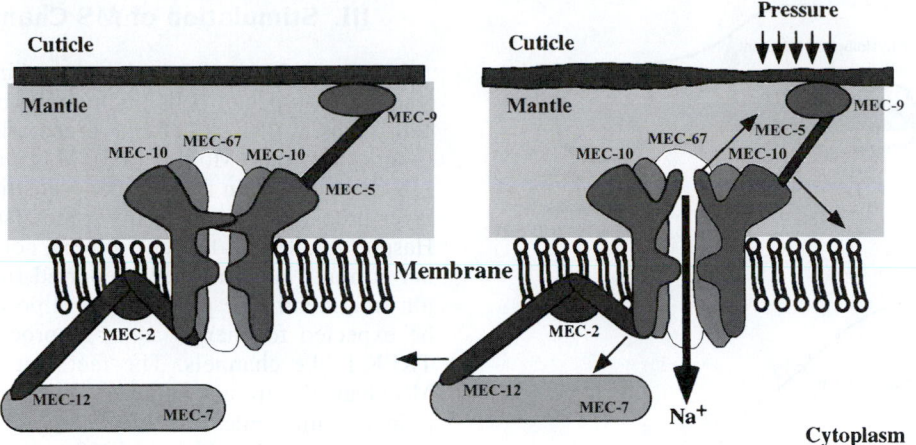

FIGURE 2. Model for a touch trandsducing complex in *C. elegans* mechanosensory neurons. MEC-number labels refer to a variety of proteins that, when mutated, can yield mechanosensory malfunctions. The degenerin MECs form the channel itself. In the absence of mechanical stimulation (left) the channel is closed. Application of a mechanical force to the body of the animal (right) stresses the assembly of interacting molecules in a way that favors the open state of the channel. (Modified from Tavernarakis and Driscoll, 1997.)

whole-organism sensory physiology. The channel in question seems to be stretch inactivated (SI) (rev. in Bourque and Oliet, 1996; Figs. 3 and 4). SI channels first aroused interest when found to coexist in growth cones with stretch-activated (SA) channels (Morris and Sigurdson, 1989, but see Morris and Horn, 1991) and then when SI channel behavior was linked to the pathophysiology of dystrophic muscle (Franco and Lansman, 1990). Though the putative osmosensory sin-

gle-channel data are tricky in their details and would benefit from repetition in other hands (see Sachs et al., 2000) the idea is that a SI cation channel mediates the osmotransducer current of hypothalamic osmosensory neurons. All-important is the concurrence between the single-channel and macroscopic input/output relations (the whole-cell osmosensory data). Because this is convoluted, it is detailed in Section VIII. Unfortunately, this important prototype sheds no light

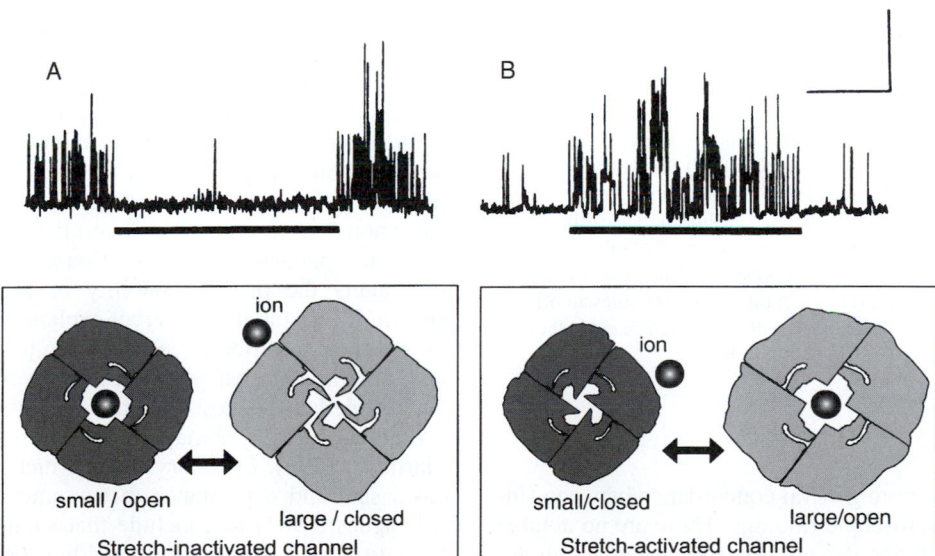

FIGURE 3. Two versions of MS channel gating: SA and SI channels. (A) Stretch-inactivated (SI) and (B) stretch-activated (SA) single-channel events recorded from molluscan neurons. Bars: application of −40 mm Hg suction. Scale: 1pA, 2 s (SI) or 4 pA, 2s (SA). (From Morris and Sigurdson, 1989.) Below each trace is a cartoon of a SI or SA channel seen face-on, suggesting how in-plane tension (conveyed from the plasma membrane itself, or from membrane skeleton, or both) could act in MS gating. At a given tension, a channel has a fixed probability of being open or closed; two-headed arrows represent stochastic equilibrium. At higher in-plane tensions, both the SI and SA channels would have higher probabilities of being in the large diameter configuration. Tension does not literally "pull" the channel open. An example: consider two tensions at which a SA channel's open probability is, say, 0.01 and 0.4. At the higher tension the open probability increases 40× yet the channel is still 2.5× more stable closed than open!

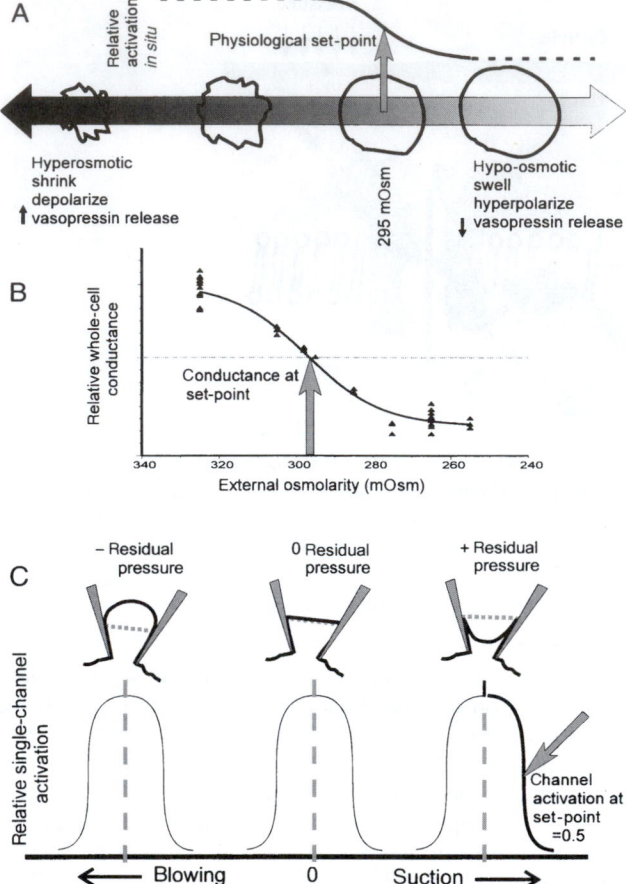

FIGURE 4. The MS channels in osmosensory neurons (OSNs); interpreting the activation curves. (A) Top, the sigmoidal activation curve for a SI, where hyperosmotic conditions (low tension) are shown to the left and hyposmotic conditions (high tension) to the right. The sigmoid has a negative slope, corresponding to data from OSNs (B). (B) Whole-cell osmosensitive conductance as a function of bath osmolarity. (Data from Bourque, with permission; a Boltzmann has been fit to the data. See Oliet and Bourque, 1996, for references.) (C) Single-channel bell-shaped activation curves. Cartoons show a pipette with a patch at two different experimental pressures. The dashed line indicates the membrane position for minimal membrane tension and the solid line indicates its position at zero experimental pressure. When the two do not coincide (i.e., when residual pressure in the pipette provides an offset), the bell curve is displaced. The bolded half of the right-hand bell curve would correspond to the sigmoids in A and B.

III. Stimulation of MS Channel Activity

MS channel currents can be elicited in patch-clamp studies of most types of cells, including a multitude of animal cells, plus wall-free cellular preparations from bacteria, fungi, and plants (Morris, 1990; Martinac, 1993). Bacterial MS channels retain full mechanosensitivity in artificial bilayer preparations and in liposomes (see Martinac, 1993; Hase *et al.*, 1995). For eukaryotes, comparable MS channel currents have not been reported for membrane skeleton-free membrane preparations (liposomes), but this can be expected to change as work proceeds, say, with the TREK-l–like channels. The fact that they, like bacterial MS channels, are activatable by membrane crenators (amphipathic molecules that cause high-curvature bending of lipid bilayers) (Patel *et al.*, 1998) suggests that tension in the bilayer suffices to change the open probability of these channels.

To obtain recordings of MS channels in animal cells, a patch of membrane several square micrometers in area is sealed at the tip of a recording pipette and pressure (usually suction) is applied to distend the membrane (see Fig. 1). On cells that lack extracellular matrix the membrane-pipette seal may occur with no treatment of the cell surface, but most cells require enzymic digestion to remove material (e.g., basement membrane, cell wall) that prevents direct access to the membrane. Sachs and colleagues have provided detailed visual and biophysical information (e.g., capacitances, elasticities) about the state of patch-clamped membrane and its underlying cytoplasmic cortex (see Sokabe *et al.*, 1991). There is no tidy mechanical stimulus for patches. Although we like to imagine that membrane tension is all that changes when suction is applied, the plasma membrane is not a simple elastic solid, so the responses of membrane patches only roughly approximate those of a Hookean (elastic) thin-walled shell. In addition to tension changes, patches under suction pressure may exhibit bilayer flow (i.e. liquidlike behavior), alterations of the structural relations between cortical cytoskeleton and the overlying membrane (i.e., viscoelastic behavior), microvesiculation, and sundry other nonideal (i.e., non-Hookean) behaviors.

The mechanism by which suction-induced membrane tension alters the open probability of MS channels is not known. It is likely that different explanations will be needed for channels in different cell types. In bacteria, experiments with amphipaths that partition asymmetrically into the bilayer's leaflets suggest that differential stress in the two leaflets of the bilayer may activate these MS channels (see Martinac, 1993). In eukaryotic cell membranes, simultaneous visual and capacitance measurements led Sokabe and colleagues (1991) to conclude that suction induces the bilayer to flow past a more unyielding protein network-like skeleton that acts as the tension-bearing (elastic) component; the channels are anchored by (and hence responsive to) unknown membrane skeletal element(s). Consistent with this view, measured membrane tether forces (which provide an estimate of membrane tension) in swelling molluscan neurons (Dai *et al.*, 1998) are considerably smaller than the tensions that gate most MS channels, which generally activate at near-lytic tensions (i.e., 1–10 mN/m). Table 1 lists

on the absence of a more general concordance between single-channel and macroscopic findings. There are no notable differences in membrane density, dynamic range, or in apparent mechanosensitivity between the putative osmotransduction SI channel and the many other MS channels that have, to date, failed to yield the expected macroscopic mechanocurrents. Encouragingly, however, a newly reported SA channel evident at single-channel and whole-cell levels in isolated epithelial cells bears many similarities (including responsiveness to swelling and cation permeability) to the putative osmosensory channel (Vanoye and Reuss, 1999).

TABLE 1 Mechanical Quantities for MS Channels

Value	Relative Size	Example
$1 \text{ N} \cong 100 \text{ g}$	An enormous force	~1 apple in earth's gravity
$1 \text{ N/m} = 10^3 \text{ dyn/cm}$	A large tension that would rupture a membrane	~1 apple dangling from a meter-wide banner
$1 \text{ N/m}^2 = 1 \text{ Pa}$	A very small experimental pressure	~1 apple per coffee table

Force[a]	Example
3 pN	Force generated by myosin molecule
7 pN	Force to pull membrane tether from a neuron
10–20 pN	Calculated force for activation of a "typical" MS channel
20 pN	Actin-gelsolin bond
50 pN	Force to pull erythrocyte membrane tethers
100 pN	Actin-actin bond
30 000 pN	Carbon-carbon bond

Tension[b]	Example	Reference
0.12 mN/m = 0.12 dyn/cm	Resting tension of plant protoplast membrane	Kell and Glaser, 1993
0.4 and 0.12 mN/m	Membrane tensions in normal and swollen molluscan neurons	Dai *et al.*, 1998
1 mN/m = 1 dyn/cm = 1 pN/nm MS	Channel activation	yeast (Gustin 1992); muscle (Sokabe *et al.*, 1991)
4 mN/m	Lytic tension for plant protoplast	Kell and Glaser, 1993
2–10 mN/m	Lytic tension for lipid bilayers and cell membranes	Sheetz and Dai, 1996

Pressure[c]
1 kPa = 7.5 mm Hg
100 mm Hg = 13.3 kPa = 133 mbar
1 mm Hg = 1.36 cm H_2O
760 mm Hg = 1 atmosphere

[a]For references, see Sheetz and Dai, 1996; Dai *et al.*, 1998; Sachs and Morris, 1998.

[b]The relevant membrane tension for MS channels is in-plane tension, not interfacial surface tension, though the units are the same. Another important quantity with the units of N/m is the spring constant (elasticity or stiffness) of Hooke's law. Tension can also have the dimensions energy/volume as in $k_B T/\text{nm}^3$. Sometimes, *forces* are referred to as "tensions" (the magnitude of a force exerted, say, via a string), but "tension" that counteracts a force is a force.

[c]±1–200 mm Hg, applied via pipettes to gate MS channels. Ideally, tension goes as Laplace's law, tension = 1/2 × radius × pressure, so the more curved a membrane, the lower the tension.

mechanical quantities and attempts to put them in a helpful context for understanding the forces acting on MS channels.

IV. Diversity of MS Channels

The designation of an ion channel as mechanosensitive (MS) is empirical—it signifies only that the channel's open probability responds to membrane deformation. In most animal cells, once a gigaohm seal is obtained during cell-

attached patch-clamp, a suction of −10 to −100 mm Hg applied through the recording electrode will generally suffice to half-maximally activate any stretch-activated MS channels present (see Table 1). Often it is difficult to saturate the effect because MS gating is elicited only as membrane tension approaches the lytic tension for the patch (e.g., Bedard and Morris, 1992; Vandorpe *et al.*, 1994). The density of MS channels is normally on the order of one channel per μm^2 of membrane. High densities such as can occur for ligand- and voltage-gated channels (say > 100 per μm^2) have never been reported for MS channels. A cell may have only one type of MS channel—for example, the SA cation channels in *Xenopus* oocytes (Hamill and McBride, 1992)—or they may have multiple types, as indicated in Table 2. Neither excision of a patch nor removal of all cytoplasmic-side calcium abolishes the mechanosensitivity of MS channels (e.g., Vandorpe *et al.*, 1994). In fact, the more stripped-down and variously traumatized a membrane patch becomes, the more readily some MS channels are activated by a standard stimulus (Wan *et al.*, 1999), suggesting that tension applied to the naked bilayer is the most effective stimulus.

There are nonselective and cation-selective (i.e., mono- and divalents) as well as K+ and Cl− or anion-selective MS channels (see Table 2). Most MS channels are stretch activated (SA) and a few are stretch inactivated (SI). The MS activity of some channels depends on the direction of membrane curvature (Bowman *et al.*, 1992; Marchenko and Sage, 1997; Maingret *et al.*, 1999). From their diversity, it is evident that MS channels belong to many different channel families. Many are activated by a variety of stimuli other than mechanical stress (Table 3), though some are known only from their MS gating (e.g., the SA cation channel in *Xenopus* oocytes). The champion so far, with four types of activation, is the smooth muscle K+ channel reported by Kirber and colleagues (1992), although this channel may really be a ligand-gated channel in disguise. Fatty acids, liberated from the membrane by stretch, appear to be the mediator of stretch activation for this channel (Ordway

et al., 1995). This possibility needs to be considered more widely.

A variety of molecularly identified (cloned) channels are now known to exhibit MS gating at the single level. Even a cursory glance at this collection reveals that the requirements for MS gating must be rather catholic. The cloned channels include MscL, the nonselective channel of nanoseimen conductance, from *E. coli* (Sukharev *et al.*, 1994); the NMDA channel, a glutamate-activated cation channel (Paoletti and Ascher, 1994); smooth muscle Ca2+-activated K+ channels (Dopico *et al.*, 1994); a G-protein-regulated K+-permeant inward rectifier, GIRK (Pleusamran and Kim, 1995; Ji *et al.*, 1998); and, most recently, several members of the TREK-like family (Maingret *et al.*, 1999). Of these, only MscL was cloned because of its mechanosensitivity. McsL forms homomultimers, probably homopentamers (Chang *et al.*, 1998), but bears no sequence resemblance to either the *heteropentameric* NMDA channel, the *heterotetrameric* GIRK, or the (presumably) *dimeric* TREK-like channels.

MS channels are defined by their responses to membrane deformation but not all are equally responsive. Dynamic ranges of several orders of magnitude are not uncommon for SA channels in, say, molluscan neurons, chick muscle, or *Xenopus* oocytes (Morris and Sigurdson, 1989; Guharay and Sachs, 1984; Hamill and McBride, 1992); this compares to the dynamic ranges of voltage- and ligand-gated channels. For some MS channels, however, mechanical stimulation only slightly changes open probability. For example, the NMDA-type glutamate channel increases its open probability

TABLE 2 Cells with Multiple Types of MS Channel

Cell	Channel(s)	Reference
Amphibian kidney cells	All demonstrable K+ channels in the basolateral membrane of amphibian proximal tubule cells are mechanosensitive	Cemerikic and Sackin, 1993
Molluscan heart	(*Lymnaea stagnalis*) SA K+ and (zero extracellular Ca2+ only) SA Na+ channels	Morris, 1990; Gardiner and Brezden, 1990
Molluscan neurons	*Lymnaea stagnalis* SA K+ and SI K+ channels; the latter are less common and have a smaller conductance *Cepaea nemoralis* SA K+ channels and (excised patches only) SA Cl− channels	Morris and Sigurdson, 1989 Bedard *et al.*, 1992
Amphibian smooth muscle	SA cation and SA K+ channels	Kirber *et al.*, 1992
Chick heart cells	5 distinct SA channels (3 K+-selective, 2 cation-selective)	Ruknudin *et al.*, 1993
Mammalian outer hair cells	The lateral walls of guinea pig outer hair cells have a SA cation channel and one that appears K+-selective	Ding *et al.*, 1992
Mammalian osteoblast-like cells	SA K+ channel, 60-pS SA cation channel, 20-pS SA channel	Davidson *et al.*, 1990
Crustacean stretch receptor neurons	SA cation channels, rectifying SA cation channels	Erxleben, 1989
E. coli	Several molecularly distinct nonselective conductance channels	Sukharev *et al.*, 1994

TABLE 3 MS Channels Whose Gating Is Also Controlled by Other Factors

Channel	Factor(s)	Reference(s)
Avian SA cation channel	Depolarization	Sachs, 1992
Amphibian smooth muscle SA K⁺ channel	Ca^{2+}, fatty acids, and membrane depolarization	Kirber *et al.,* 1992
Amphibian smooth muscle SA cation channel	Hyperpolarization	Hisada *et al.,* 1991
Crayfish rectifying SA cation channel	Hyperpolarization	Erxleben, 1989
E. coli MS channel	Depolarization	Martinac, 1993
Mammalian atrial SA K⁺ channels	Arachidonic acid and depolarization ATP-inhibited	Kim, 1992 Van Wagoner and Russo, 1992; but see Kim, 1992
Liver SA cation channel	ATP-activated (via purinergic receptors)	Bear and Li, 1991
Osteosarcoma SA cation channel	Parathyroid hormone via cytoplasmic messengers increases open probability and single-channel conductance	Duncan *et al.,* 1992
Aplysia S(erotonin)-channel	Stretch and (via FMRFamide receptor pathway) arachidonic acid metabolites	Vandorpe *et al.,* 1994; see Patel *et al.,* 1998
Mammalian neuron NMDA channel[a]	Potentiated by mechanical stimuli	Paoletti and Ascher, 1994
Cardiac GIRK[a]	Enhanced by membrane stretch	Pleusamran and Kim, 1995

[a]These channels are only known to be mechanosensitive when they are ligand bound.

approximately three-fold in response to pipette suction (Paoletti and Ascher, 1994), and the bovine epithelial Na⁺ channel ENaC only about two-fold in bilayers and not at all when expressed in *Xenopus* oocytes (Awayda *et al.,* 1998). Ideally one would reserve the term "mechanosensitive" for channels that recognize mechanical deformation as a physiological signal. Then, whether the dynamic range was 2–3 or 2–3-orders-of-magnitude, the channel would qualify as "MS" only if mechanosensitivity mattered in the life of the cell.

Various MS channels can be activated via the ligands of second messenger systems as well as by membrane stretch. They include the K⁺-selective *Aplysia* S-channel (Vandorpe and Morris, 1992), their mammalian relatives TREK-1 and TRAAK (Maingret *et al.,* 1999), and cation channels in osteoblasts (Duncan *et al.,* 1992), hepatocytes (Bear and Li, 1991), and kidney cells (Verrey *et al.,* 1995). Physiologically, the *Aplysia* S-channel (Vandorpe *et al.,* 1994) is activated by arachidonic acid metabolites and inhibited by an A-kinase, ligands controlled via the neurotransmitters FMRFamide and serotonin, respectively. This voltage-independent K⁺ channel's physiological job is to modulate electrical excitability according to neurotransmitter signalling. Under patch-clamp conditions (cell-attached or excised), however, the S-channel can be activated by stretch alone (see Fig. 1B); in fact, the maximal effect of stretch on a given patch considerably exceeds activation via second messenger paths (Vandorpe *et*

al., 1994). MS K⁺ channels are ubiquitous in neurons of *Aplysia* and other molluscs (Bedard *et al.,* 1992; Morris, 1992), but only in the identified *Aplysia* neurons are the signaling molecules, which include arachidonic acid, known. *In situ* the channels do not readily activate with mechanical stimuli (Morris and Horn, 1991; Wan *et al.,* 1995), yet "SA K⁺ channel" was until 1998 (see Patel *et al.,* 1998) the best designation available for these "S-like" channels. Many other channels may have been designated as "MS channels" simply because the primary physiological stimulus, perhaps a second messenger ligand, has not been identified. The *Xenopus* oocyte MS channel (Steffensen *et al.,* 1991) may be a case in point.

For all MS channels, tension must be conveyed to the channel via the surrounding lipids, membrane skeleton, extracellular matrix elements, or some combination. Although these structures are universal, suction does not affect gating in all channels. The classical example is the nicotinic acetylcholine channel of chick skeletal muscle. This is a cation channel with very similar permeation properties to a cation-selective SA channel in the same membrane (and a heteropentameric structure not unlike the NMDA channel), but the acetylcholine channel's kinetics are not affected by stretch (Guharay and Sachs, 1984).

Mechanotransduction at membranes need not be mediated by MS ion channels. Recently, it was shown that expression of a heterologous metabotropic ATP receptor confers mechanosensitivity on *Xenopus* oocytes. The explanation is that oocyte deformation releases ATP, which then acts on the receptor (Nakamura and Strittmatter, 1996). Here, the mechanical event that needs explanation is the stretch-induced release of ATP.

V. MS Channels in Patches: Stretch Versus Damage Plus Stretch

Because MS channels are almost ubiquitous, the question of whether their mechanosensitivity is meaningful or merely epiphenomenal is not trivial. If, *in situ*, many MS channels are **not** used for cellular mechanotransduction, why is MS gating in membrane patches so common? Perhaps for many channels anchored to the membrane skeleton, spurious (essentially pathological) MS gating can occur when the integrity of the cortical cytoskeleton in the channel's immediate environs is inadequate (Table 4). The assumption here is that an intact cortex absorbs mechanical loads that would otherwise be felt by the membrane skeleton and by attached channels. MS channels designed for use as physiological mechanotransducers would, by contrast, require an environs specifically constructed to enhance mechanical loading of the channel via the membrane skeleton or bilayer.

Patch recording is inevitably accompanied by mechanical damage to the patched membrane region. Small and Morris (1994) showed that in molluscan neurons, starting from a near-intact patch, successive rounds of mechanical stimulation augment the responsiveness of MS channels to the mechanical stimuli. "Stimulation" means "aspiration of the membrane and the underlying cytoskeleton," and it would be specious to argue that this is entirely benign. The ability of patch-induced damage to facilitate sustained MS channel responses may explain much of the widespread discordance between single-channel and whole-cell recordings (Morris and Horn, 1991; Wan *et al.*, 1999). In *Xenopus* oocytes, too, MS channel behavior is damage dependent: the rapid adaptation behavior is irreversibly destroyed by patch trauma (i.e., by successive stimuli). Interestingly,

pore (α) subunits of voltage-gated Na$^+$ channels expressed (in the absence of their auxiliary β-subunit) in *Xenopus* oocytes exhibit a physiologically nonstandard slowly inactivating gating mode that, by some unknown mechanism, irreversibly converts during membrane stretch into kinetically normal fast inactivation (Tabarean *et al.*, 1999). This irreversible effect probably says something important about the susceptibility of protein complexes to bilayer tension, but should not be taken to mean that Na$^+$ channels are mechanotransducers.

What then, is the meaning of channel mechanosensitivity? There is no unique answer, but an encapsulation of my viewpoint is: many eukaryotic channels are inherently **mechanosusceptible** to changing their transition rates among open and closed states (either reversibly or irreversibly) when mechanical stimuli increase tension in the plane of the bilayer. The channels' complex cellular environment can either act as a restraint, absorbing mechanical energy so the channel can attend to other stimuli, or it can "tune-up" the inherent mechanosusceptibility to yield a mechanotransducer of membrane tension. For channels that are not mechanotransducer specialists, mechanosusceptibility may vary as physiological and pathological conditions alter the membrane skeleton (e.g., during cell swelling, cell division, cell growth, ischemia, mechanical trauma, and nociceptive stimulation). By contrast, mechanotransducer specialist channels such as those at the tips of auditory hair cells may prove to be exceedingly **insensitive** to bilayer tension at all times, thus ensuring that they "listen" to and are gated exclusively by the relevant tension, found in their specialized accessory structures (e.g., tip links).

Whether or not it is used by the cell, mechanosusceptibility is an inherent trait in diverse integral membrane proteins. Should we marvel that this trait evolved independently as a "selected-for" feature in each protein or should we surmise that a tendency for mechanical perturbations to trigger gating transitions is, instead, a difficult to suppress trait for membrane-spanning enzymes? I favor the latter, but both postulates are valid, and deeper understanding will require molecular studies of eukaryotic channels that exhibit physiologically relevant mechanosensitivity as well

TABLE 4   Trauma, Pathology, Nociception, and MS Channel Activity

Dystrophic muscle	Lacks membrane skeleton dystrophin so spectrin may experience abnormal stress which may be transferred to channels linked to spectrin. Dystrophic cells are fragile and leaky to calcium. MS cation channels in dystrophic muscle exhibit abnormal MS gating (Franco and Lansman, 1990). There may be a causal relationship.
Brain ischemia	In the presence of agonist, NMDA-gated channels are mechanosensitive. In swelling neurons, NMDA channels generate whole-cell currents that grow larger as swelling progresses (Paoletti and Ascher, 1994). In stroke, this could exacerbate glutamate excitotoxicity.
Nociception and mechanical trauma	Neuronal SA channels are easier to activate following trauma (Small and Morris, 1994). Osteoclast MS channel mechanosensitivity is enhanced in chronically and intermittently strained cells (Duncan and Hruska, 1994). Sustained rather than transient responses of MS cation channels predominate after membrane trauma (Hamill and McBride, 1992), a phenomenon that could potentially be exploited for nociceptive signaling.
Ectopic foci in demyelinated neurons	Hypermechanosensitive (see references in Waxman, et al., 1994). Abnormalities in membrane and cortex may increase the mechanosensitivity of normally refractory channels.

as those that appear to be mechanosusceptible in spite of having no evident role as mechanotransducers.

VI. The Role of the Membrane Skeleton

When MS channels were first described (Guharay and Sachs, 1984) it was clearly demonstrated that neither tubulin nor F-actin was required for the MS gating. It was postulated that MS channels are linked into the membrane skeletal network and that when the in-parallel actin cytoskeleton is depolymerized, stress is readily transferred to the channel-linked component. What that parallel component may be has not been resolved, although it is not dystrophin since MS channels are evident in dystrophic muscle cells (Franco and Lansman, 1990). Spectrin is a candidate that has not been ruled out.

For eukaryotic systems, the bilayer may seldom experience significant stress except near lysis. Supporting this notion is the fact that the elastic constant of patches (Sokabe *et al.*, 1991) is much less than for lipid membranes. The bilayer is not, therefore, the patch's load-bearing element. That leaves the cortical cytoskeleton and the membrane skeleton. If it is generally true that the elastic constant of patches is insensitive to the actin disrupter cytochalasin (Sokabe *et al.*, 1991), that leaves, by further elimination, a cytochalasin-insensitive membrane skeletal network as the only candidate. In an intact cell, one would expect cortical actin to contribute to the elasticity of the membrane-cortex region. In patches, the fact that it does not suggests that the act of patch formation disrupts the cortical actin network, consistent with the finding of Small and Morris (1994) who showed that cytochalasin, like mechanical damage, augments rather than diminishes the responsiveness of SA channels to membrane stretch. Other membrane skeletal filaments like spectrin may constitute the load-bearing elastic elements that transfer mechanical energy to the channels.

A variety of well-characterized membrane channels (and other transporters) are tethered to the membrane skeleton, either to β-spectrin via ankyrin (Lambert and Bennett, 1993) or directly to α-spectrin (Rotin *et al.*, 1994). Additionally, some voltage- and ligand-gated channels are anchored to postsynaptic density-family proteins (Kim *et al.*, 1995). Although we surmise that eukaryotic MS channels receive mechanical energy via the membrane skeleton, there are no grounds for turning the argument around and invoking channel mechanosensitivity as a reason for linkages. Transporters may anchor to the membrane skeleton in order to stay fixed at some cellular locale. In other cases, as in erythrocytes, the transporters may serve as anchor points for a membrane skeleton whose major job is to strengthen the membrane. It is probably counterproductive for many anchored membrane enzymes to allow their activity to be influenced by mechanical stress. Thus, in asking how and if specialized channel–membrane skeleton linkages might be used by MS channels to convey force to the channel, we should also remember to ask the reverse question: what features can make such linkages **stress-proof?** (See Glogauer *et al.*, 1998.)

VII. Delay and Adaptation: Mechanically Fragile Aspects of MS Channel Behavior

The mechanical properties of patches are labile and history dependent (Hamill and McBride, 1997; Small and Morris, 1994). In fibroblasts stretch causes F-actin to depolymerize within 10 s, although when the cells are left to rest briefly, F-actin reorganizes, reappearing at greater than control levels, stiffening the cortex as it does so (see Glogauer *et al.*, 1998). Patch formation and stretch stimulation of cells may have qualitatively similar influences on the state of actin. Thixotropic properties of actin, in which mechanical stimulation of actin gels promotes the sol-state of actin (see Heidemann and Buxbaum, 1994) may affect MS channel mechanosensitivity.

In molluscan neurons and in *Xenopus* oocytes (Small and Morris, 1994; Hamill and McBride, 1992; Wan *et al.*, 1999), the behavior of SA channels following a step of applied suction has been studied in patches formed with a minimum of mechanical disruption—**gentle patches.** The responses of MS channels in the gentle patches differ dramatically from those in standard patches. The designations "gentle" and "standard" patch are operational; gentle patches inevitably become progressively more standard with successive mechanical stimuli.

In gentle oocyte patches, SA channels exhibit rapid **adaptation;** adaptation is abolished in standard patches. In gentle snail neuron patches, SA channels show a prolonged **delay** before activation; delay is abolished in standard patches. Changes in patch size and changes in cortical cytoskeleton integrity are probably not mutually exclusive, so the interpretation of input-output relations for MS channels can be fraught with difficulty. However, the mechanical fragility of the two dynamic phenomena, adaptation and delay, cannot be explained by the increases in patch size that may accompany a patch's progressive transformation from gentle to standard.

As indicated, molluscan neuron SA channel activation in response to a first hit (a first large step of suction applied to a gentle patch) proceeds only after a delay, as if the patch were viscoelastic. The delay is substantial (> 2 s at −130 mm Hg) and changes produced by treatments are easily detected. Treatments that dramatically decrease delay include repetition of mechanical stimulation, use of larger mechanical stimuli, pretreatment with cytochalasin (a drug that promotes actin depolymerization), use of recently isolated (i.e., recently disrupted) as opposed to well-established cultured neurons, hyposmotic swelling of neurons, exposure to the sulfhydryl reagent N-ethylmaleimide, and elevation of intracellular Ca^{2+} (Small and Morris, 1994; Wan *et al.*, 1999). In addition to shortening delay, these treatments increase the activation elicited for the same stimulus. In summary, the response time is long and the extent of activation is small when the cortical cytoskeleton is intact, but not when it is compromised. It is reasonable to conclude (Wan *et al.*, 1999) that the channels are not functionally mechanosensitive *in situ* when the cortex is perfectly intact.

Mechanical inflation of neurons under whole-cell clamp does not activate MS current except just before the cell rapidly expands and ruptures; a catastrophic disruption of the cortex could explain both the MS currents and the ensuing membrane rupture (Wan *et al.*, 1995) (Fig. 5). Interestingly, this echoes the situation described earlier for *E. coli* MS channels.

The fragility of MS channel dynamic responses can cause consternation and confusion but it also opens the possibility that substantial modification of MS responses may be possible through hormones that modify the cytoskeleton and by cell motility-related and volume regulation-related changes in the cytoskeleton (Schweibert *et al.*, 1994). For example, vasopressin causes both actin depolymerization and SA channel activation in kidney cells (Verrey *et al.*, 1995). Another example is during the cell cycle (Bregestovski *et al.*, 1992), when major cytoskeletal rearrangements may partly account for the observation that SA channel sensitivity varies during the cycle.

Since the formation of focal adhesions is accompanied by major cytoskeletal rearrangements, it seems likely that MS channels should show altered mechanosensitivity in connection with changed states of cell adhesivity. Hints that this might be so are beginning to accumulate. Macrophages have a MS K+ channel whose activity, monitored by single-channel recording at the upper surface of the cell, increases when the lower surface of the cell becomes adherent to a substrate (Martin *et al.*, 1995). MS channel currents in osteoblasts subjected to chronic intermittent strain are more readily activated than those in quiescent cells (Duncan and Hruska, 1994).

It is, thus, possible that the mechanical fragility of delay is telling us about variable *in situ* states of MS channels. But what about adaptation? Here there is, as yet, little insight. In phasic mechanosensory neurons, the purpose of rapid adaptation is to provide a mechanical high-pass filter, but what it might mean for amphibian oocytes, tunicate eggs, and yeast MS channels is a mystery. By contrast, the other dynamic property, delay, could in principle act like a low-pass filter; it could be useful for, say, muscle or bone cells MS channels, enabling them to ignore momentary stresses, yet inform the cell of sustained mechanical inputs and hence modify some aspect of its metabolism (volume regulation, Ca2+ loading, protein synthesis).

VIII. Physiology of MS Channels

What establishes that a putative mechanotransducer channel performs a MS physiological (or developmental) task? Suggestions about functions of the channels are legion, but experimental evidence conclusively linking MS channel activity to physiological processes is rare. A dearth of specific pharmacological tools continues to hinder progress, as does the difficulty of mechanically stimulating membrane in a calibrated and reproducible manner. Finding MS channels in patches does not mean that MS currents can readily be recorded from the channels in the intact cell.

A. Criteria for Establishing that a MS Channel Performs a Mechanical Cellular Task

The criteria that need to be met are comparable to those that have been met by ligand-gated and voltage-gated channels with established physiological roles. The following criteria summarize a list given by Morris (1992).

1. Establish the MS channel's biophysical characteristics by recording at the single-channel level.
2. Establish a way of altering channel function (e.g., a channel blocker or a mutant channel).
3. Demonstrate that some physiological or developmental aspect of the cell is MS.
4. Demonstrate that the MS cellular function is specifically impaired by the channel blocker (or by the mutation).
5. Make macroscopic recordings under conditions of minimal damage to the membrane-cortex, and demonstrate MS currents that correspond to the single-channel recordings according to (a) selectivity; (b) pharmacology; (c) noise characteristics; (d) saturability at large mechanical stimuli; (e) the expected magnitude of saturating macroscopic current given the single-channel membrane density; (f) expected MS current density variations—spatial and/or temporal—as predicted from any spatial or temporal variations noted by

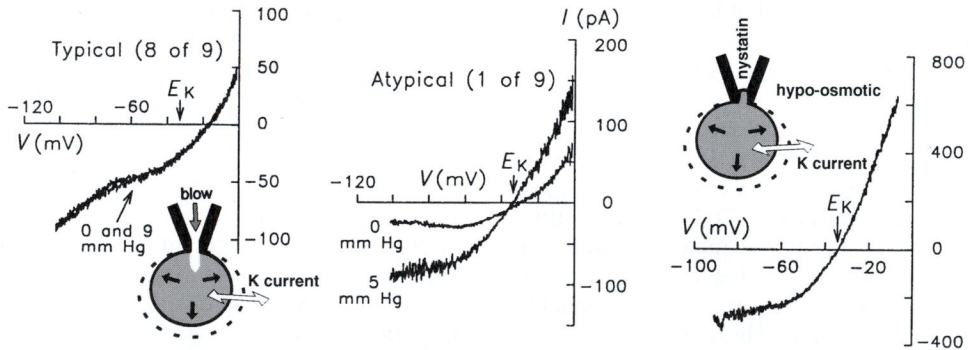

FIGURE 5. Inflation of molluscan neurons under whole-cell clamp rarely elicits MS K+ currents in spite of the abundant MS channels. Macroscopic MS currents are observed only at pressures marginally less than those that rupture the plasma membrane. The left and middle graphs are ramp-clamp currents for the stated pressures (applied via the recording pipette as shown). As expected by Laplace's law, large cells can withstand only small pressures. In the same neurons, K+ currents are elicited by osmotic swelling (right, a difference current obtained by subtracting ramp currents before swelling from those during swelling). Are these volume-activated currents the direct result of swelling (mechanical) stress? One cannot say, since swelling causes an increase in membrane capacitance (possibly recruiting new channels) and alters cytoplasmic chemistry (possibly stimulating channel activity via ligand changes) as well as increasing tension. (Modified from Wan *et al.*, 1995.)

single-channel recording; (g) degree of mechanosensitivity; and (h) sensitivity to other known physiological influences (e.g., hormones, intracellular ions).

B. A SI Cation Channel and Hypothalamic Osmotransduction

For one MS channel, many of the criteria listed above have been addressed. A subpopulation of neurons in the hypothalamus detects and responds to blood osmolarity. These magnocellular osmosensory neurons (OSNs) possess MS cation channels that may underlie the osmotransduction currents that hyperpolarize OSNs during hyposmotic stimuli and depolarize them during hyperosmotic stimuli (see Fig. 4A, B). This in turn leads to respective decreases and increases in release of the osmoregulatory hormone vasopressin from the OSNs. Under cell-attached recording conditions with physiological solutions, the ~30 pS OSN channels reverse near −40 mV, sufficiently depolarized with respect to resting potential for active channels to be excitatory. The idea that MS channels mediate osmotransduction currents is supported by the demonstration that gadolinium ions block both the macroscopic osmosensor currents and the single-channel MS currents.

Activation curves for the OSN single-channel currents in cell-attached patches are bell-shaped rather than the sigmoid curve characteristic of voltage-gated channels and of other MS channels (see Fig. 4C). This is not inconsequential; well-established physical explanations exist for sigmoid activation curves ("Boltzmanns"), but there is no precedent for "Gaussians" when it comes to activation curves. Does the bell shape reflect an inherent peculiarity of the channels? It may not. Rather, the unusual input-output relation may stem from an acknowledged systematic error in assessing the true pipette pressure. Although seeming to be data obtained at applied pressures symmetric about zero (see Fig. 4C, middle curve), the activation curves were obtained entirely with suction (negative pressures) (see Fig. 4C, right); the published bell curve peaks at "zero" only because an arithmetic offset was applied. Macroscopic currents from the same cells show a monotonic dependence on hypotonicity that fits well to a sigmoid (see Fig. 4B). If both single-channel and intact-cell currents reflect activity of the same channels, what explains the dramatic qualitative difference in input-output relations? To answer, we need to visualize how applied tensions in patch recordings would correlate with osmotic stimuli for whole-cell recording.

With whole-cell data plotted on an *x*-axis going from low-to-high osmotically induced membrane stretch (as in Fig. 4B), the Oliet and Bourque data fit a sigmoid Boltzmann relationship (akin to Hodgkin-Huxley h_∞) as expected for the intrinsic tension-activity relations of SI channels (Morris and Sigurdson, 1989; Lecar and Morris, 1993). The expected sigmoid can be found in the single-channel bell curve (see Fig. 4C, right) if the following two assumptions hold: (a) the MS channel is indeed a SI channel and (b) MS channel activity was recorded with net positive pressure in the pipette, even though suction was applied.

These assumptions may be incorrect but are not strictly *post-hoc* rationalizations—they have a precedent. For snail neuron SI channels (Morris and Sigurdson, 1989), sigmoids are more common but bell-shaped activation curves are obtained if the appropriate experimental pressure range is used. The bell corresponds to the expected h_∞ type sigmoid plus its mirror image reflected on the **zero-tension point** of the pressure axis. The *x*-axis is the **estimated** experimental pressure (see Fig. 4C), not **membrane tension;** neither the true zero pressure nor tension is known. The authors contend that the bell-shape "illusion'" arises because the membrane patch is sealed firmly in the electrode rim, and the patch is concave, flat, or convex depending on the effective pipette pressure. Tension sufficient to gate the SI channels can be achieved in both the concavely and convexly stretched patches. If residual pressure indeed persists in the pipette after gigaseal formation, the bell-shaped curve will center not at the origin but elsewhere along the axis of experimentally applied pressure, according to the sign and magnitude of the residual pressure (see Fig. 4C).

To summarize, the basis for the bell shape is critical. If the bell arises from the stimulus mechanics of patches (a bell is the SI channel congener to the U-shape of SA channels over the full stimulus range (pressures positive and negative to zero; see Sachs and Morris, 1998)) and if all the single-channel OSN channel recordings were indeed obtained starting with positive residual (not zero) pressure in the pipette, then the channel is indeed a SI channel and the microscopic and macroscopic input-output relations are not at odds.

C. Other Roles Proposed for MS Channels

This section quotes from a potpourri of papers in which mammalian or avian MS channels were tentatively linked to a wide range of cellular phenomena. The microscopic to macroscopic correspondences have yet to be established. There is a pressing need to determine if various MS channels perform *in situ* the mechanical tasks suggested for them based on their single-channel behavior.

- Bear, 1990: "It is proposed that stretch-activated channels in liver cells permit the transient influx of Ca^{2+}, which in turn acts to trigger changes in ion conductance or cytoskeletal components involved in cell **volume regulation.**" (See Bear and Li, 1991.)
- Suleymanian *et al.*, 1995: "Stretch-activated channel blockers **modulate cell volume** in cardiac ventricular myocytes . . . support[ing] the idea that [the channels] are involved in cardiac cell volume regulation."
- Puro, 1991: "Stretch-activated channels may help mediate a **compensatory response of glia to swelling** . . . [because] activation of calcium-permeable stretch-sensitive channels is associated with an increase in the activity of calcium-activated potassium channels. Activation of potassium channels to produce an efflux of potassium with a subsequent loss . . . of cell water could . . . decrease glial cell volume."
- Hansen *et al.*, 1991: "Our results indirectly implicate stretch-activated channels in the genesis of stretch-induced [cardiac] arrhythmias and provide preliminary evidence for a potential new mode of **antiarrhythmic drug action—blockade of stretch-activated channels.**"
- Kent *et al.*, 1989: "sodium entry [—an initial cellular response requisite to the growth-inducing activity of many substances—] through stretch-activated ion

channels is stimulated by deformation of the sarcolemma . . . . Streptomycin, a cationic blocker of the mechanotransducer ion channels . . . inhibited, in a dose-dependent manner, **[load-dependent] protein synthesis** otherwise observed in contracting [myocardial] muscles developing tension."

- Ding *et al.,* 1992: "SA channels . . . could affect the **motile response [of outer hair cells,** via] membrane potential or by allowing the entry of free Ca^{2+} which could lead to [cellular length changes via] actin and myosin. SA channels could also . . . [regulate the cell's osmotic pressure] thereby influencing its **electroosmotic response.**"

- Iwasa *et al.,* 1991: In auditory outer hair cells, "stretch-activated channels may play an important role in producing a mechanical feedback, an indispensible element in **cochlear tuning.**"

- Coleman and Parkington 1992: "Propagation of electrical and mechanical **activity in uterine smooth muscle:** a functional role for stretch-sensitive channels."

- Davidson *et al.,* 1990: "We propose that one or more of these [three] mechanosensitive ion channels [in osteoblast-like cells] is involved in the **response of bone to mechanical loading.**"

- Sigurdson *et al.,* 1992: "Gently prodding [tissue-cultured chick heart] cells with a pipette produced a Ca^{2+} influx that often led to **waves of calcium-induced calcium release spreading from the site of stimulation** . . . . The mechanical sensitivity probably arose from stretch-activated ion channels."

IX. Other Explorations of MS Channels

A few widely ranging examples are used to provide some flavor of potentially exciting findings. In each case, the question testifies to the need for more information about the biological impact of the readily observed mechanosensitivity.

(1) Fungal germling of bean rust: topographical sensing? Rust is a parasite whose germling needs to find the stomatal openings of host-plant leaves, a feat it achieves by sensing minute topographical features. Kung and colleagues (Zhou *et al.,* 1991) showed that germ tube protoplasts of rust have SA channels evident in both macroscopic and single-channel recordings. Saturating MS currents were readily obtained in both types of recording. Moreover, gadolinium blocks both single-channel and whole-cell currents in protoplasts and inhibits germ tube growth and differentiation *in vivo.* The rust channels are nonselective and have a very high conductance (> 0.5 nS), so that channel openings could profoundly alter the ionic composition of the minute volume of cytoplasm in the growing tips. Presumably fine control could be achieved *in vivo* by having exceedingly brief openings.

(2) Hyphal tips in fungi: MS channels and development? On the basis of numbers of SA channels per patch, Levina and colleagues (1994) report variations in the density of the channels toward hyphal tips in fungi. Based on the tip-high staining pattern of filamentous actin, and on data with cytochalasin, the authors suggest that actin causes

the channels to cluster, either by linking channels to the cytoskeleton or by increasing transport of channels to the tip. Alternately, regional variations in membrane mechanics may explain the data; channels may be present at uniform density but they may be less mechanosensitive in "old" membrane far from the tip and easy to stimulate in "new" membrane near the tip. This would echo the case of neuronal cytochalasin-sensitive SA channels which become increasingly difficult to activate as the cell culture matures (Small and Morris, 1994).

(3) Fish embryo: a role in the cell cycle? SA K^+ channels in loach embryos (2–256 cells) show dramatic changes in activity during the cell cleavage cycle, perhaps explaining the membrane voltage oscillations. At the beginning of interphase, channels tend to be inactive and less responsive to stretch, whereas at prometaphase they are highly active. Evidence is presented showing that the activity of these channels is regulated through cAMP-dependent phosphorylation (Medina and Bregestovski, 1991).

(4) Mammalian neurons: a role in neuronal development? The SA K^+ in fish embryos may well be related to both molluscan growth cone SA K^+ channels (Morris and Sigurdson, 1989) and to the TREK family. The TRAAK version of this family, found exclusively in nervous tissue, has been suggested to play a role in such processes as growth cone motility (Maingret *et al.,* 1999); while this idea was not supported for the molluscan case (Morris and Horn, 1991), it will be very worthwhile to revisit it with the more powerful tools available once a channel's molecular identity is known.

(5) Fish epithelial keratocytes: a role in modulating locomotion? During the locomotion of keratocytes, transient changes in intracellular calcium, evidently mediated by calcium-permeant SA channels, occur more frequently in cells temporarily stuck to the substratum or subjected to stretch; the increased calcium is implicated in detachment of the rear cell margin (Lee *et al.,* 1999).

(6) *Xenopus* oocytes MS channels: why the rapid adaptation? On gentle patches, a pressure step elicits a transient increase in MS channel activity—the response shows adaptation, comparable perhaps to the phasic responses of some mechanosensory cells. Is adaptation a type of slow inactivation (comparable to inactivation in voltage-gated channels) or is it associated with a change in mechanosensitivity? Double-step protocols indicate that adapted channels will reactivate with stronger stimulation, suggesting that inactivation of the MS gates has not occurred. By elimination, the transient change seems to be associated with gating sensitivity. As described earlier, adaptation is a fragile phenomenon, critically dependent on the mechanical history of the patch. A surprising aspect of adaptation is that it is not seen at depolarizing potentials. This is also true of adaptation in yeast MS channels (see Martinac, 1993; Gustin, 1992). From the point of view of potential physiological significance, these are interesting but puzzling findings because there would seem to be a Catch-22 operating. Adaptation occurs only at hyperpolarized potentials. *In vivo,* therefore, activation of the MS cation channels would depolarize the membrane, thereby precluding adaptation. Hyperpolarizing influences working in parallel with the MS channel to override their depolarizing effect (e.g., Ca^{2+}-activated K^+ channels) might, however, undo the Catch.

(7) Epithelial SA Cl⁻ channels: a role in volume regulation? In shark rectal gland cells and in renal cells, Cl⁻ channel activity depends on the state of actin (Schwiebert *et al.*, 1994; Mills *et al.*, 1994). In renal glands, filamentous actin is associated with inactive channels and depolymerized actin with active channels. Membrane stretch also stimulates channel activity. It is suggested that during swelling-induced actin depolymerization, cell swelling could be limited through activation of MS chloride channels.

(8) NMDA (glutamate) channels in mouse neurons: pathology? physiology? Membrane tension potentiates the activity of ligand-bound NMDA channels in a variety of recording configurations (Paoletti and Ascher, 1994). By contrast, activated kainate-type glutamate channels in the same patches are not mechanosensitive. The potentiation by tension is relatively small (at best, three-fold) and stretch alone will not activate the channels. Successive stimuli yield progressively larger effects, suggesting that mechanosensitivity is increased by disrupting the cytoskeleton. NMDA responses to stretch in the whole-cell mode support this view; osmosensitivity becomes evident after prolonged dialysis. The simplest explanation of this "run-up" of mechanosensitivity is that as mechanoprotection from the cytoskeleton is progressively lost, force transfer to the NMDA channels directly via the bilayer improves (consistent with newer findings that amphipaths mimic stretch in NMDA channels (Casado and Ascher, 1998)). Potentially, mechanosensitivity in NMDA-activated channels could have far-ranging implications for neuronal plasticity. If it reflects pathology, it is a phenomenon worth understanding in the context of the glutamate excitotoxicity and neuronal swelling associated with stroke.

X. Models for Gating of MS Channels

Formal models for mechanosensitive gating provide a framework for relating experimentally applied force to kinetic data, and now that channels can be crystallized, to protein structure. If a channel has two states and is in some sense a spring, then the mechanical features that should matter are the spring's position and its elasticity. There are various candidates for MS channel springs. Is the entire channel or some part of it elastic? Is an extracellular or intracellular tether (tip links in hair cells, strands of membrane skeleton or extracellular matrix) mechanically in series with the channel? Is it the collection of noncovalent bonds between lipid and protein at the channel-bilayer interface (or "the bilayer")? Is it some combination? Whatever the identity of the spring, a two-state model assumes that force acting through the spring biases a channel's open probability; the channel would have a free energy difference of several kT between open and closed states and the force helps surmount this barrier. When the larger state has a pore in the conducting configuration, then the channel is stretch activated, whereas when the larger state has a nonconducting pore, the channel is stretch inactivated.

A simple and satisfying two-state model is that developed by Hudspeth and colleagues (see Corey and Howard, 1994) for hair cell channels. It posits, on excellent empirical grounds, that an extracellular tip link acts like an elastic spring in series with a stochastic gate. An external force changes the spring's tension and thereby its mean position. Two extremes of position are helpful to visualize: a position in which the spring is fully relaxed, with its attached gate occluding the channel (the small state), and another position in which the spring is under maximal tension with the gate away from the now-open pore (the large state). If external force alters the mean position of the gate and thereby the channel's open probability, then reciprocally, thermally induced opening and closing of the channel should randomly stress and relax the gating spring. Measurements of channel noise (which indicates the rates of gating transitions) and hair cell compliance (assumed to be an indicator of gate position) in stimulated hair cells support this prediction. Note that in this model, the spring's elasticity is critical to its action, but it is regarded as a fixed quantity.

But for many eukaryotic MS channels, as for bacterial MS channels (see Batiza *et al.*, 1999) the important positional effect may be even simpler. Consider stretch-activated K^+ channels like TREK-1. It is simplest to assume they lack specialized spring structures and to postulate that, over a limited range, the bilayer is the spring around the stochastic channel. Stochastic (thermally driven) conformational transitions of the channel occur among closed and open states. If, say, a particular open conformation occupies more membrane area than all other states, it will be favored by tension because, to open, the opening channel protein has less work to do against the bilayer spring when the bilayer is under tension. Again, it is helpful to visualize two extremes of position: a position in which the spring (bilayer) is fully relaxed and its "attached" channel is closed (this is the smaller state; note that the channel protein per se is the gate) and another position with the spring (bilayer) under maximal tension and the channel (gate) open. In the low-tension condition, if the channel is to open, thermal energy in the channel will need to be sufficiently large to compress the already quite compressed bilayer. When the bilayer is expanded (i.e., under tension), however, the opening event will happen more frequently because a more frequently experienced low dose of thermal energy will suffice. Again, note the reciprocal compression and relaxation of the spring (whose elasticity constant is fixed) as the channel opens and closes. This describes stretch activation. Inversely, a scenario in which one or more of the closed states occupied more membrane area than other states would yield stretch inactivation.

The possibility that a MS channel's two major states have different **elasticities** has also been formalized. In this case (Sachs and Lecar, 1991; Lecar and Morris, 1993) the MS gate is depicted as a parabolic harmonic oscillator. At a given applied force (membrane tension) the free energies of the open and closed states depend on both mean gate position and the state-specific elasticities of the two states. Using reasonable quantities for channel transition rates and positional changes, Corey and Howard (1994) estimate, however, that differences in the second factor—elasticity—would be too small to have appreciable impact. Intriguingly, though, they point out that were the closed state significantly "softer" than the open state, the prediction would be that a channel's open probability will go through a maximum as force increases, rather than generating a sigmoid activation curve. It may be worth bearing this latter prediction in mind as further studies of SI channel mechanisms are undertaken. As discussed above, Oliet and Bourque (1993) had to assume in explaining their data that their experimental conditions generated a persistent tension offset in the osmosensory neuron membrane. Should it be found that there is no such offset and that the bell-shaped activation curves represent the true input-output relations of these MS channels, then the possibility of state-specific elasticities could be explored through detailed single-channel kinetics. Finally, some MS channels may undergo mechanosensitive open-closed transitions along a reaction coordinate whose units are not well described as a linear position (Lecar and Morris, 1993).

XI. Summary and Conclusions

MS channels are ubiquitous, abundant, and diverse. They have few unifying features other than a propensity to change open probability when stretched under patch-clamp conditions. For single-channel recordings, the name "MS channel" carries an implicit assumption that some susceptible gating structure experiences tension in the plane of the membrane and that this biases the channel's open probability.

The available information on possible mechanisms of ion channel mechanosensitivity suggest that two extremes are both important. At one extreme, the specialized receptors, cytoarchitecture is central to function. Specialized animal mechanoreceptors (e.g., hair cells in vertebrates, mechanosensory neurons in the nematode *C. elegans.*) require that mechanotransducer channels be precisely aligned with respect to special extracellular and intracellular structural proteins as well as more distant cellular structures and the whole array of hair cells. At the other extreme are bacterial MS channels, which can be reconstituted into artificial bilayers and retain their mechanosensitivity. As mechanistic details become available, definitions will need refining. A mechanically responsive channel that might **not** qualify as a MS channel would be one whose open probability changed (perhaps irreversibly) because a mechanical deformation of the membrane induced a critical lipid phase transition or altered the constituents of the bilayer or disrupted interactions among channel auxiliary subunits. It is easy to see that this could degenerate into semantics. Eventually, we may wish to reserve the term MS channel for physiological mechanotransducer channels and to speak of "mechanosusceptible" channels to describe those that show MS gating in patch recordings but not *in situ*.

In the meantime, ingenious ideas about how MS channels might contribute to cellular control processes are not in short supply, and work is underway on many fronts to test these ideas. Although MS channels detectable at the single-channel level have been on the scene for approaching two decades, they should only be afforded a status coequal with voltage-gated and ligand-gated channels once it is abundantly clear that their MS gating has a biological meaning. It should not be forgotten, however, that the inherent mechanosusceptibility of many other channels also acquires biological meaning (rather than solely biophysical meaning) if it is shown that evolution has had to contrive ways to protect cells from unwanted mechanically driven membrane currents.

Bibliography

Awayda, M. S., and Subramanyam, M. (1998). Regulation of the epithelial Na+ channel by membrane tension. *J. Gen. Physiol.* **112**, 97–111.

Bargmann, C. I. (1994). Molecular mechanisms of mechanosensation? *Cell* **78**, 729–731.

Batiza, A. F., Rayment, I., and Kung, C. (1999). Channel gate! Tension, leak and disclosure. *Structure Fold Des.* **7**, R99–R103.

Bear, C. E., and Li, C. (1991). Calcium-permeable channels in rat hepatoma cells are activated by extracellular nucleotides. *Am. J. Physiol.* **261**, C1018–C1024.

Bedard, E., Sigurdson, W. J., and Morris, C. E. (1992). Channels activated by stretch in neurons of a helix snail. *Can. J. Physiol.* **70**, 207–213.

Bourque, C. W., and Oliet, S. H. R. (1995). Mechanosensitive ion channels and osmoreception in magnocellular neurosecretory neurons. *In* Neurohypophysis: recent progress in vasopressin and oxytocin research. (T. Saito, K. Kurokawa, and S. Yoshida, Eds.) Elsevier Science, Amsterdam. pp. 205–213.

Bowman, C. L., Ding, J. P., Sachs, F., and Sokabe, M. (1992). Mechanotransducing ion channels in astrocytes. *Brain Res.* **584**, 272–286.

Bregestovski, P., Medina, I., and Goyda, E. (1992). Regulation of potassium conductance in the cellular membrane at early embryogenesis. *J. Physiol. Paris* **86**, 109–115.

Casado, M., and Ascher, P. (1998). Opposite modulation of NMDA receptors by lysophospholipids and arachidonic acid: common features with mechanosensitivity. *J. Physiol. (Lond.)* **513**, 317–330.

Cemerikic, D., and Sackin, H. (1993). Substrate activation of mechanosensitive, whole-cell currents in renal proximal tubule. *Am. J. Physiol.* **264**, F697–F714.

Chang, G., Spencer, R. H., Lee, A. T., Barclay, M. T., and Rees, D. C. (1998). Structure of the MscL homolog from *Mycobacterium tuberculosis*: a gated mechanosensitive ion channel. *Science* **282**, 2220–2226.

Corey, D. P, and Garcia-Anoveros, J. (1996). Mechanosensation and the DEG/ENac ion channels. *Science* **273**, 323–324.

Corey, D. P., and Howard, J. (1994). Models for ion channel gating with compliant states. *Biophys. J.* **66**, 1254–1257.

Dai, J., and Sheetz, M. P. (1995). Mechanical properties of neuronal growth cone membranes studied by tether formation with laser optical tweezers. *Biophys. J.* **68**, 988–996.

Dai, J., Sheetz, M. P., Wan, X., and Morris, C. E. (1998). Membrane tension in swelling and shrinking molluscan neurons. *J. Neurosci.* **18**, 6681–6692.

Davidson, R. M., Tatakis, D. W., and Auerbach, A. L. (1990). Multiple forms of mechanosensitive ion channels in osteoblast-like cells. *Pflugers Arch.* **416**, 646–651.

Ding, J. P., Salvi, R. J., and Sachs, F. (1992). Stretch-activated ion channels in guinea pig outer hair cells. *Hearing Res.* **56**, 19–28.

Dopico, A. M., Kirber, M. T., Singer, J. J., and Walsh, J. V., Jr. (1994). Membrane stretch directly activates large conductance Ca^{2+}-activated K+ channels in mesenteric artery smooth muscle cells. *Am. J. Hyperten.* **7**, 82–89.

Duncan, R. L. and Hruska, K. A. (1994). Chronic, intermittent loading alters mechanosensitive channel characteristics in osteoblast-like cells. *Am. J. Physiol.* **267**, F909–F916.

Duncan, R. L., Hruska, K. A., and Misler, S. (1992). Parathyroid hormone activation of stretch-activated cation channels in osteosarcoma cells (UMR-106.1). *FEBS Lett.* **307**, 219–223.

Erxleben, C. (1989). Stretch-activated current through single ion channels in the abdominal stretch receptor organ of the crayfish. *J. Gen. Physiol.* **94**, 1071–1083.

Franco, A., Jr., and Lansman, J. B. (1990). Calcium entry through stretch-inactivated ion channels in *mdx* myotubes. *Nature* **344**, 670–673.

French, A. S. (1992). Mechanotransduction. *Annu. Rev. Physiol.* **54**, 135–152.

Glogauer, M., Arora, P., Chou, D., Janmey, P. A., Downey, G. P., and McCulloch, C. A. (1998). The role of actin-binding protein 280 in integrin-dependent mechanoprotection. *J. Biol. Chem.* **273**, 1689–1698.

Guharay, F., and Sachs, F. (1984). Stretch-activated single ion channel currents in tissue-cultured embryonic chick skeletal muscle. *J. Physiol.* **352**, 685–701.

Gustin, M. C. (1992). Mechanosensitive ion channels in yeast. Mechanisms of activation and adaptation. *Adv. Compar. Environ. Physiol.* **10**, 19–38.

Gustin, M. C., Sachs, F., Sigurdson, W., Ruknudin, A., Bowman, C., Morris, C. E., and Horn, R. (1991). Single-channel mechanosensitive currents (Refereed Commentary and Reply). *Science* **253**, 801–802.

Hamill, O. P., and McBride, D. W., Jr. (1992). Rapid adaptation of single mechanosensitive channels in *Xenopus* oocytes. *Proc. Natl. Acad. Sci. USA* **89**, 7462–7466.

Hamill, O. P., and McBride, D. W., Jr. (1995). Mechanoreceptive membrane channels. *Am. Scientist* **83**, 30–37.

Hamill, O. P., and McBride, D. W., Jr. (1996). The pharmacology of mechanogated membrane ion channels. *Pharmacol. Rev.* **48**, 231–252.

Hamill, O. P., and McBride, D. W., Jr. (1997). Induced membrane hypo/hyper-mechanosensitivity: a limitation of patch-clamp recording. *Annu. Rev. Physiol.* **59**, 621–631.

Hase, C. C., Le Dain, A. C., and Martinac, B. (1995). Purification and functional reconstitution of the recombinant large mechanosensitive ion channel (MscL) of *Escherichia coli*. *J. Bio. Chem.* **270**, 18 329–18 334.

Heidemann, S. R., and Buxbaum, R. E. (1994). Mechanical tension as a regulator of axonal development. *Neurotoxicol.* **15**, 65–108.

Hisada, T. R. W., Ordway, R. W., Kirber, M. T., Singer, J. J., and Walsh, J. V., Jr. (1991). Hyperpolarization-activated cationic channels in smooth muscle cells are stretch-sensitive. *Pflügers Arch.* **417**, 493–499.

Hong, K., and Driscoll, M. (1994). A transmembrane domain of the putative channel subunit MEC-4 influences mechanotransduction and neurodegeneration in *C. elegans*. *Nature* **367**, 470–473.

Huang, M., Gu, G., Ferguson, E. L., and Chalfie, M. (1995). A stomatin-like protein necessary for mechanosensation in *C. elegans*. *Nature* **378**, 292–295.

Hudspeth, A. J. (1997). How hearing happens. *Neuron* **19**, 947–950.

Ji, S., John, S. A., Lu, Y., and Weiss, J. N. (1998). Mechanosensitivity of the cardiac muscarinic potassium channel. A novel property conferred by Kir3.4 subunit. *J. Biol. Chem.* **273**, 1324–1328.

Kell, A., and Glaser, R. W. (1993). On the mechanical and dynamic properties of plant cell membranes: their role in growth, direct gene transfer and protoplast fusion. *J. Theor. Biol.* **160**, 41–62.

Kim, D. (1992) A mechanosensitive K^+ channel in heart cells. Activation by arachidonic acid. *J. Gen. Physiol.* **100**, 1021–1040.

Kim, E., Niethammer, M., Rothschild, A., Jan, Y. N., and Sheng, M. (1995). Clustering of *Shaker*-type K^+ channels by interaction with a family of membrane-associated guanylate kinases. *Nature* **378**, 85–88.

Kirber, M. T., Ordway, R. W., Clapp, L. H., Walsh, J. V., Jr., and Singer, J. J. (1992). Both membrane stretch and fatty acids directly activate large conductance Ca^{2+}-activated K^+ channels in vascular smooth muscle cells. *FEBS Letts.* **297**, 24–28.

Lambert, S., and Bennett, V. (1993). From anemia to cerebellar dysfunction. A review of the ankyrin gene family. *Eur. J. Biochem.* **211**, 1–6.

Lecar, H., and Morris, C. E. (1993). Biophysics of mechanotransduction. *In* "Mechanoreception by the Vascular Wall" (G. Rubanyi, Ed.), Futura Pub. Co. Inc., Mount Kisco, NY. pp. 1–11.

Lee, J., Ishihara, A., Oxford, G., Johnson, B., and Jacobson, K. (1999). Regulation of cell movement is mediated by stretch-activated calcium channels. *Nature* **400**, 382–386.

Levina, N. N., Lew, R. R., and Heath, I. B. (1994). Cytoskeletal regulation of ion channel distribution in the tip-growing organism of *Saprolegnia ferax*. *J. Cell Sci.* **107**, 127–134.

Lewis, R. S., Ross, P. E., and Cahalan, M. D. (1993). Chloride channels activated by osmotic stress in T lymphocytes. *J. Gen. Physiol.* **101**, 801–826.

Liu, J., Schrank, B., and Waterston, R. H. (1996). Interaction between a putative mechanosensory membrane channel and a collagen. *Science* **273**, 361–364.

Maingret, F., Fosset, M., Lesage, F., Lazdunski, M., and Honore, E. (1999). TRAAK is a mammalian neuronal mechano-gated K^+ channel. *J. Biol. Chem.* **274**, 1381–1387.

Marchenko, S. M., and Sage, S. O. (1997). A novel mechanosensitive cationic channel from the endothelium of rat aorta. *J. Physiol. (Lond.)* **498**, 419–425.

Martinac, B. (1993). Mechanosensitive ion channels: biophysics and physiology. *In* "Thermodynamics of Membrane Receptors and Channels" (M. B. Jackson, Ed.), pp. 327–352. CRC Press.

McBride, D. W., Jr., and Hamill, O. P. (1992). Pressure clamp: a method for rapid step perturbation of mechanosensitive channels. *Pflügers Arch.* **421**, 606–612.

Medina, I. R., and Bregestovski, P. D. (1991). Sensitivity of stretch-activated K^+ channels changes during cell-cleavage cycle and may be regulated by cAMP-dependent protein kinase. *Proc. Roy. Soc. Lond. B* **245**, 159–164.

Mills, J. W., Schwiebert, E. M., and Stanton, B. A. (1994). Evidence for the role of actin filaments in regulating cell swelling. *J. Exp. Zool.* **268**, 111–120.

Morris, C. E. (1990) Mechanosensitive ion channels. *J. Memb. Biol.* **113**, 93–107.

Morris, C. E., (1992). Are stretch-sensitive channels in molluscan cells and elsewhere physiological mechanotransducers? *Experimenta* **48**, 852–858.

Morris, C. E., and Horn, R. (1991). Failure to elicit neuronal macroscopic mechanosensitive currents anticipated by single-channel studies. *Science* **251**, 1246–1249.

Morris, C. E., and Sigurdson, W. J. (1989). Stretch-inactivated ion channels coexist with stretch-activated ion channels. *Science* **243**, 807–809.

Nakamura, F., and Strittmatter, S. M. (1996). P2Y1 purinergic receptors in sensory neurons: contribution to touch-induced impulse generation. *Proc. Natl. Acad. Sci. USA* **93**, 10465–10470.

Oliet, S. H. R., and Bourque, C. W. (1993). Mechanosensitive channels transduce osmosensitivity in supraoptic neurons. *Nature* **364**, 341–343.

Oliet, S. H. R., and Bourque, C. W. (1996). Gadolinium uncouples mechanical detection and osmoreceptor potential in supraoptic neurons. *Neuron* **16**, 175–181.

Ordway, R. W., Petrou, S., Kirber, M. T., Walsh, J. V. Jr., and Singer, J. J. (1995). Stretch activation of a toad smooth muscle K^+ channel may be mediated be fatty acids. *J. Physiol.* **484**, 331–337.

Paoletti, P., and Ascher, P. (1994). Mechanosensitivity of NMDA receptors in cultured mouse central neurons. *Neuron* **13**, 645–655.

Patel, A. J., Honore, E., Lesage, F., Fink, M., Romey, G., and Lazdunski, M. (1999). Inhalational anesthetics activate two-pore-domain background K^+ channels. *Nat. Neurosci.* **2**, 422–2426.

Patel, A. J., Honore, E., Maingret, F., Lesage, F., Fink, M., Duprat, F., and Lazdunski, M. (1998). A mammalian two pore domain mechano-gated S-like K^+ channel. *EMBO. J.* **17**, 4283–4290.

Pleusamran, A., and Kim, D. (1995). Membrane stretch augments the cardiac muscarinic K^+ channel activity. *J. Membr. Biol.* **148**, 287–297.

Rotin, D., Bar-Sagi, D., O'Brodovich, H., Merilainen, J., Lehto, V. P., Canessa, C. M., Rossier, B. C., and Downey, G. P. (1994). An SH3 binding region in the epithelial Na⁺ channel (árENaC) mediates its localization at the apical membrane. *EMBO. J.* **13,** 4440–4450.

Sachs, F. (1992). Stretch-sensitive ion channels: an update. *In* "Sensory Transduction" (D. Corey, Ed.), pp. 242–260, Rockefeller University Press.

Sachs, F., and Lecar, H. (1991). Stochastic models for mechanical transduction (letter). *Biophys. J.* **59,** 1142–1145.

Sachs, F., and Morris, C. E. (1998). Mechanosensitive ion channels in non-specialized cells. *Rev. Physiol. Biochem. Pharmacol.* **132,** 1–78.

Sachs, F., Morris, C. E., Hamill, O., Chakfe, Y., and Bouraque, C. W. (2000). Does a stretch-inactivated cation channel integrate asmotic and peptidergic signals? (Refereed Commentary and Reply) *Nat. Neurosci.* **3,** 847–848.

Sackin, H. (1995). Mechanosensitive channels. Annu. *Rev. Physiol.* **57,** 333–353

Schwiebert, E. M., Mills, J. W., and Stanton, B. A. (1994). Actin-based cytoskeleton regulates a chloride channel and cell volume in a renal cortical collecting duct cell line. *J. Biol. Chem.* **269,** 7081–7089.

Sheetz, M. P., and Dai, J. (1996). Modulation of membrane dynamics and cell motility by membrane tension. *Trends Cell Biol.* **6,** 85–89.

Small, D. L., and Morris, C. E. (1994). Delayed activation of single mechanosensitive channels in *Lymnaea* neurons. *Am. J. Physiol.* **267,** C598–C606.

Sokabe, M., Sachs, F., and Jing, Z. (1991). Quantitative video microscopy of patch clamped membranes stress, strain, capacitance, and stretch channel activation. *Biophys. J.* **59,** 722–728.

Steffensen, I., Bates, W. R., and Morris, C. E. (1991). Embryogenesis in the presence of blockers of mechanosensitive ion channels. *Dev. Growth Differ.* **5,** 437–442.

Sukharev, S. I., Blount, P., Schroeder, M., and Kung, C. (1996). Multimeric structure of bacterial mechanosensitive channel MscL. *Biophys. J.* **70,** A366.

Sukharev, S. I., Blount, P., Martinac, B., Blattner, F. R., and Kung, C. (1994). A large conductance mechanosensitive channel in *E. coli* encoded by MscL alone. *Nature* **368,** 265–268.

Tabarean, I. V., Juranka, P., and Morris, C. E. (1999). Membrane stretch affects gating modes of a skeletal muscle sodium channel. *Biophys. J.* **77,** 758–774.

Tavernarakis, N., and Driscoll, M. (1997). Molecular modeling of mechanotransduction in the nematode *Caenorhabditis elegans.* *Annu. Rev. Physiol.* **59,** 659–689.

Vandorpe, D. H., and Morris, C. E. (1992). Stretch activation of the *Aplysia* S channel. *J. Memb. Biol.* **127,** 205–214.

Vandorpe, D. H., Small, D. L., Dabrowski, A. R., and Morris, C. E. (1994). FMRFamide and membrane stretch as activators of the *Aplysia* S-channel. *Biophys. J.* **66,** 46–58.

Vanoye, C. G., and Reuss, L. (1999). Stretch-activated single K⁺ channels account for whole-cell currents elicited by swelling. *Proc. Natl. Acad. Sci. USA* **96,** 6511–6516.

Van Wagoner, D. R., and Russo, M. (1992). Whole-cell mechanosensitive K⁺ currents in rat atrial myocytes. *Biophys. J.* **61,** A251.

Verrey, F., Groscurth, P., and Bolliger, U. (1995). Cytoskeletal disruption in A6 kidney cells: impact on endo/exocytosis and NaCl transport regulation by antidiuretic hormone. *J. Memb. Biol.* **145,** 193–204.

Wan, X., Harris, J. A., and Morris, C. E. (1995). Response of neurons to extreme osmomechanical stress. *J. Membr. Biol.* **145,** 21–31.

Wan, X., Juranka, P., and Morris, C. E. (1999). Activation of mechanosensitive currents in traumatized membrane. *Am. J. Physiol.* **276,** C318–C327.

Waxman, S. G., Kocsis, J. D., and Black, J. A. (1994). Pathophysiology of demyelinated axons. *In* "The Axon: Structure, Function and Pathophysiology" (S. G. Waxman, J. D. Kocsis, and P. K. Stys, Eds.) Oxford University Press.

Zhou, X. L., Stumpf, M. A., Hoch, H. C., and Kung, C. (1991). A mechanosensitive channel in whole cells and in membrane patches of the fungus *Uromyces.* *Science* **253,** 1415–1417.

45

Sensory Receptors and Mechanotransduction

Andrew S. French and Päivi H. Torkkeli

I. Introduction

While most living cells can detect a variety of changes in their external physical and chemical environments and respond appropriately, we usually reserve the term **sensory receptor** for those cells that transmit information about such changes to the animal's nervous system. Even this classification must be qualified. For example, the heart and other hollow organs have localized, mainly autonomous nervous systems that allow sensory receptors to produce very local functional changes. Moreover, some of the mechanisms that cells use to detect external events seem to be very similar as we move from unicellular animals to the complex sensory organs of humans. With these caveats in mind, this chapter is restricted to those cells (usually neurons) that transmit sensory information into one or more divisions of the central nervous system.

Several general principles can be applied to all sensory receptors. The concept of **modality** means that each sensory cell transmits information about only one type of environmental stimulus. Photoreceptors detect light but are insensitive to touch, while stretch receptors do not respond to odorant molecules. Within the modalities there are further specializations, so that cold temperature receptors do not respond to hot stimuli, and sweet taste receptors do not respond to bitter substances. Each type of receptor seems to produce a single specific molecular machinery for detecting just one type of stimulus. An exception to this rule is provided by the cutaneous polymodal pain receptors (Perl, 1996) which can be stimulated by mechanical, thermal, or chemical stimuli. However, it is not yet known how this multimodal behavior is produced, or if each sensory ending can detect all the different modalities.

Neurons carrying information into the nervous system form **labeled lines** that signal specific modalities and specific locations. Artificially stimulating a sensory nerve by electrical or mechanical stimulation produces a sensation that reflects the modality and location of the sensory ending, not the actual stimulus. Another general principle is that the intensity of the stimulus is encoded as the amplitude of the

electrical signal passing along the sensory neuron. In most cases, this means the action potentials propagating along the sensory axon, with more action potentials per second representing a stronger stimulus. However, there are several situations where the primary sensory cell does not produce action potentials at all, but a graded change in membrane potential that is conducted decrementally along the cell. Graded potentials can usually be propagated for only very short distances (a few millimeters at most), so such cells are generally small. Well-known examples are the rods and cones of the retina and the hair cells of the inner ear. Some invertebrate sensory axons are sufficiently short that both graded and action potential sensory signals can reach the central nervous system (Pasztor and Bush, 1982).

II. Sensory Transduction

Sensory cells respond to an external stimulus by changing their membrane potential, although several processes may occur before and after this **transduction** step. This change in membrane potential is called a **receptor potential**, and it is a graded potential, increasing and decreasing with the intensity of the stimulus. Nonsensory neurons are insensitive to external stimuli such as mechanical stress, temperature, light, and chemicals (other than synaptic transmitters), unless these reach such intensity as to threaten the cell's normal function or integrity. Therefore, sensory receptors must have specializations to detect external stimuli (Fig. 1). Conceptually, the simplest receptors are the electroreceptors, which use an external electrical current to change their own membrane potential by direct current flow. Although the concept is simple, these cells are highly specialized to maximize their electrical sensitivity. Current flows through resting membranes by the passage of ions through ion channels that are open at rest. The salt taste receptors of the tongue and mouth also rely on direct ion flow through ion channels. In this case the channels are selective for sodium and can be blocked by amiloride, so that placing a small amount of amiloride in

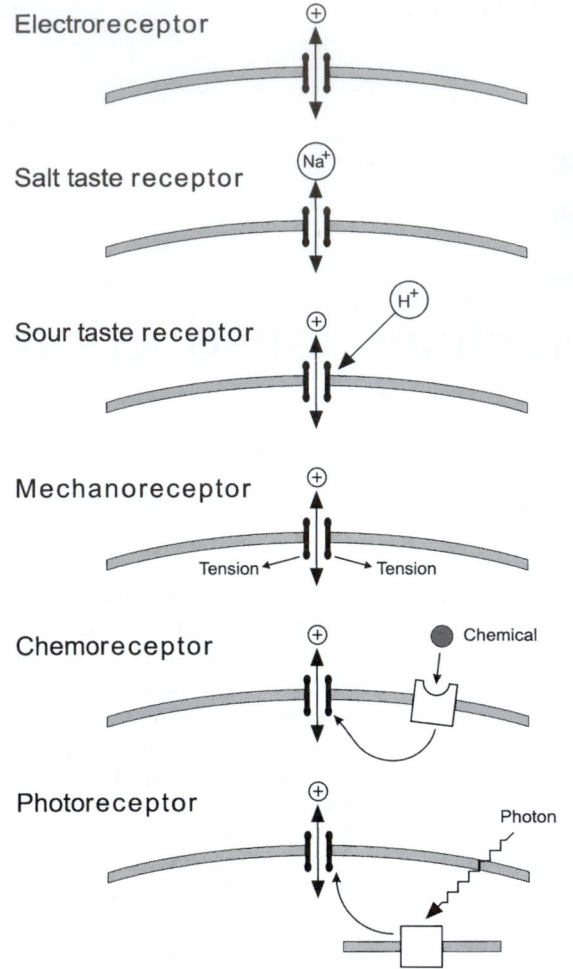

FIGURE 1. Fundamental mechanisms of sensory transduction organized by the level of complexity. In each case the cell has specialized ion channels that produce a receptor current. Modulation of the current is achieved by changing the electrochemical force for the permeant ion in the cases of electroreceptors and salt taste receptors. In other receptors the external stimulus changes the channel open probability, either by acting directly on the channel protein or via a second messenger cascade. Chemoreceptors have receptor proteins in the cell membrane. In vertebrate rods the rhodopsin receptors are in the internal disk membranes, but cones and invertebrate photoreceptors have rhodopsins in the cell membrane.

the mouth removes the ability to taste salt. Sodium diffuses through these channels to depolarize the taste cell membrane. An additional level of complexity is illustrated by sour taste receptors, where potassium-selective ion channels are closed by external hydrogen ions, reducing an outward current and depolarizing the cell.

The fundamental transduction step in mechanoreceptors is not well understood, but is assumed to involve mechanosensitive ion channels. It seems likely that these channels are opened by links to intracellular cytoskeletal components (Sachs, 1997), although this is not yet proven. If true, this is another case where the external stimulus acts directly to open or close an ion channel. Finally, there are cases where the external stimulus acts indirectly through intermediate membrane receptors. In

many chemoreceptors, specialized receptor molecules in the outer cell membrane activate membrane ion channels through a second messenger cascade when they receive an appropriate stimulus molecule. In the case of rod photoreceptors, light activates specialized receptor molecules in internal cell membranes, again producing a second messenger cascade that eventually closes ion channels in the external membrane. It is not known which types of sensory receptor mechanisms evolved first, but it is clear that several, if not all, of these processes have close parallels in other nonsensory tissues, such as the sodium channels of epithelia, the mechanically activated channels of muscle and many other tissues, and the second messenger cascades of synapses and secretory cells. It seems likely that the evolution of different sense modalities has taken advantage of preexisting cellular mechanisms, and vice-versa.

III. Sensory Adaptation

Sensory receptors often have to deal with very wide ranges of stimulus amplitudes. Therefore, they need high sensitivity to detect weak stimuli, plus the ability to reduce their sensitivity if the stimulus is strong. In some cases the maximum sensitivity approaches the limits imposed by physics or chemistry. Human rod photoreceptors and some arthropod photoreceptors can detect single photons arriving in the eye (Fein and Szuts, 1982), which is clearly the physical limit. The human ear can detect air movements of about 0.01 nm, close to the diameter of a hydrogen atom (Hudspeth, 1989). In such cases the sensory cells must amplify the initial signal considerably, which at least partly explains the complex morphology of rod photoreceptors and the human cochlea.

The reduction in sensitivity following an increase in stimulus is called **adaptation.** It may be seen as a decrease in receptor potential with time during a constant stimulus, or as an increase in the strength of stimulus required to produce a constant response. All sensory receptors adapt to some extent but there is a very wide range of adaptation speeds and amounts. At one extreme are receptors such as Pacinian corpuscles (Loewenstein and Mendelson, 1965) and spider slit sensilla (Seyfarth and French, 1994), which fire only one or two action potentials with even a strong, continuous stimulus. At the other extreme are Ruffini endings (Malinovsky, 1996), which continue to fire steadily for long periods. The type of adaptation is always appropriate to the function of the receptor. Muscle and joint receptors that signal limb position would not be useful if they adapted to silence in a few seconds, because the sense of limb position, or **kinesthesia**, would vanish if one did not keep moving. On the other hand, photoreceptors must function under a wide range of light intensities and it is essential that they can adjust their sensitivity to the ambient light level. The rapid adaptation of Pacinian corpuscles makes them ideally suited for detecting vibration, since a rapidly changing stimulus will repeatedly stimulate the receptor, while a steady stimulus will produce little response.

The mechanisms of adaptation vary widely and may occur at different stages of the process. Some involve components outside the sensory cell, such as the mechanical creep of the muscle in muscle spindles or the movements of screening pigments in insect eyes. Others may involve chemical signals within and between cells, as found in photoreceptors and olfactory receptors. Mechanical adaptation by the capsule of the Pacinian corpuscle is very well known because of the pioneering work of Loewenstein and Mendelson (1965), who described the dramatic reduction in receptor potential adaptation that can be achieved by removing most of the capsule (see Fig. 2). Unfortunately, many descriptions stop at this point, leaving the impression that mechanical adaptation dominates Pacinian corpuscle behavior, but Loewenstein and Mendelson showed that even after decapsulation the receptor will only fire one or two action potentials in response to a prolonged step stimulus (Fig. 2). They described this as electrical adaptation, and it now seems probable that many receptors use voltage- or calcium-activated ion channels in their membranes to produce electrical adaptation by raising the threshold for action potential production. This has been clearly established in several arthropod and vertebrate mechanoreceptors and is likely to occur in many other types of receptors that use action potentials to encode the sensory signal (French and Torkkeli, 1994).

IV. Information Transmission by Sensory Receptors

Nervous systems and individual nerve cells deal with information. Cells with long axons transmit information over considerable distances, and we assume that all nerve cells process information to some extent. Unfortunately, it is usually difficult to decide what the nature of the information is, and how it is being processed, except in vague and general terms. Sensory cells have been particularly important in trying to understand quantitatively how information is encoded and transmitted by nerve cells because it is usually much easier to define the input signal of a receptor neuron than other nerve cells. For example, the intensity and wavelength of light entering a photoreceptor or the length of a muscle spindle receptor can be accurately controlled while the out-

put of the receptor in terms of receptor potential or action potentials is recorded.

Quantitative investigations of information flowing through sensory neurons have been based on ideas developed by engineers for dealing with artificial communication systems, such as telephone or television signals. The recent widespread use of digital communications devices, like telephone modems, has made the concept of an **information transmission rate** (usually in bits per second, or bps) familiar to most of us. If we know the nature of the information being transmitted we can measure the actual rate of transmission, but even if nothing is known about the kind of information being transmitted, it is possible to calculate the theoretical maximum rate that could be achieved by an optimum encoding scheme. This is called the **information capacity** of system, and it is closely related to the inherent noise in the system. The basic idea is very simple. Suppose that one person (the sender) wants to send a message to another (the receiver) by changing the voltage at one end of a piece of copper wire, so that the receiver can measure the voltage change to read the message. If the copper wire were completely free of noise, the sender and receiver could agree to use voltages of any value. For example, the voltages 1 V, 2 V, and 3 V could represent the letters A, B, and C. Since any size of voltage can be used, there is no limit to the number of different signals (voltages) sent and received in any given time, and the information capacity is infinite. However, if the copper wire actually has random voltage noise of about 1 V amplitude, a signal of 1 V is difficult or impossible to read, while one of 10 V may still be detected. The higher the noise level, the more difficult it is to send and receive information. If the receiver can take time to average the output from the wire, it may still be possible to observe the 1 V signal, but this reduces how much information can be received per unit time. The temporal properties of the noise are also important, because slowly changing noise may not interfere with a quickly changing signal, and vice-versa. All these arguments can be applied, with suitable dimension changes, to other information transmission systems, such as optical fibers, satellite broadcasts, or sensory neurons. Note that when a noisy neuron fires action potentials the noise appears primarily as variability or randomness in the timing of the action potentials.

These ideas were formulated into a standard expression for calculating the information capacity from the signal-to-noise ratio by Shannon and Weaver (1949) and this approach has been used in most estimates of receptor cell information capacity. Although relatively few measurements have been made, it is clear that cells using action potentials to transmit information have lower information capacities than those that do not. Spiking mechanoreceptors in cricket cercal mechanoreceptors and spider slit sensilla had capacities of 300 bps and 200 bps respectively, compared to values of 1650 bps for fly photoreceptors and 2240 bps for the receptor potential in spider slit sensilla before encoding (Juusola and French, 1997). However, it was possible to transmit information at rates up to 500 bps for short periods using a defined encoding scheme in a cockroach mechanoreceptor (French and Torkkeli, 1998), so we must be cautious about these estimates until we understand the actual encoding schemes that neurons use.

These findings support the concept that nerve cells use action potentials to transmit information faithfully over long

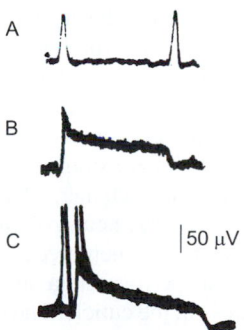

FIGURE 2. Adaptation in the mammalian Pacinian corpuscle occurs in two stages. (A) Stimulating with a mechanical step (bottom) causes rapidly adapting receptor potential responses at the start and end of the step in a normal receptor. (B) Removing most of the lamellae surrounding the sensory ending (decapsulation) eliminates most of the adaptation to reveal a slowly adapting receptor potential. (C) Suprathreshold stimulation of a decapsulated receptor still produces only two action potentials, showing that there is rapid adaptation in the conversion of receptor potential to action potentials. (Redrawn from Loewenstein and Mendelson, 1965.)

distances, but there is a significant cost in information capacity. This probably explains why some sensory systems have elaborate neural structures located peripherally, close to the primary sensory neurons. A familiar example is the vertebrate retina, with two layers of nonspiking cells and complex synaptic interactions before the spiking ganglion cells. This arrangement presumably allows important information processing to take place on the high-capacity input from the photoreceptors before it is encoded into the lower capacity ganglion cell axons.

V. Mechanoreceptors

A. Vertebrate Receptors

Vertebrate mechanoreceptors have such a wide variety of shapes and functions that classification is difficult. Many receptors bear the names of their original discoverers or rediscoverers, such as Meissner corpuscles and Ruffini endings. A variety of classification schemes have been used, based on features such as the morphology of the sensory ending and its associated tissues, rates of adaptation, and location in the body. The hair cells of the vertebrate auditory and vestibular systems have been studied extensively, and are described in Chapter (46). Vertebrate somatic mechanoreceptors have recently been classified according to the embryonic origin of the tissues associated with the sensory ending, and the complexity of the sensory ending itself (Malinovsky, 1996), as shown in Fig. 3. Type I receptors are associated with tissues of mesodermal origin (connective tissues and muscle) and include the Ruffini endings of the joints and skin, the Golgi tendon organs, and the muscle spindles. Type II receptors are transitional between Type I and Type III, and are associated with tissues of endodermal or ectodermal origin. They include the Merkel endings, where each sensory ending is closely apposed to a specialized Merkel cell. Type III receptors are clearly associated with tissues of ectodermal origin, and comprise a wide range of receptors including the bulbous endings found in hair follicles, the Meissner corpuscles, and the Pacinian corpuscles.

This classification broadly accompanies a functional shift from slowly adapting receptors (such as Ruffini endings) in the Type I group to very rapidly adapting receptors in the Type III groups (such as Pacinian corpuscles). Phylogenetically, Type I and Type II receptors are found in all classes of vertebrates from fish to primates, while Type III receptors are unknown in fish, start to occur in amphibia, and become increasingly common in higher vertebrates. A more detailed description of this classification and a history of vertebrate mechanoreceptor classification is provided by Malinovsky (1996).

B. Arthropod Receptors

Arthropods (insects, spiders, and crustaceans) have provided some important preparations for the investigation of mechanoreceptor function. Their receptors can be divided into two major groups, cuticular and multipolar, sometimes called Type I and Type II mechanoreceptors (McIver, 1985). **Cuticular receptors**, as the name implies, are generally associated with the arthropod cuticle and have their cell bodies in the periphery, close to the sensory ending. This arrangement is very different than most vertebrate and other invertebrate mechanoreceptors, which have their cell bodies in the central nervous system, distant from the sensory ending, and it allows recordings to be made close to the site of mechanotransduction. In addition, many arthropod receptor cells are relatively large, allowing penetration by microelectrodes, which is impossible in most vertebrate receptors.

There are three major groups of cuticular receptors (Fig. 4). **Hairlike** receptors are found all over the outer surfaces of most arthropods in a variety of shapes and sizes from long, thin hairs to short pegs. The hair is supported by flexible cuticle within a socket and moves relative to the skin. In insects, a single mechanoreceptor neuron is closely apposed to the base and its sensory ending contains microtubules that end in a dense tubular body. It is assumed that movement of the hair compresses the ending, with the tubular body perhaps adding a rigid structure that the compression can work against. Crustacean and spider hairs are similar, but with two or three mechanosensory neurons in each hair. All arthropod hair types can contain other sensory neurons in addition to the mechanoreceptors, such as chemoreceptors in taste hairs. **Campaniform** (bell-shaped) sensilla are found in insects, where they detect stress in the cuticle. The stress again leads to compression of a dendritic tip containing a tubular body by squeezing of the bell as it is pushed downwards. Spiders have analogous stress-detecting receptors called slit sensilla, where the neurons are located in cuticular slits. **Chordotonal** receptors are generally found further beneath the integument, although they can be connected to the integument by attachment structures. They serve a variety of functions, including hearing and joint movement detection. They generally lack tubular bodies but have dense scolopale structures surrounding the dendrite and can have multiple mechanosensory neurons.

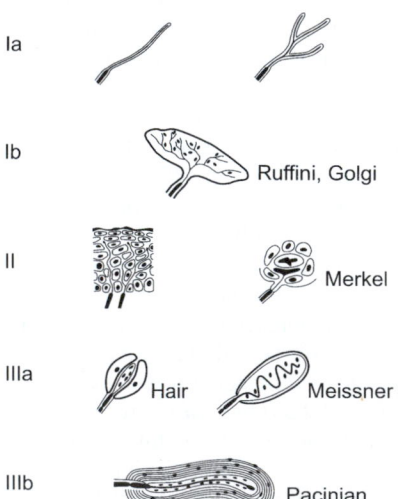

FIGURE 3.　Classification of vertebrate mechanoreceptors by embryonic origin of associated tissues and complexity of ending. Type I is associated with tissues of mesodermal origin, Type II with endodermal or epidermal, and Type III with epidermal. Subtypes (Ia, Ib, etc.) indicate morphological complexity. (Redrawn from Malinovsky, 1996.)

Hair   Campaniform   Chordotonal

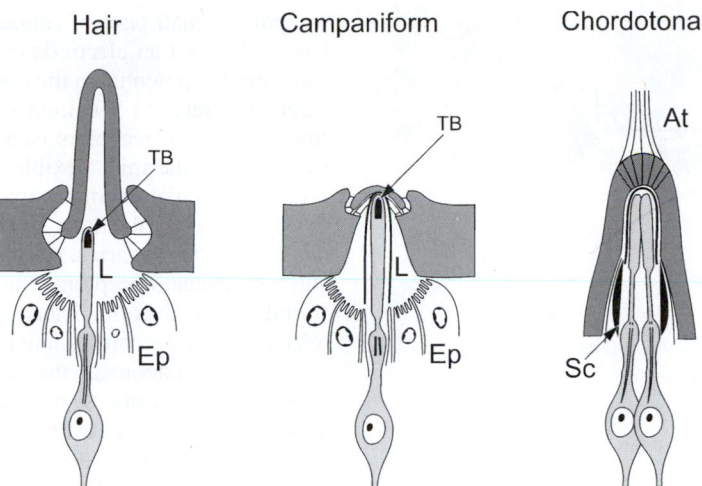

FIGURE 4. Three common types of arthropod cuticular mechanoreceptors. TB, tubular body in dendrite, formed from microtubules embedded in electron-dense material. Ep, epithelial layer, connected by tight junctions to create a lymph space (L). At, attachment from cap of chordotonal sensillum to another structure, such as a joint. Sc, scolopale surrounding a bundle of sensory dendrites.

C. Muscle Mechanoreceptors

One of the best known arthropod mechanoreceptors is the crayfish stretch receptor, which has some interesting similarities to the vertebrate muscle spindle (Fig. 5). These receptors are found on many muscles in crustacea and are multipolar or Type II arthropod mechanoreceptors. The large cell bodies are again located in the periphery, but the sensory endings consist of finely divided dendrites that cover part of the muscle surface. Unlike the muscle spindle, crayfish stretch receptors are located on the main, force-producing muscle, and so are directly excited by passive stretch of the muscle or activation of motoneurons. However, crayfish stretch receptors also have inhibitory innervation that can inhibit the sensory neuron directly, as well as the muscle.

Vertebrate muscle spindles are prominent and numerous mechanoreceptors, believed to be crucially involved in the control of muscle activation. However, they also make a major contribution to the sense of body position and movement, or kinesthesia, because appropriate artificial stimulation of muscle spindles gives sensations that the limbs are moving or in incorrect positions (Matthews, 1981). Mammalian muscle spindles are highly specialized receptors, located on their own exclusively activated intrafusal muscle fibers within the muscle, and are innervated by two types of sensory endings with different adaptation properties.

VI. Experimental Mechanoreceptor Preparations

Although many different preparations have been developed to investigate mechanoreception, only three basic methods have been used to measure the receptor current or receptor potential (Fig. 6). The **decremental conduction** method relies on insulating the sensory axon from the extracellular fluid as close as possible to the sensory ending, usually with a nonconducting substance such as paraffin wax or petroleum jelly,

Muscle spindle

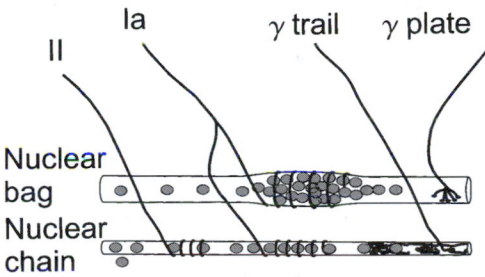

Crayfish stretch receptor

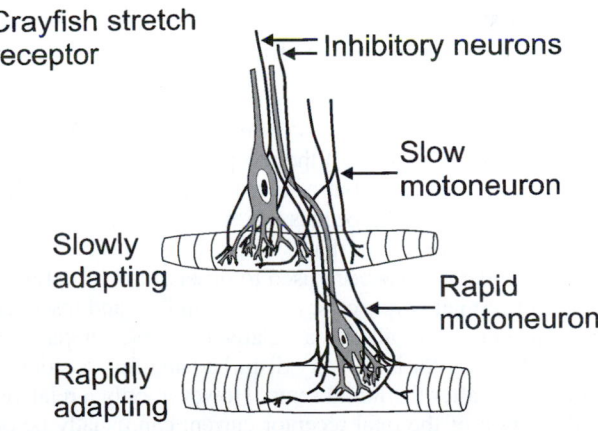

FIGURE 5. Vertebrate and invertebrate muscle mechanoreceptors. Mammalian muscle spindles contain two types of specialized muscle fibers with their own motor innervation via two types of γ-motoneurons that are activated separately from the main muscle fibers. There are also two types of sensory endings with different adaptation characteristics. Crustacean stretch receptors are multipolar mechanoreceptors with many fine dendrites embedded in muscle fibers. There are two types of sensory neurons with different adaptation properties and inhibitory neurons that synapse on the sensory neurons as well as the muscle fibers. (Redrawn from Matthews, 1981; Swerup and Rydqvist, 1992).

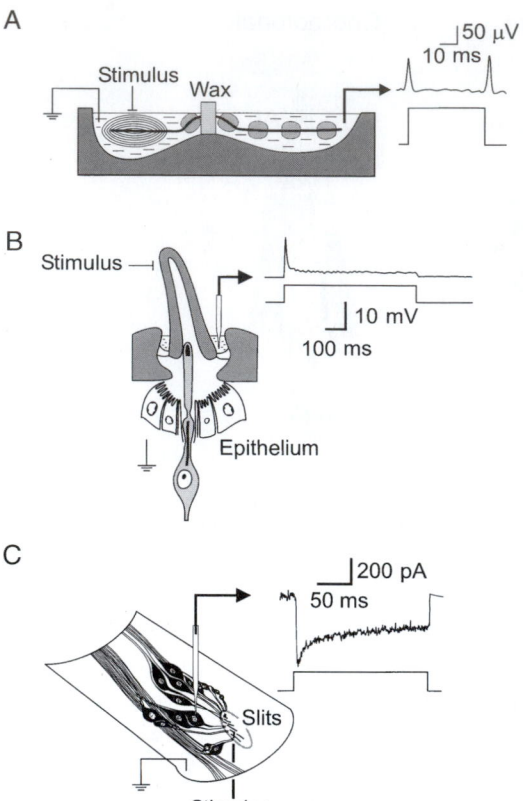

FIGURE 6. Three methods that have been used widely to observe the receptor current and potential in mechanoreceptors. (A) Decremental conduction measures the current flowing along the sensory axon from a Pacinian corpuscle by insulating the axon as close as possible to the receptor to prevent the current from reaching ground, except via the measuring bath. (B) Transepithelial experiments in an arthropod hair measure the voltage as close as possible to the lymph space surrounding the sensory ending, and rely on the high resistance of the surrounding epithelium. (C) Intracellular recording from a spider slit sensillum neuron measures the actual receptor potential during stimulation. Current measurement requires reliable voltage-clamp of the sensory ending.

before leading it to a separate conducting solution. As the receptor is stimulated, some of the receptor current flows decrementally along the axon but cannot follow its normal path to the extracellular solution because of the insulation. It can only return through the second bath, where it can easily be measured. This method has been used to observe the receptor potential in Pacinian corpuscles, muscle spindles, and insect cuticular receptors. It allows a relatively stable preparation, because the receptor is not very disturbed and fine positioning is not needed after the initial setup. However, only a relatively small fraction of the total receptor current can usually be observed, and the decremental conduction includes a substantial selective loss of high-frequency components through the membrane capacitance to ground, so that the observed receptor potential is strongly attenuated and filtered. It is generally difficult to estimate the amplitude and waveform of the original receptor potential with any accuracy.

Arthropod cuticular receptors have the sensory neurons so close to the surface that it is possible to observe part of the current flowing through the neuron from the exterior. For a hair

receptor, a small pool of conducting solution is placed in the hair socket and an electrode of some sort (glass, wire, wick) measures the potential in the pool. These **transepithelial** measurements rely on the high-resistance epithelium that surrounds cuticular receptors (see below) and the relatively low resistance of the thin, flexible socket. A similar technique is used for cuticular chemoreceptors, where it is common to cut the hair and place the tip of an electrode over the cut end to obtain even better electrical contact. However, this is more difficult for mechanoreceptors, where the hair must be moved to stimulate transduction. Transepithelial measurements allow a relatively stable recording situation, but the high-resistance access pathway attenuates the signal and emphasizes high frequencies, which can pass more easily through the cuticle. Here again, it is difficult to estimate the original receptor potential amplitude and time course with accuracy.

Intracellular recording close to the site of sensory transduction is clearly the technique of choice if it can be accomplished. However, it requires stable penetration of a small neuron while the sensory ending is moved nearby, and has only been successfully achieved in a few preparations. The real goal is **voltage-clamp** of the sensory receptor membrane, since this allows the voltage across the ion channels to be controlled while the current flowing through them is recorded. Crustacean stretch receptors are relatively easy to penetrate, even with two microelectrodes, and have been used for many pioneering studies of the receptor currents and other currents that contribute to the electrical properties of the neurons during sensory transduction (Swerup and Rydqvist, 1992). However, the receptor currents probably originate in the very fine endings that are embedded in the muscle (see Fig. 5) so that accurate voltage-clamp of the receptor current is difficult. Voltage-clamp of arthropod cuticular receptors was first performed in the cockroach tactile spine (Torkkeli and French, 1994). A more successful cuticular preparation is that of spider slit sensilla, where it is possible to clamp the receptor ending close to the sensory ending while stimulating the slits (Juusola and French, 1995). The search for improved preparations continues, and recent development of the first successful tissue-cultured mechanoreceptor system offers the promise that we may eventually be able to study single receptor cells *in vitro* as they perform mechanotransduction and sensory encoding (Torkkeli and French, 1999).

VII. Steps in Mechanoreception

Mechanoreception can be viewed as a three-stage process comprising **coupling**, **transduction**, and **encoding** (Eyzaguirre and Kuffler, 1955; Loewenstein, 1959). The stimulus is first mechanically coupled from its origin to the membrane of the sensory cell where it is transduced into a receptor potential, and if the cell is excitable this is encoded into action potentials, usually for transmission along a sensory axon.

A. Coupling

Living cells are not normally exposed to the outside surface of an animal, so most, if not all, external mechanoreceptors have some kind of tissue between them and the source of the mechanical input. In some cases these surrounding tissues are very elaborate and obviously designed to modify the input signal by affecting its amplitude and possibly its dynamic properties. Internal mechanoreceptors also have a variety of surrounding tissues with presumably similar functions. The morphology of these structures varies so strongly between different mechanoreceptors that it is widely used for classification, as described above. The Pacinian corpuscle is probably the most illustrated vertebrate mechanoreceptor, with its complex lamellar structure, but it represents only one extreme of a wide range of vertebrate mechanoreceptor structures (Malinovsky, 1996). Invertebrate mechanoreceptors also display many morphological forms, including the well-known hair receptors that cover the outside surfaces of insects and spiders, and the stretch receptors of crayfish and lobsters. Although we generally assume that these elaborate extracellular structures modify the spatial and temporal sensitivities of the receptors, there is relatively little quantitative information about these functions. An additional problem is to separate the effects of external structures from the mechanically activated ion channels themselves, since we do not yet understand the temporal properties of the channels, or how they are linked to other mechanical structures.

In most cases the external structures attenuate the stimulus, so that the displacement of the receptor cell membrane is much smaller than the original movement. For example, hairs on the human skin can be moved by several millimeters, but the resulting movements of the follicle receptors are only a few micrometers. Estimates of this attenuation have been made in some cases. For the Pacinian corpuscle, it has been suggested that the pressure reaching the inner capsule is attenuated by about two orders of magnitude (100) compared to the pressure on the outer lamellae (Bell *et al.*, 1994), but this force may already be greatly reduced compared to the initial stimulus at the skin. An attenuation of about 100 was also reported for insect hairs, based on the morphology of the sensory structure and the location of the moving elements (French and Sanders, 1979). Estimates of attenuation during coupling also allow us to estimate the **threshold,** or minimum amplitude of movement at the cell membrane that leads to sensation. For Pacinian corpuscles, movements of 1–10 nm at the capsule are probably adequate to produce action potentials at the optimum vibration frequency of about 250 Hz, and sinusoidal movements of about 0.3 nm can stimulate auditory hair cells (Hudspeth, 1989). For insect cuticular hair receptors, threshold estimates of 4 nm and 3 nm were obtained for bee (Thurm, 1965) and cockroach (French and Sanders, 1979).

The few attempts that have been made to quantify the mechanical effects of coupling over a range of input movements indicate that it can have significant temporal effects and be substantially nonlinear. The viscoelastic properties of Pacinian corpuscles are assumed to be responsible for most of the time dependence of the receptor current, but have not yet been quantified. The mechanical properties of crayfish stretch receptors have been studied very thoroughly (Swerup and Rydqvist, 1992) and require nonlinear springs plus at least one dashpot (or Voigt element) to account for the passive properties of the receptor without any muscular activity. No mechanical description is available that includes active contraction. Muscle spindles also have strongly time-dependent mechanical properties, with probably more com-

plex dynamic behavior during γ-motoneuron activation (Matthews, 1981).

B. Transduction

We assume that transduction in mechanoreceptors involves mechanically activated ion channels in the receptor cell membrane (see Chapter 44), but no single-channel recordings have yet been made that can reliably be linked to transduction in a mechanosensory neuron. Instead, most information about these channels comes from indirect measurements. Single, mechanically activated ion channels have been seen in crayfish stretch receptor neurons (Erxleben, 1989) and in tissue-cultured moth antennal mechanoreceptor neurons (Torkkeli and French, 1999), but we cannot be sure in either case that the channels were located on the fine sensory endings where transduction is thought to occur (Fig. 7). The crayfish channels had maximum conductance to potassium ions (~70 pS) but were also permeable to other monovalent and divalent cations. The moth channels were also permeable to potassium with a conductance of ~40 pS, but their selectivity has not yet been established.

Although we lack definitive single-channel measurements, we have information about some transduction channels, including the ions that flow through them, their ionic conductance, blocking chemicals, and the numbers of such channels in each receptor cell (Fig. 8). In Pacinian corpuscles, muscle spindles, and crayfish stretch receptors, development of a mechanical response depends upon the presence of sodium ions in the external solution, indicating that the channels are permeable to sodium (Loewenstein, 1971; Swerup and Rydqvist, 1992; Ito *et al.*, 1983). However, a variety of other cations can

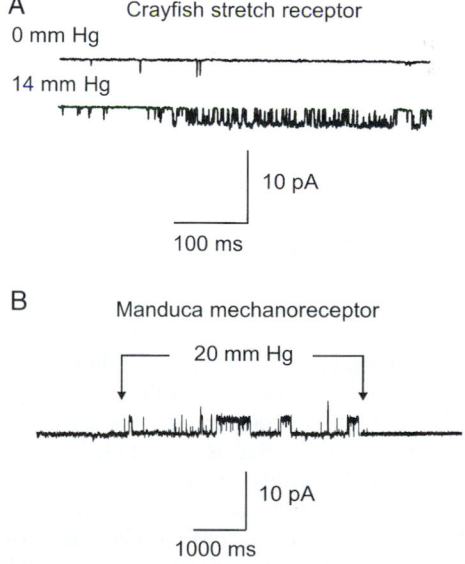

FIGURE 7. Single-channel recordings of mechanically activated ion channels in mechanoreceptor cells. (A) Channel in the membrane of a crayfish stretch receptor opens rarely at zero pressure but becomes very active with a suction of 14 mm Hg (channel opening downwards). (B) A channel in a tissue-cultured moth mechanoreceptor is activated by suction of 20 mm Hg. (Redrawn from Erxleben, 1989; Torkkeli and French, 1999).

A

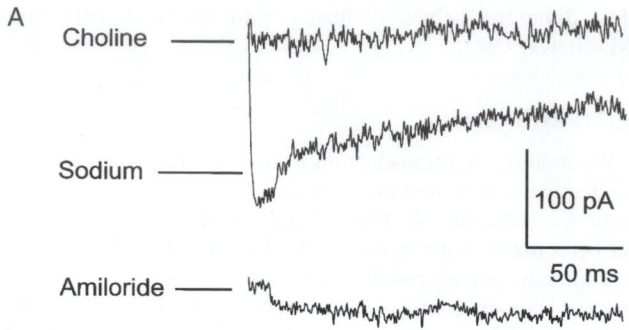

B

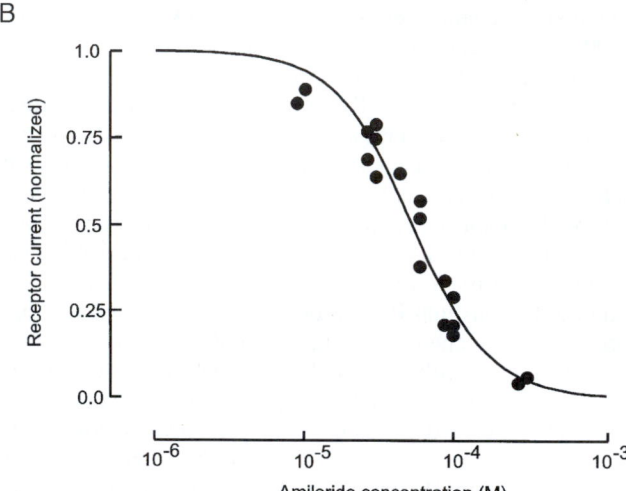

FIGURE 8.   Ionic selectivity and block of receptor current in two mechanoreceptors. (A) Step movements of a spider slit sensillum activate an inward current that is reduced to zero by replacing the sodium ions with impermeable choline, or by adding 1 mm amiloride to the bath. (B) Amiloride blocks the receptor current in mouse cochlear hair cells in a dose-dependent manner. (Redrawn from Höger *et al.,* 1997; Rüsch *et al.,* 1994).

pass through the channels of Pacinian corpuscles (Bell *et al.,* 1994) and crayfish stretch receptors (Rydqvist and Purali, 1993), and there is evidence that muscle spindle channels are also unselective (Ito *et al.,*1983). Spider slit sensilla neurons also pass sodium ions, but in this case the channels are very selective, rejecting most other cations except lithium (Höger *et al.,* 1997). In each of these cases, sodium entry during transduction causes a depolarization that leads to action potentials if the stimulus is strong enough, and each type of neuron has a receptor ending that is surrounded by an elevated concentration of sodium ions (Höger *et al.,* 1997). A different situation exists in vertebrate auditory and vestibular hair cells, where the channels are permeant to a wide range of cations, and the fluid bathing the sensory endings contains a high concentration of potassium ions (Hudspeth, 1989). Similarly, in insect cuticular mechanoreceptors, the sensory ending is surrounded by lymph that is rich in potassium ions and has a positive electrical potential relative to the normal hemolymph (French, 1988). The cases where potassium ions are used to carry the receptor current always seem to rely on separating

two regions of the sensory neurons by embedding them in an epithelial layer with tight junctions so that a potassium concentration gradient, usually combined with an electrical gradient, can drive potassium ions through the cell, depolarizing the membrane; similar arrangements exist for some of the sodium-dependent systems. Cells with nonselective receptor channels may also receive a significant calcium flux during transduction, because the inward calcium electrochemical gradient is usually high. Calcium influxes have been demonstrated in vertebrate hair cells (Ohmori, 1985) but their contributions to transduction or sensory adaptation are not clear in any mechanosensory cell.

Like many other ion channels, mechanically activated channels can be blocked or closed by several chemical agents, although it is unfortunate for mechanoreceptor researchers that no evidence for highly specific blocking molecules has been found as for voltage-activated sodium channels and a variety of ligand-gated channels. The best known blocking chemical is the trivalent positive cation of gadolinium, and sensitivity to Gd^{3+} is sometimes used as an indicator that mechanically activated channels are involved in a physiological process. Gadolinium blocked the mechanoreceptor current in the crayfish stretch receptor (Swerup *et al.,* 1991) and the spider slit sensilla (Höger *et al.,* 1997). However, Gd^{3+} is not specific for mechanically activated channels. It also blocks many calcium channels, as well as some ligand-gated and voltage-activated cation channels (Höger *et al.,* 1997). Amiloride is a well-known blocker of epithelial sodium channels, and it also blocks several types of mechanically activated ion channels (see Fig. 8). Amiloride blocks the receptor current in vertebrate hair cells (Rüsch *et al.,* 1994) and in spider slit sensilla neurons (Höger *et al.,* 1997), which has led to the suggestion that mechanoreceptor channels, and mechanically activated channels in general, may be members of the same gene family as the epithelial sodium channels. However, a direct molecular test of this theory in mammalian auditory hair cells was negative (Rüsch and Hummler, 1999). Some local anesthetics have also been used to block the receptor current in the crayfish stretch receptor (Lin and Rydqvist, 1999b) and certain antibiotics are known to block transduction in auditory hair cells (Hudspeth, 1989).

In addition to the single-channel conductance estimates described above, total mechanoreceptor currents have been estimated in some cells, and noise, or fluctuation, analysis has been used to estimate single-channel conductance and the numbers of channels per cell in several cases. **Noise analysis** relies upon the idea that the variance of the receptor current arises from receptor channels opening and closing, because all known channels regulate their average conductance by switching rapidly between open and closed states. Therefore, the variance is minimal when all of the receptor channels are either fully open or fully closed, and maximal when the mean open probability is 0.5. Assuming that the noise variance is due to the summation of currents flowing through many independent, identical channels that are randomly opening and closing, it is possible to quantify the relationships between the membrane potential, total cell membrane current variance, total current, number of channels, single-channel conductance, and mean channel open probability. In a few cases, noise analysis measurements in nonreceptor systems have been confirmed by later single-channel experiments. Noise analysis measurements are challenging because of the inherent difficulty of making good recordings of mechanoreceptor membrane currents. However, the three cases where they have been made have produced rather consistent results. In statocyst cells of the snail *Hermissenda,* noise analysis gave values of 5 pS for the single-channel conductance and 40 channels per cell (DeFelice and Alkon, 1977). In vertebrate hair cells, the corresponding values were 12 pS and 280 channels per cell (Hudspeth, 1989), and in spider slit sensilla neurons 7.5 pS and 250 channels per cell (Höger and French, 1999a). These single-channel conductances are all at the low end of the range of known mechanically activated ion channels and significantly lower than the only two single-channel measurements described above. The total membrane conductances due to sensory channels observed during these measurements agree with several estimates made in other mechanosensory neurons, and would provide enough current to depolarize the neurons to produce action potentials. Therefore, it seems likely that the number of receptor channels in each cell is limited to a few hundred at most, which will make them difficult to locate by patch-clamp methods.

Another interesting property of mechanotransduction currents is their **temperature sensitivity**. This has been measured in at least six invertebrate and vertebrate preparations with activation energy values in the range of 12–22 kcal/mol (~50–100 kJ/mol) (Höger and French, 1999b). These energy values are similar to those required to break chemical bonds, and significantly more than the energy barriers associated with ionic diffusion or conductance through ionic channels. Little is known about the activation energy of mechanically activated ion channels, but it seems likely that some crucial stage in the link between membrane tension and ion channel opening leads to this relatively high energetic barrier.

C. Encoding

Most mechanoreceptors use action potentials to transmit information to the central nervous system, and the action potentials are produced by the conventional combination of voltage-dependent sodium and potassium channels. Accurate characterization of these currents and other currents involved in the control of membrane excitability relies on voltage-clamp recordings, which have only been possible in a few cases. **Tetrodotoxin-sensitive sodium channels** are clearly responsible for the action potential upswing in Pacinian corpuscles (Loewenstein, 1959), lobster stretch receptors (Edman *et al.,* 1987a), crayfish stretch receptors (Swerup and Rydqvist, 1992), the cockroach tactile spine (Torkkeli and French, 1994), and spider slit sensilla (Seyfarth and French, 1994). The distribution of these channels in the cell membrane probably determines where the receptor potential is converted to action potentials. In the slowly adapting crayfish stretch receptor the channels are mainly located on the axon and the immediately adjacent cell body (Swerup and Rydqvist, 1992), but in the rapidly adapting receptor the channels are restricted to the axon itself (Lin and Rydqvist, 1999a). In spider slit sensilla, sodium channels are more evenly distributed, with a significant concentration on the sensory dendrite and strong excitability in the cell soma (Seyfarth *et al.,* 1995). An addi-

tional slow component of sodium inactivation was needed to model the firing behavior of lobster stretch receptors (Edman *et al.,* 1987a) and is probably present in the slowly adapting crayfish stretch receptor (Purali and Rydqvist, 1998) and the cockroach tactile spine (French and Torkkeli, 1994).

Delayed-rectifier potassium currents seem to be responsible for action potential repolarization in all of the mechanoreceptors studied, but other types of potassium currents are unevenly distributed among different receptors. Rapidly inactivating potassium currents resembling the A-type current have been described in the cockroach tactile spine (Torkkeli and French, 1994) but not in spider slit sensilla (Sekizawa *et al.,* 1999) or crayfish stretch receptors (Swerup and Rydqvist, 1992). **Calcium-activated potassium currents** were found in the cockroach tactile spine (Torkkeli and French, 1995) and vertebrate auditory hair cells (Hudspeth, 1989) but not spider slit sensilla (Sekizawa *et al.,* 1999). Single calcium-activated channels were seen in crayfish stretch receptors (Erxleben, 1993), although there have been no reports of whole-cell currents. An **inwardly rectifying potassium current** is present in the slowly adapting lobster stretch receptor (Edman *et al.,* 1987b) but has not been described elsewhere.

Calcium currents are present in auditory neurons, and negative feedback from depolarization-induced calcium entry to hyperpolarization via calcium-activated potassium current has been suggested to cause frequency tuning (Hudspeth, 1989). Calcium currents have also been measured in spider slit sensilla (Sekizawa *et al.,* 2000), although these cells do not seem to have calcium-activated potassium currents. **Electrogenic sodium pumping** is well established in crayfish stretch receptors (Sokolove and Cooke, 1971) and contributes to adaptation of action potential discharge by repolarizing the cell after sodium entry. There is also evidence for its contribution to adaptation in the cockroach tactile spine neuron (French, 1989).

The terms **rapid adaptation** and **slow adaptation** are qualitative and have been used to describe a wide range of adaptation behaviors. However, some mechanoreceptors adapt so rapidly that only one or a few action potentials are produced in response to a step stimulus (Fig. 9) and so can truly be called "rapidly adapting." It has been known for many years that the encoding stage of the Pacinian corpuscle limits its response to one or two action potentials (Loewenstein and Mendelson, 1965), although the ionic basis for the effect is not known. In the cockroach tactile spine, removal of the rapidly inactivating potassium A-current increases the overall rate of firing, but the receptor continues to adapt, while removal of the calcium-activated potassium current removes most of the adaptation (Torkkeli and French, 1994, 1995). However, blockade of these currents does not affect the shape of individual action potentials (Fig. 10) because their rapid repolarization is due to the delayed-rectifier current.

Following the encoding stage, action potentials propagate into the central nervous system along nerve axons. In vertebrates, the cell bodies of the neurons are located in the dorsal root ganglia of the spinal cord, and the axons enter via the dorsal roots. Mechanoreceptor axons are generally in the $A\alpha$, and $A\beta$ groups of myelinated fibers with conduction velocities in the range of 40–120 m/s. Bare nerve endings (Group Ia

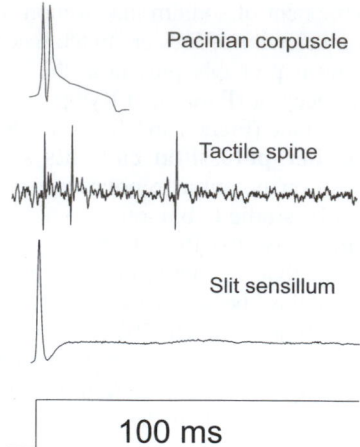

FIGURE 9.   Rapid adaptation in three mechanoreceptors. Following a step mechanical stimulus, a Pacinian corpuscle, cockroach tactile spine, and spider slit sensillum each fire a few action potentials and are then silent. While tactile spine receptors can fire longer bursts with large movements, Pacinian corpuscles and spider slit sensilla cannot normally produce more than two action potentials, however strong the stimulation. (Redrawn from French and Torkkeli, 1994).

of Fig. 3) are innervated by myelinated Aδ or unmyelinated C fibers, with conduction velocities in the range 0.5–30 m/s. The major transmitter at the first central synapse is probably the excitatory amino acid glutamate (Willis and Coggeshall, 1991). Arthropod mechanoreceptors have their cell bodies in the periphery and send axons into the central nervous system via nerve roots of the segmental ganglia, which are sometimes fused into larger structures. Conduction velocities are typically 1–5 m/s. The dominant transmitter for arthropod mechanoreceptors is acetylcholine (Burrows, 1996).

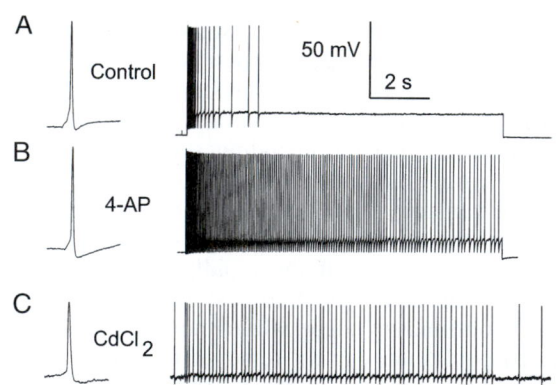

FIGURE 10.   Removal of adaptation during encoding by potassium channel block. (A) A rapidly adapting burst of action potentials in the cockroach tactile spine, following a step depolarization of the neuron. (B) 4-aminopyridine removes the rapidly inactivating potassium A-current, making the neuron fire much more, but the underlying adaptation can still be seen. (C) Cadmium blocks calcium entry and the calcium-activated potassium current, removing adaptation. Note that the shape of the individual action potentials (left) is not strongly affected by either treatment because its repolarization depends on a delayed-rectifier potassium current. (Redrawn from French and Torkkeli, 1994).

VIII. Efferent Control of Mechanoreceptors

All mechanoreceptors studied so far receive inhibitory efferent innervation close to the output synapses of their centrally located axon terminals. Well-known examples are the efferent innervation of muscle spindle and tendon organ (Frank and Fuortes, 1957). Peripheral mechanoreceptors (Semba *et al.*, 1985) and pain receptors (Schmidt and Schaible, 1998) receive similar synaptic inputs. This type of presynaptic innervation of mechanosensory neurons is remarkably similar in all vertebrate and invertebrate species studied so far (Frank and Fuortes, 1957; Semba *et al.*, 1985; Cattaert *et al.*, 1992; Pearson, 1993; Burrows, 1996). Most thoroughly studied of these synapses are GABAergic ones in insects (Burrows, 1996), crustaceans (Cattaert *et al.*, 1992), and vertebrates (Nicoll and Alger, 1979).

Evidence is now accumulating that modulation of mechanosensory neurons is more complex than the inhibitory control of axon terminals, with synaptic regulation of the sensory endings themselves and the sensory cell bodies. Transmitter receptors are found in the vertebrate dorsal root ganglion neuron somata (Robertson, 1989), and unmyelinated nerve fibers are seen in slowly adapting cutaneous mechanoreceptors and in the capsules of Pacinian corpuscles. Some of these endings are probably efferent and have been suggested to modulate sensory reception (Bell *et al.*, 1994). There is also evidence that the Merkel cell functions as a secretory cell rather than a sensory neuron and modulates the associated nerve ending (Zaccone *et al.*, 1994). Modulation of vertebrate cutaneous mechanoreceptors and pain receptors by circulating factors has also been recognized for a long time. For example, sympathetic efferents have modulatory effects on muscle spindles, Pacinian corpuscles (Akoev, 1980), several types of low-threshold skin mechanoreceptors, and pain receptors (Raja, 1995), and these neurons express adrenergic receptors (Birder and Perl, 1999). Rohon-Beard neurons, which are developmentally early amphibian and fish touch receptors, have serotonergic efferent innervation and 5-HT is believed to alter their sensitivity to mechanical stimuli by inhibiting both low- and high-voltage-activated calcium currents (Sun and Dale, 1997).

Insect mechanosensory neurons are modulated by biogenic amines, especially octopamine (Burrows, 1996). The dorsal unpaired median neurons of locust, which secrete octopamine, were recently shown to have terminals in the periphery, closely associated with the dendrites of mechanoreceptors (Bräunig and Eder, 1998). Previous findings of octopamine effects in situations where axon terminals were removed during recordings (Zhang *et al.*, 1992) also suggest that octopamine receptors are located in the neurons' peripheral parts. Although octopamine modulation of insect mechanoreceptors is probably the most thoroughly studied neuromodulatory system, there are no clear conclusions about the mechanisms involved. Octopamine may increase or decrease spiking frequency, even in the same neuron (Bräunig and Eder, 1998).

The most conclusive evidence that peripheral regions of mechanoreceptors can receive efferent innervation comes from immunocytochemical findings in arachnid and crustacean mechanoreceptors (see Fig. 5; Fig. 11A). Both types of receptors have dense efferent innervation, forming several types of synapses on all parts of the sensory neurons, includ-

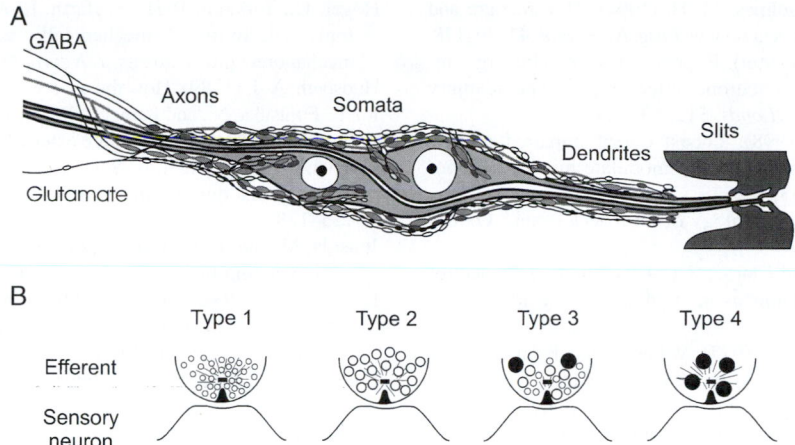

FIGURE 11. Efferent innervation of mechanosensory neurons has been mapped by immunohistochemistry in spider slit sensilla. (A) Schematized distribution of synapses surrounding a pair of neurons innervating one slit. Several fine, efferent fibers run parallel to the neurons and form numerous synapses onto their initial axon segments, somata, and dendrites, as well as onto surrounding glial cells and onto each other. The density of synapses on the dendrites is lower than other regions. At least three types of efferent fibers show immunoreactivity to GABA, and at least one efferent fiber type shows immunoreactivity to glutamate. (B) Four types of efferent synapses onto sensory neurons. Numerous other synaptic combinations between efferents, sensory neurons, and glia have been demonstrated. (Redrawn from Fabian-Fine, 1998).

ing their somata and dendrites (Foelix, 1975; Elekes and Florey, 1987; Fabian-Fine, 1998; Fabian-Fine *et al.*, 1999). The spider slit sensilla neurons and other cuticular mechanoreceptors have at least four different types of synaptic structures on the efferent terminals, based on the shapes and sizes of synaptic vesicles and immunogold labeling (Fig. 11B). These different types of synapses could have several functions, including direct inhibition or excitation of mechanosensory neurons and sensitization or desensitization. They could also control the dynamic responses of the neurons and affect their transmission of repetitive signals. It seems likely that central nervous systems are able to adjust the responsiveness of many types of mechanosensory neurons, although the functional significance of this control is not yet evident.

IX. Summary

The large variety of morphological forms of mechanoreceptors indicates that the external components surrounding the sensitive endings play important roles in controlling the receptor behavior. Most of this control is assumed to involve coupling of initial movement into deformation of the membrane containing mechanically activated channels. However, this has not been thoroughly investigated or modeled in any receptor, and there may be other functions involved, such as chemical modulation of excitability. Characterization of mechanically activated channels is proceeding, but it is important to realize that no single-channel recordings have yet been unequivocally linked to mechanotransduction in any mechanosensory neuron. Therefore, it may yet emerge that an unknown family of channel proteins is responsible for this function in true mechanoreceptors. The encoding of action potentials from receptor current and the general control of receptor excitability involves voltage- and calcium-activated ion channels that are very similar to those in other tissues, and complete models of encoding in some receptors are now available. Adaptation, and general dynamic properties of mechanoreception, occur in both the coupling and encoding stages of the process. There may also be significant dynamic behavior of the mechanically activated channels, but this will not become clear until the channels themselves are better known. Efferent control of transduction and adaptation in mechanoreceptors is probably common but generally not well understood.

Mechanoreception is a widespread and crucially important process for many physiological functions. Recent developments in electrophysiological techniques and new experimental preparations have provided important information about each stage of mechanoreception from initial deformation to action potential production, but much remains to be learned. In particular, the lack of vertebrate preparations that would allow voltage-clamp of receptor currents or detailed examination of action potential encoding is a major problem. It is also true that the molecular biological revolution has so far failed to play an important role in unraveling the machinery of mechanoreception, except insofar as providing general information about some of the families of voltage-activated ion channels that are involved in encoding. When the molecules responsible for mechanotransduction are known, it should be possible to discover how they are linked to other cellular components to confer mechanical sensitivity, and from there to move on to complete models of mechanoreception in intact cells.

Bibliography

Akoev, G. N. (1980). Catecholamines, acetylcholine and excitability of mechanoreceptors. *Prog. Neurobiol.* **15**, 269–294.

Bell, J., Bolanowski, S., and Holmes, M. H. (1994). The structure and function of Pacinian corpuscles: a review. *Prog. Neurobiol.* **42,** 79–128.

Birder, L. A., and Perl, E. R. (1999). Expression of α_2-adrenergic receptors in rat primary afferent neurones after peripheral nerve injury of inflammation. *J. Physiol. (Lond).* **512,** 533–542.

Bräunig, P., and Eder, M. (1998). Locust dorsal unpaired median (DUM) neurones directly innervate and modulate hindleg proprioceptors. *J. Exp. Biol.* **201,** 3333–3338.

Burrows, M. (1996). "The Neurobiology of an Insect Brain." Oxford University Press, Oxford.

Cattaert, D., Manira, A. E., and Clarac, F. (1992). Direct evidence for presynaptic inhibitory mechanisms in crayfish sensory afferents. *J. Neurophysiol.* **67,** 610–623.

DeFelice, L. J., and Alkon, D. L. (1977). Voltage noise from hair cells during mechanical stimulation. *Nature* **269,** 613–615.

Edman, A., Gestrelius, S., and Grampp, W. (1987a). Analysis of gated membrane currents and mechanisms of firing control in the rapidly adapting lobster stretch receptor neurone. *J. Physiol.* **384,** 649–669.

Edman, A., Gestrelius, S., and Grampp, W. (1987b). Current activation by membrane hyperpolarization in the slowly adapting lobster stretch receptor neurone. *J. Physiol.* **384,** 671–690.

Elekes, K., and Florey, E. (1987). Immunocytochemical evidence for the GABAergic innervation of the stretch receptor neurons in crayfish. *Neuroscience* **22,** 1111–1122.

Erxleben, C. F. (1989). Stretch-activated current through single ion channels in the abdominal stretch receptor organ of the crayfish. *J. Gen. Physiol.* **94,** 1071–1083.

Erxleben, C. F. (1993). Calcium influx through stretch-activated cation channels mediates adaptation by potassium current activation. *Neuroreport* **4,** 616–618.

Eyzaguirre, C., and Kuffler, S. W. (1955). Processes of excitation in the dendrites and in the soma of single isolated sensory nerve cells of the lobster and crayfish. *J. Gen. Physiol.* **39,** 87–119.

Fabian-Fine, R. (1998). Thesis: Synaptische Kontakte an sensorischen Neuronen von Spinnen: Immunocytochemische und ultrastrukturelle Untersuchungen. Erlangung des Doktorgrades, Fachbereich Biologie, Johann Wolfgang Goethe-Universität, Frankfurt am Main, 1–133.

Fabian-Fine, R., Höger, U., Seyfarth, E. A., and Meinertzhagen, I. A. (1999). Peripheral synapses at identified mechanosensory neurons in spiders: three-dimensional reconstruction and GABA immunocytochemistry. *J. Neurosci.* **19,** 298–310.

Fein, A., and Szuts, E. Z. (1982). "Photoreceptors: Their Role in Vision." Cambridge University Press, Cambridge.

Foelix, R. F. (1975). Occurrence of synapses in peripheral sensory nerves of arachnids. *Nature* **254,** 146–148.

Frank, K., and Fuortes, M. G. F. (1957). Presynaptic and postsynaptic inhibition of monosynaptic reflexes. *Fed. Proc.* **16,** 39–40.

French, A. S. (1988). Transduction mechanisms of mechanosensilla. *Annu. Rev. Entomol.* **33,** 39–58.

French, A. S. (1989). Ouabain selectively affects the slow component of sensory adaptation in an insect mechanoreceptor. *Brain Res.* **504,** 112–114.

French, A. S., and Sanders, E. J. (1979). The mechanism of sensory transduction in the sensilla of the trochanteral hair plate of the cockroach, *Periplaneta americana. Cell Tiss. Res.* **198,** 159–174.

French, A. S., and Torkkeli, P. H. (1994). The basis of rapid adaptation in mechanoreceptors. *News Physiol. Sci.* **9,** 158–161.

French, A. S., and Torkkeli, P. H. (1998). Information transmission at 500 bits/s by action potentials in a mechanosensory neuron of the cockroach. *Neurosci. Lett.* **243,** 113–116.

Höger, U., and French, A. S. (1999a). Estimated single-channel conductance of mechanically activated channels in a spider mechanoreceptor. *Brain Res.* **826,** 230–235.

Höger, U., and French, A. S. (1999b). Temperature sensitivity of transduction and action potential conduction in a spider mechanoreceptor. *Pflügers Arch.* **438,** 837–842.

Höger, U., Torkkeli, P. H., Seyfarth, E.-A., and French, A. S. (1997). Ionic selectivity of mechanically activated channels in spider mechanoreceptor neurons. *J. Neurophysiol.* **78,** 2079–2085.

Hudspeth, A. J. (1989). How the ear's works work. *Nature* **341,** 397–404.

Ito, F., Fujitsuka, N., and Kim, N. (1983). The spindle potential in the frog muscle spindle does not require external Na$^+$. *Brain Res.* **277,** 352–354.

Juusola, M., and French, A. S. (1995). Recording from cuticular mechanoreceptors during mechanical stimulation. *Pflügers Arch.* **431,** 125–128.

Juusola, M., and French, A. S. (1997). The efficiency of sensory information coding by mechanoreceptor neurons. *Neuron* **18,** 959–968.

Lin, J. H., and Rydqvist, B. (1999a). Different spatial distributors of sodium channels in the slowly and rapidly adapting stretch receptor neuron of the crayfish. *Brain Res.* **830,** 353–357.

Lin, J. H., and Rydqvist, B. (1999b). The mechanotransduction of the crayfish stretch receptor neurone can be differentially activated or inactivated by local anaesthetics. *Acta Physiol. Scand.* **166,** 65–74.

Loewenstein, W. R., (1959). The generation of electrical activity in a nerve ending. *Ann. NY Acad. Sci.* **81,** 367–387.

Loewenstein, W. R. (1971). Mechano-electric transduction in the Pacinian corpuscle: initiation of sensory impulses in mechanoreceptors. *Handbook Sensory Physiol.* **1,** 269–290.

Loewenstein, W. R., and Mendelson, M. (1965). Components of receptor adaptation in a Pacinian corpuscle. *J. Physiol.* **177,** 377–397.

Malinovsky, L. (1996). Sensory nerve formations in the skin and their classification. *Micro. Res. Tech.* **34,** 283–301.

Matthews, P. B. (1981). Evolving views on the internal operation and functional role of the muscle spindle. *J. Physiol.* **320,** 1–30.

McIver, S. B. (1985). Mechanoreception. *In* "Comprehensive Insect Physiology Biochemistry and Pharmacology" (G. A. Kerkut and L. I. Gilbert, Eds.), pp. 71–132. Pergamon Press, Oxford.

Nicoll, R. A., and Alger, B. E. (1979). Presynaptic inhibition: transmitter and ionic mechanisms. *Int. Rev. Neurobiol.* **21,** 217–258.

Ohmori, H. (1985). Mechano-electrical transduction currents in isolated vestibular hair cells of the chick. *J. Physiol.* **359,** 189–217.

Pasztor, V. M., and Bush, B. M. (1982). Impulse-coded and analog signalling in single mechanoreceptor neurons. *Science* **215,** 1635–1637.

Pearson, K. G. (1993). Common principles of motor control in vertebrates and invertebrates. *Annu. Rev. Neurosci.* **16,** 265–297.

Perl, E. R. (1996). Cutaneous polymodal receptors: characteristics and plasticity. *Prog. Brain Res.* **113,** 21–37.

Purali, N., and Rydqvist, B. (1998). Action potential and sodium current in the slowly and rapidly adapting stretch receptor neurons of the crayfish (*Astacus astacus*). *J. Neurophysiol.* **80,** 2121–2132.

Rüsch, A., and Hummler, E. (1999). Mechano-electrical transduction in mice lacking the α-subunit of the epithelial sodium channel. *Hear. Res.* **131,** 170–176.

Rüsch, A., Kros, C. J., and Richardson, G. P. (1994). Block by amiloride and its derivatives of mechano-electrical transduction in outer hair cells of mouse cochlear cultures. *J. Physiol.* **474,** 75–86.

Raja, S. N. (1995). Role of the sympathetic nervous system in acute pain and inflammation. *Ann. Med.* **27,** 241–246.

Robertson, B. (1989). Characteristics of GABA-activated chloride channels in mammalian dorsal root ganglion neurones. *J. Physiol. (Lond.)* **411,** 285–300.

Rydqvist, B., and Purali, N. (1993). Transducer properties of the rapidly adapting stretch receptor neurone in the crayfish. *J. Physiol.* **469,** 193–211.

Sachs, F. (1997). Mechanical transduction by ion channels: how forces reach the channel. *In* "Cytoskeletal Regulation of Membrane Function" (S. C. Froehner, Ed.), pp. 209–218. Rockefeller University Press, New York.

Schmidt, R. F., and Schaible, H.-G. (1998). Modulation of nociceptive information at the presynaptic terminals of primary afferent fibers. *In* "Presynaptic Inhibition and Neural Control" (P. Rudomin, R.

Romo, and L. M. Mendell, Eds.), pp. 424–439. Oxford University Press, New York.

Sekizawa, S. I., French, A. S., Höger, U., and Torkkeli, P. H. (1999). Voltage-activated potassium outward currents in two types of spider mechanoreceptor neurons. *J. Neurophysiol.* **81,** 2937–2944.

Sekizawa, S. I., French, A. S., and Torkkeli, P. H. (2000). Low-voltage activated calcium current does not regulate the firing behavior in paired mechanosensory neurons with different adaptation properties. *J. Neurophysiol.* **83,** 746–753.

Semba, K., Masarachia, P., Malamed, S., Jacquin, M., Harris, S., Yang, G., and Egger, M. D. (1985). An electron microscopy study of terminals of rapidly adapting mechanoreceptive afferent fibers in the cat spinal cord. *J. Comp. Neurol.* **232,** 229–240.

Seyfarth, E.-A., and French, A. S. (1994). Intracellular characterization of identified sensory cells in a new spider mechanoreceptor preparation. *J. Neurophysiol.* **71,** 1422–1427.

Seyfarth, E.-A., Sanders, E. J., and French, A. S. (1995). Sodium channel distribution in a spider mechanosensory organ. *Brain Res.* **683,** 93–101.

Shannon, C. E., and Weaver, W. (1949). "The Mathematical Theory of Communication." University of Illinois Press, Urbana, IL.

Sokolove, P. G., and Cooke, I. M. (1971). Inhibition of impulse activity in a sensory neuron by an electrogenic pump. *J. Gen. Physiol.* **57,** 125–163.

Sun, Q. Q., and Dale, N. (1997). Serotonergic inhibition of the T-type and high voltage-activated Ca^{2+} currents in the primary sensory neurons of *Xenopus* larvae. *J. Neurosci.* **17,** 6639–6649.

Swerup, C., and Rydqvist, B. (1992). The abdominal stretch receptor organ of the crayfish. *Comp. Biochem. Physiol. A* **103,** 423–431.

Swerup, C., Purali, N., and Rydqvist, B. (1991). Block of receptor response in the stretch receptor neuron of the crayfish by gadolinium. *Acta Physiol. Scand.* **143,** 21–26.

Thurm, U. (1965). An insect mechanoreceptor I. Fine structure and adequate stimulus. *Cold Spring Harbour Symp. Quant Biol.* **30,** 75–94.

Torkkeli, P. H., and French, A. S. (1994). Characterization of a transient outward current in a rapidly adapting insect mechanoreceptor neuron. *Pflügers Arch.* **429,** 72–78.

Torkkeli, P. H., and French, A. S. (1995). Slowly inactivating outward currents in a cuticular mechanoreceptor neuron of the cockroach (*Periplaneta americana*). *J. Neurophysiol.* **74,** 1200–1211.

Torkkeli, P. H., and French, A. S. (1999). Primary culture of antennal mechanoreceptor neurons of *Manduca sexta*. *Cell Tiss. Res.* **297,** 301–309.

Willis, W. D., and Coggeshall, R. E. (1991). "Sensory Mechanisms of the Spinal Cord." Plenum Press, New York.

Zaccone, G., Fasulo, S., and Ainis, L. (1994). Distribution patterns of the paraneuronal endocrine cells in the skin, gills and the airways of fishes as determined by immunohistochemical and histological methods. *Histochem. J.* **26,** 609–629.

Zhang, B. G., Torkkeli, P. H., and French, A. S. (1992). Octopamine selectively modifies the slow component of sensory adaptation in an insect mechanoreceptor. *Brain Res.* **591,** 351–355.

Daniel C. Marcus

46

Acoustic Transduction

I. Introduction

The detection of sound by mammals depends on a series of biological systems beginning with the collection of sound pressure waves by the external ear, followed by the mechanical transmission through the middle ear ossicles to the **cochlea** of the inner ear. The sound pressure waves in the cochlea induce motion of the basilar membrane to which the sensory organ, the **organ of Corti,** is attached. Motion of the organ of Corti with respect to another structure (tectorial membrane) causes movements of the sensory cilia (hairs) on the **hair cells** and modulation of the flow of current through these hair cells, leading to modulation of the rate of firing of the afferent auditory nerve fibers, which synapse to the base of the hair cells. The nerve fibers carry the auditory information to the brain where the signal undergoes central processing, leading to the perception of sound. Peripheral auditory processing is modulated by efferent signals originating from the brain. This chapter focuses on the cellular aspects of acoustic transduction in the auditory periphery. Much of what we know about the function of the mammalian cochlea is derived from experiments performed on preparations from the vestibular labyrinth of mammals, birds, and amphibians, and from the cochlea of birds.

II. Mammalian Inner Ear Structure

The transduction apparatus is part of an **epithelium** forming the **cochlear duct** and separating two distinct cochlear fluids, **endolymph** and **perilymph**. The composition and importance of these fluids is related later. A diagram of a cross-section of the mammalian cochlear duct is shown in Fig. 1. The organ of Corti is comprised of the sensory **inner and outer hair cells** (Fig. 11, shown later), which are surrounded by Deiters' cells, pillar cells, and Hensen's cells. A mechanically stiff apical surface of the organ of Corti, the **reticular lamina,** is formed by cuticular plates just under the apical membrane of these cells. The organ of Corti sits

on the **basilar membrane,** a fibrous sheet that transmits the acoustic stimulus.

In the medial direction from the organ of Corti, the duct is comprised of inner sulcus cells and interdental cells of the spiral limbus. The **tectorial membrane** is a gelatinous, acellular structure in the cochlear lumen apparently secreted by the interdental cells. Lateral from the organ of Corti, the cochlear duct consists of outer sulcus cells, spiral prominence cells, and **marginal cells** of the **stria vascularis.** Reissner's membrane forms the remaining wall of the triangular-shaped cochlear duct and is comprised of a thin, avascular sheet of epithelial cells.

The apical and basolateral membranes of all of these cells are separated by tight junction complexes near the endolymphatic surface, which serve to join each cell to its neighbor and to complete the barrier between endolymph and the fluid bathing the basolateral membranes. The basolateral fluid is perilymph for all cell types except strial marginal cells, as described later. The cochlear duct is closed at the apex of the cochlea and is joined at the base of the cochlea via a constriction in the epithelial lumen (ductus reuniens) to the vestibular system. Cellular physiologists have focused most of their attention on the strial marginal cells, which provide the energy source for the transduction process, and on the sensory hair cells themselves.

III. Cell Physiology of Endolymph Homeostasis

A. Composition

Even in the absence of an acoustic stimulus, the cochlea is highly active, maintaining the electrolyte composition of endolymph (Table 1) and a standing current analogous to the "dark current" of photoreceptors (Chapter 48) (Wangemann and Schacht, 1996). The transduction current from endolymph through the hair cells is carried by K^+ and so depends on the high concentration of that ion in endolymph. Gross changes in endolymph composition either by a tear in Reissner's membrane or by genetic interference in K^+ secretion lead to degeneration of the hair cells with subsequent

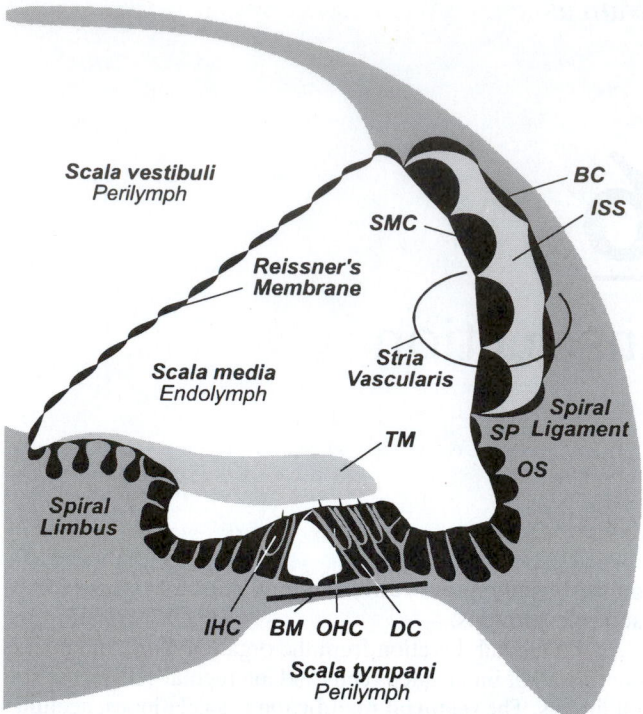

TABLE 1 Approximate Ion Composition and Electrical Potential of Cochlear Endolymph and Perilymph[a]

Ion	Endolymph	Perilymph
Potassium (mM)	157	5
Sodium (mM)	1	145
Calcium (mM)	0.02	1
Chloride (mM)	132	125
Bicarbonate (mM)	31	25
pH	7.4	7.3
Potential (mV)	+80	0

[a]From Wangemann and Schacht, 1996. Copyright 1996 by Springer-Verlag.

FIGURE 1. Diagram of the epithelial cell boundary of the cochlear duct. The triangular cross-section is about 0.3–0.5 mm on each side. The lateral wall is the site of the K^+-secretory stria vascularis, a tissue consisting of two barrier layers of strial marginal cells (SMCs) and basal cells (BCs), which enclose the intrastrial space (ISS). The organ of Corti sits on the fibrous basilar membrane (BM), forms most of the wall separating scala media and scala tympani, and consists of the inner hair cells (IHCs), outer hair cells (OHCs), and several supporting cell types including Deiters' cells (DCs). The sensory hairs at the apical membranes of the OHC project into the gelatinous tectorial membrane (TM), whereas for the most part those of the IHC do not. The spiral prominence (SP) and outer sulcus (OS) epithelial cells lie on the lateral wall between the SV and organ of Corti. The OS forms a para-sensory pathway for the absorption of cations from endolymph. The spiral ligament is the connective tissue between the stria vascularis and the bone (not shown). (Adapted with permission from Wangemann, 1997, Kalium-Ionensekretion und Entstehung des endokochlearen Potentials in der Stria vascularis. *HNO* **45,** 205–209. Copyright 1997 by Springer-Verlag.)

degeneration of the synapsing afferent auditory nerve (Vetter *et al.*, 1996). The dimensions of the gelatinous tectorial membrane are sensitive to the levels of K^+, Na^+, H^+, and Ca^{2+} as well as to unknown chemical factors (Shah *et al.*, 1995). Swelling has been associated with substitution of Na^+ for K^+, and with decreases of Ca^{2+}. Movements of water into and out of the membrane are believed to be related to the level of screening of fixed charges.

B. Stria Vascularis

The stria vascularis has two primary known functions: secretion of K^+ and generation of the lumen-positive **endocochlear potential** (EP; ca. +80 mV). The high endolymphatic K^+ concentration provides the carrier of the

transduction current. The driving force for that current consists almost exclusively of the voltage across the transduction channels in the stereocilia (Section IV) and that voltage is the sum of the intracellular potential of the hair cells and the EP. The large EP therefore heightens the sensitivity of the cochlear transduction process compared to vestibular organs, which do not have this high transepithelial electrical polarization. The cellular transport model by which the stria vascularis secretes potassium and generates the EP is shown in Fig. 2 and described next. Note that all epithelial cells that produce a vectorial transport of substances do so by virtue of different membrane properties of their apical and basolateral membranes.

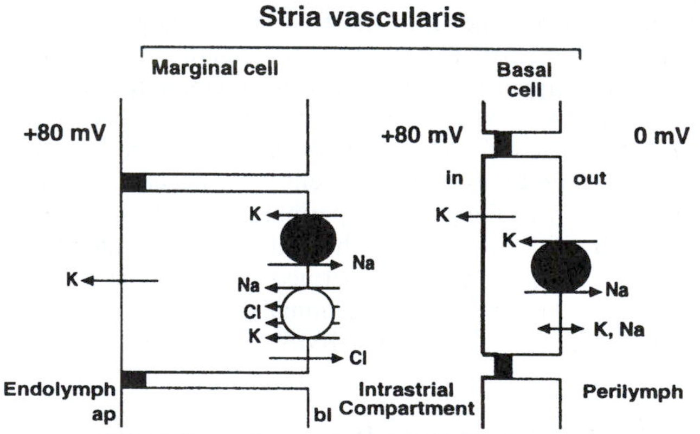

Stria vascularis

FIGURE 2. Model of ion transport by the stria vascularis. The marginal cell epithelium secretes K^+ into endolymph, and the basal cell layer produces the voltage (endocochlear potential) between endolymph and perilymph. The ion transport processes ascribed to the marginal cells have a strong experimental basis. The transport processes proposed in the model of the basal cells have not yet been experimentally demonstrated and may be distributed among basal cells, intermediate cells of the stria vascularis, and fibrocytes of the spiral ligament, which are all connected by gap junctions. Arrows indicate ion channels; open circle, ion carrier; filled circles, primary-active ion "pump." (Adapted from *Hear. Res.* **90,** Wangemann, P., Comparison of ion transport mechanisms between vestibular dark cells and strial marginal cells, pp. 149–157, Copyright 1995 with kind permission of Elsevier Science-NL, Sara Burgerhartstraat 25, 1055 KV Amsterdam, The Netherlands.)

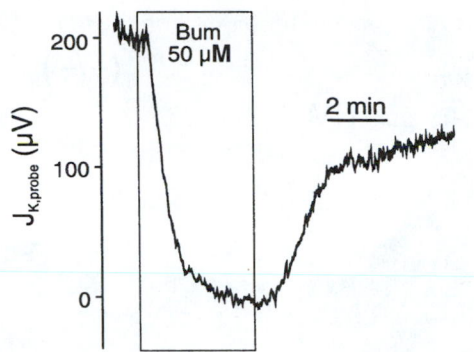

FIGURE 3. K$^+$ secretory flux from stria vascularis measured with the K$^+$-selective self-referencing probe. The ordinate reflects the difference in microvolts measured at the two positions of the probe. Flux was decreased by basolateral perfusion of the inhibitor of Na$^+$-K$^+$-Cl$^-$ cotransport, bumetanide (Bum). (Adapted from *Hear. Res.* **84**, Wangemann, P., Liu, J., and Marcus, D. C., Ion transport mechanisms responsible for K$^+$ secretion and the transepithelial voltage across marginal cells of stria vascularis *in vitro*, pp. 19–29, Copyright 1995 with kind permission of Elsevier Science-NL, Sara Burgerhartstraat 25, 1055 KV Amsterdam, The Netherlands.)

I. Division of Function between Marginal and Basal Cells

The stria vascularis consists of two barriers formed by the **marginal cells** and the **basal cells.** Each barrier consists of a continuous sheet of cells joined by tight junction complexes (Fig. 2). Between these barriers is the intrastrial space with the capillary bed for which the tissue is named and a discontinuous layer of intermediate cells. The basal cells are joined via **gap junctions** (Chapter 31) to the intermediate cells and to fibrocytes in the adjacent connective tissue, suggesting a level of cooperation among these three cell types (Kikuchi *et al.*, 1995). Unlike most sheets of epithelial cells, strial marginal cells are not coupled to each other nor to other cells by gap junctions (Kikuchi *et al.*, 1995; Sunose *et al.*, 1997). The physiological significance of this functional independence of marginal cells is not known.

There is strong evidence that there is a division of function between strial marginal cells and basal cells. The basal cell barrier produces and supports the endocochlear potential and the strial marginal cells secrete K$^+$. In spite of the separation of cellular function, the two processes are closely tied together through the composition of the intrastrial space. For example, inhibition of K$^+$ secretion by the marginal cells leads to a decline of the endocochlear potential, presumably due to a rise in intrastrial K$^+$ concentration and the consequent depolarization of the intrastrial membrane of the basal cells. Several lines of indirect evidence, including histochemical, biochemical, electrophysiological, and flux studies, have strongly suggested that the stria vascularis is responsible for secretion of K$^+$ into the cochlear lumen and for production of the EP (reviewed in Wangemann and Schacht, 1996). Recent evidence of a more direct nature is presented here.

2. K$^+$ Secretion by Strial Marginal Cell Epithelium

Active K$^+$ secretion in the cochlear duct has been demonstrated by flux measurements of radiolabeled K$^+$ intro-

duced in either the perilymphatic or vascular space and its appearance in endolymph (Konishi *et al.*, 1978; Sterkers *et al.*, 1982). These fluxes were inhibited by transport blockers and anoxia. The stria was assumed to be the site of secretion since the flux was nearly the same when the radiotracer was added to either scala vestibuli or scala tympani; possible contributions by the spiral limbus were disregarded.

More recently, the stria was isolated from the cochlear duct, a K$^+$-selective self-referencing probe (see Appendix at the end of this chapter) was placed near the tissue *in vitro* and a K$^+$ gradient was found directed away from the luminal surface (Wangemann *et al.*, 1995a) (Fig. 3). The marginal cell layer of the stria vascularis produces this K$^+$ flux using the constellation of transport processes shown in Fig. 2. The cell model is identical to that for the vestibular dark cells, the vestibular homolog of strial marginal cells (Wangemann, 1995). K$^+$ is taken up across the **basolateral membrane** from the **intrastrial space** fluid by two transporters: the Na$^+$,K$^+$-ATPase and the Na$^+$-K$^+$-Cl$^-$ cotransporter. The first is a primary-active process, which uses the energy from cytosolic adenosine triphosphate (ATP) and the second is secondary-active and uses the large Na$^+$ concentration gradient between extracellular and intracellular compartments created by the Na$^+$,K$^+$-ATPase. Na$^+$ and Cl$^-$ taken up by the cotransporter are removed from the cytosol by the Na$^+$,K$^+$-ATPase and basolateral Cl$^-$ channels, respectively (Wangemann and Schacht, 1996).

K$^+$ secretion occurs passively via channels in the apical membrane (Fig. 4). These channels are products of the genes KCNQ1 and KCNE1 for the **KvLQT1 K$^+$ channel** and its beta subunit, **IsK** (or **mink**). The resulting channel complex* and current carried by the channel are often referred to as **IKs.** The essential contribution of the IsK protein to hearing and balance is most dramatically illustrated by the development of a mouse in which the KCNE1 (*isk*) gene was knocked out (Vetter *et al.*, 1996). In these mice the entire endolymphatic space collapsed (Fig. 5) after the point in embryonic development at which endolymph secretion normally commences,

---

*Historically, the channel and gene were first named after the beta subunit before the channel itself was discovered.

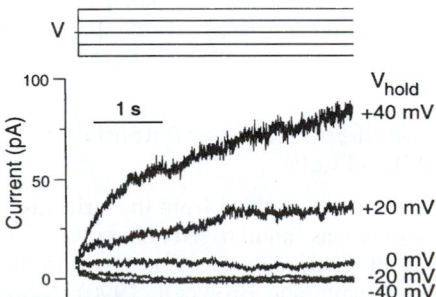

FIGURE 4. Patch-clamp currents from the apical membrane of a strial marginal cell, on-cell configuration. Sustained depolarizations activate the current slowly over several seconds, characteristic of IKs channels. (Reproduced with permission from Shen *et al.*, 1997.)

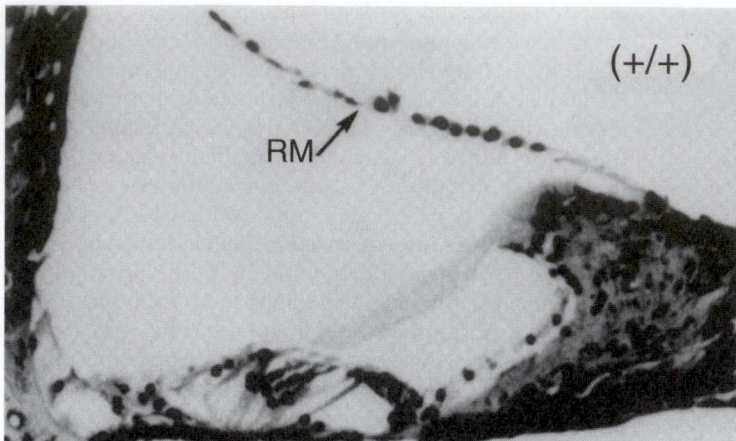
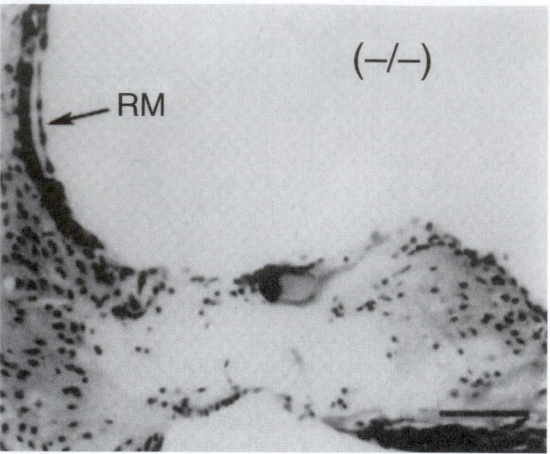

FIGURE 5. Cross-section of the cochlear duct from mice with (+/+) and without (–/–) the *isk* gene. Absence of the *isk* gene led to total collapse of the duct such that Reissner's membrane (RM) was in close apposition to the stria vascularis and tectorial membrane. (Reproduced with permission from Vetter *et al.*, 1996. Copyright 1996 by Cell Press.)

resulting in degeneration of the hair cells and the afferent nerves. The mice showed behavior that was characteristic of a lack of hearing and balance. Measurements of short circuit current (I_{sc}) under constitutive and stimulated conditions showed a lack of K⁺ secretion. The marginal cells remained present and were in other respects normal. The absence of this protein clearly has its most profound effect on the function of the auditory and vestibular periphery.

The IKs channels are characterized by slow activation (over seconds) in response to depolarization of the cell membrane and have a single-channel conductance of about 14 pS under *in vivo*-like conditions (Marcus and Shen, 1994; Shen *et al.*, 1997). These channels are believed to carry the transepithelial K⁺ flux since maneuvers that alter transepithelial K⁺ secretion, such as basolateral bumetanide or elevated bath K⁺ concentration, have the same effects on IsK currents measured from the apical membrane of strial marginal cells and vestibular dark cells in cell-attached patches (Marcus and Shen, 1994; Wangemann, 1995; Shen *et al.*, 1997). The functional IKs channel is thought to be composed of two IsK β subunits of only 130 amino acids with a single transmembrane domain and a presently unknown number of KvLQTl channel subunits (Takumi *et al.*, 1988; Wang and Goldstein, 1995; Barhanin *et al.*, 1996; Sanguinetti *et al.*, 1996). The IsK protein itself apparently modulates the activity of the associated channel (Vetter *et al.*, 1996; Tzounopoulos *et al.*, 1995; Ben-Efraim *et al.*, 1996).

3. Production of Endocochlear Potential by Strial Basal Cells

An electric current directed from the stria vascularis into the cochlear duct was found based on a voltage gradient observed when an electrode was tracked within the cochlear duct *in situ* (Zidanic and Brownell, 1990). More recently, the stria was isolated from the cochlear duct and a **voltage-sensitive vibrating probe** (see Appendix at the end of this chapter) was placed near the tissue *in vitro* (Wangemann *et al.*, 1995a). An electric field gradient was found directed

away from the luminal surface and the associated current was strongly reduced by bumetanide, a substance known to decrease the endocochlear potential *in vivo*. Evidence for the role of the basal cell layer in generation of the endocochlear potential was obtained from experiments in which the stria vascularis was mounted in a micro **Ussing chamber** to measure the voltage generated by the isolated tissue *in vitro* (Wangemann *et al.*, 1995a). It was found that the voltage was greater when the basal cell layer was intact compared to when the barrier properties had been mechanically compromised. In addition, under conditions when only the marginal cell layer contributed to the measurements, the voltage was nearly zero if the apical perfusion was high in K⁺ concentration (artificial endolymph). Under symmetrical conditions, both sides of the epithelium are perfused with the same mammalian physiological saline (high Na⁺, low K⁺ concentrations), which ensures that the transepithelial current is due completely to active cellular transport processes and not to passive chemical driving forces across the paracellular pathway (Koefoed-Johnsen and Ussing, 1958).

The proposition that the basal cell layer (perhaps in association with the intermediate cells and fibrocytes) produces the endocochlear potential is partly based on double-barreled microelectrode measurements of electrical potential and K⁺ concentration while traversing the stria *in situ* (Salt *et al.*, 1987). In the spiral ligament, the potential was taken as zero and the K⁺ concentration was a few millimolar (Fig. 6). As the electrode was advanced, a region was found where the K⁺ concentration was also low but the potential had risen to about +80 mV. This was interpreted as being located in the **intrastrial space.** Further penetration led to a jump of the K⁺ concentration to that commonly found in endolymph with no appreciable change in the potential. The voltage gradient was therefore largest across the basal cell layer and the K⁺ gradient largest across the marginal cell layer.

The **cell model** proposed for the basal cells is shown in Fig. 2. The basolateral membrane would support little potential difference due to a large nonselective cation conductance

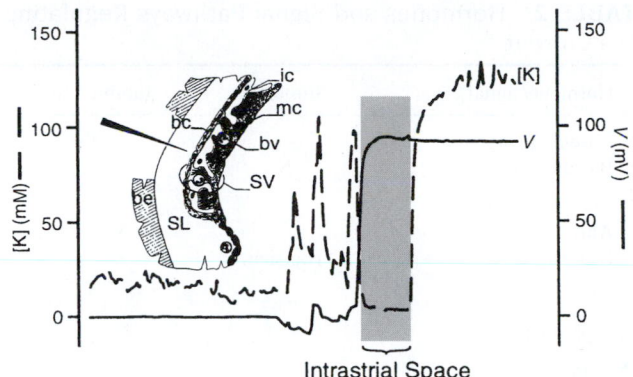

Intrastrial Space

FIGURE 6. Recording of profiles for voltage and K^+ concentration during penetration of the stria vascularis. A region was found where the voltage with respect to perilymph was highly positive but the K^+ concentration was low; this region was interpreted to be the intrastrial space. [Adapted with permission from figures in Salt *et al.*, 1987, and (inset) Dallos, P. (1996). Overview: Cochlear neurobiology. *In* "The Cochlea" (P. Dallos, A. N. Popper, and R. R. Fay, Eds.), pp. 1–43. Copyright 1996 by Springer-Verlag.]

and similar cation concentrations in the basolateral fluid (perilymph permeating the spiral ligament) and in the cytosol. The resting membrane potential of the basal cells would therefore be clamped to near zero when measured with respect to perilymph. If the membrane facing the intrastrial space has a large K^+-selective conductance, that membrane would support a large potential difference (intrastrial

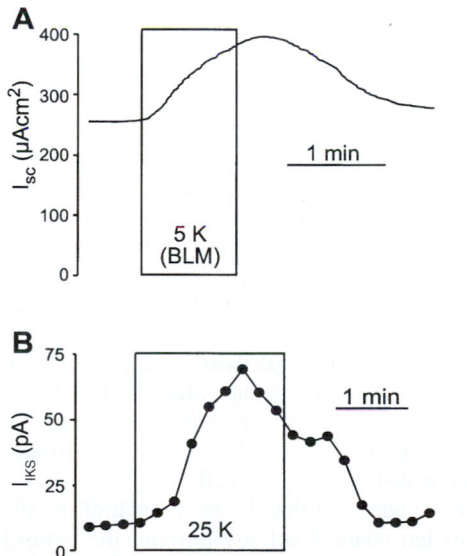

FIGURE 7. Dependence of (A) short circuit current (I_{sc}) and (B) K^+ current through the apical KvLQT1/IKs channels (I_{IKs}) on K^+ concentration in vestibular dark cells, the homologs to strial marginal cells. Basolateral $[K^+]$ was increased from 3.6 mM to 5 mM (A) or 25 mM (B). (Adapted from *Hear. Res.* **100,** Wangemann, P., Shen, Z., and Liu, J., K^+-induced stimulation of K^+ secretion involves activation of the IKs channel in vestibular dark cells, pp. 201–210. Copyright 1996 with kind permission of Elsevier Science-NL, Sara Burgerhartstraat 25, 1055 KV Amsterdam, The Netherlands.)

space positive with respect to basal cell cytosol) due to a large K^+ gradient between intra- and extracellular fluids. It is this K^+ conductance of the intrastrial membrane that generates the EP. There is little biophysical evidence yet to support experimentally the basal cell model although K^+ channels of the maxi-K^+ type have been observed on isolated basal cells (Takeuchi and Irimajiri, 1996) and K^+ channels with a pharmacology similar to that of the endocochlear potential were found to populate the membrane of the well-connected intermediate cells (Takeuchi and Ando, 1998). Some components of the model may be located in the basal cells themselves, whereas others may be associated with the intermediate cells and/or fibrocytes (Wangemann and Schacht, 1996).

4. Regulation of Ion Transport in Strial Marginal Cells and Vestibular Dark Cells

Every cell type encounters perturbations and signals by systemic variables such as osmotic strength and extracellular pH (pH_o), by hormones and/or neurotransmitters, and by cytosolic variables such as intracellular pH (pH_i). K^+ secretion in the inner ear is altered in response to many such stimuli.

The increase in flow of K^+ through hair cells in response to mechanical stimulation (sound in the cochlea, acceleration in the vestibular labyrinth) is known to increase the concentration of K^+ in the perilymph surrounding the basolateral membranes of the hair cells and the nerve endings. In strial marginal cells and vestibular dark cells, the rate of the basolateral uptake of K^+ and of its secretion through the apical IKs channel has been found to be exquisitely sensitive to the concentration of perilymphatic K^+ (Fig. 7). This K^+ sensitivity is believed to play a significant role in regulating the recirculation of K^+ back into endolymph (Wangemann *et al.*, 1996b). It has been proposed that diffusion of K^+ from the base of the hair cells to the K^+ secretory cells is facilitated by a system of supporting cells and fibrocytes connected by gap junctions (Kikuchi *et al.*, 1995).

Both **extracellular** and **intracellular acidification** transiently elevate I_{sc} in vestibular dark cells with a given decrease in pH_i causing a sevenfold greater increase in I_{sc} than the same decrease in pH_o (Fig. 8) (Wangemann *et al.*, 1995b). Fluctuations normally occur in pH_o as a result of systemic acidosis and alkalosis, whereas pH_i changes in response to varying metabolic activity of the cell. These challenges to the cellular homeostasis of pH_i are controlled by transporters such as the Na^+-H^+ exchanger and H^+-monocarboxylate transporter in vestibular dark cells and strial marginal cells (Wangemann *et al.*, 1996a; Shimozono *et al.*, 1997). In addition to controlling their own cytosolic pH, cells bordering the cochlear duct regulate the pH of endolymph (Wangemann and Schacht, 1996).

Most cells respond to a hypo-osmotic challenge by first swelling passively as water rushes in and then **regulating** their **cell volume** over several seconds or minutes by releasing KCl (**regulatory volume decrease;** RVD) by one or more of several pathways that are normally quiescent but activated by osmotic challenge (Chapter 21). In contrast, vestibular dark cells undergo RVD via K^+ and Cl^- selective pathways, which include the constitutively active IsK channel (Wangemann *et al.*, 1995c; Shiga and Wangemann,

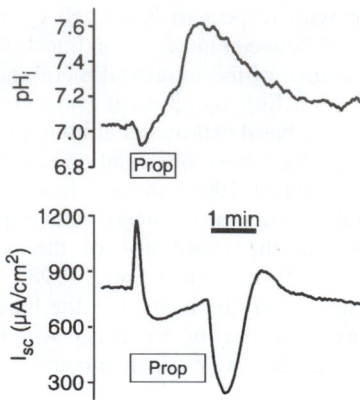

FIGURE 8. Effect of propionate (Prop) on intracellular pH (pH$_i$) and short circuit current (I_{sc}) in vestibular dark cells, the homologs to strial marginal cells. Propionate caused a biphasic change in pH$_i$ that was mirrored in the I_{sc}. (Adapted with permission from Wangemann *et al.*, 1995b. Copyright 1995 by Springer-Verlag.)

TABLE 2 Hormones and Signal Pathways Regulating IKs Current

Hormone/signal	Stimulate	Inhibit
β-Adrenergic agonist		
Basolateral	X	
Purinergic agonist		
Apical		X
Basolateral	Transient	X
cAMP	X	
Phospholipase C		X
Protein kinase C		X
Cytosolic Ca^{2+}	X	
Cytosolic pH	Transient	X
Cell swelling	X	

1995; Wangemann and Shiga, 1994). Surprisingly, a hypotonic challenge also results in sustained stimulation of K$^+$ secretion (Fig. 9).

There is no innervation of the stria vascularis, but these cells contain receptors coupled to ion transport for several local or systemic **hormones,** including catecholamines, ATP, and adrenocorticosteroid hormones (Table 2). Vaso-pressin (antidiuretic hormone) affects the EP and longitudinal K$^+$ concentration gradient (Julien *et al.*, 1994; Mori *et al.*, 1989), but the cells with vasopressin receptors have not yet been identified.

Perilymphatic perfusion of epinephrine causes no appreciable effect on the cochlear microphonic, a stimulus-dependent electrical response dependent on the EP (Klinke and Evans, 1977). However, basolateral perfusion *in vitro* of the β-adrenergic agonist isoproterenol caused maximal increases in I_{sc} of 40–75% in strial marginal cells and vestibular dark cells (Wangemann, *et al.*, 1999, 2000). β-**Adrenergic receptors** are commonly coupled via G proteins to adenylate cyclase (Chapter 12). Indeed, an increase of cytosolic cAMP in strial marginal cells and vestibular dark cells by direct stimulation of adenylate cyclase, by perfusion of a membrane-permeable **cAMP** analogue, or by inhibition of phosphodi-

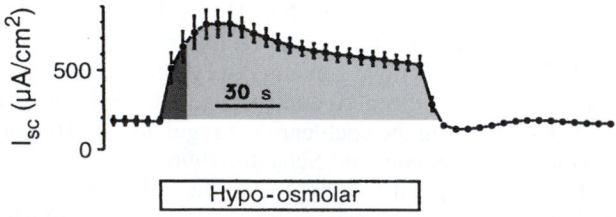

FIGURE 9. Effect of a hypo-osmotic challenge from 300 to 150 mOsm on I_{sc} in vestibular dark cells, the homologs to strial marginal cells. The dark shading indicates the increased amount of K$^+$ secretion that is sufficient to account for cell volume regulation by release of osmolyte (K$^+$). The subsequent sustained elevation of I_{sc} (light shading) indicates that a hypotonic challenge causes an upregulation of the steady-state rate of K$^+$ secretion. (Adapted with permission from Wangemann *et al.*, 1995c. Copyright 1995 by Springer-Verlag.)

esterases that catalyze the breakdown of cAMP all lead to an increase of I_{sc} (Sunose *et al.*, 1997; Wangemann *et al.*, 1999). The stimulation of K$^+$ secretion by adrenergic agonists was found to be mediated by β-adrenergic receptors through determination of the potency order of specific antagonists and by demonstration of the presence of transcripts for this subtype cochlear and in cochlear and vestibular tissue (Wangemann *et al.*, 1999, 2000).

In isolated strial marginal cell epithelium, both apical and basolateral perfusion of micromolar ATP significantly alter I_{sc} (Fig. 10) (Liu *et al.*, 1995). Apical ATP and analogues monotonically down-regulate I_{sc}, whereas basolateral ATP and analogues transiently increase I_{sc} followed by down-regulation. The sequences of agonist effects are consistent with the presence of purinergic receptors of the P2Y$_4$ or P2Y$_2$ (P2U) subtype on the apical membrane and coexisting populations of both P2Y$_1$ and P2Y$_4$ or P2Y$_2$ subtypes on the basolateral membrane. The transient increase in I_{sc} from the basolateral side is due to activation of the P2Y$_1$ receptor and the subsequent decline due to both the P2Y$_1$ and P2Y$_2$ receptors. Both P2Y receptor subtypes are known in other cells to be coupled to G proteins that stimulate phospholipase C (PLC). PLC catalyzes the breakdown of membrane phospholipid to produce both inositol trisphosphate (IP$_3$) and diacylglycerol (DAG) (Chapter 12). Although perilymphatic perfusion of high concentrations of ATP *in vivo* has little effect on the EP (Kujawa *et al.*, 1994), the basal cell layer can be expected to offer a significant barrier to diffusion from perilymph to the basolateral membrane of strial marginal cells.

It is most common that the IP$_3$ branch of the PLC pathway regulates ion channel activity through the cytosolic level of free Ca^{2+} However, it was shown that the apical P2Y$_2$ receptor down-regulates K$^+$ secretion primarily via the DAG-protein kinase C (PKC) branch (Marcus *et al.*, 1997) even though inositol phosphate production is increased in the cochlear lateral wall in response to P2Y agonists (Ogawa and Schacht, 1995). In fact, elevation of cytosolic Ca^{2+} increased the IKs channel current while activation of PKC decreased the current.

Binding sites for both **glucocorticosteroid** (Ten Cate *et al.*, 1993) and **mineralocorticosteroid** (Yao and Rarey,

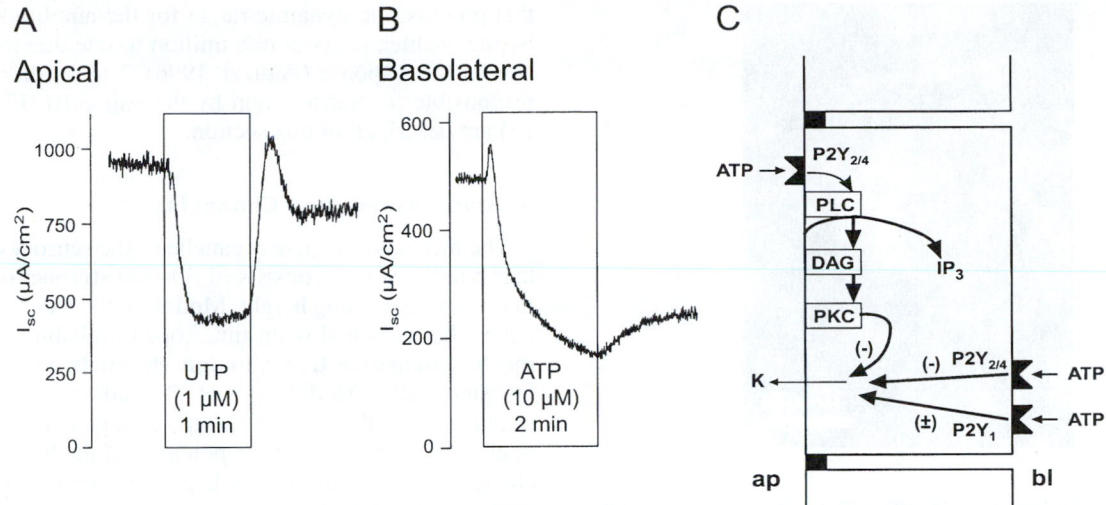

FIGURE 10. Regulation of K$^+$ secretion by extracellular ATP in strial marginal cells. (A) Representative monophasic response of short circuit current (I_{sc}) to apical perfusion of UTP, agonist of the P2Y$_2$ (P2U) receptor. (B) Representative biphasic response I_{sc} to basolateral perfusion of ATP, agonist for P2 receptors. (C) Diagram of signal pathway between activation of apical P2Y$_4$ or P2Y$_2$ receptor and down-regulation of IsK channel and of P2 receptor subtypes in basolateral membrane with effects of each on K$^+$ secretion. (A and B, Adapted with permission from Liu *et al.*, 1995.)

1996) have been demonstrated in the stria vascularis. Both corticoids control the activity of Na$^+$,K$^+$-ATPase in the stria vascularis (adrenalectomy reduced activity 60% and systemic administration of either the glucocorticosteroid dexamethasone or the mineralocorticosteroid aldosterone restored activity) (Curtis *et al.*, 1993), and the increase in Na$^+$,K$^+$-ATPase with aldosterone was not dependent on major changes in blood plasma cation concentration (Ten Cate *et al.*, 1994). However, in spite of the strong dependence of the EP on strial Na$^+$,K$^+$-ATPase, a reduction of adrenocorticosteroids by adrenalectomy did not significantly reduce the EP in the presence or absence of strong acoustic stimulation (Ma *et al.*, 1995). In addition, glucocorticoid hormones alone apparently do not stimulate the formation of Na$^+$, K$^+$-ATPase in the inner ear since no difference in the level of antibody binding to Na$^+$,K$^+$-ATPase was observed in glucocorticoid receptor knockout mice (Erichsen *et al.*, 1998).

C. Outer Sulcus Epithelium

The varying efflux of K$^+$ through the hair cells during stimulation by sound would disturb the luminal K$^+$ concentration if the only other contribution to K$^+$ homeostasis were a constant secretion of K$^+$ by the stria vascularis. A constant level of endolymphatic K$^+$ could be maintained if there were a feedback signal to the secretory process in the stria vascularis and/or to a parallel reabsorptive pathway. Although a paracrine signaling mechanism has not been unambiguously identified, it was shown that the outer sulcus epithelial cells perform a parasensory reabsorptive function and contribute to creating the extremely low Na$^+$ concentration (see Table 1). A vibrating current-density probe (see Appendix to this chapter) was used to show that outer sulcus epithelial cells generate a vectorial electric current which enters the apical side of the epithelium (Marcus and Chiba, 1999). This current was shown to be dependent on metabolic energy via the

ouabain-sensitive Na$^+$,K$^+$-ATPase in the basolateral membrane. Patch-clamp recordings demonstrated a high density of nonselective cation channels in the apical membrane that supported an inward current. Extracellular Gd^{3+} (Fig. 11) inhibited both the probe and patch-clamp currents and neither were affected by removal of Cl$^-$. The inward current

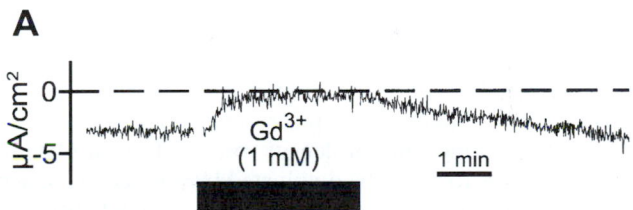

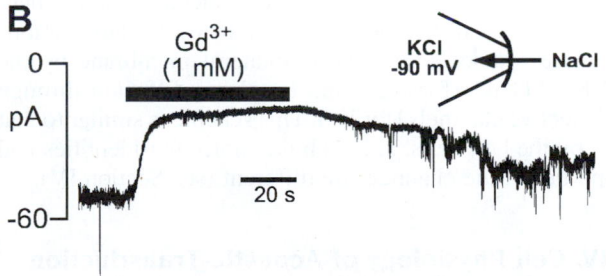

FIGURE 11. Gadolinium (Gd^{3+}) decreased the amplitude of (A) the transepithelial short circuit current measured with a current-density vibrating probe and (B) the inward apical membrane current of an excised outside-out patch. Channel density in this patch was sufficiently high that single channel events were not visible. (Reprinted from *Hear. Res.* 134, Marcus, D. C., and Chiba, T., K$^+$ and Na$^+$ absorption by outer sulcus epithelial cells, pp. 48–56, Copyright 1999, with permission from Elsevier Science.)

A

10 μm

B

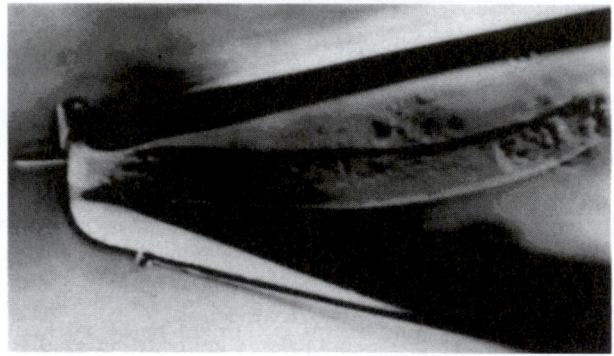

FIGURE 12. Photomicrographs of hair cells: (A) inner hair cell and (B) outer hair cell. (Reproduced with permission from (A) Kros and Crawford, 1990, (B) Evans and Dallos, 1993. Copyright 1993 National Academy of Sciences, U.S.A.)

this process, the dynamic range for the amplitude detected by the cochlea is about one million to one due to compression of the response (Patuzzi, 1996). The cellular processes responsible for transduction by the hair cells (Figs. 12 and 13) are described in this section.

A. Apical Membrane Channels

The mechanosensitive organelle of the sensory cells is the hair bundle, which consists of 30–300 **stereocilia** arranged in rows of increasing height. Motion of the hair bundle modulates the fractional open time (open probability P_o) of the **mechanosensitive transduction channels** near the tips of the stereocilia. Modulation of P_o leads to corresponding fluctuations of the current through the apical membrane and to changes in the membrane potential of the hair cell. These changes in the membrane voltage are referred to as the **receptor potential,** the primary event to modulate synaptic transmission.

Bundle movement is believed to be the result of direct interaction of the OHC stereocilia with the tectorial membrane and of the IHC stereocilia with flowing endolymph pumped back and forth in the channel under the tectorial membrane. The cytoskeleton of the stereocilia is formed by a rigid matrix of actin filaments cross-linked by fimbrin. Deflection of the hair bundle results in a rotation of the individual cilia about a pivotal region at their base, which causes a shearing motion between the ciliary tips. Shear resulting from movement of the hair bundle in the direction of the tallest stereocilia increases P_o of the transduction channels from the resting activity of about 5–15%, whereas movement of the hair bundle in the opposite direction decreases P_o (Hudspeth and Gillespie, 1994). Motion in the stimulatory direction is caused by the rarefaction phase of the external acoustic pressure wave. The

through the channels observed in on-cell patches therefore likely forms the basis of the macroscopic current observed with the vibrating current-density probe. The basolateral membrane was found to be dominated by a K^+ conductance and the basolateral membrane potential was about –90 mV.

The cell model for absorption of Na^+ and K^+ by outer sulcus epithelial cells therefore includes electrodiffusion of cations from endolymph through nonselective cation channels in the apical membrane into the cytosol. These cations leave the cytosol across the basolateral membrane by the Na^+,K^+-ATPase (for Na^+) and by electrodiffusion through K^+-selective channels (for K^+). This scheme is similar to that used by the hair cells, although the molecular identities and properties of the channels are different (see Section IV).

IV. Cell Physiology of Acoustic Transduction

The mammalian cochlea has an exquisite sensitivity, being able to detect sound pressure fluctuations of less than one millionth atmospheric pressure. At the threshold of hearing, the organ of Corti vibrates less than 1 nanometer. This sensitivity includes a gain of 100–1000 due to active mechanical amplification by cells of the organ of Corti, most likely the outer hair cells. In spite of the delicacy of

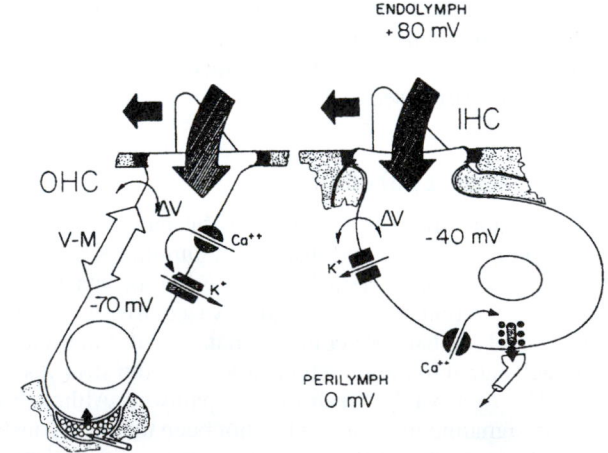

FIGURE 13. Cellular processes of acoustic transduction in inner (IHC) and outer (OHC) hair cells. Shown are the influx of K^+ through the transduction channels in the stereocilia in response to acoustic stimulation (filled horizontal arrows), voltage-gated Ca^{2+} channels in the basolateral membranes that initiate synaptic transmission in IHC and activate K^+ channels in OHC, and the voltage-driven motor elements (V-M) in the OHC. [Reproduced with permission from Dallos, P. (1992). The active cochlea. *J. Neurosci.* **12,** 4575-4585.]

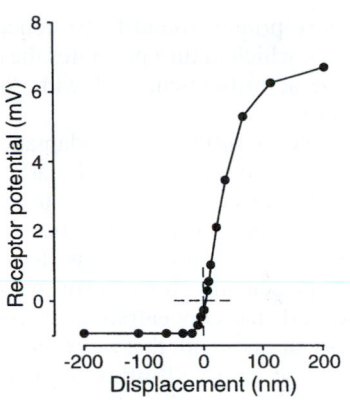

FIGURE 14. Response function of the receptor potential of OHC to mechanical displacement of the stereocilia. The asymmetrical position of the quiescent point is indicated by dashed cross-hairs. [Adapted with permission from Russell, I. J., Koessi, M., and Richardson, G. P. (1992). Nonlinear mechanical responses of mouse cochlear hair bundles. *Proc. R. Soc. London Ser. B* **250**, 217–227.]

asymmetric position on the receptor potential-stimulus transfer function of the stereocilia under unstimulated conditions (zero displacement in Fig. 14) leads to larger changes in the receptor potential with a given stimulatory displacement than for the same size inhibitory displacement. This asymmetry is a primary cause of distortion, which generates many of the nonlinear properties of the auditory system, such as distortion products and otoacoustic emissions (see later section on reverse transduction).

The shear between stereocilia is posited in the **gating-spring model** of hair cell function to transmit a mechanical force through an elastic element (the gating spring) to the mechanosensitive gate of each **transduction channel.** Increased tension on the spring increases P_o and decreased tension reduces P_o. The most widely accepted hypothesis is that the fine extracellular links (called **tip links**) that run between the tips of shorter stereocilia and the sides of taller ones constitute the elastic element of the model. This hypothesis is supported by the shared sensitivity to Ca^{2+} of the tip link structure, the transduction process, and the mechanical manifestation of the gating springs (Jaramillo, 1995).

The question of the **location of the transduction channels** has been approached by several means. The simplest and most widely accepted configuration is the presence of one channel at the insertional plaque of the tip link in the taller stereocilia of each pair. The entry of Ca^{2+} into the cytosol of the stereocilia via transduction channels was visualized with fluorescent dyes and found to be slightly below the stereocilia tip where the tallest of the two stereocilia bridged by a tip link has its insertion point. However, Ca^{2+} also enters many of the shortest stereocilia as well, suggesting that transduction channels may be located at both ends of the tip link (Denk *et al.*, 1995).

A different view is provided by evidence suggesting that the transduction channels may reside at yet another location. An antibody that recognizes an epitope on several amiloride-sensitive transport proteins binds to stereocilia not at the insertional plaque but at the point of contact be-

tween the shorter and taller stereocilia, a place below the insertion point of the tip link on the taller stereocilia (Hackney and Furness, 1995). The transduction channel is known to be inhibited by amiloride and is therefore likely the target of the antibody.

Electrophysiologic and micromechanical data from a variety of hair cells, both *in vivo* and *in vitro*, are best in accord with a model of the transduction channel in which there are transitions among one open state (O) and two closed states (C_1 and C_2) (Markin and Hudspeth, 1995):

$$C_1 \leftrightarrow C_2 \leftrightarrow O$$

and in which Ca^{2+} is required at an intracellular site to promote the transition from C_1 to C_2 (Hackney and Furness, 1995). It has been proposed that there may also be a second open state (reviewed in Kros, 1996).

Tip links are thought to be maintained under tension (see next section on adaptation), which is either increased or decreased by motion toward or away from the stimulatory direction. The time constant for response of the receptor current to a step stimulus has been found in saccular hair cells of the bullfrog to be in the range of 100–500 μs at 4 °C and to become faster at higher temperatures (Corey and Hudspeth, 1983). This time course is substantially faster than can be accounted for by typical enzymatic or second messenger pathways and points to a direct mechanical coupling between the stimulus and the gate of the transduction channel.

The **conductance of the transduction channel** has been estimated from patch-clamp records to be on the order of 300 pS under *in vivo* conditions (Kros, 1996). The channel is permeable to cations including monovalent, divalent, and small organic cations such as tetraethylammonium while selecting against anions (Corey and Hudspeth, 1979). Because potassium and calcium in endolymph are above electrochemical equilibrium with respect to the hair cell cytosol, it is primarily these two cations that flow through the transduction channel. Potassium carries the bulk of the current due to its high concentration, whereas entry of Ca^{2+} through the transduction channel is thought to control adaptation (see following section). It was recently found (Ricci and Fettiplace, 1998) that Ca^{2+} carries a surprisingly high fraction of the current through the transduction channel; about 20% (whereas the concentration of Ca^{2+} in endolymph is only about 0.01% that of K^+; c.f. Table 1). This apparent anomaly may be due to the behavior of the transduction channel like a multi-ion pore, in which the different ions interact. Sufficient Ca^{2+} enters the stereocilia to drive adaptation despite its low concentration. The channel is known to be blocked by aminoglycoside antibiotics (multivalent cations) and amiloride (a blocker of several Na^+ transport processes).

B. Adaptation

When a sustained displacement is applied to hair cells, the transduction current is first stimulated but then relaxes in the continued presence of the stimulus with a characteristic time course, a process referred to as **adaptation** (Jaramillo, 1995; Hudspeth and Gillespie, 1994). The mechanism underlying this phenomenon is thought to be an active process in which myosin motors in the insertional plaque maintain tension within the tip link through a balance of climbing and slipping (Fig. 15). The tip link slackens when there is a deflection in the negative direction but tension is soon reestablished as the motor climbs along actin filaments within the taller stereocilium of each pair. Deflection in the positive direction increases tension in the tip links, opening the mechanosensitive transduction channel. Ca^{2+} coming in through the channel is thought to bind to **calmodulin,** a

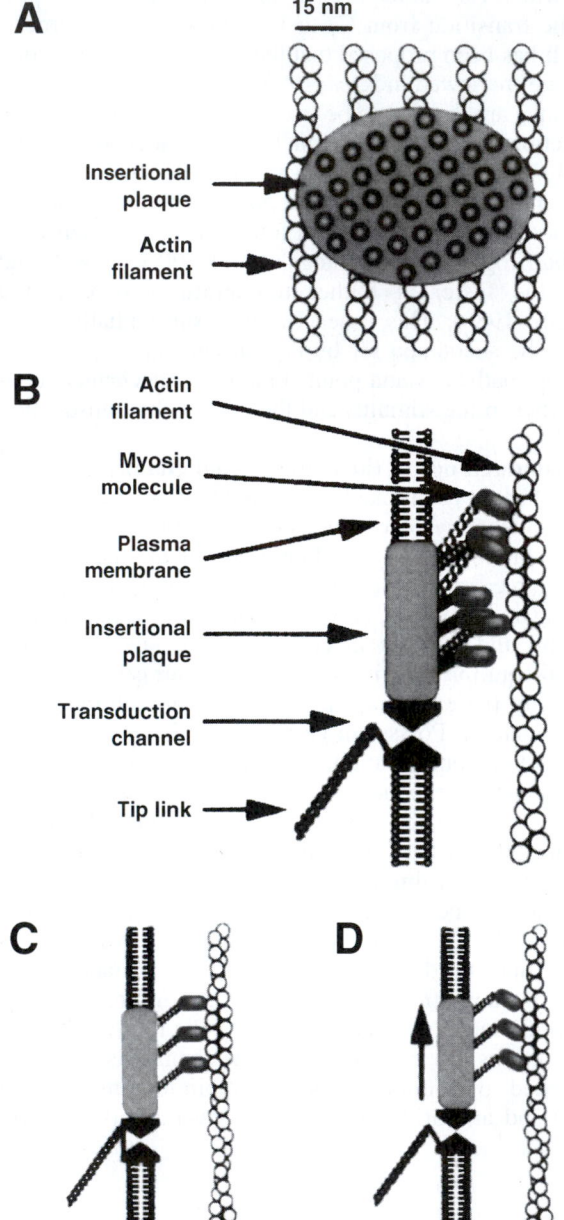

A

15 nm

Insertional plaque

Actin filament

B

Actin filament

Myosin molecule

Plasma membrane

Insertional plaque

Transduction channel

Tip link

C   **D**

FIGURE 15. Proposed mechanism of adaptation: (A) frontal view of insertional plaque on actin filaments; (B) side view of insertional plaque with associated mechanosensitive transduction channel in cell membrane; and (C and D) power stroke of actin-myosin complex. (Reproduced with permission from Hudspeth and Gillespie, 1994. Copyright 1994 by Cell Press.)

myosin regulatory protein found to be concentrated in the tips of stereocilia, which in turn promotes the dissociation of myosin from the actin filaments, allowing the plaque with the motors to slip.

The **physiologic significance** of adaptation is clear for vestibular hair cells where prolonged static displacements naturally occur during which sensitivity to transient stimuli must be maintained. However, the function of adaptation in the mammalian cochlea is less obvious (Kros, 1996). First, no sustained displacements due to acoustic stimuli are expected and, second, the concentration of Ca^{2+} in cochlear endolymph is about one-tenth that of vestibular endolymph, greatly reducing the expected influx through the transduction channels during stimulation. Third, the adaptation process has been estimated to have an upper frequency limit of about 160 Hz, limiting its potential usefulness as a response to acoustic stimuli. Physiological significance in the cochlea may lie more with homeostasis of the system in the face of small readjustments of the position and/or size of cells during normal changes in systemic osmolarity and local metabolic processes or during development.

C. Basolateral Membrane Channels

Although there are differences among hair cell types, they all contain a major basolateral **K+ conductance,** comprised of voltage-dependent and Ca^{2+}-activated K+ channels, and **voltage-gated Ca^{2+} channels** (Kros, 1996; Fuchs, 1992). The K+ channels serve to shape the receptor potential initiated by the acoustically gated transduction channels and the Ca^{2+} channels to initiate transmitter release at the presynaptic sites, respectively. The presence of voltage-gated Ca^{2+} channels alone would be expected to lead to regenerative depolarization during acoustic stimulation, however, the basolateral voltage-gated K+ channels provide an active repolarizing influence in the mature cochlea. Interestingly, immature inner hair cells have a different complement of channels including excitable Na+ channels and a relatively slow K+ channel, which lead to spontaneous and evoked action potentials. During development (Chapter 35; Developmental Changes in Ion Channels), a large, fast K+ channel is expressed that greatly speeds up the membrane time constant, preventing action potentials (Kros *et al.,* 1998).

The Ca^{2+} channels in hair cells appear to be of the L type, which are characterized by activation at depolarized membrane potentials, little or no inactivation, and block by dihydropyridines such as nifedipine. These channels differ slightly from the L-type current of neurons by a more negative voltage activation range and more rapid kinetics (Figs. 16A and B) (Fuchs, 1992). In **inner hair cells,** the Ca^{2+} currents are activated above –60 mV and peak near –20 to 0 mV (see Fig. 16A.), a range that correlates well with that of receptor potentials found in recordings made *in vivo* and with predictions from *in vitro* experiments (Kros, 1996). These Ca^{2+} channels are therefore well poised to participate in the graded release of neurotransmitter at the basolateral synapses to Type I afferent fibers in the auditory nerve. The kinetics of channel activation have been estimated to be sufficiently fast so as not to limit accurate following of nervous

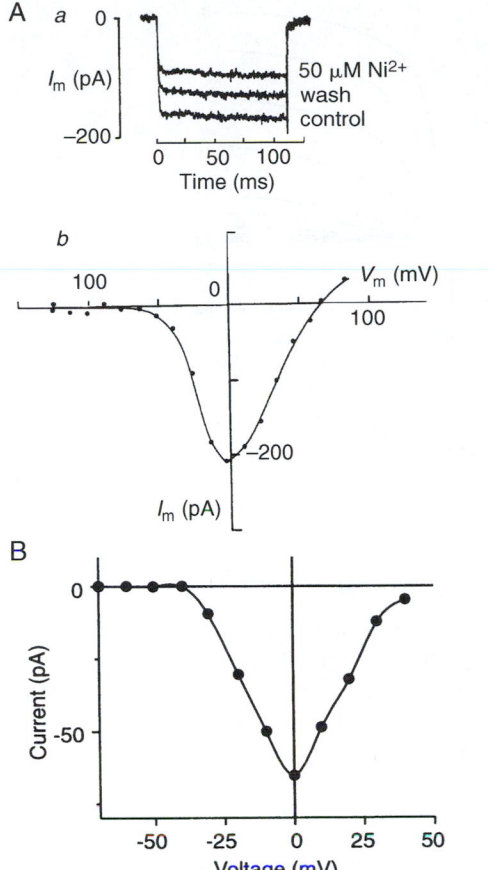

FIGURE 16. Basolateral calcium currents (whole-cell recordings): (A) the avian equivalent of inner hair cells, representative traces (a) and current-voltage relationship (b) and (B) outer hair cells. [Reprinted from (A) *Prog. Neurobiol.* **39**, Fuchs, P. A. Ionic currents in cochlear hair cells, pp. 493–505. Copyright 1992 with kind permission from Elsevier Science Ltd, The Boulevard, Langford Lane, Kidlington OX51GB, UK. and (B) from *Neurosci. Lett.* **125**, Nakagawa, T., Kakehata, S., Akaike, N., Komune, S., Takasaka, T., and Uemura, T., Calcium channel in isolated outer hair cells of guinea pig cochlea, pp. 81-84. Copyright 1991 with kind permission of Elsevier Science-NL, Sara Burgerhartstraat 25, 1055 KV Amsterdam, The Netherlands.]

TABLE 3   Inner Hair Cell Potassium Channels

Property	Fast K channel	Slow K channel
Activation time constant (ms)	0.2-0.4	2-10
Activation by membrane potential	Depol.	Depot.
Inactivation	No	No
Single-channel conductance	>200 pS	Smaller
Block by TEA	Yes	No
Block by CTX	Yes	Not tested
Block by 4-AP	No	Yes
Apamin	Not tested	No

TEA, tetraethylammonium; CTX, charybdotoxin; 4-AP, 4-amino pyridine.

cells of mammals have Ca^{2+}-activated K^+ channels that act in synergy with the voltage-gated Ca^{2+} channels to **electrically resonate** at the characteristic frequency of the cell. This forms part of the basis of frequency discrimination in these species. Electrical tuning is not, however, significantly present in mammalian cochlear hair cells. Sharp tuning of mammalian IHCs *in vivo* is thought to be the result of active mechanical feedback from other cochlear elements such as OHCs (see later section on reverse transduction).

In inner hair cells, the K^+ currents are activated by depolarization in the range of –60 to –20 mV, a range that correlates well with that of the resting and receptor potentials found in recordings made *in vivo* and with that of the voltage-gated Ca^{2+} channels described earlier. The resting potential of IHCs *in vivo* is about –40 mV, which reflects a compromise of the highly negative EMF from the basolateral K^+ channels by the depolarized EMF of the nonselective cation conductance in the apical membrane (Dallos, 1986).

The total cell K^+ current was found to be composed of both a fast and slow component (Fig. 17) The fast component is carried by channels resembling *maxi* K (or *BK*) channels that are characterized by rapid activation by membrane depolarization, block by TEA and charybdotoxin, insensitivity to 4-amino pyridine (4-AP), and single-channel conductance with symmetrical high K^+ concentration of more than 200 pS (Gitter *et al.,* 1992; Fuchs, 1992; Kros, 1996). The slow component is a delayed rectifier K^+ channel exhibiting slow activation by depolarization, no inactivation, a smaller single-channel conductance than the fast component, block by 4-AP but not TEA. Surprisingly, the fast component of the current is apparently independent of extracellular Ca^{2+} (Kros and Crawford, 1990), although the equivalent current in other types of hair cell is highly sensitive to Ca^{2+} (Fuchs, 1992).

The profile of K^+ currents in outer hair cells is different from that of IHCs (Kros, 1996; Fuchs, 1992). The primary K^+ conductance is due to $I_{K,n}$ channels, which are deactivated below –90 mV and fully activated at –50 mV, have a single-channel conductance of about 45 pS, are activated by Ca^{2+} ($P_o = 0.5$ at

discharge rates to the acoustic stimulus (Kros, 1996). By contrast, the Ca^{2+} currents in **outer hair cells** are activated only above –30 mV (see Fig. 16B), a range that is far removed from the resting membrane potential near –80 mV found *in vivo* and never reached by receptor potentials, which apparently saturate at 15 mV (Kros, 1996). (The curve would be expected to shift slightly to the left at physiological perilymphatic Ca^{2+} levels, but the channel would still not likely show significant activation at voltages observed in OHC.) There is therefore no evidence at this time for a function of these voltage-gated Ca^{2+} channels in OHCs; there are also no measurements of neural activity in the Type II afferent fibers in the auditory nerve that synapse exclusively to the OHCs.

The K^+ channels in hair cells (Table 3) vary considerably among type and species. The cochlear hair cells of many nonmammalian vertebrates and the vestibular hair

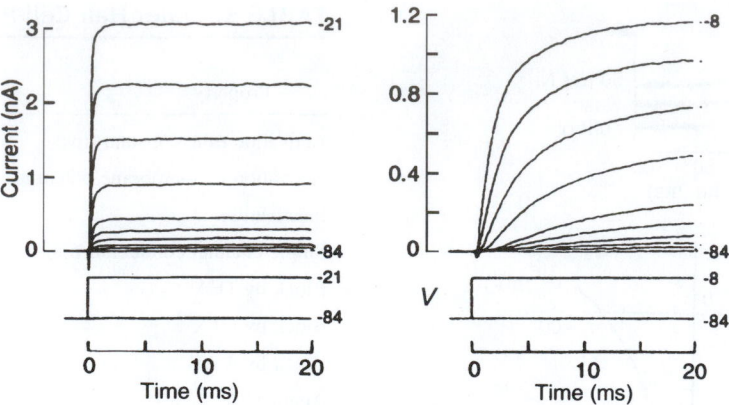

FIGURE 17. Basolateral potassium currents (whole-cell recordings) from inner hair cells. Left panel: fast component; right panel: slow component. [Reproduced with permission from Kros, C. J., and Crawford, A. C. (1989). Components of the membrane current in guinea-pig inner hair cells. *In* "Cochlear Mechanisms. Structure, Function and Models" (J. P. Wilson and D. T. Kemp, Eds.), pp. 189–195. Plenum Press, New York.]

0.2 µM), and are blocked by Ba^{2+} (Mammano *et al.*, 1995). There is also a K^+ current that activates at potentials more depolarized than –35 mV and with increasing cytosolic Ca^{2+} activity (P_o = 0.5 at 0.2 µM), has a single-channel conductance of > 200 pS, is noninactivating, is inhibited by TEA, and makes only a very small contribution to the total K^+ conductance of the basolateral membrane of OHCs (Gitter *et al.*, 1992; Housley and Ashmore, 1992). Both channels have been found to display these properties in OHCs *in situ* as well as in separated cells (Mammano *et al.*, 1995).

In addition to these two K^+ currents, the basolateral membrane contains a nonselective cation conductance that is permeable to Ca^{2+} as well as Na^+ and K^+ (Housley and Ashmore, 1992). Its function *in vivo* is not known but may be associated with a purinergic receptor (see later discussion). This conductance is clearly not dominant *in vivo* because the resting membrane potential [–70 to –80 mV (Dallos, 1986)] is not far from the Nernst potential for K^+.

D. Synaptic Release of Vesicles

Depolarization by acoustic stimulation leads to Ca^{2+}-dependent release of neurotransmitter at the 30–40 synapses with the afferent fibers at the base of each IHC (Sewell, 1996). This process can be monitored *in vitro* as changes in **membrane capacitance** (Lindau and Neher, 1988). Opening of Ca^{2+} channels during depolarization raises the local free Ca^{2+} concentration to tens or hundreds of micromolar (Roberts, 1994), which is thought to trigger synaptic exocytosis (Chapter 41). Depolarization led to increases in membrane capacitance (up to 5% of the initial value) that continued for up to 2 s. It was estimated that this was sufficient to exhaust more than five times the number of vesicles initially in close apposition to the plasma membrane at active zones, suggesting that hair cells are capable of rapidly replenishing vesicles at release sites (Parsons *et al.*, 1994). Capacitance returned during repolarization toward its prestimulus level with a time constant of about 14 s, but only in perforated-patch recordings (Chapter 26), suggesting that membrane retrieval depends on a diffusible intracellular factor.

E. Reverse Transduction

Discrimination of frequency information in acoustic signals is performed broadly at an initial level by mechanical tuning of the basilar membrane as a function of position along the cochlea. In lower vertebrates, frequency discrimination is accomplished by tuning the electrical properties of the hair cells. The exquisite sensitivity and fine frequency discrimination of mammalian hearing depends on amplification of the incoming sound within the organ of Corti. It is widely held that the OHCs sense displacements caused by sound and feedback forces, which enhance the basilar membrane motion by reducing the inherent damping of the cochlear partition. In support of this hypothesis, OHCs show membrane potential-induced length changes at acoustic rates. This process has been termed **reverse transduction** by the **cochlear amplifier**. This property of the organ of Corti leads to an epiphenomenon called **otoacoustic emission** in which mechanical fluctuations are generated by the inner ear, resulting in sounds emanating from the ear (Brownell, 1990).

Outer hair cells shorten and lengthen by up to 5% when depolarized and hyperpolarized, respectively. Longitudinal motions of the OHCs are assumed to be transmitted *in vivo* to the whole organ of Corti by the rigid reticular lamina at the top and by the Deiters' cell bodies at the base, mechanically coupled to the basilar membrane. Shortening *in vivo* would cause an upward pull on the basilar membrane and/or a downward pull on the reticular lamina. The latency is less than 0.1 ms and can be driven to frequencies in excess of 22 kHz (measurement of upper frequency limited by instrumentation) (Dallos and Evans, 1995). This response is driven by membrane potential rather than ionic currents and is independent of ATP, ruling out commonly observed cytoskeletal motors (Alberts *et al.*, 1994). It is widely believed that the motor molecules are voltage-sensitive proteins packed into the lateral membrane of the OHCs; activation of the motor proteins is thought to be associated with movements of gating charges (Fig. 18). This hypothesis is largely based on the observations that there is a congruence of the distributions of (1) densely packed particles, (2) membrane gating currents, (3) nonlin-

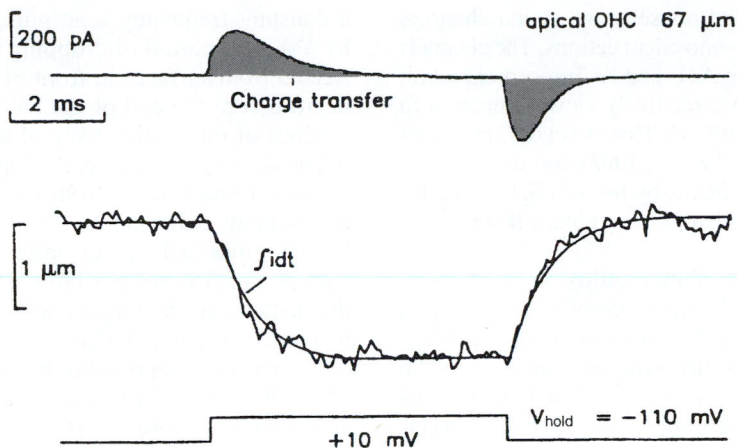

FIGURE 18. Correlation of gating charge movements (upper panel) with displacement of cell body (lower panel). Integral of the gating charge has been scaled and is shown plotted over the displacement for comparison of the kinetics. (Reprinted from *Neurosci. Res. Suppl.* **12**, Ashmore, J. F., Forward and reverse transduction in the mammalian cochlea, S39–S50. Copyright 1990 with kind permission of Elsevier Science-NL, Sara Burgerhartstraat 25, 1055 KV Amsterdam, The Netherlands.)

ear membrane capacitance, and (4) a force-generating membrane only on the lateral membrane between the level of the nucleus to near the cuticular plate. The presence of both gating current (Fig. 18) and nonlinear capacitance suggests that the motors act by conformational changes of the protein rather than by electrostatic repulsion between separate elements within the plane of the lipid bilayer (Holley, 1996). Forces estimated from distortions of the organ of Corti *in situ* due to changes of membrane potential can account for the enhanced responses to sound required by the cochlear amplifier (Mammano *et al.*, 1995).

It was initially assumed that this mechanism would function by motile responses to the OHC receptor potential. It was recently shown from measurements of the amplitude and phase relationships of the receptor potential to stereociliary displacement that those data were consistent with *attenuation* rather than amplification of the stimulus at high frequencies (Preyer *et al.*, 1996). An alternate proposed source of voltage changes that might drive the lateral motors *in vivo* is extracellular potential gradients across the OHC (Dallos and Evans, 1995).

Deiters' cells were previously thought to offer only a passive support to the OHCs and a mechanical connection to the acoustic stimulus at the basilar membrane. Recent evidence, however, suggests that these cells have a **Ca^{2+}-dependent motile response.** Increases in cytoplasmic Ca^{2+} caused an extension of the head of the phalangeal process away from the body by 0.5 to 1 μm within a few hundred milliseconds and the stiffness of the phalangeal process to increase by 28 to 51% (Dulon, 1995).

Sinusoidal fluctuations in membrane potential have been shown in lower vertebrate hair cells to cause **mechanical oscillations of the hair bundle,** but the operational frequency range is only from 20 to 320 Hz, possibly limited by the time constant of the basolateral membrane (Hudspeth and Gillespie, 1994). Myosin in the stereocilia is estimated to be sufficiently plentiful and powerful to account for the amplification produced by the organ of Corti. However, if the stereociliary motion were modulated by Ca^{2+} rather than

membrane potential, activation of stereociliar myosin by Ca^{2+} would not be restricted by the membrane's time constant. It remains to be determined whether these motors can operate at the highest frequencies detected by the mammalian cochlea.

F. Receptors

A number of receptors for neurotransmitters and neuromodulators have been identified on cochlear hair cells. Some are known to mediate synaptic transmission while the physiological significance of others has not yet been identified.

1. Outer Hair Cells

Receptors for several neurotransmitters and neuromodulators (Table 4) have been identified on outer hair cells by histochemical and electrophysiologic observations (Sewell, 1996). The efferent synapses on the outer hair cells release **acetylcholine** (ACh) and *γ-aminobutyric acid* (GABA) as neurotransmitters from two populations of fibers and may corelease ATP with ACh (Guinan, 1996; Sewell, 1996). There is evidence for the presence of both ionotropic and metabotropic receptors for ATP and ACh and for the ionotropic GABA$_A$ receptor (Housley *et al.*, 1995; Sewell, 1996). **Ionotropic receptors** are coupled directly to ion channels, whereas **metabotropic receptors** act via G protein pathways that ultimately regulate ion channel activity and/or other cellular processes. It is not yet clear whether the responses to ACh occur via a single unusual receptor isoform containing α_9-subunits or whether there are multiple isoforms of ACh receptors present in OHCs.

Activation of ACh receptors causes membrane hyperpolarization apparently through an ion channel permeable to cations including Ca^{2+}. The entry of Ca^{2+} activates Ca^{2+}-dependent K^+ channels, likely the SK type (Nenov *et al.*, 1996), leading to hyperpolarization. Ionotropic ATP receptors depolarize the hair cell by opening associated nonselective cation channels permeable to Ca^{2+} as well as monovalent cations. The channels associated with the ionotropic ATP and ACh receptors admit Ca^{2+} into the cell, leading to a relatively slow elongation (a process referred to as **slow motility**). This effect of Ca^{2+} is mediated via calmodulin (Schacht et al., 1995) and may involve phosphorylation of effector proteins by myosin light chain kinase and/or calmodulin-dependent protein kinase II (Puschner and Schacht, 1997).

In addition to its effects on slow motility, ACh has been found to control the gain and magnitude of fast motility in OHCs (Sziklai et al., 1996). ACh evokes an increase in magnitude and gain of electromotility, which is sensitive to the muscarinic blocker atropine. The effects of ACh on electromotility were found to be likely mediated via phosphorylation by PKC of undetermined proteins (King et al., 1996).

Receptors have also been found on OHCs for the peptides substance P and calcitonin gene-related peptide (CGRP). Substance P hyperpolarizes these cells by down-regulating the nonselective cation conductance in the lateral wall via a pertussis toxin-insensitive G protein (Kakehata et al., 1993). CGRP had no effect on the resting cytoplasmic Ca^{2+} concentration but potentiated by about threefold the increase in Ca^{2+} caused by ACh stimulation in chick hair cells (Shigemoto and Ohmori, 1990).

2. Inner Hair Cells

Inner hair cells have been found to have both metabotropic and ionotropic receptors for extracellular ATP (Sugasawa et al., 1996). At submicromolar ATP concentrations, the metabotropic receptors raise intracellular Ca^{2+} concentration, which hyperpolarizes IHCs via Ca^{2+}-sensitive K^+ channels. The ionotropic receptors are nonselective cation channels activated at higher ATP concentrations and which mainly have a depolarizing effect on the IHCs. Although the source of agonist for these receptors has not yet been identified, Ca^{2+}-dependent release of ATP from the organ of Corti has been observed (Wangemann, 1996).

G. Echolocation

Bats can capture flying prey by echolocation (biosonar) even among vegetation and other bats. Sounds are vocalized by the bat at frequencies near 60 kHz (the upper range of human hearing is about 20 kHz) and typically consist of a constant-frequency tone pulse for 10–100 ms followed by a shorter period of dropping frequency. The pulse is reflected from surfaces in front of the bat and is returned and heard before the end of the emitted pulse. Several characteristics of the returned sound pulse are interpreted by the bat as an acoustic image. A Doppler shift of the frequency of the returned sound from that of the emitted sound carries velocity information (the rate of closure on a target). For example, a Doppler shift from 61 to 62 kHz corresponds to 6.3 miles per hour. The delay of the echo gives the distance to the target such that a 1-ms echo delay corresponds to a target distance of 17 cm. Temporal delay acuity has been reported to be as fine as 10 ns (Simmons et al., 1990), a surprisingly short time span for recognition by a biological system. It has been suggested, however, that bats may not respond directly to processing of the time delay but rather to spectral processing, which would require only a resolution of about a few kilohertz (reviewed in Neuweiler and Schmidt, 1993, but see Simmons, 1993; Simmons et al., 1996).

The fine tuning of hearing used in echolocation is mostly achieved by mechanical specializations of the cochlea, which spread out the physical mapping of the biosonar frequencies over a full half turn of the cochlea. There is also an amplifying reverberation of the acoustic wave traveling on the basilar membrane, which is due either to reflections at a discontinuity of the basilar membrane thickness or to different radial oscillation modes of the basilar membrane (Russell and Koessi, 1995). As in other mammalian cochleae, the tuning (and the bat's special reverberation) is thought to depend on active processes in the OHCs. Tuning of auditory nerve fibers, and therefore likely of basilar membrane motion, to acoustic stimuli is amazingly high in the bat with Q10 values (center frequency/bandwidth 10 dB from the tip) as high as 610 in comparison to typical values of 1 to 10 for nonecholocating animals (Russell and Koessi, 1995). Processing by central neural pathways of the detected echos has been reviewed by Suga (1989), Suga et al., (1995), Simmons (1989), and Simmons et al., (1996). An interesting specialization of the bats' prey has been found in the dogbane tiger moth, which apparently emits clicks when a bat approaches in order to interfere with the sonar echos, thereby jamming the signal and averting becoming a meal (Fullard et al., 1994)!

V. Summary

The cellular bases of hearing in the auditory periphery were described with emphasis on the sensory hair cells and on cells of the stria vascularis. The latter is a complex structure responsible both for the generation of a lumen-positive electrical potential and for the secretion of potassium to an unusually high level in the cochlear lumen. Transduction of sound waves into neural impulses by the hair cells relies on modulation of an electric current carried by potassium flowing through the hair cells. The ion channels, carriers, and pumps in the marginal cell membranes, which account for constitutive potassium secretion, were described in this chapter. A wide variety of extracellular and cytosolic signaling pathways that regulate the rate of potassium secretion

TABLE 4 Effects of Outer Hair Cell Receptors on Fast and Slow Motility, Membrane Potential, and Calcium Concentration

	Fast	Slow	V_c	$[Ca^{2+}]_i$
ACh	x	x	Hyper.	x
ATP		x	Depol.	x
GABA			Hyper.	No
Substance P			Hyper.	No
CGRP				No, but potentiates ACh response

were also described. The specializations of hair cell physiology were discussed, including the transduction channel in the sensory cilia and the highly unusual lateral membrane motors of OHCs. The description of the physiological mechanisms employed by inner ear epithelial cells was related in this chapter to the function of the organ as a whole.

Bibliography

Alberts, B., Bray, D., Lewis, J., Raff, M., Roberts, K., and Watson, J. D. (1994). "Molecular Biology of the Cell," Third edition, Garland Publishing, New York.

Barhanin, J., Lesage, F., Guillemare, E., Pink, M., Lazdunski, M., and Romey, G. (1996). K_VLQTl and IsK (minK) proteins associate to form the I_{Ks}, cardiac potassium current. *Nature* **384**, 78-80.

Ben-Efraim, I., Shai, Y., and Attali, B. (1996). Cytoplasmic and extracellular IsK peptides activate endogenous K^+ and Cl^- channels in *Xenopus* oocytes—Evidence for regulatory function. *J. Biol. Chem.* **271**, 8768–8771.

Brownell, W. E. (1990). Outer hair cell electromotility and otoacoustic emissions. *Ear Hearing* **11**, 82–92.

Corey, D. P., and Hudspeth, A. J. (1979). Ionic basis of the receptor potential in a vertebrate hair cell. *Nature* **281**, 675–677.

Corey, D. P., and Hudspeth, A. J. (1983). Kinetics of the receptor current in bullfrog saccular hair cells. *J. Neurosci.* **3**, 962–976.

Curtis, L. M., Ten Cate, W. J., and Rarey, K. E. (1993). Dynamics of Na^+,K^+-ATPase sites in lateral cochlear wall tissues of the rat. *Eur. Arch. Otorhinolaryngol.* **250**, 265–270.

Dallos, P. (1986). Neurobiology of cochlear inner and outer hair cells: Intracellular recordings. *Hear. Res.* **22**, 185–198.

Dallos, P., and Evans, B. N. (1995). High-frequency motility of outer hair cells and the cochlear amplifier. *Science* **267**, 2006–2009.

Denk, W., Holt, J. R., Shepherd, G. M., and Corey, D. P. (1995). Calcium imaging of single stereocilia in hair cells: Localization of transduction channels at both ends of tip links. *Neuron* **15**, 1311–1321.

Dulon, D. (1995). Ca^{2+} signaling in Deiters' cells of the guinea-pig cochlea: Active process in supporting cells? *In* "Active Hearing" (A. Flock, D. Ottoson, and M. Ulfendahl, Eds.), pp. 195–208. Elsevier Science, New York.

Erichsen, S., Stierna, P., Bagger-Sjöbäck, D., Curtis, L. M., Rarey, K. E., Schmid, W., and Hultcrantz, M. (1998). Distribution of Na^+,K^+-ATPase is normal in the inner ear of a mouse with a null mutation of the glucocorticoid receptor. *Hear. Res.* **124**, 146–154.

Evans, B. N., and Dallos, P. (1993). Stereocilia displacement induced somatic motility of cochlear outer hair cells. *Proc. Natl. Acad. Sci. USA* **90**, 8347–8351.

Fuchs, P. A. (1992). Ionic currents in cochlear hair cells. *Prog. Neurobiol.* **39**, 493–505.

Fullard, J. H., Simmons, J. A., and Saillant, P. A. (1994). Jamming bat echolocation: The dogbane tiger moth Cycnw tenera times its clicks to the terminal attack calls of the big brown bat Eptesicus fuscus. *J. Exp. Biol.* **194**, 285–298.

Gitter, A. H., Froemter, E., and Zenner, H. P. (1992). C-type potassium channels in the lateral cell membrane of guinea-pig outer hair cells. *Hear. Res.* **60**, 13–19.

Guinan, J. J., Jr. (1996). Physiology of olivocochlear efferents. *In* "The Cochlea" (P. Dallos, A. N. Popper, and R. R. Fay, Eds.), pp. 435–502. Springer-Verlag, New York.

Hackney, C. M., and Fumess, D. N. (1995). Mechanotransduction invertebrate hair cells: Structure and function of the stereociliary bundle. *Am. J. Physiol.* **268**, C1–13.

Holley, M. C. (1996). Outer hair cell motility. *In* "The Cochlea" (P. Dallos, A. N. Popper, and R. R. Fay, Eds.), pp. 386–434. Springer-Verlag, New York.

Housley, G. D., and Ashmore, J. F. (1992). Ionic currents of outer hair cells isolated from the guinea-pig cochlea. *J. Physiol. London* **448**, 73–98.

Housley, G. D., Connor, B. J., and Raybould, N. P. (1995). Purinergic modulation of outer hair cell electromotility. *In* "Active Hearing" (A. Flock, D. Ottoson, and M. Ulfendahl, Eds.), pp. 221–238. Elsevier Science, New York.

Hudspeth, A. J., and Gillespie, P. G. (1994). Pulling springs to tune transduction: Adaptation by hair cells. *Neuron* **12**, 1–9.

Jaramillo, F. (1995). Signal transduction in hair cells and its regulation by calcium. *Neuron* **15**, 1227–1230.

Julien, N., Loiseau, A., Sterkers, O., Amiel, C., and Ferrary, E. (1994). Antidiuretic hormone restores the endolymphatic longitudinal K^+ gradient in the Brattleboro rat cochlea. *Pflügers Archiv.* **426**, 446–452.

Kakehata, S., Akaike, N., and Takasaka, T. (1993). Substance P decreases the non-selective cation channel conductance in dissociated outer hair cells of guinea pig cochlea. *Ann. NY Acad. Sci.* **707**,476–479.

Kikuchi, T., Kimura, R. S., Paul, D. L., and Adams, J. C. (1995). Gap junctions in the rat cochlea: Immunohistochemical and ultrastructural analysis. *Anat. Embryol. Berlin* **191**, 101–118.

King, B. F., Wang, S. Y., and Bumstock, G. (1996). P2 purinoceptor-activated inward currents in follicular oocytes of *Xenopus laeveis. J. Physiol. London* **494**, 17–28.

Klinke, R., and Evans, E. F. (1977). Evidence that catecholamines are not the afferent transmitter in the cochlea. *Exp. Brain Res.* **28**, 315–324.

Koefoed-Johnsen, V., and Ussing, H. H. (1958). The nature of the frog skin potential. *Acta Physiol. Scand.* **42**, 298–308.

Konishi, T., Hamrick, P. E., and Walsh, P. J. (1978). Ion transport in guinea pig cochlea. I. Potassium and sodium transport. *Acta Otolaryngol. Stockholm* **86**, 22–34.

Kros, C. J. (1996). Physiology of mammalian cochlear hair cells. *In* "The Cochlea" (P. Dallos, A. N. Popper, and R. R. Fay, Eds.), pp. 318–385. Springer-Verlag, New York.

Kros, C. J., and Crawford, A. C. (1990). Potassium currents in inner hair cells isolated from the guinea-pig cochlea. *J. Physiol. London* **421**, 263–291.

Kros, C. J., Ruppersberg, J. P., Rusch, A. (1998). Expression of a potassium current in inner hair cells during development of hearing in mice. *Nature* **394**, 281–284.

Kujawa, S. G., Erostegui, C., Fallon, M., Crist, J., and Bobbin, R. P. (1994). Effects of adenosine 5'-triphosphate and related agonists on cochlear function. *Hear. Res.* **76**, 87–100.

Lindau, M., and Neher, E. (1988). Patch-clamp techniques for time-resolved capacitance measurements in single cells. *Pflugers Archiv.* **411**, 137–146.

Liu, J., Kozakura, K., and Marcus, D. C. (1995). Evidence for purinergic receptors in vestibular dark cell and strial marginal cell epithelia of the gerbil. *Auditory Neurosci.* **1**, 331–340.

Ma, Y. L., Gerhardt, K. J., Curtis, L. M., Rybak, L. P., Whitworth, C., and Rarey, K. E. (1995). Combined effects of adrenalectomy and noise exposure on compound action potentials, endocochlear potentials and endolymphatic potassium concentrations. *Hear. Res.* **91**, 79–86.

Mammano, F., Kros, C. J., and Ashmore, J. F. (1995). Patch-clamped responses from outer hair cells in the intact adult organ of Corti. *Pflugers Archiv.* **430**, 745–750.

Marcus, D. C., and Chiba, T. (1999). K^+ and Na^+ absorption by outer sulcus epithelial cells. *Hear. Res.* **134**, 48–56.

Marcus, D. C., and Shen, Z. (1994). Slowly activating, voltage-dependent K^+ conductance is apical pathway for K^+ secretion in vestibular dark cells. *Am. J. Physiol.* **267**, C857–C864.

Marcus, D. C., Sunose, H., Liu, J., Shen, Z., and Scofield, M. A. (1997). P_{2U} purinergic receptor inhibits apical IsK/KvLQTl channel via protein kinase C in vestibular dark cells. *Am. J. Physiol.* **273,** C2022–C2029.

Markin, V. S., and Hudspeth, A. J. (1995). Gating-spring models of mechanoelectrical transduction by hair cells of the internal ear. *Annu. Rev. Biophys. Biomol. Struct.* **24,** 59–83.

Mori, N., Shugyo, A., and Asai, H. (1989). The effect of arginine-vasopressin and its analogues upon the endocochlear potential in the guinea pig. *Acta Otolaryngol. Stockholm* **107,** 80–84.

Nenov, A. P., Norris, C., Bobbin. R, P. (1996). Acetylcholine response in guinea pig outer hair cells. II. Activation of a small conductance Ca^{2+}-activated K^+ channel. *Hear. Res.* **101,** 149–172.

Neuweiler, G., and Schmidt, S. (1993). Audition in echolocating bats *Curr. Opin. Neurobiol.* **3,** 563–569.

Ogawa, K., and Schacht, J. (1995). P2Y purinergic receptors coupled to phosphoinositide hydrolysis in tissues of the cochlear lateral wall. *Neuroreport* **6,** 1538–1540.

Parsons, T. D., Lenzi, D., Almers, W., and Roberts, W. M. (1994). Calcium-triggered exocytosis and endocytosis in an isolated presynaptic cell: Capacitance measurements in saccular hair cells. *Neuron* **13,** 875–883.

Patuzzi, R. (1996). Cochlear micromechanics and macromechanics. *In* "The Cochlea" (P. Dallos, A. N. Popper, and R. R. Fay, Eds.), pp. 186–257. Springer-Verlag, New York.

Preyer, S., Renz, S., Hemmert, W., Zenner, H. P., and Gummer, A. W. (1996). Receptor potential of outer hair cells isolated from base to apex of the adult guinea-pig cochlea: Implications for cochlear tuning mechanisms. *Auditory Neurosci.* **2,** 145–157.

Puschner, B., and Schacht, J. (1997). Calmodulin-dependent protein kinases mediate calcium-induced slow motility of mammalian outer hair cells. *Hear. Res* **110,** 251–258.

Ricci, A. J., and Fettiplace, R. (1998). Calcium permeation of the turtle hair cell mechanotransducer channel and its relation to the composition of endolymph. *J.Physiol. London* **506,** 159–173.

Roberts, W. M. (1994). Localization of calcium signals by a mobile calcium buffer in frog saccular hair cells. *J. Neurosci.* **14,** 3246–3262.

Russell, I. J., and Koessi, M. (1995). Measurements of the basilar membrane resonance in the cochlea of the mustached bat. *In* "Active Hearing" (A. Flock, D. Ottoson, and M. Ulfendahl, Eds.), pp. 295–306. Elsevier Science, New York.

Salt, A. N., Melichar, I., and Thalmann, R. (1987). Mechanisms of endocochlear potential generation by stria vascularis. *Laryngoscope* **97,** 984–991.

Sanguinetti, M. C., Curran, M. E., Zou, A., Shen, J., Spector, P. S., Atkinson, D. L., and Keating, M. T. (1996). Coassembly of KvLQTl and minK (IsK) proteins to form cardiac I_{Ks} potassium channel. *Nature* **384,** 80–83.

Schacht, J., Fessenden, J. D., and Zajic, G. (1995). Slow motility of outer hair cells. *In* "Active Hearing" (A. Flock, D. Ottoson, and M. Ulfendahl, Eds.), pp. 209–220. Elsevier Science, New York.

Sewell, W. F. (1996). Neurotransmitters and synaptic transmission. *In* "The Cochlea" (P. Dallos, A. N. Popper, and R. R. Fay, Eds.), pp. 503–533. Springer-Verlag, New York.

Shah, D. M., Freeman, D. M., and Weiss, T. F. (1995). The osmotic response of the isolated, unfixed mouse tectorial membrane to isosmotic solutions: Effect of Na^+, K^+, and Ca^{2+} concentration. *Hear. Res.* **87,** 187–207.

Shen, Z., Marcus, D. C., Sunose, H., Chiba, T., and Wangemann, P. (1997). I_{sK} channel in strial marginal cells: Voltage-dependence, ion-selectivity, inhibition by 293B and sensitivity to clofilium. *Auditory Neurosci.* **3,** 215–230.

Shiga, N., and Wangemann, P. (1995). Ion selectivity of volume regulatory mechanisms present during a hypoosmotic challenge in vestibular dark cells. *Biochem. Biophys. Ada* **1240,** 48–54.

Shigemoto, T., and Ohmori, H. (1990). Muscarinic agonists and ATP increase the intracellular Ca^{2+} concentration in chick cochlear hair cells. *J. Physiol. London* **420,** 127–148.

Shimozono, M., Scofield, M.A., and Wangemann, P. (1997) Functional evidence for a monocarboxylate transporter (MCT) in strial marginal cells and molecular evidence for MCT1 and MCT2 in stria vascularis. *Hear. Res.* **114,** 213–222.

Simmons, J. A. (1989). A view of the world through the bat's ear: The formation of acoustic images in echolocation. *Cognition* **33,** 155–199.

Simmons, J. A. (1993). Evidence for perception of fine echo delay and phase by the FM bat, Eptesicus fuscus. *J. Comp. Physiol. A* **172,** 533–547.

Simmons, J. A., Dear, S. P., Ferragamo, M. J., Haresign, T., and Fritz, J. (1996). Representation of perceptual dimensions of insect prey during terminal pursuit by echolocating bats. *Biol. Bull.* **191,** 109–121.

Simmons, J. A., Ferragamo, M., Moss, C. F., Stevenson, S. B., and Altes, R. A. (1990). Discrimination of jittered sonar echoes by the echolocating bat, Eptesicus fuscus: The shape of target images in echolocation. *J. Comp. Physiol. A* **167,** 589–616.

Sterkers, 0., Saumon, G., Tran Ba Huy, P., and Amiel, C. (1982). K, Cl, and H_2O entry in endolymph, perilymph, and cerebrospinal fluid of the rat. *Am. J. Physiol.* **243,** F173–F180.

Suga, N. (1989). Principles of auditory information-processing derived from neuroethology. *J. Exp. Biol.* **146,** 277–286.

Suga, N., Butman, J. A., Teng, H., Yan, J., and Olsen, J. F. (1995). Neural processing of target-distance information in the mustached bat. *In* "Active Hearing" (A. Flock, D. Ottoson, and M. Ulfendahl, Eds.), pp. 13–30. Elsevier Science, New York.

Sugasawa, M., Erostegui, C., Blanchet, C., and Dulon, D. (1996). ATP activates non-selective cation channels and calcium release in inner hair cells of the guinea-pig cochlea. *J. Physiol. London* **491,** 707–718.

Sunose, H., Liu, J., Shen, Z., and Marcus, D. C. (1997). cAMP increases apical IsK channel current and K^+ secretion in vestibular dark cells. *J. Membr. Biol.* **156,** 25–35.

Sziklai, I., He, D. Z. Z., and Dallos, P. (1996). Effect of acetylcholine and GABA on the transfer function of electromotility in isolated-outer hair cells. *Hear. Res.* **95,** 87–99.

Takeuchi, S., and Ando, M. (1998). Inwardly rectifying K^+currents in intermediate cells in the cochlea of gerbils: A possible contribution to the endocochlear potential. *Neurosci. Lett.* **247,** 175–178.

Takeuchi, S., and Irimajiri, A. (1996). Maxi-K^+ channel in plasma membrane of basal cells dissociated from the stria vascularis of gerbils. *Hear. Res.* **95,** 18–25.

Takumi, T., Ohkubo, H., and Nakanishi, S. (1988). Cloning of a membrane protein that induces a slow vuliage-gai.cu potassium currciil. *Science* **242,** 1042–1045.

Ten Cate, W. J., Curtis, L. M., and Rarey, K. E. (1994). Effects of low-sodium, high-potassium dietary intake on cochlear lateral wall Na^+K^+-ATPase. *Eur. Arch. Otorhinolaryngol.* **251,** 6–11.

Ten Cate, W. J., Curtis, L. M., Small, G. M., and Rarey, K. E. (1993). Localization of glucocorticoid receptors and glucocorticoid receptor mRNAs in the rat cochlea. *Laryngoscope* **103,** 865–871.

Tzounopoulos, T., Maylie, J., and Adelman, J. P. (1995). Induction of endogenous channels by high levels of heterologous membrane proteins in *Xenopus* oocytes. *Biophys. J.* **69,** 904—908.

Vetter, D. E., Mann, J. R., Wangemann, P., Liu, J., McLaughlin, K. J., Lesage, F., Marcus, D. C., Lazdunski, M., Heinemann, S. F., and Barhanin, J. (1996). Inner ear defects induced by null mutation of the *isk* gene. *Neuron* **17,** 1251–1264.

Wang, K. W., and Goldstein, S. A. (1995). Subunit composition of minK potassium channels. *Neuron* **14,** 1303–1309.

Wangemann, P. (1995). Comparison of ion transport mechanisms between vestibular dark cells and strial marginal cells. *Hear. Res.* **90,** 149–157.

Wangemann, P. (1996). Ca^{2+}-dependent release of ATP from the organ of Corti measured with a luciferin-luciferase bioluminescence assay. *Auditory Neurosci.* **2,** 187–192.

Wangemann, P. (1997). Kalium-Ionensekretion und Entstehung des endokochlearen Potentials in der Stria vascularis. *HNO* **45,** 205–209.

Wangemann, P., Liu, J., Shimozono, M., Schimanski, S., and Scofield, M. A. (2000). K$^+$ secretion in strial marginal cells is stimulated via β_1-adrenergic receptors but not via β_2-adrenergic or vasopressin receptors. *J. Membr. Biol.* **175,** 191–202.

Wangemann, P., and Liu, J. (1996). Beta-adrenergic receptors but not vasopressin-receptors stimulate the equivalent short circuit current in K$^+$ secreting inner ear epithelial cells. *J. Gen. Physiol.* **108,** 31a.

Wangemann, P., and Schacht, J. (1996). Homeostatic mechanisms in the cochlea. *In* "The Cochlea" (P. Dallos, A. N. Popper, and R. R. Fay, Eds.), pp. 130–185. Springer-Verlag, New York.

Wangemann, P., and Shiga, N. (1994). Cell volume control in vestibular dark cells during and after a hyposmotic challenge. *Am. J. Physiol.* **266,** C1046–C1060.

Wangemann, P., Liu, J., and Marcus, D. C. (1995a). Ion transport mechanisms responsible for K$^+$ secretion and the transepithelial voltage across marginal cells of stria vascularis *in vitro. Hear. Res. 84,* 19–29.

Wangemann, P., Liu, J., and Shiga, N. (1995b). The pH-sensitivity of transepithelial K$^+$ transport in vestibular dark cells. *J. Membr. Biol.* **147,** 255–262.

Wangemann, P., Liu, J., Shen, Z., Shipley, A., and Marcus, D. C. (1995c). Hypo-osmotic challenge stimulates transepithelial K$^+$ secretion and activates apical IsK channel in vestibular dark cells. *J. Membr. Biol.* **147,** 263–273.

Wangemann, P., Liu, J. Z., and Shiga, N. (1996a). Vestibular dark cells contain the Na$^+$/H$^+$ exchanger NHE-1 in the basolateral membrane. *Hear. Res.* **94,** 94–106.

Wangemann, P., Liu, J., Shimozono, M., and Scofield, M. A. (1999). β1-adrenergic receptors but not β2-adrenergic or vasopressin receptors regulate K$^+$ secretion in vestibular dark cells of the inner ear. *J. Membr. Biol.* **170,** 67–77.

Wangemann, P., Shen, Z., and Liu, J. (1996b). IC-induced stimulation of K$^+$ secretion involves activation of the I$_{sK}$ channel in vestibular dark cells. *Hear. Res.* **100,** 201–210.

Yao, X. F., and Rarey, K. E. (1996). Localization of the mineralocorticoid receptor in rat cochlear tissue. *Acta Otolaryngol. Stockholm* **116,** 493–496.

Zidanic, M., and Brownell, W. E. (1990). Fine structure of the intracochlear potential field. I. The silent current. *Biophys. J.* **57,** 1253–1268.

Appendix: Self-Referencing Electrodes for the Measurement of Extracellular Potential and Chemical Gradients

Many physiological processes produce extracellular gradients of electric fields, ions, nutrients, and respiratory gases. Such processes include constitutive ion transport and aerobic metabolism, as well as mechanisms involved in development and neural activity. In many cases these processes can be characterized through measurements of the gradients under control and experimental conditions. In many bulk tissues, ion and oxygen fluxes can be measured with radiotracers and/or static electrodes, and ion movements in single cells can often be monitored with fluorescence microscopy and patch-clamp electrodes. There are, however, a wide variety of situations to which these techniques cannot be applied and for which the self-referencing electrode is a powerful addition to the physiologist's arsenal of techniques. Self-referencing electrodes (Smith, 1995; Smith *et al.*, 1994) are capable of making noninvasive measurements from preparations of otherwise intractable geometries (irregular shapes and/or extremely small size) and are therefore ideal for measuring fluxes from inner ear epithelia that occur in patches with dimensions on the order of 0.1 mm. Unlike most other electrophysiological approaches, including patch-clamp recordings, ion fluxes can be observed with self-referencing ion-selective probes from electroneutral as well as electrogenic transport processes.

The purpose of self-referencing the probes is to average out noise and reduce the impact of the drift normally associated with microsensing electrodes by creating a periodic signal from a source that is intrinsically static or varying only slowly and irregularly. The first such probe measurements were made of current density by detection of the associated voltage drop as the current flowed through a resistive medium, the physiological saline in which the preparation was bathed (Jaffe and Nuccitelli, 1974). High-sensitivity voltage gradient measurements are made by vibrating a metal electrode at frequencies between 100 and 1000 Hz, with signal analysis by a phase-sensitive detector (lock-in amplifier) for each plane of vibration (Fig. A-1A) (Scheffey, 1989; Jaffe and Nuccitelli, 1974). In contrast to static electrodes, which intrinsically have a noise level and stability on the order of 1 mV, the vibrating current probe can measure signal in the **nanovolt** range. The electrode is typically made from a platinum wire that has been tapered to a point and insulated to within a few micrometers of the tip. The tip is electroplated to form a "fuzzy" surface of platinum black, which greatly increases the capacitance by expanding the area of electrical contact with the bathing solution. This large contact area gives good capacitive coupling

(low impedance) between the bath and electrode at the frequencies used. The electrode can be vibrated in two orthogonal directions at different frequencies over an excursion of about 20–40 μm (Fig. A-1A). The fine spatial resolution of the probe tip (down to approximately 4–8 μm with fine-tipped electrodes) can be exploited to map the spatial distribution of current-producing processes on large and robust single cells or tissues; two-dimensional vectors representing the current flows can be constructed.

A second method (Fig. A-1B) was developed for the measurement of chemical gradients using more slowly responding sensors such as ion-selective microelectrodes (Kuhtreiber and Jaffe, 1990). An electrode is stepped between two or more positions, residing at each for about 1 s per position. Each position is between ~2–50 μm apart de-

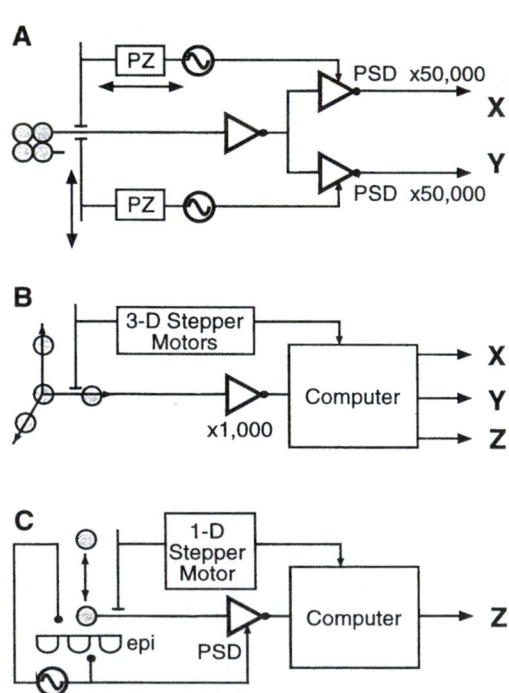

FIGURE A-1. Self-referencing probes. (A) Vibrating current probe driven in two orthogonal directions by piezoelectric (PZ) elements; signal is measured with a phase-sensitive detector (PSD). (B) Ion-selective noninvasive probe driven by stepper motors. (C) System for mapping transepithelial conductance with a combination of stepper motors and PSDs.

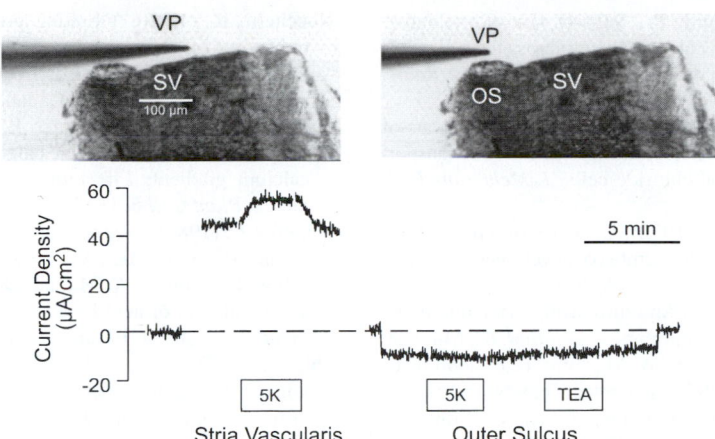

FIGURE A-2. Current density near stria vascularis (SV) and outer sulcus (OS). Upper panels: Vibrating probe (VP) near SV (*left*) and SP (*right*). Lower panel: Recording of current density showing outward (positive) current from SV and its sensitivity to change of bath K^+ concentration from 3.6 to 5mM (*left*) and the inward (negative) current from SP and its insensitivity to elevated K^+ concentration and to the maxi-K channel blocker tetraethylammonium (TEA).

pending on the experiment. Motion of the electrode between the positions disturbs the gradient being measured, so a waiting period is introduced before the signal is collected at each position after a movement, followed by a period over which the signal is collected and subsequently averaged. The averaged signal from each position is subtracted from that at the "origin." Either one-, two-, or three-dimensional paths can be defined with a trade-off between spatial and temporal resolution (Shi and Borgens, 1995). A three-dimensional path is shown in Fig. A-1. Noise reduction by this technique permits measurements of voltage gradients in the **microvolt** range.

A third method (Fig. A-1C) combines elements of the first two and is applicable to quantifying variations in transmural epithelial conductance by observing the modulation by the tissue of an imposed sinusoidal electric field (Koeckerling *et al.*, 1993). The sensor is stepped between two positions at a slow rate as in the second method. The noise reduction of the signal at each position, however, is performed by oscillating the imposed electric field at a fixed frequency and passing the signal from the sensor through a phase-sensitive detector (lock-in amplifier) referenced to that frequency. This technique is reported to give a resolution of better than 0.5 µV.

These techniques have been applied to a wide variety of biologic systems (Nuccitelli, 1990). Early measurements focused on mapping electric currents and ion gradients around eggs and in the vicinity of plant rootlets and pollen tubes (Anderson *et al.*, 1994; Pierson *et al.*, 1994; Kochian *et al.*, 1992). They have also been applied to the measurements of currents and ion fluxes across epithelia either with heterogeneous cell populations (Breton *et al.*, 1996; Durham *et al.*, 1989) or small homogeneous domains, such as the epithelia regulating the composition of endolymph in the inner ear (Chapter 46 and Fig. A-2). The probes have been found useful in studies of cell development and wound healing (Shi and Borgens, 1995; Hotary and Robinson, 1992; Borgens *et al.*, 1977) and in localizing currents in neurons (Duthie *et al.*, 1994). The probes can also be used in combination with other physiologic measurement techniques such as transepithelial voltage clamping, single-cell patch-clamp recording, and intracellular fluorescence microscopy (Foskett and Machen, 1985; Scheffey *et al.*, 1991; P. J. S. Smith, unpublished observations). With the recent addition of polarographic detection systems to the repertoire of self-referencing probes, measurements of minute oxygen gradients near respiring cells and tissues can now be recorded.

Resources are available for investigators interested in exploring the feasibility of applying these techniques to their own preparation. Self-referencing probe technology was originally developed at the National Vibrating Probe Facility (Dr. Lionel Jaffe, director) at the Marine Biological Laboratory (MBL), Woods Hole, Massachusetts. The successor to this group, the BioCurrents Research Center (Peter J. S. Smith, director), is a facility of the National Center for Research Resources (National Institutes of Health) and can host visiting scientists by application to the director (http://www.hermes.mbl.edu/labs/BioCurrents/home.html). Another laboratory providing access to probe technology is the University of Massachusetts Vibrating Probe Facility, directed by Dr. Joseph G. Kunkel. An Internet web site for this facility is at http://www.bio.umass.edu/biology/kunkel/.

Bibliography

Anderson, M., Bowdan, E., and Kunkel, J. G. (1994). Comparison of defolliculated oocytes and intact follicles of the cockroach using the vibrating probe to record steady currents. *Dev. Biol.* **162,** 111–122.

Borgens, R. B., Vanable, J. W., Jr., and Jaffe, L. F. (1977). Bioelectricity and regeneration: II. Large currents leave the stumps of regenerating newt limbs. *Proc. Nail. Acad. Sci. USA* **74,** 4528–4532.

Breton, S., Smith, P. J., Lui, B., and Brown, D. (1996). Acidiflcation of the male reproductive tract by a proton pumping (H^+)-ATPase. *Nat. Med.* **2,** 470–472.

Durham, J. H., Shipely, A., and Scheffey, C. (1989). Vibrating probe localization of acidification current to minority cells of the turtle bladder. *Ann. NY Acad. Sci.* **574,** 486–488.

Duthie, G. G., Shipley, A., and Smith, P. J. (1994). Use of a vibrating electrode to measure changes in calcium fluxes across the cell membranes of oxidatively challenged *Aplysia* nerve cells. *Free Radic. Res.* **20,** 307–313.

Foskett, J. K., and Machen, T. E. (1985). Vibrating probe analysis of teleost opercular epithelium: Correlation between active transport and leak pathways of individual chloride cells. *J. Membrane Biol.* **85,** 25–35.

Hotary, K. B., and Robinson, K. R. (1992). Evidence of a role for endogenous electrical fields in chick embryo development. *Development* **114,** 985–996.

Jaffe, L. F., and Nuccitelli, R. (1974). An ultrasensitive vibrating probe for measuring steady extracellular currents. *J. Cell Biol.* **63,** 614–628.

Kochian, L. V., Shaff, J. E., Kuhtreiber, W. M., Jaffe, L. F., and Lucas, W. J. (1992). Use of an extracellular, ion-selective, vibrating microelectrode system for the quantification of K^+, H^+ and Ca^{2+} fluxes in maize roots and maize suspension cells. *Planta* **188,** 601–610.

Koeckerling, A., Sorgenfrei, D., and Fromm, M. (1993). Electrogenic Na^+ absorption of rat distal colon is confined to surface epithelium: A voltage-scanning study. *Am. J. Physiol.* **264,** C1285–C1293.

Kuhtreiber, W. M., and Jaffe, L. F. (1990). Detection of extracellular calcium gradients with a calcium-specific vibrating electrode. *J. Cell Biol.* **110,** 1565–1573.

Nuccitelli, R. (1990). Vibrating probe technique for studies of ion transport. *In* "Noninvasive Techniques in Cell Biology" (J. K. Foskett and S. Grinstein, Eds.), pp. 273–310. Wiley-Liss, New York.

Pierson, E. S., Miller, D. D., Callaham, D. A., Shipley, A. M., Rivers, B. A., Cresti, M., and Hepler, P. K. (1994). Pollen tube growth is coupled to the extracellular calcium ion flux and the intracellular calcium gradient: Effect of BAPTA-type buffers and hypertonic media. *Plant Cell* **6,** 1815–1828.

Scheffey, C. (1989). Vibrating probe method for measuring local transepithelial current. *Ann. NY Acad. Sci.* **574,** 485.

Scheffey, C., Shipley, A. M., and Durham, J. H. (1991). Localization and regulation of acid-base secretory currents from individual epithelial cells. *Am. J. Physiol.* **261,** F963–F974.

Shi, R., and Borgens, R. B. (1995). Three-dimensional gradients of voltage during development of the nervous system as invisible coordinates for the establishment of embryonic pattern. *Dev. Dyn.* **202,** 101–114.

Smith, P. J. S. (1995). Non-invasive ion probes—Tools for measuring transmembrane ion flux. *Nature* **378,** 645–646.

Smith, P. J. S., Sanger, R. H., and Jaffe, L. F. (1994). The vibrating Ca^{2+} electrode: A new technique for detecting plasma membrane regions of Ca^{2+} influx and efflux. *Methods Cell Biol.* **40,** 115–134.

Anita L. Zimmerman

47

Cyclic Nucleotide-Gated Ion Channels

I. Introduction

Cyclic nucleotides have long been known as intracellular second messengers that regulate cell function by controlling the activity of protein kinases, which in turn control many other cellular proteins (reviewed in Greengard, 1978). However, in 1985 Fesenko and his colleagues made a startling discovery that changed our view of the physiological role of cyclic nucleotides. These investigators found that the ion channel mediating the electrical response to light in retinal rod cells was directly opened by the binding of guanosine 3',5'-cyclic monophosphate (cGMP); no phosphorylation reaction was required. Now the rod channel is considered to be a member of a special class of ion channels—the cyclic nucleotide-gated (CNG) channels. The channels are discussed in many recent reviews (Firestein and Zufall, 1994; Kaupp, 1995; Menini, 1995; Molday and Hsu, 1995; Yau and Chen, 1995; Zimmerman, 1995; Finn *et al.*, 1996; Zagotta and Siegelbaum, 1996; Wei *et al.*, 1998).

Why would Nature directly gate ion channels with cyclic nucleotides? When a cyclic nucleotide regulates a kinase, it is also in effect regulating all the proteins controlled by that kinase and by substrates of the kinase. Regulating ion channels is similar in that, like kinases, ion channels have diverse physiological effects. For example, the opening of nonselective cation channels (such as those opened by cyclic nucleotides) depolarizes the cell membrane and also allows the entry of Ca^{2+}, another important second messenger. Membrane depolarization opens Ca^{2+} channels, further increasing the entry of Ca^{2+}. Many cell functions are controlled by membrane potential and/or intracellular Ca^{2+}, including nerve impulses, muscle contraction, and the secretion of neurotransmitters and hormones. Finally, depolarization and intracellular Ca^{2+} open potassium channels, which repolarize the membrane and thereby contribute to the termination of the cellular response. Thus, there are numerous possibilities for control of cell function by cyclic nucleotide-gated ion channels. Furthermore, ion channel gating and permeation are much faster than phosphorylation reactions. Thus changes in cyclic nucleotide levels could have fast effects mediated by ion channels, followed by slower, longer lasting effects mediated by protein kinases.

II. Physiological Roles and Locations

Since their discovery in retinal rods, CNG channels have been identified in many other types of cells. In particular, CNG channels have been implicated generally in sensory transduction, as they also have been found in retinal cones, olfactory cells, invertebrate photoreceptors, cochlear hair cells, and pineal gland cells. Furthermore, CNG channels may mediate the glutamate response of retinal bipolar cells. Since CNG channels have been purified and cloned, it has become possible to screen for their presence in other types of tissue. For example, mRNA probes against the rod CNG channel have revealed its expression in cells of the heart, brain, muscle, liver, kidney, and testes. In addition, the pacemaker channel (I_h) in the heart sinoatrial node, as well as some other voltage-gated channels, have been found to contain regulatory cyclic nucleotide binding sites. The CNG channels have been studied most thoroughly in rod and olfactory cells, and therefore the CNG channels from these cells are discussed in the most detail here.

In rods and cones, CNG channels are key players in visual transduction (reviewed in Roof and Makino, 2000). It is these channels that conduct the so-called *dark current* and whose closure generates the hyperpolarizing response to light, which decreases the secretion of glutamate onto bipolar cells at the rod-bipolar synapse. The physiological second messenger in the photoreceptors is cGMP, which is at relatively high cytosolic concentration in the dark, and decreases in the light after hydrolysis by a phosphodiesterase (PDE). A similar system exists in cone visual transduction. Details of the enzyme cascade controlling the level of cGMP are given in the next section.

Rods and cones are particularly well suited to patch-clamp studies of CNG channels, since the plasma membranes of their light-sensitive outer segments contain essentially no other type of ion channel (although $Na^+/K^+/Ca^{2+}$ exchange carriers are present). Furthermore, the rod outer segment plasma membrane has an extremely high density of CNG channels—hundreds per square micrometer—allowing nanoamperes of current to be recorded from a single excised patch. Such large currents are useful in studying channel block (e.g., by divalent cations). In contrast, cone outer segments have relatively low channel densities, allowing the study of single-channel kinetics in patches containing only one channel. However, rods and cones interestingly have about the same number of total CNG channels because of the much larger plasma membrane area in cone outer segments (a consequence of the characteristic infolding of this membrane). CNG channels have also been found at low density in rod inner segments, but their functional properties appear altered there.

Olfactory receptor cells use CNG channels in sensing odorants (reviewed in Shepherd, 1992; Dionne and Dubin, 1994; Firestein and Zufall, 1994; Menini, 1995). In this system, however, there are apparently numerous receptor types; adenosine 3',5'-cyclic monophosphate (cAMP) is the physiological second messenger and the stimulus triggers cAMP production by adenylate cyclase, rather than its degradation by a PDE. Thus, in response to an odorant, the CNG channels open, and the olfactory receptor cell depolarizes, increasing the probability of generation of an action potential. Despite these differences, there is considerable similarity between the visual and olfactory transduction systems and between their respective CNG channels.

Like rods and cones, the olfactory cell has its CNG channels concentrated in a specialized region: the olfactory cilia and ciliary knob. Although the channels have been studied in excised patches from olfactory cilia (Nakamura and Gold, 1987), such experiments are extremely difficult because of the small diameter of a cilium. Luckily, the knob is larger, and some CNG channels are also located (at lower density) in the membrane of the soma. Furthermore, whole-cell patch-clamp methods have yielded considerable information on the olfactory CNG channels.

III. Control by Cyclic Nucleotide Enzyme Cascades

Like cyclic nucleotide-regulated protein kinases, CNG channels are sensors of the local concentration of cyclic nucleotides. Stimulus-induced changes in cyclic nucleotide levels are mediated by GTP-binding proteins (G proteins). The stimulus-activated receptor interacts with a G protein, causing it to release GDP and bind GTP and to dissociate into two components: an α subunit and a $\beta\gamma$ subunit complex. The α subunit of the G protein, now bound with GTP, stimulates either adenylate cyclase (in olfactory receptors) or a cGMP-specific phosphodiesterase (in photoreceptors). For photoreceptors, the stimulus that activates the receptor is a photon, whereas for olfactory cells, the stimulus is an odorant molecule that acts as a receptor ligand. This type of enzyme cascade is diagrammed in Fig. 1 for a rod photoreceptor and in Fig. 2 for an olfactory cell. The cascade in cones is similar to that in rods, except that all the membrane-associated players are located on the plasma membrane, since cones lack internal disks (see Roof and Makino, 2000).

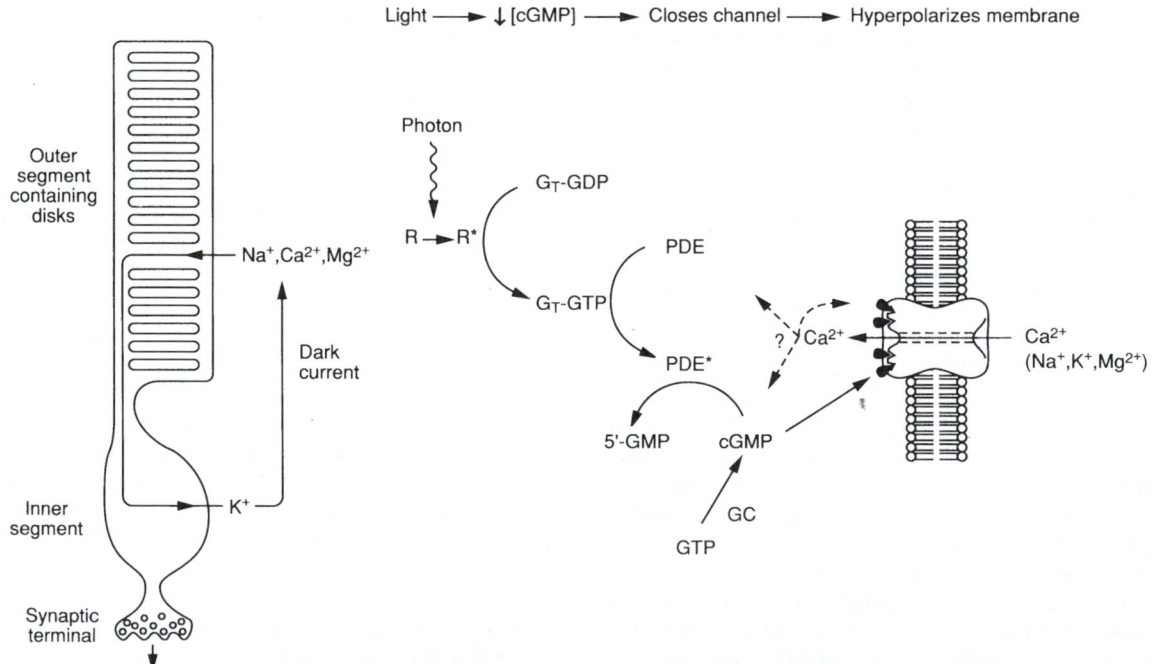

FIGURE 1. The cyclic nucleotide cascade controlling CNG channels in rods. R,R*, rhodopsin in its inactive and active forms, respectively. G_T, G protein ("Transducin"), bound to either GDP or GTP, PDE, PDE*, phosphodiesterase in its inactive and active forms, respectively. GC, guanylate cyclase. Calcium ions entering through the CNG channels are thought to modulate the function of several players in the cascade, including the channels themselves. A similar cascade operates in cones.

Odorant ⟶ ↑ [cAMP] ⟶ Opens channel ⟶ Depolarizes membrane

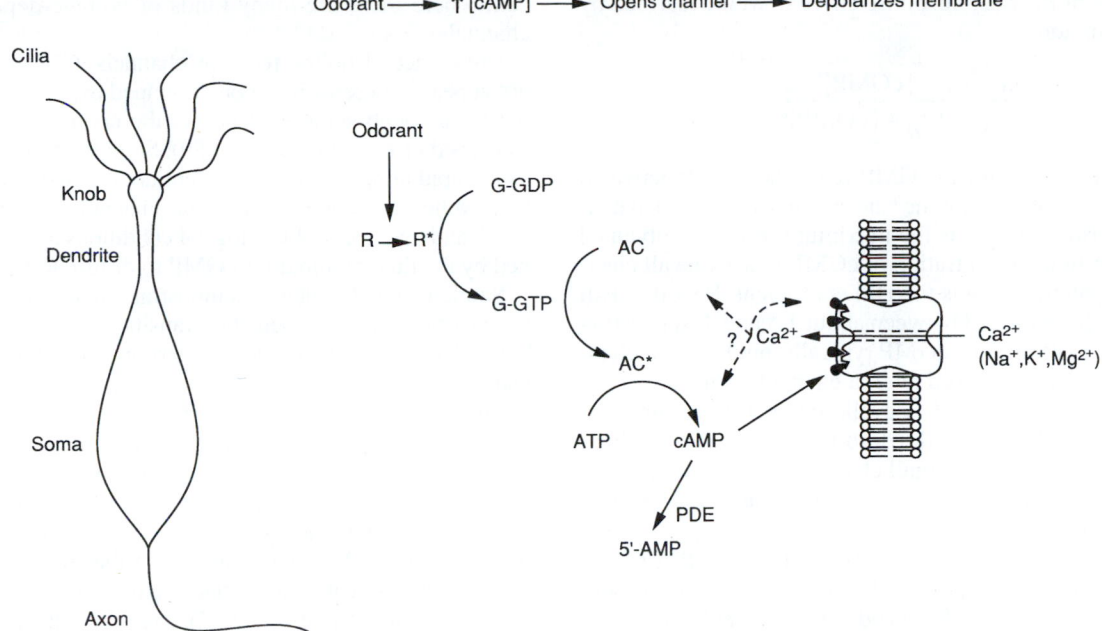

FIGURE 2. The cyclic nucleotide cascade controlling CNG channels in olfactory cells. R, R*, odorant receptor in its inactive and active (odorant-bound) forms, respectively. G, G protein, bound to either GDP or GTP. AC, AC*, adenylate cyclase in its inactive and active forms, respectively. PDE, phosphodiesterase. Here, as in photoreceptors, entering Ca^{2+} appears to modulate the cascade, including the CNG channels.

Cyclic nucleotide enzyme cascades are not fixed in their behavior. Instead, they are regulated by feedback systems, some of which involve CNG channels. For example, in rods, the Ca^{2+} that enters through CNG channels has been found to modulate the cGMP cascade. There is evidence that Ca^{2+} (in association with Ca^{2+} binding proteins) inhibits guanylate cyclase and inhibits the shutoff of rhodopsin (reviewed in Roof and Makino, 2000). In olfactory receptors, Ca^{2+} appears to be involved in both excitation and adaptation (reviewed in Menini, 1999).

In addition to such feedback regulatory systems, there are the standard shutoff mechanisms employed in cyclic nucleotide cascades (reviewed in Roof and Makino, 2000). These include phosphorylation of the receptor (e.g., the phosphorylation of rhodopsin and its binding to arrestin), GTPase activity of the G protein (converting it back to the GDP-bound inactive form), cessation of the stimulus, and competing hydrolysis or synthesis of the cyclic nucleotide. There are also hints that the ability of the channels to respond to the cyclic nucleotide may be modulated (see Section VI).

IV. Functional Properties

A. Channel Gating

CNG channels are very sensitive detectors of the local concentration of cyclic nucleotides, and they appear designed to work in the physiological concentration range of their respective agonists. Dose-response curves (e.g., Fig. 3) for activation of rod channels by cGMP give half-saturating

concentrations ($K_{1/2}$ values) ranging from about 5 to 100 µM, which is within the expected physiological concentration range. The rather wide range of values of $K_{1/2}$ may reflect functional modulation of the channels by other factors (see Section VI).

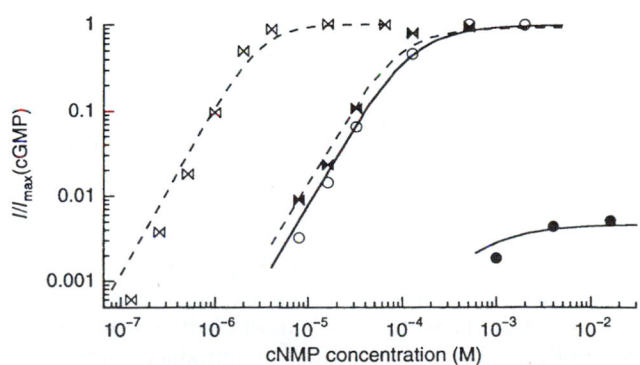

FIGURE 3. Dose-response curves for activation of cloned rod (circles) and olfactory (bows) α-homomultimeric channels. Both CNG channels show a higher apparent affinity (lower $K_{1/2}$) for cGMP (open symbols) than for cAMP (filled symbols), but for the rod channel cAMP appears to be a partial agonist, giving only a fraction of the current produced by a saturating concentration of cGMP. The smooth and dashed curves were calculated using a model in which the only difference between the results with cGMP and those with cAMP is that cGMP more effectively triggers the opening conformational change of the channels after it is bound. Currents were measured in response to voltage pulses of ±100 mV from a holding potential of 0 mV, and normalized to the current obtained in saturating cGMP (I_{max}). (Reproduced with permission from Gordon and Zagotta, 1995a. Copyright 1995 Cell Press.)

The form of the dose-response curve is well-described by the Hill equation,

$$\frac{r}{r_{max}} = \frac{[cGMP]^n}{K_{1/2}^n + [cGMP]^n}$$

where r is the response to cGMP (e.g., the cGMP-activated component of the membrane current measured in a patch-clamp experiment), r_{max} is the maximum response (obtained with a saturating concentration of cGMP to activate all channels in the patch), and n is the Hill coefficient. Reported Hill coefficients have ranged between about 1.5 and 4, suggesting that several molecules of cGMP typically bind to each channel to open it. As discussed later, a channel seems to consist of four subunits, each with a cyclic nucleotide binding site (Liu *et al.*, 1996). Because the dose-response curve for channel activation is so steep, small changes in the concentration of cAMP or cGMP produce very large changes in channel open probability.

Activation of olfactory channels is similar to that of rod and cone channels except for relative cyclic nucleotide sensitivities and efficacies. Rod and cone channels are much less sensitive to cAMP than to cGMP, with a $K_{1/2}$ for activation by cAMP of about 1.5 mM. Native olfactory CNG channels are more sensitive to both cyclic nucleotides than are photoreceptor channels, but they are only two to five times more sensitive to cGMP than to cAMP, with most $K_{1/2}$ values for activation by cGMP in the range of 1 to 5 μM (and a few as high as 20 μM—see Nakamura and Gold, 1987). Furthermore, cAMP acts as only a partial agonist for the rod channel: even at saturating concentrations, it gives only a fraction of the open probability obtained with saturating cGMP (Fig. 3). However, both cAMP and cGMP are full agonists for the olfactory channel. This difference in agonist efficacy can be explained by assuming that cGMP is a more effective agonist than cAMP for both channels, and that after agonist binding, the olfactory channel opens more easily than does the rod channel (i.e., the olfactory channel's opening conformational change is more energetically favored) (Gordon and Zagotta, 1995a).

Although some strongly voltage-dependent channels are regulated by cyclic nucleotides, channels that are primarily activated by cyclic nucleotides are only weakly voltage-dependent, with no voltage-dependent inactivation (reviewed in Zagotta and Siegelbaum, 1996). In current-voltage (*I-V*) relations from excised patches, the voltage dependence of channel gating is most obvious at low cyclic nucleotide concentrations, where it introduces significant nonlinearity (e.g., see Fig. 6D). High concentrations of cyclic nucleotides overcome the voltage dependence of gating, driving the channels (by mass action) toward high open probabilities at all voltages and linearizing the *I-V* curves. Note, however, that much of the CNG channel rectification seen in intact cells probably results from voltage-dependent channel block by Ca^{2+} and Mg^{2+}, as discussed later. Finally, the end result of both forms of nonlinearity is that current is relatively independent of voltage in the physiological voltage range. Thus sensory CNG channels are able to transduce faithfully in the face of changes in membrane potential that originate at either the transducing region or elsewhere in the

cell, where there are many kinds of voltage-dependent ion channels.

Unlike acetylcholine receptor channels, CNG channels do not appear to desensitize upon continued exposure to their agonists, although some evidence to the contrary has appeared (reviewed in Yau and Baylor, 1989). Studies on the rod channel using rapid jumps in cGMP concentration or in voltage suggest that cyclic nucleotide binding and channel opening are very rapid, and that under physiological conditions activation is limited by the time required for cGMP to diffuse to the channel.

Single-channel studies demonstrate that CNG channels have particularly fast open-shut transitions in the native membrane. This flickery behavior is striking in the cell-attached and excised-patch recordings obtained from toad rods by Matthews and Watanabe (1987) (Fig. 4). However, when purified and reconstituted, or cloned and heterologously expressed, the channels were initially found to have much slower gating kinetics (see, e.g., Kaupp *et al.*, 1989) that are more typical of many other ion channels. A similar difference was found when cGMP-gated channels in the rod outer segment were compared with those that occur at low density in the inner segment (Torre *et al.*, 1992), a region of the cell with different internal structures and functions whose plasma mem-

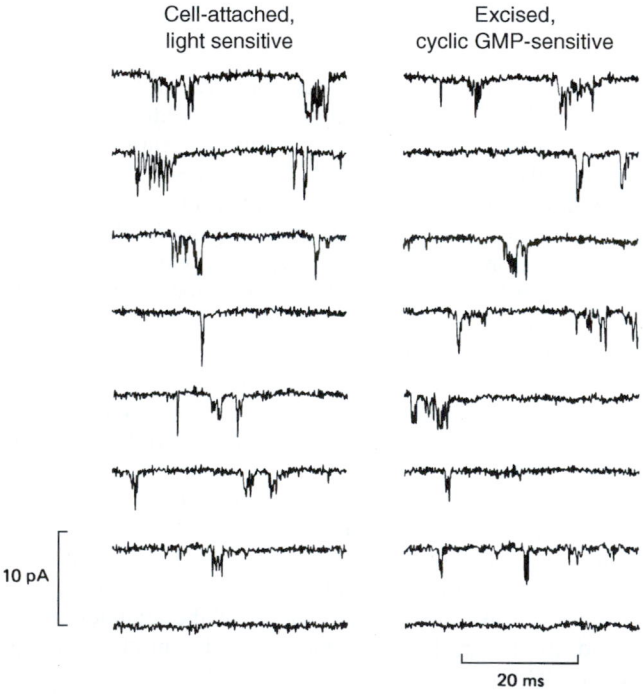

FIGURE 4. Single-channel recordings of CNG channels in native rod outer segment membranes demonstrate such rapid gating kinetics that many open-shut transitions are poorly resolved. Channel openings give downward deflections; holding potential −148 mV. Left: Cell attached patch from toad rod outer segment; the bottom trace was obtained in a saturating light, and all others in darkness. To prevent channel block, the pipette was filled with a solution lacking Ca^{2+} and Mg^{2+}. Right: The same patch after excision, with the intracellular surface bathed in either no cGMP (bottom trace) or 10 μM cGMP (all other traces); both pipette and bathing solutions lacked Ca^{2+} and Mg^{2+}. (Reproduced with permission from Matthews and Watanabe, 1987.)

brane has a different lipid and protein composition. Although Ca^{2+} and Mg^{2+} produce flicker block of these channels, the flickery gating behavior persists even in the absence of these ions. Furthermore, the transitions are too fast to reflect the binding and unbinding of cyclic nucleotides, and they also apparently do not simply reflect block by protons, which has been demonstrated in these channels (reviewed in Zagotta and Siegelbaum, 1996). Evidence from Chen *et al.* (1993) and Körschen *et al.* (1995) suggests that the flickery gating pattern of the native rod CNG channel results partly from the presence of a second channel subunit that was missing in the original reconstitution and expression studies, in which only one kind of subunit was identified (see later discussion on molecular structure). This additional subunit also gives the channel its characteristic sensitivity to L-*cis*-diltiazem. Unlike the α subunit, the β subunit (also called *subunit 2*) does not produce cyclic nucleotide-activated currents when expressed alone.

For the olfactory channel, a second subunit not only gives more flickery gating kinetics, but also increases the apparent affinity for cyclic nucleotides, giving more channel activity at low cyclic nucleotide concentrations (Fig. 5) (Liman and Buck, 1994; Bradley *et al.*, 1994). Like the rod β subunit, this subunit also does not appear to produce cyclic nucleotide-activated currents when expressed alone. However, Broillet and Firestein (1997) have been able to obtain currents by modifying the

channel with nitric oxide treatment. Finally, it now appears that the olfactory channel may have yet a third subunit (Bönigk *et al.*, 1999) whose coexpression with the first two gives behavior even more consistent with that of the native channel. With both rod and olfactory CNG channels, there still seems to be discrepancies between the behavior of the channels *in vivo* and in expression systems. Some of these differences may result from effects of channel modulators (see Section VI).

Although there remains disagreement over the detailed mechanism of activation (compare Ruiz and Karpen, 1997, 1999, with Liu *et al.*, 1998), it is clear that CNG channels can open with fewer than four ligands bound, but have highest open probabilities with all four sites occupied. Innovative methods have been very useful in dissecting the molecular mechanism of gating. These include: covalent activation (Karpen and Brown, 1996), polymer-linked cGMP dimers (Kramer and Karpen, 1998), chimeras of different channel types (e.g., Gordon and Zagotta, 1995a), and the use of modulatory substances, such as Ni^{2+}, to discern subunit interactions (Gordon and Zagotta, 1995b).

B. Permeation, Selectivity, and Block

Most CNG channels described so far are nonselective cation channels with no significant anion permeability. (Note,

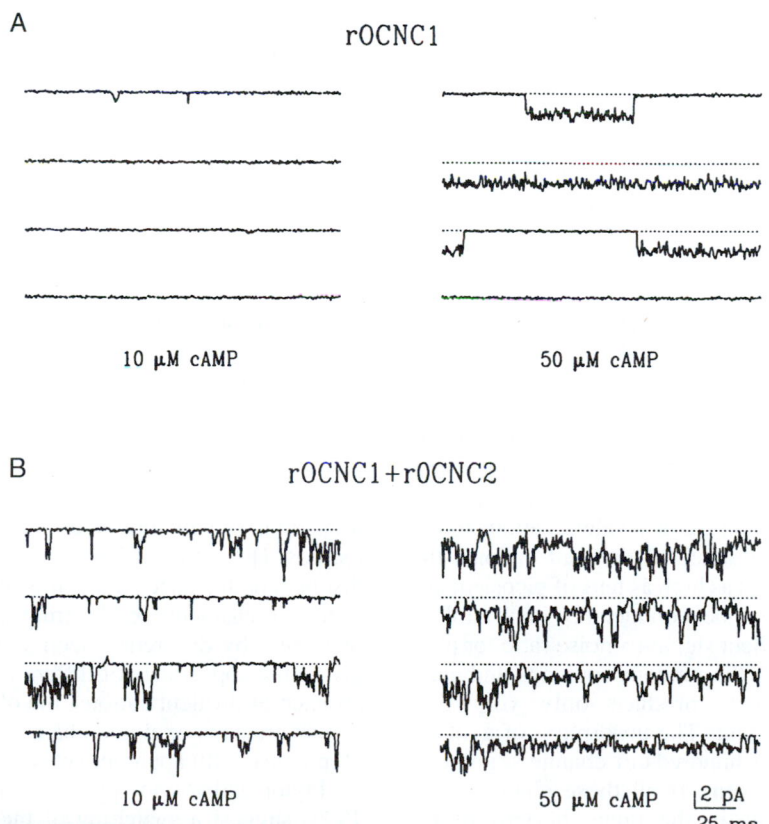

FIGURE 5. Increased cAMP sensitivity and flickery gating produced by coexpressing the second olfactory CNG channel subunit (r0CNC2) with the first (r0CNC1). When the first subunit is expressed alone (A) there are very few channel openings in 10 μM cAMP, and the openings with 50 μM cAMP are much less flickery than those seen in (B), where both subunits are expressed in the same cell. Holding potential is −80 mV. (Reproduced with permission from Liman and Buck, 1994. Copyright 1994 Cell Press.)

however, that some K^+-selective channels are regulated by direct binding of cyclic nucleotides; this is reviewed in Zimmerman (1995) and Zagotta and Siegelbaum (1996). Thus, reversal potentials for most CNG channels studied under normal ionic conditions are about +5 to +20 mV. The monovalent alkali cation permeability sequence for the rod channel has been reported to be $Li^+ \geq Na^+ \geq K^+ > Rb^+ > Cs^+$, and Ca^{2+} and Mg^{2+} ions appear to be more permeant than the monovalent cations (reviewed in Yau and Baylor, 1989; Zimmerman, 1995; Zagotta and Siegelbaum, 1996).

Cone and olfactory CNG channels also poorly discriminate among monovalent alkali cations, although their exact permeability sequences and ratios are not identical to those of the rod CNG channel. Strict comparisons are difficult because there is some variability in reported permeability ratios for each channel type. Relative permeabilities obtained may depend on whether the channels are studied in the intact cell, in excised patches, or in reconstitution or expression systems. The permeabilities may also depend on intracellular factors controlling functional modulation of the channels (see Section VI). In intact rods, the relative currents carried by monovalent and divalent cations have been found to depend on the concentration of cGMP (Cervetto *et al.,* 1988). This finding, along with the dependence of subconductance states on cGMP (below), suggests that the functional properties of CNG channels may change with the number of ligand molecules bound. If so, this may reflect a very unusual feature of these channels, and one with tremendous potential for modifying physiological responses.

In many ways, CNG channels behave as if they were single-file pores with one dominant ion-binding site that has a higher affinity for divalent cations than for monovalent cations. However, some behavior clearly indicates the presence of more than one ion-binding site, and this is confirmed by structure-function work (reviewed in Zimmerman, 1995; Zagotta and Siegelbaum, 1996). Under physiological conditions, the channel is occupied by Ca^{2+} or Mg^{2+} most of the time, and these ions prevent the passage of Na^+ and K^+, which pass through the pore much more rapidly. As a result of the very slow transport rate of the divalent cations, the mean single-channel conductance is extremely low—only about 0.1 pS in rods studied under physiological conditions. The channels can be blocked from either side, but under physiological conditions they are mostly blocked by extracellular Ca^{2+} and Mg^{2+}. To resolve single-channel currents, one must reduce the concentration of divalent cations to the micromolar range. In the absence of divalent cations, the single-channel conductance is as high as tens of picosiemens. The extremely low single-channel conductance in physiological solutions gives an excellent signal-to-noise ratio for photon detection by rods, since the random openings and closings of individual channels produce only very tiny fluctuations in the dark current. The absorption of a single photon elicits the closure of hundreds of channels, giving a smooth, stereotypical waveform. If all those channels had single-channel conductances in the range of tens of picosiemens, the rod cell would have to contend with the consequences of a large influx of Na^+ and Ca^{2+}. Cones and olfactory cells would have a similar problem, since their transducing regions also contain many CNG channels.

In the absence of divalent cations, I-V relations for CNG channels (with saturating cyclic nucleotide concentrations) are linear or nearly so (Fig. 6A and D, upper curve). Divalent cations introduce extreme nonlinearity in the I-V relations (Fig. 6B and C). When the rod channel is studied in the presence of physiological concentrations of divalent cations, its I-V relation is very outwardly rectified. Although some of this rectification is a consequence of the weak voltage dependence of channel gating described earlier, much of it results from channel block by Ca^{2+} and Mg^{2+}. Thus, at negative membrane potentials in the physiological range (about −40 to −80 mV), external Ca^{2+} and Mg^{2+} are drawn into the pore, reducing Na^+ entry and giving an approximately flat I-V relation over a wide range of voltage. This very low, voltage-independent conductance in the physiological voltage range allows light-induced outer segment voltage changes to travel relatively unattenuated to the inner segment to regulate synaptic transmission and also prevents the outer segment photosensing mechanism from fluctuating with voltage.

I-V relations of the olfactory channel, with and without divalent cations, are essentially indistinguishable from those of the rod channel, but surprisingly, the I-V relation for the cone channel is rather different. Whereas cone I-V curves from excised patches are linear in the absence of divalent cations, they are almost S-shaped when divalents are present (Fig. 6C). Although the functional significance of this difference between cone CNG channels and those in rod cells and olfactory receptors is not clear, structurally it may reflect a different location of the dominant ion-binding site within the cone channel. At subsaturating concentrations of cyclic nucleotides (which are closer to the physiological concentrations), the cone channel I-V relation is much more flat in the physiological range of membrane potential even in the absence of divalent cations (Fig. 6D, lower curve). Similarly, the rod channel I-V curve shows increasing outward rectification as the cGMP concentration is lowered (not shown).

Single-channel recordings of the rod channel in the absence of divalent cations have revealed at least two conductance states: one of about 25 to 30 pS and the other with a conductance about one-third as large. However, it has been suggested that the channel has at least one more, and perhaps many more, conductance levels. Because of the extremely rapid gating kinetics of this channel, numerous open-closed transitions are no doubt unresolved in the single-channel records. Thus, it is difficult to determine the exact number of distinct conductance states. It is also not clear whether these states are characterized by truly different ion transport rates or merely by different (incompletely resolved) open times, giving the appearance of different conductance levels. In the absence of divalent cations, the olfactory and cone channels demonstrate major single-channel conductances around 45 to 50 pS, also with apparent subconductance states. The results of Taylor and Baylor (1995) and Ruiz and Karpen (1997, 1999) suggest a switching of the rod channel from low to high conductance states with increasing ligand occupancy.

Some pharmacological agents have been tested on CNG channels. Since these channels have a strong affinity for Ca^{2+}, various calcium channel blockers (including vera-

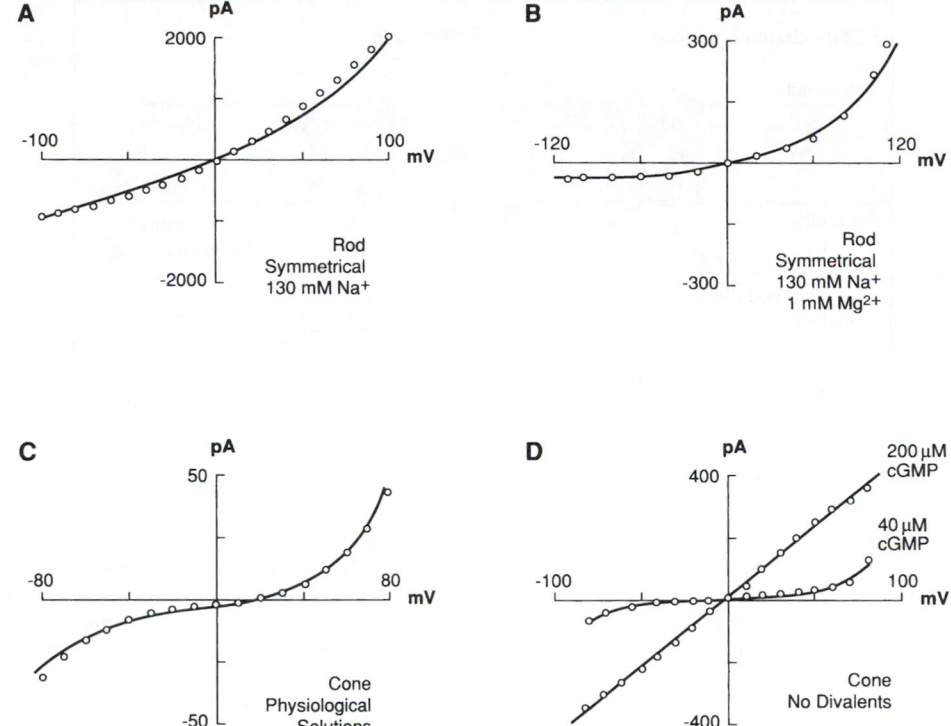

FIGURE 6. Current-voltage relations for photoreceptor cyclic GMP-activated currents from multichannel, excised patches. (A) The rod CNG channel relation is almost linear with saturating [cGMP] and no divalent cations. Olfactory CNG channels demonstrate a similar relation (not shown). (B) The addition of millimolar Mg^{2+} to both of the membrane gives strong outward rectification. Similar outward rectification has been reported for olfactory CNG channels. (C) Cone CNG channels show both outward and inward rectification in the presences of physiological (millimolar) levels of Mg^{2+} and Ca^{2+}. (D) The *I-V* relation for cone CNG channels is approximately linear with saturating [cGMP] and no divalent cations (top curve). Decreasing [cGMP] produces a nonlinear *I-V* relation (bottom curve), because gating is slightly voltage-dependent. Voltage-dependent gating of rod CNG channels also produces a very nonlinear *I-V* relation [resembling that in (B)] in the absence of divalent cations, at low [cGMP]. (Redrawn with permission from Zimmerman and Baylor, 1992 (A) and (B); from *Nature*, Haynes and Yau, 1985. Copyright 1985 Macmillan Magazines Limited. (C); and Picones and Korenbrot, 1992. (D) Reproduced from *The Journal of General Physiology*, 1992, vol. 100, pp. 647–673, by copyright permission of The Rockefeller University Press.)

pamil, dihydropyridines, and diltiazem) have been tested (reviewed for the rod channel in Yau and Baylor, 1989). Most block only weakly or not at all, but L-*cis*-diltiazem and an amiloride analogue, 3',4'-dichlorobenzamil, have been found to block effectively from the cytoplasmic surface of the membrane, with K_i values in the micromolar range. Frings *et al.* (1992) have found olfactory channels to be blocked by amiloride, D600, and diltiazem. CNG channels have also been found to be blocked by polyamines (Lu and Ding, 1999), tetracaine (Fodor *et al.*, 1997), potassium channel inactivation peptide (Kramer *et al.*, 1994), and by the snake toxin, pseudechetoxin (Brown *et al.*, 1999).

V. Molecular Structure

When the first subunit of the bovine rod CNG channel was cloned (Kaupp *et al.*, 1989), the channel was proposed to consist of multiple identical subunits, each with a cGMP-binding site located in the C-terminal region. It now seems clear that the rod channel is a tetramer (Liu *et al.*, 1996). It is thought to consist of two α and two β subunits (Shapiro and Zagotta, 1998; Shammat and Gordon, 1999), each with

its own cyclic nucleotide binding domain; however, there remains some dispute over the exact arrangement of the subunits. Second subunits have begun to be identified for other CNG channels as well, and there is now evidence that the olfactory CNG channel has a third distinct subunit (Bönigk *et al.*, 1999), although it remains to be determined how many of each subunit type constitute the channel. Figure 7 presents a general model of a CNG channel α subunit; other subunits discovered so far have been found to have the same basic structure, with slight variations.

In addition to the details of subunit composition, many other structural questions remain to be answered for CNG channels. For example, it is not known why the rod α subunit is post-translationally cleaved *in vivo*, so that before its insertion into the rod membrane, its molecular weight is reduced from 78 000 to 63 000 daltons with the loss of the end of its N-terminal tail (Bönigk *et al.*, 1993). The detailed nature of the apparent interaction between N- and C-terminal regions of olfactory CNG channels (reviewed in Zagotta and Siegelbaum, 1996) also remains to be resolved. It is this interaction that seems to be disrupted by Ca^{2+}-calmodulin, which strongly inhibits channel opening. In addition, the rod β subunit has a very long glutamic acid-rich region (GARP)

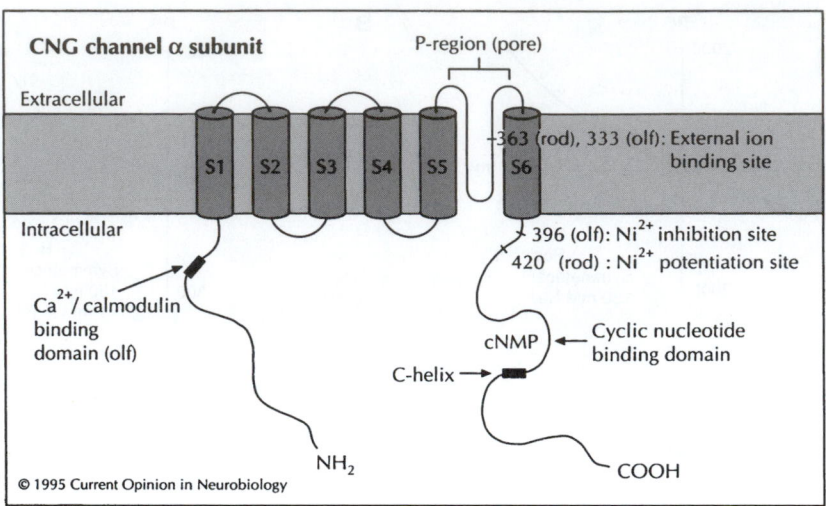

FIGURE 7.   A model for the general organization of a CNG channel α subunit. The model is based on information from many studies of rod and olfactory (olf) CNG channels. Like voltage-gated K^+ channels, CNG channels are characterized by six putative transmembrane segments (S1–S6) and a loop (P-region) that is thought to line the pore through which ions travel. Cyclic nucleotide monophosphates (cNMP), like cAMP and cGMP, are thought to bind in the cyclic nucleotide binding domain of the intracellular C-terminus. A few specific regions found to be involved in channel gating (C-helix), permeation and block (residues 363, 333) and modulation (Ca^{2+}-calmodulin binding domain) are also indicated. (Reprinted from *Current Opinion in Neurobiology* 5, Anita L. Zimmerman, "Cyclic Nucleotide-gated Channels." 296–303. Copyright 1995, with permission from Elsevier Science.)

near its N-terminus. The function of this region remains unclear, although there are recent hints that it may interact with other molecular components within the rod outer segment (Molday and Molday, 1998; Körschen *et al.*, 1999). Finally, it is not known exactly what structural changes occur in converting the binding of cyclic nucleotide at the C-terminus to the opening of the channel, or how the action of channel modulators (see following) is structurally linked to changes in channel gating. However, evidence continues to accumulate regarding interactions between domains (e.g., Varnum and Zagotta, 1997) and subunits (e.g., Varnum and Zagotta, 1996) that will no doubt be crucial in dissecting these details.

Although CNG channels are ligand-gated channels, their molecular structures more closely resemble those of voltage-gated channels. This similarity is brought out dramatically in a study in which the deletion of two amino acids in a voltage-gated K^+ channel converted it into a nonselective cation channel that was blocked by divalent cations and that demonstrated the flickery gating kinetics typical of CNG channels (Heginbotham *et al.*, 1992). Citing sequence similarities, Jan and Jan (1992) have pointed out that CNG channels probably belong to a superfamily of channels including voltage-gated Na^+, Ca^{2+}, and K^+ channels, as well as Ca^{2+}-activated K^+ channels. Also members of this superfamily are the hyperpolarization-activated, cyclic nucleotide-regulated channels that exist in the brain, heart, thalamus, and testis (DiFrancesco, 1993; Pape, 1996; Gauss *et al.*, 1998).

The members of this CNG/voltage-gated channel superfamily are characterized by a repeating structural motif, like that shown in Fig. 7, containing six putative membrane-spanning segments, located between the hydrophilic N- and C-terminal regions that project into the cytosol. Also characteristic of these channels is an S4 segment that contains many basic residues and

is thought to be a voltage-sensing region. Although CNG channels have S4 segments, they have only very weak voltage sensitivity. It has been proposed that glutamate residues in the vicinity of S4 contribute negative charges that may neutralize the effects of the positively charged arginine and lysine residues of S4 that are thought to confer voltage sensitivity to the channel (Wohlfart *et al.*, 1992; see also Tang and Papazian, 1997). The channels in this family also contain a putative pore region (P-region) in which a crucial glutamate interacts with permeating ions (reviewed in Zimmerman, 1995; Zagotta and Siegelbaum, 1996). Of course, the part of the CNG channel that best distinguishes it from the typical voltage-gated channel is its C-terminal cyclic nucleotide binding domain, which resembles that found in other cyclic nucleotide-regulated proteins, like cGMP-dependent protein kinase. The region, often referred to as the C-linker, between this binding domain and the last transmembrane segment, has recently been proposed to link the cyclic nucleotide binding event with the allosteric opening transition of the channel (Zong *et al.*, 1998; Paoletti *et al.*, 1999). Considerable evidence has accumulated to suggest that the N-terminal region is also important in gating, in part by its interaction with the C-terminus (Gordon *et al.*, 1997; reviewed in Zagotta and Siegelbaum, 1996).

VI. Functional Modulation

New information suggests that CNG channels may be modulated in ways that are only beginning to be elucidated. As mentioned previously, there are hints that the ionic selectivity of the rod CNG channel may vary with the amount of cyclic nucleotide bound. There is also now evidence that the cGMP sensitivity of the rod channel may be tuned up or down by phosphorylation (Gordon *et al.*, 1992; Molokanova *et al.*, 1997) and reduced by diacylglycerol (Gordon *et al.*, 1995a), as well as by calmodulin in the presence of Ca^{2+} (Hsu and Molday, 1993; Gordon *et al.*, 1995b). Whether other channel properties are also altered by these modulators has not been determined. Protons also modulate the rod CNG channel, increasing its open probability, while reducing its single-channel conductance (Gavazzo *et al.*, 1997; also reviewed in Zagotta and Siegelbaum, 1996). Calcium seems particularly useful in modulatory mechanisms involving feedback control since its concentration in the cell is determined partly by its entry through the CNG channels. Data from some preparations suggest that the rod channel inhibition by Ca^{2+} *in vivo* is mediated by an endogenous calcium binding protein other than calmodulin (Fig. 8) (Gordon *et al.*, 1995b; Sagoo and Lagnado, 1996; see also Bauer, 1996). Transition metal divalent cations, such as Ni^{2+}, also modulate CNG channel gating (reviewed in Zagotta and Siegelbaum, 1996); interestingly, 10 μM Ni^{2+} increases the open probability of the rod channel, but decreases the open probability of the olfactory channel. Higher concentrations of Ni^{2+} block the pores of both rod and olfactory channels. Functional modulation of the rod CNG channel may help explain the large variability in reported cGMP affinity and cooperativity and may play a role in some aspects of visual transduction. Finally, there is mounting evidence that in rods, the CNG channel interacts with $Na^+/K^+/Ca^{2+}$ exchange carriers (Bauer, 1992; Molday and Molday, 1998). The functional significance of this interaction remains to be determined.

The gating of olfactory CNG channels has been found to be modulated by intracellular Ca^{2+} (Zufall *et al.*, 1991; Kramer and Siegelbaum, 1992). Figure 9, from Zufall *et al.* (1991), shows single-channel recordings obtained from an excised patch of olfactory dendritic membrane. When the concentration of Ca^{2+} bathing the intracellular surface of the patch was increased from 0.1 to 3 μM, there was a striking increase in channel closed time, giving a decrease in open probability. Since the single-channel amplitude and mean channel open time were not affected, the mechanism of Ca^{2+} action most likely involved allosteric effects on channel gating (perhaps by stabilizing one or more closed states), rather than channel block. Kramer and Siegelbaum (1992) found that Ca^{2+} shifts the cAMP dose-response curve to the right. The channel also appears to be modulated by Ca^{2+}-calmodulin (Liu *et al.*, 1994; reviewed in Molday, 1996) and perhaps by an endogenous factor distinct from calmodulin (Balasubramanian *et al.*, 1996). This modulation is probably involved in odorant transduction and adaptation (Menini, 1999).

VII. Summary

Cyclic nucleotide-gated channels are sensitively and directly activated by the binding of cGMP and/or cAMP. Like other cyclic nucleotide-regulated proteins, CNG channels are powerful modifiers of cell function. They are found in

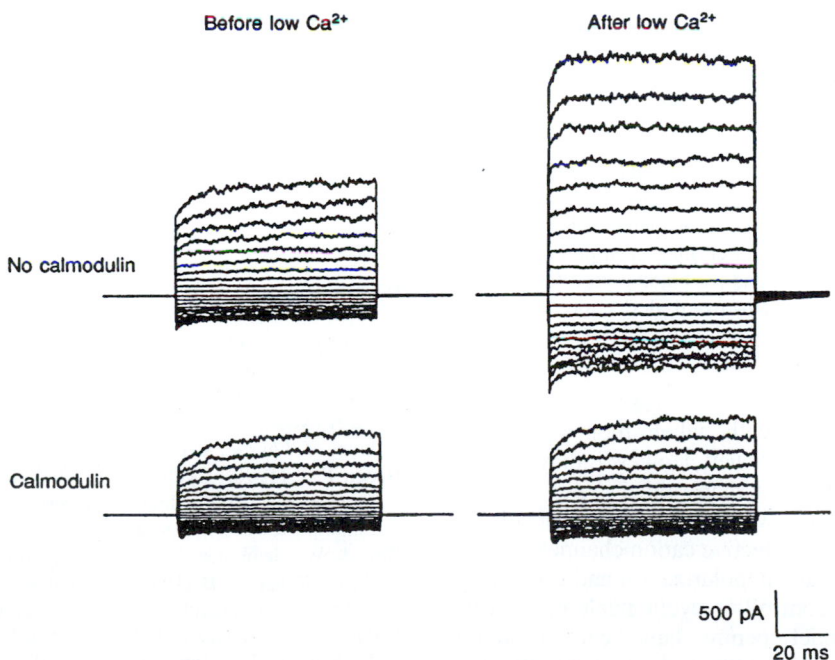

FIGURE 8. Evidence for modulation of the native rod channel by an endogenous Ca^{2+}-binding protein. The cyclic nucleotide-activated currents are slightly suppressed by Ca^{2+}-calmodulin (compare upper left panel with bottom left panel, which gives the response to 200 nM calmodulin with 1 μM Ca^{2+}). Treatment with very low Ca^{2+} (in the nanomolar range) dramatically increases the current in the absence of Ca^{2+}-calmodulin (upper right panel), as if removing an endogenous inhibitory factor. After this treatment (bottom right panel), Ca^{2+}-calmodulin reduces the current to about the same level as before the low-Ca^{2+} treatment, suggesting that the endogenous inhibitory factor and Ca^{2+}-calmodulin may be competing for the same sites. Data collected using a multichannel, excised, inside-out, rod outer segment patch exposed to 1.5 μM 8-bromo-cGMP. These difference currents are responses to voltages from −200 to +200 mV in steps of 20 mV, from a holding potential of 0 mV. Responses to saturating cyclic nucleotide were not changed by either the low-Ca^{2+} treatment or Ca^{2+}-calmodulin. (Reproduced with permission from Gordon *et al.*, 1995b.)

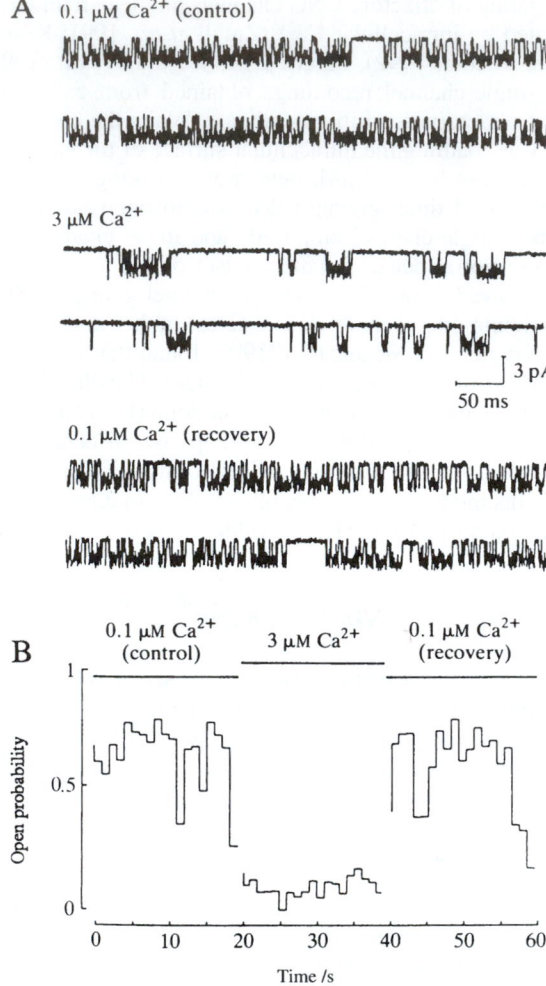

FIGURE 9. Modulation of an olfactory CNG channel by Ca^{2+}. Raising the Ca^{2+} concentration from 0.1 to 3 μM at the intracellular surface of an excised patch produced a reversible increase in channel closed time, resulting in a decrease in channel open probability. The cAMP concentration was 100 μM, and the holding potential was −60 mV. Channel opening gives a downward deflection in current. (Reproduced with permission from Zufall *et al.,* 1991.)

tant opening of CNG channels depolarizes the cell, thereby increasing its action potential firing rate.

Although it appears that the primary steps in these two cascades have been determined, many questions remain, particularly regarding feedback control mechanisms in sensory adaptation. In fact, it has become clear that many of the molecular players controlling the cascades have not yet been identified, and there is recent evidence for modulation of the CNG channels themselves. Thus, *in vivo*, CNG channel function may be modulated by phosphorylation, by Ca^{2+}, and by other factors as well. The molecular mechanisms of modulation, as well as which aspects of channel function are modulated, are currently under study.

CNG channels have striking gating and permeation properties. They are opened by the binding of cGMP or cAMP to each of four subunits, and appear to be operating in the physiological range (micromolar) of relevant cyclic nucleotide concentrations. Because the dose-response curve for channel activation is steep, small increases in [cAMP] or [cGMP] dramatically increase CNG channel open probability. Calcium and magnesium ions permeate CNG channels very slowly, so that they block sodium and potassium transport under physiological conditions, drastically reducing the mean single-channel conductance and increasing the signal-to-noise ratio in sensory transduction. CNG channel gating is very fast, typically giving very rapid flickers in single-channel currents.

Amino acid sequences of cloned CNG channels suggest that they belong to the same channel superfamily as the voltage-gated Na^+, K^+, and Ca^{2+} channels. CNG channels were initially thought to be homo-oligomers, but it is now clear that they have at least two distinct types of subunits. The correlation of CNG channel structure and function is an exciting, fast-moving area of current research which is leading to substantial insights into the mechanisms of channel gating and ion permeation.

Bibliography

Balasubramanian, S., Lynch, J. W., and Barry, P. H. (1996). Calcium-dependent modulation of the agonist affinity of the mammalian olfactory cyclic nucleotide-gated channel by calmodulin and a novel endogenous factor. *J. Memb. Biol.* **152,** 13–23.

Bauer, P. J. (1992). Association of cyclic-GMP-gated channels and Na-Ca-K exchangers in bovine retinal rod outer segment plasma membrane. *J. Physiol.* **451,** 109–131.

Bauer, P. J. (1996). Cyclic GMP-gated channels of bovine rod photoreceptors: Affinity, density and stoichiometry of Ca^{2+}-calmodulin binding sites. *J. Physiol.* **494,** 675–685.

Bönigk, W. Altenhofen, W., Müller, F., Dose, A., Illing, M., Molday, R. S., and Kaupp, U. B. (1993). Rod and cone photoreceptor cells express distinct genes for cyclic GMP-gated channels. *Neuron.* **10,** 856–877.

Bönigk, W., Bradley, J., Müller, F., Sesti, F., Boekhoff, I., Ronnett, G. V., Kaupp, U. B., and Frings, S. (1999). The native rat olfactory cyclic nucleotide-gated channel is composed of three distinct subunits. *J. Neurosci.* **19,** 5332–5347.

Bradley, J., Li, J., Davidson, N., Lester, H. A., and Zinn, K. (1994). Heteromeric olfactory cyclic nucleotide-gated channels: α subunit that confers increased sensitivity to cAMP. *Proc. Natl. Acad. Sci. USA* **91,** 8890–8894.

Broillet, M. C., and Firestein, S. (1997). Beta subunits of the olfactory cyclic nucleotide-gated channel form a nitric oxide activated Ca^{2+} channel. *Neuron* **18,** 951–958.

photoreceptors and olfactory cells, where their role in sensory transduction is well established. However, they are also beginning to be discovered in many other types of cells, where their function is not always clear. CNG channels are usually, but not always, nonselective cation channels, whose opening produces membrane depolarization and Ca^{2+} entry.

The enzyme cascades controlling cyclic nucleotide levels, and therefore CNG channel opening, have been thoroughly studied in vertebrate photoreceptors and olfactory cells. In a rod or cone, photon absorption by a pigment molecule leads to activation of a G protein, which in turn stimulates a phosphodiesterase to hydrolyze cGMP. The resultant closure of CNG channels hyperpolarizes the cell and thereby reduces synaptic transmitter release onto the next neuron in the visual circuit. In an olfactory cell, an odorant receptor with bound odorant activates a G protein, which stimulates adenylate cyclase to synthesize cAMP from ATP. The resul-

Brown, R. L., Haley, T. L., West, K. A., and Crabb, J. W. (1999). Pseudechetoxin: A peptide blocker of cyclic nucleotide-gated ion channels. *Proc. Natl. Acad. Sci. USA* **96,** 754–759.

Cervetto, L., Menini, A., Rispoli, G., and Torre, V. (1988). The modulation of the ionic selectivity of the light-sensitive current in isolated rods of the tiger salamander. *J. Physiol.* **406,** 181–198.

Chen, T. Y., Peng, Y-W., Dhallan, R. S., Ahamde, B., Reed, R. R., and Yau, K-W. (1993). A new subunit of the cyclic nucleotide-gated cation channel in retinal rods. *Nature* **362,** 764–767.

DiFrancesco, D. (1993). Pacemaker mechanisms in cardiac tissue. *Annu. Rev. Physiol.* **55,** 455–472.

Dionne, V. E., and Dubin, A. E. (1994). Transduction diversity in olfaction. *J. Exp. Biol.* **194,** 1–21.

Fesneko, E. E., Kolesnikov, S. S., and Lyubarsky, A. L. (1985). Induction by cyclic GMP of cationic conductance in plasma membrane of retinal rod outer segment. *Nature* **313,** 310–313.

Finn, J. T., Grunwald, M. E., and Yau, K.-W. (1996). Cyclic nucleotide-gated ion channels: An extended family with diverse functions. *Annu. Rev. Physiol.* **58,** 395–426.

Firestein, S., and Zufall, F. (1994). The cyclic nucleotide gated channel of olfactory receptor neurons. *Seminars in Cell Biology* **5,** 39–46.

Fodor, A. A., Gordon, S. E., and Zagotta, W. N. (1997). Mechanism of tetracaine block of cyclic nucleotide-gated channels. *J. Gen. Physiol.* **109,** 3–14.

Frings, S., Lynch, J. W., and Lindemann, B. (1992). Properties of cyclic nucleotide-gated channels mediating olfactory transduction. *J. Gen. Physiol.* **100,** 45–67.

Gauss, R., Seifert, R., and Kaupp, U. B. (1998). Molecular identification of a hyperpolarization-activated channel in sea urchin sperm. *Nature* **393,** 583–587.

Gavazzo, P., Picco, C., and Menini, A. (1997). Mechanisms of modulation by internal protons of cyclic nucleotide-gated channels cloned from sensory receptor cells. *Proc. R. Soc. Lond. Ser. B* **264,** 1157–1165.

Gordon, S. E., and Zagotta, W. N. (1995a). Localization of regions affecting an allosteric transition in cyclic nucleotide-activated channels. *Neuron* **14,** 857–864.

Gordon, S. E., and Zagotta, W. N. (1995b). Subunit interactions in the coordination of Ni^{2+} in cyclic nucleotide-gated channels. *Proc. Natl. Acad. Sci. USA* **92,** 10222–10226.

Gordon, S. E., Brautigan, D. L., and Zimmerman, A. L. (1992). Protein phosphatases modulate the apparent agonist affinity of the light-regulated ion channel in retinal rods. *Neuron* **9,** 739-748.

Gordon, S. E., Downing-Park, J., Tam, B. and Zimmerman, A. L. (1995a). Diacylglycerol analogs inhibit the rod cGMP-gated channel by a phosphorylation-independent mechanism. *Biophys. J.* **69,** 409–417.

Gordon, S. E., Downing-Park, J., and Zimmerman, A. L., (1995b). Modulation of the cGMP-gated ion channel in frog rods by calmodulin and an endogenous inhibitory factor. *J. Physiol.* **486,** 533–546.

Gordon, S. E., Varnum, M. D., and Zagotta, W. N. (1997). Direct interaction between amino- and carboxyl- terminal domains of cyclic nucleotide-gated channels. *Neuron* **19,** 431–441.

Greengard, P. (1978). Phospholated proteins as physiological effectors. *Science* **199,** 146–152.

Heginbotham, L., Abramson, T., and MacKinnon, R. (1992). A functional connection between the pores of distantly related ion channels as revealed by mutant K^+ channels. *Science* **258,** 1152–1155.

Hsu, Y.-T., and Molday, R. S. (1993). Modulation of the cGMP-gated channel of rod photoreceptor cells by calmodulin. *Nature* **361,** 76–79.

Jan, L. Y., and Jan, Y. N. (1992). Tracing the roots of ion channels. *Cell* **69,** 715–718.

Karpen, J. W., and Brown, R. L. (1996). Covalent activation of retinal rod cGMP-gated channels reveals a functional heterogeneity in the ligand binding sites. *J. Gen. Physiol.* **107,** 169–181.

Kaupp, U. B. (1995). Family of cyclic nucleotide gated ion channels. *Curr. Opin. Neurobiol.* **5,** 434–442.

Kaupp, U. B., Niidome, T., Tanabe, T., Terada, S., Bönigk, W., Stühmur, W., Cook, N. J., Kangawa, K., Matsuo, H., Hirose, T., Miyata, T., and Numa, S. (1989). Primary structure and functional expression from complementary DNA of the rod photoreceptor cyclic GMP-gated channel. *Nature* **342,** 762–766.

Körschen, H. G., Beyermann, M., Müller, F., Heck, M., Vantler, M., Koch, K.-W., Kellner, R., Wolfrum, U., Bode, C., Hofmann, K. P., and Kaupp, U. B. (1999). Interaction of glutamic-acid-rich proteins with the cGMP signalling pathway in rod photoreceptors. *Nature* **400,** 761–766.

Körschen, H. G., Illing, M., Seifert, R., Sesti, F., Williams, A., Gotezes, S., Colville, C., Müller, F., Dosé, A., Godde, M., Molday, L., Kaupp, U. B., and Molday, R. S. (1995). A 240 kDa protein represents the complete β subunit of the cyclic nucleotide-gated channel from rod photoreceptor. *Neuron* **15,** 672–636.

Kramer, R. H., and Karpen, J. W. (1998). Spanning binding sites on allosteric proteins with polymer-linked ligand dimers. *Nature* **395,** 710–713.

Kramer, R. H., and Siegelbaum, S. A. (1992). Intracellular Ca^{2+} regulates the sensitivity of cyclic nucleotide-gated channels in olfactory receptor neurons. *Neuron* **9,** 897–906

Kramer, R. H., Goulding, E., and Siegelbaum, S. A. (1994). Potassium channel inactivation peptide blocks cyclic nucleotide-gated channels by binding to the conserved pore domain. *Neuron* **12,** 655–662.

Lamb, T. D., and Pugh, E. N., Jr. (1990). Physiology of transduction and adaptation in rod and cone photoreceptors *Neurosciences* **2,** 3–13.

Liman, E. R., and Buck, L. B. (1994). A second subunit of the olfactory cyclic nucleotide-gated channel confers high sensitivity to cAMP. *Neuron* **13,** 611–621.

Liu, D. T., Tibbs, G. R., Paoletti, P., and Siegelbaum, S. A. (1998). Constraining ligand-binding site stoichiometry suggests that a cyclic nucleotide-gated channel is composed of two functional dimers. *Neuron* **21,** 235–248.

Liu, D. T., Tibbs, G. R., and Siegelbaum, S. A. (1996). Subunit stoichiometry of cyclic nucleotide-gated channels and effects of subunit order on channel function. *Neuron* **16,** 983–990.

Liu, M., Chen, T. Y., Ahamed, B., Li, J., and Yau, K.-W. (1994). Calcium-calmodulin modulation of the olfactory cyclic nucleotide-gated cation channel. *Science* **266,** 1348–1354.

Lu, Z., and Ding, L. (1999). Blockade of a retinal cGMP-gated channel by polyamines. *J. Gen. Physiol.* **113,** 35–43.

Matthews, G. and Watanabe, S.-I. (1987). Properties of ion channels closed by light and opened by guanosine 3'5'-cyclic monophosphate in toad retinal rods. *J. Physiol.* **389,** 691–715.

Menini, A. (1995). Cyclic nucleotide-gated channels in visual and olfactory transduction. *Biophys. Chem.* **55,** 185–196.

Menini, A. (1999). Calcium signaling and regulation in olfactory neurons. *Curr. Opin. Neurobiol.* **9,** 419–426.

Molday, R. S. (1996). Calmodulin regulation of cyclic-nucleotide-gated channels. *Curr. Opin. Neurobiol.* **6,** 445–452.

Molday, R. S., and Hsu, Y.-T. (1995). The cGMP-gated channel of photoreceptor cells: Its structural properties and role in phototransduction. *Behav. Brain Sci.* **18,** 441–451.

Molday, R. S., and Molday, L. L. (1998). Molecular properties of cGMP-gated channel of rod photoreceptors. *Vis. Res.* **38,** 1315–1323.

Molokanova, E., Bhavya, T., Savchenko, A., and Kramer, R. H. (1997). Modulation of rod photoreceptor cyclic nucleotide-gated channels by tyrosine phosphorylation. *J. Neurosci.* **17,** 9068–9076.

Nakamura, T., and Gold, G. H. (1987). A cyclic nuceotide-gated conductance in olfactory receptor cilia. *Nature* **325,** 442–444.

Paoletti, P., Young, E. C., and Siegelbaum, S. A. (1999). C-linker of cyclic nucleotide-gated channels controls coupling of ligand binding to channel gating. *J. Gen. Physiol.* **113,** 17–33.

Pape, H.-C. (1996). Queer current and pacemaker: The hyperpolarization-activated cation current in neurons. *Annu. Rev. Physiol.* **58,** 299–327.

Piconesl, A., and Korenbrot, J. I. (1992). Permeation and interaction of monovalent cations with the cGMP-gated channel of cone photoreceptors. *J. Gen. Physiol.* **100,** 647–673.

Roof, D. J. and Makino, C. L. (2000). The structure and function of retinal photoreceptors. *In:* The Principles and Practice of Ophthalmology, 2nd edition, vol. 3, pp. 1625–1673. D. M. Albert and F. A. Jakobiec, (Eds.) W. B. Saunders Company, Philadelphia.

Ruiz, M. L., and Karpen, J. W. (1997). Single cyclic nucleotide-gated channels locked in different ligand-bound states. *Nature* **389,** 389–392.

Ruiz, M. L., and Karpen, J. W. (1999). Opening mechanism of a cyclic nucleotide-gated channel based on analysis of single channels locked in each liganded state. *J. Gen. Physiol.* **113,** 873–895.

Sagoo, M. S., and Lagnado, L. (1996). The action of cytoplasmic calcium on the cGMP-activated channel in salamander rod photoreceptors. *J. Physiol.* **497,** 309–319.

Shammat, I. M., and Gordon, S. E. (1999). Stoichiometry and arrangement of subunits in rod cyclic nucleotide-gated channels. *Neuron* **23,** 809–819.

Shapiro, M. S., and Zagotta, W. N. (1998). Stoichiometry and arrangement of heteromeric olfactory cyclic nucleotide-gated ion channels. *Proc. Natl. Acad. Sci. USA* **95,** 14546–14551.

Shepherd, G. M. (1992). Toward a consensus working model for olfactory transduction. *In* "Sensory Transduction" (D. P. Corey and S. D. Roper, Eds.), pp. 20–37. Rockefeller Univ. Press, New York.

Tang, C. Y., and Papazian, D. M. (1997). Transfer of voltage independence from a rat olfactory channel to the *Drosophila* ether-a-go-go K+ channel. *J. Gen. Physiol.* **109,** 301–311.

Taylor, W. R., and Baylor, D. A. (1995). Conductance and kinetics of single cGMP-activated channels in salamander rod outer segments. *J. Physiol.* **483,** 567–582.

Torre, V., Straforini, M., Sesti, F., and Lamb, T. D. (1992). Different channel-gating properties of two classes of cyclic GMP-activated channel in vertebrate photoreceptors. *Proc. R. Soc. London Ser. B.* **250,** 209–215.

Varnum, M. D., and Zagotta, W. N. (1996). Subunit interactions in the activation of cyclic nucleotide-gated ion channels. *Biophys. J.* **70,** 2667–2679.

Varnum, M. D., and Zagotta, W. N. (1997). Interdomain interactions underlying activation of cyclic nucleotide-gated channels. *Science* **278,** 110–113.

Wei, J.-Y., Roy, D. S., Leconte, L., and Barnstable, C. J. (1998). Molecular and pharmacological analysis of cyclic nucleotide-gated channel function in the central nervous system. *Progr. Neurobiol.* **56,** 37–64.

Wohlfart, P. Haase, W., Molday, R. S., and Cook, N. J. (1992). Antibodies against synthetic peptides used to determine the topology and site of glycosylation of the cGMP-gated channel from bovine rod photoreceptors. *J. Biol. Chem.* **267,** 644–648.

Yau, K.-W., and Baylor, D. A. (1989). Cyclic GMP-activated conductance of retinal photoreceptor cells. *Annu. Rev. Neurosci.* **12,** 289–327.

Yau, K.-W., and Chen, T. Y. (1995). Cyclic nucleotide-gated channels. *In* "Handbook of Receptors and Channels: Ligand and Voltage-gated Ion Channels" (R. Alan North, Ed.) pp. 307–335. CRC Press, Boca Raton, FL.

Zagotta, W. N., and Siegelbaum, S. A. (1996). Structure and function of cyclic nucleotide-gated channels. *Annu Rev. Neurosci* **19,** 235–263

Zimmerman, A. L. (1995). Cyclic nucleotide gated ion channels. *Curr. Opin. Neurobiol.* **5,** 296–303.

Zimmerman, A. L., and Baylor, D. A. (1992). Cation interactions within the cyclic GMP-activated channel of retinal rods from tiger salamander. *J. Physiol.* **449,** 759–783.

Zong, X., Zucker, H., Hofmann, F., and Biel, M. (1998). Three amino acids in the C-linker are major determinants of gating in cyclic nucleotide-gated channels. *EMBO J.* **17,** 353–362.

Zufall, F., Shepherd, G. M., and Firestein, S. (1991). Inhibition of the olfactory cyclic nucleotide gated ion channel by intracellular calcium. *Proc. R. Soc. London Ser. B.* **246,** 225–230.

Anita L. Zimmerman

48

Visual Transduction

I. Introduction

The miracle of vision begins when our photoreceptors absorb light that is reflected by our surroundings. However, the photoreceptor's duty does not end with photon capture. In addition, the photoreceptor must convert the energy of the absorbed photon into an electrochemical signal to be relayed to the visual cortex of the brain. This process of visual transduction involves a G-protein-mediated second messenger system which ultimately controls membrane potential and neurotransmitter release.

This chapter gives an overview of vertebrate visual transduction. Further information can be obtained from many excellent reviews on the subject (e.g., Baylor, 1996; Yarfitz and Hurley, 1994; Yau and Baylor, 1989; Lamb and Pugh, 1990; McNaughton, 1990; Detwiler and Gray-Keller, 1992; Lagnado and Baylor, 1992; Fain et al., 1996; Koutalos and Yau, 1996; Leibrook et al., 1998; Pugh et al., 1999; Roof and Makino, 2000). Details of the cyclic nucleotide-gated ion channels that mediate the light response are covered in Chapter 47. Invertebrate visual transduction is not discussed here, but thorough descriptions are available elsewhere (e.g., Bacigalupo et al., 1990; Yarfitz and Hurley, 1994; Zuker, 1996). One interesting difference between vertebrate and invertebrate photoreceptors is that the former hyperpolarize in response to light, whereas the latter depolarize.

II. Photoreceptor Cells

The two kinds of vertebrate photoreceptors—rods and cones—are specialized for use under different conditions. The rods are employed for vision in dim light, whereas the cones are used in moderate to bright light. Thus, while a rod can reliably detect a single photon in a darkened room, its sensing mechanism shuts down completely when normal room lights are switched on. Cones, on the other hand, are not sensitive enough to detect single photons, but give us color vision and high spatial and temporal resolution when light levels are sufficiently high.

Rods and cones also occupy different parts of the retina. Cones are concentrated in the center of the retina, the fovea, which receives light from the center of our visual field. The rods, however, are almost entirely excluded from the fovea and instead dominate the peripheral retina. However, a few cones are scattered throughout the peripheral retina, and a few rods exist at the edge of the fovea. Thus, in daylight, our peripheral vision has much lower spatial resolution than our central vision because of the sparse distribution of peripheral cones. At night, our ability to locate a dim star is enhanced by looking "out of the corners of our eyes," using the abundant rods in the peripheral retina.

As shown in Fig. 1, the vertebrate photoreceptor can be divided into three regions: the outer segment, the inner segment, and the synaptic terminal. The outer segment is the region of the cell dedicated to photon capture and visual transduction. The inner segment, located closer to the front of the eye, contains the nucleus, mitochondria, and other general cell machinery. The synaptic terminal contains vesicles of neurotransmitter (thought to be glutamate) for release onto the second-order retinal neurons (the bipolar cells).

Rods and cones have very similar anatomy, except for one major feature: rods have intracellular disks, and cones have, instead, infoldings of the plasma membrane called **sacs** (Fig. 1). The reason for this difference is not known. The membranes of the disks and sacs contain photopigment molecules and various other proteins used in visual transduction. The disks or sacs are packed very tightly together: usually about 2000 in a typical outer segment whose length is about 60 μm. This parallel array of membranes serves to align the photopigment molecules in the correct orientation for optimal absorption of light, which normally travels the length of the cell from the synaptic terminal toward the tip of the outer segment. In rods, the parallel arrangement of the disks may be maintained by cytoskeletal filaments that have been found to connect the disks to each other and to the plasma membrane (Roof and Heuser, 1982).

In both rods and cones, the outer and inner segments are connected by a narrow region called the **cilium.** In some way that is not yet understood, the cilium segregates the

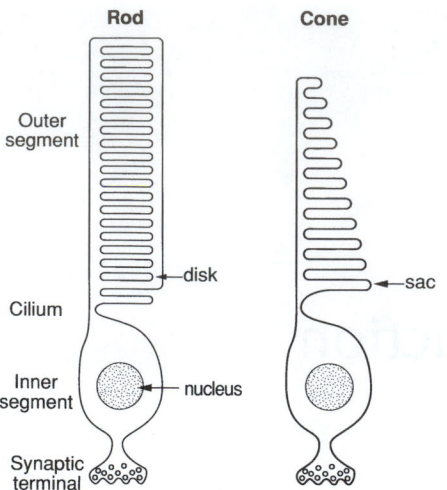

Rod   Cone

Outer segment

disk

sac

Cilium

Inner segment   nucleus

Synaptic terminal

FIGURE 1. A simplified view of rod and cone structure. The outer segment is specialized for visual transduction. Cone photopigment molecules are located on infoldings of the plasma membrane, called sacs, rather than in the membranes of intracellular disks as in rods.

inner and outer segment membrane proteins. For example, whereas the inner segment plasma membrane contains many types of ion channels (Bader *et al.*, 1982; Barnes and Hille, 1989), the outer segment plasma membrane contains essentially a pure population of light-regulated cGMP-gated channels (Baylor and Nunn, 1986). It is in the ciliary region that new disks or sacs are inserted during the outer segment regeneration process that occurs daily (Bok, 1985). Old disks or sacs are removed from the tip of the outer segment by phagocytosis by the pigment epithelial cells. These very opaque cells exist at the back of the retina, with fingerlike

cellular processes that hug the photoreceptor outer segments and also help to reduce light scatter. Because the cilium is such a narrow structure, outer and inner segments often break apart (and their membranes reseal) during cell isolation from the retina. Thus, many studies of visual transduction are conducted on isolated outer segments, which are somewhat simpler preparations than whole cells.

III. Physiology of Visual Transduction

The physiological trademark of the vertebrate photoreceptor is its light-regulated "dark current" (Fig. 2) (Hagins *et al.*, 1970). This current circulates between the outer and inner segments in the dark and is reduced upon light absorption. The dark current is carried into the outer segment mainly by Na^+ ions flowing through cGMP-gated nonselective cation channels (see Chapter 47) and out of the inner segment by K^+ ions flowing through voltage-gated K^+ channels. Since the cell has a relatively high Na^+ permeability in the dark, its resting potential is somewhat more positive than that for most cells (about -40 mV, instead of -70 mV). This resting depolarization tends to open voltage-gated Ca^{2+} channels near the synaptic terminal, allowing Ca^{2+} entry and vesicular release of neurotransmitter in the dark. When a photon is absorbed by a rhodopsin molecule in a disk membrane, a series of reactions occurs that lead to a reduction in the guanosine 3', 5'-cyclic monophosphate (cGMP) concentration, with a resulting closure of the cGMP-gated channels in the outer segment plasma membrane (discussed later). The resulting reduction in the dark current (i.e., reduction in Na^+ permeability) causes a membrane hyperpolarization, which in turn causes closure of the voltage-gated K^+ channels and Ca^{2+} channels and finally a decrease in neurotransmitter release.

Light ⟶ ↓[cGMP] ⟶ Channels close ⟶ Hyperpolarization ⟶ ↓Transmitter release

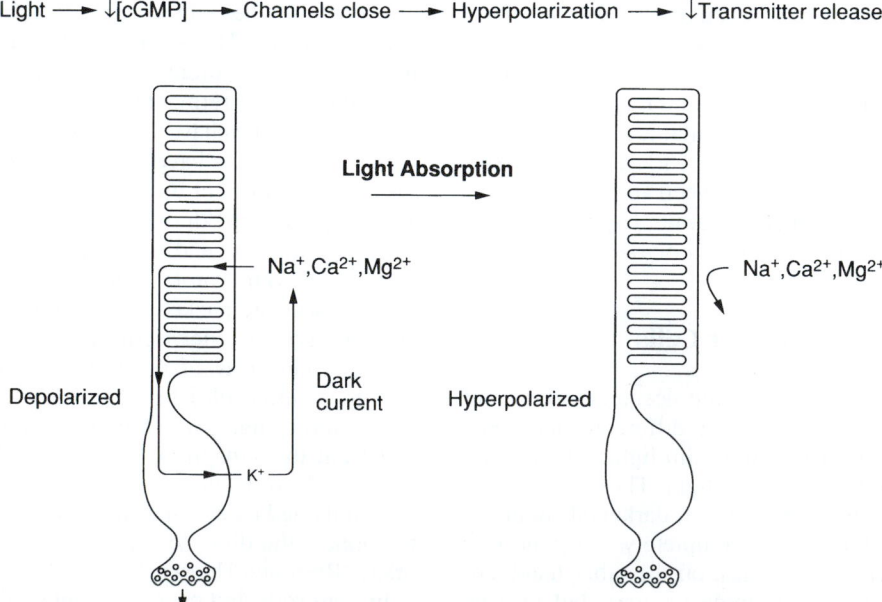

Light Absorption

Na^+, Ca^{2+}, Mg^{2+}

Na^+, Ca^{2+}, Mg^{2+}

Depolarized   Dark current   Hyperpolarized

K^+

FIGURE 2. The absorption of light shuts down the dark current that circulates between the outer and inner segments of the photoreceptor. This reduction in current hyperpolarizes the cell, reducing the release of neurotransmitter.

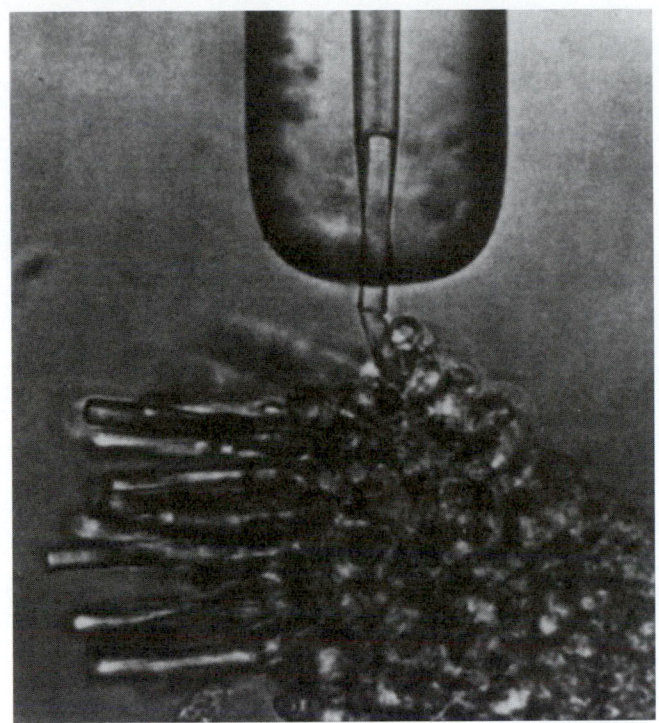

FIGURE 3. The use of a suction electrode to record the dark current flowing into the outer segment of a rod attached to a piece of toad retina. The horizontal line just below the suction electrode is the junction between the outer and inner segments. The electrode is filled with a physiological salt solution and connected to a sensitive amplifier. (Reproduced with permission from Baylor *et al.*, 1979.)

The dark current and light response can be measured in a variety of ways, including a voltage-clamp method using intracellular microelectrodes, the whole-cell patch-clamp method, and the suction electrode method. The relatively noninvasive suction electrode method (Fig. 3) works particularly well with vertebrate photoreceptors, which are usually quite small and fragile. This method is much gentler than the other techniques, yet is able to measure the dark current without interference from inner segment currents.

Families of light responses recorded with a suction electrode from a monkey rod and cone are shown in Fig. 4. Light response amplitudes increase in a graded manner with light intensity until they reach a limiting value obtained with a "saturating light" that shuts off all the inward dark current (i.e., that closes all the cGMP-gated channels in the outer segment plasma membrane). Supersaturating light intensities cannot give larger responses, but instead can give longer lasting ones. Cones are less sensitive than rods, requiring 50 to 100 times as many photons to shut down half the dark current. However, the cone responses are several times faster than those of rods, as evidenced by the shorter time to peak response. The undershoots seen in the cone responses (but not in normal rod responses) are not yet fully understood, but appear to make the cone especially sensitive to changes in illumination (Schnapf *et al.*, 1990).

Although single photon responses are too small to be measured in cones, they are about 1.0 pA in rods, and there-

fore have been studied in some detail there. Single photon responses are very reliable: they occur at least 80% of the time after absorption of a photon by a rhodopsin molecule, and they almost never occur in the absence of photon absorption (Baylor, 1987). A single absorbed photon gives a highly amplified response, shutting down 3–5% of the total dark current by closing a few hundred cGMP-gated channels in a narrow band (one to a few micrometers in width) of outer segment plasma membrane near the site of absorption. Like the dim flash responses in Fig. 4 (bottom traces of each set), the single photon response has a stereotypical waveform, with a slow, S-shaped rise and a very slow decay. These complex, slow kinetics reflect the complex enzyme cascade underlying the response (discussed later).

When membrane voltage is recorded instead of outer segment current, the waveform resembles an inverted version of the current for dim lights but not for light levels near or beyond saturation. When many photons are absorbed, the voltage recordings (Fig. 5) show a characteristic "nose" and "plateau." These features result from a shaping of the response by ion channels in the inner segment. Several types of channels contribute to the waveform, but the dominant

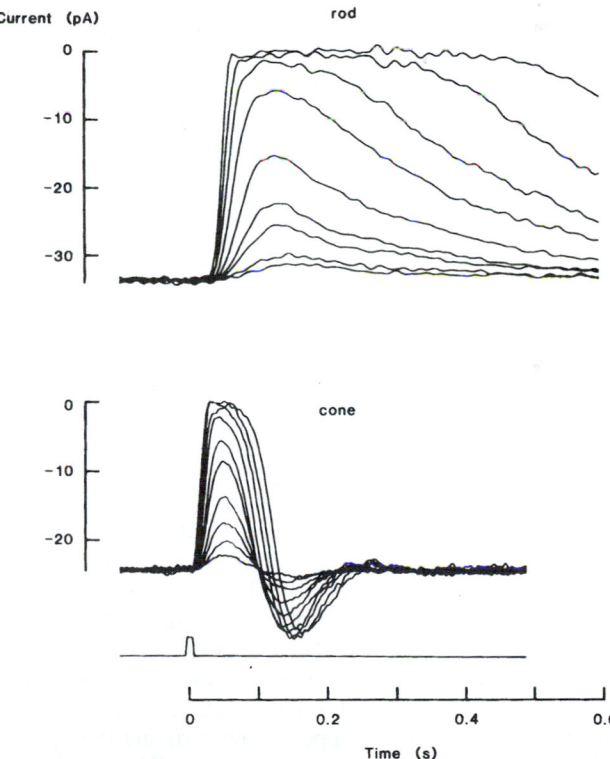

FIGURE 4. Families of photocurrents recorded by suction electrode from a rod and a cone of the monkey, *Macaca fascicularis*. Brief light flashes were given at time zero (rectangular pulse below the current recordings). Upward deflections indicate reductions in the inward dark current. From the bottom to the top of each family, flash strengths were increased by factors of 2. Expected numbers of photoisomerizations ranged from 2.9 to 860 for the rod, and from 190 to 36,000 for the cone. Saturation of the responses (top traces in each family) occurred when all dark current was shut off. (Reproduced with permission from Baylor, 1987.)

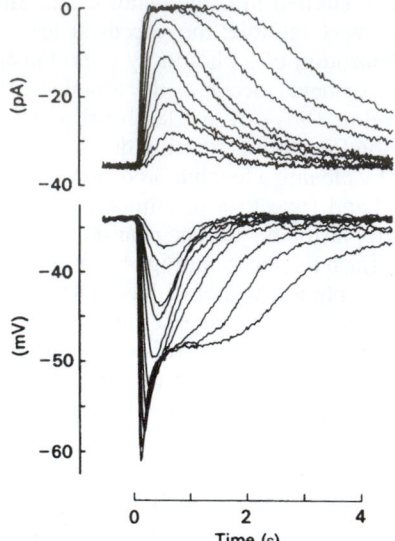

FIGURE 5. Comparison of light-evoked changes in outer segment current with those in membrane potential for a salamander rod. The currents were measured by a suction electrode, while the membrane potential was monitored by an intracellular microelectrode. The current and voltage responses to dim flashes have similar form, but a prominent "nose" and "plateau" are seen in the voltage responses to bright flashes. 500-nm, 11-ms flashes were given at $t = 0$, and photon densities increased by factors of about 2 between 1.5 and 430 photons μm^{-2}. (Reproduced with permission from Baylor and Nunn, 1986.)

channel giving rise to this more complicated shape appears to be an inner segment, nonselective cation channel that is opened by hyperpolarization. Thus, the closure of outer segment cGMP-gated channels gives the initial hyperpolarization (beginning of the nose). This hyperpolarization opens the inner segment nonselective cation channels, which in turn depolarize the membrane, giving the transition from nose to plateau. Eventually, these channels close because of the depolarization, while at the same time the outer segment channels reopen during recovery from the light response, and the membrane potential returns to its initial dark value.

Aside from the ion channels described earlier, two other types of ion transport protein are of obvious importance to the vertebrate photoreceptor. First, Na^+-K^+ pumps in the inner segment are responsible for expelling the Na^+ that enters through the cGMP-gated channels in the outer segment. Second, Na^+:Ca^{2+},K^+ exchange carriers in the outer segment expel the Ca^{2+} that enters through the cGMP-gated channels. The Na^+:Ca^{2+},K^+ carriers appear to be particularly important in light adaptation, as discussed in Section IV.B.

The ability of the visual system to detect light depends partly on its degree of adaptation to background illumination. Although much of light and dark adaptation involves pupillary responses and processes occurring in the brain and non-photoreceptor layers of the retina, some aspects of adaptation clearly occur in the rods and cones themselves. In continuous light, there is a partial recovery of the dark current after its initial suppression when the light is switched on. This sag in the photoresponse is accompanied by a decrease in light sen-

sitivity and an acceleration of response kinetics. Thus, the response to a light flash during continuous background light is smaller and briefer than that to an equally bright flash occurring in the dark. To make the amplitude of the flash response equal to that obtained in the dark, one must increase the flash intensity (Fig. 6). This light adaptation allows the photoreceptor to respond over a much larger range of light intensities than it otherwise could. In effect, the photoreceptor uses a nonlinear gain adjustment to partially overcome the response saturation that occurs as a result of having a finite number of cGMP-gated channels to close in the light.

Photoreceptor dark adaptation is a rather complicated set of processes. After exposure to a light bright enough to bleach all the photopigment, a photoreceptor ultimately cannot return to its dark state until nonexcited photopigment molecules have been regenerated (including reinsertion of the chromophore, see following). However, when the photoreceptor has been exposed to only moderately bright light, which leaves some photopigment unbleached, it is able to respond to light again after a briefer period of adaptation. This period is characterized by a lingering suppression of the dark current for many minutes after the light is switched off, a decrease in light sensitivity similar to that found in the presence of background light, and an increase in outer segment current noise (Fig. 7). These features limit visual detection and appear to derive from incomplete shutoff of the photopigment (Leibrook *et al.*, 1998).

IV. Molecular Mechanisms

A. Photopigment Activation and Shutoff

Visual pigments are integral membrane proteins called **opsins** that contain a ubiquitous light-absorbing chromophore, retinal (Fig. 8) (for reviews of visual pigments,

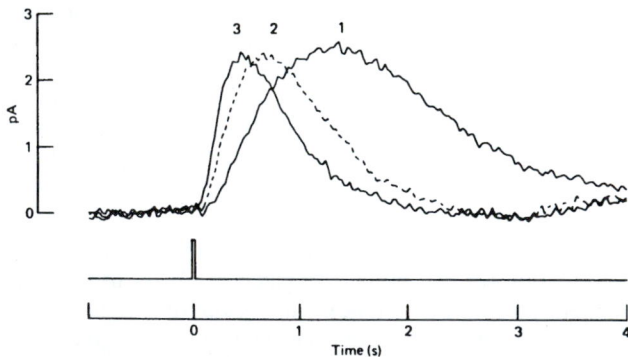

FIGURE 6. Decreased sensitivity and accelerated flash response kinetics characteristic of adaptation to background light. These outer segment currents were obtained by suction electrode recording from a toad rod that was given a test light flash in darkness (curve 1), with a dim background light (curve 2), and with a brighter background light (curve 3). In the presence of background light, the test flash intensity had to be increased to obtain responses of approximately the same amplitude. For example, the flash intensity used to obtain curve 2 was approximately five times the intensity used to obtain curve 1. (Reproduced with permission from Baylor *et al.*, 1979.)

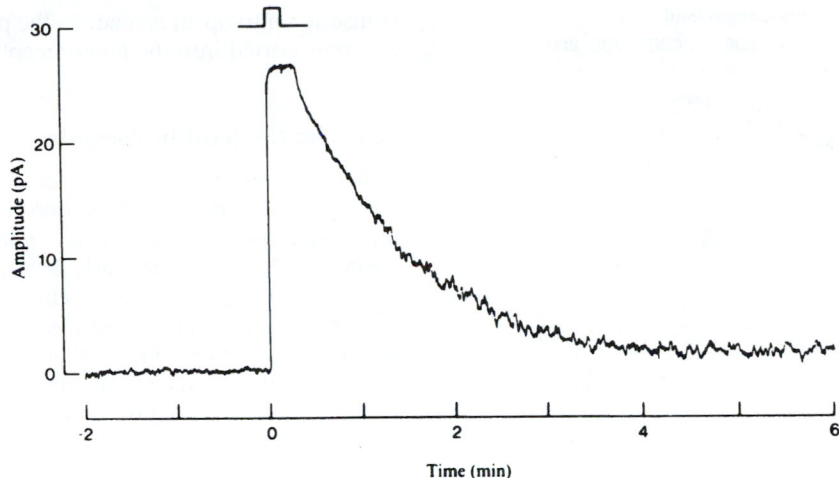

FIGURE 7. Prolonged dark current suppression and increased noise following a strong bleach. The rectangular pulse above the response indicates when the dark-adapted rod was given a bright flash of light that bleached about 0.7% of its rhodopsin (72×10^5 photons μm^{-2}, calculated to isomerize 1.8×10^7 rhodopsins). Closer examination of the lingering noise revealed a strong resemblance to that produced by the absorption of single photons. (Reprinted with permission from *Nature,* Lamb, 1980. Copyright 1980 Macmillan Magazines Limited.)

and rhodopsin in particular, see, for example, Wald, 1968; Birge, 1990; Hargrave and McDowell, 1992; Nathans, 1992). Retinal is derived from vitamin A and comes in two forms in vertebrates: $retinal_1$ and dehydroretinal, or $retinal_2$ (Dowling, 1987). Human photoreceptors use $retinal_1$. Interactions with different opsins give the retinals different spectral tuning characteristics. Thus, each type of photoreceptor has its own type of opsin, which results in a characteristic photopigment absorption spectrum, and therefore a characteristic color sensitivity of the cell. Rods contain rhodopsin (Rh) and are therefore most sensitive to blue-green light (peak wavelength around 490 nm), whereas primate cones are of three types, with characteristic peak absorbances in the short-wavelength (peak about 430 nm),

middle-wavelength (peak about 530 nm), and long-wavelength (peak about 560 nm) regions of the visible spectrum (Fig. 9). These three types of cone are sometimes referred to as blue-, green-, and red-sensitive cones, respectively, although 560 nm actually corresponds to yellow, rather than red, light. The characteristic absorption spectra determine only the probability that a photon will be absorbed by the cell's photopigment. The photoreceptor responds in exactly the same way to any absorbed photon, independent of its wavelength.

Absorption of a photon causes isomerization of retinal from the 11-*cis* to the all-*trans* form (Fig. 10), destabilizing its position within the protein. Eventually all-*trans*-retinal hops out of its binding pocket within opsin, but a chain of

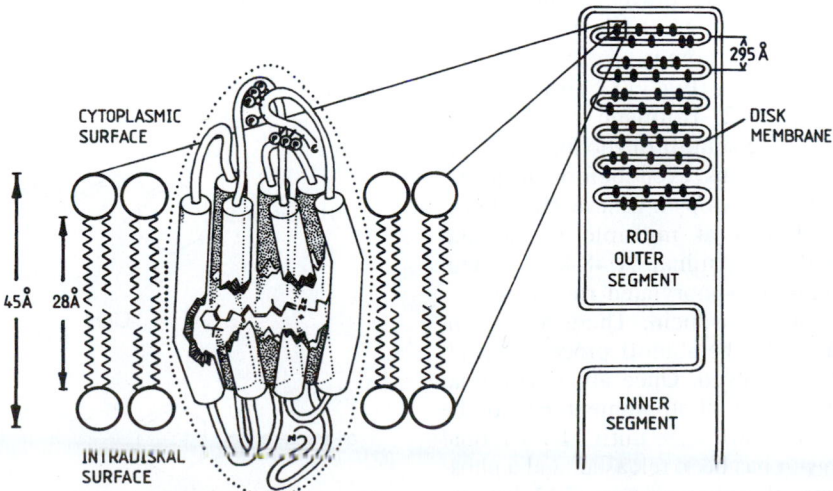

FIGURE 8. A model for the structure of rhodopsin in the disk membrane of a rod. The light-absorbing chromophore, retinal, sits in a binding pocket near the middle of the integral membrane protein, opsin. (Reproduced with permission from Dratz and Hargrave, 1983.)

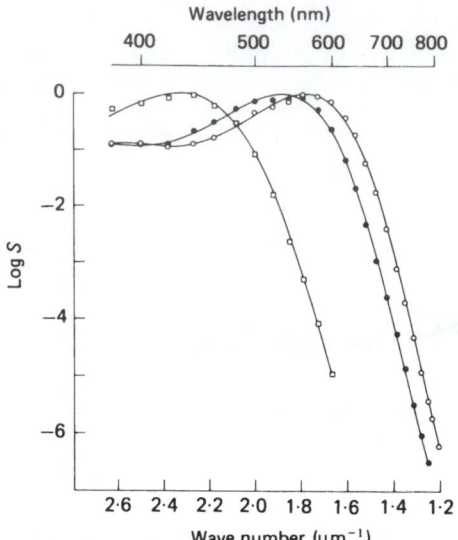

FIGURE 9. Cone spectral sensitivities, as measured by the suction electrode method. These spectra were obtained from blue-, green-, and red-sensitive monkey cones (squares, filled circles, and open circles, respectively). The spectral sensitivity, S, was derived from the cone's response to flashes of different color, and it reflects the probability of absorption of a photon of that color (or wavelength). (Reproduced with permission from Baylor, *et al.*, 1987.)

conformational intermediates is generated before this dissociation occurs. For the best-studied pigment, rhodopsin, these intermediates are (in the order that follows photon absorption) photorhodopsin, bathorhodopsin, lumirhodopsin, metarhodopsin I, metarhodopsin II, and metarhodopsin III (reviewed in Hargrave and McDowell, 1992). Similar transitions are thought to occur in other visual pigments. The metarhodopsin II (meta II) conformational state activates the G protein, transducin, as discussed later.

For the photoreceptor to reset to its dark condition after a light is switched off, the molecular players in the light response also must be switched off. Most of what we know about photopigment shutoff comes from work on rhodopsin (depicted in part of Fig. 11) (reviewed in Hargrave and McDowell, 1992; Hofmann *et al.*, 1992; Yarfitz and Hurley, 1994). The shutoff of rhodopsin's ability to activate transducin begins even before all-*trans*-retinal has left the protein. Rhodopsin kinase phosphorylates photoexcited rhodopsin at multiple serine and threonine residues near the C-terminal. A 48-kDa protein called **arrestin** then binds phosphorylated rhodopsin, reducing its interaction with transducin. There are recent hints for still other players in the shutoff process, but all the details have not been resolved. Once all-*trans*-retinal has separated from opsin, the photopigment cannot be reset to its photon-receptive dark state until 11-*cis*-retinal has been reinserted, arrestin has been released, and a phosphatase (serine/threonine phosphatase type 2A) has removed the phosphates from the C-terminal residues. This process takes several minutes. The new 11-*cis*-retinal that

is inserted into opsin is made in the pigment epithelial cells and transported into the photoreceptors.

B. Cyclic Nucleotide Cascade

A cyclic nucleotide enzyme cascade is used to translate the message of photon absorption into the language of the brain—electrical and chemical signals. Many publications written in the 1970s and early 1980s describe the Ca^{2+} hypothesis (Hagins, 1972) in which Ca^{2+} is proposed to be the second messenger mediating the response to light. There is now overwhelming evidence that the key second messenger is cGMP, rather than Ca^{2+}, and that Ca^{2+} is involved in more subtle aspects of visual transduction (discussed later). Most of our information on the molecular mechanism of visual transduction derives from studies on rods, but the process appears to be similar in cones. A large body of biochemical and physiological evidence suggests that absorption of a single photon by rhodopsin initiates the following chain of events (diagrammed in Fig. 11):

1. Photoexcited rhodopsin (Rh*) catalyzes the replacement of GDP by GTP on transducin (T). Both Rh* and the peripheral membrane protein, T, are able to diffuse laterally in the disk membrane, where they find each other and interact. Before Rh* is shut off, it activates about 500 Ts, representing the first amplification step in the transduction process.

2. The α subunit of T, now with GTP bound, dissociates from the rest of T (the β–γ subunit complex) and binds to and removes the γ inhibitory subunit of another peripheral disk membrane protein, cGMP-specific phosphodiesterase (PDE). Each Tα can activate only one PDE since the catalytic portion of PDE, PDE$\alpha\beta$, is only free to work when PDEγ is bound to Tα (and therefore unable to reassociate with PDE$\alpha\beta$).

3. Once disinhibited, one PDE molecule (PDE$\alpha\beta$) hydrolyzes about a million cGMP molecules, leading to the closure of hundreds of cGMP-gated ion channels and thereby halting entry of over a million Na^+ ions during the time course of the single photon response.

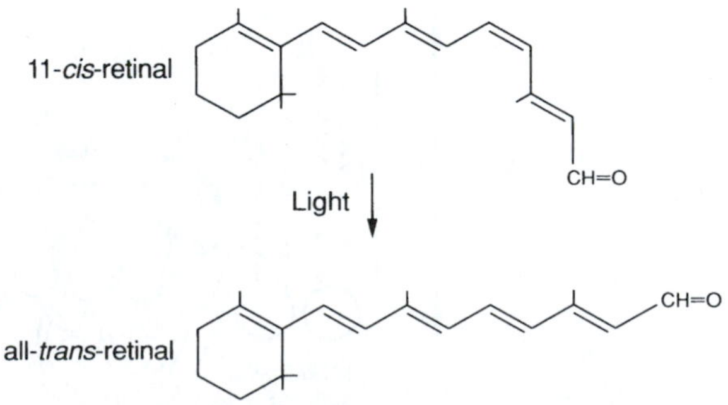

11-*cis*-retinal

Light

all-*trans*-retinal

FIGURE 10. The light-induced isomerization of retinal$_1$ from the 11-*cis* to the all-*trans* form. This is the first step in vertebrate visual transduction, and the only one that is a direct action of light.

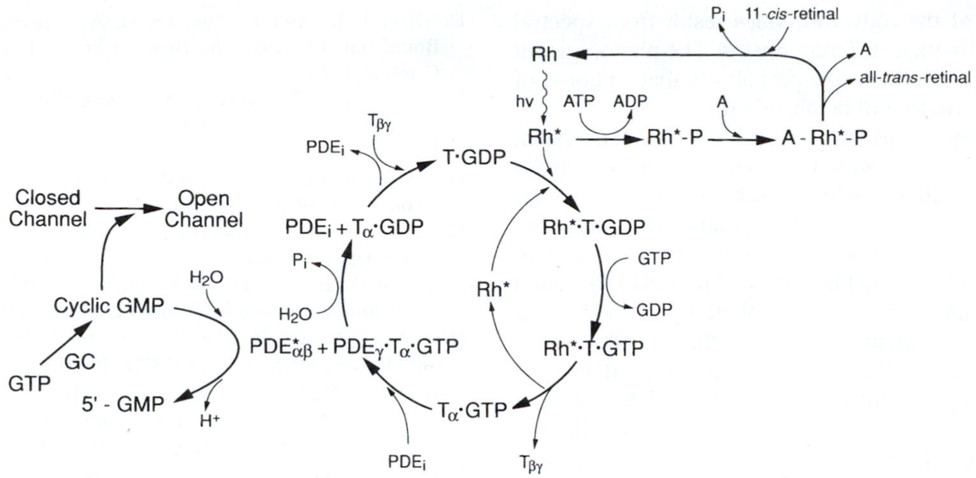

FIGURE 11. The enzyme cascade controlling visual transduction in rods. Abbreviations: A, arrestin; channel, cGMP-gated outer-segment channel; GC, guanylate cyclase; PDE, cGMP-specific phosphodiesterase (*, active form): Rh, rhodopsin (*, photoexcited, active form); T, the G protein, transducin. There is increasing evidence that in spite of its complexity, this diagram is quite incomplete. (Modified with permission from Stryer, 1991, and drawing on information from Hargrave and McDowell. 1992, and Hofmann *et al.*, 1992.)

The recovery of the dark current after photon absorption is accomplished by many processes, including these:

1. Shutoff of rhodopsin, as discussed earlier.
2. Inherent GTPase activity of $T\alpha$, converting $T\alpha$ back to the GDP-bound, inactive form that reassociates with $T\beta\gamma$.
3. Reassociation of $PDE\gamma$ with $PDE\alpha\beta$, returning PDE to its more inhibited, dark state.
4. Synthesis of new cGMP from GTP by guanylate cyclase (GC).
5. Reopening of cGMP-gated ion channels, as more cGMP becomes available.

Although these processes are the most established ones, there is recent evidence that other processes and factors may be involved in regulating the cGMP cascade, mostly through feedback mechanisms. Some of the factors that may play a role are regulatory cGMP-binding sites on PDE; calcium-binding proteins that may affect guanylate cyclase, PDE, rhodopsin, transducin, and/or the cGMP-gated channel; members of the inositol phosphate system whose function has not yet been established; and various kinases and phosphatases whose functions are also not yet clear. The field of regulation of the photoreceptor cGMP cascade is currently very unsettled and very intriguing. Some interesting findings on this subject are reviewed in Lagnado and Baylor (1992), Detwiler and Gray-Keller (1992), and Yarfitz and Hurley (1994).

One aspect of visual transduction that has been proposed to be controlled by feedback regulation is light adaptation. The following argument is suggested: coupled with calcium-binding proteins, Ca^{2+} has been found to inhibit guanylate cyclase (GC), to stimulate PDE (via inhibition of Rh phosphorylation) and to inhibit the cGMP-gated channels (see Chapter 47). In the dark, when the cGMP, concentration is relatively high, Ca^{2+} enters the cell through the cGMP-gated channels, raising intracellular $[Ca^{2+}]$ ($[Ca^{2+}]_i$). When the channels close following photon absorption, $[Ca^{2+}]_i$ decreases because its influx through the channels is decreased while its extrusion by $Na^+:Ca^{2+},K^+$ exchangers continues. The decreased $[Ca^{2+}]_i$, would be expected to stimulate guanylate cyclase and to inhibit PDE, leading to a subsequent partial recovery of the cGMP level and a reopening of some of the channels. The feedback mechanisms involving Ca^{2+} are mediated by several calcium binding proteins (reviewed in Roof and Makino, 2000).

While this scheme sounds very logical and appealing, some results suggest that it is very oversimplified. Detwiler and Gray-Keller (1992) have found that high $[Ca^{2+}]_i$ speeds the recovery from a light response, as if facilitating the return to dark levels of [cGMP] and the consequent reopening of cGMP-gated channels. This behavior is inconsistent with the argument just presented, which predicts that high $[Ca^{2+}]_i$ would *decrease* [cGMP] by stimulating PDE and inhibiting guanylate cyclase. These and other recent findings reveal that many pieces of the transduction puzzle are still missing (also see Chapter 47).

V. Summary

The process of vertebrate vision begins with photon capture and visual transduction in the rods and cones. The rods are outstanding single photon detectors that are used in dim light, whereas the cones provide excellent visual acuity and color vision in bright light. There are slight differences in the structures of rods and cones, but these photoreceptors have the same basic design for optimal photon capture and visual transduction.

The first step in the response to light is the absorption of a photon by a visual pigment molecule. The visual pigment (or photopigment) consists of a particular protein (an opsin) containing a ubiquitous chromophore, retinal. The differences in color sensitivity of the rods and cones result from spectral tuning of retinal by their different opsins. The photopigment absorption spectra determine the probability that a photon of a particular wavelength will be absorbed.

Visual transduction consists of a kind of signal conversion: the message of photon absorption is converted into a decrease in membrane conductance, which results in a hyperpolarization that decreases the release of neurotransmitter onto other retinal neurons. The light-induced decrease in membrane conductance is accomplished by the closure of nonselective cation channels in a specialized region of the photoreceptor called the outer segment. These channels conduct the so-called "dark current" that flows into the outer segment in the dark. The channels are kept open in the dark by cGMP, and they close in the light when the intracellular cGMP concentration falls.

The light-induced fall in [cGMP] occurs when a photoexcited visual pigment molecule activates a G protein, which in turn activates a phosphodiesterase to hydrolyze cGMP. There are elaborate mechanisms for returning the molecular machinery of the photoreceptor to its dark state, and for light and dark adaptation. Although many of the fundamental molecular steps in visual transduction have been determined, numerous mysteries still remain.

Bibliography

Bacigalupo, J., Johnson, E., Robinson, P., and Lisman, J. E. (1990). Second messengers in invertebrate phototransduction. *In* "Transduction in Biological Systems" (C. Hidalgo, J. Bacigalupo, E. Jaimovich, and J. Vergara, Eds.), pp. 27–45. Plenum, New York.

Bader, C. R., Bertrand, D., and Schwartz, E. A. (1982). Voltage-activated and calcium-activated currents studied in solitary rod inner segments from the salamander retina. *J. Physiol.* **331,** 253–284.

Bames, S., and Hille, B. (1989). Ionic channels of the inner segment of tiger salamander cone photoreceptors. *J. Gen. Physiol.* **94,** 718–743.

Baylor, D. A. (1987). Photoreceptor signals and vision. *Invest. Ophthalmol. Visual Sci.* **28,** 34–49.

Baylor, D. A. (1996). How photons start vision. *Proc. Natl. Acad. Sci. USA* 93, 560–565.

Baylor, D. A., and Nunn, B. J. (1986). Electrical properties of the light-sensitive conductance of rods of the salamander *Ambystoma tigrinum. J. Physiol.* **371,** 115–145.

Baylor, D. A., Lamb, T. D., and Yau, K.-W. (1979). The membrane current of single rod outer segments. *J. Physiol.* **288,** 589–611.

Baylor, D. A., Nunn, B. J., and Schnapf, J. L. (1987). Spectral sensitivity of cones of the monkey *Macaca fascicularis. J. Physiol.* **390,** 145–160.

Birge, R. R. (1990). Nature of the primary photochemical events in rhodopsin and bacteriorhodopsin. *Biochim. Biophys. Acta* **1016,** 293–327.

Bok, D. (1985). Retinal photoreceptor-pigment epithelium interactions. *Invest. Ophthalmol. Visual Sci.* **26,** 1659–1694.

Detwiler, P. B., and Gray-Keller, M. P. (1992). Some unresolved issues in the physiology and biochemistry of phototransduction. *Curr. Opin. Neurobiol.* **2,** 433–438.

Dowling, J. E. (1987). "The Retina: An Approachable Part of the Brain." pp. 195–197. The Belknap Press of Harvard Univ. Press, Cambridge, MA.

Dratz, E. A., and Hargrave, P. A. (1983). The structure of rhodopsin and the rod outer segment disk membrane. *Trends Biochem. Sci.* **8,** 128–131.

Fain, G. L., Matthews, H. R., and Cornwall, M. C. (1996). Dark adaptation in vertebrate photoreceptors. *Trends Neurosci.* **19,** 502–507.

Hagins, W. A. (1972). The visual process: Excitatory mechanisms in the primary receptor cells. *Annu. Rev. Biophys. Bioeng.* **1,** 131–158.

Hagins, W. A., Penn, R. D., and Yoshikami, S. (1970). Dark current and photocurrent in retinal rods. *Biophys. J.* **10,** 380–412.

Hargrave, P. A., and McDowell, J. H. (1992). Rhodopsin and phototransduction. *Int. Rev. Cytol.* **137B,** 49–97.

Hofmann, K. P., Pulvermuller, A., Buczylko, J., Van Hooser, P., and Palczewski, K. (1992). The role of arrestin and retinoids in the regeneration pathway of rhodopsin. *J. Biol. Chem.* **267,** 15701–15706.

Koutalos, Y., and Yau, K.-W. (1996). Regulation of sensitivity in vertebrate rod photoreceptors by calcium. *Trends Neurosci.* **19,** 73–81.

Lagnado, L., and Baylor, D. A. (1992). Signal flow in visual transduction. *Neuron* **8,** 995–1002.

Lamb, T. D. (1980). Spontaneous quantal events induced in toad rods by pigment bleaching. *Nature* **287,** 349–351.

Lamb, T. D., and Pugh, E. N., Jr. (1990). Physiology of transduction and adaptation in rod and cone photoreceptors. *Neurosciences* **2,** 3–13.

Leibrock, C. S., Reuter, T., and Lamb, T. D. (1998). Molecular basis of dark adaptation in rod photoreceptors. *Eye* **12,** 511–520.

McNaughton, P. A. (1990). Light response of vertebrate photoreceptors. *Physiol. Rev.* **70,** 847–883.

Nathans, J. (1992). Rhodopsin: Structure, function, and genetics. *Biochemistry* **31,** 4923–4931.

Pugh, E. N., Nikonov, S., and Lamb, T. D. (1999). Molecular mechanisms of vertebrate photoreceptor light adaptation. *Curr. Opin. Neurobiol.* **9,** 410–418.

Roof, D. J., and Heuser, J. E. (1982). Surfaces of rod photoreceptor disk membranes: Integral membrane components. *J. Cell Biol.* **95,** 487–500.

Roof, D. J. and Makino, C. L. (2000). The structure and function of retinal photoreceptors. *In:* The Principles and Practice of Ophthalmology, 2nd edition, volume 3, pp. 1624–1673. (D. M. Albert and F. A. Jakobiec, Eds.) W. B. Saunders Company. Philadelphia.

Schnapf, J. L., Nunn, B. J., Meister, M., and Baylor, D. A. (1990). Visual transduction in cones of the monkey *Macaca fascicularis. J. Physiol. London* **427,** 681–713.

Stryer, L. (1991). Visual excitation and recovery. *J. Biol. Chem.* **266,** 10711–10714.

Wald, G. (1968). The molecular basis of visual excitation. *Nature* **219,** 800–807.

Yau, K.-W., and Baylor, D. A. (1989). Cyclic GMP-activated conductance of retinal photoreceptor cells, *Annu. Rev. Neurosci.* **12,** 289–327.

Yarfitz, S., and Hurley, J. B. (1994). Transduction mechanisms of vertebrate and invertebrate photoreceptors. *J. Biol. Chem.* **269,** 14329–14332.

Zuker, C. S. (1996). The biology of vision in *Drosophila. Proc. Natl. Acad. Sci. USA* **93,** 571–576.

Stephen D. Roper

49

Gustatory and Olfactory Sensory Transduction

I. Introduction

Many sensory receptor cells are modified epithelial cells and consequently share properties common to epithelial tissues. These properties include (a) an ongoing renewal of the cell population and (b) the existence of two separate regions of the cell surface—**apical membrane** and **basolateral membrane.** These regions are kept separate by intercellular junctional complexes (including tight junctions) that also serve to bind adjacent cells together. Apical and basolateral cell surfaces in epithelial cells and sensory receptor cells are exposed to two vastly different milieux. This is especially the case in **gustatory** and **olfactory** receptor cells. Their apical membrane tips confront external chemical environments in the oral and nasal cavities, respectively, that can vary profoundly in composition. In contrast, their basolateral membranes are bathed in a protected, relatively constant medium—the interstitial fluid.

The exposed apical membrane of gustatory and olfactory sensory cells is the site of **chemosensory transduction,** the conversion of external chemical stimuli into intracellular electrical signals. Tight junctions confine most chemical stimuli to the specialized apical membrane and prevent the stimuli from penetrating deeper into the sensory tissue (although there may be exceptions in gustatory sensory cells, see Section II). This has two implications for gustation and olfaction. First, the molecular machinery for the initial events of signal transduction must reside in the apical membrane of the sensory cells, where chemical stimuli contact the cells. Second, because chemosensory stimulation is usually associated with a **receptor** (or **generator**) **current** (that is, a flux of ions) across the apical chemosensitive membrane, the ionic environment at the exposed apical region has a significant impact on transduction. For example, the apical chemosensitive membrane of olfactory sensory cells possesses a Ca^{2+}-dependent Cl^- conductance, as will be discussed later. This conductance exploits the low (relative to

interstitial fluid) Cl^- concentration in the external environment. When the Ca^{2+}-dependent Cl^- channels are opened during odor stimulation, Cl^- ions leave the cell, generating an **inward (depolarizing) current.** This Cl^- current augments the sensory signal. Receptor currents generated across the chemosensitive tips of olfactory and gustatory sensory cells flow into the apical membrane and out of the sensory cell across its basolateral membrane. This will be discussed in greater detail later.

Olfactory receptor cells transmit their signals directly to the brain via their axons (Fig. 1). Olfactory receptor cells are examples of primary sensory cells. In contrast, gustatory sensory cells transmit their signals synaptically to **sensory neurons** whose peripheral processes (sensory afferent fibers) innervate taste buds. The central terminals of the sensory neurons communicate with the CNS. The cells within a taste bud are also believed to communicate with each other via **chemical synapses** and **electrical synapses** (Roper, 1992). This suggests that some degree of information processing via lateral synaptic interactions, that is, cross talk, occurs in gustatory sensory organs, but the details of this are sketchy.

II. Taste Receptor Cells

A. General Comments

Peripheral sensory organs for taste consist of clusters of receptor cells, the taste buds, found throughout the oral cavity. There are **2 000–10 000 taste buds** in humans. This number varies considerably from individual to individual. Each taste bud consists of **50–100 cells.** Taste buds are embedded in the lingual epithelium in specialized mounds of tissue termed papillae (**fungiform, foliate,** and **circumvallate papillae**). Taste buds are also localized on the soft palate, uvula, epiglottis, pharynx, larynx, and have even been reported in the upper esophagus.

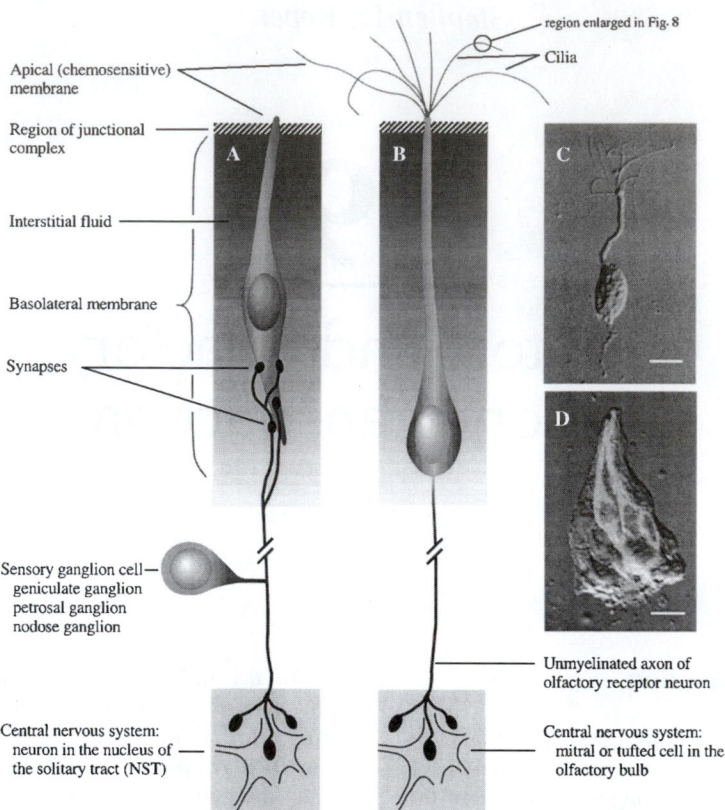

FIGURE 1. Schematic drawing comparing a gustatory sensory cell (A) and an olfactory receptor neuron (B). The apical or chemosensitive portions of the receptor cells extend above the epithelial junctional complex. The initial events in chemosensory transduction occur here. The basolateral membrane of the cells lies below the apical junctional complex and is surrounded by interstitial fluid. Gustatory sensory cells synapse onto primary sensory afferent fibers from sensory ganglion cells. Olfactory receptor neurons are primary receptor cells and send axons directly to the olfactory bulb, as shown. (C) Micrograph of a single olfactory receptor neuron isolated from a frog. (D) Micrograph of an isolated rat taste bud, containing several gustatory sensory cells. The taste bud has been immunostained for gustducin, a taste-specific G protein. This reveals three to four intensely immunostained gustatory sensory cells within this taste bud (I. Wanner and S. Roper, unpublished). Calibrations = 10 μm in C and D. (Micrograph in C modified from Kleene, S. J., and Gesteland, R.C. (1981) *Brain Res.* **229,** 536–540.)

Taste buds respond to a number of basic taste qualities, including sweet, sour, salty, bitter, **umami** (discussed later), and perhaps others. Early investigators tested whether particular regions of the tongue were selective for these different qualities. For example, there is a somewhat greater sensitivity to bitter in the posterior tongue and sweet at the tip (Fig. 2). However, these original findings have often been misinterpreted or overinterpreted in the intervening years, leading to the popular misconception that there is a discrete regional specialization of the tongue for certain tastes. The truth is that all parts of the tongue sense the basic qualities, as shown in Fig. 2.

Most of the cells in a taste bud are narrow elongate cells, extending nearly the full thickness of the lingual epithelium. The basal processes reach to the basement membrane and the apical processes stretch up to a tiny cavity in the lingual surface, the **taste pore.** The elongate, columnar morphology of gustatory sensory cells sets taste buds apart from the surrounding flat laminar tissue. That is, taste buds form tiny pockets of simple columnar epithelium within the larger sheet of stratified squamous lingual epithelium. Such an arrangement positions

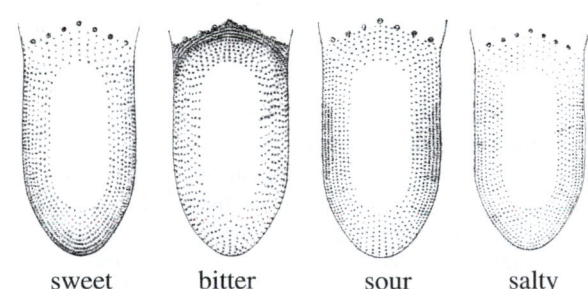

FIGURE 2. Regional distribution of sweet, bitter, sour, and salty taste on the human tongue, as originally determined by D. P. Hänig (1901). The sensitivity for a particular taste quality is represented by the density of stippling. There is a slight tendency for the anterior tip of the tongue to be more sensitive to sweet, the lateral regions to be more sensitive to sour, and the posterior to bitter. However, there is complete overlap for all these taste qualities. Also, see discussion in Lindemann (1999).

taste cells directly across the electric field generated by **transepithelial ion transport** currents. This may have implications for taste transduction mechanisms (discussed later).

The apical tips of the taste cells extend into the taste pore. **Microvilli** on the exposed tips of gustatory sensory cells are the primary sites of chemosensory transduction. The cytoplasm of taste cells varies in appearance from cell to cell. Based on the cytological and ultrastructural features of taste bud cells, some morphologists have categorized taste cells as **dark cells** and **light cells.** Other morphologists have used categories such as **Type I, II,** and **III** cells. Some ultrastructural studies have reported that all taste bud cells possess synaptic connections and thus are believed to be chemoreceptor cells. Other studies report that only one of the cell types (Type III) possesses synaptic connections and is the actual receptor cell. Some taste bud cells are believed to be **neuromodulatory cells;** they release **serotonin** to modify the chemosensitivity of the taste bud (Ewald and Roper, 1994; Delay *et al.,* 1997; Herness and Chen, 1997). Other cells may be sustentacular, or glial-like, cells and function to regulate potassium in the intercellular spaces of the taste bud (Bigani, 2001). Ongoing investigations are attempting to correlate these proposed functions of the different taste bud cells with the different cellular morphotypes (dark, light, Type I, II, III).

Taste buds also contain a population of small ovoid **stem cells** near the base of the gustatory organ. Taste cells, like the adjacent non-taste epithelium, represent a **renewing population.** Taste bud cells have an estimated life span of approximately 10 days in the mammal, slightly longer than the surrounding non-sensory epithelium. As stem cells divide and their daughter cells differentiate into mature chemoreceptive cells, they extend a long narrow process up to the taste pore. One interpretation of the morphological classes in the taste bud is that the different cell types represent different stages of maturation of gustatory sensory cells, but the data supporting this are not conclusive.

Chemosensory transduction, the focus of this chapter, is primarily concerned with events occurring at the apical chemosensitive membrane of gustatory sensory cells. However, it is important to be aware that events at the basolateral membrane can also influence gustatory signals. For example, the basolateral membrane is the site where serotonin, an important neuromodulator in taste buds, (a) acts to modify the passive (electrotonic) spread of receptor currents throughout the taste cell and (b) modulates Ca^{2+} currents that are important in synaptic transmitter release (Ewald and Roper, 1994; Delay *et al.,* 1997; Herness and Chen, 1997).

For many years, gustatory sensory cells were considered passive transducers of the chemical environment. However, microelectrode studies then revealed that these cells, like neurons, are **excitable** and possess a wide variety of **voltage-dependent ionic conductances** (Fig. 3A; Roper, 1983; Kashiwayanagi *et al.,* 1983). These conductances include voltage-dependent, tetrodotoxin (TTX)-sensitive Na^+ channels; delayed-rectifier K^+ channels; type A K^+ channels; inward-rectifier K^+ channels; Ca^{2+}-activated K^+ channels, Ca^{2+}-activated Cl^- channels; and voltage-dependent Ca^{2+} channels (reviewed in Lindemann, 1996). Certain of these ion channels are intimately involved in the taste transduction process, as described later.

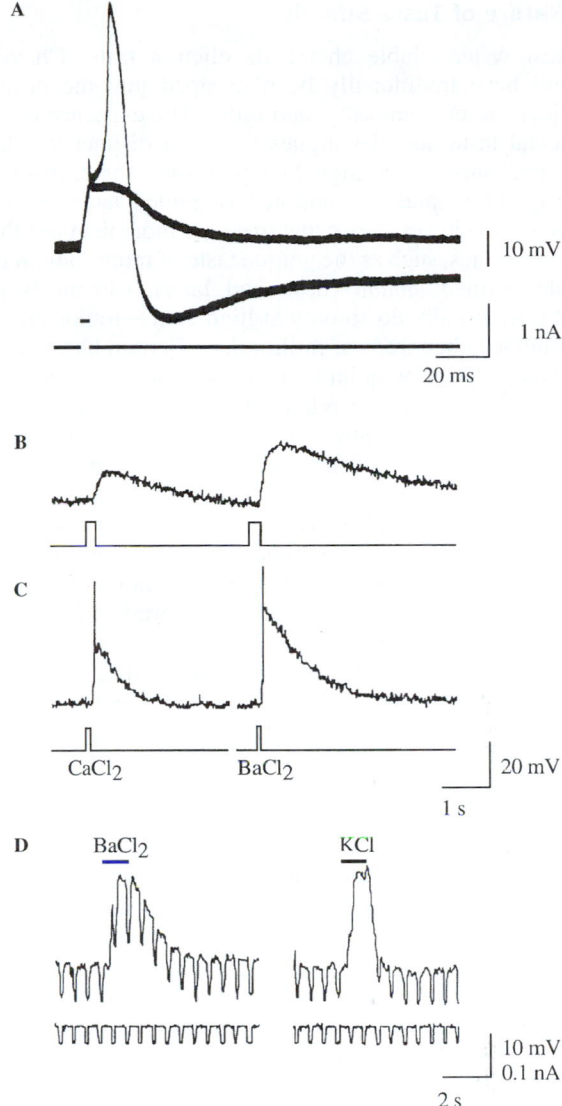

FIGURE 3. Intracellular recordings from gustatory sensory cells. (A) Gustatory sensory cells are excitable and generate action potentials when depolarized, as shown for a taste cell from the amphibian *Necturus maculosus.* Upper, two superimposed traces showing responses to brief depolarizing current pulses passed through the intracellular microelectrode. Lower, current pulse. Reprinted from Roper (1983). Copyright 1991 American Association for the Advancement of Science. (B) Subthreshold receptor potentials elicited by brief focal applications of $CaCl_2$ and $BaCl_2$, two potent taste stimuli, to the taste pore of a taste bud in *Necturus.* (C) Same as B but showing action potentials elicited when receptor potentials exceed threshold. Lower traces in B and C show the application of the chemical stimuli. (D) Intracellular receptor potentials and membrane conductance changes elicited by focal application of $BaCl_2$ and KCl to the taste pore in *Necturus.* Both stimuli elicited depolarizing receptor potentials. However, divalent cations such as Ba^{2+} decrease membrane conductance. This is shown by the increased amplitude of hyperpolarizing responses produced by a train of brief constant current pulses applied through the recording electrode. In contrast, KCl depolarized the cell, accompanied by an increased membrane conductance. Upper traces, intracellular recordings. Lower traces, current pulses applied through the recording electrode. B, C, and D modified from Bigani and Roper, 1991. Copyright 1991 American Association for the Advancement of Science.

B. Nature of Taste Stimuli

Most water-soluble chemicals elicit a taste. Chemical stimuli have traditionally been grouped into the primary qualities sweet, sour, salty, and bitter. The existence of fundamental taste qualities argues there are distinct transduction mechanisms and signaling pathways for each of the taste qualities, and this notion has guided taste research. There is growing evidence that there are more than just these four categories, such as the unique taste of monosodium glutamate, termed *umami* (described later). Chemicals that elicit taste usually do so only at high concentrations (a few millimolar to hundreds of millimolar). Certain bitter-tasting substances (such as quinine, caffeine, and strychnine) and artificial sweeteners are relatively potent stimuli and elicit responses at concentrations in the μM to low-mM range. Nonetheless, it is clear that taste cells have low stimulus sensitivity relative to the exquisite responsiveness of photoreceptors (i.e., single photons) or olfactory and pheromone chemoreceptors (i.e., picomolar stimulus concentrations, possibly single molecules). Primary functions of taste organs include regulating the intake of important nutrients and protecting against the ingestion of harmful substances. Therefore, the generally low sensitivity of gustatory sensory cells to many chemical stimuli (with the exception of bitter taste) may represent optimization of responses to physiologically relevant concentrations of nutrients. Sensitivity to bitterness may be related to the fact that many bitter-tasting chemicals are toxic and, conversely, many toxic chemicals elicit bitter taste—an aversive taste quality. Thus, it is understandable that bitter taste transduction has evolved to detect low concentrations of stimuli and to prevent the consumption of even small quantities of potentially harmful substances.

Taste stimuli are dissolved in saliva and reach only the chemosensitive microvillar (apical) surface of gustatory sensory cells. Most chemical stimuli are prevented from gaining access to the basolateral regions of the taste bud by the **tight junctions** of the junctional complex that seals off the apical membrane in the taste pore (Fig. 1). Tight junctions, however, are leaky to certain stimuli, such as Na^+, K^+, and Cl^- (Ye *et al.*, 1993). These ions can pass from the lingual surface into the intercellular spaces within the taste bud to raise the concentration there above levels normally found in interstitial fluid. It is possible that chemotransduction mechanisms for some taste stimuli, especially Na^+, K^+, Cl^-, and possibly other stimuli, exist on the basolateral membranes of taste receptor cells as well as on their exposed apical tips. Basolateral transduction mechanisms may explain **intravenous taste,** where certain chemicals that are dissolved in the blood elicit gustatory sensations. The bitter taste of some drugs injected intravenously may be an example of this phenomenon.

C. Initiation of Taste Receptor Potentials

The arrival of a taste stimulus at the apical chemosensitive membrane is signaled by the generation of a depolarizing **receptor potential** (Fig. 3B). The conversion of a gustatory chemical signal into an intracellular receptor potential is termed taste **transduction.** Unlike signal transduction in most other sensory organs, there is no single unifying hypothesis that explains how gustatory sensory cells respond to taste stimuli. The necessity of taste organs to sense a wide range of chemical substances, from simple ions to complex molecules, has led to the evolution of **multiple taste transduction mechanisms** (see Corey and Roper, 1992; DeSimone, 1991; Finger *et al.*, 2000; Kinnamon and Margolskee, 1996; Lindemann, 1996; Simon and Roper, 1993). Presumably, different gustatory sensory cells possess different transduction mechanisms and respond to different taste stimuli. Recent findings from Ca^{2+} imaging studies support this view. That is, an individual taste receptor cell may respond to a somewhat limited repertoire of chemical substances (Bernhardt *et al.*, 1996; Caicedo and Roper, 2001). However, the final answer has yet to be established. Even if an individual gustatory sensory cell is "tuned" to a subset of chemical stimuli, a taste bud is a population of gustatory sensory cells. Thus, a single taste bud with its 50–100 cells can respond to a diverse array of chemical stimuli.

The multiple signaling pathways utilized in taste transduction can be divided into three basic categories:

1. *Passive ion permeation through ion channels.* Some taste stimuli enter taste cells through ion channels in the apical membrane, thereby generating transmembrane **receptor currents** and **receptor potentials.** This is often the case when the taste stimulus is itself an ion (for example, Na^+ or K^+ salts).
2. *Block of ion channels.* As explained later, certain ions (e.g., Ca^{2+} and H^+) **block** apical membrane channels, particularly those for K^+. This also generates a receptor potential because when K^+ channels are blocked, leak conductances for other ions such as Na^+ (which tend to depolarize the cell) can then exert a greater influence on the resting potential.
3. *Receptor-ligand binding.* Interactions occur between chemical stimuli (ligands) and specific membrane-bound receptors (i.e., ligand-receptor binding) and ultimately lead to receptor currents or the release of intracellular Ca^{2+}.

Parenthetically, amphipathic stimuli such as saccharin and quinine may insert into the plasma membrane and activate taste cells independently of these three basic mechanisms. Taste transduction mechanisms involving ion permeation, channel blockage, and receptor-ligand interactions are described next.

1. Ion Permeation, Channel Blockage, and Current Flow through Apical Ion Channels

a. Na^+ Channels. Sodium salts ("salty taste") are believed to be transduced in a subset of taste cells by the direct permeation of Na^+ through passive Na^+-selective channels (as opposed to the TTX-sensitive, voltage dependent Na^+ channels that underlie action potentials) (Heck *et al.*, 1984). These ion channels resemble **amiloride-sensitive epithelial Na^+ channels (ENaC)** found in renal tubules and elsewhere. Indeed, mRNA isolated from lingual epithelium shows that all three ENaC subunits (α, β, and γ) are expressed in taste cells

(Kretz *et al.,* 1999; Lin *et al.,* 1999). Additionally, immunohistochemistry has confirmed the presence of ENaC subunits in taste cells. Yet, these channels are also expressed in lingual tissues other than taste buds, leading to speculation that non-sensory epithelial cells in the tongue somehow participate in salt taste (Li *et al.,* 1994). The proportion of α, β, and γ ENaC subunits expressed in gustatory sensory cells appears to vary between the anterior and posterior tongue, which may explain known regional differences of the tongue in salt taste. Where it has been measured (in frog taste cells), these taste Na^+ channels have small unitary conductance (1–2 pS) and are blocked by low concentrations of amiloride ($K_m = 0.2$–1 μM).

As stated previously, the apical amiloride-sensitive Na^+ channels in gustatory sensory cells are not voltage-dependent. ENaC channels are open at rest and allow an ongoing flux of Na^+, driven by this cation's electrochemical gradient. The concentration of Na^+ in human saliva (the solution bathing the apical tips of gustatory sensory cells) varies from about 3 to 63 mM, depending on the rate of salivary secretion. This results in an equilibrium potential for Na^+ (E_{Na}) of about –30 mV to +50 mV (assuming $[Na^+]_i$ is 10 mM) across the apical chemosensitive membrane. The resting potential of gustatory sensory cells is estimated to be –60 to –80 mV. Thus, there should be a small steady-state influx of Na^+ through the apical membrane at rest in those taste receptor cells that possess apical ENaC channels. During stimulation with NaCl when $[Na^+]$ in the oral cavity exceeds salivary $[Na^+]$, the electrochemical gradient driving Na^+ into the cells increases. This generates an **inward receptor current** through the apical membrane, over and above the resting Na^+ influx, and further **depolarizes** the cell. This current exits across the entire basolateral cell membrane, thus forming a return current loop by way of an extracellular (paracellular) pathway that includes the intercellular tight junctions near the taste pore. Consequently, any factors that affect the basolateral membrane conductance or the paracellular resistance, especially at the tight junctions near the taste pore, will also affect the magnitude of the receptor currents.

The density of amiloride-sensitive Na^+ channels, which is under hormonal control, determines the magnitude of Na^+ influx at rest and during salt stimulation. Specifically, **arginine-vasopressin (antidiuretic hormone, ADH)** and **aldosterone** increase the number of ENaC channels, albeit via different mechanisms. Na^+ conductance in taste cells may be tonically depressed by salivary Na^+ ("self-inhibition," Gilbertson and Zhang, 1998), but this inhibition is overridden during chemostimulation with Na^+ salts.

Accumulation of Na^+ within the taste cell is prevented by the action of the Na^+,K^+-ATPase distributed on the basolateral membrane. Na^+,K^+-ATPase pumps excess Na^+ from the cell into the surrounding interstitial spaces. Vascular elements below taste buds presumably drain off any surplus Na^+, preventing a buildup in the intercellular spaces within the taste bud. The depolarizing current produced by Na^+ influx at the apical tip may be offset in part by the electrogenic basolateral Na^+-K^+ pumps (which hyperpolarize the membrane) and partly by the compensatory efflux of K^+ through basolateral K^+ channels (i.e., an outward, repolarizing K^+ current).

Although one might anticipate that ENaC channels would be localized to the apical chemosensitive tips of the gustatory sensory cells, in fact, immunostaining for these channels reveals their presence on apical **and** basolateral membranes. The function of basolateral ENaC channels is not yet understood. A substantial resting Na^+ conductance through basolateral ENaC channels would chronically depolarize gustatory sensory cells due to the steady-state influx of Na^+ from the interstitial spaces within the taste bud. This conundrum remains to be solved.

In addition to their role in salt taste, the amiloride-sensitive Na^+ channels on gustatory sensory cells may participate in other taste qualities, particularly "sour" (discussed later). Amiloride reduces acid ("sour") taste in behavioral studies on animals and psychophysical studies on humans. Indeed, the role of ENaC channels for "saltiness" per se in human taste has been questioned. Amiloride appears to affect only the sourness, not saltiness, of NaCl solutions in human studies (Ossebaard and Smith, 1996).

Na^+ salts have an important secondary effect on taste buds apart from permeating through Na^+-selective ion channels in the apical tips of taste cells. The presence of Na^+ in saliva results in a steady **transepithelial transport of Na^+** across the non-sensory lingual epithelium, from mucosal (apical, saliva) to serosal (basolateral, interstitial fluid) spaces (Fig. 4). This Na^+ transport generates a standing voltage across the entire lingual epithelium, estimated to be about 10 mV or higher (serosal side positive). Consequently, transepithelial transport imposes an electric field across taste buds which, as described previously, represent islands of simple columnar cells embedded in a stratified squamous epithelium. In the presence of Na^+ salts, transepithelial Na^+ transport is increased and the transepithelial voltage gradient increases, thereby **increasing the electric field across the taste buds.** The transepithelial voltage drives a return current through the paracellular and transcellular pathways back to the mucosal surface (Fig. 4). The net effect of the transcellular current is to hyperpolarize the basolateral membrane of taste cells, where synapses are located, and reduce the potential difference across their apical membranes, where chemosensory transduction takes place. Hyperpolarizing the membrane near synapses might be expected to reduce transmitter release. These secondary effects oppose the direct stimulation of taste receptor cells by Na^+ salts. Furthermore, when the electrical resistance of the paracellular current pathways is increased, for example by the presence of organic anions (e.g., stimulation with sodium acetate), proportionately more of the transepithelial transport return current is shunted through the taste cells (transcellular pathway, Fig. 4). This even further hyperpolarizes the basolateral membrane and decreases the apical membrane potential. Thus, current loops generated by widespread transepithelial Na^+ transport in the tongue may **modulate taste cell activity,** independent of any direct stimulation by Na^+ of taste receptor cells *per se*. This modulation may have implications for salty taste and for taste mixtures when combinations of NaCl and other taste stimuli are present, as normally occurs during food intake. These considerations emphasize how the overall function of taste buds may be highly dependent on their histological context, that is, being situated in an intact ion-transporting epithelium.

b. K^+ and Other Cation Channels. During chemostimulation with K^+ salts ("bitter-salty taste"), K^+ presumably permeates through cation channels located in the gustatory sensory cell

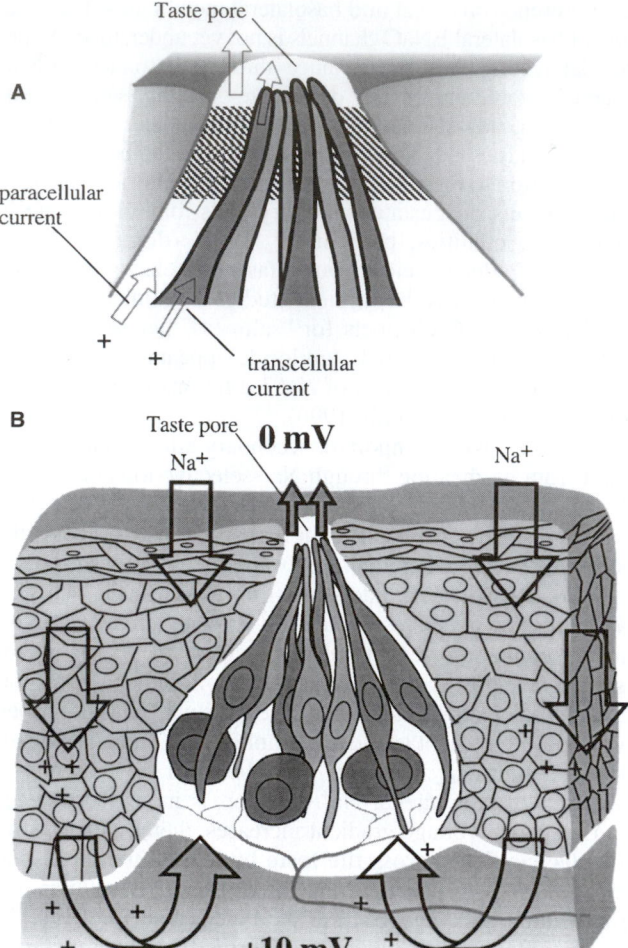

FIGURE 4. Na$^+$ transport across the non-sensory lingual epithelium produces current loops through taste buds. (A) Magnification of apical region of taste bud. (B) In lingual epithelial cells, Na$^+$ influx through passive apical channels is actively pumped out of the basolateral membranes by electrogenic Na$^+$,K$^+$-ATPase, generating a flux of Na$^+$ and transepithelial currents (open arrows). This results in a transepithelial potential of about 10 mV (serosal positive). This transepithelial voltage drives current out around taste cells (paracellular pathway) and through taste cells (transcellular pathway) (A). The current through taste cells (transcellular pathway) hyperpolarizes the basolateral membrane where synapses are located and depolarizes the apical membrane where transduction occurs. The relative contribution of current through these two pathways is determined in large part by the resistance of the junctional complex in the region of the taste pore (striped region in A) and the resistance of the taste cell membranes. Certain taste stimuli, such as organic anions, alter the resistance of the junctional complex (paracellular resistance) and thereby change the balance of current through taste buds, possibly altering synaptic transmission and/or apical transduction.

membrane. For example, when mucosal [K$^+$] is elevated, the passive influx of K$^+$ through **apical K$^+$ channels** generates a depolarizing receptor potential (Fig. 3C). Apical K$^+$ channels have been demonstrated in taste cells from amphibia. However, in taste cells from mammalian taste buds, K$^+$ channels are present mainly on the basolateral membrane (Furue and Yoshii, 1997), and it is not known whether there is an api-

cal K$^+$ conductance. Most likely, in mammalian taste buds, when stimulated with potassium salts, K$^+$ ions follow a paracellular pathway, penetrating the apical tight junction barrier and depolarizing gustatory sensory cells by permeating K$^+$ channels in the basolateral membrane.

Inwardly-rectifying K$^+$ channels that open when the cell is hyperpolarized much below E_K are also present in taste cells, and together with the delayed rectifier and other K$^+$ channels, may serve to stabilize the resting potential near E_K. Taste cells have a high input resistance (20–50 GΩ) near their resting potential. Thus, the **receptor cell at rest is in a state of maximal sensitivity.** The inward rectifier will buffer a taste cell against influences that might tend to hyperpolarize the membrane excessively and depress excitability. Such influences could include electrogenic Na$^+$ pumps in taste cells or transcellular currents generated by transepithelial transport (discussed earlier).

How acids ("sour taste") are transduced may differ from species to species. In the amphibian, *Necturus,* it has been shown that protons block apical K$^+$ channels. This blockage results in a depolarization. That is, in the absence of stimulation with K$^+$ salts (i.e., without elevated [K$^+$]$_o$), the equilibrium potential for K$^+$ (E_K) is near the resting potential and there is likely to be a small but finite ongoing efflux of K$^+$. Indeed, as detailed in Chapter 14, the resting K$^+$ conductance in large part determines a cell's resting potential. When this resting K$^+$ conductance is blocked by protons, K$^+$ efflux is curtailed and conductances for other ions (Na$^+$, Ca^{2+}, Cl$^-$) with equilibrium potentials positive with respect to resting potential determine the membrane potential. Thus, the net effect of blocking K$^+$ channels is a depolarization. Additionally, protons may activate acid-sensitive cation channels and generate an inward depolarizing receptor current. An acid-gated cation channel, MDEG1, has been reported in rat gustatory sensory cells (Ugawa *et al.,* 1998). Lastly, in hamsters it has been shown that protons themselves permeate the apical membrane, and this influx constitutes a depolarizing receptor current (Gilbertson *et al.,*1993). Interestingly, protons permeate hamster taste cells via amiloride-sensitive Na$^+$ channels (ENaC) (Fig. 5). In mixtures of Na$^+$ salts and acids, protons bind to the ion selectivity site in the ENaC channel more strongly than do Na$^+$ ions and as a result, H$^+$ conductance is less than that of Na$^+$. This also means that **protons interfere with the influx of Na$^+$.** Psychophysical studies in humans also implicate a role for amiloride-sensitive Na$^+$ channels in sour taste. (The interference between Na$^+$ and H$^+$ permeation of amiloride sensitive Na$^+$ channels may help explain the culinary wisdom that acidifying food, for example by adding vinegar or lemon juice, reduces its salty taste).

The fact that there is a common ion channel pathway for H$^+$ and Na$^+$ ions (ENaC), at least in some species, raises the question of how sour is distinguished from salty. The answer is probably that protons affect taste cells multiple ways. For example, as already mentioned, H$^+$ can block K$^+$ channels and H$^+$ can stimulate acid-gated channels. Protons may have additional less-well-characterized actions on the apical chemosensory membrane that contribute to sour taste. It remains to be determined whether different subsets of taste cells are activated by acids versus by Na$^+$ salts and whether

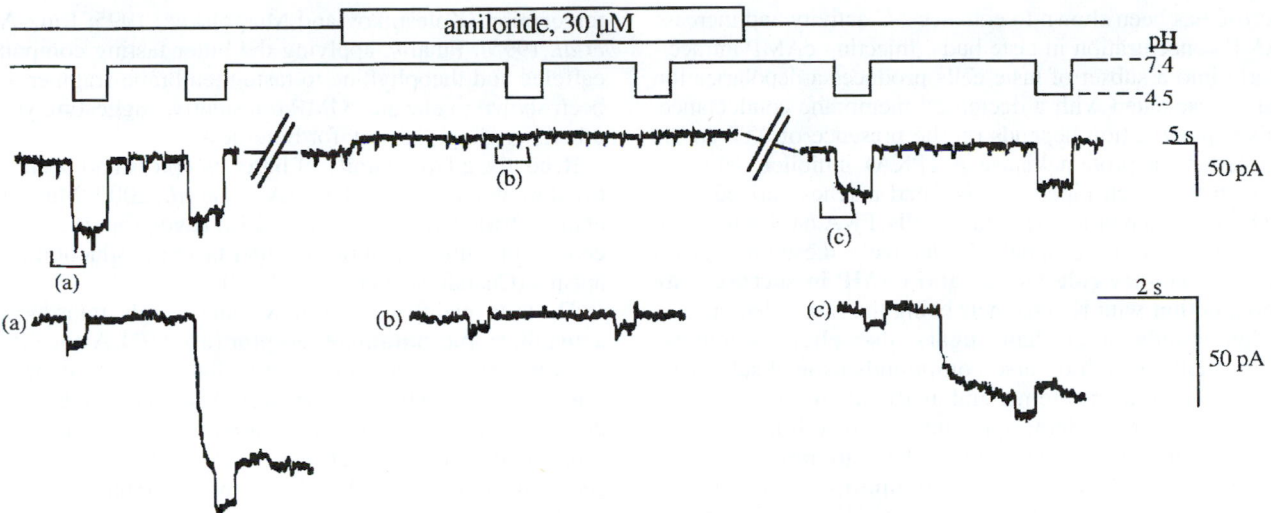

FIGURE 5. Acid taste stimuli are transduced, in part, by proton permeation of amiloride-sensitive Na$^+$ channels. This figure shows a patch-clamp record from a taste cell in an isolated fungiform taste bud from the hamster. The cell was held at −80 mV throughout the record. Acid stimulation (2.5 mM citric acid, pH 4.5) elicits an inward current (a) due to influx of H$^+$. This response is blocked by perfusing the bath with amiloride (30 μM) (b). Washing out amiloride from the bath restores the acid response (c). Top two traces indicate the periods when amiloride and citric acid were applied. Bottom traces are expanded records to show the increases in membrane conductance during citric acid stimulation more clearly. Membrane conductance was monitored by applying a train of small, constant, hyperpolarizing pulses throughout the record. From Gilbertson *et al.* (1993). Copyright 1993 by Cell Press.

any of the sour transduction mechanisms are common across species.

Other salts, such as those of the divalent cations Ca^{2+} and Ba^{2+}, are also transduced by **reducing ion permeation** through the apical membrane. To humans, these salts elicit a bitter taste. For instance, **Ca^{2+} and Ba^{2+} block apical K$^+$ channels** in amphibia (Fig. 3C), and this may constitute one mechanism for bitter taste transduction.

Gustatory sensory cells appear to detect dietary fats in the oral cavity in part by the ability of fatty acids such as linoleic acid and arachidonic acid to block K$^+$ channels (Gilbertson *et al.*, 1997). This action inhibits outward current, especially through delayed rectifier K$^+$ channels, and thereby enhances depolarizations elicited by other taste stimuli and prolongs the duration of action potentials.

Apparently, transduction mechanisms for salty, sour, certain bitter substances, and fatty acids overlap and affect many of the same ion channels. If all taste cells were to possess identical ion channels, it would be a mystery as to how taste perceptions of these substances can be differentiated. One likely answer is that different taste receptor cells may express different combinations of ion channels. Thus, different subsets of receptor cells are activated during stimulation with salts, acids, and bitter stimuli. This is a topic of ongoing investigation.

2. Membrane-Bound Receptor Events

a. Receptor-Ligand Binding. Certain taste stimuli are believed to interact with specific **membrane-bound receptors** rather than permeate or block ion channels. Sweet-tasting substances, some bitter-tasting compounds, and amino acids are examples of such stimuli. Particular amino acids, espe-

cially L-glutamate, evoke a unique taste termed **umami** that is fundamentally different from the other taste primaries **sweet, sour, salty,** and **bitter.** Consequently, the taste of L-glutamate (*umami*) is believed to have its own receptor mechanism (discussed later). *Umami* is becoming recognized as another primary taste quality, signaling the concentration of free glutamate found in many protein-rich foods such as meats, cheeses, and certain vegetables.

The most straightforward mechanism for receptor-mediated signal transduction in gustatory sensory cells is ligand-gating of ion channels (**ionotropic channels).** This is similar to the action of acetylcholine at the neuromuscular junction. In catfish taste cells, arginine and proline are believed to act this way, stimulating nonselective cation channels.

More complex receptor mechanisms involve **G-protein-coupled receptors (GPCRs).** GPCRs are members of a large family of proteins with seven transmembrane domains. The binding of a taste stimulus to a GPCR triggers intracellular second messengers and a cascade of enzymatic reactions inside the cell. These cascades amplify and/or modulate the initial signal and ultimately converge upon an **effector mechanism,** usually specific ion channels in the plasma membrane that generate a receptor potential. For instance, the taste of sucrose is believed to be transduced by GPCRs, as follows (reviewed in Lindemann, 1996). Investigators postulate that there is a GPCR for sucrose that activates G$_s$ proteins that, in turn, stimulate **adenylyl cyclase (AC)** and **increase intracellular cyclic adenosine monophosphate (cAMP) concentration.** The most plausible interpretation of studies to date is that the **increase in intracellular cAMP concentration stimulates a protein kinase that phosphorylates and blocks basolateral K$^+$ channels,** thereby indirectly depolarizing the taste cell. Stimulating taste tissue with sucrose has been shown to enhance AC activity and increase cAMP concentration in taste buds. Injecting cAMP intracellularly into a subset of taste cells produces a depolarization that is associated with a decreased membrane conductance. This depolarization depends on the presence of ATP and is blocked by a protein kinase A (PKA) inhibitor delivered through the patch pipette. In isolated patches derived from the basolateral membrane of taste cells, PKA has been shown to block certain K^+ channels. Collectively, these findings indicate a **primary role for AC and cAMP in sucrose taste transduction** with K^+ channels being the final effectors.

Compounds other than sugars also elicit sweetness. These include such diverse compounds as lead salts, certain amino acids, proteins, and artificial sweeteners (e.g., saccharin). Cross-adaptation studies and cellular analyses of taste cell responses reveal that there are multiple transduction mechanisms, and possibly multiple receptors, for sweet taste. For example, in rat taste cells, **saccharin** has been shown to stimulate the synthesis of **intracellular inositol trisphosphate (IP$_3$)** and thus **increase [Ca^{2+}]$_i$** by release from intracellular stores (Bernhardt et al.,1996). Yet, in the same cells, sucrose increases intracellular [Ca^{2+}] by influx from the extracellular medium. This influx of Ca^{2+} is consistent with activation of depolarization-gated Ca^{2+} channels following an increase in cAMP, as described previously. That is, in some gustatory sensory cells there appears to be a **dual transduction pathway for sweet,** involving the second messengers cAMP (for sucrose) and IP$_3$ (for saccharin).

Based on the involvement of the second messengers cAMP and IP$_3$, it is reasonable to infer that sucrose and saccharin are transduced via GPCRs. However, to date, no receptor has been cloned or identified that has been convincingly linked with the taste of sugars or nonsugar sweeteners.[1]

Other taste stimuli, for example certain bitter compounds, are also thought to activate membrane receptors that are coupled to G proteins. For instance, bitter compounds such as denatonium, strychnine, and caffeine **transiently generate IP$_3$** in taste cells and **stimulate Ca^{2+} release** from internal stores, presumably via specialized GPCRs. These findings are consistent with the existence of bitter receptors that activate **phospholipase C,** converting membrane **phosphatidylinositol 4,5-bisphosphate,** releasing **diacylglycerol (DAG)** and **IP$_3$** into the cytosol. An additional receptor mechanism for bitter taste has been proposed that involves a **reduction of intracellular cAMP,** possibly via a **transducin-coupled receptor** mechanism similar to that in pho-

toreception (Kolesnikov and Margolskee, 1995; Ruiz-Avila et al.,1995). Finally, applying the bitter tasting compounds caffeine and theophylline to taste membrane fragments has been shown to elevate cGMP transiently, suggesting yet another signaling pathway for bitterness.

Recently, a large family of bitter receptors had been identified in humans and rodents (Alder et al., 2000; Matsunami et al., 2000). These receptors (T2Rs) have properties that are consistent with several of the bitter taste transduction mechanisms (Chandrashekar et al., 2000).

The taste quality *umami* is now believed to be transduced by a **metabotropic glutamate receptor (mGluR).** A novel, taste-specific variant of mGluR4 (taste-mGlvR4) has been localized selectively to taste buds in lingual epithelium (Chaudhari et al., 2000). In the brain, mGluR4 is a receptor for glutamatergic synaptic transmission between neurons. However, in taste cells, a novel variant of this GPCR appears to have been put to another use—to detect dietary glutamate, a naturally-occurring component of many foods. Umami taste transduction pathways activated by taste-mGluR4 remain to be elucidated.

b. G Proteins. As implied in the preceding section, certain receptors in taste are presumed to be coupled to **G proteins.** A G α protein that is unique for taste, **α-gustducin,** has been identified in taste buds (McLaughlin et al., 1992) and is a candidate for the G α protein that couples to bitter and sweet receptors. Gene-knockout mice lacking α-gustducin show deficits in their ability to sense bitter and sweet tastes. Additionally, a G α protein found in photoreceptors, **transducin,** has also been shown to be expressed in taste cells (McLaughlin et al., 1993; Yang et al.,1999). This suggests that there may be molecular mechanisms in common between photoreception and chemoreception. Lastly, certain G protein β and γ subunits have been found in taste cells, specifically β1, β3, and γ13 (Huang et al., 1999). These βγ subunits appear to activate a particular phospholipase C (PLC β2) also found in taste cells (Rossler et al., 1998). This leads to the generation of IP$_3$ and DAG, as described for bitter taste. Thus, specific α, β, and γ G protein subunits are present in taste cells. Activating taste GPCRs can initiate transduction paths associated with G$_α$ proteins (i.e., changing the activity of AC and phosphodiesterase) as well as pathways associated with G$_{βγ}$ (i.e., stimulating phospholipase C).

The several transduction mechanisms postulated for gustatory sensory cells are summarized in Fig. 6.

D. Termination of Taste Signals

Termination of chemostimulation in taste cells is probably due to two factors: **diffusion** of the stimulus away from the apical chemosensitive tips and receptor cell **adaptation.** Very little is known about adaptation in gustatory sensory cells. However, it is plausible that the **Ca^{2+}-dependent Cl$^-$ conductance** that exists in taste bud cells contributes to sensory adaptation in taste (Taylor and Roper, 1994). An influx of Cl$^-$ across the basolateral membrane would **repolarize the gustatory sensory cell** even in the presence of a maintained chemical stimulus due to the increased [Ca^{2+}]$_i$ consequent to taste stimulation, as follows. Excitatory receptor currents result in an increase in [Ca^{2+}]$_i$ either (a) by way of

---

[1]Two putative GPCRs of unknown function, TR1 and TR2, have been cloned from rat taste tissues (Hoon et al., 1999). Based on the distribution of TR1 and TR2 expression in the tongue, it was suggested that TR1 underlies sweet taste and TR2 bitter. However, ligands for these putative GPCRs have not been identified to date and their role in taste remains to be established. Other receptors for bitter taste (T2Rs) have recently been discovered (Alder et al., 2000; Matsunami et al., 2000).

Additionally, there is suggestive evidence that saccharin (as well as quinine and other amphipathic taste stimuli) may penetrate the taste cell membrane and directly activate intracellular G proteins, bypassing any specific receptors (Naim et al.,1994) but nevertheless triggering intracellular signaling cascades.

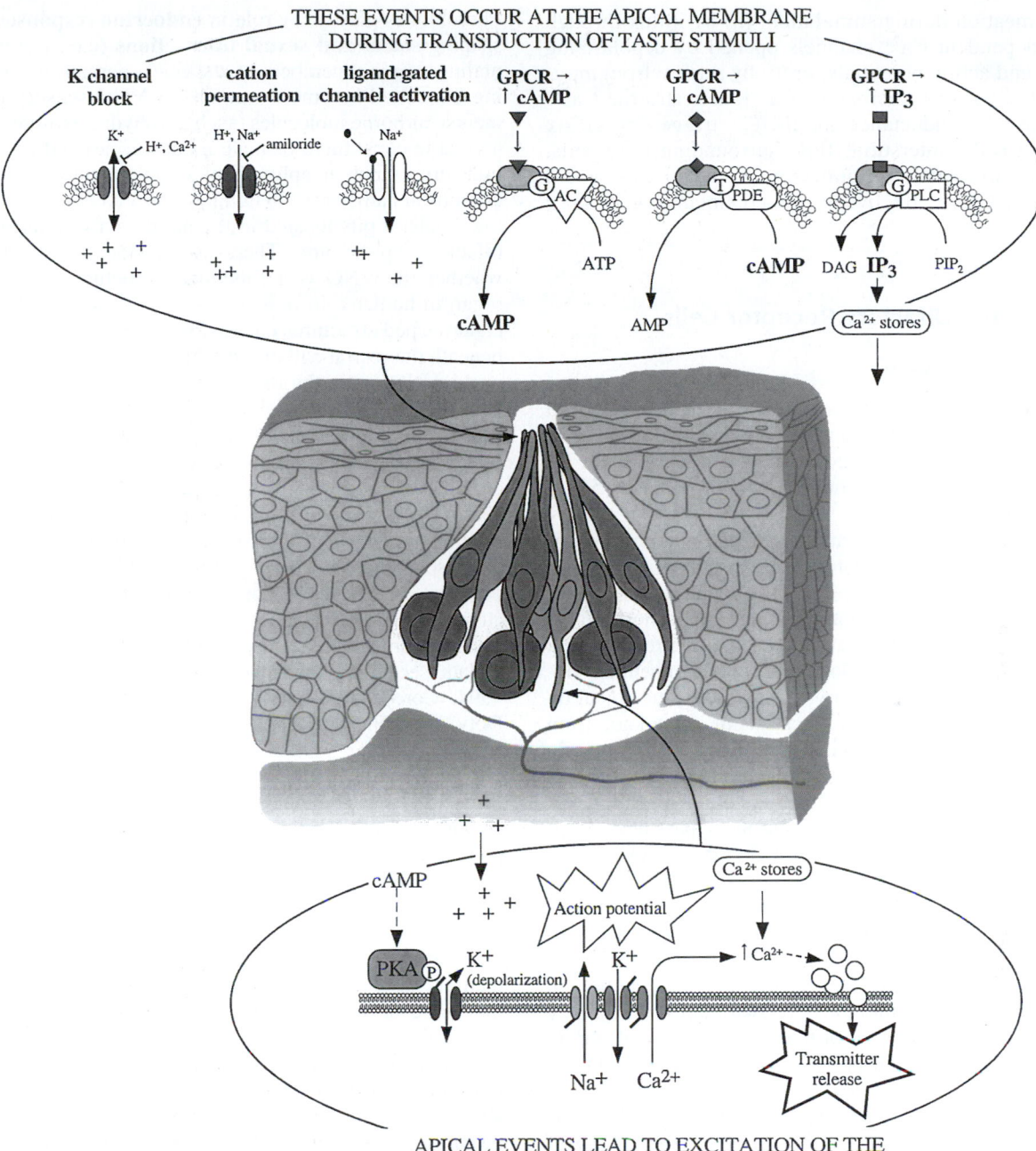

THESE EVENTS OCCUR AT THE APICAL MEMBRANE
DURING TRANSDUCTION OF TASTE STIMULI

APICAL EVENTS LEAD TO EXCITATION OF THE
BASOLATERAL MEMBRANE AND TRANSMITTER RELEASE

FIGURE 6. Summary of chemosensory transduction in taste. Events at the apical membrane of taste receptor cells, top, include (shown here from left to right) (1) block of K^+ channels, for instance by H^+ (**sour**) and Ca^{2+} (**bitter**); (2) cations (Na^+ and H^+) passing through apical amiloride-sensitive Na^+ channels (**salty** and **sour,** respectively); (3) ligand-gating of cation channels, such as proton-gated MDEG1 channels (**sour**) or L-arginine-gated channels (found in catfish gustatory sensory cells); (4) activation of G-protein-coupled receptors that stimulate adenylyl cyclase, for example, by sucrose (**sweet**); (5) activation of G-protein-coupled receptors that are negatively coupled to cAMP, for example, by monosodium glutamate (**umami**); (6) activation of G-protein-coupled receptors that stimulate phospholipase C and increase inositol trisphosphate, for example, by saccharin (**sweet**) or denatonium (**bitter**). Different cells are believed to possess different transduction mechanisms. Events illustrated on the basolateral membranes of taste receptor cells, bottom, include (1) depolarization of the membrane from a reduction of K^+ conductance due to cAMP-stimulated phosphorylation; (2) depolarization of the membrane consequent to an influx of cations at the apical membrane; (3) generation of an action potential when depolarizations reach threshold; (4) influx of Ca^{2+} ions through voltage-gated Ca^{2+} channels, or (5) intracellular release of Ca^{2+} from IP_3 gated stores; (6) release of neurotransmitter consequent to increased $[Ca^{2+}]_i$. Abbreviations: AC, adenylyl cyclase; DAG, diacyl glycerol; G, G protein; GPCR, G-protein-coupled receptor; IP_3, inositol 1,4,5-trisphosphate; PDE, phosphodiesterase; PIP_2, inositol 4,5-bisphosphate; PKA, phosphokinase A; PLC, phospholipase C.

Ca^{2+} permeation during stimulation, (b) by influx through voltage-dependent Ca^{2+} channels opened by depolarizing receptor and action potentials, or (c) by release from intracellular Ca^{2+} stores. Increased $[Ca^{2+}]_i$ activates the Ca^{2+}-dependent Cl^- conductance and thereby triggers an **influx of Cl^-** from the interstitial fluid surrounding taste cells, that is, an **outward, repolarizing current.** This would serve to reduce the effectiveness of a maintained excitatory stimulus.

III. Olfactory Receptor Cells

A. General Comments

Receptor cells for **odorants** are distributed in specialized patches of sensory epithelium embedded within the nasal respiratory epithelium. These patches comprise 1–2 cm^2 of surface in each nostril in the human, or approximately 5% of the total nasal epithelium.

Olfactory receptor cells are specialized neurons and are often termed olfactory receptor neurons. Olfactory receptor neurons extend the full thickness of the epithelium. Receptor neurons are interspersed among elongate supporting cells in the olfactory epithelium. Olfactory receptor neurons possess apical membrane specializations for receiving chemical signals. However, unlike gustatory sensory cells, the apical tip of an olfactory receptor neuron consists of a tiny knob from which extend 6 to 12 long cilia (see Fig. 1). The opposite (basal) pole of olfactory receptor neurons transforms into an axon that extends to the olfactory bulb in the CNS. Consequently, olfactory receptor neurons bypass any synaptic intervention or modulation in the periphery and transmit action potentials directly to the brain. Parenthetically, this raises the intriguing possibility that certain agents, such as drugs or toxins, can be taken up by olfactory receptor neurons and transported directly into the brain along the axons of the olfactory neurons. That is, the olfactory epithelium represents a "window" into the brain that might be exploited for pharmaceutical therapies or may be at the root of certain central nervous system disorders that are linked to environmental contaminants.

A population of **stem cells** resides at the base of the olfactory epithelium, similar to the *stratum germinativum* of epithelial tissues. Stem cells are capable of dividing and differentiating into olfactory receptor neurons. The lifespan of olfactory receptor neurons was originally thought to be approximately 30 days (Graziadei and Monti Graziadei, 1978), but more recent findings indicate that receptor neurons may live for many months (Mackay-Sim and Kittel, 1991a,b). Mitotic activity of stem cells provides a reserve population of immature olfactory receptor neurons that can rapidly differentiate and replace mature olfactory receptor neurons if necessary, but that otherwise die.

The **vomeronasal organ (VNO),** a sensory structure situated near the olfactory epithelium in many animals, detects a special category of volatile and nonvolatile chemical stimuli. These stimuli include **pheromones.** Pheromones are chemical signals discharged by one member of a species to communicate with other members of that same species.

Pheromones play a key role in **endocrine responses** and in shaping **social and sexual interactions** (e.g., territoriality, mating) among members of a species, perhaps even including humans. Pheromone stimuli for VNO sensory neurons include airborne molecules (such as dihydro-exo-brevicomin, a volatile constituent of male mouse urine) and nonvolatile molecules (such as aphrodisin, a component of vaginal secretions in hamsters).[2] In humans, the VNO consists of shallow bilateral pits located in the nasal cavity, anterior to the olfactory epithelium. There is considerable controversy whether the VNO is a functional structure or a vestigial organ in humans. In rodents, the VNO consists of bilateral cigar-shaped structures encased in a bony capsule and lying beneath the rostral end of the nasal cavity. Receptor neurons in the VNO, as in the olfactory epithelium, possess axons. However, unlike most olfactory receptor neurons, sensory neurons of the VNO possess **apical microvilli,** not cilia. The initial events of transduction are believed to occur on the microvilli. VNO sensory neurons send their axons to the **accessory olfactory bulb,** a separate portion of the olfactory bulb that is reserved for processing pheromone stimuli.

Sensory mechanisms for transducing pheromones and odorants are believed to be quite dissimilar, as will be discussed next. This chapter will focus mainly on olfactory transduction, which has been studied much more extensively to date (see reviews by Ache and Restrepo, 1999; Breer *et al.,* 1994; Corey and Roper, 1992; Dionne and Dubin, 1994; Doty, 1995; Firestein *et al.,* 1996; Gold, 1999).

B. Nature of Olfactory Stimuli

Odorants for land-dwelling animals are volatile compounds, typically diluted in large volumes of air. Odorants are only poorly soluble in aqueous solutions such as those that bathe the chemosensitive olfactory epithelium (i.e., mucus layer). Olfactory receptor neurons and VNO sensory neurons are much more sensitive to chemical stimulation than are taste receptor cells. Odorants and pheromones in the range of subnanomolar to micromolar concentrations can be detected. Many animals have a keen sense of smell; for example, dogs can detect as few as 1000 molecules/cm^3. However, reports of extreme chemosensitivity such as the ability of certain species (e.g., moths, dogs, and sharks) to detect single molecules of pheromones or odorants may be overestimates based on theoretical calculations assuming uniform diffusion of the stimulus in air or water. Yet, experiments show that the actual dispersion of odorant and pheromone molecules is more often in the form of irregular plumes rather than diffusion, somewhat like smoke rising from burning incense. Within the plume, the concentrations

---

[2]The VNO is likely to be a primary site for pheromone detection. However, some pheromones (such as 5-α-androstenone, a volatile steroid found in boar saliva) instead stimulate the olfactory system. Androstenone is also found in human male axillary sweat. Some researchers have speculated that androstenone functions as a pheromone in humans although the data are not compelling. Lastly, the VNO may serve functions in addition to pheromone detection. For instance, in snakes, the VNO is used to detect prey in addition to pheromones.

of odorants and pheromones are much higher than predicted by simple diffusion. Nonetheless, olfactory and pheromone receptors are notoriously responsive to chemical stimuli. The exquisite sensitivity of these receptor neurons might be considered an optimization of the sensory neuron to the low concentrations of biologically important chemicals that are present in the environment.

The low concentration of odorants imposes certain constraints on the possible receptor mechanisms for olfaction (chemosensory transduction in VNO sensory neurons will be discussed later). Namely, **amplifying intracellular second messenger cascades** are likely to be involved in transduction, as will be shown to be the case. Additionally, olfactory receptor neurons possess a **high membrane resistance.** The input resistance of olfactory receptor neurons is substantial (up to 30 GΩ). Consequently, tiny receptor currents—some investigators believe even those generated by the opening of a single ion channel—can produce significant voltage changes in an olfactory receptor neuron. These currents are sufficient to depolarize an olfactory receptor neuron to threshold for generating an action potential. Intracellular second messenger cascades and a high input resistance combine to increase the sensitivity of olfactory receptor neurons.

Volatile odoriferous compounds partition into the mucus layer covering the olfactory epithelial surface. A specialized carrier or transport protein has been identified in the mucus layer, **odorant binding protein** (OBP). It has been hypothesized that OBP facilitates the partitioning of odorants into the mucus. According to this hypothesis, odorants bound to OBP are carried and presented to the chemoreceptive surface on the cilia that protrude up into the mucus layer. OBP may also help rid the olfactory epithelium of odoriferous molecules.

C. Initiation of Olfactory Receptor Potentials

As far as is known, olfactory transduction involves odorant binding to specific receptors localized in the ciliary membrane of olfactory receptor neurons.

1. Membrane-Bound Receptors

The transduction pathways that have been characterized for olfaction to date involve GPCRs that are integral proteins in the ciliary membrane. When occupied by odorant, the receptor activates an intracellular G protein. Unlike gustatory sensory cells, olfactory stimuli do not permeate the plasma membrane through ion channels and carry charge (current). Volatile compounds, being lipophilic, probably do partition directly into the membrane, but this is not believed to be a principal mechanism in olfactory transduction.

Receptors for odorants have been cloned and sequenced (Buck and Axel, 1991; Raming *et al.*, 1993; Reed, 1992). **Odorant receptors** are molecules with seven membrane-spanning regions and are members of the large G-protein-coupled receptor superfamily. It has been estimated that in rodents there are **1000 odorant receptor genes.** This represents a sizeable fraction of the estimated total genome (~100 000 genes). Data suggest that any given olfactory receptor neuron

expresses a single odorant receptor. Researchers have now been able to express and demonstrate function of cloned odorant receptors in several different systems, including an insect cell line (Sf 9, Breer *et al.*, 1998), rat olfactory epithelium (Zhao *et al.*, 1998; Touhara *et al.*, 1999), and HEK-293 cells (Krautwurst *et al.*, 1998). These studies have reported responses to specific odors, thereby identifying a particular receptor with a group of odoriferous ligands.

2. G Proteins in Olfactory Receptor Neurons

Odorant binding to receptors triggers second messenger pathways, beginning with activation of G proteins. Researchers postulate that there are two separate G protein pathways for olfaction. The better characterized pathway involves G_{olf}, a G protein that is enriched in olfactory epithelium. G_{olf} is a member of the G_s family and **activates type III adenylyl cyclase** expressed in olfactory epithelium (Pace *et al.*, 1985). It has long been known that exposing isolated olfactory cilia to low concentrations (μM) of certain fruity, floral, or herbaceous odors results in an increase in intracellular cAMP. These observations were key in establishing adenylyl cyclase in the intracellular pathway for signal transduction in olfaction, especially for what are considered pleasant odors. A less well-established G protein pathway, believed by some to act as an intermediary for signaling in putrid and unpleasant odors, will be discussed later.

3. Cyclic AMP and Cyclic Nucleotide-Gated Channels in Olfactory Receptor Neurons

The effector channels for the intracellular second messengers in olfactory receptor neurons are specialized ion channels. Cyclic AMP binds to and rapidly opens cation-selective (Na^+, K^+, Ca^{2+}) **cyclic nucleotide-gated,** or **CNG, channels** from the inside of the ciliary membrane. This was demonstrated by applying cAMP and cGMP to isolated patches of membrane pulled from olfactory cilia (Nakamura and Gold, 1987) or to isolated olfactory receptor neurons (Firestein *et al.*, 1991) and observing cyclic nucleotide-gated, cation-selective conductance increases. CNG channels have little or no voltage-dependence and under physiological conditions (i.e., in the presence of extracellular Ca^{2+} and Mg^{2+}) have a small conductance, ~100 fS. The consequence of odorant stimulation and thus cAMP-gated channel activity is a **depolarizing receptor potential** (Fig. 7A). CNG channels from olfactory receptor neurons have been cloned and sequenced (Dhallan *et al.*, 1990). Olfactory CNG channels are quite similar to CNG channels found in photoreceptors. CNG channels from photoreceptors and olfactory receptor neurons differ in their sensitivity to cyclic nucleotides. The CNG channels in olfactory receptor neurons respond to cAMP and cGMP alike, with cGMP being somewhat more potent. In contrast, cAMP is far less effective than cGMP in gating photoreceptor CNG channels. It is an enigma that olfactory CNG channels are more sensitive to cGMP than cAMP because odorant stimulation is believed principally to elevate cAMP. A possible role for an indirect elevation of cGMP in olfactory adaptation (discussed later) may resolve this puzzle. Details of CNG channels are given in Chapter 47.

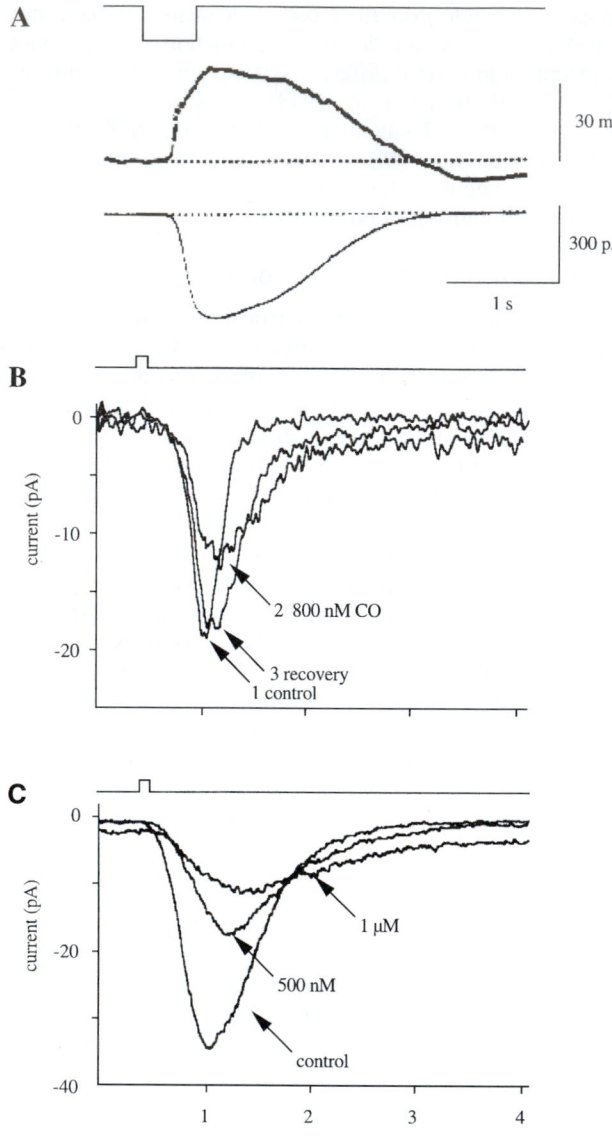

FIGURE 7. Patch-clamp recordings from olfactory receptor neurons. (A) Depolarizing receptor potential (top) and receptor current (bottom) elicited by applying an odorant (10 mM *n*-amyl acetate) to the olfactory cilia of an isolated newt olfactory receptor neuron. Modified from Kurahashi (1989). Inhibitory effects of (B) carbon monoxide, CO, and (C) cGMP on receptor currents in salamander olfactory receptor neurons after brief applications of an odorant (100 μM cineole). Traces 1 and 3 in B were recorded before and after recovery from the presence of 800 nM CO in the bathing solution. Traces in C were recorded before and during the presence of 500 nM and 1 μM 8-Br-cGMP. CO stimulates the formation of cGMP during odorant stimulation, which ultimately decreases odor responses and may underlie long-lasting olfactory adaptation. Top traces in B and C mark the application of odorant. Modified from Leinders-Zufall *et al.* (1996).

During odorant stimulation, Ca^{2+} and Na^+ enter the olfactory cilium through CNG channels. This cationic current depolarizes the sensory cell. The cation influx, in concert with a depolarizing Cl^- current (discussed next), raises the membrane potential to threshold and initiates impulses that are conducted by the axon to the olfactory bulb. A key component of the cation current through olfactory CNG channels is Ca^{2+} influx. The influx of Ca^{2+} has several important consequences in addition to supplying inward (depolarizing) current. First, Ca^{2+} influx activates a **Ca^{2+}-dependent Cl^- conductance** in the olfactory cilia. The olfactory cilia are suspended in a mucus that has a Cl^- concentration of about 55 mM. $[Cl^-]_i$ at the apical tips of olfactory receptor neurons appears to be maintained at a remarkably high level (ca. 70 mM), yielding a Cl^- equilibrium potential a few mV positive to zero and considerably more positive than resting potential. Consequently, activating the Ca^{2+}-activated anion conductance in these cells causes an **efflux of Cl^-**, that is, a **depolarizing current** that complements the influx of cations through CNG channels. This Cl^- current contributes substantially to the depolarizing receptor current. Indeed, in rat olfactory receptor neurons, the Ca^{2+}-activated Cl^- current is even greater than the initial inward Na^+ and Ca^{2+} current through CNG channels. The combined action of the cAMP-gated cation channels and Ca^{2+}-dependent Cl^- channels may exist to ensure a depolarizing receptor current even in the face of fluctuating extracellular (mucosal) cation concentrations.[3]

Second, Ca^{2+} influx through CNG channels during odorant stimulation exerts **negative feedback on the CNG channel** itself. Third, Ca^{2+} regulates the activity of a number of the enzymes involved in the second messenger cascades in olfactory signal transduction. These actions of Ca^{2+} are particularly important for signal adaptation (discussed later).

4. IP_3 in Olfactory Receptor Neurons

There is evidence, albeit quite controversial for vertebrate species, that another second messenger cascade may be activated in olfactory receptor neurons. This cascade appears to be triggered by a different set of odorants than those that are transduced by G_{olf}, adenylyl cyclase, and cAMP. The main impetus to search for alternative signal transduction pathways stems from observations that certain odorants stimulate the rapid generation of **inositol trisphosphate (IP_3)** in olfactory receptor cilia. G proteins other than those of the G_s category, described previously, are believed to activate the membrane-bound enzyme **phospholipase C (PLC).** PLC hydrolyzes a phospholipid, phosphatidyl inositol 4,5-bisphosphate, in the plasma membrane of olfactory receptor neurons. The resultant by-products, **IP_3 and diacyglycerol (DAG),** are powerful bioactive compounds. In other tissues such as

---

[3]As stated previously, activating the Ca^{2+}-dependent anion conductance in gustatory sensory cells results in the opposite effect than at the cilia of olfactory receptor neurons. Namely, in gustatory sensory cells this conductance produces an **influx** of Cl^- (i.e., a repolarizing, outward current). This is because the basolateral membrane of taste cells—where the Cl^- conductance channels are situated—is exposed to interstitial fluid and $[Cl^-]_i$ in taste cells is believed to be low. Consequently, in taste cells, the equilibrium potential for Cl^- is likely to be near the resting potential. Opening Cl^- channels, especially during depolarizations, will allow Cl^- to enter the taste cell and repolarize the membrane.

taste buds (discussed earlier), IP_3 receptors on intracellular Ca^{2+} storage compartments (e.g., endoplasmic reticulum) control the release of Ca^{2+} into the cytosol. However, Ca^{2+} stores have not been demonstrated in olfactory cilia. Instead, IP_3, if it is involved in olfactory transduction, is believed to activate specialized channels in the ciliary plasma membrane. IP_3-gated, non-selective cation channels have been demonstrated in olfactory cilia immunocytochemically and in certain species, especially invertebrates, electrophysiologically.

Both mechanisms—adenylyl cyclase→cAMP, and PLC→ IP_3—may coexist in a single olfactory receptor neuron and may be activated by different odorants. This has been clearly shown to occur in invertebrate olfactory receptor neurons, where the two pathways mediate opposing effects—excitation (IP_3-gated cation channels) versus inhibition (cAMP-gated K^+ channels) (Ache, 1994). However, if this is also the case in mammals, it might require the expression of two or more different odorant receptors in the same cell. This would seem to contradict the prevailing belief.

Lastly, although certain findings in vertebrate species support the involvement of IP_3 in olfactory signal transduction, the data are controversial. In fact, recent evidence from **gene knockout mice lacking olfactory CNG channels** indicates that odorant responses to a wide selection of odors—even odors that are known to elevate intracellular IP_3 and complex odor mixtures—are undetectable (Brunet *et al.*, 1996). The ability to eliminate odor responses by selectively eliminating CNG channels in olfactory receptor neurons seriously challenges the hypothesis that IP_3 is directly involved in mammalian olfactory transduction, at least in any straightforward scheme (reviewed by Gold, 1999).

The transduction mechanisms for olfactory receptor neurons are summarized in Fig. 8.

5. Transduction in VNO Sensory Neurons

The transduction pathways in VNO sensory neurons for pheromones and other chemical stimuli are presently under investigation. It appears that VNO sensory neurons utilize distinctive GPCRs, a different set of G proteins, and different downstream effectors than olfactory receptor neurons. For example, VNO sensory neurons express two families of GPCRs that are unrelated to the large family of odorant receptors (Dulac and Axel, 1995). In rodents, it is estimated that there are **50–100** genes in each of these two families of VNO receptors. The two families of VNO receptors are expressed in two anatomically separate regions of the VNO sensory epithelium. The G proteins G_{ao} and G_{ai2} are highly expressed in VNO sensory neurons but the olfactory-specific G_{olf} is not (Berghard and Buck, 1996). Furthermore, stimulating VNO receptors does not appear to prompt the synthesis of cAMP and cation influx through CNG channels, as occurs in olfactory receptor neurons. Instead, VNO sensory neurons express TRP2, an ion channel of the transient receptor potential (TRP) family (Liman *et al.*, 1999). Researchers currently speculate that activation of VNO receptors elevates intracellular second messengers (IP_3 and DAG) that directly gate TRP2 and depolarize the VNO sensory cell.

D. Termination of Olfactory Signals

Olfactory receptor neuron excitation is terminated by a number of mechanisms. An obvious one is the unbinding and disappearance of the odorant from the chemoreceptive surface of the cilia, perhaps aided by **odorant binding protein** (OBP, see Fig. 8). The removal of odorants is complicated by the lipophilic nature of most odoriferous compounds; they will tend to partition into the plasma membrane and thus may linger in the vicinity of the receptors. Although the details are only now unfolding, one proposed mechanism for ridding the chemosensitive surfaces of odorants is the **enzymatic modification of odorants** by broad-spectrum biotransformation enzymes found in adjacent supporting cells. These enzymes are similar to the **detoxification enzymes** found in the liver. Olfactory epithelial supporting cells adjacent to receptor neurons contain high concentrations of the detoxifying enzymes **glutathione transferase, cytochrome P-450,** and **UDP gluconosyl transferase** (Lazard *et al.*, 1991). Indeed, the lipophilic nature of many odorants and their absorption into the sensory epithelium may necessitate the existence of such degradative mechanisms.

Besides disappearance of the odorant, **sensory adaptation** is also key in terminating olfactory signals. Adaptation in olfactory receptor neurons is explained by multiple mechanisms, many of which involve Ca^{2+}, as follows. First, short-term adaptive mechanisms are believed to be set in motion by the Ca^{2+} that enters through CNG channels during odorant stimulation. Ca^{2+} inhibits CNG channels themselves (Kurahashi and Menini, 1997), a direct negative feedback on the olfactory signal. That is, an increase in Ca^{2+} in the ciliary cytosol will depress odorant responses by shutting down effector (CNG) channels.

Second, heightened intracellular Ca^{2+} also depresses the cascade of enzymes that are triggered during olfactory transduction. At high (mM) concentrations, Ca^{2+} inhibits adenylyl cyclase, thereby reducing the continued generation of cAMP during maintained stimulation. In olfactory receptor neurons, this inhibition appears to be mediated by Ca^{2+}-calmodulin-induced phosphorylation via the enzyme Ca^{2+}-calmodulin-dependent protein kinase II (CaMKII). However, at nM to µM concentrations, Ca^{2+} has the opposite effect, namely, it stimulates adenylyl cyclase. Consequently, the net effect of Ca^{2+} influx on adenylyl cyclase activity during odor stimulation is concentration-dependent. Additionally, Ca^{2+} influx stimulates phosphodiesterase (PDE), thereby accelerating the degradation of cAMP and reducing the odorant response.

Third, Ca^{2+} influx activates **Ca^{2+}-dependent K^+ channels** that initiate an **outward, repolarizing current.** Counterbalancing this, Ca^{2+} also activates a Cl^- conductance in the olfactory cilia that produces an inward depolarizing current, as described previously.

Thus, the actions of Ca^{2+} are complex. The net effect of Ca^{2+} entry through CNG channels during olfactory stimulation appears to be to reduce the responsiveness of olfactory receptor neurons perhaps as much as 20-fold, moving the stimulus-response relationship to a higher range of odorant concentrations.

Longer-lasting adaption to a maintained odor stimulus involves another mechanism. A prolonged adaptation in

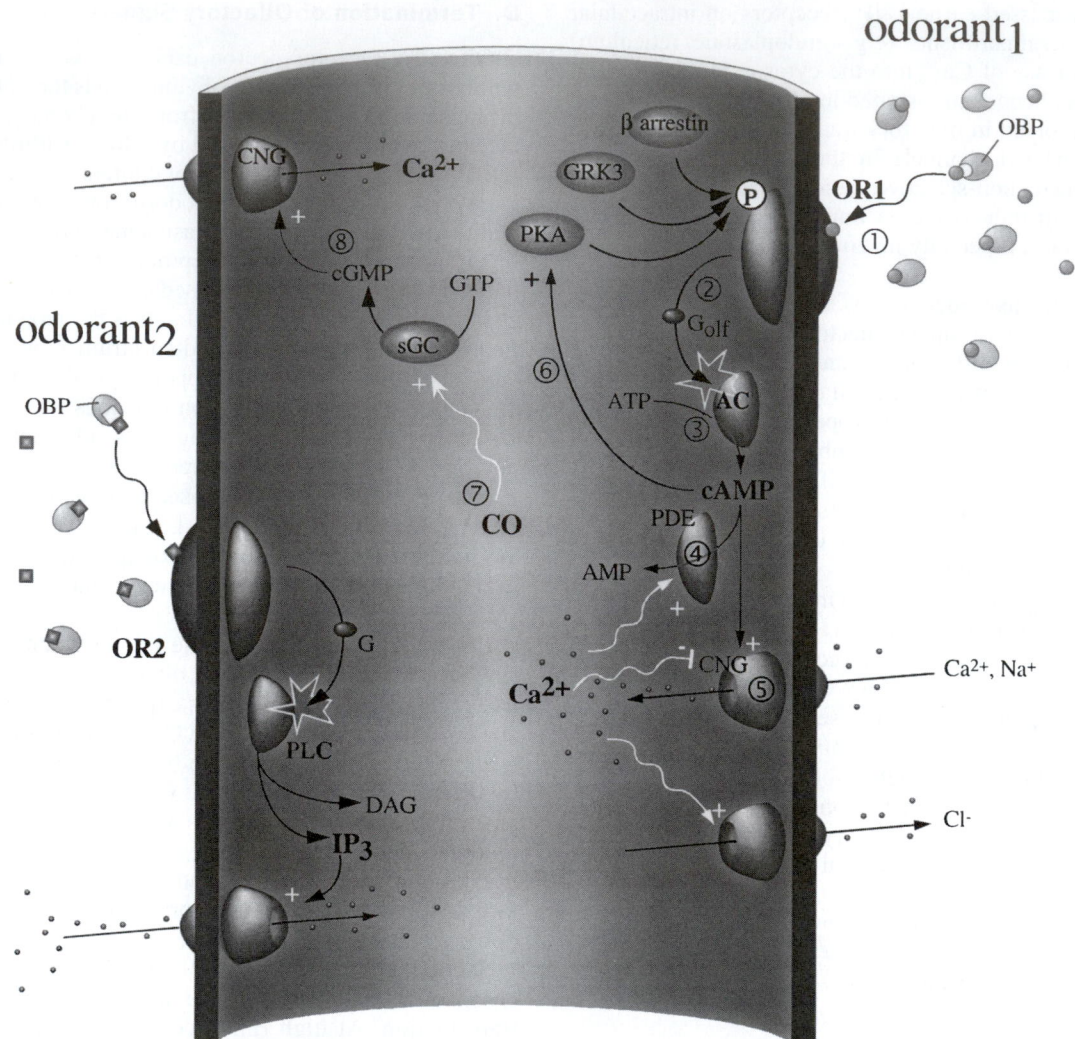

FIGURE 8.   Transduction mechanisms in olfactory receptor neurons. A portion of the shaft of an olfactory cilium is illustrated. In the mucus layer, odorants may be bound to carrier molecules, odorant binding protein (OBP). (1) Some odorants (odorant$_1$) activate receptors (OR1) that (2) stimulate a G protein (G_{olf}) and (3) activate adenylyl cyclase (AC). This converts ATP to cAMP. (4) cAMP is hydrolyzed by phosphodiesterase, PDE. (5) More importantly, cAMP activates CNG channels, allowing an influx of Ca^{2+} (also Na^+). Ca^{2+} influx opens Ca^{2+}-dependent Cl^- channels, allowing an efflux of Cl^- (i.e., a depolarizing current). (6) cAMP also stimulates protein kinases that phosphorylate and desensitize the odorant receptor, thereby quenching the response and contributing to adaptation. Intracellular Ca^{2+} decreases the long-term responsiveness of the olfactory receptor neuron, also producing adaptation. Ca^{2+} accomplishes this by exerting negative feedback on the CNG channel, positive feedback on PDE, and concentration-dependent effects on AC. Some of these actions are indicated by white wavy lines. (7) Odorant activation also generates carbon monoxide, which stimulates soluble guanylyl cyclase (sGC) to produce cGMP. (8) cGMP exerts a tonic, subthreshold excitatory influence on CNG channels, allowing a small but steady influx of Ca^{2+} into the olfactory receptor neuron.

Other odorants (odorant$_2$) may activate receptors (OR2) that are coupled via a G protein to phospholipase C (PLC). When these receptors are activated, the intracellular second messenger, inositol trisphosphate (IP_3), is formed. IP_3 opens channels in the olfactory receptor neuron membrane. The existence and significance of odorant-activated IP_3 pathways in olfactory transduction in vertebrates is currently under debate but is well-established in invertebrates.

olfactory receptor neurons, involving an intriguing role for the novel gaseous neurotransmitter **carbon monoxide (CO),** has been proposed (Leinders-Zufall *et al.,* 1996). CO, like the related neurotransmitter nitric oxide (NO), is a highly-diffusible substance and is generated during olfactory transduction via a mechanism that is not yet well-characterized. CO, in turn, activates **soluble guanylyl cyclase** (sGC) and leads to the production of **cGMP** in the

stimulated olfactory receptor neuron as well as in surrounding cells. As discussed previously, cGMP is a very effective ligand for olfactory CNG channels. At first glance, generating cGMP might seem contrary to adaptation and represent positive feedback during odorant stimulation. However, activation of olfactory CNG channels via the CO pathway produces only a **low-level, subthreshold inward current.** This maintains a persistent trickle of Ca^{2+} into the olfactory receptor neurons, **chronically reducing CNG channel activity.** The rapid transient influx of Ca^{2+} through CNG channels (the immediate effect of odorant stimulation), combined with a sustained low-level influx via cGMP activation of these channels (the CO pathway), overall leads to a long-lasting (minutes) reduction of the sensitivity of olfactory receptor neurons after the initial excitation.[4]

In addition to these sensory adaptation mechanisms, **desensitization of odorant receptors** following odorant stimulation also plays a key role in **terminating odorant responses** (Schleicher *et al.*, 1993). cAMP produced by odor stimulation opens CNG cation channels and secondarily activates **protein kinase A (PKA).** PKA and a specialized G protein-coupled receptor kinase, **GRK3**[5], phosphorylate odorant-bond receptors (Fig. 9). These protein kinases, together with regulatory proteins such as **arrestin,** render the odorant receptor inactive and **quench the subsequent transduction cascade.** Receptors are resensitized by the action of phosphatases. Thus, a **cycle of phosphorylation/dephosphoryla-tion** controls the active state of odorant receptors and ultimately, the responsiveness of olfactory receptor neurons.

Collectively, short-term adaptation, long-term adaptation, and receptor desensitization act to shape and terminate olfactory responses to prolonged odor stimulation. Sensory adaptation in many sensory systems, including olfaction, serves a dual purpose. First, over time a maintained stimulus ceases to excite responses, that is, the sensory organ adapts to the stimulus. In this sense, adaptation tends to accentuate signals from stimuli that fluctuate and emphasizes the dynamic aspects of sensory stimuli. Second, adaptation reduces responsivity to increasingly intense stimuli. This prevents the output from the sensory organs from becoming saturated by only moderately intense stimulus intensities while at the same time preserving the ability of the organ to respond well to very weak stimuli. That is, adaptation greatly extends the effective range of stimulus intensities over which a sensory organ can respond.

IV. Summary

Taste and olfaction share certain common features. For example, transduction mechanisms in both these chemical senses involve G-protein-coupled receptors and changes in intracellular cyclic nucleotides, particularly cAMP. This is the case for certain sweet, bitter, and *umami* stimuli as well as odoriferous stimuli. IP_3 mechanisms may also play a role.

---

[4]Previously, it was believed that NO played this role in olfactory adaptation. However, evolving data utilizing blockers of CO synthesis now implicate CO, not NO, as the diffusible intra- and intercellular controller of CNG channel activity in olfactory receptor neurons.

[5]GRK3 was formerly named β adrenergic receptor kinase 2, or βARK2.

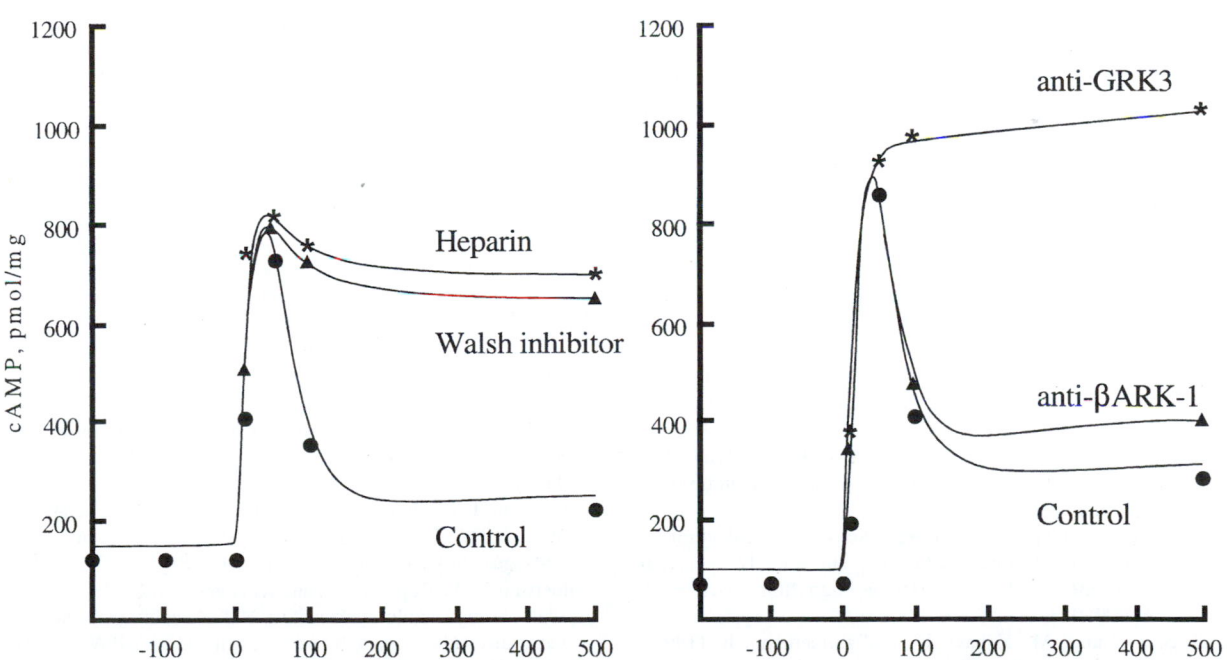

FIGURE 9. Desensitization of odorant receptors during prolonged stimulation. Olfactory second messenger signaling is quenched by phosphorylation cascades. Left, inhibiting protein kinase activity with Walsh inhibitor or heparin markedly prolongs increases in cAMP induced by odorant stimulation in cilia preparations from the rat. (●) odorant alone (1 μM citralva); (▲) odorant stimulation of cilia pretreated with Walsh inhibitor (3.8 μM); (*) odorant stimulation of cilia pretreated with heparin (1 μM). Right, pretreating the cilia preparation with antibodies to GRK3 specifically prevents desensitization of the odorant receptor and prolongs its activation(measured here as an increase in cAMP). (●) odorant alone (1 μM citralva); (▲) odorant stimulation of cilia pretreated with anti-βARK-1 antibodies (diluted 1:5000); (*) odorant stimulation of cilia pretreated with anti-GRK3 antibodies (diluted 1:5000); abscissae, time in ms. Modified from Schleicher *et al.* (1993).

In taste, IP_3 stimulates the release of Ca^{2+} from intracellular stores. In olfaction, IP_3 is believed to gate specific channels.

Transduction channels and integral membrane receptors appear to be concentrated on the exposed apical tips of both types of chemosensory receptor cells, although the basolateral membranes of taste cells may be additional sites for chemosensory transduction. In some regards, sensory transduction in taste and olfaction is similar to that in other modalities, notably photoreception. Photoreceptors, olfactory receptor neurons, and gustatory sensory cells possess ion channels that are modulated by cyclic nucleotides. Further, transducin, the G protein in photoreceptors, is also found in taste cells.

However, there are key differences between olfaction and taste. Olfactory receptor neurons are optimized for low concentrations of chemical stimuli, gustatory sensory cells for higher concentrations. Furthermore, whereas an increase in cAMP opens cation channels in olfactory receptor neurons, cAMP closes K^+ channels in taste receptor cells (by activating PKA, leading to channel phosphorylation). An additional difference is that taste cells possess ligand-gated (ionotropic) ion channels and ion channels that are either permeated or blocked by certain chemical stimuli (especially ions). These mechanisms do not participate in olfactory transduction.

Areas that are current topics of intensive research in the peripheral transduction mechanisms for taste and olfaction include, among others, extending our knowledge of the G-protein-coupled effector cascades in the receptor cells, discovering the molecular structure of olfactory and taste receptor proteins, understanding how taste and olfactory receptor neurons distinguish among the multitude of different chemical stimuli, learning how the output from the peripheral sensory organs is encoded, and determining the synaptic mechanisms that transmit signals between cells within the taste bud and on to sensory afferent fibers that innervate the taste bud.

Bibliography

Ache, B. (1994). Towards a common strategy for transducing olfactory information. *In* "Seminars in Cell Biology." (H. Breer, Ed.), pp. 55–63, Academic Press, San Diego, CA.

Ache, B. W., and Restrepo, D. (2000). Olfactory transduction. *In* "Neurobiology of Taste and Smell, 2nd edition." (T. E. Finger, D. Restrepo, and W. L. Silver, Eds.) Wiley, New York, 159–177.

Adler, E., Hoon, M. A., Mueller, K. L., Chandrashekar, J., Ryba, N. J. P., and Zuker, C. S. (2000). A novel family of mammalian taste receptors. *Cell* **100**, 693–702.

Berghard, A., and Buck, L. B. (1996). Sensory transduction in vomeronasal neurons: evidence for G_{ao}, G_{ai2} and adenylyl cyclase II as major components of a pheromone signaling cascade. *J. Neurosci.* **16**, 909–918.

Bernhardt, S. J., Naim, M., Zehavi, U., and Lindemann, B. (1996). Changes in IP_3 and cytosolic Ca^{2+} in response to sugars and non-sugar sweeteners in transduction of sweet taste in the rat. *J. Physiol.* **490**, 325–336.

Bigiani, A. (2001). Mouse taste cells with glia-like membrane properties. *J. Neurophysiol.* In press.

Bigiani, A., and Roper, S. D. (1991). Mediation of responses to calcium in taste cells by modulation of a potassium conductance. *Science* **252**, 126–128.

Breer, H., Krieger, J., Meinken, C., Kiefer, H. and Strotmann, J. (1998). Expression and functional analysis of olfactory receptors. *Ann. NY Acad. Sci.* **855**, 175–181.

Breer, H., Raming, K., and Krieger, J. (1994). Signal recognition and transduction in olfactory neurons. *Biochim. Biophys. Acta* **1224**, 277–287.

Brunet, L. J., Gold, G. H., and Ngai, J. (1996). General anosmia caused by a targeted disruption of the mouse olfactory cyclic nucleotide-gated cation channel. *Neuron* **17**, 681–693.

Buck, L., and Axel, R. (1991). A novel multigene family may encode odorant receptors: a molecular basis for odor recognition. *Cell* **66**, 175–187.

Caicedo, A., and Roper, S. D. (2001). Taste receptor cells that discriminate between bitter stimuli. *Science,* In press.

Chandrashekar, J., Mueller, K. L., Hoon, M. A., Adler, E., Feng, L., Guo, W., Zuker, C. S., and Ryba, N. J. P. (2000). T2Rs function as bitter taste receptors. *Cell* **100**, 703–711.

Chaudhari, N., Landin, A. M., and Roper, S. D. (2000). A metabotropic glutamate receptor variant functions as a taste receptor. *Nat. Neurosci.* **3**, 113–119.

Corey, D. P., and Roper, S. D. (1992). "Sensory Transduction." Rockefeller University Press, New York.

Delay, R. J., Kinnamon, S. C., and Roper, S. D. (1997). Serotonin modulates voltage-dependent calcium current in *Necturus* taste cells. *J Neurophysiol.* **77**, 2515–2524.

DeSimone, J. A. (1991). Transduction in taste receptors. *Nutrition* **7**, 146–147.

Dhallan, R. S., Yau, K.-W., Schrader, K. A., and Reed, R. R. (1990). Primary structural and functional expression of a cyclic nucleotide-activated channel from olfactory neurons. *Nature* **347**, 184–187.

Dionne, V., and Dubin, A. E. (1994). Transduction diversity in olfaction. *J. Exper. Biol.* **194**, 1–21.

Doty, R. L. (1995). "Handbook of Olfaction and Gustation." Marcel Dekker, New York.

Dulac, C., and Axel, R. (1995). A novel family of genes encoding putative pheromone receptors in mammals. *Cell* **83**, 195–206.

Ewald, D. E., and Roper, S. (1994). Bidirectional synaptic transmission in *Necturus* taste buds. *J. Neurosci.* **14**, 3791–3801.

Finger, T. E., Restrepo, D., and Silver, W. L. (2000). "Neurobiology of Taste and Smell, 2nd edition." Wiley, New York.

Firestein, S., Breer, H., and Greer, C. A. (1996). Olfaction: what's new in the nose. *J. Neurobiol.* **30**, 1–176.

Firestein, S., Darrow, B., and Shepherd, G. (1991). Activation of the sensory current in salamander olfactory receptor neurons depends on a G-protein mediated cAMP second messenger system. *Neuron* **6**, 825–835.

Furue, H., and Yoshii, K. (1997). *In situ* tight-seal recordings of taste substance-elicited action currents and voltage-gated Ba currents from single taste bud cells in the peeled epithelium of mouse tongue. *Brain Res.* **776**, 133–139.

Gilbertson, T. A., and Zhang, H. (1998). Self-inhibition in amiloride-sensitive sodium channels in taste receptor cells. *J. Gen. Physiol.* **111**, 667–677.

Gilbertson, T. A., Fontenot, D. T., Liu, L., Zhang, H., and Monroe, W. T. (1997). Fatty acid modulation of K^+ channels in taste receptor cells: gustatory cues for dietary fat. *Am. J. Physiol.* **272**, C1203–1210.

Gilbertson, T. A., Roper, S. D., and Kinnamon, S. C. (1993). Proton currents through amiloride-sensitive Na^+ channels in isolated hamster taste cells: enhancement by vasopressin and cAMP. *Neuron* **10**, 1–20.

Gold, G. H. (1999). Controversial issues in vertebrate olfactory transduction. *Annu. Rev. Physiol.* **61**, 857–871.

Graziadei, P. P. C., and Monti Graziadei, G. A. (1978). Continuous nerve cell renewal in the olfactory system. *In* "Handbook of Sensory Physiology." (M. Jacobson, Ed.).Vol. IX, pp. 55–82. Springer-Verlag, New York.

Hänig, D. P. (1901). Zur Psychophysik des Geschmackssinnes. *Philosophische Studien* **17**, 576–623.

Heck, G. L., Mierson, S., and DeSimone, J. A. (1984). Salt taste transduction occurs through an amiloride-sensitive sodium transport pathway. *Science* **223**, 403–405.

Herness, S., and Chen, Y. (1997). Serotonin inhibits calcium-activated K⁺ current in rat taste receptor cells. *Neuroreport* **8**, 3257–3261.

Hoon, M. A., Adler, E., Lindemeier, J., Battey, J. F., Ryba, N. J., and Zuker, C. S. (1999). Putative mammalian taste receptors: a class of taste-specific GPCRs with distinct topographic selectivity. *Cell* **96**, 541–551.

Huang, L., Shanker, Y. G., Dubauskaite, J., Zheng, J. Z., Yan, W., Rosenzweig, S., Spielman, A. I., Max, M., and Margolskee, R. F. (1999). Gγ13 colocalizes with gustducin in taste receptor cells and mediates IP$_3$ responses to bitter denatonium. *Nat. Neurosci.* **2**, 1055–1062.

Kashiwayanagi, M., Miyake, M., and Kurihara, K. (1983). Voltage-dependent Ca²⁺ channel and Na⁺ channel in frog taste cells. *Am. J. Physiol.* **244**, C82–88.

Kinnamon, S. C., and Margolskee, R. F. (1996). Mechanisms of taste transduction. *Curr. Opin. Neurobiol.* **6**, 506–513.

Kolesnikov, S. S., and Margolskee, R. F. (1995). A cyclic-nucleotide-suppressible conductance activated by transducin in taste cells. *Nature* **376**, 85–88.

Krautwurst, D., Yau, K. W., and Reed, R. R. (1998). Identification of ligands for olfactory receptors by functional expression of a receptor library. *Cell* **95**, 917–926.

Kretz, O., Barbry, P., Bock, R., and Lindemann, B. (1999). Differential expression of RNA and protein of the three pore-forming subunits of the amiloride-sensitive epithelial sodium channel in taste buds of the rat. *J. Histochem. Cytochem.* **47**, 51–64.

Kurahashi, T., and Menini, A. (1997). Mechanism of odorant adaptation in the olfactory receptor cell. *Nature* **385**, 725–729.

Kurahashi, T. (1989). Activation by odorants of cation-selective conductance in the olfactory receptor cell isolated from the newt. *J. Physiol.* **419**, 177–192.

Lazard, D., Zupko, K., Poria, Y., Nef, P., Lazarovits, J., Horn, S., Khen, M., and Lancet, D. (1991). Odorant signal termination by olfactory UDP glucuronosyl transferase. *Nature* **349**, 790–793.

Leinders-Zufall, T., Shepherd, G. M., and Zufall, F. (1996). Modulation by cyclic GMP of the odour sensitivity of vertebrate olfactory receptor cells. *Proc. R. Soc. Lond. [Biol.]* **263**, 803–811.

Li, X.-J., Blackshaw, S., and Snyder, S. H. (1994). Expression and localization of amiloride-sensitive sodium channel indicate a role for non-taste cells in taste perception. *Proc. Natl. Acad. Sci. USA* **91**, 1814–1818.

Liman, E. R., Corey, D. P., and Dulac, C. (1999). TRP2: a candidate transduction channel for mammalian pheromone sensory signaling. *Proc. Natl. Acad. Sci. USA* **96**, 5791–5796.

Lin, W., Finger, T. E., Rossier, B. C., and Kinnamon, S. C., (1999). Epithelial Na⁺ channel subunits in rat taste cells: localization and regulation by aldosterone. *J. Comp. Neurol.* **405**, 406–420.

Lindemann, B. (1996). Taste reception. *Physiol. Rev.* **76**, 719–766.

Lindemann, B. (1999). Receptor seeks ligand: on the way to cloning the molecular receptors for sweet and bitter taste. *Nat. Med.* **5**, 381–382.

Mackay-Sim, A., and Kittel, P. (1991a). Cell dynamics in the adult mouse olfactory epithelium: a quantitative autoradiographic study. *J. Neurosci.* **11**, 979–984.

Mackay-Sim, A., and Kittel, P. (1991b). On the life span of olfactory receptor neurones. *Eur. J. Neurosci.* **3**, 209–215.

McLaughlin, S. K., McKinnon, P. J., and Margolskee, R. F., (1992). Gustducin is a taste-cell-specific G protein closely related to the transducins. *Nature* **357**, 563.

McLaughlin, S. K., McKinnon, P .J., and Margolskee, R. F. (1993). Gustducin and transducin: a tale of two G proteins. *In* "The Molecular Basis of Smell and Taste Transduction." Ciba Foundation Symposium No. 179, pp. 186–200. Wiley, Chichester.

Matsunami, H., Montamayeur, J. P., and Buck, L. B. (2000). A family of candidate taste receptors in human and mouse. *Nature* **404**, 601–604.

Naim, M., Seifert, R., Nurnberg, B., Grunbaum, L., and Schultz, G. (1994). Some taste substances are direct activators of G-proteins. *Biochem. J.* **297**, 451–454.

Nakamura, T., and Gold, G. H. (1987). A cyclic nucleotide-gated conductance in olfactory receptor cilia. *Nature* **325**, 442–444.

Ossebaard, C. A., and Smith, D. V. (1996). Amiloride suppresses the sourness of NaCl and LiCl. *Physiol. Behav.* **60**, 1317–1322.

Pace, U., Hanski, E., Salomon, Y., and Lancet, D. (1985). Odorant-sensitive adenylate cyclase may mediate olfactory reception. *Nature* **316**, 255–258.

Raming, K., Krieger, J., and Strotmann, J. (1993). Cloning and expression of odorant receptors. *Nature* **361**, 353–356.

Reed, R. R. (1992). Signalling pathways in odorant detection. *Neuron* **8**, 205–209.

Roper, S. (1983). Regenerative impulses in taste cells. *Science* **220**, 1311–1312.

Roper, S. (1992). The microphysiology of peripheral taste organs. *J. Neurosci.* **12**, 1127–1134.

Rossler, P., Kroner, C., Freitag, J., Noe, J., and Breer, H. (1998). Identification of a phospholipase C beta subtype on rat taste cells. *Eur. J. Cell Biol.* **77**, 263–261.

Ruiz-Avila, L., McLaughlin, S. K., Wildman, D., McKinnon, P. J., Robichon, A., Spickofsky, N., and Margolskee, R. F. (1995). Coupling of bitter receptor to phosphodiesterase through transducin in taste receptor cells. *Nature* **376**, 80–85.

Schleicher, S., Boekhoff, I., Arriza, J., Lefkowitz, R. J., and Breer, H. (1993). A beta-adrenergic receptor kinase-like enzyme is involved in olfactory signal termination. *Proc. Natl. Acad. Sci. USA* **90**, 1420–1424.

Simon, S. A., and Roper, S. (1993). "Mechanisms of Taste Transduction." CRC Press, Boca Raton, FL.

Taylor, R., and Roper, S. (1994). Ca²⁺-dependent Cl⁻ conductance in taste receptor cells from *Necturus*. *J. Neurophysiol.* **72**, 475–478.

Touhara, K., Sengoku, S., Inaki, K., Tsuboi, A., Hirono, J., Sato, T., Sakano, H., and Haga, T. (1999). Functional identification and reconstitution of an odorant receptor in single olfactory neurons. *Proc. Natl. Acad. Sci. USA* **96**, 4040–4045.

Ugawa, S., Minami, Y., Guo, W., Saishin, Y., Takatsuj, K., Yamamoto, T., Tohyama, M., and Shimada, S. (1998). Receptor that leaves a sour taste in the mouth. *Nature* **395**, 555–556.

Yang, H., Wanner, I. B., Roper, S. D., and Chaudhari, N. (1999). An optimized method for in situ hybridization with signal amplification that allows the detection of rare mRNAs. *J. Histochem. Cytochem.* **47**, 431–446.

Ye, Q., Heck, G. L., and DeSimone, J. A. (1993). Voltage dependence of the rat chorda response to Na⁺ salts: implications for the functional organization of taste receptor cells. *J. Neurophysiol.* **70**, 167–178.

Zhao, H., Ivic, L., Otaki, J. M., Hashimoto, M., Mikoshiba, K., and Firestein, S. (1998). Functional expression of a mammalian odorant receptor. *Science* **279**, 237–242.

Stephen D. Roper and Michael S. Grace

Appendix: Infrared Sensory Organs

I. Introduction

Infrared radiation is an important and ubiquitous component of the diverse physical environments in which animals live. All warm objects radiate infrared energy. Naturally occurring infrared sources include terrestrial, atmospheric, and astronomical bodies. Animals, especially ectothermic species, use radiant infrared energy to assist in maintaining their body temperature and must avoid excess infrared radiation to avoid overheating. In a few species, infrared radiation is also used in a novel manner—to form spatial images of the environment and to hunt prey. These animals possess specialized infrared-receptive sensory organs, the topic of this Appendix.

Infrared-receptive organs occur in both vertebrate and invertebrate animals. Buprestid beetles, including the European fire beetle *Melanophila acuminata*, have specialized infrared-receptive organs on the ventral surfaces of their abdomens. *Melanophila* and similar beetles congregate in large numbers at forest fire sites to deposit eggs in recently burned wood. They likely use chemoreceptors and mechanoreceptors (to detect wind direction) to guide them to sites of fires, and their infrared detectors to identify potential egg-laying sites of appropriate temperature. The deep-sea shrimp *Rimicaris exoculata* also possesses an elaborate organ capable of detecting 350 °C "black smoker" thermal vents in the deep Atlantic ocean. Their infrared sensory detection system may help the shrimp maintain close proximity to life-sustaining thermal vents in the deep sea without directly contacting the steep thermocline (350 °C to 2 °C over a 2 to 3 cm lateral distance) adjacent to the thermal vent. The common vampire bat (*Desmodus rotundus*) may possess infrared radiation detectors, but its infrared sensory system has been little investigated. Behavioral experiments show that vampire bats can detect infrared radiation as low as 50 μW/cm², and thus should be able to detect mammalian prey at distances of up to 16 cm. Vampire bats may use their infrared detection system to locate suitable prey.

However, by far the best-studied infrared sensory detectors are those of snakes. As stated previously, ectothermic animals such as snakes rely on infrared radiation to regulate their body temperature. In some snakes (*Boidae*, which include boa constrictors and pythons, and *Crotalinae*, which include rattlesnakes and water moccasins), infrared radiation is also used for accurate and precise targeting of prey. These snakes preferentially feed on endothermic (warm-blooded) prey. Unlike other infrared-sensing animals, boid and crotaline snakes are the only animals known to form spatial images of their environments using extremely sensitive infrared-receptive organs.

Boid and crotaline snakes are efficient predators. While vision may be useful for some aspects of their predatory behaviors, infrared imaging alone can allow accurate prey targeting. Snakes experimentally blinded by occluding both eyes and snakes born without eyes are capable of accurately placing strikes at mammalian prey (Kardong and Mackessey, 1991). Snakes lacking one eye target their strikes as well as do normally sighted snakes. Monocularly occluded snakes do not target preferentially on the sighted side. Even in snakes with an intact visual system, vision may be relatively unimportant for prey targeting. Snakes efficiently and accurately target prey regardless of visual contrast (black versus white mice against a black background) (Grace *et al.*, 2001).

II. Nature of the Stimulus: What Is Infrared (IR) Radiation?

Radiant infrared energy is a region of the spectrum of electromagnetic radiation that lies between visible light (wavelengths of 400 to 750 nm) and microwaves (wavelengths of ~1 mm to 1 cm). The English astronomer Sir William Herschel discovered infrared radiation at the beginning of the nineteenth century. Herschel was exploring the ability of light passing through a prism to heat a thermometer that he positioned at different points in the color spectrum. Herschel noticed that visible light was capable of heating the thermometer to some extent but that this effect was much more pronounced when the thermometer was placed beyond the red end of the spectrum. Electromagnetic radiation in the infrared region is absorbed as the radiation passes through objects in its pathway. Most of this absorbed infrared energy is dissipated by the resulting increase in molecular collisions (i.e., heat). Infrared radiation penetrates body tissues deeper than does visible light, and thus IR generators produce the so-called deep heat used in physical therapy. Any object above absolute zero radiates electromagnetic waves in the infrared, the more so as it is warmed. With increasing heat, an object radiates increasingly shorter wavelengths. Objects at room temperature radiate entirely in the IR and not at all in the visible light spectrum. At 500 °C, objects begin to radiate also in the visible spectrum and they appear dull red. Endothermic (warm-blooded) animals such as rats and mice radiate maximally in the infrared region at a wavelength of approximately 10 μm. IR radiation can thus be used to detect objects in the absence of visible light (i.e., in the dark) with an object's detection being enhanced by having a temperature that contrasts with its background temperature.

Water vapor and carbon dioxide in the atmosphere absorb and severely attenuate the transmission of certain wavelengths of the IR spectrum and allow other wavelengths to pass. That is, there are windows for IR transmission in the atmosphere. One of these windows is in a region where certain IR-sensitive photographic emulsions absorb well, thereby allowing IR images of landscapes and cityscapes to be collected by satellite cameras. Another atmospheric transmission window includes the region around 10 µm, that is, the region of the IR spectrum in which endothermic animals maximally radiate. IR sensory organs in snakes are believed to have a maximum sensitivity in this region (Grace *et al.,* 1999) and thus are well suited to detect endothermic prey.

III. Infrared-Sensitive Pit Organs in Snakes

Infrared sensory organs in snakes consist of **pit organs.** Pit organs are invaginations within or between scales in the head. These invaginations are up to 3 to 4 mm wide and 3 to 4 mm deep. In boid snakes (e.g., pythons) the pit organs are arrayed in rows within labial scales of both the upper and lower jaws.[1] In crotaline snakes (e.g., rattlesnakes), there is a single facial pit organ on each side of the face between the nostril and eye (Fig. A-1). There may be additional infrared-sensitive receptor terminals in the oral cavity (especially the palate) that may be important for guiding the predatory strike when the snake's mouth is open and fangs extended (Dickman *et al.,* 1987) (Fig. A-2).

A dense plexus of infrared-receptive primary sensory afferent nerve terminals is present in pit organs. In boid snakes, these terminals lie in the epidermis just under the surface of the pit organ. Immediately below the layer of infrared-receptive neuronal terminals is a thick capillary network. In crotaline snakes, a thin membrane (15 µm) stretches across the cavity of the pit (Fig. A-1B). Infrared-receptive nerve terminals are embedded within this membrane, also over a thick capillary bed. Infrared receptor terminals and the dense capillary plexus are mainly confined to the pit organs.[2]

The anatomical configuration of the pit organ, namely that the receptive tissue is stretched across the bottom of a cavity that has a narrowed opening to the outside environment, provides the snake with a means to direct its pit organs towards prey. Thus, the pit organs can provide information about the location of an infrared-emitting source in front of the snake. A distinct image, *per se,* such as a pin-hole camera might afford, is not formed in the pit organ, but the shadows cast by the infrared radiation passing through the narrowed opening into the pit project a crude approximation of the infrared source onto the sensory terminals. This "image" is conveyed to the snake's brain, much like visual information, to establish a spatiotopic map on the optic tectum (see Section III.A).

Infrared receptor terminals form large, highly-branched, entwined clusters surrounding an aggregate of Schwann cells. Collectively, this structure is called a terminal nerve mass (~30–100 µm diameter) (Fig. A-3). Each terminal branch is densely packed with mitochondria. Clusters of electron lucent vesicles occur just beneath the plasma membrane. These specialized endings are the sites where infrared transduction takes place (see Section III.B). A complex and extensive bed of capillary loops embraces the terminal nerve masses. This capillary bed is believed to have a dual function: (1) to provide a robust oxygen supply for the mitochondria in the receptor terminals, and (2) to act as a thermal exchanger to carry away excess heat and prevent the epithelial tissues from retaining temperature increases produced by infrared radiation (Amemiya *et al.,* 1999). The thermal exchange function of the capillary bed, combined with the low heat capacity of the thin epithelial tissues in the pit organs, may ensure that the sensory receptors are more ready to respond to the transient temperature increases that are expected to occur when an infrared source (e.g., prey) passes in front of the pit organ (discussed in Section III.B).

A. Innervation and Central Nervous System Pathways of Pit Organs

Pit organs are innervated by branches of the trigeminal nerve (reviewed by Molenaar, 1992). These same nerve branches also subserve mechanoreception, nociception, and thermoreception in the facial region of reptiles, mammals, and other vertebrates. Cell bodies for the infrared-sensitive nerve terminals are located in sensory ganglia of the trigeminal nerve.

In infrared-sensitive snakes, sensory neurons in the trigeminal ganglia synapse with neurons in two distinct nuclei located in the hindbrain—the nucleus of the solitary tract (NST) and a unique group of cells termed the nucleus of the lateral descending trigeminal tract (LTTD) (see Molenaar, 1992). Neurons in the NST subserve such functions as mechanoreception, thermoreception, and nociception, as in other vertebrates. However, the LTTD is found only in infrared-sensing snakes and is responsible for processing information specifically from the pit organs (Stanford *et al.,* 1981). This nucleus is present in all infrared-sensitive snakes studied thus far, and is absent from animals that are not infrared-sensitive.

Ultimately, infrared information from the LTTD reaches the optic tectum. Spatial relationships among the axons and synapses of the infrared receptor pathway are preserved such that a systematic topographical map of the receptive membrane in the pit organ is projected onto the tectum, much like the orderly retinotopic mapping of visual information that also occurs in the optic tectum. Indeed, the infrared sensory map overlies and is aligned with the retinotopic map. Individual neurons in the optic tectum may be excited by visual and infrared stimuli arising from approximately the same point in space[3] (Hartline *et al.,* 1978). Thus,

---

[1]Boid snakes that do not have obvious pit organs possess clusters of infrared-receptive neuronal terminals in these same labial areas.

[2]The oral cavity of rattlesnakes is also sensitive to infrared and possesses dense ramifications of infrared-receptive sensory nerve terminals, similar to those in pit organs (Dickman *et al.,* 1987).

[3]The two maps—visual and infrared—do not appear to overlap well in the peripheral fields of view. The overlap is more precise in the central field, that is, directly in front of the snake.

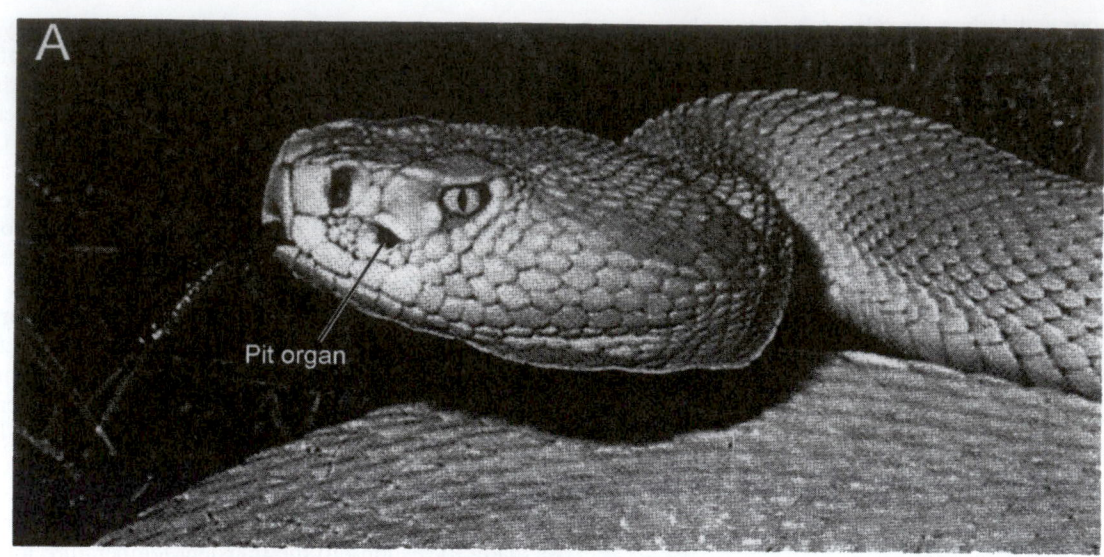

Pit organ

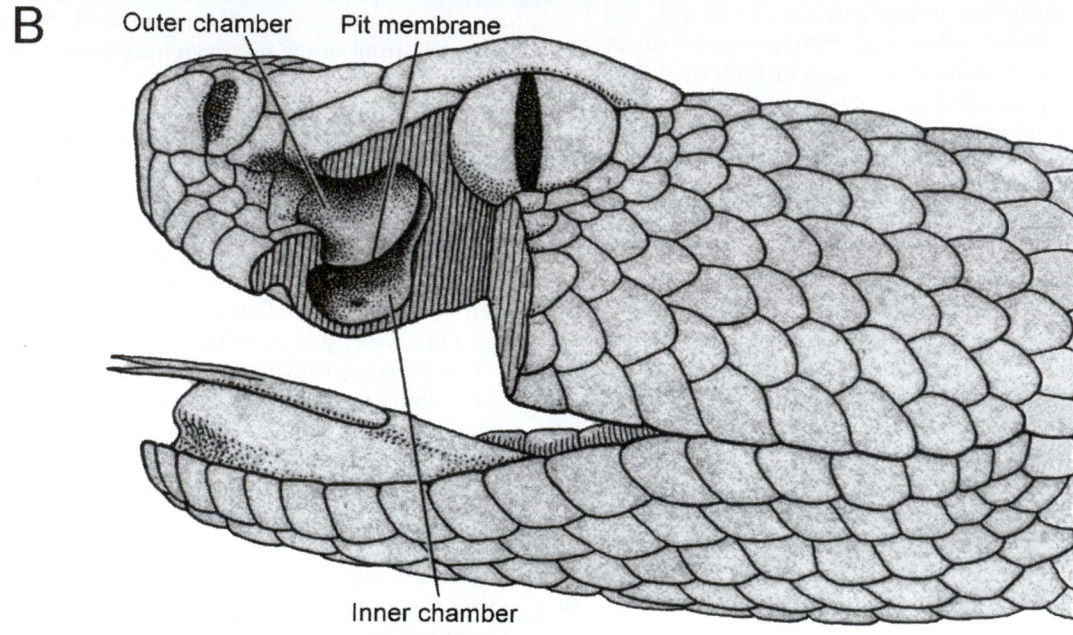

Outer chamber   Pit membrane

Inner chamber

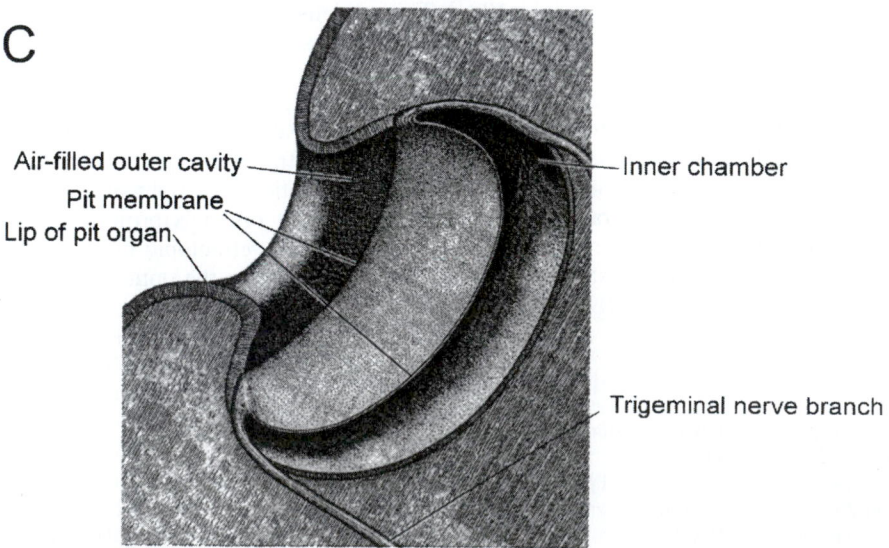

Air-filled outer cavity   Inner chamber

Pit membrane

Lip of pit organ

Trigeminal nerve branch

FIGURE A-2. Photograph of a rattlesnake striking at prey, demonstrating how infrared sensing receptors in the oral cavity may play a role in guiding the strike. Photograph by Pete Carmichael from *Florida's Fabulous Reptiles and Amphibians* (1991) World Publications, Tampa, FL.

snakes possessing infrared-sensitive pit organs view their environments simultaneously using two distinct regions of the electromagnetic spectrum.

B. Infrared Sensory Transduction

Infrared-detecting pit organs in snakes are exquisitely sensitive and selective (Bullock and Diecke, 1956; Gamow and Harris, 1973; de Cock Buning *et al.,* 1981; Newman and Hartline, 1982; Hartline, 1999). The adequate stimulus is a sharp thermal gradient, or contrast, in the surrounding environment, such as an endothermic animal (e.g., a mouse) against thermoneutral vegetation, or cool prey (e.g., a frog) against a cooler background (e.g., a pond). The temperature of the snake's body itself is not a signifi-

cant factor, nor is the average temperature of the surroundings. Air temperature is irrelevant; the pit organ responds only to radiant energy in the infrared, not conductive energy transferred by warm air. Thus, a snake is an effective hunter in the heat of midday as well as in the cool of the night. The sensitivity of pit organs is such that a pit viper can use its infrared detector to orient to and accurately strike a rat at a distance of about 1/2 meter. At this distance, a rat with a surface temperature 10 °C above background emits infrared radiation (~10 μm wavelength) that would be expected to warm the pit organ tissues by only 0.001 °C (Bullock and Diecke, 1956). This low threshold for temperature detection was also observed experimentally. Researchers infused water of different temperatures into pit organs and measured threshold nerve responses to

FIGURE A-1. Infrared pit organ in the rattlesnake. (A) Photograph of the head of a rattlesnake (*Crotalus viridis*) showing the distinctive pit organ between the nostrils and the eye. Photograph by Bill Love from *The World's Most Spectacular Reptiles and Amphibians* (1997) World Publications, Tampa, FL. (B) Schematic drawing of the internal structure of the infrared sensory organ in rattlesnake (*Crotalus viridis*). Reproduced from Gamow and Harris (1973). (C) Higher magnification schematic view of the rattlesnake pit organ. Reproduced from Newman and Hartline (1982).

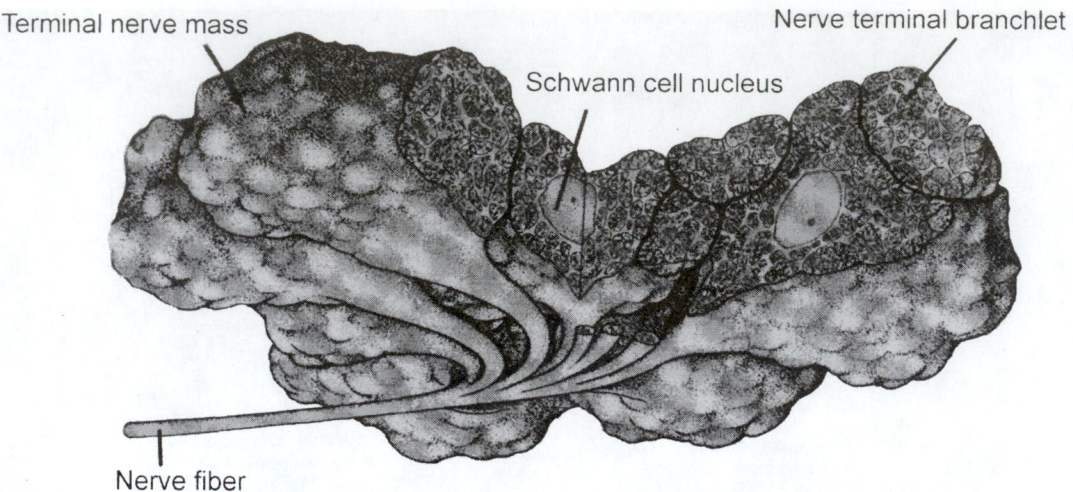

FIGURE A-3. Reconstruction of the nerve terminals for infrared sensing trigeminal sensory axons. The branches form nerve terminal masses that include Schwann cells and receptor terminals. The nerve terminals are filled with numerous mitochondria. Reproduced from Terashima *et al.* (1970).

temperature changes as small as 0.003 °C in 0.06s (Bullock and Diecke, 1956). Rapid temperature increases as small as 0.1 °C in the pit organ elicit robust responses (i.e., well above threshold detection) in the receptors (de Cock Buning *et al.*, 1981).[4]

How infrared radiation is transduced into neuronal excitation remains a mystery. It is unproven whether snake pit organ infrared receptors operate on a photochemical or thermal mechanism. That is, it is not yet known whether pit organ receptors are stimulated by specific wavelengths of infrared radiation, like photoreceptors in the retina, or whether the receptors are stimulated by temperature increases. The spectral sensitivity of snake pit organs has not been systematically investigated in much detail. Thus, the specific wavelengths that maximally excite infrared sensory terminals and therefore that might be correlated with the absorbance spectra of candidate phototransduction molecules are unknown at present. Infrared spectrometry experiments on intact pit organs in pythons show that the pit organs absorb in the ranges of 3–5 μm and 8–12 μm significantly more than do the surrounding tissues (Grace, *et al.*, 1999a). Further, the epidermal tissue overlying the IR receptor nerve terminals is essentially transparent to windows of the IR spectra centered at 4 μm and 10 μm. These results correlate with the peak emission (approximately 10 μm) of endothermic prey normally targeted by these snakes. However, the findings are not sufficiently precise to define any candidate

photoreceptive molecule(s) that absorb in a defined infrared action spectrum.

The argument against the notion that a photochemical reaction underlies infrared sensory reception is primarily that photons in the effective spectral region (~10 μm wavelength) have insufficient energy to produce molecular photoisomerization of any known transduction molecules. This may be true for retinaldehydes, which together with opsins form visual pigments. However, other molecular structures can be efficiently isomerized by infrared radiation, including wavelengths in the 10-μm region. For example, highly branched hydrocarbon dendrimers absorb infrared photons, leading to photoisomerization of an azobenzene core (Jiang and Aida, 1997). In principle, a similar mechanism could allow efficient infrared phototransduction in snake or other infrared receptors.

More likely, however, is the hypothesis that infrared receptors function as thermoreceptors. Indeed, infrared pit organ receptors are commonly termed *heat receptors* in the literature. This would imply that the sensory terminals are stimulated by all wavelengths of infrared radiation absorbed by the pit organ in proportion to the heat they generate in the tissues. The evidence that pit organs are temperature receptors includes the observation that responses to infrared stimulation can be mimicked by infusing warm water directly into the pit organ. More importantly, it is possible to calculate how much the temperature in the tissues surrounding the sensory terminals increases when infrared radiation at threshold intensity for eliciting responses is focused in the pit organ. This calculated temperature agrees well with the observed threshold temperature change determined experimentally by perfusing water of specified temperature into the pit organ (Bullock and Diecke, 1956).

Little is known about heat transduction in thermoreceptive sensory fibers. The only known thermal transduction molecules in animals are VR1, an ion channel that is activated by moderate temperatures (~43 °C, Caterina *et al.*, 1997), and

---

[4]Bullock and Diecke (1956) and de Cock Buning *et al.* (1981) alike point out that when factors such as the relative thickness of the overlying tissues, the low heat capacity of the thin pit organ membrane, and the density of innervation are taken into account, the threshold for snake pit organs is similar to that of other warm receptors, including human cutaneous thermoreceptors. One reason for the apparent higher sensitivity to small temperature increases in snakes is that the density of innervation in pit organs is very high and the epithelial tissue overlying the sensory nerve terminals is extremely thin. Thus, infrared radiation can readily and rapidly increase the temperature around a dense plexus of sensory nerve terminals.

VRL-1, a related heat-activated channel that responds to higher temperatures (52 °C, Caterina *et al.*, 1999). Low pH (acid) and capsaicin also stimulate VR1. VR1 is found in small-diameter polymodal nociceptor sensory nerve terminals. There is no indication that VR1 or VRL-1 are expressed in infrared sensory neurons, although this has not been specifically tested to date. However, given the high specificity of effective stimuli that excite infrared receptors and the extreme sensitivity of pit organs to very small thermal changes, it is unlikely that a polymodal, heat-transducing molecule such as VR1 or a high-temperature receptor like VRL-1 is involved.

Bibliography

Amemiya, F., Nakano, M., Goris, R. C., Kadota, T., Atobe, Y., Funakoshi, K., Hibiya, K., and Kishida, R. (1999). Microvasculature of crotaline snake pit organs: possible function as a heat exchange mechanism. *Anat. Rec.* **254,** 107–115.

Bullock, T. H., and Diecke, F. P. J. (1956). Properties of an infra-red receptor. *J. Physiol.* **134,** 47–87.

Caterina, M. J., Rosen, T. A., Tominaga, M., Brake, A. J., and Julius, D. (1999). A capsaicin-receptor homologue with a high threshold for noxious heat. *Nature* **398,** 436–441.

Caterina, M. J., Schumacher, M. A., Tominaga, M., Rosen, T. A., Levine, J. D., and Julius, D. (1997). The capsaicin receptor: a heat-activated ion channel in the pain pathway. *Nature* **389,** 816–824.

de Cock Buning, T., Terashima, S., and Goris, R. C. (1981). Crotaline pit organs analyzed as warm receptors. *Cell. Mol. Neurobiol.* **1,** 69–85.

Dickman, J. D., Colton, J. S., Chiszar, D., and Colton, C. A. (1987). Trigeminal responses to thermal stimulation of the oral cavity in rattlesnakes (*Crotalus viridis*) before and after bilateral anesthetization of the facial pit organs. *Brain Res.* **400,** 365–370.

Gamow, R. I., and Harris, J. S. (1973). The infrared receptors of snakes. *Sci. Am.* **228,** 94–100.

Grace, M. S., Church, D. R., Kelley, C., and Cooper, T. M. (1999a). The python pit organ: immunocytochemical and imaging analysis of a sensitive natural infrared detector. *Biosens. Bioelectron.* **14,** 53–59.

Grace, M. S., Woodward, O. M., Church, D. R., and Calisch, G. (2001). Prey targeting by the infrared-imaging snake python: effects of experimental and congenital visual deprivation. *Behav. Brain Res.* **119,** 23-31.

Hartline, P. H. (1999). Infrared sense. *In* "Encyclopedia of Neuroscience" (G. Adelman and B. H. Smith, Eds.). Elsevier, New York. pp. 957–959.

Hartline, P. H., Kass, L., and Loop, M. S. (1978). Merging of modalities in the optic tectum: infrared and visual integration in rattlesnakes. *Science* **199,** 1225–1229.

Jiang, D.-L., and Aida, T. (1997). Photoisomerization in dendrimers by harvesting of low energy photons. *Nature* **388,** 454–456.

Kardong, K., and Mackessey, G. (1991). The strike behavior of a congenitally blind rattlesnake. *J. Herpetol.* **25,** 208–211.

Molenaar, G. J. (1992). Anatomy and physiology of infrared sensitivity of snakes. *In* "Biology of the Reptilia" (C. Gans and P. S. Ulinski, Eds.). University of Chicago Press, Chicago. pp. 367–453.

Newman, E. A., and Hartline, P. H. (1982). The infrared "vision" of snakes. *Sci. Am.* **246,** 116–127.

Stanford, L. R., Schroeder, D. M., and Hartline P. H. (1981). The ascending projection of the nucleus of the lateral descending trigeminal tract: a nucleus in the infrared system of the rattlesnake (*Crotalus viridis*). *J Comp. Neurol.* **201,** 161–174.

Terashima, S., Goris, R. C., and Katsuki, Y. (1970). Structure of warm fiber terminals in the pit membrane of vipers. *J. Ultrastruct. Res.* **31,** 494–506

John G. New and Timothy C. Tricas

50

Electroreceptors and Magnetoreceptors

I. Introduction

That certain fishes could produce potent shocks has been known since antiquity, and shortly after the discovery of electricity it was determined that the shock from these animals was electrical in nature (see Wu, 1984). It was not, however, until the discovery in the 1950s that some fishes produce a much weaker electrical discharge, in the millivolt range, that the use of electrical signals for purposes other than attack or defense was considered (Lissman, 1958; Lissman and Machin, 1958). Because these weak electrical discharges offered a possible channel for communication among conspecifics, the necessity arose to determine the identity of the sensory receptors that detected them, and thus was born the study of electroreception. It has since been discovered that electroreception is more widespread among vertebrates than was initially thought, and an electrosense is possessed by many animals that lack electric discharge organs as well as by those that possess them.

Electroreception is a primitive vertebrate character, found in the common ancestor of jawless and jawed vertebrates (Bullock et al., 1983). This primitive electrosense is retained in numerous extant taxa (Fig. 1) including lampreys, chondrichthyan fishes, bichirs, sturgeon and paddlefishes, lungfishes, coelacanths, and nonanuran amphibians (salamanders and caecilians). Phylogenetic analysis of character traits indicates that the electrosenses of all of these animals are homologous, reflecting their common phylogenetic origin (Bullock et al., 1983; Bullock and Heiligenberg, 1986). Electroreception, however, has been largely lost in the teleost fish, the lineage to which most modern bony fishes belong. Most teleosts, together with their sister groups of gars and bowfin, which collectively comprise the Neopterygii, lack an electrosense. Remarkably, only two distantly related lineages of teleost fishes do possess electroreceptors, and it is apparent from character analysis that these have evolved independently. In addition, it has been recently discovered that two species of monotreme mammal, the platypus and the echidna, are also electroreceptive. Because electrosenses are not present in most other tetrapod

lineages, this clearly represents yet another "reinvention" of electroreception (Scheich et al., 1986; Gregory et al., 1989).

Electroreceptors in anamniotic vertebrates display clear affinities with the receptors of the mechanosensory lateral line. Both types of receptors probably develop from the same embryonic ectodermal placodes (Northcutt et al., 1995; Vischer, 1995), are innervated by the lateral line nerves, and possess either a kinocilium, microvilli, or both. In animals with primitive electrosensory systems, it is not known whether the electroreceptors are derived directly from lateral line receptors, or whether both were derived from a primitive mechanosensory cell (Bodznick, 1989; Jørgensen, 1989; Bullock and Heiligenberg, 1986). In teleosts, it is evident that electroreception has evolved as a specialization of the lateral line system. In contrast, the electrosense of mammals has evolved as a specialization of the trigeminal cranial nerve system since amniotes have lost the lateral line concurrent with the evolution of reproductive life histories that are not of necessity linked to aquatic environments (Andres and von During, 1988; Manger and Hughes, 1992).

All electroreceptors, whether primitive or derived, can be broadly classified as belonging to one of two categories: ampullary or tuberous (see Zakon, 1986a, 1988). **Ampullary receptors** are broadly tuned and respond to low-

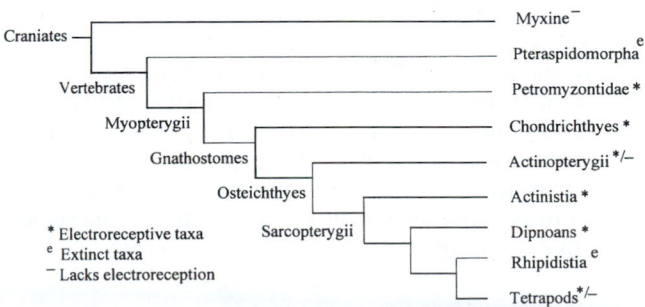

FIGURE 1. Cladogram illustrating the distribution of electroreception among craniates. Electroreception in tetrapods is found only in nonanuran amphibians and monotreme mammals.

frequency stimuli, with best frequencies generally between 0.1 and 20 Hz. This range of frequencies encompasses most bioelectric fields produced by aquatic organisms as well as those produced by nonliving physicochemical sources in aquatic environments (Peters and Bretschneider, 1972). Detection of these extrinsic sources by ampullary receptors is often referred to as electroreception in the **passive mode** (Kalmijn, 1974,1988). **Tuberous electroreceptors** are tuned to much higher frequencies, with best frequencies in the 0.1–1.0 kHz range. Tuberous organs are found only in fishes possessing weak electric organs of the sort initially described by Lissman (1958), and these receptors have best frequencies that correspond closely to the peak spectral frequency of the discharge of the animal's electric organ. The tuberous organs detect changes in the intensity and temporal pattern of the electric field produced across the body by the electric organ discharge (EOD), whether by the presence of items in the field of objects with different conductances than the water, or by the addition of the electric organ discharges of another individual. Such forms of electrodetection responding to alterations of the standing EOD are referred to as electroreception in the **active mode** (Kalmijn, 1988). In this chapter we consider the morphology and physiology of both ampullary and tuberous electroreceptors. Space limitations preclude a discussion of the central processing of electrosensory information, which has become one of the most engrossing stories in modern neuroethology. The reader is referred to several excellent recent reviews on this topic (Bullock and Heiligenberg, 1986; Zakon, 1988; Heiligenberg, 1991).

The ability to detect geomagnetic fields directly, or **magnetoreception**, has been described in a number of different taxa (for review, see Tenforde, 1989). These studies have been primarily behavioral; no studies exist concerning the physiology of magnetoreceptors or magnetoreception, and even the behavioral studies have been difficult to replicate or confirm (Moore, 1988). The direct identification of a magnetoreceptor organ in these taxa has itself been problematic; such organs are believed to be associated with localized deposits of magnetite (Fe_3O_4). Localized magnetite domains have been described in bees (Gould *et al.*, 1978), salmon (Kirschvink *et al.*, 1985), tuna (Walker *et al.*, 1984), turtles (Perry *et al.*, 1981), pigeons (Walcott *et al.*, 1979; Preston and Pettigrew, 1980), dolphins (Zoeger *et al.*, 1981), and humans (Baker *et al.*, 1983), but the organs, transduction mechanisms, and innervation have yet to be positively identified. A recent description of a putative magnetoreceptor in honeybees (Hsu and Li, 1994) has been disputed (Nesson, 1995; Nichol and Locke, 1995; Kirschvink and Walker, 1995). The exception is the description of magnetotactic behavior in magnetite-containing bacteria (Blakemore, 1975; Kalmijn and Blakemore, 1978). Magnetoreception, however, is believed to be possible in some electroreceptive taxa via the detection of electric fields induced by an animal's movement through a geomagnetic field (see discussion later).

II. Ampullary Electroreceptors

The wide phylogenetic distribution of ampullary electroreceptors among extant vertebrate taxa suggests that this class of electroreceptor has served important biological functions for hundreds of millions of years and has subsequently "re-evolved" several times (Fig. 1; see also Fig. 8). With the exception of weakly-electric fishes, which also possess tuberous electroreceptors, most species with ampullary electroreceptors lack electric organs. Thus, behaviorally relevant electric field stimuli for most species are thought to originate primarily from extrinsic sources. Ampullary electroreceptors are known to be important for the detection of prey (Kalmijn, 1971; Peters and Meek, 1973; Tricas, 1982), mates (Tricas *et al.*, 1995), and orientation to local inanimate electric fields (Kalmijn, 1982; Pals *et al.*, 1982). In addition, the ampullary electroreceptor system is theoretically capable of mediating navigation by detecting electric fields induced by movement of the animal through the earth's magnetic field (Kalmijn, 1974, 1982; Paulin, 1995), which would represent a form of electroreception in the **active mode**, as discussed earlier (Kalmijn, 1988).

A. Morphology

The ampullary electroreceptor organ in lampreys is known as an **end bud** (Fig. 2) and differs considerably in morphology from the ampullary electroreceptors found in cartilaginous and teleost fishes (Ronan and Bodznick, 1986). Each end bud consists of numerous support cells and 3 to 25 sensory cells in the epidermis that are in direct contact with the surrounding water. Individual receptor cells have numerous small microvilli on the apical surface but lack a kinocilium. Small groups or lines of end buds are distributed over the head and body surface with multiple buds being innervated by a single sensory lateral line nerve fiber (Ronan, 1986; Bodznick and Preston, 1983). Little is known about transduction in end buds, but the similarity of their responses to primitive gnathostome electroreceptors supports a homologous transduction mechanism (see discussion later). It is not known whether end buds represent the primitive electroreceptor state or whether they are a derived condition unique to the lampreys, which first appeared in their current form in the Carboniferous era considerably after the appearance of jawed fishes (Moy-Thomas and Miles, 1971).

The anatomy and physiology of primitive ampullary electroreceptor organs have been studied most extensively in the chondrichthyan fishes. Chondrichthyan fishes include the elasmobranchs (rays, skates, and sharks) and the rat fishes (Fields *et al.*, 1993), all of which possess ampullary electroreceptor organs of a similar morphology. In elasmobranch fishes, the electroreceptive unit is a prominent and highly specialized structure known as an **ampulla of Lorenzini** (Fig. 3). The ampulla proper in the marine skate is composed of multiple **alveolar sacs**, which share a common lumen (Waltman, 1966). The apex of each ampulla chamber is connected by a highly insulated **marginal zone** to a single subdermal **canal**, which is approximately 1 mm in diameter and terminates as a small epidermal pore. The canal wall is 1–2 µm thick and composed of two layers of flattened epithelial cells, which are separated by a basement membrane to which the lumenal layer is also united by tight junctions. Both the canal lumen and the ampullary chambers are filled with a K^+-rich, mucopolysaccharide, jelly-like matrix that is secreted by the superficial layer and has an ex-

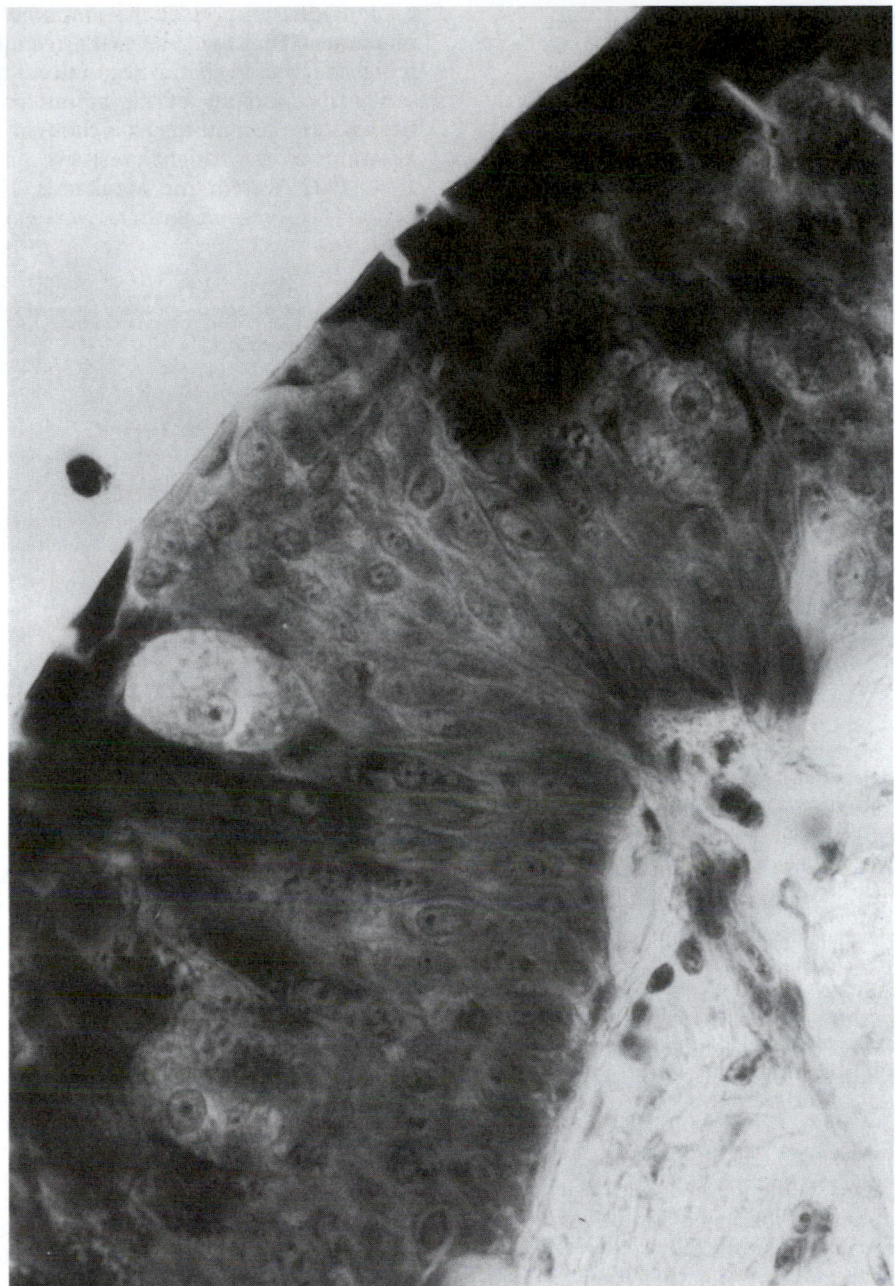

FIGURE 2. Light photomicrograph of the electrosensory end bud of the silver lamprey. (From Ronan 1986. Copyright © 1986 John Wiley & Sons. Reprinted by permission of John Wiley & Sons, Inc.)

tremely low resistivity (Murray and Potts, 1961; Waltman, 1966). These remarkable morphological and cytological features make each canal a low-resistance core conductor that connects the internal ampullary lumen with the surface pore. The sensory epithelium within the alveolus is composed of two cell types, which form a monolayer that is approximately 15 μm thick (Fig. 4). The vast majority of the alveolar surface is formed by accessory cells that are highly resistive to transmembrane currents and are bound together by tight junctions that prevent ionic leakage across the lumenal and basal surfaces of the epithelium. Interspersed among the

accessory cells are flask-shaped receptor cells (thought to be modified hair cells), which possess a single kinocilium on the apical surface and lack microvilli. This physical arrangement results in only a small fraction of the receptor cell surface being exposed to the ampullary chamber.

The basal membrane surface of the receptor cell forms a ridge seated in a postsynaptic invagination that is separated by a distance of 100–200 Å (Waltmann, 1966). A **synaptic ribbon** about 250 Å wide and 2 μm in length is located within the presynaptic ridge. A single layer of synaptic vesicles covers the ribbon, and exocytotic release of chemical

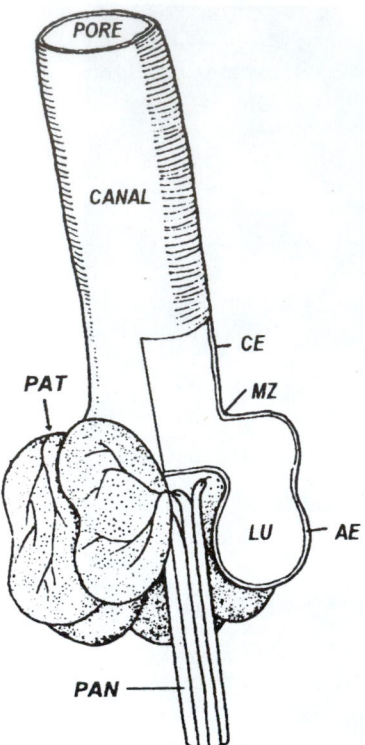

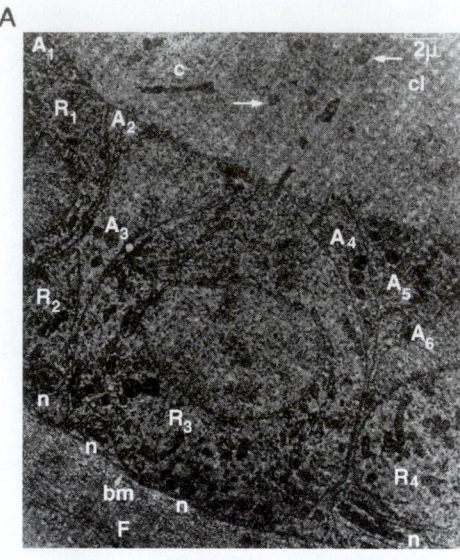

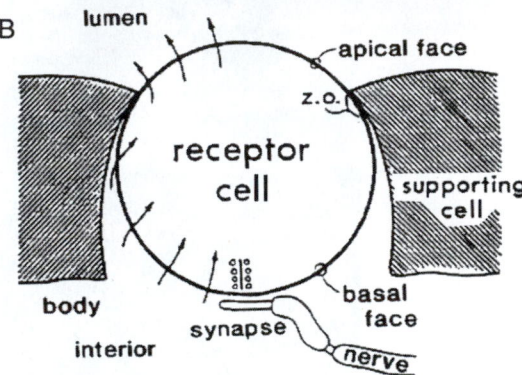

FIGURE 3. Ampulla of Lorenzini from the marine skate, *Raja*. The ampulla proper consists of multiple alveoli formed by the alveolar epithelium (AE). A high-resistance marginal zone (MZ) connects the sensory walls of the ampulla to the high-resistance canal epithelium (CE), which projects to the surface of the skin and terminates as a small pore confluent with the surrounding water. The ampulla lumen (LU) and canal are filled with a highly conductive jelly, which makes the lumen isopotential with voltages present at the pore. Myelinated primary afferent neurons (PANs) innervate the base of the ampullae and their unmyelinated primary afferent terminals (PATs) receive chemical excitation from the basal region of the sensory cells in the epithelial layer. (Modified from Waltmann, 1966.)

FIGURE 4. Receptor cell of the skate ampulla of Lorenzini. (A) Photomicrograph of flask-shaped receptor cells (R) and adjacent accessory cells (A) that are united by tight junctions to form the alveolar epithelium. A single kinocilium projects from each receptor cell into the lumen and, together with a small portion of the apical surface, is exposed to electric stimuli. Primary afferent neurons (n) innervate the basal portion of the receptors. The basement membrane (bm) and fibroblast cells (f) lie beneath the sensory epithelium. (From Waltmann, 1966.) (B) Diagrammatic representation of the receptor cell illustrating current flow during excitation. In the case of the elasmobranch, a cathodal (negative) stimulus relative to the basal region of the receptor excites the apical surface of the cell, which causes an increase in outward current flow (arrows). This in turn causes release of chemical transmitter at the cell synapse and excitation of the primary afferent. Anodal potentials in the lumen will decrease outward current flow at the apical surface and subsequently decrease the rate of transmitter release. (Reproduced from *The Journal of General Physiology*, 1972, vol. 60, pp. 534–557, by copyright permission of The Rockefeller University Press.)

neurotransmitter contained within these vesicles depolarizes the postsynaptic membrane of the innervating fibers of the anterior lateral line nerves. Unlike the hair cell receptors of the mechanosensory lateral line and octaval systems, all ampullary electroreceptors, both primitive and derived, lack efferent innervation.

Chondrichthyan fishes typically possess hundreds of ampullae that are associated in specific **regional clusters** that are closely bound by a dense matrix of connective tissue (Fig. 5A). From these clusters the subdermal canals radiate omnidirectionally and terminate in surface pores on the head, and on the enlarged pectoral disk of rays and skates. The multiple orientations of the receptor canals, and the copious distribution of the ampullary pores over the cephalic surface provide an extensive array of receptors with a high degree of spatial resolution.

The morphology of the ampullary electroreceptors in freshwater elasmobranchs is thought to reflect sensory adaptations to their highly resistive environment (Kalmijn, 1974, 1982; Raschi and Mackanos, 1989). The freshwater rays,

Pomatotrygon and *Dasyatis garouaensis*, have a hypertrophied, thick epidermis function to increase transcutaneous electrical resistance. The ampullary electroreceptors are greatly reduced in size and are referred to as **miniampullae** or **microampullae,** which are distributed individually across the skin rather than in clusters, and which have very

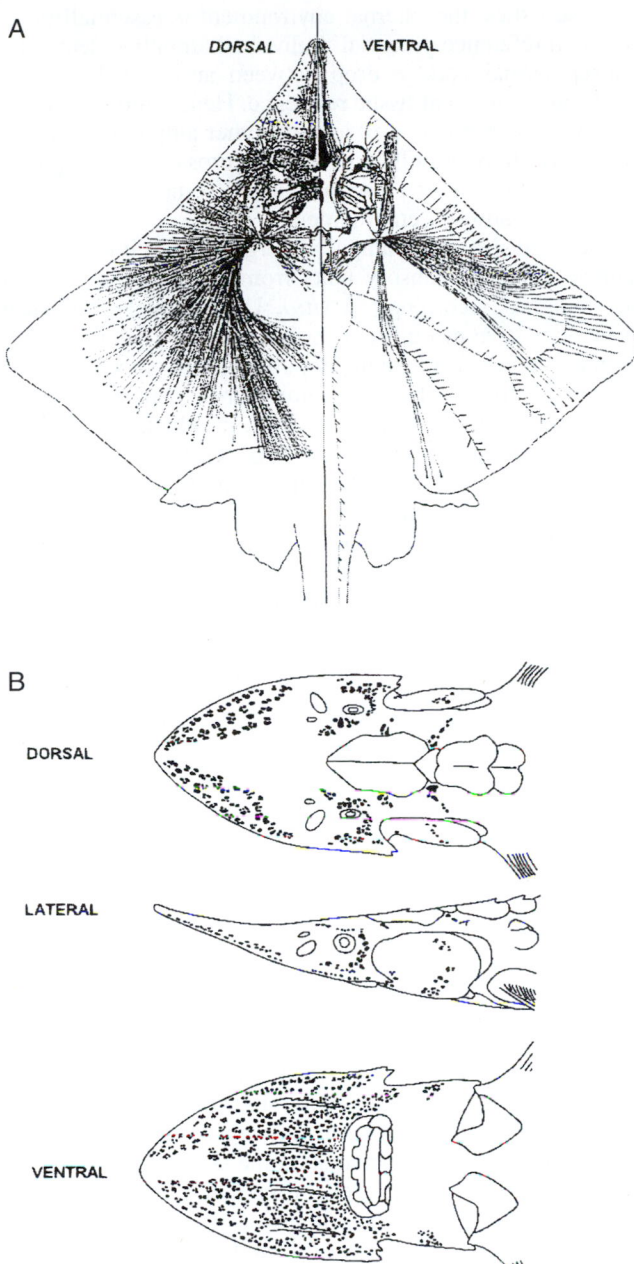

FIGURE 5. Distribution of ampullary receptors on nonteleost fishes. (A) In the skate, *Raja laevis,* ampullary canals originate at specific regional clusters and radiate in multiple directions. These canals function to measure voltage gradients that pass across the body of the animal. (Adapted from Raschi, 1986). (B) In the sturgeon, *Scaphirhynchus,* ampullary canals are extremely short and are not arranged in clusters. Instead they measure transcutaneous potential relative to the internal potential of the animal. (From Northcutt, 1986. Copyright © 1986 John Wiley & Sons. Reprinted by permission of John Wiley & Sons, Inc.)

short canals (about 0.3–2.1 mm long) that traverse the integument.

The anatomy and organization of ampullary electroreceptor organs in other nonteleost fishes and amphibians are generally similar to those of elasmobranch fishes, with which

they are believed to be homologous. Ampullary electroreceptors in chondrostean (sturgeon and paddlefishes), cladistian (bichirs), and dipnoan fishes (lungfishes) share in most respects a similar morphology among alveoli, canals, and ampullary pores. However, the ampullary organs in these fishes are most commonly arranged as single units (as opposed to large clusters) and are located immediately beneath the skin with very short (generally <0.25 mm) and small diameter (generally <0.14 mm) canals (Fig. 5B). The abundance of receptors are distributed in the head region, with the exception of the lungfishes, in which there are single ampullary electroreceptors on the head and small groups consisting of 3 to 5 ampullae scattered widely over the body (Pfeiffer, 1968; Northcutt, 1983). Ampullary electroreceptors of the head are innervated by a ramus of the anterior lateral line cranial nerve, while those on the body are innervated by a recurrent branch of the anterior lateral line nerve complex. In the bichir, *Polypterus* (Cladistia), there are about 1000 ampullae on the head region (Northcutt, 1986). Among the chondrostean fishes, electroreceptor organs in sturgeon, *Scaphirhynchus,* are arranged in about 1300 clusters of about 20 ampullae each (Northcutt, 1986), whereas in the related paddlefish, *Polyodon,* there are 50 000 to 75 000 ampullae on the elongate rostral "paddle," which are also arranged in small clusters (Nachtrieb, 1910; Jørgensen *et al.,* 1972). In the marine coelacanth, *Latimeria,* the "rostral organ" located between the eye and olfactory organ represents a complex of three principal canals that end centrally in small sensory crypts (Millot and Anthony, 1956), and is thought to be a homologous structure to the elasmobranch ampullae of Lorenzini (Bemis and Heatherington, 1982).

There is significant variability also in ampullary receptor cell morphology, particularly at the level of the apical membrane. Ampullary receptor cells in bichirs possess both a kinocilium and microvilli (Jørgensen, 1982), whereas those of chondrostean sturgeons (Teeter *et al.,* 1980) and paddlefish (Jørgensen *et al.,* 1972; Jørgensen, 1980) possess only a kinocilium as in the elasmobranchs. The receptor cells in lungfishes lack a kinocilium, but possess microvilli as in the jawless lampreys (Jørgensen, 1984). The receptor cells of the urodele amphibians (salamanders) are highly variable in morphology (Istenic and Bulog, 1984), whereas the tropical subterranean gymnophion have only microvilli (Fritsch and Münz, 1986). The primitive condition for electroreceptors is generally thought to be one possessing both kinocilium and microvilli (like other hair cells), but the reason for the loss of either kinocilium or microvilli in the various taxa and possible physiological ramifications is not known. All primitive electroreceptor cells also possess synaptic ribbons in the basal cell region, although some variation in synaptic morphology has been observed.

The fine structure of ampullary electroreceptors in teleost fishes closely resembles that of the freshwater elasmobranchs (Szabo, 1974) and other nonteleost taxa. These receptors, however, are not homologous to the primitive receptors but represent a case of parallel homoplasy, presumably the result of developmental and functional constraints necessary for the detection of extrinsic electric fields and their derivation from the hair cell receptors of the lateral line. The organs are located at the level of the basement membrane of the epidermis with a very short canal (usually about 200 μm) connecting to a pore on the skin surface. The cells of the inner walls of the ampulla and canal consist of three to five layers of flattened epithelial cells connected by tight junctions preventing current leakage across the canal wall, and the canal is filled with a conductive jelly that provides a low-resistance pathway through the lumen (Pfeiffer, 1968). The parallelism of ampullary electroreceptors among both primitive and derived groups is further demonstrated in the few existing species of electroreceptive marine teleosts. In the marine catfish, *Plotosus*, the ampullary canals have elongated, forming long subdermal tubules terminating centrally in alveolar clusters strikingly similar to the ampullae of Lorenzini in marine elasmobranchs (Obara, 1976). These teleosts can detect electric field stimuli at 80 μV/cm (Kalmijn, 1988), which is much more sensitive than the ampullary system of freshwater teleosts and approaches that of the elasmobranchs at 5 μV/cm (Kalmijn, 1982).

The ampullary receptor cells of teleosts are located in the base of the alveolus and are connected to the supporting cells via tight junctions, with only a small portion of their apical face exposed to the lumen. Teleost electroreceptor cells generally possess only microvilli; kinocilia are known only in *Xenomystus* (Notopteridae). The synaptic structure of receptors in more recently derived fishes is similar to those of the more primitive species, in which synaptic ribbons and presynaptic membrane evaginations are surrounded by a prominent postsynaptic "cup" (Szabo, 1974). Unlike primitive ampullary electroreceptor cells, ampullary receptors in teleosts may be innervated by either anterior or posterior lateral line nerves, depending upon location on the body surface. Like most other nonteleost fishes, the ampullae are distributed widely over the head, but differ in that they are usually distributed across the trunk in distinct patterns that are species specific.

B. Physiology

In marine elasmobranchs such as the skate, *Raja*, the resistance of the skin is only moderately higher than that of the body tissues (Kalmijn, 1974). Imposition of a uniform extrinsic field results in a voltage gradient across the skin and internal tissues of the body. In contrast, the ampullary canals act as **core conductors** via the highly conductive environment of the conductive jelly that is bounded by the high-resistance canal walls that lead into the ampulla. The high resistance of the alveolar surface to current leakage results in the lumen being effectively isopotential with the voltage at the ampullary pore on the skin. The electroreceptor cell potential varies as a function of the voltage drop across the apical membrane surface and the basal surface, which is at a different potential due to field gradient in the fish's body. In effect, the receptor measures the voltage drop of the field gradient along the length of the canal, and thus in a uniform field the longer the canal the greater the response of the receptor.

In freshwater elasmobranchs such as *Potamotrygon*, the resistance of the skin is relatively high compared to marine species, and the resistance of the internal tissues is relatively low, presumably as a result of osmoregulatory constraints.

In these fishes, the internal environment is essentially at a common reference potential. Individual ampullae detect the transepidermal voltage drop between an applied external field and the internal tissue reference. Hence, most freshwater elasmobranchs, as well as most other ampullary-bearing taxa, have short ampullary canals that cross only the epidermis. The majority of these taxa spend some or all of their lives in a freshwater environment.

The large size of the ampullary organ makes it possible to remove complete sensory units from the fish and conduct physiological recordings *in vitro*. Unfortunately, technical difficulties have made it impossible to obtain detailed intracellular records from single ampullary electroreceptor cells. The membrane biophysics of ampullary receptor excitation is best described for the skate, *Raja*, in which the receptor field activity was recorded as a voltage potential near the sensory epithelium within the ampulla (Obara and Bennett, 1972; Bennett and Clusin, 1978). In unstimulated electroreceptors there exists a depolarizing **bias current** caused by a standing Ca^{2+} conductance across the apical cell membrane. This resting depolarization results in a steady synaptic release of neurotransmitter at the basal cell surface and depolarization of the adjacent postsynaptic afferent nerve fiber. As a result, ampullary electrosensory primary afferent neurons are typically characterized by a regular discharge pattern. Modulation of the bias current by imposition of an extrinsic electric field results in changes in afferent fiber activity (Fig. 6). Presentation of an anodal stimulus hyperpolarizes the apical membrane and decreases the Ca^{2+}-generated bias current, inhibiting neurotransmitter release. Presentation of a relatively strong cathodal stimulus results in an increase in the Ca^{2+} conductance and a depolarization of the apical membrane. This depolarization results in the activation of a large inward Ca^{2+} conductance in the apical face, followed by activation of an inward Ca^{2+} conductance in the basal face, which increases the release of neurotransmitter and excites the primary afferent neuron. A subsequent outward K^+ current in the basal face results in repolarization of the receptor (Bennett and Clusin, 1978). Thus, changes in the polarity and intensity of the electric potential at the skin pore (and apical surface of the receptor cell) over time will modulate the resting discharge pattern of the primary afferent neuron.

Cathodal excitation and anodal inhibition of the homologous ampullary receptors has also been described in nonteleost fishes and amphibians, presumably through similar mechanisms. However, the ampullary electroreceptors in teleosts exhibit a reverse response pattern in that they are excited by anodal and inhibited by cathodal stimuli. The major difference between primitive and derived electroreceptors is that the apical surface of the teleost receptor is not excitable but has a very low electrical resistance, possibly due to a greatly increased surface area provided by the numerous microvilli. Anodal stimulation therefore results in a direct depolarization of the basal membrane, activating an inward Ca^{2+} conductance across the basal membrane and resulting in an increased release of neurotransmitter. Conversely, a negative potential applied to the apical surface of the receptor will directly hyperpolarize the receptor potential at the basal membrane and decrease the rate of transmitter release (Bennett, 1971a; Bennett and Obara, 1986).

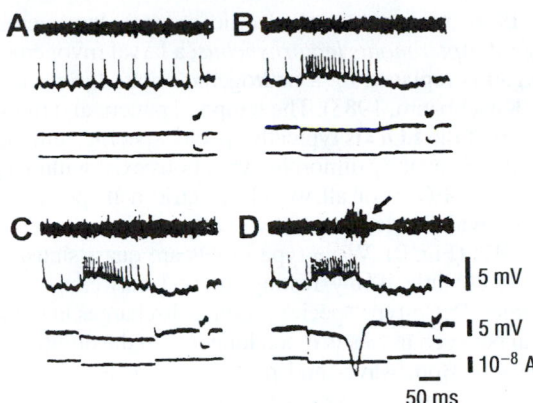

FIGURE 6. The response of ampullary receptor currents and primary afferent neurons to cathodal stimuli. The upper traces in each panel show compound action potentials recorded from the mandibular nerve. Second traces show discharges of a single primary afferent. Third traces show the potential recorded in the lumen of the ampulla and represent the microphonic record sum for all electroreceptor cells in the ampulla. Bottom traces show the excitatory current pulse injected into the canal. (A) Spontaneous discharge activity of the primary afferent without an applied extrinsic stimulus shows a regular discharge pattern. (B) Application of a weak negative current pulse evokes an excitation of the primary afferent discharge that slowly adapts. Note that the ampullary potential follows the stimulus current, the single unit activity ceases for a brief time following the end of the stimulus, and that little change is evident in the compound action potentials recorded from the whole mandibular nerve. (C) Application of a slightly stronger stimulus evokes a similar response from the single unit and a weak response from the whole nerve. (D) A similar stimulus can also produce a depolarizing **receptor spike,** which excites other ampullary receptors as seen in the whole nerve recording, but which has little additional excitatory effect on the single unit. Note that all primary afferent discharges appear to cease for a brief period during repolarization of the cell membrane. (Reproduced from *The Journal of General Physiology,* 1972, vol. 60, pp. 534–557, by copyright permission of The Rockefeller University Press.)

The high sensitivity of electrosensory primary afferent neurons was first established at a voltage gradient of about 1 µV/cm (Murray, 1962) and has recently been extended to near 20 nV/cm by Tricas and New (1997). The neural response to a prolonged, constant current field is sustained for a duration of a few seconds before it begins to adapt back to the resting discharge rate. Prolonged, constant stimulation results in a return to resting levels and accommodation of the receptor, resulting in no change in the overall sensitivity of the receptor (Bodznick *et al.,* 1993). Work on a variety of species with both primitive and derived ampullary electrosenses shows a maximum response to sinusoidal electric fields at frequencies of 1–10 Hz (Fig. 7) (Andrianov *et al.,* 1984; New, 1990; Montgomery, 1984; Peters and Evers, 1985; Tricas *et al.,* 1995). Sensitivities of primary afferent fibers innervating ampullary electroreceptors to a sinusoidal uniform field are 0.9 spikes per second per µV/cm for the little skate, *Raja erinacea* (Montgomery and Bodznick, 1993), 4 spikes per second per µV/cm for the thornback guitarfish, *Platyrhinoidis triserata* (Montgomery, 1984), and 24 spikes per second per µV/cm average for the round stingray, *Urolophus halleri* (Tricas and

New, 1998). The ampullary electroreceptors of other freshwater taxa are less sensitive to field stimuli with neural thresholds in the range of 10–100 µV (Teeter *et al.,* 1980; Teeter and Bennett, 1981).

Recordings from the lateral line nerve in the behaving dogfish (Dijkgraaf and Kalmijn, 1966) and single units from unparalyzed *in vivo* preps (Akoev *et al.,* 1967a,b) show that the regular discharge of primary afferent neurons is modulated in rhythmic bursts that are in phase with the ventilatory movements of the fish. This reafferent neuromodulation is explained by the standing (DC) bioelectric field that arises from the differential distribution of ionic charges in the animal (reviewed by Kalmijn, 1974, 1988), which in the skate is a result of both diffusion potentials and osmoregulatory ion pumping at the gills (Bodznick *et al.,* 1992). The modulation of this standing field occurs as the animal opens and closes the mouth, gills, or spiracles during the ventilatory cycle, which changes the resistance pathway between the animal's internal tissues and surrounding seawater (Bodznick *et al.,* 1992). The resultant transcutaneous potential is the source of electrosensory self-stimulation or **ventilatory reafference** (Montgomery, 1984), by which a change in the internal potential of the animal (and basal regions of the ampullary receptor cells) proportionately modulates the regular discharge of all primary afferent neurons. Thus electrosensory receptors and primary afferents exhibit common mode noise, which has important implications for central processing of electrosensory information (New and Bodznick, 1990; Bodznick and Montgomery, 1992).

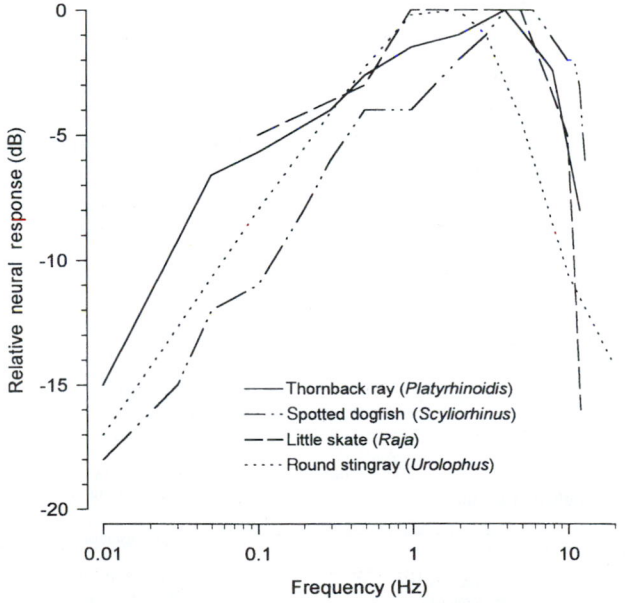

FIGURE 7. Frequency response of ampullary electrosensory primary afferent neurons in four species of elasmobranch fishes. As a group, the most sensitive frequencies range from 1 to 10 Hz, but individually some species may show distinct sensitivity peaks within this range. (Data modified for *Platyrhinoidis* from Montgomery, 1984; for *Scyliorhinus* from Peters and Evers, 1985; for *Raja* from New, 1990; and for *Urolophus* from Tricas *et al.,* 1995.)

III. Tuberous Electroreceptors

As is the case for derived ampullary electroreceptors, **tuberous electroreceptors** appear in separate and distantly related families of teleost fishes, indicating that they have evolved independently in each lineage (Bullock *et al.*, 1983; Zakon, 1986a, 1988). Among the Osteoglossomorpha, members of two families of African fishes, the Mormyridae and Gymnarchidae (superfamily Notopteroidea) (Nelson, 1994), possess both ampullary and tuberous electroreceptors; a sister group, the Xenomystinae, possesses only ampullary electroreceptors (Fig. 8) (Braford, 1986). In the superorder Ostariophysi, the siluriforms (catfishes) and South American gymnotiforms are sister groups. Catfish possess only a low-frequency ampullary system, but gymnotiforms possess both ampullary and tuberous systems. The distribution of electric senses in bony fishes supports the hypothesis that the high-frequency tuberous system is derived from the low-frequency sensitive ampullary system. Tuberous electroreceptors are found only in animals also possessing ampullary organs and weak electric organs, which have coevolved with the tuberous organs. Both taxa possessing tuberous organs have sister groups that lack tuberous organs but possess ampullary organs, which are themselves presumably derived from the neuromasts of the mechanosensory lateral line. Despite the considerable phylogenetic distance separating the two lineages, both groups show remarkable similarities in morphology and physiological response properties of the various types of tuberous receptors, presumably the result of fundamental constraints associated with the analysis of high-frequency electric organ discharges (Zakon, 1986a, 1988).

Tuberous electroreceptor organs occur only in fishes possessing **electric organs**. These specialized organs produce weak (millivolts to a few volts), high-frequency (100–1000 Hz range) discharges, to which the receptors are tuned. The electric organ-tuberous electroreceptor system therefore represents a distinct sensory channel, isolated from the low-frequency electrical signals produced by inanimate and biological sources in aquatic environments. These electric organs are usually derived from specialized muscle (see Bennett, 1971b, and

Bass, 1986, for reviews of electric organs), however in the gymnotid, *Apteronotus leptorhynchus*, a larval myogenic electric organ is replaced by a neurogenic organ during development (Kirschbaum, 1983). The temporal pattern and frequency spectrum of the EOD is typically species specific, although individual and sexually dimorphic variations exist within a given species. The EODs of all weakly electric fish species can be broadly divided into one of two categories; wave- or pulse-type EODs (Fig. 9). Wave-type EODs are quasi-sinusoidal in nature, with little if any resting interval between successive discharges. Pulse-type species produce discharges in which the interval between pulses is much longer than the duration of the pulse itself. Both wave- and pulse-type species are found in each of the lineages possessing tuberous organs, and have evolved independently within each taxon.

The EOD serves both an **electrolocation** and a **communicative** function in the animal's behavioral repertoire. The EOD generates an electric field, to which the tuberous electroreceptors are sensitive, in the water surrounding the animal. If an object of greater or lesser conductivity than the water is placed within the field, and spatial configuration of the field is altered, the resulting change in potential at the skin will be detected by the receptors (Fig. 10). Tuberous receptors allow animals to detect objects in their immediate vicinity and are especially useful in situations where other sensory systems, such as vision, are of limited usefulness in the turbid waters in which these animals generally live. In addition, the EOD produced by one animal may be detected by conspecifics; such electrocommunication plays important roles in the lives of weakly electric fishes (for reviews, see Bullock and Heiligenberg, 1986; Heiligenberg, 1977, 1991).

IV. Gymnotid Tuberous Electroreceptors

A. Anatomy

Gymnotid tuberous receptors exhibit a fairly uniform morphology despite the existence of several different functional subtypes of receptor. The organ consists of a roughly spherical chamber located in the epidermis, which communicates with the external environment via a short canal ending in an epidermal pore (Fig. 11A) (Szabo, 1965, 1974). The walls of the canal and capsule are composed of numerous (up to 50) layers of extremely flattened epithelial cells (Szamier and Wachtel, 1970). The many layers of cell membranes in series result in a decreased capacitance of the canal walls and less shunting of high frequencies across the walls of the canal and capsule than in low-frequency sensitive ampullary organs (Bennett, 1971a). Typically, the canal of gymnotid electroreceptors is filled with a plug of loosely packed epithelial cells. Directly beneath the plug and extending across the lumen of the capsule is a layer of covering cells. The cells are joined to each other and to the walls of the lumen via tight junctions (Wachtel and Szamier, 1966; Szamier and Wachtel, 1970; Szabo, 1974; Zakon, 1984a). The layer of covering cells maintains a constant ionic environment within the lumen of the capsule and also provides capacitive coupling of the receptor cells to the external environment.

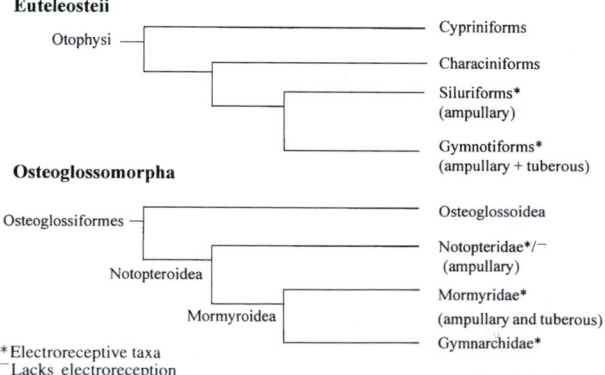

FIGURE 8. Cladogram illustrating the distribution of electroreception in the two distantly related taxa of teleost fishes possessing ampullary electroreceptors or both ampullary and tuberous electroreceptors.

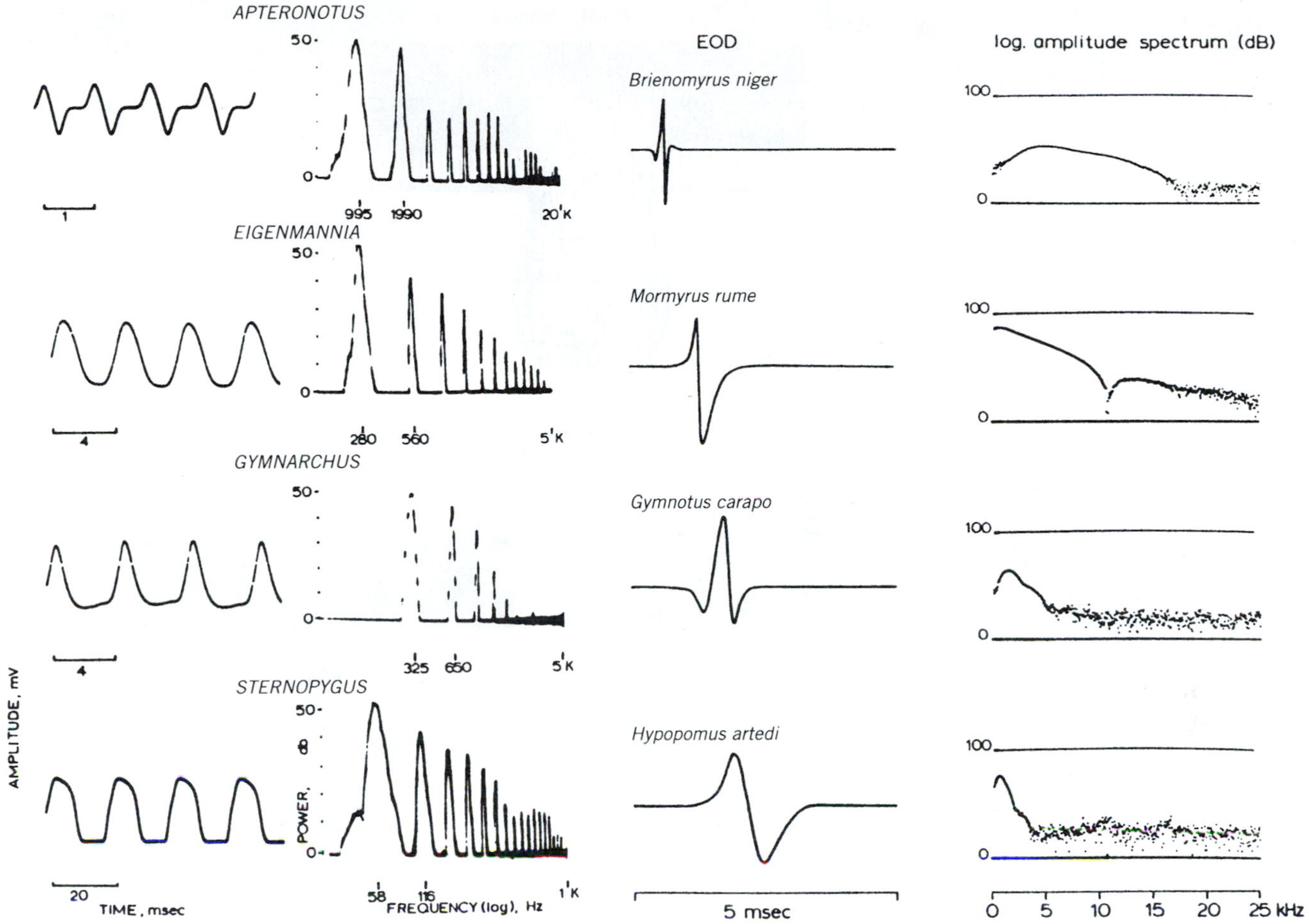

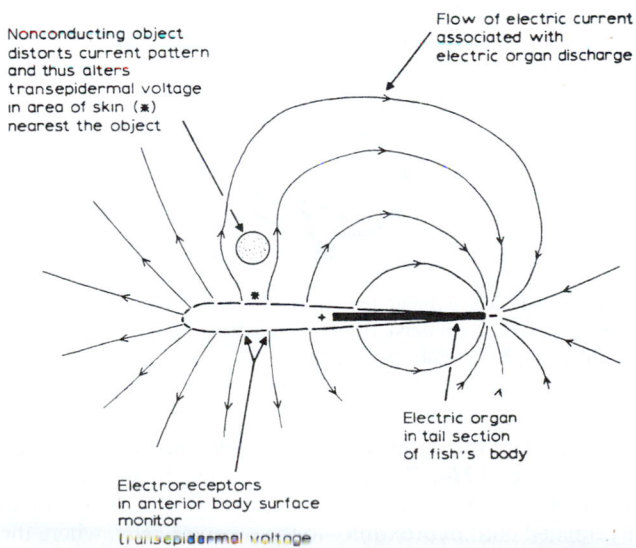

FIGURE 9. Electric organ discharges and corresponding power spectra produced by several wave- and pulse-type weakly-electric teleost fishes. EODs were recorded via a positive electrode placed in front of the head and a negative electrode behind the tail. (Reprinted from Heiligenberg, 1977, with permission.)

FIGURE 10. The principle of electrolocation. The black bar depicted within the caudal portion of the fish's body indicates the position of the electric organ. (Reprinted from Heiligenberg, 1977, with permission.)

The receptor cells themselves are located in the base of the capsule. Unlike ampullary receptors, they are attached to the base of the chamber only at the basal portion of the receptor cell's membrane, leaving most of the cell's membrane exposed to the lumen. The basal portion of the cell membrane is electrically isolated from the lumen via tight junctions with the supporting cells in the lumen (Wachtel and Szamier, 1966; Szamier and Wachtel, 1970). Generally there are 20 to 30 receptor cells per organ; however, some species may possess as many as 100 cells in certain specialized receptors (Szabo, 1965; Echague and Trujillo-Conez, 1981; Zakon, 1986b, 1987). The receptor cells are typically 20–30 μm long, with numerous apical microvilli and large numbers of mitochondria evident in the apical region of the cell (Wachtel and Szamier, 1966; Szabo, 1974). All of the receptor cells within a tuberous organ are innervated by a single afferent nerve fiber, and a given fiber may innervate either a single organ or several organs together. In the latter case, the organs form a distinct *rosette*, which is the result of division of a single organ with growth. Physiological experiments support this finding: The receptive field of an afferent fiber is centered on an individual tuberous organ (Wachtel and Szamier, 1966; Bennett, 1967; Szabo, 1974;

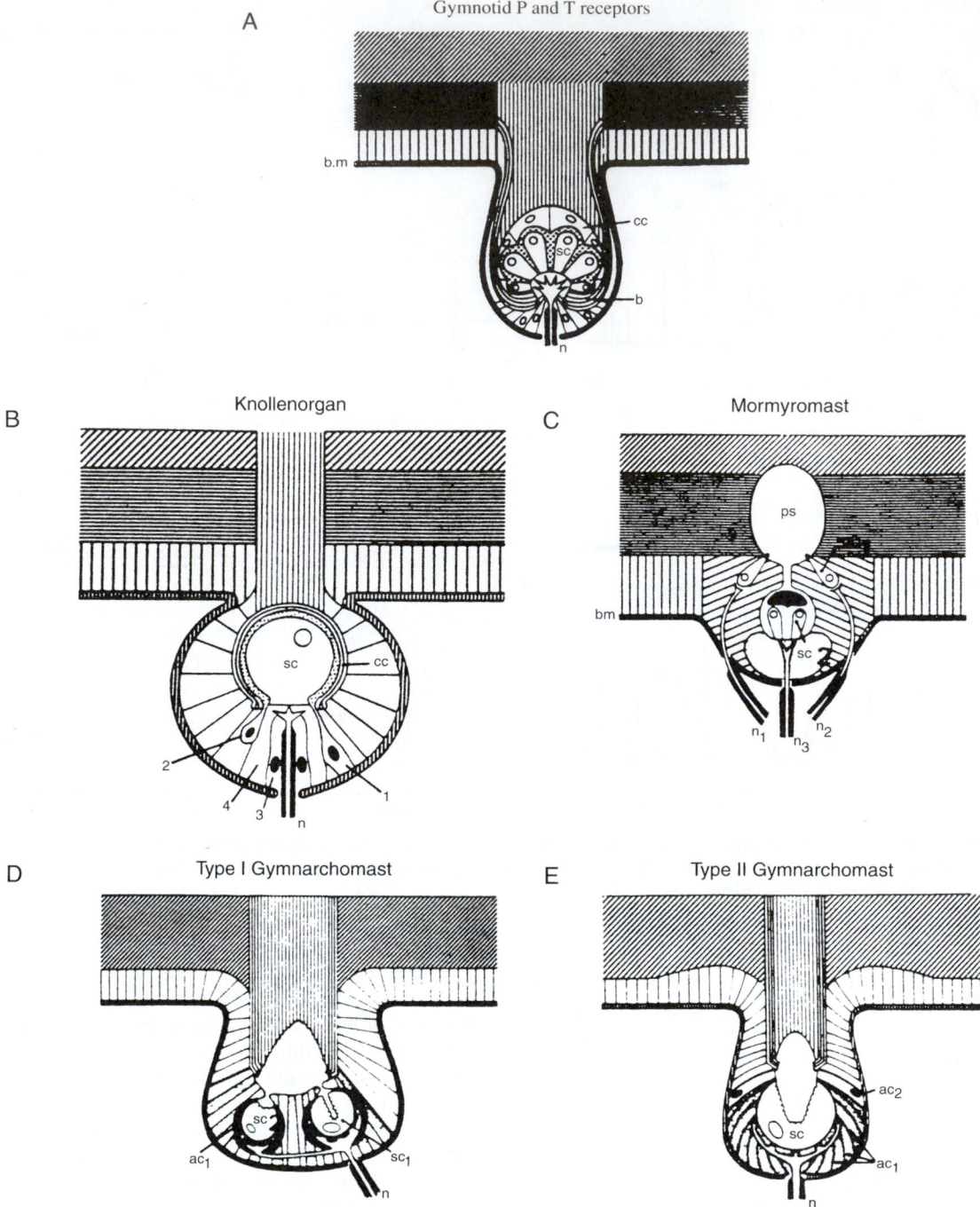

FIGURE 11. Tuberous electroreceptor organs of (A) gymnotid, (B,C) mormyrid, and (D,E) gymnarchid weakly-electric fishes. Abbreviations: ac1, ac2, accessory cells; b, capsule wall; bm, basement membrane; cc, covering cells; n#, afferent nerve fibers; ps, perisensory space; sc#, sensory cells. Numbers indicate different cell types within a given organ. (Reprinted from Szabo, 1974, with permission.)

Baker, 1980; Bastian and Heiligenberg, 1980; Zakon, 1984a, 1987). Variations in the morphology of the afferent fibers and the terminal boutons have served as a basis for anatomical classification and correspond to functional differences in the organ's physiological responses. Most tuberous organs in pulse-type fish possess an afferent fiber, which loses its myelin sheathing on entering the capsule and which innervates the individual receptor cells via numerous narrow

branches (B units, type II of Szabo) (Wachtel and Szamier, 1966; Szabo, 1974; Zakon, 1984a). Other tuberous organs are innervated by markedly thicker fibers, which remain myelinated until in proximity to the receptor cells, where the fiber swells to form a large postsynaptic knob (M units, type of Szabo) (Szabo, 1970, 1974; Szamier and Wachtel, 1970; Srivastava, 1973; Echague and Trujillo-Cenoz, 1981; Zakon, 1987). The synapses of tuberous ampullary organs in gymnotids are similar in appearance to those of ampullary receptors, with a presynaptic evagination of the receptor cell surrounded by the postsynaptic membrane of the afferent fiber. As in ampullary receptors, a synaptic ribbon is also evident at the presynaptic site in the receptor cell (Echague and Trujillo-Cenoz, 1981).

B. Physiology

Intracellular recordings from individual tuberous electroreceptors have yet to be successfully conducted, and a considerable amount remains to be learned about the physiology of transduction in these receptors. They appear to function, at least in principle, in a manner similar to that of teleost ampullary receptors, by measuring the voltage drop across the receptor cell's basal membrane surface between the internal potential of the receptor cell and the internal "reference" potential of the animal (Bennett, 1971a). The layer of covering cells that isolates the internal capsule of gymnotiform tuberous receptors from the external environment provides capacitance coupling between the two environments and acts as a high-pass filter (Fig. 12) (Bennett, 1971a). The numerous microvilli on the apical membrane surface greatly increase the surface area of the cell, which increases the input capacitance of the cell and decreases the input resistance. Bennett has suggested that the apical membrane face is electrically inexcitable and that the high blocking capacitance and low resistance of the apical face result in a membrane time constant permitting passage of frequencies in the range of the EOD. Bennett (1971a) has further suggested that for maximum sensitivity the resistance across

the excitable basal membrane face should be higher than that across the apical membrane so that the larger voltage drop is across the former.

Tuberous receptor organs generally exhibit sharper tuning characteristics than are suggested simply by the high-pass filter characteristics of their morphology alone. Active mechanisms on the basal membrane of the receptor cells have been suggested as the basis for the frequency response characteristics (Scheich *et al.,* 1973; Bastian, 1976; Scheich, 1977; Viancour, 1979a,b). Tuberous receptors produce oscillatory potentials on stimulation: The frequency of this "ringing" is the same as the best frequency of the cell and appears to be based on inward Ca^{2+} and outward K^+ currents (Zipser and Bennett, 1973; Zakon, 1984b). The tuning properties of the cell could thus be a function of the kinetics of these two currents at different stimulus frequencies.

Although the general processes of transduction in tuberous organs may be said to be similar, different populations of gymnotid tuberous organs display varying responses to elements of identical stimuli. Although a number of complex and conflicting classification schemes have been suggested to identify functional types of gymnotid tuberous electrosensory organs, we follow the classification originally proposed by Hopkins (1983) and adopted by Zakon (1986a) of two general types of tuberous organs based on their responses to EOD-like stimulation. The first group may be said to be **rapid-timing units,** which show a tightly phase-locked response to presentation of the EOD signal. The response of the receptor is remarkably precise: Single spikes are phase locked to each cycle of the EOD, exhibiting very little timing jitter (<100 µs). Rapid-timing units do not adapt and are typically sharply tuned. The second class of receptors is **amplitude-modulated units,** which demonstrate poor phase-locking of the response to the EOD but which demonstrate either an increase in the probability of a spike being produced or an increase in the number of spikes produced with increases in amplitude of the EOD stimulus. These units adapt readily and are generally more broadly tuned to lower frequencies than are rapid timing units. Thus the first category of cells encodes information on the timing of when an EOD pulse occurs, whereas the second provides information on changes in amplitude of the EOD over time. Rapid timing and amplitude-modulated units are observed in species producing both wave- and pulse-type EODS. In wave-type species such as *Eigenmannia, Sternopygus,* and *Apteronotus,* rapid-timing units have been termed **T units** (timing or phase-coding) and amplitude-modulated receptors are called **P units** (probability coding) (Fig. 13) (Scheich *et al.,* 1973; Feng and Bullock, 1977; Scheich, 1977; Bastian and Heiligenberg, 1980). The principal physiological difference between these receptor types appears to be the threshold of excitation of the receptors. Both receptor subtypes demonstrate similar widths of their dynamic ranges (approximately 20 dB) and demonstrate poor phase-locking responses at stimulus intensities that are just suprathreshold, as well as demonstrating a 1:1 following of the stimulus at intensities that are of saturating intensity. However, the stimulation threshold and saturation intensities of T units are generally 15–20 dB lower than those of P units; hence at normal EOD intensities, T units are saturated

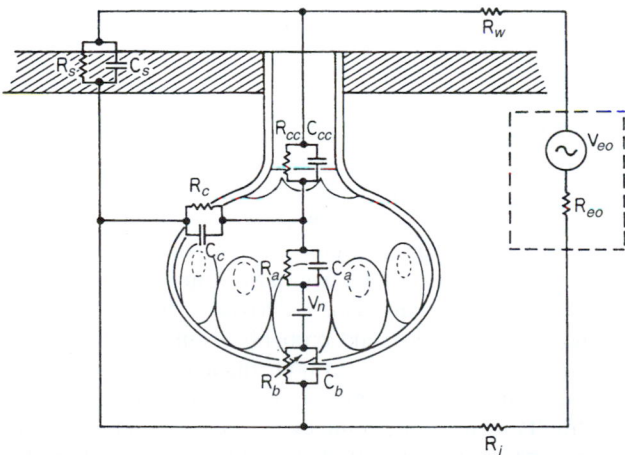

FIGURE 12. Electrical equivalent circuits of tuberous electrical organs. Abbreviations: C_a, capacitance of receptor cell apical membrane; C_b, capacitance of receptor cell basal membrane; C_c, capacitance of canal and capsule wall; C_{cc}, capacitance of covering cells; C_s, capacitance of skin; R_a, resistance of receptor cell apical membrane; R_b, resistance of receptor cell basal membrane; R_c, resistance of canal and capsule wall; R_{cc}, resistance of covering cells; R_{eo}, internal resistance of the electric organ; R_i, internal resistance of fish; R_s, resistance of skin; R_w, resistance of the water; V_{eo}, internal voltage of the electric organ; V_n, resting potential of receptor cells. Resistances and capacitances of supporting cells are not indicated and are believed to be passive. (Modified from Bennett, 1967, and Zakon, 1988.)

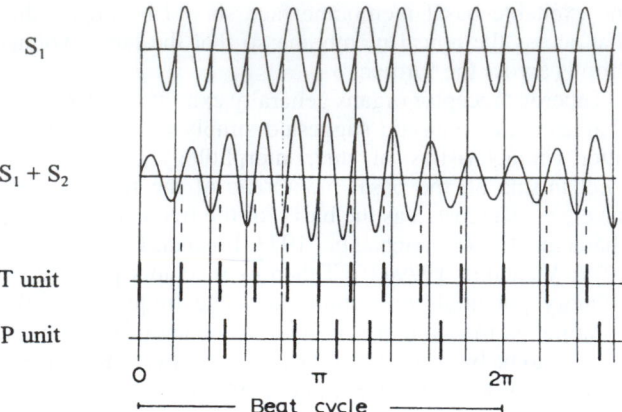

FIGURE 13. Schematic representation of the responses of gymnotid rapid-timing units (T units) and amplitude-modulated units (P units) to modulation of the EOD. The EOD of the fish (S_1) is modulated by the addition of a conspecific's EOD of a slightly different frequency ($S_1 + S_2$), creating a beat envelope. T units are tightly phase-locked to the positive zero-crossing of the modulated EOD waveform. P units are not phase-locked but exhibit an increased probability of firing as the amplitude of the EOD beat envelope increases. (Reprinted from Heiligenberg, 1986, with permission.)

and following at 1:1, whereas P units are responding in their dynamic range, generating greater or lesser spike frequencies as a function of EOD amplitude (Scheich *et al.*, 1973; Feng and Bullock, 1977). Similar units are observed in species producing pulse-type EODs such as *Hypopomus* and *Gymnotus*. **Pulse marker units** (M units) are similar to T units in their responses, producing a strongly phase-locked response to the EOD, whereas **burst duration receptors** (B units) respond to increasing EOD amplitude with monotonically increasing bursts of spikes (Hagiwara *et al.*, 1962, 1965; Hagiwara and Morita, 1963; Suga, 1967; Bastian, 1976; Hopkins and Heiligenberg, 1978; Watson and Bastian, 1979; Baker, 1980). Both types of receptors demonstrate a directional sensitivity to current flow, with optimal responses to stimuli at the optimal azimuth for transepidermal current flow (Yager and Hopkins, 1993). This directional sensitivity appears to result from the frequency-response characteristics of the receptors: The directional sensitivity can be modified by altering the stimulus frequency (McKibben *et al.*, 1993). A correlation between receptor morphology and response has been suggested in both wave- and pulse-producing species. Rapid-timing units (T, M units) are those receptors possessing thick, myelinated afferent fibers, which end in large synaptic sites at the receptors. Amplitude-modulated organs (P, B units) are those with long, thin unmyelinated afferents innervating the receptor cells (Szamier and Wachtel, 1970; Szabo, 1974).

A singular property of many gymnotid tuberous electroreceptors is their relatively narrow tuning characteristics. In general, the best frequency (BF) of an electrosensory animal's electroreceptor is closely correlated with the peak power of the EOD frequency spectrum (Fig. 14) (Scheich *et al.*, 1973; Bullock *et al.*, 1975; Bastian, 1976, 1977;

Hopkins, 1976; Zakon and Meyer, 1983). Additionally, a number of species show polymorphic EOD frequency spectra. In many species there exists a distinct sexual dimorphism of the EOD, and in wave-producing species individuals appear to have unique preferred frequencies of the EOD within the frequency range of their particular species and gender (Hopkins, 1972, 1974; Bass and Hopkins, 1983; Hagedorn and Heiligenberg, 1985). The narrow tuning of these receptors to the fish's EOD likely provides the animal with a degree of communicative and reproductive isolation from the discharges produced by other species. Furthermore, the frequency of the EOD often changes during the fish's lifetime: In *Sternopygus* the frequency of the EOD shifts downward in maturing males and increases in females. Treatment with exogenous sex steroid hormones can alter both the frequency of the EOD and the tuning of the receptors: Androgens decrease the frequency of the EOD and estrogens increase it (Meyer and Zakon, 1982; Meyer, 1983; Bass and Hopkins, 1984). Following administration of exogenous hormones, the frequencies of the EOD and receptor tuning shift at about the same rate, but with the receptor tuning lagging by a period of approximately one week (Meyer and Zakon, 1982; Meyer, 1983). Administration of steroids to animals in which the EOD has been silenced can still result in retuning of the receptors, indicating that the receptors are not merely "following" a hormonally induced shift in EOD frequency. In *Sternopygus*, tuberous receptors developing in regenerating skin segments following removal of the epidermis initially demonstrate broad tuning properties but rapidly become tuned to the BF of the fish, even if the EOD has been silenced (Zakon, 1986b).

V. Mormyroidea

A. Anatomy

Despite the fact that the tuberous electroreceptors in the Mormyroidea (*Mormyridae* and *Gymnarchidae*) are not homologous to those of the gymnotids, the receptors show striking similarities in both morphology and physiological responses, presenting a striking case of parallel homoplasy. It is likely that some of the similarities are the result of receptors in both groups being derived from hair cell octavolateralis receptors (and probably from the nonhomologous ampullary receptors in both lineages) but also likely represent the physical, physiological, and developmental constraints incumbent on the evolution of a high-frequency electrosensory system for communicative and localization purposes.

Unlike gymnotids, mormyrids possess two morphologically distinct types of tuberous electroreceptor organs: Knollenorgans (or K units) and mormyromasts (Fig. 11B, C). The **Knollenorgans** are similar in morphology to gymnotid tuberous receptors, with a number of receptor cells located in a roughly spherical chamber, the walls of which consist of flattened epithelial cells (Szabo, 1965; Harder, 1968). The capsule is connected via a canal and epidermal pore to the external environment and, as in gymnotids, the canal is loosely plugged with epithelial cells. Generally,

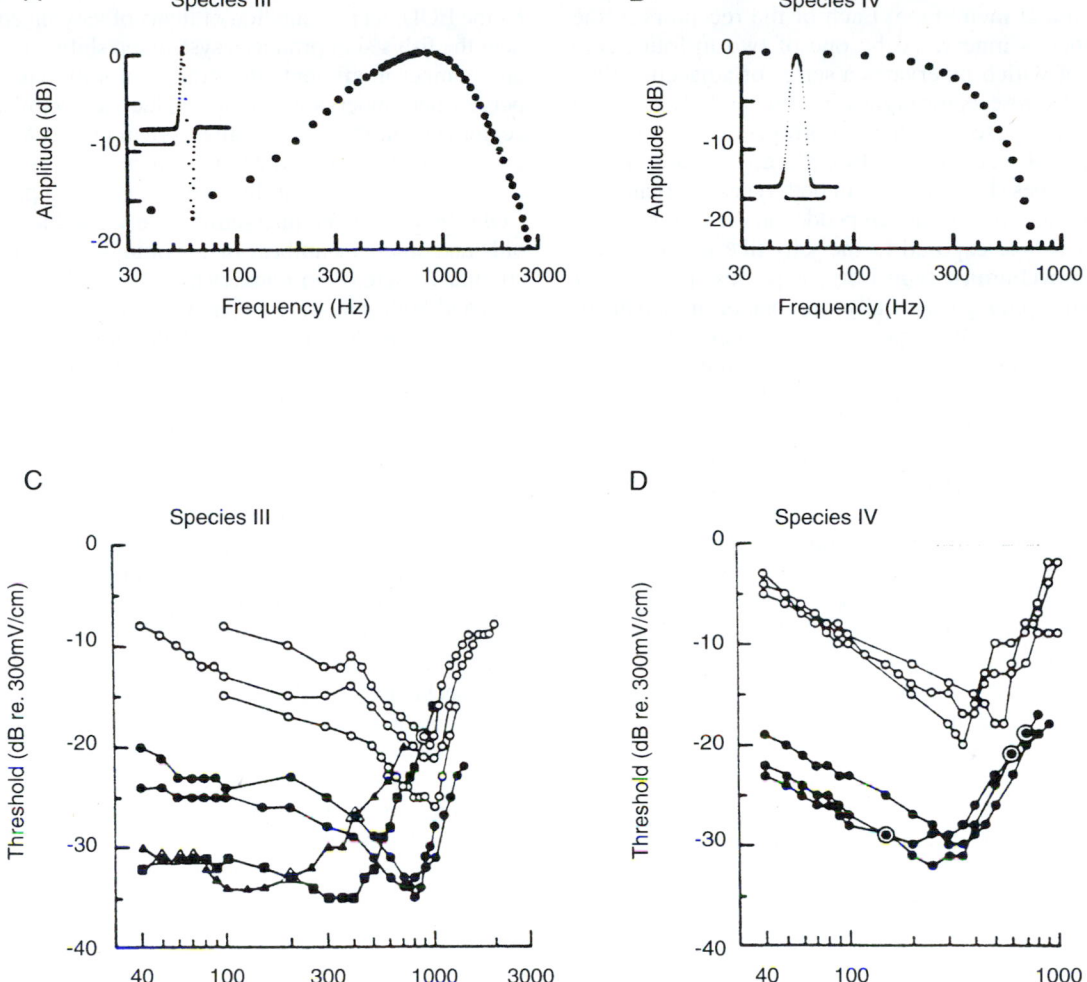

FIGURE 14. (A, B) EOD waveforms and power spectra and (C, D) corresponding frequency response functions of tuberous electroreceptors from two species of the gymnotiform *Hypopomus*. Tuning curves with open symbols are from pulse markers (M units); those with closed symbols are from burst duration coders (B units). (From Bastian, 1977.)

Knollenorgans possess a small number of receptor cells (1 to 10), although larger numbers are reported in some species. Each of the large (40–50 µm in diameter) spherical receptor cells itself is individually encapsulated by a layer of covering cells within the larger capsule of the organ. The receptor cells are attached to the walls of the capsule only at their basal membrane surface, leaving most of the receptor membrane exposed to the lumen within the covering cell capsule. The apical surface of the cell membrane is densely supplied with microvilli, and numerous mitochondria are apparent just beneath the apical membrane. Unlike most teleost electroreceptors, the membrane of Knollenorgans is believed to be electrically excitable: Stimulation of the afferent nerve fiber results in a spike that can be recorded from the organ. Although it has been suggested that an electrotonic synapse exists between Knollenorgan and afferent fiber (Bennett, 1971a; Steinbach and Bennett, 1971), only chemical synapses have been observed (Derbin and Szabo, 1968; Schessel *et al.*, 1982). Although a synaptic ribbon is present,

it does not extend into an evagination of the presynaptic membrane as is the case in gymnotids.

Mormyromast electroreceptors (D units) are more complex organs, with two distinct populations of electroreceptor cells, each innervated by different afferent nerve fibers (Szabo and Wersäll, 1970). Mormyromast receptors consist of two chambers; a superficial chamber that resembles an ampullary receptor and a deep chamber that is similar in morphology to a Knollenorgan. The deep chamber (or sensory chamber) possesses three to five small (2–3 µm) receptor cells (B cells). The receptor cells are covered with microvilli and, as in other tuberous organs, are attached to the capsule only at the basal membrane surface. A narrow canal connects the deep chamber to the superficial chamber, which is filled with a jelly-like matrix. The five to seven receptor cells of the superficial chamber (A cells) are embedded within the walls of the receptors with only the apical membrane exposed to the chamber lumen in a manner similar to ampullary electroreceptors. These cells lack microvilli, but possess a curious small dense conical structure on the apical membrane. Each of the receptors in the upper chamber is innervated by one of two to four nerve fibers, each of which innervates a series of adjacent cells.

Tuberous electroreceptor organs in gymnarchids are quite different from those of their mormyrid sister group. **Gymnarchomast** receptors are divided into two morphologically distinct types (Fig. 11D, E). In both types of organs, the cell bodies of the receptors are embedded in the capsule walls with the apical face exposed to the jelly-like matrix of the lumen. **Gymnarchomast type I** organs possess two distinct receptor cell morphologies, usually with one or more pairs of each type per organ. One type of receptor cell possesses a deep, groove-like invagination of the apical membrane surface, whereas the other possesses only a shallow indentation. Both types of receptor cells have apical microvilli, and all of the receptors within an organ are innervated by a single afferent nerve fiber. **Gymnarchomast type II** organs possess a single type of large receptor cell, with a characteristic deep invagination of the apical membrane and long apical microvilli. All of the sensory cells (7 to 12) in a type II receptor are innervated by a single afferent fiber (Szabo, 1965, 1974). Ultrastructural studies have demonstrated relatively high levels of Ca^{2+} in the cytoplasm of the receptor cells (Djebar *et al.*, 1995) presumably as a result of the inward depolarizing current produced at the membrane basal face.

B. Physiology

As in gymnotids, the tuberous electroreceptors of mormyrids can be differentiated into rapid-timing units and amplitude-coding units. The broadly tuned Knollenorgans respond to presentation of EOD-like stimuli by generating a single, highly phase-locked spike per pulse (Bennett, 1967; Szabo, 1974; Hopkins and Bass, 1981). The temporal pattern of the EOD pulse is critical to the receptor response, and only stimuli consisting of positive-negative transitions with rise-fall profiles similar to the animal's own EOD will reliably elicit a spike: Phase-shifted stimuli with power spectral peaks similar to the EOD are also poor at eliciting receptor responses (Hopkins and Bass, 1981). The Knollenorgans play a functional role in the communicative behavior of mormyrid fishes by detecting EODs produced by other conspecific individuals, since responses of Knollenorgans to the animal's own EOD are suppressed in the central nervous system. Hopkins and Bass (1981) suggested that different classes of Knollenorgan (KI and KII) may respond to different portions of the EOD cycle; the interval between the firing of the two receptor types will thus be specific for a given EOD waveform forming a basis for conspecific recognition at the level of the receptor.

Mormyromast receptors respond to the amplitude of the EOD, producing afferent fiber spikes as a function of EOD intensity. Unlike the B units of gymnotiforms, mormyromasts typically generate fewer spikes (1 to 5) per stimulus, and both response latency and interspike intervals shift with increasing stimulus amplitude (Szabo and Hagiwara, 1967). The threshold of mormyromast receptors is relatively high, indicating that these receptors are stimulated primarily by the animal's own EOD, and very little by the attenuated EODs of con-

specifics at a distance. Alteration of the electric field produced by the EOD via presentation of items of varying conductances near the fish's skin produces systematic shifts in spike latency and number in afferents innervating mormyromasts. Afferent nerve fibers innervating either of the two populations of receptor cells in the mormyromast organ project to different regions of the first-order medullary electrosensory nucleus (Bell *et al.*, 1989) and exhibit differing responses to presentation of EOD stimuli. Fibers innervating A cells exhibit higher thresholds and smaller numbers of maximum spikes per burst per stimulus, whereas those innervating B cells have lower thresholds and higher numbers of maximum spikes per burst (Bell, 1990a,b). The dynamic range of the mormyromast response intensity function is relatively limited (6 dB) suggesting that these cells are very sensitive to slight modulations of EOD amplitude within that range (Szabo and Hagiwara, 1967). The properties of the mormyromast response to EOD stimulation suggest that EOD amplitude is coded by a **latency code,** in which the intensity of the EOD is encoded by the latency of the first spike of the response, rather than by a **burst duration code,** as in the B units of gymnotids, in which the amplitude is encoded by the number of spikes generated in the burst (Szabo and Hagiwara, 1967; Bell, 1990b). Mormyromasts also exhibit extreme sensitivity to EOD phase modulation (von der Emde and Bleckmann, 1992). The different behavioral functions of Knollenorgans and mormyromasts (namely, communication versus electrolocation) in the ethogram of mormyrids is maintained throughout the central nervous system, and information from the two receptor types is segregated into two parallel pathways through the neuraxis.

The gymnarchomast receptors show striking similarities to those of gymnotids producing wave-type EODs, exhibiting either rapid-timing (S units) or amplitude-coding (O units) responses to presentation of EOD-like stimuli. S units have a low threshold and narrow dynamic range. At "normal" EOD intensities, the saturated S units show little temporal variation in their response to changes in stimulus intensity. O units have higher thresholds and dynamic ranges, and changes in EOD amplitudes are encoded as rapidly adapting bursts of afferent fiber spikes (Bullock *et al.*, 1975).

Bibliography

Akoev, G. N., Ilyinsky, O. B., and Zadan, P. M. (1976a). Physiological properties of electroreceptors of marine skates. *Comp. Biochem. Physiol.* **53A,** 201–209.

Akoev, G. N., Ilyinsky, O. B., and Zadan, P. M. (1976b). Responses of electroreceptors (ampullae of Lorenzini) of skates to electric and magnetic fields. *J. Comp. Physiol.* **106A,** 127–136.

Andres, K. J., and von During, M. (1988). The platypus bill. A structural and functional model of a pattern-like arrangement of cutaneous receptor receptors. *In* "Sensory Receptor Mechanisms" (A. Iggo, Ed.), pp. 81–89. World Scientific Publishing Company, Singapore.

Andrianov, G. N., Broun, G. R., and Ilyinsky, O. B. (1984). Frequency characteristic of skate electroreceptive central neurons responding to electrical and magnetic stimulation. *Neurophysiol.* **16,** 365–376.

Baker, C. (1980). Jamming avoidance behavior in gymnotoid electric fish with pulse-type discharges: Sensory encoding for a temporal pattern discrimination. *J. Comp. Physiol.* **136,** 165–181.

Baker, R. R., Mather, J. G., and Kennaugh, J. H. (1983). Magnetic bones in human sinuses. *Nature* **301,** 78–80.

Bass, A. H. (1986). Electric organs revisited: Evolution of a vertebrate communication and orientation organ. *In* "Electroreception" (T. H. Bullock and W. Heiligenberg, Eds.), pp. 13–70. John Wiley and Sons, New York.

Bass, A. H., and Hopkins, C. D. (1983). Hormonal control of sexual differentiation: Changes in electric organ discharge waveform. *Science* **220,** 971–974.

Bass, A. H., and Hopkins, C. D. (1984). Shifts of frequency tuning in electroreceptors in androgen treated mormyrid fish. *J. Comp. Physiol.* **155,** 713–724.

Bastian, J. (1976). Frequency response characteristics of electroreceptors in weakly electric fish (Gymnotoidei) with a pulse discharge. *J. Comp. Physiol.* **112,** 165–180.

Bastian, J. (1977). Variations in the frequency response of electroreceptors dependent on receptor location in weakly electric fish (Gymnotoidei) with a pulse discharge. *J. Comp. Physiol.* **121,** 53–64.

Bastian, J., and Heiligenberg, W. (1980). Neural correlates of the jamming avoidance response in *Eigenmannia. J. Comp. Physiol.* **136,** 135–152.

Bell, C. C. (1990a). Mormyromast receptor organs and their afferent fibers in mormyrid fish. II. Intra-axonal recordings show initial stages of central processing. *J. Neurophys.* **43,** 303–318.

Bell, C. C. (1990b). Mormyromast receptor organs and their afferent fibers in mormyrid fish III. Physiological differences between two morphological types of fibers. *J. Neurophys.* **43,** 319–332.

Bell, C. C., Zakon, H., and T. E. Finger (1989). Mormyromast receptor organs and their afferent fibers in mormyrid fish. I. Morphology. *J. Comp. Neurol.* **286,** 391–407.

Bemis, W. E., and Hetherington, T. E. (1982). The rostral organ of *Latimeria chalumnae*: Morphological evidence of an electroreceptive function. *Copeia* **1982,** 467–471.

Bennett, M. V. L. (1967). Mechanisms of electroreception. *In* "Lateral Line Detectors" (P. Cahn, Ed.), pp. 313–393. Indiana University Press, Bloomington, IN.

Bennett, M. V. L. (1971a). Electroreception. *In* "Fish Physiology" (W. S. Hoar and D. S. Randall, Eds.), Vol. 5, pp. 493–574. Academic Press, New York.

Bennett, M. V. L. (1971b). Electric organs. *In* "Fish Physiology" (W. S. Hoar and D. S. Randall, Eds.), Vol. 5, pp. 347–491. Academic Press, New York.

Bennett, M. V. L., and Clusin, W. T. (1978). Physiology of the ampulla of Lorenzini, the electroreceptor of elasmobranchs. *In* "Sensory Biology of Sharks, Skates, and Rays" (E. S. Hodgson and R. F. Mathewson, Eds.), pp. 483–505. Office of Naval Research, Arlington, VA.

Bennett, M. V. L., and Obara, S. (1986). Ionic mechanisms and pharmacology of electroreceptors. *In* "Electroreception" (T. H. Bullock and W. Heiligenberg, Eds.), pp. 157–181. John Wiley and Sons, New York.

Blakemore, R. (1975). Magnetotactic bacteria. *Science* **190,** 377–379.

Bodznick, D. (1989). Comparisons between electrosensory and mechanosensory lateral line systems. *In* "The Mechanosensory Lateral Line: Neurobiology and Evolution" (S. Coombs, P. Görner, and H. Münz, Eds.), pp. 655–680. Springer-Verlag, New York.

Bodznick, D., and Montgomery, J. C. (1992). Reafference in the elasmobranch electrosensory system: Medullary neuron receptive fields support a common-mode rejection mechanism. *J. Exp. Biol.* **171,** 127–137.

Bodznick, D., Hjelmstad, G., and Bennett, M. V. L. (1993). Accommodation to maintained stimuli in the ampullae of Lorenzini: How an electroreceptive fish achieves sensitivity in a noisy world. *Japan J. Physiol.* **43,** Suppl. 1, S231–S237.

Bodznick, D., Montgomery, J. C., and Bradley, D. J. (1992). Suppression of common-mode signals within the electrosensory system of the little skate, *Raja erinacea. J. Exp. Biol.* **171,** 107–125.

Bodznick, D., and Preston, D. G. (1983). Physiological characterization of electroreceptors in the lampreys, *Ichthyomyzon unicuspis* and *Petromyzon marinus. J. Comp. Physiol.* **152,** 209–217.

Braford, M. R. (1986). African knifefishes: The Xenomystinae. *In* "Electroreception" (T. H. Bullock and W. Heiligenberg, Eds.), pp. 453–464. John Wiley and Sons, New York.

Bullock, T. H., Behrend, K., and Heiligenberg, W. (1975). Comparison of the jamming avoidance response in gymnotoid and gymnarchid electric fish: A case of convergent evolution of behavior and its sensory basis. *J. Comp. Physiol.* **103,** 97–121.

Bullock, T. H., Bodznick, D. A., and Northcutt, R. G. (1983). The phylogenetic distribution of electroreception: Evidence for convergent evolution of a primitive vertebrate sense modality. *Brain Res. Rev.* **6,** 25–46.

Bullock, T. H., and Heiligenberg, W. (1986). "Electroreception." John Wiley and Sons, New York.

Derbin, C., and Szabo, T. (1968). Ultrastructure of an electroreceptor (Knollenorgan) in the mormyrid fish, *Gnathonemus petersii Int. J. Ultrastruct. Res.* **22,** 469–484.

Dijkgraaf, S., and Kalmijn, A. J. (1966). Versuche zur biologischen Bedeutung der Lorenzinischen Ampullen an Haifischen. *Z. Vgl. Physiol.* **47,** 438–456.

Djebar, B., Bensouilah, M., and Denizot, J.-P. (1995). Ultrastructural distribution of calcium in cutaneous electroreceptor organs of teleost fish. *Biotech. Histochem.* **70,** 81–89.

Echague, A., and Trujillo-Cenoz, O. (1981). Innervation patterns in the tuberous organs of *Gymnotus carapo. In* "Advances in Physiological Sciences," Vol 31: "Sensory Physiology of Aquatic Lower Vertebrates" (T. Szabo and G. Czeh, Eds.), pp. 29–40. Akademiai Kiado, Budapest.

Feng, A. S., and Bullock, T. H. (1977). Neuronal mechanisms for object discrimination in the weakly electri fish, *Eigennmania viriscens. J. Exp. Biol.* **66,** 141–158.

Fields, R. D., Bullock, T. H., and Lange, O. D. (1993). Ampullary sense organs, peripheral, central and behavioral electroreception in chimeras (*Hydrolagus*, Holocephali, Chondrichthyes). *Brain Behav. Evol.* **41,** 269–289.

Fritzsch, B., and Münz, H. (1986). Electroreception in amphibians. *In* "Electroreception" (T. H. Bullock and W. Heiligenberg, Eds.), pp. 483–496. John Wiley and Sons, New York.

Gould, J. L., Kirschvink, J. L., and Deffeyes, K. F. (1978). Bees have magnetic resonance. *Science* **201,** 1026–1028.

Gregory, J. E., Iggo, A., McIntyre, A. K., and Proske, U. (1989). Responses of electroreceptors in the snout of the echidna. *J. Physiol. (Lond.)* **414,** 521–538.

Hagedorn, M., and Heiligenberg, W. (1985). Court and spark: Electric signals in the courtship and mating of gymnotoid fish. *Anim. Behav.* **33,** 254–265.

Hagiwara, S., Kusano, K., and Negishi, K. (1962). Physiological properties of electroreceptors of some gymnotids. *J. Neurophys.* **25,** 430–449.

Hagiwara, S., and Morita, H. (1963). Coding mechanisms of electroreceptor fibers in some electric fish. *J. Neurophys.* **26,** 551–567.

Hagiwara, S., Szabo, T., and Enger, P. S. (1965). Electroreceptor mechanisms in a high-frequency weakly electric fish, *Sternarchus albifrons. J. Neurophys.* **28,** 784–799.

Harder, W. (1968). Die Beziehungen zwischen Elektrorezeptoren, elektrischem Organ, Seitenlinienorganen und Nervensystem bei den Mormyridae (Teleostei, Pisces). *Z. Vgl. Physiol.* **59,** 272–318.

Heiligenberg, W. (1977). "Principles of Electrolocation and Jamming Avoidance Response in Electric Fish." Springer, New York.

Heiligenberg, W. (1986). Jamming avoidance responses. *In* "Electroreception" (T. H. Bullock and W. Heiligenberg, Eds.), pp. 613–649. John Wiley and Sons, New York.

Heiligenberg, W. F. (1991). Neural nets in electric fish. MIT Press, Cambridge.

Hopkins, C. D. (1972). Sex differences in signaling in an electric fish. *Science* **176**, 1035–1037.

Hopkins, C. D. (1974). Electric communication: Functions in the social behavior of *Eigenmannia viriscens. Behaviour* **50**, 270–305.

Hopkins, C. D. (1976). Stimulus filtering and electroreception: Tuberous electroreceptors in three species of gymnotoid fish. *J. Comp. Physiol.* **103**, 171–207.

Hopkins, C. D. (1983). Functions and mechanisms in electroreception. *In* "Fish Neurobiology," Vol. 1: "Brain Stem and Sense Organs" (R. G. Northcutt and R. E. Davis, Eds.), pp. 215–259. University of Michigan Press, Ann Arbor, MI.

Hopkins, C. D., and Bass, A. H. (1981). Temporal coding of species recognition signals in an electric fish. *Science* **212**, 85–87.

Hopkins, C. D., and Heiligenberg, W. F. (1978). Evolutionary designs for electric signals and electroreceptors in gymnotoid fishes of Surinam. *Behav. Ecol. Sociobiol.* **3**, 113–134.

Hsu, C.-Y., and Li, C.-W. (1994). Magnetoreception in honeybees. *Science* **265**, 95–97.

Istenic, L., and Bulog, B. (1984). Some evidence for the ampullary organs in the European cave salamander *Proteus anguineus* (Urodela Amphibia). *Cell Tissue Res.* **235**, 393–402.

Jørgensen, J. M. (1980). The morphology of the Lorenzinian ampullae of the sturgeon *Acipenser ruthenus* (Pisces: Chondrostei). *Acta Zool.* **61**, 87–92.

Jørgensen, J. M. (1982). Fine structure of the ampullary organs of the bichir *Polypterus senegalus* Cuvier, 1829 (Pisces: Brachiopterygii) with some notes on the phylogenetic development of electroreceptors. *Acta Zool.* **63**, 211–217.

Jørgensen, J. M. (1984). On the morphology of the electroreceptors of the Australian and one African lungfish species. *Vidensk. Meddr. Dansk Naturh. Foren.* **145**, 77–85.

Jørgensen, J. M. (1989). Evolution of octavolateralis sensory cells. *In* "The Mechanosensory Lateral Line: Neurobiology and Evolution" (S. Coombs, P. Görner, and H. Münz, Eds.), pp. 115–146. Springer-Verlag, New York.

Jørgensen, J. M., Flock, Å., and Wersäll, J. (1972). The Lorenzinian ampullae of *Polyodon spathula. Z. Zellforsch.* **130**, 362–377.

Kalmijn, A. J. (1971). The electric sense of sharks and rays. *J. Exp. Biol.* **55**, 371–383.

Kalmijn, A. J. (1974). The detection of electric fields from inanimate and animate sources other than electric organs. *In* "Handbook of Sensory Physiology" (A. Fessard, Ed.), Vol. III/3, pp. 147–200. Springer-Verlag, New York.

Kalmijn, A. J. (1982). Electric and magnetic field detection in elasmobranch fishes. *Science* **218**, 915–918.

Kalmijn, A. J. (1988). Detection of weak electric fields. *In* "Sensory Biology of Aquatic Animals" (J. Atema, R. R. Fay, A. N. Popper, and W. N. Tavogla, Eds.), pp. 151–186. Springer-Verlag, New York.

Kalmijn, A. J., and Blakemore, R. (1978). The magnetic behavior of mud bacteria. *In* "Animal Migration, Navigation and Homing" (K. Schmidt-Koenig and W. T. Keeton, Eds.), pp. 347–353. Springer-Verlag, New York.

Kirschbaum, F. (1983). Myogenic electric organ precedes the neurogenic organ in apteronotid fish. *Naturwissenschaften* **70**, 205.

Kirschvink, J. L., and Walker, M. M. (1995). Honeybees and magnetoreception. *Science* **269**, 1889.

Kirschvink, J. L., Walker, M. M., Chang, S. B., Dizon, A. E., and Peterson, K. A. (1985). Chains of single-domain magnetite particles in chinook salmon. *J. Comp. Physiol. A* **157**, 375–381.

Lissman, H. W. (1958). On the function and evolution of electric organs in fish. *J. Exp. Biol.* **35**, 156–191.

Lissman, H. W., and Machin, K. E. (1958). The mechanisms of object location in *Gymnarchus niloticus* and similar fish. *J. Exp. Biol.* **35**, 451–486.

Manger, P. R., and Hughes, R. L. (1992). Ultrastructure and distribution of epidermal sensory receptors in the beak of the echidna, *Tachyglossus aculeatus. Brain Behav. Evol.* **40**, 287–296.

McKibben, J. R., Hopkins, C. D., and Yager, D. D. (1993). Directional sensitivity of tuberous electroreceptors: Polarity preferences and frequency tuning. *J. Comp. Physiol.* **173**, 415–424.

Meyer, J. H. (1983). Steroid influences upon the discharge frequency of a weakly electric fish. *J. Comp. Physiol.* **153**, 29-38.

Meyer, J. H., and Zakon, H. H. (1982). Androgens alter the tuning of electroreceptors. *Science* **217**, 635–637.

Millot, J., and Anthony, J. (1956). L'organe rostral de *Latimeria* (Crossoptérygien Coelacanthidè). *Ann. Sci. Nat. Zool., 11e Série*, **18**, 381–387.

Montgomery, J. C. (1984). Frequency response characteristics of primary and secondary neurons in the electrosensory system of the thornback ray. *Comp. Biochem. Physiol.* **79A**, 189–195.

Montgomery, J. C., and Bodznick, D. (1993). Hindbrain circuitry mediating common-mode suppression of ventilatory reafference in the electrosensory system of the little skate, *Raja erinacea. J. Exp. Biol.* **183**, 203–315.

Moore, B. R. (1988). Magnetic fields and orientation in homing pigeons: Experiments of the late W. T. Keeton. *Proc. Natl. Acad. Sci. USA* **85**, 4907–4909.

Moy-Thomas, J. A., and Miles, R. S. (1971). "Paleozoic Fishes." W. B. Saunders, Philadelphia.

Murray, R. W. (1962). The response of the ampullae of Lorenzini of elasmobranches to electrical stimulation. *J. Exp. Biol.* **39**, 119–128.

Murray, R. W., and Potts, T. W. (1961). The composition of the endolymph and other fluids of elasmobranches. *Comp. Biochem. Physiol.* **2**, 65–75.

Nachtrieb, H. F. (1910). The primitive pores of *Polyodon spathula* (Walbaum). *J. Exp. Zool.* **9**, 455–468.

Nelson, J. S. (1994). "Fishes of the World," 3rd ed. John Wiley and Sons, New York.

Nesson, M. H. (1995). Honeybees and magnetoreception. *Science* **269**, 1889–1890.

New, J. G. (1990). Medullary electrosensory processing in the little skate. I. Response characteristics of neurons in the dorsal octavolateralis nucleus. *J. Comp. Physiol.* **167**, 285–294.

New, J. G., and Bodznick, D. (1990). Medullary electrosensory processing in the little skate. II. Suppression of electrosensory reafference via a common-mode rejection mechanism. *J. Comp. Physiol.* **167**, 295–307.

Nichol, H., and Locke, M. (1995). Honeybees and magnetoreception. *Science* **269**, 1888–1889.

Northcutt, R. G. (1983). The primary lateral line afferents in lepidosirenid lungfishes. *Soc. Neurosci. Abstr.* **9**, 1167.

Northcutt, R. G. (1986). Electroreception in nonteleost bony fishes. *In* "Electroreception" (T. H. Bullock and W. Heiligenberg, Eds.), pp. 257–285. John Wiley and Sons, New York.

Northcutt, R. G., Brändle, K., and Fritzsch, B. (1995). Electroreceptors and mechanosensory lateral line organs arise from single placodes in axolotls. *Dev. Biol.* **168**, 358–373.

Obara, S. (1976). Mechanisms of electroreception in ampullae of Lorenzini of the marine catfish *Plotosus. In* "Electrobiology of Nerve, Synapse and Muscle" (J. P. Reuben, D. P. Purpura, M. V. L. Bennett, and E. R. Kandel, Eds.), pp. 128–147. Raven Press, New York.

Obara, S., and Bennett, M. V. L. (1972). Mode of operation of ampullae of Lorenzini of the skate, *Raja. J. Gen. Physiol.* **60**, 534–557.

Pals, N., Valentijn, P., and Verwey, D. (1982). Orientation reactions of the dogfish, *Scyliorhinus canicula*, to local electric fields. *Neth. J. Zool.* **32**, 495–512.

Paulin, M. G. (1995). Electroreception and the compass sense of sharks. *J. Theor. Biol.* **174**, 325–339.

Perry, A., Bauer, G. B., and Dizon, A. E. (1981). Magnetite in the green turtle. *Trans. Am. Geophys. Union* **62**, 850.

Peters, R. C., and Bretschneider, F. (1972). Electric phenomena in the habitat of the catfish, *Ictalurus nebulosus* LeS. *J. Comp. Physiol.* **81**, 345–362.

Peters, R. C., and Evers, H. P. (1985). Frequency selectivity in the ampullary system of an elasmobranch fish (*Scyliorhinus canicula*). *J. Exp. Biol.* **118**, 99–109.

Peters, R. C., and Meek, F. (1973). Catfish and electric fields. *Experientia* (*Basel*) **29**, 299–300.

Pfeiffer, W. (1968). Die Fahrenholzschen Organe der Dipnoi und Brachiopterygii. *Z. Zellforsch.* **90**, 127–147.

Preston, D., and Pettigrew, J. D. (1980). Ferromagnetic coupling to muscle receptors as a basis for geomagnetic field sensitivity in animals. *Nature* **285**, 99–101.

Raschi, W. (1986). A morphological analysis of the ampullae of Lorenzini in selected skates (Pisces-rajoidei). *J. Morphol.* **189**, 225–247.

Raschi, W., and Mackanos, L. A. (1989). The structure of the ampullae of Lorenzini in *Dasyatis garouaensis* and its implications on the evolution of freshwater electroreceptive systems. *J. Exp. Zool.* **2**, 101–111.

Ronan, M. C. (1986). Electroreception in cyclostomes. *In* "Electroreception" (T. H. Bullock and W. Heiligenberg, Eds.), pp. 209–224. John Wiley and Sons, New York.

Ronan, M. C., and Bodznick, D. (1986). End buds: Non-ampullary electroreceptors in adult lampreys. *J. Comp. Physiol.* **158**, 9–16.

Scheich, H. (1977). Neural basis of communication in the high frequency electric fish, *Eigenmannia viriscens* (jamming avoidance response). III. Central integration in the sensory pathway and control of the pacemaker. *J. Comp. Physiol.* **113**, 229–255.

Scheich, H., Bullock, T. H., and Hamstra, R. H. (1973). Coding properties of two classes of afferent fiber: High frequency electroreceptors in the electric fish, *Eigenmannia*. *J. Neurophys.* **36**, 39–60.

Scheich, H., Langner, G., Tidemann, C., Coles, R. B., and Guppy, A. (1986). Electroreception and electrolocation in platypus. *Nature* **319**, 401–402.

Schessel, D. A., Ginzburg, R. D., and Highstein, S. M. (1982). Ultrastructural and physiological evidence for chemical synaptic transmission between the vestibular type I hair cell and its primary nerve chalice. *Anat. Rec.* **202**, 168.

Srivastava, C. B. L. (1973). Peripheral nerve ending types in tuberous electroreceptors of a high frequency gymnotid, *Sternarchus albifrons*. *J. Neurocytol.* **2**, 77–83.

Steinbach, A. B., and Bennett, M. V. L. (1971). Effect of divalent ions and drugs on synaptic transmission in phasic electroreceptors in a mormyrid fish. *J. Gen. Physiol.* **58**, 580–598.

Suga, N. (1967). Coding in tuberous and ampullary organs of a gymnotid electric fish. *J. Comp. Neurol.* **131**, 437–453.

Sugawara, Y. (1989). Two Ca⁺⁺ current components of the receptor current in the electroreceptors of the marine catfish *Plotosus*. *J. Gen. Physiol.* **93**, 365-380.

Sugawara, Y., and Obara, S. (1989). Receptor Ca⁺⁺ current and Ca⁺⁺-gated K⁺ current in tonic electroreceptors of the marine catfish *Plotosus*. *J. Gen. Physiol.* **93**, 343–364.

Szabo, T. (1965). Sense organs of the lateral line system in some electric fish of the Gymnotidae, Mormyridae and Gymnarchidae. *J. Morphol.* **117**, 229–250.

Szabo, T. (1970). Morphologische und funktionelle Aspekte bei Elektrorezeptoren. *Verh. Dtsch. Zool. Ges.* **64**, 141–148.

Szabo, T. (1974). Anatomy of the specialized lateral line organs of electroreception. *In* "Handbook of Sensory Physiology" (A. Fessard, Ed.), Vol. III/3, pp. 13–58. Springer-Verlag, New York.

Szabo, T., and Hagiwara, S. (1967). A latency-change mechanism involved in sensory coding of electric fish (mormyrids). *Physiol. Behav.* **2**, 331–335.

Szabo, T., and Wersäll, J. (1970). Ultrastructure of an electroreceptor (mormyromast) in a mormyrid fish. *Gnathonemus petersii*. *J. Ultrastruct. Res.* **30**, 473–490.

Szamier, R. B., and Wachtel, A. W. (1970). Special cutaneous receptor organs of fish. VI. Ampullary and tuberous organs of Hypopomus. *J. Ultrastruct. Res.* **30**, 450–471.

Teeter, J. G., and Bennett, M. V. L. (1981). Synaptic transmission in the ampullary electroreceptor of the transparent catfish, *Kryptopterus*. *J. Comp. Physiol.* **142**, 371–377.

Teeter, J. G., Szamier, R. B., and Bennett, M. V. L. (1980). Ampullary electroreceptors in the sturgeon *Scaphirhynchus platorynchus* (Rafinesque). *J. Comp. Physiol.* **138**, 213–223.

Tenforde, T. S. (1989). Electroreception and magnetoreception in simple and complex organisms. *Biomagnetics* **10**, 215–221.

Tricas, T. C. (1982). Bioelectric-mediated predation by swell sharks, *Cephaloscyllium ventriosum*. *Copeia* **1982**, 948–952.

Tricas, T. C., and New, J. G. (1998). Sensitivity and response dynamics of elasmobranch electrosensory primary afferent neurons to near threshold fields. *J. Comp. Physiol.* **182,** 89-101.

Tricas, T. C., Michael, S. W., and Sisneros, J. A. (1995). Electrosensory optimization to conspecific phasic signals for mating. *Neurosci. Lett.* **202**, 129–132.

Viancour, T. (1979a). Electroreceptors of a weakly electric fish. I. Characterization of tuberous receptor tuning. *J. Comp. Physiol.* **133**, 317–325.

Viancour, T. (1979b). Electroreceptors of a weakly electric fish. II. Individually tuned receptor oscillations. *J. Comp. Physiol.* **133**, 327–338.

Vischer, H. A. (1995). Electroreceptor development in the weakly electric fish *Eigenmannia*: A histological and ultrastructural study. *J. Comp. Neurol.* **360**, 81–100.

von der Emde, G., and Bleckmann, H. (1992). Extreme phase sensitivity of afferents which innervate mormyromast electroreceptors. *Naturwissenschaften* **79**, 131–133.

Wachtel, A. W., and Szamier, R. B. (1966). Special cutaneous receptor organs of fish: The tuberous organs of *Eigenmannia*. *J. Morphol.* **119**, 51–80.

Walcott, C., Gould, J. L., and Kirschvink, J. L. (1979). Pigeons have magnets. *Science* **201**, 1027–1029.

Walker, M. M., Kirschvink, J. L., Chang, S. B. R., and Dizon, A. E. (1984). A candidate magnetic sense organ in the yellowfin tuna, *Thunnus albacares*. *Science* **224**, 751–753.

Waltmann, B. (1966). Electrical properties and fine structure of the ampullary canals of Lorenzini. *Acta Physiol. Scand.* **66**, Suppl. 264, 1–60.

Watson, D., and Bastian, J. (1979). Frequency response characteristics of electroreceptors in the weakly electric fish, *Gymnotus carapo*. *J. Comp. Physiol.* **134**, 191–202.

Wu, C. H. (1984). Electric fish and the discovery of animal electricity. *Am. Sci.* **72**, 598–607.

Yager, D. D., and Hopkins, C. D. (1993). Directional characteristics of tuberous electroreceptors in the weakly electric fish, Hypopomus (Gymnotiformes). *J. Comp. Physiol.* **173**, 401–414.

Zakon, H. H. (1984a). Postembryonic changes in the peripheral electrosensory system of a weakly electric fish: Addition of receptor organs with age. *J. Comp. Neurol.* **228**, 557–570.

Zakon, H. H. (1984b). The ionic basis of the oscillatory receptor potential of tuberous electroreceptors in Sternopygus. *Soc. Neurosci. Abstr.* **10**, 193.

Zakon, H. H. (1986a). The electroreceptive periphery. *In* "Electroreception" (T. H. Bullock and W. Heiligenberg, Eds.), pp. 103–156. John Wiley and Sons, New York.

Zakon, H. H. (1986b). The emergence of tuning in newly-generated tuberous electroreceptors. *J. Neurosci.* **6**, 3297–3308.

Zakon, H. H. (1987). Variation in the mode of receptor cell addition in the electrosensory system of gymnotiform fishes. *J. Comp. Neurol.* **262**, 195–214.

Zakon, H. H. (1988). The electroreceptors: Diversity in structure and function. *In* "Sensory Biology of Aquatic Animals" (J. Atema, R. R. Fay, A. N. Popper, and W. N. Tavolga, Eds.), pp. 813–850. Springer, New York.

Zakon, H. H., and Meyer, J. H. (1983). Plasticity of electroreceptor tuning in the weakly electric fish *Sternopygus dariensis. J. Comp. Physiol.* **153**, 477–487.

Zipser, B., and Bennett, M. V. L. (1973). Tetrodotoxin resistant electrically excitable responses of receptor cells. *Brain Res.* **62**, 253–259.

Zoeger, J., Dunn, J. R., and Fuller, M. (1981). Magnetic material in the head of the common Pacific dolphin. *Science* **213**, 892–894.

Harvey M. Fishman

Appendix: The Biophysics of Electroreception in Ampullary Organs of Elasmobranch Fishes

Specialized organs, the ampullae of Lorenzini, make electroreception possible in marine elasmobranch fishes (sharks, skates, and rays) and other animals that possess this unusual sensory modality (see Chapter 50). These organs transduce variations in electric field intensity into modulations of the basal spike-firing rate in the primary afferent nerve, which innervates the ampullae and conveys this signal to the brain. Most animals, including humans, do not have an electroreceptive sense, but ampullary epithelial cells and mechanosensitive hair cells of the inner ear have some common morphological and functional properties (see Chapter 46). Electroreception in marine elasmobranchs[1] also ranks among the most sensitive sensory systems (for a comparison see Block, 1992). Recently, the mechanism that underlies this acute sensitivity has become of interest with respect to understanding how weak electromagnetic fields could possibly affect humans (Hafemeister, 1998). Based on behavioral responses, sharks and rays respond to electric field intensities as low as, and perhaps lower than, 0.5 μV/m (Kalmijn, 1966). This acute electrical sense apparently enables them to find prey and to orient themselves with respect to the earth's magnetic field (Kalmijn, 1988).

In this brief appendix, we consider the biophysical properties of ampullary organs that enable transduction of electric field intensities into neural signals, which are perceived in the brain. Some of these properties were deduced from voltage-clamp and complex admittance measurements (Lu and Fishman, 1994) using the isolated, single-organ preparation[2] of Clusin and Bennett (1977) excised from skates. Such whole-organ preparations allow determination of an organ's output (neural spike activity) to various electrical stimuli applied to the ampullary epithelium (Fig. A-1). Based on these isolated organ data, we describe the evidence that supports the following four assertions. (1) The behavioral responses of animals to low-intensity electric fields are primarily due to the properties of ampullary organs. (2) The electric-field orientation (stimulus polarity) that produces an excitatory output (i.e., increased afferent nerve activity) is due to the conduction properties of the specific ion channels in apical and basal membranes of ampullary epithelial cells. (3) Signal amplification is achieved by interaction of physically separated and different ion conductances in apical and basal membranes of the ampullary epithelial cells. (4) The acute electric-field sensitivity is achieved by a convergent system that enhances the signal relative to background noise in the signal generation and amplification processes of the sensory epithelium.

When marine elasmobranchs are in their seawater environment, electrically conductive paths through seawater and their low skin resistance (Murray, 1967) shunt all ampullary organs. Consequently, the high shunt resistance of "open-circuit" (current-clamp) measurements of transorgan voltages in response to currents applied to an electrically isolated organ do not correspond to the operational conditions of an ampullary organ *in situ* in an animal (Lu and Fishman, 1994). In contrast, the low shunt resistance or "short-circuit" (voltage–clamp) measurements of ampullary transepithelial current in response to control of transepithelial voltage more closely approximate the electrical "load" on an organ in an animal swimming in seawater. Thus, voltage-clamp responses are more likely to reflect organ operational characteristics and to provide the electrical data that are essential for understanding ampullary electroreceptor function.

From such voltage-clamp data (Lu and Fishman, 1994), the relationship between the spike-firing rate of the afferent nerve (organ output) and applied step-voltage changes across an epithelium (input) was determined. This transfer curve (Fig. A-2) shows that an organ provides a discernible signal to the brain of an animal, in response to applied transepithelial voltages of a few microvolts or less! Furthermore, when the transepithelial voltage (luminal with respect to basal side) is more negative than −100 μV (Fig. A-2), no additional increase in the rate of afferent spike firing occurs (i.e., the output becomes saturated). Thus, the saturated output for inputs more negative than −100 μV is indicative of the upper limit of the excitatory response of an organ. With respect to accounting for marine elasmobranch sensitivity (Kalmijn, 1966; Kalmijn, 1982), a 0.5 μV/m field applied to a 0.2-m length canal should produce an ampullary transepithelial voltage drop of 0.1 μV. Although this transepithelial voltage is still an order of magnitude lower than the lowest level (1 μV) step clamp response

[1]Here we consider only the biophysical properties of ampullary organs in marine elasmobranchs, which are the most sensitive with respect to behavioral responses to exogenous electric fields. Some details of ampullary organ function in freshwater elasmobranchs, such as the electric field orientation for excitation versus inhibition of afferent nerve activity, are significantly different (see Chapter 50).

[2]Each excised organ preparation consists of a 4-cm length of ampullary canal attached to an ampulla, lined with epithelial cells, which are innervated by a 1-cm length of primary afferent nerve. The canal is filled with a highly conductive gel and lined with nonconducting epithelial cells.

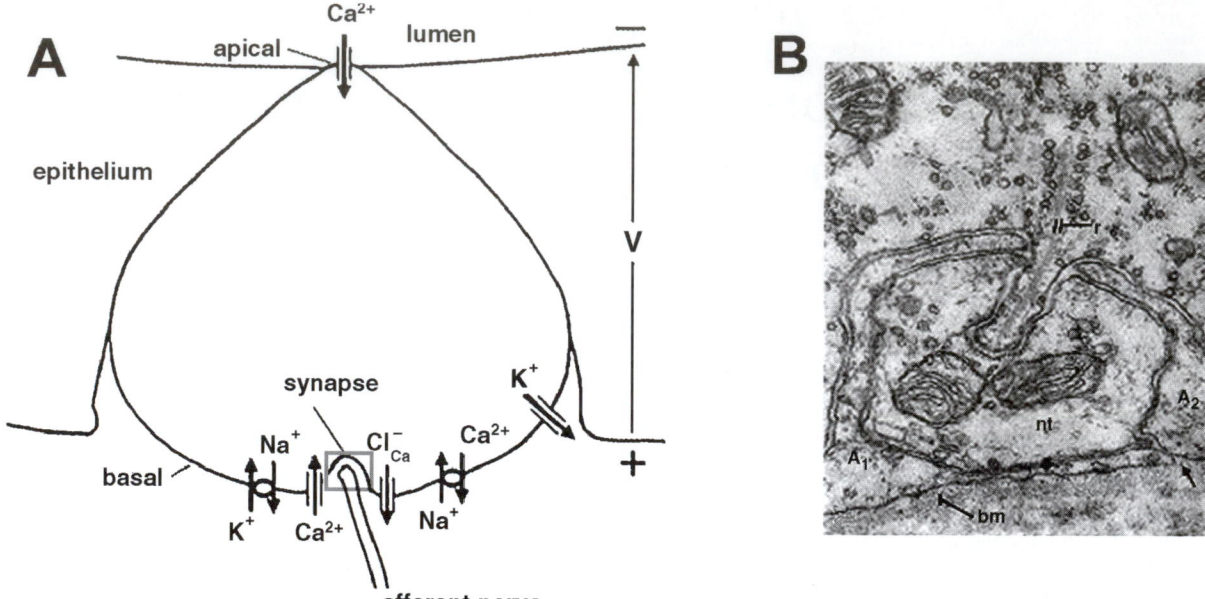

FIGURE A-1. (A) Diagram showing the location of various ion channels and transporters in the ampullary epithelium as identified by the effect of specified antagonists, agonists, or ion substitutions on admittance data from isolated organs (from Lu and Fishman, 1995a). Conduction by Ca^{2+} channels in apical membranes constitutes the negative conductance reflected in electrophysiological data (Fig. A-3) in that they provide the inwardly directed current (arrow) in response to depolarization of apical membranes (i.e., luminal negative-polarity voltages). Conduction by ion channels (Cl_{Ca}^{-} (calcium-activated Cl^{-}), Ca^{2+}, and K^{+}) in basal membranes produces an oscillation involved in the release of neurotransmitter at presynaptic membrane (see Lu and Fishman, 1995b). The ion transporters (Na^{+}-K^{+} and Na^{+}-Ca^{2+}) maintain a highly regulated membrane voltage and intracellular Ca^{2+} concentration to enable detection of fields that produce sub-microvolt changes in transepithelial voltage. (B) Electron micrograph of the ribbon synapse (boxed region in A) formed at the basal membrane of an ampullary epithelial cell with the afferent nerve, which conveys electroreceptor signals to the brain. From Waltman (1996).

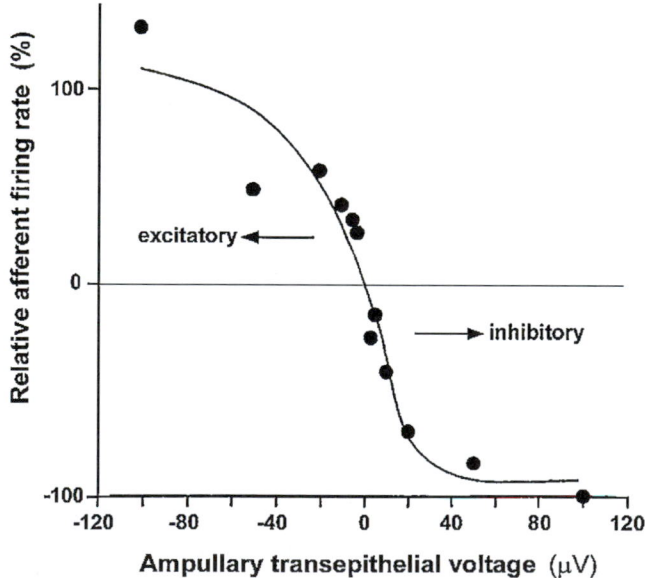

FIGURE A-2. Transfer characteristic curve of an isolated ampullary organ excised from the skate *Raja erinacea*. Data (filled circles) were obtained by measuring the afferent spike-firing rate relative to the basal firing rate (35 spikes/s at a holding potential of 0 mV) in response to transepithelial step voltage clamps from the holding potential. Applied step changes in ampullary transepithelial voltage polarized the apical (luminal) side relative to the basal side (see Fig. A-1A), Increasingly negative luminal voltages (<0) are excitatory stimuli (increase afferent activity) up to −100 mV, beyond which (data not shown) the output saturates. A voltage step of 3 µV produced the smallest discernable change in afferent spike rate. Solid line is a curve fit of a logistic function to the data. Modified from Lu and Fishman (1994).

that has been observed in an isolated ampullary organ (Fig. A-2), it is two or more orders better than any other biological electrotransducer. Consequently, when enhancements of signal-to-noise performance (discussed later) are considered, such as integration of inputs from multiple organs together with integration of inputs from multiple cells in each organ, the detectable response of an isolated ampullary organ appears to be sufficient to account for the observed behavioral responses.

To aid in the description of properties that underlie electrical signal generation and processing within the epithelium of an ampullary organ, the driving-point complex admittance[3] of an isolated organ is useful (Lu and Fishman, 1994). The locus of the complex admittance determined at 400 frequencies (ranging from 0.05 to 20 Hz) shows (Fig. A-3A) that the **real part** of the admittance is negative (i.e., all data points are in the left half of the complex plane). Such admittance behavior is analogous to a negative conductance. That is, when the polarity of the lumen is negative with respect to the basal side, the current flows

---

[3]The complex admittance, $Y(jf)$, is a function of frequency, where $j = \sqrt{-1}$, defined as

$$Y(jf) = G(f) + jB(f)$$

where $G(f)$ is the **real part** and $B(f)$ is the **imaginary part** of the admittance. The admittance is plotted as a locus of frequency points in the complex plane $B(f)$ versus $G(f)$).

inward (from lumen into the epithelial cells) instead of outward for a positive conductance. A luminal negative-polarity stimulus also depolarizes apical membranes of ampullary epithelial cells. Measurement of the ampullary transepithelial current-voltage relationship for increasingly lumen-positive step voltage clamps from various holding potentials (HP) shows (Fig. A-3B) a negative slope (conductance) region, within which an organ operates, in accordance with the admittance data. Pharmacological identification of ion channels (Fig. A-1A) in apical membranes provides an explanation of inwardly directed current for apical membrane depolarizations (Lu and Fishman, 1995a). For a luminal negative stimulus, the inward current through depolarized apical membranes (i.e., the negative conductance-like behavior) is due to conduction through Ca^{2+} channels. In contrast, conduction by the ion channels in basal membranes (Fig. A-1A) results in a positive membrane conductance, and the current flow is directed outwardly (from within epithelial cells toward the basal side). Because of apical membrane Ca^{2+} channel conduction, the presynaptic terminals become depolarized whenever apical membranes are depolarized. Basal membrane depolarization then initiates a sequence of processes in the presynaptic membrane (Fig.1B) that results in neurotransmitter release (see Chapter 41) and a subsequent increase in afferent nerve activity. Hence, a luminal negative-polarity stimulus is excitatory.

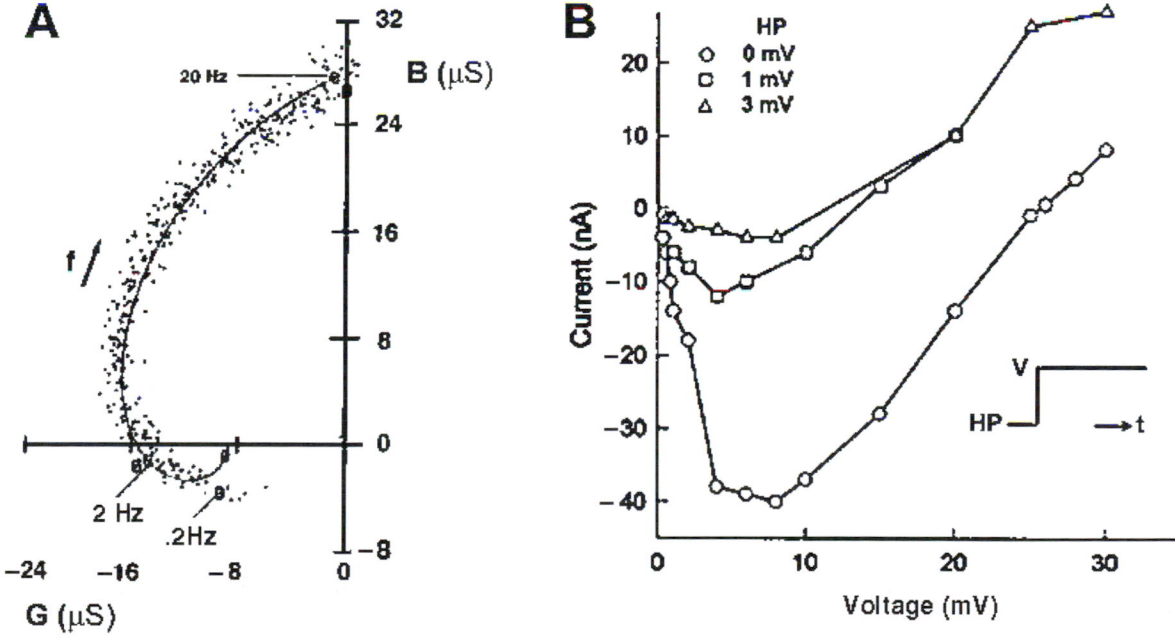

FIGURE A-3. (A) Locus of the admittance function of complex frequency jf, Y(jf) = G(f) +jB(f), of an excised ampullary organ from the skate *Raja erinacea* plotted in the complex plane (B(f) versus G(f)) from determinations at 400 frequencies ranging from 0.05 to 20 Hz. The dots are the data; the solid line is a best fit (minimumization of mean square error criterion) of the complex admittance of an electrical model to the data. The *real part,* G(f), of the admittance os negative (<0), indicative of conduction behavior similar to a negative conductance. (B) Current-voltage relationship of an isolated ampullary transepithelial current (I) (current densities were not possible due to the unknown area of the epithellium) in response to step voltage clamps (inset) from various holding potentials (HPs). I<0 is a current directed from the basal to apical side of the ampullary epithelium for a lumen-positive polarity step clamp The negative slope (conductance) of the *I-V* data is indicative of the negative conductance behavior inferred from the admittance locus in (A). Due to the unknown surface area of the epithelium Modified from Lu and Fishman (1994).

The first synapse in the electroreception-signaling pathway is located in the basal membranes of ampullary epithelial cells (Fig. A-1A). At sites of innervation of basal membranes, the afferent nerve forms "ribbon-type" synapses (Fig. A-1B). To release enough neurotransmitter to increase postsynaptic activity, most presynaptic membranes must be depolarized by at least several **millivolts.** Moreover, as discussed previously (Fig. A-2), a change in spike activity of afferents that innervate an ampullary epithelium is measurable for **microvolt** changes in transepithelial voltage. This 1000-fold or more difference between a small input (transepithelial voltage) and the change in presynaptic membrane voltage necessary for the release of neurotransmitter suggests that an important function of the ampullary epithelium must be to provide signal amplification.

To explain how amplification could be achieved in an ampullary epithelium (the first stage of signal processing (Fig. A-4A), consider the interplay of a negative conductance ($g_a < 0$), generated by Ca^2 channels in apical membranes, and a positive conductance ($g_b > 0$), generated in basal membranes. A negative conductance in the "voltage divider" circuit (Fig. A-4A) transforms attenuation of the transepithelial voltage (input signal) at basal membranes ($V_b/V_A < 1$) into amplification ($V_b/V_A > 1$). The "gain" or amplification (Fig. A-4B) of V_A at presynaptic membranes (V_b) increases as the difference in the magnitude of apical and basal membrane conductance diminishes. That is, if $g_a < 0$ and $g_b > 0$, then $(g_a + g_b) < g_a$ such that

$$V_b/V_A = g_a/(g_a + g_b) > 1.$$

Thus, the interaction of the different types of ion channels in apical (Ca^{2+}) versus those in basal membranes ($K^+ + Cl^- + Ca^{2+}$) in the two membrane faces of ampullary epithelial cells might serve to amplify the transepithelial signal. Until actual measures of intracellular voltages in ampullary epithelial cells confirm that $V_b/V_A > 1$, this explanation is tentative. This amplification scheme also provides a rationale for epithelial cells to function as the primary transducers of the input signal. That is, epithelial cells within an epithelium typically have the requisite, spatially separated membranes, apical and basal, each with different types of ion channels and conductance characteristics.

Finally, even with sufficient amplification, sensitivity to low-intensity fields can only be achieved by greatly enhancing the signal-to-noise ratio (S/N). For many years, the ability of sharks and rays to respond to weak electric fields was assumed to imply a special detection mechanism other than electric field-gated conduction through the ion channels of ampullary epithelia. However, several hundred ampullae in an animal each contribute to a confluent system in which the outputs of 10 000 epithelial cells in each ampulla (Murray, 1974) converge on a few afferent nerves. The noise contributed by individual ampullae and epithelial cells within an ampulla is incoherent, whereas the signal from these sources is coherent. If the signal and noise were averaged from several hundred ampullae, the S/N ratio would be enhanced by a factor of $\sqrt{n}$ (where n is the number of ampullae) compared to a single-stage of amplification (Fishman, 1987; Pickard, 1988). This would enhance the S/N ratio by ten-fold or more. Extrapolating from the discernible response of a single iso-

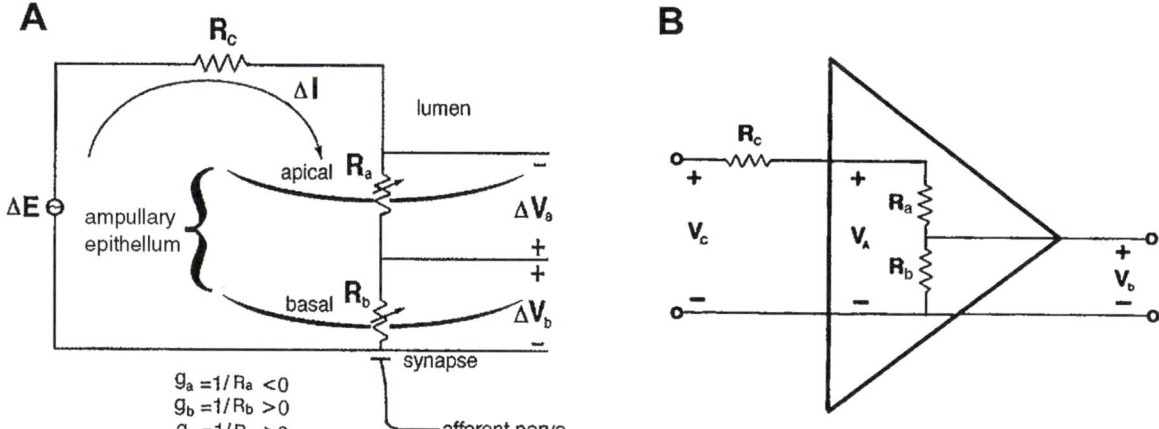

FIGURE A-4. An electrical circuit that reflects the conduction properties of an isolated ampullary organ. The highly conductive gel-filled canal ($g_C = 1/R_C > 0$) leads to the lumen of the ampullary epithelium whose conduction properties are dominated by Ca^{2+} channels in apical membranes ($g_a = 1/R_a < 0$) and basal membrane ion channels ($g_b = 1/R_b > 0$). The ampullary epithelium responds to the small-signal voltage, ΔE, with a steady state inwardly directed (apical to basal) current (ΔI) and by *changes* in both apical and basal membrane voltages, as indicated by the polarities of ΔV_a and ΔV_b (the common point between R_a and R_b corresponds to the interior of the epithelial cells). Batteries corresponding to ion driving forces are not included in the diagram because the information was derived from incremental (admittance) analyses. Portions of basal membranes constitute the presynaptic membrane of synapses formed with the afferent nerve that conveys encoded information about the transampullary epithelial voltage to the animal's brain. (B) Signal amplification ($V_b/V_A > 1$) can occur because of the separate and different conduction properties of apical and basal membranes ($R_a < 0$ and $R_b > 0$) of epithelial cells in an ampulla. Input signal, V_A, can be amplified to produce an output, V_b, affecting neurotransmitter release and postsynaptic afferent nerve activity. See text for details. (From Lu and Fishman, 1995).

lated organ to a change in transepithelial voltage of 1 μV, the detectable response from an ensemble of ampullae could be 0.1 μV. While this value is an order higher than the transepithelial voltage (10 nV) estimated for 5-cm length canals in dogfish (Petracchi and Cercignani, 1998), it nevertheless indicates that the acute sensitivity to electric fields is primarily due to the ampullary organ. Furthermore, recent considerations (Petracchi and Cercignani, 1998) suggest that S/N enhancement is accomplished by conventional means, and that the gating of ampullary epithelial-cell ion channels is still the most likely process that generates the signal.

Bibliography

Block, S . M. (1992). Biophysical principles of sensory transduction. *Soc. Gen. Physiol. Series* **47,** 1–17. Rockefeller Univ. Press, New York.

Clusin, W. T., and Bennett, M. V. L. (1977). Calcium-activated conductance in skate electroreceptors: current clamp experiments. *J. Gen. Physiol.* **69,** 121–143.

Fishman, H. M. (1987). On the responsiveness of elasmobranch fishes to weak electric fields. *In* "Mechanistic Approaches to Interactions of Electric and Electromagnetic Fields with Living Systems." (M. Blank and E. Findl, Eds.), pp. 431–436. Plenum, New York.

Hafemeister, D. (1998). "Biological Effects of Low-Frequency Electromagnetic Fields." American Association of Physics Teachers, College Park, MD.

Kalmijn, A. J. (1966). Electroreception in sharks and rays. *Nature* **212,** 1232–1233.

Kalmijn, A. J. (1982). Electric and magnetic field detection in elasmobranch fishes. *Science* **218,** 916–918.

Kalmijn, A. J. (1988). Electromagnetic orientation: a relativistic approach. *Prog. Clin. Biol. Res.* **257,** 23–45.

Lu, J., and Fishman, H. M. (1994). Interaction of apical and basal membrane ion channels underlies electroreception in ampullary epithelia of skates. *Biophys. J.* **67,** 1525–1533.

Lu, J., and Fishman, H. M. (1995a). Ion channels and transporters in the electroreceptive ampullary epithelium from skates. *Biophys. J.* **69,** 2467–2475.

Lu, J., and Fishman, H. M. (1995b). Localization and function of the electrical oscillation in electroreceptive ampullary epithelium from skates. *Biophys. J.* **69,** 2458–2466.

Murray, R. W. (1967). The function of the ampullae of Lorenzini of elasmobranchs. *In* "Lateral Line Detectors." P. Cahn, Ed. pp. 277–293. Indiana University Press, Bloomington, IN.

Murray, R. W. (1974). The ampullae of Lorenzini. *In* "Handbook of Sensory Physiology." A. Fessard, Ed. Berlin, Springer-Verlag. **3:** 125–148.

Petracchi, D., and Cercignani, G. (1998). A comment on the sensitivity of fish to low electric fields. *Biophys. J.* **75,** 217–218.

Pickard, W. F. (1988). A model for the acute electrosensitivity of cartilaginous fishes. *IEEE Trans. Biomed. Eng.* **35,** 243–249.

Waltman, B. (1966). Electrical properties and fine structure of the ampullary canals of Lorenzini. *Acta Physiol. Scand.* **66,** Suppl. 264, 1–60.

SECTION

VI

Muscle and Other Contractile Systems

Nicholas Sperelakis and Hugo Gonzalez-Serratos

51

Skeletal Muscle Action Potentials

I. Introduction

The normal contraction of twitch-type skeletal muscle cells is always preceded by an action potential (AP) or depolarization of the sarcolemma beyond the membrane potential (E_m) level at which contraction is triggered, that is, the mechanical threshold (see Chapter 54 on excitation-contraction coupling in skeletal muscle). This is the first step in the chain of events triggered by the initial excitatory process in the sarcolemma, linking it to the final mechanical response. This chain of events is known as the excitation-contraction (E-C) coupling process. In this chapter, we will study the sarcolemmal electrophysiological properties that are the bases of the AP generation. Action potential generation and excitability in neurons were covered in an earlier chapter. Most of the general electrophysiological principles discussed there also apply to skeletal muscle fibers, and so are only briefly reviewed and summarized in this chapter. In most respects, the electrogenesis of the APs in nerve axons and skeletal muscle fibers is quite similar, except for their types of afterpotentials, saltatory propagation in myelinated axons, and high Cl^- conductance in skeletal muscle fibers. Both are long cables and have very brief and fast-rising APs whose inward current is carried by Na^+ ions through fast Na^+ channels. Skeletal muscle fibers, however, have the added complexity of an extensive internal transverse (T) tubular system formed by a periodic invagination of the surface cell membrane (Fig. 1), forming an orderly three-dimensional array of tubules that propagate excitation from the cell surface into the deep interior of the fiber for purposes of excitation-contraction coupling.

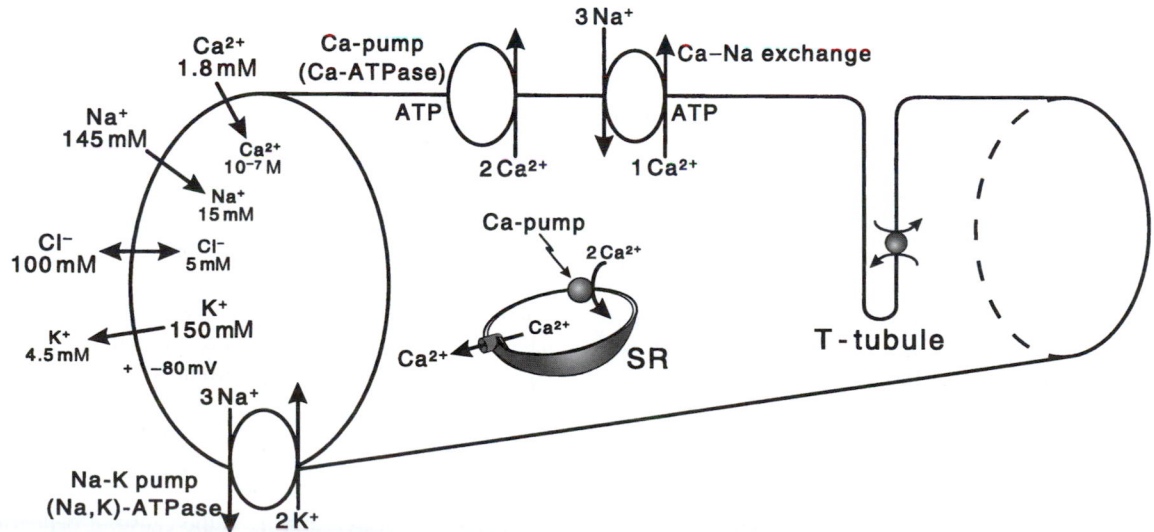

FIGURE 1. Intracellular and extracellular ion distributions in vertebrate skeletal muscle fibers. Also shown are the polarity and magnitude of the resting potential. Arrows indicate direction of the net electrochemical gradient. The Na^+-K^+ pump and Ca^{2+}-Na^+ exchange carrier are located in the cell surface and in the T-tubule membranes. A calmodulin-dependent Ca^{2+}-ATPase and Ca^{2+} pump, similar to that in the sarcoplasmic reticulum (SR), is located in the cell surface and T-tubule membranes.

The skeletal AP is considerably different from the AP of cardiac muscle cells which has a very long duration with a pronounced plateau and a substantially lower rate of rise and propagation velocity. The myocardial cells are short, and there is a slight delay in propagation at each cell-to-cell junction. Like skeletal muscle fibers, myocardial cells have a fast I_{Na} responsible for the rapid upstroke of the AP, but the delayed rectifier K^+ conductance is turned on slowly and there is a substantial inward I_{Ca} during the entire plateau.

The APs of smooth muscles are markedly different from those of skeletal muscle fibers, in that the resting (takeoff) potential is lower (more depolarized), the rate of rise of the APs is much slower, the AP overshoot is much less, and the AP duration (APD) is considerably longer. Propagation velocity is much slower, and the smooth muscle cells are short and small in diameter. The inward current for the APs in smooth muscle cells is primarily a slow Ca^{2+} current carried through L-type Ca^{2+} channels, but some cells do possess some functioning fast Na^+ channels.

II. General Overview of Electrogenesis of the Action Potential

The ion distributions and ion pumps and exchangers found in skeletal muscle fibers are similar to those of other types of cells, as described in the chapter on the resting potential (Fig. 1). The APs in vertebrate skeletal muscle twitch fibers consist of a **spike** followed by **depolarizing ("negative") afterpotential** (Figs. 2 and 3). A large fast inward Na^+ current, passing through fast Na^+ channels, is responsible for electrogenesis of the spike, which rises rapidly (400–700 V/s). Subsequently, a small slow inward Ca^{2+} current, passing through kinetically slow channels, may be involved in excitation-contraction coupling. The skeletal muscle cell membrane has at least two types of voltage-dependent K^+ channels (Fig. 4). One type allows K^+ ions to pass more readily inward than outward, the so-called **inward-going rectifier**. This channel is responsible for **anomalous rectification** (i.e., decrease in g_K with depolarization). There is a quick decrease in K^+ conductance on depolarization and increase in K^+ conductance with repolarization. The second type of K^+ channel is similar to the usual K^+ channel found in squid giant axon membrane, the so-called **delayed rectifier**. Its conductance turns on more slowly than g_{Na} on depolarization. This channel allows K^+ to pass readily outward down the electrochemical gradient for K^+. The activation of this channel produces the large increase in total g_K that helps to terminate the AP (see Fig. 3).

The AP amplitude is about 120 mV, from a resting potential of –80 mV to a peak overshoot potential of about +40 mV (see Figs. 2 and 3). The duration of the AP (at 50% repolarization, or APD_{50}) ranges between 3 and 6 ms, depending on the species and temperature. The **threshold potential** (V_{th}) for triggering of the fast Na^+ channels is about –55 mV; thus, a **critical depolarization** of about 25 mV is required to reach V_{th}. The turn-on of the fast g_{Na} (fast I_{Na}) is very rapid (within 0.2 ms), and E_m is brought rapidly toward E_{Na} (Figs. 2 and 3). There is an explosive (positive exponential initially) increase in g_{Na}, caused by a positive feedback relationship between g_{Na} and E_m.

From the current versus voltage curves, the maximum inward fast Na^+ current occurs at an E_m of about –20 mV. The current decreases at more depolarized E_m levels because of the diminution in electrochemical driving force as the membrane is further depolarized, even though the conductance remains high. At the reversal potential (E_{rev}) for the current, the current goes to zero; I_{Na} then reverses direction with greater depolarization.

As E_m depolarizes, it crosses V_{th} for slow Ca^{2+} channels located in the transverse tubules, which is about –35 mV. Turn-on of the Ca^{2+} conductance (g_{Ca}) and I_{Ca} is slower. The peak I_{Ca} is considerably smaller than the peak I_{Na}.

III. Ion Channel Activation and Inactivation

As discussed in the chapter on nerve excitability, the fast Na^+ channels (and the slow Ca^{2+} channels) have a double gating mechanism; one gate is the **inactivation gate** (I-gate) and the second gate is the **activation gate** (A-gate). For a channel to be conducting, both the A-gate and I-gate must be open; if either one is closed, the channel is nonconducting. The A-gate is closed at the resting E_m and opens rapidly on depolarization, whereas the I-gate is open at the resting E_m and closes slowly on depolarization. In the Hodgkin-Huxley (1952) analysis, the opening of the A-gate requires simultaneous occupation of three negatively charged sites by three positively charged m^+ particles. Therefore

$$g_{Na} = \bar{g}_{Na} m^3 h \qquad (1)$$

where m is the activation variable, h is the inactivation variable, and $\bar{g}_{Na}$ is the maximum conductance. A small gating current (I_g) has been measured that corresponds to the movement of the charged m particles (or rotation of an equivalent dipole). The outward I_g leads into the inward I_{Na}.

The fast I_{Na} lasts only for 1–2 ms because of the spontaneous voltage **inactivation** of the fast Na^+ channels, that is, they inactivate quickly, even if the membrane were to remain depolarized. Inactivation is produced in the ion channels by the voltage-dependent closing of the inactivation gate (I-gate).

The voltage dependency of inactivation is given by the h_∞ versus E_m curve. The Na^+ conductance (g_{Na}) at any time is equal to the maximal value ($\bar{g}_{Na}$) times $m^3 h$. Therefore, when $h = 0$, $g_{Na} = 0$, and when $h = 1.0$, $g_{Na} = \bar{g}_{Na}$ (if $m = 1.0$). At the normal resting potential, h_∞ is nearly 1.0 and diminishes with depolarization, becoming nearly zero at about –30 mV. The maximal rate of rise of the AP (max dV/dt) is directly proportional to the net inward current or I_{Na}, which is directly proportional to g_{Na} and can be expressed as

$$\max dV/dt \ \propto \ \frac{I_{Na}}{C_m} = \frac{\bar{g}_{Na} \, m^3 \, h \left(E_m - E_{Na} \right)}{C_m} \qquad (2)$$

Therefore, the decrease in h_∞ is the cause of decrease in max dV/dt. At about –30 mV, $h_\infty \to 0$, and there is nearly complete inactivation of the fast Na^+ channels. Thus, depolarization by any means (e.g. elevated $[K^+]_o$ or applied current pulses) decreases max dV/dt, and excitability disappears at about –50 mV.

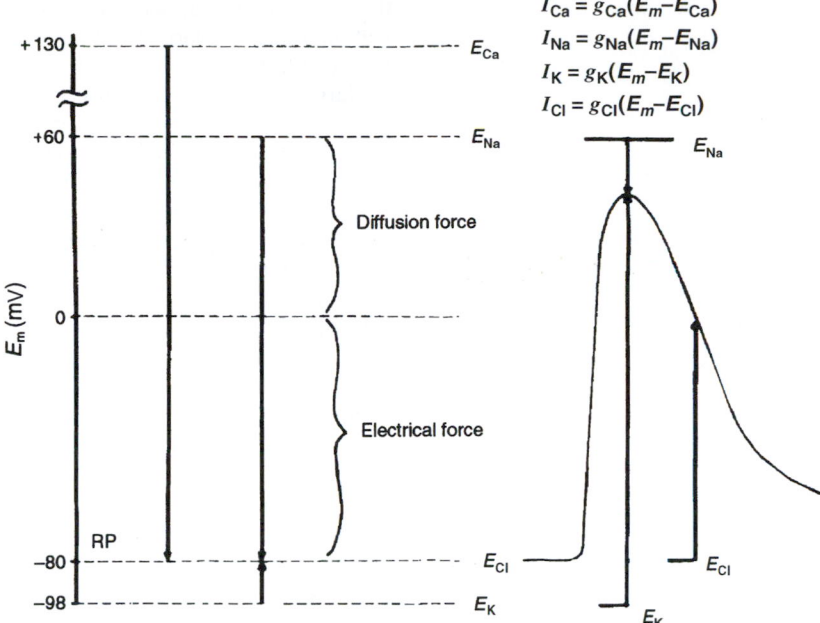

$$I_{Ca} = g_{Ca}(E_m - E_{Ca})$$
$$I_{Na} = g_{Na}(E_m - E_{Na})$$
$$I_{K} = g_{K}(E_m - E_{K})$$
$$I_{Cl} = g_{Cl}(E_m - E_{Cl})$$

FIGURE 2.   Representation of the electrochemical driving forces for Na^+, Ca^{2+}, K^+, and Cl^- at rest (left diagram) and during the AP in a skeletal muscle fiber (right diagram). Equilibrium potentials for each ion (e.g., E_{Na}) are positioned vertically according to their magnitude and sign; they were calculated from the Nernst equation for a given set of extracellular and intracellular ion concentrations. Measured resting potential is assumed to be –80 mV. Electrochemical driving force for an ion is the difference between its equilibrium potential (E_i) and the membrane potential (E_m), that is, ($E_m - E_i$). Thus, at rest, the driving force for Na^+ is the difference between E_{Na} and the resting E_m; if E_{Na} is +60 mV and resting E_m is –80 mV, the driving force is 140 mV. The driving force is then the algebraic sum of the diffusion force and the electrical force, and is represented by the length of the arrows in the diagram. Driving force for Ca^{2+} (about 210 mV) is even greater than that for Na^+, whereas that for K^+ is much less (about 18 mV). Direction of the arrows indicates the direction of the net electrochemical driving force, namely, the direction for K^+ is outward, whereas that for Na^+ and Ca^{2+} is inward. If Cl^- is passively distributed, then for a cell sitting a long time at rest, $E_{Cl} = E_m$ and there is no net driving force. The driving forces change during the AP, as depicted. The equations for the different ionic currents are given in the upper right-hand portion of the figure. (Adapted from Sperelakis, 1979.)

The **slow Ca^{2+} channels** behave much the same way as the fast Na^+ channels with respect to inactivation, with one main difference being the voltage range over which the slow channels inactivate: –45 to –10 mV for the slow channels, compared to –80 to –30 mV for the fast Na^+ channels. Another major difference is that the slow Ca^{2+} conductance inactivates much more slowly than the fast Na^+ conductance; that is, they have a long inactivation time constant (τ_{inact}). (The h variable for the slow channel is sometimes referred to as the f variable, and the m variable as the d variable.) One other difference is that, because slow Ca^{2+} channels are located in the T-tubular system, their inactivation is affected by tubular Ca^{2+} depletion caused by the Ca^{2+} ions that flow into the myoplasm during I_{CaS}. The recovery process for the slow Ca^{2+} channels is slow compared to 1–2 ms for fast Na^+ channels.

The K^+ channel (delayed rectifier) may have only an activation gate, because it does not inactivate quickly. In the Hodgkin–Huxley analysis of squid giant axon, the A-gate opens when four positively charged n^+ particles simultaneously occupy four favorable positions (negatively charged sites). If n is the probability that one site is occupied, then n^4 is the probability that all four sites are occupied. Therefore,

$$g_{K} = \bar{g}_{K} n^4 \qquad (3)$$

The fourth power to which n must be raised causes a delay (sigmoidal foot) in turn-on of the K^+ conductance.

IV. Mechanisms of Repolarization

The skeletal AP is terminated partly by the turn-on of the K^+ conductance (g_K) (see Fig. 3). The increase in g_K acts to bring E_m towards E_K (about –98 mV), since the membrane potential at any time is determined primarily by the ratio of g_{Na}/g_K. This type of g_K channel is activated by depolarization and turned off by repolarization. Therefore, this g_K channel is **self-limiting**, in that it turns itself off as the membrane is repolarized by its action.

In addition to the g_K turn-on, turnoff of g_{Na} occurs (see Fig. 3) (contributing to repolarization) for two reasons: (1) spontaneous inactivation of fast Na^+ channels that had been activated, that is, closing of their I-gate (inactivation τ of 1–3 ms); and (2) reversible shifting of activated channels directly back to the resting state (deactivation), because of the rapid repolarization occurring due to the g_K increase (Fig. 5). Theoretically, it would be possible to have an AP that

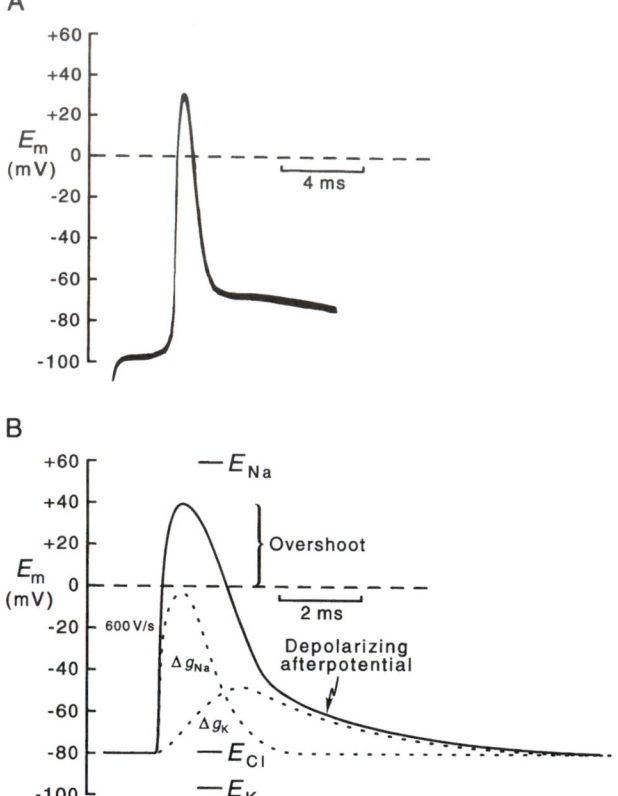

FIGURE 3. (A) Action potential (AP) recorded with an intracellular microelectrode in a skeletal muscle fiber of frog semitendinosus muscle bathed in normal frog Ringer's solution. Note the prominent depolarizing afterpotential. Shock artifact is at left of spike. (Modified from Sperelakis *et al.,* 1973.) (B) Diagrammatic representation of the relative conductance changes for Na^+ and K^+ during an AP. The rising phase of the AP is caused by an increase in g_{Na}, which brings the membrane potential (E_m) toward the Na^+ equilibrium potential (E_{Na}). The falling phase in the AP is due to the rise in g_K and to the decrease in g_{Na} and to an outward Cl^- current. The depolarizing afterpotential is explained in part by the contribution of the AP traveling down the transverse tubular network and in part by the fact that the delayed rectifier K^+ channel is less selective for K^+ (30:1 over Na^+) than is the resting channel (100:1).

would repolarize (but more slowly) even if there were no g_K mechanism, because the g_{Na} channels would spontaneously inactivate, and so the g_{Na}/g_K ratio and E_m would be slowly restored to their original resting values.

In skeletal muscle, there is an important third factor involved in repolarization of the AP: the **Cl^- current** (see Fig. 2). The Cl^- permeability (P_{Cl}) and conductance (g_{Cl}) are very high in skeletal muscle. In fact, P_{Cl} of the surface membrane is much higher than P_K, the P_{Cl}/P_K ratio being about 3–7. As discussed in the chapter on resting potential, the Cl^- ion is passively distributed, or nearly so, and thus cannot determine the resting potential under steady-state conditions. However, net Cl^- movements inwards (hyperpolarizing) or outward (depolarizing) do affect E_m transiently until reequilibration occurs and there is no further net movement. Although there is no net electrochemical driving force for Cl^- current (I_{Cl}) at

the resting potential, since $E_m = E_{Cl}$, during the AP depolarization, there is a larger and larger driving force for outward I_{Cl} (i.e., Cl^- influx), since $I_{Cl} = g_{Cl} (E_m - E_{Cl})$. In other words, the large electric field that was keeping Cl^- out (i.e., $[Cl^-]_i \ll [Cl^-]_o$) diminishes during the AP, and so Cl^- ion enters the fiber. This Cl^- entry is hyperpolarizing, and so tends to repolarize the membrane more quickly than would otherwise occur. That is, repolarization of the AP is sharpened by the Cl^- mechanism. There may be voltage-dependent g_{Cl} channels, as well as a high g_{Cl} in the membrane at rest.

To further illustrate some of the preceding points on the role of Cl^-, when skeletal muscle fibers are placed into Cl^--free Ringer solution (e.g., methanesulfonate substitution), depolarization and spontaneous APs and twitches occur for a few minutes until most or all of the $[Cl^-]_i$ is washed out. After equilibration, the resting E_m returns to the original value ca. –90 mV for frog skeletal muscle and –80 mV for mammalian, clearly indicating that Cl^- does not determine the resting potential and that net Cl^- efflux produces depolarization. Readdition of Cl^- to the bath produces a rapid large hyperpolarization, for example, to –120 mV, due to net Cl^- influx; the E_m then slowly returns to the original value (–90 mV) as Cl^- reequilibrates, that is, redistributes itself passively. These same effects occur in cardiac muscle, smooth muscle, and nerve, but to a lesser extent, because in these tissues P_{Cl} is much lower (for example, P_{Cl}/P_K ratio is only about 0.5 in vascular smooth muscle).

The importance of the Cl^- current in repolarization in skeletal muscle fibers is illustrated by one type of **myotonia** in which an abnormally low P_{Cl} causes repetitive APs to occur. Because g_{Cl} is abnormally low, total membrane conductance G_m is also low. From the relationship between membrane current (I_m) and G_m: $I_m/G_m = E_m$, it can be deduced that, because G_m is low in these muscle cells, only a smaller than normal outward depolarizing capacitative current I_m is necessary to reach threshold E_m for an AP. Since g_{Cl} is abnormally low, making G_m also low, the membrane resistance R_m will be abnormally high, making the space constant λ (see Eq. 5) larger than normal. Consequently, APs will propagate at a faster velocity while travelling along the muscle cell and into the T-tubular membranes (see Eq. 8 and Section XIV.A in this chapter.) Also, because g_{Cl} is abnormally low, the Cl^- influx during AP repolarization is much less than normal, and so the repolarization process is slowed, thus increasing the duration of the AP. As a consequence of the above, when depolarization occurs, the AP threshold is easily and quickly reached, and the generated APs spread easily and fast along the sarcolemma and into the T-tubular membranes. These factors make the whole system unstable, and oscillations trigger repetitive discharge of APs in the skeletal muscle fibers. That is, the muscle fibers lose their tight control by the motor neurons, and so contraction becomes partly involuntary. For example, persons with myotonia find it difficult to release a handshake or to remove their hand from a drinking glass. There are several causes of myotonia, including genetic abnormalities in ion channels as well as drug-induced conditions. Any agent that greatly lowers P_{Cl} or g_{Cl} will have the same effect. It has been shown that simply decreasing g_{Cl} causes repetitive firing in equivalent circuit models of skeletal

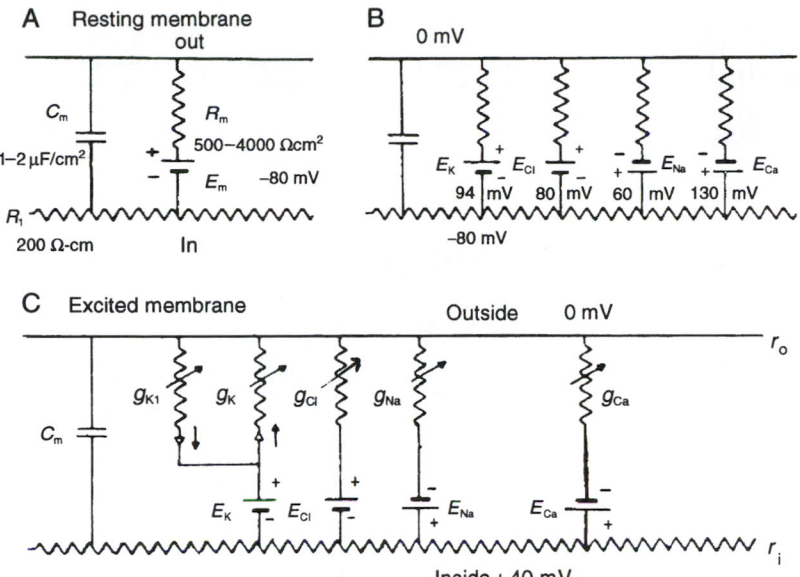

FIGURE 4.   Electrical equivalent circuits for a skeletal muscle fiber cell membrane (A and B) at rest and (C) during excitation. (A) Membrane as a parallel resistance-capacitance circuit, the membrane resistance (R_m) being in parallel with the membrane capacitance (C_m). Resting potential (E_m) is represented by an 80-mV battery in series with the membrane resistance, the negative pole facing inward. (B) Membrane resistance is divided into four component parts, one for each of the four major ions of importance: K^+, Cl^-, Na^+, and Ca^{2+}. Resistances for these ions (R_K, R_{Cl}, R_{Na}, and R_{Ca}) are parallel to one another and represent totally separate and independent pathways for permeation of each ion through the resting membrane. These ion resistances are depicted as their reciprocals, namely, ion conductances (g_K, g_{Cl}, g_{Na}, and g_{Ca}). Equilibrium potential for each ion (e.g., E_K), determined solely by the ion distribution in the steady state and calculated from the Nernst equation, is shown in series with the conductance path for that ion. Resting potential of –80 mV is determined by the equilibrium potentials and by the relative conductances. (C) Equivalent circuit is further expanded to illustrate that, for the voltage-dependent conductances, there are at least two separate K^+-conductance pathways (labeled here g_{K1} and g_K). In series with the K^+ conductances are rectifiers pointing in the direction of least resistance to current flow. There is one Na^+ conductance pathway, the kinetically fast Na^+ conductance (g_{Na}). In addition, there is a kinetically slow pathway that allows Ca^{2+} to pass through. Arrows drawn through the resistors indicate that the conductances are variable, depending on membrane potential and time. (Adapted from Sperelakis, 1979.)

muscle fibers. In addition, K^+ ions tend to accumulate in the lumen of the T-tubules under normal conditions (see Section VIII on late depolarizing afterpotentials), and this accumulation is exaggerated with the prolonged APs, and so tends to partially depolarize the fibers and increase their excitability. Some forms of myotonia are produced by **abnormal fast Na^+ channels**; namely, a small fraction of these channels does not inactivate as quickly as usual (i.e., their I-gates do not close normally), and so causes a prolonged small depolarization after the AP and repetitive discharge.

V. Voltage-Dependent Cl⁻ Channels

As discussed previously, the Cl^- conductance (g_{Cl}) of skeletal muscle fibers is very high normally, and it is important for producing a sharp repolarization of the AP. As stated, one type of myotonia is caused by an abnormally low Cl^- conductance, and therefore low Cl^- current (I_{Cl}), which leads to repetitive firing of APs. Also as mentioned previously, the P_{Cl}/P_K ratio is about 3–7 in skeletal muscle, thus indicating that the dominant permeability or conductance (g_{Cl}/g_K ratio) is Cl^-.

This high g_{Cl} is due to a large number of voltage-dependent gated Cl^- channels, which are outwardly recti-

fying. These Cl^- channels are located primarily on the surface sarcolemma in frog skeletal muscle, there being only a few in the T-tubules; whereas in rat or mouse, these Cl^- channels are primarily in the T-tubules, there being very few on the surface membrane. Denervation of mammalian fibers causes g_{Cl} to decrease almost to zero.

In frog, there are several subtypes of Cl^- channels that have single-channel conductances ranging between 40 and 70 pS, and each channel may exhibit several subconductance states. Those that do, often have a main gate that opens or closes the entire channel. In fetal mammalian fibers, Cl^- channels with conductances of about 40, 60, and 300 pS have been observed. Myoballs cultured from muscle biopsies of patients having one form of myotonia had a reduced (ca. 50%) single-channel conductance for the Cl^- channel, which would contribute to the myotonia (Fahlke *et al.*, 1993). In primary cultures of rat skeletal muscle, the fast Cl^- channel showed a behavior consistent with six closed states and two open states (Weiss and Magleby, 1992). The Cl^- channel in myoblasts and myotubes of the L6 cell line derived from rat skeletal muscle had a high conductance of about 330 pS (Hurnak and Zachar, 1992). Voltage-gated Cl^- channels have also been found in the SR membrane of skeletal muscle.

Some Cl^- channels described for other tissues include (1) Ca^{2+}-dependent Cl^- channels, (2) stretch-activated Cl^- channels,

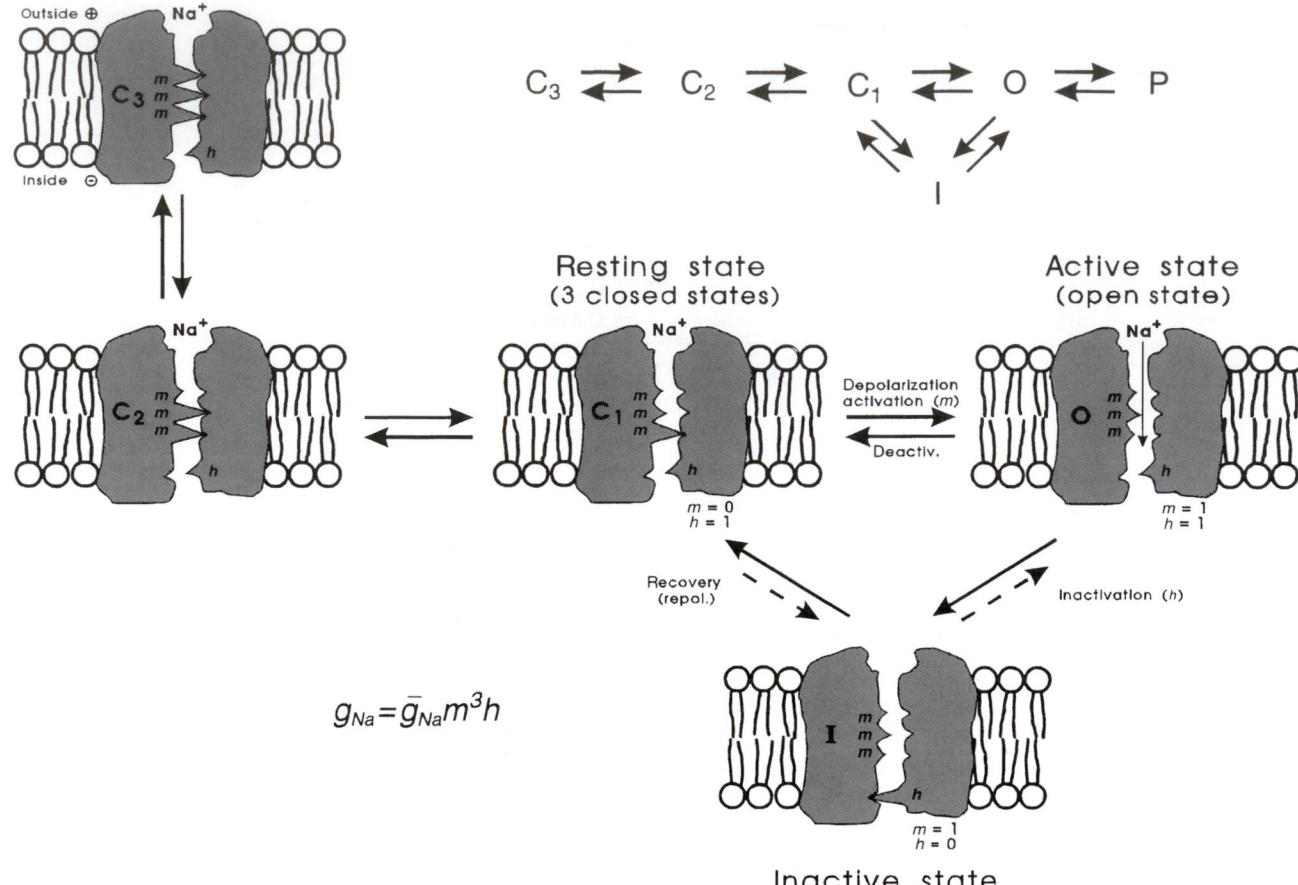

$$C_3 \rightleftarrows C_2 \rightleftarrows C_1 \rightleftarrows O \rightleftarrows P$$

$$g_{Na} = \bar{g}_{Na} m^3 h$$

FIGURE 5.   Illustration of the hypothetical states of the fast Na$^+$ channel. The three states patterned after the Hodgkin-Huxley view were modified to reflect the fact that there is evidence for three closed states. As depicted, in the most closed state (C_3), all three m gates (or particles) are in the closed configuration. In the mid-closed state (C_2), two m gates are closed and one is open. In the least closed state (C_1), one gate is closed and two are open. In the resting state, the activation gate (A) is closed and the inactivation gate (I) is open: $m = 0$, $h = 1$. Depolarization to the threshold activates the channel to the active state, the A-gate opening rapidly and the I-gate still being open: $m = 1$, $h = 1$. The activated channel spontaneously inactivates to the inactive state due to closure of the I-gate: $m = I$, $h = 0$. The recovery process on depolarization returns the channel from the inactive state back to the resting state, thus making the channel again available for reactivation. Na$^+$ ion is depicted as being bound to the outer mouth of the channel and poised for entry down its electrochemical gradient when both gates are in the open configuration. The reaction between the resting state and the active state is readily reversible, and there is some reversibility of the other reactions. The fast Na$^+$ channel is blocked by tetrotodoxin (TTX) binding to the outer mouth and plugging it.

and (3) cyclic AMP-stimulated Cl$^-$ channels. The receptor-operated Cl$^-$ channels apparently have a G-protein (e.g., G_s or G_i) as intermediate for coupling.

The voltage-dependent Cl$^-$ channels can be blocked relatively selectively by several methods, including acidosis and use of compounds such as the stilbene derivatives (DIDS and SITS) and 9-anthracene carboxylic (9-AC) acid. The Cl$^-$ channels in frog skeletal muscle are relatively insensitive to 9-AC acid, whereas those in adult mammalian muscle are highly sensitive. The anion selectivity sequence for some voltage-dependent Cl$^-$ channels is I$^-$ > Br$^-$ > Cl$^-$ > F$^-$.

VI. ATP-Dependent K$^+$ Channels

ATP-dependent K$^+$ (K_{ATP}) channels are characterized by the inhibition of channel openings by ATP. In addition to being ligand sensitive, these K_{ATP} channels are voltage-dependent; the open state probability increases with depolarization. Most of the studies of K_{ATP} channels have been performed using patch-clamp of isolated inside-out patches, including skeletal muscle (Spruce *et al.*, 1987). The unitary conductance of this channel varies with [K$^+$]$_o$ that ranges from 15 pS in 2.5 mM [K$^+$]$_o$ to 42 pS in 60 mM [K$^+$]$_o$. K_{ATP} channels are closed when the membrane is polarized and when the ATP concentration in the intracellular myoplasm is in the range of 1.0 mM or higher, which is the normal physiological concentration. A decrease in ATP concentration or depolarization activates the channel. The half-maximum inhibition of this channel opening by ATP, measured at a constant [K$^+$]$_o$/[K$^+$]$_i$, is 0.135 mM at pH 7.2 (Fig. 6). Hydrolysis of ATP (into ADP + P$_i$) is not required for ATP to close K_{ATP} channels (Spruce *et al.*, 1987).

Although ATP is the most specific ligand to close K_{ATP} channels, there are other ligands that modulate these channels. The metabolites produced during contraction, like ADP,

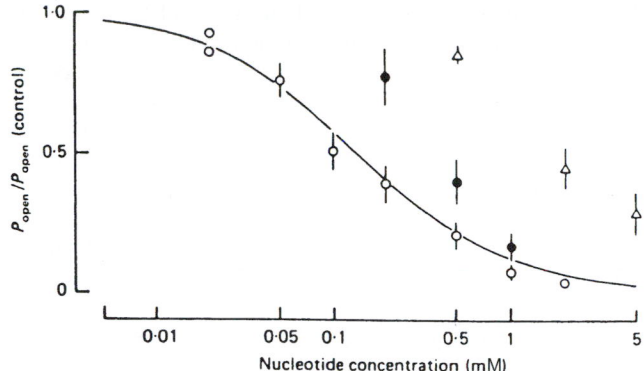

FIGURE 6. Effect of ATP (open circles), ADP (filled circles) and AMP (open triangles) on closing the ATP-regulated K⁺ channels. Ordinate: open-state probability (P_o) of the channel relative to its value in nucleotide-free solution (P_{open} control). (Reproduced with permission from Spruce *et al.*, 1992)

Mg^{2+}, and H^+ (see Chapter 56 on contractility of muscle) also modulate K_{ATP} channels. The two adenine nucleotides, ADP and AMP, retain the ability to modulate these channels. ADP and AMP (in the absence of ATP) also can block K_{ATP} channels in a dose-dependent manner. However, they are less effective than ATP in closing the channels; the affinity of ADP and AMP are approximately 3 and 16 times lower than for ATP, respectively (Fig. 6). ATP analogs like GTP, ITP, XTP, CTP, and UTP (in which the ribose or the adenosine moieties are replaced) also have reduced effectiveness (about tenfold) in closing these channels. In the presence of ATP, 0.3 mM ADP reduces the K_{ATP} channel current at ATP concentrations below 1 µM. At higher ATP concentrations, ADP increases the currents. Furthermore, ADP shifts the ATP dose-response curve, increasing the maximum half-inhibition concentration by about tenfold (Spruce *et al.*, 1987; Vivaudou *et al.*, 1991). These results are consistent with competition between ATP and ADP for the K_{ATP} channel.

A decrease in pH at the cytoplasmic surface (pH_i) reduces the degree of K_{ATP} channel inhibition caused by ATP. The ATP concentration for half-inhibition is 2.5 times and 15 times greater at pH_i 6.8 and 6.3, respectively, as compared to that at pH_i 7.2. Therefore, the curve that relates K_{ATP} channel opening to ATP concentration is shifted to the right. Thus, during acidosis, a smaller decrease in ATP concentration will lead to a larger opening of K_{ATP} channels. In other words, more K_{ATP} channels are activated by a decrease in pH_i at the same ATP concentration. Perhaps proton binding to the channel prevents ATP binding (Davies *et al.*, 1992). Mg^{2+} has a similar effect to protons. Increases of cytosolic Mg^{2+} (millimolar level) reduces the ability of ATP to close K_{ATP} channels. The ATP concentration for half-inhibition is 77 µM in the presence of 0.1 mM cytoplasmic Mg^{2+} concentration and 203 µM in 3 mM (Vivaudou *et al.*, 1991). This inhibitory effect may be caused by the ability of Mg^{2+} to bind to ATP. In addition, intracellular Mg^{2+} may reduce the K_{ATP} current at positive membrane potentials by binding inside the channels and plugging them (Woll *et al.*, 1989).

K_{ATP} channels have been studied in excised patches from surface membrane blebs from muscles of frogs and adult

mammals. These channels are, therefore, likely present in the sarcolemma. It is not known whether they are present in the transverse tubular system. From the high number of K_{ATP} channels (3–10) found in the patches from frog skeletal muscle, the K_{ATP} channel density has been estimated to be as high as 10 channels per µm² of surface membrane (Spruce *et al.*, 1985), a density equal to or higher than that for K⁺ delayed rectifier channels (Standen *et al.*, 1985).

At present, there is no clear understanding of the physiological role played by K_{ATP} channels in skeletal muscle. Under physiological conditions, the intracellular ATP concentration is about 5 mM at rest (Dawson *et al.*, 1978, 1980). This ATP concentration is much higher than the concentration necessary to close half the K_{ATP} channels. Therefore, at the resting ATP concentration, the K_{ATP} channels should be inactive (Spruce *et al.*, 1987; Davies *et al.*, 1992). Even during repetitive contractions that lead to muscle fatigue, the ATP concentration is maintained at near normal levels by the action of creatine kinase and creatine phosphate (Carlson and Siger, 1960; Nassar-Gentina *et al.*, 1978).

It has been proposed that K_{ATP} channels may be associated with the decrease in force development underlying muscle fatigue. A drug (SR44866) that opens K_{ATP} channels (closed by ATP) in sarcolemmal membrane patches from frog skeletal muscles also reduces the AP duration, the early afterpotential, and the peak twitch (without affecting the resting potential) in intact frog muscles (Sauviat *et al.*, 1991). The high K⁺ permeability found in muscle fibers that have undergone a permanent contracture (rigor) produced by repetitive stimulation when poisoned with cyanide and iodocetate (Fink and Lüttgau, 1976), may be caused by the decreased ATP concentration opening K_{ATP} channels. As described in Chapter 56 ATP breaks the bond between myosin cross-bridges and actin filaments, permitting cross-bridge cycling of attachment and detachment to actin. Thus, the loss of ATP leads to a permanent attachment of myosin and actin (rigor). In metabolically-exhausted frog semitendinosus muscle fibers (nonphysiological), the addition of tolbutamide and glyburide, two K_{ATP} channel antagonists, significantly reduces the K⁺ efflux rate (Castle and Haylett, 1987). As said above, even under fatigue induced by prolonged repetitive stimulation, the decrease in intracellular ATP is small (Nassar-Gentina *et al.*, 1978), and fatigued muscle fibers can contract when Ca^{2+} is released directly from the intracellular SR stores (Gonzalez-Serratos *et al.*, 1978; Garcia *et al.*, 1991) (see Chapter 55 for a discussion of Ca^{2+} release from sarcoplasmic reticulum in muscle). In intact animals, some of the skeletal muscle fatigue results from synaptic fatigue, including synaptic fatigue at the neuromuscular junction.

An attractive candidate for regulation of K_{ATP} channel activity is pH_i. As stated previously, a decrease in pH_i (e.g., from lactic acid production) reduces the inhibitory effect of ATP on K_{ATP} channels (Davies *et al.*, 1992), leading to an increase in the probability of K_{ATP} channels opening when compared to the same ATP concentration at a higher pH_i. That is, a decrease in pH_i increases the ATP concentration required for half-inhibition of channel activity. It has been shown that during exercise, with fatigue development, pH_i may decrease by about one unit (Renaud, 1989). A possible functional role of K_{ATP} channels is that during prolonged repetitive mechanical activation of skeletal muscle, the small decrease in ATP concentration is accompanied by an increase in ADP, H^+, and Mg^{2+} concentrations, and that the overall combination of these chemical changes may then lead to the activation of K_{ATP} channels. As K_{ATP} channels open, they contribute to the increased K^+ efflux found during repetitive muscle contraction. Because of the restricted diffusion out of the T-tubular system (see Chapter 54) and the closeness of the intercellular fiber spacing, extracellular K^+ concentration increases, especially inside the tubular system. This may cause a decrease in cell excitability and generation of APs, which may be reflected as decreased force development. Even though speculative, this mechanism may protect skeletal muscle cells from large ATP depletions that would have deleterious effects.

VII. Slow Delayed Rectifier K+ Current

Two types of K^+ delayed rectifier currents occur in skeletal muscle. A slow I_K was first described by Adrian and co-workers (1970a,b) in voltage-clamped frog sartorius fibers (at 30 °C); the slow component of outward I_K reached a maximum in about 3 s (at -30 mV) and declined with a time constant of about 0.5 s (at -100 mV). In voltage-clamped frog toe muscle, Lynch (1978) observed that, with depolarizing clamps more positive than the threshold of -55 mV for activation of the outward K^+ current (about 5 mS/cm^2), most fibers had both fast and slow components of the outward I_K. The voltage dependences of both K^+ currents were shifted equally in the depolarizing direction by elevated $[Ca^{2+}]_o$ or $[H^+]_o$, presumably due to altering the net negative outer surface charge of the membrane, thereby hyperpolarizing (see Chapter 14 on resting potential). Acidosis also increased the rate of turn-on of the slow delayed rectifier ($pK_a = 5.8$). The fast delayed current was relatively selectively blocked by TEA or by a sulfhydryl reagent, whereas the slow delayed current was selectively depressed by a histidine reagent. It was estimated that about 25% of the delayed rectifier channels are in the T-tubular membrane. The functional significance of the slow outward I_K is unknown, although it may be partly responsible for the delayed depolarizing afterpotential (see following section).

VIII. Electrogenesis of Depolarizing Afterpotentials

As mentioned previously in this chapter, the AP spike in skeletal muscle fibers is followed by a prominent **depolarizing afterpotential** (also called a negative afterpotential based on the old terminology from external recording) (see Fig. 3). In addition to this early depolarizing afterpotential (i.e., emerging from the spike), there is a late depolarizing afterpotential that follows a tetanic train of spikes (e.g., 10 spikes). The electrogenesis of the early and late afterpotentials are different. The early afterpotential is due to a T-tubular membrane conductance change, whereas the late afterpotential is due primarily to K^+ accumulation in the T-tubules.

The **early depolarizing afterpotential** of frog skeletal fibers is about 25 mV in amplitude immediately after the spike component, and gradually decays to the resting potential in 10–20 ms. It was shown by Adrian *et al.* (1970a) that this decay results from the fact that the delayed rectifier K^+ channel that opens during depolarization to terminate the spike is less selective for K^+ (ca. 30: 1, K^+ : Na^+) than is the K^+ channel in the resting membrane (ca. 100:1). Therefore, the constant-field equation (see Chapter 14) predicts that the membrane should be partly depolarized when the membrane potential is dominated by this K^+ conductance that is turned on during the AP. Thus, the early depolarizing afterpotential is partly due to the persistence and slow decay of this less-selective K^+ conductance. Based on the electrophysiological characteristics and ionic conductances of the surface membrane, Adrian and Peachey (1973) reconstructed the surface-recorded AP from striated muscle. The time course of the AP, shape of the early afterpotential depolarization, and the conduction velocity could be best reconstructed by giving a value to the access resistance of the T-tubular system, activatable Na^+ and K^+ tubular membrane currents, and velocity of the tubular AP. Thus, the early depolarizing afterpotential reflects, in part, the tubular AP. This is also confirmed as shown by the disappearance of the early afterpotential recorded from muscles in which the T-tubular system has been disrupted and disconnected from the surface membrane by the glycerol osmotic shock method[2] (Eisenberg and Gage, 1969).

The **late depolarizing afterpotential** of frog skeletal fibers may result from accumulation of K^+ ions in the T-tubules (Adrian and Freygang, 1962). During the AP depolarization and turn-on of g_K (delayed rectifier), there is a large driving force for K^+ efflux from the myoplasm coupled with a large K^+ conductance, resulting in a large outward K^+ current [$I_K = g_K \cdot (E_m - E_K)$] across all surfaces of the fiber, namely across the surface sarcolemma and T-tubule walls. The K^+ efflux at the fiber surface membrane can rapidly diffuse away and mix with the relatively large interstitial fluid (ISF) volume, whereas the K^+ efflux into the T-tubules (TT) is trapped in this restricted diffusion space. The resulting high $[K^+]_{TT}$ decreases E_K across the T-tubule membrane, and thereby depolarizes this membrane. Because of cable properties, part of this depolarization is transmitted to the surface sarcolemma, and is recorded by an intracellular microelectrode. The K^+ accumulation in the T-tubules can only be dissipated relatively slowly by diffusion out of the mouth of the T-tubules, and by active pumping back into the myoplasm (across the T-tubule wall) by the Na^+-K^+ pump sites located in the T-tubular membrane. Thus the decay of the late afterpotential will be a function of these two processes.

The amplitude and duration of the late depolarizing afterpotential of skeletal muscle fibers is a function of the number of spikes in the train and their frequency. That is, the greater the spike activity, the greater the amplitude and duration of the late afterpotential. If the train consists of 20 spikes at a frequency of 50/s, a typical value for the amplitude of the late afterpotential in frog fibers is about 20 mV. When the diameter of the T-tubules is increased by placing the fibers in hypertonic solu-

---

[1] To produce glycerol osmotic shock, about 300 mOsm glycerol is added to Ringer's solution. The glycerol rapidly permeates into the fiber interior (so the fiber shrinks only transiently) and equilibrates. But when the gylcerol is washed out, there is a great hypotonic shock that disrupts the T-tubules.

tions[2] the amplitude of the late afterpotential decreases as expected because of the greater dilution of the K^+ ions accumulating in the T-tubule lumen. When the T-tubular system is disrupted and disconnected from the surface membrane by the **glycerol osmotic shock method**[1] the late afterpotentials disappear together with the early afterpotentials.

An alternative explanation for the late depolarizing afterpotential is that it may be due to the slow delayed rectifier g_K change described above (Adrian *et al.*, 1970b). The equilibrium potential for the slow I_K is –83 mV, and the sign (direction) of the late afterpotential reverses when the fiber is depolarized below –80 mV. Hence, the late afterpotential may arise from the slow relaxation of a component of the K^+ conductance increase, which is less selective for K^+ than the K^+ channels open in resting membrane. That is, in this view, the electrogenesis of the late afterpotential would be similar to that for the early afterpotential, but a different K^+ channel is involved.

All depolarizing afterpotentials, regardless of whether early or late, have physiological importance because they alter excitability and the propagation velocity of the fiber. A depolarizing afterpotential should enhance excitability (lower threshold) to a subsequent AP. This is because the **critical depolarization** required to reach the **threshold potential** would be decreased. A large late depolarizing afterpotential, such as that due to K^+ **accumulation** in the T-tubules, can, under certain pathological conditions, trigger repetitive APs. The effect of depolarizing afterpotentials on velocity of propagation in skeletal muscle fibers involves two opposing factors: (1) the decrease in critical depolarization required and (2) the decrease in maximal rate of rise of the AP (max dV/dt), which is a function of the takeoff potential (h_∞ versus E_m curve). Therefore, what factor dominates will depend on the degree of depolarization and the shape of the h_∞ curve. When frog skeletal fibers are depolarized slightly by elevating $[K^+]_o$, only a decrease in propagation velocity is observed (Sperelakis *et al.*, 1970).

IX. Ca^{2+}-Dependent Slow Action Potentials

Slow APs are recorded under conditions in which the fast Na^+ current is blocked by Na^+-deficient solution, TTX, or voltage inactivation of the fast Na^+ channels in high $[K^+]_o$. Under these conditions, the only carrier of inward current available to produce an AP is Ca^{2+} ion. Spontaneously occurring slow APs were first observed in frog sartorius fibers equilibrated in Cl^--free solution containing TTX (Sperelakis *et al.*, 1967). Upon addition of Ba^{2+} ion (e.g., 0.5 mM), which is a potent blocker of K^+ channels and P_K, the fibers partially depolarize, and spontaneously discharge slowly rising (e.g., 1–10 V/s) overshooting APs of long duration (e.g., several seconds), having a prominent plateau component (resembling a cardiac AP in shape). Ba^{2+} depolarizes rapidly in Cl^--free solution, because the voltage-clamping effect of the Cl^- distribution (E_{Cl}), due to the large P_{Cl}, is circumvented. In addition, since Cl^--free solu-

tion raises the resistance of the cell membrane about sevenfold, an intracellular microelectrode can better detect the potential changes occurring across the T-tubule wall because of less short-circuiting by the surface membrane.

By using two intracellular microelectrodes, one for applying intracellular current to stimulate the impaled frog skeletal muscle fiber and the other for recording voltage a short distance away in the same fiber ($[K^+]_o$ of 25 mM to depolarize the fibers to about –45 mV and thereby voltage-inactivate the fast Na^+ channels, and $[Na^+]_o$ reduced to zero so that there could be no inward fast Na^+ current), application of small hyperpolarizing current pulses during the slow AP indicates that membrane resistance increases progressively during the plateau component, leading to an abrupt repolarization terminating the AP (Kerr and Sperelakis, 1982). The AP responses fatigued with repetitive stimulation. The rate of rise, overshoot, and duration of the slow APs are a function of $[Ca^{2+}]_o$ (Beaty and Stefani, 1976; Vogel *et al.*, 1978; Kerr and Sperelakis, 1982). For example, the AP duration at 50% amplitude (APD_{50}) was about 8 s when $[Ca^{2+}]_o$ was 6 mM. The amplitude of the slow AP plotted against log $[Ca^{2+}]_o$ gave a straight line with a slope of 28 mV/decade, which is close to the theoretical 29 mV/decade (at 21 °C) from the Nernst relationship for a situation in which only Ca^{2+} ion carried the inward current. The slow APs were depressed and blocked by the Ca^{2+}-antagonistic and slow-channel-blocking drugs, verapamil and bepridil, with an ED_{50} of about 5×10^{-8} M. It was shown that the slow AP arises from the T-tubular system of the skeletal muscle fiber (Vogel *et al.*, 1978; Kerr and Sperelakis, 1982), based on their disappearance when the T-tubules were disrupted and disconnected from the surface membrane by the glycerol osmotic shock method (Eisenberg and Gage, 1969). The normal fast APs are not affected by the glycerol treatment. These results indicate that the slow Ca^{2+} channels giving rise to the slow APs are located primarily in the tubular system.

What is the mechanism that causes the slow APs? Isotope flux measurements have shown that there is a net Ca^{2+} influx during contractions of phasic skeletal muscle fibers. This finding has led to the proposition that an influx from the extracellular space may initiate the contraction (Bianchi and Shanes 1959, Curtis, 1966). Additionally, it was discovered that voltage-clamped phasic skeletal muscle fibers have slow inward Ca^{2+} currents (I_{Ca}) (Stanfield, 1977; Sanchez and Stefani, 1978). Furthermore, like for slow APs, elevation of $[Ca^{2+}]_o$ increased I_{Ca}, and I_{Ca} was depressed by the slow Ca^{2+} channel blockers D-600, nifedipine, and Ni^{2+} (Stanfield, 1977; Sanchez and Stefani, 1978; Almers *et al.*, 1981). Detubulation by the glycerol osmotic shock method abolishes I_{Ca} (Nicola-Siri *et al.*, 1980; Potreau and Raymond, 1980). These results support the conclusion that I_{Ca} produces slow APs. The various conformational states that the Ca^{2+} slow channels undergo during excitation are depicted in Fig. 7. These states are similar to those of the fast Na^+ channels (Fig. 5), except there are only two closed states.

Do slow inward calcium currents (I_{CaS}), and thereby slow APs, play a role in coupling excitation with contraction? A substantial contraction, of between 20 and 50% of the normal twitch tension, accompanies the slow APs (Vogel *et al.*, 1978), suggesting that the voltage-dependent Ca^{2+} channels in the

---

[2] In hypertonic solutions, skeletal muscle fibers shrink (fiber diameter decreases) like a perfect osmometer with an osmotically inactive volume of about 32%), but T-tubules swell.

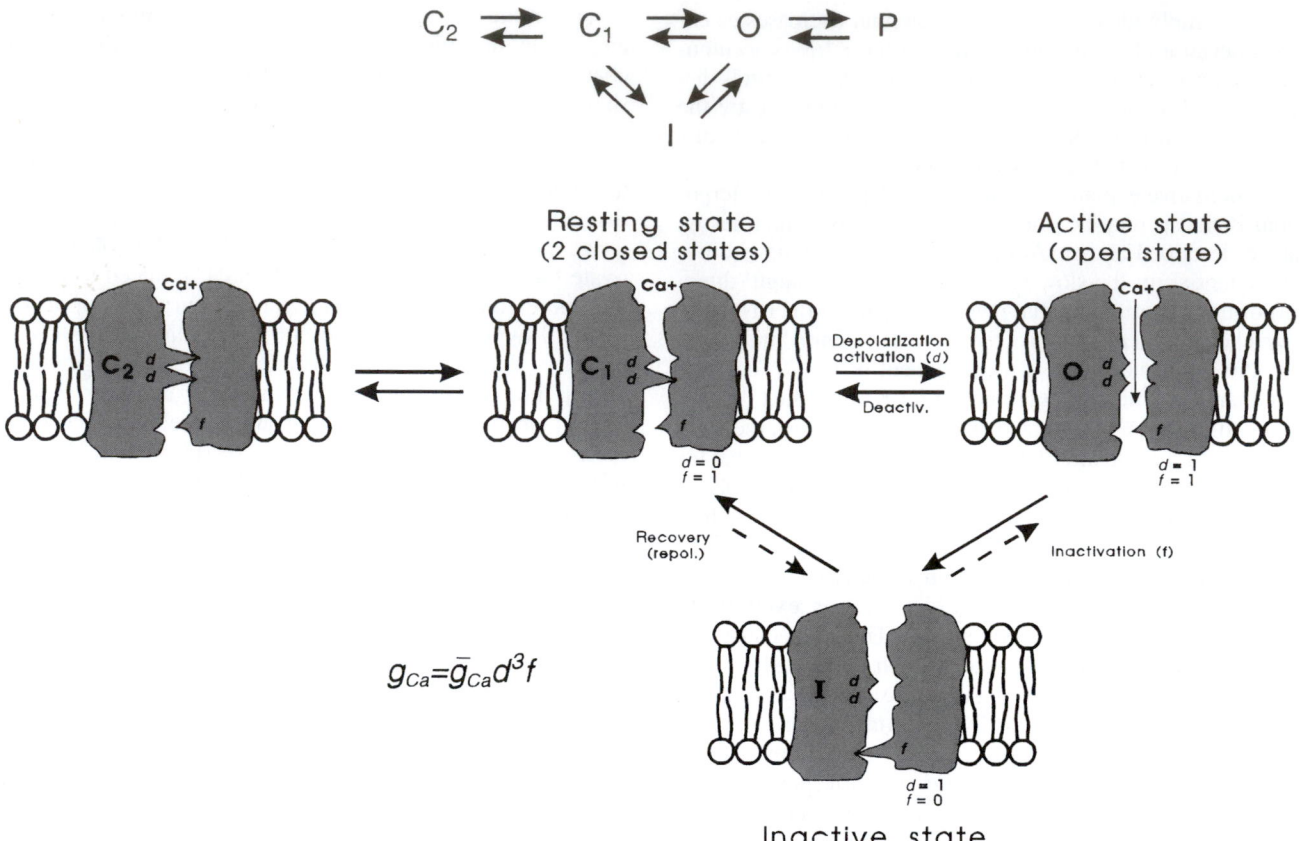

$$g_{Ca} = \bar{g}_{Ca} d^3 f$$

FIGURE 7. Illustration of the four hypothetical states of the slow Ca^{2+} channel. There is evidence for two closed states. As depicted, in the most closed state (C_2), both d gates (or particles) are in the closed configuration. In the least closed state (C_1), one gate is closed and one is open. In the resting closed states (C_2, C_1), the activation gate (A) is closed and the inactivation gate (I) is open: $d = 0, f = 1$. Depolarization to the threshold activates the channel to the active state, the A-gate opening rapidly and the I-gate still being open: $d = 1, f = 1$. The activated channel spontaneously inactivates to the inactive state due to closure of the I-gate: $d = 1, f = 0$. The recovery process on repolarization returns the channel from the inactive state back to the resting state, thus making the channel again available for reactivation. Ca^{2+} ion is depicted as being bound to the outer mouth of the channel and poised for entry down its electrochemical gradient when both gates are in the open configuration. The reaction between the resting state and the active state is readily reversible, and there is some reversibility in the other reactions. The slow channels behave similarly to the fast channels, except that their gates appear to move more slowly on a population basis; that is, the slow channels activate, and recover more slowly. (Although the gates of any individual slow channel may move quickly, the stochastic behavior of the population of channels is such that their summed conductance changes slowly.) The slow channel gates operate over a different voltage range than the fast channels (i.e., less negative, more depolarized). TTX does not block the slow channels, but drugs such as nifedipine do block by binding to the channel.

tubular system may play a role during excitation-contraction (E-C) coupling. However, phasic skeletal muscle cells contract for several minutes after $[Ca^{2+}]_o$ is lowered to $< 10^{-8}$ M (Armstrong *et al.*, 1972). Gonzalez-Serratos et al. (1982) observed that blocking I_{Ca} with the organic channel blocker diltiazem did not depress twitch or tetanic force development. On the contrary, diltiazem and D-600, another organic calcium blocker, potentiate twitch tension at a concentration that blocks I_{Ca}. These experiments indicate that I_{Ca} plays no obvious role in E-C coupling in normal amphibian phasic muscle fibers. The twitch potentiation observed with these Ca^{2+} channel blockers is caused by an inhibitory effect of these drugs on Ca^{2+} uptake by the SR (Ortega *et al.*, 1997). Nevertheless, in dysgenic mice in which contraction of skeletal muscles is weak, the Ca^{2+} channels in the T-tubules are few or absent.

Slow APs were also recorded from mouse skeletal muscle fibers equilibrated in a solution that was Cl⁻-free, low Na⁺ (10 mM), and high K⁺ (20 mM) (Kerr and Sperelakis, 1982). As with frog muscle, the slow APs were abolished after detubulation. The rate of rise, amplitude, and duration of the slow AP increased as a function of $[Ca^{2+}]_o$, the maximal rate of rise (max dV/dt) being about 0.5 V/s (in 8 mM $[Ca^{2+}]_o$). The slow APs also were blocked by verapamil, bepridil, Mn^{2+}, and La^{3+}.

Given these facts, what is the biological role of I_{Ca} in phasic skeletal muscle? External stimulation of one-day-old cultured embryonic skeletal muscle cells from *Xenopus laevis* elicit local contractions (Cordoba-Rodriguez *et al.*, 1996) at the time when the T-system and the SR in other types of muscle cells have not yet fully developed (Flucher *et al.*, 1993) and may not be functional. During the first five days in culture, these embryonic skeletal muscle cells need extracellular Ca^{2+} to contract when stimulated. This is in contrast with adult phasic skeletal muscle fibers that can produce normal twitches for many minutes after extracellular Ca^{2+} concentration is lowered to $<10^{-8}$ M (Armstrong et al., 1972). Consequently, an inflow of Ca^{2+} from the extracellular space may be required in embrionic cells as the means to produce contraction. The cytosolic Ca^{2+} concentration increase that causes contraction in the young myocytes may be produced by an inward flow of Ca^{2+} via voltage-dependent Ca^{2+} channels (i.e., I_{Ca}). In whole-cell voltage-clamp studies, currents with the characteristics of I_{Ca} have been observed in embryonic and neonatal skeletal muscles (Moody-Corbett et al.,1989; Cognard et al., 1992; Cordoba-Rodriguez et al., 1997, Gonzalez-Serratos et al., 1996). One-day-old cultured myocytes show inward currents that increase in magnitude as the ionic carrier is increased in concentration, and are blocked by external Co^{2+}. The current density increased from 1.7 pA/pF, to 3.3 pA/pF, and to 7.9 pA/pF at 1, 5, and 15 days in culture, respectively. One-day-old cells develop local contractures only when they are exposed to 20 mM caffeine, in contrast to 10–day-old cells that contract like fully developed phasic skeletal muscle fibers with only 1 mM caffeine. These results indicate that the T-tubules and SR are poorly developed or not functional in early stages of skeletal muscle development, and that I_{Ca} may be an important mechanism by which Ca^{2+} flows into the cells to trigger the contraction.

In summary, there are voltage-dependent slow Ca^{2+} channels located primarily in the T-tubules of amphibian and mammalian skeletal muscle fibers that have properties similar to those of the slow Ca^{2+} channels found in heart muscle and smooth muscle. Thus, the slow Ca^{2+} channels may play an important role in E-C coupling. This Ca^{2+} influx could trigger the release of more Ca^{2+} from the nearby TC-SR via the Ca^{2+}-trigger Ca^{2+}-release mechanism (Fabiato, 1982). Although contractions can still be evoked when the surface membrane is depolarized to E_{Ca} to prevent Ca^{2+} entry (Miledi et al., 1977), this does not exclude a possible role for the Ca^{2+} entry across the T-tubule wall because it is the ΔE_m across the T-tubule that would affect Ca^{2+} influx at this location. Because the time course of the slow AP is much longer than that of a twitch contraction, it was suggested that the inward Ca^{2+} current may play a role in K^+ contracture, in tetanic contraction, or in long-term regulation of contraction, perhaps by increasing the Ca^{2+} concentration in the SR, and thereby increasing the amount of internal Ca^{2+} available for release on subsequent activation (Nicola-Siri et al., 1980). $[Ca^{2+}]_{SR}$ does increase following tetanic stimulation (Gonzalez-Serratos et al., 1982). In the early stages of development, the function of the T-system and SR are not well developed, and I_{Ca} may be an important mechanism to trigger contraction.

X. Developmental Changes in Membrane Properties

The cell membranes of most excitable cells apparently pass through similar stages of differentiation during development. For example, young (2- to 3-day-old) embryonic chick heart (tubular) has few or no functional fast Na^+ channels, but has a high density of slow (Na^+ and Ca^{2+}) channels and fires slowly rising TTX-insensitive APs. Fast Na^+ channels then appear and progressively increase in number, reaching the maximal (adult) level at late embryonic development (e.g., day 20). The P_{Na}/P_K ratio is high in young hearts, due to a low P_K, and accounts for the low resting potential and automaticity in nearly all the cells.

Skeletal muscle fibers and neurons also undergo developmental changes in membrane electrical properties (e.g., Spector and Prives, 1977; Spitzer, 1979) (see Chapter 35). In general, fast Na^+ channels are absent in the young, less differentiated cells, but they do possess excitability because of a large number of slow channels. The AP is TTX insensitive, slowly rising, and of long duration, resembling a slow AP in cardiac muscle. Later during development, fast Na^+ channels make their first appearance, and the fast Na^+ channels and slow channels coexist. At that period, TTX does not abolish the APs, but reduces max dV/dt (i.e., slow APs remain). At a later stage, the slow channels in the sarcolemma are lost (or greatly reduced in number), and the fast Na^+ channels progressively increase in density. The APs become fast rising and of short duration and are completely abolished by TTX. As discussed previously, some functional slow Ca^{2+} channels remain in the T-tubular system.

XI. Electrogenic Na^+-K^+ Pump Stimulation

The Na^+,K^+-ATPase pump is **electrogenic** in skeletal muscle fibers (both mammalian and amphibian). The pump is electrogenic, producing a net outward current, because 3 Na^+ ions are pumped out to every 2 K^+ ions pumped in. The **electrogenic pump potential** contribution to the resting potential (see Chapter 14) is about 12–16 mV in rat skeletal muscle fibers (Sellin and Sperelakis, 1978). The **net pump current** can be stimulated by increasing the number of pump sites per unit area of cell membrane or by increasing the turnover rate of each pump site. β-adrenergic agonists (e.g., isoproterenol) rapidly (within 5 min) hyperpolarize (e.g., 7–9 mV) skeletal muscle fibers, and insulin more slowly (e.g., peak reached by 10 min) hyperpolarizes to a smaller degree (e.g., 5–7 mV) (Iannaccone et al., 1989). Since cAMP also hyperpolarizes, the action of β-agonists is believed to be mediated by elevation of cAMP and phosphorylation of the Na^+-K^+ pump (or an associated regulatory protein) by protein kinase A (PKA) . The action of insulin is thought to be mediated by the incorporation of spare membrane in an internal pool, which contains Na^+-K^+ pumps, into the cell membrane.

The pump current (I_p) can be directly measured in single fibers (cultured skeletal myotubes rounded by use of colchicine, a microtubule disrupter) by doing whole-cell voltage clamp under conditions in which all the ionic conductances are blocked. When this is done, the pump current can be measured at different voltages and normalized for unit membrane capacitance and membrane area. Values of about 1 pA/pF or 1.0 μA/cm^2 were obtained, with a reversal or zero current potential of about –140 mV (Li and Sperelakis, 1994).

When $[K^+]_o$ is lowered below the normal physiological level, for example, from 4.5 mM to about 0.1 mM, a large depolarization occurs in mammalian skeletal muscle fibers (Fig. 8). This depolarization is caused, in part, by inhibition of the Na^+-K^+ pump current. The K_m value for $[K^+]_o$ for the Na^+,K^+-ATPase is about 2 mM, and the relationship between Na^+,K^+-ATPase activity and $[K^+]_o$ is very steep. Therefore, inhibition of the Na^+-K^+ pump occurs. The contribution of the electrogenic pump to the RP is relatively large in mammalian skeletal muscle fibers.

XII. Slow Fibers

One type of skeletal muscle fibers, known as **slow fibers**, subserves tonic functions including posture. Slow fibers should not be confused with "slow twitch fibers." The true slow fibers do not fire APs, whereas all types of twitch fibers do. The slow fibers are usually smaller in diameter than twitch fibers, and they exhibit a less distinct myofibrillar arrangement. Slow fibers have been found in a number of vertebrate muscles, for example, in the frog rectus abdominus muscle, frog ileofibu-

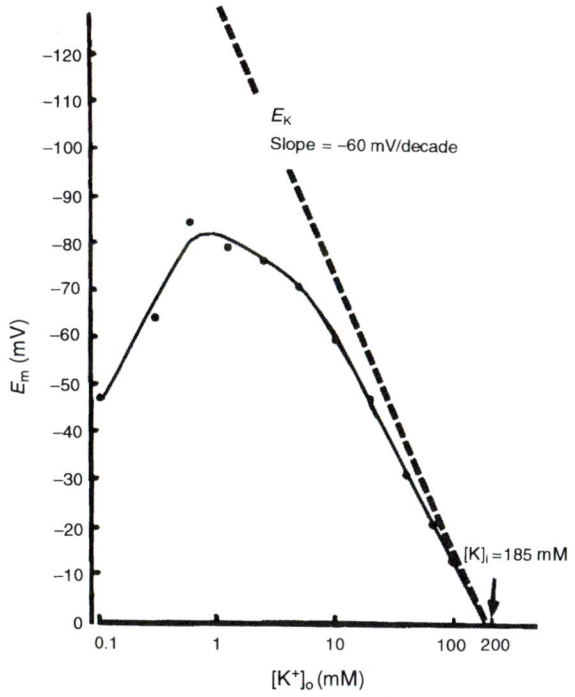

FIGURE 8. The mean resting membrane potential (E_m) of normal mouse skeletal muscle plotted as a function of the extracellular K^+ concentration ($[K^+]_o$) on a logarithmic scale. The straight line drawn through the data points for 20 mM $[K^+]_o$ and above has a slope of 50 mV/decade. Extrapolation of this line to zero potential gives the intracellular K^+ concentration ($[K^+]_i$) of 185 mM. The dashed line gives the calculated E_K values (slope of 61 mV/decade). Note the "fold-over" of the E_m curve at $[K^+]_o$ levels below 1 mM, presumably due to inhibition of the electrogenic pump potential (V_p) and to a decrease in P_K and g_K at low $[K^+]_o$ levels. (Reproduction from Sellin and Sperelakis, 1978.)

laris muscle, and mammalian extraocular muscles. It is probable that careful searching will reveal some slow fibers in other mammalian muscles.

The slow fibers have **multiple innervation** by a series of motor endplates (spaced about 1 mm apart) from a single motoneuron. As with twitch fibers, acetylcholine (ACh) is the synaptic transmitter. The force of contraction of the slow fibers is controlled by graded **end plate potentials** (EPPs). That is, an increase in frequency of impulses in the motoneuron produces a larger EPP (by temporal summation) and this, in turn, produces a greater contraction in the vicinity of the end plate. Since the end plates are spaced closely together—at a distance of about one length constant—the entire fiber becomes nearly uniformly depolarized, even though there are no propagated APs. Therefore, the entire length of the slow fiber contracts almost uniformly.

The slow fibers do possess T-tubules and triadic junctions with the terminal cisternae of the SR (TC-SR). Therefore, the T-tubules may act as **passive conduits** in the slow fibers to bring the depolarization (produced in the surface membrane by the EPP) deep into the fiber interior. Thus, depolarization of the T-tubule occurs by their passive cable properties. This depolarization, in turn, could bring about the influx of Ca^{2+} by activation of voltage-dependent slow Ca^{2+} channels located in the T-tubules.

APs normally cannot be induced to occur in vertebrate slow fibers under a variety of experimental conditions. However, denervation of frog slow fibers does allow an AP-generating mechanism to appear (Miledi *et al.*, 1971). APs can be induced in slow fibers of invertebrates [e.g., crustacean skeletal muscles and horseshoe crab (*Limulus*) heart] (Fatt and Ginsborg, 1958). In the neurogenic Limulus heart, which normally is activated by summating excitatory postsynaptic potentials, propagating (ca. 5 cm/s) and overshooting spontaneous APs can be rapidly induced by Ba^{2+} (0.1–10 mM) (Rulon *et al.*, 1971). These slowly rising (ca. 1.0 V/s) APs are resistant to TTX, and these voltage-dependent slow channels can pass Ba^{2+}, Sr^{2+}, and Ca^{2+}.

XIII. Conduction of the Action Potential

When the EPP, generated at the neuromuscular junction, reaches threshold for eliciting an AP in the skeletal muscle fiber, an AP is propagated down the muscle fiber in both directions from the end plate. (In some muscle fibers, there is a second end plate innervated by a motoneuron exiting the spinal cord at another level.) The AP is overshooting (to about +40 mV), and propagates at a constant velocity of about 5 m/s over the surface sarcolemma. As discussed in the chapter on propagation in axons, propagation occurs by means of the **local-circuit currents** that accompany the propagating impulse. The **radial (transmembrane) currents** are triphasic: (1) the first phase is outward across the membrane and is of moderate intensity; (2) the second phase is inward and is most intensive (at the current "sink" region); and (3) the third phase is outward and is the least intensive. The first outward phase corresponds to the passive exponential foot of the AP due to the **passive cable spread** of voltage and current. The second inward phase corresponds to the large increase in g_{Na} and fast inward Na^+ current during the later portion of the rising phase and peak of the AP. The third outward phase corresponds to the small increase in g_K and net outward K^+ current and the less steep repolarizing phase of the AP spike.

The **longitudinal (axial) currents**, both inside and outside the fiber, are biphasic: (1) the first phase is most intensive and is in the direction of propagation of the impulse (forward); and (2) the second phase is less intensive and is in the backward direction. The external longitudinal currents, of course, must equal the internal axial currents, but they are in the opposite directions. The internal axial currents are confined to the myoplasm, whereas the external longitudinal currents can use the entire ISF space or so-called **volume conductor** (since current takes the path of least resistance). It is this latter fact that allows the **electromyogram** (EMG) to be recorded from the skin overlying an activated skeletal muscle. The amplitude of the EMG potentials becomes larger when more fibers within the muscle are activated (**fiber summation**), because of summation of the IR voltage drops produced by each fiber activated simultaneously. The frequency of the EMG potentials reflects the frequency and asynchrony of activation of the muscle.

The skeletal muscle fibers, composed of **myoblast cells** fused end to end and multinucleated, behave as semi-infinite cables. That is, an AP can propagate from one end of the fiber to the other, uniformly and unimpeded. The space or **length constant** (λ) of the fiber cable is about 1.5 mm for frog sartorius fibers (Sperelakis *et al.*, 1967) and about 0.76 mm for the rat EDL muscle (Sellin and Sperelakis, 1978). The length constant is the distance over which the potential impressed at one region would decay to $1/e$ ($1/2.717 = 0.37$) or 37% of the initial value. That is, in a passive cable, voltage decays exponentially with a certain length constant as given by

$$V_x = V_o e^{-x/\lambda} \tag{4}$$

where V_x is the voltage at the distance x, and V_o is the voltage at the origin ($x = 0$). λ is given by

$$\lambda = \sqrt{\frac{r_m}{r_i + r_o}} \tag{5}$$

$$cm = \sqrt{\frac{\Omega \cdot cm}{\frac{\Omega}{cm} + \frac{\Omega}{cm}}} = \sqrt{cm^2}$$

Assuming that r_o (the outside longitudinal resistance) is negligibly small compared to r_i (this would be true for a superficial fiber in a bundle immersed in a large bath)

$$\lambda = \sqrt{\frac{r_m}{r_i}} = \sqrt{\frac{R_m}{R_i} \frac{a}{2}} \tag{6}$$

$$cm = \sqrt{\frac{\Omega \cdot cm}{\Omega/cm}} = \sqrt{\frac{\Omega \cdot cm^2}{\Omega \cdot cm} cm} = \sqrt{cm^2}$$

where r_m ($\Omega \cdot cm$) and r_i (Ω/cm) are the membrane resistance and the internal longitudinal resistance normalized for unit length of fiber, R_m ($\Omega \cdot cm^2$) is the membrane resistance normalized for both fiber radius and length, R_i ($\Omega \cdot cm$) is the resistivity of the myoplasm (normalized for length

and cross-sectional area), and a (cm) is the fiber radius. R_m is often loosely called membrane resistivity or specific resistance, but this is not accurate because for true membrane resistivity (ρ_m) there must be correction for membrane thickness δ:

$$\rho_m = \frac{R_m}{\delta} \tag{7}$$

$$\Omega \cdot cm = \frac{\Omega \cdot cm^2}{cm}$$

For a derivation of these equations, and for the interconversions between r_i and R_i, and r_m and R_m, the reader is referred to the Appendix of this book (Chapter 70).

The factors that determine active **velocity of propagation** (θ_a) include the intensity of the local-circuit current, threshold potential, and the passive cable properties, λ and τ_m. As discussed previously, the greater the rate of rise of the AP, the greater the intensity of the local-circuit current, hence the greater the θ_a. In addition to its dependence on the density of the fast Na^+ channels (determinant of the maximum Na^+ conductance, $\bar{g}_{Na}$), the kinetic properties of the channel gating, and the threshold potential (V_{th}), max dV/dt is determined also by the resting (takeoff) potential (related to the h_∞ versus E_m curve), as discussed previously (Fig. 9). In addition, because cooling decreases max dV/dt ($Q_{10} \simeq 3$), θ_a is slowed accordingly. R_m also is increased by cooling, the Q_{10} for R_K in frog sartorius fibers being about 2.8 (ion diffusion in free solution has a Q_{10} of about 1.2) (Sperelakis, 1969).

In a passive cable, such as a skeletal muscle fiber, the **passive propagation velocity** (θ_p) is directly proportional to the length constant [3] and inversely proportional to the time constant

$$\theta_p = \frac{\lambda}{\tau_m} \tag{8}$$

$$\frac{cm}{s} = \frac{cm}{s}$$

$$\theta_p = \frac{\sqrt{\frac{R_m}{R_i} \frac{a}{2}}}{R_m C_m} \tag{8a}$$

$$\frac{cm}{s} = \frac{\sqrt{\frac{\Omega \cdot cm^2}{\Omega \cdot cm} \frac{cm}{}}}{\Omega \cdot cm^2 \frac{F}{cm^2}} = \frac{\sqrt{cm^2}}{s}$$

$$\theta_p = \frac{\sqrt{a}}{\sqrt{R_m R_i 2} C_m} \tag{8b}$$

$$\frac{cm}{s} = \frac{\sqrt{cm}}{\sqrt{\Omega \cdot cm^2} \sqrt{\Omega \cdot cm} \frac{F}{cm^2}} = \frac{\sqrt{cm}}{cm \sqrt{cm} \frac{s}{cm^2}} = \frac{cm}{s}$$

Therefore, propagation velocity is directly proportional to the square root of the fiber radius (a), and inversely proportional

---

[3] The length constant for sinusoidally-varying applied currents (λ_{ac}) is shorter than λ_{dc}, depending on the ac frequency.

to membrane capacitance (C_m) and to the square root of R_i and the square root of R_m. For example, propagation velocity is greater in large-diameter muscle fibers.

The relationship between propagation velocity and membrane current density (I_m) is given by

$$I_m = \frac{a}{2}\frac{1}{R_i}\frac{1}{\theta^2}\frac{d^2V}{dt^2} \tag{9}$$

$$\frac{\text{amp}}{\text{cm}^2} = (\text{cm})\left(\frac{1}{\Omega\cdot\text{cm}}\right)\left(\frac{1}{\text{cm}^2/\text{s}^2}\right)\left(\frac{\text{V}}{\text{s}^2}\right) = \frac{\text{amp}}{\text{cm}^2}$$

where d^2V/dt^2 is the second time derivative of the AP. As indicated, membrane current is proportional to d^2V/dt^2, whereas the longitudinal current (I_l) or the capacitative current (I_c) is proportional to dV/dt

$$I_c = C_m\frac{dV}{dt} \tag{10}$$

XIV. Excitation Delivery to Fiber Interior

A. Conduction into the T-Tubular System

The experiments of Huxley and Taylor (1958) were the first to provide evidence that there was some structure, located at the level of the Z-lines in frog skeletal muscle fibers, which is involved in EC coupling. This structure allows fast conduction of the excitatory process (AP) from the surface membrane to the center of the muscle cells. These investigators applied current pulses at different points along the length of the sarcomeres in isolated fibers and found that when the microelectrode tip was opposite the Z-line, graded contractions of the two half-sarcomeres occurred. The greater the current, the greater was the inward spread of the contraction. In addition, they discovered that there were **sensitive spots** located around the perimeter of the fiber at the Z-line level; that is, the membrane was not uniformly sensitive. At about the same time, it was discovered by electron microscopists that transverse (T-) tubules were located at the level of the Z-lines in amphibian skeletal muscle (and at the level of the A-I junctions of the sarcomere in mammalian skeletal muscle). Thus, the T-tubules probably represent the morphological conduit for the findings of Huxley and Taylor. The morphological arrange-

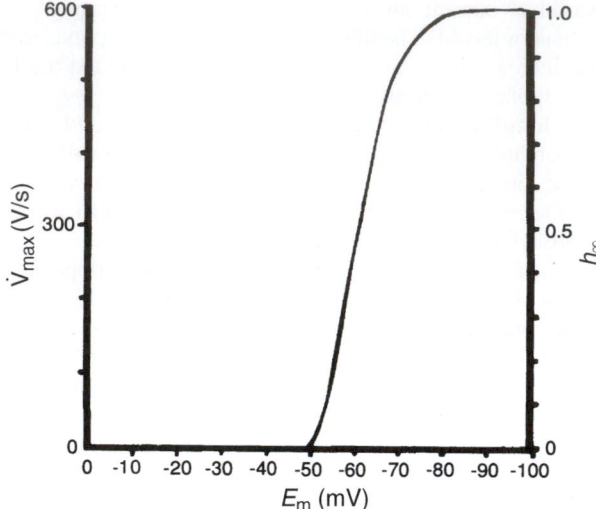

FIGURE 9. Graphic representation of the maximal rate of rise of the action potential (max dV/dt) as a function of resting E_m or takeoff potential. Max dv/dt is a measure of the inward current intensity (membrane capacitance being constant), which is dependent on the number of channels available for activation; h is the inactivation factor of Hodgkin-Huxley as $g_{Na} = \bar{g}_{Na}m^3h$, where g_{Na} is the Na^+ conductance, $\bar{g}_{Na}$ is the maximal conductance, and m and h are variables; h_∞ represents h at $t = \infty$ or steady state (practically, after 20 ms). The fast Na^+ channels begin to inactivate at about -75 mV, and nearly complete inactivation occurs at about -30 mV (h_∞ low). Therefore, max dV/dt, decreases because h_∞ decreases.

ments of the sarcotubular system of skeletal muscle fibers are illustrated in Fig. 10.

Diffusion of some substance from the surface membrane into the skeletal muscle fiber interior is much too slow to account for the relatively short latent period of about 1–3 ms between the beginning of the AP and the beginning of contraction. That is, the diameter of the fibers (mean value of about 70 μm in frog sartorius fibers) is much too large for a diffusion mechanism from the fiber surface to be responsible. **Diffusion time** (for 95% equilibration) increases by the square of the distance, and would require about 2.5 s for a small molecule freely diffusing across a cell radius of 50 μm; estimates for Ca^{2+} diffusion time are considerably longer than this

FIGURE 10. Sarcotubular system of skeletal muscle fibers from (A,C) tibialis anterior muscle of mouse and (B) iliotibialis muscle of lizard. (A) Longitudinal section showing the sarcomere structure of several myofibrils: A-band, I-band, Z-line. The network sarcoplasmic reticulum (N-SR), also known as the longitudinal SR, appears as a torn sleeve surrounding the surface of each myofibril. The N-SR is continuous with the junctional SR (J-SR) that abuts close to the transverse tubules (TT). The TT membranes are invaginations of the cell surface membrane at the level of the A-I junctions in mammalians (or at the level of the Z-line in lizards and amphibians). The J-SR and TT form the complex coupling known as a triad (*). The N-SR is continuous across the I-band, but this is obscured in this section by the presence of paired mitochondria (Mit) over the I-bands. The TT and SR are both selectively filled with osmium tetroxide precipitate, causing their profiles to be more electron opaque than the other structures. Scale bar at lower right represents 1.0 μm. (B) Higher magnification of a triad to show more detail. As shown, the triad consists of a single T-tubule sandwiched between two cisternae of the J-SR. Scale bar = 0.1 μm. (C) High magnification of a triadic junction to illustrate the array of regularly spaced junctional processes or SR foot processes (several indicated by arrowheads) that project between the TT membrane and the J-SR membrane. There are dense granules within the lumen of the J-SR cisternae (*). Scale bar = 0.1 μm. (Electron micrographs provided courtesy of Dr. Mike Forbes, University of Virginia.)

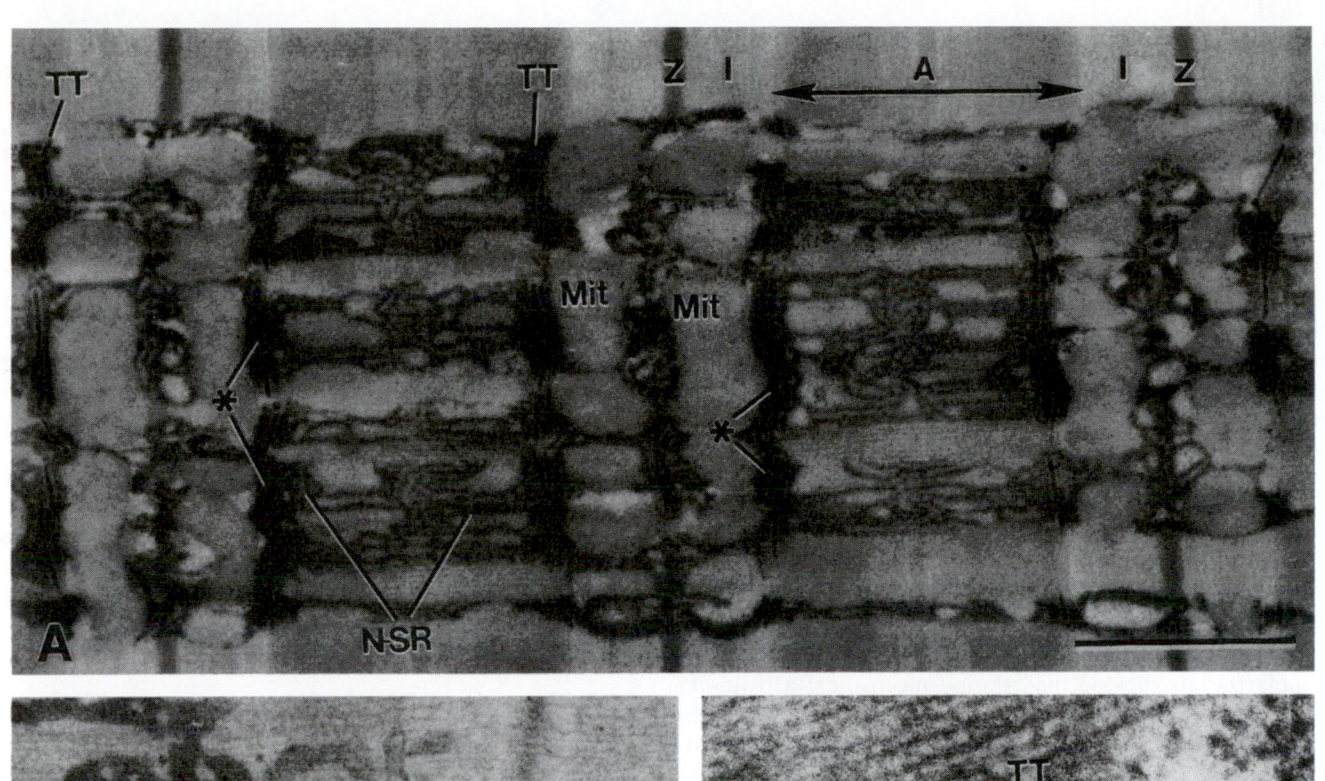

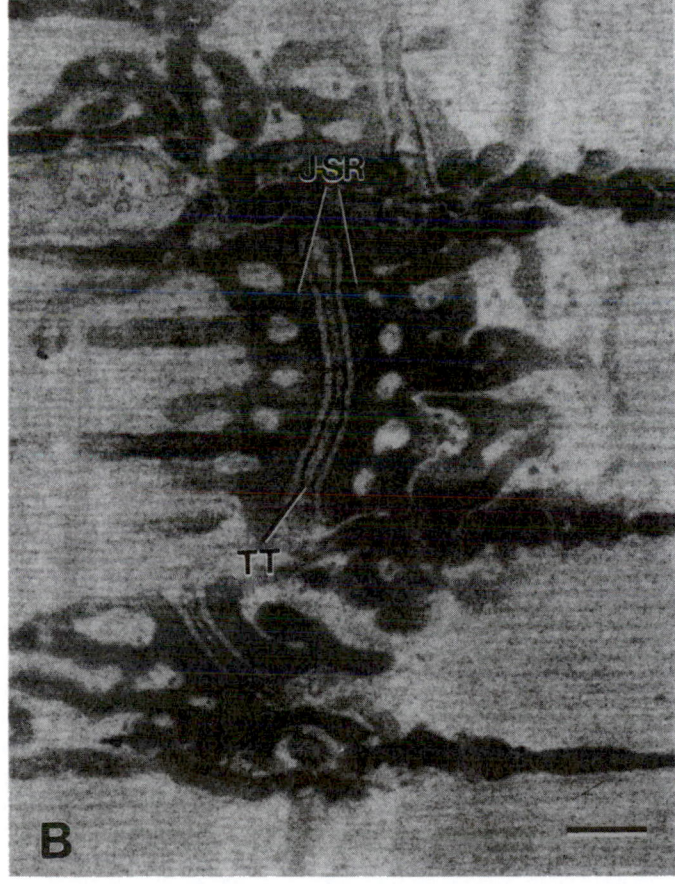

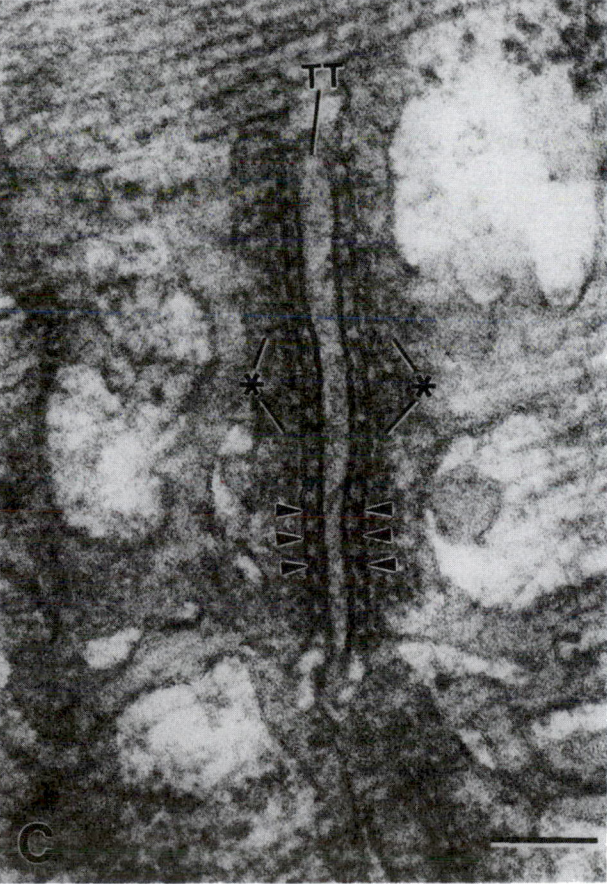

(Podolsky and Costantin, 1964). Therefore, the T-tubular system serves to bring excitation deep into the fiber interior rapidly and thereby reduces the required diffusion distance to an average value of about 0.7 μm (Sperelakis and Rubio, 1971). Disruption of the T-tubules (with their disconnection from the surface membrane by the glycerol osmotic-shock method) uncouples contraction from excitation (Eisenberg and Gage, 1969), thus further underscoring the essential role of the T-tubules in E-C coupling.

Estimates of the **length constant** of the T-tubules (λ_{TT}), assuming the T-tubule membrane has about the same resistivity (R_m) as the surface membrane $\lambda_{TT} = \sqrt{R_m/R_i}\,\sqrt{a/2}$, give values of about 50 μm. Because the resistivity of the T-tubule membrane of frog muscle is probably higher than that of the surface sarcolemma because of lower g_{Cl} (Hodgkin and Horowicz, 1959; Adrian and Freygang, 1962; Sperelakis and Schneider, 1968), this would give a longer λ_{TT} value. Therefore, it is possible for the T-tubules to serve as **passive conduits** to bring the depolarization from the surface membrane (during its AP) into the fiber interior.

However, direct microscopic observations of the degree of myofibril activation across the fibers caused by raised $[K^+]_o$ depolarization showed that, during mechanical activation, λ_{TT} is smaller than the fiber radius (Gonzales-Serratos, 1975). Electrophysiological experiments revealed that when the tubular membrane is depolarized exponentially (like with the foot of the action potential) λ_{TT} is also smaller than the fiber radius. Therefore, during the foot of the AP only a small outer ring of the tubular network membrane depolarizes beyond the threshold to trigger mechanical activation (Hodgkin and Nakajima, 1972). These results imply that in order for all the myofibrils in the cross-section of a fiber to be activated, which had been demonstrated previously (Gonzalez-Serratos, 1971), there must be a T-tubular AP (TAP).

There is evidence that the T-tubules actually fire APs, that is, they **actively propagate impulses** inward, and so bring large depolarization deep into the fiber interior. The evidence for this includes the observation of a threshold for sudden initiation of localized contraction by Costantin and colleagues (1967, 1973). The T-tubule AP is sensitive to TTX and is Na+ dependent, and therefore is apparently similar in nature to the surface membrane AP. By use of high-speed cinemicrography to measure sequential activation of the myofibrils in a radial direction, Gonzalez-Serratos (1971) estimated the propagation velocity of the T-tubule AP (θ_{TT}) to be about 10 cm/s, with a Q_{10} of 2.2 which is similar to the Q_{10} of the surface membrane AP conduction velocity. Although θ_{TT} is about 15 times slower than propagation down the fiber longitudinally (about 1.6 m/s), it is sufficient to account for the short latent period before contraction begins.

Further evidence of a Na+-dependent TAP came from observations in which muscles placed in low $[Na^+]_o$, and stimulated briefly at high frequency, initially develop normal tetanic tension that rapidly falls and steadily decreases to a lower sustained tension simultaneous with the central myofibrils becoming inactive. These results are due to Na+ depletion in the T-tubule network, particularly in the deeper parts far from the orifice at the fiber surface. Na+ depletion occurs because g_{Na} increases during the TAP generation,

causing an inward Na+ flow into the myoplasm (Bezanilla et al., 1972). The fatigue occurs more rapidly in fibers pre-equilibrated in low $[Na^+]_o$ (e.g., 60 mM). It is thought that the Na+ influx (the inward fast Na+ current) with each AP in the T-tubule produces a progressive decline in $[Na^+]_{TT}$, which slows propagation velocity down the T-tubules and eventually leads to loss of excitability when $[Na]_{TT}$ drops below some critical level (e.g., 30 mM). Na+ depletion should occur more rapidly deep in the T-tubule network of the fiber because there would be less diffusion of Na+ in from the mouth of the T-tubule to replenish the Na+ loss. Active Na+-K+ pumping in the T-tubules may not occur fast enough to keep up with the Na+ loss into the fiber myoplasm.

There are also voltage-dependent slow Ca2+ channels in the T-tubule membrane, and slow APs that arise from the T-tubule can be recorded under appropriate conditions (Sperelakis et al., 1967; Vogel and Sperelakis, 1978). The evidence for the existence of this type of channel and some of its properties was discussed in Section IX. The Ca2+ influx into the myoplasm through these Ca2+ channels could play a role in EC coupling.

B. Evidence for T-Tubule Communication with the Sarcoplasmic Reticulum across the Triadic Junction under Some Conditions

Ca2+ for contraction in skeletal muscle is primarily released from the TC-SR (Winegrad, 1968), and there is an *internal cycling* of Ca2+ ion. Changes in $[Ca^+]_o$ of the bathing solution take a relatively long time (e.g., 30–60 min) before exerting a large effect on the force of contraction in skeletal muscle. In contrast, in cardiac muscle, the effect of lowered $[Ca^{2+}]_o$ is obvious within a few seconds, indicating that the primary determinant of the force of contraction is the Ca2+ influx across the sarcolemma through the slow Ca2+ channels. Therefore, in skeletal muscle, excitation propagates actively down the T-tubules and Ca2+ is released from the TC-SR, but it is not known how the signal is transferred from the T-tubule to the TC-SR across the triadic junction.

Electron-opaque **tracer molecules**, like horseradish peroxidase (HRP) (ca. 60 Å diameter), enter into the T-tubules and from there can enter into some of the TC-SR of frog skeletal muscle (Rubio and Sperelakis, 1972; Kulczycky and Mainwood, 1972) (Fig. 11A and B). Exposure of the fibers to hypertonic solutions facilitates the entry of HRP into the TC-SR, so that nearly 100% of the TC-SR becomes filled (Fig. 11C and D). Thus, there may be a functional connection between the SR and the extracellular space (Sperelakis et al., 1973). If so, there may be lumen-to-lumen continuity between the T-tubules and TC-SR during excitation, allowing the AP in the T-tubules to invade directly into the TC-SR to depolarize and bring about the release of Ca2+. The depolarization of the TC-SR could activate voltage-dependent slow Ca2+ channels, allowing Ca2+ influx into the myoplasm down an electrochemical gradient.

If the longitudinal SR (L-SR) were electrically isolated from the TC-SR by a substantial resistance (e.g., zippering between the two SR compartments, described later), this would account for the **fiber capacitance** measured being relatively low (Mathias et al., 1980). The effect of this

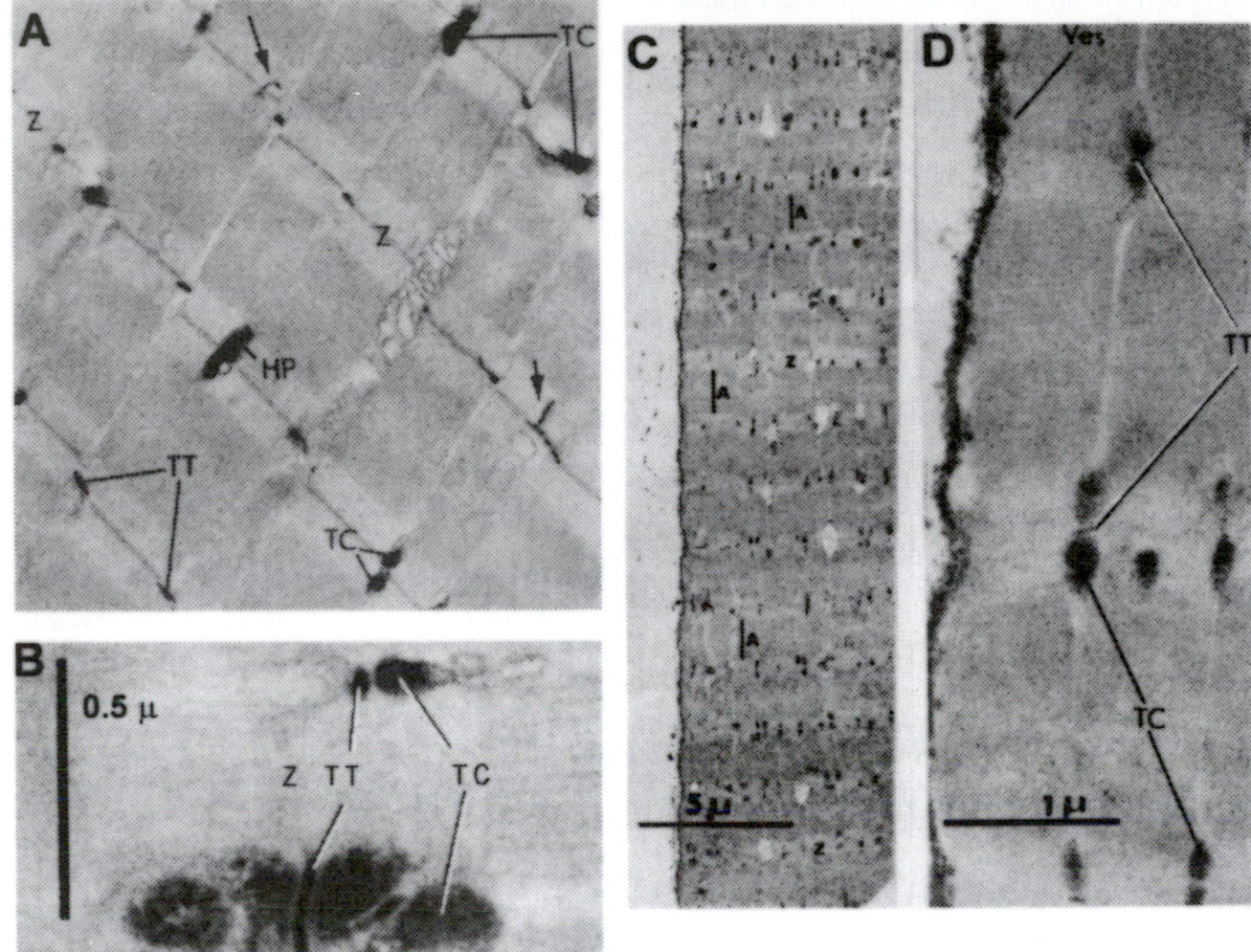

FIGURE 11. Evidence that large molecules of horseradish peroxidase (HP) can enter into the terminal cisternae (TC) of the SR via the transverse tubules (TT) of frog sartorius fibers. Electron micrographs of longitudinal sections. (A,B) Fiber was exposed to HP under isosmotic conditions. (A) Section through several myofibrils showing presence of HP activity (as a dense electron-opaque material) in the TT and in some of the TC at triadic junctions. In amphibian muscle, the TT occur at the level of the Z-lines (Z) of the sarcomeres. Arrows point to two branches of the TT running longitudinally. (B) Higher magnification of two triads, one with both cisternae filled with HP and the other with only one cisterna filled. (C,D) Fiber was exposed to HP under hypertonic condition (3 X isotonic, using NaCl), showing that almost all cisternae were filled with peroxidase. (C) Section at low magnification. (D) Portion of same section as in Part C shown at higher magnification. The surface vesicles (Ves) also became filled with HP. [Modified from Figs. 2 and 4 of Rubio, R., and Sperelakis, N. (1972). *Z. Zellforsch.* **124**, 57–72.]

would be to remove the very large membrane surface area of the L-SR and hence greatly reduce the capacitance that would be measured.

It has been suggested that the SR is depolarized during the release of Ca^{2+} in EC coupling. For example, optical signals (e.g., birefringence and fluorescence changes) can be recorded from the SR membranes during contraction (e.g., Baylor and Oetliker, 1975; Bezanilla and Horowicz, 1975). In addition, Natori (1965) demonstrated that propagation of contraction (1–3 cm/s) triggered by electrical stimulation can occur in muscle fiber regions that had been denuded (skinned) of their sarcolemma, the propagation of excitation presumably occurring by means of the SR membranes.

Investigators (Sperelakis *et al.,* 1973) have demonstrated that EC uncoupling could be produced by exposing frog skeletal muscle fibers to Mn^{2+} (1 mM) or La^{3+} (1 mM) while in hypertonic solution (to facilitate entry of the blockers into the TC-SR). After the fibers were returned to normal Ringer's solution, normal fast APs could be elicited, but there were no contractions accompanying them; that is, a "permanent" EC uncoupling was produced. These results were interpreted as suggesting that Mn^{2+} and La^{3+} entered into the lumen of the TC-SR and blocked the slow Ca^{2+} channels. A similar exposure of frog sartorius fibers to Mn^{2+}, La^{3+}, or to Ca^{2+}-free solution blocked the caffeine-induced contracture as well (Rubio and Sperelakis, 1972).

Thus, from these physiological and ultrastructural studies, it was suggested that the lumen of the SR is continuous with that of the T-tubule under conditions of hypertonicity or excitation, and that substances can enter into the TC-SR to exert an effect on Ca^{2+} release into the myoplasm.

Compartmental analysis of skeletal muscle has also suggested that the SR is open to the ISF. [In contrast, in cardiac muscle, there is no evidence that the SR is open to the ISF (Rubio and Sperelakis, 1971).] For example, Conway (1957), Harris (1963), and Keynes and Steinhardt (1968) concluded that Na^+ in frog skeletal muscle fibers is distributed in two separate compartments. Harris (1963) suggested that the Na^+, K^+, and Cl^--concentrations in one compartment (presumably the SR) were about equal to those of the ISF, and Rogus and Zierler (1973) concluded that the Na^+ concentration in the SR of rat skeletal muscle approximates that of the ISF. The volume of the SR compartment was 14.3% of fiber volume, and in hypertonic solution, the SR volume increased and the washout of the SR compartment was faster. Tasker *et al.* (1959) also had reported a large sucrose space of 26.5% for frog sartorius fibers.

Other researchers (Birks and Davey, 1969) have demonstrated that the volume changes of the SR of skeletal muscle in hypertonic (sucrose) and hypotonic solutions were always opposite of those occurring within the myoplasmic compartment. They concluded that sucrose must enter into the SR, pulling water osmotically from the myoplasm, to produce the marked swelling of the SR that occurred in hypertonic solutions. Vinogradova (1968) concluded from the distribution of nonpenetrating sugars in frog sartorius muscle that the SR compartment is continuous with the ISF; **the inulin space** was 19.0% and increased in hypertonic solution and decreased in hypotonic solution and in glycerol-treated fibers (for disruption of the T-tubules).

The total [³H] **sucrose space** of frog sartorius muscles was found to be 18.0% in isotonic solution and 22.6% in twofold hypertonic solution (Sperelakis *et al.*, 1978). The relative SR volume (including the small T-tubule volume) was 12.4% and 17.0% of fiber volume, respectively. This value for SR volume of frog skeletal muscle is close to that measured by ultrastructural techniques (Peachey, 1965; Mobley and Eisenberg, 1975). Evidence that the TC-SR and L-SR may not be freely connected to one another under resting conditions comes from the observations that (1) the L-SR did not fill with HRP, whereas the TC-SR did (Rubio and Sperelakis, 1972), and (2) there is a **zippering** of the membranes connecting these two components of the SR in mouse and frog skeletal muscle (Howell, 1974; Wallace and Sommer, 1975; Forbes and Sperelakis, 1979).

In ⁴⁵Ca washout experiments on frog muscles, Kirby *et al.* (1975) found three compartments, similar to the three **sucrose compartments** described previously, except the half-times were about two- to threefold shorter. They suggested that the first compartment was the ISF space, the second was the T-tubule plus the TC-SR, and the third was the L-SR. Bianchi and Bolton (1974) also found a transient increase in ⁴⁵Ca efflux and a marked loss of muscle Ca^{2+} from frog sartorius muscles exposed to hypertonic solutions (twice isotonicity), and suggested that hypertonicity produces transient communication between the TC-SR and the T-tubules,

thus allowing their Ca^{2+} to be lost to the ISF. In addition, it has been reported in a human muscle disease, polymyositis, that the T-tubules are spatially continuous with the SR, as visualized with lanthanum tracer, and that enzymes leak from the TC-SR into the T-tubules and ISF (Chou *et al.*, 1980).

Frog skeletal muscle fibers have an **osmotically inactive volume** of about 32% when placed into Ringer's solution made hypertonic with sucrose or other nonpenetrating solutes; that is, fiber diameter does not shrink to the theoretical value expected if it were a perfect **osmometer** (Sperelakis and Schneider, 1968; Sperelakis *et al.*, 1970). For example, in twofold hypertonic solution, there should be a decrease in fiber volume to one-half and fiber radius to 0.707 ($1/\sqrt{2}$) of the original value. The observed change is to only 0.81 of the original diameter. Because the SR volume increases in hypertonic solution (Huxley *et al.*, 1963; Sperelakis and Schneider, 1968; Birks and Davey, 1969), it is likely that the osmotic inactive volume is due to the SR. The swollen SR would prevent the fiber volume from decreasing to one-half in twofold hypertonic solution, even if the volume of the myoplasm proper were to decrease to one-half.

In cardiac muscle, an osmotically inactive volume is not present (Sperelakis and Rubio, 1971), electron-opaque tracers do not enter the SR (Sperelakis *et al.*, 1974), and the SR volume does not increase with hypertonicity (Sperelakis and Rubio, 1971).

This brief section was included to let the student know that there are data that do not fit with currently accepted hypotheses.

XV. Summary

The resting potential of skeletal muscle twitch fibers is about –80 mV (mammalian) or –90 mV (amphibian), and the AP overshoots to about +40 mV. The maximal rate of rise of the AP is very fast, being about 500–700 V/s, and is due to a large inward fast Na^+ current (I_{Na}) which brings E_m up close to E_{Na}. The APD_{50} is brief, being about 1–3 msec, and the falling phase of the AP is produced by several repolarizing factors: (1) Na^+ channel inactivation, (2) increase in the delayed rectifier K^+ conductance, (3) I_{Na} deactivation due to some repolarization, and (4) Cl^- influx (outward I_{Cl}).

The skeletal muscle fibers are long cables, formed by fusion of myoblast cells, and therefore produce a multinucleated long myotube or fiber, and have a fast propagation velocity of about 5 m/s Each skeletal muscle twitch fiber is normally closely controlled by the motor innervation, there being one or two motor end plates (neuromuscular junctions) located near the midregion of each fiber. Excitation spreads in both directions from the neuromuscular junction.

The twitch fibers undergo developmental changes similar to those in cardiac muscle and neurons. In early development, there are few or no fast Na^+ channels, and the AP upstroke is slow and produced by an inward current through slow Ca^{2+} channels. The AP duration is also long because the delayed rectifier K^+ conductance is not fully developed. During subsequent development, fast Na^+ channels are gained, the resting potential is greater (more negative), and the AP shortens to a brief spike.

The skeletal muscle AP spike is immediately followed by a large and prominent early depolarizing afterpotential, that slowly decays over 10–20 ms. This early afterpotential is caused in part by the persistence and slow decay of the delayed rectifier K^+ conductance (that was turned on by the Na^+ influx-caused depolarization), which has a Na^+:K^+ selectivity or P_{Na}/P_K ratio higher (e.g., 1/30) than that of the resting membrane (e.g., 1/100). Since this delayed rectifier K^+ conductance dominates, E_m is held for a time more depolarized than the normal resting potential.

After (and during) a tetanic burst (train) of AP spikes, a large prominent late depolarizing afterpotential is produced. This late afterpotential is caused by a cumulative K^+ accumulation in the T-tubules that acts to depolarize them due to the decrease in E_K, and thereby depolarize the surface sarcolemma passively. In addition, a slow component of the delayed rectifier K^+ conductance may persist during the train and slowly decay after the train, and as discussed previously, this K^+ conductance is less selective for K^+ than is the resting K^+ conductance.

The AP invades the T-tubules and propagates inward at a slow velocity and thereby serves to bring excitation deep into the fiber interior. The depolarization of the T-tubules activates slow (L-type) Ca^{2+} channels and the voltage sensors that are present in them, and this serves as a critical step in EC coupling in skeletal muscle fibers. In some circumstances, the lumen of the TC of the SR may communicate directly with the lumen of the T-tubule, perhaps allowing excitation to invade directly into the TC-SR, from which Ca^{2+} is released into the myoplasm to bring about contraction. The chapter on EC coupling provides a detailed discussion of a mechanism that involves the Ca^{2+} channels acting as voltage sensors that somehow are coupled to and open the Ca^{2+} release channels in the TC-SR (surface facing the T-tubule).

Some skeletal muscles also contain a fraction of fibers that are nontwitch slow muscle fibers, which normally do not fire APs. They are multiply innervated by the motor neuron, with numerous motor end plates spaced about 1 mm apart along the entire length of the fiber. Graded contraction of each fiber is produced by varying the frequency of axon APs that increase the amplitude of the EPPs by temporal summation. The membrane potential change produced by the summed EPPs is carried passively into the T-tubules to bring about contraction.

Bibliography

Adrian, R. H., and Freygang, W. H. (1962). The potassium and chloride conductance of frog muscle membrane. *J. Physiol. London.* **163**, 61–103.

Adrian, R. H., Chandler, W. K., and Hodgkin, A. L. (1970a). Voltage clamp experiments in striated muscle fibers. *J. Physiol. London.* **208**, 607–644.

Adrian, R. H., Chandler, W. K., and Hodgkin, A. L. (1970b). Slow changes in potassium permeability in skeletal muscle. *J. Physiol. London* **208**, 645–668.

Adrian, R. H., and Peachey, L. D. (1973). Reconstruction of the action potential of frog sartorius muscle. *J. Physiol. London.* **235**, 103–131.

Almers, W., Fink, R., and Palade, P. T. (1981). Calcium depletion in frog muscle tubules: The decline of calcium depletion in frog mus-

cle tubules: The decline of calcium current under maintained depolarization. *J. Physiol. London* **312**, 177–207.

Armstrong, C. M., Bezanilla, F. M., and Horowicz, P. (1972). Twitches in the presence of ethylene glycol bis(β-aminoethyl ether)-N, N'-tetraacetic acid. *Biochem. Biophys. Acta.* **267**, 605–608.

Baylor, S. M., and Oetliker, H. (1975). Birefingence experiments on isolated skeletal muscle fibres suggest a possible signal from the sarcoplasmic reticulum. *Nature* **253**, 97–101.

Beaty, G. N., and Stefani, I. (1976). Calcium dependent electrical activity in twitch muscle fibers of the frog. *Proc. R. Soc. London (Biol.)* **194**, 141–150.

Bezanilla, F., and Horowicz, P. (1975). Fluorescence intensity changes associated with contractile activation in frog muscle stained with Nile Blue A. *J. Physiol. London* **246**, 709–735.

Bezanilla, F., Caputo, C., Gonzalez-Serratos, H., and Venosa, R. A. (1972). Sodium dependence of the inward spread of activation in isolated twitch muscle fibres of the frog. *J. Physiol. London* **223**, 507–523.

Bianchi, C. P., and Bolton, T.C. (1974). Effect of hypertonic solutions and glycerol treatment on calcium and magnesium movements of frog skeletal muscle. *J. Pharmacol. Exp. Ther.* **188**, 536–522.

Bianchi, C. P., and Shanes, A. M. (1959). Calcium influx in skeletal muscle at rest, during activity, and during potassium contracture. *J. Gen. Physiol.* **42**, 803–815.

Birks, R. I., and Davey, D. F. (1969). Osmotic responses demonstrating the extracellular character of sarcoplasmic reticulum. *J. Physiol. London* **21**, 171–188.

Carlson, E. D., and Siger, A. (1960). The mechanochemistry of muscular contraction. I. The isometric twitch. *J. Gen. Physiol.* **44**, 33–60.

Castle, N. A., and Haylett, D. G. (1987). Effect of channel blockers on potassium efflux from metabolically exhausted frog skeletal muscle. *J. Physiol.* **383**, 31–43.

Chou, S. M., Nonaka, I., and Voice, G. F. (1980). Anastomoses of transverse tubules with terminal cisternae in polymyositis. *Arch. Neurol.* **37**, 257–266.

Cognard, C., Rivet-Bastide, M., Constantin, B., and Raymond, G. (1992). Progressive predominance of 'skeletal' versus 'cardiac' types of excitation-contraction coupling during *in vitro* skeletal myogenesis. *Pflügers Archiv.* **422**, 207–209.

Conway, E. J. (1957). Nature and significance of concentration relations of potassium and sodium ions in skeletal muscle. *Physiol. Rev.* **37**, 84–132.

Cordoba-Rodriguez, R., Gonzalez-Serratos, H., Matteson, D. R., and Rozycka, M. (1997). Ica is important in E-C coupling in developing cultured embryonic amphibian skeletal muscle cells. *Biophy. J.* **72**, A119, H5.

Cordoba-Rodriguez, R., Matteson, D. R., and Gonzalez-Serratos, H. (1996). Embryonic E-C coupling in cultured skeletal muscle cells. *Biophy. J.* **70**, A390.

Costantin, L. L., and Podolsky, R. J. (1967). Depolarization of the internal membrane system in the activation of frog skeletal muscle. *J. Gen. Physiol.* **50**, 1101–1124.

Costantin, L. L., and Taylor, S. R. (1973). Graded activation in frog muscle fibers. *J. Gen. Physiol.* **61**, 424–443.

Curtis, R. A. (1966). Ca fluxes in single twitch muscle fibers. *J. Gen. Physiol.* **50**, 225–267.

Davies, N. W., Standen, N. B., and Stanfield, P. R. (1992). The effect of intracellular pH on ATP-dependent potassium channels of frog skeletal muscle. *J. Physiol.* **445**, 549–568.

Dawson, M. J., Gadian, D. G., and Wilkie, D. R. (1978). Muscular fatigue investigated by phosphorus nuclear magnetic resonance. *Nature* **274**, 861–866.

Dawson, M. J., Gadian, D. G., and Wilkie, D. R. (1980). Mechanical relaxation rate and metabolism studied in fatiguing muscle by phosphorus nuclear magnetic resonance. *J. Physiol.* **299**, 465–484.

Eisenberg, R.S., and Gage, P. W. (1969). Ionic conductances of the surface and transverse tubular membranes of frog sartorius fibers. *J. Gen. Physiol.* **53**, 279–297.

Fabiato, A. (1982). Mechanism of calcium-induced release of calcium from the sarcoplasmic reticulum of skinned cardiac cells studied with potential-sensitive dyes. *In* "The Mechanism of Gated Calcium Transport Across Biological Membranes" (S. T. Ohnishi and M. Endo, Eds.), pp. 237–255. Academic Press, New York.

Fahlke, C., Zachar, E., and Rudel, R. (1993). Chloride channels with reduced single-channel conductance in recessive myotonia congenita. *Neuron* **10**, 225–232.

Fatt, P., and Ginsborg, B. L. (1958). The ionic requirements for the production of action potentials in crustacean muscle fibers. *J. Physiol. London* **142**, 156–543.

Fink, R., and Luttgau, H. C. (1976). An evaluation of the membrane constants and the potassium conductance in metabolically exhausted muscle fibers. *J. Physiol.* **263**, 215–238.

Flucher, B. E., Takekura, H., and Franzini-Armstrong, C. (1993). Development of the excitation-contraction coupling apparatus in skeletal muscle: Association of sarcoplasmic reticulum and transverse tubules with myofibrils. *Developmental Biol.* **160**, 135–147.

Forbes, M. S., and Sperelakis, N. (1979). Ruthenium red staining of skeletal and cardiac muscles. *Z. Zellforsch. Cell Tissue Res.* **200**, 367–382.

Garcia, M. del C., Gonzalez-Serratos, H., Morgan, J. P., Perreault, C. L., and Rozycka, M. (1991). Differential activation of myofibrils during fatigue in phasic skeletal muscle cells. *J. Musc. Res. Cell Motil.* **12**, 412–424.

Gonzalez-Serratos, H. (1971). Inward spread of activation in vertebrate muscle fibers. *J. Physiol. London* **212**, 777–799.

Gonzalez-Serratos, H., (1975). Graded activation of myofibrils and the effect of diameter on tension development during contractures in isolated skeletal muscle fibers. *J. Physiol. London* **253**, 321–339.

Gonzalez-Serratos, H., Cordoba-Rodriguez, R., Matteson, D. R., and Rozycka, M. (1996). Role of calcium currents in excitation-contraction coupling in developing cultured embryonic amphibian skeletal muscle cells. *J. Physiol.* **494P**.

Gonzalez-Serratos, H., Somlyo, A. V., McClellan, G., Shuman, H., Borrero, L. M., and Somlyo, A. P. (1978). Composition of vacuoles and sarcoplasmic reticulum in fatigued muscle: Electron probe analysis. *Proc. Natl. Acad. Sci.* **75**, 1329–1333.

Gonzalez-Serratos, H., Valle-Aguilera, R., Lathrop, D. A., and Garcia, M. del. (1982). Slow inward calcium currents have no obvious role in muscle excitation-contraction coupling. *Nature* **298**, 292–294.

Harris, E. J. (1963). Distribution and movement of muscle chloride. *J. Physiol. London* **166**, 87–109.

Hodgkin, A. L., and Horowicz, P. (1959). The influence of potassium and chloride ions on the membrane potential of single muscle fibers. *J. Physiol. London* **148**, 127–160.

Hodgkin, A. L., and Huxley, A. F. (1952). Currents carried by sodium and potassium ions through the membrane of the giant axon of *Loligo, J. Physiol. London* **116**, 449–472.

Hodgkin, A. L., and Nakajima, S. (1972). Analysis of the membrane capacity in frog muscle. *J. Physiol. London* **221**, 121–136.

Howell, J. N. (1974). Intracellular binding of ruthenium red in frog skeletal muscle. *J. Cell Biol.* **62**, 242-247

Hurnak, O., and Zachar, J. (1992). Maxi chloride channels in L6 myoblasts. *Gen. Physiol. Biophys.* **11**, 389–400.

Huxley, A. F., and Taylor, R. E. (1958). Local activation of striated muscle fibres. *J. Physiol. London* **144**, 426–441.

Huxley, H. E., Page, S., and Wilkie, D. R. (1963). Appendix. An electron microscopic study of muscle in hypertonic solutions. (M. Dydynsk and D. R. Wilkie, Eds.). *J. Physiol. London* **169**, 312–329.

Iannaccone, S. T., Li, K. -X., Sperelakis, N., and Lathrop, D. A. (1989). Insulin-induced hyperpolarization in mammalian skeletal muscle. *Am. J. Physiol.* **256**, C368–C374.

Kerr, L. M., and Sperelakis, N. (1982). Effects of the calcium antagonists verapamil and bepridil (CERM-1978) on Ca^{2+}-dependent slow action potentials in frog skeletal muscle, *J. Pharmacol. Exp. Ther.* **222**, 80–86.

Keynes, R. D., and Steinhardt, R.A. (1968). The components of the sodium efflux in frog muscle. *J. Physiol. London* **198**, 581–599.

Khan, A. R. (1981). Influence of ethanol and acetaldehyde on electromechanical coupling of skeletal muscle fibres. *Acta Physiol. Scand.* **111**, 425–430.

Kirby, A. C., Lindley, B. D., and Picken, J. R. (1975). Calcium content and exchange in frog skeletal muscle. *J. Physiol. London* **253**, 37–52.

Kulczycky, S., and Mainwood, G. W. (1972). Evidence for a functional connection between the sarcoplasmic reticulum and the extracellular space in frog sartorius muscle. *Can. J. Physiol. Pharmacol.* **50**, 87–98.

Li, K.-X., and Sperelakis, N. (1994). Electrogenic Na-K pump current in rat skeletal myoballs. *J. Cell. Physiol.* **159**, 181–186.

Lynch, C., III (1978). Kinetic and biochemical separation of potassium currents in frog striated muscle. Ph.D. thesis. University of Rochester, New York.

Mathias, R. T., Levis, R. A., and Eisenberg, R. S. (1980). Electrical models of excitation-contraction coupling and charge movement in skeletal muscle. *J.Gen. Physiol.* **76**, 1–31.

Miledi, R., Stefani, E., and Steinbach, A. B. (1971). Induction of the action potential mechanism in slow muscle fibres of the frog. *J. Physiol. London* **217**, 737–754.

Miledi, R., Parker, R. I., and Schalow, G. (1977). Measurement of calcium transients in frog muscle by the use of arseno III. *Proc. R. Soc. London (Biol.)* **198**, 201–210.

Mobley, B. A., and Eisenberg, B. R. (1975). Sizes of components in frog skeletal muscle measured by methods of stereology. *J. Gen. Physiol.* **66**, 31–45.

Moody-Corbett, F., Gilbert, R., Akbarali, H., and Hall, J. (1989). Calcium current in embryonic *Xenopus* muscle cells in culture. *Canadian J. Physiol. Pharmacol.* **67**, 1259–1264.

Nassar-Gentina, V., Passonneau, J. V., Vergara, J. L., and Rapoport, S. I. (1978). Metabolic correlates of fatigue and of recovery from fatigue in single frog muscle fibers. *Gen. Physiol.* **72**, 593–606.

Natori, R. (1965). Propagated contractions in isolated sarcolemma-free bundle of myofibrils. *Jikeidai Med. J.* **12**, 214–221.

Nicola-Siri, L., Sanchez, J. A., and Stefani, E. (1980). Effect of glycerol treatment on calcium current of frog skeletal muscle. *J. Physiol. London* **305**, 87–96.

Ortega, A., Gonzalez-Serratos, H., and Lepock, J. (1997). Effect of organic calcium channel blocker D-600 on sarcoplasmic reticulum calcium uptake in skeletal muscle. *Am. J. Physiol. (Cell Physiol. 41)* **272**, C310–C317.

Peachey, L. D. (1965). The sarcoplasmic reticulum and transverse tubules of the frog's sartorius. *J. Cell. Biol.* **25**, 209–231.

Podolsky, R. J., and Costantin, L. L. (1964). Regulation by calcium of the contraction and relaxation of muscle fibers. *Fed. Proc.* **23**, 933–939.

Potreau, D., and Raymond, G. (1980). Calcium-dependent electrical activity and contraction of voltage-clamped frog single muscle fibers. *J. Physiol. London* **307**, 9–22.

Renaud, J. M. (1989). The effect of lactate on intracellular pH and force recovery of fatigued sartorius muscles of frog, *Rana pipiens. J. Physiol.* **416**, 31–47.

Rogus, E., and Zierler, K. L. (1973). Sodium and water contents of sarcoplasm and sarcoplasmic reticulum in rat skeletal muscle: Effects of anisotonic media, ouabain, and external sodium. *J. Physiol. London* **233**, 227–270.

Rubio, R., and Sperelakis, N. (1971). Entrance of colloidal ThO_2 tracer into the T-tubules and longitudinal tubules of the guinea pig heart. *Z. Zellforsch.* **116**, 20–36.

Rubio, R., and Sperelakis, N. (1972). Penetration of horseradish peroxidase the terminal cisternae of frog skeletal muscle fibers and blockade of caffeine contracture by Ca^{++} depletion. *Z. Zellforsch.* **124**, 57–71.

Rulon, R., Hermsmeyer, K., and Sperelakis, N. (1971). Regenerative action potentials induced in the neurogenic heart of *Limulus polyphemus. Comp. Biochem. Physiol.* **39A**, 333-335.

Sanchez, J. A., and Stefani, E. (1978). Inward calcium current in twitch muscle fibers of the frog. *J. Physiol. London* **283**, 197–209.

Sauviat, M.-P., Ecault, E., Faivre, J.-F., and Finlay, I. (1991). Activation of ATP-sensitive K channels by a K channel opener (SR 44866) and the effect upon electrical and mechanical activity of frog skeletal muscle. *Pflügers Arch.* **418**, 261–265.

Sellin, L. C., and Sperelakis, N. (1978). Decreased potassium permeability in dystrophic mouse skeletal muscle. *Exp Neurol.* **62**, 609–617.

Spector, I., and Prives, J. M. (1977). Development of electrophysiological and biochemical membrane properties during differentiation of embryonic skeletal muscle in culture. *Proc. Natl. Acad. Sci. USA* **74**, 5166–5170.

Sperelakis, N. (1969). Changes in conductance of frog sartorius fibers produced by CO_2, ReO_4, and temperature. *Am J. Physiol.* **217**, 1069–1075.

Sperelakis, N. (1979). Origin of the cardiac resting potential. *In* "Handbook of Physiology , the Cardiovascular System, Vol. 1: The Heart" (R. Berne and N. Sperelakis, Eds.), Chap. 6, pp. 187–267. American Physiological Society, Bethesda, MD.

Sperelakis, N., and Rubio, R. (1971). Ultrastructural changes produced by hypertonicity in cat cardiac muscle. *J. Mol. Cell. Cardiol.* **3**, 139–156.

Sperelakis, N., and Schneider, M. F. (1968). Membrane ion conductances of frog sartorius fibers as a function of tonicity. *Am. J. Physiol.* **215**, 723–729.

Sperelakis, N., Forbes, M. S., and Rubio, R. (1974). The tubular systems of myocardial cells: Ultrastructure and possible function. *In* "Recent Advances in Studies on Cardiac Structure and Metabolism" (N. S. Dhalla and G. Rona, Eds.), Myocardial Biology, Vol. 4, pp. 163–194. University Park Press, Baltimore.

Sperelakis, N., Mayer, G., and Macdonald, R. (1970). Velocity of propagation in vertebrate cardiac muscles as functions of tonicity and $[K^+]_0$. *Am. J. Physiol.* **219**, 952–963.

Sperelakis, N., Schneider, M. F., and Harris, E. J. (1967). Decreased K$^+$ conductance produced by Ba^{++} in frog sartorius fibers. *J. Gen. Physiol.* **50**, 1565–1583.

Sperelakis, N., Shigenobu, K., and Rubio, R. (1978). ^{3}H-Sucrose compartments in frog skeletal muscle relative to sarcoplasmic reticulum. *Am. J. Physiol.* **234**, C181–C190.

Sperelakis, N., Valle, R., Orozco, C., Martinez-Palomo, A., and Rubio, R. (1973). Electromechanical uncoupling of frog skeletal muscle by possible change in sarcoplasmic reticular content. *Am J. Physiol.* **225**, 793–800.

Spitzer, N. C. (1979). Ion channels in development. *Annu. Rev. Neurosci.* **2**, 363–397.

Spruce, A. E., Standen, N. B., and Stanfield, P. Ṙ. (1985). Voltage-dependent ATP-sensitive potassium channels of skeletal muscle membrane. *Nature* **316**, 736–738.

Spruce, A. E., Standen, N. B., and Standfield, P. R. (1987). Studies of the unitary properties of adenosine-5'-triphospate-regulated potassium channels of frog skeletal muscle. *J. Physiol.* **382**, 213–236.

Standen, N. B., Stanfield, P. R., and Ward, T. A. (1985). Properties of single potassium channels in vesicles formed from the sarcolemma of frog skeletal muscle. *J. Physiol.* **364**, 339–358.

Stanfield, P. R. (1977). A calcium dependent inward current in frog skeletal muscle fibers. *Pflugers Arch.* **368**, 267–270.

Stephenson, E. W. (1981). Activation of fast skeletal muscle: Contributions of studies on skinned fibers. *Am J. Physiol.* **240**, C1-19.

Tasker, P., Simon, S. E., Johnstons, B. M., Shankly, K. H., and Shaw, F. H. (1959). The dimensions of the extracellular space in sartorius muscle. *J. Gen. Physiol.* **43**, 39–53.

Vinogradova, N. A. (1968). Distribution of nonpenetrating sugars in the frog's sartorius muscle under hypo- and hypertonic conditions. *Tsitologiya* **10**, 831–838.

Vivaudou, M. B., Arnoult, C., and Villaz, M. (1991). Skeletal muscle ATP-sensitive K$^+$ channels recorded from sarcolemmal blebs of split fibers: ATP inhibition is reduced by magnesium and ADP. *J. Membrane Biol.* **122**, 165–175.

Vogel, S., and Sperelakis, N. (1978). Valinomycin blockade of myocardial slow channels is reversed by high glucose. *Am J. Physiol.* **235**, H46–H51.

Vogel, S., Harder, D., and Sperelakis, N. (1978). Ca^{++} dependent electrical and mechanical activities in skeletal muscle. *Fed. Proc.* **37**, 517.

Wallace, N., and Sommer, J. R. (1975). Fusion of sarcoplasmic reticulum with ruthenium red. *Proc. Electron Microsc. Soc. A.* **33rd**, 500–501.

Weiss, D. S., and Magleby, K. L. (1992). Voltage-dependent gating mechanism for single fast chloride channels from rat skeletal muscle. *J. Physiol. London* **453**, 279–306.

Winegrad, S. (1968). Intracellular calcium movements of frog skeletal muscle during recovery from tetanus. *J. Gen. Physiol.* **51**, 65–83.

Woll, K. H., Lonnendonker, U., and Neumcke, B. (1989). ATP-sensitive potassium channels in adult mouse skeletal muscle: Different modes of blockage by internal cations, ATP, and tolbutamide. *Pflügers Arch.* **414**, 622–628.

Gordon M. Wahler

52

Cardiac Action Potentials

I. Introduction

The electrophysiological behavior of heart cells, individually and as a functional syncytium, subserve the function of the heart; namely to pump blood to the body. Heart cells are similar to other cell types in that their internal ionic composition is quite different from the extracellular ionic environment. For example, measurement of the intracellular versus extracellular ionic composition shows that the intracellular ionic composition is low in sodium ions (Na^+) and high in potassium ions (K^+), while the reverse is true of the extracellular ionic composition. These concentration differences, together with the selective permeability characteristics of the cell membrane, generate a potential difference of between 80 and 90 mV across the cell membrane of the resting cardiac cell, with the inside being negative with respect to the outside. Additionally, heart cells are excitable cells. That is, they are capable of generating all-or-none electrical responses known as **action potentials** (APs). The cardiac AP is caused by the complex interaction of a number of different ionic currents. This chapter reviews the characteristics of the cardiac AP, with emphasis on the major currents responsible for the various components or phases of the APs, as well as the regional differences in cardiac APs.

The electrical activity of cardiac cells has been studied for several decades by impaling the cells with high-resistance microelectrodes. The more recent developments of methods to isolate viable single adult cardiac cells (e.g., Powell *et al.,* 1980), together with the development of the patch-clamp technique (Hamill *et al.,* 1981) for recording single-channel (microscopic) currents and whole-cell (macroscopic) currents from single cells, has led to an explosion of information on the currents responsible for generation of the cardiac AP. While most of the information on cardiac ion currents has been obtained from cardiac cells isolated from experimental animals, the small number of studies on human cardiac cells (e.g., Coraboeuf and Nargeot, 1993; Ravens *et al.,* 1996) suggest that the currents in the human heart are gen-

erally similar to those observed in other animals, especially other mammals.

II. Resting Membrane Potential

In cardiac cells at rest, the membrane is quite permeable to K^+ and relatively impermeable to Na^+ and other ions. Thus, K^+ flows out of the cell down the concentration gradient, resulting in rapid buildup of a negative potential inside the cell. As the electric potential buildup increases in magnitude, it becomes sufficient to counterbalance the chemical driving force generated by the concentration gradient. At this potential, called the **equilibrium potential,** the net ion flux is zero. Note that this does not mean there is no flux of the ion, only that the inward and outward fluxes of the ion are equal. The Nernst equation describes the relationship between the intracellular and extracellular concentrations of a single ion and its equilibrium potential. The equilibrium potential for K^+ (E_K) is calculated by the Nernst equation

$$E_K = -\frac{RT}{z\mathscr{F}} \ln \frac{\left[K^+\right]_i}{\left[K^+\right]_o}$$

where T is the absolute temperature (in kelvins), z is the valence (or charge) of the ion (for K^+ it is +1), R is the universal gas constant, $\mathscr{F}$ is the Faraday constant, $[K^+]_i$ is the K^+ ion concentration inside the cell, $[K^+]_o$ is the K^+ ion concentration outside the cell, and ln is the natural logarithm. The intracellular and extracellular concentrations of K^+ are such that E_K is approximately -90 mV at 37 °C.

Since the ventricular cell at rest is very permeable to K^+, and not very permeable to other ions, the cell resting potential should be close to the calculated E_K. Indeed, this is the case. For example, in Fig. 1, the resting potential is approximately -87 mV. The resting potential does not quite reach E_K because there is a small, but finite, permeability to Na^+ ions and, in addition, there are other electrogenic ion transport

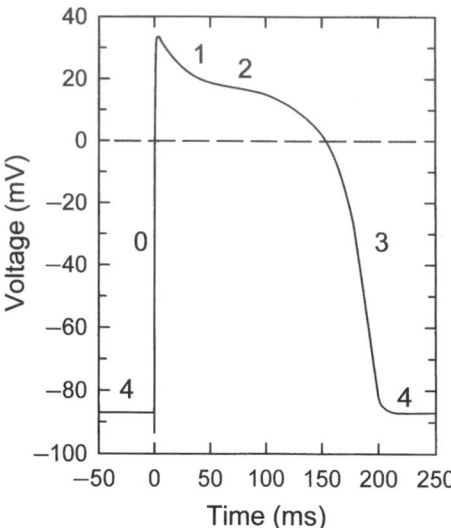

FIGURE 1. Ventricular action potential. This is a diagrammatic representation of a typical adult mammalian ventricular AP. The five phases of the ventricular AP are labeled (0–4).

systems (primarily the Na^+,K^+-ATPase) which contribute slightly to the resting potential.

III. Currents During Phases of the Action Potential

A. Overview

Cardiac cells have certain properties in common with other excitable cells, such as nerve and skeletal muscle cells. However, the behavior of cardiac cells also differs from the behavior of nerve and skeletal muscle cells in some important respects. The nerve and skeletal muscle APs are relatively brief, consisting primarily of a rapid depolarization phase (when the inside of the cell becomes more positive) followed immediately by a rapid repolarization phase (when the inside of the cell returns to a more negative potential). The depolarization phase of nerve and skeletal muscle cells is caused by the rapid influx of positive sodium ions into the cell, down the electrochemical gradient for Na^+, which makes the cell interior more positive. The subsequent repolarization is caused by the efflux of positive potassium ions from the cell, down the electrochemical gradient for K^+, which returns the cell interior to its original more negative resting potential. The entire process of the nerve or skeletal muscle APs is largely complete within a few milliseconds. In cardiac cells, the APs are more complex and generally much longer in duration. Thus, for example, a typical cardiac ventricular AP may be at least 100–200 ms in duration (Fig. 1). Additionally, unlike for nerve and skeletal muscle cells, APs from different regions of the heart vary substantially in shape.

There are two primary types of cardiac cells. One cell type is found in the working cells of the atria, ventricles, and the specialized conduction cells of the His-Purkinje net-

work. These cells have a high resting K^+ permeability between APs. The APs in these cells are generated by a fast Na^+ current and are known as fast APs. The prolonged duration of the AP in ventricular cells and Purkinje fibers is caused by an approximate balance of an inward Ca^{2+} current and outward K^+ currents, which results in a prominent plateau phase. Atrial cells have a less prominent plateau.

The second type of cardiac cell is found in the sinoatrial and atrioventricular nodes. These cells have a low K^+ permeability between APs and are automatic (i.e., spontaneously active). The upstroke of these APs is generated by a Ca^{2+} current, which is smaller and slower than the fast Na^+ current. Because of this, and the comparatively low density of Ca^{2+} channels, conduction in nodal cells is much slower than conduction in regions having Na^+-dependent APs. These APs are known as slow APs. Slow APs (a.k.a. slow responses) may sometimes also occur in cells that normally have fast APs. This occurs when I_{Na} is blocked pharmacologically or pathophysiologically by tetrodotoxin or voltage inactivation by depolarization with high K^+. The investigator-initiated generation of slow responses in ventricular cells has been used extensively in the past to indirectly study I_{Ca}. While this method has largely been supplanted by patch-clamp methodology, it does have the advantage of examining the cells under more physiological conditions (e.g., in the presence of sympathetic nerve terminals). Many earlier discoveries about the physiological and pharmacological modulation of I_{Ca} were made using this methodology (e.g., Wahler and Sperelakis, 1985). The use of perforated-patch techniques to study the currents under more physiological conditions and molecular biology techniques to clone and express channel subunits is currently revolutionizing our understanding of the precise nature of various currents. However, each of these methodologies has its limitations as well. For example, it is often not clear which combinations of cloned and expressed subunits are the physiological form of the channel (e.g., Tseng, 1999).

The configuration of the fast cardiac AP can be divided into several phases. The following description describes the five phases of the AP found in ventricular cells (phases 0 through 4), and indicates the current or currents that are primarily responsible for each phase. The phases and primary currents in the ventricle are summarized in Table 1. Phase 0 is the upstroke of the AP, phase 1 is the early repolarization phase, phase 2 is the plateau phase, phase 3 is the primary repolarization phase, and phase 4 is the resting potential phase of the ventricular AP. Additional information about regional differences in the AP phases and currents will be also presented for sinoatrial nodal cells, atrial cells, atrioventricular nodal cells, and Purkinje fibers in Section V.

B. Phase 0: Sodium Current

Phase 0 is the rapid upstroke of the AP. In ventricular cells, as well as in cells of the atria and His-Purkinje network, this upstroke is dependent on a fast Na^+ current quite similar to current responsible for the upstroke of the nerve or skeletal muscle APs. This current is designated as fast because it exhibits very rapid activation and inactivation kinetics, in contrast to other currents. Because of the fast upstroke and fast

TABLE 1 Phases of the Ventricular Action Potential and Primary Currents Responsible[a]

Phase	Current
Phase 0	I_{Na}
Phase 1	I_{to}
Phase 2	I_{Ca}, I_K, I_{to}
Phase 3	$I_K (I_{Ks}, I_{Kr}), I_{Kur}?, I_{K1}$
Phase 4	I_{K1}

[a]I_{Na} and I_{Ca} are inward currents; the potassium currents (I_{to}, I_K, I_{K1}, etc.) are outward currents.

conduction of the ventricular APs, the APs of these cells (and those of atrial and Purkinje fiber cells) are referred to as fast APs, as noted earlier. Figure 2 illustrates the kinetics and voltage-dependence of the Na$^+$ current. Following a large depolarization, I_{Na} reaches a peak in less than 1 ms. Following this peak activation of I_{Na}, the amplitude of this current spontaneously decreases. This decay of I_{Na} is due to closure (inactivation) of the fast Na$^+$ channels; thus, I_{Na} is nearly zero after only a few milliseconds. The fast activation and large magnitude of I_{Na} has caused difficulties in accurately recording this current in voltage-clamped cardiac cells. Thus, I_{Na} can generally only be recorded in adult cardiac cells at reduced $[Na^+]_o$ levels and/or at low temperatures. Alternatively, I_{Na} can be studied in very small cells, such as the embryonic chick ventricular cell (e.g., Fig. 2) in which I_{Na} (and other currents) are at least an order of magnitude smaller than in adult cells.

For the rapid upstroke of phase 0 to occur, the cell needs to be depolarized to the voltage necessary to open some of the fast Na$^+$ channels (approximately −70 mV). When the

Na$^+$ channels begin to open, Na$^+$ flows down the electrochemical gradient into the cell. This causes the cell to depolarize further (i.e., the inside becomes more positive), which opens additional Na$^+$ channels. Therefore, once sufficient Na$^+$ channels open, the process becomes self-perpetuating, resulting in rapid depolarization (i.e., the membrane potential rapidly moves toward the equilibrium potential for Na$^+$). The voltage at which a sufficient number of Na$^+$ channels open to initiate the AP is the threshold for firing of the AP. As the cell begins to depolarize further beyond the threshold, I_{Na} increases as more Na$^+$ channels are activated. Eventually, with even greater depolarization, I_{Na} begins to decline as the membrane potential approaches E_{Na}. Thus, the peak I_{Na} occurs between −30 to −20 mV. The membrane potential never reaches E_{Na} for several reasons: (1) As the membrane potential gets closer to E_{Na}, the driving force for Na$^+$ influx is diminished. (2) The Na$^+$ channels close shortly after opening (beginning after about 1 ms); thus, some Na$^+$ channels are already closing during the latter part of the upstroke. (3) Repolarizing currents are beginning to activate during the latter portion of the upstroke. Thus, the maximum positive membrane potential normally attained is approximately +35 mV (Fig. 1). Nevertheless, the upstroke causes a substantial voltage change (110–120 mV) within 1–2 ms in fast cardiac cells.

C. Phase 1: Transient Outward Current

Phase 1 of the cardiac AP is the transient and relatively small repolarization phase which immediately follows the upstroke of the AP. The size of phase 1 repolarization varies between species and also between different regions of the heart within a given species. Thus, APs recorded from the outer (epicardial) layer of ventricular cells display a more prominent phase 1, whereas APs recorded from the inner (endocardial) layer of ventricular cells, display a small phase

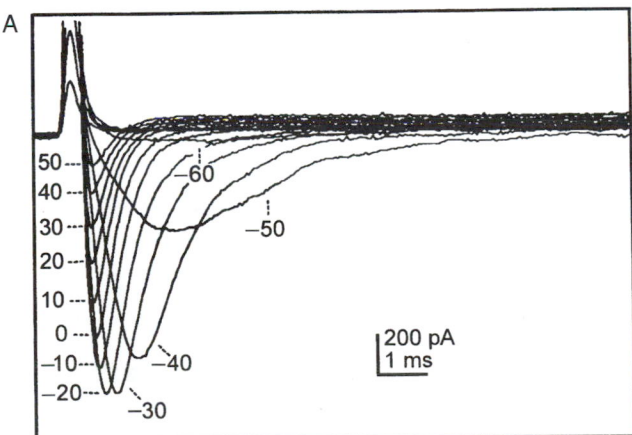
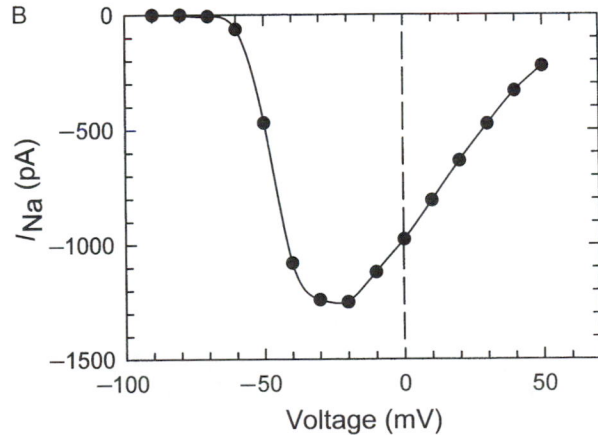

FIGURE 2. Sodium current (I_{Na}) recorded from a small embryonic chick ventricular cell. I_{Na} in these cells is much smaller than in adult cells, due to the very small size of these cells. Additionally, the small spherical shape of these cells and lack of t-tubules also contribute to better voltage control. (A) Shown are the original I_{Na} currents obtained upon stepwise depolarization from −100 mV to the indicated voltages. The current peaks in less than 1 ms at very depolarized potentials, and then rapidly inactivates. (B) The current-voltage curve of the peak current at each voltage. Note that the threshold for I_{Na} is approximately −70 mV and the current peaks at approximately −25 mV.

1 repolarization (Liu *et al.*, 1993). Phase 1 is also very large in Purkinje fibers and in atrial cells, but is largely absent in nodal cells.

Identification of the specific current or currents responsible for phase 1 repolarization has been controversial. It is now clear that the primary ion responsible for most of the phase 1 repolarization is K^+ (Kenyon and Gibbons, 1979), although a Cl^- component does appear to contribute to phase 1 repolarization (Bouron *et al.*, 1991) (discussed later).

Phase 1 repolarization is largely due to a transient outward current (I_{to}). I_{to} turns on rapidly with depolarization (i.e., beginning during the final portion of the AP upstroke), and is only active at very depolarized potentials; the threshold for activation is approximately -30 mV (Fig. 3). Thus, I_{to} has a characteristic transient shape—the rapid activation of this current is followed by inactivation during the AP plateau. Because of its voltage dependence and time course, I_{to} significantly overlaps (and opposes) the inward Ca^{2+} current (which is the primary depolarizing current during the plateau phase, see later discussion).

This current (I_{to}) is actually composed of at least two separate currents (I_{to1} and I_{to2}) which are carried through two physically distinct channels (Tseng and Hoffman, 1989). One of the currents (I_{to1}) is a K^+ current that is independent of the internal Ca^{2+} concentration ($[Ca^{2+}]_i$) and is sensitive to the K^+ channel blocker 4-aminopyridine (4-AP). This component of I_{to} is very similar to the I_A current recorded in nerve fibers. The second component of I_{to} (I_{to2}) is Ca^{2+}-dependent and less sensitive to 4-AP but is more sensitive to another K^+ channel blocker, tetraethylammonium ion (TEA$^+$). It is thought that I_{to2} is, at least in part, a Ca^{2+}-activated Cl^- channel (Harvey, 1996).

Under physiological conditions, the first I_{to} component, I_{to1}, is by far the larger of the two components. Thus, irrespective of the exact nature of I_{to2}, efflux of K^+ (through I_{to1} channels) appears responsible for the vast majority of phase 1 repolarization. The second component (I_{to2}) may become more important when intracellular levels of Ca^{2+} become too high. Thus, under Ca^{2+}-overload conditions, I_{to2} would be activated and shorten the AP duration, thereby indirectly abbreviating the duration of Ca^{2+} current, resulting in a reduced Ca^{2+} influx. Thus, activation of I_{to2} by intracellular Ca^{2+} likely acts as a negative feedback mechanism to reduce calcium overload.

D. Phase 2: Calcium Current

Phase 2 is commonly called the plateau phase. It follows the early repolarization phase (phase 1), and is a period of time in which the membrane potential remains relatively constant, that is, it does not rapidly repolarize. The presence of this prominent plateau phase is responsible for the long AP duration in cardiac cells, which is the major difference between cardiac cells and nerve or skeletal muscle fibers. The plateau is caused by an approximate balance of positive inward current (which would tend to cause depolarization) and positive outward currents (which would tend to cause repolarization). The primary inward current is a Ca^{2+} current. The primary outward K^+ current during the plateau phase of ventricular cells (particularly in the latter part of the plateau), is the slowly activating K^+ current known as the delayed rectifier, which is described in greater detail under phase 3 repolarization. Additionally, I_{to} contributes to the early plateau phase in those cells that have a substantial I_{to}. In addition to the Ca^{2+} and K^+ currents, there is a small contribution of the fast Na^+ current to the plateau. Thus, while I_{Na} largely inactivates within a few milliseconds after depolarization, a very small fraction of the I_{Na} inactivates slowly (see Fig. 1), causing a late sodium current known as the **sodium window current,** which is sufficient to affect the plateau of the AP, and, hence, AP duration (Attwell *et al.*, 1979).

The inward Ca^{2+} current (I_{Ca}) exhibits activation and inactivation much like I_{Na}, but on a slower time scale (Fig. 4). This second inward current is carried by Ca^{2+} and peaks

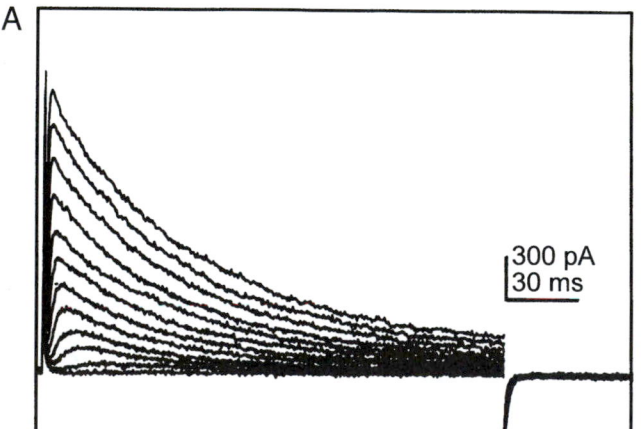

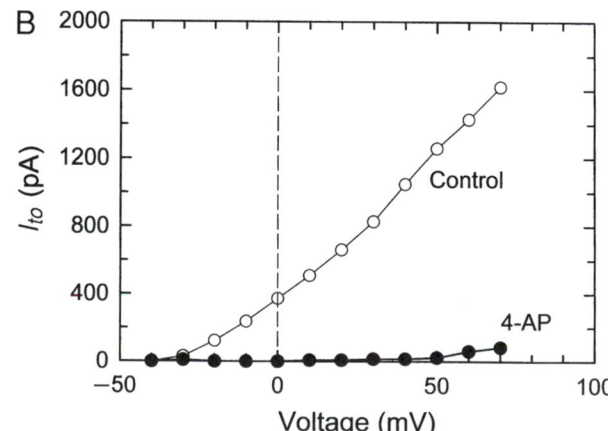

FIGURE 3. Transient outward current (I_{to}) recorded from a 21-day-old rat ventricular cell. (A) Shown are the original currents obtained upon stepwise depolarization from -80 mV. I_{to} activates rapidly and then inactivates. Currents were recorded in the presence of tetrodotoxin to eliminate I_{Na} and cadmium to eliminate overlapping I_{Ca}. (B) The current-voltage curve shows increasing activation of this current at voltages above approximately -30 mV (indicated by open circles). This current is the Ca^{2+}-independent, 4-aminopyridine-sensitive component of I_{to} (i.e., I_{to1}) and is therefore readily blocked by 4-aminopyridine (4-AP, indicated by closed circles).

A

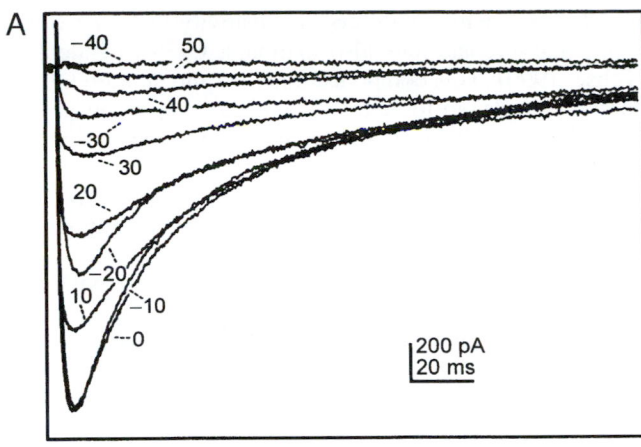

B

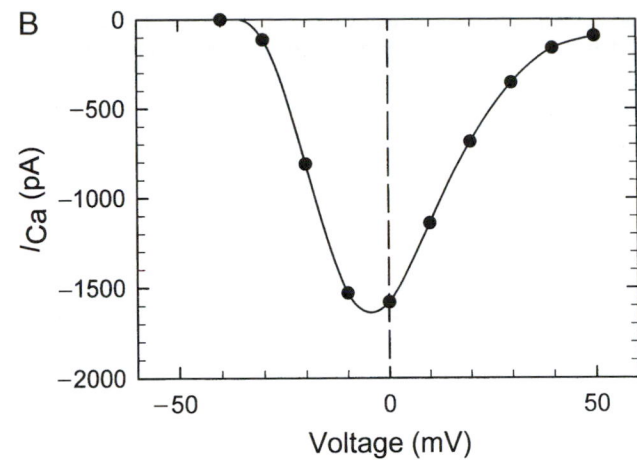

FIGURE 4. L-type calcium current ($I_{Ca(L)}$) recorded from an adult guinea pig ventricular cell. (A) Shown are the original currents obtained upon stepwise depolarization from –80 mV to the indicated voltages. I_{Na} was inactivated by briefly pulsing to –40 mV prior to applying the indicated test pulse. The Ca^{2+} currents show the same general shape as the Na^+ currents (Fig. 2.), i.e., they activate upon depolarizing steps and inactivate during the voltage pulse. However, both activation and inactivation are much slower than for I_{Na} (note difference in time scale). (B) The current-voltage curve shows a similar shape to the I_{Na} current-voltage relationship; however, both the threshold voltage and peak voltage are shifted approximately 30 mV in the depolarizing direction. $I_{Ca(L)}$ is roughly the same size as I_{Na} shown in Fig. 2 only because this cell is so much larger than the embryonic chick cell used in Fig. 2. I_{Na} in this guinea pig cell would be approximately 20 nA or more, and could not be voltage-clamped under these conditions.

within a few ms, but requires a few hundred ms to completely inactivate. I_{Ca} is at least one order of magnitude smaller than I_{Na} in a given cell. The threshold for activation of I_{Ca} is approximately –40 mV, the current is maximal near 0 mV, and the E_{Ca} is around +100 mV. Thus, I_{Ca} is active over a more positive (depolarized) potential range than I_{Na}. This classical Ca^{2+} current is now commonly referred to as the **L-type calcium current** ($I_{Ca(L)}$), to distinguish it from the novel transient **T-type Ca^{2+} current** ($I_{Ca(T)}$) described later.

The L-type Ca^{2+} current is central to many aspects of cardiac function. For example, it is the primary link in excitation-contraction coupling in the heart. Thus, the influx of Ca^{2+} ions during the plateau of the AP is what links the electrical events of the AP to the mechanical events, namely contraction. Ca^{2+} influx via $I_{Ca(L)}$ stimulates Ca^{2+} release from internal stores (which is the major source of contractile Ca^{2+} in most adult mammalian hearts), replenishes the internal Ca^{2+} stores available for subsequent release (for more detail, see Chapter 54), and, to some extent, directly activates the contractile proteins. $I_{Ca(L)}$ is also very important in automaticity and conduction, due to the Ca^{2+}-dependent nature of the nodal APs.

$I_{Ca(L)}$ is a major regulatory site in control of cardiac electrical activity and contraction by neurotransmitters, hormones, intracellular ions, etc. Perhaps the most important regulator of $I_{Ca(L)}$ in the heart is the autonomic nervous system. Thus, release of norepinephrine from cardiac sympathetic nerves or release of epinephrine from the adrenal gland stimulates the beta-adrenergic receptors of the cardiac cell. A cascade of events ensues, which involves production of cyclic AMP (adenosine 3´,5´-cyclic monophosphate) and stimulation of the cyclic AMP-dependent protein kinase. This protein kinase directly phosphorylates the Ca^{2+} chan-

nel, thereby enhancing the channel activity. The net result of this cascade is that $I_{Ca(L)}$, and thereby force of contraction, is stimulated. For a detailed review of this process, see the chapter on phosphorylation. The parasympathetic neurotransmitter, acetylcholine (ACh), inhibits cyclic AMP formation. Additionally, ACh stimulates cyclic GMP (guanosine 3´,5´-cyclic monophosphate) production. Cyclic GMP is a second messenger similar to cyclic AMP, but which reduces $I_{Ca(L)}$ (see Wahler and Dollinger, 1995).

E. Phase 3: Delayed Rectifier Current

Phase 3 is the late or final repolarization phase following the AP plateau. It is similar to the repolarization observed in nerve and skeletal muscle cells. It is primarily caused by the unbalance of the currents that were relatively balanced during phase 2. $I_{Ca(L)}$ decreases with time (due to inactivation) and the delayed rectifier current (I_K) increases (due to slow activation). This eventually leads to the outward current (I_K) overwhelming the inward current ($I_{Ca(L)}$). I_K is the primary repolarizing current in most ventricular preparations. The classical delayed rectifier activates slowly, compared to most other currents, and does not inactivate with time. Thus, I_K increases gradually during sustained depolarization at voltages around the plateau level. This current is similar to the delayed rectifier in nerve cells, although slower. I_K can be clearly distinguished from I_{to} by slow activation, lack of inactivation, and different pharmacology. Due to technical reasons, I_K is often greatly underestimated in patch-clamp recordings. For example, in canine atria I_K was found in 4% of cells isolated using a "chunk" method versus 99% of cells isolated using a "perfusion" method (Yue *et al.*, 1996).

There are several components of I_K; thus, many investigators divide I_K into a very slowly activating component (I_{Ks})

and a more rapidly activating component (I_{Kr}) (Sanguinetti and Jurkiewicz, 1990; Lindblad *et al.*, 1996). The sensitivity of the two I_K components to pharmacological agents differs, as does the shape of their current-voltage curves and their activation threshold voltage, suggesting that these two components of I_K are carried through distinct channels.

Mutations in the genes encoding I_{Kr} and I_{Ks} have both been associated with long QT syndrome, which is characterized by slowed repolarization, increased incidence of Torsades de Pointes arrhythmias, and a higher risk for sudden death (Veldkamp, 1998; Vizgirda, 1999). Interestingly, the APs are longer in females than males, resulting in a greater QT interval. This suggests that the repolarizing currents may be smaller in females than males, which may explain the greater incidence of Torsades de Pointes arrhythmias in females (Vizgirda, 1999).

The beta-adrenergic-cAMP cascade can stimulate I_K, similar to its ability to stimulate $I_{Ca(L)}$ (e.g., Walsh and Kass, 1988), which would tend to shorten the AP. Thus, beta-adrenergic stimulation tends to lengthen the AP duration by enhancing I_{Ca} and at the same time tends to shorten AP duration by enhancing I_K. The overall effect of beta-adrenergic stimulation on AP duration, thus, is determined by the relative contribution of the changes in these two currents (and perhaps also the contribution of I_{Cl}, see later discussion). The slower component of I_K, I_{Ks}, is the component of I_K which is enhanced in conditions of beta-adrenergic stimulation (Sanguinetti and Jurkiewicz, 1990).

Another potentially important repolarizing current has been demonstrated in human atrial myocytes. This **ultrarapid-activating delayed rectifier K$^+$ current** (I_{Kur}) shows slow deactivation and is insensitive to TEA, Ba^{2+}, and dendrotoxin, but is sensitive to 4-AP (Feng *et al.*, 1998). Thus, I_{Kur} is a K$^+$ current that can be distinguished from both I_{to1} (from which it differs by voltage dependence) and the clas-

sical I_K (from which it differs by sensitivity to 4-AP). Some ventricular preparations also exhibit a Ca^{2+}-dependent I_K (Tohse, 1990). This component may help to shorten the AP duration similar to the effect of the Ca^{2+}-dependent I_{to} described above. However, the relative magnitude and importance of these currents is still unclear.

As the membrane potential continues to repolarize and approaches the resting membrane potential, largely due to the effects of I_K, the inwardly rectifying K$^+$ current (I_{K1}, which is the primary determinant of the resting potential) also begins to contribute to repolarization. Thus, it is clear that a number of potassium currents may contribute to phase 3 of the action potential.

F. Phase 4: Inward Rectifier Current

In ventricular cells, and most other cardiac cells, phase 4 is the resting potential. The resting potential is defined as the stable, negative potential that occurs between APs in nonspontaneous cells. The resulting negative resting potential is maintained by the Na$^+$-K$^+$ pump. It is very near E_K due to a relatively high permeability of the resting ventricular cell membrane to K$^+$ ions, and a very low permeability to other ions. Thus, the resting potential of nonautomatic cardiac cells is similar to the resting potential in skeletal muscle cells. The resting potential is determined largely by a K$^+$ current known as the **inward** (or **anomalous**) **rectifier**, I_{K1}.

The most notable characteristic of I_{K1} is that it displays inward rectification. Thus, it passes current more readily in the inward direction than in the outward direction. This characteristic is evident in Fig. 5. This inward rectification of the I_{K1} channel current is thought to be due to blockade of the channels by intracellular Mg^{2+} (Matsuda *et al.*, 1987) and polyamines (Lopatin *et al.*, 1994).

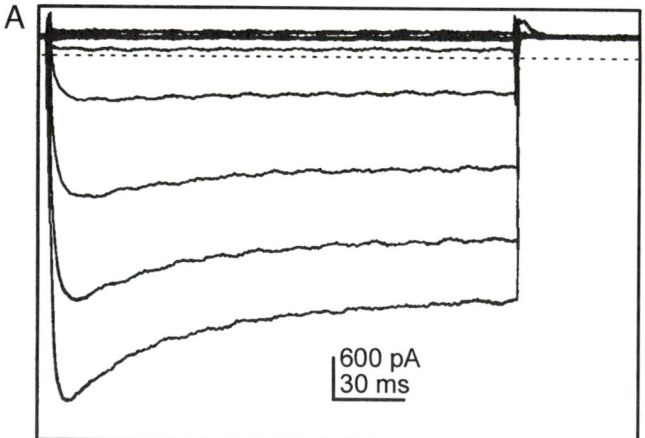
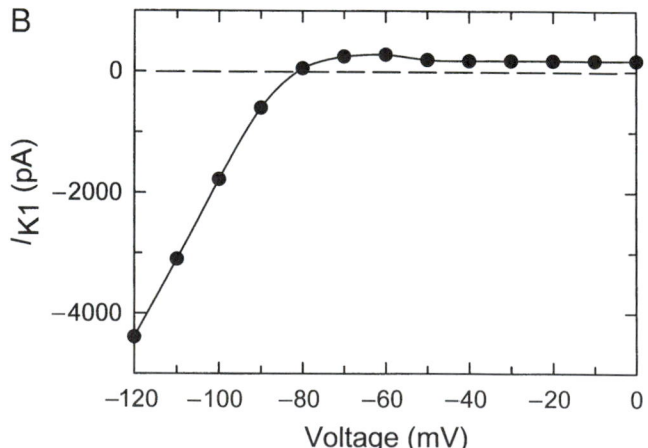

FIGURE 5.   Inwardly-rectifying K$^+$ current (I_{K1}) recorded from an adult guinea pig ventricular cell. (A) Shown are the original barium-subtracted currents obtained upon stepwise hyperpolarization and depolarization from a holding potential of −40 mV. The current displays the typical inward rectification. That is, the current is smaller in the outward (physiological) direction than the inward direction. (B) The current-voltage curve for I_{K1}. The current also displays a negative slope region between approximately −60 mV and −30 mV. Thus, larger depolarizations actually result in a decrease in outward K$^+$ flux during this voltage range, such that the current at −30 mV is less than half as large as at −60 mV. The net result is that during repolarization (phase 3) the K$^+$ current increases as the membrane potential approaches the resting potential.

Of course, the outward K^+ current is the only one occurring physiologically, since the membrane potential does not normally hyperpolarize beyond E_K. In most preparations, I_{K1} also demonstrates a negative slope region in the outward direction. That is, as the cell is depolarized from near its resting potential to progressively more depolarized potentials, the outward current first increases and then decreases with further depolarization, with the outward current dropping to very low levels at voltages positive to -40 mV. Thus, the contribution of I_{K1} to repolarization tends to be less at potentials near the plateau, and greater as the membrane potential approaches the resting potential, that is, at the time that I_K is declining.

IV. Additional Currents Contributing to the Action Potential

A. Pump Current

The Na^+,K^+-ATPase, or pump, is the primary transport system that maintains the ionic imbalance between the cell exterior and interior. That is, each pump cycle extrudes three Na^+ ions out of the cell and transports two K^+ ions into the cell, thus building up $[K^+]_i$ and reducing $[Na^+]_i$. Because of the exchange of three positive for two positive ions, each pump cycle generates a net loss of one positive charge, which generates a pump current (I_{pump}). At the resting potential, I_{pump} is an outward current which may hyperpolarize the membrane potential slightly. In addition, I_{pump} can cause considerable shortening of the action potential when $[Na^+]_i$ increases pathologically (Gadsby, 1984). Additionally, blockade of the pump with toxic concentrations of digitalis may lead to significant increase in intracellular $[Na^+]$, which may in turn activate a Na^+-dependent K^+ current ($I_{K(Na)}$) (Luk and Carmeliet, 1990). However, the primary role of the Na^+,K^+-ATPase is to set up and maintain the ionic gradients that generate the electrochemical driving forces for the currents responsible for the action potential.

B. Na^+-Ca^{2+} Exchange Current

A Na^+-Ca^{2+} exchanger also exists in the cardiac sarcolemma. Working in the "normal mode" the Na^+-Ca^{2+} exchanger exchanges intracellular Ca^{2+} for extracellular Na^+; thus, the exchanger is an important mechanism whereby Ca^{2+} is removed from the cytoplasm. The exchanger may also work in the "reverse" mode, exchanging intracellular Na^+ for extracellular Ca^{2+}. Under these conditions, the exchanger can contribute Ca^{2+} influx for excitation-contraction coupling. The exchanger transports three Na^+ for each Ca^{2+} under most conditions, leading to the net movement of one positive charge. Thus, the exchanger generates a current which can also contribute to the action potential. The equilibrium potential for Na^+-Ca^{2+} exchange current is generally slightly negative to 0 mV; therefore, near the resting potential the Na^+-Ca^{2+} exchanger works in the normal mode and generates an inward current. During the initial portion of the plateau, the Na^+-Ca^{2+} exchanger transiently works in the reverse mode and briefly generates

an outward current prior to returning to the normal mode. Thus, the Na^+-Ca^{2+} exchange current may contribute to the shape of the AP (for review, see Janvier and Boyett, 1996).

C. cAMP-Stimulated Chloride Current

Under basal conditions, the Cl^- current (I_{Cl}) in the heart is relatively small and probably does not contribute a great deal to the configuration of the AP. However, when cAMP levels are stimulated (as with sympathetic nerve stimulation), a significant time-independent Cl^- current develops. Activation of this current by beta-adrenergic stimulation can cause a small depolarization of the resting potential and significant shortening of the action potential (Harvey et al., 1990).

D. ATP-Sensitive Potassium Current

When the oxygen supply declines, as occurs during ischemia, AP duration shortens. The shortening of the AP duration accelerates inactivation of I_{Ca}, thereby reducing contractility. The reduced contractility greatly decreases the energy demands of the cell, thereby sparing ATP. This mechanism contributes to the survival of the myocardial cell during temporary ischemia. However, in addition to this beneficial effect, regional shortening of the AP can also lead to arrhythmias, due to the dispersion of refractory periods.

The shortening of the AP during ischemia is largely caused by activation of a unique outward K^+ current, which is inhibited by intracellular ATP. The decreased oxygen supply reduces ATP levels in the cell, and I_{K1} is inhibited. At the same time, another K^+ current ($I_{K(ATP)}$) is activated (Noma, 1983). This ATP-sensitive K^+ current is inhibited by physiological levels of intracellular ATP (Fig. 6) and thus appears to contribute little to the AP configuration under conditions of adequate oxygenation. However, during inadequate oxygenation, the AP is shortened as $I_{K(ATP)}$ replaces I_{K1}. This is in large part because $I_{K(ATP)}$ displays less inward rectification than I_{K1}, and thus has a greater effect on repolarization. The physical characteristics of the $I_{K(ATP)}$ channel (e.g., ATP sensitivity and/or degree of rectification) are altered in some pathophysiological states, such as hypertrophy (Cameron et al., 1988) or diabetic cardiomyopathy (Smith and Wahler, 1996; Shimoni et al., 1998). This may be an important factor in the abnormal responses to ischemia in these conditions.

The intracellular ATP levels reached during acute ischemia are generally higher than required to open a substantial number of $I_{K(ATP)}$ channels. However, the decreasing intracellular pH and increasing lactate accumulation also can promote $I_{K(ATP)}$ channel opening (Fan and Makielski, 1993) in addition to a fall in intracellular ATP levels. Furthermore, opening of only a small fraction of $I_{K(ATP)}$ channels may be sufficient to substantially shorten the AP (Faivre and Findlay, 1990).

Other channels may also be activated by ischemia, in addition to $I_{K(ATP)}$ channels; for example, lysophosphatidylcholine and long-chain acylcarnitine resulting from membrane phospholipid metabolism rapidly accumulate in early ischemia and may activate the arachidonic acid–activated K^+

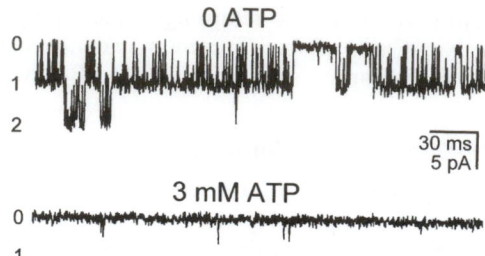

FIGURE 6. ATP-sensitive K$^+$ current $I_{K(ATP)}$ recorded from an inside-out patch from an adult rat ventricular cell. Single-channel currents shown were recorded in the absence of ATP (upper traces) and in the presence of physiological levels of ATP (3 mM, lower traces). Channel openings are downward. In the absence of ATP, channel activity is high. Thus, up to two channels are open simultaneously (number of open channel levels are indicated at the side). In the presence of 3 mM ATP, the channels are rarely open. In the presence of ATP, the openings are extremely brief, such that they do not appear to reach the normal open level (1).

channel ($I_{K(AA)}$) and the phosphatidylcholine-activated K$^+$ channel ($I_{K(PC)}$) (Montsuez, 1997).

V. Regional Differences in Action Potentials

A. Overview

The normal pathway for electrical activation of the heart is the following: sinoatrial (SA) node, atria, atrioventricular (AV) node, bundle of His, Purkinje fibers, ventricles. The APs differ from region to region, reflecting the different roles played by the different cell types. The following description characterizes the APs for each region, and indicates how each differs from the ventricular AP.

B. Sinoatrial Node

The SA node contains specialized cells that generate APs that are quite different from the ventricular APs described above (Fig. 7). Unlike ventricular APs, these cells do not have a true resting potential; that is, the membrane potential between APs is not stable but rather exhibits a slow spontaneous depolarization known as **phase 4 depolarization,** or the **pacemaker potential.** Since there is no resting potential in these cells, the most negative potential the cell reaches between APs is called the **maximum diastolic potential.** This potential is less negative than the resting potential of ventricular cells, due to a lower K$^+$ permeability (caused by a lack of I_{K1} in these cells). The maximum diastolic potential ranges from approximately −55 mV for true primary pacemaker cells, to approximately −70 mV for transitional cells on the border between the SA node and atria. The pacemaker potential takes the nodal cell from the maximum diastolic potential to the threshold for generating an AP in these cells (approximately −40 mV). Thus these cells are spontaneously active, and the slope of the phase 4 depolarization is an important determinant of the rate of AP generation, and, thereby, heart rate.

Once the phase 4 depolarization brings the cell to the threshold for AP firing, an AP occurs, as in ventricular cells. However, in nodal cells, the upstroke (phase 0) is quite different from the upstroke in ventricular (and nerve or skeletal muscle) cells: it is a much slower upstroke and is Ca^{2+}-dependent rather than Na$^+$-dependent. That is, I_{Na} is negligible in SA node cells and does not contribute significantly to phase 0 (Irisawa *et al.*, 1993). Thus, the upstroke in nodal cells is generated by the inward Ca^{2+} current, $I_{Ca(L)}$. Since the speed of the upstroke largely determines the speed of conduction, this slow phase 0 is very important in cardiac function, since it results in a slow conduction in nodal cells. There is generally no phase 1 and a brief plateau (phase 2) in nodal cells. Phase 3 repolarization returns the cell to the maximum diastolic depolarization. Since nodal cells only repolarize to the maximum diastolic potential of approximately −55 to −70 mV, the lack of I_{K1} in these cells does not significantly slow repolarization, because I_{K1} contributes to repolarization primarily at potentials nearer to E_K.

C. Atria

Atrial cells have APs that are similar in many aspects to the ventricular APs described above. Thus, the resting potential (phase 4) is approximately −85 mV, and there is a fast upstroke (phase 0) generated by I_{Na}. The most distinguishing feature of the atrial AP is that it has a more triangular appearance than the ventricular AP. This more triangular appearance seems to be due to a prominent phase 1 in atrial cells. Thus, in atrial cells phases 1, 2, and 3 tend to run together resulting in a triangular shape, with a distinct plateau not always apparent. This is likely due to a large I_{to} in atrial cells.

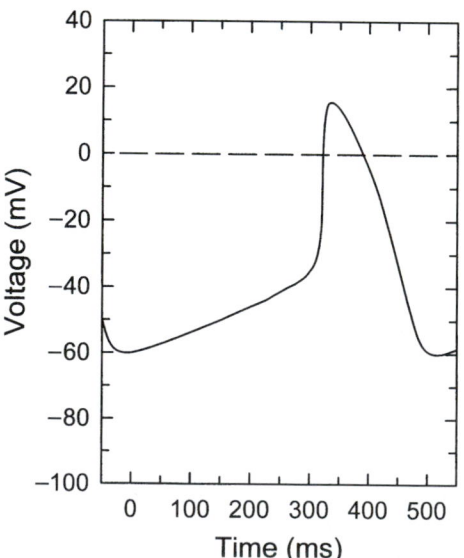

FIGURE 7. Sinoatrial node AP. This is a diagrammatic representation of a typical sinoatrial node AP in the mammalian heart. Note the pacemaker potential and the less negative membrane potential, and the slower upstroke compared to the fast AP of the ventricular cell (see Fig. 1).

D. Atrioventricular Node

The cells of the AV node generate APs that are quite similar to the APs of the SA node. Thus, these cells fire Ca^{2+}-dependent APs and also display spontaneous phase 4 depolarization (i.e., automaticity). However, the rate of the phase 4 depolarization in AV nodal cells is much slower than the rate of phase 4 depolarization in SA nodal cells. Thus, the SA node cells fire APs before the AV node cells fire, which is why the SA node cells are the normal pacemaker cells of the heart.

E. Purkinje Fibers

Purkinje APs are in most respects similar to the ventricular APs. Thus, these cells have a negative resting potential between APs (phase 4) and a very rapid upstroke (phase 0) generated by I_{Na}. Purkinje fibers do differ from ventricular cells in that they have a more prominent phase 1 repolarization and a longer plateau (phase 2). The plateau is followed by a phase 3 repolarization that is virtually identical to phase 3 in ventricular cells. Cells in the Bundle of His appear to have APs similar to those in the Purkinje fibers; however, in general, Bundle of His cells have not been studied in detail.

Additionally, Purkinje cells may exhibit automaticity, especially when the extracellular K^+ concentration is low. Thus, under some conditions, they exhibit phase 4 depolarization. Purkinje cells have all the currents found in ventricular cells, described above (I_{Na}, I_{to}, I_{Ca}, I_K, I_{K1}). Purkinje cells also have some additional currents, which are absent or very small in ventricular cells, that are related to the latent pacemaker function of these cells. These additional currents are described in the following section on automaticity.

VI. Automaticity

A. Overview

Automaticity refers to the ability of some cardiac cells to depolarize and fire repetitive APs spontaneously. Thus, as noted above, automatic cells (e.g., in the SA node) do not have a stable resting potential between APs, but rather have a maximum diastolic potential followed by a spontaneous phase 4 depolarization known as the pacemaker potential. The spontaneous depolarization to threshold generates the AP in that cell type. Following repolarization of the AP, the spontaneous phase 4 depolarization occurs again. The slope of the pacemaker potential (i.e., rate of spontaneous phase 4 depolarization) largely determines the rate of AP firing. Because the rate of firing of APs is normally fastest in the primary pacemaker cells of the SA node, this region acts as the normal pacemaker of the heart. Once this region fires an AP, the wave of depolarization is propagated to other regions of the heart, ultimately leading to contraction of the heart. Automatic cells in other regions of the heart (e.g., AV node) normally do not have the opportunity to spontaneously fire before the wave of depolarization arising from the SA node drives them to threshold.

B. Mechanisms of Automaticity

Automaticity is a property of several cell types in the heart under physiological conditions. Under pathophysiological conditions, even normally nonspontaneous cells (e.g., ventricular cells) may exhibit automaticity. Physiologically, the most important pacemaker potential is that of the normal pacemaker cells in the SA node. However, the very small size of the SA node cells makes all the currents difficult to measure accurately; additionally, the very high input resistance (due to lack of I_{K1} in the SA node) means that extremely small currents may be significant contributors to the pacemaker potential in the SA node cells. These technical difficulties have limited our understanding of the pacemaker process in SA node cells. Therefore, there is considerable controversy regarding the precise mechanism of automaticity in SA node cells (for reviews, see Baumgarten and Fozzard, 1992; Campbell *et al.,* 1992). Because of these limitations, much of what we know about automaticity is experimentally derived from other automatic cell types (e.g., Purkinje cells) and then extrapolated to SA node cells using various mathematical models. There are limitations to this approach, since there are considerable differences between automatic cell types. Thus, for instance, the pacemaker potential range is much more depolarized in SA node cells compared to Purkinje cells, which is bound to influence the relative contribution of various currents.

1. Automaticity in Purkinje Fibers

Under physiological conditions, Purkinje fibers can have an extremely slow phase 4 pacemaker potential. In contrast to SA node cells, the mechanisms for automaticity in Purkinje cells are fairly well understood. As noted above, the Purkinje cells have all the currents that ventricular cells have, plus some additional currents not found to any significant degree in adult ventricular cells. The large I_{K1} current in Purkinje cells tends to clamp the membrane potential near E_K, and therefore a large depolarizing current is needed to overcome this clamping effect. The primary current responsible for the pacemaker potential in Purkinje has some unusual properties, earning it the designation as the "funny current" (I_f).

I_f is a slowly activating inward depolarizing current activated by hyperpolarization, which is present in automatic cells. I_f is largely absent in adult ventricular cells (or it may be present but nonfunctional in ventricular cells due to voltage inactivation, see Yu and Cohen, 1993). It is a nonselective cation current, that is, it is carried by a mixture of both Na^+ and K^+ ions. In Purkinje cells, I_f is responsible for most of the depolarizing current which generates the pacemaker potential.

In addition to I_f, there may be a small contribution of the T-type Ca^{2+} current to the latter stages of the pacemaker potential in Purkinje cells (see later description of automaticity in nodal cells), as well as a small contribution of I_{Na} to the final portion of the pacemaker potential in these cells. Once the Purkinje cell is depolarized to the threshold for I_{Na} (approximately −70 mV), Na^+ channels will open. The inward flux of Na^+ may contribute to the final phase of the pacemaker potential, as the membrane potential approaches the threshold for AP generation. Thus, the Purkinje fiber will fire an AP which is generated by I_{Na}, as described above for ventricular cells. The remainder of the AP is generated by essentially the same mechanisms as previously described for ventricular cells.

2. Automaticity in Nodal Cells

Several factors contribute to the pacemaker potential in nodal cells. Because of the very low density of I_{K1} channels in nodal cells, the resting K$^+$ permeability is much lower in nodal cells than in ventricular cells. The large resting K$^+$ permeability in ventricular cells generated by I_{K1} tends to keep the interior of the cells negative, opposing depolarization of the cell toward threshold by "clamping" the membrane potential near E_K. A much smaller current is sufficient to depolarize the nodal cells, due to this much lower resting K$^+$ permeability. Thus currents which may be too small to accurately measure using present electrophysiological techniques (small background currents or currents produced by various electrogenic transport mechanisms) could produce sufficient current to affect the pacemaker potential. Because of this limitation, the analysis of the relative contribution of various currents to the pacemaker potential in nodal cells is much less clear than for Purkinje cells. Therefore, investigators often use mathematical models of nodal electrophysiology which are largely based on the characteristics of various currents measured in other automatic cell types (e.g., Purkinje fibers, sinus-venosus of amphibian hearts) in which the measurements are more readily made. However, the mathematical models of nodal cells at present are inadequate to accurately describe changes in automaticity under a variety of physiological conditions (see Dokos *et al.*, 1996).

The major depolarizing current during the pacemaker potential of Purkinje cells, I_f, is also present in nodal cells and probably contributes somewhat to the pacemaker potential in the SA node. As noted earlier, I_f is an unusual depolarizing current in that it is activated by hyperpolarization (DiFrancesco, 1993). Thus, it is likely to be quantitatively less important to normal pacemaker activity in nodal cells as in comparison to Purkinje cells, since nodal cells normally operate at much more depolarized voltages. In fact, in studies on SA node cells using perforated-patch recording, in which the intracellular environment should be more physiological, blockade of I_f had only little effect on SA node automaticity (Liu *et al.*, 1998). It may be that I_f is not very important for normal automaticity, but may be a protective mechanism, maintaining automaticity when the maximum diastolic potential becomes much more hyperpolarized (e.g., with strong vagal stimulation, resulting in the opening of many $I_{K(ACh)}$ channels and considerable hyperpolarization— see later discussion).

In addition to I_f, the interaction between a depolarizing background current (I_b) and the decay of the major repolarizing current, the delayed rectifier (I_K) may also provide a depolarizing current capable of contributing to the pacemaker potential in nodal cells. Since the relative contribution of these currents to the pacemaker potential in nodal cells cannot be accurately determined experimentally, there has been considerable controversy over which depolarizing current (I_f or I_b) plays the greater role in determining the slope of the pacemaker potential in these cells.

This small inward background current (I_b), present in nodal cells is a cation current carried primarily by Na$^+$ ions (Hagiwara *et al.*, 1992). I_b contributes a constant small depolarizing current. Due to the small size of this current, relatively little is known about its magnitude and characteristics in mammalian SA node cells; however, indirect evidence suggests that it may be a very important component in determining automaticity (see Campbell *et al.*, 1992; Dokos *et al.*, 1996). The role of a constant I_b in generating a variable pacemaker potential likely stems from the interaction of I_b with a variable I_K. I_K is the primary current responsible for repolarization in nodal cells, as in other cardiac cells. As in ventricular cells, I_K has also been shown to consist of at least two components (I_{Ks} and I_{Kr}) (see Dokos *et al.*, 1996). I_K displays essentially no inactivation during a prolonged depolarizing pulse, but displays a slow decay upon repolarization toward E_K. The time course of the I_K decay is very slow at membrane potentials in the voltage range of the pacemaker potential in nodal cells. The depolarizing action of I_b is opposed by I_K. Thus, the depolarization due to a constant background current I_b increases progressively with time due to a gradual reduction of opposing repolarizing current I_K. This is thought to be a significant factor in development of the pacemaker potential in nodal cells.

Other small currents may also contribute to and/or modulate the pacemaker potential in nodal cells. For example, the electrogenic sodium-potassium ATPase pump current, I_{pump}, and/or the electrogenic Na$^+$-Ca^{2+} exchange current (Noble, 1984; Dokos *et al.*, 1996).

As noted earlier, the upstroke (phase 0) of nodal cells is generated by an L-type Ca^{2+} current rather than an Na$^+$ current. This Ca^{2+} current appears identical to the classic long-lasting L-type Ca^{2+} current ($I_{Ca(L)}$) which is largely responsible for the plateau in both nodal and working cardiac cells. In addition to $I_{Ca(L)}$, nodal cells (and other pacemaker cells) have a second type of Ca^{2+} current which is activated at more negative potentials and has a much more rapid inactivation (Hagiwara *et al.*, 1988). This second, rapid type of Ca^{2+} current has been named the transient, or T-type Ca^{2+} current ($I_{Ca(T)}$), in contrast to the classical, slowly inactivating long-lasting, or L-type, Ca^{2+} current ($I_{Ca(L)}$). This T-type Ca^{2+} current is small to nonexistent in adult ventricular cells. It appears that both T- and L-type Ca^{2+} currents may contribute to the latter part of the pacemaker potential in nodal cells (Doerr *et al.*, 1989). Since the T-type channels are active at more negative potentials than the L-type channels, the presence of T-channels effectively lowers the threshold for $I_{Ca(L)}$. That is, the T-type current may not only contribute significantly to the pacemaker potential, but opening of T-type channels can depolarize the cells further toward the threshold for the L-type channels.

C. Modulation of Automaticity

I_f and $I_{Ca(L)}$ are both enhanced by the sympathetic neurotransmitter, norepinephrine (NE), and inhibited by the parasympathetic neurotransmitter, acetylcholine (ACh). Thus, NE increases the slope of the phase 4 depolarization, the threshold is reached sooner, and heart rate increases. In contrast, ACh decreases the slope of the phase 4 depolarization, the threshold is reached more slowly, and heart rate decreases. In addition to the effect on I_f and $I_{Ca(L)}$, ACh also activates another specific K$^+$ current, $I_{K(ACh)}$, which hyperpolarizes the cell, i.e., driving the maximum diastolic potential further from threshold. Thus, when $I_{K(ACh)}$ is activated, it takes a longer time to reach threshold and, also, the rate of phase 4 depolarization (slope) is decreased; thus the heart rate is decreased. The pacemaker cells of the SA node are richly innervated by sympathetic and parasympathetic nerves. These actions of NE and ACh on I_f, $I_{Ca(L)}$, and $I_{K(ACh)}$ in SA node cells are the basis of the stimulating and inhibiting effect on heart rate of sympathetic or parasympathetic nerve stimulation.

VII. Summary

APs in the heart are generated by the complex time-dependent interaction of several currents carried primarily by Na$^+$, K$^+$, and Ca^{2+} ions. Some cardiac cells display automaticity, due to the presence of a cyclical spontaneous depolarization called the pacemaker potential. The cells of the sinoatrial node normally exhibit the fastest spontaneous depolarization of automatic cells in the heart and are therefore the normal pacemaker cells. Automaticity in nodal cells is caused by the interaction of several currents, including a hyperpolarization-activated cation current (the "funny current"), a small background Na$^+$ current, the delayed rectifier K$^+$ current, and Ca^{2+} currents.

The sympathetic and parasympathetic nerves play important (and opposite) roles in regulating the force of myocardial contraction and heart rate. Thus, the sympathetic nerves stimulate the Ca^{2+} current, which enhances the force of contraction of the heart. The parasympathetic nerves inhibit the Ca^{2+} current, which decreases the force of contraction of the heart. The sympathetic nerves enhance the rate of spontaneous phase 4 depolarization in the sinoatrial node, resulting in an increased heart rate. The parasympathetic nerves decrease the rate of phase 4 depolarization, and also hyperpolarize sinoatrial node cells, resulting in a decreased heart rate. Thus, actions of neurotransmitters on cardiac ionic currents are a major site for modulating cardiac function.

Bibliography

Attwell, D., Cohen, I., Eisner, D., Ohba, M., and Ojeda, C. (1979). The steady-state TTX-sensitive ("window") sodium current in cardiac Purkinje fibers. *Pfluegers Archiv.* **379**, 137–142.

Baumgarten, C. M., and Fozzard, H. A. (1992). Cardiac resting and pacemaker potentials. *In* "The Heart and Cardiovascular System: Scientific Foundations" (H. A. Fozzard, H. Haber, R. B. Jennings, A. M. Katz, and H. E. Morgan, Eds). Vol. 1, Raven Press, New York. pp. 963–1001.

Bouron, A., Potreau, D., and Raymond, G. (1991). Possible involvement of a chloride conductance in the transient outward current of whole-cell voltage-clamped ferret ventricular myocytes. *Pfluegers Archiv.* **419**, 534–536.

Cameron, J. S., Kimura, S., Jackson-Burns, D. A., Smith, D. B., and Bassett, A. L. (1988). ATP-sensitive K$^+$ channels are altered in hypertrophied ventricular myocytes. *Am. J. Physiol.* **255**, H1254–H1258.

Campbell, D. L., Rasmussen, R. L., and Strauss, H. C. (1992). Ionic current mechanisms generating vertebrate primary cardiac pacemaker activity at the single cell level: an integrative view. *Annu. Rev. Physiol.* **54**, 279–302.

Coraboeuf, E., and Nargeot, J. (1993). Electrophysiology of human cardiac cells. *Cardiovasc. Res.* **27**, 1713–1725.

DiFrancesco, D. (1993). Pacemaker mechanisms in cardiac tissue. *Annu. Rev. Physiol.* **55**, 455–472.

Doerr, T., Denger, R., and Trautwein, W. (1989). Calcium currents in single SA nodal cells of the rabbit heart studied with the action potential clamp. *Pfluegers Archiv.* **413**, 599–603.

Dokos, S., Celler, B., and Lovell, N. (1996). Ion currents underlying sinoatrial node pacemaker activity: a new single cell mathematical model. *J. Theor. Biol.* **181**, 245–272.

Faivre, J.-F., and Findlay, I. (1990). Action potential duration and activation of ATP-sensitive potassium current in isolated guinea-pig ventricular myocytes. *Biochim. Biophys. Acta* **1029**, 167–172.

Fan, Z., and Makielski, J. C. (1993). Intracellular H$^+$ and Ca^{++} modulation of trypsin modified ATP sensitive K$^+$ channels in rabbit ventricular myocytes. *Circ. Res.* **72**, 715–722.

Feng, J., Xu, D., Wang, Z., and Nattel, S. (1998). Ultrarapid delayed rectifier current inactivation in human atrial myocytes: Properties and consequences. *Am. J. Physiol.* **275**, H1717–H1725.

Gadsby, D. C. (1984). The Na$^+$/K$^+$ pump of cardiac cells. *Annu. Rev. Biophys. Bioeng.* **13**, 373–378.

Hagiwara, N., Irisawa, H., and Kameyama, M. (1988). Contribution of two types of calcium currents to the pacemaker potential of rabbit sino-atrial node cells. *J. Physiol.* **395**, 233–254.

Hagiwara, N., Irisawa, H., Kasanuki, H., and Hosoda, S. (1992). Background current in sino-atrial node cells of the rabbit heart. *J. Physiol.* **448**, 53–72.

Hamill, O. P., Marty, A., Neher, A., Sakmann, B., and Sigworth, F. J. (1981). Improved patch-clamp techniques for high-resolution current recording from cells and cell-free membrane patches. *Pfluegers Archiv.* **391**, 85–100.

Harvey, R. D., (1996). Cardiac chloride currents. *NIPS* **11**, 175–181.

Harvey, R. D., Clark, C. D., and Hume, J. R. (1990). Chloride current in mammalian cardiac myocytes. Novel mechanism for autonomic regulation of action potential duration and resting membrane potential. *J. Gen. Physiol.* **95**, 1077–1102.

Irisawa, H., Brown, H. F., and Giles, W. (1993). Cardiac pacemaking in the sinoatrial node. *Physiol. Rev.* **73**, 197–227.

Janvier, N. C., and Boyett, M. R. (1996). The role of Na-Ca exchange current in the cardiac action potential. *Cardiovasc. Res.* **32**, 69–84.

Kenyon, J. L., and Gibbons, W. R. (1979). Influence of chloride, potassium, and tetraethylammonium on the early outward current of sheep cardiac Purkinje fibers. *J. Gen. Physiol.* **73**, 117–138.

Lindblad, D. S., Murphey, C. R., Clark, J. W., and Giles, W. R. (1996). A model of the action potential and underlying membrane currents in a rabbit atrial cell. *Am. J. Physiol.* **271**, H1666–H1696.

Liu, D. W., Gintant, G. A., and Antzelevitch, C. (1993). Ionic bases for electrophysiological distinctions among epicardial, midmyocardial, and endocardial myocytes from the free wall of the canine left ventricle. *Circ. Res.* **72**, 671–687.

Liu, Y. M., Yu, H., Li, C. Z., Cohen, I. S., and Vassalle, M. (1998). Cesium effects on I_f and I_K in rabbit sinoatrial node myocytes: implications for SA node automaticity. *J. Cardiovasc. Pharmacol.* **32**, 783–790.

Lopatin, A. N., Makhina, E. N., and Nichols, C. G. (1994). Potassium channel block by cytoplasmic polyamines as the mechanism of intrinsic rectification. *Nature* **372**, 368–369.

Luk, H. N., and Carmeliet, E. (1990). Na$^+$ activated K$^+$ current in cardiac cells: rectification, open probability block and role in digitalis toxicity. *Pfluegers Archiv.* **416**, 766–768.

Matsuda, H., Saigusa, A., and Irasawa, H. (1987). Ohmic conductance through the inwardly rectifying K channel and blocking by the integral Mg^{++}. *Nature* **325**, 156–159.

Montsuez, J.-J. (1997). Cardiac potassium currents and channels. Part I: basic science aspects. *Internat. J. Cardiol.* **61,** 209–219.

Noble, D. (1984). The surprising heart: a review of recent progress in cardiac electrophysiology. *J. Physiol. (London)* **353,** 1–50.

Noma, A. (1983). ATP-regulated K$^+$ channels in cardiac muscle. *Nature* **305,** 147–148.

Powell, T., Terrar, D. A., and Twist, V. W. (1980). Electrical properties of individual cells isolated from adult rat ventricular myocardium. *J. Physiol. (London)* **302,** 131–153.

Ravens, U., Wettwer, E., Ohler, A., Amos, G. J., and Mewes, T. (1996). Electrophysiology of ion channels of the heart. *Fundam. Clin. Pharmacol.* **10,** 321–328.

Sanguinetti, M. C., and Jurkiewicz, N. K. (1990). Two components of cardiac delayed rectifier K$^+$ current. Differential sensitivity to block by class III antiarrhythmic agents. *J. Gen. Physiol.* **96,** 195–215.

Shimoni, Y., Light, P. E., and French, R. J. (1998). Altered ATP sensitivity of ATP-dependent K$^+$ channels in diabetic rat hearts. *Am. J. Physiol.* **275,** E568–E576.

Smith, J. M., and Wahler, G. M. (1996). ATP-sensitive potassium channels are altered in ventricular myocytes from diabetic rats. *Mol. Cell. Biochem.* **158,** 43–51.

Tohse, N. (1990). Calcium-sensitive delayed rectifier potassium current in guinea pig ventricular cells. *Am. J. Physiol.* **258,** H1200–H1207.

Tseng, G. N. (1999). Molecular structure of cardiac I channels: Kv4.2, Kv4.3, and other possibilities? *Cardiovasc. Res.* **41,** 16–18.

Tseng, G. N., and Hoffman, B. F. (1989). Two components of transient outward current in canine ventricular myocytes. *Circ. Res.* **64,** 633–647.

Veldkamp, M. W. (1998). Is the slowly activating component of the delayed rectifier, Iks, absent from undiseased human ventricular myocardium? *Cardiovasc. Res.* **40,** 433–435.

Vizgirda, V. M. (1999). The genetic basics for cardiac dysrhythmias and the Long QT Syndrome. *J. Cardiovasc. Nurs.* **13,** 34–45.

Wahler, G. M., and Dollinger, S. J. (1995). Nitric oxide donor SIN-1 inhibits mammalian cardiac calcium current through cGMP-dependent protein kinase. *Am. J. Physiol.* **268,** C45–C54.

Wahler, G. M., and Sperelakis, N. (1985). Intracellular injection of cGMP depresses cardiac slow action potentials. *J. Cycl. Nucl. Protein Phosph. Res.* **10,** 83–95.

Walsh, K. B., and Kass, R. S. (1988). Regulation of a heart potassium channel by protein kinase A and C. *Science* **242,** 67–69.

Yu, H., Chang, F., and Cohen, I. S. (1993). Pacemaker current exists in ventricular myocytes. *Circ. Res.* **72,** 232–236.

Yue, L., Feng, J., Li, G. R., and Nattel, S. (1996). Transient outward and delayed rectifier currents in canine atrium: properties and role of isolation methods. *Am. J. Physiol.* **270,** H2157–H2168.

Nancy J. Rusch and R. Kent Hermsmeyer

53

Smooth Muscle Action Potentials and Electrical Profiles

I. Introduction

Action potentials (APs) were first defined in nerve fibers as propagating electrical impulses with uniform conduction velocity and amplitude. Similarly, in long skeletal muscle fibers that show uniform geometry throughout their length, electrical **propagation** (conduction) is accomplished by APs of constant velocity and uniform amplitude and overshoot. In contrast, APs in smooth muscle tissues are variable in amplitude, shape, pattern, and conduction velocity. This diversity of electrical patterns distinguishes APs in smooth muscle cells from those in nerve and skeletal muscle fibers. In fact, it may be more appropriate to refer to the diverse AP patterns in smooth muscle tissues as electrical **spikes** rather than action potentials, with this more generic term recognizing the diverse types of depolarizing patterns that form the basis for electrical excitability in smooth muscle tissues.

At least several key differences exist between smooth muscle cells and the cells that compose skeletal and cardiac muscle tissues. By definition, smooth muscle cells lack **striations** (aligned dark and light bands at the light microscope level), and this is a unifying property of this cell type. Another important difference is the very small size of smooth muscle

cells compared to cardiac or skeletal muscle cells, and hence, their dependence upon **multicellular interaction** to control smooth muscle tone even at the local level. Smooth muscle tissues consist of an assembly of short (e.g., 100–300 μm in length) and very narrow (e.g., 3–10 μm in diameter) cells with differing biophysical properties (Table 1). These cells interact by multiple coordination mechanisms to produce relatively long or tonic contractions associated with the function of the hollow organs, such as the blood vessels, intestine, uterus, and urinary bladder in which they appear. A final distinction of smooth muscle cells is the tremendous complexity of their electrical patterns, and the stimuli that trigger electrical excitability. Some of the smooth muscle tissues function through propagation of spikes by electrical coupling, whereas others rely primarily on tonic contractions maintained by chemical or mechanical mechanisms to regulate the level of cell and organ excitability. Thus the best generalization is that, through a variety of electrical and non-electrical coupling mechanisms, smooth muscle cells produce **coordinated contractions or relaxations** to carry out the functions of their host organs. As an example, Fig. 1A shows that vascular smooth muscle cells encircle the lumen of small arteries to form the **media** (muscle layer) of these hollow vessels. Within

TABLE 1 Electrical Properties of Some Smooth Muscles (P = 0–15 cm/s)

Muscle	Resting E_m	Spike amplitude	dV/dt max	Cell length	S	C_m TF
Vascular muscle	−55 mV	40 mV	5 V/s	100 μm	150 μm	1.9
Taenia coli (GP)	−50 mV	55 mV	7 V/s	150 μm	1500 μm	1.7
Intestinal (longitudinal)	−60 mV	65 mV	5 V/s	150 μm	2000 μm	1.7
Uterine (pregnant)	−70 mV	75 mV	5 V/s	200 μm	1500 μm	1.5
Vas deferens	−60 mV	65 mV	5 V/s	150 μm	2100 μm	1.5

Data are from Prosser *et al.,* 1960; Tomita, 1970; and Hernandez-Nicaise, *et al.,* 1980.

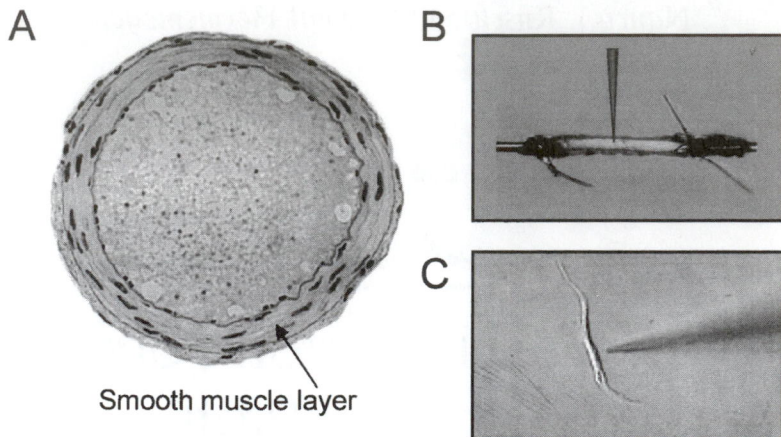

FIGURE 1. (A) The smooth muscle cell layer in a small mesenteric artery. (B) The electrophysiological profile of these smooth muscle cells can be studied in isolated, cannulated vessels using intracellular microelectrode impalements to record membrane potential in the smooth muscle cells. (C) Alternatively, single smooth muscle cells can be enzymatically isolated from the artery, and fire-polished patch-clamp pipettes can be positioned on the plasma membrane to measure and identify ion channel currents (Unpublished photos by J. DeBruin and J. H. Lombard, and by K. M. Gauthier, A. M. Runnells and N. J. Rusch, Medical College of Wisconsin).

this arrangement, excitatory input may induce a coordinated contraction of these smooth muscle cells to reduce arterial diameter and thereby attenuate blood flow to distal organs. Conversely, relaxation of the same smooth muscle cells will result in vessel dilation and a subsequent increase in blood flow to the distal tissues. Figure 1B shows that the underlying electrical patterns mediating smooth muscle contractile and dilator responses can be monitored using sharp microelectrodes to impale the smooth muscle cells of isolated, cannulated arteries. In addition, Fig. 1C shows that a living single vascular muscle cell may be approached by a polished pipette tip used to record ion channel currents in the plasma membrane. To accomplish this, individual smooth muscle cells are enzymatically isolated from arteries, and ion channel currents are measured by patch-clamp techniques in the membrane patches encircled by the pipette tip. In the past two decades, the microelectrode and patch-clamp techniques, which permit the correlation between membrane potential properties in smooth muscle cells and the native ion channel populations, have provided significant insight into the ionic basis of smooth muscle cell excitability.

Given the diverse function of blood vessels and internal organs for which smooth muscle cells are the contractile element, there have been numerous attempts to provide appropriate classifications for different smooth muscle cell types. The principal classification system used in the last 50 years has been the **unitary** and **multi-unit** categories of Bozler (1938, 1948). This classification, referred to in many textbooks, is based on whether the sphere (or cylinder) that contains a wall of smooth muscle cells acts in unison as a single electrical unit, or whether the smooth muscle cells are separately controlled through innervation, thus forming multiple discrete functional units in the organ. However, classifying smooth muscle tissues as primarily unitary or multi-unit smooth muscle structures is often precluded by the highly variable electrophysiological profiles of smooth muscle cells. Within an organ, and at times within even a small part of an organ, there will be a variety of

cells contracting, relaxing, and releasing signals to carry out function, including reduction or enlargement of the dimensions of that organ. Additionally, within a single blood vessel or visceral organ, there are **spontaneously active** cells that show pacemaker properties capable of generating spontaneous electrical spikes. These spikes, in turn, are conducted for distances that range from a few cells to several thousand cells. In contrast, other cells will show no spontaneous electrical activity, and may even remain **quiescent** during attempts to electrically stimulate them. Yet, these cells may contribute to the overall contraction of the organ when they are exposed to the multiplicity of stimulatory inputs that characterize the surrounding environment. Evidence from cellular studies show that the chemical modulation of smooth muscle function dominates the transitions between relaxation and contraction. Depending on the environmental conditions, a single smooth muscle cell may show tonic contraction or spontaneous activity associated with the generation of action potentials. Indeed, smooth muscle cells have multiple receptors, activation mechanisms, and inhibitors. They can be stimulated or inhibited by transmitters released from nerve endings, by local or circulating hormones, by neighboring cells, or by drugs. Electrophysiological analyses of the rapid phasic contractions of smooth muscle cells have demonstrated a tight temporal correlation between membrane depolarization and smooth muscle contraction, implying cause and effect between electrical signals and mechanical responses. Also within a given organ, there appear to be mechanisms for sustaining contractions for long periods of time with relatively low energy cost, a concept referred to as **smooth muscle tone**. The electrophysiological specialization for such a long-term maintenance of the **state of activation** (reduced diameter) correlates closely in some cells with membrane potential, and in others, most strongly with chemical mediators. Thus, diverse populations of smooth muscle cells coupled to their regulation by a wide variety of stimuli, provide rapid and appropriate changes in contractile tone to optimize the function of the vascular system, the digestive tract, and other organ systems that depend upon the regulation of smooth muscle tone to perform their function.

II. Determinants of Membrane Potential

The phospholipid bilayer that forms the plasma membrane of smooth muscle cells is impermeant to negatively-charged proteins and phosphorylated compounds found in the cell cytoplasm, and the sequestration of these substances confers a negative voltage to the cell interior called the **resting membrane potential**. The level of resting membrane potential is apparently the predominant influence on contraction in most smooth muscle cells. Excitatory stimuli that trigger **depolarization** (change in membrane potential toward 0 mV) cause increasing levels of contraction, because membrane depolarization activates voltage-gated Ca^{2+} channels to permit Ca^{2+} influx, and also results in the release of Ca^{2+} ions from intracellular stores to increase free Ca^{2+} ions available to bind to the contractile proteins. On the other hand, **hyperpolarization** (change to more negative potential inside the cell) results in a decrease of intracellular free Ca^{2+}, and is associated with smooth muscle relaxation. Regulation of contraction and of function in many types of cells by voltage-sensitive ion channels explains why changes in membrane potential and therefore ion conductances (e.g., the electrical current mediated by ion channel opening) are of great importance.

In smooth muscle cells, the negative level of resting membrane potential is primarily maintained by the steady-state efflux of K^+ ions from the cell mediated by the opening of K^+ channels. Notably, as reviewed elsewhere in this text, major ion species are distributed between the intracellular and extracellular spaces according to their propensity for membrane permeation and their electrochemical gradients. The extracellular space contains high concentrations of Na^+ and Cl^- ions, whereas K^+ is conserved in high concentrations inside the cell which permits its efflux when K^+ channels open at resting membrane potential. Notably, the concentration of Ca^{2+} is more than 10^4 times higher outside than inside the cell, providing a steep chemical gradient for Ca^{2+} influx when voltage-gated Ca^{2+} channels are activated during cell excitation to mediate smooth muscle contraction. In most smooth muscle cells, Na^+ and Cl^- channels appear to contribute less than K^+ and Ca^{2+} channels to the regulation of resting membrane potential. Indeed, the interaction between K^+ and Ca^{2+} channels is thought to be the primary ion channel interplay that determines the level of excitation-contraction coupling.

The activity of the **Na^+,K^+-ATPase pump** in the plasma membrane of the smooth muscle cell also helps to establish a negative level of membrane potential. This large polypeptide spans the cell membrane, and uses energy derived from ATP to transport Na^+ and K^+ against their concentration gradients. The Na^+,K^+-ATPase pump is composed of two subunits: a main functional α subunit protein that contains the binding sites for Na^+, K^+, and ATP, and a regulatory β subunit that influences the expression and activity of the α subunit protein. Traditionally, it is thought that 3 Na^+ ions are pumped out of the cell and 2 K^+ ions are pumped in. This unequal exchange results in the generation of a hyperpolarizing current that contributes to the negativity of the cell interior. The contribution of the Na^+,K^+-ATPase pump to resting membrane potential represents between 5 and 20 percent of the negative charge that characterizes the resting potential of smooth muscle cells. Thus, its hyperpolarizing influence reinforces the primary contribution of membrane K^+ channels to negative membrane potential levels.

Because different types of smooth muscle cells express variable levels and subtypes of ion channel and transport proteins, their electrical profiles vary greatly. Unlike neurons or skeletal muscle cells, smooth muscle cells lack or show low expression levels of Na^+ channels, and instead predominantly rely on the inhibition of resting K^+ efflux to depolarize the membrane. The resting membrane potential level of –35 mV to –65 mV in smooth muscle cells is considerably less negative than the values of –70 mV to –90 mV generally recorded from neuronal or cardiac cells. In effect, this more positive potential level may enable the steady-state activation of voltage-gated Ca^{2+} channels to mediate Ca^{2+}-dependent vascular tone in smooth muscle cells. Furthermore, whereas neurons and cardiac myocytes rely on spontaneously electrical activity for the repetitive secretion of neurotransmitters and cardiac pacemaker activity, respectively, many types of smooth muscle cells do not demonstrate spontaneous or repetitive electrical activity. Rather, they rely on graded changes in membrane potential to provide activator Ca^{2+} for tonic contraction. This tonic activation of the smooth muscle cells provides for graded changes in muscle tension so that tissues and organs are activated commensurate with their functional demands.

III. Voltage-Gated Ion Channels

A. Potassium Channels

Potassium channels are the primary determinants of resting membrane potential, and at least four K^+ channel superfamilies are co-expressed in smooth muscle cells. The **voltage-gated K^+ (K_V) channels** appear to be ubiquitously expressed in smooth muscle membranes, where they participate in setting the resting membrane potential and the regulation of contraction. The selective transcription and translation of specific K_V channel subtypes from at least eight gene families is proposed as a possible source of tissue diversity in electrical excitability. Structurally, the K_V α subunits contain six hydrophobic transmembrane segments (S1–S6) flanked by hydrophilic amino- and carboxy-terminal sequences located in the cell interior (Fig. 2A). Domain S4 appears to possess an intrinsic voltage sensor, which is associated with channel gating. Because the channel pore represents a heterotetrameric α subunit complex (Fig. 2A, inset), which may be composed of similar or different α subunits originating from the same K_V gene family, the "mixing and matching" of K_V α subunits provides for a heterogeneous population of channels in smooth muscle cell membranes. An additional level of complexity is introduced by the K_V β subunits, which may interact with the α subunit protein to enhance channel inactivation or modify cell surface expression. Not surprisingly, a complex pattern of site-specific expression and co-assembly of K_V channels is thought to confer electrical heterogeneity within and between different types of smooth muscle cells (Horowitz *et al.*, 1999; Farrugia, 1999; Berger and Rusch, 1999).

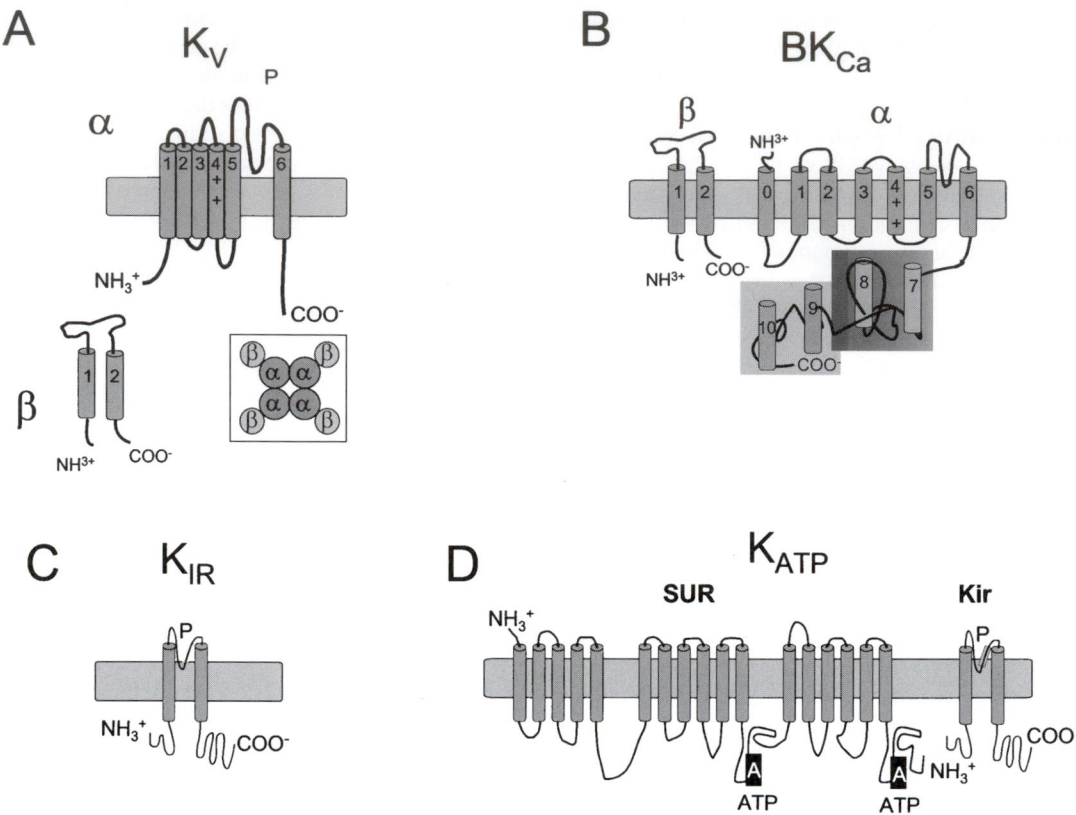

FIGURE 2. Proposed topology of K^+ channels expressed in smooth muscle membranes, including: (A) voltage-gated K^+ (K_V) channels, (B) high conductance Ca^{2+}-activated K^+ (BK_{Ca}) channels, (C) inward rectifier K^+ (K_{IR}) channels, and (D) ATP-sensitive K^+ (K_{ATP}) channels. The inset in (A) indicates that four α subunits assemble to form the functional pore of the K_V channel, with a 1:1 association with regulatory β subunits.

High-conductance, Ca^{2+}-activated K^+ channels, called BK_{Ca} channels because of their big (B) unitary conductance (150 to 300 pS), also are densely expressed in smooth muscle membranes. Similar to K_V channels, they contribute to the resting membrane potential and buffer depolarizing responses and smooth muscle contraction. Unlike K_V channels which originate from multiple gene families, BK_{Ca} channel α subunits arise from a single gene family, but channel diversity is still generated by a high level of alternative splicing of the common primary transcript. Notably, the N-terminal region of the pore-forming α subunit of the BK_{Ca} channel shares partial homology with the K_V channels, showing six transmembrane domains (S1–S6) and a highly conserved pore region between S5 and S6 (Fig. 2B). The Ca^{2+}-sensitivity of the BK_{Ca} channel appears to be conferred by an extra four transmembrane domains (S7–S10) at the C-terminal region of the α subunit, and by close association with a regulatory β subunit (Meera *et al.,* 1997).

Additionally, **inward rectifier K^+ (K_{IR}) channels** and **adenosine 5′-triphosphate-sensitive K^+ (K_{ATP}) channels** also may contribute to the resting membrane potential in smooth muscle cells. The pore-forming α-subunits of K_{IR} and K_{ATP} channels are both subtypes of the inward rectifier K^+ channel family (K_{IR}) that show the common property of inward rectification, in that they conduct inward K^+ current

more readily than outward current. The K_{IR} channel pore also is formed by four α subunits, but it has a unique membrane topology of only two transmembrane (M1, M2) regions (Fig. 2C). The K_{ATP} channels, which are inhibited by intracellular ATP, represent a K_{IR} channel coupled to a sulphonylurea receptor (SUR) that possesses **ATP binding cassettes** (Fig. 2D). The functional role of the K_{ATP} channels has been studied most intensely in vascular smooth muscle. In this smooth muscle tissue, these channels are thought to open when cytosolic ATP levels become suppressed during conditions of metabolic stress, thereby mediating hyperpolarization and vasodilation to enhance blood flow to compromised tissues (Quayle *et al.,* 1997; Farrugia, 1999).

The tight regulation of resting membrane potential is required for normal smooth muscle function, and the parallel and redundant pathways for K^+ efflux enabled by the four K^+ channel superfamilies provide the basis for this regulation in smooth muscle cells. For example, Fig. 3 shows that the depolarization of smooth muscle membranes activates both K_V and BK_{Ca} channels in the smooth muscle membrane of small arteries, implying that these two channel families act in unison as a repolarizing force to restore resting membrane potential to the smooth muscle cells. Although the dependence of the resting potential on K^+ conductance appears to be universal, each type of smooth muscle cell expresses its own unique popula-

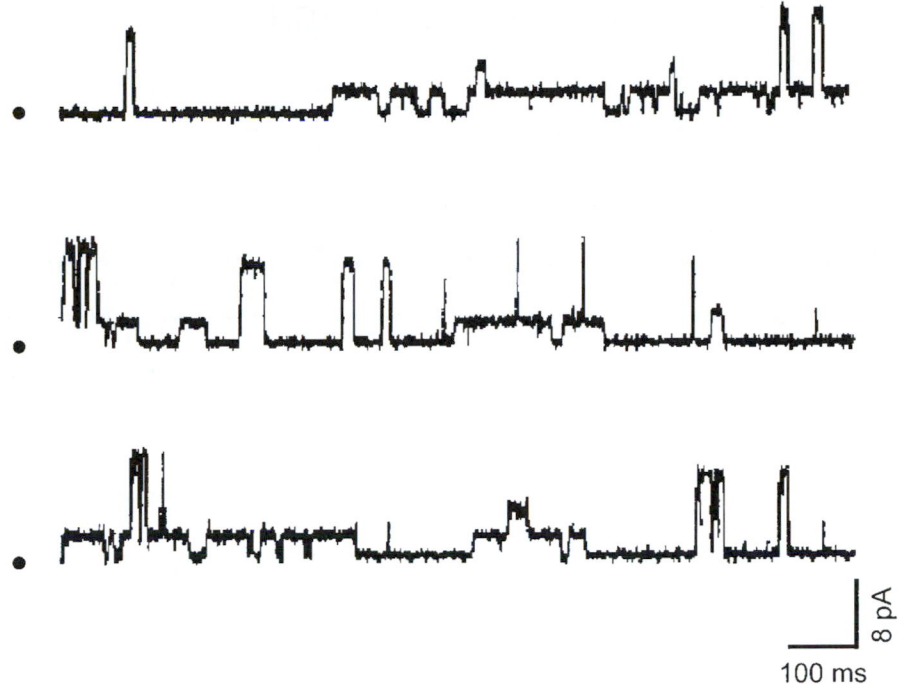

FIGURE 3. Single-channel recording of low-amplitude K_V currents and high-amplitude BK_{Ca} currents in inside-out membrane patches from smooth muscle cells of the rat mesenteric artery. Simultaneous opening of two K_V channels in the membrane patch sometimes resulted in stacked currents of low amplitude. The patch was depolarized to a membrane potential of +40 mV to elicit frequent openings of the channels. (Unpublished recording by P. Li and N. J. Rusch, Medical College of Wisconsin.)

tion of K^+ channels, which in turn, provides the appropriate level of resting membrane potential tailored to its integrated function. The small size and high input resistance of smooth muscle cells, coupled to the high density of K^+ channels expressed in their plasma membrane, provide the smooth muscle tissue with a highly favorable environment in which to regulate electrical excitability and coordinate contraction.

B. Ca^{2+} Channels

Smooth muscle cells express at least two types of voltage-activated Ca^{2+} channels. The **transient (T-type) Ca^{2+} channels** are activated and inactivated at negative potentials near or negative to the resting membrane potential level of smooth muscle cells. The opening of these channels mediates a short (transient) burst of Ca^{2+} influx. In contrast, the **long-lasting (L-type) Ca^{2+} channels** activate and inactivate at more positive membrane potentials, and show sustained activity at the membrane potential values typical of smooth muscle cells (Sturek & Hermsmeyer, 1986; Farrugia, 1999). Within a single smooth muscle cell, single-channel currents through both types of voltage-gated Ca^{2+} channels can be detected by patch-clamp techniques, confirming their co-expression in the plasma membrane, as shown in Fig. 4 (Benham *et al.*, 1987). Only the L-type Ca^{2+} channel is sensitive to block by the several classes of Ca^{2+} **antagonists** that are in common medical use as smooth muscle relaxants to treat hypertension, coronary artery spasm, asthma, urinary bladder dysfunction, and other smooth muscle diseases associated with abnormal contractile tone. Furthermore, whereas the T-type Ca^{2+} channel

has been implicated as a depolarizing force in some pacemaker cells, the L-type Ca^{2+} channel appears to be the predominant pathway for voltage-gated Ca^{2+} influx in most smooth muscle cells. The sustained activation of the L-type Ca^{2+} channel underlies the graded increases in smooth muscle tone during membrane depolarization that is required for the normal function of the smooth muscle tissues embedded in the walls of hollow organs (Nelson *et al.*, 1990; Rich *et al.*, 1993; Shimigol *et al.*, 1998).

Structurally, the L-type Ca^{2+} channel in smooth muscle has been identified as a splice variant of the related cardiac L-type Ca^{2+} channel. As depicted in Fig. 5, it minimally represents a pore-forming α_{1C} subunit associated with regulatory $\alpha_2\delta$ and β Ca^{2+} channel subunits. In contrast to the heteromultimer structure of the K^+ channel pore that is assembled from four discrete α subunits, the pore-forming structure of the L-type Ca^{2+} channel is a single polypeptide consisting of four repeat domains (I, II, III, IV) each containing six transmembrane spanning segments (S1–S6). Pore formation and voltage sensing are conferred by this central subunit, and the α_{1C} subunit also contains the binding sites for Ca^{2+} channel blocking drugs. Patch-clamp studies in vascular smooth muscle cells indicate that the biophysical properties of the L-type Ca^{2+} channel are consistent with its role in cell excitation. Modeling of L-type Ca^{2+} channels indicates that at least several thousand of these pore-forming subunits are expressed in a single cell, and hence, voltage-gated Ca^{2+} influx permitted by a small fraction of these channels is sufficient for tonic contraction of smooth muscle cells (Nelson *et al.*, 1990). The L-type Ca^{2+} channels, which have been detected in all types of smooth muscle cells

OUTSIDE-OUT PATCH

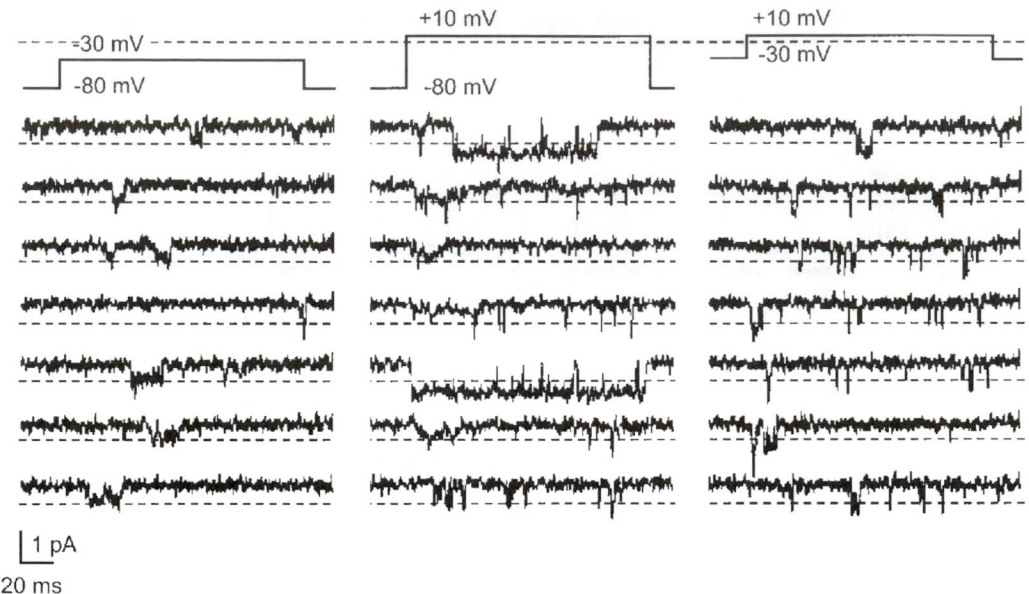

FIGURE 4. Single-channel recording of T-type and L-type Ca^{2+} channels in isolated patches of vascular smooth muscle membranes. (left panel) Low-amplitude T-type Ca^{2+} currents were elicited by depolarizing the membrane patch from −80 mV to −30 mV. (middle panel) More powerful depolarizing pulses from −80 mV to +10 mV elicited both T-type currents and high threshold L-type Ca^{2+} channel currents. The latter are identified by their higher inward amplitudes. (right panel) Depolarizing pulses to +10 mV initiated at the more positive membrane potential of −30 mV elicited only the high-threshold L-type Ca^{2+} channel currents. [Reproduced with permission from Benham *et al.* (1987).]

that have been studied, mediate low-amplitude single-channel Ca^{2+} currents that provide for the fine regulation of Ca^{2+} influx during membrane depolarization. Conversely, hyperpolarization of the smooth muscle membrane closes a fraction of opened L-type Ca^{2+} channels, which results in the reversal of the contractile process and smooth muscle relaxation. Thus the generation of tonic smooth muscle tone relies on the steady-state activation of L-type Ca^{2+} channels and the subsequent intracellular signaling cascade that mediates smooth muscle contraction (Nelson *et al.*, 1990; Rich *et al.*, 1993; Shimigol *et al.*, 1998; Farrugia, 1999).

C. Chloride Channels

Whereas the structures of the L-type Ca^{2+} channel and many types of K^+ channels are partially delineated, less is known about the structure or functional role of other types of potentially important voltage-gated ion channels expressed in smooth muscle cells. For example, recent findings suggest that volume- or Ca^{2+}-sensitive Cl^- channels, which are activated by membrane stretch or by increases in intracellular Ca^{2+}, may contribute to the membrane depolarization observed during cell excitation by mediating the efflux of Cl^- ions. Two different genes corresponding to volume- and Ca^{2+}-sensitive Cl^- channels have been identified in vascular smooth muscle, and Cl^- selective channels have been detected in several types of smooth muscle cells by patch-clamp methods (Farrugia, 1999). At least one of these Cl^- channels appears to have 12 transmembrane-spanning regions, suggesting a double-

barreled structure that is unique among ion channel pores. Although the opening of Cl^- channels to mediate Cl^- efflux may help to depolarize and contract smooth muscle cells, the blocking drugs that are used to inhibit Cl^- flux also may affect other transport processes. Thus the structure of Cl^- channels in smooth muscle membranes and their functional role with respect to electromechanical coupling remain unclear.

IV. Receptor Modulation of Membrane Potential

The **multiplicity of receptors** on a given smooth muscle cell is noteworthy because the resulting array of possibilities for integration of separate excitatory and inhibitory inputs offers diversity in end-organ electrical responses. The additional complexity added by multi-input modulation of the activity of individual cells provides a further diversity that virtually defies monolithic classification. Thus the importance of integration of multiple inputs and consequently diverse mechanisms for electrical modulation of contraction cannot be overemphasized (Table 2). Since each cell has an array of **receptors** (high affinity drug binding sites) and modulations, there are many permutations and combinations. Thus, attempts to categorize smooth muscle cells require a multi-criterion system.

For example, smooth muscle cells are not easily categorized by innervation, which is a useful criterion in skeletal muscle. While smooth muscle tissues are coordinated by motor nerves of the **autonomic** (involuntary) nervous sys-

L-type Ca²⁺ Channel

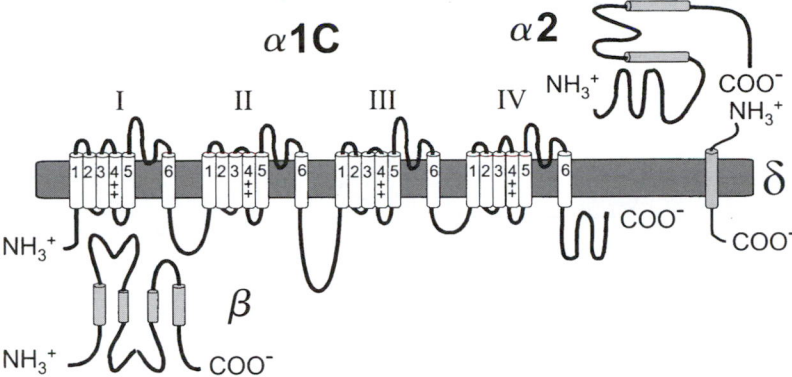

FIGURE 5. Proposed topology of the L-type Ca²⁺ channel showing its subunit components. [Adapted from Harder *et al.* (1985) with permission.]

tem, the distribution of receptor input is typically more diffuse, with release of **neurotransmitters** such as norepinephrine, acetylcholine, serotonin, angiotensin, ATP, substance P, calcitonin gene-related peptide (CGRP), adenosine, and still other important regulatory substances, often influencing hundreds or even thousands of cells. Indeed, the function of the nerves in smooth muscle tissues is more often to modulate rather than to initiate contraction. Smooth muscle cells are slower to contract than cardiac or skeletal muscle, and may maintain long-duration contractions in response to depolarizing stimuli. As depicted in Fig. 6, the sustained depolarization and contraction of small arteries during stimulation by the adrenergic neurotransmitter, norepinephrine, is a characteristic example of this response (Hermsmeyer, 1982). In addition to neurotransmitters, the electrical and contraction profile of smooth muscle can be

modulated by stretch, changes in oxygen tension, pH, osmolality, locally sensed chemical substances, ionic imbalances, and other substances found in abundance in the host organ. For example, Fig. 7 shows that although small pulmonary arteries are electrically quiescent under normal conditions, the presence of hypoxia triggers depolarization, the generation of rhythmic activity, and spike potentials in these vessels which results in vasoconstriction of the pulmonary circulation (Harder *et al.*, 1985). The excitatory spikes of the

TABLE 2 Receptor Types on Smooth Muscles

Type	Transmitters	Receptors
Adrenergic	Norepi, epi	α, β
Cholinergic	Acetylcholine	ACh
Serotonergic	Serotonin (5–HT)	5–HT
Nucleotide	ATP, adenosine	ATP
Angiotensin	Angio	AT
Vasopressin	Vasopressin (ADH)	VP
Endothelin	Endothelin	ET
Prostanoid	Prostaglandins E, F, I	PG
Leukotrienes	Leukotriene series	L
Insulin	Insulin	I
Interleukin	IL series	IL
Histamine	Histamine	H
Thromboxane	Thromboxane, PGH2	Tx-A$_2$

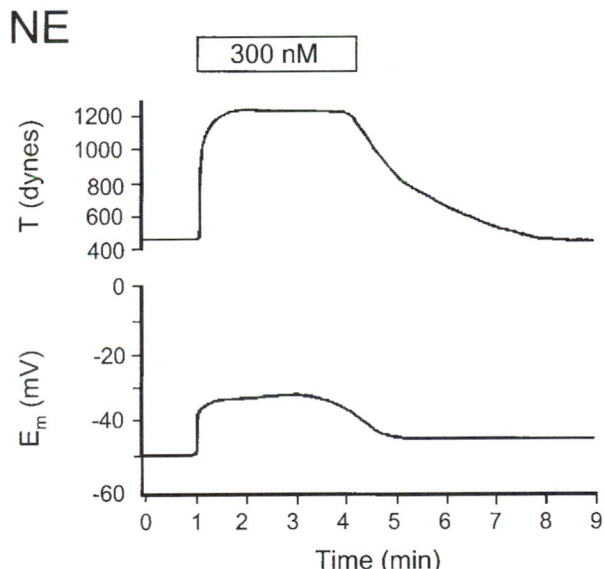

FIGURE 6. Depolarization and corresponding tension allow a steady state to be maintained in a rat caudal artery during stimulation by 300 nM norepinephrine. These drawings are traced from original records and illustrate the positive correlation between membrane potential and tension that is typical of small arteries. Both membrane potential and tension levels can remain relatively constant for many minutes during stimulation by vasoconstrictors, and may explain a major part of how tone is developed and maintained in blood vessels. (Reproduced with permission from Hermsmeyer, 1982.)

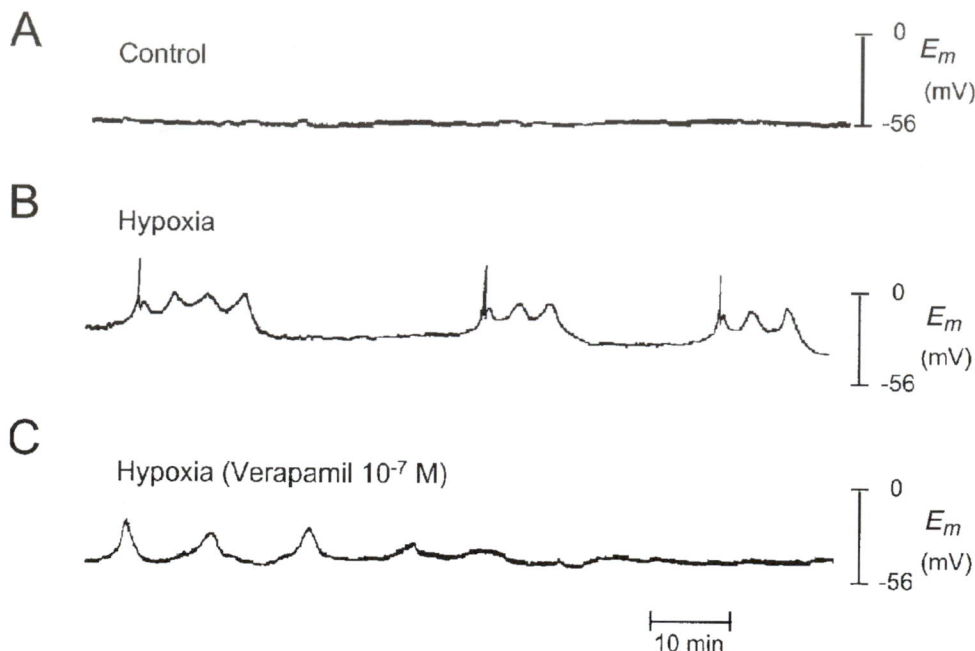

FIGURE 7. Microelectrode recording of membrane potential in a cannulated, pressurized (80 mmHg) pulmonary resistance artery in: (A) control oxygenated conditions (pO_2 = 300 Torr), (B) the presence of hypoxia (pO_2 = 30 Torr), and (C) in the continued presence of hypoxia after the addition of the L-type Ca^{2+} channel blocking drug, verapamil (10^{-7} M). The presence of hypoxia triggered depolarization and generation of action potentials in pulmonary smooth muscle cells, which were eliminated by verapamil. [Adapted from Harder *et al.* (1985) with permission.]

smooth muscle cells, which may be related to the inhibition of K^+ channels, are inhibited by drugs that block L-type Ca^{2+} channels, indicating a key role for voltage-gated Ca^{2+} influx in spike generation. Importantly, the contractile response of the pulmonary arteries to hypoxia reduces blood flow to poorly ventilated areas of the lung, but enhances blood flow to ventilated alveoli, where effective gas exchange can take place to support normal respiration.

At least two general families of Ca^{2+} permeant channels, the **receptor-operated (nonselective) cation channels** and the voltage-gated L-type Ca^{2+} channels, appear to act in concert to mediate cell excitation and depolarization in response to receptor activation. Receptor-operated cation channels conduct inward Na^+ and Ca^{2+} currents in response to ligand binding to G protein-coupled receptors, which induces a conformational change in the pore-forming protein (Bolton *et al.*, 1999). Under some conditions, this influx of cations is thought to provide the initial depolarizing signal that activates L-type Ca^{2+} channels to trigger smooth muscle contraction. The molecular structure of nonselective channels in smooth muscle tissues remains unclear, but the channel complex is thought to consist of both receptor-associated and channel domains, and may have only two transmembrane-spanning segments, similar to the K_{IR} channel. These channels may arise from multiple gene families, because they show variable properties of ion selectivity, voltage-sensitivity, and regulatory profiles in various smooth muscle cells. One of the first nonselective cation channels to be detected by patch-clamp techniques in smooth muscle membranes was the **ATP-gated channel** in vascular smooth muscle, depicted in Fig. 8 (Benham and Tsien,

1987). This channel opens in the presence of the ATP ligand molecule to conduct Na^+ as well as Ca^{2+} ions through the channel pore. In any given smooth muscle tissue, **electromechanical (voltage-gated)** and **pharmacomechanical (ligand-gated, voltage-insensitive) coupling** may interact to determine the final level of contraction.

Finally, receptor activation may permit the direct signaling of intracellular Ca^{2+} release, without the opening of surface membrane Ca^{2+} channels. In this case, the membrane-delineated enzyme **phospholipase C** causes the conversion of a principal substituent of the membrane, phosphatidyl inositol into the metabolite, inositol 1,4,5-trisphosphate (**InsP$_3$**), which causes localized (but highly effective) release of intracellular Ca^{2+} (Berridge and Irvine, 1984; Bolton *et al.*, 1999). Intracellular Ca^{2+} release is primarily from membranous structures thought to function analogously to the **sarcoplasmic reticulum** (SR), the prominent Ca^{2+} release organelle in skeletal muscle. One portion of the SR that regulates intracellular Ca^{2+} by release in response to membrane signals is located immediately inside the surface membrane, and functions as the superficial buffer barrier (Daniel *et al.*, 1995). This sub-sarcolemmal SR is important in Ca^{2+} homeostasis in vascular muscle cells because Ca^{2+} entering through sarcolemmal Ca^{2+} channels must pass through this region of SR, giving rise to the buffer barrier concept. Intracellular stores are modulated by the Ca^{2+} content of the superficial buffer barrier, but at least in localized sites, the Ca^{2+} concentration may increase more than one-hundred fold from its resting level of 100–300 nM when Ca^{2+} is released from the larger central component of the SR. Diacylglycerol (**DAG**) is also formed by the phospholi-

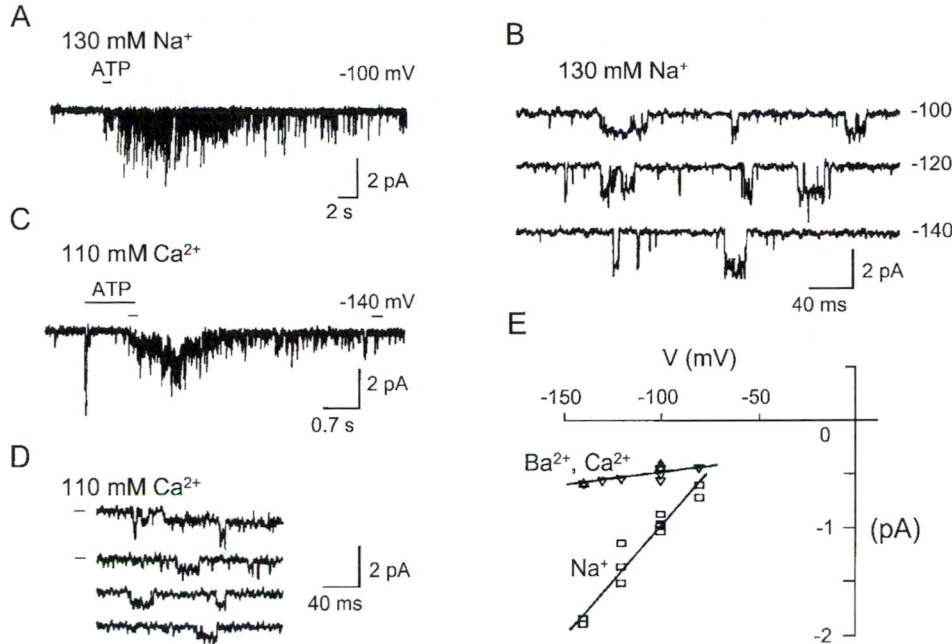

FIGURE 8. Individual currents activated by ATP in outside-out membrane patches torn from a cell are shown here. (A) Currents carried by Na⁺, and in (B), the same single channel events are spread out for individual inspection. Values to the right of each trace indicate the membrane potential at which the trace was obtained. (C, D) The same kind of membrane patch, but with Ca^{2+} as the current carrier. From these records, the amplitude of current for a given potential was determined and plotted, as shown in (E). From such plots, the conductance (Δcurrent/Δvoltage) was determined as 25 picosiemens (pS) for the divalent ion current and 200 pS for the Na⁺ currents. Reprinted by permission from *Nature* [Benham and Tsien (1987) copyright 1987 Macmillan Magazines Ltd.]

pase C-induced metabolism of phosphatidyl inositol, and DAG causes the activation of protein kinase C (**PKC**). The PKC signal triggers contraction through multiple mechanisms. Activation of PKC causes intracellular Ca^{2+} release, phosphorylation of the contractile proteins leading to an increased contraction for a given level of Ca^{2+}, and a markedly prolonged contraction. This combination of activation responses triggered by PKC results in strong, long-lasting contractions that correlate poorly, or at times not at all, with the level of depolarization, especially at later times during the contraction. Thus InsP₃ and PKC are other examples of substances that can induce **pharmacomechanical coupling**, that is, changes in contraction without requisite participation of membrane potential change (Somlyo and Somlyo, 1990).

In summary, once membrane excitation has begun, a complex set of membrane-related processes amplify the original discrete set of signals. The intracellular concentration of Ca^{2+} rises as a result of Ca^{2+} influx or release from intracellular stores (or both). In many types of smooth muscle cells, the key event appears to be Ca^{2+} entry through ion channels, with perhaps even the opening of a single Ca^{2+} channel beginning the messenger cascade leading to contraction. In addition, some smooth muscle activators also are believed to produce contraction through a direct action on the intracellular signaling processes, whether or not membrane potential is allowed to change. The latter event is often triggered by the receptor-linked production of InsP₃ from phosphatidyl inositol during the activation of phospholipase C in the cell membrane. This process is a biochemical signaling cascade, capable of causing activation of multiple intracel-

lular processes, through amplification by intracellular messengers, that ultimately result in smooth muscle contraction. Interference with this process can be at several points, but Ca^{2+} entry and release are particularly notable and vulnerable. Hence, Ca^{2+} channel blocking drugs that interfere with voltage-gated Ca^{2+} influx and the Ca^{2+} signal process are very effective smooth muscle relaxants.

V. Heterogeneous Electrical Properties of Smooth Muscle Cells

The electrical properties of different smooth muscle cells, including resting potentials and electrical spikes, have many fundamental similarities to other types of excitable cells. Among smooth muscles, there are common characteristics such as the co-expression of different ion channel types, but it is difficult to generalize about them. In fact, smooth muscle cells, like striated muscles and perhaps even more like neurons, have a diverse array of electrical properties that correspond with functional differences in various organs, and even parts of an organ. For example, the resting membrane potential of smooth muscles varies over a range between about –35 mV and –65 mV. This variation reflects real differences in the ionic conductances of the cells being studied. In fact, within a smooth muscle organ, there is a diversity of resting membrane potentials distributed even within a very finite distance, suggesting different subpopulations of smooth muscle cells within an organ. The range of resting potentials in arteries has been reported by many investigators to span more than 10 mV. Studies of visceral smooth muscle show a similar range, possibly contributing to the composite characteristics of an organ. Figure 9 emphasizes the vast differences in spontaneous contractile activity found in circular smooth muscle isolated from different layers of the human colon, which reflect the localized differences in electrical patterns for smooth muscle excitation within this tissue (Rae *et al.*, 1998). Thus, one way to think of an organ, such as the intestinal tract, is as a composite of smooth muscle cells consisting of subpopulations of cells which have a range of electrical properties that, when regarded together, constitute the electrical characteristics of that organ.

At the basis of electrical heterogeneity is the diverse population of K^+ channels expressed in smooth muscle cells. Because the opening of K^+ channels sets the membrane potential level, it also influences the availability of voltage-gated Ca^{2+} and Na^+ channels that mediate the amplitude of the electrical spike. The K^+ channels also can modify the form (shape and duration) of the electrical spike, because they mediate the flow of electrical current (K^+ efflux) which repolarizes the cells after excitation. Thus, the outward current generated by K^+ channels contributes to the resting membrane potential, the pacemaker mechanism, and spike repolarization. If the K^+ channels are blocked by such drugs as tetraethylammonium or by the cation, barium, there is depolarization of the cells (usually producing contraction), and slowed repolarization of the spikes (that often leads to enhanced electrical activity, especially in organs showing rhythmic contractions).

The ion that carries the inward (depolarizing) current that causes the electrical spikes found in smooth muscle tissues usually is calcium (Ca^{2+}). In cells that are known for their

spontaneous spiking activity, there is a higher expression of T-type Ca^{2+} channels. However, the same spontaneously active cells also express L-type Ca^{2+} channels in high density, and generate electrical spikes which are inhibited by Ca^{2+} antagonists (blockers of L-type Ca^{2+} channels). Thus the role of the T-type Ca^{2+} channel in spike generation may be permissive rather than mandatory. Examples of cells showing prominent Ca^{2+}-dependent action potentials or spikes include the azygos vein of newborn animals, the hepatic portal vein in all animals, numerous smooth muscle cells in the intestine, and the pacemaker cells of the progesterone-dominated uterus. Under some conditions, the mechanism for smooth muscle spikes includes Na^+ channels, which can be found in certain smooth muscle cells, especially those that show prominent pacemaker activity. The function of Na^+ channels in smooth muscle cells that have a Ca^{2+} channel pacemaker which functions independently of Na^+ channels is enigmatic. One useful way to think of a function for Na^+ channels in smooth muscle is the role of an adjunct to the pacemaker mechanism (Sturek and Hermsmeyer, 1986; Bolton *et al.*, 1999; Farrugia, 1999). If the Ca^{2+} pacemaker should fail to initiate spontaneous spikes, the Na^+ pacemaker is present and available for initiating spikes. Thus, it would provide a safety function for maintenance of pacemaker function.

In comparison to neuronal and cardiac tissues which rely on the dense expression of Na^+ channels to mediate the upstroke of the action potential, Ca^{2+} channels are expressed in lower densities in the plasma membrane of smooth muscle cells, and have a slower rate of **activation** (onset of channel opening). Thus, the electrical spikes in smooth muscle cells show a slower upstroke velocity and lower amplitude than the action potentials observed in neuronal cells or cardiac myocytes. Even more importantly, the **inactivation** (turn-off of current) of Ca^{2+} channels occurs over a time course that is at least a thousandfold as long as for Na^+ channels. These channel kinetics permit intracellular Ca^{2+} and contractile force to reach higher levels, and for contractions to be sustained much longer in smooth muscles than in cardiac cells expressing Na^+ channels. The prolonged process of inactivation also allows for a diversity of function, and for custom tailoring of the excitation event by hormones and regulatory substances that act through variables of time and space.

Thus pacemaker activity in smooth muscle cells depends on a balance of ion currents. The dynamic balance between inward and outward currents determines whether a cell will depolarize to reach the threshold voltage that triggers an electrical spike. A model of pacemaker activity based on Ca^{2+} and K^+ channels is shown in Fig. 10 (Hermsmeyer and Akbarali, 1989). In pacemaker cells, the spontaneous depolarization caused by the influx of Ca^{2+} and/or Na^+ is balanced by an outflow of K^+ current that tends to repolarize the membrane. If there is a slight increase in outward K^+ channel current, there will be hyperpolarization and suppression of pacemaker activity. If K^+ conductance slightly decreases, the cell will depolarize sufficiently to activate voltage-gated Na^+ or Ca^{2+} channels and generate spontaneous electrical activity. It is only under conditions where K^+ conductance is in an optimum range and there is a balance of currents, that tireless, repetitive cycles of sponta-

Circular Smooth Muscle of Human Colon

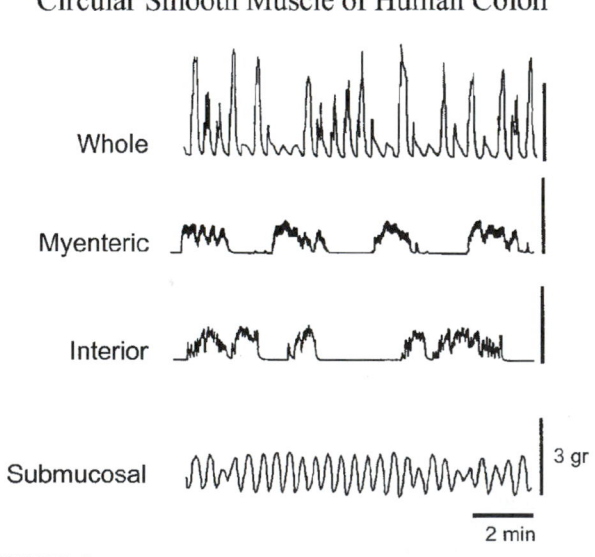

FIGURE 9. The highly variable patterns of spontaneous contractile activity observed in different circular smooth muscle layers of the human colon. The contraction patterns of the isolated smooth muscle strips were examined using isometric tension recording. [Adapted from Rae *et al.* (1998) with permission.]

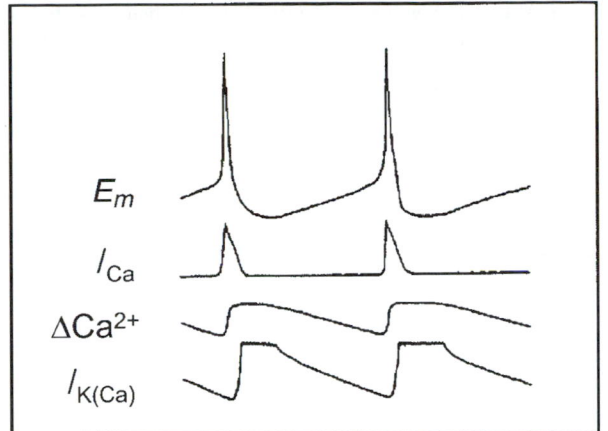

FIGURE 10. Pacemaker potentials are modeled by this diagram of membrane potential (E_m), Ca^{2+} current (I_{Ca}), intracellular Ca^{2+} concentration (ΔCa^{2+}), and Ca^{2+}-dependent K^+ current ($I_{K(Ca)}$). In this hypothetical vascular muscle cell, pacemaker potentials are evident as the slow upward drift of membrane potential toward zero to an inflection point, at which the slow upward movement suddenly bends to a nearly vertical rise that is the electrical spike. The spike is brief because of inactivation of Ca^{2+} channels that is evident in the return to baseline of Ca^{2+} current. The increase in intracellular Ca^{2+} (ΔCa^{2+}) enhances outward current through the Ca^{2+}-sensitive K^+ channels ($I_{K(Ca)}$), which accelerate the repolarizing phase and allow the downsweep to the most negative E_m at the beginning of the pacemaker depolarization. [Adapted from Hermsmeyer and Akbarali (1989) with permission.]

neous activity can occur. Because of the multiplicity of regulators that modulate ion channels, spontaneous activity is somewhat irregular in smooth muscle cells, distinguishing it from the more clocklike timing mechanism found in the pacemaker cells of the heart.

VI. Summary

Smooth muscle cells are small cells arranged in groups to form the contractile wall of hollow organs and blood vessels. In most cases, the muscle cell function is specialized for slow, maintained contraction (tone) rather than speed. Smooth muscle spikes show a high degree of diversity which translates into varied excitation. This diversity is in contrast to the rapid (but inflexible) signals that result from fast Na^+ channels in nerve and skeletal muscle. In contrast, the variability in the electrical signal in smooth muscles, generally based on the dynamic balance between outward K^+ current and inward Ca^{2+} current, provides a range of gradation of activity for internal organs that permits normal function. Smooth muscle pacemakers, although relatively less studied than those in the heart, provide vital functions for blood vessels and visceral organs, and are important for understanding regulatory mechanisms and drug actions. The frequency and configuration of action potentials and electrical spikes are regulated by a multiplicity of stimulatory and inhibitory inputs to a diverse array of receptors on a single cell. Coordination through chemicals released by nerve fibers, secretory cells, or other local factors influences an array of ion conductances that allows for long-duration con-

tractions and continuously graded regulation. With the multiplicity of receptors and response mechanisms found in the deceptively small smooth muscle cell, the complexity of signaling mechanisms possible with the simple smooth muscle spike should be regarded as a compact masterpiece of multiple responsiveness.

Bibliography

Benham, C. D., and Tsien, R. W. (1987). A novel receptor-operated Ca^{2+}-permeable channel activated by ATP in smooth muscle. *Nature* **328**, 275–278.

Benham, C. D., Hess, P., and Tsien, R. W. (1987). Two types of calcium channels in single smooth muscle cells from rabbit ear artery studied with whole-cell and single-channel recordings. *Circ. Res.* **61**, 10–16.

Berger, M. G., and Rusch, N. J. (1999). Voltage and calcium-gated potassium channels: functional expression and therapeutic potential in the vasculature. *Perspect. Drug Disc. Design* **15/16**, 313–332.

Berridge, M. J., and Irvine, R. F. (1984). Inositol trisphosphate, a novel second messenger in cellular signal transduction. *Nature* **312**, 315–321.

Bolton, T. B., Prestwich, S. A., Zholos, V., and Gordienko, D. V. (1999). Excitation-contraction coupling in gastrointestinal and other smooth muscles. *Annu. Rev. Physiol.* **61**, 85–115.

Bozler, E. (1938). Electric stimulation and conduction of excitation in smooth muscle. *Am. J. Physiol.* **122**, 614–623.

Bozler, E. (1948). Conduction, automaticity, and tonus of visceral muscles. *Experientia* **4**, 213–218.

Daniel, E. E., Van Breemen, C., Schilling, W. P., and Kwan, C. Y. (1995). Regulation of vascular tone: Cross-talk between sarcoplasmic reticulum and plasmalemma. *Can. J. Physiol. Pharmacol.* **73**, 551–557.

Farrugia, G. (1999). Ionic conductances in gastrointestinal smooth muscles and interstitial cells of *Cajal. Annu. Rev. Physiol.* **61**, 45–84.

Harder, D. R., Madden, J. A., and Dawson, C. A. (1985). Hypoxic induction of Ca^{2+}-dependent action potentials in small pulmonary arteries of the cat. *J. Appl. Physiol.* **59**, 113–118.

Hermsmeyer, K. (1982). Electrogenic ion pumps and other determinants of membrane potential in vascular muscle. *Physiologist* **25**(6), 454–465.

Hermsmeyer, K., and Akbarali, H. (1989). Cellular pacemaker mechanisms in vascular muscle. *Prog. Appl. Microcirc.* **15**, 32–40.

Hernandez-Nicaise, M.-L., Mackie, G. O., and Meech, R. W. (1980). Giant smooth muscle cells of *Beroë.* Ultrastructure, innervation, and electrical properties. *J. Gen. Physiol.* **75**, 79–105.

Horowitz, B., Ward, S. M., and Sanders, K. M. (1999). Cellular and molecular basis for electrical rhythmicity in gastrointestinal muscles. *Annu. Rev. Physiol.* **61**, 19–43.

Meera, P., Wallner, M., Song, M., and Toro, L. (1997). Large conductance voltage- and calcium-dependent K^+ channel, a distinct member of voltage-dependent ion channels with seven N-terminal transmembrane segments (S0–S6), an extracellular N terminus, and an intracellular (S9–S10) C terminus. *Proc. Natl. Acad. Sci. USA* **94**(25), 14066–14071.

Nelson, M. T., Patlak, J. B., Worley, J. F., and Standen, N. B. (1990). Calcium channels, potassium channels, and voltage-dependence of arterial smooth muscle tone. *Am. J. Physiol.* **259**, C3–C18.

Prosser, C. L., Burnstock, G., and Kahn, J. (1960). Conduction in smooth muscle: Comparative structural properties. *Amer. J. Physiol.* **199**, 545–552.

Quayle, J. M., Nelson, M. T., and Standen, N. B. (1997). ATP-sensitive and inwardly rectifying potassium channels in smooth muscle. *Physiol. Rev.* **77**, 1165–1231.

Rae, M. G., Fleming, N., McGregor, D. B., Sanders, K. M., and Keef, K. D. (1998). Control of motility patterns in the human colonic circular muscle layer by pacemaker activity. *J. Physiol.* **510**, 309–320.

Rich, A., Kenyon, J. L., Hume, J. R., Overturf, K., Horowitz, B., and Sanders, K. M. (1993). Dihydropyridine-sensitive calcium channels expressed in canine colonic smooth muscle cells. *Am. J. Physiol.* **264**, C745–C754.

Shmigol, A. V., Eisner, D. A., and Wray, S. (1998). Properties of voltage-activated $[Ca^{2+}]$ transients in single smooth muscle cells isolated from pregnant rat uterus. *J. Physiol.* **511**, 803–811.

Somlyo, A. P., and Somlyo, A. V. (1990). Flash photolysis studies of excitation-contraction coupling, regulation, and contraction in smooth muscle. *Ann. Rev. Physiol.* **52**, 857–874.

Sturek, M., and Hermsmeyer, K. (1986). Calcium and sodium channels in spontaneously contracting vascular muscle cells. *Science* **233**, 475–478.

Tomita, T. (1970). Electrical properties of mammalian smooth muscle. *In* "Smooth Muscle" (E. Bülbrin, A.F. Brading, A.W. Jones, and T. Tomita, Eds). The Williams and Wilkins Company, Baltimore. pp. 197–243.

Judith A. Heiny

54

Excitation-Contraction Coupling in Skeletal Muscle

I. Introduction

Skeletal muscle is activated to contract by a sequence of fast, electrically driven events that are collectively termed **excitation-contraction coupling** (EC coupling). EC coupling is a highly organized process of signal transduction that utilizes specialized membranes, membrane junctions, and ion channels on both the exterior and interior of the cell. This chapter describes the cellular and molecular processes that mediate electrical excitation of the outer membranes and transduce it into a signal for intracellular Ca^{2+} release, focusing on vertebrate fast-twitch skeletal muscle which is the best characterized. Chapter 55 discusses molecular mechanisms for control of Ca^{2+} release and reuptake by the sarcoplasmic reticulum (SR). Chapter 56 discusses mechanisms of force generation by the contractile proteins actin and myosin.

Skeletal muscle is the fastest contracting of the three major muscle types and is related to cardiac and smooth muscle types evolutionarily and embryonically. The proteins that mediate excitation-contraction in cardiac, smooth, and skeletal muscle share a high degree of homology. Nonetheless, each muscle type has evolved highly differentiated mechanisms of using these proteins to serve its specialized functions. While cardiac and smooth muscle primarily mediate enteric contractile processes that are under autonomic and humoral control, skeletal muscle primarily mediates rapid, willed bodily movements that are under the control of the central nervous system. Consequently, the defining features of skeletal muscle activation are speed and voluntary control, as needed for fine control of body movement.

II. Overview of EC Coupling

The sequence of EC coupling in a vertebrate fast-twitch skeletal muscle fiber is shown schematically in Fig. 1. Activation begins when an action potential from a **motor neuron** arrives at the **neuromuscular junction,** resulting in the release of the neurotransmitter **acetylcholine** (ACh) into the synaptic clefts (see Fig. 1 of Chapter 41). Binding of ACh to ACh receptors on the adjacent end plate (see Chapter 40) locally depolarizes the postsynaptic membrane to threshold for exciting an action potential on the muscle **sarcolemma.** The action potential rapidly propagates the depolarization to the entire sarcolemma by a mechanism similar to action potential generation in the nervous system (see Chapters 24 and 25). The action potential also propagates into the fiber interior via a specialized system of tubular membranes (**transverse-tubules** or **T-tubules**) which are continuous with and invaginate transversely from the sarcolemma at periodic intervals. The T-tubules provide the conduit for the action potential to reach the fiber interior, and also bring the outer membranes into close proximity with the internal **sarcoplasmic reticulum** (SR) at specialized intracellular junctions called **triads.** The triad junction mediates rapid communication between the extracellular sarcolemma/T-tubules and the intracellular SR. At the triad junctions, **voltage-dependent Ca^{2+} channels** in the T-tubular membrane (also called **voltage sensors** because they respond to the action potential depolarization, or **dihydropyridine receptors** (DHPRs) because of their sensitivity to the dihydropyridine class of Ca^{2+} channel blocking drugs) detect the depolarization and transduce it into a signal for opening **Ca^{2+} release channels** (also called the **ryanodine receptors** (RyRs) because they bind this plant alkaloid with high affinity) on the closely opposed SR membrane. The molecular interactions between these two distinct types of Ca^{2+} channel are not completely known but it is likely that they form a macromolecular complex that interacts allosterically. It is hypothesized that membrane depolarization drives a conformational change on charged, intramembrane domains of the T-tubule Ca^{2+} channel which, in turn, drives a conformational change on the SR Ca^{2+} release channel, which results in its opening. Whether this

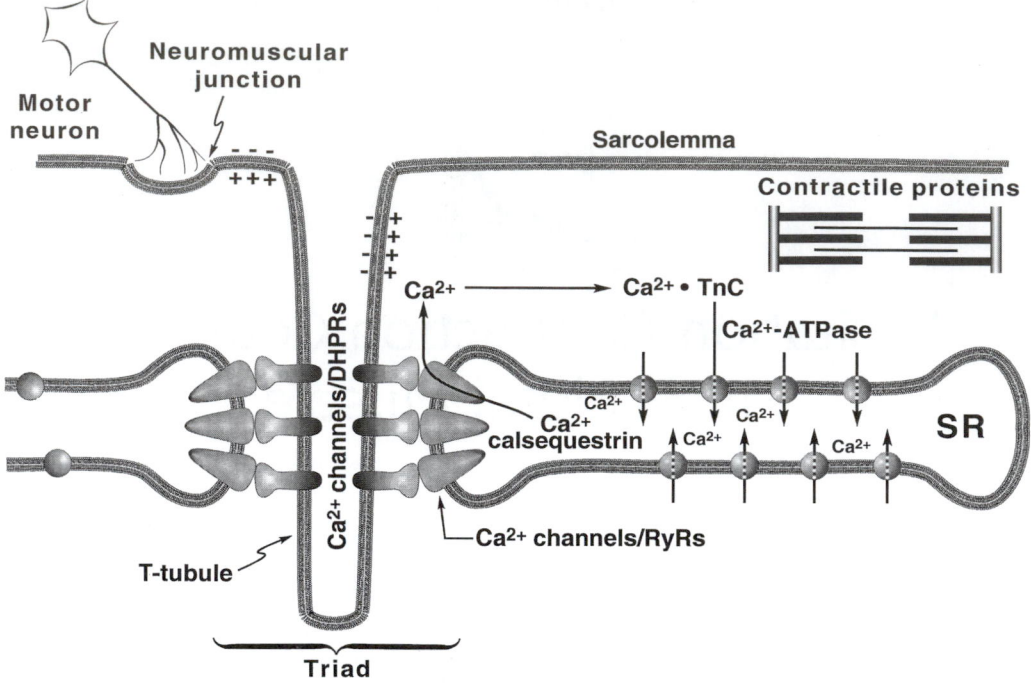

1. **Resting [Ca^{2+}]$_i$ ~0.1 µM**
2. **Neuromuscular transmission**
3. **Action potential propagation along sarcolemma and into T-tubules**
4. **Signal transduction from Ca^{2+} channels/DHPRs to Ca^{2+} release channels/RyRs at triad junctions**
5. **Ca^{2+} release from SR**
6. **Cytosol [Ca^{2+}] reaches 1–10 µM**
7. **Diffusion and binding of Ca^{2+} ions to TnC**
8. **Removal of troponin inhibition of the contractile proteins**
9. **Contractile proteins shorten to generate force**
10. **Reuptake of Ca^{2+} into SR by Ca^{2+}-ATPase**
11. **Inactivation of Ca^{2+} release channels**
12. **Binding of Ca^{2+} to calsequestrin**

FIGURE 1. Sequence of muscle activation and relaxation.

interaction requires accessory proteins as part of a larger complex remains to be established. The SR, related evolutionarily to the endoplasmic reticulum, is a highly specialized membrane compartment for controlling cytosolic Ca^{2+}. In a resting muscle fiber, the SR actively maintains cytosolic [Ca^{2+}] at submicromolar levels, typically 0.1 µM, by active transport mediated by Ca^{2+}-ATPase molecules present at high density along its length. Upon receiving a stimulus from the T-tubule the SR rapidly releases Ca^{2+} into the cytosol at rates approaching 100 µM/ms to raise cytosolic [Ca^{2+}] to micromolar levels (typically 1–10 µM). Ca^{2+} efflux occurs through RyRs localized on the junctional regions of the SR (**jSR**) where the SR membrane nearest the T-tubules form cisternae (**terminal cisternae**). Released Ca^{2+} ions diffuse and bind to **troponin-C** (TnC), the regulatory subunit of troponin, thereby removing troponin's inhibitory effect on the **contractile proteins,** actin and myosin, which shorten to generate force (see Chapter

56). Force output is proportional to cytosolic Ca^{2+} concentration (Fig. 2). Upon repolarization, Ca^{2+} release terminates and multiple mechanisms act in concert to return cytosolic Ca^{2+} to resting levels. High-density **Ca^{2+}-ATPase** pumps on the SR membrane rapidly pump Ca^{2+} back into the SR where it is largely bound to the Ca^{2+} binding protein **calsequestrin.** Cytosolic Ca^{2+}-binding proteins with rapid kinetics buffer Ca^{2+} transiently while the slower SR pumps return it to the SR. Some Ca^{2+} is also extruded extracellularly by active Ca^{2+} transport mechanisms in the sarcolemma. The RyR is itself inactivated by a Ca^{2+}-dependent mechanism. Force terminates when cytosolic Ca^{2+} returns to resting levels.

Thus, cytosolic [Ca^{2+}] is the central link between membrane excitation and activation of the contractile proteins (Fig. 3). Contraction is inhibited at low resting Ca^{2+} levels and proceeds when Ca^{2+} is elevated transiently in response to membrane excitation.

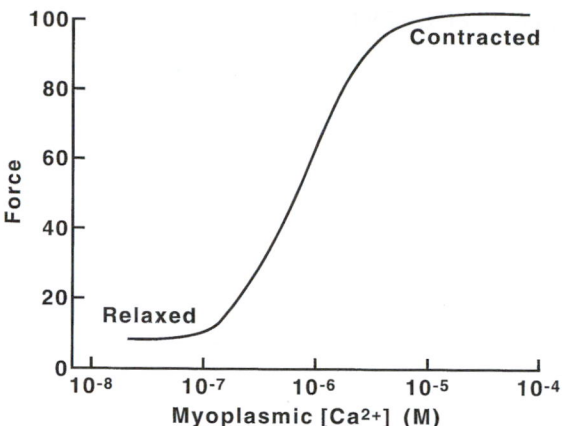

FIGURE 2. Relationship between force and myoplasmic Ca^{2+} concentration.

III. Speed of Skeletal Muscle Activation

Activation of skeletal muscle is rapid despite the fact that skeletal muscle cells are among the largest of mammalian cells. Skeletal muscle fibers are multinucleated cells that form during embryogenesis by fusion of precursor myoblasts. They are cylindrically shaped with diameters from 50–150 μM and lengths from millimeters to centimeters. To produce effective mechanical force, all parts of a muscle fiber must develop tension simultaneously. It was recognized early that a trigger process other than one involving a diffusion-limited second messenger must be involved (Hill, 1948). For example, a small ion entering through the outer membrane of a 100-μm muscle fiber would require 100–200 ms to diffuse and bind to putative activating sites distributed throughout the cross-section. By comparison, a fast-twitch muscle fiber is activated in 20–30 ms.

Synchronous activation of skeletal muscle is achieved via a rapidly propagating action potential that spreads the depolarization along the sarcolemma and into the cell interior along the specialized conducting membranes of the T-tubules. The final trigger, Ca^{2+}, diffuses at most a few micrometers from its release site on the SR to the activator sites on TnC. Figure 4A shows the typical delay between membrane excitation and contraction for a frog fast-twitch muscle at 18 °C. Tension starts to develop within 3–5 ms of the action potential and peaks in 20–30 ms. Table 1 summarizes the speed of the key events following neuromuscular transmission that contribute to this delay in a frog muscle fiber at 18 °C. For example, propagation of the action potential laterally along the sarcolemma from the middle of the fiber to both tendons introduces a delay of 5 ms. Signal transduction into the fiber interior, including all events from sarcolemma depolarization to activation of the innermost contractile proteins, requires 4–5 ms. Each of these events is faster in mammalian muscle at 37 °C. Figure 4B shows the time course of the cytosolic Ca^{2+} change and tension in a mammalian fast-twitch muscle at 37 °C, stimulated with a single action potential or at tetanic frequency to produce a maintained contraction.

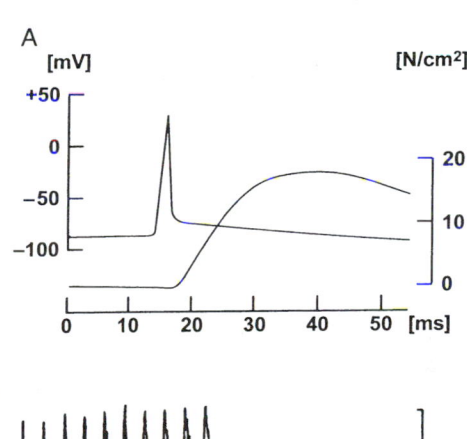

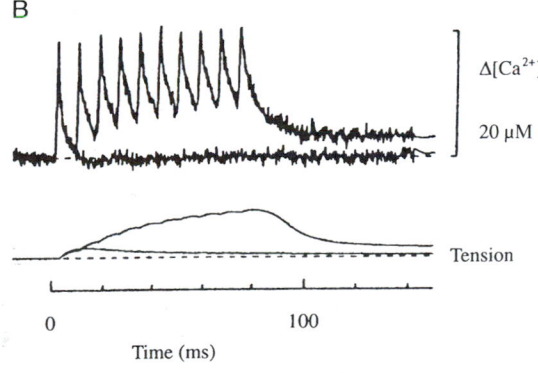

FIGURE 4. Speed of EC coupling in skeletal muscle. (A) Speed of EC coupling in a fast-twitch frog skeletal muscle fiber at 18 °C, showing the action potential (top trace) and tension (lower trace). (Modified from Hodgkin and Horowicz, 1957.) (B) Speed of EC coupling in a mouse skeletal muscle fiber at 37 °C. The top traces show the myoplasmic Ca^{2+} change measured with a Ca^{2+} indicator dye. The lower trace shows tension. The fiber was stimulated at time zero with a single action potential, and with repetitive action potentials at high frequency, to simulate a muscle tetanus. (Reproduced from *The Journal of General Physiology,* 1996, **108,** 455–469 by copyright permission of The Rockefeller University Press.)

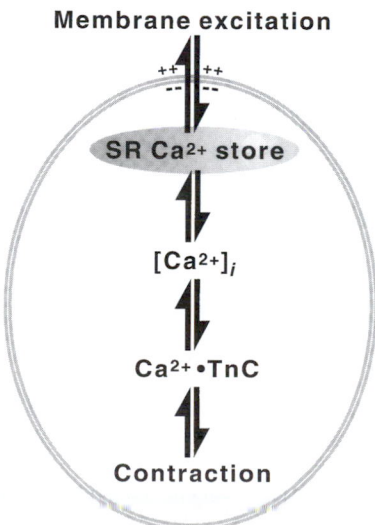

FIGURE 3. A rise in cytosolic $[Ca^{2+}]$ links membrane excitation to activation of contractile proteins.

TABLE I   Duration of Key Steps in the Activation of a Fast-Twitch
Skeletal Muscle

EC coupling steps	Duration[a]
Action potential propagation along sarcolemma	5–10 ms
Action potential propagation to center of fiber along T-tubules	~ 0.7 ms
Signal transduction at triad junction, from T-tubule depolarization to activation of RyR on SR	~ 0.5 ms
Peak rate of Ca^{2+} release to peak Ca^{2+} binding to TnC (start of tension)	2–3 ms
Peak myoplasmic Ca^{2+} change to peak tension	15–25 ms

[a]Durations were calculated for a hypothetical frog fiber of 50 μm diameter and 5 cm length, having a central end plate. Literature values for conduction velocity and duration of intermediate steps (Gonzales-Serratos, 1971; Vergara and Delay, 1986; Jong *et al.*, 1996) were adjusted to 18 °C.

IV. Membrane Architecture of EC Coupling

Synchronous activation depends on effective, rapid communication between the specialized extracellular membranes of the sarcolemma/T-tubules and the intracellular SR. To achieve this, the membranes of skeletal muscle are highly organized to bring into close physical proximity the proteins involved in each of the key events: membrane excitation, Ca^{2+} release, and force generation (rev. in Franzini-Armstrong and Jorgensen, 1994; Franzini-Armstrong and Protasi, 1997). Figure 5 illustrates the three-dimensional architecture of the sarcolemma, T-tubules, triad junctions, and SR membranes of a fast-twitch skeletal muscle fiber, and their physical relationship to the contractile proteins. The sarcolemma and T-tubule

membranes are organized to conduct the action potential rapidly to all parts of the fiber. The T-tubules compose the majority of the sarcolemma membrane, representing 50–80% of the total sarcolemma area (Peachey, 1965). They invaginate from the sarcolemma in a transverse plane approximately twice every 1–2 μM (at two planes per sarcomere in mammalian muscle). Within this plane they branch extensively to form a network that covers the entire cross-sectional area of the muscle cell, as shown in Fig. 6. Up to 80% of the tubular membrane is associated with the SR at triad junctions (Peachey, 1965; Dulhunty, 1984). Given this membrane architecture, no part of the sarcolemma is more than a few tenths of a micrometers and at most a few milliseconds signal transduction time from the Ca^{2+} release channels of the SR.

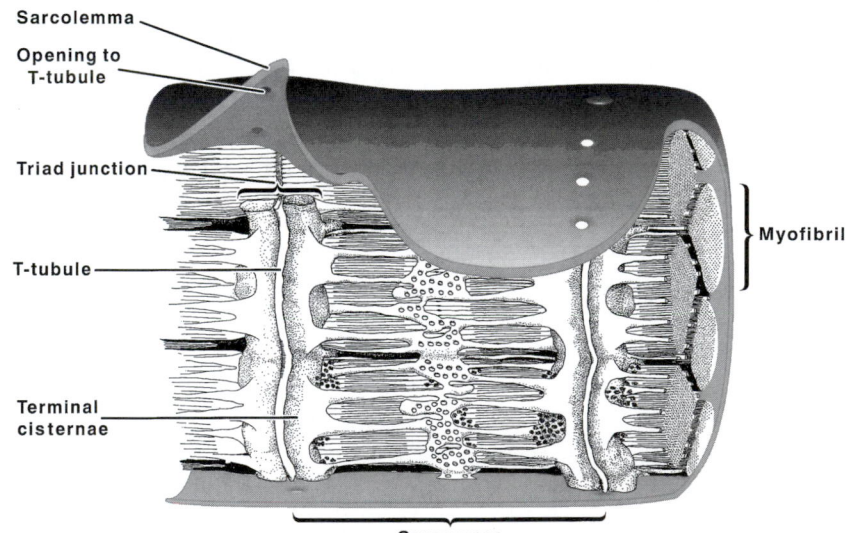

FIGURE 5.   Three-dimensional reconstruction of a longitudinal section of a skeletal muscle fiber from the frog. The sarcolemma surrounds bundles of contractile proteins called myofibrils. T-tubules invaginate transversely from the sarcolemma at periodic intervals and form junctions with the SR along their entire length. At the triad junctions, the longitudinally oriented SR membranes widen into sacs called terminal cisternae, which closely oppose the T-tubules and contain the Ca^{2+} release channels/RyRs. The terminal cisternae also closely overlay the activating Ca^{2+} sites on TnC. The functional unit of force generation is the sarcomere, consisting of overlapping actin and myosin filaments anchored at each end. The longitudinally oriented tubular regions of SR contain the Ca^{2+}-ATPase which takes up released Ca^{2+} all along the sarcomere. (Reproduced from *The Journal of Cell Biology*, 1965, **25**, 209–232 by copyright permission of The Rockefeller University Press.)

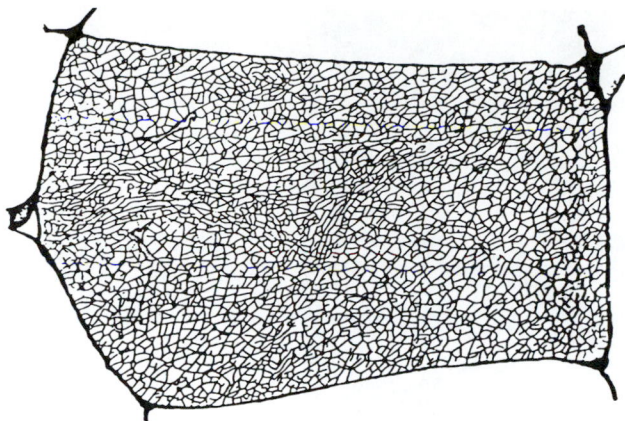

FIGURE 6. Reconstruction of the T-tubule network over the whole cross-section of a frog twitch muscle fiber. The reconstruction was made by tracing electron micrographs of transverse serial slices of muscle perfused with the electron-dense and membrane-impermeant molecule peroxidase, which diffused into the T-tubules. (From Peachey & Eisenberg, 1978.)

Likewise the triad junctions are optimized for rapid signal transduction from the T-tubule to the SR. Figure 7A shows the morphology of the triad junction, at which discrete junctional domains of the SR interact with specialized junctional domains of the T-tubule. Figure 7B shows the spatial organization of the two key proteins of the triad junction, the DHPRs and the RyRs, which associate at the junctions to form a functional signal transduction complex. At the junctions, the T-tubule and SR membranes flatten and face each other across a narrow gap of about 10 nm. The junctional surface of the terminal cisternae of the SR contains two rows of proteins corresponding to the Ca^{2+} release channels/RyRs. The RyRs are packed in highly ordered arrays, possibly touching at the corners, in a skewed pattern with a center-to-center spacing of about 30 nm. The RyRs are also termed **foot proteins** because of their unique appearance as dense, bridging structures in transverse electron micrographs of the triad. Each RyR/Ca^{2+} release channel is composed of four identical subunits. Each subunit has a membrane-spanning domain and a large cytosolic domain that extends across the junctional gap and comes to within 1 nm of the T-tubule membrane. In a top view, the large cytosolic domains of the tetramer roughly resemble four spheres assembled in a four-leaf clover or quatrefoil pattern. The large cytoplasmic domain is identified with the foot structure observed in electron micrographs of the triad junction. In a side view, the tetramer assumes a mushroom shape with a smaller central region composed of four equal lobes that presumably insert into the jSR membrane to form the ion-conducting pore of the molecule. The junctional T-tubule membranes also contain a high density of proteins that are aligned with a regular periodicity in parallel rows of four-particle arrays, termed **junctional tetrads** (Block et al., 1988). The tetrads have been identified as groups of four voltage sensors/DHPRs and are arranged in a pattern approximately corresponding to the outer corners of the cytosolic spheres of the RyR tetramer. Tetrads are asso-

ciated with RyRs in an alternate disposition such that one tetrad is associated with every other RyR (indicated by dotted lines in Fig. 7B).

Figure 7B was taken from a fish skeletal muscle fiber, in which tetrads line up opposite every other RyR in a 1:2 ratio. A similar arrangement of tetrads is suggested in triads and peripheral couplings of other skeletal muscles from the frog, rat, chicken, mouse, and human (Franzini-Armstrong and Jorgensen, 1994; Franzini-Armstrong and Kish, 1995). However, the ratio determined from binding studies may be less than that inferred from electron micrographs, so the 1:2 ratio of DHPR tetrads to RyRs may be a lower limit (Anderson et al., 1994).

Variations on this architecture theme for the intracellular junctions occur in different species. These include **dyads,** in which a junctional region of terminal cisterna is apposed to a single short segment of T-tubule, and **peripheral couplings,** in which a junctional region of terminal cisterna is apposed to an approximately circular junctional domain of the surface membrane. Triads, dyads, and peripheral couplings are structurally and functionally equivalent. Each is the intracellular site at which the interaction between DHPRs and RyRs links electrical excitation of the outer membrane to intracellular Ca^{2+} release. Collectively, these intracellular junctions are called **calcium release units** (CRUs; Franzini-Armstrong et al., 1999). A further functional subdivision of the system, termed a **couplon** has been applied to a functional grouping of RyRs and DHPRs that act in concert during EC coupling (Stern et al., 1997). Accordingly, the frog triad junction contains two couplons and a peripheral coupling contains a single couplon.

A. Proteins of the Triad Junction

Figure 8 shows the molecular structure and membrane topology of DHPRs and RyRs at the triad junction.

The DHPR/T-tubule Ca^{2+} channel is a voltage-gated ion channel belonging to the L-type family of Ca^{2+} channels (rev. in DeWaard et al., 1996; Perez-Reyes and Schneider, 1995; see also Chapters 27 and 33). Multiple subtypes of the L-type calcium channel with tissue-specific distribution have been identified, including skeletal muscle, heart, brain, and endocrine isoforms. The L-type Ca^{2+} channels belong to the superfamily of voltage-gated ion channels that share common structural and functional motifs. All L-type calcium channels are multisubunit complexes composed of at least three subunits associated in an equimolar ratio: α, α_2-δ, and β. The skeletal channel has a fourth γ subunit. The alpha isoforms are distributed in a highly tissue-specific manner and are the product of different genes and/or alternative splicing mechanisms. Apparently all L-type channels contain α_2-δ, which is encoded by a single gene. The β isoforms are tissue specific and result from different genes and alternative splicing. The skeletal-specific γ subunit is a single gene product. The complete skeletal muscle calcium channel consists of α_{1s}, α_2-δ, β_{1a}, and γ subunits (Fig. 9A). These have been purified, sequenced, cloned, and expressed. Expression of α_{1s} alone or in various combinations with α_2-δ, β_{1a} and γ can reconstitute the Ca^{2+} current. However, all four subunits are required for the formation of

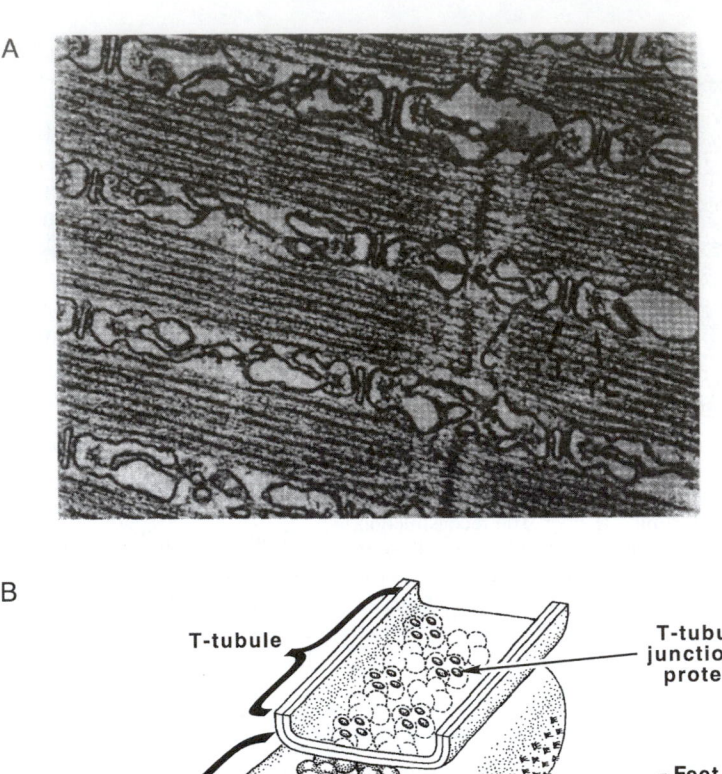

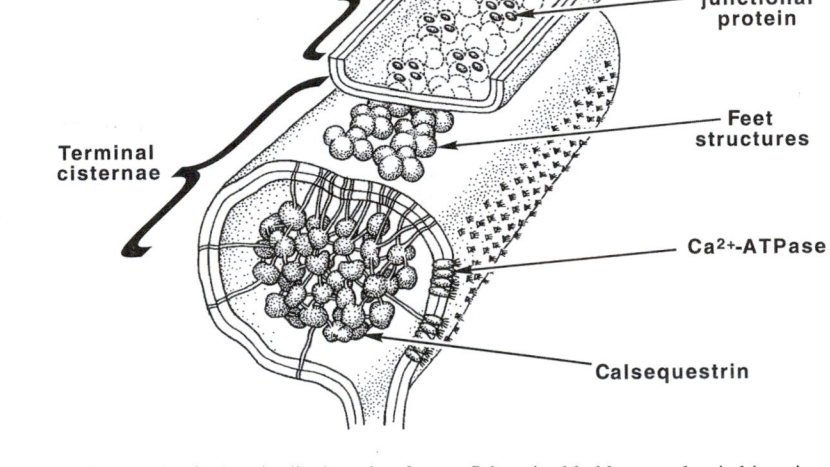

FIGURE 7. (A) Electron micrograph of a longitudinal section from a fish swim bladder muscle triad junction. (From Franzini-Armstrong and Peachey, 1981.) (B) Three-dimensional reconstruction of a half-triad showing the organization of the junctional membranes and proteins involved in signal transduction. T-tubule and SR membranes face each other across a gap of about 10 nm. The junctional surface of the terminal cisternae of the SR bears two rows of proteins corresponding to the RyR/Ca^{2+} release channels. RyRs project into the gap and come to within 1 nm of the T-tubule membrane. The junctional T-tubule membrane contains parallel rows of four-particle arrays, termed junctional tetrads, identified as DHPRs. In fast-twitch fish muscle, tetrads are arranged in a zig-zag spacing that corresponds exactly to the position of every other RyR on the adjacent SR (indicated by dotted lines). Nonjunctional and longitudinal regions of the SR are continuous with the terminal cisternae and contain densely packed Ca^{2+}-ATPase molecules. The lumen of the SR contains calsequestrin, a high-capacity Ca^{2+} binding protein. (Reproduced from *The Journal of Cell Biology,* 1988, **107,** 2587 by copyright permission of The Rockefeller University Press.)

a functional DHP receptor with properties of the native channel (Suh-Kim *et al.*, 1996).

The primary α_{1s} subunit contains the major functional domains of the channel. It is a large integral membrane protein (212 kDa) that contains the pore-forming and voltage-sensing regions as well as the binding site for dihydropyridines and other calcium channel blockers. The $\alpha 2$ and δ chains (125 and 27 kDa, respectively) are linked through a disulfide bond to form a single integral membrane glycoprotein (125 kDa). The δ chain contains a transmembrane segment that may anchor the α_2-δ complex to the membrane. The α_2-δ subunit may play a role in determining the spatial distribution and membrane localization of the Ca^{2+} channel. The β_{1a} subunit is a smaller cytoplasmic protein (58 kDa) that associates specifically with the skeletal isoform of alpha at an interaction domain on the I–II cytoplasmic linker. This subunit increases the activation kinetics of α_{1s} and is required for the proper expression and membrane targeting of α_{1s} (Neuhuber *et al.*, 1998; Gregg *et al.*, 1996). It also participates in regulation of the voltage-sensing and EC coupling functions of the Ca^{2+} channel (Beurg *et al.*, 1999). The γ subunit is a small glycosylated membrane

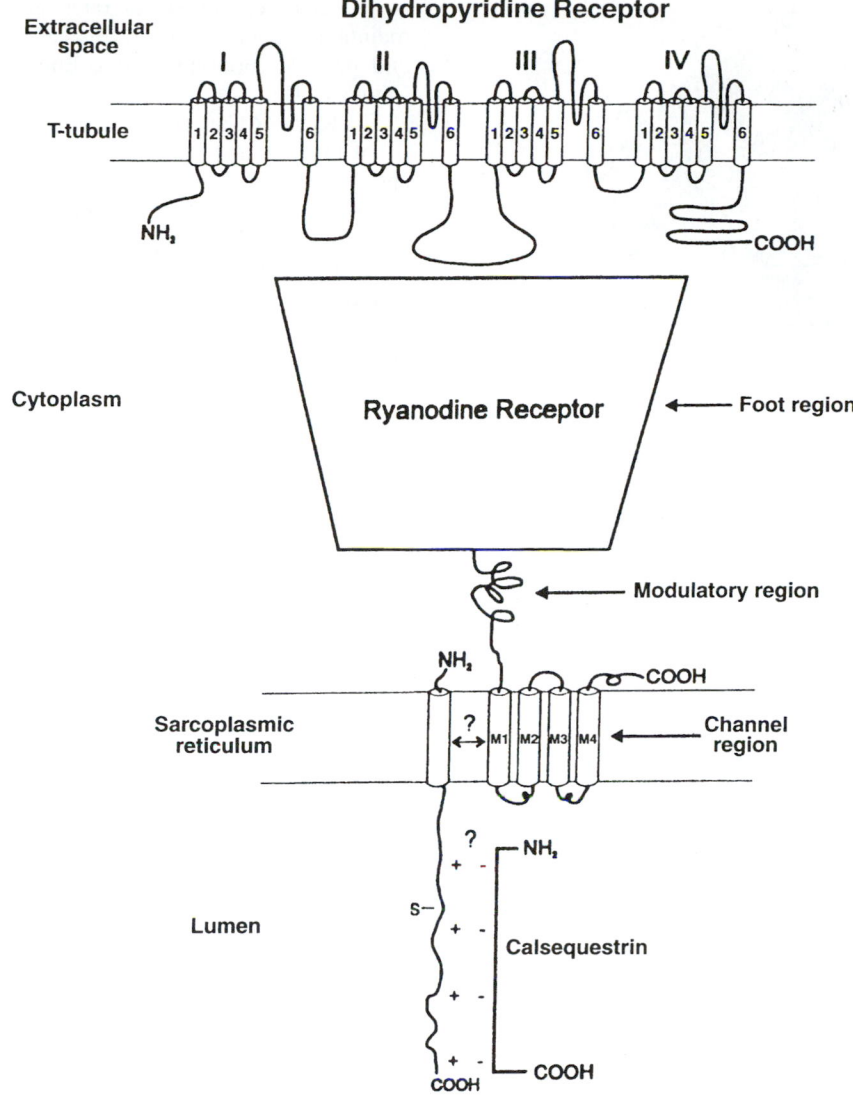

FIGURE 8. Key proteins of the triad junction, the DHPR (top) and RyR (bottom), showing primary sequence and presumed secondary membrane topology. For simplicity, only the α_1 subunit of the DHPR and only one subunit of the ryanodine receptor homotetramer are shown. DHPRs are composed of one each of four subunits (α_1, α_2-δ, β, γ). The DHPR α_1 subunit contains the key functional domains including the voltage sensor, pore, and DHP-binding site. It consists of four homologous hydrophobic domains (I–IV) connected by cytosolic linker sequences. Each of the four domains in turn is composed of eight transmembrane helices (S1–S6 alpha helices are drawn schematically as cylinders plus two shorter intramembrane helices that fold into the bilayer to form the central pore). Alternative models of the RyR suggest as many as 12 transmembrane domains. A question mark indicates the interaction between triadin and calsequestrin. Alternatively, the cytosolic carboxy end of triadin may fold parallel to the lumenal face of the SR membrane and bind calsequestrin. (Adapted from McPherson and Campbell, 1993.)

protein (25 kDa) of unknown function that associates with the skeletal α_{1S} isoform but not with heart or brain alpha subunits. It contains four transmembrane domains and two N-linked glycosylation sites.

The alpha subunit (Fig. 9B) is composed of four hydrophobic internal repeats, termed domains I, II, III, and IV, that share a high degree of homology. Each domain consists of eight transmembrane alpha-helical segments, denoted S1,

S2, S3, S4, S5, S5–6 linker (or P-linker), and S6. Domains I–IV are connected by longer segments of more hydrophilic amino acids (denoted I–II, II–III and III–IV linkers). The linkers as well as N- and C-terminals are cytosolic. The domains assemble to form a transmembrane protein with a central pore.

The major functional domains on α_{1S} have been identified. The P-linker helices of each domain fold into the membrane

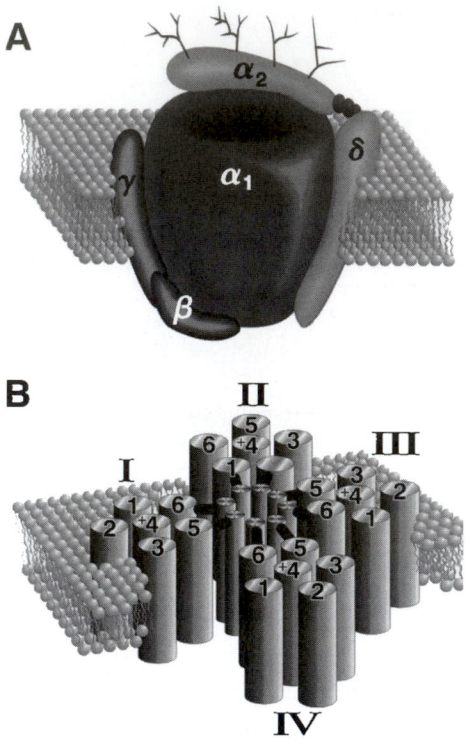

FIGURE 9. (A) Subunit structure of the skeletal muscle calcium channel showing the four subunits, α_{1s}, α_2-δ complex, β_{1a} and γ. (B) Assembly of the primary α_{1s} subunit, showing the four hydrophobic domains that fold into the bilayer to form an integral membrane protein with a central ion-conducting pore.

to form the pore for ion flux. The S4 transmembrane helix within each domain is highly charged, containing a positively charged residue on every third amino acid. S4 is presumed to form the voltage sensor that responds to the depolarization during the action potential. Upon depolarization, all four S4 helices move in a cooperative fashion to activate both the DHPR and the allosterically coupled RyR. Movement of the S4 segments is presumed to generate the macroscopic intramembrane charge movement currents that are detected electrically during EC coupling.

The skeletal muscle Ca^{2+} channel is capable of functioning both as a voltage sensor for EC coupling and as a Ca^{2+} ion channel. As a Ca^{2+} channel, it is characterized by high-voltage activation, brief openings, and low single-channel ion flux. Activation of the skeletal Ca^{2+} channel is slow compared with the closely related cardiac channel. The S4 charge movements may be followed by slower, less steeply voltage-dependent conformational changes leading to opening. It is thought that the early rapid movement of S4 domains, and not the Ca^{2+} current that follows after a delay of several hundred milliseconds, has been exploited evolutionarily by skeletal muscle to control RyR activation. Due to the slow kinetics of opening, Ca^{2+} influx through the channel pore is not significant during a normal action potential of a few milliseconds duration. However, it may build up to measurable levels during a tetanus, when a skeletal muscle

is activated repetitively at rates of 50–100 Hz to elicit a maintained contraction. In this case Ca^{2+} entry may add to and perhaps modulate intracellular Ca^{2+} release (Oz and Frank, 1991).

The RyR/SR Ca^{2+} release channel belongs to the superfamily of ligand-gated ion channels (rev. in Franzini-Armstrong and Protasi, 1997; see also Chapter 55). RyRs share significant sequence and structural homology with the inositol 1,4,5-trisphosphate (IP$_3$) receptors that release Ca^{2+} from internal stores in other cell types. RyRs are larger molecules with a significantly greater Ca^{2+} conductance than IP$_3$ receptors and are found in cells such as muscle that generate large, fast changes in intracellular Ca^{2+} concentration.

Three main types of RyR have been identified with tissue-specific distributions. The predominant isoform in mammalian skeletal muscle is RyR1. RyR2 is the major isoform in heart and is also widely distributed in the brain. RyR3 is present in brain, epithelium and selected muscles of some species. The RyR isoforms are products of different genes and share 65–70% homology. Alternatively spliced transcripts have also been detected.

The skeletal RyR1 isoform is associated either directly or indirectly with a number of other proteins and therefore must be viewed as a larger complex when considering its functional interactions. RyR1 interacts primarily with the DHPR on the cytoplasmic side and with calsequestrin on the luminal side of the jSR. In addition, it interacts with other components of the jSR including junctin, FKBP12 protein (FK506 binding protein), triadin, and a novel 90-kDa jSR protein that may associate with RyR1 in fast-twitch skeletal muscle (Froemming *et al.,* 1999). **Calsequestrin** is localized on the luminal jSR membrane, closely associated with the RyR. Its main function is to increase the total capacity of the SR for Ca^{2+}. **Junctin** is a small calsequestrin-binding protein that is enriched in the jSR and may anchor calsequestrin to the membrane. **FKBP12 protein** (FK506 binding protein) is stably bound to the corners of the cytosolic domain of the RyR in a 1:1 ratio and may mediate an end-to-end interaction. **Triadin** is present in high abundance in jSR and also binds the RyR in a 1:1 ratio. Its functional role is not known.

The RyRs are extremely large molecules (~5000 amino acids and ~560 kDa). The tetramer assembles to form a molecule of approximately $30 \times 30 \times 12$ nm size with a four-leaf clover–shaped extracellular domain and a smaller intramembrane domain with a central pore. Figure 10 shows a three-dimensional reconstruction of RyR1 in closed and open conformations.

Sequence analysis indicates a large hydrophilic N-terminal region thought to constitute the cytoplasmic or foot domain and a smaller mostly hydrophobic C-terminal region predicted to form the intramembrane channel. Both N- and C-termini are predicted to be cytoplasmic. The foot region contains four repeat motifs that occur in two tandem pairs. The membrane-crossing region is highly conserved and has strong similarity with the same region of the IP$_3$ receptor. The number of membrane crossings of the C-terminal region has not been established, but is proposed to be between 4 and 10.

The skeletal RYR1 is capable of dual modes of activation, both allosteric and ligand gating. It can be allosteri-

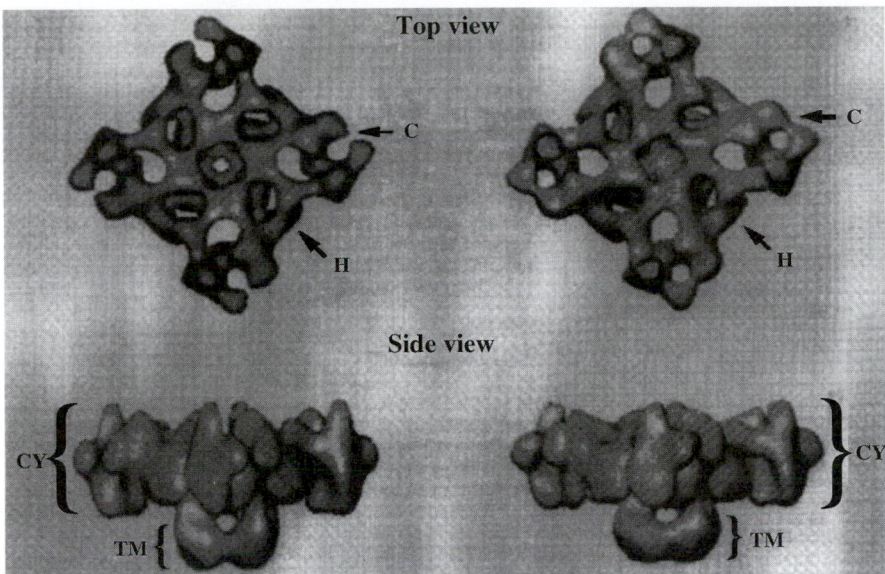

FIGURE 10. The open and closed states of the skeletal muscle RyR/Ca^{2+} release channel are shown in three different views: top, side, and bottom. The putative transmembrane and cytoplasmic (TM and CY) parts of the protein resemble the stem and cap of a mushroom-shaped protein. The cytoplasmic side of the tetramer contains clamp-shaped and handle (C and H) domains which alternately touch and overlap in the open and closed conformations. The cytoplasmic side of the tetramer contains the regions of spatial superposition with the T-tubule tetrads/DHPRs. The transmembrane region of the channel appears open towards the SR, whereas in the closed state a central opening is not seen in this region. (Reproduced from *The Journal of General Physiology*, 1996, **106**, 659–704, by copyright permission of The Rockefeller University Press.)

cally opened by an associated skeletal DHPR or opened by direct Ca^{2+} binding when not associated with a skeletal DHPR. When isolated and purified in the absence of triad junctions or DHPRs, the skeletal RYR1 behaves as a ligand-gated ion channel with a high single-channel ion flux. Ca^{2+} and adenine nucleotides at mM levels open the pore, and mM Mg^{2+} inhibits it. In this respect, the skeletal RyR1 is similar to cardiac RyR2, although its Ca^{2+} sensitivity is lower. Studies in which DHPRs have been inserted into dysgenic myotubes that congenitally lack DHPRs (see also Section V.C) have demonstrated that RyR1 can be activated by Ca^{2+} entering through DHPRs when the inserted DHPR contains key cardiac-specific sequences. On the other hand, RyR1 can be opened allosterically and mediate a skeletal-type EC coupling (independent of extracellular Ca^{2+}) only when associated with a skeletal-type DHPR. Thus, the complete skeletal DHPR-RyR1 complex is required to initiate Ca^{2+} release by the physiologically relevant mode in fast-twitch skeletal muscle.

V. Mechanisms of Interaction between DHPRs and RyRs

At the triad junction a unique intracellular signal transduction occurs whereby a voltage-gated Ca^{2+} channel on an extracellular membrane controls the opening of a different, normally ligand-gated Ca^{2+} channel on an intracellular membrane. The essential steps in signal transduction at the triad junctions are detection of the action potential depolarization by voltage-sensing molecules/DHPRs in the T-tubule, communication from the voltage sensor/DHPR to the Ca^{2+} release channel/RyR1 of the SR, and Ca^{2+} efflux from the SR.

A. Voltage Sensing

It was recognized early that EC coupling in fast skeletal muscle of higher vertebrates is directly controlled by membrane potential, without a requirement for extracellular Ca^{2+}. This contrasts with EC coupling in skeletal muscle of some lower invertebrates and cardiac muscle, in which Ca^{2+} influx across the sarcolemma/through sarcolemmal DHPRs is absolutely required for contraction. In cardiac muscle the Ca^{2+} that enters through the sarcolemmal DHPR functions as a second messenger to activate the cardiac RyR2, a process termed **calcium-induced calcium release** (CICR). In a classic experiment, Armstrong and colleagues (1972) demonstrated that a frog skeletal muscle fiber continues to twitch in the absence of extracellular Ca^{2+} when stimulated with action potentials. Similarly, voltage-clamp experiments demonstrated that a skeletal muscle fiber can develop tension provided only that the membrane is depolarized to a minimum potential and duration, termed the **mechanical threshold.** That is, when the Na$^+$, K$^+$, and Cl$^-$ ion channels that mediate the action potential are blocked, all that is required to activate contraction is a suprathreshold change in membrane potential. The relationship between tension and membrane potential is very precise in skeletal muscle as shown in Fig. 11.

Subsequently, Schneider and Chandler (1973) discovered the presence in skeletal muscle membranes of mobile

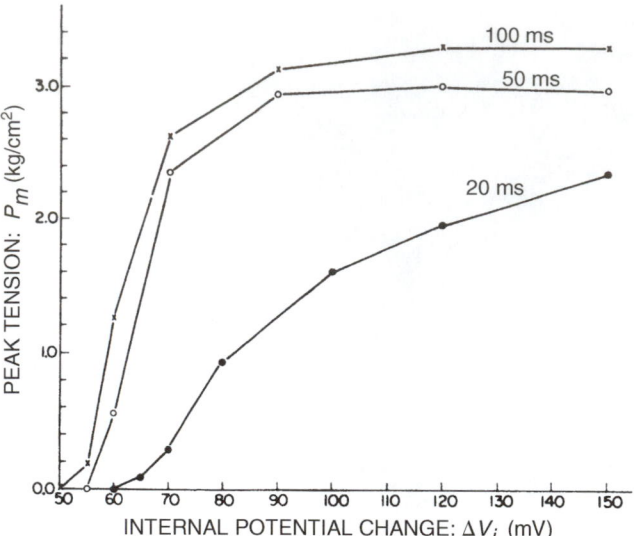

FIGURE 11. Relationship between peak tension and membrane potential in the absence of an action potential. The fibers were depolarized by holding the membrane potential constant for 50 ms with a voltage-clamp. Tension was measured simultaneously with a transducer attached to one tendon. (Reproduced from *The Journal of General Physiology*, 1984, **84,** 133–154, by copyright permission of The Rockefeller University Press.)

intramembrane charges whose movement could be detected electrically as a voltage-dependent dielectric current, termed **charge movement.** Based on the similarity of the kinetics and voltage dependence of charge movement to mechanical activation (Fig. 12), they proposed that it reflected the movement of charged intramembrane domains of a molecule in the T-tubules that functioned as the essential **voltage sensor** for EC coupling. They further suggested that the voltage sensor might interact directly with

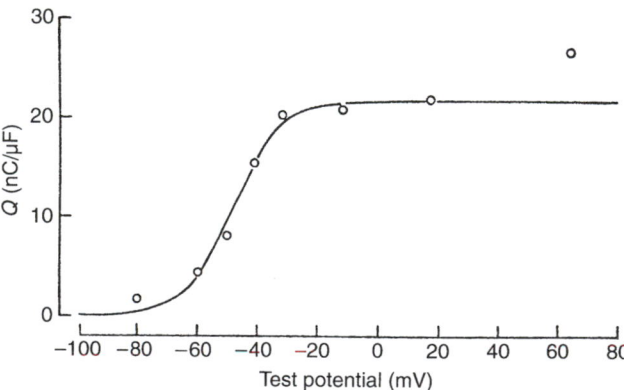

FIGURE 12. Relationship between charge movement (nC/μF) and membrane potential (mV). Charge moved at each potential was obtained by integrating the voltage-dependent capacity current, normalized to the fiber linear capacitance near the resting potential. Symbols represent the mean values of charge moved at the "on" and "off" of the pulse. The continuous curve is a fit of the data to a two-state Boltzmann function, with parameters: $V_{mid} = -47.7$ mV, $k = 8$ mV, and $Q_{max} = 21.5$ nC/μF. (From Chandler *et al.,* 1976.)

a hypothetical Ca^{2+}-conducting protein on the SR to cause its opening. The mechanical or allosteric hypothesis of EC coupling (electromechanical coupling) proposes that movement of charged voltage-sensing domains of the skeletal DHPR allosterically alters an activation domain of RyR1, either directly or via an intermediate protein. In this model, the triad junction exists to bring together the key proteins in a macromolecular complex. This early allosteric model of EC coupling, formulated before the identity of the proteins of the triad junction had been established, has formed the framework for subsequent investigations into the nature of the molecular interactions at the triad junction complex.

Several lines of evidence led to the identification of the voltage sensor as the DHPR/T-tubule Ca^{2+} channel which was shown to be localized in skeletal muscle T-tubules at one of the highest densities found in nature (rev. in Rios and Pizarro, 1991; Schneider, 1994; Melzer *et al.,* 1995). A definitive identification came following the cloning of the skeletal muscle Ca^{2+} channel (Tanabe *et al.,* 1987) and its use in experiments on dysgenic mice (rev. in Adams and Beam, 1990). Muscular dysgenesis arises from a single lethal mutation in the gene for the DHPR α-subunit, which results in reduction of DHPR protein, absence of slow Ca^{2+} currents and charge movement, and failure of EC coupling. Tanabe and colleagues (1988) demonstrated that Ca^{2+} currents and EC coupling could be restored in dysgenic muscle fibers transfected with the cDNA for the α-subunit of the DHPR, confirming the essential role of the DHPR as the voltage sensor in signal transduction to the SR.

B. Molecular Interactions between the DHPR and RyR

The molecular mechanism by which the Ca^{2+}-conducting pore of the RyR moves from a closed to open configuration under control of the T-tubule Ca^{2+} channel is the subject of intense research (rev. in Meissner and Lu, 1995; Leong and MacLennan, 1998a). The unique structure of the RyR—a Ca^{2+} channel domain in the SR bilayer and a huge cytosolic domain that spans most or all of the junctional gap—suggests that opening and closing of the pore could be remotely controlled by a cytosolic region that makes contact with the DHPR across the junctional gap. For example, the DHPR in its resting state could exert an inhibitory influence on the RyR/Ca^{2+} release channel that could be removed upon depolarization, or the RyR could be activated by conformational movements of the DHPR during depolarization. Inhibition of the RyR by the DHPR in the resting state and its removal by depolarization is attractive because of the inherent speed of this type of on-but-inhibited switching mechanism and because positively charged transmembrane helices in the S4 domain of the DHPR would move outward in the T-tubule membrane upon depolarization, away from the RyR. The following sections and Fig. 13 summarize the major domains of the skeletal DHPR α subunit that have been shown to mediate EC coupling interactions with the RyR.

C. Information from Genetic and Transgenic Models

Results from studies with dysgenic mice first identified specific domains of the DHPR that can activate RyR1 in in-

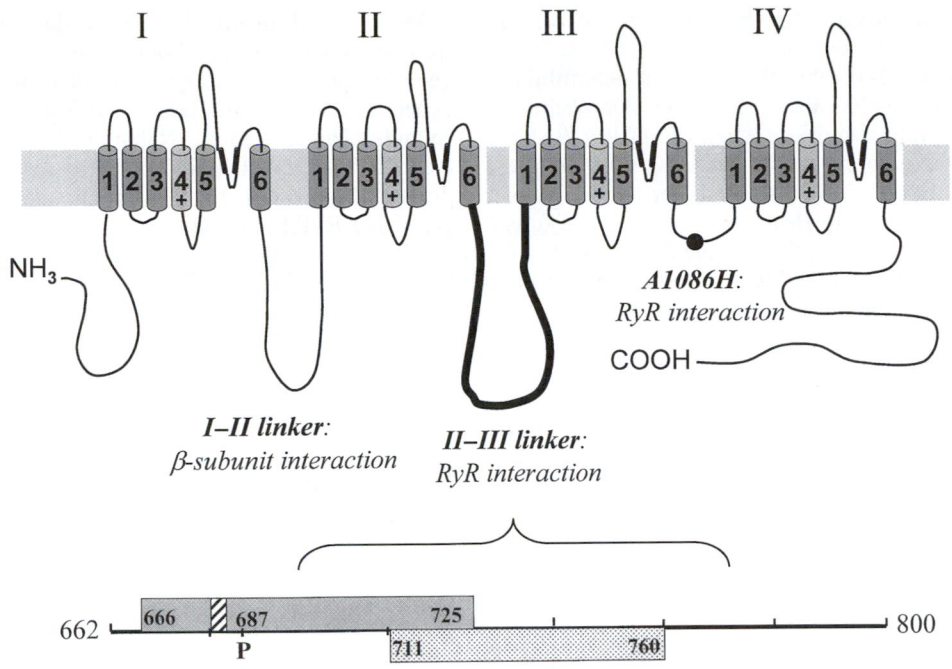

FIGURE 13. Sequences of the DHPR identified with functional roles in EC coupling. S4 voltage-sensing domains are indicated with a positive charge. Pore sequences between transmembrane segments S5 and S6 are indicated in bold. The 138-amino-acid cytosolic linker between domains II–III is a key interaction domain with the RyR and is essential for activating skeletal-type EC coupling. An Arg residue in the III–IV cytosolic linker of the human DHPR is critical for controlling Ca^{2+} release from the skeletal RyR1 and causes malignant hyperthermia when mutated. The I–II linker is the site of interaction with the β-subunit and influences EC coupling. Inset: Within the II–III linker a subdomain from approximately residues 666–726 is important for activation of RyR1 by the skeletal DHPR and contains a cluster of five basic residues (striped) that are required for the interaction. Residue Ser 687 inhibits RyR1 activation when phosphorylated. A second overlapping region from approximately residues 711–760 is important for both activation and blocking of RyR1 by the DHPR, and includes a region that receives a retrograde signal from RyR1 to enhance the function of DHPRs as calcium channels. Numbering corresponds to the rabbit skeletal muscle DHPR sequence.

tact muscle under physiological conditions. The α subunits of the L-type Ca^{2+} channel/DHPR of skeletal and cardiac muscle share a high degree of homology and both play a key role in triggering Ca^{2+} release. The main differences in primary structure between the skeletal and cardiac isoforms reside in the large cytosolic regions at the N- and C-termini, and the linker sequences II–III and III–IV. A key functional difference is that Ca^{2+} entry through the channel is required to elicit Ca^{2+} release in cardiac but not in skeletal muscle. Beam and collaborators (Tanabe *et al.*, 1990) exploited these differences to identify molecular regions of the α_1 subunit of the DHPR that are critical for initiating a skeletal-type EC coupling. When the skeletal muscle α_{1s} subunit was expressed in dysgenic myotubes, a Ca^{2+} release independent of Ca^{2+} entry could be elicited in response to electrical depolarization, whereas Ca^{2+} entry was required to elicit Ca^{2+} release when the cardiac α_{1c} subunit was expressed. By constructing and expressing chimeric DHPRs composed of various cardiac and skeletal domains, Tanabe and collaborators (1990) demonstrated that the 138-amino-acid cytoplasmic linker between domains II and III is a critical determinant of skeletal-type EC coupling. When a chimeric DHPR having the cardiac protein backbone but the skeletal II-III linker was expressed (see Fig. 13), contraction could be

elicited by electrical stimulation even when Ca^{2+} entry was blocked. This indicates that the II–III linker is involved in signal transmission from the skeletal DHPR to RyR1 on the SR. Interestingly, expression of both the wild-type skeletal and cardiac DHPRs as well as the chimeric channels all led to the restoration of charge movement. This indicates that the voltage-sensing step in EC coupling does not depend on the nature of the transmission mechanism, and likely involves separate molecular domains of the α subunit.

Other cytosolic regions on the α subunit of the skeletal DHPR have also been implicated in direct interactions with the RyR1. A mutation in the III–IV cytosolic linker of the human DHPR (A1086H) is associated with susceptibility to one form of malignant hyperthermia (MH), a hereditary skeletal muscle disorder that is characterized by an uncontrolled, sustained Ca^{2+} release (Monnier *et al.*, 1997). This suggests that the III–IV linker also participates in control of RyR1 activation by the skeletal DHPR, or, alternatively, it may participate in turning off Ca^{2+} release under control of the skeletal DHPR. Earlier studies have shown that Ca^{2+} release, which is initiated by membrane depolarization, is also terminated upon membrane repolarization, under tight control of the voltage sensor (rev. in Rios and Pizzaro, 1991; Suda, 1995). The mutation in MH may delay or prevent closure of RyR1 upon repolarization (Leong and MacLennan, 1998a).

A minor role of the I–II cytosolic linker in determining the mode of EC coupling has also been shown (De Waard *et al.*, 1996). EC coupling is blocked when the DHPR β subunit gene, which binds to the I–II linker of skeletal DHPR, was knocked out. Expression of β_{1a} but not the predominantly cardiac β_{2a} isoform in β-null myotubes could fully restore a skeletal-type EC coupling, suggesting that the interaction of β_{1a} with skeletal DHPR is required for a skeletal, Ca^{2+}-independent type of EC coupling.

D. Evidence from *in Vitro* Studies

Information from *in vitro* studies supports the existence of protein-protein interactions between the skeletal DHPR and RyR (rev. in Meissner and Lu, 1995; Leong and MacLennan, 1998a). Early biochemical evidence for a direct association between proteins in the T-tubule and the SR came from the observation that purified fractions of T-tubule membrane could form junctions with SR membranes and that triad fractions could be isolated intact, despite stringent fractionation procedures (Caswell *et al.*, 1979). Strong protein-protein associations of some kind at the triad junction would be required to preserve the structural integrity of the triad junction as it sustains the strong forces of contraction.

Further studies have confirmed the importance of the II–III cytoplasmic linker in conferring skeletal-type EC coupling. Peptides composed of either the skeletal or cardiac α subunit II-III linker sequence can activate the skeletal RyR1 by increasing open channel probability and the affinity of ryanodine binding (Lu *et al.*, 1994). The peptides did not activate the cardiac RyR2 Ca^{2+} release channel. This suggests that the skeletal isoforms of both the RyR and the DHPR are required to achieve a skeletal-type EC coupling. Studies with shorter peptides have identified subdomains of the II–III linker which are essential to activate Ca^{2+} release (see Fig. 13, inset). Amino acid segments within the region 666–725 confer skeletal-type EC coupling (666–726: Lu *et al.*, 1995; 671–690: El-Hayek *et al.*, 1995; 1998). A 25-amino-acid segment (Glu 666–Pro 691) within this region of α_{1s}, but not the corresponding cardiac sequence, is able to activate the skeletal ryanodine receptor. A cluster of five basic residues within this segment (681RKRRK685) is absolutely required to activate the skeletal RyR1 Ca^{2+} release channel (Zhu *et al.*, 1999). A second overlapping region from amino acids 711–760 appears to be important both for activation and blocking of RyR1 activation by the DHPR (711–742: Nakai *et al.*, 1998b; 725–742: Saiki *et al.*, 1999).

Conversely, as expected for direct protein-protein interactions, regions on RyR1 that interact specifically with the skeletal DHPR have also been identified. An N-terminal sequence on RyR1 (Leu 922–Asp 1112) interacts strongly with the II–III linker and weakly with the III–IV linker of the skeletal DHPR, suggesting that the DHPR linkers bind to contiguous and possibly overlapping sites on RyR1. A 37-amino-acid segment within this region of RyR1, but not the corresponding cardiac sequence on RyR2, binds specifically to the II–III linker of the skeletal DHPR (Leong and MacLennan, 1998b).

Much useful information on the skeletal RyR1 has come from studies using mice homozygous for a disrupted RyR1 gene (dyspedic mice). Dyspedic mice lack EC coupling, as expected, and in addition have a thirty-fold lower density of calcium channel current (Nakai *et al.*, 1996). Expression of the cDNA for RyR1 in dyspedic myotubes restores EC coupling and enhances the calcium channel current. This suggests that RyR1, in addition to receiving the EC coupling signal from the DHPR, also transmits a retrograde signal that enhances the function of skeletal DHPRs as calcium channels. The interactions between the DHPR and the RyR leading to EC coupling are specific for the skeletal RyR1, indicating that the identity of the RyR isoform may determine the type of EC coupling. Expression of the predominant cardiac isoform, RyR2, did not restore skeletal-type gating of Ca^{2+} release by the DHPR or retrograde signaling to enhance calcium channel activity (Nakai *et al.*, 1997). By expressing chimeras of the skeletal RyR1 and cardiac RyR2, Nakai and collaborators (1998a) identified distinct regions of RyR1 that are involved in the reciprocal interactions of RyR1 with the skeletal DHPR. Residues 1635–2636 of RyR1 mediated skeletal-type EC coupling and enhanced calcium channel function, whereas a chimera containing adjacent RyR1 residues (2659–3720) was only able to enhance calcium channel function. Residues 720–765 within the II–III linker of α_{1s} are required for receiving the feedback signal from RyR1 that leads to enhancement of Ca^{2+} current (Grabner *et al.*, 1999).

E. Evidence from *in Vivo* Studies

Early functional studies demonstrated that manipulations that alter T-tubule Ca^{2+} channel function influence SR Ca^{2+} release, and conversely that manipulations that affect Ca^{2+} release alter the function of the DHPR. Such bidirectional cross-talk between the DHPR and RyR, involving both forward and retrograde interactions, is expected for an allosteric interaction along a macromolecular complex. In studies using intact skeletal muscle cells, conformational transitions of the DHPR are detected as charge movement and SR Ca^{2+} release is detected by monitoring myoplasmic Ca^{2+} transients using Ca^{2+} indicator dyes introduced into the myoplasm.

Much useful information on the kinetics and voltage-dependence of the voltage-sensing and Ca^{2+} release steps under physiological conditions has come from these studies (rev. in Rios and Pizarro, 1991; Melzer *et al.*, 1995). Additional information has come from investigating the actions of pharmacologic agents that are known to alter EC coupling in skeletal muscle, including caffeine and perchlorate (ClO_4^-). ClO_4^- is a chaotropic anion that is a potent and specific potentiator of EC coupling in skeletal muscle. It shifts the voltage dependence of charge movement by 15 to 30 mV to more negative voltages, slows the kinetics of both on and off charge movements, and reduces the minimum threshold charge required to activate Ca^{2+} release. It was demonstrated early that several of the actions of ClO_4^- could be explained by a simple model in which the two proteins interact allosterically as shown in Fig. 14 (Rios *et al.*, 1993). In this model, the RyR/Ca^{2+} release channel is a homotetramer

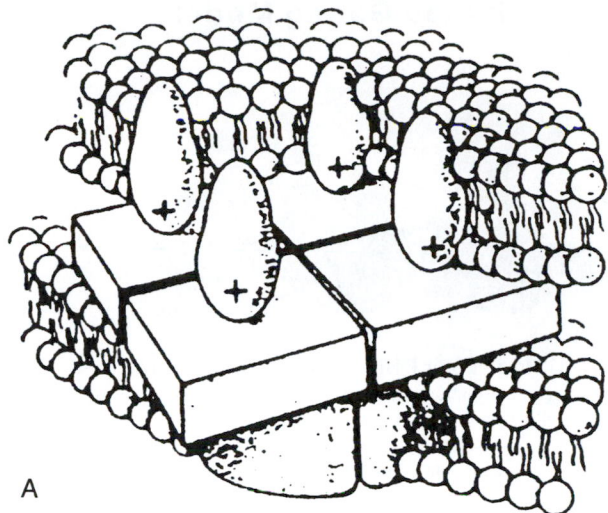

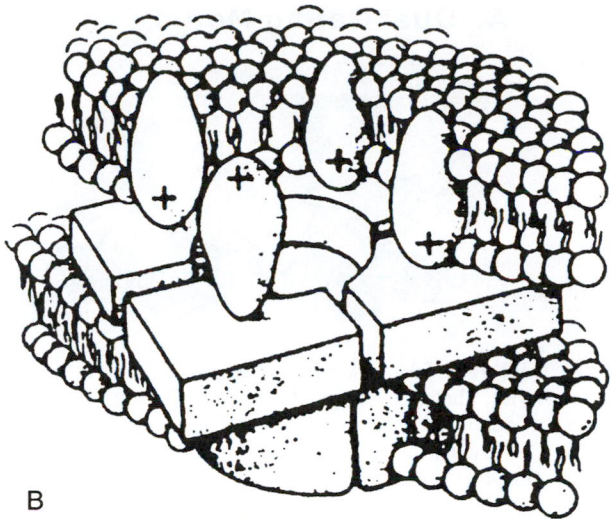

FIGURE 14. Allosteric model of EC coupling. Four independent voltage-sensor molecules/DHPRs in the T-tubule membrane compose a tetrad that is aligned with and contacts one RyR/Ca^{2+} release channel, depicted as a homotetramer. Two states of the release channel (closed and open) are pictured, as well as two of five possible dispositions of the dihydropyridine receptor tetrad. (A) All DHPRs in the inactive conformation. (B) One DHPR in the active conformation and three in the inactive state. (Reproduced from *The Journal of General Physiology*, 1993, **102**, 449–481, by copyright permission of The Rockefeller University Press.)

that can assume two conformations, open and closed. The four DHPRs aligned above it (one tetrad) act as heterotropic ligands to modulate this transition. Each shift of one voltage sensor/DHPR to its fully active state increases the probability that the RyR will open. In this model, ClO$_4^-$ acts primarily on the RyR/Ca^{2+} release channel to increase its probability of opening and secondarily on the DHPR via allosteric feedback. As part of this feedback to the DHPR, one kinetic component of charge movement, termed Q_γ charge, could be simulated. A related hypothesis in this model is that released Ca^{2+} ions in the triadic gap modify the voltage-sensing process in a positive-feedback manner to generate additional charge movement. Thus, both allosteric interactions between the RyR and the DHPR, and Ca^{2+} ions released into the triadic gap may contribute to the generation of the Q_γ charge movements detected electrically from intact muscle.

Other studies have likewise shown that Ca^{2+} release can influence the DHPR and vice-versa, but have led to a somewhat different model of the interaction (rev. in Jong *et al.*, 1996). An alternate proposal is that the steep voltage dependence of Ca^{2+} release is an intrinsic function of the voltage sensor in the membranes of the T-system, without requiring additional positive feedback from Ca^{2+} release. In this model, the SR Ca^{2+} content or release or a related process directly alters the kinetics of charge movement, but does not generate a separate species of charge. Both of these findings suggest a close association between the DHPR and the SR Ca^{2+} release channel.

Although Ca^{2+} entry is not required for contraction in skeletal muscle, the fact that Ca^{2+} release and/or the RyR through retrograde allosteric interactions can modify the slow Ca^{2+} current also supports the idea of a close interaction between RyR1 and the skeletal DHPR. Intracellular BAPTA,

a rapid Ca^{2+} buffer that is capable of buffering released Ca^{2+} ions in the triadic space near the release sites (Jong *et al.*, 1996) eliminates the slow Ca^{2+} current (Feldmeyer *et al.*, 1993).

F. Overall Control and Integration of Ca^{2+} Release Events

Although there is general agreement that the initial event leading to activation of calcium release is a depolarization-driven allosteric interaction between a skeletal DHPR tetrad and one RyR1 (electromechanical coupling), the mechanism by which this signal is communicated to adjacent RyRs and perhaps amplified or synchronized with other RyRs, is less understood.

The position of DHPR tetrads alternating with every other RyR1, and the packing of RyR1s in highly ordered arrays possibly touching at the corners suggests a dual control of RyR1 activation in which RyR1s linked to a skeletal DHPR are activated allosterically by the DHPR while neighboring RyRs are ligand gated by binding of released Ca^{2+} (Fig. 15A). The latter is analogous to Ca^{2+}-induced Ca^{2+} release in heart muscle, with the source of activator Ca^{2+} being intracellular rather than extracellular. According to the **dual control theory**, each RyR is opened independently either by allosteric or ligand gating and the global myoplasmic change arises from the summation of thousands of individual single Ca^{2+} release events from both populations of RyRs.

An alternate model of RyR gating, termed **coordinated gating**, proposes that clusters of physically linked RyRs open simultaneously in a coordinated fashion (Fig. 15B; Marx *et al.*, 1998). In the later model, clusters of RyRs operate synergistically as all-or-none Ca^{2+} release units (all

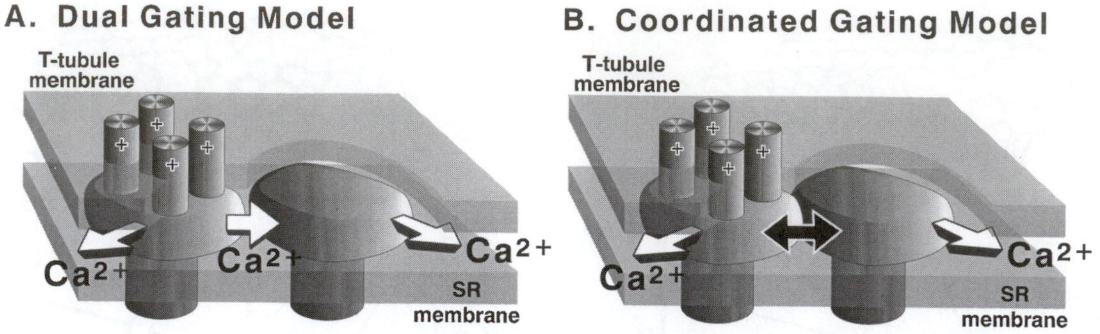

A. Dual Gating Model

B. Coordinated Gating Model

FIGURE 15. Two models of integration of Ca^{2+} release from RyRs. A tetrad of four DHPRs is shown aligned above one RyR1. The adjacent RyR is not associated with a DHPR tetrad and is not necessarily attached to the first RyR. (A) Dual gating model of RyR1 activation. Upon depolarization, one DHPR tetrad allosterically activates its associated RyR; the resulting local Ca^{2+} change activates the unassociated RyR. Thus, one population of RyR1s is gated allosterically while the second population of RyR1s is ligand gated by released Ca^{2+}. The overall myoplasmic Ca^{2+} change results from the summation of individual Ca^{2+} release events from both populations of RyR1s. (B) Coordinated gating model. A tetrad of four DHPRs is aligned above one RyR1. The DHPR-associated RyR and adjacent RyRs are linked to each other in such a way that one DHPR and a cluster of RyRs form a functional macromolecular complex. Upon depolarization, one DHPR tetrad allosterically activates its associated RyR and that allosteric change is transmitted down the complex to activate simultaneously all RyRs in the cluster. In this model, both populations of RyR1s, those associated and those unassociated with DHPR tetrads, are activated by the DHPR without intermediate Ca^{2+} ligand. The overall myoplasmic Ca^{2+} change results from the coordinated gating of linked RyR1s.

channels in a cluster open or close together). Opening of one RyR1 by its linked DHPR results in simultaneous opening of all contiguous RyRs in the cluster. RyRs operating as functional units would be expected to generate faster local Ca^{2+} release events and a more synchronous global change in myoplasmic Ca.

Small, local Ca^{2+} release events, termed Ca^{2+} **sparks,** have been detected intracellularly in skeletal and cardiac muscle using Ca^{2+} indicator dyes (Cheng *et al.,* 1993; Tsugorka *et al.,* 1995). It is generally thought that the Ca^{2+} spark is the elementary unit of SR Ca^{2+} release. The Ca^{2+} spark, which exhibits a stereotypical amplitude and duration, may arise from the stochastic opening of several RyRs or may represent the coordinated opening of a functional RyR cluster.

G. Modulation of EC Coupling

There is growing evidence that EC coupling can be modulated under physiological conditions, and a number of potential sites and mechanisms have been identified. The skeletal DHPR α subunit contains a number of consensus phosphorylation sites including a serine residue, Ser 687 within the critical II–III loop that is rapidly phosphorylated by a cAMP-dependent kinase. Phosphorylation of this residue is required for binding of the II–III loop to RyR1 (but not RyR2) but only the dephosphorylated II–III loop can activate the RyR (Lu *et al.,* 1995). In studies using dysgenic myotubes, substitution of Ser 687 with alanine, the corresponding cardiac residue, did not abolish skeletal-type EC coupling (Nakai *et al.,* 1998b). Together, these data suggest that phosphorylation of the DHPR II–III linker may be a potential regulatory mechanism for EC coupling. The α_{1s} subunit contains sites that can inhibit as well as activate Ca^{2+} release from the RyR (residues 671–690 in the II–III

linker: El-Hayek *et al.,* 1995; residues 1487–1506 in the C-terminus: Slavik *et al.,* 1997). Similarly, subdomains of RyR1 have been identified that participate in the modulation, inhibition, or inactivation of Ca^{2+} release. In addition to the regions of RyR1 involved in retrograde enhancement of DHPR function as a Ca^{2+} channel, it has been shown that the C-terminal one-fourth of RyR1 has a major role in both the activation and inactivation of RyR1 by Ca^{2+} (Nakai *et al.,* 1999; Du and MacLennan, 1999).

Other proteins of the triad may also modulate EC coupling as well as participate in the structural architecture of the junction. The triad region of skeletal muscle forms supramolecular membrane complexes that include, in addition to the DHPR and RyR1, FKBP12, calsequestrin, junctin, triadin, and possibly other proteins. Finally, a long-term regulation of EC coupling may occur at the level of gene expression. Studies have shown that the expression of α_{1s} can be altered during adaptation to physiological demands, including altered patterns of nerve activity, exercise, changes in the load-bearing state and metabolic or endocrine status (Ray *et al.,* 1995; Kandarian *et al.,* 1996; Pereon *et al.,* 1997).

VI. Summary

Excitation-contraction coupling in skeletal muscle is a fast signal transduction process by which an action potential on the sarcolemmal/T-tubule membranes is transduced into a biochemical trigger for the contractile proteins to shorten and generate force. The immediate intracellular trigger is Ca^{2+} released from intracellular stores. Contraction is inhibited when cytosolic Ca^{2+} is at resting, submicromolar levels, and proceeds when Ca^{2+} rises to micromolar levels.

The key transduction steps that link membrane depolarization to Ca^{2+} release from the SR occur at specialized junctions that bring together the sarcolemmal and intracellular SR membranes. At the junctions, a sarcolemmal Ca^{2+} channel/DHPR senses the depolarization during the action potential and communicates it to the Ca^{2+} release channels/RyRs on the SR, resulting in the release of stored Ca^{2+}. The molecular mechanisms of interaction between these two distinct Ca^{2+} channels involves an allosteric process. DHPRs and RyRs associate at the junctions to form a functional macromolecular complex that is capable of distinct molecular interactions at identified sites. Additional associated junctional proteins may participate in or modulate this interaction. The result of this signal transduction is a fast, coordinated activation of muscle contraction.

Bibliography

Adams, B. A., and Beam, K. G. (1990). Muscular dysgenesis in mice: a model system for studying excitation-contraction coupling. [Review]. *FASEB J.* **4**, 2809–2816.

Anderson, K., Cohn, A. H., and Meissner, G. (1994). High-affinity [^{3}H]PN200-110 and [^{3}H]ryanodine binding to rabbit and frog skeletal muscle. *Am. J. of Physiol.* **266**, C462–C466.

Armstrong, C. M., Bezanilla, F. M., and Horowicz, P. (1972). Twitches in the presence of ethylene glycol bis(-aminoethyl ether)-N,N'-tetracetic acid. *Biochim. Biophys. Acta* **267**, 605–608.

Beam, K. G., Tanabe, T., and Numa, S. (1989). Structure, function, and regulation of the skeletal muscle dihydropyridine receptor. [Review]. *Ann. NY Acad. Sci.* **560**, 127–137.

Beurg, M., Sukhareva, M., Ahern, C.A., Conklin, M.S., Perez-Reyes, E., Powers, P.A., Gregg, R.G., Coronado, R. (1999). Differential regulation of skeletal muscle L-type Ca^{2+} current and excitation-contraction coupling by the dihydropyridine receptor beta subunit. *Biophys. J.* **76**, 1744–1756.

Block, B. A., Imagawa, T., Campbell, K. P., and Franzini-Armstrong, C. (1988). Structural evidence for direct interaction between the molecular components of the transverse tubule/sarcoplasmic reticulum junction in skeletal muscle. *J. Cell Biol.* **107**, 2587–2600.

Caputo, C., Bezanilla, F., and Horowicz, P. (1984). Depolarization-contraction coupling in short frog muscle fibers. A voltage clamp study. *J. Gen. Physiol.* **84**, 133–154.

Caswell, A.H., Lau, Y.H., Garcia, M., and Brunschwig, J.P. (1979). Recognition and junction formation by isolated transverse tubules and terminal cisternae of skeletal muscle. *J. Biol. Chem.* **254**, 202–208.

Chandler, W. K., Rakowski, R. F., and Schneider, M. F. (1976). A nonlinear voltage-dependent charge movement in frog skeletal muscle. *J. Physiol.* **254**, 243–283.

Cheng, H., Lederer, W. J., and Cannell, M. B. (1993). Calcium sparks: elementary events underlying excitation-contraction coupling in heart muscle. *Science* **262**, 740–744.

De Waard, M., Gurnett, C. A., and Campbell, K. P. (1996). Structural and functional diversity of voltage-activated calcium channels. *Ion Channels* **4**, 41–87.

Du, G. C. and MacLennan, D. H. (1999). Ca(2+) inactivation sites are located in the COOH-terminal quarter of recombinant rabbit skeletal muscle Ca(2+) release channels (ryanodine receptors). *J. Biol. Chem.* **274**, 26 120–26 126.

Dulhunty, A. F. (1984). Heterogeneity of T-tubule geometry in vertebrate skeletal muscle fibres. *J. Muscle Res. Cell Motility* **5**, 333–347.

El-Hayek, R., and Ikemoto, N. (1998). Identification of the minimum essential region in the II–III loop of the dihydropyridine receptor alpha 1 subunit required for activation of skeletal muscle-type excitation-contraction coupling. *Biochem.* **37**, 7015–7020.

El-Hayek, R., Antoniu, B., Wang, J., Hamilton, S. L., and Ikemoto, N. (1995). Identification of calcium release–triggering and blocking regions of the II–III loop of the skeletal muscle dihydropyridine receptor. *J. Biol. Chem.* **270**, 22 116–22 118.

Feldmeyer, D., Melzer, W., Pohl, B., and Zollner, P. (1993). A possible role of sarcoplasmic Ca^{2+} release in modulating the slow Ca^{2+} current of skeletal muscle. *Pflügers Arch.* **425**, 54–61.

Franzini-Armstrong, C., and Jorgensen, A. O. (1994). Structure and development of E-C coupling units in skeletal muscle. [Review]. *Ann. Rev. Physiol.* **56**, 509–534.

Franzini-Armstrong, C., and Kish, J. W. (1995). Alternate disposition of tetrads in peripheral couplings of skeletal muscle. *J. Muscle Res. Cell Motility* **16**, 319–324.

Franzini-Armstrong, C., and Peachly, L. D. (1981). Striated muscle—contractile and control mechanisms. [Review]. *J. Cell Biol.* **91**, 166s–186s.

Franzini-Armstrong, C., and Protasi, F. (1997). Ryanodine receptors of striated muscles: a complex channel capable of multiple interactions. *Physiol. Rev.* **77**, 699–729.

Franzini-Armstrong, C., Protasi, F., and Ramesh, V. (1999). Shape, size and distribution of Ca^{2+} release units and couplons in skeletal and cardiac muscles. *Biophys. J.* **77**, 1528–1539.

Froemming, G. R., Pette, D., and Ohlendieck, K. (1999). The 90-kDa junctional sarcoplasmic reticulum protein forms an integral part of a supramolecular triad complex in skeletal muscle. *Biochem. Biophys. Res. Commun.* **261**, 603–609.

Gonzalez, A., and Rios, E. (1993). Perchlorate enhances transmission in skeletal muscle excitation-contraction coupling. *J. Gen. Physiol.* **102**, 373–421.

Gonzalez-Serratos, H. (1971). Inward spread of activation in vertebrate muscle fibres. *J. Physiol. (Lond.)* **212**, 777–799.

Grabner, M., Dirksen, R. T., Suda, N., and Beam, K. G. (1999). The II–III loop of the skeletal muscle dihydropyridine receptor is responsible for the bi-directional coupling with the ryanodine receptor. *J. Biol. Chem.* **274**, 21 913–21 919.

Gregg, R. G., Messing, A., Strube, C., Beurg, M., Moss, R., Behan, M., Sukhareva, M., Haynes, S., Powell, J. A., Coronado, R., Powers, P. A. (1996). Absence of the beta subunit (cchb1) of the skeletal muscle dihydropyridine receptor alters expression of the alpha 1 subunit and eliminates excitation-contraction coupling. *Proc. Natl. Acad. Sci. USA* **93**, 13 961–13 966.

Hill, A. V. (1948). On the time required for diffusion and its relation to processes in muscle. *Proc. R. Soc. B* **135**, 446–453.

Hodgkin, A. L., and Horowicz, P. (1957). The differential action of hypertonic solutions on the twitch and action potential of a muscle fibre. *J. Physiol.* **136**, 17–18.

Hollingworth, S., Zhao, M., and Baylor, S. M. (1996). The amplitude and time course of the myoplasmic free [Ca^{2+}] transient in fast-twitch fibers of mouse muscle. *J. Gen. Physiol.* **108**, 455–469.

Jong, D. S., Pape, P. C., Geibel, J., Chandler, W. K. (1996). Sarcoplasmic reticulum calcium release in frog cut muscle fibers in the presence of a large concentration of EGTA. *Soc. Gen. Physiol. Ser.* **51**, 255–268

Kandarian, S. C., Peters, D. G., Favero, T. G., Ward, C. W., and Williams, J. H. (1996). Adaptation of the skeletal muscle calcium-release mechanism to weight-bearing condition. *Am. J. Physiol.* **270**, C1588–C1594.

Leong, P., and MacLennan, D. H. (1998a). Complex interactions between skeletal muscle ryanodine receptor and dihydropyridine receptor proteins. *Biochem. Cell Biol.* **76**, 681–694.

Leong, P., and MacLennan, D. H. (1998b). A 37-amino-acid sequence in the skeletal muscle ryanodine receptor interacts with the cytoplasmic loop between domains II and III in the skeletal muscle dihydropyridine receptor. *J. Biol. Chem.* **273**, 7791–7794.

Lu, X., Xu, L., and Meissner, G. (1994). Activation of the skeletal muscle calcium release channel by a cytoplasmic linker of the dihydropyridine receptor. *J. Biol. Chem.* **269**, 6511–6516.

Lu, X., Xu, L., and Meissner, G. (1995). Phosphorylation of dihydropyridine receptor II–III linker peptide regulates skeletal muscle calcium release channel function. Evidence for an essential role of the beta-OH group of Ser687. *J. Biol. Chem.* **270,** 18 459–18 464.

Marx, S. O., Ondrias, K., and Marks, A. R. (1998). Coupled gating between individual skeletal muscle Ca^{2+} release channels. *Science* **281,** 818–821.

Mcpherson, P. S., and Campbell, K. P. (1993). The ryanodine receptor/Ca^{2+} release channel. *J. Biol. Chem.* **268,** 13 765–13 768.

Meissner, G., and Lu, X. (1995). Dihydropyridine receptor-ryanodine receptor interactions in skeletal muscle excitation-contraction coupling. *Biosci. Rep.* **15,** 399–407.

Melzer, W., Herrmann-Frank, A., and Luttgau, H. C. (1995). The role of Ca^{2+} ions in excitation-contraction coupling of skeletal muscle fibres. [Review]. *Biochim. Biophys. Acta* **1241,** 59–116.

Monnier, N., Procaccio, V., Stieglitz, P., and Lunardi, J. (1997). Malignant-hyperthermia susceptibility is associated with a mutation of the alpha 1-subunit of the human dihydropyridine-sensitive L-type voltage-dependent calcium-channel receptor in skeletal muscle. *Am. J. Hum. Genet.* **60,** 1316–1325.

Nakai, J., Gao, L., Xu, L., Xin, C., Pasek, D. A., and Meissner, G. Evidence for a role of C-terminus in Ca(2+) inactivation of skeletal muscle Ca(2+) release channel. (1999). *FEBS Lett.* **459,** 154–158.

Nakai, L. J., Dirkson, R. T., Nguyen, H. T., Pessah, I. N., Beam, K. G., and Allen, P. D. (1996). Enhanced dihydropyridine receptor channel activity in the presence of ryanodine receptor. *Nature* **380,** 72–75.

Nakai, J., Ogura, T., Protase, F., Franzini-Armstrong, C., Allen, P. D., and Beam, K. G. (1997). Functional nonequality of the cardiac and skeletal ryanodine receptors. *Proc. Natl. Acad. Sci. USA* **94,** 1019–1022.

Nakai, J., Sekiguchi, N., Rando, T. A., Allen, P. D., and Beam, K. G. (1998a). Two regions of the ryanodine receptor involved in coupling with L-type Ca^{2+} channels. *J. Biol. Chem.* **273,** 13 403–13 406.

Nakai, J., Tanabe, T., Konno, T., Adams, B., and Beam, K. G. (1998b). Localization in the II–III loop of the dihydropyridine receptor of a sequence critical for excitation-contraction coupling. *J. Biol. Chem.* **273,** 24 983–24 986.

Neuhuber, B. Gerster, U., Mitterdorfer, J., Glossmann, H., Flucher, B. E. (1998). Differential effects of Ca^{2+} channel beta1a and beta 2a subunits on complex formation with alpha1S and on current expression in tsA201 cells. *J. Biol. Chem.* **273,** 9110–9118.

Orlova, E. V., Serysheva, I. I., Van Heel, M., Hamilton, S. L. & Chiu, W. (1996). Two structural configurations of the skeletal muscle calcium release channel. *J. Gen. Physiol.* **106,** 659–704.

Oz, M., and Frank, G. B. (1991). Decrease in the size of tetanic responses produced by nitrendipine or by extracellular calcium ion removal without blocking twitches or action potentials in skeletal muscle. *J. Pharmacol. Exper. Ther.* **257,** 575–581.

Paul, R. J., Ferguson, D. G., and Heiny, J. A. (1996). Muscle physiology: molecular mechanisms. *In* "Essentials of Physiology", (N. Sperelakis and R. O. Banks, Eds.) pp. 203–216. Little, Brown and Co., Boston.

Peachey, L. D. (1965). The sarcoplasmic reticulum and transverse tubules of the frog's sartorius. *J. Cell Biol.* **25,** 209–232.

Peachey, L. D., and Eisenberg, B. R. (1978). Helicoids in the T system and striations of frog skeletal muscle fibers seen by high voltage electron microscopy. *Biophys. J.* **22,** 145–154.

Pereon, Y., Navarro, J., Hamilton, M., Booth, F. W., and Palade, P. (1997). Chronic stimulation differentially modulates expression of mRNA for dihydropyridine receptor isoforms in rat fast twitch skeletal muscle. *Biochem. Biophys. Res. Commun.* **235,** 217–222.

Perez-Reyes, E. and Schneider, T. (1995). Molecular biology of calcium channels. *Kidney International* **48,** 1111–1124.

Ray, A., Kyselovic, J., Leddy, J. J., Wigle, J. T., Jasmin, B. J., and Tuana, B. S. (1995). Regulation of dihydropyridine and ryanodine receptor gene expression in skeletal muscle. Role of nerve, protein kinase C, and cAMP pathways. *J. Biol. Chem.* **270,** 25 837–25 844.

Rios, E., and Pizarro, G. (1991). Voltage sensor of excitation-contraction coupling in skeletal muscle. [Review]. *Physiol. Rev.* **71,** 849–908.

Rios, E., Karhanek, M., Ma, J., and Gonzalez, A. (1993). An allosteric model of the molecular interactions of excitation-contraction coupling in skeletal muscle. *J. Gen. Physiol.* **102,** 449–481.

Saiki, Y., El-Hayek, R., and Ikemoto, N. (1999). Involvement of the Glu724–Pro760 region of the dihydropyridine receptor II–III loop in skeletal muscle-type excitation-contraction coupling. (1999). *J. Biol. Chem.* **274,** 7825–7832.

Schneider, M. F. (1994). Control of calcium release in functioning skeletal muscle fibers. [Review]. *Ann. Rev. Physiol.* **56,** 463–484.

Schneider, M. F. & Chandler, W. K. (1973). Voltage-dependent charge movement in skeletal muscle: a possible step in excitation-contraction coupling. *Nature* **242,** 244–246.

Slavik, K. J., Wang, J. P., Aghdasi, B., Zhang, J. Z., Mandel, F., Malouf, N., and Hamilton, S. L. (1997). A carboxy-terminal peptide of the alpha 1-subunit of the dihydropyridine receptor inhibits Ca(2+)-release channels. *Am. J. Physiol.* **272,** C1475–C1481.

Stern, M. D., Pizarro, G., and Rios, E. (1997). Local control model of excitation-contraction coupling in skeletal muscle. *J. Gen. Physiol.* **110,** 415–440.

Suda, N. (1995). Involvement of dihydropyridine receptors in terminating Ca^{2+} release in rat skeletal myotubes. *J. Physiol. (Lond.)* **486,** 105–112.

Suh-Kim, H., Wei, X., Klos, A., Pan, S., Ruth, P., Flockerzi, V., Hofmann, F., Perez-Reyes, E., and Birnbaumer, L. (1996). Functional interaction among alpha 1, beta, gamma and alpha 2 delta subunits. *Receptors Channels* **4,** 217–225.

Tanabe, T., Beam, K. G., Adams, B., Niidome, T., and Numa, S. (1990). Regions of the skeletal muscle dihydropyridine receptor critical for excitation-contraction coupling. *Nature* **346,** 567–569.

Tanabe, T., Beam, K. G., Powell, J., and Numa, S. (1988). Restoration of excitation-contraction coupling and slow calcium current in dysgenic muscle by dihydropyridine receptor complementary DNA. *Nature* **336,** 134–139.

Tanabe, T., Takeshima, H., Mikami, A., Flockerzi, V., Takashiki, H., Kangawa, K., Kojima, M., Matsuo, H., Hirose, T., and Numa, S. (1987). Primary structure of the receptor for calcium channel blockers from skeletal muscle. *Nature* **328,** 313–318.

Tsugorka, A., Rios, E., and Blatter, L. A. (1995). Imaging elementary events of calcium release in skeletal muscle cells. *Science* **269,** 1723–1726.

Vergara, J. and Delay, M. (1986). A transmission delay and the effect of temperature at the triadic junction of skeletal muscle. *Proc. R. Soc. Lond. B Biol. Sci.* **229,** 97–110.

Zhu, X., Gurrola, G., Jiang, M. T., Walker, J. W., and Valdivia, H. H. (1999). Conversion of an inactive cardiac dihydropyridine receptor II–III loop segment into forms that activate skeletal ryanodine receptors. *FEBS Lett.* **450,** 221–226.

Gerhard Meissner

55

Ca^{2+} Release from Sarcoplasmic Reticulum in Muscle

I. Introduction

In excitable cells, the release of Ca^{2+} ions from intracellular membrane compartments is triggered by an excitatory electrical signal, or it occurs via a chain of voltage-independent steps that involve the agonist-induced formation of inositol 1,4,5-trisphosphate (IP$_3$) and subsequent activation of an intracellular membrane receptor/Ca^{2+} channel complex, the **IP$_3$ receptor** (Berridge, 1993). The voltage-dependent mechanism, commonly referred to as **excitation-contraction (EC) coupling,** is the focus of this chapter. It has been most thoroughly studied in vertebrate striated muscle.

Figure 1 illustrates the major components involved in EC coupling: **A transverse (T-) tubule** membrane system of invaginations through which muscle contraction is triggered and an intracellular Ca^{2+} storing and Ca^{2+} releasing membrane system, the **sarcoplasmic reticulum** (SR), which contains a Ca^{2+} pump and a Ca^{2+} release channel. Lumenal ionized SR Ca^{2+} has been estimated at about 1.0 mM and the ionized cytosolic Ca^{2+} at about 0.1 μM, yielding an approximately ten-thousand-fold Ca^{2+} gradient across the SR membrane. Sequestration of Ca^{2+} into the SR against the Ca^{2+} gradient is mediated by a Ca^{2+}-ATPase that transports two Ca^{2+} into the SR at the expense of one ATP. The release channels span the narrow gap where the SR (the Ca^{2+} store) and the T-tubule (the conduit of the action potential) are within ~15 nm of each other (for reviews, see Fleischer and Inui, 1989; Franzini-Armstrong and Protasi, 1997). The SR Ca^{2+} release channels are also known as **feet** or **junctional processes,** and as **ryanodine receptors** (RyRs) because they have the ability to bind the plant alkaloid **ryanodine** with high affinity and specificity. They are often viewed, at least in skeletal muscle, as directly linked to four particles located in the T-tubule membrane and presumed to represent another Ca^{2+} channel (L-type), also known as the **dihydropyridine receptor** (DHPR). The relative number of RyRs and DHPRs in striated muscle varies greatly, ranging from about 10 RyRs per DHPR in cardiac muscle to a 1:1 stoichiometry in fast-twitch mammalian skeletal muscle.

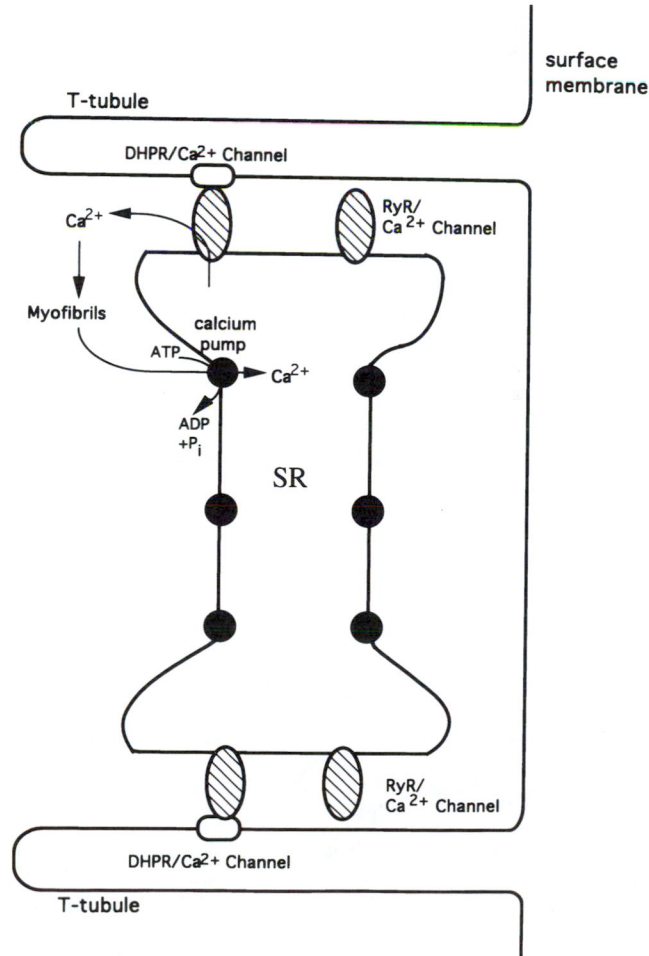

FIGURE 1. Schematic representation of a segment of a mammalian skeletal muscle cell illustrating the major components involved in EC coupling.

Accordingly, not all RyRs may be linked or closely apposed to DHPRs in striated muscle.

In addition to the Ca^{2+} release channel, the SR of skeletal and cardiac muscle contains monovalent ion-selective channels: (1) a K^+ channel, (2) a Cl^- channel, and (3) a H^+ (OH) permeable pathway. Movement of ions through the monovalent cation- and anion-selective channels has been proposed to compensate charge movements across the SR membrane, and thereby to support rapid Ca^{2+} release and reuptake (for review, see Meissner, 1983).

II. Mechanisms of EC Coupling

Different excitation-contraction coupling mechanisms exist in skeletal and cardiac muscles (for reviews, see Fabiato, 1983; Rios and Pizarro, 1991). A distinguishing feature is that EC coupling in **cardiac muscle** is dependent on extracellular Ca^{2+}, whereas **skeletal muscle** EC coupling is not (in the short run). In cardiac muscle, dihydropyridine-sensitive, L(long-lasting)-type Ca^{2+} channels located in the surface membrane and T-tubule mediate the influx of Ca^{2+} during an action potential by functioning as voltage-dependent Ca^{2+} channels (Fig. 2). The resulting rise in intracellular Ca^{2+} concentration triggers the massive release of Ca^{2+} by opening SR Ca^{2+} release channels. This process is known as **calcium-induced Ca^{2+} release** (CICR). Intracellular Ca^{2+} transients have been suggested to arise as the sum of localized Ca^{2+} release events called **Ca^{2+} sparks.** The opening of a single T-tubule Ca^{2+} channel is apparently sufficient to evoke a Ca^{2+} spark by activating a functional Ca^{2+} release unit that may consist of one or more Ca^{2+} release channels (Santana *et al.*, 1996). Ca^{2+} also plays a major role in regulating ryanodine-sensitive SR Ca^{2+} release channels in **mammalian smooth muscle** and **neurons.**

Mammalian heart cells contain two additional Ca^{2+} entry pathways, T(transient)-type Ca^{2+} channels and the Na^+-Ca^{2+} exchanger; however, there is little evidence that they have a significant role in initiating SR Ca^{2+} release in normal hearts.

In vertebrate skeletal muscle, a different mechanism of EC coupling is present. A unique mechanism, referred to as the **mechanical coupling mechanism,** has been formulated and suggests that the SR Ca^{2+} release channel is opened via a direct physical interaction with a voltage-sensing molecule in the T-tubule (see Fig. 2). Biophysical and pharmacological evidence, as well as molecular expression studies, suggest that the T-tubule DHPR is the **voltage sensor** for EC coupling in vertebrate skeletal muscle (Rios and Pizarro, 1991).

The mammalian skeletal muscle DHPR is composed of five subunits: α_1, α_2, β, γ, and δ (Catterall, 1995). α_1 alone forms a Ca^{2+} channel, binds Ca^{2+} channel antagonists, and belongs to a voltage-sensitive ion channel superfamily that includes Na^+ and K^+ channels. Co-immunoprecipitation and cross-linking of skeletal muscle DHPRs and RyRs as well as morphological evidence suggest a well-defined interaction between the two receptors. Clusters of four particles, **tetrads** presumed to represent four DHPRs, are located opposite four subunits of every other RyR (Franzini-Armstrong and Protasi, 1997). The formation of DHPR tetrads is dependent on the presence of the skeletal muscle RyR (Protasi *et al.*, 1998). Other SR junctional proteins that may assist in the formation of the triad are triadin and junctin (for review, see Meissner and Lu, 1995).

The existence of two populations of skeletal muscle RyRs, one coupled to tetrads and one not, raises the question of how unlinked RyRs are activated. One suggestion is that Ca^{2+} released by DHPR-linked RyRs activates DHPR-unlinked RyRs in skeletal muscle by a Ca^{2+}-induced mechanism resembling that in cardiac muscle (Rios and Pizarro, 1991). An alternative

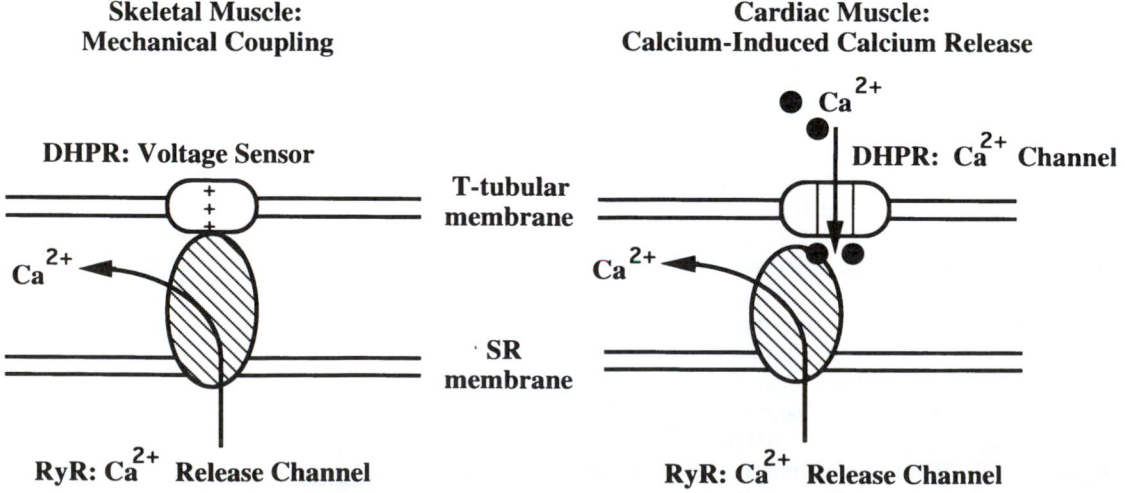

Muscle Excitation-Contraction Coupling

Skeletal Muscle:
Mechanical Coupling

Cardiac Muscle:
Calcium-Induced Calcium Release

DHPR: Voltage Sensor

T-tubular membrane

Ca^{2+}

DHPR: Ca^{2+} Channel

Ca^{2+}

SR membrane

RyR: Ca^{2+} Release Channel

RyR: Ca^{2+} Release Channel

FIGURE 2. Models of skeletal and cardiac muscle EC coupling. Depicted are the mechanical coupling model for vertebrate skeletal muscle, and the calcium-induced calcium release (CICR) model for cardiac muscle.

mechanism is that neighboring skeletal muscle RyRs are physically linked, leading to simultaneous channel opening and closing, termed **coupled gating** (Marx *et al.*, 1998).

We shall discuss below studies carried out to determine the structural and functional features of the SR RyR/Ca²⁺ release channel. These studies have shown that mammalian tissues, including muscles, express three RyR/Ca²⁺ release channel isoforms that are encoded by three different genes: *ryr1*, *ryr2*, and *ryr3*. The three RyR isoforms are known as the skeletal muscle (RyR1), cardiac muscle (RyR2), and brain (RyR3) RyRs. All three share a high-affinity binding site for [³H]ryanodine and a high-conductance pathway for Ca²⁺ and monovalent cations, but display isoform- and species-dependent differences in their *in vitro* regulation by Ca²⁺ and other effector molecules.

III. Isolation of Membrane Fractions Enriched in RyR/Ca²⁺ Release Channels

The molecular properties of the RyR/Ca²⁺ release channels have been extensively studied with rabbit skeletal muscle as the source for RyRs. Fragmentation of the SR during homogenization and subsequent fractionation by differential and density-gradient centrifugation yields a "heavy" SR vesicle fraction enriched in Ca²⁺ release channels and [³H]ryanodine-binding activity and that corresponds to the junctional region of the SR, **junctional SR** (see Fig. 1). Another important advance allowing study of the coupling between T-tubule depolarization and SR Ca²⁺ release in skeletal muscle (Ikemoto *et al.*, 1985) has been the isolation of membrane fractions composed of a T-tubule segment sandwiched between two junctional SR vesicles (Fleischer and Inui, 1989). These junctional complexes are known as **triads**.

Microsomal membrane fractions enriched in ryanodine-sensitive Ca²⁺ release channels have been also isolated from other excitable tissues, including cardiac muscle, smooth muscle, and brain. In these cases, however, the membrane fractions are typically of a lower purity than those from skeletal muscle.

IV. Isolation and Structure of RyRs

The isolation and structural determination of the SR Ca²⁺ release channel has been greatly facilitated by the identification of ryanodine as a channel-specific ligand. Ryanodine is a neutral plant alkaloid that is obtained from the stems of the South American shrub, *Ryania speciosa,* and is composed of two major compounds: ryanodine and 9,21-didehydroryanodine (Fig. 3). Ryanodine is a highly toxic compound. Its pharmacological effects have been most clearly shown in muscle, where depending on muscle type and activity, it can cause either contracture or a decline in contractile force (Jenden and Fairhurst, 1969). Ca²⁺ flux studies with isolated SR vesicles and single-channel recordings have shown that ryanodine specifically affects the SR Ca²⁺ release channel. Ryanodine activates the channel at low (nanomolar) concentrations, but inhibits the channel at high (micromolar) concentrations (see Section VI. C, Figs. 9 and 10). Because the drug binds with high specificity and dissociates slowly from the high-affinity site of the receptor (ei-

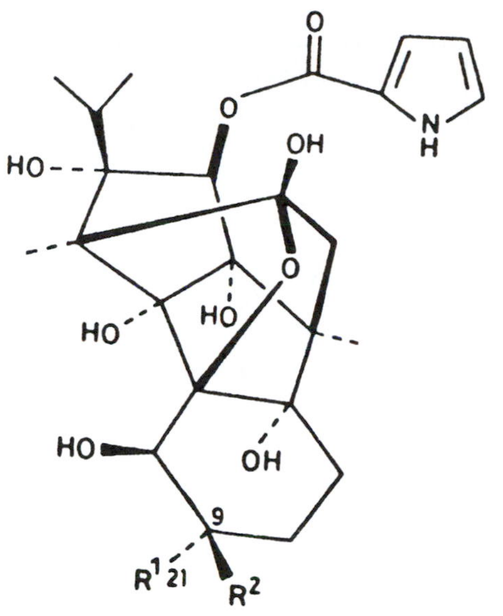

FIGURE 3. Structure of ryanodine ($R_1 = CH_3$, $R_2 = H$) and 9,21-didehydroryanodine (R_1, $R_2 = CH_2$).

ther membrane bound or detergent solubilized), [³H]ryanodine has been found to be an ideal probe in the isolation of the RyR from a variety of tissues and species.

The RyR was first isolated from striated muscle because its role in regulating free cytoplasmic Ca²⁺ levels was originally recognized in muscle, and relatively large amounts of membranes enriched in [³H]ryanodine-binding activity can be obtained from skeletal muscle and cardiac muscle. In most studies, the membrane-bound Ca²⁺ release channels were solubilized in the presence of [³H]ryanodine, using the zwitterionic detergent Chaps and high ionic strength (1.0 M) NaCl; they were then purified by sequential column chromatography (Inui *et al.*, 1987). Purification of a functional receptor can be also obtained in essentially one step by immunoaffinity chromatography (Smith *et al.*, 1988) or density gradient centrifugation through a linear sucrose gradient (Lai *et al.*, 1988). Figure 4A shows the [³H]ryanodine pattern on the sucrose gradients and sedimentation profile of the proteins associated with junctional SR vesicles isolated from rabbit skeletal muscle. A single peak of bound radioactivity, comigrating with a small protein peak possessing an apparent sedimentation coefficient of 30S, is observed in the lower half of the gradients. Binding to the small protein peak is specific since no radioactivity is present in the lower half of the gradients when the membranes are incubated with an excess of cold ryanodine. A sedimentation coefficient of 30S suggests that the RyR is a very large protein complex with a molecular weight (M_r) in excess of 1 000 000.

The sucrose gradient centrifugation procedure is relatively simple and straightforward. The procedure results in efficient separation of the large 30S RyR complex from the other solubilized smaller SR proteins because of its faster sedimentation rate. This method has been used to isolate a functional 30S RyR from several species and tissues including skeletal muscle, cardiac muscle, smooth muscle, and brain (Meissner, 1994).

A

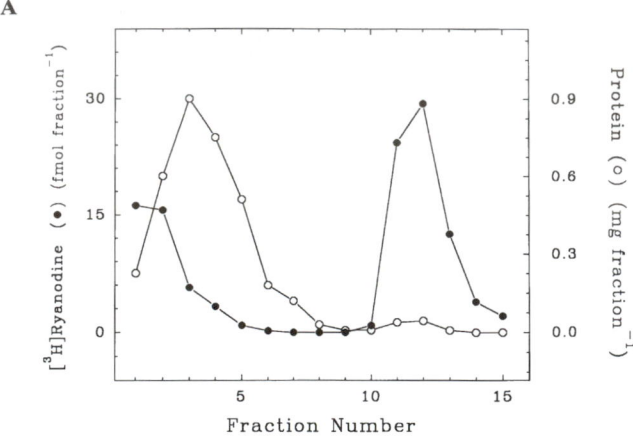

B

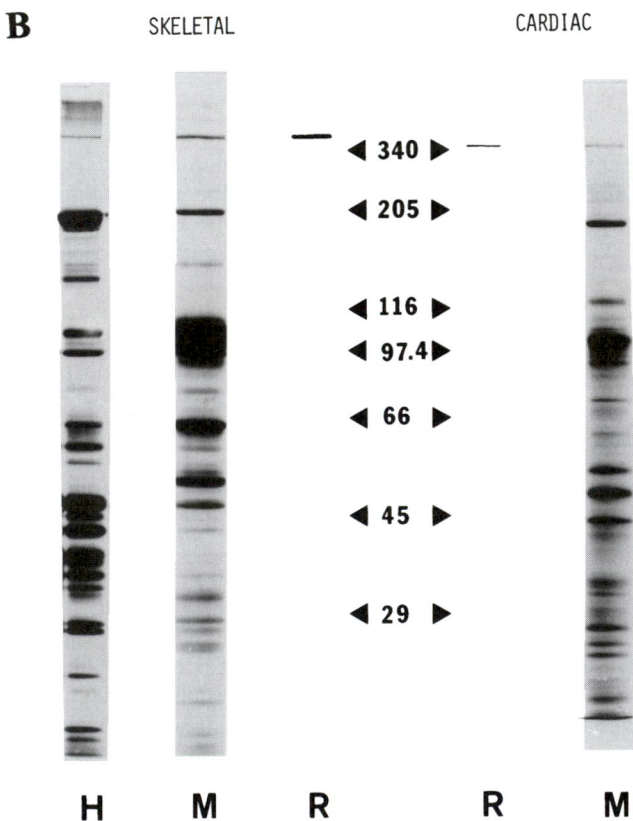

FIGURE 4. Sedimentation profile and SDS gel electrophoresis of rabbit skeletal muscle and canine cardiac RyRs. (A) Junctional skeletal muscle SR vesicles were solubilized in Chaps, centrifuged through a linear sucrose gradient, and fractionated. Fractions were analyzed for protein and ^{3}H-radioactivity. Unbound [^{3}H]ryanodine and the majority of the solubilized proteins sedimented near the top of the gradient (fractions 1–6), whereas the [^{3}H]ryanodine-labeled RyR comigrated with a small protein peak to the bottom of the gradient (fractions 11–13). (In modified form from Lai *et al., Nature* **331,** 315–319, 1988. Copyright 1988 by Macmillan Magazines Limited.) (B) Silver-stained SDS-polyacrylamide gel of whole rabbit skeletal muscle homogenate (H) and rabbit skeletal and canine cardiac muscle heavy SR membranes (M) and purified RyRs (R). Sizes of molecular weight standards are shown ($\times 10^{-3}$). (From Meissner *et al., Mol. Cell. Biochem.* **82,** 59–65, 1989.)

SDS polyacrylamide gel electrophoresis has shown that the 30S protein complexes of mammalian skeletal and cardiac muscles are composed of a major high–molecular-weight RyR polypeptide and isoform-specific low–molecular-weight immunophilins (FK506 binding protein) which migrate with an apparent $M_r > 340\,000$ (Fig. 4B) and $M_r \sim 12{,}000$ (Timerman *et al.*, 1996; not visible on the gels in Fig. 4B), respectively. Cloning and sequencing of the complementary DNA of the mammalian skeletal and cardiac muscle RyR isoforms have revealed an open reading frame of about 15 kb encoding RyR polypeptides of $M_r \sim 560{,}000$ (rev. in Sutko and Airey, 1996; Ogawa *et al.*, 1999). Purified RyR preparations from mammalian brain, crustacean skeletal muscle, and the nematode *Caenorhabditis elegans* also displayed a single protein band of high molecular weight on SDS gels. In contrast, the presence of two immunologically distinct high–molecular-weight RyR protein bands (corresponding to the mammalian RyR1 and RyR3) has been described for the main skeletal muscles of chicken, frog, and fish (Sutko and Airey, 1996). The two RyR isoforms are present as discrete homooligomers in amphibian and avian skeletal muscles. The appearance of the RyR1 isoform alone in some very fast-contracting muscle of fish suggests that this isoform is selectively expressed when rapid contraction is required in nonmammalian vertebrate muscles (O'Brien *et al.*, 1993).

Electron microscopy has revealed that the purified mammalian RyRs have a morphology nearly identical to that of the protein bridges (junctional feet) that span the T-tubule–SR junctional gap in vertebrate skeletal muscles (Franzini-Armstrong and Protasi, 1997). The skeletal muscle RyR is composed of four polypeptides of $M_r \sim 560\,000$ as evidenced by (a) the four-leaf clover–like (quatrefoil) appearance of negatively stained samples (Fig. 5), (b) a high apparent sedimentation coefficient of 30S (see Fig. 4A), and (c) cross-linking studies (Lai *et al.*, 1989). Three-dimensional reconstruction of images from electron microscopy indicates that the skeletal and cardiac muscle RyRs consist of a large loosely packed cytosolic foot region and a smaller transmembrane region that extends ~ 7 nm toward the SR lumen and is thought to contain the Ca^{2+} channel pore (for review, see Wagenknecht and Radermacher, 1997).

V. Molecular Cloning and Expression of RyRs

Mammalian tissues express three types of RyRs that are encoded by three different genes. The three RyR isoforms are also known as the skeletal muscle (RyR1), cardiac muscle (RyR2), and brain (RyR3) RyR because they were first identified and isolated from skeletal muscle, cardiac muscle, and brain, respectively (Sutko and Airey, 1996; Franzini-Armstrong and Protasi, 1997; Ogawa *et al.*, 1999). Each of the large RyR polypeptides in RyR1, RyR2, and RyR3 is comprised of ~ 5000 amino acids with a predicted molecular mass of ~ 560 kDa and amino acid sequence identity of 65–70%. RyR1 and RyR2 are the predominant isoforms in skeletal muscle and cardiac muscle, respectively; however, both isoforms are also expressed in brain and other tissues at low levels. In turn, the brain RyR is expressed as a minor

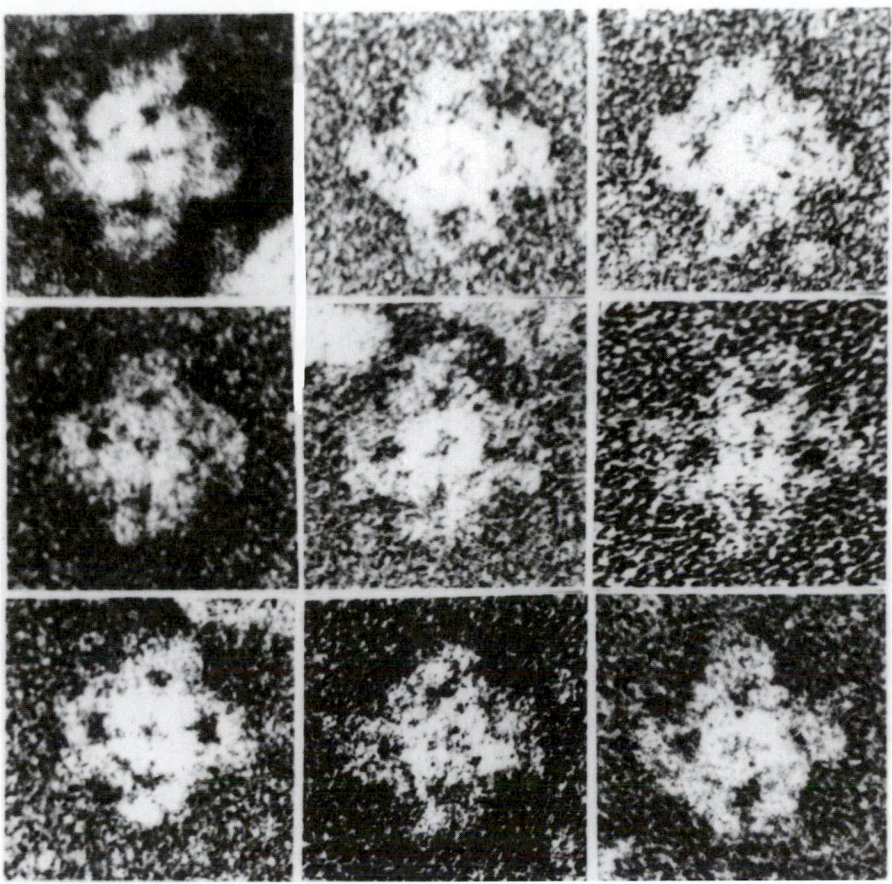

FIGURE 5.　Negative-stain electron micrograph of the purified rabbit skeletal muscle RyR. Shown is a selected panel of particles displaying the characteristic four-leaf clover (quatrefoil) structure of the 30S RyR complex. Dimensions of the quatrefoils are 34 nm from the tip of one leaf to the tip of the opposite one, with each leaf 14 nm wide. The central electron-dense region has a diameter of 14 nm with the central hole of a diameter of 1–2 nm (from Lai *et al.*, *Nature* **331**, 315–319, 1988. Copyright 1988 by Macmillan Magazines Limited).

component in skeletal and cardiac muscles. cDNAs encoding the amphibian, avian, and fish homologs of mammalian RyR1 and RyR3 have been isolated and sequenced. Isolation and sequencing of a gene for RyR in the fruit fly (*Drosophila melanogaster*) and *Caenorhabditis elegans* revealed 40–47% identity between the amino acid sequences of the *Drosophila* and *C. elegans* RyRs and the three mammalian RyRs (Sakube *et al.*, 1997).

Hydropathy plots of the amino acid sequences of RyRs have suggested that the M_r ~560 000 polypeptides have two major structural regions: (1) a C-terminal pore region that exhibits a high extent of similarity in amino acid sequence and traverses the membrane at least four times (hence 16 or more transmembrane segments per tetrameric RyR); and (2) a large, more variable extramembrane region, which is thought to correspond to the cytoplasmic foot structure. The four membrane-spanning segment model (Takeshima *et al.*, 1989) is supported by studies with SR lumenal site-directed antibodies and single-channel recordings with tryptic fragments and deletion mutants. Studies with site-directed antibodies suggest that the N- and C-termini of RyR1 are cytoplasmically localized, with the C-terminus being important for the expression of a functional RyR1 complex (Gao *et al.*, 1997). Expression studies in cul-

tured cells and *Xenopus* oocytes have suggested that the M_r 560 000 polypeptides are sufficient to form Ca^{2+} channels sensitive to ryanodine, caffeine, and Ca^{2+}.

Primary sequence analysis suggests the presence of several phosphorylation sites and sites of Ca^{2+}, ATP, and calmodulin binding on the large cytoplasmic foot structure. Experimental evidence for one phosphorylation site (Witcher *et al.*, 1991), three calmodulin binding sites (Guerrini *et al.*, 1995), and several Ca^{2+}-sensitive channel domains has been reported (see Section VII, Fig. 12). RyRs constitute a potential target for reactive nitrogen and oxygen species because they contain a large number of sulfhydryls (~100/RyR subunit) whose oxidation modulates function (for review, see Zucchi and Ronca-Testoni, 1997). In favor of a regulatory role of sulfhydryls, the cardiac channel is S-nitrosylated endogenously, and S-nitrosylation and oxidation by exogenous NO-generating molecules lead to channel activation and inactivation (rev. in Eu *et al.*, 1999).

RyR gene knockout mice were constructed to address the question of a functional requirement of the predominant (RyR1) and minor (RyR3) RyR isoforms for skeletal muscle EC coupling. Mutant mice lacking RyR1 died perinatally and the skeletal muscle fibers failed to show a contractile response to electrical stimulation under physiological conditions. RyR3 knockout mice showed an impairment of contraction in neonatal but not adult skeletal muscle (Takeshima *et al.*, 1996; Sonnleitner *et al.*, 1998). Deletion of RyR2 resulted in embryonic lethality and altered cardiomyocytes (Takeshima *et al.*, 1998).

VI. RyRs Are High-Conductance Ligand-Gated Channels

Function of the Ca^{2+} release channels has been extensively studied *in vitro* in SR vesicle Ca^{2+} flux and single-channel measurements. A third, more indirect method relies

on the use of [3H]ryanodine, a plant alkaloid that is widely used as an indicator of channel activity due to its preferential binding to open channels.

A. SR Vesicle Ca^{2+} Efflux Measurements

In skeletal and cardiac muscles, the SR releases its Ca^{2+} stores in milliseconds in response to an action potential. To measure similarly rapid Ca^{2+} fluxes *in vitro*, rapid mixing and filtration devices must be employed. The released Ca^{2+} can be measured using Ca^{2+} indicator dyes such as fura-2 or a radioisotope of Ca^{2+} ($^{45}Ca^{2+}$).

Figure 6 shows a typical $^{45}Ca^{2+}$ efflux experiment. A rapid mixing device (Fig. 6A) is used to measure $^{45}Ca^{2+}$ re-

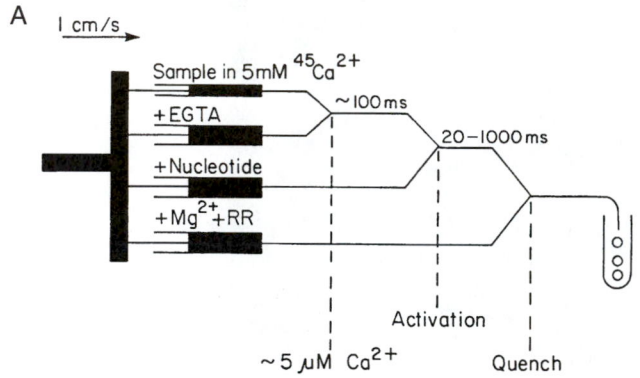

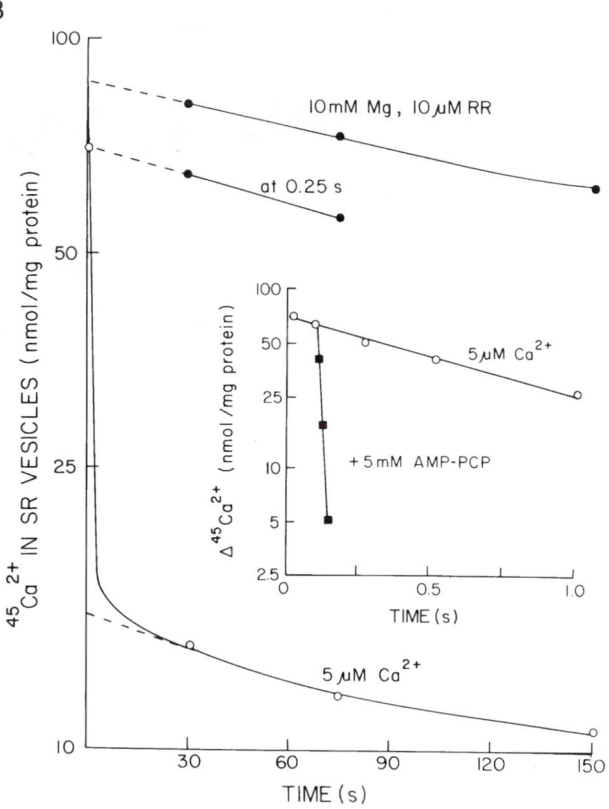

FIGURE 6. Measurement of rapid $^{45}Ca^{2+}$ efflux rates. (A) Diagrammatic representation of rapid $^{45}Ca^{2+}$ efflux quench experiments. A major part of the rapid mixing apparatus is a ram that simultaneously pushes four syringes filled with SR vesicles and three mixing solutions. In the example depicted, vesicles are passively loaded with 5 mM $^{45}Ca^{2+}$ by prolonged incubation (2 h at room temperature) before being placed in the top syringe. In the first mixing chamber, the extravesicular Ca^{2+} concentration is reduced to a more physiological concentration of 5 μM or less by dilution into a solution containing a Ca^{2+} chelating agent (EGTA). A nucleotide such as ATP or AMP-PCP can then be added in a second mixing chamber to induce very rapid $^{45}Ca^{2+}$ release. In the third mixing chamber, $^{45}Ca^{2+}$ release is inhibited by the addition of Mg^{2+} and ruthenium red (RR), which, in combination, strongly inhibit the skeletal muscle RyR. After the final mixing step, vesicles are collected, placed on a filter under vacuum, and washed. The radioactivity remaining with the vesicles is determined by liquid scintillation counting. (B) Time course of $^{45}Ca^{2+}$ efflux from heavy rabbit skeletal muscle SR vesicles that were passively loaded with 5 mM $^{45}Ca^{2+}$ by incubation at 23 °C for 2 h and then diluted into isosmolal, unlabeled release media. Rapid $^{45}Ca^{2+}$ efflux was inhibited by the addition of a quench solution (10 mM Mg^{2+} and 10 μM RR). $^{45}Ca^{2+}$ remaining with the vesicles was determined by placing them on a filter and washing with the quench solution. Amounts of $^{45}Ca^{2+}$ initially trapped by all vesicles (87 nmol/mg protein) and not readily released by the subpopulation of vesicles lacking the Ca^{2+} release channel (15 nmol/mg protein) were estimated by extrapolating back to the time of vesicle dilution (broken lines). Inset: The time course of $^{45}Ca^{2+}$ efflux from the vesicle population containing the Ca^{2+} release channel was obtained by subtracting the amount not readily released (15 nmol/mg protein). (From Meissner, *Meth. Enzymol.* **32**, 417–437, 1988).

TABLE 1 Comparison of $^{45}Ca^{2+}$ Efflux from SR Vesicles of Rabbit Skeletal Muscle and Canine Cardiac Muscle

Free Ca²⁺ (M)	Additions to efflux medium (M)	$^{45}Ca^{2+}$ efflux rate $(t_{1/2}, s)^a$	
		Rabbit skeletal	Canine cardiac
< 10⁻⁸	- - -	8	25
	5 × 10⁻³ ATP	0.06	12
~ 10⁻⁵	- - -	0.6	0.02
	5 × 10⁻³ AMP-PCP	0.01	0.01
	2 × 10⁻⁶ calmodulin	1.2	0.1
	10⁻³ Mg²⁺	15	0.25
10⁻³	- - -	10	0.1

^{a}SR vesicles were passively loaded with 1 mM $^{45}Ca^{2+}$ and diluted into efflux media containing the indicated concentrations of free Ca²⁺ and other additions. $t_{1/2}$ indicates the time required for the Ca²⁺-permeable vesicles to release half their $^{45}Ca^{2+}$ content.

lease by a passively loaded vesicle fraction derived from the junctional SR of rabbit skeletal muscle. $^{45}Ca^{2+}$ efflux is slow when the vesicles are directly diluted into a medium containing two inhibitors (Mg²⁺ and ruthenium red) of the RyR/Ca²⁺ release channel (Fig. 6B). This allows the determination of the amounts of $^{45}Ca^{2+}$ trapped by all vesicles at zero time. In release media containing micromolar concentrations of free Ca²⁺ (5 μM in Fig. 6B), the RyR Ca²⁺ channel is partially activated, resulting in the release of a majority of the trapped $^{45}Ca^{2+}$ in just a few seconds. Some radioactivity remains with the vesicles for longer times (> 30 s) because not all of them contain the RyR. In the figure inset, $^{45}Ca^{2+}$ release is stopped by the addition of Mg²⁺ and ruthenium red at varying time intervals (ranging from 25–1000 ms). In the presence of 5 μM free Ca²⁺, the vesicles release half their $^{45}Ca^{2+}$ stores within 0.7 s, corresponding to a first-order rate constant of about 1.0 s⁻¹. Figure 6B further shows that AMP-PCP (a nonhydrolyzable ATP analog) increases the initial release rate about fifty-fold, resulting in nearly complete release in 50 ms.

Table 1 compares the Ca²⁺ release properties of skeletal muscle and cardiac muscle SR vesicles that were diluted into media containing different concentrations of Ca²⁺, Mg²⁺, adenine nucleotide, and calmodulin. In both membrane fractions, millimolar ATP and micromolar Ca²⁺ activate $^{45}Ca^{2+}$ release. In the presence of micromolar Ca²⁺ and millimolar AMP-PCP, skeletal and cardiac vesicles release half their $^{45}Ca^{2+}$ stores in about 10 ms. Millimolar Ca²⁺ and Mg²⁺ and micromolar calmodulin have an inhibitory effect.

Major differences in the *in vitro* regulation of the skeletal muscle and cardiac muscle Ca²⁺ release channels are observed. At 10⁻⁵ M extravesicular (cytoplasmic) Ca²⁺, the half-time of $^{45}Ca^{2+}$ release is about 20 ms for cardiac vesicles, as compared with 600 ms for skeletal vesicles (see Table 1). In contrast, adenine nucleotides are more effective

in stimulating $^{45}Ca^{2+}$ release from skeletal than from cardiac vesicles. In addition, $^{45}Ca^{2+}$ efflux from cardiac vesicles is only partially inhibited by 1 mM Ca²⁺ or 1 mM Mg²⁺, whereas the Ca²⁺-activated release channel of skeletal muscle is nearly fully inhibited.

B. Single-Channel Planar Lipid Bilayer Measurements

The RyR/Ca²⁺ release channel is a cation-selective channel that displays an unusually large ion conductance for monovalent cations (~750 pS with 250 mM K⁺ as the current carrier) and divalent cations (~150 pS with 50 mM Ca²⁺). The existence of a large Ca²⁺ conductance was originally demonstrated in single-channel recordings with native skeletal and cardiac muscle SR vesicles fused with planar lipid bilayers, and subsequently in studies with purified RyRs incorporated in lipid bilayers (Meissner, 1994). Since RyR/Ca²⁺ release channels conduct monovalent cations such as Na⁺, K⁺, or Cs⁺ more efficiently than Ca²⁺, many of the reported studies have been performed using monovalent cations rather than Ca²⁺ as the conducting ion. As illustrated in Fig. 7, the 30S RyR complex, purified in the absence of [³H]ryanodine, can be recorded in planar lipid bilayers in the presence of a symmetric 250 mM KCl solution. The Chaps-solubilized and purified RyR can be directly incorporated into a lipid bilayer; however, more reproducible results are obtained when the Chaps-solubilized, purified RyR is first reconstituted into lipid bilayer vesicles by removal of the detergent by dialysis. The lipid vesicles are then fused with a planar lipid bilayer in the presence of an asymmetric ion gradient. Further fusion is prevented by equalizing the ion concentration in both bilayer chambers. In most experiments, the channel complex incorporates into the bilayer with the cytoplasmic (foot) region of the channel facing the *cis* side of the bilayer (i.e., the side to which the complex is added) (Xu and Meissner, 1998). Regulation of the channel can therefore be conveniently studied by varying the Ca²⁺, Mg²⁺, or ATP concentration in the *cis* chamber of the bilayer

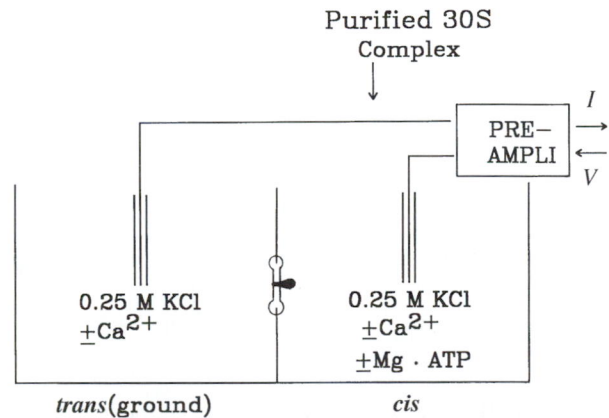

FIGURE 7. Diagrammatic representation of planar lipid bilayer apparatus used for recording purified 30S Ca²⁺ release channel complexes. The *trans* and *cis* sides of the planar lipid bilayer are equivalent to the lumenal and cytoplasmic sides of the SR membrane, respectively.

apparatus. If desired, a Ca^{2+} conducting channel can be obtained by perfusing the *trans* chamber with a $CaCl_2$ buffer. Cl^- can be used in these studies because the channel does not conduct anions.

In Fig. 8, the purified skeletal muscle and cardiac muscle RyR/Ca^{2+} release channels were recorded in planar lipid bilayers in a symmetrical KCl medium. The single-channel traces show that the activity of both channels is dependent on the Ca^{2+} concentration at the *cis* (cytoplasmic) side of the bilayer. The activity of the skeletal muscle channel is increased by increasing the *cis* Ca^{2+} concentration from ~1 µM to ~10 µM. Further increase in *cis* Ca^{2+} to 1 mM inhibits channel openings. Cardiac channels are activated to a greater extent than skeletal channels when Ca^{2+} is the sole activating ligand and require higher Ca^{2+} concentrations for inhibition of channel activity than skeletal channels. Single-channel recordings such as those shown in Fig. 8, as well as vesicle-ion flux and $[^3H]$ryanodine-binding measurements, have suggested that the mammalian skeletal and cardiac Ca^{2+} release channels possess cytosolic high-affinity activating and low-affinity inhibitory Ca^{2+} binding sites.

In addition to cytosolic Ca^{2+}, SR lumenal Ca^{2+} regulates the skeletal and cardiac muscle RyRs at two potential channel sites: Ca^{2+} binds to lumenal channel sites (for review, see Sitsapesan and Williams, 1997) or accesses cytosolic Ca^{2+} activation and inactivation sites following ion flux to the cytosolic receptor side (Xu and Meissner, 1998). At present, the reasons for the different results are not clear but may reflect a predominance of one of the

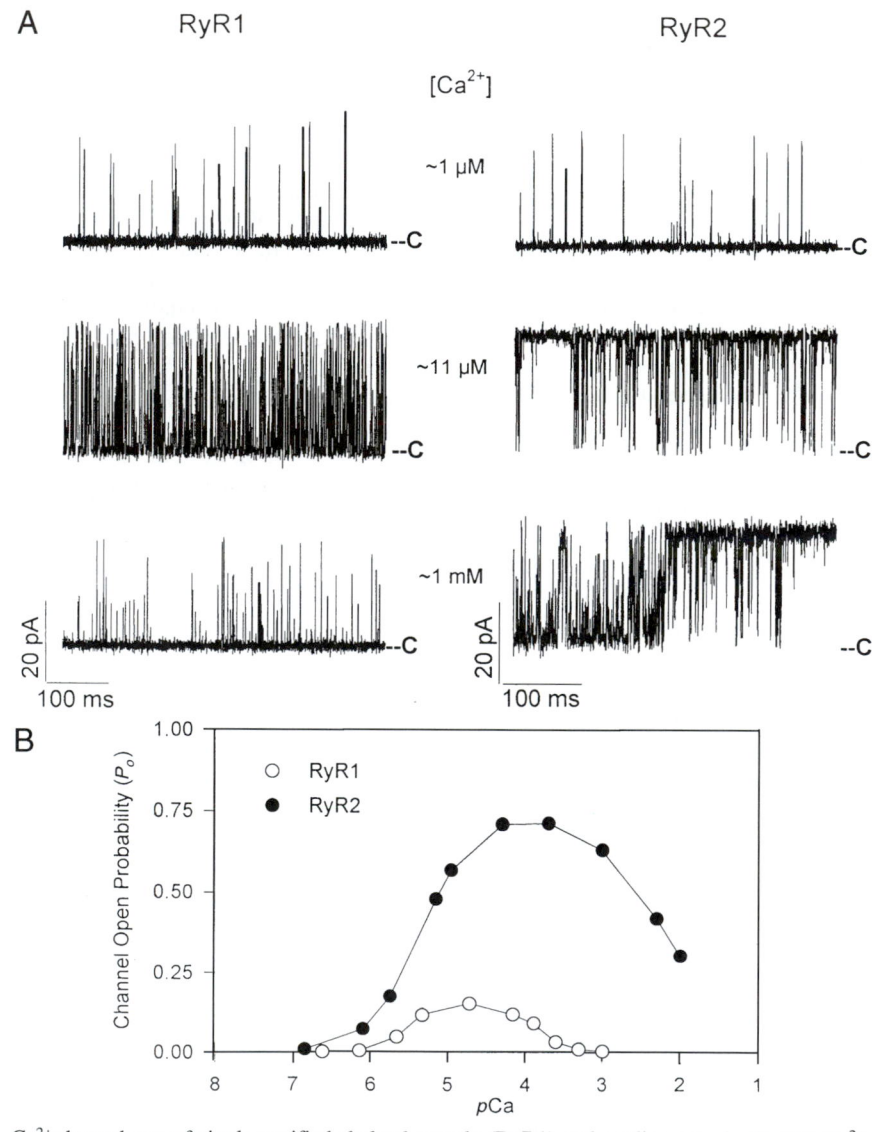

FIGURE 8. Ca^{2+} dependence of single purified skeletal muscle (RyR1) and cardiac muscle (RyR2) Ca^{2+} release channels. (A) Single-channel currents were recorded at +40 mV (RyR1) and +35 mV (RyR2) in symmetrical 250 mM KCL, 20 mM KPipes, pH 7.35 media containing the indicated concentrations of free cytosolic Ca^{2+}. *Trans* (SR lumenal) free Ca^{2+} concentration was ≤ 10 µM. Bars on the right (—) represent the closed (C) channels. (B) Channel open probability (P_o) as a function of cytosolic free Ca^{2+} concentration. (From Xu *et al., Ann. NY Acad. Sci.* **853,** 130–148, 1998.)

two mechanisms, depending on the experimental conditions. Other possible Ca^{2+}-dependent mechanisms include channel adaption (Gyorke and Fill, 1993), SR Ca^{2+} depletion (Baylor and Hollingworth, 1988), and regulation of RyR by SR lumenal (calsequestrin, Szegedi *et al.*, 1999) and cytosolic (calmodulin, Tripathy *et al.*, 1995) binding proteins.

C. Effects of Ryanodine and Caffeine

Among the large number of drugs that have been identified to affect SR Ca^{2+} release (Zucchi and Ronca-Testoni, 1997), ryanodine and caffeine are two of the best known because of their extensive use in the assessment of SR function by controlling cytoplasmic Ca^{2+} concentrations. Measurement of $^{45}Ca^{2+}$ flux in SR vesicles shows that ryanodine has a dual effect in that nanomolar concentrations lock the channel into an open configuration, whereas at concentrations above 10 μM, ryanodine completely closes the channel (Fig. 9). Single-channel recordings allow researchers to directly study the functional consequences of the interaction of ryanodine with its receptor. In turn, in the planar lipid bilayer measurements, the channel's highly characteristic modification by ryanodine provides a reliable means of distinguishing between the RyR/Ca^{2+} release channel and other types of ion channel currents. Figure 10A shows that several minutes after the addition of micromolar concentrations of ryanodine, the channel enters into a subconductance state with a channel open probability close to unity. In this study, a relatively high ryanodine concentration of 30 μM was used to reduce the time required to observe the otherwise very slow binding of ryanodine to the channel. Upon the addition of mM *cis*-ryanodine, the chan-

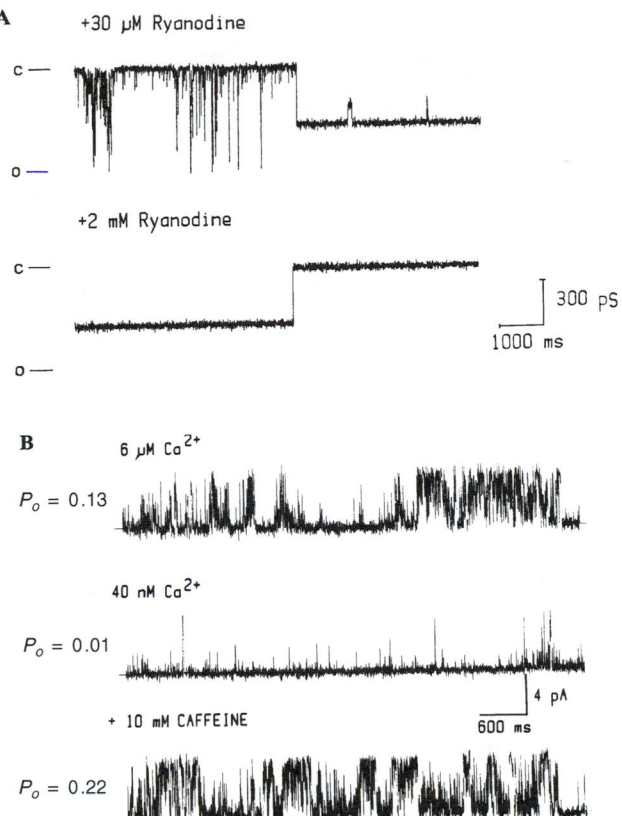

FIGURE 10. Effects of ryanodine and caffeine on single skeletal muscle RyRs. 30S purified channel complexes were reconstituted in their soluble form into planar lipid bilayers (see Fig. 7). (A) The upper trace shows the appearance of a subconducting channel state with an open probability (P_o) of ~1, following several minutes after the addition of 30 μM *cis* ryanodine. An additional, infrequent substate is also observed. The lower trace illustrates the sudden transition from the subconductance state to a fully closed state 1 min after addition of 2 mM *cis* (cytoplasmic) ryanodine. The bars on the left represent the closed (c) channel. (From Lai *et al.*, *J. Biol. Chem.* **264,** 16 776–16 785, 1989.) (B) Single-channel activity with 6 μM free *cis* Ca^{2+} (upper trace), after *cis* free Ca^{2+} was decreased to 40 nM by the addition of a Ca^{2+} buffer (EGTA) (middle trace), and after the addition of 10 mM caffeine to the 40 nM free *cis* Ca^{2+} medium (bottom trace). In contrast to ryanodine, caffeine activates the channel without changing its conductance. (From Rousseau *et al.*, *Arch. Biochem. Biophys.* **267,** 75–86, 1988.)

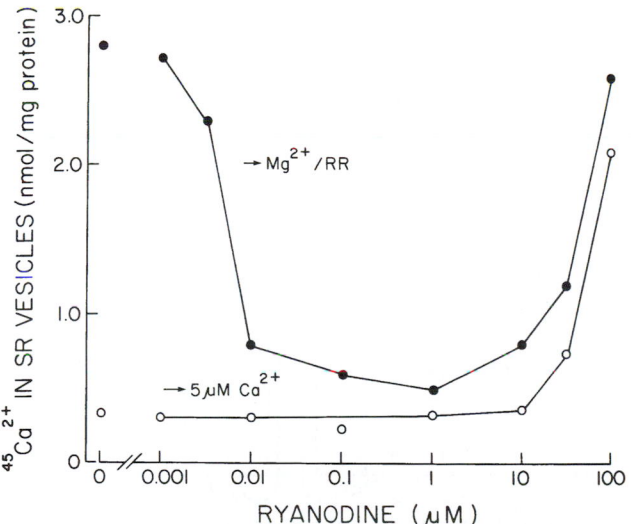

FIGURE 9. Dependence of $^{45}Ca^{2+}$ permeability of SR vesicles on ryanodine concentration. Heavy rabbit skeletal muscle SR vesicles were incubated for 45 min at 37 °C with 0.1 mM $^{45}Ca^{2+}$ and the indicated concentrations of ryanodine. Amounts of $^{45}Ca^{2+}$ retained by the vesicles in Ca^{2+} release channel–inhibiting (10 mM Mg^{2+}, 10 μM ruthenium red (RR); •) and –activating (5 μM free Ca^{2+}; o) media are shown. (From Meissner, *J. Biol. Chem.* **261,** 6300–6306, 1986.)

nel's subconductance state generally disappears and the channel enters into a fully closed state. A characteristic property of ryanodine-modified channel states is their insensitivity to regulation by Ca^{2+}, Mg^{2+}, and ATP, all of which greatly affect the gating behavior of the unmodified channel. [³H]Ryanodine-binding studies provide independent evidence for a complex interaction of the drug with the Ca^{2+} release channel (Meissner, 1994; Fig. 11). The existence of multiple interacting high- and low-affinity sites (Lai *et al.*, 1989), as well as the dependence of [³H]ryanodine binding on Ca^{2+}, Mg^{2+}, ATP and monovalent cations and anions has been described (Liu *et al.*, 1998). As a general rule, conditions that open the channel such as the presence of μM Ca^{2+}, mM ATP, or high ionic strength, have

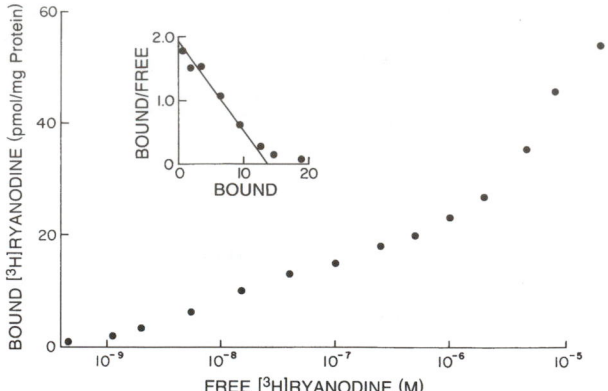

FIGURE 11. High- and low-affinity [³H]ryanodine binding to heavy rabbit skeletal muscle SR membranes. Bound [³H]ryanodine was determined as the difference between total [³H]ryanodine and unbound [³H]ryanodine before and after removal of SR membranes by centrifugation, respectively. Inset: Scatchard analysis reveals a curvilinear slope indicating the presence of both high-affinity ($K_D = 7$ nM) and low-affinity sites. (Adapted from Lai *et al., J. Biol. Chem.* **264,** 16 776–16 785, 1989).

been found to favor the interaction of ryanodine with the channel by increasing the rate of association and affinity of [³H]ryanodine binding.

In contrast to ryanodine, caffeine activates the SR Ca²⁺ release channel by increasing channel open probability without significantly affecting single-channel conductance (see Fig. 10B). Another important difference is that caffeine activates the Ca²⁺ release channel without loss of sensitivity to regulation by Mg²⁺ and ATP. Some xanthines (pentifylline, 1,7-dimethylxanthine) are more potent than caffeine (1,3,7-trimethylxanthine) in activating the skeletal RyR (Liu and Meissner, 1997; Xu *et al.,* 1998).

VII. Identification of Functional Regions of Skeletal Muscle RyR

The isolation and primary structure determination as well as functional expression of the three mammalian RyR isoforms have opened the way to a systematic study of the molecular properties of a key component in Ca²⁺ signaling in cells. The relationship between RyR protein structure and function has been probed by preparing bacterial fusion proteins and small synthetic peptides with sequences from the DHPRs and RyRs. Chimeric constructs single-site and deletion mutants have been expressed in myotubes lacking the α_1-subunit of the DHPR (**dysgenic** myotubes) or RyR1 (**dyspedic** myotubes) and in mammalian cell culture systems (human embryo kidney (HEK) cells, Chinese hamster ovary (CHO) cells). These studies have yielded preliminary information on the location of amino acids involved in the formation of the conductance pathway and the regulation of the skeletal muscle RyR by the DHPR and Ca²⁺ (Fig. 12).

A. RyR Conductance Pathway

The lumenal loop region between the two most C-terminal membrane segments of RyRs (M3 and M4; see Fig. 12) has sequence similarities to segments of the related inositol 1,4,5-trisphosphate receptors (IP₃Rs) and an unrelated class of ion channels, the voltage-gated K⁺ channels. A recent x-ray analysis of the structure of the K⁺ channel from *Streptomyces lividans* detailed the structure of the pore region (Doyle *et al.,* 1998). An important finding was the identification of a conserved VGYG motif that comprised the ion selectivity filter. A sequence related to the K⁺ channel VGYG motif is GGIG in the lumenal loop between transmembrane regions 3 and 4 of RyRs. Conservation of the VGYG motif in the K⁺ channels and the related GGIG motif in all isoforms of the RyR and GGI(V)G motif in the related inositol trisphosphate receptors implies that the lumenal loop between transmembrane regions 3 and 4 of RyRs may fold back into the membrane (Balshaw *et al.,* 1999), with GIG (aa 4896–4898 in RyR1) in the membrane as illustrated in Fig. 12. In support of the model is that the mutation G4824A in the cardiac RyR (equivalent to G4894A in RyR1) in the lumenal loop between transmembrane regions 3 and 4 alters channel conductance without affecting channel activity (Zhao *et al.,* 1999). Single-site mutations of amino acid residues flanking the GIG motif that are highly conserved among the RyR and IP₃R families result in loss of high-affinity [³H]ryanodine binding and an altered ion conductance, providing additional support that the M3–M4 loop is important in determining channel function (Gao *et al.,* 2000).

B. DHPR-RyR Interactions in Skeletal Muscle

Different coupling mechanisms exist in cardiac muscle and skeletal muscle. As discussed earlier, a unique property

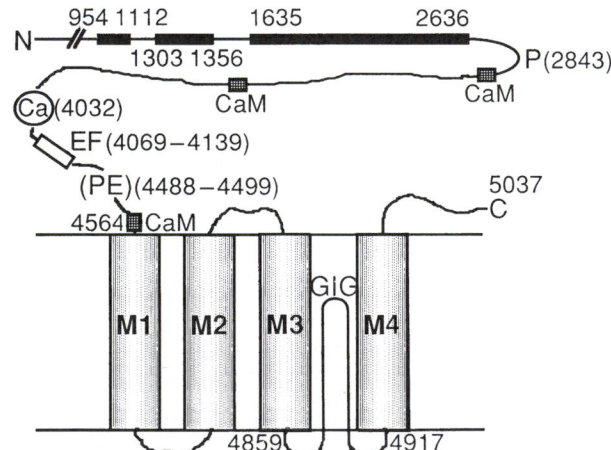

FIGURE 12. Working model of RyR1. The model proposes that the lumenal loop between transmembrane regions 3 and 4 forms part of the ion conductance pathway of RyR1. Also shown are three regions (amino acid residues 954–1112, 1303–1356, and 1635–2636) reported to interact with DHPR, a phosphorylation (P) site (S2843), three calmodulin (CaM)-binding sites (3042–3058, 3617–3634, 4540–4557), and three putative Ca²⁺ regulatory sites (E4032, 4069–4139, and 4488–4499). For references see text.

of the mammalian skeletal muscle RyR is its direct linkage to the skeletal muscle DHPR/Ca^{2+} channel isoform, which acts in EC coupling as a voltage-sensing molecule rather than as a Ca^{2+} channel (see Fig. 2). Compelling evidence for the different function of the DHPR in cardiac and skeletal muscle EC coupling—as a Ca^{2+} channel in cardiac muscle and as a voltage-sensing molecule in skeletal muscle—has been obtained through the elegant work of Tanabe and Beam. These investigators used an animal model lacking the skeletal muscle DHPR α_1-subunit to show that one of the putative cytoplasmic loop regions (II–III region) of the α_1-subunit plays a major role in determining the type of EC coupling that exists in striated muscle (Tanabe et $al.$, 1990). Lu and colleagues (1994, 1995) obtained evidence for a direct functional interaction between the DHPR and RyR by showing that a peptide expressed in $E.$ $coli$ and derived from the II–III loop region of the skeletal muscle DHPR α_1-subunit activates the purified skeletal muscle RyR/Ca^{2+} release channel in planar lipid bilayers (Fig. 13). In Fig. 13A, a single skeletal muscle RyR channel was recorded in the presence of a suboptimally activating Ca^{2+} concentration in the cis (cytoplasmic) chamber. Fig. 13B shows that addition

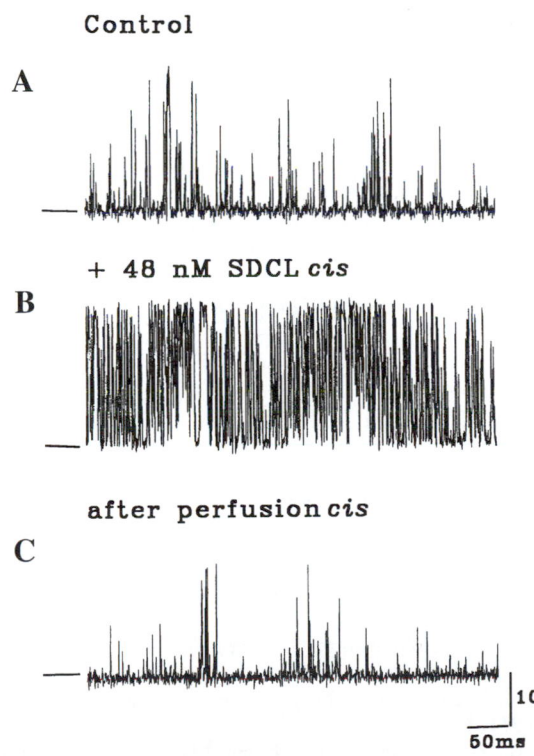

Control

A

+ 48 nM SDCL cis

B

after perfusion cis

C

10pA

50ms

FIGURE 13. Effect of skeletal muscle DHPR α_1-subunit II–III loop peptide on single-channel activity of purified skeletal muscle Ca^{2+} release channel. Proteoliposomes containing the purified 30S channel complex were fused with a planar lipid bilayer (see Fig. 7). Single-channel currents, shown as upward deflections, were recorded in symmetric 0.25 M KCl media with 6 μM and 50 μM free Ca^{2+} in the cis and $trans$ chambers, respectively. (A) Control, P_o = 0.04. (B) 3 min after the addition of 48 nM peptide to the cis chamber, P_o = 0.71. (C) After perfusion of cis chamber with 0.25 M KCl buffer containing 6 μM free Ca^{2+}, P_o = 0.003. Holding potential = 30 mV. (From Lu et $al.$, $J.$ $Biol.$ $Chem.$ **269**, 6511–6516, 1994.)

of the peptide to the cis chamber resulted in an increase in the activity of the RyR channel. Removal of the peptide by perfusion (Fig. 13C) decreased single-channel activity to the control level, showing that the activation of the channel by the peptide was reversible. Additional studies showed that a peptide derived from the putative II–III loop region of the cardiac DHPR α_1-subunit is as effective as the skeletal peptide in activating the skeletal muscle RyR/Ca^{2+} release channel. Both the skeletal muscle and cardiac muscle peptide bind to the skeletal muscle RyR, but neither peptide is able to bind or activate the cardiac RyR. These results imply that, in addition to the DHPR, the RyR has an important role in determining the EC coupling mode in muscle.

Progress has been made in identifying the regions in the II–III cytoplasmic loop of the DHPR that interact with RyR1. Microinjection of skeletal muscle and cardiac DHPR chimeric cDNAs into dysgenic myotubes indicates that a 42-amino-acid region is a major determinant of EC coupling (skeletal or cardiac coupling) (Nakai et $al.$, 1998b). A 10-amino-acid region in the II–III cytoplasmic loop region interacts with RyR1 (El-Hayek and Ikemoto, 1998); however, this region is not critical for skeletal muscle EC coupling in the expression studies with dysgenic myotubes. Two possible explanations are that some regions in the II–III cytoplasmic loop of the DHPR have primarily a structural role, or that because of sequence similarity, skeletal/cardiac chimeras fail to detect these regions as being specific for a functional interaction between the DHPR and RyR1. The C-terminal sequence of the DHPR α_1-subunit (Slavik et $al.$, 1997) and the β-subunit of the DHPR (Gregg et $al.$, 1996) have been reported to contribute to the functional coupling of the two receptors.

Similar approaches have been used in identifying regions of RyR1 that interact with the skeletal muscle DHPR. Expression of deletion and chimeric RyRs in myotubes lacking RyR1 (dyspedic myotubes) indicates that residues 1303–1356 (Yamazawa et $al.$, 1997) and 1635–2636 (Nakai et $al.$, 1998a) interact with DHPR. The use of RyR1-derived peptides (aa 954–1112) suggests additional interactions between the skeletal muscle RyR and DHPR (Leong and MacLennan, 1998).

C. Ca²⁺ Binding Sites

RyR1 is activated by micromolar Ca^{2+} and inhibited by millimolar Ca^{2+} in the absence of other channel effectors. It is likely that this biphasic behavior is a consequence of at least two classes of Ca^{2+} binding sites: a high-affinity activation site and a low-affinity inactivation site. Three lines of evidence suggest that the C-terminal one-third of RyR1 has a critical role in Ca^{2+} activation. First, antibodies raised against a negatively charged sequence (residues 4489–4499: PEPEPEPEPEPE) (see Fig. 12) blocked Ca^{2+}-dependent activation of the channel (Chen et $al.$, 1993). Second, the mutation E3885A in RyR3 (equivalent to E4032A in RyR1) forms a functional channel with normal conductance but with greatly decreased Ca^{2+} sensitivity (Chen et $al.$, 1998). Third, a deletion mutant encoding the C-terminal 1377 amino acids of RyR1 (3661–5037) could be activated by μM Ca^{2+} (Bhat et $al.$, 1997).

TABLE 2 Properties of Mammalian RyR/Ca^{2+} Release Channel Complexes

Property	Skeletal muscle (RyR1)	Cardiac muscle (RyR2)	Brain (RyR3)
Apparent sedimentation coefficient	30S	30S	30S
Morphology	Quatrefoil	Quatrefoil	Quatrefoil
High-affinity ryanodine binding (K_D)	< 10 nM	< 10 nM	< 10 nM
Single-channel conductance (γ_{max})			
in 0.25 M KCl	~775 pS	~775 pS	~775 pS
in 0.05 M Ca^{2+} (with 0.25 M KCl *cis*)	~145 pS	~145 pS	~135 pS
Regulation			
activation by Ca^{2+}	µM	µM	µM
activation by mM ATP and			
mM caffeine	Yes	Yes	Yes
inhibition by mM Ca^{2+} or Mg^{2+}	Yes	Yes	Yes
Modification by ryanodine			
to subconductance state	Yes	Yes	Yes
to closed state	Yes	Yes	—–

Less is known about the location of the Ca^{2+} inactivation site(s). Unlike the full-length RyR1, truncated RyR1 ($\Delta 1$–3660) failed to close at high Ca^{2+} concentrations, which suggests that the N-terminal foot structure has a role in Ca^{2+} regulation (Bhat *et al.*, 1997). Consistent with this finding, single amino acid mutations in the N-terminal and central regions of RyR1 link to a rare muscle disorder known as malignant hyperthermia, which is characterized by elevated Ca^{2+} release from the SR (Phillips *et al.*, 1996). At variance with the above results, chimeric constructs with a RyR2 C-terminus showed a reduced channel inactivation at elevated Ca^{2+} concentrations, comparable to that of RyR2 (Du and MacLennan, 1999; Nakai *et al.*, 1999). The reasons for the different results are not clear but they may be due to limitations inherent in using deletion mutants and chimeric constructs. The RyR is a cooperatively coupled tetramer. Therefore, one possibility is that deletion of a RyR region or replacement with a region from another RyR isoform does not directly affect the Ca^{2+} inactivation site(s). Rather, the properties of the RyR may be altered through global conformational changes, accounting for the loss of Ca^{2+} inactivation.

VIII. Summary

The identification of [^{3}H]ryanodine as a specific probe has enabled the isolation and subsequent cloning of a family of large intracellular channel complexes composed of four ~560-kDa polypeptides. The mammalian RyRs are encoded by three different genes and show a tissue-specific expression. They share several properties including an apparent sedimentation coefficient of 30S, the presence of a high-affinity ryanodine binding site, and a high-conductance mono- and divalent cation pathway that can be modified by ryanodine in an essentially identical manner (Table 2).

RyR/Ca^{2+} release channels constitute a rich target for controlling cellular functions. A large number of endogenous effectors regulate RyR function, including Ca^{2+}, Mg^{2+}, ATP, and calmodulin. RyRs also contain phosphorylation sites and reactive thiols, which suggests that receptor phosphorylation and reactive nitrogen and oxygen species have a role in the *in vivo* regulation of channel activity.

Acknowledgments

I would like to thank Xiangyang Lu for providing Figures 1 and 2 and Daniel Pasek for Figure 12. Support from United States Public Health Service grants AR18687 and HL27430 is gratefully acknowledged.

Bibliography

Balshaw, D., Gao, L., and Meissner, G. (1999). Commentary: lumenal loop of the ryanodine receptor: a pore-forming segment? *Proc. Natl. Acad. Sci. USA* **96**, 3345–3347.

Baylor, S. M., and Hollingworth, S. (1988). Fura-2 calcium transients in frog skeletal muscle fibers. *J. Physiol.* **403**, 151–192.

Berridge, M. J. (1993). Inositol tris-phosphate and calcium signalling. *Nature* **361**, 315–325.

Bhat, M. B., Zhao, J., Takeshima, H., and Ma, J. (1997). Functional calcium release channel formed by the carboxy-terminal portion of ryanodine receptor. *Biophys. J.* **73**, 1329–1336.

Catterall, W. A. (1995). Structure and function of voltage-gated ion channels. *Annu. Rev. Biochem.* **64**, 493–531.

Chen, S. W. R., Ebisawa, K., Li, X., and Zhang, L. (1998). Molecular identification of ryanodine receptor Ca^{2+} sensor. *J. Biol. Chem.* **273**, 14 675–14 678.

Chen, S. R. W., Zhang, L., and MacLennan, D. H. (1993). Antibodies as probes for Ca^{2+} activation sites in the Ca^{2+} release channel (ryanodine receptor) of rabbit skeletal muscle sarcoplasmic reticulum. *J. Biol. Chem.* **268**, 13 414–13 421.

Doyle, D. A., Cabral, J. M., Pfuetzner, R. A., Kuo, A., Gulbis, J. M., Cohen, S. L., Chait, B. T., and MacKinnon, R. (1998). The structure of the potassium channel: molecular basis of K⁺ conduction and selectivity. *Science* **280,** 69–76.

Du, G. G., and MacLennan, D. H. (1999). Ca²⁺ inactivation sites are located in the COOH-terminal quarter of recombinant rabbit skeletal muscle Ca²⁺ release channels (ryanodine receptors). *J. Biol. Chem.* **274,** 26 120–26 126.

El-Hayek, R., and Ikemoto, N. (1998). Identification of the minimum essential region in the II–III loop of the dihydropyridine receptor α_1 subunit required for activation of skeletal muscle–type excitation-contraction coupling. *Biochemistry* **37,** 7015–7020.

Eu, J. P., Xu, L., Stamler, J. S., and Meissner, G. (1999). Regulation of ryanodine receptors by reactive nitrogen species. *Biochem. Pharmacol.* **57,** 1079–1084.

Fabiato, A. (1983). Calcium-induced release of calcium from the cardiac sarcoplasmic reticulum. *Am. J. Physiol.* **245,** C1–C14.

Fleischer, S., and Inui, M. (1989). Biochemistry and biophysics of excitation-contraction coupling. *Annu. Rev. Biophys. Biophys. Chem.* **18,** 333–364.

Franzini-Armstrong, C., and Protasi, F. (1997). Ryanodine receptors of striated muscles: a complex channel capable of multiple interactions. *Physiol. Rev.* **77,** 699–729.

Gao, L., Tripathy, A., Lu, X., Meissner, G. (1997). Evidence for a role of C-terminal amino acid residues in skeletal muscle Ca²⁺ release channel (ryanodine receptor) function. *FEBS Lett.* **412,** 223–226.

Gao, L., Balshaw, D., Xu, L., Tripathy, A., Xin, C., and Meissner, G. (2000). Evidence for a role of the lumenal M3-M4 loop in skeletal muscle Ca²⁺ release channel (ryanodine receptor) activity and conductance. *Biophys. J.* **79,** 828–840.

Gregg, R. G., Messing, A., Strube, C., Beurg, M., Moss, R., Behan, M., Sukhareva, M., Haynes, S., Powell, J. A., Coronado, R., and Powers, P. A. (1996). Absence of the beta subunit (cchb1) of the skeletal muscle dihydropyridine receptor alters expression of the alpha 1 subunit and eliminates excitation-contraction coupling. *Proc. Nat. Acad. Sci. USA* **93,** 13 961–13 966.

Guerrini, R., Menegazzi, P., Anacardio, R., Marastoni, M., Tomatis, R., Zorzato, F., and Treves, S. (1995). Calmodulin binding sites of the skeletal, cardiac and brain ryanodine receptor Ca²⁺ channels: modulation by the catalytic subunit of cAMP-dependent protein kinase? *Biochemistry* **34,** 5120–5129.

Gyorke, S., and Fill, M. (1993). Ryanodine receptor adaptation: control mechanism of Ca²⁺-induced Ca²⁺ release in heart. *Science* **260,** 807–809.

Ikemoto, N., Antoniu, B., and Meszaros, L. G. (1985). Rapid flow chemical quench studies of calcium release from isolated sarcoplasmic reticulum. *J. Biol. Chem.* **260,** 14 096–14 100.

Inui, M., Saito, A., and Fleischer, S. (1987). Isolation of the ryanodine receptor from cardiac sarcoplasmic reticulum and identity with the feet structures. *J. Biol. Chem.* **262,** 15 637–15 642.

Jenden, D. J., and Fairhurst, A. S. (1969). The pharmacology of ryanodine. *Pharmacol. Rev.* **21,** 1–25.

Lai, F. A., Erickson, H. P., Rousseau, E., Liu, Q. Y., and Meissner, G. (1988). Purification and reconstitution of the calcium release channel from skeletal muscle. *Nature* **331,** 315–319.

Lai, F. A., Misra, M., Xu, L., Smith, H. A., and Meissner, G. (1989). The ryanodine receptor–Ca²⁺ release channel complex of skeletal muscle sarcoplasmic reticulum. Evidence for a cooperatively coupled, negatively charged homotetramer. *J. Biol. Chem.* **264,** 16 776–16 785.

Leong, P., and MacLennan, D. H. (1998). The cytoplasmic loops between domains II and III and domains III and IV in the skeletal muscle dihydropyridine receptor bind to a contiguous site in the skeletal muscle ryanodine receptor. *J. Biol. Chem.* **273,** 29 958–29 964.

Liu, W., and Meissner, G. (1997). Structure-activity relationship of xanthines and skeletal muscle ryanodine receptor/Ca²⁺ release channel. *Pharmacology* **54,** 135–143.

Liu, W., Pasek, D. A., and Meissner, G. (1998). Modulation of Ca²⁺-gated cardiac muscle Ca²⁺ release channel (ryanodine receptor) by mono- and divalent ions. *Am. J. Physiol.* **274,** C120–C128.

Lu, X., Xu, L., and Meissner, G. (1994). Activation of the skeletal muscle calcium release channel by a cytoplasmic loop of the dihydropyridine receptor. *J. Biol. Chem.* **269,** 6511–6516.

Lu, X., Xu, L., and Meissner, G. (1995). Phosphorylation of dihydropyridine receptor II–III loop peptide regulates skeletal muscle calcium release channel function. Evidence for an essential role of the β-OH group of Ser⁶⁸⁷. *J. Biol. Chem.* **270,** 18 459–18 464.

Marx, S. O., Ondrias, K., and Marks, A. R. (1998). Coupled gating between individual skeletal muscle Ca²⁺ release channels (ryanodine receptors). *Science* **281,** 818–821.

Meissner, G. (1983). Monovalent ion and calcium ion fluxes in sarcoplasmic reticulum. *Mol. Cell. Biochem.* **55,** 65–82.

Meissner, G. (1988). Ionic permeability of isolated muscle sarcoplasmic reticulum and liver endoplasmic reticulum vesicles. *Meth. Enzym.* **157,** 417–437.

Meissner, G. (1994). Ryanodine receptor/Ca²⁺ release channels and their regulation by endogenous effectors. *Annu. Rev. Physiol.* **56,** 485–508.

Meissner, G., and Lu, X. (1995). Dihydropyridine receptor–ryanodine receptor interactions in skeletal muscle excitation-contraction coupling. *Biosci. Rep.* **15,** 399–408.

Meissner, G., Rousseau, E., Lai, F. A., Liu, Q. Y., Anderson, and K. A. (1988). Biochemical characterization of the Ca²⁺ release channel of skeletal and cardiac sarcoplasmic reticulum. *Mol. Cell. Biochem.* **82,** 59–65.

Nakai, J., Gao, L., Xu, L., Xin, C., Pasek, D. A., and Meissner, G. (1999). Evidence for a role of C-terminus in Ca²⁺ inactivation of skeletal muscle Ca²⁺ release channel (ryanodine receptor). *FEBS Lett.,* **459,** 154–158.

Nakai, J., Tanabe, T., Konno, T., Adams, B., and Beam, K. G. (1998b). Localization in the II–III loop of the dihydropyridine receptor of a sequence critical for excitation-contraction coupling. *J. Biol. Chem.* **273,** 24 983–24 986.

Nakai, J., Sekiguchi, N., Rando, T. A., Allen, P. D., and Beam, K. G. (1998a). Two regions of the ryanodine receptor involved in coupling with L-type Ca²⁺ channels. *J. Biol. Chem.* **273,** 13 403–13 406.

O'Brien, J., Meissner, G., and Block, B. A. (1993). The fastest contracting muscles of nonmammalian vertebrates express only one isoform of the ryanodine receptor. *Biophys. J.* **65,** 2418–2427.

Ogawa, Y., Kurebayashi, N., and Murayama, T. (1999). Ryanodine receptor isoforms in excitation-contraction coupling. *Adv. Biophys.* **36,** 27–64.

Phillips, M. S., Fujii, J., Khanna, V. K., DeLeon, S., Yokobata, K., de Jong, P. J., and MacLennan, D. H. (1996). The structural organization of the human skeletal muscle ryanodine receptor (RYR1) gene. *Genomics* **34,** 24–41.

Protasi, F., Franzini-Armstrong, C., and Allen, P. D. (1998). Role of ryanodine receptors in the assembly of calcium release units in skeletal muscle. *J. Cell Biol.* **140,** 831–842.

Rios, E., and Pizarro, G. (1991). Voltage sensor of excitation-contraction coupling in skeletal muscle. *Physiol. Rev.* **71,** 849–908.

Rousseau, E., LaDine, J., Liu, Q. Y., and Meissner, G. (1988). Activation of the Ca²⁺ release channel of skeletal muscle sarcoplasmic reticulum by caffeine and related compounds. *Arch. Biochem. Biophys.* **267,** 75–86.

Sakube, Y., Ando, H., and Kagawa, H. (1997). An abnormal ketamine response in mutants defective in the ryanodine receptor gene *ryr1* (unc-68) of *Caenorhabditis elegans*. *J. Mol. Biol.* **267,** 849–864.

Santana, L. F., Cheng, H., Gomez, A. M., Cannell, M. B., and Lederer, W. J. (1996). Relation between the sarcolemmal Ca²⁺ current and Ca²⁺ sparks and local control theories for cardiac excitation-contraction coupling. *Circ. Res.* **78,** 166–171.

Sitsapesan, R., and Williams, A. J. (1997). Regulation of current flow through ryanodine receptors by luminal Ca^{2+}. *J. Membrane Biol.* **159**, 179–185.

Slavik, K. J., Wang, J. P., Aghdasi, B., Zhang, J. Z., Mandel, F., Malouf, N., and Hamilton, S. L. (1997). A carboxy-terminal peptide of the alpha1-subunit of the dihydropyridine receptor inhibits Ca^{2+}-release channels. *Am. J. Physiol.* **272**, C1475–C1481.

Smith, J. S., Imagawa, T., Ma, J., Fill, M., Campbell, K. P., and Coronado, R. (1988). Purified ryanodine receptor from rabbit skeletal muscle is the calcium-release channel of sarcoplasmic reticulum. *J. Gen. Physiol.* **92**, 1–26.

Sonnleitner, A., Conti, A., Bertocchini, F., Schindler, H., and Sorrentino, V. (1998). Functional properties of the ryanodine receptor type 3 (RyR3) Ca^{2+} release channel. *EMBO J.* **17**, 2790–2798.

Sutko, J. L., and Airey, J. A. (1996). Ryanodine receptor Ca^{2+} release channels: does diversity in form equal diversity in function? *Physiol. Rev.* **76**, 1027–1071.

Szegedi, C., Sarkozi, S., Herzog, A., Jona, I., and Varsanyi, M. (1999). Calsequestrin: more than "only" a luminal Ca^{2+} buffer inside the sarcoplasmic reticulum. *Biochem. J.* **337**, 19–22.

Takeshima, H., Ikemoto, T., Nishi, M., Nishiyama, N., Shimuta, M., Sugitani, Y., Kuno, J., Saito, I., Saito, H., Endo, M., Iino, M., and Noda, T. (1996). Generation and characterization of mutant mice lacking ryanodine receptor type 3. *J. Biol. Chem.* **271**, 19 649–19 652.

Takeshima, H., Komazaki, S., Hirose, K., Nishi, M., Noda, T., and Iino, M. (1998). Embryonic lethality and abnormal cardiac myocytes in mice lacking ryanodine receptor type 2. *EMBO J.* **17**, 3309–3316.

Takeshima, H., Nishimura, S., Matsumoto, T., Ishida, H., Kangawa, K., Minamino, N., Matsuo, H., Ueda, M., Hanaoka, M., Hirose, T., and Numa, S. (1989). Primary structure and expression from complementary DNA of skeletal muscle ryanodine receptor. *Nature* **339**, 439–445.

Tanabe, T., Beam, K. G., Adams, B. A., Niidome T., and Numa, S. (1990). Regions of the skeletal muscle dihydropyridine receptor critical for excitation-contraction coupling. *Nature* **346**, 567–569.

Timerman, A. P., Onoue H., Xin, H. B., Barg, S., Copello, J., Wiederrecht, G., and Fleischer, S. (1996). Selective binding of FKBP12.6 by the cardiac ryanodine receptor. *J. Biol. Chem.* **271**, 20 385–20 391.

Tripathy, A., Xu, L., Mann, G., and Meissner, G. (1995). Calmodulin activation and inhibition of skeletal muscle Ca^{2+} release channel (ryanodine receptor). *Biophys. J.* **69**, 106–119.

Xu, L., and Meissner, G. (1998). Regulation of cardiac muscle Ca^{2+} release channel by sarcoplasmic reticulum lumenal Ca^{2+}. *Biophys. J.* **75**, 2302–2312.

Xu, L., Tripathy, A., Pasek, D. A., and Meissner, G. (1998). Potential for pharmacology of ryanodine receptor/calcium release channels. *Ann. NY Acad. Sci.* **853**, 130–148.

Yamazawa, T., Takeshima, H., Shimuta, M., and Iino, M. (1997). A region of the ryanodine receptor critical for excitation-contraction coupling in skeletal muscle. *J. Biol. Chem.* **272**, 8161–8164.

Wagenknecht, T., and Radermacher, M. (1997). Ryanodine receptors: structure and macromolecular interactions. *Curr. Opin. Struct. Biol.* **7**, 258–265.

Witcher, D. R., Kovacs, R. J., Schulman, H., Cefali, D. C., and Jones, L. R. (1991). Unique phosphorylation site on the cardiac ryanodine receptor regulates calcium channel activity. *J. Biol. Chem.* **266**, 11 144–11 152.

Zhao, M., Li, P., Li, X., Zhang, L., Winkfein, R. J., and Chen, W. S. R. (1999). Molecular identification of the ryanodine receptor pore-forming segment. *J. Biol. Chem.* **274**, 25 971–25 974.

Zucchi, R., and Ronca-Testoni, S. (1997). The sarcoplasmic reticulum Ca^{2+} channel/ryanodine receptor: modulation by endogenous effectors, drugs and disease states. *Pharm. Rev.* **49**, 1–51.

Richard J. Paul

56

Contraction of Muscles

I. Introduction

The generation of force and movement by muscle is an area of physiology and biophysics that has fascinated scientist and layman alike since the dawn of scientific inquiry. The history of the study of muscle is elegantly chronicled by Dorothy Needham in *Machina Carnis* (1971). Because of its highly organized and repeating structure, skeletal muscle has proven more amenable to structural analysis (such as X-ray diffraction) than most biological tissues. Thus it has served as a paradigm for unraveling relationships between function and structure. This has become particularly exciting in conjunction with the techniques of molecular biology, which offer the potential for altering particular amino acids and molecular manipulators, such as laser tweezers. These tools offer the potential for directly testing the links between structure at nanometer resolution and function. The focus of this chapter is on the nature of the mechanochemical energy conversion. Muscle is one of the most efficient energy converters known, and studies in this area couple classical enzyme kinetics and muscle mechanics.

Although we will focus on the molecular and cellular level, one should be aware of the functions and consequences of muscle activity at the organ and whole-animal levels. Many important effects at the whole-organism level are related simply to muscle mass. Skeletal muscle constitutes approximately 40% of human body mass. If we include cardiac and smooth muscle, the total muscle mass reaches the 50% level. One consequence of this is that approximately 30% of basal metabolism is related to muscle and as much as 90% of a person's total metabolism during strenuous exercise can be related to meeting the energy requirements of muscle. This chemical activity can in turn produce a significant heat load for the organism. Other functions that are associated with the large muscle mass are the storage and mobilization of metabolites (primarily glucose and amino acids). Also, there are significant consequences to alterations in muscle electrolyte metabolism, since it is a major storage site for ions such as H^+, K^+, and Mg^{2+}. Some of these ramifications, as related to molecular processes in

muscle, are presented. The major emphasis of this chapter, however, is on the relationships between muscle structure and function at the cellular and subcellular levels. (Other nonmuscle motile systems relating to cilia and flagella are covered in the following two chapters.)

The study of the relationships between structure and function can be divided into two primary areas. The first area involves investigation of mechanisms underlying the generation of macroscopic force and shortening. We first consider how force is developed and how the mechanical behavior of muscle is quantitated, a field known collectively as *muscle mechanics*. This is then integrated with the current picture of muscle structure as a first step in constructing theories of muscle function. Next, we consider the thermodynamic rules governing energy conversion and the constraints they place on models proposed for muscle contraction. We relate energetics and mechanics to the kinetics of the myosin ATPase as a basis for crossbridge cycling mechanisms. Muscle metabolism and the matching of adenosine triphosphate (ATP) demand with ATP synthesis will complete the picture of mechanochemical energy conversion.

The second major area involves study of the mechanisms underlying the regulation of muscle contraction. The control of intracellular Ca^{2+} concentration, a key intracellular messenger, forms an area of study known as *excitation-contraction coupling* and is the focus of Chapters 54 and 55. The intracellular receptors transducing the Ca^{2+} signal reflect the great diversity of types of muscle that have evolved in response to a wide variety of functional needs. These muscles also have much in common. All muscle contains the proteins actin and myosin, which are the locus of the mechanochemical energy conversion. Chemical energy in the form of ATP hydrolysis is the immediate driving reaction for all muscle energy transduction. Another feature common to all muscle types is that calcium ions, at micromolar concentrations, are the primary second messenger in the regulatory mechanisms. We first focus on these common aspects, using a generalized striated muscle as the model. Then, with an understanding of these common mechanisms, we consider the different muscle types.

II. The Mechanisms of Force Production and Shortening: Muscle Mechanics

Studies of the mechanical behavior of muscle have played a central role in our understanding of muscle and have also formed an integral part of the language of muscle physiology (Hill, 1965). These studies before the late 1950s were primarily phenomenological, though they were also important in characterization of muscle performance (Jewel and Wilkie, 1958). Such studies remain important in characterization of muscle myopathies and in current mechanistic studies (for example, in describing the functional consequences of changing muscle protein isoforms in transgenic animals).

There are two arbitrary but natural divisions in studies of mechanics. The first division involves **steady-state** relationships. This information was crucial to the development of the sliding-filament theory and was extensively investigated in the 1960s (Gordon *et al.*, 1966). Studies of **transients** form a second division. They assumed a more central importance during the early 1970s (Huxley and Simmons, 1971) with the growing realization that information at the level of individual crossbridges could be gained from such mechanical studies on single fibers. These studies of muscle responses to rapid changes in mechanical constraints are even more valuable to unraveling crossbridge behavior when coupled with the recently improved temporal resolution of X-ray diffraction of muscle (Huxley, 1996).

A. Steady-State Relations between Force and Length and the Sliding Filament Theory

A first step in muscle mechanics involves a description of muscle behavior in terms of relationships between the mechanical variables of force and length. Apparatus for transduction and recording of these variables have evolved considerably over this century, but the historical apparatus are still responsible for much of the language of muscle physiology. Since it was easier to control force than length, by hanging a fixed load on the muscle, terms such as **preload** and **afterload** entered this vocabulary (see later section on force-velocity relationships). However, in view of what is now known about structure, it is conceptually easier to use length as the independent variable.

Figure 1 shows a schematic of an experimental apparatus for measurement of muscle force-length relationships. In this setup, length is controlled, and the steady-state force at various lengths is measured. In developing these relationships, we consider the performance of an isolated muscle (or single muscle cell) known as a **muscle fiber**. After mounting in the apparatus, muscle length is varied, then held isometric at a specific length (constant total length) while the steady-state force is measured. The relationship between the unstimulated force (often designated as passive force) and muscle length is designated the passive **length-tension relationship**. This relationship can be characterized as an exponential spring, $F = A_1 + A_2 \exp(A_3 X)$, whose behavior is similar to that of a rubber band. This relationship can show a dependence on the direction of the imposed length changes,

known as **hysteresis**, but deviations are small in a true steady state. Some form of passive force is common to all muscles, but an exact anatomical assignment of the structures underlying passive force is dependent on both the type of muscle and the preparation studied.

The next page of this analysis involves a similar protocol but includes stimulation of the muscle to identify the parameters associated with activated muscle. With active muscle, the language that evolved reflected the state of understanding and the experimental apparatus. The response to a single electrical shock is known as a **twitch contraction**. At one time, this was believed to be some form of elemental or quantal behavior of muscle, hence its historical importance. Increasing the frequency of stimulation leads to a summation in time of the individual twitch responses, known as **temporal summation**. Beyond a certain frequency (depending on muscle type and temperature) the force response becomes a smooth, fused curve, called an **isometric tetanus**. These responses to stimulation are ultimately related to the Ca^{2+} handling underlying activation of the contractile proteins. For our present purposes, we consider only isometric tetani, so that the mechanical behavior of the fully activated contractile apparatus can be considered, without complications arising from behavior attributable to non–steady-state Ca^{2+} signaling.

Tetanic stimulation adds an additional increment of force to the passive force present at a given length. The passive force plus active (stimulated) force, measured as a function of muscle length, is shown in Fig. 1. This relationship is known as the **total force-length curve**. The total force-length relationship varies considerably from muscle to muscle, though the component passive and active force-length relationships are qualitatively similar. The differences are largely ascribable to the relative amount of passive force developed at the length at which active force is optimal.

For the understanding of mechanism, the relationship between the additional active force generated when a muscle is stimulated and muscle length is paramount. This relationship, the **active force-length curve**, is unusual in that it decreases to zero at both long and short muscle lengths. For most materials, including polymers like rubber, force increases as length is increased. This typical behavior is also seen for the passive force as shown in Fig. 1. The observation that active force decreased at long muscle lengths was critical to eliminating theories that involved folding of continuous muscle filaments as the basis for the generation of force upon activation. To understand the relationship between active force and muscle length, it is necessary to consider muscle structure.

B. The Structure of Muscle: Interdigitating Filament Systems

Muscle cells are composed of a filament system underlying their mechanical properties and an internal membrane system related to their control functions. The filament structure repeats on both the transverse and longitudinal directions as shown in Fig. 2. A muscle fiber is composed of myofibrils whose fundamental longitudinal repeating unit is the sarcomere. The sarcomere consists of two interdigitating filament systems—thick (14-nm) myosin-containing filaments

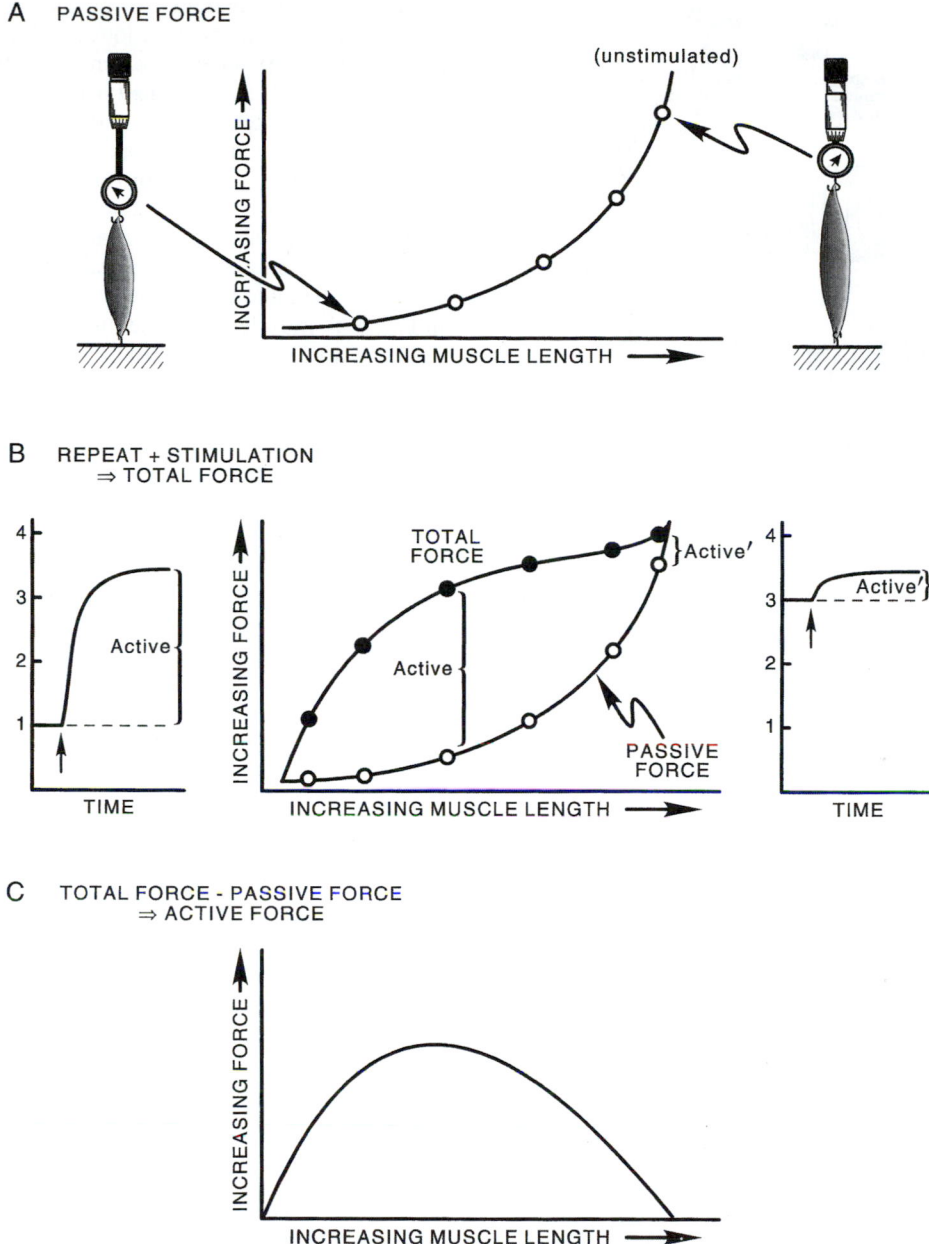

FIGURE 1. Measurement and operational definitions of muscle force-length relationships. (A) Relation between isometric force under unstimulated conditions, passive force, and muscle length. (B) Relation between isometric force and length under stimulated conditions, total force. (C) Active force operationally defined as the difference between total force and passive force is shown as a function of muscle length.

and thin (7-nm) actin-containing filaments—which underlie the banding seen under optical microscopy, characteristic of striated or striped muscle. The optical properties of the sarcomere due to the overlapping filaments gave rise to the nomenclature for the banding regions. The I-band contains only thin filaments and is optically **isotropic**, whereas the A-band contains both filament types and is **anisotropic**. The constancy of A-band dimensions, independent of total muscle length (Huxley and Niedergerke, 1954), was a key experimental finding, leading to the concept that filament length was constant. Constant filament lengths and interdigitating filaments are best observed at the electron microscope level (Fig. 3), where interpretation of the changing banding pattern with muscle length was first elucidated.

C. Sliding Filament Theory

We can now better understand the tenets of the sliding-filament theory (the most generally accepted theory of muscle contraction). This theory postulates that muscle force is gener-

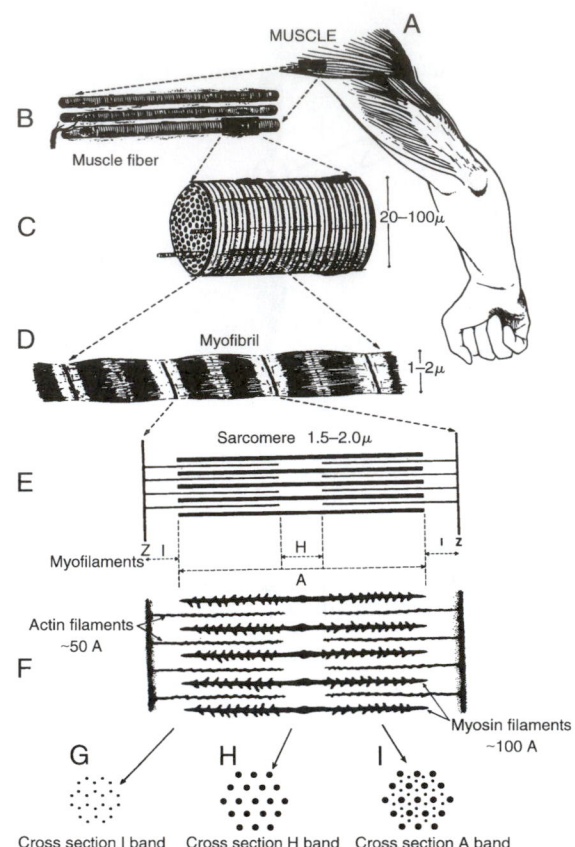

FIGURE 2. Skeletal muscle structure. The whole muscle is constructed of fundamental repeating units. A muscle cell or fiber is composed of myofibrils, and its fundamental longitudinal repeating unit is the sarcomere. All views are longitudinal except the cross-sections of the sarcomere shown in the bottom left drawings. (From Bloom and Fawcett, 1994, *Textbook of Histology,* 11th ed. New York: Chapman and Hall.)

ated by the interaction between thick and thin filaments of constant length, which are free to interdigitate and slide past one another. In this theory, the ability to generate isometric force is proportional to the extent of overlap between the filaments. This mechanical correlate of the proposed sliding-filament structure was tested in the classic work of Gordon *et al.* (1966) and is summarized in Fig. 4. The clearest correlation between the level of active isometric force and extent of filament overlap is in the region of decreasing force between sarcomere lengths of 2.2 and 3.6 μm, that is, the descending limb of the active force-length curve. The interpretation of the ascending limb of the force-length curve (e.g., 1.2 to 2.2 μm) is not as straightforward. It is complicated by possible changes in activation at short lengths. The loss of force at these lengths can also be attributed to double filament overlap and compression of the contractile elements while interdigitating. Sliding filaments are widely accepted, but the basis of force generation remains an active area of research (see Section II.E on crossbridge theory).

D. Relationships among Force, Velocity, Work, and Energy Utilization

Together with the relationships between force and length, the relationship between force and shortening velocity is a

fundamental characteristic of muscle performance. Studies of work production under various conditions (to define muscle performance) have an extensive history. To eliminate considerations due to transient activation, only contractions under maximum tetanic stimulation are presented here.

With current technology, one would use an ergometer to impose a constant velocity or force, and measure the resulting force or velocity as a function of time. However, it is instructive to consider the apparatus used historically because some of the nomenclature prevalent in muscle physiology can be attributed to these protocols. Such apparatus is shown in Fig. 5. A muscle is attached to a lever system so that addition of a weight or **preload** can freely stretch the muscle. The importance of the preload, in this context, is that it sets the initial length of the muscle and, consequently, the active force, as governed by the force-length relationship. A mechanical stop placed on the lever system allows additional weights to be added without changing length. The total load now on the muscle is called the **afterload**, as the muscle can support this load only after stimulation, when its force development exceeds that of the afterload. Afterloaded isotonic contractions are depicted in Fig. 5, which shows the effects of increasing afterload on muscle length as a function of time after stimulation. Although the change in length is curvilinear with time, reflecting changes in force due to altered filament overlap, the initial change in length is constant with time and thus has a steady velocity. To isolate the relationship between force and velocity from changes in force due to different degrees of thick and thin filament overlap, experiments are designed to exploit the plateau region of force-length relationship in which the crossbridge number is constant. Furthermore, muscles in which the passive force was negligible in this region (e.g., the frog semitendinosis) are commonly used to avoid complications in interpretation of force-velocity data due to passive force. The relationship between afterload and velocity under these conditions is hyperbolic (Fig. 5). Several equations have been proposed to fit this relationship between velocity (V) and afterload (F). However, the most widely used is an equation attributed to A. V. Hill, which is expressed as

$$(F + a)(V + b) = a(V_{max} + b) = b(F_o + a) \qquad (1)$$

where a and b are constants and $a/F_o = b/V_{max}$ are dimensionless constants that determine the curvilinearity of the Hill equation and which are approximately 0.25 for skeletal muscle. The mechanistic origin of this parameter is controversial. Although low values are often associated with efficient muscle performance, a generally accepted theoretical basis remains to be found. The parameter a was initially thought to be related to a constant derived from heat measurements, called the **shortening heat**. However, subsequent experimentations proved that the shortening heat was not a constant but a function of load as well. Although other equations can fit the relationship between force and velocity, the scientific stature of Hill and the apparent link between heat and mechanical measurements led to the dominance of the Hill equation in the literature.

The Hill equation empirically provides a reasonable guide to muscle performance. A consequence of the hyperbolic nature is that the power output of muscle shows a relatively

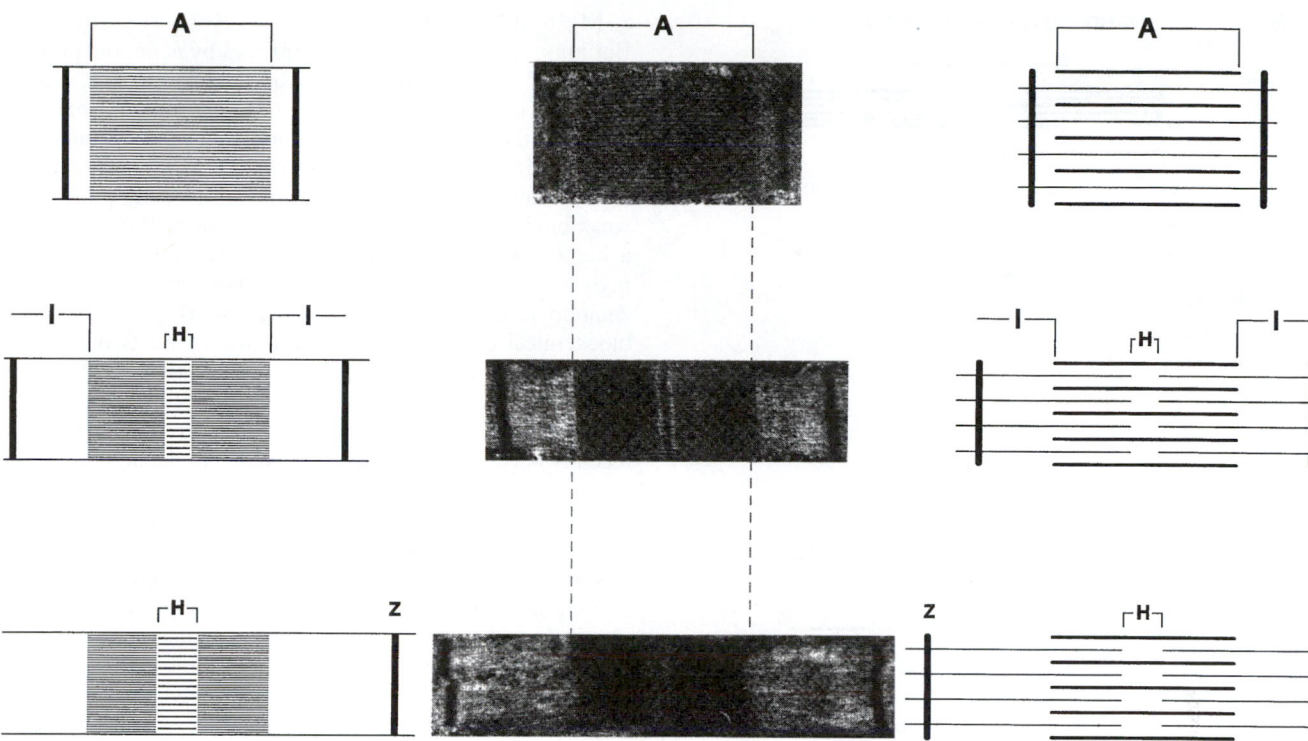

FIGURE 3. Electron micrographs and diagrams of sarcomeres at various muscle lengths. The sliding-filament theory is based on the constancy of the thick and thin filaments. (A) A-bands; (I) I-bands; (Z) Z-bands; (H) H-zone.

broad range around a maximum. An advantage of the Hill formulation over others is that it is symmetrical. This can be seen best in its normalized form,

$$\left(\xi + \theta\right)\left(\nu + \theta\right) = \theta\left(\theta + 1\right) \tag{2}$$

where ξ is the normalized force (F/F_o), ν is normalized velocity (V/V_{max}), and $\theta = a/F_o = b/V_{max}$. As power is given by the product of force times velocity, Eq. 1 or 2 can be readily used to derive a relationship between power and load or velocity. The optimal load or velocity for maximum power output can also be derived from these equations and is equal to $\sqrt{\theta^2 + \theta} - \theta$. For skeletal muscle this equals about 0.3 times the maximum load or velocity. These empirical characteristics are useful in the design of ergonometric devices to optimize performance.

The Hill equation is most commonly used as a characteristic for testing models of muscle mechanical behavior. The most widely known of these models is the kinetic scheme of A. F. Huxley (1957). In this model, rate constants for crossbridge formation-breakage are functions of position. This model (Fig. 6) can explain the relationship between force and velocity, but it is not adequate for transitions between states, such as those seen (see Fig. 10) when rapid step changes are imposed (see Section II.F on tension transients). It is interesting that a major constraint, within which the parameters of Huxley's kinetic scheme were developed, was based on data of heat production as a function of shortening. With hindsight, we now know that these data probably do not simply represent ATP breakdown at the crossbridge level as Huxley assumed. Direct chemical measurement of ATP breakdown has

improved substantially with development of rapid-freezing apparatus and application of nuclear magnetic resonance (NMR) analysis to intact muscle. However, the dependence of the ATP hydrolysis rate attributable to crossbridges on mechanical conditions known with certainty (see Section III).

E. Crossbridge Theory

The most prevalent theory for force generation involves the concept of cyclic interactions between the heads of myosin molecules (crossbridges) projecting from the thick filaments and the actin molecules comprising the thin filaments. Evidence from electron micrographs indicates that these crossbridges, in various conformations, bridge the gap between thick and thin filaments. Based on both micrographs and X-ray diffraction data, these projections occur at intervals of 14.3 nm. The best evidence (from mass comparisons) is that 3 myosin molecules are located at each site, with an identical repeat at about 43 nm. The structure of the thick filament is shown in Fig. 7.

The evidence for cyclic interaction is partly based on structural considerations. Striated muscle can generate force and shorten over a range of about 50–150% of its rest length. Since a muscle is composed of identical sarcomeres, each sarcomere could also operate over the same range. For a sarcomere of 2.2 μm, this operating range would be about 1.1–3.3 μm. Since the thin filaments of each half of the sarcomere move toward the center, a sarcomere shortening from 3.3 to 1.1 μm (Δ of 2.2 μm) would require a movement of thin filaments on each side of the sarcomere of one half that

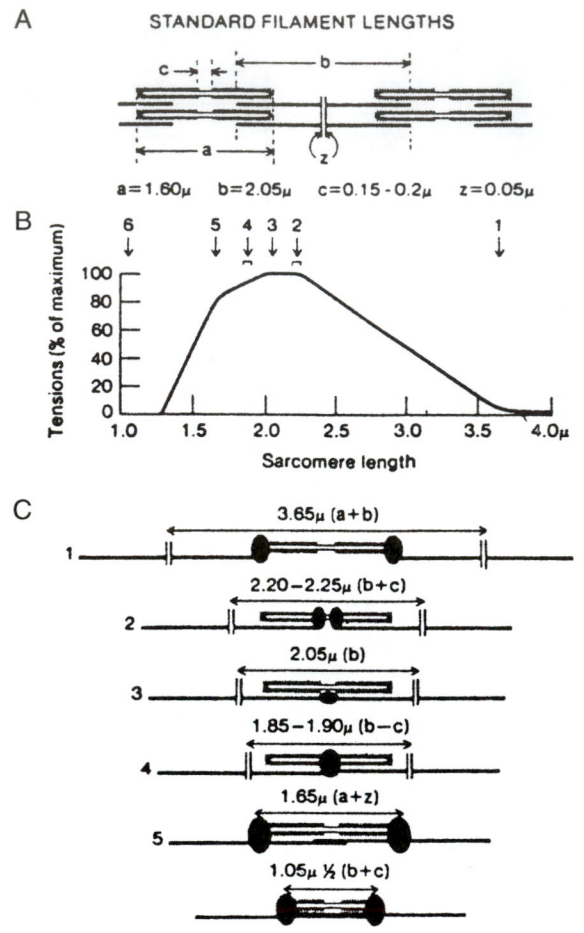

FIGURE 4. Structural basis for the active isometric force-length relationship. (From Gordon *et al.*, 1966.)

distance, or 1.1 μm relative to the thick filament. This distance is considerably longer than the crossbridge spacing (0.014 μm) and is, in fact, longer than a single myosin molecule (0.150 μm). It is thus difficult to envision a myosin crossbridge remaining attached to actin in a structure with interdigitating filaments of constant length and relative sliding of more than 1 μm. Hence some form of cyclic interactions between the myosin crossbridge and actin are envisioned to permit the observed degree of shortening. For example, if the working range of a cross bridge is 10 nm, then 100 repeated cycles of attachment and detachment would be required for a relative filament movement of 1 μm.

A second basis for the crossbridge theory arises from biochemical studies on isolated muscle proteins. The myosin molecule consists of a long rodlike region important in assembly into thick filaments and a globular head region, which contains the ATPase activity. The globular head, or S1 region (Fig. 7), has dimensions consistent with the projections identified with the crossbridge in electron micrographs. Studies of the kinetics of the ATP hydrolysis by myosin, involving stopped-flow apparatus and other techniques for the measurement of rapid time courses, have provided a framework for cyclic interaction of the S1 myosin head with actin. The kinetic scheme is shown in Fig. 8. The physiological ATPase activity (known as Mg^{2+}-ATPase activity) of purified myosin is relatively low. But importantly, it is activated 200-fold by actin, the principal protein of the thin filament. In the absence of ATP, purified actin and myosin bind strongly. In intact muscle, this is the cause for the stiffness of muscle in the absence of nucleotide associated with "rigor." Addition of ATP to a "rigor complex" of actin and myosin dissociates the complex by producing a weak binding state because the binding of ATP to myosin has a higher affinity (Fig. 8, steps 3 to 4). ATP hydrolysis by myosin alters the binding characteristics and crossbridge orientation, leading to a strong binding state (Fig. 8, step 2). The biochemical cycle is completed as the products of ATP hydrolysis, first inorganic phosphate (P_i) and then ADP, dissociate from the myosin head. One ATP molecule is hydrolyzed per crossbridge cycle. The structural details of this cycle have recently been significantly upgraded due to the high resolution X-ray diffraction data of the myosin head by Rayment (1996) and colleagues (Rayment *et al.*, 1996). As shown in the insets in Fig. 8, these data suggest that binding of ATP to myosin alters the conformation of the myosin head and consequently its binding to actin. Correlating this crossbridge cycle with mechanical steps in force generation is a major focus of muscle physiology. To understand how the mechanical behavior of an intact macroscopic muscle can be interpreted in terms of individual crossbridge behavior, we must consider the studies of mechanical transients.

F. Transient Mechanical Behavior and the Crossbridge Cycle

Analysis of the mechanical properties of any material generally involves the imposition of known perturbations (for example, step or sinusoidal changes in length or force) while recording the response, of the system. When modeling the time course of the response, combination of two types of components (springlike and viscous elements) can account for the behavior of most materials. The behavior of spring-like material is characterized by an instantaneous relationship between force and length, whereas for a viscous element, a resistive force is proportional to the rate of imposed length change. Combination of these elements can approximate the behavior of many materials, including muscle to a degree (see Fig. 9).

The steady-state behavior of muscle can be adequately described by two parallel elements: a springlike element, representing the passive force-length characteristics; and a contractile apparatus, representing the active force characteristics. For skeletal muscle at L_o, a length of optimal filament overlap, little passive force exists. Thus the behavior we are describing is that of the contractile element alone. Imposition of a rapid step shortening leads to a rapid drop in active force, followed by a redevelopment of tension. This suggests that muscle behavior can be modeled by two components: a series elastic spring, which instantaneously responded to the length change; and a contractile element, whose behavior could be described by a viscous-like relationship between force and velocity (see later section). Step changes in force yield a rapid, in-phase letter change followed by a slower shortening phase, also consistent with this two-component model.

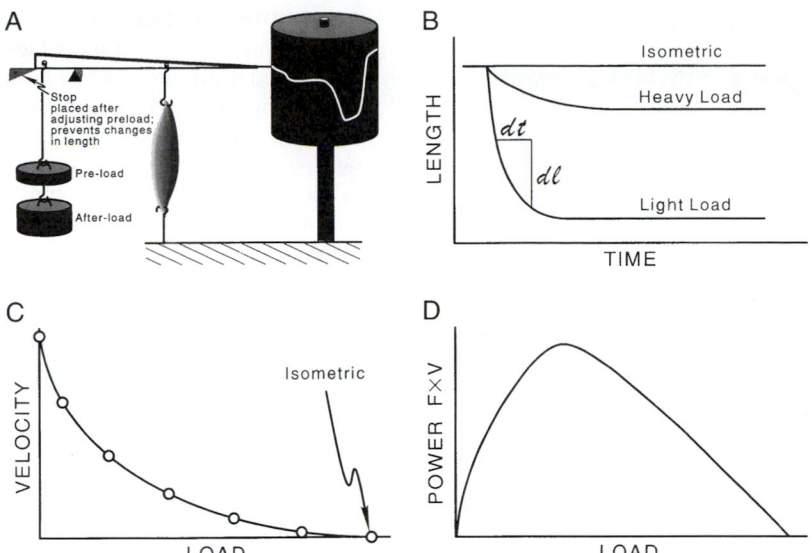

FIGURE 5. Relationships among velocity of shortening, power output, and load. (A) The experimental apparatus used to generate afterloaded, isotonic conditions. (B) The experimental results yielded by various loads. (C and D) The derived relationships between velocity and load, and power and load, respectively. dl = change in length; dt = change in time; $F \times V$ = force × velocity = power.

Although this modeling can explain mechanical behavior, its value lies in its potential for associating anatomical elements with model elements. A localization for the series elastic component (SEC) has evolved and has considerably altered our view of muscle mechanics. The SEC can be characterized in terms of the extent of shortening required to discharge the maximal isometric force (F_o). Studies on intact, whole muscle suggested that a change in the length of muscle

of 3% L_o was sufficient to transiently reduce active force to zero. This elasticity is ascribed to connective elements such as tendons and similar passive structures. A 3% change in L_o can be translated to a relative filament motion of 33 nm. This is significantly larger than the crossbridge spacing (14.3 nm), reinforcing the concept that such behavior is external to the contractile component and in series with crossbridges. The two-component model is phenomenologically useful, and it

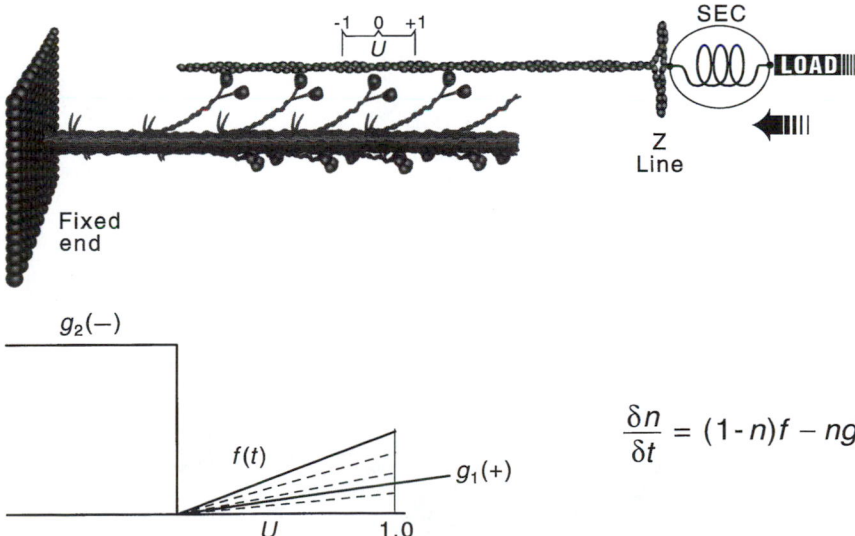

$$\frac{\delta n}{\delta t} = (1 - n)f - ng$$

FIGURE 6. Diagram representing a two-element model of muscle and Huxley's (1957) mathematical model. Top: Half-sarcomere containing a myosin thick filament and actin-containing thin filament, which constitutes a contractile component, coupled in series with a series elastic component (SEC). Bottom left: Functional form of rate constants for crossbridge attachment (f) and detachment (g_1 and g_2) as a function of U; the crossbridge position coordinate shown at top. Bottom, right: Differential equation describing the change in number of crossbridges. Force generated in this model is equal to the number of attached crossbridges, n, times the force of an individual crossbridge.

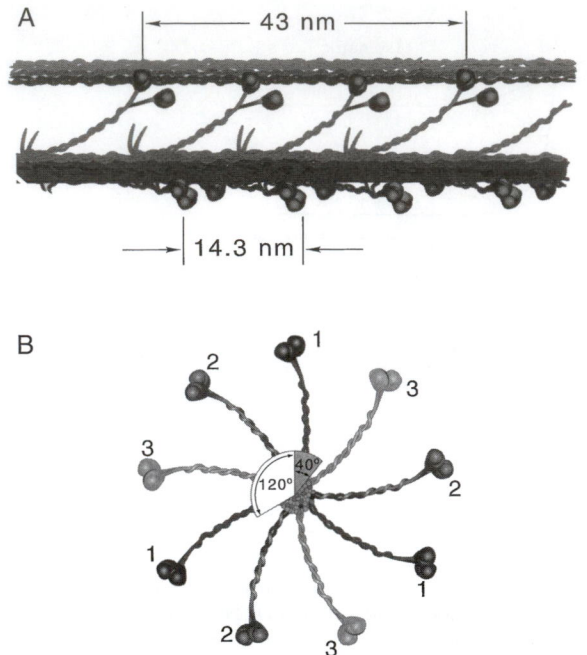

FIGURE 7. Thick filament. (A) A rendition of the thick-filament structure based on current evidence. (B) The filament viewed from the end.

can predict behavior of many whole tissues, particularly those with large intrinsic SEC, such as some smooth muscle. However, as improvements in the temporal resolution of mechanical measurements advanced, this model was found to be inadequate for the behavior of single skeletal muscle fibers.

In the 1970s, techniques for imposition of step changes greatly improved as both transducers and recording apparatus achieved millisecond resolution. At the same time, techniques for working with single muscle fibers and devices for control of sarcomere length, rather than overall muscle length, were developed. These new measurements indicated that the true SEC extent was much smaller than that previously measured in whole muscle. (Previous measurements apparently included an artifact of the slow response time of the recording system.) Current estimates of the SEC for striated muscle are less than 0.5% L_o, approximately 6 nm per half-sarcomere, clearly in a range to be potentially associated with crossbridges themselves. Moreover, the time courses of responses (Fig. 10) were not consistent with an instantaneous spring connected in series with a contractile element, which also was characterized by an instantaneous force-velocity relationship. Huxley and Simmons (1971) provided the first evidence that an instantaneous spring-like behavior could be attributed to the crossbridges. They exploited the fact that the number of crossbridges could be varied by taking advantage of the force-length relationship. As shown in Fig. 10, if the SEC is intrinsic to the crossbridge, then the shortening step required to discharge the instantaneous elasticity is also intrinsic to the crossbridge. Thus the change in length to discharge force would be independent of the number of crossbridges and, consequently, independent of force. Alternatively, if the instantaneous elasticity was external to the crossbridges, it would be governed by its own relationship between force and extension. Reduction of force (and number of crossbridges) by changing the initial muscle length in this case would reduce the step shortening required. Huxley and Simmons (1971) provided evidence

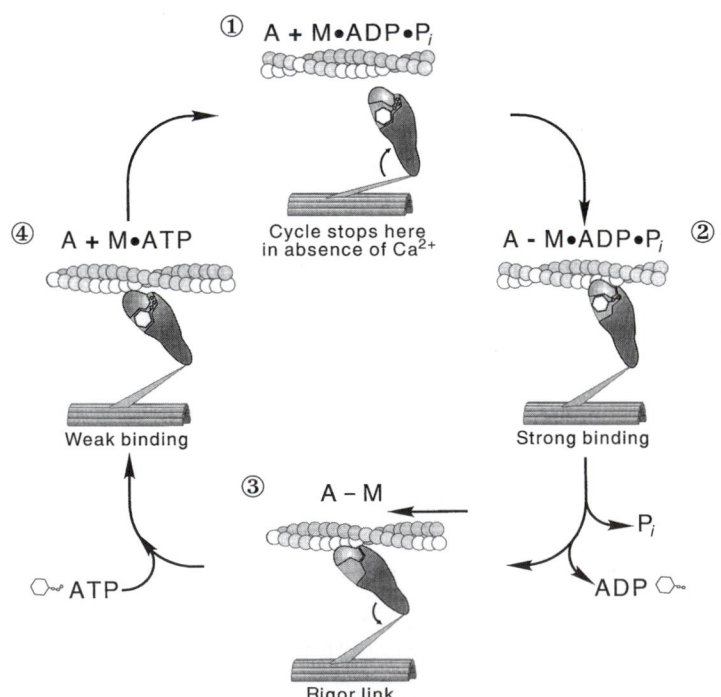

FIGURE 8. The crossbridge cycle. This depiction represents a crossbridge cycle based on currently accepted information from muscle structure, mechanics, and biochemical kinetics of the actin-activated myosin ATPase.

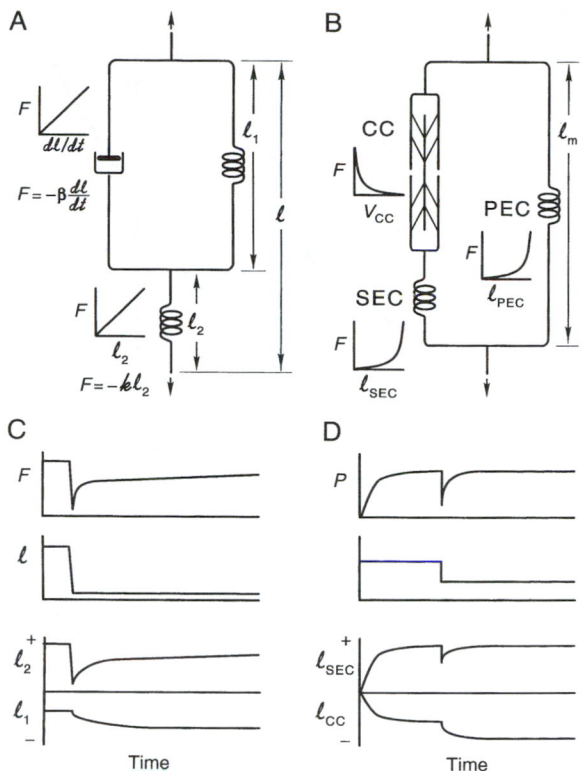

FIGURE 9. Models forming the basis of classic analysis of muscle mechanics. (A) Three-element Voight model containing two linear spring elements (l_1 and l_2) combined with a "dashpot," a linear viscous element. (B) Three-element model of muscle containing a contractile component (CC), a parallel elastic component (PEC), and a series elastic component (SEC). Graphs indicate the behavior of each element. (C) Time course of the response of the Voight model and its individual components to imposition of a rapid step change in length. (D) Similar responses of the muscle model to a rapid step change (note that the initial muscle length here is chosen such that the PEC does not play a role in this response).

that showed that altering active force by changing the initial muscle length did not alter the size of the step shortening required to discharge the maximum isometric force at any initial length. These results supported the concept that the instantaneous spring-like behavior is an intrinsic crossbridge property. Moreover, it also provides evidence that the crossbridges act as independent force generators. An important corollary of these studies is that the instantaneous stiffness can be used as an index of the number of crossbridges operating at any moment. This corollary has been used in numerous studies to assess the effects of various interventions on crossbridge number.

Recent mechanical (Higuchi *et al.*, 1995) and X-ray diffraction studies (Huxley, 1996) suggest that a significant fraction of the instantaneous stiffness is attributable to thin filaments. Thus one should exercise some caution in the interpretation of stiffness measurements.

To summarize, force generation is due to cyclic interaction of myosin heads projecting from the thick filament with actin of the thin filament. These crossbridges act as independent sites for force generation. Total active force is proportional to the number of activated crossbridges whose number is geometrically constrained by the extent of filament overlap. The mechanism of crossbridge force generation is still the subject of intense investigation. A current model based on biochemical kinetics and crossbridge mechanics studies is shown in Fig. 8. A shift in orientation of the crossbridge in the strong binding states (steps 3–4) is associated with the force-generating step in this model.

III. Muscle Energetics

Studies of muscle energetics parallel those of mechanics and biochemical kinetics and have the ultimate goal of understanding the mechanism of mechanochemical energy transduction at the crossbridge level. Historically, muscle energetics has been associated with the measurements of muscle heat production. Heat production is the first indication of the chemical reactions in muscle, and the temporal resolution of measurements of heat far exceeds that of direct chemical determinations. Moreover, heat measurements were made long before the chemical reactions were known. In fact, heat measurements and thermodynamic analysis were essential to ultimately identifying the chemical reactions underlying muscle activity. Thus it is not surprising that a tremendous literature, complete with its own terminology, has arisen. The essential principle comes from an application of the first law of thermodynamics. The change in enthalpy, ΔH, is equal to the sum of heat (Q) plus work (W) of the muscle. The change in enthalpy, in turn, can be related to the sum of the changes in number of moles of each species (Δn_i) multiplied by its specific enthalpy (h_i):

$$\Delta H = \sum_i \Delta n_i \times h_i = Q + W \qquad (3)$$

Thus, by measuring muscle heat and work, some indication of the underlying chemical reactions can be obtained.

One of the first applications of this type of study was carried out by Wallace Fenn (1924) in the laboratory of A. V. Hill. A theory, known as the **viscoelastic model**, proposed that stimulation elicited a fixed or quantal extent of some unknown reaction(s), whose energy was transduced to a fixed amount of heat plus work. A muscle shortening against a load would produce work and, according to this theory, less heat than a muscle under isometric conditions not producing work. Fenn's studies showed that the $Q + W$ in a work-producing contraction exceeded the heat produced in an isometric contraction (in which work is minimal). This "Fenn effect" was important in that theories of muscle energy conversion had to be configured in terms of chemical reactions that were closely coupled to mechanical events.

The studies of the nature of the chemical reactions coupled to muscle activity, both those immediately coupled to mechanical events and those involved in the subsequent recovery by intermediary metabolism, have been closely associated with muscle energetics. The history of the search for the identity of the reactions immediately coupled to muscle mechanical output parallels the history of biochemistry itself. Because of increased understanding of the biochemistry of fermentation, lactate was first believed to be this

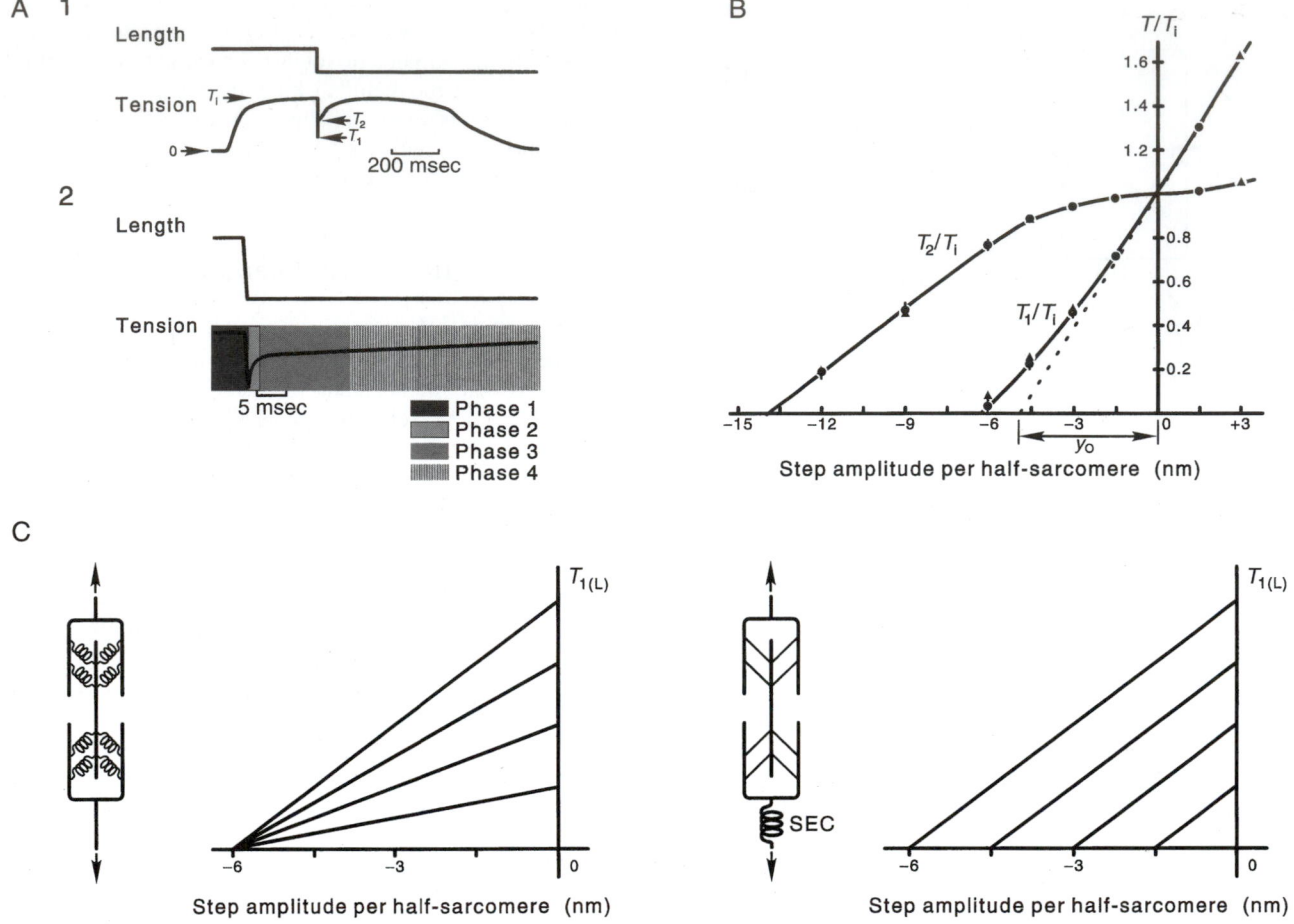

FIGURE 10. Huxley-Simmons (1971) experiments. (A) Transient force responses to very rapid (<1 ms) step changes in length, which led to revision of classical mechanics. (B) Dependence of phase 1 and phase 2 force amplitudes on muscle length. (C) Dependence of T_1 if the series elasticity resided within the crossbridge itself (left), and if the series elasticity was located in a classic SEC (right). Huxley and Simmons data support left panel of part C.

driving reaction. However, Lundsgaard (1938) showed that muscle treated with iodoacetate (which blocks glycolysis and thus lactate production) could still contract. Fiske and SubbaRow (1927) showed that inorganic phosphate (P_i) is liberated during contraction, arising not from a glycolytic intermediate, but from a new "phosphagen" phosphocreatine (PCr). Myosin was known to be an ATPase, but there was little change in ATP concomitant with contractile activity. The connection was made by Lohmann (1934), who discovered an enzyme (creatine kinase) that catalyzes the reversible transphosphorylation:

$$PCr + ADP \leftrightarrow ATP + Cr \qquad (4)$$

From these observations grew the energetic schema that ATP is the immediate energy source coupled to the myosin ATPase. In intact muscle, PCr rapidly rephosphorylates ADP via this Lohmann reaction (or creatine kinase reduction). Thus breakdown of PCr to Cr and P_i is the only net reaction occurring and accounts for muscle heat production during contraction. Quantitative tests of this scheme awaited the development of rapid-freezing technology and microchemical analysis of muscle extracts, which occurred during

the late 1960s. These "energy balance" studies (Woledge *et al.*, 1985) tested the validity of Eq. 3. Under conditions in which metabolic resynthesis was minimized, the measured breakdown of PCr multiplied by the partial molar enthalpy (34 kJ/mole) would have been equal to $Q + W$ if this had been the only chemical reaction occurring. As shown in Fig. 11, this equality was not valid as $Q + W$ was significantly greater than that accounted for by PCr breakdown. The implication was that there was a "missing" reaction associated with an "unexplained" enthalpy production occurring during contraction. The implications of these observations were critical to the validity of the energetic dogma. Moreover, heat records to this day are still interpreted largely in terms of ATP breakdown (coupled to PCr breakdown) in most current kinetic models, starting with the classic Huxley (1957) formulation. Thus energetics studies were focused on identification of the nature of the unexplained enthalpy and a potential missing reaction associated with contraction.

In a parallel fashion, "biochemical" balance studies (Kushmerick, 1983) were also aimed at approaching the question of a missing reaction, but with chemical rather than physical techniques. The rationale was that if PCr had been

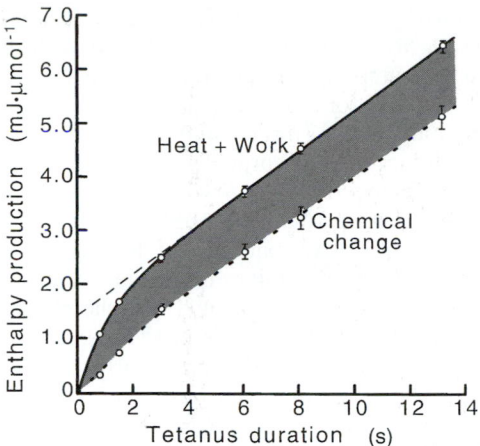

FIGURE 11. Skeletal muscle "energy balance." Enthalpy (heat + work) production as a function of tetanus duration is given as the solid line. The expected enthalpy (chemical change times the enthalpy of PCr breakdown, 34 kJ/mol) is plotted as the dotted line. Shaded area represents the unexplained enthalpy. [Adapted from Kushmerick, M. J., 1983. Energetics of muscle contraction. *In* "Handbook of Physiology (Section 10)—Skeletal Muscle" (Peachy, Lee, and Adrian, Richard H., Eds.), pp. 189–236. Copyright 1983 by Oxford University Press.]

the only net reaction occurring during contraction, and if resynthesis (which is governed by theoretical biochemical stoichiometry) had occurred during recovery, then the amount of oxygen consumed during recovery would have been equal to 1/6.5 times the PCr broken down during contraction. Again as shown in Fig. 12, an imbalance was observed, with the extent of recovery metabolism associated with a resynthesis of ATP that exceeded that accounted for by PCr breakdown during contraction. Paul (1983a), using both "energy balance" techniques, showed that oxidation of substrate was the only net reaction occurring. The combinations of techniques yield data that indicated that the chemical imbalance is attributable to additional breakdown of ATP after contraction.

The resolution of the "missing" reaction question apparently resides in the discovery of the protein parvalbumin and its role as a Ca^{2+}-binding site (Gillis *et al.*, 1982). Parvalbumin binds a significant amount of the Ca^{2+} released by the sarcoplasmic reticulum (SR) during contraction. This binding is associated with a substantial evolution of heat. There is also evidence that Ca^{2+} is returned to the SR after the contraction ceases, which would require activity of the Ca^{2+} pump and concomitant hydrolysis of ATP. This explanation is perhaps the most widely held, but it is not universal. For example, the biochemical balance studies would suggest that the postcontractile ATP breakdown is proportional to the duration of stimulus, whereas the enthalpy balances apparently indicate that the unexplained enthalpy attains a plateau value. Moreover, some unexplained enthalpy is associated with mammalian striated muscle, which does not contain parvalbumin. Nevertheless, the schema involving ATP as the immediate driving reaction, with PCr as the source of chemical energy for rapid resynthesis in intact muscle, appears valid. This is further supported by (1) ther-

modynamic data indicating that the free energy provided by ATP during contraction is sufficient to account for the work produced (Woledge *et al.*, 1985) and (2) many studies using permeabilized muscle indicating that ATP is the only chemical energy source required.

IV. Muscle Metabolism

Mechanochemical energy transformation and the breakdown of high-energy phosphagens under various conditions form one branch of study known as muscle energetics. Energy

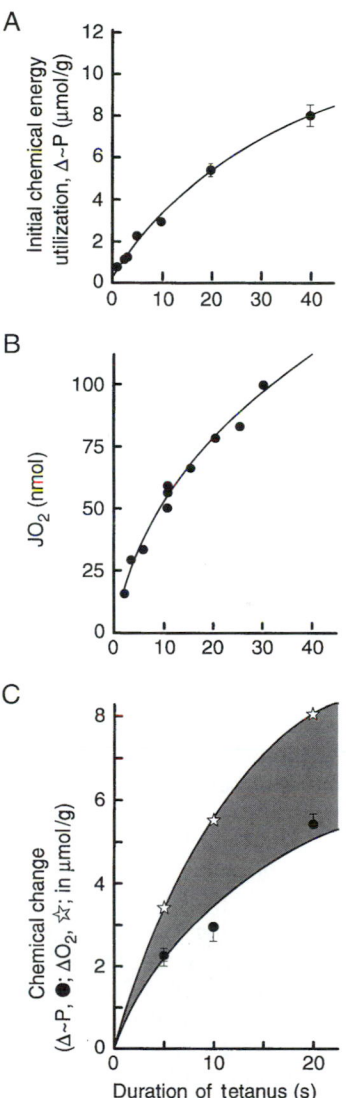

FIGURE 12. Skeletal muscle "biochemical energy balance." (A) The initial high-energy phosphagen (PCr + ATP) breakdown during contraction as a function of tetanus duration. (B) Total recovery oxygen consumption elicited by the contraction durations plotted on the ordinate. (C) The total phosphagen resynthesized based on theoretical stoichiometry and the measured O_2 consumption (open stars) and the measured phosphagen breakdown during contraction (filled circles). The shaded area represents the excess of ATP synthesis over the phosphagen breakdown measured during contraction.

metabolism and intermediary metabolism are two terms often used to describe the other major branch of energetics, that of the study of the metabolic provision of phosphagen in support of muscle activity and the mechanisms of coordination of metabolism with contractility. Muscle mechanical activity is supported by ATP provided by both oxidative phosphorylation and glycolysis. Historically, the biochemistry of these pathways often mirrors our understanding of muscle, as this tissue was often the major model for biochemical studies. Although general metabolism and energy production were covered in Chapter 7, the historical aspects (Needham, 1971) and biochemical details are beyond the scope of this text; the treatise of Needham is highly recommended for further reading.

Perhaps of most recent interest in this field is the reexamination of the theories whereby contractile ATP use is coordinated with metabolic resynthesis of ATP. The most prevalent theory is that of acceptor- or ADP-limited respiration. In the case of isolated mitochondria in the presence of substrate, their rate of oxidative phosphorylation (state III) is limited by the availability of ADP as the substrate for rephosphorylation (Kushmerick, 1983). Because the hydrolysis of ATP concomitant with contractile activity is increased, one might anticipate that an increase in ADP could naturally couple the increased usage with synthesis. The creatine kinase reaction, coupled with the amount of PCr in striated muscle, buffers the ATP concentration. Under the assumption that this reaction is in equilibrium, the level of free (i.e., not bound) ADP can be calculated to be in the region of 10 µM. This is also in the range for the K_m for ADP for mitochondrial oxidative phosphorylation. This is apparently the major mechanism for skeletal muscle (Kushmerick, 1983), for which significant changes in PCr, and thus ADP, occur during contractile activity and correlate well with the increased oxidative metabolism. On the other hand, significant increases in cardiac (Heineman and Balaban, 1990) and smooth muscle oxidative metabolism occur in the absence of changes in PCr or ATP, indicating that increases in energy metabolism can match the contractile ATP utilization in the absence of changes in ADP. There are a number of hypotheses in this area. The most prominent ones propose regulation of oxidative metabolism by $[Ca^{2+}]_i$ and mechanisms involving a substrate level "push," for example, by mobilization of substrate, leading to increased NADH levels.

A second area of growing interest on the metabolism side of muscle energetics is the effect of microcompartmentation of metabolism and its subserved function. Instead of being distributed in a random fashion in solution in the cytosol, many enzymes important in energy metabolism are apparently localized within the cell, often in close apposition to energy-dependent processes. Perhaps the most widely studied system is that of creatine kinase (Ishida et al., 1994), which is localized at the m-line of striated muscle as well as in membrane structures and the mitochondria. In permeabilized muscle, endogenous phosphocreatine alone can support contractile activity, suggesting a close relationship between the myosin ATPase and creatine kinase. Similarly, phosphocreatine has been shown to support Ca^{2+} uptake by sarcoplasmic reticulum isolated from skeletal muscle. The

enzymes involved with glycolysis are also localized, primarily on the thin filaments and associated with the plasmalemma and sarcoplasmic reticulum. There are increasing numbers of studies for skeletal (James et al., 1999), cardiac (Weiss and Lampe, 1987), and smooth muscle (Ishida et al., 1994), indicating that ATP provided by glycolysis supports functions different from those dependent on oxidative metabolism. This is particularly striking in smooth muscle, in which the oxidative and glycolytic components of metabolism can be of similar magnitude (Paul, 1983b). ATP-dependent processes associated with the plasma membrane, such as the Na^+-K^+ pump and ATP-dependent potassium channels, are apparently particularly dependent on ATP generated by aerobic glycolysis, whereas the ATP requirements for the force generating actin-myosin interaction are apparently more strongly correlated with oxidative metabolism. The bases and consequences of this cellular organization are not fully understood, but have been suggested to be related to either greater energy transduction efficiency or the need for independent regulation of different cellular functions.

V. Comparative Muscle Physiology

A. Striated Muscle

The general features just described are common to muscle function; however, there are a great variety of muscle types adapted to specialized conditions. Among skeletal muscle, perhaps the most familiar adaptation is the difference between red and white muscle. The nomenclature has a long history and many different schemes exist. Histologically and functionally three major classifications predominate, termed fast glycolytic (FG), slow oxidative (SO), and fast oxidative-glycolytic (FOG), reflecting their gross properties. Most whole muscle contains various mixtures of these three basic muscle fibers and appears pale compared to red muscle, which is composed primarily of SO fibers. The red color is due to the presence of myoglobin, a protein that facilitates the diffusion of oxygen.

These fiber types are adapted for different power requirements. The FG fibers provide large forces that can be rapidly activated and relaxed, but are susceptible to fatigue. Their metabolism depends heavily on glycolytic production of ATP, which can respond rapidly to large changes in energy requirements associated with muscle contraction but has relatively limited capacity. For low-level but continuous muscular activity, such as that required by postural muscles, the SO fibers are adapted with a high oxidative metabolic capacity. A lower actin-activated myosin ATPase of the SO fibers relative to FG fibers facilitates a more economical maintenance of force, however, this is accompanied by a slower velocity of shortening.

In primates, FOG fibers are relatively rare and highly specialized muscles (some ocular muscles appear to be composed primarily of this fiber type). Their twitch duration and shortening speed are intermediate between those of SO and FG fibers and are characterized by high levels of both glycolytic and oxidative metabolism. These differences in adaptation also extend to speed of activation, which is approximately three times faster in FG than SO fibers. Cardiac muscle has its own special adaptations and in many ways shares similarities with SO skeletal muscle fibers. (These are considered elsewhere in this text). A third muscle type, smooth muscle, is significantly different from striated muscle and is discussed in detail in the following section.

These gross differences in muscle fiber type have long been studied. Applying advanced molecular biology techniques, we now know that most muscle proteins exist as isoforms. Thus recent interest has focused on the isoform-function relationships as well as the regulation of the expression of these various isoforms. What is of particular interest is that the number of expressed isoforms—and the theoretical possible variants—far exceed the three gross skeletal fiber types. Detailed information on isoforms is beyond the scope of this text and is available in specialized treatises (Mahdavi *et al.*, 1986; Pette and Staron, 1990).

B. Smooth Muscle

Historically, smooth muscle has been of interest because of its specialized function. It lines the hollow organs, such as blood vessels and the gastrointestinal (GI) tract. Its structure is less organized than that of striated muscle and thus somewhat less amenable to biophysical experimentation. For many years, its properties have been simply extrapolated from striated muscle. However, it is now clear that while many similarities exist, there are also significant differences. Studies of smooth muscle have significantly increased over the past decade. This can be attributed largely to its clinical relevance in terms of the diseases of industrialized society such as hypertension, asthma, and GI motility disorders. However, much of the current interest has arisen from the discovery that the regulatory mechanisms at the contractile filament level are radically different from those of striated muscle.

Smooth muscle can develop isometric forces per cross-sectional area that are equal to or greater than those generated by striated muscle. This is all the more surprising in that the myosin content of smooth muscle is considerably less (about one-fifth) than that of striated muscle. These large forces can be maintained with an ATP use that is 100- to 500-fold lower than the corresponding rate in skeletal muscle. The trade-off for high forces and economy of force maintenance is apparently shortening velocity, which is much slower (up to 1000-fold) than that in striated muscle. The efficiency, in terms of work per ATP, is also somewhat lower (approximately one-fifth that of skeletal muscle); however, the data here are less extensive. The key question then is how does smooth muscle accomplish these specialized functions given that its contractile apparatus uses basically similar actin and myosin components?

1. Smooth Muscle Structure and Its Relationship to Function

The structure of smooth muscle is shown in Fig. 13. Smooth muscle cells are smaller than skeletal muscle, with maximum diameters of the spindle-shaped cells in the range

10–20 μm. Filament structure is less organized, with no distinct sarcomeric or banding structure. There are no Z-bands, but actin filaments are organized through attachment to specialized cytoskeletal regions called dense bodies or patches. These areas contain α-actinin, a Z-band protein, which further strengthens this analogy. It is also likely that these structures mediate transmission of force between cells. The ratio of thin to thick filaments is much higher in smooth muscle (~15:1) than in striated muscle (2:1).

This latter observation raises the question of whether structural factors could account for these functional differences. The answer clearly depends on the mechanism of contraction in smooth muscle. It is assumed, by analogy to striated muscle, that a sliding-filament mechanism is involved in smooth muscle contraction. There is very limited evidence to date, but the available data are consistent with this theory. Primarily, the evidence in intact tissue is limited to a dependence of active force on length, which is qualitatively similar to that of striated muscle. The loose organization of myofilaments is not readily amenable to direct structural data regarding the extent of thin-thick filament overlap for comparison with force-length behavior. Other supportive evidence can be gained from the more recent "motility assays." In these experiments, actin filaments are observed to move on a myosin-coated surface, or myosin-coated beads move on actin filament networks (Harris and Warshaw, 1993b). One can infer that folding of filaments is not essential to motion, consistent with a sliding-filament model. Moreover, the only difference between striated and smooth muscle myosin in these experiments is a slower velocity observed with smooth muscle myosin.

Assuming that an analogous sliding-filament mechanism operates in smooth muscle, what types of structure could account for the differences in function? Force is proportional to the number of crossbridges in parallel, whereas shortening velocity and the maximal shortened length are related to the number of fundamental units in series. Models of arrangement of cells or sarcomeres yielding an increase in force and holding economy while reducing velocity are shown in Fig. 14. While increasing the number of cells in parallel (Fig. 14A) works in the appropriate direction (large force, low tension cost), it falls short of true smooth muscle behavior in that the amount of total tissue shortening is very limited. Smooth muscle can shorten to relatively short lengths; some 20–30% of initial tissue length is not uncommon. In the model of Fig. 14A, the absolute length change would be that attributable to shortening of only one cell and would not account for the shortening of any macro size tissue.

In Fig. 14B, different mechanical properties are associated with the assembly of the sarcomere. Longer myosin filaments and consequently longer sarcomeres are associated with more parallel crossbridges and higher forces. However, for a given myosin content, this arrangement has fewer sarcomeres in series and hence a slower overall velocity. Moreover, if the individual myosin crossbridge ATPase is not altered, one would have a similar ATP use for both models, and thus the longer sarcomere version would have a greater **economy** (force maintained per rate of ATP hydrolysis) or a lower **tension cost** (reciprocal of economy). These

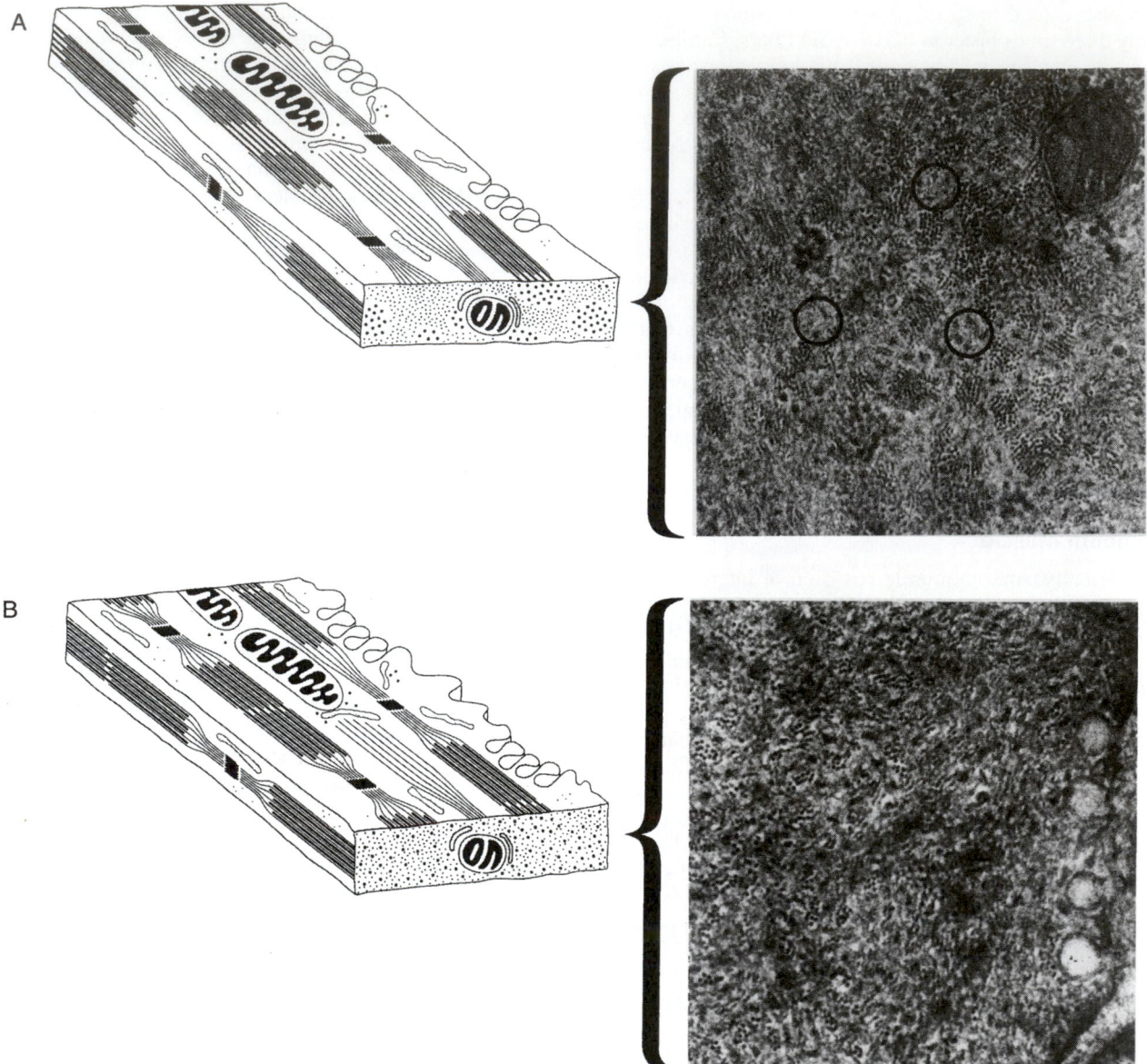

FIGURE 13.   Morphology of relaxed and contracted smooth muscle. (A) This diagram represents a portion of a relaxed smooth muscle cell, highlighting the arrangement of the actin-containing thin filaments and the myosin-containing thick filaments. This rendering is greatly simplified from the actual arrangement. From the longitudinal orientation, thin filaments arise at dense bodies and project to interact with thick filaments, forming the contractile apparatus. In the cross section, the filaments are distributed in a nonuniform manner, with thick filaments forming clusters surrounded by groups of thin filaments. This arrangement is shown in the electron micrograph (circled), which is a cross-section of a relaxed visceral smooth muscle cell. (B) A contracted smooth muscle cell. The notable differences from the relaxed cell shown in part A are that, in the longitudinal views, the thin filaments overlap considerably more of the thick filaments, drawing the dense bodies closer together and shortening the cell. In cross section, the thin and thick filaments are randomly distributed to form a uniform pattern. The electron micrograph is a cross section of a contracted visceral smooth muscle cell, which shows more uniform distribution of the thin and thick filaments. (Diagrams are modified from Heumann, 1973.)

changes are in the direction that distinguishes smooth muscle from skeletal muscle. The question then is how much of the difference between skeletal and smooth muscle can be related to simply the "mechanical advantage" of longer sarcomeres? Although no obvious sarcomeric structure exists for smooth muscle, the length of the myosin filament of smooth muscle relative to striated answers this question.

There is evidence that some smooth muscles, particularly in invertebrates such as the scallop or mussel, whose sarcomere lengths may be up to 10-fold greater than mammalian skeletal muscle, make use of this mechanical advantage. However, for mammalian tissues, myosin filament length has been estimated at approximately 2.2 μm (Somlyo, 1980), a figure not significantly different from that of stri-

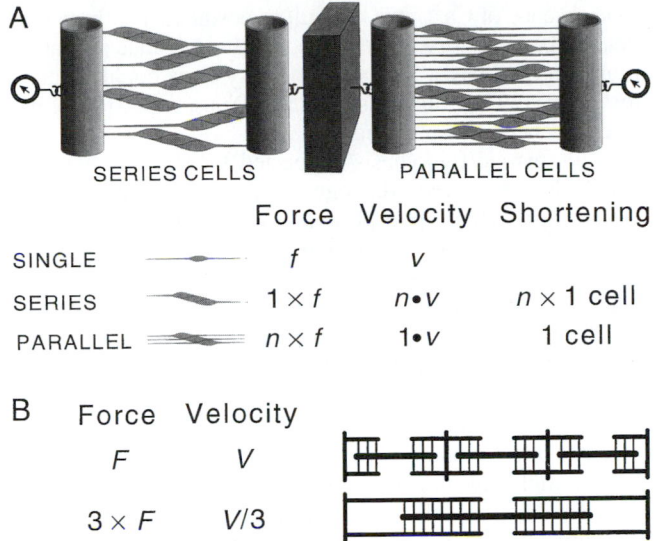

FIGURE 14. (A) Parallel and series models of force transmission in smooth muscle-containing tissue. Parallel arrangement increases total force, but with a decrease in velocity and total shortening compared with cells coupled in series. (B) Parallel and series model at the level of the sarcomere. Both sliding filament models have the same myosin content, but it is arranged in short (top) and long sarcomeres (bottom).

ated myosin (1.6 μm), and differences here are not primarily ascribable to structural features. The differences in contractile properties between smooth and skeletal muscle would thus appear to reside largely in the nature of the smooth muscle myosin molecule and its intrinsically lower ATPase.

2. Regulation of Smooth Muscle Contractility

Regulation of contractility can be divided into (1) mechanisms for control of $[Ca^{2+}]_i$ and (2) mechanisms for transduction of the Ca^{2+} signal to activation of the contractile apparatus. Regulation of $[Ca^{2+}]_i$ is considered elsewhere, and the focus here is on transduction mechanisms.

Up to the mid-1970s, mammalian smooth muscle, based largely on analogy to striated muscle, was considered to be a "thin-filament" regulated system, that is, one in which the troponin is the Ca^{2+} receptor, and tropomyosin is a transducing component of the regulatory system (see Chapter 54). Several lines of evidence led to the current, most widely accepted view that phosphorylation-dephosphorylation of the 20-kDa light chain of myosin (MLC-P_i) is the primary transduction site of the Ca^{2+} signal (see Hartshorne, 1987; de Lanerolle and Paul, 1991). Although tropomyosin is known to be a major smooth muscle protein, troponin is not present in smooth muscle. However, proving a protein absent is difficult, and that proof came slowly. Smooth muscle actomyosin also proved considerably different from that of striated muscle in that, as the purity increased, the ATPase activity of smooth muscle actomyosin decreased, the opposite of that found for striated actomyosin. This suggested that smooth muscle myosin required activation, whereas striated muscle myosin required de-inhibition. This was tested more rigorously by "competition" experiments. In

these experiments, the Ca^{2+} sensitivity of the actin-activated Mg^{2+}-ATPase of myosin was measured using "unregulated" thin filaments (i.e., purified actin without troponin and tropomyosin). Striated muscle myosin showed little Ca^{2+} sensitivity, whereas vertebrate smooth muscle myosin retained Ca^{2+} sensitivity, indicating that the site of its regulation was on the myosin itself. The regulatory site was identified at the isolated protein level when it was discovered that the actin activation of the Mg^{2+}-ATPase of smooth muscle myosin was dependent on phosphorylation of the 20-kDa light chain.

Several lines of evidence supported this thick filament regulatory mechanism at the cellular level. One strategy was to alter MCL-P_i, independent of $[Ca^{2+}]_i$ through manipulation of the kinase or phosphatase. Inhibition of MLCK by various inhibitors or use of calmodulin-binding peptides leads to relaxation and dephosphorylation in the presence of $[Ca^{2+}]_i$. Alternatively, proteolysis of MLCK yields a fragment that is constitutively active, independent of Ca^{2+}. This enzyme added to permeabilized smooth muscle generates a contraction, mechanically indistinguishable from that induced by Ca^{2+}. Pressure injection of these constitutively active MLCK into intact smooth muscle cells also leads to a contraction. This is important, for one could argue that making tissue permeable could lead to loss of regulatory elements. Taken in total, these results suggest that myosin light chain phosphorylation is the major regulatory site and is necessary for activation.

Whether myosin light chain phosphorylation is sufficient as a regulatory mechanism or indeed has additional roles is open to question. During the past several years a number of experimental observations suggest that smooth muscle regulation may be more complex than initially envisioned. Figure 15 summarizes the time courses of several parameters after stimulation. Isometric force increases monotonically and then maintains a plateau value, whereas MLC-P_i rapidly increases to a maximum achieved early in the contraction and then decreases. The extent of the decline in the MLC-P_i is controversial, and this is likely attributable to different smooth muscle types, dependence on the nature of the stimulus, and technique for measurement of MLC-P_i. The decline in MLC-P_i with maintained or increasing force suggests that other regulatory factors may play a role. Moreover, near maximal forces are attained at MLC-P_i levels considerably lower than 100%, which further questions its role as a simple switch. The decline in MLC-P_i is similar to that of $[Ca^{2+}]_i$ and, importantly, to the decline in shortening velocity. In view of these observations, Murphy and colleagues (Hai and Murphy, 1988, 1989) suggested that MLC-P_i might be a regulator of contractile velocity. They coined the expression **latch** for the state of maintained force with reduced MLC-P_i and slower contraction speeds. **Latch bridges,** or crossbridges, in this state were initially postulated to be non- or slowly cycling, and thus they reduced the overall tissue velocity by acting as a type of internal load on the more rapidly cycling bridges. As force retained its dependence on Ca^{2+} in the latch state, it was postulated that latchbridges may be regulated by a system different from MLC-P_i. This is a rapidly changing and intensely investigated area.

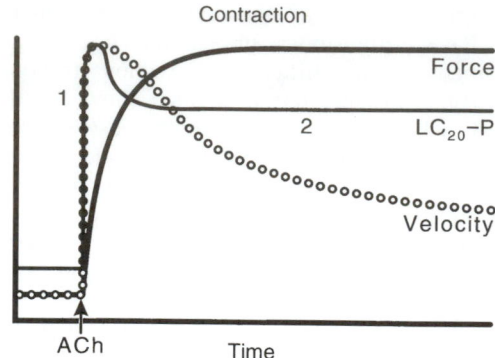

FIGURE 15. Relations between isometric force, maximum short-ening velocity, and myosin regulatory light chain phosphorylation (LC_{20}-P), and the duration of stimulation. Data are for tracheal smooth muscle (de Lanerolle and Paul, 1991).The decrease in velocity at times when force is maintained is the basis for the **latch bridge** theories.

Currently there are two major classes of theories for latch behavior, which are not necessarily mutually exclusive. Thin-filament regulatory mechanisms based on several actin-binding proteins, namely, leiotonin, caldesmon, and calponin, have been suggested. A common feature is that they are all pro-posed to be inhibitory, and for the latter two, the interaction with actin is sensitive to Ca^{2+}-calmodulin. Although all are promising in terms of the ability to inhibit myosin ATPase ac-tivity in the test tube, the relevance to intact smooth muscle has yet to be unequivocally demonstrated. As smooth muscle myosin requires activation, these systems could only be ancil-lary. Based on the ability of MLC-P_i to activate smooth muscle

in the absence of Ca^{2+}, it is difficult to postulate that these sys-tems can be inhibitory in the sense that they regulate activation on an "on-off" basis. However, these systems could regulate crossbridge cycling and thus velocity (Obara *et al.*, 1996), but more evidence of this is needed.

An alternative hypothesis suggested by Hai and Murphy (1989) does not require regulatory components beyond myosin light chain phosphorylation-dephosphorylation. A version of their model is shown in Fig. 16. They postulate that if a phosphorylated crossbridge were dephosphorylated while attached, its detachment rate would be significantly slower than if it were to remain phosphorylated. The kinetic scheme of Fig. 16 is sufficient to fit the available data on the time courses of force and MLC-P_i. It can also predict the time course of ATP utilization, which also decreases similarity to velocity, with stimulation time. However, this scheme predicts that a strongly curvilinear relationship exists between the rate of ATP utilization and force and that ATP utilization due to the "futile" cycle of myosin light chain phosphorylation-dephosphorylation is the dominant (~85%) source of ATP uti-lization. This view has been challenged by Paul (1990) based on the available smooth muscle energetics data. Moreover, he shows that simply a high ratio of attachment to detachment rate constants, as proposed several decades earlier by Bozler (1977), can fit the available data without invoking a special "latch" detachment cycle. Hai and Murphy (1992), using the model of Fig. 16 coupled with the kinetic scheme of Huxley (1957) for crossbridge cycling, were able to circumvent some of the objections to this model.

Recent experiments show that myosin phosphorylation-dephosphorylation can be an appreciable energy cost when levels of MLC-P_i are high (Wingard *et al.*, 1997), as during

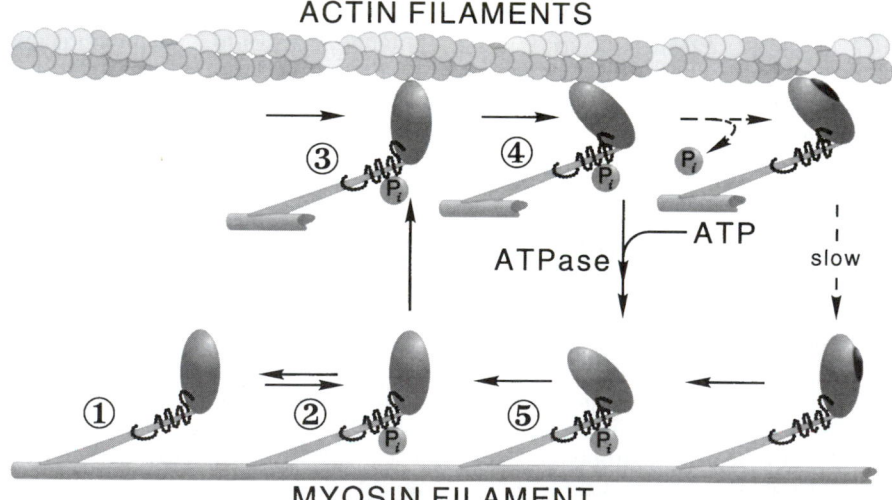

FIGURE 16. A schematic model of the smooth muscle myosin crossbridge interaction with actin filaments is depicted. The transition from state 1 to state 2 represents the Ca^{2+}-calmodulin–dependent phosphorylation/dephosphorylation activation mech-anism proposed for smooth muscle regulation. The cycle of interaction (states 2–5) hydrolyzed 1 ATP molecule and generates 1 quantum of tension with each pass. Only the angulated attached myosin crossbridge (state 4), however, generates isometric force. The inherent speed of the cycle is much slower in smooth muscle than in striated, with the ATP-dependent dissociation step (state 4–5) apparently being slower. The dashed arrows indicate formation of a proposed dephosphorylated actomyosin crossbridge (Hai and Murphy, 1989), which then dissociates only very slowly and may be responsible for the high holding economy of vascular smooth muscle.

the early stages of contraction. Thus in addition, longer "sarcomeres" and a low intrinsic actomyosin ATPase, modulation of the crossbridge cycle rate, and the cost of this "regulatory" mechanism contribute to the high economy of tension maintenance in smooth muscle. The roles of **latch bridges** or thin filament proteins in the regulation or modulation of smooth muscle contractility is yet to be fully resolved and universally accepted.

Much recent attention has been focused on mechanisms of alteration of the "Ca^{2+}-sensitivity" of smooth muscle. It is well documented that under conditions of constant or fixed $[Ca^{2+}]$, the level of force generated by smooth muscle can be modulated. Several mechanisms for Ca^{2+} sensitization/desensitization have been put forth, with much attention given to regulation of myosin phosphatase and C-kinase pathways. Inhibition of myosin phosphatase activity via activation of the monomeric GTP-binding protein RhoA and Rho-kinase has been proposed by the Somlyos *et al.* (1999). While evidence for protein kinase C and actin-based regulatory mechanisms championed by the Morgan Laboratory (Horowitz *et al.*, 1996). This will continue to be an area of very active research in the smooth muscle field.

VI. Summary

This chapter focuses largely on muscle structure in relation to contractility. Mechanisms of regulation of intracellular Ca^{2+} and excitation-contraction coupling are treated in more detail elsewhere (Chapters 53 and 54). Muscle structure is first developed in conjunction with mechanics, leading to the formulation of the sliding-filament theory. Muscle contractility continues with further characterization of the relationships among force, velocity, work, and energy utilization. The mechanism at the molecular level, that is, the crossbridge cycle, is similarly developed, matching structural and biochemical knowledge with transient mechanical studies. The energy requirements of the crossbridge cycle lead to presentation of energy use in intact muscle, which is considered in detail in muscle energetics. Energy use is then followed by a discussion of muscle metabolism and the route for synthesis of ATP necessary for contractile activity. The chapter then shifts from considerations of muscle in general to comparative muscle physiology and behavior of different fiber types. In particular, smooth muscle is treated in detail, and its regulation, mechanical properties, and energetics are considered and contrasted to those of striated muscle.

Bibliography

Bloom, W., and Fawcett, D. W. (1968). "A Textbook of Histology." W. B. Saunders, Philadelphia.

Bozler, E. (1977). Introduction: Thermodynamics of smooth muscle contraction. *In* "The Biochemistry of Smooth Muscle" (N. L. Stevens, Ed.), pp. 3–14. University Park Press.

Dillon, P. F., Aksoy, M. O., Driska, S. P., and Murphy, R. A. (1981). Myosin phosphorylation and the cross-bridge cycle in arterial smooth muscle. *Science* **211**, 495–497.

Fenn, W. O. (1924). The relation between the work performed and the energy liberated in muscular contraction. *J. Physiol.* **58**, 373–395.

Fiske, C. H., and SubbaRow, Y. (1927). The nature of inorganic phosphate in voluntary muscle. *Science* **65**, 401–403.

Gillis, J. M., Thomason, D., Lefèvre, J., and Kretsinger, R. H. (1982). Parvalbumins and muscle relaxation: A computer simulation study. *J. Muscle Res. Cell Mot.* **3**, 377–398.

Gordon, A. M., Huxley, A. F., and Julian, F. J. (1966). The variation in isometric tension with sarcomere length in vertebrate muscle fibres. *J. Physiol.* **184**, 170–192.

Hai, C.-M., and Murphy, R. A. (1988). Regulation of shortening velocity by cross-bridge phosphorylation in smooth muscle. *Am. J. Physiol.* **255**, C86–C94.

Hai, C.-M., and Murphy, R. A. (1989). Ca^{2+}, crossbridge phosphorylation, and contraction. *Annu. Rev. Physiol.* **51**, 285–298.

Hai, C.-M., and Murphy, R. A. (1992). Adenosine 5'-triphosphate consumption by smooth muscle as predicted by the coupled fourstate crossbridge model. *Biophys. J.* **61**, 530–541.

Harris, D. E., and Warshaw, D. M. (1993a). Smooth and skeletal muscle actin are mechanically indistinguishable in the *in vitro* motility assay. *Circ. Res.* **72**, 219–224.

Harris, D. E., and Warshaw, D. M. (1993b). Smooth and skeletal muscle myosin both exhibit low duty cycles at zero load *in vitro. J. Biol. Chem.* **268**, 14765–14768.

Hartshorne, D. J. (1987). Biochemistry of the contractile process in smooth muscle. *In* "Physiology of the Gastrointestinal Tract" (L. R. Johnson, Ed.), Vol. 1, pp. 423–482. Raven Press, New York.

Heineman, E. W., and Balaban, R. S. (1990). Control of mitochondrial respiration in the heart *in vivo. Annu. Rev. Physiol.* **52**, 523–542.

Heumann, H. G. (1973). Smooth muscle: Contraction hypothesis based on the arrangements of actin and myosin filaments in different states of contraction. *Philos. Trans. R. Soc. London (Biol.)* **B265**, 213.

Higuchi, H., Yanagida, T., and Goldman, Y. E. (1995). Compliance of thin filaments in skinned fibers of rabbit skeletal muscle. *Biophys. J.* **69**, 1000–1010.

Hill, A. V. (1965). "Trails and Trials in Physiology." Arnold, London.

Horowitz, A., Menice, C. B., Laporte, R., and Morgan, K. G. (1996). Mechanisms of smooth muscle contraction. *Physiol. Rev* **76**, 967–1003.

Huxley, A. F. (1957). Muscle structure and theories of contraction. *Prog. Biophys. Chem.* **7**, 255–318.

Huxley, A. F., and Niedergerke, R. (1954). Interference microscopy of living muscle fibres. *Nature* **173**, 971–973.

Huxley, A. F., and Simmons, R. M. (1971). Proposed mechanism of force generation in striated muscle. *Nature* **233**, 533–538.

Huxley, H. E. (1966). A personal view of muscle and motility mechanisms. *Annu. Rev. Physiol.* **58**, 1–19.

Ishida, Y., Reisinger, I., Walliman, T., and Paul, R. J. (1984). Compartmentation of ATP synthesis and utilization in smooth muscle: Roles of aerobic glycolysis and creatine kinase. *Mol. Cell. Biochem.* **133/134**, 39–50.

James, J. H., Wagner, K. R., King, J. K., Leffler, R. E., Upputuri, R. K., Balasubramaniam, A., Friend, L. A., Shelly, D. A., Paul, R. J., and Fischer, J. E. (1999). Stimulation of both aerobic glycolysis and Na(+)-K(+)-ATPase activity in skeletal muscle by epinephrine or amylin. *Am. J. Physiol.* **277**, E176–186.

Jewell, B. R., and Wilkie, D. R. (1958). An analysis of the mechanical components in frog's striated muscle. *J. Physiol.* **143**, 515–540.

Krisanda, J. M., and Paul, R. J. (1984). Energetics of isometric contraction in porcine carotid artery. *Am. J. Physiol.* **246**, C510–C519.

Kushmerick, M. J. (1983). Energetics of muscle contraction. *In* "Handbook of Physiology (Section 10)—Skeletal Muscle" (L. Peachy and R. H. Adrian, Eds.), pp. 189–236, Oxford University, New York.

de Lanerolle, P., and Paul, R. J. (1991). Myosin phosphorylation/dephosphorylation and the regulation of airway smooth muscle contractility. *Am. J. Physiol.* **261**, L1–L14.

Lohmann, K. (1934). Über die enzymatische Aufspaltung der Kreatinephosphorsaüre; zugleich ein Beitrag zum Chemismus der Muskelkontraktion. *Biochem. Z.* **271**, 264–277.

Lundsgaard, E. (1938). The biochemistry of muscle. *Annu. Rev. Biochem.* **7**, 377–398.

Mahdavi, V., Streher, E. E., Periasamy, M., Wieczorek, D. F., Izumo, S., and Nadal-Ginard, B. (1986). Sarcomeric myosin heavy chain gene family: Organization and pattern of expression. *Med Sci. Sports Ex.* **18**, 299–308.

Murray, J. M., and Weber, A. (1974). The cooperative action of muscle proteins. *Sci. Am.* **230**, 58.

Needham, D. M. (1971). "*Machina carnis*. The Biochemistry of Muscular Contraction in Its Historical Development," pp. 1–782. University Press, Cambridge.

Obara, K., Szymanski, P. T., Tao, T., and Paul, R. J. (1996). Effects of calponin on isometric force and shortening velocity in permeabilized *taenia coli* smooth muscle. *Am. J. Physiol.* **270**, C481–487.

Paul, R. J. (1983a). Physical and biochemical energy balance during an isometric tetanus and steady state recovery in frog sartorius at 0° C. *J. Gen. Physiol.* **81**, 337–354.

Paul, R. J. (1983b). Aerobic glycolysis and ion transport in vascular smooth muscle. *Am. J. Physiol.* **244**, C399–C409.

Paul, R. J. (1990). Smooth muscle energetics and theories of cross-bridge regulation. *Am. J. Physiol.* **258**, C369–C375.

Pette, D., and Staron, R. S. (1990). Cellular and molecular diversities of mammalian skeletal muscle fibers. *Rev. Physiol. Biochem. Pharmacol.* **116**, 2–47.

Rayment, I. (1996). The structural basis of the myosin ATPase activity. *J. Biol. Chem.* **271**, 15850–15853.

Rayment, I., Smith, C., and Yount, R. G. (1996). The active site of myosin. *Annu. Rev. Physiol.* **58**, 671–702.

Somlyo, A. V. (1980). Ultrastructure of vascular smooth muscle. *In* "Handbook of Physiology (Section 2)—The Cardiovascular System, (Vol. II) Vascular Smooth Muscle," pp. 33–67. American Physiological Society, Bethesda, MD.

Somlyo, A. P., Wu, X., Walker, L. A., and Somlyo, A. V. (1999). Pharmacomechanical coupling: the role of calcium, G proteins, kinases and phosphatases. *Rev. Physiol. Biochem. Pharmacol.* **134**, 201–234.

Warshaw, D. M., Desrosiers, J. M., Work, S. S., and Trybus, K. M. (1990). Smooth muscle myosin cross-bridge interactions modulate actin filament sliding velocity *in vitro*. *J. Cell Biol.* **111**, 453–463.

Weiss, J. N., and Lamp, S. T. (1987). Glycolysis preferentially inhibits ATP-sensitive K^+ channels in isolated guinea pig cardiac myocytes. *Science* **238**, 67–69.

Wingard, C. J., Paul, R. J., and Murphy, R. A. (1997). Energetic cost of activation processes during contraction of swine arterial smooth muscle. *J. Physiol. (Lond.)* **501**, 213–223.

Woledge, R. C., Curtin. N. A., and Homsher, E. (1985). "Energetic Aspects of Muscle Contraction," pp. 1–357. Academic Press, London/New York.

Edna S. Kaneshiro, Michael J. Sanderson, and George B. Witman

57

Amoeboid Movement, Cilia, and Flagella

I. Introduction

Movement is characteristic of life, whether it be of whole organisms, individual cells, or cytoplasmic components. For a system to cause movement, a force must be exerted on a mass over some distance; thus movement requires expenditure of energy. Several basic mechanisms for the generation of force have evolved (Table 1), the most well-characterized of which are: (1) actin-filament–based systems, which utilize the energy from adenosine triphosphate (ATP) (e.g., muscle); (2) microtubule-based systems, which utilize the energy from ATP (e.g., eukaryote cilia and flagella); (3) spasmin-based systems, which utilize energy from the Ca^{2+} chemical potential (e.g., spasmonemes); (4) prokaryotic rotary motors, which utilize energy from a transmembrane ion gradient (e.g., bacterial flagella). Muscle contraction was discussed in the preceding chapter. The present chapter covers actin-filament–based movement in nonmuscle cells as well as microtubule-based movements. Spasmonemes and bacterial flagella are discussed in the next chapter.

II. Amoeboid Movement and Actin-Based Systems

Several cellular functions are attributed to actin-based systems and include pseudopod formation, vesicle transport, and exocytosis and endocytosis in cells such as macrophages, neutrophils, and retinal pigmented epithelial cells (Table 2). Temperature-independent precipitation or patching of antibodies bound to cell surfaces is followed by actin-based aggregation of the patches called **capping**, which requires metabolic energy and is temperature dependent. The contractile ring in the cortex of cells undergoing cytokinesis contains actin, myosin, and other actin-binding proteins. Thus, actin-based motility also contributes to the basic cellular events involved in cell proliferation.

Actin-based motility can be divided into two basic forms: (1) changes in cell shape or position can be achieved by polymerizing or depolymerizing actin filaments and (2) polymer-

ized actin filaments can interact with myosin to form a contractile unit.

A. Actin Polymerization

One of the most dramatic examples of rapid actin polymerization was described by Tilney in the sperm of the sea cucumber, *Thyone*. The acrosome vacuole of unstimulated sperm is filled with actin monomers and profilin. The term **profilactin** (profilamentous actin) was originally coined to describe the unpolymerized state of precursors below the acrosome vacuole; thus this term should not be confused with profilin. Upon activation, the acrosome process forms as the result of polymerization of actin monomers into filaments. Filaments about 60–70 μm long are assembled within 10 s.

Actin monomers bind ATP, which is subsequently hydrolyzed during polymerization. Polymerization can occur by the addition of actin monomers at the plus (fast-growing) end as well as at the minus (slow-growing) end of actin filaments. The plus and minus ends correspond to the barbed and pointed ends, respectively, of heavy meromyosin (HMM)-decorated filaments. When associated with membranes, it is the plus end that is usually found adjacent to the membrane.

Inhibitors of actin polymerization include the cytochalasins, which are compounds that bind to the ends of actin filaments, resulting in the inhibition of monomer addition. At high concentrations, cytochalasins can even sever filaments. Cellular functions, such as cell locomotion, cytokinesis, and cell movements during morphogenesis, are inhibited by these alkaloid drugs. In contrast, the highly toxic compound phalloidin stabilizes actin filaments and inhibits actin depolymerization.

In recent years, dozens of proteins associated with actin-based contractile systems have been described in many different cell types. These actin-binding proteins sever, stiffen, nucleate, bundle, cap, cause sliding between, cross-link, polymerize, and depolymerize actin filaments, and others interact with actin monomers (Table 3). Some actin-binding proteins can function in more than one capacity; for example, they can cap and sever filaments. These molecules have at least one actin-binding site. Many, such as those involved

Copyright © 2001 by Academic Press. All rights of reproduction in any form reserved.

TABLE 1   Comparison of Systems for Force Generation

Property	Actin-filament–based	Microtubule-based	Spasmin-based	Eukaryotic rotary motor
Major protein	Actin	Tubulin	Spasmin	Flagellin
Molecular mass	42 kDa	50 kDa	18–20 kDa	50 kDa
Isoforms[a]	4 Muscle (α): cardiac, skeletal, vascular, and enteric; 2 nonmuscle (β and γ): cytoplasmic	2 Microtubule-building blocks (α, β); 2 microtubule-organizing centers (γ and δ)	Spasmin A (18 kDa) Spasmin B (20 kDa)	
Isoelectric point[a]	α 5.41 β 5.44 γ 5.47	α 4.75 β 4.69 γ 5.55	Spasmin A 4.7 Spasmin B 4.8 α Centrin 4.9 β Centrin 4.8	
Unpolymerized form	Monomer	Dimer	?	Monomer
Nucleotide required for polymerization	ATP (1/monomer)	GTP (2/dimer)	?	None
Polymer structure	Double helical filament	Hollow tube of 13 protofilaments (14 nm lumen)	Fibrous filament	Hollow cylinder (5 nm lumen)
Polymer diameter	8 nm	25 nm	4 nm	14–20 nm
Subunits/μm polymer	370 monomers	1600 dimers	?	2000 monomers
Subunits/turn	15	13	?	5.5

[a]Differences occur among various species.

in gel-sol transformations, also have binding sites for Ca^{2+} or the signal-transducing lipid phosphatidylinositol 4,5-bis-phosphate (PIP_2).

Profilins are proteins that can regulate actin polymerization. In the soil amoeba *Acanthamoeba* and in yeast, profilin has two actin-binding sites. It caps the free ends of actin polymers by forming dimers with the actin monomers, thus stopping fil-ament growth. In vertebrates, profilin has only one actin-binding site, and hence cannot cap actin filaments. Profilin and actin-depolymerizing factor inhibit actin polymerization. After an external stimulus, actin must first dissociate from such molecules before cell surface membrane extensions can proceed. The dissociation may be controlled by inositol lipids in the membrane.

TABLE 2   Functions of Actin-Based Motility Systems

Function	Examples
Locomotion	Vertebrate smooth, cardiac and skeletal muscles, amoeboid cells, neutrophils
Cytoplasmic streaming (cyclosis)	*Nitella, Physarum*
Feeding and endocytosis	*Amoeba*, macrophages, heliozoans, forams
Exocytosis and secretion	Neurotransmitter release
Cell proliferation	Contractile ring formed during cytokinesis
Reproduction	Invertebrate sperm acrosome process
Transport and absorption	Intestinal and kidney brush border microvilli
Sensory reception and transduction	Vertebrate stereocilia, squid photoreceptor
Tissue morphogenesis	Neurulation

B. Amoeboid Locomotion

1. Crawling of Amoeboid Cells

The mechanism of amoeboid movement was among the first nonmuscle actin-myosin systems to receive the attention of early cell biologists. Studies were first done mainly on large free-living amoebae such as *Amoeba proteus*. The amoeba cell body is differentiated into three regions: the endoplasm (sol) at the core of the cell occupied by organelles; a cortical ectoplasm (gel); and a shear zone separating the ectoplasm from the endoplasm (Fig. 1). Organelles are excluded from a thin clear hyaline layer adjacent to the cell surface membrane. This layer is thicker at the pseudopodial tip, where it is called the **hyaline cap**. *Amoebae* move by extension of the pseudopodium, attachment of the pseudopodial membrane to the substratum, and detachment of the membrane from the substratum at the posterior uroid, thus allowing the cell to translocate itself to an anteriorad location. Concomitantly, there is a movement of solated cytoplasm in the endoplasm anteriorly into the newly forming pseudopod and movement of cytoplasm in the gel of the cortex posteriorly. Similarly, rapid cytoplas-

TABLE 3 Some Actin-Binding Proteins

Protein	Subunit molecular mass (kDa)	Activity
Actin-binding protein (ABP)	250	Gelating
Actin depolymerizing factor (ADF)	19	Monomer binding, severing
α-Actinin	90–100	Bundling, gelating
β-Actinin	34–37	Capping
Actophorin-depactin	17	Monomer binding
Band 4.9	48–52	Bundling
Caldesmon	87	Bundling
Calpactin	36	Bundling, membrane binding
Capping protein	29–31	Capping
Cofilin	21	Polymer binding
DNAse-1	35	Monomer binding
Fascin	58	Bundling
Filamin	250	Gelating
Fimbrin	68	Bundling
Fragmin, severin	42	Capping, severing, nucleating
Gelactins I-IV	23–28	Gelating
Gelsolin	90	Capping, severing, nucleating
Myosin I	110–130	Sliding, membrane binding
Myosin II	175–200	Sliding
Ponticulin	17	Membrane binding
Profilin	12–15	Monomer binding, capping
Radixin	82	Capping
Spectrin	220–260	Gelating
Synapsin I	76–78	Bundling
Tropomyosin	35	Polymer binding
Troponin I	23	Polymer binding
Villin	95	Capping, bundling, severing, nucleating
120-kDa protein	120	Gelating

mic streaming (cyclosis) commonly occurs and is easily observed in large plant cells (such as *Nitella*) and in fungi (such as *Dictyostelium* and *Physarum*).

During locomotion, cell surface membrane is added at the forming pseudopodium and internalized at the posterior uroid of the cell. Large free-living amoebae have numerous large Golgi bodies (dictyosomes), correlating with the demand for active membrane turnover and recycling. The addition of new membrane material to the cell surface during locomotion places a high demand on the cell. Hence, when an amoeba is actively crawling along a substratum, it stops feeding and drinking (phagocytosis and pinocytosis), which are endocytotic processes known to also require actin, myosin, and ATP. Thus, amoebae apparently are unable to walk and eat at the same time.

Microinjection of the Ca^{2+}-sensitive bioluminescent protein aqueorin shows that the Ca^{2+} concentration is elevated at the tip of extending pseudopodia and at the posterior

uroid region. The pseudopod tip membrane is apparently more electrically active than the rest of the cell. Ionic currents, probably carried by Ca^{2+}, have been recorded by the sensitive vibrating microprobe technique (which measures extracellular currents concentrated near "hot spots").

2. Contractile Proteins

Ultrastructural studies of extracted amoebae indicate that the cortical region is filled with actin filaments (Fig. 2). These actin filaments can be decorated with heavy meromyosin (HMM) from vertebrate muscle. Actin associated with the cell membrane was thus recognized as a mechanism for explaining surface extensions and other cell shape changes that occur in amoeboid cells during crawling, feeding, and dividing. The first convincing biochemical indication that amoeboid movement was based on molecules and mechanisms similar to those of muscle was obtained by Korn and Pollard on the small soil

A

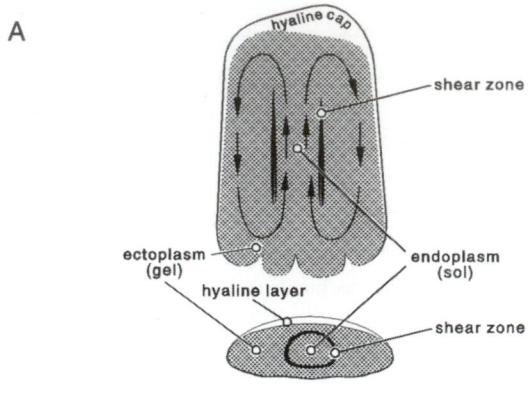

B

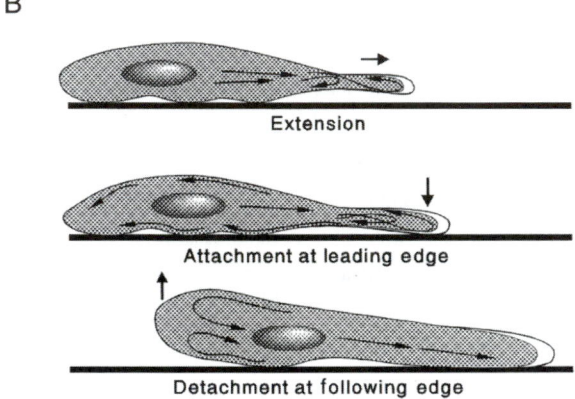

C

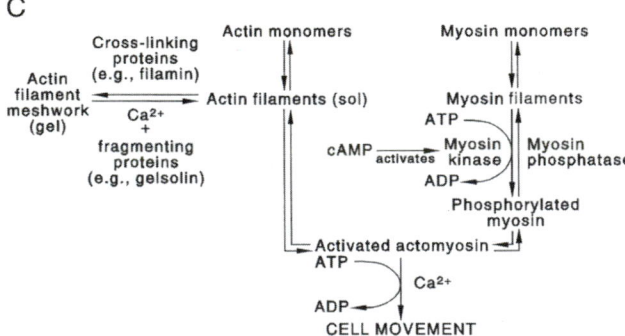

FIGURE 1. Amoeboid locomotion and cytoplasmic streaming. (A) The cytoplasm of a large free-living amoeba such as *Amoeba proteus* is differentiated into the solated endoplasm, the gelated ectoplasm at the cortex, and the clear region next to the cell surface membrane called the *hyaline layer,* which is enlarged at the tip of the pseudopod, where it is called the *hyaline cap.* Organelles move in the endoplasm toward the extending pseudopod. In the central endoplasm, actin filaments are presumably short and not highly cross-linked, whereas in the ectoplasm, where organelles move posteriad, they are long and in a gelatin state. (B) Locomotion of the cell involves three events: (1) The extension of the pseudopod. This extension involves the assembly of actin filaments by addition of monomers to actin filaments attached to the cell membrane. Movement of other molecules to the advancing pseudopod may be achieved by actin-myosin I interaction. (2) Attachment of the extended pseudopod membrane to the substratum is correlated with the aggregation of actin in this region. (3) The membrane at the posterior becomes detached from the substratum, thus allowing the cell to locomote. The cortical network of actin in the posterior region probably contracts by myosin II action, generating the force for the flow of cytoplasmic organelles in the

amoeba, *Acanthamoeba*, which can be grown to high densities in culture. Prior to these studies, two models for amoeboid movement were hotly debated: (1) the rear contraction model (posterior cortex contracts, pushing the cytoplasm forward) and (2) the fountain head model (molecules at the anterior fold, resulting in transformation of sol to gel and pulling of endoplasm forward). With recent biochemical, molecular, and immunochemical data, these models are being integrated into a more comprehensive view of this type of cellular movement.

Contractile proteins purified from *Acanthamoeba* indicate that actin has been highly conserved in different organisms. Amoeba myosin molecules, however, were found to be much smaller and less complex than vertebrate muscle myosin. Two forms of myosin have now been characterized in *Acanthamoeba*: (1) myosin II, which forms bipolar filaments like myosin in striated muscle; and (2) myosin I, which has a single head and a short tail and does not self-assemble into bipolar filaments as does myosin II. Myosin I, which can bind to actin and to cell membranes, causes cell shape changes in many nonmuscle cells by forces generated at the cell surface (Fig. 3).

It was demonstrated that even after the cytoplasm is removed from the amoeba (by suction or by allowing the cytoplasm to move into a glass capillary tube), cytoplasmic streaming continued; thus, the machinery and energy for this movement reside within the cytoplasm. The mechanism controlling the direction and rate of cytoplasmic streaming involves the cell surface. It is now generally accepted that extensions and retractions of cell surface in various types of pseudopods (as in microspikes and lamellopodia and filopodia of platelets, macrophages, and fibroblasts) is regulated by transmembrane signaling events such as those involving inositol lipids and G proteins. Stimulation of the cell surface results in the polymerization of actin by addition of monomers to filaments attached to the cell surface membrane. Hormones and growth factors stimulate cell surface ruffling activity in tissue culture systems. Thus, it is likely that transmembrane signals activate second messengers, cyclic nucleotides, and protein phosphorylations by protein kinases.

3. Sol–Gel Transformation

A current model for explaining the reversible transformation from sol to gel states in free-living amoeba cytoplasm proposes that the viscous gel contains cross-linked actin filaments. Cytoplasm isolated from cells that exhibit active streaming contains high concentrations of actin-binding proteins (such as myosin) and cross-linking and severing proteins. Proteins such as α-actinin and filamin can link actin at both ends, and thereby cross-link the filaments. Cross-linking produces a gel state of the actin network in the cytoplasm (endoplasm). If highly cross-linked, the gel is more rigid and less able to contract.

Proteins such as gelsolin and severin tightly bind actin, resulting in the displacement of adjacent monomers in the filament. The displacement causes the fragmentation or disintegration of the filament, resulting in shorter actin filaments. In

solated endoplasm. (C) Regulation of actin and myosin interactions for amoeboid locomotion is schematically diagrammed. (Adapted from Eckert, Randall, and Augustine, "Animal Physiology; Mechanisms and Adaptations.")

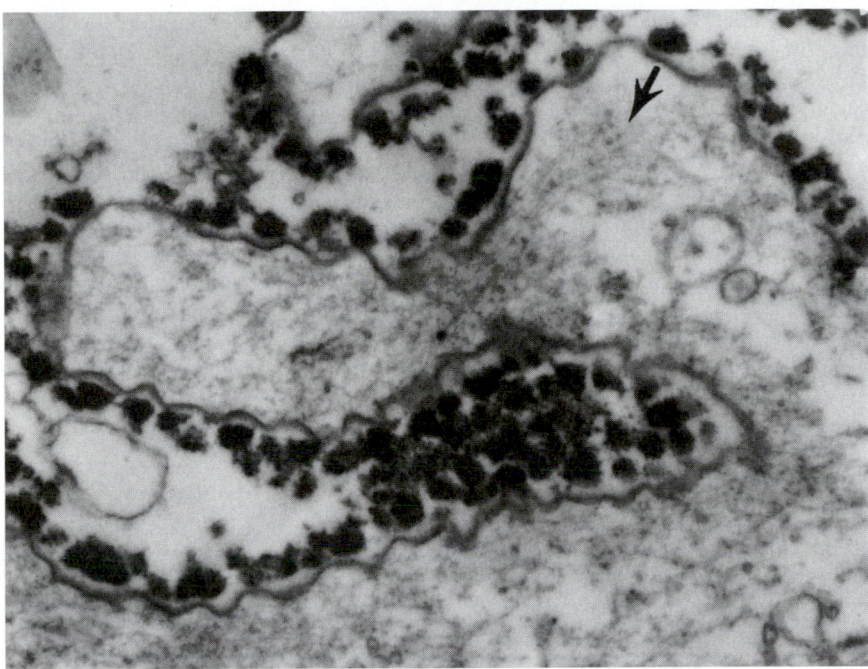

FIGURE 2. Electron micrograph of actin filament networks. The free-living amoeba *Chaos carolinensis* exhibits typical amoeboid locomotion. An extensive network of actin filaments (arrow) is detected by electron microscopy of the gelated ectoplasm cortex under the cell surface membrane. The cell surface membrane is clearly identified by its association with the stain Alcian blue, which binds surface glycocalyx mucopolysaccharides (dark granules). Magnification, ×49,000 (Courtesy of Vivian Nachmias.)

response to increased intracellular Ca^{2+} and/or H^+ concentration, the cross-links between actin filaments are broken, and the actin filaments are now present as short fragments and transform the cytoplasm from the gel to the sol state.

4. Movement of Cytoplasm and Organelles

Amoeboid locomotion is accompanied by cytoplasmic streaming involving reversible sol-to-gel transformations of actin filament networks. However, myosin causes cytoplasm to move by providing the motor for the sliding of actin filaments (see Fig. 1). In the gel state, cytoplasmic actin filaments exist in a network or mesh with phosphorylated myosin. In the sol state, the short actin filaments interact with the small myosin molecules and contraction occurs using the energy derived from ATP hydrolysis. The forces generated by the contraction in these regions are then transferred to surrounding gel regions. Myosin I provides this motor function, and

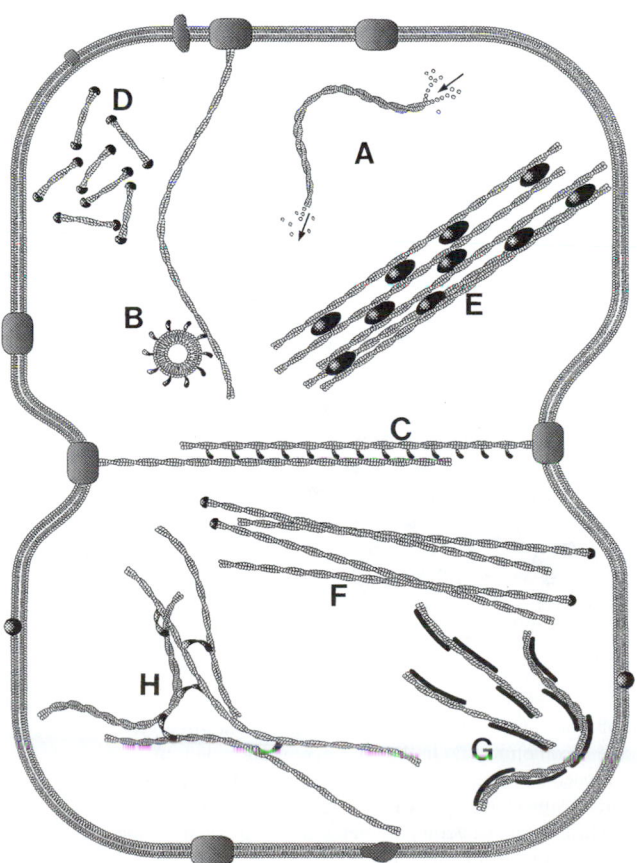

FIGURE 3. Mechanisms of actin functions and the role of actin-binding proteins. A number of actin-binding proteins serve to change the state of actin filaments and their aggregation. (A) Polymerization of monomers into filaments and depolymerization of the filaments can produce movement and cell shape changes. (B) A motor molecule attached to a vesicle membrane can cause movement of the organelle along actin filaments. (C) Sliding of actin filaments past each other in the contractile ring is aided by the action of myosin motor molecules during cytokinesis. Cell shape changes can also occur by motor molecules attached to the cell surface membrane interacting with membrane-attached actin filaments. (D–H) Various actin-binding proteins cause fragmenting (D), bundling (E), capping (F), stiffening (G), and gelating (H) of actin filaments.

myosin II may function in the generalized contraction of the cell cortex. The distribution of myosins has been described in *Acanthamoeba* by antibody binding studies. Myosin II is concentrated in the posterior uroid, whereas myosin I is concentrated in the anterior pseudopod where the cell membrane forms local ruffling and microspikes. *Acanthamoeba* myosin I can interact directly with membrane lipids, especially acidic lipids such as phosphatidylserine and inositol phospholipids, which may enable linking of the regulatory mechanism at the membrane with movement in the cytoplasm. Myosin bound to specific lipids in the membrane may be the motor for movement of the cell surface and organelles along actin filaments (see Fig. 3). The actin network of many different cell types contains actin-binding proteins, such as the cross-linking protein filamin and the Ca^{2+}-sensitive proteins severin, gelsolin, and fragmin. Some of these proteins bind to integral membrane proteins, some link actin with membrane-associated proteins (e.g., erythrocyte spectrin), and others (e.g., calpactin and lipocortin) bind to membrane phospholipids. Because PIP_2 binds to myosin I, α-actinin, profilin, and gelsolin, this lipid may be directly involved in transduction of signals across the membrane to the cytoplasmic actin network.

"Unconventional" (nonmuscle) myosins are also involved in the movement of organelles and vesicles along actin filaments. More than a dozen unconventional myosins have been identified, suggesting that different myosins have specialized functions within the cell.

5. Chemotaxis

The amoeboid stage of the slime mold *Dictyostelium* exhibits chemotactic behavior. It moves up a cyclic adenosine mono-phosphate (cAMP) concentration gradient (Fig. 4). Upon receptor binding by the chemoattractant at the cell surface membrane, increases occur in cortical actin polymerization, cell surface ruffling, and microspike formation. When a pseudopod extends, this establishes a polarity to the cell and the direction of cytoplasmic flow. These foci of membrane activity may be sites of actin polymerization. It is expected that the part of the cell closest to the higher concentration of attractant molecules would have higher receptor occupancy and local transmembrane signals, ion currents, second messengers, and protein phosphorylations. Thus, the signals resulting from cAMP binding are likely to enhance the polymerization of actin filaments near the cell membrane, resulting in pseudopodial extension in that region and movement up the chemoattractant gradient.

C. Contractile Ring

Cytokinesis is accomplished by the pinching of the cell into daughter cells by forces in the cell cortex. The cleavage furrow of dividing cells contains actin filaments recruited from other parts of the cell and arranged parallel to the membrane in a belt called the **contractile ring**. The location of the contractile ring is controlled by and develops in the plane bisecting the mitotic spindle. The contractile ring also contains myosin, α-actinin, and filamin (Fig. 5). In smooth muscles and in various types of nonmuscle cells, myosin light chains are regulated by phosphorylation catalyzed by Ca^{2+}-calmodulin-associated myosin light chain kinases. Phosphorylation allows the myosin to interact with actin, resulting in contraction; dephosphorylation causes dissociation of actin and myosin. Similarly, the division of a cell by the contractile ring within the cell cortex involves

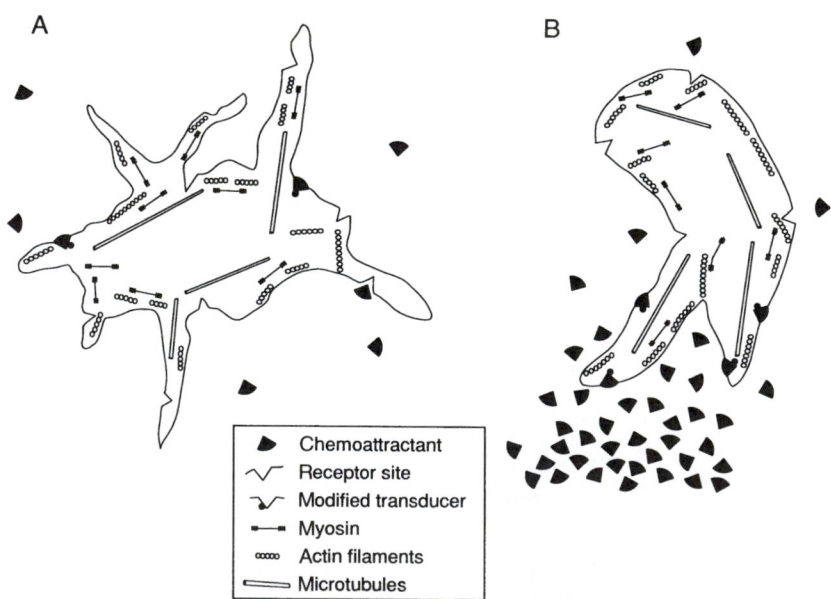

FIGURE 4. Chemotactic responses in amoeboid cells. (A) In the absence of an attractant or repellent concentration gradient, the chemotactic cell exhibits normal random pseudopod extensions and locomotion. (B) In the presence of a gradient of chemoattractant, more receptor sites on the cell membrane facing the gradient become occupied, thus changing the conformation of the receptor or modifying the transducer. Transducer-induced transmembrane signals result in cascade reactions involving intracellular second messengers and an increase in cyclic nucleotides, which in turn stimulate actin polymerization in that part of the cell cortex. The resulting pseudopod extension causes the cell to turn and move up the attractant gradient.

a cascade of protein phosphorylations. After cell division, the components of the contractile ring disperse.

D. Embryogenesis

Movement associated with whole tissues has long been recognized during a number of developmental events, such as neurulation (Fig. 6). Neurulation involves epithelial cells with apical actin filament bundles arranged parallel to the cell surface. Cells in the epithelial sheet are cemented together at the apical pole by belt desmosomes. Thus, contraction at the apexes of the cells and the folding of the epithelial sheet are highly coordinated. Contraction presumably results from sliding of actin filaments caused by interaction with myosin.

E. Microvilli

Brush borders in intestinal and kidney epithelia consist of extensive 1- to 2-μm-long outfoldings of the cell surface membrane called **microvilli**. Microvilli contain bundles of actin filaments cross-linked by the actin-associated proteins fimbrin and villin (Fig. 7). Fimbrin aligns actin filaments in a parallel paracrystalline array. Villin has three actin-binding sites and, like fimbrin, also bundles actin; but at high Ca^{2+} concentrations, villin severs the filament and caps the plus end at the break. The actin filaments are linked laterally to the cell membrane by myosin I associated with the Ca^{2+}-binding protein calmodulin. Movement of the cell membrane relative to the actin bundle may be related to the high turnover and sloughing of intestinal brush border membranes.

At the tips of microvilli, the filaments end in an electron-dense amorphous structure. The tip of the microvillus corresponds to the actin filament plus end. At the bases of microvilli, the filaments are anchored in a terminal web composed of the cytoskeletal elements myosin II, tropomyosin, spectrin, and caldesmon (another actin-binding protein). Villin is absent from

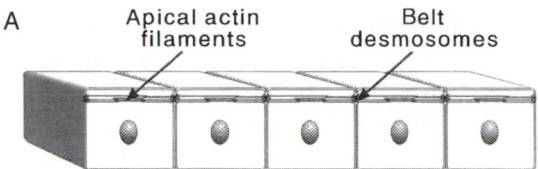
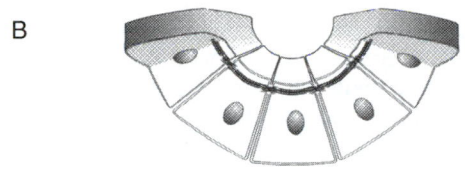

Folding of epithelial sheets

FIGURE 6. Mechanism of epithelial folding during embryogenesis. (A) Actin filament networks are formed at the apical pole of epithelial cells in a sheet. They are aligned parallel to the apical cell membrane. Belt desmosomes serve as welding structures between the cells. (B) When the apical actin network contracts, presumably involving myosin interaction with the filaments, the entire epithelial sheet undergoes synchronous folding.

the terminal web. Myosin II and tropomyosin may interact with actin in the terminal web by mechanisms similar to that in muscle. The terminal web complex may be responsible for maintaining the stiff cell surface microvillus extensions, thereby ensuring the maximum absorptive area of the epithelial cell.

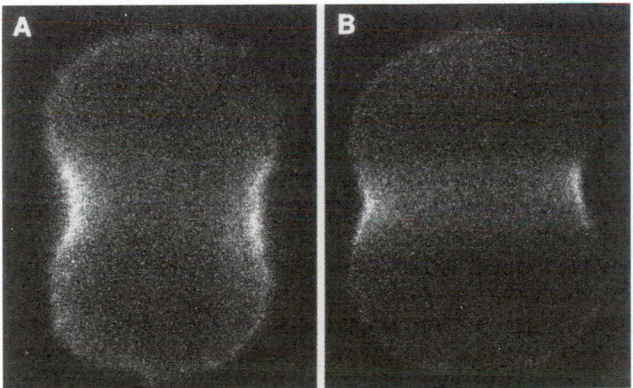

FIGURE 5. Contractile ring in the cortex of a dividing human cell. (A) Antibodies directed against purified actin were tagged with a fluorescent label and reacted with a HeLa cell at early telophase of mitosis. The antibody bound the actin filaments concentrated in the cleavage furrow. (B) Similar to (A) but using anti-myosin II antibodies demonstrating the concentration of myosin in the contractile ring. Radixin and α-actinin are also concentrated in the cleavage furrow. (Courtesy of Pamela Maupin and Thomas Pollard.)

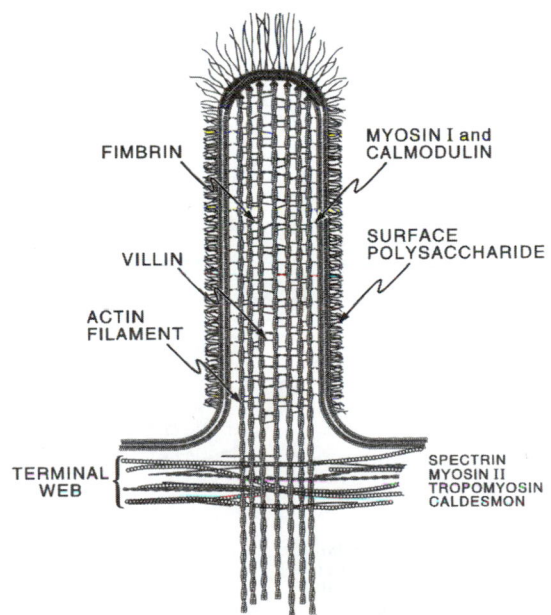

FIGURE 7. Actin in microvilli. The intestinal microvillus is packed with parallel-bundled actin filaments. The filaments are held together by actin-binding bundling molecules such as villin and fimbrin. The filaments are also linked to the membrane through actin-binding myosin I that is associated with calmodulin. In the terminal web, other actin-binding proteins, caldesmon and spectrin, tropomyosin, and myosin II, anchor the actin bundle into the cytoplasm, attach the terminal web to the cell surface membrane, and generate tension to maintain the stiffness of the microvillus structure.

F. Sensory Reception

Among vertebrates, hair cells are found in the organs for hearing (cochlea), balance and equilibrium (semicircular canals), and the lateral line system of fish and amphibians. When present, hair cells contain one true cilium, the **kinocilium**. Since adult mammals lack a kinocilium in the cochlea, the kinocilium may not be critical for hair cell function. The actual mechanoreceptors of the hair cells are the **stereocilia**, which are actin-filament–filled microvilli unrelated to true cilia. During development, the filaments grow distally (plus end) to form the microvilli and then firmly anchor into the cytoplasm by monomer addition; this probably occurs at the minus end to form an extension of the actin bundle deep into the cytoplasm. Stereocilium length increases from one side of the cell toward the other. The stereocilia are arranged in a teepee-like fashion with their tips touching the tips of neighboring stereocilia (Fig. 8). Thin bridging material links the tip of one stereocilium to what may be ion channels in the membrane at the tip of adjacent stereocilia. Displacement of the tuft in the direction of the longest stereocilia results in depolarization of the hair cell, whereas displacement toward the shortest stereocilia results in a hyperpolarizing receptor potential. Depolarizing stimuli produce greater potential changes than equivalent hyperpolarizing stimuli. These processes are discussed in more detail in Chapter 36.

Microvilli in the squid photoreceptor contain rhodopsin (visual pigment) and a single actin filament. Because light disrupts the filament, it may function with the visual pigment in transducing light energy.

III. Eukaryote Cilia and Flagella

Eukaryote cilia and flagella should not be equated to prokaryote (bacterial) flagella (described in the next chapter) nor to the stereocilia (an unfortunate misnomer). As described previously, stereocilia are microvillar structures. Prokaryote flagella are extracellular organelles with a distinct substructure, biochemical composition (see Table 1), and mechanism of force generation. Eukaryote cilia and flagella are intracellular organelles that are surrounded by the cell surface membrane. Force generation uses the energy released from ATP hydrolysis and involves interaction between microtubules and dynein ATPases.

A. Beat Patterns

Cilia were first observed in about 1675 by van Leeuwenhoek, who described them as "little legs" of *Paramecium*. Eukaryote cilia and flagella have similar substructures and their functions are based on similar mechanisms: thus these terms will be used interchangeably. Although cilia and flagella are identical in diameter (ca. 0.2 μm), cilia are usually shorter (5–15 μm) and often occur in large numbers and in close proximity to each other. Cilia exhibit asymmetrical beat cycles (effective or power stroke and recovery stroke), causing fluid to move parallel to the cell surface in the direction of the effective stroke (refer to Table 4 for functions). The movement of cilia propels small organisms through the medium, or moves the medium or mucus along tissue surfaces. During the effective stroke, the cilium is relatively straight with a bend near the base. During the recovery stroke, the bend propagates from the base to the tip, thus offering minimal resistance to the medium (see Fig. 17).

In contrast to cilia, flagella usually are longer (up to 100–200 μm or more), occur singly or in low numbers, and exhibit relatively symmetrical waves that propagate along their lengths (either base-to-tip or tip-to-base), causing fluid to move parallel to the flagellar long axis. However, these distinctions are not absolute; for example, some flagella can beat with either a ciliary or flagellar type of waveform.

Flagella occur in all of the major eukaryotic phylogenetic groups, with the possible exception of some lower fungi, which have lost them during evolution. Nematodes, crustaceans, and adult insects lack motile cilia, but they do have sensory cilia.

B. Structure

Familiarity with the ultrastructure of the cilium is essential for understanding how the organelle functions. The major regions of the cilium, the shaft, the transition zone, and the basal body (kinetosome), are diagrammed as observed by longitudinal sections and cross-sections in Fig. 9. Thin section freeze-fracture and other ultrastructural techniques such as deep etching have helped elucidate substructures, such as those of the ciliary membrane and dynein arms.

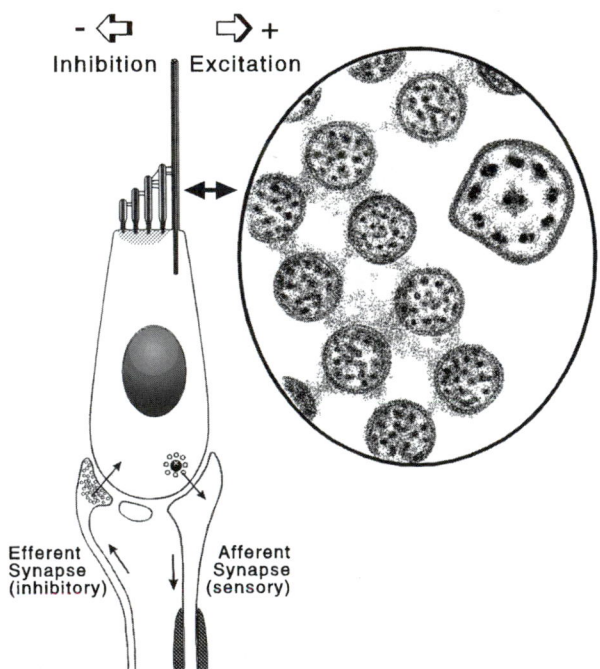

FIGURE 8. Stereocilia and a kinocilium of a hair cell. The stereocilia are actually actin-filled microvilli. Linkers attach the tips of adjacent stereocilia. The single kinocilium is a typical cilium with a 9 + 2 axoneme pattern (inset, right). Mechanostimulation that bends stereocilia in different directions causes increased or decreased rates of firing of the afferent second order neuron. (Adapted from Eckert, Randall, and Augustine, "Animal Physiology; Mechanisms and Adaptations.")

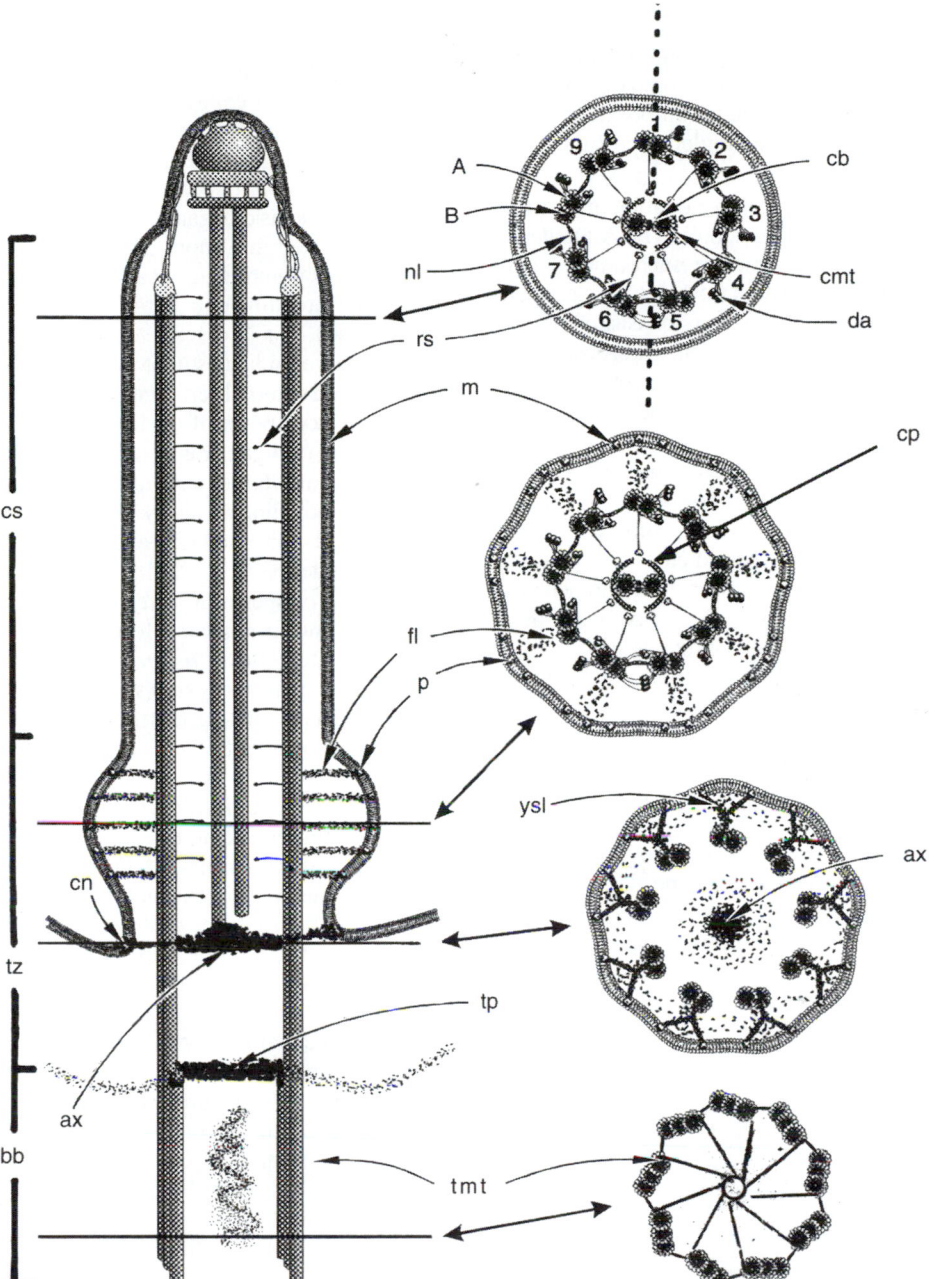

FIGURE 9.   Longitudinal and cross-sectional views at different levels of the cilium. A longitudinal section illustrates the different regions of the cilium: the basal body (bb), the transition zone (tz), and the ciliary shaft (cs). The structure of the tip depends on species. The ciliary membrane (m) is continuous with the cell surface membrane. The outer doublet microtubules in the ciliary shaft (shown in cross-section as viewed from base to tip) are each composed of an A (complete) microtubule and B (incomplete) microtubule. The outer dynein arms (with three large globular heads) (da) and the inner dynein arms (with two large globular heads) are attached to the A microtubule. At the 5–6 bridge, the dynein arms are permanently linked to the adjacent doublet. Doublet microtubules are joined by nexin links (nl). Radial spokes (rs) extend centrally toward the pair of central singlet microtubules (cmt), which are linked by bridges (cb); the enlarged spoke heads are in close proximity to the pair of central microtubule projections (cp). The plane (dotted line) that passes through the 5–6 bridge and doublet microtubule No. 1 is the plane of the effective power stroke of ciliary action. In the transition zone, fibrous linkers (fl) connect the doublet microtubules and the ciliary membrane at a level where the membrane bulges out and assumes a scalloped outline in cross-section. Within the membrane, intramembranous particles that form the ciliary membrane plaques (p) are in register with these linkers (also see Fig. 14). Proximally, at about the level of the axosome (ax), another intramembranous particle array, the ciliary necklace (cn) is in register with linking material that appears Y-shaped (ysl) in cross-sections. One of the central microtubules is embedded in the axosome, whereas the other originates distal to the axosome. The transition zone ends proximally at the terminal plate (tp) of the basal body. In the basal body (centriole), a C microtubule is added; the resulting triplet microtubules (tmt) form a characteristic pattern as viewed in cross-section.

1. Shaft

The common ultrastructural pattern of the axoneme (microtubules and associated cytoplasmic structures) is described as 9 + 2. That is, the axoneme is composed of 9 outer doublet microtubules and 2 central singlet microtubules.

a. Microtubules and Tubulins. Each outer doublet microtubule is composed of an A tubule and a B tubule. The A tubules, as well as the central singlets, are complete microtubules, composed of 13 protofilaments that extend the length of the microtubule. The B tubule is incomplete and may consist of 10 or 11 protofilaments (Fig. 10). The protofilaments are observed as fibrils that splay out at the tips of damaged microtubules, indicating strong longitudinal and weak lateral bonds between subunits. Protofilaments are composed largely of the protein tubulin. Tubulin exists as a 110-kDa heterodimer of α and β monomers. The protofilament diameter is 4 nm and the tubulin dimer repeat occurs longitudinally every 8 nm. The tubulin dimer dimension determines the periodicities of axonemal structures (note in the following discussion that nanometer distances between repeats of structures along the axoneme are all divisible by 8). In addition to tubulin, proteins called *tektins* form intermediate-filament-like filaments that are associated with three protofilaments of the A-tubule wall and extend the entire length of the microtubule.

By convention, micrographs of cross-sections are published as viewed from the base looking toward the tip of the cilium (see Fig. 9). Thus, cross-sections of the shaft (with the typical 9 + 2 axonemal pattern) show inner and outer dynein arms extending clockwise from the A tubule of one doublet microtubule toward the B tubule of the adjacent doublet microtubule. Convention adopted in the numbering of the outer doublets comes from their relative positions with respect to the direction of the effective stroke and structures located asym-

metrically within the axoneme. The effective stroke occurs in a plane (shown in Fig. 9) that passes through the permanent bridge between outer doublets 5 and 6 (5–6 bridge), through the central microtubules, and through the outer doublet opposite the 5–6 bridge. That doublet has been designated outer doublet 1, and numbering proceeds clockwise. In axonemes lacking the 5–6 bridge, the doublet at the leading edge of the effective stroke is designated doublet 1. Individual doublets are also designated in a more general way as doublet N with its dynein arms pointing at the adjacent doublet, $N + 1$.

Other microtubule patterns exist; these are described as 9 + 0, 9 + 1, 3 + 0, 6 + 0, etc. A comparative study of sperm ultrastructure indicates that there is a correlation between species with external fertilization and the typical 9 + 2 pattern. These include sperm of coelenterates, echinoderms, and protists. Sperm of organisms with internal fertilization, such as insects, some molluscs, and mammals, have large variations in structure, including a variety of accessory structures (Fig. 11).

Numerous inherited defects that affect axonemal structure have been described in humans. These defects cause abnormal ciliary or sperm motility, leading to chronic respiratory problems and, in males, infertility. Abnormal axonemes are also correlated with *situs inversus*, a condition characterized by displacement of organs (such as the heart) to the opposite side of the body. The abnormal body symmetry results from defective movement of the nodal cilia during embryonic development.

The tip of the cilium varies in different cell types, but in general, the central and A tubules extend further than do the B tubules (see Fig. 9). The central microtubules have caps and the A tubules contain plugs. The cap and plugs have linking structures that anchor the microtubules to the ciliary membrane. The cap material shares cross-reactive epitopes with kinetochores, structures associated with centromeres of chromosomes (see following), suggesting a common mechanism for anchoring microtubules in these two structures.

The central microtubules are joined by the central pair bridge. The central tubule projections (formerly referred to as the central sheath) are complex structures that extend from each central microtubule. Some of these projections have a periodicity of 16 nm along the axoneme. The projections on the two microtubules differ in size, which distinguishes one central microtubule from the other. In some organisms the central pair appears to rotate once per beat cycle. This has given rise to the hypothesis that the central microtubules serve as a distributor that activates outer doublet sliding.

b. Dynein Arms and ATPases. Axonemal ATPase activity was demonstrated in experiments conducted by Gibbons (1965), who combined biochemical methods with ultrastructural analyses of *Tetrahymena* cilia. He extracted cilia with a low-ionic-strength buffer containing EDTA and noted that the arms on microtubule doublets were removed. Along with the loss of doublet arms, the loss of ATPase activity of the cilia was observed. These armless cilia were then incubated with fractions of cilia extracts containing ATPase activity, an experimental manipulation that restored the arms to the microtubule doublets. Based on these observations, he coined the term *dynein* (force protein) for material in the arms, and

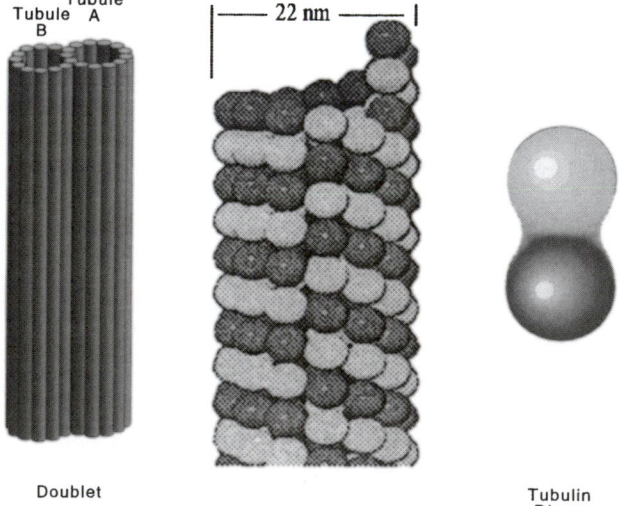

FIGURE 10. Subunit structure of microtubules. A complete microtubule is composed of 13 protofilaments (longitudinal arrangement of dimers). In a doublet, tubule A is complete and shares three protofilaments with tubule B, which is incomplete and composed of 10–11 protofilaments. Soluble tubulin occurs as a heterodimer of α- and β-tubulin.

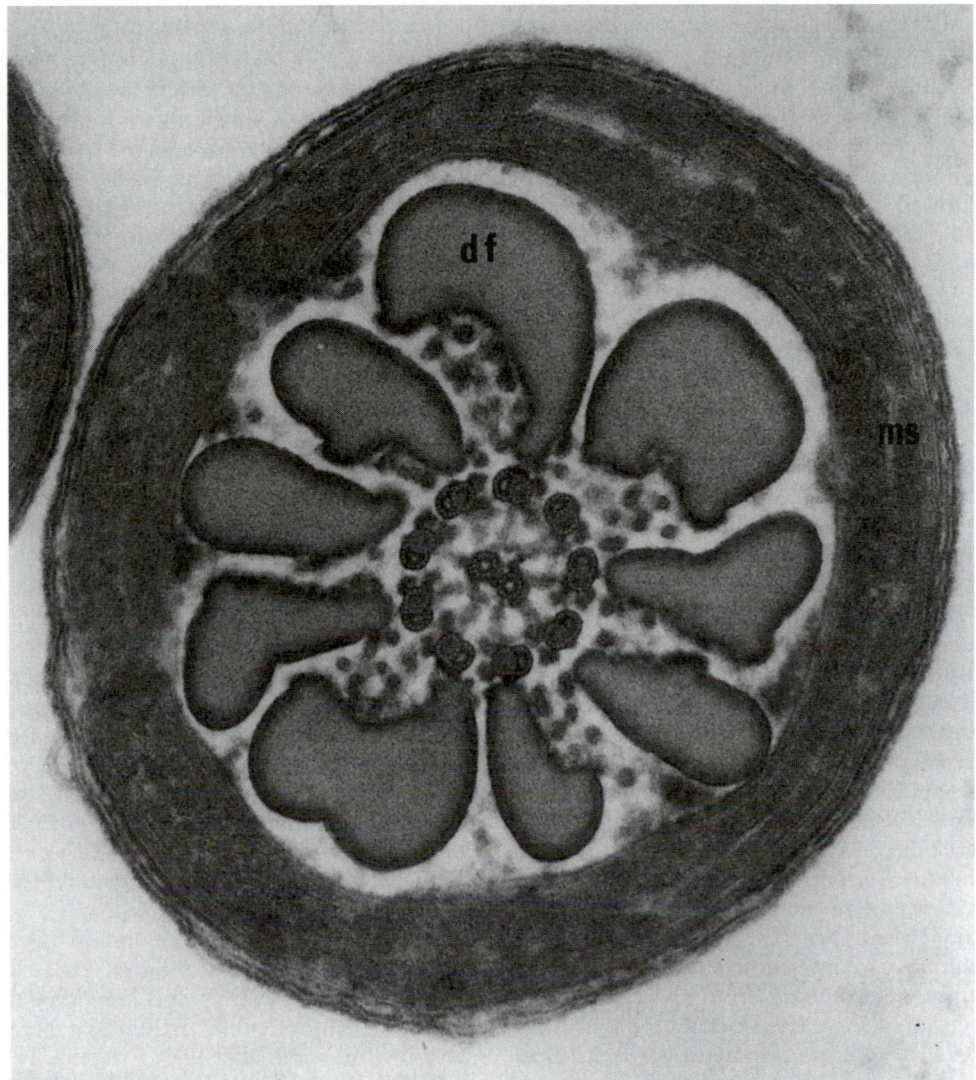

FIGURE 11.   Accessory structures typical of sperm flagella. This electron micrograph of the flagellum of a Chinese hamster spermatozoa shows dense fibers (df) and the mitochondrial sheath (ms) found in mammalian sperm cells. (Courtesy of David Phillips.)

postulated that the enzyme was essential for releasing energy from ATP to generate force for ciliary motility.

Most cilia contain two categories of arms, outer and inner arms. The outer dynein arms have a longitudinal periodicity of 24 nm along the microtubule doublets. Two or three large polypeptides (~500 000 MW) with ATPase activity correspond to the two or three heads (depending on species) seen in the arm's substructure; about a dozen smaller proteins also contribute to the arm's structure. Beat frequency of *Chlamydomonas* mutant flagella lacking outer arms is one-third that exhibited by wild-type cells. Thus, the outer arms provide much of the power for flagellar movement. There are at least two types of inner arms: dyads and triads with two and three globular heads, respectively. These arms repeat longitudinally at 96-nm intervals along the A tubules, with each repeat unit composed of one triad and two dyads. The inner arms contain at least six large polypeptide ATPases in addition to several intermediate- and low-molecular-weight

chains. Beating of *Chlamydomonas* inner arm mutants indicates that inner arms may regulate waveform symmetry. Actin and centrin (discussed later) have been identified at the base of the inner arms. These proteins may regulate the activity of these structures.

c. Linkers and Accessory Structures. Nexin (interdoublet links) connect the A tubules to the B tubules of the adjacent doublet microtubules. These structures have a longitudinal periodicity of 96 nm.

Radial spokes extend from A tubules toward the center of the axoneme. Each spoke ends in an enlargement called the *spoke head*, which interacts with the projections of the central microtubules. Spokes occur along the length of the doublet microtubules in groups of either two or three that have a longitudinal periodicity of 96 nm.

Within the flagellar shaft, accessory structures are commonly found between the axoneme and membrane. For example, in many sperm cells, prominent dense fibers are associated with each outer doublet microtubule (Fig. 11), and a fibrous sheath is found external to the dense fibers. In the midpiece, located between the principal piece (shaft) and the cell body of the sperm cell, extensive mitochondrial profiles are visible surrounding the axoneme and dense fiber structures.

In the phytoflagellate *Euglena* and in the trypanosomes, an accessory structure termed the *paraxial rod* runs along the length of the flagellum (Fig. 12). In *Euglena*, this structure and the paraxial ribbon are sites of mastigoneme (see following) attachment. Mutants with defects in the paraxial rod have impaired motility, indicating that the rod has an important function in flagellar movement. The substructure of the paraxial rod consists of fibrous helices with a pitch of 45° and a longitudinal periodicity of 54 nm.

d. Ciliary Membrane. The ciliary membrane is closely apposed to the axoneme along the length of the shaft. Membrane-to-microtubule linkers occur along the axoneme shaft, especially at the tips. Links joining the outer dynein arm and the membrane have been described in some cilia.

The outer surfaces of mucus-transporting cilia are decorated at their tips with fibrillar material called *crowns*; the crowns may enhance interaction with the mucus layer. Hemidesmosomes are formed at points along the trypanosome flagellar membrane (Fig. 13) where the parasite attaches to host tissue. In many flagellated protists, the site of flagellar emergence from the flagellar pocket (reservoir)

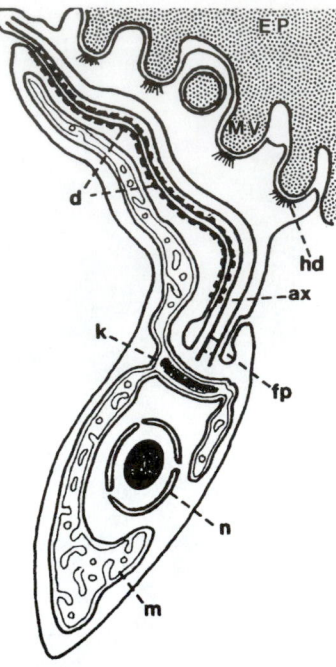

FIGURE 13. Flagellar membrane of a parasitic protozoan. *Trypanosoma brucei* attaches to the microvillous (MV) surface of the tsetse fly salivary gland epithelium (EP). The parasite anchors by forming hemidesmosomes (hd) to its host. Desmosomes (d) are commonly observed between the flagellar membrane and flagellar pocket (fp) at the site of flagellar emergence from the cell body. Desmosomes also occur along the entire length of the flagellum, linking it with the cell body. The kinetoplast (k), axoneme (ax), nucleus (n), and mitochondrion (m) are indicated. (From Vickerman and Tetley, 1990.)

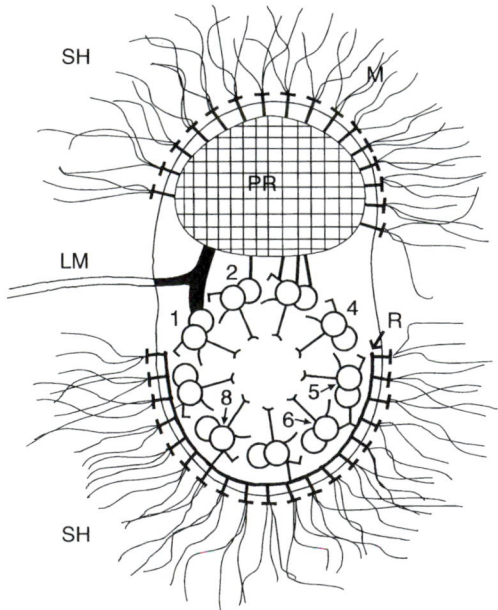

FIGURE 12. The *Euglena* flagellum has many accessory structures. The flagellum contains a paraxial rod (PR). Two half-sheaths (SH) partially enclose the flagellar membrane; one associated with the paraxial rod, and the other attached to a paraxial ribbon (R). The half-sheaths are sites of mastigoneme (M) attachment and of anchoring to the membrane and underlying cytoplasmic structures. A unilateral row of long mastigonemes (LM) also exhibits anchoring through the membrane. (Adapted from Bouck *et al.*, 1990.)

is characterized by desmosomes between the flagellar membrane and the membrane of the flagellar pocket.

Scales and mastigonemes (flagellar hairs) decorate the flagellar exterior in some unicellular organisms (see Fig. 12). These external structures have species-specific architecture useful for taxonomic assignments. The arrangement of mastigonemes can be classified as acroneme (single hair at the tip), pantoneme (many hairs surrounding the flagellum shaft), or stichoneme (hairs only on one side of the flagellum). Scales and mastigonemes, which are composed primarily of glycoproteins, are assembled in the Golgi complex and are seen within vesicles that eventually undergo exocytosis. Scales are easily removed from the membrane, indicating electrostatic bonding, whereas mastigonemes are firmly attached to the axoneme (or accessory structures) by linking material that extends through the membrane.

2. Transition Zone

a. Axoneme. The proximal region of the cilium, usually at the level of the cell surface membrane, contains unique structures not present in the shaft (see Fig. 9). The outer doublet microtubules continue in this zone, but here they lack dynein arms, nexin links, and radial spokes. Cilia of different species vary with respect to some structures; however, the proximal end of this region is normally delineated by the terminal plate of the basal body (kinetosome, centriole), which is always present regardless of whether the basal body is associated with a cilium. The number of transverse plates in this zone can vary, and there seems to be little consistency in the names used to refer to each one. The most distal has been called the *axosome plate*. A thin intermediate plate is also present in some cilia.

Embedded in the axosome plate is the axosome, an amorphous structure from which one of the two central microtubules arises. The other central microtubule begins at a distance above the axosome plate.

The transition zone of some flagella contains two central cylinders (proximal and distal) that have an H shape when viewed in longitudinal sections. Cross-sections of this region demonstrate fibers connecting the cylinders to the A tubules, forming a nine-pointed stellate pattern. Flagellar scission has been shown to occur along the plane of the nine-pointed stellate structure (see centrin following).

b. Transition Zone Membrane. The ciliary membrane bulges out in the transition zone and is associated with structures not present elsewhere. Intramembranous particles (IMPs) are organized in two distinct arrays, the ciliary necklace and the ciliary plaques (patches) (Fig. 14). Other IMP arrays are found in various species. A single plaque usually consists of three longitudinal rows containing three to six 10-nm IMPs. Nine plaques encircle the cilium in register with the nine peripheral

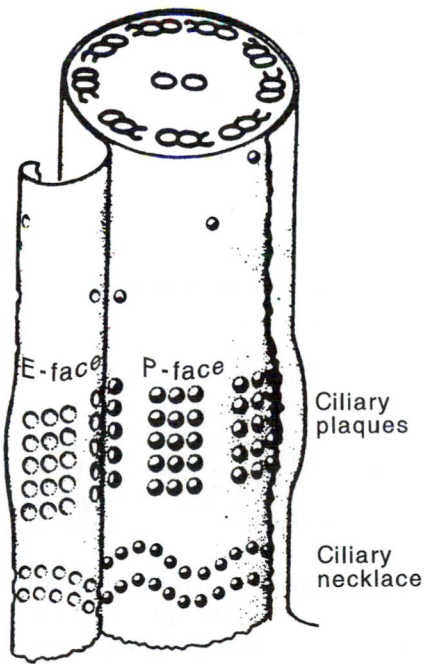

FIGURE 14. Intramembranous particles within ciliary membranes. The membrane is fractured between the lipid bilayer revealing particles on the protoplasmic face (P-face) with imprints on the external face (E-face). Characteristic arrays called the ciliary plaques occur as nine groups in register with the nine outer doublets of the axoneme. Each plaque consists of three longitudinal rows of three to six particles. The ciliary necklace is proximal to the ciliary plaques and consists of two or more zigzag rows of particles, depending on species. (Adapted from Sleigh, 1989.)

doublets in the axoneme and are attached to the doublets by linking structures. The rectangular plaques are located distal to the axosome plate. Proximal to the plaques, the ciliary necklace, consisting of one or more rows of IMPs, surrounds the cilium in a scalloped pattern. The scalloped pattern corresponds to the rims of champagne-glass-shaped linkers (Y-shaped in thin sections) that attach to the doublet microtubules. The plaque and/or necklace particles may be sites of Ca^{2+} entry into the cilium or may have Ca^{2+}-ATPase activity responsible for active pumping of this divalent cation out from the cilium against a concentration gradient. Studies on deciliation in the ciliate *Tetrahymena* demonstrate that the plaques are removed with the excised cilia and the necklace remains with the cell body. Thus, scission occurs where there are no microtubule-membrane links, dyneins, nexin links, IMPs, central microtubules, or radial spokes.

3. Basal Body

The basal body (kinetosome, centriole) is characterized by triplet microtubules connected by linkers between the A and C tubules of adjacent triplet microtubules (see Fig. 9). The A and B tubules are continuous with those of the transition zone; the C tubule is added in the basal body. In the proximal regions of many basal bodies, triplets and the intertriplet and radial linkers form a distinctive cartwheel pattern in cross-sections of the structure.

The basal body is essential for ciliary and centriolar growth and replication, probably serving as a template. Cilia appear and grow only at sites first occupied by basal bodies. How this occurs is not known, but new basal bodies characteristically develop at right angles to the existing basal body during replication. However, centrioles appear *de novo* during spermiogenesis in several plant species and the amoeboflagellate *Naegleria*. The basal body anchors the flagellum to the cell body, but is not essential for flagellar movement *per se*.

Several types of accessory structures can be found associated with basal bodies: cross-striated rootlets, basal feet, microtubular ribbons and bands, kinetodesmal fibers, and microfibrillar bundles. These appear to anchor the cilium to the cell body and help maintain proper orientation or polarity of cortical structures as well as the rest of the cell. Fibrous linkers connect the cell surface membrane to basal bodies of mechanoreceptor cilia of nudibranch statocysts. This relationship suggests a mechanism of signal transduction from cilium to membrane.

C. Flagellar Assembly and Intraflagellar Transport

Growth or regeneration of axonemes occurs at the tips of cilia, as first demonstrated by Rosenbaum and Child. When cells of the unicellular biflagellate alga *Chlamydomonas* are deflagellated, synthesis of flagellar proteins is stimulated. Autoradiograms of cells exposed to radiolabeled amino acids during flagellar regrowth showed a concentration of radioactivity at the flagellar tips, indicating that newly synthesized material is added to the tip.

Movement of axonemal precursors to the flagellar tip appears to be an active process. When flagella of *Chlamydomonas* are viewed by differential interference contrast microscopy, particles are observed moving from the base of the flagellum to the tip, and then returning to the base. Electron microscopy indicates that these particles are moving in the space between the axonemal microtubules and the flagellar membrane. This movement, termed **intraflagellar transport** (IFT), is powered by motors that move along the doublet microtubules (Fig. 15). Movement in the base-to-tip, or anterograde, direction is generated by a microtubule motor termed kinesin II. Movement in the opposite, or retrograde, direction is generated by a type of dynein termed cytoplasmic dynein 1b. Kinesin II has been shown to be necessary to move the inner dynein arms to the tip of the flagellum, and mutations in kinesin II block flagellar assembly. Cytoplasmic dynein 1b is also essential for flagellar assembly; defects in this dynein lead to accumulation of the IFT particles at the tip of the flagellum. The particles are multisubunit protein complexes containing at least 15 different polypeptides. IFT motors and particle proteins have been identified in the ciliated sensory neurons of the nematode *Caenorhabditis elegans* and in the connecting cilium of retinal photoreceptors, so IFT is probably involved in the assembly and maintenance of a broad range of cilia and flagella, including modified sensory cilia.

D. Mechanism of Ciliary Motility

1. Sliding of Doublet Microtubules

Early experiments by Hoffman-Berling showed that glycerinated striated muscle (models) that had lost small, soluble molecules after damage to the cell surface membrane contracted in the presence of ATP and Mg^{2+}. Sperm models prepared in a similar manner began swimming upon subsequent exposure to reactivation solutions containing ATP and Mg^{2+}. This indicates that flagellar motility, like muscle contraction, is also based on a system using ATP to generate force. As described earlier, Gibbons (1965) identified ATPase

activity in cilia extracts and correlated it with the presence of dynein arms on the outer doublets.

The sliding microtubule hypothesis was proposed by Satir (1968), who noted that at the tips of straight cilia all doublets ended at the same level, but in bent cilia the outer doublets along the convex outer side did not extend as far as those on the concave inner side. This suggested that the microtubules themselves did not shorten during force generation, and established the concept that the mechanism for cilia movement involves the sliding of doublet microtubules past each other, analogous to sliding of thick filaments past thin filaments in striated muscles.

Definitive evidence that microtubules can slide past each other was obtained by Summers and Gibbons (1971), who partially digested demembranated flagella with trypsin to destroy the linkers that normally resist interdoublet sliding. Upon the addition of ATP and Mg^{2+}, the axonemes elongated by a telescopic action of sliding microtubules. The maximum length of telescoping axonemes sometimes reached nine times the original length before disintegration occurred. Trypsin-treated axonemes of *Chlamydomonas* paralyzed mutants that lack central microtubules and radial spokes also undergo sliding disintegration in reactivation solutions. Thus, sliding involves doublet-doublet interactions and does not require central microtubules or radial spokes.

The direction in which the arms on one doublet microtubule apply force to the adjacent doublet microtubule was analyzed by Sale and Satir (1977) in trypsin-treated axonemes undergoing ATP-induced sliding. They found that the arms always pushed the adjacent doublet $N + 1$ in a base-to-tip direction. This polarity of force generation appears to be the same in all organisms and under all conditions examined.

2. Conversion of Sliding to Bending

If all microtubule doublets slide past each other with each doublet N pushing $N + 1$ tipward, ciliary bends or flagellar waves cannot be formed. Since every dynein arm apparently works in the same manner and all doublets slide in the same way, if all dyneins around the axoneme operated at the same time, the resultant forces would cancel each other's effects, resulting in no ciliary movement. Therefore, there must be regulation of local transient resistance to, and activation of, inter-doublet sliding for coordinated activity. Separate sliding and control systems must operate in the cilium.

Bending is produced in cilia by local and temporary restriction to sliding (Fig. 16). How this comes about is not certain. One possibility is that binding of the spoke heads to the central microtubule projections causes resistance to sliding. Along the length of the flagellum, radial spokes are perpendicular to microtubules in straight regions but are tilted in bent regions. It has been proposed that at the leading edge of a propagating bend, attachment causes resistance to doublet sliding, and at the trailing edge, spoke heads become detached, releasing the resistance to sliding and allowing that region to straighten. However, the presence of motile flagella in mutants without normal central microtubules or radial spokes implies that other structures cause resistance to sliding and bend formation. Thus, restriction of sliding may involve nexin links or the binding of dynein arms to neighboring microtubule doublets. The radial spoke-central microtubule interaction may control wave symmetry.

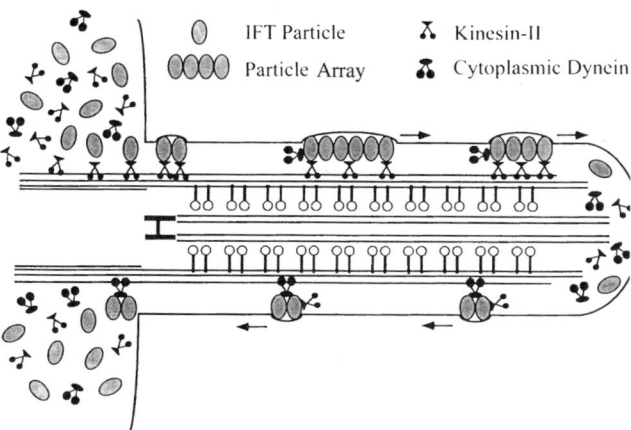

FIGURE 15. Model of intraflagellar transport (IFT) in a *Chlamydomonas* flagellum. Linear arrays of IFT particles are transported by the motor molecule kinesin-II distally along the flagellum in the space between the doublet microtubules and the flagellar membrane. The particles are returned to the base of the flagellum by the action of cytoplasmic dynein isoform 1b. A large pool of IFT particles and motors is located at the base of the flagellum. (Adapted from Cole *et al.*, 1999.)

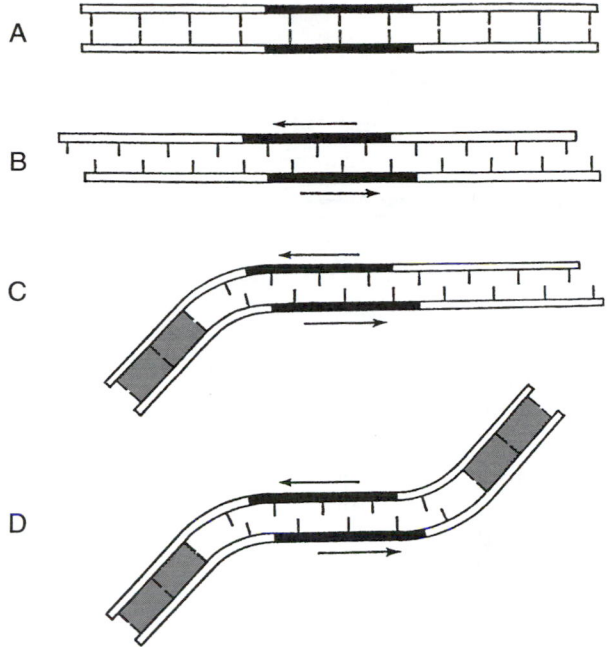

FIGURE 16. Conversion of sliding into bending. (A) Outer doublet microtubules with associated (inactive) linking structures. (B) If there is no resistance to the forces generated by dynein arm action (darkened region), doublet microtubules slide freely. (C) Local transient resistance to sliding caused by activation of linking structures in one region of the cilium (stippled area) causes bend formation during sliding in an adjacent region (darkened region). (D) Resistance to sliding (stippled regions) on both sides of the active force generation site (darkened region) causes two bends to form in opposite directions, resulting in the formation of a wave. (Reprinted by permission from *Nature*, vol. 265, pp. 269–270. Copyright © 1977 Macmillan Magazines Limited.)

3. Coordination of Sliding during the Ciliary Beat Cycle

Unlike the one-dimensional sliding of thick and thin filaments of muscle, ciliary motility involves movement in two or three dimensions. Although internal local restriction of interdoublet sliding can result in bend formation, propagation of bends and the different movements in the effective and recovery strokes in the ciliary beat cycle require coordination of events in different parts of the cilium. Computer simulations of the coordinated movements of *Paramecium* cilia suggest that dynein arms of only doublet numbers 1, 2, 3, and 4 on one side of the cilium are active during the effective stroke (Fig. 17). During the recovery stroke, activity switches to doublet numbers 6, 7, 8, and 9 on the other side. All doublets in each half do not act synchronously but are activated progressively from doublet N to doublet $N + 1$. Activity along the length of the cilium is also not synchronous but begins at the base with activity spreading tipward.

There is evidence that in at least some organisms the pair of central microtubules rotates continuously in one direction during ciliary activity, with one complete rotation being made per beat cycle. These observations make the central microtubules attractive candidates for controlling the coordination of doublet activation during the beat cycle by functioning as a distributor. If this is indeed true, the obvious

transducing structures in the axonemes are the radial spokes. However, as mentioned above, there are mutants with motile flagella that lack radial spokes. Therefore, if the radial spokes normally activate the dynein arms, there must be an additional control mechanism located in the doublet microtubules themselves.

4. Modulation and Control of Ciliary Activity

Changes in beat frequency, arrest of movement, and reversal of the direction of the effective stroke (ciliary reversal) are modulations superimposed on the dynein-ATP-Mg^{2+}-microtubule system. These control mechanisms, which impart meaningful behavior such as taxis and the avoidance reaction, primarily involve regulation by the intraciliary level of Ca^{2+}, which in turn is controlled by the ciliary membrane. Neuronal, hormonal, and purely physical interactions of adjacent cilia within a group are other modulation factors. Three basic parameters of the beat—orientation (changes in the direction of the effective stroke), frequency, and waveform—are modified by stimuli.

a. Ciliary Reversal and the Avoidance Reaction. The physiological basis for the control of ciliary activity has been particularly well studied in the protozoan *Paramecium*. A *Paramecium* cell usually swims about 1–2 mm/s in a forward left spiral pattern with the effective stroke of the cilia directed toward the posterior of the cell body (Fig. 18). When the ciliate encounters an appropriate negative stimulus (mechanical, electrical, or chemical stimulation at the anterior end), it reverses the direction of the effective stroke (ciliary reversal). Thus the cell rapidly swims backward, away from and avoiding the stimulus source. Then, during a renormalization period (partial ciliary reversal), some cilia revert back to normal beat while others are still reversed, or individual cilia exhibit circling movements, all of which cause the cell to gyrate or spin while remaining in the same spot. After all effective strokes of cilia are directed posteriorly, the ciliate returns to forward swimming in a new direction, completing the avoidance reaction.

The resting membrane potential of *Paramecium* is about −40 mV (inside negative). Early in the 1930s, Kamada's laboratory observed that the membrane of *Paramecium* exhibited depolarizations that correlated with reversals of the effective strokes of somatic cilia, indicating that electrophysiologic events at the membrane were involved in the modulation of ciliary activity. The definitive evidence that Ca^{2+} regulated this activity was provided by Naitoh and Kaneko (1973), who found that detergent-extracted models of *Paramecium*, reactivated in solutions containing ATP and Mg^{2+}, swam in a forward left spiral pattern, but that those reactivated in ATP, Mg^{2+} plus Ca^{2+} swam backward. After a series of experiments on *Paramecium*, Eckert and Naitoh formulated the Ca^{2+} hypothesis for ciliary reversal (Fig. 19).

The intracellular Ca^{2+} concentration is normally $<10^{-6}$ M, whereas extracellular concentrations are usually in the millimolar range. Upon stimulation to the anterior end of *Paramecium*, receptor potentials that activate voltage-sensitive Ca^{2+} channels located in the ciliary membrane are generated. The open Ca^{2+} channels allow extracellular Ca^{2+} to enter the

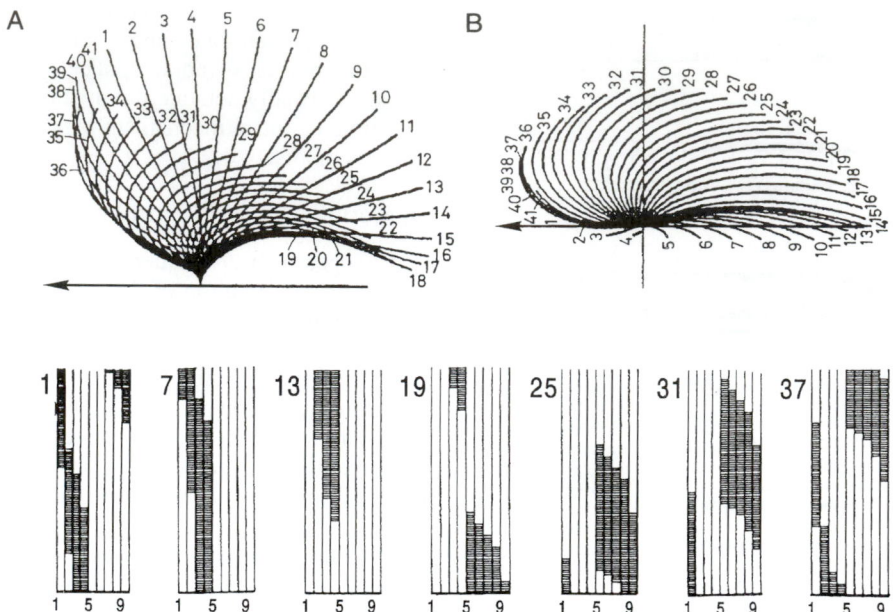

FIGURE 17.   Computer simulation of regions of dynein arm activity during a ciliary beat cycle. (A) Side view of the activity of a *Paramecium* somatic cilium during one cycle of effective and recovery strokes. The positions of the cilium at different times during the cycle are numbered. The arrow indicates the direction of forward swimming (toward the cell's anterior); the effective stroke is directed posteriorly. (B) Similar to (A) but showing the top view of the effective and recovery stroke of a *Paramecium* somatic cilium. The effective stroke begins at the position numbered 1; the cilium rapidly and relatively stiffly bends posteriorly, offering high resistance to the medium. During the recovery stroke the cilium exhibits a greater bend, moves relatively slowly, and is directed more laterally, thus offering less resistance to the surrounding medium. (C) To obtain the movements of the ciliary cycle, activation of the individual outer doublets (hatched regions) changes with different positions of the cilium. During the effective stroke (e.g., positions 1, 7, 13), activation of doublets 1–4 dominates, whereas during the recovery stroke (e.g., positions 19, 25, 31, 37), activation of doublets 5–9 dominates. Adapted from Sugino and Naitoh (1982). Reprinted by permission from *Nature* **295,** 609–611, copyright 1982 Macmillan Magazines Ltd.

cilia by diffusion down its concentration gradient, causing a regenerative Ca^{2+} current that lasts 10–20 ms. This depolarization is graded with stimulus intensity; the spike upstroke is about 10 times slower than those of neurons. The inward Ca^{2+} current at depolarizing voltages has been measured in voltage-clamped cells. These voltage-sensitive Ca^{2+} channels also allow Sr^{2+} or Ba^{2+} into the cilium, which is accompanied by all-or-none membrane action potentials and backward jerks by the cell called the **barium dance** (Fig. 20).

The location of Ca^{2+} channels in *Paramecium* cilia was determined by experiments in which cells were deciliated, resulting in the loss of the Ca^{2+} response in stimulated cells. During regeneration, the Ca^{2+} response returned and the amplitude increased as cilia grew. Thus it is probable that Ca^{2+} channels are distributed along the length of the ciliary membrane.

The Ca^{2+} influx through channels results in increased intraciliary Ca^{2+} concentration, which activates the mechanism that causes the reversal of the effective stroke. Thus, the ciliary effective stroke is directed anteriorly, and the ciliate swims backward. The latent period between stimulus and ciliary reversal lasts several milliseconds. In some ciliated epithelia (e.g., in the bivalve mussel gill), increased intracellular Ca^{2+} concentration causes beat arrest instead of ciliary reversal. The mechanism of reversal is not understood.

The depolarization-activated Ca^{2+} channels in the ciliary membrane are inactivated by the Ca^{2+} that enters the or-

ganelle. The inactivation of the Ca^{2+} channels and an outflux of K^+ via delayed depolarization-activated K^+ channels in the somatic membrane restore the membrane's resting potential. During the renormalization period of the avoidance reaction, Ca^{2+}-ATPases known to be present in ciliary membranes are thought to pump Ca^{2+} out, thus lowering its intraciliary concentration. Several distinct Ca^{2+}-ATPase activities specifically found in the ciliary membrane have been characterized; these are distinct from those of the somatic membrane. Following the decrease of intracellular Ca^{2+} to a threshold value, the control mechanism restores the direction of the effective stroke toward the posterior and the ciliate resumes forward swimming. Although Ba^{2+} and Sr^{2+} can enter cells through Ca^{2+} channels, they apparently are not effectively pumped out by Ca^{2+}-ATPase pumps and are also not as effective as Ca^{2+} in inactivating the activated Ca^{2+} channels.

Hundreds of locomotory behavioral mutants of *Paramecium* that are useful for dissecting events at the membrane and at the axoneme have been isolated. Kung first described *pawn* mutants that fail to exhibit the avoidance reaction (so named because pawn chess pieces are not allowed to be moved backward). In a series of studies with Eckert and Naitoh, Kung demonstrated that extracted models of the mutant swam backward when Ca^{2+} was present in the reactivation solution, indicating that the defect was not in the axoneme. Electrophysiologic data showed no Ca^{2+}

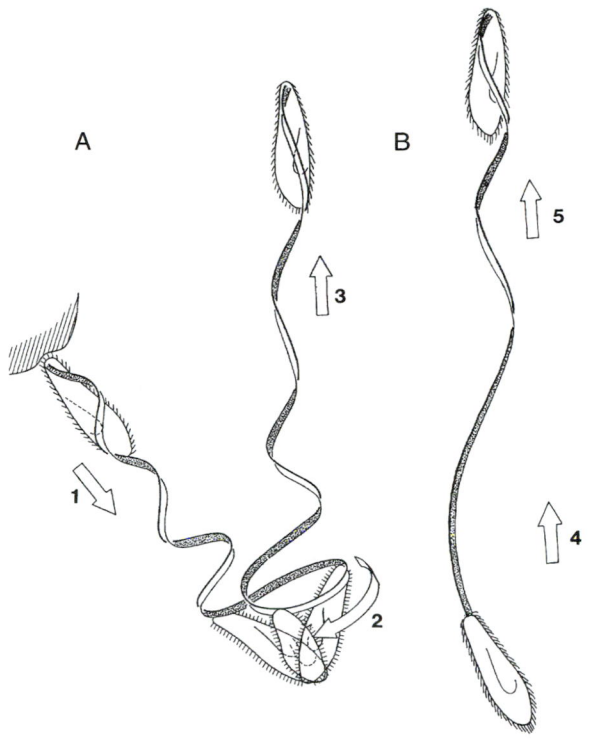

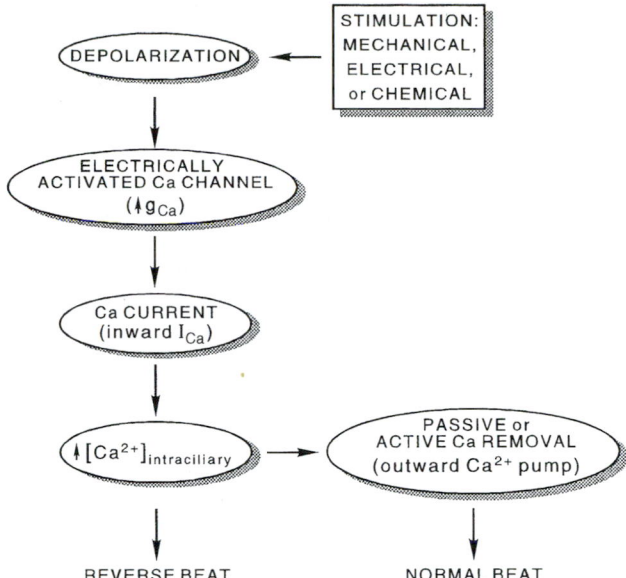

FIGURE 18. Locomotory behavior in *Paramecium*. (A) The avoidance reaction occurs when the cell is stimulated (e.g., bumps against a solid object with its anterior end). The cell reacts by switching the direction of the effective stroke of the somatic cilia toward the cell's anterior, resulting in backward swimming (1). The cell then pivots about its posterior pole during a renormalization period (2), then resumes normal front left spiral forward swimming, in which the effective power stroke of the somatic cilia is redirected toward the cell's posterior (3). (B) The escape reaction is initiated by stimulation of the posterior pole of the organism. The cell responds by increasing the velocity of swimming in a front left spiral forward direction (4), then resumes its normal rate of forward swimming (5). (Adapted from Naitoh and Sugino, 1984, with permission.)

FIGURE 19. The calcium hypothesis of reversal of the effective stroke of ciliary beat. In the avoidance reaction of a ciliated protozoan, a stimulus causes depolarization of the cell surface membrane, including the ciliary membrane. The depolarization of the membrane activates voltage-sensitive Ca^{2+} channels in the ciliary membrane, resulting in the membrane increasing its Ca^{2+} conductance. The inward Ca^{2+} current reflects the movement of the divalent cation down its concentration gradient into the cell (cilia), resulting in increased intraciliary Ca^{2+} concentration. The increased intraciliary Ca^{2+} causes a modification of the ciliary beat by changing the direction of the effective stroke. During the renormalization period, Ca^{2+} is removed from the cilia, probably aided by outward-directed Ca^{2+} pumps (ATPase) in the ciliary membrane, leading to the resumption of the normal direction of the ciliary effective stroke.

spike in response to depolarizing current stimuli and no inward Ca^{2+} current under voltage-clamp conditions, thus identifying the voltage-sensitive Ca^{2+} channels as the lesion site in this mutant (Fig. 21). The channel (or gate) molecules have been reconstituted into artificial lipid membranes by fusion with ciliary membrane vesicles, and single-channel conductances have been examined in patch-clamp experiments. Several types of channels, one of which behaves like the voltage-dependent Ca^{2+} channels of intact cells, have been thus detected. Recently, molecular genetic techniques have been used to clone some of the pawn genes.

b. Beat Frequency and the Escape Reaction. Beat frequency of cilia in nonstimulated *Paramecium* is about 15 Hz, which can change to 50 Hz in response to a stimulus. In the avoidance reaction, the beat frequency during ciliary reversal increases to a maximum of 40–50 Hz and then decreases with cilia beat reorientation.

Paramecium responds to mechanical stimulation at its posterior by increasing cilia beat frequency (ciliary augmentation), and swimming forward faster, described as the **es-**

cape reaction (see Fig. 18). Beat frequency is also controlled by intraciliary Ca^{2+} levels (Fig. 22). Hyperpolarization of the membrane follows posterior stimulation and probably involves Ca^{2+} influx and voltage- and Ca^{2+}-dependent K^+ fluxes, which are not understood as well as those associated with the avoidance reaction. There is evidence indicating that hyperpolarization activates adenylate cyclase, leading to increased intraciliary cAMP, which in turn stimulates cAMP-dependent protein kinase(s) that phosphorylates several ciliary proteins (Fig. 23). Protein kinases may phosphorylate dyneins or other proteins associated with the arms, including a low-molecular-weight light chain.

Calmodulin, adenylate and guanylate cyclases, protein kinases, cAMP and cGMP, cAMP- and cGMP-dependent protein kinases, and phosphoprotein phosphatase (calcineurin) have been detected in *Paramecium* cilia. Hence Ca^{2+} regulation of cilia activity probably involves cascade systems involving second messengers and cyclic nucleotides; the substrates that become phosphorylated are now being examined. It has been suggested that guanylate cyclase activity regulates the avoidance reaction, but the rate of increase in cGMP may not be sufficiently rapid to account for this behavioral response. In sperm, it has been demonstrated that cAMP is required to activate flagellar motility of fully

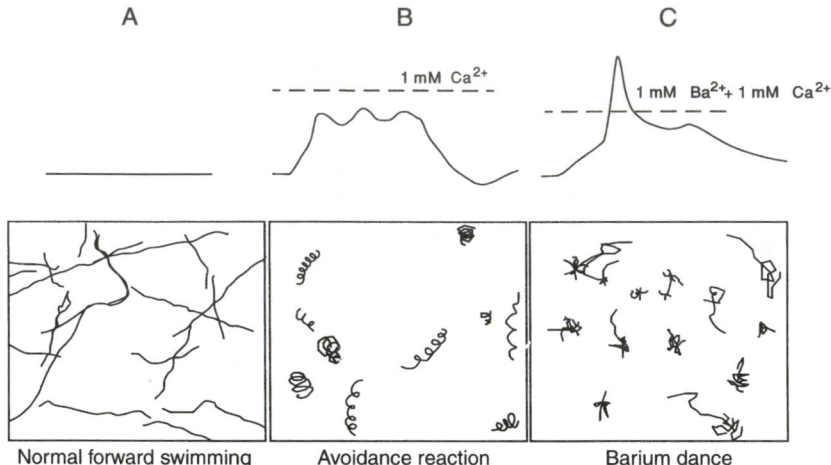

FIGURE 20. Electrical properties of the cell surface membrane and swimming behavior. (A) Top: In the absence of a stimulus, the membrane exhibits a stable resting potential of about −40 mV. Bottom: Swim tracks of the organism reflect normal front left spiral forward locomotion. (B) Top: Stimulation is followed by the Ca^{2+} response, a spike graded with respect to stimulus strength. Bottom: The behavioral correlate is illustrated by the swim tracks reflecting backward swimming in a tight spiral pattern. (C) Top: The ciliary membrane voltage-sensitive Ca^{2+} channel also accommodates Ba^{2+} and Sr^{2+}. In the presence of Ba^{2+}, all-or-none action potentials occur. Bottom: The cell exhibits a characteristic barium dance of rapid backward jerks.

assembled but nonmotile sperm upon release from the male reproductive tract.

In ciliated epithelial systems, acetylcholine and serotonin cause an increase in beat frequency, whereas epinephrine causes a decrease. Increased intraciliary Ca^{2+} (experimentally induced by Ca^{2+} ionophores) causes the arrest of cilia beat in mussel gills. This response also occurs *in situ* when the nerve is stimulated. The involvement of transmitter substances in the control of ciliary activity further implicates second messenger and cyclic nucleotide cascade reactions in ciliary reversal, ciliary augmentation, or both.

c. M*etachrony.* The coordinated beating of a group of cilia forming large waves over the tops of cilia on the cell surface is called a **metachronal wave** (Fig. 24). The waves result from purely physical viscous coupling between neigh-

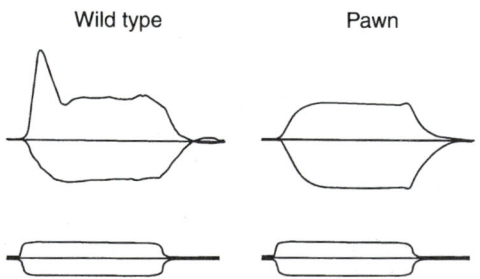

FIGURE 21. Behavioral mutants. Locomotory mutants of various types have been correlated with ciliary and flagellar abnormalities. In *Paramecium,* a membrane-defective pawn mutant was isolated by its inability to swim backward. Electrophysiological analysis demonstrated that when stimulated, the mutant membrane did not exhibit the Ca^{2+} spike. Top traces are the responses (mV) to depolarizing and hyperpolarizing stimuli; the bottom traces are the corresponding current stimuli (I_s). (Adapted from Kung and Eckert, 1972, with permission of the author.)

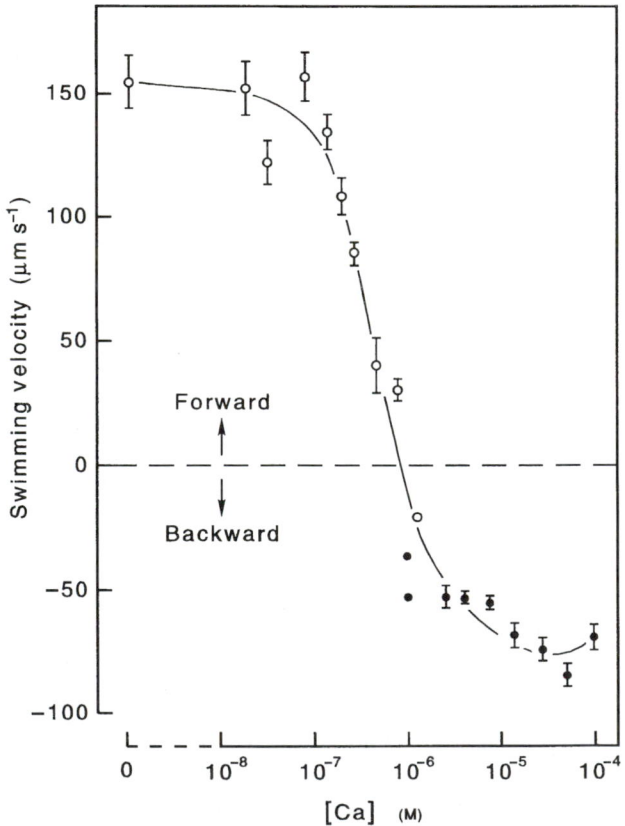

FIGURE 22. Intraciliary calcium concentration regulates the velocity of backward and forward swimming. Correlation with swimming velocity was determined by analyzing the swimming rate of ATP-Mg^{2+}-reactivated Triton-extracted models of *Paramecium* in the presence of various concentrations of Ca^{2+}. (From Naitoh and Sugino, 1984, with permission.)

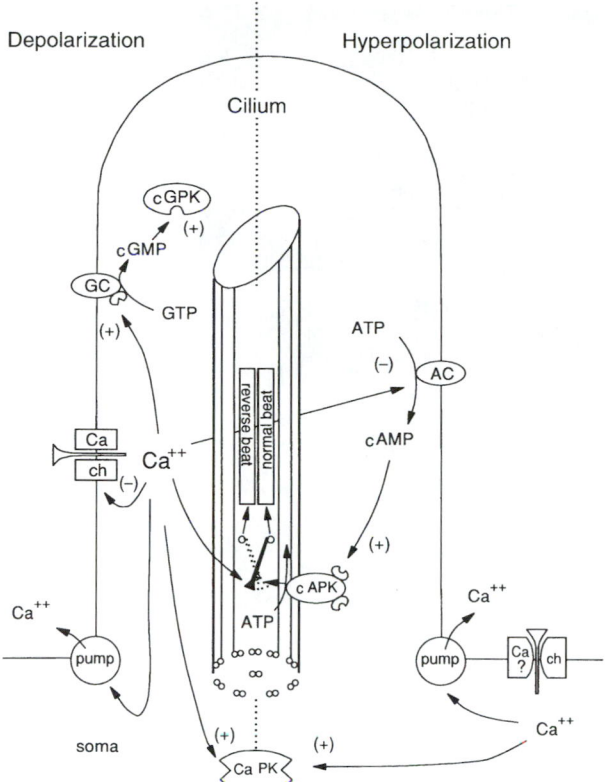

FIGURE 23. Control of *Paramecium* ciliary activity. Ciliary reversal (avoidance reaction, depolarization) and augmentation (escape reaction, hyperpolarization) involves Ca²⁺ and cyclic nucleotides. Upon depolarization, ciliary voltage-dependent Ca²⁺ channels open, increasing the intraciliary Ca²⁺ concentration that activates the switch controlling reversal of the effective stroke and backward swimming. The Ca²⁺ channels are inactivated by the increase in intraciliary Ca²⁺, and the concentration is restored by sequestration and/or extrusion by pumps, resulting in the resumption of forward swimming. The Ca²⁺ response activates a protein kinase (CaPK) and guanylate cyclase (GC) that causes an increase in cGMP, which in turn activates a cGMP-dependent protein kinase (cGPK). Membrane hyperpolarization is accompanied by the activation of adenylate cyclase (AC) and increased cAMP. The increase in intraciliary cAMP activates a cAMP-dependent protein kinase (cAPK) that leads to increased ciliary beat frequency. (From Preston and Saimi, 1990.)

boring cilia transmitted through the surrounding fluid. The coupled-oscillator hypothesis developed as the result of experimental evidence refuting the idea that protozoan cells have intracellular neural systems and that ciliary coordination was controlled by such neuromotor networks. Microsurgical cuts of fibrillar structures in the cortex of ciliates did not alter metachrony; waves continued over small gaps made in the cell cortex. Also, detergent-extracted models were shown to exhibit metachrony.

E. Sensory Receptors

The association of cilia with sensory reception is common among animals. Usually these cilia lack dynein arms and have a 9 + 0 axoneme pattern. The following examples of sensory cilia demonstrate that they are commonly involved in sensory reception and transduction and may also serve additional functions in these cells.

The major division during metazoan evolution apparently has separated photoreceptors into two groups. The photoreceptors of most protostomes (annelids, arthropods, and mollusks), such as the cells of squid retina, have actin-based microvilli substructures, whereas most deuterostomes (echinoderms and chordates) have microtubule-based photoreceptors.

Vertebrate photoreceptor cells have a specialized region between their inner and outer segments that contains a single short connecting cilium that is analogous in structure to the transition zone. The outer segment is a modified cilium. During development of the retina, just before the outer segment forms, the cilium grows from a basal body–centriole complex. The membrane of the distal region of the cilium expands, forming the photosensitive disk membranes. The immature cilium is differentiated into proximal and distal regions. The proximal region contains Y-shaped membrane-microtubule doublet linkers, and an IMP array, comparable to the ciliary necklace. This configuration is also true of the mature connecting cilium. Involvement of the distal region in the transport of opsin into the developing and mature outer segment has been indicated by immunocytochemical evidence. These regions bind anti-opsin antibodies, as does the mature connecting cilium. Thus, in addition to separating the photosensory outer segment domain of the cell from the release of synaptic transmitter by the inner segment, the connecting cilium may function in the delivery and turnover of the disk membrane.

In olfactory receptors, evidence for odorant receptors within the ciliary membrane has been provided by the loss of electrophysiologic responses after experimental deciliation of the neurons. Receptor occupancy probably causes a voltage drop across the membrane, which then spreads to the cell body. As in motile cilia, cyclic nucleotides are involved in olfactory sensory transduction. Some cilia-based mechanoreceptors may involve the deformation of the ciliary membrane, causing the opening of ion channels such as the Na⁺ channels in nudibranch statocysts. The change in ion conductance then elicits generator potentials in the sensory cell.

IV. Other Microtubule Systems

A. Occurrence of Nonaxonemal Microtubules

Nonaxonemal microtubules are widespread. They occur as single microtubules in the cytoplasm, and they form complex structures such as the mitotic apparatus of dividing cells (Table 4). Cytoplasmic microtubules may be linked together by bridges to form elaborate arrays such as the heliozoan axopodium (Fig. 25). Cytoplasmic microtubules can function like railroad tracks, guiding vesicles and other organelles to specific sites in the cell. They may also do work by polymerization and depolymerization.

B. Microtubule Polymerization and Depolymerization

Microtubule polymerization and depolymerization, as well as tubulin pools and posttranslational modifications, are highly regulated in a cell. A dynamic and dramatic example of rapid microtubule polymerization and depolymerization

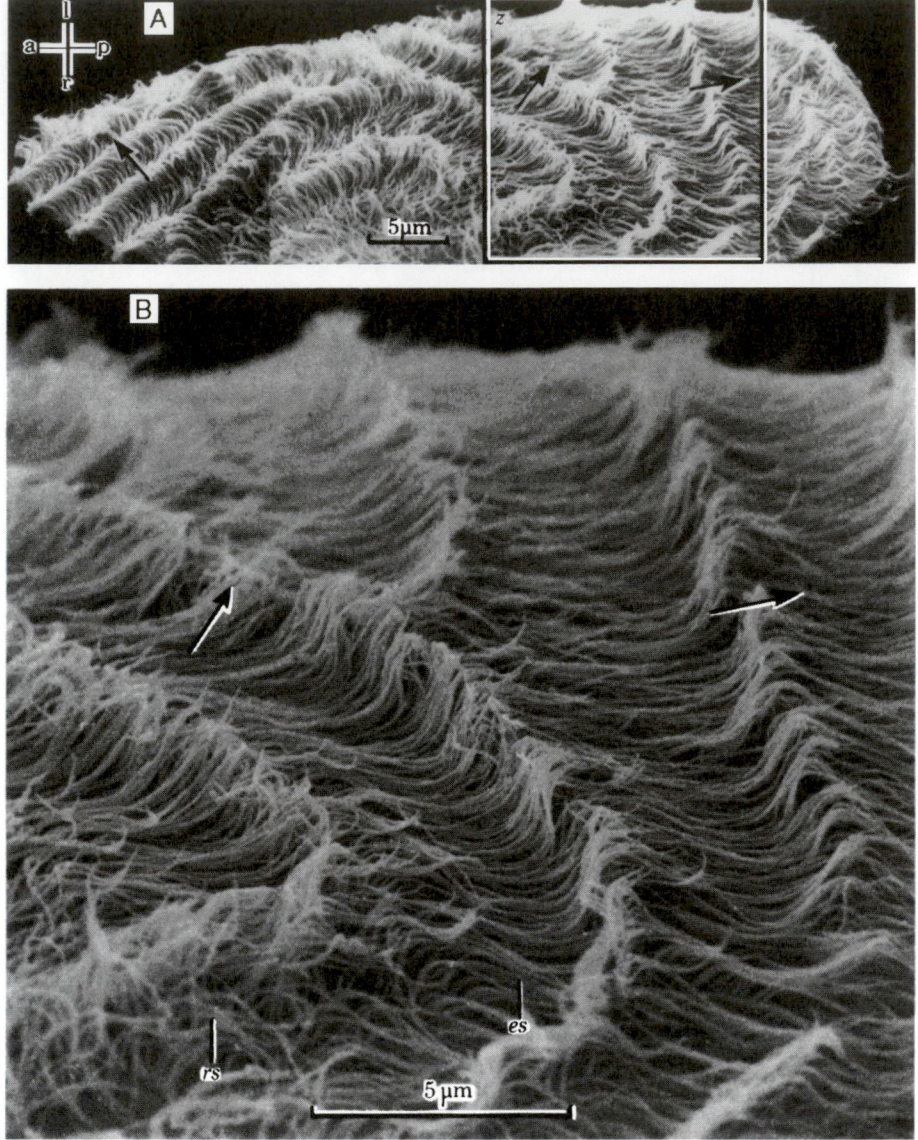

FIGURE 24. Metachronal waves. Coordinated activity of many closely packed cilia and flagella form large metachronal waves over the surface of the cell. (A) Scanning electron micrograph of *Opalina* that was rapidly fixed to preserve the metachronal waves. (B) Enlargement in which individual flagella are resolved. The coordination of ciliary beat resulting in metachrony is attributed to physical viscous coupling interactions between neighboring cilia. (Reprinted with permission from Tamm and Horridge, 1970.)

can be seen in heliozoans (Fig. 26). The axopodia of these protozoans are stiff pseudopodia radiating from the cell in a starburst pattern. These structures collapse when a prey organism is taken into the cell body. This disintegration and re-extension can happen so rapidly that special fixation procedures are necessary for ultrastructural visualization of the structure. The axopod contains a stiff bundle of microtubules extending to the nucleus. When viewed in cross-sections, the microtubules of some axopodia are seen to form an elaborate pattern that looks like two microtubule sheets rolled up together and maintained by extensive links. The processes of polymerization and depolymerization can occur in seconds, as rapidly as actin polymerization in the acrosome reaction of sea cucumber sperm described previously. The rapid disinte-

gration and reextension of axopodia may be controlled by changes in Ca^{2+} levels, as Ca^{2+} causes microtubule disassembly.

Microtubule polymerization is usually initiated at structures termed **microtubule organizing centers** (MTOCs), which include spindle pole bodies in yeast and basal bodies and centrosomes in animal cells. A third isoform of tubulin, γ-tubulin, is found at MTOCs and is involved in the initiation of microtubule polymerization and determination of microtubule subunit polarity. The location and orientation of the MTOC determines the location and orientation of the microtubular structure that it organizes.

Studies of microtubule assembly and disassembly *in vitro* have revealed that microtubules grow more rapidly at one end

TABLE 4 Functions of Microtubular Systems

Function	Examples
Cilia and flagella	
Locomotion	Ciliates; flagellates; flatworms; sperm cells; aquatic metazoan larvae
Feeding (generation of water currents that move suspended or mucus-trapped food toward the mouth and/or toward phagocytic cells)	Ciliates; flagellates; sponge choanocytes; sea anemone tentacles and gastrovascular cavities; mucociliary feeding in gastropods, bivalves, crinoids, and annelids
Respiration (move water over respiratory surface)	Aquatic mollusk ctenidia; sea star epidermis and tube feet
Excretion and osmoregulation	Protonephridial flame cells of flatworms and rotifers; ciliated funnels of annelid nephridia
Surface cleaning	Ciliated epithelia of vertebrate tracheal and bronchial airways; ependymal surfaces of cerebrospinal cavity; Eustachian tube and middle ear
Mating (cell-cell recognition, gamete agglutination)	Mating-reactive flagellates and ciliates; invertebrate and vertebrate gametes
Transport	Passage of egg along mammalian oviduct; movement of ingested material along alimentary tracts of annelids, mollusks, echinoderms, and tunicates
Adhesion and anchoring	Parasitic trypanosomes and leishmanias to host tissues; cilia of mammalian respiratory tract to bacterial pathogens
Circulation	Coelomic fluid of invertebrates; cerebrospinal fluid of vertebrates
Sensory reception and transduction	Connecting cilia in photoreceptor cells; kinocilia of fish and amphibian lateral line systems; nudibranch, jellyfish and crayfish statocysts; insect and nematode sensory receptors; vertebrate photoreceptor cells
Morphogenesis and development	Rotation of archenterons (*situs inversus* correlated with immotile cilia); mirror-image and other global cortical abnormalities in ciliates
Cytoplasmic singlet microtubules	
Cell division	Chromosomal and polar fibers of mitotic apparatus of eukaryotic cells
Transport and intracellular trafficking guidance	Movement of organelles toward and away from neuron soma; viral transport; transport of lens precursors during development of squid eye; transport of endosomes and lysozomes; movement of discoid vesicles to ciliate oral region; transport of rhodopsin in rod inner segment
Structural integrity and architecture	Cytoskeletal network; axopods of heliozoan protozoa
Cell shape changes	Fish photoreceptor cells
Organelle positioning	Endoplasmic reticulum; Golgi apparatus; nuclear migration

(the plus end). This end is usually distal to the MTOC. In a Ca^{2+}-free solution, one dimer of tubulin binds two GTP molecules, one of which is hydrolyzed with a slight delay after polymerization. Thus, a rapidly growing microtubule is composed mostly of GDP-tubulin but has a cap of GTP-tubulin at its plus end. This cap protects the end from disassembly; however, if the cap is lost, exposing GDP-tubulin at the plus end, then rapid depolymerization ensues. As a result, a population of microtubules generally will be made up of both rapidly growing and rapidly shrinking microtubules. This continuous growing and shrinking of microtubules is known as *dynamic instability*. In the cell, the minus end of the microtubule is usually associated with an MTOC, which protects the microtubule from depolymerization

at that end. Dynamic instability provides a mechanism by which the microtubules continuously probe the cell for sites that will bind their plus ends. When the plus ends are captured, as for example by a kinetochore (see next section), the microtubules are stabilized. However, the protein katanin (from the Japanese word for sword) can sever microtubules, thus creating new ends and destabilizing the microtubule.

C. Mitotic Spindle

As in cilia and flagella, mitotic spindles form by growth of microtubules nucleated at centrosomes and by addition of tubulin distally. At the end of prophase, the nuclear envelope in most

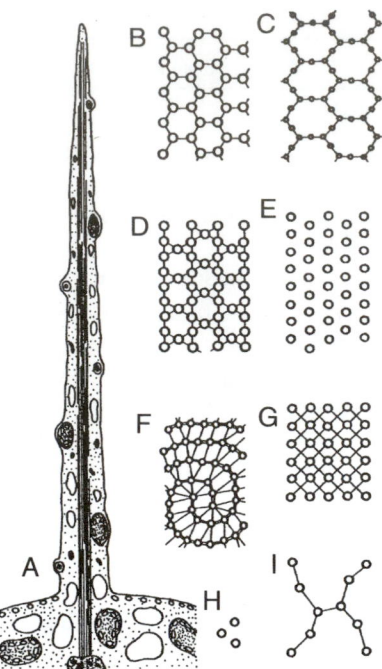

FIGURE 25. Microtubules are important cytoskeletal elements. (A) Axiopods are cell extensions supported by microtubular arrays. (B)–(I) Illustration of the diversity of microtubular arrays found in various protozoa. (From Sleigh, 1989.)

cells disintegrates, and three types of microtubules grow from the spindle poles with plus ends distal to the poles: (1) polar microtubules that extend from the pole in the direction of the opposite pole, (2) astral microtubules that extend away from the pole and may reach as far as the cell membrane, and (3) kinetochore microtubules that attach to chromosomes (Fig. 27). The **kinetochore** is a trilaminar disk, conspicuous at metaphase when chromatid pairs are seen with a constriction called the *centromere*. In higher eukaryotic cells, the chromosomal centromeres are associated with kinetochores. The microtubules growing from the pole bind to the kinetochore in prometaphase.

At anaphase, two movements are associated with the mitotic spindle: (1) chromosome movement toward the poles (anaphase A) and (2) separation of the poles (anaphase B) by apparent lengthening of the polar fibers. Polar movement of chromosomes in anaphase A results from depolymerization at the kinetochore. Depolymerization at the spindle pole may also occur. Yeast chromosomal fibers have only one microtubule at the kinetochore, which demonstrates that depolymerization of a single microtubule can move chromatids to the poles. The pushing apart of mitotic poles during anaphase B does not result from tubulin polymerization. Sliding of overlapping polar microtubules driven by kinesin or dynein (see later) in the central zone of the spindle may be responsible for this movement.

Compounds such as colchicine bind tubulin and inhibit polymerization. Other microtubule inhibitors include colcemid, podophyllotoxin, and nocodazole. Vinblastine and vincristine inhibit microtubules by precipitation of tubulin and the

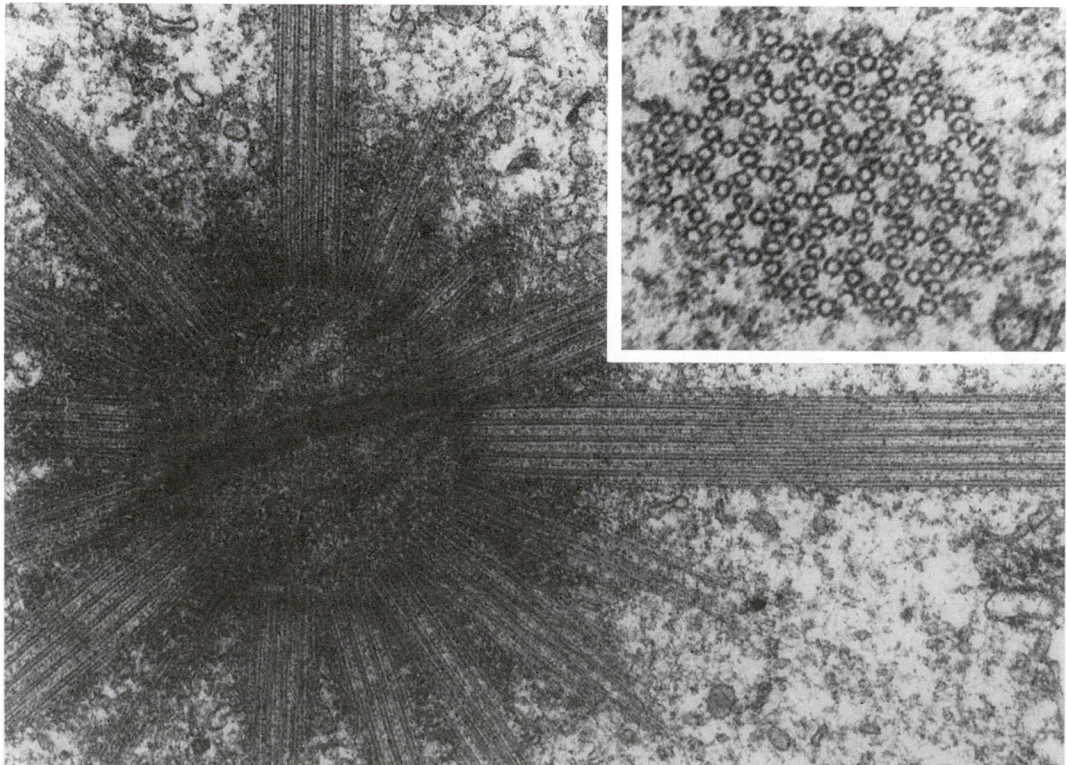

FIGURE 26. Electron micrograph of a heliozoan showing microtubules that support the organism's axopod. The microtubular bundles radiate from the centroplast, which contains a centriolar structure. The inset shows a cross-sectional view of one of the microtubule bundles. (Reproduced from Bardele, 1977, Courtesy of The Company of Biologists Ltd., Cambridge, UK.)

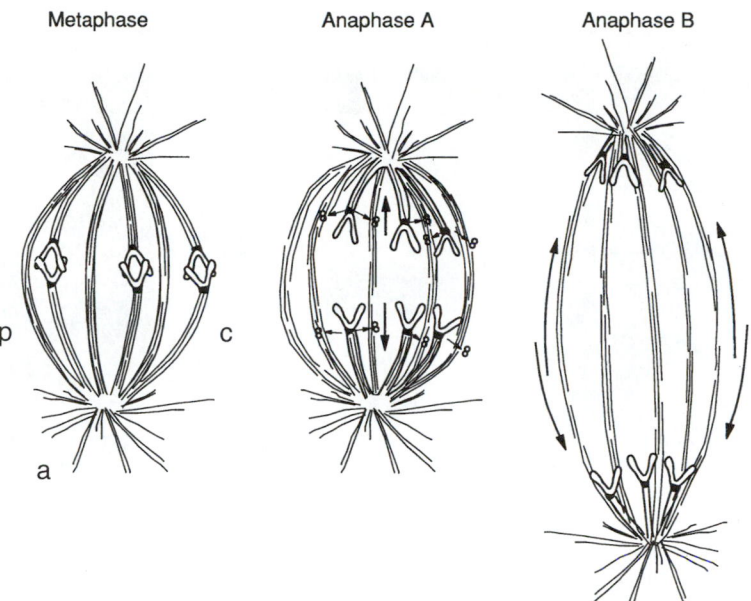

Metaphase   Anaphase A   Anaphase B

FIGURE 27. Mitotic spindle activity. Microtubular bundles occur in the mitotic spindle as astral rays (a), chromosomal fibers (c), and polar fibers (p). During anaphase A, the chromosomal microtubules depolymerize at the end attached to the kinetochore, resulting in shortening of the microtubules and movement of chromosomes toward the poles. During anaphase B, the distance between the two poles increases and the entire spindle apparatus becomes elongated, presumably by sliding of polar microtubules past each other by motor molecule action.

formation of crystalline aggregates. Taxol, which is used as an anticancer drug, favors polymerization and stabilizes microtubules, leading to cellular accumulation of microtubules.

Amino acid sequences of tubulins from organisms representing broad phylogenetic backgrounds indicate that the genes for tubulin are highly conserved. However, microtubules from various sources differ in sensitivity to cold, colchicine, ionic strength, pH, and detergents. Some of these differences may be due to posttranslational modifications of tubulin, or to association with specific microtubule-associated proteins (MAPs). Widespread posttranslational modifications involve tubulin acetyltransferase (and deacetylase) and tubulin detyrosinase. Enzymatic acetylation occurs at the axonemal tip as microtubule subunits are added during growth of cilia. The enzyme for deacetylation of tubulin is present in the cytoplasm; hence, tubulin subunits may be exchangeable between axonemal and cytoplasmic pools. Tubulin synthesis, which is controlled by negative feedback mechanisms, is finely regulated in the cell. An increase in cellular free tubulin decreases the half-life of tubulin mRNA.

D. Movement of Particles and Organelles along Microtubules

Two major classes of molecular motors use the energy of ATP to move components along cytoplasmic microtubules. The first class is represented by cytoplasmic dynein 1a. Just as cytoplasmic dynein 1b moves IFT particles along doublet microtubules toward the base of the flagellum, cytoplasmic dynein 1a transports viruses, membrane-bound vesicles, and other cell components toward the minus end of cytoplasmic microtubules. The second class of microtubule motor is repre-

sented by kinesin (Fig. 28), the ATPase responsible for fast anterograde transport of vesicles in axons. In some organisms, dozens of kinesin-like proteins have been identified, suggesting that these proteins have very specific roles in cells. Kinesin and many other members of this family move toward the plus end of the microtubule. However, some kinesin family members, such as ncd in *Drosophila* and Kar3 in *S. cerevisiae*, move toward the minus end. Therefore, as in the case of IFT, components may be transported in one direction by plus-end-directed kinesins, and in the opposite direction by minus-end-directed kinesins or cytoplasmic dynein. For example, kinesin and cytoplasmic dynein 1a are believed to be responsible for fast anterograde and retrograde axonal transport, respectively.

Whether plus-end- or minus-end-directed motors dominate in the movement of a particular cell organelle is tightly controlled by the cell. For example, in fish melanophores (Fig. 29) pigmented granules move along microtubules that radiate from

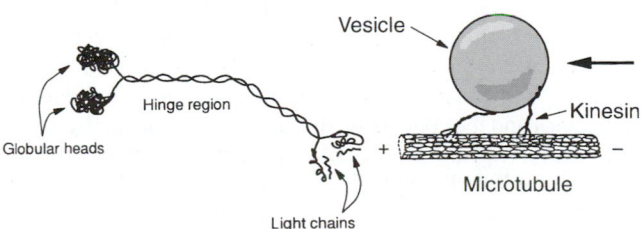

FIGURE 28. Kinesin, a motor molecule. Kinesin is a relatively small motor molecule that interacts with microtubules. Interaction of kinesin with microtubules and membrane-bound organelles is responsible for movement of organelles within the cytoplasm. The large arrow indicates the direction in which the vesicle moves relative to the plus (+) and minus (−) ends of a microtubule. (From Bray, 1992.)

A   B   C

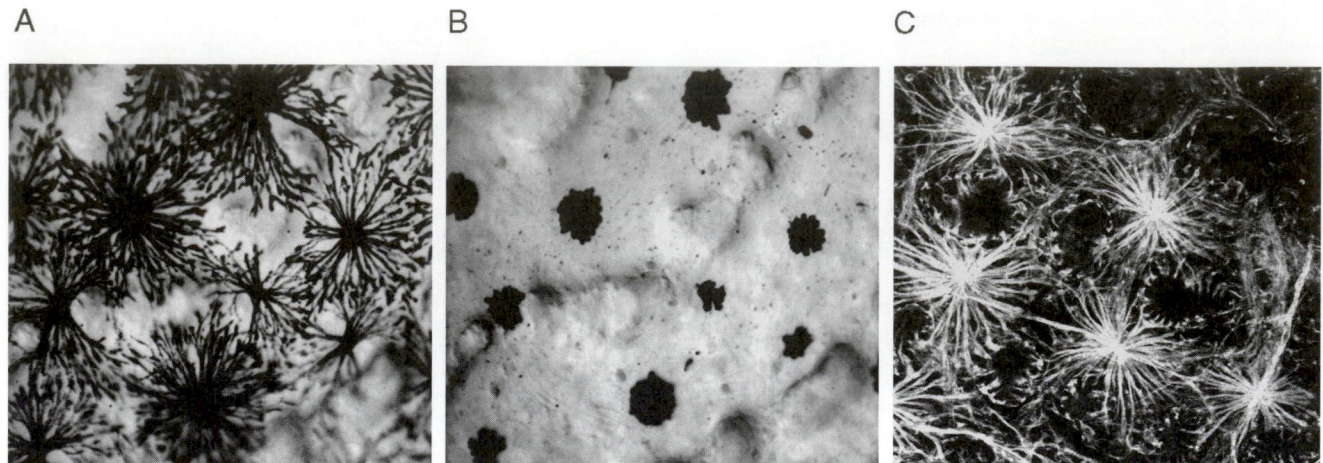

FIGURE 29 Brightfield micrographs of a field of living fish melanophores in which the pigmented granules are (A) dispersed or (B) aggregated. (C) Immunofluorescence micrograph of another field of fish melanophores that have been fixed and stained with anti-tubulin antibody to reveal the distribution of microtubules. During dispersion the pigment granules are moved outward along the microtubules by a kinesin family member, whereas during aggregation the granules are moved inward by cytoplasmic dynein. (Courtesy of Dr. Leah Haimo.)

the center of the cell. This movement is controlled by neurostimulation. *In vitro* studies indicate that aggregation of granules in the center of the cell, which depends on cytoplasmic dynein, occurs in response to dephosphorylation of protein substrates by a protein phosphatase. In contrast, dispersion of granules, which is driven by a kinesin family member, is dependent upon phosphorylation of substrates by cAMP-dependent protein kinase. The targets of the kinases and phosphatases may be subunits of the molecular motors themselves.

All microtubule motors have two major domains, a globular head and a tail (Fig. 28). The head contains the site of ATPase activity and is responsible for force generation, whereas the tail appears to determine the cell component to which the molecule binds. Most microtubule motors also contain smaller protein subunits that mediate binding to the cargo. The interaction of cytoplasmic dynein 1a with membranous structures is mediated by dynactin, itself a multisubunit complex.

V. Summary

Both actin-filament- and microtubule-based systems for cell motility use the energy from ATP hydrolysis, and these two mechanisms often complement each other during cell shape changes and organelle transport. Studies of striated muscle structure and function have identified the nature of relevant molecules and the concept of sliding as a mechanism by which contraction can be achieved. Hence, amoeboid locomotion of nonmuscle cells, cytoplasmic streaming, and cell shape changes can now be explained by similar actin-based systems in which myosin molecules provide the generation of force.

Although ciliary and flagellar movement is based on a microtubule-dynein system, it shares features with the actomyosin system; that is, motility is based on a sliding mechanism. Furthermore, motors that interact with microtubules cause such cell motility as organelle transport, similar to the

actin filament system. In some cases actin-filament- and microtubule-based organelle movement cannot be explained by sliding, for example, shortening of microtubules during movement of chromosomes to spindle poles. Polymerization and depolymerization of actin filaments or microtubules apparently constitute the underlying mechanism for force generation in these cases.

Three-dimensional ciliary and flagellar movements involve several levels of regulation. The generation of force is achieved by dynein ATPases using the energy released by Mg^{2+}-ATP hydrolysis. Conversion of sliding to bending results from local transient resistance to sliding of doublet microtubules. Reversal of the ciliary effective stroke and cessation of beat of some epithelial cilia are controlled by intraciliary Ca^{2+} levels. The intraciliary Ca^{2+} level also controls beat frequency. Cyclic nucleotides also participate in the transduction of signals during changes in the rate of ciliary beat and the direction of the effective stroke.

Bibliography

Adamek, G. D., Gestland, R. C., Mair, R. G., and Oakley, B. (1984). Transduction physiology of olfactory receptor cilia. *Brain Res.* **310**, 87–97.

Bardele, C. F. (1977). Comparative study of axopodial microtubule patterns and possible mechanisms of pattern control in the centrohelidian heliozoa *Acantocystis, Raphidiophrys*, and *Heterophrys. J. Cell Sci.* **25**, 205–232.

Bloodgood, R. A. (Ed.) (1990). "Ciliary and Flagellar Membranes." Plenum, New York.

Bouck, B., Rosiere, T. K., and Levasseur, P. J. (1990). *Euglena gracilis*: a model for flagellar surface assembly, with reference to other cells that bear flagellar mastigonemes and scales. *In* "Ciliary and Flagellar Membranes" (R. A. Bloodgood, Ed.), pp. 65–90. Plenum, New York.

Bray, D. (1992). "Cell Movements." Garland Publ. Inc., New York.

Brokaw, C. J. (1985). Cyclic AMP-dependent activation of sea urchin and tunicate sperm motility. *Ann. NY Acad. Sci.* **438**, 132–141.

Cole, D. G., Diener, D. R., Himelblau, A. L., Beech, P. L., Fuster, J. C., and Rosenbaum, J. L. (1998). *Chlamydomonas* kinesin-II-dependent intraflagellar transport (IFT): IFT particles contain proteins required for ciliary assembly in *Caenorhabditis elegans* sensory neurons. *J. Cell Biol.* **141**, 993–1008.

Conrad, G. W., and Schroeder, T. E. (Eds.) (1990). "Mechanisms of Furrow Formation During Cell Division," *Ann. NY Acad. Sci.* Vol. 582. New York Academy of Science, New York, 325 pages.

Doughty, M. J., and Kaneshiro, E. S. (1983). Divalent cation-dependent ATPase activities associated with cilia and other subcellular fractions of *Paramecium*: an electrophoretic characterization on Triton polyacrylamide gels. *J. Protozool.* **30**, 565–573.

Eckert, R., and Brehm, P. (1979). Ionic mechanisms of excitation in *Paramecium. Annu. Rev. Biophys. Bioenerg.* **8**, 353–383.

Eckert, R., Randall, D., and Augustine, G. (1988). "Animal Physiology. Mechanisms and Adaptation." Freeman, New York.

Gibbons, I. R. (1965). Chemical dissection of cilia. *Arch. Biol.* **75**, 317–352.

Gibbons, I. R. (1981). Cilia and flagella of eukaryotes. *J. Cell Biol. Suppl.* **91**, 107–124.

Hirokawa, N. (1998). Kinesin and dynein superfamily proteins and the mechanism of organelle transport. *Science* **279**, 519–526.

Inoue, S., and Stephens, R. E. (Eds.) (1975). "Molecules and Cell Movement." Raven Press, New York.

Kaneshiro, E. S. (1984). Symposium—the structure and function of cilia and flagella. *J. Protozool.* **31**, 7–40.

Klumpp, S., Steiner, A. L., and Schultz, J. E. (1983). Immunocytochemical localization of cyclic GMP, cGMP-dependent protein kinase, calmodulin and calcineurin in *Paramecium tetraurelia. Eur. J. Cell Biol.* **32**, 164–170.

Kung, C., and Eckert, R. (1972). Genetic modification of electric properties in an excitable membrane. *Proc. Natl. Acad. Sci. USA* **69**, 93–97.

Lefebvre, P. A., and Rosenbaum, J. L. (1986). Regulation of the synthesis and assembly of ciliary and flagellar proteins during regeneration. *Annu. Rev. Cell Biol.* **2**, 517–546.

Mermall, V., Post, P. L., and Mooseker, M. S. (1998). Unconventional myosins in cell movement, membrane traffic, and signal transduction. *Science* **279**, 527–533.

Naitoh, Y., and Eckert, R. (1969). Ionic mechanisms controlling behavioral responses of *Paramecium* to mechanical stimulation. *Science* **164**, 963–965.

Naitoh, Y., and Kaneko, H. (1973). Control of ciliary activities by adenosine-triphosphate and divalent cations in Triton-extracted models of *Paramecium caudatum. J. Exp. Biol.* **58**, 657–676.

Naitoh, Y., and Sugino, K. (1984). Ciliary movement and its control in *Paramecium. J. Protozool.* **31**, 31–40.

Omoto, C. K., and Witman, G. B. (1981). Functionally significant central-pair rotation in a primitive eukaryotic flagellum. *Nature* **290**, 708–710.

Omoto, C. K., Gibbons, I. R., Kamiya, R., Shingyoji, C., Takahashi, K., and Witman, G. B. (1999). Rotation of the central pair microtubules in eukaryotic flagella. *Mol. Biol. Cell* **10**, 1–4.

Preston, R. R., and Saimi, Y. (1990). Calcium ions and the regulation of motility in *Paramecium. In* "Ciliary and Flagellar Membranes" (R. Bloodgood, Ed.), pp. 173–200. Plenum, New York.

Rosenbaum, J. L., Cole, D. G., and Diener, D. R. Intraflagellar transport: the eyes have it. *J. Cell Biol.* **144**, 385–388.

Sale, W. S., and Satir, P. (1977). Direction of active sliding of microtubules in *Tetrahymena* cilia. *Proc. Natl. Acad. Sci. USA* **74**, 2045–2049.

Satir, P. (1968). Studies on cilia. III. Further studies on the cilium tip and a sliding filament model of ciliary motility. *J. Cell Biol.* **39**, 77–94.

Satir, P. (1984). The generation of ciliary motion. *J. Protozool.* **31**, 8–12.

Shinghoji, C., Murakami, A., and Takahashi, K. (1977). Local reactivation of Triton-extracted flagella by iontophoretic application of ATP. *Nature (London)* **265**, 269–270.

Sleigh, M. A. (Ed.) (1974). "Cilia and Flagella." Academic Press, New York.

Sleigh, M. (1989). "Protozoa and Other Protists." Arnold, London.

Sugino, K., and Naitoh, Y. (1982). Simulated cross-bridge patterns corresponding to ciliary beating in *Paramecium. Nature* **295**, 609–611.

Summers, K. E., and Gibbons, I. R. (1971). Adenosine triphosphate-induced sliding of tubules in trypsin-treated flagella of sea urchin sperm. *Proc. Natl. Acad. Sci. USA* **68**, 3092–3096.

Tamm, S. L., and Horridge, G. A. (1970). The relation between the orientation of the central fibrils and the direction of the beat in cilia of *Opalina. Proc. Roy. Soc. Lond. B* **175**, 219–233.

Thaler, C. D., and Haimo, L. T. (1996). Microtubules and microtubule motors: mechanisms of regulation. *Int. Rev. Cytol.* **164**, 269–327.

Vickerman, K., and Tetley, L. (1990). Flagellar surfaces of parasitic protozoa and their role in attachment. *In* "Structure and Function of Ciliary and Flagellar Surfaces" (R. A. Bloodgood, Ed.), pp. 267–304. Plenum, New York.

Warner, F. D., Satir, P., and Gibbons, I. R. (Eds.) (1989). "Cell Movement," Vol. 1. Liss, New York.

Warrick, H. M., and Spudich, J. A. (1987). Myosin structure and function in cell motility. *Annu. Rev. Cell Biol.* **3**, 379–421.

Witman, G. B. (1990). Introduction to cilia and flagella. *In* "Ciliary and Flagellar Membranes" (R. A. Bloodgood, Ed.), pp. 1–30. Plenum, New York.

Yanagimachi, R. (1988). Mammalian fertilization. *In* "The Physiology of Reproduction" (E. Knobil, J. Neill, L. L. Ewing, G. S. Greenwald, C. L. Markert, and D. W. Pfaff, Eds.), pp. 135–185. Raven Press, New York.

Edna S. Kaneshiro

58

Centrin-Based Contraction and Bacterial Flagella

I. Spasmonemes and Centrin-Containing Structures

The peritrich ciliates include solitary (e.g. *Vorticella*) and colonial (e.g. *Zoothamnium*) forms that have stalks, some of which contract rapidly when the organism is mechanically, chemically, or electrically stimulated (Fig. 1). Contractile peritrich stalks can contract to 30% of their original lengths in an all-or-none response. In some colonial forms, the stalks of the individual organisms contract independently, whereas in others, stimulation results in a synchronous contraction of all stalks, branches, and the trunk. The tubular stalk is formed by extracellular secretions of the cell when the nonstalked, free-swimming telotroch stage settles to the substratum. Secretion of the stalk transforms the telotroch to the sessile form.

A. Correlation of Contractility with Spasmonemes and Myonemes

Contractility of peritrich stalks is absolutely correlated with the presence of spasmonemes; stalks of peritrichs without spasmonemes are not contractile. The spasmoneme,

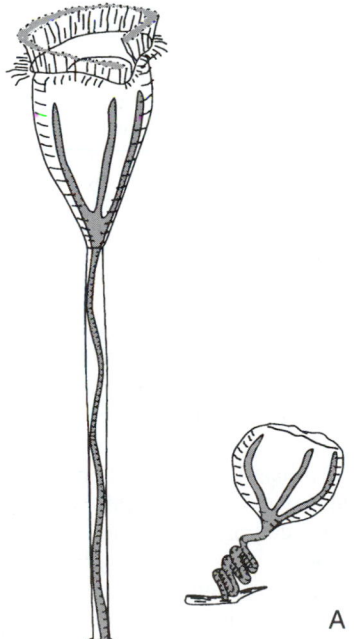
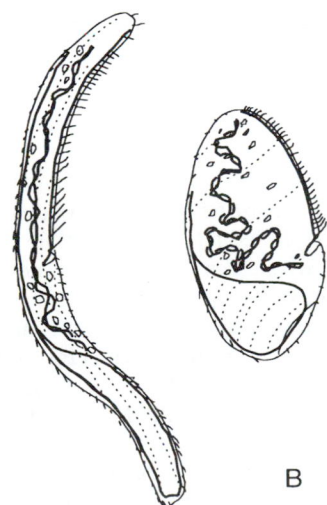

A

B

FIGURE 1. Contraction of *Vorticella* and *Spirostomum*. (A) Both the stalk and the cell body of *Vorticella* contract. The stalk coils during contraction by spasmoneme activity. Similar structures, called myonemes, in the cell body are responsible for the shortening of the cell body. (B) Myonemes in the cortex of *Spirostomum* cause contraction of the cell along its longitudinal axis. (Modified from *The Journal of General Physiology*, 1970, vol. 56, pp. 168–179, by copyright permission of The Rockefeller University Press.)

readily visible by light microscopy, is present in the cytoplasmic extension of the cell body within the stalk. Spasmonemes are commonly about 10 μm in diameter; however the giant spasmoneme found in the colonial peritrich *Zoothamnium* is 30 μm in diameter and 1 mm long. The stalk also contains stiffening fibers that serve to reextend the contracted stalk. When contracted spasmonemes within stalks break, the stalk sheath rapidly recoils and the stalk straightens and extends.

Recognizing that spasmonemes differ from the actinomyosin-based contractile systems (Table 1 in Chapter 57), early workers preferred the term *spasmoneme*. The term *myoneme* implied too close a relationship with muscle; however, the structures in cell bodies that behave like spasmonemes are still called myonemes. Myonemes are joined to spasmonemes in peritrichs. Similar myonemes are present in stalkless ciliates such as the large ciliate *Spirostomum* (Fig 1), which can contract to 50% of its original length, and the trumpet-shaped *Stentor*, which can contract its cell body down to 13% of its original length.

1. Ultrastructure

a. Spasmonemes. Spasmonemes have a trabecular texture containing 4-nm microfibrils arranged roughly along the long axis of the spasmoneme. By negative staining, extracted spasmoneme microfibrils show a 3.5-nm longitudinal periodicity. Numerous mitochondria and a membranous system (endoplasmic reticulum, ER) of saccules and tubules are associated with the microfibrils (Fig. 2). The ER vesicles or cisternae surround and penetrate microfibrillar bundles. This enclosed membrane system is believed to be analogous to the sarcoplasmic reticulum of muscle fibers and is thought to sequester and release Ca^{2+} during spasmoneme contraction and extension. This notion is supported by electron microscopy of oxalate-treated myonemes that show deposits of calcium precipitates within the vesicles.

b. Myonemes. The myonemes of the stalkless, giant ciliates *Stentor* and *Spirostomum* are similar in ultrastructure to those in the cell bodies of stalked peritrich ciliates. Huang and Pitelka studied *Stentor* that had been pretreated with EGTA. The myonemes in these treated cells were relaxed and electron microscopy showed a meshwork of the 4-nm microfibrils. In cells stimulated to contract during fixation, the microfibrils coiled, forming short 10- to 12-nm-diameter tubes with a 4- to 5-nm wall consisting of 4 to 6 globular subunits (Fig. 3). However, these coiled tubular conformations are not seen in all contracted myonemes of all species, nor are they seen in spasmonemes of stalked peritrichs; therefore, this configuration may represent a supercoiled state of the microfibrils.

Myonemes in *Stentor* are associated with cross-linked microtubular bundles (kinetodesmal fibers), which was suggested by Huang and Pitelka to be the Mg^{2+}-ATP-requiring mechanism for reextension of the cell body. The microtubular bundles extend from ciliary basal bodies. The number of overlapping microtubules at any given anterior-to-posterior level of the cell is greater in contracted cells than in extended cells. During contraction, cross-links between microtubules detach, suggesting that this allows passive sliding of microtubules when myonemes contract. Sliding of micro-

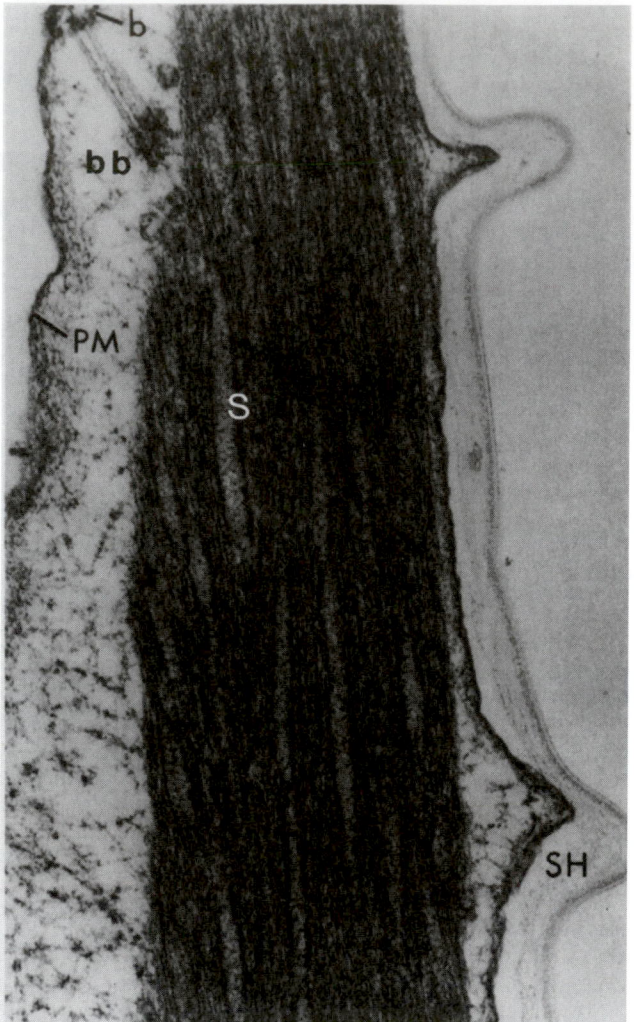

FIGURE 2. Electron microscopy of the *Vorticella* spasmoneme. The spasmoneme in a peritrich stalk consists of 4-nm fibrils arranged parallel to the long axis of the organelle, and the stalk. Within the organelle, membrane-bound saccules (S), sites of calcium sequestration, are seen interspersed between the fibrils. The stalk sheath (SH), the cell surface membrane (PM), and a modified basal body (bb) associated with short bridges (b) are indicated. (Reproduced with permission from Allen, 1973.)

tubules in the opposite direction may accompany reextension and therefore may serve as the antagonist of myoneme contraction, but this idea has not been tested.

c. The Linkage Complex. Allen described a structure associated with the microfibrils and called it the linkage complex. The linkage complex is oriented in various directions with respect to the spasmoneme or myoneme long axis. It consists of a midpiece, thin structures that cross perpendicular to that midpiece, rails that run parallel to the midpiece, and long, thin, striated microfilaments that connect the midpiece with structures at the outer cortex of the cell or stalk (Fig. 4). At the cortex, the striated filaments make close associations with basal bodies or modified basal bodies and can even be seen passing through the lumen of modified basal bodies.

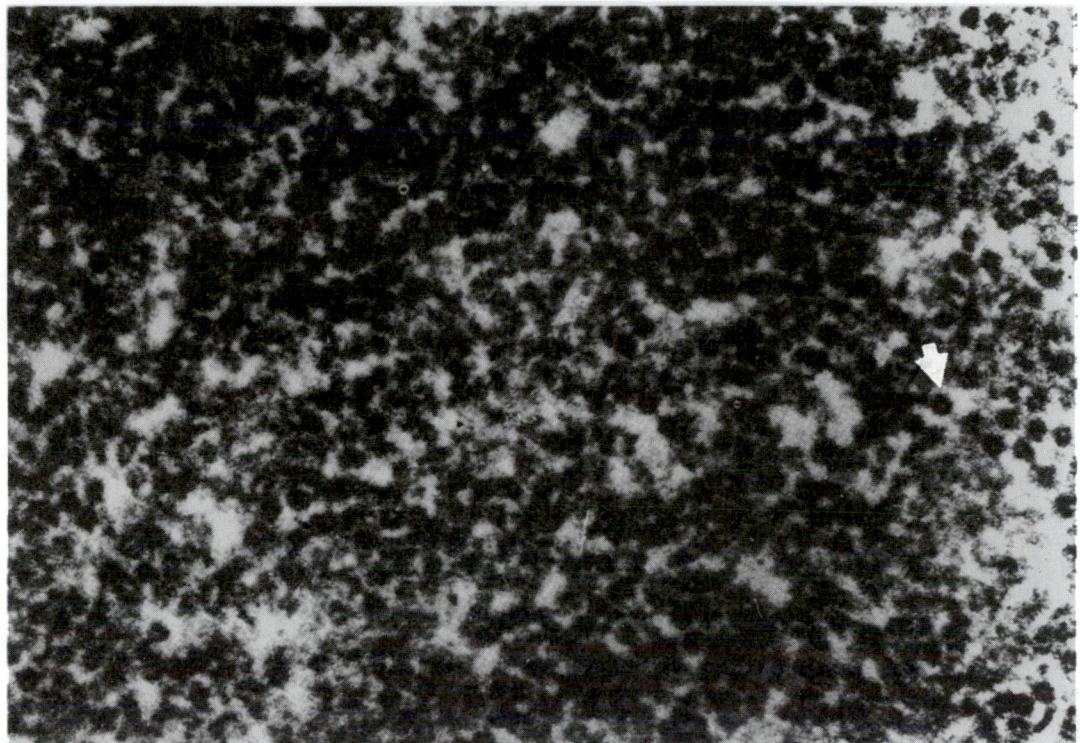

FIGURE 3. Contracted myoneme of *Stentor*. The 4-nm fibrils in myonemes of a contracted *Stentor* cell form supercoils. They appear doughnut-shaped when viewed by electron microscopy in cross-sections of the myoneme. [Reprinted with permission from Huang, B., and Mazia, D. (1975). Microtubules and filaments in ciliate contractility. *In* "Molecules and Cell Movement" (S. Inoué and R. E. Stephens, Eds.), pp. 389–409. Raven Press, New York.]

2. Properties of Spasmonemes and Myonemes

Many of the early significant advances in the understanding of spasmoneme structure and function came from a group that included Weiss-Fogh and Amos (see, for example, Routledge *et al.*, 1976). Difficulty in obtaining large numbers of purified spasmonemes for biochemical analyses and the lack of a suitable method for growing mass cultures of peritrichs have slowed progress in understanding the mechanism by which these organelles contract. Nonetheless, considerable information has been assimilated by some imaginative approaches used by these and other investigators.

a. Speed of Contraction. Spasmoneme and myoneme contractions differ from those in the actomyosin–ATP systems in many respects. Myonemes do not bind HMM. Following a latent period of 1–4 ms, spasmonemes contract at a speed of 11–21 cm/s, or about 200 lengths/s, whereas the fastest known striated muscle (in mouse fingers) contracts at only 22 lengths/s. In peritrichs, the contraction begins at the cell body–stalk junction, which has also been shown to have a lower stimulus threshold compared with distal regions. Contraction propagates at a maximum velocity of 21–60 cm/s in long stalks and 1.4–5.6 cm/s in small stalks. In fatigued stalks, conduction velocity decreases to 3 cm/s, with fatigue appearing first at the distal end. The duration of contraction is 2–20 ms. In contrast, reextension of contracted spasmonemes takes several seconds. The Q_{10} of extension is 2.5,

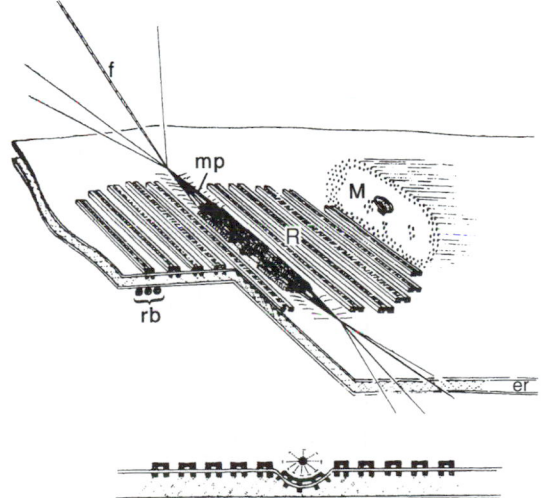

FIGURE 4. Linkage complex of spasmonemes and myonemes of *Vorticella*. The linkage complex is associated with the myoneme and the rough ER (er); the ribosomes of the ER are shown. The complex consists of rails (R), the midpiece (mp), and several filaments (f) that fan out from the tips of the midpiece (rb), ribosomes, A cross-sectional view of the linkage complex is shown at the bottom. (Reproduced with permission from Allen, 1973).

suggesting that enzymatic processes are involved during this slower reextension process.

b. Role of Ca²⁺. Injection of Ca^{2+} into cells with myonemes results in contraction, and stimulation of cells preloaded with the Ca^{2+}-sensitive bioluminescent protein aequorin results in the emission of light before contraction (Fig. 5). Glycerinated or detergent-extracted models of peritrich stalks contract and extend repeatedly when sequentially exposed to a Ca^{2+} solution and then to an EGTA solution; ATP or Mg^{2+} is not required. In extracted models, Ca^{2+} can be substituted by Sr^{2+} and less so by Ba^{2+}, but Mg^{2+} does not initiate contraction. The threshold for Ca^{2+} is estimated at about 10^{-7} g ions/L. Contraction and extension of spasmoneme models can be repeated for at least two days at room temperature. This activity is not affected by the mercury-containing inhibitor salyrgan (mersalic acid) nor by parachloromercurobenzoate, cyanide, dinitrophenol, or fluorodinitrobenzene. It is also not affected by the detergents digitonin, saponin, or Tween 80.

c. Optical Properties. An important difference between spasmonemes and striated muscles lies in their optical properties. Like the A-band of muscle, spasmonemes exhibit positive birefringence (A-band, 2.3×10^{-3}; spasmonemes, 4×10^{-3}), which indicates that substructures are ordered with their long axis in line with the long axis of the fibers; therefore, spasmonemes and muscle A-bands are comparable. The positive birefringence of spasmonemes is largely due to form (substructures), rather than to intrinsic (molecular conformations) birefringence. During contrac-

tion, the birefringence of the striated muscle A-band remains the same, indicating no dramatic molecular conformational changes. However, the positive birefringence of spasmonemes decreases to almost zero during contraction (Fig. 6). This indicates that during spasmoneme contraction there is a decrease in order or folding of polymeric molecules. Since the birefringence appears to be due mainly to form birefringence, the optical properties of spasmonemes suggest a mechanism of microfibril coiling, in agreement with the observation that microfibrils in *Stentor* myonemes coil or supercoil during contraction. Unstrained *Zoothamnium* giant spasmonemes are apparently isotropic but they increase their birefringence when stretched.

d. Energetics. With respect to energetics, the tension developed by spasmonemes is similar to that of striated muscles operating at the optimum tension for performance of work. During isometric tension, the work done by striated muscle in a single twitch is equivalent to 10^5-10^6 N·m⁻² (N = 10^5 dyn). Spasmonemes develop $4-8 \times 10^4$ N·m⁻² tension per unit cross-sectional area, and the work done in a single twitch has been calculated at 11 J/kg wet wt. Therefore, spasmonemes are equivalent to, or develop somewhat lower tension than, striated muscle. But, during contraction with different loads, striated muscles exhibit high shortening velocity with low load, which is associated with very little force and thin-thick filament cross-bridging. On the other hand, in spasmonemes (as measured by changing the viscosity of the medium), force does not vary with speed. This is similar to the energetics of a rubber band or metal spring. The calculated instantaneous power is 2.7 kW/kg wet wt, compared with that of the most

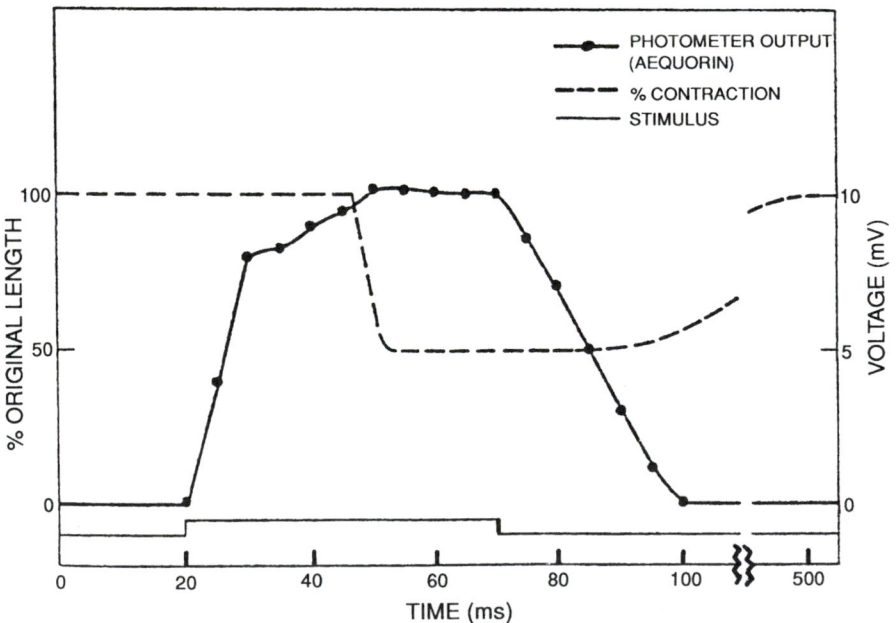

FIGURE 5. Excitation-contraction coupling in *Spirostomum*. Temporal relationship between excitation, increase in free calcium, and the contraction of a *Spirostomum* cell preinjected with aequorin, a Ca^{2+}-dependent luminescent protein. Contraction begins when aequorin-induced photon emission is maximal. Cessation of electrical stimulation is followed by a decrease in photon emission and the cell relaxes. (Reproduced from *The Journal of General Physiology*, 1970, vol. 56, pp. 168–179, by copyright permission of The Rockefeller University Press.)

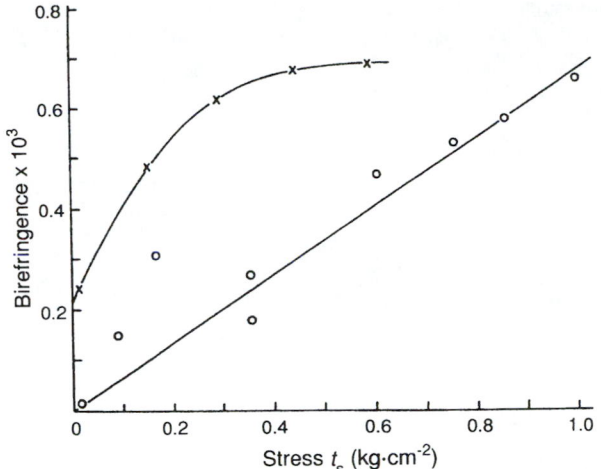

FIGURE 6. Birefringence changes of spasmonemes. In the presence of high calcium (o) the spasmoneme birefringence is directly proportional to stress (t_s = tensile stress). Thus in the high calcium state, the spasmoneme behaves optically and mechanically as rubber. However, in the low calcium state at 10^{-8} M concentration (X), the relationship between birefringe and stress is not the same. In the low calcium medium, the spasmoneme actively elongates and exhibits a pushing force, and the unstressed organelle becomes birefringent. [Reprinted with permission from Amos, W. B. (1975). Contraction and calcium binding in the vorticellid ciliates, *In* "Molecules and Cell Movement" (S. Inoué and R. E. Stephens, Eds.), pp. 411–436. Raven Press, New York.]

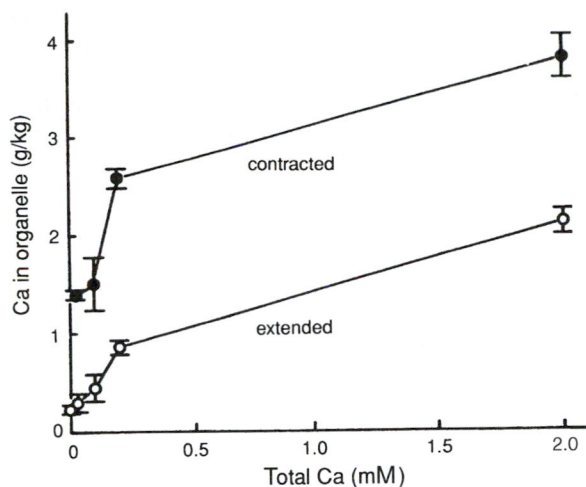

FIGURE 7. Electron microprobe analysis of the calcium content of isolated glycerinated *Zoothamnium* giant spasmonemes. Contracted spasmonemes were prepared by treatment with 10^{-6} M Ca^{2+}. Extended spasmonemes were treated with 10^{-8} MCa^{2+}, except at the ordinate, at which a 20 mM EDTA solution was used [Reprinted with permission from Amos, W. B. (1975). Contraction and calcium binding in the vorticellid ciliates. *In* "Molecules and Cell Movement" (S. Inoué and R. E. Stephens, Eds.), pp. 411–436. Raven Press, New York.]

energetic known muscle (insect flight muscle), which is only 0.05–0.2 kW/kg wet wt.

Spasmonemes have a high degree of passive extensibility and can be stretched to four times their resting lengths and then return to their normal lengths. At low Ca^{2+} concentrations, a measurable pushing force and increased birefringence develop. If the stalk is prevented from elongating, the spasmoneme is seen elongating and bending within the stalk. This pushing force observed during spasmoneme extension is not thoroughly understood, but it may involve ATP hydrolysis, which is consistent with the presence of abundant mitochondria closely associated with spasmonemes.

Elemental analysis (calcium K_α peak) by electron microscopic microprobe of glycerinated *Zoothamnium* giant spasmonemes indicates that the organelle has 1.5 and 0.36 g bound calcium/kg dry wt during contraction and extension, respectively (Fig. 7). The change in chemical potential of calcium ($\Delta\mu_{Ca} = RT \ln [Ca^{2+}]_{high} / [Ca^{2+}]_{low}$) was examined to determine whether Ca^{2+} alone can account for work done by spasmonemes during contraction. The value of 10^4 J/mol calcium was calculated for a change from 10^{-8} to 10^{-6} M; therefore, for each mole of calcium bound to the spasmoneme, 10^4 J of energy would be available for work. As mentioned previously, the work done in a single twitch of the spasmoneme is 11 J/kg wet wt. Therefore, to produce the estimated energy output, the spasmoneme must bind at least 11×10^{-4} moles of Ca^{2+}, or 44 mg calcium/kg wet wt. The microprobe elemental analyses indicated values of 1.5 (contracted spasmoneme) and 0.36 (extended spasmoneme) g calcium bound/kg wet wt. Therefore, the chemical potential of calcium can account for the work done by spasmoneme contraction.

3. Spasmins and the Mechanism of Contraction

The major proteins of spasmonemes, spasmins, are acidic proteins with pI values of 4.7 and 4.8. They constitute 50–60% of the total stainable proteins of the isolated organelles separated by SDS-PAGE. Their migrations, which are not affected by β-mercaptoethanol (no –S–S– bonds), indicate that their masses are about 18 kDa (spasmin A) and 20 kDa (spasmin B). In EGTA–Ca^{2+}-buffered gels, their mobilities (anodally) decrease in the presence of high Ca^{2+} (10^{-6} M), suggesting a decrease in net negative charge, an increase in the Stokes radius of the molecules, or both. However, in alkaline urea gels with Ca^{2+}, spasmin B exhibits increased mobility, which is not consistent with a decrease in net negative charge due to the binding of calcium. This observation provided strong evidence against an earlier hypothesis, the electrostatic model of spasmoneme contraction. This hypothesis was based on the idea that Ca^{2+} binding neutralized repulsive negative charges, resulting in the folding of polymeric molecules. Using the microprobe data for bound calcium in contracted and extended spasmonemes, it was calculated that Ca^{2+} binds to spasmin with stoichiometry of one or two Ca^{2+} per molecule. Spasmins are high in serine, aspartate + asparagine, and glutamate + glutamine. Unlike troponin C and parvalbumin, spasmins lack cysteine.

Coiling of microfibrils is a plausible hypothesis for myoneme contraction. An alternative hypothesis is the entropic rubber model, which generally equates spasmonemes with material such as elastin or rubber that possesses oriented isoprene units. In this model, contraction results by the release of stored elastic energy.

4. Excitation-Contraction Coupling

Less examined and understood is the nature of the coupling of excitation with contraction (EC coupling). Since the activation time is on the order of 2–3 ms in *Zoothamnium*, and its giant spasmoneme is 30 μm in diameter, the signal is transduced 10 times faster than can be explained by simple diffusion of Ca^{2+} through the surface. Currently, the linkage complex associated with spasmonemes appears the most likely candidate for EC coupling (see Fig. 4). Allen has suggested that the striated filament that links the midpiece of the linkage complex with cortical structures near the surface membrane (e.g., basal bodies) may transduce the electrical signal to the midpiece. The midpiece is physically coupled to the ER and hence could trigger Ca^{2+} release from the cisternae. Furthermore, the rails may be involved in the movement of Ca^{2+} in and out of the ER and may be the sites of ATPase activity for concentrating Ca^{2+} into the ER.

B. Striated Rootlets

Striated rootlets (rhizoplasts), well-developed in unicellular organism such as flagellated algal cells (e.g., *Tetraselmis*), are organelles that are associated with the flagellar basal bodies (Fig. 8). They extend, branch, and link to adjacent basal bodies. In some cases they extend and pass along the nuclear envelope and anchor near the cell membrane opposite the flagella. The anchor site has been termed laminated oval, half-desmosome, or rhizankyra. The rootlet exhibits cross-striations of greater than 160-nm periodicity and is composed of 3- to 8-nm microfibrils aligned parallel to the long axis of the organelle. That these were contractile organelles was clearly demonstrated by fixation of *Tetraselmis* cells in 0.1–5 mM Ca^{2+}. Salisbury (1989) found that, compared with organelles in cells fixed without Ca^{2+}, the organelle was 65% shorter, 30% wider, and had a reduced cross-striation periodicity (see Fig. 8). Other studies of striated rootlets demonstrate that regions proximal to the flagellar basal bodies have smaller periodicities than distal regions, suggesting a polarity in the organelle. However, this observation may reflect the direction in which contraction propagates. In living cells, it was observed that the rootlets undergo cyclic contractions and extensions with millimolar concentrations of Ca^{2+} and ATP.

Much like spasmonemes, neither muscle HMM nor myosin S-1 fragments decorate extracted rootlets. Also, striated rootlets and other related organelles contract rapidly within less that 20 ms; reextension is slow (a few seconds to about 1 hour.) Contraction is initiated by elevated intracellular Ca^{2+} and is independent of ATP hydrolysis. When contracted, the microfibrils in striated rootlets assume a supercoiled configuration. Although the direct requirement for ATP in the reextension of spasmonemes is not clearly established, there is good evidence that ATP is required by striated rootlets during their reextension process.

The major proteins of striated rootlets, centrins (called caltractin in *Chlamydomonas*), make up greater than 60% of the total proteins of the isolated organelle. Centrins are low-molecular-mass (20-kDa) acidic proteins (α centrin, pI 4.9; and β centrin, pI 4.8).

Metabolic radiolabeling with $^{32}PO_4$ demonstrates that the β isoform is phosphorylated whereas the α isoform is not. The role of phosphorylation of these contractile proteins is currently unclear. However, it has been suggested that the phosphorylated β and dephosphorylated α isoforms are

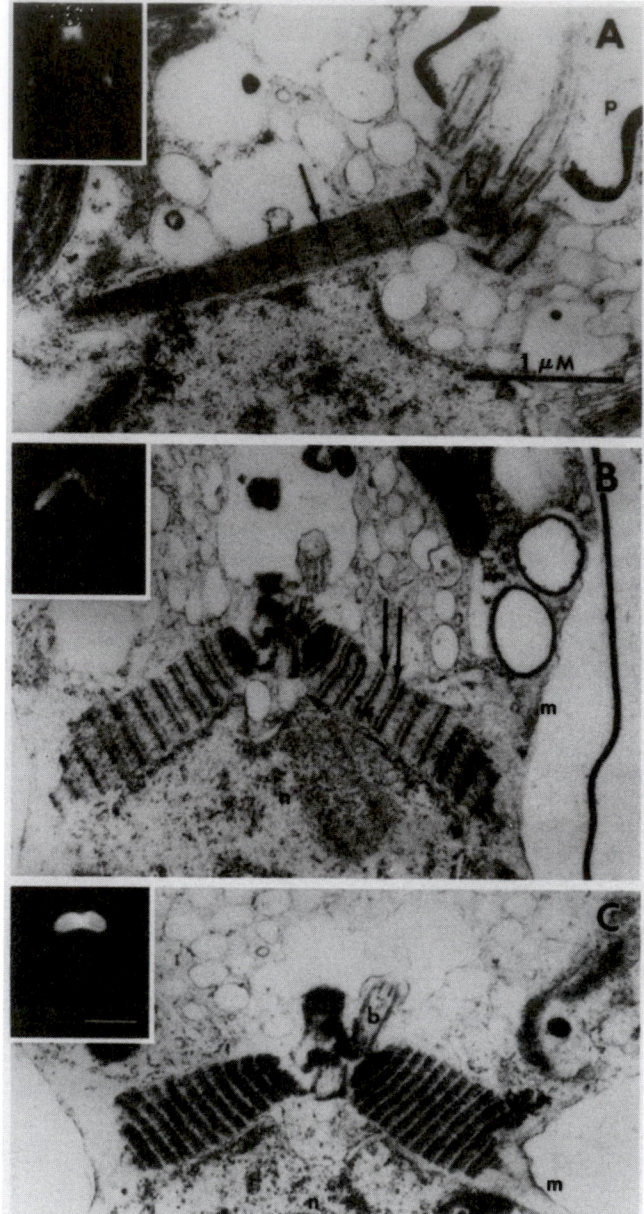

FIGURE 8. Striated rootlet of the marine flagellate, *Tetraselmis*. Cells were fixed for electron microscopy under various conditions. The insets show the binding patterns of flourescence-labeled anti-centrin antibodies. (A) Fixation with low Ca^{2+} illustrates the normal width of the organelle and the periodicity of its cross-striations. (B) Fixation with high Ca^{2+} plus ATP may be equivalent to the reextension phase after contraction. (C) Fixation with high Ca^{2+} (0.1–5mM) caused a dramatic shortening of the striated rootlets. The organelle in this state is characterized by increased thickness and reduced periodicity of the cross-striations. (Courtesy of Jeffrey Salisbury.)

correlated with the extended and contracted states, respectively. Both isoforms of centrin exhibit Ca^{2+}-sensitive mobilities in electrophoresis gels. However, in alkaline urea gels, the mobilities decrease in the presence of Ca^{2+}, similarly to the mobility of spasmins in SDS-PAGE under denaturing conditions. These observations on centrin may be

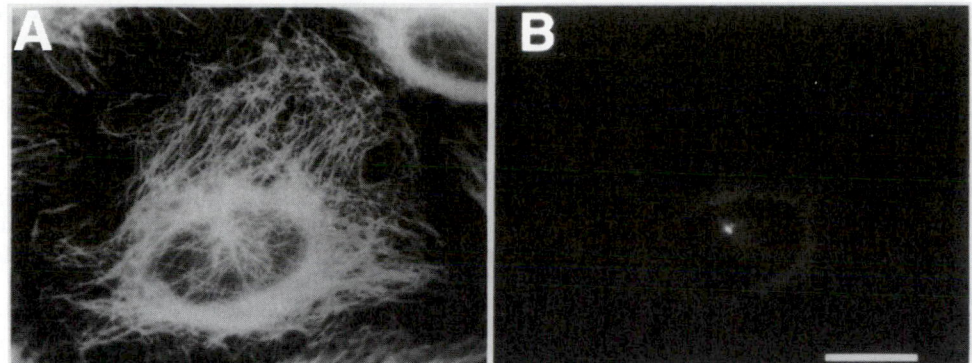

FIGURE 9. Anti-centrin antibody binding to cells. The wide occurrence of centrin-like molecules is demonstrated by a HeLa cell double-stained with fluorescence-labeled anti-tubulin and anti-centrin antibodies. (A) Anti-tubulin binding pattern. (B) Anti-centrin binds to the centrosomes of cells from broad phylogenetic backgrounds. (Courtesy of Jeffrey Salisbury.)

related to those made with troponin C, which binds Ca^{2+} in urea gels, causing an increase in mobility. This has been interpreted as a change to a more compact molecular conformation, analogous to the interaction of troponin I and troponin C when the two proteins interact and form a slower migrating complex.

C. Occurrence and Functions of Centrin and Related Molecules

Monospecific antibodies to flagellar rootlet centrin proteins have been reported to bind to centrioles and centrosomes of mitotic spindle poles, as well as to the mitotic spindle matrix (Fig. 9). These observations, made on a wide range of cell types, revealed the ubiquitous occurrence of centrin-like molecules and inspired Salisbury (1989) to coin the term centrin for the protein. These findings indicate that centrin may have a universal function in cells, since it is found in cell division structures.

The anti-centrin antibodies also bind to a nine-pointed stellate structure in the transition zone between the flagellar shaft and the basal body of some flagella. The antibody binding occurs at the plane of scission during the deflagellation process. The stellate structure was shown to contract during Ca^{2+}-induced deflagellation, implicating that structure in the generation of force required for autotomy (Fig. 10). However, autotomy occurs in many cilia and flagella that normally do not have the nine-pointed stellate structure,

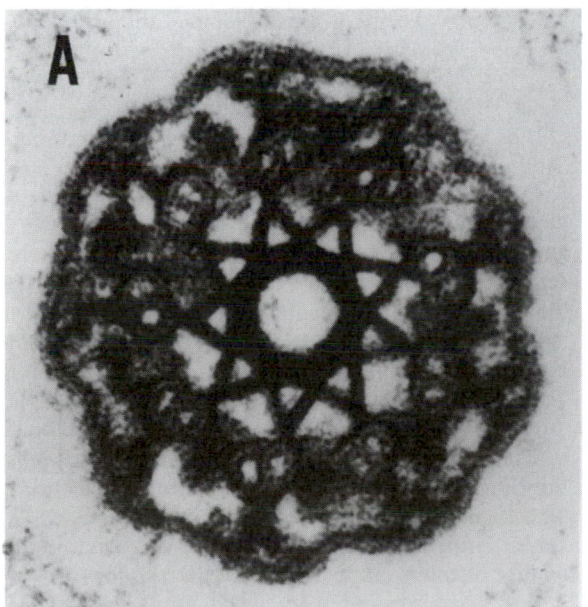
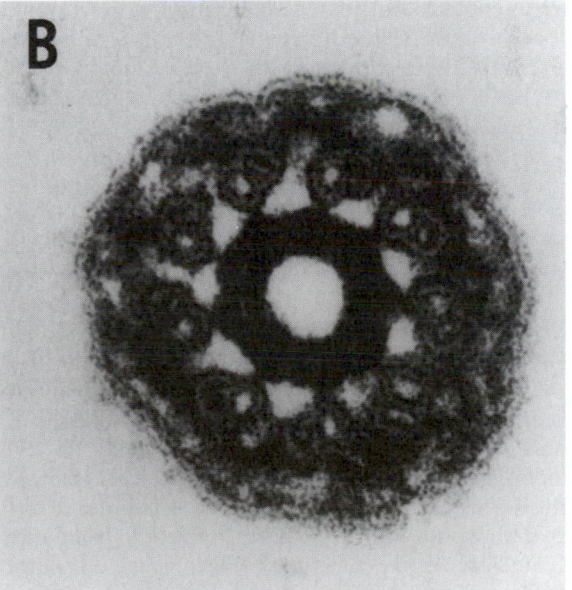

FIGURE 10. Transition zone of *Chlamydomonas* flagella at the nine-pointed stellate structure. (A) The configuration in a normal flagellated cell. (B) The nine-pointed stellate structure appears contracted in a cell treated with high concentrations of Ca^{2+}, a treatment that induces autotomy. Constriction of the stellate structure may be the direct cause of weakening at this plane leading to severing of the flagella. (Reproduced from *The Journal of Cell Biology*, 1989, vol. 108, pp. 1751–1760, by copyright permission of The Rockefeller University Press.)

and recently Jarvik has described autotomy in a *Chlamydomonas* mutant lacking this structure.

The centrin-related proteins are related to the E-F hand superfamily of calcium-binding proteins. Huang isolated and sequenced *Chlamydomonas* caltractin (centrin) cDNA and demonstrated that the protein contains four E-F hands. Sequence identity with calmodulin is 45–48%, and sequence identity with the yeast spindle pole *CDS31* gene is 50%. Anti-caltractin antibodies, however, do not cross-react with calmodulin.

An alternative hypothesis for the function of centrin has been proposed for the biflagellated alga *Spermatozopsis*, whose rhizoplast (rootlet) does not exhibit cross-striations. A connecting fiber joins the two basal bodies. During forward swimming, the two flagella of the cell exhibit asymmetrical breast stroke movements. The organism displays a photophobic response to light (avoidance reaction), resulting in backward swimming. During this response, the flagella undulate and the basal bodies reorient by contraction of the connecting fiber. This switch for basal body reorientation is independent of the motor that drives flagellar beating.

Anti-centrin antibodies also bind to a structure called the paraxial rod, which is found within the transverse flagellum of some dinoflagellates. The paraxial rod may cause a rapid and strong contraction of the dinoflagellates transverse flagellum that causes a stop reaction of the swimming cell (equivalent to an avoidance reaction). Thus, centrin may function as a modulator of motile structures.

II. Prokaryote Locomotion

Locomotion, characteristic of some bacteria, serves as a useful taxonomic criterion. Motility assays are standardly employed in identifying bacterial isolates. Two types of locomotion are recognized in prokaryotic cells: (1) flagella driven swimming and (2) gliding. Gliding prokaryotes and eukaryotes is discussed later in Section III.

A. Bacterial Flagella

Flagellated bacteria swim by the rotation of rigid helical flagella driven by motors at their bases. Bacterial flagella are thin (ca. 20 nm in diameter) extracellular organelles that extend about 15–20 μm from the surface cells (Fig. 11). They can occur as a single flagellum (monopolar) or a bundle of several flagella (lophotrichous) at one end of the cell. When bundles occur at both poles, the pattern is called amphitrichous. Cells with randomly distributed flagella are peritrichous. In spirochetes, flagella do not protrude from the cell surface. Instead, an axial filament consisting of two sets of fibrils that extend from pole to pole, enclosed within the outer layer of the cell surface, is responsible for the flexing movements exhibited by these organisms.

B. Structure

When stained, bacterial flagella are visible by light microscopy, but details of their structure can be elucidated only by special microscopic and spectroscopic techniques. The bacterial flagellum is composed of three parts: the basal body, hook, and filament (Fig. 12).

1. Basal Body

The basal body is relatively small, but is a complex structure made up of a rod and rings. The rod is analogous to a transmission shaft. Basal body rings differ in gram-negative and gram-positive bacteria, and these substructures correlate with the differences in cell wall structures. In gram-negative species such as *Escherichia coli* and *Salmonella typhimureum*, there are usually four rings: (1) the M ring resides within the cell membrane; (2) the S ring is located within the periplasmic space, (3) the P ring is in the peptidoglycan layer; and (4) the L ring is within the lipopolysaccharide-containing outer layer. The outer two rings probably serve as the motor's bushing and allow the rod to pass through the outer layers of the cell envelope. The inner two are involved in rotation of the motor. How rotation is generated is not yet clear. It has been suggested that the M ring may serve as the stator and the P ring as the rotor. Alternatively, some have argued that a stator needs a large mass offering sufficient stability into which it can anchor, since the filament needs to overcome high viscous drag. The most likely structure for greater mass is the peptidoglycan layer. In this case, the rotation would then be generated with the cell membrane and the M ring directly involved in the generation of rotation. In gram-positive species that lack the outer layer, only two rings are found in the basal body: one is the equivalent of the M ring within the cell membrane, and the other, the outer ring, is associated with the teichoic acid component of the cell wall.

2. Hook

The hook functions like an elastic universal joint between the basal body and filament. Its 90-nm length is genetically predetermined and involves hook-associated proteins (HAP) at the hook-filament junction. Mutants lacking HAP continuously secrete unassembled filament protein into the medium.

3. Filament

The filament, which acts like a propeller, is a rigid, brittle helical polymer of a single 51-kDa protein, flagellin. Some bacterial flagella (such as in *Caulobacter*) are composed of two types of flagellin molecules. The flagellin subunits determine the final shape of the filament structure by non-equivalent interactions between the subunits. Left-handed helices are more common than right-handed helices. Flagellin self-assembles *in vitro*, forming a cylindrical wall of about 11 subunits per 2 turns. A mixture of two different flagellins also self-assembles *in vitro*. Unlike that of actin and tubulin, polymerization of flagellin does not require energy from nucleotide triphosphates. Flagellins are among the most antigenic proteins known, which is an adaptation of the vertebrate immune system for effective elimination of bacterial infections.

C. Assembly of Flagellar Substructures

Much like that observed in growing eukaryote flagellar axonemes, the assembly of the growing flagellum occurs by the addition of individual protein subunits from the proximal to

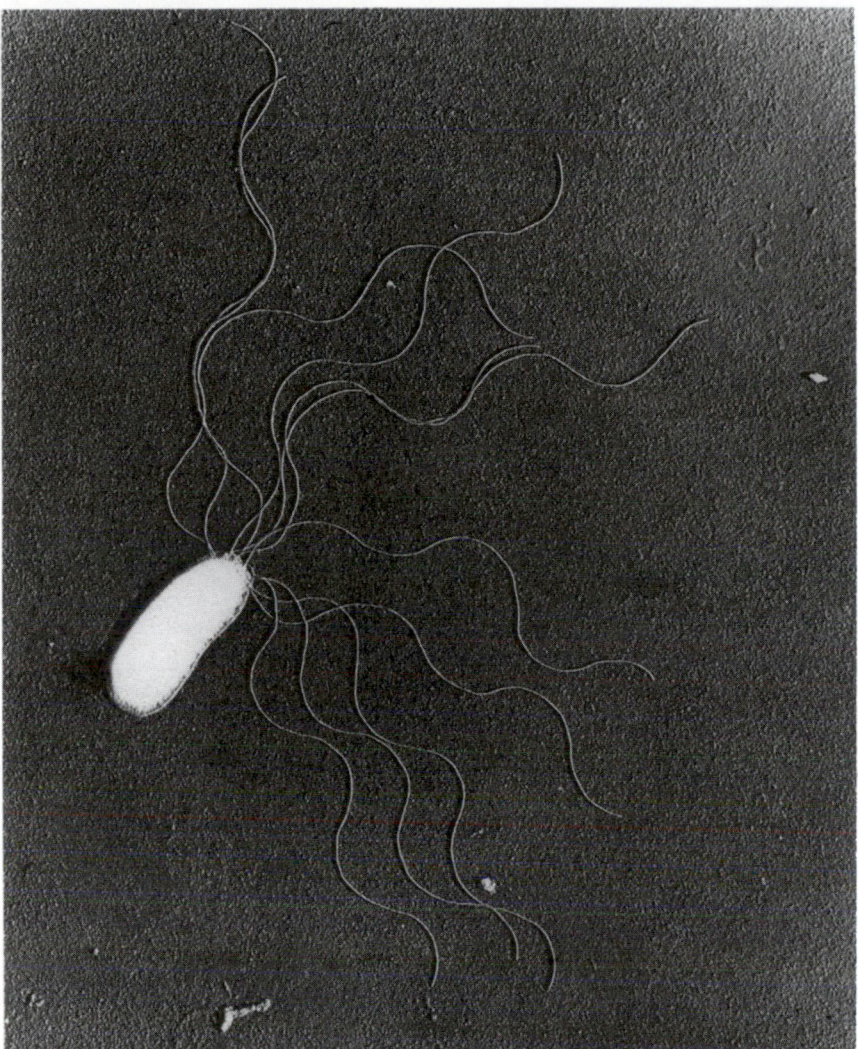

FIGURE 11.   Electron micrograph of *Pseudomonas marginalis*. Flagella emerge from one pole of the bacterial cell. (Courtesy of Arthur Kelman.)

distal end; the tip is the last to be added. This base-to-tip mechanism probably evolved in bacteria, since assembly occurs primarily outside the cell membrane and the export and addition of simple subunits to a single structure are easier than the export and addition of larger, more complex modular structures. Also, since most of the torsional load is at the proximal end, and because the motor rotates while the organelle is being elongated, any proximal separation in the nascent structure for insertion of new material would be mechanically difficult. The precursors, therefore, must be exported, single-file, through the growing structure; they do not reach the tip on the external surface. The information for the entire flagellum structure is probably inherent in the structure of each subunit, since self-assembly *in vitro* reconstitutes the different components very close in size and sequence to the intact organelle. Moreover, all rod, hook, HAP, and filament proteins do not have cleaved signal peptides, strongly suggesting that the subunits include topological targeting information. Assembly, therefore, is closely regulated by the sequence of synthesis

and the export of the subunits. A central channel of 5-nm diameter has been detected in bacterial flagella by x-ray fiber diffraction techniques. This channel is sufficiently large to serve as a conduit for single subunits to get to the growing tip.

D. Energy Source

The energy for rotation of the motor does not come from the hydrolysis of high-energy phosphate bonds of ATP. It is derived from the dissipation of an ionic transmembrane electrochemical gradient (ion-motive force). In most neutrophilic species, such as *E. coli*, the ionic species that has been actively pumped out of the cell by the electron transport chain in the bacterial membrane during respiration is H^+ (proton-motive force). The coupling of energy released in the respiratory chain requires solute and ion transport, not ATP, as an intermediate; therefore, ATP synthesis and flagellar rotation are alternative links with the respiratory chain. When H^+ is the ion actively pumped out, forming a transmembrane proton chemical and

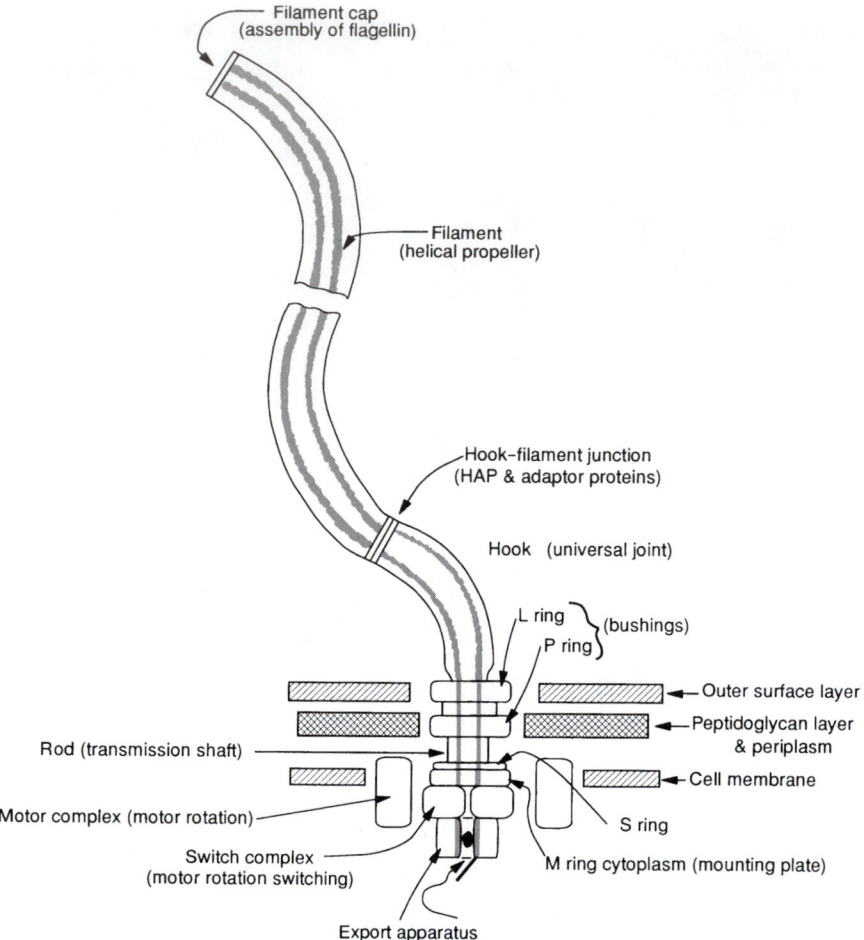

FIGURE 12.   Structure of the flagellum of a gram-negative bacterium. The filament is flagellin polymer that is rigid and helical. The hook, L ring, P ring, and rod lie distal to the cell surface membrane. The S ring and M ring are embedded in the membrane bilayer and are associated with the complexes that serve as the motor and the motor switch. The export apparatus allows molecules into a central channel for flagellar growth at the site of assembly at the tip. (Modified from MacNab, 1990, with permission of Cambridge University Press.)

electrical gradient, the proton-motive force becomes available to drive flagellar rotation. Under anaerobic conditions, the translocation of protons is achieved by proton ATPase activities. In marine and alkalophilic bacteria, the ion is Na^+, which has been pumped out of the cell by a Na^+-H^+ antiporter or by the electron transport chain. Since the energy can come from the electrochemical gradients of either H^+ or Na^+, the rotation of the motor must involve electrostatic interactions and events involving association and dissociation of ions.

Although no morphological structures have been identified, genetic and biochemical evidence indicates that there must be structures involved in energy transduction for converting ion-motive force to motor rotation. Since it is difficult to control specific membrane potentials and pH gradients experimentally in gram-negative bacteria, energy coupling has been best examined in gram-positive bacteria such as *Streptococcus* and *Bacillus subtilis*. It has been clearly established that rotation is coupled to the inward current of protons and that the speed of rotation is proportional to the proton-motive force. A flux of 1200 protons per revolution, independent of load or velocity, has been calculated. Membrane potentials or pH gradients of opposite polarity also drive the rotation of the motor. Gliding motility in some species of bacteria has also been shown to be powered by proton-motive force.

E. Swimming Behavior

In general, peritrichous bacteria swim slower and rotate in a relatively straight line; those with flagella restricted to poles swim in many directions and spin rapidly. During swimming, the flagella of peritrichous bacteria rotate as a coordinated bundle (Fig. 13). In amphitrichous cells, the bundle at one end rotates clockwise (CW) and the one at the other end rotates counterclockwise (CCW) to propel the cell in one direction. The rotary nature of the flagellar motor can be visualized in tethered cells with their flagella adhered onto the substratum. Tethered cells rotate; the shorter the cell, the less the hydrodynamic drag, and thus the faster the cell rotates.

The force created by the dissipation of the transmembrane potential propels the cell at a speed of 20–80 μm/s or

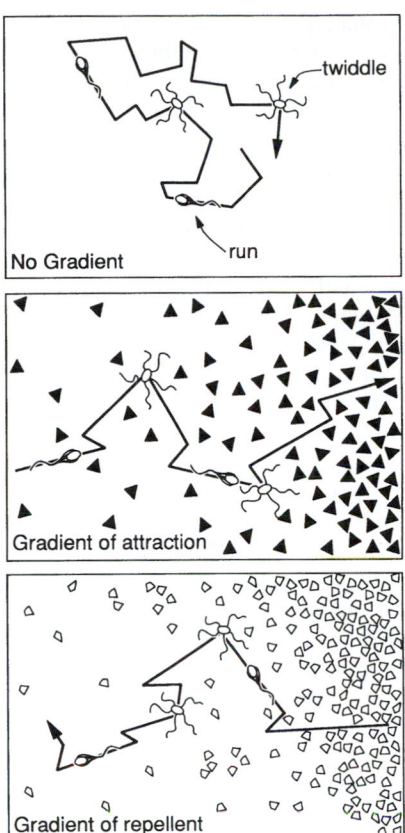

FIGURE 13. Swim tracks illustrating chemotaxis in bacteria. Top: The cell exhibits spontaneous runs and tumbles (twiddles) with random orientations when attractant or repellent molecules are uniformly distributed in the medium. During runs, flagella form a bundle and rotate together. Tumbling results from the flagella splaying out of the bundle. Middle: In the presence of a gradient of an attractant, a random walk changes to a biased walk characterized by longer runs toward the region of higher attractant concentration. Bottom: The cell exhibits longer runs away from higher repellent concentration.

≥ 10 lengths/s. In cells like *E. coli,* swimming is a three-dimensional random walk with gentle curves called runs. The flagellar bundle revolves CCW at a rate of 12 000 rpm. The current consumption of each flagellum has been calculated at 10^{-5} A. Considering that bacteria commonly live in dilute aqueous solutions, with Reynolds numbers of 10^4–10^5, inertia is insignificant. That is, if the cell stopped swimming, it would come to a halt estimated at $<10^{-5}$ of its cell length. Swimming patterns are well characterized and they result from the control of flagellar activity. Runs are interrupted by brief breaks called *tumbles* or *twiddles* during which individual flagella in bundles splay out causing a pause. The flagella in monopolar cells or flagellar bundles in peritrichous cells reverse to a CW rotation and the cell swims backward.

F. Chemotaxis

1. Responses to Attractants and Repellents

In the presence of attractants such as amino acids and sugars or repellents such as phenol, organic acids, and heavy met-

als, random walk becomes biased (see Fig. 13). This is unlike chemotaxis of *Dictyostelium* amoebae, which move directly toward or away from attractants and repellents, respectively. Taxis is possible because the flagellar motor has an all-or-none switch. The morphological identity of the motor switch has yet to be determined, but there is good genetic and physiological evidence for its existence. When *E. coli* is placed in a spatial chemical gradient of an attractant, suppression of tumbling leads to net progress toward the attractant; repellents increase tumbling. Actually, bacteria do not sense a spatial gradient; that is they cannot detect differences in concentrations between their anterior and posterior parts. Berg found that chemotactic behavior is the result of a response to a temporal gradient that develops as the cell moves through the medium. Thus, bacteria must be able to measure the attractant concentration, store that information, and compare it with a value measured at a later time. To do this, they must have memory. The time between stimulus and response (response latency) is on the order of 200 ms. In signal transduction, the cell can integrate information from one type of receptor or it can integrate information from different receptors; the effect of an attractant (CCW flagellar rotation) is canceled by the effect of a repellent (CW flagellar rotation). Chemotactic responses are transient; thus the cell exhibits adaptation.

2. Receptor Binding and Transduction of Signal to Switch

An important breakthrough in understanding bacterial chemotaxis was the discovery in Koshland's and Adler's laboratories that methylation of proteins was involved in the behavioral response. Since then, use of the powerful tools of molecular genetics and biochemistry has advanced this field at a remarkable rate. In *E. coli* and *S. typhimurium,* the components of most flagellar structures are biochemically and genetically identified; almost all flagellar genes have now been cloned and sequenced.

The chemotactic system of *E. coli* involves four receptor-transducer proteins, six cytoplasmic proteins, and three components of the flagellar switch. Different attractants bind to different receptors, called Tsr, Tar, Trg, and Tap. These receptor–transducer proteins are involved in measuring past and current concentrations of the ligand and in comparing the two values. The ligand-binding site of the transducer is in the periplasmic domain. In the cytoplasmic domain the transducer molecules contain methyl-accepting sites. Ligand binding at the periplasmic domain is fast and leads to covalent modifications of the cytoplasmic domain; the reactions involved in modification of the cytoplasmic methyl-accepting sites are slower.

In three dimensions (Fig. 14), a receptor-transducer protein is thought to contain four helical bundles in the periplasmic space and two helices that contain the ligand-binding sites. Two transmembrane helices connect the periplasmic and cytoplasmic domains. In the cytoplasm, there are four helical bundles with four to five methyl-accepting sites. Occupancy of ligand-binding sites represents the measurement of the current ligand concentration, and modification by methylation of the cytoplasmic domain is involved in storage of information about prior concentrations (Fig. 15). In the transduction

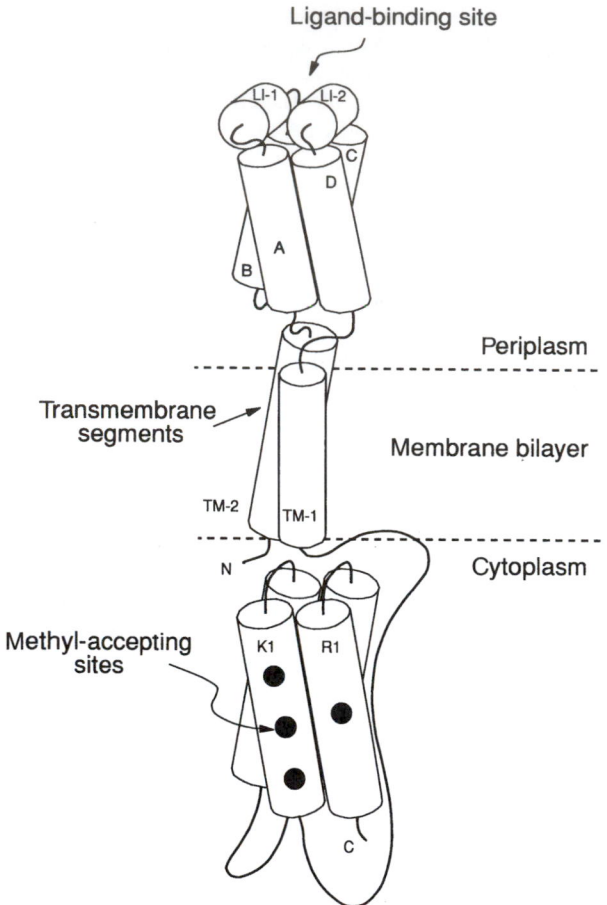

FIGURE 14. The transmembrane receptor-transducer molecule. The receptor sites (LI-1 and LI-2) that bind ligands lie in the periplasmic space of gram-negative bacteria cell surfaces. There are segments (TM-1 and TM-2) that span the bilayer membrane and methylaccepting sites (solid dots) are on α-helical cylindrical structures (K1 and R1) within the cytoplasm. The C- and N-termini of the polypeptide are indicated. (From Hazelbauer *et al.*, 1990; reprinted with permission of the Cambridge University Press.)

cascade, the initial source of methyl groups is S-adenosylmethionine (SAM), and the initial source of phosphate groups is adenosine triphosphate (ATP).

In the unstimulated state the ligand-binding site of the receptor is unoccupied. In this state, the methyl-accepting sites on the cytoplasmic domain of the transducer are continuously methylated and demethylated at a steady state of about one methyl group per protein.

In the stimulated state the ligand-binding site is occupied, which leads to a bias of the switch to CCW rotation. This is translated to suppression of tumbling (longer runs) as the cell moves up the attractant gradient. The information at the binding site in the periplasmic domain is transduced through the cell membrane domain to the cytoplasmic domain, resulting in the activation of methyl-accepting sites. In the cytoplasm, phosphotransfer reactions involving the proteins CheA, CheW, CheY, CheZ, CheR, and CheB are activated. CheA is a protein kinase that is phosphorylated by ATP and then transfers phosphate groups to CheY and CheB. Phosphorylated CheY and CheB are unstable compounds. Phospho-CheY may interact directly with the flagellar motor to introduce CCW rotation. Experiments using cell envelopes with depleted cytoplasmic contents demonstrate that flagellar motors are locked in the excited CCW rotation state.

Methylation of the cytoplasmic domain of the transducer offsets the effects of ligand binding, leading to adaptation; thus, the cytoplasmic components of the chemotactic system receive a null signal. The cytoplasmic protein CheR is an enzyme that catalyzes the methylation of the transducer.

Loss of the ligand on the periplasmic domain represents a negative stimulus. The effects of methylation of the cytoplasmic domain are unbalanced and cause a bias of the flagellar switch to CW rotation. The cytoplasmic domain of the transducer becomes activated for demethylation. The cytoplasmic protein CheB is an enzyme that catalyzes the demethylation of the transducer. CheZ is a phosphatase that acts on phospho-CheY, increasing its hydrolysis rate.

The result of demethylation of the cytoplasmic domain balances the effect of ligand loss at the periplasmic domain as the transducer assumes the adapted state. Again, the cytoplasmic components of the chemotactic system receive a null signal. The system then returns to the unstimulated state in the absence of a chemotactic gradient.

III. Gliding and Other Movements

There are several cell motility systems that may not belong to one of the four systems described in this and the previous chapter. Examples of unique movements that have been reported are described here.

A. Gliding

Gliding of an organism is characterized by active movement in contact with a solid substratum, with no obvious locomotory organelle responsible for this movement nor any distinct change in shape of the organism.

1. Bacteria

Extensive investigations for over a century have yet to bring consensus on the mechanism of bacterial gliding. It is only agreed that gliding requires cell contact with the substratum and that extracellular slime, mucilage, or mucoid material is probably involved.

Particles such as polystyrene latex beads placed on the surface of *Cytophaga* adhere and move longitudinally in both directions, as well as clockwise and counterclockwise. Hence the entire surface of the cell is capable of gliding motility, and the force for this movement is generated in all directions lateral to the cell surface. The cell can turn at a rate of 100 rpm and move as fast as 10 μm/s.

Gliding mutants were grouped by Pate (1988) into two types: (1) mutants with defective machinery or motors that are paralyzed or nonmotile and (2) mutants with defective systems that interact with the motor to translocate the cell over the surface. Mutants that can glide and move beads

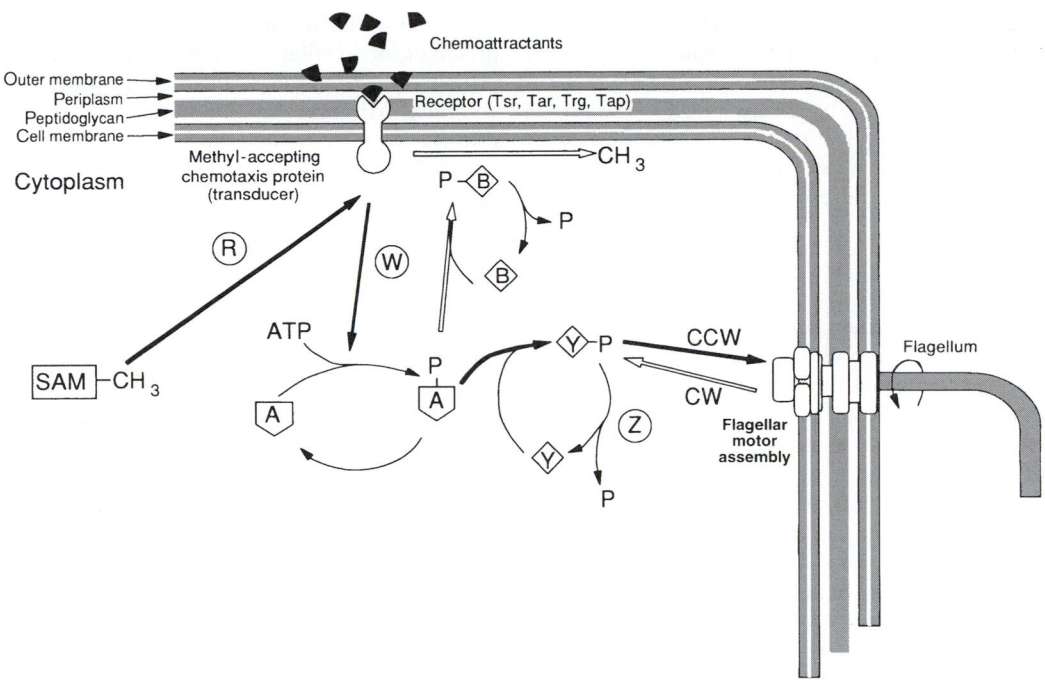

FIGURE 15. Bacterial chemotactic behavior. The attractant molecules bind to receptors located in the periplasm of gram-negative bacteria. This induces a signal that is transduced through the region spanning the membrane bilayer to the methyl-accepting portion in the cytoplasm. Activation of the cytoplasmic portion of the receptor-transducer methyl-accepting protein results in methylation involving S-adenosylmethionine (SAM), ATP, and the enzyme CheR. A protein kinase (CheA) is phosphorylated, which in turn phosphorylates CheY (which interacts with the flagellar motor switch to induce CCW rotation) and CheB (an enzyme that demethylates the transducer). The phosphatase CheZ dephosphorylates CheY; dephosphorylated CheY interacts with the flagellar switch to induce CW rotation.

over their surfaces, but whose colonies fail to spread, may be defective in accessory systems.

All truly nonmotile mutants are resistant to phages that infect wild-type cells and have several membrane properties that differ from those of wild-type membranes. One of several major cell envelope proteins can be missing from different nonmotile strains; therefore, these may be components of the surface machinery or motor. Leadbetter has identified an interesting biochemical difference between wild-type and nonmotile mutants that do not exhibit colony spreading, gliding, or movement of latex beads over their surfaces. The mutants are deficient in a specific group of unusual lipids called sulphonolipids (Fig. 16). He further demonstrated that restoration of normal sulphonolipids in the cell by supplementation of mutant cultures with a sulfonolipid precursor also restored all types of movement present in wild-type cells. Several other changes in surface polypeptides and carbohydrates have been observed in both classes of mutants, but it is still unclear whether these biochemical lesions are primary or secondary to the machinery at the cell surface driving gliding motility.

The fluidity of the membrane bilayer can modulate gliding, as shown by a rapid temperature downshift that quickly stops bacteria from moving. If bacteria are kept at the lower temperature, gliding resumes after an adaptation period during which the fatty acid composition of the bacteria lipids has changed with an increase in the ratio of unsaturated to

saturated fatty acids, as well as an increase in the ratio of branched-chain to straight-chain fatty acids. The cell surface membrane lipids probably undergo thermal phase transition from liquid crystalline to crystalline gel, and metabolic compensations resulting in more fatty acids that increase bilayer fluidity allow the transition point to be lowered.

There is evidence that the membrane potential or proton-motive force, not ATP, is probably the source of energy for

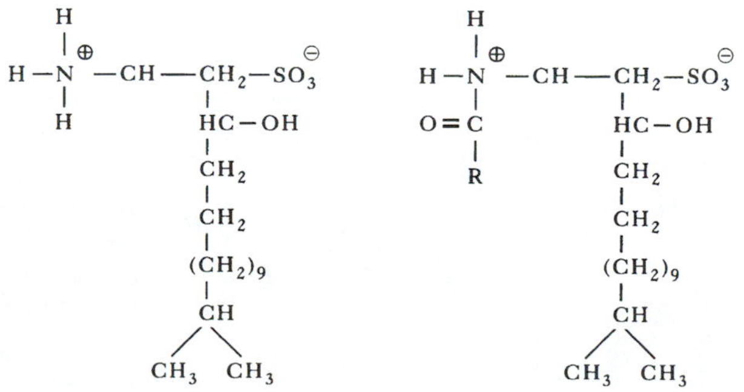

Capnine

N-acylcapnine

FIGURE 16. Gliding motility in some bacteria is correlated with the presence of sulphonolipids. Sulphonolipids are lipids not commonly found in organisms. They contain a fatty acid (R–CO–) moiety.

bacterial gliding. The presence of many rotary motors distributed over the entire cell surface may explain this form of locomotion. These motors are thought to be similar to the motors at the proximal end of bacterial flagella (see preceding) but are obscure because they do not have obvious external structures. Extracts of cells examined by electron microscopy reveal ring structures that resemble flagellar motors of bacteria.

2. Unicellular Eukaryotes

Gregarines are large protozoans that live as parasites in invertebrates. They exhibit smooth gliding locomotion but have no cilia or other external organelles that can account for this movement. The cell surface has an unusual ultrastructure that may be related to its motility, or it may simply represent an efficient absorptive surface for nutrients. The cell surface is highly folded with parallel longitudinally arranged ridges (Fig. 17). The organism secretes copious amounts of mucus to which small particles adhere and are moved posteriad. The ridges undulate, which may cause mucus transport and/or gliding. Although myosin has been identified in the organism, it is still not clear whether it functions in gliding and/or other movements such as dilating, contracting, and flexing of the cell body.

Gliding is exhibited by unicellular algae such as diatoms. It is thought that extracellular mucoid material at the suture between the two parts of the cells is responsible for movement of the cell.

Desmids, another group of unicellular algae, are composed of adjoining semicells joined by an isthmus. They exhibit end-over-end flipping, which probably involves extracellular mucoid secretions released through pores at each end of the cell. The structure of the mucilage is described as prismatic or fibrillar and may therefore contain contractile proteins. Like the gregarines, the desmids' surfaces are also complex. Microfibrillar networks are present in the inner, primary layer of the cell wall, and microfibril bundles present in the external, secondary layer form flat, longitudinal ridges or ribbons.

3. Movement along Surfaces of Eukaryote Flagella

Euglena, Chlamydomonas, and other protists exhibit cellular movements that are not related to other movements involving known motility mechanisms. When *Chlamydomonas* flagella make contact with a solid substratum, they can glide at a velocity of 2 μm/s. Bloodgood demonstrated that beads move up and down the flagellar surface in a saltatory fashion. Nongliding mutants, and cells treated with local anesthetics (lidocaine), calmodulin inhibitors (trifluoperazine), and low extracellular Ca^{2+} ($<10^{-6}$ M), fail to exhibit latex bead movement. Thus Ca^{2+} is apparently required for flagellar surface movements. The motor for this movement has not been identified or located.

B. The Rotary Axostyle

The guts of termites and wood-eating roaches are filled with many different protozoan species. One group, the devescovinid flagellates, has complex morphologies that serve as excellent landmarks for the investigator (Fig. 18). Internal landmarks include the nucleus and the Golgi apparatus, which wraps around the posterior part of the nucleus and part of the axostyle. These organelles form a complex coupled during axostyle rotation. External landmarks include an asymmetrical papilla at the tip and several different species of exosymbiotic bacteria with distinct distributions over the

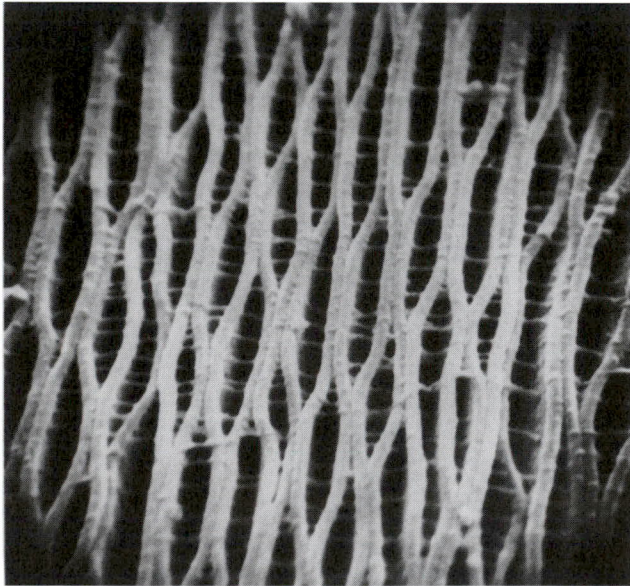

FIGURE 17. Gliding motility in gregarine protozoa is not understood. Scanning electron microscopy of the surface of a gregarine cell suggests that deep cortical folds undulate. The undulations may move secreted mucus posteriad, thus causing the gliding locomotion of the organism. (Courtesy of Eugene Small.)

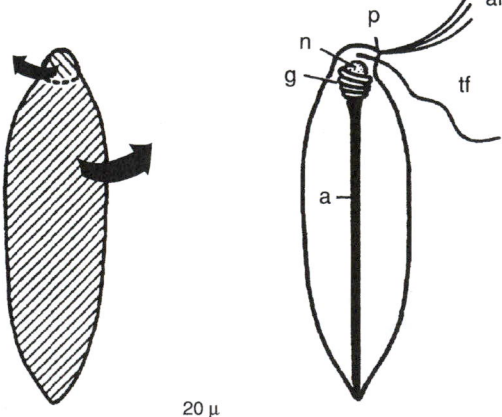

FIGURE 18. Rotary motor of some devescovinid flagellates. The anterior pole ("head") of the cell spins in one direction relative to the cell body. The axostyle (a) in the cytoplasm of the cell also rotates in one direction with the "head." This organelle is associated with the nucleus (n), which is encircled by the Golgi complex (g) at the anterior of the cell. Three anterior (af) and one trailing (tf) flagella arise from the papilla (p). The cytoplasmic organelles and flagella serve to visualize the rotation of the anterior pole. (Adapted from Tamm, 1976, with permission.)

surface of individual devescovinid flagellate cells. Some of the devescovinid flagellates rotate by mechanisms that probably involve the flagella or other external motor mechanism. At least one species has a rotary axostyle.

The axostyle extends from the anterior to the posterior poles of the cell. Axostyles are composed of numerous microtubules arranged similarly to the axostyles in heliozoan axopodia. Here the microtubules are arranged by cross-links in a spiral pattern when viewed in cross-section (Fig.19). Surrounding the axostyle is a sheath containing microfilaments aligned perpendicular to the axostyle axis. Outside this sheath is a region called the girdle, consisting of a finely granular inner layer and an outer layer filled with vesicles and inclusions. The outer layer of the girdle remains stationary during axostyle rotation.

Flagella that emerge from the anterior papilla exhibit normal flagellar waves. Independent of those flagellar movements, the

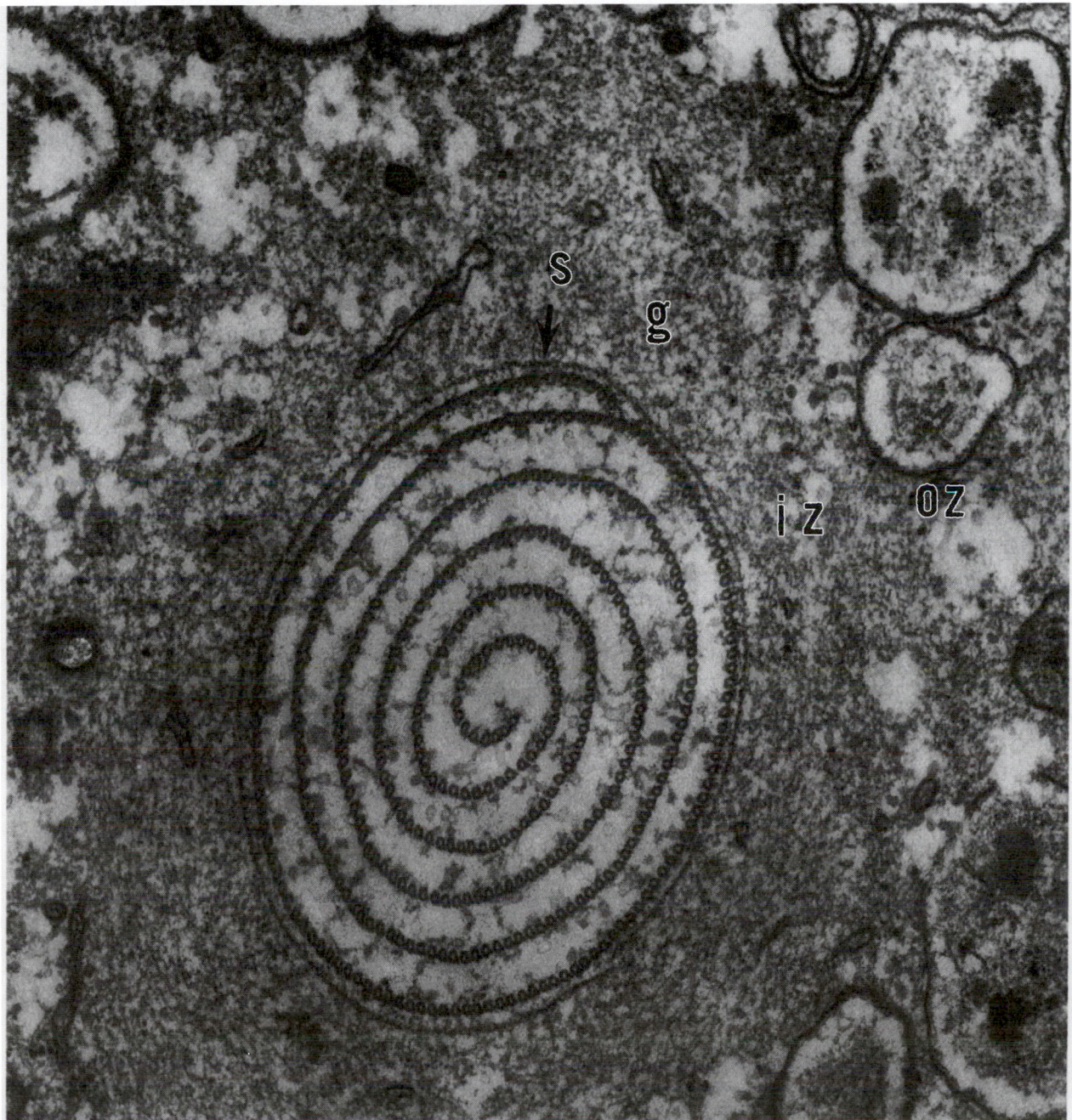

FIGURE 19. The axopod of the devescovinid flagellate serves as a shaft for the rotary motor. Electron micrograph of a cross-section of the axopod illustrates its microtubular structure. The microtubules are organized within a sheath (s), which is surrounded by a region called the girdle (g). The granular inner zone (iz), of the girdle is distinct from the outer zone (oz), where vesicles and other inclusions are present. It is not known how the girdle-sheath interaction gives rise to rotation of the axostyle. (Courtesy of Sidney Tamm.)

cells exhibit another type of movement that involves the rotation of the anterior pole (head) relative to the rest of the cell (body). The parts of the cell that rotate include the cell surface membrane at the anterior pole, the flagella, the nucleus-Golgi complex, the axostyle, the sheath, and the inner layer of the girdle. Rotation rate can be as high as 1.5 s per turn. The drive shaft is the rigid axostyle, which does not contract or undulate as do axostyles in some other flagellates. This devescovinid axostyle rotates only in a clockwise direction, looking at the cell from the anterior. If the head is tethered and prevented from moving, the body rotates. The cell surface membrane between the head and body is called the shear zone. Here membrane domains continuously move past each other and must therefore be highly fluid and free of ectosymbiotes. Paralysis of flagellar activity by pH shock does not stop the rotational movements, thus, the force for rotation does not come from flagellar action.

Tamm graphically demonstrated that the axostyle generates a torque along its length. Following destruction of segments along the axostyle by a laser microbeam, the posterior half continues rotating and does so at higher speeds. The head rotates more slowly or stops. Thus, the piece of axostyle attached to the head, which represents a load, provides less force for rotation than does the intact axostyle. The axostyle is analogous to the armature of an electric motor, and the surrounding cytoplasm represents the field coils. Torque is probably developed between the sheath and the inner layer of the girdle since the sheath rotates with the axostyle. The absence of a sheath and girdle is correlated with the lack of rotary motors in devescovinids.

Experiments on glycerinated cell models demonstrate that reactivation of the rotary motor requires ATP and Mg^{2+}; Ca^{2+} apparently does not affect motor activity. It has been suggested, but not proved, that force generation may be actin-based, arranged in one direction similarly to cytoplasmic streaming in *Nitella*.

IV. Summary

The Ca^{2+}-spasmin/centrin system does not use ATP for contraction, but it may be responsible for reextension. The mechanism for spasmoneme and centrin-based contraction cannot be explained by sliding; coiling of 4-nm filaments or more basic molecular conformational changes may be involved in this type of contraction. Spasmins and centrins are polymers that appear to require only the binding of Ca^{2+} to change their lengths. The role of phosphorylation of centrin function is currently unknown. Much less is understood about this motility system compared with the actin-myosin and tubulin-dynein systems, but the demonstration of its distribution in diverse organisms makes it highly probable that increased attention will given to this motility system in the future.

The energy for the rotation of bacterial flagella is derived from a transmembrane ion gradient (e.g., proton-motive force) initially established by input of energy (e.g., pumps) to set up the gradient. Although energy from ATP hydrolysis may be involved in setting up an electrochemical gradient across the cell membrane, ATP does not directly act on the rotation of bacterial flagella. Cell motility using the energy of ion gradients has yet to be demonstrated in eukaryotic cells. The molecular genetics of bacterial flagella and proteins involved in chemotaxis are well elucidated but only in a few species of bacteria. Unlike eukaryotic cilia and flagella, bacterial flagella are rigid helical structures that are rotated in one direction by the rotary motor associated with the cell membrane. Details on the mechanism by which the motor is driven by the ion-motive force are still unclear.

There are cell movements (e.g., gliding) that are not explained by the four recognized motility systems discussed in this and the previous chapter. It may be that many of these will be shown to be governed by one of the known mechanisms, but there remains the possibility for the discovery of novel systems.

Bibliography

Abbanat, D. R., Leadbetter. E. R., Godchaux III, W., and Escher, A. (1986). Sulfonolipids are molecular determinants of gliding motility. *Nature* **324**, 367–369.

Allen R. D. (1973). Contractility and its control in peritrich ciliates. *J. Protozool.* **20**, 25–36.

Amos, W. B. (1975). Contraction and calcium binding in the vorticellid ciliates. *In* "Molecules and Cell Movement" (S. Inoué and R. E. Stephens, Eds.). pp. 411–436. Raven Press, New York.

Cachon, J., and Cachon, M. (1985). Non-actin filaments and cell contraction in *Kofoidinium* and other dinoflagellates. *Cell Motil.* **5**, 1–15.

Ettiene, E. M. (1970). Control of contractility in *Spirostomum* by dissociated calcium ions. *J. Gen. Physiol.* **56**, 168–179.

Hazelbauer, G. L., Yaghmai, R., Burrows, G. G., Baumgartner, J. W., Dulton, D. P., and Morgan D. G. (1990). Transducers: Transmembrane receptor proteins involved in bacterial chemotaxis. *In* "Biology of the Chemotactic Response" (J. P. Armitage and M. M. Lackie, Eds.), pp.107–134. Cambridge University Press, London, New York.

Huang, B., and Mazia, D. (1975). Microtubules and filaments in ciliate contractility. *In* "Molecules and Cell Movement" (S. Inoué and R. E. Stephens, Eds.), pp. 389–409. Raven Press, New York.

Huang, B., Mengersen, A., and Lee, V. D. (1988). Molecular cloning of cDNA for caltractin, a basal body-associated Ca^{2+}-binding protein: Homology in its protein sequence with calmodulin and the yeast CDC31 gene product. *J. Cell Biol.* **107**, 133–140.

MacNab, R. M. (1990). Genetics, structure and assembly of the bacterial flagellum. *In* "Biology of the Chemotactic Response" (J. P. Armitage and J. M. Lackie, Eds.), pp. 77–106. Cambridge University Press, London, New York.

Manson, M. D. (1992). Bacterial motility and chemotaxis. *In* "Advances in Microbial Physiology" (A. H. Rose, Ed.), pp. 277–346, Academic Press, New York.

Melkonian, M., Beech, P. L., Katsaros, C., and Schulze, D. (1992). Centrin-mediated cell motility in algae. *In* "Algal Cell Motility" (M. Melkonian, Ed.), pp. 179–221. Chapman and Hall, New York.

Pate, J. L. (1988). Gliding motility in procaryotic cells. *Can. J. Microbiol.* **34**, 459–465.

Roberts, T. M. (1987). Fine (2–5nm) filaments: New types of cytoskeletal structures. *Cell Motil. Cytoskel.* **8**, 130-142.

Routledge, L. M., Amos, W. B., Yew, F. F., and Weis-Fogh, T. (1976). New calcium binding contractile proteins. *In* "Cell Motility" (R. Goldman, T. Pollard, and J. Rosenbaum, Eds.), pp.93–114. Cold Spring Harbor Laboratory Press, Cold Spring Harbor, New York.

Salisbury, J. L. (1989). Algal centrin: Calcium sensitive contractile organelles. *In* "Algae as Experimental Systems" (A.W. Coleman, L. J. Goff, and J. R. Stein-Taylor, Eds.), pp. 19–37. Liss, New York,

Salisbury, J. L., and Floyd, G. L. (1978). Calcium induced contraction of the rhizoplast of the quadriflagellate green algae. *Science* **202**, 975–977.

Sanders, M. A., and Salisbury, J. L. (1989). Centrin-mediated microtubule severing during flagellar excision in *Chlamydomonas rheinhardii. J. Cell Biol.* **108**, 1751–1760.

Tamm, S. L. (1976). Properties of a rotary motor in eukaryotic cells. *In* "Cell Motility," Book C (R. Goldman, T. Pollard, and J. Rosenbaum, Eds.), pp. 949–967. Cold Spring Harbor Laboratory Press, Cold Spring Harbor, N.Y.

Tamm, S. L. (1978). Laser microbeam study of a rotary motor in termite flagellates. Evidence that the axostyle complex generates torque. *J. Cell. Biol.* **78**, 76–92.

Alister G. Macdonald

59

Effects of High Pressure on Cellular Processes

I. Introduction

Hydrostatic pressure is a force acting in all directions, in the air we breathe, in water, or in body fluids. The hydrostatic pressure generated by a 10-m depth of water is, in round figures, equal to normal atmospheric pressure, or 100 kPa (Table 1). By convention the 100 kPa at 10-m depth implies "in excess of atmospheric pressure," but for some purposes it may be expressed as two atmospheres, or 200 kPa, absolute (total) pressure. Hydrostatic pressure is a thermodynamic intensity parameter, affecting equilibria and reaction rates in a fashion analogous to the way temperature (heat) contributes to the energy involved in physical and chemical reactions.

While the ways in which temperature affects physiological and biochemical processes are familiar, it is less well known that hydrostatic pressure also exerts remarkable and interesting effects on many life processes and, like temperature, is an important environmental factor. When a cell or an aquatic organism is subjected to a high hydrostatic pressure, as for example in an experimental chamber or the ocean depths, pressure is transmitted throughout the system by the molecules, mostly water, impinging on each other. Nothing is squashed, but molecular interactions and the conformations of macromolecules are certainly affected. The solvent water, which constitutes the bulk phase, is hydrostatically compressed, reducing its volume about 4% at the maximum pressure in the deep oceans, 100 MPa^{-1} (Table 1).

Realizing what hydrostatic pressure is not will help in the understanding of how it acts. Hydrostatic pressure is not directional. It is not involved in baroreceptors for example, which generally detect the stretch caused by the differential pressure acting across a tissue. Osmotic pressure is also quite separate from hydrostatic pressure.

When the hydrostatic pressure acting on a gas is doubled, the volume is halved (Boyle's Law) and the density is approximately doubled. In the case of a gas mixture such as air (79% N_2, 21% O_2 by volume), the partial pressure of the constituents is doubled and so too is their tendency to dissolve in a solvent, such as water. Henry's Law states that the amount of gas dissolved is the product of the partial pressure (pp) and the characteristic solubility coefficient (sc) for the gas/solvent combination.

$$\text{Quantity of gas dissolved} = \text{pp} \times \text{sc} \qquad (1)$$

Now, consider a beaker of water or the surface of seawater. In each case the water, in equilibrium with the air above, will contain nitrogen and oxygen dissolved at a partial pressure of, respectively, 79% and 21% of atmospheric pressure, that is, 79 and 21 kPa. Now the interesting question is, what happens to the partial pressures of the dissolved gases when the water in the beaker or the seawater

TABLE 1 Hydrostatic Pressure Units and Occurrence

Units	Occurrence
16 kPa	Peak aortic blood pressure, humans
100 kPa	Pressure at 10 m depth of water
2 MPa (2×10^6 Pa)	At 200 m depth; e.g., edge of continental shelf
20–40 MPa	In typical oil well reservoirs
100 MPa	Approximately, deepest ocean trench
100–5000 MPa (0.1–5 GPa)	Used in high-pressure chemistry and industrial food sterilization
> GPa	Pressures in geo- and astrophysics

1 pascal (Pa) = 1 newton per m^2 (N/m^2), force per unit area. 10^3 Pa = 1 kPa. 100 kPa corresponds to normal atmospheric pressure (actually 0.98697 atm) and also 1 bar and 750.064 mm Hg. Note: unless stated otherwise, pressure is expressed as "gauge" pressure, that is, in excess of atmospheric pressure.

High hydrostatic pressures can be generated in an ultra centrifuge, according to $w^2 Po (r^2 - ro^2/2)$. w is angular velocity, r^o the radius of rotation at the top of the liquid column, r the radius of rotation at the depth of interest and po the density of the solution at atmospheric pressure (Mertens-Strijthagen et al., 1985). Multiplying the product by 10^{-4} converts dyn·cm^{-2} to kPa. More usually high hydrostatic pressure is generated by a special hydraulic pump and confined in steel cylinders.

is subjected to an increase in purely hydrostatic pressure? The seawater can be imagined to be carried to depth in a current and the beaker of water hydrostatically pressurized in a special oil-filled chamber. In both cases a gas phase is now absent, so it would be reasonable to conclude that the pN_2 and pO_2 remain unchanged. However, there is a second order effect, as the hydrostatic pressure does very slightly increase the partial pressure of the dissolved gases, by approximately 14% 10 MPa^{-1}. Intuitively this can be imagined to arise from the bulk compression of the water, providing fewer "holes" for the gases to occupy. Formally it may be expressed as a reduction in the Henry's Law solubility coefficient. From Eq. 1, if the solubility coefficient is reduced while the quantity of gas dissolved remains constant, then the partial pressure must increase.

This chapter will explain how high hydrostatic pressure affects molecules of biological interest, and their interactions; the ways in which pressure affects cellular processes; and how animals including humans are affected by pressure.

II. Molecular Effects of Pressure and Temperature

A. Equilibrium Processes in Aqueous Solution

High pressure generally dissociates salts, weak acids, bases, and multimeric proteins in aqueous solution. Why is this? The dissociation processes are reversible and often simple, involving only weak bonds and no covalent bonds. The degree of dissociation is expressed as a ratio, the equilibrium constant K. For a weak acid,

$$HA \rightleftharpoons A^- + H^+ \qquad K = \frac{[A^-][H^+]}{[HA]} \qquad (2)$$

For a protein multimer,

$$P_n \rightleftharpoons nP \qquad K = \frac{[P]^n}{[P_n]} \qquad (3)$$

In accordance with Le Chatelier's principle, high pressure shifts these equilibria in the direction of least volume. K is increased when the difference in volume across the equilibrium is negative ($-\Delta V$), that is, a net volume decrease occurs in the direction left to right. In Eq. 2, for example, pressure increases $[H^+]$ (decreases pH) according to

$$pH = pK_a + \log\frac{[A^-]}{[HA]} \qquad (4)$$

Changes in pK_a with pressure closely match changes in pH when the ratio $[A^-]/[HA]$ is constant. In the case of a phosphate buffer the pH is decreased by 0.4 unit by 100 MPa.

The volume change, ΔV, relates to the system as a whole: solvent water, hydrated solutes, and nonhydrated solutes. In Eq. 2, the $-\Delta V$ arises from the electrical constriction of the water of hydration surrounding the ions. The hydration of hydrophobic groups incurs a $-\Delta V$, if the locally ordered water is more dense than bulk water. In Eq. 3, protein multimers often dissociate through numerous other $-\Delta V$ processes, in-

cluding ionic and hydrophobic interactions with the solvent. Such volume changes can be directly measured in a dilatometer (a precision volume measuring device) or a densitometer (a device for determining density).

The change in an equilibrium constant with increase in pressure at constant temperature is given by

$$\left(\frac{\partial \ln K}{\partial P}\right)_T = -\frac{\Delta V}{RT} \qquad (5)$$

where P is pressure, T is temperature, and R is the gas constant. The sign of ΔV determines the direction in which pressure shifts the equilibrium and its magnitude determines the extent of the shift. A convenient equation for calculating ΔV (ml·mol^{-1}, + or −) from experimentally determined values of K is

$$-\Delta V = RT \, 2.3 \frac{\log_{10} K_1 - \log_{10} K_2}{P_1 - P_2} \qquad (6)$$

In analogous fashion, changes in temperature affect an equilibrium constant at constant pressure according to

$$\left(\frac{\partial \ln K}{\partial T}\right)_P = \frac{\Delta H}{RT^2} \qquad (7)$$

where ΔH is the change in enthalpy (heat content). When ΔH is positive, the equilibrium shifts to the right when the temperature is increased (that is, the process is endothermic), and vice versa. For example, the pH of Tris buffer is reduced by the considerable amount of 0.028 pH units per °C. This buffer is often used in high-pressure work because its pH is virtually unaffected by pressure (Disteche, 1972).

Pressure and temperature are thus two important thermodynamic intensity parameters which contribute to the fundamental energy change underlying equilibrium processes, the Gibbs free energy change (Laidler, 1978). At constant temperature and pressure, the equilibrium state is that at which the Gibbs energy, G, is zero. This is Le Chatelier's principle, mentioned earlier. If we imagine the equilibrium shifting, then the Gibbs energy change (ΔG) is

$$\Delta G = \Delta H - T\Delta S \qquad (8)$$

where ΔS is the entropy change and T is the temperature, as before.

When ΔG is negative the shift will be spontaneous. It is not often explained in textbooks that the ΔH term includes a pressure/volume term and U, the internal energy:

$$\Delta H = \Delta U + P\Delta V$$

hence

$$\Delta G = (\Delta U + P\Delta V) - T\Delta S \qquad (9)$$

Additionally, the fundamental relationship between the standard Gibbs energy change, ΔG^0 in an equilibrium, the equilibrium constant K, and temperature is

$$\Delta G^0 = -RT \ln K \qquad (10)$$

Many multimeric proteins, of which the ubiquitous actin is a good example, can exist in equilibrium with their monomeric form. Thus globular, monomeric actin (G) polymerizes to the filamentous (F) form of actin

$$G \rightleftharpoons F \qquad K = \frac{[F]}{[G]} \qquad (11)$$

Applying Eq. 5, Fig. 1. shows that a plot of the natural log of K against pressure gives a straight line at lower pressures, the ΔV is positive and becomes smaller at higher pressures. The former ΔV is 107 ml·mol^{-1}, and as ΔH and ΔS are also positive, the self-assembly is described as entropy-driven, consistent with hydrophobic interactions being involved. The positive ΔS reflects the increase in disordered solvent, released from the ordered state surrounding the monomers, as polymerization proceeds. The overall effect of pressure is to favor the monomeric state and to shift the simple equilibrium (Eq. 11) to the left. This is generally true of protein multimers (Table 2). However, it does not follow that the actin or other protein multimers in living cells always become dissociated to monomers when the cells are subject to high pressure. Secondary covalent bonds can readily stabilize the multimers, but nevertheless the equilibria shifts seen *in vitro* may occur *in vivo* (Section III). Furthermore these simple monomer-multimer equilibria are susceptible to pressures of less than 100 MPa. Higher pressures, often greater by an order of magnitude or more, denature proteins. This is an unfolding, opening up of protein structure, increasing the hydration of hydrophobic groups within the interior. However the sum of the latter $-\Delta V$ seems to be largely offset by other changes. Pressure denaturation is used to kill microorganisms, that is, to sterilize (Table 2).

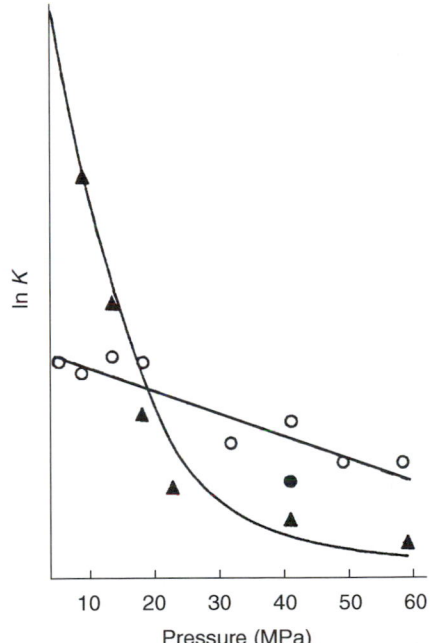

FIGURE 1. Effect of pressure on the G-F equilibrium of actin extracted from chicken muscle (▲). Pressure reduces the equilibrium constant $K = [F]/[G]$, that is, it favors the monomeric G form (Eqs. 5 and 6). At low pressure $\Delta V = 107$ ml·mol^{-1}, but this decreases at higher pressures. The G-F equilibrium of actin extracted from the muscle of a deep-sea fish (O) (*Coryphaenoides armatus*) is radically different. See Section V. Reprinted with permission from Swezy and Somero (1985), copyright (1985) American Chemical Society.

TABLE 2 Stability of Proteins and Supramolecular Structures *in Vitro* at High Pressure

Dissociated to subunits by pressure	ΔV (ml·mol^{-1})	Reference
Enzymes		
Hexokinase	−120	Silva and Weber (1993)
Ca^{2+}-ATPase	−167	Siebenaller and Somero (1989)
Lactate dehydrogenase	−170 to −220	Seibenaller and Somero (1989)
Phosphorylase A	200	Silva and Weber (1993)
Structural and motor proteins		
Tubulin	−90	Siebenaller and Somero (1989)
Actin	up to −139	Siebenaller and Somero (1989)
Myosin	[a]	Tumminia *et al.* (1989)
Fibrin	[a]	Collen *et al.* (1970)
Heme-containing and oxygen-carrying proteins		
Cytochrome c-cytochrome b5	−122	Rodgers and Sligar (1991)
Cytochrome c	−87	Kornblatt *et al.* (1988)
Hemoglobin and hemocyamin	[a]	Silva *et al.* (1996)
Supramolecular structures		
70G ribosomes	−242	Siebenaller and Somero (1989)
Tobacco mosaic virus and many others	[a]	Silva *et al.* (1996)

[a]Highly dependent on conditions.

Typically pressures much higher than those used for the above denature proteins and can also kill bacteria and inactivate viruses (Silva and Weber, 1993; Silva *et al.*, 1996; Jonas and Jonas, 1994; Ludwig *et al.*, 1996; Heremans and Smeller, 1998).

B. Lipid Bilayers Under Pressure

While high pressure, surprisingly perhaps, dissociates molecules in aqueous solution, it has the intuitively expected effect on lipid bilayers, compressing and ordering their structure. In this respect it mimics the effects of a decrease in temperature (Chapter 3). An important feature of bilayers is their anisotropic structure (Chapter 3), which, under high pressure, undergoes an anisotropic compression. The bulk compressibility of bilayers is about 5.6×10^{-7} kPa^{-1}. For comparison the compressibility of water is 4.5×10^{-7} kPa^{-1}, and that of hexane is 16.6×10^{-7} kPa^{-1}. In certain model bilayers, DOPC, DPPC, DPPC/cholesterol, DMPC/cholesterol (both 70:30 molar ratio), the thickness compression is negative, that is the bilayer becomes thicker at high hydrostatic pressure due to the marked lateral compression. (i.e., in the plane of the bilayer). This has been elegantly demonstrated in experiments in which large bilayer vesicles of egg yolk PC are pressurized in pure water and their volume measured by optical microscopy. Figure 2 shows that the volume of vesicles decreases at high pressure by 16%, which is more than can be accounted for by the bulk compression of water (10%). This arises from the lateral compression of the bilayer reducing the area, and hence the volume, of the vesicle. Water moves out of the vesicle to accommodate the change. Associated with the marked lateral compression of the bilayer structure is a reduction in the lateral diffusion of, for example, fluorescent probes.

Pure lipid bilayers undergo phase transition when they are passed through a sufficient temperature range (Chapter 3). A well known case is the liquid-crystalline → gel state transition of DPPC bilayers that occurs on cooling to ~ 41 °C (T_m), with a ΔH of –87 kcal mol^{-1}. At T_m, a quasi-freezing point, heat is given off at a constant temperature as the change from the liquid-crystalline to gel state takes place. There also occurs a $-\Delta V$, which can be directly measured (Fig. 3), which suggests that the process should be affected by high pressure. This indeed is the case; high pressure causes bilayers to pass through a transition to the gel state at temperatures at which the bilayer would normally be in the liquid crystalline state. Thus pressure raises T_m, without affecting ΔV or ΔH in the Clausius-Clapeyron equation.

$$\frac{dT_m}{dp} = \frac{T_m \Delta V}{\Delta H} \tag{12}$$

Transitions can be detected by many methods, most of which have been used at high pressure to show that T_m is increased by 0.2 °C/MPa in a variety of lipid bilayers (Fig. 3). The effect is also seen in cases where a transition occurs in a natural membrane bilayer. Generally, however, transitions are not common in natural membranes because of their heterogeneous lipid composition; hence the effects of pressure

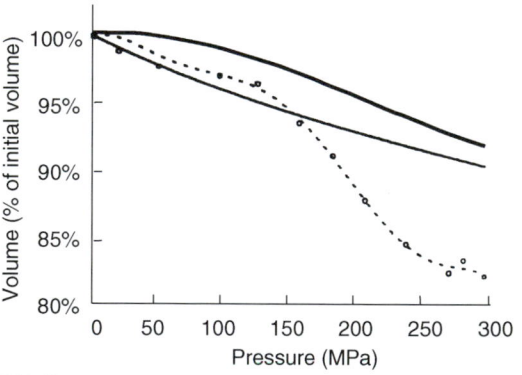

FIGURE 2. Lateral compression of lipid bilayers by hydrostatic pressure. The volume of vesicles made of egg yolk phosphatidylcholine bilayers was derived from microscopic observations and is plotted as a function of pressure. Vesicles without cholesterol (dotted line) undergo marked lateral compression; their volume is reduced much more than an equivalent volume of water (lower continuous line). The volume of vesicles whose membrane contains cholesterol (upper continuous line) is reduced by an amount similar to water showing that the lateral compression of the bilayer is reduced by cholesterol. (From Beney et al., 1997).

there are largely confined to ordering the bilayer. Examples of the latter include rat liver mitochondria, human erythrocytes, and fish synaptic and brain myelin vesicles, in which the ordering effect of pressure can be measured by the increase in temperature required to offset it, 0.13–0.21 °C MPa^{-1} (Macdonald, 1987). (The implications for deep sea organisms are discussed in Section V).

Very high pressures induce a variety of phase states in lipid bilayers, (Czeslik et al., 1998). In the case of natural membranes, the ordered, pressurized state causes some proteins to be shed from the bilayer (Plager and Nelsestuen, 1992), and it appears to augment the "surface presentation" of the major histocompatibility complex in certain cells (Eisenthal et al., 1996).

The partial molar volume of lipid bilayers and proteins is the sum of the constitutive volume (i.e., of atoms and bonds), the structural void volumes, and the interfacial volume, that is, change in the solvent volume, arising from the hydration of the accessible surfaces. In the present context only the second and third of these are affected by pressure.

C. Rates of Chemical Reactions

The thermodynamics of equilibrium processes in aqueous solution, and the phase state of lipid bilayers at high pressure, inevitably fall short of explaining how pressure affects *rates* of reaction. For this we need to appreciate, on the one hand, transition state theory, and on the other, the complexity of the reaction system as a whole. Consider the elementary reaction, A converting to B. Transition state theory holds that underlying the overall reaction there are two separate processes:

$$\mathrm{A} \to \mathrm{B}$$
$$\mathrm{A} \rightleftharpoons \mathrm{A^*} \to \mathrm{B} \tag{13}$$

A has to become activated (A*), gaining Gibbs energy (Fig. 4), and may then either revert to A or proceed to form B, at a rate determined by [A*]. The equilibrium constant for the activation step ($K^\#$) is therefore an important factor in determining the overall rate of reaction. The other main contribution is derived from the absolute temperature (T) and two fundamental constants: Boltzmann's constant (k_B) and Planck's constant (h), in the form $k_B T/h$. This yields a factor with the units of frequency per second, which relates to molecular motions. Combining these two factors gives the rate constant (k) for the elementary reaction.

$$k = \frac{k_B T}{h} K^\# \tag{14}$$

The equilibrium constant for the activation process, $K^\#$, is given by

$$\ln K^\# = -\frac{\Delta G^\#}{RT} \tag{15}$$

similar to Eq. 10. $\Delta G^\#$ is the standard free (Gibbs) energy change in the activation step (Fig. 4). As we have seen, ΔG is composed of three terms (ΔU, ΔV, ΔS); hence

A

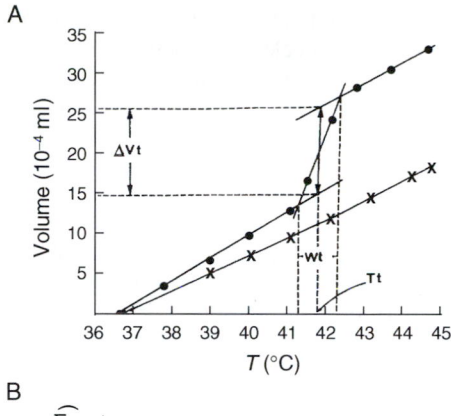

B

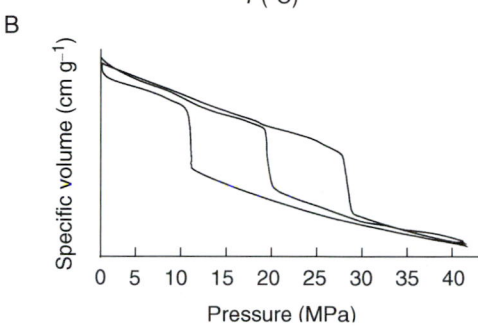

C

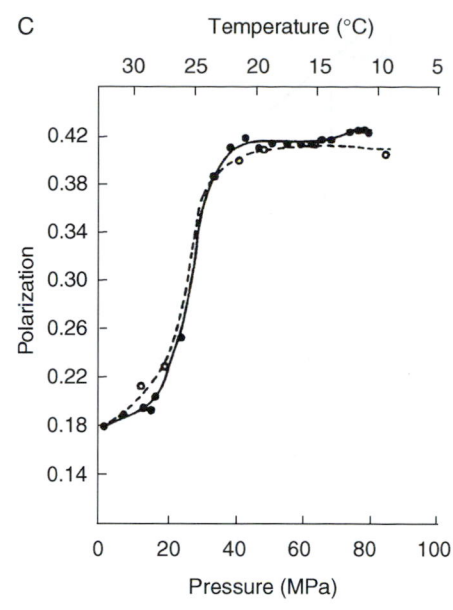

FIGURE 3. The volume and fluidity changes at the main liquid-crystalline-to-gel-phase transition in lipid bilayers at high pressure. (A) The ΔV_T in DPPC liposomes passing through the phase transition (●) is directly observed in a dilatometer. Subtracting the equivalent volume of water (x) from the liposome data gives $\Delta V_T = 25$ ml·mol^{-1}. At pressures up to 30 MPa the liposome curve translates to the right, ΔV_T remaining constant. (B) Phase transition in DMPC liposomes in a 5 wt % suspension apparent as a change in specific volume as a function of pressure, measured with a bellows dilatometer at (from left to right) 26 °C, 28 °C, 30 °C. (C) The fluidity of DMPC liposome bilayers, measured by the fluorescence polarization of diphenylhexatriene, as a function of temperature at atmospheric pressure (●), upper *x* axis, and as a function of pressure at 30.5 °C (●), lower *x* axis. The steep change in polarization indicates the phase transition, with fluidity decreasing as the bilayer passes to the gel state. [(A) From Macdonald 1978; (B) Reprinted from *Zeitschr. Physikalische Chemie*, **184**, Böttner, M., Ceh, D., Jacobs, U., and Winter, R., High pressure volumetric measurements on phospholipid bilayers, 5205–5218, copyright (1994), with permission from Elsevier Science. (C) Reprinted with permission from Chong and Weber (1983), copyright (1983) American Chemical Society.]

$$\Delta G^{\#} = (\Delta U + P\,\Delta V)^{\#} - T\,\Delta S^{\#} \qquad (16)$$

$\Delta S^{\#}$ is the change in entropy and ΔU the change in internal energy. $(\Delta U + P\Delta V)^{\#}$ is usually abbreviated as $\Delta H^{\#}$, the activation enthalpy, but is expanded here to emphasize how pressure will influence $\Delta G^{\#}$ through the $P\Delta V^{\#}$ term.

We can now consider how pressure and temperature af-

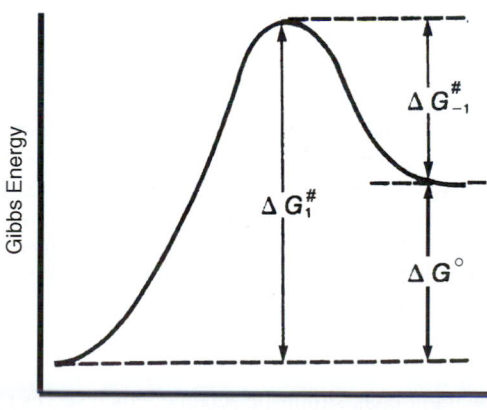

FIGURE 4. Energy diagram. Activation involves a gain in Gibbs energy $\Delta G_1^{\#}$, and it decays with a loss of $\Delta G_{-1}^{\#}$. (From Laidler, 1978).

fect the rate constant k. First, for historical reasons, many temperature effects are analyzed using the Arrhenius equation

$$\ln k = \ln A - \frac{E_a}{RT} \qquad (17)$$

where E_a is the activation energy, that is, the thermal energy required to activate the process, and A is the frequency factor for the reaction. A plot of ln k against $1/T$ (K^{-1}) gives a straight line of slope $-E_a/R$, the so-called Arrhenius plot. The magnitude of E_a determines the sensitivity of the reaction to temperature and it is related to $\Delta H^{\#}$ by $E_a = \Delta H^{\#} + RT$. Figure 5 illustrates an Arrhenius plot; the steeper the slope, the greater the effect temperature has in accelerating the reaction, by enabling activation to take place.

Pressure affects k through the $\Delta V^{\#}$ term

$$\left(\frac{\partial \ln k}{\partial P}\right)_T = -\frac{\Delta V^{\#}}{RT} \qquad (18)$$

A plot of ln k against P will yield a straight line if $\Delta V^{\#}$ is constant, and the slope is proportional to $\Delta V^{\#}$. An equation analogous to Eq. 6 may be used to calculate $\Delta V^{\#}$. Whereas the Arrhenius plots (Fig. 5) generally show negative slopes, the pressure equivalent show positive or negative slopes. This genuinely reflects the fact that high pressure is as likely to

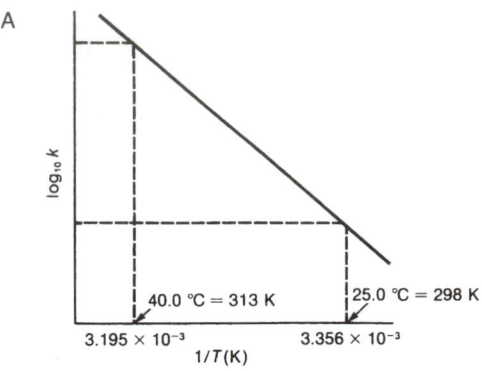

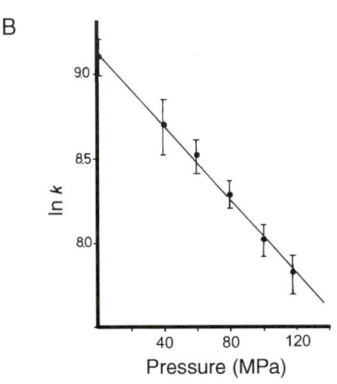

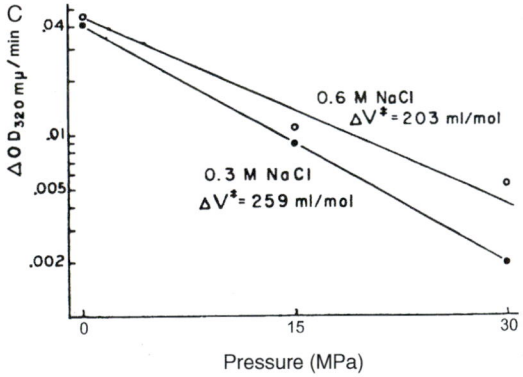

FIGURE 5. Effects of temperature and pressure on the rates of simple reactions. (A) A generalized plot of the logarithm of rate (k) against the reciprocal of the temperature. Such Arrhenius plots are obtained for simple and enzyme-catalyzed reactions and also for complex processes, such as heart rate. The slope indicates the activation energy (E_a) for the rate-limiting step so $\Delta H^{\#} = (E_a) + RT$. See text, and Laidler (1978). (B) The pressure equivalent of (A) here referring to the "flip rate" (s^{-1}) (k) of a phenylalanine aromatic ring in the protein basic pancreatic trypsin inhibitor. The $\Delta V^{\#}$ is 83 ml·mol^{-1} or 50 Å^3 per molecule. Reprinted from *FEBS Lett.*, 112, Wagner, G., Activation volumes for the rotational motion of interior aromatic rings in globular proteins determined by high resolution ^{1}HNMR at variable pressure, 280–284, copyright (1980), with permission from Elsevier Science. (C) The thermal, entropy-driven aggregation of a protein (poly-L-valyl-ribonuclease) is slowed by pressure in a way which is modified by ionic strength. Log Δ optical density (OD) min^{-1} is a measure of the reaction rate. (From Kettman *et al.*, 1966).

accelerate reactions as it is to slow them, while an increase in temperature generally accelerates reactions. Intuitively this is not difficult to understand. The molecular changes underlying $\Delta V^{\#}$, (conformational changes, hydration, etc,) are just as likely to cause a net volume increase as they are to cause a net volume decrease. An increase in temperature in raising the reactants to the activation, transition state, invariably feeds energy into the system to speed the reaction.

These considerations apply to both intramolecular reactions and intermolecular reactions. For example, the dynamic structure of proteins manifests thermal motion. In the case of a protein called basic pancreatic trypsin inhibitor, BPTI, tyrosines and phenylalanines provide aromatic rings whose rotational flip rate has been measured as a function of temperature and pressure by an NMR technique which need not concern us here. Figure 5 shows how pressure retards the flip rate of the phenylalanine ring in the 45 position; the calculated $\Delta V^{\#}$ is 50 Å^3 (83 ml·mol^{-1}). This activation volume is imagined to relate to the void which has to arise for the ring to flip unimpeded. Clearly high pressure slows this component of the protein's motion. The effect of temperature is to accelerate the flip rate, yielding a linear plot as in Fig. 5A, with $\Delta H^{\#} = 17$ kcal·mol^{-1} (Wagner, 1980).

As an example of an intermolecular reaction, consider the thermally driven aggregation of polyvalyl ribonuclease, a protein artificially decorated with hydrophobic groups on its outer surface. Using an optical pressure chamber to record the turbidity increase as the aggregates formed, Kettman *et al.* (1966) found that high pressure reversibly slowed the rate of aggregation (Fig. 5). The $\Delta V^{\#}$ was +259·ml·mol^{-1} in low salt concentration, and was reduced in high salt concentration. The spontaneous aggregation ($-\Delta G^{\#}$) at room temperature and the $+\Delta H$ mean that the reactions are entropy driven, since

$$\Delta G^{\#} = (\Delta U + P\Delta V)^{\#} - T\Delta S^{\#}$$

The aggregated state represents a loss of entropy, but this is more than offset by the entropy gain in the solvent, here imagined as the release of dense, ordered water in the activation stage leading to the hydrophobic interaction between the valyl groups. The positive $\Delta V^{\#}$ arises from this loss of ordered water.

From such elementary reactions, which are rarely encountered in physiology, we can now proceed to consider reactions catalyzed by enzymes. The simplest reaction scheme to envisage is

$$E + S \underset{k_{-1}}{\overset{k_1}{\rightleftharpoons}} ES \overset{k_2}{\rightarrow} P + E \qquad (19)$$

The enzyme, E, and the substrate, S, combine to form an enzyme-substrate complex, ES, which has a fleeting existence before decaying to product, P, or dissociating to E and S (Fig. 6). The overall reaction velocity is determined by the slowest, rate-limiting step. If the substrate concentration is high, then the rate of reaction is the product of the total enzyme concentration, $[E]_o$ and k_2 irrespective of the rates k_1, k_{-1} (Laidler, 1951).

$$V_{\max} = k_2[E]_o \qquad (20)$$

This is the traditional, simplistic scheme which serves to introduce the phenomena. Northrop (1996); Balny (1996); and before them, Morild (1981); and Laidler and Bunting (1973) have discussed the more detailed analysis of high pressure kinetics.

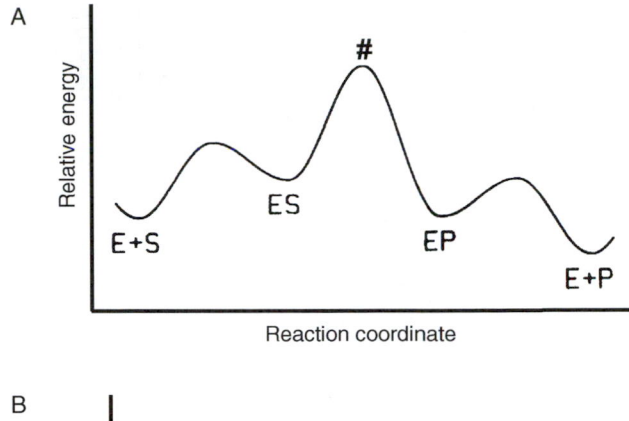

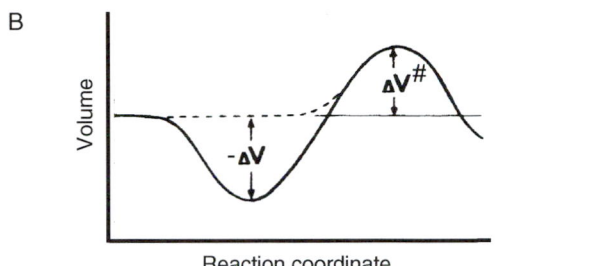

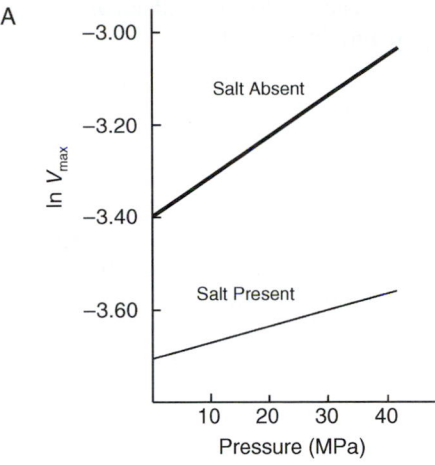

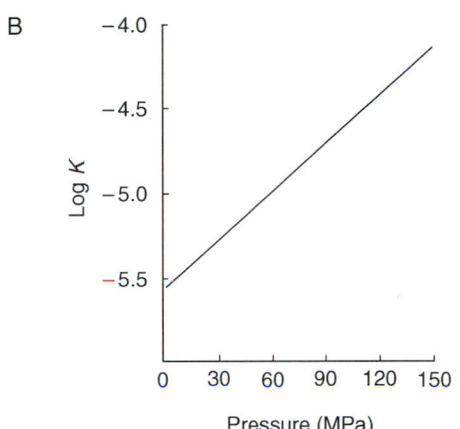

FIGURE 6. (A) Energy diagram for the simplistic enzyme reaction scheme described in the text, Eq. 19. (B) shows the ΔV component of ΔG in the same scheme, with, for purposes of illustration, a $-\Delta V$ shown in the first, binding step followed by $+\Delta V^{\#}$ for the catalytic stage. [(A) from Northrop, 1996; (B) from Morild (1981)]

The first part of the reaction scheme in Eq. 19 is an enzyme-substrate binding equilibrium and the second part is a catalytic reaction. Each passes through its activation (transition) stage; thus the $\Delta V^{\#}$ relating to k_2 is the activation volume in forming the transition stage ES*, en route to releasing products from the enzyme-substrate complex. In high substrate concentration, the reaction velocity at high pressure is limited by the $\Delta V^{\#}$ of k_2; hence the slope of plot of ln K against pressure will reveal $\Delta V^{\#}$.

An example of the effect of pressure on the ES → P + E step is given in Fig. 7 A. The enzyme alkaline phosphatase catalyses the release of phosphate from P-nitrophenyl phosphate. High substrate concentrations ensure the reaction velocity is maximal, V_{max}, and the data enable $\Delta V^{\#}$ to be calculated at -21.6 ml·mol^{-1}, that is, pressure accelerates the catalysis. If the reaction is run in the presence of the neutral salt KI, $-\Delta V^{\#}$ is much reduced by, it is argued, affecting the hydration changes which attend the activation process in the ES → P + E stage. (Low and Somero, 1975).

An example of pressure affecting the E + S step, that is, the binding of an enzyme and its substrate, is shown in Fig. 7B. As the enzyme phophofructokinase binds to its substrate, fructose 6-phosphate, it undergoes a decrease in intrinsic fluorescence. Thus the substrate binding equilibrium can be measured optically at high pressure. In Fig. 7B the log of K, the dissociation constant, the inverse of the binding constant, plotted against pressure, shows a linear positive slope, of ΔV 45 ml·mol^{-1} for the binding reaction. The enzyme's regulatory ligands, phosphoenol pyruvate (PEP), which inhibits, and MgADP, which activates, also bind to

FIGURE 7. Effects of high pressure on enzyme-catalyzed reactions. (A) Alkaline phosphatase catalyzes the release of phosphate from P-nitrophenyl phosphate. V_{max} obtained at high substrate concentrations is plotted against pressure, and the slope reflects the $-\Delta V^{\#}$ of 21.6 ml·mol^{-1} (salt absent) and -8.6 ml mol^{-1} (salt present) for the ES* activation stage. See text. Reprinted with permission from Greaney and Somero (1979), copyright (1979) American Chemical Society. (B) The binding of the enzyme phosphofructokinase to its substrate fructose-6-phosphate may be monitored by fluorescence. The dissociation constant $K=([E][S]/[ES])$ for the binding process is increased by pressure, ΔV 45 ml·mol^{-1}. (From Johnson and Reinhart, 1996).

the enzyme with $+\Delta V$, ($+40$ and $+44$ ml mol^{-1} respectively). In all these cases the $+\Delta V$ of binding (not a $V^{\#}$) is likely to arise from the release of electrostricted water from the binding surfaces. If substrate concentrations are low, the rate determining step may not be a single reaction, the $\Delta V^{\#}$ of substrate binding can be involved, and interpreting $\Delta V^{\#}$, the slope of a plot as in Fig. 7A, is more complicated.

These examples show that the reaction conditions, *in vitro*, the concentration of enzyme, substrate, and regulatory ligands, and the nature of the ionic environment and the molecular changes involved in particular reactions, all influence the effect which pressure has on enzyme-catalyzed reaction. If we consider such reactions *in vivo*, there are other important variables, such as the structured environment in which catalysis takes place (lipid bilayer, ribosome), which add to the complexity of relating the overall effect of pressure on a reaction to the molecular volume changes underlying the rate limiting step.

III. Cellular Effects

When cells and tissues are subjected to hydrostatic pressure, numerous processes and reactions are affected, usually reversibly. This section should be read with due reference to Chapters 6, 13–18, 24–27, and 51–56, and with two particular questions in mind: (1) Are the effects caused directly or indirectly by pressure? (2) What useful and illuminating generalizations can be drawn from these effects?

A. Cell Structures

Pioneer experiments with fertilized sea urchin eggs and other oocytes, carried out by Marsland and colleagues, showed that high hydrostatic pressure (20–60 MPa) reduced the gel strength of the cortical cytoplasm, simultaneously blocking and reversing the contractile cleavage furrow (Marsland, 1970). More recent experiments (Begg *et al.,* 1983) have shown that filament bundles of actin in microvilli are disorganized, but recover on decompression. Generally the cytoplasm of eukaryote cells is reversibly "liquefied" by high pressure, largely caused by the depolymerization of actin stress fibers. As a result, cells round up and amoeboid movement and cytokinesis are reversibly blocked. Microtubules, polymers of tubulin (Chapter 6), are variously susceptible to high pressure (Fig. 8). In the projecting cytoplasmic spines of the protozoan *Actinophrys,* the microtubules are somewhat differently organized but similarly susceptible to pressure (Kitching, 1970), whereas the microtubules in HeLa cells, and osteosarcoma cells are relatively resistant (Crenshaw *et al.,* 1996). Also in HeLa cells, intermediate filaments composed of cytokeratin (Chapter 36) are readily depolymerized by pressure, but myosin II filaments are not. Although pressure generally depolymerizes proteins *in vitro* (Section II), intracellular structures such as actin filaments, microtubules, and cytokeratin intermediate filaments are now thought to be indirectly affected by pressure. There are at least two reasons for this. As we have seen, the same protein, for example, tubulin, polymerized into microtubules in different cells, shows radically different susceptibilities to high pressure, presumably reflecting the state of the polymerizing process *in vivo.* Furthermore, the state of cells in culture influences the susceptibility of their intracellular structures. In particular HeLa cells growing in contact with each other fail to round up at high pressure (30 MPa), whereas isolated cells round up and lack actin stress fibers (Crenshaw *et al.,* 1996).

A specific attempt to measure cytoplasmic [Ca²⁺] in mouse fibroblasts caused to round up by 40 MPa failed to reveal any significant change. Nevertheless, other protein-regulating mechanisms such as phosphorylation could be responsible for the structural changes (Crenshaw and Salmon, 1996).

If indeed pressure acts indirectly then the apparent ΔV cited in the case of oocyte microtubules (Fig. 8) is merely an arbitrary numerical index and does not directly relate to tubulin polymerization. The widespread disruption of cytoskeletal

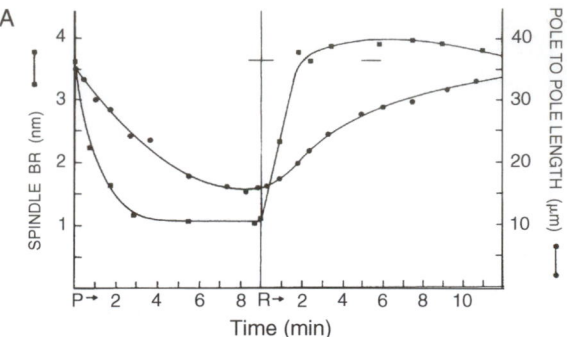

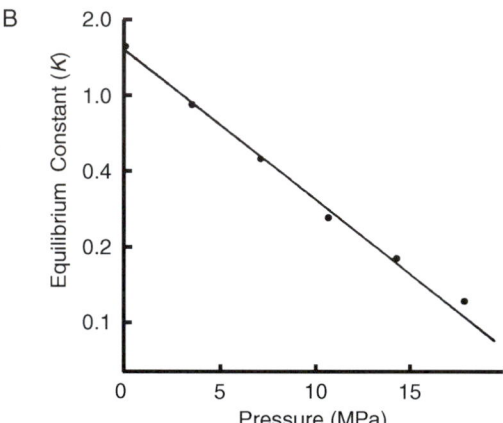

FIGURE 8. High pressure causes microtubules to depolymerize. (A.) Birefringence retardation, (BR, left hand axis (■) a measure of the concentration of orientated tubulin, and length of the metaphase spindle (right hand scale (●) in an oocyte of *Chaetopterus,* subjected to 13.6 MPa, 22 °C, at P and decompressed to atmospheric pressure 9 minutes later at R. The approach to the equilibrium birefringence value is rapid (■). (B) The polymer-monomer equilibrium constant (K) is $[B]/[A_o\text{-}B]$, in which A_o is the maximum birefringence (fully polymerized state) and B the value under given conditions. K is plotted on a log scale as a function of pressure, and from Eq. 6 an apparent ΔV of 400 ml·mol⁻¹ at 22 °C is calculated. (From Inoué *et al.,* 1975).

proteins which high pressure causes, directly or indirectly, may well affect cellular processes which seem somewhat remote from cell structure per se. For example, the distribution and activity of ion channels are intimately associated with the state of cytoskeletal proteins (Chapter 36) and so too are transport and permeability processes in epithelia. Furthermore, the association of mRNA with the cytoskeletal proteins is affected (for example, actin mRNA) (Zimmerman *et al.,* 1994). This is one of several aspects of protein synthesis which high pressure inhibits. Another is the pressure dissociation of ribosomes (Gross and Jaenicke, 1994).

The pressure blockade of cell cleavage has enabled the ploidy of cells to be manipulated for research and commercial purposes. In fish farming the production of large quantities of triploid salmon and trout are achieved by pressurizing the fertilized oocyte to block the shedding of a polar body (Lakra and Das, 1998). Triploid female fish do not sexually mature, and their marketable quality does not decline upon reaching maturity.

B. Muscle

Muscle, a tissue whose cells are conspicuously filled with polymerized proteins, is affected by high pressure in ways determined by the contraction cycle. Some of these were discovered by the first physiologists to apply high pressure techniques to eukaryote cells (Table 3). Generally pressure has no depolymerizing effect on muscle proteins *in vivo*. Consider first the mechanical responses of skeletal muscle fibers (rabbit psoas) whose plasma membrane has been removed. Such "skinned" fibers generate tension which can be initiated by the experimenter controlling the perfusing solution (Chapter 56). The tension in relaxed fibers is unaffected by 10 MPa, but fibers maximally activated by Ca^{2+} show a reduction in tension. In the latter condition, cross-bridge cycling is inhibited by pressure, whereas in the relaxed state cross bridges are absent. A third state, rigor, is interesting because cross bridges are formed but not cycling. In this state, high pressure increases tension by the bulk compression of the protein matrix, in much the same way as it affects collagen (Ranatunga *et al.*, 1990).

In intact skeletal and cardiac muscle, twitch tension is accelerated and relaxation slowed by high pressure (Fig. 9). A major cause of this is an enhanced $[Ca^{2+}]_i$ transient following the action potential. The $[Ca^{2+}]_i$ transient is the net result of Ca^{2+} release from, and simultaneous transport back into, the sarcoplasmic reticulum, and pressure's main effect appears to be the inhibition of Ca^{2+} transport. The enhanced $[Ca^{2+}]_i$ more than offsets the simultaneous and lesser inhibition of cross-bridge cycling. This latter effect is manifest in the reduced tetanic tension which is produced at high pressure (Fig. 9). However, when the tetanic tension is markedly reduced in severely fatigued muscle, pressure again increases the tension because the tension-limiting process switches from cross-bridge cycling to Ca^{2+}-activation (Vawda *et al.*, 1996). Thus positive and negative inotropic effects of pressure can be revealed by manipulating the limiting reaction.

This was elegantly demonstrated by Brown (1936) in brilliant pioneering work. He applied 40 ms pulses of hydrostatic pressure at different times in the twitch process. A pressure pulse applied just before or at the same time as the electrical stimulus caused the same increase in tension as is seen during continuously applied pressure. The same pressure pulse applied 40 ms after the stimulus, had no effect. These pressure jump experiments, recently confirmed by Vawda *et al.* (1995), showed that pressure enhanced twitch tension by affecting an early stage in excitation-contraction coupling. This is now known to be the increase in $[Ca^{2+}]_i$. In vertebrate cardiac muscle, pressure also increases the force generated but two important differences from skeletal muscle are noted. The first is that an isolated heart shows spontaneous, rhythmic contractions. Pressure reduces the rate of spontaneous beating; for example, in the isolated rat atria, 10 MPa reduces the beat frequency by 22%, and increases the force of contraction by approximately 50% (Ornhagen and Sigurdsson, 1981). Part of the latter is caused by a prolonged action potential, which can mobilize more calcium to initiate contraction (see later discussion). If the beat frequency is electrically paced, however, the same pressure

causes a 20-30% increase in force of contraction, which probably arises from the way cardiac muscle regulates $[Ca^{2+}]_i$, the second distinctive feature to consider. The idea here is that pressure acts like ouabain, inhibiting the Na^+-K^+ pump. The result is that $[Na^+]_i$ increases and causes $[Ca^{2+}]_i$ to increase, by biasing the Na^+-Ca^{2+} exchange mechanism. (Chapter 52; and Hogan and Besch, 1993).

C. Membrane Permeability and Transport

Simple processes, such as the diffusion of gases or water in aqueous solution, are little affected by the pressure range of interest here. However, the diffusion of K^+ across liposome bilayers is a more complicated process, involving adsorption and translocation, and is much more sensitive to pressure. When facilitated by valinomycin, the $\Delta V^\#$ for K^+ flux through a liposome bilayer is 40 ml·mol^{-1} compared to ~20 ml·mol^{-1} in the absence of carrier. The translocation of the K-valinomycin complex across a planar bilayer is limited by the viscosity of the interior which pressure increases and the resultant activation volume $\Delta V_T^\#$ is 12 ml·mol^{-1} (Benz and Conti, 1986). The formation of the K-valinomycin complex, which probably occurs at the bilayer water interface, appears to involve a small positive ΔV, and its dissociation, a negative ΔV. Overall the positive activation volume of translocation predominates, and K^+ flux is reduced at high pressure. When the carrier-mediated mechanisms of erythrocytes are blocked, the resultant passive diffusion of K, Na, Cs, and Rb is accelerated by pressure ($\Delta V^\#$ –50 to –90 ml·mol^{-1}) (Hall *et al.*, 1993). This clearly indicates an unexpected underlying complexity. Additionally there is a KCl co-transport pathway which Godart and Ellory (1996), and Gibson and Hall (1995) have shown to be indirectly activated by pressure, which favors dephosphorylation. Reassuringly perhaps, the passive flux of some amino acids (e.g. alanine, glycine) in red cells is reduced by high pressure to an extent similar to liposome fluxes.

Primary active transport processes, which use metabolic energy directly, are susceptible to high pressure through a variety of mechanisms. As with enzyme kinetics we may distinguish changes in K_m (a measure of the affinity of the transporter for the molecule it transports) and changes in the maximum rate of transport (i.e., V_{max}). The ubiquitous Na^+-K^+ pump is inhibited by pressure in erythrocytes (human, fish, frog skin, and fish gill) (Hall *et al.*, 1993).

In human erythrocytes, 40 MPa halves the K_m and V_{max} for K^+ influx and Na^+ efflux at 37 °C, hence $[Na^+]_i$ increases. The three main reaction steps (Chapter 17) ion binding, occlusion (translocation through the bilayer), and release are each possible sites of action. Another is the uncoupling of the ATPase, energy capturing process, from the pump. Insofar as pressure (15 MPa) activates the ATPase yet inhibits the pump in human erythrocytes, uncoupling is an attractive possibility (Goldinger *et al.*, 1980). The uncoupled ATPase, isolated from dog kidney and fish gill, is inhibited by pressure (40 MPa) which also reduces fluidity of the bilayers which support it (Chong *et al.*, 1985; Gibbs, 1997). The Ca^{2+} pump in the sarcoplasmic reticulum of mammalian muscle (Chapter 18) has been studied by Hasselbach and colleagues (Stephan and Hasselbach,

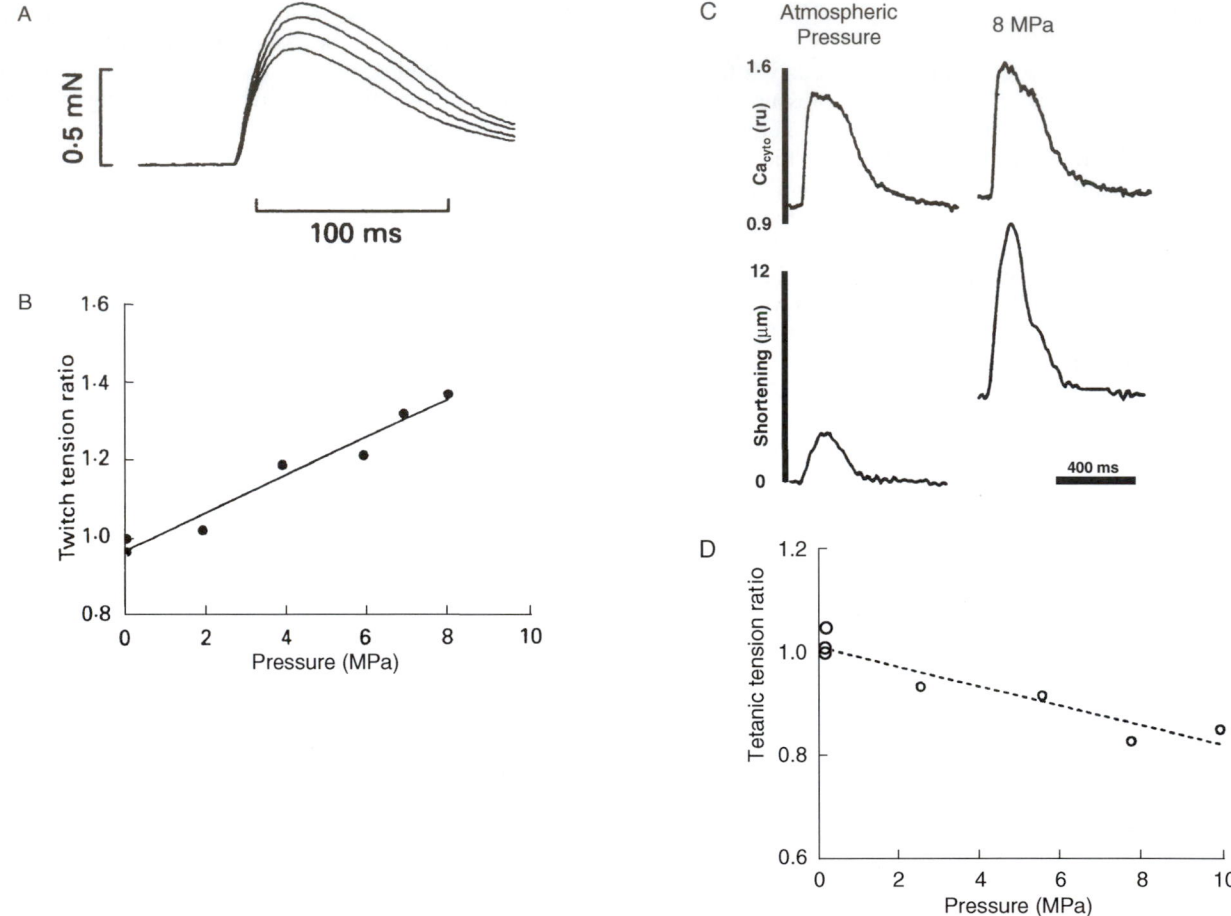

FIGURE 9.   Positive and negative inotropic effects of high hydrostatic pressure. (A) Twitch contractions (tension, mN) in a bundle of fibers from rat skeletal muscle are accelerated and enhanced at 2, 4, and 6 MPa. Relaxation is slowed. The lowest curve was obtained at normal atmospheric pressure. (B) Pressure dependence of the twitch tension, as in A, normalized to the atmospheric pressure value. (C) The electrically stimulated rise in $[Ca]_i$ (in relative units) in an individual cardiac myocyte (guinea pig) is enhanced at 8 MPa. These data were obtained with an optical high pressure apparatus, enabling both cell length and fluorescence ($[Ca]_i$), to be simultaneously monitored. (D) The negative inotropic action of pressure is manifest in the reduced tetanic tension of fiber bundles normalized to the atmospheric pressure value. [(A), (B), and (D) from Ranatunga and Geeves, (1991), and (C) from Besch and Hogan (1996).]

1991), who have shown that its inhibition by pressure probably causes the increased $[Ca^{2+}]_i$ transient in cardiac muscle, mentioned above. Both this Ca^{2+} pump and the equivalent one in erythrocyte membranes appear to undergo subunit dissociation at very high pressure (Verjovski-Almeida *et al.,* 1986; Coelho-Sampaio *et al.,* 1991).

Transport across the amphibian epithelia provides an interesting example in which pressure affects the passive permeability to sodium (Hong *et al.,* 1984, Brouha *et al.,* 1970) similar, perhaps, to the pressure diuresis in human divers (Takeuchi *et al.,* 1995). It also affects the insertion of aquaporins (Coluccio *et al.,* 1983).

D. Excitable Membranes

The hyperexcitable effects of pressure, first seen in marine animals in the 19th century and then rediscovered in deep simulated dives with laboratory mammals and humans, has stimulated an interest in electrophysiology at high pressure.

The pressure reversal of anesthesia in whole animals has also been another stimulus. Virtually all electrical recording techniques have been adapted for use at high pressure (Table 3) but our understanding of how pressure "excites" animals (Section IV) is far from clear.

1. Squid Giant Axon

Spyropoulos (1957a) was the first to record the prolonged time course of the action potential in the squid axon at high pressure, and Table 3 shows the development of voltage-clamp experiments in the giant axon and other neurons. The most refined recordings show pressure slows both the activation and inactivation of the sodium current (Fig. 10). The $\Delta V^{\#}$ for the opening of the m gate is 58 ml·mol^{-1}, and for the equivalent potassium channel n gate it is 37 ml·mol^{-1} (Chapter 25). The $\Delta V^{\#}$ of the gating current for sodium channels is much smaller (17 ml·mol^{-1}), reflecting the movement of the voltage sensitive part of the channel prior

TABLE 3 Electrophysiological Techniques Used with Simple Preparations at High Pressures

Method	Experimental description	Reference
I Extracellular recording	1. Single motor nerve fibres, amphibian, 108 MPa	Spyropoulos (1957b)
	2. Motor nerve, amphibian, 68 MPa	Gershfeld and Shanes (1958)
	3. Vagus nerve, deep-sea fish, 40 MPa, <10 °C	Harper *et al.* (1987)
II Intracellular recording	1. Cannulated squid giant axon, 108 MPa	Spyropoulos (1957a)
(a) simple recording	2. Crab neuromuscular junction, micrometer drive to insert microelectrode outside pressure vessel, 50 MPa	Campenot (1975)
	3. Micrometer drive to insert microelectrode outside vessel, *Helix* neurons.	Harper *et al.* (1975)
	4. External micromanipulation of microelectrode, helium compression to 20 MPa	Henderson *et al.* (1975)
(b) voltage clamp	1. Squid giant axon, cannulated, helium compression to 20 MPa	Henderson and Gilbert (1975) also Henderson *et al.* (1975)
	2. *Helix* neurons, 2 electrode voltage clamp. 50 MPa	Harper *et al.* 1981 and Wann *et al.* (1979)
	3. Vaseline gap voltage-clamp, node of Ranvier amphibian nerve and lobster axon, helium compression to 20 MPa	Kendig (1984) Grossman and Kendig (1984)
	4. 2 electrode voltage-clamp, oocytes with perfusion, helium compression 40 MPa	Daniels *et al.* (1991)
	5. 2 electrode voltage-clamp, oocytes, 60 MPa	Schmalwasser *et al.* (1998)
III Patch-clamp recording	1. G-ohm patch-clamp Ach R channels, rat muscle cells, 60 MPa	Heinemann *et al.* (1987a)
	2. G-ohm seal whole cell capacitance recording of exocytosis, as above	Heinemann *et al.* (1987b)
	3. G-ohm patch-clamp, channels reconstituted in liposome bilayers, 90 MPa	Macdonald & Martinac (1999)
	4. G-ohm seal patch-clamp, patch equilibrated with experimental gas (O_2) in 10 mins	Macdonald and Vjotosh (1999)
	5. M-ohm patch-clamp, locust muscle glutamate-R channels repeated seals formed at high pressure, helium compression 50 MPa	Macdonald *et al.* (1989)
IV Planar bilayer	Alamethicin channels in planar bilayer, helium compression to 100 MPa	Bruner and Hall (1983)
V Epithelia	Frog skin, electrogenic Na^+ transport potential	Brouha *et al.* (1970)
VI Tissue slices	Perfused brain slices, remote manipulation of microelectrodes, helium compression to 20 MPa	Fagni *et al.* (1987)

In general the first published of each particular technique or variant thereof is listed. Unless stated otherwise compression is hydraulic, that is, in the absence of gas.

Historical: Regnard (1887); Ebbecke (1914); Grundfest and Cattell (1935); Cattell (1936)

to opening (Conti *et al.*, 1984). These positive $\Delta V^{\#}$ are, of course, the cause of the slow and prolonged action potential at high pressure.

2. Single-Channel Recording

Heinemann *et al.* (1987a) were the first to extend patch-clamp recording to high pressure (Table 3), and in Fig. 10 the similarity between sodium channel kinetics in the squid axon and in chromaffin cells at high pressure is explained. The molecular basis for these effects of pressure are still obscure. The slowing of the Na^+ and K^+ channel kinetics probably does not reflect the more ordered state of the lipid environment at high pressure.

Furthermore, Na^+ channel inactivation is thought to involve a "ball and chain" mechanism, in which a tethered peptide, otherwise freely diffusing in an aqueous phase, binds to the interior of the channel pore. This idea has been extended to account for the rapid inactivation of K^+ channels (N-type inactivation) which Meyer and Heinemann (1997) have compared with the slower C-type inactivation in Shaker B K^+ channels expressed in oocytes. They found that N-type inactivation was slowed by pressure ($+\Delta V^{\#}$) and C-type inactivation was accelerated ($-\Delta V^{\#}$). Simultaneous temperature experiments showed that N-type inactivation was entropy-driven, and C-type inactivation was enthalpy-driven (Table 4). The data for N-type inactivation are wholly consistent with the binding of a largely hydrophobic

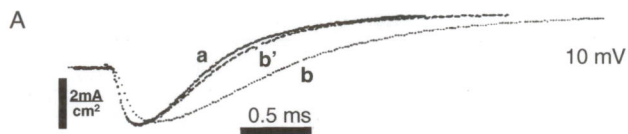

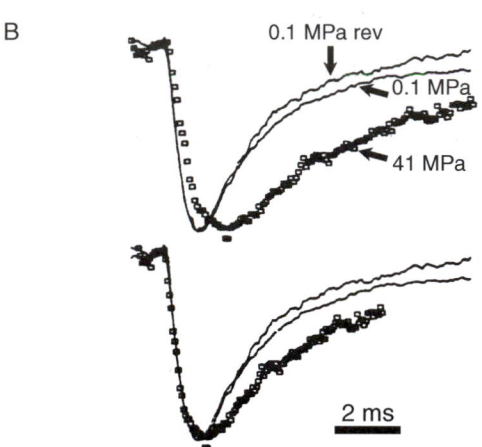

FIGURE 10. Pressure slows the kinetics of the fast Na⁺ channel in the squid giant axon (A) and in chromaffin cells from the bovine adrenal gland (B). (A) Voltage-clamp recording of the current flowing when an axon is depolarized to 10 mV from −80mV, 15 °C (a) at atmospheric pressure, and (b) at 40 MPa. Curve b is replotted as b′, scaled by a factor of 1.05 (amplitude) and 1.61 (time and delayed 25 μs), showing the main effect of pressure to be on the time course of activation (downward curve) and inactivation (upward curve). Higher pressures required a higher time scaling factor. The time scaling factor provides a measure of the characteristic time constant of activation, and using Eq. 6, $\Delta V^\#$, the apparent activation volume for channel activation was calculated to be 32 ml·mol⁻¹, at 15 °C and somewhat larger at lower temperatures. The curve b′ fails to superimpose on the upward curve, the inactivation phase, hence pressure slows inactivation rather more than activation especially at voltages more depolarized than −20 mV. (B) Upper curves show whole cell Na⁺ currents in a chromaffin cell at a test potential of 5 mV, 18 °C. The trace at normal pressure (0.1 MPa), before exposure to pressure is unscaled and the post-pressure trace at normal pressure (0.1 MPa rev) is scaled by a factor of 1.43 (amplitude). The trace at 41 MPa is scaled by a factor of 1.67 (amplitude). The lower set of traces shows how, with a scaling factor of 1.4 (time), the rising phase of the traces superimpose but the falling (inactivation) phase does not. It requires a scaling factor of 2 to match the controls. Thus Na⁺ channel inactivation is more pressure sensitive than activation (Apparent $\Delta V^\#$ at 17 °C; activation, 20 ml · mol⁻¹; $\Delta V^\#$ inactivation, approximately 37 ml·mol⁻¹) [(A) From Conti *et al.*, (1982); (B) Heinemann *et al.*, (1987b)].

peptide ball, and contrast with those for C-type inactivation, imagined to be a structural change in the channel entrance.

At present the most illuminating generalization to be made about ion channels under high pressure is that their kinetics are variously affected but their conductance is not. For example the voltage-sensitive Ca²⁺ channels in bovine chromaffin cells are unaffected by pressures up to 40 MPa (Heinemann *et al.*, 1987a) but the BK$_{ca}$ channel in the same cell (at negligible [Ca²⁺]) is activated by pressures

monotonically up to 90 MPa with no change in channel conductance (Macdonald, 1997). Ligand-gated channels are similarly diverse in their responses to pressure (Kendig *et al.*, 1993).

E. Chondrocytes

Chondrocytes are cells which synthesize the load-bearing extracellular matrix which covers the articulating surfaces of joints. Their mechanical environment is complex, comprising dynamic loading, shear, compression with accompanying fluid movement, and ionic changes. There is a purely hydrostatic pressure component which fluctuates, for example, in human hip joints, in the 3–18 MPa region. Osteocytes also experience complex mechanical stresses which include hydrostatic pressure.

The effects of pressure on chondrocytes have been studied using cells in culture, slices of cartilage (extracellular matrix containing chondrocytes), and intervertebral discs. High pressures not found *in vivo* are to be distinguished from lower pressures which would normally be experienced *in vivo*. For example, in chondrosarcoma cells the former (50 MPa) reduced the expression of proteoglycan core protein mRNA while the latter (1–5 MPa) enhanced it (Takahashi *et al.*, 1998). Intervertebral discs (Ishihara *et al.*, 1996) and articular cartilage respond to "physiological" cyclic compression (Parkkinen *et al.*, 1993). Transport and ultrastructural aspects of chondrocytes have also been studied at high pressure, creating a plausible picture of selective adaptation to their distinctive high pressure environment.

Chondrocytes and some other cell types also respond to very small hydrostatic pressures, (20 kPa). These responses may be physiological but they are difficult to understand and their elucidation may cause a reappraisal of some of the high pressure effects described in this chapter (see Section IV. A.).

IV. Effects of Hydrostatic Pressure on Animals and Humans

A. Aquatic Animals

First consider the response of invertebrates such as crustacea, and certain fish, to small increments of hydrostatic pressure, similar to or less than would arise from tidal depth changes. The shore crab *Carcinus moenas*, for example, increases its motor activity when its pressure is cyclically raised and lowered some 100–1000 Pa. Remarkably small pressures (1 kPa = 10 cm water) can elicit responses in a variety of animals, all lacking an obvious compressible, that is, gas, compartment. These normal behavioral responses imply that hydrostatic pressure is a sensory modality. Such hydrostatic pressures are too small to be detected (transduced) by the thermodynamics of solute-solvent interactions outlined in Section II, (Macdonald and Fraser, 1999). Transducing very small hydrostatic pressures may be a fundamental cellular property of which we are largely ignorant (see Section III.E), and its relationship to higher pressure effects is not at all clear. In the case of the crab, the "micro" pressure transduction

TABLE 4 Thermodynamic Activation Parameters for the Inactivation Process in K⁺ Channels

K⁺ channels expressed in *Xenopus* oocytes	$\Delta G^{\#}$ kJ/mol	$\Delta H^{\#}$ kJ/mol	$T \cdot \Delta S^{\#}$ kJ/mol	$\Delta V^{\#}$ ml·mol⁻¹
Shaker B channel; N-type inactivation	57	75	18	63
Mutant ShΔT449A; C-type inactivation	74	66	−8	−75

Data are for 20 °C and +50 mV, from Meyer and Heinemann (1997).

mechanism appears to involve mechano-activated elements in the animal's statocyst.

In contrast, when animals with a macroscopic gas phase, such as that found in the fish swim bladder, are subjected to small hydrostatic pressure changes, the compressible gas undergoes a volume change sufficient to activate stretch receptors in the bladder wall. In these cases the transduction of small hydrostatic pressures appears to be readily accounted for. Consider now aquatic animals subjected to larger pressure increments in an observational pressure vessel. Normal locomotor activity is increased by pressure acting directly on the neuromuscular system. It then becomes increasingly uncoordinated, culminating in whole body convulsions. These were seen by the first physiologist to study high pressures, Paul Regnard (1891). The pressure which elicits convulsions is used as an index of the tolerance of animals to pressure. Generally pressure convulsions are not lethal, although an immobilized state occurs at a higher pressure, at which stage death is likely. In those aquatic crustacea that cannot tolerate hypoxia, paralysis of the pleopods (that ventilate the gills) leads to death. Fish also become hyperexcitable and eventually convulse at similar pressures. Additionally, pressure affects heart rate and metabolism; Fig. 11 shows how oxygen consumption in the eel is stimulated during the initial hyperexcitable phase but thereafter is somewhat depressed.

B. Buoyancy

The buoyancy of aquatic animals subjected to significant pressure is interesting because buoyancy requires the creation of a void or region of low-density, working against the ambient pressure. The main buoyancy mechanisms accumulate either (a) low-density solutions, that is, solutes with a large partial molar volume, or (b) lipids and hydrocarbons, or (c) gas (Pelster, 1997).

Certain cephalopod molluscs, *Sepia* (cuttlefish), *Spirula*, and *Nautilus* (and by implication, ammonites known only from fossil remains), are buoyed by a pressure-resistant shell which contains a gas similar to air but with a pressure slightly less than atmospheric. The cuttlefish shell, which lies within its body, can withstand a pressure of 2.4 MPa, and the external shells of *Nautilus* and *Spirula* implode at around 6 MPa and 17 MPa, or 600 m and 1700 m depth, respectively (Fig. 12 illustrates the *Nautilus* shell). Despite their superficial dissimilarities, the shells of all three animals are deposited by living tissue which retains an interface with a restricted region of the gas-filled shell interior. Mechanical resistance to external pressure is achieved by the strength of the shell and by the toughness, and restricted dimensions, of the soft living parts which would otherwise be forced into the gas cavity (Fig. 12). The same force however, does cause a hydraulic flow of water into the cavity, raising the question, what keeps the water out? The answer is osmosis. Active transport sets up an osmotic pressure gradient, in some cases greater than is available from normal sea water (2.4 MPa = 240 m depth) with which the body fluids of these mollusks are in equilibrium (Denton, 1971; Greenwald *et al.*, 1980). Thus, energy metabolically sustains the osmotic removal of water from the gas cavities in the shell, which would otherwise sink the animal.

The fish swim bladder, mentioned earlier, typically occupies 5% of the animal's volume, sufficient to create neutral buoyancy. The gas within is at the same pressure as the ambient sea water and fish's tissues. The composition of the gas is air enriched a little with carbon dioxide, and more so with oxygen, the proportions of the latter increasing with depth. The vascular supply to the bladder is complex, comprising an interface with the bladder (gas gland) and a rete, or system of closely packed linear blood vessels in which the flow in neighboring vessels is mutually opposed. There exists a steep partial pressure gradient, from the gas in the bladder to the gas partial pressure in the perfusing blood, which, in the gills, equilibrates with the gas dissolved in the ambient seawater. There is an ingenious active transport mechanism which "pumps" up the partial pressures of the swim bladder gases, ensuring that in the bladder they combine to mechanically match the ambient hydrostatic pressure. (Pelster, 1997; Mylvaganam *et al.*, 1996; Perutz, 1996).

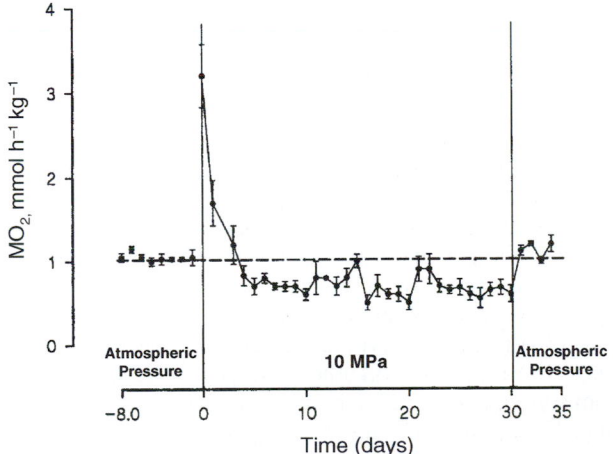

FIGURE 11. The oxygen consumption of the eel *Anguilla anguilla* kept in a high pressure aquarium at 10 MPa for 30 days. The means (SEM) of four separate experiments are shown. The hyperactivity of the animals on compression is apparent from the high oxygen consumption which declined to a subnormal level. (Data of Simon, Sebert, and Barthelemy, 1989, in Sebert, 1997).

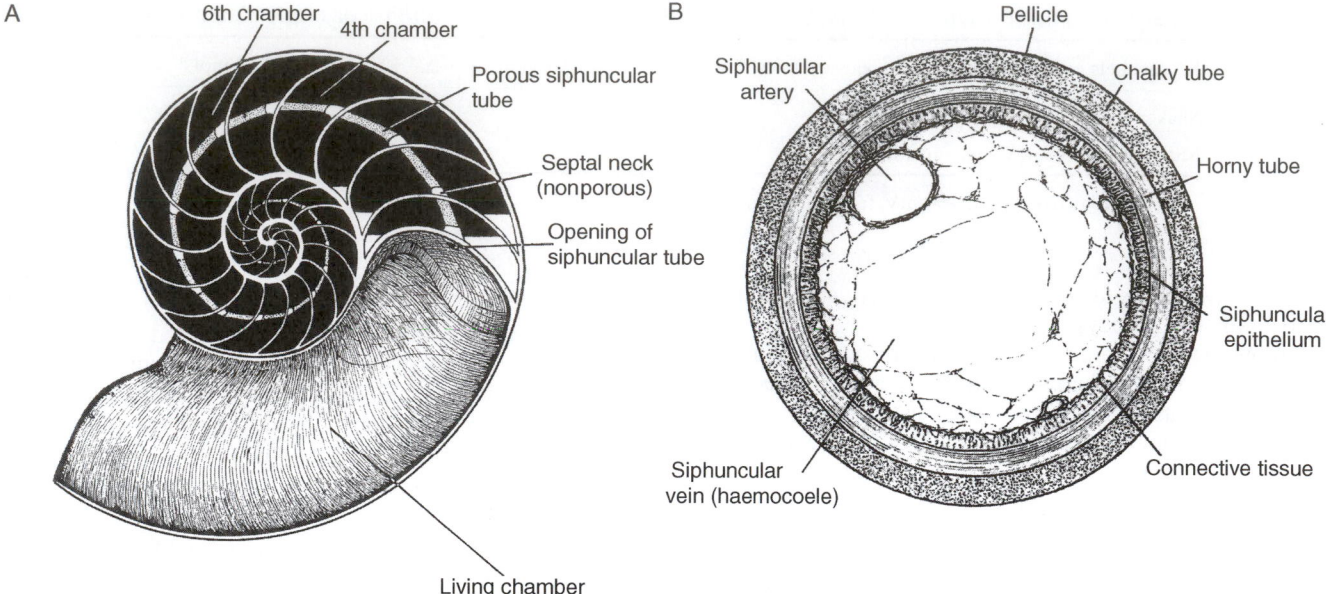

FIGURE 12. The buoyancy shell of *Nautilus*, an oceanic Cephalopod confined to depths of less than 300 m. (A) A vertical section shows the terminal, living chamber, which is occupied by the animal. The siphuncle, an extension of its mantle, runs up the center of the other chambers, which contain gas at subatmospheric pressure. Water is pumped osmotically out of these buoyancy chambers, the osmotic gradient being set up in the siphuncle. (B) Cross-section of the siphuncle. Active transport renders the fluid on the gas side hypotonic and therefore subject to osmotic flux into the siphuncle from which it is transported along ducts in the epithelium. The depth/pressure capability of this system is determined by the mechanical strength of the shell (it implodes at 6 MP = 600 m) or the osmotic gradient in the siphuncle. (From Denton and Gilpin-Brown, 1966 quoted in Denton, 1971).

C. Breathold Diving

Various species of air breathing vertebrates undertake breathold dives, during which their lung gases are compressed and their tissues experience significant hydrostatic pressure (Fig. 13). Some diving mammals, notably the seals *Mirounga leonina*, Southern Elephant seal; *M. angustirostris*, Northern Elephant seal; and *Leptonychotes weddellii*, Weddell seal, routinely dive to a depth of several hundred meters, and sometimes to beyond 1000 m. (Kooyman and Ponganis, 1998; Hindell and Lea, 1998; Schreer and Testa, 1996; Castellini, 1991). The deepest diver, however, appears to be the sperm whale *Physeter catadon* which has safely attained 2000 m depths (20 MPa) (Clarke, 1978). Research has focused on the mechanisms of collapsing the lungs, economizing on oxygen consumption, and coping with the hypoxia and thermal stress. However, rapid compression to hydrostatic pressures of 5–20 MPa is a profound and unfamiliar stress which, in human diving, has emerged as the main depth limiting factor (see Section IV.D.), and as a consequence is slowly being recognized in diving mammals. Human breathold diving, which has been practiced for centuries, has revealed some interesting systems physiology (Holm *et al.,* 1998; Elsner *et al.,* 1996; Anderson and Schagatay, 1998).

D. Scuba Diving

Diving with self-contained underwater breathing apparatus (SCUBA) for sport, or the equivalent techniques used by military and commercial divers, enables humans to experi-

ence "wet" hydrostatic pressures while lung volumes remain normal (Fig. 13). The SCUBA diver experiences hydrostatic pressure in three distinct ways. First, there is the mechanical effect on lung gases, and second, there is the all-pervading effect on tissues and the molecular processes therein, just as in purely aquatic animals. Third, hydrostatic pressure may be used to introduce a carefully selected partial pressure of an inert gas anesthetic to counteract some of the deleterious effects of high pressure.

Consider a sports diver using a compressed air supply. At a depth of 20 m the ambient pressure is 200 kPa gauge or 300 kPa absolute and thus the pO_2 of the air entering the diver's airway is 63 kPa (air is 21% O_2; 0.21 × 300 kPa = 63 kPa; water vapor pressure is ignored). This is three times the normal pO_2, and given sufficient time, will adversely affect the diver's tissues, beginning with the lungs. The pN_2 at 20 m depth will be 273 kPa, which is somewhat below the pN_2 which affects the CNS in ways similar to weak general anesthetics or ethanol. At greater depths (>50 m) adverse effects of pO_2 and pN_2 are avoided by using the inert gas helium as a diluent for oxygen. A commercial diver working at 300 m (3 MPa) might breathe a safe pO_2 of 25 kPa obtained from a mixture of 99.17% helium and 0.83% oxygen by volume. Using such mixtures it is possible to pressurize air breathing animals to very high pressure (~20 MPa) without significant effects from the pO_2 or pHe. In these conditions hydrostatic pressure can affect excretion (Section III.C.), heart rate (see later discussion), and other facets of human physiology. The conspicuous changes in the activity of laboratory mammals have been studied carefully and compared

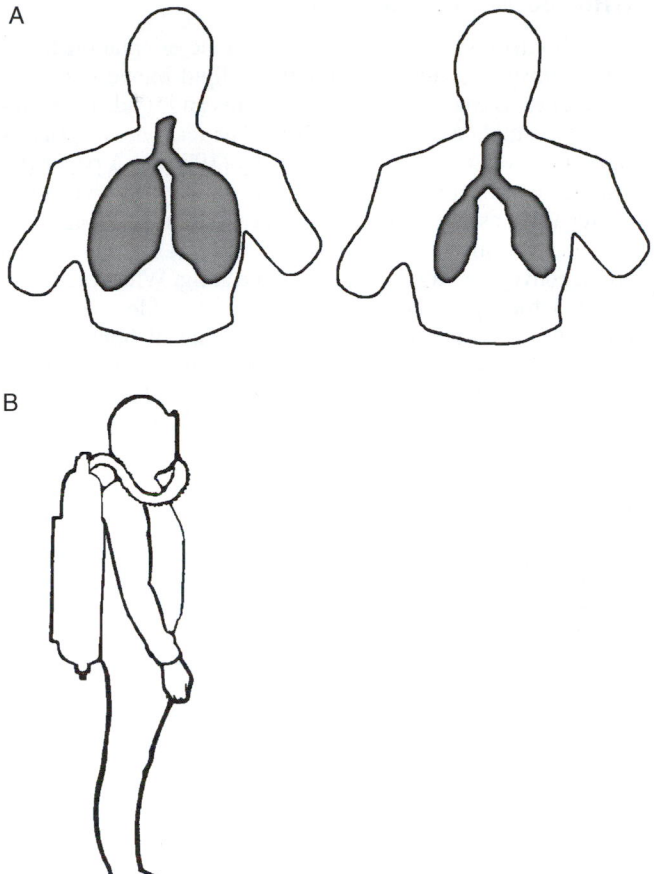

FIGURE 13. Divers. (A) Breathold diving in humans entails the compression, and on ascending the potentially hazardous expansion, of lung gas. Diving mammals such as seals accomplish this readily. (B) Self-contained underwater breathing apparatus (SCUBA) is used by sports divers breathing air; the commercial diver uses air or oxygen-helium mixtures to operate at greater depths, usually from a diving bell. The breathing equipment provides gas "on demand" at the diver's ambient pressure, and the diver's lungs are of a normal volume but contain a dense (compressed) gas. Commercial "wet" dives have attained depths of 500 m and "dry" chamber dives have exposed humans to a pressure of up to 7.1 MPa and baboons to 11 MPa. (See Bennett and Elliott, 1993; Rostain, 1993; Lafay *et al.*, 1995).

with, on the one hand, the effects seen in pressurized aquatic animals, and on the other, with the more limited pressures to which primates have been safely exposed. The main conclusion is that high hydrostatic pressure causes hyperexcitability in mammals and humans, just as it does in aquatic animals. Three examples will illustrate the extent of this excitability. First, there is a commonly observed "diving bradycardia" caused not by face immersion, which elicits a bradycardia in diving mammals, but by the ambient high hydrostatic pressure. This bradycardia is not caused by a direct effect of pressure on cardiac muscle (Section III.B.) but by pressure affecting the autonomic control of heart rate, most probably increasing parasympathetic activity via the vagus nerve. However other, reflex influences may also contribute. Thus bradycardia reflects hyperactivity in a part of the CNS

(Yamazaki *et al.*, 1998). In practical deep diving this bradycardia is not particularly important, being offset by other regulating mechanisms. Another example of excitability is the way stretch reflexes in humans (e.g., knee-jerk reflex) are enhanced by the pressure experienced in simulated deep dives (Harris and Bennett, 1983). Finally there is the High Pressure Neurological Syndrome (HPNS), the grand term describing a multiplicity of neurological changes seen in humans at 1.5 MPa or higher. The syndrome includes finger tremor, complex EEG changes and, in primates and laboratory mammals, seizures. It is these ominous changes in the CNS which limit man's safe exposure to high pressure (Rostain, 1993), although inert gas anaesthetics ameliorate the symptoms.

E. Hydrostatic Pressure - Inert Gas Interactions

Consider a diver breathing a gas mixture which provides a safe pO_2 and a pN_2 of 400 kPa. The resultant concentration of nitrogen dissolved in the tissues, especially the CNS, leads to a condition similar to that induced by general anesthetics, or the inebriation caused by ethanol. At low dose there is a loss of behavioral inhibition, but at higher doses analgesia occurs, followed by a reversible loss of consciousness. The nitrogen molecules affect certain central synaptic pathways and hence impair consciousness without entering into chemical reactions. That is, they act through weak bonds and not by forming covalent bonds. Nitrogen shares this property with other chemically inert gases, such as nitrous oxide and xenon, and with volatile anesthetics, for example, chloroform and halothane. Their anesthetic potency is closely correlated with their physical properties, most notably their solubility in solvents such as olive oil or benzene. Helium has the lowest solubility and in the diving context it does not act as an anesthetic but merely as a hydrostatic fluid which can be breathed safely, diluting oxygen and filling the lung as it hydrostatically pressurizes the diver or animal. Now if a laboratory mammal, which is "anesthetized" with nitrogen or a clinical anesthetic, is pressurized with helium, it recovers consciousness, that is, wakes up, and behaves relatively normally. This is the pressure reversal of anesthesia, first discovered by Johnson and Flagler (1951) who, having anesthetized tadpoles by dissolving ethanol in their water, revived them with 13–20 MPa hydrostatic pressure. Figure 14 shows the results of a recent study.

To return to the helium experiments, if higher pressures are applied to a laboratory mammal, then hyperexcitability ensues and eventually convulsions occur. However if an anesthetic dose of nitrogen is present, the convulsions occur at a pressure much higher than in the absence of the nitrogen. This is an inert gas or general anesthetic counteracting the hyperexcitable effects of pressure, and in extremely high pressure dives, nitrogen or the weaker inert gas hydrogen is incorporated in the breathing mixture, to subdue the HPNS symptoms. (Rostain, 1993; Bennett and Elliott, 1993). The mutual antagonism between pressure and general anesthetics is not confined to the hyperexcitable or anesthetized states of mammals. It is apparent in cold-blooded animals (Fig. 14), and in dividing cells and lipid bilayer properties. How do we explain this phenomenon? One idea is that anesthetics dissolve in

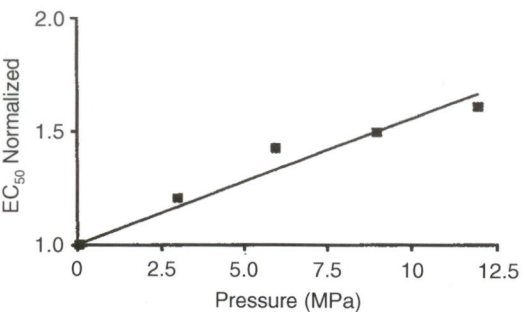

FIGURE 14. The reversal of anesthesia induced by propofol (2,6-diisopropylphenol). EC_{50} refers to the experimental concentration of propofol which abolishes the righting reflex in 50% of a group of tadpoles (*Rana pipiens*) in a water-filled observational pressure vessel. Rotating the vessel inverts the narcotized animals whose righting reflex is thus scored. Increasing hydrostatic pressure raises EC_{50} and thus opposes the anesthetic action of propofol. (From Tonner *et al.*, 1992).

their "targets," increasing their volume which, in view of the correlation between anesthetics, potency and their degree of hydrophobicity, is entirely plausible. Conversely, pressure would reduce the volume of the same target, hence restoring its original volume, and function. This "critical volume" model has been presented in elegant physical terms by Miller and colleagues (1973), but remains somewhat abstract (reviewed in Wann and Macdonald, 1988). An alternative view is that anesthetics bind to their molecular targets by weak readily reversible interactions, and these incur a $+\Delta V$, which pressures of 5–10 MPa oppose. Thus anesthesia might be reversed by pressure dissociating the anesthetic from its site of action. An example consistent with this view is the way N_2O reduces the open probability, P_o, of a glutamate-gated ion channel, and the addition of 10 MPa helium pressure restores it to normal (Macdonald and Ramsey, 1995).

$$N_2O + \text{channel} \underset{-\Delta V}{\overset{+\Delta V}{\rightleftharpoons}} N_2O-\text{channel complex} \qquad (21)$$
$$(\text{normal } P_o) \qquad\qquad (\text{reduced } P_o)$$

A third view is that pressure does not affect general anesthetics at their molecular site of action (antagonism) but acts physiologically by stimulating the CNS in ways which counteract the depressive effect of anaesthetics. The converse process, the "anaesthesia" of pressure-activated pathways in the control of the HPNS, gives this view appeal.

V. Adaptation to High Pressure

Animals and bacteria live at high pressures in the deep sea, down to the greatest depths where the pressure is approximately 100 MPa. The oceans stimulated the interest of 19th century biologists and oceanographers in high pressure science (Regnard, 1891) but the difficulties of gaining access to suitable organisms has limited experimental work right up to the present day. Nevertheless, a number of studies have demonstrated adaptations to high pressure in deep sea organisms.

A. Homeoviscous Adaptation

The homeoviscous theory of membrane adaptation holds that organisms regulate the fluidity of lipid bilayers in their membranes to enable membrane proteins and lipids to participate in normal cellular metabolism. The term *homeoviscous adaptation* was first used by Sinensky (1974) in a paper describing the phenomenon in bacteria. At the time, viscosity was regarded as an appropriate physical parameter to apply to bilayers, but since then the convenient term fluidity is preferred, conveying vaguely the same meaning. When precision is called for, spectroscopic terms such as order parameter should be used. The theory is an extension of homeostasis, which is usually concerned with the regulation of ionic-water balance, $pO_2/pCO_2/pH$ balance, excretion of nitrogenous waste, and, in endotherms in particular, body temperature. Homeoviscous regulation of the composition of the lipid bilayer stabilizes the fluidity of the bilayer against adverse changes imposed by the diet or the temperature of the environment. Thus the fluidity of mammalian membranes at 37 °C matches that of homologous membranes in fish living at arctic temperatures. The composition of the phospholipids in the bilayers is the main variable; an increased proportion of phospholipids containing double bonds, that is, unsaturated fatty acids, is found in low temperature membranes, counteracting the low temperature-ordering of the bilayer. The double bond introduces a 30 °C bend in the carbon backbone of the acyl chains, increasing the bilayers' free volume and fluidity.

Homeoviscous adaptation of membrane lipids is seen in animals, plants, and bacteria as a correlation of bilayer fluidity with living conditions, most often temperature. Precisely how the fluidity of the bilayer provides the required kinetic conditions for the multiplicity of membrane processes is poorly understood. Furthermore we do not understand how the cell "knows" that its membrane bilayer is the "wrong" fluidity, and instigates corrective proceedings. Nevertheless the deep ocean presents a clear test for homeoviscous theory, because the high pressure and invariable low temperature both order bilayers. Homeoviscous theory predicts that compensatory changes in the bilayer composition of deep sea organisms ensures a normal, functional bilayer fluidity. This prediction has been tested and confirmed (Macdonald and Cossins, 1985). Deep sea, high pressure adaptation takes the form of modifying the chemical composition of the bilayer lipids to offset the decrease which pressure brings about in the amplitude of the motion.

The boundary lipid of the Na^+,K^+-ATPase extracted from the gills of deep sea fish has a fluidity adapted to its normal high pressure environment (see Section III.C.) (Gibbs, 1997). Furthermore, deep sea prokaryotes also manifest homeoviscous adaptation to pressure. Certain pressure tolerant (barophilic) bacteria isolated from abyssal depths and cultured in the laboratory have a proportion of unsaturated fatty acids in their membrane lipids which increases with culture pressure (De Long and Yayanos, 1985; Wirsen *et al.*, 1987; Yano *et al.*, 1998). The deep sea archaeon *Methanococcus jannaschii* grows at high pressure, producing a mixture of lipids which do not conform to homeoviscous expectations, but then the lipids of archaebacteria are distinctively different from those in eubacteria or eukaryotes (Koga *et al.*, 1993). Interestingly, the pressure-ordering effect on the fluidity of lipids extracted from *M. jannaschii* is similar to that seen in phospholipid bilayers (Section II. B.) (Kaneshiro and Clark 1995).

B. Proteins

The proteins of deep sea animals and bacteria might be expected to be adapted to their normal high ambient pressure. Good evidence that this is the case comes from experiments developed by Hochachka, Somero, and co-workers, largely with enzymes and structural proteins extracted from deep sea fish. For example, lactic dehydrogenase (LDH), which catalyzes the conversion of pyruvate to lactate, requires to bind to NADH, a pressure-sensitive component of the overall reaction (Somero, 1990).

$$\text{pyruvate} + H^+ + NADH \rightarrow NAD + lactate \qquad (22)$$

The K_m of the enzyme for NADH is increased by pressure, which favors the dissociated state. LDH from the muscle of shallow water fish shows this clearly, but the same enzyme from deep sea species has a K_m which is little affected by pressure. The implications are that in shallow water fish the above reaction would be retarded by pressure, and that the binding reaction (NADH-LDH) involves a significant $+\Delta V$. Deep sea LDH appears to have a modified binding mechanism which is not susceptible to pressure, hence it would not rate-limit the reaction at pressure (Fig. 15). In contrast, the K_m for the substrate pyruvate is little affected by pressure in either deep or shallow species. Thus the deep sea NADH-LDH binding reaction is pressure-adapted and the K_m for NADH remains close to levels seen in other species, all of which are close to the intracellular concentration. In other words, the K_m values are a conserved kinetic property.

Structural proteins in deep sea animals also appear to be

adapted to high pressure. Actin, whose G-F equilibrium is described in Section II.A, polymerizes with a $+\Delta V$ which in chicken muscle is large. Actin extracted from the muscle of the deep sea fish *Coryphaenoides armatus*, which lives at 20–50 MPa, polymerizes with a much smaller $+\Delta V$, and one that is unaffected over the pressure range tested (Fig. 1). The idea is that the subunit interactions are different to those of chicken actin, and less pressure labile (Swezey & Somero, 1985).

VI. Summary

High hydrostatic pressure affects equilibria and reaction rates through the molar volume changes involved in the reaction system as a whole. Many biochemically important processes are susceptible to pressure: pH and ionic equilibria, protein monomer-multimer equilibria, lipid bilayer phase transitions, and various ligand-binding reactions. At a cellular level most physiological processes are thus affected, for example, the stability of intracellular structures and the motive force produced, muscular contraction, transport processes, excitability, and metabolic processes. At the level of the whole organism, in humans also, sublethal pressures cause hyperexcitability, and higher pressures cause immobilization. Aquatic animals show elegant adaptations to pressure, in their ability to sense it, and in evolving ingenious buoyancy mechanisms. Deep sea organisms, living at pressures as high as 100 MPa, are adapted at the molecular level through subtle changes in their enzymes, structural proteins, and membrane lipids.

Several technological applications have arisen from high pressure biology: the manipulation of ploidy, and of membrane proteins; sterilization and inactivation of microorganisms; food processing; and safe human diving.

This chapter intends to stimulate interest not only in high pressure, but more generally in a particular approach to physiology. The study of high pressure effects in molecular processes provides a way of exploring their thermodynamics and kinetics. The effects of pressure on cells and organisms are manifestations of fundamental perturbations. Some of these effects are direct, the result of pressure affecting the observed process, while others are indirect, the result of pressure affecting a regulatory process, thus revealing the complex coordination within cells and organisms. The study of adaptations of life processes to high pressure environments demonstrates how the evolutionary selection of functional, integrated systems is to be understood in thermodynamic and kinetic terms.

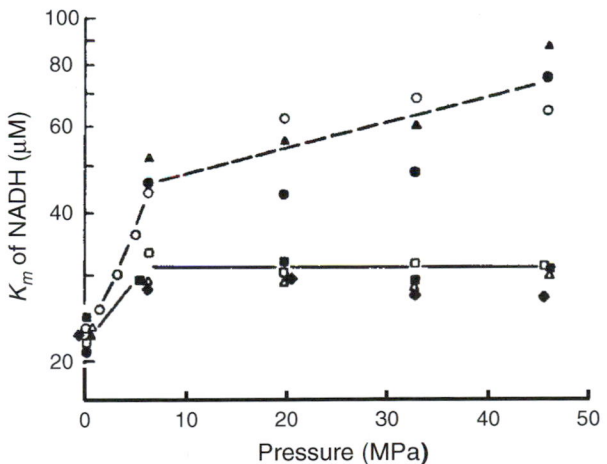

FIGURE 15. Adaptation of enzyme-ligand binding in deep sea fish. The K_m of NADH fish muscle lactate dehydrogenase is plotted against pressure. The K_m of NADH is markedly increased by pressure in the shallow water enzyme but is unaffected in the deep sea enzyme. This is interpreted as enzyme adaptation to high pressure. Shallow species: dashed line, ●, *Pagothenia borchgrevinki*, ▲, *Scorpaena guttata*; O, *Sebastolobus alascanus*. Deep sea species: continuous line, □, *Sebastolobus altivelis*; Δ, *Antimora rostrata*; ■, *Coryphaenoides acrolepis*; ◆, *Halosauropsis machrochir*. (From Siebenaller and Somero, 1979).

Acknowledgment

I thank Professor Horst Ludwig for commenting on the manuscript.

Bibliography

Anderson, J., and Schagatay, E. (1998). Effects of lung volume and involuntary breathing movements on the human diving response. *Eur. J. Appl. Physiol.* **77**, 19–24.

Balny, C. (1996). Transient enzyme kinetics at high pressure. In "High Pressure Effects in Molecular Biophysics and Enzymology" (J. L. Markley, D. B. Northrop, and C. A. Royer, Eds., Oxford University Press. New York/Oxford. pp. 196–210.

Begg, D. A., Salmon, E. D., and Hyatt, H. A. (1983). The changes in structural organization of actin in the sea urchin egg cortex in response to hydrostatic pressure. J. Cell. Biol. **97**, 1795–1805.

Behan, M. K., Macdonald, A. G., Jones, G. R., and Cossins, A. R. (1992). Homeoviscous adaptation under pressure : The pressure dependence of membrane order in brain myelin membranes of deep sea fish. Biochim. Biophys. Acta. **1103**, 317–323.

Beney, L., Perrier-Cornet, J. M., Wayert, M., and Gervais, P. (1997). Modification of phospholipid vesicles induced by high pressure: Influence of bilayer compressibility. Biophys J. **72**, 1258–1263.

Bennett, P. B., and Elliott, D. H. (Eds.) (1993). "The Physiology and Medicine of Diving." W. B. Saunders Co. Ltd., London.

Benz, R., and Conti, F. (1986). Effects of hydrostatic pressure on lipid bilayer membranes II. Activation and reaction volumes of carrier mediated ion transport. Biophys. J. **50**, 99–107.

Besch, S. R., and Hogan, P. M. (1996). A small chamber for making optical measurements on single living cells at elevated hydrostatic pressure. Undersea Hyperbaric Medicine **23**, 175–184.

Böttner, M., Ceh, D., Jacobs, U., and Winter, R. (1994). High pressure volumetric measurements on phospholipid bilayers. Zeitschr. Physikalische Chemie **184**, S205–S218.

Brouha, A., Pequex, A., Schoffeniels, E., and Disteche, A. (1970). The effects of high hydrostatic pressure on the permeability of characteristics of the isolated frog skin. Biochim. Biophys. Acta **219**, 455–462.

Brown, D. E. S. (1936). The effect of rapid compression upon events in the isometric contraction of skeletal muscle. J. Cell. Comp. Physiol. **8**, 141–157.

Bruner, L. J., and Hall, I. E. (1983). Pressure effects on alamethicin conductance in bilayer membranes. Biophys J. **44**, 39–47.

Campenot, R. B. (1975). The effect of high hydrostatic pressure on transmission at the crustacean neuromuscular junction. Comp. Biochem. Physiol. **52B**, 133–140.

Castellini, M. A. (1991). The biology of diving mammals: Behavioral, physiological and biochemical limits. In "Advances in Comparative and Environmental Physiology." (R. Gilles, Ed.), Vol. 8, Springer-Verlag, New York/Berlin. pp. 105–138.

Cattell, M. (1936). The physiological effects of pressure. Biol. Rev. **11**, 441–475.

Chong, P. L. G., and Weber, G. (1983). Pressure dependence of 1,6-Diphenyl-1, 3, 5-hexatriene fluorescence in single compartment phosphatidylcholine liposomes. Biochemistry **22**, 5544–5550.

Chong, P. L. G., Fortes, P. A. G., and Jameson, D. M. (1985). Mechanisms of inhibition of Na,K–ATPase by hydrostatic pressure studied with fluorescent probes. J. Biol.Chem. **260**, 14484–14490.

Clarke, M. R. (1978). Buoyancy control as a function of the spermaceti organ in the sperm whale. J. Mar. Biol. Ass. U.K. **58**, 27–71.

Coelho-Sampaio, T., Ferreira, S. T., Benaim, G., and Vieyra, A. (1991). Dissociation of purified erythrocyte Ca-ATPase by hydrostatic pressure. J. Biol. Chem. **266**, 22266–22272.

Collen, D., Vandereycken, G., and De Maeyer, L. (1970). Influence of hydrostatic pressure on the reversible polymerization of fibrin monomers. Nature **228**, 669–671.

Coluccio, L. M., Brady, R. J., and Parsons, R. H. (1983). Pressure effects on the ADH-induced initiation of water flow in toad bladder. Am. J. Physiol. **244**, (Renal Fluid Electrolyte Physiol 13) F547–553.

Conti, F., Fioravanti, R., Segal, J.R., and Stühmer, W. (1982). Pressure dependence of sodium currents of squid axon. J. Membr. Biol. **69**, 23–34.

Conti, F., Inoue, I., Kukita, F., and Stühmer, W. (1984). Pressure dependence of sodium gating currents. Eur. J. Biophys. **11**, 137–147.

Cossins, A. R., and Macdonald, A. G. (1986). Homeoviscous adaptation under pressure. III. The fatty acid composition of liver mitochondrial phospholipids of deep sea fish. Biochim. Biophys. Acta **860**, 325–335.

Crenshaw, H. C., and Salmon, E. D. (1996). Hydrostatic pressure to 400 atm does not induce changes in the cytosolic concentration of Ca^{++} in mouse fibroblasts: Measurements using Fura-2 fluorescence. Expt. Cell Res. **227**, 277–284.

Crenshaw, H. C., Allen, J. A., Skeen, V., Harris, A., and Salmon, E. D. (1996). Hydrostatic pressure has different effects on the assembly of tubulin, actin, myosin II, vinculin, talin, vimentin and cytokeratin in mammalian tissue cells. Exper. Cell Res. **227**, 285–297.

Czeslik, C., Reis, O., Winter, R., and Rapp, G. (1998). Effects of high pressure on the structure of dipalmitoylphosphatidylcholine bilayer membranes: A synchroton-X-ray diffraction and FT-IR spectroscopy study using the diamond anvil technique. Chem. Phys. Lipids. **91**, 135–144.

Daniels, S., Zhao, D. M., Inman, N., Price, D. J., Shelton, C. J., and Smith, E. B. (1991). Effects of general anaesthetics and pressure on mammalian excitatory receptors expressed in Xenopus oocytes. Ann. N.Y. Acad. Sci. **625**, 108–115.

De Long, E. F., and Yayanos, A. A. (1985). Adaptation of the membrane lipids of a deep sea bacterium to changes in hydrostatic pressure. Science **228**, 1101–1103.

Denton, E. J. (1971). Examples of the use of active transport of salts and water to give buoyancy in the sea. Phil. Trans. Roy. Soc. Lond. B. **262**, 277–287, Cambridge University Press.

Disteche, A. (1972). Effects of pressure on the dissociation of weak acids. Symp. Soc. Exper. Biol. **26**, 27–60.

Ebbecke, U. (1914). Wirkung allseitiger kompression aug den fraschmuskel. Archiv für Physiologie **157**, 79–116.

Eisenthal, A., Matsaev, A., Gelfand, A., Kohn, P., Lifschitz-Mercer, B., Skormick, Y., and Shinitzky, M. (1996). Surface projection of murine major histocompatibility determinants induced by hydrostatic pressure and cytokines. Pathobiology **64**, 142–149.

Elsner, R., Shiraki, K., and Mohri, M. (1996). Seals and ama, animal and human diving experts. In "Physiological Basis of Occupational Health : Stressful Environments." (K. Shiraki, S. Sagara, and M. K. Yousef, Eds.), SPB Academic Publishing, The Netherlands. pp. 127–135.

Fagni, L., Hugon, M., Folco, A., and Imbert, G. (1987). A versatile chamber for microphysiologic studies with gases under high pressure. Undersea Biomed. Res. **14**, 161–168.

Gerschfeld, N. L., and Shanes, A. M. (1958). The influence of high hydrostatic pressure on cocaine and veratrine action in a vertebrate nerve. J. Gen. Physiol. **42**, 647–653.

Gibbs, A. G. (1997). Biochemistry at depth. In "Deep Sea Fishes" (D. J. Randall, and A. P. Farell, Eds.), Academic Press, New York/London. pp. 239–277.

Gibson, J. S., and Hall, A. C. (1995). Stimulation of KCl co-transport in equine erythrocytes by hydrostatic pressure: Effects of kinase/phosphate inhibition. Pflügers Archiv **429**, 446–448.

Godart, H., and Ellory, J. C. (1996). KCl cotransport activation in human erythrocytes by high hydrostatic pressure. J. Physiol. **491**, 423–434.

Goldinger, J. M., Kang, B. S., Choo, Y. E., Paganelli, C. V., and Hong, S. K. (1980). Effect of hydrostatic pressure on ion transport and metabolism in human erythrocytes. J. Appl. Physiol. Respir. Environ. Exercise Physiol. **49**, 224–231.

Greaney, G. S., and Somero, G. N. (1979). Effects of anions on the activation thermodynamics and fluorescence emission spectrum of alkaline phosphatase : Evidence for enzyme hydration changes during catalysis. Biochemistry **18**, 5322–5332.

Greenwald, L., Ward, P. L., and Greenwald, O. E. (1980). Cameral liquid transport and buoyancy control in chambered nautilus (Nautilus macromphalus). Nature **286**, 55–56.

Gross, M., and Jaenicke, R. (1994). Proteins under pressure. Eur. J. Biochem. **221**, 617–630.

Grossman, Y., and Kendig, J. J. (1984). Pressure and temperature: Time-dependent modulation of membrane properties in a bifurcating axon. *J. Neurophysiol.* **52**, 692–708.

Grundfest, H., and Cattell, M. (1935). Some effects of hydrostatic pressure on nerve action potentials. *Am. J. Physiol.* **113**, 56–57.

Hall, A. C., Pickles, D. M., and Macdonald, A. G. (1993). Aspects of eukaryotic cells. *In* "Effects of High Pressure on Biological Systems" (A. G. Macdonald, Ed.), Springer Verlag-Heidelberg, New York/Berlin. pp. 30–85.

Harper, A. A., Macdonald, A. G., and Wann, K. T. (1975). Apparatus for intracellular recording from excitable cells subjected to high hydrostatic pressure. *J. Physiol.* **251**, 7–8P.

Harper, A. A., Macdonald, A. G. and Wann, K. T. (1981). The action of high hydrostatic pressure on the membrane currents of *Helix* neurones. *J. Physiol.* **311**, 325–339.

Harper, A. A., Macdonald, A. G., Wardle, C. S., and Pennec, J. P. (1987). The pressure tolerance of deep sea fish axons. Result of Challenger cruise 6B/85. *Comp. Biochem. Physiol.* **88A**, 647–653.

Harris, D. J., and Bennett, P. B. (1983). Force and duration of muscle twitch contractions in humans at pressures up to 70 bar. *J. Appl. Physiol: Respir. Environ. Exercise Physiol.* **54**, 1209–1215.

Heinemann, S. H., Conti, F., Stühmer, W., and Neher, E. (1987a). Effects of hydrostatic pressure on membrane processes. Sodium channels, calcium channels and exocytosis. *J. Gen. Physiol.* **90**, 765–778.

Heinemann, S. H., Stühmer, W., and Conti, F. (1987b). Single acetylcholine receptor channel currents reduced at high hydrostatic pressures. *Proc. Natl. Acad. Sci. USA.* **84**, 3229–3233.

Henderson, J. V., and Gilbert, D. L. (1975). Slowing of ionic currents in the voltage-clamped squid axon by helium pressure. *Nature* **258**, 351–352.

Henderson, J. V., Morin, R. A., and Lanphier, E. H. (1975). Apparatus for intracellular electrophysiological measurements at 200 ATA. *J. Appl. Physiol* **38**, 353–355.

Heremans, K., and Smeller, L. (1998). Protein structure and dynamics at high pressure. *Biochem. Biophys. Acta* **1386**, 353–370.

Hindell, M. A., and Lea, M.-A. (1998). Heart rate swimming speed, and estimated oxygen consumption of a free-ranging Southern Elephant Seal. *Physiol. Zool.* **71**, 74–84.

Hogan, P. M., and Besch, S. R. (1993). Vertebrate skeletal and cardiac muscle. *In* "Effects of High Pressure on Biological Systems" (A.G. Macdonald, Ed.), Springer-Verlag, New York/Berlin. pp. 125-146.

Holm, B., Schagatay, E., Kobayashi, T., Masuda, A., Ohdaira, T., and Honda, Y. (1998). Cardiovascular changes in elderly male breathhold divers (ama) and their socio-economical background at Chikura in Japan. *Appl. Human Sci.* **17**, 181–187.

Hong, S. K., Duffy, M. E., and Goldinger, J. M. (1984). Effect of high hydrostatic pressure on sodium transport across toad skin. *Undersea Biomed. Res.* **11**, 37–47.

Inoué, S., Fuseler, J., Salmon, E. D., and Ellis, G. W. (1975). Functional organization of mitotic microtubules. *Biophys. J.* **15**, 725–744.

Ishihara, M., McNally, D. S., Urban, J. P G., and Hall, A. C. (1996). Effects of hydrostatic pressure on matrix synthesis in different regions of the intervertebral disc. *J. Appl. Physiol.* **80**, 839–846.

Johnson, F. H., and Flagler, E. A. (1951). Activity of narcotized amphibian larvae under hydrostatic pressure. *J. Cell. Comp. Physiol.* **37**, 15–25.

Johnson, J. L. and Reinhart, G. D. (1996). Effects of high pressure on the allosteric properties of phosphofructokinase from *Escherichia coli*. *In* "High Pressure Effects in Molecular Biophysics and Enzymology" (J. L. Markley, D. B. Northrop, and C. A. Royer, Eds). Oxford University Press, New York/Oxford. pp. 242–255.

Jonas, J., and Jonas, A. (1994). High pressure NMR spectroscopy of proteins and membranes. *Annu. Rev. Biophys. Biomol. Struct.* **23**, 287–318.

Kaneshiro, S. M., and Clark, D. S. (1995). Pressure effects on the composition and thermal behavior of lipids from the deep sea thermophile *Methanococcus jannaschii*. *J. Bacteriol.* **177**, 3668–3672.

Kendig, J. J., (1984). Ionic currents in vertebrate myelinated nerve at hyperbaric pressure. *Am. J. Physiol.* **246**, C84–C90.

Kendig, J. J., Grossman, Y., and Heinemann, S. H. (1993). Ion channels and nerve function. *In* "Effects of High Pressure on Biological Systems" (A. G. Macdonald, Ed.), Springer-Verlag, New York/Berlin. pp. 87–124.

Kettman, M. S., Nishikawa, A. H., Morita, R. Y., and Becker, R. R. (1966). Effects of hydrostatic pressure on the aggregation reaction of poly-L-valyl-ribonuclease. *Biochem. Biophys. Res. Comm.* **22**, 262–267.

Kitching, J. A., (1970). Some effects of high pressure on protozoa. *In* "High Pressure Effects on Cellular Processes" (A. M. Zimmerman, Ed.), Academic Press, New York/London. pp. 155–177.

Koga, Y., Nishihara, M., Morri, H., and Akagawa-Matsushita, M. (1993). Ether polar lipids of methanogenic bacteria: Structures, comparative aspects, and biosynthesis. *Microbiol. Rev.* **57**, 164–182.

Kooyman, G. L., and Ponganis, P. J. (1998). The physiological basis of diving to depth: Birds and mammals. *Annu. Rev. Physiol.* **60**, 19–32.

Kornblatt, J. A., Bon Hoa, G., and Heremans, K. (1988). Pressure induced effects on cytochrome oxidase: The aerobic steady state. *Biochemistry* **27**, 5122–5128.

Lafay, V., Barthelemy, P., Comet, B., Frances, Y., and Jammes, Y. (1995). ECG changes during the experimental human dive HYDRA 10 (71 atm / 7,200 kPa). *Undersea and Hyperbaric Medicine* **22**, 51–60.

Laidler, K. J., (1951). The influence of pressure on the rates of biological reactions. *Arch. Biochem. Biophys.* **30**, 226–236.

Laidler, K. J. (1978). "Physical chemistry with biological applications." Benjamin Cummings Pub. Menlo Park, California.

Laidler, K. J., and Bunting, P. S. (1973). "The Chemical Kinetics of Enzyme Action." Clarendon, Oxford.

Lakra, W. S., and Das, P. (1998). Genetic engineering in aquaculture. *Indian J. Animal Sciences* **68**, 873–879.

Low, P. S., and Somero, G. N. (1975). Protein hydration changes during catalysis: A new mechanism of enzyme rate-enhancement and ion activation/inhibition of catalysis. *Proc. Nat. Acad. Sci. USA.* **72**, 3305–3309.

Ludwig, H., Scigalla, W., and Sojka, B. (1996). Pressure and temperature induced inactivation of microorganisms. *In* "High Pressure Effects in Molecular Biophysics and Enzymology." (J. L. Markley, D. B. Northrop, and C. A. Royer, Eds.), Oxford University Press, New York/Oxford. pp. 346–363.

Macdonald, A. G. (1976). Locomotor activity and oxygen consumption in shallow sea and deep sea invertebrates exposed to high hydrostatic pressure and low temperature. *In* "Underwater Physiology V" (C. J. Lambertson, Ed.). FASEB, Bethesda, MD. pp. 405–420.

Macdonald, A. G. (1978). A dilatometric investigation of the effects of general anesthetics, alcohols and hydrostatic pressure on the phase transition in smectic mesophases of dipalmitoyl phosphatidylcholine. *Biochim. Biophys. Acta* **507**, 26–37.

Macdonald, A. G. (1987). The role of membrane fluidity in complex processes under high pressure. *In* "Current Perspectives in High Pressure Biology" (H. W. Jannasch, R. E. Marquis, and A. M. Zimmerman, Eds.), Academic Press, New York/London. pp. 209–223.

Macdonald, A. G. (1997). Effect of high hydrostatic pressure on the BK channel in bovine chromaffin cells. *Biophys. J.* **73**, 1866–1873.

Macdonald, A. G. and Cossins, A. R., (1985). The theory of homeoviscous adaptation of membranes applied to deep sea animals. *Symp. Soc. Exper. Biol.* **39**, 300–322.

Macdonald, A. G., and Fraser, P. J. (1999). The transduction of very small hydrostatic pressures. *Comp. Biochem. Physiol.* A **122**, 13–36.

Macdonald, A. G., Ramsey, R. L., Shelton, C. J., and Usherwood, P. N. R. (1989). An apparatus for single-channel patch recording at high pressure. *J. Physiol.* **409**, 2P.

Macdonald, A. G. and Ramsey, R. L. (1995). The effects of nitrous oxide on a glutamate-gated ion channel and their reversal by high pressure; A single channel analysis. *Biochim. Biophys. Acta* **1236**, 135–141.

Macdonald, A. G., and Martinac, B. (1999). Effect of high hydrostatic pressure on the porin OmpC from *Escherichia coli*. *FEMS Microbiol. Letters.* **173**, 327–334.

Macdonald, A. G., and Vjotosh, A. N. (1999). Patch clamp recording of BK_{Ca} channels in hyperbaric oxygen. *J. Physiol.* **518P**, 111P.

Marsland, D. (1970). Pressure-temperature studies on the mechanisms of cell division. *In* "High Pressure Effects on Cellular Processes" (A. M. Zimmerman, Ed.), Academic Press, New York/London. pp. 260–312.

Mertens-Strijthagen, J., de Schryver, C., Wattiaux-de Coninck, S., and Wattiaux, R. (1985). The effect of pressure on fetal and neonatal liver mitochondrial membranes. *Archiv. Biochem. Biophys.* **236**, 825–831.

Meyer R., and Heinemann, S. H. (1997). Temperature and pressure dependence of Shaker K^+ channel N- and C- type inactivation. *Eur. Biophys. J.* **26**, 433–445.

Miller, K. W., Paron, W. D. M., Smith, R. A., and Smith, E. B. (1973). The pressure reversal of general anesthesia and the critical volume hypothesis. *Molec. Pharmacol.* **9**, 131–143.

Morild, E. (1981). The theory of pressure effects on enzymes. *Adv. Protein Chem.* **34**, 93–166.

Mylvaganam, S. E., Bonaventura, C., Bonaventura, J., and Getzoff, E. (1996). Structural basis for the Root effect in hemoglobin. *Nature Struct. Biol.* **3**, 275–283.

Northrop, D. B. (1996). Steady state enzyme kinetics at high pressure. *In* "High Pressure Effects in Molecular Biophysics and Enzymology" (J. L. Markley, D. B. Northrop, and C. A. Royer, Eds.), Oxford University Press, New York/London. pp. 211–241.

Ornhagen, H. C., and Sigurdsson, S. B. (1981). Effects of high hydrostatic pressure on rat atria muscle. *Undersea Biomed. Res.* **8**, 113–120.

Parkkinen, J. J., Ikonen, J., Lammi, M. J., Laakonen, J., Tammi, M., and Helminen H. J. (1993). Effects of cyclic hydrostatic pressure on protoglycan synthesis in cultured chondrocytes and articular cartilage explants. *Arch. Biochem. Biophys.* **300**, 458–465.

Pelster, B. (1997). Buoyancy at depth. *In* "Fish Physiology" (W. S. Hoar, D. J. Randall, and A. P. Farrell, Eds.), Vol. 16 Academic Press, London/New York. pp. 195–237.

Perutz, M. F. (1996). Cause of the Root effect in fish hemoglobins. *Nature Structural Biol.* **3**, 211–212.

Plager, D. A., and Nelsestuen, G. L. (1992). Dissociation of peripheral protein membrane complexes by high pressure. *Protein Science* **1**, 530–539.

Ranatunga, K. W., Fortune, N. S., and Geeves, M. A. (1990). Hydrostatic compression in glycerinated rabbit muscle fibres. *Biophys. J.* **58**, 1401–1410.

Ranatunga, K. W., and Geeves, M. A. (1991). Changes produced by increased hydrostatic pressure in isometric contractions of rat fast muscle. *J. Physiol.* **441**, 423–431.

Regnard, P. (1887). Les phenomenes de la vie sous les hautes pressions—la contraction musculaire. *Compte Rendu Soc Biol. Paris* **39**, 265–409.

Regnard, P. (1891). "Recherche Experimentales sur les Conditions Physique de la Vie dans les Eaux." Masson, Paris.

Rodgers, K. K., and Sligar, S. G. (1991). Mapping electrostatic interactions in macromolecular associations. *J. Mol. Biol.* **221**, 1453–1460.

Rostain, J.-C. (1993). The nervous system: Man and laboratory mammals. *In* "Effects of High Pressure on Biological Systems" (A. G. Macdonald, Ed.), Springer-Verlag, New York/Berlin. pp. 198–238.

Schmalwasser, H., Neef, A., Elliott, A. A., and Heinemann, S. H. (1998). Two-electrode voltage clamp of *Xenopus* oocytes under high hydrostatic pressure. *J. Neurosci. Methods* **81**, 1–7.

Schreer, J. F. and Testa, J. W. (1996). Classification of Weddell seal diving behavior. *Marine Mammal Sci.* **12**, 227–250.

Sebert, P. (1997). Pressure effects on shallow water fishes. *In* "Deep Sea Fishes" (D.J. Randall, and A. P. Farrell, Eds.), Academic Press, New York/London. pp. 279–323

Siebenaller, J. F., and Somero, G. N. (1979). Pressure-adaptive differences in the binding and catalytic properties of muscle-type (M4) lactate dehydrogenases of shallow and deep-living marine fish. *J. Comp. Physiol. B.* **129**, 295–300.

Siebenaller, J. F., and Somero, G. N. (1989). Biochemical adaptation to the deep sea. *CRC Critical Reviews in Aquatic Sciences* **1**, 1–25.

Silva, J. L., and Weber, G. (1993). Pressure stability of proteins. *Annu. Rev. Phys. Chem.* **44**, 89–113.

Silva, J. L., Fogual, D., Da Poian, A. T., and Prevelige, P. E. (1996). The use of hydrostatic pressure as a tool to study viruses and other macromolecular assemblages. *Curr. Opin. Struct. Biol.* **6**, 166–175.

Sinensky, M. (1974). Homeoviscous adaptation—a homeostatic process that regulates the viscosity of membrane lipids in *E. coli*. *Proc. Natl. Acad. Sci. USA* **71**, 522–525.

Somero, G. N. (1990). Life at low volume change: Hydrostatic pressure as a selective factor in the aquatic environment. *Am. Zool.* **30**, 123–135.

Spyropoulos, C. S. (1957a). The effect of hydrostatic pressure upon the normal and narcotized nerve fiber. *J. Gen. Physiol.* **40**, 849–857.

Spyropoulos, C. S. (1957b). Response of single nerve fibres at different hydrostatic pressures. *Am. J. Physiol.* **189**, 214–218.

Stephan, L., and Hasselbach. W. (1991). Volume changes in high affinity calcium binding of the sarcoplasmic reticulum calcium-transport enzyme. *Eur. J. Biochem.* **202**, 551–557.

Swezey, R. R., and Somero, G. N. (1985). Pressure effects on actin self-assembly: Interspecific differences in the equilibrium and kinetics of the G to F transformation. *Biochemistry* **24**, 852–860.

Takahashi, K., Kubo, T., Anai, Y., Kitajima, I., Takigawa, M., Imanishi, J., and Hirasewa, Y. (1998). Hydrostatic pressure induces expression of interleukin-6 and tumour necrosis factor and mRNAs in a chondrocyte-like cell line. *Ann. Rheum. Dis.* **57**, 231–236.

Takeuchi, H., Mohri, M., Shiraki, K., Lin, Y.C., Claybaugh, J. R., and Hong, S. K. (1995). Diurnal renal responses in man to water loading at sea level and 31 atm ab. *Undersea and Hyperbaric Medecine* **22**, 61–71.

Tonner, P. H., Poppers, D. M., and Miller, K. W. (1992). The general anaesthetic potency of propofol and its dependence on hydrostatic pressure. *Anaesthesiology* **77**, 926–931.

Tumminia, S. J., Koretz, J. F., and Landau, J. V. (1989). Hydrostatic pressure studies of native and synthetic thick filaments in vitro myosin aggregates at pH7.0 with and without G proteins. *Biochim. Biophys. Acta* **999**, 300–312.

Vawda, F., Ranatunga, K. W., and Greeves, M. A. (1995). Pressure induced changes in the isometric contractions of single intact frog muscle fibres at low temperatures. *J. Muscle Res. Cell Motility* **16**, 412–419.

Vawda, F., Ranatunga, K. W., and Geeves, M. A. (1996). Effects of hydrostatic pressure on fatiguing frog muscle fibres. *J. Muscle Res. Cell Motility* **17**, 631–636.

Verjovski-Almeida, S., Jurtenbach, E., Amorim, A. F., and Weber, G. (1986). Pressure-induced dissociation of solubilized sarcoplasmic reticulum ATPase. *J. Biol. Chem.* **261**, 9872–9878.

Wagner, G. (1980). Activation volumes for the rotational motion of interior aromatic rings in globular proteins determined by high resolution ^{1}HNMR at variable pressure. *FEBS Lett.* **112**, 280–284.

Wann, K. T., and Macdonald, A. G. (1988). Actions and interactions of high pressure and general anesthetics. *Prog. Neurobiol.* **30**, 271–307.

Wann, K. T., Macdonald, A. G., Harper, A. A., and Wilcock, S. E. (1979). Electrophysiological measurements at high hydrostatic pressure: Methods for intracellular recording from isolated ganglia and for extracellular recording *in vivo. Comp. Biochem. Physiol.* **64A**, 141–147.

Wirsen, C. O., Jannasch, H. W., Wakeham, S. G., and Cannel, E. A. (1987). Membrane lipids of a psychrophilic and barophilic deep sea bacterium. *Curr. Microbiol.* **14**, 319–322.

Yamazaki, F., Shiraki, K., Sagawa, S., Endo, Y., Torri, R., Yamaguchi, H., Mohril, M., and Lin, Y.-C. (1998). Assessment of cardiac auto-nomic nervous activities during heliox exposure at 24 atm abs. *Aviat. Space Environ. Med.* **69**, 643–646.

Yano, Y., Nakayama, A., Ishihara, K., and Saito, H. (1998). Adaptive changes in membrane lipids of barophilic bacteria in response to changes in growth pressure. *Appl. Environ. Microbiol.* **64**, 479–485.

Zimmerman, A. M., Symington, A. L., and Zimmerman, S. (1994). Pressure induced disruption of the cytoskeleton and the protein synthesizing machinery. *In* "Basic and Applied High Pressure Biology" (P. B. Bennett, and R. E. Marquis, Eds.), University of Rochester Press, Rochester, NY, pp. 23–30.

60

Electrocytes of Electric Fish

I. Introduction

Electric fish such as the marine **electric ray** (genus *Torpedo*) and the freshwater **electric eel** (*Electrophorus electricus*) are capable of generating powerful electrical discharges that can be measured in the water surrounding these animals. These fish use the production of bioelectricity as an effective mechanism to stun prey and ward off predators. Electrical discharges are generated by electric cells, called **electroplax** or **electrocytes**, that produce end-plate potentials and action potentials that are remarkably similar to the membrane potentials of neurons and myocytes. In fact, the membrane receptors, ion channels, and ATPases responsible for electric tissue electrophysiology are biochemically and functionally identical to those of mammalian muscle and nerve. For this reason, electrocytes have been used extensively as a specialized and appropriate model system for the study of excitable cell membrane electrophysiology and biochemistry. Due to the specialized nature of electric tissue, it has also been used as an enriched source of membrane proteins for biochemical studies. Previous chapters have described in detail the generation of acetylcholine (ACh)-mediated muscle end-plate potentials and the propagation of action potentials of nerve and muscle (see Chapters 24, 25, 28, 41, and 51). This chapter examines the anatomy and cellular morphology that electric fish have evolved in order to produce powerful electrical discharges. An electrophysiological and biochemical comparison is made between the electrocytes of the freshwater electric eel and the marine electric ray. The major contributions that electric tissue has made to the understanding of the electrophysiology and biochemistry of excitable membranes are also reviewed.

The shocking sensations produced by electric fish were undoubtedly experienced by mankind long before the recording of scientific phenomena. Some of the first recorded reports of unusual effects produced by electrical discharges of electric fish were of the Nile river catfish, *Malapterus electricus*. Nile river fishermen reported unpleasant sensations when handling live *Malapterus*, or even the water-soaked nets containing the fish. Godigno, a seventeenth century Jesuit father, noted that dead fish could be induced to move when a live *Malapterus* was thrown among them (Grundfest, 1957). At the time when Ben Franklin and other investigators were experimenting with static electricity of the Leyden jar, the electric eel provided insight into the basic conductive properties of electricity. In 1775, John Walsh conducted numerous experiments, one of which involved 10 people holding hands in a circle where the first and last "subjects" touched the opposite ends of a moderate-sized eel. All 10 people received a severe shock. The relative conductivities of various materials, including glass, wood, silk, brass chains, and iron rods, were then determined by holding these materials between two of the investigators and noting the severity of the electrical discharge. Although these experiments were likely to be very convincing to Walsh and his assistants, others doubted the electrical nature of the discharge from *Electrophorus* and *Torpedo*. The bioelectric nature of the discharge had not gained widespread acceptance until Du Bois-Raymond demonstrated that nerve and muscle were electrogenic (Grundfest, 1957). Since that time, the usefulness of *Electrophorus*, *Torpedo*, and other electric fish as models for excitable membranes has been realized.

II. Anatomy of *Electrophorus* and Mechanism of the Electrical Discharge

Powerful electric fish possess a specialized anatomy and cellular morphology devoted to the production of electrical discharges. The electric eel is an excellent example of this specialization. It has been well characterized on the cellular and biochemical level and is used here to describe the production of bioelectricity. Figure 1A depicts the location of the electric organs within *Electrophorus*. The viscera are crowded into the rostral 20% of the animal; the remaining 80% is comprised predominantly of electric tissue and swimming muscles. The electric organs are confined to the ventral portion of this caudal region, whereas most of the swimming muscles, major blood vessels, and spinal cord are situated in the dorsal one-third (Fig. 1B). The eel also possesses a tubular swim

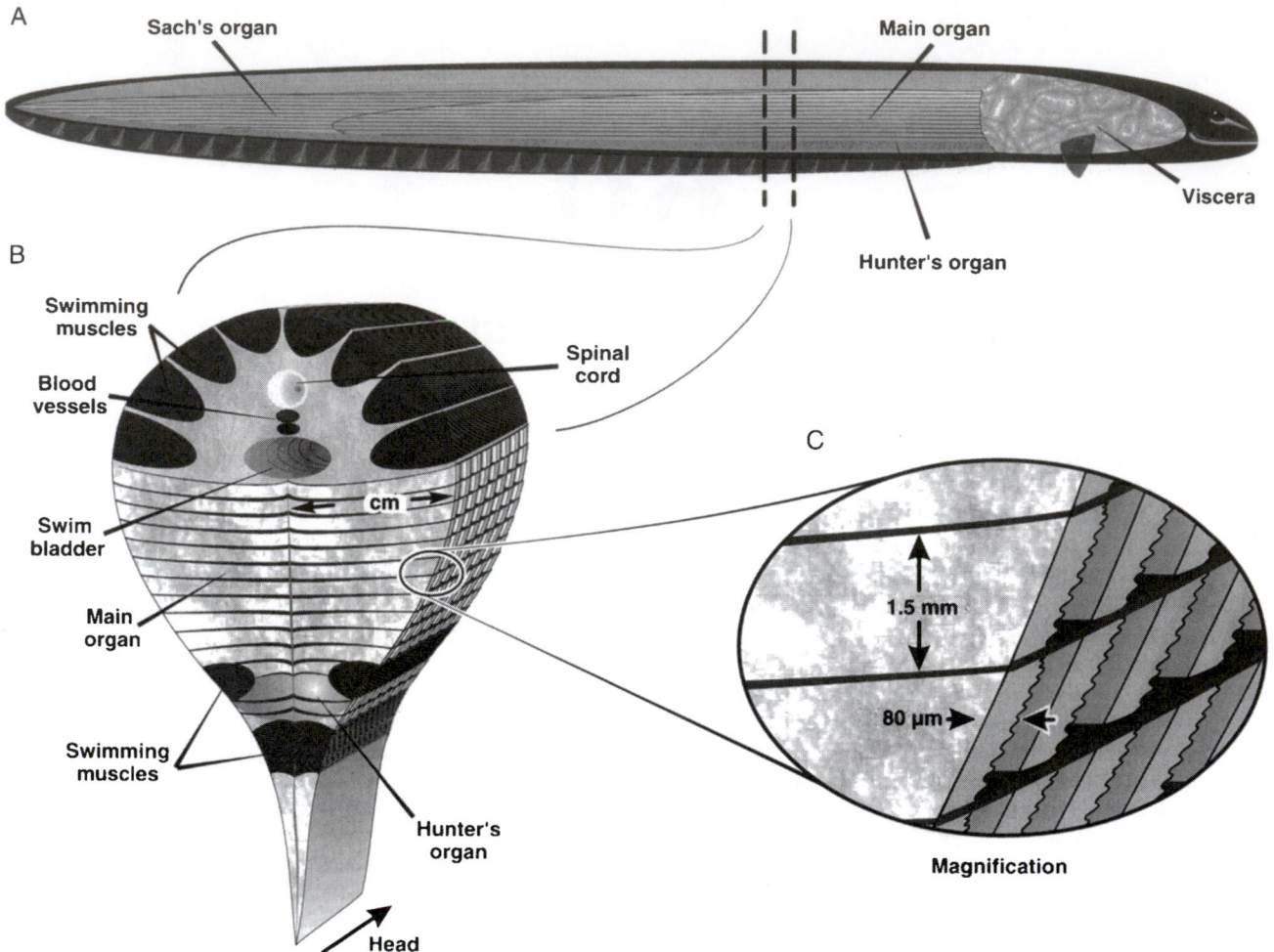

FIGURE 1.   Anatomy of the electric eel. (A) Diagram illustrating the anatomical orientation of electric organs. (B) A section through the middle portion of the eel, drawn such that the anterior surface is nearest the reader. (C) Columns of electrocytes extend the length of the electric organ. In this panel, the flatter, caudal surface of each electrocyte would be innervated by numerous electromotor neurons (not shown).

bladder that extends the length of the fish, and is positioned dorsal to the main electric organ and ventral to the spinal cord. The central nervous system consists of a small brain typical of teleost fish and a spinal cord that extends down the length of the animal. The electrical discharge is coordinated in the central control nucleus in the medulla. Axons from these neurons of the brain project caudally and synapse on neurons of the electromotor nucleus of the spinal cord. **Electromotor neurons** radiate into the electric organ, innervating individual electrocytes (Bennett and Sandri, 1989). To generate the whole-animal electrical discharge, each electrocyte of the entire electric organ must be stimulated simultaneously. In other words, action potentials reaching proximal electrocytes of the electric organ must be delayed to varying degrees relative to more distal regions. Neurons innervating proximal electrocytes are smaller in diameter and conduct action potentials more slowly. Some of these neurons wind their way to these rostral electrocytes, thereby slowing stimulation of this part of the electric organ, aiding in synchronous activation (Bennett, 1971). The delay may also occur in the electromotor nucleus

of the spinal cord where central neurons synapse on electromotor neurons. Presumably, synaptic transmission is slower to electromotor neurons innervating proximal portions of the electric organ, whereas faster signaling occurs between neurons that innervate electrocytes near the tail (Szabo, 1961).

As depicted in Fig. 1, *Electrophorus* has three well-defined electric organs. The main organ is the largest and is responsible for voluntarily generating powerful high-voltage discharges. The main organ extends from behind the peritoneal cavity of the viscera down the tail of the eel, where it eventually gives rise to Sach's organ. This organ, along with Hunter's organ, generates repetitive low-voltage discharges and is thought to be involved in electrolocation of objects in the eel's environment. In the cross-sectional view of Fig. 1B, Hunter's organ is seen to be partially delineated from the main organ by two columns of skeletal muscles. Electric tissue develops from these columns of skeletal muscle tissue in immature eels, and is thought to arise from embryonic myocyte precursor cells (Keynes, 1961). As shown later, the membranes of electrocytes are biochemically and functionally very similar to skeletal muscle sarcolemma.

Electrocytes of both the main electric organ and Sach's organ are large ribbon-shaped cells. Each electrocyte extends laterally from the midline of the electric organ to the skin, a distance of up to 4 cm. They have a width of up to 1.5 mm and thickness of 80 μm. As seen in Fig. 1C, electrocytes are positioned one after another along their flat axis to give rise to long rectangular columns of cells running along the longitudinal axis of the eel. These columns are delineated and electrically insulated from one another by **connective tissue septa**, which help to maintain the physical structure of the electric organ. This stacked arrangement is a common feature of electric organs in electric fish, and enables greater voltages to be produced (see later). When viewed under light microscopy, electrocytes are seen as multinucleated syncytiums, similar to the skeletal muscle myocytes from which they are derived. Electrocytes are seen in cross-section to have one relatively flat posterior membrane relative to the other more undulated anterior membrane. The flat caudal membrane is innervated by ACh-releasing electromotor neurons that form synapses that are morphologically similar to motor end plates of skeletal muscle cells (Chapter 41).

Immunofluorescent localization microscopy and electrophysiological experiments have led to the understanding of how the electrical discharge is generated at the cellular level. Electrocytes use membrane receptors and ion channels polarized to either the innervated or noninnervated membrane in order to produce **transcellular potentials** that give rise to the discharge of the electric organ (Fig. 2A). The caudal innervated membrane contains acetylcholine receptors (AChRs), inward and outward rectifying K$^+$ channels, a high density of voltage-gated Na$^+$ channels, and only trace amounts of Na$^+$,K$^+$-ATPase. This membrane is both chemically and electrically excitable. That is, this surface of the electrocyte produces **action potentials** (APs) in response to artificial stimulation with AChR agonists or direct electrical stimulation. The noninnervated membrane, on the other hand, does not respond to these manipulations, since it has no AChRs or voltage-gated Na$^+$ channels. Instead, this membrane contains a high concentration of Na$^+$,K$^+$-ATPase and ion channels responsible for maintaining the **resting potential** of −70 to −85 mV. Evidence suggests that the resting current of *Electrophorus* electrocytes is carried predominantly by K$^+$ channels (Lester, 1978), but the contribution of a Cl$^-$ conductance cannot be excluded (Nakamura *et al.*, 1965). In *Torpedo*, the resting current has been shown to be carried at least partially by Cl$^-$ (Miller and White, 1980). In skeletal muscle myocytes, 30–70% of the resting current is carried by this anion (Chapter 51). Since *Electrophorus* electric tissue is derived from skeletal muscle, it is possible that the resting current of these cells is also carried by Cl$^-$.

APs arriving at the nerve termini of electromotor neurons cause ACh to be released onto the innervated membrane of electrocytes (Fig. 2A). **End-plate potentials** (EPPs) produced by AChRs surpass the threshold for Na$^+$ channel activation, and trigger the production of APs that propagate very short distances between electromotor junctions and have overshoots of +40 to +65 mV. Meanwhile, the potential of the noninnervated membrane remains at the resting value of up to −85 mV, due to the

abundance of resting current channels and the absence of voltage-gated Na$^+$ channels. When the AP peaks on the innervated membrane, a net **transcellular potential** of up to 150 mV results across the electrocyte (+65 mV of the innervated membrane minus −85 mV of the noninnervated membrane). This transcellular potential difference is accompanied by a net flow of positive current moving in the innervated membrane-to-noninnervated membrane direction (left to right as shown in Fig. 2). Insulating septa that form electrocyte columns effectively insulate the extracellular regions on either side of the electrocyte. These connective tissue structures prevent current from flowing around the outside of the electrocyte, which would short-circuit the transcellular potential difference (Fig. 2B). This arrangement allows each electrocyte to act as a simple battery having an electrical potential of up to 150 mV. Since each electrocyte of a column is stimulated simultaneously, the potentials of each electrocyte battery within a column summate to generate a large voltage, as predicted by Ohm's law. In Fig. 2B, the potentials of three electrocytes summate to give a potential of 450 mV. In large eels where the potentials of many thousands of electrocytes summate, the net electric discharge can reach 700 V.[1] Connective tissue septa channel the current down the longitudinal axis of the electric organ toward the head of the eel. The current leaves the eel through low-resistance regions of the skin, and is conducted through the water, producing an electric field, the magnitude of which diminishes with the square of the distance from the animal. Objects, like other fish or the human hand, experience a potential difference in this electric field, and currents sufficient to excite muscles, nerves, and sensory endings flow through them, producing a shocking sensation. The circuit of the discharge is closed by current flowing through the skin of the tail region of the eel back into electrocyte columns from which the electromotive force (EMF) originated (Bennett, 1971).

III. Electrocyte Membrane Electrophysiology

A. Membrane and Extracellular Potentials

Electrocyte membranes contain many of the same protein elements found in myocytes and neurons. In fact, the individual APs of the innervated membrane of *Electrophorus* electrocytes are quite similar to those of other excitable cells. However, electric cells are different, in that potential changes are polarized to a particular membrane, resulting in the generation of transcellular potentials and an asymmetric flow of current. Also, to produce whole-animal electrical discharges having the maximum possible voltage or current output, electrocytes have evolved to express exaggerated amounts of key excitable membrane proteins.

In electric tissue preparations where the flat innervated electrically excitable surface of electrocytes is exposed, APs can be triggered with extracellular stimulating electrodes situated close to the membrane surface. Membrane potentials and **transcellu-**

---

[1] A discharge of this magnitude requires that at least 4700 electrocytes be stimulated simultaneously. That is, 4700 electrocytes × 0.15 V per electrocyte = 705 V. This situation is analogous to a flashlight, where more batteries aligned in series produce a brighter light source.

A

Resting

Stimulated

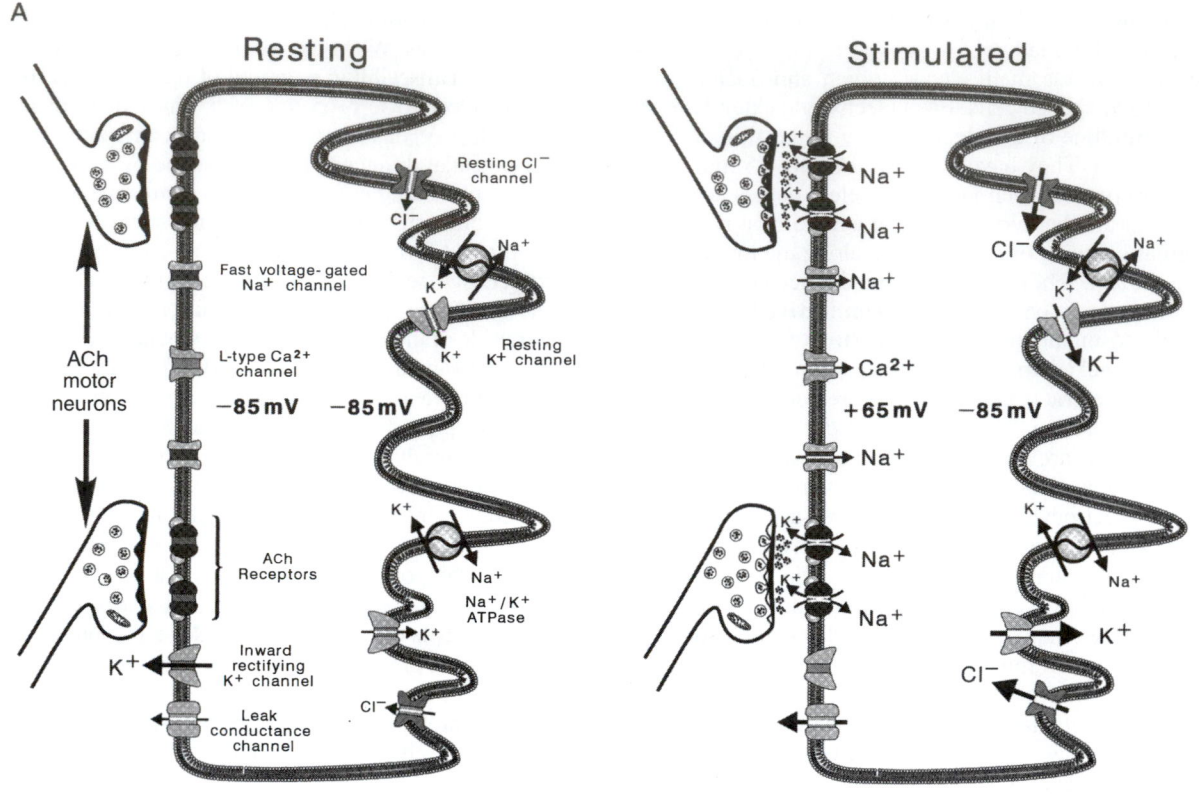

B

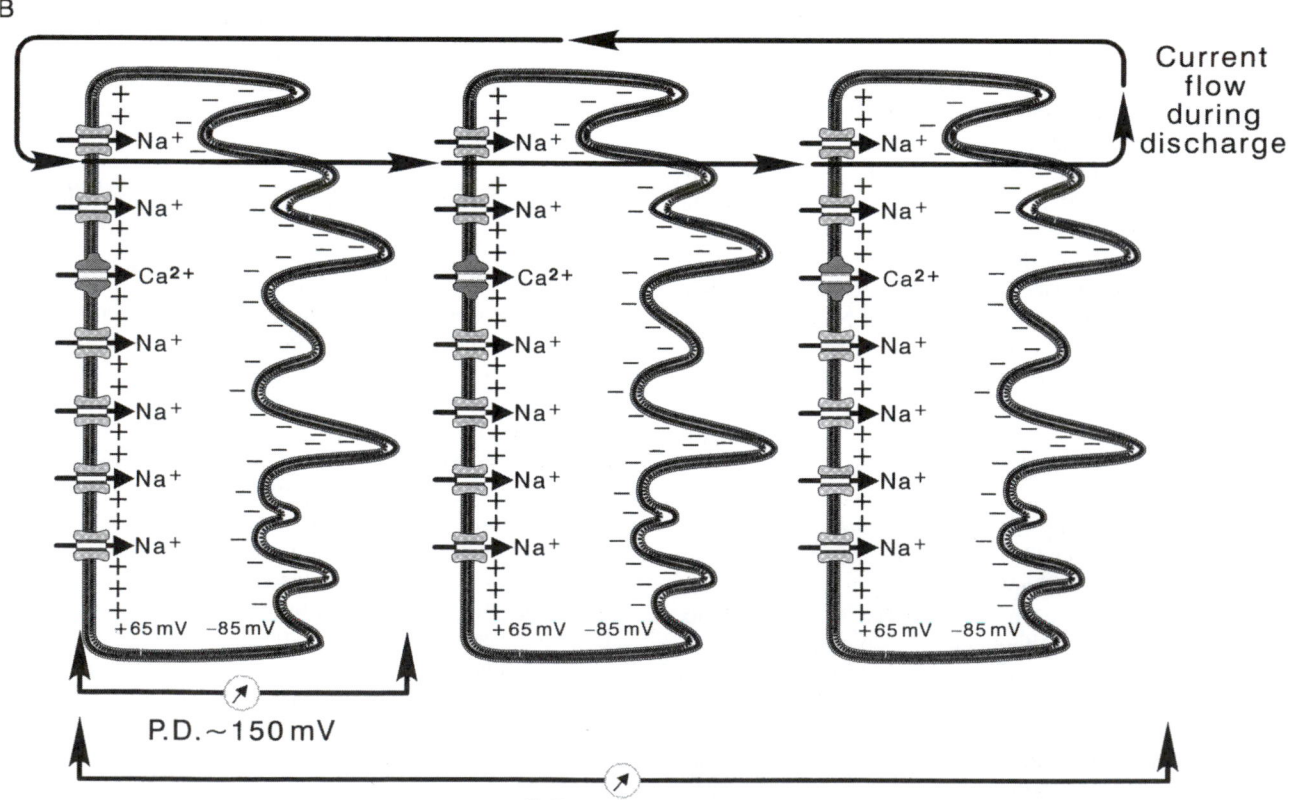

lar potentials can then be measured through recording electrodes lowered across the innervated membrane or through the entire cell, respectively (Fig. 3). As the recording electrode approaches the innervated membrane and a stimulus is applied, a negative deflection is recorded (Fig. 3A). Because the recording electrode measures the potential difference between the region just outside the membrane compared to the reference electrode placed in the bath, the AP here is recorded as a negative deflection. After the electrode is advanced through the innervated membrane, an AP that has propagated from the site of stimulation to the recording electrode is detected. The characteristics of electrocyte membrane potentials vary considerably from cell to cell, but are typically similar to those measured on the myocyte sarcolemma or neuronal axolemma. As seen in Fig. 3B, a typical resting potential is about -75 mV, and ranges from -65 to -85 mV. The electrocyte AP seen in Fig. 3B is typical in its 3.5-ms duration and +50-mV overshoot. Generally, the duration ranges from 2 to 4 ms and the overshoot between +35 and +65 mV. These values for the overshoot are considerably larger than that of other excitable cells and are due to an extraordinarily high density of voltage-gated Na^+ current, and a relatively low level of outward rectifying K^+ current (see discussion later). When the recording electrode is lowered even further until it completely penetrates the electrocyte, transcellular APs are recorded. Because the recording electrode is once again in the extracellular space, the resting potential here is measured as 0 mV. However, the interstitium where the recording electrode is positioned is electrically insulated by connective tissue septa from the reference electrode located in the bath solution. When the electrocyte is stimulated, an AP is recorded that is identical to the intracellular AP, except that it initiates at 0 mV. It has a peak equal to the total amplitude of the intracellular AP, in this case about 125 mV. This transcellular AP arises because the noninnervated nonexcitable membrane does not fire an AP that would cancel out the spike of the innervated membrane. Insulating connective tissue septa prevent the potential difference from being short-circuited around the outside of the electrocyte. If the stimulus is increased such that the electrocyte beneath the recording electrode is also stimulated, a negative deflection in the extracellular potential is recorded just behind the 125-mV transcellular AP (Fig. 3C, right). This negative deflection corresponds to the resulting AP of the lower electrocyte, which is stimulated later than the electrocyte on the surface. In the intact electric organ, these electrocytes would be stimulated simultaneously by the eel's nervous system such that these APs would be occurring at the same time. In this way, the transcellular potentials summate to yield a powerful electrical discharge.

Ionic currents responsible for *Electrophorus* electrocyte resting potentials, as well as EPPs and APs are similar to other excitable cells, and the reader is referred to previous chapters where their mechanisms have been described in detail. However, some differences between electrocyte membranes and those of neurons and myocytes are worthy of mention. The noninnervated membrane of electrocytes has a very low resistance of about 0.1 Ω/cm^2, which is one to two orders of magnitude less than that typically found for nerve or muscle (Nakamura *et al.*, 1965). Physiologically, the eel needs this high K^+ and Cl^- current to clamp the noninnervated membrane at the resting potential in order to set up transcellular potentials like those observed in Fig. 3C. On the other hand, the innervated membrane at rest has a resistance of 3–6 Ω/cm^2 (Nakamura *et al.*, 1965). On stimulation, this resistance decreases due to a large voltage-activated Na^+ current. In fact, Shenkel and Sigworth (1991) were able to measure macroscopic Na^+ currents in excised patches of the innervated membrane corresponding to a density of as much as 1300 channels/μm^2. This high density of Na^+ channels is accompanied by relatively few outwardly rectifying K^+ channels. Because of this distribution of channels in the innervated membrane, the electrocyte AP has a large overshoot that nearly reaches the Na^+ equilibrium potential, E_{Na}. The repolarization phase is therefore due primarily to the inactivation of Na^+ channels, and secondarily to delayed rectifier K^+ channels and the resting current of the innervated membrane.

Skeletal muscle-like EPPs that trigger APs on the surface of electrocytes are generated by AChR-mediated currents. Basically, macroscopic and single-channel currents conducted by the eel AChR are very similar to those seen on skeletal muscle sarcolemma. The permeability of the eel receptor to both Na^+ and K^+ is nearly equal since the reversal potential is the midpoint between E_{Na} and E_K (Sheridan and Lester, 1977; Pasquale *et al.*, 1986). In other words, the peak of the EPP moves toward a value of approximately -10 mV in order to activate Na^+ channels for an AP. Eel AChRs elicit single-channel opening events that are similar to the receptor from mammalian sources in that their mean open time is dependent on membrane potential, temperature, and the AChR agonist used. Single-channel conductances through individual receptors do not depend on the ligand used. However, these preparations of the eel AChR are different from other excitable cells in that single-channel open times can be fitted to a single exponential compared to the more complex distributions found for other sources of the receptor. This indicates that the eel expresses only one isoform of each of the receptor subunits, yielding a receptor with a single unique conductance.

B. Equivalent Circuits

From what is known of electrocyte electrophysiology, equivalent circuits can be derived that explain the production

---

FIGURE 2. Diagrammatic representation of electrocytes. The left surface of each cell represents the posterior innervated membrane. (A) At rest, both the innervated and noninnervated membrane exhibit a potential of -85 mV. When stimulated, activated AChRs generate EPPs, triggering Na^+ channel-mediated APs peaking at +55 mV on the innervated membrane. The noninnervated membrane contains no voltage-gated Na^+ channels and maintains the -85-mV resting potential. The result is a transcellular potential difference of approximately 150 mV. (B) Because each cell is stimulated simultaneously, electrocyte transcellular potentials summate. The potentials of three electrocytes culminate to produce 450 mV. Currents generated by stimulated electrocytes flow down electrocyte columns in the posterior-to-anterior direction. The circuit is closed by current flowing out the head of the eel, through the water, and back into the tail region.

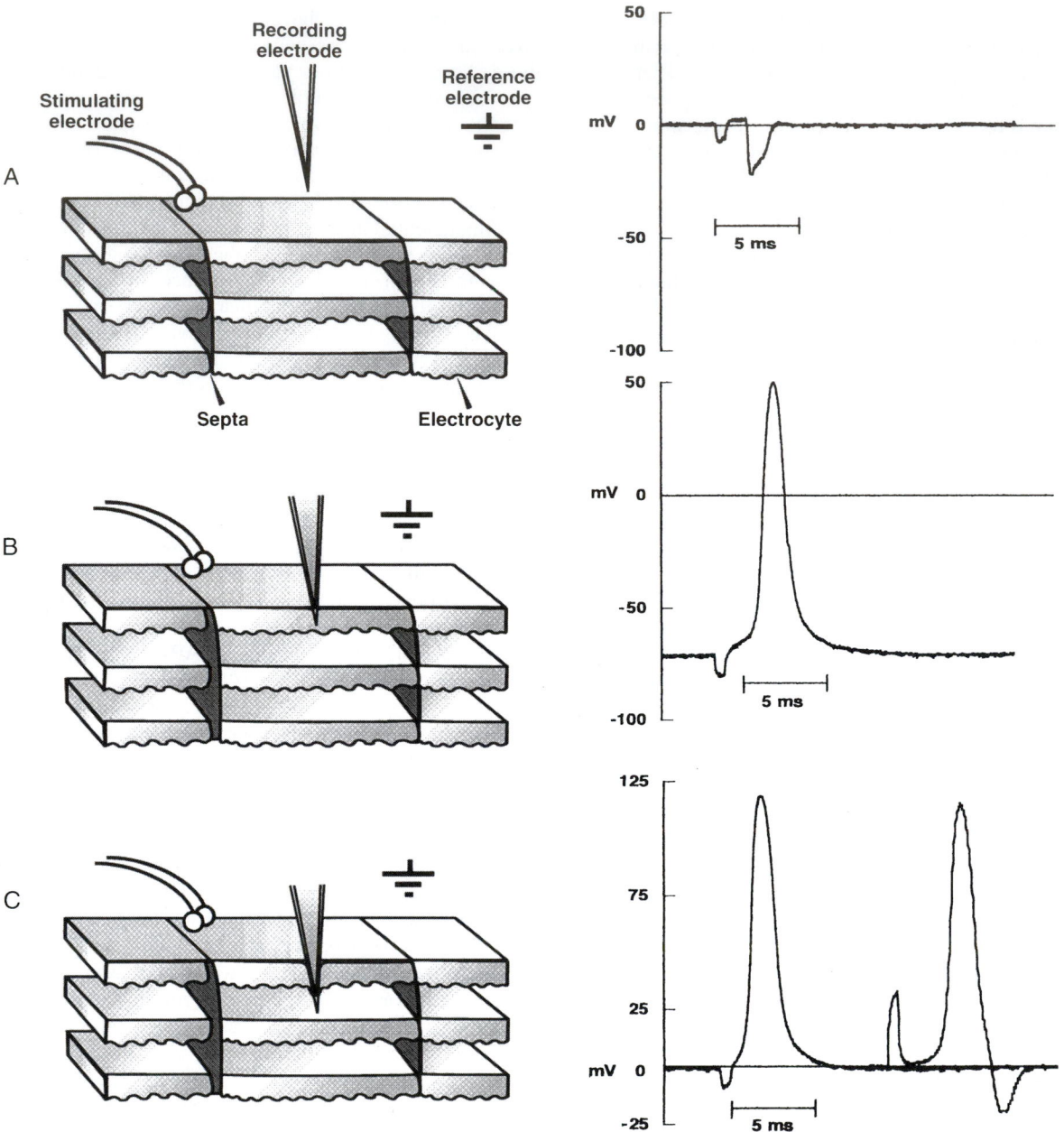

FIGURE 3.   Extracellular potentials, APs, and transcellular potentials of *Electrophorus* electrocytes in an electric tissue slice preparation. Diagrams to the left depict a slice of electric tissue in cross-section where electrocytes are oriented such that the innervated membrane is uppermost. Columns of electrocytes run in the vertical direction and are delineated by insulating connective tissue septa. Potentials recorded by the recording electrode in the indicated positions are shown on the right. (A) The recording electrode near the innervated membrane records a negative deflection in the extracellular potential. (B) The recording electrode penetrating the innervated membrane records an intracellular AP. (C) Transcellular potentials measured after the recording electrode has penetrated the entire electrocyte. The second potential recording on the right shows the negative deflection of the electrocyte beneath the recording electrode when a higher intensity stimulus is applied.

of transcellular potentials that arise during the AP. Because electrocytes are large flat cells comprised of essentially two parallel membranes having a uniform potential across their entire surfaces, whole-cell potentials can be described with two equivalent circuits, pertaining to the innervated and non-innervated membranes, connected by a resistor representing the resistance of the cytoplasm (R_{cyt}). At rest, the permeabil-

ity of both membranes to K^+ and Cl^- is high, so that their equilibrium potentials are expressed more than that of Na^+, resulting in an E_m of about -85 mV (for a detailed description, see Chapter 14). The equivalent circuit for the electrocyte at rest can then be reduced to that seen in Fig. 4A (right), where both membranes have composite E_m values that drive an outward flow of positive current. Notice that the equiva-

A

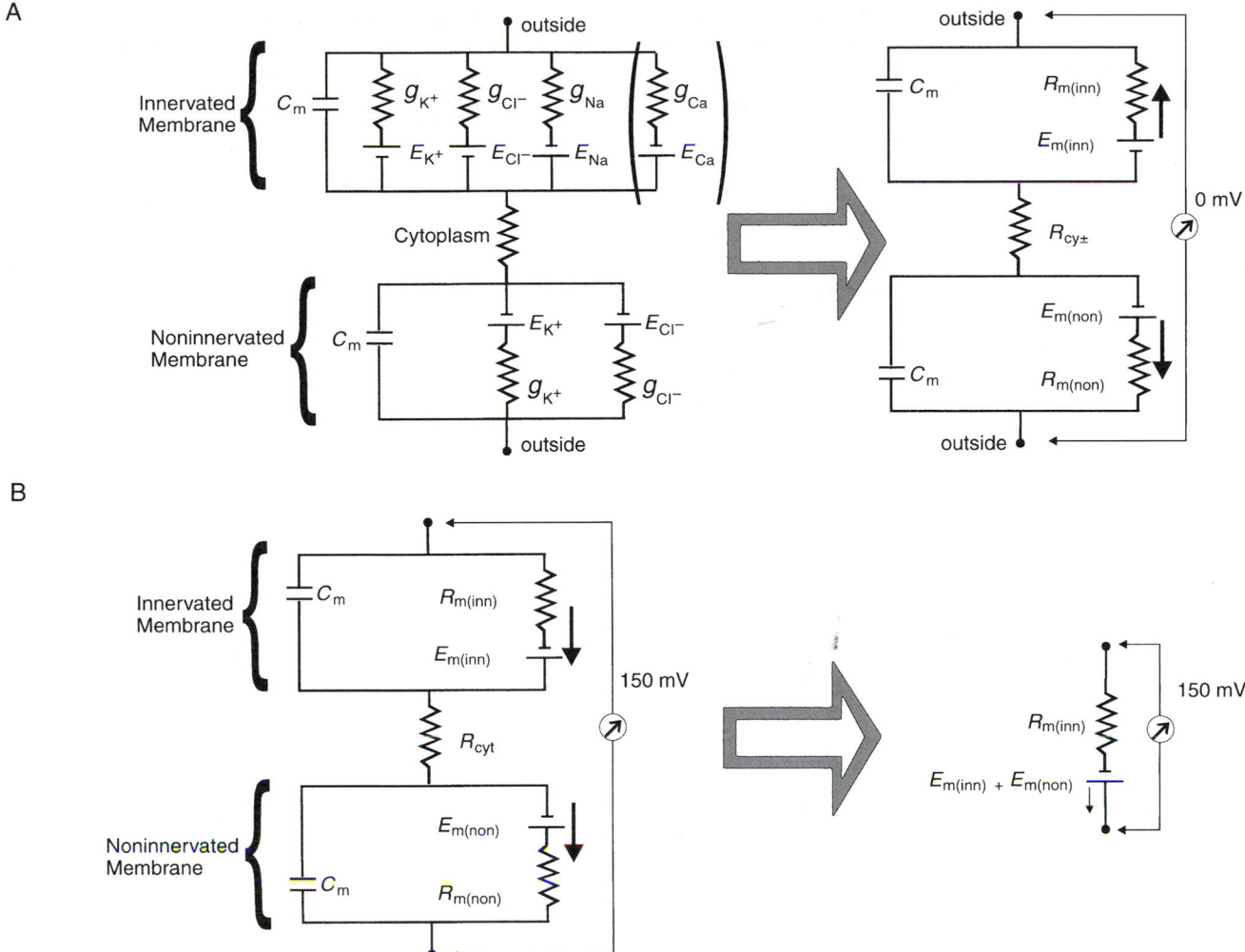

FIGURE 4. Electrical equivalent circuit diagrams for both the innervated and noninnervated membranes of an electrocyte. The innervated membrane is uppermost in both A and B. Both membranes are represented by parallel resistance-capacitance circuits, and are connected by the cytoplasmic resistance, R_{cyt}. C_m is the membrane's capacitance. The conductances for K^+, Cl^-, Na^+, and Ca^{2+} across the membranes are represented by g_K, g_{Cl}, g_{Na}. and g_{Ca}, respectively, and are inversely proportional to the resistance of the membrane for these ions. Nernst potentials for each of these ions across the membranes are E_K, E_{Cl}, E_{Na}, and E_{Ca}. The Ca^{2+} leg of the circuit is in brackets, since the existence of a selective Ca^{2+} conductance in the electrocyte has not yet been demonstrated. (A) At rest, the innervated membrane reduces to a circuit where $R_{m(inn)}$ represents the total membrane resistance of this face of the cell, and is a composite of the resistances of the membrane to each of the ions listed to the left. The $E_{m(inn)}$ and $E_{m(non)}$ are the resting membrane potentials for the innervated membrane and noninnervated membrane, respectively. At rest, E_K and E_{Cl} are expressed the most, since the conductance of the membrane to both these ions is greatest. In this state, the circuit diagrams are mirror images of one another, no net current flows across the cell, and the transcellular potential is 0 mV. (B) At the peak of the AP, the conductance of the membrane to Na^+ (and possibly Ca^{2+}) increases dramatically, and E_{Na} and E_{Ca} is expressed more than E_K and E_{Cl}. The polarity of the $E_{m(inn)}$ battery is now reversed, such that there is a net flow of positive current in the direction of the noninnervated membrane, and a transcellular 150-mV potential results. Given that R_{cyt} and $R_{m(non)}$ are negligible compared to $R_{m(inn)}$, the equivalent circuit can be reduced to the "electrocyte unit" on the right.

lent circuits for both membranes are mirror images of one another, both having a symmetrical outward movement of current that cancels to give a transcellular potential of 0 mV. At the peak of the AP, however, the permeability of the innervated membrane to Na^+ increases dramatically, such that the membrane potential is influenced primarily by E_{Na}. The composite E_m of the innervated membrane has now reversed its polarity, so that current across this membrane moves inward (Fig. 4B, right). Because the noninnervated membrane has no Na^+ conductance, the polarity of the potential here is the same as at rest, and a net outward current continues to flow. Now, the total driving force for both membranes is in the same direction, so that current flows from the innervated membrane to and through the noninnervated membrane. If one considers that R_{cyt} and $R_{m(non)}$ are negligible compared to $R_{m(inn)}$, the equivalent circuit for the electrocyte can be reduced to a resistor and battery in series, where the resistance is equal to $R_{m(inn)}$ and the battery represents the composite potentials of $E_{m(\text{inn})} + E_{m(\text{non})}$. During stimulation, the potential across this unit, then, equals the transcellular potential and has a magnitude equal to that of the amplitude of the innervated membrane's AP, or approximately 150 mV.

In the electric organ of *Electrophorus*, electrocytes are stacked one after another in very long columns. Electrically, this arrangement is represented by many electrocyte resistor-battery units connected in series, as shown in Fig. 5A. Each unit contributes an additional 150 mV to the overall electrical discharge. However, Kirchhoff's first law dictates that the current measured at every point along an unbranching leg of a circuit, such as a series of batteries, is constant.[2] In other words, the value of the overall current output of a column of electrocytes does not depend on the number of cells in series, but on the electrocyte that has the greatest resistance to the flow of current. (For a comprehensive description of Ohm's law and Kirchoff's laws applied to biological equivalent circuits, see Sperelakis, 1979.) In order for an electric organ to increase the current of an electrical discharge, it must have additional electrocytes arranged in parallel. This amplification is accomplished in electric fish by having numerous electrocyte columns situated alongside one another. The electric eel, being a long slender animal, has fewer electrocyte columns arranged in parallel relative to some other electric fish. Because of this anatomical arrangement, the eel produces discharges of very high voltage with less current compared to other electric fish, such as *Torpedo*, the electric ray.

The electric ray is a marine elasmobranch with two large electric organs positioned laterally on either side of its flattened head (Fig. 6). Hexagonal-shaped columns of electrocytes run in the vertical direction, and conduct current up away from the ocean floor, around the edge of the lateral fin and back into the bottom of the electric organ. *Torpedo* has a short but wide electric organ accommodating large numbers of electrocyte columns situated next to one another. As shown in Fig. 5B, Kirchoff's law predicts that electrocyte resistor-battery units in parallel will produce electrical discharges of low voltage and high current, proportional to the total number of electrocyte units in parallel. This is indeed the case with the electric ray, in which discharges of up to 16 A and 60 V, totaling up to 1 kW, have been measured (Grundfest, 1960).

IV. Comparative Physiology of *Electrophorus* and *Torpedo*—Models for Mammalian Excitable Cells

A. *Electrophorus*

1. Na⁺ Channel

Because electric tissue is specialized for membrane excitability and carries out its function with membrane proteins common to mammalian tissues, electrocytes of electric fish provide excellent models for mammalian excitable cell membranes. Compared to electrocytes of *Torpedo* and other electric fish, *Electrophorus* electrocytes express more of the membrane proteins common to mammalian excitable tissues and therefore provide a more general model for excitable membranes. For example, both voltage-gated Na⁺ and AChR-mediated currents have been measured from the innervated membranes of these cells. Early preparations took advantage of the large size of these cells by sealing single electrocytes over windows in Lucite chambers (Schoffeniels, 1961). Because the resistance of the noninnervated membrane is negligible relative to the innervated membrane, electrocytes in this configuration are treated as a single membrane without having to thread a space clamping electrode down the middle of the cell. With this method, Nakamura *et*

---

[2] Specifically, Kirchoff's first law states that the current entering a point along a circuit is equal to the sum of the currents of all the branches leaving that point. Therefore, if the circuit is unbranching, then the current at every point is constant.

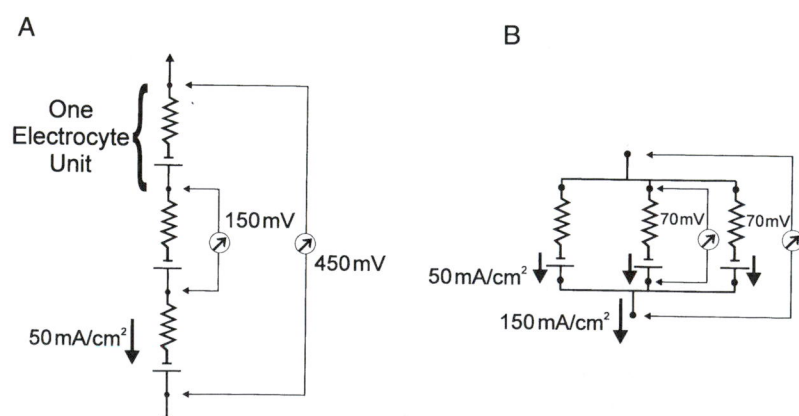

FIGURE 5.   Electrocyte units in series and in parallel. (A) When connected in series, the potentials of each 150-mV electrocyte unit summate, while the current remains constant. In this case, three electrocyte units are shown to summate their potentials to yield a total of 450 mV. This tends to be the case in the electric organ of *Electrophorus,* where transcellular potentials of 150 mV and currents of 50 mA/cm² are measured. (B) The currents of electrocytes in parallel summate while the potential remains constant. Three electrocyte units, each having a current of 50 mA/cm², are shown to produce a 150 mA/cm² current when arranged in parallel. This is the case in *Torpedo,* where many electrocytes are situated next to one another in parallel. The transcellular potential produced by *Torpedo* electrocytes is 70 mV.

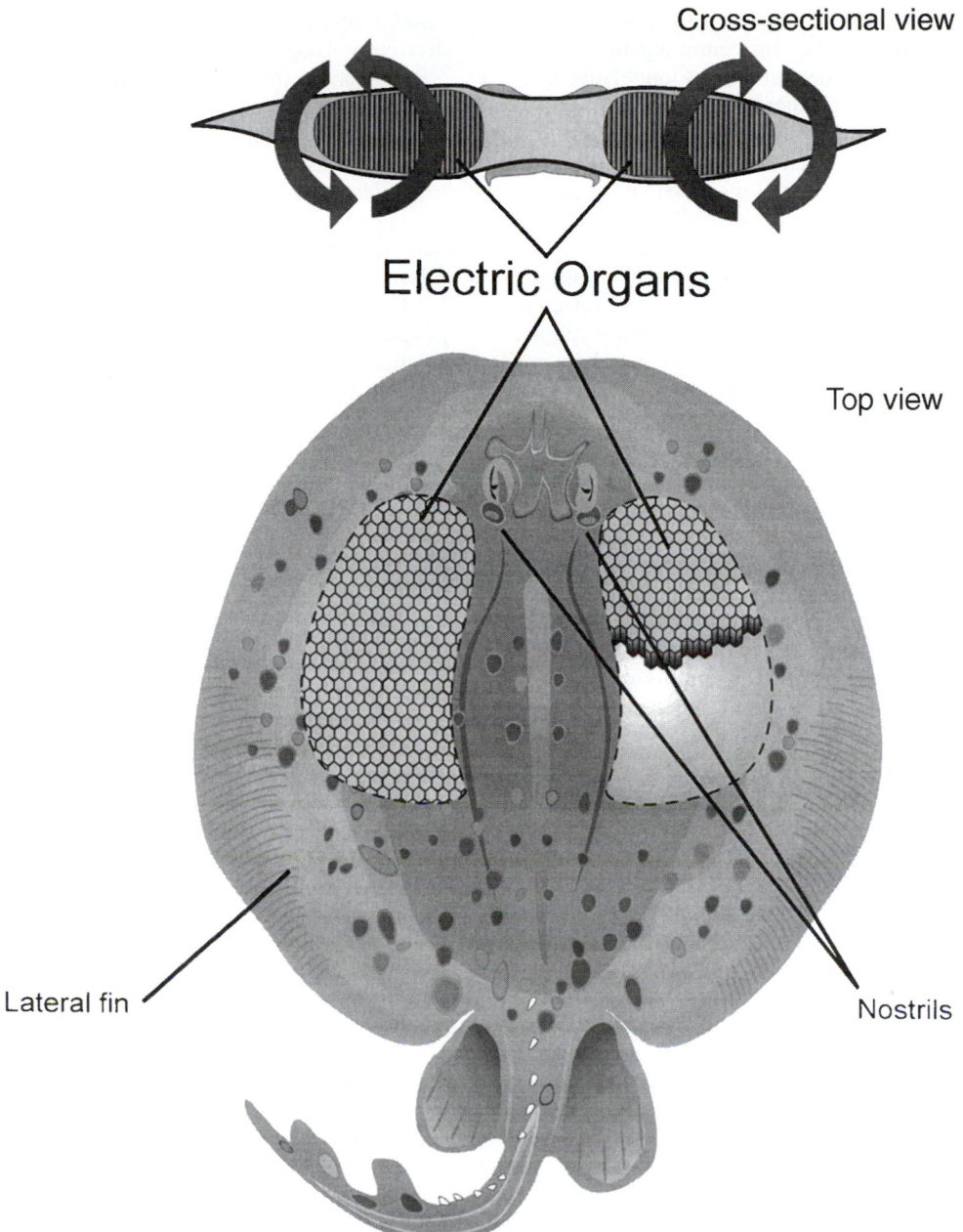

Cross-sectional view

Electric Organs

Top view

Lateral fin

Nostrils

FIGURE 6. Drawing of an electric ray (genus *Torpedo*), depicting the location of its two lateral electric organs. The inset shows the direction of the flow of current around the fish while producing an electrical discharge.

al. (1965) were able to measure Na$^+$ currents responsible for the rising phase of eel electrocyte APs, and to construct a current-voltage (*I-V*) relation for the *Electrophorus* Na$^+$ channel. These researchers also described an abundant inward rectifying K$^+$ current that has an *I-V* relation similar to those seen in myocytes. However, since delayed K$^+$ currents are not seen in these preparations, the repolarization of the eel electrocyte AP is apparently due to Na$^+$ channel inactivation. Much of what is known about the electrogenesis of the AP of electrocytes and other excitable cells has come from studies such as this. The innervated membrane of *Electrophorus* electrocytes was also **patch-clamped** to examine macroscopic Na$^+$ cur-

rents in order to determine minute charge movements associated with channel opening, as well as to determine the permeability of the channel to K$^+$ relative to Na$^+$. Upon membrane depolarization, the Na$^+$ channel undergoes a conformational change that is associated with the movement of positively charged amino acids within the protein. This movement of positive charge is thought to be a consequence of the opening of the channel gate, and can be measured in a population of Na$^+$ channels as a small outward current. Opening of the eel Na$^+$ channel is associated with the movement of approximately 1.5 charges upon channel opening, a value similar to nerve and muscle preparations (Shenkel and

Bezanilla, 1991; for a comprehensive description of gating mechanisms, see Hille, 1992). On examining the selective permeability of the eel Na^+ channel for this ion relative to K^+, Shenkel and Sigworth (1991) found P_{Na}/P_K ratios of 8 to 43. This range represents a substantial variation that was even seen in membrane patches taken from the same cell. Given that *Electrophorus* electrocytes are known to express just one isoform of the channel protein, these results suggested that this variability might arise from posttranslational modifications such as glycosylation or even **phosphorylation**. This possibility was substantiated by experiments that showed the eel Na^+ channel to be modulated by exogenously applied protein kinase A (Emerick *et al.*, 1993; also see Chapter 33 of this book). These studies contributed to our basic understanding of the electrophysiological function of the Na^+ channel protein.

The first Na^+ channel ever purified, and later sequenced, came from electric tissue of *Electrophorus*. The purified 260-kDa protein is heavily glycosylated and consists of a single functional α-subunit (Agnew *et al.*, 1978; Miller *et al.*, 1983). Na^+ channels from mammalian brain and muscle express additional β-subunits. Since it has been suggested that these auxiliary subunits may regulate channel gating, preparations of the eel Na^+ channel are advantageous in that they eliminate the possibly complicating influence of β-subunits. An *Electrophorus* electric tissue cDNA library was used to clone and sequence the channel for the first time (Noda *et al.*, 1984). Its structure includes four homologous repeats that each contain six membrane-spanning helices. The fourth transmembrane segment contains a cluster of positively charged amino acids thought to be involved in gating of the channel. Movement of these positive amino acids during opening of the channel is thought to be responsible for the minute currents measured in patch-clamp experiments such as those discussed earlier. Compared to sequences of Na^+ channels of mammalian muscle and brain, the eel Na^+ channel shows the greatest homology with the muscle protein, as expected since electrocytes develop ontogenetically from myocytes. Both of these channels, however, lack a 202-amino-acid segment located between the first and second homologous repeat domains of the brain Na^+ channel (for complete discussions of Na^+ channel purification, structure, and diversity, see Chapter 27 in this book and Hille, 1992).

2. Acetylcholine Receptor

As mentioned in a previous section, the large size of eel electrocytes has facilitated their dissection in order to measure single-channel AChR conductances. Other excitable cells express numerous isoforms of the subunits that make up the receptor, and produce single-channel recordings that show complex mean open-time distributions and variable conductance values. The eel AChR yields more homogeneous values owing to its simplified subunit composition (Pasquale *et al.*, 1986). The simple mean open-time distributions for the eel AChR were found for both the main electric organ as well as for Sach's organ, suggesting that AChR diversity evolved in order to meet varying physiological needs of myocytes and neurons. The eel receptor also desensitizes to a lesser extent in the presence of sustained concentrations of agonists (Pallotta and Webb, 1980). Simple subunit composition and lack of agonist-

induced desensitization make the eel AchR advantageous for electrophysiological and biochemical studies.

Recognizing the cholinergic nature and the specialization of *Electrophorus* electrocytes, biochemists utilized electric tissue as a source for some of the first purifications of AChRs. Using various separation techniques, including differential centrifugation to separate membrane fractions and affinity chromatography to selectively purify the AChR from other membrane proteins, a 260-kDa macromolecule was isolated (Olsen *et al.*, 1972; Biesecker, 1973). Not knowing *a priori* that the receptor is a pentameric protein made up of α-, β-, γ-, and δ-subunits (in a respective ratio of 2:1:1:1), the initial isolation and identification of the peptides that make up the whole receptor were arduous. After some debate, the 44-kDa α-subunit was established as the ligand-binding portion of the protein. The 50- to 65-kDa β-, γ-, and δ-subunits, along with the α-subunit, are arranged symmetrically around a central axis to make up the ion channel pore of the protein (Karlin and Cowburn, 1973; Chang, 1974).

3. Na^+,K^+-ATPase

As we have seen, electrocytes express massive quantities of membrane receptors and ion channels in order to carry out their specialized function. To maintain resting potentials in the face of currents associated with EPPs and APs that dissipate Na^+ and K^+ gradients, electrocytes need to express large amounts of **Na^+,K^+-ATPase**. For this reason, *Electrophorus* electric tissue has been used as a source for the purification of this enzyme for structure-function studies. The eel protein, like that from other tissues, consists of a 94-kDa α-subunit and a glycosylated 47-kDa β-subunit. During purification, the ATPase is solubilized from electrocyte membranes with various detergents. To analyze its functional characteristics, the protein was reconstituted into liposomes where its ATP-driven translocation of radiolabeled Na^+ and K^+ has been found to be similar to preparations of native electrocyte membranes containing the ATPase (Yoda *et al.*, 1984). These preparations of the eel protein have been invaluable in determining the reaction mechanisms of the ATPase involved in its function. Drugs that target and inhibit different partial reactions of the protein's translocation mechanism have also been investigated and used to examine the pump's function in cell physiology. Drugs, such as the cardiac glycoside digoxin, have also been used clinically specifically to inhibit Na^+,K^+-ATPase function in the heart. This treatment dissipates the membrane Na^+ gradient, which indirectly augments intracellular Ca^{2+} within cardiac myocytes. This results in an increase in the heart's force of contraction, which can alleviate some forms of heart disease. (See Chapter 17; for literature review, see Lingrel and Kuntzweiler, 1994.)

4. Calmodulin

Electrophorus electric tissue also expresses large quantities of the calcium-binding protein **calmodulin**. (For a detailed discussion of Ca^{2+}-binding protein function, see Chapter 10.) In fact, calmodulin makes up roughly 2%, by weight, of electrocyte protein (Munjaal *et al.*, 1986). Once again, electric tissue was used as a source for the purification of this 17-kDa soluble protein (Childers and Siegel,

1975). Unlike the membrane proteins discussed earlier, the function of calmodulin within electrocytes remains elusive, even though some of the functions of this Ca^{2+}-mediator protein in intracellular signaling mechanisms is well documented in other electrically excitable cells. Figure 7 shows the intracellular location of calmodulin within electrocytes. Calmodulin is present throughout the cytoplasm of electrocytes, but is particularly concentrated near both the innervated and noninnervated membranes. In light of this membrane localization, along with the fact that calmodulin is so abundant in this tissue specialized for membrane excitability, a role for this protein in membrane function is likely. Determining the role of calmodulin in electrocyte function will undoubtedly lend insight into the role of this protein in membrane function of other excitable cells.

B. *Torpedo*

1. Comparative Electrophysiology

The marine electric ray (genus *Torpedo* and various species: *marmorata, californica, nobilianae, occidentalis*) can produce high-amperage electrical discharges by virtue of numerous electrocyte columns arranged in parallel, as discussed briefly earlier. In this way it differs from *Electrophorus*, which produces high-voltage discharges owing to numerous electrocytes arranged in series. The basic arrangement of *Torpedo* electrocytes within electric organ columns is remarkably similar to that of *Electrophorus*, considering that these two fish belong to different orders and the existence of electric tissue in both orders of fish represents convergent evolution. Although *Torpedo* electrocytes are smaller and pancake shaped (10–30 μm × 5 mm in diameter) relative to wafer-shaped *Electrophorus* electrocytes, they are still stacked one after another in columns delineated by electrically insulating connective tissue septa. *Torpedo* electrocytes also display membrane polarity similar to that depicted in the diagrams of Fig. 2.

Some basic differences in the membrane biochemistry and electrophysiology exist between electrocytes of these two fishes, however. Like the eel, *Torpedo* electrocytes can be stimulated to produce EPPs in response to nervous stimulation. That is, these electrocytes are chemically excitable. *Torpedo* electrocytes, however, do not fire APs in response to EPPs or artificially applied electrical stimuli and are therefore electrically nonexcitable. This is due to a lack of voltage-dependent Na^+ channels on the innervated membrane that would generate and propagate APs. Instead, these cells are richly innervated by ACh-releasing electromotor neurons, and have an abundance of postsynaptic AChRs. These ligand-gated channels conduct EPPs of 5-ms duration that peak just below 0 mV, halfway between E_{Na} and E_K. Like *Electrophorus* electrocytes, the noninnervated membrane of these cells has a large resting current, partially carried by Cl^-. Transcellular potentials measured across *Torpedo* electrocytes then, have an amplitude equal to that of the EPP, or 70 to 85 mV. The equivalent circuits diagrammed in Figs. 4A and B also apply to these electrocytes, except that the potential produced by each electrocyte unit equals 70 to 85 mV, instead of 150 mV for *Electrophorus* electrocytes.

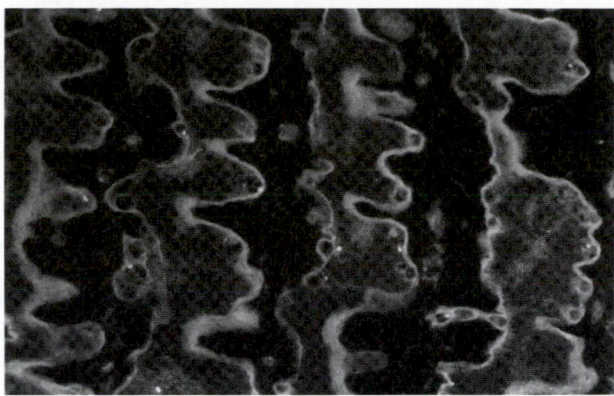

FIGURE 7. Immunofluorescent localization of calmodulin within main organ electrocytes. Paraffin-embedded 4-μm sections were probed with anti-calmodulin sheep antibodies. The location of primary antibodies was visualized with fluorescein-labeled rabbit anti-sheep secondary antibodies, and photographed using epifluorescence microscopy.

2. Acetylcholine Receptor

Without voltage-gated Na^+ channels to propagate an AP, EPPs decay exponentially with the distance traveled from the **electromotor end-plate**. However, very little EPP decay is actually measured on the innervated membrane of *Torpedo* electrocytes, because the density of end-plates is so great. In fact, one might describe the innervated membrane of these cells as one large electromotor end-plate. These cells, therefore, provide a very specialized model for the motor end-plate. Like *Electrophorus* electric tissue, *Torpedo* electric tissue has been used for the purification of the AChR, which has supplied a wealth of knowledge about the biochemical properties of the receptor, as described in the previous section. The first sequences ever to be determined for each of the subunits of the receptor were obtained by screening *Torpedo* electric tissue cDNA libraries (Noda *et al.*, 1982, 1983a,b; Claudio *et al.*, 1983). The mRNAs encoding each of the receptor subunits were together injected into *Xenopus* oocytes in order for the functional protein to be expressed on the membrane of these cells. Interestingly, an increase in the ACh-induced conductance could be measured after microinjection (Mishina *et al.*, 1984). By knowing the sequence of the receptor's subunits, a three-dimensional model of the receptor has been constructed and continually amended in light of ongoing biochemical research. (A model of the AChR appears in Chapter 41.) These studies established the basic protein structure of the nicotinic AChR, the findings of which have only been slightly modified to apply to the receptor of mammalian muscle and nerve.

3. Acetylcholinesterase

AChE is an important enzyme found in the postsynaptic membrane of cholinergic synapses of neurons, motor endplates of myocytes, and the electromotor end-plates of electrocytes. It catalyzes the hydrolysis of ACh to choline and acetate, thereby terminating ligand-gated activation of the AChR. Some pesticides and chemical warfare agents contain **anti-AChE agents** that cause acute and chronic alterations in central nervous system and neuromuscular function. Anti-AChE drugs have also been developed to alleviate symptoms of glaucoma, Alzheimer's dementia, and myesthenia gravis—diseases marked by attenuated postsynaptic AChR density or compromised ACh release (for review, see Millard and Broomfield, 1995). Obviously, great care is needed when administering these drugs because overmedication can cause side effects similar to exposure to harmful anti-AChE agents.

Since the *Torpedo* electrocyte represents an exaggerated cholinergic system, it was used as a source for the purification and subsequent structural analysis of AChE. AChE is an 80-kDa protein that self-associates into tetramers, octamers, and dodecamers, and is anchored in the postsynaptic membrane through a phospholipid linkage (Parker *et al.*, 1978; Ratman *et al.*, 1986). *Torpedo* electric tissue provided massive enough quantities of AChE for the protein to be crystallized for subsequent X-ray diffraction studies. The resulting diffraction pattern obtained from X-rays shot through these crystals was analyzed to construct a three-dimensional structure of the protein, localizing atoms within the enzyme to within 2.8 Å (Sussman *et al.*, 1991). These experiments with the *Torpedo* enzyme will continue to be invaluable to the development of new drugs aimed at the treatment of cholinergic diseases and for therapies for individuals exposed to toxic anti-AChE agents.

4. Cl⁻ Channel

Perhaps the most dramatic contribution that electric tissue has made to recent membrane biochemistry and physiology has been toward elucidating the structure and function of Cl⁻ channels. Recall that in order for electrocytes to produce large transcellular potentials, the noninnervated membrane must have a tremendous resting current. This current clamps the noninnervated membrane potential at highly negative resting potentials even while the innervated membrane depolarizes dramatically. In *Torpedo* electrocytes, the resting current is carried at least partially by Cl⁻. Electric tissue of the electric ray, therefore, has been used to isolate the channel protein for physiological studies, and as a source for mRNA used in cloning and sequencing of the channel.

In the course of developing the planar lipid bilayer method for measuring ion channel conductances, Miller and White (1980) found that vesicles derived from the noninnervated membrane of *Torpedo* electrocytes contained Cl⁻ channels having novel **"double-barreled" gating kinetics**. With depolarization, individual Cl⁻ channel complexes acted as two channels with two separate but equal conductances. When one channel of the complex was open, the other was more likely to be subsequently activated as well. Single-channel recordings showed periods of inactivity until one channel of the complex was opened, after which a second equal conductance would superimpose on the first. Purification of the *Torpedo* Cl⁻ channel confirmed that the protein was a homodimer consisting of two 90-kDa polypeptides. When the purified protein was incorporated into planar lipid bilayers for single-channel recording, the same double-barreled gating kinetics were observed (Middleton *et al.*, 1994).

The first Cl⁻ channel ever to be sequenced came from *Torpedo* electric tissue and has greatly expanded the field of Cl⁻ channel molecular biology. Using the expression cloning technique, the mRNA responsible for the *Torpedo* Cl⁻ conductance was identified, and its corresponding cDNA sequenced. The encoded protein was predicted to consist of 805 amino acids, and to have a molecular weight similar to that of the purified protein (Jentsch *et al.*, 1990). When mRNA for this channel, termed ClC-0, was injected into *Xenopus* oocytes, Cl⁻ conductances having double-barreled gating kinetics were expressed (Bauer *et al.*, 1991). Recognizing that electric tissue is a model for skeletal muscle membranes, Steinmeyer *et al.* (1991b) screened a rat muscle cDNA library with oligonucleotide sequences derived from the *Torpedo* Cl⁻ channel. In this way, the sequence for the major Cl⁻ channel of mammalian skeletal muscle (called ClC-1) was obtained. It was later found that genetic aberrations in the mammalian ClC-1 gene result in symptoms of skeletal muscle myotonia (Steinmeyer *et al.*, 1991a). These findings confirmed the results of Bryant and Morales-Aguilera (1971) that showed this disease to be associated with compromised Cl⁻ conductance (see Chapter 39). Once the sequences were known for both ClC-0 and ClC-1, investigators began screening libraries derived from virtually every mammalian tissue. Numerous Cl⁻ channel sequences have now been determined, and have been implicated in various physiological functions from neuronal membrane excitability to epithelial solute transport (for a review, see Fong and Jentsch, 1995). More Cl⁻ channels having even greater diversity and function will undoubtedly be uncovered in the future.

V. Summary

Both the freshwater electric eel and the saltwater electric ray produce extraordinarily powerful electrical discharges with membrane ion channels, receptors, and pumps common to other excitable cells. These fish have separately evolved a specialized anatomy and cellular morphology designed for this function. Because of their specialized membrane asymmetry, APs and EPPs generated on the innervated membrane are not reproduced on the noninnervated membrane, thereby setting up an asymmetrical flow of current across the cell. The arrangement allows transcellular potentials to be generated, which is essentially the basis for the generation of bioelectricity within the electric organs of these fish. Connective tissue septa that delineate columns of electrocytes prevent transcellular potentials from being short-circuited around the outside of individual electrocytes, and also channel the resulting current along the electric organ.

The membrane potentials used by electrocytes to produce transcellular potentials are remarkably similar to those of other excitable cells, such as myocytes and neurons. The electrophysiology, therefore, can be explained by currents conducted through ligand-gated receptors and channels having known characteristics. Currents that give rise to electrocyte membrane potentials can even be represented by equivalent circuits similar to those of other excitable cells. However, two major differences exist between electrocytes and other excitable cells: (1) Electrocytes express exaggerated amounts of key excitable membrane proteins, such as the Na⁺ channel of

Electrophorus and the AChR of *Torpedo*. These proteins that exist in high density tend to produce greater currents and peak potentials than what is customarily seen on other excitable cells. (2) Membrane proteins are polarized to particular sides of the cell to facilitate the production of transcellular potentials. In the past, however, researchers have taken advantage of these differences to utilize these fish as useful model systems.

Electrophorus electrocytes provide a general model system for excitable cells such as neurons and myocytes, since they contain common membrane receptors, channels, and ATPases. They are also large and easy to dissect in order to perform potential recording, voltage-clamp analysis, and patch-clamp measurements. Since it expresses large quantities of proteins such as the Na$^+$ channel, the Na$^+$,K$^+$-ATPase, AChR, and calmodulin, eel electric tissue has been used as a source for the purification of these proteins for molecular and functional analysis.

Torpedo electrocytes, on the other hand, are richly innervated with ACh-releasing electromotor neurons and are electrically inexcitable. Therefore, they provide a very specialized model for the motor end-plate. Because of the exaggerated cholinergic nature of *Torpedo* electric tissue, it has been used as a rich protein and mRNA source for the AChR and AChE. The expression of Cl$^-$ channels on the noninnervated membrane of these cells has led researchers to use this tissue as a source for ClC-0 protein and mRNA as well.

Investigations with electric tissue of both the electric eel and the electric ray have opened wide avenues of study in electrophysiology, protein biochemistry, and clinical research. Electrophysiological techniques have been used, refined, and in some cases developed while using electrocytes as model systems. Like the squid giant axon, these cells have been instrumental in defining and confirming the ionic currents responsible for excitable cell membrane potential changes. Biochemically, *Electrophorus* and *Torpedo* electric tissue has supplied abundant quantities of key excitable membrane proteins that exist in only trace amounts in mammalian tissues. Since electric tissue develops from skeletal muscle, the biochemical properties and three-dimensional structures of these proteins are similar, if not identical, to those of mammalian skeletal muscle and other excitable cells. These discoveries will continue to further our understanding of the mechanisms by which membrane potentials of excitable cells are generated and regulated, as well as the understanding and the treatment of disease.

Bibliography

Agnew, W. S., Levinson, S. R., Brabson, J. S., and Raftery, M. A. (1978). Purification of the tetrodotoxin-binding component associated with the voltage-sensitive sodium channel from *Electrophorus electricus* electroplax membranes. *Proc. Natl. Acad. Sci. USA* **75**, 2606–2610.

Bauer, C. K., Steinmeyer, K., Schwartz, J. R., and Jentsch, T. J. (1991). Completely functional double-barreled chloride channel expressed from a single *Torpedo* cDNA. *Proc. Natl. Acad. Sci. USA* **88**, 11052–11056.

Bennett, M. V. L. (1971). Electric organs. *In* "Fish Physiology" (W. S. Hoar and D. J. Randall, Eds.), pp. 347–491. Academic Press, New York.

Bennett, M. V. L., and Sandri, C. (1989). The electromotor system of the electric eel investigated with horseradish peroxidase as a retrograde tracer. *Brain Res.* **488**, 22–30.

Biesecker, G. (1973). Molecular properties of the cholinergic receptor purified from *Electrophorus electricus*. *Biochemistry* **12**, 4403–4409.

Bryant, S. H., and Morales-Aguilera, A. (1971). Chloride conductance of normal and myotonic goat fibres and the action of monocarboxylic aromatic acids. *J. Physiol. (London)* **219**, 367–382.

Chang, H. W. (1974). Purification and characterization of acetylcholine receptor-I from *Electrophorus electricus*. *Proc. Nat. Acad. Sci. USA* **71**, 2113–2117.

Childers, S. R., and Siegel, F. L. (1975). Isolation and purification of a calcium-binding protein from electroplax of *Electrophorus electricus*. *Biochim. Biophys. Acta* **455**, 99–108.

Claudio, T., Ballivet, M., Patrick, J., and Heinemann, S. (1983). Nucleotide and deduced amino acid sequences of *Torpedo californica* acetylcholine receptor γ subunit. *Proc. Natl. Acad. Sci. USA* **80**, 1111–1115.

Emerick, M. C., Shenkel, S., and Agnew, W. S. (1993). Regulation of the eel electroplax Na channel and phosphorylation of residues on amino- and carboxyl-terminal domains by cAMP-dependent protein kinase. *Biochemistry* **32**, 9435–9444.

Fong, P., and Jentsch, T. J. (1995). Molecular basis of epithelial Cl channels. *J. Membrane Biol.* **144**, 189–197.

Grundfest, H. (1957). The mechanisms of discharge of the electric organs in relation to general and comparative electrophysiology. *Prog. Biophys.* **7**, 3–74.

Grundfest, H. (1960). Electric organ. *McGraw-Hill Encycl. Sci. Technol.* **8**, 427–433.

Hille, B. (1992). In "Ionic Channels of Excitable Membranes" (B. Hille, Ed.). Sinauer Associates, Sunderland, MA.

Jentsch, T. J., Steinmeyer, K., and Schwarz, G. (1990). Primary structure of *Torpedo marmorata* chloride channel isolated by expression cloning in *Xenopus* oocytes. *Nature* **348**, 510–514.

Karlin, A., and Cowburn, D. (1973). The affinity-labeling of partially purified acetylcholine receptor from electric tissue of *Electrophorus*. *Proc. Natl. Acad. Sci. USA* **70**, 3636–3640.

Keynes, R. D. (1961). The development of the electric organ in *Electrophorus electricus*. *In* "Bioelectrogenesis" (C. Chagas and A. Paes De Carvalho, Eds.), pp. 14–19. Elsevier, New York.

Lester, H. (1978). Analysis of sodium and potassium redistribution during sustained permeability increases at the innervated face of *Electrophorus* electroplaques. *J. Gen. Physiol.* **72**, 847–862.

Lingrel, J. B., and Kuntzweiler, T. (1994). Na$^+$,K$^+$-ATPase. *J. Biol. Chem.* **269**, 19659–19662.

Middleton, R. E., Pheasant, D. J., and Miller, C. (1994). Purification, reconstitution, and subunit composition of a voltage-gated chloride channel from *Torpedo* electroplax. *Biochemistry* **33**, 13189–13198.

Millard, C. B., and Broomfield, C. A. (1995). Anticholinesterases: Medical applications of neurochemical principles. *J. Neurochem.* **64**, 1909–1918.

Miller, C., and White, M. M. (1980). A voltage-dependent chloride conductance channel from *Torpedo* electroplax membrane. *Ann. N.Y. Acad. Sci.* **80**, 534–551.

Miller, J. A., Agnew, W. S., and Levinson, S. R. (1983). Principal glycopeptide of the tetrodotoxin/saxitoxin binding protein from *Electrophorus electricus*: Isolation and partial chemical and physical characterization. *Biochemistry* **22**, 462–470.

Mishina, M., Kurosaki, T., Tobimatsu, T., Morimoto, Y., Noda, M., Yamamoto, T., Terao, M., Lindstrom, J., Takahashi, T., Kuno, M., and Numa, S. (1984). Expression of functional acetylcholine receptor from cloned cDNAs. *Nature* **307**, 604–608.

Munjaal, R. P., Conner, C. G., Turner, R., and Dedman, J. R. (1986). Eel electric organ: Hyperexpressing calmodulin system. *Molec. Cell. Biol.* **6**, 950–954.

Nakamura, Y., Nakajima, S., and Grundfest, H. (1965). Analysis of spike electrogenesis and depolarizing K inactivation in electroplaques of *Electrophorus electricus*, L. *J. Gen. Physiol.* **49**, 321–349.

Noda, M., Shimizu, S., Tanabe, T., Takai, T., Kayano, T., Ikeda, T., Takahashi, H., Nakayama, H., Kanaoka, Y., Minamino, N., Kangawa, K., Matsuo, H., Raftery, M. A., Hirose, T., Inayama, S.. Hayashida, H., Miyata, T., and Numa, S. (1984). Primary structure of *Electrophorus electricus* sodium channel deduced from cDNA sequence. *Nature* **312**, 121–127.

Noda, M., Takahashi, H., Tanabe, T., Toyosato, M., Furutani, Y., Hirose, T., Asai, M., Inayama, S., Miyata, T., and Numa, S. (1982). Primary structure of α-subunit precursor of *Torpedo californica* acetylcholine receptor deduced from cDNA sequence. *Nature* **299**, 793–797.

Noda, M., Takahashi, H., Tanabe, T., Toyosato, M., Kikyotani, S., Furutani, Y., Hirose, T., Takashima, H., Inayama, S., Miyata, T., and Numa, S. (1983a). Structural homology of *Torpedo californica* acetylcholine receptor subunits. *Nature* **302**, 528–532.

Noda, M., Takahashi, H., Tanabe, T., Toyosato, M., Kikyotani, S., Hirose, T., Asai, M., Takashima, H., Inayama, S., Miyata, T., and Numa, S. (1983b). Primary structures of β- and δ-subunit precursors of *Torpedo californica* acetylcholine receptor deduced from cDNA sequences. *Nature* **301**, 251–255.

Olsen, R. W., Meunier, J.-C., and Changeux, J.-P. (1972). Progress in the purification of the cholinergic receptor protein from *Electrophorus electricus* by affinity chromatography. *FEBS Lett.* **28**, 96–100.

Pallotta, B. S., and Webb, G. D. (1980). The effects of external Ca^{++} and Mg^{++} on the voltage sensitivity of desensitization in *Electrophorus* electroplaques. *J. Gen. Physiol.* **75**, 693–708.

Parker, K. K., Chan, S. L., and Trevor, A. J. (1978). Purification of native forms of eel acetylcholinesterase: Active site determination. *Arch. Biochem. Biophys.* **187**, 322–327.

Pasquale, E. B., Udgaonkar, J. B., and Hess, G. P. (1986). Single-channel current recording of acetylcholine receptors in electroplax isolated from the *Electrophorus electricus* main and Sachs' electric organs. *J. Membrane Biol.* **93**, 195–204.

Ratman, M., Sargent, P. B., Sarin, V., Fox, J. L., Nguyen, D. L., Rivier, J., Criado, M., and Lindstrom, J. (1986). Location of antigenic determinants on primary sequences of subunits of nicotinic acetylcholine receptor by peptide mapping. *Biochemistry* **25**, 2621–2632.

Schoffeniels, E. (1961). The flux of cations in the single isolated electroplax of *Electrophorus electricus* (L.). *In* "Bioelectrogenesis" (C. Chagas and A. Paes De Carvalho, Eds.), pp. 147–165. Elsevier, New York.

Shenkel, S., and Bezanilla, F. (1991). Patch recordings from the electrocytes of *Electrophorus*. Na channel gating currents. *J. Gen. Physiol.* **98**, 465–478.

Shenkel, S., and Sigworth, F. J. (1991). Patch recordings from the electrocytes of *Electrophorus electricus*. Na currents and P_{Na}/P_K variability. *J. Gen. Physiol.* **97**, 1013–1041.

Sheridan, R. E., and Lester H. A. (1977). Rates and Equilibria at the acetylcholine receptor of *Electrophorus* electroplaques. A study of neurally evoked postsynaptic currents and of voltage-jump relaxations. *J. Gen. Physiol.* **70**, 187–219.

Sperelakis, N. (1979). Origin of the cardiac resting potential. *In* "Handbook of Physiology." Vol. 1 "The Cardiovascular System" (R. M. Berne and N. Sperelakis, Eds.), pp. 187–267. American Physiological Society, Bethesda, MD.

Steinmeyer, K., Klocke, R., Ortland, C., Gronemeier, M., Jockusch, H., Grunder, S., and Jentsch, T. J. (1991a). Inactivation of muscle chloride channel by transposon insertion in myotonic mice. *Nature* **454**, 304–308.

Steinmeyer, K., Ortland, C., and Jentsch, T. J. (1991b). Primary structure and functional expression of a developmentally regulated skeletal muscle chloride channel. *Nature* **354**, 301–304.

Sussman. J. L.. Harel, M., Frolow, F., Oefner, C., Goldman, A.,Toker, L., and Silman I. (1991). Atomic structure of acetylcholinesteras, from *Torpedo californica*: A prototypic acetylcholine-binding protein. *Science* **253**, 872–879.

Szabo, T. H. (1961). Anatomo-physiologie des centres nerveux specifiques de quelques organes electriques. *In* "Bioelectrogenesis" (C. Chagas and A. Paes De Carvalho, Eds.), pp. 185–201. Elsevier, New York.

Walsh, J. (1775). Experiments and observations on the *Gymnotus electricus*, or electric eel. *Philos. Trans.* **65**, 94–101.

Yoda. A., Clark, A. W., and Yoda, S. (1984). Reconstitution of $(Na^+ + K^+)$-ATPase proteoliposomes having the same turnover rate as the membraneous enzyme. *Biochim. Biophys. Acta* **778**, 332–340.

SECTION
VII

Protozoa and Bacteria

Michael Levandowsky and Thomas E. Gorrell

61

Physiological Adaptations of Protists

I. Introduction

This chapter introduces the reader to the great diversity to be found in the physiology of the **single-celled eukaryotes**, or **protists**. These include a variety of groups, some **autotrophic** or plant-like, some **phagotrophic** or **osmotrophic** and thus animal-like, and many with a combination of these traits.

First, a word about terminology and classification. Traditionally, these organisms comprised the **algae** and the **protozoa**. The early classifications of protozoa divided them into three groups based on their locomotion (**ciliates, flagellates, amoebae**) and a parasitic group (**sporozoa**). The algae were classified largely on the basis of pigments (**red algae, brown algae, golden-brown algae, green algae**) and obvious structural differences (**cryptophytes, dinoflagellates**). Many of the flagellated groups appeared in both classifications. These characteristics are clearly important, and the early classifications and terminology tend to persist in informal usage. However, subsequent work with the electron microscope, biochemical advances, and molecular approaches have changed many of our views on phylogenetic relationships, and revised classification schemes have been proposed (Cavalier-Smith, 1993; Patterson 1994).

In this chapter, the term **Protista**, or simply **protists**, is used to describe all of these groups. Figure 1 gives an informal indication of current thinking about relationships among protistan and other eukaryotic groups, as indicated by molecular evidence, particularly rRNA homology (Sogin, 1989, 1991) as well as ultrastructural information.

Molecular and other evidence suggests that many protist lines may have originated over a billion years ago. Thus, the genetic divergence (in the sense of divergence in nucleic acid sequences) among the various protists is at least comparable to that separating the animal, plant, and fungal kingdoms.

Although they are certainly phylogenetically diverse, the protists do have a certain physiological unity, based on common problems confronted by unicellular organisms.

While, in general, they are typical eukaryotic cells and follow the principles described elsewhere in this book, a number of evolutionary solutions have appeared here that are not present in the cells of multicellular organisms. An example is the **light antenna** ("eyespot") present in many photoautotrophic flagellates. This structure, which appears to have evolved independently in several protist groups, is not found in the multicellular animals and plants.

A feature of protists, just starting to be appreciated, is the apparent ease with which many cells "capture" and use physiological units of other cells. A prime example is the sequestration and use of **prey organelles** such as **chloroplasts** by phagotrophic protists. Indeed, the incorporation of symbionts is a pervasive feature of protistan physiology, occurring in most if not all groups. There are a variety of endosymbiotic phenomena, ranging from the temporary uptake, endocytosis, and use of foreign cells or organelles, to established symbioses in which the endosymbiont has become a required part of the host cell. A dramatic example of this was the appearance of an intracellular bacterial infection in a laboratory culture of *Amoeba*. The infecting bacterium, initially deleterious to the host cell, was transformed from an **endoparasite** to an obligate **endosymbiont** over the course of a few years' culture in the laboratory, and the host and its endosymbiont are now mutually dependent (Jeon, 1987, 1995). This nearly universal tendency of protists to form temporary or permanent endosymbiotic complexes is a feature that the physiologist comes to expect, and a major theme of this chapter.

The relationships shown in Fig. 1 come from molecular and other evidence relating to the basic genome in these groups. However, if one considers plastids and other organelles, it is clear that many of the groups shown as separate are also linked by past endosymbiotic events. Thus, some (but not all) of the dinoflagellates are photosynthetic, and it is clear that their chloroplasts were obtained endosymbiotically from various other protistan sources. Euglenids, on the other hand, though distantly related to the dinoflagellates, apparently got their chloroplasts endosymbiotically from a green

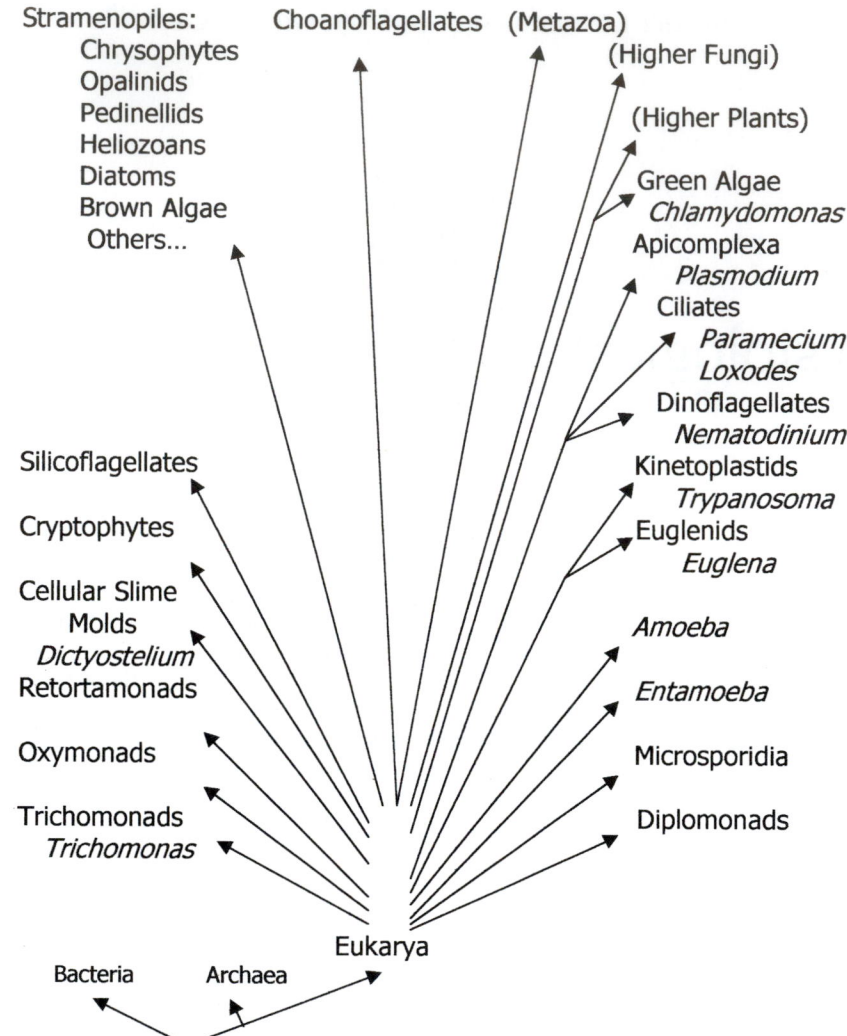

FIGURE 1. An informal diagram showing probable phylogenic relations among some of the more well-known protistan groups, based on molecular and ultrastructural evidence. Many groups are not shown. Genera mentioned in this chapter appear in italics. For detailed descriptions and discussion, see Sogin (1989), Cavalier-Smith (1993), Patterson (1994), or visit http://phylogeny.arizona.edu/tree/phylogeny.html.

algal source. These and other complexities of protistan evolutionary relationships do not appear in simple branching "family tree" diagrams such as Fig. 1.

Also included in this chapter are descriptions of some striking protistan structures whose functions are not clear. Thus, this chapter catalogs the many types of **extrusomes**, microprojectiles that are found in many protistan groups, but whose function is in most cases still a matter of speculation.

A comprehensive review of protistan physiology is not given, because that would duplicate much of the treatment of basic cell physiology in other chapters. Rather, the focus is on unique or unusual features that have evolved in this group, and that are not found in the cells of multicellular organisms. These features are adaptations to the various free-living and parasitic niches filled by this very diverse group of organisms. In particular, emphasis is given here to understudied phenomena.

II. Biophysical Constraints of Scale: The Example of Filter-Feeding

The dimensions of most protists and their appendages, the values of relative velocity in water, and the viscosity of water all combine to yield **Reynolds numbers**[1] much smaller than one. Thus, inertial forces are generally completely excluded as a factor in protistan biophysics. As an illustration of the practical implications of this, **filter-feeding** in flagellates and ciliates is examined. This has been worked

[1]Reynolds number is defined as $R = dv/n$, where d is a spatial dimension, v is velocity, and n is the kinematic viscosity. Essentially, it can be thought of as the ratio of inertial to viscous forces acting on a body moving through a fluid. Thus, for a large, fast-moving organism in water, such as a fish, the ratio is much greater than one and inertial forces dominate. For microbes, on the other hand, the ratio is usually much less than one, and they live in a world dominated by viscous forces.

out theoretically and experimentally by Fenchel (1986a, b) for a number of species, and his discussion is followed here.

Filter-feeding is widespread among protists that consume bacteria, other protists, or other particles. Suspended food particles may be "captured" passively by diffusion, where the particle reaches the protist by chance, through Brownian motion or swimming, and sticks to an adhesive surface. This happens with many protistan cells with amoeboid tentacles (e.g., heliozoa). Particles may be caught raptorially by direct interception and ingestion during active swimming (many dinoflagellates, euglenids, ciliates). Finally, they may be obtained by filtering a current of water, which is often produced by the protist. This latter method is examined here.

Following Fenchel, let $U(x)$ be an **uptake function**, where x is the particle concentration. **Clearance**, F, is defined as the volume of water cleared per unit time. The relation between these two quantities is given by $F = U(x)/x$, where U is assumed to be nearly proportional to x at low particle concentrations, but becomes saturated at high concentrations. It is assumed that the rate of retention of particles by the filter is proportional to concentration, and that ingestion of retained particles takes a finite time, t, during which no other particles are ingested. These assumptions lead to a hyperbolic equation:

$$U(x) = F_m x (1 - Ut) \qquad (1)$$

Substituting variables and rearranging yields the equation:

$$U(x) = U_m x / (U_m / F_m + x) \qquad (2)$$

where $U_m = 1/t$ (the maximum rate of ingestion as $x \to \infty$), F_m is the maximum clearance realized as $x \to 0$, and U_m/F_m is a constant (dimension L^{-3}). From Eq. 2, we can see that the kinetics of filter-feeding is formally identical to the familiar Michaelis-Menten kinetics of enzymology, with the constant U_m/F_m corresponding to the half-saturation constant.

Equation 2 has been verified experimentally for protistan uptake of latex particles and of bacterial food, and F_m varies greatly with the particle size (Fig. 2). When the size for which F_m is maximal is determined experimentally, that size is assumed to be retained with an efficiency approaching 100%; thus, that value of F_m can be taken as a direct measure of the rate of flow of water through the filter. This can then be compared for different flow fields of water and for different species' filter designs.

To understand the biological significance of clearance, it is useful to express it as volume-specific clearance, that is,

$$F_m / (\text{cell volume})$$

As an example of the application of this theory, consider a situation in which the volume fraction of bacteria in a seawater sample is 10^{-6} (e.g., 2×10^6 cells/mL $\times$ 0.5 μm^3 cell volume). A typical heterotrophic flagellate with a specific clearance of 10^5/h (it filters 10^5 times its volume of water per hour) will need 10 h to ingest its own volume of bacteria. With a 50% growth efficiency, it would thus be able to divide every 20 h. Growth efficiency, or **yield**, of protists is nearly invariant over a large range of growth rates (Fenchel, 1986a, b). For balanced (i.e., log phase) growth,

$$u(x) = U(x)Y \qquad (3)$$

where $u(x)$ is the instantaneous growth rate at a food concentration x, and Y is a yield constant. Thus, and as verified experimentally, $u(x)$ and $U(x)$ have similar functional forms, so that growth data can be used to estimate clearance.

Fenchel (1986b) also investigated the flow field generated by various filter-feeding flagellates and ciliates both theoretically and empirically. It is here that the assumption of low Reynolds number is used in calculating the theoretical flow field produced in the medium relative to the cell by ciliary motion. An example is shown in Fig. 3, where the predicted theoretical flow field is compared to the observed paths of suspended particles through the filter apparatus of a ciliate. From these flow lines one can see that a larger volume is sampled than might be expected simply from the area of the filter itself.

III. Nutrition and Excretion

A. Endocytosis, Digestion, and Defecation

As with other cells, protists take in nutrients and other materials by a variety of passive and active mechanisms. Of particular interest here are specialized processes of phagocytosis and digestion not found elsewhere. Many ciliates and some flagellates have complex feeding structures. The best known case is *Paramecium*, the subject of many studies by Allen and Fok (Allen, 1984; Fok and Allen, 1990). This cell has a fixed **cytopharynx** (or **buccal cavity**), a tube-shaped cavity where food particles are taken in. At the end of this is the **cytostome**, where the **food vacuoles** (FV) (also **phagosomes**, or **digestive vacuoles**) are formed. The cytostomal membrane, which pinches off to form the FV, is distinct from the cell membrane antigenically and in structure. During formation of the FV, it grows by fusion with disk-shaped vesicles, which are transported to it along microtubular ribbons.

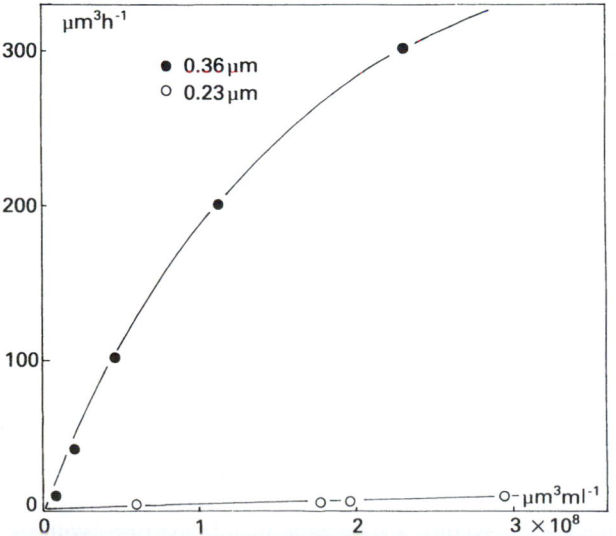

FIGURE 2. Volume uptake of two sizes of latex beads as functions of environmental concentration by *Cyclidium* and with the data fitted to Eq. 2. (Adapted with permission from Fenchel, 1986a, b.)

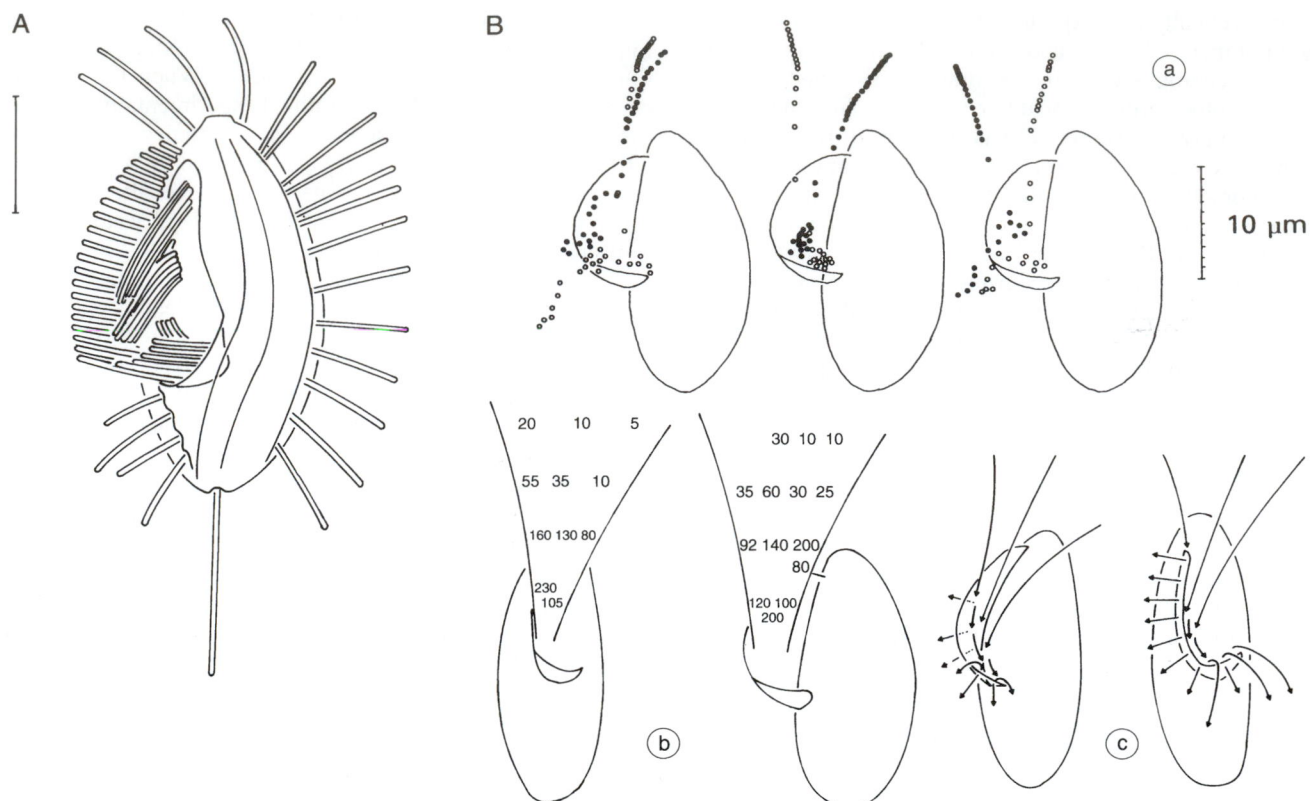

FIGURE 3.   (A) Diagrammatic rendition of *Cyclidium*, showing filtration apparatus. Bar is 5 µm. (B) *a*. Position of 1.1-µm latex particles at 0.02-s intervals along six flow lines during nitration by *Cyclidium*. *b*. Schematic presentation of the critical flow lines and approximate velocities (µm/s). c. Schematic presentation of the flow lines. (Reprinted with permission from Fenchel, 1986a.)

The movement is thought to be due to cytoplasmic dynein in a microtubule-based motor. The vesicles are essentially recycled membrane from earlier FVs, returning to the cytostome by a defined path.

After forming, the FV follows a defined path through the cytoplasm. As it progresses, vesicles called **acidosomes** also move along the cytopharyngeal microtubular ribbons and fuse with the vacuole just after it pinches off, lowering the pH of its interior. Whether the acidosomes are filled with acid or simply deliver proton pumps is not resolved yet. The low pH kills the prey and favors the activity of digestive enzymes. The release of the FV from the cytostome and its subsequent movement involve actin. As the acidosomes fuse with it, the original membrane of the FV pinches off to form small vesicles that recycle back to the cytostome, so that the FV is totally reconstituted. Next, lysosomes migrate from the golgi apparatus and fuse synchronously with the FV, bringing acid hydrolases, including acid phosphatase. The membrane may be completely replaced again at this stage. Digestion takes about 20 m. At the end of this phase, vesicles form and are pinched off from the FV, probably to be used again in lysosomes. The remaining FV membrane binds to microtubules that guide it to the **cytoproct,** or **cytopyge,** where the FV membrane fuses with the cell membrane to form a pore. Its contents (undigested residue), are then egested (**defecation**) and the FV collapses. The cytoproct (cytopyge) is a specific location

where the **plasmalemma,** a layer containing infraciliature and other structures, has a gap allowing the food vacuole to reach the cell membrane and fuse with it. After defecation, the membrane is recycled in discoidal vesicles to the cytostome, completing the cycle. This last step appears to involve an actin-based system that may be calcium regulated.

B. The Contractile Vacuole

Protists without cell walls that live in hypotonic media (freshwater species) have **contractile vacuoles** (CV), which periodically excrete fluid. In the best studied case, the ciliate *Paramecium*, this consists of a central vacuole, a surrounding complex of **ampullae,** and a network, or **spongiome,** of tubules (Fig. 4). Certain tubules in this complex are decorated with peg-like elements that are vacuolar-type proton pumps. These are found in both the cellular slime mold *Dictyostelium* and the ciliate *Paramecium* (Heuser *et al.,* 1993; Allen, 1997).

Excretion occurs through a cycle. During **diastole,** the vacuole forms and grows by the fusion of smooth-membrane vesicles. During this period fluid travels from the spongiome to the CV. In **systole,** the vacuole membrane fuses with the cell membrane at one site to form a pore, and the fluid is excreted. As this happens, the vacuole contracts as its membrane fragments and forms vesicles again. Actin has not been

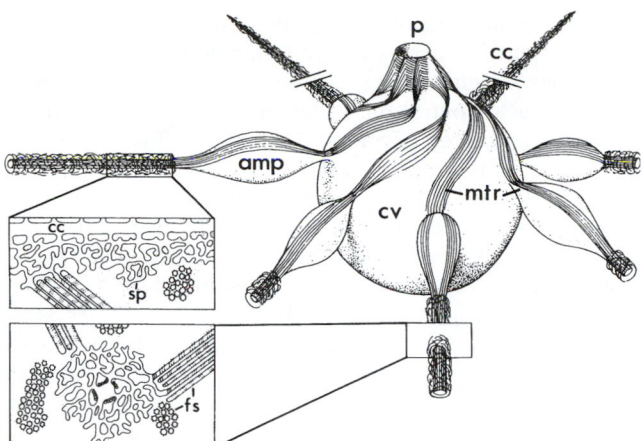

FIGURE 4. Contractile vacuole complex in *Paramecium*, showing ampullae (amp), collecting canal (cc), contractile vacuole (cv), pore (pv), spongiomal tubules (sp), fluid segregation organelles (fs), and microtubular ribbons (mtr). (Adapted with permission from Hausmann and Hülsmann, 1996.)

detected near the CV, and contraction is thought to be due to cellular pressure after the appearance of the pore opening to the exterior. At one time it was thought that the connections of the CV to the spongiome were interrupted during systole, preventing backflow.

Recent work, however, indicates that the connections persist throughout the cycle. The narrowness of the tubular connections presents great resistance, so that considerable pressure would be required for a rapid backflow during systole.

A major function of the CV is clearly osmoregulation. The cycle ceases in cells placed in a hypertonic medium, resuming after a time when the cell adapts and increases internal tonicity. Fluid in the CV is high in K and Na, relative to the cytoplasm, also suggesting a role in maintenance of ionic balance.

IV. Energetic Adaptations: Fermentative Microbodies

Although aerobic protists have mitochondria similar to those in other cells, the earliest protists may have lacked mitochondria (Martin and Müller, 1998). Many modern species, both parasitic and free-living, are either **microaerophiles** or **anaerobes** (facultative or obligate), and specialized organelles have evolved in response to energy requirements in these groups.

Studies of parasitic protists reveal a pair of organelles: **glycosomes** and **hydrogenosomes.** Glycosomes contain enzymes and have been found in the **kinetoplastida,** a group that includes many important parasites (*Trypanosoma, Leishmania*) as well as free-living organisms. Hydrogenosomes are hydrogen-producing organelles found in the parasitic trichomonad flagellates, as well as in certain anaerobic ciliates and fungi.

Both hydrogenosomes and glycosomes belong to the category of organelles called **microbodies.** Most microbodies have a single surrounding membrane and function quite differently from mitochondria and chloroplasts (the "energy organelles"). **Peroxysomes** and **glyoxysomes** are classic examples of microbodies in higher plants and animals, as well as in many protists. Glycosomes and hydrogenosomes, however, perform functions that more closely resemble the energy organelles. Each facilitates formation of adenosine triphosphate (ATP) by a fermentative process. Glycosomes contain the first seven enzymes of glycolysis (Opperdoes, 1987), while hydrogenosomes metabolize pyruvate to H_2, CO_2, and acetate (Müller, 1993). Glycosomes and hydrogenosomes function in separate groups of organisms for energy production and use organic molecules as terminal electron acceptors. ATP formation occurs in these microbodies by a **substrate-level phosphorylation** mechanism, in which a phosphorylated intermediate of glycolysis serves to phosphorylate adenosine diphosphate (ADP). Both types of organelle show unusual aerobic and anaerobic modes of metabolism. Moreover, the glycosomes are found in parasites that have a complex life cycle, including separate stages adapted to life in the bloodstream of vertebrates and in insect tissues.

Both glycosomes and hydrogenosomes are of considerable interest as potential targets for development of new drugs for serious parasitic infections such as African sleeping sickness, Chagas's disease, leishmaniasis, and trichomoniasis. Biogenesis of glycosomes and hydrogenosomes requires that their proteins have a characteristic signal sequence that targets proteins to the organelle.

A. Glycosomes

Glycosomes were initially discovered in 1977 in the bloodstream form of the cattle parasite, *Trypanosoma brucei*, and later in other members of the kinetoplastida. The latter are a group of mostly parasitic flagellates that have a single mitochondrion with a characteristic DNA-enriched structure, the **kinetoplast.** Besides the bloodstream form, trypanosomes can have several other morphological forms, including the **procyclic** form found in the invertebrate host. Procyclics are often used in comparative studies because they are easy to culture. Bloodstream forms (also known as slender forms) can be obtained by harvesting from infected vertebrate hosts, or in recent times by **axenic** culture (pure culture, without other cell types) (Hirumi and Hirumi, 1989).

Information about glycosomal function is derived mainly from studies of overall reactions and the subcellular location of enzymes. The cell membrane of bloodstream forms is dominated by a variable surface glycoprotein, which prevents host antibodies from reaching invariant essential proteins such as those used for glucose transport (Borst and Fairlamb 1998). The plasma membrane of the bloodstream form has a more positive potential than in insect forms, and this potential is more easily depolarized by several ATPase inhibitors. The response of the glycosomal membrane to these conditions is not understood. Much remains to be learned about specific transport mechanisms (Borst and Fairlamb, 1989).

The bloodstream form of the parasite shows high rates of glucose catabolism and enrichment of glycolytic enzymes in

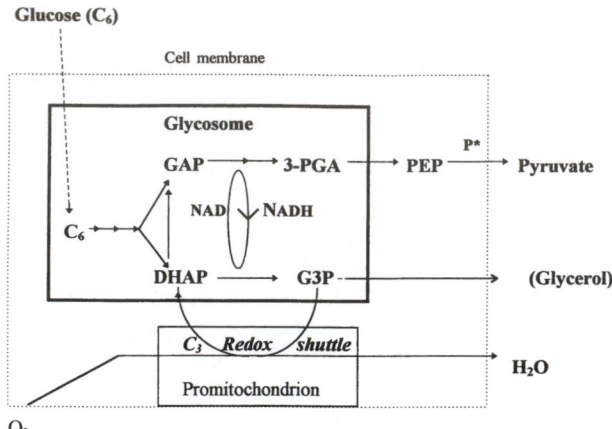

FIGURE 5. Summary of glycosomal metabolism in bloodstream forms of *Trypanosoma brucei*. Redox shuttle: glycosomally produced glycerol-3-phosphate (G3P) is oxidized in the promitochondrion to dihydroxyacetone phosphate (DHAP), and this oxidized three-carbon unit returns to the glycosomes. The P* denotes where net formation of ATP occurs under aerobic conditions. Under anaerobic conditions, glycerol is produced in equal amounts to pyruvate. Additional abbreviations: glyceraldehyde-3-phosphate, GAP; 3-phosphoglycerate, 3-PGA; phosphoenolpyruvate, PEP. (Adapted with permission from Fairlamb and Opperdoes, 1986.)

the glycosome (Opperdoes, 1987). Under aerobic conditions, pyruvate is the sole product of the cells, but oxygen is consumed by the **promitochondria**, a process quite distinct from that found in normal mitochondria (see Chapter 8). Without oxygen, no Pasteur effect is seen (i.e., the rate of glucose consumption does not change), but now glycerol is produced with pyruvate (Fig. 5). Glycerol production occurs through a selective permease in the cell membrane. This process ensures removal of glycerol and thus helps overcome an unfavorable thermodynamic gradient for formation of ATP by a glycerol kinase in the cytosol.

Most interestingly, under aerobic conditions the amount of ATP consumed in activating the six-carbon sugars is balanced by formation of 2 mol of 3-phosphoglycerate (3-PGA). Thus, the glycosome under these conditions does not provide net formation of ATP for the cell, despite the considerable turnover of ATP.

The excretion of pyruvate by the cell leaves unanswered the question of how the glycosome oxidizes reduced nicotinamide adenine dinucleotide (NADH), a rather typical problem for a fermentative pathway. Studies of enzyme activities and their subcellular distribution lead to the conclusion that an unusual C_3 redox shuttle connects the glycosome with the mitochondrion of the cell (see Fig. 5). In the shuttle, dihydroxyacetone phosphate (DHAP) reduction to glycerol-3-phosphate (G3P) results in the regeneration of NAD from NADH for the pathway leading to pyruvate production. The G3P moves to the mitochondrion, where an alternative oxidase regenerates DHAP to resupply the glycosome.

The mitochondrion in bloodstream forms is regarded as a promitochondrion since it lacks several components of the mitochondrion. Bloodstream forms do not obtain ATP by the alternative oxidase.

Glycosomal metabolism is controlled to a large extent by glucose transport across the cell membrane. Under certain conditions of glucose excess, several glycolytic enzymes appear to also control the metabolism of glucose. Entry of glucose into the cell occurs by facilitated diffusion with different transport proteins used by bloodstream forms and procyclic forms. These proteins belong to a superfamily of glucose transporters such as Glut1 of the mammalian erythrocyte, another cell that lacks mitochondria. The significance of the glycosomal membrane is not understood. This membrane presents a significant barrier to metabolites, since treatment of glycosome enriched fractions with triton is required in order to measure the enzyme activities. Glucose and the products of the organelle appear to have free access. An antiport may function to exchange G3P and DHAP, but two antiport transporters cannot be excluded (Borst and Fairlamb, 1998). Pyruvate and glycerol leave the cell by energy-independent permeases, and at least for pyruvate, transport occurs by facilitated diffusion.

While glycosomes do function in bloodstream forms to degrade glucose, several other functions also occur in the organelle, depending on the species and the stage of the life cycle. Procyclic forms rely on respiration by a complete mitochondrion. Glycosomes function in CO_2 fixation. In addition, they show key enzyme activities for both fatty acid oxidation and ether lipid peroxidation. These enzyme activities, along with the presence of catalase (in a few strains) and protein sequences of glycosomal enzymes, suggest a close relationship between glycosomes and peroxysomes (Opperdoes, 1987; Sommer and Wang, 1994).

The glycosome is unique for the presence of the glycolytic enzymes. How this compartmentation benefits the cell is not clear, however. More detailed studies are needed to clarify the transport mechanisms and how the transport processes are regulated, especially during transformation between the vertebrate and insect forms.

B. Hydrogenosomes

Hydrogenosomes have been found in several unrelated groups, including parasitic flagellates and free-living protists (Müller, 1993, Yarlett, 2000). Besides these protozoa, at least one rumen fungus has hydrogenosomes. They were first discovered in 1973 in the cattle parasite *Tritrichomonas foetus* and in the human parasite *Trichomonas vaginalis*. Hydrogenosomes represent about 5% of the cell protein in the trichomonads. No mitochondria are present in these cells, and they lack the ATPase characteristic of oxidative phosphorylation (Müller, 1993).

The organelle has a double membrane and was named for its ability to produce hydrogen gas while oxidizing pyruvate, produced by cytosolic glycolysis (Fig. 6). Hydrogenosomes are morphologically distinct from mitochondria since for most hydrogenosomes no inner membranes are present except for the organelle in a free-living ciliate. Electron microscopy reveals that the organelles elongate and divide, or can be consumed by autophagy (Benchimol *et al.*, 1996). Inhibition of polyamine metabolism in *Tritrichomonas foetus* by the putrescine analog 1,4-diamino-2-butanone results in inhibition of cell growth and a degraded appearance of the

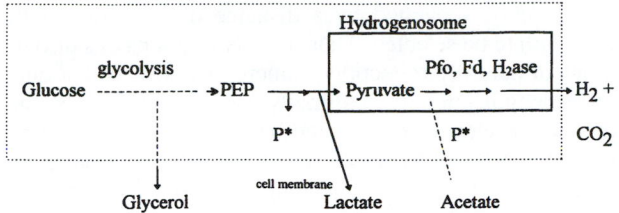

FIGURE 6. Summary of *T. vaginalis* hydrogenosomal metabolism. Gly-colysis occurs by the Embden-Meyerhoff pathway in the cytoplasm. Oxidation of pyruvate occurs in the hydrogenosome by the sequential action of pyruvate: ferredoxin oxidoreductase (Pfo), ferredoxin (Fd), and hydrogenase (H_2ase). Net formation of ATP is designated by P* and phosphoenolpyruvate by PEP. The dashed line indicates that more than one step occurs in the formation of acetate and of ATP. (Adapted with permission from Müller, 1993.)

hydrogenosomes, yet the cells retain motility (Reis *et al.*, 1999).

The ability to produce hydrogen gas is a highly unusual feature for a eukaryote cell. Cells ferment significant amounts of pyruvate through this organelle with ATP formation occurring by a substrate level phosphorylation step (Fig. 6). The iron-sulfur proteins that metabolize the pyruvate are now used in molecular approaches to identify hydrogenosomes when only a few cells can be obtained.

Removal of pyruvate by the hydrogenosome eliminates its use as an electron acceptor for reoxidizing glycolytically reduced NAD. It was discovered recently that *T. vaginalis* oxidizes the NADH under anaerobic conditions by the production of glycerol. In contrast to trypanosomes, trichomonads lose energy by producing glycerol. Trichomonads have a G3Pase rather than a glycerol kinase.

Trichomonads prefer growth under anaerobic or microaerophilic conditions, but do have the ability to consume oxygen (Müller, 1993). Likewise, the isolated hydrogenosomes consume oxygen, but in contrast to whole cells this activity is rapidly lost. The cytosol of trichomonads contains a soluble NADH oxidase that directly reduces oxygen to water; this would protect the oxygen-labile components of the hydrogenosome, such as the two iron-sulfur enzymes. Oxygen does not change the rate of glucose consumption, in contrast to aerobic eukaryotic cells, and cells gain no ATP by oxygen reduction. (Oxygen stops hydrogen production by intact cells, but acetate and CO_2 production continue.)

The presence of the hydrogenosomal pathway is not essential for growth of trichomonads since several strains of *T. foetus* and *T. vaginalis* have been obtained by long-term culture in the presence of metronidazole. This drug is reductively activated to a cytotoxic form by the oxidation of pyruvate. The cells grow at a decreased rate and show an increased rate of glucose consumption, however (Kulda *et al.*, 1993).

Free-living ciliates with hydrogenosomes show an unusual symbiotic relationship by having established methanogenic bacteria as endosymbionts (see Fig. 12 Section V.B). The hydrogenosomes in these ciliates have an unusual membrane since it forms invaginations, and thus the organelle structurally resembles a mitochondrion. The methane production pulls reoxidation of NADH against a standard midpoint potential difference of -100 mV by (presumably) removing H_2 produced via NADH.

The presence of hydrogenosomes in such diverse anaerobic organisms as flagellates, free-living and symbiotic ciliates, and rumen fungi provides for many variations in structure and function. Additional functions for hydrogenosomes, such as accumulation of Ca and formation of a transmembrane potential, have also been reported for some species (Müller, 1993).

The hydrogenosome can be regarded as a fermentative analog of a mitochondrion, since both organelles degrade pyruvate and form ATP as a major function. The hydrogenosome is believed to have originated in an endosymbiotic event. DNA has not been found in hydrogenosomes, except for a report that hydrogenosomes of the ciliate *Nyctotherus* were stained with DNA-specific antibodies (Akhmanova *et al.*, 1998).

It has been suggested that the Eukaryotes arose through symbiotic association of an anaerobic, strictly hydrogen-dependent, strictly autotrophic Archaebacterium host with a Eubacterium symbiont that was able to respire, and which generated molecular hydrogen as a waste product (Martin and Müller, 1998).

V. Sensory Adaptations, Membrane Potentials, and Ion Channels

Like other cells, protists exhibit sensitivity and respond in various ways to environmental stimuli. The basic physiological and biochemical mechanisms, as they are known, are similar to those found in metazoan systems. *Paramecium*, for example, has been referred to informally as a "swimming neuron." In this section several examples of photoreceptors and gravireceptors are examined, and then recent work on the underlying transduction mechanisms is discussed.

A. Photoreceptors

Ambient light is a source of energy to autotrophic protists. In addition, for motile species, it can serve as a directional signal. Many flagellates, and probably all the motile photosynthetic species, exhibit a positive phototaxis. For this purpose a directional receptor is needed. Directional receptors, or "eyespots," fall into two main categories: (1) receptors with opaque screens, which detect direction by the screen's shadow, and (2) receptors with **antennae** based on interference and diffraction to detect direction, essentially the same optical principle as that seen in a reflection hologram. In addition to these directional detectors, one group of predatory dinoflagellates has developed a system of **ocelloids** with intracellular lenses. In this section we discuss the antennae and the ocelloids, two unique but understudied systems.

1. Receptors with Light Antennae

The interferometric **light antenna** is found in a phylogenetically diverse array of flagellates, including many genera of green algae, dinoflagellates, and cryptophytes (Foster and

Smyth, 1980; Melkonian and Robenek, 1984; Smyth *et al.*, 1988). In the cases where experiments have been done, these antennae appear to be selective with regard to wavelength and light direction. The presence of essentially similar organelles in a wide array of groups considered on many other grounds, including molecular homology, to be phylogenetically diverse raises a question: are these all cases of convergent evolution? If not, have these organelles spread through endosymbiotic events?

The antennae operate through an interference mechanism similar to that which produces structural color in iridescent objects. The essential structure consists of a number of parallel reflective layers that act as partial mirrors. These are spaced evenly, separated by a distance of a quarter of the wavelength to be selected. Thus, incident light passes through a series of partially reflecting pigment layers spaced at quarter-wavelength intervals. Normally incident light of the appropriate wavelength is reflected back at each partial mirror. These reflected light rays are in phase with each other and with the incoming radiation, leading to positive reinforcement of this light signal in front of the eyespot, where the receptor pigment is located. The basic principle is illustrated in Fig. 7. Light from other angles, and light of other wavelengths, also passes through this filter, but the reflections are out of phase with each other, and the resultant interference minimizes their contribution.

The most intensive studies of light antennae were done with the green flagellate *Chlamydomonas* (Fig. 8) (Foster and Smyth, 1980), but the antennae are also found in dinoflagellates, chrysophytes, and many other light-responsive flagellates.

2. Intracellular Lenses in Dinoflagellates

Perhaps the most complex and surprising photoreceptors are the ocelloids (ocelli), eyespots found in a family of nonphotosynthetic dinoflagellates, the Warnowiaceae. A typical ocelloid consists of a lens, a retinoid, and an opaque pigment cup (Fig. 9). The pigment cup, or **melanosome**, consists of a wall of pigment granules, and it surrounds the paracrystalline **retinoid**. In front of this is the lens, or **hyalosome**, consisting of a peripheral corneal zone and a central crystalloid body. The entire organelle is approximately 24 μm long and 15 μm wide (Greuet, 1978, 1987).

Francis (1967) measured the refractive index of the lens in *Nematodinium* by three methods, and found it to be approximately 1.52. He determined the focal plane by tracing rays parallel to the optic axis and found that light was focused in the retinoid layer. The field of view in these species was determined to be about 30° but may be wider in other species. From these observations we conclude that the ocelloids could probably act as directional light receptors.

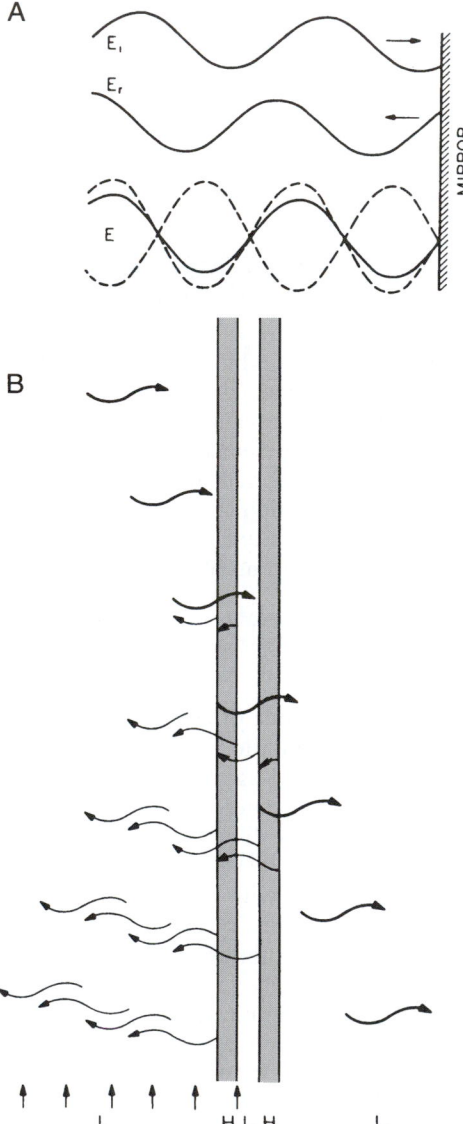

FIGURE 7.   Reflection. (A) Perfect mirror. Incident wave E_i, reflected wave E_r, and the sum $E = E_i + E_r$ a standing wave: the dashed lines show the extrema. (B) Interference reflector. A segment of a wave one wavelength long at successive instants, showing reflections from partial mirrors spaced quarter-wavelengths apart. The reflected waves are in phase. (From Foster and Smythe, 1980.)

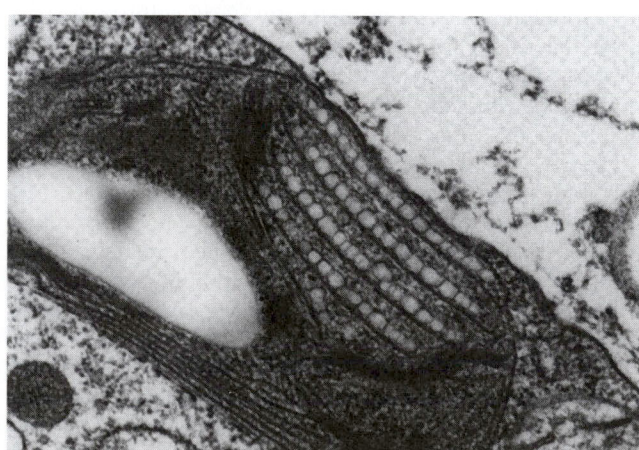

FIGURE 8.   Sectional view of the eyespot of *Chlamydomonas*, showing reflecting pigment layers forming a quarter-wavelength stack. (From Foster and Smyth, 1980.)

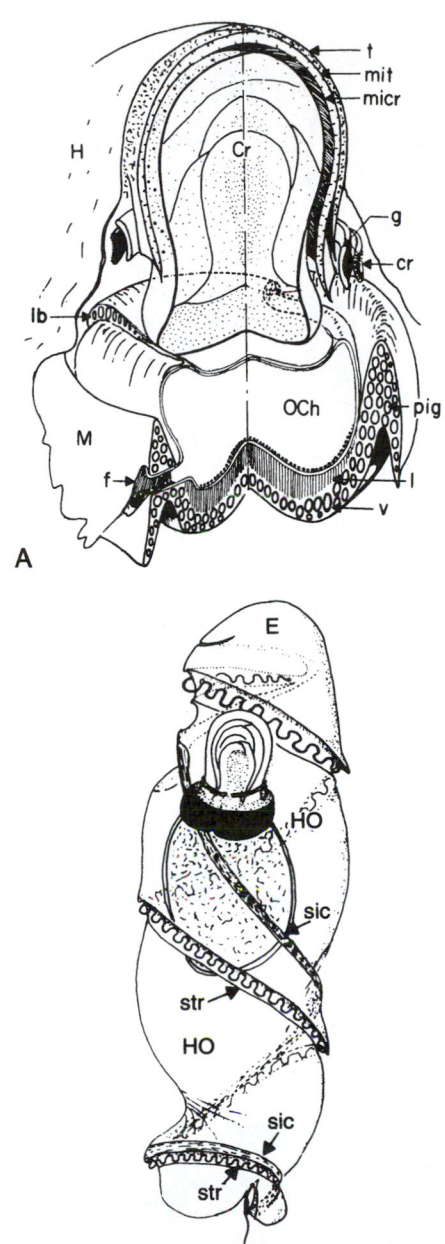

FIGURE 9. (A) Diagram of structure of *Nematodinium* ocellus, showing interior structure: crystalline body (Cr), ocelloid channel (f), constricting ring (cr), periocelloid gallery (g), hyalosome (H), lamellae of retinal body (1), microcrystalline layer (micr), mitochondrion (mit), ocelloid chamber (OCh), pigmentary ring (pig), basal plate (lb), microtubular layer (t), vesicular layer of the retinal body (v). Magnification: 3900X. (B) Position of ocellus in the *Nematodinium* cell, showing parts of the cell: episome (E), hyposome (HO), intercingular sulcus (sic), cingular sulcus with transverse flagellum (str), posterior flagellum (fl). (From Greuet, 1978.)

Given the presence of **nematocysts** (see Section VI.C) in these same dinoflagellate species, and the fact that they are predators rather than autotrophs, it has been suggested that the ocelloids might act as "range finders," leading to discharge of nematocysts when the contrast of the focused image on the retinoid is maximal. While certainly plausible, such behavior has not been reported yet. Unfortunately, these species have not been cultured, and there appear to be no studies of ocelloid function, so that all we have at this time are speculations based on the structure.

B. Gravity Receptors in Ciliates

Many single-celled organisms orient with respect to the earth's gravitational field while swimming. In some, such as the well-studied ciliate *Paramecium*, this is thought to be a purely hydrodynamic effect due to the shape and the distribution of mass in the cell: as the organism swims, the center of mass pulls the rear of the cell downward, orienting it so that it swims upward (Roberts, 1981). In this case, then, there seems to be no need for a gravity receptor. However, many protists can migrate either up or down, switching between a positive and a negative geotaxis in response to external signals or to a circadian or a tidal rhythm, suggesting a more complex type of response.

In particular, a group of ciliates containing the genera *Loxodes* and *Remanella* has a gravity response that depends on the dissolved oxygen level. In nature, these ciliates collect at the interface between anaerobic and aerobic zones in the water column. In oxygen-containing water they swim downward, while in an anoxic environment they swim upward. They all contain a characteristic organelle, the Müller body, which acts as a gravity receptor. This is a fluid-filled vesicle, in which a membrane-covered mineral body is suspended (Fig. 10). The mineral body, consisting of a strontium or barium salt, acts as a statolith, and its position on the interior surface of the vesicle appears to inform the cell of its orientation in the gravitational field (Fenchel and Finlay, 1986). The vesicle is anchored to the system of connected cilia of the cell, and it is suggested, by analogy to other mechanoreceptors, that oriented mechanical stress on the cell membrane at this point of connection may lead to a depolarization, which in turn would affect ciliary beating and determine the orientation of the swimming motion (Fenchel and Findlay, 1986) (see also Section VIII).

C. Sensory Transduction: Membrane Potentials, Ion Channels, and Intracellular Components

1. Membrane Potentials, Calcium, and Behavior

As with many aspects of protistan physiology, membrane potentials are best known in the large ciliate *Paramecium*. A classic feature of ciliate behavior is the **avoidance reaction**, in which the cell stops, swims backward for a short time, then swims forward again, usually in a somewhat different direction. Early workers observed prolonged backward swimming in cells subjected to various stresses: high temperatures, sudden changes in pH or osmotic pressure, exposure to solvents and other deleterious chemicals. The common denominator in all these proved to be that the cell membrane became "leaky" or permeable to Ca^{2+} and other ions. Using the methods developed by Szent-Gyorgy and others for muscle tissues, *Paramecium* cells were exposed to

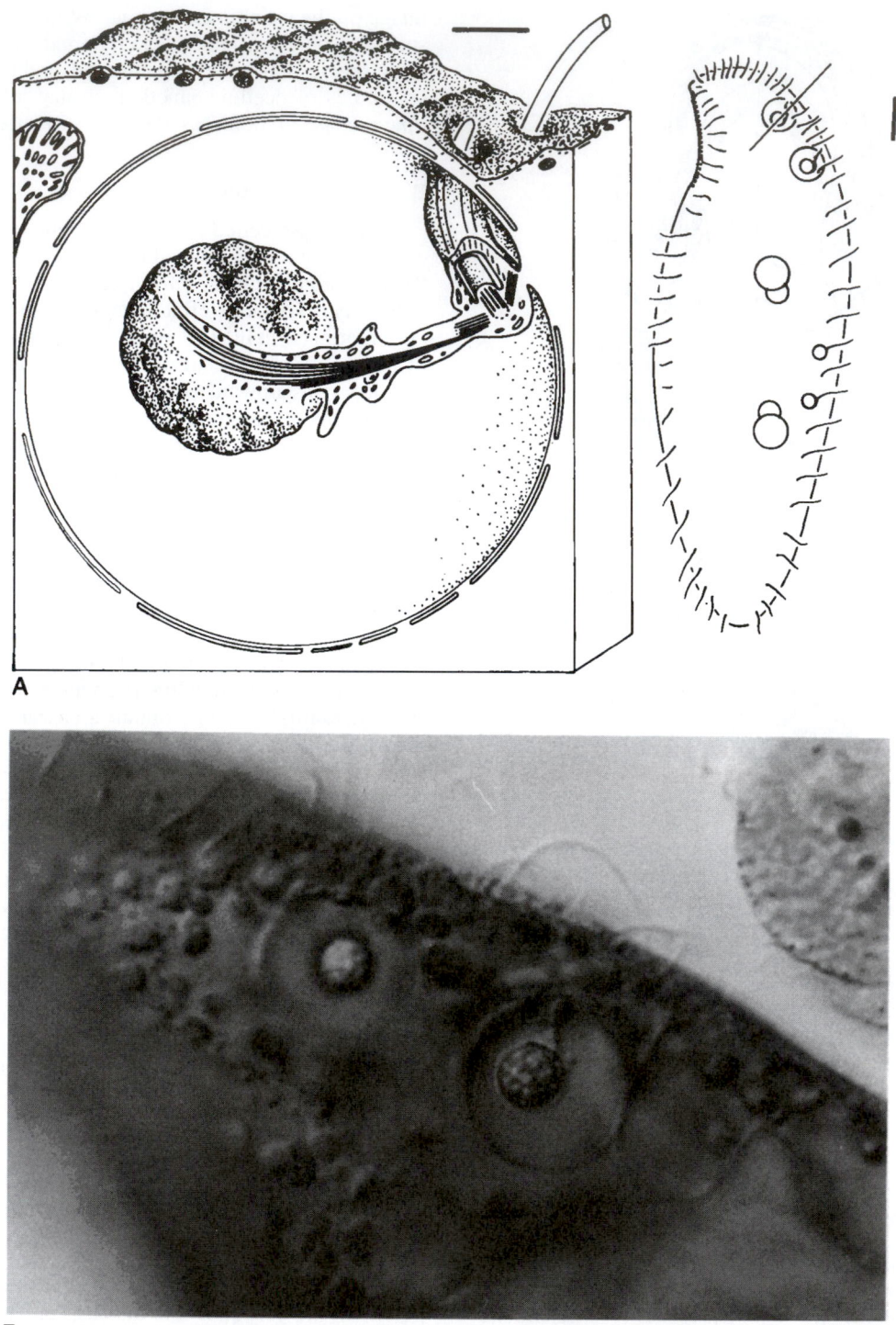

FIGURE 10.   (A) Müller body, gravity receptor. Drawing showing position in *L. striatus* cell (scale bar: 10 μm), and a three-dimensional reconstruction from serial sections (scale bar: 1 μm). (B) Müller body in living cell of *L. striatus*. (From Fenchel and Finlay, 1986.)

detergents such as Triton X-100 to render the cell membrane permeable. Such model extracted-cell preparations would swim forward if provided with ATP and Mg^{2+} in a medium containing $[Ca^{2+}] < 10^{-7}$ M. At higher Ca^{2+} concentrations the cells swam backward (Naitoh and Kaneko, 1973).

Electrophysiological studies with intracellular electrodes in normal cells have given the following general picture: During forward swimming, a Ca^{2+}-dependent membrane potential is maintained. Appropriate mechanical, chemical, or electrical stimuli cause an action potential in which calcium channels open in the membrane, causing depolarization and backward swimming. The membrane potential is restored by a calcium pump after a short time, and forward swimming resumes. Cells are depolarized by repellents (stimuli that cause backward swimming) and hyperpolarized by attractants (stimuli that inhibit backward swimming).

2. Behavioral Mutants

Kung and his colleagues obtained a number of behavioral mutants in which swimming behavior could be correlated with abnormal electrophysiological properties of the *Paramecium* membrane. Examples include the following:

1. *Pawn* (several types), in which voltage-gated Ca^{2+} channels are affected, depolarization does not occur, and the cell cannot swim backward.
2. *Pantophobiac*, where Ca^{2+}-dependent K^+ currents are affected, and prolonged responses occur to all stimuli.
3. *Paranoiac* (several types), affecting Ca^{2+}-dependent Na^+ channels. Prolonged responses occur in Na^+ solutions.

Many other mutants are known (Saimi and Kung, 1987).

3. Ion Channel Types and Membrane Excitation

Electrophysiological and behavioral genetic studies have revealed eight distinct types of ion channels in *Paramecium* and at least three types of membrane excitation that govern behavioral responses, as follows (Hinrichsen and Schultz, 1988):

1. Two mechanically induced currents: a depolarizing, Ca^{2+}-based current in the cell's anterior and a hyperpolarizing, K^+-based current in the posterior.
2. A rectifying K^+ current that can cause regenerating hyperpolarization during the action potential.
3. A second Ca^{2+}-dependent K^+ current, which may play a role in maintaining the resting potential.

4. Second Messengers and Transduction Pathways

Second messengers and internal biochemical events following mechanical or chemical stimuli have been the subject of study in various protists. In the case of the *Pantophobiac* mutants of *Paramecium*, for example, it was found that normal behavior could be restored by microinjection of wild-type calmodulin (CaM). This led to the discovery that these mutants were specific point mutations leading to

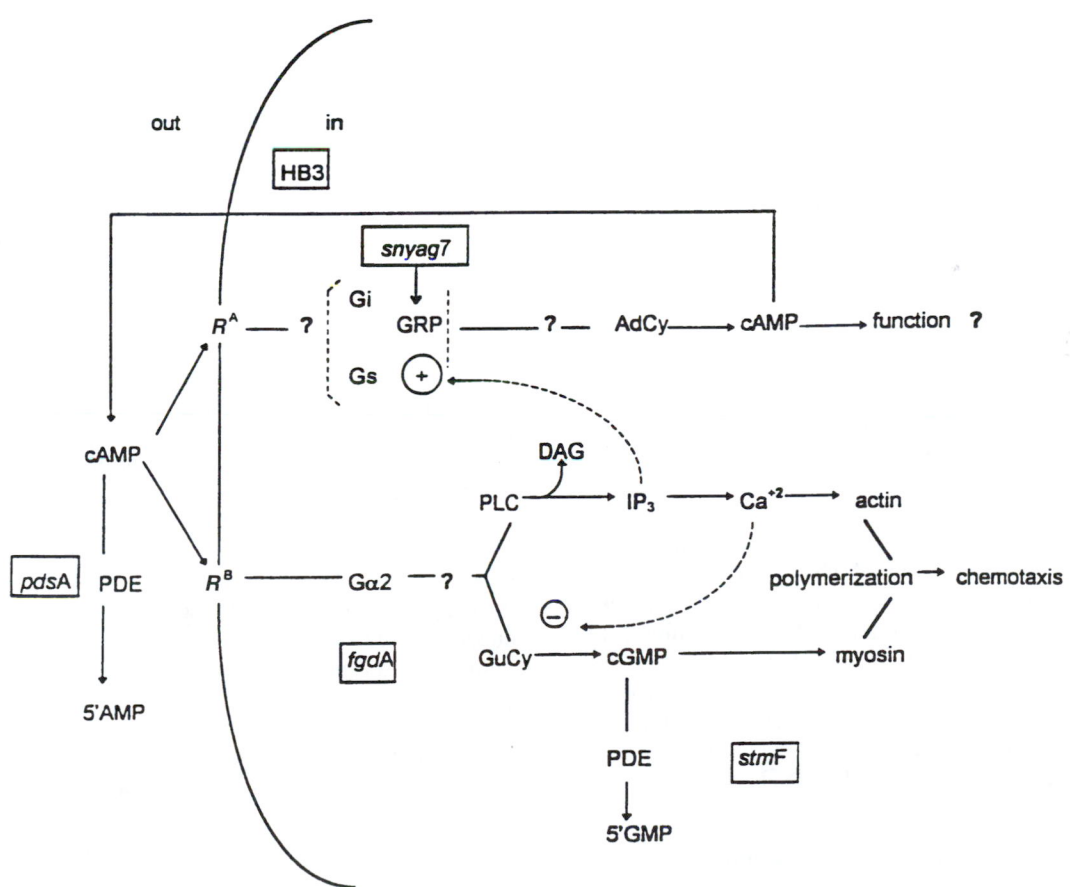

FIGURE 11. Model for sensory transduction in *D. discoideum*. Boxes denote mutants with defects in the nearby step of the pathway. Abbreviations: cAMP, cyclic adenosine monophosphate; cGMP, cyclic guanosine monophosphate; PDE, phosphodiesterase; R^A, R^B, receptors; PLC, phospholipase C; IP_3, inositol trisphosphate; G, G-proteins; GRP, GTP-reconstituting protein; GuCy, guanylyl cyclase; AdCy, adenylyl cyclase; DAG, diacylglycerol. (Adapted with permission from Van Haastert, 1991.)

amino acid substitutions at specific CaM sites. Further studies showed that the Ca^{2+}-CaM complex regulates calcium-dependent Na^+ channels by direct interaction and is also required for the functioning of K^+ channels.

Progress at this level has been greatest with the cellular slime mold *Dictyostelium discoideum*. In this organism there appear to be several chemosensory transduction pathways, similar in general to that found in animal cells such as leukocytes but differing in some aspects. Figure 11 shows a model summarizing current information and theory for this species. Greater diversity is expected to be found as more protistan groups are examined (Van Houten, 1992; Van Haastert, 1991).

VI. Incorporation of Physiological Units from Other Cells

Symbiotic relationships, including cellular endosymbiosis, are widespread in biology. In the protists, this tendency appears to be accentuated, and many new physiological opportunities have been produced by endosymbiotic combinations. Thus, many protists have prokaryotic or eukaryotic endosymbionts. In some cases these associations appear to be more or less permanent, as in the green ciliate *Paramecium bursaria*, which always has an algal *Chlorella* endosymbiont. Sometimes the association is more intimate, and parts of another cell have been permanently incorporated into the host cell. There are also more transient associations, in which the host "cultures" and uses all or part of an ingested cell in its cytoplasm for a period of time. Examples of all of these possibilities are presented in this section.

A. Intracellular Capture and Culture of Foreign Organelles

Early reports of photosynthesis by certain natural populations of chlorophyll-containing ciliates suggested that these supposedly heterotrophic organisms might harbor algal symbionts. Electron microscopy, however, revealed that some ciliates contain isolated fragments of algal cells, which appear to be functional. The photosynthetic marine ciliate *Mesodinium rubrum*, which sometimes forms "red tides" (massive blooms that color the sea), contains endosymbiotic algal organelles in the form of chloroplast-mitochondrial complexes (Taylor *et al.*, 1971). Other species of planktonic ciliates have been found to contain intact algal chloroplasts, pyrenoids, and even eyespots. A significant proportion of the planktonic ciliates in various marine habitats has been found to contain "captured" chloroplasts, which remain functional within the ciliate for extended periods, eventually being consumed by the host (Stoecker *et al.*, 1987).

B. Xenosomes: Bacterial Endosymbionts

The term **xenosome** was used by Soldo (1987) to describe certain bodies in marine ciliates. These were later identified as endosymbiotic bacteria, and Corliss (1985) suggested that the term be defined to include all DNA-containing, membrane-bound intracellular bodies or organisms. By this definition, mitochondria and chloroplasts would be included. In practice, however, the term has been used largely for bacterial endosymbionts of marine ciliates.

1. Bacterial Endosymbionts in Ciliates

Some endosymbiotic bacteria, such as the *Kappa* and *Alpha* particles in *Paramecium* and the *Omicron* particles in Euplotes, are classical objects of study (Görtz 1996). More recently there have been a number of studies by Soldo and colleagues of the xenosomes from marine ciliates (Soldo, 1987; Soldo *et al.*, 1992). Originally found in the species *Parauronema acutum*, they are infective to 12 strains of this species, and also to the phylogenetically distant species *Uronema marinum*, but not to strains from five other marine ciliate genera, or the freshwater genera *Paramecium* and *Tetrahymena*. They are toxic to some marine ciliates, such as *Uronema nigricans*, and the toxic effect is abolished by proteolytic enzymes.

These xenosomes contain DNA, RNA, proteins, and lipids in amounts typical of small bacteria, and are selectively destroyed by a number of antibiotics, including penicillin, ampicillin, tetracycline, and chloramphenical, but not neomycin or cycloheximide. They are small, gram-negative rods, present in numbers ranging from 100 to 200 per host cell, and they divide in synchrony with the host. When released by gentle mechanical rupture of the host, they swim with darting motions or spin like propellers. They average two flagella per cell, inserted at random sites in the cell wall. Within the host cytoplasm they have a fairly typical gram-negative double outer membrane and single inner membrane, with a layer of peptidoglycan between these. They are not enclosed in vacuoles but contact the host cytoplasm directly.

Xenosome chromosomal DNA is multicopy, consisting of 9–14 circularly permuted duplex molecules of about 515 kb (kilobase pairs). In addition there are several plasmids. Analysis of restriction sites revealed that all the adenines of GATC sequences in the plasmids are methylated, whereas those in chromosomal sequences are not, suggesting that there are two replicons, one controlling chromosomal and the other plasmid replication.

Surprisingly, 30% of protein in xenosomes is from the host cell. This is, however, consistent with the relatively small genome. Assuming 1.3 kb of DNA is needed to encode a protein of average M_r 66 000, then the xenosome can encode fewer than 400 proteins, less than one-tenth the number for a typical free-living species such as *Escherichia coli*.

Xenosomes *in vitro* consume oxygen at a very low rate, comparable to that observed in Rickettsia and *Kappa* symbionts, and preferentially use succinate as an energy source. Symbiont-containing ciliates, on the other hand, consume oxygen at a rate 20–30% higher than symbiont-free ciliates, and host glycogen is consumed at a significantly higher rate in the latter than in the former.

The affinities and origins of these organisms present an intriguing problem. Ribosomal RNA analysis of marine ciliate xenosomes indicates some homology to the μ and π

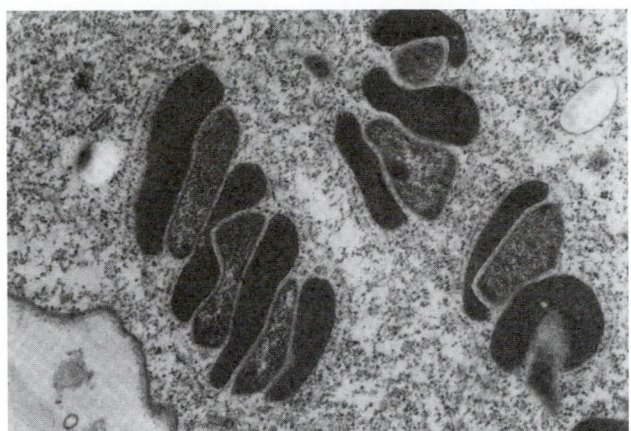

FIGURE 12. Hydrogenosomes and methanogen symbionts in the ciliate *P. frontata*. The darker bodies are the hydrogenosomes and the lighter bodies are the symbionts. Magnification: 20000×. (From Fenchel and Finlay, 1991.)

xenosomes of *P. aurelia*, but little or no homology to known free-living species, suggesting that these endosymbionts may have a long history of association with ciliates (Soldo *et al.*, 1992).

2. Methanogenic Endosymbionts in Anaerobic Ciliates

Most of the free-living anaerobic ciliates examined have endosymbiotic and/or ectosymbiotic bacteria (Fenchel *et al.*, 1977), and the role of some of the former have been studied in some detail. In particular, a number of species harbor methanogenic bacteria. In the cases where these have been investigated, the bacteria are intimately associated with a series of hydrogenosomes. The arrangement of bacteria and organelles has been compared to a stack of coins, with bacteria and hydrogenosomes alternating, and the latter on the ends (Fig. 12). This complex is highly organized in some species, and possibly a permanent feature of the host cell. In the ciliate *Plagiopyla frontata* the bacteria divide synchronously with the host. The number of methanogens per host cell (about 3000) remains constant until a late stage in the host cell cycle, when it doubles. Thus, the ciliate apparently controls reproduction of the symbiont to maintain a stable population density.

When the methanogen inhibitor 2-bromoethanesulfonic acid was used to inactivate the methanogenic symbionts of three ciliate species, *P. frontata*, *Metopus contortus*, and *M. palaeoformus*, growth rate and yield were reduced in the first two species, but not in the last. It is suggested that the energetic advantage conferred by the symbiont in the first two species may be due to the secretion of organic material by the bacteria (Fenchel and Finlay, 1991). The advantage to the methanogens of consumption of H_2 and acetate from the adjacent hydrogenosomes seems clear. This could be a significant advantage, especially in marine or other sulfate-rich environments, where free-living methanogenic bacteria would compete with the more efficient sulfate-reducing bacteria for H_2.

3. Bacterial Endosymbionts in *Amoeba proteus*

Many endosymbionts have been reported in large free-living amoebae, such as *Amoeba proteus* (Jeon, 1995). Particularly dramatic is the case of a bacterial endosymbiont that has been studied from its initial appearance as a contaminant in laboratory culture, through the coevolution of host and endosymbiont, to a mutually dependent symbiotic relationship (Jeon, 1987). The bacterium, which first appeared in the culture in 1966, is termed the **X-bacterium**. It infected a culture of the D strain of *A. proteus*, which already contained other symbiont-like particles of unknown origin. The X-bacterium is described as a gram-negative rod, with an ultrastructure similar to *E. coli*, but is not otherwise identified taxonomically as yet. Initially most of the infected amoebae died, but a few survived, and the X-bacteria gradually lost virulence. The number of bacteria per amoeboid cell, originally greater than 100000, stabilized at about 42000. Within a few years, the host cell, now called the xD strain, became dependent on the presence of the endosymbiont. X-bacteria can be transferred into other D strain cells by microinjection or by induced phagocytosis. The bacteria are enclosed in host-generated vesicles, or **symbiosomes**, and when observed in freeze-fracture preparations are found to be embedded in a matrix of fibrous material. The symbiosomes do not fuse with lysosomes, and during infection the X-bacteria seem to be somewhat resistant to lysozymes, since about 10% of them avoid digestion in the phagolysosomes. Two kinds of plasmids were found in X-bacteria, and isolated X-bacteria treated with ethidium bromide or acridine orange for 3 h failed to infect amoebae.

Several molecules produced by the symbiont have been studied. One, the Xd29 protein, appears to be a peripheral membrane protein that is constantly shed into the host cytoplasm, passing readily through the symbiosome membrane. Symbiont-produced lipopolysaccharides (LPS) have also been identified and were shown by immunostaining to be present on the cytoplasmic side of the symbiosomes. Injected antibodies to the LPS abolished the fusion-avoiding properties of symbiosomes, causing them to fuse with lysosomes. A 96-kDa protein from the symbiont is also present on the symbiosome membrane, and is suspected of playing a role in preventing lysosomal fusion. X-bacteria contain a large amount of 67-kDa heat-shock protein (HSP, GroEx), but since there are no free-living X-bacteria cultures for comparison, it is not known whether this is an indication of stress. In other intracellular infective bacteria (e.g., *Legionella*) the GroEx protein in the intracellular bacteria is more than seven times greater than in free-living cells. The complete nucleotide sequence of the *GroEx* operon of the X-bacteria was determined and it has a high degree of homology with those of other endoparasitic or symbiotic bacteria, such as *Legionella* and *Coxiella*.

Some polypeptide bands detected by gel electrophoresis of the amoebae cytosol are no longer present after prolonged endosymbiosis. One protein that disappears after symbiosis is a SAM (S-adenosyl-L-methionine) synthetaselike protein. A genomic library of *A. proteus* has been prepared in an effort to detect genetic changes as a result of long-term symbiosis with X bacteria.

The following changes are reported in experimentally infected cells:

1. Accelerated cell growth during the initial phase of experimental infection for up to 12 months
2. Increased sensitivity to starvation
3. Newly acquired temperature sensitivity above 26 °C, one degree higher than their optimum growth temperature
4. Increased sensitivity to overfeeding
5. Increased sensitivity to crowding in culture vessels
6. A symbiont-synthesized 29-kDa protein in the host cytoplasm
7. Development of dependence on the symbiotic X-bacteria

Clearly, the analysis of this evolving system is only beginning. With regard to future research directions, Jeon (1995) lists the following unanswered questions:

1. Why are X-bacteria dependent on amoebae?
2. Why are host amoebae dependent on their symbionts for survival?
3. How do symbiosomes avoid fusing with amoebic lysosomes?
4. Are any chromosomal and/or plasmid genes of X-bacteria involved in rendering the bacteria resistant to digestion by lysosomal enzymes?
5. What are the precursor bacteria from which the X-bacteria arose? What are the basic differences between them and the X-bacteria?
6. Is there a permanent genomic change of amoebae caused by symbiosis?

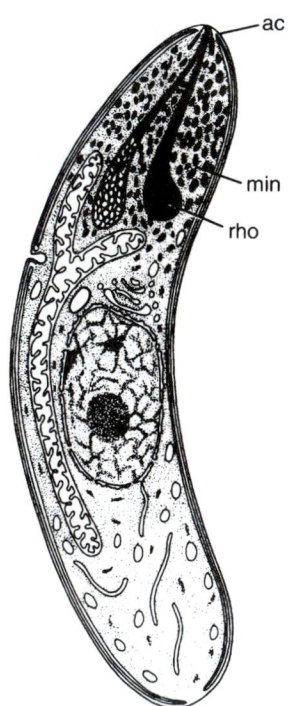

FIGURE 13. Diagram of apicomplexid cell, showing rhoptries (rho), micronemes (min), and apical complex (ac). (Adapted from Chobotar and Scholtyseck, 1982.)

VII. Structures with Unknown Functions

Next, emphasizing the opportunities for future research, a number of more or less prominent structures are examined that are not understood at the basic level of function (though some have been the subject of interesting speculations).

A. Rhoptries

A large group of medically important parasitic protists, the **Apicomplexa** (formerly the **Sporozoa**), is defined by a structure called the **apical complex**. This group includes such important parasites as *Plasmodium*, which causes malaria. All the apicomplexa are obligate intracellular parasites at some stage in their life cycle, and the apical complex is thought to be an instrument of invasion. A prominent part of this complex are the **rhoptries**, secretory organelles containing lipids and proteins (Fig. 13) that originate in the **Golgi system**, filled with enzymes. For many years the general assumption has been that rhoptry and microneme secretions must play a role in the invasion of the host cell, but it has proven difficult to identify actual function (Sam-Yellowe, 1996). In some apicomplexid species the parasite is contained in a **parasitophorous vacuole** after invasion, but in others the vacuole disappears, and the parasite is in direct contact with the host cytoplasm. Rhoptries contain dense protein granules, and epitopes corresponding to these have been identified in the host cell membrane, its cytoskeleton, and also in the parasitophorous vacuole membrane, where present. Their role, however, is not clear.

B. Apicoplastids

Plasmodium and other parasitic apicocomplexids have a structure that appears to be homologous to the chloroplast of green algae (Köhler *et al.*, 1997). An organelle surrounded by four membranes contains a 35-kb circular fragment of DNA, which is shown by cluster analysis to be most closely related to DNA in green algal chloroplastids. Localization of the fragment was done by an *in situ* hybridization technique. The organelle divides by binary fission and is introduced into daughter cells early in replication. The genome is transcribed, and transcription products have been identified. The presence of four membranes enclosing this organelle suggests that it originated endosymbiotically, following ingestion of an algal protist that contained a plastid (Dzierzinski *et al.*, 1999).

The function of this organelle, which has been named the **apicoplastid**, is not known, but it clearly must have a function. The new organelle has been found in all apicocomplexids examined so far, and has generated much excitement as a potential target in the design of new drugs for malaria and other apicocomplexid diseases. Preliminary results with several drugs that inhibit apicoplast replication or metabolism are most promising (Fichera and Roos, 1997; Jomaa *et al.*, 1999).

C. The Paraflagellar Rod

The **paraflagellar rod** (PFR) is found in parasitic kinetoplastids and also in the free-living euglenids and dinoflagel-

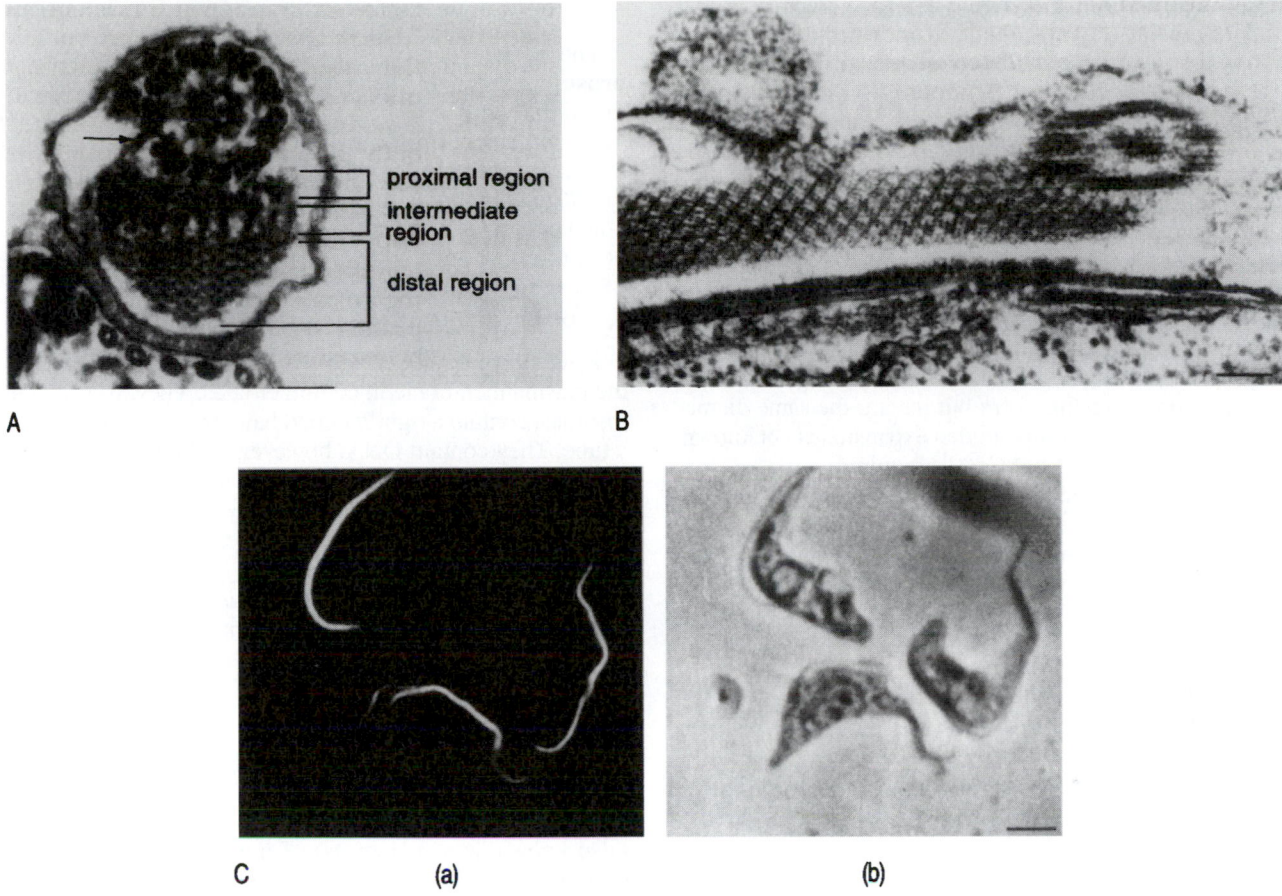

FIGURE 14. (A) Cross-section of the flagellum and paraflagellar rod of *Trypanosoma brucei* (scale bar: 74 nm). (B) Longitudinal section (scale bar: 256 nm). (C) Immunofluorescent staging of the PFR with monoclonal antibody ROD-1. Fluorescence picture (a) and phase image (b) (scale bar: 5.5 μm). (From Sherwin, T. and Gull, K. *Phil. Trans. Roy. Soc. Lond. Ser B* 323: 575–588 (1989).)

lates. The PFR is a complex, lattice-like structure running parallel to the flagellum (Fig. 14). In both euglenids and kinetoplastids there have been many ultrastructural, biochemical, and molecular studies of this organelle (Hyams, 1982; Woodward *et al.*, 1994).

Comparison of proteins from flagella of *Euglena* (with PFR) and *Chlamydomonas* (without PFR) indicated a pair of major proteins as PFR components. Subsequently, similar protein pairs were found in various kinetoplastid species. Cross-reactivity of kinetoplastid and euglenid PFR was established, using antibodies to PFR of the kinetoplastid *Crithidia*. In addition to the two major proteins, a number of minor protein components have been found.

Regarding function, in the euglenids the PFR is located adjacent to the eyespot, and is thought to be the photoreceptor. In the parasitic kinetoplastids, current opinion favors a role in attachment to host cells during infection. Direct evidence for these rather different roles is lacking, however[2] (Bastin *et al.*, 1996).

[2]Interestingly, the only group of kinetoplastids that lacks a PFR is a group of monogenetic parasites that have bacterial endosymbionts of the genus *Bordetella*.

D. Extrusomes

A feature of many protists is the ability to extrude preformed structures. **Extrusomes** are vesicles that contain some organized substance or apparatus, which usually changes its form when released to the exterior in exocytosis. In some cases these organelles are clearly related to the feeding activities of the cell, while in others they may be a defensive adaptation, but in many cases their function is as yet unknown. They are extruded in response to chemical, mechanical, or electrical stimuli. While some metazoa eject structures, such as the **cnidocysts** (also called **nematocysts**) of **Hydra** and other coelenterates, such functions are usually performed by differentiated cells. In the unicellular protists specialized organelles have evolved instead.

The terminology and organization of this section follows that of Hausmann and Hülsmann (1996). The classification is based largely on morphology, and it is not clear whether, for example, **trichocysts** in ciliates and flagellates are homologous or represent parallel evolutionary developments. In the case of the **ejectosomes,** a truly remarkable question of homology arises: the *Kappa* particles in the ciliate *Paramecium* (bacterial endosymbionts, see earlier discussion) contain **R-bodies**, which are morphologically indistinguishable from the ejectosomes of certain flagellates (see later section), and these have even been observed to unroll and form tubes, as ejectosomes do. Did the ejectosome originate as part of a bacterial endosymbiont?

1. Spindle Trichocysts

These are probably the best studied of the extrusomes. Spindle trichocysts of *Paramecium*, a favorite demonstration in introductory biology courses, are found in the cortex of the cell, just under the plasma membrane. In the resting state, the trichocyst is a spindle-shaped or rhomboid paracrystalline protein body. In response to various chemical and/or mechanical stimuli, this unfolds in a few milliseconds to form an expanded, thread-shaped filament, about eight times as long as the former resting form but having the same diameter. The driving force for this sudden expansion is not known. It is independent of ATP, but Ca^{2+} is required.

It is thought that the function of trichocysts is to repel predators, though the evidence for this is not very clear. Trichocysts are found in ciliates, and in dinoflagellates and other flagellate groups (Hausmann and Hülsmann, 1996).

2. Mucocysts

Like trichocysts, **mucocysts** are found just under the plasma membrane, and consist of paracrystalline filamentous bodies. Expansion to the exterior takes place in three dimensions and lasts for several seconds. They are found in various ciliates, flagellates, and the ameboid actinopods, where they may be responsible for the sticky surface used in capturing food organisms. Otherwise, they are thought to have a protective function. They resemble the cortical granules of sea urchin eggs, which are involved in the formation of the fertilization membrane.

3. Discobolocysts

Discobolocysts are found in certain flagellates (**chrysophytes** and others). In the intracellular resting state they are almost spherical, with a disk in the part next to the plasma membrane. On ejection, the disk is unaltered, but the remainder is changed into long filamentous material. The function of these organelles is not known.

4. Toxicysts

These are found in ciliates and some phagotrophic flagellates. They function in the capture of prey, and perhaps in defense against predators. The toxicyst capsule contains a long tube, which during extrusion is either telescoped or everted. This enters the prey and is used to inject a toxin, which kills or paralyzes the prey. In the suctorian ciliates, they discharge on contact, and serve also to hold the prey until it can be taken in and consumed. The nature of the toxin(s) does not appear to have been investigated.

5. Rhabdocysts

These are rod shaped and occur in one group of ciliates (the **karyorelictids**). As with toxicysts, they discharge telescopically—an event that has been compared to the discharge of an arrow from a blowpipe. Their function is unknown.

6. Ejectosomes

These occur in certain flagellates (**cryptophytes** and **prasinophytes**). In the intracellular resting state they are like tightly coiled ribbons. Extruded, they unroll and form very long tubes. This is said to be an escape reaction. The remarkable similarity of this organelle with the R-bodies in the *Kappa* particle endosymbionts of *Paramecium* was noted earlier, and raises the possibility of an endosymbiotic origin for this organelle.

7. Epixenosomes

Epixenosomes occur tightly bound to the outer surface of the plasma membrane in certain ciliates. They, like the ejectosomes, contain a tightly coiled band that unrolls and forms a tube. They contain DNA, however, and thus are epibionts, not organelles, and on the basis of several structural features it has been suggested that they may represent a primitive type of organism somewhere between the prokaryotes and the eukaryotes. In any case they are found in all specimens of different species of the genus *Euplotidinium*, from different geographic regions. Thus, this appears to be a tight, ancient symbiotic relation.

8. Nematocysts

These are found in certain dinoflagellates, and are thought to function in predation, though this has not been observed. As with toxicysts, they strongly resemble the **cnidocysts** (also known as nematocysts) of multicellular coelenterates such as Hydra. They are capsules containing a coiled tube, which evaginates on extrusion.

E. Crystalline Bodies

Many protists contain crystals, usually of calcium and phosphorus salts, and with small amounts of magnesium, chloride, or organic material, in some cases. Classically, these were considered simply waste products, since they seemed to be somewhat dependent on diet. More recently it has been suggested that they may be reservoirs of ions needed in metabolism. They are often formed when the cells are grown in optimal conditions, and when removed experimentally they tend to be replaced quickly.

VIII. Protistan Responses to Gravity and to Gradients of Oxygen and Light: An Example from Physiological Ecology

In this section an example is presented of the integration of protists in their environment, based on some of the physiological capacities discussed in earlier sections. The example is taken from a study of a nutrient-rich pond (Berninger *et al.*, 1986). Figure 15 shows the distribution of two kinds of ciliates in relation to several critical physicochemical factors. The situation depicted is typical of summer when temperatures are high and there is not much wind. In these circumstances, bacterial metabolism depletes oxygen in the lower part of the water column, and in the absence of mixing by wind, the pond

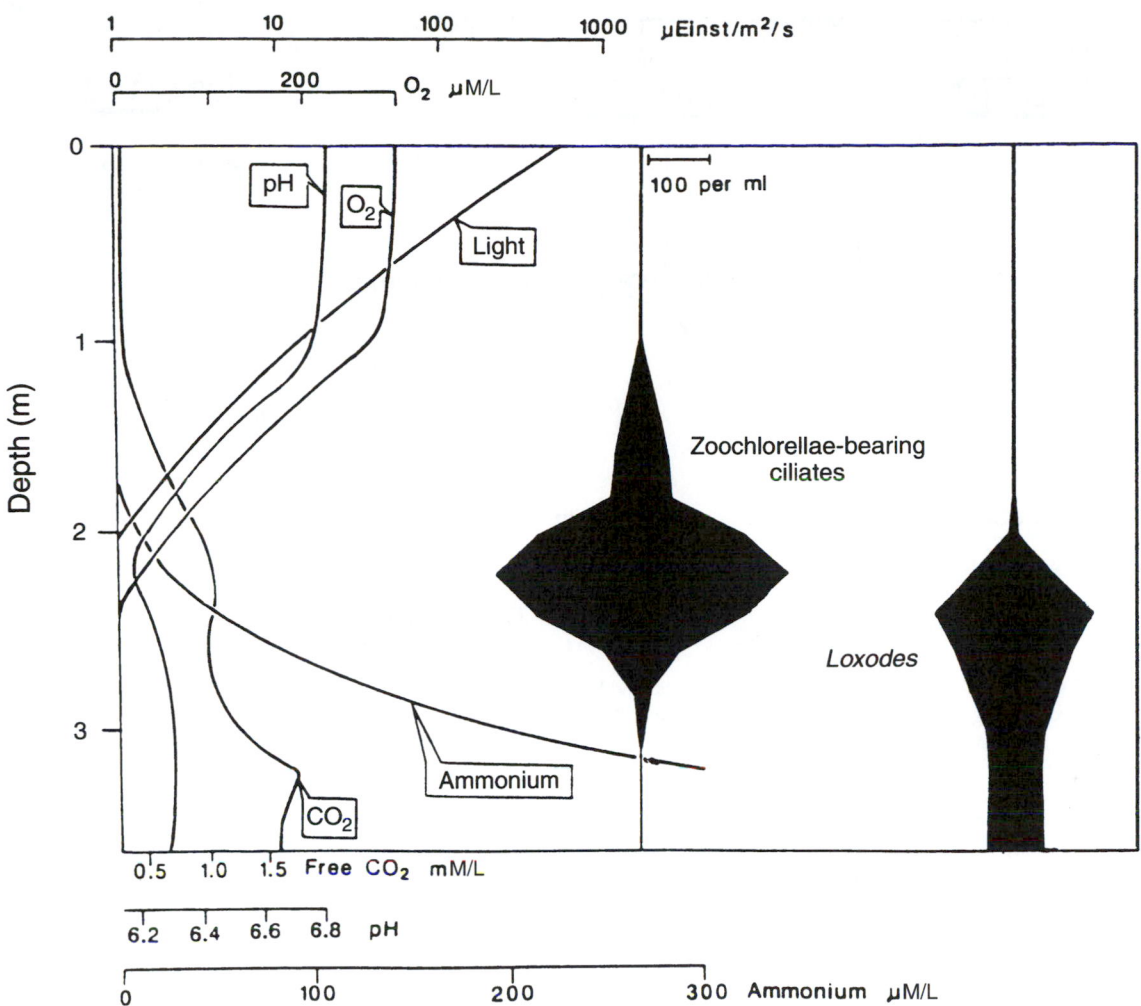

FIGURE 15. Vertical distribution of microaerophilic zoochlorellae (algal symbiont)-bearing ciliates, and the ciliate *Loxodes* in a small productive pond, with profiles of some relevant physical and chemical factors. (Adapted from Berninger *et al.*, 1986, and Fenchel and Finlay, 1986.)

becomes stratified. The depth profiles of two ciliate populations are shown. The first consists of three species of microaerophilic zoochlorellae-bearing ciliate. These remain in a low-oxygen zone, where major predators have difficulty following them. Their lower limit is set by the light requirement of their symbionts, which produce the oxygen that enables them to survive in this zone. The second group consists of a species of the ciliate *Loxodes*, mentioned earlier (Section V.B) in connection with the gravity receptor, or Müller body.

Loxodes responds to three interacting factors: oxygen, blue light, and the force of gravity. As noted earlier, in the dark, cells accumulate in regions of low oxygen tension (ca. 5% of saturation). If then exposed to light, they swim into the dark or into anaerobic water. When exposed to high O_2 and light simultaneously, after initial episodes of backward swimming (the **avoiding reaction**), they swim downward (a **positive geotaxis**). If placed in anaerobic water, especially in the dark, they respond by swimming upward (**negative geotaxis**). The end result of all this is that, in the pond, they accumulate by geotaxis in a zone in the **oxycline** (oxygen gradient) that is optimal. Typically this is in the dark, just

below the ciliates with zoochlorellae symbionts. The cells are sensitive to very low light levels, so that levels as low as 10 W m^{-2} will cause them to swim down into anaerobic water. *Loxodes* can use nitrate as a terminal electron acceptor, which is unusual in ciliates. Meanwhile, by swimming down into the anaerobic zone, the cells avoid most of their potential predators—zooplankters, planktonic larvae, juvenile fish—which are restricted to the aerobic zone.

Figure 16 summarizes what is known or suspected regarding the physiological basis for this adaptive behavior. Obviously, much remains to be done before the physiological basis for some of the arrows in the figure is understood. Photoreception remains to be worked out in detail, but it is clear that a blue light receptor is involved, possibly a flavin, and one product of its excitation in the presence of oxygen is superoxide radical (Finlay and Fenchel, 1986). However, *Loxodes* has only low levels of superoxide dismutase and catalase. Thus, superoxide, or a product of its dismutation, for example, hydrogen peroxide, might be the internal signal for oxygen perception, binding to cytochrome oxidase or perhaps reducing another component of the electron transport system

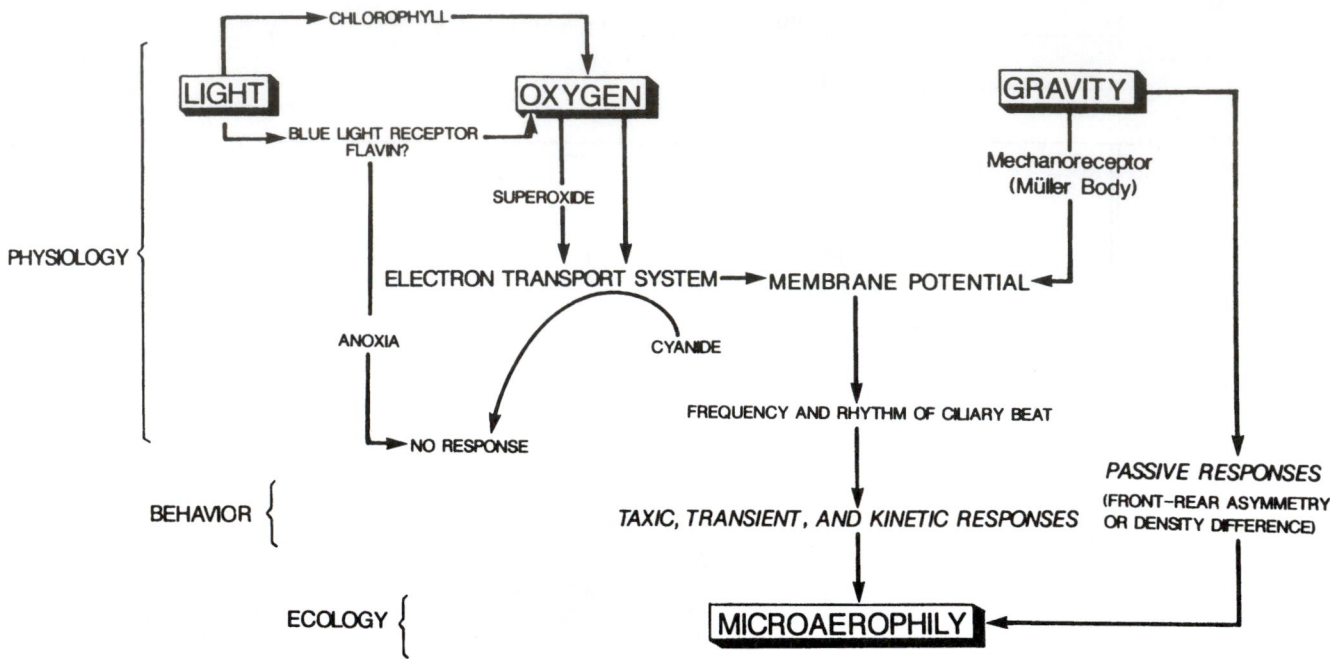

FIGURE 16. Model of the probable physiological and behavioral responses to the three cardinal factors (light, oxygen, and gravity) controlling microaerophily in ciliated protozoa. (Adapted with permission from Finlay, 1990.)

(ETS), such as cytochrome *c*. A drawback of this speculation is that the ETS is in the mitochondrion, while the pigment granules are in the cell membrane. Thus, a change in the ETS might influence the membrane potential of the mitochondrion, but it is not clear how this signal might be transmitted to the cell membrane to influence ciliary beating and change the swimming behavior (Finlay *et al.*, 1986).

In any case the linked responses to light, oxygen tension, and gravity seem very understandable given the ecology of this organism. What is attractive in this system is the possibility of linking the organism's physiology and its ecology.

IX. Summary: Protistan Diversity

The theme of this chapter has been the physiological diversity that has evolved among the many protistan groups, as seen in some of the better-studied examples. However, even these examples are as yet only poorly understood, and many groups remain unstudied or little studied. the diversity of known physiological systems and mechanisms is likely to increase enormously as these groups are explored. Let us consider this at several different levels of organization.

A. Molecular Diversity

There is great diversity at the level of molecular genetics, not treated in this chapter. Ciliates, for example, display a slight variation on the otherwise universal genetic code: the universal stop codons, TAA and TGA, instead code for glutamine in some (but not all) ciliates. (It has been suggested that this may serve as a barrier to viruses, which have not been detected in ciliates.) The presence of two nuclei in cil-

iates is also different: the **micronucleus** serves as the genetic archive, for storage and recombination of the genetic pattern, whereas the **macronucleus** is the regulator of cellular function. In trypanosomes and other kinetoplastids, the **kinetoplast** is a large network of DNA associated with a single giant mitochondrion. This **kinetoplast DNA** (kDNA) consists of a network of about 50 **maxicircles**, which carry the mitochondrial genes, and about 10,000 **minicircles**. It was noticed that many genes in the maxicircles appeared nonfunctional, since they lacked conventional punctuation and contained frameshifts. However, mRNA with functional coding was produced. It turned out that the minicircle DNA was used to produce "guide RNA," which edited the pre-mRNA from the maxicircles. Since this discovery, similar RNA editing has been found in other organisms, including a mammalian nucleus.

Indeed, protists have been a fruitful source of discovery in fundamental biology. Early in the century it was noted that some ciliates could divide approximately 50 times asexually, and then had to undergo sexual genetic recombination, anticipating Heyflick's epochal discovery of a similar mortality in mammalian cell lines by several decades. The mechanism involved here, change in length of the telomeres with each division, was first discovered in a ciliate, *Tetrahymena* (Blackburn, 1992; Blackburn and Greider, 1995). Studies with this ciliate also revealed for the first time that RNA could be an enzyme (Cech, 1987).

B. Organellar Diversity

Figure 1, based largely on nucleic acid homologies, gives a view of the evolutionary sequence in the development of protists, based to a large degree on molecular evidence, especially rRNA homology. However, the ever-branching tree-like structure in Fig. 1 may be misleading. For example, the earliest eukaryotes, possibly represented by certain relict contemporary anaerobic organisms, lacked both mitochondria and chloroplasts, while another early group, the red algae (Rhodophyta), has the latter organelles but lacks the axoneme and thus the eukaryotic flagellum. If one assumes that the absence of the latter organelles is a primitive feature (i.e., not the result of secondary evolutionary losses), then logic indicates that either (1) these organelles evolved more than once or (2) an endosymbiotic event occurred, in which a cell containing mitochondria and chloroplasts combined with one having the eukaryotic flagellum with an axonemal structure. (Note that many biologists think that mitochondria and chloroplasts, and perhaps also flagella, originated from symbiotic prokaryotes.)

As seen in this chapter, though, endosymbiotic events do occur often among protists. Protists are not necessarily prisoners of their phylogenetic past. If a useful invention appears in one evolutionary line, it can cross to another by a symbiotic transfer. Many dinoflagellates are photosynthetic, but ultrastructural evidence indicates that their chloroplasts originated in other evolutionary lines (the cryptophytes and the chrysophytes). The ciliates, distant relatives of the dinoflagellates, also lack chloroplasts, but, as mentioned in this chapter, planktonic species often "capture" and use chloroplasts from their algal prey, and in some cases these symbioses appear to have become more or less permanent.

In summary, Fig. 1 reflects the phylogeny of sRNA and other nucleic acids in the cell, but, in organellar organization, one evolutionary line can borrow or take from another. At this level, a flowchart is more appropriate than a tree.

C. Cellular Diversity

Finally, protistan cells can vary in space and time. Some have complex life cycles, or can form multicellular structures in which there is specialization and differentiation. Perhaps most well known here are the cellular slime molds. At one stage these function as individual small amoebae. Under certain conditions, however, the individual cells send chemical signals to each other and come together to form multicellular "slugs." These move about for a time and then metamorphose into a more or less complex multicellular reproductive structure. Curiously, in one group of ciliates essentially the same thing occurs—presumably a parallel evolution.

This chapter started with the description of filter-feeding by protists—an important ecological process that is illustrative of the problems faced by cells that are also free-living organisms. This was followed by a more detailed discussion of feeding: phagotrophy and digestion in the well-studied case of the ciliate *Paramecium*. Since mitochondria in the protists, though varying somewhat in structure, are essentially similar to those in other cells, their metabolism was not examined here. Instead the unique glycosomes and hydrogenosomes of anaerobic and microaerophilic protists were discussed.

Returning to the view of protists as organisms, several sensory organelles unique to these cells were examined, fol-lowed by a consideration of their sensory transduction. Here there are some exciting similarities and differences with vertebrate and invertebrate neurons and sensory cells. In particular, a calcium potential dominates in those cells that have been studied, and calmodulin-based pathways are prominent aspects of transduction.

The widespread occurrence of endosymbiosis in protists was illustrated with well-documented examples. Next, to emphasize the opportunities for breaking new ground with protistan physiology, a number of striking organelles of unknown function were described. Finally, returning to the original emphasis on the cells as organisms, an example from the physiological ecology of a freshwater ciliate in a high-nutrient freshwater pond was described.

The record suggests that, by studying the diversity of protistan adaptations, cell physiologists can find new answers to old questions, and also new questions.

Acknowledgments

We thank Nigel Yarlett, Miklos Müller, Tom Fenchel, F. J. R. Taylor, and Kenneth Foster for helpful discussions.

Bibliography

Akhmanova, A., Vonken, F., van Alen, T., van Hoek, A., Boxma, B., Vogels, G., Veenhulst, M., and Hackstein, J. (1998). A hydrogenosome with a genome. *Nature* **396,** 527–528.

Allen, R. D. (1984). *Paramecium* phagosome membrane: From oral region to cytoproct and back again. *J. Protozool.* **31,** 1–6.

Allen, R. D. (1997). Membrane tubulation and proton pumps. New ideas in cell biology. *Protoplasma* **189,** 1–8.

Bastin, P., Matthews, K. R., and Gull, K. (1996). The paraflagellar rod of Kinetoplastida: Solved and unsolved questions. *Parasit. Today* **12,** 302–307.

Benchimol, M., Johnson, P. J., and de Souza, W. (1996). Morphogenesis of the hydrogenosome: an ultrastructural study. *Biol. Cell* **87,** 197–205.

Berninger, U.-G., Finlay, B. J., and Canter, H. M. (1986). The spatial distribution and ecology of zoochlorellae-bearing ciliates in a productive pond. *J. Protozool.* **33,** 557–563.

Blackburn, E. H. (1992). Telomerases. *Annu. Rev. Biochem.* **61,** 113–129.

Blackburn, E. H., and Greider, C. W. (Eds.) (1995). "Telomeres." Cold Spring Harbor Laboratory Press, Cold Spring Harbor, New York.

Borst, P., and Fairlamb, A. (1998). Surface receptors and transporters of *Trypanosoma brucei. Annu. Rev. Microbiol.* **52,** 745–778.

Cavalier-Smith, T. (1993). Kingdom Protozoa and its 18 phyla. *Microbiol. Rev.* **57,** 953–994.

Cavalier-Smith, T., and Lee, J. J. (1985). Protozoa as hosts for endosymbioses and the conversion of symbionts into organelles. *J. Protozool.* **32,** 376–379.

Cech, T. (1987). The chemistry of self-splicing RNA and RNA enzymes. *Science* **236,** 1532–1539.

Chobotar, W., and Scholtyseck, E. (1982). Ultrastructure. *In* "The Biology of the Coccidia." (D. M. Hammond and P. L. Long, Eds.), pp. 10–37. University Park Press, Baltimore.

Corliss, J. 0. (1985). Concept, definition, prevalence and host interactions of xenosomes (cytoplasmic and nuclear endosymbionts). *J. Protozool.* **32,** 373–376.

Dzierzinski, F., Popescu, O., Toursel, C., Slomianny, C., Yahiaoui, B., and Tomavo, S. (1999). The protozoan parasite *Toxoplasma gondii* expresses two functional plant-like glycolytic enzymes. Implications for evolutionary origin of apicomplexans. *J. Biol. Chem.* **274,** 24888–24895.

Fairlamb, A. H., and Oppordoes, F. R. (1986). Carbohydrate metabolism in African trypanosomes, with special reference to the glycosome. *In* "Carbohydrate Metabolism in Cultured Cells" (M. J. Morgan, Ed.), pp. 183–224. Plenum, New York.

Fenchel, T. (1986a). "Ecology of Protozoa." Springer-Verlag, New York.

Fenchel, T. (1986b). Protozoan filter feeding. *Progr. Protistol.* **1,** 65–113.

Fenchel, T., and Finlay, B. J. (1986). The structure and function of Müller vesicles in loxodid ciliates. *J. Protozool.* **33,** 69–76

Fenchel, T., and Finlay, B. J. (1991). Endosymbiotic methanogenic bacteria in anaerobic ciliates: Significance for the growth efficiency of the host. *J. Protozool.* **38,** 18–22.

Fenchel, T., and Finlay, B. J. (1995). "Ecology and Evolution in Anoxic Worlds." Oxford University Press, Oxford, UK.

Fenchel, T., Perry, T., and Thane, A. (1977). Anaerobiosis and symbiosis with bacteria in free-living ciliates. *J. Protozool.* **24,** 154–163,

Fichera, M., and Roos, D. (1997). A plastid organelle as a drug target in apicocomplexid parasites. *Nature* **390,** 407–409.

Finlay, B. J. (1990). Ecology of free-living protozoa. *Adv. Microb. Ecol.* **11,** 1–36.

Finlay, B. J., and Fenchel, T. (1986). Physiological ecology of the ciliated protozoon *Loxodes. Rep. Freshwater Biol. Ass.* **54,** 73–96.

Finlay, B. J., Fenchel, T., and Gardner, S. (1986). Oxygen perception and O_2 toxicity in the freshwater ciliated protoozoon *Loxodes J. Protozool.* **33,** 157–165.

Fok, A. K., and Allen, R. D. (1990). The phagosome–lysosome membrane system and its regulation in *Paramecium. Int. Rev. Cytol.* **123,** 61–94.

Foster, K. W., and Smyth, R. D. (1980). Light antennas in phototactic algae. *Microbiol. Rev.* **44,** 572–630.

Francis, D. (1967). On the eyespot of the dinoflagellate, *Nematodinium. J. Exp. Biol.* **47,** 495–501.

Gallo, J. M., and Schrevel, J. (1985). Homologies between paraflagellar rod proteins from trypanosomes and euglenoids revealed by a monoclonal antibody. *Eur. J. Cell Biol.* **36,** 163–168.

Görtz, H. D. (1996). Symbiosis in ciliates. *In* "Ciliates" (K. Hausmann and P. Bradbury, Eds.), pp. 441–462. Gustav Fischer, Stuttgart.

Greuet, C. (1978). Organization ultrastructurale de l'ocelloide de *Nematodinium.* Aspect phylogenetique du photorecepteur de Peridiniens Warnowiidae Lindemann. *Cytobiologie* **17,** 114–136.

Greuet, C. (1987). Complex organelles. *In* "The Biology of Dinoflagellates." (F. J. R. Taylor, Ed.), pp. 119–142. Blackwell, Oxford.

Hausmann, K., and Hülsmann, N. (1996). "Protozoology." Georg Thieme Verlag, Stuttgart, Germany.

Heuser, J., Zhu, Q., and Clarke, M. (1993). Proton pumps populate the contractile vacuoles of *Dictyostelium* amoebae. *J. Cell Biol.* **121,** 1311–1327.

Hinrichsen, R. D., and Schultz, J. (1988). *Paramecium*: A model system for the study of excitable cells. *Trends Neur. Sci.* **11,** 27–32.

Hirumi, H., and Hirumi, K. (1989). Continuous cultivation of *Trypanosoma brucei* bloodstream forms in a medium containing a low concentration of serum protein without feeder cell layers. *J. Parasitol.* **75,** 985–989.

Hyams, J. (1982). The *Euglena* paraflagellar rod: Structure, relationship to other flagellar components and preliminary biochemical characterization. *J. Cell Sci.* **55,** 199–210.

Jeon, K. W. (1987). Change of cellular pathogens into required cell components. *Ann. NY Acad. Sci.* **503,** 359–371.

Jeon, K. W. (1995). The large, free-living amoebae: Wonderful cells for biological studies. *J. Euk. Microbiol.* **42,** 1–7.

Joiner, K. A. (1991). Rhoptry lipids and parasitophorous vacuole formation: a slippery issue. *Parasitol. Today* **7,** 226–227.

Jomaa, H., Wiesner, J., Sanderbrand, S., Altincicek, B., Weidemeyer, C., Hintz, M., Turbachova, I., Eberl, M., Zeidler, J., Lichtenthaler, H., Soldati, D., and Beck, E. (1999). Inhibitors of the nonmevalonate pathway of isoprenoid biosynthesis as antimalarial drugs. *Science* **285,** 1573–1576.

Köhler, S., Delwiche, C. F., Denny, P., Tilney, L. G., Webster, P., Wilson, R. J. M., Palmer, J. D., and Roos, D. S. (1997). A plastid of probable green algal origin in Apicomplexan parasites. *Science* **275,** 1485–1489.

Kulda, J., Tachezy, J., and Čerkasovova, A. (1993). *In vitro* induced anaerobic resistance io metronidazole in *Trichomonas vaginalis. J. Euk. Microbiol.* **40,** 262–269.

Martin, W., and Müller, M. (1998). The hydrogen hypothesis for the first eukaryotes. *Nature* **392,** 15–16.

Melkonian, M., and Robenek, H. (1984). The eyespot apparatus of flagellated green algae: a critical review. *Progr. Phycol. Res.* **3,** 193–268.

Müller, M. (1993). The hydrogenosome. *J. Gen. Microbiol.* **139,** 2879–2889.

Naitoh, Y., and Kaneko, H. (1973). Control of ciliary activities by adenosine triphosphate and divalent cations in Triton-extracted models of *Paramecium caudatum. J. Exp. Biol.* **58,** 657–676.

Opperdoes, F. R. (1987). Compartmentation of carbohydrate metabolism in trypanosomes. *Annu. Rev. Microbiol.* **41,** 127–151.

Patterson, David J. (1989). Stramenopiles: chromophytes from a protistan perspective. *In* "The Chromophyte Algae: Problems and Perspectives," pp. 357–379. Clarendon Press, Oxford. (J. Green, B. Leadbeater and W. Diver, Eds.).

Patterson, David J. (1994). Protozoa: evolution and systematics. *In* "Progress in Protozoology," pp. 1–14. Gustave Fischer Verlag, Stuttgart. (K. Hausmann and N. Hülsmann, Eds.).

Reis, I., Martinez, M., Yarlett, N., Johnson, P., Silva-Filho, F., and Vannier-Santos, M. (1999). Inhibition of polyamine synthesis arrests trichomonad growth and induces destruction of hydrogenosomes. *Antimicrob. Agents and Chemother.* **43,** 1919–1923.

Roberts, A. M. (1981). Hydrodynamics of protozoan swimming. *In* "Biochemistry and Physiology of Protozoa" 2nd ed.. Vol. 4, pp. 5–66. Academic Press, New York. (M. Levandowsky and S. H. Hutner, Eds).

Saimiu, Y., and Kung, C. (1987). Behavioral genetics of *Paramecium. Ann. Rev. Genet.* **21,** 47–65.

Sam-YelIowe, T. Y. (1996). Rhoptry organelles of the apicomplexa: Their role in host cell invasion and intracellular survival. *Parasitol. Today* **12,** 308–316.

Sherwin, T., and Gull, K. (1989). The cell cycle of *Trypanosoma brucei brucei*: timing of event markers and cytoskeleton modifications. *Phil. Trans. Roy. Soc. London Ser B* **323,** 575–588.

Smyth, R. D., Saranak, J., and Foster, K. W. (1988). Algal visual systems and their photoreceptor pigments. *Progr. Phycol. Res.* **6,** 254–286.

Sogin, M. L. (1989). Evolution of eukaryotic microorganisms and their small subunit ribosomal RNAs. *Am. Zool.* **29,** 487–497.

Sogin, M. L. (1991). Early evolution and the origin of eukaryotes. *Curr. Opin. Gen. Devel.* **1,** 457–463.

Soldo, A. T. (1987). *Parauronema* and its xenosomes: A model system. *J. Protozool.* **34,** 447–451.

Soldo, A. T., Brickson, S. A., and Vazquez, D. (1992). The molecular biology of a bacterial endosymbiont. *J. Protozool.* **39,** 196–198.

Sommer, J. M., and Wang, C. C. (1994). Targeting proteins to the glycosomes of African trypanosomes. *Annu. Rev. Microbiol.* **48,** 105–138.

Stoecker, D. K., Michaels, A. E., and Davis, L. H. (1987). Large proportion of marine planktonic ciliates found to contain functional chloroplasts. *Nature* **326,** 790–792.

Taylor, F. J. R., Blackbourne, D. J., and Blackbourne, J. (1971). The redwater ciliate *Mesodinium rubrum* and its "incomplete symbionts": A review including new ultrastructural observations. *J. Fish. Res. Bd. Can.* **28**, 391–407.

Van Haastert, P. (1991). Sensory transduction in eukaryote cells. *Eur. J. Biochem.* **195**, 289–303.

Van Houten, J. (1992). Chemosensory transduction in eukaryotic microorganisms. *Annu. Rev. Physiol.* **54**, 639–663.

Woodward, R., Garden, M. J., and Gull, K. (1994). Molecular characterisation of a novel, repetitive protein of the paraflagellar rod in *Trypanosoma brucei. Mol. Biochem. Parasitol.* **67**, 31–39.

Yarlett, Nigel (2000). Trichomonads. *In* "Encyclopedia of Life Science." Macmillan, London, http://www.els.net. (Printed version to appear in 2001.)

Dennis W. Grogan

62

Physiology of Prokaryotic Cells

I. Introduction

As electron microscopy first revealed structural detail within cells, a distinction between two very different classes of cells emerged. Plant, animal, fungal, and protozoan cells were always found to contain a variety of internal membranous organelles, the most prominent of which was the nucleus. In contrast, corresponding structures were not observed within bacterial cells. Plants, animals, fungi, and protozoa were thus collectively termed **eukaryotes**, that is, having a true nucleus, whereas bacteria were termed **prokaryotes**, reflecting a presumably primitive form of life predating eukaryotic cells.

A summary of prokaryotic physiology must emphasize two seemingly dissonant themes. On one hand, the cellular structures and functions of one prokaryote, *Escherichia coli,* have been defined in such exquisite molecular detail that it has become the most thoroughly understood living organism. On the other hand, this knowledge does not predictably translate into a detailed understanding of other prokaryotic cells because prokaryotes do not constitute an evolutionarily or biochemically coherent group of organisms. Prokaryotic cells arose extremely early in evolution and have diverged to cover a tremendously broad spectrum of metabolic and molecular alternatives. Putting prokaryotic physiology in perspective thus requires an appreciation of the distant relationships among these apparently simple organisms.

The nucleotide sequence of the small-subunit ribosomal RNA provides a highly conserved molecular chronometer that places all cellular organisms into one of three discrete natural groupings. Two of these groupings, the **bacteria** and the **archaea**, consist of prokaryotes that have diverged from each other about as much as either has diverged from eukaryotes. The bacteria consist of several major sublineages (Table 1), but for many purposes they are simply divided into two groups based upon their response to the **Gram stain**. Bacterial species that stain in this procedure are termed **gram-positive** and have a different cell-surface architecture from most **gram-negative** bacteria (Section II. E). The Gram reaction is thus a useful way to distinguish

two fundamentally different cell types among the bacteria. The archaea are typically divided into three groups according to metabolic properties: the methanogens, the extreme halophiles, and the extreme thermophiles. Table 1 illustrates how various groups of bacteria, archaea, and eukaryotes compare to each other by the criterion of ribosomal RNA sequence.

These relationships provide the necessary perspective for discussing physiological features of prokaryotic cells. They imply, for example, that bacterial cells differ greatly from eukaryotic cells at the molecular level, which has been borne out by decades of research. However, at least two eukaryotic organelles, the **mitochondrion** and the **chloroplast**, seem to have descended from bacteria which entered into symbiotic relationships with a progenitor of modern eukaryotic cells (Schwartz and Dayhoff, 1978). Thus the composite nature of the eukaryotic cell results in significant similarities between eukaryotic organelles and bacterial cells. Molecular taxonomy also implies that archaea differ greatly from bacteria, despite the fact that both groups consist of prokaryotes. This is consistent with the available metabolic and structural data, but the full extent of these differences remains to be discovered.

For convenience, this chapter presents the prokaryotic world in a concentric fashion. *Escherichia coli*, which belongs to the "purple" group of gram-negative bacteria (named after its pigmented, photosynthetic members) is described in greatest detail. Gram-negative bacteria that differ from *E. coli* in significant respects are discussed where appropriate. In turn, gram-positive bacteria and other structurally distinct bacteria are contrasted with the gram-negative bacteria. Finally, archaea, insofar as their physiological features have been established, are discussed in comparison to all bacteria.

II. Prokaryotic Cytology

The prokaryotic cell is the smallest and simplest form of life capable of an independent existence. Compared to the typical eukaryotic cell, it is remarkable for what it lacks. A

TABLE 1 Major Groups of Organisms According to Molecular Taxonomy

I. Bacteria

    A. Deinococci

        1. *Thermus*

        2. *Deinococcus*

    B. Green, nonsulfur bacteria

    C. Cyanobacteria

    D. "Purple" bacteria

        1. *Escherichia*

        2. *Azotobacter*

        3. *Thiobacillus*

        4. *Desulfovibrio*

    E. Gram-positive bacteria

        1. *Clostridium*

        2. *Lactobacillus*

        3. *Streptococcus*

II. Archaea

    A. Thermophiles

        1. *Sulfolobus*

        2. *Pyrolobus*

        3. *Thermoproteus*

        4. *Archaeoglobus*

    B. Methanogens

    C. Extreme halophiles

        1. *Haloferax*

        2. *Halobacterium*

III. Eukaryotes

    A. Microsporidia

    B. Flagellates

    C. Slime molds

    D. Ciliates

    E. Green plants

    F. Fungi

    G. Animals

The scheme above summarizes the relatedness of major groups of organisms according to ribosomal RNA comparisons (Woese *et al.*, 1990). All groups at the same level in the outline (capital letters, e.g.) have comparable evolutionary status; several major groups have been omitted for the sake of clarity.

typical prokaryotic cell measures less than 1 μm in diameter (Fig. 1). It has no cytoskeleton, no identifiable mitotic apparatus, and no internal membranous organelles, yet it carries out all the basic functions of life, including nutrient uptake, catabolism, energy conversion, biosynthesis, genetic regulation, genome replication, cell division, and adaptive responses to environmental stimuli. The spatial and functional

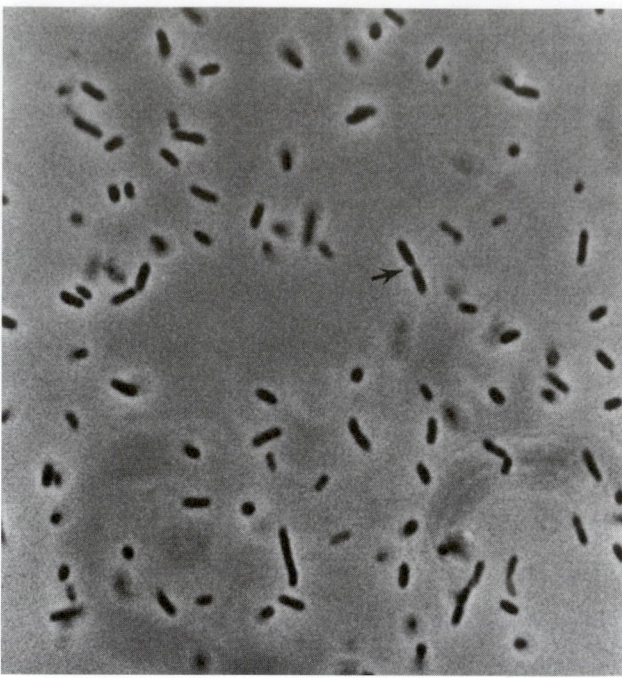

FIGURE 1. *Escherichia coli* cells suspended in growth medium. The average width of the cells is about 0.5 μm. The arrowheads show site of constriction (septation) in a cell undergoing division. This light micrograph (phase contrast) illustrates the limited structural information obtainable by optical methods due to the small size of prokaryotic cells.

organization required for these processes occurs at the level of molecular assemblies; only the larger of these assemblies are readily apparent in the electron microscope.

All prokaryotic cells have a **nucleoid**, a **cytoplasm**, and a **cytoplasmic membrane**. In addition to these three essential structures, most bacteria and some archaea have a **cell wall**. Gram-negative bacteria have an additional membrane outside the cell wall called the **outer membrane** (Fig. 2). The following discussion of prokaryotic cytology centers on gram-negative bacteria, which include the greatest number of the basic cellular structures normally found in bacteria.

A. Nucleoid

The prokaryotic chromosome consists of a single circular DNA of several million base-pairs. This DNA is neither confined within a nucleus nor evenly dispersed throughout the cytoplasm. It occurs instead as a defined yet convoluted region near the center of the cell, called the **nuclear region** or **nucleoid**. The nucleoid replicates and partitions into daughter cells during growth and cell division, and bears a functional analogy to the eukaryotic nucleus.

The *E. coli* cell is a short rod about 1 μm long, yet its DNA is about 1 mm, or 10^3 μm in length. Thus, although bacteria do not have nucleosomes or true histones, their DNAs are compacted in the nucleoid. Cations, including Mg^{2+}, spermidine and other polyamines, and small histone-like proteins help condense the DNA *in vivo*. Longer-range organization occurs in the form of 50–100 topologically constrained domains or "loops." Lysis of cells under appropriate conditions

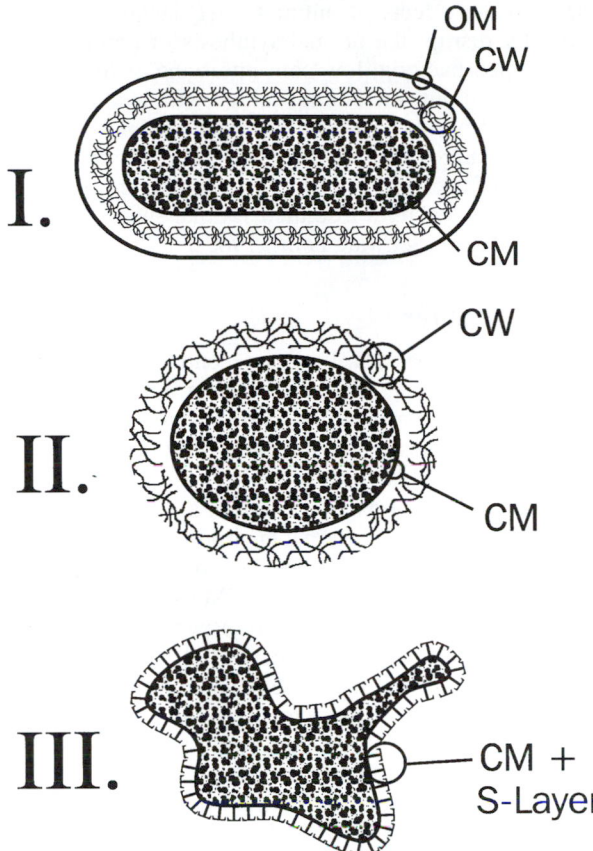

FIGURE 2. Basic cell structures of prokaryotes. The diagrams summarize the three most common types of prokaryotic cell structure. I: A gram-negative bacterium has a thin cell wall sandwiched between the cell (i.e., cytoplasmic) membrane (CM) and outer membrane (OM). II: A gram-positive bacterium has a thick cell wall (CW) immediately outside the cell membrane (CM). III: many archaea have only an S-layer, composed of glycoprotein subunits, as a structural support for the cell membrane, although others (not depicted) have cell walls (Section II.H.5).

(nonionic detergent and high salt concentration) releases nucleoids in apparently intact form, and allows the loops to be visualized by electron microscopy (Drlica, 1987).

B. Cytoplasm

All other components of the cell interior are collectively termed the cytoplasm. Often described as a fluid (hence the equivalent term *cytosol*), the **cytoplasm** is highly concentrated and probably quite viscous. It consists primarily of metabolic enzymes and ribosomes, reflecting the central importance of rapid growth for prokaryotic organisms. An *E. coli* cell adapted to growth in simple glucose medium at 37 °C contains about 19 000 ribosomes. The ribosomes, transfer RNA, and various translation factors together account for nearly half of the total cell mass; this high ribosomal content enables the *E. coli* cell to reproduce itself every 40 min under these conditions (Neidhardt *et al.*, 1990).

Its lack of internal compartmentalization (Fig. 3) contributes to a prokaryote's ability to grow and regulate its

gene expression quickly. Prokaryotic cells have no diffusional barriers segregating the sources of energy, raw material, and sequence information needed for DNA, RNA, and protein synthesis, and the small dimensions of the cell ensure that diffusion is extremely rapid. Furthermore, transcription and translation are temporally and spatially coupled in bacteria; a series of ribosomes begins "reading" the mRNA before the mRNA itself has been completed and released from the transcription complex. This coupling has functional significance beyond simply allowing rapid growth, as demonstrated by the phenomenon called **attenuation**. This form of genetic regulation uses disruption of the normally tight coupling between ribosome and RNA polymerase to abort the transcription of certain amino acid biosynthetic genes when the supply of the corresponding amino acids is adequate (Yanofsky and Crawford, 1987).

C. Cytoplasmic Membrane

The prokaryotic cell interior is chemically separated from the external environment by a unit membrane called the **cytoplasmic (cell) membrane** (CM) (Fig. 2) which is the only lamellar structure common to all prokaryotic cells (Fig. 1). The molecular composition of the bacterial CM generally conforms to the fluid mosaic model of biological membranes (see Chapter 6 on ultrastructure of cells). Its phospholipid bilayer is intrinsically permeable to gases and water but impermeable to ionic or large polar molecules. This impermeability is essential, allowing the cell to maintain ion potentials and retain metabolites (see later discussion). However, the cell also requires provision for chemical exchange with its surroundings, including nutrient uptake and environmental sensing. This problem is solved by a large number of different integral and peripheral proteins

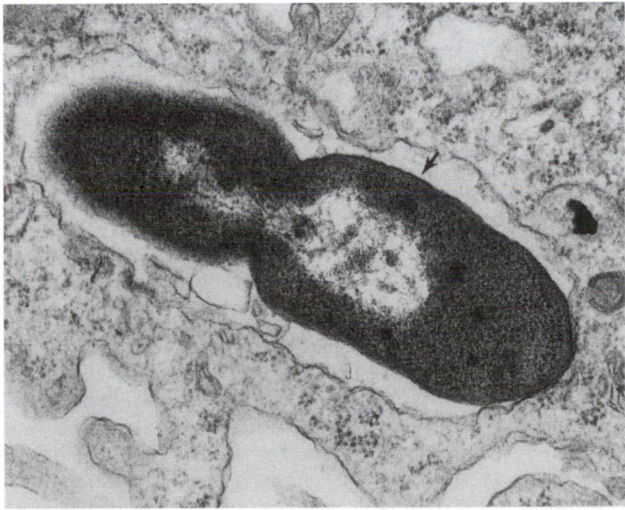

FIGURE 3. Electron micrograph (ultrathin section) of a bacterial cell. An *E. coli* cell is shown that has been engulfed by a macrophage and is in the late stages of cell division. The nucleoid is visible as a clear zone in each of the nascent daughter cells. The arrowhead shows a region in which the three-layer structure of the gram-negative cell envelope can be seen. (Photograph courtesy of A. J. Mukkada.)

which breach the phospholipid bilayer with various solute-specific pumps and gates. By means of these proteins, which comprise about 70% of the CM by weight, the prokaryotic cell maintains transmembrane gradients and controls transmembrane fluxes of critical solutes.

D. Cell Wall

With few exceptions (Section II.D.4), bacteria have a **cell wall** immediately outside the cytoplasmic membrane (Fig. 2). The cell wall is composed of a polymer, called **peptidoglycan** or **murein**, that is unique to bacteria. Peptidoglycan has a repeating unit consisting of two amino sugars linked via β-1,4 glycosidic bonds. The resulting linear glycan chains are cross-linked at frequent intervals by peptide bridges (Fig. 4). This cross-linked network results in one huge, baglike macromolecule of high tensile strength that completely envelopes the mechanically fragile cell membrane (Fig. 2).

Nearly all prokaryotic cells live in environments whose water activity exceeds that of their cytoplasms. The cell wall provides the mechanical resistance necessary to counteract the resulting osmotic (turgor) pressure (see Chapter 21 on osmosis and regulation of cell volume) which can be as high as 20 atm. As a result, the bacterial cell wall preserves the integrity of the CM, determines cell shape, and plays a major role in cell division. Evidence of these functions can

External environment

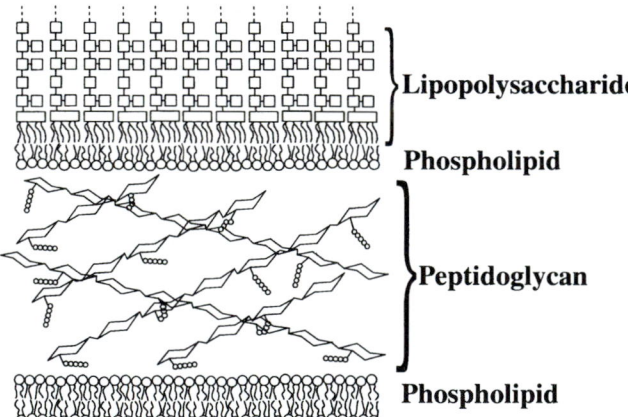

Bacterial cytoplasm

FIGURE 4. Molecular organization of the cell envelope of gram-negative bacteria. The outer membrane (OM) consists of one leaflet of lipopolysaccharide (LPS) and one of phospholipid. Each LPS molecule has a large polysaccharide chain (linked squares) whose precise structure depends on the bacterial species. The cell wall (CW) is composed of peptidoglycan, whose repeating unit consists of a dimer of two modified sugars: N-acetyl muramic acid and N-acetyl glucosamine. Each repeating unit has a short peptide (circles) attached to the N-acetyl muramic acid; the peptides of adjacent glycan chains become joined by enzymes in the periplasm to form the periodic cross-links found in the mature peptidoglycan polymer. The cytoplasmic membrane (CM) is composed of phospholipid. In addition to lipid, both the OM and CM contain protein species, which are not depicted in the figure.

be seen in the effects of antibiotics (β-lactams, e.g.) that specifically disrupt the normal synthesis of peptidoglycan. In rod-shaped bacteria, low concentrations of these antibiotics block cell division but not cell growth, leading to elongated cells of normal diameter. At moderate antibiotic concentrations, bulges (rather than constrictions) form at the normal site of bacterial cell division. At high antibiotic concentrations, the bacteria lose their rodlike shape, swell into spheres, and lyse (Schwarz *et al.*, 1969).

E. Outer Membrane

Gram-negative bacteria have a second lipid bilayer surrounding the cell wall, called the **outer membrane**. This forms a protected compartment, bounded by the cytoplasmic and outer membranes, called the **periplasm**. Thus the wall in a gram-negative cell is sandwiched between two membranes (Fig. 2, Fig. 4), in contrast to the wall of a gram-positive cell, which is exposed to the external environment (Fig. 2).

The biochemical composition of the bacterial outer membrane (OM) differs from that of the CM with respect to both lipid and protein constituents. With respect to lipid, the OM is a hybrid membrane. Its inner leaflet, which faces the periplasm, incorporates the same phospholipids as does the CM. The outer leaflet, however, consists of much higher molecular weight class of lipid known as **lipopolysaccharide**, or **LPS** (Fig. 4). With respect to protein, the OM has a relatively simple protein composition dominated by only a few protein species (Nikaido and Vaara, 1985), which contrasts with the heterogeneity of CM proteins.

These two compositional differences from the CM underly the primary physiological function of the OM, which is to shield the gram-negative cell from toxic, hydrophobic compounds in the environment. An intact LPS leaflet greatly impedes diffusion of soluble hydrophobic (i.e., lipophilic) molecules across the OM. This presumably reflects the permeability properties of the dense array of cooperatively interacting charged and polar sugar residues at the membrane surface, through which nonpolar solutes must pass (Fig. 4). Experimental evidence of the importance of this screening function can be seen in the greater sensitivities of gram-positive bacteria, as well as of LPS-depleted gram-negative cells, to toxic dyes, detergents, and hydrophobic antibiotics (Nikaido and Vaara, 1985).

As in the case of the CM, the OM cannot completely insulate the cell from its external environment. The necessary exchange of nutrients is mediated by the few, abundant protein species found in the OM. These proteins, termed **porins**, trimerize to form relatively nonspecific solute channels, or **pores**, through the OM. The pores exclude solutes with molecular weights of about 700 Da and above, and tend to further discriminate against anionic solutes (Nikaido and Vaara, 1985).

F. Appendages

Two basic types of external filaments may be anchored in the CM. **Flagella** are helical protein filaments that extend several cell lengths (2–20 μm). Depending on the bacterial species, they may be single, numerous, clustered, dispersed over the cell, or absent altogether. Each flagellum is rotated by a protein complex at its base, and acts like a propeller to push the cell through its liquid medium (see Chapter 58). Thus, bacterial flagella have a cellular function (i.e., motility) analogous to that of eukaryotic cilia and flagella, but no structural, molecular, or mechanistic homology to them.

Pili and **fimbriae** are structurally similar to each other and distinct from flagella. Both pili and fimbriae consist of straight protein fibers that protrude less than about 1 μm from the cell, giving it a bristled appearance in the electron microscope. Both types of appendages appear to mediate attachment of a bacterium to another cell. Pili attach a "donor" bacterium to a recipient cell for the subsequent transfer of DNA (conjugation), whereas fimbriae typically attach a pathogenic bacterium to specific cells of the host. In some bacteria, pili mediate a type of slow, surface-dependent motility whose mechanistic basis is not thoroughly understood (Strom and Lory, 1993).

Flagella, pili, and fimbriae all form via spontaneous self-assembly of small protein subunits. Bacterial flagella grow by adding protein subunits to the distal tip, which requires the subunits (monomers) to travel from the cytoplasm to the growing tip of the flagellum, several cell lengths away. This is accomplished by using the hollow core of the flagellum itself as a conduit. In contrast, pili and fimbriae appear to polymerize at the base of the fiber where it attaches to the CM (Neidhardt *et al.*, 1990).

G. Capsule

Many bacteria secrete hydrophilic polymers (usually polysaccharides) that form a gelatinous matrix around the cell. If this matrix remains attached to the cell and forms a defined zone around it, it is called a **capsule**. Capsules have been observed to inhibit the ingestion of bacteria by human phagocytes or by protozoa. Capsules may also help free-living bacteria survive desiccation or starvation. Even if it does not form a defined capsule, secreted polysaccharide aids the nonspecific adherence of bacteria to wetted solid surfaces, leading to the formation of **biofilms**. Bacteria in biofilms tend to tolerate much higher concentrations of antibiotics than do bacteria suspended in fluids. These films therefore pose serious health risks if they form on surfaces of heart valves or implanted devices (Costerton *et al.*, 1987).

H. Other Cell-Surface Architectures

1. Gram-positive Bacteria

The gram-positive bacteria lack an outer membrane and have thicker cell walls than do gram-negative bacteria. These cell walls are chemically more complex; in addition to peptidoglycan, they contain various anionic polymers (teichoic, lipoteichoic, and teichuronic acids). The resulting cell wall does not readily decolorize in the Gram staining procedure. Table 2 summarizes the major physiological differences between gram-positive and gram-negative bacterial cells.

2. Acid-fast Bacteria

Some environmentally and clinically important bacteria (*Mycobacterium* and *Nocardia* spp.) have a unique type of cell envelope that includes very high-molecular weight lipids

TABLE 2 Properties of Gram-positive and Gram-negative Bacteria

Property	Gram-positive bacteria	Gram-negative bacteria
Cell wall:		
Chemical constituents	Peptidoglycan teichoic, lipoteichoic and teichuronic acids	Peptidoglycan
Typical thickness	20-80 nm	2-3 nm
Outer membrane	(Absent)	Composed of LPS and phospholipid
Intrinsic resistance to detergents, dyes, and certain antibiotics	Low	High
Desication resistance	High	Low
Examples	*Clostridium Lactobacillus*	*Escherichia Thiobacillus*

called **mycolic acids**. These waxy lipids are complexed to an additional polysaccharide layer surrounding the cell wall of an otherwise gram-positive cell. X-ray diffraction data suggest that the mycolic acids form a very thick lipid membrane analogous to the OM of gram-negative cells (Nikaido *et al.*, 1993). The resulting cell envelope makes the cells very difficult to stain and destain.[1] The unusual inertness and impenetrability of these cell envelopes to solutes probably explains other notable properties of *Mycobacterium* and *Nocardia* spp., including slow growth rates and a general resistance to antibiotics and to the immune system (Nikaido *et al.*, 1993).

3. Surface Layers

Either gram-positive or gram-negative bacteria may bear regular paracrystalline arrays of protein subunits on their surfaces. These **surface (S-) layers** give whole cells or cell envelope preparations a cobblestone pattern when metal-shadowed for electron microscopy. A number of functions have been proposed for these protein layers, including sieving effects, mediation of cell-cell interactions, and modification of the cell's net surface charge. Bacterial S-layers are not essential for bacteria growing under laboratory conditions, however, and tend to be lost during prolonged cultivation in the laboratory (Messner and Sleytr, 1992).

4. Bacteria Without Cell Walls

A few bacteria, most notably *Mycoplasma* spp., have no cell wall. These cells have relatively strict requirements for the composition and osmolarity of their growth medium. As expected from the absence of a bacterial cell wall (see previous discussion), the cells have irregular shape, and when transferred into hypotonic medium, the cells swell and lyse.

[1]Cells of these bacteria, stained by heating in carbol-fuschin dye, fail to decolorize even in acidic phenol, yielding the empirical designation "acid-fast."

5. Archaea

Archaeal cell membranes differ from all other cell membranes with respect to their lipids. Archaea contain only *sn*-(2,3-di-O-alkyl)-glycerol membrane lipids, which occur in no other group of organisms. The hydrocarbon chains of these lipids are C20 or C40 saturated isoprenoids attached to the glycerol backbone via ether linkages. To date, no archaea are known to have a second lipid membrane analogous to the OM of gram-negative bacteria. Some archaea (a few species of methanogens and extreme halophiles) have cell walls, but none of them is composed of peptidoglycan.

Most archaea lacking true cell walls have some type of S-layer attached to the CM. The location of these S-layers is analogous to that of a cell wall, however, which suggests an analogous function (Fig.1). It should be noted that archaeal S-layers differ from bacterial S-layers (Section II H 3) in several respects. Unlike bacterial S-layers, 1) their subunits are usually glycosylated, 2) they seem to be essential, and are not lost during laboratory cultivation, 3) they are usually anchored to the cytoplasmic membrane (Fig. 5), and 4) they can define cell shape (Baumeister *et al.*, 1989). It should be also noted that some archaea, such as *Thermoplasma* spp., lack both an S-layer and a cell wall; these cells are osmotically sensitive.

III. Metabolic Strategies

The prokaryotic world is metabolically diverse. Many prokaryotes depend on metabolic processes unknown in eukaryotes.

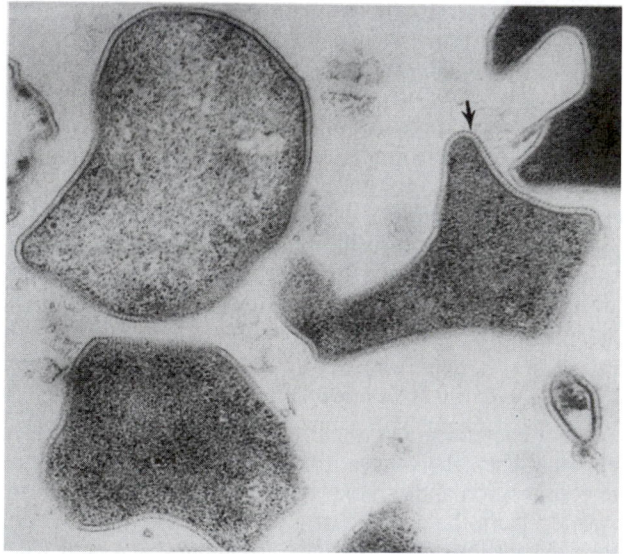

FIGURE 5. Cells of *Sulfolobus acidocaldarius*, an archaeon from geothermal environments (Section VII). The ultrathin sections show the irregular cell shape and regular repeating structure of the glycoprotein S-layer. The S-layer is held at a fixed distance from the cytoplasmic membrane by some form of spacer (Baumeister *et al.*, 1989); this results in a "picket fence" boundary around cells or cell ghosts (Grogan, 1996a) in thin sections.

A. Autotrophy

Many prokaryotes can derive all of their carbonaceous cell material from CO_2. This property, termed **autotrophy**, requires external sources of energy and reducing equivalents (i.e., electrons).

Several different families of bacteria (collectively called **photoautotrophs**) use light as the energy source, whereas various inorganic and organic compounds in the environment serve as sources of electrons. Photoautotrophic bacteria exhibit great diversity with regard to the biochemistry of photosynthesis. Among them, only the **cyanobacteria** carry out **oxygenic photosynthesis**. Cyanobacteria contain **chlorophyll b** and have two connected **photosystems**, one of which (PS2) oxidizes water to supply electrons to the other (PS1) (see Chapter 64 on Photosynthesis). This is the scheme also used by plant chloroplasts, to which cyanobacteria are evolutionarily related (Schwartz and Dayhoff, 1978).

The other photoautotrophic bacteria perform **anaerobic (anoxygenic) photosynthesis**. They have only one photosystem, which can operate in a cyclic manner to produce a proton potential (Section IV.B). In these cases, the electron equivalents for the reduction of CO_2 come from H_2S, H_2, or dissolved organic compounds, depending on the type of bacterium and the resources available to it. The photosynthetic pigments of the anoxygenic photoautotrophs, called **bacteriochlorophylls,** differ from cyanobacterial and plant chlorophylls in having intense absorbance maxima in the infrared region of the spectrum (White, 1995). In many of these organisms, photoautotrophic growth represents only one among several metabolic options, and production of photosynthetic pigments occurs only in the absence of oxygen.

Another form of autotrophy, **chemoautotrophy,** is unique to prokaryotes. Chemoautotrophs derive energy by mediating the oxidation and reduction of inorganic compounds in their environments. Aerobic chemoautotrophs include organisms that oxidize H_2, CO, NH_4^+, NO_2^-, elemental S, H_2S, or Fe^{2+}, using O_2. Anaerobic chemoautotrophs include organisms that derive energy from the reduction of CO_2 to CH_4, or of SO_4^{2-} to H_2S.

Among both photo- and chemo-autotrophs, at least three metabolic pathways have been identified by which CO_2 is "fixed" (i.e., reduced and incorporated into some common intermediary metabolite): the **Calvin cycle,** the **acetyl CoA pathway,** and **the reductive tricarboxylic acid cycle** (Caldwell, 1995). Most photoautotrophic bacteria (and all green plants) use the Calvin cycle: carboxylation of ribulose bisphosphate to yield two molecules of 3-phosphoglycerate, followed by a complex series of reactions to regenerate ribulose bisphosphate (see Chapter 64 on photosynthesis). The acetyl-CoA pathway, used by methanogenic archaea and sulfate-reducing bacteria, involves the differential reduction of two CO_2 molecules. One is reduced to form a methyl group, whereas another is reduced to form a carbonyl unit, which is condensed to the methyl group. The resulting acetyl unit is transferred to coenzyme A for assimilation into cell material. The reductive tricarboxylic acid pathway begins with carboxylation of succinyl CoA to yield 2-oxo-glutarate, which is in turn carboxylated to yield isocitrate. The transformations of the TCA cycle then continue in reverse, leading to the cleavage of citrate by an ATP-dependent citrate lyase to yield acetyl CoA and oxaloacetate. Prokaryotes using this pathway include the "green" photosynthetic bacteria.

B. Nitrogen Fixation

The only organisms that can convert N_2 into ammonia are prokaryotes. This capability, though relatively rare, is widely dispersed among prokaryotic groups. It occurs in various aerobic, anaerobic, facultatively anaerobic bacteria (i.e., those able to grow either aerobically or anaerobically, including both heterotrophs and autotrophs), and in methanogenic archaea. N_2 reduction is carried out by a large enzyme complex, **nitrogenase**. It is composed of an iron-containing component and a molybdenum- and iron-containing component, and utilizes ATP plus a reduced low-potential ferredoxin or flavoprotein as the source of electrons. Nitrogenases are intrinsically O_2-sensitive, and N_2-fixing prokaryotes use a variety of strategies to protect these enzymes from oxidative inactivation. For example, facultative anaerobes such as *Klebsiella* spp. produce nitrogenase only under anaerobic conditions. Filamentous cyanobacteria, which generate O_2 during photosynthesis, sequester nitrogenase in specialized cells called **heterocysts**, in which the oxygenic photosystem II does not operate. *Azotobacter* spp. are obligate aerobes that can fix nitrogen; they appear to scavenge intracellular O_2 by maintaining very high respiration rates.

IV. Energetics of Bacterial Cells

With a few notable exceptions, the endergonic cellular processes of bacteria are driven by being enzymatically coupled to the hydrolysis of ATP. Life thus requires a steady resupply of this energy currency. As illustrated by an *E. coli* cell growing aerobically using glucose, prokaryotes employ two fundamentally different strategies of ATP regeneration: **substrate-level phosphorylation** and **chemiosmotic coupling**.

A. Substrate-level Phosphorylation

As in the eukaryotic cytosol, soluble enzymes in the *E. coli* cytoplasm convert one mole of glucose to two moles of pyruvic acid via the Embden-Meyerhof-Parnas pathway (glycolysis), which involves phosphorylation of 2 moles of ADP to form ATP and the reduction of 2 moles NAD^+ to NADH. If the cell has no exogenous electron acceptors available (Section IV.B), the two moles ATP represent the cell's sole energy harvest, whereas the reduced cofactor (NADH) represents unusable electrons. If the electrons are not transferred to some other molecule, the oxidized cofactor (NAD^+) will not be regenerated and a second mole of glucose cannot be metabolized. Bacteria (and eukaryotic cells) solve the latter problem by transferring the electrons to pyruvate or its metabolites, thereby forming various end-products which the cell excretes; examples include lactic, acetic, propionic, or butyric acids, and ethanol, butanol, or acetone. Substrate-level phosphorylation, coupled to reduction of metabolites derived from the growth substrate, constitute the biochemical definition of **fermentation**.

B. Chemiosmotic Coupling

The general features of this mechanism resemble those of mitochondria and chloroplasts, which are described elsewhere in this volume (see chapters on ultrastructure of cells;

physiology of the mitochondrion; energy production and metabolism; and photosynthesis). The process has two distinct stages, each of which involves vectoral enzymatic processes at the CM. The first stage transfers electrons (taken from carbon compounds during glycolysis and the tricarboxylic acid cycle) to some electron acceptor (such as O_2) via a series of enzymatically catalyzed, strongly exergonic, oxidation-reduction reactions. This respiratory electron transport is coupled at certain points to the extrusion of protons from the cytoplasm, creating a **proton potential** or **protonmotive force** (PMF) across the CM. The second stage of chemiosmotic coupling converts the PMF into ATP; it uses an $\mathbf{F_1F_0}$ **proton-translocating ATPase** to couple the entry of three protons from outside the cell to phosphorylation of an ADP molecule (Maloney, 1987). As a result of these two processes, the yield of ATP per glucose molecule is many times higher for a bacterium that is respiring than for one that is fermenting.

Respiration can take many forms in prokaryotic cells, reflecting the fact that oxygen represents only one of several terminal electron acceptors which prokaryotes can use to form a PMF. Table 3 lists the major types of **anaerobic respiration** carried out by prokaryotes. *E. coli*, for example, can use four electron acceptors other than oxygen (Table 3), but it produces the necessary oxido-reductases only according to a complex regulatory hierarchy. This hierarchy ensures that the energetically most favorable electron acceptor is used, should more than one be available (Gunsalus, 1992).

The PMF of a respiring bacterium represents an energetic intermediate in the formation of ATP by the oxidation of carbon compounds. It should be noted, however, that the PMF appears to be a basic energetic feature of all bacterial cells whether or not they are respiring. The **lactic acid bacteria**, for example, have no electron-transport chain and cannot respire; they generate ATP only by substrate-level phosphorylation. They nevertheless have an F_1F_0 ATPase, and use it in the reverse sense of respiring bacteria, that is, to maintain a PMF at the expense of ATP hydrolysis. This may reflect the fact that certain basic prokaryotic processes use the PMF directly as their energy source. *E. coli*, for example, uses direct

TABLE 3 Respiratory Strategies of Prokaryotes

Electron acceptor	Reduced product	Organism
Oxygen	Water	*E. coli*, other aerobes
Nitrate	Nitrite	*E. coli*, other bacteria
Fumarate	Succinate	*E. coli*, other bacteria
Dimethyl sulfoxide	Dimethyl sulfide	*E. coli*, other bacteria
Trimethylamine oxide	Trimethylamine	*E. coli*, other bacteria
Sulfate	Hydrogen sulfide	*Desulfovibrio*, *Archaeoglobus**
Sulfur	Hydrogen sulfide	*Thermoproteus**
Carbon dioxide	Methane	*Methanobacterium**

* denotes archaea; all others are bacteria

coupling to H$^+$ influx to transport several nutrients (see later discussion) and to drive flagellar rotation.

As discussed in Chapter 7, Energy Production and Metabolism, the PMF consists of two components: the electrical potential ($\Delta\Psi$), and the chemical potential (ΔpH). In those bacteria which live at pH values near 7, both components contribute significantly to the PMF. Although the small size of prokaryotic cells precludes direct electrical measurement of these potentials, they can be estimated by chemical probes. The $\Delta\Psi$ can be measured by the fluorescence yield of triphenylmethyl- or tetraphenylphosphonium ions, or by the equilibrium distribution of radioactive K$^+$ across the CM in the presence of valinomycin. The ΔpH can be estimated by the distribution of radioactive weak acids such as acetic or benzoic acids across the CM, and the overall PMF can be independently estimated by the maximal accumulation of lactose by cells able to transport, but not metabolize, this sugar (Maloney, 1987).

According to these methods, which generally agree, a typical E. coli cell in medium at pH 6.5 has a $\Delta\Psi$ of about 100 mV (inside negative) and a ΔpH that corresponds to an additional 100 mV (Maloney, 1987). Extreme acidophiles, however, may maintain an electrical potential which is of opposite polarity as the normal $\Delta\Psi$, that is, inside positive. This helps counteract the very large ΔpH (which can be more than 4 pH units) across the cytoplasmic membrane. In contrast, extreme alkaliphiles, which grow at external pH values of 10–12, have a negative ΔpH across the CM, and accordingly, a very large $\Delta\Psi$ (White, 1995). Some prokaryotes, particularly those living in marine or alkaline environments, can use the more abundant Na$^+$ ion to drive ATP synthesis (Skulachev, 1994).

V. Solute Transport

Prokaryotes have efficient transport systems for a wide range of nutrients and inorganic ions. The reason for these systems is straightforward: prokaryotes must scavenge nutrients from their environments in order to survive and grow. Table 4 lists the basic transport mechanisms used by bacterial cells.

A. Facilitated Diffusion

This mechanism is rare among bacterial transport systems. Although mechanistically simple and energetically cheap, it is also relatively ineffective. One of the best studied examples of a facilitated diffusion through a bacterial membrane is glycerol uptake in E. coli. In this case, extracellular glycerol diffuses passively through a polyol-specific membrane channel encoded by the glpF gene.

B. Group Translocation

In an elaboration of facilitated diffusion, the solute becomes chemically modified upon entering the cytoplasm. The chemical modification, or **group translocation** (typically phosphorylation or phosphoribosylation) converts an uncharged solute molecule into a form that cannot diffuse out through the same carrier. In this way, the uncharged species actually transported by the channel can remain at a lower steady-state concentration inside the cell than outside, whereas the modified solute can reach high intracellular concentrations, due to the investment of chemical energy in the modification process. In most cases, this "trapping reaction" doubles as a necessary step in metabolizing the solute.

C. Active Transport

Active transport in bacteria is defined by the following features (Neidhardt et al., 1990): 1) specific steric recognition occurs between the solute and a membrane-bound transport protein, 2) the solute is released into the cytoplasm in its unmodified form, 3) accumulation occurs against a solute concentration gradient, and 4) energy is expended.

Two types of active transport are distinguished from each other by the immediate source of energy used (Table 4). The first type requires the hydrolysis of ATP or an equivalent high-energy phosphoryl bond. In these systems, the membrane-bound carriers typically consist of several subunits, one of which contains an ATP-binding site (Furlong, 1987). In gram-negative bacteria, such a transporter often utilizes a nonmembrane (i.e., soluble) protein located in the periplasm. Transport requires these periplasmic solute-binding proteins, as demonstrated by the fact that cold hypotonic shock treatment of bac-

TABLE 4 Solute Transport Systems in Bacteria

Basic type	Mechanistic features	Examples in E. coli
I. Facilitated diffusion	Selective channel, no net accumulation of transported species	OM pores, glycerol uptake
II. Group transfer	Chemical modification of transported species	Phosphotransferase system for sugars
III. Active transport	Solute interacts specifically with a cognate, membrane-bound carrier	
	Transported species accumulates against a concentration gradient	
A. ATP-driven	Requires ATP hydrolysis	Histidine uptake
B. PMF-driven	Coupled to ion flux	Lactose uptake

terial cells, which releases only periplasmic proteins, also destroys transport capability. Studies of bacterial mutants and of cytoplasmic membrane vesicles confirm that the membrane-bound transporter interacts specifically, not with external solute in its free form, but with the solute complexed to its cognate binding protein (Furlong, 1987).

The second type of active transport utilizes ionic potentials directly. In some cases the solute enters via symport with a proton, in other cases, with a Na^+ ion. Alternatively, certain ions, such as Ca^{2+}, are pumped out of the cells by proton-coupled antiport systems (Maloney, 1987).

VI. Stress Responses

Prokaryotic cells must routinely respond to changing environmental conditions in order to optimize their growth or to simply survive. These homeostatic responses, though diverse, all involve changes in the general biochemical composition of the prokaryotic cell.

Some responses are mediated directly by activation of existing proteins, as exemplified by the effects of osmotic stress. An increase in the osmotic strength of the growth medium triggers a response in which the osmotic strength of the bacterial cytoplasm increases (Neidhardt *et al.*, 1990). In *E. coli*, this increase is primarily due to net K^+ uptake by transporters in the CM, which, in turn, seem to respond directly to cell turgor, perhaps as signaled by the state of compression of membrane lipid. The physiological stress imposed by extremely high external ionic strength, however, can apparently not be effectively countered by K^+ accumulation alone. Under these conditions, *E. coli* and other bacteria accumulate "compatible solutes" in the cytoplasm, of which the two best studied examples are L-proline and glycine betaine. Typically, these compounds are not synthesized by the cell, but must be available in the medium. Their accumulation in the cytoplasm permits growth at otherwise inhibitory external salt concentrations.

Two conclusions have been drawn from these observations. First, high concentrations of compatible solutes in the cytoplasm are innocuous to cellular function, and may even protect the bacterial enzymes against the deleterious effects of high solute concentrations. Second, the conservation and sophistication of osmotic stress responses among bacteria imply that turgor pressure *per se* is essential for the bacterial cell. Turgor pressure is thought to provide the physical force for expanding the cell, thus permitting growth.

Other stress responses involve changes (often complex) in the pattern of gene expression (Neidhardt *et al.*, 1990). This strategy takes advantage of the relatively large biosynthetic capacity of the growing prokaryotic cell, its simple, streamlined organization, and a very short half-life of messenger RNA, which in *E. coli* (under standard conditions) averages 1.3 min. Studies of such regulatory responses in bacteria have revealed an array of elegant and diverse molecular mechanisms for controlling gene expression, a full discussion of which lies outside the scope of this chapter. However, one example warrants mention here because it involves transmembrane signaling and appears to be shared among a variety of environmental responses in bacteria. Another regulatory response is described because it relates to general nutritional stress and is highly conserved among bacteria.

A. Two-Component Regulatory Systems

When gram-negative bacteria begin to starve for inorganic phosphate, they activate the transcription of several families of genes, whose products help the cell scavenge phosphate (Wanner, 1993). The process, which typifies a mechanism underlying many signal transduction systems in bacteria, involves the specific interaction and phosphorylation of two proteins, named PhoR and PhoB after their respective genes in *E. coli*.

The phosphorylation state of the PhoB protein determines expression of the various genes of the phosphate-starvation response. The phosphorylated form of PhoB (here designated "*PhoB") activates transcription of these genes, whereas the unphosphorylated form does not. The relative abundance of *PhoB is in turn determined by the conformational state of a CM protein, PhoR (Fig. 6). In its activated conformation ("*PhoR"), this protein acts as a PhoB kinase, whereas its inactive conformation ("PhoR"), acts as a *PhoB phosphatase. Ultimately, the conformational state of PhoR is determined by the availability of inorganic phosphate in the external medium. By a mechanism which is not entirely clear, PhoR can sense the absence of phosphate-periplasmic binding protein complexes in the periplasm and assumes under these conditions its active conformation. The resulting *PhoR phosphorylates PhoB, forming *PhoB, which in turn stimulates the transcription of *phoA* and other phosphate-scavenging genes by the bacterial RNA polymerase (Fig. 6). The result is a rapid increase in certain enzymatic activities which serve to physiologically adapt the cell to a specific change in the environment.

Protein pairs resembling PhoR/PhoB mediate a wide variety of regulatory responses of bacteria to various types of environmental change (Tables 5 and 6). The PhoR protein, a transmembrane signal transducer, typifies the **component I** or **sensor-kinase** of these systems. With a few exceptions, this is a cytoplasmic membrane protein that senses the status of some environmental parameter and accordingly adopts either of two conformations. In its active conformation, the component I has two characteristic phosphotransferase activities: (1) it phosphorylates one of its own histidine residues, using ATP, and (2) it transfers the phosphate to an aspartyl residue on its cognate component II. These two activities result in the net transfer of a phosphate group from ATP to the cognate **component II** or **response regulator**, which in most cases activates transcription for a certain set of genes, but only in its phosphorylated form.

The similarities in interactions and biochemical properties of various components I and II (Table 6) are underscored by regions of sequence homology. Although they respond to diverse signals, all components I have two highly conserved sequence motifs at their carboxy termini. Similarly, all components II have sequence homology near their amino termini, and various sub-families may share regions of additional homology elsewhere in the amino acid sequence. The conserved regions of components I presumably recognize and interact with the conserved regions of

PHOSPHATE STARVATION

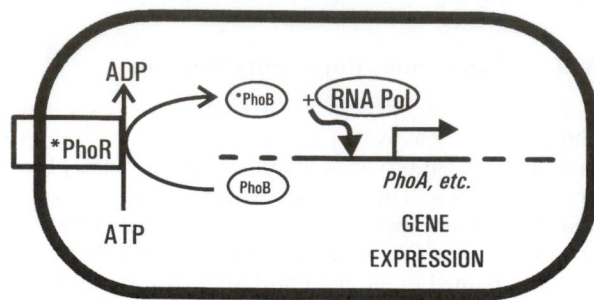

PHOSPHATE ABUNDANCE

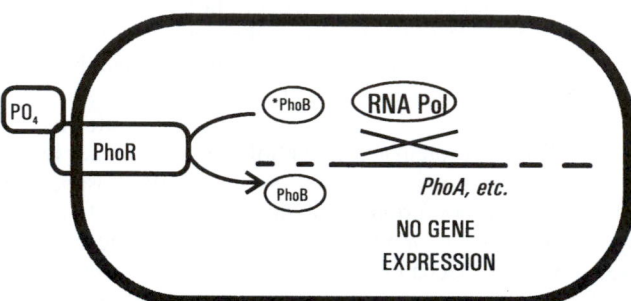

FIGURE 6. Basic features of a two-component regulatory system. A simplified version of phosphate-regulated gene expression is shown. The PhoR protein senses the extracellular availability of phosphate, either directly through an externally exposed domain, or indirectly via interaction with the high-affinity phosphate transporter (Wanner, 1993; Neidhardt *et al.*, 1990). When the external phosphate concentration is low, PhoR becomes activated; it autophosphorylates at the expense of intracellular ATP and then phosphorylates the PhoB protein. When the external phosphate concentration is high, PhoR becomes deactivated; in this form it removes phosphate groups from phosphorylated PhoB protein. PhoB is a soluble transcription factor for a specific set of genes, but stimulates transcription only when phosphorylated.

components II as an essential part of the signal transmission process. The nonconserved regions among the components I and II presumably define their specificity for the external stimulus detected, and for the particular cellular response, respectively.

B. The Stringent Response

When cells growing in complete media are suddenly deprived of exogenous amino acids, they quickly and specifically halt the transcription of ribosomal- and transfer-RNA genes. This switching-off, called the **stringent response**, benefits a starving cell by stopping the production of new ribosomes at a time when the cell has simultaneously 1) inadequate biosynthetic resources to meet its future growth demands and 2) an excess of ribosomes in proportion to the growth rate it can maintain (Cashel and Rudd, 1987). The mechanism by which stable RNA gene transcription responds to general starvation is uniquely bacterial and involves the ribosome. When uncharged tRNAs bind to the ribosomes (a symptom of severe amino acid depletion), a ribosome-associated protein is stimulated to form an unusual nucleotide, **ppGpp**, from ATP and GDP. This **guanosine tetraphosphate** appears to interact with the bacterial RNA polymerase in a way that decreases its affinity for rRNA and tRNA promotors (Cashel and Rudd, 1987).

VII. Prokaryotes Living in Extreme Environments

The physiological diversity of prokaryotes is dramatically illustrated by those organisms which not only tolerate chemical and physical extremes, but actually require them for normal cellular function. This heterogenous collection of prokaryotes includes many new species of archaea that have been discovered only in recent years, in part, because their natural habitats and growth conditions are so unusual.

The **extreme halophiles** are archaea that normally grow in concentrated brines. Well-studied species include the rod-shaped *Halobacterium salinarum* (=*halobium*) and the irregularly shaped *Haloferax volcanii*. *H. salinarum* grows best in media containing 3–4 M NaCl and will not grow at salt concentrations less than 1.5 M. The cell maintains an internal K^+ concentration of up to 5 M and contains Cl^- as the major counterion. Accordingly, most *H. salinarum* enzymes function best in extremely high salt concentrations, and denature at low salt concentrations (Kushner, 1985). The native structure of these halophilic enzymes tends to incorporate an unusually high number of acidic amino acid residues on the protein surface. Other notable features of the *H. salinarum* cell include small,

TABLE 5 Examples of Two-Component Regulatory Systems

Mnemonic	Environmental stimulus	Cellular response	Component I	Component II
*Pho*sphate	Low external [PO_4]	Induction of various PO_4-metabolizing enzymes	PhoR	PhoB
*N*itrogen *r*egulation	Low external [NH_4^+]	Induction of glutamine synthetase	NR_I	NR_{II}
*O*uter *m*embrane *p*rotein	External osmolarity	Regulation of alternate porin synthesis	EnvZ	OmpR
*A*erobic *r*espiratory *c*ontrol	Oxygen concentration	Regulation of central metabolic enzymes	ArcB	ArcA
*N*itrate reductase	External NO_3^- present, oxygen absent	Induction of nitrate reductase	NarX	NarL
*Che*motaxis	Temporal gradients of attractants and repellents	Change in swimming interval	CheA	CheY

TABLE 6   General Features of Two-Component Regulators

Feature	Component I	Component II
Cellular location	Membrane	Cytoplasm
Cellular function	Sense and signal an external condition	Control gene expression
Biochemical activities	Autophosphorylation at histidine residue	Binding to regulatory regions of genes
	Phosphorylation of cognate component II	Stimulating transcription of specific genes
	May dephosphorylate cognate component II	May inhibit transcription of other specific genes
Sequence homology	Two highly conserved regions near carboxyl end	One highly conserved region near amino end
		(subfamilies have additional regions of homology near carboxyl end)

proteinaceous **gas vesicles** in the cytoplasm, and patches of a special retinal protein, **bacteriorhodopsin**, in the cytoplasmic membrane. The gas vesicles make the cells buoyant, and thus help keep them in contact with two sources of energy: O_2 and light. Oxidation of organic compounds in the environment normally supports growth, but if oxygen becomes limiting, extra energy can be supplied by bacteriorhodopsin. This is a membrane-bound retinal protein that functions as a light-driven ion pump. When the cell is illuminated, it pumps protons from the cytoplasm; this contributes to the cell's PMF and thus to the production of ATP during O_2 deprivation (Kushner, 1985).

Extreme **acidophiles** (organisms which require low pH for optimal cell growth) have been identified among the bacteria and archaea. Certain gram-negative bacteria of the genus *Thiobacillus* derive energy from the oxidation of reduced sulfur compounds. The oxidized end product is often H_2SO_4, so these bacteria acidify their environment in the normal course of their metabolism. At least one species, *Tb. ferrooxidans*, can also oxidize Fe^{2+}, whose auto-oxidation proceeds slowly at low pH values. Some *Thiobacillus* cells grow optimally at about pH 2 but maintain a cytoplasmic pH near 6.5.

A moderately thermophilic archaeon, *Picrophilus oshimae*, represents the most extreme acidophile known to date. This organism, isolated from the soil of a Japanese geothermal field, grows optimally in dilute sulfuric acid at about pH 0.9. *P. oshimae* cells can grow at pH 0 but not at pH values above 3.5. At pH values greater than 5, the cells lyse (Schleper *et al.,*1995).

Bacteria that require low temperatures are termed **psychrophiles**, and typically occur in marine, arctic, and antarctic habitats. Enzymes of these organisms may denature, and the cells may die, at room temperature. In several cases, the minimum temperature for growth has not been determined, but lies below about –10 °C. The membrane phospholipids of these bacteria contain unusually high proportions of unsaturated and short-chain fatty acids (Morita, 1975).

Some bacteria exhibit extreme resistance to ionizing radiation. *Deinococcus radiodurans* was first discovered in a heavily gamma-irradiated can of food that subsequently spoiled. The radiation doses necessary to kill 63% of a population of *D. radiodurans* are about 1800 krad of gamma radiation and about 900 J/m^2 of 254-nm UV. These doses can be compared those required to kill 63% of a population of *E. coli*, that is, about 5 krad of X radiation and 30 J/m^2 254-nm UV (Gutman *et al.,* 1993; see Chapter 69, Effects of Ionizing Radiation). The extremely

high radiation resistance of *D. radiodurans* has been attributed to various DNA repair enzymes, multiple copies per cell of the circular chromosome, and a particularly active DNA recombinase. These features allow a *D. radiodurans* cell to reconstruct an intact chromosome from at least 100 individual fragments formed by double-stranded breaks (Gutman *et al.*, 1993).

Prokaryotes that grow optimally at 80 °C or higher temperatures are termed **extreme thermophiles** or **hyperthermophiles** (Stetter *et al.*, 1990). This group consists almost entirely of archaea isolated from geothermal habitats such as terrestrial hot springs, and submarine thermal vents. Aerobic hyperthermophiles include *Sulfolobus* spp., which require acidic conditions as well as high temperatures. The combination of high temperature and low pH required for optimal growth (about 80 °C and pH 3) is extremely effective at denaturing proteins. The *Sulfolobus* cell helps protect its cytoplasmic proteins, which appear to be generally heat-stable but not acid-stable (Grogan, 1996b), by maintaining its cytoplasmic pH at about 6 (Schäfer *et al.*, 1990). Physical properties of the unique lipids of *Sulfolobus* spp. probably help maintain the resulting ΔpH. The ether-linked phytanyl chains of these lipids span the entire membrane, forming a stable monolayer. Such membranes exhibit very low rates of proton leakage at high temperature, compared to ester-linked phospholipid bilayers (Van De Vossenberg *et al.*, 1995).

The most thermophilic organisms known are anaerobic archaea isolated from submarine vents, where hydrostatic pressure permits liquid water to be superheated. *Pyrolobus fumarii*, the most thermophilic organism yet cultivated, grows optimally at about 106 °C (Blöchl *et al.*, 1997). The very existence of cells with such growth requirements raises fundamental questions regarding their need to stabilize nucleic acids, proteins, and other molecules necessary for all life. The thermostability of enzymes of most hyperthermophiles appears to be intrinsic, that is, not due to counterions or other solutes. Biophysical analyses indicate that several classes of interactions of the amino acid residues contribute to this intrinsic structural stability (Jaenicke and Böhm, 1998). The DNA of hyperthermophiles, in contrast, has no discernable mechanism for intrinsic thermostability, and is subject to chemical decomposition at these temperatures. The need for special stabilization or repair of the DNA of hyperthermophiles, and the biochemical mechanisms which may be employed, remain unclear and relatively unexplored (Grogan, 1998).

VIII. Summary

Prokaryotes represent the smallest and simplest living cells. Some mediate the global cycling of elements, others cause disease, and still others provide some of the best-defined systems available for elucidating cellular structure and function in molecular terms. The "minimal cell", that is, that basic set of structures common to all prokaryotic cells, consists of a nucleoid and cytoplasm enclosed in a cytoplasmic membrane. However, most free-living prokaryotes also have some type of cell wall, and many bacteria have a second lipid membrane outside the cell wall. The apparently simple substructures of prokaryotic cells are functionally complex. They mediate all of the cell's diverse functions, including nutrient uptake, energy conversion and conservation, growth, secretion, genome replication, cell division, and regulation of gene expression in response to environmental change.

Prokaryotes employ a wide range of metabolic strategies, including many not represented among eukaryotes. Several alternatives of CO_2 fixation and photosynthesis are represented among bacteria and archaea, of which only one, that of the cyanobacteria, has been adopted by green plants. Some prokaryotes grow chemoautotrophically, some can fix N_2, and some can "respire" anaerobically, using electron acceptors other than O_2 in the environment.

Bacteria depend on solute-specific transport proteins located in the CM to accumulate nutrients and to move inorganic ions into and out of the cell. Those proteins involved in active transport display a high affinity for the external solute and expend energy to accumulate it to high intracellular levels. A common type of active transport in gram-negative bacteria utilizes soluble proteins which bind the solute in the periplasm and deliver it to the membrane-bound transporter.

Some membrane transporters mediate homeostatic responses to environmental stress. Others mediate "two-component" regulation of bacterial gene expression, by which bacteria change their macromolecular compositions in response to external conditions. This form of adaptation can be accomplished rather quickly in many bacteria, due to their high biosynthetic capacity relative to cell mass.

The functional diversity of prokaryotic cells becomes apparent in extreme environments. Certain bacteria require low external pH or low temperature to survive and grow, and others can cope with very high levels of DNA damage. Certain archaea require extremely high salt concentration or extremely high temperature for normal cellular function. Because these archaea differ radically from *E. coli* and other well-studied microorganisms, pinpointing the physiological and biochemical properties of their cells relevant to their harsh environments remains a challenging area of research.

Bibliography

Baumeister, W., Wildhaber, I., and Phipps, B. (1989). Principles of organization in eubacterial and archaebacterial surface proteins. *Can. J. Microbiol.* **35**, 215–227.

Blöchl, E., Rachel, R., Burggraf, S., Hafenbrandl, D., Jannasch, H., and Stetter, K. O. (1997). *Pyrolobus fumarii,* gen. and sp. nov., represents a novel group of archaea, extending the upper temperature limit for life to 113 °C. *Extremophiles* **1**, 14–21.

Caldwell, D. R. (1995). "Microbial Physiology and Metabolism." William C. Brown Publishers, Dubuque, IA.

Cashel, M., and Rudd, K. E. (1987). The stringent response. *In* "*Escherichia coli* and *Salmonella typhimurium*: Cellular and Molecular Biology" (F. C. Neidhardt, Ed.), American Society for Microbiology, Washington, DC. pp. 1410–1438.

Costerton, J. W., Cheng, K. J., Geesey, G. G., Ladd, T. I., Nickel, J. C., Dasgupta, M., and Marrie, T. J. (1987). Bacterial biofilms in nature and disease. *Annu. Rev. Microbiol.* **41**, 435–464.

Csonka, L. N. (1989). The physiological and genetic responses of bacteria to osmotic stress. *Microbiol. Rev.* **53**, 121–147.

Drlica, K. (1987). The nucleoid. *In* "*Escherichia coli* and *Salmonella typhimurium*: Cellular and Molecular Biology" (F. C. Neidhardt, Ed.), American Society for Microbiology, Washington, DC. pp. 91–103.

Dutton, P. L. (1986). Energy transduction in anoxygenic photosynthesis. *In* "Photosynthesis III: Photosynthetic Membranes and Light-Harvesting Systems" (L. A. Staehelin, and C. J. Arntzen, Eds.), Springer-Verlag, Berlin. pp 197–237.

Furlong, C. (1987). Osmotic-shock-sensitive transport systems. *In* "*Escherichia coli* and *Salmonella typhimurium*: Cellular and Molecular Biology" (F. C. Neidhardt, Ed.), American Society for Microbiology, Washington, DC. pp. 768–796.

Grogan, D. W. (1996a). Isolation and fractionation of cell envelope from the extreme thermo-acidophile *Sulfolobus acidocaldarius. J. Microbiol. Meth.* **26**, 35–43.

Grogan, D. W. (1996b). Organization and interactions of cell envelope proteins of the extreme thermoacidophile *Sulfolobus acidocaldarius. Can. J. Microbiol.* **42**, 1163–1171.

Grogan, D. W. (1998). Hyperthermophiles and the problem of DNA instability. *Molec. Microbiol.* **28**, 1043–1050.

Gunsalus, R. P. (1992). Control of electron flow in *Escherichia coli*: Coordinated transcription of respiratory pathway genes. *J. Bacteriol.* **174**, 7069–7074.

Gutman, P. D., Fuchs, P., Ouyang, L., and Minton, K. W. (1993). Identification, sequencing, and targeted mutagenesis of a DNA polymerase gene required for the extreme radioresistance of *Deinococcus radiodurans. J. Bacteriol.* **175**, 3581–3590.

Jaenicke, R., and Böhm, G. (1998). The stability of proteins in extreme environments. *Curr. Op. Struct. Biol.* **8**, 738–748.

Kushner, D. J. (1985). The Halobactericeae. *In* "The Bacteria" Vol. VIII, "Archaebacteria" (J. R. Socatch, and L. N. Ornston, Eds.), Academic Press, Orlando. pp. 171–214.

Maloney, P. (1987). Coupling to an energized membrane: Role of ion-motive gradients in the transduction of metabolic energy. *In* "*Escherichia coli* and *Salmonella typhimurium*: Cellular and Molecular Biology" (F. C. Neidhardt, Ed.), American Society for Microbiology, Washington, DC. pp. 222–243.

Messner, P., and Sleytr, U. (1992). Crystalline bacterial cell-surface layers. *Adv. Microb. Physiol.* **33**, 212–275.

Morita, R.Y. (1975). Psychrophilic bacteria. *Bacteriol. Rev.* **89**, 144–167.

Neidhardt, F. C., Ingraham, J. L., and Schaechter, M. (1990). "Physiology of the Bacterial Cell." Sinauer Associates, Sunderland, MA.

Nikaido, H., and Vaara, M. (1985). Molecular basis of bacterial outer membrane permeability. *Microbiol. Rev.* **49**, 1–32.

Nikaido, H., Kim, S.-H., and Rosenberg, E. Y. (1993). Physical organization of lipids in the cell wall of *Mycobacterium chelonae. Molec. Microbiol.* **8**, 1025–1030.

Raetz, C. (1987). Biosynthesis of lipid A in *Escherichia coli*. *In* "*Escherichia coli* and *Salmonella typhimurium*: Cellular and Molecular Biology" (F. C. Neidhardt, Ed.), American Society for Microbiology, Washington, DC. pp. 498–503.

Schäfer, G., Anemüller, S., Moll, R., Meyer, W., and Lübben, M. (1990). Electron transport and energy conservation in the archaebacterium *Sulfolobus acidocaldarius. FEMS Microbiol. Rev.* **75**, 335–348.

Schleper, C., Pühler, G., Holz, I., Gambacorta, A., Janekovic, D., Santarius, U., Klenk, H.-P., and Zillig, W. (1995). *Picrophilus* gen. nov., fam. nov.: A novel heterotrophic, thermoacidophilic genus and family comprising archaea capable of growth around pH 0. *J. Bacteriol.* **177**, 7050–7059.

Schwartz, R. M., and Dayhoff, M. O. (1978). Origin of prokaryotes, eukaryotes, mitochondria, and chloroplasts. *Science* **199**, 395–403.

Schwarz, U., Asmus, A., and Frank, H. (1969). Autolytic enzymes and cell division of *Escherichia coli. J. Mol. Biol.* **41**, 419–429.

Skulachev, V. P. (1994). The latest news from the sodium world. *Biochim. Biophys. Acta* **1187**, 216–221.

Stetter, K. O., Fiala, G., Huber, G., Huber, R., and Segerer, A. (1990). Hyperthermophilic microorganisms. *FEMS Microbiol. Rev.* **75**, 117–124.

Strom, M., and Lory, S. (1993). Structure-function and biogenesis of the type IV pili. *Annu. Rev. Microbiol.* **47**, 565–596.

Van De Vossenberg, J., Ubbink-Kok, T., Elferink, M., Driessen, A., and Konings, W. N. (1995). Ion permeability of the cytoplasmic membrane limits the maximum growth temperature of bacteria and archaea. *Molec. Microbiol.* **18**, 925–932.

Wanner, B. L. (1993). Gene regulation by phosphate in enteric bacteria. *J. Cell. Biochem.* **51**, 47–54.

Westerhoff, J. V., and Welch, G. R. (1992). Enzyme organization and the direction of metabolic flow: Physicochemical considerations. *In* "Current Topics in Cellular Regulation" (E. R. Stadtman, and P. B. Chock, Eds.), Vol. 33, Academic Press, New York.

White, D. (1995). "The Physiology and Biochemistry of Prokaryotes." Oxford University Press, New York.

Woese, C. R., Kandler, O., and Wheelis, M. (1990). Towards a natural system of organisms: Proposal for the domains Archaea, Bacteria, and Eucarya. *Proc. Nat. Acad. Sci. U.S.A.* **87**, 4576–4579.

Yanofsky, C., and Crawford, I. P. (1987). The tryptophan operon. *In* "*Escherichia coli* and *Salmonella typhimurium*: Cellular and Molecular Biology" (F. C. Neidhardt, Ed.), American Society for Microbiology, Washington, DC. pp. 1453–1472.

SECTION VIII

Plant Cells, Photosynthesis, and Bioluminescence

Richard G. Stout and Lawrence R. Griffing

63

Plant Cell Physiology

I. Introduction

Superficially, plants and animals appear to be completely different forms of life. Sedentary flowering plants may seem alien to motile warm-blooded mammals like us. However, if we look at the cellular level, we see that animals and plants are not so different and unrelated. Plant cells and animal cells are connected evolutionarily, sharing a common ancestor, but separated by over half a billion years of evolution.

Despite this length of time, plant and animal cells are remarkably similar with regard to fundamental cellular processes such as chemiosmotic adenosine triphosphate (ATP) synthesis, DNA replication, and protein synthesis. Membranes and mem-

brane-bound organelles, such as mitochondria, nuclei, Golgi, and endoplasmic reticulum (ER), appear similar in both plant and animal cells, although differences can be detected at the biochemical and molecular levels. The differences between plant and animal cells become more evident at the higher order cellular functions, such as cell replication, transport, defense, and intercellular communication.

II. Plant Cell Ultrastructure

The major organelle systems of plant cells are illustrated in Fig. 1. Throughout the plant's life, new cells are typically pro-

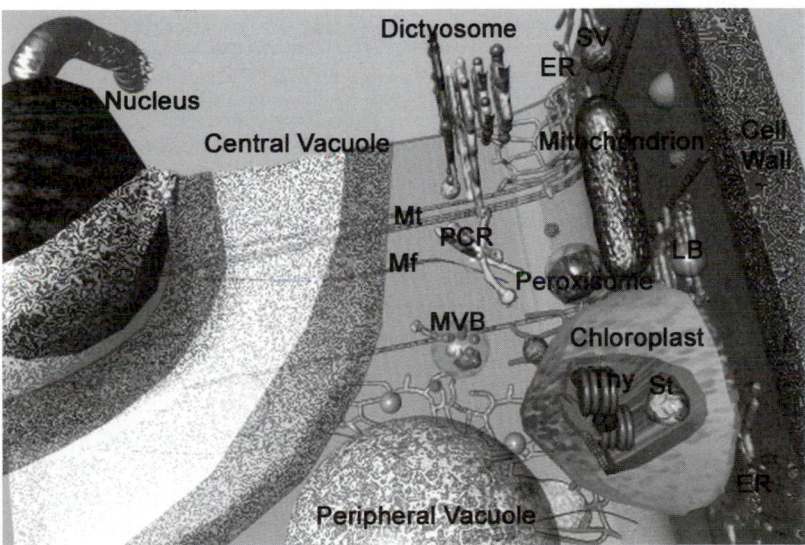

FIGURE 1. An illustration in 3D perspective showing the most important organelle systems of plants. The relative sizes of the organelles are accurate. The plasma membrane of the cell is juxtaposed to the cell wall. Adjacent to the plasma membrane is the cortical endoplasmic reticulum (ER). Microtubules (Mt) and microfilaments (Mf) are also associated with the cell cortex. The membrane (tonoplast) of the large central vacuole wraps around the nucleus residing in a trans-vacuolar strand. Another distinct vacuole type is labeled as a peripheral vacuole. The Golgi complex is represented by the dictyosome and the secretory vesicle (SV). The chloroplast is shown in cut-away view revealing the interior thylakoids (Thy), starch grains (St) and lipid droplets. The endolysosomal pathway is represented by the multivesicular body (MVB) and the partially-coated reticulum (PCR). The organelles of oxidative metabolism, the mitochondrion, and the peroxisome, are represented near a lipid body (LB).

duced in localized regions of relatively high mitotic activity called **meristems.** Meristematic regions are usually found at the tips of shoots and roots and at the base of branch shoots. Compared with meristematic cells, mature plant cells are distinguished by a large central vacuole, Fig. 2. During growth, this vacuole expands through the uptake of water, ions, and other solutes. Proper regulation of water and ion transport properties of this vacuole is necessary for normal plant growth and development. However, it has recently been "rediscovered" that not all vacuoles in a plant cell are the same. Different vacuoles exist side-by-side together in the same cell (Okita and Rogers, 1996). While some expand during growth (large central vacuole in Fig. 2A and B), others are the sites of storage, nutrient cycling, or other lytic events (small peripheral vacuole in Fig. 2A and B).

Along with the vacuole, the plastids (including chloroplasts) and the cell wall are the main structural features of plant cells that distinguish them from animal cells. The cell wall affects virtually every aspect of plant cell physiology, including development, transport, defense, and osmoregulation (Burgess, 1985; Tolbert, 1980). Morphological and physiological specialization of plant cells occurs primarily through modifications of the cell wall and plastids.

A. Cell Wall Structure and Biosynthesis

The walls of plant cells consist of a skeleton of interwoven microfibrils embedded in an amorphous matrix (Bacic

A

Meristematic Cell

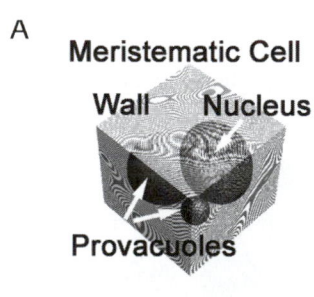

B

Mature Cell

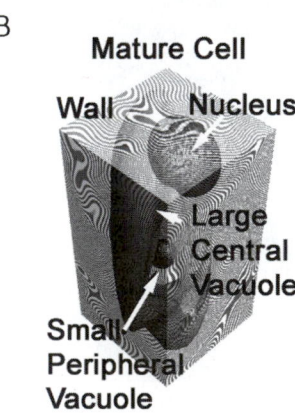

FIGURE 2. A meristematic cell (A) undergoes elongation to form a mature cell (B) through wall extension and uptake of water into the large central vacuole. One of the provacuoles (diagrammatically shown here as spherical) takes up the water and enlarges. The nucleus is included for reference to show that many organelles do not change in size during cell maturation.

et al., 1988). Cellulose (β1,4-linked poly-D-glucose) is the microfibrillar component of plant cell walls (Fig. 3A). Cellulose molecules are linear ribbonlike polymers of 500 to 2000 glucose residues, and in some cases achieving polymerization of as many as 23 000 glucose residues. Cellulose microfibrils generally consist of 40 to 70 cellulose molecules, running in parallel, overlapping, and oriented with the same polarity. The linear nature of the cellulose molecule is due to the β1-4 linkage between the glucose units. Hydrogen bonding occurs among the linear cellulose chains, but van der Waals forces are also important in promoting the formation of a highly ordered crystalline aggregate (Delmer, 1999).

The cellulose microfibrils are cross-linked in the plant cell wall by the matrix polysaccharides, the hemicelluloses, and the pectins. The hemicelluloses (Fig. 3B) are a mixture of branched heterogeneous polymers of glucose, xylose, galactose, arabinose, and other sugars. Structurally, the hemicelluloses are typified by a backbone of β1-4-linked glucose or xylose, from which relatively short side chains extend. The long backbone of the hemicellulose molecules may hydrogen bond to the surface of a cellulose microfibril, with the side chains protruding perpendicular to the surface of the microfibril.

A heterogeneous complex family of branched polymers, the pectins (Fig. 3C), is enriched in galacturonic acid residues. Their major function is probably to cross-link cellulose microfibrils by attaching to the microfibril-bound hemicelluloses (Fig. 4).

In addition to hemicelluloses and pectins, the cell wall matrix contains both structural glycoproteins and enzymes. Cell wall proteins have been classified based on repeating amino acid motifs of glycine (the glycine-rich proteins), proline (the proline-rich proteins), and hydroxyproline (the hydroxyproline-rich proteins), or by the presence of covalently-attached oligosaccharides containing arabinogalactans (the arabinogalactan proteins). Some wall proteins cannot be classified in these terms, being instead a mosaic of these protein domains (Carpita et al., 1996). The structural hydroxyproline-rich glycoprotein extensin is often compared to the animal structural protein collagen, which is also rich in hydroxyproline, and is found in the extracellular matrix. Woven into and covalently bound to the polysaccharide components of the plant cell wall, extensin serves to structurally reinforce the wall (Varner and Lin, 1989). The rodlike extensin molecules may also be cross-linked to one another (via tyrosine residues), catalyzed by peroxidases present in the cell wall.

The relative proportion of these cell wall components depends on the type of cell wall. The polysaccharide composition of monocot cells differs from that of dicot cells. Monocots have hemicelluloses with relatively little xyloglucan, using glucuronoarabinoxylans instead; they also have different pectins. Cell walls also change during development of the cell. The initial cross-wall of dividing cells is structurally different, in some cells containing more callose (Samuels et al., 1995). As the wall develops, cellulose is made at the newly-formed plasma membrane of the cell plate and matrix polysaccharides are deposited. Regions of

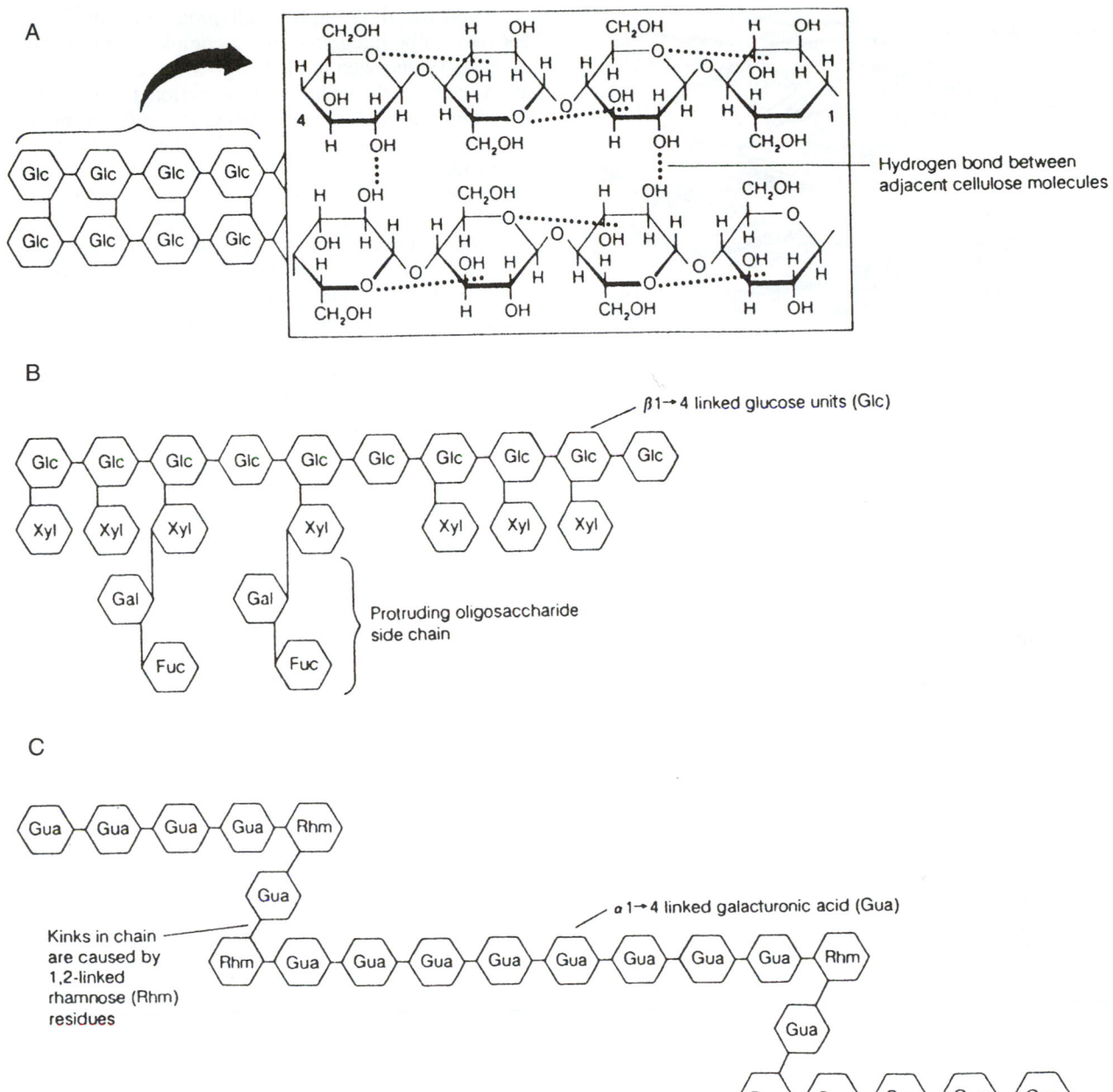

FIGURE 3. Structures of (A) cellulose, (B) hemicellulose, and (C) pectin. (From Taiz, L., and Zeiger, E. (1991). *Plant Physiology,* Copyright © 1991 Benjamin/Cummings Publishing Company and Dr. Eldon Newcomb, Department of Botany, University of Wisconsin, Madison, WI.)

specialization within the wall occur, with proteins and pectins moving outwards from the cell to form the middle lamella, the region between the walls of adjacent plant cells that may be dissolved to form intercellular air spaces. Air spaces appear between cells as a natural consequence of cellular enlargement.

During growth, the wall is considered to be a primary cell wall. After growth and division stops, the secondary wall forms. Cells may continue to deposit cellulose at their surface, but the wall is often rigidified through impregnation and cross-linking with **lignins,** polymers of aromatic alcohols. Lignins contribute to cell wall rigidity and greatly increase resistance to cell wall degradation. Cork cell walls, for example, contain very high levels of lignins.

Cellulose microfibrils, synthesized by enzymes in the plasma membrane, appear in the electron microscope to be pressed against the outer surface of the plasma membrane. A microfibril can sometimes be traced along the membrane surface until it ends at a structure called a **terminal complex** or **rosette.** Many plant cell biologists think these are cellulose synthases (Delmer, 1999). Multiple cellulose chains (36)

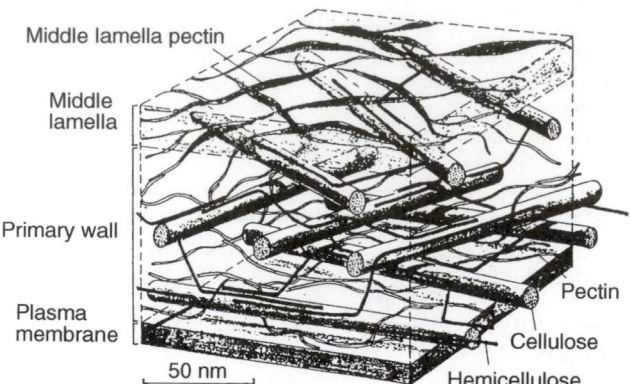

FIGURE 4. A simplified illustration of how the three main components of the primary cell wall of higher plants may be spatially arranged in three dimensions. (From McCann and Roberts, 1991, by permission.)

may be synthesized by a single rosette, and the structure of the rosette appears to be consistent with six subunits, each making six chains. These terminal complexes presumably are laterally mobile in the fluid plasma membrane, and as cellulose microfibrils are made, the complexes are pushed around the surface of the plasma membrane, as shown in Fig. 5 (Giddings and Staehelin, 1991). The cellulose synthases likely use cytoplasmic UDP-glucose as the building blocks for cellulose.

The deposition of cellulose microfibrils at the cell surface is not random. The morphology of a plant cell is determined by the orientation of these cellulose microfibrils in the cell wall. A good correspondence between microtubule alignment and the alignment of cellulose microfibrils near the plasma membrane exists. These observations have given rise to the hypothesis that the cellulose synthase complexes may somehow be guided as they move laterally through the lipid bilayer matrix of the plasma membrane by one or two microtubule "tracks," see Fig. 5.

To generate the complex wall, proteins synthesized in the ER have to be integrated into a complex with the polysaccharides synthesized in the Golgi. Much of the final associations apparently occur after secretion takes place, with movement through the wall being controlled somehow by the cell (Stout and Griffing, 1993). There is probably some organization of the synthesis and secretion of polysaccharides within the Golgi membranes. Immunocytochemical work from sycamore suspension culture cells (Zhang and Staehelin, 1992) and *Vicia* root hairs (Sherrier and VandenBosch, 1994) indicate that xyloglucan is synthesized in the *trans* Golgi cisterna, while methyl-esterified pectin is first detectable in the medial Golgi cisterna, Fig. 6. Double labeling shows that vesicle matrix polysaccharides co-exist in the same vesicle (Sherrier and VandenBosch, 1994). Labeling occurs throughout the stack when antibodies to glycoprotein oligosaccharides are used. The organization of the plant Golgi stack, at least in the scale-forming algae that make cellulosic scales in the Golgi, provides the paradigm that Golgi cisternae mature, with the *cis* cisternae maturing into medial and then *trans* cisternae (Mollenhauer and Morré, 1980). Clear evidence for cisternal maturation has recently been found in animal cells, which until recently had been widely thought to transfer Golgi products only through vesicle shuttles between the cisternae (Bonfanti *et al.*, 1998).

B. Cell Wall Functions

Cell walls play roles in addition to serving as scaffolding for the society of plant cells that compose a living plant. For example, cell walls provide a physical barrier to insects and microbial pathogens.

In addition to acting as a suit of armor, the cell wall allows plants to take up water. Plant cells take up water passively, that is, by osmosis. Thus, to remain hydrated, the cell's water potential must be lower (more negative) than its surroundings. Plant cells must be hypertonic to their immediate environment. In this condition, the cells would eventually

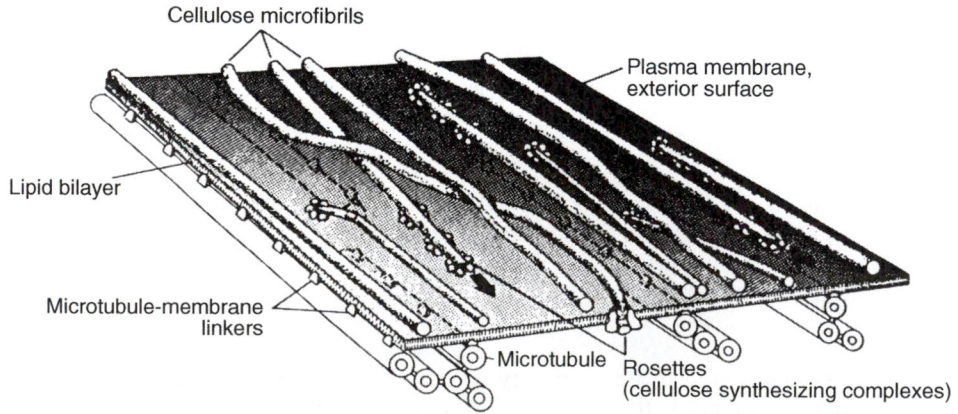

FIGURE 5. A model depicting microtubule-directed cellulose microfibril deposition during primary cell wall formation. (From Staehelin and Giddings, 1982. Copyright © John Wiley and Sons, Inc. Reprinted by permission of John Wiley and Sons, Inc.)

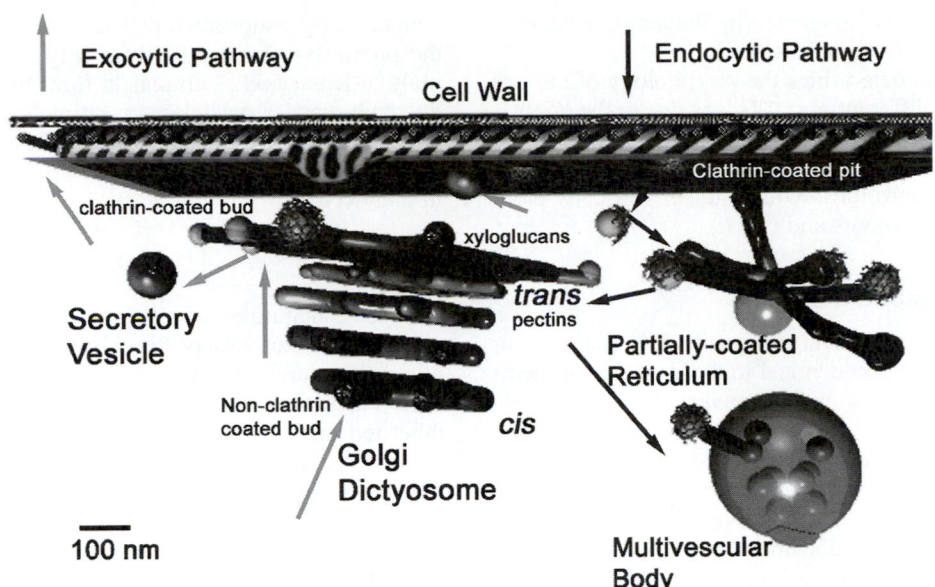

FIGURE 6. A 3D representation of the organelles involved in the pathway of polysaccharide exocytosis and membrane endocytosis during plasma membrane recycling. Clathrin-coated buds are associated with the plasma membrane and the trans-Golgi as well as the partially-coated reticulum and tubular extensions of the multivesicular body. Matrix polysaccharides are assembled in the Golgi, with pectin assembly starting first. Pectins and xyloglucans are the secreted to the plasma membrane via secretory vesicles from the *trans* cisterna. Endocytosis from coated pits moves through the partially-coated reticulum to the trans-Golgi, then the other cisternae and the multivesicular body. Some endocytosed molecules are recycled while others are deposited in vacuoles.

swell and burst without rigid cell walls. Since the cell wall prevents swelling, a relatively small amount of water uptake leads to a large increase in hydrostatic pressure (turgor pressure) within the cell. As turgor pressure increases, the cell approaches osmotic equilibrium with its surroundings (Taiz and Zeiger, 1998).

Cell turgor pressure is critical to plants. For example, it is responsible for the distention of leaves. Moreover, reversible changes in cell turgor pressure allow the regulation of gas exchange between the leaf and the environment through pores on the leaf epidermis called **stomata.** Stomata consist of two kidney-shaped cells, called guard cells, that change shape with changes in their turgor pressure. This reversible change in shape leads to the opening and closing of the pores. The most important role of plant cell turgor pressure is, however, to drive cell enlargement. Without plant cell turgor pressure plants would not be able to grow.

C. Developmental Aspects of the Cell Wall

Plant cell division cannot occur without the formation of a new cell wall. In plants, the final phase of cell division (cytokinesis) is characterized by the formation of a cell plate between daughter cells. Unlike animal cells, the site of future cell plate formation is predicted by a band of cortical microtubules that occurs in preprophase. Hence, cytoplasmic, as well as spindle, microtubules may serve to direct the orientation of cell plate formation along the plane of cell division. The cell plate results from the concerted fusion of Golgi-derived, and perhaps plasma membrane-derived, vesicles and tubules in the region of the cytoplasm bisecting the spindle (Samuels *et al.*, 1995).

Soon after a plant cell divides, it begins to enlarge. Most of a plant's size comes from the growth of its cells. In general, plant cells enlarge much more than animal cells. Plant cells may grow in volume by 10- to 50-fold. Most of this increase in volume is due to water uptake into the central vacuole.

For uptake to occur, either the osmotic solute concentration in the cell must increase or the cell's turgor pressure must decrease. In growing plant cells, the turgor pressure decreases because of a special process called **stress relaxation.** Because water is not compressible, only a slight "give" in the cell wall is sufficient to reduce the cell's turgor pressure and, simultaneously, the stress on the wall. With this drop in turgor pressure, the cell's water potential is lowered (more negative) relative to its surroundings, and water uptake occurs. For plant cell growth to be sustained, the cell strikes a balance between water uptake (tends to increase turgor pressure) and cell wall extension (tends to decrease turgor pressure). During this growth phase, the plant cell accumulates enough solutes to maintain the osmotic concentration as the cell volume increases.

In most nongrowing plant cells, the cell wall is rigid, except for a slight elasticity. For the cells to enlarge, the cell wall must irreversibly extend in response to the cell's turgor pressure. It must become less rigid throughout, that is, "loosened." This increased wall extensibility depends on the rates of complex biochemical processes in the cell wall. Several mechanisms have been proposed for this. One is called the acid growth theory (Rayle and Cleland, 1992). According to this hypothesis, the growing plant cell actively extrudes protons into the cell wall space (probably via ATPase proton pumps in the plasma membrane). The resulting lowered pH may activate enzymes called expansins (Taiz, 1994; Carpita *et al.*, 1996) present in the cell wall that cleave load-bearing bonds. "Acid growth" is augmented by the active deposition of new cell wall material via exocytosis.

The cell wall also determines the morphology of the cell during growth and differentiation. This is due to the orientation of the cellulose microfibrils around the cell (see Fig. 6 and Section II.A). For example, a horizontal barrel-hoop arrangement of the microfibrils will result in vertically elongated cells typical of shoots and roots.

D. Plant Endomembrane Systems

The physical properties and composition of plant membranes are similar, hut not identical to those of animal membranes. For example, although both membrane systems have sterols, plants have little or no cholesterol in their membranes. Plant and animal cells also generally resemble each other when it comes to fundamental membrane-related processes, but may differ considerably in detail. For example, the plasma membrane of plant cells is constructed, replenished, and recycled as it is in animal cells, except during the process of cell plate formation, a process with few analogs in the animal kingdom. Generally, Golgi-derived vesicles fuse with the plasma membrane, adding lipids and proteins to it in addition to depositing their contents in the cell wall region. As described above, Golgi vesicles contain a wide range of cell wall matrix polysaccharides as well as structural and enzymatic glycoproteins. The glycoproteins are synthesized by ER and their oligosaccharides are modified in the Golgi, while the polysaccharides are synthesized by Golgi enzymes. In contrast, the Golgi apparatus in animal cells does not synthesize polysaccharides, but packages and secretes mostly glycoproteins that are differently glycosylated from plant cell glycoproteins. For example, plant cell glycoproteins commonly contain xylose and fucose modifications, while animal cell glycoproteins often contain sialic acid modifications.

Just as Golgi contents differ in plants and animals, so do the shape, cellular location, and method of replication of the Golgi apparatus. The Golgi apparatus in plants is composed of one to hundreds of individual dictyosomes, Figs. 1 and 6, often distributed throughout the cell. In contrast, most animal cells contain one to ten connected Golgi stacks per cell found in the perinuclear or pericentriolar region only. The Golgi stacks in animal cells are thought to break down during the later part of the cell cycle, allowing uniform inheritance of Golgi membranes by daughter cells. In single-celled algae with only one dictyosome, the Golgi divides once, with a single daughter Golgi stack going to each daughter cell. In plants with multiple dictyosomes, Golgi replication occurs, but no apparent breakdown of the stacked structures occurs, as it does in animal cells (Griffing, 1991).

The plant cell plasma membrane is recycled in a manner reminiscent of animal cells. Clathrin-coated pits have been observed in plant cells, Fig. 6 (Griffing, 1991), and evidence suggests that the macromolecular components of the plasma membrane are recycled via endocytosis (Griffing, 1991; Battey et al., 1999). Clathrin-coated membranes as well as uncoated membranes are involved in the deposition of proteins in the vacuole. The degree of similarity between the endolysosomal compartment in animal cells and that in plants has only been explored on the structural level. Plants contain early endosomes that are partially clathrin-coated, the **partially-coated reticulum** (Figs. 1 and 6), as well as multivesicular bodies similar in form to those found as endosomal carrier vesicles in animals, (Figs. 1 and 5). Different routes of delivery of content to the vacuoles may give rise to the different vacuole populations that are found in plant cells.

E. Vacuoles

In some mature plant cells, the central vacuole may constitute more than 90% of the cell volume. Meristematic and other immature cells may have many small, distinct vacuoles called **provacuoles** (see Fig 2A), that could either be inherited or derived from the Golgi apparatus. Maturation of the vacuole is a complicated phenomenon. Some individual provacuoles may coalesce as the cell grows to maturity, helping to build the large central vacuole. However, vacuoles also grow by absorption of water and direct addition of membrane precursors from the Golgi. Vacuoles often have complicated forms, ranging from simple tubes and spheres to branched, anastomosing globoid shapes. Cytoplasmic strands traverse the vacuole in channels necessarily bounded by the vacuole membrane, contributing to the complexity of form. It is clear that there are several populations of vacuoles in cells (Griffing and Fowke, 1985, Okita and Rogers, 1996).

The vacuole is bounded by a lipid bilayer membrane called the **tonoplast,** and it contains inorganic ions, metabolites (i.e., organic acids, sugars, amino acids), phenolic and alkaloid compounds, and hydrolytic enzymes. The acidic vacuolar pH (4 to 6) is often several pH units below the cytoplasmic pH (ca. pH 7.0).

The vacuole is the cell's principal storage site (except for oils, see next section). Vacuoles also may (1) help to defend the cell, (2) play a role in the sexual reproduction of plants, (3) function as a cytoplasmic homeostat, (4) recycle macromolecular components of the cell, and (5) most importantly, enable cells to grow. Different vacuoles may be specialized for these different functions.

As a large compartment separated from the cytoplasm, the vacuole serves as both a short- and long-term storage site. In the seed, different populations of vacuoles are made via different pathways to accommodate the different storage proteins (Okita and Rogers, 1996). Plant secondary metabolites or inorganic ions such a Na^+ that may be potentially harmful to cytoplasmic functions are sequestered by one or more different vacuoles in the cell. Organic acids such as malate may be temporarily stored in the vacuole, depending on the rate of photosynthesis. Toxic compounds produced by cells are permanently stored in the vacuole. These noxious compounds may discourage insects and other herbivores from eating such plant tissues. Vacuoles may also help to attract beneficial insects. Colorful pigments such as anthocyanins may be stored in the vacuole of flower petal cells, attracting insect pollinators.

The metabolic and signal transduction processes in the cell cytoplasm are very sensitive to changes in the H^+ and Ca^{2+} concentration. The vacuole serves as both source and sink for these two ions. Transport systems located in the tonoplast (discussed later in Section IV) actively mediate this dynamic ebb and flow of H⁺ and Ca²⁺.

Sometimes referred to as the cell's garbage dump, the vacuole is better described as the cell's recycling center, behaving like a giant lysosome. The hydrolytic enzymes in the vacuole digest proteins, RNA, lipids, and other molecules that have outlived their usefulness. The resulting breakdown products may be reused by the cell or, in the case of dying cells, may be transported to other parts of the plant.

As previously mentioned, plant cells can grow much larger than animal cells because of the vacuole. The vacuole is much less complex and protein-rich than the cytoplasm, which costs more to support in terms of energy and mineral input (particularly nitrogen). For most plants, nitrogen is in short supply in the soil. Thus, by increasing their volume through water uptake primarily into the vacuole, plant cells can enlarge without having to support a larger, nitrogen-rich cytoplasm.

F. Lipid Bodies

Lipid bodies (Fig. 1), also called oleosomes, spherosomes, or oil bodies, are commonly found in most plant cells. They store triglycerides in the cell and are very abundant in oil seeds. Unlike intracellular fat bodies in animal adipocytes, the lipid body has a peculiar unilamellar outer membrane enriched in oleosin. The origin of lipid bodies is unknown, although they are thought to arise, at least in part, from membrane components originating in the ER (Huang, 1992). Plants have many important desaturases of lipids that are not found in animals, making these products valuable in industry and medicine (Topfer *et al.*, 1995).

G. Plastids

Chloroplasts are complex organelles that contain an outer and an inner membrane and a complex internal membrane system composed of thylakoids and lipid and starch inclusions (Fig. 1). They are only one example of a group of plant cell organelles called **plastids** (Burgess, 1985). Other plastids include proplastids, amyloplasts ("starch-containing"), and multicolored chromoplasts (found in fruits and flowers). The membrane system within the plastids, as well as their contents, is what defines the different types of plastids. Immature undifferentiated plastids, the proplastids (four to five times smaller than chloroplasts), are most commonly found in meristematic cells. These proplastids give rise to the other plastids.

The nature of the plastids found in a cell is determined by the cell type and the developmental stage of the tissue. Most plastid proteins (ca. 90%) are encoded in the cell nucleus. These genes are translated by cytoplasmic ribosomes, and the resulting polypeptides are targeted to the plastids and actively imported by them. Differential expression of these nuclear genes determines the kind of plastids present in a given cell (Mullet, 1988).

H. Microbodies, Glyoxysomes, and Peroxisomes

Microbodies, oxidative organelles of the plant cell, vary in form depending on their function. In mature leaves they are known as peroxisomes and are involved in photorespiration. While performing this role, they are often in close apposition to both the chloroplast and the mitochondrion. In the seed, where they are known as glyoxysomes, they are involved in beta-oxidation of lipids from storage triglycerides in lipid bodies. In root nodules, purine metabolism is their specialty. All of the microbodies can convert to other types, as protein turnover and targeting of new proteins to the organelle occur as the cell develops (Olsen and Harada, 1995).

I. Cytoskeleton

The plant cell cytoskeleton appears to be a dynamic structure composed of microtubules, actin filaments, and intermediate filaments (Lloyd, 1991). As previously discussed, the orientation of cortical microtubules may direct the deposition of cellulose microfibrils in the cell wall (see Section II.A). The orientation of the filaments that constitute the cytoskeleton may be affected by light, gravity, and plant hormones.

Cortical and transvacuolar actin filaments choreograph the motion of the cell's cytoplasm. This phenomenon, called **cytoplasmic streaming,** is commonly observed as a unidirectional streaming of cytoplasm in large vacuolated cells. It is also commonly observed as bidirectional rapid movements of organelles through transvacuolar strands or in thickened regions of the conical cytoplasm. Cytoplasmic organelles, such as mitochondria and chloroplasts, have been observed traveling along bundles of actin filaments at up to 75 μm/s in giant algal cells (Williamson, 1986). The mechanism of cytoplasmic streaming is the subject of some debate. From analysis of streaming in the giant *Characean* algae, a model has been put forward by which the ER movement along actin generates bulk flow that moves all of the other organelles in the cytoplasm by viscous drag (Kachar and Reese, 1988). However, many organelles have been shown to be able to stream independently on the actin-based cortex of the *Characean* alga, *Nitella,* and so must have their own cytoplasmic myosins associated with them (Kohno *et al.*, 1990). If these organelles have their own myosin-actin links, they may not be influenced by bulk flow, but track independently of other organelles.

The cytoskeleton also directs intracellular vesicular traffic. Although membrane traffic in plant cells is relatively insensitive to anti-microtubule drugs, it is very sensitive to drugs that disrupt the actin cytoskeleton (Griffing, 1991). Even though the actin cytoskeleton is intimately linked to streaming events, streaming movements are probably regulated differently from directed movement of organelles to their target compartment.

J. Plasmodesmata

Most plasmodesmata are established during the final stage of cell division. Strands of tubular ER become entrapped in the developing cell plate, leading to the formation of hundreds of tiny conduits in the wall separating the two new daughter cells. Under special circumstances, secondary plasmodesmata (having multiple branches united via a central cavity) may develop in nondividing cells.

As illustrated in Fig. 7, plasmodesmata are structurally complex (Robards and Lucas, 1990). They are lined with plasma membrane, thus linking the plasma membranes of adjacent cells. A single plasmodesma contains an axial component called the **desmotubule,** which passes through the plasmodesma and often connects the ER of adjacent cells. Indeed, the desmotubule may be an extension of the ER. In between the desmotubule and the plasma membrane is a sleeve of cytoplasm. The two neck regions of a plasmodesma may be ringed by nine or ten "5-nm subunits" which may govern aperture size (analogous to a sphincter).

III. Cell-to-Cell Communication

A. Intercellular Transport via Plasmodesmata

Plasmodesmata mediate transport between adjacent plant cells, as gap junctions facilitate transport between animal cells (Robards and Lucas, 1990). However, gap junctions and plasmodesmata possess no structural similarities.

Recent evidence has dramatically changed our view of plasmodesmata. Once considered mere conduits for the passive diffusion of metabolites and ions (typically 1 kDa or smaller), plasmodesmata have the ability to mediate the cell-to-cell transport of proteins and nucleic acids (Lucas, 1999). The demonstrated capacity of plasmodesmata to transport both plant and viral proteins, as well as the ability of some of these proteins to mediate the transport of their own mRNA, has led to the hypothesis that plants may function as supracellular organisms, in contrast to the multicellular mode adapted by animals (Lucas, 1999).

B. Plant Hormones

Like animals, plants depend on chemical signals for intercellular communication. Five plant hormones—auxin, cytokinin, gibberellin, abscisic acid, and ethylene—play many

roles in plant growth and development (Davies, 1995; Taiz and Zeiger, 1998). They are all relatively small molecules (Fig. 8), which allows them to easily pass through cell walls and plasmodesmata, and they are effective at low concentrations. Plant hormones often act at points remote from their site of synthesis. Recent evidence suggests that plants may also have steroid hormones (Hooley, 1996).

Each of these hormones may affect several different physiological processes during the course of a plant's life (see Table 1). A hormone may elicit a particular biological response depending on several factors. These include its point of action within the plant, the stage of development of the tissue, the concentration of the hormone, and the presence or absence of one or more of the other hormones.

C. Defensive Signals

Plant cells have the ability to defend themselves against attack by herbivores or pathogens (Maleck and Dietrich, 1999). As noted previously, vacuoles may contain metabolites, such as alkaloids, as constitutive chemical deterrents to attack. Cells also have inducible defensive responses, including (1) the biosynthesis of antibiotic compounds, collectively known as phytoalexins, toxic to some microorganisms; (2) the activation of genes coding for protease inhibitor proteins that inhibit the digestive serine proteases of microbial pathogens or insect herbivores; and (3) the increased deposition of extensin and lignins into the cell wall region to fortify the wall.

These inducible defensive responses can be elicited locally and systematically by signal molecules produced at the site of pathogen or insect attack. Such wounding results in the rapid accumulation of phytoalexins and protease inhibitors, not only in the wounded tissues but also in neighboring and distal unwounded parts of the plant, thereby indicating that a signal, or signals, is released from wounded cells to travel throughout the plant.

Specific cell wall fragments (oligosaccharides) of fungi and plants (Fig. 9) can induce defensive responses in cells (Ryan and Farmer, 1991). Because oligogalacturonides have

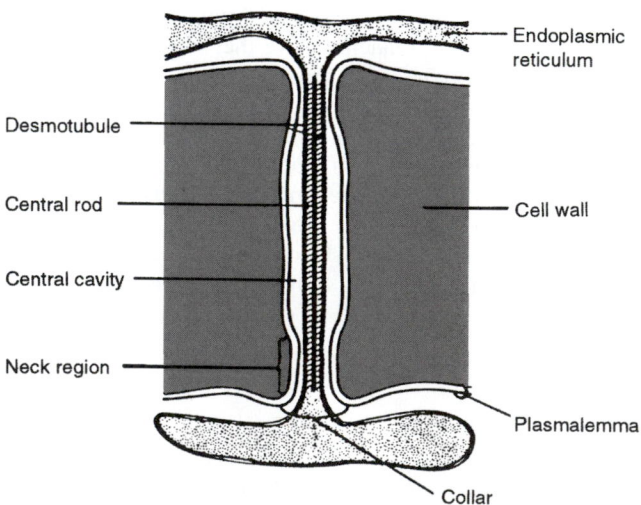

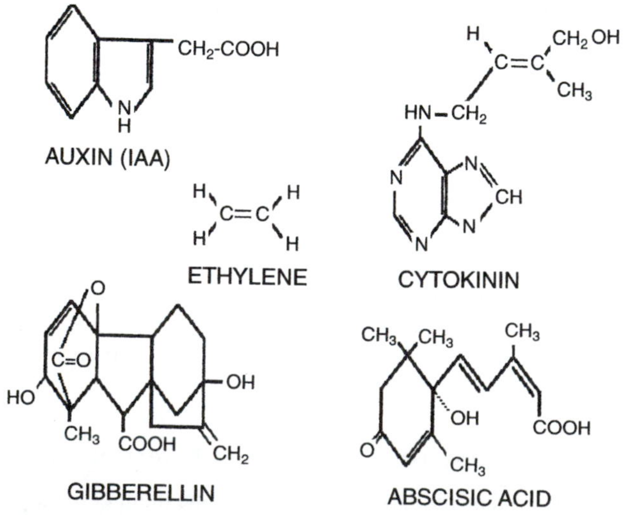

FIGURE 7. Diagram of the components of a simple plasmodesma as seen in longitudinal section. [Reproduced from Robards, A. W. (1976). Plasmodesmata in higher plants. *In* "Intercellular Communication in Plants: Studies on Plasmodesmata" (B. E. S. Gunning and A. W. Robards, Eds.), p. 31. Springer-Verlag, New York/Berlin, with permission.]

FIGURE 8. Structures of plant hormones.

TABLE 1 Primary Physiological Roles of Plant Hormones

Hormone	Main functions
Auxin	Promotes stem elongation, bud dormancy, and fruit development; stimulates formation of vascular tissue; promotes cell division
Ethylene	Promotes fruit ripening and lateral growth in stems
Cytokinin	Promotes cell division; delays senescence
Gibberellin	Promotes shoot elongation, flowering, and seed germination
Abscisic acid	Inhibits growth; promotes stomatal closure; promotes desiccation tolerance

been shown to move only a relatively short distance from a wound site, cell wall fragments may function chiefly as localized intercellular wound signals. Other, more mobile signals have also been reported.

An 18-amino-acid polypeptide, systemin (perhaps the first polypeptide hormone found in plants), is able to induce the synthesis of wound-inducible protease inhibitor proteins in several plant species (McGurl *et al.*, 1992). Systemin is relatively mobile within the plants tested to date.

Abscisic acid (ABA), ethylene, salicylate (structurally related to aspirin), and methyl jasmonate are other possible defensive signal molecules in plants. Both salicylate (Raskin, 1992) and methyl jasmonate, a lipid-derived molecule found in plants, trigger the production of protease inhibitors (Farmer and Ryan, 1990). Because of their volatile nature, methyl jasmonate and ethylene may function not only in intercellular communication but also in plant-plant communication (Staswick, 1992; Pieterse and van Loon, 1999).

D. Interactions between Plants and Other Organisms

Complex interactions exist between plants and other living organisms. A glimpse of this complexity is provided by the current understanding of how plant cells interact with two soil-borne bacteria.

One of these bacteria, *Rhizobium,* forms a symbiotic relationship with leguminous plants such as soybean, pea, and alfalfa (Long, 1989). This symbiosis provides the plant with a valuable mineral nutrient, namely, nitrogen. Plant cells cannot directly use nitrogen from the atmosphere. Plants must rely on mineral nitrogen (primarily NO_3^-) from the soil, and nitrate availability often limits plant growth and productivity. Some plants can obtain nitrogen derived from atmospheric N_2 by forming symbiotic relationships with prokaryotic organisms that fix nitrogen. This ability to incorporate or fix atmospheric N_2 into organic form is limited to only certain prokaryotes, including some cyanobacteria and bacteria in the genus *Rhizobium.* The plant receives a usable form of nitrogen from the *Rhizobium,* and the bacteria derive energy in the form of carbohydrates from the plant.

Legumes that thrive in nitrogen-poor soil likely have small, pea-size nodules that are packed with symbiotic nitrogen-fixing *Rhizobium* on their roots. The steps involved in establishing such a relationship are numerous and complex. The coordi-

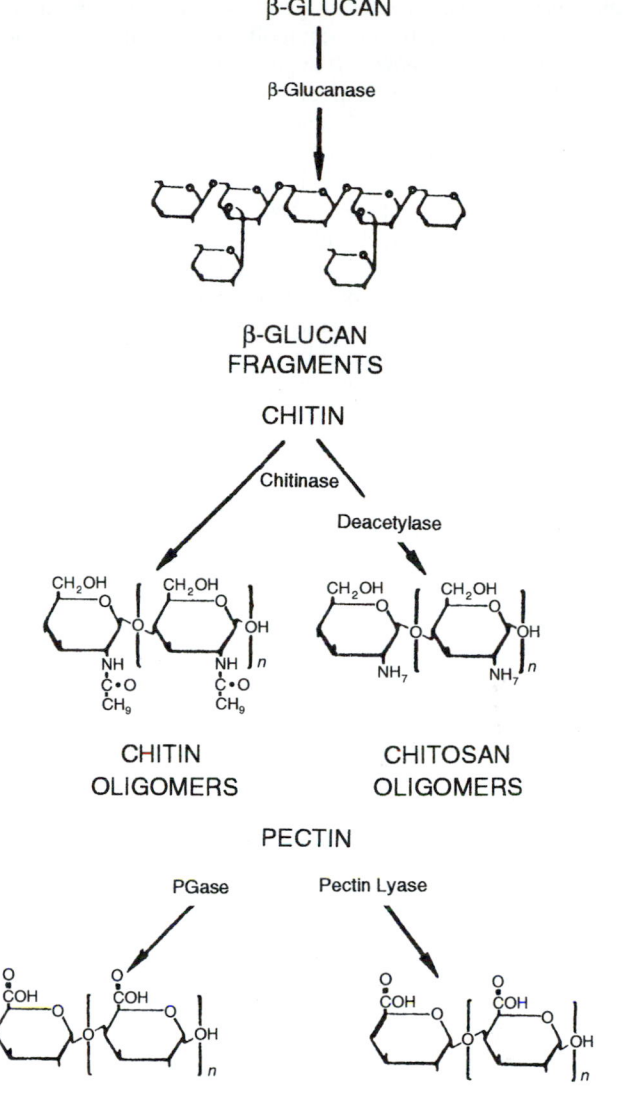

FIGURE 9. Enzymatic degradation of fungal and plant cell walls produces oligosaccharides that may function as localized defensive signals in plants. (Reprinted with permission from Ryan, 1988. Copyright 1988 American Chemical Society.)

nated expression of both bacterial and plant genes is critical for the successful formation and maintenance of this symbiosis. Recent advances in the molecular genetics of *Rhizobium* have revealed that most symbiotic genes are located on a large plasmid called the Sym plasmid. Sym plasmid genes involved in host-specific infection and nodule formation are referred to as nodulation (*nod*) genes. Some of these bacterial *nod* genes may be required for the production of a signal that leads to the expression in the host plant of a family of proteins, called nodulins, involved in various stages of nodule formation.

The release into the soil of chemical signals produced by the plants may be the first step toward symbiosis. For example, alfalfa root cells release flavones and other related compounds that preferentially attract the species *R. meliloti*. These compounds not only mediate this chemotactic response, but may also regulate bacterial genes (Geurts and Franssen, 1996). One of these flavones, luteolin, induces the expression of *nod* genes in *R. meliloti*. It apparently does so by activating a 33-kDa protein, constitutively expressed in these bacteria and coded for by the *nodD* gene (Fisher and Long, 1992).

Several of the *nod* genes are responsible for the production of signal molecules critical for the establishment of legume/*Rhizobium* symbiosis. These are the Nod factors, lipid-linked sulfated oligosaccharides that can modulate the plant response at extremely low concentrations (Spaink, 1995). The sensitivity and specificity of Nod factor interactions with host legumes speaks in favor of some extremely sensitive reception mechanism perhaps mediated by a receptor protein. Nod factors trigger predetermined morphogenic pathways in the roots of receptive legume plants leading to the formation of root nodules (Cohn *et al.*, 1998; Schultze and Kondorosi, 1998).

The other soil-borne bacterium, *Agrobacterium tumefaciens,* causes crown gall disease in plants. It transfers genes to infected plant cells, transforming them into tumor cells. The transformed cells grow and divide abnormally to form large galls on the plant.

Agrobacterium infects the plant only at wound sites. Ironically, wounded plant cells release a chemical that signals *Agrobacterium* to activate the infection process. This signaling molecule specifically activates *Agrobacterium* genes that mediate the infection process. These *vir* (virulence) genes, as well as the DNA transferred to the cell (T-DNA), are parts of a large plasmid of *Agrobacterium,* the Ti (tumor-inducing) plasmid. The T-DNA portion of the Ti plasmid is copied within *Agrobacterium,* transferred to the cell, and integrated into the cell nuclear genome. The expression of the T-DNA in the cells results in their transformation into tumor cells (Beijersgergen *et al.*, 1992).

The ability of *Agrobacterium* to transfer and stably integrate T-DNA into cells is the basis for most plant genetic engineering. Recombinant DNA technology can be used to replace tumor-causing genes from within the T-DNA with "desirable" genes. Then the T-DNA can be used as a vector for the genetic transformation of plants.

IV. Membrane Transport

A. Electrogenic Proton Pumps

Unlike animal cells, the cells of plants, fungi, and bacteria can live in an environment of little nutritional value. The ability to accumulate nutrients such as mineral ions, sugars, and amino acids, which may be present extracellularly at only minute levels, may be explained in two ways. First, these organisms produce a substantial resting membrane potential, and second, they rely on the ubiquitous proton, rather than Na^+, to generate this potential.

The primary active transport mechanisms in cell membranes generate electrochemical H^+ gradients to drive secondary active transport. The average plasma membrane potential in cells ranges from −120 to −180 mV (cytoplasmic side negative), with the electrogenic pump component being −60 to −120 mV. Electrogenic H^+ transport occurs not only in plants but also in bacteria and fungi. Some animal cells, such as those of the gas-

tric epithelium, also possess proton pumps. In plant cells, this electrogenic component is produced by proton-translocating ATPases (H^+-ATPases), which pump protons from the cytoplasm, across the plasma membrane, into the cell wall space (Sze *et al.*, 1999).

Recent efforts to clone and sequence plant plasma membrane H^+-ATPase genes have revealed differences from and similarities to animal P-type ATPases. The enzyme has eight membrane-spanning regions, with both the C- and the N-terminal regions present on the cytoplasmic side of the plasma membrane. The sequences of several plant P-type H^+-ATPases, except for certain hydrophilic regions, do not show great overall homologies with analogous enzymes in animal cells. One conserved region (containing aspartate) may be associated with the formation of the phosphorylated intermediate; another may be the binding site for ATP (Sze *et al.*, 1999).

There are multiple forms of the P-type H^+-ATPase genes in plants, and their expression may be cell- or tissue-specific (Palmgren and Harper, 1999). There is evidence that the plasma membrane H^+-ATPase exists as multigene families in oat, tobacco, and tomato. In *Arabidopsis thaliana,* characterized by a relatively small genome, at least 10 putative isoforms of plant P-type H^+-ATPase have been identified. At least one of these isoforms is expressed principally in roots.

Regulation of the plasma membrane H^+-ATPases may involve an autoinhibitor domain located within the C-terminal region. Reversible phosphorylation of certain amino acids of this domain, interaction of this domain with a 14-3-3 protein, or both, may stabilize the H^+-ATPase in a high-activity state (Palmgren and Harper, 1999; Sze *et al.*, 1999).

The plasma membrane may contain components of a redox chain involved in H^+/e^- transport across the membrane (Crane and Barr, 1989). Although it is not clear whether this system is electrogenic, it likely increases H^+ extrusion through exporting negative charge, thus stimulating the H^+-ATPases by reducing the "backpressure" of the charge gradient on the proton pumps. This redox system may be modulated by light.

A proton-motive force also drives transport at the vacuolar membrane (tonoplast), though this membrane potential (⁻90 mV, cytoplasmic side negative) is typically less than that of the plasma membrane. The primary active transport system in the tonoplast is a V-type H^+-ATPase that pumps protons into the vacuole (Sze *et al.*, 1999). These V-type H^+-ATPases are distinguished in several ways from the P-type ATPases in the plasma membrane. Structurally more complex, the vacuolar H^+-ATPases are large (450- to 650-kDa) oligomeric proteins consisting of 8 to 13 subunits. A comparison of vacuolar H^+-ATPases from several plant species indicates considerable variation in subunit composition. As with the plasma membrane H^+-ATPases, the cell genome may contain multiple genes coding for the individual subunits of the vacuolar H^+-ATPases. Regulation of the vacuolar V-type H^+-ATPase is poorly understood. Factors that affect the H^+-ATPases include Ca^{2+} concentration and cytoplasmic pH.

These factors also affect another type of electrogenic proton pump in the tonoplast. A proton translocating pyrophosphatase (PPase) may significantly contribute to the electrogenic component of the vacuolar membrane potential. Pyrophosphate (PP_i) may be generated by a variety of metabolic reactions in cells.

B. Ion and Water Channels

The applications of patch-clamp techniques to plants, as well as recent advances in the cloning, sequencing, and expression of putative ion channel genes, have demonstrated the presence of channels conducting K^+, Cl^-, Ca^{2+}, and water in the plasma membrane and vacuolar membrane of plant cells (Chrispeels *et al.*, 1999; Gzempinski *et al.*, 1999).

Present at relatively high concentrations in cells (typically 75-100 mM), K^+ activates cytoplasmic enzymes and serves as a major osmotic solute. In most plant cells, K^+ passively enters the cytoplasm down the relatively large electrical gradient across the plasma membrane (inside negative). Major pathways for this K^+ uptake by cells are probably K^+-influx channels. These are activated by membrane potentials more than −100 mV, generated by the action of the plasma membrane H^+-ATPases.

K^+ efflux is mediated by channels that are also regulated by the membrane potential. In the case of K^+-efflux channels, however, K^+ efflux is activated by the depolarization of the membrane potential to values more positive than −40 mV. The physiological roles for the K^+ efflux channels include rapid turgor/volume regulation in stomatal guard cells (see Section II.B) and motor tissue cells in plants as well as repolarization of action potentials (APs) in both plants and algae.

Plant cell K^+ channels differ both physiologically and biophysically from animal cell K^+ channels. In plant cells, the two types of K^+ channels mediate the long-term K^+ transport into and out of the cells. As in animal cells, K^+ channels help to reset the membrane potential close to the equilibrium potential for K^+ after the induction of short-term potential changes by other channels. Biophysically, plant voltage-dependent K^+ channels activate 10 to 100 times more slowly than those of animal cells. Despite these functional differences, plant and animal K^+ channels may have some structural similarities.

Several other plasma membrane ion channels have been reported in plants (Fig. 10). These include anion channels (particularly Cl^-) and nonselective stretch-activated channels (Tyerman, 1992). The latter may function as turgor sen-

sors and mechanosensors (Cosgrove and Hedrich, 1991). Signal-regulated Ca^{2+}-permeable ion channels may also be present in both the plasma membrane and the tonoplast (see separate discussion of Ca^{2+} transport later).

Patch-clamp studies of the vacuolar membrane indicate the presence of voltage-dependent ion channels that are modulated by cytosolic free Ca^{2+}. One is the SV-type channel, so named because the kinetics of activation of these currents are slow, hence the term "slow-vacuolar" (SV-type) currents, and the other is the FV-type channel, characterized by relatively "fast" kinetics. The SV-type channels are permeable to both cations and anions and are activated by an increase in cytoplasmic Ca^{2+} (>0.3 μM). The gating of the Ca^{2+}-dependent channels depends on both Ca^{2+} concentration and voltage across the membrane. The FV-type channels may be chiefly responsible for the physiological passage of anions into the vacuole. Another ion channel in the vacuolar membrane, distinct from the SV and FV channels, may transport malate. An abundant anion in vacuoles, malate serves as the major anionic solute involved in charge balance with inorganic cations such as K^+ and Ca^{2+}.

Recent evidence supports the idea that membrane proteins called aquaporins, present in both the tonoplast and plasma membrane, facilitate the permeation of water across plant membranes and tissues (Chrispeels *et al.*, 1999; Tyerman *et al.*, 1999). Aquaporins belong to a highly conserved group of membrane proteins called the major intrinsic proteins (MIPs) found in both animals and plants.

C. Carriers

In addition to ion-selective (or nonselective) channels that create pores in the lipid bilayer, plant cell membranes have carriers or porters. Carrier-mediated transport of sugars, amino acids, and ions across the plasma membrane and tonoplast is coupled to proton transport down the H^+ gradients across these membranes. These proton cotransporters may be either symporters (H^+ and solute travel in same direction) or antiporters (H^+ and solute exchange). In the plasma membrane, H^+-coupled symporters may transport sucrose, amino acids, and anions such as nitrate or phosphate. H^+-coupled antiporters may include Na^+ carriers in the plasma membrane and Na^+, Ca^{2+}, Cd^{2+}, and sucrose carriers in the vacuolar membrane (Chrispeels *et al.*, 1999).

Sugar-proton cotransport carriers may exist in both the plasma membrane and the tonoplast. The superfamily of sugar transporter genes reported for bacteria and mammalian cells supports the idea that cell sugar transporters may be part of this family. Nitrogen-containing substances, such as nitrate and amino acids, are probably taken up by cells by proton-coupled symporters. Electrical measurements and genetic studies support the idea of carrier-mediated NO_3^- uptake at the plasma membrane. An H^+-NO_3^- antiporter may also exist in the vacuolar membrane. There is also evidence for the proton-coupled uptake of inorganic phosphate at the plant cell plasma membrane (Chrispeels *et al.*, 1999)

D. Ca^{2+} Transport Mechanisms

In plant cells, as in animal cells, Ca^{2+} is an important second messenger in intracellular signal transduction (Sanders *et al.*,

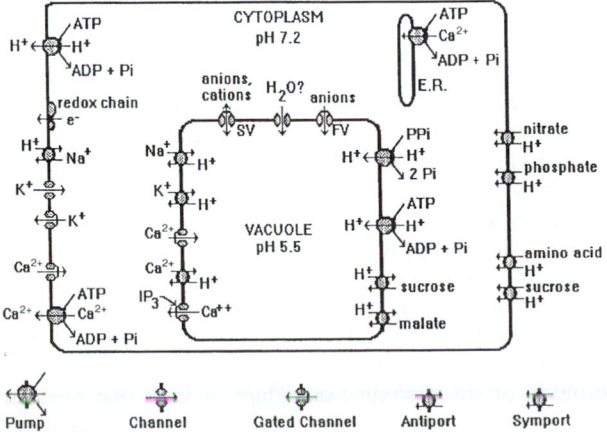

FIGURE 10. A model summarizing the membrane transport systems in plant cells.

1999). For this and other reasons, the regulation of Ca^{2+} levels in the cytoplasm is critical for cell physiology. Although extracellular and intravacuolar Ca^{2+} concentrations may be in the millimolar range, the cytoplasmic concentration of free Ca^{2+} is about 10 000 times lower (100 nM in resting conditions). Obviously, transport mechanisms must exist to maintain such steep Ca^{2+} concentration gradients.

Ca^{2+}-ATPases are associated with the plasma membrane, the tonoplast, and the ER. The plasma membrane Ca^{2+}-ATPases act as primary Ca^{2+}-efflux pumps, moving Ca^{2+} into the cell wall space against a large electrochemical gradient. Ca^{2+}-ATPases pump Ca^{2+} into the lumen of the ER and into the vacuole, which are major sites of intracellular storage. These Ca^{2+}-ATPases are P-type ATPases and are stimulated by calmodulin (similar to the Ca^{2+}-ATPase of red blood cells). The Ca^{2+}-ATPase is a dimer, consisting of two identical polypeptides of 130–140 kDa each.

A secondary active transport mechanism for Ca^{2+} may exist on the vacuolar membrane. In mature plant cells, the largest intracellular pool of Ca^{2+} is the vacuole. Calcium ion transport into the vacuole occurs against both a chemical and an electrical potential difference. The H^+ gradient across the tonoplast may drive Ca^{2+} uptake into the vacuole via H^+-Ca^{2+} antiporters.

Ca^{2+} channels also mediate transmembrane Ca^{2+} movement. Ca^{2+} actively pumped out of the cell can passively reenter the cell through Ca^{2+} channels in the plasma membrane. Ca^{2+} influx channels in the plasma membrane have been implicated in the establishment (and perhaps maintenance) of cell polarity. For example, in the zygote of the brown alga *Fucus,* asymmetry in the environment (e.g., low light, gravity, sperm entry, or low pH) leads to localization (or local activation) of Ca^{2+}-influx channels and Ca^{2+}-efflux pumps in the cell membrane. This leads to a Ca^{2+} flux out at one side of the cell and in at the opposite side. This intracellular Ca^{2+} gradient may promote structural asymmetries, such as cytoskeletal orientation and cell wall deposition, leading to cell polarity.

There are also at least two Ca^{2+} channels in the vacuolar membrane. One is activated (gated) by inositol 1,4,5-trisphosphate (IP_3) and may increase cytoplasmic free Ca^{2+} through releasing Ca^{2+} from the vacuolar store. A voltage-dependent Ca^{2+} channel has also been recently reported in vacuoles. This channel may be activated after the release of Ca^{2+} into the cytoplasm by the IP_3-stimulated Ca^{2+}-release channels. IP_3-stimulated Ca^{2+} release from the vacuole may polarize the vacuolar membrane potential, which, in turn, would activate the inward-rectifying Ca^{2+} channels. These channels may assist the Ca^{2+}-ATPases and the H^+-Ca^{2+} antiporters in reestablishing the low cytoplasmic resting levels of calcium. As discussed in the following section, changes in the cytoplasmic free Ca^{2+} concentration may play a critical role in stimulus response coupling.

V. Signal Perception and Response

A. Photoperception

Light has profound effects on many aspects of plant physiology, especially development. Sunlight, absorbed primarily by the green pigment chlorophyll, is the energy source for the life-giving process of photosynthesis. By activating specific nuclear genes coding for proteins involved in chlorophyll biosynthesis and chloroplast development, sunlight also acts as the environmental cue to developing cells to begin photosynthesis. Light does not directly alter gene expression, of course, but acts through a mediator, a photoreceptor known as phytochrome (von Arnim and Deng, 1996). In addition to mediating the light-triggered greening of plant tissue, phytochrome is involved in many other light-sensitive processes, including seed germination, shoot elongation, circadian rhythms, leaf development, and flower induction.

Phytochrome has been well characterized. The native phytochrome molecule is a dimer of approximately 240 kDa. Each monomer (ca. 120 kDa) contains a single covalently bound chromophore, an open-chain tetrapyrrole. A water-soluble protein, phytochrome can exist in either of two conformations. The conversion from one form to the other is triggered by specific wavelengths of light and is fully reversible. The P_r form of phytochrome absorbs red light (650–680 nm) and is converted to the P_{fr} form, which absorbs light of longer wavelengths, that is, far-red light (710–740 nm). In most cases, P_{fr} (produced by red light) is the active form of phytochrome. The P_{fr} form initiates a transduction process that leads to altered expression of specific genes. The resulting differential gene expression culminates in modified cell growth or physiology appropriate to the prevailing light environment.

The intracellular signal transduction mechanism probably involves Ca^{2+}. Since there is some evidence that the P_{fr} form associates with membranes (and the P_r form is soluble), P_{fr} may act by altering the cytoplasmic Ca^{2+} concentration through interaction with one or more of the membrane-bound Ca^{2+}-transport mechanisms discussed earlier.

Phytochrome is involved in a wide range of light-sensitive processes. From recent cloning and sequencing studies of phytochrome, the smallest known flowering plant genome, *Arabidopsis,* was found to contain as many as five different phytochrome genes. This supports earlier evidence for the existence of more than one type of phytochrome in plants. Different phytochrome isoforms may mediate different red light-sensitive responses.

Another class of photoreceptors in plant cells is activated by blue light. Blue light is involved in the control of a variety of physiological and developmental processes. The most well studied of these include phototropism (e.g., the growing of a shoot toward light), stomatal opening, and inhibition of shoot elongation. The blue-light photoreceptors that mediate these responses are likely plasma membrane-bound flavoproteins.

Blue light apparently directly activates a kinase moiety on a plasma membrane-bound flavoprotein to cause rapid phosphorylation of a class of plasma membrane proteins (Kaufman, 1993). This may represent an early step in the signal transduction chain for phototropism and for the blue-light inhibition of shoot elongation. Thus, at least one blue-light photoreceptor in cells may be a photoactivated protein kinase and may initiate a phosphorylation cascade culminating in the inhibition of elongation.

Evidence also indicates that a protein phosphorylated by a blue-light-mediated photoreaction may be associated with cytoskeletal elements. This suggests that the cell cytoskeleton may play a functional role in the inhibition of cell growth by blue light.

B. Gravity Perception

Both shoots and roots correct their direction of growth if they are displaced from the vertical. The growing regions of a plant, especially the zones of cell elongation in primary roots and shoots, are particularly responsive to gravity, though roots and shoots respond in opposite ways. When a young shoot is displaced from the vertical, it will curve upward after several hours. Conversely, a root in the same situation will curve downward. Both responses are a result of unequal rates of cell elongation on the upper side of the organ compared with the lower side. As the curving root or shoot regains a vertical orientation, the elongation rates of the cells become equal.

Gravity perception by plants probably involves the movement of free-falling bodies within specialized cells (Volkmann and Sievers, 1979). These gravity-sensitive bodies are called **statoliths.** In gravity-sensing cells (statocytes), the statoliths are probably starch grains located within amyloplasts, which are specialized plastids. The amyloplasts sediment through the cytosol and come to rest on the lowermost part of the cell (Fig. 11). This result somehow redirects the transport of the plant growth substance auxin, leading to unequal elongation growth across the root or shoot (McClure and Guilfoyle, 1989; Hasenstein and Evans, 1988).

C. Mechanosensory Mechanisms

Most plants display altered development as a result of mechanical stimulation. For example, plants growing in windy landscapes tend to be shorter, with thicker stem diameters, than the same species sheltered from the wind. To be shorter and sturdier in windy environments is obviously an adaptive advantage. This change in plant growth and development in response to mechanical stimulation is called **thigmomorphogenesis.** In the laboratory, thigmomorphogenesis can be simulated by touch. These touch responses are different from touch-induced rapid movements in plants (e.g., the Venus flytrap or the touch-sensitive *Mimosa*) in that the plant movements occur through rapid and reversible turgor pressure changes in motor cells. Thigmomorphogenesis involves irreversible alterations in cell development. Recent reports of mechanosensitive plasma membrane ion channels (Cosgrove and Hedrich, 1991), known to exist in animal cells, may offer a clue to how mechanical stimulation is perceived by plant cells. Patch-clamp analyses of stomatal guard cells (see Section II.B) have revealed at least three types of stretch-activated (SA) ion channels (anion efflux, K^+ efflux, and Ca^{2+} influx) in the plasma membrane. These SA channels are gated, presumably by strain on the membrane, and may act primarily in the regulation of guard cell volume and turgor (see Section II.B). SA channels in actively growing cells may also serve as growth-rate sensors so that cell wall expansion can be coordinated with the plasma membrane material via Golgi-derived vesicular fusion. Stretch-activated Ca^{2+}-influx channels in the plasma membrane may modulate the process of cell enlargement through changes in cytoplasmic Ca^{2+} concentration. This Ca^{2+} signal would be amplified by a variety of Ca^{2+}-modulated proteins, including calmodulin (CaM).

As shown in Fig. 12, activated CaM may amplify the signal in several ways: first, through CaM-modulated protein kinases that subsequently phosphorylate other enzymes; second, through alterations in the microtubules and microfilaments as well as through increased production of the plant hormone ethylene, increasing the lateral growth of cells (e.g., leading to shorter, thicker stems). Finally, Ca^{2+}-activated CaM also may stimulate the expression of calmodulin genes, serving as a positive feedback mechanism. Changes in cytosolic Ca^{2+} may also activate anion efflux channels in the plasma membrane, which may account for action potentials (see Section IV.B) that have been observed in algae and plants following mechanical stimulation (Braam and Davis, 1990a).

D. Receptor Proteins

As previously discussed, chemical signals, such as plant hormones and cell wall fragments, affect cell development

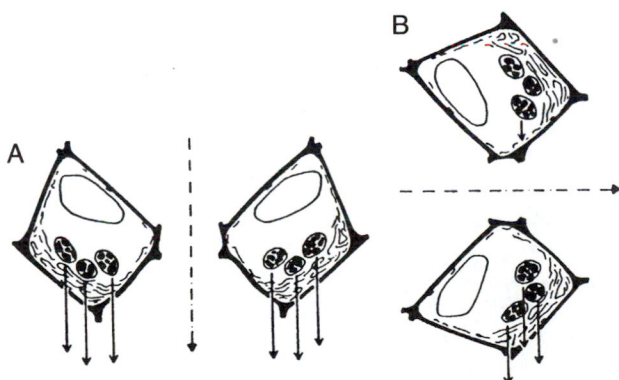

FIGURE 11. A diagram illustrating the reorientation of amyloplasts in statocytes in response to the direction of gravity (denoted by the solid arrows). The dashed arrows point to the root tip. (A) Vertical orientation of the root. (B) Horizontal orientation of the root. [Reproduced from Volkmann, D., and Sievers, A. (1979). Graviperception in multicellular organs. *In* "Encyclopedia of Plant Physiology, New Series" (W. Haupt and M. E. Feinleib, Eds.), Vol. 7. pp. 573–600. Springer-Verlag, Berlin/New York, with permission.]

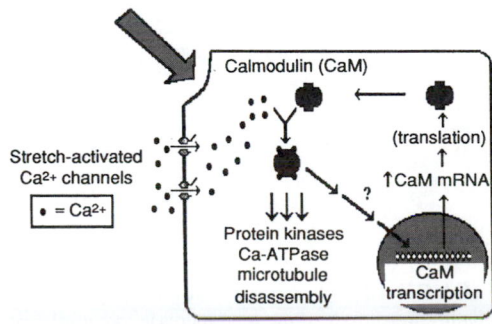

FIGURE 12. A hypothetical model of mechanosensory perception and response at the cellular level in plants. (Modified from Braam and Davis, 1990b, by permission.)

and physiology. Most hormone receptor proteins in animal cells are located on the cell surface, that is, in the plasma membrane. The exception to this is steroid hormone receptors, which are located intracellularly. Similarly, most of the putative receptor proteins identified in plant cells are associated with the plasma membrane.

Although putative receptors exist for most of the plant hormones (see Fig. 8), the two most well characterized are for auxin and ethylene. A probable auxin receptor has been located in elongating cells. This 22-kDa protein is apparently bound to the plasma membrane, with the auxin-binding site on the extraprotoplasmic surface. Physiological evidence suggests that the auxin receptor may somehow modulate ion transport activity at the plasma membrane (Venis and Napier, 1995). Plants sense ethylene by a protein kinase cascade (Bleeker and Schaller, 1996).

Specific receptors for oligosaccharide signal molecules, such as beta-glucans (see Section III.C), may also reside in the plasma membrane. The receptors for these and other extracellular effector molecules that elicit defensive responses in cells (e.g., systemin, methyl jasmonate, and salicylic acid) presumably trigger signal transduction pathways, culminating in the transcriptional activation of defense-related genes.

E. Signal Transduction Mechanisms

Many of the signal perception mechanisms discussed previously invoke the participation of Ca^{2+}, inositol trisphosphate (IP_3), and protein kinases in intracellular signaling (Palme, 1992). Two of the most important intracellular second messengers in animal cells are cyclic AMP (cAMP) and Ca^{2+}. Although there is little reliable evidence that cAMP functions as such in plants, there is ample experimental evidence that Ca^{2+} does. Changes in cytosolic Ca^{2+} occur during stimulus-response coupling in plants. Calcium-modulated proteins, including calmodulin and calcium-dependent protein kinases, are present in plant cells. Several CaM-regulated proteins have been identified, notably, Ca^{2+}-ATPase, NAD kinase, and protein kinase. The recent identification of annexins in plants, which are involved in the regulation of exocytosis in animal cells, provides evidence that Ca^{2+}-binding proteins may serve to link increases in cytosolic Ca^{2+} concentration with the stimulation of exocytosis in plant cells (Xu et al., 1992).

At least two breakdown products of phosphatidylinositol 4,5-bisphosphate (PIP_2), IP_3 and diacylglycerol (DAG), serve as intracellular signaling molecules in animal cells. These products of PIP_2 hydrolysis are known to exist in plants (Trewavas and Gilroy, 1991), but unequivocally demonstrating their participation in signal transduction has been difficult. The enzyme that hydrolyzes PIP_2, namely, phospholipase C, has been localized to the cytoplasmic surface of plasma membranes. Thus, the enzymatic basis for signal transduction via PIP_2 hydrolysis is apparently present in cells (Roberts and Harmon, 1992). A conceptualization of events in cells involving stimulus-evoked turnover of phosphatidylinositol in the plasma membrane and the resulting Ca^{2+} release is summarized in Fig. 13.

In animal cells, GTP-binding regulatory proteins (G proteins) serve as intermediates between plasma membrane receptors and the enzymes or ions channels they affect. G-protein-linked receptors typically activate a chain of events that alter the intracellular concentration of second messengers such as cAMP, Ca^{2+}, and IP_3. They also transduce extracellular signals into membrane-mediated events, including regulation of ion channels. GTP-binding proteins have been reported to exist in plants. Antibodies to animal G proteins have been shown to cross-react with plasma membrane proteins from a variety of plants, and a G-protein gene has been

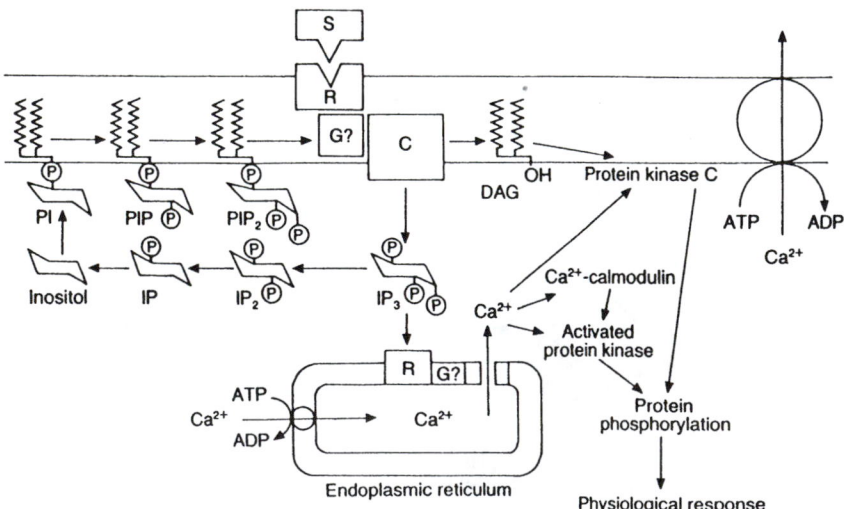

FIGURE 13. A summary of events in plant cells involving the stimulus-induced turnover of plasma membrane-bound phosphatidylinositol, leading to an increase in cytoplasmic Ca^{2+} concentration. S, stimulus; R, receptor; G, G protein; C, phospholipase C; DAG, diacylglycerol; IP_3, inositol 1,4,5-trisphosphate; IP_2, inositol 1,4-bisphosphate; IP, inositol 1-phosphate; PI, phosphatidylinositol; PIP, phosphatidylinositol 4-phosphate; PIP_2, phosphatidylinositol 4,5-bisphosphate. The G protein may stimulate the gating of ion channels. (From Trewavas and Gilroy, 1991, by permission.)

isolated from *Arabidopsis*. Although there is evidence that a G protein may regulate inward K⁺ channel current in stomatal guard cells (Fairly-Grenot and Assman, 1991), G proteins have yet to be shown to mediate second messenger production in plant cells.

VI. Summary

The cell wall, plastids, and the vacuole are the three main structural features that distinguish plant cells from animal cells. Consisting of cellulose microfibrils embedded in a polysaccharide and protein matrix, the cell wall of plants both supports and protects the protoplast. Chloroplasts serve as the photosynthetic organelles in green plant tissues, while other types of plastids function in carbohydrate storage (amyloplasts) and in reproduction (chromoplasts). Most mature plant cells are characterized by a large central vacuole, which permits plant cells to enlarge to a greater degree than animal cells and serves as an intracellular storage area for metabolites and for hydrolytic enzymes involved in the breakdown and recycling of cellular components. In contrast to these three structural features, plant and animal cell membrane and cytoskeletal systems display fundamental similarities.

Plant cells use chemical signals to communicate with each other as well as with other organisms. Such molecules may travel from cell to cell via plasmodesmata, small conduits through the cell walls of adjacent cells, which normally allow the passage of relatively small molecules (<800 kDa). The coordination and regulation of plant development and physiology are primarily accomplished by only five different plant hormones. These, as well as other chemical signals such as specific oligosaccharides and oligopeptides, can trigger systematic defensive responses in plants. Complex interactions, such as differential gene expression, between microbial symbionts or pathogens and plant cells are mediated by chemical signals produced by both the plant and the microorganisms.

Plant cells maintain a substantial resting potential (−120 to −180 mV) across their plasma membrane primarily through electrogenic, proton-translocating ATPases. This primary active transport mechanism drives most of the other secondary transport systems in plasma membrane. An electrochemical proton gradient, maintained by proton-transport ATPases and pyrophosphatases, also drives transport at the vacuolar membrane. Gated channels for potassium and calcium ions, for example, are present in both the plasma and vacuolar membranes. Many of the transport proteins in plant cells structurally resemble analogous proteins in animal cells.

Plants can perceive subtle changes in light quality and quantity through at least two different kinds of photoreceptors, phytochrome (red/far-red light) and a blue-light photoreceptor. Gravity perception is probably achieved through falling bodies in the cytoplasm called statoliths. Mechanical stimulation may be mediated by plasma membrane bound, stretch-activated ion channels. Both chemical and environmental signal responses in plant cells may be mediated by second-messenger systems similar to those found in animal cells, that is, calcium ions, phosphoinositides, and G proteins.

Bibliography

Bacic, A., Harris, P. J., and Stone, B. A. (1988). Structure and function of plant cell walls. *In* "Biochemistry of Plants: A Comprehensive Treatise" (J. Preiss, Ed.), Vol. 14, pp. 298–371. Academic Press, San Diego.

Battey, N. H., James, N. C. , Greenland, A. J. , and Brownlee, C. (1999). Exocytosis and endocytosis, *Plant Cell* **11**, 643–659.

Beijersgergen, A., Dulk-Ras, A. D., Schilperoort, R. A., and Hooykaas, P. J. J. (1992). Conjugative transfer by the virulence system of *Agrobacterium tumefaciens*. Science **256**, 1324–1327.

Bleeker, A. B., and Schaller, G. E. (1996). The mechanism of ethylene perception. *Plant Physiol.* **111**, 653–660.

Bonfanti, L., Mironov, A.A. Jr., Martinez-Menárguez, J.A., Martella, 0., Fusella, A., Baldassarre, M., Buccione, R., Geuze, H.J., Mironov, A. A., and Luini, A. (1998). Procollagen traverses the Golgi stack without leaving the lumen of cisternae: Evidence for cisternal maturation. *Cell* **95**, 993–1004.

Braam, J., and Davis, R. W. (1990a). Rain-, wind-, and touch-induced expression of calmodulin and calmodulin-related genes in *Arabidopsis*. *Cell* **60**, 357–364.

Braam, J., and Davis, R. W. (1990b). The mechanosensory pathway in *Arabidopsis*: touch-induced regulation of expression of calmodulin and calmodulin-related genes and alterations of development. *Curr. Top. Plant Biochem. Physiol.* **9**, 85–100.

Burgess, J. (1985). "An Introduction to Plant Cell Development." Cambridge University Press, London/New York.

Carpita, N., McCann, M., and Griffing, L. R. (1996). The plant extracellular matrix: news from the cell's frontier. *Plant Cell* **8**, 1451–1463.

Cheong, J.-J., and Hahn, M. G. (1991). A specific, high-affinity binding site for the hepta-beta-glucoside elicitor exists in soybean membranes. *Plant Cell* **3**, 137–147.

Chrispeels, M J., Crawford. N M and Schroeder, J. I. (1999). Proteins for transport of water and mineral nutrients across the membranes of plant cells. *Plant Cell* **11**, 661–675.

Cohn. J., Day. R. B., and Stacey, G. (1998). Legume nodule organogenesis. *Trends Plant Sci.* **3**, 105–110.

Cosgrove, D. J., and Hedrich, R. (1991). Stretch-activated chloride, potassium, and calcium channels coexisting in plasma membranes of guard cells in *Vicia faba* L. *Planta Med.* **186**, 143–153.

Crane, F. L., and Barr, R. (1989). Plasma membrane oxidoreductases. *Crit. Rev. Plant Sci.* **8**, 273–307.

Czempinski, K., Gaedeke, N., Zimmermann, S., and Müller-Röber, B. (1999). Molecular mechanisms and regulation of plant ion channels. *J. Exp. Bot.* **50**, 955–966.

Davies, P. J. (Ed.) (1995). "Plant Hormones: Physiology, Biochemistry and Molecular Biology." Kluwer Academic Publishers, Dordrecht, The Netherlands.

Delmer, D. P. (1999). Cellulose biosynthesis: exciting times for a difficult field of study. *Annu. Rev. Plant Physiol. Plant Mol. Biol;.* **50**, 245–276.

Delmer, D. P., and Amor, Y. (1995). Cellulose biosynthesis. *Plant Cell* **7**, 987–1000.

Ehrhardt, D. W., Wais, R., and Long, S. R. (1996). Calcium spiking in plant root hairs responding to rhizobium nodulation signals. *Cell* **85**, 673–681.

Fairly-Grenot, K., and Assman, S. M. (1991). Evidence for G-protein regulation of inward K⁺-channel current in guard cells of Fava Bean. *Plant Cell* **3**, 1037–1047.

Farmer, E. E., and Ryan, C. A. (1990). Interplant communication: airborne methyl jasmonate induces the synthesis of proteinase inhibitors in plant leaves. *Proc. Natl. Acad. Sci. USA* **87**, 7713–7716.

Fisher, R. F., and Long, S. L. (1992). *Rhizobium*-plant signal exchange. *Nature* **357**, 655–660.

Geurts, R., and Franssen, H. (1996). Signal transduction in *Rhizobium*-induced nodule formation. *Plant Physiol.* **112**, 447–453.

Giddings, T. H., Jr., and Staehelin, L. A. (1991). Microtubule-mediated control of microfibril deposition: a re-examination of the hypothesis. *In* "The Cytoskeletal Basis of Plant Growth and Form" (C. W. Lloyd, Ed.), pp. 85–100. Academic Press, London.

Griffing, L. R. (1991 Comparisons of Golgi structure and dynamics in plant and animal cells. *J. Electron. Microsc. Tech.* **17,** 179–199.

Griffing. L. R., and Fowke, L. C. (1985). Cytochemical localization of peroxidase in soybean suspension culture cells and protoplasts: intracellular vacuole differentiation and presence of peroxidase in coated vesicles and multivesicular bodies. *Protoplasma* **128,** 22–30.

Hasenstein, K. H., and Evans, M. L. (1988). Effects of cations on hormone transport in primary roots of *Zea mays. Plant Physiol.* **86,** 890–894.

Hayashi. T. (1989). Xyloglucans in the primary cell wall. *Annu. Rev. Plant Physiol. Plant Mol. Biol.* **40,** 139–168.

Hooley, R. (1996). Plant steroid hormones emerge from the dark. *Trends Genet.* **12,** 281–283.

Huang, A. C. (1992). Oil bodies and oleosins in seeds. *Annu. Rev. Plant Physiol. Plant Mol.* **43,** 77–100.

Kachar, B., and Reese, T. S. (1988). The mechanism of cytoplasmic streaming in *Characean* algal cells: sliding of endoplasmic reticulum along actin sheets. *J. Cell Biol.* **106,** 1545–1552.

Kaufman, L. S. (1993). Transduction of blue-light signals. *Plant Physiol.* **102,** 333–337.

Kohno. T., Chaen, S., and Shimmen, T. (1990). Characterization of the translocator associated with pollen tube organelles. *Protoplasma* **154,** 179–183.

Lloyd, C. W. (Ed.)(1991). "The Cytoskeletal Basis of Plant Growth and Form." Academic Press, London.

Long, S. R. (1989). *Rhizobium*-legume nodulation: life together in the underground. *Cell* **56,** 203–214.

Lucas, W. J. (1999). Plasmodesmata and the cell-to-cell transport of proteins and nucleoprotein complexes. *J. Exp. Bot.* **50,** 979–987.

Maleck, K., and Dietrich, R. A. (1999). Defense on multiple fronts: how do plants cope with diverse enemies? *Trends Plant Sci.* **4,** 215–219.

McCann, M. C., and Roberts, K. (1991). Architecture of the primary cell wall. *In* "The Cytoskeletal Basis of Plant Growth and Form" (C. W. Lloyd, Ed.), pp.109–130. Academic Press, London.

McClure, B. A., and Guilfoyle, T. (1989). Rapid redistribution of auxin-regulated RNAs during gravitropism. *Science* **243,** 91–93.

McGurl, B., Pearce, G., Orozoco-Cardenas. M., and Ryan, C. A. (1992). Structure, expression, and antisense inhibition of the systemin precursor gene. *Science* **255,** 1570–1573.

Mollenhauer, H. H., and Morré, D. J. (1980). The Golgi apparatus. *In* "The Biochemistry of Plants: A Comprehensive Treatise" (P. K. Stumpf. and E. E. Conn., Eds.) pp. 437–488, Academic Press, Inc., New York.

Mullet, J. E. (1988). Chloroplast development and gene expression. *Annu. Rev. Plant Physiol. Plant Mol. Biol.* **39,** 475–502.

Okita, T. W., and Rogers, J. C. (1996). Compartmentation of proteins in the endomembrane system of plant cells. *Annu. Rev. Plant Physiol. Plant Mol. Biol.* **47,** 327–350.

Olsen, L. J., and Harada, J. J. (1995). Peroxisomes and their assembly in higher plants. *Annu. Rev. Plant Physiol. Plant Mol. Biol.* **46,** 123–146.

Palme, K. (1992). Molecular analysis of plant signalling elements: relevance of eukaryotic signal transduction models. *Crit. Rev. Cytol.* **132,** 223–283.

Palmgren, M. G., and Harper, J. F. (1999). Pumping with plant P-type ATPases. *J. Exp. Bot.* **50,** 883–893.

Pierson, E. S., Miller. D. D., Callaham D. A., Shipley, A. M., Rivers, B. A., Cresti, M., and Hepler, P. K. (1994). Pollen tube growth is coupled to the extracellular calcium ion flux and the intracellular calcium gradient: effect of BAPTA-type buffers and hypertonic media. *Plant Cell* **6,** 1815–1828.

Pieterse, C M. J., and van Loon, L. C. (1999). Salicylic acid-independent plant defence pathways. *Trends Plant Sci.* **4,** 52–58.

Raskin, I. (1992). Salicylate, a new plant hormone. *Plant Physiol.* **99,** 799–803.

Rayle, D. L., and Cleland, R. E. (1992). The acid growth theory of auxin-induced cell elongation is alive and well. *Plant Physiol.* **99,** 1271–1274.

Robards, A. W. (1976). Plasmodesmata in higher plants. *In* "Intercellular Communication in Plants: Studies on Plasmodesmata" (B. E. S. Gunning and A. W. Robards, Eds.), p. 31. Springer-Verlag, New York/Berlin.

Robards, A. W., and Lucas, W. J. (1990). Plasmodesmata. *Annu. Rev. Plant Physiol. Plant Mol. Biol.* **41,** 369–419.

Roberts, P. M., and Harmon, A. C. (1992). Calcium-modulated proteins: targets of intracellular calcium signals in higher plants. *Annu. Rev. Plant Physiol. Plant Mol. Biol.,* **43,** 375–414.

Robinson, P. G., and Depta, H. (1988). Coated vesicles. *Annu. Rev. Plant Physiol. Plant Mol. Biol.* **39,** 53–99.

Ryan, C. A. (1988). Oligosaccharides as recognition signals for the expression of defensive genes in plants. *Biochemistry* **27,** 8879–8883.

Ryan, C. A., and Farmer, E. E. (1991). Oligosaccharide signals in plants: a current assessment. *Annu. Rev. Plant Physiol. Plant Mol. Biol.* **42,** 651–674.

Samuels, A. L., Giddings. T. H., Jr., and Staehelin, L. A. (1995). Cytokinesis in tobacco BY-2 and root tip cells: a new model of cell plate formation in higher plants. *J. Cell Biol.* **130,** 1345–1357.

Sanderfoot, A. A., and Raikhel, N. V. (1999). The specificity of vesicle trafficking: coat proteins and SNAREs. *Plant Cell* **11,** 629–664.

Sanders D., Brownlee, C., and Harper, J. F. (1999). Communicating with calcium. *Plant Cell* **11,** 691–706.

Schultze, M., and Kondorosi, A. (1998). Regulation of symbiotic root nodule development. *Annu. Rev. Genet.* **32,** 33–57.

Shérrier, D. J., and VandenBosch, K. A. (1994). Secretion of cell wall polysaccharides in *Vicia* root hairs. *Plant J.* **5,** 185–196.

Spaink, H. P. (1995). The molecular basis of infection and nodulation by rhizobia the ins and outs of sympathogenesis. *Annu. Rev. Phytopathol.* **33,** 345–368.

Staehelin, L. A., and Giddings, T. H. (1982). Membrane control of microfibrillar order. *In* "Developmental Order: Its Origin and Regulation" (S. Subtelney and P. B. Green, Eds.), p.144. Alan R. Liss. New York.

Staswick, P. F. (1992). Jasmonate, genes, and fragrant signals. *Plant Physiol.* **99,** 804–807.

Stout, R. G., and Griffing, L. R. (1993). Transmural secretion of a highly-expressed cell surface antigen of oat root cap cells. *Protoplasma* **172,** 27–37.

Sze, H., Li, X., and Palmgren, M. G. (1999). Energization of plant cell membranes by H^+-pumping ATPases: regulation and biosynthesis. *Plant Cell* **11,** 677–689.

Taiz, L. (1994). Expansins: proteins that promote cell wall loosening in plants. *Proc. Natl. Acad. Sci. USA* **91,** 7387–7389.

Taiz, L., and Zeiger, E. (1998). "Plant Physiology." Sinauer Associates, Inc., Sunderland, MA.

Tolbert, N. F. (Ed.) (1980). "The Plant Cell." Academic Press, New York.

Topfer, R. Martini, N., and Schell, H. (1995). Modification of plant lipid synthesis. *Science* **268,** 681–686.

Trewavas, A., and Gilroy, S. (1991). Signal transduction in plant cells. *Trends Genet.* **7,** 356–361.

Tyerman, S. D. (1992). Anion channels in plants. *Annu. Rev. Plant Physiol. Plant Mol. Biol.* **43,** 351–373.

Tyeman, S. D., Bohnert, H. J., Maurel, C., Steudle, E., and Smith, J. A. C. (1999). Plant aquaporins: their molecular biology, biophysics and significance for plant water relations. *J. Exp. Bot.* **50,** 1055–1071.

Varner, J. F., and Lin, L.-S. (1989). Plant cell wall architecture. *Cell* **56,** 231–239.

Venis, M. A., and Napier, R. M. (1995). Auxin receptors and auxin binding proteins. *Crit Rev. Plant Sci.* **14,** 27–47.

Volkmann, D., and Sievers, A. (1979). Graviperception in multicellular organs. *In* "Encyclopedia of Plant Physiology, New Series" (W. Haupt and M. F. Feinleib, Eds.), Vol. 7, pp. 573–600. Springer-Verlag, Berlin/New York.

von Arnim, A., and Deng, X.-W. (1996). Light control of seedling development. *Annu. Rev. Plant Physiol. Plant Mol. Biol.* **47,** 215–243.

Williamson, R. F. (1986). Organelle movement along actin filaments and microtubules. *Plant Physiol.* **82,** 631–634.

Xu, P., Lloyd, C. W., Staiger, C. J., and Drobak, B. K. (1992). Association of phosphatidylinositol-4-kinase with the plant cytoskeleton. *Plant Cell* **4,** 941–951.

Zhang, G. F., and Stachelin. L. A. (1992). Functional compartmentation of the Golgi apparatus of plant cells. *Plant Physiol.* **99,** 1070–1080.

Darrell Fleischman

64

Photosynthesis

I. Introduction

A. Historical Aspects

It must have been apparent from the earliest times that animals can survive only by eating plants or other animals, whereas plants seem to subsist on nothing but sunshine and a little water and earth. Julius Mayer, who introduced the concept of conservation of energy, suggested in 1845 that green plants can capture the energy of light and store it in a chemical form. His insight was correct. Almost all of the free energy used by living organisms is derived from energy of sunlight, which has been stored in photosynthetically produced molecules.

Clues about how plants might store free energy were obtained early (Loomis, 1960). In 1771, the English philosopher Joseph Priestly began a series of experiments that demonstrated that plants produce a gaseous substance, oxygen, which animals require for respiration. By the end of the 19th century it was recognized that oxygen is produced only by the green parts of the plant and that its production requires light, carbon dioxide, and water. The first visible products of photosynthesis were grains of starch, a polymer of glucose.

We now know that photons are captured by **chlorophyll,** the green pigment found in leaves (Fig. 1). Plants use the energy of the photons to extract electrons from water (leaving oxygen as a by-product), and use them to convert inorganic molecules such as carbon dioxide to organic molecules such as glucose:

$$6\,CO_2 + 6\,H_2O \xrightarrow{\text{light}} C_6H_{12}O_6 + 6\,O_2$$

The products of the above process contain more free energy per mole of glucose formed than do the reactants, that is,

$$\Delta G^0 = 686 \text{ kcal/mol}$$

Nitrogen and sulfur also can be fixed, that is, converted from inorganic to organic form, by the photosynthetic system. Nitrogen and sulfur that are contributed by inorganic starting materials (including dinitrogen (N_2), nitrate, and sulfate) are incorporated into amino acids and other organic molecules. In this chapter we discuss in detail only carbon fixation.

B. Photosynthetic Microorganisms

Higher plants are not the only organisms that perform **photosynthesis**. Algae also do so, as do many bacteria, including

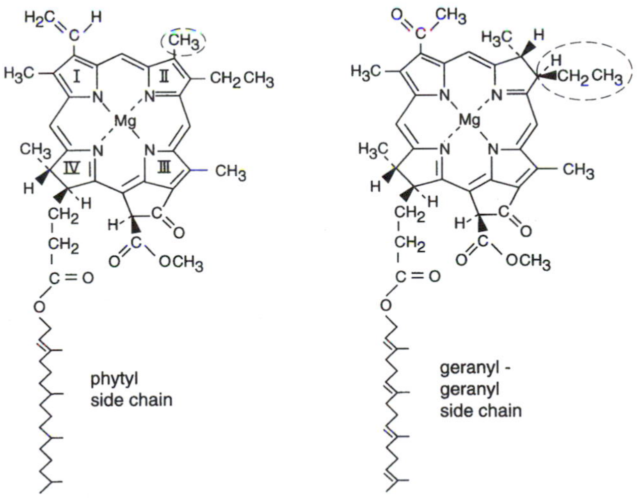

Chlorophyll *a*   Bacteriochlorophyll *a*

FIGURE 1. The structures of chlorophyll *a* (Chl *a*) and bacteriochlorophyll *a* (BChl *a*). Differences between the structures of the various forms of chlorophyll contribute to differences between the absorption spectra of chlorophyll proteins, allowing them to capture light in different regions of the spectrum. The conjugated system in BChl has one fewer double bond (in ring II) than that of Chl. Single resonance forms of the conjugated systems of double bonds are shown. In Chl *b*, –HC=O replaces the circled methyl substituent on ring II of Chl *a*. In BChl *b*, =CH–CH₃ replaces the circled substituents on ring II of BChl *a*. In some bacteria, phytyl rather than geranylgeranyl serves as the side chain of bacteriochlorophyll *a* or bacteriochlorophyll *b*. Other forms of chlorophyll are found in some organisms. Pheophytins resemble the respective chlorophylls, but the Mg^{2+} is replaced by two protons.

the **cyanobacteria** (blue-green algae), the **purple and green photosynthetic bacteria**, heliobacteria and *Chloroflexus* (see Chapter 62). As much as a third of the earth's photosynthesis is performed by such microorganisms in the oceans. Algae and cyanobacteria use water as their source of electrons and evolve oxygen, as higher plants do. Most other photosynthetic bacteria cannot derive electrons in this manner, and instead extract electrons from donors such as elemental sulfur, sulfide, thiosulfate, organic acids, alcohols, and hydrogen. The bacterium *Halobacterium halobium* performs photosynthesis in a quite different way. Rather than chlorophyll, it uses **bacteriorhodopsin**, a carotenoid-containing protein resembling the visual pigment rhodopsin, to capture the energy of light in a process that does not involve electron transfer (Stoeckenius, 1999; Subramaniam and Henderson, 2000).

C. Evolution of Photosynthesis

It is believed that the earliest photosynthetic bacteria appeared more than 3.7 billion years ago at a time when the earth's atmosphere contained almost no oxygen. They might even have originated from chemotrophic bacteria near deep ocean hydrothermal vents (Yurkov *et al.*, 1999). Like present-day purple and green photosynthetic bacteria, these primordial bacteria could not use water as an electron donor. Then, perhaps as early as 3.7 billion years ago, there appeared a bacterium that, in the light, could generate an oxidant strong enough to oxidize water (Blankenship, 1992, 1998). The new bacterium had access to a seemingly limitless source of electrons—water. The bacterium formed oxygen as the byproduct of water oxidation. At first the oxygen may have been consumed in chemical reactions with reductants such as Fe^{2+}, but by 2.3 billion years ago it was accumulating in the atmosphere. The oxygen gave rise to the ozone layer and so permitted the development of higher organisms by shielding them from ultraviolet radiation.

Once oxygen was present, respiration and oxidative phosphorylation became possible. Using oxygen as an electron acceptor, bacteria could reoxidize photosynthetically synthesized molecules and regain some of the energy stored in them:

$$C_6H_{12}O_6 + 6\,O_2 \longrightarrow O_2 + 6\,H_2O$$

$$\Delta G^0 = -686 \text{ kcal/mol}$$

Many of the proteins photosynthetic bacteria used for photosynthesis could also be used for oxidative phosphorylation. Several lines of evidence suggest that purple photosynthetic bacteria might have been the ancestors of nonphotosynthetic bacteria that perform oxidative phosphorylation, and that the mitochondria of eukaryotic cells evolved from such bacteria, which had formed symbiotic associations with primitive eukaryotic cells (Woese, 1987). Thus, it is no coincidence that the mechanisms of photosynthesis and respiration are similar in many respects. Much of our understanding of oxidative phosphorylation has come from studies of photosynthesis, which have exploited the experimental advantages of a system that can be driven by light.

In algae and higher plants, photosynthesis takes place in organelles known as **chloroplasts**. There is evidence that chloroplasts evolved from oxygen-evolving photosynthetic bacteria such as cyanobacteria or *Prochloron*, which had formed symbiotic associations with eukaryotes. This process appears to have occurred several times.

The evolution of photosynthesis is discussed in excellent on-line photosynthesis tutorials (Whitmarsh and Govindjee, 2000; Cramer, 1999). A shorter tutorial is offered by Kramer (2000). Links to a number of very useful photosynthesis resources, many of which include interactive graphics and animations, are presented by Orr and Govindjee (2000) and Roberts (2000).

II. Chloroplasts

In higher plants, most chloroplasts are found in the **mesophyll cells** of leaves (Fig. 2). The chloroplast consists of a collection of flattened membranous vesicles called **thylakoids**, which are enclosed in an **envelope** formed by a double membrane (Fig. 3). The outer membrane contains channels formed by the protein **porin** and is freely permeable to substances whose molecular mass is below about 10 kDa. The inner membrane is selectively permeable and contains a number of metabolite transporters, which mediate the interaction between metabolism in the plant cytosol and within the chloroplast (Flügge and Heldt, 1991). The volume surrounding the thylakoids is known as the **stroma**. The thylakoid membranes contain the pigments that capture light, and, like the cristae of mitochondria, they contain an electron transport system coupled to an adenosine triphosphate (ATP) synthase. The soluble enzymes responsible for carbon dioxide assimilation are found in the stroma. In many chloroplasts of land plants and green algae, the thylakoid membranes are stacked tightly together in some regions to form structures called **grana**, which look like green grains under the light microscope. The grana are connected by nonstacked thylakoid extensions, the **stroma lamellae**.

Chloroplasts still retain genes, which are located on a chromosome found in the stroma, for many of the proteins involved in photosynthesis and for transfer and ribosomal RNA (Rochaix *et al.*, 1998). Proteins coded by chloroplast genes are synthesized on ribosomes located in the stroma. Genes for other chloroplast proteins are now located on nuclear chromosomes of the plant cell. Remarkably, many multi-subunit proteins of the chloroplast thylakoid, including the ATP synthase, contain both subunits coded on nuclear genes and synthesized on cytosolic ribosomes and subunits coded on chloroplast genes and synthesized on stromal ribosomes. How the plant coordinates the synthesis of these subunits and the assembly of the proteins presents an intriguing research problem. Chloroplast proteins that are synthesized in the plant cytosol are initially formed with amino terminal extensions, termed **transit peptides**, which allow them to bind to receptors on the outer envelope membrane and then enter the chloroplast. The process requires GTP, ATP, and molecular chaperones (Schnell, 1995). After the proteins have entered the chloroplast stroma, the transit peptides are removed by a highly specific protease. Proteins destined for the membrane or lumen of the thylakoid are formed with an additional presequence, which is proteolytically cleaved after it has allowed the protein to enter the thylakoid membrane or lumen.

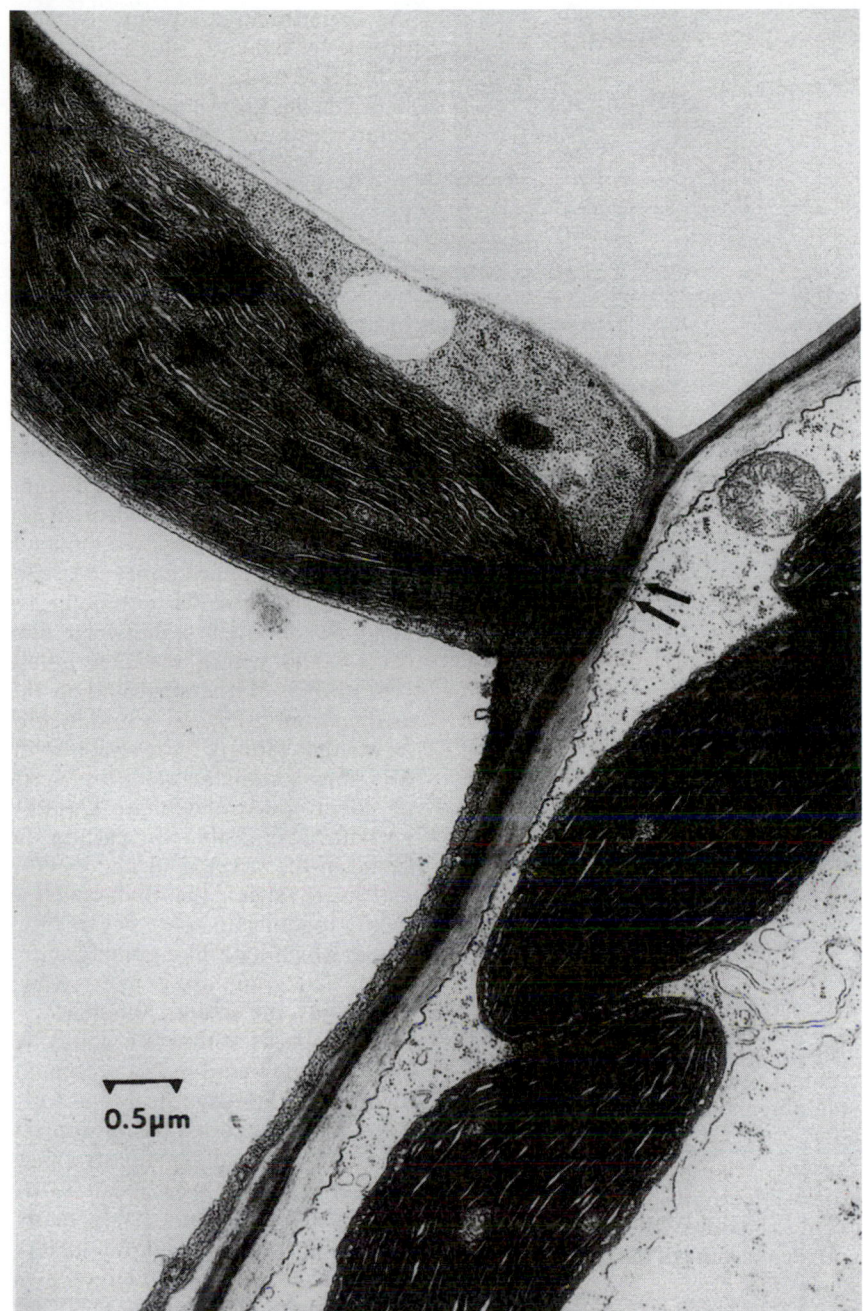

FIGURE 2. Transmission electron micrograph of corn (*Zea mays*) chloroplasts. Upper left: chloroplast within mesophyll cell. Lower right: chloroplast within bundle sheath cell. Arrows indicate plasmodesmata through which metabolites are exchanged between the cells. The mesophyll cells lie near the leaf surface. The bundle sheath cells lie near the leaf vascular bundles. Note the grana in the mesophyll chloroplast. (Micrograph courtesy of Iain M. Miller.)

III. Biochemistry of Carbon Assimilation

A. The Reductive Pentose Phosphate Cycle

To determine the sequence of reactions involved in the conversion of carbon dioxide to sugars, in the early 1950s Melvin Calvin and his associates exposed algae to $^{14}CO_2$ and light for different periods. After treatment with boiling alcohol, the carbon compounds that had been formed were separated by paper chromatography. The manner in which the distribution of ^{14}C among the carbon atoms of the various carbon compounds changed with time revealed the operation of a cyclic process, which has become known as the **reductive pentose phosphate cycle**, or **Calvin cycle** (Fig. 4).

In the first step, CO_2 combines with ribulose 1,5-bisphosphate to form two molecules of 3-phosphoglycerate. The

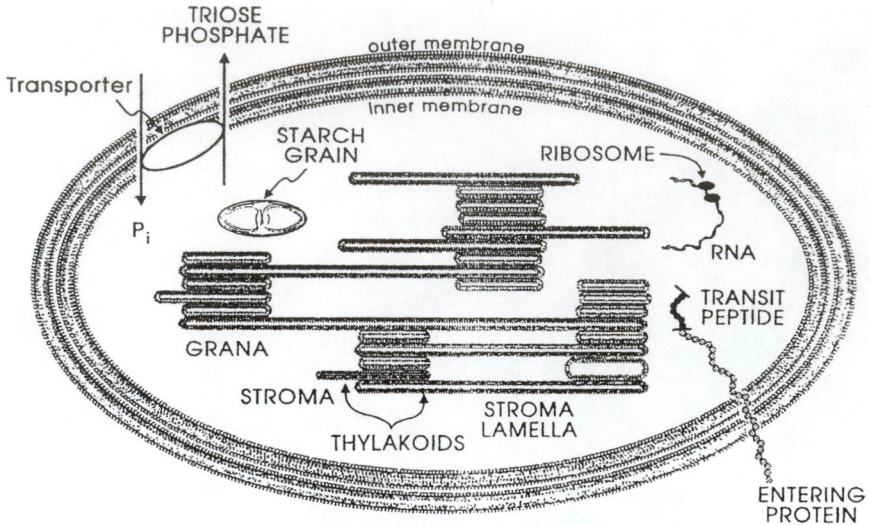

FIGURE 3. Diagram of a chloroplast.

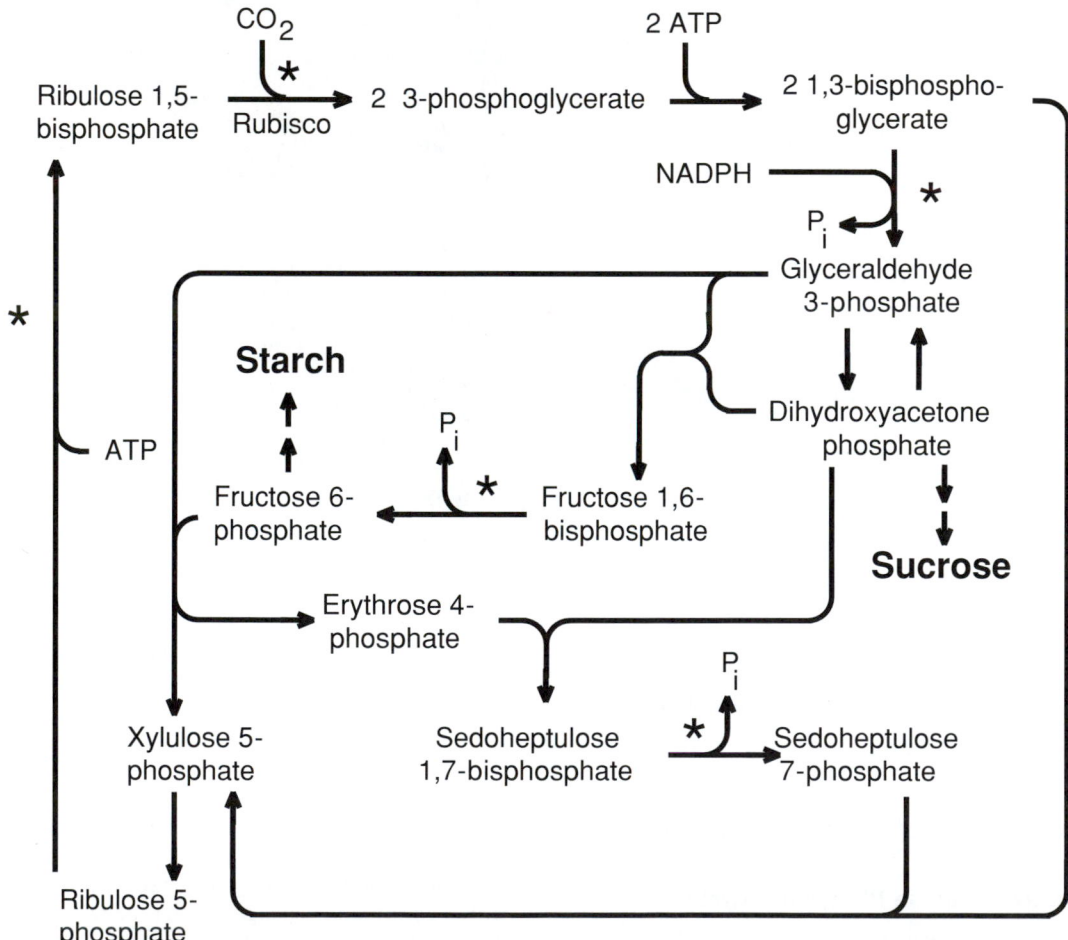

FIGURE 4. The reductive pentose phosphate or Calvin cycle. Stoichiometries are shown only for the first two steps. Erythrose, xylulose and sedoheptulose are sugars that contain 4, 5, and 7 carbons, respectively. Stars indicate steps catalyzed by light-regulated enzymes.

reaction is catalyzed by ribulose 1,5-bisphosphate carboxylase/oxygenase, often referred to as "**rubisco.**" This enzyme has a rather low affinity for CO_2. To compensate, it is present in extremely high concentrations in the chloroplast and is probably the most abundant protein on earth. Next, the 3-phosphoglycerate is phosphorylated to form 1,3-bisphosphoglycerate, which in turn is reduced to glyceraldehyde 3-phosphate by nicotinamide adenine dinucleotide phosphate (NADPH). In the remaining steps of the cycle, ribulose 1,5-bisphosphate is regenerated by a pathway that includes a complicated series of rearrangements catalyzed by transketolases and aldolases.

Intermediates in the cycle, including fructose 6-phosphate, glyceraldehyde 3-phosphate, and dihydroxyacetone phosphate, serve as precursors for the starch, sucrose, and amino acids that are the final products of photosynthesis.

ATP is required for the conversion of 3-phosphoglycerate to 1,3-bisphosphoglycerate and for the conversion of ribulose 5-phosphate to ribulose 1,5-bisphosphate. NADPH is required for the reduction of 1,3-bisphosphoglycerate to glyceraldehyde 3-phosphate. For each molecule of CO_2 fixed, three molecules of ATP and two molecules of NADPH are needed. It is the function of the light-driven reactions in the thylakoid membranes to furnish this ATP and NADPH.

B. Photorespiration and C_4 Plants

Not only does ribulose 1,5-bisphosphate carboxylase/oxygenase have a low affinity for CO_2. It also catalyzes a competing reaction, in which O_2 rather than CO_2, is added to ribulose-1,5-bisphosphate (Fig. 5). The products are 3-phosphoglycerate and phosphoglycolate. This reaction seems to serve no useful purpose. Many plants grow much faster in a CO_2-enriched atmosphere in which the carboxylation reaction can compete more effectively with the oxygenation reaction. It appears that in the 2 billion years since plants began to fill the atmosphere with oxygen while removing CO_2 from it, they have been unable to modify the enzyme so that its affinity for CO_2 is increased or its affinity for O_2 is decreased significantly. Molecular biologists are now trying to accomplish that task.

Part of the carbon appearing in phosphoglycolate is rescued. In a series of reactions occurring in **peroxisomes** and in mitochondria, two molecules of phosphoglycolate are converted to one molecule of glycerate, which is returned to the chloroplast and rephosphorylated. But one carbon atom is lost as CO_2, and ATP and O_2 are consumed. This process is known as **photorespiration.** It is not coupled to ATP formation, and the net result is waste of ATP and fixed carbon.

A number of plants, including corn, sugarcane, and crabgrass, partially avoid photorespiration by concentrating CO_2 in the cells that contain the Calvin cycle enzymes, so that carboxylation can compete more effectively with oxygenation. In these plants, the Calvin cycle enzymes are located in the chloroplasts of the **bundle sheath cells**, which surround the vascular bundles deep within the leaves (Fig. 6). CO_2 is first captured in the mesophyll cells, which lie near the leaf surface, by carboxylation of phosphoenolpyruvate (PEP):

$$PEP + CO_2 \xrightarrow{\text{PEP carboxylase}} \text{oxaloacetate} + P_i$$

The carboxylation is catalyzed by PEP carboxylase, an enzyme that has a high affinity for CO_2 and does not catalyze an oxygenation reaction. Oxaloacetate is next reduced to malate (or transaminated to form aspartate in some plants). The malate or aspartate is transferred to the bundle sheath cells through fibers known as **plasmodesmata.** Some of these can be seen at the arrows in Fig. 2. Malate is oxidatively decarboxylated to pyruvate in the bundle sheath cells in a reaction that also generates NADPH. The CO_2 that is released enters the Calvin cycle in the bundle sheath cells. Pyruvate returns to the mesophyll cells, where it is again transformed to PEP. The net result is the transfer of CO_2 and NADPH to the bundle sheath cells.

The CO_2-concentrating mechanism consumes energy, since ATP is converted to adenosine monophosphate (AMP) and phosphate. Nevertheless, plants that use it are often referred to as efficient plants because avoidance of photorespiration more than compensates for this expense. Such plants are usually known as C_4 plants because of the involvement of four-carbon acids. Many C_4 plants are native to tropical areas, where the C_4 pathway is especially advantageous since high temperatures and bright sunlight encourage photorespiration.

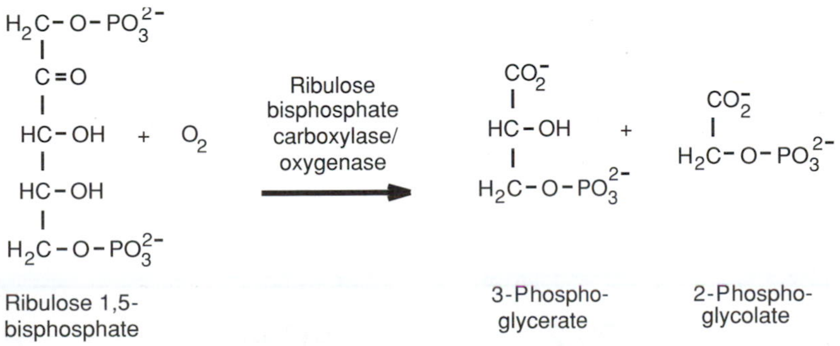

FIGURE 5. The oxygenation reaction catalyzed by ribulose 1,5-bisphosphate carboxylase.

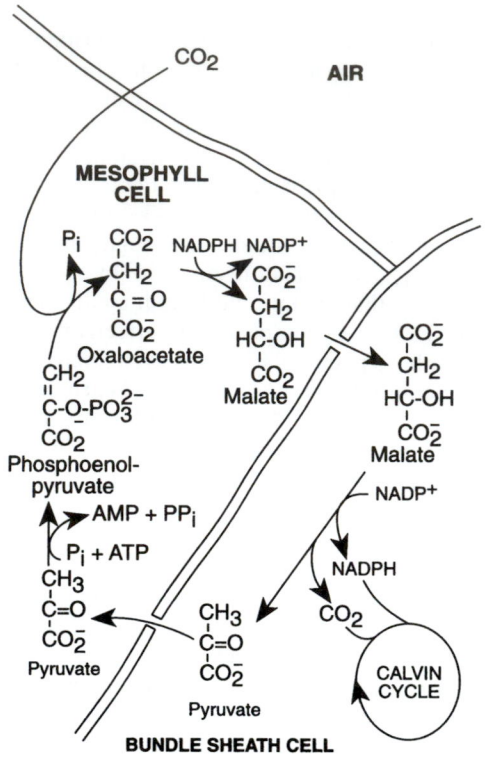

FIGURE 6. The mechanism used by C_4 plants to concentrate CO_2 and NADPH in bundle sheath cells. CO_2 is captured in the mesophyll cells by carboxylation of phosphoenolpyruvate. Malate is transferred to the bundle sheath cells, where CO_2 is released and NADPH is formed to enter the Calvin cycle. Pyruvate returns to the mesophyll cells.

IV. Formation of ATP

A. Light-Driven Electron Transport and ATP Synthesis

Carbon dioxide fixation requires ATP and NADPH. It seemed reasonable to suspect that the role of light is to provide the energy necessary for their formation. Photosynthetic membranes contain electron transport chains much like those of mitochondria, and light can drive electron transport along the chains (Figs. 7 and 8). As in mitochondria, ATP formation is coupled to the electron transport.

How can electron flow through the electron carriers shown in Fig. 7 cause ATP formation? Mitchell (1961) suggested that electron transport and ATP formation are both coupled to the movement of protons across membranes. His **chemiosmotic hypothesis** helps us understand the reason for the arrangement of the electron carriers within photosynthetic membranes (Mitchell, 1979).

B. The Chemiosmotic Hypothesis

It was known that electron transport in mitochondria and chloroplasts is accompanied by the movement of protons across the cristae or thylakoid membranes. Mitchell suggested that the transmembrane pH difference (often called a proton gradient) that is formed is an intermediate in phosphorylation. Electron transport would drive proton translocation, and the proton gradient in turn would drive ATP synthesis.

Two experiments involving photosynthetic material played a critical role in convincing scientists that the chemiosmotic hypothesis must be taken seriously. Chloroplasts will synthesize ATP even if they are illuminated in the absence of ADP and P_i and then left in the dark

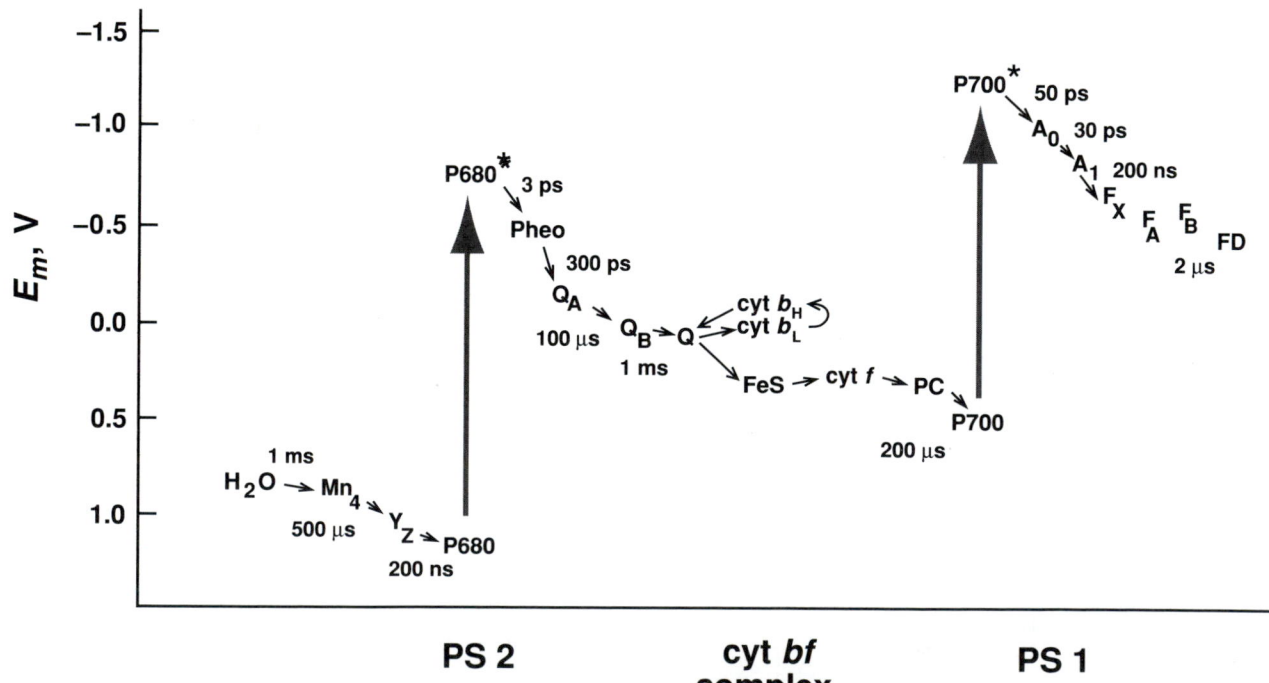

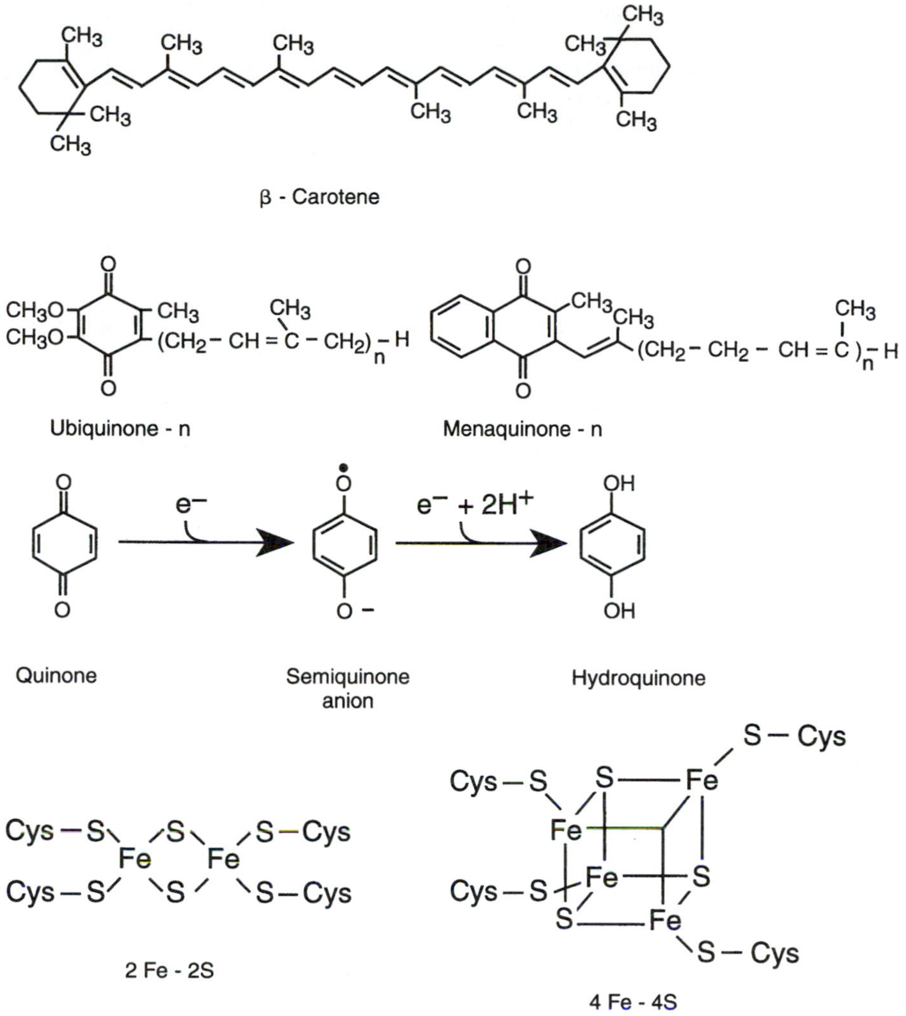

β - Carotene

Ubiquinone - n   Menaquinone - n

Quinone   Semiquinone anion   Hydroquinone

2 Fe - 2S

4 Fe - 4S

FIGURE 8. Structures of some of the molecules and cofactors involved in photosynthesis. Other carotenoids and quinones are found in many photosynthetic organisms. The quinone, semiquinone anion, and hydroquinone forms of the quinone six-membered ring are shown.

for several minutes before these substrates are added. A relatively stable "high energy intermediate" has been formed in the light and can later drive ATP formation in the dark. Andre Jagendorf and Ernest Uribe learned that the illumination step was most efficient when the pH of the medium was about 4, while the ATP synthesis step was most efficient at pH 8. Therefore they routinely illuminated the chloroplasts in medium buffered at pH 4 with succinate, then increased the pH to 8 before adding ADP and P_i. But ATP was formed by this procedure even if the chloroplasts were not illuminated! The quick increase in pH had created a proton gradient across the thylakoid membranes; the lumen pH had remained near 4 while the external pH had increased to 8. The simplest explanation was that the proton gradient had driven the ATP formation.

Halobacterium halobium is a red bacterium found in lakes that have a high salt content. It can form ATP in the light in a quite unusual way. The membrane of the bacterium contains

FIGURE 7. The electron transport system of higher plant chloroplasts. Midpoint redox potentials (E_m) of the components are indicated. P680 and P700 are chlorophyll *a* dimers, which, when excited by light (large arrows), transfer an electron to the acceptors pheophytin *a* (Pheo) and chlorophyll *a* (A_0), respectively. Typical electron transfer times are shown for the electron transfer steps, which are indicated by arrows. ps: picoseconds; μs: microseconds; ms: milliseconds; Mn_4, a complex containing 4 manganese nuclei that accepts electrons from water; Y_Z, tyrosine; Q_A, Q_B, and Q, plastoquinone; cyt b_H and cyt b_L, the high and low potential hemes, respectively, of cytochrome *b*; FeS, the 2 Fe-2S center of the Rieske protein; cyt *f*, cytochrome *f*; PC, plastocyanin, a water-soluble copper-containing protein; A_1, phylloquinone; F_X, F_A, F_B, 4 Fe-4 S clusters; FD, ferredoxin, a water-soluble protein containing a 4 Fe-4 S cluster.

purple domains in which molecules of the transmembrane protein bacteriorhodopsin are packed together in a hexagonal lattice. Each molecule of bacteriorhodopsin contains a molecule of all-*trans*-retinal attached as a Shiff base to a lysine residue. Upon absorbing light, the retinal is isomerized to the 13-*cis* isomer, initiating a cycle of events resulting in the translocation of a proton from the inside to the outside of the cell and regeneration of all-*trans*-retinal (Subramaniam and Henderson, 2000; Stoeckenius, 1999). The electrochemical proton gradient that is established is used to drive ATP formation.

ATP synthesis in mitochondria is catalyzed by a membrane-associated ATP synthase. Photosynthetic ATP formation is catalyzed by a similar protein. Ephraim Racker and Walter Stoeckenius incorporated purified bacteriorhodopsin into artificial phospholipid vesicles. When the vesicles were illuminated, protons were pumped into the vesicles. Protons were pumped inward rather than outward because the bacteriorhodopsin incorporated into the membranes in an orientation that was opposite to that in the bacterial membrane. When purified mitochondrial ATP synthase was also incorporated into the vesicles, light caused ATP to be formed from ADP and P_i. This experiment showed that the mitochondrial ATP synthase can catalyze ATP formation in the absence of any other mitochondrial protein, including those of the electron transport chain. Only an electrochemical proton gradient is required.

How might the pH gradient be established? Some electron carriers, such as cytochromes and iron-sulfur proteins, cannot cross membranes. Others, such as quinones, are soluble in membranes and, as illustrated in Fig. 8, may bind protons after they have accepted electrons.

Mitchell proposed that the electron carriers are arranged asymmetrically in the membrane, as in the upper part of Fig. 9. Carriers that bind protons would alternate in the electron transport chain with those that do not. Electron transfers that result in H^+ binding (A → B) would occur on one side of the membrane (the stromal side in Fig. 9), while electron transfers that result in H^+ release (B → C) would occur on the opposite side (the luminal side). In one direction (from the stromal side to the luminal side), electrons and protons would always move together across the membrane. In Fig. 9, BH serves as the carrier of an electron and a proton. Electrons would always move alone across the membrane in the opposite direction (C → D). In this way electron transport would be accompanied by the net transfer of protons across the membrane. Such electron transport is sometimes described as **vectorial**, since it has a defined direction within the membrane, unlike electron transfer that occurs in solution.

Protons would flow back across the membrane through the **ATP synthase**. This proton flow would drive ATP formation. How it does so will be discussed in IV.D.

C. Storage of Free Energy in an Electrochemical Gradient

When an uncharged molecule moves from a region where its concentration is C_1 to a region where its concentration is C_2 there is a free energy change

$$\Delta G = RT \ln C_2/C_1$$

per mole transferred. Since

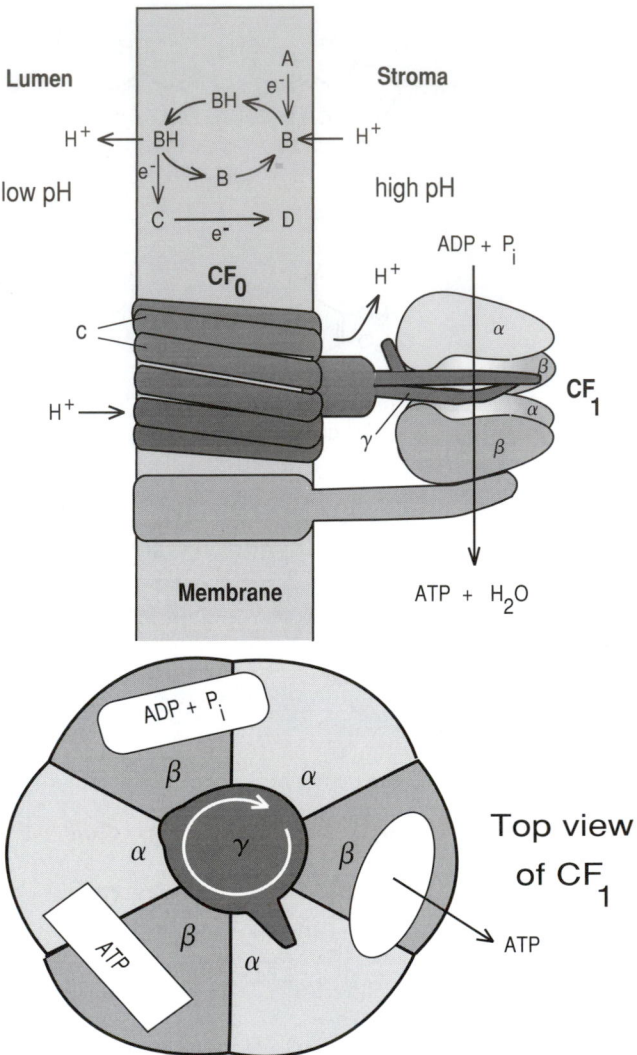

FIGURE 9. An illustration of how electron transport and ATP formation might be coupled. Top: Example of how electron and proton transport might be coupled. A, B, C, and D are electron carriers that cannot move through the membrane. B is a molecule, such as a quinone, which becomes protonated when it accepts an electron, diffuses across the membrane and releases the proton when it is reoxidized. Center: The ATP synthase, illustrating current conceptions of how proton transport might be coupled to ATP formation. The α and β subunits of CF1 are held fixed, while H+ flow through CF_0 causes rotation of the c and γ subunits. As projections of γ contact each $\alpha\beta$ pair in succession, they cause conformational changes that allow binding of ADP and phosphate, ATP formation, and ATP release. CF_1 is shown as a section through the γ subunit. Bottom: top view of CF_1, illustrating how rotation of the γ subunit may induce successive steps in ATP formation.

$$\Delta G = \Delta H - T\Delta S$$

the free energy change is due to the change in entropy, ΔS, that accompanies the change in distribution of the molecules. When a charged particle such as a proton moves between regions whose electrical potentials differ, there is an additional free energy change per mole transferred,

$$\Delta G = z \mathscr{F} \Delta \psi$$

where z is the number of charges on the molecule, $\mathscr{F}$ is Faraday's constant (23 kcal/V·mol), and $\Delta \psi$ is the electrical potential difference in volts. Such a difference in both concentration and electrical potential is often called an **electrochemical gradient**. There can be an electrochemical proton gradient between the chloroplast stroma and the thylakoid lumen. This gradient is created by the transfer of protons and electrons across the membrane during electron transport. For the transfer of protons ($z = +1$) from the stroma to the lumen

$$\Delta G = RT \ln \left(\left[H^+ \right]_{lumen} / \left[H^+ \right]_{stroma} \right) + \mathscr{F} \left(\psi_{lumen} - \psi_{stroma} \right)$$

$$\Delta G = -2.30 \, RT \, \Delta pH + \mathscr{F} \Delta \psi$$

This free energy difference is sometimes known as **proton-motive force** and abbreviated Δp, $\Delta \mu_{H^+}$, or PMF. Protons can be transported across the thylakoid membrane against their electrochemical gradient because the transfer is coupled to exergonic electron transfer. The proton translocation shown in Fig. 9 can occur if the sum of ΔG for proton translocation and ΔG of electron transfer from A to D is less than zero.

D. ATP Synthase

The chloroplast ATP synthases are found in the stroma lamellae and in the unappressed membranes of the grana, but rarely in appressed granal membranes (Fig. 10). The synthase has a structure similar to that of the mitochondrial ATP synthase (Fig. 9). It consists of a transmembrane domain (CF$_0$), and a catalytic domain (CF$_1$) located on the stromal surface of the thylakoid membrane. Each includes several kinds of subunit.

Biochemical (Boyer, 1997) and crystallographic (Abrahams *et al.*, 1994) studies of the ATP synthase have suggested that proton efflux through F$_0$ may be coupled to ATP formation in the following way. CF$_1$ includes three α and three β subunits, which form a ring of $\alpha\beta$ dimers (Fig. 9, bottom). Each β subunit contains a catalytic site. At each of the three catalytic sites, the following steps occur. (1) ADP and phosphate bind loosely. (2) After a conformation change, the binding becomes tight and ATP is formed. (3) After a second conformation change, ATP is released. It is this step, rather than ATP formation, that requires the greatest energy input. (4) After a third conformational change, the site is ready to bind ADP and phosphate loosely again. The steps occur at each catalytic site in sequence and cooperatively, e.g., as one site is binding ADP and phosphate, another is converting them to ATP, and the third is releasing ATP. What then causes the conformational changes? F$_0$ includes 9–12 copies of subunit c, a small transmembrane protein. It has been suggested that the c subunits may form a ring in the membrane. Each may be able to bind a proton from the luminal side of the membrane and release it on the stromal side, but only after the ring of c subunits has rotated, rather like the function of a waterwheel. The c subunits are attached to the γ subunit, a rod-shaped protein that extends into the hole formed by the $\alpha\beta$ ring. Proton efflux through the c ring would thus cause the γ subunit to rotate inside the $\alpha\beta$ ring. As projections on the γ ring contact each $\alpha\beta$ dimer in turn, they would induce the conformational changes that lead to ATP formation (Leslie *et al.*, 1999). Translocation of 9–12 protons and formation of three ATPs would accompany each full rotation of the c-γ complex, consistent with the observation that translocation of 3–4 protons is required for each ATP formed. Thus ATP formation is thermodynamically possible when μ_{H^+} is more than one-third or one-fourth of ΔG of ATP formation. Links to beautiful graphic illustrations of ATPase function can be found at Roberts (1999).

The ATP synthase is intricately regulated, presumably to prevent the reversal of ATP synthesis in the dark. It is activated by the presence of a sufficiently large $\Delta \mu_{H^+}$. Until it is activated, it

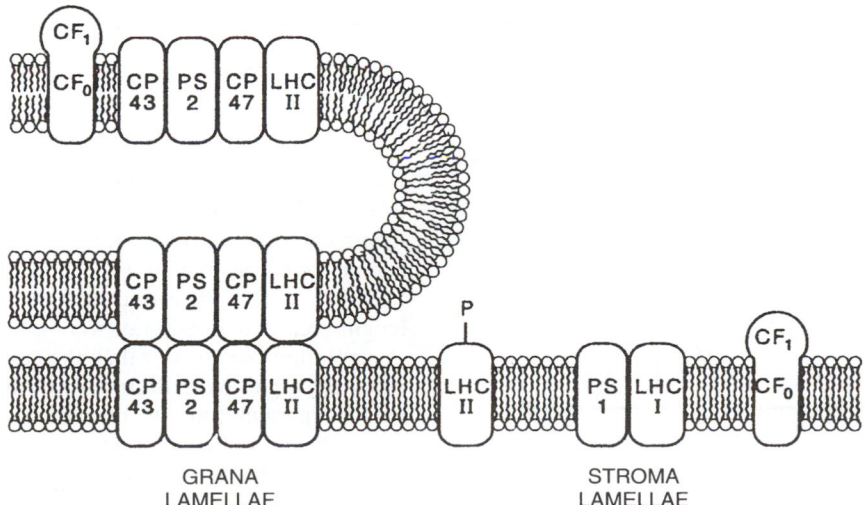

FIGURE 10. Distribution of photosynthetic complexes between the grana and stroma lamellae. Photosystem 2 (PS 2) is found primarily in the grana lamellae, whereas photosystem 1 (PS 1) and ATP synthase are primarily in stroma lamellae and unappressed granal membranes. When PS 2 activity exceeds PS 1 activity, LHC II complexes become phosphorylated and move from the grana lamellae to the stroma lamellae.

will neither synthesize nor hydrolyze ATP. Reduction of a CF_1 disulfide bridge by reduced thioredoxin, which is formed when chloroplasts are illuminated, enhances the efficiency of the activated ATP synthase when $\Delta\mu_{H^+}$ is small.

V. Photosynthetic Electron Transport

A. The Interaction of Light with Molecules

A light wave consists of an electric field and a magnetic field whose directions are perpendicular to each other. The wave moves through space at a velocity

$$v = nc$$

where n is the refractive index of the medium through which the light is moving, and c is the speed of light in a vacuum, 3.0×10^8 m/s (Fig. 11). The wavelength of the light, λ, is the distance between the crests of the waves. Visible light has a wavelength between 400 nm and 700 nm. The frequency of the light, ν, is the frequency with which crests of the waves pass a given point in space, and

$$\nu = c/\lambda$$

According to the principles of quantum theory, light can be absorbed or emitted only in discrete units called **quanta**. The energy, E, of such a quantum depends upon the frequency of the light according to the equation

$$E = h\nu$$

where h is Planck's constant, 2.86×10^{-37} kcal·s.

Just as the electrons of atoms occupy atomic orbitals having discrete energies, the electrons in molecules occupy molecular orbitals having discrete energies. Molecular orbitals may extend over several nuclei. Each molecular orbital can contain two electrons of opposite spin. Absorption of light can promote (excite) an electron to an orbital of higher energy (Fig. 12) if the energy difference between the orbitals is equal to the energy of a quantum of the light. Several competing processes may occur after light is absorbed. The excited electron may drop back to its ground molecular orbital. As it does so, the excitation energy may be reemitted as light (**fluorescence**), it may be lost as heat, or it may be transferred to a neighboring molecule, causing excitation of one of the neighbor's electrons to a higher orbital. Finally, the excited

electron may be transferred to an empty orbital of a neighboring molecule. When this occurs, part of the energy of the absorbed light is conserved as chemical free energy of the electron donor-acceptor pair.

The tendency of a biological molecule to donate electrons is usually expressed as its midpoint redox potential at pH 7, E_{m7}. The **midpoint potential**, E_m, is the electrical potential difference that would exist between the standard hydrogen half cell and a solution in which the oxidized and reduced forms of the molecule are present in equal concentrations. In the standard hydrogen half cell, the pH is 0 and the partial pressure of H_2 is 1 atm. The E_{m7} values for members of the chloroplast electron transport system are indicated in Fig. 7. A molecule in which an electron has been excited to a higher orbital has a more negative E_m than does the molecule in its ground state, by an amount about equal numerically to the energy difference between the ground and excited orbitals (in electron volts). Thus, it is a better electron donor. Note the difference between the E_m's of P680 and P680* (and between P700 and P700*) in Fig. 7.

The standard free energy change of an electron transfer is related to the difference between the E_m values of the electron donor and acceptor

$$\Delta G^o = -n\mathcal{F}\Delta E_m$$

where n is the number of electron equivalents transferred per mole. Thus, excitation of a molecule lowers the free energy change accompanying the transfer of an electron to another molecule. This equation can be used to calculate the free energy changes accompanying the light-induced charge separations and the subsequent electron transfers shown in Fig. 7.

Especially lucid explanations of the photochemical processes involved in photosynthesis have been given by Clayton (1970, 1980).

B. The Presence of Two Photochemical Reactions in Chloroplasts

The most careful measurements of the **quantum yield** of oxygen evolution indicate that at least 8 light quanta must be absorbed for each molecule of oxygen that is evolved. Because the oxidation of two water molecules to form one O_2 molecule requires the removal of only 4 electrons,

$$2\,H_2O \rightarrow O_2 + 4\,H^+ + 4\,e^-$$

it appeared that two quanta are necessary for the extraction of each electron from water. Robert Emerson illuminated the green alga *Chlorella* with light of different wavelengths and observed the rates of O_2 evolution. He found that when light of wavelength less than 680 nm and light of wavelength greater than 680 nm were given together, the rate of O_2 evolution was greater than the sum of the rates observed when the algae were illuminated with light of each wavelength separately; that is, the two wavelengths acted synergistically. This phenomenon, which has come to be known as **Emerson enhancement**, suggested that efficient O_2 evolution requires the cooperation of a system which absorbs long-wavelength light and a system which absorbs short-wavelength light. L. N. M. Duysens found that 680-nm light caused the oxidation of cytochrome f, whereas 562-nm light caused its re-reduction. In 1960 R. Hill and F. Bendall

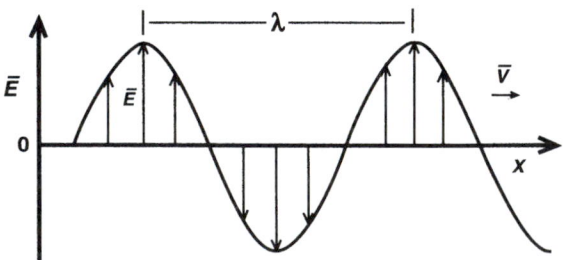

FIGURE 11. The electric field of a light wave. The magnetic field (not shown) is perpendicular to the electric field and to the direction of motion of the wave.

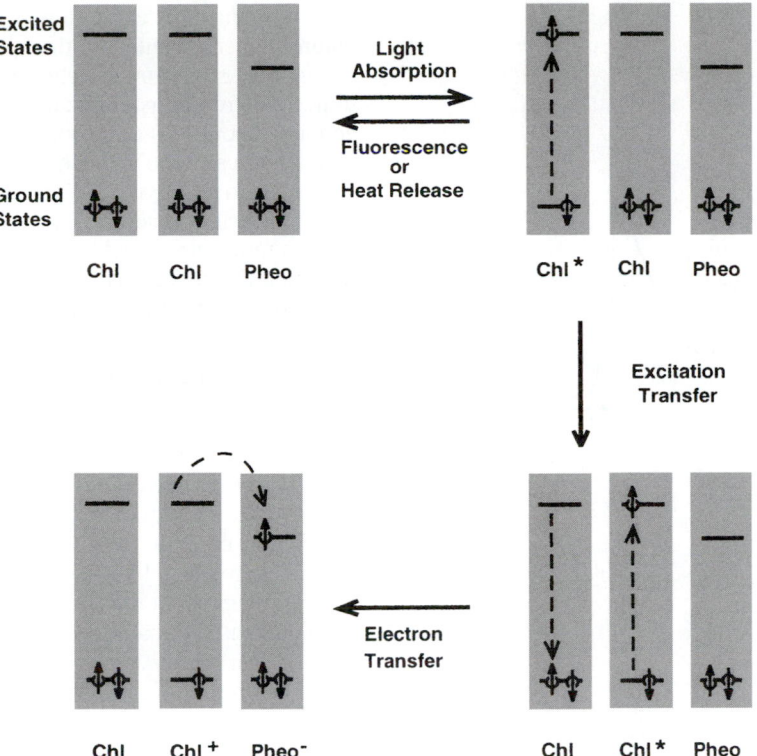

FIGURE 12. Light absorption and emission, excitation transfer, and photochemical electron transfer. Chlorophyll (Chl) and pheophytin (Pheo) molecules are used as examples. Circles with upward and downward arrows represent electrons having opposite spins. Molecules have many ground and excited orbitals. Only one of each is shown. Chl*, electronically excited Chl; Chl⁺, oxidized Chl; Pheo⁻, reduced Pheo.

suggested the **Z scheme**, in which a photosystem driven by short-wavelength light (**photosystem 2, PS 2**) and a photosystem driven by long-wavelength light (**photosystem 1, PS 1**) are connected in series. Light absorbed by PS 2 would cause the transfer of an electron from H_2O to a chain of electron carriers between the photosystems (cytochrome *f* is one of these). Light absorbed by PS 1 would cause the transfer of the electron from this chain to NADP⁺ to form NADPH:

$$H_2O \rightarrow PS2 \rightarrow cytochrome f \text{ (etc.)} \rightarrow PS1 \rightarrow NADP^+$$

Absorption of two quanta, one by each photosystem, would be required for the transfer of an electron all the way from H_2O to NADP⁺. Cytochrome *f* would be reduced by PS 2 and oxidized by PS 1. The complete Z-scheme as it is now understood is presented in Fig. 7.

The light-absorbing pigments associated with PS 1 and PS 2 have somewhat different absorption spectra. The PS 2 pigments have very little absorption beyond about 680 nm, so light in this wavelength region drives PS 1 almost exclusively.

C. Primary Electron Donors and Light-Harvesting Pigments

The absorption spectra of a typical plant cell and of Chl *a* and Chl *b* are shown in Fig. 13. Bessel Kok noticed that

when chloroplasts are illuminated, there is a small decrease in their optical absorbance ("bleaching") at 700 nm. He suggested that the change was due to the light-induced oxidation of a chlorophyll molecule, resulting in the loss of its optical absorbance at that wavelength. It was suggested that the primary light reaction is the transfer of an electron from a special chlorophyll molecule, named **P700**, to an electron acceptor (Fig. 12). In bacteria, illumination caused cytochrome oxidation (detected as a change in optical absorbance at wavelengths where cytochrome absorbs light) as well as bacteriochlorophyll oxidation. To determine whether bacteriochlorophyll or cytochrome oxidation occurs first, William Parson illuminated bacteria with very short flashes from a Q-switched laser. Immediately after the flash, he observed oxidation of bacteriochlorophyll. The bacteriochlorophyll became reduced again in a few milliseconds and, as it did, the cytochrome became oxidized. The chlorophyll had been oxidized first; the cytochrome had then donated an electron to the oxidized chlorophyll.

The laser flash had been so brief that it ended before the electron lost by the bacteriochlorophyll had been replaced by an electron from the cytochrome, so the flash had caused the transfer of only a single electron from the bacteriochlorophyll. Such **single turnover flashes** have been a powerful tool for the study of photosynthetic electron transport. Lasers capable of producing flashes femtoseconds (10^{-15} s) in duration have made it possible to measure the

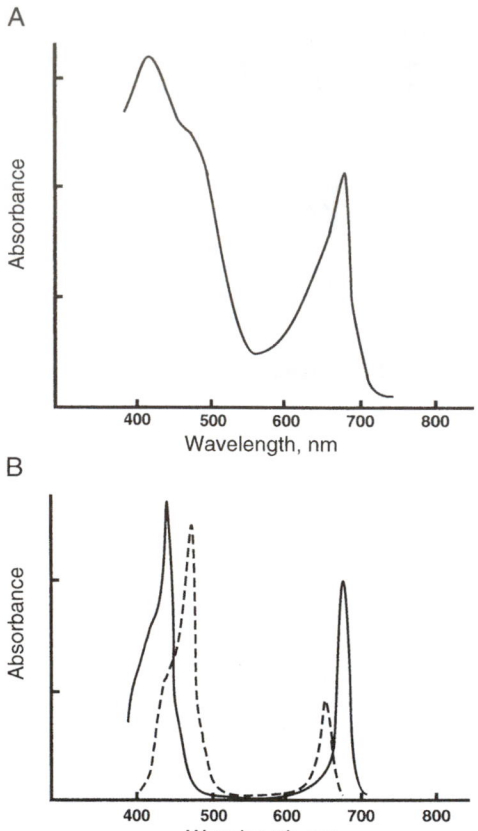

FIGURE 13. (A) Absorption spectrum of a typical plant chloroplast. (B) Absorption spectra of Chl *a* (solid line) and Chl *b* (dashed line). Each absorbance band corresponds to excitation of chlorophyll from the ground state to a different excited state. In the chloroplast, both Chl *a* and Chl *b* absorb light between 400 nm and 500 nm and between 600 nm and 700 nm. In addition, β-carotene absorbs light between 400 nm and 500 nm.

rates of early electron transfer steps, some of which occur in picoseconds (10^{-12} s).

P700 is the primary electron donor in PS 1. The primary donor of PS 2 absorbs at 680 nm and is known as **P680**. Oxidized chlorophyll contains an unpaired electron (see Fig. 12). Such molecules can be studied by the technique of **electron paramagnetic resonance**, in which the absorption of microwaves by the unpaired electrons is measured. Study of P680 and P700 by this technique, as well as other experiments, suggested that they are *dimers* of chlorophyll. After they have lost an electron, the remaining unpaired electron is shared between the two chlorophyll molecules that comprise the dimers.

The details of many biophysical techniques, including optical spectroscopy and magnetic resonance, which are employed in photosynthesis research are described in Amesz and Hoff (1996).

Most of the chlorophyll in cells serves only to capture light and does not participate in photochemistry. Each photochemically active chlorophyll dimer (P680, P700) is associated with an aggregate of several hundred light-harvesting chlorophyll molecules, known as a **photosynthetic unit**. After any chlorophyll molecule in the unit absorbs a quan-

tum of light, the excitation energy migrates among the chlorophyll molecules in the aggregate (Figs. 12 and 14). When the excitation reaches the photochemically active dimer, electron transfer occurs. A typical photosynthetic unit contains about 300 chlorophylls molecules. **Beta carotene** and other carotenoids also serve as light-harvesting pigments, as do tetrapyrrole pigments known as **phycobilins** in cyanobacteria and some algae. The pigments are attached to protein molecules. Such accessory pigments allow the capture of light in regions of the spectrum where chlorophyll does not absorb light strongly.

D. The Structure of Photosynthetic Reaction Centers

It has been possible to isolate many of the oligomeric protein complexes involved in photosynthesis by dissolving the membranes in detergent and purifying the complexes through chromatography and other techniques. The complexes that contain the photochemically active chlorophyll dimers are referred to as **reaction centers.** It has become increasingly apparent that PS 1, PS 2, and the reaction centers of photosynthetic bacteria resemble each other much more closely than had been suspected (Nitschke and Rutherford, 1991; Büttner *et al.,* 1992). They may have evolved from a

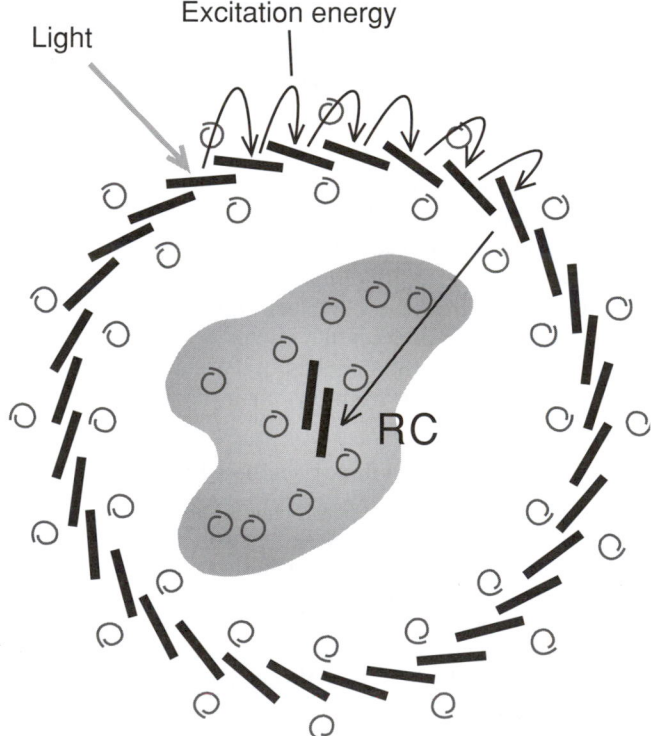

FIGURE 14. Light absorption and the transfer and capture of excitation energy. Light is absorbed by any chlorophyll molecule (dark rectangles) in a large array. Excitation energy is transferred among the chlorophyll molecules and finally captured by a special chlorophyll pair in the reaction center (RC). The figure is based on the structure of the bacterial light-harvesting complex (Cogdell *et al.,* 1999). Protein dimers to which BChl is bound form a ring around the reaction center. (Higher plant light harvesting pigments have a quite different structure.)

common ancestor (Blankenship, 1992). PS 1 bears an especially strong resemblance to the reaction centers of the green bacteria and heliobacteria (Olson, 1996), while PS 2 resembles those of the purple bacteria and *Chloroflexus*. Roderick Clayton and Dan Reed, and George Feher succeeded in isolating reaction centers from purple bacteria. Hartmut Michel, Johann Deisenhofer, and Robert Huber devised a way to crystallize reaction centers from the purple bacterium *Rhodopseudomonas viridis* and determined their structure by x-ray diffraction (Deisenhofer and Michel, 1989, Deisenhofer and Norris, 1993). Many of the current ideas about the structure and function of PS 1 and PS 2 are based on the belief that their structures probably resemble that of the *Rp. viridis* reaction center. In addition, the reaction center structure provides important insights into the general properties to be expected of integral membrane proteins, which have been extremely difficult to crystallize. The *Rp. viridis* reaction center consists of 4 subunits (Fig. 15). Subunits L and M each contain five α-helices that cross the membrane. The co-factors that participate in electron transport are bound to the L and M subunits. Their arrangement is sketched in Fig. 15 and can be seen in three dimensions in Fig. 16. The primary electron donor, P985, is indeed a dimer, in this case of bacteriochlorophyll *b*. The reaction center also includes two additional molecules of bacteriochlorophyll *b*; two of bacteriopheophytin *b*; one molecule

each of menaquinone, ubiquinone, and carotenoid; and a nonheme ferrous iron ion. Surprisingly, the reaction center has approximate two-fold symmetry. It has been shown that electrons move down only the L side of the reaction center after light has been absorbed. The electrons move from P985 to the pheophytin associated with subunit L, to Q_A (menaquinone), and finally to Q_B (ubiquinone). Why they move only along this pathway, what function the other cofactors may have, and whether the bacteriochlorophyll *b* monomer associated with the L subunit participates in electron transfer are currently being investigated. The reaction center also includes a cytochrome subunit containing 4 hemes, which is located on the periplasmic surface (facing the outside of the cell) of the membrane. The hemes donate electrons to oxidized P985. There is also an H subunit, which forms a sort of cap on the cytoplasmic side of the L and M subunits and is anchored to the membrane by a single membrane-spanning helix. Its function is not known. All aspects of bacterial photosynthesis are discussed in Blankenship *et al.*(1995).

E. The Structure and Function of the Chloroplast Electron Transport System

We can now summarize the way the chloroplast electron transport system is believed to operate (Fig. 17; Ort *et al.*,

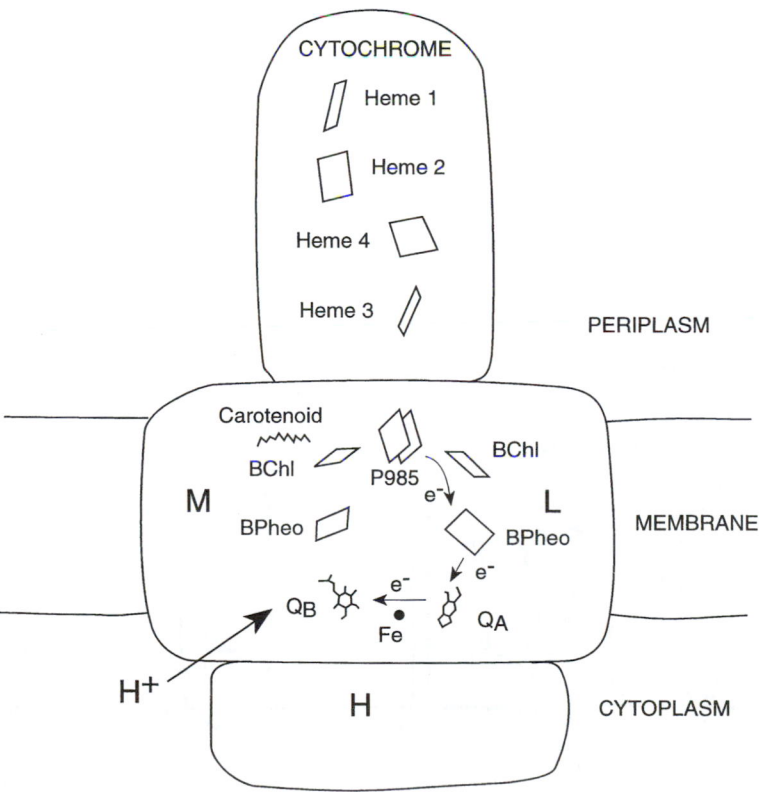

FIGURE 15. The arrangement of the cofactors in the *Rhodopseudomonas viridis* reaction center. The reaction center has approximate two-fold (C_2) symmetry, but electrons move only along the path indicated. After Q_B accepts 2 electrons, 2 protons from the cytoplasmic side of the membrane are bound. P985, the photochemically active bacteriochlorophyll *b* dimer; Bchl, bacteriochlorophyll *b*; BPheo, bacteriopheophytin *b*; Q_A, menaquinone; Q_B, ubiquinone; Fe, ferrous iron.

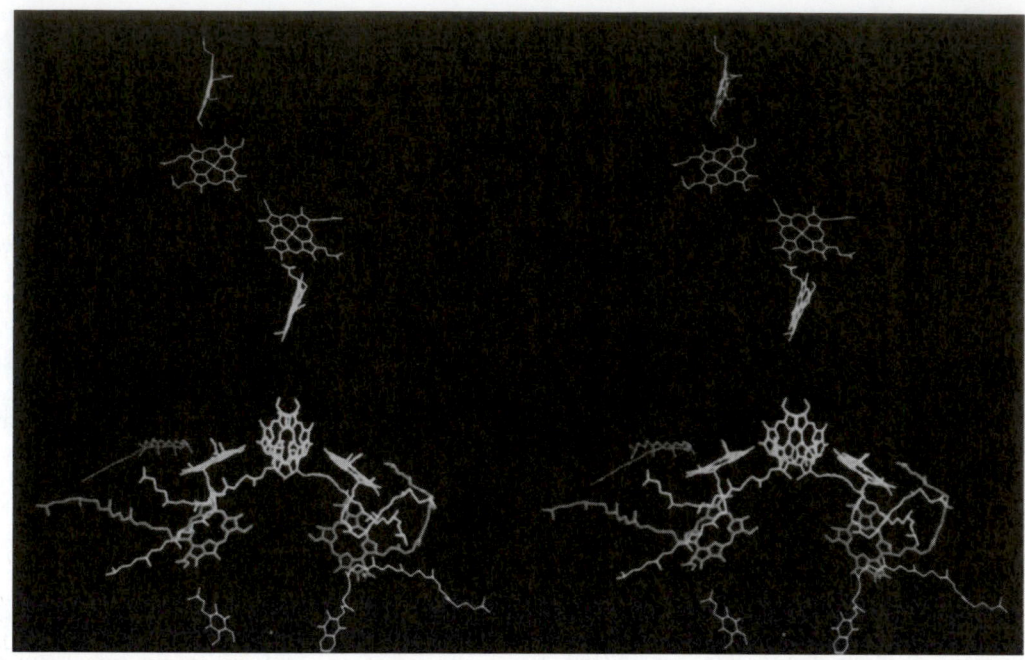

FIGURE 16. Stereo drawing of the *Rhodopseudomonas viridis* reaction center cofactors. To view the drawing in three dimensions, focus on a spot midway between the figures and allow their images to become superimposed. (From EMBO J. **8,** 2149–2170 (1989), (c) 1989 The Nobel Foundation.)

1996; He and Malkin, 2000; Nugent, 1996; Singhal *et al.,* 1999; Cramer, 1999; Ke, 2000). The components have redox potentials that allow electrons to move forward along the chain; they are arranged in the membrane and oriented in such a way that forward electron transfer is fast, and electron flow results in translocation of protons across the membrane.

PS 2 contains two subunits, D1 and D2, whose amino acid sequences partially resemble those of purple bacterial reaction center subunits L and M. Like the L and M sub-

units, D1 and D2 bind the electron transport cofactors. PS 2 contains several additional proteins that are not discussed here.

Formation of one O_2 molecule requires the removal of four electrons from two molecules of H_2O. Kok suggested that each photochemical act removes one electron from an oxygen-evolving complex (OEC). When four electrons have been removed, oxygen molecules are released (Fig. 18). The OEC contains four manganese nuclei. Determining its structure is a

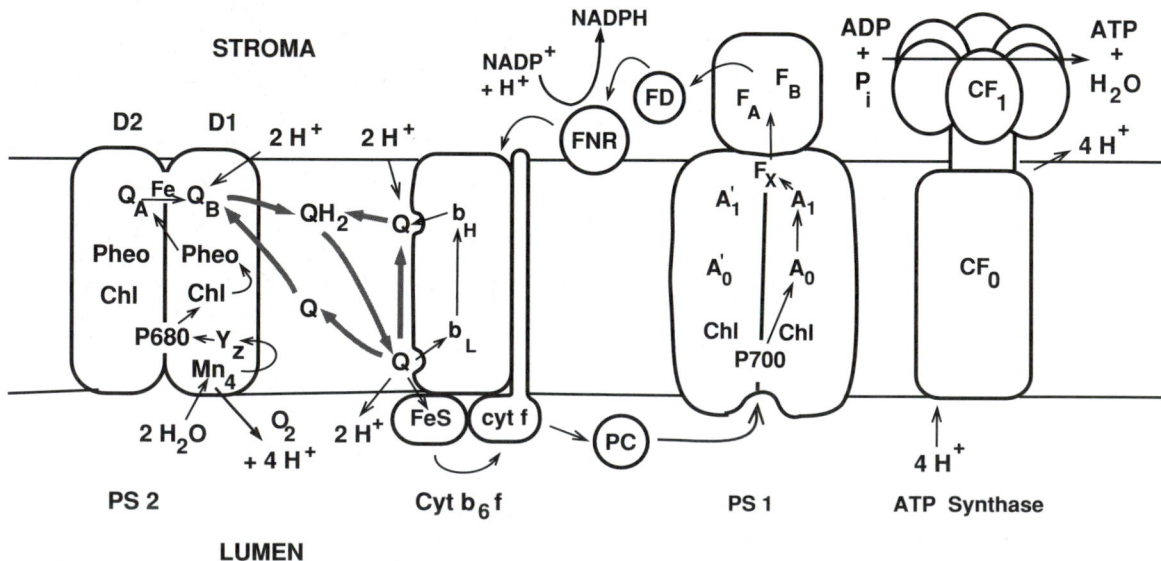

FIGURE 17. Electron and proton transport and ATP formation in chloroplasts. Symbols are defined in the legend of Fig. 7.

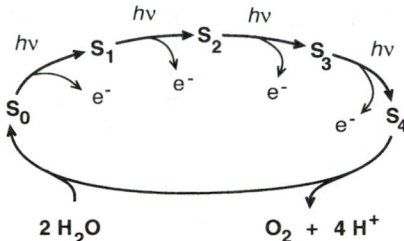

FIGURE 18. Storage of positive charges in the OEC during a series of flashes. S_0, S_1, S_2, S_3, and S_4 represent states of the OEC containing 0, 1, 2, 3, and 4 stored positive charges, respectively. O_2 is released after S_4 is formed.

major objective of current research. The following sequence of events is now believed to occur (see Fig. 7). Upon excitation, P680 transfers an electron to a special pheophytin a molecule (pheophytin a is similar to Chl a, but the Mg^{2+} is replaced by two protons). An electron is then transferred to P680$^+$ from Y_Z, a tyrosine residue, which is part of D1. Next, Y_Z^+ receives an electron from the OEC, leaving a positive charge stored in the OEC. This sequence of events is repeated until four positive charges have been stored in the OEC and O_2 is released. During the process, four protons are released into the chloroplast lumen.

The first electron that had been transferred to pheophytin a moves to Q_A (plastoquinone) and then to Q_B (also plastoquinone) to form a plastosemiquinone anion. The rates of these electron transfers are shown in Fig. 7. After a second photon is absorbed, a second electron moves down the chain to Q_B, reducing it to Q_B^{2-}. Q_B^{2-} then binds two protons from the stroma to form plastohydroquinone. The structures of quinone, semiquinone anion, and hydroquinone are shown in Fig. 8; the mechanism that may be involved in proton binding is discussed by Shinkarev and Wraight (1993) and Okamura and Feher (1995). At this point two protons have been taken up from the stroma and four protons have been released into the lumen. The herbicides atrazine and diuron act by competing with plastoquinone for binding to the Q_B site on D1.

The plastohydroquinone is released from its binding site on PS 2 and diffuses within the membrane (as QH_2 in Fig. 17) until it reaches the cytochrome bf complex (Crofts and Berry, 1998; Yu $et\ al.$, 1998). Here it binds to a site near the luminal surface of the membrane.

At this point a series of reactions is initiated that will result in the transfer of additional protons across the membrane. A possible mechanism of the proton translocation, which was suggested by Mitchell, is known as a **Q cycle** (Crofts, 2000). The bound QH_2 donates an electron to a 2 Fe-2 S center known as the **Rieske Fe-S center**. The electron moves from there to cytochrome f. The cytochrome bf complex also contains two cytochrome b-type hemes, b_H and b_L. QH_2 as a reductant is not strong enough to donate an electron to these hemes, but the plastosemiquinone that was formed when QH_2 transferred the electron to the Fe-S center is, and does so. In effect, the transfer of the first electron from QH_2 to the Fe-S center has driven the transfer of the second electron to b_L. QH_2 has been oxidized to Q, releasing two protons into the lumen. A second QH_2 molecule is oxidized by a similar mechanism, and a second electron

is donated to the hemes. The two heme electrons are donated to a plastoquinone (Q) molecule, which is bound near the stromal surface. It then binds two protons from the stroma. In the net process occurring at the cytochrome bf complex, one QH_2 has been oxidized, two protons have been removed from the stroma and four protons have been released into the lumen.

The electrons that were transferred to cytochrome f move to **plastocyanin**, a soluble, copper-containing protein. The plastocyanin diffuses within the lumen until it reaches the PS 1 complex, where it transfers the electron to P700$^+$. P700 is a chlorophyll a dimer.

Along with several other proteins, PS 1 contains a protein dimer, the PsaA and PsaB proteins, to which electron transfer cofactors are bound. The crystal structure of PS 1 reveals that the PsaA-PsaB dimer has approximate two-fold rotational symmetry, as does the purple bacterial reaction center (Schubert $et\ al.$, 1997). Excited P700 donates an electron to A_0, a chlorophyll a monomer (remember that the acceptor in PS 2 is pheophytin a) (Chitnis, 1996). From A_0, the electron moves to A_1, which is phylloquinone, then through the [4 Fe-4 S] cluster F_X to the [4 Fe-4 S] clusters F_A and F_B, and finally to **ferredoxin**, a soluble [4 Fe-4 S] protein. From ferredoxin the electron is transferred to (NADP$^+$) by way of the flavoprotein **ferredoxin-NADP oxidoreductase, FNR**. On reduction the NADP binds a proton in the stroma. Protons return from the lumen to the stroma through the ATP synthase and ATP is formed.

Some of the ferredoxin molecules may transfer their electrons back to oxidized quinone in the membrane, by a mechanism that is not well understood, so that they can be used to generate more ATP. This process is known as **cyclic photophosphorylation** (Bendall and Manasse, 1995).

VI. Regulation of Photosynthesis

Virtually every step in the photosynthetic process is tightly regulated. The precision of this regulation is illustrated in Fig. 19. During illumination of spinach leaves with light whose intensity was varied sinusoidally, the net rate of CO_2 fixation perfectly paralleled the light intensity except between hours 4 and 10, when it was saturated. As the light intensity changed, the rates of the reactions leading to CO_2 uptake changed in such a way that efficiency of the use of the light remained constant. The final output of photosynthesis, export of fixed carbon from the leaves, remained almost perfectly constant throughout the cycle.

If photosynthesis is to work efficiently, PS 1 and PS 2 must transfer electrons at the same rate. Chloroplasts regulate the rates of the photosystems by regulating the transfer of excitation energy to P680 and P700. Both PS 1 and PS 2 contain chlorophyll molecules bound to proteins that are a permanent part of the PS 1 and PS 2 complexes. CP 43, CP 47 and LHC I shown in Fig. 10 are among these. Chlorophyll is also bound to the PsaA and PsaB proteins of PS 1. Such chlorophyll is known as **antenna chlorophyll**. Chloroplasts thylakoids also contain a protein known as **LHC II** (for light-harvesting complex II), to which are typically attached 8 molecules of chlorophyll a, 7 of chloro-

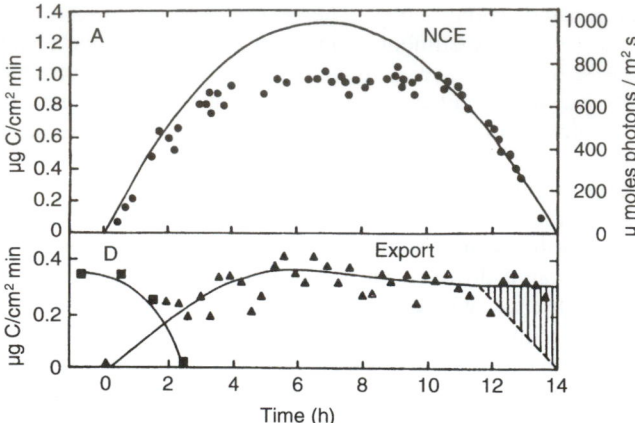

FIGURE 19. Illustration of how rigorously photosynthesis is regulated. Spinach leaves were illuminated with sinusoidally varying light to simulate the variation of light intensity during the day. Top: Net CO_2 exchange. Solid line, light intensity; closed circles, net uptake of CO_2. Bottom: Export of fixed carbon from the leaf. Triangles, newly-fixed carbon; squares, carbon originating from stored starch. (From Servaites *et al.*, 1989, by permission of the American Society of Plant Physiologists.)

phyll *b,* and 1 or 2 of xanthophyll per peptide (Allen, 1992). Under some conditions LHC II is tightly attached to PS 2, to which it transfers excitation energy. If PS 2 is working faster than PS 1, the electron carriers between PS 2 and PS 1 become overreduced. When this happens, a kinase is activated. The kinase catalyzes the phosphorylation of some of the LHC II. The phosphorylated molecules dissociate from PS 2 and diffuse into the stroma lamellae (Fig. 10). As a result, they no longer transfer excitation energy to PS 2, and so its rate slows. It is suspected that the some of the phosphorylated LHC II may associate with PS 1, which is found predominantly in the stroma lamellae, and transfer excitation energy to it instead. The kinase activity is thought to be regulated by the oxidation state of the cytochrome *bf* complex and perhaps also by the oxidation state of plastoquinone. Once the electron carriers between PS 2 and PS 1 are no longer overreduced, the kinase is inactivated and a phosphatase removes the phosphate from the LHC II. It returns to the grana lamellae and reassociates with PS 2.

A regulatory mechanism that deserves special mention involves the protein **thioredoxin**, which serves a signal that it is daytime. Thioredoxin contains a disulfide bridge which can be reduced by ferredoxin in a reaction catalyzed by ferredoxin-thioredoxin reductase. In the light, ferredoxin is reduced by PS 1 and in turn reduces thioredoxin. Thioredoxin then activates a number of Calvin cycle enzymes and increases the sensitivity of the ATP synthase to $\Delta\mu_{H^+}$, by disulfide exchange. As a result, the Calvin cycle and the ATP synthase are active when there is enough light to drive electron transport.

VII. Summary

Plants, algae, and photosynthetic bacteria capture the energy of sunlight and store it in molecules such as starch and

sucrose. Such photosynthetically synthesized molecules are the source of almost all of the free energy used by living organisms.

Light is first absorbed by antenna and light-harvesting pigments such as chlorophyll and β-carotene, raising them to electronically excited states. The excitation energy is then transferred to a special pair of chlorophyll molecules in the photosynthetic reaction center.

Excitation energy is converted to chemical energy in the reaction center as the excited special chlorophyll pair transfers an electron to an acceptor, which is either a pheophytin molecule or a chlorophyll molecule. The electrons proceed along an electron transport chain. The photosynthetic reaction centers of some purple photosynthetic bacteria have been isolated and crystallized and their structures determined by x-ray diffraction.

In higher plants, photosynthesis takes place in organelles called chloroplasts. Chloroplasts contain two photosystems, each having a reaction center. The photosystems are connected in series, forming an electron transport system known as the Z scheme. As electrons move along this electron transport chain from water to NADP+, protons are bound on the stromal side of the chloroplast thylakoid membrane and released on the luminal side, forming a proton gradient. As the protons move back across the membrane through the ATP synthase, ATP is formed.

The ATP and NADPH are used for the synthesis of carbohydrates in a cyclic series of reactions known as the reductive pentose phosphate cycle.

Bibliography

Abrahams, J. P., Leslie, A. G. W., Lutter, R., and Walker, J. E. (1994). Crystal structure at 2.8-angstrom resolution of F_1 ATPase from bovine heart mitochondria. *Nature* **370**, 621–628.

Allen, J. F. (1992). Protein phosphorylation in regulation of photosynthesis. *Biochim. Biophys. Acta* **1098**, 275–335.

Amesz, J., and Hoff, A. J. (Eds.). (1996). "Biophysical Techniques in Photosynthesis." Kluwer Academic Publishers, Dordrecht, Boston, London.

Bendall, D. S., and Manasse, R. S. (1995). Cyclic photophosphorylation and electron transport. *Biochim. Biophys. Acta* **1229**, 23–38.

Blankenship, R. E. (1992). Origin and early evolution of photosynthesis. *Photosynthesis Research* **33**, 91–111.

Blankenship, R. E. (1998). The origin and evolution of oxygenic photosynthesis. *Trends Biochem. Sci.* **23,** 94–97.

Blankenship, R. E., Madigan, M. T., and Bauer, C. E. (Eds.). (1995). "Anoxygenic Photosynthetic Bacteria". Kluwer Academic Publishers, Dordrecht.

Boyer, P. D. (1997). The ATP synthase: a splendid molecular machine. *Annu. Rev. Biochem.* **66**, 717–749

Büttner, M., Xie, D., Nelson, H., Pinther, W., Hauska, G., and Nelson, N. (1992). Photosynthetic reaction center genes in green sulfur bacteria and in photosystem 1 are related. *Proc. Natl. Acad. Sci. USA* **89**, 8135–8139.

Chitnis, P. R. (1996). Photosystem I. *Plant Physiol.* **111**, 661–669.

Clayton, R. K. (1970). "Light and Living Matter, Volume 1: the Physical Part." McGraw-Hill Book Company, New York.

Clayton, R. K. (1980). "Photosynthesis: Physical Mechanisms and Chemical Patterns." Cambridge University Press, Cambridge.

Cogdell, R. J., Isaacs, N. W., Howard, T. D., McLuskey, K., Fraser, N. J., and Prince, S. M. (1999). How photosynthetic bacteria harvest solar energy. *J. Bacteriol.* **181**, 3869–3879.

Cramer, W. A. (1999). Bioenergetics and Photosynthesis. http://www.biosci.umn.edu/biophys/OLTB/bioenerg-photo.html.

Crofts, A. R., and Berry, E. A. (1998). Structure and function of the cytochrome bc_1 complex of mitochondria and photosynthetic bacteria. *Curr. Opinions in Struct. Biol.* **8**, 501–509.

Crofts, A. R. (2000). Structure and mechanism in photosynthesis and bioenergetics. http://www.life.uiuc.edu/crofts/ahab/home.html.

Deisenhofer, J., and Michel, H. (1989). The photosynthetic reaction centre from the purple bacterium *Rhodopseudomonas viridis*. *EMBO J.* **8**, 2149–2169.

Deisenhofer, J., and Norris, J. R. (Eds.). (1993). "The Photosynthetic Reaction Center," Vols. I and II. Academic Press, San Diego.

Flügge, U., and Heldt, H. W. (1991). Metabolite translocators of the chloroplast envelope. *Annu. Rev. Plant Physiol. Plant Mol. Biol.* **2**, 129–144.

Geiger, D. R., Servaites, J. C., and Shieh, W.-J. (1992). Balance among parts of the source-sink system: a factor in crop productivity. *In* "Crop Photosynthesis: Spatial and Temporal Determinants" (N. Baker and H. Thomas, Eds.), pp. 155–192. Elsevier, Amsterdam.

He, W.-Z., and Malkin, R. (2000). Photosystem II and Photosystem I. *In* "Photosynthesis: A Comprehensive Treatise" (A. S. Raghavendra, Ed.), Cambridge University Press, Cambridge (in press).

Ke, B. "Photosynthesis: Photobiochemistry, and Photobiophysics." Kluwer Academic Publishers, Dordrecht (in press).

Kramer, D.M. (2000). Bioenergetics. http//ibc.wsu.edu/faculty/dkother/index.html.

Leslie, A. G. W., Abrahams, J. P., Braig, K., Lutter, R., Menz, R. I., Orriss, G. L., Van Raaij, M. J., and Walker, J. E. (1999). The structure of bovine mitochondrial F_1-ATPase: an example of rotary catalysis. *Biochem. Soc. Trans.* **27**, 37–42.

Loomis, W. E. (1960). Historical introduction. *In* "Encyclopedia of Plant Physiology" (W. Ruhland, Ed.), Vol. V, pp. 85–114. Springer-Verlag, Berlin.

Mitchell, P. (1961). Coupling of phosphorylation to electron and hydrogen transfer by a chemi-osmotic mechanism. *Nature* **191**, 144–148.

Mitchell, P. (1979). Keilin's respiratory chain concept and its chemiosmotic consequences. *Science* **206**, 1148–1159.

Nakamoto, R. K., Ketchum, C. J., and Al-Shawi, M. (1999). Rotational coupling in the F_0F_1 ATP synthase. *Annu. Rev. Biophys. Biomol. Struct.* **28** (in press).

Nitschke , W., and Rutherford, A. W. (1991). Are all the different types of photosynthetic reaction centers variations of a common structural theme? *Trends Biochem. Sci.* **16**, 241–245.

Nugent, J. H. A. (1996). Oxygenic photosynthesis. Electron transfer in photosystem I and Photosystem II. *Eur. J. Biochem.* **237**, 519–531.

Okamura, M. Y., and Feher, G. (1995). Proton-coupled electron transfer reactions of Q_B in reaction centers from photosynthetic bacteria.

In "Anoxygenic Photosynthetic Bacteria" (R. E. Blankenship, M. T. Madigan, and C. E. Bauer, Eds.), pp. 594–597. Kluwer Academic Publishers, Dordrecht.

Olson, J. M. (1996). Iron-sulfur-type reaction centers. Introduction. *Photochem. Photobiol.* **64**, 1–4.

Orr, L. and Govindjee (2000). Photosynthesis and the World Wide Web. http://www-new.life.uiuc.edu/govindjee/photoweb/.

Ort, D. R., Yocum, C. F., and Heichel, I. F. (Eds.) (1996). "Photosynthesis: The Light Reactions". Kluwer Academic Publishers, Dordrecht.

Roberts, A. (2000). Art's biotech resource. http://www.arc.unm.edu/~aroberts/.

Rochaix, J.-D., Goldschmidt-Clermont, M., and Merchant, S. (Eds.) (1998). "The Molecular Biology of Chloroplasts and Mitochondria in Chlamydomonas" Kluwer Academic Publishers, Dordrecht.

Saenger, A. G. (2000). Photosystem I: x-ray structure analysis. http://userpage.chemie.fu-berlin.de/fb_chemie/ikr/ag/saenger/projects/phosys/Welcome.html.

Schnell, D. J. (1995). Shedding light on the chloroplast protein import machinery. *Cell* **83**, 521–524.

Schubert, W.-D., Klukas, O., Krauss, N., Saenger, W., Fromme, P., and Witt, H. T. (1997). Photosystem I of *Synechococcus elongatus* at 4 A resolution: comprehensive structure analysis. *J. Mol. Biol.* **272**, 741–769.

Servaites, J. C., Fondy, B. R., Li, B., and Geiger, D. R. (1989). Source of carbon for export from spinach leaves throughout the day. *Plant Physiol.* **90**, 1168–1174.

Shinkarev, V., and Wraight, C. A. (1993). Electron and proton transport in the acceptor quinone complex of reaction centers of phototropic bacteria. *In* "The Photosynthetic Reaction Center" (Deisenhofer, J., and Norris, J. R., Eds.), Vol. I , pp. 194–255. Academic Press, San Diego.

Singhal, G. S., Renger, G, Irrang, K-D., and Govindjee (Eds) (1999). "Concepts in Photobiology. Photosynthesis and Photomorphogenesis." Kluwer Academic Publishers, Dordrecht and Narosa Publishing House, Hingham.

Stoeckenius, W. (1999). Bacterial rhodopsins: evolution of a mechanistic model for the ion pumps. *Protein Science* **8**, 447–459.

Subramaniam, S., and Henderson, R. (2000). Molecular mechanisms of vectorial proton translocation by bacteriorhodopsin. *Nature* **406**, 653–657

Whitmarsh, J., and Govindjee (2000). The photosynthetic process. http://www.life.uiuc.edu/govindjee/paper/gov.html.

Woese, C. R. (1987). Bacterial evolution. *Microbiol. Rev.* **51**, 221–271.

Yu, C.-A., Xia, D., Kim, H., Deisenhofer, J., Zhang, L., Kachurin, A. M., and Yu, L. (1998). Structural basis of the functions of the mitochondrial cytochrome bc_1 complex. *Biochim. Biophys. Acta* **1365**, 151–158.

Yurkov, V. Y., Krieger, S., Stackebrandt, E., and Beatty, J. T. (1999). *Citromicrobium bathomarinum*, a novel aerobic bacterium isolated from deep sea hydrothermal vent plume waters that contains photosynthetic pigment-protein complexes. *J. Bacteriol.* **181**, 4517–4525.

· J. Woodland Hastings

65

Bioluminescence

I. Introduction

Although the emission by living organisms of light that is visible to other organisms is a rather rare occurrence, and in that sense a curiosity of nature, it has several different and fascinating functions. Bioluminescence is also a unique tool for investigating and understanding numerous different basic physiological processes, both cellular and organismic. It allows one to explore in one system a gamut of questions that confront biologists, ranging from gene expression and its regulation to enzymology, bioenergetics, physiology, function, ecology, and evolution (Hastings, 1983). Lastly, luciferases and associated proteins have recently been developed for use as reporters of gene expression. Such proteins may be visualized noninvasively from the same cell *in vivo,* and over an extended time course, for example, during development (Hastings *et al.,* 1997; Chalfie and Kain, 1997).

The phenomenon is not only rare; in the different groups that do emit light, the biochemical and physiological mechanisms responsible for it are very different, as are their specific functional roles. Indeed, bioluminescence is *not* an evolutionarily conserved function; in the different groups of organisms the genes and proteins involved are mostly unrelated and evidently originated and evolved independently. How many times this may have occurred is difficult to say, but it has been estimated that present-day luminous organisms come from as many as 30 different evolutionarily distinct origins (Harvey, 1952; Herring, 1978; Hastings, 1983; Hastings and Morin, 1991).

II. Physical and Chemical Mechanisms

Bioluminescence does not come from or depend on light absorbed by the organism. It derives from an enzymatically catalyzed chemiluminescence, a highly exergonic (energy-yielding) reaction in which chemical energy is transformed into light energy (McCapra, in Herring, 1978; Campbell, 1988; Wilson, 1985, 1995; Wilson and Hastings, 1998).

Thus in the reaction of substance A with substance B, one of the reaction products is formed in an electronically excited state (D*), which then emits a photon ($h\nu$).

$$A + B \rightarrow C + D*$$

$$D* \rightarrow D + h\nu$$

Chemiluminescence is a special case of the more general phenomenon of luminescence, in which energy is specifically channeled to a molecule; excited state production is not dependent on the temperature of the molecule. Other kinds of luminescence include fluorescence and phosphorescence, in which the excited state is created by the prior absorption of light, or triboluminescence and piezoluminescence, involving crystal fracture and electric discharge, respectively. The color is a characteristic of the excited molecule, independent of how it was excited.

Luminescence is contrasted with incandescence, in which excited states are produced by virtue of the thermal energy. An example is the light bulb, in which a filament is heated, and the color of the light depends on the temperature ("red hot" reflecting a lower temperature than "white hot"). The energy (E) of the photon is related to the color or frequency of the light, and is given by the equation $E = h\nu$, where h is Planck's constant and ν is the frequency. In the visible light range, E is very large in relation to most biochemical reactions. Thus, the energy released by a mole of photons (6.02×10^{23}) in visible wavelengths is about 50 kcal, which is much more than the energy from the hydrolysis of a mole of ATP—about 7 kcal. A visible photon is thus able to do a lot of work (e.g., photosynthesis) or a lot of damage (mutation; photodynamic action, which can kill.) Conversely, it takes a highly exergonic reaction to create a photon.

A question of fundamental importance, then, is what kind of chemical process possesses enough energy, and evidently in a single step (an important point), to "populate" an excited state. A clue is the fact that chemiluminescences in solution generally require oxygen, which in its reaction with a substrate, forms an organic peroxide. The energy from the breakdown of such peroxides—which can generate up to

1115

100 kcal per mole—is ample to account for a product in an electronically excited state.

A model of the reaction mechanism in such chemiluminescent reactions is referred to as chemically initiated electron exchange luminescence (CIEEL) (Schuster, 1979; Catalani and Wilson, 1989). In this mechanism, peroxide breakdown involves electron transfer with chemiexcitation. It is initiated by an electron transfer from a donor species (D) to an acceptor (A), which is the peroxide bond in this case. After electron transfer, the weak O–O bond cleaves to form products B and C^-. The latter is a stronger reductant than A^-, so the electron is transferred back to D with the concomitant formation of a single excited state D* and emission. Thus,

$$A + D \rightarrow A^- + D^+ \quad D^* \rightarrow D + h\nu$$
$$\downarrow \qquad\qquad$$
$$B + C^- \quad C$$

The mechanisms of different bioluminescence reactions can possibly be accommodated within this general scheme.

A useful way to think of chemiluminescence is to regard it as the reverse of a photochemical reaction, in which the excited state created by the absorption of a photon gives rise to chemical species capable of further reaction. But such species can also react reversibly to repopulate the excited state. Photosynthesis is a good example: the primary chemical states formed in photosynthesis are comparable to C^- and D^+, which may be the penultimate states in bioluminescence.

In the primary step of photosynthesis, the energy of the excited state of chlorophyll (Chl*) gives rise to an electron transfer, with the consequent formation of a primary oxidant and a primary reductant (see Chapter 64). With A as the electron acceptor, these steps can be represented as

$$Chl + h\nu \rightarrow Chl^*$$

$$Chl^* + A \rightleftharpoons Chl^+ \cdot + A^- \cdot$$
$$\downarrow \qquad \downarrow$$
$$\text{stable oxidant} \quad \text{stable reductant}$$

Most of these redox species give rise to stable products and ultimately to CO_2 fixation. However, some species recombine and reemit a "delayed light." This is essentially the reverse of the reaction, with the formation of the singlet excited state of chlorophyll and its subsequent emission of a photon.

In bioluminescence, the substrates and enzymes, though chemically different in different organisms, are all referred to as luciferin and luciferase (DeLuca and McElroy, 1981). These are thus generic terms, and to be correct and specific, each should be identified with the organism (Tables 1 and 2), thus firefly luciferin or bacterial luciferase. The luciferases, as well as the structures and reaction intermediates of the different known luciferins, will be discussed in connection with the individual groups of organisms.

III. Luminous Organisms: Abundance, Diversity, and Distribution

Bioluminescence is indeed rare as measured by the total number of luminous species, but it is phylogenetically diverse, being found in more than 13 phyla (Herring, in Herring, 1978;

Herring, 1987); only the major ones are considered here (Table 1). These include bacteria, unicellular algae, and fungi, as well as animals ranging from jellyfish, annelids, and mollusks to shrimp, fireflies, echinoderms, and fish. Luminescence is unknown in higher plants and in vertebrates above the fish (Cormier, in Herring, 1978). It is also absent in several invertebrate phyla. In some phyla or taxa, a substantial population of the genera are luminous (e.g., ctenophores, ~50%; cephalopods, >50%; echinoderms and annelids, ~4%). Commonly, all members of a luminous genus emit light, but in some cases there are both luminous and nonluminous species.

The fact that the enzymes and substrates, as well as the physiological and functional aspects of bioluminescence, differ in the several major taxa is indicative of their independent evolutionary origins. In fact, there are chemically different systems found in different taxa within some phyla, so the total number of evolutionarily independent groups may be 30 or more (Hastings, 1983). Fewer than half of these have been studies in detail, and knowledge of the luciferins and luciferases is available for only about 10.

Although luminescence is prevalent in the deep sea (Herring, 1985a), it is not associated especially with organisms that live in total darkness. There are no known luminous species either in deep freshwater bodies, such as Lake Baikal, Russia, or in the darkness of terrestrial caves. There are luminous dipteran larvae (*Arachnocampa*) that live near the mouths of caves in New Zealand and Australia (Fig. 1), but they also occur in culverts and the undercut banks of streams, where there is considerable daytime illumination. Although insect displays of bioluminescence are among the most spectacular, bioluminescence is relatively rare in the terrestrial environment (<0.2% of all genera). Some other terrestrial luminous forms are millipedes, centipedes, earthworms, and snails, but in none of these is the display very bright.

For reasons that are still obscure, bioluminescence is most prevalent in the marine environment (Kelly and Tett, in Herring, 1978); it is greatest at midocean depths (200–1200 m), where daytime illumination fluxes range between $\sim 10^{-1}$ and 10^{-12} $\mu W \cdot cm^{-2}$. In these locations, bioluminescence in fish may occur in over 95% of the individuals and in 75% of the species; similar percentages are tabulated for shrimp and squid. The midwater luminous fish *Cyclothone* is considered to be the most abundant vertebrate on the planet. Where high densities of luminous organisms occur, their emissions can exert a significant influence on the communities and may represent an important component in the ecology, behavior, and physiology of these organisms. Above the below midocean depths, luminescence decreases to <10% of all individuals and species. It may be somewhat higher (~20%) at abyssal depths, whereas among coastal species, fewer than 2% are bioluminescent.

IV. Functions of Bioluminescence

The functional importance of bioluminescence and its selection in evolution are believed to be based largely on its being detected by another organism; the response of that organism then favors in some way the luminous individual

TABLE 1   Luminous Organisms: Biochemical Mechanisms and Biological Functions

Type of organism	Representative genera	Luciferins and other factors (Emission max, nm)	Displays and functions
Bacteria	*Photobacterium* *Vibrio* *Xenorhabdus*	Reduced flavin and long-chain aldehyde (475–535) Some with accessory emitters	Steady bright glow Autoinduction of luciferase Function as symbionts
Mushrooms	*Panus, Armillaria* *Pleurotus*	Unknown (535)	Steady dim glow; function unknown
Dinoflagellates	*Gonyaulax* *Pyrocystis* *Noctiluca*	Linear tetrapyrrole pH change (470)	Short (0.1 s) bright flashes function to frighten or deter
Cnidaria			
Jellyfish	*Aequorea*	Ca^{2+}, coelenterazine	Bright flash or train of flashes; function to
Hydroid	*Obelia*	Imidazo pyrazine nucleus	frighten or deter
Sea pansy	*Renilla*	(460–510), some accessory emitters	
Ctenophores	*Mnemiopsis* *Beroe*	Ca^{2+}, coelenterazine (460)	Bright flashes; function to frighten or deter
Annelids			
Earthworms	*Diplocardia*	*N*-Isovaleryl-3 amino propanal	Cellular exudates, sometimes very bright;
Marine polychaetes	*Chaetoptorus*	Unknown	function to divert and to deter; others unknown
Syllid fireworm	*Odontosyllis*	Unknown (480)	
Mollusks			
Limpet	*Latia*	Aldehyde	Exuded luminescence in all three; photophores
Clam	*Pholas*	Clam luciferin, but structure is unknown,	and symbiotic bacteria in some squid; diversion,
Squid	*Heteroteuthis*	Cu^{2+}	decoy
Crustacea			
Ostracod	*Vargula*	Imidazolopyrazine nucleus (465)	Squirts enzyme and substrates; diversion, decoy
Shrimp (euphausids) Copepods	*Meganyctiphanes*	Linear tetrapyrrole (470)	Photophores; camouflage
Insects			
Coleopterids (beetles)			
Firefly	*Photinus, Photuris*	Benzothiazole, ATP, Mg^{2+}	Flashes, specific kinetic patterns
Click beetles	*Pyrophorus*	Similar chemistry in all coleoptera	Communication: courtship, mating
Railroad worm	*Phengodes, Phrix-othrix*		
Diptera (flies)	*Arachnocampa*	Biochemistry unknown	Lure to attract prey
Echinoderms			
Brittle stars	*Ophiopsila*	Biochemistry unknown	Trains of rapid flashes; frighten, divert predators
Sea cucumbers	*Laetmogone*	Biochemistry unknown	Unknown
Chordates			
Tunicates	*Pyrosoma*	Organelles evolved from bacteria (480–500)	Brilliant trains of flashes stimulated by light and other factors
Fish			
Cartilaginous	*Squalus*	Biochemistry unknown	Unknown
Bony			
Ponyfish	*Leiognathus*	Symbiotic luminous bacteria (~490)	Camouflage, ventral luminescence
Flashlight fish	*Photoblepharon*	Symbiotic luminous bacteria (~490)	To attract and capture prey
Angler fish	*Cryptopsaras*	Symbiotic luminous bacteria (~490)	
Midshipman	*Porichthys*	Self luminous, Vargula type luciferin, nutritionally obtained	Camouflage? courtship display?
Midwater fish			
	Cyclothone	Self-luminous, biochemistry unknown	Many photophores, ventral and lateral
	Neoscopelus	Self-luminous, biochemistry unknown	Photophores: lateral, on tongue
	Tarletonbeania	Self-luminous, biochemistry unknown	Sexual dimorphism; males have dorsal (police car) photophores

TABLE 2 Luciferases

	kDa	E.C. No.
Bacterial	~80 (α, 41; β, 39)	1.14.14.3
Dinoflagellate	~135	
Coelenterate	~35	1.13.12.5
Mollusk		1.14.99.21
Firefly	~60	1.13.12.7
Crustacean	~68	1.13.12.6

FIGURE 1. Luminous dipteran larvae (*Arachnocampa*) on the ceiling of a cave in New Zealand.

(Buck, in Herring, 1978; Hastings, 1983; Buck, 1938; Herring, 1990). In higher animals luminescence is generally controlled neurally (Anctil, 1987).

Bioluminescence is interesting biologically because it is a clear and well-documented example of a function that, while not metabolically essential, confers an advantage on the individual. Although bioluminescence has evidently arisen independently many times, it may also have been lost many times in different evolutionary lines, particularly where it was not a truly important function.

Bioluminescence may be thought of as a bag of tricks: the light can be used in different ways and for different functions. Most of the perceived functions of bioluminescence may be classified under three main rubrics: defense, offense, and communication (Table 3).

Important defensive strategies associated with bioluminescence are to frighten, to serve as a decoy, to provide camouflage, and to aid in vision. Organisms may be frightened or diverted by flashes, which are typically bright and brief (0.1 s); light is emitted in this way by many organisms, and experimental studies confirm that flashes can indeed frighten (Morin, 1983).

On the other hand, a glowing object in the ocean often appears to attract feeders or predators. Although a luminous organism would evidently be at risk by virtue of this attraction, the fact can be used defensively if an organism creates a decoy light to attract the predator, and then slips off under the cover of darkness. This is exactly what quite a number of organisms do. A luminous squid in darkness squirts luminescence instead of ink; ink would be useless in such a case. Some organisms sacrifice more than light; in scaleworms and brittle stars, a part of the body may be automized (broken off) and left behind as a luminescent decoy to attract the predator. In these cases the body part flashes while still attached but glows after detachment, exemplifying that a flash deters whereas a glow attracts.

A unique method for evading predation from below is to camouflage the silhouette by emitting light that matches the

TABLE 3 Functions of Bioluminescence

Function	Strategy	Method
Defense	Frighten, startle	Bright, brief flashes
	Decoy, diversion	Glow, luminous cloud, sacrificial lure
	Camouflage	Ventral luminescence during the day, disrupting or concealing the silhouette seen from below
Offense	Frighten, startle	Bright flash may temporarily immobilize prey
	Lure	Glow to attract, then capture prey
	Vision	To see and capture prey
Communication	Courtship, mating	Specific flashing signals; patterns of light emission recognized by opposite sex
Dispersal, propagation	Glow to attract feeders	Bacteria ingested by feeder pass through gut tract alive and are therefore dispersed

color and intensity of the downwelling background light. By analogy with countershading in reflected light, this has been called counterillumination. Imagine a plane in the sky during the day. If it could emit light from its bottom surface matching the sky behind, it would be invisible from below. Actually, it is not necessary for the entire surface to emit light; emission by only a part would mean that the object would no longer look like a plane. This can be called **disruptive illumination,** and many luminous marine organisms, including fish, apparently use this to aid in escaping detection (McFall-Ngai and Morin, 1991). Another novel defensive strategy has been dubbed the burglar alarm: Dinoflagellates flash when grazed upon, which may enhance predation on the grazers, and thus reduce grazing on dinoflagellates (Abrahams and Townsend, 1993).

There are also several ways in which luminescence can aid in predation. Several of these, such as helping in vision, may be of value for both offense and defense. For example, flashes, which are more typically used defensively, can be used offensively to temporarily startle or blind prey. A glow can also be used offensively; it can serve as a lure. The organism is attracted to the light but is then captured by the organism that produced the light. Camouflage may also be used offensively, allowing the luminous predator to approach its prey undetected. Vision is certainly useful offensively; prey may be seen and captured under conditions that are otherwise dark, as practiced by the flashlight fish *Photoblepharon* (Fig. 2).

Communication involves information exchange between individual members of a species, and luminescence is used for this in several organisms, including annelids, crustaceans, insects, squid, and fishes. The most common such use of light is for courtship and mating, as in fireflies (Buck and Buck, 1976; Lloyd, 1977, 1980). But there are numerous examples in the ocean (Herring 1990). In the syllid fireworm *Odontosyllis* a truly extraordinary display occurs as the animals engage in mating, which occurs just post-twilight a few days after the full moon. Readily observed in many parts of the world (e.g., Bermuda), the females come to the surface

FIGURE 3. Luminous mushrooms.

and swim in a tight circle. A male streaks from below and joins the female, with eggs and sperm shed in the ocean in a luminous circle. Another example occurs over shallow reefs in the Caribbean: male ostracod crustaceans produce complex species-specific trains of secreted luminous material—ladders of light—which attract females (Morin and Cohen, 1991).

There remain some luminous organisms for which it is difficult to determine what the function of light emission may be. In the luminous fungi (Wassink, in Herring, 1978) (Fig. 3), both the mycelium and mushrooms emit a continuous light, day and night, but is typically very dim. Moreover, the mycelium itself is almost never exposed: it is underground or inside a decaying tree. The role of luminescence in this and other such cases remains to be understood.

V. Bacterial Luminescence

A. Occurrence and Functions

Luminous bacteria (Fig. 4) occur ubiquitously in the oceans and can be isolated from most seawater samples from the surface to depths of 1000 m or more. A primary habitat where most species abound is in association with another (higher) organism, dead or alive, where growth and propagation occur. Planktonic forms are readily isolated, but they do not grow in seawater, as it is a poor growth medium, so they may be viewed as having overflowed from primary habitats (Nealson and Hastings, 1991).

The most exotic specific associations involve specialized light organs (e.g., in fish and squid) in which a pure

FIGURE 2. The flashlight fish (*Photoblepharon*), showing the light organ (harboring luminous bacteria) just below the eye.

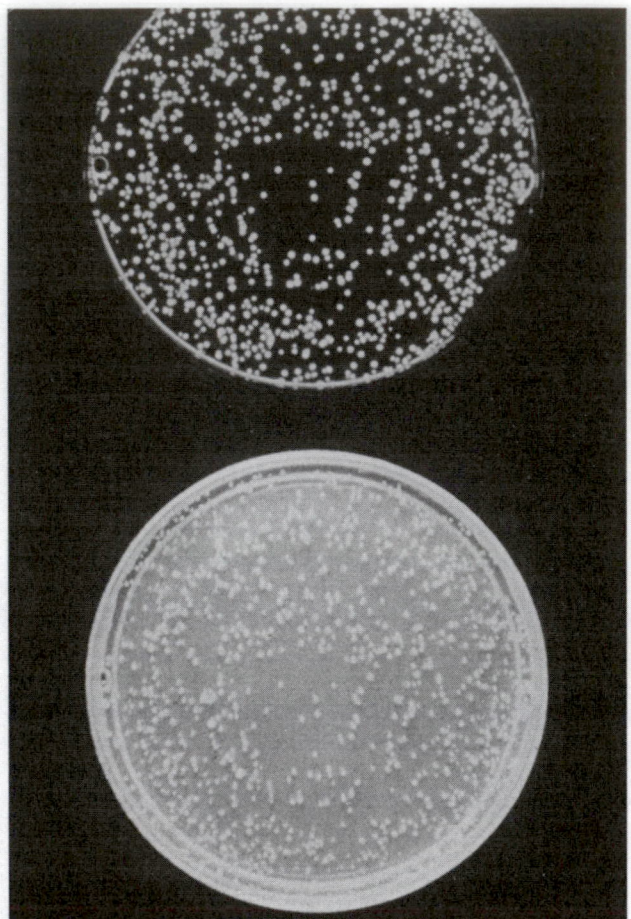

FIGURE 4. Luminous bacteria, photographed by their own light (top) and by room light (bottom). Note the dark mutant near the center.

culture of luminous bacteria is maintained at a high density and at high light intensity. In teleost fishes, some 11 different groups carrying such bacteria are known (Fig. 5). Exactly how symbioses are achieved—the initial infection, exclusion of contaminants, nutrient supply, restriction of

growth but bright light emission—is not understood (Hastings *et al.,* 1987). In such associations, the host receives the benefit of the light and may use it for one or more specific purposes; the bacteria in return receive a niche and nutrients.

Direct (nonspecific) associations include parasitizations and commensals. Intestinal bacteria in marine animals, notably fish, are often luminous, and heavy pigmentation of the gut tract is often present, presumably to prevent the light from betraying the location of the fish to predators. Indeed, the light emission in these cases is thought to benefit the bacteria more directly, for example, by attracting predators and thereby promoting bacterial dispersion and propagation. Luminous bacteria growing on a substrate, be it a parasitized crustacean, the surface of a dead fish, or a fecal pellet, can produce a light bright enough to attract other organisms, presumably to feed on the material. Attraction to promote dispersal and thus propagation of bacteria may thus be viewed as yet another function of bioluminescence, although, because the luminous organism must survive ingestion, it is probably limited to bacteria.

Terrestrial luminous bacteria are rare. The best known are those harbored by nematodes that are parasitic on insects such as caterpillars. The nematode carries the bacteria as symbionts and injects them into the host along with its own fertilized eggs. The bacteria grow and the developing nematode larvae feed on them. The dead but now luminous caterpillar (Fig. 6) attracts predators, which serves to disperse the nematode offspring, along with the bacteria.

Luminous bacteria are also examples of organisms that can exploit—or at least survive in—different habitats, for example, in a light organ or free in seawater, in the planktonic environment. This versatility also includes the capacity to turn the luminescent system on and off, at both physiological and genetic levels (Hastings, 1987). Where advantageous, it is expressed at high levels; where not, the genes are repressed and energy is conserved.

B. Biochemistry

Luminous bacteria typically emit a continuous light, usually blue-green. When strongly expressed, a single bacterium

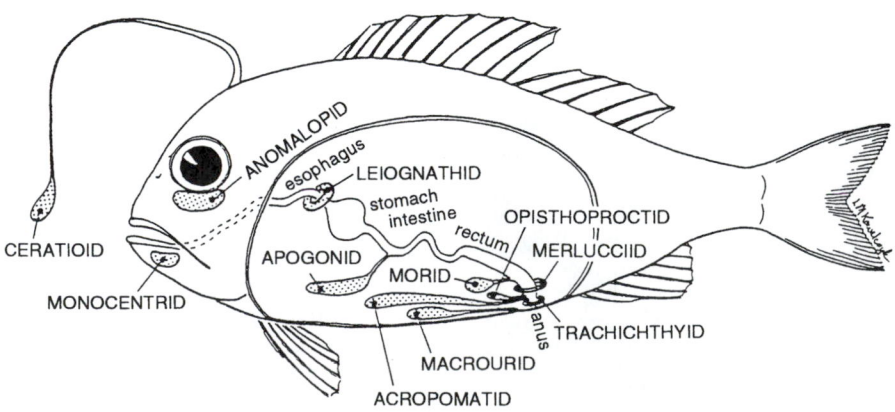

FIGURE 5. The "ichthylicht." A diagrammatic fish is used to indicate the approximate locations, sizes, and configurations of the light organs in the several groups of luminous fishes that culture luminous bacteria as a source of light for the organ.

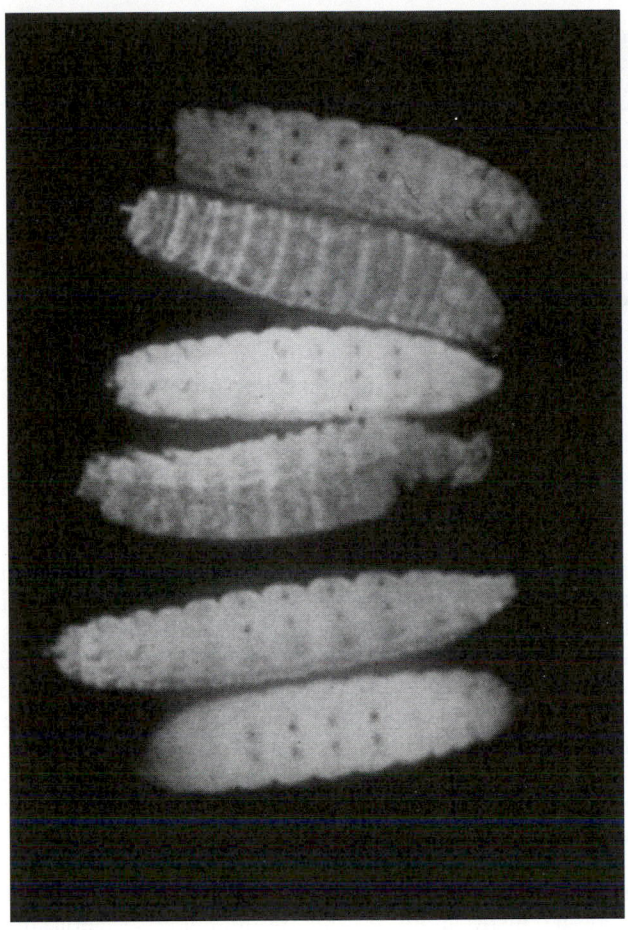

FIGURE 6. Luminous caterpillars; caused by parasitic luminous bacteria.

may emit 10^4 or 10^5 photons s^{-1}. The system is biochemically unique and is diagnostic for a bacterial involvement in the luminescence of a higher organism, as endosymbionts, for example. The pathway itself (Fig. 7) constitutes a shunt of cellular electron transport at the level of flavin, and reduced flavin mononucleotide is the substrate (luciferin) that reacts with oxygen in the presence of bacterial luciferase to produce an intermediate peroxy flavin (Wilson and Hastings, 1998; Meighen and Dunlap, 1993; Baldwin and Ziegler, 1992; Hastings *et al.,* 1985). This intermediate then reacts with a long-chain aldehyde (tetradecanal) to form the acid and the luciferase-bound hydroxy flavin in its excited state. Although there are two substrates in this case, the flavin can claim the name luciferin on etymological grounds, since it forms (bears) the emitter. The bioluminescence quantum yield has been estimated to be about 30%.

The enzyme is an external flavin monooxygenase (EC 1.14.14.3). Curiously, no other enzymes of this general type have been found to emit light, even at very low quantum yields. The light-emitting steps have been modeled in terms of an electron exchange mechanism (see Section II), and the experimental evidence is consistent with this.

There are enzyme systems that serve to maintain the supply of myristic aldehyde, and genes coding for these enzymes are part of the *lux* operon (Meighen, 1991; Fig. 8). The luciferases themselves are homologous heterodimeric (α-β) proteins (~80 kDa) in all species. They possess a single active center per dimer, mostly associated with the α-subunit. Structurally, they appear to be relatively simple; that is, no metals, disulfide bonds, prosthetic groups, or non-amino acid residues are involved.

The luciferase and the mechanism of the bacterial reaction have been studied in great detail. An interesting feature of this luciferase reaction is its inherent slowness: at 20 °C, the time required for a single catalytic cycle is about 20 s.

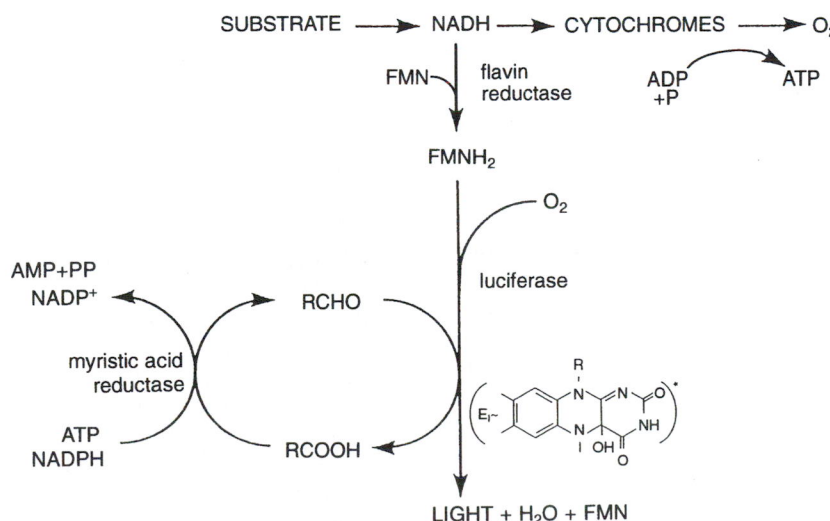

FIGURE 7. The luciferase reaction in bacteria. In the electron transport pathway, adenosine triphosphate (ATP) is generated; luciferase shunts electrons at the level of reduced flavin (FMNH$_2$) directly to molecular oxygen. In the mixed function, oxidation with long-chain aldehyde, hydroxy-FMN is produced in its excited state (*) along with long-chain acid. The FMN product is reduced again and recycles, and the aldehyde is regenerated from the acid.

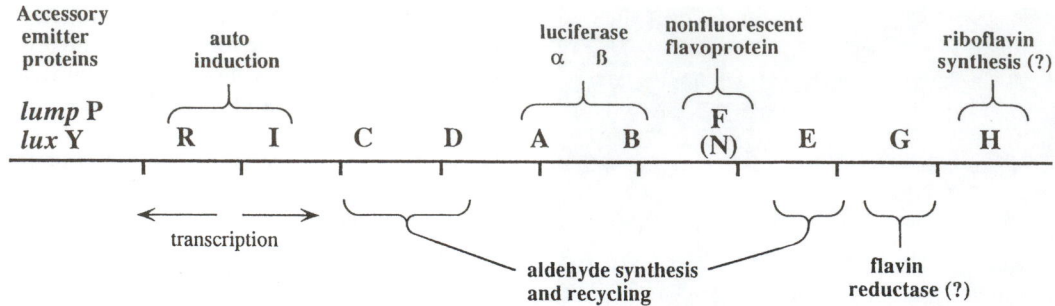

FIGURE 8.   Organization of the *lux* genes in *Vibrio fischeri*. The operon on the right, transcribed from the 5' to the 3' end, carries genes for synthesis of autoinducer (*lux* I), for luciferase α and β peptides (*lux* A and B), and for aldehyde production (*lux* C, D, and E). The operon on the left encodes for a gene (*lux* R), which encodes for a receptor molecule that binds autoinducer; the complex controls the transcription of the right operon. Other genes, *lux* F (N), G and H (right), are associated with the operon but with still uncertain functions, genes for accessory emitter proteins also occur (left).

The luciferase peroxy flavin itself has a long lifetime; at low temperatures (0 to −20 °C) it has been isolated, purified, and characterized. It can be further stabilized by aldehyde analogs such as long-chain alcohols and amines, which bind at the aldehyde site.

C. Regulation of Bacterial Luminescence

Typically, bacteria are unable to regulate emission on a fast time scale (ms, s), as in organisms that emit flashes. However, bacteria are generally able to control the development and expression of luminescence at both physiological and genetic levels. The most unique of these mechanisms is "autoinduction," in which the transcription of the luciferase and aldehyde synthesis genes of the *lux* operon is regulated by a gene product of the operon itself. A substance produced by the cells called autoinducer (Fig. 9) is a product of the *lux* I gene (Nealson and Hastings, 1991). The ecological implications are evident: in planktonic bacteria, a habitat where luminescence has no value, autoinducer cannot accumulate, and no luciferase synthesis occurs. However, in the confines of a light organ, high autoinducer levels are reached and the luciferase genes are transcribed. Interestingly, it has recently been discovered that similar mechanisms control the expression of other specific genes in bacteria, and this has been dubbed quorum sensing (Hastings and Greenberg, 1999; Fuqua *et al.*, 1994).

There are a number of other control mechanisms that serve to regulate the transcription of the *lux* operon, including glucose (catabolic repression), nutrient levels, iron, and oxygen. Each of these factors represents a different mechanism for the physiological control of gene expression, and each has implications concerning the ecology of luminous

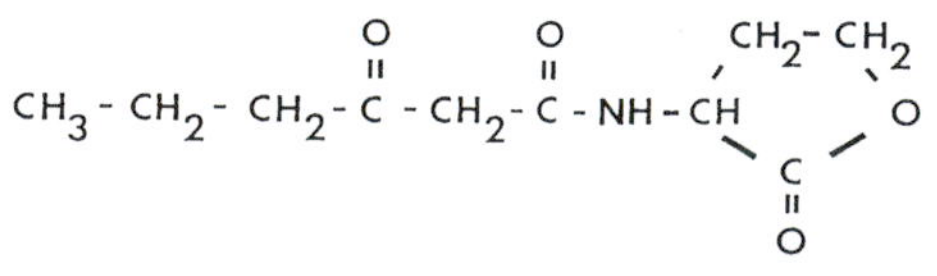

bacteria and the function of their luminescence. But there is also control at the genetic level: in some species of bacteria there are "dark" (very, very dim) mutants that may arise spontaneously (see Fig. 4). In these the synthesis of the luminescent system fails altogether to occur, irrespective of conditions, and this is an inheritable property. However, the *lux* genes are not lost and revertants occur. Thus, the organism can compete under conditions where luminescence is not advantageous, yet be able to produce luminous forms and populate the appropriate habitat when and where it is found. Indeed, the bacterial *lux* genes may occur in many bacterial strains but not be expressed. This means that there may be many more potentially luminous bacteria than would be deduced from the luminescence of colonies on plates.

VI. Dinoflagellate Luminescence

A. Occurrence and Function

Dinoflagellates occur ubiquitously in the oceans as planktonic forms and contribute substantially to the so-called "phosphorescence" commonly seen at night (especially in summer) when the water is disturbed. They occur primarily in surface waters and many species are photosynthetic. In the phosphorescent bays (e.g., in Puerto Rico and Jamaica), high densities of a single species (*Pyrodinium bahamense*) usually occur. The so-called "red tides" are blooms of dinoflagellates.

About 6% of all dinoflagellate genera contain luminous species, but since there are no luminous dinoflagellates among the freshwater species, the proportion of luminous forms in the ocean is higher. As a group, dinoflagellates are important as symbionts, notably for contributing photosynthesis and carbon fixation in animals, but unlike bacteria, no luminous dinoflagellates are known from symbiotic niches.

Since dinoflagellates are stimulated to emit light when predators (e.g., crustacea) are active, predators on the crustacea might be alerted, resulting in a reduced predation on dinoflagellates (Abrahams and Townsend, 1993). Predation on

FIGURE 9.   Structure of autoinducer from *Vibrio fischeri*.

dinoflagellates may also be impeded more directly, since the flash could startle or divert the predator. The response time to stimulation (ms) is certainly fast enough to have this effect.

B. Biochemistry, Cell, and Molecular Biology

Luminescence in dinoflagellates is emitted from many small (~0.5 μm) cortical structures, identified as a new type of organelle, termed the **scintillon** (flashing unit) (Hastings and Dunlap, in DeLuca and McElroy, 1986). They occur as outpocketings of the cytoplasm into the cell vacuole, like a balloon, with the neck remaining connected (Fig. 10). Scintillons contain only dinoflagellate luciferase and luciferin (with its binding protein), other cytoplasmic components being somehow excluded. The genes for luciferase and luciferin binding protein have been cloned and sequenced; the first ~100 N-terminal amino acids of the two exhibit about 50% identity; neither has sequence similarity to other known proteins. The luciferase is unusual in that after the first ~100 residues it comprises three contiguous intramolecularly conserved domains, each with a catalytic site, thus three active centers in a single molecule (Li *et al.*, 1997; Wilson and Hastings, 1998). Scintillons can be identified by immunolabeling with antibodies raised against the luminescence proteins (Nicolas *et al.*, 1987), and visualized by their bioluminescent flashing after stimulation, as well as by the fluorescence of luciferin (Fig. 11). Dinoflagellate luciferin is a novel tetrapyrrole related to chlorophyll (Fig. 12).

Activity can be obtained in extracts made at pH 8 simply by shifting the pH from 8 to 6; it occurs in both soluble and particulate (scintillon) fractions, suggesting that during extraction some scintillons are lysed, whereas others seal off at

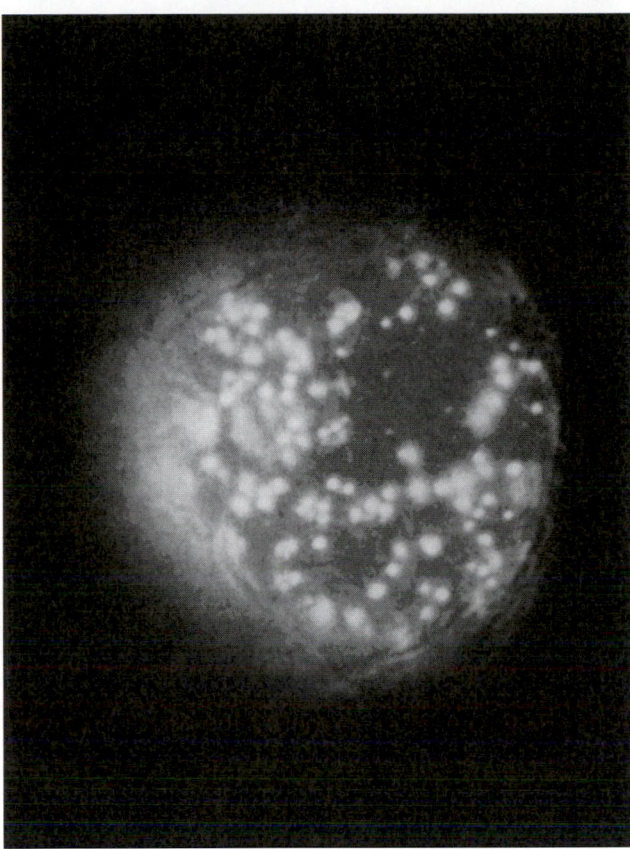

FIGURE 11. A *Gonyaulax* cell visualized by fluorescence microscopy, showing scintillons by the fluorescence of dinoflagellate luciferin.

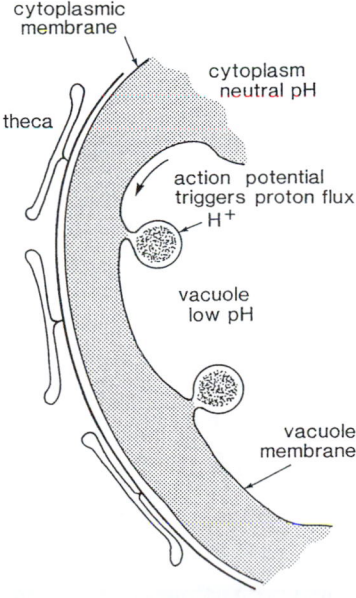

FIGURE 10. Scintillons of dinoflagellates represented as organelles formed as cytoplasmic outpocketings hanging in the acidic vacuole.

the neck and form closed vesicles. With the scintillon fraction, the *in vitro* activity occurs as a flash (~100 ms), very close to that of the living cell, and the kinetics are independent of the dilution of the suspension. For the soluble fraction, the kinetics are dependent on dilution, as in an enzyme reaction. A distinctive feature is that at pH 8 the luciferin is bound to the luciferin binding protein, which thereby prevents it from reaction with luciferase, but it is free at pH 6 (Fig. 12B).

The flashing of dinoflagellates *in vivo* is postulated to result from a transient pH change in the scintillons, triggered by an action potential in the vacuolar membrane which, while sweeping over the scintillon, opens ion channels that allow protons from the acidic vacuole to enter (see Fig. 10 and Section III).

C. Control of Dinoflagellate Luminescence: The Circadian Clock

The composition of the medium and nutrient conditions apparently have little effect on the development and expression of bioluminescence in dinoflagellates. However, in *Gonyaulax polyedra* and some other dinoflagellates, luminescence is regulated by day-night light-dark cycles and an internal circadian biological clock mechanism (Morse *et al.*,

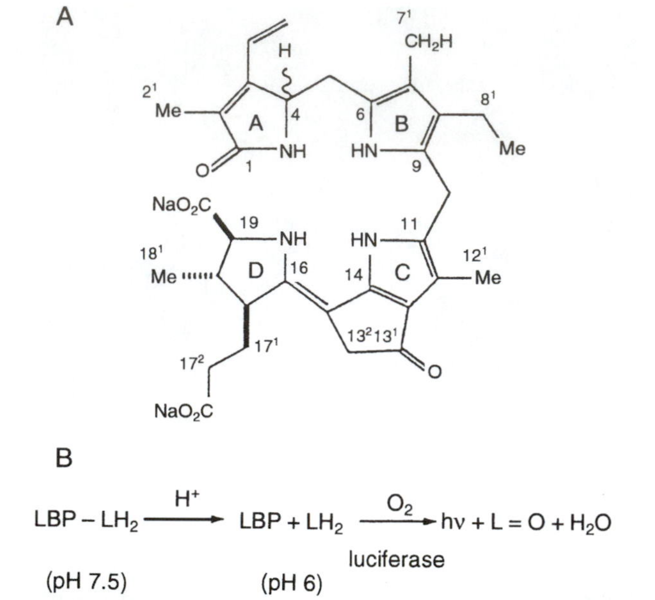

A

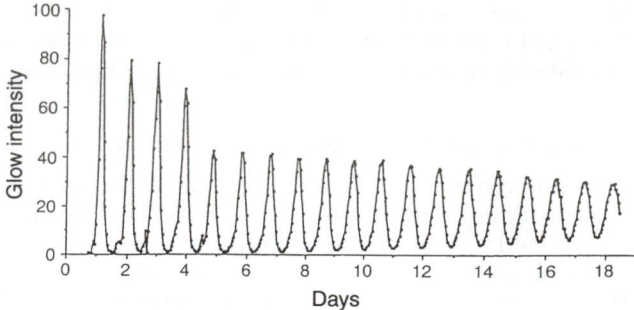

FIGURE 13. The circadian rhythm of the steady glow of bioluminescence in *Gonyaulax*. A culture grown in a 24-hour light-dark cycle
was transferred at zero time to constant conditions (19 °C; dim white
light); measurements were made about once every hour for 18 days.
The average period length was 22.8 hours.

B

$$LBP - LH_2 \xrightarrow{\;H^+\;} LBP + LH_2 \xrightarrow[\text{luciferase}]{\;O_2\;} h\nu + L = O + H_2O$$

(pH 7.5)   (pH 6)

FIGURE 12. (A) Structure of dinoflagellate luciferin, a tetrapyrrole (fluorescence maximum, ~475 nm). (B) Reaction steps for
Gonyaulax bioluminescence. Luciferin (LH$_2$) is attached to luciferin
binding protein (LBP) at pH 8, but is free at pH 6 and thus able to be
oxidized by luciferase; the product has a carbonyl at position 13^2.

1990). The flashing response to mechanical stimulation is
far greater during the night than during the day, and a steady
low-level emission (glow) exhibits a peak toward the end of
the night phase. The regulation is attributed to an endogenous mechanism; cultures maintained under constant conditions (light, temperature) continue to exhibit rhythmicity for
weeks (Fig. 13), but with a period that is not exactly 24
hours; it is only about (circa) one day (diem), thus the origin
of the term.

The nature of this circadian clock remains one of the real
enigmas in physiology. In humans and other higher animals,
where it regulates the sleep-wake cycle and many other
physiological processes, the mechanism involves the nervous system (Hastings *et al.,* 1991). But it also occurs in
plants and unicellular organisms, including *Euglena,*
Chlamydomonas, and *Paramecium.* In the case of *G. polyedra* it is known that daily changes occur in the cellular concentrations of luciferase, luciferin, and its binding protein;
the proteins are synthesized and destroyed each day. Hence,
the biological clock exerts control at a very basic level by
controlling gene expression.

VII. Coelenterates and Ctenophores

A. Occurrence and Function

Luminescence is common and widely distributed in these
groups (Cormier, in Herring, 1978; Herring, in Herring,
1978). In the ctenophores (comb jellies), luminescent organisms constitute over half of all genera, whereas in the
coelenterates (cnidaria) the figure is about 6%. Luminous
hydroids, siphonophores, sea pens, and jellyfish, among oth-

ers, are well known. The organisms are mostly sessile or
sedentary, and upon stimulation emit light as flashes (Fig.
14A). Bioluminescence is absent in sea anemones and
corals.

Hydroids occur as plantlike growths, typically adhering
to rocks below low tide level in the ocean. When they are
touched, a sparkling emission is conducted along the

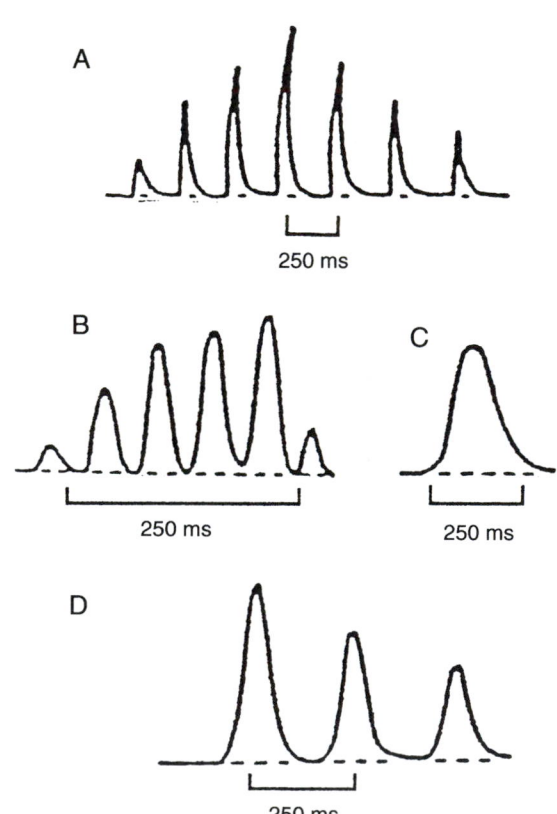

FIGURE 14. Bioluminescent flashes. (A) A train of flashed in the
hydrozoan *Obelia geniculata* following a single electrical stimulus.
Spontaneous flashes from three species of fireflies: (B) *Photinus
evanescens* (C) *P. marginellis,* and (D) *P. versicolor.*

colony; repetitive waves from the origin may occur. Luminous jellyfish (such as *Pelagia noctiluca*) are well known; the bright flashing comes from photocytes along the edge of the umbrella at the base of the tentacles. *Aequorea*, a hydromedusan that is very abundant during the summer in the ocean off the northwest United States (San Juan Islands region), has been the material used for much of the research on the biochemistry of the system (Shimomura, 1985). The sea pansy, *Renilla*, which occurs near shore on sandy bottoms, has also figured importantly in the elucidation of the biochemistry of coelenterate luminescence (Cormier, 1981).

B. Biochemistry, Cell Biology, and the Control of Flashing

Photocytes occur as specialized cells located singly or in clusters in the endoderm. They are commonly controlled by epithelial conduction in hydropolyps and siphonophores and by a colonial nerve net in anthozoans. The light may be emitted as one to many flashes per stimulus. The putative neurotransmitter involved in luminescence control in *Renilla* is adrenaline or a related catecholamine.

The luciferin from coelenterates, coelenterazine, possesses an imidazolopyrazine skeleton (Fig. 15A). It is notable for its widespread phylogenetic distribution, but whether the reason is nutritional or genetic (hence, possible evolutionary relatedness) has not yet been elucidated. In some cases (e.g., *Renilla*), the sulfated form of luciferin may occur as a precursor or storage form and is convertible to active luciferin by sulfate removal with the cofactor 3',5'-diphosphadenosine. The active form may also be sequestered by a Ca^{2+}-sensitive binding protein, analogous to the dinoflagellate binding protein. In this case Ca^{2+} triggers the release of luciferin and flashing.

Another, more novel type of control of the reaction occurs in other cnidaria (e.g., *Aequorea*); this involves a luciferase-peroxyluciferin intermediate poised for the completion of the reaction. The photoprotein aequorin (Shimomura, 1985; Charbonneau *et al.*, 1985), isolated from the jellyfish *Aequorea* (in the presence of EDTA to chelate calcium), emits light simply upon the addition of Ca^{2+}, which is presumably the trigger *in vivo* (Blinks *et al.*, 1982; Cormier *et al.*, 1989). This luciferin and the luciferase (EC 1.13.12.5) react with oxygen to form the peroxide in a calcium-free compartment (the photocyte), where it is stored. An action potential allows Ca^{2+} to enter and bind to the protein, changing its conformational state and allowing the reaction to continue, but without the need for free oxygen at this stage. An enzyme-bound cyclic peroxide, a dioxetanone, is a postulated intermediate; it breaks down with the formation of an excited emitter, along with a molecule of CO_2. Coelenterate luciferase possesses homology with calmodulin (Lorenz *et al.*, 1991).

In several coelenterates, the light emission occurs at a longer wavelength *in vivo* than *in vitro*. This is attributed to energy transfer from the excited luciferase-bound emitter to the fluorophore of the accessory green fluorescent protein (GFP), now widely used as a reporter of gene expression (Chalfie and Kain, 1997; Hastings *et al.*, 1997).

It had been reported in the early literature that coelenterates could emit bioluminescence without oxygen. The explanation is now evident: the animal contains the luciferase-bound peroxyluciferin (analogous to the bacterial flavin peroxide) in a stored and stable state, and only calcium is needed for the light-emitting step.

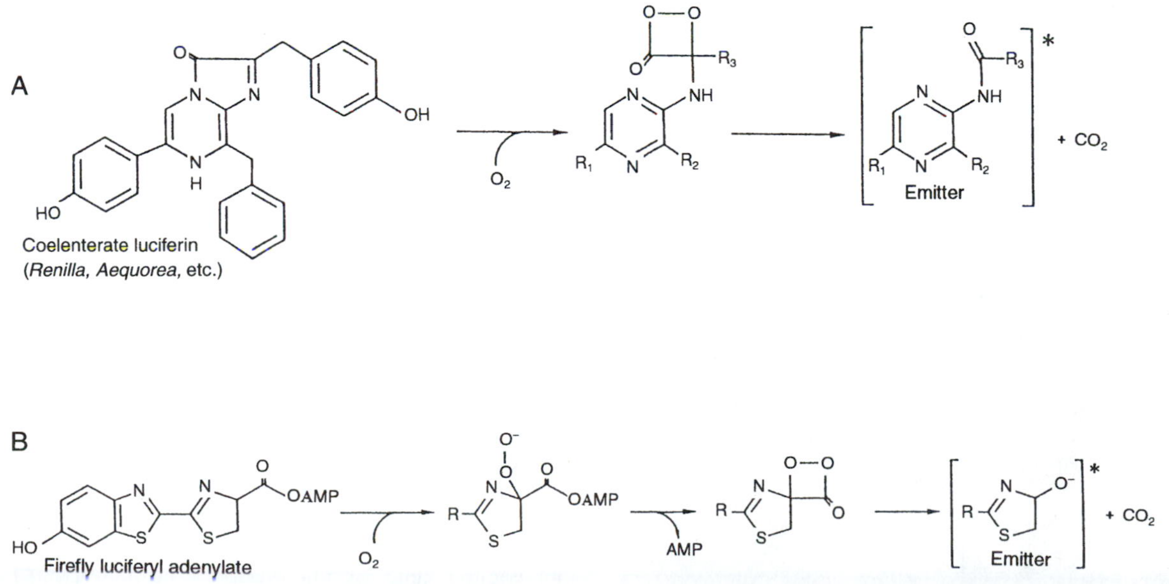

FIGURE 15. (A) The structure of coelenterate luciferin (coelenterazine) and the light-emitting reaction, showing the postulated cyclic peroxide intermediate and excited state in the light-emitting reaction. (B) The structures of firefly luciferin, showing the postulated cyclic peroxide intermediate and excited state.

VIII. Fireflies

A. Occurrence and Function

Only about 100 genera of insects are classed as luminous out of a total of approximately 70000 insect genera (Lloyd, in Herring, 1978). But when it occurs, the luminescence is impressive, most notably in the many species of beetles, fireflies, and their relatives. Fireflies themselves possess ventral light organs on posterior segments (Fig. 16). The South American railroad worm, *Phrixothrix*, has paired green lights on the abdominal segments and red head lights, while the click and fire beetles, *Pyrophorini*, have both running lights (dorsal) and landing lights (ventral). The dipteran cave glow worm (see Fig. 1) exudes beaded strings of slime from its ceiling perch, serving to entrap minute flying prey, which are attracted by the light emitted by the animal.

The variety of different fireflies, with their different habitats and behaviors, is impressive. The major function of light emission in fireflies is for communication during courtship, in which one sex emits a flash as a query, to which the other responds, usually in a species-specific pattern (Case, 1984; Lloyd, 1977, 1980). Some flashing patterns are shown in Figs. 14B–14D. Two signal system types have been distin-

guished. In the first, one sex (usually female) is stationary and emits light or a flashing signal to which the other sex is attracted. In the second, one sex (usually the flying male), emits a species-specific flash, while the other emits a species-specific flash response. The time delay between the two may be a signaling feature; for example, it is 2 s in some North American species. But the flashing pattern (e.g., trains distinctive in duration and/or intensity) is also important in some cases, as is the kinetic character of the individual flash (duration; onset and decay kinetics). In some species, flickering occurs in the flashes, sometimes at frequencies higher than those detectable by the human eye (~40 Hz).

In other cases, the communication patterns and their interpretations are not as clear or as readily classified. Fireflies in Southeast Asia are particularly noteworthy in this respect, especially the synchronous flashing of *Pteroptyx* spp. These fireflies form congregations of many thousands in single trees, where the males produce an all-night-long display, with flashes every 1–4 s, dependent on species (Buck and Buck, 1976). This appears to serve to attract females to the tree.

B. Biochemistry

The firefly system was the first in which the biochemistry was well characterized. It had been known since before 1900 that cell-free extracts could continue to emit light for several minutes or hours, and that after the complete decay of the light, emission could be restored by adding a second extract, prepared using boiling water to extract the cells (then cooled). The enzyme luciferase was assumed to be in the first (cold water) extract (with all the luciferin substrate being used up during the emission), whereas since the enzyme was denatured by the hot water extraction, some substrate was left intact. This was referred to as the **luciferin-luciferase reaction,** and it was already known in the first part of this century that luciferins and luciferases from the different major groups would not cross-react, indicative of their independent evolutionary origins (Harvey, 1952).

In 1947 it was discovered that the addition of adenosine triphosphate (ATP), a "high-energy" intermediate, to an "exhausted" cold water extract resulted in an enormous bioluminescence response. This response showed that luciferin had not actually been consumed in the cold water extract; ATP could not be the emitter, since it did not have the appropriate fluorescence. For some time, ATP was believed to be providing the energy for light emission. But, as noted already, the energy available from ATP hydrolysis is only about 7 kcal per mole, whereas the energy of a visible photon is 50 kcal or more per mole. It was soon discovered that firefly luciferin, later shown to be a unique benzothiazole (Fig. 15B), was still present in large amounts in the exhausted cold water extract and that it was ATP that was consumed, but available, in the hot water extract. ATP was shown to be required to form the luciferyl adenylate intermediate, which then reacted with oxygen to form a cyclic luciferyl peroxy species, which broke down to yield CO_2 and an excited state of the carbonyl product (McElroy and DeLuca, in Herring, 1978). Luciferase catalyzes both the luciferin activation and the subsequent steps leading to the excited product.

FIGURE 16.   Ventral light organs of fireflies.

In reactions in which luminescence has decreased to a low level (this may continue for days), it was long ago found that emission is greatly increased by coenzyme A, but the reason for this was obscure. The discovery that long-chain acyl-CoA synthetase (EC 6.2.1.3) has homologies with firefly luciferase (EC 1.13.12.7) both explains this observation and indicates the evolutionary origin of the gene.

Firefly luciferase has been cloned and expressed in other organisms, including *Escherichia coli* and tobacco. In both cases, luciferin must be added exogenously; tobacco plants "light up" when the roots are immersed in luciferin (Ow *et al.,* 1986). There are some beetles in which the light from different organs is a different color. The same ATP-dependent luciferase reaction with the same luciferin occurs in the different organs, but the luciferases are slightly different, coded by different (but homologous) genes (Wood *et al.,* 1989). They are presumed to differ with regard to the site that binds the luciferin, which could thereby alter the emission wavelength.

C. Cell Biology and Regulation of Flashing

The firefly light organ comprises a series of photocytes arranged in a rosette, positioned radially around a central tracheole, which supplies oxygen to the organ (Ghiradella, 1998). The organ itself comprises a series of such rosettes, stacked side by side in many dorsoventral columns. Photocyte granules or organelles containing luciferase have been identified with peroxisomes on the basis of immunochemical labeling.

It is still not known how flashing is controlled in fireflies (Case and Strause, in Herring, 1978). Although flashing is initiated by a nerve impulse that travels to the light organ, most of the nerve terminals in the light organ are not on photocytes but on special tracheolar cells, which control the supply of oxygen. This accounts for the considerable time delay between the arrival of the nerve impulse at the organ and the onset of the flash. The possibility that the flash is somehow directly triggered by an action potential thus seems unlikely. Also, none of the ions typically gated by membrane potential changes (Na^+, K^+, and Ca^{2+}) appear to be likely candidates for controlling luminescence chemistry. An alternate theory is that the flash is controlled by the availability of oxygen, which is required in the luminescence reaction. A comparison of different species shows a strong positive relationship between the extent of the tracheal supply system in the adults and the flashing ability. On the other hand, the mechanism must account for the rapid kinetics, complex waveforms, multiple flashes, and high-frequency flickering, all of which seem unlikely to be regulated by a gas in solution. However, although oxygen might diffuse slowly, it reacts very rapidly chemically in this system. The half rise time of luminescence with the anaerobic enzyme system (luciferase-luciferyl adenylate) is about 10 ms. The mechanism is still being actively investigated.

IX. Other Organisms: Other Chemistries

The four systems already described are known best, but several others have been studied in some detail. Several of these are briefly described below.

A. Mollusks

Snails (gastropods), clams (bivalves), and cephalopods (squid) have bioluminescent members (Young and Bennett, 1988). The squid luminous systems are by far the most numerous and diverse, in both form and function, rivaling the fishes in these respects. As is also true for fishes, some squid use symbiotic luminous bacteria (Ruby, 1996), but most are self-luminous, indicating that bioluminescence had more than one evolutionary origin within the class.

Many squid possess photophores, which may be used in spawning and other interspecific displays (communication). Photophores are compound structures with associated optical elements such as pigment screens, chromatophores, reflectors, lenses, and light guides. They may emit different colors of light and are variously located near the eyeball, on tentacles, on the body integument, or associated with the ink sac or other viscera. In some species, luminescence intensity has been shown to be regulated in response to changes in ambient light, indicative of a camouflage function.

Along the coasts of Europe a clam, *Pholas dactylus*, inhabits compartments that it makes by boring holes into the soft rock. When irritated, these animals produce a bright cellular luminous secretion, squirted out through the siphon as a blue cloud. This animal and its luminescence have been known since Roman times, and the system was used by DuBois in his discovery and description of the luciferin-luciferase reaction in the 1880s (see Section VIII). Well ensconced in its rocky enclosure, the animal presumably uses the luminescence to somehow deter or thwart would-be predators.

The *Pholas* reaction has been studied extensively, but the structure of the luciferin, which involves a protein-bound chromophore, remains unknown. The luciferase is a copper-containing large (>300-kDa) glycoprotein (Henry *et al.,* 1975). It can serve as a peroxidase with several alternative substrates, indicating the involvement of a peroxide in the light-emitting pathway; the superoxide ion is apparently involved in the reaction.

There are luminous species in several families of gastropods; a New Zealand pulmonate limpet, *Latia neritoides*, is notable as the only known luminous eukaryote that can be classed as a truly freshwater species. It also secretes a bright luminous slime (green emission, $\lambda_{max} = 535$ nm), whose function may be similar to that of the *Pholas* emission. Its luciferin is an enol formate of an aldehyde, but the emitter and products in the reaction are unknown. In addition to its luciferase ($M_r \sim 170$ kDa; EC 1.14.99.21), a "purple protein" ($M_r \sim 40$ kDa) is required, but only in catalytic quantities, suggesting that it may be somehow involved as a recycling emitter.

B. Annelids

The annelids also include many luminous species, both marine and terrestrial (Herring, in Herring 1978). *Chaetopterus* are marine polychaetes that construct and live in U-shaped tubes in sandy bottoms; they also exude luminescence upon stimulation, but the chemistry of the reaction has eluded researchers. Other marine polychaetes include the Syllidae, such as the Bermuda fireworm mentioned earlier, and the polynoid worms, which shed their luminous scales as decoys. Extracts of the latter have been shown to emit light upon the addition of superoxide ion.

More but still limited knowledge is available concerning the biochemistry of the reaction in terrestrial earthworms, some of which are quite large, over 60 cm in length (Wampler in DeLuca and McElroy, 1981). Upon stimulation they exude coelomic fluid from the mouth, anus, and body pores. This exudate contains cells that lyse to produce a luminous mucus, emitting in the blue-green region. However, the exudate from animals deprived of oxygen does not emit, but will do so after the admission of molecular oxygen to the free exudate. In *Diplocardia longa* the cells responsible for emission have been isolated; luminescence in extracts involves a copper-containing luciferase $M_r \sim 300$ kDa), and the luciferin (N-isovaleryl-3-amino-1-propanal). The *in vitro* reaction requires H_2O_2, not free O_2.

C. Crustaceans

Many crustaceans are luminescent (Herring, 1985b). The cypridinid ostracods such as *Vargula* (formerly *Cypridina*) *hilgendorfii* are small organisms that possess two glands with nozzles from which the luciferin and luciferase (EC 1.13.12.6) are squirted into the seawater, where they react and produce a spot of light, useful either as a decoy or for communication (Morin and Cohen, 1991).

Cypridinid luciferin and its reaction have differences and similarities to those of the coelenterazine system (Cormier, in Herring, 1978). The luciferin in both is a substituted imidazopyrazine nucleus that reacts with oxygen to form an intermediate cyclic peroxide, which then breaks down to yield CO_2 and an excited carbonyl. However, the cypridinid luciferase gene has been cloned and appears to have no homologies with the gene for the corresponding coelenterate proteins, and calcium is not involved in the cypridinid reaction. The two different luciferases reacting with similar luciferins have apparently had independent evolutionary origins, indicative of convergent evolution at the molecular level.

Euphausiid shrimp possess compound photophores with accessory optical structures and emit a blue, ventrally directed luminescence. The system is unusual because both luciferase and luciferin cross-react with the dinoflagellate system. This cross-taxon similarity indicates another possible exception to the rule that luminescence in distantly related groups had independent evolutionary origins. The shrimp might obtain luciferin nutritionally, but the explanation for the occurrence of functionally similar proteins is not evident. One possibility is lateral gene transfer; convergent evolution is another. Analyses of gene structures for homologies should provide insight into this question.

D. Fish

Bioluminescence in fish is highly diverse and occurs in both teleost (bony) and elasmobranch (cartilaginous) fish. Partly because animals have not been so readily available,

relatively little is known about their physiology and biochemistry, but many have been described (Herring, 1982).

As noted in the section on bacteria, many fish obtain their light-emitting ability by culturing luminous bacteria in special organs (see Figs. 2 and 5), but most are self-luminous. Self-luminous species include *Porichthys,* the midshipman fish, so-called because of the array of its photophores distributed linearly along the four pairs of lateral lines, much as are the buttons on a military uniform. Because it occurs close to shore, it has been the object of considerable study, and more is known about the physiological control of luminescence in *Porichthys* than in any other fish. Biochemically it is less well characterized, but its luciferin and luciferase cross-react with the cypridinid ostracod crustacean system already described. This was an enigma until it was discovered that Puget Sound fish have photophores but are nonluminous; however, they can emit if injected with cypridinid luciferin or fed the animals, showing that luciferin may be obtained nutritionally. Did the luciferase in this fish originate independently to make use of the available substrate, or was the ability to synthesize luciferin lost secondarily from a complete system? If the latter, this would be analogous to the loss of the ability to synthesize vitamins in mammals.

Open sea and midwater species include the sharks (elasmobranchs), some of which may have several thousand photophores. The teleosts include the gonostomatids such as *Cyclothone,* with simple photophores, and the hatchet fish, having compound photophores with elaborate optical accessories; emission is directed exclusively downward, indicative of a camouflage function of the light.

Fish often possess different kinds of photophores located on different parts of the body, particularly ventrally and around the eyes, evidently with different functions. One interesting arrangement, known in both midwater squid and myctophids, makes use of a special photophore positioned to shine on the eye or a special photoreceptor. Its intensity parallels that of the other photophores, so it provides information to the animal concerning its own brightness, thus allowing it to match the intensity of its own counterillumination to that of the downwelling ambient light. Another clear case of functional use is in *Neoscopelus;* in addition to the many photophores on the skin, they also occur on the tongue, allowing it to attract prey to just the right location.

Sexual dimorphism is also frequent in fish luminescence. Males and females of the myctophid *Tarletonbeania* were originally thought to be different species. Only one (now known to be the male) has caudal luminous organs, and the occurrence of these fish was known only from stomach contents of predator fish; only the female could be captured in nets, but never together with a male. The apparent explanation is that when a predator attacks, the males dart off in all directions with their dorsal lights flashing, like a police car, leading the predators on a wild chase (and sometimes getting caught), but leaving the females, who remain in place, safe from the predator in the cover of darkness, yet easy to catch in a net.

A number of self-luminous fish eject luminous material; in the searsid fishes this is cellular in nature, but it is not bacterial, and its biochemical nature is not known. Such animals may also possess photophores.

X. Applications of Bioluminescence

Instrumentation for measuring light emission is very sensitive and free of background and interference characteristic of many other analytical techniques (Wamplar, in Herring, 1978; Kricka and Whitehead, 1984). A typical photon counting instrument can readily detect an emission of about 10^4 photons s^{-1}, which corresponds to the transformation of 10^5 molecules s^{-1} (or 6×10^6 min^{-1}) if the quantum yield is 10%. (Those bioluminescent reactions that have been measured have quantum yields of this or higher; the firefly yield is about 90%.) A substance at a concentration of 10^{-9} M could readily be detected in a single cell 10 μm in diameter by this technique.

For these and other reasons, luminescence has come into widespread use (Roda *et al.*, 1999; Hastings *et al.*, 1997). Luminescent tags have been developed that replace radioactivity with as high a sensitivity. Since the different luciferase systems have different specific chemistries, quantitative determinations of many different specific substances can be accomplished. One of the first and still widely used assays uses firefly luciferase to detect ATP. Many different enzymes use or produce ATP, so their activities may be followed using this assay. With bacterial luciferase, any reaction that produces or utilizes NAD(H), NADP(H), or long-chain aldehyde, either directly or indirectly, can be coupled to this light-emitting reaction.

A photoprotein from the scaleworm has been used to detect superoxide ion; the purified photoprotein aequorin is widely used to detect intracellular Ca^{2+} and its changes under various experimental conditions (e.g., during muscle contraction). The protein is relatively small, nontoxic, and it may be readily injected into cells, reporting Ca^{2+} over the range of 3×10^{-7} to 10^{-4} M. The luminescence of bacteria has also been used as a very sensitive test for oxygen, making use of the fact that the K_m for O_2 in that reaction is extremely low. An oxygen electrode–incorporating luminous bacteria has been developed recently.

Luciferase genes have also been exploited as reporters for many different specific genes. Such systems are noninvasive and nondestructive, and the relevant activity can be measured in the living cell and in the same cell over the course of time. Recent studies of circadian rhythms have made use of the expression of the bacterial *lux* gene for the study of rhythms in cyanobacteria (Kondo *et al.*, 1993), the firefly *luc* gene for rhythms in higher plants (Millar *et al.*, 1995), and the *aequorin* gene for tracking intracellular calcium rhythms in *Arabidopsis* (Johnson *et al.*, 1995). As indicated earlier, the *gfp* gene is in wide use in many different types of applications (Chalfie and Kain, 1997).

These many diverse applications illustrate the fact that studies of bioluminescence, sometimes thought to be of little importance, have contributed to knowledge in ways not foreseen at the time the investigations were being made. Basic knowledge is a powerful tool in more than one respect.

XI. Summary

The emission of visible light by living organisms is an unusual phenomenon, both in terms of its relative rarity and with respect to the biochemical and regulatory mechanisms involved. But where it does occur, bioluminescence is sometimes spectacular and can usually be inferred to have functional importance—a consequence of the fact that another organism detects and responds to the light.

The uses of the light may be classified under three headings: defense, offense, and communication. Light may be used defensively to startle or frighten (flashes), to divert predators, as a decoy, or to provide camouflage. Offensively, light may be used as a lure, to attract and convert would-be predators into prey. Communication occurs in courtship and mating displays.

Biologically, the single most striking aspect of bioluminescence is its wide and diverse phylogenetic distribution and the independent evolutionary origin of different systems. The ability to emit light occurs in some 13 different phyla, and apparently originated many times, perhaps 30 or more. This is reflected not only in the gene and protein structures, but also in its biological, biochemical, and functional diversity, as well as its sporadic phylogenetic distribution.

Another unusual and unexplained fact is that bioluminescence is primarily a marine phenomenon. Although there are terrestrial forms, it is virtually absent in fresh water; only one such species is known. It is also not confined to or especially prevalent in animals that live in complete darkness (caves, the deep ocean).

Bioluminescence is an enzymatically catalyzed chemiluminescence, a chemical reaction that emits light. The enzymes involved are all referred to generically as luciferases, somewhat unfortunately, because they are not conserved evolutionarily, and are thus structurally different in different groups of organisms. The genes coding for several of the luciferases have been cloned and sequenced, confirming that they possess no homologous regions in common. The substrates, generally called luciferins, react with molecular oxygen to form intermediate luciferase-bound peroxides, which break down to give a product in an excited state, which subsequently emits light.

In the marine environment, luminous bacteria are ubiquitous as planktonic forms in seawater, and they are also responsible for the light emission of many species of higher organisms, usually as symbionts. Terrestrial forms are not common, but do occur as symbionts in nematodes, as an agent in the nematode's parasitization of insects. All species use the same biochemical system, a shunt of the electron transport pathway, in which reduced flavin and aldehyde are oxidized by molecular oxygen to give a luciferase-bound flavin intermediate in an excited state. Luminous species are versatile with respect to the alternate uses of light emission in different situations, the exploitation of alternate habitats, and the capacity to turn the synthesis of the system on and off at both the physiological and genetic levels. The genes involved may be widely distributed and mobile.

Dinoflagellates are unicellular algae; in the ocean these organisms are largely responsible for the sometimes brilliant sparkling "phosphorescence" seen at night when the water is disturbed, and also for "red tides." Their luminescent flashing originates from novel cellular organelles called scintillons, formed as spherical outpocketings of the cytoplasm into the vacuole. They contain dinoflagellate luciferase and luciferin (a novel reduced linear tetrapyrrole), the latter bound to a second protein. Flashing is triggered by an action potential in the vacuolar membrane that causes a transient pH change in the scintillon, releasing the luciferin from its binding protein. Luminescence in some dinoflagellates is controlled by a cellular circadian biological clock, which causes the synthesis and destruction of the components to occur on a daily cycle.

There are many luminous comb jellies, hydroids, siphonophores, sea pens, and jellyfish. Upon stimulation, light is generally emitted as brief bright flashes or trains of flashes. The luminescence originates in specialized cells called photocytes, triggered by a conducted epithelial or nerve action potential. The luciferin is a substituted imidazole called coelenterazine, which, the name notwithstanding, occurs in several other phyla; its oxidation results in light emission. The reaction, catalyzed by coelenterate luciferase, is regulated by Ca^{2+}. This is a case where the evolutionary origin of the luciferase is known; it bears homology with calmodulin.

Fireflies typically emit light as flashes, which are used as species-specific signals for communication in courtship. The light organ is a complex structure with photocytes arranged in a rosette pattern, invested with tracheoles to transport the required molecular oxygen directly to the cells. It may be that oxygen ultimately regulates flashing, though a nerve impulse initiates the process. The firefly reaction is unique in having a requirement for ATP, which serves to "activate" the luciferin (a benzothiazole). The luciferyl adenylate is thus the "true" substrate that reacts with oxygen, forming an intermediate cyclic peroxide whose breakdown results in light emission. Firefly luciferase also shows homologies with another protein, namely, long-chain acyl-CoA synthetase.

Other major luminous groups include the mollusks (snails, clams, squid), annelid worms (both marine and terrestrial), crustacea (shrimp and ostracods), echinoderms (brittle stars, starfish, sea cucumbers), and fish, both cartilaginous (sharks) and teleost (bony fishes). Of all the groups, fish and squids have the greatest variety of luminous systems; some make use of symbiotic luminous bacteria as a source of light, whereas others are self-luminous. Luminous organisms are most abundant at midwater depths (500–1000 m) in the open ocean.

Bibliography

Abrahams, L. V., and Townsend, L. D. (1993). Bioluminescence dinoflagellates: A test of the burglar alarm hypothesis. *Ecology* **74**, 258–260.

Anctil, M. (1987). Neural control of luminescence. *In* "Nervous Systems in Invertebrates" (M. A. Ali, Ed.), pp. 573–602. Plenum, New York.

Baldwin, T., and Ziegler, M. (1992). The biochemistry and molecular biology of bacterial bioluminescence. *In* "Chemistry and Biochemistry of Flavoenzymes" (F. Müller, Ed.), Vol. 3, pp. 467–530. CRC Press, Boca Raton, FL.

Blinks, J. R., Wier, W. G., Hess, P., and Prendergast, F. G. (1982). Measurement of Ca^{2+} concentrations in living cells. *Prog. Biophys. Mol. Biol.* **40**, 1–114.

Buck, J. B. (1938). Synchronous rhythmic flashing of fireflies. II. *Q. Rev. Biol.* **63**, 265–289.

Buck, J. B., and Buck, E. (1976). Synchronous fireflies. *Sci. Am.* **234**, 74–85.

Campbell, A. K., (1988). "Chemiluminescence: Principles and Applications in Biology and Medicine." Verlag Chemie, Weinheim.

Case, J. (1984). Firefly behavior and vision. *In* "Insect Communication" (T. Lewis, Ed.), pp. 195–222. Royal Entomological Society of London, published for them by Harcourt Brace Jovanovich, New York.

Catalani, L. H., and Wilson, T. (1989). Electron transfer and chemiluminescence. Two inefficient systems: 1,4-Dimethoxy-9,10-diphenylanthracene peroxide and diphenoylperoxide. *J. Am. Chem. Soc.* **111**, 2633–2639.

Chalfie, M., and Kain, S. (Eds.) (1997). "GFP: Green Fluorescent Protein Strategies and Applications." John Wiley & Sons, New York.

Charbonneau, H., Walsh, I., McCann, R., Prendergast, F., Cormier, M., and Vanaman, T. (1985). Amino acid sequence of the calcium-dependent photoprotein aequorin. *Biochemistry* **24**, 6762–6771.

Cody, C. W., Prasher, D. C., Westler, W. M., Prendergast, F. G., and Ward, W. W. (1993). Chemical structure of the hexapeptide chromophore of the *Aequorea* green-fluorescent protein. *Biochemistry* **32**, 1212–1218.

Cormier, M. J. (1981). *Renilla* and *Aequorea* bioluminescence. *In* "Bioluminescence and Chemiluminescence" (M. DeLuca and W. D. McElroy, Eds.), pp. 225–233. Academic Press, New York.

Cormier, M. J., Prasher, D. C., Longiaru, M., and McCann, R. O. (1989). The enzymology and molecular biology of the Ca^{2+}-activated photoprotein, aequorin. *Photochem. Photobiol.* **49**, 509–512.

DeLuca, M., and McElroy, W. D. Eds., (1981). "Bioluminescence and Chemiluminescence." Academic Press, New York.

DeLuca, M., and McElroy, W. D. (1986). *Methods Enzymol.* 133.

Fuqua, W. C., Winans, S. C., and Greenberg, E. P. (1994). Quorum sensing in bacteria: The LuxR-LuxI family of cell density-responsive transcriptional regulators. *J. Bact.* **176**, 269–275.

Ghiradella, H. (1998). The anatomy of light production: the fine structure of the firefly lantern. *In* "Microscopic Anatomy of Invertebrates" **11A**, 363–381. Wiley-Liss, Inc.

Grober, M. S. (1988). Brittle star bioluminescence functions as an aposematic signal to deter crustacean predators. *Anim. Behav.* **36**, 493–501.

Harvey, E. N. (1952). "Bioluminescence." Academic Press, New York.

Hastings, J. W. (1983). Biological diversity, chemical mechanisms and evolutionary origins of bioluminescent systems. *J. Mol. Evol.* **19**, 309–321.

Hastings, J. W. (1987). Why luminous bacteria emit bright light—and sometimes do not. *In* "Perspectives in Microbial Ecology," (F. Megusar and M. Ganter, Eds.), pp. 249–252. Ljubljana, Yugoslavia.

Hastings, J. W., and Greenberg, E. P. (1999). Quorum sensing: the explanation of a curious phenomenon reveals a common characteristic of bacteria. *J. Bacteriol.* **181**, 2667–2668.

Hastings, J. W., and Morin, J. G. (1991). Bioluminescence. *In* "Neural and Integrative Animal Physiology" (C. L. Prosser, Ed.), pp. 131–170. Wiley-Interscience, New York.

Hastings, J. W., Potrikus, C. J., Gupta, S., Kurfürst, M., and Makemson, J. C. (1985). Biochemistry and physiology of bioluminescent bacteria. *Adv. Microb. Physiol.* **26**, 235–291.

Hastings, J. W., Makemson, J., and Dunlap, P. V. (1987). How are growth and luminescence regulated independently in exosymbionts? *Symbiosis* **4**, 3–24.

Hastings, J. W., Boulos, Z., and Rusak, B. (1991). Circadian rhythms. *In* "Neural and Integrative Animal Physiology" (C. L. Prosser, Ed.), pp. 435–546. Wiley-Interscience, New York.

Hastings, J. W., Kricka, L., and Stanley, P., Eds. (1997). "Bioluminescence and Chemiluminescence: Molecular Probes." John Wiley & Sons, Chichester.

Henry, J. P., Monny, C., and Michelson, A. M. (1975). Characterization and properties of *Pholas* luciferase as metalloglycoprotein. *Biochemistry* **14**, 3458–3466.

Herring P. J. (Ed.) (1978). "Bioluminescence in Action." Academic Press, New York.

Herring P. J. (1982). Aspects of bioluminescence in fishes. *Oceanogr. Mar. Biol. Annu. Rev.* **20**, 415–470.

Herring P. J. (1985a). How to survive in the dark: Bioluminescence in the deep sea. *In* "Physiological Adaptations of Marine Animals" (M. S. Laverack, Ed.), Soc. Exp. Biol. Symp. No. 39, pp. 323–350. The Company of Biologists Ltd., Cambridge, UK.

Herring P. J. (1985b). Bioluminescence in the crustacea. *J. Crustacean Biol.* **5**(4), 557–573.

Herring P. J. (1987). Systematic distribution of bioluminescence in living organisms. *J. Biolumin. Chemilumin.* **1**, 147–163.

Herring P. J. (1990). Bioluminescent communication in the sea. *In* "Light and Life in the Sea" (P. J. Herring, A. K. Campbell, M. Whitfield, and L. Maddock, Eds.), pp. 245–264. Cambridge University Press, London/New York.

Johnson, C. H., Knight, M., Kondo, T., Masson, P., Sedbrook, J., Haley, A., and Trewavas, A. (1995). Circadian oscillations of cytosolic and chloroplastic free calcium in plants. *Science* **269**, 1863–1865.

Kondo, T., Strayer, C., Kulkarni, R., Taylor, W., Ishiura, M., Golden, S., and Johnson, C. (1993). Circadian rhythms in prokaryotes: Luciferase as a reporter of circadian gene expression in cyanobacteria. *Proc. Natl. Acad. Sci. USA* **90**, 5672–5676.

Kricka, L., and Whitehead, T. P. (Eds.) (1984). "Analytical Applications of Bioluminescence and Chemiluminescence." Academic Press, New York.

Li, L., Hong, R., and Hastings, J. W. (1997). Three functional luciferase domains in a single polypeptide chain. *Proc. Natl. Acad. Sci.* **94**, 8954–8958.

Lloyd, J. E. (1977). Bioluminescence and communication. *In* "How Animals Communicate" (T. A. Sebeok, Ed.), pp. 164–183. Indiana University Press, Bloomington.

Lloyd, J. E. (1980). Firefly signal mimicry. *Science* **210**, 669–671.

Lorenz, W. W., McCann, R. O., Longiaru, M., and Cormier, M. J. (1991). Isolation and expression of a cDNA encoding *Renilla reniformis* luciferase. *Proc. Natl. Acad. Sci. USA* **88**, 4438–4442.

McFall-Ngai, M., and Morin, J. W. (1991). Camouflage by disruptive illumination in leiognathids, a family of shallow-water, bioluminescent fishes. *J. Exp. Biol.* **156**, 119–137.

Meighen, E. A. (1991), Molecular biology bacterial bioluminescence. *Microbiol. Rev.* **55**, 123–142.

Meighen, E. A., and Dunlap, P. V. (1993). Physiological, biochemical and genetic control of bacterial bioluminescence. *Adv. Microb. Physiol.* **34**, 1–67.

Millar, A. J., Carré, I. A., Strayer, C. A., Chua, N.-H., and Kay, S. A. (1995). Circadian clock mutants in Arabidopsis identified by luciferase imaging. *Science* **267**, 1161–1163.

Morin, J. (1983). Coastal bioluminescence: patterns and functions. *Bull. Mar. Sci.* **33**, 787–817.

Morin, J. G., and Cohen, A. C. (1991). Bioluminescent displays, courtship and reproduction in ostracodes. *In* "Crustacean Sexual Biology" (R. Bauer and J. Martin, Eds.), pp. 1–16. Columbia University Press, New York.

Morse, D., Fritz, L., and Hastings, J. W. (1990). What is the clock? Translational regulation of circadian bioluminescence. *Trends Biochem. Sci.* **15**, 262–265.

Nealson, K., and Hastings, J. W. (1991). The luminous bacteria. In "The Prokaryotes" 2nd ed., Volume I, Part 2, Chap. 25 (A. Balows, H. G. Trüper, M. Dworkin, W. Harder and K. H. Schleifer, Eds.), pp. 625–639. Springer-Verlag, New York.

Nicolas, M.-T., Nicolas, G., Johnson, C. H., Bassot, J.-M., and Hastings, J. W. (1987). Characterization of the bioluminescent organelles in *Gonyaulax polyedra* (dinoflagellates) after fast-freeze freeze fixation and antiluciferase immunogold staining. *J. Cell Biol.* **105**, 723–735.

Ow, D. W., Wood, K. V., DeLuca, M., de Wet, J. R., Helinski, D. R., and Howell, S. H. (1986). Transient and stable expression of the firefly luciferase gene in plant cells and transgenic plants. *Science* **234**, 856–859.

Roda, A., Pazzagli, M., Kricka, L. J., and Stanley, P. E. (eds.) (1999). *Bioluminescence and Chemiluminescence: Perspectives for the 21st Century*. John Wiley and Sons, Chichester.

Ruby, E. G. (1996). Lessons from a cooperative, bacterial-animal association: The *Vibrio fischeri-Euprymna scolopes* light organ symbiosis. *Annu. Rev. Microbiol.* **50**, 591–624.

Schuster, G. B. (1979). Chemiluminescence of organic peroxides. *Acc. Chem. Res.* **12**, 366–373.

Shimomura, O. (1985). Bioluminescence in the sea: Photoprotein systems. *Soc. Exp. Biolo. Symp.* **39**, 351–372.

Wilson, T. (1985). Mechanism of chemiluminescence. *In* "Singlet Oxygen" (A. Frimer, Ed.), Vol. 2, pp. 37–57. CRC Press, Boca Raton, FL.

Wilson, T. (1995). Comments on the mechanisms of chemi-and bioluminescence. *Photochem. Photobiol.* **62**, 601–606

Wilson, T., and Hastings, J. W. (1998). Bioluminescence *Annu. Rev. Cell Devel. Bio.* **14**, 197–230.

Wood, K. V., Lam, Y. A., Seliger, H. H., and McElroy, W. D. (1989). Different cDNAs elicit bioluminescence of different colors. *Science* **244**, 700–702.

Young, R. E., and Bennett, T. M. (1988). Cephalopod luminescence. *In* "The Mollusca" (M.R. Clarke and E. R. Trueman, Eds.), Vol. 12, pp. 241–251. Academic Press, New York.

SECTION
IX

Cell Division and Programmed Cell Death

66

Regulation of Cell Division in Higher Eukaryotes

Francisco Martínez, Alma D. Chávez, Diana González, and Andrés A. Gutiérrez

I. Introduction

The cells of a living organism must choose one of three pathways: live and reproduce, live without reproducing, or die. Cell division has been found to be tightly controlled in association with other fundamental events in the cell's life, such as, **differentiation, senescence,** and **programmed cell death** (e.g., apoptosis). Deregulation of these processes has been involved in the origin of pathological conditions, such as malignancy and autoimmune disorders. How and by which pathways the cell cycle is controlled have been central questions since the **cell theory** was proposed more than a century ago.

Research in the past has been mainly oriented towards the phase of the cycle in which the cell divides into two replicas of itself, that is, **mitosis.** In contrast, the other part of the cell cycle, known as **interphase,** remained largely unexplored. Late in the 1960s, the biochemical description of the **maturation promoting factor** (MPF) in amphibian eggs became the first clue as to what controls the cell cycle passage from interphase to mitosis (rev. in Norbury and Nurse, 1992). It is now known that this factor is a dimer composed of a **protein kinase unit, Cdc2,** and its **regulatory subunit,** known as **cyclin.** Further studies revealed the existence of homologous protein kinases that were responsible for regulating all the phases of the cell cycle by binding to different cyclins at different periods of time (e.g., Cdc28 in budding yeast).

Fortunately, the cell cycle machinery of the eukaryotic cells studied to date, from yeasts to humans, has been well conserved throughout evolution. This similarity has greatly facilitated the characterization of, for example, presumed human cyclin genes in reconstitution assays by their ability to rescue mutant strains of defective yeasts. These and other powerful biochemical and genetic approaches have contributed to fully characterize these genes and their products (Table 1) (Dunphy, 1997).

To date, a large number of proteins have been cloned and characterized for each one of the phases of the eukaryotic cell cycle. These proteins include various **cyclins, cyclin-dependent-kinases** (CDKs), **protein kinase inhibitors,** and **transcription factors** such as the products of tumor suppressor genes, p53 and pRB. All of these molecules have been shown to interact in critical ratios to determine the passage from one phase to another in the cell cycle.

TABLE 1 Common Strategies to Identify and Characterize Key Molecules of the Cell Cycle

Gene cloning strategies	*In vitro* assays	*In vivo* assays
Yeast two-hybrid screens	Kinase (phosphorylating) activity Substrates: Histone H1, pRB,[a] p107	Reconstitution assays in defective mutant yeasts and insects
Positional cloning	Immunoprecipitations of complexes (e.g., cyclin-CDK-CDK inhibitors)	Cells in culture DNA synthesis–inhibitory activity (e.g., thymidine incorporation) Cell cycle analysis (e.g., FACS[b]) Genetic overexpression and down-regulation (e.g., Ab[c] and antisense)
Subtractive hybridization	Ubiquitination assays Fluorescence *in situ* hybridization (FISH) Crystallographic analysis	Transgenic mice Gain of function or knockout

[a] Retinoblastoma protein.
[b] Fluorescence-activated cell sorter.
[c] Antibody.

Furthermore, the past three years have witnessed the description of an enlarging list of molecules involved in yet another important mechanism in control of the cell cycle, **ubiquitin-mediated proteolysis.**

This chapter, therefore, presents novel concepts regarding this proteolytic system and updates the roles of the other systems regulating the mitotic cell cycle of higher eukaryotic cells. Major reviews have been updated in each area but, as in the previous edition, neither the meiotic cell cycle nor studies in other eukaryotic cells (e.g., yeasts) are analyzed in detail.

II. General Overview

The **mitotic cell cycle** of higher eukaryotic cells consists of four phases (Fig. 1A) (1) **gap phase 1,** or **G_1 phase,** when the cell grows and prepares to synthesize DNA; (2) **S phase,** when DNA synthesis takes place and replicates the whole set of chromosomes; (3) a second gap period, **G_2 phase,** in which the cell prepares for division; and (4) the mitotic **M phase,** in which the actual division of the original cell into two daughter cells takes place. Mitosis also consists of various phases: (1) **prophase,** when chromosome condensation takes place; (2) **prometaphase-metaphase,** in which two members of each pair of sister chromatids attach to microtubules; (3) **anaphase,** when sister chromatids are separated and move along the microtubules; and (4) **telophase,** with the reformation of nuclei and the decondensation of the chromosomes (Fig. 1B). At the exit from mitosis, and depending upon the circumstances, the new daughter cells (1) reenter G_1 phase and keep cycling, (2) remain temporarily quiescent in a stage named **G_0 phase,** or (3) are terminally differentiated and unable to reenter the cell cycle.

Eukaryotes have developed a complicated network of molecules that act in a finely coordinated manner during the cell cycle. Central participants in the cell cycle machinery are **CDK complexes,** which contain a catalytic subunit, the CDK, and a regulatory subunit, a cyclin. Once the complexes have been activated, they are able to phosphorylate diverse substrates and, directly or indirectly, control gene expression and the progression of the cell cycle. As a feedback mechanism, cyclin-CDK complexes are negatively regulated by two different families of **CDK inhibitors:** CDK-inhibitor proteins, **Cip/Kip,** and inhibitors of kinases,

Ink (see Section III.B). Further control of the cell cycle progression is exerted through **ubiquitin-mediated proteolysis** of cyclins (e.g., D, E, A, and B), CDK inhibitors (e.g., p27), transcription factors (e.g., E2F), anaphase inhibitors, glue proteins, and so on (see Section III.E).

This complicated network has allowed cells to count on at least three **checkpoints** to safeguard the normal progression of the cell cycle, at G_1-S, at G_2-M, and at the exit of mitosis. The term "checkpoint" was originally used by Weinert and Hartwell (1989) in relation to the gene products that negatively regulate the cell cycle in response to DNA damage in yeasts. In their experiments in yeasts, cell cycle arrest occurred in response to chromosome damage by x-rays, and this arrest was mediated by the expression of the RAD 9 gene. Thus, the term was used in the sense of a border that could not be crossed in the presence of DNA damage. In this case, checkpoints could be considered monitors of the physical integrity of chromosomes that coordinate cell cycle transitions. However, other authors have also used the term "checkpoint" in relation to biochemical pathways that ensure that the initiation of particular cell cycle events depends upon the successful completion of others.

To help the reader through this general overview, Fig. 2 shows the most important mechanisms that control the mitotic cell cycle in higher eukaryotic cells. Tables 2 and 3 list the number of amino acid residues, the NCBI accession numbers,[1] and the main activities of the participants in charge of regulating this process.

A. Regulation of the G_1-S Phase Transition

The driving force that makes cells cycle comes from an intricate network of negative and positive signals. When the ratio of growth stimulatory signals to growth inhibitory signals increases, the cell cycle is switched on. For simplicity, these signals have been referred to as **mitogens** in Fig. 2.

---

[1]NCBI accession numbers refer to the gi number of the protein sequences at the National Center for Biotechnology Information, NIH, USA. These sequences are compiled from different databases (e.g., GenBank, EMBL, PIR, etc.), and the complete list can be obtained from the, NCBI-NIH URL: http://www.ncbi.nlm.nih.gov.

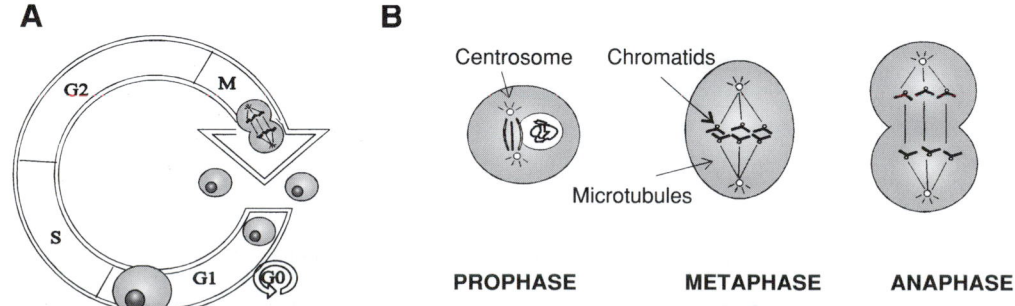

FIGURE 1. Phases of the cell cycle: (A) Interphase consists of three phases known as G_1, S, and G_2, in which the cell prepares for cell division. The latter takes place during mitosis. (B) Phases of mitosis. See text for details.

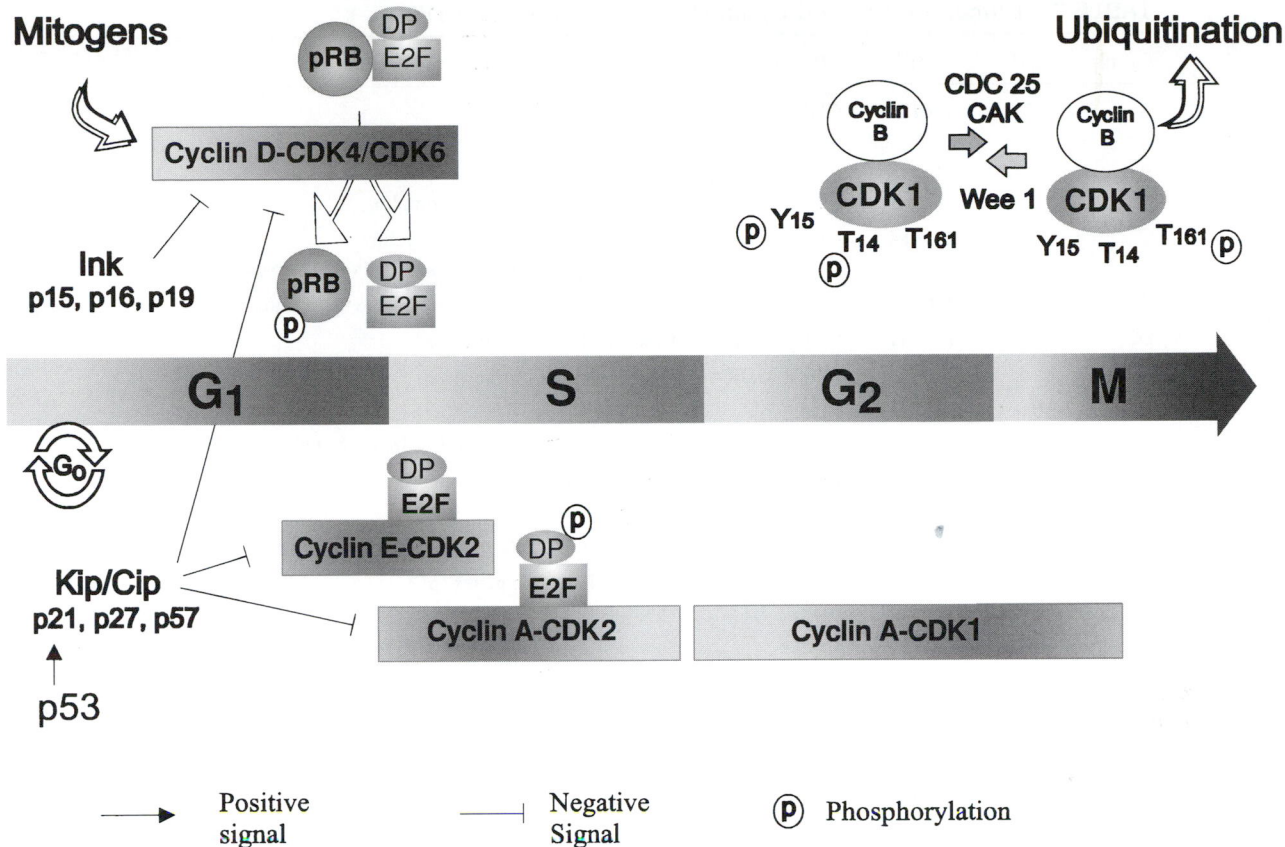

FIGURE 2. Regulation of the mitotic cell cycle in higher eukaryotic cells. The sequential formation, activation, and inactivation of various cyclin-CDK complexes govern the progression of the cell cycle. The timing of the events herein depicted and the patterns of interaction among the different members of this machinery are representative and may differ for some models. Negative regulation of the cell cycle occurs at checkpoints G_1-S, G_2-M, and exit of mitosis. CDK, cyclin-dependent kinase; E2F-DP, transcription factor; pRB, retinoblastoma protein; Ink, inhibitors of kinase; Kip/Cip, CDK-inhibitor protein; CAK, CDK-activating kinase; Cdc25, dual specificity phosphatase; Wee1, phosphorylase; T14, Thr 14; Y 15, Tyr 15; T161, Thr 161. Of note, pRB has more than a dozen phosphorylation sites on either serine or threonine residues. For further details, see Section II.

Mitogens induce the synthesis of **D-type cyclins** (D1, D2, D3), which are short-lived proteins ($t_{1/2} < 25$ min). The highest cyclin D levels are reached in the late G_1 phase, and they rapidly decay after the cell has moved into the S phase. Thereafter, D cyclins are no longer needed for the completion of the cell cycle (Sherr, 1995). The importance of these cyclins is that they become transducers of signals from the environment by acting as regulator subunits for cyclin-dependent kinases CDK2, CDK4, and CDK6.

As previously mentioned, the activation of CDKs regulates progression through critical transitions in the cell cycle (Lania *et al.*, 1999). Two distinct types of CDKs control entry into S phase: CDK4/CDK6 and CDK2. In addition to cyclin binding, CDK4 and CDK2 require phosphorylation by a **CDK-activating-kinase** (CAK) to become fully active. In contrast, inactivation of either kinase activity leads to cessation of proliferation and withdrawal from the mitotic cycle.

As D cyclins accumulate, they associate with CDK4 and CDK6 and form active holoenzyme complexes, that is, cyclin D–CDK4 and cyclin D–CDK6. These complexes are critical because they facilitate the exit from G_1 by phosphorylating a master regulatory protein of the cell cycle at the

G_1-S transition: the **retinoblastoma protein** (p105Rb). Indeed, cyclin D–kinase activity is the most important mechanism by which pRB is phosphorylated in the mid to late G_1 phase.

pRB is the product of a tumor suppressor gene and a member of a family referred to as **pocket proteins,** which includes **p107** and **p130.** Like pRB, p107 and p130 are phosphorylated by various CDKs (Grana *et al.*, 1998). When pRB proteins are phosphorylated by CDKs, they become inactive and release normally bound transcription factors known as **E2F-DP heterodimers.** The latter, in turn, mediate progress toward S phase. Because pRB phosphorylation occurs 1 to 2 hours before the start of S phase, a vast array of S-phase genes is induced by the E2F-DP heterodimers during this period (Weinberg, 1996).

But the activation of CDKs (and their effects) depends not only on cyclin abundance (e.g., expression and proteolysis) and on the phosphorylation state of their kinase subunits. As mentioned above, their activity is also controlled by the presence of negative regulators that belong to one of two families: Inks and Cip/Kips (Sherr and Roberts, 1999). These two families of kinase inhibitors differ in structure and specificity. Three members of the **Cip/Kip family** have now been

TABLE 2　Human Cyclins and Cyclin-Dependent Kinases (CDKs)

Cyclin	aa[a]	gi[b]	Binds	Synonyms/Function
A	432	116169	CDK1, CDK2, p107	G_2 = M-specific cyclin G_2 = M cyclin abruptly destroyed at mitosis
B	433	116176	CDK1	G_2 = M-specific cyclin MPF = Cyclin B–CDK1 complex
C	303	1117984	CDK1, CDK2, CDK4	CCNC gene. G_1-S-specific cyclin Cyclin PRAD1 encoded by bcl-1 linked oncogene
D1	295	116152	CDK6	CCND gene [11q13].[d] Controls the G_1-S transition
D2	289	231741	CDK2, CDK4/6	Controls the G_1-S transition
D3	292	231743	CDK2, CDK4/6	Controls the G_1-S transition
E1	395	116154	CDK2 CDK3	CCNE gene Nuclear protein essential for the G_1-S transition
E2	404	3769614	CDK2	CDK2 partner in late G_1 and S
F	786	1082313	?	CCNF gene [16p13.3]. Nuclear protein Activity during interphase with peak levels in G_2
G1	295	1236913	?	Damage-inducible genes (p53 dependent) Nuclear Protein
G2	344	1236915	?	Damage-inducible genes (p53 dependent) Cytoplasmic Protein
H	323	532561	CDK7 CDK2 TFIIB	CAK (MO15) regulatory subunit along with MAT1
I	377	1183162	?	Height-constraint levels in post-mitotic tissues Damage-inducible genes (p53 dependent)
J	391	1276897	CDK2?	Cyclin-dependent kinase interactor
K	357	4502625	CDK9?	Plays a role in regulating basal transcription
T1	726	2981196	CDK9	Cyclin T1/CDK9 = RNA polymerase II transcription factor Strongly enhances the affinity and specificity of the 　　tat: TAR RNA interaction in HIV-1 transcription
T2A	663	2981198	CDK9	Cyclin subunits—transcription elongation factor B-human
T2B	730	2981200	CDK9	Cyclin subunits—transcription elongation factor B-human

Other CDK Regulators

p35	307	1090763	CDK5	Neuronal-specific noncyclin regulator of CDK 5
PCNA	261	129694	CDK2, CDK4 CDK6	Proliferating cell nuclear antigen Auxiliary protein of DNA pol δ Controls eukaryotic DNA replication
p50	378	1421821	CDK4	p50/Cdc 37 protein kinase homolog Targets Hsp90 and both stabilize CDK 4
CKS1	79	461746	CDK2	p9. CDK regulatory subunit 1
CKS2	79	461747	CDK2	p9. CDK regulatory subunit 2 CKS1 and 2 bind to the catalytic subunit of CDKs Involved in CDK oligomerization ($\cong$ 6 Kinase subunits)
MAT1	309	1989848	CDK7 H	*Ménage à trois* 1. CDK7–cyclin H assembly factor p36. – RING finger protein
Wee1	646	1085293	CDK1	Tyrosine protein kinase. Phosphorylates CDK1 Negative regulator of entry into mitosis at G_2-M

described, known as **p21, p27,** and **p57,** according to their protein molecular weight. These molecules are considered universal inhibitors since they can bind to and inhibit the kinase activity of various CDK2, CDK4, and CDK6 complexes. In this respect, the Cip/Kip family can be considered more promiscuous than the Ink family, since the latter binds only to cyclin D–CDK4 and cyclin D–CDK6 complexes. Members of the **Ink family** include the proteins **p15, p16, p18,** and **p19.**

In addition to these events, G_1-S transition requires the presence of **cyclin E–CDK2** and **cyclin A–CDK2 com-**

TABLE 2 (*cont.*)   Human Cyclins and Cyclin-Dependent Kinases (CDKs)

Cyclin	aa[a]	gi[b]	Binds	Synonyms/Function
CDKs[d]				
CDK1	297	115922	A, B, p9	p34-Cdc 2. T-loop contains Thr 161 Phosphorylates the C-terminus of RNA pol II Regulates the formation of the mitotic spindle Essential for entry into S phase and mitosis
CDK2	298	116051	A, E, D2, D3, E1A, Skp1/skp2	p33. T-loop contains Thr 160 Adeno. E1A-associated kinase Maximum activity during S and G_2 phases
CDK3	305	231726	E	Putative role in G_1 exit by activating E2F 1, 2, and 3
CDK4	303	1168867	D1-D3 P50/Hsp90	PSJK3. Phosphorylates pRB Essential for progression at the G1-S transition Constant levels throughout the cell cycle
CDK5	292	4033704	p35 D?, p25	PSSALRE. Homology to the CDK1/Cdc28 family Role of CDK5/p35 in neuronal differentiation
CDK6	326	266423	D1-D3	PLSTIRE Phosphorylates pRB Essential for progression at G_1-S transition
CDK7	346	1705722	MAT1 H	CDK-activating kinase, CAK (MO15) = CDK7-cyclin H-MAT-1 Phosphorylates CDK2-(cyclin E), CDK4-(cyclin D) and Thr 161 in CDK1 kinase
CDK8	464	1000491	D-type and C	Interacts with D-type and C cyclins
CDK9	372	4502747	T1, T2a, T2b CDK2	PITALRE Nuclear Cdc 2-related protein kinase that phosphorylates the pRB protein *in vitro*
CDK10	360	4502731		PISSLRE, Cdc 2-related protein kinase Its effect is exerted at G_2-M Possible tumor suppressor gene

[a] Number of amino acid residues.

[b] gi accesion number, NCBI locus.[1]

[c] Brackets indicate chromosomal location (when available).

[d] Serine/threonine kinases that belong to the Cdc2/Cdc28 family.

plexes. Just like cyclin D–kinase activity (cyclin D–CDK4/6), cyclin E–CDK2 complexes also participate in pRB phosphorylation. The levels of cyclin E–CDK2 complexes correlate with their kinase activity; that is, they are first observed in late G_1, peak at the G_1-S transition, and disappear in early S. Thus, their activity precedes the S-phase role of cyclin A.

Since one of the genes regulated by E2F is cyclin E, it has been hypothesized that cyclin D controls cyclin E activity through pRB and E2F (Botz *et al.*, 1996). According to this model, cyclin D–CDK4/CDK6 complexes carry out the initial phosphorylation of pRB that leads to the induction of small amounts of cyclin E (Lundberg and Weinberg, 1998). Thereafter, cyclin E–CDK2 complexes could participate in pRB phosphorylation too. This hypothesis could explain why the ectopic expression of cyclin E is able to rescue the phenotype generated in cyclin D1 knockout mice (see Section IV).

Cyclin A appears around the G_1-S transition, and its level gradually increases and peaks at prophase. Then, cyclin A is abruptly degraded by ubiquitin-mediated proteolysis at the metaphase-anaphase transition of mitosis (see Section III.E). This cyclin is found in the nucleus, and its expression occurs earlier than that of cyclin B. In contrast to the latter, cyclin A binds both CDK2 and CDK1 kinases. Thus, cyclin A–CDK2

complexes are formed as the synthesis of cyclin A progresses, reaching maximum concentration during S phase.

Once cells enter into S phase, they become fully committed to divide and the cell cycle continues without the presence of further external signals. The most important point of control in S phase is the beginning of replication, marked by the origins' fire. Multiple replication origins exist per chromosome. These sites are where the replication forks are established through the unwinding of the DNA strands and the synthesis of the first primers for replicative chain elongation. In metazoan cells in general, more than defined sequences, these replication origins are circumscribed to a structural context (Delgado *et al.*, 1998). In fact there is a temporal control of the origins' fire during the S phase, for example, the chromosome regions associated with the maintenance of the chromosome structure (i.e., telomeric and centromeric regions) fire later than the chromosome regions associated with a high transcription rate (De Pamphilis, 1999; Greider, 1999).

The consensus view that has emerged is that there are two stages of chromatin during the cell cycle: pre-replication and post-replication. The **pre-replication complex** assembles during the G_1 phase and makes chromatin competent for replication. This complex consists of orc (1–6) proteins, Cdc6, and

TABLE 3 Human CDK Inhibitors and Pocket Proteins[a]

Name	aa[b]	gi[c]	Synonyms/Function[d]
p14	138	639716	ARF. Product of the human CDKN2A beta transcript Binds to MDM2 and induces p53 cell cycle arrest CDK4/CDK6 inhibitor
p15	148	1168868	CDK4 inhibitor B (p14-Ink4B)(p15-Ink4B)(MTS2) [9p21] Effector of TGFβ-induced G_1 arrest Inhibits CDK4/6 interaction with cyclins D and their function
p16	138	6397716	CDK4 inhibitor A (CDK4I)(p16-Ink4A)[9p21] Multiple tumor suppressor 1(MTSI) Inhibits CDK4/6 interaction with cyclins D and their function Negative regulator of the cell cycle at G_1
p18	168	1168870	CDK6 inhibitor (p18-Ink6; Ink4C). Weak interaction with CDK4 Arrests cell cycle with a correlated dependence on pRB Contains ANK repeat [1p32]
p19	166	1705730	CDK inhibitor specific to CDK4 and CDK6(p19-Ink4D) Inhibitor of cyclin D–dependent kinases [19p13]
Cip/Kip Family			
p21	164	729143	CDK inhibitor 1 (CDKN1A, Cip 1, Sdil, Wafl, Cap20) Potent "universal" inhibitor of G_1CDKs Induction by DNA damage (p53 dependent), IFN, hypoxia Terminal differentiation and senescence
p27	198	1168871	CDK inhibitor p27Kip1 [12p-12p13.1] Potent inhibitor of G_1CDKs N'-terminal (CDK-binding/inhibitory domain) similar to p21 Inactivates cyclin E/A–CDK2 and cyclin D CDK4 complexes Mediates extracellular antimitogenic signals
p28	209	1222531	Cyclin-dependent kinase inhibitor (African clawed frog)
p35	307	1082315	Neural-specific regulatory subunit of CDK5 Neural migration and development of mammalian cortex
p57	316	790248	p57Kip2. Inhibitor of G_1 CDKs [11p15.5] Not a significant suppressor in adult cells N-terminal (inhibitory domain) similar to p21
Pocket Proteins			
p105	928	132164	pRB, retinoblastoma tumor suppressor gene [13q14] Blocks cells in G_1 phase. Phosphorylated by cyclin D–CDK4/6 Inhibits E2F-DP activity and RNA pol I–III transcription Binds various cellular transcription factors and viral proteins
p107	1068	292371	RB-related p107 protein Phosphorylated by cyclins A/E–CDK2 complexes and D kinases Similar but nonredundant pRB actions
p130	1139	132164	RB-related p130 protein [16q12.2-13] Same as in p107

[a] Human CDK inhibitors are classified in one of two families: Ink4 or Cip/Kip. Pocket proteins include the retinoblastoma suppressor gene and its related proteins, p107 and p130.
[b] Amino acid number.
[c] gi accession number. NCBI locus.[1]
[d] The names by which the same protein might be known and the chromosome mapping in brackets.

Mcm (2–7) proteins. The functional interaction between Cdc6 and Orc proteins is essential for the loading of Mcm proteins. The latter, in turn, form hexameric complexes that have ATP-dependent helicase activity. Chromatin, thus "licensed" for replication, is guided into the S phase by the activation of cell-cycle-regulated protein kinases. Upon entry into S phase, the pre-replication complex partially dissolves, first by dissocia-

tion of Cdc6 and then by the gradual release of Mcm proteins. These events are accompanied by a recruitment of chain elongation factors and the establishment of replication forks.

The **post-replication stage** is characterized by the disassembly of the first complex. Cdc6 is released just before, or at the beginning of the S phase and it is replaced by the protein Cdc45, which in turn serves to attract DNA polymerase α-primase and the single-stranded-specific DNA-binding protein, RPA, as a first step in establishing replications forks (Mimura and Takisawa, 1998). Then, as the S phase proceeds, Mcm proteins are gradually dissociated from chromatin and, since replicated chromatin is devoided of Mcm proteins, the chromatin is no longer competent for replication. Thus dissociation of Cdc6 and of Mcm proteins guarantees that the genome is normally replicated once and only once per cell cycle (reviewed by Ritzi and Knippers, 2000).

Furthermore, cyclin A–CDK2 holoenzymes are responsible for the phosphorylation of: (a) Cdc45, (b) Cdc6 and its release from chromatin, and (c) Mcm4 that inactivates its DNA helicase activity (Ishimi et al., 2000). Other S phase regulatory kinases include protein kinase A for the release of Cdc6 proteins and Cdc7 kinases for the phosphorylation of Mcm proteins.

B. Regulation of the G$_2$-M Phase Transition

The G$_2$-M transition is one of the best characterized phases of the cell cycle. In this phase, **cyclin A/B–CDK1** (i.e., cyclin A/B–Cdc2) complexes become the dominant players. The role of cyclin A has already been described, so only that of cyclin B will be referred to in this section. The expression of this cyclin is transcriptionally regulated throughout the cell cycle. As shown in Fig. 2, it appears in S phase (mainly in the cytoplasm), and it is destroyed along with cyclin A at metaphase. Interestingly, cyclin B is imported into the nucleus just before nuclear envelope breakdown.

As cyclin B accumulates, cyclin B–CDK1 complexes are formed. Originally described as **maturating promoter factors,** these complexes require further posttranslational modifications to become fully active. In higher eukaryotes, this activation requires phosphorylation of CDK1 at Thr 161 and dephosphorylation at Tyr 15–Thr 14 (see Fig. 2 and Table 2). The inhibitory phosphorylation at Thr 14 and at Tyr 15 is carried out by a **kinase, Wee1,** while a dual specificity **phosphatase, Cdc25 homolog,** carries out their dephosphorylation. Interestingly, both enzymes also require up-regulatory phosphorylation by a protein kinase that is active at M phase, probably CDK1. As the cell enters into mitosis, Wee1 decreases its kinase activity while Cdc25 increases its phosphatase activity. At this time, **CDK-activating kinase (CAK)** exerts its up-regulatory activity by phosphorylating the Thr 161 residue in the T-loop of CDK1.

At this transition the mammalian genome is protected by a multiplicity of G$_2$-M checkpoints. For example, in the presence of DNA damage, Cdc25 phosphatases are phosphorylated or inactivated by proteins of the families **Chk1/14.3.3,** while Wee1/Mik proteins are activated by **Cds1** (Blasina et al., 1999). The overall result is the phosphorylation/inactivation of the CDK1 subunit of the cyclin B–CDK1 complex. Apparently, the expression of another regulator called **Gadd45** also arrests the cell cycle at this transition through a p53-dependent inhibitory effect on the cyclin B1–CDK1 complex (Wang et al., 1999). See Sections III.D.1–2 for further discussion.

Once the cyclin A/B–CDK1 complexes are active, they promote M-phase progression. It is during this phase that the actual division of the cell into two replicas of itself takes place. Evidence from years of descriptive and experimental work has established the existence of different phases in mitosis: prophase, prometaphase-metaphase, and transition to anaphase. Thus, there is an overwhelming number of studies on diverse events that occur during each one of these phases, including membrane partitioning (Warren, 1993), mitotic chromosome condensation, mitotic spindle formation, and chromosome movement (Nigg et al., 1996). One of the areas of major interest at present is the control and function of the mitotic spindle. This spindle consists of a dynamic array of microtubules and associated proteins (e.g., kinesins, survivin, etc.) and it is responsible for segregation of chromosomes during mitosis. Various **microtubule motor proteins** (e.g., **kinesins**) have been found within the spindle, so they are generally considered controllers of the segregation process. However, the importance of kinesins and related proteins in this process remains undetermined (reviewed in Sack et al., 1999).

C. Exit from Mitosis

At the end of the cell cycle, during metaphase, cyclin A and B levels decrease abruptly through **ubiquitin-mediated proteolysis.** This mitotic cyclin proteolysis is essential for the exit from mitosis. As described later, cyclin A and B contain a motif within the N'-termini, the **destruction box,** that confers regulation by ubiquitination pathways. Once the cyclin-CDK1 complexes are disrupted by destruction of their cyclin subunits, CDK1 monomers are rapidly and completely dephosphorylated.

Although the disruption of cyclin A/B–CDK1 complexes has been largely recognized as a major event for the exit from mitosis, recent studies have shown that anaphase entry is controlled by yet another mechanism in which cyclin destruction is not required, **sister chromatid separation** at kinetochores (reviewed in Townsley and Ruderman, 1998; Koepp et al., 1999). This process seems to be orchestrated by **E3 holoenzymes,** which conduct the ubiquitination of proteins whose degradation is needed for sister chromatid separation (Gorbsky, 1997). Therefore, ubiquitination and proteolysis of these proteins at **proteasomes** seem to be important events for progression from metaphase to anaphase. Interestingly, E3 complexes also mediate cyclin destruction during G$_1$ phase (see Section III.E).

III. Participants in the Cell Cycle

A. Cyclins and Cyclin-Dependent Kinases (CDKs)

As mentioned previously, CDKs are recipients of **growth signals** that regulate passage through sequential cell cycle transitions. In monomeric form, CDKs are inactive. Thus, CDKs are regulated by association with positive regulatory subunits (cyclins), with negative regulators known as CDK inhibitors, and by phosphorylation. The balance of these factors controls CDK activity, and the levels of all these proteins are tightly controlled both transcriptionally and posttranslationally. The latter is mainly achieved by ubiquitin-mediated proteolysis.

The complexity of the cell cycle machinery in higher eukaryotes is unparalleled in other organisms. So, to have a quick reference of most molecules involved in this process, Table 2 shows the protein sequences, NCBI accession numbers,[1] and

---

[1]NCBI accession numbers refer to the gi. number of the protein sequences at the National Centers for Biotechonlogy Information, NIH, USA. These sequences are compiled from different databases (e.g., GenBank, EMBL, PIR, etc.), and the complete list can be obtained from the NCBI-NIH URL: http://www.nbci.nlm.nih.gov.

functions of most cyclins and CDKs known to date. The phylogenetic trees of these and other cell cycle regulators are also shown in Fig. 3.

1. Cyclins

As mentioned before, the cyclins are a group of proteins well conserved through evolution and among different species (see Fig. 3). Cyclins A to I are structurally related to each other by 11.5–60.6% similarity. This is mainly apparent at the so-called **cyclin box,** a 100–150-amino-acid region that is the putative site for interaction with CDKs. Cyclins also contain a **destruction box** (RxxLxxxxN) in the N'-terminal region, or a consensus sequence **PEST** at the C'-terminus. The former is important for timing-triggered degradation, for example, in cyclins A, B, F, and I. The latter is rich in proline, glutamic acid, aspartic acid, serine, and threonine residues, which are necessary for the rapid turnover of these proteins, for example, in cyclins C, D, E, F, G_2, and I.

Because cyclin oscillations are critical throughout the cycle, they must be finely regulated by various mechanisms,

including (1) the transcriptional control of mRNA production, (2) specific proteolysis mediated by special ubiquitin ligases that turn on and off at particular times in the cell cycle (see Section III.E), and (3) external events such as a diminution of nutrients (in yeasts) or growth factors (in mammalian cells).

The functions of these molecules have already been described in different phases of the cell cycle, and they are summarized in Table 2. Some others are described in relation to CDKs and CDK inhibitors. However, the recent description of four new types of cyclins (i.e., F, G, I, and T) deserves a brief comment. **Cyclin F** is a nuclear protein ubiquitously expressed in human tissues. Its expression dramatically oscillates during the cell cycle, peaking in G_2 phase and decreasing prior to cyclin B destruction in mitosis. Cyclin F controls cell cycle transitions through pathways that are different from those reported for other cyclins (see Table 2).

Another group of human cyclins is composed of **cyclins G1 and G2,** whose expression differs in various tissues. Cyclin G1 levels can be found constitutively expressed ei-

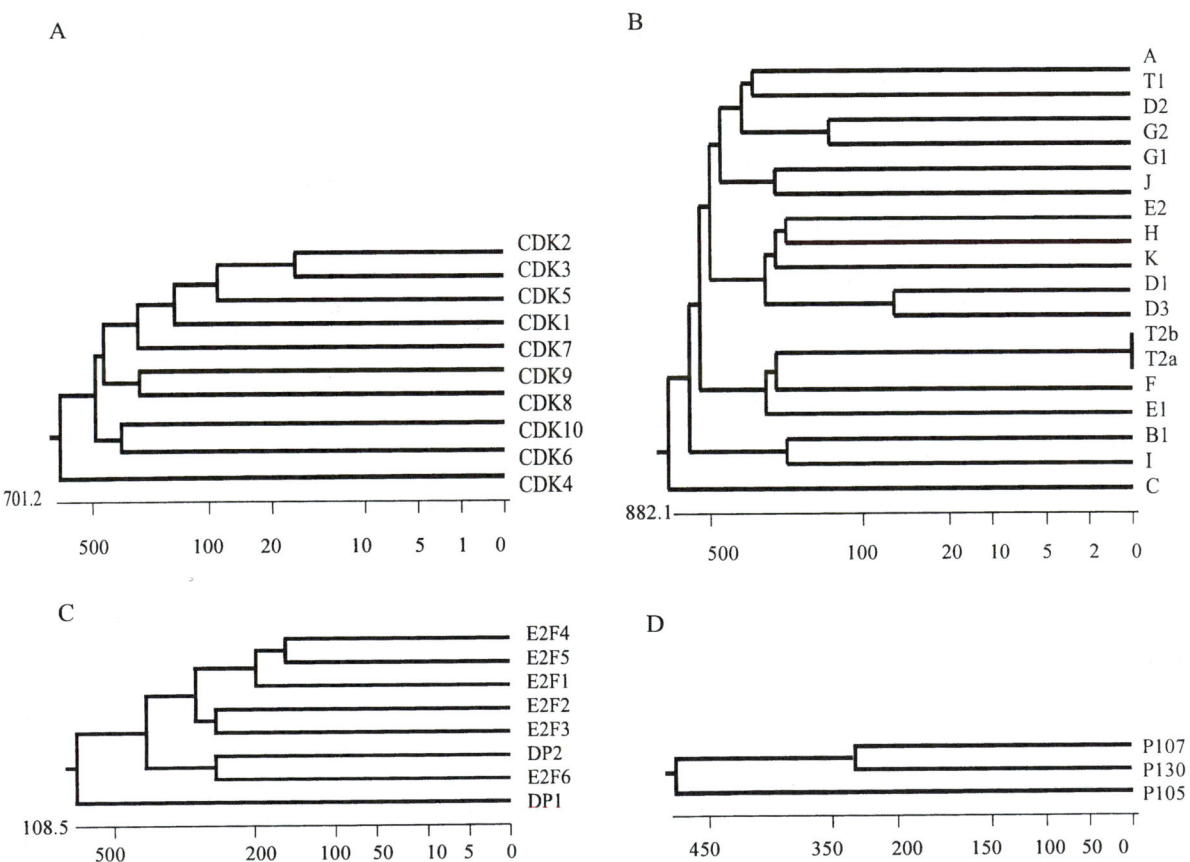

FIGURE 3. Phylogenetic trees of human cell cycle regulators. The dendrograms illustrate the relative similarity between the members of four different groups of human cell cycle regulators. The scale beneath the tree measures the distance between sequences and the units indicate the number of substitution events, that is, horizontal branch lengths are inversely proportional to the degree of similarity between the sequences. The multiple sequence alignment of these proteins was created by the Megalign program from the DNASTAR package, using the Clustal method with PAM250 residue weight table (licensed to GGA, INNSZ, Mexico). Except for E2F-DP proteins, the sequence accession numbers used for this analysis are shown in Tables 2 and 3. (A) Cyclin-dependent kinases, CDKs., (B) Cyclins., (C) E2F and DP transcription factors. E2F1, 181918; E2F2, 1082847; E2F3, 738758; E2F4, 1095443; E2F5, 1095444; E2F6, 4503439; DP1, 424317; DP2, 604479. (D) Pocket proteins.

ther throughout the cell cycle or exclusively elevated during G_1 phase, depending on the type of cell under study. In contrast, cyclin G2 levels oscillate during the cell cycle, with peak expression in late S phase. Cyclin G protein contains a typical cyclin box at the N'-terminus but no apparent destruction box or PEST sequence like those conserved in other cyclin family proteins. The function of G cyclins is unknown, but they seem to be transcriptionally activated by p53 in the case of DNA damage, arresting the cell cycle at the G_2-M transition (Shimizu et al., 1998).

Finally, **cyclin I** mRNA is primarily expressed in human postmitotic tissues. Because cyclin I is constantly expressed during the cell cycle, it is possible that its functions are independent of cell cycle control (see Table 2). Similarly, **cyclin T** (i.e., T1 and T2) is a new family of regulators that are ubiquitously expressed. Cyclin T1 has been identified as a regulatory subunit of CDK9 and as a component of the transcription elongation factor P-TEFb. Cyclin T1–CDK9 complexes phosphorylate RNA polymerase II (RNAP II) in vitro and strongly enhance the affinity and specificity of the tat:TAR RNA interaction in HIV-1 transcription (Garriga et al., 1998; Wimmer et al., 1999). Therefore, it has been considered that cyclin T1–CDK9 complexes are required HIV-1 host cellular cofactors generated during T-cell activation.

2. Cyclin-Dependent Kinases

Proteins with putative CDK activity have been identified and isolated through their ability to complement mutant strains of yeast or insects defective for a particular CDK. These reconstitution assays test the capacity of a particular gene to replace the activity of the defective mutant. For example, CDK1, CDK2, and CDK3 genes have been found to complement and rescue Cdc28 mutants of *Saccharomyces cerevisiae*.

All members of this group belong to a family of serine/threonine protein kinases that can phosphorylate histone H1 or other substrates in vitro (Lania et al., 1999). CDK1 to CDK8 are related to one another by 20.3–75.5% similarity and the phylogenetic tree of these proteins is shown in Fig. 3. Interestingly, CDK1, CDK2, and CDK3 share a **PSTAIRE motif** at their N'-terminal region, while the rest of the CDKs differ in this motif, that is, CDK4, PISTVRE; CDK5, PSSALRE; CDK6, PLSTIRE; CDK7, NRTALRE; CDK8, SMSACRE; and CDK9, PITALRE. In CDK2, for example, the PSTAIRE motif has been found in a helix known to be important for recognition of cyclin A. In the inactive state, the helix is displaced in CDK2 so the residues for binding ATP are wrongly disposed. Once cyclin A binds to the catalytic cleft of CDK2, it induces conformational changes in its activation segment and PSTAIRE helix. These changes, in turn, activate the kinase function of CDK2 (Brown et al., 1999). The significance of these motifs in other CDKs is under study.

Just as in the case of cyclins, the activity of CDKs is also tightly controlled by various mechanisms: (1) cyclin abundance, or synthesis and proteolysis rates; (2) levels of cyclin-CDK complexes formed, that is, activation; (3) presence of CDK inhibitors; (4) degree of proteolysis of various cell cycle regulators (see Section III.E); and (5) CDK subunit phosphorylation state. The latter might be, for example, increased (activatory) or decreased (inhibitory) depending on the balance of CAK and Wee1 activities, respectively (see Fig. 1). The intricate network has been recently reviewed by Morgan and colleagues (1998), and the following paragraphs describe some findings on this subject.

a. CDK1 (Cd*c*2). Although the mechanisms of activation of Cdc2 are now well understood and its in vitro substrates recognized (e.g., histone H1), the identification of its in vivo substrates has remained elusive. Certainly, CDK1 plays an important role by triggering, directly or indirectly, various processes at the onset of mitosis. For example, it is known that CDK1 can activate/phosphorylate the C'-terminus of RNA pol II. Moreover, studies in HeLa cells have shown that cyclin B–CDK1 complexes are also able to activate/ phosphorylate a protein, Eg5, required for the efficient separation of centrosomes and the assembly of the bipolar mitotic spindles. Of relevance, this process occurs in a cell cycle–dependent manner, with CDK1 phosphorylating a Ser residue during S phase and another critical one, Thr 297, during M phase (Blangy et al., 1995).

b. CDK2. As mentioned in Section II.A, the activity of cyclin A–CDK2 complexes peaks at late G_1 and S phases in both normal and malignant cells. These complexes promote the passage through S phase. For example, in normal fibroblasts, the binary complexes are really quaternary ones, composed of cyclin A–CDK2 bound to **proliferating cell nuclear antigen** (PCNA), and the CDK inhibitor, p21 (see Section III.B.2.a). In malignant cells, as the tumor suppressor gene p53 is lost, so goes the level of p21 protein found in these complexes. This could explain why, in these type of cells, a substantial fraction of cyclin A–CDK2 complexes associate to other proteins of 9, 19, and 45 kDa (i.e., $p9^{Cks1/Cks2}$, $p19^{Skp1}$, $p45^{Skp2}$). Binding of p19 to cyclin A–CDK2 complexes requires p45, and these proteins are essential for kinase activity (see Section III.E.2.a). Moreover, the p45 protein bound to cyclin A–CDK2 complexes is required for execution of DNA replication in both normal and malignant cells (Yu, Z. K. et al., 1998). Because p45 levels have been found greatly increased in various tumor cells, this could be another important player in the tumorigenic process.

c. CDK3. CDK3 has been proposed as a regulator of G_1 exit. Little is known about the pathway(s) by which this CDK might mediate this effect. One hypothesis is that CDK3, through its binding to the DP subunit of the transcription factors E2F1 and E2F2, might be licensing events necessary for S-phase entry. However, the mechanism by which the activation of E2F factors takes place and its exact timing, whether this occurs prior to or after the release of E2F from the pRb, remain to be elucidated.

d. CDK4 *and* CDK6. CDK4 and CDK6 both bind selectively to D-type cyclins during G_1 phase (see Table 2). As mentioned in Section II.A, the cyclin D–CDK4/CDK6 complexes are the main regulators that phosphorylate pRB prior to G_1-S phase. Both CDKs are also specifically down-regulated by Ink proteins and, nonselectively, by Cip/Kip inhibitors. In addition, CDK4 phosphorylation by CAK up-regulates its functions. CAK is composed of three subunits: CDK7, MAT 1, and cyclin H (see Section III.A.2.f).

e. CDK5. CDK5 stands apart from the rest of the known CDKs. Apparently, CDK5 does not play an important role in the control of the cell cycle in terminally differentiated cells. Its levels are low in most adult tissues, except in the brain where it is associated with a **noncyclin activator, p35.** Intriguingly, the CDK5/p35 kinase has been found to be involved in neuronal differentiation during embryogenesis and in the control of cytoskeletal functions in postmitotic neurons (Nikolic *et al.*, 1996). How p35 activates CDK5 is unknown. However, studies in embryo brain extracts and Mv1Lu cells have demonstrated that p35 can inhibit the recognition of CDK5 by the p27 inhibitor (Lee *et al.*, 1996). This protective function could partially explain the positive regulatory activity that p35 exerts on CDK5.

f. CDK7. Formerly known as MO15, CDK7 is the kinase subunit of the CAK complex. The other subunits of this complex are cyclin H and MAT-1 (*ménage à trois*). The name MAT-1 was given to this protein because it actually acts as the assembly factor between CDK7 and cyclin H. However, the activation of the CAK complex is not complete without the phosphorylation of CDK7 in the T loop (Thr 170). Once CAK becomes active, it phosphorylates CDK1, CDK2, and CDK4 at their Thr residues in the T loop (i.e., Thr 160/161), up-regulating the functions of these molecules. Thus, a single CAK activity can control major cyclin kinase activities throughout the cell cycle.

g. CDK9. Previously named PITALRE, this Cdc2-related kinase protein is a serine-threonine kinase involved in many physiological processes. CDK9 is not cell cycle–regulated and acts preferentially in differentiation. Its cyclin partners, cyclins of the T family, recently have been isolated. As mentioned in Section III.A.1., CDK9 has been shown to associate with the HIV-Tat protein, suggesting a possible involvement in AIDS (de Falco and Giordano, 1998).

3. Crystallographic Studies

Standing as landmarks in our understanding of cyclin-CDK interactions are the pioneering determinations of the crystal structures of cyclin A, CDK2, and the partially activated human cyclin A–CDK2 complex (Brown *et al.*, 1999). These studies have revealed how cyclin A binding induces the conformational changes in CDK2 that allow the formation of a catalytic center. In brief, cyclin A binding to CDK2 produces structural arrangements at the activation segment and at the PSTAIRE helix of the kinase subunit. At the same time, these changes expose and allow the phosphorylation of a Thr 160 residue, which is the homologous T-loop phosphorylatable Thr 161 in CDK1. As a result, the phosphorylated cyclin A–CDK2 complex has a seventeen-fold increase

in activity in comparison to the nonphosphorylated form. Surprisingly, as the complexes become active, they also become susceptible to inhibition by Cip/Kip inhibitors. This mechanism, therefore, could act as the negative feedback of cyclin A–CDK2 activation. It is impossible to predict if all CDKs are activated as CDK2 is, but this indeed could be the case.

Recent crystal structures reported include (1) the proliferation cell nuclear antigen, PCNA, compelled with a 22-residue peptide derived from the C-terminus of the protein p21; (2) the 310 helix region of p57 in CDK2 inhibition; (3) yeast cyclin G; (4) the CDK5 activation domain; (5) cyclin H; (6) p27 bound to cyclin A–CDK2; and (7) p21 bound to CDK2.

B. Cyclin-CDK Inhibitors

Recent work has identified two families of proteins that act as negative regulators of CDKs: **inhibitors of CDK4/6** (Ink4) and **CDK-inhibitor proteins** (Cip/Kip). The former specifically target the CDK4 and CDK6 kinases and bind to these proteins, preventing their interaction with cyclins D and/or inhibiting the kinase activity of formed complexes. In contrast Kip, binds preferentially to cyclin-CDK complexes and either prevents their activation by CDK activating kinase (CAK) or inhibits their kinase activity. In addition, and unlike Inks, Kips are promiscuous and interact with most G_1-CDKs (see Fig. 2). For example, they can effectively target CDK2 in complexes with cyclins A and E, as well as CDK4 and CDK6 in complexes with cyclins D1, D2, and D3. Therefore, Cip/Kip family members have been referred to as universal CDK inhibitors (rev. in Sherr and Roberts, 1999). A summary of the names, NCBI accession numbers,[1] and functions of these inhibitors is given in Table 3.

1. Ink4 Family

p16[Ink4a] (p16) was the first member described of the Ink family. This inhibitor was initially observed as a CDK4-associated protein in human cells, and subsequently found to be a specific cyclin D–CDK4/6 inhibitor. Since the original description of p16 in 1993, three other members of this family, **p15**[Ink4b] (p15), **p18**[Ink4c] (p18), and **p19**[Ink4d] (p19), have been isolated (Sherr and Roberts, 1999). They all share structural and biological properties, although each one has a different transcriptional control. For example, by using two different promoters, the same locus, Ink4a, gives rise to two different transcripts: p16[Ink4a] and p19[ARF] (Quelle *et al.*, 1995). Both transcripts have common exons (E2 and E3) but differ at the 5′ exon: E1α for p16[Ink4a] and E1β for p19[ARF] (note that the latter is not the same as p19[Ink4d]). Interestingly, the locus that encodes p15, Ink4b, is in proximity to the Ink4a locus of the chromosome 9p21.

A characteristic of this family is that all its members specifically bind CDK4 and CDK6. The only exception is p18, which binds weakly to CDK4. It is also known that p15, p16, and p18 have ankyrin repeats (Ank repeats), a protein interaction motif, that could mediate the action of these inhibitors. It is thought that Ink proteins arrest the cell cycle through these interactions.

Another characteristic of Ink proteins is that they are inducible. For example, p15 is induced by TGFβ, while E2F proteins and terminal differentiation induce p16. However, the ectopic overexpression of either one of these proteins arrests the cell cycle at G_1. Intriguingly, p19^{ARF} overexpression arrests the cycle both in G_1 and G_2 phases, but apparently through mechanisms not involving direct inhibition of known cyclin-CDK complexes.

Another important aspect of this family is that p16 and p18 rely on the presence of functional pRB in order to arrest the cycle (this fact is unknown for p19). Since the ability of Ink proteins to arrest G_1 progression is restricted to pRB-positive cells, this implies that loss of these proteins, like loss of pRB, could predispose toward tumor development.

In fact, this hypothesis has proved to be right in the case of p16, which is also known as **multiple tumor suppressor gene 1** (MTS 1). First, many human cancers have been shown to harbor deletions and mutations at the Ink4a locus (Table 4). These abnormalities encode for functional aberrant p16 proteins that apparently allow unrestricted cell proliferation. On the other hand, Ink4a -/- (null) mice have been found to develop tumors at an early age (Table 5), and this fact clearly corroborates the role of p16 as a tumor suppressor gene (Serrano *et al.*, 1996). Since the p15 locus is found in physical proximity to the p16 locus, it has also been considered a tumor suppressor gene and it was named **multiple tumor suppressor gene 2** (MTS 2). However, its role in tumorigenesis has not been established as in the case of p16 and p19^{ARF} mutations, which coexist in up to 50% of human tumors.

2. Cip/Kip Family

The three members of this family are **p21**, **p27**, and **p57** (see Table 3). As a general characteristic, these proteins can inhibit all the cyclin-CDK complexes essential for G_1 progression and S-phase entry. Indeed, when either one of these inhibitors is ectopically overexpressed in cells *in vitro*, these arrest in G_1.

It has been recently found that the sequence WNFD-FXXXXPLEGXXXWXXV is a common amino acid motif at the N'-terminal region of p21 and p27. This sequence seems to be the active CDK-binding/inhibitory domain of both proteins (Nakanishi *et al.*, 1995). A more detailed analysis of this region in p21 has revealed that (1) the amino acids 42 to 71 are important for growth inhibition, (2) the minimum amino acid sequence required for inhibition of DNA synthesis spans amino acids 22 to 71, and (3) amino acids 49 to 71 are involved in the actual binding/inhibition of CDK2 activity. Of note, this sequence similarity in members of the Cip/Kip family is not present in p16 nor in p15, so there must be at least two distinct mechanisms for CDK inhibition.

In contrast, Cip/Kip inhibitors are more divergent at their C-terminal region. Nonetheless, both p21 and p27 share a PCNA-binding domain by which they could inhibit its replicative activity. The function of the C-terminal region of p57 remains to be elucidated.

a. The p21 and the p53 tumor suppresor gene. In response to DNA damage, eukaryotic cells use a system of checkpoint control to arrest the cell cycle progression. These checkpoints give time for DNA repair prior to its replication in S phase (G_1-S) and prior to segregation of chromatids in M phase (G_2-M). In the first case, **p53** induces G_1 arrest through the transcriptional induction of p21 (G_1-S checkpoint) (Fig. 4). In contrast, the G_2-M arrest has been largely considered a p53-independent process, controlled by the inhibitory phosphorylation of CDK1. Recent studies, however, have found that p53 also plays a critical role at the G_2-M transition by (1) decreasing the intracellular levels of cyclin B1 protein and therefore inhibiting the CDK1 kinase activity for mitotic initiation (Innocente *et al.*, 1999), (2) inducing the expression of 14.3.3σ, a p53-regulated inhibitor of G_2-M progression (Hermeking *et al.*, 1997), and (3) providing a p53-dependent **spindle checkpoint** (Cross *et al.*, 1995). In p53-mutated bladder cancer cells, for example, the sole ectopic expression of wild-type p53 arrests the cells at the G_2-M transition and restores their radiosensitivity (Fig. 5). In summary, p53 and ATM proteins are guardians of the genome that mediate cell cycle perturbation in response to DNA damage and that play a role in cell death, genetic stability, and cancer susceptibility (see Section III.D.3).

In DNA-damaged cells, the expression and activation of p53 is induced by DNA breaks and phosphorylation. Double-strand breaks and single-strand gaps greater than 30 nucleotides, for example, are sufficient to induce the expression of p53 and arrest the cell cycle at G_1. This delay, in turn, allows the repair of the DNA damage or promotes the cells toward apoptosis (Schwartz and Retter, 1998). On the

TABLE 4 Cell Cycle Players in Tumorigenesis

Gene	Genetic changes	Human tumors
p16[a]	Homozygous deletions	NSC lung cancer,[b] ALL,[c] renal, bladder, prostate, head and neck cancers
	Point mutation + small deletions	Pancreas, esophagus, biliary tract, f. melanoma[d]
	Methylation	Breast, colon
D1	Rearrangement	Parathyroid adenoma
	Amplification/ overexpression	B cell lymphoma, breast, colorectal, squamous cell (SC) cancers
pRB[a]	Germline mutations	Hereditary retinoblastoma
	Mutation/ deletion of both alleles	Sporadic retinoblastoma osteosarcoma, SC lung and breast cancers
p53[a]	Mutation/deletion	Wide range of tumors

[a] Tumor suppressor genes
[b] Non-small cell
[c] Acute lymphocytic leukemia
[d] Familial melanoma

TABLE 5 Transgenic Mice

Gene	Phenotype	Reference(s)
p53 −/−	Viable and normal development Females with exencephaly (some) T cell lymphomas and sarcomas Mammary tumors	(Donehower et al., 1995)
E2F1−/−	Viable and fertile Testicular atrophy Exocrine gland dysplasia Thymic and lymph node enlargement Reproductive tract sarcomas, lung adenocarcinomas and lymphomas	(Field et al., 1996; Yamasaki et al., 1996)
pRB −/−	Lethal phenotype (<16th embryonic day) Neuronal cell death, abnormal erythropoiesis	(Lee et al., 1992)
+/−	Viable and fertile Pituitary and thyroid tumors Bronchial epithelium hyperplasia	(Jacks et al., 1992)
D1 −/−	Viable and fertile/dwarfism Severe retinopathy Hypoplasia of mammary glands	(Fantl et al., 1995)
++/++	Selective overexpression in stratified epithelia Normal differentiation and function Epidermal hyperproliferative response Severe thymic aplasia	(Robles et al., 1996)
++/++	Selective overexpression in mammary gland leads to abnormal proliferation/adenocarcinoma	(Wang et al., 1994)
D1 E[a]	Rescues all phenotypic manifestations of cyclin D1 deficiency	(Geng et al., 1999)
p21 −/−	Viable and normal development Nontumorigenic	(Deng et al., 1995)
++/++	Selective overexpression in hepatocytes halts postnatal liver development and regeneration	(Wu et al., 1996)
p27 −/−	Normal prenatal development; viable Enlarged body size; female sterility Hyperplasia of thymus, adrenals and gonads Pituitary tumors and disorganization of retina	(rev. in Raff, 1996)
p16 −/−	Viable Extramedullar hematopoiesis Soft tissue sarcomas and lymphomas	(Serrano et al., 1996)

[a] Replacement of cyclin D1 with cyclin E.

other hand, DNA damage-induced p53-phosphorylation at Ser 15 is carried out by many kinases but particularly by the ATM–Rad 3 protein family. The latter not only phosphorylates p53 at Ser 15, but it also does it at Ser 37 *in vitro* (Tibbetts *et al.*, 1999). Once phosphorylated, p53 induces the transcription of many genes including gadd45, mdm2, mck, bax, IGF—BP3, BTG2, p21, cyclin G, and 14.3.3 proteins (Hermeking *et al.*, 1997).

Two other members of the mammalian p53 family, **p73** and **p51**, have been recently described too. Both proteins share elevated sequence homology with p53 and they also activate p53-responsive promoters and induce apoptosis. However, p73 is not induced by DNA damage and it is not targeted by oncoviral proteins such as SV40 and adenovirus E1B. The role of p73 and p51 in cell growth and carcinogenesis, therefore, has not been established (Kalein, 1999).

The **p21 gene** was first cloned as an inhibitor of DNA synthesis that was overexpressed in terminally nondividing senescent human fibroblasts (SDI1), and later as a p53-transactivated gene (WAF1) and a CDK-interacting protein (CIP1, p21) that inhibited cyclin-dependent kinase activity. Now it is known that p21 is the only member of this family induced by p53. Other inducers of the p21 inhibitor include various growth factors (e.g., TGFβ) and transcription factors (e.g., MyoD and C/EBPα).

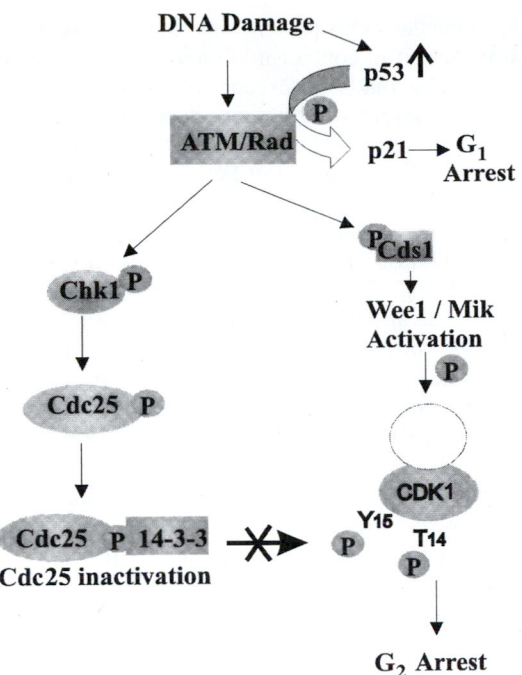

FIGURE 4. Checkpoints induced in DNA-damaged cells. This scheme summarizes the pathways by which genotoxic DNA damage induces arrest of the cell cycle at G_1 and G_2. The expression and subsequent phosphorylation of p53 by ATM, for example, transcriptionally activates p21 and induces G_1 arrest. In contrast, two independent pathways induce G_2 arrest through the phosphorylation/inactivation of CDK1 in cyclin B–CDK1 complexes. However, p53 also controls the expression of cyclin B and 14.3.3 proteins in the presence of DNA damage. See Sections III.B.2.a and III.D for further details.

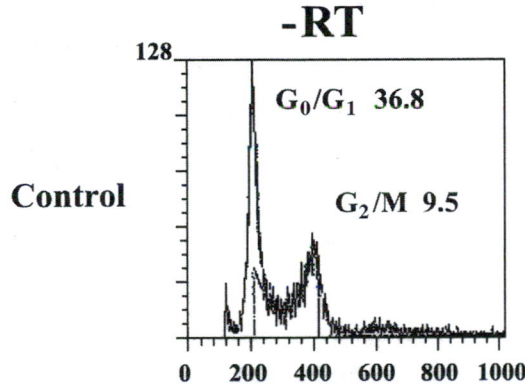

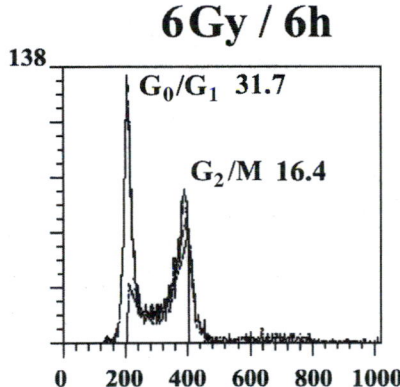

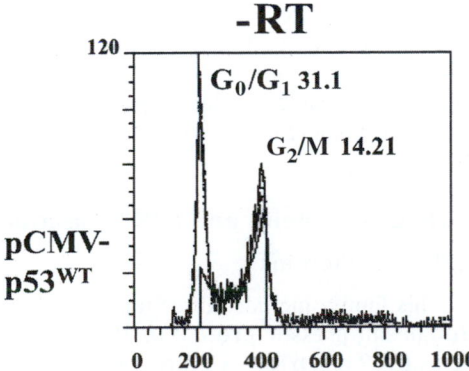

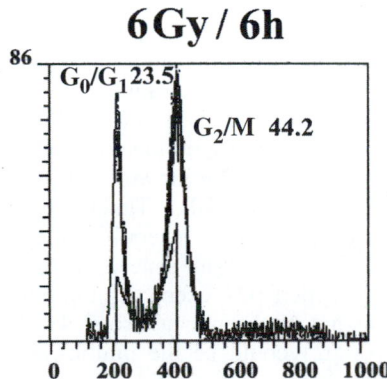

FIGURE 5. G_2-M arrest induced by wild-type p53. The p53-mutated transitional bladder cancer cell line, T24, was transfected with a pCMV-p53[wt] plasmid and the effects on DNA content were analyzed before and after irradiation *in vitro* (6Gy). The DNA histogram shows that p53 overexpression, especially when associated with radiation, induces G_2 arrest in these cells. Most transfected cells die within 24 hours after radiation, whereas nontransfected cells remain unaffected (data not shown). (The pCMV-p53[wt] plasmid was a kind gift of Dr. B. Vogelstein, JHU.)

p21 is considered a universal inhibitor, and cells might utilize it in various ways: (1) in proliferating cells in G_1 phase, it may contribute to cell cycle transitions timing or G_1 arrest; (2) in S phase, it may block proliferating cell nuclear antigen, PCNA, and slow down DNA synthesis to facilitate repair processes (Umar *et al.,* 1996); and (3) in development, it may contribute to cell cycle arrest in terminally differentiating cells. Finally, the ectopic expression of p21 in human diploid fibroblasts and mouse NIH3T3 has been shown to inhibit DNA synthesis and, in tumor cells, growth inhibition.

It has been shown that p21 contains specific cyclin-binding motifs that are important for its activity (Chen *et al.,* 1996). Immunoprecipitations with antibodies against different cyclins and CDKs have demonstrated that p21 preferentially associates with cyclin-CDK complexes *in vivo*. In normal fibroblasts, for example, it can be found associated with CDK2 bound to cyclins A and E, and with CDK4 or CDK6 bound to D-type cyclins. Furthermore, the majority of cyclin-CDK molecules in proliferating, nontransformed cells are assembled in heterodimeric complexes containing stoichiometric amounts of both PCNA and p21. In contrast, p21 usually disappears from these complexes in transformed cells, apparently due to the lack of p53-antioncogene function (see Section III.A.2b).

This is because p21 is a p53-inducible inhibitor (its promoter has two consensus sites for p53). Thus, if p53 function is lost in tumor cells, p21 expression is also stopped. On the contrary, low DNA precursor pools, radiation, hypoxia, and other DNA-damaging agents induce p53-mediated G_1 arrest through the transcriptional activation of p21(Linke *et al.,* 1996). According to the current cell cycle model, DNA damage in p53 +/+ cells induces p21 activation and inhibits cyclin-kinase activity. These changes, in turn, block pRB phosphorylation and the expression of S phase genes by keeping E2F bound to pRB. Thereafter, the cells follow one of two pathways: (1) repair DNA damage and continue in the cell cycle, or (2) die. This mechanism, then, is an essential regulator that coordinates cell cycle progression with transcriptional activity, DNA repair mechanisms, and cell death (e.g., apoptosis).

But p21 can also be induced by p53-independent pathways. For example, p21 is elevated in the postmitotic state in terminally differentiated cells, during embryogenesis, and in senescent cells. In these cases, p21 elevation is not necessarily p53 dependent. For example, p21 is elevated in developing mouse embryo muscle by the transcriptional activator of the myogenic MyoD, and during the growth and differentiation of mouse preadipocytes by C/EBPα (Timchenko *et al.,* 1996).

While p21 levels increase reversibly in quiescent cells, they irreversibly increase in senescent cells. The exact mechanism by which p21 exerts its inhibitory effects in senescence has not yet been established. However, one putative target of p21 could be the inhibition of the SAPK group of MAP kinases. Since SAPK phosphorylation of transcription factors is important in stress-activated signaling cascades, p21 could be the converging point for the regulation of cellular stress, the cell cycle, tumor suppression, and senescence. It has been suggested that this inhibition may reside in the N-terminal domain of p21 too.

b. p27 inhibitor. This protein was first identified as a CDK inhibitor present in contact-inhibited cells and in mink lung epithelial cells made quiescent by the antimitogenic cytokine transforming growth factor beta (TGFβ). In the latter case, the cells contained cyclin E–CDK2 complexes associated with a 27κDa protein. Further studies demonstrated that p27 cooperates with p15 to cause the G_1 arrest induced by TGFβ. In other models, p27 also mediates G_1 arrest induced by serum ablation and cAMP. In contrast, it can be down-regulated by growth-stimulatory signals, such as interleukin-2, IL-2, and adenovirus E1A proteins.

It is now known that p27 predominantly inhibits cyclin D–CDK4 kinase activity and also inactivates cyclin E/A–CDK2 complexes in G_1. Through the former action, p27 inhibits pRB phosphorylation/inactivation, the master control at the G_1-S transition. Thus, if p27 expression is down-regulated with antisense molecules, the fraction of cells in S phase increases and that in G_1 decreases. It has also been reported that p27 has the ability to induce a proteolytic activity that cleaves cyclin A, yielding a truncated cyclin A lacking the mitotic destruction box (Bastians *et al.,* 1998). Nevertheless, this truncated cyclin is still destroyed by the proteasome.

Recent evidence from three independent groups has confirmed the importance of p27 in cell growth control (rev. in Raff, 1996). In these studies, both copies of the p27 gene were inactivated in mice (p27 nullizygous mice –/–) by targeted gene disruption, and in all cases the animals grew more rapidly and bigger than the controls. Since these abnormalities could not be attributed to other factors (e.g., endocrinopathy), p27 inactivation was implicated as the main cause of these results. Thus, p27 probably limits cell proliferation in various tissues, by inhibiting G_1 CDK activities (Table 5).

c. p57 inhibitor. As previously mentioned, the p57 inhibitory domain is homologous to the one in the N-terminal region of p21. This inhibitor, as well as the others, induces G_1 arrest. Nonetheless, its contribution for the maintenance of the nonproliferative state in adult tissues is marginal. However, due to the identification of imprinting of the p57 gene on chromosome 11p15, it has been considered a putative tumor-suppressor gene in embryonal carcinomas and in a cancer-prone syndrome known as Beckwith-Weidemand (i.e., BWS). The importance of this finding resides in being the first cell cycle regulatory protein controlled by imprinting (Matsuoka *et al.,* 1996).

C. Pocket Proteins and E2F Transcription Factors

1. Pocket Proteins

This family includes three members: the **retinoblastoma tumor suppressor protein** (p105 RB) and the related proteins **p107** and **p130.** A fourth member has already been described in *Drosophila*, RBF, but its human homolog has not been identified yet (Du *et al.,* 1996). **Pocket proteins** share many structural and functional features. They display high sequence homology in two discontinuous regions (CR1 and CR2) that make up the **pocket domain** (i.e., E1A/T-binding domain). This region is particularly important in complex formation and modulation of the activity of several cellular and viral proteins. For example, all three proteins bind E2F-DP heterodimers at the pocket region and promote transcriptional repression. Apparently, complexes containing E2F, a pocket protein, and the histone deacetylase enzyme, HDAC1, regulate E2F activity in G_1 (Ferreira *et al.*, 1998).

As shown in Fig. 6, the pRB pocket also interacts with (1) other transcription factors such as MyoD, BRG$_1$, UBF; (2) nuclear RNA polymerases pol I, pol II, and pol III; (3) viral proteins such as HPV-16, simian virus large T antigen, and adenovirus E1A; and (4) chaperone molecules, e.g., hsp 75. Furthermore, pRB protein contains an HPV-E7 binding domain at the C-terminal that is able to inhibit the binding of E2F to pRB [with the sole union of E7.] Thus, viral proteins, whether bound to the pocket or to the E7 domain, can displace and modify the function of cellular proteins normally bound to the pocket. The consequence of these interactions, for example, is that free E2F-DP heterodimers become transcriptionally active and promote cell cycle progression.

The mechanism by which pocket proteins regulate the cell cycle progression and arrest the cells in G_1 phase has not been fully understood. First, depending upon the cell under study, these proteins exhibit different growth-suppressive properties and exert control at various phases of the cell cycle. Thus, despite acting in a similar way, they are not functionally redundant. As described later, pocket proteins also have preferential binding to certain transcription factors, for example, pRB binds to E2F1, E2F2, and E2F3, while p107 and p130 bind to E2F4 and E2F5.

The activation/inactivation of the RB family members is also puzzling. Both pocket protein phosphorylation and pocket protein E2F-complex formation are cell cycle dependent. As mentioned earlier, pRB is phosphorylated by cyclin D–CDK4/6 and cyclin E–CDK2 complexes in mid to late G_1 phase. Once that pRB is phosphorylated/inactivated, it dissociates from E2F–DP heterodimers and allows cell cycle

progression. Similarly, p107 and p130 stably interact *in vivo* with cyclin E/A–CDK2 complexes that appear in late G_1 phase. These complexes seem to be responsible for the existence of various phosphorylated forms of p107 and p130 throughout the different phases of the cell cycle. However, there is evidence that their phosphorylation by cyclin E/A–CDK2 complexes does not necessarily lead to dissociation of pocket proteins from E2F-DP heterodimers (i.e., transcriptional activation). For example, cyclin E/A–CDK2 complexes phosphorylate p107-E2F4 complexes, but this phosphorylation does not dissociate p107 from E2F4. Instead, only cyclin D–kinase activity seems to be able to dissociate them, resulting in the relief of the p107 G_1 exit block (Xiao *et al.*, 1996). Thus, the timing of activation of the various complexes formed by pocket proteins and the activities of these complexes are numerous and intricate (Grana *et al.*, 1998).

However, it is important to point out the tumor suppressor activity of pRB. Mutations of pRB (e.g., deletions, duplications, point mutations) are frequently found in human tumors, and the inactivation of both copies of the gene appears to be the rate-limiting event in the genesis of retinoblastoma and osteosarcoma. Moreover, pRB nullizygous transgenic mice (pRB –/–) are not viable, whereas heterozygous adult mice (pRB +/–) are highly predisposed to pituitary and thyroid carcinomas (see Table 5). These findings have reinforced the essential role of pRB in growth control and development and, despite the fact that p107 and p130 exert related functions, mutations of these proteins have never been described in human tumors. For this reason, it is thought that p107 and p130 only serve in a limited fashion in the inhibition of tumor formation. An emerging concept is that the tumor suppressor functions of pRB not only correlate with the suppression of S phase, but also with the inhibition of RNA pol I, pol II, and pol III transcriptional activity. Since the transcription by RNA pol III is higher in S and G_2 phases, its activity must certainly be important in protein synthesis and cell growth. Therefore, if pol activities are

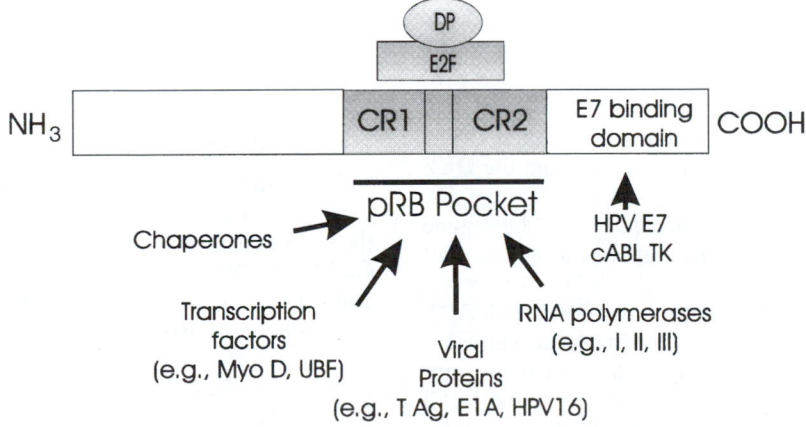

FIGURE 6. RB protein domains and regulators. The pRB protein is the master regulatory protein at the G_1-S transition. Its main function is to inhibit the activity of the E2F family of transcription factors. However, other important participants in the cell cycle, such as polymerases, transcription factors, and chaperones, also interact with it. In contrast, mutations and binding of viral proteins can inactivate the pRB gene and its protein, respectively.

also repressed by pRB, this could explain its critical role in arresting unrestrained proliferation and tumorigenesis.

2. E2F and DP (DRTF1) Proteins

E2F-DP heterodimers are a group of cell cycle–regulated transcription factors that coordinate the expression of a variety of viral and cellular genes (Johnson and Schneider-Broussard, 1998). While E2F-DP activity has been traditionally considered an essential step for the G_1-S transition, studies have also suggested a role for these heterodimers in abnormal growth proliferation and in the control of programmed cell death. This is because E2F-DP binding sites are present in the promoters of a wide number of genes that encode proteins necessary for the progression of these processes.

Structural analysis has shown that these heterodimers consist of two distinct proteins: E2F and DP (i.e., E2F–DP). The **E2F subunit** is encoded by a family of genes consisting of at least six members (i.e., E2F1–E2F6). The E2F1 to E2F5 proteins are related to one another by 20.9–61.4% similarity (see Fig. 3). Despite being highly conserved and structurally related proteins, E2F1, E2F2, and E2F3 prefer liaisons with pRB, while E2F4 and E2F5 associate with other pocket proteins, p107 and p130 (Weinberg, 1996; Grana *et al.*, 1998). Another important difference is that E2F1, E2F2, and E2F3 present a cyclin A–binding domain nonexistent in other E2F proteins (see later discussion). Moreover, a novel E2F6 has been recently reported (Cartwright *et al.*, 1998). This member has homology to E2F1–5 proteins within the DNA-binding, heterodimerization, and marked box domains but overall is more homologous to DP than to E2F proteins (see Fig. 3). This fact could explain why (1) E2F6 expression inhibits transcription from promoters possessing E2F recognition sites rather than activating transcription and (2) E2F6 delays the exit from S phase rather than inducing S phase as the other E2F proteins do.

The formation of diverse E2F-DP–pocket protein complexes occurs at different times during the cell cycle. So far, two human **DP proteins** have been identified as heterodimeric partners of E2F in these complexes: DP1 and DP2. DP1 protein (410 aa) results from the ubiquitous expression of a single message. Instead, the DP2 gene encodes at least five DP2 RNAs from which three DP2-related proteins are expressed *in vivo* (55, 48, and 43 kDa) (Rogers *et al.*, 1996). DP proteins are conserved proteins in their DNA-binding domains, which, in the C-termini, contain the **DEF box**. This domain is essential for the dimerization of DP with E2F proteins. It is worth mentioning that a DP3 gene has been cloned from mouse, but its human homolog has not yet been described.

Despite both types of phosphoproteins, DP1 and DP2, being unable to bind pRB directly (they do not have an E2F-like pRB-binding domain), either one can first form transcriptionally active complexes with E2F1, E2F2, and E2F3 through the DEF box, and then associate to pRB (see Fig. 6).

Late in G_1 phase, cyclin D–CDK4/6 complexes phosphorylate the retinoblastoma protein and these, in turn, release E2F-DP heterodimers from their pockets. Thus, E2F-DP becomes transcriptionally active and binds to promoters in

diverse type of genes whose products control the cell cycle progression. The latter include (1) transcription genes: dihydrofolate reductase (DHFR), polymerase alpha, thymidine kinase (TK); (2) CDK activity: cyclins D, E, A, and CDK1; (3) nuclear oncogenes: *c-myc, N-myc, B-myb;* (4) pocket proteins; and (5) the E2F1 gene itself. Of note, in *Drosophila melanogaster* embryos, cyclin E acts as an upstream activator of E2F, and E2F-directed transcription cannot proceed in cyclin E–mutated strains (Duronio *et al.*, 1996).

Later in the cycle, during S, the function of DP1 is partially controlled by phosphorylation. Thus, it has been shown that cyclin A–CDK2 complexes initially bind to a cyclin A–binding domain in the amino-terminal region of E2F (see Fig. 2). Subsequently, the kinase activity of this complex phosphorylates DP1 and inhibits E2F-DP activity. Interestingly, when these DP1 heterodimers are phosphorylated *in vitro* by cyclin A–CDK2 complexes (but not cyclin E–CDK2), E2F1 DNA-binding and transactivation activities are suppressed. This could explain why some promoters that are transcriptionally activated by E2F during G_1 are found inactive during S phase.

However, when E2F1 mutants defective in cyclin A–CDK2 binding are expressed in fibroblasts, these cells undergo S phase delay/arrest followed by regrowth or apoptosis depending on the transactivation activity of DNA-bound E2F (Krek *et al.*, 1995). The connection between a continuous E2F activity and apoptosis has also been shown in *Drosophila* imaginal discs, where the overexpression of E2F activates programmed cell death (Asano *et al.*, 1996). Since the current theory considers E2F an up-regulator of cell proliferation, these results were completely unexpected.

The findings in **E2F1 null mice** (E2F1 –/–) have been puzzling too. These animals not only were viable but, contrary to what was expected, developed tumors. For this reason, the E2F gene has been proposed as the first of a class that could exert oncogene and antioncogene functions (Weinberg, 1996). Many theories have been developed in trying to understand those confusing results.

Altogether, these facts make obvious that normal cell cycle progression requires a fine balance between cyclin-CDK activity, transcriptional control, and DNA replication.

D. Other Cell Cycle Regulators

1. Cdc25 Proteins

Three distinct Cdc25 phosphatases (i.e., A, B, and C) are expressed in mammalian cells. Only Cdc25A plays a role at the G_1-S transition; Cdc25B and Cdc25C act primarily at the G_2-M transition (Nishijima *et al.*, 1997). However, the exact contribution of isoforms B and C in the progression of the cell cycle in mitosis remains to be established.

2. Negative Regulators of the Cyclin B-CDK Complex: Chk1, Cds1, and 14.3.3

Three new families of cyclin B-CDK complex negative regulators, **Cds1** (Chk2/Rad53), **Chk** (Chk1 [p56]), and the **14.3.3** proteins (Rad24/Rad25), have been recently described in mammals (see Fig. 4). Both Cds1 and Chk are kinases that phosphorylate and inactivate Cdc25 proteins in response to genotoxic DNA damage (Blasina *et al.,* 1999; Furnari *et al.,* 1999). These kinases, therefore, have been considered to be involved in the S-M replication checkpoint that prevents mitosis when DNA is incompletely replicated.

Apparently, Chk1 is phosphorylated and activated by kinases of the ATM/Rad3 family (see Fig. 4). Thereafter, Chk1 phosphorylates Cdc25 dual phosphatases at specific serine residues (Ser 216 and Ser 287 in humans and *Xenopus laevis* oocytes, respectively). These phosphorylated residues, in turn, act as binding sites for 14.3.3 proteins. The latter are a family of proteins that, along with Chk1, inhibit the Cdc25 phosphatase activity and arrest the cell cycle at G_2-M phase (Hermeking *et al.,* 1997).

Binding by 14.3.3 inactivates Cdc25 and maintains the checkpoint-induced G_2 arrest. Interestingly, 14.3.3 proteins do not modify the activity of Cdc25 assayed *in vitro,* but rather alter the subcellular localization of Cdc25. At least in yeasts and *Xenopus,* it has been shown that the activation of the DNA-damage checkpoint causes the net nuclear export of Cdc25 by a mechanism that depends on Chk1, 14.3.3 proteins, and the nuclear export machinery (López-Girona *et al.,* 1999; Yang *et al.,* 1999). This function enables 14.3.3 proteins to inhibit the access of Cdc25 to the cyclin B–CDK1 substrates and, subsequently, induces G_2 arrest.

Cds1 is also phosphorylated/activated by kinases of the ATM/Rad family (see Fig. 4). Once phosphorylated, Cds1 induces the Cdc25-dependent dephosphorylation of CDK1 in cyclin B–CDK1 complexes by increasing the activity of Wee1 and Mik1 kinases. Therefore, Cds1 and Chk1 seem to act in different checkpoint responses that phosphorylate/inactivate the cyclin B–CDK1 complex.

3. ATM

Ataxia telangiectasia (AT) is a rare human autosomal recessive disorder with pleiotropic phenotypes including neuronal degeneration, immune dysfunction, growth retardation, premature aging, and increased cancer risk. The responsible mutated gene, **ATM,** encodes a protein containing, at its C-terminus, a phosphatidylinositol 3-kinase catalytic domain, and at its central part, a RAD3-homologous domain. It also contains a proline-rich region and a leucine zipper, both of which are implicated in protein signal transduction. Proline-rich regions, for example, have been shown to bind C-Abl.

ATM protein is involved in cell cycle control and DNA repair and recombination in response to DNA damage. On the one hand, ATM is required to optimally induce the p53-dependent G_1 cell cycle checkpoint (see Fig. 4). In this case, the interaction between ATM protein and p53 involves two regions in ATM, one at the amino-terminus and the other one at the carboxy-terminus corresponding to the PI3-kinase domain. In the presence of DNA damage, ATM binds and phosphorylates p53 on Ser 15 and Ser 37, contributing to the activation and stabilization of p53 (Khanna *et al.,* 1998). Thus, it is not surprising that the G_1 checkpoint is defective in AT cells. The role of ATM in cell cycle control at G_2-M has already been discussed in relation to two human kinases, Cds1 and Chk1 (see Section III.D.2).

E. Protein Ubiquitination and Destruction

1. Structure and Function of the System

The **ubiquitin-mediated proteolytic system** plays a central role between the different phases of the cell cycle. This system targets many naturally short-lived proteins including cell cycle proteins (e.g., cyclins), transcription factors, cell growth modulators, and signal inducers (Table 6).

In this proteolytic system, ubiquitin is conjugated with target proteins through the sequential activation of three enzymes: (1) **ubiquitin-activating enzyme,** E1, uses ATP to form a thioester bond between itself and ubiquitin; this pathway is initiated by the ATP-dependent activation of ubiquitin's C'-terminus by E1; (2) E1 transfers ubiquitin to a **ubiquitin-conjugating enzyme,** UBC or E2-Cdc34, through a thioester transfer reaction; and finally (3) ubiquitin is transferred to the substrate directly by an E2 enzyme or by E2 acting together with an **ubiquitin-protein ligase,** E3 (Fig. 7). In the latter scenario, the E3 complex would only facilitate the transfer of ubiquitin from E2 to the ϵ-amino groups of lysine residues in substrates. For example, it has not been demonstrated that APC forms thioester bonds with ubiquitin, so it is unlikely that APC/C ubiquitinates its substrates through a thioester cascade.

In either case, once the protein has been ubiquitinated, it is prone to ATP-dependent proteolysis at the **proteasome.** The latter is a large protease (2000 kDa) also known as the 26S proteasome complex. This complex results from the interaction of a 20S proteasome with the proteasome activator 700 (i.e., PA700). It is through this association that the complex obtains ubiquitin dependency, although many other substrates of the ubiquitin system have been described.

TABLE 6 Regulatory Proteins Degraded by the Proteasome

Transcriptional regulators	Cell cycle proteins	Oncogenic products and tumor suppressors
NK-κB(p105)	Cyclins (mitotic cyclins: A & B; G_1 cyclins: D & E)	c-Jun
IκB	Cyclin-dependent kinase (CDK) inhibitors (p27, p21, etc.)	c-Fos
ATF2 (activating transcription factor 2)		c-Mos
ICER (inducible cAMP early repressor)		p53

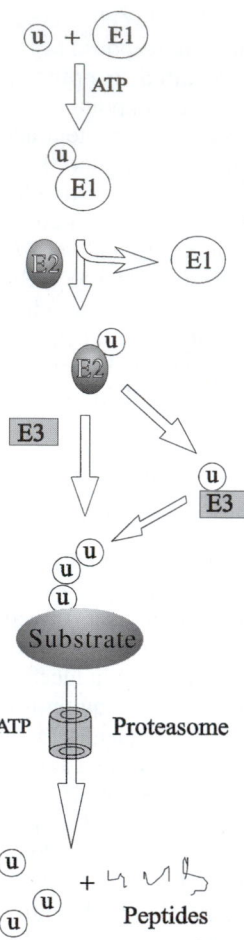

FIGURE 7. Ubiquitin-mediated proteolysis. The ubiquitination and subsequent proteolysis of a target protein is shown. E1–E3 holoenzyme complexes ubiquitinate proteins in an energy-dependent process. E3 holoenzymes (e.g., SCF and APC) can transfer the ubiquitin from E2 to the target protein or, in other cases, the ubiquitin is transferred first from E2 to E3 in a thioester cascade, and then to the target protein. Once ubiquitinated, the target substrates are degraded by the proteasome. See Section III.E for further details.

So far, 13 E2s have been described in *Saccharomyces cerevisiae* and some homologs from *Xenopus* (**UBCx**) and humans (**UBCH10**) have also been cloned. Two E3 holoenzymes, the **cyclosome/anaphase-promoting complex** (APC) and the **SCF complex,** have been partially characterized (Fig. 8). It is now accepted that APC, SCF, and the von Hippel-Lindau complexes belong to the same category of SCF-like complexes because they share homology in common constitutive proteins: (1) Skp family: Skp1, elongin; (2) Cullin family: Cul, Cdc53, Hrt1, Hrt 2; and (3) R-box proteins: Rbx1/Roc1, Apc 11, and so on (reviewed in Koepp *et al.,* 1999; Peters, 1998; and Townsley and Ruderman, 1998).

a. SCF Complex. The SCF is a large collection of modular E3s that are responsible for the ubiquitination of various proteins (see Fig. 8). This complex targets the proteolysis of molecules involved at the G_1-S transition (i.e., cyclins D and E, Cip 27, E2F1) and at the mitotic entry after completion of

DNA replication (i.e., Wee1). The SCF was named for three of its core components: **Skp1** (p19^{Skp1}), **Cdc53/cullin,** and an **F-box-containing protein** (e.g., p45^{Skp2}). The fact that F-box-containing SCF components are unstable, suggests a mechanism of regulating SCF function through ubiquitination and proteolysis of F-box components (Zhou and Howley, 1998). Specific F-box proteins (Fbps) recruit different substrates to the SCF and many Fbps have been identified in mammals (Latres *et al.,* 1999).

Moreover, another protein named **Rbx1/Roc1** has also been found to be a component of the SCF complex. This protein, which contains a Ring H2 finger domain termed the R-box, was originally found as a protein that bound the von Hippel-Lindau (VHL) tumor suppressor complex (Kamura *et al.,* 1999). Since VHL also contains a Skp1 homolog (elongin 1) and a Cdc53 homolog (Cul2), it has been considered structurally analogous to SCF.

Data from numerous studies have revealed the way SCF complexes work. First, p19^{Skp1} and F-box proteins (e.g., p45^{Skp2}) interact through the F-box motif. Thereafter, Cdc53 proteins function as the bridge that joins Skp1–F-box proteins complexes with E2-Cdc34 (Seol *et al.,* 1999). The activation of SCF complexes, in addition, occurs when R-box proteins recruit E2-Cdc34 into the SCF complexes and induce E2-autopolyubiquitination (see Fig. 8). Therefore, the presence of R-box proteins in cullin complexes seems to play a very important role for the transfer of ubiquitin or ubiquitin-like proteins to the substrate. The latter requires phosphorylation by protein kinases in order to bind SCF complexes.

The timing of SCF-mediated ubiquitination is regulated by substrate phosphorylation in contrast to the cell cycle–regulated APC ubiquitination activity.

b. APC Complex. The anaphase-promoting complex/cyclosome (APC) is a ubiquitin-protein ligase composed of at least 10 subunits in *Xenopus laevis* and mammals (Grossberger *et al.,* 1999). The APC system controls sister chromatid separation (i.e., progression to anaphase) and exit from mitosis (i.e., telophase into G_1) (Yu *et al.,* 1998). Cyclin A, cyclin B, and the APC regulator Cdc55 are the main targets in mammals. Other targets, in yeasts for example, include anaphase inhibitors (e.g., Pds1 and Cut2) and molecules whose destruction is required for the disassembly of the mitotic spindle during cytokinesis (e.g., Ase1). In yeasts, for example, cyclin A is degraded just before metaphase and the anaphase inhibitors are degraded for the metaphase-anaphase transition; later on, completion of anaphase, inactivation of Cdk1, and exit from mitosis are achieved by degrading cyclin B1 and B3, in that order. At the end, Ase1 is degraded for disassembly of the mitotic spindle.

The mechanisms by which APC complexes regulate the timing of substrate destruction are being elucidated. Studies in yeasts and *Drosophila melanogaster* have shown that APC complexes are activated in a cell cycle–dependent sequence. During mitosis, they are first activated by *fizzy*/**Cdc20** proteins and, thereafter, by *fizzy*-related (*fzr*) proteins for exit into G_1 (Fig. 8). *Fizzy* and *fzr* proteins have seven C-terminal WD40 repeats that form families distinct

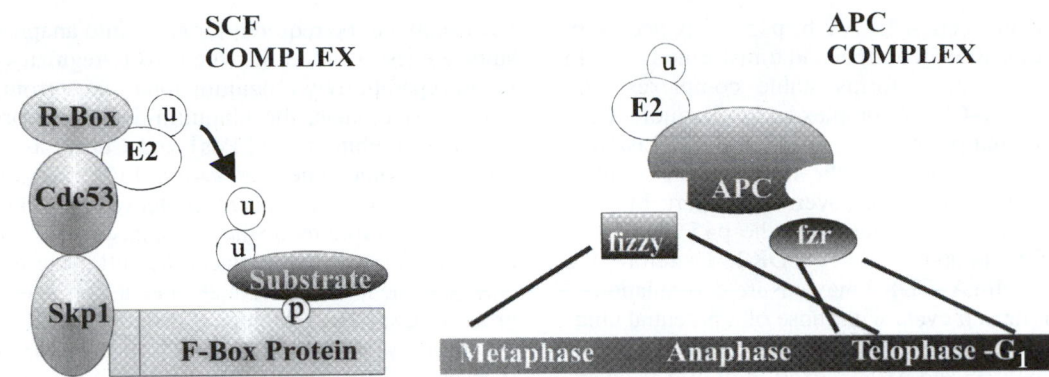

FIGURE 8. E3 complexes SCF and APC. E3 holoenzymes regulate the ubiquitination of target proteins during the cell cycle. SCF binds phosphorylated substrates and, similar to APC, helps in the transfer of ubiquitin from E2 to the substrates. SCF complexes act mainly at the G_1-S transition, whereas APC complexes act at mitosis. However, the proper timing of proteolysis by APC requires the sequential binding of *fizzy* and *fizzy*-related (*fzr*) proteins. See Section III.E for details.

from other proteins that contain WD40 repeats. Homologs of *fizzy*/Cdc20 have been cloned from *Saccharomyces cerevisiae*, *Saccharomyces pombe* (*slp1+*), *Xenopus*, and humans (p55 Cdc). On the other hand, homologs of *fzr* are known for *Drosophila* and *S. cerevisiae* (Cdh1/Hct1). Apparently, the proteolytic activation by *fizzy* proteins is D-box dependent while *fzr* activity shows a much more relaxed specificity for the D-box. Proteins that have a conserved nine-residue motif destruction box (D-box) include cyclins, Pds 1 in budding yeast, and Cut2 in *S. pombe*.

The exact mechanism(s) by which *fizzy* and *fzr* exert the activation of APC complexes at specific phases of the cell cycle is under study. In yeasts, for example, *Fizzy*/Cdc20 mutants arrest at metaphase and fail to degrade cyclin A and anaphase inhibitors (Pds1, Cut2), suggesting that *fizzy* function is required for the destruction of substrates during the metaphase-anaphase transition. On the other hand, *fzr* function is directed mainly to the proteolysis of cyclins B1, B3, and Ase1 at the exit from mitosis. Therefore, it has been suggested that *fizzy* and *fzr* proteins activate APC/C complexes at mitosis and G_1, respectively.

Furthermore, *fizzy* and *fzr* proteins are finely regulated by yet another set of proteins. In the presence of spindle damage, for example, the expression of **Mad/Bub proteins** acts as a negative regulator of Cdc20 activity. Thus, Mad/Bub proteins down-regulate the activity of APC complexes and induce metaphase arrest. Homologs of these proteins have been found in yeasts (Mad2p), mice (Bub1), and humans (Bub3) (Taylor *et al.*, 1998). On the other hand, *fzr* (Cdh1) proteins are either activated by a Cdc14 phosphatase or inactivated by CDK-dependent phosphorylation. However, Cdc14 phosphatase also induces cyclin B destruction and mitotic exit by mechanisms independent of *fzr* dephosphorylation (Shou *et al.*, 1999).

2. Major Cycle Transitions and Proteolysis

Currently, three major cell cycle transitions are known to require the degradation of specific proteins by the ubiquitin-proteasome pathway: (1) entry into S phase, (2) separation of sister chromatids, and (3) exit from mitosis. Some examples have already been given but others are mentioned in the following paragraphs.

a. Proteolysis and Entry into S Phase. The SCF complex plays a predominant role in this part of the cycle. By promoting proteolysis, the SCF complex exerts a posttranslational control not only on G_1 cyclins (i.e., cyclin D1 and E), but also on p27, E2F1, and Wee1.

As already stated, cyclin abundance is often rate-limiting for CDK activity and, by controlling rates of cyclin synthesis and degradation, the cell is able to regulate cell cycle transitions. The human p19[Skp1]/ p45[Skp2]/Cul1 complex, for example, acts as a ubiquitin E3 ligase that regulates the G_1-S transition *in vivo*. This complex selectively targets **cyclin D** and **p21** for ubiquitin-dependent proteolysis, but leaves p27 protein levels unaffected (Yu, Z. K. *et al.*, 1998).

In contrast, the intracellular level of **p27** is rapidly reduced at the G_1-S transition phase by two other mechanisms (Shirane *et al.*, 1999). The first one involves a ubiquitin-dependent proteolysis that eliminates the cyclin-binding domain at its N-terminus and produces a 22-kDa intermediate. In this system the phosphorylation of Thr 187 is essential for p27 degradation and it is possible that p45[Skp2] specifically targets p27 for phosphorylation and degradation during the cell cycle progression. The p22 intermediate has virtually no CDK2 kinase inhibitory activity.

The second pathway by which p27 proteolysis takes place is a ubiquitin-independent mechanism. Experiments using caspase inhibitors strongly suggest that p23 results from the proteolysis of p27 by a caspase-3-like protease. This cleavage in p27 does not alter the binding properties of p23 to CDK2 and CDK4 kinases but produces cytosolic sequestration of this intermediate (Loubart *et al.*, 1999). Therefore, p23 is still able to inhibit cyclin/CDK2 activity and to induce G_1 arrest in various models.

SCF complexes also play a role in the proteolysis of **E2F** and **Wee1**. For example, E2F1 binds to p45[Skp2] and is likely to be ubiquitinated by this complex. On the other hand, the E2-dependent degradation of Wee1 in *Xenopus* is required for mitotic entry. The importance of these mechanisms remains to be elucidated.

As mentioned in Section III.A.2.b, p45^{Skp2} is necessary for the S phase transition in normal and transformed cells. In human cells, this protein forms stable complexes with p19^{Skp1} and cyclin A–CDK2 complexes. Furthermore, Cul1 and E2 are additional partners of p45^{Skp2} *in vivo* (Lisztwan *et al.*, 1998). The formation of the multiprotein complex p45^{Skp2}-Cul 1-p19^{Skp1} could be governed, in part, by periodic S phase-specific accumulation of the p45^{Skp2} subunit and by the p45^{Skp2}-bound cyclin A–CDK2. Therefore, the dependency on cyclin A–CDK2 may ensure coordination of the activities of the cell cycle with those of a potential ubiquitin conjugation pathway. Nevertheless, the exact mechanism by which p45^{Skp2} and p19^{Skp1} exert their effects on cyclin A–CDK2 complexes remains unclear. For example, Yam and colleagues (1999) found that Skp2 inhibits the kinase activity of cyclin A-CDK2 *in vitro*, but not that of CDK1 or CDK5. Furthermore, overexpression of Skp2 in mammalian cells did not promote the S-phase transition, but rather induced G$_1$-S arrest.

b. Proteolysis and Separation of Sister Chromatids. Sister chromatids in early mitotic cells are held together mainly by interactions between centromeres. The separation of sister chromatids at the transition between the metaphase and the anaphase stages of mitosis depends on the anaphase-promoting complex (APC) (see Fig. 8). Mutations in certain subunits of the APC block the ubiquitination and destruction of mitotic cyclins and also arrest cells in metaphase, suggesting that the APC is required for the segregation of sister chromatids during anaphase (Zachariae *et al.*, 1996). As mentioned in Section II.C, sister chromatids segregation (i.e., anaphase entry) is not simply a consequence of **cyclin B** destruction because nondegradable forms of cyclin B lacking the D-box block exit from mitosis, but anaphase occurs as normal. These and other data suggest that sister chromatid segregation depends on the destruction of proteins other than cyclin B.

The targets of APC in this case could be either from a family of **glue proteins** that link sister chromatids together or, alternatively, from a group of proteins that act as **anaphase inhibitors** for regulating glue proteins. A candidate for a physical sister chromatid glue protein degraded during mitosis has been recently identified in *S. cerevisiae*. This protein, Scc1/Mcd1, was identified in association with Smc1, a chromosomal protein involved in chromosome condensation. From various studies, it has been proposed that Scc1p-Mcd1p loss from chromatin is sufficient for the loss of cohesion that triggers sister chromatid segregation at anaphase (Townsley and Ruderman, 1998). Scc1p/Mcd1p is rarely detectable in early G$_1$ and accumulates in phase S. It is possible, therefore, that Scc1p/Mcd1p is degraded at S-G$_2$ and at late M-G$_1$ by separate pathways.

Another interesting protein involved in anaphase entry in *Xenopus* is a subunit of the APC called APC-alpha (Jorgensen *et al.*, 1998). Some homologs of this protein have already been identified including APC1, Cut4, BIME, and Tsg24. The mammalian isoform Tsg24, for example, has been identified in two forms. One is a soluble form that associates with other components of the APC complex; the second is a centromere-associated protein. Since Tsg24 has

been shown to be required for entry into anaphase, these results suggest a model in which APC regulates sister chromatid separation by ubiquitination of a centromere protein.

In yeast models, the ubiquitin-dependent proteolysis of anaphase inhibitors (i.e., Pds1 and Cut2) has an important role for chromosome segregation. In *S. cerevisiae*, the ubiquitination and subsequent proteolysis is carried out by *fizzy*/Cdc20 proteins at the metaphase-anaphase transition. For this process to take place, the cell cycle regulatory cyclin-dependent kinase Cdc28 and a conserved associated protein, Cks1/Suc1, interact genetically, physically, and functionally with components of the 26S proteasome (Kaiser *et al.*, 1999).

Many other proteins control the mitotic spindle and centrosome cycle in mammals including p19^{Skp1} (Gstaiger *et al.*, 1999), Aurora2/Aik (Farruggio *et al.*, 1999), and the Mad/Bub family (Taylor and McKeon, 1997; Taylor *et al.*, 1998). In murine models, for example, Bub1 is required for checkpoint response to spindle damage and restrains progression through a normal mitosis (Taylor and McKeon, 1997). Thus, Bubs exert a feedback mechanism that delays the onset of anaphase until all the chromosomes are correctly aligned on the mitotic spindle. A human homolog of Bub1, Bub3, has also been found as a 37-kDa protein with four WD repeats. Like Bub1, Bub3 localizes to kinetochores before chromosome alignment (Taylor *et al.*, 1998).

In an attempt to prevent chromosome missegregation, the activation of the mitotic checkpoint in response to mitotic spindle damage delays the exit from mitosis. Another component of this pathway, hsMad2 in mammalian cells, physically associates with components of APC when the checkpoint is activated, thereby preventing the degradation of inhibitors of the mitotic exit machinery. The inhibitory association between Mad2 and the APC component Cdc27 also takes place transiently during the early stages of a normal mitosis and is lost before mitotic exit, suggesting a model for the regulation of the APC by Mad2 and the control of mitosis under normal growth conditions (Wassmann and Benezra, 1998).

c. Proteolysis and Exit from Mitosis. A key mechanism of CDK inactivation is ubiquitin-mediated cyclin proteolysis. This process is carried out by the late mitotic activation of APC. However, activation of the APC requires its association with substoichiometric activating *fzr* subunits (Cdh1/Hct1). The latter are constitutively expressed in the cell and only bind the APC during the mitotic exit and G$_1$. In yeasts, for example, the APC-*fzr* complex targets cyclins B1, B3, and Ase1 but requires the dephosphorylation of *fzr* proteins by the phosphatase Cdc14 to accomplish this task (Jaspersen *et al.*, 1999). Thus, in yeasts, APC-*fzr* complexes play their major role at the mitotic exit, whereas the APC-*fizzy* complexes act at the metaphase-anaphase transition (i.e., proteolysis of cyclin A, Pds1, and Cut2). Nevertheless, APC-*fizzy* complexes probably participate in both phases in *Drosophila*.

Other novel molecules for exit from mitosis in yeasts include Cdc15, Cdc5, Cdc14, Dbf2, and Tem1. All of these cooperate in the activation of the APC in late mitosis (Jaspersen *et al.*, 1998).

3. Deubiquitinating Enzymes

Similar to protein phosphorylation, protein ubiquitination is a dynamic process, involving enzymes that add ubiquitin (ubiquitin conjugating enzymes) and enzymes that remove ubiquitin (deubiquitinating enzymes). To date, considerable progress has been made in the understanding of ubiquitin conjugation and its role in regulating protein degradation. However, the regulation that occurs at the level of deubiquitination has not been characterized.

It is now known that deubiquitinating enzymes are cysteine proteases that specifically cleave ubiquitin from ubiquitin-conjugated protein substrates. Many candidate deubiquitinating enzymes have been identified, making them the largest family of enzymes in the ubiquitin system. Deubiquitinating enzymes have significant sequence diversity and may have a broad range of substrate specificities during the cell cycle (D'Andrea and Pellman, 1998; Cai *et al.,* 1999).

IV. Transgenic Mice

Once a putative molecule has been cloned and shown to be of relevance in the cell cycle in *in vitro* assays, the final goal is to test its potential regulatory activity in *in vivo* conditions. This has been achieved by the creation of transgenic mice for each one of the molecules of interest. For this purpose, pluripotential embryonic stem cells (ES) in culture are genetically modified prior to their return to mouse, where they will contribute to all tissues of a mouse including the germline. ES cells are modified to either (1) carry and over-express foreign DNA for **gain of function** or (2) gene targeted, to create loss-of -function mutations in the mouse, that is, **knockout** mice.

A partial list of the transgenic mice specifically created for evaluating the activity of cell cycle regulator proteins is shown in Table 5.[2] Although the phenotype of some of these chimeric animals fully corroborated the putative activity of certain regulators in *in vitro* experiments (e.g., pRB, p53, p16), in other cases the phenotype has significantly differed from the expected outcomes (e.g., E2F1, cyclin D1, p21). Some of these findings have already been discussed in relation to specific proteins in Section III.

Except for the **pRB nullizygous mice** (pRB –/–), none of the other knockout mice present a lethal phenotype. Indeed, pRB is not necessary for cell division or for differentiation up to day 13–14 of gestation. However, this lethal phenotype can be partially potentiated in pRB –/–, p107 –/– mice. In contrast, **pRB heterozygous** animals (pRB +/–) are viable and develop tumors of the thyroid and pituitary glands after 8 months of age. This is in agreement with its role as a tumor suppresor gene, but these heterozygous animals do not develop retinoblastomas or osteosarcomas as humans do.

Interestingly, pRB +/– mice and **p27 nullizygous mice** (p27 –/–) share the feature of generating pituitary tumors. This has given further support to the concept of an integrated pRB–cyclin D–CDK4–p27 pathway. Furthermore,

[2]Further information regarding available transgenic mice can be accessed at http://www.jax.org/tbase.

p27 –/– mice grow faster and present with generalized organomegaly in comparison to their controls. Since these effects could not be attributed to another factor (e.g., endocrinopathy), p27 has been implicated as an essential regulator of growth in multiple cell types (Raff, 1996).

As expected, **p53 nullizygous** (p53 –/–) and **p16 nullizygous** (p16 –/–) mice present normal development and tumor formation early in life. These findings are in agreement with their known tumor suppressor activity *in vitro* and with their genetic abnormalities in human cancers. However, it was also expected that **p21 nullizygous mice** (p21 –/–) would develop tumors, but this was not the case. Nonetheless, both p53 –/– thymocytes and p21 –/– embryonic fibroblasts are (1) deficient in their ability to arrest in G_1 in response to DNA damage, and (2) achieve a high saturation density due to a significant alteration in growth control *in vitro*. In contrast, p21 overexpression in liver halts cell cycle progression and postnatal liver development and regeneration, suggesting a critical role of p21 in normal development.

Another unexpected finding occurred in **mice lacking E2F1** (i.e., E2F1 –/–). These animals were viable, developed normally, and presented a broad spectrum of tumors. Thus, although overexpression of E2F1 in tissue culture cells promotes growth proliferation and oncogenesis, the lack of this gene in E2F1 –/– mice can also induce oncogenesis *in vivo*. For this reason, the E2F1 gene has been considered the first of a class acting both as an oncogene and as a tumor suppressor gene (Weinberg, 1996).

Puzzling too was the case of **cyclin D1 nullizygous mice.** It is known that the sole overexpression of cyclin D1 deregulates cell proliferation and induces tumorigenesis in the tissues where it is selectively expressed. Thus, cyclin D1 has been considered an essential component in promoting cell proliferation and cyclin D1 –/– mice were expected to present a lethal phenotype. However, these animals developed to term with a reduced body size. They also lacked normal development in retina and in mammary gland. It was concluded, then, (1) that D1 kinase activity is important for G_1 progression, but that it is dispensable for the proliferation of the majority of tissues in mammals; and (2) that other D-type cyclins could partially replace cyclin D1 activity in these animals.

However, the later statement has been recently challenged in a paper where the replacement of the coding sequence of mouse cyclin D1 with that of human cyclin E in the mouse germline (**cyclin D-1E**) rescued the phenotypic manifestations of cyclin D1 deficiency (Geng *et al.,* 1999). Of note, knockin cyclin E did not induce an additional phenotype in these mice despite close homology of the sequence of cyclin E and those of M-phase cyclins, in contrast to those of cyclin D types (see Fig. 3). Altogether these data suggest that (1) cyclin E bypasses cyclin D functions in mammalian cells and (2) this cyclin could be a rate-limiting target in these cells for cyclin D. The exact mechanisms by which these events take place are under study.

V. Summary

The mitotic cell cycle of higher eukaryotes consists of four phases: G_1, S, G_2, and mitosis. In the first three phases, the cell prepares for cell division, which actually takes place during mitosis. To guarantee the high-fidelity transmission of genetic information, eukaryotic cells have developed a complicated network of molecules that act in a finely coordinated manner. Central participants in this process include cyclins, CDKs, CDK inhibitors, transcription factors such as the products of tumor suppressor genes (e.g., pRB, p53), and the ubiquitin-mediated proteolytic system. Cyclins bind to CDKs and act as regulatory subunits of these catalytic kinases. The activation of these cyclin-CDK complexes and their effects depend on various factors: (1) cyclin abundance, (2) phosphorylation state of their kinase subunits, (3) presence of negative regulators known as CDK inhibitors; and (4) proteolysis of positive and negative regulators of the cell cycle.

Once CDKs become active, they phosphorylate diverse substrates that, directly or indirectly, control gene expression and the progression of the cell cycle. Twelve groups of cyclins (i.e., A type to T type) and ten types of CDKs (i.e., 1, 2, etc.) are known to date, some of which have various isoforms. Established interactions include (1) cyclin D–CDK4 and cyclin D–CDK6 during G_1 phase, (2) cyclin E–CDK2 and cyclin A–CDK2 at the G_1-S transition and into S phase, and (3) cyclin A–CDK1 and cyclin B–CDK1 from G_2 until mitosis.

Cyclin D–CDK4/CDK6 complexes phosphorylate pRB protein, the product of the tumor suppressor gene pRB and the master regulatory protein of the cell cycle at the G_1/S transition. Once phosphorylated, pRB protein liberates E2F–DP heterodimers, which are active transcription factors that promote the passage from G_1 to S phase and allow the progression of the cell cycle. For this transition to occur, the activities of cyclin E/A–CDK2 complexes are also required. Once in G_2 phase, cyclin A/B–CDK1 complexes appear and promote M-phase progression. Interestingly, these complexes dissociate at metaphase (i.e., one of the phases of mitosis) due to cyclin degradation via ubiquitin-mediated proteolysis. This process is essential for the exit from mitosis.

As a negative feedback, CDK inhibitors down-regulate the function of CDK complexes. These inhibitors belong to one of two families, Ink4 or Kip/Cip. Members of the former family include proteins p15, p16, p18, and p19 and exclusively inhibit cyclin-CDK4/CDK6 complexes. In contrast, the Kip/Cip family proteins p21, p27, and p57 are considered universal inhibitors since they can down-regulate the activities of cyclin-CDK4/CDK6 and cyclin-CDK2 complexes.

As previously discussed, the proteolytic system also controls the activity of many proteins during the cell cycle. To date, two E3 holoenzymes are known (i.e., SCF and APC) that mediate the ubiquitination of target proteins that subsequently undergo proteolysis at the proteasome. SCF complexes, for example, preferentially ubiquitinate G_1 cyclins (i.e., cyclins D1, E), p27, E2F1, and Wee1. In contrast, APC complexes target proteins during mitosis. In mammals, APC mediates the ubiquitination of cyclin A and cyclin B but, in yeasts, the target list includes anaphase inhibitors (e.g., Pds1 and Cut2) and molecules whose destruction is required for the disassembly of the mitotic spindle during cytokinesis (e.g., Ase1).

Altogether, positive and negative regulators are very important in the control of the cell cycle during various physi-

ological (e.g., embryogenesis, terminal differentiation, senescence) and pathological conditions (e.g., p53-mediated G_1 arrest through the transcriptional activation of p21 in the presence of DNA damage), including the cell death pathways (e.g., apoptosis).

Bibliography

Asano, M., Nevins, J. R., and Wharton, R. P. (1996). Ectopic E2F expression induces S phase and apoptosis in *Drosophila* imaginal discs. *Genes and Dev.* **10**, 1422–1432.

Bastians, H., Townsley, F. M., Ruderman, J. V. (1998). The cyclin-dependent kinase inhibitor p27 (Kip1) induces N-terminal proteolytic cleavage of cyclin A. *Proc. Natl. Acad. Sci. USA* **95**, 15 374–15 381.

Blangy, A., Lane, H. A., d´Herin, P., Harper, M., Kress, M., and Nigg, E. A. (1995). Phosphorylation by p34 Cdc2 regulates spindle association of human Eg5, a kinesin-related motor essential for bipolar spindle formation *in vivo*. *Cell* **83**, 1159–1169.

Blasina, A., Van de Weyer, I., Laus, M. C., Luyten, W. H. M. L., Parker, A. E., and McGowan, C. H. (1999). A human homologue of the checkpoint kinase Cds1 directly inhibits Cdc25 phosphatase. *Curr. Biol.* **9**, 1–10.

Botz, J., Zerfass-Thome, K., Spitkovsky, D., Delius, H., Vogt, B., Eilers, M., Hatzigeorgiou, A., and Janssen-Durr, P. (1996). Cell cycle regulation of the murine cyclin E gene depends on an E2F site in the promoter. *Mol. Cell. Biol.* **16**, 3401–3409.

Brown, N. R., Noble, M. E., Lawrie, A. M., Morris, M. C., Tunnah, P., Divita, G., Johnson, L. N., and Endicott, J.A. (1999). Effects of phosphorylation of threonine 160 on cyclin-dependent kinase 2 structure and activity. *J. Biol. Chem.* **274**, 8746–8756.

Brugarolas, J., Chandrasekaran, C., Gordon, J. I., Beach, D., Jacks, T., and Hannon, G. J. (1995). Radiation-induced cell cycle arrest compromised by p21 deficiency. *Nature* **377**, 552–557.

Cai, S. Y., Babbitt, R. W., and Marchesi, V. T. (1999). A mutant deubiquitinating enzyme (Ubp-M) associates with mitotic chromosomes and blocks cell division. *Proc. Natl. Acad. Sci. USA* **96**, 2828–2833.

Cartwright, P., Muller, H., Wagener, C., Holm, K., and Helin, K. (1998). E2F-6: a novel member of the E2F family is an inhibitor of E2F-dependent transcription. *Oncogene* **17**, 611–623.

Chen, J., Saha, P., Kornbluth, S., Dynlacht, B. D., and Dutta, A. (1996). Cyclin-binding motifs are essential for the function of p21[CIP1]. *Mol. Cell. Biol.* **16**, 4673–4682.

Cross, S. M., Sanchez, C. A., Morgan, C. A., Schimke, M. K., Ramel, S., Idzerda, R. L., Raskind, W. H., and Reid, B. J. (1995). A p53-dependent mouse spindle checkpoint. *Science* **267**, 1353–1356.

D'Andrea, A., and Pellman, D. (1998). Deubiquitinating enzymes: a new class of biological regulators. *Crit. Rev. Biochem. Mol. Biol.* **33**, 337–352.

de Falco, G., and Giordano, A. (1998). CDK9 (PITALRE): a multifunctional Cdc2-related kinase. *J. Cell. Physiol.* **177**, 501–506.

Delgado, S., Gómez, M., Bird, A., Antequera, F. (1998). Initiation of DNA replication at CpG islands in mammalian chromosomes. *EMBO J* **17**, 2426–2435.

Deng, C., Zhang, P., Harper, J. W., Elledge, S .J., and Leder, P (1995). Mice lacking p21[CIP1/WAF1] undergo normal development, but are defective in G1 checkpoint control. *Cell* **82**, 675–684.

De Pamphilis, M. L. (1999). Replication origins in metazoan chromosomes: facts or fiction? *Bioessays* **21**, 5–16.

Donehower, L. A., Godley, L. A., Marcelo Aldaz, C., Pyle, R., Shi, Y. P., Pinkel, D., Gray, J., Bradly, A., Medina, D., and Varmus, H. E. (1995). Deficiency of p53 accelerates mammary tumorigenesis in Wnt-1 transgenic mice and promotes chromosomal instability. *Genes Dev.* **9**, 882–895.

Du, W., Vidal, M., Xie, J.-E., and Dyson, N. (1996). RBF, a novel RB-related gene that regulates E2F activity and interacts with cyclin E in *Drosophila*. *Genes Dev.* **10**, 1206–1218.

Dunphy, W. G. (Ed) (1997). "Methods in Enzymology," Vol. 283. "Cell Cycle Control." Academic Press, New York.

Duronio, R .J., Brook, A., Dyson, N., and O'Farrel, P. H. (1996). E2F-induced S phase requires cyclin E. *Genes Dev.* **10**, 2505–2513.

Fantl, V., Stamp, G., Andrews, A., Rosewell, I., and Dickson, C. (1995). Mice lacking cyclin D1 are small and show defects in eye and mammary gland development. *Genes Dev.* **9**, 2364–2372.

Farruggio, D.C., Townsley, F. M., and Ruderman, J. V. (1999). Cdc20 associates with the kinase aurora2/Aik. *Proc. Natl. Acad. Sci. USA* **96**, 7306–7311.

Felsenfeld, G. (1996). Chromatin unfolds. *Cell* **86**, 13–19.

Ferreira, R., Magnagni-Jaulin, L., Robin, P., Harel-Bellan, A., and Trouche, D. (1998). The tree members of the pocket proteins family share the ability to repress E2F activity through recruitment of a histone deacetylase. *Proc. Nat. Acad. Sci. USA* **95**, 10 493–10 498.

Field, S. J., Tsai, F-Y., Kuo, F., Zubiaga, A. M., Kaelin, W. G., Livingston, D. M., Orkin, S. H., and Greenberg, M. E. (1996). E2F-1 functions in mice to promote apoptosis and suppress proliferation. *Cell* **85**, 549–561.

Furnari, B., Blasina, A., Boddy, M. N., McGowan, C. H., Russell, P. (1999). Cdc25 inhibited in *in vivo* and in vitro by checkpoint kinases Cds1 and Chk1. *Mol. Biol. Cell.* **10**, 833–845.

Garriga, J., Peng, J., Parreno, M., Price, D. H., Henderson, E. E., and Grana, X. (1998). Upregulation of cyclin T1/CDK9 complexes during T cell activation. *Oncogene* **17**, 3093–3102.

Geng, Y., Whoriskey, W., Park, M. Y., Bronson, R. T., Medema, R. H., Li, T., Weinberg, R.A., and Sicinski, P. (1999). Rescue of cyclin D1 deficiency by knockin cyclin E. *Cell* **97**, 767–777.

Gorbsky, G. J. (1997). Cell cycle checkpoints: arresting progress in mitosis. *Bioessays* **19**, 193–197.

Grana, X., Garriga, J., and Mayol, X. (1998). Role of the retinoblastoma protein family, pRB, p107 and p130 in the negative control of cell growth. *Oncogene* **17**, 3365–3383.

Greider, C. W. (1999). Telomeres do D-loop-T-loop. *Cell* **97**, 419–422.

Grossberger, R., Gieffers, C., Zachariae, W., Podtelejnikov, A. V., Schleiffer, A., Nasmyth, K., Mann, M., and Peters, J. M. (1999). Characterization of the DOC1/APC10 subunit of the yeast and the human anaphase-promoting complex. *J. Biol. Chem.* **274**, 14 500–14 507.

Gstaiger, M., Marti, A., and Krek, W. (1999). Association of human SCF(Skp2) subunit p19(Skp1) with interphase centrosomes and mitotic spindle poles. *Exp. Cell. Res.* **247**, 554–562.

Hartwell, L. H., and Weinert, T. A. (1989). Checkpoints: controls that ensure the order of cell cycle events. *Science* **246**, 629–634.

Hermeking, H., Lengauer, C., Polyak, K., He, T-C., Zhang, L., Thiagalingam, S., Kinzler, K. W., and Vogelstein, B. (1997). 14.3.3, is a p53-regulated inhibitor of G_2-M progression. *Mol. Cell* **1**, 3–11.

Innocente, S. A., Abrahmson, J. L., Cogswell, J. P., and Lee, J. M. (1999). p53 regulates the G_2 checkpoint through cyclin B1. *Proc. Natl. Acad. Sci. USA* **96**, 2147–2152.

Ishimi, Y., Komamura-Hohno, Y., You, Z., Omori, A., and Kitagawa, M. (2000). Inhibition of Mcm 4 6 7 helicase activity by phosphorylation with cyclin A/CDK2. *J. Biol. Chem.* **275**, 16 235–16 241.

Jacks, T., Fazeli, A., Schmitt, E. M., Bronson, R. T., Goodell, M. A., and Weinberg, R. A. (1992). Effects of Msp 31 and *Rb* mutation in the mouse. *Nature* **359**, 295–230.

Jaspersen, S. L., Charles, J. F., and Morgan, D. O. (1999). Inhibitory phosphorylation of the APC regulator Hct1 is controlled by the kinase Cdc28 and the phosphatase Cdc14. *Curr. Biol.* **9**, 227–236.

Jaspersen, S. L., Charles, J. F., Tinker-Kulberg, R. L., and Morgan, D. O. (1998). A late mitotic regulatory network controlling cyclin destruction in *Saccharomyces cerevisiae. Mol. Biol. Cell.* **9**, 2803–2817.

Johnson, D. G., and Schneider-Broussard, R. (1998) Role of E2F in cell cycle control and cancer. *Front. Biosci.* **3**, d447–d448.

Jorgensen, P. M., Brundell, E., Starborg, M., and Hoog, C. (1998). A subunit of the anaphase-promoting complex is a centromere-associated protein in mammalian cells. *Mol. Cell. Biol.* **18**, 468–476.

Kaiser, P., Moncollin, V., Clarke, D. J., Watson, M. H., Bertolaet, B. L., Reed, S. I., and Bailly, E. (1999). Cyclin-dependent kinase and CksSuc1 interact with the proteasome in yeast to control proteolysis of M-phase targets. *Genes. Dev.* **13**, 1190–1202.

Kalein, W. G., Jr. (1999). The emerging p53 family. *J. Natl. Cancer Inst.* **91**, 594–598.

Kamura, T., Sato, S., Haque, D., Liu, L., Kaelin, W. G., Elledge, S. J., Conaway, R. C., Harper, J. W., and Conaway, J. W. (1999). Rbxl, a component of the Vh1, tumor suppressor complex and SCF ubiquitin ligase. *Science* **284**, 657–661.

Khanna, K. K., Keating, K. E., Koslov, S., Scott, S., Gatei, M., Hobson, K., Taya, K., Gabrielli, B., Chan, D., Less-Miller, S. P., and Lavin, M. F. (1998). ATM associates with and phosphorylates p53: mapping the region of interaction. *Nat. Genet.* **20**, 398–400.

King, R. W., Peters, J. M., Tugendreich, S., Rolfe, M., Hieter, P., and Kirschner, M. W. (1995) A 20S complex containing Cdc27 and Cdc16 catalyzes the mitosis-specific conjugation of ubiquitin to cyclin B. *Cell* **81**, 279–288.

Koepp, D.M., Harper, J. W., and Elledge, S. J. (1999). How the cyclin became a cyclin: regulated proteolysis in the cell cycle. *Cell.* **97**, 431–434.

Krek, W., Xu, G., and Livingston, D. M. (1995). Cyclin A-kinase regulation of E2F-1 DNA binding function underlies suppression of an S phase checkpoint. *Cell* **83**, 1149–1158.

Lania, L., Majello, B., Napolitano, G. (1999). Transcriptional control by cell-cycle regulators: a review. *J. Cell. Physiol.* **179**, 134–141.

Latres, E., Chiaur, D. S., and Pagano, M. (1999). The human F-box protein beta-Trcp associates with the Cul1-Skp1 complex and regulates the stability of beta-catenin. *Oncogene* **18**, 849–854.

Lee, E. Y.-H. P., Chang, C-Y., Hu, N., Wang, Y.-C. J., Lai, C.-C., Herrup, K., Lee, W.-H., and Bradley, A. (1992). Mice deficient for *Rb* are nonviable and show defects in neurogenesis and hematopoiesis. *Nature* **359**, 288–294.

Lee, M. H., Nikolic, M., Baptista, C., Lai, E., Tsai, L. H., and Massague, J. (1996). The brain specific activator allows CDK5 to escape inhibition by p27Kip1 in neurons. *Proc. Natl. Acad. Sci. USA* **93**, 3259–3263.

Linke, S. P., Clarkin, K. C., DiLeonardo, A., Tsou, A., and Wahl, G. M. (1996). A reversible, p53-dependent G_0/G_1 cell cycle arrest induced by ribonucleotide depletion in the absence of detectable DNA damage. *Genes Dev.* **10**, 934–947.

Lisztwan, J., Marti, A., Sutterluty, H., Gstaiger, M., Wirbelauer, C., and Krek, W. (1998). Association of human Cul-1 and ubiquitin-conjugating enzyme Cdc34 with the F-box protein p45(Skp2): evidence for evolutionary conservation in the subunit composition of the Cdc34-SCF pathway. *EMBO. J.* **17**, 368–383.

López-Girona, A., Furnari, B., Mondesert, O. and Russell P. (1999). Nuclear localization of Cdc25 is regulated by DNA damage and a 14.3.3 protein. *Nature* **397**, 172–175.

Loubart, A., Rochet, N., Turchi, L., Rezzonico, R., Far, D. F., Auberger, P., Rossi, B., and Ponzio, G. (1999). Evidence for a p23 caspase-cleaved form of p27[Kip1] involved in G_1 growth arrest. *Oncogene* **18**, 3324–3333.

Lundberg, A. S., and Weinberg, R. A. (1998). Functional inactivation of the retinoblastoma protein requires sequential modification by at least two distinct cyclin-CDK complexes. *Mol. Cell. Biol.* **18**, 753–761.

Matsuoka, S., Huang, M., and Elledge, S. J. (1998). Linkage of ATM to cell cycle regulation by the CHK2 protein kinase. *Science* **282**, 1893–1897.

Matsuoka, S., Thompson, J. S., Edwards, M. C., Barletta, J. M., Grundy, P., Kalikin, L. M., Harper, J. W., Elledge, S. J., and Feinberg, A. P. (1996). Imprinting of the gene encoding a human cyclin-dependent kinase inhibitor, p57 Kip2, on chromosome 11p15. *Proc. Natl. Acad. Sci USA.* **93**, 3026–3030.

Mimura, S., and Takisawa, H. (1998). *Xenopus* Cdc45-dependent loading of DNA polymerase alpha onto chromatin under the control of S-phase Cdk. *EMBO J* **17**, 5699–5707.

Morgan, D. O., Fisher, R. P., Espinoza, F. H., Farrell, A., Nourse, J., Chamberlin, H., and Jin, P. (1998). Control of eukaryotic cell cycle progression by phosphorylation of cyclin-dependent kinases. *Cancer J. Sci. Am.* **4 Suppl. 1**,. S77–S83.

Nakanishi, M., Robetorye, R. S., Adami, G. R., Pereira-Smith, O. M., and Smith, J. R. (1995). Identification of the active region of the DNA synthesis inhibitory gene p21 Sdi1/Cip1/WAF1. *EMBO J.* **14**, 555–563.

Nigg, E. A., Blangy, A., Lane, H. A. (1996). Dynamic changes in nuclear architecture during mitosis: on the role of protein phosphorylation in spindle assembly and chromosome segregation. *Exp. Cell Res.* **229**, 174–180.

Nikolic, M., Dudek, H., Kwon, Y. T., Ramos, Y. F. M., and Tsai, L. H. (1996). The CDK5/p35 kinase is essential for neurite outgrowth during neuronal differentiation. *Genes Dev.* **10**, 816–825.

Nishijima, H., Nishitani, H., Seki, T., and Nishimoto, T. (1997). A dual-specificity phosphatase Cdc25B is an unstable protein and triggers p34 Cdc2/cyclin B activation in hamster BHK21 cells arrested with hydroxyurea. *J. Cell. Biol.* **138**, 1105–1116.

Norbury, C., and Nurse, P. (1992). Animal cell cycles and their control. *Annu. Rev. Biochem.* **61**, 441–470.

Peters, J. M. (1998). SCF and APC: the Yin and Yang of cell cycle regulated proteolysis. *Curr. Opin. Cell. Biol.* **10**, 759–768.

Quelle, D. E., Zindy, F., Ashmun, R. A., and Sherr, C. J. (1995). Alternative reading frames of the INK4a tumor suppressor gene encode two unrelated proteins capable of inducing cell cycle arrest. *Cell* **83**, 993–1000.

Raff, M. C. (1996). Size control: the regulation of cell numbers in animal development. *Cell* **86**, 172–175.

Reed, S. I. (1992). The role of p34 kinases in the G_1 to S-phase transition. *Annu. Rev. Cell Biol.* **8**, 529–561.

Ritzi, M. and Knippers, R. (2000). Initiation of genome replication: assembly and disassembly of replication-competent chromatin. *Gene* **245**, 13–20.

Robles, A. I., Larcher, F., Whalin, R .B., Murillas, R., Richie, E., Gimenez-Conti, I. B., Jorcano, J. L., and Conti, C. J. (1996). Expression of cyclin D1 in epithelial tissues of transgenic mice results in epidermal hyperproliferation and severe thymic hyperplasia. *Proc. Natl. Acad. Sci. USA* **93**, 7634–7638.

Rogers, K. T., Higgins, P. D. R., Milla, M. R., Phillips, R. S., and Horowitz, J. M.(1996). DP-2, a heterodimeric partner of E2F: identification and characterization of DP-2 proteins expressed *in vivo*. *Proc. Natl. Acad. Sci. USA* **93**, 7594–7599.

Sabbatini, P., Lin, J., Levine, A. J., and White, E. (1995). Essential role for p53-mediated transcription in E1A-induced apoptosis. *Genes Dev.* **9**, 2184–2192.

Sack, S., Kull, F. J., and Mandelkow, E. (1999) Motor proteins of the kinesin family. Structures, variations, and nucleotide binding sites. *Eur. J. Biochem.* **262**, 1–11

Schwartz, D., and Rotter, V. (1998). p53-dependent cell cycle control: response to genotoxic stress. *Semin. Cancer Biol.* **8**, 325–336.

Seol, J. H., Feldman, R. M., Zachariae, W., Shevchenko, A., Correll, C. C., Lyapina, S., Chi, Y., Galova, M., Claypool, J., Sandmeyer, S., Nasmyth, K., Deshaies, R. J., Shevchenko, A., and Deshaies, R. J.(1999). Cdc53/Cullin and the essential Hrt1 RING-H2 subunit of SCF define a ubiquitin ligase module that activates the E2 enzyme Cdc34. *Genes Dev.* **13**, 1614–1626.

Serrano, M., Lee, H.-W., Chin, L., Cordon-Cardo,C., Beach, D., and DePinho, A.(1996). Role of the *INK4a* locus in tumor suppression and cell mortality. *Cell* **85**, 27–37.

Sherr, C. J. (1995). D-type cyclins. *Trends. Biochem. Sci.* **20**, 187–190.

Sherr, C. J., and Roberts, J. M. (1999). CDK inhibitors: positive and negative regulators of G_1-phase progression. *Genes. Dev.* **13**, 1501–1512.

Shimizu, A., Nishida, J.-I., Ueoka, Y., Kato, K., Hachiya, T., Kuriaki, Y., and Wake, N. (1998). Cyclin G contributes to G_2-M arrest of cells in response to DNA damage. *Biochem. Biophys. Res. Comm.* **242**, 529–533.

Shirane, M., Harumiya, Y., Ishida, N., Hirai, A., Miyamoto, C., Hatakeyama, S., Nakayama, K., and Kitagawa, M. (1999). Down-regulation of p27(Kip1) by two mechanisms, ubiquitin-mediated degradation and proteolytic processing. *J. Biol. Chem.* **274**, 13 886–13 893.

Shou, W., Seol, J. H., Shevchenko, A., Baskerville, C., Moazed, D., Chen, Z. W., Jang, J., Shevchenko, A., Charbonneau, H., and Deshaies, R.J. (1999). Exit from mitosis is triggered by Tem1-dependent release of the protein phosphatase Cdc14 from nucleolar RENT complex. *Cell* **97**, 233–244.

Taylor, S. S., and McKeon, F. (1997). Kinetochore localization of murine Bub1 is required for normal mitotic timing and checkpoint response to spindle damage. *Cell* **89**, 5727–5735.

Taylor, S. S., Ha, E., and McKeon, F. (1998). The human homolog of Bub3 is required for kinetochore localization of Bub1 and a Mad3/Bub1-related protein kinase. *J. Cell. Biol.* **142**, 1–11.

Tibbetts, R. S., Brumbaugh K. M., Williams, J. M., Sarkaria, J. N., Cliby, W. A., Sieh, S. Y., Taya, Y., Prives, C., and Abraham, R. T. (1999). A role for ATR in the DNA damage-induced phosphorylation of p53. *Genes Dev.* **13**, 152–157.

Timchenko, N. A., Wilde, M., Nakanishi, M., Smith, J. R., and Darlington, G. J. (1996). CCAAT/enhancer-binding protein α (C/EBP α) inhibits cell proliferation through the p21 (WAF-1/CIP-1-SDI-1) protein. *Genes Dev.* **10**, 804–815.

Townsley, F. M., and Ruderman, J. V. (1998). Proteolytic ratchets that control progression through mitosis. *Trends Cell Biol.* **8**, 338–344.

Tsvetkov, L. M., Yeh, K. H., Lee, S. J., Sun, H., Zhang, H. (1999). p27(Kip1) ubiquitination and degradation is regulated by the SCFSkp2 complex through phosphorylated thr187 in p27. *Curr. Biol.* **9**, 661–664.

Umar, A., Buermeyer, A. B., Simon, J. A., Thomas, D. C., Clark, A. B., Liskay, R. M., and Kunkel, T. A. (1996). Requirement for PCNA in DNA mismatch repair at a step preceding DNA resynthesis. *Cell* **87**, 65–73.

Wang, T. C., Cardiff, R. D., Zukerberg, L., Lees, E., Arnold, A., and Schmidt, E. V. (1994). Mammary hyperplasia and carcinoma in MMTV-cyclin D1 transgenic mice. *Nature* **369**, 669–671.

Wang, X. W., Zhan, Q., Khan, M. A., Khonty, H. U., Yu, L., Hollander, M. C., O'Connor, P. M., Fornace, A. J., Harris, C. C. (1999). GADD45 induction of a G_2-M cell cycle checkpoint. *Proc. Natl. Acad. Sci. USA* **96**, 3706–3711.

Warren, G. (1993). Membrane partitioning during cell division. *Annu. Rev. Biochem.* (1993). **62**, 323–348.

Wassmann, K., and Benezra, R. (1998). Mad2 transiently associates with an APC/p55Cdc complex during mitosis. *Proc. Natl. Acad. Sci. USA* **95**, 11 193–11 198.

Weinberg, R. A. (1996). E2F and cell proliferation: a world turned upside down. *Cell* **85**, 457–459.

Wimmer, J., Fujinaga, K., Taube, R., Cujec, T. P., Zhu, Y. Peng J., Price, D. H., and Perterlin, B. M. (1999). Interactions between Tat and Tar and human immunodeficiency virus replication are facilitated by human cyclin T1 but not cyclin T2a or T2b. *Virology* **255**, 182–189.

Wu, H., Wade, M., Krall, L., Grisham, J., Xiong, Y., and VanDyke, T. (1996). Targeted *in vivo* expression of the cyclin-dependent kinase inhibitor p21 halts hepatocyte cell-cycle progression, postnatal liver development and regeneration. *Gene Dev.* **10**, 245–260.

Xiao, Z.-X., Ginsberg, D., Ewen, M., and Livingston, D. M. (1996). Regulation of the retinoblastoma protein-related protein p107 by G_1 cyclin associated kinases. *Proc. Natl. Acad. Sci. USA* **93**, 4633–4637.

Yam, C. H., Ng, R. W., Siu, W. Y., Lau, A. W., Poon, R. Y. (1999). Regulation of cyclin A-CDK2 by SCF component Skp1 and F-box protein Skp2. *Mol. Cell Biol.* **19,** 635–645.

Yamasaki, L., Jack, T., Bronson, R., Goillot, E., Harlow, E., and Dyson, N. J. (1996). Tumor induction and tissue atrophy in mice lacking E2F-1. *Cell* **85,** 537–548.

Yang, J., Winkler, K., Yoshida, M., and Kornbluth, S. (1999). Maintenance of G$_2$ arrest in the *Xenopus* oocyte: a role for 14.3.3-mediated inhibition of Cdc25 nuclear import. *EMBO J.* **18,** 2174–2183.

Yu, H., Peters, J. M., King, R. W., Page, A. M., Hieter, P., Kirschner, M. W. (1998). Identification of a Cullin homology region in a sub-unit of the anaphase-promoting complex. *Science.* **279,** 1219–1222.

Yu, Z. K., Gervais, J .L., and Zhang, H. (1998). Human CUL-1 associ-ates with the Skp1/Skp2 complex and regulates p21(Cip1/WAF1) and cyclin D proteins. *Proc. Natl. Acad. Sci. USA* **95,** 11 324–11 329.

Zachariae, W., Shin, T. H., Galova, M., Obermaier, B., and Nasmyth, K. (1996). Identification of subunits of the anaphase-promoting complex of *Saccharomyces cerevisiae. Science* **274,** 1201–1204.

Zhou, P., and Howley, P. M. (1998). Ubiquitination and degradation of the substrate recognition subunits of SCF ubiquitin-protein ligases. *Mol. Cell.* **2,** 571–580.

67

Cancer Cell Properties

I. Introduction

Understanding the process of neoplastic transformation involves understanding the physiology of the transformed cell in comparison to its normal counterpart. Although sometimes obscure in their beginnings, the characteristics of transformed cells eventually include a general loss of responsiveness to signals that promote growth arrest, differentiation, or cell death. The study of how normal processes are altered during neoplastic transformation will enable the development of newer and better therapies to treat a variety of tumors.

One approach to defining the physiological changes in the transformed cell is through the use of genetics. Such techniques have led to the identification of genes altered during tumorigenesis, and **disease genes**—genes whose aberrant alleles are responsible for defined clinical syndromes that predispose to cancer (Fig. 1). These genetic clues have introduced us to the pathways and processes that are normally required for growth control and have allowed the definition of physiological systems within the cell.

It is widely accepted that as a cell becomes neoplastic, it acquires mutations in a variety of genes, and that these mutations lead to both losses and gains of functions. These genes encode the regulatory components of systems normally required to maintain the cell in equilibrium with its environment. Through mutation of one or more of these genes, the transformed cell can evade normal inhibitory signals from the environment and become dysplastic or ultimately invasive and/or metastatic. Such physiological systems that are targets of mutation include those that regulate the cell cycle and the transcription of genes by signal transduction pathways, that maintain genomic integrity, that control the ability of cells to interact with their neighbors, and that induce programmed cell death or apoptosis under specific circumstances.

II. The Cell Cycle

The life of a growing cell is cyclical in nature. It progresses through four major phases during which specific cellular functions are performed. In S-phase, the cell replicates its genome, producing two copies of its chromosomes. The cell ultimately divides into two daughter cells during M-phase, with each daughter inheriting one set of chromosomes. Following M, and prior to S, is a growth phase, G_1, in which the cell grows and prepares for S-phase. Similarly, following S and prior to M, is a second growth phase, G_2, in which the cell grows further and prepares for M. This cycle is tightly controlled, assuring one S-phase is completed prior to every M-phase.

The cell commits to one complete cycle at a specific point during the G_1 phase. Beyond this point, referred to as START in yeast or the restriction point in mammalian cells, the cell no longer will be influenced by external signals until it has completed the full cycle. Prior to START, the cell receives information from its external environment in the form of positive regulators such as growth factors, or negative regulators such as contact inhibition or other signals that induce differentiation and exit from the cell cycle. These positive and negative signals are communicated to the cell cycle machinery of the nucleus through a cascade of proteins that begins at the cell surface. Cell-surface receptors receive external signals and set off a wave of protein-protein interactions in the cytoplasm which eventually conduct signals to the nucleus. The result is transcription of specific genes involved in cell cycle regulation (reviewed in Grunicke, 1990).

Cells induced to differentiate exit the cell cycle and enter a quiescent state referred to as G_0. The decision to enter G_0 or to proceed through the cell cycle is determined by a series of cell cycle control proteins. The major proteins of the cell cycle machinery are a family of kinases called the cyclin dependent kinases, or CDKs. These kinases are activated by pairing with partner cyclin proteins, enabling them to phosphorylate downstream targets and propel the cell through the cell cycle. An additional class of cell cycle proteins, the cyclin-CDK inhibitors, or CDIs, inhibit progression through the cell cycle by interfering with the cyclin activation of a CDK.

In addition to regulating the transition at START, different members of these protein families control transitions prior to S- and M-phases. These transition points are referred to as cell cycle checkpoints. Disruptions of these checkpoints were first

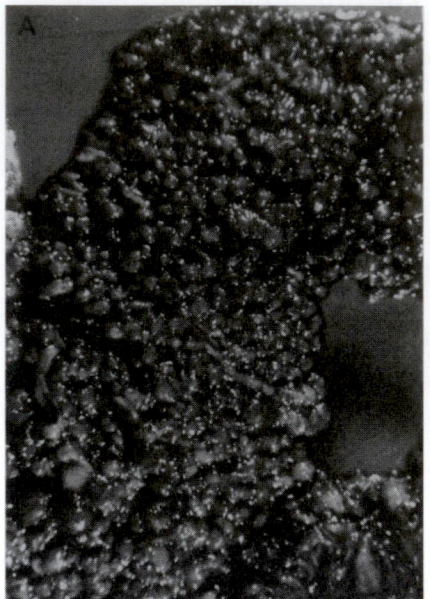
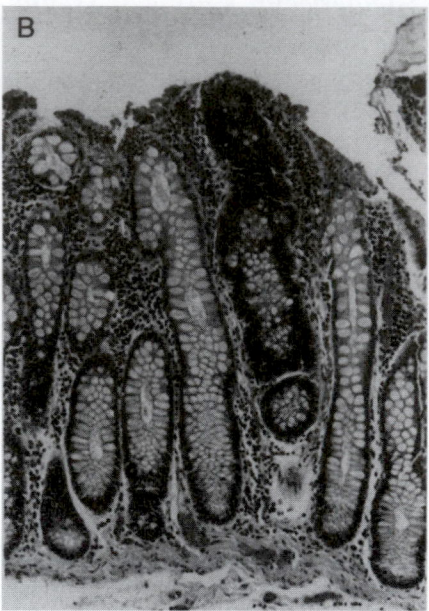
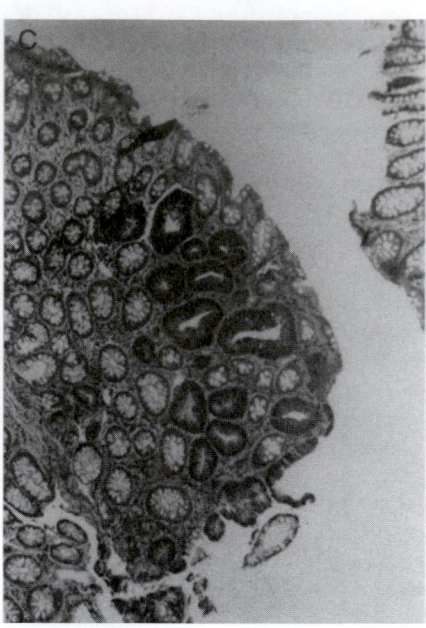

FIGURE 1. Disruptions of the tumor suppressor gene *APC* are associated with abnormal proliferation of the colonic epithelium. With the identification of genes involved in hereditary cancers, researchers have begun the study of different physiological processes in normal cells, such as those that control cell growth. When the *APC* gene is mutated in the germline, numerous adenomas form in the colon and rectum of the affected individual. (A) A section of colon from a patient with adenomatous polyposis coli that is carpeted with numerous adenomas. (B) A hematoxylin-and-eosin (H&E)-stained section of the colon from a different patient, illustrating the change in the epithelium of a single-crypt adenoma as compared to the normal crypts surrounding it. A second *APC* mutation has most likely occurred in these cells that have become neoplastic. (C) An H&E-stained section from a larger adenoma, showing the increase in epithelial dysplasia. (A, B, and C are courtesy of Cecelia Fenoglio-Presier M. D., University of Cincinnati.)

identified in specific tumor types and in cells infected with tumor viruses. In addition, these checkpoints are regulated by tumor suppressor proteins, of which the retinoblastoma gene product, or RB, is the best example (Fig. 2).

The identification of the RB tumor suppressor stemmed from studies of a pediatric tumor of the eye, retinoblastoma, that occurs in either a sporadic or a familial form. The work of Knudson (1971) first suggested that the familial form was due to the inheritance of one mutant allele at a tumor suppressor locus, and that the age of onset and tumor number was due to the frequency with which the second or normal allele acquired an inactivating mutation. This paradigm has been applied to understanding the function of most tumor suppressors associated with inherited predisposition to cancer, where loss of function at both alleles of a locus is associated with loss of growth control in a cell, and where the inheritance of one mutant allele accelerates the incidence of tumor formation. Subsequent cloning of the gene and the analysis of its gene product shows that RB is a nuclear protein that is phosphorylated in a cell cycle-specific manner.

The RB tumor suppressor is a key member of a pathway of proteins that function at the restriction point checkpoint. During G_1, the RB protein is hypophosphorylated and arrests cells in G_1 by binding members of the E2F family of transcription factors. Genes essential for the progression of cells into S-phase are activated or derepressed. Upon phosphorylation of RB, the E2F proteins are released and cells begin to cycle (reviewed in Hinds and Weinberg, 1994). The phosphorylation of RB is mediated by the D-type cyclins in

association with their partner CDKs (CDK4 and CDK6) (reviewed in Hunter and Pines, 1994). The CDI, p16^{INK4}, which binds to CDK4 or CDK6 and prevents the binding of cyclin D, inhibits this reaction. The CDIs represent an additional level at which cell cycle progression can be regulated (reviewed in Hunter and Pines, 1994). This pathway of proteins, termed the cyclin D/RB pathway, ultimately results in the release of free E2Fs. The E2Fs override growth arrest caused by serum withdrawal (Johnson *et al.*, 1993) or RB overexpression (Qin *et al.*, 1995). In addition, E2F family members prevent G_1 arrest by p16^{INK4a} overexpression, or when cells are treated with antibodies to neutralize cyclin D/CDK4 (DeGregori *et al.*, 1995; Lukas *et al.*, 1996). At the same time, a G_1 arrest caused by neutralization of cyclinD/CDK4 or overexpression of p16^{INK4a} requires RB (Lukas *et al.*, 1995a; Lukas *et al.*, 1995b; Medema *et al.*, 1995). These data indicate that the E2F transcription factors function downstream of RB and the cyclin D/CDK4 or CDK6 complex.

Many tumors contain mutations in at least one member of the cyclin D/RB pathway, thus disrupting regulation of G_1/S progression. DNA amplification occurs commonly at the **cyclin D1** locus in adenocarcinomas of the breast, head and neck, esophagus, bladder, and lung (Hall and Peters, 1996), while the *CDK4* locus is often amplified in certain sarcomas and gliomas (Hall and Peters, 1996). *p16^{INK4a}* is mutated in pancreatic adenocarcinoma, non-small cell lung cancer, and both familial and sporadic melanoma (Sherr, 1996); tumors mutated at the *RB* locus include retinoblastoma, small

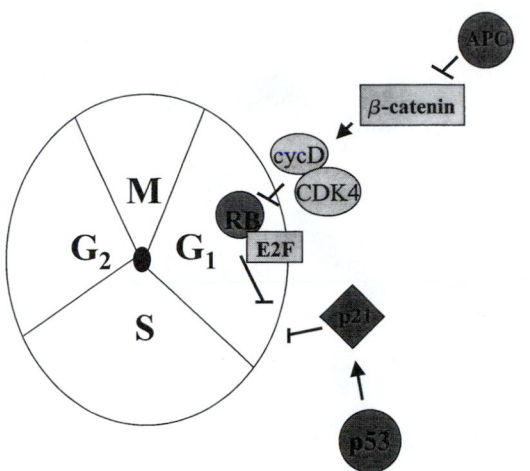

FIGURE 2. **Tumor suppressors control cell cycle checkpoints.**
The four major phases of the cell cycle are displayed. Tumor suppressors such as RB and p53 have roles in regulating cell cycle checkpoints during G_1. The APC tumor suppressor also plays a role in G_1 regulation via its ability to regulate β-catenin-induced gene transcription. Mutations of these tumor suppressor genes result in the loss of checkpoint regulators and thus, uncontrolled growth.

cell lung cancer, breast adenocarcinoma, and osteosarcoma (Benedict *et al.*, 1990). The placement of these gene products in the same pathway is also supported by the observation that tumors with *RB* mutations generally have reduced cyclin D1 and high p16^{INK4a} (Sherr, 1996).

A tumor suppressor whose gene product also functions at the G_1/S checkpoint is *APC*. *APC* was identified as the gene responsible for a hereditary predisposition to colon cancer known as familial adenomatous polyposis coli (Groden *et al.*, 1991, Nishishio *et al.*, 1991). Mutations of the *APC* gene are found in over 80% of all colorectal tumors, both familial and sporadic, and are believed to be early if not initiating events in colorectal tumorigenesis (Powell *et al.*, 1992). Transfection of *APC* into colon cancer cell lines lacking wild-type APC induces a G_1 arrest accompanied by reduced expression of cyclin D1 and reduced phosphorylation of RB (Heinen *et al.*, 2000). APC also regulates β-catenin-mediated transcription (Behrens *et al.*, 1996; Huber *et al.*, 1996; Molenaar *et al.*, 1996) with this activity crucial to cell cycle arrest. Interestingly, two genes recently identified as targets of β-catenin have direct implications in cell cycle control: *c-MYC* (He *et al.*, 1998) and *cyclin D1* (Tetsu and McCormick, 1999). Thus, APC functions in the cyclin D/RB pathway by inhibiting the ability of β-catenin to activate key genes for G_1 progression. This also explains the observation that colorectal tumors do not carry other mutations in the cyclin D/RB pathway genes (Hall and Peters, 1996), as these tumors disrupt G_1-regulation through loss of APC function.

Another well-characterized cell cycle checkpoint occurs at the G_1/S transition, where cells arrest in response to DNA damage. This delay is mediated by the p53 tumor suppressor, germline mutations of which lead to an increase in childhood sarcomas and/or the familial cancer disorder known as Li-Fraumeni syndrome. Experiments by

Kuerbitz *et al.* (1992) revealed a cell cycle block at G_1 in cells with normal p53 in response to radiation-induced DNA damage. The G_1 block was not present in cells lacking functional p53. When DNA damage is detected, the p53 protein is upregulated at the G_1/S boundary (Kastan *et al.*, 1991), which leads to an upregulation of the CDI p21 (reviewed in Sherr, 1994). The p21 protein interferes with a number of G_1 cyclin-CDK complexes as well as the proliferating cell nuclear antigen (PCNA) which functions in DNA replication and repair. This pauses cell cycle progression, presumably to correct the damaged DNA prior to replication. Additionally, extensive DNA damage results in p53-dependent apoptosis. Failure to induce appropriate cell cycle arrest or apoptosis in damaged cells results in the propagation of DNA mutations (reviewed in Sherr, 1994).

III. Genome Stability

The likelihood of developing a tumor might seem rare when considering the requirement for multiple mutations in tumor development and the slow rate of mutation in human cells, estimated at 1.4×10^{-10} mutations per nucleotide per cell generation. Therefore, researchers have hypothesized that the mutation rate of tumor cells must exceed that of normal cells (Loeb *et al.*, 1974) and that tumors might require early mutational events that not only confer a growth advantage to the cell, but also lead to an increase in genomic instability.

Studies of *p53* have revealed increased levels of genomic instability when these genes are mutant. Using cells from Li-Fraumeni patients which carry one mutation of *p53*, it was observed that gene amplification occurred at a much higher frequency in cells that lose the remaining *p53* allele than those with at least one functional *p53* gene (Livingstone *et al.*, 1992; Yin *et al.*, 1992).

The breast cancer susceptibility genes *BRCA1* and *BRCA2* also play a role in maintenance of the genome. These proteins form a complex with RAD51 (Scully *et al.*, 1996; Wong *et al.*, 1997), the homolog of bacterial RecA required for double-strand DNA break repair (Malkova *et al.*, 1996; Sung and Robberson, 1995). *Brca1$^{-/-}$* ES cells are sensitive to oxidative DNA damage (Gowen *et al.*, 1998), while fibroblasts from *Brca2$^{-/-}$* embryos frequently contain chromosome structure abnormalities accompanied by growth arrest (Bertwistle *et al.*, 1997; Patel *et al.*, 1998). Tumors from *Brca2*-deficient animals often contain mutations in genes that regulate the spindle assembly checkpoint at G_2/M, such as *p53*, *Bub1*, and *Mad3L* (Lee *et al.*, 1999). Thus, cooperativity between mitotic checkpoint inactivation and BRCA2 deficiency may promote tumorigenesis.

Additional candidates for genes involved in maintaining genome stability are those involved in DNA replication or repair. The strongest evidence for this has been the discovery of a link between hereditary nonpolyposis colon cancer (HNPCC) and genes involved in the mismatch repair system (reviewed in Kinzler and Vogelstein, 1996). Linkage analysis identified five genes responsible for HNPCC: *hMLH1*, *hMSH2*, *hMSH6*, *PMS1*, and *PMS2*. The protein products of these five genes are involved in the human mismatch repair system which identifies and corrects DNA replication errors.

One feature of HNPCC tumors is the presence of instability within homopolymeric sequences and at small repeat sequences called **microsatellites.** Microsatellite instability results from slippages of DNA polymerase as it replicates these tracts of DNA and leads to deletion or insertion of additional repeat units (reviewed in Kinzler and Vogelstein, 1996). Microsatellite instability has been detected in a variety of tumor types and also in the earliest detectable stages of colon tumor development, indicating that mutations of the DNA mismatch repair system are early events in tumorigenesis (Shibata *et al.,* 1994; Augenlicht *et al.,* 1996; Heinen *et al.,* 1996).

Finally, the chromosome breakage syndromes such as ataxia telangiectasia, Bloom's syndrome, Fanconi anemia, Werner's syndrome, and xeroderma pigmentosum are autosomal recessive diseases that are characterized by increases in both chromosome instability and cancer predisposition. The discovery of the disease genes for these disorders has led to the identification of numerous proteins necessary for cells to replicate or repair their DNA. These proteins include helicases and kinases that respond to DNA damage, and proteins necessary for nucleotide excision repair. Mutation of these genes and the genes of the DNA mismatch repair system act in a recessive manner in cells (similar to that of tumor suppressors), although their absence does not directly affect growth regulation. Mutation of these genes results in a mutator phenotype within the cell which leads to an increased frequency of further mutation, the targets of which may then be those genes directly involved in growth control (Fig. 3).

Therefore, inherited predisposition to human cancer can occur by one of two genetic mechanisms. The first is by the inheritance or acquisition of a germline mutation in one of the genes involved in a growth control pathway for a particular cell type. The second is by the inheritance or acquisition of a germline mutation in a gene that controls the ability of a cell to replicate or repair its DNA, and whose disruption consequently is associated with an increased mutation frequency throughout the genome. The disruption of such genes can be recognized clinically by an increased incidence of a particular tumor type in affected individuals or by an increased incidence of many tumor types in affected individuals.

IV. Cell Adhesion and Motility

A cell that has lost growth control and formed a tumor eventually becomes restricted from further growth due to the size of the cell mass and the lack of nutrients in the immediate area. There are three possible outcomes of such circumstances: the tumor remains benign and growth ceases; the tumor induces angiogenesis, such that vascularization allows it to grow larger; or tumor cells undergo further mutations that allow them to invade surrounding normal tissue and metastasize to secondary sites. Cells of this last group also acquire alterations in adhesive and motile properties and often express proteases inappropriately. Such changes in cellular adhesion are poorly understood during tumorigenesis, although it is clear that cellular adhesion plays a fundamental role during normal development and is required for the maintenance of tissue integrity in the adult. Almost all cells *in vivo* are in contact with surrounding cells, either of the same or different tis-

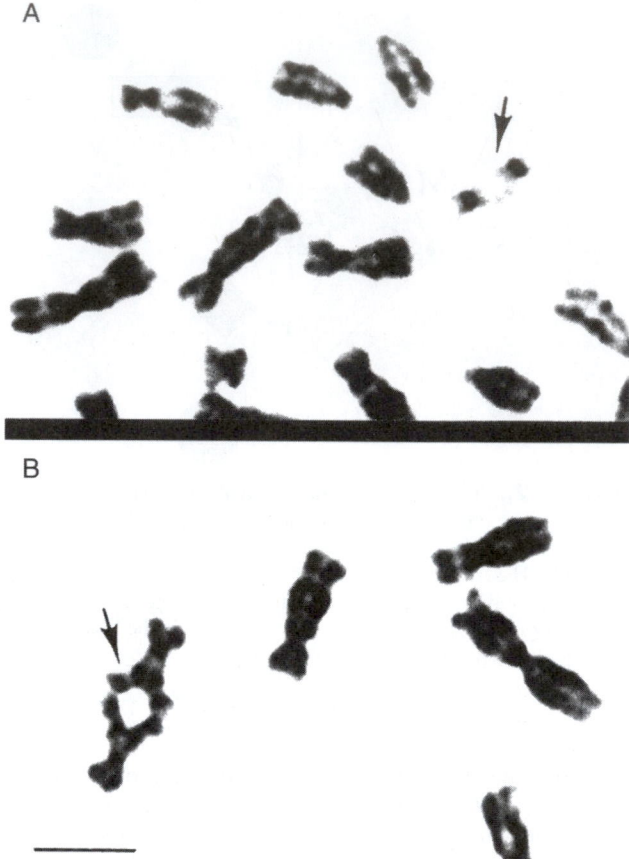

FIGURE 3. **Symmetrical quadriradial configurations (Qrs) from a Bloom's syndrome lymphocyte are examples of genomic instability.** Arrows point to Qrs in G-banded chromosome preparations. (A) Qr is due to an exchange between the long arms of chromosome 19. (B) Qr is due to an exchange between the long arms of chromosome 9 (bar scale: 5 μm). The increased genomic instability in Bloom's syndrome cells may explain the increased predisposition to cancer in these patients. (Courtesy of Steven Schonberg and James German, The New York Blood Center.)

sue type, and/or surrounding extracellular matrix (ECM). Cellular adhesion can be considered in two categories: cell-cell adhesions and cell-matrix adhesions.

The specificity of cellular adhesion is at the molecular level. Cell adhesion molecules or CAMs facilitate adhesion between a cell and another substrate, typically a cell surface or matrix. Adhesion between identical molecules is known as **homophilic adhesion,** whereas adhesion between distinct molecules is termed **heterophilic adhesion.** As discussed later, CAMs perform more than just an adhesive role (reviewed in Kirkpatrick and Peifer, 1995), and are arranged into families that share homology and functional domains but can also be categorized according to their binding specificities. The four major families are the immunoglobulin superfamily, the cadherins, the integrins, and the selectins (Fig. 4). These families are increasing in size, although they are becoming less distinct as functional domains are often shared between more than one family of molecules. For example, the protein tyrosine phosphatases function both as adhesion receptors and

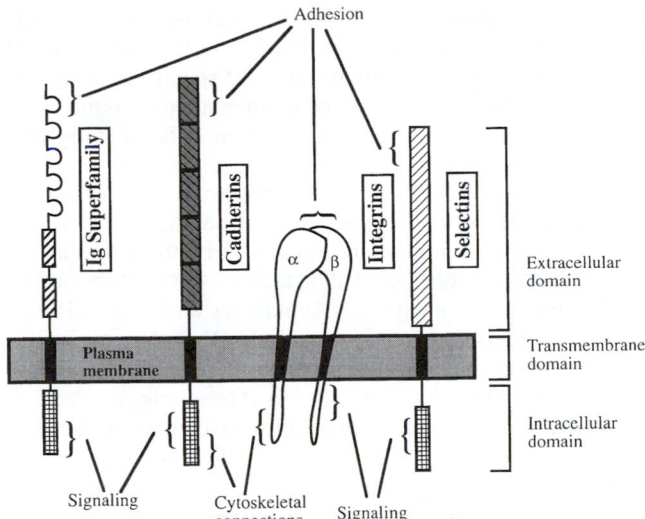

FIGURE 4. Structure and function of cell adhesion molecules. Four major families of CAMs, the immunoglobulin (Ig) superfamily, the cadherins, the integrins, and the selectins are diagrammed. Note that each CAM molecule contains an extracellular domain, a transmembrane domain, and a cytoplasmic domain. While there are common functional domains found *within* each family, note the common functions found *between* families. CAMs serve as adhesion molecules by way of extracellular domains. The CAMs also transduce signals from the outside of the cell to the inside of the cell via activation of regions in the cytoplasmic domain. These regions then bind cytoskeletal elements, thereby forming a line of communication between the outside of the cell and the motile apparatus.

signaling molecules (Brady-Kalnay *et al.,* 1995). CAMs contain an extracellular domain, a transmembrane domain, and a cytoplasmic domain. Regulation of CAM function occurs at the level of protein expression, conformational alteration of the protein, or biochemical modification of the CAM. This regulation is crucial in determining when a cell is adhesive.

E-cadherin is a classic example of a CAM whose function is modified in many tumors. Decreased E-cadherin expression is correlated with invasiveness in many types of malignant epithelial tumors (Christofori and Semb, 1999). Alternatively, molecules that regulate cadherin function, for example, the catenins, can be absent due to mutational inactivation or down-regulation, providing tumor cells with the same advantage for motility (Morton *et al.,* 1993; Sommers *et al.,* 1994; Breen *et al.,* 1995; Vermeulen *et al.,* 1995).

During development, many embryonic cells down-regulate certain adhesion molecules prior to their migration through the embryonic tissue. Once the cells arrive at their destination, those same molecules are up-regulated, to allow cells to coalesce and form mature tissue. Neural crest cells, a transient population of migratory embryonic cells, down-regulate N-CAM prior to their exit from the neural tube (Akitaya and Bronner-Fraser, 1992). As cell-cell adhesion is reduced, neural crest cells separate from one another and, following specific guidance cues, migrate through the developing embryo. Once these cells reach their destination, N-CAM is again up-regulated and the cells readhere in order to develop adult tissue such as the dorsal root ganglia of the peripheral nervous system. Function-blocking antibodies against N-CAM inhibit neural crest cell migration, demonstrating that it is involved in cell-cell adhesion and also in the cell-matrix adhesion required for migration (Bronner-Fraser *et al.,* 1992). Many early studies on neural crest cell migration are considered a template for current studies on metastasis. The basic questions are: what controls the detachment of cells from an originating population? What mechanisms are involved in their invasion of surrounding tissue and/or migration? What determines the final resting places where a neural crest cell will differentiate, or a neoplastic cell will proliferate and form a new tumor?

For migration to occur, a cell engages its motile apparatus by making connections from the ECM to the cellular cytoskeleton. CD44 is a transmembrane glycoprotein that binds the ECM component collagen extracellularly, and cytoskeletal components, such as ankyrdin, intracellularly. This binding is modified by phosphorylation, acylation, and GTP binding. Intracellular regulation of CD44 controls the adhesive properties of the extracellular domain of CD44, forming a line of communication between cell adhesion and the cell motility apparatus. Some metastatic carcinomas express isoforms of CD44 that do not interact with ankyrin in the correct manner, allowing the cells to lose their spatial restriction (Bourrguignon *et al.,* 1995). In addition, expression of the integrin class of ECM receptors is often misregulated during tumor formation (Keely *et al.,* 1998), resulting in the disruption of key ECM/cytoskeletal connections.

Reorganization of the cellular cytoskeleton is necessary for cell motility. Members of the Rho family of small GTPases, including Rho, Rac, and Cdc42, regulate the formation of cytoskeletal structures such as stress fibers, lamellipodia, and filopodia that are involved in migration. GTPase activation, then, plays a role in the ability of tumor cells to acquire a motile, invasive phenotype. For example, constitutively active Rac or Cdc42 induces breast epithelial cell motility and invasion through an ECM substrate (Keely *et al.,* 1997).

To facilitate migration, a cell secretes proteases to digest the local ECM and to induce surrounding stromal cells to produce matrix-degrading proteases. Many advanced tumors express increased levels of matrix metalloproteinases (MMPs) as compared to normal tissues or noninvasive tumors, suggesting an important role for these enzymes in invasion and metastasis (Chambers and Matrisian, 1997). *In vivo* studies of malignant breast cancer demonstrated elevated levels of the membrane type MMP (MT-MMP), stromelysin-3, and gelatinase A mRNA in advanced tumors as compared to normal breast tissue and noninvasive tumors (Heppner *et al.,* 1996). Conversely, tissue inhibitors of metalloproteinases (TIMPs) are negative regulators of MMPs, and their down-regulation correlates with increased invasiveness of glioblastomas and astrocytomas (Mohanam *et al.,* 1995). Additionally, inhibition of MMPs with synthetic inhibitors suppresses tumor invasion and metastasis in animals (Chirivi *et al.,* 1994; Watson *et al.,* 1995). Other proteases are similarly implicated in tumor cell motility. The expression of serine proteases, such as plasmin and urokinase-type plasminogen activator, cysteine proteases, such as cathepsin B, and aspartate proteases, such as cathepsin D, have been correlated with a more invasive tumor phenotype (Andreasen *et al.,* 1997; Sivaparvathi *et al.,* 1995; Rochefort and Liaudet-Coopman, 1999).

The presence of antiadhesive molecules, such as proteoglycans, mucins, laminins, and thrombospondins, has been well documented in the developing nervous system (Chiquet-Ehrismann, 1995), where they prohibit the migration of certain cell types. These molecules may play a role in restricting invading tumor cells to a more limited terrain. On the other hand, if such antiadhesive molecules are inappropriately down-regulated, invasion may increase.

The combined effects of cellular adhesion, motility, and protease secretion must be considered when studying tumor progression and metastasis. The events involved in the progression to the metastatic state can be considered as a recapitulation of normal developmental events, although one difference is that changes in a developing cell are preprogrammed while the changes in a cancer cell are the result of mutation (Fig. 5). In teleological terms, the tumor cells undergo similar changes to normal cells, given that they have equivalent objectives: to migrate to a distant site, to settle there, and to divide and populate the new area.

V. Apoptosis

The size of cell populations is determined not only by the rate of cell proliferation, but also by the rate of cell death. Thus, tumor formation can result from the deregulation of proper growth control, as well as from the deregulation of apoptosis, or programmed cell death. Apoptosis is a tightly regulated process of cell death necessary for proper development and cell turnover (Kerr et al., 1972). It occurs in all multicellular organisms, its timing and specificity essential to both normal development and maintenance of tissue function. Apoptosis is biochemically and morphologically distinct from necrosis or cell death due to tissue injury.

The biochemical changes in an apoptotic cell begin with an external or internal trigger for cell death. External triggers for apoptosis include removal of growth factors or the presence of a death signal that activates receptors on the cell surface (for review see Ashkenazi and Dixit, 1998). Internal triggers for apoptosis include DNA damage or improper microtubule formation (for review see Evan and Littlewood, 1998; Li et al., 1998). These triggers lead to the activation of signal transduction cascades that activate proteases known as caspases (for review see Thornberry and Lazebnik, 1998). Caspases cleave specific protein substrates in the cell and can be used experimentally as markers for apoptotic cell death (Fig. 6). Caspase activation produces a variety of distinct morphological changes in the cell. For example, during apoptosis, chromatin compacts and associates with the nuclear envelope. The cell and the nucleus then convolute extensively, forming discrete, membrane-enclosed cellular fragments. These cell fragments are then phagocytosed by surrounding cells and degraded inside lysosomal vacuoles. No tissue inflammation is present and apoptosis can occur on a single-cell basis without affecting the status of neighboring cells.

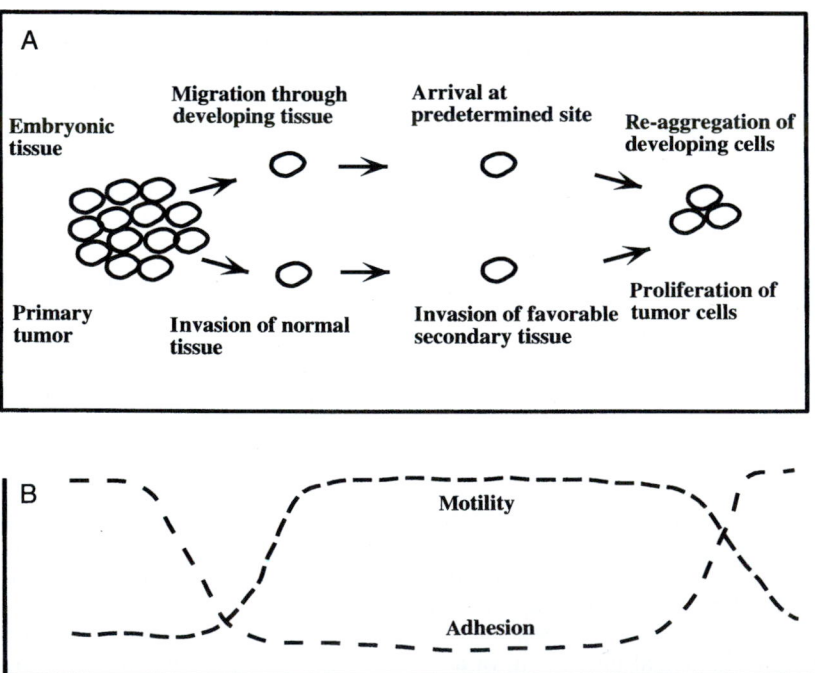

FIGURE 5. Tumor metastasis is a recapitulation of development. There are parallels between cellular events that take place during development and metastasis. (A) depicts a population of cells that represents embryonic cells in the upper portion of the figure and primary tumor cells in the lower portion. In both scenarios, the migrating cell will undergo similar changes in the relative levels of adhesion and motility (B). In order for the cell to detach from its original mass, it must down-regulate adhesion molecules. In order for the cell to migrate, it must up-regulate molecules involved in motility. Once the embryonic cell reaches its destination, or the tumor cell finds itself in a suitable environment, the opposite occurs: adhesion is up-regulated and motility is down-regulated.

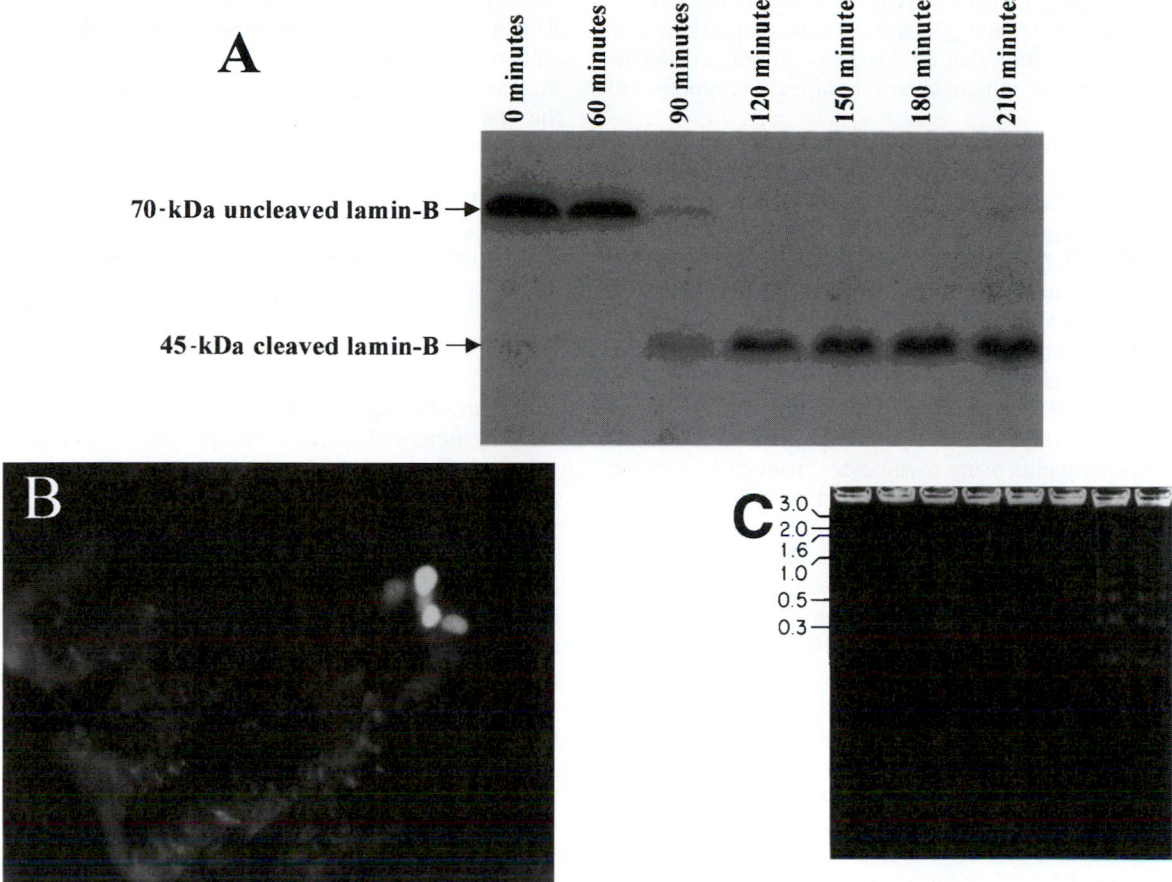

FIGURE 6. **Specific caspase-mediated protein cleavage events are characteristic of the apoptotic process.** (A) Western blot showing cleavage of the 70 kDa nuclear protein, lamin-B, into a 45 kDa fragment. Lamin-B is cleaved during apoptosis by caspases-3 and -6. Lamin-B is from a normal mouse liver nuclei source incubated with *Xenopus* egg extract competent for apoptosis (Newmeyer *et al.,* 1994). Time in minutes is listed above each lane. Caspase-mediated cleavage is visible at 90 minutes. (B) Cross-section of normal mouse colon showing brightly staining epithelial cells undergoing apoptosis at the end of their life span. Cells are stained using the terminal deoxynucleotidyl transferase-mediated dUTP nick end labeling (TUNEL) method which adds a fluorescent tag to double-strand DNA breaks and single-strand DNA nicks present in high numbers in apoptotic cells. (C) Ethidium bromide-stained agarose gel showing degradation of genomic DNA into nucleosomal fragments. DNAs are from a transformed cell line that is dependent on interleukin-3 (IL-3) for growth. When IL-3 is removed from the culture medium, the cell line undergoes apoptosis. Evidence for this process is seen in lanes 7 and 8. Size markers are shown on the left in kilobases. This DNA laddering is the result of caspase-mediated cleavage and inactivation of a DNase inhibitor, allowing increased DNase activity in the dying cell. (Courtesy of David Askew, Ph.D., University of Cincinnati College of Medicine).

During normal development of multicellular organisms, subsets of cells undergo apoptosis as an obligate part of the developmental process to ensure proper cell number and body shape (Sanders and Wride, 1995). Apoptosis is also essential in the developing immune system for elimination of T-cells in the thymus both during positive and negative selection. In the adult, apoptosis maintains a constant cell number in continuously renewing cell populations such as hematopoietic and epithelial cells.

When apoptotic mechanisms are disrupted, the ability to remove unwanted cells is lost. Thus the down-regulation of apoptosis is associated with a number of different tumor types. A cytogenetic hallmark of chronic myeloid leukemia (CML) is a balanced translocation, t(9;22), known as the Philadelphia chromosome. This translocation creates a chimeric gene encoding the fusion protein BCR-ABL, which suppresses apoptosis *in vitro*. Inhibition of BCR-ABL using antisense strategies reverses this effect. The relative rates of cell proliferation are not increased in CML (Strife and Clarkson, 1988), suggesting that the primary mechanism by which BCR-ABL increases cell number is through the suppression of apoptosis (for review see Deininger and Goldman, 1998).

In a second example, most non-Hodgkin's B-cell lymphomas are associated with a t(14;18) translocation. These translocations disrupt normal apoptotic function by joining the *BCL-2* gene with the immunoglobulin heavy-chain locus and increasing the expression of *BCL-2*. Normally, BCL-2 will suppress apoptosis (Vaux *et al.,* 1988). In tumors, increased BCL-2 increases cell number by decreasing cell death. Additionally, increased BCL-2 *in vitro* confers resistance to killing by some glucocorticoids (Miyashita and Reed, 1995), suggesting that the apoptotic characteristics of a tumor can dictate which chemotherapeutic regimes are more effective.

Apoptosis plays a role in maintaining consistent epithelial cell populations. Therefore, loss or down-regulation of apoptotic function can be involved in the formation of carcinomas. An important example is the loss of apoptotic activity associated with the transformation of normal colorectal epithelium to carcinoma (Bedi *et al.*, 1995). Gross rearrangements of *BCL-2*, like the lymphoma-related *BCL-2* translocations, are not generally observed in carcinomas, suggesting that there are alternative mechanisms for dysregulation of apoptosis in these tumor types. One such mechanism is the alteration of p53 function. The importance of p53 in apoptosis was demonstrated by experiments using thymocytes from *p53*-knockout mice; comparisons of $p53^{-/-}$ and normal thymocytes treated with ionizing radiation revealed that the $p53^{-/-}$ cells were resistant to radiation-induced apoptosis. The normal cells, meanwhile, underwent massive apoptosis (Lowe *et al.*, 1993; Clark *et al.*, 1993). It is now well accepted that p53 functions in response to DNA damage, either by arresting the cell cycle at the G_1/S boundary to facilitate DNA repair, or by its ability to initiate apoptosis and eliminate cells that have acquired DNA damage. The loss of p53 function in many cancers is consistent with the idea that tumors acquire mutations that result in the inhibition of apoptotic pathways.

Normal p53 protein mediates apoptosis by several different mechanisms. It is a transcription factor that down-regulates the expression of *BCL-2 in vitro* via a 195 bp segment in the 5' untranslated region of *BCL-2*. This down-regulation of *BCL-2* expression results in an up-regulation of apoptosis. p53 up-regulates the expression of *BAX*, a BCL-2-related protein that is a dominant-inhibitor of BCL-2 (Miyashita and Reed, 1995). The *BAX* gene is regulated at the level of expression via p53 binding to the *BAX* promoter via four p53-binding motifs (Miyashita and Reed, 1995). Radiation-induced DNA damage up-regulates *BAX* expression through p53 (Miyashita *et al.*, 1994). In this case, by up-regulating a pro-apoptotic protein, the net effect of p53 activity is to up-regulate apoptosis. Normal p53 protein also up-regulates apoptosis by increasing expression of the FAS receptor, known as one of the death receptors (for review see Gottleib and Oren, 1998). This occurs through a p53 responsive element within the first intron of the *FAS* gene. The resultant increase in FAS receptor provides for FAS-mediated apoptosis if a positive death-inducing signal is received by the cell via binding of FAS-ligand. Conversely, apoptosis can be induced by the absence of anti-apoptotic signals provided by growth factors. p53 up-regulates expression of the insulin-like growth factor 1 binding protein 3 (IGF-BP3) which inhibits signaling by the insulin-like growth factor 1. This growth factor normally provides an anti-apoptotic signal to the cell. Again, the net effect on the cell as a result of p53 signaling is pro-apoptotic. These four examples of downstream effector pathways of p53 are just a subset of p53-dependent mechanisms of apoptosis.

Colorectal carcinomas usually carry mutations in both copies of the *p53* gene. In addition, the earliest mutation in the formation of most of these carcinomas is mutation of the *APC* gene. *APC* mutations increase the number of epithelial cells in an intestinal crypt via a decrease in apoptosis or an increase in cell growth. Gene transfer studies have shown that the introduction of *APC* to tumorigenic cells in culture decreases cell number by both increasing apoptosis and decreasing cell growth (Morin *et al.*, 1996; Heinen *et al.*, 2000). Therefore, APC may be part of a normal developmental process in the colonic crypts, where apoptosis of the luminal epithelial cells is constantly occurring.

VI. Summary

Proteins that regulate the cell cycle have been identified by studying inherited diseases that predispose to cancer and by genetic analyses of different tumor types. Processes that regulate DNA replication and cell division are also intimately involved with the maintenance of the genome. Gene mutations that increase genomic instability allow cells to accumulate further mutations at a rapid rate. Such mutator phenotypes may be essential characteristics of many tumors, facilitating the mutations that a cell must acquire before becoming tumorigenic. Malignant tumors are also characterized by the ability to invade surrounding normal tissues. Cells then metastasize to new sites through subsequent changes in the expression of proteins that regulate cell adhesion and motility. Finally, unregulated tumor growth can result from the inability of tumor cells to induce appropriate cell death.

As cells replicate their DNA and undergo cell division, they must coexist and cooperate with neighboring cells in order to compose different tissues. Therefore, the growth and life span of each cell are tightly regulated, with a precise balance between cell division and its inhibition. When the regulatory systems that maintain this balance are disrupted, the scales tilt in favor of growth with a subsequent increase in cell number. Such altered cells ultimately produce clonal populations of cells, or what we see clinically as tumors. Comparative studies of normal and transformed cells have identified many of the proteins and processes that maintain the normal physiology of the cell. Ultimately, this information will lead to new therapies that can modulate the systems that are disrupted in tumors.

Bibliography

Akitaya, T., and Bronner-Fraser, M. (1992). Expression of cell adhesion molecules during initiation and cessation of neural crest cell migration. *Dev. Dyn.* **194,** 12–20.

Andreasen, P. A., Kjoller, L., Christensen, L., and Duffy, M. J. (1997). The urokinase-type plasminogen activator system in cancer metastasis: A review. *Int. J. Cancer* **72,** 1–22.

Ashkenazi, A., and Dixit, V. M. (1998). Death receptors: Signaling and modulation. *Science* **281,** 1305–1308.

Augenlicht, L. H., Richards, C., Corner, G., and Pretlow, T. P. (1996). Evidence for genomic instability in human colonic aberrant crypt foci. *Oncogene* **12,** 1767–1772.

Bar-Sagi, D., and Feramisco, J. R. (1985). Micro-injection of the *ras* oncogene protein into PC12 cells induces morphological differentiation. *Cell* **42,** 841–848.

Bedi, A., Pasricha, P., Akhtar, A., Barber, J., Bedi, G., Giardiello, F., Zehnbauer, B., Hamilton, S., and Jones, R. (1995). Inhibition of apoptosis during development of colorectal cancer. *Cancer Res.* **55**, 1811–1816.

Behrens, J., von Kries, J. P., Kuhl, M., Bruhn, L., Wedlich, D., Grosschedl, R., and Birchmeier, W. (1996). Functional interaction of β-catenin with the transcription factor LEF-1. *Nature* **382**, 638–642.

Benedict, W. F., Xu, H. J., Hu, S. X., and Takahashi, R. (1990). Role of the retinoblastoma gene in the initiation and progression of human cancer. *J. Clin. Invest.* **85**, 988–993.

Bertwistle, D., Swift, S., Marston, N. J., Jackson, L. E., Crossland, S., Crompton, M. R., Marshall, C. J., and Ashworth, A. (1997). Nuclear location and cell cycle regulation of the BRCA2 protein. *Cancer Res.* **57**, 5485–5488.

Bokoch, G. M., and Der, C. J. (1993). Emerging concepts in the *Ras* superfamily of GTP-binding proteins. *FASEB* **7**, 750–759.

Bourrguignon, L. Y., Iida, N., Welsh, C. F., Zhu, D., Krongrad, A., and Pasquale, D. (1995). Involvement of CD44 and its variant isoforms in membrane-cytoskeleton interaction, cell adhesion and tumor metastasis. *J. Neurooncol.* **26**, 201–208.

Brady-Kalnay, S. M., and Tonks, N. K. (1995). Protein tyrosine phosphatases as adhesion receptors. *Curr. Opin. Cell. Biol.* **7**, 650–657.

Breen, E., Steele, G., Jr., and Mercurio, A. M. (1995). Role of the E-cadherin/alpha-catenin complex in modulating cell-cell and cell-matrix adhesive properties of invasive colon carcinoma cells. *Ann. Surg. Oncol.* **2**, 378–385.

Bronner-Fraser, M., Wolf, J. J., and Murray, B. A. (1992). Effects of antibodies against N-cadherin and N-CAM on the cranial neural crest and neural tube. *Dev. Biol.* **153**, 291–301.

Chambers, A. F., and Matrisian, L. M. (1997). Changing views of the role of matrix metalloproteinases in metastasis. *J. Natl. Cancer Inst.* **89**, 1260–1270.

Chiquet-Ehrismann, R. (1995). Inhibition of cell adhesion by anti-adhesive molecules. *Curr. Opin. Cell. Biol.* **7**, 715–719.

Chirivi, R. G. S., Garofalo, A., Crimmin, M. J., Bawden, L. J., Stoppacciaro, A., Brown, P. D., and Giavazzi, R. (1994). Inhibition of the metastatic spread and growth of B16-BL6 murine melanoma by a synthetic matrix metalloproteinase inhibitor. *Int. J. Cancer* **58**, 460–464.

Christofori, G., and Semb, H. (1999). The role of the cell-adhesion molecule E-cadherin as a tumour-suppressor gene. *Trends Biochem. Sci.* **24**, 73–76.

Clarke, A. R., Purdie, C. A., Harrison, D. J., Morris, R. G., Bird, C. C., Hooper, M. L., and Wyllie, A. H. (1993). Thymocyte apoptosis induced by p53-dependent and independent pathways. *Nature* **362**, 849–852.

DeGregori, J., Leone, G., Ohtani, K., Miron, A., and Nevins, J. R. (1995). E2F-1 accumulation bypasses a G1 arrest resulting from the inhibition of G1 cyclin-dependent kinase activity. *Genes Dev.* **9**, 2873–2887.

Deininger, M. W., and Goldman, J. M. (1998). Chronic myeloid leukemia. *Curr. Opin. Hematol.* **5**, 302–308.

Evan, G., and Littlewood, T. (1998). A matter of life and cell death. *Science* **281**, 1317–1322.

Goldfarb, M., Shimizu, K., Perucho, M., and Wigler, M. (1982). Isolation and preliminary characterization of a human transforming gene from T24 bladder carcinoma cells. *Nature* **296**, 404–409.

Gottleib, T. M., and Oren, M. (1998). p53 and apoptosis. *Cancer Biol.* **8**, 359–368.

Gowen, L. C., Avrutskaya, A. V., Latour, A. M., Koller, B. H., and Leadon, S. A. (1998). BRCA1 required for transcription-coupled repair of oxidative DNA damage. *Science* **281**, 1009–1012.

Groden, J., Thliveris, A., Samowitz, W., Carlson, M., Gelbert, L., Albertson, H., Joslyn, G., Stevens, J., Spirio, L., Robertson, M., Sargeant, L., Krapcho, K., Wolff, E., Burt, R., Hughes, J. P., Warrington, J., McPherson, J., Wasmuth, J., Le Paslier, D., Abderrahim, H., Cohen, D., Leppert, M., White, R. (1991). Identification and char-

acterization of the familial adenomatous polyposis coli gene. *Cell* **66**, 589–600.

Groden, J., Joslyn, G., Samowitz, W., Jones, D., Bhattacharyya, N., Spirio, L., Thliveris, A., Robertson, M., Egan, S., and Meuth, M. (1995). Response of colon cancer cell lines to the introduction of APC, a colon-specific tumor suppressor gene. *Cancer Res.* **55**, 1531–1539.

Grunicke, H. (1990). Signal transduction mechanisms in cancer. *Biochem. Soc. Trans.* **18**, 67–69.

Hall, M., and Peters, G. (1996). Genetic alterations of cyclins, cyclin-dependent kinases, and CDK inhibitors in human cancer. *Adv. Cancer Res.* **68**, 67–108.

He, T. C., Sparks, A. B., Rago, C., Hermeking, H., Zawel, L., da Costa, L. T., Morin, P. J., Vogelstein, B., and Kinzler, K. W. (1998). Identification of c-MYC as a target of the APC pathway. *Science* **281**, 1509–1512.

Heinen, C. D., Heppner Goss, K., Cornelius, J. R., Babcock, G. F, Knudsen, E. S., Kowalik, T., and Groden, J. (2000). The APC tumor suppressor controls entry into S-phase through its ability to regulate β-catenin and the cyclin D/RB pathway. Manuscript submitted.

Heinen, C. D., Shivapurkar, N., Tang, Z., Groden, J., and Alabaster, O. (1996). Microsatellite instability in aberrant crypt foci from human colons. *Cancer Res.* **56**, 5339–5341.

Heppner, K. J., Matrisian, L. M., Jensen, R. A., and Rodgers, W. H. (1996). Expression of most matrix metalloproteinase family members in breast cancer represents a tumor-induced host response. *Am. J. Pathol.* **149**, 273–282.

Hinds, P. W., and Weinberg, R. A. (1994). Tumor suppressor genes. *Curr. Opin. Genet. Dev.* **4**, 135–141.

Huber, O., Korn, R., McLaughlin, J., Ohsugi, M., Herrmann, B. G., and Kemler, R. (1996). Nuclear localization of β-catenin by interaction with transcription factor LEF-1. *Mech. Dev.* **59**, 3–10.

Hunter, T., and Pines, J. (1994). Cyclins and cancer II: Cyclin D and CDK inhibitors come of age. *Cell* **79**, 573–582.

Johnson, D. G., Schwarz, J. K., Cress, W. D., and Nevins, J. R. (1993). Expression of transcription factor E2F1 induces quiescent cells to enter S phase. *Nature* **365**, 349–352.

Kastan, M. B., Onyekwere, O., Sidransky, D., Vogelstein, B., and Craig, R. W. (1991). Participation of p53 protein in the cellular response to DNA damage. *Cancer Res.* **51**, 6304–6311.

Keely, P. J., Westwick, J. K., Whitehead, I. P., Der, C. J., and Parise, L.V. (1997). Cdc42 and Rac1 induce integrin-mediated cell motility and invasiveness through PI(3)K. *Nature* **390**, 632–636.

Keely, P., Parise, L., and Juliano, R. (1998). Integrins and GTPases in tumour cell growth, motility and invasion. *Trends Cell. Biol.* **8**, 101–106.

Kerr, J. F. R., Wyllie, A. H., and Currie, A. R. (1972). Apoptosis: Basic biological phenomenon with wide-ranging implications in tissue kinetics. *Br. J. Cancer* **26**, 239–257.

Kinzler, K. W., and Vogelstein, B. (1996). Lessons from hereditary colorectal cancer. *Cell* **87**, 159–170.

Kinzler, K. W., Nilbert, M. C., Su, L. K., Vogelstein, B., Bryan, T. M., Levy, D. B., Smith, K. J., Preisinger, A. C., Hedge, P., and McKechnie, D. (1991). Identification of FAP locus genes from chromosome 5q21. *Science* **253**, 661–665.

Kirkpatrick, C., and Peifer, M. (1995). Not just glue: Cell-cell junctions as cellular signaling centers. *Curr. Opin. Genet. Dev.* **5**, 56–65.

Knudson, A. G. (1971). Mutation and cancer: Statistical study of retinoblastoma. *Proc. Natl. Acad. Sci. USA* **68**, 820–823.

Kuerbitz, S. J., Plunkett, B. S., Walsh, W. V., and Kastan, M. B. (1992). Wild-type p53 is a cell cycle checkpoint determinant following irradiation. *Proc. Natl. Acad. Sci. USA* **89**, 7491–7495.

Lee, H., Trainer, A. H., Friedman, L. S., Thistlethwaite, F. C., Evans, M. J., Ponder, B. A. J., and Venkitaraman, A. R. (1999). Mitotic checkpoint inactivation fosters transformation in cells lacking the breast cancer susceptibility gene Brca2. *Mol. Cell* **4**, 1–10.

Li, F., Ambrosini, G., Chu, E. Y., Plescia, J., Tognin, S., Marchisio, P. C., and Altieri, D. C. (1998). Control of apoptosis and mitotic spindle checkpoint by survivin. *Nature* **396**, 580–584.

Livingstone, L. R., White, A., Sprouse, J., Livanos, E., Jacks, T., and Tlsty, T. D. (1992). Altered cell cycle arrest and gene amplification potential accompany loss of wild-type p53. *Cell* **70**, 923–935.

Loeb, L. A., Springgate, C. F., and Battula, N. (1974). Errors in DNA replication as a basis of malignant change. *Cancer Res.* **34**, 2311–2321.

Lowe, S. W., Schmitt, E. M., Smith, S. W., Osborne, B. A., and Jacks, T. (1993). p53 is required for radiation-induced apoptosis in mouse thymocytes. *Nature* **363**, 847–849.

Lukas, J., Bartkova, J., Rohde, M., Strauss, M., and Bartek, J. (1995a). Cyclin D1 is dispensable for G1 control in retinoblastoma gene-deficient cells independently of CDK4 activity. *Mol. Cell. Biol.* **15**, 2600–2611.

Lukas, J., Parry, D., Aagaard, L., Mann, D. J., Bartkova, J., Strauss, M., Peters, G., and Bartek, J. (1995b). Retinoblastoma-protein-dependent cell-cycle inhibition by the tumour suppressor p16. *Nature* **375**, 503–506.

Lukas, J., Petersen, B. O., Holm, K., Bartek, J., and Helin, K. (1996). Deregulated expression of E2F family members induces S-phase entry and overcomes p16^{INK4a}-mediated growth suppression. *Mol. Cell. Biol.* **16**, 1047–1057.

Malkova, A., Ivanov, E. L., and Haber, J. E. (1996). Double-strand break repair in the absence of RAD51 in yeast: A possible role for break-induced DNA replication. *Proc. Natl. Acad. Sci. USA* **93**, 7131–7136.

Medema, R. H., Herrera, R. E., Lam, F., and Weinberg, R. A. (1995). Growth suppression by p16^{INK4} requires functional retinoblastoma protein. *Proc. Natl. Acad. Sci. USA* **92**, 6289–6293.

Miyashita, T., and Reed, J. C. (1995). Tumor suppressor p53 is a direct transcriptional activator of the human bax gene. *Cell* **80**, 293–299.

Miyashita, T., Harigai, M., Hanada, M., and Reed, J. C. (1994). Identification of a p53-dependent negative response element in the bcl-2 gene. *Cancer Res.* **54**, 3131–3135.

Mohanam, S., Wang, S. W., Rayford, A., Yamamoto, M., Sawaya, R., Nakajima, M., Liotta, L. A., Nicolson, G. L., Stetler-Stevenson, W. G., and Rao, J. S. (1995). Expression of tissue inhibitors of metalloproteinases: Negative regulators of human glioblastoma invasion *in vivo*. *Clin. Exp. Metastasis*. **13**, 57–62.

Molenaar, M., van de Wetering, M., Oosterwegel, M., Peterson-Maduro, J., Godsave, S., Korinek, V., Roose, J., Destree, O., and Clevers, H. (1996). XTcf-3 transcription factor mediates β-catenin-induced axis formation in *Xenopus* embryos. *Cell* **86**, 391–399.

Morin P. J., Vogelstein, B., and Kinzler, K. W. (1996). Apoptosis and APC in colorectal tumorigenesis. *Proc. Natl. Acad. Sci. USA* **93**, 7950–7954.

Morton, R. A., Ewing, C. M., Nagafuchi, A., Tsukita, S., and Isaacs, W. B. (1993). Reduction of E-cadherin levels and deletion of the alpha-catenin gene in human prostate cancer cells. *Cancer Res.* **53**, 3585–3590.

Newmeyer, D. D., Farschon, D. M., and Reed, J. C. (1994). Cell-free apoptosis in *Xenopus* extracts: Inhibition by Bcl-2 and requirement for an organelle fraction enriched in mitochondria. *Cell* **79**, 353–364.

Nishisho, I., Nakamura, Y., Miyoshi, Y., Miki, Y., Ando, H., Horii, A., Koyama, K., Utsunomiya, J., Baba, S., Hedge, P., Markham, A., Krush, A. J., Peterson, G. Hamilton, S. R., Nilbert, M. C., Levy, D. B., Bryan, T. M., Preisinger, A., Smith, K. J., Su, L. K., Kinzler, K. W., Vogelstein, B. (1991). Mutations of chromosome 5q21 genes in FAP and colorectal cancer patients. *Science* **253**, 665–669.

Patel, I. C. J., Vu, V. P., Lee, H., Corcoran, A., Thistlethwaite, F. C., Evans, M. J., Colledge, W. H., Friedman, L. S., Ponder, B. A., and

Venkitaraman, A. R. (1998). Involvement of BRCA2 in DNA repair. *Mol. Cell* **1**, 347–357.

Powell, S. M., Zilz, N., Beazer-Barclay, Y., Bryan, T. M., Hamilton, S. R., Thibodeau, S. N., Vogelstein, B., and Kinzler, K. W. (1992). APC mutations occur early during colorectal tumorigenesis. *Nature* **359**, 235–237.

Qin, X. Q., Livingston, D. M., Ewen, M., Sellers, W. R., Arany, Z., and Kaelin, W. G., Jr. (1995). The transcription factor E2F-1 is a downstream target of RB action. *Mol. Cell. Biol.* **15**, 742–755.

Rochefort, H., and Liaudet-Coopman, E. (1999). Cathepsin D in cancer metastasis:A protease and a ligand. *APMIS* **1999**, 86–95.

Sanders, E. J., and Wride, M. A. (1995). Programmed cell death in development. *Int. Rev. Cytology* **163**, 105–173.

Scully, R., Chen, J., Plug, A., Xiao, Y., Weaver, D., Feunteun, J., Ashley, T., and Livingston, D. M. (1996). Association of BRCA1 with Rad51 in mitotic and meiotic cells. *Cell* **88**, 265–275.

Sherr, C. J. (1994). G1 phase progression: Cycling on cue. *Cell* **79**, 551–555.

Sherr, C. J. (1996). Cancer cell cycles. *Science* **274**, 1672–1677.

Shibata, D., Peinado, M. A., Ionov, Y., Malkhosyan, S., and Perucho, M. (1994). Genomic instability in repeated sequences is an early somatic event in colorectal tumorigenesis that persists after transformation. *Nature Genet.* **6**, 273–281.

Sivaparvathi, M., Sawaya, R., Wang, S. W., Rayford, A., Yamamoto, M., Liotta, L. A., Nicolson, G. L., and Rao, J. S. (1995). Overexpression and localization of cathepsin B during the progression of human gliomas. *Clin. Exp. Metastasis* **13**, 49–56.

Sommers, C. L., Gelmann, E. P., Kermler, R., Cowin, P., and Byers, S. W. (1994). Alterations in beta-catenin phosphorylation and plakoglobin expression in human breast cancer cells. *Cancer Res.* **54**, 3544–3552.

Strife, A., and Clarkson, B. (1988). Biology of chronic myelogenous leukemia: Is discordant maturation the primary defect? *Semin. Hematol.* **25**, 1–19.

Sung, P., and Robberson, D. L. (1995). DNA strand exchange mediated by a RAD51-ssDNA nucleoprotein filament with polarity opposite to that of RecA. *Cell* **82**, 453–461.

Suzuki, H., Harpaz, N., Tarmin, L., Yin, J., Jiang, H-Y., Bell, J. D., Hontanosas, M., Groisman, G. M., Abraham, J. M., and Meltzer, S. J. (1994). Microsatellite instability in ulcerative colitis-associated colorectal dysplasias and cancers. *Cancer Res.* **54**, 4841–4844.

Tetsu, O., and McCormick, F. (1999). β-catenin regulates expression of cyclin D1 in colon carcinoma cells. *Nature* **398**, 422–426.

Thornberry, N. A., and Lazebnik, Y. (1998). Caspases: Enemies within. *Science* **281**, 1312–1316.

Vaux, D. L., Cory, S., and Adams, J. M. (1988). Bcl-2 gene promotes haemopoietic cell survival and cooperates with c-myc to immortalize pre-B cells. *Nature* **335**, 440–442.

Vermeulen, S. J., Bruyneel, E. A., Bracke, M. E., De Bruyne, G. K., Vennekens, K. M., Vleminckx, K. L., Berx, G. J., van Roy, F. M., and Mareel, M. M. (1995). Transition from the noninvasive to the invasive phenotype and loss of alpha-catenin in human colon cancer cells. *Cancer Res.* **55**, 4722–4728.

Watson, S. A., Morris, T. M., Robinson, G., Crimmin, M. J., Brown, P. D., and Hardcastle, J. D. (1995). Inhibition of organ invasion by the matrix metalloproteinase inhibitor batimistat (BB-94) in two human colon carcinoma metastasis models. *Cancer Res.* **55**, 3629–3633.

Wong, A. K. C., Pero, R., Ormonde, P. A., Tavtigian, S. V., and Bartel, P. L. (1997). RAD51 interacts with the evolutionarily conserved BRC motifs in the human breast cancer susceptibility gene BRCA2. *J. Biol. Chem.* **272**, 31941–31944.

Yin, Y., Tainsky, M. A., Bischoff, F. Z., Strong, L. C., and Wahl, G. M. (1992). Wild-type p53 restores cell cycle control and inhibits gene amplification in cells with mutant p53 alleles. *Cell* **70**, 937–948.

Agustín Guerrero and Juan Manuel Arias

68

Apoptosis

I. Introduction

Apoptosis is a program for cell deletion triggered by either physiological or noxious signals. For many reasons, apoptosis, or **programmed cell death,** has been the subject of intense work in recent years. First, it has been recognized that the quantity of cells in a tissue depends on the balance between cell death and cell division. Second, important biochemical and molecular advances have occurred in the elucidation of the pathways triggering and regulating cell death. Third, the underlying cause of some diseases can be the consequence of alterations in the cell death machinery; for instance tumoral growth, which can be generated by either a reduced rate of cell death or an increased rate of cell division. Conversely, degenerative disorders can arise from a reduced production of cells or an enhanced rate of cell death. Furthermore, physiological cell death is an essential process in the ontogeny of multicellular organisms. In this respect, it has been proposed that normal cell death is involved in different embryonic processes such as the development of the central nervous system, the lumina of tubular structures, the shaping of limbs, etc. It has been suggested that physiological cell death is programmed particularly during development, implying that some internal molecular clock is involved in triggering cell death. Nevertheless, there are also many reports of apoptosis being triggered by either external signals or the absence of trophic factors. It is quite possible that both internal and external signals are involved in regulating the fate of a given cell. Interestingly, a picture is emerging where the signals for cell growth and cell death are so closely intertwined that they probably represent different manifestations of a common theme. Examples of that are the **nerve growth factor (NGF)** and the **T-cell receptor (TCR).** The former induces either growth or cell death in neurons; the latter induces either differentiation or cell death in thymocytes. How the same signal or receptor could produce such divergent results, is just one of many areas of study in the field of apoptosis.

II. Morphological Characterization of Cell Death

Physiological cell death was described for the first time based on morphological grounds only, and was given the name of **apoptosis** (Kerr *et al.,* 1972). This was defined as a physiological cell death occurring sporadically in tissues, that did not generate an inflammatory reaction, and which could be either programmed or induced, but it was morphologically different from necrosis. It was also proposed that apoptosis was an active process because messenger RNA (mRNA) and protein synthesis appeared to be required for cell death to occur. At that time, it was appreciated that apoptosis could be an essential element in tissue homeostasis by balancing cell division with cell death. It was also suggested that apoptosis could be important in ontogeny and in some pathological processes (Kerr *et al.,* 1972).

Strictly speaking then, apoptosis depicts only the morphological changes of dying cells in a tissue. Therefore, to name a given cell death event apoptosis, the event has to fulfill the morphological characterization described later. This, together with the absence of *in vitro* models (that is, before the appearance of molecular and biochemical markers of apoptosis), retarded elucidation of the mechanisms involved in physiological cell death. Such models for studying apoptosis *in vitro* have emerged by combining morphological, biochemical, and molecular approaches to the point that apoptosis can now be studied in cell-free extracts. Despite this progress, there is still an incipient classification of the different types of cell deletion. Necrosis and apoptosis are two different expressions of cell death that are clearly recognized, but there may be more forms of cell deletion.

A. Necrosis

Necrosis is considered a nonphysiological cell death. This occurs apparently because of an extremely toxic stimulus or a massive cell injury. Cell death by necrosis is characterized by cellular swelling, most importantly of the mitochondria, and the absence of morphological alterations in the nucleus. The integrity of the plasma membrane is lost at the end, producing leakage of the internal contents into the extracellular milieu and vice versa. The disappearance of the plasma membrane as a selective barrier seems to be a differentiating characteristic in necrosis. Presumably, this is why necrosis produces a very strong inflammatory reaction, which is not the case for apoptosis (Wyllie *et al.*, 1980). The fact that the same noxious stimulus could produce either necrosis or apoptosis, depending on the concentration and the duration of exposure, has made it difficult to separate the pathways characteristic of apoptosis from those for necrosis (Lemasters, 1999).

B. Apoptosis

Apoptosis was described as a particular set of transformations at the microscopic level associated with cell death (Kerr *et al.*, 1972). Apparently, one of the earliest steps a cell takes when it is committed to die in a tissue is to cease communicating with its neighbors (Fig. 1). This is evident as the dying cell detaches from the adjacent ones and rounds up (Wyllie *et*

al., 1980; Ruoslahti and Reed, 1994). Condensation of the chromatin at the nuclear membrane and fragmentation of the nucleus are manifested. The cell shrinks due to cytoplasmic condensation, probably in response to cross-linking of proteins and loss of water (Fig. 1). Organelles preserve their normal ultrastructure except for the endoplasmic reticulum, which becomes slightly dilated. An increase in the activity of the cell surface is another event characteristic of apoptosis, where the plasma membrane becomes ruffled and blebbed. Eventually this activity separates the cell into a number of membrane-bound fragments of different size, termed apoptotic bodies (Kerr *et al.*, 1972). They could contain cytoplasm only or fragments of nucleus and other organelles, depending on the size of such bodies. Membrane integrity of apoptotic bodies is not lost before being readily engulfed by either surrounding macrophages or, most frequently, the healthy neighboring cells. An intact plasma membrane that impedes the contact of the cytoplasm with the immune system seems to be the reason for the absence of inflammatory reaction in apoptosis, making this process practically imperceptible for the organism.

The morphological hallmark of apoptosis is the condensation of the nuclear chromatin in either crescents around the periphery of the nucleus or a group of condensed spherical fragments. These changes can be seen easily with permeable fluorescent DNA-binding dyes, such as Hoechst 33342, by fluorescence microscopy. Apoptosis also involves the activation of endonucleases; this enzyme produces the characteristic ladder pattern of DNA fragmentation (Wyllie *et al.*, 1984), which is now considered one of the biochemical hallmarks of apoptosis (Fig. 2).

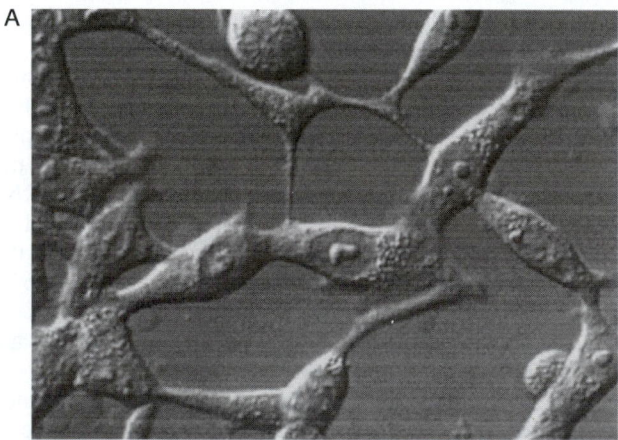

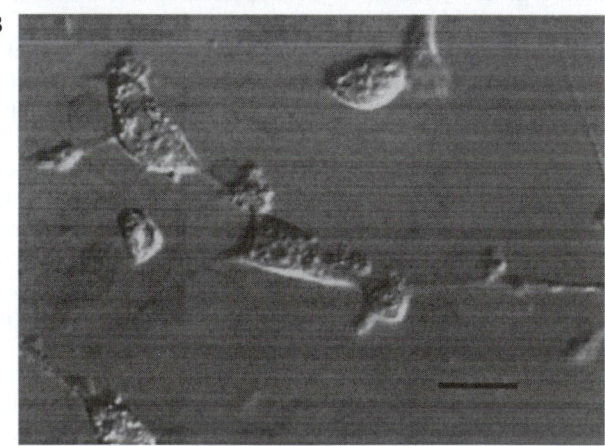

FIGURE 1. Nomarski imaging of resting and apoptotic cells. LNCaP cells were fixed and imaged with an optical microscope (Nomarski optics) after being cultured for 8 hours in the presence of either 0.1% DMSO (A) or 1 μM staurosporine (B). Notice how apoptosis (B) is associated with cell shrinkage and loss of both nuclear integrity and cellular contacts. Bar represents 20 μm.

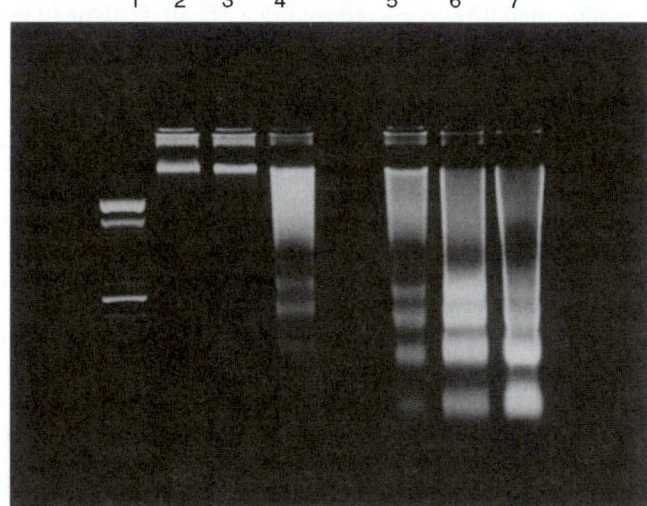

FIGURE 2. Typical ladder pattern of DNA degradation in apoptosis. LNCaP cells were exposed to ionomycin (a Ca^{2+} ionophore) for 14 hours in the presence (2–4 lanes) or the absence (5–7 lanes) of serum. DNA was extracted and electrophoresed in a 2% agarose gel and stained with ethidium bromide. Lanes 2 and 5 are in the absence of ionomycin and show the effect of serum removal. Lanes 3 and 6 are DNA from cells incubated with 1 μM ionomycin, and lanes 4 and 7 are from cells incubated with 10 μM ionomycin. Lane 1 has the molecular weight markers. Notice how DNA is fragmented in multiples of approximately 200 bp, generating the typical ladder pattern associated with apoptosis.

Apoptosis seems to be an active process implying that the cell participates in its own demise ("suicide program"). This is, in principle, an important difference with necrosis where the cell is just a victim of a strong harmful condition. Initially, it was suggested there was a strict requirement for RNA and protein synthesis (Kerr, *et al.*, 1972; Wyllie *et al.*, 1984). However, recent studies aimed at reevaluating the role of protein synthesis in thymocyte apoptosis found their inhibition does not always block DNA fragmentation (Chow *et al.*, 1995). Furthermore, the cytoplasmic changes typical of apoptosis can also be seen in enucleated cells, suggesting that physiological cell death is regulated by cytoplasmic factors. This also indicates that the characteristic nuclear changes in apoptosis and the nuclear activity do not determine the fate of the cell (Jacobson *et al.*, 1994). The apoptosis machinery is present in the cytoplasm of healthy cells, but it is not functional. Energy is required in apoptosis, most probably to generate those signals that will start the death machinery but not necessarily to keep it working (Jacobson *et al.*, 1993; Jacobson *et al.*, 1994).

Physiological cell death could be considered a programmed event mainly during ontogeny, meaning that some internal molecular clock determines the time when a given cell should die. This is ultimately believed to be associated with specific genes being either turned on or deactivated to ignite the cell death machinery. This has originated a strong effort in characterizing the battery of genes involved in this process, as is reviewed later. However, it seems that apoptosis is not always programmed, and important signal transduction events are triggered by external signals. Programmed cell death can also be considered in more general terms to mean that there is a "program" or a sequence of steps associated with cell death, known as the **execution phase**.

III. Regulation of Programmed Cell Death

Signal transduction in apoptosis is one of the rapidly evolving areas in the study of physiological cell death. It is now evident that there is a cell suicide mechanism or program that involves receptors, regulators, and effectors. Some of the important questions in apoptosis include these: Is there a single checkpoint where all different inducers of cell death converge? What are the nature and mechanisms of physiological regulators of cell death? How does a cell "know" when to die? Unraveling the mechanisms involved in controlling cell death will be an important step for coping with the pathological consequences of alterations in the apoptosis program.

A. Inducers of Apoptosis

Apoptosis can be induced by a large variety of conditions that, a priori, do not seem to have anything in common (Fig. 3). Interestingly, the same signal could produce either apoptosis or necrosis. This is apparently related to the level of stress imposed by the noxious signal. Lower levels will induce apoptosis, where the cell is still able to activate its own demise, but for higher insults, necrosis will occur instead. Apparently, a drastic depletion of energy is

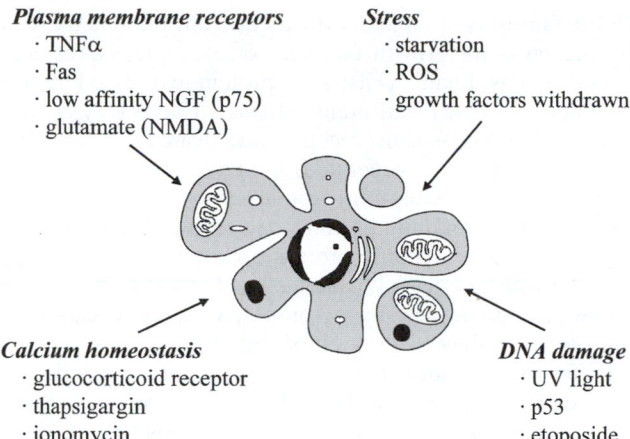

FIGURE 3. Inducers of apoptosis. A schematic separation of inducers of apoptosis based on similar mechanisms for inducing cell death. Nevertheless, there is an important overlap among the different groups. For instance, reactive oxygen species (ROS) can induce cell death by increasing Ca^{2+} influx, and elevated $[Ca^{2+}]_i$ increases ROS production. The four groups imply four different mechanisms for recognizing signals that induce cell death, but the execution phase is probably the same for all.

the turning point (Samali *et al.*, 1996). This is because apoptosis requires normal levels of ATP, and a sudden drop in ATP concentration can only lead to necrosis. Examples of this dual outcome are oxidative stress and excitotoxicity. *In vitro*, apoptosis is always followed by secondary necrosis, that is, plasma membrane disintegration.

We have subdivided the different inducers of apoptosis into four different groups (see Fig. 3), mainly to facilitate discussion of the different transducing mechanisms involved; however, keep in mind that many of them overlap quite importantly.

1. Plasma Membrane Receptors

Plasma membrane limits a cell, and it is endowed with a multitude of receptors that recognize external signals and transduce their messages by fostering the generation of second messengers or a cascade of interacting proteins. These molecules in turn trigger a new cell behavior by activating different effectors. Apoptosis is not an exception to that picture of signal transduction. Different plasma membrane receptors are involved in triggering cell death. The family of **Tumor Necrosis Factor Receptor (TNFR)** and p75 (a low affinity version of the NGF receptor) are examples of plasma membrane receptors coupled to apoptosis.

Receptors of the TNFR family (TNF-R1, Fas, DR3, DR4, and DR5) are characterized by having extracellular cysteine-rich regions and a single membrane-spanning domain, their activation requires trimerization, and their transduction mechanism is based on protein-protein interaction. In a broad sense, the TNFR family is more closely related to the receptors for trophic factors than to those coupled to G proteins, highlighting the similarity between cell death and cell growth. Nevertheless, the transduction mechanism of the

TNFR family is completely different from the one used by the receptors for growth factors. For example, Fas is activated by Fas ligand (FasL), a protein expressed in the plasma membrane of different activated T-cells (Nagata and Golstein, 1995). Once the receptor is activated, a chain of interacting proteins is generated, which is mediated by the death domain of coupling proteins (i.e. FADD/MORT1). Some of these coupling proteins can recruit, via their death effector domains, specific proteases (i.e., procaspase 8) associated with cell death (Ashkenazi and Dixit, 1998). Due to a low proteolytic activity of procaspases, these can auto-activate when they are brought together.

Different mechanisms of protection against the action of these death receptors have been discovered. Decoy receptors can bind agonists of apoptosis, but because they lack the death domain, the apoptosis signal cannot be transduced in activation of caspase 8. Another form of protection involves the presence of SODD (silencer of death domains), a protein that impedes the spontaneous activation of TNF-R1 (Jiang et al., 1999). TNF-R1 and DR3 can also increase cell proliferation. This occurs because different coupling proteins, such as TRAF2 and RIP, trigger activation of nuclear factor κB (NF-κB) and c-Jun. The latter transcription factors could be up-regulating genes encoding apoptosis-inhibitory proteins (Wang et al., 1996). Actually, the capability to kill a cell by TNF-R1 is better expressed by combining TNF with inhibitors of transcription or blockers of protein synthesis, in agreement with the capacity of TNF-R1 to activate NF-κB. Interestingly, besides TNF-R1, ionizing radiation and cancer chemotherapeutic compounds also activate NF-κB. This might explain their failure in eliminating cancer cells (Wang et al., 1996). NGF is a well-known inducer of neuronal survival during development, and it does that by activating a specific tyrosine kinase receptor (trkA). Surprisingly, NGF can also produce cell death; in this case a different type of receptor is involved known as p75 (Frade et al., 1996). This is a low-affinity receptor for NGF without an intrinsic tyrosine kinase activity, but it contains a death domain. The presence of a death domain explains why NGF can induce cell death as well.

Integrins form another family of plasma membrane proteins that have been associated with apoptosis of normal epithelial or endothelial cells. When these cells are detached from the extracellular matrix or forced to grow in suspension, they undergo apoptosis. This effect can be specifically prevented by attaching these cells with anti-integrin antibodies, suggesting that integrins produce apoptotic signals when they are not interacting with the extracellular matrix (Rhouslati and Reed, 1994).

Interestingly, all these studies have documented the existence of plasma membrane receptors linked to apoptosis, but have also shown that the same receptors can alter cell proliferation. This shows that cell death and cell division are closely intertwined, suggesting the presence of common points in these two supposedly divergent events.

2. DNA Damage

Apoptosis can also be seen as a safeguard mechanism because it is involved in deleting damaged cells. p53 is a tumor suppressor gene normally activated by DNA damage and is the most frequently mutated gene in human tumors (Yonish-Rouach, 1996). Activation of p53 blocks cell cycle in G1 phase and, it is presumed, this gives time to the DNA-repairing machinery to restore DNA before replication. On activation of p53, the cell could undergo either growth arrest or apoptosis. Growth arrest induced by p53 depends on its role as transcriptional activator, which is located in the N-terminus of the protein. On the other hand, apoptosis induced by DNA damage, which has been shown to depend strictly on p53 (Lowe et al., 1993), does not seem to require gene expression (Caelles et al., 1994; Haupt et al., 1995). It seems the ability of p53 to induce cell death or growth arrest resides in different parts of the peptide sequence. Similarly to the TNFR family, the induction of cell death by p53 does not require protein synthesis. This is in agreement with the cell death machine being present in the cytoplasm of healthy cells, but under constant repression.

3. Stress

Withdrawal of growth factors or serum deprivation has been shown to induce cell death by apoptosis. However, this effect depends on the stage of the cell cycle, because there are points where cells can be arrested and survive (Farinelli and Greene, 1996). This seems to be linked to the presence of cyclin-dependent kinase (CDK) inhibitors, such as p21 or p16 (Wang and Walsh, 1996). The former produces growth arrest at the G_1 phase of the cell cycle, allowing the cell to survive. This could be explained as the presence of these inhibitors impeding the interaction of conflicting signals (proliferation in harsh conditions) that otherwise would eventually trigger apoptosis.

By studying the mechanism of action of Bcl-2 (an important blocker of apoptosis, discussed later), it was found that this protein protects cells against oxidative insults. This raised the possibility that reactive oxygen species (ROS) could be important mediators of cell death. However, it was shown later that apoptosis can occur in the absence of ROS, implying that they are not universal signals for apoptosis. Nevertheless, antioxidants and ROS scavengers (Slater et al., 1995) can block the effect of different inducers of apoptosis. Further, Bcl-2 protects cells from the lethal effects of H_2O_2 or t-butyl hydroperoxide, two well-known inducers of oxidative stress (Korsmeyer et al., 1995). Interestingly, mutations in a Cu/Zn superoxide dismutase, an enzyme that catalyzes the dismutation of the toxic superoxide anion (O_2) to O_2 and H_2O_2, have been associated with loss of motor neurons in some cases of familial amyotrophic lateral sclerosis (Rosen et al., 1993). All of these studies point to the importance of ROS as inducers of cell death and argue for the relevance of having mechanisms of protection against oxidative insults.

4. Ca^{2+} Homeostasis

Apoptosis is also induced by disturbing the homeostasis of intracellular Ca^{2+} concentration ($[Ca^{2+}]_i$). Typical nuclear changes associated with apoptosis are triggered by either reducing or increasing $[Ca^{2+}]_i$ (Table 1). Reducing external $[Ca^{2+}]$ or blocking endoplasmic reticulum Ca^{2+} pumps with thapsigargin, for instance, will deplete internal Ca^{2+} pools.

TABLE 1 The Role of External Ca^{2+} in Nuclear Fragmentation in LNCaP Cells Exposed to Different Inducers of Apoptosis

Treatment	Intact nuclei (%)	Fragmented nuclei (%)
Control	97	3
5 mM BAPTA	63	37
5 mM BAPTA + 5 mM CaCl$_2$	91	9
1 μM staurosporine	62	38
1 μM staurosporine + 1 mM Bapta	81	19
0.1 μM thapsigargin	63	37
0.1 μM thapsigargin + 1 mM Bapta	66	34

LNCaP cells were treated with the indicated chemicals for 48 hours in fetal calf serum supplemented RPMI medium ([Ca^{2+}] = 0.4 mM). Cells were harvested and their nuclei stained with the DNA-binding dye, Hoechst 33258 (2 μg·ml^{-1}), and counted with a fluorescence microscope (excitation wavelength 360 nm). 1 mM Bapta (a calcium chelator) did not induce cell death (not shown). Reducing external Ca^{2+} prevented staurosporine-induced apoptosis, however, complete removal of external Ca^{2+} triggered apoptosis on its own (5 mM Bapta). Apoptosis due to thapsigargin was not affected by reducing external Ca^{2+}.

This may impair the role of luminal Ca^{2+}-sensitive molecular chaperones (Gill et al., 1996), affecting the quality control mechanism of protein folding in endoplasmic reticulum. Furthermore, it has been shown that depletion of nuclear Ca^{2+} stores stops traffic between the nucleus and the cytoplasm, blocking protein synthesis (Perez-Terzic et al., 1996). Recently, it has been proposed that a protease associated with apoptosis (caspase 12) is released from the endoplasmic reticulum when this organelle is stressed by different conditions (Nakagawa et al., 2000). It is conceivable that this protease is responsible for the apoptosis induced by depletion of internal Ca^{2+} stores.

Ca^{2+}-dependent apoptosis is the consequence of a sustained elevation in [Ca^{2+}]$_i$, which is due to a continued Ca^{2+} influx through the plasma membrane (Dowd, 1995). However, the nature of the Ca^{2+} channels involved and their regulation are not known. Inducers of apoptosis such as ionomycin and thapsigargin are well known for their ability to deplete internal Ca^{2+} stores; other inducers, such as dexamethasone and H$_2$O$_2$, have been shown to lower Ca^{2+} content in internal stores. A reduction in the luminal [Ca^{2+}] from the endoplasmic reticulum activates what are called store-operated Ca^{2+} (SOC) channels, also known as depletion-activated channels (Berridge, 1995). Accordingly, it has been proposed that inducers of apoptosis increase [Ca^{2+}]$_i$ via SOC channels by depleting internal Ca^{2+} stores. The poor pharmacological and molecular characterization of SOC channels has precluded the corroboration of this hypothesis.

Although SOC channels might play a critical role in apoptosis, there is also evidence for the involvement of other types of Ca^{2+} channels in this process. It has been shown that a Ca^{2+} permeable channel (type-3 inositol 1,4,5-trisphosphate receptor, IP$_3$R) is expressed in the plasma membrane of lymphocytes undergoing apoptosis (Khan et al., 1996). Significantly, when the expression of this channel is inhibited, so is apoptosis for this type of cells (Khan et al.,

1996). This was the first demonstration that apoptosis in lymphocytes depends on expression of a Ca^{2+}-permeable channel. Although the importance of IP$_3$R in cell death is definitive (Marks, 1997), it is not clear what role IP$_3$R localization plays in apoptosis (i.e., plasma membrane or endoplasmic reticulum). In neurons, cell death has been associated with another kind of Ca^{2+}-permeable channel, the NMDA type of glutamate receptors. The increased activity of this channel produces a Ca^{2+} overload, which has been associated with cell death by excitotoxicity. This condition can result in either necrosis or apoptosis, depending on the metabolic state of the cell (Ankarcrona et al., 1995).

As discussed earlier, oxidative stress can induce apoptosis, and the same type of insult has been shown to directly activate Ca^{2+}-permeable, nonselective cation channels in endothelial cells (Koliwad et al., 1996), suggesting a connection between Ca^{2+} influx and apoptosis induced by reactive oxygen species. Apoptosis in prostate cells is also dependent on Ca^{2+} influx (Furuya et al., 1994) where a new type of Ca^{2+}-permeable channel, not operated by internal Ca^{2+} stores, has been recently identified and associated with apoptosis (Gutiérrez et al., 1999). Interestingly, cationic channels not permeable to Ca^{2+}, encoded by mdeg genes, produce cell death by swelling, characteristic of necrosis (Ellis et al., 1991; Waldmann et al., 1996). This suggests that cationic channels do not produce apoptosis unless they are permeable to Ca^{2+}. From this limited number of studies on the role of Ca^{2+} channels in apoptosis, it appears that the capacity to increase [Ca^{2+}]$_i$ is more important than the type of Ca^{2+} channel involved in elevating [Ca^{2+}]$_i$. Furthermore, it is now evident that such increase in [Ca^{2+}]$_i$ is not necessarily reflected in an elevated bulk [Ca^{2+}]$_i$, but in a localized mitochondrial Ca^{2+} overload (Szalai et al., 1999).

A rise in [Ca^{2+}]$_i$ is a widely used second messenger for different physiological events (Fig. 4). It depends on different Ca^{2+}-binding proteins to transduce Ca^{2+}-rising external

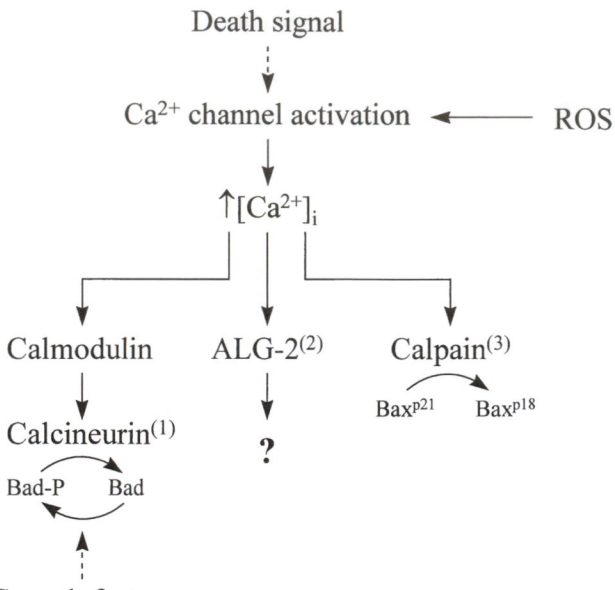

FIGURE 4. Working model of the role of Ca^{2+} as second messenger in apoptosis. Death signals, like reactive oxygen species (ROS), could directly activate Ca^{2+}-permeable channels or indirectly do so by depleting internal Ca^{2+} stores. An increase in $[Ca^{2+}]_i$ is recognized by apoptosis-specific Ca^{2+}-binding proteins that activate different kinds of cell death effectors. Solid lines indicate direct activation. Dashed lines represent a pathway of multiple stages. The rise in $[Ca^{2+}]_i$ that triggers apoptosis does not need to be a global increment in $[Ca^{2+}]_i$; in this case it will be more difficult to detect such increment. (1) Wang *et al.*, 1999; (2) Vito *et al.*, 1996; (3) Wood *et al.*, 1998.

messages in specific cell behaviors. In this regard, different groups have demonstrated an increment in calmodulin expression induced by cell death agonists, despite a general reduction in protein synthesis (Furuya *et al.*, 1994; Dowd, 1995). Calmidazolium, a potent inhibitor of calmodulin, blocks apoptosis in different systems (Dowd, 1995), indicating that this Ca^{2+}-binding protein could be transducing the elevated $[Ca^{2+}]_i$ in an active cell death program. Recently, ALG-2 was identified as a new kind of Ca^{2+}-binding protein in thymocytes (Vito *et al.*, 1996). This protein has two canonical Ca^{2+}-binding domains, and apoptosis triggered by different, unrelated inducers, is blocked when the expression of ALG-2 is specifically abrogated by an antisense approach. The expression of this protein is not limited to thymocytes, because it is also present in hepatic and kidney cells. This suggests that ALG-2 could be involved in mediating cell death in different types of cells.

An increased $[Ca^{2+}]_i$ activates different Ca^{2+}-dependent enzymes such as endonucleases, proteases, kinases, phosphatases, and tissue transglutaminase, where some of them seem to be involved in cell death. In this regard, activation of **calcineurin**, a Ca^{2+}-dependent phosphatase, readily induces cell death in the presence of low levels of serum. Furthermore, the overexpression of a constitutively active form of calcineurin does induce apoptosis without a Ca^{2+} increase, implying that the activation of this phosphatase is enough to trigger programmed cell death (Shibasaki and McKeon, 1995). Dephosphorylation of Bad, a pro-apoptotic

protein of the Bcl-2 family, appears to explain the apoptogenic activity of calcineurin. In the absence of phosphate, Bad promotes caspase activation, apparently by overcoming the anti-apoptotic effect of Bcl-x_L at the level of Apaf-1 (see Fig. 5). On the other hand, growth factors via several kinases, including PKB/Akt, phosphorylate Bad, and block apoptosis (Wang *et al.*, 1999).

Bcl-2 is a membrane protein that blocks apoptosis, as discussed later. This protein modifies regulation of $[Ca^{2+}]_i$ in different type of cells. The main effect of the overexpression of Bcl-2 is the preservation of the endoplasmic reticulum Ca^{2+} pool and an increased mitochondrial Ca^{2+} buffering power (Baffy *et al.*, 1993; Szalai *et al.*, 1999). Bcl-2 seems to block preferentially the apoptosis-associated rise in intranuclear $[Ca^{2+}]$, suggesting that this is how Bcl-2 inhibits DNA degradation (Marin *et al.*, 1996). Whether this effect is directly due to Bcl-2 or is just a consequence of blocking cell damage is not clear yet.

Ca^{2+} is a very important second messenger in a variety of physiological processes; the evidence reviewed here suggests that it is also involved in inducing cell death. Whether this occurs because Ca^{2+} is directly activating Ca^{2+}-dependent enzymes involved in the execution path of apoptosis (see Section III.C) or indirectly because alterations in Ca^{2+} regulation block cell growth, which in turn triggers cell death, is not yet clear. Understanding how Bcl-2 affects regulation of $[Ca^{2+}]_i$ and how proteins like ALG-2 transduce cell death will undoubtedly shed light on the role of Ca^{2+} in apoptosis.

B. Regulators of Apoptosis

Apoptosis is finely regulated, and the discovery of genes involved in this task has made the study of them one of the quickly developing areas in programmed cell death. The importance of these genes, for instance the *bcl-2* family, has been widely demonstrated. Nevertheless, there are other proteins unrelated to the bcl-2 family that seem to modulate apoptosis as well.

1. Bcl-2 Family of Proteins

Bcl-2 is a membrane protein considered a **proto-oncogene.** It was found that up to 85% of human follicular B-cell lymphomas express Bcl-2. This was the consequence of a t(14; 18) chromosomal translocation that places bcl-2 (18q21) under the transcriptional control of the Ig H-chain locus (14q32). This explains its abundance in lymphoma cells (Korsmeyer *et al.*, 1995).

It was later discovered that Bcl-2 produces lymphomas without increasing the rate of cell proliferation, as was expected for an oncogene, but by slowing down cell death (Korsmeyer *et al.*, 1995). Bcl-2 then became the foundation of a new family of regulators of cell death. The importance of this family was more evident when it was realized that *ced 9* (an inhibitory gene of cell death in the worm *C. elegans*) is a functional and structural homologue of bcl-2. This suggests that programmed cell death has been highly conserved in evolution from invertebrates to humans (Häcker and Vaux, 1995).

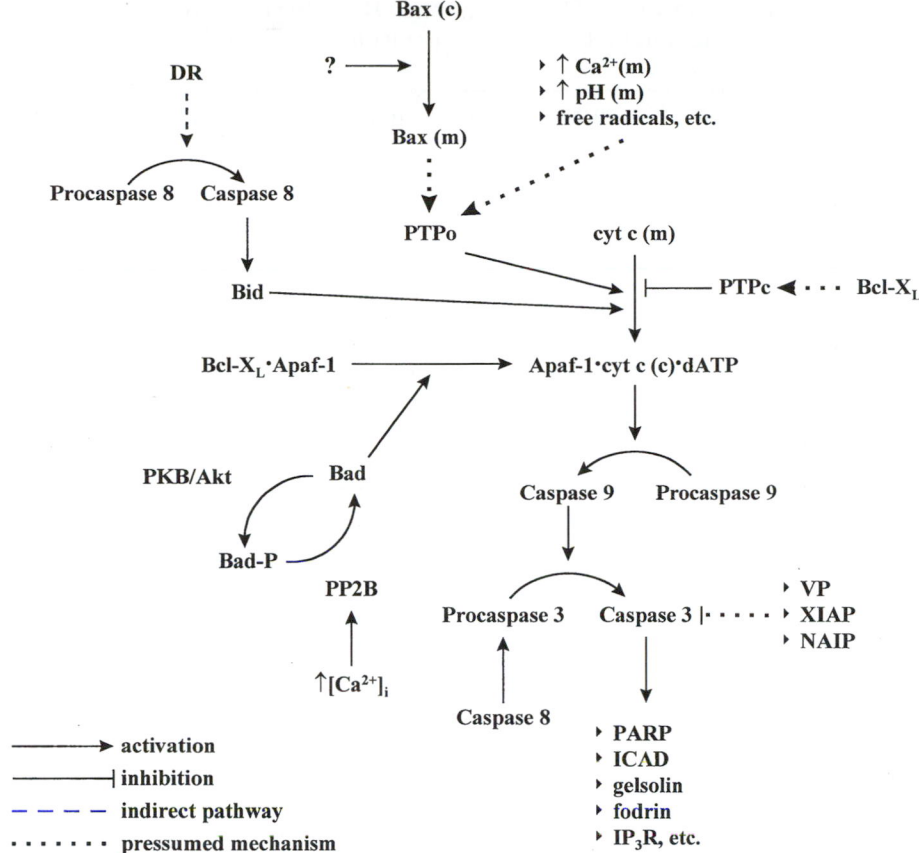

FIGURE 5. Biochemical pathways in apoptosis. The diagram shows how different inducers of apoptosis converge in activating a caspase cascade due to the translocation of cytochrome c from mitochondria (cyt c(m)) to the cytoplasm (cyt c(c)) and generating a cytoplasmic apoptosome. The latter is constituted by cytochrome c, Apaf-1, dATP/ATP, and procaspase 9. Cytochrome c release can be induced either by plasma membrane death receptors (DR) through activation of caspase 8 and cleavage of Bid or by opening of a mitochondrial permeability transition pore (PTPo). Conceivably, antagonists of apoptosis such as Bcl-2 or Bcl-x$_L$, block cytochrome c release by closing PTP (PTPc). The apoptosome can also be affected by altering the interactions between Bcl-x$_L$ and Apaf-1, for instance by calcineurin (PP2B) mediated dephosphorylation of Bad. There are different proteins, not related to the Bcl-2 family, which can inhibit apoptosis by affecting caspase activity. These can be derived from viruses (VP) or mammalian cells (i.e., XIAP).

Bcl-2 is a 25-kDa membrane protein localized to the external membrane of mitochondria, nuclear envelope, and endoplasmic reticulum. Bcl-2 insertion in intracellular membranes requires the hydrophobic carboxy-terminus of the protein. This is called Bcl-2α, which is the most abundant isoform. There is a splice variant (Bcl-2β) that lacks the anchoring region but that binds to Bcl-2α isoform. Deletion of the hydrophobic region of Bcl-2α reduces but does not abolish its function. This suggests that the active site is located in its cytoplasmic part and the hydrophobic C terminus targets the protein to the different organelles. It also implies that the location of Bcl-2 plays a role in protecting cells from apoptosis.

Most of the organs in mice lacking Bcl-2 (*bcl-2 –/–*) complete development and survive, indicating that other proteins must be involved in regulating cell death (Korsmeyer *et al.,* 1995). It is now evident that Bcl-2 is only one member of a large family of regulators of apoptosis (Table 2). Remarkably, this family comprises proteins that are present in some viruses

as well. The Bcl-2 family of proteins is divided in two main groups, agonists and antagonists, based on their ability to favor or block apoptosis, respectively (Table 2).

Bcl-2 is the archetypal antagonist. *In vitro,* this protein is able to block or retard apoptosis triggered by a wide and diverse variety of inducers (Table 3), although *in vivo* its function seems to be more limited, as indicated earlier. This is not the case for another antagonist, Bcl-x$_L$. This protein is highly expressed in central nervous system, kidney, and marrow. In this case *bcl-x* knockout embryos were not viable. Massive apoptosis was evident in the immature nervous system and in hematopoietic cells. This is in agreement with the abundance of Bcl-x$_L$ in those same cells, implying a very important role for *bcl-x* gene in development (Motoyama *et al.,* 1995). Interestingly, long (Bcl-x$_L$) and short (Bcl-x$_S$) forms of Bcl-x have been identified. The former is an antagonist, but the latter counteracted the antagonistic action of Bcl-2, suggesting that this form produces cell death.

TABLE 2 Presence of Bcl-2 Sequence Homologies (BH)
and Transmembrane Domain for Agonist and Antagonist
of Apoptosis

Protein	BH1	BH2	BH3	BH4	TD	Cell death
Bcl-2	+	+	+	+	+/–	A
Bcl-X_L	+	+	+	+	+	A
Ced-9	+	+	–[a]	+	+	A
BHRF1	+	+	–	–	+	A
Bax	+	+	+	–	+/–	P
Bad	–	–	+	–	–	P
Bcl-X_S	–	–	+	+	+	P
Bik	–	–	+	–	+	P
EGL-1	–	–	+	–	–	P

Note: The presence (+) or absence (–) of Bcl-2 sequence homology regions
(BH1 to BH4) is shown, based on the references listed here. The transmembrane
domain (TD) is either present (+) or not (–); (+/–) represents splice variants lacking the TD. The cell death column indicates whether the proteins antagonizes (A)
or promotes (P) cell death. [a]The presence of this domain is not clear.
References: Adams and Cory, 1998. Reed, 1998. Complete references are found
in the bibliography at the end of the chapter

Bax was the first protein identified to be an agonist of apoptosis; this protein was isolated in association with Bcl-2.
Activated Bax is an integral membrane protein of 21 kDa
which shares 21% identity and 43% similarity with Bcl-2.
Bcl-2 homology domains are clustered in four regions, termed
BH domains. They are found in the following order: BH4,
BH3, BH1, and BH2, from the amino to the carboxy end of the
protein (see Table 2). Bax-deficient mice show alterations due
to uncontrolled accumulations of only lymphocytes and germ
cells (Knudson *et al.,* 1995). Two conclusions arise from this
study. First, Bax is effectively an agonist of cell death. Second,
there must be more agonists of cell death, since only a few
cells were affected by the absence of Bax (see Table 2).

BH1 domain is crucial in the antagonistic action of the
Bcl-2 family of proteins, and BH3 has been associated with
their agonistic effect (see Table 2). Point mutations in a
highly conserved glycine in BH1 completely abrogate the
antagonistic action of Bcl-2 and Bcl-x_L. The hypothesis that
BH3 is involved in triggering cell death stems from the existence of pure agonists of apoptosis that have only BH3,
like Bcl-x_S, Bik, and Bid (see Table 2). However, BH3 is
also present in antagonists such as Bcl-2 and Bcl-x_L, indicating that there must be other factors besides these domains
to explain their mechanism of action.

The three-dimensional structure of a monomer of Bcl-x_L
has been resolved (Muchmore *et al.,* 1996). This study
showed that BH1, BH2, and BH3 form a hydrophobic groove
inside of which the highly conserved glycine of BH1 is located. Apparently, this hydrophobic cleft allows the interaction with Bax, which has been suggested to be necessary for
the antagonistic actions of Bcl-2 and Bcl-x_L. Accordingly, increasing the side chain of the highly conserved glycine in
BH1 would generate a steric hindrance that would diminish
the strength of the interaction with Bax. Although homodimerization and heterodimerization among the different pro-

teins of the Bcl-2 family are well documented, the importance of this interaction on their physiological roles is not
clear-cut (McDonnell *et al.,* 1996). Thus, Muchmore *et al.*
(1996) have highlighted the structural similarity between
Bcl-x_L and the membrane insertion domain of diphtheria
toxin and colicin. Accordingly, it has been shown that both
agonists (Bax) and antagonists (Bcl-x_L, Bcl-2) of apoptosis
produce nonselective ion channels in lipid bilayers. A priori,
the characteristics of the ion channels do not help to explain
the anti-apoptotic activity of Bcl-2. Nevertheless, it has been
shown that alpha helices 5 and 6, which are involved in making the ionic pore, are essential for the antagonistic action of
Bcl-2 (Matsuyama *et al.,* 1998). Interestingly, Bax seems to
have two independent ways of killing a cell, one that involves
the pore-forming α helices and the other mediated by the
BH3 domain (Matsuyama *et al.,* 1998).

Importantly, Bcl-2 inhibits the release by mitochondria of
apoptogenic factors (see later). Still, this does not seem to
totally explain the action of Bcl-2 (Rossé *et al.,* 1998).
Whatever the Bcl-2 mechanism of action is, it must be
downstream and very close to a common effector in the execution phase of programmed cell death. This is because
Bcl-2 is able to abrogate or retard the effect of widely diverse inducers of apoptosis or even necrosis that a priori do
not have anything else in common (see Table 3).

2. Mitochondria

Mitochondria appear to be the checkpoint where different
inducers of apoptosis converge to trigger cell death (Fig. 5).
The central role of mitochondria in apoptosis is supported by
the presence in the intermembrane space of different inducers of apoptosis (cytochrome c and Apoptosis-Inducing
Factor (AIF)), as well as other proteins that are involved in
apoptosis (procaspases and calpain-like proteases). A diverse

TABLE 3 Overexpression of Bcl-2 Blocks Apoptosis Induced by a Wide Variety of Cell Death Stimuli

Apoptosis inducers	Reference
Physical	
Gamma- and UV-radiation	2
Heat shock	7
Chemical	
Chemotherapeutic drugs	3
Reactive oxygen species	7
Azide	7
Growth factors withdrawal	
Glucose	7
IL-3	5
Androgens	6
Neurotrophic factors	2
Others	
Calcium	4
p53	7
c-Myc	2
Viruses	7
CD3/TCR	8
Dexamethasone	3
Glutamate	1

This table does not intend to be an exhaustive list of apoptosis inducers, but rather to highlight their diversity in nature. (1) Ankarcrona *et al.*, 1995; (2) Boise *et al.*, 1995; (3) Cidlowski *et al.*, 1996; (4) Dowd, 1995; (5) Hockenberry *et al.*, 1990; (6) Raffo *et al.*, 1995; (7) Reed, 1994; (8) Sentman *et al.*, 1991. Complete references are found in the bibliography at the end of the chapter.

variety of apoptogenic signals promote the release of these proteins to the cytoplasm, which in turn activate the **execution phase** of apoptosis.

It is not clear how all these proteins are released from mitochondria. The movement of cytochrome c during apoptosis is one of the best studied at both cellular and subcellular levels. It has been proposed that either mitochondrial permeability transition pore (PTP) or a component of it (VDAC) is involved in releasing cytochrome c to the cytoplasm. Opening of VDAC by Bax and closing of it by Bcl-x_L could explain their respective roles in apoptosis (Shimizu *et al.*, 1999). This, however, poses the following dilemma. On one hand, ATP is clearly a requisite for apoptosis, and synthesis of ATP by mitochondria depends on the inner mitochondrial membrane potential. On the other hand, the evidence suggests that PTP is required for releasing cytochrome c, but opening PTP strongly depolarizes mitochondria and halts ATP synthesis. Recent reports suggest the presence of two different mechanisms for cytochrome c release, one involving PTP and mitochondria depolarization and the other a still unidentified mechanism that does not produce mitochondria depolarization. The latter mechanism is proposed to be the target of

BH3-only proteins such as Bid and Bik (Shimizu and Tsujimoto, 2000). It is plausible that the presence of two different mechanisms for cytochrome c release solves the dilemma.

Once cytochrome c is in the cytoplasm, it triggers apoptosis by activating a cytoplasmic apoptosome. Cytochrome c, Apoptosis Protease-Activating Factor (Apaf-1), dATP or ATP, and procaspase 9 form the latter (Fig. 5). Apaf-1 (the mammalian homolog of CED-4, a protein essential for apoptosis in *C. elegans*) is a 130-kDa ATPase with a caspase recruitment domain (CARD), a CED-4 homologous domain, and 13 to 14 WD-40 repeats, which are involved in protein-protein interactions. Apaf-1 is the cytoplasmic receptor of cytochrome c. Oligomerization of Apaf-1, which is promoted by the hydrolysis of dATP/ATP and the presence of cytochrome c, recruits procaspase 9. This allows procaspase 9 its own processing in a fully active caspase 9. Once caspase 9 is released from the apoptosome, it initiates a caspase cascade responsible for many, if not all, of the cell transformations typical of apoptosis (see Fig 5).

Studies in mice lacking Apaf-1 stress the importance of this protein in mitochondrial apoptosis (Yoshida *et al.*, 1998). Interestingly, Fas-induced apoptosis showed no requirement of Apaf-1 in thymocytes. In those cells where mitochondria is involved in Fas-mediated cell death (i.e., hepatocytes), it can be explained by Fas-induced activation of caspase 8, which in turn hydrolyzes Bid (a BH3-only protein). The latter protein is able to trigger cytochrome c release and the ensuing caspase cascade (see Fig 5).

3. Other Regulators of Apoptosis

Although the Bcl-2 family of proteins is the most studied one in terms of regulation of apoptosis, other proteins can affect apoptosis. For instance, the gene encoding the **neuronal apoptosis inhibitory protein (NAIP)** does not show any resemblance to bcl-2, yet it protects neurons from apoptosis. Deletions of NAIP gene have been correlated with type I spinal muscular atrophy, a fatal autosomal recessive disorder characterized by an important neuronal cell death (Chinnaiyan and Dixit, 1996).

Studies on the mechanism of action of **Tumor Necrosis Factor (TNF)** found that a sphingomyelinase is activated during TNF-induced apoptosis. This enzyme is a sphingomyelin-specific phospholipase C that generates phosphocholine and **ceramide** by hydrolysis of sphingomyelin. The use of cell-permeable ceramides suggested that natural ceramide could be involved in triggering cell death, among other cellular responses, probably by modulating the activity of different transcription factors (Hannun, 1996). Subsequently, it has been demonstrated that activation by TNF-R1 of plasma membrane–neutral sphingomyelinase is not linked to the death domain of this receptor. However, the death domain of TNF-R1 is required for the indirect activation of an acidic sphingomyelinase. In conclusion, it appears that TNF-R1 activates both neutral and acidic sphingomyelinases; ceramide produced by the former is involved in cell survival, but the ceramide produced by the latter is linked to apoptosis (Testi, 1996). Ceramides are insoluble, so they are restricted to those membranes where they are generated. That, combined with segregation of the effector system sensitive to ceramide, could explain the paradoxical roles of ceramide in cell physiology.

C. Effectors in Apoptosis

Apoptosis is characterized by specific morphological changes in the nucleus, plasma membrane, and endoplasmic reticulum, among others. These changes are very similar in different type of cells, implying that a conserved set of enzymes exists and produces those modifications when activated during the **execution phase** of cell death. These enzymes determine the fate of the cell, mainly because their effects are irreversible. Caspases (cysteine-dependent proteases), endonucleases, and transglutaminase are some of the enzymes activated during the execution phase of apoptosis.

1. Proteases

The discovery that ced-3 (an essential gene for cell death in the worm *C. elegans*) was similar to an Interleukin-1β-converting enzyme (ICE/caspase 1) highlighted the importance of **proteases** in the execution phase of apoptosis. Those proteases involved in apoptosis are substrate-specific, nonlysosomal enzymes that have either cysteine or serine in their active site.

The **caspase family** is the best-studied family of proteases involved in apoptosis. Caspases are cysteine-dependent proteases characterized by cleaving at aspartic residues in P1 with different consensus sequences for cleaving substrates: caspase 1 recognizes (W/Y/F)EHD, so it generates interleukin-1β from its precursor by cleaving after aspartic (D). Another enzyme, caspase 3, recognizes DEXD (Wolf and Green, 1999), cleaving and inactivating poly(ADP-ribose) polymerase (one of the enzymes involved in DNA repair), among others. The importance of caspases in apoptosis has been demonstrated by three different criteria: the use of permeable and specific inhibitory tetrapeptides, viral proteins that are specific and potent inhibitors of these proteases, and last, the targeted disruption of caspase genes in mice (Hakem *et al.,* 1998). In general, caspases are localized in the cytoplasm and present as inactive proenzymes that activate on proteolysis, in some cases by autocatalysis. It appears that only one caspase (CED-3) exists in the worm *C. elegans*; however, up to 14 caspases have been identified in mammals. They have been divided in two subfamilies. One includes caspases 1, 4, 5, 11, 12, 13, and 14. Here, caspases 1 and 11 are associated with the inflammatory response (processing of IL-1) more than cell death. The other subfamily is formed by caspases 2, 3, 6, 7, 8, 9, and 10, and is directly involved in apoptosis.

The latter subfamily can be further subdivided in two groups, one known as initiator caspases (2, 8, 9, and 10) characterized by long prodomains. The other group is the executioner caspases (3, 6, and 7) characterized by short prodomains (Wolf and Green, 1999). Two apoptosomes have been identified for the activation of initiator caspases, one at the plasma membrane for activation of caspase 8 via recruitment through the death effector domain, and the other in the cytoplasm for the activation of procaspase 9 via homologous CARD-

mediated interaction with Apaf-1 (see Fig. 5). Once the initiator caspases are activated, they generate active executioner caspases, like caspase 3, by cleaving their corresponding procaspases. In turn, the executioner caspases cut the different proteins responsible for the cell transformation associated with apoptosis. Interestingly, different viral proteins (p35, CrmA) block apoptosis by inhibiting the activity of different caspases. Furthermore, there are inhibitors of apoptosis (IAP) present in mammalian cells (xIAP, cIAP, etc.) that appear to block apoptosis by interfering with caspase function. Caspase activation is a requisite for apoptosis; however, their activation appears to be part of a feedforward loop. This kind of situation is very clear in the release of cytochrome c. The latter can produce activation of caspase 3, but this caspase promotes further release of cytochrome c (Kirsch *et al.,* 1999).

Conversely, we know much less for serine-dependent proteases in cell death. Granzyme B is a serine protease specific for aspartic acid P1 residue, which seems to be required for the cytotoxic T-cell-induced apoptosis (Chinnaiyan and Dixit, 1996). It has been shown that granzyme B can activate caspase 3, which is an important effector in apoptosis. Presumably, the introduction by perforin of granzyme B into targeted cells will trigger apoptosis by activation of caspase 3. Another enzyme activated in apoptosis is a Ca^{2+}-dependent serine protease involved in the degradation of the nuclear scaffold (McConkey, 1996), and it is one of the candidates producing the characteristic apoptotic nuclear alterations.

Changes in the shape and volume of the cell and nucleus are very prominent characteristics of apoptosis. These represent important modifications in the cytoskeleton (Zhivotovsky *et al.,* 1996). Proteolytic modification of different cytoskeletal proteins has been documented: α-fodrin, actin, and vimentin, among others. Proteins with roles in nuclear function and shape are also substrates for apoptosis-activated proteases. Enzymes involved in DNA repair, such as Poly (ADP-ribose) polymerase and the DNA-dependent protein kinase, are also cleaved by caspases. Proteins associated with the matrix attachment region of the nucleus, like lamins and the U1 small ribonucleoprotein, and many others, like histone H1 and topoisomerase I and II, are other examples of proteins cleaved during apoptosis. The specific cleavage of these proteins conceivably disrupts the normal nuclear integrity and function, preparing the cell for a complete shutdown.

It is clear that understanding regulation of the different proteases in apoptosis should allow a significant control of the outcome in cell death.

2. Endonucleases

Fragmentation of the nucleus and DNA degradation into multimers of 200 bp are well established characteristics of apoptosis. It has been shown that these two changes depend very importantly on activation of caspase 3. Fragmentation of the nucleus involves a recently described protein, Acinus, which induces chromatin condensation after cleavage by caspase 3 and another protease (Sahara *et al.,* 1999).

Recently, the endonuclease involved in degrading DNA during apoptosis has been identified (Enari *et al.,* 1998) as a caspase-activated deoxyribonuclease (CAD). This enzyme appears to be present in the cytoplasm and contains a nuclear-localization signal. In the absence of apoptogenic signals, CAD is associated with an inhibitor (ICAD). The latter protein contains two putative cleavage sequences for caspase 3. This protease relieves the inhibition on CAD by cleaving ICAD at aspartic 117; this in turn frees CAD, so it can now translocate to the nucleus and generate the typical ladder pattern of DNA digestion (Enari *et al.*, 1998).

3. Transglutaminase

Limited cytoplasmic leakage is another characteristic of apoptosis, which seems to be due to a shell formed around the cell. It has been proposed that tissue transglutaminase (tTG) could be involved in generating such a stabilizing structure. This enzyme is a cytoplasmic, Ca^{2+}-dependent, protein-glutamine γ-glutamyltransferase that cross-links proteins, leading to their polymerization (this probably confers cyto-architectural stability). A substantial increase in tTG mRNA occurs on induction of apoptosis in different types of cells (Fesus *et al.*, 1996; Piacentini, 1995). Overexpression in some cell lines of tTG cDNA increased the susceptibility to apoptosis and, conversely, transfection with antisense tTG cDNA increased resistance to apoptotic stimuli by diminishing expression of tTG. Although these experiments suggest that tTG can induce apoptosis, further evidence suggests that the role of tTG in apoptosis, albeit not universal, is that of an effector and not a regulator of cell death. Interestingly, overexpression of tTG cDNA can transform necrotic cell death in typical apoptosis, stressing the importance of this enzyme in the expression of programmed cell death (Fesus *et al.*, 1996).

4. Other Effectors

Significant reductions in cell volume are associated with apoptosis. Budding of the endoplasmic reticulum or alterations in the cell volume regulatory mechanisms could carry the cell shrinkage. The former may involve the release in the extracellular space of endoplasmic reticulum-derived vesicles that were not allowed to condensate. The latter may be due to an increased efflux of KCl in response to an elevated activity of Ca^{2+}-dependent K^+ and Cl^- channels, among other ion transporters.

Programmed cell death is characterized by "silent" removal of apoptotic cells from the organism. Alterations in the plasma membrane signal neighboring cells and surrounding phagocytes for engulfing and deletion of the dying cell. There are different mechanisms for the recognition of target cells by phagocytes (Hart *et al.*, 1996). One of these mechanisms is the **externalization of phosphatidylserine.** Phosphatidylserine (PS) is a negatively charged phospholipid that is asymmetrically present in the internal leaflet of the plasma membrane in normal conditions. This asymmetry is rapidly lost by externalization of PS on induction of apoptosis. Inhibition of ICE-like proteases abolishes the reduction in PS asymmetry, placing the activation of proteases upstream to the externalization of PS. This also means that proteins are responsible for keeping lipid asymmetry in normal cells. In this regard, cytoskeletal proteins such as α-fodrin or

enzymes, such as translocases, have been implicated in regulating this asymmetry, and they could be the targets of apoptosis proteases.

A plasma membrane receptor present in phagocytes is involved in PS recognition. The best candidate is CD68 or macrosialin; however, it seems that this protein is only part of a more complex structure involved in the recognition of apoptotic and senescent cells.

IV. Roles of Physiological Cell Death

It is now recognized that not all physiological cell death occurs by apoptosis. Different forms of cell death could arise because there is one single program (highly conserved from worms to mammals, e.g., the caspase family of proteases) with different downstream effectors. The other possibility is the existence of more than one program. The latter seems to be supported by the observation that ciliary ganglion neurons can die either by apoptosis or not, depending on the stimulus (Schwartz *et al.*, 1993). The different functions of physiological cell death are considered here, regardless of whether it is manifested as apoptosis or not.

Physiological cell death is presumably present in multicellular organisms only, with prominent roles in development and homeostasis (see Fig. 6). Accordingly, it has been proposed that cell death can be divided into five different categories, depending on the roles of the dying cell in the organism (Ellis *et al.*, 1991). First, we consider **futile cells,** those that have no apparent function. In general, they are removed in very early stages of development, and it has been proposed that they represent evolutionary vestiges. Second, we have **surplus** or **redundant cells.** One case is the nervous system,

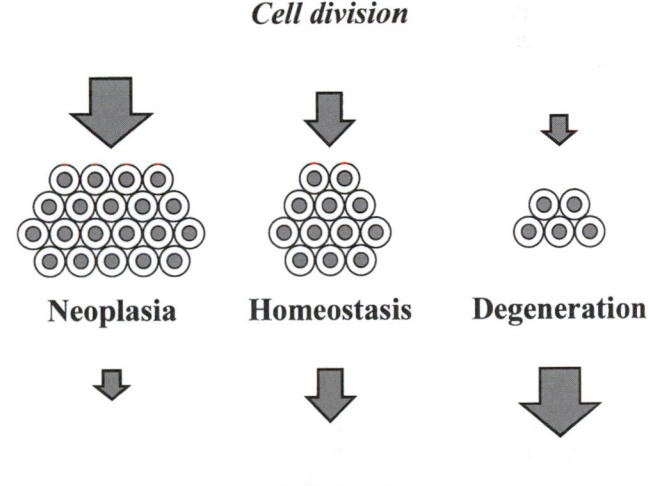

Cell division

Neoplasia   **Homeostasis**   **Degeneration**

Cell death

FIGURE 6. Tissue homeostasis depends on the balanced rates of cell division and cell death. The combination of an increased cell division with a reduced cell death in the generation of "proliferative" disorders is depicted. The opposite is the case for degenerative diseases. The scheme is trying to stress the idea that both processes work in concert to produce dramatic changes in tissue homeostasis. However, it is possible to have alterations in cell death independently of cell division.

where an excess of cells is produced, and this is suggested to help innervation of target sites, those neurons that fail to make contact are removed later. Third, **mistaken** or **damaged cells** occur when cells develop improperly or sustain genetic damage, respectively. Fourth, **obsolete cells** are those that have completed their function, either in development or homeostasis of a given organ. Fifth, **harmful cells** include, for example, the negative selection of autoreactive T lymphocytes in the thymus. These cells of the immune system, unless deleted, represent a risk of autoimmune diseases.

Alterations of the cell death program will inevitably generate important disorders, due to the relevance of cell deletion in development and homeostasis in multicellular organisms (Thompson, 1995). Diseases such as cancer, autoimmune disorders, and viral infections are some examples in which diminished cell death has been implicated. However, neurodegenerative disorders, such as Alzheimer's and Parkinson's diseases, as well as osteoporosis and AIDS, are cases where an accelerated rate of cell deletion has been involved in their pathophysiology.

V. Frontiers in the Study of Apoptosis

Given the importance of cell death in both normal and altered biological functions, this suggests that understanding apoptosis will have important repercussions in medicine. Currently, apoptosis has become synonymous with regulation of proteolytic enzymes (e.g., caspases). Unraveling those mechanisms involved in controlling caspases in apoptosis and defining the diversity and isoforms of the execution phase will permit the design of strategies for therapeutically regulating cell death.

Bcl-2 is a pluripotent antagonist of apoptosis *in vitro*. However, we do not know how this is achieved. The fact that diverse and unrelated inducers of apoptosis are blocked by Bcl-2 suggests that a single, intracellular homeostasis sensor would activate the execution path on alterations in cell physiology. This is probably the target being repressed by the different antagonists of cell death. The mitochondrion appears as the most plausible homeostasis sensor, but further studies are needed to fully understand the role of this organelle in apoptosis, for instance during development. Alternatively, Bcl-2 has more than one mechanism of action to explain its role in apoptosis.

VI. Summary

Physiological cell death, apoptosis, and programmed cell death may not, strictly speaking, be synonyms, However, the wealth of information on cell deletion emphasizes the importance of cellular machinery, which, for the most part, is constitutively present in the cytoplasm and readily executes a cell death program. This program seems to be the same regardless of the nature of the inducer triggering apoptosis. Nevertheless, it is also evident that there is redundancy in this machinery, that is, there is more than one way of activating the same protease or there is more than one protease to cleave the same substrate. Furthermore, the execution phase seems

to be working in a self-amplification loop, making it difficult to identify the critical regulators in apoptosis. Conceivably, the activation of the execution phase requires coincidence of different signals to avoid spontaneous or spurious activation. Therefore, understanding the regulation of proteolytic enzymes in apoptosis is a necessary step in order to design therapeutic approaches to eliminate cancer cells specifically or to stop cell death in degenerative disorders.

RNA transcription and protein synthesis do not appear to be an essential characteristic for the action of the execution machinery in apoptosis, as opposed to previous thinking. This indicates that cell death effectors are constitutively present, but repressed. Nevertheless, other stages of cell death, like signaling or determination, could be under transcriptional or translational control. Thus, the role of protein synthesis might be more evident during development.

Although we do not yet understand how Bcl-2 works, this protein has already revolutionized our understanding of cell death mechanisms. First, we know that the so-called "proliferative" disorders could also originate by a reduction in cell death, independently of an increment in cell proliferation (see Fig. 6). Second, the finding that Bcl-2 can partially protect against necrotic insults in addition to apoptosis implies that both processes could have more in common than previously thought.

Interestingly, the two ends, cell death and cell proliferation, seem to come into close contact. This is evident when we consider the relationship between apoptosis and cell cycle and the paradoxical effect of some inducers of cell death in cell proliferation. This has made difficult the understanding and identification of apoptosis-specific regulatory mechanisms. Nevertheless, important advances have been carried out in the molecular identification of the cell death machinery, for regulators and for effectors as well. The similarity of this program from worms to mammals has allowed the combination of forces for a faster understanding of cell deletion, and it highlights the importance of apoptosis in cell function.

Acknowledgments

We thank Ricardo Mondragón for Figure 1, and Jose A. Arias and Jose Segovia for critical reading of the manuscript.

Bibliography

Adams, J. M., and Cory, S. (1998). The Bcl-2 protein family: arbiters of cell survival. *Science* **281**, 1322–1326.

Ankarcrona, M., Dypbukt, J. M., Bonfoco, E., Zhivotovsky, B., Orrenius, S., Lipton, S. A., and Nicotera, P. (1995). Glutamate-induced neuronal death: a succession of necrosis or apoptosis depending upon mitochondrial function. *Neuron* **15**, 961–973.

Ashkenazi, A., and Dixit, V. M. (1998). Death receptors: signaling and modulation. *Science* **281**, 1305–1308.

Baffy, G., Miyashita, T., Williamson, J. R., and Reed, J. C. (1993). Apoptosis induced by withdrawal of Interleukin-3 (IL-3) from an IL-3-dependent hematopoietic cell line is associated with repartitioning of intracellular calcium and is blocked by enforced Bcl-2 oncoprotein production. *J. Biol. Chem.* **268**, 6511–6519.

Berridge, M. J. (1995). Capacitative calcium entry. *Biochemical J.* **312**, 1–11.

Boise, L. H., Gottschalk, A. R., Quintáns, J., and Thompson, C. B. (1995). Bcl-2 and Bcl-2 related proteins in apoptosis regulation. *Curr. Topics in Microbiol. and Immunol.* **200**, 107–121.

Caelles, C., Helmberg, A., and Karin, M. (1994). *p53*-dependent apoptosis in the absence of transcriptional activation of *p53*-target genes. *Nature* **370**, 220–223.

Chinnaiyan, A. M., and Dixit, V. M. (1996). The cell-death machine. *Current Biol.* **6**, 555–562.

Chow, S. C., Peters, I., and Orrenius, S. (1995). Reevaluation of the role of *de novo* protein synthesis in rat thymocyte apoptosis. *Exp. Cell Research* **216**, 149–159.

Cidlowski, J. A., King, K. L., Evans-Storm, R. B., Montague, J. W., Bortner, C. D., and Hughes, F. M. (1996). The biochemistry and molecular biology of glucocorticoid-induced apoptosis in the immune system. *Recent Progress in Horm. Res.* **51**, 457–491.

Dowd, D. R. (1995). Calcium regulation of apoptosis. *In* "Advances in second messenger and phosphoprotein research" (Means A. R. ed.), vol. 30, pp. 255–281. Raven Press, Ltd., N. Y.

Ellis, R. E., Yan, J., and Horvitz, H. R. (1991). Mechanisms and functions of cell death. *Ann. Rev. Cell Biol.* **7**, 663–698.

Enari, M., Sakahira, H., Yokoyama, H., Okawa, K., Iwamatsu, A., and Nagata, S. (1998). A caspase-activated DNase that degrades DNA during apoptosis, and its inhibitor ICAD. *Nature* **391**, 43–50.

Farinelli, S. E., and Greene, L. A. (1996). Cell cycle blockers mimosine, ciclopirox, and deferoxamine prevent the death of PC12 cells and postmitotic sympathetic neurons after removal of trophic support. *J. Neurosci.* **16**, 1150–1162.

Fesus, L., Madi, A., Balajthy, Z., Nemes, Z., and Szondy, Z. (1996). Transglutaminase induction by various cell death and apoptosis pathways. *Experientia* **52**, 942–949.

Frade, J. M., Rodriguez-Tebar, A., Barde, Y. A. (1996). Induction of cell death by endogenous nerve growth factor through its p75 receptor. *Nature* **383**, 166–168.

Furuya, Y., Lundmo, P., Short, A. D., Gill, D. L., and Issacs, J. T. (1994). The role of calcium, pH, and cell proliferation in the programmed (apoptotic) death of androgen-independent prostatic cancer cells induced by thapsigargin. *Cancer Res.* **54**, 6167–6175.

Gill, D. L., Waldron, R. T., Rys-Sikora, K. E., Ufret-Vincenty, C. A., Graber, M. N., Favre, C. J., and Alfonso, A. (1996). Calcium pools, calcium entry, and cell growth. *Biosci. Rep.* **16**, 139–157.

Gutiérrez, A. A., Arias, J. M., Garcia, L., Mas-Oliva, J., and Guerrero-Hernandez, A. (1999). Activation of a Ca^{2+} permeable cation channel by two different inducers of apoptosis in a human prostatic cancer cell line. *J. Physiol (London)* **517**, 95–107.

Häcker, G., and Vaux, D. L. (1995). Apoptosis. A sticky business. *Current Biol.* **5**, 622–624.

Hakem, R., Hakem, A., Duncan, G. S., Henderson, J. T., Woo, M., Soengas, M. S., Elia, A., de la Pompa, J. L., Kagi, D., Khoo, W., Potter, J., Yoshida, R., Kaufman, S. A., Lowe, S. W., Penninger, J. M., and Mak, T. W. (1998). Differential requirement for caspase 9 in apoptotic pathways *in vivo*. *Cell* **94**, 339–352.

Hannun, Y. A. (1996). Functions of ceramide in coordinating cellular responses to stress. *Science* **274**, 1855–1859.

Hart, S. P., Haslett, C., and Dransfield, I. (1996). Recognition of apoptotic cells by macrophages. *Experientia* **52**, 950–956.

Haupt, Y., Rowan, S., Shaulian, E., Vousden, K. H., and Oren, M. (1995). Induction of apoptosis in HeLa cell by trans-activation-deficient *p53*. *Genes Dev.* **9**, 2170–2183.

Hockenbery, D., Nuñez, G., Milliman, C., Scheiber, R. D., and Korsmeyer, I. (1990). Bcl-2 is an inner mitochondrial membrane protein that blocks programmed cell death. *Nature* **348**, 334–336.

Jacobson, M. D., Burne, J. F., King, M. P., Miyashita, T., Reed, J. C., and Raff, M. C. (1993). Bcl-2 blocks apoptosis in cells lacking mitochondrial DNA. *Nature* **361**, 365–369.

Jacobson, M. D., Burne, J. F., and Raff, M. C. (1994). Programmed cell death and Bcl-2 protection in the absence of nucleus. *EMBO J.* **13**, 1899–1910.

Jiang, Y., Woronicz, J. D., Liu, W., and Goeddel, D. V. (1999). Prevention of constitutive TNF receptor 1 signaling by silencer of death domains. *Science* **283**, 543–546.

Kerr, J. F. R., Wyllie, A. H., and Currie, A. R. (1972). Apoptosis: a basic biological phenomenon with wide-ranging implications in tissue kinetics. *Br. J. Cancer* **26**, 1790–1794.

Khan, A. A., Soloski, M. J., Sharp, A. H., Schilling, G., Sabatini, D. M., Li, S. H., Ross, C. A., and Snyder, S. H. (1996). Lymphocyte apoptosis: Mediation by increased type-3 Inositol 1, 4, 5-trisphosphate receptor. *Science* **263**, 503–507.

Kirsch, D. G., Doseff, A., Chaun, B. N., Lim, D. S., de Souza-Pinto, N. C., Hansford, R., Kastan, M. B., Lazebnik, Y. A., and Hardwick, J. M. (1999). Caspases-3-dependent cleavage of Bcl-2 promotes release of cytochrome c. *J. Biol. Chem.* **274**, 21155–21161.

Knudson, C. M., Tung, K. S., Tourtellotte, W. G., Brown, G. A., and Korsmeyer, S. J. (1995). Bax-deficient mice with lymphoid hyperplasia and male germ cell death. *Science* **270**, 96–99.

Koliwad, S. K., Kunze, D. L., and Elliot, S. J. (1996). Oxidant stress activates a nonselective cation channel responsible for membrane depolarization in calf vascular endothelial cells. *J. Physiol.* **491**, 1–12.

Korsmeyer, S. J., Yin, X. M., Oltvai, Z. N., Veis-Novack, D. J., and Linette, G. P. (1995). Reactive oxygen species and the regulation of cell death by Bcl-2 gene family. *Biochim. Biophys. Acta* **1271**, 63–66.

Lemasters, J. J. (1999). Mechanisms of hepatic toxicity. V. Necroapoptosis and the mitochondrial permeability transition: shared pathways to necrosis and apoptosis. *Am. J. Physiol.* **276**, G1–G6.

Lowe, S. W., Schmitt, E. M., Smith, S. W., Osborne, B. A., and Jacks T. (1993). *p53* is required for radiation-induced apoptosis in mouse thymocytes. *Nature* **362**, 847–849.

Marin, M. C., Fernandez, A., Bick, R. J., Brisbay, S., Buja, L. M., Snuggs, M., McConkey, D. J., Eschenbach, A. C., Keating, M. J., and McDonnel, T. J. (1996). Apoptosis suppression by Bcl-2 is correlated with the regulation of nuclear and cytosolic Ca^{2+}. *Oncogene* **12**, 2259–2266.

Marks, A.R. (1997). Intracellular calcium-release channels: regulators of cell life and death. *Am. J. Physiol.* **272**, H597–H605.

Matsuyama, S., Schendel, S. L., Xie, Z., and Reed, J. C. (1998). Cytoprotection by Bcl-2 requires the pore-forming α5 and α6 helices. *J. Biol. Chem.* **273**, 30995–31001.

McConkey, D. J. (1996). Calcium-dependent, interleukin 1β-converting enzyme inhibitor-insensitive degradation of lamin B_1 and DNA fragmentation in isolated thymocyte nuclei. *J. Biol. Chem.* **271**, 22398–22406.

McDonnell, T. J., Beham, A., Sarkiss, M., Andersen, M., and Lo, P. (1996). Importance of Bcl-2 family in cell death regulation. *Experientia* **52**, 1008–1017.

Motoyama, N., Wang, F., Roth, K. A., Sawa, H., Nakayama, K., Negishi, I., Senju, S., Zhang, Q., Fuji, S., and Loh, D. Y. (1995). Massive cell death of immature hematopoietic cells and neurons in Bcl-x-deficient mice. *Science* **267**, 1506–1510.

Muchmore, S. W., Sattler, M., Liang, H., Meadows, R. P., Harlan, B. S., Yoon, H. S., Nettesheim, D., Chang, B. S., Thompson, C. B., Wong, S. L., Ng, S. C., and Fesik, S. W. (1996). X-ray and NMR structure of human Bcl-x_L, an inhibitor of programmed cell death. *Nature* **381**, 335–341.

Nagata, S., and Golstein, P. (1995). The Fas death factor. *Science* **267**, 1449–1455.

Nakagawa, T., Zhu, H., Morishima, N., Li, E., Xu, J., Yankner, B.A., and Yuan, J. (2000). Caspase-12 mediates endoplasmic-reticulum-specific apoptosis and cytotoxicity by amyloid-beta. *Nature* **403**, 98–103.

Perez-Terzic, C., Pyle, J., Jaconi, M., Stehno-Bittel, L., and Clapham, D. E. (1996). Conformational states of the nuclear pore complex induced by depletion of nuclear Ca^{2+} stores. *Science* **273**, 1875–1877.

Piacentini, M. (1995). Tissue transglutaminase: a candidate effector element of physiological cell death. *Curr. Topics in Microbiol. and Immunol.*, **200**, 163–175.

Raffo, A. J., Perlman, H., Chen, M. W., Day, M. L., Streitman, J. S., and Buttyan, R. (1995). Overexpression of Bcl-2 protects prostate cancer cells from apoptosis *in vitro* and confers resistance to androgen depletion *in vivo*. *Cancer Res.* **55**, 4438–4445.

Reed, J. C. (1994). Bcl-2 and the regulation of programmed cell death. *J. Cell Biol.* **124**, 1–6.

Reed, J. C. (1998). Bcl-2 family proteins. *Oncogene* **17**, 3225–3236.

Rosen, D. R., Siddique, T., Patterson, D., Figlewicz, D. A., Sapp, P., Hentati, A., Donaldson, D., Goto, J., O'Regan, J. P., Deng, H. X, *et al.* (1993). Mutations in Cu/Zn superoxide dismutase gene are associated with familial amyotrophic lateral sclerosis. *Nature* **362**, 59–62.

Rossé, T., Olivier, R., Monney, L., Rager, M., Conus, S., Fellay, I., Jansen, B., and Borner, C. (1998). Bcl-2 prolongs cell survival after Bax-induced release of cytochrome c. *Nature* **391**, 496–499.

Ruoslahti, E., and Reed, J. C. (1994). Anchorage dependence, integrins, and apoptosis. *Cell* **77**, 477–478.

Sahara, S., Aoto, M., Eguchi, Y., Imamoto, N., Yoneda, Y., and Tsujimoto, Y. (1999). Acinus is a caspase-3-activated protein required for apoptotic chromatin condensation. *Nature* **401**, 168–173.

Samali, A., Gorman, A. M., and Cotter, T. G. (1996). Apoptosis—the story so far . . . *Experientia* **52**, 933–941.

Schwartz, L. M., Smith, S. W., Jones, M. E. E., and Osborne, B. A. (1993). Do all programmed cell deaths occur via apoptosis? *Proc. Natl. Acad. Sci. USA* **90**, 980–984.

Sentman, C. L., Shutter, J. R., Hockenber, D., Kanagawa, O., and Korsmeyer, S. J. (1991). Bcl-2 inhibits multiple forms of apoptosis but not negative selection in thymocytes. *Cell* **67**, 879–888.

Shibasaki, F., and McKeon, F. (1995). Calcineurin functions in Ca^{2+}-activated cell death in mammalian cells. *J. Cell Biol.* **131**, 735–743.

Shimizu, S., Narita, M., and Tsujimoto, Y. (1999). Bcl-2 family proteins regulate the release of apoptogenic cytochrome c by the mitochondrial channel VDAC. *Nature* **399**, 483–487.

Shimizu, S., and Tsujimoto, Y. (2000). Proapoptotic BH3-only Bcl-2 family members induce cytochrome c release, but not mitochondrial membrane potential loss, and do not directly modulate voltage-dependent anion channel activity. *Proc. Natl Acad. Sci. USA* **97**, 577–582.

Slater, A. F. G., Nobel, C. S. I., and Orrenius, S. (1995). The role of intracellular oxidants in apoptosis. *Biochim. Biophys. Acta* **1271**, 59–62.

Szalai, G., Krishnamurthy, R., and Hajnóczky, G. (1999) Apoptosis driven by IP3-linked mitochondrial calcium signals. *EMBO J.* **18**, 6349–6361.

Testi, R. (1996). Sphingomyelin breakdown and cell fate. *Trends in Biochem. Sci.* **21**, 468–471.

Thompson, C. B. (1995). Apoptosis in the pathogenesis and treatment of disease. *Science* **267**, 1456–1462.

Vito, P., Lacaná, E., and D'Adamio, L. (1996). Interfering with apoptosis: Ca^{2+}-binding protein ALG-2 and Alzheimer's disease gene ALG-3. *Science* **271**, 521–525.

Waldmann, R., Champigny, G., Voilley, N., Lauritzen, I., and Lazdunski, M. (1996). The mammalian degenerin MDEG, an amiloride-sensitive cation channel activated by mutations causing degeneration in Caenorhabditis elegans. *J. Biol. Chem.* **271**, 10433–10436.

Wang, J., and Walsh, K. (1996) Resistance to apoptosis conferred by Cdk inhibitors during myocyte differentiation. *Science* **273**, 359–361.

Wang, C. Y., Mayo, M. W., and Baldwin, A. S., Jr. (1996). TNF- and Cancer therapy-induced apoptosis: potentiation by inhibition of NF-κB. *Science* **274**, 784–787.

Wang, H-G., Pathan, N., Ethell, I. M., Krajewski, S., Yamaguchi, Y., Shibasaki, F., McKeon, F., Bobo, T., Franke, T. F., and Reed, J. C. (1999). Ca^{2+}-induced apoptosis through calcineurin dephosphorylation of BAD. *Science* **284**, 339–343.

Wolf, B. B., and Green, D. R. (1999). Suicidal tendencies: Apoptotic cell death by caspase family proteinases. *J. Biol. Chem.* **274**, 20049–20052.

Wood, D. E., Thomas, A., Devi, L. A., Berman, Y., Beavis, R. C., Reed, J. C., and Newcomb, E. W. (1998). Bax cleavage is mediated by calpain during drug-induced apoptosis. *Oncogene* **17**, 1069–1078.

Wyllie, A. H., Kerr, J. F. R., and Currie, A. R. (1980). Cell death: the significance of apoptosis. *Int. Rev. Cytol.* **68**, 251–306.

Wyllie, A. H., Morris, R. G., Smith, A. L., and Dunlop, D. (1984). Chromatin cleavage in apoptosis: association with condensed chromatin morphology and dependence on macromolecular synthesis. *J. Pathol.* **142**, 67–77.

Yonish-Rouach, E. (1996). The *p53* tumour suppressor gene: a mediator of a G1 growth arrest and of apoptosis. *Experientia* **52**, 1001–1007.

Yoshida, H., Kong, Y. Y., Yoshida, R., Elia, A. J., Hakem, A., Hakem, R., Penninger, J. M., and Mak, T. W. (1998). Apaf1 is required for mitochondrial pathways of apoptosis and brain development. *Cell* **94**, 739–750.

Zhivotovsky, B., Burgess, D. H., and Orrenius, S. (1996). Proteases in apoptosis. *Experientia* **52**, 968–978.

Eric J. Hall *and* James G. Kereiakes

69

Effects of Ionizing Radiation on Cells

I. Introduction

Life on earth has developed with an ever-present background of radiation. It is not something new, invented by the wit of man; radiation from natural sources has always been here in the form of, for example, radioactivity in the ground and in food, as well as cosmic rays from outer space. An issue often debated is whether life has evolved in spite of the potential deleterious effects of radiation—the winner in a constant battle—or whether the ability of radiation to cause mutations has been a vital factor in the continued upward evolution of biological species. No one is sure at present which is the case, and it is probable that the answer will never be known with any certainty.

What is new, what is man-made, is the extra radiation to which we are subjected from medical X-rays in the hospital or dentist's office, from journeys in high-flying aircraft, from the fallout of nuclear weapons testing, from nuclear reactors built to generate electrical power, and from low-level radioactive waste. There can be no denying that a man-made component of radiation is being continuously added to the natural background level. This is a cause of great concern to the public, and must ultimately implicate the whole of society because of the critical choices and issues involved. However, the use of radiation has become an integral part of modern life. From the X-ray picture of a broken limb to the treatment of cancer, the medical applications of radiation are now accepted as commonplace, and X-ray facilities are available in every community hospital.

II. Types of Radiation

Radiation is ionizing if it is able to disrupt the chemical bonds of molecules of which living things are made and thus cause biologically important changes. Visible light, radio waves, and the radiant heat from the sun are also forms of radiation, but are not able to produce damage in this way by ionization, though, of course, they too can cause biological effects if larger amounts are involved.

Ionizing radiations come in several varieties. First, there are rays, such as **X-rays** and **gamma rays.** These represent energy transmitted in a wave without the movement of any material, just as heat and light from the sun cross the vast emptiness of space to reach the earth. X- and gamma rays do not differ from one another in nature or in properties; the only difference between the two is where they originate. X-rays in general, are made in an electrical device, such as may be seen in any dentist's office, whereas gamma rays are emitted by unstable or radioactive isotopes. The electromagnetic spectrum is illustrated in Fig. 1.

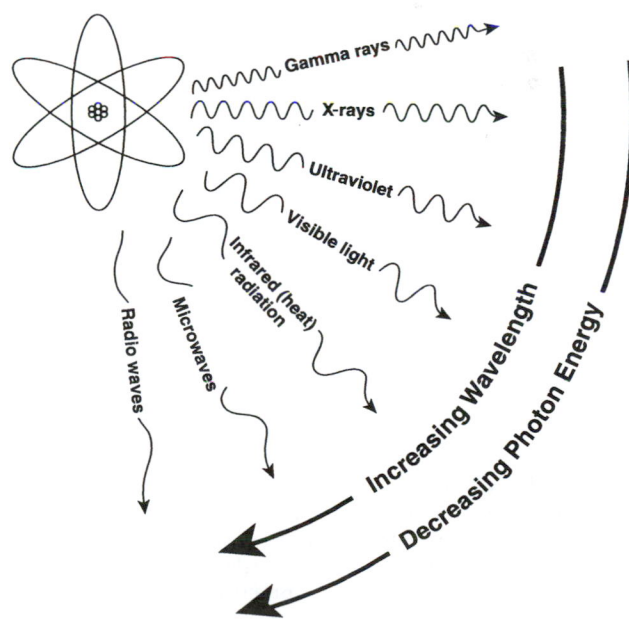

FIGURE 1. Illustrating the electromagnetic spectrum. X-rays and gamma rays have the same nature as visible light, radiant heat, and radiowaves; however, they have a shorter wavelength and consequently a larger photon energy—more energy per "packet."

Other types of ionizing radiations are fast-moving particles of matter. Some carry a charge of electricity, some do not. The penetrability of different types of radiation is illustrated in Fig. 2.

Neutrons are the only uncharged particles of any consequence, and they are an important form of ionizing radiation because they are generally associated with atomic bombs and nuclear reactors. Neutrons are particles with a mass similar to that of the proton, but they carry no electrical charge. Because they are electrically neutral, they are very penetrating in material of all sorts, including living tissues. Neutrons constitute one of the fundamental particles of which the nucleus of most atoms is built. Neutrons are emitted as a by-product when heavy radioactive atoms such as uranium undergo fission, that is, split up to form two smaller atoms. They can also be produced artificially by large accelerators in physics research laboratories.

Electrons are small negatively charged particles, which are found in all normal atoms. Electrons are often given off when radioactive materials break down and decay, in which case they are called **beta rays.** They can also be produced artificially in the laboratory and accelerated in electrical devices.

Protons are positively charged particles found in the nucleus of every atom. They have a mass approximately equal to that of the neutron and are almost 2000 times heavier than an electron. Protons are not usually given off by radioactive isotopes on earth, but they are found in great abundance in outer space and may constitute a hazard to astronauts.

Alpha particles are nuclei of helium atoms; that is, helium atoms with the planetary electrons stripped off. An alpha particle consists of two protons and two neutrons stuck together. They have a net positive charge and are relatively massive. These particles are commonly emitted when heavy radioactive isotopes, such as uranium or radium, decay and break down.

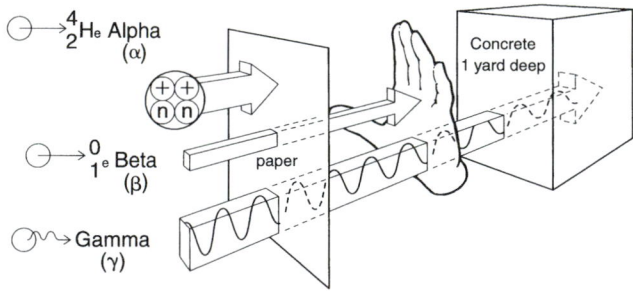

FIGURE 2. Illustration of some of the different types of ionizing radiations that were labeled α, β, and γ before their true nature was known or understood. Gamma (γ) rays form part of the electromagnetic spectrum and are of the same nature as heat or light. They can be very penetrating and pass through thick barriers. Beta (β) rays are comprised of a stream of electrons, tiny negatively charged particles. Beta rays can pass through a hand, but unless they are of very high energy, they can be stopped by a modest barrier. Alpha (α) rays are relatively massive positively charged particles: they are in fact helium nuclei and each is made up of two neutrons and two protons in close association. In general, alpha rays can be stopped by a thin barrier—even a sheet of cardboard.

Heavy ions are the nuclei of any atoms that are stripped of their planetary electrons and are moving at high speed. Ions of almost all of the known elements are present in space and constitute one of the problems of space flight. These particles move with great speed and have enormous energy, and it is virtually impossible to design spacecraft to fully protect their occupants from all of the heavy ions.

III. Interactions of Radiation with Matter

When such radiations pass through living things, they give up energy to the tissues and cells of which all biological material is made. The energy is not spread out evenly, but is deposited or dumped very unevenly in discrete "packets." A lot of energy is given to some parts of some cells, and little, if any, to others.

This uneven pattern of energy deposition accounts for the special consequences of ionizing radiation. The total amount of energy involved may be small; however, some cells of the living material may be adversely affected because it is deposited so unevenly. The smallness of the amount of energy involved may be illustrated in a number of ways. For example, the dose of X-rays that would undoubtedly kill a person if given to the whole body may be compared with heat energy; it would be less than that absorbed by drinking a cup of warm coffee, or sunbathing for a few minutes on a hot day. Energy in the form of heat is absorbed uniformly and evenly, and much greater quantities of energy in these forms are required to produce damage in living things.

The situation may be illustrated by an example that, although trivial, contains the essence of the difference between ionizing and nonionizing radiations. Imagine throwing a kilogram of material at a running rabbit. One may choose to throw sand, in which case millions of particles would make up one kilogram, or one may choose to throw a single rock weighing one kilogram. The same total amount of energy would be involved in throwing either projectile a given distance. In the case of the sand, the energy would be divided into such small individual packets that no damage would result to the rabbit from the impact of any of them. If a single rock were chosen instead, the chance of scoring a hit at all would be greatly reduced, but biological damage would undoubtedly follow in the event that a hit occurred. Ionizing radiations constitute large discrete packets of energy. To absorb a dose of ionizing radiation is to be hit by a rock, not by many particles of sand!

Radiations may be directly or indirectly ionizing. All of the charged particles previously discussed including α particles and heavy ions are **directly ionizing;** that is, provided the individual particles have sufficient kinetic energy, they can directly disrupt the atomic structure of the absorber through which they pass, and produce chemical and biological changes. Electromagnetic radiations (X- and gamma rays) are **indirectly ionizing.** They do not produce chemical and biological damage themselves, but when they are absorbed in the material through which they pass they give up their energy to produce fast-moving electrons.

The process by which X-ray photons are absorbed depends on the energy of the photons concerned and the chemical composition of the absorbing material. At high energies, the Compton process predominates, while at lower energies the photoelectric process is more important; but in either case the end result is the same: The photon energy is converted into kinetic energy of a fast-moving secondary electron. In practice, when an X-ray beam is absorbed by tissue, a vast number of photons interact with a vast number of atoms to produce a large number of fast electrons, many of which can ionize other atoms of the absorber, break vital chemical bonds, and initiate the chain of events that ultimately is expressed as biological damage.

Neutrons are uncharged particles. For this reason they are highly penetrating compared with charged particles of the same mass and energy. They are indirectly ionizing and are absorbed by elastic or inelastic scattering. Fast neutrons differ basically from X-rays in the mode of their interaction with tissue. **X-ray photons** interact with the **orbital electrons** of atoms of the absorbing material and set in motion **fast electrons. Neutrons,** however, interact with the **nuclei** of atoms of the absorbing material and set in motion **fast recoil protons, α particles,** and **heavier nuclear fragments.**

In the case of intermediate fast neutrons, **elastic scattering** is the dominant process. The incident neutron collides with the nucleus of an atom of the absorber (which in the case of tissue is predominantly hydrogen). Part of its kinetic energy is transferred to the nucleus and part is retained by the deflected neutron, which may go on to make further collisions. The recoil protons that are set in motion lose energy by excitation and ionization as they pass through the biological material.

At energies above about 5 MeV, **inelastic scattering** begins to take place and assumes increasing importance as the neutron energy rises. The neutron may interact with a carbon nucleus to produce three α particles. These are the so-called "spallation" products, which become very important at higher energies. The α particles produced in this way represent a relatively modest proportion of the total absorbed dose, but they are densely ionizing and have an important effect on the biological characteristics of the radiation.

IV. Measuring Radiation

The amount or quantity of radiation, or as it is usually called, the **dose,** is measured in terms of the energy absorbed in the tissues. The unit of absorbed dose is the Gray (Gy) defined to be an energy absorption of 1 joule per kilogram.

In some instances, the dose is very much less than one Gray and then the units milligray (mGy) or microgray (μGy) are used. There are 1000 mGy or one million μGy in 1 Gy. To give an idea of the way in which these units are used, two examples are quoted from later sections. A total body dose of 5 Gy to a human being would most probably be fatal; the average natural background radiation to which we are exposed is about 3 mGy per year.

Equal doses of different types of radiation do not necessarily produce equal biological effects. For example, 0.5 Gy of neutrons is more effective than 0.5 Gy of X-rays. In general, X-rays, gamma rays, and electrons are least effective for a given dose, while heavy ions are the most damaging. Neutrons fall somewhere between.

For general discussions of radiation effects, a different quantity is used, namely, **equivalent dose,** which is the absorbed dose (in Gray) multiplied by a factor known as the **radiation weighting factor** (W_r), which allows for the relative effectiveness of the particular type of radiation involved. If dose is measured in Gray, the equivalent dose will be in Sievert (Sv). X-rays and gamma rays are regarded as the standard, and for these types of radiation Gray and Sievert are interchangeable. A dose of 1 Gy of X-rays is, by definition, 1 Sv. Alpha particles, however, are roughly 20 times as effective as X-rays for the same absorbed dose and are therefore assigned a radiation weighting factor (W_r) of 20. Consequently, a dose of 1 Gy of α particles represents an effective dose of 20 Sv. In some instances, the equivalent dose may be much less than 1 Sv or μSv, in which case smaller subunits are used. There are 1000 millisieverts (mSv) and 1 million microsieverts (μSv) in 1 Sievert.

Next, there is the **collective equivalent dose.** This is obtained by multiplying the equivalent dose by the number of individuals exposed. Thus, for example, in the incident at Three Mile Island, an estimate of the collective equivalent dose received by the 2 million people who lived within 50 miles turned out to be 32 person-sieverts. Some individuals received up to 1 mSv, others less than 0.01 mSv.

V. DNA Damage and Chromosome Breaks

Many lines of evidence support the view that DNA is the principal target for the biological effects of ionizing radiations. For example, using a polonium-tipped microneedle that emits short-range α particles, it has been shown that irradiating the cytoplasm of the cell to huge doses in excess of 250 Gy does not affect proliferation, whereas small doses to the nucleus prove to be lethal. Further, the incorporation of the short-range beta emitter tritiated thymidine into DNA is lethal, and incorporation of halogenated pyrimidines into DNA results in increased radiosensitivity, which is directly proportional to the amount incorporated.

The interaction of ionizing radiation with DNA has classically been described as resulting from a *direct* or an *indirect* interaction. In the direct interaction, a photon is absorbed by the medium, resulting in a secondary fast-moving electron that directly hits and breaks the DNA strand. In the indirect interaction the fast electron hits a water molecule (tissue is mostly water) producing a hydroxyl radical (OH·). A free radical is an atom or molecule with an unpaired electron in the outer orbit, a state that is associated with a high degree of chemical reactivity. This free radical then diffuses to the DNA causing a strand break. Experiments using free radical scavengers have led to the estimation that two-thirds of DNA damage from photon irradiation results from the indirect action. For α particles and neutrons the balance shifts in favor of the direct effect, which becomes dominant.

DNA consists of two strands that form a double helix, with each strand composed of deoxynucleotides, the sequence of which must be complementary, A pairing with T, and C with G. A break in a single strand, caused by either the direct or indirect effects discussed earlier, has little

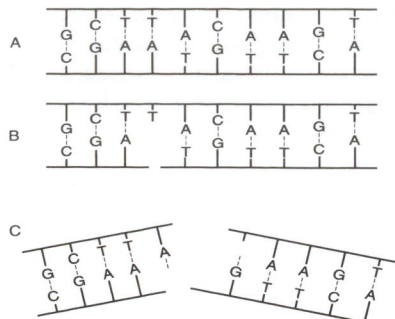

FIGURE 3. Diagrams of single- and double-strand DNA breaks caused by radiation. (A) Two-dimensional representation of the normal DNA double helix. The base pairs carrying the genetic code are complementary (i.e., adenine pairs with thymine, guanine pairs with cytosine). (B) A break in one strand is of little significance because it is readily repaired, using the opposite strand as a template. (C) If breaks occur in both strands and are directly opposite or separated by only a few base pairs, this may lead to a double-strand break where the chromatin snaps into two pieces. (Courtesy of John Ward, 1993.)

biological consequence since the break is rapidly repaired using the opposite strand as a template. However, if breaks occur in both strands that are directly opposite, or separated by only a few base pairs, then a double-strand break (DSB) occurs and the piece of chromatin snaps in two. This is illustrated in Fig. 3.

There is good evidence to believe that most of the important biological effects of ionizing radiations are a direct consequence of the rejoining of two DSBs. Figure 4 illustrates two ways in which a DSB in each of two chromosomes may rejoin. In the lower half of the figure, the two broken chromosomes rejoin in such a way that a dicentric and an acentric fragment are formed. This exchange type aberration is lethal to the cell; the fragment with no centromere will be lost dur-

ing cell division, whereas the aberrant chromosome with two centromeres will make a normal mitosis impossible. Aberrations of this type represent the principal mechanism whereby radiation kills cells. Note that breaks in *two* chromosomes are necessary for an exchange type aberration. The two breaks can be produced by a single electron track (see Fig. 5) in which case the probability of an interaction will be proportional to dose—that is, doubling the dose doubles the yield of aberrations. However, the two chromosome breaks may result from two separate electrons, in which case the probability of an interaction increases with the $(\text{dose})^2$—that is, doubling the dose makes an aberration *four* times as likely.

In practice, both possibilities occur, so that the yield of chromosomal aberrations is a linear-quadratic function of dose, that is,

$$\text{Aberration yield} = \alpha D + \beta D^2 \qquad (1)$$

This linear-quadratic relationship is ubiquitous in radiation biology for all biological effects of gamma and X-rays. Scoring dicentric aberrations can be used as a biological dosimeter. Suppose someone is suspected of being accidentally exposed to a large dose of radiation. A sample of blood can be taken, the peripheral lymphocytes stimulated to divide, and the incidence of dicentric aberrations scored. The lowest dose that can be assessed conveniently is about 25 cGy.

Referring back to Fig. 4, it is possible for the breaks in the two chromosomes to rejoin as shown in the upper half of the figure. This leads to a symmetrical translocation that is quite compatible with cell viability. However, it represents a rearrangement of genetic material, and in a few instances leads to a malignant change. For example, Burkitt's lymphoma and some types of leukemia result from a translocation that moves an oncogene from a quiescent to an active chromosomal site. This is one of the most likely mechanisms for radiation-induced carcinogenesis.

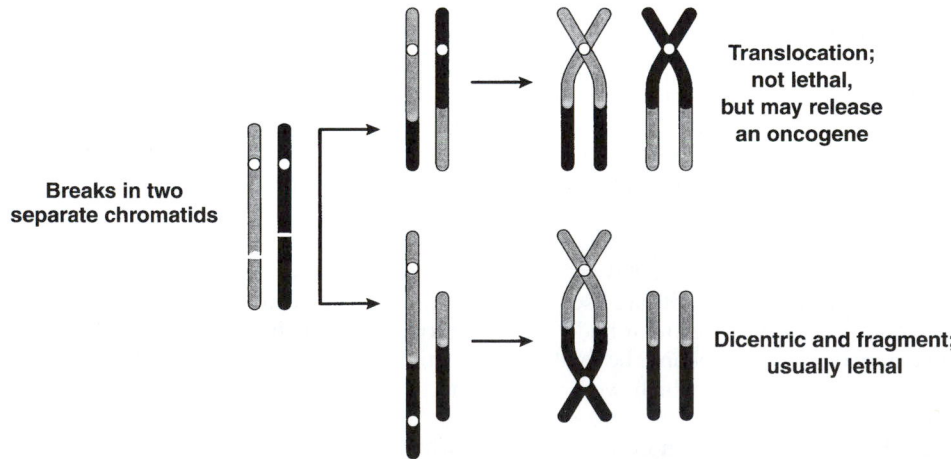

FIGURE 4. Most biological effects of radiation are due to the incorrect joining of breaks in two chromosomes. For example, the two broken chromosomes may recombine to form a dicentric (a chromosome with two centromeres) and an acentric fragment (a fragment with no centromere). This is a lethal lesion resulting in cell death. Alternatively, the two broken chromosomes may exchange broken ends. This is a called a symmetrical translocation. It does not lead to the death of the cell, but in a few special cases activates an oncogene by moving it from a quiescent to an active site.

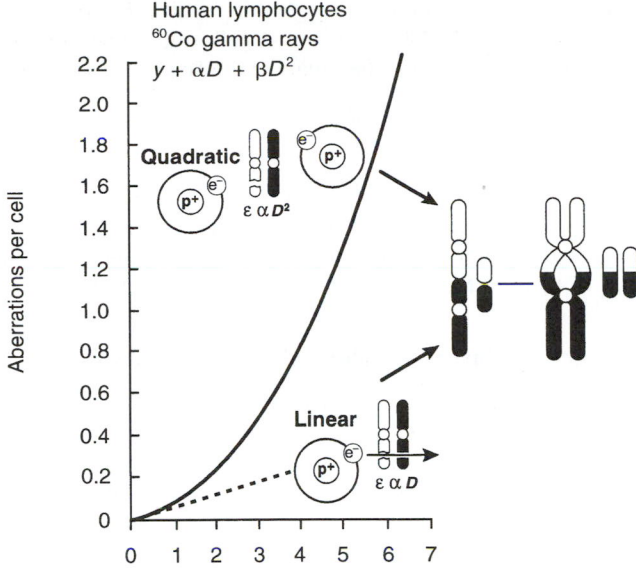

Human lymphocytes
^{60}Co gamma rays

$y + \alpha D + \beta D^2$

Quadratic

$\varepsilon \propto D^2$

Linear

$\varepsilon \propto D$

FIGURE 5. The frequency of chromosomal aberrations (dicentrics and rings) is a linear-quadratic function of dose because the aberrations are the consequence of the interaction of two separate breaks. At low doses, both breaks may be caused by the same electron; the probability of an exchange aberration is proportional to the square of the dose (D^2) (Redrawn from Hall, 1994.)

VI. Cell Survival Curves

A. Clonogenic

A **cell survival curve** describes the relationship between the radiation dose and proportion of cells that survive. What is meant by "survival"? Survival is the opposite of death! For differentiated cells that do not proliferate, such as nerve, muscle, or secretory cells, death can be defined as the loss of a specific function. For proliferating cells, such as hematopoietic stem cells or cells growing in culture, loss of the capacity for sustained proliferation—that is loss of **reproductive integrity**—is an appropriate definition. This is sometimes called **reproductive death** and a cell that survives by this definition is said to be **clonogenic** because it can form a clone or colony.

This definition is generally relevant to the radiobiology of whole animals and plants and their tissues. It has particular relevance to the radiotherapy of tumors. For a tumor to be eradicated, it is only necessary that cells be "killed" in the sense that they are rendered unable to divide and cause further growth and spread of the malignancy.

B. Why Cells Die

The classic mode of cell death following exposure to radiation is "mitotic death." Cells die while attempting to divide because of chromosomal damage, such as the formation of a ring or a dicentric that causes loss of genetic material and prevents the clean segregation of DNA into the two daughter cells. Death does not necessarily occur at the first mitosis following irradiation; cells can often manage to

complete several divisions, but death is inevitable if chromosomal damage is severe.

The other form of cell death is programmed cell death or **apoptosis.** This is an important form of cell death during the development of the embryo, and is implicated, for example, in the regression of the tadpole tail during metamorphosis. Is it also important in many facets of biology including cell renewal systems and hormone-related atrophy. Apoptosis is characterized by a sequence of morphological events; cells condense and the DNA breaks up into pieces before the cells are phagocytosed and removed. In radiation biology, apoptosis is a dominant mode of radiation-induced cell death in cells of lymphoid origin, but of variable importance in most other cell types.

C. The Shape of a Cell Survival Curve

Most cell survival curves have been obtained by growing cells *in vitro* in petri dishes. Many cell lines have been established from malignant tumors and from normal tissues taken from humans or laboratory animals.

If cells are seeded as single cells, allowed to attach to the surface of a petri dish, and provided with culture medium, and appropriate conditions, each cell will grow into a macroscopic colony that is visible by eye in a period of a few weeks. If, however, the same number of cells is placed into a parallel dish and exposed to a dose of radiation just after they have attached to the surface of the dish, some cells will grow into colonies indistinguishable from those in the unirradiated dish, but others will form only tiny abortive colonies because the cells die after a few divisions. This is illustrated in Fig. 6. The **surviving fraction** is the number of macroscopic colonies counted in the irradiated dish divided by the number on the unirradiated dish. This process is repeated so that estimates of survival are obtained for a range of doses; surviving fraction is plotted on a logarithmic scale against dose on a linear scale in Fig. 7.

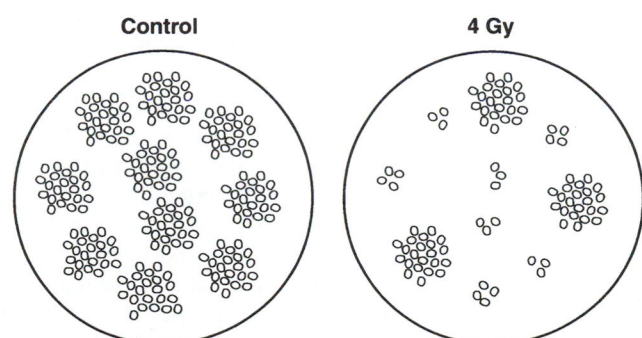

Control　　　**4 Gy**

FIGURE 6. In the left-hand dish, 10 cells were seeded and grew up into 10 macroscopic colonies, each containing hundreds or thousands of cells. The same number of cells were seeded in the right-hand dish but it was exposed to a dose of 4 Gy of X-rays as soon as the cells had attached. In this case, 3 cells grew into colonies indistinguishable from those on the unirradiated dish, while three formed abortive colonies. The surviving fraction is 3/10, or 0.3. In practice, of course, much larger numbers of cells are used to achieve better precision.

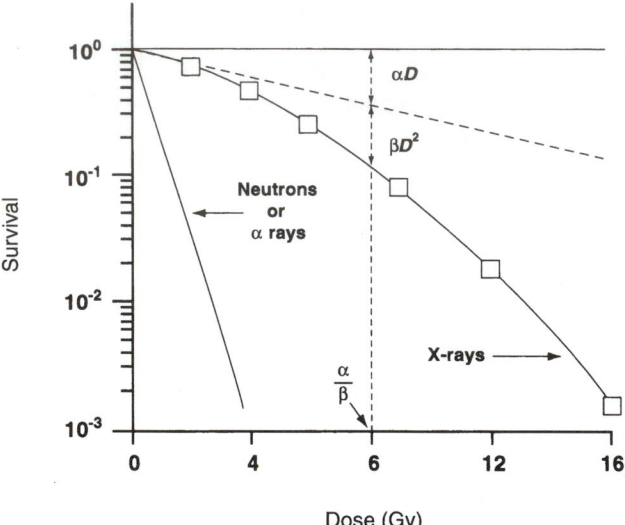

FIGURE 7. Shape of survival curve for mammalian cells exposed to radiation. The fraction of cells surviving is plotted on a logarithmic scale against dose on a linear scale. For α particles or low-energy neutrons (said to be densely ionizing) the dose-response curve is a straight line from the origin (i.e., survival is an exponential function of dose). The survival curve can be described by just one parameter, the slope. For X- or gamma rays (said to be sparsely ionizing), the dose-response curve has an initial linear slope, followed by a shoulder; at higher doses, the curve tends to become straight again. The experimental data for X-rays are fitted to a linear-quadratic function. There are two components of cell killing: One is proportional to dose (αD); the other is proportional to the square of the dose (βD^2). The dose at which the linear and quadratic components are equal is the ratio α/β. The linear-quadratic curve bends continuously but is a good fit to experimental data for the first few decades of survival. (Redrawn from Hall, 1994.)

Qualitatively, the shape of the survival curve can be described in relatively simple terms. At "low doses" for sparsely ionizing radiations, such as X-rays, the survival curve starts out straight on the log-linear plot with a finite initial slope; that is, the surviving fraction is an exponential function of dose. At higher doses, the curve bends. This bending or curving region extends over a dose range of a few Gy. At very high doses the survival curve often tends to straighten again; in general, this does not occur until doses are in excess of 10 Gy. By contrast, for densely ionizing (high linear energy transfer) radiations, such as α particles or low-energy neutrons, the cell survival curve is a straight line from the origin; that is, survival approximates to an exponential function of dose.

The linear quadratic model is currently the model of choice for cell survival curves and is fitted to the data in Fig. 7. This model assumes that there are two components to cell killing by radiation, one that is proportional to dose and one that is proportional to the square of the dose. The notion of a component of cell inactivation that varies with the square of the dose goes back to the early work with chromosomes in which many chromosome aberrations are clearly the result of two separate breaks, as previously described.

By this model the expression for the cell survival curve is

$$S = e^{-\alpha D - \beta D^2} \qquad (2)$$

where S is the fraction of cells surviving a dose D, and α and β are constants. The components of cell killing that are proportional to dose and the the square of the dose are equal when

$$\alpha D = \beta D^2 \qquad (3)$$

or

$$D = \alpha/\beta \qquad (4)$$

This is an important point that bears repeating: The linear and quadratic contributions to cell killing are equal at a dose that is equal to the ratio of α/β. This is illustrated in Fig. 7.

D. Survival Curves for Normal Tissues in Vivo

A great deal of ingenuity has been shown in devising systems to determine clonogenic survival curves for the cells of normal tissues *in vivo*. A compilation of many of these is shown in Fig. 8. There is a range of radiosensitivity with bone marrow stem cells being the most radiosensitive; these cells are characterized by a survival curve that has little shoulder, probably because these cells are prone to die an apoptotic rather than a mitotic death.

E. Radiosensitivity of Various Organisms

Figure 9 is a compilation of survival data to compare the radiosensitivity of various organisms. Its purpose is to illustrate that mammalian cells are exquisitely radiosensitive to radia-

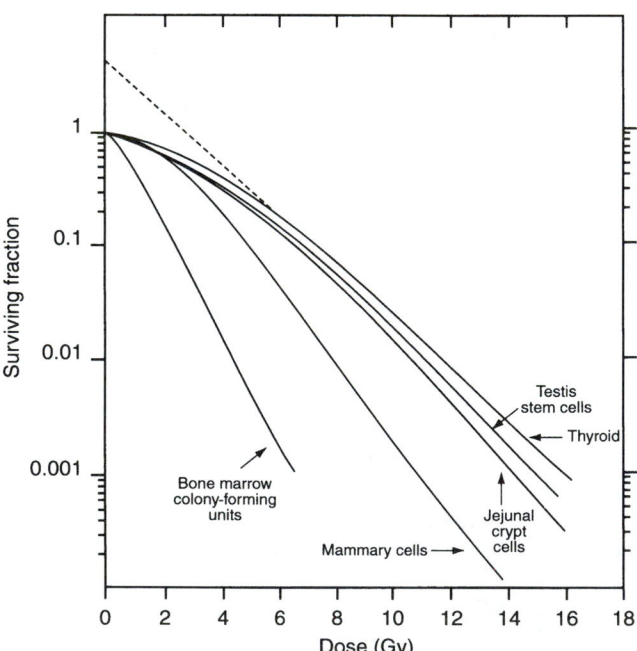

FIGURE 8. Summary of survival curves for clonogenic assays of cells from normal tissues. The bone marrow colony-forming units, together with the mammary and thyroid cells, represent systems in which cells are irradiated and assayed by transplantation into a different tissue in recipient animals. The jejunal crypt and testis stem cells are examples of systems in which cells are assayed for regrowth *in situ* after irradiation.

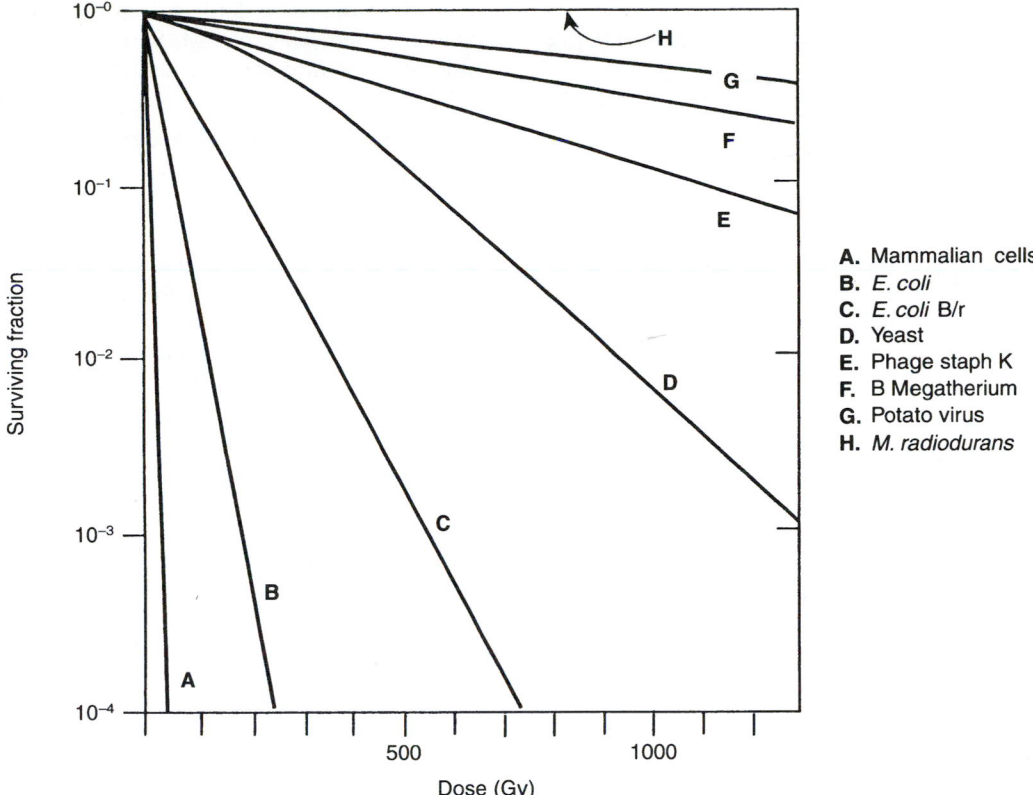

FIGURE 9. A comparison of the radiosensitivity of various organisms. Mammalian cells are exquisitely sensitive compared with bacteria and yeast, principally because of their larger DNA content.

tion compared with microorganisms. Bacteria and yeast are much more resistant to radiation. *Escherichia coli* B/r is a resistance mutant of *E. coli* that has particularly efficient repair mechanisms. Top of the league is *Micrococcus radioduras;* there is no measurable killing even after a dose of 100 Gy! This dramatic variation of radiosensitivity is largely a function of DNA content—a large DNA content leads to radiosensitivity. This figure also illustrates why a dose of tens of thousands of Gray is required when radiation is used to sterilize surgical devices; common bacteria are **very** resistant.

VII. Sensitivity and Phase of the Cell Cycle

A. The Cell Cycle

Mammalian cells propagate by mitosis. When a cell divides, two daughter cells are produced, each of which carries a chromosome complement identical to that of the mother cell. After an interval of time has elapsed, each of the daughter cells may undergo a further division. The time between successive divisions is known as the **mitotic cycle time** or, as it is commonly called, the **cell cycle time.**

When a population of dividing cells is observed with a conventional light microscope, the only event in the entire cell cycle that can be distinguished is mitosis itself. Just before the cell divides to form two daughter cells, the chromosomes

(which are diffuse and scattered in the cell in the period between mitoses) condense into clearly distinguishable forms. In addition, in monolayer cultures of cells, just before mitosis, the cells round up and become loosely attached to the surface of the culture vessel. This whole process of mitosis—in preparation for which the cell rounds up, the chromosome material condenses, the cell divides into two, and then stretches out again and attaches to the surface of the culture vessel—lasts only about 1 hour. The remainder of the cell cycle, interphase, occupies all of the intermitotic period. No events of interest can be identified with a conventional microscope during this time.

Because cell division is a cyclic phenomenon, repeated in each generation of the cells, it is usual to represent it as a circle, as shown in Fig. 10. The circumference of the circle represents the full mitotic cycle time for the cells (T); the period of mitosis is represented by M. The remainder of the cell cycle can be further subdivided by the use of **autoradiography**. This technique was first introduced by Howard and Pelc (1953) and has revolutionized the study of cell biology.

The basis of the technique is to feed the cells thymidine, a basic building block for making a new set of chromosomes, which has been labeled with radioactive tritium. Cells that are actively synthesizing new DNA as part of the process of replicating their chromosome complement will incorporate the radioactive thymidine. The surplus radioactive thymidine is then flushed from the system, and the preparation of cells is coated with a very thin layer of nuclear (photographic) emulsion. Beta

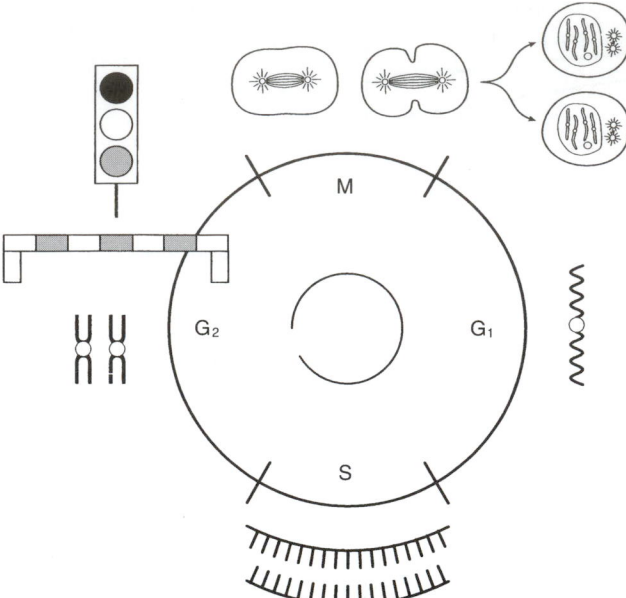

FIGURE 10. The stages of the mitotic cycle for actively growing mammalian cells (M, mitosis; S, DNA synthetic phase; G_1 and G_2, "gaps" or periods of apparent inactivity between the major discernible events in the cycle). Also shown are the site of action and function of the molecular checkpoint gene. Cells exposed to any DNA damaging agent, including ionizing radiation, are arrested in G_2 phase. The function of the pause in cell cycle progression is to allow for a check of chromosome integrity before the complex task of mitosis is attempted. Cells in which the checkpoint gene is inactivated are more sensitive to killing by gamma rays or ultraviolet light.

particles from cells that have incorporated radioactive thymidine pass through the nuclear emulsion and produce a latent image. When the emulsion is subsequently developed and fixed, the area through which a β particle has passed appears as a black spot. It is then a comparatively simple matter to view the preparation of cells and to observe that some of the cells have black spots or "grains" over them, which indicates that they were actively synthesizing DNA at the time of radioactive thymidine was made available. Other cells do not have any grains over their nuclei; this is interpreted to mean that the cells were not actively making DNA when the radioactive label was made available to them. When this is done, it becomes obvious that cells incorporate thymidine, that is, make DNA, only during a discreet part of the cycle, labeled S in Fig. 10. The interval between mitosis and S was called by Howard and Pelc "the first gap in activity," or G_1; the interval between S and M represents the second gap or G_2. Because of the problems related to disposal of radioactive waste, tritiated thymidine has been largely replaced by 5-bromodeoxyuridine, which is not radioactive and is identified by fluorescence.

All proliferating mammalian cells, whether in culture or growing normally in a tissue, have a mitotic cycle consisting of mitosis (M), followed by G_1, a period of DNA synthesis (S), and G_2, after which mitosis occurs again. The overall length of the cell cycle may vary from about 10 h for a hamster cell grown in culture to hundreds of hours for stem cells in some self-renewal tissues. This is due almost entirely to a

dramatic variation in the length of G_1. The remaining components of the cell cycle, M, S, and G_2, vary comparatively little between different cells in different circumstances.

B. Synchronously Dividing Cells

A study of the variation of the radiosensitivity with the position or age of the cell in the cell cycle was only made possible by the development of techniques to produce synchronously dividing cell cultures—populations of cells in which all of the cells occupy the same phase of the cell cycle at a given time. The most satisfactory way to produce a synchronously dividing cell population is to use the **mitotic harvest** technique, first described by Terasima and Tolmach (1963). This technique can only be used for cultures that grow in monolayers attached to the surface of the growth vessel. It exploits the fact that when such cells are close to mitosis, they round up and become loosely attached to the surface. If at this stage the growth medium over the cells is subjected to gentle motion (by shaking), the mitotic cells become detached from the surface and float in the medium. If this medium is then removed from the culture vessel and plated out into new petri dishes, the population consists almost entirely of mitotic cells. Incubation of these cell cultures at 37 °C then causes the cells to move together synchronously in step through their mitotic cycle. By delivering a dose of radiation at various times after the initial harvesting of mitotic cells, one can irradiate cells at various phases of the cell cycle.

C. Radiosensitivity and the Cell Cycle

Using the mitotic harvest technique, complete survival curves at a number of discrete points during the cell cycle were measured by Sinclair. The results are shown in Fig. 11.

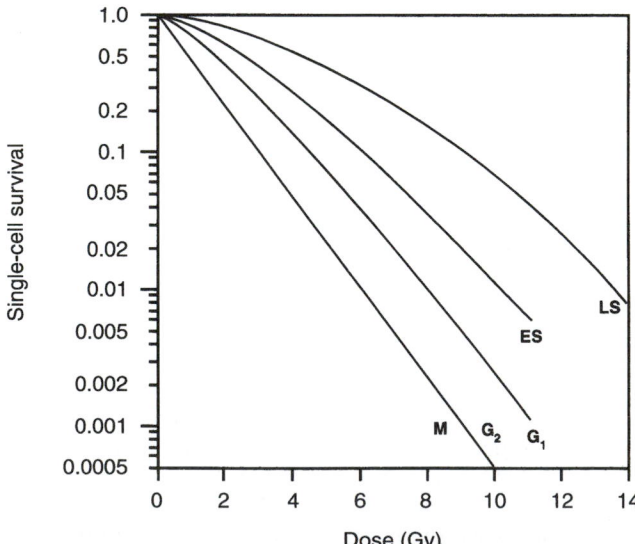

FIGURE 11. Cell survival curves for Chinese hamster cells at various stages of the cell cycle. The survival curve for cells in mitosis is steep and has no shoulder. The curve for cells in late S phase is shallower and has a large initial shoulder. G_1 and S phases are intermediate in sensitivity. (Redrawn from the data of Sinclair and Morton, 1966.)

Survival curves are shown for mitotic cells (M), for cells in G_1 and G_2, and for cells in early and late S. It is at once evident that the most sensitive cells are those in M and G_2, which are characterized by a survival curve that is steep and has no shoulder. At the other extreme, cells in the latter part of S phase exhibit a survival curve that is less steep, but the essential difference is that the survival curve has a very broad shoulder. The other phases of the cycle, such as G_1 and early S, are intermediate in sensitivity between the two extremes.

The experiments of Sinclair and Morton (1966), shown in Fig. 11, were performed with a rapidly dividing cell having a very short G_1. In other cell lines with a long G_1, which may be closer to the situation for most proliferating stem cells *in vivo*, there is a second relatively radioresistant period in early G_1.

The following is a summary of the main characteristics of the variation of radiosensitivity with cell age in the mitotic cycle:

1. Cells are most sensitive at or close to mitosis.
2. Resistance is usually greatest in the latter part of the S phase.
3. If G_1 has an appreciable length, a resistant period is evident early in G_1, followed by a sensitive period toward the end of G_1 phase.
4. G_2 is usually sensitive, perhaps as sensitive as M.

VIII. Molecular Checkpoint Genes

Cell cycle progression is controlled by a family of genes known as **molecular checkpoint genes.** In radiation biology the most important checkpoint appears to be the G_2 phase of the cycle where cells are temporarily halted after a dose of radiation (see Fig. 10). The importance of this function is to halt cells and allows chromosomal damage to be repaired before the complex task of mitosis is attempted. The checkpoint gene has been cloned and sequenced in some strains of yeast. Mutant cell lines in which the checkpoint gene is defective are very sensitive to radiation or, in fact, to any DNA damaging agent.

IX. Repair of Radiation Damage

Radiation damage to mammalian cells may be (1) **lethal damage,** which is irreversible, irreparable, and by definition, leads irrevocably to cell death; or (2) **sublethal damage,** which under normal circumstances can be repaired unless additional sublethal damage is added (e.g., from a second dose of radiation) with which it can interact to form lethal damage. **Sublethal damage repair** is the operational term for the increase in cell survival that is observed when a given radiation does is split into two fractions separated by a time interval.

Figure 12 shows data obtained in a split-dose experiment with cultured Chinese hamster cells. Figure 12A refers to cells that are not moving through the cycle because they were maintained at 24 °C, a nonphysiological temperature, between doses.

A single dose of 15.58 Gy leads to a surviving fraction of 0.005. When the dose is divided into two equal fractions,

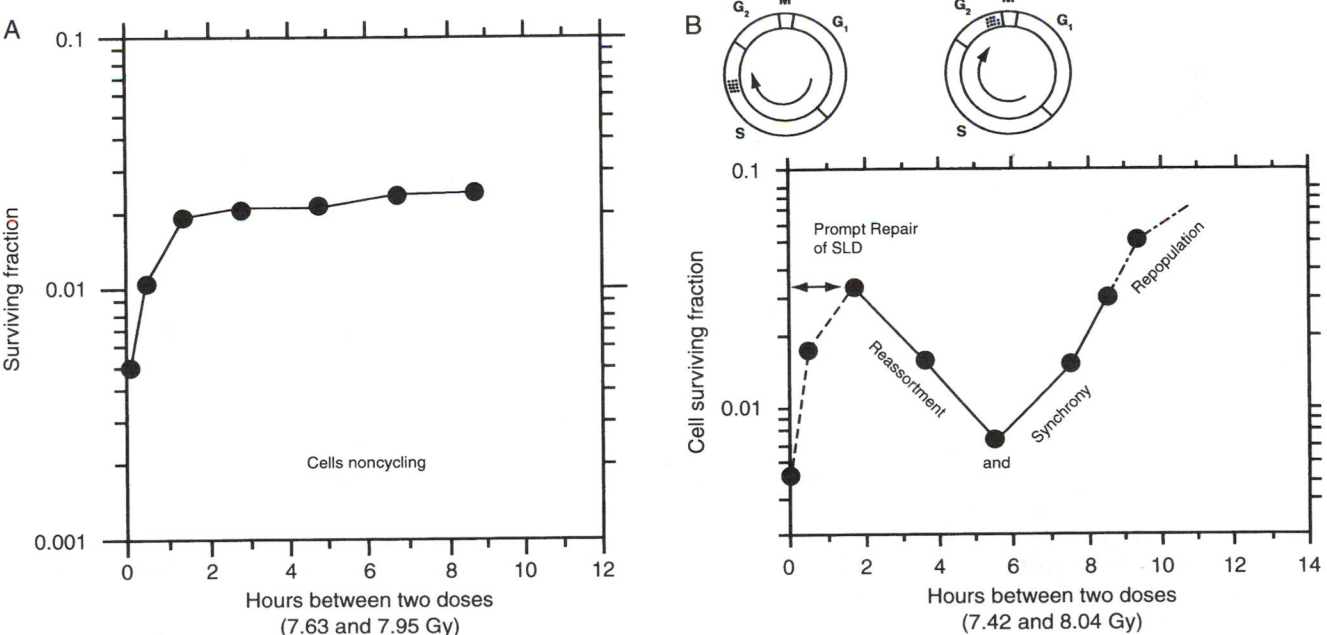

FIGURE 12. (A) Survival of Chinese hamster cells exposed to two fractions of X-rays and incubated at room temperature for various time intervals between the two exposures. (B) Survival of Chinese hamster cells exposed to two fractions of X-rays and incubated at 37 °C for various time intervals between the two doses. The survivors of the first dose are predominantly in a resistant phase of the cell cycle (late S). When the interval between doses is about 6 h, these resistant cells will have moved to the G_2/M phase, which is sensitive. (Redrawn from Elkind *et al.,* 1965.)

separated by 30 min, the surviving fraction is already appreciably higher than for a single dose. As the time interval is extended, the surviving fraction increases until a plateau is reached at about 2 h, corresponding to a surviving fraction of 0.02. This represents about four times as many surviving cells as for the dose given in a single exposure. A further increase in the time interval between the dose fractions is not accompanied by any additional increment in survival. The increase in survival in a split-dose experiment is due to the repair of sublethal radiation damage. This illustrates the phenomenon of repair without the complication of movement of the cell through the cell cycle.

Figure 12B shows the results of the parallel experiment in which cells were exposed to split doses while being maintained at their normal growing temperature of 37 °C. The pattern of repair seen in this case differs from that observed for cells kept at room temperature. In the first few hours prompt repair of sublethal damage is again evident, but at longer intervals between the two split doses the surviving fraction of cells decreases again. An understanding of this phenomenon is based on the age-response function described earlier. When an asynchronous population of cells is exposed to a large dose of radiation, more cells are killed in the sensitive than in the resistant phases of the cell cycle. The surviving population of cells, therefore, tends to be partly synchronized. In Chinese hamster cells, most of the survivors from a first dose are located in the S phase of the cell cycle. If about 6 h are allowed to elapse before a second dose of radiation is given, this cohort of cells will progress around the cell cycle and will be in G_2 or M, a sensitive period of the cell cycle, at the time of the second dose. If the increase in radiosensitivity in moving from late S to the G_2/M period exceeds the effect of repair of sublethal damage, the surviving fraction will fall.

The pattern of repair shown in Fig. 12B is, therefore, a combination of three processes occurring simultaneously. First, there is the prompt repair of sublethal radiation damage. Second, there is progression of cells through the cell cycle during the interval between the split doses, which has been termed **reassortment.** Third, there is an increase in surviving fraction due to cell division, or repopulation, when the interval between the split doses is 10–12 h because this exceeds the length of the cell cycle of these rapidly growing cells.

A simple experiment, performed *in vitro,* illustrates the three "R's" of radiobiology: **repair, reassortment,** and **repopulation.** It should be emphasized that the dramatic dip in the split-dose curve at 6 h, caused by reassortment, and the increase in survival by 12 h, because of repopulation, are seen only for rapidly growing cells. Hamster cells in culture have a cycle time of only 9 or 10 h. The time sequence of these events would be longer in more slowly proliferating normal tissues *in vivo.* Repair of sublethal radiation damage has been demonstrated in just about every biological test system for which a quantitative endpoint is available.

Figure 13 shows that when a dose of radiation is split into several fractions, each separated by a time interval sufficiently long for sublethal damage to be repaired, more cells survive than for the same total dose given in a single fraction, because the shoulder of the curve must be repeated

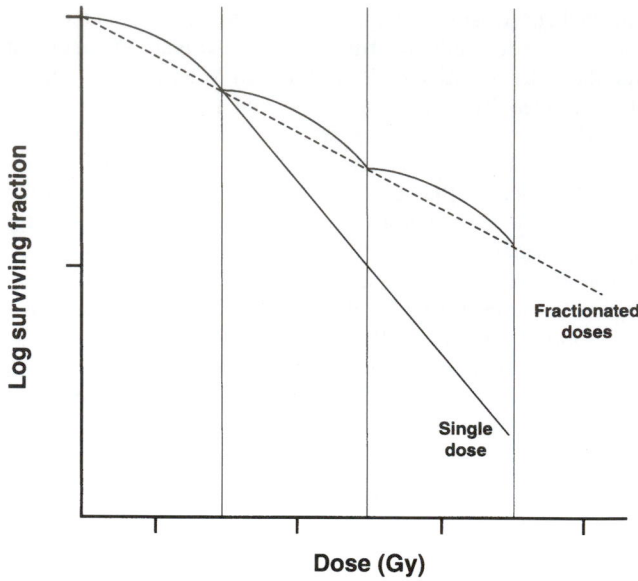

FIGURE 13. The effect of fractionation when radiation is delivered in a series of fractions. With a time interval between fractions sufficiently long for sublethal damage to be repaired, fewer cells are killed than for the same dose delivered in a single exposure.

with each fraction. In general, there is a good correlation between the extent of repair of sublethal damage and the size of the shoulder of the survival curve. This is not surprising, since both are manifestations of the same basic phenomenon: the accumulation and repair of sublethal damage. Some mammalian cells are characterized by a survival curve with a broad shoulder, and split-dose experiments then indicate a substantial amount of sublethal damage repair. Other types of cells show survival curves with a minimal shoulder, and this is reflected in more limited repair of sublethal damage. In the terminology of the linear-quadratic (α/β) description of the survival curve, it is the quadratic component (β) that causes the curve to bend and results in the sparing effect of a split dose.

X. The Mechanism of Sublethal Damage Repair

As previously mentioned, there is a good correlation between cell killing and the production of asymmetrical chromosomal aberrations, such as dicentrics. This in turn is a consequence of an interaction between two (or more) double-strand breaks in the DNA. On this interpretation, the repair of sublethal damage is simply the repair of DSBs. When a dose is split into two parts separated by a time interval, some of the DSBs produced by the first dose are rejoined and repaired before the second dose. The breaks in two chromosomes that must interact to form a dicentric may be formed by (1) a single track breaking both chromosomes or (2) separate tracks breaking the two chromosomes. The component of cell killing that results from single-track damage will be the same whether the dose is given in a single exposure or fractionated. The same is not true of multiple-track damage. If the dose is given in a single exposure (i.e., two fractions with $t = 0$ between them), all breaks produced by separate electrons can interact to form dicentrics. On the other hand, if the two dose fractions, D/2, are separated by (for example) 3 h, breaks produced by the first dose may be repaired before the second dose is given. Consequently, there will be fewer interactions between broken chromosomes to form di-centrics and more cells will survive. On this simple interpretation, the repair of sublethal damage reflects the repair and rejoining of DSBs before they can interact to form lethal lesions. This interpretation readily accounts for the repair of radiation damage in cells where mitotic death dominates. In cells that die an apoptotic death, survival tends to be an exponential function of dose, that is, the survival curve is straight with no shoulder on the usual semilog plot. In this case, there is no sparing from a split-dose exposure.

A. Repair and Radiation Quality

For a given biological test system, the size of the initial shoulder on the acute survival curve and, therefore, the amount of sublethal damage repair indicated by a split-dose experiment varies with the type of radiation used. The shoulder is largest for X-rays, smaller for neutrons, and nonexistent for densely ionizing α particles, where survival is an exponential function of dose. The amount of sublethal damage repair follows the same pattern; largest for X-rays, smaller for neutrons, and nonexistent for α particles.

B. The Dose-Rate Effect

For X- or gamma-ray doses, rate is one of the principal factors that determines the biological consequences of a given absorbed dose. As the dose rate is lowered and the exposure time extended, the biological effect of a given dose is generally reduced. Continuous low dose-rate irradiation may be considered to be an infinite number of infinitely small fractions; consequently the survival curve under these conditions would also be expected to have no shoulder and to be shallower than for single acute exposures.

The magnitude of the dose-rate effect from the repair of sublethal damage varies enormously between different types of cells. Cells characterized by a survival curve for acute exposures that has a small initial shoulder usually exhibit a modest dose-rate effect. This is to be expected, since both are expressions of the cell's capacity to accumulate and repair sublethal radiation damage. This is generally true of cells for which apoptosis is an important mechanism of cell death. Cell lines characterized by a survival curve for acute exposures, which has a broad initial shoulder, exhibit a dramatic dose-rate effect. An example is shown in Fig. 14 for Chinese hamster cells; in this cell line, mitotic death dominates and apoptosis is unimportant.

XI. The Oxygen Effect

A. Nature of Oxygen Effect

Many chemical and pharmacologic agents that modify the biological effect of ionizing radiations have been discov-

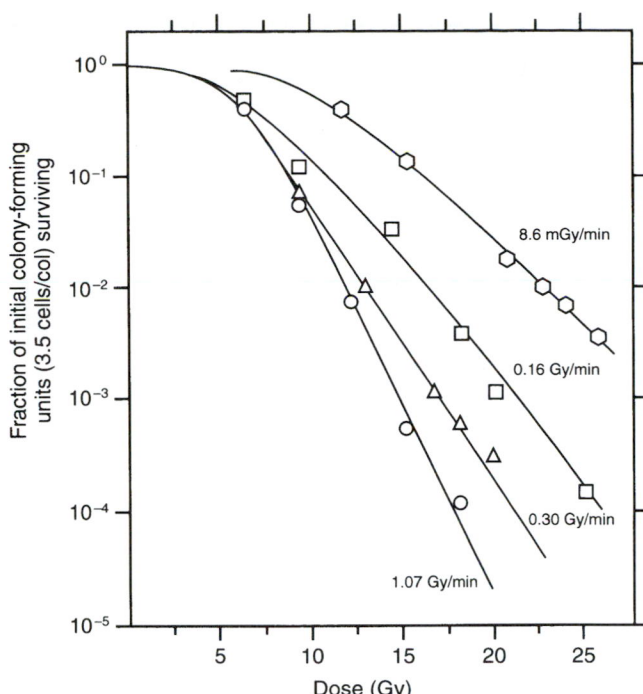

FIGURE 14. Dose-response curves for Chinese hamster cells (CHL-F line) grown *in vitro* and exposed to cobalt-60 gamma rays at various dose rates. At high doses a substantial dose late effect is evident even between 1.07, 0.3, and 0.16 Gy/min. The decrease in cell killing becomes even more dramatic as the dose rate is further reduced. (Redrawn from Bedford *et al.*, 1973).

ered. None is simpler than oxygen, and none produces such a dramatic effect. The oxygen effect was observed by Schwarz (1910), who noted that the skin reaction produced on his forearm by a radium applicator was reduced if the applicator was pressed hard onto the skin. In England, Mottram (1936) explored the question of oxygen in detail and was the first to discuss the possible importance of the oxygen effect in the radiotherapy of human cancer.

Survival curves for mammalian cells exposed to X-rays in the presence and absence of oxygen are illustrated in Fig. 15. The ratio of hypoxic to aerated doses needed to achieve the same biological effect is called the **oxygen enhancement ratio** (OER). For sparsely ionizing radiations, such as X- and gamma rays, the OER at high doses has a value of between 2.5 and 3. The OER has been determined for a wide variety of chemical and biological systems with different endpoints, and its value for X-rays always tends to fall in this range.

B. Mechanism of Oxygen Effect

In practical terms, oxygen must be present during irradiation for its sensitizing effect to be observed. In fact, sophisticated experiments have shown that oxygen still sensitizes if if is added a few *microseconds* after a brief pulse of radiation. This is the clue to the mechanism of the oxygen effect since microseconds correspond to the lifetime of the free radicals formed when X-rays interact with water.

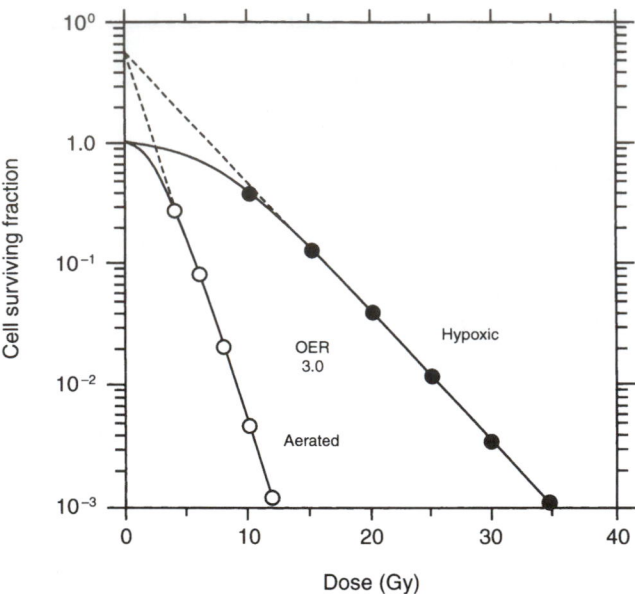

FIGURE 15. Survival curves for mammalian cells exposed to X-rays under aerated and hypoxic conditions produced by passing a stream of pure nitrogen over the cells. These data are typical of many in the literature. Oxygen is a dose-modifying factor; that is, at all levels of cell survival, the dose required under hypoxic conditions is three times greater than that required under aerated conditions to produce the same biological effect. This ratio of doses is known as the oxygen enhancement ratio (OER).

Oxygen is essential to "fix" the damage to DNA resulting from radiation-produced free radicals. In the absence of oxygen, radical induced damage is readily repaired, but if oxygen is present, the damage is made permanent, that is, fixed. (The word *fix* here is used in the European sense of making something permanent, as in fixing a film, not in the American sense of repairing, as in fixing a flat tire.) Consequently, oxygen modifies only that component of radiation damage mediated by free radicals—the indirect damage. Oxygen has no effect on the direct damage caused by ionization of DNA.

C. Oxygen Effect for Different Types of Radiation

The oxygen enhancement ratio for X- or gamma rays is approximately 3. It is this large because two-thirds of the damage produced by X-rays is mediated by free radicals. Densely ionizing α particles kill cells primarily by direct action, that is, by ionization of the DNA, a process not dependent on oxygen. Consequently, the OER for low energy (2-MeV) α particles is 1.0. The OER for neutrons has an intermediate value of about 1.6, because while the direct action is dominant there is still a contribution from free radical damage.

In summary, the oxygen effect is large and important in the case of sparsely ionizing radiations, such as X-rays; is absent for densely ionizing radiations, such as α particles; and has an intermediate value for fast neutrons.

D. Concentration of Oxygen Required

Only a small amount of oxygen is required to produce the dramatic and important oxygen effect characteristic of X-rays. An oxygen concentration of 3 mm Hg (about 1/2%) results in a radiosensitivity halfway between complete hypoxia and 100% oxygen. By a concentration of 30 mm Hg, characteristic of most normal tissues in the body, the radiosensitivity of cells is indistinguishable from that in air or even in 100% oxygen.

E. Oxygen Effect in Radiotherapy of Human Cancer

Malignant tumors tend to outgrow their blood supply so that areas of necrosis (dead cells) are a common feature of most cancers. Oxygen diffusing from blood vessels has a limited range because it is used up by rapidly growing and respiring cancer cells. Consequently, there are pockets of cells distant from capillaries that are hypoxic—still with sufficient oxygen to be viable, but at a low enough oxygen concentration to be resistant to killing by X-rays. This diffusion-limited hypoxia is known as "chronic" hypoxia. Regions of "acute" hypoxia develop in tumors as a result of the temporary closing of a particular blood vessel. If this blockage were permanent, the cells downstream would, of course, eventually die and would be of no further consequence. There is, however, good evidence that tumor blood vessels open and close in a random fashion so that different regions of the tumor become hypoxic intermittently. At the moment when a dose of radiation is delivered, a proportion of the tumor cells may by hypoxic, and consequently resistant to killing by X-rays. The problem of hypoxia, both chronic and acute, is largely overcome in practice by delivering radiotherapy in a large number of small fractions, typically 30 to 40, over a period of 6–8 weeks. Other attempts to eliminate the problem of hypoxia cells has been to use neutrons rather than X-rays (because of their smaller OER) and to develop chemicals that specifically sensitize or kill hypoxic cells.

XII. Radiation Quality and Biological Effects

A. Deposition of Radiant Energy

When radiation is absorbed in biological material, ionizations and excitations occur that are not distributed at random but tend to be localized along the tracks of individual charged particles in a pattern that depends on the type of radiation involved. For example, photons of X-rays give rise to fast electrons, particles carrying unit electric charge and having very small mass; neutrons, on the other hand, give rise to recoil protons, particles again carrying unit electric charge but having a mass nearly 2000 times greater than that of the electron. Alpha particles carry two electric charges on a particle four times as heavy as a proton. The charge-to-mass ratio of α particles therefore differs from that for electrons by a factor of about 8000. As a result, the spatial distribution of the ionizing events produced by different particles vary enormously. For X-rays, the primary ionizing events are well separated in space and for this reason X-rays are said to be "sparsely ionizing." The ionizing events produced along the track of a low-energy α particle form a dense column, and for this reason α particles are referred to as "densely ionizing."

B. Linear Energy Transfer

Linear energy transfer (LET), a term introduced by Zirkle (1940), is the energy transferred per unit length of the track. The special unit usually used for this quantity is kiloelectron volt per micron (keV/μm) of unit density material. The International Commission of Radiological Units (1962a, 1962b) defined this quantity as follows: the linear energy transfer (L) of charged particles in medium is the quotient of dE/dl, where dE is the average energy locally imparted to the medium by a charged particle of specified energy in traversing a distance of dl. That is,

$$L = dE/dl \qquad (5)$$

Since most radiations in practice consist of a wide spectrum of energies, LET can only be an average quantity. Typical LET values for commonly used radiations are listed in Table 1. Note that for a given type of charged particle, the higher the energy, the *lower* the LET and therefore the lower its biological effectiveness. For example, gamma rays and X-rays both give rise to fast secondary electrons; therefore, 1.1-MV cobalt-60 gamma rays have a lower LET than 250-kV X-rays and are less effective biologically by about 10%. By the same token, 150-MeV protons have a lower LET than 10-MeV protons and are, therefore, slightly less effective biologically.

C. Relative Biological Effectiveness

Figure 16 illustrates the survival curves obtained for three different types of radiation, namely, X-rays, 15-MeV neutrons, and α particles. As the LET increases from about 2 keV/μm for X-rays up to 150 keV/μm for α particles, the survival curve changes in two important respects. First, the survival curve becomes steeper. Second, the extrapolation number tends toward unity; that is, the shoulder of the curve becomes progressively smaller as the LET increases.

It is evident from Fig. 16 that *equal* doses of different types of radiation do not result in *equal* biological effects. One Gray

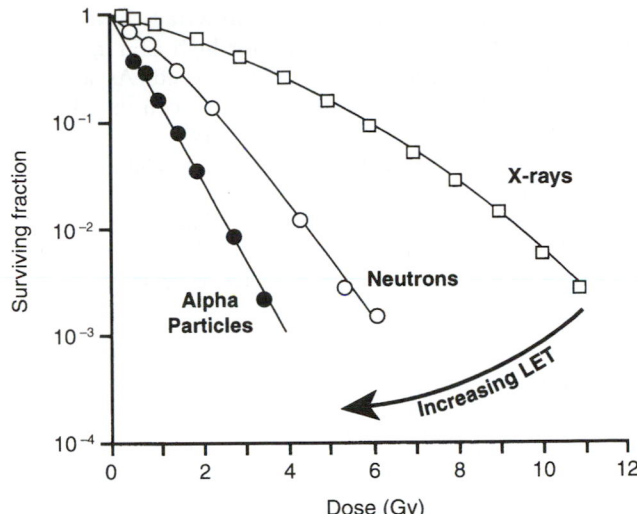

FIGURE 16. Survival curves for cells of human origin exposed to 250-kV X-rays, 15-MeV neutrons, and 4-MeV α particles. As the LET of the radiation increases, the slope of the survival curves gets steeper and the size of the initial shoulder gets smaller. (Redrawn from Broerse *et al.,* 1968.)

of α particles produces more cell killing than 1 Gy of X-rays. The key to the difference lies in the pattern of energy deposition at the microscopic level. In comparing different radiations, it is customary to use X-rays as the standard. The formal definition of **relative biological effectiveness** (RBE) is as follows: The RBE of some test radiation (r) compared with X-rays is defined by the ratio D_x/D_r where D_x and D_r are, respectively, the doses of X-rays and the test radiation required for equal biological effect.

To determine the RBE of some test radiation, say, α particles, one chooses a biological system in which the effect of radiation may be scored quantitatively, and one compares α particles with the standard radiation, that is, X-rays. The data in Fig. 16 are an example. To achieve a surviving fraction of 10^{-2} requires a dose of about 9.5 Gy or X-rays or 2.5 Gy of α particles. The RBE of α particles relative to X-rays is the ratio of doses required to produce the same biological effect, that is, 9.5/2.5 = 4.6. Neutrons have a somewhat lower RBE; the dose to result in a surviving fraction of 10^{-2} in 4.3 Gy, so that the RBE is about 9.5/4.3, or 2.2.

It is important to note that, because the survival curves for X-rays (low LET) and neutrons or α particles (high LET) have a different shape, with the survival curve for X-rays having an appreciable shoulder, RBE does not have a unique value. At lower doses, corresponding to higher levels of survival, the RBE will assume a larger value; the maximum RBE will correspond to the ratio of the initial slopes for X-rays and for the test radiation.

D. Relative Biological Effectiveness for Different Cells and Tissues

Even for a given total dose, the RBE varies significantly according to the tissue or endpoint used to measure it. Survival

TABLE 1 Typical LET Values

Radiation	Track avg.	LET (keV/μm)	Energy avg.
Cobalt-60 X-rays		0.2	
250-kV X-rays		2.0	
10-MeV protons		4.7	
150-MeV protons		0.5	
14-MeV neutrons	12		100
2.5-MeV α particles		166	
2-GeV Fe ions		1000	

curves for mammalian cells exposed to X-rays have a large but variable shoulder region, while for high LET radiations the shoulder region is reduced or eliminated. As a consequence, the RBE will be different for each cell line. In general, cells characterized by an X-ray survival curve with a large shoulder, indicating that they can accumulate and repair a large amount of sublethal radiation damage, will show a large RBE for neutrons or α particles. Conversely, cells for which the X-ray survival curve has little if any shoulder will exhibit small RBE values for α particles or neutrons.

E. Relative Biological Effectiveness as a Function of Linear Energy Transfer

Figure 16 shows data for just three types of radiation to illustrate the way in which the shape of the survival curve changes as the density of ionization (LET) increases. Studies have been performed with many types of radiation covering a wide range of LET values. Figure 17 is a plot of the RBE as a function of LET. As the LET increases, the RBE increases slowly at first, then more rapidly as the LET increases beyond 10 keV/μm. Between 10 and 100 keV/μm, the RBE increases rapidly with increasing LET and in fact reaches a maximum of about 100 keV/μm. Beyond this value for the LET, the RBE falls to lower values.

The LET at which the RBE reaches a peak is much the same (about 100 keV/μm) for a wide range of mammalian cells, from mouse to human, and is the same for all biological endpoints. It reflects the "target" size and is related to the DNA content, which is similar for all mammalian cells. The

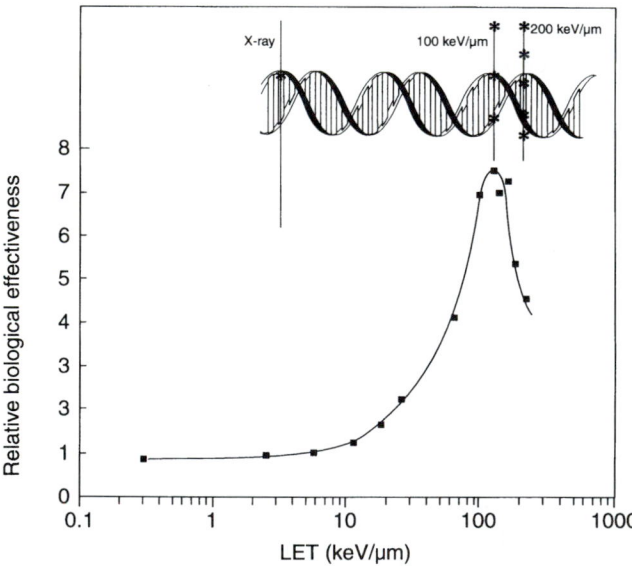

FIGURE 17. Variation of RBE with LET. RBE rises to a maximum at an LET of about 100 keV/μm and subsequently falls for higher LET values. The inset diagram illustrates that radiation with an LET of 100 keV/μm is most biologically effective because the average separation between ionizing events coincides with the diameter of the DNA double helix (2 nm); consequently a single-track of this quality can produce double-strand breaks with greatest efficiency. (Graph redrawn from Barendsen, 1968.)

reason why radiation with a LET of 100 keV/μm is optimal in terms of producing a biological effect can be understood from the upper panel of Fig. 17. At this density of ionization, the *average* separation between ionizing events just about coincides with the diameter of the DNA double helix (20 Å or 2 nm). Radiation with this density of ionization is most likely to cause a double-strand break by the passage of a single charged particle, and DSBs are the basis of most biological effects, as discussed earlier. This is illustrated in the top panel of Fig. 17. In the case of X-rays, which are more sparsely ionizing, the probability of a single track causing a double-strand break is low, and in general more than one track will be required to produce a double-strand break. As a consequence, X-rays have a low biological effectiveness. At the other extreme, much more densely ionizing radiations (with a LET of 200 keV/μm, for example) will readily produce DSBs but energy will be "wasted" because the ionizing events are too close together. Since RBE is the ratio of *doses* to produce equal biological effect, this more densely ionizing radiation will have a lower RBE than the optimal LET radiation. The more densely ionizing radiation will be just as effective *per track,* but less effective per unit dose. It is possible, therefore, to understand why RBE reaches a maximum value in terms of the production of DSBs, since the interaction of two DSBs to form an exchange type aberration is the basis of most biological effects. Radiation having this optimal LET includes neutrons of a few hundred kiloelectron volts, as well as low-energy protons and α particles.

F. Radiation Weighting Factors

It is evident from the preceding discussion that different types of radiation differ in their biological effectiveness per unit of absorbed dose. The complexities of RBE are too difficult to apply in specifying dose limits in radiation protection; it is necessary to have a simpler way to consider differences in biological effectiveness of different radiations. One cGy of neutrons, for example, is more hazardous than 1 cGy of X-rays. The term **radiation weighting factor** (W_r) has been introduced for this purpose. The quantity produced by multiplying the absorbed dose by the weighting factor is called the **equivalent dose.** When dose is expressed in Gray (Gy), the equivalent dose is in Sievert (Sv). Radiation weighting factors are chosen by the International Commission on Radiological Protection (1991), based on a consideration of experimental RBE values, biased for biological endpoints relevant to radiation protection at low dose and low dose rate. There is a considerable element of judgment involved. The W_r is set at unity for all low LET radiations (X-rays, gamma rays, and electrons), with a value of 20 for maximally effective neutrons and α particles. Thus, using this system, an absorbed dose of 0.1 Gy of radiation with a radiation weighting factor of 20 would result in an equivalent dose of 2 Sv.

XIII. Radioprotectors

In the late 1940s it was discovered that either cysteine or cysteamine could protect mice from the lethal effects of total body X-irradiation if administered in large amounts prior to exposure. The structures of these compounds are:

$$\underset{\text{COOH}}{\overset{\overset{\displaystyle NH_2}{|}}{SH-CH_2-CH}} \qquad \text{cysteine} \qquad (6)$$

$$SH-CH_2-CH_2-NH_2 \qquad \text{cysteamine} \qquad (7)$$

Animals receiving these radioprotectors could tolerate almost double the radiation dose. The ratio of the radiation dose to produce the same biological effect in the presence and absence of the drug is known as the **dose reduction factor** (DRF). Many similar compounds have been tested and found to be effective as radioprotectors. The most efficient tend to have certain structural features in common: a free SH group (or potential SH group) at one end of the molecule and a strong basic function such as amine or guanidine at the other end, separated by a straight chain of two or three carbon atoms. Sulfhydryl compounds are efficient radioprotectors against sparsely ionizing radiations, such as X- or gamma rays. The mechanism of action involves the scavenging of free radicals.

X-ray photons give up their energy to produce fast electrons as described earlier. These electrons may hit the DNA and cause a break directly (direct action) or interact with a water molecule to form a hydroxyl radical (OH·), which diffuses to the DNA and causes a break (indirect action). The protective effect of sulfhydryl compounds stems from their ability to "scavenge" free radicals. This process is illustrated in Fig. 18.

The protective effect of sulfhydryl compounds tends to parallel the oxygen effect, being maximal for sparsely ionizing radiations (e.g., X- or gamma rays) and minimal for densely ionizing radiations (e.g., low-energy α particles). It might be predicted that with effective scavenging of all free radicals the largest possible value of DRF would equal the oxygen enhancement ratio, with a value of 2.5 to 3.0.

It is not surprising that the discovery in 1948 of a compound that offered protection against radiation excited the interest of the U.S. Army, since the memory of Nagasaki and Hiroshima was vivid in the years immediately after World War II. Although cysteine is a radioprotector, it is also toxic and induces nausea and vomiting at the dose levels required for radioprotection. Consequently, the Walter Reed Army Hospital in Washington, D.C., synthesized more than 3000 compounds in an attempt to find the perfect radioprotector, one that would protect against radiation without debilitating side effects. At an early stage, the important discovery was made that the toxicity of the compound could be greatly reduced if the sulfhydryl group was covered by a phosphate group. This is illustrated in Table 2. The LD_{50} of the compound in animals can be doubled and the protective effect in terms of the DRF greatly enhanced if the SH group in cysteamine is covered by a phosphate. This tends to reduce systemic toxicity. Once in the cell, the phosphate group is stripped, and the SH group begins scavenging free radicals. Many compounds have been synthesized, but only two have been put to practical use. The structure and effectiveness of these is summarized in Table 3. The first compound, WR-638, called Cystaphos, was said to be carried routinely in the field pack of Russian infantry in Europe during

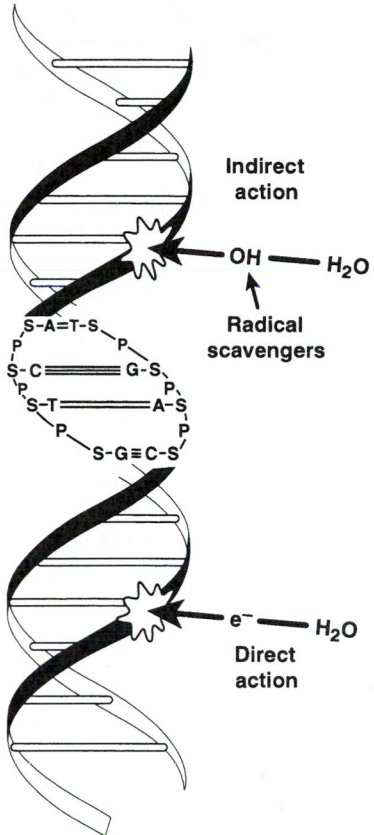

FIGURE 18. Photons of X-rays either interact directly with DNA to produce a strand break (direct action) or indirectly via the production of a free radical (indirect action). Radioprotectors that are radical scavengers "intercept" the direct action. (Redrawn from Hall, 1994.)

the era of the Cold War for use in the event of a nuclear conflict. Its usefulness would have been largely psychological, since the compound was carried as a tablet to be administered orally, when in fact these sulfhydryl compounds break down in stomach acid and are only effective when administered intravenously or intraperitoneally. A further factor, of course, is that such compounds will protect only from sparsely ionizing radiation; consequently, they would offer little protection against the prompt release of neutrons produced by the detonation of a nuclear device. They would be effective only against the gamma rays from the resulting fallout.

The second compound, WR-2721, now known as amifostine, is perhaps the most effective of those synthesized in the

TABLE 2 Effect of Adding a Phosphate-Covering Function on the Free Sulfhydryl of β-Mercaptoethylamine

Drug	Formula	LD_{50} in mice	DRF
MEA	$NH_2-CH-CH_2-SH$	343 (323–364)	1.6 at 200 mg/kg
MEA-PO$_3$	$NH_2-CH_2-CH-SH_2PO_3$	777 (700–864)	2.1 at 500 mg/kg

TABLE 3 Two Protectors in Practical Use

Compound	Structure	Drug dose (mg/kg)	DRF 7 days	DRF 30 days
WR-638	$NH_2CH_2CH_2SPO_3HNa$	500	1.6	2.1
WR-2721	$NH_2(CH_2)_3NHCH_2CH_2SPO_3H_2$	900	1.8	2.7

Walter Reed series (Kligerman *et al.,* 1992). It gives good protection to the blood-forming organs, as can be seen by the DRF for 30-day death in mice, which approaches the theoretical maximum value of 3. It was probably the compound carried by the U.S. astronauts on their trips to the moon to be used if a solar event occurred. On these missions, when the space vehicle left earth's orbit and began coasting toward the moon, the astronauts were committed to a 14-day mission, since they did not have sufficient fuel to turn around without first orbiting the moon and using its gravitational field. If there had been a major solar event in that period, the astronauts would have been exposed to a shower of high-energy protons, resulting in an estimated dose of several hundred Gray. The availability of a radioprotector with a DRF of between 2 and 3 would have been very important in such a circumstance. As it turned out, no major solar event ever occurred during any manned lunar mission.

Amifostine also has a potential in radiotherapy. One application would be for local topical use, to reduce for example the mucosal reaction that occurs during radiotherapy of head and neck cancers. Another possibility is to protect all normal tissues by delivering the drug systemically; there is some evidence that the drug gets into tumors more slowly than into normal tissues, so that if the radiation is delivered within minutes of the drug, a differential effect is obtained with more protection of the normal tissues than of the tumor.

XIV. Summary

Life on earth has always been exposed to radiation, but there is now an additional exposure from medical X-rays, nuclear power, and journeys in high-flying aircraft. Of concern are ionizing radiations, so called because they have a sufficient photon energy to knock electrons out of the atoms through which they pass, break chemical bonds, and cause a variety of biological changes.

Ionizing radiations come in a variety of types. X-rays are usually generated in an electrical device. Gamma rays are emitted by a radioactive material during decay. X- or gamma rays are electromagnetic waves, similar to radiant heat and visible light, but having a shorter wavelength. Neutrons are uncharged particles, often associated with nuclear fission. High-energy high-z charged particles are a hazard to astronauts in long-duration space flights.

The principal target for all biological effects of ionizing radiations is the DNA. The basic lesion is a double-strand break, caused when a charged particle breaks both strands of the DNA double helix, causing the chromatin strand to snap. Chromosomal aberrations represent one of the earliest detectable biological changes; if DSBs occur in two separate chromosomes, the broken ends may rejoin in incorrect and bizarre ways. These rearrangements can often be seen at the first metaphase after irradiation.

Following irradiation, cells may die from one of two mechanisms:

1. Mitotic cell death, due to the failure of cells to cope with severely damaged chromosomes
2. Apoptosis, or programmed cell death, similar to the way that cells die and are removed during embryogenesis

The quantity of radiation, or dose, is measured in terms of the energy absorbed per unit mass. The current mass is the Gray (Gy), defined to be an energy absorption of one joule per kilogram.

A number of factors influence the biological effects of a given absorbed dose of radiation. First, the response of cells to radiation depends critically on their position in the division cycle at the time that they are exposed. In general, cells are most radiosensitive when they are close to mitosis. Second, cells vary greatly in their ability to repair radiation damage. In addition, a given dose of radiation is much less effective when spread out over a long period of time, because much of the radiation damage can be repaired during a prolonged exposure. Third, in the case of X- and gamma rays (but less so for neutrons and α particles), the biological consequences of a given dose can be modified by chemical means. For example, cells are more radiosensitive if molecular oxygen is present than if it is absent. On the other hand, compounds that scavenge free radicals are often found to be radioprotectors, and such compounds were used by the astronauts during lunar missions.

Fourth, the biological effect of a given absorbed dose depends on the "quality" of the radiation. X- and gamma rays are said to be "sparsely ionizing" because the ionizing events are well separated along the tracks of the electrons set in motion. Consequently, they are less effective than neutrons and α particles that are said to be "densely ionizing" because the ionizing events are clustered together and more likely to produce a DSB in a chromosome.

Bibliography

Barendsen, G. W. (1968). Responses of cultured cells, tumors and normal tissues to radiations of different linear energy transfer. *Curr. Top. Radiat. Res.* **4,** 293.

Bedford, J. S., and Mitchell, J. B. (1973). Dose-rate effects in synchronous mammalian cells in culture. *Radiat. Res.* **54,** 316–327.

Broerse, J. J., Barendsen, G. W., and Van Kersen, G. R. (1968). Survival of cultured human cells after irradiation with fast neutrons of different energies in hypoxic and oxygenated conditions. *Int. J. Radiat. Biol.* **13,** 559–572.

Elkind, M. M., Sutton-Gilbert, H., Moses, W. B., Alescio, T., and Swain, R. B. (1965). Radiation response of mammalian cells in culture: V. Temperature dependence of the repair of x-ray damage in surviving cells (aerobic and hypoxic). *Radiat. Res.* **25,** 359–376.

Hall, E. J. (1994). "Radiobiology for the radiologist." (4th ed.) J. B. Lippincott Company, Philadelphia.

Howard, A., and Pelc, S. R. (1953). Synthesis of deoxyribonucleic acid in normal and irradiated cells and its relation to chromosome breakage. *Heredity* **6**, Suppl., 261.

International Commission on Radiological Protection. (1991) "Recommendations," Report No. 60. Pergamon Press, New York.

International Commission on Radiological Units and Measurements. (1962a). "Radiation Quantities and Units," Report 10a, Handbook 84. National Bureau of Standards, Washington, DC.

International Commission on Radiological Units and Measurements. (1962b). "Physical Aspects of Irradiation," Report 10b, Handbook 85. National Bureau of Standards, Washington, DC.

Kligerman, M. M., Liu, T., Liu, Y., Scheffler, B., He, S., and Zhang, S. (1992). Interim analysis of a randomized trial of radiation therapy of rectal cancer with/without WR-2721. *Int. J. Radiat. Oncol. Biol. Phys.* **22**, 799–802.

Mottram, J. C. (1936). Factor of importance in radiosensitivity of tumors. *Br. J. Radiol.* **9**, 606–614.

Schwarz, W. (1910). *Wein. Klin. Wochenschr. Nr.* **11S**, 397.

Sinclair, W. K., and Morton, R. A. (1966). X-ray sensitivity during the cell generation cycle of cultured Chinese hamster cells. *Radiat. Res.* **29**, 450–474.

Terasima, R., and Tolmach, L. J. (1963). X-ray sensitivity and DNA synthesis in synchronous populations of HeLa cells. *Science* **140**, 490–492.

Zirkle, R. E. (1940). The radiobiological importance of the energy distribution along ionization tracks. *J. Cell. Comp. Physiol.* **16**, 221.

Appendix

Nicholas Sperelakis

70

Review of Electricity and Cable Properties

I. Introduction

An appreciation of some elementary principles of electricity is essential to the understanding of many aspects of cell physiology and biophysics and the electrophysiology of excitable cells. The purpose of this section is to give a brief review of the most relevant aspects of electricity and electric circuits that pertain to the cell membrane.

II. Definition of Circuit Elements and Ohm's Law

The simple electric circuit shown schematically in Fig. 1 consists of a battery, a resistor, an ammeter, a switch, and copper wire. Closing the switch allows electrical current to flow from the positive side of the battery, through the resistor and ammeter, and back to the negative side of the battery. Electric current is considered, by convention, to flow from **positive to negative** through the external circuit, even though the **electrons**—the negatively charged particles—actually flow in the opposite direction, that is, from negative to positive. One can consider that holes, that is, the absence of an electron in a space that previously contained one, flow in the direction opposite that of the electrons, that is, from positive to negative around the external circuit. Within the battery, the flows are just the opposite; namely, electrons flow from positive to negative, and current flows from negative to positive.

In the circuit of Fig. 1, the relationship between the applied voltage, the total resistance, and the current that flows is given by **Ohm's law**. This relationship, first formulated by George S. Ohm, states that the current (I) that flows is equal to the applied voltage (V) divided by the resistance (R):

$$I = \frac{V}{R} \tag{1a}$$

$$\text{amperes} = \frac{\text{volts}}{\text{ohms}}$$

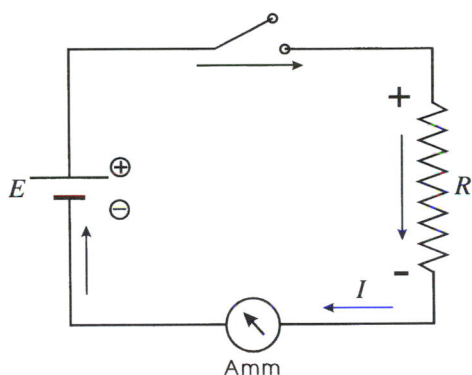

FIGURE 1. A simple electric circuit containing a battery (E) and a resistor with resistance R. When the switch is closed, electrical current (I) flows in the direction indicated by the arrows, from the positive pole of the battery to the negative pole in the external circuit, and can be measured by the ammeter (Amm) placed in series with the resistor. Current through the battery flows from the negative pole to the positive pole. The relationship between, E, R, and I is given by Ohm's law: $I = E/R$.

An **ampere** (A) of current is the movement past a point of one **coulomb** (C) of electrical charge per second (1 A = 1 C/s). The unit of **resistance** is the **ohm** (Ω), which may be operationally defined as that value of resistance through which there is a current of 1 A when the applied potential is 1 V ($R = V/I$). Resistance is a measure of the opposition to current flow; namely, for a given applied voltage, the greater the resistance the less the flow of current (inverse proportionality). The **volt** can be operationally defined from Ohm's law as that potential difference (PD) necessary to produce 1 A of current flow through a resistance of 1 Ω:

$$V = IR \tag{1b}$$

$$\text{volts} = (\text{amperes})(\text{ohms})$$

Two points are at a PD of 1 V when it takes 1 **joule** (J) of energy to transport 1 C of charge from one point to the other, given as

$$1 \text{ V} = 1 \text{ J/C} \tag{2a}$$

Note that energy equals voltage times the charge:

$$\text{joules} = \text{volts} \times \text{coulombs} \tag{2b}$$

(Thus the electron-volt unit of atomic physics is a unit of energy, the charge on an electron being 1.60×10^{-19} C.)

A coulomb of charge is equal to that represented by 6.24×10^{18} electrons or univalent ions (reciprocal of 1.60×10^{-19} C/electron).

$$1 \text{ C} = 6.24 \times 10^{18} \text{ univalent ions} \tag{3}$$

Multiplication of the charge on an electron (q_e) by **Avogadro's number** (N_A), which is the number of molecules per mole, gives the number of coulombs in an equivalent (eq) of charge or mole of univalent ions

$$\mathscr{F} = q_e \times N_A \tag{4}$$

$$\frac{C}{eq} = \left(1.60 \times 10^{-19} \frac{C}{\text{univalent ion}}\right)\left(6.02 \times 10^{23} \frac{\text{ions}}{eq}\right)$$

$$\cong 96\,500 \frac{C}{eq}$$

This derived value is known as the **Faraday constant** ($\mathscr{F}$), used in electrochemistry in equations such as the Nernst equation.

Multiplying the Faraday constant by the valence of the ion (eq/mol) then gives the number of coulombs per mole:

$$z\mathscr{F} = \frac{eq}{mol} \times \frac{C}{eq} = \frac{C}{mol} \tag{5}$$

Conductance (G) is the reciprocal of the resistance:

$$G = \frac{1}{R} \tag{6}$$

The greater the resistance, the lower the conductance, and vice versa. Therefore, Ohm's law can also be written as

$$I = \frac{V}{R} = \frac{1}{R} V = GV \tag{7}$$

This states that the current is a product of conductance and voltage. This form is useful for electrophysiology, since conductances in parallel simply add. The unit of conductance is the **mho** (ohm spelled backwards) or Ω^{-1} or **siemens** (S).

III. Resistors and Conductances in Series and In Parallel

A. Resistors In Series

For resistors in series (Fig. 2A.), the total resistance (R_T) is a simple sum of the values of the resistances in series:

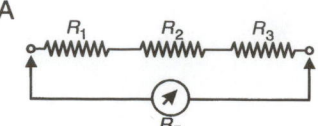

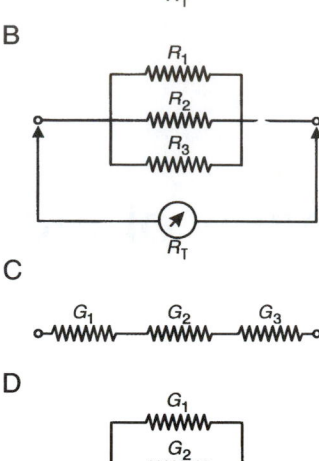

FIGURE 2. (A) Resistors placed in series–the total resistance (R_T) is the sum of the resistances: $R_T = R_1 + R_2 + R_3$. (B) Resistors placed in parallel—the reciprocal of the total resistance is equal to the sum of the reciprocals of the individual resistances: $1/R_T = 1/R_1 + 1/R_2 + 1/R_3$. (C) Reciprocal of the total conductance ($1/G_T$) of conductances in series is equal to the sum of the reciprocals of the individual conductances: $1/G_T = 1/G_1 + 1/G_2 + 1/G_3$. (D) Total conductance of conductances in parallel is equal to the sum of the conductances: $G_T = G_1 + G_2 + G_3$.

$$R_T = R_1 + R_2 + R_3 + \dots \tag{8}$$

If all the resistors are of the same value ($R_1 = R_2 = R_3$), then the total resistance is

$$R_T = R_1 N \tag{9}$$

where N is the number of resistors in series. Because myelin consists of many wrappings of membrane around the axon membrane (membranes in series), it should be obvious why myelination effectively raises membrane resistance.

B. Resistors In Parallel

For resistors in parallel (Fig. 2B), the reciprocal of the total resistance ($1/R_T$) is equal to the sum of the reciprocals of the resistances:

$$\frac{1}{R_T} = \frac{1}{R_1} + \frac{1}{R_2} + \frac{1}{R_3} + \dots \tag{10}$$

If all the resistors are the same value ($R_1 = R_2 = R_3$), then the total resistance is equal to any one resistance divided by the total number of resistors in parallel:

$$R_T = \frac{R_1}{N} \tag{11a}$$

If there are only two resistances in parallel, then Eq. 10 can be reduced algebraically to

$$\frac{1}{R_T} = \frac{1}{R_1} + \frac{1}{R_2} = \frac{R_1 + R_2}{R_1 R_2} \qquad (11b)$$

Taking the reciprocal gives

$$R_T = \frac{(R_1 \times R_2)}{(R_1 + R_2)} \qquad (11c)$$

Equation 11c states that the total resistance of two resistors in parallel is equal to the product of the resistances divided by the sum (often called the **rule of product over the sum**). Note that R_T is always less than the lowest R for resistors in parallel.

C. Conductances In Series

Conductances in series (Fig. 2C) add like resistances in parallel; namely, the reciprocal of the total conductance $(1/G_T)$ is equal to the sum of the reciprocals of each conductance:

$$\frac{1}{G_T} = \frac{1}{G_1} + \frac{1}{G_2} + \frac{1}{G_3} \qquad (12)$$

D. Conductances In Parallel

Conductances in parallel (Fig. 2D) add like resistances in series; namely, the total conductance (G_T) is simply equal to the sum of each conductance:

$$G_T + G_1 + G_2 + G_3 + \ldots \qquad (13a)$$

That this is true is demonstrated by substituting $G = 1/R$ into Eq. 13a and obtaining

$$\frac{1}{R_T} = \frac{1}{R_1} + \frac{1}{R_2} + \frac{1}{R_3} + \ldots \qquad (13b)$$

which is identical to Eq. 10 for resistances in parallel. It is convenient to use conductances in the study of the electrical properties of membranes, since the conductance pathways for the various species of ions are in parallel to each other, and therefore can simply be added to give the total conductance.

IV. Kirchhoff's Laws

Kirchhoff's two laws may be stated in the following manner: (1) The current going toward a point is equal to that leaving it. For example, in a simple circuit composed of three resistors in series and a battery in series (Fig. 3A), the current is constant throughout the circuit; that is, an ammeter placed anywhere in the circuit gives the same reading. In a simple parallel circuit composed of a network of three resistors in parallel connected to a series battery (Fig. 3B), the current is not constant throughout the circuit. The sum of the currents in each branch is equal to the current in the main branch; in other words, the sum of the currents converging toward a branch point is equal to that leaving it:

$$I_T = i_1 + i_2 + i_3 \qquad (14)$$

(2) The algebraic sum of the voltage drops (IR) across each of the resistors in any closed network is equal to the source electromotive force (emf), e.g., battery. For the series circuit

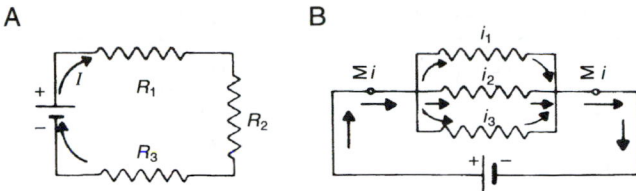

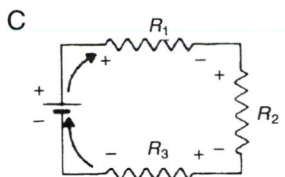

FIGURE 3. Diagrammatic representation of Kirchhoff's two laws: (1) Current approaching a point is equal to that leaving the point. Thus in the simple series circuit (A) the current is constant throughout, and in the simple parallel circuit (B), the sum of the currents in each leg is equal to the total current (I_T): $I_T = i_1 + i_2 + i_3$. (2) The sum of the voltage drops (IR) in the external circuit must equal the source emf. Thus in the simple series circuit (C), the sum of the IR drops is equal to the source emf (battery voltage): source emf = $IR_1 + IR_2 + IR_3$.

shown in Fig. 3C,

$$\text{source emf} = IR_1 + IR_2 + IR_3 \qquad (15)$$

The IR or voltage drop across each resistor is always in the polarity depicted; namely, the proximal side with respect to the direction of the current flow is always positive, and the distal side is negative. Current is always considered to flow from positive to negative in the external circuit. By convention, a positive potential is higher (or greater) than a negative one; hence, the potential is progressively dropped or lowered around the external circuit.

V. Nature of Capacitors

A capacitor is a device that can store electric charge or electricity. It consists of two parallel plates (conductors) separated by a dielectric. All effective dielectrics have a high electrical resistance, but the reverse is not necessarily true; namely, all high resistors are not necessarily very good dielectrics (e.g., air). The **dielectric constant** (ϵ) of a material is a relative index of its ability to function as an efficient dielectric compared to a vacuum, which has an ϵ value of 1.000. The dielectric constants of some selected materials are given in Table 1. The greater the **dipolar nature** of the molecules, the greater the dielectric constant. For example, water is an efficient dielectric because of its highly dipolar nature. The dipoles must be free to flip-flop and orient with changes in the electric field, much like the water dipole orients around small charged ions (e.g., Na^+, K^+) to form the hydration layers. Polar phospholipids in the cell membrane should have a higher dielectric constant than less polar oils.

It is seen from **Coulomb's law,** for the force $(F,$ in **dynes**) of interaction between point charges, that a large dielectric constant reduces the force of interaction between electrical

TABLE 1   Dielectric Constants of Some Materials

Material	Dielectric constant (ϵ)
Vacuum	1.00000
Air	1.0006
Oils	3–5
Water (distilled)	80
Cell membrane	3–6
Solid insulators	2–10
Paraffin wax	2–3
Quartz	4–5
Mica	6–7
Glass	5–7
Thin layers of oxidized metal	>30

charges, that is, the dielectric shields or buffers (see Section IX):

$$F = \frac{q_1 q_2}{\epsilon d^2} \tag{16}$$

where q_1 and q_2 are the magnitudes of the charges (in statcoulomb), d is the distance between the charges, and ϵ is the dielectric constant. The distance squared term in Coulomb's law reflects the geometrical factor: the surface area of a sphere equals $4\pi a^2$ or πd^2, in which a is the radius and d is the diameter. Therefore, the surface area over which the total force is distributed increases with the square of the radius (or diameter) of the sphere. From Eq. 16, a **statcoulomb** may be defined as that charge which, when placed 1 cm from an equal and like charge in a vacuum, repels the charge with a force of 1 dyne (dyn). A **dyne** is the force necessary to give a mass of 1 g an acceleration of 1 cm/s^2. Because the gravitational constant is 980 cm/s^2, 1 dyn is equal to approximately 1 mg of force (1 g/980).

The magnitude of a **capacitance** (C) is equal to

$$C = \frac{\epsilon A}{d}\frac{1}{4\pi k} \tag{17a}$$

in which A is the area of each plate (in cm^2), d is the distance (in cm) between the plates (i.e., thickness of the dielectric), and k is a constant (9.0×10^{11} cm/F). The unit of capacitance is the **farad** (F), named after Michael Faraday. Thus, the greater the dielectric constant, the greater the capacitance (Eq. 17a).

The greater the area, the greater the capacitance (Eq. 17a). Therefore, the total capacitance of several capacitors in parallel should be the simple sum (see Section VI.A) because the area of the plates is effectively increased. For example, the capacitance of the **tuning capacitor** (dial) of a radio is varied by altering the degree of overlap between two sets of interdigitating plates: the more the overlap, the greater the capacitance, because the area (A) is greater; air is the dielectric in this case. (Commercial capacitors may use substances such as air, oil, paper, or mica for the dielectric; the area can be made very large by many wrappings of double sheets of aluminum foil separated by paper or mica). The area term is often incorporated with the capacitance to give the specific capacitance (C/A), that is, the capacitance normalized to unit area (e.g., 1 cm^2):

$$\frac{C}{A} = \frac{\epsilon}{d}\frac{1}{4\pi k} \tag{17b}$$

The shorter the distance between the plates, the greater the capacitance (Eq. 17a). Therefore, the cell membrane has a very high capacitance, because the distance between the plates is only about 3.5–7.0 nm or 35–70 Å (the nonpolar phospholipid tail length of the membrane bilayer). For example, if the measured specific capacitance of the cell membrane (C_m) is 1.0 µF/cm^2, and assuming a dielectric constant of 5, then from Eq. 17b

$$d = \frac{\epsilon}{(C/A)}\frac{1}{4\pi k} \tag{17c}$$

$$= \frac{5}{1 \times 10^{-6}\,\text{F/cm}^2}\frac{1}{(12.56)(9.0 \times 10^{11}\,\text{cm/F})}$$

$$= 44.2 \times 10^{-8}\,\text{cm}$$

$$= 44\,\text{Å}$$

$$= 4.4\,\text{nm}$$

All capacitors have a **breakdown voltage;** that is, at more than a critical voltage level, the dielectric material cannot withstand the PD impressed across it. Therefore the capacitor breaks down, and electric current can now flow through the dielectric material from one plate to the other. The capacitor can no longer function as a capacitance, because the two plates are almost shorted to one another. A typical breakdown voltage of a commercial radio capacitor may be 200 V. In contrast, the capacitance of the nerve or muscle membrane has a breakdown voltage of only about 200 mV. Because the two plates of the membrane capacitor are much closer (about 50 Å or 0.5×10^{-6} cm), however, the voltage gradient necessary for breakdown is 200 mV/0.5×10^{-6} cm or 400 000 V/cm. At a resting potential of -80 mV, the voltage gradient is about 160 000 V/cm. Thus, the cell membrane is indeed a very effective capacitor and dielectric.

VI. Capacitors in Parallel and Series

A. Capacitors In Parallel

For capacitances in parallel (Fig. 4A), the total capacitance (C_T) is a simple sum of the capacitances

$$C_T = C_1 + C_2 + C_3 + \dots \tag{18}$$

The reason for this should be clear from Eq. 17a, which demonstrates that capacitance is directly proportional to the area of the plates; hence placing capacitors in parallel effectively increases the area of the plates, as can be deduced from Fig. 4A. If the number (N) of capacitors in parallel all have the same value ($C_1 = C_2 = C_3$), then the total capacitance is

$$C_T = C_1 N \tag{19}$$

Thus, capacitors in parallel add like resistors in series. Invaginations of the skeletal muscle fiber or ventricular myocardial cell surface membranes to form transverse (T)

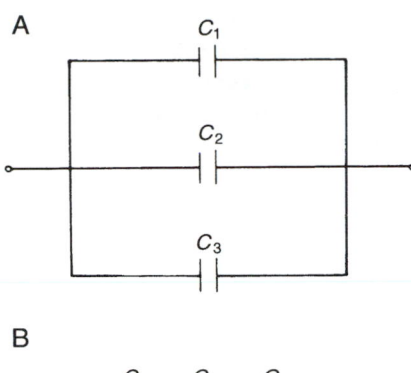

FIGURE 4. (A) Capacitors placed in parallel. The total capacitance is the sum of the individual capacitances: $C_T = C_1 + C_2 + C_3$. (B) Capacitors placed in series. The reciprocal of the total capacitance is equal to the sum of the reciprocals of the individual capacitances: $1/C_T = 1/C_1 + 1/C_2 + 1/C_3$.

tubules increase the total measured fiber capacitance, because the increased membrane area effectively adds capacitors in parallel.

B. Capacitors In Series

For capacitances in series (Fig. 4B), the reciprocal of the total capacitance ($1/C_T$) equals the sum of the reciprocals of each capacitance:

$$\frac{1}{C_T} = \frac{1}{C_1} + \frac{1}{C_2} + \frac{1}{C_3} + \cdots \qquad (20)$$

If there are several capacitors in series that are all of equal value ($C_1 = C_2 = C_3 = ...$), then the total capacitance is equal to any capacitance divided by N:

$$C_T = \frac{C_1}{N} \qquad (21a)$$

If there are only two capacitors in series, regardless of values, the total capacitance equals the product over the sum:

$$C_T = \frac{C_1 C_2}{C_1 + C_2} \qquad (21b)$$

Note that C_T is always less than the lowest C. Because myelin consists of many (20–200) wrappings of membrane around the axon membrane, it should be obvious why myelination reduces the total capacitance of a nerve fiber. Capacitors in series effectively increase the distance between the plates (d), and thereby lower the total capacitance (see Eq. 17a).

VII. Capacitive Reactance

Capacitors have almost infinite resistance to direct current (DC). Their impedance (Z, in Ω) or so-called "AC resistance" to sinusoidal alternating current (AC) is less, however, and is a function of the frequency (f) of the AC. The

higher the frequency, the lower the **capacitive reactance** (X_C, in Ω). The relationship is given by

$$X_C = \frac{1}{2\pi f C} \qquad (22a)$$

$$= \frac{1}{\omega C}$$

in which C is capacitance and the **angular velocity** (ω) in radians/s is equal to $2\pi f$. For DC, $f = 0$ and $X_C = \infty$. At $f = \infty$, $X_C = 0$. For a given f, the larger C is, the lower X_C is.

For a nerve or muscle membrane, if C is 1.0, μF/cm^2 and f is 1000 cycles/s, then

$$X_C = \frac{1}{(6.28)(1000/s)(1 \times 10^{-6}\,\mathrm{F/cm^2})} \qquad (22b)$$

$$= 0.159 \times 10^3\,\frac{\mathrm{s \cdot cm^2}}{\mathrm{F}}$$

$$= 159\ \Omega \cdot \mathrm{cm^2}$$

VIII. Membrane Impedance

A. Parallel RC Circuit

Because the cell membrane has a capacitive component (because of the continuous phospholipid bilayer matrix) in parallel with a resistive component (because of protein ion channels embedded in the lipid bilayer matrix), membrane impedance (Z) should diminish as the AC frequency is increased (Fig. 5A). The relationship between frequency and relative impedance for a typical cell membrane is given in Fig. 5B. When f is extremely low, X_C is extremely high, and Z approaches R. As f is increased, X_C diminishes, and more and more of the current passes through C. Current does not actually pass through the dielectric of C, but it appears so because of the phenomenon of **induced charge** that occurs (see Section IX).

When f is extremely high, X_C approaches zero, and therefore Z approaches zero. This is analogous to two resistors in parallel, one of whose impedance decreases as a function of the applied frequency. In diathermy, a very high frequency AC is used to heat damaged tissues without producing excitation of the nerve and muscle membranes; this is possible because of the capacitive component of the membrane, which causes Z_m to be extremely low and which does not allow a significant membrane potential change to be produced.

At the frequency at which $X_C = R$, however, Z is not equal to $0.5 \times R$, but is actually $0.707 \times R$, as determined by the following equation for a resistor and capacitor in parallel (note its similarity to that for two resistors in parallel, except for the squared function):

$$\left(\frac{1}{Z}\right)^2 = \left(\frac{1}{X_C}\right)^2 + \left(\frac{1}{R}\right)^2 \qquad (23a)$$

$$Z = \sqrt{\frac{X_C^2 R^2}{X_C^2 + R^2}} \qquad (23b)$$

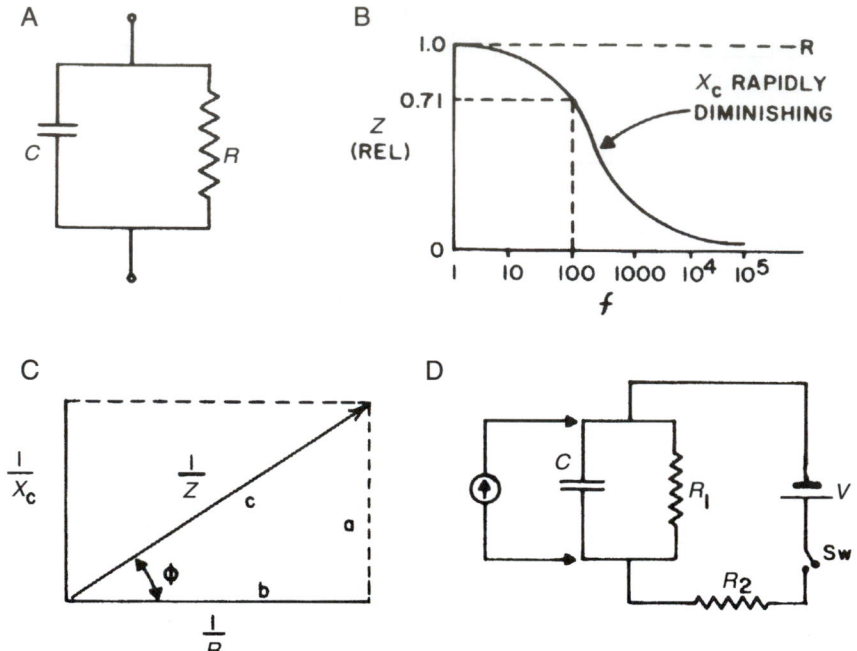

FIGURE 5.   (A) Parallel resistance–capacitance (RC) network model of the cell membrane. (B) Impedance (Z) of such a parallel resistance–capacitance network decreases as the sinusoidal frequency (f) is increased: a plot of Z versus log frequency is an inverse sigmoidal shape. Impedance decreases because the capacitive reactance (X_C) decreases according to $X_C = 1/2\pi fC$. (C) Parallelogram method of obtaining the impedance of a parallel resistance–capacitance circuit. Parallelogram constructed of $1/X_C$ against $1/R$ gives the vector that is equal to $1/Z$; the phase angle (ϕ) between current and voltage is also obtained. (ϕ approaches $90°$ as resistance (R) approaches infinity; i.e., the circuit becomes purely capacitive.) Also, as can be reasoned, the larger the capacitance (for a given resistance), the lower the X_C for a given f and hence the larger the ϕ. Thus the circuit behaves as the most purely capacitive ($\phi \to 90°$), the larger the capacitance and the larger the resistance. (D) Circuit diagram illustrates why current leads the voltage (by a certain phase angle) in a resistance–capacitance circuit. The instant the switch (Sw) is closed the frequency appears to be infinite, and so X_C is zero; therefore no voltage appears across the capacitance, although the (induced) current through the capacitance is maximum; that is, current leads the measured voltage.

The impedance (Z) and the **phase angle** (ϕ) between current and voltage can also be measured graphically by the **parallelogram method,** as illustrated in Fig. 5C. In this method, the reciprocal of the capacitive reactance is plotted on the ordinate against the reciprocal of the resistance on the abscissa, and the parallelogram is completed by drawing lines parallel to the two axes. The diagonal line is equal to the reciprocal of the impedance, and the angle between the diagonal and the abscissa gives the phase angle (ϕ) between voltage and current. In a purely resistive circuit, the current and voltage are exactly in phase ($\phi = 0°$). In a purely inductive circuit, current lags the voltage by $90°$, whereas in a purely capacitive circuit, **current leads the voltage** by $90°$. In a resistance-capacitance circuit, parallel or series, ϕ is between $0°$ and $90°$, depending on the relative values of R and X_C.

The reason that current leads the voltage in a resistance-capacitance circuit may be discerned from Fig. 5D. At the instant the switch is closed, the frequency appears to be infinite (step change), and so $X_C = 0$. Therefore, there can be no potential difference across the capacitor, and the entire voltage is dropped across R_2. Thus, the current through the capacitor is maximum, while the voltage across the capacitor is minimum, that is, the current leads the voltage by $90°$.

The identity of the parallelogram method with Eq. 23a is given by the **Pythagorean theorem** of trigonometry,

$$c^2 = a^2 + b^2 \qquad (24)$$

where c is the length of the hypotenuse of a right triangle, and a and b are the lengths of the other two sides. By substituting for a, b, and c from Fig. 5C, the following expression, which is identical to Eq. 23a is obtained:

$$\left(\frac{1}{Z}\right)^2 = \left(\frac{1}{X_C}\right)^2 + \left(\frac{1}{R}\right)^2 \qquad (23a)$$

B. Series RC Circuit

For a series resistance–capacitance network (Fig. 6A), the corresponding analysis is

$$Z^2 = X_C^2 + R^2 \qquad (25a)$$

or

$$Z = \sqrt{X_C^2 + R^2} \qquad (25b)$$

which is similar to having two resistors in series, except for the squared function. When $X_C = R$, then $Z = 1.414R$ (not $2 \times R$). At $f = \infty$, $X_C = 0$, and $Z = R$. At $f = 0$ (DC), $X_C = \infty$, and $Z = \infty$. The parallelogram analysis is shown in Fig. 6B.

A

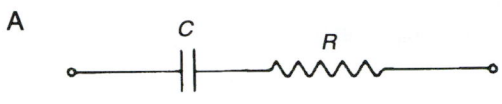

B

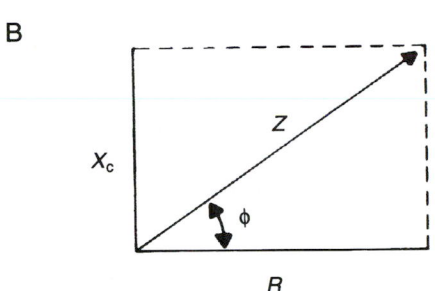

FIGURE 6. (A) Series resistance–capacitance (*RC*) network. (B) Parallelogram method of obtaining the impedance (*Z*) of a series resistance–capacitance circuit. Parallelogram constructed with capacitative reactance (*X_C*) on the ordinate and resistance (*R*) on the abscissa gives the vector which is equal to *Z*. The phase angle (ϕ) between current and voltage approaches 90° as *R* approaches zero.

IX. Capacitive Charge and Capacitive Current

The total **charge** (*Q*, in coulombs) stored on the plates of a capacitor is equal to the product of the capacitance and applied voltage:

$$Q = CV \tag{26}$$

$$C = (F)(V)$$

Thus, if a certain charge is placed on a capacitor (air dielectric) and then the distance between the plates is decreased so that capacitance increases, it follows that the PD across the capacitor must decrease in proportion. It also follows that because an increase in dielectric constant increases capacitance, an increase in dielectric constant must also decrease the PD for a given charge separation; that is, it reduces the force acting between the separated charges.

The electric charge stored in a capacitor can be used to do work. Kinetic energy is needed to charge a capacitor, the charged capacitor representing potential energy (like a battery), and discharge of the capacitor converts the potential energy back into kinetic energy. From Eq. 26, and because electric current flow (*I*) is a charge flowing past a point per unit of time (*I* = *Q*/*t*), the **capacitive current** flow (*I_C*, in amps) during charge or discharge of a capacitor is equal to the capacitance (in farads) times the rate of change of the PD:

$$I_C = C \frac{dV}{dt} \tag{27}$$

$$\frac{C}{s} = F \frac{V}{s}$$

Thus for a given capacitance, the greater the rate of change of voltage, the greater the capacitive current is. In a myocardial cell membrane, for example, the maximum rate of change in

PD across the membrane during the rising phase of the AP is relatively large (e.g., 100 mV/0.5 ms or 200 V/s); therefore capacitive current is also relatively large. There is little or no change in the value of the membrane capacitance during an AP.

As stated in Section VIII, current does not actually flow through the capacitor, but only appears to do so because of the phenomenon of **induced charge.** When electrons flow into one plate of the capacitor, the negative charges on this plate repel the electrons from the other adjacent plate, since like charges repel one another. Thus, as diagrammed in Fig. 7, a flow of electrons into the lower plate of the capacitor leads to a flow of electrons out of the upper plate. Hence it appears as though the flow of electrons was directly through the dielectric. The deficiency of electrons in the upper plate makes the upper plate become positive, whereas the lower plate is negatively charged.

X. Membrane Time Constant

When current is through a simple resistance or series of resistors, the PD across the network is developed instantaneously. This is not true for a biological membrane or for the parallel resistance-capacitance network shown in Fig. 8A. Instead, the PD builds in a nearly exponential (negative) manner until the final maximal value is attained, as illustrated in Fig. 8B. The slow charging occurs because of the presence of the capacitance. The time it takes for **63% of the final PD** to be reached is, by definition, the **time constant** (τ). The equation describing this PD buildup is

$$V_t = V_{max} (1 - e^{-t/\tau}) \tag{28}$$

A
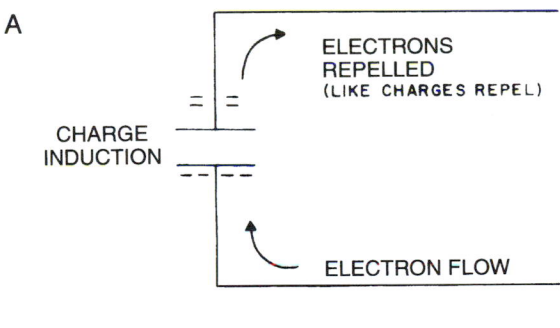

B
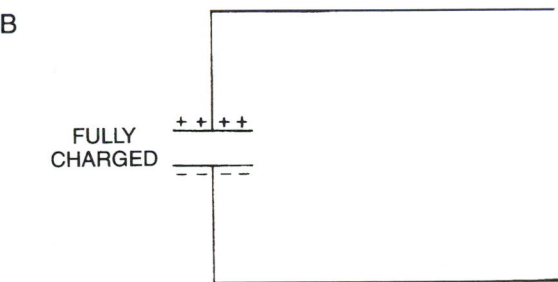

FIGURE 7. Illustration of the apparent flow of current through the dielectric of a capacitor. Current does not actually flow through the capacitor but appears to because of the phenomenon of charge induction. (A) When one plate of the capacitor is made negative, this plate repels the electrons from the other plate nearby, because like charges repel. (B) Thus the second plate has an electron deficiency that gives the plate a positive charge.

where V_t is the voltage across the network at any time (t) and V_{max} is the maximum or final PD when the steady state is reached. When $t = \tau$, then $e^{-t/\tau}$ becomes

$$e^{-1} = \frac{1}{e} = \frac{1}{2.717} = 0.37 \qquad (29a)$$

Therefore,

$$V_t = V_{max}(1 - 0.37) \qquad (29b)$$
$$= 0.63 V_{max}$$

For all other values of t, Eq. 28 is solved by taking the logarithms and obtaining

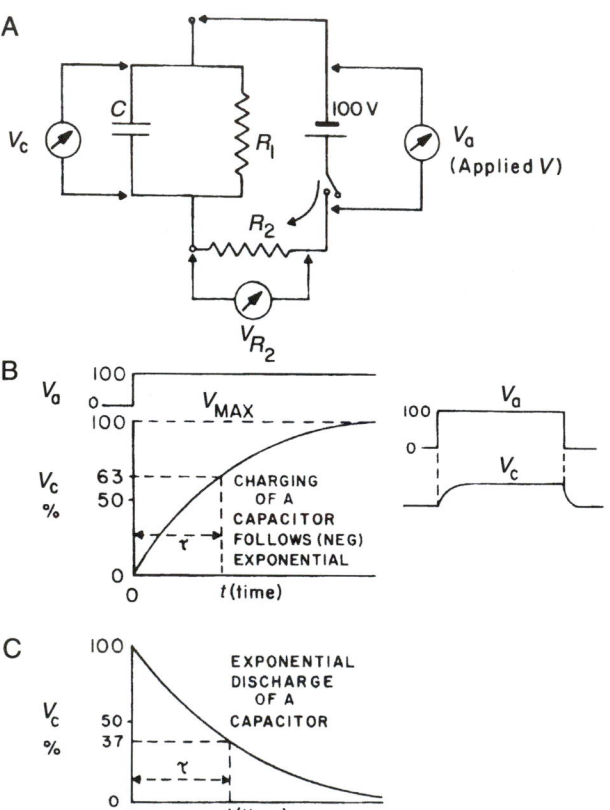

$$V_t = V_{max} - \text{antiln}\left[\ln V_{max} - \frac{t}{\tau}\right] \qquad (30)$$

$$V_t = V_{max} - \text{antilog}\frac{2.303 \log V_{max} - (t/\tau)}{2.303} \qquad (31)$$

On discharge of the resistance–capacitance network, the voltage across the capacitor (V_C) decreases exponentially, as indicated in Fig. 8C. Again, the time it takes for 63% of the final PD to be reached (which is to the level of **37% of the initial value**) is the time constant (τ). The equation describing the decay in PD as a function of time (analogous to that for the decay in voltage as a function of distance along a cable) is

$$V_t = V_{max} e^{-t/\tau} \qquad (32)$$

where V_{max} is the maximum or initial PD at $t = 0$. When $t = \tau$,

$$V_t = 0.37 V_{max} \qquad (33)$$

For all other values of t, Eq. 32 is solved by taking the logarithms

$$\ln V_t = \ln V_{max} - \frac{t}{\tau} \qquad (34a)$$

$$V_t = \text{antiln}\left[\ln V_{max} - \frac{t}{\tau}\right] \qquad (34b)$$

$$V_t = \text{antilog}\frac{2.303 \log V_{max} - (t/\tau)}{2.303} \qquad (34c)$$

In the circuit illustrated in Fig. 8A, upon closing the switch, the capacitor (C) charges exponentially with a time constant about equal to $R_1 C$ (if R_1 is much greater than R_2). At the first instant after the switch is closed, there is a step change in the applied potential that appears as an infinite frequency. Therefore, because $X_C = 1/2\pi fC$, $X_C \cong 0$. According to Kirchhoff's laws pertaining to the relative voltage drops across series impedances, the voltage drop is in proportion to the relative impedance (i.e., the higher impedance has the greater voltage drop across it), and the sum of all voltage drops must equal the source emf. Thus, all the emf of the battery should be dropped across R_2 ($V_{R_2} = 100$ V) and none across R_1 ($V_C = 0$ V).

At the final steady-state value (plateau of applied voltage pulse), the applied voltage pulse is steady, and therefore the frequency is zero; hence $X_C \cong \infty$, and from analysis of the impedance of a parallel resistance–capacitance network, $Z = R_1$. At this time, nearly all of the battery emf should be dropped across C (or R_1) because $R_1 \gg R_2$. In this condition, the current continues to flow through R_1 and R_2.

At all intermediate instants, $V_C > 0$ V but <100 V. The capacitor charges exponentially. One way to view why a capacitor charges and discharges exponentially is to assume that the negative plate of a capacitor is a compartment that can fill up with electrons. If so, the **rate of fill-up (charging)** at any moment should be a function of the actual degree of fill-up at that moment, according to the laws of **first-order kinetics.** As the concentration builds, the rate of further buildup becomes slower and slower (a negative exponential). The process

FIGURE 8. (A) Exponential (negative) charge and discharge of a capacitor for the parallel resistance–capacitance (R_1C) circuit in series with a second resistance (R_2). Applied voltage is from a 100-V battery upon closing of the switch and is measured by voltmeter V_a. Voltmeter V_C measures the voltage across the capacitor, and voltmeter V_{R_2} measures the voltage across R_2. (B) Upon closing the switch, V_a suddenly jumps from 0 to 100 V. Potential recorded by V_C, however, slowly increases exponentially; the time constant (τ) is the time it takes for V_C to build to 63% ($1 - e^{-1}$) of its final (or maximal) value. The instant that the switch is closed, the frequency effectively is infinite, and so capacitive reactance (X_C) is zero; therefore no voltage appears across the parallel resistance–capacitance network, although the induced current through the capacitor is maximum. The entire voltage drop of the battery thus appears across R_2. After a long time, that is, when the circuit reaches steady state, $f = 0$ and X_C is infinite; therefore the voltage across the parallel resistance–capacitance network is equal to $(R_1/R_1 + R_2) \times 100$ V. This charging of a capacitor can be viewed as a compartment filling with electrons and obeying first-order kinetics. (C) Upon opening of the switch, the capacitor discharges exponentially through R_1. Time constant is the time it takes for discharge to 37% (e^{-1}) of the initial (maximal) value.

reaches equilibrium when the maximum number of electrons (maximum density or concentration of electrons) that can be stored on the plate for a given impressed voltage is reached.

Similarly, the **rate of emptying** of the compartment at any instant depends only on the number of electrons available at that instant. As concentration decays, the rate of further decay becomes slower and slower. The time to build up or to decay to one-half (50%) of the final value is called the **half-time**, $t_{1/2}$. At one half-time, the rate should be one-half what it was initially, because only half of the original number of electrons are now available. At the second half-time, the rate should be one-half what it was at the first half-time or one-fourth of what it was initially, and so on. About seven half-times are necessary to reach more than 99% of the final steady-state value.

The **time constant** is the time required to build to 63% of the final steady-state value or to decay to 37% (100% − 63%) of the initial value. About five time constants are necessary to reach more than 99% of the final value. The time constant and half-time values for a particular exponential process can be interconverted by the relationship

$$\tau = \frac{t_{1/2}}{\ln 2} = \frac{t_{1/2}}{0.693} = 1.44 t_{1/2} \tag{35}$$

Thus, if the membrane time constant were 1.0 ms, the half-time would be 0.69 ms. The time constant (in seconds) and first-order **rate constant** (k, in s^{-1}) are reciprocally related as

$$k = \frac{1}{\tau} \tag{36}$$

In a nerve or skeletal muscle fiber cable, the change in membrane potential during the application of electrotonic constant-current pulses is actually an **error function**, because the longitudinal resistance of the intracellular fluid is not negligible. An error function curve builds more rapidly than does an exponential curve. The equivalent time constant for such an error function curve corresponds to a buildup to 84% of steady-state value, rather than to 63%.

XI. Specific Resistance and Specific Capacitance

A. Specific Resistance

The resistance (R) actually measured for any material depends on the nature of the material (its **specific resistance** or **resistivity**, ρ) and on the amount and shape of the material used. If discussion here is confined to uniform elongated cylinders or cubes of material (Fig. 9A), then

$$R = \rho \frac{L}{A_X} \tag{37a}$$

$$\Omega = (\Omega \cdot cm)\left(\frac{cm}{cm^2}\right)$$

In Eq. 37a, A_X is the cross-sectional area (πa^2) and L is the length. As seen from this equation, the greater the resistivity of the material, the greater the resistance measured for any given length or cross-sectional area of material. The resistivity, by definition, is equal to the resistance of a material when the ma-

terial is of unit cross-sectional area (1 cm^2) and of unit length (1 cm), hence of unit volume (1 cm^3). The units of resistivity are in $\Omega \cdot cm$. Rearranging Eq. 37a gives

$$\rho = R \frac{A_X}{L} \tag{37b}$$

$$\Omega \cdot cm = \Omega \left(\frac{cm^2}{cm}\right)$$

Mammalian Ringer's solution has a resistivity of about 50 $\Omega \cdot cm$. The resistivity of such electrolyte solutions varies with the concentration of each of the salts present, and with the mobilities of the various species of ions (e.g., K^+, Na^+, and Ca^{2+}). The resistivity (reciprocal of conductivity or specific conductance) of solutions decreases with temperature elevation (whereas that of metallic conductors increases).

B. Resistivity of Axoplasm or Myoplasm

The notation is confusing because physicists use ρ to denote resistivity, whereas physiologists use R (see Fig. 9A and B). Thus R_i is the resistivity of the internal cytoplasm (usual values of 100–300 $\Omega \cdot cm$ for axoplasm and myoplasm), and R_o is the resistivity of the outside or external solution, which is normally the interstitial fluid (value of about 50 $\Omega \cdot cm$, comparable to that of Ringer's solution). Hence, we are forced to use r (instead of R) to represent the absolute resistance (in Ω), and Eq. 37b becomes

$$R_i = r \frac{A_X}{L} \tag{38}$$

$$\Omega \cdot cm = \Omega \frac{cm^2}{cm}$$

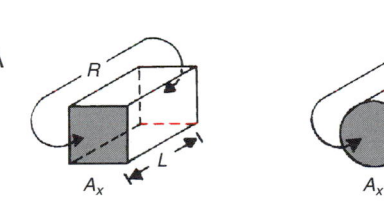

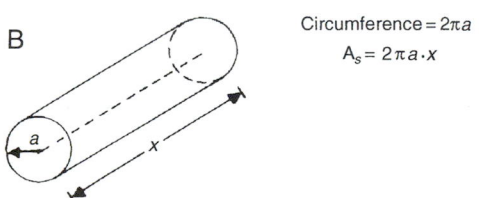

FIGURE 9. Resistivity measurements on a material. (A) For material such as myoplasm or axoplasm, the longitudinal resistance (R) depends on the resistivity of the material (ρ_i), the cross-sectional area (A_x or πa^2, where a is the radius), and length (L): $R = \rho L / \pi a^2$. Box shape and cylindrical shape are illustrated. (B) For skin material such as cell membrane, the transverse resistance (R) depends on the so-called "resistivity" of the membrane (R_m) and on the surface area (A_s or $2\pi a \cdot x$): $R = R_m / 2\pi a x$.

It is convenient to substitute r_i for r/L,

$$R_i = r_i A_X \qquad (39a)$$

$$\Omega \cdot cm = \frac{\Omega}{cm} cm^2$$

where r_i is the internal resistance normalized for unit length, but not for unit cross-sectional area. That is, r_i is numerically equal to the internal resistance of a 1-cm length of a given cable, such as a skeletal muscle fiber or nerve axon. Because $A_X = \pi a^2$, Eq. 39a can be modified to

$$R_i = r_i \pi a^2 \qquad (39b)$$

which is the final useful form of the equation when making biological measurements (see Table 2).

C. Specific Resistance and Resistivity of the Cell Membrane

In dealing with the resistance of the cell surface membrane, we may proceed starting from the basic Eq. 37b and rearranging it to give

$$\rho L = R A_X \qquad (37c)$$

Because the length being considered is the thickness of the membrane (δ) and because the area being considered is the surface area (A_S) of the cylinder (Fig. 9B), we must make the appropriate substitutions into Eq. 37c. Letting $\delta = L$,

$$\rho \delta = R A_X \qquad (40)$$

Letting $r = R$ and $A_S = A_X$

$$\rho \delta = r A_S \qquad (41a)$$

Because A_S = circumference ($2\pi a$) times length (x) = $2\pi a x$,

$$\rho \delta = r 2\pi a x \qquad (41b)$$

Letting $R_m = \rho \delta$,

$$R_m = r 2\pi a x \qquad (42a)$$

Rearranging gives

$$R_m = (rx)(2\pi a) \qquad (42b)$$

Letting $r_m = rx$,

$$R_m = r_m (2\pi a) \qquad (43)$$

$$\Omega \cdot cm^2 = (\Omega \cdot cm)(cm)$$

where r_m is the membrane resistance normalized for unit length, but not for unit surface area. That is, r_m is numerically equal to the transverse membrane resistance of a 1-cm length of a given cable (e.g., frog sartorius fiber). The r_m is converted to R_m by multiplying by the circumference. These interconversions are summarized in Table 2.

Although R_m is often called the membrane resistivity or specific resistance, the term is not accurate, because as just seen, the units for R_m are $\Omega \cdot cm^2$, not $\Omega \cdot cm$. That is, the true resistivity has not been normalized for the unit length (thickness of 1 cm). Thus, if R_m is 1000 $\Omega \cdot cm^2$ and, δ is 100 Å (10^{-6} cm), then the **true membrane resistivity** (ρ_m) would be

$$\rho_m = \frac{R_m}{\delta} \qquad (44)$$

$$= \frac{1000 \, \Omega \cdot cm^2}{1 \times 10^{-6} \, cm}$$

$$= 1 \times 10^9 \, \Omega \cdot cm$$

This resistivity value makes the cell membrane an effective insulator indeed, comparable to the best commercial insulators. The cell membrane, however, is extremely thin (the animal cannot afford the space to have a 1-cm thick insulation for each nerve and muscle fiber); therefore, the actual nerve cable has a great deal of energy loss (cable decrement). Note that r_m is also not the true resistivity even though it has the proper units.

D. Specific Capacitance

In dealing with the capacitance of the cell surface membrane, we may proceed in a similar manner using an analysis analogous to that for R_m just given, and taking into account the fact that the total resistance of resistors added in parallel becomes smaller, whereas the total capacitance of capacitors added in parallel becomes larger. Therefore, the measured capacitance must be divided by the surface area—rather than multiplied, as for membrane resistivity—to obtain the specific capacitance. Thus, the units for membrane specific capacitance (C_m) are F/cm^2, compared to $\Omega \cdot cm^2$ for

TABLE 2 Summary of the Interconversions Used in the Measurements of Properties of Biological Cables

Orientation, with respect to fiber axis	Actual value measured	Operation			Normalized for unit length of fiber	Operation			Normalized for radius of fiber (A_x or circumference)
Longitudinal	$r(\Omega)$	÷	×	=	$r_i(\Omega/cm)$	×	πa^2	=	$R_i(\Omega \cdot cm)$
Transverse	$r(\Omega)$	×	×	=	$r_m(\Omega \cdot cm)$	×	$2\pi a$	=	$R_m(\Omega \cdot cm^2)$
Transverse	$c(F)$	÷	×	=	$c_m(F/cm)$	÷	$2\pi a$	=	$C_m(F/cm^2)$

Note: A_x, cross-sectional area; r, resistance; x, fiber length; r_i, internal resistance per unit length of fiber; a, fiber radius; R_i, internal resistivity; r_m, membrane resistance for a unit length of fiber; R_m, membrane resistivity; c, capacitance; c_m, membrane capacitance per unit length of fiber; C_m, membrane specific capacitance.

membrane specific resistance (R_m). In addition, as with R_m, C_m is not prorated for 1-cm thickness, but is given for the actual thickness of the membrane; hence, the units are F/cm^2 (not F/cm). Thus, the equation analogous to that for membrane resistivity (Eq. 42a) is

$$C_m = \frac{c}{A_S} = \frac{c}{2\pi a x} \tag{45}$$

where c is the absolute capacitance in farads.

Letting $c_m = c/x$,

$$C_m = \frac{c_m}{2\pi a} \tag{46}$$

$$\frac{\text{F}}{\text{cm}^2} = \frac{\text{F/cm}}{\text{cm}}$$

where c_m (measured in F/cm) is the membrane capacitance normalized for unit length but not for unit surface area. The c_m is converted to C_m (F/cm^2) by dividing by the circumference (Table 2). Thus, C_m is normalized both for unit length of the fiber and for unit circumference (radius).

XII. Biological Cable Decrement

Nerve and muscle fibers make relatively poor cables, in comparison with commercial cables (such as a transoceanic submarine cable). There is a great deal of **energy loss** in the transmission of information along these biological cables. If the cable remained passive (no active impulse), the signal would become greatly attenuated and distorted after traveling only a relatively short distance, e.g., 1 mm. Hence biological signal transmission in long fibers is an active energy-consuming process, the energy input being distributed along the entire length of the fiber. The **biological cable is poor** for the following reasons: (1) R_i is approximately 10^7 times that for copper wire; (2) R_m is approximately 10^{-6} times that for an effective insulator such as rubber. (But, as stated in the preceding section, the true resistivity of the cell membrane is comparable to that of an efficient commercial insulator; that is, the cell membrane would be as efficient as a commercial insulator if both were equally thick); and (3) C_m is high.

The biological cable, however, is the best possible, given the biological circumstances and requirements. In addition, the nerve cable has been vastly improved by an evolutionary development in the vertebrates (namely, **myelination**); myelination increases R_m and decreases C_m by more than 100-fold. There are two consequences that result from improvement of the nerve cable by myelination: (1) The **velocity of propagation** is greatly increased; and (2) the **energy cost** of signaling is greatly decreased.

A cable consists of two conductors arranged in parallel (usually concentrically) separated by an insulating material, for example, a copper wire in the center core surrounded by insulation and a sleeve of copper, as illustrated in Fig. 10A. The circuit diagram for an ideal cable is given in Fig. 10B. In a submarine cable, the ocean can serve as the outer conductor. In the biological cable, the tissue fluids serve as the ocean, providing a fairly effective outside conductor. The in-

side conductor is not as effective, however, as discussed previously (because of high R_i and small cross-sectional area); hence the internal resistance is not negligible. Therefore the circuit diagram shown in Fig. 10C more closely approximates the biological cable.

Furthermore, efforts are made to minimize any stray (unwanted) capacitance in the submarine cable so that AC signals are not attenuated and distorted. As stated in the previous paragraph, the biological cable has a relatively large capacitance (but that capacitance is greatly reduced in myelinated fibers). Hence the biological cable is more accurately represented by the circuit diagram (Fig. 10D), which includes the large c_m, the large r_i, and the relatively low r_m. These resistance and capacitance parameters are distributed along the length of the cable.

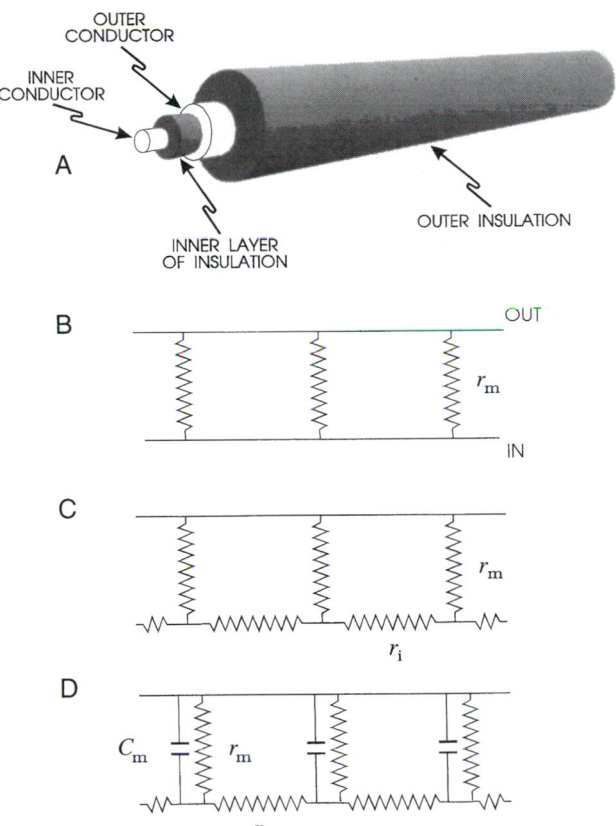

FIGURE 10. Diagrams illustrating cables. (A) Cable composed of copper wire coated with insulation and surrounded concentrically by a copper sleeve, with insulation over the latter. (B) Simple circuit diagram for an idealized cable, which consists of two good conductors separated by a material of high resistance (R). (C) For a biological cable, such as nerve fiber or skeletal muscle fiber, the inner conductor is not zero or low resistance (as would nearly be true for a copper wire), and therefore a resistance must be drawn. Even in the biological cable, however, the transverse membrane resistance (r_m) is much higher than the longitudinal internal resistance (r_i). In the biological cable, the outside resistance (r_o) can be drawn as a straight line, since r_o is relatively low in resistance compared with r_i, because the external current makes use of the entire interstitial fluid space (i.e., many resistors in parallel). (D) Biological cable with the large membrane capacitance (C_m) included.

A potential applied at one end of a cable diminishes exponentially with distance (x), with a certain **length** (or **space**) **constant** (λ) as given by

$$V_x = V_o\, e^{-x/\lambda} \tag{47}$$

where λ, by definition, is the distance at which the potential has fallen to $1/e$ or 37% of the initial value. This exponential decay can be determined experimentally using a cable network or can be calculated using Ohm's and Kirchhoff's laws as the principal analytic aids.

Equation 47 is for an infinitely long fiber cable (i.e., one whose length is at least 10 times the length constant λ). In such cases, there is an exponential fall of potential in both directions from the site of current injection ($x = 0$). Cable properties are discussed in more detail in Chapter 24. However, if the fiber is short (i.e., a **truncated cable**, shorter than the hypothetical λ), then Eq. 47 does not apply. Instead the following equation applies:

$$V_x = V_0\, \frac{\cosh\left(\dfrac{L - x}{\lambda_h}\right)}{\cosh\left(\dfrac{L}{\lambda_h}\right)} \tag{48}$$

where cosh represents the hyperbolic cosine, L is the length of the cell, x is the distance from the site where current is injected (e.g., at one end of the fiber) and V_x is measured, λ_h is the hypothetical length constant, and V_0 is the voltage change at the site of current injection ($x = 0$). This equation indicates that when $x = 0$, the cosh term becomes 1.0, and so $V_x = V_0$. When x is not zero, but is at the middle or far end of the fiber, then V_x is less than V_0, in accordance with Eq. 48. The actual values calculated from this equation are given in Table 3.

Cardiac muscle cells and smooth cells are short, so they behave as truncated cables. If there are no gap junction channels (connexons) between contiguous cells lying end to end, then Eq. 48 applies. The cable is considered to be "open" at both ends, that is, terminated in nearly infinite resistance (like an open electrical wall switch), because of the high-resistance cell membrane at the ends. Applied current is considered to reflect from the high-resistance end, thus causing the voltage falloff over the length of the cell to be very small, as indicated in Table 3. The reader is referred to Appendix 1 of Chapter 24 for evidence and references showing that there is actually very little voltage decay within a single myocardial cell. If there were many connexon channels providing low-resistance pathways between the cells, then Eq. 48 would not apply.

The values given in Table 3 are for a cylindrical myocardial cell having a length (L) of 150 µm, and V_0 is assumed to be 100 mV. The V_x values were calculated at two points in the cell: at midpoint ($x = 75$ µm) and at the far end ($x = 150$ µm). The calculations were done for a wide range of assumed λ_h values. As can be seen, the larger the λ_h value, the less voltage decay along the length of the cell.

To measure λ_h requires two microelectrodes to be placed within one cell at a known distance apart, for the voltage decay to be measured between the two electrodes. The first electrode

TABLE 3 Calculations of Voltage Falloff at Two Distances within a Myocardial Cell Behaving as a Truncated Cable and for Which Eq. 48 Applies

λ_h (µm)	V_x (mV)	
	$x = 75$ µm	$x = 150$ µm
150	73.1	64.8
200	82.7	77.2
300	91.4	88.7
500	96.8	95.7
1000	99.2	98.9
3000	99.9	99.9

Note: Calculations for cell length (L) of 150 µm, V_0 of 100 mV, and assumed hypothetical λ_h values ranging between 150 and 3000 µm.

is used to apply current (I_0) and to measure simultaneously the voltage change (V_1 or V_0) by use of a balanced Wheatstone bridge circuit. From these data, λ_h can be calculated. However, the procedure is difficult and inaccurate because of bridge imbalance that often occurs due to the microelectrode resistance changing with applied current, pressure, and intracellular ionic milieu; in addition, the exact interelectrode distance between the microelectrode tips is hard to determine, and the placing of two microelectrodes into one short cell often damages the cell.

Because of these difficulties, it is simpler and more accurate to measure the **input resistance** of an isolated single cell, and from this calculate the resistance of the cell membrane (the resistance of the myoplasm is negligibly low). Values for R_{in} of 10–80 MΩ have been reported (Sperelakis *et al.*, 1960; Tarr and Sperelakis, 1964; Sperelakis and Lehmkuhl, 1966; Josephson *et al.*, 1984). Then the membrane specific resistance R_m can be calculated using the measured surface area of the cell. R_m is the product of R_{in} and the surface area (A_s) of the cell ($A_s = \pi 2aL$):

$$R_m = R_{in}\,(\pi 2aL) \tag{49}$$

For an R_{in} of 30 MΩ:

$$R_m = \left(30 \times 10^6\,\Omega\right)\left(7.54 \times 10^{-5}\,\text{cm}^2\right)$$

$$= 2260\,\Omega\cdot\text{cm}^2$$

For a higher or lower R_{in} value, R_m becomes proportionally higher or lower.

The resistivity of the myoplasm R_i is a relatively constant value, and can be assumed to be 200 Ω·cm. Therefore, from the measured radius (a) of the cell, the λ_h value can be calculated using the following relationship:

$$\lambda_h = \sqrt{\frac{R_m}{R_i}\,\frac{a}{2}} \tag{50}$$

This is the same as Eq. 3 of Chapter 24. Table 4 gives the calculated λ_h values for a myocardial cell with a radius of 8 µm for various assumed R_m values ranging between 100 Ω·cm² and 30 000 Ω·cm². The calculations are given for two different assumed R_i values. As expected, λ_h is a direct function of the square root of R_m. Values of λ_h of 320–375 μm have been reported (Josephson *et al.*, 1984). If so, then R_m must be about 500 Ω·cm² (see Table 4), and there is only about a 10% decay of potential over the length of the cell (see Table 3).

XIII. Inductance, Inductive Reactance, and Oscillations

A. Nature of Inductance

An inductor consists of a coil of wire with multiple turns. When current in any conductor is changed in magnitude, there is an **opposing emf induced** in any neighboring conductor, for example, in the adjacent turns of the coil. The greater the rate of change of the current, the larger the opposing emf induced. The opposing emf is in the opposite polarity of the applied voltage, and therefore acts to oppose the change in current. This is why an inductor is also known as a choke, because the inductor reacts against AC, causing a high impedance (or high so-called "AC resistance") to rapidly varying currents. When the current through the coil is increasing, the **induced emf** opposes the current flow, whereas when the current is decreasing, the induced emf facilitates and prolongs the current. Thus, inductance (L) tends to keep current constant when the voltage is varied.

The induced emf is due to the magnetic field that surrounds a wire that is carrying current. When current increases, the magnetic field expands, and when current decreases, the magnetic field shrinks. Thus, if the conductor is in the form of a coil, the expanding and shrinking magnetic field cuts across the neighboring turns of the wire. Whenever a moving magnetic field cuts across a conductor, an electromotive force (emf) is generated in the conductor. [This is the principle of the electric generator, in which the magnetic field is usually fixed (permanent magnet) and the conductor (rotor armature) is moved.] An **expanding field** induces the emf in one polarity (opposing), and the **shrinking field** induces the emf in the opposite polarity (facilitating). For example, a large voltage and current (spark) is produced when a circuit containing an inductance is opened (broken).The unit of inductance is the

Henry (H), named for the research by Joseph Henry published in 1831.

The phenomenon of induction of an opposing emf and a facilitating emf in a simple coil is known as **self-inductance**. The self-inductance of any given coil is a constant value, depending on the number of turns in the coil. A coil has a self-inductance of 1 H when 1 V of opposing emf is caused by a dI/dt of 1 A/s:

$$L = \frac{V}{dI/dt} \qquad (51)$$

Inductance is analogous to mechanical inertia or a mass that resists an increase or decrease in velocity. The energy (in joules) stored in the magnetic field of an inductance is equal to $0.5\,LI^2$ (analogous to 0.5 mass × velocity²). **Mutual inductance** occurs between two closely spaced parallel coils, for example, in a transformer.

B. Inductive Reactance and Impedance

In contrast to capacitive reactance, the **inductive reactance** (X_L, in ohms) varies directly with frequency and magnitude of the inductance.

$$X_L = 2\pi f L \qquad (52)$$

$$= \omega L$$

Hence, for a given inductance, the higher the frequency, the greater the reactance. The inductive reactance is zero for DC ($f = 0$). In a purely inductive circuit, the current lags the voltage by a **phase angle** of 90°. The current through the inductance (I_L) is equal to the applied voltage (V) divided by X_L (according to Ohm's law).

The impedance for series and parallel resistance–inductance circuits is similar to those for the corresponding resistance-capacitance circuits. For a series resistance–inductance circuit

$$Z^2 = X_L^2 + R^2 \qquad (53)$$

For a parallel resistance–inductance circuit

$$\left(\frac{1}{Z}\right)^2 = \left(\frac{1}{X_L}\right)^2 + \left(\frac{1}{R}\right)^2 \qquad (54)$$

The same **parallelogram analysis** also applies, except by convention X_L is drawn downward (instead of upward, as in the case of X_C) to represent the fact that the current lags the voltage.

C. Resonance and Oscillations

If an inductance is placed in parallel with a capacitance (parallel inductance–capacitance or so-called "**tank circuit**"), then the impedance follows a peculiar pattern as the AC frequency is increased. As f is increased, X_L increases, whereas X_C decreases; hence Z passes through a rather sharp maximum (approaches infinite), and the current is minimum (approaches zero), at some intermediate frequency called the **resonant frequency** (f_0). In a series inductance–capacitance circuit, at the resonant frequency, Z is minimum and the current is maximum. These facts are used in many electronic tuning circuits and filter circuits.

TABLE 4 Calculations of the Hypothetical Length Constant λ_h from Eq. 50 for a Myocardial Cell Having Radius of 8 μm and for Various Assumed R_m Values

	λ_h (mm)	
R_m (Ω · cm²)	$R_i = 200\ \Omega \cdot$ cm	$R_i = 100\ \Omega \cdot$ cm
100	0.141	0.200
300	0.245	0.346
500	0.316	0.447
1 000	0.447	0.632
3 000	0.775	1.095
10 000	1.414	2.000
30 000	2.449	3.464

Note: Calculations made for two different assumed R_i values.

The resonant frequency for either the parallel or the series inductance–capacitance circuit may be calculated from

$$f_0 = \frac{1}{2\pi} \sqrt{\frac{1}{LC}} \qquad (55)$$

At low frequencies, $X_C > X_L$ and current leads voltage; at high frequencies, $X_L > X_C$ and current lags voltage. At f_0, the current is in phase with the applied voltage and $X_C = X_L$. Thus,

$$\frac{1}{\omega C} = \omega L \qquad (56a)$$

or

$$\omega^2 = \frac{1}{LC} \qquad (56b)$$

and

$$\omega = \sqrt{\frac{1}{LC}} \qquad (56c)$$

which is the same as Eq. 55, since $\omega = 2\pi f$. If a significant resistance (in series with the inductive branch, as for the excitable membrane) is present in the parallel $(L + R)C$ circuit, then a **damping factor** (R^2/L^2) must be introduced, and f_0 is now given by

$$f_0 = \frac{1}{2\pi} \sqrt{\frac{1}{LC} - \frac{R^2}{L^2}} \qquad (57)$$

One characteristic of the parallel inductance–capacitance tank circuit is that when suitable energy is supplied to such a circuit, **oscillations** in potential tend to occur at

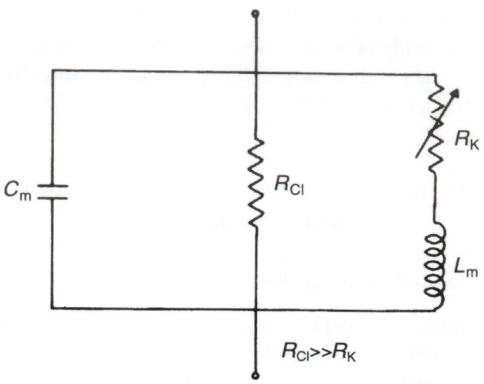

FIGURE 11. Circuit diagram (K. S. Cole circuit) depicting the apparent inductance of biological membranes (L_m). L_m is in parallel to the membrane capacitance (C_m) and is due to the peculiar behavior of the K^+ resistance (R_K); that is, there is no physical basis for the inductance as there is for the capacitance (namely, the lipid bilayer matrix). When Cl^- resistance (R_{Cl}) is high compared to R_K, as is true for most myocardial cells and neurons, the parallel inductance–capacitance circuit tends to oscillate at a resonant frequency (f_0): $f_0 = 1/2\pi\sqrt{1/LC}$. The arrow through R_K indicates that it is a variable resistance, depending on membrane potential. When the membrane is depolarized, R_K increases (because of the K^+ channels responsible for so-called "anomalous rectification"). Because $I_K = (E_m - E_K)/R_K$, the increase in driving force $(E_m - E_K)$ for outward K^+ current (I_K) is counterbalanced by the increase in R_K, the latter acting to hold the K^+ current constant. This is behavior typical of an inductance and gives rise to the apparent membrane inductance.

TABLE 5 Range of the Electromagnetic Spectrum

	Frequency	(Hz)	Wavelength (m)
Longwave	1×10^4	(10 kHz)	30 000 (30 km)
Radio broadcast			
AM	1×10^6	(1,000 kHz)	300 (0.3 km)
Shortwave	1×10^7	(10 MHz)	30
FM	3×10^8	(300 MHz)	1.0
TV broadcast	3×10^8	(300 MHz)	1.0
Microwave	3×10^{10}	(30 GHz)	0.01 (1.0 cm)
Light			
Infrared	3×10^{12}		1×10^{-4} (0.1 mm)
Visible	3×10^{14}		1×10^{-6} (1.0 μm)
Ultraviolet	3×10^{16}		1×10^{-8} (0.01 μm; 10 nm)
X-rays	3×10^{18}		1×10^{-10} (1.0 Å; 0.1 nm)
Gamma rays	3×10^{20}		1×10^{-12} (0.01 Å)

Source: Now You're Talking, edited by L. D. Wolfgang, J. Kearman, and J. P. Kleinman. Published by the American Radio Relay League, Newington, CT, 1996. Frequencies listed are for about the midrange of each spectrum. The corresponding wavelengths were calculated from the following equation, where c is the velocity of light, f is the frequency, and L is the wavelength. The sample calculation is for an AM radio wave of 1000 kHz.

$$L = \frac{c}{f} = \frac{m/s}{cycles/s} = \frac{m}{cycle}$$

$$= \frac{300\,000\,000 \text{ m/s}}{1000 \text{ kHz}} = \frac{3 \times 10^8 \text{ m/s}}{1 \times 10^6 \text{ cycles/s}} = 3 \times 10^2 \text{ m}$$

the resonant (or natural) frequency. In effect, at one instant, all the **energy is stored in the magnetic field** around the inductor; this field collapses when the current through the inductor goes to zero. This effect induces an emf that drives current in the opposite direction and charges the capacitor. When fully charged, all the **energy is stored as electrical charge on the plates of the capacitor.** When the capacitor is fully charged and the current in the circuit goes to zero, the inductance now appears as a zero resistance ($f = 0$) across the capacitor, causing the capacitor to discharge into the inductor. This changing current induces a magnetic field in the inductor, and the cycle is repeated over and over. The oscillations produced are **sinusoidal** in nature. A small amount of energy needs to be put into the system; otherwise the oscillations dampen out because of the small resistance of the components and the resultant power (I^2R) loss as heat.

Oscillations of membrane potential, namely, the **pacemaker potentials,** are known to occur in many nerve and muscle cells under natural conditions and can be made to occur in all nerve and muscle membranes under experimental conditions. The nerve membrane (e.g., squid giant axon) has been demonstrated by Cole, Mullins, and others to also contain an inductive component (see Cole, 1968). Estimates of the **apparent inductance** for biological membranes (L_m) range from 0.5 to 10 H·cm^2. This inductance does not have a simple physical molecular basis, as do the resistance and capacitance components; hence it is described as an apparent inductance only. The L_m seems to be intimately associated with the K$^+$ resistive component of the membrane (R_K), as depicted-in the circuit diagram of Fig. 11. As indicated by the arrow through the K$^+$ resistance, R_K varies as a function of the membrane potential (E_m); that is, there is K$^+$ rectification. In K$^+$ anomalous rectification, when the membrane is depolarized, thereby increasing the driving force ($E_m - E_K$) for outward K$^+$ current, g_K decreases; this decrease in g_K tends to keep the outward

K$^+$ current constant, since $I_K = g_K (E_m - E_K)$, thereby mimicking the behavior of an inductance.

XIV. Electromagnetic Spectrum

The electromagnetic spectrum from radio waves and light waves through gamma rays is summarized in Table 5. As Table 5 shows, the higher the frequency of the electromagnetic wave, the shorter its wavelength. The equation relating wavelength (L) to frequency (f) and to relativistic velocity (c) is given in the footnote to Table 5. Note the similarity of this relationship to that for calculating the wavelength of a nerve impulse discussed in Chapter 24. The information given in this table has relevance to Fig. 1 of Chapter 69.

Bibliography

Cole, K. S. (1968). "Membranes, Ions, and Impulses." University of California Press, Berkeley.

Josephson, I. R., Sanchez-Chapula, J., and Brown, A. M. (1984). A comparison of calcium currents in rat and guinea pig single ventricular cells. *Circ. Res.* **54**, 144–156.

Sperelakis, N. (1979). Origin of the cardiac resting potential. *In* "Handbook of Physiology, The Cardiovascular System, Vol. 1: The Heart" (R. M. Berne and N. Sperelakis, Eds.), pp. 187–267, American Physiological Society, Bethesda, MD.

Sperelakis, N. (1995). "Electrogenesis of Biopotentials." Kluwer Academic Publishers, New York.

Sperelakis, N., and Lehmkuhl, D. (1966). Ionic interconversion of pacemaker and nonpacemaker cultured chick heart cells. *J. Gen. Physiol.* **49**, 867–895.

Sperelakis, N., Hoshiko, T., and Berne, R. M. (1960). Non-syncytial nature of cardiac muscle: membrane resistance of single cells. *Am. J. Physiol.* **198**, 531–536.

Tarr, M., and Sperelakis, N. (1964). Weak electrotonic interaction between contiguous cardiac cells. *Am. J. Physiol.* **207**, 691–700.

Index

A

Acetylcholine, inhibition of calcium channel, 568–569

Acetylcholine receptor channel
  acoustic transduction, 787–788
  dysfunction in myasthenia gravis, 668–669
  electrocytes, 1029, 1034–1037
  neuronal channels, 682–683
  nicotinic, *see* Nicotinic acetylcholine receptor/channel
  structure, 679–683

Acetylcholinesterase, *Torpedo* electrocyte, 1035–1036

Acoustic transduction, *see also* Cochlea; Echolocation; Endolymph; Hair cell; Stria vascularis
  cell physiology, 782–788

Acquired neuromyotonia, potassium channel mutation, 668

Actin, *see also* F-actin
  amoeboid movement, 959–966
  binding proteins, 708–709, 959–961
  cell motility, 959–960
  comparison to other motility proteins, 960
  crossbridge interaction with myosin, 945–949, 953, 955–956
  depolymerization of stress fibers, 1010
  filament, pH dependence, 368
  ion channels, 602, 616
  microvilli, 965
  neuronal cytoskeleton, 480–481
  polymerization, 959–960

Action potential (AP), *see also* Action potential propagation; Afterpotential
  accommodation, 421–422
  anodal-break excitation, 422–423
  calcium channel activation and inactivation, 428–432
  calcium-dependent, 873–875
  cardiac, 585–586, 887–897
  compound recording, 405–406
  electrocytes, 1025, 1027, 1029, 1030
  feedback loops, 414
  frequency-modulated signaling, 395–396
  gating currents of ion channels, 429–430
  generation, overview, 427–428
  local circuit currents, 400–401, 417–419
  local excitatory state, 418
  neuron developmental changes, 594–595
  patch clamp analysis, 426
  phases in external recording, 404–405, 419

  potassium channel activation and inactivation, 432–434
  refractoriness, 420–421
  repolarization, 434–435
  resting potential effects, 436
  skeletal muscle, 591–592, 865–883
  smooth muscle, 899–909
  sodium channels, 409, 428–429, 430–434
  strength-duration curve, 421
  threshold of excitation, 414, 418, 420
  voltage clamp analysis, 424–427
  wavelength determination, 403–404

Action potential propagation
  alternating current, 415–416
  biological fiber as cable, 396, 412–416
  conduction velocity calculation, 414
  determinants of velocity, 401
  electric field model, 407–410
  fiber diameter effects, 395
  input resistance, 399–400, 415
  length constant, 396–398
  linear cable theory, 413–416
  local potentials, 400
  membrane properties, 412–413
  myelination effects, 401–403
  saltatory conduction, 401–403
  time constant, 398–399
  velocities, 395, 412–416

Active transport
  cotransporters, 250
  countertransporters, 250
  nutrients, 128
  primary versus secondary, 250

Activin, development of electrical activity in skeletal muscle, 594

Activity coefficient, ions, 8, 228–229

Adenosine triphosphate, *see* ATP

Adenylate cyclase
  cyclic AMP synthesis, 155–156
  regulation by G proteins, 155–156

β-Adrenergic receptor, stria vascularis, 780

Afterpotential, *see also* Action potential
  early depolarizing, 436, 872
  early hyperpolarizing, 437
  electrogenesis, 436–438
  late depolarizing, 437, 872
  late hyperpolarizing, 437, 872–873
  physiological significance, 437–438

Agatoxin, 637–639

Agonist-activated channel, 446–447

Allosteric regulation, metabolic enzymes, 123

Amiloride-sensitive sodium channel
  gene locus, 486–487
  hormonal stimulation, 485–486
  ion selectivity, 485
  structure, 486
  taste transduction, 818–821
  tissue distribution, 485

Amino acid
  acid-base properties, 28–29
  degradation for energy production, 126
  essential, 126
  interactions in a protein chain, 23–24
  structures, 20

γ-Aminobutyric acid (GABA) receptor
  conductance, 683
  desensitization, 683
  drug modulation, 684
  structure, 683–685

Ammonium chloride prepulse technique, alteration of intracellular pH, 375–376

Amoeboid movement
  actin-based systems, 959–966
  chemotaxis, 964
  contractile proteins, 961–962
  cytoplasmic streaming, 962, 963
  mechanism, 960–961
  myosin distribution, 962–964
  organelle movement, 964
  sol-gel transformation, 962–963

AMPA (α-amino-3-hydroxy-5-methyl-4-isoxazole propionic acid) receptor
  postsynaptic density, 604, 606–607
  properties, 684–686

Ampulla of Lorenzini, *see* Ampullary electroreceptor

Ampullary electroreceptor
  active mode, 840
  bias current, 844
  biophysics, 857–861
  chondrichthyan (cartilaginous) fishes, 840–843, 844, 857–861
  clustering, 842–843
  elasmobranchs, 840–842, 844, 857–861
  frequency response, 839–840, 845, 859
  lampreys, 840
  morphology, 840–844
  passive mode, 840
  physiology, 844–845
  sensitivity, 844, 845, 860–861
  species distribution, 839–840, 843–844